2014 International Power Electronics Conference

(IPEC-Hiroshima 2014 ECCE-ASIA)

Hiroshima, Japan
18-21 May 2014

Pages 2413-3189

IEEE Catalog Number: CFP14CPB-POD
ISBN: 978-1-4799-2706-7

Copyright © 2014 by the Institute of Electrical and Electronic Engineers, Inc
All Rights Reserved

Copyright and Reprint Permissions: Abstracting is permitted with credit to the source. Libraries are permitted to photocopy beyond the limit of U.S. copyright law for private use of patrons those articles in this volume that carry a code at the bottom of the first page, provided the per-copy fee indicated in the code is paid through Copyright Clearance Center, 222 Rosewood Drive, Danvers, MA 01923.

For other copying, reprint or republication permission, write to IEEE Copyrights Manager, IEEE Service Center, 445 Hoes Lane, Piscataway, NJ 08854. All rights reserved.

***This publication is a representation of what appears in the IEEE Digital Libraries. Some format issues inherent in the e-media version may also appear in this print version.**

IEEE Catalog Number: CFP14CPB-POD
ISBN 13:   978-1-4799-2706-7

Additional Copies of This Publication Are Available From:

Curran Associates, Inc
57 Morehouse Lane
Red Hook, NY 12571 USA
Phone: (845) 758-0400
Fax: (845) 758-2633
E-mail: curran@proceedings.com
Web: www.proceedings.com

TABLE OF CONTENTS

A NOVEL CONTROL SCHEME FOR THREE-LEVEL FULL-BRIDGE CONVERTER ACHIEVING LOW THD OUTPUT VOLTAGE............66
Liu, Jilong ; Xiao, Fei ; Chen, Wei ; Yang, Guorun

PARALLEL CONNECTED THREE PHASE INVERTERS BASED ON MODULAR DESIGN AND DISTRIBUTED CONTROL............72
Xiao, Fei ; Chen, Wei ; Liu, Jilong ; Wang, Hengli

EFFICIENCY INVESTIGATIONS OF A 3KW T-TYPE INVERTER FOR SWITCHING FREQUENCIES UP TO 100 KHZ............78
Anthon, Alexander ; Zhang, Zhe ; Andersen, Michael A.E. ; Franke, Toke

MINIATURIZATION OF THE BOOST-UP TYPE ACTIVE BUFFER CIRCUIT IN A SINGLE-PHASE INVERTER............84
Watanabe, Hiroki ; Koiwa, Kazuhiro ; Itoh, Jun-ichi ; Ohnuma, Yoshiya ; Miyawaki, Satoshi

TESTING FACILITY USING LARGE CAPACITY INVERTER............92
Ishimaru, Yusuke ; Adachi, Mitsuo ; Tsukakoshi, Masahiko ; Nakamura, Ritaka ; Masuda, Hiroyuki ; Ogashi, Yoshihiro ; Tsuboi, Yuichi

PERFORMANCE EVALUATION UNDER THE ACTUAL OPERATING CONDITION OF A LARGE CAPACITY VSI INVERTER FOR STEEL MILL APPLICATIONS............97
Mamun, Mostafa ; Yoshizawa, Daisuke ; Mukunoki, Makoto

A SOFT-SWITCHING SINGLE-PHASE UNIFIED POWER QUALITY CONDITIONER............105
Jiang, Maoh-Chin ; Chang, Kai-Chi ; Lu, Kao-Yi ; Shih, Bing-Jyun ; Liu, Tai-Chun

NOVEL THREE-PHASE PWM AC-AC CONVERTERS SOLVING COMMUTATION PROBLEM............110
Khan, Ashraf Ali ; Shin, Hyunhak ; Cha, Honnyong ; Kim, Heung-Geun

EXPERIMENTAL INVESTIGATION OF NORMALLY-ON TYPE BIDIRECTIONAL SWITCH FOR INDIRECT MATRIX CONVERTERS............117
Sung, Kyungmin ; Iijima, Ryuji ; Nishizawa, Shinichi ; Norigoe, Isami ; Ohashi, Hiromichi

VISUALIZATION OF PWM WAVEFORMS OF OUTPUT VOLTAGE AND INPUT CURRENT FOR A DIRECT MATRIX CONVERTER............123
Asai, Inami ; Takeshita, Takaharu

SPACE VECTOR MODULATION BASED ON VIRTUAL INDIRECT CONTROL FOR HIGH FREQUENCY AC-LINKED MATRIX CONVERTER............130
Inoue, Keita ; Shioda, Masashi ; Katade, Motohumi ; Goto, Akira ; Morishita, Shin ; Itoh, Junichi ; Koiwa, Kazuhiro

A FUNDAMENTAL VERIFICATION OF A SINGLE-PHASE TO THREE-PHASE MATRIX CONVERTER WITH A PDM CONTROL BASED ON SPACE VECTOR MODULATION............138
Nakata, Yuki ; Itoh, Jun-ichi

STEADY STATE CHARACTERISTICS OF THE BOOST-TYPE MATRIX CONVERTER FOR STAND-ALONE POWER SOURCE............146
Nagano, Y. ; Yamamura, N. ; Ishida, M. ; Hirokado, K.

DESIGN PROCEDURE FOR OUTPUT CURRENT CONTROL AND DAMPING CONTROL OF MATRIX CONVERTER............152
Takahashi, Hiroki ; Itoh, Jun-ichi

A NOVEL LCL FILTER PARAMETER DESIGN METHOD BASING ON RESONANT FREQUENCY OPTIMIZATION OF THREE-LEVEL NPC GRID CONNECTED INVERTER............160
Li, Ning ; Wang, Yue ; Niu, Ruigen ; Guo, Wei ; Lei, Wanjun ; Wang, Zhao'An

DESIGN AND ANALYSIS OF ISOLATED BI-DIRECTIONAL DC/DC CONVERTER USING QUASI-RESONANT ZVS............166
Noh, Yong-Su ; Won, Chung-Yuen ; Oh, Min-Seok ; Jeon, Jin-Yong ; Jung, Yong-Chae

AN ACTIVE-CLAMPING ZVS FLYBACK CONVERTER WITH INTEGRATED TRANSFORMER............172
Lin, Jing-Yuan ; Lo, Yu-Kang ; Chiu, Huang-Jen ; Wang, Chao-Fu ; Lin, Chien-Yu

PFM AND PWM HYBRID CONTROLLED LLC CONVERTER............177
Yamamoto, Junichi ; Zaitsu, Toshiyuki ; Abe, Seiya ; Ninomiya, Tamotsu

DISCUSSIONS ON VARIOUS VOLTAGE EQUALIZERS FOR EDLCS USING CW CIRCUIT............183
Khant, Hlaing Kyi Pyar ; Matsui, Keiju ; Hasegawa, Masaru ; Yasubayashi, Mikio ; Umeno, Masayoshi ; Ooishi, Eiji

ISOLATION SYSTEM WITH WIRELESS POWER TRANSFER FOR MULTIPLE GATE DRIVER SUPPLIES OF A MEDIUM VOLTAGE INVERTER............191
Kusaka, Keisuke ; Orikawa, Koji ; Itoh, Jun-ichi ; Morita, Kazunori ; Hirao, Kuniaki

STUDY AND IMPLEMENTATION OF A 15-W POWER AMPLIFIER FOR PIEZOELECTRIC ACTUATOR............199
Lo, Yu-Kang ; Chiu, Huang-Jen ; Liu, Yu-Chen ; Lin, Chung-Yi ; Cheng, Shih-Jen ; Yang, CS

ISOLATED VOLTAGE-BOOSTING CONVERTER............204
Hwu, K.I. ; Jiang, W.Z. ; Shieh, Jenn-Jong

HIGH VOLTAGE CONVERSION RATIO CASCADE BOOST CONVERTER WITH DC SNUBBER............208
Lee, Yuang-Shung ; Yu, Ling-Chia ; Chou, Tzu-Han

DESIGN-ORIENTED ANALYSIS OF RESONANCE DAMPING AND HARMONIC COMPENSATION FOR LCL-FILTERED VOLTAGE SOURCE CONVERTERS............216
Wang, Xiongfei ; Blaabjerg, Frede ; Loh, Poh Chiang

STATE-SPACE AVERAGE MODELING OF BIDIRECTIONAL DC-DC CONVERTER FOR BATTERY CHARGER USING LCLC FILTER..224
Moon, Sang-Ho ; Jou, Sung-Tak ; Lee, Kyo-Beum

A NEW SVPWM STRATEGY FOR INPUT SWITCHED MULTILEVEL CONVERTER................................230
Xiong, Li ; Prasanna, U.R. ; Bilal, Akin ; Rajashekara, Kaushik

ESD RELIABILITY INFLUENCE OF A 60 V POWER LDMOS BY THE FOD-BASED (& DOTTED-OD) DRAIN..236
Chen, Shen-Li ; Lee, Min-Hua

ENHANCED TRANSVERSE-FLUX MOTOR WITH TORUS COILS..240
Tanaka, Junya ; Sakai, Kazuto

THE INFLUENCE OF MAGNETIC PROPERTIES OF PERMANENT MAGNET ON THE PERFORMANCE OF IPMSM FOR AUTOMOTIVE APPLICATION..246
Yoshioka, S. ; Morimoto, S. ; Sanada, M. ; Inoue, Y.

CHARACTERISTICS OF INTERIOR PERMANENT MAGNET SYNCHRONOUS MOTOR WITH IMPERFECT MAGNETS..252
Shinagawa, Syuhei ; Ishikawa, Takeo ; Kurita, Nobuyuki

STUDY OF STATOR STRUCTURE TO IMPROVE RELUCTANCE TORQUE FOR IPMSM WITH CONCENTRATED WINDING..258
Morikawa, R. ; Sanada, M. ; Morimoto, S. ; Inoue, Y.

DEVELOPMENT AND VERIFICATION OF ENERGY-ACCURATE SIMULATION MODELS FOR PERMANENT MAGNET SYNCHRONOUS MOTORS IN AUTOMATION SYSTEMS................................264
Blank, Frederic ; Roth-Stielow, Jorg

COMPARISON OF THE RESISTANCE- AND INDUCTANCE-BASED SALIENCY OF A PMSM DUE TO A SHORT-CIRCUITED ROTOR WINDING..270
Graus, Johannes ; Rambetius, Alexander ; Hahn, Ingo

DESIGN AND OPTIMIZATION OF HIGH-SPEED SWITCHED RELUCTANCE MOTOR USING SOFT MAGNETIC COMPOSITE MATERIAL..278
Gaing, Zwe-Lee ; Kuo, Kuan-Yi ; Hu, Jia-Sheng ; Hsieh, Min-Fu ; Tsai, Ming-Hsiao

INFLUENCE OF PULSE WIDTH MODULATION (PWM) ON THE IRON LOSSES OF ELECTRICAL STEEL................283
Boehm, Andreas ; Hahn, Ingo

INVESTIGATION ON IRON LOSS CHARACTERISTICS IN STAR-CONNECTION AND DELTA-CONNECTION UNDER THREE PHASE PWM INVERTER EXCITATION..289
Odawara, Shunya ; Fujisaki, Keisuke ; Fukuhara, Shuhei

OPTIMIZATION ON ARRANGEMENT OF PERMANENT MAGNETS FOR MAGNETIC LEVITATION SYSTEM FOR THIN STEEL PLATE (FUNDAMENTAL CONSIDERATION ON LEVITATION PROBABILITY)..294
Ishii, Hirotaka ; Hasegawa, Shinya ; Narita, Takayoshi ; Oshinoya, Yasuo

EFFECT OF A MAGNETIC FIELD FROM THE HORIZONTAL DIRECTION ON A MAGNETICALLY LEVITATED STEEL PLATE (FUNDAMENTAL CONSIDERATIONS ON THE SHAPE ANALYSIS OF ULTRATHIN STEEL PLATE)..299
Kurihara, Takeshi ; Hasegawa, Shinya ; Narita, Takayoshi ; Oshinoya, Yasuo

NOVEL MAGNETIC STRUCTURE OF INTEGRATED DIFFERENTIAL-MODE AND COMMON-MODE INDUCTORS TO SUPPRESS DC SATURATION..304
Umetani, Kazuhiro ; Tera, Takahiro ; Shirakawa, Kazuhiro

A NOVEL CONTROL METHOD IN FLUX-WEAKENING REGION FOR EFFICIENT OPERATION OF INTERIOR PERMANENT MAGNET SYNCHRONOUS MOTOR..312
Ueda, K. ; Morimoto, S. ; Inoue, Y. ; Sanada, M.

IMPLEMENTATION OF THE MTPA AND MTPV CONTROL WITH ONLINE PARAMETER IDENTIFICATION FOR A HIGH SPEED IPMSM USED AS TRACTION DRIVE................................318
Nguyen, Quoc Khanh ; Petrich, Matthias ; Roth-Stielow, Jorg

CORRECTION OF REFERENCE FLUX FOR MTPA CONTROL IN DIRECT TORQUE CONTROLLED INTERIOR PERMANENT MAGNET SYNCHRONOUS MOTOR DRIVES................................324
Shinohara, Atsushi ; Inoue, Yukinori ; Morimoto, Shigeo ; Sanada, Masayuki

VOLTAGE REGULATION AND MAXIMUM OUTPUT POWER TRACKING OF A 4.5KW PERMANENT-MAGNET SYNCHRONOUS GENERATOR..330
Chang, Yuan-Chih ; Chang, Hsiu-Feng ; Dai, Wei-Fu ; Wu, Chun-Wei

A NOVEL FLUX-WEAKENING CONTROL METHOD BASED ON SINGLE CURRENT REGULATOR FOR PERMANENT MAGNET SYNCHRONOUS MOTOR..335
Fang, Xiaocun ; Hu, Taiyuan ; Lin, Fei ; Yang, Zhongping

PREDICTIVE CURRENT CONTROL METHOD IN INDUCTION MOTOR SPEED SENSORLESS DRIVE................341
Wei, Sun ; Yong, Yu ; Dianguo, Xu ; Jin, Xu ; Li, Ding

REAL-TIME IMPLEMENTATION OF AN ONLINE MODEL PREDICTIVE CONTROL FOR IPMSM USING PARALLEL COMPUTING ON FPGA..346
Leuer, Michael ; Bocker, Joachim

AN INTEGRAL SLIDING-MODE CONTROLLER FOR ENERGY EFFICIENCY IMPROVEMENT IN AC POWER SOURCE SUPPLIED AC MACHINE DRIVES..351
Shieh, Hsin-Jang ; Chen, Ying-Zuo

PERFORMANCE IMPROVEMENT OF ULTRA-HIGH-SPEED PMSM DRIVE SYSTEM BASED ON DTC BY USING SIC INVERTER..356
Togashi, Ryo ; Inoue, Yukinori ; Morimoto, Shigeo ; Sanada, Masayuki

MATHEMATICAL MODEL FOR HIGH-EFFICIENCY CONTROL OF PERMANENT-MAGNET SYNCHRONOUS MOTOR IN STATOR FLUX LINKAGE SYNCHRONOUS FRAME...................363
Inoue, Tatsuki ; Inoue, Yukinori ; Morimoto, Shigeo ; Sanada, Masayuki

WIDE-SPEED-RANGE OPERATION OF DTC-BASED PMSM DRIVE SYSTEM USING MTPF CONTROL...........370
Inoue, Yukinori ; Ichiya, Takahiro ; Morimoto, Shigeo ; Sanada, Masayuki

AN INDUSTRIAL LOW-VOLTAGE INVERTER FOR PRM CONTROL.....................376
Nakamura, M. ; Oka, T. ; Oishi, K.

OPTIMAL PULSE PATTERN DETERMINATION BASED ON PULSE HARMONIC MODULATION....................383
Furukawa, Kimihisa ; Ajima, Toshiyuki ; Miyazaki, Hideki

METHOD FOR AUTO-TUNING OF CURRENT AND SPEED CONTROLLER IN IPMSM DRIVE SYSTEM BASED ON PARAMETER IDENTIFICATION 390
Tadokoro, D. ; Morimoto, S. ; Inoue, Y. ; Sanada, M.

COMPARATIVE STUDY OF PWM STRATEGIES FOR THREE-PHASE OPEN-END WINDING INDUCTION MOTOR DRIVES....................395
Zhu, B. ; Prasanna, U.R. ; Rajashekara, K. ; Kubo, H.

10MW,3.3MWH ENERGY STORAGE SYSTEM CONSISTING OF 4000 FLYWHEELS CONTROLLED BY ICT NETWORK FOR SHORT CYCLE POWER FLUCTUATION COMPENSATION 403
Kato, Koji ; Ishigma, Satoru ; Nakajima, Yoichiro ; Arai, Haruki ; Ueda, Tetsuya ; Iwata, Tetsuki ; Ito, Yoichi ; Sugao, Kazumi

VERSATILE POWER TRANSFER STRATEGIES OF PV-BATTERY HYBRID SYSTEM FOR RESIDENTIAL USE WITH ENERGY MANAGEMENT SYSTEM....................409
Choi, Seong-Chon ; Sin, Min-ho ; Kim, Dong-Rak ; Won, Chung-Yuen ; Jung, Yong-Chae

HIGH-EFFICIENCY AND COST-MINIMIZATION METHOD OF ENERGY STORAGE SYSTEM WITH MULTI STORAGE DEVICES FOR GRID CONNECTION 415
Haga, Hitoshi ; Shimao, Toshihiro ; Kondo, Seiji ; Kato, Koji ; Itoh, Youichi ; Arimatsu, Kenji ; Matsuda, Katsuhiro

BIDIRECTIONAL DC-DC CONVERTER WITH MULTIPLE SWITCHED-CAPACITOR CELLS 421
Lee, Yuang-Shung ; Huang, Hsin-Wei ; Chou, Tzu-Han

SWITCHED-CAPACITOR CHARGE EQUALIZATION CIRCUIT FOR SERIES-CONNECTED BATTERIES 429
Hsieh, Yao-Ching ; Cai, Zheng-Xiu ; Wu, Wen-Zhe

PERFORMANCE ANALYSIS OF UNITL-H6 INVERTER WITH SIC MOSFETS 433
Barater, Davide ; Buticchi, Giampaolo ; Concari, Carlo ; Franceschini, Giovanni ; Gurpinar, Emre ; De, Dipankar ; Castellazzi, Alberto

MAXIMUM POWER POINT TRACKING OF GRID-TIED PHOTOVOLTAIC POWER SYSTEMS 440
Lee, Ya-Ting ; Chiu, Chian-Song ; Chiu, Tse-Wei

A NEW VOLTAGE TYPE MAGNETICALLY COUPLED T-SOURCE INVERTER....................446
Tran, Q.V. ; Low, K.S.

A HIGH EFFICIENCY HYBRID 7-LEVEL INVERTER WITH SINGLE DC SOURCE 452
Yanhong, Zhang ; Kazuya, Ogura ; Oi, Kazunobu

OPTIMAL IDLING CONTROL STRATEGY FOR THREE-PORT FULL-BRIDGE CONVERTER 458
Jiang, Yongjie ; Liu, Fuxin ; Ruan, Xinbo ; Wang, Lipeng

FILTER DESIGN FOR THREE-LEVEL GRID-CONNECTED INVERTER WITH LOW SWITCHING FREQUENCY....................465
Ren, Kangle ; Zhang, Xing ; Wang, Fusheng ; Tu, Yunwu ; Wang, Lingxiang ; Deng, Lirong

A NOVEL EFFICIENT T TYPE THREE LEVEL NEUTRAL-POINT-CLAMPED INVERTER FOR RENEWABLE ENERGY SYSTEM....................470
Wu, Wenlong ; Wang, Fei ; Wang, Yong

A NOVEL NEUTRAL POINT VOLTAGE AUTOMATIC BALANCING CARRIER-BASED MODULATION STRATEGY OF THREE-LEVEL NPC CONVERTER....................475
Li, Ning ; Wang, Yue ; Niu, Ruigen ; Guo, Wei ; Lei, Wanjun ; Wang, Zhao'An

A HIGH VOLTAGE GAIN SWITCHED-COUPLED-INDUCTOR QUASI-Z-SOURCE INVERTER....................480
Ahmed, Furqan ; Cha, Honnyong ; Kim, Su-Han ; Kim, Heung-Geun

A NOVEL CONTROL STRATEGY TO SUPPRESS DC CURRENT INJECTION TO THE GRID FOR THREE-PHASE PV INVERTER....................485
Zhang, Tao ; He, Guofeng ; Chen, Min ; Xu, Dehong

CLC FILTER DESIGN OF A FLYBACK-INVERTER FOR PHOTOVOLTAIC SYSTEMS 493
Shin, Yesl ; Lee, June-Hee ; Lee, June-Seok ; Lee, Kyo-Beum

THREE-PHASE INVERTER TOPOLOGIES FOR GRID-CONNECTED PHOTOVOLTAIC SYSTEMS....................498
Ozkan, Ziya ; Hava, Ahmet M.

A THREE-PORT TOPOLOGY COMPARISON FOR A LOW POWER STAND-ALONE PHOTOVOLTAIC SYSTEM....................506
Mira, Maria C. ; Knott, Arnold ; Andersen, Michael A.E.

EFFECT OF CONVENTIONAL GRID-VOLTAGE FEEDFORWARD ON THE OUTPUT IMPEDANCE OF A THREE-PHASE PHOTOVOLTAIC INVERTER....................514
Messo, T. ; Jokipii, J. ; Suntio, T.

POWER AMPLIFIER SUITABLE FOR PHOTOVOLTAIC CELL BOOSTER....................522
Kohama, Teruhiko ; Sogawa, Yuki ; Tsuji, Satoshi

REALIZATION STUDY OF INTERLEAVED PV MICROINVERTER BY QUADRATURE-PHASE-SHIFT SPWM CONTROL 526
Hsieh, Hung-I ; Hsieh, Guan-Cyun ; Hou, Jiaxin

CURRENT SENSORLESS MPPT METHOD FOR A PV FLYBACK MICROINVERTERS USING A DUAL-MODE 532

Lee, June-Hee ; Lee, June-Seok ; Lee, Kyo-Beum

A NOVEL METHOD OF SUPPRESSING INRUSH CURRENTS OF SQUIRREL-CAGE INDUCTION MACHINE USING MATRIX CONVERTER IN WIND POWER GENERATION SYSTEMS 538

Yamada, Hiroaki ; Hanamoto, Tsuyoshi

NONLINEAR PITCH CONTROL DESIGN FOR LOAD REDUCTION ON WIND TURBINES 543

Xiao, Shuai ; Yang, Geng ; Geng, Hua

DEVICE LOADING OF MODULAR MULTILEVEL CONVERTER MMC IN WIND POWER APPLICATION 548

Popova, L. ; Pyrhonen, J. ; Ma, K. ; Blaabjerg, F.

A NOVEL OPTIMAL DESIGN OF DFIG CROWBAR RESISTOR DURING GRID FAULTS 555

Hu, Sheng ; Zou, XuDong ; Kang, Yong

DC-VOLTAGE REGULATION OF A FIVE LEVELS NEUTRAL POINT CLAMPED CASCADED CONVERTER FOR WIND ENERGY CONVERSION SYSTEM 560

Merahi, Farid ; Mekhilef, Saad ; Berkouk, El Madjid

A REACTIVE POWER SHARING METHOD BASED ON VIRTUAL CAPACITOR IN ISLANDING MICROGRID 567

Xu, Haizhen ; Zhang, Xing ; Liu, Fang ; Shi, Rongliang ; Yu, Changzhou ; Zhao, Wei ; Yu, Yong ; Cao, Wei

STORAGE CAPACITY PERFORMANCE FOR HYBRID PV/DIESEL SYSTEM IN SABAH MALAYSIA 573

Hidayat, Nabil M ; Kari, Mat Nasir ; Mohd Arif, Mohd Johari

NEW TECHNIQUES FOR MEASURING ISLANDED MICROGRID IMPEDANCE CHARACTERISTICS BASED ON CURRENT INJECTION 577

Hou, Lixiang ; Liu, Baoquan ; Shi, Hongtao ; Yi, Hao ; Zhuo, Fang

A GENERAL FRAMEWORK TO DESIGN OPERATION MODES OF DC MICROGRIDS WITHOUT COMMUNICATION LINKS 582

Pan, Miao ; Shen, Na ; Yang, Geng ; Morita, Kazunori ; Ogura, Kazuya ; Wu, Weiyang

IMPLEMENTATION DESIGN OF THE CONVERTER-BASED GALVANIC ISOLATION FOR LOW VOLTAGE DC DISTRIBUTION 587

Mattsson, A. ; Vaisanen, V. ; Nuutinen, P. ; Kaipia, T. ; Lana, A. ; Peltoniemi, P. ; Silventoinen, P. ; Partanen, J.

PEAK DETECTION METHOD USING TWO-DELTA OPERATION FOR SINGLE VOLTAGE SAG 595

Lee, Woo-Cheol ; Lee, Taeck-Kie

LINE LOSS MINIMIZATION IN RADIAL DISTRIBUTION SYSTEM USING MULTIPLE STATCOMS AND STATIC CAPACITORS 601

Miyazaki, Kensuke ; Takeshita, Takaharu

A NOVEL CONTROL METHOD FOR INDIVIDUAL DC VOLTAGE BALANCING IN H-BRIDGE CASCADED STATCOM 609

Xu, Rong ; Yu, Yong ; Yang, Rongfeng ; Qu, Lizhi ; Sun, Wei ; Xu, Dianguo

RESEARCH ON THE CONTROL STRATEGY OF STATCOM BASED ON MODULAR MULTILEVEL CONVERTER 614

Zhang, Wei ; Gao, Qiang ; Su, Bonan ; Jin, Miaoxin ; Xu, Dianguo ; Liu, Jianyu

FAULT DIAGNOSIS IN LARGE FORMAT LIFEPO4 ESS APPLICATION THROUGH DWT-BASED MRA 619

Kim, Jonghoon

COMPARISON OF DIFFERENT IGBT BASED DESIGNS OF POWER ELECTRONIC TRANSFORMER 624

Wang, Xinyu ; Ouyang, Shaodi ; Liu, Jinjun ; Meng, Fei ; Javed, Riffat

SEMI-ADAPTIVE HARMONIC CONTROL FOR POWER BALANCING DEVICE FOR AC TRACTION 629

Akagi, Masataka ; Tsuruta, Hironori ; Oso, Hiroshi

RESEARCH OF EFFICIENT MAIN POWER EQUIPMENT USING SIC POWER DEVICE 634

Shinbo, Mitsuo ; Sonoda, Hideki ; Ishida, Takahito ; Abiko, Hiroshi ; Shibanuma, Kenichi ; Chiba, Yoshinori

A HIGH PERFORMANCE CONTROL STRATEGY FOR THREE-LEVEL NPC EMU CONVERTERS 640

Song Kejian ; Wu Mingli ; Wang Hui ; Agelidis, Vassilios Georgios

A DESIGN OF INRUSH CURRENT IDENTIFICATION SYSTEM FOR HIGH-SPEED TRAIN'S TRACTION TRANSFORMER 647

Yu, Weikai ; Liu, Xiankai ; Zhang, Yuzhuo ; Cao, Yuan ; Ma, Weigang ; Hei, Xinhong ; Huang, Zhenhui ; Jiang, Dawang

CURRENT SOURCE INVERTER BASED CASCADED SOLID STATE TRANSFORMER FOR AC TO DC POWER CONVERSION 651

Roy, Sudhin ; De, Ankan ; Bhattacharya, Subhashish

EVALUATION OF HIGH VOLTAGE 15 KV SIC IGBT AND 10 KV SIC MOSFET FOR ZVS AND ZCS HIGH POWER DC -DC CONVERTERS 656

Moballegh, Shiva ; Madhusoodhanan, Sachin ; Bhattacharya, Subhashish

THE DIRECT YAW-MOMENT CONTROL TO FOLLOW THE NEUTRAL STEERING PATH REGARDLESS OF VELOCITY 664

Jang, Young-Jin ; Nam, Kwang-Hee

NEXT-GENERATION IGBT MODULE STRUCTURE FOR HYBRID VEHICLE WITH HIGH COOLING PERFORMANCE AND HIGH TEMPERATURE OPERATION 671

Morozumi, Akira ; Gohara, Hiromichi ; Momose, Fumihiko ; Saito, Takashi ; Nishimura, Yoshitaka ; Mochizuki, Eiji ; Takahashi, Yoshikazu

INTEGRATION OF PLUG-IN ELECTRIC VEHICLES IN POWER SYSTEMS USING CHARGING MODE SWITCHING 677

Wen-Tai Li ; Wen, Chao-Kai ; Chen, Jung-Chieh ; Teng, Jen-Hao ; Ting, Pangan

A NOVEL COMPENSATION METHOD FOR A MOTOR PHASE CURRENT SENSOR OFFSET ERROR VARIED DURING A VSI-MOTOR DRIVE 682

Tamura, Hiroshi ; Noto, Yasuo ; Ajima, Toshiyuki ; Itoh, Jun-ichi

INVESTIGATION OF CALCULATION METHOD OF LOSSES IN PWM INVERTER WITH VOLTAGE BOOSTER USING BOTH DC LINK VOLTAGE CONTROL AND FLUX WEAKENING CONTROL 689

Imakiire, Akihiro ; Hikita, Masayuki ; Yamamoto, Kichiro ; Yonemori, Ryo

DYNAMIC AND STEADY-STATE BEHAVIOR OF A PARALLELING THREE-PHASE AC-TO-DC CONVERTER WITH REDUCED DC BUS CAPACITOR 694

Kamnarn, Uthen ; Kanthaphayao, Yutthana ; Chunkag, Viboon

REACTIVE POWER LOSS OPTIMIZATION METHOD FOR BI-DIRECTIONAL ISOLATED DC-DC CONVERTERS 702

Wen, Huiqing

POWER SUPPLY FOR A WIRELESS SENSOR NETWORK: AIRLINER FLIGHT TEST CASE STUDY 707

Durand Estebe, P. ; Boitier, V. ; Bafleur, M. ; Dilhac, J-M. ; Berhouet, S.

A CONFIGURABLE THREE-PHASED INVERTER FOR TEACHING POWER ELECTRONICS 712

Kern, Ansgar

A BACHELOR-STUDENT PROJECT: BUCK-BOOST OPERATION OF AN INTEGRATED H-BRIDGE FOR VARIABLE-SPEED ENERGY STORAGE SYSTEMS USING MEASUREMENT COILS IN THE STATOR OF A DC-MACHINE 718

De Belie, Frederik ; Darba, Araz ; Melkebeek, Jan

DEVELOPMENT OF A WEB-BASED REMOTE EXPERIMENT SYSTEM FOR ELECTRICAL MACHINERY LEARNERS 724

Ishibashi, Makoto ; Fukumoto, Hisao ; Furukawa, Tatsuya ; Itoh, Hideaki ; Ohchi, Masashi

DEVELOPMENT OF POWER MEASUREMENT SYSTEM IN SIMULATED MICRO GRID SYSTEM FOR EDUCATION 730

Hira, Yuki ; Furukawa, Tatsuya ; Yakabe, Seichiro ; Fukumoto, Hisao ; Itoh, Hideaki ; Ohchi, Masashi

POWER ELECTRONIC TECHNOLOGIES FOR FLEXIBLE DC DISTRIBUTION GRIDS 736

De Doncker, Rik W.

2.5KV, 200KW BI-DIRECTIONAL ISOLATED DC/DC CONVERTER FOR MEDIUM-VOLTAGE APPLICATIONS 744

Matsuoka, Yuji ; Wada, Keiji ; Nakahara, Mizuki ; Takao, Kazuto ; Kyungmin Sung ; Ohashi, Hiromichi ; Nishizawa, Shinichi

POWER-LOSS BREAKDOWN OF A 750-V, 100-KW, 20-KHZ BIDIRECTIONAL ISOLATED DC-DC CONVERTER USING SIC-MOSFET/SBD DUAL MODULES 750

Akagi, Hirofumi ; Yamagishi, Tatsuya ; Tan, Nadia M.L. ; Kinouchi, Shin-ichi ; Miyazaki, Yuji ; Koyama, Masato

DESIGN CONSIDERATIONS OF A 15KV SIC IGBT ENABLED HIGH-FREQUENCY ISOLATED DC-DC CONVERTER 758

Tripathi, Awneesh ; Mainali, Krishna ; Patel, Dhaval ; Kadavelugu, Arun ; Hazra, Samir ; Bhattacharya, Subhashish ; Hatua, Kamalesh

COMMON-MODE CURRENTS IN MULTI-CELL SOLID-STATE TRANSFORMERS 766

Huber, Jonas E. ; Kolar, Johann W.

SINGLE-STAGE RECONFIGURABLE DC/DC CONVERTER FOR WIDE INPUT VOLTAGE RANGE OPERATION IN HEVS 774

Zeljkovic, Sandra ; Reiter, Tomas ; Gerling, Dieter

A TWO STAGE DC/DC CONVERTER WITH WIDE INPUT RANGE FOR EV 782

Peng Wen ; Changsheng Hu ; Haitao Yang ; Longlong Zhang ; Cheng Deng ; Yashun Li ; Dehong Xu

INTERMEDIATE AND LIGHT LOAD EFFICIENCY IMPROVEMENT OF A HIGH-POWER DENSITY BIDIRECTIONAL DC-DC CONVERTER IN HYBRID ELECTRIC VEHICLES WITH MR FLUID GAP INDUCTOR 790

Ahmed, Furqan ; Su-Han Kim ; Cha, Honnyong ; Kim, Dong-Hun ; Heung-Geun Kim

REGENERATIVE CONTROL OF BI-DIRECTIONAL DC-DC CONVERTER CONTROLLING VARIABLE DC-LINK FOR FCEV 796

Il-Kuen Won ; An-Yeol Ko ; Do-Yun Kim ; Chung-Yuen Won ; Young-Ryul Kim

LARGE DRIVING RANGE INCREASE OF SERIES CHOPPER BASED POWER TRAIN USING MOTOR TEST BENCH 801

Hosoyamada, Yu ; Takeda, Masashi ; Motoi, Naoki ; Kawamura, Atsuo

THE POWER ELECTRONICS PROGRAM AT BEIJING JIAOTONG UNIVERSITY 807

Fei Lin ; Zhongping Yang ; Zheng, T.Q.

EFFORTS FOR POWER ELECTRONICS EDUCATION IN A START-UP COMPANY 811

Hattori, Fumiya ; Imaoka, Jun ; Ishitobi, Manabu ; Nagai, Shinichiroh ; Yamamoto, Masayoshi

EDUCATION FOR THE ENGINEERS OF TRACTION POWER SUPPLY DIVISION IN EAST JAPAN RAILWAY COMPANY 817

Takino, Toshiaki ; Iwakami, Tetsuro

SUCCESSFUL ONLINE EDUCATION - GECKOCIRCUITS AS OPEN-SOURCE SIMULATION PLATFORM 821

Musing, Andreas ; Kolar, Johann W.

AN ELECTRIC VEHICLE PROJECT FOR ECO-RUN RACE 829

Yamagata, Shinichi ; Oda, Yoshinori ; Tanai, Masanobu ; Sung, Kyungmin

MULTI-LOOP CONTROLLER DESIGN FOR DIODE-ASSISTED BUCK-BOOST VOLTAGE SOURCE INVERTER 835

Yan Zhang ; Jinjun Liu ; Xiaolong Ma ; Junjie Feng

VOLUME 2

REAL-TIME SIMULATION OF WIND TURBINE CONVERTER-GRID SYSTEMS 843
Shah, Shahil ; Vieto, Ignacio ; Nian Heng ; Sun, Jian

TECHNOLOGIES FOR MITIGATING FLUCTUATION CAUSED BY RENEWABLE ENERGY SOURCES 850
Katoh, Shuji ; Ohara, Shinya ; Itoh, Tomomichi

RELIABILITY-ORIENTED ENERGY STORAGE SIZING IN WIND POWER SYSTEMS 857
Zian Qin ; Liserre, Marco ; Blaabjerg, Frede ; Poh Chiang Loh

A MULTI-LEVEL VIRTUAL CONDUCTOR AS A BACKBONE OF A DC POWER ROUTING SYSTEM 863
Ramadan, Husam A. ; Imamura, Yasutaka ; Kawachi, Konosuke ; Yang, Sihun ; Shoyama, Masahito

SEMI-NUMERICAL METHOD FOR LOSS-CALCULATION IN FOIL-WINDINGS EXPOSED TO AN AIR-GAP FIELD 868
Leuenberger, D. ; Biela, J.

LOSS REDUCTION OF LAMINATED CORE INDUCTOR USED IN ON-BOARD CHARGER FOR EVS 876
Tera, Takahiro ; Taki, Hiroshi ; Shimizu, Toshihisa

FEASIBLE EVALUATIONS OF COUPLED MULTILAYER CHIP INDUCTOR FOR POL CONVERTER 883
Imaoka, Jun ; Kimura, Shota ; Itoh, Yuki ; Yamamoto, Masayoshi ; Suzuki, Michiaki ; Kawano, Kenji

OPTIMAL INDUCTOR DESIGN FOR 3-PHASE VOLTAGE-SOURCE PWM CONVERTERS CONSIDERING DIFFERENT MAGNETIC MATERIALS AND A WIDE SWITCHING FREQUENCY RANGE 891
Burkart, Ralph M. ; Uemura, Hirofumi ; Kolar, Johann W.

COMPARATIVE ANALYSIS OF INDUCTOR CONCEPTS FOR HIGH PEAK LOAD LOW DUTY CYCLE OPERATION 899
Leibl, Michael ; Kolar, Johann W.

INITIAL POSITION ESTIMATION FOR IPMSMS USING COMB FILTERS AND EFFECTS ON VARIOUS INJECTED SIGNAL FREQUENCIES 907
Suzuki, Toshiki ; Tomita, Mutuwo ; Hasegawa, Masaru ; Doki, Shinji

ADAPTIVE SIGNAL INJECTION METHOD COMBINED WITH EEMF BASED POSITION SENSORLESS CONTROL OF IPMSM DRIVES 914
Ohnuma, Takumi ; Makaino, Yuki ; Saitoh, Ryoh

STUDY OF LOW SPEED SENSORLESS DRIVES FOR SPMSM BY CONTROLLING ELLIPTICAL INDUCTANCE 919
Maekawa, Sari ; Hinata, Toshifumi ; Suzuki, Nobuyuki ; Kubota, Hisao

SUPPRESSION OF INJECTION VOLTAGE DISTURBANCE FOR HIGH FREQUENCY SQUARE-WAVE INJECTION SENSORLESS DRIVE WITH REGULATION OF INDUCED HIGH FREQUENCY CURRENT RIPPLE 925
Dongouk Kim ; Yong-Cheol Kwon ; Seung-Ki Sul ; Jang-Hwan Kim ; Rae-Sung Yu

APPLICATION TREND OF SALIENCY-BASED SENSORLESS DRIVES 933
Yamazaki, Akira ; Ide, Kozo

SWITCHING-LEVEL SIMULATION MODEL OF MMC-BASED BACK-TO-BACK CONVERTER FOR HVDC APPLICATION 937
Byung Moon Han ; Jong kyou Jeong

POWER-CELL SWITCHING-CYCLE CAPACITOR VOLTAGE CONTROL FOR THE MODULAR MULTILEVEL CONVERTERS 944
Wang, Jun ; Burgos, Rolando ; Boroyevich, Dushan ; Bo Wen

A COMPARISON OF MODULAR MULTILEVEL ENERGY CONVERSION PROCESSES: DC/AC VERSUS DC/DC 951
Kish, Gregory J. ; Lehn, Peter W.

A NOVEL TOPOLOGY OF WIND POWER PLANT SUITABLE FOR DC POWER TRANSMISSION SYSTEMS 959
Nishikata, Shoji ; Tatsuta, Fujio ; Suzuki, Katsumi

AN IMPEDANCE-BASED APPROACH TO HVDC SYSTEM STABILITY ANALYSIS AND CONTROL DEVELOPMENT 967
Liu, Hanchao ; Shah, Shahil ; Sun, Jian

TOPOLOGY EVALUATION OF SLOTLESS BEARINGLESS MOTORS WITH TOROIDAL WINDINGS 975
Steinert, Daniel ; Nussbaumer, Thomas ; Kolar, Johann W.

WINDING ARRANGEMENT IN SINGLE-DRIVE BEARINGLESS MOTOR WITH RADIAL GAP 982
Sugimoto, Hiroya ; Tanaka, Seiyu ; Chiba, Akira ; Rahman, M.A.

DEVELOPMENT OF A ONE-AXIS ACTIVELY REGULATED BEARINGLESS MOTOR WITH A REPULSIVE TYPE PASSIVE MAGNETIC BEARING 988
Asama, Junichi ; Watanabe, Daisuke ; Oiwa, Takaaki ; Chiba, Akira

CONTROL CHARACTERISTICS OF 8/10 AND 12/14 BEARINGLESS SWITCHED RELUCTANCE MOTOR 994
Zhenyao Xu ; Dong-Hee Lee ; Jin-Woo Ahn

BASIC CHARACTERISTIC OF A TWO-UNIT OUTER ROTOR TYPE BEARINGLESS MOTOR WITH CONSEQUENT POLE PERMANENT MAGNET STRUCTURE 1000
Takemoto, Masatsugu

VOLTAGE RIPPLE ELIMINATION IN INDUCTOR-LESS AC-TO-AC CONVERTERS FOR MULTI-POLE PERMANENT MAGNET SYNCHRONOUS GENERATORS 1006

Tanaka, Koutaro ; Fujita, Hideaki

A NEW SVM METHOD TO REDUCE COMMON-MODE VOLTAGE IN DIRECT MATRIX CONVERTER 1013

Huu-Nhan Nguyen ; Hong-Hee Lee

EXPERIMENTAL VERIFICATION OF HIGH FREQUENCY LINK DC-AC CONVERTER USING PULSE DENSITY MODULATION AT SECONDARY MATRIX CONVERTER 1021

Itoh, Jun-ichi ; Oshima, Ryo ; Takahashi, Hiroki

LOSS ANALYSIS AND DESIGN METHOD FOR HIGH EFFICIENCY MATRIX CONVERTER 1028

Koiwa, Kazuhiro ; Goh Teck Chiang ; Itoh, Jun-ichi

CAPACITOR CLAMPED MULTI-LEVEL MATRIX CONVERTER 1036

Raju, Siddharth ; Mohan, Ned

EUROPEAN TRENDS AND TECHNOLOGIES IN TRACTION 1043

Drofenik, Uwe ; Canales, Francisco

CO-PHASE POWER SUPPLY SYSTEM FOR HSR 1050

Qunzhan Li ; Wei Liu ; Zeliang Shu ; Shaofeng Xie ; Fulin Zhou

THE APPLICATION OF ELECTRONIC FREQUENCY CONVERTER TO THE SHINKANSEN RAILYARD POWER SUPPLY 1054

Shimizu, Toshimasa ; Kunomura, Ken ; Kai, Masahiko ; Onishi, Mitsuru ; Masuzawa, Hiroshi ; Miyajima, Hiroki ; Otsuki, Midori ; Tsuruma, Yoshinori

APPLICATION EXAMPLES OF ENERGY SAVING MEASURES IN JAPANESE DC FEEDING SYSTEM 1062

Suzuki, Takashi ; Hayashiya, Hitoshi ; Yamanoi, Takashi ; Kawahara, Keiji

LITHIUM ION BATTERY APPLICATION IN TRACTION POWER SUPPLY SYSTEM 1068

Teshima, Masato ; Takahashi, Hirotaka

INTEGRATED ISOLATION AND VOLTAGE BALANCING LINK OF 3-PHASE 3-LEVEL PWM RECTIFIER AND INVERTER SYSTEMS 1073

Boillat, David O. ; Kolar, Johann W.

VOLTAGE STEP-UP CONVERTER BASED ON MULTISTAGE STACKED BOOST ARCHITECTURE (MSBA) 1081

Rufer, Alfred ; Barrade, Philippe ; Steinke, Gina

COMPARISON OF CASCADED MULTILEVEL CONVERTER TOPOLOGIES FOR AC/AC CONVERSION 1087

Ilves, Kalle ; Bessegato, Luca ; Norrga, Staffan

EVALUATION OF ISOLATED THREE-PHASE AC-DC CONVERTER USING MODULAR MULTILEVEL CONVERTER TOPOLOGY 1095

Nakanishi, Toshiki ; Itoh, Jun-ichi

SELF-DECOUPLED DUAL PICK-UP COILS WITH LARGE LATERAL TOLERANCE FOR ROADWAY POWERED ELECTRIC VEHICLES 1103

Choi, Su Y. ; Lee, Sung W. ; Lee, Eun S. ; Jeong, Seog Y. ; Gu, Beom W. ; Rim, Chun T.

CONTACTLESS POWER TRANSFER SYSTEM SUITABLE FOR LOW VOLTAGE AND LARGE CURRENT CHARGING FOR EDLCS 1109

Kudo, Takahiro ; Toi, Takahiro ; Kaneko, Yasuyoshi ; Abe, Shigeru

EXCITATION SYSTEM BY CONTACTLESS POWER TRANSFER SYSTEM WITH THE PRIMARY SERIES CAPACITOR METHOD 1115

Nozawa, Ryosuke ; Kobayashi, Ryota ; Tanifuji, Hikaru ; Kaneko, Yasuyoshi ; Abe, Shigeru

DESIGN OF FERRITE CORES OF INDUCTIVE POWER COLLECTION COILS FOR MOVING VEHICLES 1122

Shimode, Daisuke ; Murai, Toshiaki ; Sawada, Tadashi

TORQUE/CURRENT RATIO IMPROVEMENT AND VIBRATION REDUCTION OF SWITCHED RELUCTANCE MOTORS USING MULTI-STAGE STRUCTURE 1128

Matsui, Ryota ; Nakao, Noriya ; Akatsu, Kan

IMPROVEMENT OF EFFICIENCY BY STEPPED-SKEWING ROTOR FOR SWITCHED RELUCTANCE MOTORS 1135

Sugiura, Makoto ; Ishihara, Yuji ; Ishikawa, Hiroki ; Naitoh, Haruo

A SINGLE PHASE SRM DRIVEN BY COMMERCIAL AC POWER SUPPLY 1141

Aiso, Kohei ; Nakao, Noriya ; Akatsu, Kan

FAST ANALYTICAL MODEL OF SWITCHED RELUCTANCE MACHINE 1148

Smaka, Senad ; Masic, Semsudin ; Cosovic, Mirsad

DETAILED ANALYSIS AND A GENERAL DESIGN PROCEDURE OF DAMPED LCL FILTERS IN THREE PHASE VOLTAGE SOURCE CONVERTERS 1155

Baoquan Liu ; Shaohui Zhong ; Yixin Zhu ; Hao Yi ; Fang Zhuo

70 KHZ, 15 KW SILICON-CARBIDE MOSFET INVERTER FOR INDUSTRIAL INDUCTION HEATING SYSTEMS 1160

Komeda, Shohei ; Tsuboi, Yoshiki ; Fujita, Hideaki

A STUDY ON EFFICIENCY IMPROVEMENT OF HIGH-FREQUENCY CURRENT OUTPUT INVERTER BASED ON IMMITTANCE CONVERSION ELEMENT 1166

Suzuki, Shun ; Shimizu, Toshihisa

HIGH-SPEED SWITCHING METHOD OF MOSFET USING VOLTAGE BOOST AUXILIARY CIRCUIT FED BY GATE DRIVE POWER SUPPLY 1173

Noguchi, Toshihiko ; Murata, Munehiro

OPERATING STRATEGY FOR BI-DIRECTIONAL LLC RESONANT CONVERTER WITH SEAMLESS OPERATION 1179

Abe, Seiya ; Yamamoto, Junichi ; Zaitsu, Toshiyuki ; Ninomiya, Tamotsu

NEGATIVE SEQUENCE CURRENT INJECTION CONTROL ALGORITHM COMPENSATING FOR UNBALANCED PCC VOLTAGE IN MEDIUM VOLTAGE PMSG WIND TURBINES 1185

Jayoon Kang ; Daesu Han ; Suh, Yongsug ; Byoungchang Jung ; Jeongjoong Kim ; Jonghyung Choi

OPTIMIZATION OF AN OFF-GRID HYBRID SYSTEM FOR SUPPLYING OFFSHORE PLATFORMS IN ARCTIC CLIMATES 1193

Kalogera, Maria ; Bauer, Pavol

ACTIVE DAMPING CONTROL OF LLCL FILTERS FOR THREE-LEVEL T-TYPE GRID CONVERTERS 1201

Alemi, Payam ; Lee, Dong-Choon

DEVELOPING A NEW TOPOLOGY FOR THE DC-DC CONVERTER USED IN FUEL CELL-ELECTRIC DOUBLE LAYER CAPACITOR HYBRID POWER SOURCE SYSTEM FOR MOBILE DEVICES 1207

Tosaka, Shuhei ; Yamanaka, Tatsuya ; Katayama, Noboru ; Hayase, Masanori ; Dowaki, Kiyoshi ; Kogoshi, Sumio

MULTIPLE OUTPUT CHARGER BASED ON PHASE SHIFT FULL BRIDGE CONVERTER WITH NOVEL TIME DIVISION MULTIPLE CONTROL TECHNIQUE 1214

Van-Long Tran ; Woojin Choi

DC-BREAKER FOR A MULTI-MEGAWATT BATTERY ENERGY STORAGE SYSTEM 1220

Demetriades, Georgios D. ; Hermansson, Willy ; Svensson, Jan R ; Papastergiou, Konstantinos ; Larsson, Tomas

ENERGY MANAGEMENT METHOD USING THE IIR FILTER FOR PEMFC-SUPERCAPACITOR HYBRID POWER SOURCE 1227

Yamanaka, Tatsuya ; Katayama, Noboru ; Tosaka, Shuhei ; Kogoshi, Sumio

ADVANCED TORQUE AND CURRENT CONTROL TECHNIQUES FOR PMSMS WITH A REAL-TIME SIMULATOR INSTALLED BEHAVIOR MOTOR MODEL 1234

Tanabe, Ryo ; Akatsu, Kan

COMPENSATION OF THE CURRENT MEASUREMENT ERROR WITH PERIODIC DISTURBANCE OBSERVER FOR MOTOR DRIVE 1242

Yamaguchi, Takashi ; Tadano, Yugo ; Hoshi, Nobukazu

RAPID AND STABLE SPEED CONTROL OF SPMSM BASED ON CURRENT DIFFERENTIAL SIGNAL 1247

Kitajima, Jun ; Ohishi, Kiyoshi

PARALLEL CONNECTED MULTIPLE DRIVE SYSTEM USING SMALL AUXILIARY INVERTER FOR NUMBERS OF PMSM 1253

Nagano, Tsuyoshi ; Itoh, Jun-chi

A TRANSFORMER INRUSH REDUCTION TECHNIQUE FOR LOW-VOLTAGE RIDE-THROUGH OPERATION OF RENEWABLE CONVERTERS 1261

Hsin-Chih Chen ; Ping-Heng Wu ; Cheng, Po-Tai

A CELL CAPACITOR ENERGY BALANCING CONTROL OF MODULAR MULTILEVEL CONVERTER CONSIDERING THE UNBALANCED AC GRID CONDITIONS 1268

Jung, Jae-Jung ; Shenghui Cui ; Kim, Sungmin ; Sul, Seung-Ki

FAULT CURRENT LIMITATION USING THYRISTOR BASED DEVICES 1276

Komatsu, Wilson ; Giaretta, Antonio Ricardo ; de Miranda, Rubens Domingos ; Jardini, Jose Antonio ; Casolari, Ronaldo Pedro ; Vasquez-Arnez, Ricardo Leon ; Hojo, Toshiaki ; Carvalho, Eden Luiz ; Maezono, Paulo Koiti

DC-DC BOOST CONVERTER BASED MSHE-PWM CASCADED MULTILEVEL INVERTER CONTROL FOR STATCOM SYSTEMS 1283

Law, Kah Haw ; Dahidah, Mohamed S.A.

NOVEL PRINCIPLE FOR FLUX SENSING IN THE APPLICATION OF A DC + AC CURRENT SENSOR 1291

Schrittwieser, L. ; Mauerer, M. ; Bortis, D. ; Ortiz, G. ; Kolar, J.W.

UTILIZING VOLTAGE MEASUREMENT OF FET SWITCH FOR MPPT OF DC ENERGY SOURCE 1299

Kimura, Noriyuki ; Niijima, Koji ; Morizane, Toshimitsu ; Omori, Hideki

HIGH FREQUENCY TRANSFORMER BASED ON A COUPLED INDUCTOR TOPOLOGY WITH DIELECTRIC ISOLATION 1303

Amanci, Adrian Z. ; Dawson, Francis P. ; Ruda, Harry E.

CONCEPT AND EXPERIMENTAL EVALUATION OF A NOVEL DC- 100MHZ WIRELESS OSCILLOSCOPE 1309

Lobsiger, Yanick ; Ortiz, Gabriel ; Bortis, Dominik ; Kolar, Johann W.

INTRODUCTION AND EFFECTIVENESS OF STATCOM TO THE INDEPENDENT POWER SYSTEM OF JR EAST 1317

Omi, Masataro ; Kotegawa, Ryo ; Ando, Masato ; Masui, Takeshi ; Horita, Yasuhisa

THE ANALYSIS OF TIME-VARYING RESONANCES IN THE POWER SUPPLY LINE OF HIGH SPEED TRAINS 1322

Chu, Xi ; Lin, Fei ; Yang, Zhongping

FUZZY FEED-FORWARD CHARGE/DISCHARGE CONTROL OF STATIONARY ENERGY STORAGE SYSTEMS FOR DC ELECTRIC RAILWAYS 1328

Kikuchi, Takuya ; Taga, Hironori ; Takagi, Ryo

TRAIN GROUP CONTROL FOR ENERGY-SAVING DC-ELECTRIC RAILWAY OPERATION 1334

Watanabe, Shoichiro ; Koseki, Takafumi

TRANSFORMER-LESS UNIFIED POWER FLOW CONTROLLER USING THE CASCADE MULTILEVEL INVERTER 1342

Fang Zheng ; Shao Zhang ; Shuitao Yang ; Gunasekaran, Deepak ; Karki, Ujjwal

A NEW POWER FLOW CONTROLLER USING SIX MULTILEVEL CASCADED CONVERTERS FOR DISTRIBUTION SYSTEMS..1350

Tsuruta, Ryoji ; Hosaka, Tatsuya ; Fujita, Hideaki

A PROPOSAL OF MODULAR MULTILEVEL CONVERTER APPLYING THREE WINDING TRANSFORMER..1357

Tamada, Shunsuke ; Nakazawa, Yosuke ; Irokawa, Shoichi

BACK-TO-BACK SYSTEM FOR FIVE-LEVEL CONVERTER WITH COMMON FLYING CAPACITORS................1365

Hasegawa, Isamu ; Urushibata, Shota ; Kondo, Takeshi ; Hirao, Kuniaki ; Kodama, Takashi ; Hui Zhang

HARMONIC MODELING OF A VEHICLE TRACTION CIRCUIT TOWARDS THE DC BUS................................1373

Haghbin, Saeid ; Karvonen, Andreas ; Thiringer, Torbjorn

AC/DC CONVERTER BASED ON INSTANTANEOUS POWER BALANCE CONTROL FOR REDUCING DC-LINK CAPACITANCE..1379

Tokumasu, Akira ; Taki, Hiroshi ; Shirakawa, Kazuhiro ; Wada, Keiji

MODULAR CONVERTER ARCHITECTURE FOR MEDIUM VOLTAGE ULTRA FAST EV CHARGING STATIONS: DUAL HALF-BRIDGE-BASED ISOLATION STAGE..1386

Vasiladiotis, Michail ; Bahrani, Behrooz ; Burger, Niklaus ; Rufer, Alfred

NEW INTERLEAVED CURRENT-FED RESONANT CONVERTER WITH SIGNIFICANTLY REDUCED HIGH CURRENT OUTPUT FILTER FOR EV AND HEV APPLICATION..1394

Moon, Dongok ; Park, Junsung ; Choi, Sewan

15 PHASE INDUCTION MOTOR DRIVE WITH 1:3:5 SPEED RATIOS USING POLE PHASE MODULATION................1400

Umesh B S ; Sivakumar K

MATHEMATICAL MODEL OF NOVEL WOUND-FIELD SYNCHRONOUS MOTOR SELF-EXCITED BY SPACE HARMONICS..1405

Aoyama, Masahiro ; Noguchi, Toshihiko

DUAL PURPOSE NO VOLTAGE WINDING DESIGN FOR THE BEARINGLESS AC HOMOPOLAR AND CONSEQUENT POLE MOTORS..1412

Severson, Eric ; Nilssen, Robert ; Undeland, Tore ; Mohan, Ned

HARVESTING ENERGY FROM SHIP ROLLING USING AN ECCENTRIC DISK REVOLVING IN A HULA-HOOP MOTION..1420

Yu-Jen Wang

LOAD-INDEPENDENT CURRENT OUTPUT OF INDUCTIVE POWER TRANSFER CONVERTERS WITH OPTIMIZED EFFICIENCY..1425

Zhang, Wei ; Wong, Siu-Chung ; Tse, Chi K. ; Chen, Qianhong

VOLTAGE CONTROL OF INDUCTIVE CONTACTLESS POWER TRANSFER SYSTEM WITH COAXIAL CORELESS TRANSFORMER FOR DC POWER DISTRIBUTION..1430

Miiura, Yushi ; Ojika, Satoshi ; Ise, Tomofumi

CONTACTLESS HIGH POWER TRANSFORMER TECHNOLOGIES FOR RAILWAY VEHICLES................1438

Kondo, Keiichiro ; Yamamoto, Kohei ; Kitazawa, Satochi

TWO-SWITCH VOLTAGE EQUALIZER BASED ON HALF-BRIDGE CONVERTER WITH MULTI-STACKED CURRENT DOUBLERS FOR SERIES-CONNECTED BATTERIES..1444

Uno, Masatoshi ; Kukita, Akio

OPTIMAL ENERGY STORAGE SYSTEM PLANNING FOR MICROGRIDS WITH CONTRACT CAPACITY CONSTRAINT..1452

Shu-Hung Liao ; Jen-Hao Teng ; Yung-Ching Huang ; Dong-Jing Lee

OPTIMAL ZERO SEQUENCE INJECTION IN MULTILEVEL CASCADED H-BRIDGE CONVERTER UNDER UNBALANCED PHOTOVOLTAIC POWER GENERATION..1458

Yu, Yifan ; Konstantinou, Georgios ; Hredzak, Branislav ; Agelidis, Vassilios G.

SIMPLE METHOD FOR MEASURING OUTPUT IMPEDANCE OF A THREE-PHASE INVERTER IN DQ-DOMAIN..1466

Jokipii, Juha ; Messo, Tuomas ; Suntio, Teuvo

ANALYSIS AND DESIGN OF POWER MANAGEMENT SCHEME FOR AN ON-BOARD SOLAR ENERGY STORAGE SYSTEM..1471

Jiang, W. ; Yu, F.Y. ; Lin, Z.Y. ; Wu, G.F. ; Chen, H. ; Hashimoto, S

LVRT CONTROL STRATEGY OF CSC-DPMSG-WGS UNDER UNBALANCED GRID FAULTS................1476

Meiqin Mao ; Yong Ding ; Shiting Weng ; Liuchen Chang

A NEW CURRENT CONTROL DROOP STRATEGY FOR VSI-BASED ISLANDED MICROGRIDS................1482

Shoeiby, B. ; Davoodnezhad, R. ; Holmes, D.G. ; McGrath, B.P.

POWER EXCHANGE USING PFC FOR MICRO GRID..1490

Sakai, Tomoyasu ; Takeda, Takashi ; Yukita, Kazuto ; Goto, Yasuyuki ; Ichiyanagi, Katsuhiro ; Morita, Hiroshi

DETERMINATION OF ROTOR TEMPERATURE FOR AN INTERIOR PERMANENT MAGNET SYNCHRONOUS MACHINE USING A PRECISE FLUX OBSERVER..1501

Specht, Andreas ; Wallscheid, Oliver ; Bocker, Joachim

MONITORING CRITICAL TEMPERATURES IN PERMANENT MAGNET SYNCHRONOUS MOTORS USING LOW-ORDER THERMAL MODELS..1508

Huber, Tobias ; Peters, Wilhelm ; Bocker, Joachim

ROBUST CURRENT CONTROL INSENSITIVE TO GAIN DEVIATION AND OFFSET OF INVERTER DC-LINK CURRENT SENSOR FOR SPMSM..1516

Matsuura, Kei ; Ando, Itaru ; Ohishi, Kiyoshi ; Matsuhashi, Masataka

AUTO-TUNING METHOD OF INDUCTANCES FOR PERMANENT MAGNET SYNCHRONOUS MOTORS................1522

Nomura, Naofumi ; Higuchi, Shinichi

AN IMPEDANCE-BASED STABILITY ANALYSIS METHOD FOR PARALLELED VOLTAGE SOURCE CONVERTERS 1529
Wang, Xiongfei ; Blaabjerg, Frede ; Loh, Poh Chiang

DYNAMIC CHARACTERISTICS AND STABILITY COMPARISONS BETWEEN VIRTUAL SYNCHRONOUS GENERATOR AND DROOP CONTROL IN INVERTER-BASED DISTRIBUTED GENERATORS 1536
Jia Liu ; Miura, Yushi ; Ise, Toshifumi

EMBEDDED LIMITATIONS AND PROTECTIONS FOR DROOP-BASED CONTROL SCHEMES WITH CASCADED LOOPS IN THE SYNCHRONOUS REFERENCE FRAME 1544
D'Arco, Salvatore ; Guidi, Giuseppe ; Suul, Jon Are

VIRTUAL SYNCHRONOUS GENERATOR CONTROL WITH DOUBLE DECOUPLED SYNCHRONOUS REFERENCE FRAME FOR SINGLE-PHASE INVERTER 1552
Hirase, Yuko ; Noro, Osamu ; Yoshimura, Eiji ; Nakagawa, Hidehiko ; Sakimoto, Kenichi ; Shindo, Yuji

CONTACTLESS DC CONNECTOR BASED ON GAN LLC CONVERTER FOR NEXT GENERATION DATA CENTERS 1560
Hayashi, Yusuke ; Toyoda, Hajime ; Ise, Toshifumi ; Matsumoto, Akira

ANALYSIS OF MIS-INTERRUPTION OF SEMICONDUCTOR BREAKER IN DC POWER FEEDING SYSTEM 1567
Murai, Kensuke ; Kanai, Yasuyuki ; Asakimori, Koki ; Babasaki, Tadatoshi

A RELIABLE ELECTRONIC CHOKE WITH NO NEED OF GAIN ADJUSTMENT FOR WIRE COMMUNICATION SYSTEM 1575
Katsuki, Akihiko ; Nakamura, Tatsuya ; Mizuki, Tatsuya ; Shibahara, Kohei ; Abe, Tomohiko ; Ikeda, Tomohiko ; Maeyama, Shigetaka

DESIGN OF NEW CONTROL STRATEGIES FOR A FOUR-LEG THREE-PHASE INVERTER TO ELIMINATE THE NEUTRAL CURRENT UNDER UNBALANCED LOADS 1580
Zhao-Qin Guo ; Panda, Sanjib Kumar ; Prasanna, I.V.

RESEARCH TRENDS OF MODULAR MULTILEVEL CASCADE INVERTER (MMCI-DSCC)-BASED MEDIUM-VOLTAGE MOTOR DRIVES IN A LOW-SPEED RANGE 1586
Okazaki, Yuhei ; Matsui, Hitoshi ; Hagiwara, Makoto ; Akagi, Hirofumi

AN INPUT SWITCHED MULTILEVEL INVERTER FOR OPEN-END WINDING INDUCTION MOTOR DRIVE 1594
Zhu, B. ; Jia, Y. ; Prasanna, U.R. ; Rajashekara, K. ; Kubo, H.

VARIABLE CARRIER FREQUENCY MIXED PWM TECHNIQUE BASED ON CURRENT RIPPLE PREDICTION FOR REDUCED SWITCHING LOSS 1601
Kubo, Hajime ; Yamamoto, Yasuhiro

SLIDING MODE PWM FOR EFFECTIVE CURRENT CONTROL IN SWITCHED RELUCTANCE MACHINE DRIVES 1606
Manolas, Iakovos ; Papafotiou, Georgios ; Manias, Stefanos N.

EXPERIMENTAL VERIFICATION OF AN EMC FILTER USED FOR PWM INVERTER WITH WIDE BAND-GAP DEVICES 1613
Itoh, Jun-ichi ; Araki, Takahiro ; Orikawa, Koji

PACKAGING FOR SIC POWER DEVICE 1621
Funaki, Tsuyoshi

SOLID STATE TRANSFORMER AND MV GRID TIE APPLICATIONS ENABLED BY 15 KV SIC IGBTS AND 10 KV SIC MOSFETS BASED MULTILEVEL CONVERTERS 1626
Madhusoodhanan, Sachin ; Tripathi, Awneesh ; Patel, Dhaval ; Mainali, Krishna ; Kadavelugu, Arun ; Hazra, Samir ; Bhattacharya, Subhashish ; Hatua, Kamalesh

VOLUME 3

GENERALIZED MODULAR MULTILEVEL CONVERTER AND MODULATION 1634
Hui Liu ; Loh, Poh Chiang ; Blaabjerg, Frede

AVERAGE POWER CONTROL OF DC BUS VOLTAGES OF CASCADED H-BRIDGE MULTILEVEL CONVERTERS 1639
Lee, Chia-Tse ; Chen, Hsin-Chih ; Ching-Wei Wang ; Ching-Hsiang Yang ; Cheng, Po-Tai

ANALYSIS AND COMPARISON OF HIGH POWER SEMICONDUCTOR DEVICE LOSSES IN 5MW PMSG MV WIND TURBINES 1646
Kihyun Lee ; Kyungsub Jung ; Seunghoo Song ; Suh, Yongsug ; Changwoo Kim ; Hyoyol Yoo ; Sunsoon Park

APPLICATION OF MODULAR MATRIX CONVERTER TO WIND TURBINE GENERATOR 1654
Inomata, Kentaro ; Hara, Hidenori ; Morimoto, Shinya ; Fujii, Junji ; Takeda, Kotaro ; Yamamoto, Eiji

FREE MOTION MECHANICAL POWER FACTOR; COMPARISON BETWEEN ROBOTS IN DIFFERENT STRUCTURE AND COORDINATE 1660
Mizoguchi, Takahiro ; Nozaki, Takahiro ; Ohnishi, Kouhei

ANALYSIS OF SETTLING BEHAVIOR AND DESIGN OF CASCADED PRECISE POSITIONING CONTROL IN PRESENCE OF NONLINEAR FRICTION 1665
Ruderman, Michael ; Iwasaki, Makoto

FIELD AND BENCH TEST EVALUATION OF RANGE EXTENSION CONTROL SYSTEM FOR ELECTRIC VEHICLES BASED ON FRONT AND REAR DRIVING-BRAKING FORCE DISTRIBUTIONS 1671
Fujimoto, Hiroshi ; Harada, Shingo ; Goto, Yuichi ; Kawano, Daisuke ; Sato, Koji ; Matsuo, Yusuke

VIBRATION SUPPRESSION OF INTEGRATED RESONANT AND TIME DELAY SYSTEM BY REFLECTED WAVE REJECTION 1679
  Saito, Eiichi ; Oboe, Roberto ; Katsura, Seiichiro

THRUST CHARACTERISTICS IMPROVEMENT OF A CIRCULAR SHAFT MOTOR FOR DIRECT-DRIVE APPLICATIONS 1685
  Omura, Mototsugu ; Shimono, Tomoyuki ; Fujimoto, Yasutaka

DESIGN OF A BEARINGLESS FLUX-SWITCHING SLICE MOTOR 1691
  Gruber, Wolfgang ; Radman, Karlo ; Schob, Reto.T.

PROPOSAL OF A PERMANENT MAGNET HYBRID TYPE AXIAL MAGNETICALLY LEVITATED MOTOR 1697
  Kurita, Nobuyuki ; Ishikawa, Takeo ; Takada, Hiromu ; Suzuki, Genri

COMPARISON OF HIGH SPEED BEARINGLESS DRIVE TOPOLOGIES WITH COMBINED WINDINGS 1701
  Mitterhofer, Hubert ; Mrak, Branimir ; Gruber, Wolfgang

HIGH-SPEED MAGNETICALLY LEVITATED REACTION WHEEL DEMONSTRATOR 1707
  Zwyssig, Christof ; Baumgartner, Thomas ; Kolar, Johann W.

STABILIZED SUSPENSION CONTROL CONSIDERING ARMATURE REACTION IN A D-Q AXIS CURRENT CONTROL BEARINGLESS MOTOR 1715
  Ooshima, Masahide ; Kumakura, Yoshito

ANALYSIS AND DESIGN OF A HIGH-FREQUENCY ISOLATED DUAL-TANK LCL RESONANT AC-DC CONVERTER 1721
  Du, Yimian ; Bhat, Ashoka K.S.

VERIFICATION OF LLC RESONANT CONVERTER APPLIED A CURRENT-BALANCING HIGH-FREQUENCY TRANSFORMER WITH MULTI-OUTPUT WINDINGS 1728
  Araki, Jun ; Shinozaki, Ikki ; Funato, Hirohito ; Ogasawara, Satoshi ; Murakami, Daichi ; Hirota, Yukitsugu ; Mihara, Teruyoshi ; Mouri, Masayuki ; Okazaki, Fumihiro

LIGHT-LOAD EFFICIENCY IMPROVEMENT STRATEGY FOR LLC RESONANT CONVERTER UTILIZING A STEP-GAP TRANSFORMER 1734
  Huang, Wen-Nan ; Lee, Shiu-Hui ; Chen, Ching-Guo

A NOVEL ACCURATE PRIMARY SIDE CONTROL (PSC) METHOD FOR HALF-BRIDGE (HB) LLC CONVERTER 1738
  Jae-Bum Lee ; Kim, Chong-Eun ; Jae-Hyun Kim ; Cheol-O Yeon ; Young-Do Kim ; Moon, Gun-Woo

A SIMPLE CONTROL SCHEME FOR IMPROVING LIGHT-LOAD EFFICIENCY IN A FULL-BRIDGE LLC RESONANT CONVERTER 1743
  Kim, Jae-Hyun ; Kim, Chong-Eun ; Lee, Jae-Bum ; Young-Do Kim ; Han-Shin Youn ; Moon, Gun-Woo

POWER CONDITIONER FOR STABILIZING POWER DISTURBANCE CAUSED OF WIND TURBINE GENERATOR SYSTEM 1748
  Saga, Yasunao ; Fujii, Kansuke ; Yoda, Kazuyuki

A FRONT-TO-FRONT (FTF) SYSTEM CONSISTING OF MULTIPLE MODULAR MULTILEVEL CASCADE CONVERTERS FOR OFFSHORE WIND FARMS 1761
  Sasongko, Firman ; Hagiwara, Makoto ; Akagi, Hirofumi

MODELLING, DESIGN AND CONTROL OF GRID CONNECTED CONVERTER FOR HIGH ALTITUDE WIND POWER APPLICATION 1775
  Adhikari, Jeevan ; Rathore, Akshay K. ; Panda, S K

PRACTICAL STUDY OF A HIGH STEP-DOWN CONVERTER 1781
  Jinno, Masahito ; Su, Hong-Wei ; Tsai, Jiung-Lin ; Matsuo, Hirofumi

GENERALIZED MODELING AND OPTIMIZATION OF A BIDIRECTIONAL DUAL ACTIVE BRIDGE DC-DC CONVERTER INCLUDING FREQUENCY VARIATION 1788
  Jauch, Felix ; Biela, Jurgen

BALANCED DISCHARGING OF POWER BANK WITH BUCK-BOOST BATTERY POWER MODULES 1796
  Moo, Chin-Sien ; Wu, Tsung-Hsi ; Hou, Chih-Hao ; Hsieh, Yao-Ching

Y-SOURCE IMPEDANCE-NETWORK-BASED ISOLATED BOOST DC/DC CONVERTER 1801
  Siwakoti, Yam P. ; Town, Graham E. ; Loh, Poh Chiang ; Blaabjerg, Frede

MULTI-PHASE DC-DC CONVERTER WITH RIPPLE-LESS OPERATION FOR THERMO-ELECTRIC GENERATOR 1806
  Kimura, Noriyuki ; Niijima, Koji ; Morizane, Toshimitsu ; Omori, Hideki

POSITION SENSORLESS START-UP METHOD OF SURFACE PERMANENT MAGNET SYNCHRONOUS MOTOR USING NONLINEAR ROTOR POSITION OBSERVER 1811
  Hanamoto, Tsuyoshi ; Yamada, Hiroaki ; Okuyama, Yoshihiro

SENSORLESS CONTROL OF PMSM FOR THE WHOLE SPEED RANGE USING TWO-DEGREE-OF-FREEDOM CURRENT CONTROL AND HF TEST CURRENT INJECTION FOR LOW SPEED RANGE 1816
  Seilmeier, Markus ; Piepenbreier, Bernhard

ELLIPSE-TRAJECTORY-ORIENTED VECTOR CONTROL FOR ENERGY EFFICIENT/WIDE-SPEED-RANGE DRIVES OF SENSORLESS PMSM 1824
  Shinnaka, Shinji ; Amano, Yuki

DEVELOPMENT OF POSITION SENSORLESS CONTROL FOR PERMANENT-MAGNET SYNCHRONOUS GENERATOR DRIVE 1832
  Chang, Yuan-Chih ; Lin, Chia-Yu ; Dai, Wei-Fu ; Wu, Chun-Wei

CONTROL OF A 750KW PERMANENT MAGNET SYNCHRONOUS MOTOR 1837
  Liping Zheng ; Dong Le

REGIONAL SMART GRID OF ISLAND IN CHINA WITH MULTIFOLD RENEWABLE ENERGY 1842
Xu Cai ; Zheng Li

STABILIZING SMALL ISLAND POWER SYSTEM WITH RENEWABLES BY USE OF POWER CONDITIONING SYSTEMS - JAPANESE ISLAND SYSTEM CASE - 1849
Baba, Jumpei

POWER ELECTRONICS SOLUTIONS APPLIED TO A VARIETY OF DEMONSTRATIVE MICROGRID PROJECTS 1855
Ueda, Yoshinobu

MOVING TOWARDS THE SMART GRID: THE NORWEGIAN CASE 1861
Fosso, Olav B. ; Molinas, Marta ; Sand, Kjell ; Coldevin, Grete H.

POWER ELECTRONICS TECHNOLOGY IN SMART GRID PROJECTS -APPLICATIONS AND EXPERIENCES- 1868
Kobayashi, Takenori

EV AND HEV MOTOR DEVELOPMENT IN TOSHIBA 1874
Arata, Masanori ; Kurihara, Yoshihiro ; Misu, Daisuke ; Matsubara, Masakatsu

MOTOR STATOR WITH THICK RECTANGULAR WIRE LAP WINDING FOR HEVS 1880
Ishigami, Takashi ; Tanaka, Yuichiro ; Homma, Hiroshi

COMPARISON STUDY OF VARIOUS MOTORS FOR EVS AND THE POTENTIALITY OF A FERRITE MAGNET MOTOR 1886
Matsuhashi, Daiki ; Matsuo, Keisuke ; Okitsu, Takashi ; Ashikaga, Tadashi ; Mizuno, Takayuki

OPTIMAL FIELD EXCITATION CONTROL OF A CLAW POLE MOTOR FOR HYBRID ELECTRIC VEHICLE 1892
Azuma, M. ; Hazeyama, M. ; Morita, M. ; Kuroda, Y. ; Daikoku, A. ; Inoue, M.

A WIDE SPEED RANGE HIGH EFFICIENCY EV DRIVE SYSTEM USING WINDING CHANGEOVER TECHNIQUE AND SIC DEVICES 1898
Takatsuka, Yushi ; Hara, Hidenori ; Yamada, Kenji ; Maemura, Akihiko ; Kume, Tsuneo

PERFORMANCE COMPARISON OF A GAN GIT AND A SI IGBT FOR HIGH-SPEED DRIVE APPLICATIONS 1904
Tuysuz, Arda ; Bosshard, Roman ; Kolar, Johann W.

WIDE-BAND GAP DEVICES IN PV SYSTEMS - OPPORTUNITIES AND CHALLENGES 1912
Sintamarean, C. ; Eni, E. ; Blaabjerg, F. ; Teodorescu, R. ; Wang, H.

POWER ELECTRONICS EQUIPMENTS APPLYING NOVEL SIC POWER SEMICONDUCTOR MODULES 1920
Mino, Kazuaki ; Yamada, Ryuji ; Kimura, Hiroshi ; Matsumoto, Yasushi

EMI PREDICTION METHOD FOR SIC INVERTER BY THE MODELING OF STRUCTURE AND THE ACCURATE MODEL OF POWER DEVICE 1929
Maekawa, Sari ; Tsuda, Junichi ; Kuzumaki, Atsuhiko ; Matsumoto, Shuhei ; Mochikawa, Hiroshi ; Kubota, Hisao

SYSTEM INTEGRATION OF GAN TECHNOLOGY 1935
Ferreira, J.A. ; Popovic, J. ; van Wyk, J.D. ; Pansier, F.

POWER LOSSES OF MULTILEVEL CONVERTERS IN TERMS OF THE NUMBER OF THE OUTPUT VOLTAGE LEVELS 1943
Kashihara, Yugo ; Itoh, Jum-ichi

A LARGE CAPACITY 3-LEVEL IEGT INVERTER 1950
Yoshizawa, Daisuke ; Mukunoki, Makoto ; Omote, Kenichiro ; Hayashi, Makoto ; Isida, Takashi

VIBRATION SUPPRESSING CONTROL METHOD OF ANGULAR TRANSMISSION ERROR OF CYCLOID GEAR FOR INDUSTRIAL ROBOTS 1956
Yoshioka, Takashi ; Hirano, Yosei ; Ohishi, Kiyoshi ; Miyazaki, Toshimasa ; Yokokura, Yuki

AN ADVANCED POSITION CONTROL OF OVERHEAD CRANE BY SWAY SUPPRESSION METHOD EMULATING NATURAL DAMPING 1962
Kurabayashi, Toshiyuki ; Yang Chuan ; Murakami, Toshiyuki

A ROBOTIC CANE FOR WALKING ASSISTANCE 1968
Shimizu, Kyohei ; Smadi, Issam ; Fujimoto, Yasutaka

HAND POSITION ESTIMATION IN BINOCULAR VISUAL SPACE USING LINEAR APPROXIMATION OF KINEMATICS 1974
Komada, Satoshi ; Turpin, Santiago ; Hashimoto, Kento ; Yashiro, Daisuke ; Hirai, Junji

CONTACT STATE RECOGNITION BASED ON HAPTIC SIGNAL PROCESSING FOR ROBOTIC TOOL USE 1978
Matsuzaki, Ryohei ; Okuma, Jun ; Sakaino, Sho ; Tsuji, Toshiaki

RECENT TECHNICAL TRENDS IN MAGNETIC MATERIALS 1984
Wajima, Kiyoshi ; Toda, Hiroaki ; Kosaka, Takashi ; Marukawa, Yasuhiro ; Ishihara, Chio

MULTI-DOMAIN CO-SIMULATION WITH NUMERICALLY IDENTIFIED PMSM INTERWORKING AT HILS FOR ELECTRIC PROPULSION 1990
Park, Gyeong-Jae ; Jung, Hochang ; Kim, Yong-Jae ; Jung, Sang-Yong

RECENT TECHNICAL TRENDS IN PMSM 1997
Morimoto, Shigeo ; Asano, Yoshinari ; Kosaka, Takashi ; Enomoto, Yuji

RECENT TECHNICAL TRENDS IN SRM AND FSM 2004
Kano, Yoshiaki

RECENT TECHNICAL TRENDS IN VARIABLE FLUX MOTORS 2011
Toba, Akio ; Daikoku, Akihiro ; Nishiyama, Noriyoshi ; Yoshikawa, Yuichi ; Kawazoe, Yosuke

A GENERAL DISCRETE TIME MODEL TO EVALUATE ACTIVE DAMPING OF GRID CONVERTERS WITH LCL FILTERS 2019

Parker, S.G. ; McGrath, B.P. ; Holmes, D.G.

ANALYSIS AND REDUCTION OF POWER LOSSES IN PV CONVERTERS FOR GRID CONNECTION TO LOW-VOLTAGE THREE-PHASE THREE-WIRE SYSTEMS 2027

Amma, Ryosuke ; Fujita, Hideaki

DESIGN OF GRID CONNECTED PWM CONVERTERS CONSIDERING TOPOLOGY AND PWM METHODS FOR LOW-VOLTAGE RENEWABLE ENERGY APPLICATIONS 2034

Kantar, Emre ; Hava, Ahmet M.

PERFORMANCE OF DEAD TIME COMPENSATION METHODS IN THREE-PHASE GRID-CONNECTION CONVERTERS 2042

Mannen, Tomoyuki ; Fujita, Hideaki

D-S DIGITAL CONTROL FOR THREE-PHASE BI-DIRECTIONAL INVERTERS 2050

Wu, T.-F. ; Chang, C.-H. ; Lin, L.-C.

EXPECTATIONS OF NEXT-GENERATION POWER DEVICES FOR HOME AND CONSUMER APPLIANCES 2058

Kanouda, Akihiko ; Shoji, Hiroyuki ; Shimada, Takae ; Okubo, Toshikazu

APPLICATION TREND AND FORESIGHT OF SIC POWER DEVICES TO AIR CONDITIONERS 2064

Kamikura, Mamoru ; Murata, Yuichiro ; Kutsuki, Tomohiro ; Saito, Katsuhiko

RECENT TECHNICAL TRENDS AND FUTURE PROSPECTS OF IGBTS AND POWER MOSFETS 2068

Ogura, Tsuneo

RECENT DEVELOPMENT AND FUTURE PROSPECTS OF POWER SIC DEVICES 2074

Nakamura, T. ; Nakano, Y. ; Aketa, M. ; Hanada, T.

RECENT ADVANCES AND FUTURE PROSPECTS ON GAN-BASED POWER DEVICES 2075

Ueda, Tetsuzo

SCALING AND BALANCING OF MULTI-CELL CONVERTERS 2079

Kasper, Matthias ; Bortis, Dominik ; Kolar, Johann W.

HYBRID MODULATED UNIVERSAL SOFT-SWITCHING CURRENT-FED DC/DC CONVERTER FOR WIDE VOLTAGE REGULATION FOR PV/FUEL CELLS/BATTERY APPLICATIONS 2087

Moorthy, Radha Sree Krishna ; Rathore, Akshay Kumar

HIGH EFFICIENCY POWER CONVERTERS FOR BATTERY ENERGY STORAGE SYSTEMS 2095

Kawakami, Noriko ; Iijima, Yukihia ; Li, Haiqing ; Ota, Satoru

IMPLEMENTATION OF BRIDGELESS CUK POWER FACTOR CORRECTOR WITH POSITIVE OUTPUT VOLTAGE 2100

Yang, Hong-Tzer ; Chiang, Hsin-Wei

A NOVEL SYNCHRONOUS RECTIFIER METHOD FOR A LLC RESONANT CONVERTER WITH VOLTAGE-DOUBLER RECTIFIER 2108

Murata, Koji ; Kurokawa, Fujio

LATEST DEVELOPMENTS IN INCREASING THE POWER DENSITY OF TRACTION DRIVES 2113

Bakran, Mark-M. ; Marz, Andreas ; Laska, Bernd ; Krafft, Eberhard ; Korner, Olaf ; Nagel, Andreas

CATENARY AND STORAGE BATTERY HYBRID SYSTEM FOR ELECTRIC RAILCAR SERIES EV-E301 2120

Kono, Y. ; Shiraki, N. ; Yokoyama, H. ; Furuta, R.

TECHNOLOGY FOR ENERGY-SAVING RAILWAY OPERATION THROUGH POWER-LIMITING BRAKES—A CASE STUDY AT AN URBAN RAILWAY 2126

Koseki, Takafumi ; Watanabe, Shoichiro ; Hamazaki, Yasuhiro ; Kondo, Keiichiro ; Hasegawa, Tomonori ; Mizuma, Takeshi

AN OVERVIEW ON BRAKING ENERGY REGENERATION TECHNOLOGIES IN CHINESE URBAN RAILWAY TRANSPORTATION 2133

Yang, Zhongping ; Xia, Huan ; Wang, Bin ; Lin, Fei

TRACTION INVERTER THAT APPLIES COMPACT 3.3 KV / 1200 A SIC HYBRID MODULE 2140

Ishikawa, Katsumi ; Yukutake, Seigo ; Kono, Yasuhiko ; Ogawa, Kazutoshi ; Kameshiro, Norifumi

POWER ELECTRONIC-BASED PROTECTION FOR DIRECT-CURRENT POWER DISTRIBUTION IN MICRO-GRIDS 2145

Tseng, K.J. ; Luo, Guomin

A CONCEPT OF HIGH POWER DC/DC CONVERTER WITH DOUBLE LOW POWER OUTPUTS 2152

Hojo, Masahide ; Nishioka, Tomoya ; Yamanaka, Kenji

PERFORMANCE EVALUATION FOR GRID IMPEDANCE BASED ISLANDING DETECTION METHOD 2156

Liu, Ning ; Aljankawey, A.S. ; Diduch, C.P. ; Chang, L. ; Mao, Meiqin ; Yazdkhasti, Pegah ; Su, Jianhui

IDENTIFYING NATURAL DEGRADATION/AGING IN POWER MOSFETS IN A LIVE GRID-TIED PV INVERTER USING SPREAD SPECTRUM TIME DOMAIN REFLECTOMETRY 2161

Li, Qian ; Khan, Faisal H.

CONTROL METHOD FOR INDUCTIVE POWER TRANSFER WITH HIGH PARTIAL-LOAD EFFICIENCY AND RESONANCE TRACKING 2167

Bosshard, R. ; Kolar, J.W. ; Wunsch, B.

STANDARD MODELS FOR SMART GRID SIMULATIONS 2175

Noda, Taku ; Nagashima, Tomohiro ; Sekisue, Takayuki ; Kabasawa, Yuichiro ; Kato, Shinji ; Sekiba, Yoichi ; Tokuda, Hirokazu ; Kounoto, Masaaki

MODEL DEVELOPMENT FOR MOTOR DRIVE SYSTEM SIMULATIONS 2183

Ishikawa, Hiroki ; Abe, Takashi ; Kato, Toshiji ; Kubota, Yutaka ; Shimomura, Junichi ; Kohno, Yusuke ; Ikeda, Masahiro ; Umeda, Nobuhiro ; Kimura, Noriyuki ; Shigematsu, Koichi ; Inoue, Yukinori

PRACTICAL SIMULATION EXAMPLES OF AUTOMOTIVE AND POWER SUPPLY SYSTEMS...........2189
Abe, Takashi ; Fukushima, Kentaro ; Sekisue, Takayuki ; Shigematsu, Koichi ; Ichihara, Junichi ; Kato, Toshiji ; Ishikawa, Hiroki ; Kouno, Yusuke ; Konoto, Masaaki ; Saito, Ryoji ; Nishida, Yasuyuki

ADMITTANCE MATRICES OF VOLTAGE SOURCE CONVERTERS FOR DISTRIBUTED GENERATORS...........2195
Lian, K.L. ; Huang, T.D.

FPGA-BASED SIMULATION OF POWER ELECTRONICS USING ITERATIVE METHODS...........2202
Zhang, Huiguo ; Sun, Jian

GALLIUM ARSENIDE IC TECHNOLOGY FOR POWER SUPPLIES ON CHIP...........2208
Pala, Vipindas ; Peng, Han ; Hella, Mona ; Chow, T.Paul

SILICON ON NANOCRYSTALLINE AND MICROCRYSTALLINE DIAMOND STACKING STRUCTURE FOR POWER SUPPLY ON CHIP...........2212
Yamada, Takatoshi ; Hasegawa, Masataka

A NOVEL LOAD REGULATION TECHNIQUE FOR POWER-SOC WITH PARALLEL CONNECTED POLS...........2216
Abe, Seiya ; Matsumoto, Satoshi ; Hidaka, Akira ; Rikitake, Jungo ; Ninomiya, Tamotsu

MATRIX-POL ARCHITECTURE FOR INTEGRATED POWER SUPPLY...........2222
Ishizuka, Yoichi ; Shibahara, Ryota ; Ninomiya, Tamotsu ; Tanaka, Kiminori ; Abe, Seiya

ON-CHIP BUCK CONVERTER WITH SPIRAL FERRITE INDUCTOR AND REDUCING IR DROP IN 3D STACKED INTEGRATION...........2228
Fuketa, Hiroshi ; Shinozuka, Yasuhiro ; Ishida, Koichi ; Takamiya, Makoto ; Sakurai, Takayasu

DCM ANALYSIS OF A SINGLE SIC SWITCH BASED ZVZCS TAPPED BOOST CONVERTER...........2232
Choi, Bo H. ; Lee, Eun S. ; Kim, Ji H. ; Rim, Chun T.

EFFECT OF INPUT AND OUTPUT TERMINAL SOURCES ON DYNAMIC BEHAVIOR OF SWITCHED-MODE CONVERTERS...........2240
Suntio, T. ; Viinamaki, J. ; Jokipii, J. ; Messo, T. ; Sitbon, M. ; Kuperman, A.

A FULLY SOFT-SWITCHED MULTIPHASE DC-DC CONVERTER WITH REDUCED SWITCH COUNT FOR HIGH POWER APPLICATION...........2247
Kim, Minjae ; Yang, Daeki ; Choi, Sewan

A STATIC CHARACTERISTIC ANALYSIS OF PROPOSED BI-DIRECTIONAL DUAL ACTIVE BRIDGE DC-DC CONVERTER...........2252
Nagata, Shun ; Takasaki, Mika ; Furukawa, Yutaka ; Hirose, Toshiro ; Ishizuka, Yoichi

HYBRID BATTERY CHARGING SYSTEM COMBINING OBC WITH LDC FOR ELECTRIC VEHICLES...........2260
Kim, Seonghye ; Kang, Feel-soon

TRANSIENT BEHAVIOR OF THE DUAL ACTIVE BRIDGE CONVERTER IN HIGH EFFICIENT ENERGY CONVERSION SYSTEM...........2266
Aoyama, Kohei ; Motoi, Naoki ; Tsuruta, Yukinori ; Kawamura, Atsuo

STATE-OF-CHARGE ESTIMATION FOR LITHIUM-ION BATTERY PACK USING RECONSTRUCTED OPEN-CIRCUIT-VOLTAGE CURVE...........2272
Chun, Chang Yoon ; Seo, Gab-Su ; Yoon, Sung Hyun ; Cho, Bo-Hyung

SYSTEM DESIGN OF ELECTRIC ASSISTED BICYCLE USING EDLCS AND WIRELESS CHARGER...........2277
Itoh, Jun-ichi ; Noguchi, Kenji ; Orikawa, Koji

STUDY ON LOW-LOSS GATE DRIVE CIRCUIT FOR HIGH EFFICIENCY SERVER POWER SUPPLY USING NORMALLY-OFF SIC-JFET...........2285
Katoh, Kaoru ; Ishikawa, Katsumi ; Hatanaka, Ayumu ; Ogawa, Kazutoshi ; Akiyama, Satoru ; Ogawa, Takashi ; Yokoyama, Natsuki ; Maru, Naoki ; Takahashi, Osamu ; Nishisu, Koji

A SHORT CIRCUIT PROTECTION METHOD BASED ON A GATE CHARGE CHARACTERISTIC...........2290
Horiguchi, Takeshi ; Kinouchi, Shin-ichi ; Nakayama, Yasushi ; Oi, Takeshi ; Urushibata, Hiroaki ; Okamoto, Shoji ; Tominaga, Shinji ; Akagi, Hirofumi

HIGHLY RELIABLE 1200-V P-TYPE MOSFET FOR LEVEL-SHIFT CIRCUIT USED IN DRIVER IC...........2297
Sakurai, Naoki ; Hakutou, Takuma ; Yura, Masashi

A NEW LEVEL UP SHIFTER FOR HVICS WITH HIGH NOISE TOLERANCE...........2302
Akahane, Masashi ; Jonishi, Akihiro ; Yamaji, Masaharu ; Kanno, Hiroshi ; Tanaka, Takahide ; Nishio, Haruhiko ; Sumida, Hitoshi

OUTPUT RIPPLE MINIMIZATION OF SINGLE-STAGE POWER FACTOR CORRECTED BI-DIRECTIONAL BUCK AC/DC CONVERTER...........2310
Veerasamy, Balaji ; Kitagawa, Wataru ; Takeshita, Takaharu

THREE-PHASE ISOLATED FULL-BRIDGE BOOST PFC WITH FLYBACK PASSIVE AUXILIARY CONVERTER...........2318
Meng, Tao ; Yu, Shuai ; Ben, Hongqi ; Wei, Guo ; Sun, Shaohua

CONTROL AND EXPERIMENT OF A MODULAR PUSH-PULL PWM CONVERTER FOR A BATTERY ENERGY STORAGE SYSTEM...........2323
Hagiwara, Makoto ; Akagi, Hirofumi

ACTIVE FRONT-END TOPOLOGY FOR 5 LEVEL MEDIUM VOLTAGE DRIVE SYSTEM WITH ISOLATED DC BUS...........2330
Oka, Toshiaki ; Kusunoki, Hironobu ; Tsukakoshi, Masahiko ; Kleinecke, John ; Daskalos, Mike

A DUAL ACTIVE BRIDGE DC-DC CONVERTER WITH OPTIMAL DC-LINK VOLTAGE SCALING AND FLYBACK MODE FOR ENHANCED LOW-POWER OPERATION IN HYBRID PV/STORAGE SYSTEMS...........2336
Poshtkouhi, Shahab ; Trescases, Olivier

NOVEL MODULAR MULTIPLE-INPUT BIDIRECTIONAL DC-DC POWER CONVERTER (MIPC)...........2343
Hintz, Andrew ; Prasanna, Udupi.R. ; Rajashekara, Kaushik

SINGLE-SWITCH PWM CONVERTER INTEGRATING VOLTAGE EQUALIZER FOR PHOTOVOLTAIC MODULES UNDER PARTIAL SHADING............2351

Uno, Masatoshi ; Kukita, Akio

NEW DC RAIL SIDE SOFT-SWITCHING PWM DC-DC CONVERTER WITH VOLTAGE DOUBLER RECTIFIER FOR PV GENERATION INTERFACE............2359

Sayed, Khairy ; Kwon, Soon-Kurl ; Nishida, Katsumi ; Nakaoka, Mutsuo

MODELING METHOD OF STRAY MAGNETIC COUPLINGS IN AN EMC FILTER FOR A SIC SOLAR INVERTER............2366

Masuzawa, Takashi ; Hoene, Eckart ; Hoffmann, Stefan ; Lang, Klaus-Dieter

DC BUS VOLTAGE EMI MITIGATION IN THREE-PHASE ACTIVE RECTIFIERS USING A VIRTUAL NEUTRAL FILTER............2372

Parker, S.G. ; Segaran, D.S. ; Holmes, D.G. ; McGrath, B.P.

EFFECTS OF TRANSFORMER STRUCTURES ON THE NOISE BALANCING AND CANCELLATION MECHANISMS OF SWITCHING POWER CONVERTERS............2380

Hsieh, Hung-I ; Shih, Sheng-Fang

A NOVEL TECHNIQUE FOR REDUCING LEAKAGE CURRENT BY APPLICATION OF ZERO-SEQUENCE VOLTAGE............2385

Ayano, Hideki ; Murakami, Kouhei ; Matsui, Yoshihiro

AC-CHOPPERS USING INSTANTANEOUS VOLTAGE CONTROL TECHNIQUE TO SOLVE VOLTAGE SAG PROBLEMS............2392

Khomfoi, Surin

VOLTAGE REGULATION IN DISTRIBUTION SYSTEM USING THE COMBINED DVR............2400

Nakamura, Sota ; Aoki, Mutsumi ; Ukai, Hiroyuki

NONLINEAR CONTROL OF THREE-PHASE FOUR-WIRE DYNAMIC VOLTAGE RESTORERS FOR DISTRIBUTION SYSTEM............2406

Jeong, Seon-Yeong ; Nguyen, Thanh Hai ; Lee, Dong-Choon ; Kim, Jang-Mok

VOLUME 4

DISTURBANCE CALCULATION BASED ON SPACE VECTOR DOT PRODUCT: APPLICATIONS TO COMPENSATORS............2413

de Carvalho, Kelly Caroline Mingorancia ; Ama, Naji Rajai Nasri ; Komatsu, Wilson ; Martinz, Fernando Ortiz ; Figueredo, Ricardo Souza ; Matakas, Lourenco

PROPOSAL OF 6TH RADIAL FORCE CONTROL BASED ON FLUX LINKAGE............2421

Kanematsu, Masato ; Miyajima, Takayuki ; Fujimoto, Hiroshi ; Hori, Yoichi ; Enomoto, Toshio ; Kondou, Masahiko ; Komiya, Hiroshi ; Yoshimoto, Kantaro ; Miyakawa, Takayuki

AIR GAP CONTROL OF MULTI-PHASE TRANSVERSE FLUX PERMANENT MAGNET LINEAR SYNCHRONOUS MOTOR BY USING INDEPENDENT VECTOR CONTROL............2427

Hwang, Seon-Hwan ; Bang, Deok-Je ; Kim, Ji-Won

MODIFIED DIRECT INSTANTANEOUS TORQUE CONTROL OF SWITCHED RELUCTANCE MOTOR WITH HIGH TORQUE PER AMPERE AND REDUCED SOURCE CURRENT RIPPLE............2433

Suryadevara, Rohit ; Fernandes, B.G.

CONTROL OF WOUND FIELD SYNCHRONOUS MOTOR INTEGRATED WITH ZSI............2438

Tajima, G. ; Kosaka, T. ; Matsui, N. ; Tonogi, K. ; Minoshima, N. ; Yoshida, T.

A NOVEL IPMSM MODEL FOR ROBUST POSITION SENSORLESS CONTROL TO MAGNETIC SATURATION............2445

Matsumoto, Atsushi ; Hasegawa, Masaru ; Doki, Shinji

MOTOR DRIVE SYSTEM USING NONLINEAR MATHEMATICAL MODEL FOR PERMANENT MAGNET SYNCHRONOUS MOTORS............2451

Iwaji, Yoshitaka ; Nakatsugawa, Junnosuke ; Sakai, Toshifumi ; Aoyagi, Shigehisa ; Nagura, Hirokazu

SENSORLESS-ORIENTED DESIGN OF IPMSM............2457

Kano, Yoshiaki

NOISE REDUCTION METHOD BY INJECTED FREQUENCY CONTROL FOR POSITION SENSORLESS CONTROL OF PERMANENT MAGNET SYNCHRONOUS MOTOR............2465

Taniguchi, Shun ; Yasui, Kazuya ; Yuki, Kazuaki

FORCE SENSORLESS BILATERAL CONTROL USING A DYNAMICAL ASYMMETRIC COMPENSATOR............2470

Hama, Ryota ; Imai, Jun ; Takahashi, Akiko ; Funabiki, Shigeyuki

DESIGN OF M-IPD CONTROLLER OF MULTI-INERTIA SYSTEM USING DIFFERENTIAL EVOLUTION............2476

Ikeda, Hidehiro ; Tsuyoshi, Hanamoto

A GUIDE TO DESIGN DISTURBANCE OBSERVER BASED MOTION CONTROL SYSTEMS............2483

Sariyildiz, Emre ; Ohnishi, Kouhei

IDENTIFICATION OF TWO-MASS MECHANICAL SYSTEMS USING TORQUE EXCITATION: DESIGN AND EXPERIMENTAL EVALUATION............2489

Saarakkala, Seppo E. ; Hinkkanen, Marko

INDUCTOR LOSS CALCULATION OF COUPLED INDUCTORS FOR HIGH POWER DENSITY BOOST CONVERTER............2497

Itoh, Yuki ; Kimura, Shota ; Imaoka, Jun ; Yamamoto, Masayoshi

1.2KW DUAL-ACTIVE BRIDGE CONVERTER USING SIC POWER MOSFETS AND PLANAR MAGNETICS............2503

De, D. ; Castellazzi, A. ; Lamantia, A.

ANALYSIS OF HYSTERESIS AND EDDY-CURRENT LOSSES FOR A MEDIUM-FREQUENCY TRANSFORMER IN AN ISOLATED DC-DC CONVERTER..2511
Nakahara, Mizuki ; Wada, Keiji

EXPERIMENTAL VERIFICATION OF CAPACITIVE POWER TRANSFER USING ONE PULSE SWITCHING ACTIVE CAPACITOR FOR PRACTICAL USE...2517
Kitabayashi, Tatsuaki ; Funato, Hirohito ; Kobayashi, Hiroya ; Yamaichi, Katsuya

A SINGLE-STAGE HIGH-PF DRIVER FOR SUPPLYING A T8-TYPE LED LAMP............................2523
Cheng, Chun-An ; Chang, Chien-Hsuan ; Cheng, Hung-Liang ; Chung, Tsung-Yuan

ELIMINATION OF ELECTROLYTIC CAPACITOR IN AC-DC SYSTEM OF LED DRIVER................2529
Mustapa, Rijalul Fahmi ; Hidayat, Nabil M ; Tukiman, Rahayu

A NOVEL BRIDGELESS BOOST HALF-BRIDGE ZVS-PWM SINGLE-STAGE UTILITY FREQUENCY AC-HIGH FREQUENCY AC RESONANT CONVERTER FOR DOMESTIC INDUCTION HEATERS................2533
Mishima, Tomoakzu ; Nakagawa, Yuki ; Nakaoka, Mutsuo

APPLICATION OF VIRTUAL VALIDATION SYSTEM FOR INVERTER HEAT PUMP SYSTEM............2541
Kanamori, Masaki ; Noda, Koji ; Endo, Takahisa ; Suzuki, Nobuyuki

TEST SETUP FOR ACCELERATED TEST OF HIGH POWER IGBT MODULES WITH ONLINE MONITORING OF VCE AND VF VOLTAGE DURING CONVERTER OPERATION................................2547
de Vega, Angel Ruiz ; Ghimire, Pramod ; Pedersen, Kristian Bonderup ; Trintis, Ionut ; Beczckowski, Szymon ; Munk-Nielsen, Stig ; Rannestad, Bjorn ; Thogersen, Paul

DESIGN OF HIGH-SPEED IGBT-BASED SWITCHING MODULES FOR PULSED POWER APPLICATIONS................2554
Kluge, Andreas ; Goehler, Lutz ; Gueldner, Henry ; Trompa, Thomas ; Mory, David ; Segsa, Karl-Heinz

COMPARATIVE SUITABILITY EVALUATION OF REVERSE-BLOCKING IGBTS FOR CURRENT-SOURCE BASED CONVERTER................................2562
De, Ankan ; Roy, Sudhin ; Bhattacharya, Subhashish

NEW REVERSE-CONDUCTING IGBT (1200V) WITH REVOLUTIONARY COMPACT PACKAGE................2569
Takahashi, K. ; Yoshida, S. ; Noguchi, S. ; Kuribayashi, H. ; Nashida, N. ; Kobayashi, Y. ; Kobayashi, H. ; Mochizuki, K. ; Ikeda, Y. ; Ikawa, O.

AN IMPROVED MODULATED CARRIER CONTROL OF SINGLE-PHASE CCM BOOST PFC CONVERTER................2575
Kim, Hyejin ; Cho, Bo-Hyung ; Choi, Hangseok

MODIFIED INTERLEAVED CURRENT SENSORLESS CONTROL FOR THREE-LEVEL BOOST PFC CONVERTER WITH ASYMMETRIC LOADS................2580
Chen, Hung-Chi ; Liao, Jhen-Yu

A NOVEL CRITICAL-CONDUCTION-MODE BRIDGELESS INTERLEAVED BOOST PFC RECTIFIER................2587
Cao, Guoen ; Kim, Hee-Jun

ANALYSIS AND DESIGN OF A PUSH-PULL SINGLE-STAGE FLYBACK POWER FACTOR CORRECTOR................2593
Lo, Yu-Kang ; Chiu, Huang-Jen ; Liu, Yu-Chen ; Lin, Chung-Yi ; Cheng, Shih-Jen ; Yang, CS

LINEAR OVER-MODULATION STRATEGY FOR CURRENT CONTROL IN PHOTOVOLTAIC INVERTER................2598
Park, Yongsoon ; Sul, Seung-Ki ; Hong, Ki-Nam

DESIGN OF DECENTRALIZED VOLTAGE CONTROL FOR PV INVERTERS TO MITIGATE VOLTAGE RISE IN DISTRIBUTION POWER SYSTEM WITHOUT COMMUNICATION................2606
Lee, Tzung-Lin ; Yang, Shih-Sian ; Hu, Shang-Hung

STABILITY ANALYSIS AND ACTIVE DAMPING FOR LLCL-FILTER BASED GRID-CONNECTED INVERTERS................2610
Huang, Min ; Blaabjerg, Frede ; Loh, Poh Chiang ; Wu, Weimin

INTEGRATED COMMON AND DIFFERENTIAL MODE FILTER APPLIED TO A SINGLE-PHASE TRANSFORMERLESS PV MICROINVERTER WITH LOW LEAKAGE CURRENT................2618
Figueredo, Ricardo Souza ; de Carvalho, Kelly Caroline Mingorancia ; Matakas, Lourenco

DESIGN AND INTEGRATION OF INTERPHASE INDUCTORS FOR INTERLEAVED THREE PHASE VOLTAGE-SOURCE-INVERTERS IN DC-FED MOTOR DRIVE SYSTEMS................2626
Zhang, Xuning ; Boroyevich, Dushan ; Burgos, Rolando

A NOVEL TRANSFORMER MODEL USING MAGNETIC CIRCUIT................2632
Nakamurame, Fuminori ; Ise, Toshifumi

HARDWARE-IN-THE-LOOP SIMULATION OF A MACHINE MODEL WITH REAL-TIME ANIMATION................2638
Xiaojie Zhuang ; Hibino, Shinya ; Harakawa, Masaya ; Terabe, Ryosuke ; Ozaki, Takayuki ; Nagano, Tetsuaki

DEVELOPMENT OF REAL TIME DIGITAL SIMULATOR FOR SELF-COMMUTATED SVC TO SUPPRESS VOLTAGE FLICKER................2644
Terao, Yutaka ; Shishida, Yasuhiro ; Tsuruma, Yoshinori ; Ishizuka, Tomotsugu ; Aoyama, Fumio ; Yoshino, Teruo ; Kato, Yutaka ; Belanger, Jean

OPERATIONAL ASPECTS AND POWER ARCHITECTURE DESIGN FOR A MICROGRID TO INCREASE THE USE OF RENEWABLE ENERGY IN WIRELESS COMMUNICATION NETWORKS................2649
Kwasinski, Alexis ; Kwasinski, Andres

P+ MULTIPLE RESONANT CONTROL FOR OUTPUT VOLTAGE REGULATION OF MICROGRID WITH UNBALANCED AND NONLINEAR LOADS................2656
Kyungbae Lim ; Jaeho Choi ; Juyoung Jang ; Junghum Lee ; Jaesig Kim

130MVA-STATCOM FOR TRANSIENT STABILITY IMPROVEMENT................2663
Imanishi, Takao ; Nagatomo, Yoshinobu ; Iwasaki, Shinya ; Masaki, Kenji ; Fujii, Toshiyuki ; Ieda, Jun

IMPROVED DROOP CONTROLLER FOR MICROGRID INVERTER CONSIDERING THE LINE IMPEDANCE MISMATCHING................2668
Du Yan ; Liuchen Chang ; Meiqin Mao ; Jianhui Su ; Ning Liu

SUPPRESSION CONTROL METHOD FOR IRON LOSS OF MATRIX MOTOR UNDER FLUX WEAKENING UTILIZING INDIVIDUAL WINDING CURRENT CONTROL 2673

Hijikata, Hiroki ; Akatsu, Kan ; Miyama, Yoshihiro ; Arita, Hideaki ; Daikoku, Akihiro

PERFORMANCE ANALYSIS OF A NEW CONCENTRATEDWINDING INTERIOR PERMANENT MAGNET SYNCHRONOUS MACHINE UNDER FIELD ORIENTED CONTROL 2679

Nguyen, D. ; Dutta, R. ; Fletcher, J. ; Rahman, F. ; Lovatt, Howard

ONLINE PARTICLE SWARM OPTIMIZATION FOR SENSORLESS IPMSM DRIVES CONSIDERING PARAMETER VARIATION 2686

Song, Z.Q. ; Xiao, D. ; Rahman, M.F.

A DTC-PWM CONTROL SCHEME OF PMSM BASED ON 12-SECTORS DIVISION AND SPEED INFORMATION 2693

Yunchang Kwak ; Jin-Woo Ahn ; Dong-Hee Lee

CONTROL OF POWER FLOW BETWEEN THE WIND GENERATOR AND NETWORK 2700

Stumpf, Peter ; Nagy, Istvan ; Vajk, Istvan

ADVANCES IN NANOGRID TECHNOLOGY AND ITS INTEGRATION INTO RURAL ELECTRIFICATION IN INDIA 2707

Mishra, Santanu ; Ray, Olive

STUDY AND IMPLEMENTATION OF SEVEN-LEVEL INVERTER USING COUPLED INDUCTOR AND SWITCHED-CAPACITOR 2714

Yi-Chun Lin ; Jiann-Fuh Chen ; Wen-Chien Hsu ; Sheng-Kai Kao

CASCADED MULTILEVEL CONVERTER BASED BIDIRECTIONAL INDUCTIVE POWER TRANSFER (BIPT) SYSTEM 2722

Bac Xuan Nguyen ; Vilathgamuwa, D.M. ; Foo, Gilbert ; Ong, Andrew ; Sampath, Prasad K. ; Madawala, Udaya K.

UNDERSAMPLING CONTROL OF A BIDIRECTIONAL CASCADED BUCK+BOOST DC-DC CONVERTER 2729

Rosekeit, Martin ; Joebges, Philipp ; Lelie, Markus ; Sauer, Dirk Uwe ; De Doncker, Rik W.

SUB-MICROSECOND RESPONSE DIGITAL CONTROLLER FOR POL 2737

Nonaka, Hirotaka ; Ishizuka, Yoichi ; Mii, Kenji ; Takenami, Fumiaki ; Kanemoto, Daisuke

GAIN CONTROLLED HIGH EFFICIENCY POWER FACTOR CORRECTION CIRCUIT 2745

Yonezawa, Yu ; Nakao, Hiroshi ; Sasaki, Tomotake ; Matsui, Yoshinobu ; Nakashima, Yoshiyasu ; Kaneko, Junji ; Shimamori, Hiroshi ; Yoshino, Yukio ; Hisato, Hosoyama ; Atsushi, Manabe ; Motizuki, Shun ; Yamashita, Shigeharu

DESIGN OF QUASI-RESONANT FLYBACK CONVERTER CONTROL IC WITH DCM AND CCM OPERATION 2750

Kai-Hui Chen ; Tsorng-Juu Liang

LOAD TRANSIENT RESPONSE IMPROVEMENT BASED ON PID CONTROL 2754

Yau, Y.T. ; Hwu, K.I.

AN ACTIVE-CLAMPING FORWARD CONVERTER WITH NON-LINEAR STEP-DOWN CONVERSION 2758

Jing-Yuan Lin ; Yu-Kang Lo ; Huang-Jen Chiu ; Chao-Fu Wang ; Chien-Yu Lin

SWITCHING LOSS MINIMIZATION OF 3-PHASE INTERLEAVED BIDIRECTIONAL DC-DC CONVERTER 2763

Eui-Cheol Nho ; Jae-Hun Jung ; Hak-Soo Kim ; In-Dong Kim ; Heung-Geun Kim ; Tae-Won Chun

MODIFIED THREE-PHASE THREE-LEVEL DC-DC CONVERTER -ADOPTING ASYMMETRICAL DUTY CYCLE CONTROL 2768

Yue Chen ; Xuling Chen ; Liu, Fuxin ; Ruan, Xinbo

DEADBEAT CONTROL OF POWER LEVELING UNIT WITH BIDIRECTIONAL BUCK/BOOST DC/DC CONVERTER 2775

Hamasaki, Shin-ichi ; Mukai, Ryosuke ; Yano, Yoshihiro ; Tsuji, Mineo

DESIGN OF OPTIMIZED ON-OFF CONTROL TO IMPROVE EFFICIENCY OF PARALLELED CONVERTER SYSTEM 2781

Kohama, Teruhiko ; Sogawa, Yuki ; Tsuji, Satoshi

EFFICIENCY IMPROVEMENTS IN A SINGLE ACTIVE BRIDGE MODULAR DC-DC CONVERTER WITH SNUBBER CAPACITANCE OPTIMISATION 2787

Ting, Yeh ; de Haan, Sjoerd ; Ferreira, Jan A.

A WIRELESS POWER TRANSFER SYSTEM OPTIMIZED FOR HIGH EFFICIENCY AND HIGH POWER APPLICATIONS 2794

Bani Shamseh, Mohammad ; Kawamura, Atsuo ; Yuzurihara, Itsuo ; Takayanagi, Atsushi

NON-ITERATIVE LCL FILTER DESIGN FOR THREE-PHASE TWO-LEVEL VOLTAGE-SOURCE PWM CONVERTERS 2802

Byung-Geuk Cho ; Seung-Ki Sul

DSP-BASED INTERLEAVED BUCK POWER FACTOR CORRECTOR 2810

Yu-Chen Liu ; Tsan Chen ; Po-Jung Tseng ; Yu-Kang Lo ; Huang-Jen Chiu

THE AVERAGE MODEL OF A THREE-PHASE THREE-STAGE POWER ELECTRONIC TRANSFORMER 2815

Shaodi Ouyang ; Liu, Jinjun ; Wang, Xinyu ; Wang, Xiaojian ; Fei Meng ; Riffat, Javid

A MULTI-CARRIER PWM FOR AC-DC-AC CONVERTER WITHOUT DC LINK ELECTROLYTIC CAPACITOR 2821

Chung-Chuan Hou ; Hsin-Ping Su

A DECOUPLING OFFSET-BASED PWM CONTROL FOR A MULTILEVEL INVERTER UNDER DC VOLTAGE UNBALANCE 2826

Nho Van Nguyen ; Tam Khanh Tu Nguyen ; Lee, Hong-Hee

?-? PARETO OPTIMIZATION OF 3-PHASE 3-LEVEL T-TYPE AC-DC-AC CONVERTER COMPRISING SI AND SIC HYBRID POWER STAGE...2834
Uemura, Hirofumi ; Krismer, Florian ; Okuma, Yasuhiro ; Kolar, Johann W.

PRACTICAL INVESTIGATION OF THE GATE BIAS EFFECT ON THE REVERSE RECOVERY BEHAVIOR OF THE BODY DIODE IN POWER MOSFETS..2842
Lindberg-Poulsen, Kristian ; Petersen, Lars Press ; Ouyang, Ziwei ; Andersen, Michael A.E.

AN ONLINE VCE MEASUREMENT AND TEMPERATURE ESTIMATION METHOD FOR HIGH POWER IGBT MODULE IN NORMAL PWM OPERATION...2850
Ghimire, Pramod ; de Vega, Angel Ruiz ; Beczkowski, Szymon ; Munk-Nielsen, Stig ; Rannested, Bjorn ; Thogersen, Paul Bach

EVALUATION ON IRON LOSS CHARACTERISTICS IN SERIES CONNECTION AND PARALLEL CONNECTION OF LOADS WITH INVERTER EXCITATION..2856
Odawara, Shunya ; Fujisaki, Keisuke

LOSS AND THERMAL MODEL FOR POWER SEMICONDUCTORS INCLUDING DEVICE RATING INFORMATION...2862
Ma, K. ; Bahman, A.S. ; Beczkowski, S.M. ; Blaabjerg, F.

IMPROVING RELIABILITY OF IGBT SURFACE ELECTRODE FOR 200 C OPERATION.........................2870
Nishimura, Tomohiro ; Ikeda, Yoshinari ; Hokazono, Hiroaki ; Mochizuki, Eiji ; Takahashi, Yoshikazu

INFLUENCE OF CARRIER FREQUENCY ON IRON LOSS TAKING ACCOUNT OF DEAD TIME EFFECT.......2874
Kogi, Ryosuke ; Odawara, Shunya ; Fujisaki, Keisuke

DECREASE OF SIC-BJT DRIVER LOSSES BY ONE-STEP COMMUTATION...2881
Barth, Henry ; Hofmann, Wilfried

POWER PROFILE BASED SELECTION AND OPERATION OPTIMIZATION OF PARALLEL-CONNECTED POWER CONVERTER COMBINATIONS...2887
Vogt, T. ; Peters, A. ; Frohleke, N. ; Bocker, J. ; Kempen, S.

A NOVEL POWER LOSS CALCULATION METHOD FOR IGBTS IN POWER CONVERTERS VIA CHAOTIC SPWM CONTROL...2893
Boyu Wang ; Li, Hong ; Xiaojie You ; Trillion Zheng

LOSS ANALYSIS AND SOFT-SWITCHING CHARACTERISTICS OF FLYBACK-FORWARD HIGH GAIN DC/DC CONVERTER WITH GAN FET...2899
Zhang Yajing ; Zheng, Trillion Q. ; Li Yan

INSULATED METAL SUBSTRATE FOR POWER MODULES USING ANODIC OXIDE FILM OF ALUMINUM..2904
Tokuyama, Takeshi ; Kusukawa, Jumpei ; Nakatsu, Kinya

A FAST-TRANSIENT-RESPONSE BUCK CONVERTER WITH SPLIT-TYPE III COMPENSATION AND CHARGE-PUMP CIRCUIT TECHNIQUE...2910
Chen, Jiann-Jong ; Wei-Ting Hsu ; Jih-Hua Yu ; Hwang, Yuh-Shyan ; Cheng-Chieh Yu

ADVANTAGES OF LOW PARASITIC INDUCTANCE PACKAGES OF POWER MOSFET FOR SERVER POWER APPLICATIONS...2914
Wonsuk Choi ; Dongkook Son ; Dongwook Kim

MODULAR INTEGRATION OF A MATRIX CONVERTER...2920
Solomon, Adane Kassa ; Skuriat, Robert ; Castellazzi, Alberto ; Wheeler, Pat

A MODULAR NANOSECOND PULSE GENERATION SYSTEM FOR PLASMA-ASSISTED IGNITION.............2926
Peng Gao ; Fletcher, John ; O'Byrne, Sean

DEVELOPMENT OF A SINGLE SWITCH CELL FOR MODULAR NANOSECOND PULSE GENERATION SYSTEMS..2932
Peng Gao ; Fletcher, John ; O'Byrne, Sean

ADVANTAGE OF SUPER JUNCTION MOSFET FOR POWER SUPPLY APPLICATION.................................2939
Tabira, K. ; Watanabe, S. ; Shimatou, T. ; Watashima, T. ; Takenoiri, S.

STUDY ON AN ACCURATE CALCULATION OF THE CONDUCTED EMI NOISE OF THE POWER CONVERTERS...2944
Omata, Shinpei ; Shimizu, Toshihisa

AN EXACT DISCRETE-TIME MODEL CONSIDERING DEAD-TIME NONLINEARITY FOR AN H-BRIDGE GRID-CONNECTED INVERTER...2950
Xie, Ruiliang ; Hao, Xiang ; Yang, Xu ; Chen, Wenjie ; Huang, Lang ; Chao Wang

THEORETICAL ANALYSIS OF THE DUALITY PRINCIPLE APPLIED TO INTERLEAVED TOPOLOGIES....................2954
Caris, M.L.A. ; Huisman, H. ; Duarte, J.L.

A NEW IMPEDANCE MEASUREMENT METHOD BASED ON HIGH FREQUENCY COMPENSATION............................2960
Yue, Xiaolong ; Zhuo, Fang ; Hao Yi

NUMERICAL AND EXPERIMENTAL INVESTIGATION OF PARASITIC EDGE CAPACITANCE FOR PHOTOVOLTAIC PANEL...2967
Wenjie Chen ; Xiaomei Song ; Hao Huang ; Xu Yang

VEHICLE INTERIOR NOISE CONTROL OF ULTRA-COMPACT ELECTRIC VEHICLE (FUNDAMENTAL CONSIDERATION USING RECTANGULAR ENCLOSURE)...2972
Kato, Taro ; Kato, Hideaki ; Oshinoya, Yasuo ; Suzuki, Ryosuke ; Hasegawa, Shinya

CONSIDERATION FOR THE PROPAGATION PATH OF CONDUCTIVE NOISE IN AIR CONDITIONERS......................2977
Tokiwa, Tsuyoshi ; Kanamori, Masaki ; Endo, Takahisa ; Iida, Mikiya ; Ogasawara, Satoshi ; Yizhanyi Tang

IRON LOSS EVALUATION OF IRON POWDER CORE SUITABLE FOR INDUCTOR USED IN POWER CONVERTERS...2983
Mori, Tomohiro ; Igarashi, Kazunori ; Kanagawa, Kinji ; Yamashita, Nobuyuki ; Shimizu, Toshihisa ; Bizen, Yosio

OPTIMIZED TUNING METHOD OF STATIONARY FRAME PROPORTIONAL RESONANT CURRENT CONTROLLERS 2988

Martinz, Fernando Ortiz ; de Carvalho, Kelly Caroline Mingorancia ; Ama, Naji Rajai Nasri ; Komatsu, Wilson ; Matakas, Lourenco

INSTANTANEOUS POWER THEORY APPLIED TO POWER CONDITIONING UNDER DISTORTED MAINS VOLTAGES: A MATLAB/SIMULINK APPROACH...2996

Nicolae, Petre-Marian ; Popa, Lucian-Dinut ; Nicolae, Marian-Stefan ; Nicolae, Ileana-Diana

THE RESEARCH ON RELIABILITY AND REAL-TIME OF THE SCHEME OF PROCESS LAYER GOOSE NETWORK IN SMART SUBSTATION BASED ON ARTIFICIAL COBWEB TOPOLOGY STRUCTURE 3002

Liu, Xiaosheng ; Zhu, Honglin ; Xu, Dianguo ; Li, Yanxiang

EFFICIENCY IMPROVEMENT OF A SELF-START TYPE PERMANENT MAGNET SYNCHRONOUS MOTOR..3007

Saikusa, H. ; Arikawa, S. ; Higuchi, T. ; Yokoi, Y. ; Abe, T.

CONSIDERATION OF OPTIMAL NUMBER OF POLES AND FREQUENCY FOR HIGH-EFFICIENCY PERMANENT MAGNET MOTOR 3012

Misu, Daisuke ; Matsushita, Makoto ; Takeuchi, Katsutoku ; Oishi, Koji ; Kawamura, Mitsuhiro

BASIC STUDY ON THE SUITABLE STRUCTURE OF A PERMANENT MAGNET SYNCHRONOUS MOTOR WITH A POWDER MAGNETIC CORE 3018

Hashimoto, Shizuka ; Sanada, Masayuki ; Morimoto, Shigeo ; Inoue, Yukinori

CHARACTERISTICS OF A HALF-WAVE RECTIFIED BRUSHLESS SYNCHRONOUS GENERATOR...............................3024

Hirakawa, Yuki ; Higuchi, Tsuyoshi ; Yokoi, Yuichi ; Abe, Takashi

MODELING OF WOUND ROTOR SYNCHRONOUS MACHINES CONSIDERING HARMONICS, GEOMETRIC SALIENCIES AND SATURATION INDUCED SALIENCIES 3029

Rambetius, Alexander ; Luthardt, Sven ; Piepenbreier, Bernhard

DESIGN AND COMPARISON OF HIGH FREQUENCY TRANSFORMERS USING FOIL AND ROUND WINDINGS...3037

Iyer, Kartik V ; Robbins, William P ; Mohan, Ned

A METHOD TO CALCULATE THE PERFORMANCE OF LINEAR INDUCTION MOTORS USING SIMPLE TWO-PHASE MODEL 3044

Hirahara, Hideaki ; Yamamoto, Shu ; Ara, Takahiro ; Shimizu, Toshihisa

AN ESP DOWNHOLE PARAMETERS MONITORING SYSTEM BASED ON CURRENT LOOP TRANSMISSION METHOD 3050

Jin Miaoxin ; Zhang Wei ; Gao Qiang ; Xu Dianguo

BENDING MAGNETIC LEVITATION CONTROL FOR THIN STEEL PLATE (EXPERIMENTAL CONSIDERATION USING SLIDING MODE CONTROL)...3055

Yonezawa, Hikaru ; Narita, Takayoshi ; Oshinoya, Yasuo ; Marumori, Hiroki ; Hasegawa, Shinya

TRANSFORMER WINDING LOSSES WITH ROUND CONDUCTORS FOR DUTY-CYCLE REGULATED SQUARE WAVES 3061

Iyer, Kartik V ; Robbins, William P ; Basu, Kaushik ; Mohan, Ned

SIMULATION OF RESIN MOLDED TYPE SENSOR IN POLE SWITCH FOR POWER DELIBERY SYSTEMS 3067

Furukawa, Tatsuya ; Muta, Shoichiro ; Fukumoto, Hisao ; Itoh, Hideaki ; Ohchi, Masashi

ROBUST STARTUP CONTROL OF SENSORLESS PMSM DRIVES WITH SELF-COMMISSIONING 3072

Lin, Chiao-Chien ; Tzou, Ying-Yu

POSITION SENSORLESS CONTROL OF PMSM WITH A LOW-FREQUENCY SIGNAL INJECTION 3079

Nimura, Tomohiro ; Doki, Shinji ; Fujitsuna, Masami

A COMPARISON OF DIFFERENT SENSORLESS POSITION ACQUISITION METHODS AT LOW SPEEDS FOR A PERMANENT MAGNET SYNCHRONOUS MACHINE IN VEHICLE APPLICATIONS..3085

Lehmann, Oliver ; Zehelein, Matthias ; Schuster, Johannes ; Roth-Stielow, Jorg

STABILITY COMPARISON OF IPMSM SENSORLESS VECTOR CONTROL SYSTEMS USING EXTENDED EMF 3093

Tsuji, Mineo ; Mizusaki, Hiroshi ; Hamasaki, Sin-ichi

INDUCTION MACHINE BASED FLYWHEEL SPEED ESTIMATION AT STAND-BY MODE...3099

Liu, Rongqiang ; Xu, David

SYMMETRICAL SIGNALING SYSTEM FOR SENSOR-LESS SRM DRIVE 3106

Yamamoto, Kenji ; Takahashi, Hisashi ; Ushiro, Nobumasa ; Shirasawa, Koki

DIGITAL INTEGRATORS FOR CONDITION MONITORING: A DC AND MULTITONE SIGNAL ANALYSIS 3111

Peretti, L.

AUDIBLE NOISE REDUCTION METHOD IN IPMSM POSITION SENSORLESS CONTROL BASED ON HIGH-FREQUENCY CURRENT INJECTION 3119

Tauchi, Yuki ; Kubota, Hisao

A NOVEL DESIGN FOR INDUCTION MOTOR FLUX ESTIMATION USING IMPULSIVE OBSERVER..............................3124

Peng Wang ; Yan Li ; Jianwen Zhang ; Xu Cai ; Zhengzhi Han

LOAD TORQUE AND INERTIA SIMULATION BASED ON DOUBLE-STATOR PERMANENT-MAGNET SYNCHRONOUS MOTOR 3129

Zhe Wang ; Mingyan Wang ; Ben Guo ; Chai Feng

INDEPENDENT SPEED AND POSITION CONTROL OF TWO PERMANENT MAGNET SYNCHRONOUS MOTORS FED BY A FOUR-LEG INVERTER 3134

Kubo, Yuji ; Moroi, Takayuki ; Kouki, Matsuse ; Kubota, Hisao ; Rajashekara, Kaushik

MINIMIZATION OF STATOR CURRENTS FOR MONO INVERTER DUAL PARALLEL PMSM DRIVE SYSTEM 3140
Yongjae Lee ; Ha, Jung-Ik

PERFORMANCE COMPARISON OF INVERTER AND DRIVE CONFIGURATIONS WITH OPEN-END AND STAR-CONNECTED WINDINGS 3145
Neubert, Markus ; Koschik, Stefan ; De Doncker, Rik W.

INPUT CURRENT HARMONICS REDUCTION CONTROL FOR ELECTROLYTIC CAPACITOR LESS INVERTER BASED IPMSM DRIVE SYSTEM 3153
Abe, Kodai ; Ohishi, Kiyoshi ; Haga, Hitoshi

NONCONTACT GUIDE SYSTEM FOR TRAVELING ELASTIC STEEL PLATES (THEORETICAL STUDY ON THE SHAPE OF TRAVELING STEEL PLATE) 3159
Sakaba, Kouichi ; Hasegawa, Shinya ; Narita, Takayoshi ; Oshinoya, Yasuo

ACTIVE SEAT SUSPENSION FOR ULTRA-COMPACT VEHICLE (FUNDAMENTAL CONSIDERATION ON ELECTROMYOGRAM WHEN FALL FROM THE BUMP) 3162
Mashino, Masahiro ; Sunaga, Keita ; Hasegawa, Shinya ; Ishida, Masaki ; Kato, Hideaki ; Oshinoya, Yasuo

ADAPTIVE CURRENT TRACKING OF THREE-PHASE ACTIVE POWER FILTER USING BACKSTEPPING CONTROL 3168
Yunmei Fang ; Juntao Fei ; Shixi Hou ; Weili Dai

FAST IDENTIFICATION OF RESONANCE CHARACTERISTIC FOR 2-MASS SYSTEM WITH ELASTIC LOAD 3174
Ming Yang ; Liang Hao ; Dianguo Xu

AUTONOMOUS NAVIGATION SYSTEM BASED ON COLLISION DANGER-DEGREE FOR UNMANNED GROUND VEHICLE 3179
Yasuno, Takashi ; Tanaka, Daiki ; Kuwahara, Akinobu

A HIGH-PERFORMANCE BIDIRECTIONAL DC-DC CONVERTER FOR DC MICRO-GRID SYSTEM APPLICATION 3185
Shu-Wei Kuo ; Yu-Kang Lo ; Huang-Jen Chiu ; Shih-Jen Cheng ; Chung-Yi Lin ; Yang, CS

VOLUME 5

IMPROVEMENT IN EFFICIENCY OF LED LIGHTING SYSTEM 3190
Hwu, K.I. ; Jiang, W.Z. ; Jenn-Jong Shieh

COMPARISON AND EVALUATION OF VIBRATION-BASED PIEZOELECTRIC POWER GENERATORS 3194
Basari, Amat A. ; Awaji, Sosuke ; Hashimoto, Seiji ; Kasai, Makoto ; Suto, Kenji ; Kumagai, Shunji ; Kasai, Makoto ; Suto, Kenji ; Wei Jiang ; Shuren Wang

BATTERY SELECTION FOR HYBRID ENERGY SYSTEMS AND THERMAL MANAGEMENT IN ARCTIC CLIMATES 3200
Kalogera, Maria ; Bauer, Pavol

100KW PV PCS WITH NATURAL CONVECTION COOLING FOR OUTDOOR INSTALLATION 3207
Jin, Yasuhiro ; Matsuoka, Kazumasa ; Takahashi, Takehiro ; Takahashi, Nobuhiro

A NEW PLL BASED ON FAST POSITIVE AND NEGATIVE SEQUENCE DECOMPOSITION ALGORITHM WITH MATRIX OPERATION UNDER DISTORTED GRID CONDITIONS 3213
Shaohua Sun ; Hongqi Ben ; Tao Meng ; Jinyong Zhang

PERFORMANCE IMPROVEMENT OF PHOTOVOLTAIC POWER GENERATION SYSTEMS USING ON-OFF CONTROL METHODS 3218
Kenji, Matsumoto ; Nomura, Shinichi

LOW VOLTAGE PV POWER INTEGRATION INTO MEDIUM VOLTAGE GRID USING HIGH VOLTAGE SIC DEVICES 3225
Chattopadhyay, Ritwik ; Bhattacharya, Subhashish ; Foureaux, Nicole C. ; Silva, Sidelmo M. ; Braz Cardoso, F. ; de Paula, Helder ; Pires, Igor A. ; Cortizio, Porfirio C. ; Moraes, Lenin ; de S.Brito, Jose A.

A NOVEL GLOBAL MAXIMUM POWER POINT TRACKING METHOD FOR PHOTOVOLTAIC GENERATION SYSTEM OPERATING UNDER PARTIALLY SHADED CONDITION 3233
Jing-Hsiao Chen ; Yu-Shan Cheng ; Shun-Chung Wang ; Huang, Jia-Wei ; Liu, Yi-Hua

AN APPLICATION OF Z-SOURCE CONVERTER TO BATTERIES CHARGE WITH A PHOTOVOLTAIC SYSTEM 3239
Razik, H. ; Zitouni, Y. ; Maret, C.

PCS WITH SCANNING-TYPE MPPT CONTROL FOR INDUSTRIAL GRID-CONNECTED PV POWER GENERATION SYSTEM 3244
Itako, Kazutaka

FEASIBLE METHOD OF CALCULATING LEAKAGE REACTANCE OF 9-WINDING TRANSFORMER FOR HIGH-VOLTAGE INVERTER SYSTEM 3249
Fukumoto, Hisao ; Furukawa, Tatsuya ; Itoh, Hideaki ; Ohchi, Masashi

HIGH POWER HVDC-DC CONVERTERS FOR THE INTERCONNECTION OF HVDC LINES WITH DIFFERENT LINE TOPOLOGIES 3255
Schon, Andre ; Bakran, Mark-M.

CHARACTERIZATION OF A CURRENT SHUNT AND AN INDUCTIVE VOLTAGE DIVIDER FOR PMU CALIBRATION 3263
Kon, Saytaro ; Yamada, Tatsuji

DISTRIBUTED SERIES/HYBRID-SHUNT COMPENSATION FOR HARMONIC MITIGATION IN COMMERCIAL FACILITIES 3270

Diniz, Rogerio Azevedo ; Pires, Igor A. ; Franca, Gleisson J. ; Cardoso, Braz J.

ROBUST CONTROL DESIGN FOR THE VOLTAGE TRACKING LOOP OF A DVR 3278

Ferrari, Bruno Augusto ; Ama, Naji Rajai Nasri ; de Carvalho, Kelly Caroline Mingorancia ; Martinz, Fernando Ortiz ; Matakas, Lourenco

MULTI-PORT SOLID STATE TRANSFORMER FOR INTER-GRID POWER FLOW CONTROL 3286

Roy, Sudhin ; De, Ankan ; Bhattacharya, Subhashish

REACTIVE POWER CONTROL STRATEGY BASED ON DC CAPACITOR VOLTAGE CONTROL FOR ACTIVE LOAD BALANCER IN THREE-PHASE FOUR-WIRE DISTRIBUTION SYSTEMS 3292

Tint Soe Win ; Hisada, Yoshihiro ; Tanaka, Toshihiko ; Hiraki, Eiji ; Okamoto, Masayuki ; Lee, Seong Ryong

VOLTAGE SAG RIDE-THROUGH PERFORMANCE OF VIRTUAL SYNCHRONOUS GENERATOR 3298

Alipoor, Jaber ; Miura, Yushi ; Ise, Toshifumi

CONTROL OF DISTRIBUTED GENERATION SYSTEMS UNDER UNBALANCED VOLTAGE CONDITIONS 3306

Kabiri, R. ; Holmes, D.G. ; McGrath, B.P.

STABILITY ANALYSIS OF GRID-CONNECTED INVERTERS WITH LCL-FILTER BASED ON HARMONIC BALANCE AND FLOQUET THEORY 3314

Jing Bian ; Hong Li ; Zheng, Trillion Q.

COMPARATIVE EVALUATION OF PASSIVE DAMPING TOPOLOGIES FOR PARALLEL GRID-CONNECTED CONVERTERS WITH LCL FILTERS 3320

Beres, Remus ; Wang, Xiongfei ; Blaabjerg, Frede ; Bak, Claus Leth ; Liserre, Marco

STUDY AND IMPLEMENTATION OF A SEPIC LED DRIVER WITH ADJUSTABLE OUTPUT VOLTAGE 3328

Po-Jung Tseng ; Yu-Chen Liu ; Yu-Kang Lo ; Chiu, Huang-Jen ; Yun-Chu Chiu

AN INTERLEAVED SINGLE-STAGE LLC RESONANT CONVERTER USED FOR MULTI-CHANNEL LED DRIVING 3333

Chang, Chien-Hsuan ; Cheng, Chun-An ; Jinno, Masahito ; Cheng, Hung-Liang

A NOVEL TYPE OF WIRELESS V2H SYSTEM WITH BIDIRECTIONAL RESONANT SINGLE-ENDED INVERTER 3341

Fukuoka, Hiroki ; Iga, Yuichi ; Omori, Hideki ; Morizane, Tosimitsu ; Kimura, Noriyuki ; Nakaoka, Mutuo

DESIGN AND IMPLEMENTATION OF AN INTERLEAVED BCM BOOST PFC CONTROL IC 3346

Kuan-Hsien Chou ; Tsorng-Juu Liang ; Kai-Hui Chen ; Ji-Shiang Lee

LOW CAPACITIVE INDUCTORS FOR FAST SWITCHING DEVICES IN ACTIVE POWER FACTOR CORRECTION APPLICATIONS 3352

Hernandez, Juan C. ; Petersen, Lars P. ; Andersen, Michael A.E.

TEMPERATURE-ROBUST LC3 LED DRIVER WITH LOW THD, HIGH EFFICIENCY, AND LONG LIFE 3358

Lee, Eun S. ; Choi, Bo H. ; Cheon, Jun P. ; Kim, Bong C. ; Rim, Chun T.

OPTIMIZING REPULSIVE LORENTZ FORCES FOR A LEVITATING INDUCTION COOKER 3365

Zingerli, Claudius M. ; Nussbaumer, Thomas ; Kolar, Johann W.

DESIGN OF A MODULAR RESONANT CONVERTER FOR 25KV-8A DC POWER SUPPLY OF RF CAVITIES 3371

Siemaszko, Daniel ; Pittet, Serge ; Aguglia, Davide ; de Mallac, Louis

A NOVEL TRANSFORMER-LESS INTERLEAVED FOUR-PHASE HIGH STEP-DOWN DC CONVERTER WITH LOW SWITCH VOLTAGE STRESS 3379

Ching-Tasi Pan ; Chen-Feng Chuang ; Chia-Chi Chu ; Hao-Chien Cheng

EFFICIENCY IMPROVEMENT OF POWER SUPPLY WITH TRANSIENT CURRENT CIRCUIT USING DIGITAL CONTROL 3386

Takashita, Haruomi ; Shoyama, Masahito ; Yonezawa, Yu ; Nakashima, Yoshiyasu

ULTRA HIGH STEP-DOWN CONVERTER 3392

Yau, Y.T. ; Hwu, K.I.

DIGITAL CONTROL OF PWM INVERTER USING ULTRA HIGH SPEED NETWORK FOR FEEDBACK SIGNALS WITH COMMUNICATION DISTURBANCE OBSERVER BASED ON ROCKET I/O PROTOCOL 3397

Saito, Ryo ; Tsuchida, Kazuo ; Yokoyama, Tomoki

100 KHZ DC CHOPPER DIGITALLY GATE CONTROLLED WITH PARTIAL TURN- OFF SWITCHING USING SIC-MOSFET AND FPGA 3403

Tsuruta, Yukinori ; Kawamura, Atsuo

VARIABLE CARRIER DEADBEAT CONTROL WITH DIGITAL HYSTERESIS METHOD USING SOC-FPGA FOR UTILITY INTERACTIVE INVERTER 3410

Ohashi, Shunsuke ; Yoshida, Morito ; Yokoyama, Tomoki

A SPACE VECTOR MODULATION STRATEGY FOR THREE-LEVEL OPERATION BASED ON DUAL TWO-LEVEL VOLTAGE SOURCE INVERTERS 3417

Kumsuwan, Yuttana ; Srirattanawichaikul, Watcharin

INVESTIGATION ON THE PARALLEL OPERATION OF ALL-GAN POWER MODULE AND THERMAL PERFORMANCE EVALUATION 3425

Cheng, Stone ; Po-Chien Chou

FULL SILICON CARBIDE BOOST CHOPPER MODULE FOR HIGH FREQUENCY AND HIGH TEMPERATURE OPERATION 3432

Pettersson, Sami ; Kicin, Slavo ; Holm, Toni ; Bianda, Enea ; Canales, Francisco

DEVELOPMENT OF ULTRAHIGH VOLTAGE SIC POWER DEVICES 3440

Fukuda, Kenji ; Okamoto, Dai ; Harada, Shinsuke ; Tanaka, Yasunori ; Yonezawa, Yoshiyuki ; Deguchi, Tadayoshi ; Katakami, Shuji ; Ishimori, Hitoshi ; Takasu, Shinji ; Arai, Manabu ; Takenaka, Kensuke ; Fujisawa, Hiroyuki ; Takei, Manabu ; Matsumoto, Kazushi ; Ohse, Naoyuki ; Ryo, Mina ; Ota, Chiharu ; Takao, Kazuto ; Mizukami, Makoto ; Kato, Tomohisa ; Izumi, Toru ; Hayashi, Toshihiko ; Nakayama, Koji ; Asano, Katsunori ; Okumura, Hajime ; Kimoto, Tsunenobu

HIGH SWITCHING PERFORMANCE OF 1.7KV, 50A SIC POWER MOSFET OVER SI IGBT FOR ADVANCED POWER CONVERSION APPLICATIONS 3447

Hazra, Samir ; De, Ankan ; Bhattacharya, Subhashish ; Lin Cheng ; Palmour, John ; Schupbach, Marcelo ; Hull, Brett ; Allen, Scott

CONTROL METHOD FOR FIVE LEVEL CONVERTER WITH COMMON FLYING CAPACITORS TO AVOID VOLTAGE LEVEL SKIP 3455

Wei Yan ; Hui Zhang ; Ogura, Kazuya ; Urushibata, Shota

LOW-COMPLEXITY ANALYTICAL APPROXIMATIONS OF SWITCHING FREQUENCY HARMONICS OF 3-PHASE N-LEVEL VOLTAGE-SOURCE PWM CONVERTERS 3460

Burkart, Ralph M. ; Kolar, Johann W.

DYNAMIC VOLTAGE BALANCING ALGORITHM FOR MODULAR MULTILEVEL CONVERTER WITH THREE-LEVEL FLYING CAPACITOR SUBMODULES 3468

Dekka, Apparao ; Wu, Bin ; Zargari, Navid R.

MODULAR MEDIUM VOLTAGE DRIVE FOR DEMANDING APPLICATIONS 3476

Dujic, Drazen ; Wahlstroem, Jonas ; Marrero Sosa, Juan Alberto ; Fritz, Dominik

ASYMMETRICAL FAULT RIDE-THROUGH OF THREE-PHASE PV SYSTEMS USING FOUR-WIRE DC-AC CONVERTERS 3482

Iyer, Shivkumar ; Bin Wu ; Yunwei Li ; Singh, B.N.

OPERATION MODE ANALYSIS FOR SOLVING THE PARTIAL SHADOW IN A NOVEL PV POWER GENERATION SYSTEM 3489

Qi Zhang ; Xiangdong Sun ; Yanru Zhong ; Lie Guo ; Matsui, Mikihiko

ANALYSIS OF PARTIAL POWER PROCESSING DISTRIBUTED MPPT FOR A PV POWERED ELECTRIC AIRCRAFT 3496

Marzouk, Ahmad Diab ; Fournier-Bidoz, Sebastien ; Yablecki, Jessica ; McLean, Kenneth ; Trescases, Olivier

IMPACTS OF RECTIFIER CIRCUIT LOADS ON ISLANDING DETECTION OF PHOTOVOLTAIC SYSTEMS 3503

Yoshida, Yoshiaki ; Suzuki, Hirokazu

INDUCTION MOTOR MADE OF SMC 3509

Morimoto, Masayuki ; Inamori, Mamiko

ESTIMATION AND COMPARISON OF THE WINDAGE LOSS OF A 60 KW SWITCHED RELUCTANCE MOTOR FOR HYBRID ELECTRIC VEHICLES 3513

Kiyota, Kyohei ; Kakishima, Takeo ; Chiba, Akira

DEVELOPMENT OF HIGH-POWER PMASYNRM USING FERRITE MAGNETS FOR REDUCING RARE-EARTH MATERIAL USE 3519

Sanada, Masayuki ; Morimoto, Shigeo ; Inoue, Yukinori

CONSIDERATION OF 10KW IN-WHEEL TYPE AXIAL-GAP MOTOR USING FERRITE PERMANENT MAGNETS 3525

Sone, Kodai ; Takemoto, Masatsugu ; Ogasawara, Satoshi ; Takezaki, Kenichi ; Hino, Wataru

POWER CONTROL METHOD FOR MULTI-PARALLEL DC DISTRIBUTION SYSTEM THROUGH THE EQUIVALENT CIRCUIT MODEL 3532

Seok-Jin Hong ; Soo-Cheol Shin ; Hee-Jun Lee ; Chung-Yuen Won ; Taeck-Kie Lee

A COMMUNICATION-LESS DISTRIBUTED VOLTAGE CONTROL STRATEGY FOR A MULTI-BUS AC ISLANDED MICROGRID 3538

Wang, Yanbo ; Yongdong Tan ; Chen, Zhe ; Wang, Xiongfei ; Tian, Yanjun

AN ENHANCED LOAD POWER SHARING STRATEGY FOR LOW-VOLTAGE MICROGRIDS BASED ON INVERSE-DROOP CONTROL METHOD 3546

Yixin Zhu ; Fang Zhuo ; Baoquan Liu ; Hao Yi

ADDING VIRTUAL RESISTANCE IN SOURCE SIDE CONVERTERS FOR STABILIZATION OF CASCADED CONNECTED TWO STAGE CONVERTER SYSTEMS WITH CONSTANT POWER LOADS IN DC MICROGRIDS 3553

Mingfei Wu ; Lu, Dylan D.C.

EXPANSION OF OPERATING RANGE AND IMPROVEMENT OF TORQUE RESPONSE OF PMSM DRIVE BY USING MODEL PREDICTIVE CONTROL 3557

N/A

NONLINEAR MODEL PREDICTIVE TORQUE CONTROL OF A LOAD COMMUTATED INVERTER AND SYNCHRONOUS MACHINE 3563

Almer, Stefan ; Besselmann, Thomas ; Ferreau, Joachim

MODEL PREDICTIVE CURRENT CONTROL FOR PMSM CONSIDERING NUMBER OF SWITCHING OPERATIONS 3568

Zanma, Tadanao ; Yasumura, Yuji ; Liu, KangZhi

PREDICTIVE INDIRECT MATRIX CONVERTER FED TORQUE RIPPLE MINIMIZATION WITH WEIGHTING FACTOR OPTIMIZATION 3574

Uddin, Muslem ; Mekhilef, Saad ; Rivera, Marco ; Rodriguez, Jose

HIGH-POWER DENSITY HYBRID CONVERTER TOPOLOGIES FOR LOW-POWER DC-DC SMPS 3582

Radic, Aleksandar ; Ahssanuzzaman, S.M. ; Mahdavikhah, Behzad ; Prodic, Aleksandar

COUPLED INDUCTOR BASED CURRENT-FED SWITCHED INVERTER FOR LOW VOLTAGE RENEWABLE INTERFACE 3587
Nag, Soumya Shubhra ; Mishra, Santanu Kumar

A SEMI-ISOLATED MULTI-INPUT CONVERTER FOR HYBRID PV/WIND POWER CHARGER SYSTEM 3592
Cheng-Wei Chen ; Kun-Hung Chen ; Chen, Yaow-Ming

HFL PV MICRO-INVERTER WITH FRONT-END CURRENT-FED CONVERTER AND HALF-WAVE CYCLOCONVERTER 3598
Nayanasiri, D.R. ; Vilathgamuwa, D.M. ; Maskell, D.L.

COMPREHENSIVE STUDY ABOUT STABILITY ISSUES OF MULTI-MODULE DISTRIBUTED SYSTEM 3604
Liu, Fangcheng ; Liu, Jinjun ; Zhang, Haodong ; Xue, Danhong ; Dou, Qinyun

CHARACTERISTICS STUDY OF NEURAL NETWORK AIDED DIGITAL CONTROL FOR DC-DC CONVERTER 3611
Maruta, Hidenori ; Motomura, Masashi ; Kurokawa, Fujio

ZERO CURRENT SWITCHING CURRENT-FED PARALLEL RESONANT PUSH-PULL (CFPRPP) CONVERTER 3616
Moorthy, Radha Sree Krishna ; Rathore, Akshay Kumar

CHARACTERISTICS OF TRANSMISSION CARRIER IN A NEW WIRE COMMUNICATION SYSTEM BY THE USE OF HIGH-RIPPLE DC-DC CONVERTER 3624
Katsuki, Akihiko ; Mizuki, Tatsuya ; Shibahara, Kohei ; Morita, Kosuke ; Masutomo, Kazufumi ; Maeyama, Shigetaka

5MHZ PWM-CONTROLLED CURRENT-MODE RESONANT DC-DC CONVERTER USING GAN-FETS 3630
Hariya, Akinori ; Yanagi, Hiroshige ; Ishizuka, Yoichi ; Matsuura, Ken ; Tomioka, Satoshi ; Ninomiya, Tamotsu

DESIGN AND PERFORMANCE EVALUATION OF DIGITAL CONTROL FOR LLC SERIES RESONANT DC-TO-DC CONVERTERS 3638
Pidaparthy, Syam Kumar ; Choi, Byungcho ; Jang, Jinhaeng

EXPERIMENTAL VERIFICATION OF NOISELESS SAMPLING FOR BUCK CHOPPER CIRCUIT WITH CURRENT CONTROL 3646
Takeuchi, Shun ; Wada, Keiji

CONTROL CHARACTERISTICS IMPROVEMENT OF FULL-BRIDGE DC-DC CONVERTER WITH SNUBBER CAPACITOR 3652
Domoto, Kazuhide ; Ishizuka, Yoichi ; Abe, Seiya ; Ninomiya, Tamotsu

DCM CONTROL METHOD OF BOOST CONVERTER BASED ON CONVENTIONAL CCM CONTROL 3659
Le Hoai Nam ; Orikawa, Koji ; Itoh, Jun-ichi

TECHNICAL ASSESSMENT OF LOAD COMMUTATION SWITCH IN HYBRID HVDC BREAKER 3667
Hassanpoor, Arman ; Hafner, Jurgen ; Jacobson, Bjorn

CONTROL OF HEXAGONAL MODULAR MULTILEVEL CONVERTER FOR 3-PHASE BTB SYSTEM 3674
Hamasaki, Shin-ichi ; Okamura, Kazuki ; Tsubakidani, Takashi ; Tsuji, Mineo

A SYNTHESIZED CAPACITORS VOLTAGE CONTROL FOR MODULAR MULTILEVEL CONVERTER IN HVDC APPLICATION 3680
Rongfeng Yang ; Shunke Sui ; Binbin Li ; Wei Wang ; Dianguo Xu

OPERATING PHASE AND FREQUENCY SELECTION OF LOW FREQUENCY AC TRANSMISSION SYSTEM USING CYCLOCONVERTERS 3687
Achara, Pichetjamroen ; Ise, Toshifumi

FAST ACTING DC CIRCUIT BREAKER FOR HVDC TRANSMISSION LINE BASED ON DC/DC CHOPPER 3695
Liangyi Tang ; Bin Wu ; Yaramasu, Venkata ; Weirong Chen ; Athab, Hussain S.

1700V SI-IGBT AND SIC-SBD HYBRID MODULE FOR AC690V INVERTER SYSTEM 3702
Haining Wang ; Ikawa, O. ; Miyashita, S. ; Nishimura, T. ; Igarashi, S.

SWITCHING SIMULATION OF SIC HIGH-POWER MODULE WITH LOW PARASITIC INDUCTANCE 3707
Yamamoto, Takashi ; Hasegawa, Kohei ; Ishida, Masaaki ; Takao, Kazuto

SWITCHING PERFORMANCE OF PARALLEL-CONNECTED POWER MODULES WITH SIC MOSFETS 3712
Colmenares, Juan ; Peftitsis, Dimosthenis ; Nee, Hans-Peter ; Rabkowski, Jacek

BUILT-IN RELIABILITY DESIGN OF A HIGH-FREQUENCY SIC MOSFET POWER MODULE 3718
Jianfeng Li ; Gurpinar, Emre ; Lopez-Arevalo, Saul ; Castellazzi, Alberto ; Mills, Liam

EXPERIMENTAL SWITCHING FREQUENCY LIMITS OF 15 KV SIC N-IGBT MODULE 3726
Kadavelugu, Arun ; Bhattacharya, Subhashish ; Ryu, Sei-Hyung ; Van Brunt, Edward ; Grider, Dave ; Leslie, Scott

SELECTION OF SUITABLE CARRIER-BASED PWM METHOD FOR MODULAR MULTILEVEL CONVERTER 3734
Ciftci, Baris ; Erturk, Feyzullah ; Hava, Ahmet M.

CONTROL AND EXPERIMENT OF A 380-V, 15-KW MOTOR DRIVE USING MODULAR MULTILEVEL CASCADE CONVERTER BASED ON TRIPLE-STAR BRIDGE CELLS (MMCC-TSBC) 3742
Kawamura, Wataru ; Hagiwara, Makoto ; Akagi, Hirofumi

A POWER ELECTRONIC TRANSFORMER WITH SINUSOIDAL VOLTAGES AND CURRENTS USING MODULAR MULTILEVEL CONVERTER 3750
Sahoo, Ashish Kumar ; Mohan, Ned

VARYING AND UNEQUAL CARRIER FREQUENCY PWM TECHNIQUES FOR MODULAR MULTILEVEL CONVERTERS 3758
Konstantinou, Georgios ; Darus, Rosheila ; Pou, Josep ; Ceballos, Salvador ; Agelidis, Vassilios G.

COMPARISON OF PHASE-SHIFTED AND LEVEL-SHIFTED PWM IN THE MODULAR MULTILEVEL CONVERTER 3764
Darus, Rosheila ; Konstantinou, Georgios ; Pou, Josep ; Ceballos, Salvador ; Agelidis, Vassilios G.

A SINGLE-PHASE POWER CONDITIONER WITH A BUCK-BOOST-TYPE POWER DECOUPLING CIRCUIT 3771
Yamaguchi, Shota ; Shimizu, Toshihisa

A NOVEL ASYMMETRICAL FLC-BASED MPPT TECHNIQUE FOR PHOTOVOLTAIC GENERATION SYSTEM 3778
Yi-Hsun Chiu ; Yu-Shan Cheng ; Yi-Hua Liu ; Shun-Chung Wang ; Zong-Zhen Yang

A NOVEL CURRENT LINK DISTRIBUTED MPPT PV SYSTEM - OVERALL SYSTEM PROTOTYPING AND EVALUATION 3784
Mikihiko ; Toru ; Akira ; Xiang-Dong Sun ; Byung-Gyu Yu

POWER FLOW CONTROL AND MPPT PARAMETER SELECTION FOR RESIDENTIAL GRID-CONNECTED PV SYSTEMS WITH BATTERY STORAGE..3789
Chokchai, Chuenwattanapraniti

A MAXIMUM POWER POINT TRACKING METHOD WITH RIPPLE CURRENT ORIENTATION 3796
Moo, Chin-Sien ; Wu, Gwo-Bin

OUTPUT CHARACTERISTICS OF A SURFACE PERMANENT MAGNET-TYPE VERNIER MOTOR - COMPARISON OF TEST RESULTS AND CALCULATION 3801
Kataoka, Yasuhiro ; Takayama, Masakazu ; Anazawa, Yoshihisa ; Matsushima, Yoshitarou

TOPOLOGY OPTIMIZATION FOR SKEW OF SPMSM BY USING MULTI-STEP PARALLEL GA 3809
Kitagawa, Wataru ; Takeshita, Takaharu

LOSS MINIMIZATION DESIGN USING MAGNETIC EQUIVALENT CIRCUIT FOR A PERMANENT MAGNET SYNCHRONOUS MOTOR 3815
Sato, Daisuke ; Itoh, Jun-ichi

THE PROPOSAL OF A NEW MOTOR WHICH HAS A HIGH WINDING FACTOR AND A HIGH SLOT FILL FACTOR 3823
Makita, Shinji ; Ito, Yasuhide ; Aoyama, Tomohiro ; Doki, Shinji

VARIABLE LEAKAGE FLUX INTERIOR PERMANENT MAGNET SYNCHRONOUS MACHINE FOR IMPROVING EFFICIENCY ON DUTY CYCLE 3828
Minowa, Masanao ; Hijikata, Hiroki ; Akatsu, Kan ; Kato, Takashi

HISTORY AND TRENDS OF CONVERTER TECHNOLOGY FOR DC AND AC TRANSMISSION IN JAPAN 3834
Yoshino, Teruo

ACCURATE OUTPUT POWER CONTROL OF CONVERTERS FOR MICROGRIDS BASED ON LOCAL MEASUREMENT AND UNIFIED CONTROL..3842
Meiqin Mao ; Zheng Dong ; Yong Ding ; Liuchen Chang

IMPEDANCE-BASED ANALYSIS OF ACTIVE FREQUENCY DRIFT ISLANDING DETECTION METHOD FOR GRID-TIED INVERTER SYSTEM 3850
Wen, Bo ; Boroyevich, Dushan ; Burgos, Rolando ; Shen, Zhiyu ; Mattavelli, Paolo

DEVELOPMENT OF 200-MVAR CLASS THYRISTOR SWITCHED CAPACITOR SUPPORTING FAULT RIDE-THROUGH 3857
Ohtake, Asuka ; Fei Zhang ; Fujimoto, Takafumi ; Nakayama, Naoyuki

DETAILED ANALYSIS AND DESIGN OF A THREE-PHASE PHASE-MODULAR ISOLATED MATRIX-TYPE PFC RECTIFIER 3864
Cortes, Patricio ; Fassler, Lukas ; Bortis, Dominik ; Kolar, Johann W. ; Silva, Marcelo

AN ENERGY SAVING DRIVE METHOD OF AN INDUCTION MOTOR WITH THE SUPPRESSION OF SUDDEN ACCELERATION AND DECELERATION 3872
Asano, Yuji ; Inoue, Kaoru ; Kotera, Keito ; Kato, Toshiji

FIELD ORIENTED CONTROL OF SENSORLESS LINEAR INDUCTION MOTOR USING MATRIX CONVERTER..3877
Sayed, Mahmoud A. ; Mohamed, Essam Ebaid ; Mohamed, Tarek Hassan ; Takeshita, Takaharu

A STATOR-EQUATION-BASED REDUCED-ORDER OBSERVER FOR POSITION-SENSORLESS VECTOR CONTROL SYSTEM OF DOUBLY-FED INDUCTION MACHINES 3885
Smiththisomboon, Somrat ; Suwankawin, Surapong

INPUT CURRENT RIPPLE ANALYSIS OF INVERTER FED DUAL THREE-PHASE AC MOTORS 3893
Dahono, Pekik Argo ; Satria, Andri

OFFLINE EXTRACTION OF INDUCTION MACHINE PARAMETERS FOR CONTROL STRATEGY SYNTHESIS 3898
Koschik, Stefan ; Bauer, Florian ; De Doncker, R.W.

HIGH CURRENT PLANAR TRANSFORMER FOR VERY HIGH EFFICIENCY ISOLATED BOOST DC-DC CONVERTERS 3905
Pittini, Riccardo ; Zhe Zhang ; Andersen, Michael A.E.

HIGH VOLTAGE-GAIN INTERLEAVED BOOST DC-DC CONVERTER DISCARDED ELECTROLYTIC CAPACITOR 3913
Nha, Quang Trong ; Huang-Jen Chiu ; Yu-Kang Lo ; Pham Phu Hieu

PARALLEL BI-DIRECTIONAL DC-DC CONVERTER FOR ENERGY STORAGE SYSTEM...3920
Ouchi, Takayuki ; Kanoda, Akihiko ; Takahashi, Naoya

CHARGING SCENARIO OF SERIAL BATTERY POWER MODULES WITH BUCK-BOOST CONVERTERS 3928
Jhen-Yu Jian ; Chu-Shen Chang ; Moo, Chin-Sien ; Hau-Chen Yen

COMPARATIVE THERMAL PERFORMANCE EVALUATION OF SIC MOSFETS AND SI MOSFET FOR 1.2 KW 300 KHZ DC-DC BOOST CONVERTER AS A SOLAR PV PRE-REGULATOR 3933
Taekyun Kim ; Minsoo Jang ; Agelidis, Vassilios G.

TOLERANCE ANALYSIS OF A CONSTANT-ON TIME CURRENT-MODE VOLTAGE REGULATOR WITH ADAPTIVE VOLTAGE POSITION FEATURE 3938

Chih Wei Chen ; Dan Chen ; Shin Shiung Wang

FPGA-BASED DIGITAL-CONTROLLED POWER CONVERTER DESIGNED WITH UNIVERSAL INPUT MEETING 80 PLUS PLATINUM EFFICIENCY CODE AND STANDBY POWER CODE FOR SEVER POWER APPLICATIONS 3942

Lai, Yen-Shin ; Ho, Kung-Min

STATIC AND DYNAMIC ANALYSES OF DIGITAL PEAK CURRENT MODE DC-DC CONVERTER 3950

Kajiwara, Kazuhiro ; Kurokawa, Fujio ; Shibata, Yuichiro

EXTENDED DISCRETE CONTROL OF CLASS E AMPLIFIER IN ORDER TO ACHIEVE NOMINAL OPERATION 3955

Suetsugu, Tadashi ; Xiuqin Wei ; Kuga, Shotaro

ADAPTIVE POWER EFFICIENCY CONTROL BY COMPUTER POWER CONSUMPTION PREDICTION USING PERFORMANCE COUNTERS 3959

Kawaguchi, Shinichi ; Yachi, Toshiaki

Author Index

The 2014 International Power Electronics Conference

Disturbance calculation based on space vector dot product: applications to compensators

Kelly Caroline Mingorancia de Carvalho, Naji Rajai Nasri Ama, Wilson Komatsu, Fernando Ortiz Martinz,
Ricardo Souza Figueredo, Lourenço Matakas Junior

Electrical Energy and Automation Department
Polytechnic School of the University of São Paulo
São Paulo, Brazil
kellymingorancia@gmail.com, matakas@pea.usp.br

Abstract-The ability to individually extract the many disturbances of the AC line currents or voltages is desirable for reference signal calculation algorithms employed in power conditioners, allowing improved utilization of the power converter. This paper proposes a novel algorithm based on the dot product of three dimensional space vectors in the *abc* coordinate system, for real time calculation of the instantaneous positive, negative and zero sequence components for each individual harmonic. It is presented the functionalities for compensating fundamental reactive power, selected current harmonics and current unbalances. This method is computationally simple and requires no coordinate transformation. The proposed algorithms are validated by simulation and experiments.

Keywords— Amplitude and phase estimation, phase-locked loops (PLL), power quality, space vectors.

I. INTRODUCTION

Many strategies are employed for the real time calculation of the reference signals for disturbance active compensators. One desirable feature for these reference calculation algorithm is its ability to separate the disturbance in individual relevant parts, since this characteristic has impact on the rating and on the operation of the converter. It is known that reactive power and unbalance compensation require high power, low frequency converter. On the other hand, harmonics compensation require high bandwidth for the converter controller, high switching frequency and a converter operating at lower power ratings. For high power applications, current specific harmonics compensation may be also desirable.

Among these strategies, it can be noted that instantaneous pq theory [1] does not require a phase locked loop (PLL) and is convenient to compensate a set of disturbances. However it is not possible to extract its individual components, and harmonic polluted voltages will require a PLL. Instantaneous Conservative Power theory [2] proposes a decomposition in a number of terms with physical meaning. However even a reactive compensation will need a high bandwidth converter to inject current harmonics that appear for distorted voltages.

Calculations based on the Fourier Series are shown in [3], and methods based on the dq transformation are also widely used [4]. The use of multiple dq frames rotating at different speeds and directions [5][6] enables the

extraction of the individual positive and negative sequence harmonics.

Methods based on the Fortescue Transformation [7] usually do not allow the separation of the many disturbances. Authors of [8] propose a new definition of instantaneous sequence components, but it cannot separate the individual harmonics. Reference [9] proposes a method based on the inner (dot) product to obtain the instantaneous fundamental positive sequence.

This paper proposes a novel strategy to extract the instantaneous positive, negative and zero sequences for each harmonic, based on the dot product (DP) of three dimensional stationary frame space vectors. Additionally, a graphical explanation based on *abc* and *αβ*0 facilitates the comprehension of this method. Three application examples are studied and numerically simulated, including the compensation of fundamental reactive power, current specific harmonics and current unbalances. The reactive current and unbalance compensation cases are experimentally validated.

Compared to existing methods, the proposed algorithm is simpler, has lower computational complexity, and requires no coordinate transformation. Moreover, it can also be applied to compensate several voltage and current disturbances such as specific harmonics, unbalances, or power factor correction and voltage sags and swells.. This strategy can also be extended to three phase four wire systems.

II. DISTURBANCE CALCULATION BASED ON SPACE VECTOR DOT PRODUCT

Considering an orthonormal basis *abc* formed by the unit length vectors $\vec{a}, \vec{b}, \vec{c}$ of $\Re^3$ (Fig. 1).

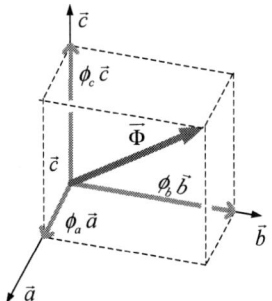

Fig. 1 Space vector in the *abc* system ($\Re^3$).

978-1-4799-2706-7/14 $31.00 © 2014 IEEE

With a set of time varying coordinates $\phi_a(t), \phi_b(t), \phi_c(t)$, one can define the instantaneous space vector $\vec{\Phi}$ (Fig. 1) by:

$$\vec{\Phi} = \phi_a(t)\vec{a} + \phi_b(t)\vec{b} + \phi_c(t)\vec{c} = [\phi_a \ \phi_b \ \phi_c]^T \tag{1}$$

where T is the transposed vector. The coordinates ϕ_a, ϕ_b, ϕ_c can be the line to neutral voltages v_a, v_b, v_c defining the vector $\vec{V}$ or the line currents i_a, i_b, i_c defining the vector $\vec{I}$. The proposed algorithm uses only variables in the abc coordinate system, but its explanation is more conveniently done by using the $\alpha\beta0$ system defined in [1],[2],[3] and shown in Fig. 2.

Considering $\vec{V}$, the $\alpha\beta$ plane is the locus of vector with null instantaneous zero sequence, defined by:

$$v_a(t) + v_b(t) + v_c(t) = 0 \tag{2}$$

An arbitrary space vector $\vec{\Phi}$ can be decomposed in a zero sequence vector $\vec{\Phi}^0$ and $\vec{\Phi}_{\alpha\beta}$ according to Fig. 3.

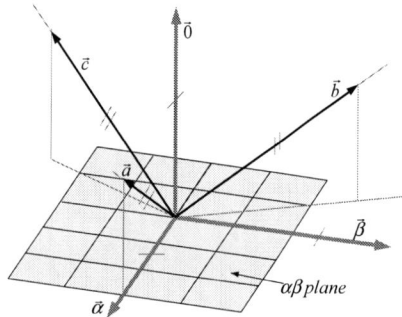

Fig. 2. abc and $\alpha\beta0$ $\Re^3$ coordinate systems.

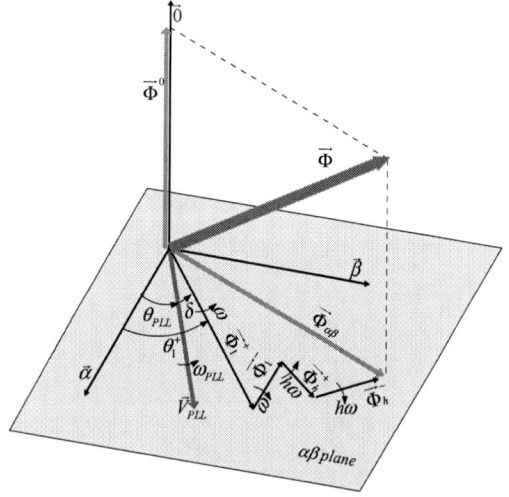

Fig.3. Vector $\vec{V}_{PLL}$ and $\vec{\Phi}$ (decomposed into $\vec{\Phi}_{\alpha\beta}, \vec{\Phi}_1^+, \vec{\Phi}_1^-, \vec{\Phi}_h^+, \vec{\Phi}_h^-, \vec{\Phi}^0$).

The vector $\vec{\Phi}_{\alpha\beta}$ lies on the $\alpha\beta$ plane and contains the vectors corresponding to the hth order positive and

negative sequence harmonic components $\vec{\Phi}_h^+$ and $\vec{\Phi}_h^-$ respectively, as shown in Fig. 3. $\vec{\Phi}_h^+$ rotates in counter clockwise direction with $h\omega$ speed and $\vec{\Phi}_h^-$ rotates in the clockwise direction with $h\omega$ speed.

A. Calculation of hth order positive sequence harmonic components $\vec{\Phi}_h^+$

The dot product (DP) of two vectors rotating at the same speed and direction is constant. Otherwise the DP will be an oscillating signal with zero mean value. This property can be used to calculate the hth order positive sequence harmonic components $\vec{\Phi}_h^+ = [\phi_{ah}^+ \ \phi_{bh}^+ \ \phi_{ch}^+]^T$. In this way, the dot products $\vec{\Phi} \bullet \vec{V}_{PLL\ h\parallel}^+$ and $\vec{\Phi} \bullet \vec{V}_{PLL\ h\perp}^+$ are directly evaluated using abc coordinates, where the PLL positive sequence vectors $\vec{V}_{PLL\ h\perp}^+$ and $\vec{V}_{PLL\ h\parallel}^+$ (Fig. 4) rotate in the counter clockwise direction with speed $h\omega$ on the $\alpha\beta$ plane and are defined by.

$$\begin{bmatrix} \vec{V}_{PLL\ h\perp}^+ = \begin{bmatrix} v_{PLLah\perp} \\ v_{PLLbh\perp} \\ v_{PLLch\perp} \end{bmatrix} = \begin{bmatrix} \cos\left(h(\theta_{PLL})\right) \\ \cos\left(h(\theta_{PLL} - 2\pi/3)\right) \\ \cos\left(h(\theta_{PLL} + 2\pi/3)\right) \end{bmatrix} \\ \vec{V}_{PLL\ h\parallel}^+ = \begin{bmatrix} v_{PLLah\parallel} \\ v_{PLLbh\parallel} \\ v_{PLLch\parallel} \end{bmatrix} = \begin{bmatrix} \cos\left(h(\theta_{PLL} + \pi/2)\right) \\ \cos\left(h(\theta_{PLL} - 2\pi/3 + \pi/2)\right) \\ \cos\left(h(\theta_{PLL} + 2\pi/3 + \pi/2)\right) \end{bmatrix} \end{bmatrix} \tag{3}$$

For h=1 the vectors $\vec{V}_{PLL\ h\perp}^+$ and $\vec{V}_{PLL\ h\parallel}^+$, are synchronized to positive sequence fundamental frequency component $\vec{V}_1^+$ of the mains vector $\vec{V}$. This is done by obtaining the angle θ_{PLL}, which is generated by a three phase PLL[1] so as to force the vector $\vec{V}_{PLL\ 1\parallel}^+$, calculated by imposing $h = 1$ in (3), to be instantaneously parallel to the positive sequence vector of the mains fundamental frequency ($\vec{V}_1^+$). The strategy used in this paper is shown in appendix I.

As the zero sequence component is instantaneously orthogonal to $\vec{V}_{PLL\ h\perp}^+$ and $\vec{V}_{PLL\ h\parallel}^+$, and remembering that the DP of two orthogonal vectors is null, one can obtain (4):

$$\begin{bmatrix} \vec{\Phi} \bullet \vec{V}_{PLL\ h\parallel}^+ = (\vec{\Phi}_{\alpha\beta} + \vec{\Phi}^0) \bullet \vec{V}_{PLL\ h\parallel}^+ = \vec{\Phi}_{\alpha\beta} \bullet \vec{V}_{PLL\ h\parallel}^+ \\ \vec{\Phi} \bullet \vec{V}_{PLL\ h\perp}^+ = (\vec{\Phi}_{\alpha\beta} + \vec{\Phi}^0) \bullet \vec{V}_{PLL\ h\perp}^+ = \vec{\Phi}_{\alpha\beta} \bullet \vec{V}_{PLL\ h\perp}^+ \end{bmatrix} \tag{4}$$

Where:

$$\vec{\Phi}_{\alpha\beta} = \vec{\Phi}_1^+ + \vec{\Phi}_1^- + \vec{\Phi}_h^+ + \vec{\Phi}_h^- \tag{5}$$

---

[1] Any three phase PLL that tracks on the fundamental positive sequence of the input signal can be used [3][4][5][6]. This paper uses strategies implemented in the abc coordinate system [3][4][5] (See appendix I).

The geometrical analysis of the algorithm to extract the hth order positive sequence harmonic components can be done by using the vector diagram shown in Fig. 4 in the $\alpha\beta$ plane.

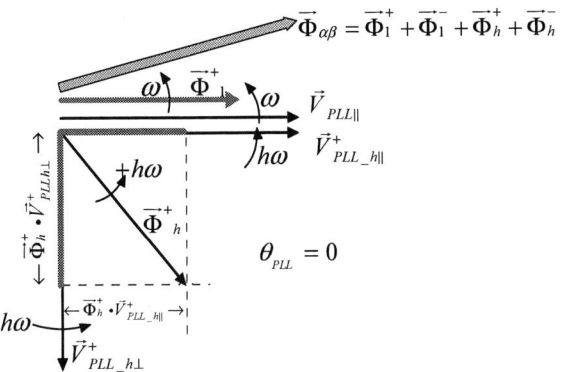

Fig.4 Vectors $\vec{\Phi}_1^+$, $\vec{\Phi}_h^+$, $\vec{\Phi}_{\alpha\beta}$, $\vec{V}_{PLL\,\|}$, $\vec{V}_{PLL\,h\|}^+$ e $\vec{V}_{PLL\,h\perp}^+$ in the $\alpha\beta$ plane.

Fig. 5 presents the positive sequence extraction algorithm as a single wire block diagram, where $\vec{\Phi}\cdot\vec{V}_{PLL\,h\|}^+$ is low pass filtered, originating $\vec{\Phi}_h^+\cdot\vec{V}_{PLL\,h\|}^+$, which followed by a gain $2/3$ results in the magnitude of the projection of $\vec{\Phi}_h^+$ on the vector $\vec{V}_{PLL\,h\|}^+$ (Fig. 4), named as $\Phi_{h\|}^+$. The instantaneous values of the parallel terms are finally obtained by $\Phi_{h\|}^+\,\vec{V}_{PLL\,h\|}^+$. The same reasoning is done for the orthogonal component $\vec{\Phi}\cdot\vec{V}_{PLL\,h\perp}^+$, resulting in the extracted positive sequence components of the hth harmonics:

$$\vec{\Phi}_h^+ = [\phi_{ah}^+\ \phi_{bh}^+\ \phi_{ch}^+] = \Phi_{h\|}^+\,\vec{V}_{PLL\,h\|}^+ + \Phi_{h\perp}^+\,\vec{V}_{PLL\,h\perp}^+ \quad (6)$$

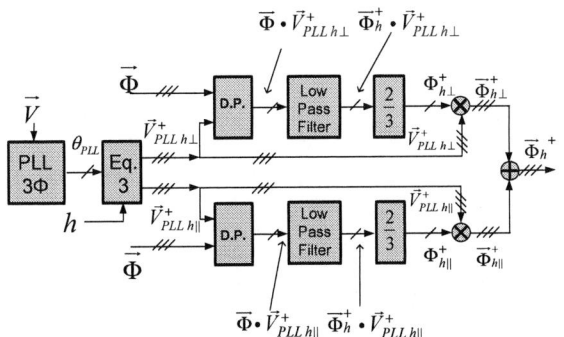

Fig. 5 Positive sequence ($\vec{\Phi}_h^+$) extraction algorithm - single wire block diagram.

B. Calculation of hth order negative sequence harmonic components $\vec{\Phi}_h^-$

The negative sequence harmonics $\vec{\Phi}_h^- = [\phi_{ah}^-\ \phi_{bh}^-\ \phi_{ch}^-]^T$ can be obtained by substituting the pair $\vec{V}_{PLL\,h\|}^+$, $\vec{V}_{PLL\,h\perp}^+$ by a new pair of negative sequence orthogonal vectors

$\vec{V}_{PLL\,h\|}^-$ and $\vec{V}_{PLL\,h\perp}^-$ that rotates with speed $h\omega$ in the clockwise direction, as shown in Fig. 6. These two vectors are easily obtained by swapping the signals corresponding to phases b and c in (3), as it is implemented in the single wire diagram in Fig.7 by the "phase swap" block.

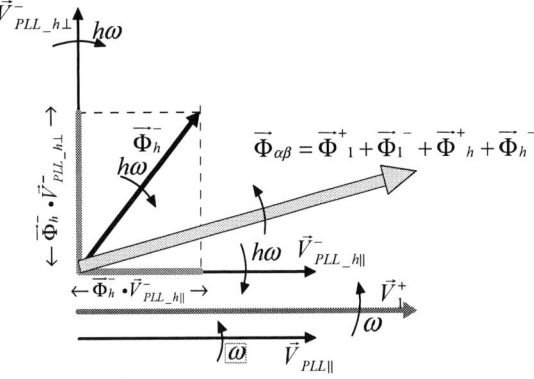

Fig. 6 Vectors $\vec{V}_1^+$, $\vec{\Phi}_h^-$, $\vec{\Phi}_{\alpha\beta}$, $\vec{V}_{PLL\,\|}^+$, $\vec{V}_{PLL\,h\|}^-$ and $\vec{V}_{PLL\,h\perp}^-$ in the $\alpha\beta$ plane.

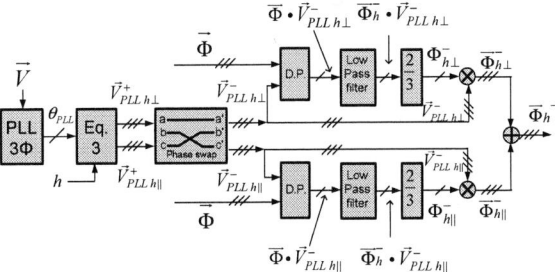

Fig. 7 Negative sequence ($\vec{\Phi}_h^-$) extraction algorithm - single wire block diagram.

C. Calculation of hth order zero sequence harmonic components $\vec{\Phi}_h^0$

The calculation of $\vec{\Phi}_h^0 = [\phi_{ah}^0\ \phi_{bh}^0\ \phi_{ch}^0]^T$ is based in the dot products $\vec{\Phi}\cdot\vec{V}_{PLL\,h\|}^0$ and $\vec{\Phi}\cdot\vec{V}_{PLL\,h\perp}^0$, where $\vec{V}_{PLL\,h\|}^0$ and $\vec{V}_{PLL\,h\perp}^0$ are zero sequence vectors with $h\omega$ frequency:

$$\begin{bmatrix} \vec{V}_{PLL\,h\|}^0 = v_{PLLah\|}[1\ 1\ 1]^T = \cos\!\big(h(\theta_{PLL})\big)[1\ 1\ 1]^T \\ \vec{V}_{PLL\,h\perp}^0 = v_{PLLah\perp}[1\ 1\ 1]^T = \cos\!\big(h(\theta_{PLL}+\pi/2)\big)[1\ 1\ 1]^T \end{bmatrix} \quad (7)$$

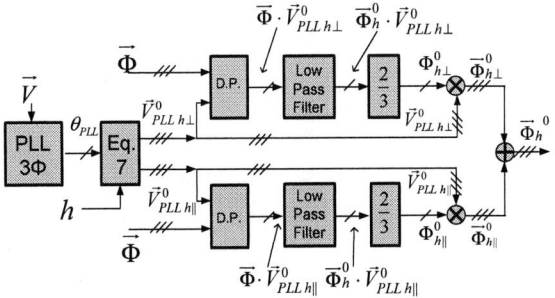

Fig. 8 Zero sequence ($\vec{\Phi}_h^0$) extraction algorithm - single wire block diagram.

In order to avoid unnecessary calculation, the block diagram in Fig. 8 can be simplified to the one shown in Fig. 9. The instantaneous phase amplitudes of a zero component signal are equal; therefore, the product of only one term of the vector $\vec{\Phi}^0 = [\phi^0 \; \phi^0 \; \phi^0]^T$ is needed to calculate the zero h^{th} order sequence component by:

$$\phi^0 = \frac{\phi_a + \phi_b + \phi_c}{3} \qquad (8)$$

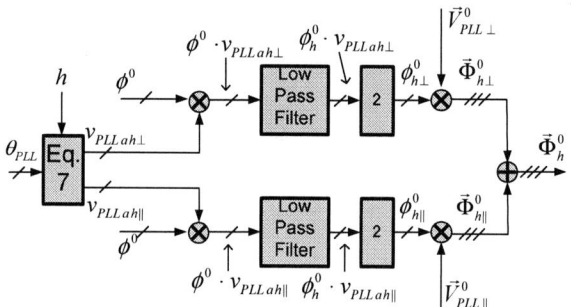

Fig. 9 Simplified zero sequence ($\vec{\Phi}_h^0$) extraction algorithm - single wire block diagram.

III. APPLICATION TO PARALLEL DISTURBANCE COMPENSATORS - SIMULATION RESULTS

Fig. 10 shows the grid voltages, a delta connected RL switched load (Load 1, $R_1 = 33\Omega$ and $L_1 = 400mH$), a non linear load composed by a three-phase diode rectifier (Load 2, $R_2 = 370\Omega$, $L_2 = 1.7\,mH$, $C_2 = 47\,\mu F$) and a three phase voltage source converter which compensates a set of chosen load disturbances. The converter is rated for 220V line to line RMS voltage, 3.5A RMS line current, 350V DC voltage, using line inductor filters of 10mH, and 800µF DC side capacitor.

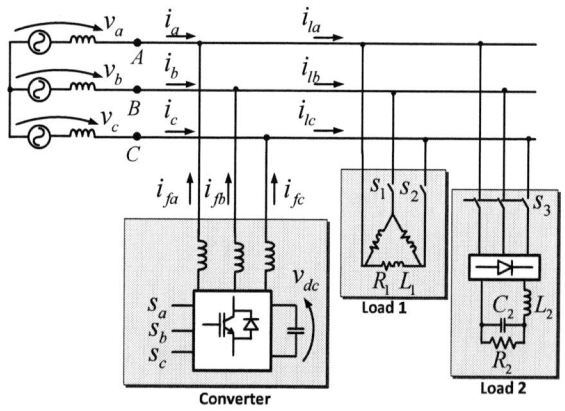

Fig. 10 Simulation / Experimental Setup: VSC converter and loads.

Closing only switch S_1 of Load 1 results in a single phase RL load connected between phases A and B (unbalanced load with reactive power). Simultaneous closure of S_1 and S_2 results in a balanced three-phase load with reactive power. Switch S_3 of Load 2 connects

to the grid a diode rectifier with LC filter operating in discontinuous current mode.

Fig. 11 presents the control diagram for the converter of Fig. 10, including the reference current calculator, the DC voltage control loop, the current tracking loop, the PLL and the PWM block. To maintain coherence with the proposed algorithms for disturbance calculation, all the blocks in Fig. 11 are implemented in *abc* system, with no coordinate transformation. Current loop uses three individual digital PI controllers with the anti windup algorithm presented in [10], one for each phase. To avoid the well known non controllability of three phase, three wire converters using three independent controllers, this paper subtracts the instantaneous zero sequence component, given by $(i_{fa} + i_{fb} + i_{fc})/3$, from the measured filter currents i_{fa}, i_{fb}, i_{fc}. The current loop digital PI controller parameters were calculated to set the damping factor of the closed loop poles to 0.4, resulting in $K_p = 40.4$ and $T_i = 402\mu s$. Besides that, a feed forward of the grid voltage is also included, to improve the disturbance and transient response. Three individual single update sampled PWM are employed. Instantaneous zero sequence is added to the PWM reference signal to emulate Space Vector PWM behavior [11]. Carrier and sampling frequency were set to 12KHz and grid frequency is 60Hz.

The PLL block is based on dot product PLL [12], also known as pPLL. The Proportional Integral controller gains were adjusted as described in [13], resulting in k_{PPLL}=139.426 and k_{IPLL}=0.574. DC voltage controller uses a PI compensator, and regulates v_{dc}^2 instead of v_{dc} for exact linearization of the DC loop transfer function. Setting $K_p = 5.94\times10^{-5}$ and $T_i = 49ms$, the resulting settling time is equal to 9 cycles of mains frequency.

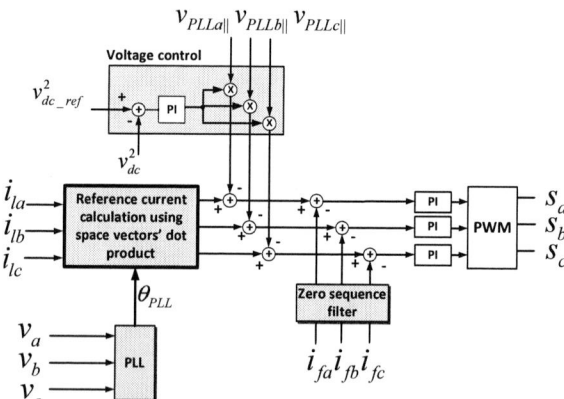

Fig. 11. Block Diagram of the compensator control loops.

The proposed method to calculate positive and negative sequences of voltage and currents is applied to the three different disturbance compensators described below:

- Fundamental Reactive Currents Compensation;
- Fundamental Negative Sequence Compensation;
- Current specific Harmonics Compensation.

A. Fundamental Reactive Currents Compensation

The calculation of the reference signals for the fundamental reactive current compensator, described in the block diagram of Fig. 12, can be derived from Fig. 5 by:
- imposing the load currents as input vector $\vec{\Phi} = \vec{I}_l = [i_{la}\ i_{lb}\ i_{lc}]^T$;
- imposing $h = 1$;
- using only the lower signal path corresponding to the current components orthogonal to the positive sequence of the fundamental grid voltages;

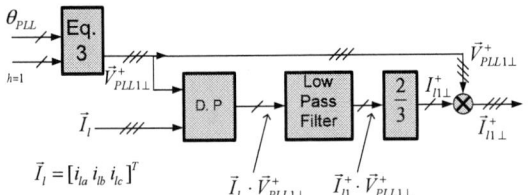

$$\vec{I}_l = [i_{la}\ i_{lb}\ i_{lc}]^T \qquad \vec{I}_l \cdot \vec{V}_{PLL1\perp}^+ \quad \vec{I}_{l1}^+ \cdot \vec{V}_{PLL1\perp}^+$$

Fig. 12. Reference Calculation Block of Fig. 11 for the Fundamental Reactive Current Compensator.

Moving average low pass filter with a sampling frequency of 12kHz and a window width of $T/2$ (where T is the grid voltage period) was employed for the three compensators. Simulation results were obtained using the PSIM software. Fig. 13 shows line to neutral voltages, grid currents and load currents for the fundamental reactive power compensation. Till 0.2s it presents steady state operation with non linear Load 1 (S_3 On). At 0.2s switches S_1 and S_2 are turned on, including Load 2 (linear balanced RL load).

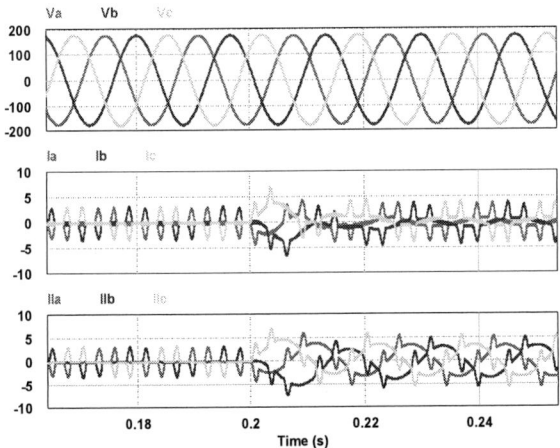

Fig. 13. Simulated results for fundamental reactive power compensator: top - line to neutral voltages (v_a, v_b, v_c), middle: line currents (i_a, i_b, i_c), bottom- load currents (i_{la}, i_{lb}, i_{lc}).

Table I summarizes the calculated values of the positive sequence of the load and the grid fundamental currents (peak values) for $t < 0.2s$ and $t \gg 0.2s$. Grid voltage positive sequence is $179.63 \underline{|0^o}$ V (peak).

TABLE I
CALCULATED VALUES OF FUNDAMENTAL POSITIVE SEQUENCE OF GRID AND LOAD CURRENTS - REACTIVE POWER COMPENSATION - SIMULATION

	Grid (A - peak)	Load (A - peak)
$t < 0.2s$	$0.936\underline{|-0.7^\circ}$	$0.936\underline{|0.1^\circ}$
$t \gg 0.2s$	$1.712\underline{|2.0^\circ}$	$3.711\underline{|-63.4^\circ}$

Reactive fundamental currents are adequately compensated. Time response is expected to be equal to the moving average filter width ($T/2$). Longer transient times in Fig, 13 are explained by the RL load large time constant ($\tau = L_1 / R_1 \sim 12ms$), reaching steady state in about 2 cycles. Current harmonics do not disturb the operation of the reactive power compensator. This is especially important for high power converters, which will be able to inject only mains frequency currents, and operate with lower switching frequencies.

B. Fundamental negative sequence compensation

The calculation of the reference signals for the fundamental negative sequence current compensator, described in the block diagram of Fig. 14, can be obtained from Fig. 7 by:
- imposing the load currents as input vector $\vec{\Phi} = \vec{I}_l = [i_{la}\ i_{lb}\ i_{lc}]^T$;
- imposing $h = 1$.

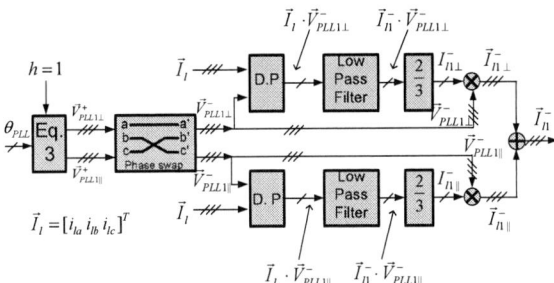

Fig. 14 Reference Calculation Block of Fig. 11 for the Fundamental Negative Sequence Current Compensator.

Fig. 15 shows line to neutral voltages, grid currents and load currents for the Fundamental Negative Sequence Current Compensator. Till 0.2s it presents steady state operation with unbalanced linear load (Load 2 with S_1 On). At 0.2s the load becomes a balanced inductive load (S_1 and S_2 are turned on).

Table II presents the calculated values of the positive and negative sequences of the load and the grid fundamental currents (peak values) for $t < 0.2s$ and $t \gg 0.2s$. Grid voltage positive sequence is $179.63 \underline{|0^o}$ V (peak).

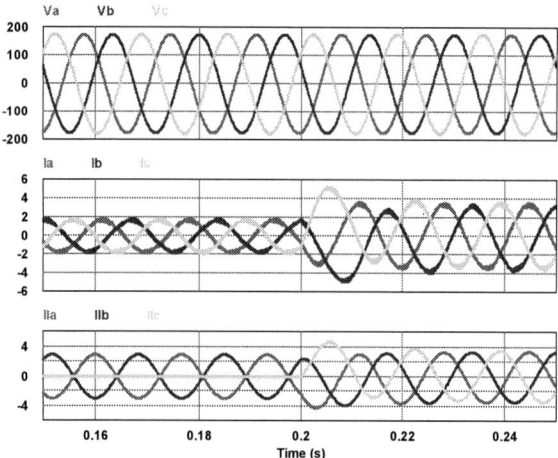

Fig. 15. Simulated results for fundamental negative sequence compensator: top - line to ground voltages (v_a, v_b, v_c), middle: line currents (i_a, i_b, i_c), bottom- load currents (i_{la}, i_{lb}, i_{lc}).

TABLE II
CALCULATED VALUES OF FUNDAMENTAL SYMMETRICAL COMPONENTS
FOR GRID AND LOAD CURRENTS - NEGATIVE COMPONENT
COMPENSATION- SIMULATION

		I_{grid}	I_{load}
$t < 0.2s$	positive	$1.709\lfloor{-77.2°}$	$1.707\lfloor{-77.7°}$
	negative	$0.040\lfloor{-172.3°}$	$1.707\lfloor{-17.7°}$
$t \gg 0.2s$	positive	$3.427\lfloor{-77.5°}$	$3.415\lfloor{-77.7°}$
	negative	0.000	0.000

For $t < 0.2s$ the negative sequence is adequately compensated. The long settling time for $t > 0.2s$ is again explained by the RL load transient.

C. Current specific harmonics compensation

The calculation of the reference signals for current specific harmonics compensation described in the block diagram of Fig. 16, considering only the 5th and 7th harmonics, can be obtained from Fig. 5 and Fig. 7 by:

- imposing the load currents as input vector $\vec{\Phi} = \vec{I}_l = [i_{la} \ i_{lb} \ i_{lc}]^T$;

- considering that the three phase load currents have the same waveform and are equally displaced, this results in negative sequence 5th harmonic and positive sequence 7th harmonic. As a consequence Fig. 16 simultaneously uses Fig. 7 block algorithm with $h = 5$, and Fig. 5 with $h = 7$.

As the PI controller with sampling frequency of 12kHz presents excessive tracking error for the 5th and 7th harmonics, the harmonics compensator employs three proportional plus resonant controllers tuned at 60Hz, 300Hz and 420Hz respectively, according to (9):

$$40.4\left(1 + \frac{(2 \cdot 2\pi 60)s}{s^2 + (2\pi 60)^2} + \frac{(2 \cdot 5 \cdot 2\pi 60)s}{s^2 + (5 \cdot 2\pi 60)^2} + \frac{(2 \cdot 7 \cdot 2\pi 60)s}{s^2 + (7 \cdot 2\pi 60)^2}\right)$$
(9)

Using Tustin with pre-warping [14] method in (9) results in the discrete transfer function:

$$40.4 + \frac{1.269(1 - z^{-2})}{1 - 1.999z^{-1} + z^{-2}} + \frac{6.320(1 - z^{-2})}{1 - 1.975z^{-1} + z^{-2}} + \frac{8.884(1 - z^{-2})}{1 - 1.952z^{-1} + z^{-2}}$$
(10)

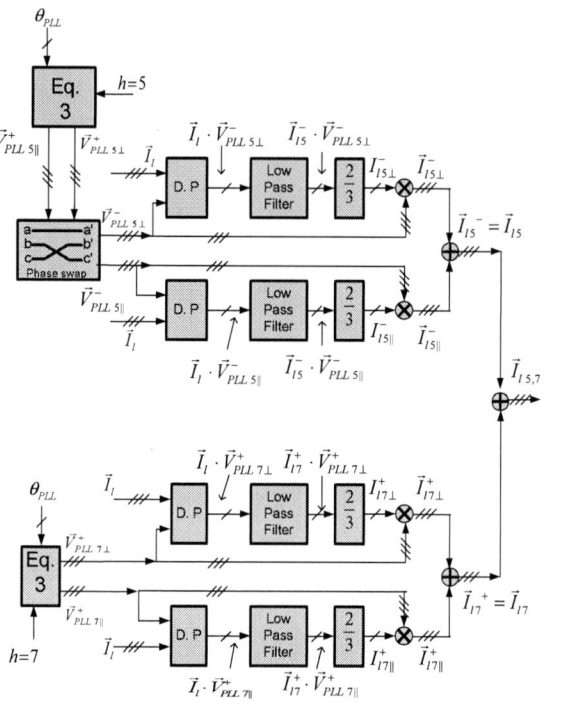

Fig. 16. Reference Calculation Block of Fig. 11 for the compensation of 5th and 7th harmonics.

Fig. 17 presents simulated line and load current waveforms for the compensation case of Fig. 16. For $t < 0.2s$ S_3 is on, and the non linear load is connected to the grid. For $t > 0.2s$ S_3 is off.

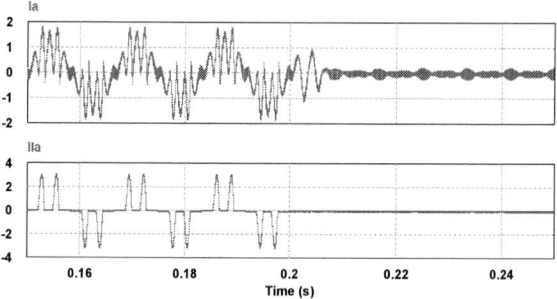

Fig. 17 Simulated results for current harmonics : top - line current (i_a), bottom- load currents (i_{la}).

Table III shows calculated values of the 1st, 5th and 7th harmonics for the grid and load phase a current, for $t < 0.2s$.

TABLE III
CALCULATED 1ST, 5TH AND 7TH CURRENT COMPONENTS (A PEAK) FOR
CURRENT SPECIFIC HARMONICS COMPENSATION - SIMULATION

	I_{grid}	I_{load}
1	0.952	0.947
5	0.002	0.840
7	0.003	0.742
11	0.501	0.500
13	0.376	0.376

The 5th and 7th harmonics are adequately attenuated. Fig. 17 shows that the settling time is around half of mains period, which corresponds to the employed moving average filter window size. For $t > 0.2s$ grid harmonic components presented in table III are null, since there is no load connected to the grid.

IV. EXPERIMENTAL RESULTS

Control loops and PWM blocks of the experimental setup were implemented in a TMS320F28335 DSP. Waveforms were obtained using Agilent DSO6014 oscilloscope, current probes Tektronix A6302, and then data were processed by using MATLAB. The performed experiments have used the same loads and the same controllers described in section III and the setup of Fig. 10.

Fig. 18 shows the line to neutral voltages corresponding to the fundamental reactive current compensation using Load 1 (*S1* and *S2* On) and Load 2 (*S3* On) in steady state operation.

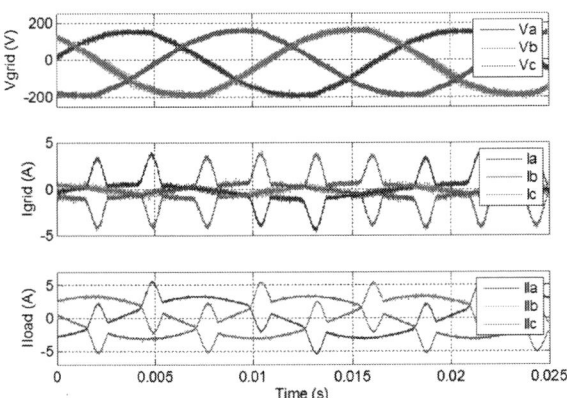

Fig. 18. Experimental results for fundamental reactive power compensator: top - line to ground voltages (v_a, v_b, v_c), middle: line currents (i_a, i_b, i_c), bottom- load currents (i_{la}, i_{lb}, i_{lc}).

Table IV shows the fundamental symmetrical components of the grid voltage, the compensated grid current and the load current.

TABLE IV
CALCULATED SYMMETRICAL COMPONENTS FOR REACTIVE POWER
COMPENSATION- EXPERIMENTAL

	V_{grid} (Vpeak)	I_{grid} (Apeak)	I_{load} (Apeak)
positive	$176.15\underline{|-0.5^o}$	$1.62\underline{|-3.5^o}$	$3.42\underline{|-62.6^o}$
negative	$0.82\underline{|-149.6^o}$	$0.02\underline{|-40.8^o}$	$0.04\underline{|-41.1^o}$
zero	$1.25\underline{|-106.8^o}$	0.00	$0.01\underline{|-62.6^o}$

Reactive currents were compensated as expected and the results are similar to the ones obtained in Table I for the simulated case.

Fig 19 shows the fundamental negative sequence compensation using Load 1 (*S1* On and *S2* Off). The waveforms were taken in steady state. Table V summarizes the calculated symmetric components of the measured values, showing good agreement with the results presented in Table II for the simulated case.

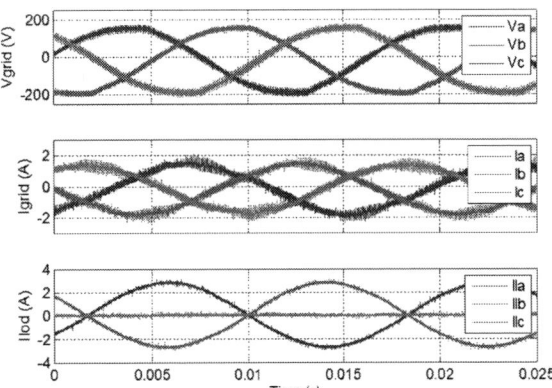

Fig. 19. Experimental results for fundamental negative sequence compensator waveforms: top - line to ground voltages (v_a, v_b, v_c), middle: line currents (i_a, i_b, i_c), bottom- load currents (i_{la}, i_{lb}, i_{lc}).

TABLE V
CALCULATED SYMMETRICAL COMPONENTS FOR FUNDAMENTAL
NEGATIVE COMPONENT COMPENSATION- EXPERIMENTAL

	V_{grid}	I_{grid}	I_{load}
positive	$175.82\underline{|0.3^o}$	$1.57\underline{|-73.0^o}$	$1.59\underline{|-76.8^o}$
negative	$0.63\underline{|-80.4^o}$	$0.03\underline{|-3.1^o}$	$1.59\underline{|-16.7^o}$
zero	$0.29\underline{|-137.8^o}$	$0.01\underline{|-109.7^o}$	$0.02\underline{|-127.9^o}$

V. CONCLUSION

This paper has presented a method to calculate the instantaneous positive, negative and zero sequences of any current or voltage harmonic component, which is based on the dot product of three dimensional space vectors in the stationary frame. Simulated data have shown the compensation of current harmonics, reactive power, and unbalances. The two last cases were experimentally verified, validating the algorithm. This paper has demonstrated that the proposed strategies allow separation of the individual disturbances, in a computationally simpler and faster way than those presented in literature, with no need to perform coordinate transformations.

APPENDIX I – PLL

The implemented PLL block is shown in Fig. 20.

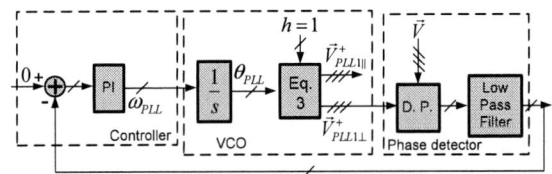

Fig. 20. PLL block diagram.

VCO block generates two sets of positive sequence signals, $\vec{V}^+_{PLL1\perp}$ and $\vec{V}^+_{PLL1\parallel}$, with unit amplitude and frequency ω_{pll}. The phase detector block calculates the dot product (D.P.) between $\vec{V}$ and $\vec{V}^+_{PLL1\perp}$ and filters its ripple by a low pass filter. The PI controller adjusts ω_{PLL}

to force $\vec{V}^+_{PLL\,1\perp}$ to be in quadrature with the fundamental positive sequence of $\vec{V}$, and $\vec{V}^+_{PLL\,1\|}$ to be in phase with the fundamental positive sequence of $\vec{V}$. The PI controller is adjusted according to [13].

ACKNOWLEDGMENT

The authors are grateful to Texas Instruments Academic Program, for donating the DSP Evaluation Kit and respective software, and to CNPq, FAPESP and FDTE for the financial support.

REFERENCES

[1] Akagi et al "Instantaneous Power Theory and Applications to Power Conditioning", John Wiley and Sons, 2007, 380p.

[2] Tenti et al. "Conservative Power Theory, a Framework to Approach Control and Accountability Issues in Smart Microgrids, "*Power Electronics, IEEE Transactions on*, vol.26, no.3, pp. 664-,673, March 2011

[3] Tong Xiangqian et al., "Single order harmonic detection with incremental algorithm of DFT, "*Power Electronics and Motion Control Conference, 2009. IPEMC '09. IEEE 6th International*, vol. , no. , pp. 2402-2405, May 2009

[4] Xu Jin; Hu Min; Chen Haiyun; Fang Wei, "A Detection Approach for Random Harmonic Current in Single-Phase Circuit," *Intelligent System Design and Engineering Application (ISDEA), 2012 Second International Conference on* , vol. , no. , pp. 1346-1350, Jan. 2012

[5] Asiminoaei et al., "Detection is key - Harmonic detection methods for active power filter applications," *Industry Applications Magazine, IEEE*, vol. 13, no. 4, pp. 22-33, July-Aug. 2007

[6] Su Shiping et al."A Specific Harmonic Detection Method for Three-Phase Four-Wire Circuits Based on Multi-frequency Rotating Coordinate Transformation, "*Electrical and Control Engineering (ICECE), 2010 International Conference on*, vol. , no. , pp. 261-264, June 2010

[7] Cutri, R.; Matakas, L., "A Generalized Instantaneous Method for Harmonics, Positive and Negative Sequence Detection/Extraction," *Power Electronics Specialists Conference, 2007. PESC 2007. IEEE* , vol. , no. , pp. 2294-2297, June 2007.

[8] Tenti, et al. "Generalized symmetrical components for periodic non-sinusoidal three-phase signals." *Electrical Power Quality and Utilisation Journal* 13.1 (2007): 9-15.

[9] Pádua et al. "Frequency-Adjustable Positive Sequence Detector for Power Conditioning Applications," *Proceedings of IEEE Power Electronics Specialists Conference - PESC'05*, pp. 1928–1934, June 2005.

[10] Buso, S.; Mattavelli, P., *Digital Control in Power Electronics*. 1st ed. San Rafael: Morgan & Claypool, 2006, 151p. ISBN 978-1598291124

[11] D. G. Holmes and T. A. Lipo, *"Pulse width modulation for power converters: principles and practice,"* IEEE Press Series Power Eng., 2003.

[12] Matakas, L.; Komatsu, W.; Martinz, F.O., "Positive sequence tracking Phase Locked Loops: A unified graphical explanation," *Power Electronics Conference (IPEC), 2010 International*, vol., no., pp.1273-1280, June 2010.

[13] Naji Rajai Nasri AMA, Wilson KOMATSU, Fuad KASSAB JUNIOR, Lourenco MATAKAS JUNIOR, "Adaptive single phase moving average filter PLLs: analysis, design, performance evaluation and comparison", Przeglad Elektrotechniczny, Issue 3, 2014.

[14] A.G. Yepes, F.D. Freijedo, J. Doval-Gandoy, O. Lopez, J. Malvar, and P. Fernandez-Comesana, "Effects of Discretization Methods on the Performance of Resonant Controllers", *IEEE Transactions on Power Electronics*, vol. 25, no. 7, pp. 1692-1712, 2010.

Proposal of 6th Radial Force Control Based on Flux Linkage
– Verification on Load Condition –

Masato Kanematsu, Takayuki Miyajima
Hiroshi Fujimoto, and Yoichi Hori
The University of Tokyo
5-1-5 Kashiwanoha,
Kashiwa, Chiba, 277-8561 Japan
kanematsu, miya@hflab.k.u-tokyo.ac.jp
fujimoto, hori@k.u-tokyo.ac.jp

Toshio Enomoto, Masahiko Kondou, Hiroshi Komiya
Kantaro Yoshimoto, and Takayuki Miyakawa
Nissan Motor Co., Ltd.
1-1, Morinosatoaoyama, Atsugi-shi
Kanagawa, 243-0123, Japan
enomoto, m-kondo, h-komiya@mail.nissan.co.jp
ka-yoshimoto, takayuki_miyakawa@mail.nissan.co.jp

Abstract— **IPMSMs (Interior Permanent Magnet Synchronous Motors) are widely used for many industrial applications. However, IPMSMs cause large noise and vibration due to torque ripple and radial force. In this paper, 6th radial force modelling is constructed based on flux linkage and 6th radial force is controlled on the ground of this modelling. Firstly, we make some assumptions to lead 6th radial force modelling. Secondly, 6th radial force modelling is derived. Simulation results and Experimental results verify the validity of 6th radial force model. Finally, 6th radial force control is performed by injecting dq-axis harmonic current.**

I. INTRODUCTION

In many industrial applications, IPMSMs (Interior Permanent Magnet Synchronous Motors) are often selected as drive motors. In these applications, IPMSMs face strong demands about the reduction of noise and vibration. Specially the noise and vibration problems in the inside of cars remain to be one of the problems which should be solved. Furthermore lower acoustic noise and vibration is desirable to enhance the value of products[1].

The analysis technology on the vibration of electromagnetic force is investigated in [2][3][4] and [5]. The Reference [6] and [7] propose the designing method to reduce radial force vibration. It is natural to apply skew method for reducing the vibration of radial force [8][9][10]. However, the methods with structal designing usually increase the cost of the products. We therefore focus on the method to reduce radial force with current control. The Reference [11] proposes the control method to suppress harmonic radial force with harmonic current. However, this method needs iterative calculations of electromagnetic field analysis in each drive condition and how to lead optimal reference to suppress radial force is unclear.

It is known that in IPMSMs electrical 2nd and 6th radial forces usually cause serious noise and vibration. It is known that there exist phase differences between the 2nd radial force

on U, V and W-phase tooth as:

$$F_{rU}(\theta) = F_{r2}\cos 2\theta \tag{1}$$

$$F_{rV}(\theta) = F_{r2}\cos 2\left(\theta - \frac{2}{3}\pi\right) \tag{2}$$

$$F_{rW}(\theta) = F_{r2}\cos 2\left(\theta - \frac{4}{3}\pi\right) \tag{3}$$

where F_{r2} is the amplitude of 2nd radial force. This causes Pth order annular elastic deformation, where P denotes pole pairs. 2nd radial froce modelling and control method have been proposed in [15]. On the other hand, 6th radial forces on U, V and W-phase tooth are in phase as:

$$F_{rU}(\theta) = F_{r6}\cos 6\theta \tag{4}$$

$$F_{rV}(\theta) = F_{r6}\cos 6\left(\theta - \frac{2}{3}\pi\right) \tag{5}$$

$$F_{rW}(\theta) = F_{r6}\cos 6\left(\theta - \frac{4}{3}\pi\right) \tag{6}$$

Therefore the transfer characteristics from radial force to acceleration has very high amplitude because 6th radial force excites spatially 0th annular mode. Fig. 1 shows the concept of 2nd and 6th radial force and typical natural frequency mode of the stator.

In this paper, 6th component of the radial force on a tooth is expressed mathematically based on flux linkage. This expression clarifies the relationship between dq-axis harmonic current and 6th radial force and enables us to control 6th radial force. The simulation and experiment are performed to validate the utility of the modelling and suppression control method.

II. ASSUMPTION OF APPROXIMATION

In this chapter, assumption of approximation are shown. Based on these assumptions, approximation model of 6th radial force is derived. JMAG (electromagnetic field analysis software) produced by JSOL Corporation is utilized for this analysis.

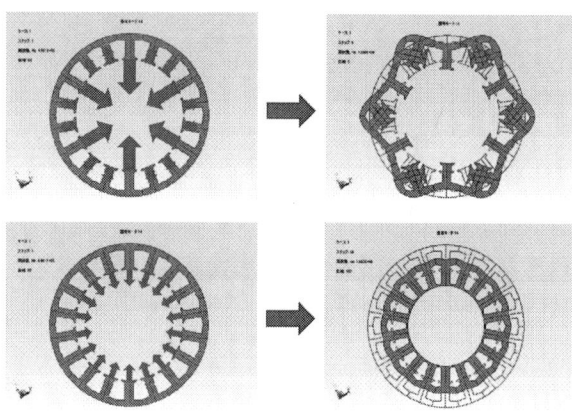

Fig. 1. Typical natural frequency mode of stator(12P18S)

A. The relationship between current, flux linkage and radial force

Flux linkage through a U-phase tooth $\phi_u(t)$ is expressed as

$$\phi_u(t) = \frac{\psi_u(t)}{N} \tag{7}$$

where, N is turn number per a phase and ψ_u is flux linkage on U-phase tooth.

We assume the tangential flux distribution $B_\theta(t)$ is small, and all flux linkage is generated by the radial flux distribution $B_r(t)$.

$$\phi_u(t) = \int B_r(t)dS \tag{8}$$

Radial force on a U-phase tooth $f_u(t)$ is calculated based on Maxwell stress.

$$f_u(t) = \int \frac{B_r(t)^2}{2\mu_0}dS \tag{9}$$

where S is a tooth area facing air region. This paper has the assumption that flux is distributed equally over the tooth area S. With this assumption, (8) and (9) are rewritten as (10) and (11).

$$\phi_u(t) = B_r(t)S \tag{10}$$

$$f_u(t) = \frac{B_r^2(t)}{2\mu_0}S \tag{11}$$

In this paper, we consider constant speed condition and radial force is regarded as a function for electrical angle θ. Substituting (7) and (10) into (11), equation (12) is obtained.

$$f_u(\theta) = \frac{\psi_u^2(\theta)}{2\mu_0 SN^2} = A\psi_u^2(\theta) \tag{12}$$

$$A := \frac{1}{2\mu_0 SN^2} \tag{13}$$

(12) is called the approximation of radial force in this paper.

B. Assumption on flux linkage

It is also assumed that flux linkage caused by permanent magnet $\psi_{um}(\theta)$ and flux linkage caused by current $\psi_{ui}(\theta)$

TABLE I
PARAMETERS OF IPMSM

turn number N	120
a pair of poles P	6
teeth area S [m^2]	4.13×10^{-4}
ψ_{m1}[mWb]	36.2
ψ_{m5}[mWb]	0.811
ψ_{m7}[mWb]	-0.114
L_d[mH]	0.866
L_q[mH]	1.31

satisfy linear independency.

$$\psi_u(\theta) = \psi_{um}(\theta) + \psi_{ui}(\theta) \tag{14}$$

This paper considers 12 poles 18 slots IPMSM. The winding pattern is concentrated winding. To consider 6th radial force, $\psi_{um}(\theta)$ is defined as:

$$\psi_{um}(\theta) = \psi_{1m}\cos\theta + \psi_{5m}\cos5\theta + \psi_{7m}\cos7\theta \tag{15}$$

5th and 7th flux linkage ψ_{5m} and ψ_{7m} have negative value when they have opposite phase against fundamental flux linkage. On ground of the symmetry, flux linkage on U-phase tooth is considered. The parameter of IPMSM is shown in Table I. dq-axis current reference i_d, i_q is defined as :

$$i_d := I_{d0} + i_{d6} \tag{16}$$

$$i_{d6} := I_{d6}\cos(6\theta - \theta_{d6}) \tag{17}$$

$$i_q := I_{q0} + i_{q6} \tag{18}$$

$$i_{q6} := I_{q6}\cos(6\theta - \theta_{q6}) \tag{19}$$

III. THE INFLUENCE OF d-AXIS HARMONIC CURRENT

Flux linkage on U-phase caused by d-axis harmonic current ψ_{uih} is shown as:

$$\begin{bmatrix} \psi_{uih} \\ \psi_{vih} \\ \psi_{wih} \end{bmatrix} = \boldsymbol{C}_{dq}^{uvw} \begin{bmatrix} L_d I_{d6}\cos(6\theta - \theta_{d6}) \\ 0 \end{bmatrix}$$

$$= \sqrt{\frac{1}{6}} L_d I_{d6} \begin{bmatrix} \cos(5\theta - \theta_{d6}) + \cos(7\theta - \theta_{d6}) \\ \cos(5\theta - \theta_{d6} + \frac{2}{3}\pi) + \cos(7\theta - \theta_{d6} - \frac{2}{3}\pi) \\ \cos(5\theta - \theta_{d6} + \frac{4}{3}\pi) + \cos(7\theta - \theta_{d6} - \frac{4}{3}\pi) \end{bmatrix} \tag{20}$$

Adding harmonic flux linkage ψ_{uih}, all flux linkage on U-phase is written by :

$$\psi_u(\theta) = \sqrt{\frac{2}{3}}(\Psi_{1m} + L_d I_{d0})\cos\theta - \sqrt{\frac{2}{3}}L_q I_{q0}\sin\theta$$

$$+ \psi_{5m}\cos5\theta + \sqrt{\frac{1}{6}}L_d I_{d6}\cos(5\theta - \theta_{d6})$$

$$+ \psi_{7m}\cos7\theta + \sqrt{\frac{1}{6}}L_d I_{d6}\cos(7\theta - \theta_{d6}) \tag{21}$$

where, $\Psi_{1m} := \sqrt{\frac{3}{2}}\psi_{1m}$. In this paper, 2-phase/3-phase transform is absolute transformation. Substituting (21) into (12), the 6th order component generated by d-axis harmonic

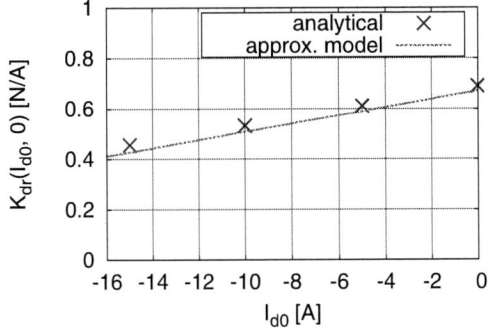

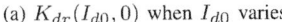

(a) $K_{dr}(I_{d0}, 0)$ when I_{d0} varies

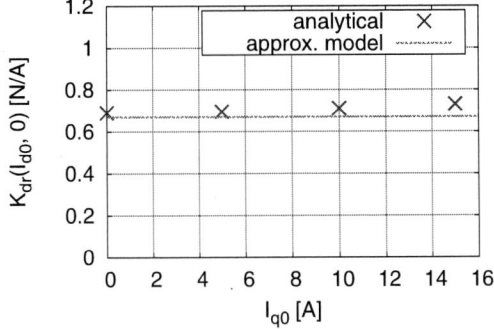

(b) $K_{dr}(0, I_{q0})$ when I_{q0} varies

Fig. 2. The comparison between $K_{dr}(I_{d0}, I_{q0})$ and FEA results

current is extracted as :

$$f_{i_{d6}}(I_{d0}, I_{q0}, I_{d6}, \theta_{d6}) = \frac{A}{3}(\Psi_{1m} + L_d I_{d0})L_d I_{d6}\cos(6\theta - \theta_{d6})$$

$$= K_{dr}(I_{d0}, I_{q0})i_{d6} \quad (22)$$

$$K_{dr}(I_{d0}, I_{q0}) := \frac{A}{3}(\Psi_{1m} + L_d I_{d0})L_d \quad (23)$$

where $f_{i_{d6}}(I_{d0}, I_{q0}, I_{d6}, \theta_{d6})$ is 6th radial force caused by i_{d6}. To verify the accuracy of (22), FEA is performed on the condition that $I_{d6} = 1$A and $\theta_{d6} = 0$deg. The FEA result is shown in Fig.2. Although a lot of assumptions have been made to lead (22), we can see that 6th radial force model (22) differs very little from FEA result.

IV. THE INFLUENCE OF q-AXIS HARMONIC CURRENT

In the same way, flux linkage generated by q-axis harmonic current $\psi_{uih}(\theta)$ is calculated as:

$$\begin{bmatrix} \psi_{uih}(\theta) \\ \psi_{vih}(\theta) \\ \psi_{wih}(\theta) \end{bmatrix} = C_{dq}^{uvw} \begin{bmatrix} 0 \\ L_q I_{q6}\cos(6\theta - \theta_{q6}) \end{bmatrix}$$

$$= \sqrt{\frac{1}{6}}L_q I_{q6} \begin{bmatrix} \sin(5\theta - \theta_{q6}) - \sin(7\theta - \theta_{q6}) \\ \sin(5\theta - \theta_{q6} + \frac{2}{3}\pi) - \sin(7\theta - \theta_{q6} - \frac{2}{3}\pi) \\ \sin(5\theta - \theta_{q6} + \frac{4}{3}\pi) - \sin(7\theta - \theta_{q6} - \frac{4}{3}\pi) \end{bmatrix}$$

$$(24)$$

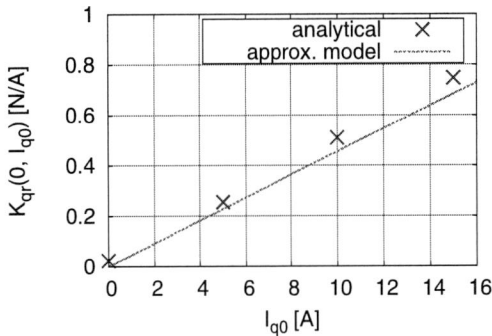

Fig. 3. $F_{i_{q6}}(0, I_{q0})$ when I_{q0} varies

Substituting (24) into (12), the 6th order component generated by q-axis harmonic current is extracted as:

$$f_{i_{q6}}(I_{d0}, I_{q0}) = \frac{A}{3}L_q I_{q0} L_q I_{q6}\cos(6\theta - \theta_{q6}) \quad (25)$$

$$= K_{qr}(I_{d0}, I_{q0})i_{q6} \quad (26)$$

$$K_{qr}(I_{d0}, I_{q0}) := \frac{A}{3}L_q^2 I_{q0} \quad (27)$$

Fig.3 shows the FEA results which are performed on the condition that $I_{q6} = 1$A, $\theta_{q6} = 0$A, and $I_{d0} = 0$A. Fig.3 shows that Eq. (27) can predict 6th radial force caused by q-axis harmonic current well.

A. 6th radial force control

The transfer characteristics between 6th harmonic current and 6th radial force have been obtained. All 6th radial force is expressed as

$$f_{r6}(i_d, i_q) = f_{\text{base}}(I_{d0}, I_{q0}) + K_{dr}(I_{d0}, I_{q0})i_{d6} + K_{qr}(I_{d0}, I_{q0})i_{q6} \quad (28)$$

$$f_{\text{base}}(I_{d0}, I_{q0}) := F_{\text{base}}\cos(6\theta - \theta_{\text{base}}) \quad (29)$$

where $f_{\text{base}}(I_{d0}, I_{q0})$ is 6th radial force caused by harmonic inductance and harmonic magnetic flux. $f_{\text{base}}(I_{d0}, I_{q0})$ varies according to I_{d0} and I_{q0}. $f_{\text{base}}(I_{d0}, I_{q0})$ can be obtained by calculating 6th radial force without harmonic current in FEA. With $F_{\text{base}}(I_{d0}, I_{q0})$ and $\theta_{\text{base}}(I_{d0}, I_{q0})$, 6th harmonic current reference is obtained as

$$i_{d6:\text{opt}} = -\frac{F_{\text{base}}(I_{d0}, I_{q0})}{K_{dr}(I_{d0}, I_{q0})}\cos(6\theta - \theta_{\text{base}}(I_{d0}, I_{q0})) \quad (30)$$

$$i_{q6:\text{opt}} = -\frac{F_{\text{base}}(I_{d0}, I_{q0})}{K_{qr}(I_{d0}, I_{q0})}\cos(6\theta - \theta_{\text{base}}(I_{d0}, I_{q0})) \quad (31)$$

Fig. 4 shows simulation result of 6th radial force control by d-axis harmonic current. 4th and 8th radial forces are affected by 6th d-axis current because 5th and 7th harmonic fluxes generated by 5th and 7th harmonic current produce 4th and 8th radial forces. However, considering transfer characteristics of 4th and 8th radial forces, which has Pth annular mode, it is said that 4th and 8th radial forces cause little noise and

978-1-4799-2706-7/14 $31.00 © 2014 IEEE

The 2014 International Power Electronics Conference

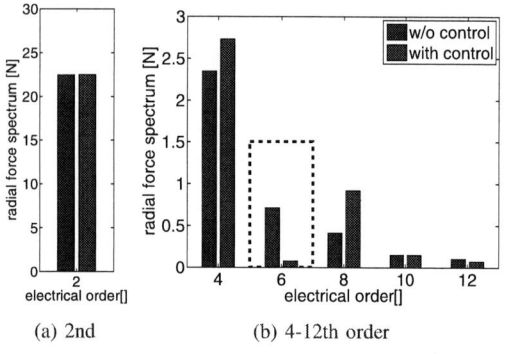

(a) 2nd   (b) 4-12th order

Fig. 4. 6th radial force control by d-axis harmonic current(simulation result)

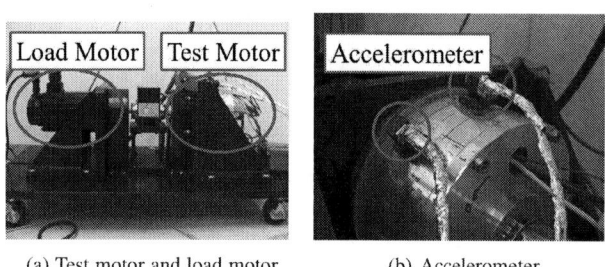

(a) Test motor and load motor   (b) Accelerometer

Fig. 5. Experimental environment

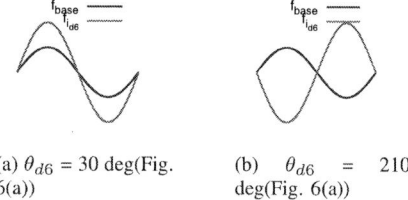

(a) $\theta_{d6} = 30$ deg(Fig. 6(a))   (b) $\theta_{d6} = 210$ deg(Fig. 6(a))

Fig. 7. The phase relationship between f_{base} and $f_{i_{d6}}$ in Fig. 6(a)

vibration.

V. EXPERIMENTAL RESULTS

In experiment, radial acceleration on the surface of test motor is evaluated intead of radial force. The speed of the test motor is controlled as 800rpm by load motor. Fig. 5 shows test motor, load motor and accelerometer. The current controller is designed as Perfect Tracking Control and Pole zero cancellation PI Control[16]. 6th radial acceleration $a_{r6}(\omega)$ is expressed theoretically as

$$a_{r6}(\omega) = H(\omega)f_{r6}(i_d, i_q). \tag{32}$$

where $H(\omega)$ is transfer characteristics between radial force and vibratoin at the speed of ω.

A. The validation of approximation model

To validate the approximation model (22) and (26), some expreriments are performed. At first, 6th radial acceleration is shown in 6(a) under the condition that $I_{d6} = 2.5$A, $I_{q6} = 0$A, $I_{d0} = 0$A and I_{q0}, $\theta_{i_{d6}}$ vary. It can be seen in Fig. 6(a) that the phase of 6th radial force is varied by shifting the phase of d-axis harmonic current. In the result, when

f_{base} is opposite to $\theta_{i_{d6}}$ 6th radial acceleration decreases and when f_{base} corresponds to $\theta_{i_{d6}}$ 6th radial acceleration increases in Fig. 6(a). Fig. 7 shows the situation when I_{q0} is 5A and θ_{d6} is 30 and 210deg in Fig. 6(a). θ_{base} is identified as 30deg from Fig. 6(a) in this drive condition. In Fig. 6(b), the result of d-axis 6th harmonic current effect to acceleration is shown. The fact that 6th radial acceleration caused by d-axis harmonic current $a_{i_{d6}}$ is independent of θ_{d6} and I_{q0} validates the approximation model (22).

The transfer characteristics of radial acceleration from d-axis 6th harmonic current $H(\omega_0)K_{dr}(I_{d0}, I_{q0})$ is identified as

$$(H(\omega_0)K_{dr}(0,5))' = 1.35 \times 10^{-2} [\text{m/s}^2/A] \tag{33}$$

because in Fig. 6(b) the average acceleration of $a_{i_{d6}}$ is $3.37 \times 10^{-2} m/s^2$, which is generated by I_{d6} of 2.5A. In Eq. (33), the symbol of $()'$ denotes the parameter indentified by the experiment.

In the same way, Fig. 6(c) shows the experimental results when $I_{d6} = 0$A, $I_{q6} = 2.5$A, $I_{d0} = 0$A and I_{q0}, $\theta_{i_{q6}}$ vary. In Fig. 6(d), the result of q-axis 6th harmonic current effect to acceleration is shown. Eq. (26) indicates that the average acceleration at $I_{q0} = 10$A is twice as large as the average acceleration at $I_{q0} = 5$A. In Fig. 6(d) the average acceleration at $I_{q0} = 10$A is one point three larger than that at $I_{q0} = 5$A.

The transfer characteristics of radial acceleration is obtained as

$$(H(\omega_0)K_{qr}(0,5))' = 2.26 \times 10^{-2} [\text{m/s}^2/A]. \tag{34}$$

B. The result of 6th radial force control

In this chapter, 6th radial force control is realized under the condition that the rotation speed is 800rpm, I_{d0} is 0A, and I_{q0} is 5A. In this section, to maximize the effect of 6th radial force control, the current references of 6th radial force control are identified by experiment.

At first, 6th radial acceleration $a_{r6}(I_{d0}, I_{q0})$ is measured with keeping harmonic current zero.

$$(H(\omega_0)F_{\text{base}}(I_{d0}, I_{q0}))' = a_{r6}(I_{d0}, I_{q0}) \tag{35}$$

Experimental results is shown in Fig. 8(a) 8(d) 8(g) 8(j) 8(m), where to make discussions the output of torque meter and theta acceleration are shown. $(H(\omega_0)F_{\text{base}}(0,5))'$ is identified by 6th radial acceleration in Fig. 8(a) as:

$$(H(\omega_0)F_{\text{base}}(0,5))' = 2.97 \times 10^{-2} [\text{m/s}^2] \tag{36}$$

With Eq. (33) and (34), dq-axis harmonic current references are calculated as:

$$i_{d6:\text{opt}} = -\frac{(H(\omega_0)F_{\text{base}}(0,5))'}{(H(\omega_0)K_{dr}(0,5))'} \cos(6\theta - \theta_{\text{base}}) \tag{37}$$

$$i_{q6:\text{opt}} = -\frac{(H(\omega_0)F_{\text{base}}(0,5))'}{(H(\omega_0)K_{qr}(0,5))'} \cos(6\theta - \theta_{\text{base}}) \tag{38}$$

$$i_{d6:\text{opt}} = -2.20 \cos\left(6\theta - \frac{1}{6}\pi\right) \tag{39}$$

978-1-4799-2706-7/14 $31.00 © 2014 IEEE

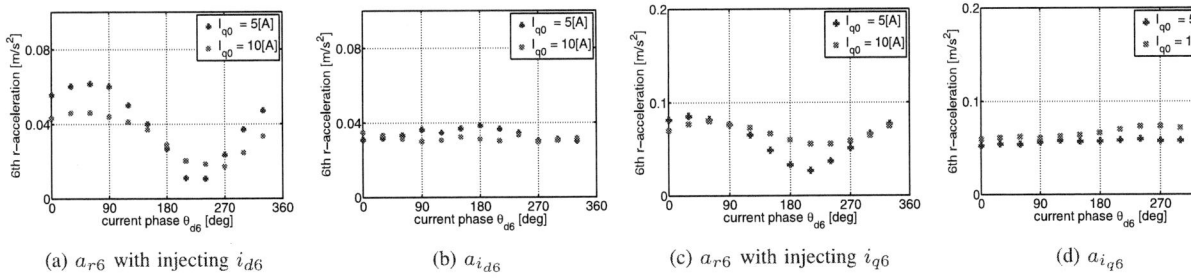

(a) a_{r6} with injecting i_{d6} (b) $a_{i_{d6}}$ (c) a_{r6} with injecting i_{q6} (d) $a_{i_{q6}}$

Fig. 6. The validation of approximation model(Experimental result)

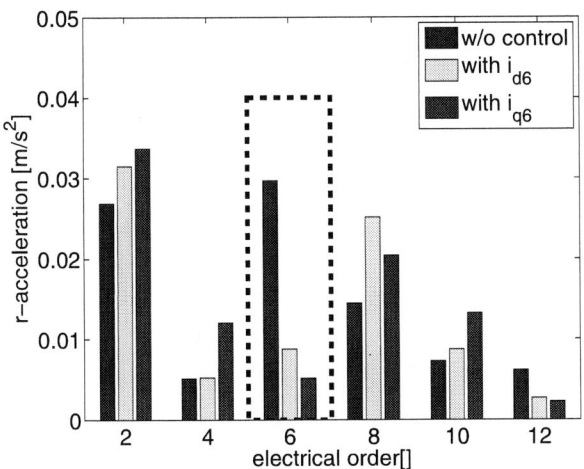

Fig. 9. The comarison of 6th radial force controls(Experimental Result)

$$i_{q6:opt} = -1.31 \cos\left(6\theta - \frac{1}{6}\pi\right) \qquad (40)$$

Fig. 8 shows the result of 6th radial force control by $i_{d:opt}$ and $i_{q:opt}$.

6th radial accelerations caused by 6th radial force are largely suppressed in Fig. 8(b) and 8(c). when $i_{q6:opt}$ are selected to suppress radial force, 6th circumferential acceleration caused by torque ripple increases in Fig. 8(f). On the other hand, there are little deterioration of 6th circumferential acceleration with $i_{d6:opt}$ in Fig. 8(e). Fig. 9 shows the comparison of radial acceleration with 6th radial force control. It is notable that the decrease of 6th radial acceleration is larger than the increase of 4th and 8th radial accelerations.

VI. CONCLUSION

In this paper, we proposed the modelling of 6th radial force caused by dq-axis harmonic current based on flux linkage. This model enables us to evaluate the harmonic radial force generated by dq-axis harmonic current. The validity of the proposed model were verified by FEA and experiment. Based on this modelling, 6th radial force was suppressed. In our future works, 6th radial force control combined with dq-axis harmonic current will be proposed to suppress both circumferential and radial acceleration.

REFERENCES

[1] Kondo Keiichiro,Kubota Hisao: "Innovative Application Technologies of AC Motor Drive Systems" IEEJ Trans. Industry Applications, Vol.1, No.3 pp.132-140(2012)

[2] M.Islam, R.Islam, T.Sebastian : "Noise and vibration characteristics of permanent magnet synchronous motors using electromagnetic and structural analyses", IEEE International Conference on Energy Conversion Congress and Exposition (ECCE), Vol.17, No.22, pp.3399-3405,2011

[3] H.Y.Issac Du, Lei Hao, and Hejie Lin: "Modeling and analysis of electromagnetic vibrations in fractional slot PM machines for electric propulsion," Energy Conversion Congress and Exposition (ECCE), pp.5077-5084(2013)

[4] J.F. Gieras, Chong Wang, Joseph. C.S.Lai, Nesimi Ertugrul: "Analytical Prediction of Noise of Magnetic Origin Produced by Permanent Magnet Brushless Motors," IEEE International Conference on Electric Machines Drives Conference(IEMDC), Vol.1, No.5, pp.148-152(2007)

[5] M. Boesing, R.W. De Doncker: "Exploring a Vibration Synthesis Process for the Acoustic Characterization of Electric Drives" IEEE Trans. Ind. Appl., vol.48, no.1, pp.70-78(2012)

[6] T.Kobayashi, Y.Takeda, M.Sanada, and S.Morimoto: "Vibration Reduction of IPMSM with Concentrated Winding by Making Holes", IEEJapan Trans. D, Vol. 124-D, No.2, pp. 202-207, 2004(in Japanese).

[7] Sang-Ho Lee, Jung-Pyo Hong, Sang-Moon Hwang, Woo-Taik Lee, Ji-Young Lee, Young-Kyoun Kim: "Optimal Design for Noise Reduction in Interior Permanent-Magnet Motor", IEEE Trans. Ind. Appl., vol.45, no.6, pp.1954-1960, Nov.-dec. 2009

[8] D. C. Hanselman: "Effect of skew, pole count and slot count on brushless motor radial force, cogging torque and back EMF", Inst. Elect. Eng. Proc. mdash,Elect. Power Appl., Vol.144, No.5, pp.325-330, 1997

[9] Jae-Woo Jung; Do-Jin Kim; Jung-Pyo Hong; Geun-Ho Lee; Seong-Min Jeon: "Experimental Verification and Effects of Step Skewed Rotor Type IPMSM on Vibration and Noise", IEEE Trans. Magn., vol.47, no.10, pp.3661-3664, 2011

[10] A. Cassat, C. Espanet, R. Coleman, L. Burdet, E. Leleu,D. Torregrossa, J. M' Boua, A. Miraoui: "A Practical Solution to Mitigate Vibrations in Industrial PM Motors Having Concentric Windings", IEEE Trans. Ind. Appl., vol.48, no.5, pp.1526-1538, 2012

[11] W. Zhu, B. Fahimi, and S. Pekarek, gA field reconstruction method for optimal excitation of surface mounted permanent magnet synchronous machines,h IEEE Trans. Energy Convers., vol. 21, no. 2, pp. 303-313, Jun. 2006.

[12] Jae-Woo Jung, Do-Jin Kim, Jung-Pyo Hong, Geun-Ho Lee, Seong-Min Jeon: "Experimental Verification and Effects of Step Skewed Rotor Type IPMSM on Vibration and Noise," IEEE Trans. Magnetics, Vol.47, No.10, pp.3661-3664(2011)

[13] Tao Sun, Ji-Min Kim, Geun-Ho Lee, Jung-Pyo Hong, Myung-Ryul Choi: "Effect of Pole and Slot Combination on Noise and Vibration in Permanent Magnet Synchronous Motor", IEEE Trans. Magnetics, Vol.47, No.5, pp.1038-1041(2011)

[14] Jiao Guandong, C.D.Rahn : "Field weakening for radial force reduction in brushless permanent-magnet DC motors" IEEE Trans. Magnetics, Vol.40, No.5, pp. 3286- 3292(2004)

[15] M.Kanematsu, T.Miyajima, H.Fujimoto, Y.Hori, T.Enomoto, M.Kondou, H.Komiya, K.Yoshimoto, T.Miyakawa: "Suppression Control of Radial Force Vibration due to Fundamental Permanent-Magnet Flux in IPMSM", IEEE Energy Conversion Congress and Exposition(ECCE), pp.2812-2816 (2013)

The 2014 International Power Electronics Conference

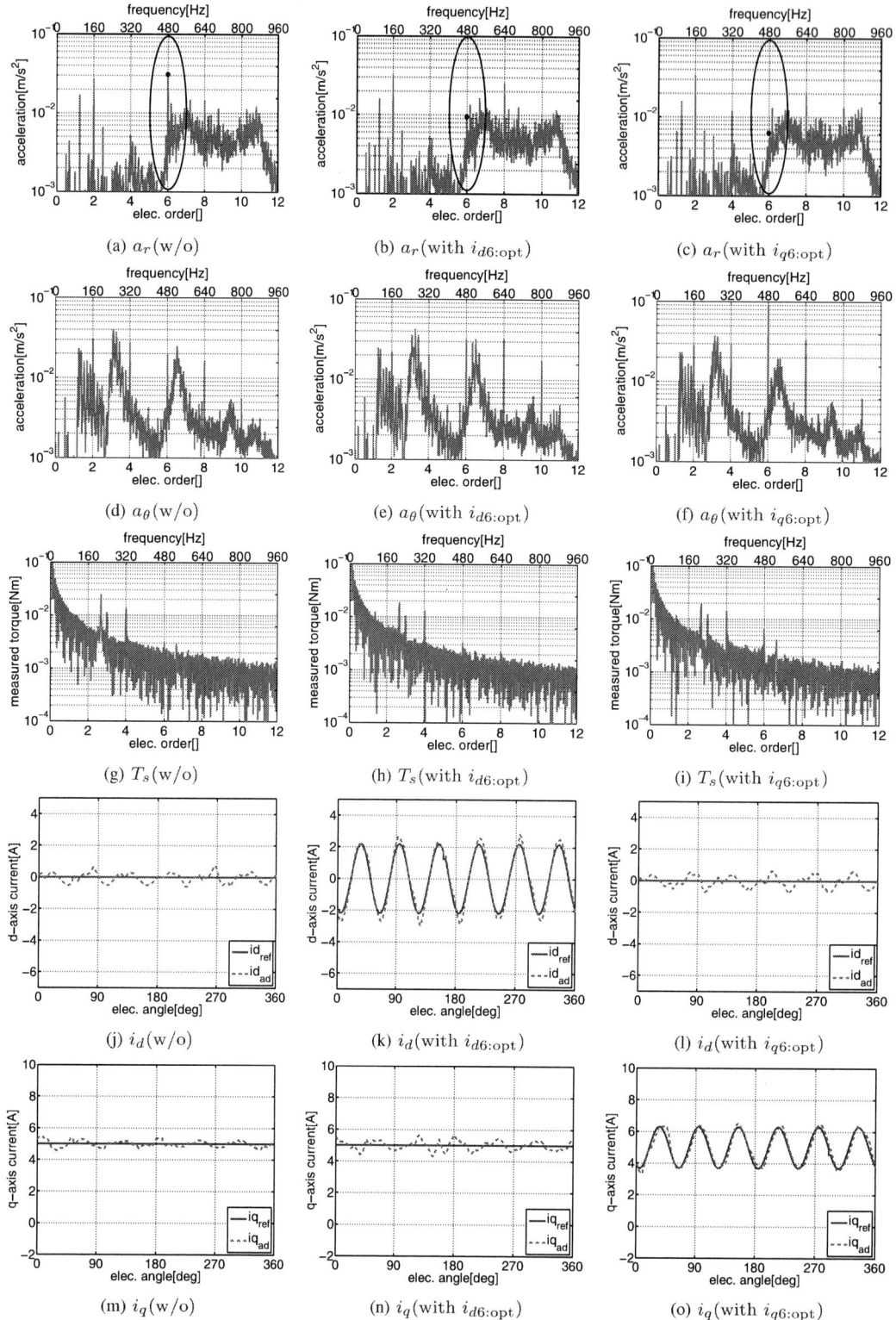

Fig. 8. 6th radial force controliExperimental Resultj

[16] K.Nakamura, H.Fujimoto, M.Fujitsuna: "Torque ripple suppression control for PM motor with current control based on PTC", Power Electronics Conference (IPEC), 2010 International , pp.1077-1082, June 2010

Air Gap Control of Multi-phase Transverse Flux Permanent Magnet Linear Synchronous Motor by using Independent Vector Control

Seon-Hwan Hwang
Department of Electrical Engineering
Kyungnam University
Changwon, Korea
seonhwan@kyungnam.ac.kr

Deok-Je Bang, Ji-Won Kim
KERI
Changwon, Korea
djbang@keri.re.kr, jwkim@keri.re.kr

Abstract— **This paper presents an air gap control method of a multi-phase transverse flux permanent magnet linear synchronous motor (MP-TFPMLSM) based on independent vector control. Especially, the MP-TFPMLSM is composed of symmetrical multi-phase and multiple-module structures, which are basically three-phase configurations. Hence, in this paper, a *d*-axis current control applying the *d-q* transformation and the independent vector control are proposed for the air gap control between two symmetric stators and mover of the MP-TFPMLSM. The control performance and characteristics of the MP-TFPMLSM, which is based on two basic three-phase structures of the air gap control and the position control of multi-phase linear machines using a concept of vector control are analyzed. As a result, the air gap control by using the independent vector control is a suitable method to efficiently control the interaction between stator and mover of the MP-TFPMLSM. The proposed method is easily implemented and has less computation time to design controllers for operating the multi-phase machines. The effectiveness of the proposed independent vector control algorithm is verified through several experimental results.**

Keywords— *Independent vector control, MP-TFPMLSM, air gap control, d-axis control.*

I. INTRODUCTION

Permanent magnet (PM) linear synchronous motors are beginning to find widespread industrial applications such as transportation system, automation system, and machine tool positioning. The main features of permanent magnet linear synchronous motors are the high force density achievable and the high positioning precision and accuracy associated with the mechanical system simplicity. Especially, the transverse flux permanent magnet linear synchronous motors (TFPMLSMs) in this paper have not been used widely in industrial applications due to the nonlinear characteristics. Therefore, the novel control algorithm can be applied for improving the performance of these machines [1]-[3].

In particular, the linear motor drives need to develop not only a contactless levitation system but also a contact-free propulsion system to drive the transportation system. In general, the linear motors produce a high normal force compared to the thrust force developed by

the motor itself [2], [3]. In addition, the use of a magnetic levitation system is steadily increasing in high-speed motor drives, generators like a flywheel drive, medical equipments, and harsh environments such as low temperature, high temperature, and vacuum. Moreover, the magnetic levitation drive offers many advantages such as high velocity, no wear and no maintenance. In general, the magnetic levitation system needs the additional windings which are used to produce suspension force of the linear machines.

In this paper, however, there is no additional winding to generate the normal force for controlling the air gap. In addition, this paper focuses on the continuous air gap and position controls considering the special structures of the developed MP-TFPMLSM [4], [5].

As a result, the independent vector control applying transformation matrix and the symmetric three-phase configuration are proposed for producing the thrust and the levitation forces of the MP-TFPMLSM. The control performance and characteristics of the proposed control algorithms are analyzed and compared. In addition, the independent vector control is more appropriate for the MP-TFPMLSM compared to the stationary PI current regulator for controlling the air gap and mover position. The proposed algorithm is easily implemented by using the vector control concept and less computation efforts. Thus, the feasibility and effectiveness of the proposed algorithm are verified through the several experiments.

II. DESCRIPTION OF MP-TFPMLSM

A. MP-TFPMLSM Structure

Fig. 1 shows the structures of the MP-TFPMLSM. From Fig. 1 (a), the manufactured drive system has the flexible moving part and stationary parts to control the air gap and position of the linear machine. The used MP-TFPMLSM has symmetrical independent multi-phase windings based on the 3-phase winding with 2 set which is installed in left and right sides of its mover as shown in Fig. 1 (b). In addition, the machine consists of multiple-module and multi-phase structures having separated magnetic and electric circuits per each phase. Each stator

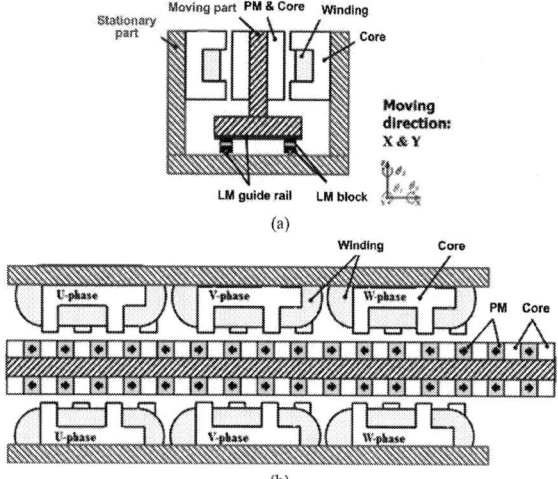

Fig. 1. Structures of MP-TFPMLSM. (a) Upper view. (b) Front view.

phase winding is separated by the 120° electrical degrees [4], [5]. In particular, there are no additional winding in order to produce the magnetic levitation unlike the conventional systems. Finally, it can be seen that only the coils facing the mover PMs will produce the magnetic levitation and the thrust force as shown in Fig. 1 (b).

B. Mathmatical Model of MP-TFPMLSM

From Fig. 1 (b), the each single-phase voltage equation with a racetrack-shaped winding can be expressed as:

$$v_x = R_s i_x + \frac{d\lambda_x}{dt} \quad , \quad x = 1, 2, \cdots, N \tag{1}$$

where v_x is the stator voltage in each phase. R_s is the stator resistance. i_x and λ_x is the stator current and flux linkage.

From (1), the MP-TFPMLSM has the independent three-phase configuration without a neutral point unlike the conventional three-phase Y-connected ac machines as shown in Fig. 2 [4], [6].

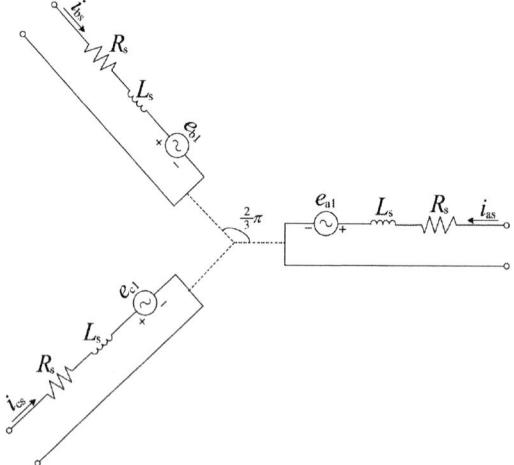

Fig. 2. Three-phase equivalent circuit of MP-TFPMLSM.

Therefore, the independent three-phase voltage equations of the MP-TFPMLSM can be easily expressed as follows:

$$
\begin{aligned}
v_{as} &= R_s i_{as} + \frac{d\lambda_{as}}{dt} \\
v_{bs} &= R_s i_{bs} + \frac{d\lambda_{bs}}{dt} \\
v_{cs} &= R_s i_{cs} + \frac{d\lambda_{cs}}{dt}
\end{aligned}
\tag{2}
$$

where v_{as}, v_{bs}, and v_{cs} are the three-phase phase voltages. λ_{as}, λ_{bs}, and λ_{cs} are the flux linkages per phase. i_{as}, i_{bs}, and i_{cs} are the stator phase currents.

In addition, the mutual inductance between each phase winding can be considered as zero due to the independent magnetic circuit of each phase [4], [5]. Therefore, the dq-axes voltage equations considering the mutual inductance in the synchronous reference frame can be represented by

$$
\begin{aligned}
v_{ds}^e &= R_s i_{ds}^e + L_s \frac{di_{ds}^e}{dt} - \frac{\pi}{\tau} v_r L_q i_{qs}^e \\
v_{qs}^e &= R_s i_{qs}^e + L_s \frac{di_{qs}^e}{dt} + \frac{\pi}{\tau} v_r L_d i_{ds}^e + \frac{\pi}{\tau} v_r \lambda_f
\end{aligned}
\tag{3}
$$

where v_{ds}^e and v_{qs}^e are the dq-axes voltages in the synchronous reference frame. i_{ds}^e and i_{qs}^e are the dq-axes currents in the synchronous reference frame. λ_f and L_s is the flux linkage and stator inductance. v_r is the velocity of linear motor. τ is pole pitch.

The generated thrust force from the independent three-phase MP-TFPMLSM can be given by

$$F_y = \frac{3}{2} \frac{\pi}{\tau} \left(\lambda_f i_{qs}^e + \left(L_d - L_q \right) i_{ds}^e i_{qs}^e \right) \tag{4}$$

The dynamic equation of the MP-TFPMLSM can be represented by

$$F_y = M \frac{dv_r}{dt} + F_d + B v_r \tag{5}$$

From Eq. (4), the q-axis current is directly related to the position control the MP-TFPMLSM which is generated by the thrust force when the L_d is the same as L_q.

III. INDEPENDENT VECTOR CONTROL OF MP-TFPMLSM

As mentioned previously, the independent vector control concept can be applied to control the MP-TFPMLSM. Therefore, the normal force in order to control air gap between the stator and the mover focuses on the paper. The normal force related to the air gap variation is regulated to levitate the mover of the MP-

978-1-4799-2706-7/14 $31.00 © 2014 IEEE

TFPMLSM. The unbalanced flux density in each air gap causes the unbalanced normal force. Namely, the field weakening control is operated and then the air gap flux density is decreased. On the contrary it is increased [7]-[9]. Finally, the air gap can be adjusted in the arbitrary direction by controlling the d-axis current on the vector control strategy.

In order to derive the relation with the d-axis current and the normal force, Fig. 3 shows the thrust and normal forces in the Y- and X-axes direction in the two axes coordinates. In Fig. 3 (b), first quadrant represents the force, F_r caused by the dq-axes currents on the right side of the MP-TFPMLSM. The force, F_l is also the total force generated from the thrust and normal forces of the left side stators.

Fig. 4 shows the block diagram of the independent vector control for the MP-TFPMLSM. The d-axis current command is generated from the PID gap controller through displacement sensor. The thrust force for tracking the mover position control is produced by the position and speed controllers as shown in Fig. 4. The each phase current is transformed into d-q rotating frame by the transformation matrix.

IV. EXPERIMENTS

Fig. 5 shows the overall block diagram of the MP-TFPMLSM drive system to verify the proposed control algorithm. As shown in Fig. 5, the MP-TFPMLSM used in the experiments is based on the symmetric configuration with the independent 3-phases. In addition, the single-phase bi-directional AC/DC/AC converter modules are implemented by the IGBT modules with a switching frequency of 10 kHz. The total control algorithm is performed by a digital signal processor (DSP) control system. The displacement sensor for controlling the air gap is installed at the stationary part.

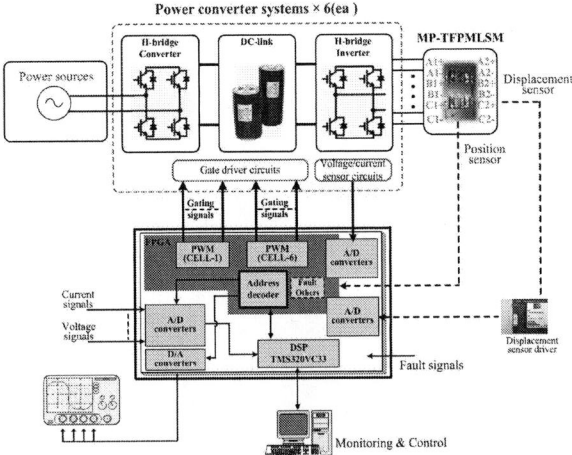

Fig. 5. Block diagram of MP-TFPMLSM.

The detailed specifications of the MP-TFPMLSM are shown in Table 1.

Fig. 6 depicts the photo of the actual MP-TFPMLSM, which includes single-phase power converters, a control board, and a displacement sensor.

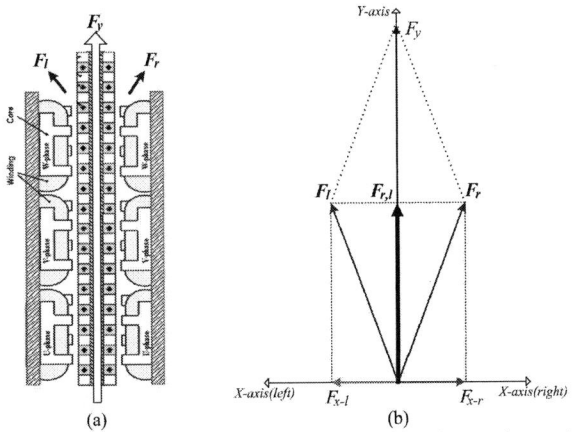

Fig. 3. Total thrust force and each normal force. (a) Thrust and normal forces in the MP-TFPMLSM. (b) Each thrust and normal forces in two-axes coordinates.

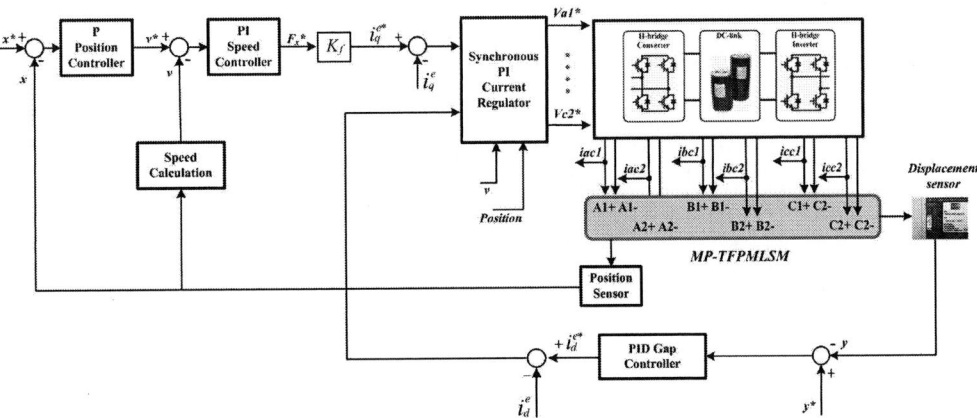

Fig. 4. Overall control block diagram of independent vector control for MP-TFPMLSM.

TABLE I
SPECIFICATIONS OF THE MP-TFPMLSM

	Parameters	Values	Unit
Rated MP-TFPMLSM	MMF	6,000	[AT]
	Thrust force	250	[N], Per phase
Mover PM	Material	Ferrite	
	Thickness	8	[mm]
Stator	Material	S20c	
	Pole pitch	20	[mm]
	Thickness	16	[mm]
	Width	30	[mm]
	Height	80	[mm]
Mover	Material	S20c	
	Width	110	[mm]
	Height	35	[mm]

In order to simultaneously drive the air gap control and the position control, the independent three-phase windings and the proposed vector control algorithm are used without the any additional winding to control air gap length.

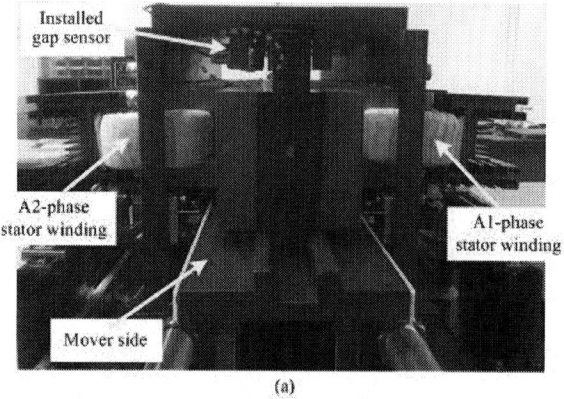

(a)

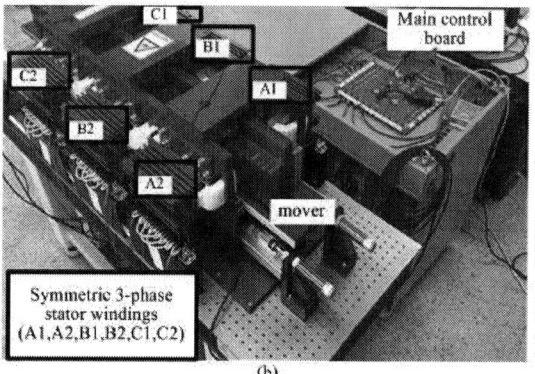

(b)

Fig. 6. Pictures of MP-TFPMLSM drive system. (a) Front view. (b) Upper view.

Figs. 7 and 8 show the experimental results of the independent vector control for the MP-TFPMLSM. When the MP-TFPMLSM is standstill and the air gap command is 3 [mm], the actual air gap is well tracking with the air gap command as shown in Fig. 7. In addition, the d-axis currents of the each stator side are symmetrically generated.

Fig. 8 describes the experimental results under the air gap command variation and the mover position is 0.15[m]. In spite of the air gap command variation from

2.5 [mm] to 3.5 [mm], the actual air gap is also well controlled by the proposed independent vector control with the d-q transformation. The d-axis currents are the same as Fig. 7.

Fig. 9 shows the experimental waveforms when the air gap command is changed from 1.2 [mm] to 3.0 [mm] and the mover position also varies from 0.01 [m] to 0.15 [m], respectively. The actual mover position and air gap are controlling with each command as shown in Fig. 9.

Figs. 10 and 11 shows the experimental results under the constant air gap command and the mover stroke is from 0.05 [m] to 0.19 [m]. According to the proposed algorithm, the actual air gap is well controlled below 100 [μm]. Each a-phase current is shown in Fig. 12.

Fig. 12 represents the experimental waveforms under the mover position is changed from 0.05 [m] to 0.2 [m] and the air gap command is from 2.5 [mm] to 3.5 [mm]. As a result, the real air gap is following the air gap command as shown in Fig. 12.

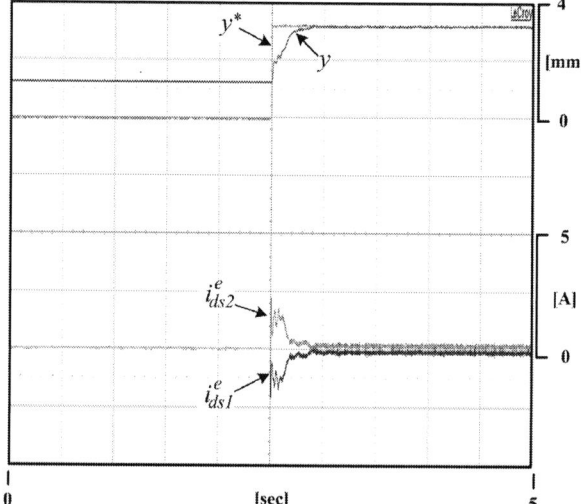

Fig. 7. Experimental waveform of air gap control under standstill.

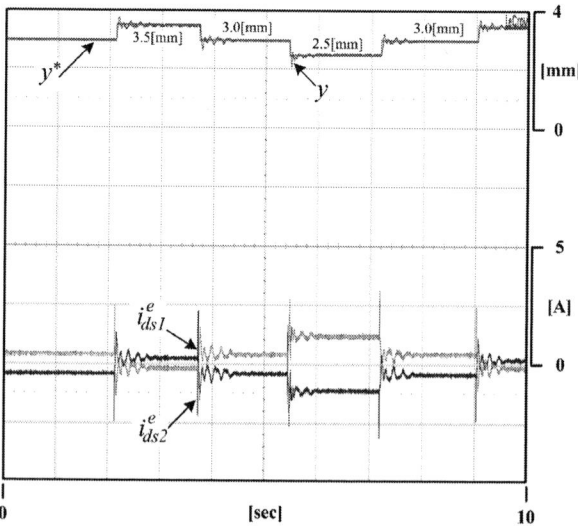

Fig. 8. Experimental waveform of air gap control under the air gap command variation.

The 2014 International Power Electronics Conference

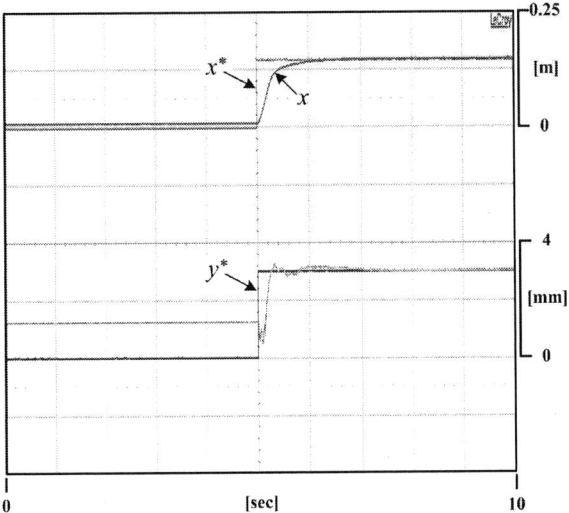

Fig. 9. Experimental waveform according to simultaneous mover position and air gap command changes.

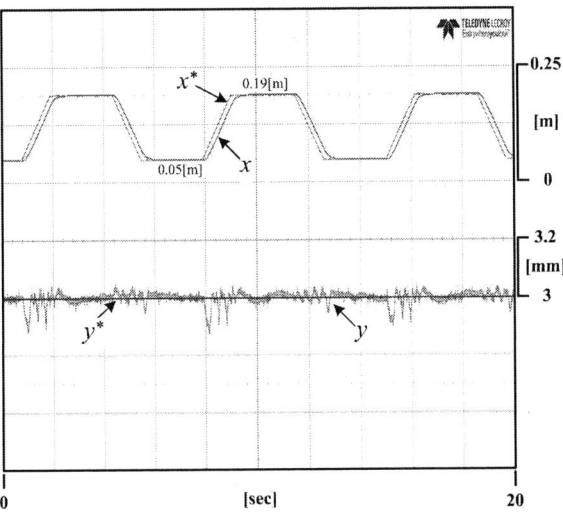

Fig. 10. Experimental waveform of constant air gap command at constant mover stroke operation.

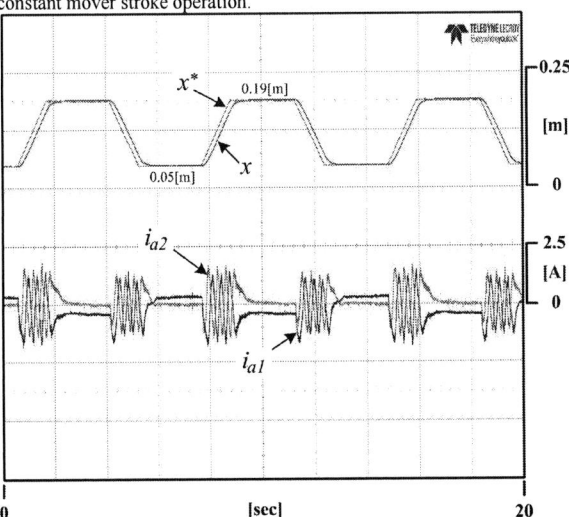

Fig. 11. Experimental waveform of each a-phase current at constant mover stroke operation.

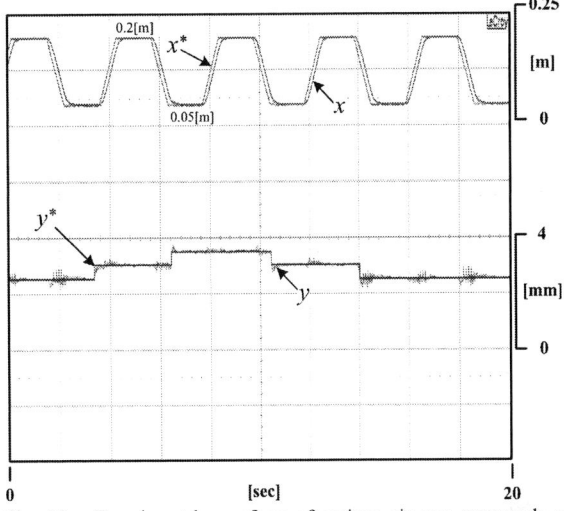

Fig. 12. Experimental waveform of various air gap commands at constant mover stroke operation.

V. CONCLUSIONS

This paper proposes the air gap and position control methods based on the independent vector control for the developed MP-TFPMLSM without the additional windings in order to generate the normal force which is the magnetic levitation force.

The mover position can be controlled by the q-axis current of the independent vector control. Especially, the air gap between the mover and the stators can be regulated by the d-axis current. As a result, the air gap and position control based on the multi-phase and multiple-module structures can be easily implemented by using the proposed independent vector control having less computational burden compared to the stationary PI current regulator. In addition, it is possible to facilitate extremely the system extension of the MP-TFPMLSM with multiple-modules. The usefulness of the proposed vector control method was verified through the experimental results.

ACKNOWLEDGMENT

This research was partly supported by Basic Science Research Program through the National Research Foundation of Korea (NRF) funded by the Ministry of Education, Science and Technology (No. 2013R1A1A1013670) and partly supported by the KERI Primary Research Program of MSIP/ISTK (No. 13-12-N0101-67)

REFERENCES

[1] D. H. Kang, "A study on the design of transverse flux linear motor with high power density", *IEEE International Symposium on*, vol. 2, pp. 707-711, Aug. 2002.

[2] W. Y. Kim, J. M. Lee, and S. J. Kim, "Rolling motion control of a levitated mover in a permanent-magnet-type bearingless linear motor", *IEEE Trans. Magn.*, vol. 46, no. 6, pp. 2482-2485, June 2010.

[3] W. G. Kim, S. J. Cho, "Control of transverse flux linear motor to the linear and curve section by using low-cost position sensors", *Industrial Electronics, 2007. ISIE 2007 IEEE International Symposium on*, pp. 1322-1326, Nov. 2007.

[4] S. H. Hwang, H. Li, J. W. Park, J. M. Kim, and D. J. Bang, "Vector control of multiple-module transverse flux PM generator for large-scale direct-drive wind turbines," *2011 IEEE, Energy Conversion Congress and Exposition*, pp. 2365-2372, Sept. 2011.

[5] D. Bang, H. Polinder, J. A. Ferreira, and S. S. Hong, "Structural mass minimization of large direct-drive wind generators using a buoyant rotor structure", *Energy Conversion Congress and Exposition, 2010 IEEE*, pp. 3561-3568, Sept. 2010.

[6] O. S. Park, J. W. Park, C. B. Bae, and J. M. Kim, "A dead time compensation algorithm of independent multi-phase PMSM with three-dimensional space vector control," *Journal of Power Electronics*, vol. 13, no. 1, pp. 77-85, 2013.

[7] S. Kobayashi, M. Ooshima, and M. Nasir Uddin, "A radial position control method of bearingless motor based on d-q axis current control", *IAS, 2011 IEEE*, pp. 1-8, Oct. 2011.

[8] M. Ooshima, S. Kobayashi, and M. Nasir Uddin, "Magnetic levitation tests of a bearingless motor based on d-q axis current control", *IAS, 2012 IEEE*, pp. 1-7, Oct. 2012.

[9] Z. Guan, D. H. Lee, J. W. Ahn, and F. Zhang, "A compensation strategy of suspending force in hybrid type stator pole bearingless switched reluctance motor", *ICEMS, 2011 International Conference on*, pp. 1-6, Aug. 2011.

Modified Direct Instantaneous Torque Control of Switched Reluctance Motor with High Torque per Ampere and Reduced Source Current Ripple

Rohit Suryadevara, B. G. Fernandes
Department of Electrical Engineering
Indian Institute of Technology Bombay
Mumbai, India
rohitsurya@ee.iitb.ac.in, bgf@ee.iitb.ac.in

Abstract—**This paper presents a modified version of direct instantaneous torque control (DITC) for minimizing torque ripple in switched reluctance motor (SRM). In DITC, it is observed that the phase current profiles and torque ripple characteristics depend on the turn-on instant. For current profiles with low RMS value, an overshoot in torque is observed during the commutation period, which increases the torque ripple. This overshoot in torque can be limited by delaying the turn-on instant. Consequently, the resultant current profiles have higher source current ripple, higher RMS value and thereby, lower torque per ampere ratio. In order to address these limitations, a modified version of DITC is proposed for a three-phase SRM, which simultaneously improves the torque per ampere ratio and prevents torque overshoot. In addition, the proposed strategy also aims at improving the source current ripple, by reducing its peak values during the commutation period. Finally, the improved performance with modified DITC is verified by simulation studies.**

Index Terms—**Direct instantaneous torque control, source current ripple, switched reluctance motor, torque ripple minimization**

I. INTRODUCTION

Switched reluctance motor (SRM) is gaining importance due to its simple, rugged and low-cost construction, absence of permanent magnets, fault-tolerant capability and reliability. Presence of torque ripple is one of its main drawbacks and minimization of torque ripple is an ongoing area of research in SRM. Torque ripple can be minimized by improving the magnetic design of the motor and by using electronic control techniques [1]. These control techniques can be broadly classified into two categories: indirect torque control techniques and direct torque control techniques [2].

Indirect torque control techniques control the torque by converting it into equivalent phase current references. Alternatively, direct torque control techniques do not require any torque-to-current conversions and therefore, avoid precomputation or storage of current profiles [2] - [10]. Direct instantaneous torque control (DITC) is one among the available direct torque control techniques. The prominent features of DITC are its simplicity, requirement of low-precision position sensor, instantaneous torque estimation solely from terminal

quantities and robustness against variations in input quantities such as torque, voltage and speed [5].

In DITC, it is observed that the phase current profiles and torque ripple characteristics depend on the turn-on instant. For current profiles with low RMS value, an overshoot in torque is observed during the commutation period, which increases the torque ripple. This overshoot in torque can be limited by delaying the turn-on instant. Consequently, the resultant current profiles have higher source current ripple, higher RMS value and thereby, lower torque per ampere ratio.

In order to address the aforementioned limitations with DITC, a modified version of DITC is proposed for a three-phase SRM, which simultaneously improves the torque per ampere ratio and at the same time, prevents torque overshoot. In addition, the proposed strategy also aims at improving the source current ripple, by reducing its peak values during the commutation period. Detailed simulation studies are performed and the results are verified for a 3-phase, 12/26 pole, 600 rpm, 1.5 kW SRM [11].

II. DIRECT INSTANTANEOUS TORQUE CONTROL - DITC

This section gives a brief description of DITC for SRM. The nonlinear model of SRM used for demonstrating the simulation results is developed in MATLAB-Simulink environment [12]. The necessary motor characteristics for this model are obtained through FEM simulations using MagNet 2D package. Asymmetric H-bridge converter is used for this scheme [13]. Different switching states possible with this converter are shown in Fig. 1.

In DITC, a torque-hysteresis controller generates switching signals for the converter based on error between reference torque and motor torque. Torque is regulated within two hysteresis bands. During single-phase conduction, active phase generates the required motor torque and it is regulated using the inner band. During commutation period, motor torque is shared by the incoming and outgoing phases as single phase alone cannot meet the complete torque requirement. The outer and inner bands control the switching of outgoing and incoming phases respectively.

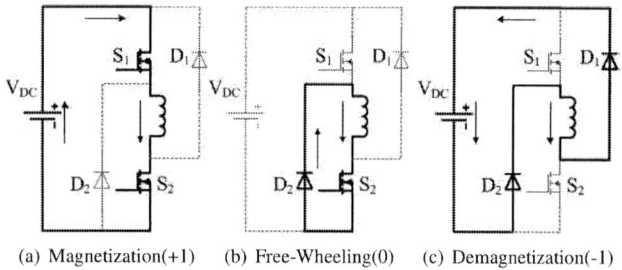

(a) Magnetization(+1) (b) Free-Wheeling(0) (c) Demagnetization(-1)

Fig. 1: Switching states of H-bridge Asymmetric Converter for one phase of SRM.

The turn-on of incoming phase initiates the commutation process. The outgoing phase is set into free-wheeling mode. Few instants later, when the total torque falls below the lowermost limit, the outgoing phase is turned-on again. When the incoming phase can regulate the total torque, the outgoing phase is set back to free-wheeling mode. When the total torque reaches upper-most limit, the incoming phase is set to free-wheeling mode and outgoing phase is demagnetized.

A. Limitations with existing DITC

In a three-phase SRM, inductance overlap among the phases is small. During the commutation period, the value of $dL/d\theta$ is low in both the commutating phases due to saturation. Therefore, high value of current is required to meet the torque requirement and this results in peaky currents during the commutation period.

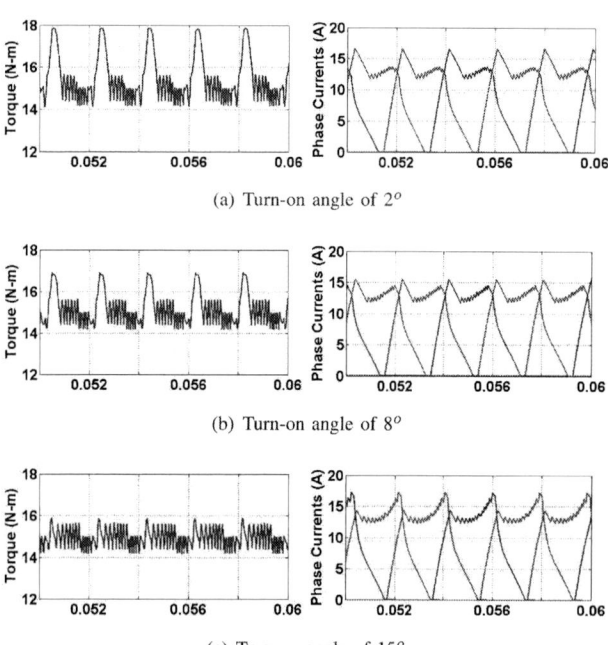

Fig. 2: Comparison of torque ripple and phase current profiles for various turn-on angles in DITC.

With the existing DITC strategy, it is observed that the phase current profiles and torque ripple characteristics depend on the turn-on instant. A comparison of phase current profiles and their corresponding torque performance for three different turn-on angles is shown in Fig. 2. Their peak-peak torque ripple and phase current RMS values at a speed of 400 rpm are listed in Table I.

It can be inferred that few current profiles can have a detrimental effect on torque performance of the machine. Detailed waveforms with DITC for turn-on angles of 2° and 15° are shown in Figs. 3 and 4 respectively. It is important to note that the turn-on and turn-off angles corresponding to low RMS value are approximately determined through iterative study. Precise determination of optimal current profiles is limited by the lack of accurate mathematical model of SRM, resulting from non-linear behaviour.

TABLE I: Comparison of torque ripple and phase current RMS values in DITC

Reference Torque (N-m)	Turn-on angle (Deg.-electrical)	Peak-peak Torque Ripple (N-m)	RMS value of phase current (A)
15	2	3.6	8.61
15	8	2.6	8.56
15	15	1.6	9

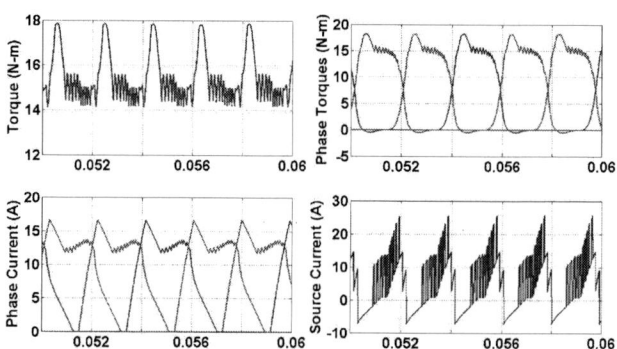

Fig. 3: Waveforms with DITC for turn-on angle of 2°.

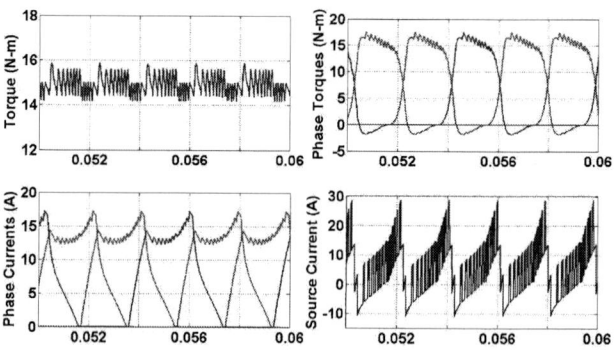

Fig. 4: Waveforms with DITC for turn-on angle of 15°.

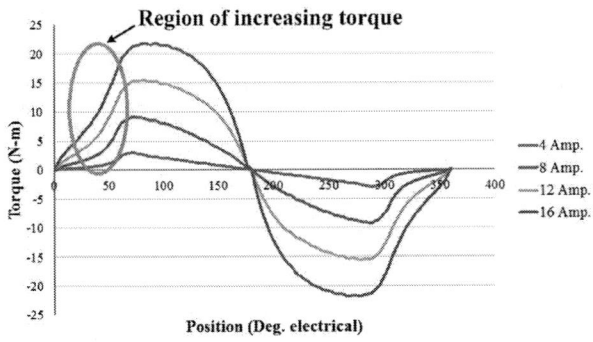

Fig. 5: Phase torque characteristics of three-phase 12-26 SRM.

An overshoot occurs in the torque with turn-on angles of 2° and 8° because at that instant, the phase torque characteristics of incoming phase are rising sharply with respect to rotor position. The torque characteristics (T-i-θ) for the chosen three-phase SRM are shown in Fig. 5.

When torque crosses the upper-most limit, the incoming phase is set into free-wheeling mode and outgoing phase is demagnetized as per existing DITC strategy. However, during free-wheeling operation of incoming phase, the current falls at a relatively slow rate and is unable to limit the torque within specified limits. Therefore, slow rate of fall of current coupled with the rising nature of phase torque characteristics in the incoming phase is responsible for torque overshoot.

To maintain good torque performance, the existing control strategy must delay the turn-on angle and follow a current profile similar to (c) of Fig. 2. The drawbacks in maintaining such a current profile are listed as follows:

- A large current peak near aligned position will take longer time to reach zero. This increases the duration of current in negative torque region. To overcome this negative torque, the active phase has to generate higher torque, increasing the RMS value of phase current.
- Moreover, large current peak during the turn-off period results in high rate of fall of current near the aligned position. This increases the radial forces on the machine and hence, vibrations and acoustic noise [14].
- Due to absence of direct control over current, there is a possibility of phase current overshoot due to very low dL/dθ near the aligned position, as shown in Fig. 6.
- Increase in source current ripple imposes higher burden on input source.

In order to address the aforementioned limitations with existing DITC, modified DITC is proposed in following section.

III. MODIFIED DITC

In the proposed strategy, when torque crosses the upper-most limit of hysteresis band, rate of fall of current in the incoming phase is increased by applying $-V_{DC}$ across it, instead of free-wheeling. This increase in rate of fall of current in incoming phase limits the torque overshoot.

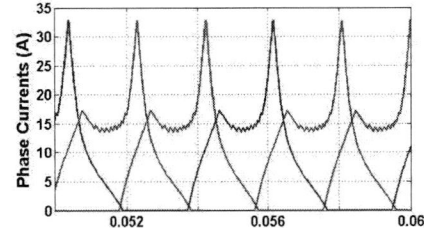

Fig. 6: Phase current overshoot.

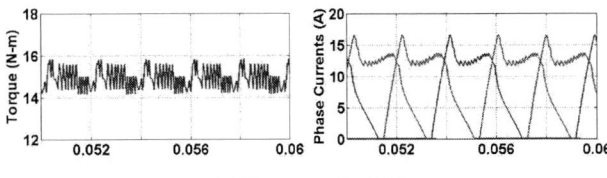

(a) Turn-on angle of 2°

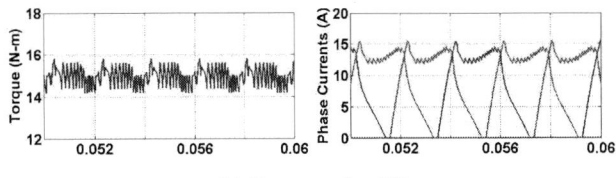

(b) Turn-on angle of 8°

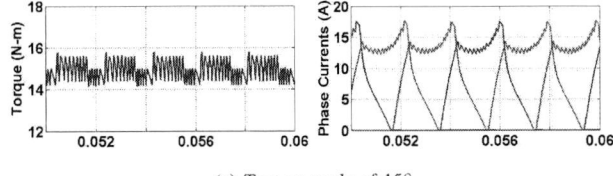

(c) Turn-on angle of 15°

Fig. 7: Comparison of torque ripple and phase current profiles for various turn-on angles in Modified DITC.

TABLE II: Comparison of torque ripple and phase current RMS values in Modified DITC

Reference Torque (N-m)	Turn-on angle (Deg.-electrical)	Peak-peak Torque Ripple (N-m)	RMS value of phase current (A)
15	2	1.6	8.4
15	8	1.6	8.5
15	15	1.6	9

A comparison of phase current profiles and their corresponding torque performance for three different turn-on angles is shown in Fig. 7. Their peak-peak torque ripple and phase current RMS values at a speed of 400 rpm are listed in Table II. With the proposed modification, the issue of torque overshoot with current profiles of low RMS value is addressed.

Detailed waveforms with modified DITC for turn-on angle of 2°, which has low RMS value, are shown in Fig. 8. However, it can be observed that the source current profile has deteriorated. Table III gives a comparison of existing DITC and modified DITC schemes at a speed of 400 rpm.

978-1-4799-2706-7/14 $31.00 © 2014 IEEE

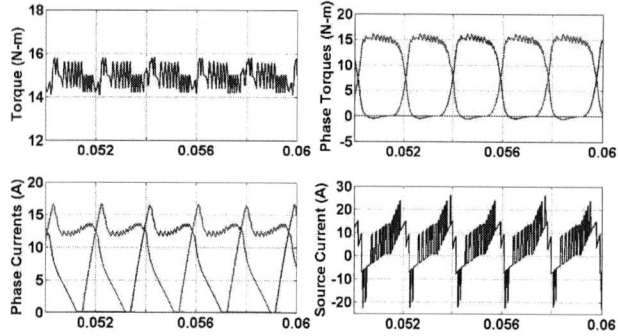

Fig. 8: Waveforms with Modified DITC for turn-on angle of $2°$.

TABLE III: Comparison of DITC and Modified DITC

Control Strategy	Torque (N-m)	Phase RMS Current (A)	Torque per Ampere (N-m/A)	Source Current Peak Values (A)
DITC	15	9	1.667	28.5, -11
Modified DITC	15	8.4	1.785	26, -22

IV. REDUCTION OF SOURCE CURRENT RIPPLE

A comparison between peak values of source currents of existing DITC and modified DITC is shown in Table III. The main reason for increase in negative peak of source current in modified DITC is the simultaneous turn-off of outgoing and incoming phases during commutation period. To minimize this negative peak, simultaneous turn-off of commutating phases should be avoided.

From the waveforms shown in Figs. 3 and 4, it can be inferred that the incoming phase is mainly responsible for producing the torque overshoot. Therefore, controlling the phase torque of incoming phase can effectively limit the total torque within specified limits. When $-V_{DC}$ is applied across the incoming phase (-1) to reduce the torque, the outgoing phase can either be set into free-wheeling mode (0) or turned-on (+1). The current directions for these two states are shown in Fig. 9, where phase-A is assumed to be the incoming phase and phase-B is assumed to be the outgoing phase.

A. Outgoing phase is set into free-wheeling mode

When the torque reaches upper-most limit, $-V_{DC}$ is applied across the incoming phase and the outgoing phase is set into free-wheeling mode. As a result, current flowing into the source is only the current present in incoming phase, as shown in Fig. 9(a). This switching state is denoted by (-1,0). Simulation is done for a reference torque of 15 N-m at two speeds: 400 rpm and 50 rpm and the results are tabulated in Tables IV and V respectively. The simulated waveforms at 400 rpm speed are shown in Fig. 10.

B. Outgoing phase is turned-on

When the torque reaches upper-most limit, $-V_{DC}$ is applied across the incoming phase and the outgoing phase is turned-on. The current in the incoming phase is now divided into two parts: a part of it flows into the outgoing phase and the remaining part flows into the source. As a result, current flowing into the source is only a part of incoming phase current, as shown in Fig. 9(b). This switching state is denoted by (-1,+1). Simulation is done for a reference torque of 15 N-m at two speeds: 400 rpm and 50 rpm and the results are tabulated in Tables IV and V respectively. The simulated waveforms at 400 rpm speed are shown in Fig. 11.

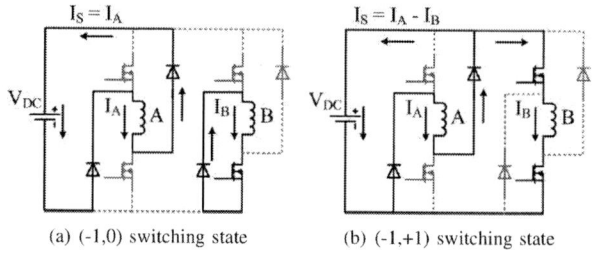

(a) (-1,0) switching state   (b) (-1,+1) switching state

Fig. 9: Modified DITC switching states to limit torque overshoot and reduce source current ripple.

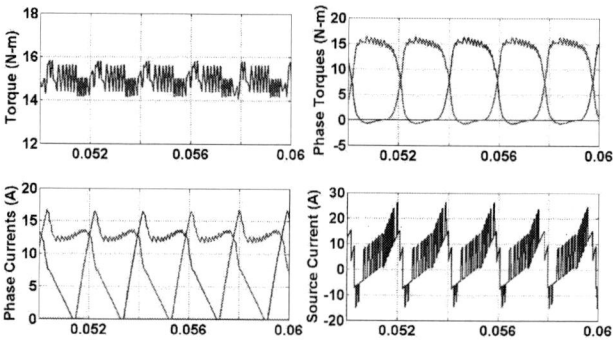

Fig. 10: Waveforms with Modified DITC using (-1,0) switching state.

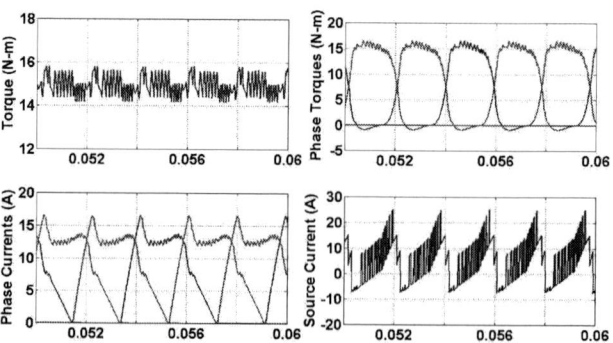

Fig. 11: Waveforms with Modified DITC using (-1,+1) switching state.

TABLE IV: Comparison of DITC and proposed Modified DITC at 400 rpm

Control Strategy	Torque (N-m)	Peak-Peak Torque ripple (N-m)	Phase RMS Current (A)	Torque per Ampere (N-m/A)	Copper Loss (W)	Reduction in Copper loss	Source Current Peak Values (A)
DITC	15	1.6	9	1.667	58.32	-	28.5, -11
Modified DITC using (-1,0)	15	1.6	8.5	1.765	52	10.83%	26, -15
Modified DITC using (-1,+1)	15	1.6	8.56	1.752	52.75	9.55%	26, -7.5

TABLE V: Comparison of DITC and proposed Modified DITC at 50 rpm

Control Strategy	Torque (N-m)	Peak-Peak Torque ripple (N-m)	Phase RMS Current (A)	Torque per Ampere (N-m/A)	Copper Loss (W)	Reduction in Copper loss	Source Current Peak Values (A)
DITC	15	2.2	8.56	1.752	52.75	-	26.5, -20
Modified DITC using (-1,0)	15	1.6	8.13	1.845	47.59	9.78%	20, -16
Modified DITC using (-1,+1)	15	1.6	8.13	1.845	47.59	9.78%	20, -15

V. RESULTS AND DISCUSSION

Tables IV and V give a comparison between existing DITC and the proposed modified DITC, in terms of phase RMS current, torque per ampere ratio, copper loss reduction and source current ripple. For a reference torque of 15 N-m, the results are presented for speeds of 400 rpm and 50 rpm respectively, to compare the performances at different speeds.

In the proposed commutation strategy with (-1,0) and (-1,+1) states, the turn-off duration of outgoing phase is slightly increased as it is not demagnetized continuously. Therefore, when the phase RMS currents of modified DITC in Tables III and IV are compared, there is a slight increase in the RMS value of phase current. This is the necessary trade-off for minimizing source current ripple.

At low speeds, modified DITC with (-1,0) and (-1,+1) switching states yields similar results because the available time for commutation is relatively more, as compared at high speeds. At the time instant when incoming phase is demagnetized to limit torque overshoot, the outgoing phase current is already reduced to a low value due to more available time for turn-off. Therefore, significant variation in RMS values and source current ripple values is not observed at low speeds with (-1,0) and (-1,+1) switching states in modified DITC.

VI. CONCLUSION

This paper presents a modified version of direct instantaneous torque control (DITC) for torque ripple minimization in switched reluctance motor (SRM). For current profiles with low RMS value, torque overshoot is observed during the commutation period in existing DITC, which increases the torque ripple. This overshoot in torque can be limited by delaying the turn-on instant. Consequently, the resultant current profiles have higher source current ripple, higher RMS value and thereby, lower torque per ampere ratio.

To address the aforementioned limitations, modified direct instantaneous torque control strategy is proposed in this paper. The proposed strategy simultaneously improves the torque per ampere ratio and prevents torque overshoot. In addition, source current ripple is also reduced by suitable modification of switching states during the commutation period.

REFERENCES

[1] I. Husain, "Minimization of torque ripple in SRM drives," *IEEE Trans. Ind. Electron.*, vol. 49, no. 1, pp. 28-39, Feb. 2002.

[2] S. K. Sahoo, S. K. Panda, and J. X. Xu, "Direct torque controller for switched reluctance motor drive using sliding mode control," in *Proc. PEDS*, 2005, pp. 1129-1134.

[3] P. Jinupun and P. C.-K. Luk, "Direct torque control for sensorless switched reluctance motor drives," in *Proc. 7th Int. Conf. Power Electron. Variable Speed Drives*, 1998, pp. 329-334.

[4] A. D. Cheok and Y. Fukuda, "A new torque and flux control method for switched reluctance motor drives," *IEEE Trans. Power Electron.*, vol. 17, no. 4, pp. 543-557, Jul. 2002.

[5] R. B. Inderka and R. W. A. A. De Doncker, "DITC-direct instantaneous torque control of switched reluctance drives," *IEEE Trans. Ind. Appl.*, vol. 39, no. 4, pp. 1046-1051, Jul./Aug. 2003.

[6] N. H. Fuengwarodsakul, M. Menne, R. Inderka, and R. W. De Doncker, "High dynamic four-quadrant switched reluctance drive based on DITC," *IEEE Trans. Ind. Appl.*, vol. 41, no. 5, pp. 1232-1242, Sep./Oct. 2005.

[7] N. H. Fuengwarodsakul, S. E. Bauer, J. Krane, C. P. Dick, and R. W. De Doncker, "Sensorless direct instantaneous torque control for switched reluctance machines," in *Proc. Eur. Conf. Power Electron. Appl.*, Dresden, Germany, Sep. 11-14, 2005.

[8] C. R. Neuhaus, N. H. Fuengwarodsakul, and R. W. De Doncker, "Predictive PWM-based direct instantaneous torque control of switched reluctance drives, in *Proc. 37th IEEE Power Electron. Spec. Conf.*, Jeju, Korea, Jun. 2006, pp. 1-7.

[9] H. J. Brauer, M. D. Hennen, and R. W. De Doncker, "Control for polyphase switched reluctance machines to minimize torque tipple and decrease ohmic machine losses," *IEEE Trans. Power Electron.*, vol. 27, no. 1, pp. 370-378, Jan. 2012.

[10] S. K. Sahoo, S. Dasgupta, S. K. Panda, and J. X. Xu, "A Lyapunov function based robust direct torque controller for switched reluctance motor drive system," *IEEE Trans. Power Electron.*, vol. 27, no. 2, pp. 555-564, Feb. 2012.

[11] S. P. Nikam, V. Rallabandi, and B. G. Fernandes, "A high-torque-density permanent-magnet free motor for in-Wheel electric vehicle application," *IEEE Trans. Ind. Appl.*, vol. 48, no. 6, pp. 2287-2295, Nov./Dec. 2012.

[12] F. Soares and P. J. Costa Branco, "Simulation of a 6/4 switched reluctance motor based on Matlab/Simulink environment," *IEEE Trans. Aerosp. Electron. Syst.*, vol. 37, no. 3, pp. 989-1099, Jul. 2001.

[13] Rik De Doncker, Duco W. J. Pulle, and André Veltman, *Advanced Electrical Drives: Modeling, Analysis and Control*. Springer, 2011.

[14] R. Krishnan, *Switched Reluctance Motor Drives: Modeling, Simulation, Analysis, Design and Applications*. Boca Rotan, FL: CRC Press, 2001.

Control of Wound Field Synchronous Motor Integrated with ZSI

G. Tajima, T. Kosaka and N. Matsui
Dept. of Computer Science and Engineering
Graduate School of Engineering
Nagoya Institute of Technology
Nagoya, Japan
kosaka@nitech.ac.jp

K. Tonogi, N. Minoshima and T. Yoshida
Toyota Industries Corporation
8, Chaya, Obu, Aichi, Japan
kazuki.tonogi@mail.toyota-shokki.co.jp

Abstract— **This paper proposes a novel variable speed motor drive, wound field synchronous motor (WFSM) drive integrated with Z-source inverter (ZSI). The proposed WFSM is non-permanent magnet machine and employs two field coils as dc field mmf sources. On the other hand, ZSI has an impedance-network which is composed of two dc reactors and two capacitors. In the proposed drive, the two field coils of WFSM act not only as dc field mmf sources, but also as dc reactors in ZSI, resulting in size and cost reductions of the drive system. In order to achieve both the voltage boost-up and the field current controls simultaneously, its control algorithm is examined. Simulation and experimental results show that the proposed drive under the proposed control works properly.**

Keywords— *boost-up/field current control, integrated motor and drive circuit system, wound field synchronous motor, z-source inverter*

I. INTRODUCTION

At present, the application of permanent magnet motors with the use of rare-earth permanent magnets is growing to improve efficiency while reducing size and weight. Supply and price instabilities of rare earth materials, however, become a problem in recent years. Therefore, researches and developments of less- or non-rare-earth permanent magnet motors are received much attention from industries.

In response, the authors have also proposed wound field synchronous motor (WFSM) as a candidate of non-permanent magnet motors [1]. In WFSM, field fluxes are generated by two field toroidal coils which are wound on stationary field pole cores located at both ends of motor in the axial direction. By this structural nature, it enables to feed current into the field coils without any brushes or slip rings. However, one of disadvantages is the increases in motor size and cost because the field pole cores and the field coils are net additional components against conventional motors.

On the other hand, in case of a variable speed drive system in which a motor is designed as high speed machine under a low-voltage battery drive, it often employs DC-DC boost-up converter inserted between a battery and an inverter [2]. In this case, it is expected to reduce motor size and weight while improving its efficiency. However, it is indispensable task to design the system taking into account of efficiency degradation and cost-up due to an addition of DC-DC converter.

In this paper, a variable speed drive system shown in Fig. 1 is proposed, where WFSM is integrated with Z-source inverter (ZSI) [3]. ZSI has an impedance-network, which consists of two dc reactors L_f and two capacitors C and is inserted between a DC voltage source and a three-phase inverter. ZSI is functioning not only as three-phase inverter, but also as DC-DC boost-up converter. One of disadvantages of ZSI is size and weight of bulky dc reactors. This integrated system can solve the drawbacks of WFSM and ZSI simultaneously. Specifically, two dc reactors of ZSI are replaced by two dc field coils of WFSM as the sharing component in motor and converter and hence, the proposed drive system is expected as a solution for size, weight and cost reductions as well as for the loss minimization of the total drive system. In the proposed drive system, however, a difficulty lies in its control by which the voltage boost-up and field current regulation required for a given operating condition are satisfied harmoniously. Therefore, the control algorithm of the proposed system is examined as a main objective in this paper. Some simulation results are conducted and show that the proposed drive system works properly.

II. BASIC WORKING PRINCIPLE AND MODEL OF PROPOSED INTEGRATED DRIVE SYSTEM

A. Structure and Basic Working Principle of WFSM

Fig. 2 illustrates a general view of 20-pole/24-slot WFSM employing fractional-slot concentrated windings. The main machine part in the middle of the figure is

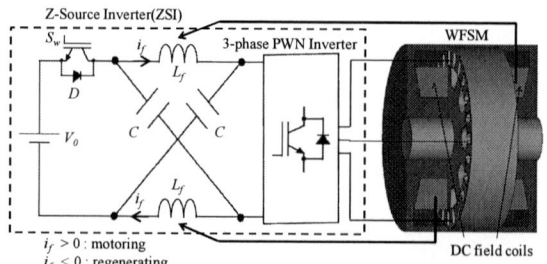

Fig. 1. Proposed drive system of WFSM integrated with ZSI.

978-1-4799-2706-7/14 $31.00 © 2014 IEEE

composed of laminated rotor core with 10 salient poles and laminated stator core with three-phase armature concentrated windings accommodated in 24 slots. The back yoke core is made of soft magnetic composites (SMC) and forms axis field flux paths flowing into the field pole SMC cores located at the both ends of motor. Each field pole SMC core located at each end of motor in the axial direction has the toroidal field coil. Fig. 3 depicts the field flux paths when the field coils are only energized by a certain DC current excitation. Although the airgap flux density is unipolar, it varies with the rotor position displacement due to the doubly salient-pole machine nature. The spatial distribution of gap permeance with respect to the rotor position has AC components. Thus, the three-phase symmetrical AC induced voltage waveforms arises in the three-phase symmetric armature windings by the modulation of the field flux with respect to the rotor position displacement. As a result, WFSM generates torque by feeding three-phase armature currents with an adequate phase angle against the three-phase induced voltage.

B. Motor Model

Given that the AC component of the airgap flux density distribution is a sinusoidal, a *d-q* axis model can be introduced similar to that in conventional AC motors under three-phase sinusoidal current drive.

Assuming that the direction of the field current i_f shown in Fig. 3 is positive and the origin of rotor position is defined as the one where the flux linkage in the U-phase winding reaches its positive maximum, the voltage equation of WFSM can be expressed in the following

fashion through the conventional d-q transformation.

$$\begin{bmatrix} v_d \\ v_q \end{bmatrix} = \begin{bmatrix} R + pL_d & -\omega L_q \\ \omega L_d & R + pL_q \end{bmatrix} \begin{bmatrix} i_d \\ i_q \end{bmatrix} + \sqrt{\frac{3}{2}} \omega K i_f \begin{bmatrix} 0 \\ 1 \end{bmatrix} \quad (1)$$

The first term expression follows the custom in interior permanent magnet motors. The second term Ki_f in the expression is the amplitude of AC component of the flux linkage in each phase winding generated by field mmfs.

The input power can be calculated from (1) and the torque equation can be obtained in the following manner,

$$\tau = \sqrt{\frac{3}{2}} P K i_f i_q + P(L_d - L_q) i_d i_q . \quad (2)$$

The first term expresses the torque produced by the field flux and the q-axis current similar to the magnet torque in permanent magnet motors. The second term is the reluctance torque. As WFSM shown in Figs. 2 and 3 employs the fractional concentrated windings, the reluctance torque expressed in the second term is negligibly lower than the torque in the first term. Paying attention to the torque expressed in the first term, it turns out that the polarity of the torque is determined by the product of the polarities of the field current i_f and the q-axis current i_q.

III. Operating Modes of ZSI

The operating modes of ZSI are divided into three states in accordance with the switching pattern of the three-phase PWM inverter and OFF or ON of the switch S_w shown in Fig. 1 for motoring and regenerating modes, respectively. Here after, it is noted that the inductance of each field coil is L_f, the resistance of each field coil is R_f, the capacitance of the capacitors is C, each voltage across each impedance-network component is $L_f di_f/dt$, $R_f i_f$ and v_c, respectively. Furthermore, the voltage and the current of the DC voltage source are V_0 and i_{in}, respectively, and the field current is i_f, the capacitor current is i_c, the voltage across the inverter leg is v_i, the current flowing to the inverter is i_i, respectively. Each operating mode can be expressed as corresponding voltage equation regardless of motoring or regenerating and therefore, the behaviors of the field current and the capacitor voltage are determined accordingly. The behavior of each operating mode during motoring in ZSI examined in following section is analyzed on the assumption that the PWM carrier frequency of the inverter is enough high compared with the resonance frequency of the impedance-network.

A. Active State

Fig. 4 shows the equivalent circuit at *active state* in motoring mode. The *active state* is only the state among three states, where the inverter outputs non-zero voltage vector and supplies power to WFSM. The load side is approximated as a constant current source I_i and therefore, the current fed to the inverter i_i should be equal to I_i ($i_i = I_i$). With regard to time t, the beginning time of the *active state* is defined as $t = 0$.

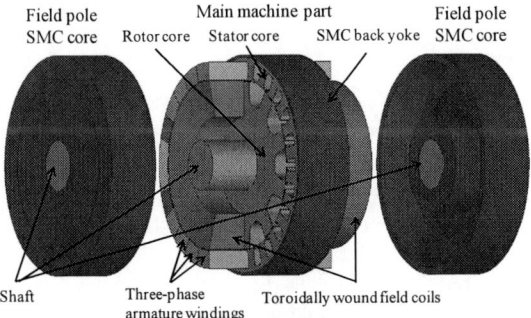

Fig. 2. General view of WFSM.

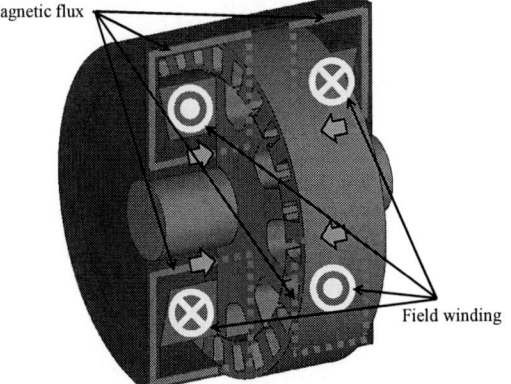

Fig. 3. Field flux paths in WFSM.

From the voltage equation of impedance-network side at the *active state*, the time response of field current can be expressed as,

$$\frac{di_f}{dt} = \frac{V_0 - v_c - R_f i_f}{L_f}. \tag{3}$$

The time response of capacitor voltage can be similarly expressed in the following equation.

$$\frac{dv_c}{dt} = \frac{i_f - i_i}{C} = \frac{i_f - I_i}{C} \tag{4}$$

On the other hand, the current of DC voltage source i_{in} can be expressed in the following fashion according to the relationship between the currents in ZSI.

$$i_{in} = 2i_f - I_i \tag{5}$$

In addition, the following condition has to be satisfied because the switch S_w is always in the OFF in motoring mode.

$$2i_f - I_i > 0 \tag{6}$$

If (6) is not satisfied, the voltage across the inverter leg v_i becomes zero because a part of the current fed to the motor flows through some freewheeling diode of some switching devices. As a result, this state is identical with *shoot through state* mentioned later and thus, the inverter is unable to output non-zero voltage vector. This state is called *freewheeling diode shoot through state* [4] to distinguish from an intentionally executed *shoot through state*. In order to avoid this state, one of the control restrictions of ZSI is that the field current i_f has to control to meet (6). As long as the field current i_f is controlled to meet (6), the voltage across the inverter leg v_i meets the following relation.

$$v_i = 2v_c - V_0 \tag{7}$$

B. Active Zero State

The equivalent circuit at *active zero state* in motoring mode is shown in Fig. 5. As the inverter outputs zero voltage vector in the *active zero state*, the state of the inverter is disconnected from DC side ($i_i = 0$). With regard to time t, the changing time to the *active zero state* is redefined as $t = 0$. In this case, as the voltage equation of impedance-network side is identical with the

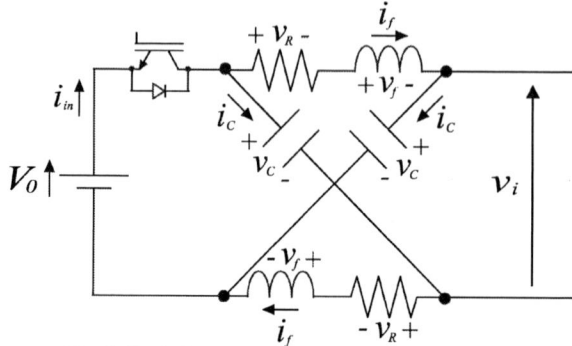

Fig. 5. Equivalent circuit at *active zero state* (motoring).

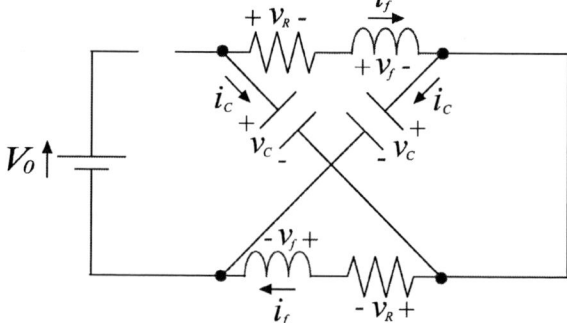

Fig. 6. Equivalent circuit at *shoot through state* (motoring).

previously mentioned *active state*, the time response of field current can be expressed similar to (3). The time response of capacitor voltage is similarly expressed in (4), but it is obtained from the following equation by considering the inverter disconnection.

$$\frac{dv_c}{dt} = \frac{i_f}{C} \tag{8}$$

The voltage across the inverter leg v_i can be also similarly expressed in (7).

C. Shoot Through State

The equivalent circuit at *shoot through state* in motoring is shown in Fig. 6. The *shoot through state* is the state of some inverter arm in short circuit. With regard to time t, the changing time to the *shoot through state* is redefined as $t = 0$.

From the voltage equation of impedance-network side, the time response of field current can be expressed as,

$$\frac{di_f}{dt} = \frac{v_c - R_f i_f}{L_f}. \tag{9}$$

The time response of capacitor voltage is obtained from the following equation.

$$\frac{dv_c}{dt} = -\frac{i_f}{C} \tag{10}$$

Needless to say, the voltage across the inverter leg at this state can be expressed as $v_i = 0$.

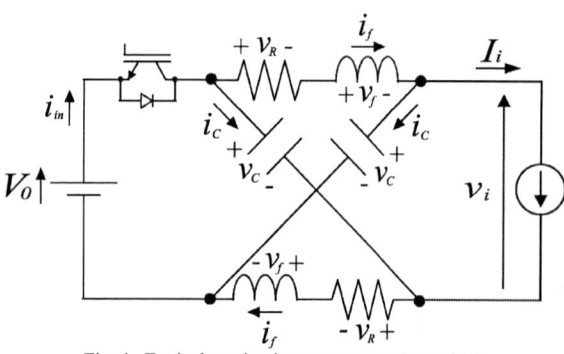

Fig. 4. Equivalent circuit at *active state* (motoring).

IV. Control Algorithm of Field Current and Capacitor Voltage

A. Control manners for Field Current and Capacitor Voltage in the Proposed Drive System

The basic control principle of ZSI is examined from the time response of field current i_f and capacitor voltage v_c expressed by each operating state described in the preceding chapter.

The duty ratios d_A, d_Z, and d_S of *active state*, *active zero state*, and *shoot through state* have the following relation.

$$d_A + d_Z + d_S = \frac{T_A}{T} + \frac{T_Z}{T} + \frac{T_S}{T} = 1 \tag{11}$$

The time response of field current at *active state* and *active zero state* is determined by (3), and that at *shoot through state* is obtained from (9). The total time response of the field current during one control interval T is derived from (3), (9) and (11) and given in,

$$\frac{di_f}{dt} = \frac{(2v_c - V_0)d_S + V_0 - v_c - R_f i_f}{L_f} \tag{12}$$

(12) means that the field current can be controlled by *shoot through state* duty ratio d_S based on a feedback of the field current. Similarly, the total time response of capacitor voltage during one control interval T is derived from (4), (8), (10) and (11) and given in,

$$\frac{dv_c}{dt} = \frac{-2i_f d_S - I_i d_A + i_f}{C} \tag{13}$$

(13) implies that the capacitor voltage can be controlled by *active state* duty ratio d_A based on a feedback of the capacitor voltage under the given *shoot through state* duty ratio d_S decided by the field current controller. On the other hand, *active state* duty ratio d_A also has to be determined so as to meet the requirement for the power control in WFSM simultaneously. In the proposed drive system, therefore, the capacitor voltage v_c and the motor power have to be controlled by only one control input d_A. In general, this control might be impossible and therefore, we have to make another independent control input for the capacitor voltage control. In the right side of (13), the inverter load current I_i corresponds to the input power of WFSM for the given torque and speed condition. As the input power includes not only the output, but also the inverter, copper and iron losses, the inverter load current I_i can be adjusted by the copper loss which is controlled by the d-axis current with negligible torque change.

B. Proposed Capacitor Voltage Control

First of all, both the field and the q-axis current controllers are designed so as to respond faster than the capacitor voltage and the d-axis current controllers because these current responses directly dominate over the torque response. Under this condition, the controller for achieving the capacitor voltage and the motor output controls simultaneously is examined based on AC Small Signal Modeling Method [5].

Firstly, the relation between the input power and the motor output is examined. The input power can be expressed in $v_i i_i d_A$. To explain the control algorithm simply, it can be assumed that the inverter and the iron losses are negligible compared to the copper loss. As a result, the following relation is obtained from (1), (2) and (7).

$$(2v_c - V_0)i_i d_A = R\{i_d^2 + i_q^2\} + \omega\tau \tag{14}$$

Based on the AC small signal modeling, the variables v_c and i_d in the above equation can be expressed by using the target value (x^*) and the deviation ($\hat{x}$), and given in the following fashions.

$$v_c = v_c^* + \hat{v}_c \quad , \quad i_d = i_d^* + \hat{i}_d \tag{15}$$

In addition, the field and the q-axis currents can be also expressed in $i_f = i_f^*$ and $i_q = i_q^*$, respectively. Finally, the transfer function from $\hat{i}_d$ to $\hat{v}_C$ is obtained from (2), (12), (13) and (14) by substituting the variables expressed in (15) and given in,

$$G_{v_c} = \frac{\hat{v}_c(s)}{\hat{i}_d(s)} = -\frac{2Ri_d^* + \omega P(L_d - L_q)i_q^*}{sC(2v_c^* - V_0)} \tag{16}$$

(16) implies that the deviation of the capacitor voltage $\hat{v}_c$ is controllable by the fine tune control of motor output with the deviation of the d-axis current $\hat{i}_d$ and hence, it can be seen that to achieve both the capacitor voltage and the motor output controls simultaneously is possible.

C. Control System Configuration of the Proposed Drive System

Fig. 5 illustrates the control block diagram configured based on the examination in the preceding section. The field current controller employs a PI controller based on the error between its reference and detected values. The capacitor voltage controller is also composed of a PI controller based on the deviation between its reference and detected values. In addition, the output is connected with the d-axis current controller so that the independent control of the capacitor voltage and the motor output can be achieved. WFSM controller takes control of each axis current with PI controller by employing the decoupling control as with conventional AC machines. The proposed system has to be controlled by the appropriate control reference parameters i_f^*, v_c^*, i_d^*, i_q^* taking into account of control constraint conditions under the given operating point (τ^*, ω). For this reason, the sets of control reference parameters for the operating region are stored in a look-up table in advance.

V. Control Constraint Conditions and Control Reference Parameter Search

This chapter explains how to search for the control reference parameters to control the field current, the capacitor voltage and the motor output simultaneously at given operating points in the proposed system.

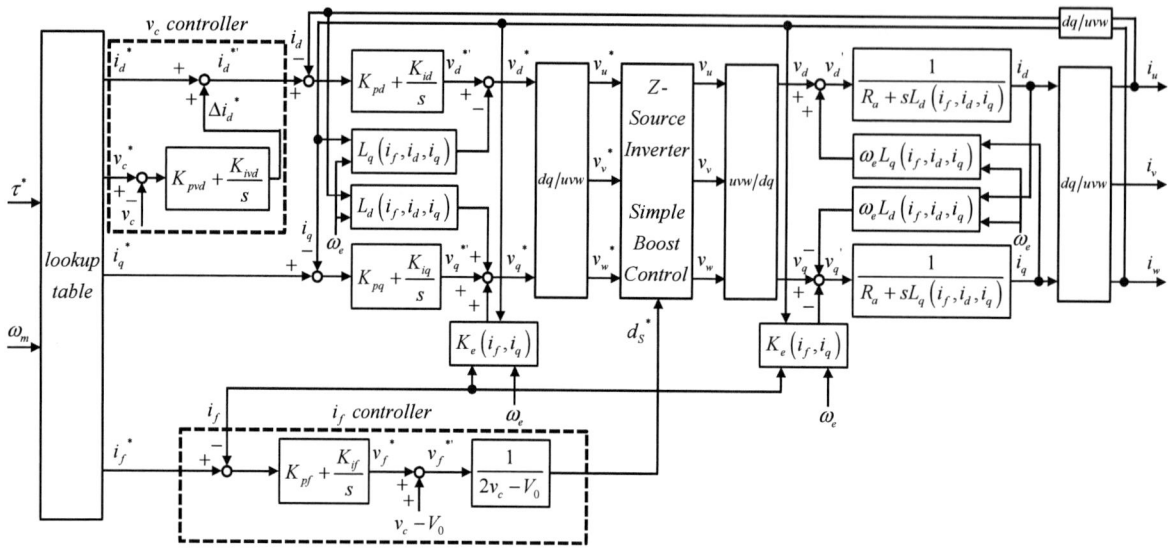

Fig. 7. Block diagram for the proposed controller.

A. Constraint of Input-output Balance Equation in ZSI

The steady-state input-output relation in ZSI is expressed in the following equation with the control reference parameters of given operating points from energy conservation.

$$V_0 i_f^* = 2R_f i_f^{*2} + R(i_d^{*2} + i_q^{*2}) + \omega\tau^* \tag{17}$$

The left-hand side of Eq. (17) expresses the input from the DC power source, the first term on the right side expresses the copper loss of the field coils, the second term expresses the copper loss of the armature windings, and the third term expresses the motor output.

B. Constraint of the Field Current Reference

The field current i_f and the current fed to the inverter bridge I_i at the *Active State* have to satisfy the following equation derived from (6) for avoiding the *freewheeling diode shoot through state*.

$$2i_f^* > I_i \tag{18}$$

C. Constraint of the Operable Duty Ratio

ZSI is controlled by the duty ratio of the three states. The control reference parameters have to be chosen such that the sum of the duty ratios at operating points is not exceeded 1. The following equation is obtained by rearranging (11).

$$m^* = d_A^* + d_{Z\_min}^* \leq 1 - d_S^* \tag{19}$$

where $d_{Z\_min}^*$ is the minimum duty ratio reference of the *active zero state*. m^* is the sum of d_A^* and $d_{Z\_min}^*$ and expressed by another equation below for the reason that the duty ratio equals the area ratio as shown in Fig. 8.

$$m^* = \frac{2\pi}{3\sqrt{3}} d_A^* \tag{20}$$

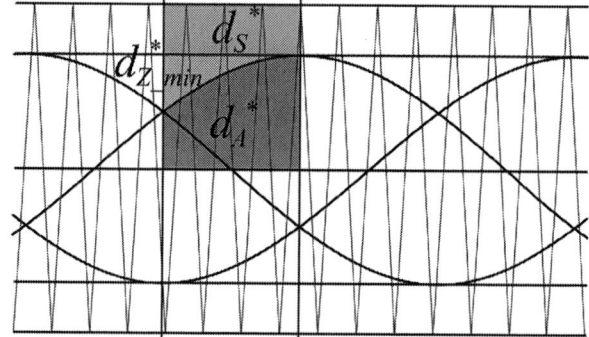

Fig. 8. Duty ratio of each operating state.

Rearranging (19) and (20), the following equation is derived.

$$d_A^* \leq \frac{3\sqrt{3}}{2\pi}(1 - d_S^*) \tag{21}$$

Thus, ZSI has to be controlled such that the duty ratios d_A^* and d_S^* satisfy (21).

D. Control Parameter Search at Motor Operating Points

The flowchart of control reference parameters search at operating points of the drive system of WFSM integrated with ZSI is shown in Fig. 9. The current reference i_f^* and the armature current reference i_d^*, i_q^* of the control reference parameters satisfying the constraint conditions of (17), (18) and (21) is searched at the torque reference and the number of revolutions. They and the capacitor voltage reference v_c^* is eventually stored in a look-up table.

The 2014 International Power Electronics Conference

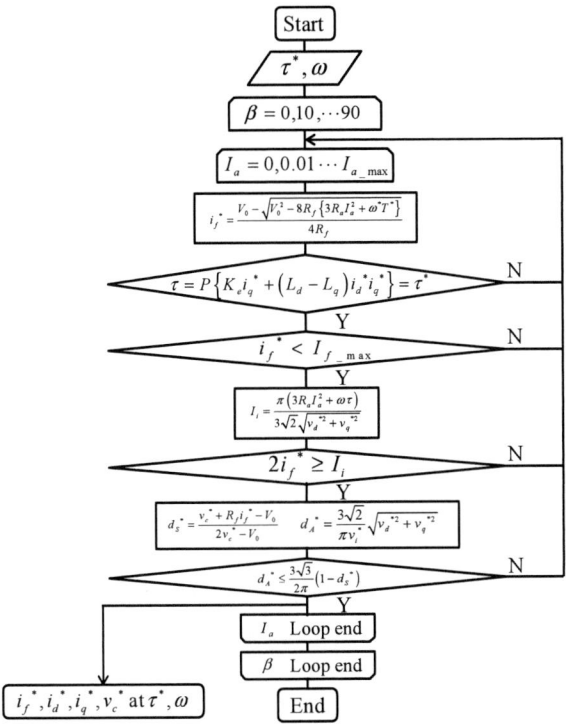

Fig. 9. Flowchart of control reference parameters search.

VI. SIMULATION RESULTS

As the first step of validation of the derived control algorithm for the proposed drive system, the drive simulation is conducted using a circuit simulator (PSIM Professional Version 9.3.2). The parameters of WFSM and ZSI appear in Table I. As the first step, all IGBTs and diodes are regarded as ideal devices which have no conduction and switching losses. The iron loss of WFSM is also out of consideration.

Fig. 10 demonstrates the simulation result under the condition that the motor speed is constant at 250r/min. The control reference parameters at this operating point appear in Table II. It can be seen that all the controlled variables converge to the references and hence, the validity of the proposed control algorithm can be confirmed.

VII. EXPERIMENTAL RESULTS

The experiment under the same control reference parameters as simulation is conducted by test WFSM and test ZSI shown in Figs. 11 and 12.

Fig. 13 demonstrates the experimental result under the condition that the DC bus voltage V_0 is set to 60V. The reason why the DC bus voltage higher than that used in the simulation is that the real experimental system includes losses of switching devices, iron losses and other stray losses without considering in the simulation. In addition, the reference of the capacitor voltage v_c^* is set to 115V so that the voltage across the inverter leg v_i is almost same with that in the simulation. Although the

TABLE I
PARAMETERS OF WFSM AND ZSI

WFSM parameters		ZSI parameters	
R [Ω]	0.36	C [uF]	9.9
L_d [mH]	Non-linear element	R_f [Ω]	1.13
L_q [mH]	Non-linear element	L_f [mH]	Non-linear element

TABLE II
CONTROL REFERENCE PARAMETERS USED FOR SIMULATION

Control reference parameters			
τ^* [Nm]	1.0	i_f^* [A]	8.0
ω [r/min]	250	i_d^* [A]	10.8
V_0 [V]	30	i_q^* [A]	8.9
		v_c^* [V]	100

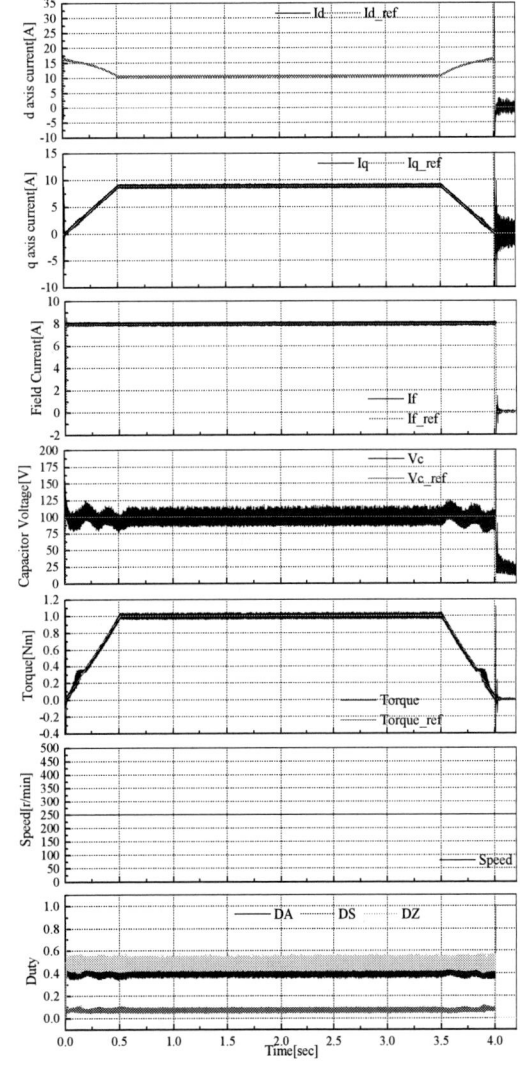

Fig. 10. Simulation result operating at 1Nm- 250r/min(V_o=30V).

armature current i_d and i_q have relatively large current ripple due to the ripple of capacitor voltage v_c^*, It can be seen that all the controlled variables converge to the references and hence, the validity of the proposed control algorithm can be confirmed by the experiment.

SMC core and field coil (load-side)   SMC core and field coil (anti-load-side)

Stator   Rotor

Fig. 11. Photographs of tested WFSM.

Fig. 12. Photograph of fabricated ZSI for test.

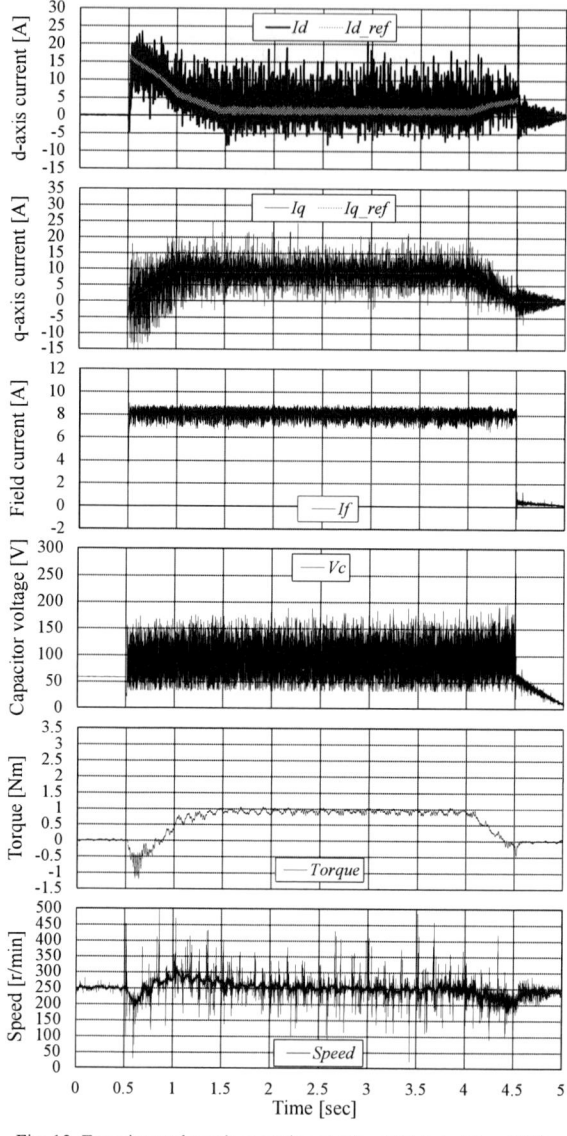

Fig. 13. Experimental result operating at 1Nm-250r/min ($V_0 = 60$V).

VIII. CONCLUSIONS

This paper has proposed the variable motor drive system using WFSM integrated with ZSI. The basic working principle of the proposed integrated drive system has been explained. The validity of the proposed control algorithm can be confirmed by the simulation and the experiment. It was also realized the necessity of having the loss of WFSM integrated with ZSI under consideration in simulation.

REFERENCES

[1] T. Kosaka, T. Hirose, and N. Matsui, "Brushless Synchronous Machines with Wound-Field Excitation using SMC Core Designed for HEV Drives," *IPEC-Sapporo 2010.*

[2] M. Kamiya, "Development of Traction Drive Motors for the Toyota Hybrid Systems," *IEEJ Trans. on IA.*, vol. 126, no. 4, pp. 473-479, 2006.

[3] F. Z. Peng, "Z-Source Inverter," *IEEE Trans. on Industry Applications*, vol. 39, no. 2, 2003.

[4] M. Shen and F. Z. Peng, "Operation Modes and Characteristics of the Z-Source Inverter With Small Inductance of Low Power Factor," *IEEE Trans. on Industrial Electronics*, vol. 55, no. 1, 2008.

[5] Jingbo Liu, Jiangang Hu, and Longya Xu, "Dynamic Modeling and Analysis of Z Source Converter-Derivation of AC Small Signal Model and Design Oriented Analysis," *IEEE Trans. on Power Electronics*, vol. 22, no. 5, 2007.

A Novel IPMSM Model for Robust Position Sensorless Control to Magnetic Saturation

Atsushi Matsumoto
Department of Electrical Engineering
and Computer Science
Nagoya University
Aichi, Japan 464-8603
Telephone: +81–52–789–2777
Fax: +81–52–789–3140
Email: atsushi.matsumoto@nagoya-u.jp

Masaru Hasegawa
Department of Electrical Engineering
Chubu University
Aichi, Japan 487-8501
Telephone: +81–568–51–1111
Fax: +81–568–51–1219
Email: mhasega@isc.chubu.ac.jp

Shinji Doki
Department of Electrical Engineering
and Computer Science
Nagoya University
Aichi, Japan 464-8603
Telephone: +81–52–789–2777
Fax: +81–52–789–3140
Email: doki@nagoya-u.jp

Abstract—**This paper presents a novel interior permanent synchronous motors (IPMSMs) model for position sensorless control. The proposed model has robustness with respect to magnetic saturation. Furthermore, this model can approximately estimate the maximum torque control (MTC) frame, which is possible to realize simple and robust sensorless control system. In addition, the sensitivity of inductance parameters in this model has also been discussed. Moreover, a position sensorless control system based on this model has been shown.**

Keywords—*IPMSM, Position Sensorless Control, Magnetic Saturation, Maximum Torque Per Ampere Control.*

I. INTRODUCTION

Interior permanent magnet synchronous motors (IPMSMs) are widely applied in various fields, such as household appliance, industry field, and so on, because IPMSMs arealize high efficiency and contribute to system compactness. To the IPMSMs for high efficiency control, detection of the rotor position is necessary. However, position sensors for detection of this signal decrease reliability and restrict the environment for installation. Therefore, position sensorless control has been required, and various strategies have been proposed so far[1], [2].

The maximum torque control per ampere (MTPA) control, which optimizes the overall torque including the reluctance torque, is often utilized as the most efficient drive of IPMSMs[3], [4]. In order to implement this control, generally, accurate parameters are required. Inductances dramatically vary, however, due to magnetic saturation, which has been one of the most important problems in recent years.

On the other hand, the maximum torque control (MTC) frame, which stands for a new coordinate aligned with the current vector at the MTPA control has been proposed, in which the MTC frame is directly estimated based on the modified extended electromotive force (EEMF), and the position sensorless control technique on this frame is realized [5], [6], [7]. This approach seems to be considerably questionable since the modified EEMF estimation is achieved based on "virtual inductance" model, which is defined by setting intentional

parameter mismatches in L_q. This approach is discussed in literature [5] is suitable for the motor which has characteristics of a large magnetic saturation and low saliency ratio. However, the scope of application of this approach is not shown.

This paper presents a novel mathematical model and a position sensorless control method for IPMSMs. The proposed position sensorless control system uses only the d-axis inductance which depends relatively little on magnetic saturation. That is, the proposed method does not need the q-axis inductance, thus providing position sensorless control robust to magnetic saturation, which occurs mainly along the q-axis. Thus, the proposed method provides easier parameter setting. First, we introduce a novel mathematical model, and explain its physical interpretation. Next, we explain that the proposed model can approximately estimate the MTC frame, and show the scope of application of the proposed position sensorless control system based on this model. Finally, we carry out some experiments to verify the effectiveness of the proposed methods.

II. PROPOSED MODEL AND POSITION SENSORLESS CONTROL SYSTEM

A. Derivation of proposed flux model

First, a novel mathematical model for position sensorless control of IPMSMs is derived. The voltage equation of IPMSMs on the rotating coordinates ($d - q$ coordinates) is given by

$$\begin{bmatrix} v_d \\ v_q \end{bmatrix} = \begin{bmatrix} R + pL_d & -\omega_{re}L_q \\ \omega_{re}L_d & R + pL_q \end{bmatrix} \begin{bmatrix} i_d \\ i_q \end{bmatrix} + \begin{bmatrix} 0 \\ \omega_{re}K_E \end{bmatrix}. \quad (1)$$

where
R resistance;
L_d d-axis inductance;
L_q q-axis inductance;
p differential operator;
K_E EMF constant;
ω_{re} angular speed at electrical angle;
θ_{re} rotor position at electrical angle.

This paper aims to unify all inductance of the impedance matrix in (1) to L_d. Equation (1) is transformed as follows:

$$\begin{bmatrix} v_d \\ v_q \end{bmatrix} = \begin{bmatrix} R + pL_d & -\omega_{re}L_d \\ \omega_{re}L_d & R + pL_d \end{bmatrix} \begin{bmatrix} i_d \\ i_q \end{bmatrix}$$
$$+ \begin{bmatrix} 0 \\ p(L_q - L_d)i_q \end{bmatrix} + \omega_{re} \begin{bmatrix} -(L_q - L_d)i_q \\ K_E \end{bmatrix}. \quad (2)$$

In this paper, the proposed flux $\boldsymbol{\lambda}$ is defined as

$$\boldsymbol{\lambda} = \begin{bmatrix} \lambda_d \\ \lambda_q \end{bmatrix} = \begin{bmatrix} K_E \\ (L_q - L_d)i_q \end{bmatrix}. \quad (3)$$

where λ_d and λ_q are components of the flux vector on the $d - q$ coordinates. By using (3), (2) is transformed as follows:

$$\begin{bmatrix} v_d \\ v_q \end{bmatrix} = \begin{bmatrix} R + pL_d & -\omega_{re}L_d \\ \omega_{re}L_d & R + pL_d \end{bmatrix} \begin{bmatrix} i_d \\ i_q \end{bmatrix}$$
$$+ (p\boldsymbol{I} + \omega_{re}\boldsymbol{J}) \begin{bmatrix} \lambda_d \\ \lambda_q \end{bmatrix}. \quad (4)$$

The mathematical model on the fixed coordinates ($\alpha - \beta$ coordinates) can be derived as

$$\begin{bmatrix} v_\alpha \\ v_\beta \end{bmatrix} = \begin{bmatrix} R + pL_d & 0 \\ 0 & R + pL_d \end{bmatrix} \begin{bmatrix} i_\alpha \\ i_\beta \end{bmatrix} + p \begin{bmatrix} \lambda_\alpha \\ \lambda_\beta \end{bmatrix}, \quad (5)$$

where λ_α and λ_β are components of the transformed flux vector in (3) on the $\alpha - \beta$ coordinates. The above equation can mathematically be derived without any approximation. In this paper, (5) is defined as "L_d model". It should be noted that L_d model is insensitive to L_q and K_E which tends to widely vary according to load and thermal conditions. In addition, it can be seen from (5) that this mathematical model is expressed with only L_d as inductance parameter, which almost maintains constant generally. Therefore, inductance variation due to magnetic saturation hardly affects this mathematical model, yielding realization of the robust flux estimation. However, the proposed flux $\boldsymbol{\lambda}$ does not necessarily align with d-axis. The phase relationship of the proposed flux $\boldsymbol{\lambda}$ will be described in the following subsection.

The physical meaning of $\boldsymbol{\lambda}$ can be interpreted as shown in Fig. 1. Figure 1(a) illustrates the general physical model of an IPMSM. This paper intentionally divides L_q into the stator-side inductance L_d and the virtual rotor-side inductance $L_q - L_d$ to define the proposed flux. As a result, Fig. 1(b) can be obtained and the rotor-side flux vector $\boldsymbol{\lambda}$ can be expressed as (3). This physical model can be regarded as a Surface Permanent Magnet Synchronous Motor (SPMSM) since the stator-side inductances are all unified to L_d as shown in Fig. 1(b). Hence, the mathematical model of Fig. 1(b) can be derived on $\alpha - \beta$ coordinates by (5).

B. Phase relation ship between proposed flux and current for torque maximization

This subsection clarifies that the phase relationships between the proposed flux and the current for torque maximization.

The phase relationships between $d - q$ coordinates and the proposed flux ($\gamma - \delta$ coordinates) are shown in Fig. 2, where

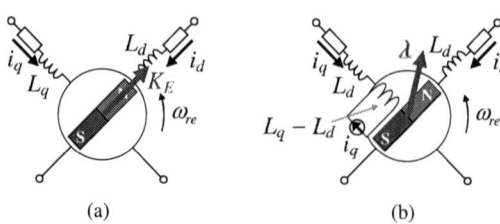

Fig. 1. Physical models of IPMSM.

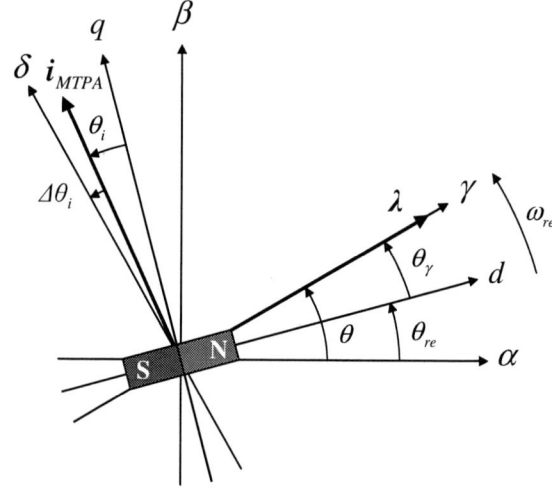

Fig. 2. Definition of coordinates.

θ_γ, θ_i and $\Delta\theta_i$ stand for δ-axis and the current phases at MTPA control from with d-axis, q-axis and the MTPA axis respectively. θ stands for δ-axis phase from with α-axis, and is given by

$$\theta = \tan^{-1}\left(\frac{\lambda_\beta}{\lambda_\alpha}\right). \quad (6)$$

Generally, the current phase θ_i under the MTPA control is expressed as the follow:

$$\theta_i = \tan^{-1}\left(\frac{-i_d}{i_q}\right)$$
$$= \tan^{-1}\left(\frac{-K_E + \sqrt{K_E{}^2 + 4\{(L_q - L_d)i_q\}^2}}{2(L_q - L_d)i_q}\right)$$
$$= \tan^{-1}\left(\frac{-\lambda_d + \sqrt{\lambda_d{}^2 + 4\lambda_q{}^2}}{2\lambda_q}\right)$$
$$= \tan^{-1}\left(\frac{\frac{\lambda_q}{\lambda_d}}{\frac{1 + \sqrt{1 + \left(2\frac{\lambda_q}{\lambda_d}\right)^2}}{2}}\right). \quad (7)$$

On the other hand, from (3), θ_γ is given by

$$\theta_\gamma = \tan^{-1}\left(\frac{\lambda_q}{\lambda_d}\right) \quad (8)$$
$$= \tan^{-1}\left(\frac{(L_q - L_d)i_q}{K_E}\right).$$

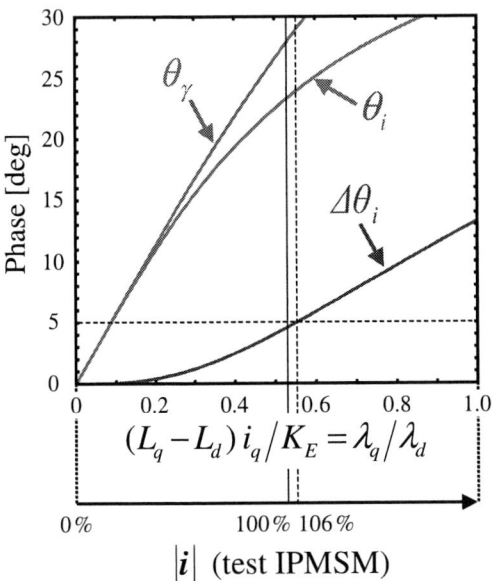

Fig. 3. Analysis results of flux and current phase characteristics with $(L_q - L_d)i_q/K_E$.

TABLE I. PARAMETERS OF TEST IPMSM

Parameters	Value
Rated Power	1.5 kW
Rated Speed	3600 min^{-1}
Rated Phase Current	5.0 A
Rated Line Voltage	200 V
R	0.55 Ω
L_d	8.31 mH
L_q	22 – 35 mH
K_E	0.206 V· s/rad
Pole Pairs	2

It should be noted that the MTC frame can be estimated by the proposed flux coordinate if $\theta_\gamma \simeq \theta_i$, yielding the MTPA control can be realized by the regulating current amplitude on the proposed flux coordinate.

Figure 3 shows that the analysis results of $\Delta\theta_i$ characteristics between the proposed flux and the current for torque maximization with increase of $(L_q - L_d)i_q/K_E$. From this figure, $(L_q - L_d)i_q/K_E$ of less than 0.556 gives rise to five degrees or less in $\Delta\theta_i$. Therefore, if IPMSMs corresponding to the above condition, it is possible to estimate approximately the MTC frame by using this model.

The test IPMSM with parameters listed in Table I satisfies relation of $(L_q - L_d)i_q/K_E = 0.556$ at the rated current. Therefore, the proposed flux estimation makes it possible to estimate the MTC frame, which enables to eliminate the MTPA control algorithm with some parameters.

C. MTPA control method based on proposed flux model

The proposed MTPA control method implemented in the MTC frame, similarly to literature [5]. The detail of the proposed method is described as follows. First, a flux observer is configured based on (5) to estimate the proposed flux (λ_α and λ_β). Next, the angle θ between the α-axis and the γ-axis is estimated using (6). By using this theta, L_d model of (5) is

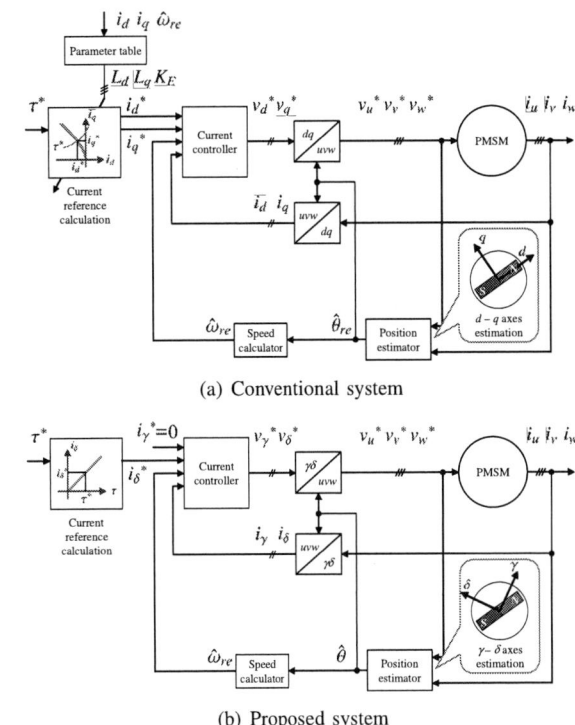

(a) Conventional system

(b) Proposed system

Fig. 4. The block diagrams of the position sensorless control system.

transformed into $\gamma - \delta$ coordinates as the follow:

$$
\begin{bmatrix} v_\gamma \\ v_\delta \end{bmatrix} = \begin{bmatrix} \cos\theta & -\sin\theta \\ \sin\theta & \cos\theta \end{bmatrix} \begin{bmatrix} v_\alpha \\ v_\beta \end{bmatrix}
$$
$$
= \begin{bmatrix} R + pL_d & -\omega_{re}L_d \\ \omega_{re}L_d & R + pL_d \end{bmatrix} \begin{bmatrix} i_\gamma \\ i_\delta \end{bmatrix} + \begin{bmatrix} p\lambda_\gamma - \omega_{re}\lambda_\delta \\ p\lambda_\delta + \omega_{re}\lambda_\gamma \end{bmatrix} \quad (9)
$$

where λ_γ and λ_δ are components of the flux vector on the $\gamma - \delta$ coordinates. From the definition of $\gamma - \delta$ coordinates, the proposed flux occurs only γ-axis; that is, $\lambda_\delta = p\lambda_\delta = 0$ and $\lambda_\gamma = |\boldsymbol{\lambda}|$. Therefore, (9) is transformed as the follow:

$$
\begin{bmatrix} v_\gamma \\ v_\delta \end{bmatrix} = \begin{bmatrix} R + pL_d & -\omega_{re}L_d \\ \omega_{re}L_d & R + pL_d \end{bmatrix} \begin{bmatrix} i_\gamma \\ i_\delta \end{bmatrix} + \begin{bmatrix} p|\boldsymbol{\lambda}| \\ \omega_{re}|\boldsymbol{\lambda}| \end{bmatrix}, \quad (10)
$$

where $|\boldsymbol{\lambda}| = \sqrt{\lambda_d^2 + \lambda_q^2} = \sqrt{\lambda_\alpha^2 + \lambda_\beta^2}$. It can be seen from (10), L_d model on $\gamma - \delta$ coordinates can be interpreted equivalently as a SPMSM, because the inductance components in impedance matrix are all unified to L_d. Based on the above, the current control system is configured based on (10). Finally, the MTPA control can be implemented approximately via $i_\gamma = 0$ control. The block diagrams of the position sensorless control system are shown in Fig. 4. As a result, the proposed position sensorless control system does not need to generate the d-axis current reference like conventional MTPA control. Therefore, the proposed method makes it possible to achieve the simple and robust MTPA control, although some $\Delta\theta_i$ remains, as shown in Fig.3.

D. Design method of current controller in the proposed system

It can be seen from (10), the voltage equation of IPMSMs based on L_d model is the different equation from conventional

978-1-4799-2706-7/14 $31.00 © 2014 IEEE

one. Equation (10) is transformed the following state equation:

$$p \begin{bmatrix} i_\gamma \\ i_\delta \end{bmatrix} = \begin{bmatrix} -\frac{R}{L_d} & \omega_{re} \\ -\omega_{re} & -\frac{R}{L_d} \end{bmatrix} \begin{bmatrix} i_\gamma \\ i_\delta \end{bmatrix} + \begin{bmatrix} \frac{1}{L_d} & 0 \\ 0 & \frac{1}{L_d} \end{bmatrix} \begin{bmatrix} v_\gamma \\ v_\delta \end{bmatrix}$$
$$+ \begin{bmatrix} -p|\lambda| \\ -\frac{1}{L_d}\omega_{re}|\lambda| \end{bmatrix} \quad (11)$$

Therefore, the decoupling controller in $\gamma - \delta$ coordinates can be configured as follows:

$$\begin{bmatrix} v_\gamma{}^* \\ v_\delta{}^* \end{bmatrix} = \begin{bmatrix} v_\gamma{}' \\ v_\delta{}' \end{bmatrix} + \omega_{re} \begin{bmatrix} -L_d i_\delta \\ L_d i_\gamma + \left|\hat{\lambda}\right| \end{bmatrix}, \quad (12)$$

where, $v_\gamma{}'$ and $v_\delta{}'$ are the output voltages of the current controller configured in $\gamma - \delta$ coordinates. In addition, " ^ " means estimated value. Here, since $p|\lambda|$ is difficult to determine in linear state equations, it is neglected as a modeling error (disturbance). The above decoupling controller does not include L_q and K_E. Therefore, the proposed method provides the robust current control system with respect to magnetic saturation.

The current controller based on the proposed system is configured based on PI regulator and decoupling controller. The transfer function of PI regulator can be described as $G(s) = K_p + \frac{K_i}{s}$, where K_p ans K_i are the proportional and integral gain of the regulator, respectively. If the proportional and integral gain designed as (13), the transfer function of the closed-loop system can be set as the first-order low-pass filter.

$$\begin{aligned} K_p &= \omega_{cc} L_d \\ K_i &= \omega_{cc} R, \end{aligned} \quad (13)$$

where ω_{cc} is bandwidth of current controller. Because the L_d model can be interpreted equivalently as a SPMSM, gamma-axis and delta-axis proportional gains are designed as same value.

E. Sensitivity of L_d in proposed position sensorless control system

1) Estimated phase characteristics: This paper analyses the flux and phase estimation error caused by L_d setting error based on the method described in literature [8]. It is suppose that inductances vary as shown in following equation with respect to their set values $\tilde{L}_d$ and $\tilde{L}_q$.

$$L_d = \tilde{L}_d + \Delta L_d \quad (14)$$
$$L_q = \tilde{L}_q + \Delta L_q, \quad (15)$$

where ΔL_d and ΔL_q are the inductances setting error on the $d - q$ coordinates, which are negative at magnetic saturation. By using the above equations, the error vector $\epsilon_{\Delta L}$ caused by inductance setting error can be derived as the follow:

$$\epsilon_{\Delta L} = \Delta L_d \boldsymbol{i}, \quad (16)$$

where $\boldsymbol{i}$ is the current vector. Figure 5 shows the flux vector diagram in the case of the inductances setting error. Figure 5(a) shows the error vector and the phase estimation error in the case that the inductance decreases due to magnetic saturation. By using the figure, the phase estimation error $\Delta\theta_{(\Delta L_d)}$ which

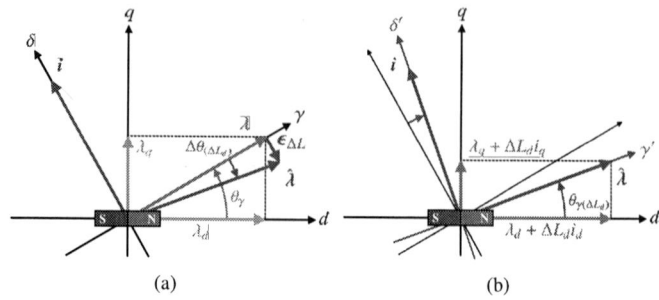

Fig. 5. Flux vector diagram in the case of inductances setting error.

caused by the inductances setting error is expressed as the follow:

$$\begin{aligned} \Delta\theta_{(\Delta L_d)} &= \tan^{-1}\left(\frac{\Delta L_d \boldsymbol{i}}{|\boldsymbol{\lambda}|} \right) \\ &= \tan^{-1}\left(\frac{\Delta L_d \boldsymbol{i}}{\sqrt{K_E{}^2 + \{(L_q - L_d)i_q\}^2}} \right) \end{aligned} \quad (17)$$

As can be seen from Fig. 5(a), the phase of the estimated flux vector lags behind than the phase of the original flux vector. By using the figure, the estimated phase $\theta_{\gamma(\Delta L_d)}$ is expressed as the follow:

$$\theta_{\gamma(\Delta L_d)} = \tan^{-1}\left(\frac{(L_q - L_d + \Delta L_d)i_q}{K_E + \Delta L_d i_d} \right) \quad (18)$$

Therefore, the current control system which includes the MTPA control is reconstructed in the lagged coordinate as shown in Fig. 5(b). Here, we define $\gamma' - \delta'$ coordinates as shown in Fig. 5(b). As shown in Fig. 3, the phase of the original flux vector leads than the phase of the current vector for the MTPA control, and therefore when a lagged error occurs in the phase estimation, the estimated phase eventually approaches the current phase at MTPA control. Figure 6 shows the analysis results of the estimated phase characteristics with respect to the inductance variation in the test IPMSM with parameters listed in Table I. As can be seen from the figure, in the test motor, the proposed method allows for the robust phase estimation even if the inductances setting error occurs 25 %. From the above, the proposed method has also robustness to the L_d variation due to magnetic saturation.

2) Current control characteristics: If the L_d setting error is occurred, the impedance matrix of (1) is transformed as follows:

$$\begin{aligned} \begin{bmatrix} v_d \\ v_q \end{bmatrix} &= \begin{bmatrix} R + p\tilde{L}_d & -\omega_{re}\tilde{L}_d \\ \omega_{re}\tilde{L}_d & R + p\tilde{L}_d \end{bmatrix} \begin{bmatrix} i_d \\ i_q \end{bmatrix} \\ &+ \begin{bmatrix} p(L_d - \tilde{L}_d)i_d \\ p(L_q - \tilde{L}_q)i_q \end{bmatrix} + \omega_{re} \begin{bmatrix} -(L_q - \tilde{L}_q)i_q \\ K_E + (L_d - \tilde{L}_d)i_d \end{bmatrix}. \end{aligned} \quad (19)$$

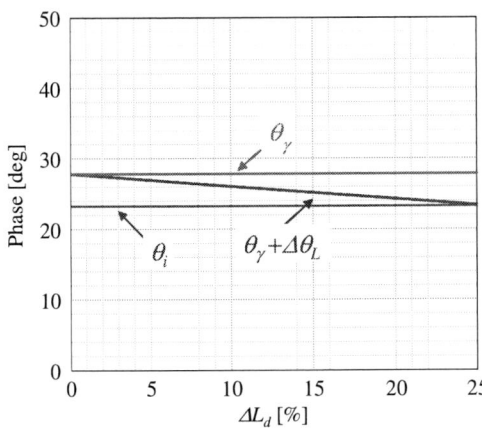

Fig. 6. Analysis results of estimated phase characteristics with respect to the inductance variation in test IPMSM.

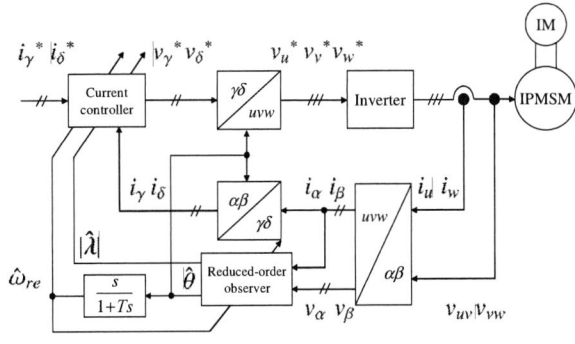

Fig. 7. Position sensorless control system.

Accordingly, the definition of the flux is transformed as follows:

$$\boldsymbol{\lambda}(\tilde{L}_d) = \begin{bmatrix} \lambda_d(\tilde{L}_d) \\ \lambda_q(\tilde{L}_d) \end{bmatrix} \tag{20}$$

$$= \begin{bmatrix} K_E + (L_d - \tilde{L}_d)i_d \\ (L_q - \tilde{L}_d)i_q \end{bmatrix}$$

$$= \begin{bmatrix} K_E + \Delta L_d i_d \\ (L_q - L_d + \Delta L_d)i_q \end{bmatrix}. \tag{21}$$

Using the above equation, (19) can be rewritten as follows:

$$\begin{bmatrix} v_d \\ v_q \end{bmatrix} = \begin{bmatrix} R + p\tilde{L}_d & -\omega_{re}\tilde{L}_d \\ \omega_{re}\tilde{L}_d & R + p\tilde{L}_d \end{bmatrix} \begin{bmatrix} i_d \\ i_q \end{bmatrix}$$

$$+ p \begin{bmatrix} \lambda_d(\tilde{L}_d) \\ \lambda_q(\tilde{L}_d) \end{bmatrix} + \omega_{re} \begin{bmatrix} -\lambda_q(\tilde{L}_d) \\ \lambda_d(\tilde{L}_d) \end{bmatrix}. \tag{22}$$

Therefore, the L_d model which includes the ΔL_d on $\gamma' - \delta'$ coordinates can be derived as follows:

$$\begin{bmatrix} v_{\gamma'} \\ v_{\delta'} \end{bmatrix} = \begin{bmatrix} R + p\tilde{L}_d & -\omega_{re}\tilde{L}_d \\ \omega_{re}\tilde{L}_d & R + p\tilde{L}_d \end{bmatrix} \begin{bmatrix} i_{\gamma'} \\ i_{\delta'} \end{bmatrix} + \begin{bmatrix} p|\boldsymbol{\lambda}(\tilde{L}_d)| \\ \omega_{re}|\boldsymbol{\lambda}(\tilde{L}_d)| \end{bmatrix}. \tag{23}$$

From the above, the estimated flux is varied by L_d setting error. However, the structure of (23) is the same as L_d model. As a result, the proposed current controller is reconstructed based on (23). Therefore, it can be considered that the L_d setting error (which includes the inductance variation) hardly affects the proposed current control system if the estimated flux is used. However, the proposed current control system is influenced by the estimated flux accuracy, because this system uses the estimated flux. Hence, the important problem of this system is to improve the performance of the position estimator, it is future work.

III. EXPERIMENTAL RESULTS

We carried out to show the effectiveness of the MTC frame estimation. The configuration of the position sensorless control system is shown in Fig. 7. A 1.5 kW concentrated winding IPMSM was used as the test machine; its parameters are given in Table I. The magnetic saturation characteristic

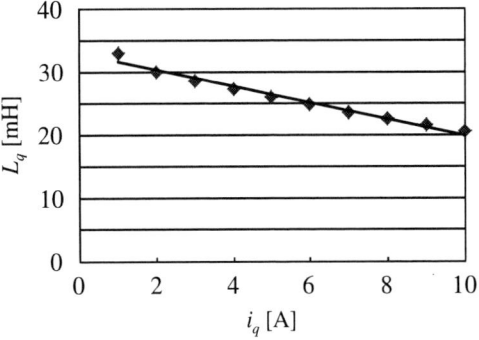

Fig. 8. Measured values of L_q.

of L_q which measured according to literature [8] is shown in Fig. 8. In following experiments, IPMSM was operated by position sensorless vector control with the proposed method. In addition, the proposed flux is estimated by the reduced-order flux observer [9] as shown in Fig. 9. The estimated speed $\hat{\omega}_{re}$ is substituted with differential operation of $\hat{\theta}$ and through the first order low-pass filter (1000 rad/s). The reduced-order flux observer, the current controller, and the coordinate transformer were executed with DSP (TI:TMS320C6713B), and the pulse width modulation of the voltage reference was made by FPGA (Altera:EPF10K20TC144-4). The estimation period and the control period were 100 μs. Hence, the carrier frequency of the PWM inverter was 10 kHz.

Figure 10 shows the MTC frame estimation results at 1800 min^{-1}, in which $\hat{\theta}_\gamma$ stands for the estimated MTC current phase by using L_d model. In these figures, solid lines represent the developed torque measurement results under constant current amplitude $|i| = 4.33$ A (50 % of the rated current, $L_q \simeq 27$ mH) and 8.66 A (the rated current, $L_q \simeq 22$ mH), respectively. It turns out from Fig. 10 that errors of $\hat{\theta}_\gamma$ are approximately suppressed to 4° regardless current amplitude i, so that reduction of the developed torque is hardly visible. Therefore, these experimental results conclude that L_d model realizes the robust MTC frame estimation to magnetic saturation.

The 2014 International Power Electronics Conference

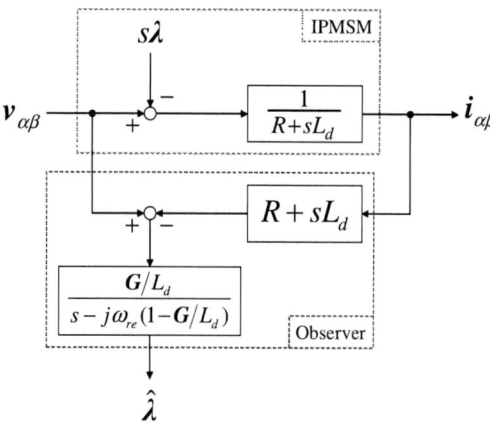

Fig. 9.   Structure of reduced-order flux observer.

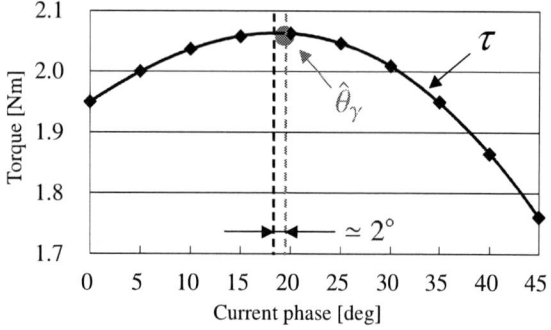

(a) $|i| = 4.33$ A (50 % of the rated current)

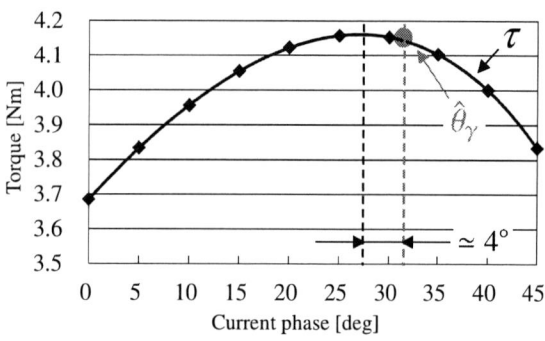

(b) $|i| = 8.66$ A (the rated current)

Fig. 10.   Current phase estimation results at 1800 min^{-1}.

IV. CONCLUSION

In this paper, the novel mathematical model and the position sensorless control method for IPMSMs has been examined. This paper derived the proposed model with clear physical meaning and described the simple MTPA control method. From the analysis results, this paper showed that the proposed model is robust to magnetic saturation. In addition, we verified the effectiveness of the proposed method in experiments.

From the above, we confirmed that the proposed position sensorless control system has robustness to magnetic saturation.

REFERENCES

[1] Z. Chen, M. Tomita, S. Doki and S. Okuma, "An Extended Electromotive Force Model for Sensorless Control of Interior Permanent-Magnet Synchronous Motors," *IEEE Trans. on Industry Electronics*, vol. 50, no. 2, pp. 288–295, Apr. 2003.

[2] M. Hasegawa and K. Matsui, "IPMSM Position Sensorless Drives Using Robust Adaptive Observer on Stationary Reference Frame," *IEEJ Trans. on Electrical and Electronics Engineering*, vol. 3, no. 1, pp. 120–127, Jan. 2008.

[3] T. M. Jahns, G. B. Kliman and T. W. Neumann, "Interior Permanent-Magnet Synchronous Motors for Adjustable-Speed Drives," *IEEE Trans. on Industry Applications*, vol. IA-22, no. 4, pp. 738–747, July 1986.

[4] S. Morimoto, K. Hatanaka, Y. Tong, Y. Takeda and T. Hirasa, "Servo Drive System and Control Characteristics of Salient Pole Permanent Magnet Synchronous Motor," *IEEE Trans. on Industry Applications*, vol. 29, no. 2, pp. 338–343, Mar./Apr. 1993.

[5] H. Hida, Y. Tomigashi and K. Kishimoto, "Novel Sensorless Control for PM Synchronous Motors Based on Maximum Torque Control Frame," *Proc. EPE2007*, no. 175, (10pages), 2007

[6] K. Tobari, K. Sakamoto, D. Maeda and T. Endo, "Maximum Torque Control Technique Suitable for Sensorless Permanent Magnet Synchronous Motor Drives," *Proc. 2006 IEE Japan Industry Applications Society Annual Conference*, no. 64, pp. 389–392, Apr. 2006 (in Japanese).

[7] S. Shinnaka and K. Sano: "A New Unified Analysis of Estimate Errors by Model-Matching Phase-Estimation Methods for Sensorless Drive of Permanent Magnet Synchronous Motors and New Trajectory-Oriented Vector Control, Part I", *IEEJ Trans. on Industry Applications*, Vol.127, No.9, pp.950–961, Sept. 2007 (in Japanese).

[8] S. Ichikawa, M. Tomita, S. Doki and S. Okuma, "Sensorless Control of Synchronous Motors Based on an Extended EMF Model and Inductance Measurement in the Model," *Proc. EPE-PEMC 2004*, Sep. 2004.

[9] Y. Imaeda, S. Doki, M. Hasegawa, K. Matsui, M. Tomita and T. Ohnuma, "PMSM position sensorless control with extended flux observer," *Proc. IECON2011*, pp. 4721–4726, Nov. 2011.

Motor Drive System Using Nonlinear Mathematical Model for Permanent Magnet Synchronous Motors

Yoshitaka Iwaji, Junnosuke Nakatsugawa,

Toshifumi Sakai and Shigehisa Aoyagi

Hitachi, Ltd., Hitachi Research Lab
Hitachi-shi, Ibaraki, Japan

Hirokazu Nagura

Hitachi, Ltd., Infrastructure Systems Company
Hitachi-shi, Ibaraki, Japan

Abstract— **We have developed a mathematical model for permanent magnet synchronous motors (PMSMs). The nonlinear characteristics of this model are expressed using a total of eight constants. The magnet flux characteristics of PMSMs can be obtained by using this simplified model. This model can also be transformed into a reverse function, and motor drive simulation including nonlinear characteristics can easily be implemented. Here, we present the simulation results using our nonlinear model. We also demonstrated the effectiveness of the model by simulating estimates of rotor positions based on magnet flux nonlinear characteristics.**

Keywords— *Permanent magnet synchronous motor, magnetic saturation, vector control, position sensorless control*

I. INTRODUCTION

A simplified nonlinear model (N-model) for permanent magnet synchronous motors (PMSMs) has already been proposed by Nakatsugawa et al. [1], [2]. The N-model is constructed of simple fractional equations using a total of eight constants. The N-model is well approximated by using these constants with magnetic flux characteristics by taking magnetic saturation and cross coupling effects with the *d*- and *q*-axes into account.

Methods of position sensorless control have been widely used in PMSMs, and this trend has recently expanded to include methods based on nonlinear characteristics [4]–[7]. Nonlinear PMSM models are increasingly needed in the design of position sensorless controllers.

This paper introduces a simulation model using the N-model. Some kinds of nonlinear sensorless methods were simulated using this model. We also present some experimental results demonstrating the validity of the N-model simulator.

II. NON-LINEAR PMSM MODEL

A. N-model

Figures 1 (a) and (b) show the characteristics of the magnetic flux of a PMSM. The linear model in Fig. 1 (a)

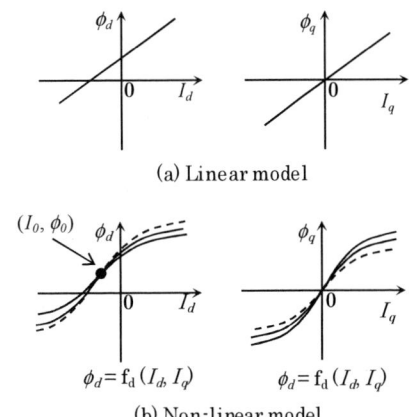

(a) Linear model

(b) Non-linear model

Fig. 1. Characteristics of magnetic flux of PMSM.

is used in most PMSM drives to estimate rotor positions. In this case, the magnetic flux characteristics are obtained as

$$\phi_d = L_d I_d + \phi_m \text{ and } (1)$$
$$\phi_q = L_q I_q. \qquad (2)$$

However, nonlinearity characteristics appear in actual PMSMs. Nonlinearity generally includes two features, as can be seen from Figure 1 (b). The first involves the saturation curves of the magnetic flux in both axes. The second involves cross coupling effects with the *d*- and *q*-axes.

These features can be expressed by using the N-model as

$$\phi_d(I_d, I_q) = \frac{K_{Ld}(I_d + I_0)}{1 + K_{Sd}|I_d + I_0| + K_{Sdq}|I_q|} + \phi_0 \text{ and } (3)$$

$$\phi_q(I_d, I_q) = \frac{K_{Lq} I_q}{1 + K_{Sqd}|I_d + I_0| + K_{Sq}|I_q|}. \qquad (4)$$

The 2014 International Power Electronics Conference

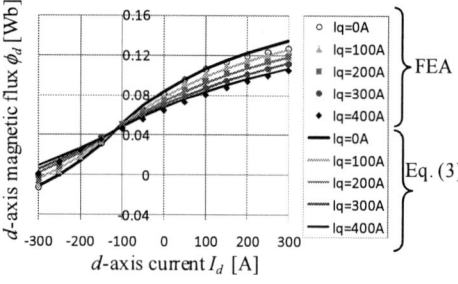

(a) ϕ_d

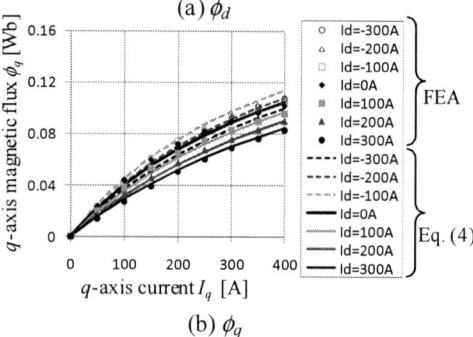

(b) ϕ_q

Fig. 2. Results from analysis of magnetic fluxes ϕ_d and ϕ_q.

The K_{Ld}, K_{Sd}, K_{Sdq}, K_{Lq}, K_{Sqd}, K_{Sq}, I_0, and ϕ_0 in these equations are the N-model constants. Nonlinearities are expressed by these constants. The four physical meanings of these constants are:

(i) K_{Ld} and K_{Lq} mean overall inductances of PMSM. These correspond to the inductances at the point of the largest $d\phi/di$.

(ii) K_{Sd} and K_{Sq} mean the saturation characteristics of the magnet flux of PMSM. If these values are larger, the saturation characteristics strengthen.

(iii) K_{Sdq} and K_{Sqd} mean the cross coupling effects with the d- and q-axes. If these values are large, the magnetic fluxes are strongly influenced by the other axis current.

(iv) I_0 and ϕ_0 correspond to the cross point of I_d-Φ_d curves shown in Figure 1 (b). This point is a singular point that is not influenced by the I_q values.

Figures 2 (a) and (b) plot the magnetic flux characteristics obtained by conducting finite element analysis (FEA) (plotted points in Figure 2). The results from curve fitting when using the N-model also appear in these figures as continuous lines. The N-model can well approximate nonlinear PMSM characteristics, as seen in Figure 2.

B. N-model constants

Some examples of N-model constants are discussed here. Table 1 summarizes the specifications of the analyzed motors. Table 2 lists the fitting results for the N-model. These values are scattered over a wide range, as we can see from Table 2. The order of the values depends on the specifications of the motors. Therefore, it was difficult to only compare non-linearities by using constants.

Table 1. Specifications for analyzed motors.

	Motor A	Motor B	Motor C
Rated power [kW]	22.0	22.0	1.0
Type of rotor	IPM	SPM	IPM
Number of poles	8	8	4
Rated torque [Nm]	70	70	3
Rated current [Arms]	105	84	4.7
Rated speed [r/min]	3,000	3,000	3,180

Table 2. Constants of N-model (before normalization).

	K_{Ld}	K_{Sd}	K_{Sdq}	K_{Lq}	K_{Sq}	K_{Sqd}	I_0	Φ_0
Motor A	0.0004	0.0024	0.0022	0.0006	0.0026	0.0018	119	0.044
Motor B	0.0003	0.0024	0.0009	0.0003	0.0005	0.0009	205	0.053
Motor C	0.0171	0.0609	0.0366	0.0160	0.0328	0.0312	6.6	0.077

Table 3. Base value for normalization.

	Unit	Motor A	Motor B	Motor C	
Torque	[Nm]	70	70	3	Rated torque τ_m
Current	[A]	148	119	6.6	I_q at τ_m
Magnetic flux	[Wb]	0.079	0.098	0.152	ϕ_d at τ_m

Table 4. Constants of N-model (after normalization).

	K_{Ldn}	K_{Sdn}	K_{Sdqn}	K_{Lqn}	K_{Sqn}	K_{Sqdn}	I_{0n}	Φ_{0n}
Motor A	0.82	0.36	0.33	1.10	0.38	0.26	0.80	0.55
Motor B	0.42	0.28	0.11	0.32	0.06	0.11	1.72	0.54
Motor C	0.74	0.40	0.24	0.69	0.22	0.21	1.00	0.51

Hence, normalization was introduced to the N-model parameters. The base values for torque, current, and magnetic flux were defined according to the values in Table 3. All constants were normalized with these base values. The normalized N-model parameters are listed in Table 4. The parameters approached these values due to this normalization.

It is obvious from Table 4 that motor B has small K_{Sd}, K_{Sdq}, K_{Sq}, and K_{Sqd}. This means that motor B has greater linearity than the others. We expected that motor A would have the strongest nonlinearity of these.

C. N-model for time domain simulation

Motor currents I_d and I_q in conventional time domain simulation can be calculated with Eqs. (1) and (2). I_d and I_q are easily obtained by using the PMSM linear model as seen in Figure 3. This linear model corresponds to the reverse functions of Eqs. (1) and (2).

A non-linear simulation model can similarly be obtained with Eqs. (3) and (4). The reverse function of the N-model can easily be derived because the N-model only consists of simple primary functions.

Figure 4 outlines the simulation model using the reverse N-model. Motor currents I_d and I_q are obtained by using the reverse functions of Eqs. (3) and (4):

$$I_d\left(\phi_d,\phi_q\right)=$$
$$-I_0+\frac{\left(\phi_d-\phi_0\right)\left(K_{Ld}+\left|\phi_q\right|\left(K_{Sdq}-K_{Sq}\right)\right)}{\left(K_{Ld}-\left|\phi_d-\phi_0\right|K_{Sd}\right)\left(K_{Lq}-\left|\phi_q\right|K_{Sq}\right)-\left|\phi_d-\phi_0\right|\left|\phi_q\right|K_{Sdq}K_{Sqd}} , \quad (5)$$

978-1-4799-2706-7/14 $31.00 © 2014 IEEE   2452

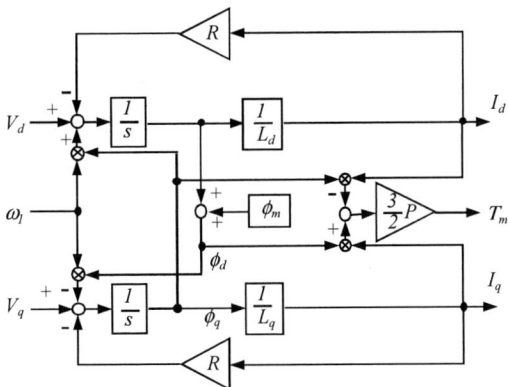

Fig. 3. Conventional PMSM linear model.

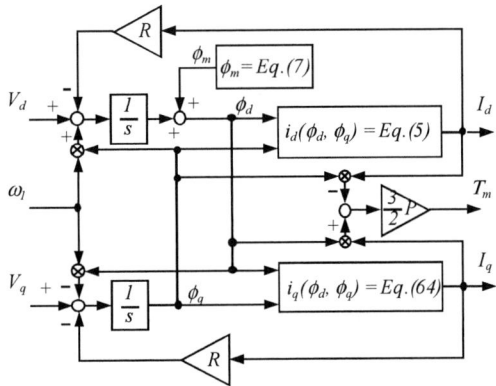

Fig. 4. PMSM non-linear model (N-model).

$$I_q(\phi_d,\phi_q)=$$

$$\frac{\phi_q\big(K_{Ld}-|\phi_d-\phi_0|(K_{Sd}-K_{Sqd})\big)}{\big(K_{Ld}-|\phi_d-\phi_0|K_{Sd}\big)\big(K_{Lq}-|\phi_q|K_{Sq}\big)-|\phi_d-\phi_0|\phi_q|K_{Sdq}K_{Sqd}} , \text{ and } (6)$$

$$\phi_m=\phi_d(0,0)=\frac{K_{Ld}I_0}{1+K_{Sd}I_0}+\phi_0 \quad (7).$$

The motor torque is expressed by

$$T_m=\frac{3}{2}P\big(\phi_d i_q-\phi_q i_d\big) , (8).$$

III. SENSORLESS METHOD OF CONTROL BY USING N-MODEL

A non-linear model to simulate motors can easily be built by using the N-model, as was previously mentioned. Figure 5 outlines the configuration for a position sensorless controller with the nonlinear PMSM model. A sensorless method based on estimating rotor position error [3] is introduced in this controller. The rotor position error, $\Delta\theta_{dc}$, is defined as the difference between the d-axis (the real d-axis) and dc-axis (the assumed d-axis in the controller). The $\Delta\theta_{dc}$ is obtained as

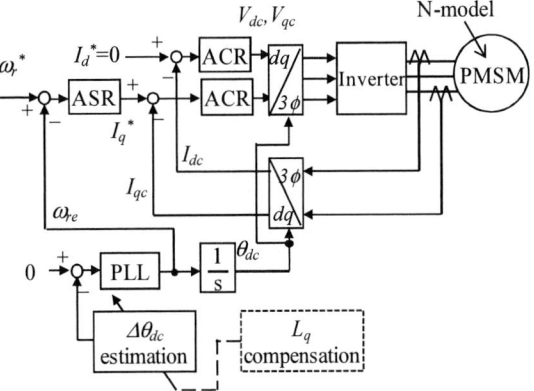

Fig. 5. Configuration for position sensorless controller.

$$\Delta\theta_{dc}=\tan^{-1}\frac{V_{dc}-R_1 I_{dc}+\omega_1 L_q I_{qc}}{V_{qc}-R_1 I_{qc}-\omega_1 L_q I_{dc}}, \quad (9)$$

where ω_l is the rotor angular velocity, L_q is the q-axis inductance, and V_{dc} and V_{qc} are the dc-qc axis voltages. Here I_{dc} and I_{qc} are the dc-qc axis currents and R_1 is the stator winding resistance. These estimates were carried out in the $\Delta\theta_{dc}$ estimation block in Figure 5. Furthermore, the estimated rotation speed, ω_{re}, was obtained as a result of phase-locked loop (PLL) control.

The $\Delta\theta_{dc}$ is estimated with Eq. (9) with the conventional method [3]. The $\Delta\theta_{dc}$ is regulated to zero by the PLL controller. The premise is that L_q is constant, as shown in Eq. (9). However, actual motors have relatively nonlinear characteristics. Hence, L_q might be changed according to certain conditions.

The nonlinearity of PMSMs generally strengthens in proportion to the power density of motors. Therefore, the calculation error in Eq. (9) increases for motors with high power density. Hence, the transient response including torque disturbance will be lowered for high power density motors.

Figure 6 plots the relationship between power density and the tolerance limit of torque disturbance with conventional position sensorless methods. The tolerance limit of the disturbance torque step decreases according to the increase in power density.

The L_q is adjusted from a nominal value to a stabilized value for this system to prevent this phenomenon. This adjustment is often conducted in actual systems.

However, the L_q in Eq. (9) can dynamically be changed depending on conditions by using the N-model. The L_q for Eq. (4) is obtained as:

$$L_q=\frac{K_{Lq}}{1+K_{Sqd}|I_d+I_0|+K_{Sq}|I_q|} . (10)$$

Equation (10) is transformed into the dc-qc axis by $\Delta\theta_{dc}$ as:

The 2014 International Power Electronics Conference

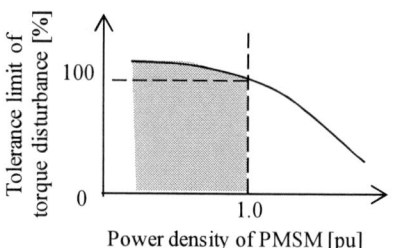

Fig. 6. Relationship between power density and tolerance of torque disturbance.

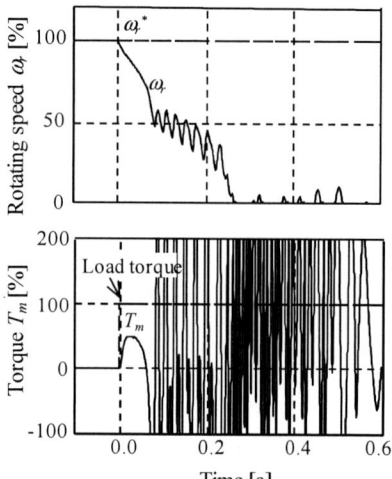

Fig. 7. Simulation results for conventional method.

Fig. 8. Simulation results for proposed method.

$$L_q = \frac{K_{Lq}}{1 + A_1 + A_2} \cdot \quad (11)$$

Here,

$$A_1 = K_{Sqd}\left|I_{dc}\cos\Delta\theta_{dc} - I_{qc}\sin\Delta\theta_{dc} + I_0\right|,$$
$$A_2 = K_{Sq}\left|I_{dc}\sin\Delta\theta_{dc} + I_{qc}\cos\Delta\theta_{dc}\right|$$

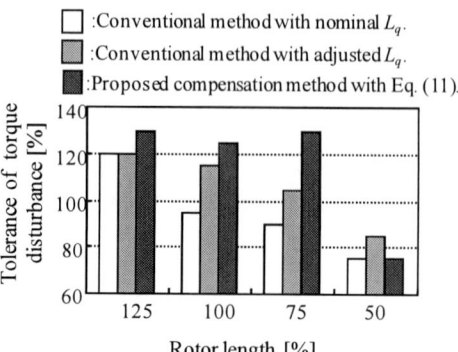

Fig. 9. Simulation results for torque disturbance tolerance.

Figure 7 presents the simulation results with a nominal L_q value, where the motor torque response was very low so that the position sensorless method failed.

Figure 8 plots the simulation results with L_q compensation by using Eq. (11), where the motor torque rapidly generated.

Figure 9 has the simulation results for tolerance to torque disturbance. Four kinds of PMSMs were designed to change the power density, and the N-model parameters were calculated for each. The number of turns of stator windings were adjusted according to the rotor length to maintain the basic specifications.

The tolerance limit of torque disturbance decreased depending on power density (rotor length) with the conventional method, as can be seen from Figure 9. Some improvements could be achieved by adjusting L_q. However, the tolerance limit of torque disturbance was mainly improved by introducing Eq. (11).

IV. ESTIMATION OF INITIAL POSITION

The N-model includes saturation characteristics, as was previously mentioned. Hence, some position sensorless methods based on magnetic saturation could easily be simulated. Sections IV and V explain simulation analysis with these methods.

Figure 10 clarifies the principles underlying the estimation of initial rotor position [5], where voltage pulses were applied to PMSMs. Positive and negative currents were generated in each phase by these pulses.

The differences between the negative and the positive current peak values expressed by Eq. (12) change depending on magnetic saturation.

$$\Delta P_u = \left||I_{u1}| - |I_{u2}|\right|, \ \Delta P_v = \left||I_{v1}| - |I_{v2}|\right|, \ \Delta P_w = \left||I_{w1}| - |I_{w2}|\right|. \quad (12)$$

The rotor initial position can be estimated by:

$$\Delta P_\alpha = \frac{2}{3}\left\{\Delta P_u - \frac{1}{2}\Delta P_v - \frac{1}{2}\Delta P_w\right\},$$
$$\Delta P_\beta = \frac{2}{3}\left\{\frac{\sqrt{3}}{2}\Delta P_v - \frac{\sqrt{3}}{2}\Delta P_w\right\} \text{ and } (13)$$
$$\theta_{est} = \tan^{-1}\frac{\Delta P_\beta}{\Delta P_\alpha}. \quad (14)$$

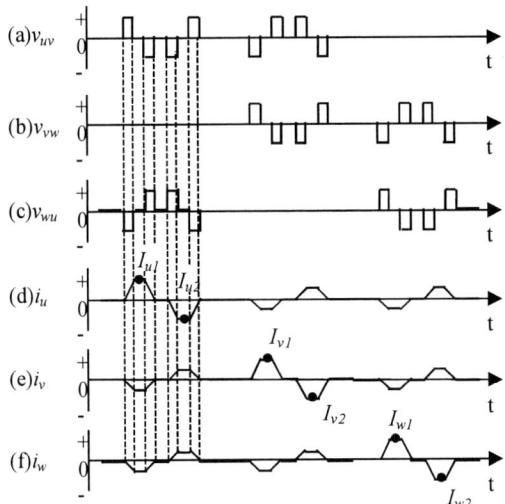

Fig. 10. Principles underlying estimation of initial rotor position.

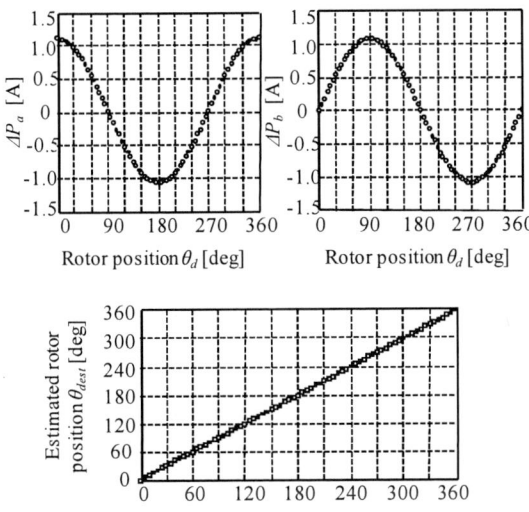

Fig. 11. Simulated estimation of initial rotor position.

Figures 11 and 12 plot the simulation and the experimental results obtained from estimating the initial positions. Both estimates were considerably close. The ΔP_α and ΔP_β are strongly distorted in Fig. 12, which was caused by space harmonics in the motor. Space harmonics was not considered in the N-model. However, the detected values of ΔP_α and ΔP_β were almost the same as the simulation results. This indicated that the N-model was very useful for designing estimates of the initial position.

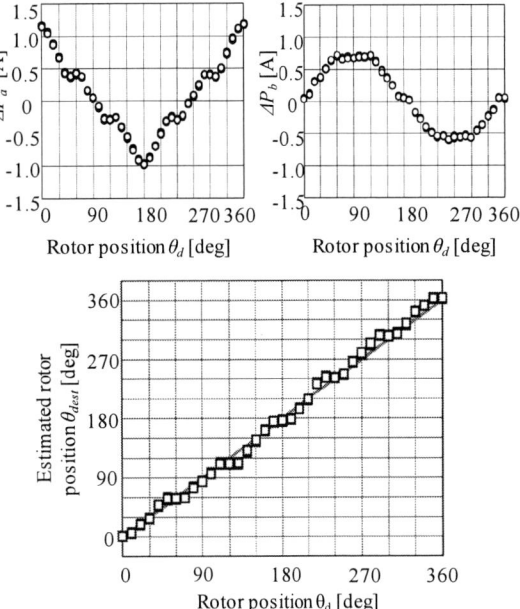

Fig. 12. Experimental results for estimation of initial rotor position.

V. LOW SPEED POSITION SENSORLESS METHOD FOR NON-SALIENCY PMSMs

A position sensorless method for non-salient PMSMs has previously been reported [5] and [6], where PMSM is driven by block commutation and open phase voltage is detected. The generated voltage in the open phase is called induced voltage caused by magnetic saturation (IVMS).

Figure 13 shows the circuits used for the IVMS measurements. When a V-W pulse was applied to the motor, open phase voltage E_u, IVMS, changed according to the rotor position.

Figure 14 shows the generation principle of IVMS. This figure expresses the relationship between the rotor magnet position and the windings of each phase. The origin of θ_d corresponds to the phase-U winding position. At θ_d=60[deg], the magnet flux of the rotor is on the phase-W winding. In this case, the phase-W inductance L_w slightly decreases because of the magnet flux saturation. Hence, the open phase voltage E_u, IVMS, is also decreased to a minus value (Fig. 14 (a)). Similarly, E_u becomes a positive at θ_d=120[deg] (Fig. 14 (b)). In this way, IVMS changes in accordance with the rotor position [5]-[7]. Hence, the rotor position can be estimated using IVMS value.

Figures 15 (a) and (b) plot the simulation using N-model and experimental results for the IVMS value. Both results estimated about the same IVMS values. Nonlinear based sensorless drives could be evaluated by using N-model simulation, as shown in the figure. Therefore, the detection circuit for IVMS could be designed in advance without actual measurements.

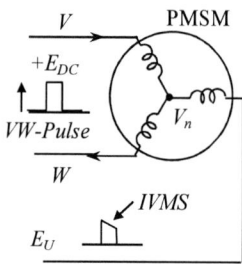

Fig. 13. Measured circuit of IVMS.

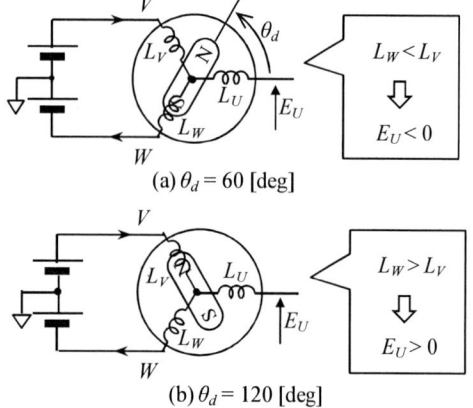

(a) $\theta_d = 60$ [deg]

(b) $\theta_d = 120$ [deg]

Fig. 14. Principle of generation for IVMS.

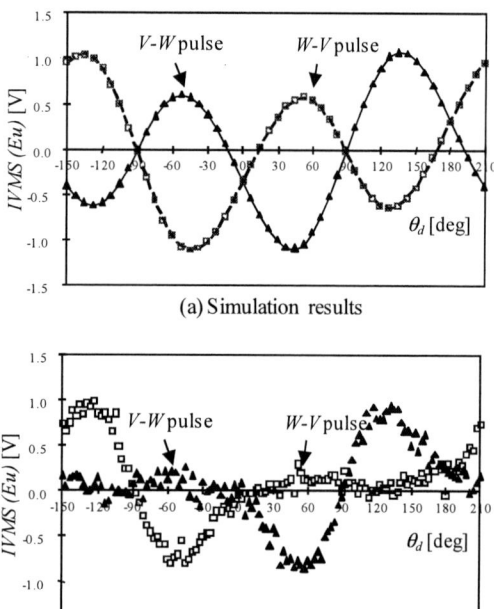

(a) Simulation results

(b) Experimental results

Fig. 15. Generated IVMS values.

VI. CONCLUSIONS

PMSM drive simulation using a simplified nonlinear model, the N-model, was demonstrated. Nonlinear characteristics of PMSM could easily be evaluated by using the N-model.

For example, the tolerance limit for torque disturbances was evaluated with the N-model simulator. Furthermore, the effectiveness of the method of inductance L_q compensation for nonlinear PMSMs was demonstrated.

Some kinds of sensorless methods based on magnetic saturation were discussed by using the N-model.

The simulation results corresponded well with those from experiments and verified that N-model based simulation and system design were extremely effective for position sensorless PMSM drives.

REFERENCES

[1] J. Nakatsugawa, N. Iwasaki, H. Nagura, and Y. Iwaji, "Proposal of Mathematical Modeling for Permanent Magnet Synchronous Motors Considering Magnetic Saturation and Cross Coupling Effects", Proceedings of the 2009 Japan Industry Applications Society Conference, Vol. 1, Nos. 1–150, pp. 715–720 (2009) (in Japanese).

[2] H. Nagura, Y. Iwaji, J. Nakatsugawa, and N. Iwasaki, "New Vector Controller for PM Motors which Modeled the Cross-Coupling Magnet Flux Saturation", Proceedings of The 2010 International Power Electronics Conference (IPEC-Sapporo 2010), No. 23E1-2, pp. 1064–1070 (2010).

[3] K. Sakamoto, Y. Iwaji, T. Endo, and Y. Takakura, "Position and Speed Sensorless Control for PMSM Drive Using Direct Position Error Estimation", in Proceedings of IECON'01, pp. 1680–1685.

[4] D. Kaneko, Y. Iwaji, K. Sakamoto, T. Endo, and T. Okubo: "An Estimation Method of Initial Position for PMSMs based on Magnetic Saturation", SPC-04-143, IEA-04-61 (2004) (in Japanese).

[5] Y. Iwaji, Y. Kokami, and M. Kurosawa, "Position Sensorless Control Method at Low Speed for Permanent Magnet Synchronous Motors Using Induced Voltage Caused by Magnetic Saturation", Proceedings of IPEC-Sapporo 2010, No. 24G1-1, Jun. 2010.

[6] Y. Iwaji, S. Aoyagi, R. Takahata, K. Tobari, I. Suzuki, and M. Hano, "Position Sensorless Control Method at Zero Speed Region for Permanent Magnet Synchronous Motors Based on Block Commutation Drive", Proceedings of the 2013 IEEE IEMDC, TS12-3, pp. 525–531, May 2013.

[7] Y. Iwaji, R. Takahata, T. Suzuki, H. Tokoi and Y. Enomoto, "Low-Speed Position Sensorless Drive for Highly Efficient Permanent Magnet Synchronous Motor without Rare-Earth Metals", proceedings of the 2013 IEEE ECCE, pp. 532-539, Sep. 2013.

Sensorless-Oriented Design of IPMSM

Yoshiaki Kano

Dept. of Information and Computer Engineering
Toyota National College of Technology
Toyota, Japan
kano@toyota-ct.ac.jp

Abstract— **This paper deals with the design of a concentrated-winding interior permanent magnet synchronous motor (IPMSM) under saliency based-sensorless drive with a high-frequency signal injection for a general industrial application. Compared with distributed-winding IPMSMs, the serious disadvantages of concentrated winding IPMSMs are the narrow sensorless operating region and large torque ripple. To solve the problem, a novel rotor flux barrier design is proposed. The proposed rotor include its V-shape PM arrangements and circular flux-barriers at the rotor-yoke in the center of the N-pole and S-pole. The IPMSM designed is built and tested experimentally. As a result, it is demonstrated that the test machine realizes low-torque ripple under the frequent operating conditions while satisfying the required maximum torque under the sensorless drive.**

Keywords—High-frequency injection, position sensorless control, torque ripple, cross-coupling, magnetic saturation, incremental inductance, industrial drive

I. INTRODUCTION

In recent years, distributed winding interior permanent magnet (IPM) motors are replacing induction motors in general industrial applications due to high efficiency, high torque-to-volume ratio, high power factor, and sensorless rotor position detection capability [1]. The use of fractional-slot (FS) concentrated-winding will make the IPM motors more attractive because of their advantages [2],[3]: reduced Joule losses and efficiency improved as a consequence of short end winding length, reduced machine material and manufacturing costs. However, FS windings are characterized by several parasitic effect [4] that are mainly due to the local magnetic saturation and the high harmonic content in the magneto- motive force (MMF). As a result, the maximum torque under saliency based-sensorless drive with a high-frequency signal injection is limited; this makes it difficult to reduce size of the motor by using concentrated-winding. In addition, the serious disadvantages of concentrated-winding IPMSMs are the torque ripple and acoustic noise. Therefore, the continuous research and development of sensorless-oriented design method of FS winding IPMSM to realize the wide range of torques as well as low torque ripple would be very important [5-7].

This paper presents the design of a concentrated-winding IPMSM under saliency-based sensorless drive with a high-frequency signal injection for a general industrial application. A 45Nm-5.5kW IPMSM is designed by the proposed approach so as to meet torque ripple requirements with wide sensorless operating region. Special features of the resulting motor include its V-shape PM arrangements and circular flux-barriers at the rotor-yoke in the center of the N-pole and S-pole. This makes significant reduction of the torque ripple under heavy load conditions. The IPMSM designed is built and tested experimentally. As a result, it is demonstrated that the test machine realizes low torque ripple under the frequent operating conditions while satisfying the required maximum torque under the sensorless drive.

II. DESIGN RESTRICTIONS AND REQUIREMENTS FOR INDUSTRIAL DRIVE APPLICATIONS

The design restrictions and target specifications for industrial drive applications appear in table I. The required torque-speed characteristic is shown in Fig.1. The stator outer diameter D_o and stack length L are 134mm and 120mm, respectively. This motor size is

TABLE I
DESIGN RESTRICTIONS AND SPECIFICATIONS FOR INDUSTRIAL DRIVE APPLICATIONS.

Dimensional & material constraints	Stator outer diameter D_o	134 mm
	Stack length L	120 mm
	Shaft diameter D_{sh}	46 mm
	Air gap length g	0.4 mm
	Type of lamination steel	50A600
	Type of PM	NEOMAX38VH
	Shape of PM	Flat
	Coil-filling factor	0.55
Limitations on power circuit	Inverter bus voltage V_{dc}	190 V
	Rated and max. phase current	23.0, 34.5 A_{rms}
Requirements	Rated and max. torque(≤1750rpm)	30, 45Nm
	Torque ripple ratio (50%~100% rated torque conditions)	<10%
	Cogging torque (peak-to-peak)	0.82 Nm

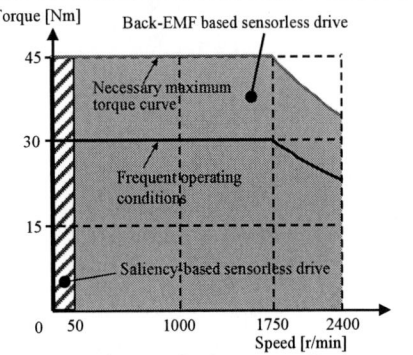

Fig. 1. Requirements for the target application.

achieved 40% reduction of motor-volume compared to the commercialized distributed-winding IPMSM (D_o=175 mm, L=120mm). The stator yoke width is 9.94 mm to ensure the mechanical strength. The air gap length and the coil-filling factor are chosen as the practical values taking account of manufacturability. Sintered neodymium-iron-boron (N_dF_eB) magnets are selected because of high remanence flux-density. In the target application, the d-axis current i_d=0 control is applied to drive the motor under saliency-based sensorless drive.

Under these restrictions, the motor designed must satisfy the following requirement as an alternative to the current distributed winding IPM motor:

1. The desired maximum torque is 45Nm under the saliency-based sensorless drive in the shaded area shown in Fig.1.
2. The desired torque ripple ratios $\Delta T/T_{ave}$ (ΔT: peak-to-peak value of torque ripple, T_{ave}: average torque) are less than 10% from 50% to 100% of the rated torque.
3. The desired peak-to-peak value of cogging torque is less than 0.82Nm.
4. The magnet volume must be kept lower than 77.8cm^3 similar with the baseline motor.
5. The stator configuration shown in Fig.2 is used to the design study.

Commercial FEA package, JMAG-Studio ver.10.0, released by JSOL corporation is used as 2D-FEA solver for this design and analysis.

III. DESIGN EXAMPLE WITH MAXIMUM TORQUE CAPABILITY UNDER SENSORLESS DRIVE

A. Design example [7]

To realize the wide range of torques, authors have already established and verified the following design procedure to increase the maximum torque under the sensorless drive [7] :

1) The rotor tooth opening is designed by using the inset PM rotor.
2) The depth of the embedded magnet has to be designed by fixed rotor tooth opening.
3) The combination of the magnet width and length should be designed with the maximum torque requirement.

The design procedure is applied to 45Nm-5.5kW 6-pole-9-slot industrial drive IPM motor to fulfill the maximum torque requirement [7]. According to the optimum design result, the prototype IPM motor has been constructed and tested. Fig.2 shows the measured dimensions of the prototype. Fig.3 shows the sensorless operating point trajectories (at d-q plane) with a contour map of the torque-ripple ratio. In the sensorless drive, the estimation of the rotor position is performed by the conventional signal injection-based sensorless method which forces the high frequency component in the q-axis current to be zero[8]. In Fig.3, L_{dif} ($=L_{qh}$- L_{dh}) and L_{dqh} are the difference and cross-coupling incremental inductances, respectively.

In the ideal case, the operating point trajectory coin-

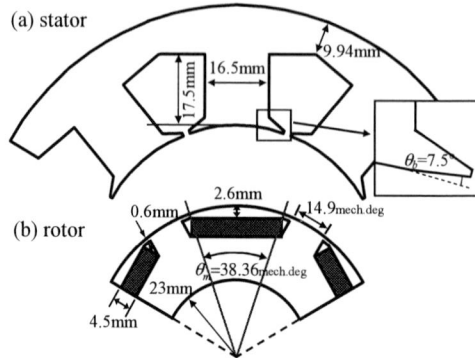

(a) stator

(b) rotor

Fig.2. Configuration of prototype. (saliency ratio=2.17)

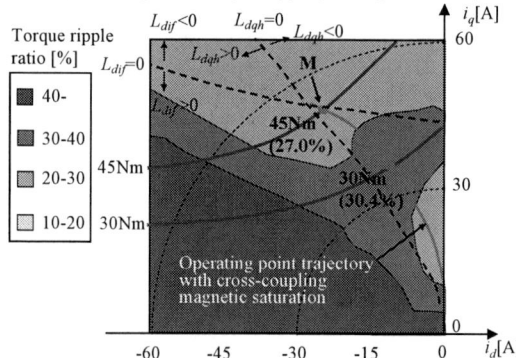

Fig. 3. Sensorless operating point trajectories of prototype.

cides with i_q-axis. However, the actual operating point trajectory swings counterclockwise in the direction of the negative d-axis current as a result of cross-coupling magnetic saturation. Thus, the position error arises by cross-coupling magnetic saturation and the negative d-axis current is applied to the motor. Consequently, it is possible to use reluctance torque whereas the i_d=0 control is applied to drive the motor. In this case, the operating limit at point M reaches more than 45Nm, and therefore the prototype satisfy the maximum torque capability. On the other hand, the operating limit M reaches very close to the condition with L_{dif}=L_{dqh}=0. There is only 5% of the torque difference between the operating limit and the condition with L_{dif}=L_{dqh}=0. Because of the insignificant difference, using the torque at L_{dif}=L_{dqh}=0 as a substitute for the maximum torque capability in the designing studies.

From Fig.3, it is found that the torque ripple ratio is more than 20% for all operating points. To develop the industrial drive motor, the torque ripple ratio must be less than 10%. Generally, the torque ripple under the sinusoidal current drive arises from permeance variations in the air gap due to the slotting effect (cogging torque), harmonics in the EMF, saturation in the magnetic circuit of the machine, and harmonics in the current induced by the inverter. In the prototype motor, the measured peak-to-peak cogging torque is 0.71Nm. The corresponding torque ripple ratio is 2.4% at the rated torque.

Fig.4(a) shows the measured back-EMF waveform at 1,750 r/min and Fig.4(b) shows the main harmonic components included in the measured waveform. As apparent from Fig.4, the test motor has a large amount of

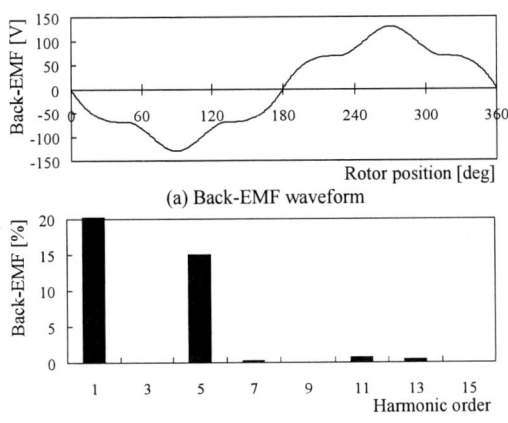

(a) Back-EMF waveform

(b) Main harmonic components

Fig. 4. Measured back-EMF waveform at 1750r/min.

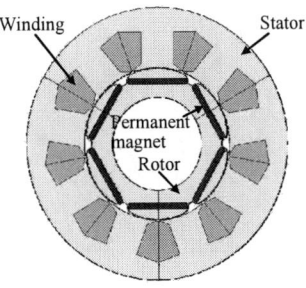

Fig. 5. Construction of test motor with sinusoidal profiling rotor surface.

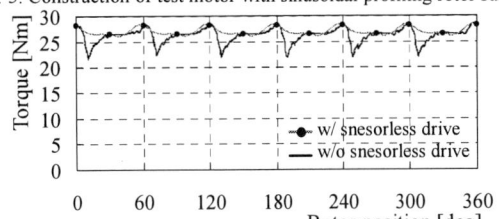

Fig. 6. Measured torque waveforms at 20r/min with and without sensorless drive.

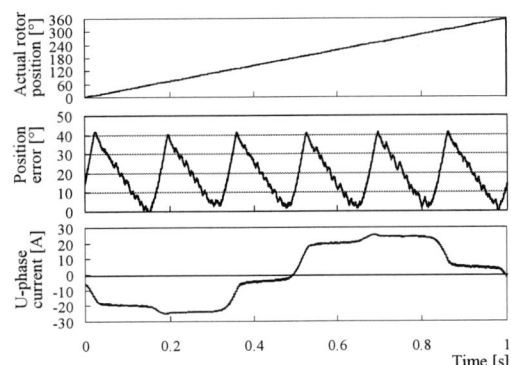

Fig. 7. Measured current and position error under the sensorless drive.

the harmonics of back-EMF waveform, the torque ripple ratio due to 5th and 7th harmonics in the back-EMF is 22.1% at the rated torque. From this result, the torque harmonic of 6th order is evident. Since the torque ripple ratio is 30.4% at the rated torque, the torque fluctuation of the test motor is mainly caused by the 5th and 7th harmonics of back-EMF. Therefore, the reduction of harmonics of back-EMF is very important to satisfy the torque ripple requirements.

B. Examination of torque ripple reduction for several techniques

Many researchers proposed the several torque ripple minimization techniques for FS winding IPMSMs with slot/pole/phase=1/2.

In [9], it has been shown that a reduction of the torque ripple can be achieved by using a rotor step-skewing. The rotor is split into two or more parts, each of them is skewed with respect to the others. It is applied to the targeted 45Nm-5.5kW 6-pole-9-slot industrial drive IPMSM. As a result, the torque ripple ratio is dramatically decreased from 30.4% to 5.7% at the rated torque. However, the maximum torque under the sensorless drive is decreased from 45Nm to 31.9Nm by the reduction of the operating region L_{dif}>0. The reason is that the q-axis inductance can be reduced with q-axis flux-pass by using the rotor step-skewing. To satisfy the torque requirements, the PM length must be increase for the expansion of the sensorless operating region. Drawbacks of the approach is the increase of the PM volume and motor cost.

In [10], it has been shown that a reduction of the torque ripple can be achieved by using a sinusoidal profiling rotor surface. In this way, the torque ripple due to harmonics in the EMF can be reduced. This technique is applied to the targeted industrial drive IPM motor. Fig.5 shows the construction of test motor and Fig.6 shows the measured torque ripple waveforms with and without sensorless drive. Fig.7 shows the measured U-phase current and position error at the 87% load torque under the sensorless drive. From Fig.6, the torque ripple ratio is dramatically increased from 8.1% to 26.4% by the sensorless drive. The increase of torque ripple caused by the harmonics in the phase current as shown in Fig.7. The

current harmonics is induced by the position error fluctuation. Moreover, the maximum torque under the sensorless drive is 26Nm. It is 42.3% smaller than the desired value of 45Nm. Therefore, a sinusoidal profiling rotor surface is not suitable for the target applications.

IV. DESIGN OF ROTOR GEOMETRY IN V-SHAPE PM ARRANGEMENT

In Section III, it was shown that the harmonics of EMF must be reduced to satisfy the torque ripple requirement. The magnet pole-arc θ_m is a particularly important parameter in regard to the harmonics of EMF. In this section, the V-shape PM arrangement shown in Fig.8 are examined to reduce torque ripple.

A. Investigation of Optimum Magnet Pole-Arc

The most influential torque ripple components is normally the 6th harmonic, followed by the 12th, and the 18th harmonics. The sum of these three torque harmonics when fed with purely sinusoidal current is given by

$$\sum_{n=1}^{3} T_{6n} = \frac{3}{2\omega_m} \cdot \sum_{n=1}^{3} \left(\widehat{E}_{6n+1} - \widehat{E}_{6n-1} \right) \cdot \widehat{I}_1 \qquad (1)$$

where ω_m is the angular speed of the rotor, $\widehat{E}_n$ is the amplitude of the nth time harmonic of the phase EMF,

The 2014 International Power Electronics Conference

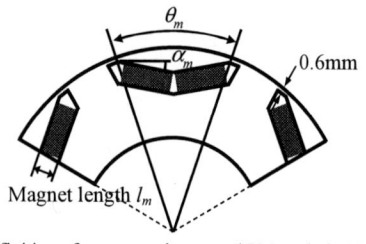

Fig. 8. Definition of magnet pole-arc and PM angle in V-shape PM arrangement.

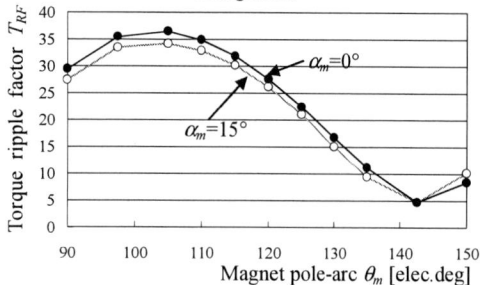

Fig. 9. Torque ripple factor vs. magne pole-arc.

and $\hat{I}_1$ is the fundamental amplitude of the phase current, respectively. The torque ripple factor T_{RF} is defined as the ratio of the peak-to-peak torque ripple to the average torque, which is given as [4]

$$T_{RF} = \frac{\Delta T_{pp}}{T_{ave}} \approx \frac{2\sqrt{\sum_{n=1}^{3} T_{6n}^2}}{T_{ave}} \qquad (2)$$

$$= \frac{2\sqrt{(\hat{E}_7 - \hat{E}_5)^2 + (\hat{E}_{13} - \hat{E}_{11})^2 + (\hat{E}_{19} - \hat{E}_{17})^2}}{\hat{E}_1}$$

Fig.9 shows the torque ripple factor T_{RF} versus magnet pole-arc θ_m for various PM angle. The magnet volume (103cm^3) is the same in all the models, thus the magnet length l_m depends on θ_m and α_m.

From Fig.9, the factor T_{RF} is not much changing even if the α_m is changed. The optimum magnet pole-arc θ_{m_opt} that minimize the torque ripple factor can be found $\theta_{m\_opt}=142.5°$.

As far as the torque ripple is concerned, 44-number of analysis models which have different pole-arc and PM angle are considered to determine optimum model. To evaluate the torque ripple due to harmonics in the EMF and saturation in the magnetic circuit of the machine, the instantaneous torque waveforms under $i_d=0$ control have been simulated for various load torques. In the torque computation, the effect of the cogging torque on the torque ripple is excluded by using the cogging torque calculation technique presented in [11]. The load torques have been set to 25%, 50%, and 100% of the rated load torque, respectively. Fig.10 shows the results of the computation.

Since the magnetic saturation does not occur at the light-load condition(25% rated-load torque), the variation of torque ripple ratio for magnet pole-arc are similar to the result of torque ripple factor as shown in Fig.9. In contrast, the torque ripple ratios are gradually changed with the load torque increase because of magnetic

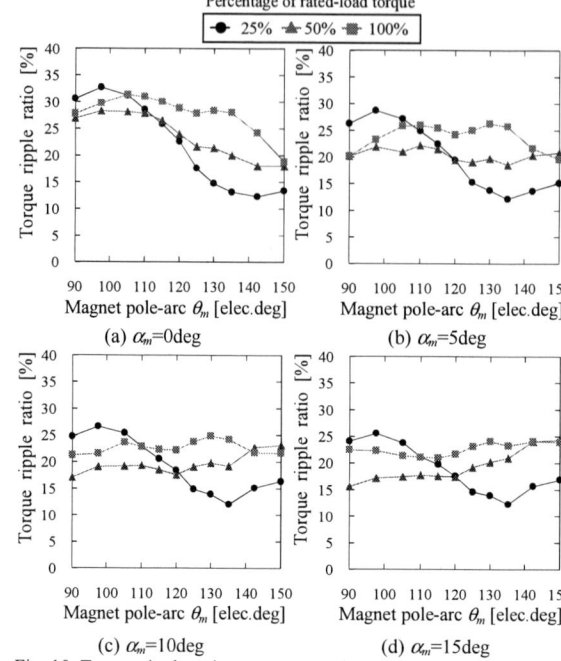

Fig. 10. Torque ripple ratio vs. magnet pole-arc under i_d=0 control.

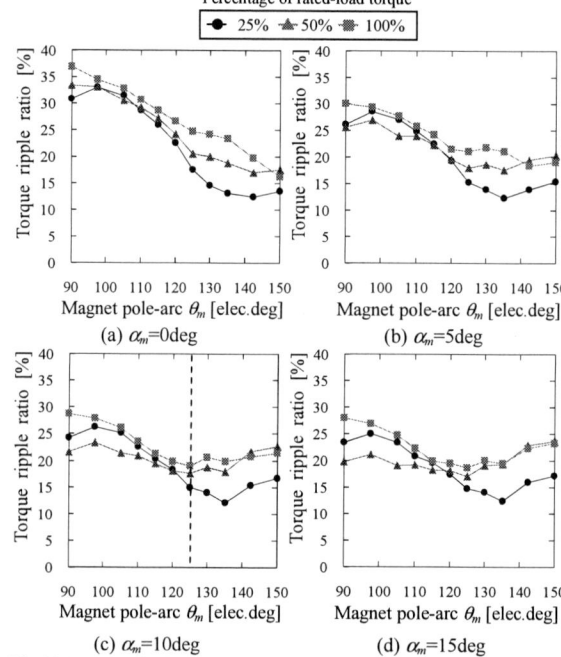

Fig.11. Torque ripple ratio vs. magnet pole-arc under sensorless control.

saturation in the core. Specifically, remarkable partial magnetic saturation occur at the inner surface of the stator teeth and the outer surface of the rotor for larger magnet pole-arc (θ_m>120elec.deg). This saturation cause the increase of torque ripple. For smaller magnet pole-arc (θ_m<115elec.deg), partial magnetic saturation occur at the rotor yoke. It should be noted that this magnetic saturation at the rotor yoke makes the decrease of torque ripple. This is because the magnetic saturation cause the reduction of the harmonics of phase EMF at load.

Figs.11 and 12 show the characteristics of torque ripple

978-1-4799-2706-7/14 $31.00 © 2014 IEEE

The 2014 International Power Electronics Conference

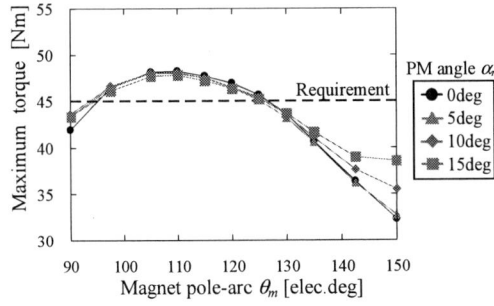

Fig. 12. Maximum Torque vs. magne pole-arc.

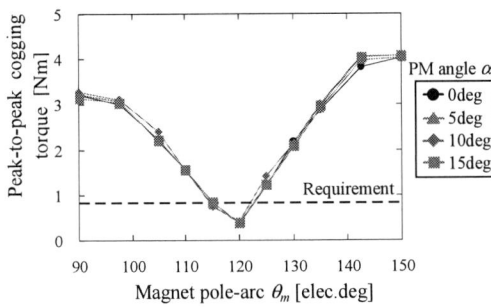

Fig. 13. Cogging torque vs. magne pole-arc.

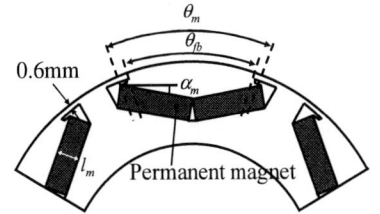

Fig.14. Definition of θ_{fb} in V-shape PM arrangement.

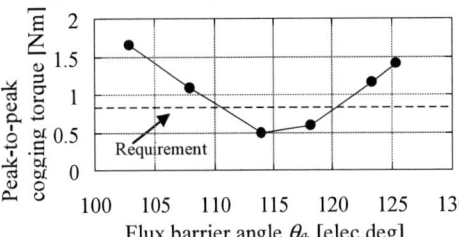

Fig. 15. Peak-to-peak cogging torque variation with flux-barrier angle.

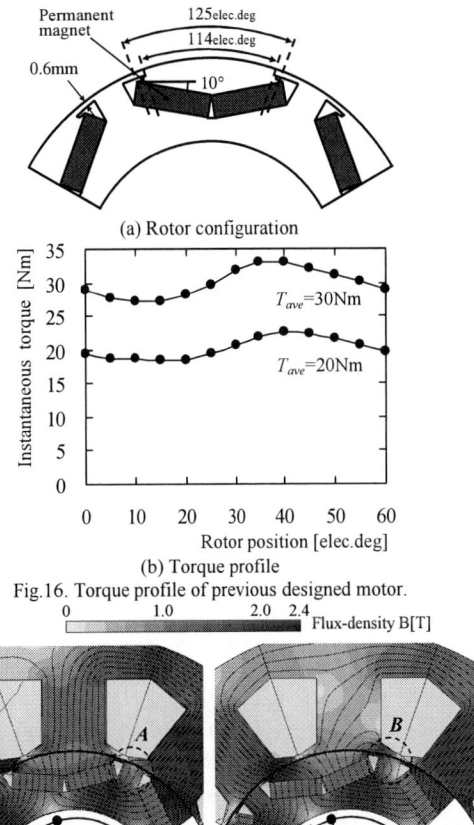

(a) Rotor configuration

(b) Torque profile

Fig.16. Torque profile of previous designed motor.

(a) θ=12.5°   (b) θ=40.0°

Fig.17. Flux plot of previous designed motor.

ratio and the maximum torque under the sensorless drive. Fig.13 shows the peak-to-peak cogging torque versus magnet pole-arc θ_m for various PM angle.

In the sensorless drive, the negative d-axis current is applied to the motor because of position error caused by cross-coupling effect. As a result, the change of torque ripple ratio caused by the above partial magnetic saturation decreases as shown in Fig.11.

Consequently, the optimum combination between pole-arc θ_{mo} and PM angle α_{mo} that minimize the torque ripple ratio can be found θ_{mo}=125elec.deg and α_{mo}=10°. In this case, the maximum torque is 45.5Nm, the torque ripple ratio at the rated torque is about 17.0%, and the peak-to-peak value of cogging torque is 1.41Nm. It is 0.59Nm larger than the cogging torque requirement. To satisfy the cogging torque requirement, the flux-barrier design shown in Fig.14 will be investigated in the next sub-section.

B. Design of flux barrier angle

Fig.15 shows the peak-to-peak cogging torque variation for changing the flux-barrier angle θ_{fb} from 103 to 125elec.deg. The 4.1mm magnet length and magnet opening angle θ_m=125elec.deg are the same in all the models.

From Fig.15, the optimum flux-barrier angle θ_{fb_opt} that minimize the peak-to-peak value of cogging torque can be found θ_{fb_opt}=114elec.deg. In the following design, the θ_{fb_opt}=114elec.deg is treated as the optimum value. However, the above design is not enough to fulfill the torque ripple requirements. In the following section, we propose the new flux-barrier design to reduce the torque ripple while keeping the maximum torque under the sensorless drive.

V. A NEW FLUX-BARRIER DESIGN TO REDUCE TORQUE RIPPLE [12]

A. Effect of proposed circular flux-barrier in rotor-yoke

Fig.16(a) is the previous designed motor and the torque profile is shown in Fig.16(b). Fig.17 shows the flux plot at the rotor positions θ =12.5° and 40°, which correspond to the minimum and maximum torque conditions in Fig.16(b). To reduce the torque fluctuations,

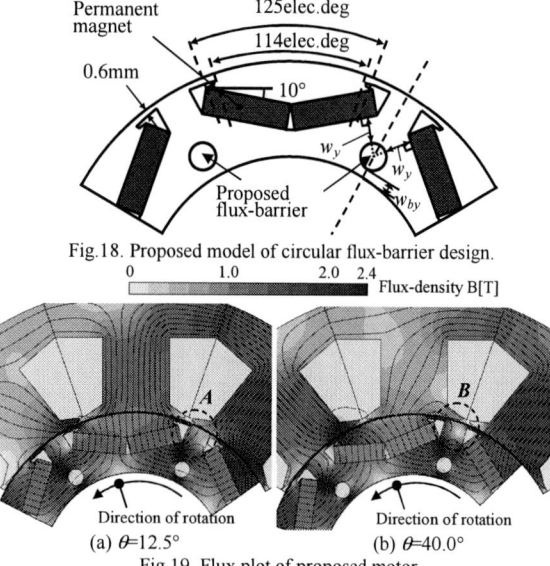

Fig.18. Proposed model of circular flux-barrier design.

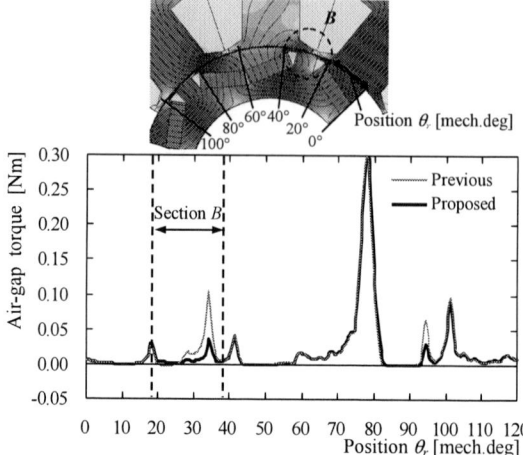

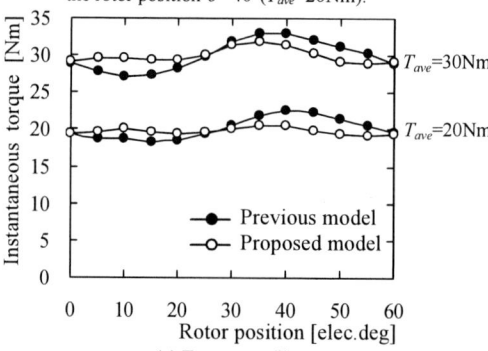

Fig.19. Flux plot of proposed motor.

Fig.20. Effect of proposed flux-barrier on air-gap torque distribution at the rotor position $\theta=40°$ ($T_{ave}=20$Nm).

it is necessary to increase the minimum torque around θ $=12.5°$ and decrease the maximum torque around $\theta=40°$. To solve the problem, the negative torque at the section "A" and the positive torque at the section "B" in Fig.17 have to be reduced. Thus, the flux passing through the rotor pole should be reduced at each position. From the viewpoint, we finally set up the circular flux-barrier at the rotor-yoke in the center of the N-pole and S- pole.

Fig.18 shows the proposed new flux-barrier design. The flux-barrier should be decided by the rotor-yoke width w_y and w_{by} in Fig.18. As an example, Fig.19 shows the flux plot of the proposed model with $w_y=4.5$mm and $w_{by}=1.5$mm at the rotor position $\theta=12.5°$ and $40°$. Fig.20 shows the air-gap torque distributions of both the previous designed motor and the proposed model under the 67% rated torque of 20Nm at the rotor position $\theta=40°$. The calculation of air-gap torque is based on the Maxwell Stress Tensor.

It is obvious from Fig.19, that the flux at the sections "A" and "B" are reduced compared with that of the previous designed motor. Because using the proposed flux-barrier to get the iron saturation in between the V-shape-magnet and the barrier that effectively works well. It is to make reduction of the flux passing through the rotor-pole and the air-gap torque at the section "B" as shown in Fig.20.

Fig 21 shows the effect of the proposed flux-barrier on the magnet torque-ripple and the operating region $L_{dif}>0$ under the sensorless drive. The torque ripple due to harmonics in the EMF is significantly reduced by the proposed flux-barrier. In addition, the proposed flux-barrier makes the large operating region because the d-axis inductance can be reduced while keeping the q-axis inductance.

B. Circular flux-barrier design

Fig.22 shows the torque ripple ratio for changing the rotor yoke width w_y from 3.0mm to 6.0mm. In the study, the α_m changes from 8.7° to 17.1°. The maximum value

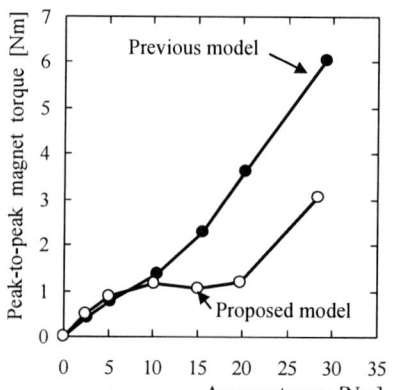

(a) Torque profile

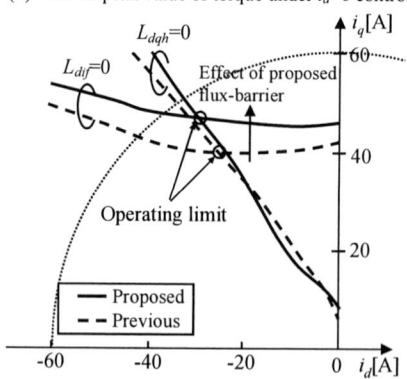

(b) Peak-to-peak value of torque under $i_d=0$ control

(c) Operating region $L_{dif}>0$

Fig.21. Effect of proposed flux-barrier on torque performance.

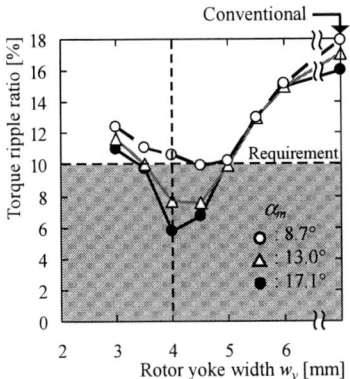

Fig.22. Torque ripple ratio vs. rotor yoke width w_y.

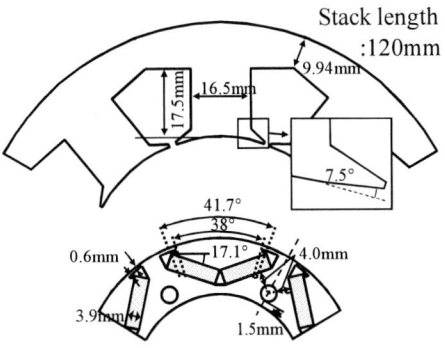

Fig. 23. Measured dimensions of prototype.

TABLE II
DESIGNED MOTOR SPECIFICATIONS

Rotor outer diam. D_r [mm]	74.2	Shaft diameter [mm]	46
Stator outer diameter [mm]	134	Stack length L [mm]	120
Stator yoke width [mm]	9.94	Stator teeth width [mm]	16.5
PM length l_m [mm]	3.85	PM width / piece [mm]	13.59
Magnet opening angle θ_m [°]	41.7	Flux barrier angle α_{fb} [°]	38.0
Rotor yoke width w_{by} [mm]	1.5	PM angle α_m [°]	17.1
Wdg.resistance/phase R [Ω]	0.167	Rotor yoke width w_y[mm]	4.0
Back-emf const. K_E [V/rad/s]	0.26	Max. torque [Nm]	45.9
d-axis inductance L_d [mH]	5.03	Rated torque [Nm]	30
q-axis inductance L_q [mH]	9.33	Cogging torque $\tau_{p\text{-}p}$ [Nm]	0.3
Torque ripple ratio [%]	<6.5	Saliency ratio	1.86

of PM angle is 17.1°, which is selected to ensure the mechanical strength in the rotor yoke width at the center of V-shape magnet. The w_{by} is 1.5mm.

From Fig.22, larger PM angle is better for the torque ripple reduction by using the proposed flux-barrier. In case of α_m=17.1°, the torque ripple ratios are minimized at the rotor yoke width w_y= 4.0mm. To fulfill all the requirements, 17.1° of the PM angle and 4.0mm of the rotor yoke width are chosen as the optimum values.

VI. EXPERIMENTAL VERIFICATIONS

A. Prototype Motor Specifications and Experimental Setup

According to the optimum design results, a prototype 6-pole-9-slot IPMSM has been constructed. Fig.23 shows the measured dimensions of the prototype motor. Table II lists the prototype motor specifications. The measured air gap of 0.32mm is 20% smaller than the designed value of 0.4mm because the shrink fit of the stator core into frame brought about a reduction in the size of the stator core. However, the other measured dimensions are mostly identical with those of the optimum designed values.

Fig.24 shows the control system configurations. In the control system, the estimated speed $\hat{\omega}$ is compared to the speed reference ω^*. The speed error is processed in the PI speed regulator to obtain the torque component of current reference $i_q{}^*$, while the d-axis current reference $i_d{}^*$ is zero. The estimations of the rotor position $\hat{\theta}$ and speed $\hat{\omega}$ are performed by the conventional sensorless method [8]. In the drive system, the injected signal is 20V, 300Hz. The carrier frequency of the pulse width modulation IGBT inverter is 6.0kHz and the control loop cycle is 166μs. The digital signal processor (DSP) controller implements the sensorless algorithm.

B. Back-EMF

Fig.25 shows the measured U-phase back-EMF voltage compared to the FEA result. The measured and calculated harmonic spectra of the back EMF waveform are shown in Fig.26. It can be seen that the measured and calculated back-EMF waveforms consist almost entirely of the fundamental and 5th harmonic components. The measured 5th harmonic component is slightly lower than the calculated one. This result suggests that the prototype

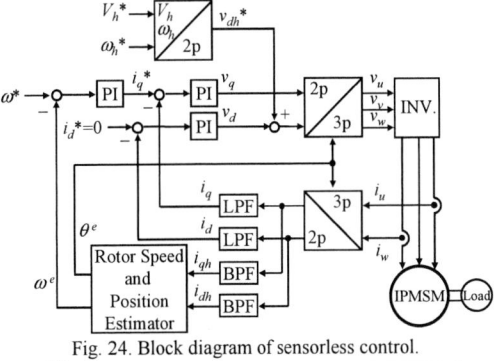

Fig. 24. Block diagram of sensorless control.

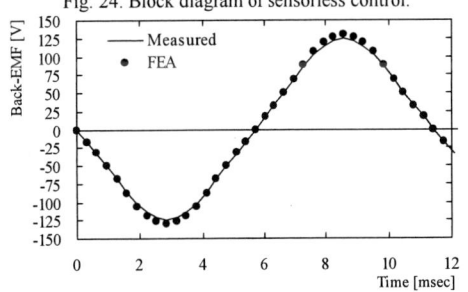

Fig. 25. Comparison of measured and calculated back-EMF voltage waveforms at 1750r/min.

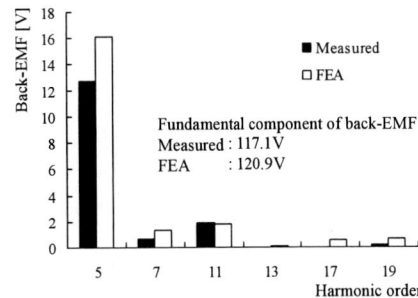

Fig. 26. Harmonic spectrum of the measured and calculated back-EMF voltage waveforms at 1750r/min.

978-1-4799-2706-7/14 $31.00 © 2014 IEEE

The 2014 International Power Electronics Conference

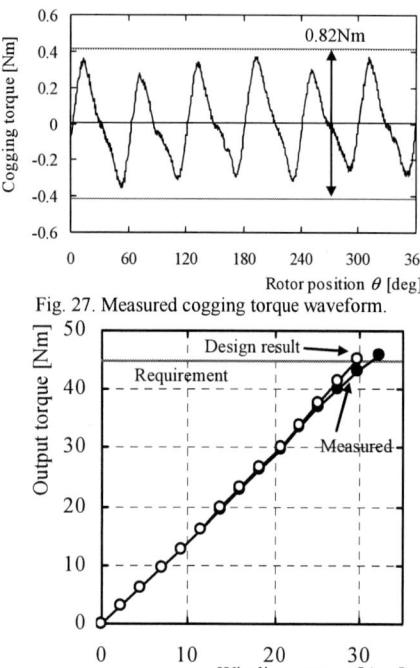

Fig. 27. Measured cogging torque waveform.

Fig. 28. Measured torque vs. current characteristics.

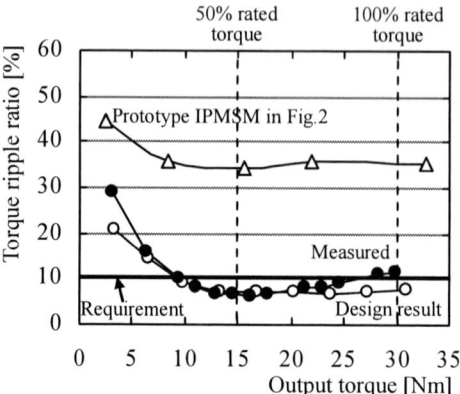

Fig. 29. Measured torque ripple ratio.

motor will have a lower torque ripple than predicted.

C. Cogging Torque

Fig.27 shows the measured cogging torque waveforms of the prototype motor. The measured peak-to-peak value of cogging torque is 0.72 Nm. The designed motor realizes the necessary torque requirements.

D. Maximum Torque under the Sensorless Drive

Fig.28 shows the measured torque versus current characteristics under the sensorless drive at 20r/min. In this figure, the calculated torque vs. current characteristics is also shown. It is found that the measured and calculated torque agree well. The maximum torque is 45.8Nm. The prototype fulfills the maximum torque desired in the target application. The developed motor realizes the increase of 1.67 times the torque density compared to the commercialized IPMSM.

E. Torque Ripple under the Sensorless Drive

Fig.29 shows the measured torque ripple ratio of the prototype motor. From Fig.29, it is found that the measured torque ripple ratio is slightly larger than the calculated one. The reason may be that the shrink fit causes the degradation of the permeability in the lamination steel. However, the prototype motor almost satisfies the target torque ripple requirements.

VII. CONCLUSIONS

This paper has presented a design of concentrated-winding IPMSM under the saliency-based sensorless drive for a 5.5kW general industrial application. In order to achieve the wide sensorless operating region and low torque ripple, the novel rotor flux barrier has been presented and designed. The designed rotor configuration

has been built and tested experimentally. Consequently, the test machine has fulfilled all the requirements to the target application.

REFERENCES

[1] M. Motoike, "Permanent Magnet Type Synchronous Motor -EDM Series-," Toyo technical review, no.106, pp.29-33, 2000.

[2] A.EL-Refaie, "Fractional-slot concentrated-windings synchronous permanent magnet machines: Opportunities and challenges," *IEEE Transactions on Industrial Electronics*, vol.57, no.1, pp.107-121, jan. 2010.

[3] J.Cros and P.Viarouge, "Synthesis of high performance PM motors with concentrated windings," *IEEE Transactions on Energy Conversion*, vol.17, no.2, pp.248-253, jun. 2002.

[4] F. Magnussen and H. Lendenmann, "Parasitic Effects in PM Machines With Concentrated Windings," *IEEE Transactions on Industry Applications*, vol.43, no.5, pp.1223–1232, Sep./Oct. 2007.

[5] J.M. Park, S.I. Kim, J.P. Hong, J.H. Lee, "Rotor Design on Torque Ripple Reduction for a Synchronous Reluctance Motor With Concentrated Winding Using Response Surface Methodology," *IEEE Transactions on Magnetics*, vol. 42, no. 10, pp. 3479–3481, Oct. 2006.

[6] U.J. Seo, Y.D. Chun; J.H. Choi, P.W. Han, D.H. Koo, J. Lee, "A Technique of Torque Ripple Reduction in Interior Permanent Magnet Synchronous Motor," *IEEE Transactions on Magnetics*, vol. 47, no. 10, pp. 3240–3243, Oct. 2011.

[7] Y.Kano, T.Kosaka, N.Matsui, and T.Nakanishi: "Design of Saliency-Based Sensorless Drive IPM Motors for General Industrial Applications ", Conference Record of the 2008 IEEE Industry Applications Conference 43rd Annual Meeting, CD-ROM (ISBN 978-1-4244-2279-1), (2008.10)

[8] Y.Ohmori, S.Hagiwara, "Control of Permanent Magnet Synchronous Motor," Toyo technical review, no.111, pp.13-21, 2005.

[9] Y.Kawase, T.Yamaguchi, T.Yano, M.Igata, K.Ide, Y.Kataoka, and A.Yamagiwa, "Three-Dimensional Electromagnetic Force Analysis of an Interior Permanent Magnet Synchronous Motor With Skewed Rotor," The Papers of Joint Technical Meeting on Static Apparatus and Rotating Machinery, IEE Japan, SA-05-31, RM-05-31(2005)(in Japanese)

[10] S.Kim, J.Bhan, J.Hong, and K.Lim, "Optimization technique for improving torque performance of concentrated winding interior PM synchronous motor with wide speed range," *in Conf. Rec. of the 2006 IEEE IAS Annual Meeting*, Tampa, pp.1933-1200, Oct.2006.

[11] Ki-Chan Kim, and Seung-Ha Jeon, "Analysis on Correlation Between Cogging Torque and Torque Ripple by Considering Magnetic Saturation," IEEE Transactions on Magnetics, vol.49, no. 5, May 2013

[12] Y.Kano, T.Kosaka, N.Matsui, and T.Nakanishi, " A New Technique of Torque Ripple Reduction in Saliency-Based Sensorless Drive IPM Motors for General Industrial Applications," Proceedings of 13rd European Conference on Power Electronics and Applications, CD-ROM (2009)

Noise Reduction Method by Injected Frequency Control for Position Sensorless Control of Permanent Magnet Synchronous Motor

Shun Taniguchi, Kazuya Yasui, and Kazuaki Yuki

TOSHIBA Corporation

1 Toshiba-cho, Fuchu-City, Tokyo, Japan

Abstract— **This paper proposes a novel noise reduction method by injected frequency control for position sensorless control of permanent magnet synchronous motor (PMSM). The proposed method reduces acoustic noise caused by injected high-frequency voltage without reducing the accuracy of rotor position estimation. The proposed method is verified by numerical simulations and experiments using a 0.8kW-PMSM.**

Keywords— Permanent Magnet Synchronous Motor, Position Sensorless Control, Acoustic Noise, Random PWM

I. INTRODUCTION

Recently, permanent magnet synchronous motors (PMSMs) have been applied to many applications owing to their high efficiency and high power density.

A rotor position is usually measured by a rotational position sensor, such as encoders or resolvers, for precise current (torque) control according to a rotational angle. The sensor reduces the overall reliability and increases cost, size, and weight.

Therefore, position sensorless control, which requires no rotational position sensor, has been discussed to be applied during the last decades [1][2][3]. At low speed range, high-frequency voltage injection method is applied to rotor position estimation [1]. However, the method sometime causes magnetic noise issue in some application such as railway applications because the injected high-frequency voltage increases the audio-frequency components of the motor current.

Several noise reduction methods have been reported [2][3]. One of these methods estimates rotor position by only Pulse Width Modulation (PWM) ripples [2]. The method reduces acoustic noise because it does not inject high-frequency voltage. However, the method reduces accuracy of rotor position estimation. On the other hand, a signal injection technique whose frequency is PWM switching frequency has been proposed [3]. If PWM switching frequency is above audible frequency range, acoustic noise by injected signal can be also remarkably reduced. However, in some applications such as railway applications, the method provides little benefit when PWM switching frequency is approximately 1 [kHz] or 2 [kHz].

On the other hand, for reducing motor electric noise caused by PWM, random PWM (RPWM) has been applied to many applications [4][5][6]. The method can also reduce motor electric noise by the injected high-frequency voltage when the injected high-frequency voltage is synchronized with PWM carrier. However, the method also reduces the accuracy of rotor position estimation.

To cope with this problem, this paper proposes a novel noise reduction method based on RPWM. The proposed method can reduce acoustic noise by the injected high-frequency voltage without decreasing estimation performance. The features of the proposed method are 2 points as follows.

1. Injected frequency controller can distribute current spectrums widely in the range that the injected frequency can be selected.

2. Injected voltage controller can maintain the accuracy of rotor position estimation when the injected frequency is changed.

The proposed method is verified by numerical simulations and experiments using a 0.8kW-PMSM. The proposed method is also expected to increase the efficiency of PMSMs.

II. CONVENTIONAL METHOD

Fig.1 shows a block diagram of conventional method. An error of rotor position estimation $\Delta\theta$ is defined as $\Delta\theta = \theta - \theta_{est}$ (where θ is actual rotor position and θ_{est} is estimated rotor position) and stator voltage equations of a PMSM in estimated rotor reference frame 'dcqc' are given by equation (1).

$$
\begin{bmatrix} v_{dc} \\ v_{qc} \end{bmatrix} = R_m \begin{bmatrix} i_{dc} \\ i_{qc} \end{bmatrix} + \begin{bmatrix} L_{dc} & L_{dqc} \\ L_{dqc} & L_{qc} \end{bmatrix} \frac{d}{dt} \begin{bmatrix} i_{dc} \\ i_{qc} \end{bmatrix} \\
+ \omega_{est} \begin{bmatrix} -L_{dqc} & -L_{qc} \\ L_{dc} & L_{dqc} \end{bmatrix} \begin{bmatrix} i_{dc} \\ i_{qc} \end{bmatrix} + \Phi_f \omega \begin{bmatrix} -\sin\Delta\theta \\ \cos\Delta\theta \end{bmatrix}
\tag{1}
$$

Where, (v_{dc}, v_{qc}) are dc-qc axis stator input voltages, (i_{dc}, i_{qc}) are dc-qc axis stator currents, R_m is stator resistances, Φ_f is magnet flux linkage, ω is actual rotor electrical angular velocity, ω_{est} is estimated rotor electrical angular velocity, and $(L_{dc}, L_{qc}, L_{dqc})$ are parameters defined as equation (2).

$$\begin{cases} L_{dc} = \dfrac{1}{2}\left(L_d + L_q\right) + \dfrac{1}{2}\left(L_d - L_q\right)\cos 2\Delta\theta \\[2mm] L_{qc} = \dfrac{1}{2}\left(L_d + L_q\right) - \dfrac{1}{2}\left(L_d - L_q\right)\cos 2\Delta\theta \\[2mm] L_{dqc} = \dfrac{1}{2}\left(L_d - L_q\right)\sin 2\Delta\theta \end{cases} \quad (2)$$

Where, (L_d, L_q) are d-q axis inductances.

In equation (1), in lower speed range, the third term and the fourth term can be neglected with smaller ω. Moreover, if high-frequency voltage is injected, the second term is more dominant than the first term. Thus, equation (3) is derived from equation (1).

$$\begin{bmatrix} v_{dch} \\ v_{qch} \end{bmatrix} = \begin{bmatrix} L_{dc} & L_{dqc} \\ L_{dqc} & L_{qc} \end{bmatrix} \frac{d}{dt} \begin{bmatrix} i_{dch} \\ i_{qch} \end{bmatrix} \quad (3)$$

Where, (v_{dch}, v_{qch}) are dc-qc axis injected high-frequency voltages, and (i_{dch}, i_{qch}) are dc-qc axis high-frequency currents.

When injected high-frequency voltage references, $v_{dh}{}^*$, $v_{qh}{}^*$ are given by equation (4), high-frequency motor currents are given by equation (5).

$$\begin{cases} v_{dch}{}^* = V_{dh}{}^* \cos 2\pi f_h \\ v_{qch}{}^* = 0 \end{cases} \quad (4)$$

$$\begin{bmatrix} i_{dch} \\ i_{qch} \end{bmatrix} = \frac{V_{dh}{}^* \sin 2\pi f_h}{L_d L_q 2\pi f_h} \begin{bmatrix} \dfrac{L_d + L_q}{2} - \dfrac{L_d - L_q}{2}\cos 2\Delta\theta \\[2mm] -\dfrac{L_d - L_q}{2}\sin 2\Delta\theta \end{bmatrix} \quad (5)$$

The amplitude of i_{qch} is given by Fourier expansion as shown in equation (6).

$$I_{qh} = \frac{1}{\pi} \int_{-\pi}^{\pi} i_{qc} \sin 2\pi f_h \, dt \quad (6)$$

Where, I_{qh} is the amplitude of i_{qch}. As shown in equation (5), I_{qh} is proportional to $\sin 2\Delta\theta$. Therefore, θ_{est} can be estimated by phase locked loop as shown in Fig.1.

III. Proposed Method

The proposed method is equipped with injected frequency controller and injected amplitude controller in addition to the conventional method. That is, the proposed method replaces injected voltage controller shown in Fig.1 with that in Fig.2. This section explains the injected voltage controller as follows.

A. Injected frequency controller

Fig.3 shows a block diagram of proposed injected frequency controller. Fig.4 shows a state transition diagram of the proposed injected frequency controller. Where, f_{low} is lower frequency of injected voltage, f_{high} is higher frequency of injected voltage, P_{lh} is probability of transition from f_{low} to f_{high}, P_{hl} is probability of transition from f_{high} to f_{low}, S_{freq} is status of injected frequency. The proposed method, which uses only two injected frequencies, controls current spectrums by transferring probability of the two injected frequencies, f_{low} and f_{high}. We explain basic principle of the proposed method below.

Random signal generator produces random number in the range from 0 to 1. Transition selector compares the random number and state transition probability produced

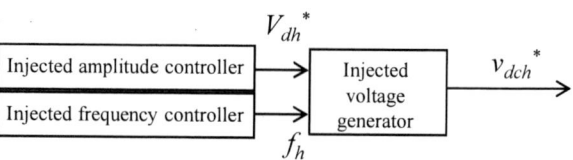

Fig. 2. Block diagram of proposed injected voltage controller.

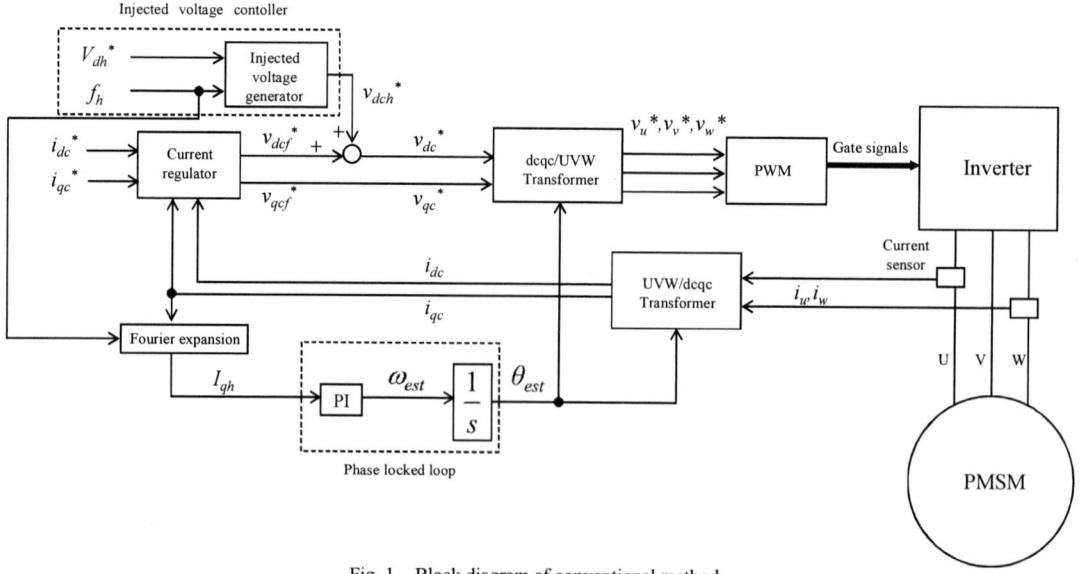

Fig. 1   Block diagram of conventional method.

by probability selector. If the random number is smaller than the state transition probability, transition signal is switched on, and injected frequency is changed. In this way, injected frequencies are controlled by state transition probability.

The proposed method has only four modes. That is a) continuing f_{low}, b) continuing f_{high}, c) transition from f_{low} to f_{high}, d) transition from f_{high} to f_{low}. Each peak frequencies of the current spectrums are given as follows.

a) continuing f_{low}

If an injected frequency is continued to be selected, the continuing frequency becomes a peak frequency of current spectrums. Therefore, a peak frequency f_{peak} is given in (7).

$$f_{peak} = f_{low} \qquad (7)$$

b) continuing f_{high}

By the same token, a peak frequency of current spectrums is given in (8).

$$f_{peak} = f_{high} \qquad (8)$$

c) transition from f_{low} to f_{high}
d) transition from f_{high} to f_{low}.

We discuss both c) and d) at the same time. At the case of the transition from f_{high} through f_{low} to f_{high}, a peak frequency of current spectrums is given in (9).

$$f_{peak} = \frac{2}{\dfrac{1}{f_{low}} + \dfrac{1}{f_{high}}} = \frac{2 f_{low} f_{high}}{f_{low} + f_{high}} \qquad (9)$$

At the case of the transition from f_{low} through f_{high} to f_{low}, a peak frequency of current spectrums is given in (10).

$$f_{peak} = \frac{2}{\dfrac{1}{f_{low}} + \dfrac{1}{f_{high}}} = \frac{2 f_{low} f_{high}}{f_{low} + f_{high}} \qquad (10)$$

Therefore, current spectrums of the proposed method consist of three spectrums shown in Fig.5. That is a) the spectrums at the case of continuing f_{low}, b) the spectrums at the case of continuing f_{high}, and c), d) the spectrums at the case of transition between f_{low} and f_{high}.

Each spectral peaks of the three spectrums are given in (11), (12), and (13).

$$Spectral\ peak\ of\ f_{low} = \frac{f_{high}}{f_{low}} \frac{P_{hl}(1 - P_{lh})}{f_{low} P_{lh} + f_{high} P_{hl}} C \qquad (11)$$

$$Spectral\ peak\ of\ f_{high} = \frac{f_{low}}{f_{high}} \frac{P_{lh}(1 - P_{hl})}{f_{low} P_{lh} + f_{high} P_{hl}} C \qquad (12)$$

$$Spectral\ peak\ of\ \frac{2 f_{low} f_{high}}{f_{low} + f_{high}}$$
$$= \frac{(f_{low} + f_{high})^2}{2 f_{low} f_{high}} \frac{P_{hl} P_{lh}}{f_{low} P_{lh} + f_{high} P_{hl}} C \qquad (13)$$

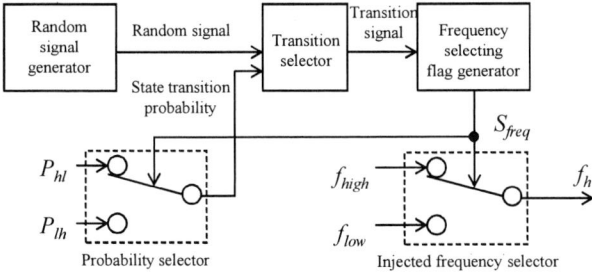

Fig. 3.   Block diagram of proposed injected frequency controller.

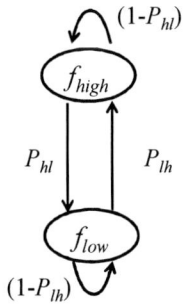

Fig. 4.   State transition diagram of proposed method.

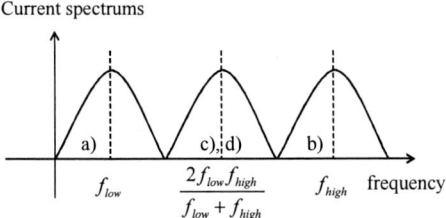

Fig. 5.   Simplified diagram of current spectrums caused by proposed method.

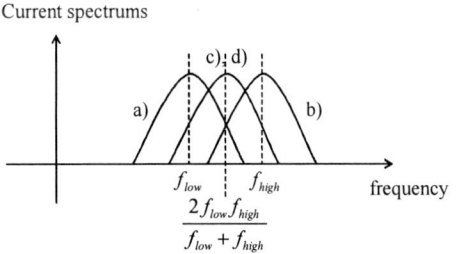

Fig. 6.   Current spectrums at the case of the range injected frequency can be selected is narrow.

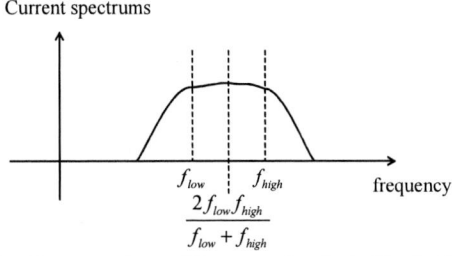

Fig. 7.   Current spectrums of the proposed method with adjusting P_{hl} and P_{lh}.

Where, C is common value that is changed according to modulation factor, f_{high}, f_{low}, and so on.

Generally, the range is restricted as follows, that the injected frequency can be selected.

1. f_{low} must be set as high as the control delay is permitted.

2. f_{high} must be set as low as the processing time is permitted.

For this reason, the current spectrums caused by proposed method overlap as shown in Fig.6. Therefore, the spectral peak can be reduced by setting probabilities, P_{hl} and P_{lh}, lower so that the spectrums become flat as shown in Fig.7.

As above, harmonics peaks become lower and are distributed in wider frequency area by the proposed injected frequency controller.

B. Injected amplitude controller

Fig.8 shows a block diagram of the injected amplitude controller. The proposed method selects the injected voltage amplitude according to the status of the injected frequency.

As shown in equation (5), I_{qh} is inversely proportional to f_h. On the other hand, I_{qh} is proportional to V_{dh}^*. If V_{dh}^* is set so as to be inversely proportional to f_h, I_{qh} becomes constant as shown in Fig.9. It can maintain the accuracy of rotor position estimation when the injected frequency is changed. Moreover, current sampling time is also changed according to f_h. It can realize real-time Fourier

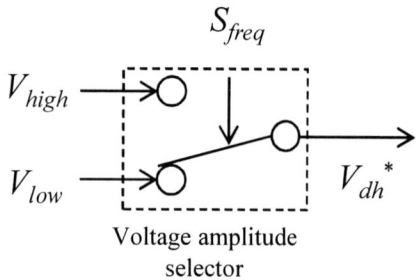

Fig. 8. Block diagram of proposed injected amplitude controller.

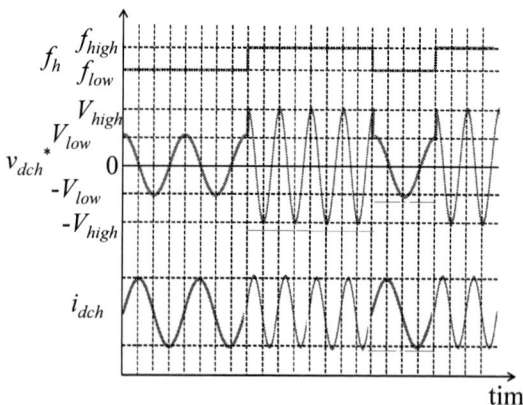

Fig. 9. Relationship among injected frequency, injected voltage, and high-frequency motor current by injected amplitude controller.

expansion by equation (6).

IV. NUMERICAL SIMULATIONS AND EXPERIMENTS

The proposed method has been verified by numerical simulations and experiments using a 0.8kW-PMSM.

The experimental parameters are given in Table 1, assuming railway applications in Ex.1, and automotive applications in Ex.2.

Fig.10 shows simulation result of conventional method. Fig.11 shows simulation result of proposed method without injected amplitude controller. Fig.12 shows simulation result of proposed method with injected amplitude controller. Where, T is torque. As shown in Fig.10-12, the injected amplitude controller can maintain the accuracy of rotor position estimation when the injected frequency is changed.

Fig.13 shows the noise measurement results of the conventional method and the proposed method (Ex.1). As shown in Fig.13, the spectral peak of the proposed method is about 46.2 [dBA], and that of the conventional method is about 53.0 [dBA]. Fig.14 shows the noise measurement results of the conventional method and the proposed method (Ex.2). As shown in Fig.14, the spectral peak of the proposed method is about 51.7 [dBA], and that of the conventional method is about 59.1 [dBA]. Therefore, it can be seen that the proposed method reduces spectral peak of noise.

As above, it can be seen that the proposed method reduces acoustic noise caused by injected high-frequency voltage without reducing the accuracy of rotor position estimation.

V. CONCLUSION

This paper has proposed a novel noise reduction method by injected frequency control for PMSM. The proposed method is equipped with injected frequency controller and injected amplitude controller in addition to the conventional method. The proposed method reduces

Table 1. Experimental parameters.

Item	Simbol	Value
Rated output [kW]	P	0.8
Pole pairs	N	3
Inverter DC link voltage [V]	V_{dc}	100
Fundamental motor current [Arms]	i_f	1.4
Rotor electric frequency [Hz]	f_r	3
Modulation factor [%]	M	10
High-frequency motor current [Arms]	i_h	0.28
Carrier frequency [Hz]	f_c	750 (ex.1) 4000 (ex.2)
Lower injected frequency [Hz]	f_{low}	325 (ex.1) 450 (ex.2)
Higher injected frequency [Hz]	f_{high}	425 (ex.1) 550 (ex.2)
Probability of the transition from f_{low} to f_{high} [%]	P_{lh}	23 (ex.1) 10 (ex.2)
Probability of the transition from f_{high} to f_{low} [%]	P_{hl}	23 (ex.1) 10 (ex.2)

978-1-4799-2706-7/14 $31.00 © 2014 IEEE

magnetic noise caused by injected high-frequency voltage without reducing the accuracy of rotor position estimation. The proposed method has been verified by numerical simulations and experiments using a 0.8kW-PMSM. The proposed method is also expected to increase the efficiency of PMSMs.

REFERENCES

[1] K.Yasui, Y.Nakazawa, O.Yamazaki, and I.Yasuoka, "Development of Rotor Position Sensorless Control for PRM Applied to Railway Traction Drive", *IPEC-Niigata 2005*, Japan S50-4, pp. 1671-1675, (April 2005)

[2] Y.Shibano, and K.Kubota, "Pole Position Estimation Method of IPMSM at Low Speed without High Frequency Components Injection", *Applied Power Electronics Conference and Exposition, 2009. APEC 2009. Twenty-Fourth Annual IEEE* Page(s): 233 - 239 (2009)

[3] S.Kimu, YC. Kwon, SK.Sul, J. Park and SM. Kim, "Position sensorless operation of IPMSM with near PWM switching frequency signal injection", *Power Electronics and ECCE Asia (ICPE & ECCE), 2011 IEEE 8th International Conference on* Page(s): 1660 - 1665 (2011)

[4] Y.Nakazawa, "A Position Sensorless Control for Permanent Magnet Reluctance Motor", *The Papers of Technical Meeting IEE Japan, VT-02-12*, pp.67-72 (2002-3)(in Japanese)

[5] C.B.Jacobina, A.M.N.Lima, E.R.C.Silva, and A.M.Trzynadlowski, "Current control for induction motor drives using random PWM", *IEEE Trans. on Inductrial Electnics*, Vol.45 No.5 (1998.10)

[6] K.S.Kim, Y.G.Jung, and Y.C.Lim, "Shaping the Spectra of the Acoustic Noise Emitted by Three-Phase Inverter Drives based on the New Hybrid Random PWM Technique", *in Conf. Rec. of IEEE Power Electronics Specialist Conference*, PESC'06, 2006, pp.545-550 (2006).

[7] S.Tanicuchi, Y.Kamijo, K.Yasui, M.Matsushita, K.Yuki, S.Noda, "New PWM Carrier Distribution Technique for Reducing Motor Electromagnetic Noise", *The Papers of Technical Meeting IEE Japan*, Vol.MD-13, No.26-31,33-35,37-41, pp.55-60 (2013-7.11),(in Japanese)

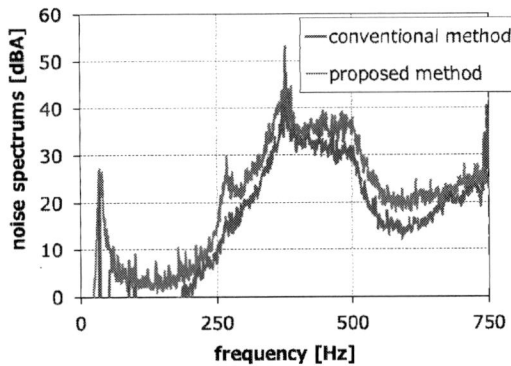

Fig. 13. Noise measurement results of conventional method and proposed method (Ex.1).

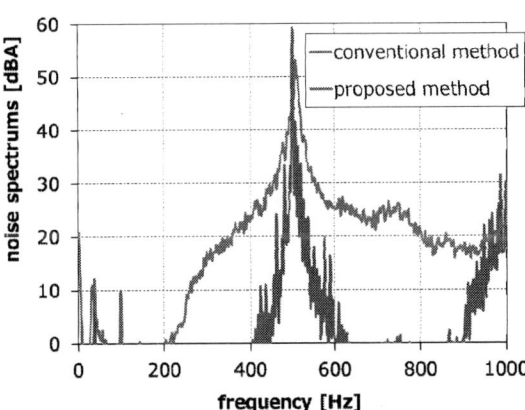

Fig. 14. Noise measurement results of conventional method and proposed method (Ex.2).

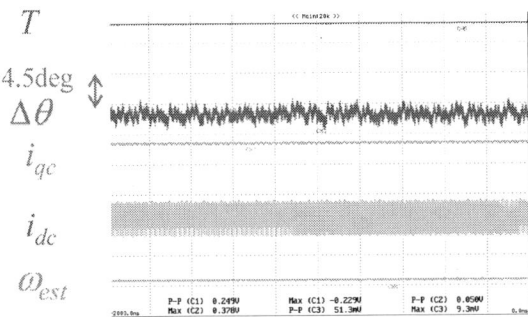

Fig. 10. Simulation result of conventional method.

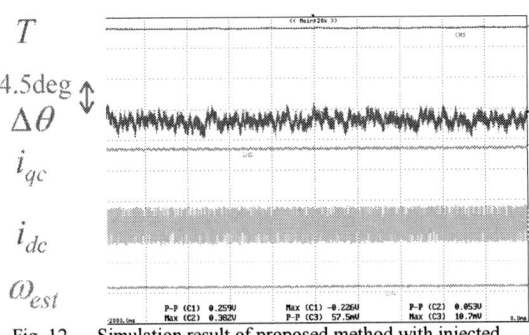

Fig. 11. Simulation result of proposed method without injected amplitude controller.

Fig. 12. Simulation result of proposed method with injected amplitude controller.

Force sensorless bilateral control using a dynamical asymmetric compensator

Ryota Hama, Jun Imai, Akiko Takahashi, and Shigeyuki Funabiki

Department of Electrical and Electronic Engineering
Okayama University
Okayama, JAPAN
Email: hama.ryota@s.okayama-u.ac.jp

Abstract— Among other bilateral control systems, those of symmetric type have advantage of simple structure, force sensorless and readiness to stabilize. However, a symmetric type bilateral system is known to have poor haptic performance. In this paper, we propose a bilateral system design without force sensor, based on a generalized system structure of master slave systems. Force feedback and position tracking performance are taken into account by H2 performance at frequency range of the desired haptic transfer, and influence of the measurement noise is suppressed by H-infinity condition. Such a multi-objective controller is designed by solving a semidefinite programming problem. The derived controller tends to have large condition number; we truncate balanced realization of the controller to obtain feasible feedback systems. Effectiveness of the proposed approach is illustrated by numerical studies in time as well as frequency domains, and also experiments.

Keywords— Bilateral control, Force sensorless, asymmetrric compensator, semidefinite programming problem

I. INTRODUCTION

To operate a remote robot in the location where people cannot approach, a master slave system is employed in various fields. Usually, an operator gives commands via the master to let the slave work in right way. But then, just by looking at slave's motion, it is hard to operate properly as desired. If the operating environment is dangerous or fragile, the operator is also required to adjust his/her force. Various bilateral control methods for master slave systems have been proposed thus far(e.g. [1] [2]).

The bilateral master slave system shows the operator the external force which the slave received in working. A drawback of the bilateral control is that the control system structure tends to be complex and therefore it may causes instability of the system. And more, it requires force sensors which are expensive and it can be dangerous in case of some failure. For symmetric methods, such problems do not occur while it is difficult to give accurate haptic to operator.

Recently, bilateral control methods using force observer, instead of force sensor, have been proposed. Especially in [3], a model based force observer is presented. A drawback of the method is that not only a controller but also an observer are necessary to design;

further, how to determine observer gain is not readily clarified. On the other hand, a switching controller has been proposed which is based on cyclic algorithm [4]. However, design method for the control gain is complicated and unknown environment is assumed to be modeled by a mass. Further, little considerations have been given to the frequency band regarding force sensing.

In this paper, we propose a force-sensorless bilateral system that has better haptic performance than the conventional symmetric one. The controller design problem is reduced to a linear matrix inequality problem. In experimental study, we employ two direct drive motors of the same type as a 1-DOF master slave system as shown in Fig. 1. To avoid complexity of the control system structure, we employ the one without force observers here. Instead of using force sensor, the dynamic output feedback controller estimates the external force as disturbances.

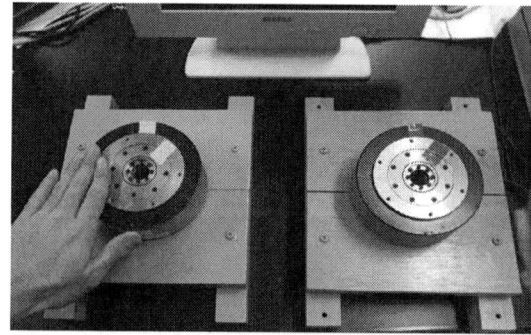

Fig. 1. 1-DOF master slave system

II. STRUCTURE OF THE BILATERAL MASTER SLAVE SYSTEM

The 1-DOF master slave system using two DD motors of the same type is modeled by (1), (2):

$$J\ddot{\theta}_m(t) + D\dot{\theta}_m(t) = F(t) - \tau_m(t) \qquad (1)$$

$$J\ddot{\theta}_s(t) + D\dot{\theta}_s(t) = -F_g(t) + \tau_s(t) \qquad (2)$$

where J is the moment of inertia, D the viscous friction, $\theta_m(t)$ and $\theta_s(t)$, the position of master and slave, respectively, $F(t)$ the human force, $F_g(t)$ the external force, τ_m and τ_s the control input torque of master and slave, respectively.

978-1-4799-2706-7/14 $31.00 © 2014 IEEE

The symmetric method is a basic bilateral control one. The block diagram is shown in Fig. 2. The positions of the master and slave are observed. Typically PD-type controller is adopted here. This method has relatively high stability. Designing such a controller is not so difficult. Moreover it is not necessary to implement extra force sensors. The symmetric controller provides torque depending on the position deviation. On the contrary, it can presents only inaccurate force based on position deviation. For example, an operator manipulate the master when the slave is free, no torque on the master side is required to be generated. This wasted power is preventing the exact force presentation.

In this paper, we propose an asymmetric bilateral control based on the structure of the symmetric ones. The block diagram is shown in Fig. 3. In this method, master and slave are modeled as a simple state-space model. The controller observes the position and velocity of the master and slave. If the system is modeled faithfully, disturbance applied to the master can be regarded as human force, and disturbance applied to the slave can be regarded as an external force. When this problem setting is incorporated in semidefinite programing problem, the designed controller performs bilateral control while estimating information of each force indirectly. This method has a higher degree of freedom than symmetric method. When the slave is free, the controller operates as a unilateral controller. And when the slave receives an external force, the controller presents the estimated force to the master and makes the slave follow the movement of the master.

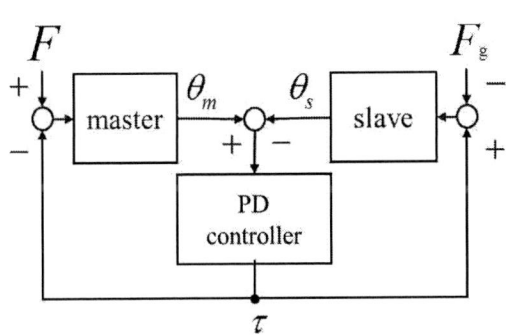

Fig. 2. Symmetric method

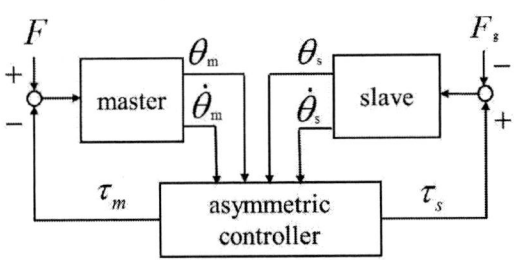

Fig. 3. Proposed method

III. CONTROLLER DESIGN OF THE PROPOSED METHOD

A. Generalized plant

In order to achieve good performance, errors of position and force must be decrease rapidly. However, observed signals are contaminated by noise, it is necessary to taking the noise effect into the controller design. Therefore, we consider the evaluation outputs of multi-objective control satisfying such control requirements. Here we define

$$z_2 = \begin{bmatrix} \theta_e & F_e \end{bmatrix} = \begin{bmatrix} W_\theta (\theta_m - \theta_s) & W_F (F_g - \tau_m) \end{bmatrix}^T \quad (3)$$

$$z_\infty = W_n \begin{bmatrix} \tau_m & \tau_s \end{bmatrix}^T \quad (4)$$

where W_θ, W_F and W_n are frequency weight. W_θ and W_F are band-pass filter specifying frequency range of the haptic. W_n is a high-pass filter in order to reduce effect of high frequency noise. Other variables are defined as follows:

$$x = \begin{bmatrix} \theta_m & \dot{\theta}_m & \theta_s & \dot{\theta}_s & x_\theta & \theta_e & x_F & F_e & x_{u1} & x_{u2} \end{bmatrix}^T \quad (5)$$

$$w_2 = \begin{bmatrix} F & F_g \end{bmatrix}^T \quad (6)$$

$$w_\infty = \begin{bmatrix} F & F_g & n \end{bmatrix} \quad (7)$$

$$y = \begin{bmatrix} \theta_m & \dot{\theta}_m & \theta_s & \dot{\theta}_s \end{bmatrix}^T \quad (8)$$

where $x_\theta, x_F, x_{u1}, x_{u2}$ are state variable contained in the frequency weight W_θ, W_F and W_n, n is observation noise. Then generalized plant of the master slave system is as follows:

$$P : \begin{cases} \dot{x} = Ax + B_{w_\infty} w_\infty + B_{w_2} w_{w_2} + Bu \\ z_2 = C_{z_\infty} x + D_{z_\infty w_\infty} w_\infty + D_{z_\infty u} u \\ z_\infty = C_{z_2} x + D_{z_2 w_2} w_2 + D_{z_2 u} u \\ y = Cx + D_{yw_\infty} w_\infty \end{cases} \quad (9)$$

The block diagram of the generalized plant is shown in Fig. 4.

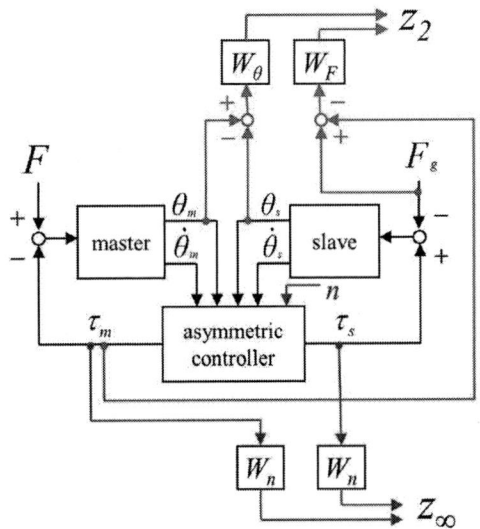

Fig. 4. Generalized plant

B. Semidefinite programing problem

For the generalized plant (9), we consider the following dynamic output feedback controller:

$$K : \begin{cases} \dot{x}_K(t) = A_K x_K(t) + B_K y(t) \\ u(t) = C_K x_K(t) + D_K y(t) \end{cases} \quad (10)$$

where x_K is the state variable of the controller, A_K, B_K, C_K and D_K are matrices of proper size. These controller parameters are to be determined via linear matrix inequalities. In case of H_2 and/or H_∞ controller design, optimal solution can be practically obtained due to high speed computation of suboptimal problems [5]. Generally, multi-objective design problem is non-convex, and therefore we derive suboptimum solutions based on common lyapunov solution. A semidefinite programming problem for the controller of the proposed method is described as follows:

$$\inf \left\| T(P,K)_{z_2, w_2} \right\|_2 \text{ subject to } \left\| T(P,K)_{z_\infty, w_\infty} \right\|_\infty < 1 \quad (11)$$

where $\| T(P,K)_{zn,wn} \| (n=2, \infty)$ is a H_n norm from w_n to z_n. Solving this SDP, robustness for noise is sustained while tracking and force sensing performance for specified frequency range can be optimized.

C. Order reduction of the controller

The order of the controller designed above, is 9 which corresponds to the order of the generalized plant. The complexity of the controller might cause a numerically bad condition, and the whole system tends to be unstable [6]. In order to avoid the problem, we reduce the controller order via truncated balanced realization.

Using a nonsingular matrix T, a new state variable of the controller z_K is defined as follows:

$$z_K := T^{-1} x_K \quad (12)$$

Then the dynamic output feedback controller (10) is transformed as follows:

$$\hat{K} : \begin{cases} \dot{z}_K(t) = T^{-1} A_K T z_K(t) + T^{-1} B_K y(t) \\ u(t) = C_K T z_K(t) + D_K y(t) \end{cases} \quad (13)$$

In this state space representation, the observability grammian and the controllability grammian are diagonal and equivalent. A singular value of grammian means the influence of the control system. It is possible to reduce the controller order while minimizing the deterioration of control performance which is designed in (11) by truncating the state variable which has a mall singular value of grammian. By truncating the variables that do not affect the control performance, the condition number has been drastically improved.

IV. SIMULATION

A. Set up

Effectiveness of the proposed method is illustrated with Matlab/SIMULINK. The sampling time is set to 0.1 ms. Frequency band of W_θ and W_F are from 0.01 through 50 Hz and of W_n is more than 1000 Hz. As a result of these choices, attained H_2 norm is 6.0. Then the condition number of the controller is more than 9.5×10^{12}. This is

improved to 3.0×10^3 by reducing to a 6 order controller. Performance comparison between the original controller K and the reduced controller K_r is shown in fig. 5.

The motor parameter is $J=0.0025$, $D=0.0013$. The slave motor is bound by the spring of $K=1$. The human force strength is set to pulse wave of maximum $F=0.2$. For comparison, with the symmetric method is simulated under the same condition. The proportion gain k_p and the derivative gain k_d are chosen as $k_p=10$ and $k_d=0.1$.

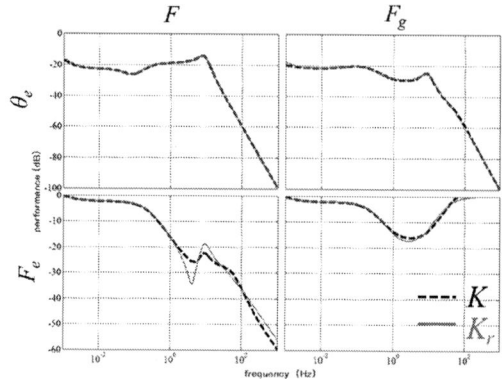

(a) H_2 performance

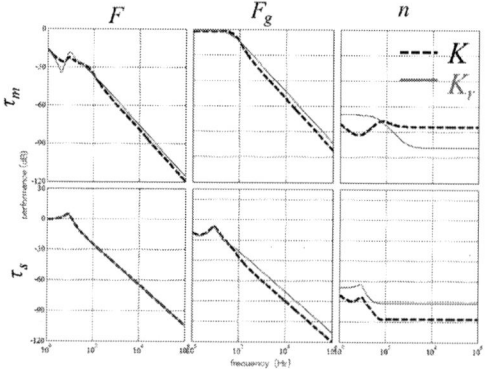

(b) H_∞ performance

fig. 5. Performance differences

B. Results and discussion

This simulation result is shown in Fig. 6 and 7. Position tracking performance is satisfactory for both methods. Regarding force deviation, the proposed method is better than the symmetric one. However, it is not possible to determine a "good force" only in this result. What is important is that the force represents the feeling of the spring. Next these waveforms are compared to the reaction force us applied to the single motor bound by the same spring. This result is shown in Fig. 8. It is observed that the proposed method is able to present the feel of spring more accurately. The frequency domain result indicates that the proposed method presents a wide frequency range haptic. It means that this method is difficult to present the haptic of flexible object.

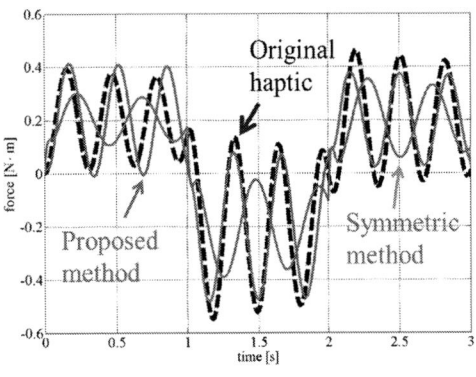

(b) Proposed method

fig. 7. Haptic performance

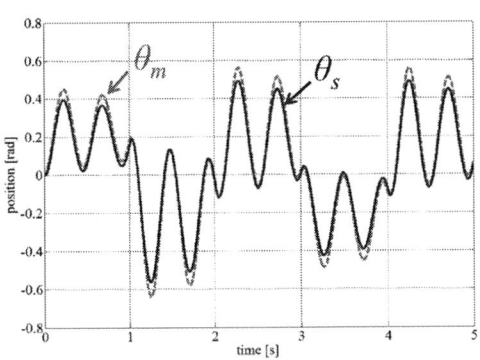

(a) Symmetric method

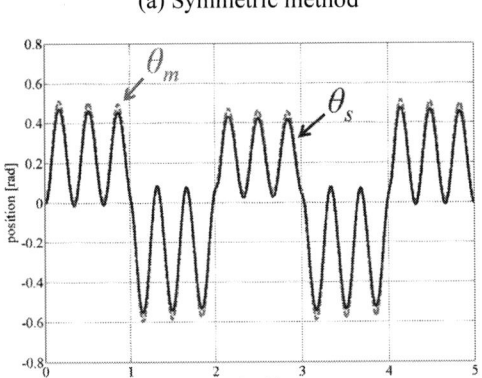

(b) Proposed method

fig. 6. Position tracking performance

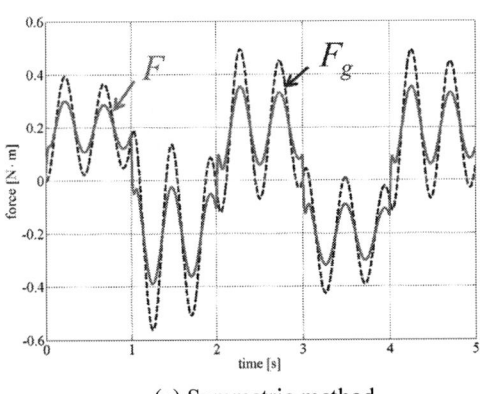

(a) Symmetric method

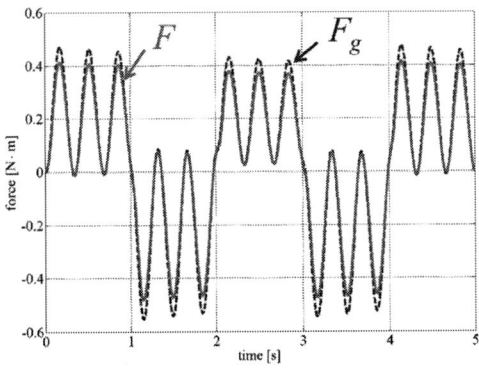

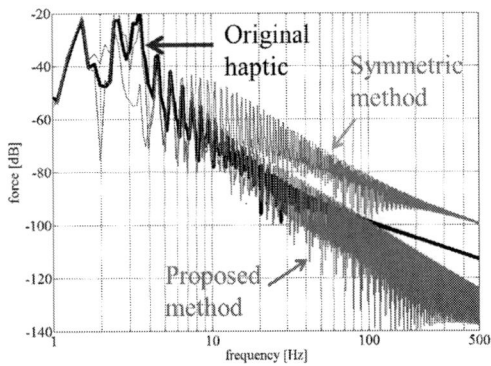

(a) time domain

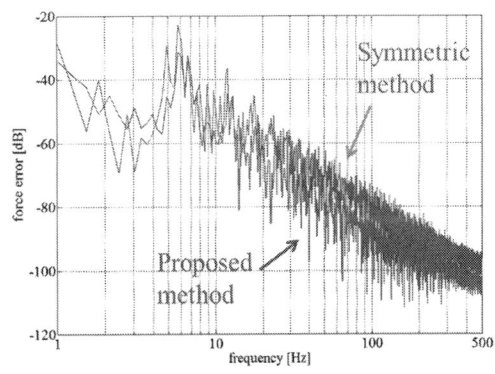

(b) frequency domain

(c) force error

fig. 8. Comparison with Original haptic

V. EXPERIMENT

A. Set up

In this section, we illustrate the effectiveness of the proposed method in the experiment using Matlab/SIMULINK and Real-Time Workshop. As DD motors of the master and the slave, SGMCS-02B made by YASKAWA Electric Corporation is adopted. These equipments mount a encoder which can measure the angle. The angular velocity is derived by the time difference values of the angle. All parameters and problem settings are the same as the simulation.

978-1-4799-2706-7/14 $31.00 © 2014 IEEE

B. Results and discussion

This experimental result is shown in Fig. 9 and 10. Position tracking performance is satisfactory for both the symmetric and the proposed methods as well as simulation. In the deviation between the haptic force and the external force, the proposed method is slightly better than the symmetric method. Comparison with original haptic is shown in Fig. 11. Force error is improved around the high bandwidth in the proposed method, but it is unfavorably compared to the simulation result. From the difference in the behavior of the motor in the simulations and experiments, this reason is considered to be a modeling error of the master and slave system.

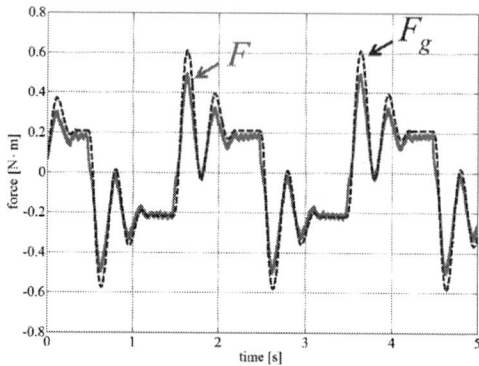

(b) Proposed method

fig. 10. Haptic performance

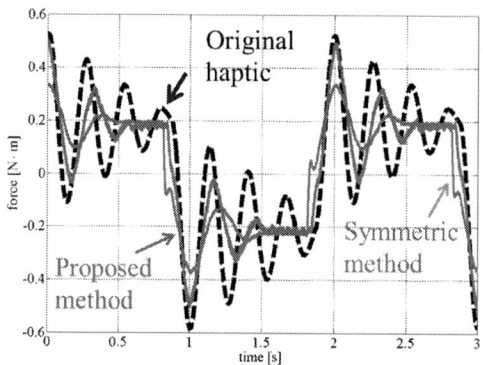

(a) time domain

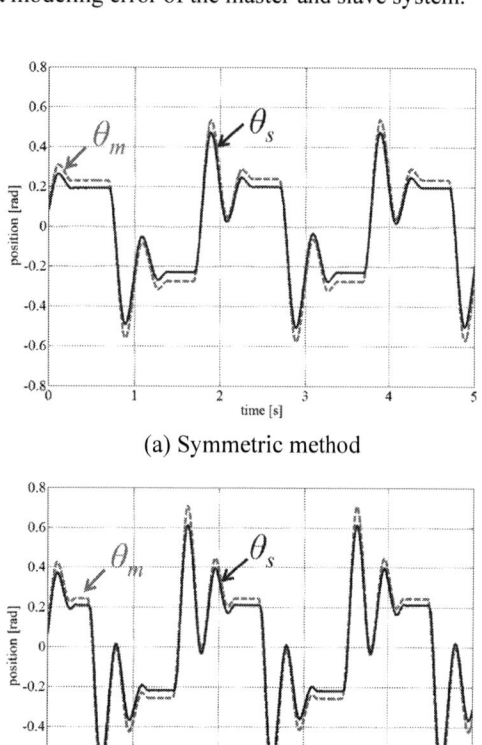

(a) Symmetric method

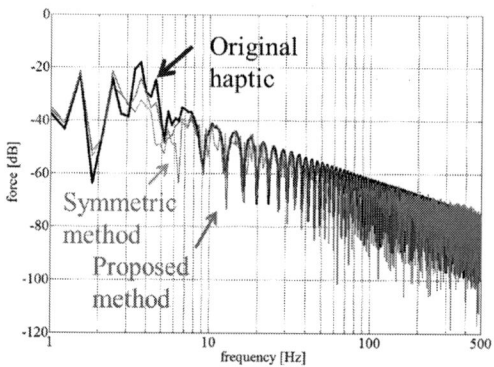

(b) frequency domain

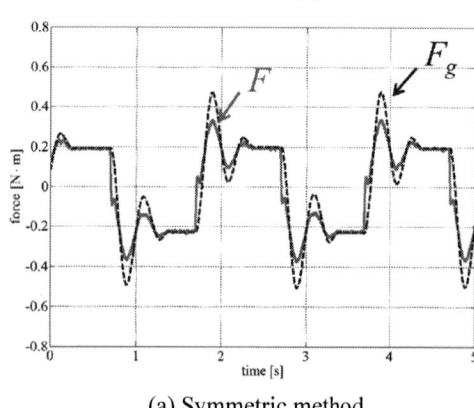

(b) Proposed method

fig. 9. Position tracking performance

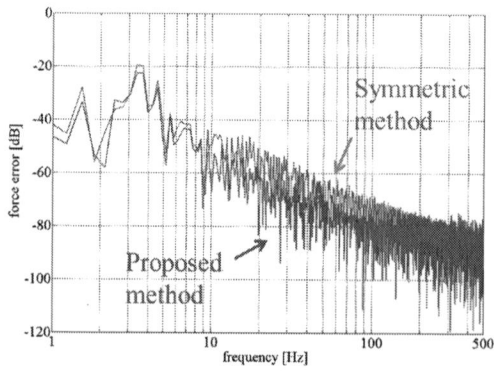

(c) force error

fig. 11. Comparison with Original haptic

(a) Symmetric method

VI. CONCLUSIONS

In this paper, we proposed a force sensorless bilateral control design with a dynamical asymmetric compensator. This method does not need to design separate controller and force observer, and the position tracking performance and the haptic performance are obtained for a required frequency band. Furthermore, it has flexibility to generate the torque of the master and the slave separately unlike a symmetric method. It was demonstrated that better haptic performance is obtained than the controller from the symmetric one. The drawback of the symmetric methods that the frequency band of the haptic is narrow was also revealed by simulation result. In order to further improve control performance, it is necessary to reduce the conservatism in semidefinite program and truncation of the balanced realized controller for implementation. In addition, it is necessary to consider the robustness under the modeling error. In future work, Performance limitation of the proposed controller shall be discussed analytically and experimentally.

REFERENCES

[1] J. Cui, S. Tosunoglu, C. Moore and W. Repperger, "A Review of Teleoperation System Control," *Florida Conference on Recent Advances in Robotics*, Florida Atlantic University, Boca Raton, Florida, May 8-9, 2003.

[2] N. Kobayashi, R. Masuda and K. Komoriya, "Practical Control of Robot Manipulators," *CORONA publishing co*, pp. 228-241, 1999. (in Japanese)

[3] Stefan Lichiardopol, Nathan van de Wouw, and Henk Nijmeijer, "Bilateral Teleoperation for Linear Force Sensorless 3D Robots,"Informatics in Control, Automation and Robotics, LNEE 89, pp.197-210,2011

[4] A. Alcocer, A. Robertsson, A. Valera and R. Johansson,"Force estimation and control in robot manipulators," Proceedings the 7th Symposium on Robot Control (SYROCO 2003), pp31-36, 2003

[5] C. Scherer, P. Gahinet and M. Chilali, "Multiobjective Output-Feedback Control via LMI Optimization," *IEEE Trans. Automatic Control*, Vol. 41, No. 7, pp. 896-911, 1997.

[6] G. Ohinata and B. Anderson, "Control System Design," Asakura publishing co, pp. 8-19, 1999. (in Japanese)

Design of m-IPD Controller of Multi-Inertia System using Differential Evolution

Hidehiro Ikeda
Department of Electrical Electronics Engineering
Nishi-Nippon Institute of Technology
Kanda-machi, Fukuoka, Japan
E-mail: ikeda@nishitech.ac.jp

Hanamoto Tsuyoshi
Graduate School of Life Science and Systems Engineering
Kyushu Institute of Technology
Kitakyushu, Japan
Email: hanamoto@nishitech.ac.jp

Abstract— **In this paper, a new design method of vibration suppression controller for multi-mass (especially, 2-mass) resonance systems is proposed. The controller consists of a digital modified-IPD controller for speed loop and a digital PI controller for current minor loop. The six controller gains are determined by Differential Evolution algorithm. The Differential Evolution is one of optimization techniques and a kind of evolutionary computation technique. In this paper, we have applied the *DE/rand/*1/*bin* strategy to design swift the optimal controller parameters. Furthermore, we consider the parameter identification method of 2-inertia experiment system by the Differential Evolution. Finally, the effectiveness of the proposal methods has been confirmed by the computer simulations and the experiments.**

Keywords— *2-inerita system, vibration suppression, Differential Evolution, Parameter Identification*

I.   INTRODUCTION

High precision and fast response motor drive systems are widely used in many industrial applications (e.g. Steel Rolling Mill, Blue-ray Disc Drive, Hard Disk Drive, Robot Manipulator, Electrical Vehicle and etc.). Advances of the control theory and the actuator technology have made it possible to widen the bandwidth of the control system for faster responses.

On the other hand, modern mechanical systems tend to lack stiffness due to miniaturization and weight reduction because constructions of those systems have become complicated.   Therefore, the motor drive systems generally are multi-mass systems with several inertia moments, gears and springs. It can be analyzed by an approximate 2-inertia system. More effective control methods to suppress vibrations of the 2-inertia resonance systems have been proposed: e.g. resonance ratio control, full state feedback control, H $_\infty$ control method, coefficient diagram method, Pole Placement Method and Fractional Order PID$_k$ Control [1]-[3]. Ikeda, et al. have proposed a fuzzy control the 2-inertia systems with the scaling factors determined by Genetic Algorithm [4]. Furthermore, some researchers have proposed the vibration suppression control methods for more then 3-inertia system [5]-[7].

In this paper, we propose two new design methods for the 2-inertia resonance system using Differential Evolution Algorithm (DE) [8]-[11].

DE algorithm is one of the evolutionary algorithms. Unlike simple GA that uses binary coding for representing problem parameters, DE uses real-valued vectors. The crucial idea behind DE is a scheme for generating trial parameter vectors. There are several variants of DE design. In this paper, we utilize the DE combination *DE/rand/*1/*bin* strategy.

First, we consider the parameter identification method of the 2-inertia experiment system by the DE. In general, the experimental system of the 2-inertia system is consists of a power supply, a motor, a power electronics circuit (such as a dc chopper and an inverter) and controller. Especially, the motor contains uncertainties and disturbances such as the parameter variation to the nominal value and the nonlinear friction torque like a Coulomb torque. Here, we identify the experimental parameters using the DE, which are the armature resistance R_a, the armature inductance L_a, the inertia of the motor J_M, the inertia of the load J_L, the stiffness of the shaft K_s, the back EMF constant K_e and the coulomb torque T_f.

Next, a new design method of vibration suppression controller is proposed. The controller consists of modified-I-PD controller for the speed loop and a PI controller for the current minor loop. The six controller gains are determined by DE algorithm. Comparing with the conventional design algorithm such as the GA or the PSO, the proposed technique is able to shorten the time of the controller design to a large extent and to obtain accurate results.

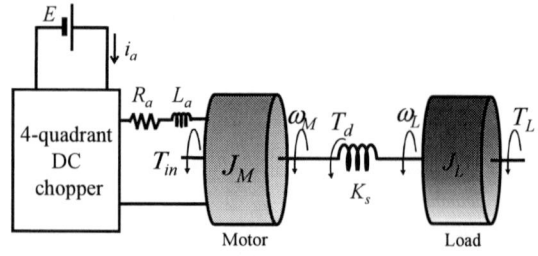

Fig. 1. 2-inertia model with the current characteristic.

978-1-4799-2706-7/14 $31.00 © 2014 IEEE

Finally, we confirmed the effectiveness of the proposal method by the computer simulation and experiment by an experimental setup.

II. VIBRATION SUPPRESION CONTROLLER

A. 2-Inertia Model

The 2-inertia model, which consists of two rigid inertias with a torsional shafts, is shown in Fig. 1, where ω_M is the angular speed of the motor, ω_L is the angular speed of load, T_{in} is the input torque, T_L is the disturbance torque, J_M is the motor inertia, J_L is the inertia of load, T_{dis} is the torsional torque of shaft and K_s is the stiffness of shaft. In this research, we consider the current loop for the high speed torque control. Here, R_a is the armature resistance, L_a is the armature inductance and E is the DC voltage of the DC power supply. And we have neglected the viscous frictions.

The continuous state equation of the 2-inertia model is given by equation (1).

$$\frac{d}{dt}\begin{bmatrix} i_a \\ \omega_M \\ \omega_L \\ T_{dis} \end{bmatrix} = \begin{bmatrix} -R_a/L_a & -K_e/L_a & 0 & 0 \\ K_t/J_M & 0 & 0 & -1/J_M \\ 0 & 0 & 0 & 1/J_L \\ 0 & K_s & -K_s & 0 \end{bmatrix} \begin{bmatrix} i_a \\ \omega_M \\ \omega_L \\ T_{dis} \end{bmatrix} \quad (1)$$

$$+ \begin{bmatrix} 1/L_a \\ 0 \\ 0 \\ 0 \end{bmatrix} E + \begin{bmatrix} 0 \\ 0 \\ -1 \\ 0 \end{bmatrix} T_L$$

Thus, the block diagram of the 2-inertia model is given in Fig. 2. From the equation of the 2-inertia model, the transfer function of the input torque T_{in} to the motor angular velocity ω_M is described as follows:

$$\frac{\omega_M}{T_{in}} = \frac{s^2 + \omega_a^2}{J_M s \left(s^2 + \omega_r^2\right)} \quad (2)$$

The resonance frequency ω_r and the anti-resonance frequency ω_a are calculated as

$$\omega_r = \sqrt{\frac{K_s}{J_M} + \frac{K_s}{J_L}} \quad (3)$$

$$\omega_a = \sqrt{\frac{K_s}{J_L}} \quad (4)$$

Here, the nominal parameters of the 2-inertia experimental model in this paper are listed in TABLE I. Fig. 3 shows the frequency response of the nominal 2-inertia model. In this figure, a maximum peak of gain characteristic and a minimum peak of gain characteristic are observed.

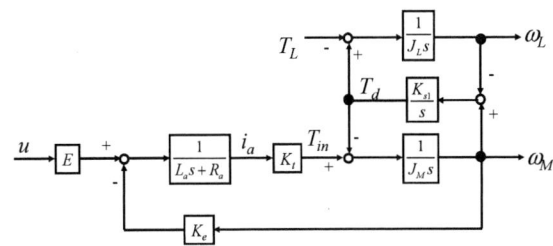

Fig. 2. Block diagram of the 2-inertia model.

TABLE I
NOMINAL PARAMETER OF THE 2-INETIA MODEL

Symbol	Specification
J_M	2.744×10^{-4} [kgm^2]
J_L	2.940×10^{-4} [kgm^2]
K_s	70.7 [Nm/rad]

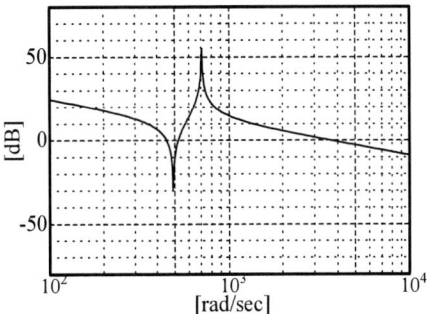

Fig. 3. Frequency response of 2-inertia model.

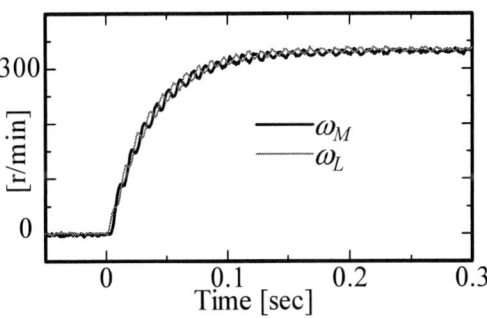

Fig. 4. Step response of dc voltage input.

There are the resonance frequency ω_r and the anti-resonance frequency ω_a using the nominal parameters, respectively. Then, ω_r is 705.8 [rad/sec] (112.3 [Hz]) and ω_a is 490.4 [rad/sec] (78.0 [Hz]) in this paper. The purpose of the vibration suppression control is to reduce the peak of ω_r.

In particular, Fig. 4 shows the experimental result with the angular speed step response of DC voltage input. From this figure, we can see the resonance vibrations. The resonance frequency is about 100[Hz] which is different from above the nominal value. Hence, it is necessary to identify the real value of the experimental system in order to design the vibration suppression controller exactly.

B. Digital m-IPD Controller

In this paper, we apply the classical PID controller to the vibration suppression of 2-inertia speed control system to simplify the controller structure. The proposed the speed controller and the current controller are described below, respectively:

$$i_r = \left(\frac{K_i}{s} \left(\omega_{ref} - \omega_M \right) - \left(K_p + K_d s \right) \omega_M \right) \frac{1}{T_d s + 1} \quad (5)$$

$$u = \left(K_{ap} + \frac{K_{ai}}{s} \right) \left(i_r - i_a \right) \quad (6)$$

Since the structure of the resonance system is complex, it is difficult to suppress the vibrations using only the simple PID controller. In this paper, we use the modified-I-PD speed controller which is consist 2-DOF control, considering response speed, and in addition, we use the first lag element in order to increase degree of freedom of the control design. Where, i_r is the control input of the current minor loop, u is the control input of the 4-quadrant DC chopper and ω_{ref} is the speed reference. Then, K_p, K_i, K_d and T_d are the modified-I-PD speed controller gains and K_{ap} and K_{ai} are the PI current controller gains.

Furthermore, considering the application to the apparatus, we construct the digital control system which has discrete controller. Fig. 5 shows the block diagram of the proposed digital control system.

III. DESIGN OF THE CONTROL SYSTEM BY DIFFERENTIAL EVOLUTION

In general, since determinations of the controller gains are a rule of thumb, it takes a long time to determine the controller gains by trial and error or some other methods. In this paper, we apply the Differential Evolution for the determination of six controller gains.

In addition, we utilize the Differential Evolution to identify the DC motor driving system of the experimental setup. The experimental system is generally contained the parameter error comparing with the nominal value. For this reason, we should estimate the real parameters to evaluate the proposed control system correctly.

A. Differential Evolution

Differential Evolution algorithm (DE) is one of the evolutionary algorithms. Unlike simple GA that uses binary coding for representing problem parameters, DE uses real-valued vectors. DE algorithm is one of the evolutionary algorithms. Unlike simple GA that uses binary coding for representing problem parameters, DE uses real-valued vectors. The crucial idea behind DE is a scheme for generating trial parameter vectors.

The design procedure using the DE consists of 4 steps (Initial Population, Mutation, Crossover and Selection). Fig. 6 shows the flow of DE algorithm.

There are several variants of DE design. In this paper, we utilize the DE combination *DE/rand/1/bin* strategy.

A set of D optimization parameters is called an individual. It is represented by D-dimensional parameter vector. A population consists of NP parameter vector $x_{i,G}$. Where, i=1, 2, ..., NP, NP denotes the number of members in one population and G indicates one generation. We have one population for each generation. Here, the initial population vector is determined randomly.

In this DE optimization, F is the scaling factor and CR is the crossover rate. The scaling factor F works for creating a mutation vector $v_{i,G}$. For each target vector $x_{i,G}$, the mutation vector v_i are generated according to

$$v_{i,G+1} = x_{r1,G} + F \left(x_{r2,G} - x_{r3,G} \right) \qquad r_1 \neq r_2 \neq r_3 \neq i \quad (7)$$

where, r_1, r_2 and r_3 are distinct. In the crossover, the target vector $x_{i,G}$ mixed with the mutation vector $v_{i,G+1}$, using following scheme for j = 1, 2, ... D,

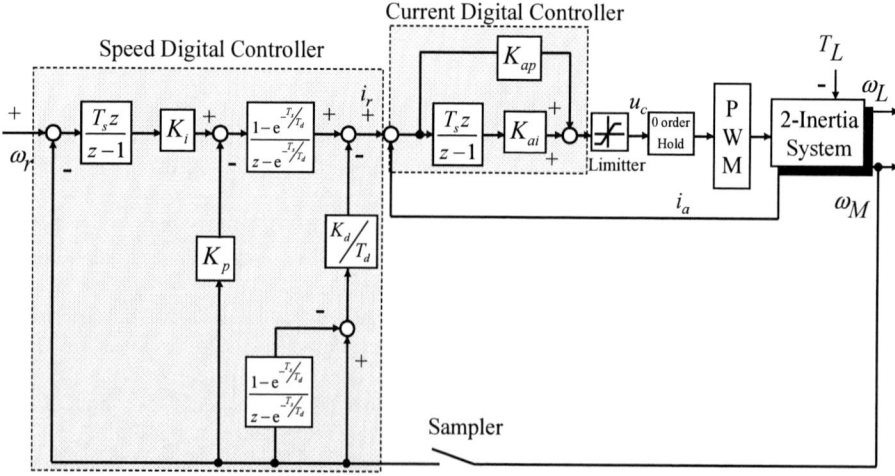

Fig. 5. Proposed control system.

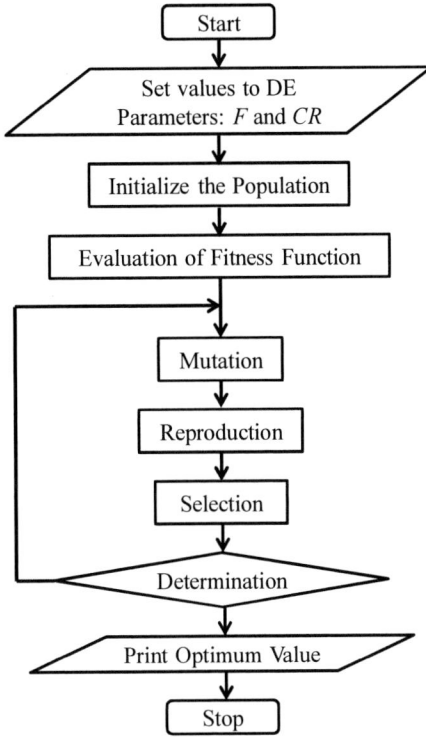

Fig. 6. Flow of DE algorithm.

$$u_{ji,G+1} = \begin{cases} v_{ji,G+1}, & \text{if } rand() \le CR \text{ or } j=\text{start point} \\ x_{ji,G}, & \text{if } rand() > CR \text{ or } j \ne \text{start point} \end{cases} \quad (8)$$

$u_{i,G+1}$ is the trial vector. $rand()$ is the j-th evaluation of uniform random generator number. The start point (1, 2, ... , D) is a randomly chosen index which ensures that $u_{i,G+1}$ gets at least one element from $v_{i,G+1}$.

A selection algorithm is utilized

$$x_{i,G+1} = \begin{cases} u_{i,G+1}, & \text{if } y(u_{i,G+1}) > y(x_{i,G}) \\ & \text{for maximization problems} \\ x_{i,G}, & \text{otherwise} \end{cases} \quad (9)$$

for $j = 1, 2, ..., D$.

B. Off-line Prameter Identification by DE

In this paper, we propose the off-line parameter identification method for the seven parameters of experimental system utilizing DE. Here, the seven parameters to identify are the armature resistance R_a, the armature inductance L_a, the inertia of the motor J_M, the inertia of the load J_L, the stiffness of the shaft K_s, the back EMF constant K_e and the coulomb torque T_f.

First, the experiment of simple step response is carried out. The data is the load angular speed $\omega_{L, ex}$ and the armature current $i_{a, ex}$. Next, we calculate the identified simulation model $\omega_{L, sim}$ and $i_{a, sim}$ by DE. In this paper, we use the error area for the index function in order to identify. Fig. 7 shows the conceptual diagram of the

index function and equation (10) is the index function y in this paper. Where, t is the simulation time and w is the design parameter.

$$y = \frac{1}{\sum \sqrt{t \left(\left| \omega_{L,\text{ex}} - \omega_{L,\text{sim}} \right|^2 + w \left| i_{a,\text{ex}} - i_{a,\text{sim}} \right|^2 \right)}} \quad (10)$$

In the DE, the number of problem dimension is 7 (R_a, L_a, J_M, J_L, K_s, K_e and T_f), the population size is 2000, the scaling factor F is 0.5 and the crossover rate CR is 0.9.

Fig. 8 shows the waves of the identification results using the proposed method. From this figure, we can observe the same waves between the experiment and the identification. Therefore, it is obvious that the proposed identification method is worked very well. Fig. 9 and Fig. 10 show the transition of the maximum index value and the K_e vectors, respectively. And TABLE. II shows the identification parameters by DE.

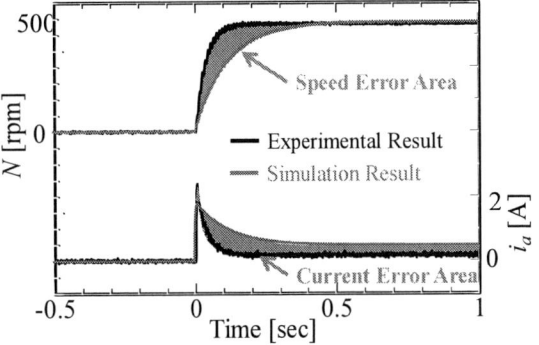

Fig. 7. Index function of parameter identification.

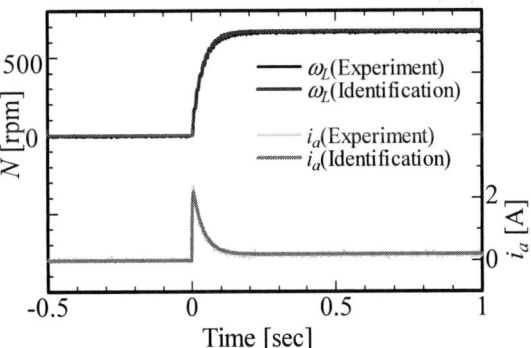

Fig. 8. Identification results by DE.

TABLE II
NOMINAL PARAMETER OF THE 2-INETIA MODEL

Symbol	Value
R_a	$2.7729 \times 10^{-4}[\text{kgm}^2]$
L_a	$4.3806[\Omega]$
J_M	$2.7729 \times 10^{-4}[\text{kgm}^2]$
J_L	$3.2042 \times 10^{-4}[\text{kgm}^2]$
K_s	$54.112[\text{Nm/rad}]$
K_e	$0.26365[\text{Vsec/rad}]$
T_f	$0.04459[\text{Nm}]$

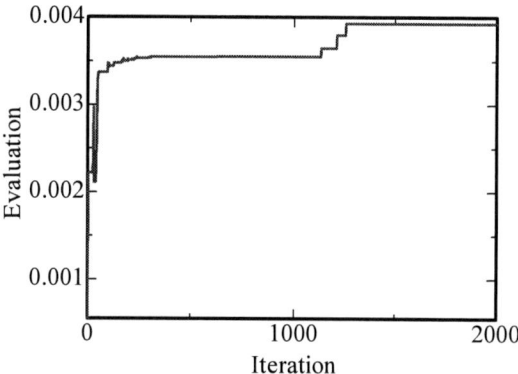

Fig. 9. Flow of DE algorithm.

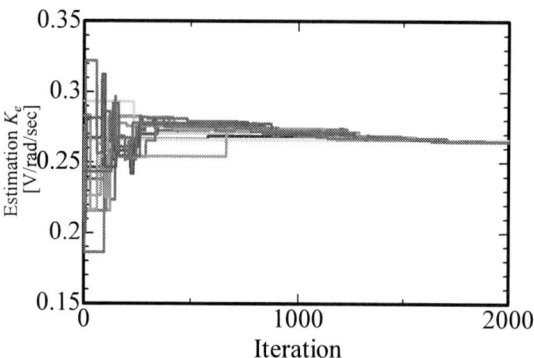

Fig. 10. Flow of DE algorithm.

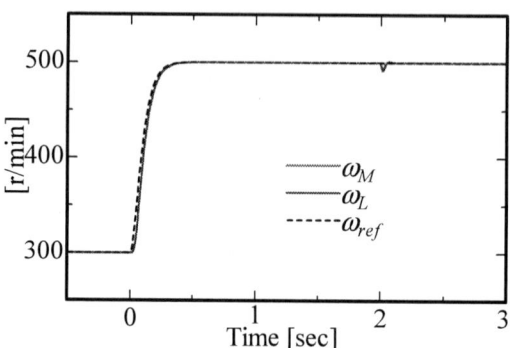

Fig. 11. Simulation Result of DE Design.

TABLE III
DESIGN CONTROLLER GAINS BY DE

Symbol	Value
K_p	4.3
K_i	300.0
K_d	7.53×10^{-4}
T_d	1.691×10^{-4}
K_{ap}	99.6
K_{ai}	1.004

C. Controller Design by DE

In general, since determinations of the controller gains are some complex design methods or a rule of thumb, it takes a long time to determine these gains by trial and error or some other methods. In this paper, we apply the DE for the determination of six gains in order to design quickly..

Therefore, the number of problem dimension is six (K_p, K_i, K_d, T_d, K_{ap} and K_{ai}), the population size is 200, the scaling factor F is 0.5 and the crossover rate CR is 0.9. Equation (11) is the index function, where ω_M is the motor angular speed and ω_{ref} is the angular speed reference command which is the shape of the step response of the second lag element described as equation (12), where ω_n is natural angular frequency.

$$y = \int_0^\infty t \left| \omega_{ref} - \omega_M(k) \right| dt \qquad (11)$$

$$\omega_{ref} = \omega_{step\,width} \left(1 - e^{-\omega_n t} \left(\omega_n t + 1 \right) \right) + \omega_{offset} \qquad (12)$$

Fig. 11 shows the simulation result using the proposed design method, where $\omega_{step\,width}$ is 200[r/min], ω_{offset} is 300 [r/min] and ω_n is 20 [rad/sec]. It can be seen that the proposed method works very well. Here, TABLE. III is the controller gains designed by the proposed method.

IV. EXPERIMENTAL RESULTS

The performance of the proposed vibration suppression controller has been tested in an experimental system. The system is shown in Fig. 12, where consists of a 300 [W] DC motor, a 300 [W] DC generator, a four quadrant DC chopper and a DSP board.

The DSP board (PE-PRO/F28335 STARTER KIT) which is made by Myway Plus Corporation, is consists the DSP TMS320F28335PGFA, Digital I/O, ABZ counters, AD converters and DA converters. The motor and load angles are detected by 5000 [ppr] optical rotary encoders. The armature current is detected by current sensors and is sent to 12 bit AD Converters.

The control period is 1 [msec] and the detected period of the current and the angular speeds is 10 [µsec]. Then, the design language is C. The proposed digital controller has been realized as the difference equation.

The disturbance torque is inputted to the load terminal by the constant current load of the electrical load equipment.

Fig. 13 shows the apparatus of the 2-Inetia model in this experimental setup.

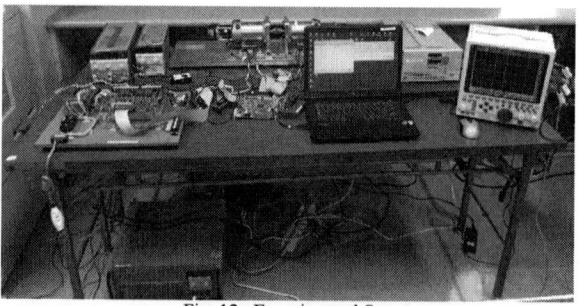

Fig. 12. Experimental Setup.

Fig. 13. 2-Inerita Model.

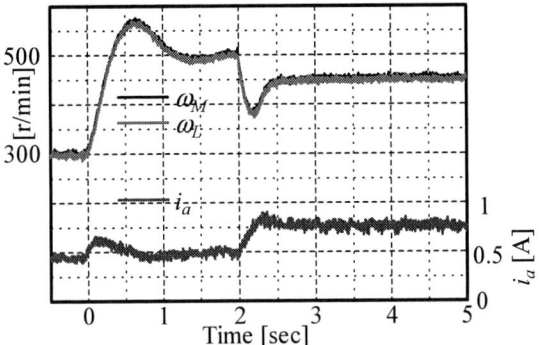

Fig. 14. Experimental Results using conventional PI Speed Controller.

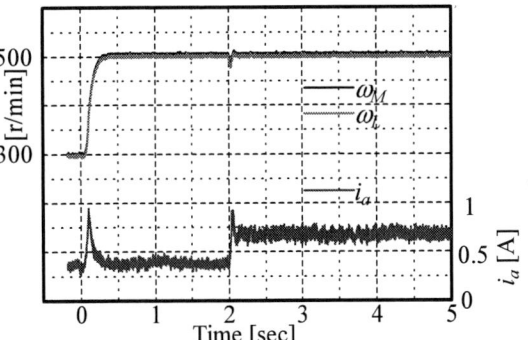

Fig. 16. Experimental Results using the proposed method (ω_n=20 [rad/sed]).

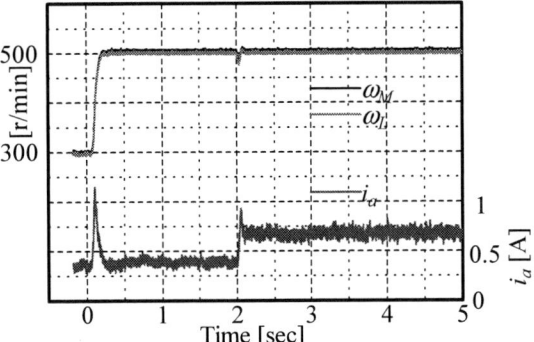

Fig. 17. Experimental Results using the proposed method (ω_n=30 [rad/sed]).

Fig. 14 shows the experimental results using the conventional PI speed controller. Here, the speed reference command input ω_{ref} is changed from 300 to 500 [rpm] at t = 0 [sec], and the disturbance input T_L is changed from 0 [Nm] to 10 [%] at t = 2 [sec]. From this figure, we can observe the overshoot at the speed step response and the steady state error at the disturbance input. Meanwhile, Fig. 15 shows the results using the proposed method as ω_{ref} (ω_n=10 [rad/sec]) using equation (10). It is observed that the speed errors reduced very well and the resonance vibration also suppressed very well. Then, Fig. 16 and Fig. 17 show the results for the variation of the response time, ω_n=20 [rad/sec] and ω_n=30 [rad/sec] respectively. As indicated in these figures, according to the reference shape, the response time has changed satisfactorily.

Moreover, we proposed the design method of six controller gains for 2-inertia system using the differential evolution in order to design in short time. And we concurrently proposed the off-line identification method of the experimental system by the differential evolution. Experimental results demonstrated the validity of the proposed methods.

V. CONCLUSIONS

In this paper, we proposed the vibration suppression controller for 2-inertia system. The controller is consists of the digital m-I-PD speed controller and the digital PI current controller.

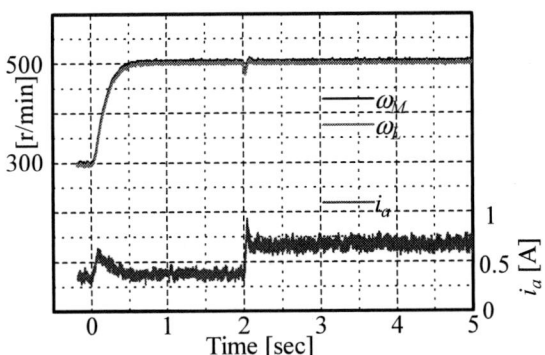

Fig. 15. Experimental Results using the proposed method (ω_n=10 [rad/sed]).

REFERENCES

[1] Y.Hori, H. Sawada and Y. Chun, "Slow resonance ratio control for vibration suppression and disturbance rejection in torsional system," *IEEE Trans. on Ind. Electron.*, vol. 46, no. 1, pp. 162-168, 1999.

[2] K. Szabat and T. O.-Kowalska, "Vibration Suppression in a Two-Mass Drives System Using PI Speed Controller and Additional Feedbacks – Comparative Study," *IEEE Trans. on Industrial Electronics*, vol. 54, no. 2, pp. 1193-1206, 2007.

[3] K. Erenturk, "Fractional-Order PI$^\lambda$D$^\mu$ and Active Disturbance Rejection Control of Nonlinear Two-Mass Drive System," *IEEE Trans. on Industrial Electronics*, vol. 60, no. 9, pp. 3806-3813, 2013.

[4] H. Ikeda, T. Hanamoto, Y. Tanaka and T. Tsuji, "Fuzzy Control of 2-Inertia System with Scaling Factors Determined by GA," *Trans. of IEE Japan*, vol. 121-D, no.9, pp. 996-997, 2001.

[5] G. Zhang and J. Furusho, "Control of Three-inertia System by PI/PID Control," *Trans of IEE Japan.*, vol. 119-D, no. 11, pp. 1386-1392, 1999.

[6] H. Ikeda, T. Hanamoto and T. Tsuji, "Design of Multi-Inertia Digital Speed Control System Using Taguchi Method," *Proc. of ICEM 2008*, Paper ID 1167, PB.3.9, pp. 1-6, 2008.

[7] H. Ikeda, S. Ueda and T. Hanamoto, "Design of Vibration Suppression Controller for High Order Resonance System by

Mutation-Type Grouping PSO," *Proc. of PEDS 2013*, Paper ID 9270, 6 pages, 2013.

[8] M. S. Saad, H. Jamaluddin and I. Z. M. Darus, "Implementation of PID controller tuning using differential evolution and genetic algorithm," *ICIC International*, vol. 8, no. 11, pp. 7761-7779, 2012.

[9] J. S. Yadav, N. P. Patidar and J. Singhai, "Controller Design of Discrete System by Order Reduction Technique Employing Differential Evolution Optimization Algorithm," *International Journal of IMS*, 6:1, pp. 43-49, 2010.

[10] S. Yamaguchi, "An Automatic Control Parameters Tuning Method for Differential Evolution," *Trans of IEE Japan*, Vol. 128-C, No.11, pp. 1696-1703, 2008.

[11] J. Brest, et al, "Self-Adapting Control Parameters in Differential Evolution: A Comparative Study on Numerical Benchmark Problems," *IEEE Trans. on Evolutionary Computation*, Vol. 10, No. 6, pp. 646-657, 2006.

A Guide to Design Disturbance Observer based Motion Control Systems

Emre SARIYILDIZ, Kouhei OHNISHI
Department of System Design Engineering
Keio University, Yokohama, Japan
emre@sum.sd.keio.ac.jp, ohnishi@sd.keio.ac.jp

Abstract- **This paper proposes new practical design tools for the robust motion control systems based on disturbance observer (DOB). Although DOB has long been used in several motion control applications, it has insufficient analysis and design tools. The paper proposes a new practical robustness constraint, which improve the robustness at high frequencies, on the bandwidth of a DOB and nominal inertia. Although increasing the bandwidth of a DOB and nominal inertia improves the performance and stability, they are limited by the robustness constraint. Besides, a novel stability analysis method is proposed for reaction force observer (RFOB) based robust force control systems. It is shown that not only the performance, but also the stability changes significantly by the imperfect identification of inertia and torque coefficient. The robustness and stability of a DOB based motion control system are improved by proposing new design tools. The validity of the proposals are verified by experimental results.**

Index Terms: Disturbance Observer, Motion Control Systems, Reaction Force Observer, Robustness and Stability

I. INTRODUCTION

Three decades before, DOB was proposed by Ohnishi et al. in the first IPEC conference [1]. After that, it has been used in several robust motion control applications such as robotics, industrial automation, automotive, and so on [2-4]. A DOB estimates external disturbances and system uncertainties, e.g., friction, inertia variation etc.; and the robustness of the motion control system is achieved by feeding back the estimated disturbances in an inner-loop [5]. To achieve control goals, e.g., position, force or admittance control goals, an outer-loop controller is designed by considering the nominal plant parameters, since a DOB can nominalize an uncertain plant [5].

A low-pass-filter (LPF), which satisfies the properness in the inner-loop, and the nominal inertia and torque coefficient of a motor are used in the design of a DOB. The dynamic characteristics of the DOB's LPF and the ratio between the uncertain and nominal plant parameters change the stability and robustness of a DOB based motion control system, significantly [6]. It is a well-known fact that the bandwidth of the DOB's LPF is desired to set as high as possible to estimate/compensate disturbances in a wide frequency range. However, it is limited by practical and robustness constraints [7]. Besides that the order of the DOB's LPF changes the robustness and performance significantly. As it is increased, the performance of the system improves, yet the robustness

deteriorates [7]. The stability and performance of a DOB based robust motion control system can be improved by using higher/lower nominal inertia/torque coefficient; however, its upper/lower bound has not been derived yet [8].

A RFOB, which was proposed by Murakami et al., is an application of a DOB and is used to estimate environmental impedance [9]. It has several superiorities over a force sensor such as sensorless force control, higher force control bandwidth, etc.; and they have been shown in the literature experimentally [10, 11]. A RFOB is designed easily by subtracting the external disturbances and system uncertainties from the input of a DOB. Therefore, a DOB and a RFOB are quite similar, however only the latter has a model based control structure. It is the main challenging issue in a RFOB based robust force control system, since imperfect system identification may change the stability and performance significantly. So far, only the performance of a RFOB based robust force control system has been considered in the literature due to the oversimplified analysis methods. However, not only the performance, but also the stability changes significantly by the design parameters of a DOB and a RFOB.

In this paper, new design constraints are proposed for DOB based robust position and force control systems. It is shown that the bandwidth of a DOB and nominal inertia have robustness constraints due to imperfect velocity measurement in practice. The proposed robustness constraint is more dominant than noise, so the bandwidth of a DOB should be determined by considering the proposed robustness constraint. The stability of a DOB based motion control system is improved by increasing/decreasing nominal inertia/torque coefficient, yet the robustness deteriorates. Therefore, there is a trade-off, which is adjusted by the ratio of the uncertain and nominal plant parameters, between the stability and robustness of a DOB based robust motion control system. A novel stability analysis method is proposed for RFOB based robust force control systems. It is shown that the stability of a RFOB based robust force control system deteriorates as the identified inertia that is used in the design of a RFOB is increased. Besides, it is shown that a DOB and a RFOB can be designed as a phase lead-lag compensator in the robust force control systems if their bandwidths are set to the different values.

The rest of the paper is organized as follows. In section II, a DOB and a RFOB are presented briefly. In sections III and IV,

The 2014 International Power Electronics Conference

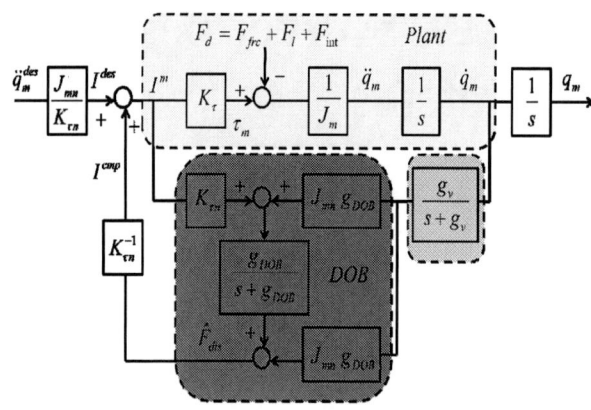

(a) Disturbance Observer

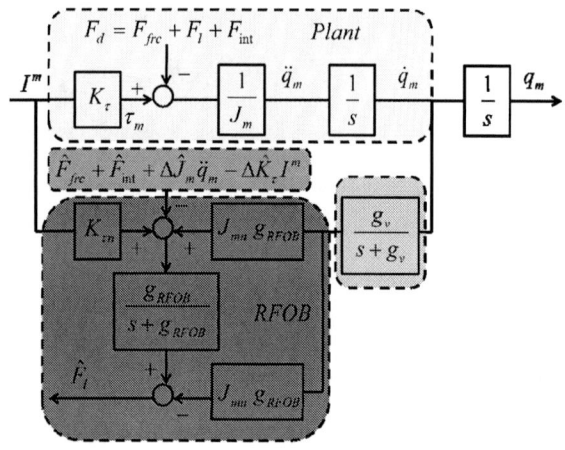

(b) Reaction Force Observer

Fig. 1 Block diagrams of a DOB and a RFOB

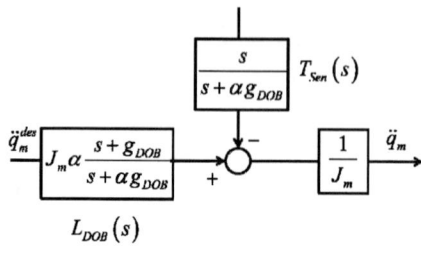

(a) Ideal velocity measurement

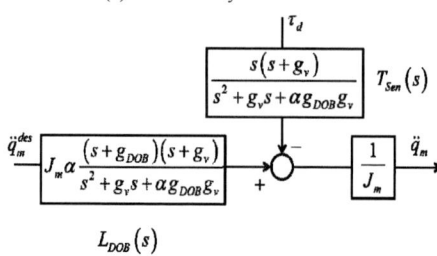

(b) Imperfect velocity measurement

Fig. 2 Simplified block diagrams of a DOB

ΔK_τ and $\Delta \hat{K}_\tau$	Torque coefficient variation and its estimation;
F_l	Loading force;
F_{frc}	Nonlinear friction force;
F_{int}	Interactive force;
$F_d = F_{int} + F_l + F_{frc}$	Total external disturbance;
$F_{dis}, \hat{F}_{dis}$	Total disturbance and its estimation;

As shown in Fig. 1, although a DOB and a RFOB are quite similar, only the latter has a model based control structure. Since, not only the nominal, but also the uncertain plant parameters are required in the design of a RFOB.

Without any approximation, Fig. 1a can be simplified as shown in Fig. 2, in which $\alpha = \dfrac{J_{mn} K_\tau}{J_m K_{\tau n}}$. Fig. 2a is drawn when ideal velocity measurement is achieved, i.e., g_v is infinite; however, Fig. 2b is drawn when g_v is finite. It is a well-known fact that a DOB requires precise velocity measurement [12]. Therefore, in practice, g_v should be finite to suppress noise and obtain precise velocity measurement in a determined bandwidth.

The simplified block diagrams given in Fig. 2 show the open loop $\left(L_{DOB}(s)\right)$ and sensitivity $\left(T_{Sen}(s)\right)$ transfer functions of a DOB based motion control system. The open loop transfer functions show that a DOB can be designed as a phase lead-lag compensator that is adjusted by α. The stability and performance can be improved by using a DOB as a phase lead compensator, i.e., $\alpha > 1$. The dynamic characteristics of the sensitivity function changes significantly at high frequencies when g_v is finite. Therefore, although it has never been considered so far, the robustness of a DOB changes significantly when a low-pass-filter is used in velocity

new practical design constraints are proposed for DOB and RFOB based position and force control systems, respectively. In section V, experimental results are given. The paper ends with conclusion given in the last section.

II. DISTURBANCE AND REACTION FORCE OBSERVERS

Block diagrams of a DOB and a RFOB are shown in Fig. 1. Hereinafter, force and torque are used interchangeably. In this figure

J_m, J_{mn}	Uncertain and nominal inertias;
$K_\tau, K_{\tau n}$	Uncertain and nominal torque coefficients;
I^m, I^{des}, I^{cmp}	Total, desired and compensate motor currents;
$q_m, \dot{q}_m, \ddot{q}_m$	Angle/position, velocity and acceleration;
g_{DOB}	Cut-off frequency of a DOB;
g_{RFOB}	Cut-off frequency of a RFOB;
g_v	Cut-off frequency of velocity measurement;
ΔJ_m and $\Delta \hat{J}_m$	Inertia variation and its estimation;

978-1-4799-2706-7/14 $31.00 © 2014 IEEE

The 2014 International Power Electronics Conference

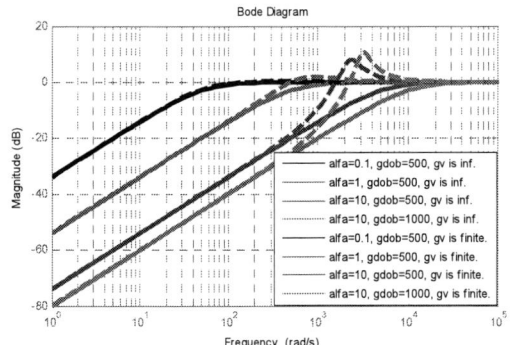

Fig. 3. Sensitivity function $\left(T_{Sen}\right)$ frequency responses of the inner-loop

measurement. A new robustness design constraint for a DOB based motion control system is derived as follows:

Let us consider the sensitivity function that is given in Fig. 2 b and apply $g_v = \kappa\, g_{DOB}$.

$$T_{Sen} = \frac{s\left(s + \kappa g_{DOB}\right)}{s^2 + \kappa g_{DOB}s + \alpha \kappa g_{DOB}^2} \tag{1}$$

The characteristic function of (1) can be designed by using

$$w_n = \sqrt{\alpha \kappa}\, g_{DOB} \quad\text{and}\quad \xi = 0.5\sqrt{\frac{\kappa}{\alpha}} \tag{2}$$

where w_n and ξ denote natural frequency and damping coefficient of a general second order characteristic polynomial, respectively. To suppress the peak of the frequency response of T_{Sen} , if it is assumed that $\xi \geq 0.707$, then

$$\alpha g_{DOB} \leq \frac{g_v}{2} \tag{3}$$

Eq. 3 indicates that α and/or g_{DOB} cannot be increased freely due to the robustness constraint. As a result, there is a trade-off between the stability and robustness in the design of a DOB. Fig.3 shows the robustness constraint and the trade-off between

the stability and robustness of a DOB based motion control system. Although the robustness of a DOB is improved as α is increased when g_v is infinite, the peak of the sensitivity function increases, which deteriorates the robustness, as α is increased when g_v is finite.

III. DOB BASED ROBUST POSITION CONTROL SYSTEMS

Fig. 4 shows a block diagram for a DOB based robust position control system. In this figure, q_m^{ref}, $\dot{q}_m^{ref}$ and $\ddot{q}_m^{ref}$ denote angle/position, velocity and acceleration reference inputs, respectively; $\ddot{q}_m^{des}$ denotes the desired acceleration; and K_p and K_D denote the proportional and derivative gains of the outer-loop controller, respectively. A DOB provides the robustness of the position control system in the inner-loop, and the performance goals are achieved by using an acceleration based controller in the outer-loop. The transfer functions between $\ddot{q}_m^{ref}$ and $\ddot{q}_m$ can be derived from Fig. 4 directly as follows:

$$\frac{\ddot{q}_m}{\ddot{q}_m^{ref}} = \frac{\alpha\left(s + g_{DOB}\right)\left(s^2 + K_D s + K_p\right)}{s^2\left(s + \alpha g_{DOB}\right) + \alpha\left(s + \alpha g_{DOB}\right)\left(K_D s + K_p\right)} \tag{4}$$

when g_v is infinite; and

$$\frac{\ddot{q}_m}{\ddot{q}_m^{ref}} = \frac{\alpha\left(s + g_v\right)\left(s + g_{DOB}\right)\left(s^2 + K_D s + K_p\right)}{s^2\left(s^2 + g_v s + \alpha g_v g_{DOB}\right) + \alpha\left(s + g_v\right)\left(s + g_{DOB}\right)\left(K_D s + K_p\right)} \tag{5}$$

when g_v is finite.

If the stabilities of the transfer functions given in (4) or (5) are analyzed, for instance the Routh-Hurwitz theorem can be used, then it can be shown that the stability of the position control system is improved by increasing α , i.e., using higher nominal inertia or lower nominal torque coefficient improve the stability of a DOB based position control system. However, Eq. 3 shows that α cannot be increased freely due to the robustness constraint in practice.

IV. RFOB BASED ROBUST FORCE CONTROL SYSTEMS

Fig. 5 shows a block diagram for a RFOB based robust force control system. In this figure, C_f denotes the force control gain; F_l^{ref} denotes the force reference; and F_l and $\hat{F}_l$ denote external force and its estimation, respectively. The other parameters are same as defined above. The stability of a RFOB based robust force control system can be analyzed by deriving the open-loop transfer function as follows:

Environmental contact model can be described effectively by using a simple lumped spring damper model as follows:

$$F_l = D_{env}\left(\dot{q}_m - \dot{q}_e\right) + K_{env}\left(q_m - q_e\right) \tag{6}$$

where D_{env} and K_{env} denote environmental damping and stiffness coefficients, respectively; and q_e and $\dot{q}_e$ denote the position and

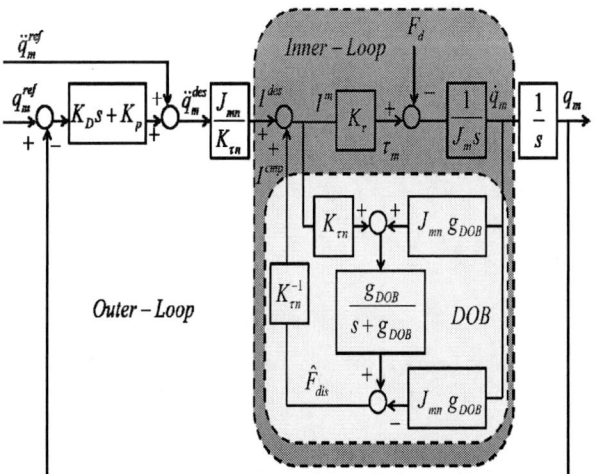

Fig. 4. A Block diagram for a DOB based robust position control system

The 2014 International Power Electronics Conference

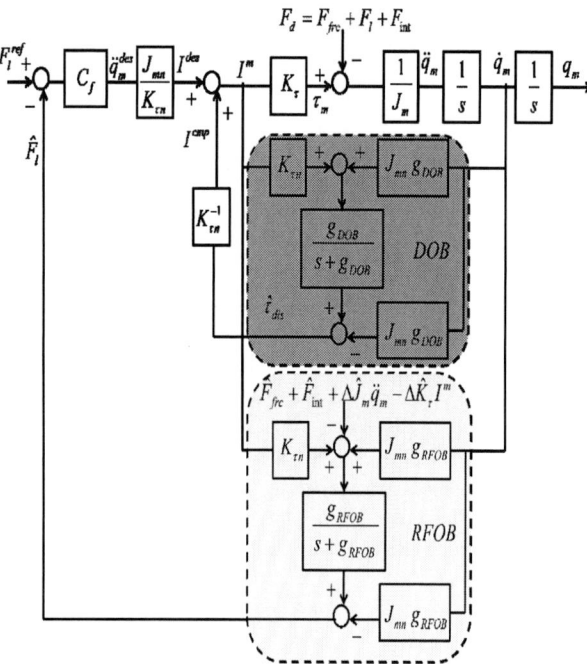

Fig. 5. A Block diagram for a RFOB based robust force control system

velocity of environment at equilibrium, respectively. The open loop transfer function of a RFOB based robust force control system is derived from the Fig. 5 as follows:

$$L_{RFOB}(s) = C_f \frac{g_{RFOB} \frac{J_{mn}}{K_{\tau n}} (s + g_{DOB}) \varphi(s)}{s\{J_m s(s + \alpha g_{DOB}) + (D_{env} s + K_{env})\}(s + g_{RFOB})} \quad (7)$$

where $\varphi(s) = (J_m \hat{K}_\tau - \hat{J}_m K_\tau) s^2 + \hat{K}_\tau D_{env} s + \hat{K}_\tau K_{env}$. Eq. 7 shows that if $J_m \hat{K}_\tau < \hat{J}_m K_\tau$, then the open loop transfer function has a zero at right half plane zero. Therefore, the stability and performance of a RFOB based robust force control system may deteriorate by the imperfect identification of inertia and torque coefficient, significantly. Consequently, a RFOB should be designed by satisfying $J_m \hat{K}_\tau > \hat{J}_m K_\tau$ to improve the stability of the system. Besides, Eq. 7 shows that the relative degree of $L_{RFOB}(s)$ is one, so the root loci have asymptotes, at π rad.

If a RFOB is designed by using perfect system identification, i.e., $\hat{K}_\tau = K_\tau$ and $\hat{J}_m = J_m$, then

$$L_{RFOB}(s) = C_f \frac{(s + g_{DOB})}{(s + g_{RFOB})} \frac{g_{RFOB} J_m \alpha (D_{env} s + K_{env})}{s\{J_m s(s + \alpha g_{DOB}) + (D_{env} s + K_{env})\}} \quad (8)$$

Eq. 8 shows that a DOB and a RFOB can be designed as a phase lead-lag compensator, and the stability of the robust force control system can be improved by using $g_{DOB} < g_{RFOB}$. When perfect system identification is achieved, the relative degree of $L_{RFOB}(s)$ is two, so the root loci have asymptotes, at

$\pm\pi/2$ rad. Eq. 7 and Eq.8 show that the asymptotic behaves of the root loci deteriorate by the perfect inertia and torque coefficient identification.

In general, the bandwidths of a DOB and a RFOB are set to the same value in the robust force control systems. If $g_{DOB} = g_{RFOB} = g$, then the open-loop transfer function is

$$L_{RFOB}(s) = C_f \frac{g J_m \alpha (D_{env} s + K_{env})}{s\{J_m s(s + \alpha g) + (D_{env} s + K_{env})\}} \quad (9)$$

The relative degree of $L_{RFOB}(s)$ is two, so the root loci have asymptotes, at $\pm\pi/2$ rad. However, the phase lead-lag compensator cannot be used in the design of the robust force control systems.

Eq. 7, Eq. 8 and Eq. 9 show that each of the open loop transfer functions have a pole at the origin, so there is no a steady state error in the DOB based robust force control systems.

Table I
Specifications of the experimental setup

$m_m \cong 0.62\,kg$	Mass of dc motor
$K_\tau = 33\,N/A$	Force coefficient
$g_v = 1000\,rad/s.$	Cut-off frequency of vel. meas.
$g_{DOB} = 250\,rad/s.$	Cut-off frequency of DOB.
$g_{RTOB} = 750\,rad/s.$	Cut-off frequency of RFOB
$K_p = 1200$	Proportional position control gain
$K_D = 90$	Derivative position control gain
$C_f = 1$	Proportional force control gain

V. SIMULATION AND EXPERIMENT

In this section, simulation and experimental results will be presented. In the experiments, a linear DC motor which is shown in Fig. 6 is used. Specifications of the experimental set-up are shown in Table-I. The sampling time is 0.1 ms, and KYOWA LUR-A-50NSA1 force sensor is used to verify the performance of RFOB in force control.

Fig. 7 and Fig. 8 show the root loci of a DOB based position

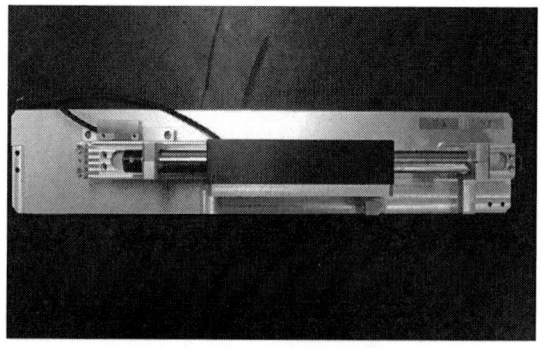

Fig. 6. Linear DC motor

978-1-4799-2706-7/14 $31.00 © 2014 IEEE

The 2014 International Power Electronics Conference

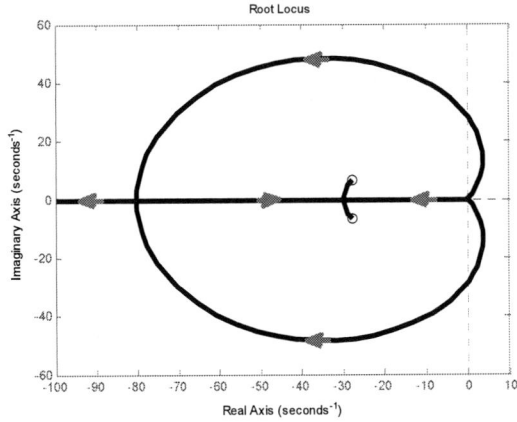

Fig. 7. Stability of the robust position control system

and force control systems, respectively. The root locus of the position control system is plotted with respect to α in Fig. 7. It is clear from the figure that the stability of the robust position control system improves as α is increased. In Fig. 8, the root-loci of the robust force control system are plotted with respect

to the force control gain C_f. Fig. 8a shows that the stability of the robust force control system deteriorates as C_f is increased; increasing the bandwidth of a RFOB improves the stability. Fig. 8b shows that the stability of the robust force control system deteriorates due to right half plane zero if $\hat{J}_m > J_m$. To improve the stability, $\hat{J}_m \leq J_m$ should be guaranteed in the design of a RFOB.

Fig. 9 shows the position control responses of the DC motor when a DOB is implemented. In this experiment, a sinusoidal position reference input is applied between 1 to 10 seconds, and the position control responses are observed by changing the nominal inertia in the design of the DOB. The Fig 9 shows that the stability of the position control system is improved by increasing the nominal inertia. Although the robustness deteriorates in the inner-loop when α is increased, the outer loop controller improves the robustness of the position control system.

Fig. 10 shows the force control responses of the DC motor when a hard environment (aluminum box) is used in the contact motion. In this experiment, the hard environment is at 0.01 m. initially, and a step force control reference is applied at 1 second. The Fig. 10 clearly shows that the stability of the robust force control system changes significantly by the design parameters of DOB and RFOB, and the stability is improved

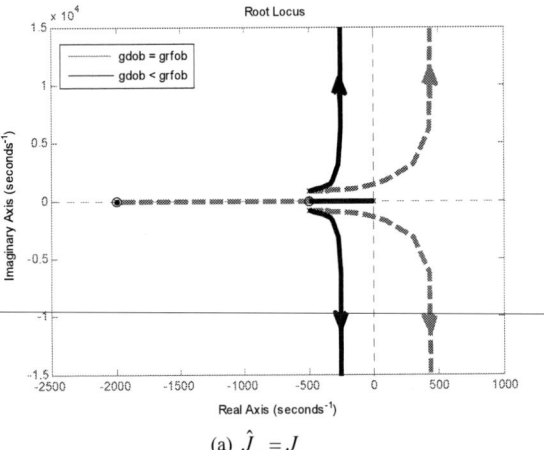

(a) $\hat{J}_m = J_m$

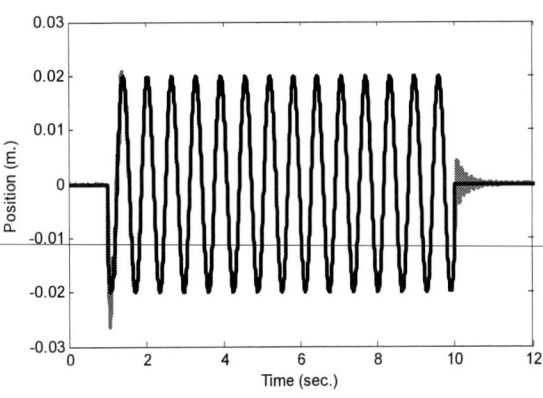

(a) Small nominal mass

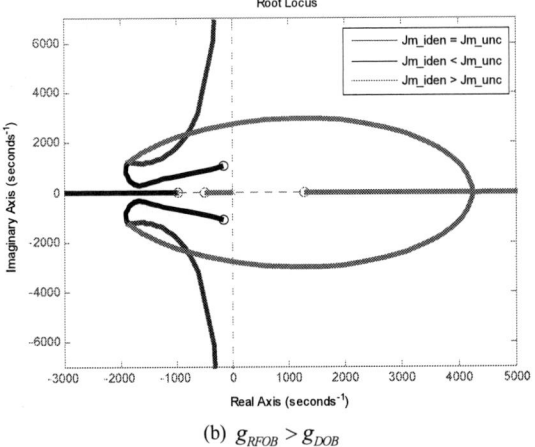

(b) $g_{RFOB} > g_{DOB}$

Fig. 8 Stability of the robust force control system

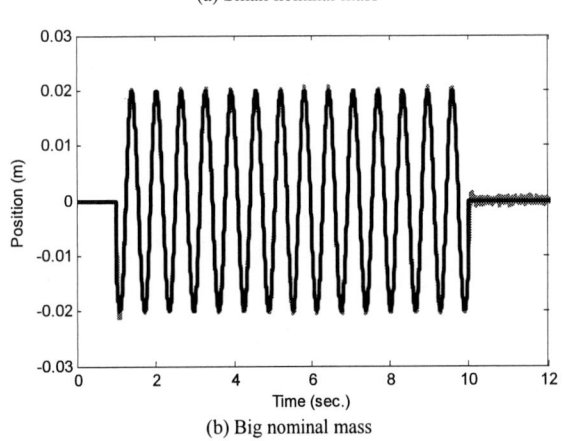

(b) Big nominal mass

Fig. 9. Position control responses

978-1-4799-2706-7/14 $31.00 © 2014 IEEE

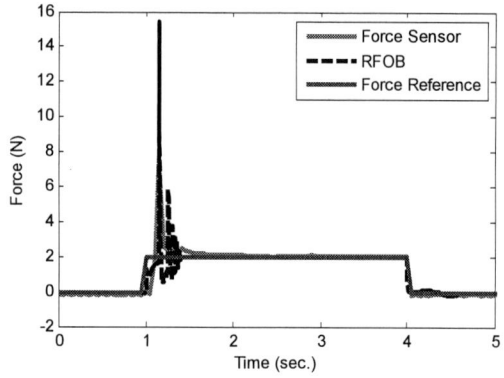

(a) $J_m \cong \hat{J}_m$ and $g_{RFOB} = g_{DOB}$

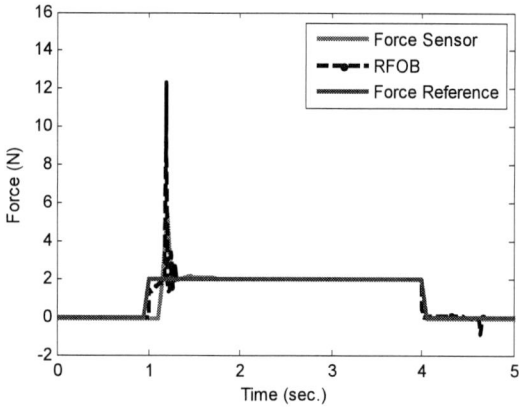

(b) $J_m > \hat{J}_m$ and $g_{RFOB} > g_{DOB}$

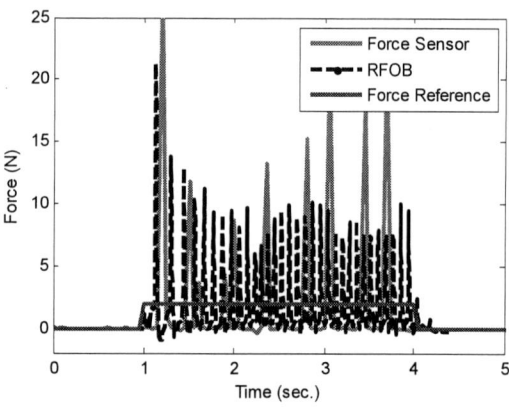

(c) $J_m < \hat{J}_m$ and $g_{RFOB} > g_{DOB}$

Fig. 10. Force control responses

by designing $J_m > \hat{J}_m$ and $g_{RFOB} > g_{DOB}$.

CONCLUSION

This paper proposes new design tools for the DOB and RFOB based robust position and force control systems, respectively. The proposed design tools improve the stability,

robustness and performance of the robust motion control systems. By using the proposed design tools, a DOB based robust motion control system can be designed easily without requiring any pre-experiences on DOB.

The paper clarifies that velocity measurement has significant effect in the stability, robustness and performance of a DOB; and not only the performance but also the stability changes significantly by the design parameters of a DOB and a RFOB in the robust force control systems.

To improve the stability, robustness and performance of a DOB based motion control system, although the exact motor inertia is not required, it should be identified to design a DOB and a RFOB by using the proposed constraints. Although increasing the identified torque coefficient $\hat{K}_r$ improves the stability of the robust force control system, the performance deteriorates significantly. Therefore, the torque coefficient should be identified precisely.

ACKNOWLEDGEMENT

This research was supported in part by the Ministry of Education, Culture, Sports, Science and Technology of Japan under Grant-in-Aid for Scientific Research (S), 25220903, 2013.

REFERENCES

[1] K. Ohishi, K. Ohnishi, K. Miyachi. "Torque-speed regulation of DC-motor based on load torque estimation method". In Proceedings of IPEC-Tokyo pp. 1209–1218, 1983

[2] W. H. Chen, D. J. Ballance, P. J. Gawthrop, J. O'Reilly, " A Nonlinear Disturbance Observer for Robotic Manipulators", Industrial Electronics, IEEE Transactions on, vol. 47, no. 4, pp. 932-938, August 2000,

[3] B. A. Guvenc, L. Guvenc and S. Karaman, "Robust MIMO disturbance observer analysis and design with application to active car steering ", Int. J. Robust Nonlinear Control, vol. 20, no. 8, pp.873–891, 2010

[4] C. J. Kempf and S. Kobayashi, "Disturbance observer and feed-forward design of a high-speed direct-drive positioning table," IEEE Trans. on Control Systems Tech., vol. 7, no. 5, Sep., 1999

[5] K. Ohnishi, M. Shibata, and T. Murakami, "Motion control for advanced mechatronics," IEEE/ASME Trans.Mechatronics, vol. 1, no. 1, pp. 56–67,Mar. 1996.

[6] E. Sariyildiz, K. Ohnishi, Bandwidth constraints of disturbance observer in the presence of real parametric uncertainties, Eur. J. of Control, vol.19, pp. 199-205, 2013

[7] E. Sariyildiz, K. Ohnishi, "A Guide to Design Disturbance Observer", ASME Trans. J. Dyn. Sys., Meas., Control, vol.136, no.2, 2013 021011-021011-10, doi:10.1115/1.4025801

[8] H. Kobayashi, S. Katsura, K. Ohnishi, "An analysis of parameter variations of disturbance observer for motion control," IEEE Transactions on Industrial Electronics, vol. 54, no. 6, pp. 3413-3421, December, 2007.

[9] T. Murakami, F. Yu, and K. Ohnishi, "Torque sensorless control in multi-degree-of-freedom manipulator," IEEE Trans. Ind. Electron., vol. 40, no. 2, pp. 259–265, Apr. 1993.

[10] S. Katsura, Y. Matsumoto, K. Ohnishi,"Analysis and Experimental Validation of Force Bandwidth for Force Control", IEEE Transactions on Industrial Electronics, vol. 53, no. 3, pp. 922-928, June 2006.

[11] S. Katsura, Y. Matsumoto, K. Ohnishi, "Modeling of Force Sensing and Validation of Disturbance Observer for Force Control," IEEE Transactions on Industrial Electronics, vol. 54.no. 1, pp. 530-538, February 2007.

[12] T. Tsuji, T. Hashimoto, H. Kobayashi, M. Mizuochi, K. Ohnishi, "A Wide-Range Velocity Measurement Method for Motion Control," IEEE Transactions on Industrial Electronics, vol. 56, no. 2, pp. 510-519, February 2009.

Identification of Two-Mass Mechanical Systems Using Torque Excitation: Design and Experimental Evaluation

Seppo E. Saarakkala and Marko Hinkkanen
Aalto University School of Electrical Engineering
P.O. Box 13000, FI-00076 Aalto, Helsinki, Finland

Abstract—**This paper deals with methods for parameter estimation of two-mass mechanical systems in electric drives. Estimates of mechanical parameters are needed in the start-up of a drive for automatic tuning of model-based speed and position controllers. A discrete-time output error (OE) model is applied to parameter estimation. The resulting pulse-transfer function is transformed into a continuous-time transfer function, and parameters of the two-mass system model are analytically solved from the coefficients of this transfer function. An open-loop identification setup and two closed-speed-loop identification setups (direct and indirect) are designed and experimentally compared. The experiments are carried out at nonzero speed, making the closed-loop identification setup easier to apply. It was found out that all the identification setups are applicable for the parameter estimation of two-mass mechanical systems.**

Index Terms—**Electric drives, parameter estimation, resonant mechanical load, torsional oscillation.**

I. Introduction

High-performance ac electric drives are replacing pneumatic and hydraulic actuators or dc motor drives in modern machineries—such as injection molding machines [1], machine tools [2], industrial robots [3]—due to their energy efficiency, compact size, and flexible control algorithms. These machineries often consist of several moving or rotating masses, which are coupled together with flexible mechanical transmissions (e.g., belts, gearboxes, long shafts), leading to mechanical resonances. In order to achieve high dynamic performance, motion control of the drive systems with resonant mechanical loads should be based on higher-order mechanical models. The model-based automatic controller tuning typically relies on the knowledge of mechanical parameters and some performance specifications (e.g., closed-loop bandwidth) [4], [5]. However, datasheets of the mechanical components are not often available or the calculation of the mechanical parameters can be a highly complex task. Hence, to enable model-based automatic tuning of the motion controllers, the mechanical parameters should be automatically identified during the start-up of a drive [6] or during the drive operation [7]. Moreover, the identification of the mechanical system may offer a possibility to diagnose mechanical faults. As an example, a method to detect a rolling-bearing damage is proposed in [8].

The identification routines, proposed for parameter estimation of two- or multi-mass mechanical systems, can be roughly divided to parametric methods [3], [6], [9]–[11] and nonparametric methods [12]–[14]. The nonparametric methods use the frequency-domain characteristics of the system, while,

in the parametric methods, the parameters of the two-mass system transfer-function polynomials are estimated in the time domain. When the identification is completed offline, e.g., during the start-up of a drive, the parameters of the mechanical model can be estimated either in open loop or using closed-loop speed control [15]. It is desirable to reduce the effect of nonlinear friction phenomena on parameter estimates by operating at nonzero speed. However, when using the open-loop method, it may be difficult to find a suitable value for the offset torque without causing the system to rush. On the other hand, in closed-loop identification, the drive can be easily operated at desired (nonzero) speed. Closed-loop identification methods can be divided into direct and indirect methods [15]. In the case of direct methods, the input signal is affected via the feedback loop. Hence, a correct noise model is needed. In the case of indirect methods, the closed-loop system is first identified, and the open-loop model is then solved using the known control law.

The excitation signal should contain all frequencies evenly distributed, and the variance of the excitation signal should be as large as possible. White noise is normally utilized in stochastical identification. In electric drives, the torque and the speed are limited to their maximum allowed values. With limited input signals, the largest variance is obtained by binary signals, which have only two possible values (e.g., -1 and 1). A pseudo-random binary signal (PRBS) fulfills the previously stated requirements, and it can be easily formed with a shift register. The statistical properties of the PRBS are studied in [16].

In this paper, the mechanical system is excited using the PRBS, which is superimposed on the electromagnetic torque by means of field-oriented control. The rotor-speed response of the driving motor is measured. Because the rotor-speed response is noisy, the discrete-time OE model is used in identification, in accordance with [3]. The main contributions of this paper are: 1) An indirect closed-loop method is proposed for identification of two-mass mechanical system. According to the authors' knowledge, indirect methods have not been applied in this context before (except in the preliminary study in [11])[1]; 2) The effect of the speed controller gain

---

[1]The main differences between this paper and the preliminary study are: 1) the parameters of the OE model are estimated using a straightforward iterative method; 2) the continuous-time transfer function parameters are analytically derived from the discrete-time pulse-transfer function parameters; and 3) the method is applied to estimate the mechanical parameters of an experimental two-mass system, whereas a two-mass system emulator was used in [11].

978-1-4799-2706-7/14 $31.00 © 2014 IEEE

on identifiability is analyzed by means of simulations and experiments; 3) The proposed indirect identification method is experimentally compared with the open-loop identification method, the direct identification method, and the frequency-response based method proposed in [13].

If the identified system is highly nonlinear, the linear identification methods presented in this paper could be augmented with methods that can estimate the nonlinear elements, such as backlash or friction [17]–[19]. If the load inertia and the coupling stiffness vary during the drive operation [20], the proposed identification methods could be used to estimate the parameters of the mechanical system in various operating points, and then construct a look-up table of parameter values as a function of operating point.

II. MODEL OF A TWO-MASS MECHANICAL SYSTEM

The mechanical dynamics of the resonating two-mass system are given as

$$J_M\ddot{\theta}_M = T_M - T_S - b_M\dot{\theta}_M \tag{1a}$$

$$J_L\ddot{\theta}_L = T_S - T_L - b_L\dot{\theta}_L \tag{1b}$$

$$T_S = K_S(\theta_M - \theta_L) + c_S(\dot{\theta}_M - \dot{\theta}_L) \tag{1c}$$

where the angular positions of the motor and the load are θ_M and θ_L, respectively. The motor electromagnetic torque, the loading torque, and the shaft torque are T_M, T_L, and T_S, respectively. The motor speed is denoted by $\omega_M = \dot{\theta}_M$ and the load speed by $\omega_L = \dot{\theta}_L$. The moments of inertias of the motor and the load are denoted as J_M and J_L, respectively. The torsional stiffness and the damping of the shaft are K_S and c_S, respectively. The friction is modelled as viscous damping both on the motor and load sides, denoted as b_M and b_L, respectively. From (1), the open-loop transfer function from the torque $T_M(s)$ to the speed $\omega_M(s)$ is obtained as

$$G(s) = \frac{B(s)}{A(s)} \tag{2}$$

where

$$B(s) = J_L s^2 + (c_S + b_L)s + K_S$$
$$A(s) = J_M J_L s^3 + (J_M c_S + J_L c_S + J_L b_M + J_M b_L)s^2$$
$$+ (J_M K_S + J_L K_S + c_S b_M + c_S b_L + b_M b_L)s + K_S(b_M + b_L)$$

If $b_M = 0$, $b_L = 0$, and $c_S = 0$ are assumed, the antiresonance frequency and the resonance frequency are

$$f_{ares} = \frac{1}{2\pi}\sqrt{\frac{K_S}{J_L}} \qquad f_{res} = \frac{1}{2\pi}\sqrt{K_S\frac{J_M + J_L}{J_M J_L}} \tag{3}$$

respectively.

III. PARAMETER ESTIMATION

First, three different identification setups are introduced. Then, parameters of the continuous-time mechanical model are linked with parameters of the discrete-time OE model. Further, factors affecting the accuracy of the parameter estimation are discussed.

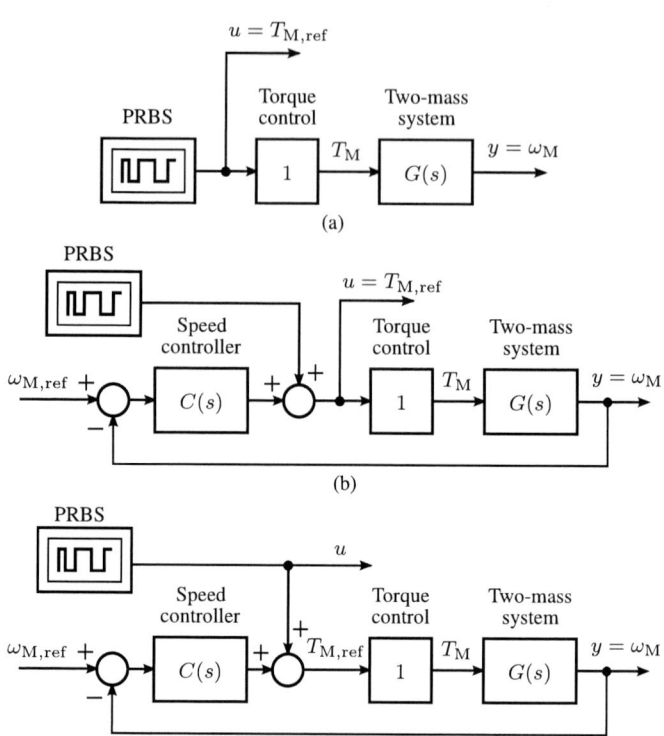

Fig. 1. Identification setups: (a) open loop; (b) direct closed loop; (c) indirect closed loop. Ideal torque control is assumed. Input and output identification signals are denoted by u and y, respectively.

A. Identification Setups

Three identification setups shown in Fig. 1 are considered. The PRBS torque excitation is applied in all setups. Typical torque-control bandwidths in ac servo drives are from several hundred hertz up to a few kilohertz, while dominant resonance frequencies of mechanical systems are lower. Hence, the torque-control loop is usually significantly faster than the mechanical system. If the sampling frequency of the parameter estimation is set significantly below the torque-control bandwidth, the effect of torque control cannot be seen in the identification signals and thus it can be omitted. In this paper, ideal torque control is assumed, i.e. $T_M = T_{M,ref}$.

An open-loop setup is shown in Fig. 1(a). The open-loop transfer function (2) can be directly estimated from the excitation signal u and the output signal y, i.e. $y(s) = G(s)u(s)$. If the excitation signal has a zero average, identification will be performed in the vicinity of zero speed. In this case, friction phenomena are highly nonlinear and can cause bias to the parameter estimates. Open-loop identification could also be performed during an acceleration test or a deceleration test. In these cases, the trend of the speed signal should be removed. Similar open-loop setups have been studied in [13] and [10].

Fig. 1(b) shows a direct closed-loop setup, where the excitation signal were superimposed on the torque reference obtained from the speed controller.[2] The identification procedure is similar as in the open-loop case. However, the identification

---

[2]An equivalent identification setup could be formed, if the excitation signal is superimposed on the speed reference and a proportional speed controller is used. In this case, the amplitude of the excitation signal would depend on the gain of the speed controller.

input signal u and the noise (not shown in the figure) are now correlated due to the speed controller, which may lead to biased parameter estimates. Similar direct setups have been considered in [13] and [10].

Fig. 1(c) shows an indirect closed-loop identification setup. The excitation signal is superimposed on the torque reference as in the direct setup. The identification input signal u, however, is now the PRBS, which is not affected by the speed controller. The transfer function from the input $u(s)$ to the output $y(s)$ is

$$\frac{y(s)}{u(s)} = \frac{G(s)}{1 + G(s)C(s)} \tag{4}$$

This closed-loop transfer function to be estimated contains the speed controller $C(s)$, whose effect on parameter estimates must be removed afterwards. Therefore, the method is called indirect [21]. For simplicity, a proportional (P) speed controller is used, i.e., $C(s) = k_\mathrm{p}$. Hence, the order of the transfer function to be identified is the same in all three setups. It is worth noticing that the indirect method can be applied for parameter estimation even if the speed controller output is not accessible (i.e., the direct closed-loop estimation method cannot be used), assuming that the speed controller gain is known a priori.

B. Mechanical Parameters

The discrete-time OE model applied in parameter estimation is

$$y(k) = \frac{B_\mathrm{d}(z)}{A_\mathrm{d}(z)} u(k) + e(k) \tag{5}$$

where z is the time-shift operator, $y(k)$ and $u(k)$ are the discrete samples corresponding to the signals y and u, respectively, shown in Fig. 1, and $e(k)$ is the output noise in the system. The pulse-transfer function to be identified is given as

$$\frac{B_\mathrm{d}(z)}{A_\mathrm{d}(z)} = \frac{\theta_1 z^2 + \theta_2 z + \theta_3}{z^3 + \theta_4 z^2 + \theta_5 z + \theta_6} \tag{6}$$

where $\theta_1 \ldots \theta_6$ are the six parameters to be estimated.

The output of the pulse-transfer function (6) can be expressed as

$$y(k) = \boldsymbol{\phi}^\mathrm{T}(k)\boldsymbol{\theta} \tag{7}$$

where the predictor vector and the parameter vector are

$$\boldsymbol{\phi}(k) = \begin{bmatrix} u(k-1) \\ u(k-2) \\ u(k-3) \\ -y(k-1) \\ -y(k-2) \\ -y(k-3) \end{bmatrix} \qquad \boldsymbol{\theta} = \begin{bmatrix} \theta_1 \\ \theta_2 \\ \theta_3 \\ \theta_4 \\ \theta_5 \\ \theta_6 \end{bmatrix} \tag{8}$$

respectively. When the noise component is summed to the output of the system, solving the parameter vector using (7) and (8) leads to biased parameter estimates [21]. Here, a straightforward iterative method is applied to reduce the bias in the parameter estimates [22]. In this method, the input and output signals are filtered using the estimated system polynomial $\hat{A}_\mathrm{d}(z)$ from the previous iteration. The output of the adaptive filtered system is given as

$$y_\mathrm{f}(k) = \boldsymbol{\phi}_\mathrm{f}^\mathrm{T}(k)\boldsymbol{\theta} \tag{9}$$

where the filtered predictor vector and output are

$$\phi_\mathrm{f}(k) = \frac{1}{\hat{A}_\mathrm{d}(z)}\phi(k) \qquad y_\mathrm{f}(k) = \frac{1}{\hat{A}_\mathrm{d}(z)}y(k) \tag{10}$$

respectively. When estimating the parameter vector $\boldsymbol{\theta}$, the adaptive filtered output vector $\boldsymbol{y}_\mathrm{f}$ and the predictor matrix $\boldsymbol{\Phi}_\mathrm{f}$ are given as

$$\boldsymbol{\Phi}_\mathrm{f} = \begin{bmatrix} \phi_\mathrm{f}(3) & \phi_\mathrm{f}(4) & \cdots & \phi_\mathrm{f}(N) \end{bmatrix}$$
$$\boldsymbol{y}_\mathrm{f} = \begin{bmatrix} y_\mathrm{f}(3) & y_\mathrm{f}(4) & \cdots & y_\mathrm{f}(N) \end{bmatrix}^\mathrm{T} \tag{11}$$

where N is the total number of samples used in the parameter estimation. Then, the matrices are used in an iterative least-squares algorithm to solve the parameter vector

$$\boldsymbol{\theta} = \left(\boldsymbol{\Phi}_\mathrm{f}^\mathrm{T}\boldsymbol{\Phi}_\mathrm{f} \right)^{-1} \boldsymbol{\Phi}_\mathrm{f}^\mathrm{T}\boldsymbol{y}_\mathrm{f} \tag{12}$$

which can be used to find $\hat{A}_\mathrm{d}(z)$ for the next iteration:

$$\hat{A}_\mathrm{d}(z) = z^3 + \theta_4 z^2 + \theta_5 z + \theta_6. \tag{13}$$

During the first iteration, the filtering polynomial $\hat{A}_\mathrm{d}(z) = 1$. The iterations are continued until the estimated parameters converge to the final values. It is important to check that the roots of $\hat{A}_\mathrm{d}(z)$ are inside the unit circle after each iteration.

The pulse-transfer function (6) is then converted to a zero-pole matching equivalent continuous-time transfer function

$$\frac{y(s)}{u(s)} = \frac{b_1 s^2 + b_2 s + b_3}{s^3 + a_1 s^2 + a_2 s + a_3} \tag{14}$$

where the parameters $b_1 \ldots b_3$ and $a_1 \ldots a_3$ are given in the Appendix. When comparing (4) and (14), the following system of equations is obtained:

$$b_1 = \frac{1}{J_\mathrm{M}} \qquad b_2 = \frac{c_\mathrm{S} + b_\mathrm{L}}{J_\mathrm{M} J_\mathrm{L}} \qquad b_3 = \frac{K_\mathrm{S}}{J_\mathrm{M} J_\mathrm{L}} \tag{15a}$$

$$a_1 = \frac{(J_\mathrm{M} + J_\mathrm{L})c_\mathrm{S} + J_\mathrm{L}b_\mathrm{M} + J_\mathrm{M}b_\mathrm{L} + k_\mathrm{p}J_\mathrm{L}}{J_\mathrm{M} J_\mathrm{L}} \tag{15b}$$

$$a_2 = \frac{(J_\mathrm{M} + J_\mathrm{L})K_\mathrm{S} + (b_\mathrm{M} + b_\mathrm{L})c_\mathrm{S} + b_\mathrm{M}b_\mathrm{L} + k_\mathrm{p}(c_\mathrm{S} + b_\mathrm{L})}{J_\mathrm{M} J_\mathrm{L}} \tag{15c}$$

$$a_3 = \frac{K_\mathrm{S}(b_\mathrm{M} + b_\mathrm{L} + k_\mathrm{p})}{J_\mathrm{M} J_\mathrm{L}} \tag{15d}$$

From (15), the mechanical parameters J_M, J_L, b_M, b_L, K_S, and c_S can be solved. In the open-loop and direct closed-loop identification setups, $k_\mathrm{p} = 0$ is substituted into (15).

If the dominant resonance frequencies of the mechanical system were near the bandwidth of the torque control, the parameter estimates from (15) would be biased. If the bandwidth of the torque control is known, the dynamics of the torque-control loop could be included in the identification setups shown in Fig. 1. This inclusion would lead to a different system of equations to be solved.

C. Sampling Frequency and the Number of Samples

According to the Nyquist-Shannon sampling theorem, the sampling frequency of the discrete-time system should be at least twice the highest frequency in the original continuous-time signal. In most cases, the system response should also

Fig. 2. Experimental setup.

be modeled slightly above the resonance frequency to see if there are some additional dynamics at higher frequencies. However, if the sampling frequency is selected too high, numerical sensitivity issues can appear and cause the loss of identifiability. A high sampling frequency also causes the model fit to concentrate at high frequencies. A rule of thumb is to select the sampling frequency ten times the bandwidth of the process [21].

If the model structure is chosen correctly, increasing the number of samples N should decrease the effect of disturbance noise and enhance the accuracy of parameter estimates. However, increasing the number of samples increases the requirements for memory and processing capacity.

D. Model Validation

Model validation is an essential part of the identification procedure. The designer needs to know whether the selected model structure and the identification setup offers good enough information from the real system. A common tool in validation is residual analysis. Residual analysis is based on the statistical properties of the residuals $\varepsilon(k) = y(k) - \hat{y}(k)$. The simulated-system output is denoted as $\hat{y}(k) = G_d(z)u(k)$, where $G_d(z)$ represents the zero-pole equivalent discretization of the continous-time time transfer function (2), which is obtained using the estimated system parameters.

The autocorrelation

$$R_\varepsilon(\tau) = \frac{1}{N}\sum_{k=1}^{N}\varepsilon(k)\varepsilon(k-\tau) \qquad (16)$$

of the residuals should ideally resemble that of white noise. Furthermore, the cross-correlation

$$R_{\varepsilon,u}(\tau) = \frac{1}{N}\sum_{k=1}^{N}\varepsilon(k)u(k-\tau) \qquad (17)$$

between the input signal and the residuals should ideally be zero [22]. For the OE model, the emphasis in the residual analysis is in the cross-correlation since a noise model is not included in the OE structure [23]. If possible, the residual analysis should be performed using a different input-output

dataset than the one which is used for the parameter estimation. Moreover, the identified model can be validated through the comparison of time- and frequency-domain responses of the identified and the measured (real) systems.

IV. RESULTS

The identification methods described in Section III are evaluated by means of simulations and experiments. First, the effect of the speed controller gain on the parameter estimates is studied by means of simulations. Then, the mechanical parameters of the experimental system are estimated and compared with the parameter estimates obtained using a frequency-response based identification method proposed in [13]. Finally, the results are validated using correlation and frequency-domain analyses.

A. Experimental System

The experimental setup is shown in Fig. 2. The setup consists of two mechanically coupled permanent-magnet synchronous motors (PMSMs). An inverter-fed 4-kW 2400-rpm PMSM, controlled with a dSPACE DS1104 board, is used as a driving motor. The driving motor is connected to a 4-kW loading servo motor using a flexible toothed belt. In order to vary coupling stiffness, different belts can be used. An additional inertia disk can be added to the shaft of the load motor.

The experiments were carried out using two mechanical configurations, referred to as Configurations A and B. The load-side inertia equals the motor-side inertia in Configuration A, while the additional inertia disk increasing the load-side inertia is applied in Configuration B. Furthermore, a stiffer belt is applied in Configuration B.

Mechanical parameters of both configurations were calculated based on the datasheet values of mechanical components. These parameters are given in Tables I and II for Configurations A and B, respectively. The antiresonance and resonance frequencies of Configuration A, calculated using (3) with the datasheet values, are close to each other ($f_{ares} = 60$ Hz and $f_{res} = 84$ Hz), while the antiresonance and resonance frequencies of Configuration B are far away from each other ($f_{ares} = 27$ Hz and $f_{res} = 79$ Hz). The datasheet values for c_S were approximated using

$$c_S = \frac{K_S}{2\pi f_{res}Q_k} \qquad (18)$$

where $Q_k = 10$ was used for flexible couplings [24].

Torque control is accomplished through field-oriented control. The torque-control loop operates at 10-kHz sampling frequency and the torque-control bandwidth is 350 Hz. The speed-control loop operates at 1-kHz sampling frequency. The sampling frequency of the parameter estimation is 333 Hz, and the torque-control loop is ignored in parameter estimation. The number of samples is $N = 1620$. The excitation signal is a PRBS with values -2 Nm and 2 Nm (the rated torque being 17 Nm).

The rotor speed ω_M of the driving motor is measured using an incremental encoder. The angular speed is calculated

The 2014 International Power Electronics Conference

TABLE I
DATASHEET VALUES AND ESTIMATED MECHANICAL PARAMETERS FOR CONFIGURATION A

| Parameter | Datasheet | Parameters estimated using (15) | | | Frequency-response method [13] |
		Open loop	Direct	Indirect	Open loop
J_M (kgm^2)	0.005	0.0058	0.006	0.0058	0.005
J_L (kgm^2)	0.005	0.0047	0.0044	0.0044	0.004
K_S (Nm/rad)	700	656	651	646	591
c_S (Nms/rad)	0.13	0.08	0.04	0.03	0.05
b_M (Nms/rad)	0	−0.01	−0.16	−0.15	−0.33
b_L (Nms/rad)	0	0.03	0.17	0.16	0.34

TABLE II
DATASHEET VALUES AND ESTIMATED MECHANICAL PARAMETERS FOR CONFIGURATION B

| Parameter | Datasheet | Parameters estimated using (15) | | | Frequency-response method [13] |
		Open loop	Direct	Indirect	Open loop
J_M (kgm^2)	0.005	0.0074	0.0069	0.0066	0.0057
J_L (kgm^2)	0.038	0.049	0.045	0.039	0.032
K_S (Nm/rad)	1100	1495	1414	1374	1231
c_S (Nms/rad)	0.22	1.14	1.27	1.12	1.68
b_M (Nms/rad)	0	−1.28	−1.5	−1.39	−2.12
b_L (Nms/rad)	0	1.3	1.58	1.4	2.12

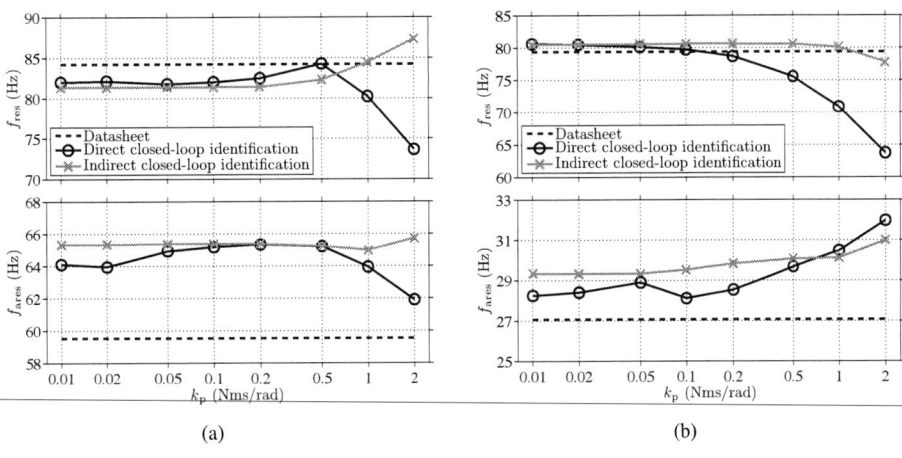

Fig. 3. Estimated resonance and antiresonance frequencies as a function of the speed controller gain: (a) Configuration A; (b) Configuration B. The results are obtained from simulations.

from the measured angular position difference within the fixed sampling interval of 1 ms. This sampling scheme leads to a significant quantization noise especially at low rotational speeds which also favours the use of the OE model structure in the identification.

B. Benchmark Method for Experimental Comparison

The identification methods described in Section III are experimentally compared with a frequency-response based open-loop method proposed in [13]. The identification setup shown in Fig. 1(a) is applied. The experimental frequency-response function $G_\mathrm{e}(\mathrm{j}\omega)$ is evaluated in $M = 166$ data points between the frequencies of 1 Hz and 166 Hz by means of the Welch method, having the Hamming-window length of 540. Then, the parameter values of the analytical frequency-response function $G(\mathrm{j}\omega)$ [obtained from (2)] are varied, and the best fit is iteratively searched by minimizing the error function

$$J(\boldsymbol{\vartheta}) = \sum_{i=1}^{M} |G_\mathrm{e}(\mathrm{j}\omega_i) - G(\mathrm{j}\omega_i, \boldsymbol{\vartheta})|^2 \qquad (19)$$

where the parameter vector is $\boldsymbol{\vartheta} = [J_\mathrm{M}, J_\mathrm{L}, K_\mathrm{S}, c_\mathrm{S}, b_\mathrm{M}, b_\mathrm{L}]$. The initial values of the parameter vector, needed in the first iteration, are selected according to the datasheet values given in Tables I and II.

C. Simulation Results

The speed controller is a P controller. The effect of the speed controller gain k_p on the estimated antiresonance and resonance frequencies is examined by means of simulations. A white-noise signal with variance of 1 rad^2/s^2 is added to the simulated motor speed. Numerical values for the resonance frequencies are calculated using (3), based on the estimated system parameters.

Fig. 3 shows the estimated resonance and antiresonance frequencies as a function of the speed controller gain for both the configurations. The system resembles the open-loop setup at low gain values ($k_\mathrm{p} < 0.1$ Nms/rad). In both closed-loop setups, high controller gains increase the effect of the measurement noise due to the feedback. The high controller gains ($k_\mathrm{p} > 1$ Nms/rad) also speed up the system, in which

The 2014 International Power Electronics Conference

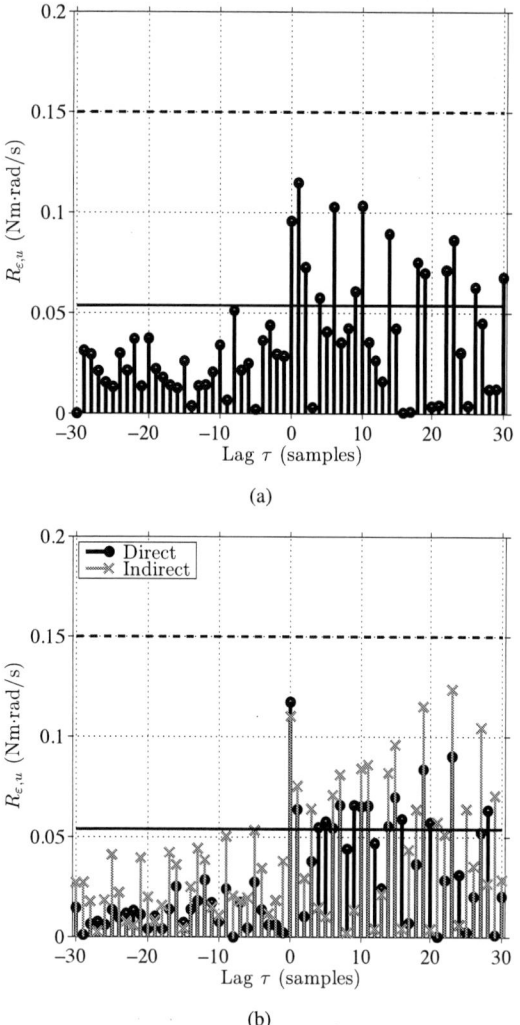

(a)

(b)

Fig. 4. Cross-correlation between input and residuals of Configuration A: (a) open loop; (b) closed loop. The 97% confidence limit is indicated as solid black line and the practical confidence limit as dashed black line.

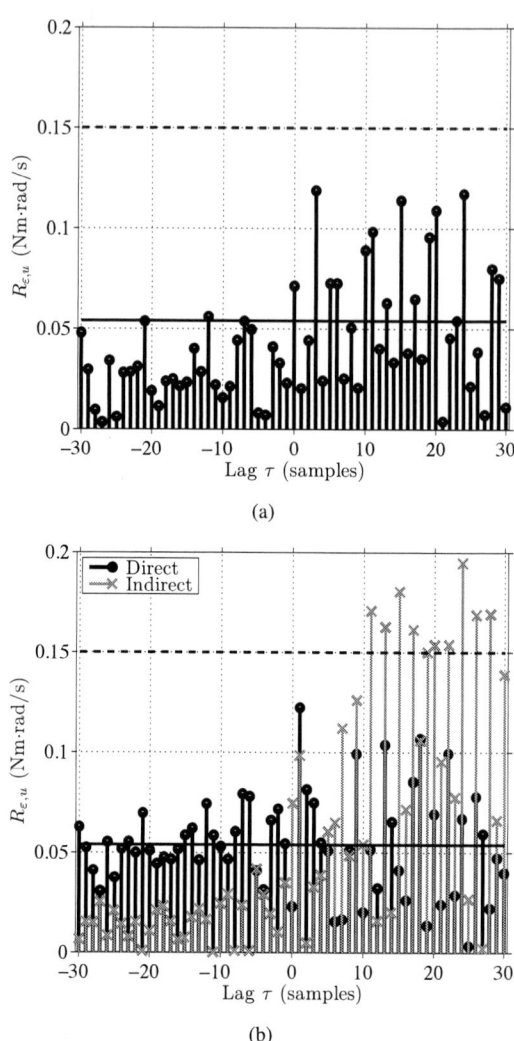

(a)

(b)

Fig. 5. Cross-correlation between input and residuals of Configuration B: (a) open loop; (b) closed loop. The 97% confidence limit is indicated as solid black line and the practical confidence limit as dashed black line.

case the sampling frequency should be increased. This can be seen as biased resonance frequency estimates. It can be seen that the proposed indirect method is less sensitive to the selection of the speed controller gain than the direct method.

D. Experimental Results

In the closed-loop identification setups, the speed controller gain $k_p = 0.2$ Nms/rad is selected and the parameters are estimated at the rotational speed of 200 r/min. In order to operate approximately at the speed of 200 r/min also in the open-loop setup, a constant offset torque is applied. The offset values are removed both from the input and output signals before identification. For model validation, a separate input-output dataset is measured in closed loop using the speed controller gain of $k_p = 0.05$ Nms/rad.

The parameter estimates of Configurations A and B are given in Tables I and II, respectively. It can be seen that all the estimated parameters agree well with the datasheet values. In all the cases, an estimate of the motor-side damping b_M is negative. However, the sum $b_M + b_L$ of the viscous

damping estimates is positive. Furthermore, when substituting the obtained parameter values back to the open-loop transfer function (2), all the coefficients of the transfer function are positive (i.e., the obtained poles and zeros are stable). When comparing the open-loop parameter estimates obtained using (15) with those of the frequency-response method, it can be seen that the frequency-response method gives smaller values for the inertia moments and for the coupling stiffness. Moreover, it was observed that the window length has an impact on the parameter estimates obtained using the frequency-response method. When the window length was reduced to 270, the parameter estimates of both the configurations were reduced almost by an average of 15%.

The parameter estimates obtained from (15) are first analyzed by means of stochastical analysis. The cross-correlation between the input signal and the residuals is evaluated using (17). Fig. 4 shows the results of the cross-correlation analysis for Configuration A and Fig. 5 for Configuration B. A 97% confidence limit and a practical confidence limit are introduced in the figures [22]. The 97% confidence limit is calculated as

$2.17/\sqrt{N}$, where N is the number of samples used in the estimation. Cross-correlation values remaining below the practical confidence limit will indicate that stochastically acceptable parameter estimates are obtained. It can be seen in Figs. 4 and 5 that the cross-correlations between the input and the residuals remain mostly below the practical confidence limit in all the identification cases.

The frequency responses, obtained using the datasheet parameter values and the estimated parameter values, are compared. Fig. 6 shows the frequency responses of both the configurations obtained using the open-loop identification setups. It can be seen that the estimated amplitude is higher at lowest frequencies when the frequency-response method is applied because the inertia estimates are too low. When analyzing solely the locations of the antiresonance and resonance frequencies, it can be seen that the estimated resonance frequencies agree well with the datasheet-based resonance frequencies. In the case of the frequency-response method, the antiresonances appear at too high frequencies, because the load-inertia estimate is too low. These observations agree also with the numerical values given in Tables I and II. Fig. 7 shows the frequency responses of both the configurations obtained using the closed-loop identification setups. It can be seen that the estimated frequency responses agree well with the datasheet-based frequency responses. It should be noted that the estimated amplitudes at the resonance frequencies are not directly comparable with the datasheet-based amplitudes, because $b_M = b_L = 0$ are assumed in the case of datasheet values. Furthermore, the datasheet values of c_S are only rough approximations, obtained using (18).

V. Conclusion

This paper proposes an indirect closed-loop method for identification of two-mass mechanical system. Based on the simulation results, the proposed method is less sensitive to the selection of the speed controller gain than the direct method (when the simple OE model structure is used). The proposed indirect identification method was experimentally compared with the open-loop identification method, the direct identification method, and the frequency-response method. Based on the validation results, it can be concluded that all the identification setups are applicable for the parameter estimation of two-mass mechanical systems. The most biased estimate was the sum of the viscous friction coefficients, which is rarely needed in motion controller tuning.

Appendix
Continuous-Time Transfer Function Parameters

The denominator of the pulse-transfer function (6) can be expressed as a combination of the first-order pole and the second-order complex-conjugate poles:

$$\frac{B_d(z)}{A_d(z)} = \frac{\theta_1 \left(z^2 + \beta_2 z + \beta_3\right)}{(z + \alpha_1)\left(z^2 + \alpha_2 z + \alpha_3\right)} \tag{20}$$

where

$$\beta_2 = \theta_2/\theta_1 \qquad \beta_3 = \theta_3/\theta_1$$
$$\alpha_1 + \alpha_2 = \theta_4 \qquad \alpha_1\alpha_2 + \alpha_3 = \theta_5 \qquad \alpha_1\alpha_3 = \theta_6$$

The pulse-transfer function (20) is converted to a continuous-time zero-pole matching equivalent transfer function using the relation $s = \frac{1}{h}\ln(z)$, where h is the sampling interval [25]. The transfer function

$$\frac{y(s)}{u(s)} = \frac{b_1'\left(s^2 + 2b_2's + b_3'^2 + b_3'^2\right)}{(s + a_1')\left(s^2 + 2a_2's + a_2'^2 + a_3'^2\right)} \tag{21}$$

is obtained, where the parameters are

$$b_2' = -\frac{1}{h}\ln\left(\sqrt{\beta_3}\right) \tag{22a}$$

$$b_3' = \frac{1}{h}\arctan\left(\sqrt{4\beta_3/\beta_2^2 - 1}\right) \tag{22b}$$

$$a_1' = -\frac{1}{h}\ln\left(-\alpha_1\right) \qquad a_2' = -\frac{1}{h}\ln\left(\sqrt{\alpha_3}\right) \tag{22c}$$

$$a_3' = \frac{1}{h}\arctan\left(\sqrt{4\alpha_3/\alpha_2^2 - 1}\right) \tag{22d}$$

$$b_1' = \frac{a_1'\beta_1\left(1 + \beta_2 + \beta_3\right)\left(a_2'^2 + a_3'^2\right)}{\left(1 + \alpha_1 + \alpha_2 + \alpha_1\alpha_2 + \alpha_3 + \alpha_1\alpha_3\right)\left(b_2'^2 + b_3'^2\right)} \tag{22e}$$

Using (22), the parameters in (14) are given as

$$b_1 = b_1' \qquad b_2 = 2b_1'b_2' \qquad b_3 = b_1'\left(b_2'^2 + b_3'^2\right) \tag{23a}$$

$$a_1 = a_1' + 2a_2' \qquad a_2 = 2a_1'a_2' + a_2'^2 + a_3'^2 \tag{23b}$$

$$a_3 = a_1'\left(a_2'^2 + a_3'^2\right) \tag{23c}$$

Acknowledgment

The authors gratefully acknowledge ABB Oy for the financial support.

References

[1] K. Ohishi and R. Furusawa, "Actuators for motion control: Fine actuator force control for electric injection molding machines," *IEEE Ind. Electron. Mag.*, vol. 6, no. 1, pp. 4–13, Mar. 2012.

[2] C. Hu, B. Yao, and Q. Wang, "Coordinated adaptive robust contouring controller design for an industrial biaxial precision gantry," *IEEE/ASME Trans. Mechatronics*, vol. 15, no. 5, pp. 728–735, Oct. 2010.

[3] M. Östring, S. Gunnarsson, and M. Norrlöf, "Closed-loop identification of an industrial robot containing flexibilities," *Control Engineering Practice*, vol. 11, no. 3, pp. 291–300, 2003.

[4] S. N. Vukosavic and M. R. Stojic, "Suppression of torsional oscillations in a high-performance speed servo drive," *IEEE Trans. Ind. Electron.*, vol. 45, no. 1, pp. 108–117, Feb. 1998.

[5] L. Harnefors, S. E. Saarakkala, and M. Hinkkanen, "Speed control of electrical drives using classical control methods," *IEEE Trans. Ind. Appl.*, vol. 49, no. 2, pp. 889–898, 2013.

[6] H.-B. Beck and D. Turschner, "Commissioning of a state-controlled high-powered electrical drive using evolutionary algorithms," *IEEE/ASME Trans. Mechatronics*, vol. 6, no. 2, pp. 149–154, June 2001.

[7] M. A. Valenzuela, J. M. Bentley, and R. D. Lorenz, "Dynamic online sensing of sheet modulus of elasticity," *IEEE Trans. Ind. Appl.*, vol. 46, no. 1, pp. 108–120, 2010.

[8] H. Zoubek, S. Villwock, and M. Pacas, "Frequency response analysis for rolling-bearing damage diagnosis," *IEEE Trans. Ind. Electron.*, vol. 55, no. 12, pp. 4270–4276, Sep. 2008.

[9] Y. Guo, L. Huang, and M. Muramatsu, "Research on inertia identification and auto-tuning of speed controller for AC servo system," in *IEEE Power Conversion Conf.*, vol. 2, Osaka, Japan, Apr. 2002, pp. 896–901.

[10] I. Eker and M. Vural, "Experimental online identification of a three-mass mechanical system," in *IEEE CCA Conf.*, vol. 1, Istanbul, Turkey, Jun. 2003, pp. 60–65.

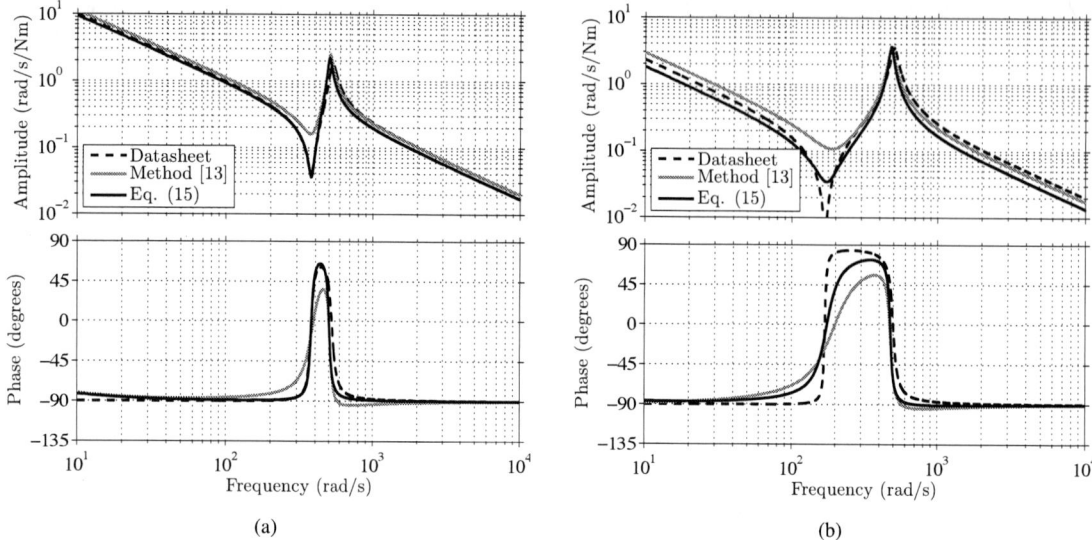

Fig. 6. Frequency responses obtained using open-loop identification method: (a) Configuration A; (b) Configuration B.

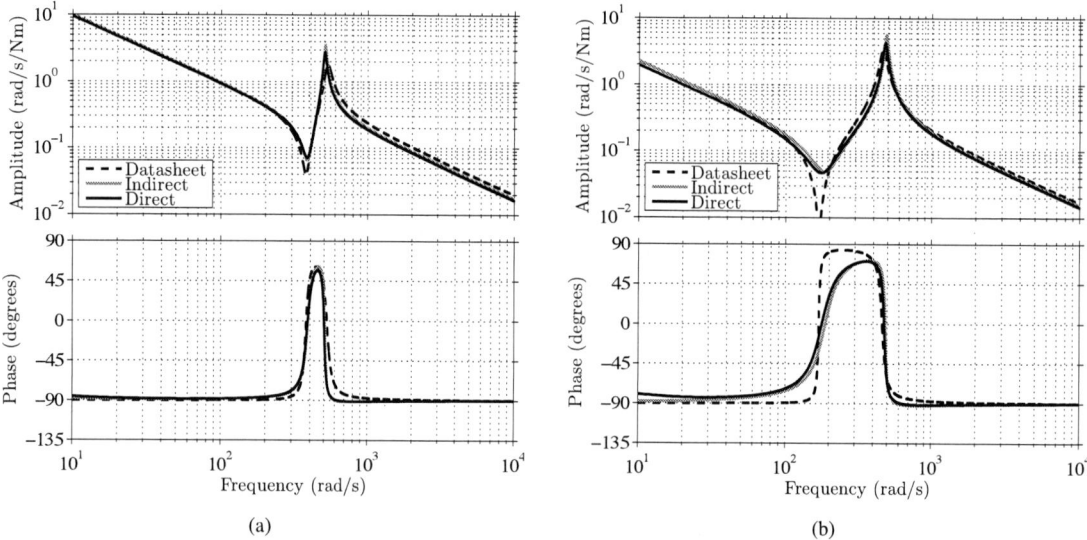

Fig. 7. Frequency responses obtained using closed-loop identification methods: (a) Configuration A; (b) Configuration B.

[11] S. E. Saarakkala, T. Leppinen, M. Hinkkanen, and J. Luomi, "Parameter estimation of two-mass mechanical loads in electric drives," in *Advanced Motion Control Workshop*, Sarajevo, Bosnia and Hertsegovina, Mar. 2012.

[12] S. Beineke, F. Schütte, H. Wertz, and H. Grotstollen, "Comparison of parameter identification schemes for self-commissioning drive control of nonlinear two-mass systems," in *IEEE IAS Conf.*, vol. 1, New Orleans, LA, Oct. 1997, pp. 493–500.

[13] S. Villwock and M. Pacas, "Application of the Welch-method for the identification of two- and three-mass-systems," *IEEE Trans. Ind. Electron.*, vol. 55, no. 1, pp. 457–466, Jan. 2008.

[14] Y. Yoshioka and T. Hanamoto, "Estimation of a multimass system using the LWTLS and a coefficient diagram for vibration-controller design," *IEEE Trans. Ind. Appl.*, vol. 44, no. 2, pp. 566–574, 2008.

[15] U. Forssell and L. Ljung, "Closed-loop identification revisited," *Automatica*, vol. 35, no. 7, pp. 1215–1241, Jul. 1999.

[16] S. Villwock, A. Baumuller, M. Pacas, F.-R. Gotz, B. Liu, and V. Barinberg, "Influence of the power density spectrum of the excitation signal on the identification of drives," in *IEEE IECON Conf.*, Orlando, FL, Nov. 2008, pp. 1252–1257.

[17] C. T. Johnson and R. D. Lorenz, "Experimental identification of friction and its compensation in precise, position controlled mechanisms," *IEEE Trans. Ind. Appl.*, vol. 28, no. 6, pp. 1392–1398, 1992.

[18] D.-H. Lee and J.-W. Ahn, "Dual speed control scheme of servo drive system for a nonlinear friction compensation," *IEEE Trans. Power Electron.*, vol. 23, no. 2, pp. 959–965, 2008.

[19] S. Villwock and M. Pacas, "Time-domain identification method for detecting mechanical backlash in electrical drives," *IEEE Trans. Ind. Electron.*, vol. 56, no. 2, pp. 568–573, Feb. 2009.

[20] M. Jokinen, S. Saarakkala, M. Niemela, R. Pollanen, and J. Pyrhonen, "Physical drawbacks of linear high-speed tooth belt drives," in *International Symposium on Power Electronics, Electrical Drives, Automation and Motion (SPEEDAM)*, Ischia, Italy, June 2008, pp. 872–877.

[21] L. Ljung, *System Identification: Theory for the User*, 2nd ed. Troy, NY: Prentice-Hall, 1999.

[22] I. D. Landau and G. Zito, *Digital Control Systems: Design, Identification and Implementation*, 1st ed. Germany: Springer, 2006.

[23] L. Ljung, *System Identification Toolbox: User's Guide*. Natick, MA: Mathworks, 2010.

[24] A. Frei, A. Grgic, W. Heil, and A. Luzi, "Design of pump shaft trains having variable-speed electric motors," in *International Pump Symposium*, Houston, TX, 1986, pp. 33–44.

[25] G. F. Franklin, J. D. Powell, and M. L. Workman, *Digital Control of Dynamic Systems*, 3rd ed. Menlo Park, CA: Addison-Wesley, 1997.

Inductor Loss Calculation of Coupled Inductors for High Power Density Boost Converter

Yuki Itoh, Shota Kimura, Jun Imaoka, Masayoshi Yamamoto

Shimane University
1060 Nishikawatsu, Matsue, Shimane 690-8504, Japan
yamamoto@ecs.shimane-u.ac.jp

Abstract— **The interleaved boost converter with coupled inductors which achieves high power density has gained great attention in automotive applications. However, the temperature of the inductor tends to be high because loss density is increased by the high power density. Therefore, it is important to calculate the inductor losses when the inductors in high power density converters are designed. In this paper, firstly, a core loss calculation method for the coupled inductor is proposed. Then, accuracy of the core loss calculation method is confirmed by experimental validation. Finally, magnetic properties of the used cores are shown and then the usefulness of the coupled inductors is discussed from different viewpoints of inductor losses and volume comparing single-phase, two-phase and two-phase converter with coupled inductors methods by using the imaginary inductors theory. As a result, inductor loss of the coupled inductors can be reduced compared to conventional methods, and a superior core material is suggested.**

I. INTRODUCTION

Power converters for EV and HEV have been required to miniaturization and high power density in order to increase the interior space in vehicles and improve the driving performance [1]-[3]. Interleaved boost converter with coupled inductors are known as one of the topologies to satisfy these demands. This is because higher frequency operation can be achieved for capacitive components by using interleaved technique [4]-[5]. And also, the coupled inductors are applied in an interleave converter in order to miniaturize the magnetic components. However, the temperature increase of these components becomes a problem since high power density of converter means that the thermal density of these components increases. As for magnetic components, the saturation flux density of the core is reduced in accordance with the temperature rise. As a result, a performance of the inductor is deteriorated. Recently, however, the core material with different magnetic properties began to be provided by many magnetic material manufacturers, the selection of the core corresponding to the application has been enabled [6]-[12]. The design of the magnetic component is required considering the reduction of the saturation magnetic flux density due to temperature rise of the core in driving the power converter. It is thought that a major factor of the temperature rise is the core loss.

It had been studied for general core loss of inductors for a long time as typified by Steinmetz equation [13]. In

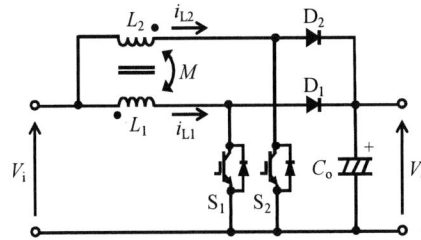

Fig. 1. Interleaved boost converter with coupled inductors.

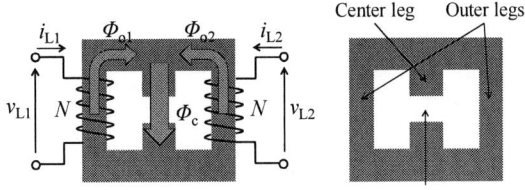

Fig. 2. Coupled inductors and core structure.

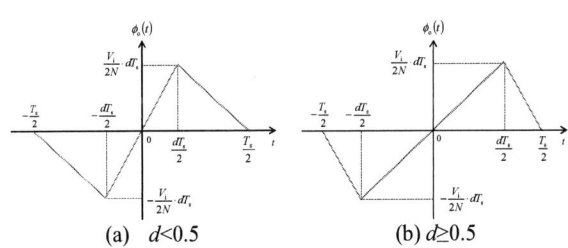

(a) $d<0.5$ (b) $d\geq0.5$
Fig. 3. Flux waveforms in outer leg.

(a) $d<0.5$ (b) $d\geq0.5$
Fig. 4. Flux waveforms in center leg.

978-1-4799-2706-7/14 $31.00 © 2014 IEEE

TABLE I
CIRCUIT PARAMETERS

Input voltage	V_i	26 V
Output voltage	V_o	84 V
Output power	P_o	1 kW
Duty ratio	d	0.692
Switching frequency	f_s	50 kHz
Inductor average current	I_{Lave}	19.2 A
Ratio of inductor ripple current	i_{Lpp}/I_{Lave}	0.2
Inductor ripple current	i_{Lpp}	3.85 A

TABLE II
INDUCTOR DESIGN AND CORE PARAMETERS

Core		PC40 EC90 (TDK Co.,Ltd)
Appearance of core		
Number of winding turns	N	13 turns
Mutual inductance	M	409 μH
Leakage inductance	L_{lk1}, L_{lk2}	28 μH
Air gap length	$l_g/2$	2.2 mm
Cross-sectional area of outer leg	A_o	2.85×10^{-4} m^2
Cross-sectional area of center leg	A_c	7.07×10^{-4} m^2
Core valume of outer leg	V_o	7.464×10^{-5} m^3
Core volume of center leg	V_c	4.483×10^{-5} m^3
Mass density	ρ_c	4.8×10^3 kg/m^3
Specific heat	c	600 J/(kg·K)

recent years, the study of iron loss calculation and evaluation method based on the conditions used in switching converters had been conducted. There are the cases that were discussed for filter inductors in single-phase PWM inverter and DC/DC converter so far [14]-[16]. However, as for the coupled inductor with higher frequency of the inductor ripple current and magnetic flux in the core, it has not been examined yet.

This paper proposes a core loss calculation method for the coupled inductors. This core loss calculation method is a method which is calculated from a core loss characteristic graph on datasheet of a core by applying the Fourier series expansion method to flux waveforms is magnetized under square wave voltage magnetizing condition. Then, the validity of the core loss calculation method is discussed from experimental viewpoints. In addition, the materials provided by the manufacture show magnetic property and the relationship of inductor loss and volume are discussed using the imaginary inductor theory in several core materials.

II. MODELING OF FLUX DENSITY AND FREQUENCY

Interleaved boost converter with coupled inductors is shown in Fig. 1. The coupled inductors and magnetic core structure are shown in Fig. 2. It is necessary for core

loss calculation to determine ripple flux density and switching frequency. In the coupled inductors core, the flux characteristics of outer and center leg are different. Fig. 3 and Fig. 4 show flux the waveforms of the outer and center legs of the coupled inductors. These AC flux is generate in the core by applying voltage to the inductor windings based on Faraday's law. From Fourier series of the flux in the outer and center legs, the ripple flux density at each frequency component ΔB_{ok}, ΔB_{ck} can be modeled as follows:

$$\begin{cases} \Delta B_{ok} = \dfrac{V_i \cdot \sin(d \cdot k \cdot \pi)}{N \cdot f_s \cdot (1-d) \cdot k^2 \cdot \pi^2 \cdot A_o} \\ \Delta B_{ck\_d<0.5} = \dfrac{V_i \cdot T_s \cdot \sin(2d \cdot k \cdot \pi)}{2 \cdot N \cdot (1-d) \cdot k^2 \cdot \pi^2} \\ \Delta B_{ck\_d\geq0.5} = \dfrac{V_i \cdot T_s \cdot \sin(2d-1) \cdot k \cdot \pi}{2 \cdot N \cdot (1-d) \cdot k^2 \cdot \pi^2} \end{cases} \quad (1)$$

Where, f_s is switching frequency, d is duty ratio, N is number of winding turns, k is the order of the Fourier series. Also, the frequencies of the flux density at each term f_{ok}, f_{oc} are given by:

$$\begin{cases} f_{ok} = k \cdot f_s \\ f_{ck} = 2 \cdot k \cdot f_s \end{cases} \quad (2)$$

III. CORE LOSS CALCULATION

Circuit parameters are shown in Table I. Inductor design and core parameters are given in Table II. The circuit parameters are a 1/60 equivalent model of a design specification of a boost converter which is mounted in PRIUS (TOYOTA, HEV). The core of coupled inductors is PC40 EC90 (TDK Co., Ltd: material ferrite). The maximum flux density value in the coupled inductor is 250mT.

From core loss characteristic graph on datasheet of PC40, the core loss calculation equation when calculating the core loss is given by the following formula:

$$P_{cv} = 1.95 \times 10^{-11} \cdot \Delta B^{2.33} \cdot f^{1.24} \text{ [W/m}^3\text{]} \quad (3)$$

This is the equation that the core loss characteristic is linearly approximated in order to calculate the core loss at any ripple flux density and frequency. But, the equation is the core loss per unit the core volume. Hence, the core losses at each frequency can be calculated from (1), (2) and (3). Core loss is calculated from the circuit constants of Table I. The core loss of the coupled inductors is 0.57W from the calculation results.

IV. EXPERIMENTAL VERIFICATION

We experimentally verified the accuracy of the calculation results of the core loss calculation method. The circuit and inductor design parameters are the same as it is shown in Table I and Table II. In general, the power dissipation P_c of the core can be determined as a function of the temperature rise ΔT of the core, the mass density ρ_c and the core volume V_{core} with:

$$P_{core} = c \cdot V_{core} \cdot \rho_c \cdot \Delta T \quad (4)$$

Where, c is the specific heat capacity of the core. Therefore, by measuring the temperature rise of the core during driving, the core loss can be back-calculated from (4). Fig. 7 shows the experimental waveforms. The power converter was driven for three hours in this state, and the measured points of temperature are the three points of the outer leg, the center leg and the winding of the coupled inductors as shown Fig. 8. The measured results of the temperatures are shown in Fig. 9. The temperatures of the outer leg, the center leg and the winding were 39.6℃, 39.8℃, 42.1℃, respectively after three hours. Therefore, the core loss of the coupled inductors was 0.56W from (4).

V. COMPARATIVE CORE LOSS VALUES

Table III shows the core loss comparative results. The core loss values in the calculated result and the experimental result are 0.57W and 0.56W, respectively. The deviation between the calculation and experimental results was less than 3%. As for the calculated result, core loss is almost identical, but DC bias component of the magnetic flux is not considered. On the other hand, as for the experimental result, the temperature increase of the core is thought to be affected by the heat generation of the windings.

VI. MAGNETIC MATERIALS CHARACTERISTICS

This chapter explains the characteristics of the core material that is provided. In this analysis, we used PC40 (Ferrite) from TDK , 10JNEX900 (Silicon Steel) from JFE , FINEMET (Nanocrystalline) from Hitachi Metals, 2605SA1 (Fe-Amorphous) from Metglas, and a selection of pressed powdered cores of different permeability, Kool Mu 125, MPP 550, High Flux 160 from Magnetics Inc. Magnetism properties   of these materials are shown in Table IV [8]-[11]. Magnetic properties of the material are dependent on the preparation method of the core and the basic material used.

PC40 (Ferrite) is a ceramic whose main component is iron oxide. It has high electrical resistance and low loss in the high frequency range. The characteristic of 10JNEX900 is a core with a high saturation magnetic flux density and high permeability, which solves the problem of the steel that becomes hard and brittle at 3.5% or more of silicon content. The precursor of FINEMET is amorphous ribbon obtained by rapid quenching at one million °C/second from the molten metal consisting of mainly Fe, Si, B and small amounts of Cu and Nb. This core material has a relatively high saturation magnetic flux density, high permeability, and low loss. 2605SA1 of iron-based amorphous metal has high saturation flux density, and the rate of temperature increase of core is low. Thus, this core material is capable of being used at high temperatures, and a compact design is possible. Powdered cores are produced from powder of various materials by coagulation with an epoxy resin. Basically, magnetic properties of the powdered cores are strongly dependent on the iron powder. However, by combining

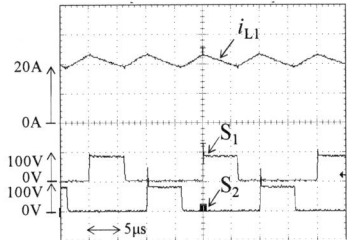

Fig. 7. Experimental Waveforms.

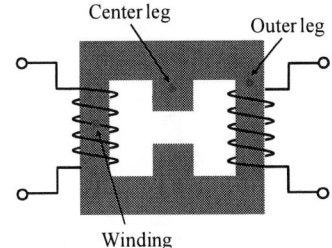

Fig. 8. Measured points of temperatures.

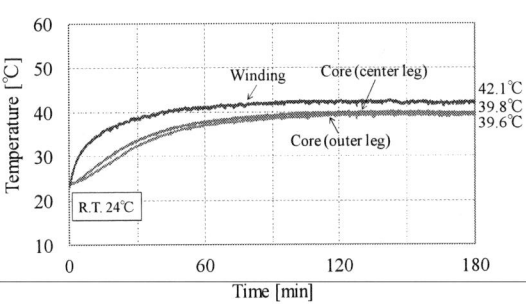

Fig. 9. Temperature measured results

TABLE III
CORE LOSS COMPARATIVE RESULTS

	Core loss
Calculated result	0.57 W
Experimental result	0.56 W

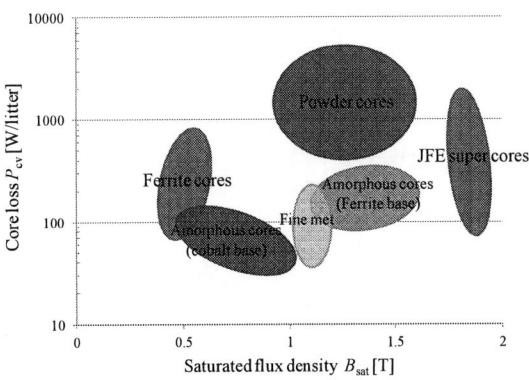

Fig. 10. Relationship between core loss and saturation flux density

TABLE IV
Magnetic Material Properties

Magnetic Material Type	Manufacturer	Material	B_{sat} [T] @25°C	Rel.Permeability	Curie temp. [°C]	Thermal conductivity [W/mK]	Electrical Resistivity [μΩ·m]	Density[g/cm³]	Core Loss @0.1T, 100kHz [W/litter]
Ferrite	TDK	PC40	0.5	2300	200	5	6.5×10^6	4.8	164
Silicon Steel	JFE	10JNEX900	1.8	23000	700	19.6	0.82	7.49	2066
Nanocrystalline	Hitachi metals	FINE MET	1.23	70000	570	7.1	1.3	7.4	445
Fe-Amorphous	Metglas	2605SA1	1.56	1200	395	10	1.37	7.18	889
Powder Core	Magnetics	Kool Mu 125	1.05	125	500	8	TBD	5.5	1051
Powder Core	Magnetics	MPP 550	0.75	550	460	8	TBD	8.0	1168
Powder Core	Magnetics	High Flux 160	1.5	160	500	8	TBD	7.6	1482

TBD – to be determined.

various materials, powdered cores achieve lower power losses than powdered cores of iron, and which have different permeability.

Material comparison chart shows core loss vs saturation flux density in the Fig.10. The higher core loss is reported in the 10JNEX900 and powder cores.

As shown in the Table IV, 10JNEX900 has the highest saturation flux, followed by 2605SA1, High Flux 160, FINE MET, Kool Mu 125, MPP 550, and PC40.

The core maximum operating temperature is limited by the Curie temperature and the thermal capability limits of laminated structure and coating.

Core loss is compared using different materials. Fig.11 shows the relationship between core loss and magnetic flux density in the case of frequency of 5kHz and 100kHz. From Fig.11, Core losses vary depending on the material of the core, and core loss is different for each material by the magnitude of the magnetic flux density and the frequency band. The lowest core loss is reported in the PC40 and FINEMET. The core loss of powder cores is relatively high. The highest core loss is presented by 10JNEX900. Consequently, materials need to be selected by operating frequency and magnitude of the magnetic flux density ripple.

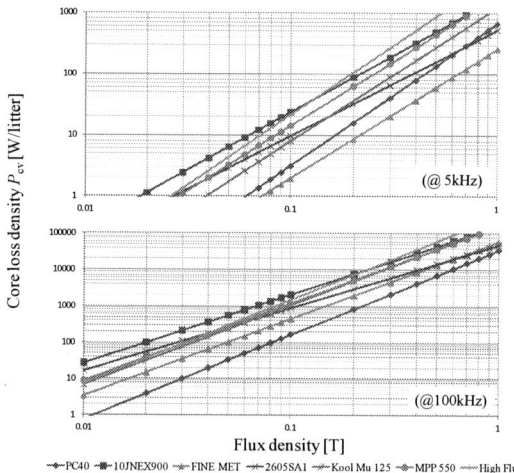

Fig. 11. Core loss and flux density changing at 5kHz and 100kHz

VII. INDUCTOR LOSS COMPARISON USING IMAGINARY INDUCTORS

In the previous sections, a core loss calculation method and magnetic material properties were introduced. This section shows comparative data of inductor loss and volume between single-phase, two-phase converter with non-coupled inductors and coupled inductors methods using imaginary cores in various materials. The inductor losses are total loss of core loss and copper loss.

A. Calculation model of core volume

Imaginary cores of non-coupled and coupled inductors are given in Fig. 11 and Fig. 12. The window shape and the cross-section of the imaginary cores are all defined as the square for simplicity. Flux density ripple, frequency and core volume are needed for calculating core loss. When a core size is discussed, important factors are window and cross-sectional area of the core. The window area A_w can be calculated as:

$$A_w = \frac{N \cdot A_{winding}}{k} \qquad (5)$$

Where N is the number of winding turns $A_{winding}$ is the sectional area of winding and k is space factor.

The sectional areas of a core can be determined from the condition that peak flux core Φ_p does not exceed the defined maximum flux Φ_{max}. The sectional area A_{core} of non-coupled inductor cores in single and two-phase methods is expressed as the following equation:

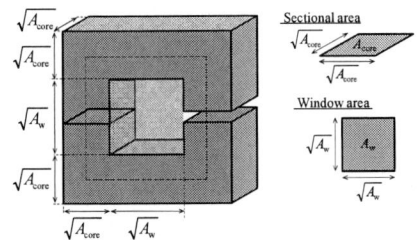

Fig. 11. Imaginary core of non-coupled inductor.

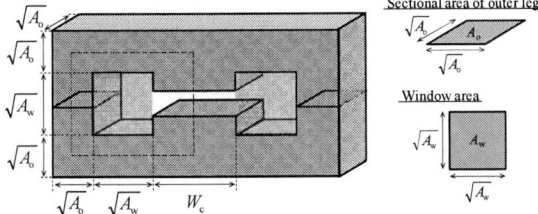

Fig. 12. Imaginary core of coupled inductors.

$$2 \cdot i_{Lpp} \cdot B_{max} \cdot A_{core} \cdot N - \left(2 \cdot I_{Lave} + i_{Lpp}\right) \cdot V_i \cdot d \cdot T_s \geq 0 \quad (6)$$

On the other hand, in case of $d \geq 0.5$, the outer leg sectional area A_o of coupled inductor core in two-phase transformer linked method is determined by:

$$\frac{2 \cdot i_{Lpp} \cdot B_{max} \cdot A_o}{V_i \cdot d \cdot T_s} \cdot \frac{d}{2d-1} \cdot N^3 - \left(I_{Lave} + i_{Lpp} \cdot \frac{d}{2d-1}\right) \cdot N^2$$

$$+ \left(1 - \frac{2d}{2d-1}\right) \cdot \frac{3\sqrt{A_w} + 4\sqrt{A_o}}{\mu_0 \cdot \mu_r \cdot A_o} \cdot B_{max} \cdot A_o \cdot N$$

$$+ \left(\frac{d}{2d-1} - \frac{1}{2}\right) \cdot \frac{3\sqrt{A_w} + 4\sqrt{A_o}}{\mu_0 \cdot \mu_r \cdot A_o} \cdot V_i \cdot d \cdot T_s \geq 0 \quad (7)$$

By setting N for (5), (6) and (7), core size can be calculated. The core loss will be calculated similarly as the section III.

B. Copper loss calculation model

The copper loss P_{copper} can be calculated from the DC resistance R_{dc}, AC resistance R_{ac} of the winding and the effective value I_{Lrms} of inductor ripple and average current I_{Lave}.

$$P_{copper} = R_{dc} \cdot I_{Lave}^2 + R_{ac} \cdot I_{Lrms}^2 \quad (8)$$

Here, when defining the section of winding as the square, DC resistance R_{dc} is expressed as (10) by the total length of windings l_L of (9) and the electrical resistivity ρ.

$$l_L = \left(4 \cdot \sqrt{A_{winding}} + 4 \cdot \sqrt{A_{core}}\right) \cdot N \quad (9)$$

$$R_{dc} = \rho \cdot \frac{l_L}{A_{winding}} \quad (10)$$

In consideration of the skin effect, AC resistance R_{ac} is expressed by the following formula.

$$R_{ac} = \rho \cdot \frac{l_L}{4 \cdot \delta \cdot \left(\sqrt{A_{winding}} - \delta\right)} \quad (11)$$

Where δ is the skin depth of winding.

Copper loss is calculated from the equation above.

TABLE V
Circuit Parameters.

Input voltage	V_i	202 V
Output voltage	V_o	650V
Output power	P_o	60 kW
Duty ratio	d	0.692
Switching frequency	f_s	50 kHz
Ratio of inductor current ripple	i_{Lpp}/I_{Lave}	0.2
Maximum flux density	B_{max}	500~1800mT
Turn number of winding	N	10 ~ 30 turns
Space factor	k	0.8
Current density of winding	j	5 A/mm^2

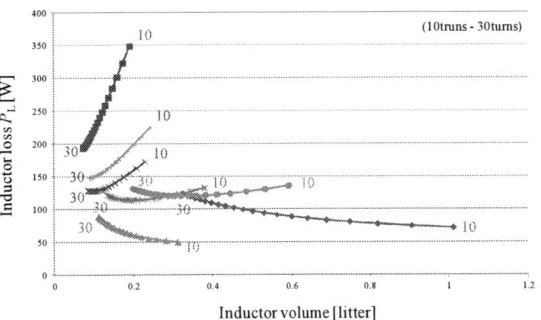

Fig.13. Relationship between the inductor volume and inductor loss in the case of changing the number of turns in the single boost converter.

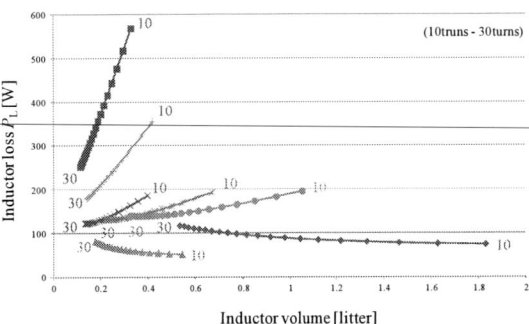

Fig.14. Relationship between the inductor volume and inductor loss in the case of changing the number of turns in the two-phase boost converter.

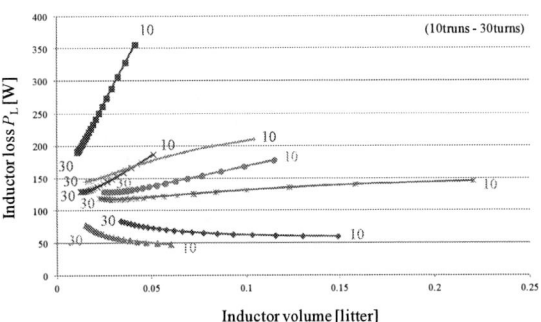

Fig.15. Relationship between the inductor volume and inductor loss in the case of changing the number of turns in the two-phased boost converter with coupled inductors.

C. *Comparison of inductor loss and volume*

The circuit parameters that are used in this analysis are shown in Table V. The circuit specification is the boost converter that is installed in the current Hybrid Vehicle Prius (TOYOTA). In the single-phase, two-phase methods with non-coupled inductors and coupled inductors, calculation results of inductor volume and inductor loss in the case of changing the number of turns from 10turns to 30turns are shown in Fig.13, Fig.14 and Fig.15.

Firstly, when focusing on the change of the winding turns *N*. From the calculation results, inductor volume can be miniaturized by increasing the number of turns, and reduction of inductor loss is possible. However number of turns is increased, inductor loss increases because copper loss becomes predominant

Then, when focusing on Fig.13 to Fig.15 and Table IV in terms of the material, inductor loss is understood that it is very different by core materials. 10JNEX900 has the highest inductor loss. Inductor loss of Powder cores is relatively high, PC40 and FINE MET is relatively low. Additionally, in terms of inductor volume, 10JNEX900 that has the largest saturation magnetic flux density is suitable for miniaturization in core materials of Table I. However, inductor loss per unit volume of 10JNEX900 is very high and specific heat of which is also high. Consequently, it is expected that the core of 10JNEX900 produces high temperature. From the calculation result, a superior core material is thought FINEMET which achieved the miniaturization of inductor volume and the decrease of inductor loss.

Inductor volume and inductor losses are compared in the different circuit method. Two-phase boost converter using mutual coupled inductors is more suitable for miniaturization than single and two-phase boost converter. This is because the current ripple of coupled inductors operates at the twice frequency from the effects of mutual induction as compared with the non-coupled inductor. The inductor loss of two-phase boost converter using mutual coupled inductors can also be reduced compared with the single-phase converter and two-phase converter.

However, since the inductor losses of the coupled inductors are higher than non-coupled inductor, the inductor loss per unit volume become high, in this way, temperature rise in the core is concerned. In particular, the material selection is important in the coupled inductors.

From the above results, it is shown that the inductor loss and inductor volume are changed by the number of turns, the inductor volume and inductor loss are largely determined by the magnetic properties of the material, and coupled inductors are superior in terms of miniaturization, but the inductor loss per unit volume is increased.

VIII. CONCLUSION

In this paper, a core loss calculation method for integrated magnetic components is proposed. Even when the flux characteristics in each part of the core are different as coupled inductors, these core losses can separately be calculated by this method. Therefore, it is considered that this method is useful for integrated magnetic components. The consistency is validated by experimental tests. Furthermore inductor loss and volume are compared between single, two-phase methods and two-phase transformer-linked method using imaginary cores in various materials. As a result of the comparison, transformer-linked method is effective for downsizing the inductor volume and reducing the inductor loss compared to the conventional methods. Depending on the material of the core, the difference of the inductor volume and the difference of inductor loss are shown. The calculation result of the above suggest that FINEMET is useful in the inductor loss and inductor volume

REFERENCES

[1] M. Hirakawa, M. Nagano, Y. Watanabe, K. Andoh, S. Nakatomi and S. Hashino, T. Shimizu: "High Power Density interleaved DC/DC Converter using a 3-phase integrated Close-Coupled Inductor set aimed for electric vehicle", IEEE Energy Conversion Congress and Exposition (ECCE2010), pp.2451-2457, (2010).

[2] A. Fratta, P. Casasso, G. Griffero, P. Guglielmi, S. Nieddu, G.M. Pellegrino: "New Design Concepts and Realisation of Hybrid DC/DC Coupling Reactors for light EVs", 29th Annual Conference of the IEEE Industrial Electronics Society (IECON '03), Vol.3, pp.2877-2882, (2003).

[3] K. Hartnett, J. Hayes, and M. Egan: "Novel CCTT-Core Split-Winding Integrated Magnetic for High-power DC-DC Converters", IEEE Energy Conversion Congress and Expo (ECCE2011), pp. 598-605, (2011)

[4] S.-Y Tseng, C.-L.Ou, S.-T. Peng and J.-D.Lee, "Interleaved coupled-inductor Boost Converter with Boost Type Snubber for PV system", the 1st IEEE Energy Conversion Congress and Exposition (ECCE2009), pp.1860-1867, (2009).

[5] J. Imaoka, M. Yamamoto: "Novel integrated Magnetic core structure suitable for Transformer-Linked Interleaved Boost chopper circuit" IEEE Energy Conversion Congress and Expo (ECCE2012), pp. 3279- 3284,(2012).

[6] Rylko, M.S. Univ. Coll. Cork, Cork ; Hartnett, K.J. ; Hayes, J.G. ; Egan, M.G.Magnetic "Material Selection for High Power High Frequency Inductors in DC-DC Converters" Applied Power Electronics Conference and Exposition,(APEC 2009),pp.2043-2049,(2009).

[7] B.J. Lyons, J.G. Hayes, M.G. Egan, "Magnetic material comparisons for high-current inductors in low-medium frequency dc-dc converters," IEEE Applied Power Electronics Conference, 2007, pp. 71-77.

[8] www.tdk.co.jp

[9] www.metglas.com

[10] www.hitachi-metals.co.jp

[11] www.mag-inc.com

[12] www.jfe-steel.co.jp

[13] C. P. Steinmetz, "On the law of hysteresis," *Proc. IEEE*, vol. 72, pp.196-221, 1984.

[14] Seiji Iyasu, T. Shimizu and K. Ishii, "A Novel Inductor Loss Calculation Method on Power Converters Based on Dynamic Minor Loop", *IEEJ Trans.* on *I. A.*, Vol. 126, No. 7 pp. 1028-1034 (2006-7)

[15] T. Shimizu, K. Kakazu, H. Matsumori, K. Takano and H. Ishii, "Iron Loss Evaluation of Filter Inductor used in PWM Inverters" Proc. of the 1st IEEE Energy Conversion Congress and Exposition (ECCE 2011), pp.606-613 (2011).

[16] J. Muhlethaler, J. Biela, J. Kolar and A. Ecklebe: "Improve Core-loss Calculation for Magnetic Components Employed in Power Electronic Systems", IEEE Trans on Power Electronics, Vol.27, No.2, pp.964-973 (2012).

978-1-4799-2706-7/14 $31.00 © 2014 IEEE

1.2kW Dual-Active Bridge Converter using SiC Power MOSFETs and Planar Magnetics

D. De, A. Castellazzi
Power Electronics and Machines Control Group,
University of Nottingham,
Nottingham, UK
Email: alberto.castellazzi@nottingham.ac.uk

A. Lamantia
Operations Department,
BLU Electronic,
Desio, Italy
Web: www.bluelectronic.com

Abstract— **This paper proposes the development of a compact high frequency bi-directional dc-dc converter, particularly designed for the case of future avionic applications, where one DC power bus at 270V needs to be interconnected with a 28V one. The selection of the topology (dual active bridge) in comparison with other isolated high step down topologies is presented with state-of-the-art power device technology. The topology is implemented with new semiconductor device technology to maximize the switching frequency and deliver a solution with contained volume and weight. Planar magnetic are employed to further optimize the converter shape. An experimental evaluation and detailed loss separation based on simulation study are presented.**

Keywords— *Dual active bridge converter, SiC power MOSFETs, planar magnetics.*

I. INTRODUCTION

Future avionic applications foresee a substantial electrification of the on-board energy generation and distribution infrastructure. In particular, the need will arise for interfacing a DC power bus at 270 V with a DC power bus at 28 V with bi-directional power transfer capability between the two. To that aim, highly efficient dc-dc converters are needed, which will be however viable for this application only if they can offer contained volume and weight. In this work we focused on a 1.2 kW rated converter and went through a process of topology selection, detailed design, prototyping and testing. Each phase is presented in detail in a dedicated section.

II. TOPOLOGY SELECTION AND DESIGN

A. Selection

Many different topologies can in principle be considered for the above discussed requirements. Here, however, we further restricted our attention to topologies offering:

✓ *fixed frequency operation*, for predictable EMI performance and easier control;

✓ *resonant-transition soft-switching capability* (particularly, zero-voltage switching ZVS), for circuit simplicity and better long term performance stability;

✓ *low-order dynamics,* for ease of implementation and option for parallel operation.

The following isolated topologies were initially considered and benchmarked against the above listed requirements:

➢ Two-Stage Isolated DC-DC converter (TSI) [1];

➢ Cyclo-Converter based DC-DC converter (CCB) [2];

➢ Dual Active Bridge converter (DAB) [3].

In particular, for the dual-active bridge converter we considered two implementation options: one with full-bridge switch arrangements on both the primary and secondary sides [4], shown in Fig. 1 and one with a center-tapped transformer secondary winding [5], shown in Fig.2. Indeed, the latter offers the possibility in principle to reduce the number of active switches for the same efficiency or to improve efficiency for the same number of switches (using two parallel MOSFETs for each switch on the output) against some complication in the design of the transformer.

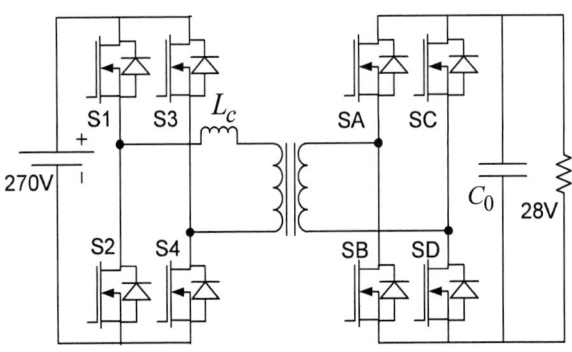

Fig. 1. Topology 3 – Dual active bridge topology with full bridge at either side of transformer.

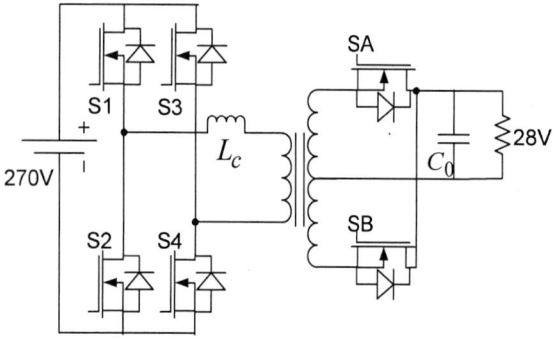

Fig. 2. Topology 4 – Dual active bridge topology with center-tapped transformer at secondary side.

Beyond the power and voltage rating, the initial design aims were as summarized in Table I, while the benchmark criteria were efficiency overall volume and input current harmonics content.

TABLE I
BENCHMARK DESIGN TARGETS

Output voltage ripple (peak-to-peak)	100 mV
Inductor current ripple (where applicable)	10% of load current
Switching frequency	$\geq$100 kHz

The initial study was based on extensive electro-thermal circuit simulation based on realistic physics-based compact models of the semiconductor devices [6, 7]. Both directions of power transfer were analyzed and two cases were considered: first, ideal components were assumed, with all parasitic inductance values set to zero; then, realistic values were taken into account for the stray inductance of the transformer and device package terminals. Efficiency benchmark results across the various topologies are summarized in Fig. 3 (a) and (b) for the two cases, respectively, for a nominal power output of 1.2 kW; it is interesting to note that, the DAB with center-tapped secondary and parallel output MOSFETs performs best when ideal conditions are assumed. As can be seen, when the presence of unavoidable leakage inductance of the transformer and interconnections is realistically taken into account, the most efficient topology is the DAB and the implementation with two full-bridge cells is more efficient that the secondary-center-tapped one, even when two devices are paralleled on the 28 V (i.e., the high current) side.

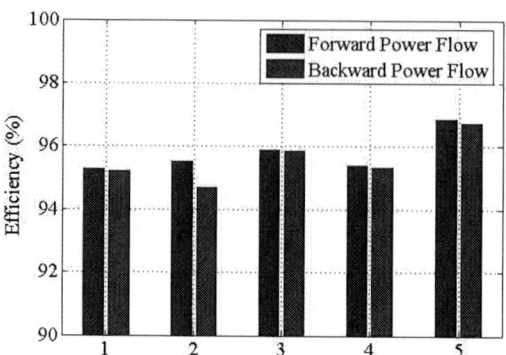

1: Two stage isolated DC-DC
2: Cycloconverter based DC-DC,
3: DAB with Full bridge,
4: DAB with Center-tap at secondary
5: DAB with Center-tap and with 2 devices in parallel at secondary

(a)

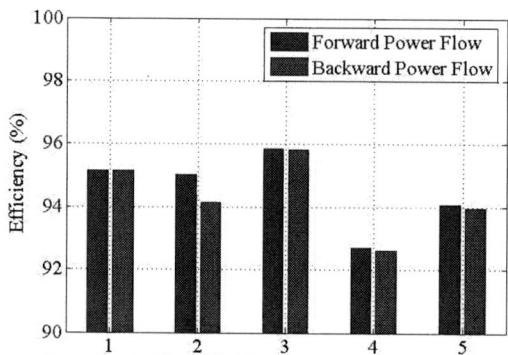

1: Two stage isolated DC-DC
2: Cycloconverter based DC-DC,
3: DAB with Full bridge,
4: DAB with Center-tap at secondary
5: DAB with Center-tap and with 2 devices in parallel at secondary

(b)

Fig. 3. Simulated efficiency at full load for various converter topologies in both power-flow directions (here, forward is power transfer from the 270V to the 28 V side).

In terms of volume and weight, the size of the transformer and inductor for the DAB implemented with two full-bridge cells is found to be the minimum (Fig. 4 and Fig. 5), as well as the overall size after adding the contribution of the capacitors (in principle negligible in comparison to that of the magnetic components). Indeed, the CCB topology enables to minimise the size of the EMI filters, but the DAB is still estimated to be superior in terms of overall size.

The 2014 International Power Electronics Conference

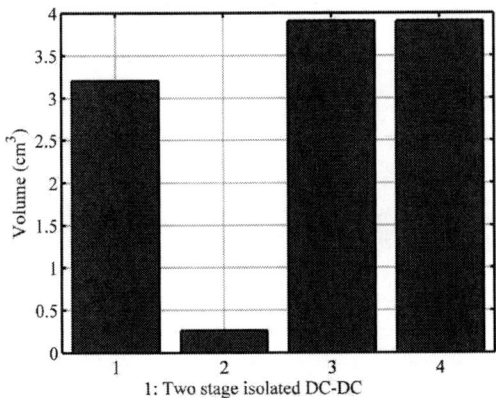

Fig. 4. Estimated output capacitor volume of various converter topologies for a 1.2 kW delivery with nominal 270V and a 28V input/output voltage levels.

1: Two stage isolated DC-DC
2: Cycloconverter based DC-DC,
3: DAB with Full bridge,
4: DAB with Center-tap at secondary

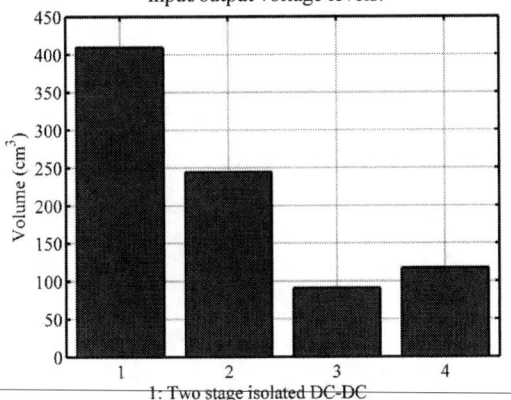

1: Two stage isolated DC-DC
2: Cycloconverter based DC-DC,
3: DAB with Full bridge,
4: DAB with Center-tap at secondary

Fig. 5. Estimated magnetic volume of various converter topologies for a 1.2 kW delivery with nominal 270V and a 28V input/output voltage levels.

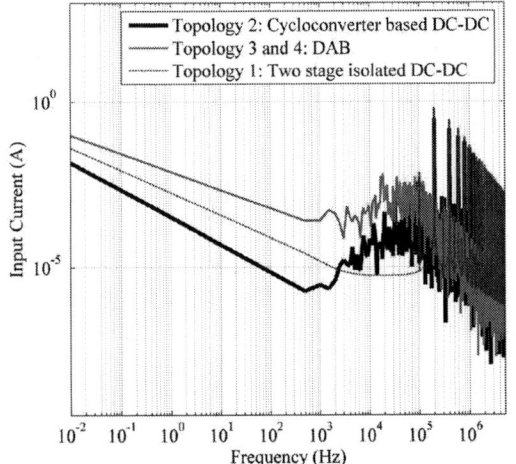

Fig. 6. Input current frequency components spectrum for various converter topologies with same input capacitance.

Also, from a control point of view the DAB is an ideal solution as the plant has first order dynamics and acts as a controlled current generator, thus enabling relative ease of paralleling for higher power demands. Moreover, the primary inductor could be in principle implemented as leakage inductance of the transformer, with enhanced integration potential. In summary, though additional pros and cons could be discussed in more detail for each of the investigated topologies, in view of all the above considerations, the DAB was retained as the most suitable topology.

B. Design

Fig. 7 shows the circuit schematic. Although the converter is intended for full bi-directional operation, in the following reference is made to the 270V-side as the primary and to the 28V-side as the secondary side.

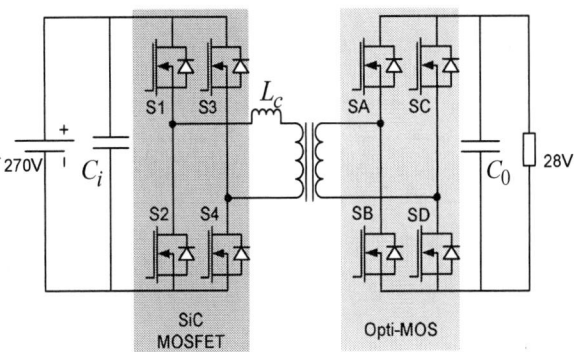

Fig. 7. Dual-active bridge dc-dc converter interfacing a 270V and a 28V power bus.

The circuit basic operation is broadly reported in literature and is here summarised by the waveforms in Fig. 8, which shows the device switching sequence and representative steady state waveforms of the converter for a given phase shift between the bridges (the voltage across L_c and the transformer primary current I_{Lc}). Between the commutation of the switch pairs S1-S4 and S2-S3 on the primary and switch pairs SA-SD and SB-SC on the secondary side, a dead-time is inserted to enable resonant-transition turn-on zero voltage switching (ZVS) of the converter. Depending on the value of L_C, this feature can be kept over a signifcant load range [8]. Typically, by increasing the converter switching frequency the size of the magnetic components and the capacitive filters can be reduced; however, this is achieved at the expense of increased switching losses and the associated cooling requirements. Recent advancements in power device technology, in particular the availability of wide-band-gap (SiC) power devices, enable higher switching frequencies without degradation of performance and without additional cooling requirements. In the present application SiC MOSFETs rated at 600V are used in the primary side (270V) and Infineon Opti-MOS transistors are used in the secondary side (28V), as indicated in Fig. 7. It is worth pointing out that both transistor types enable for the use of the body-

978-1-4799-2706-7/14 $31.00 © 2014 IEEE   2505

diode as a freewheeling element for the current in L_C, without the need for additional anti-parallel diodes. Moreover, both transistor types are avalanche rugged, capable of easily withstanding the dissipation of the typical energy levels stored in parasitic inductances prior to commutation and thus allow for reliable fully snubberless design of the converter [9].

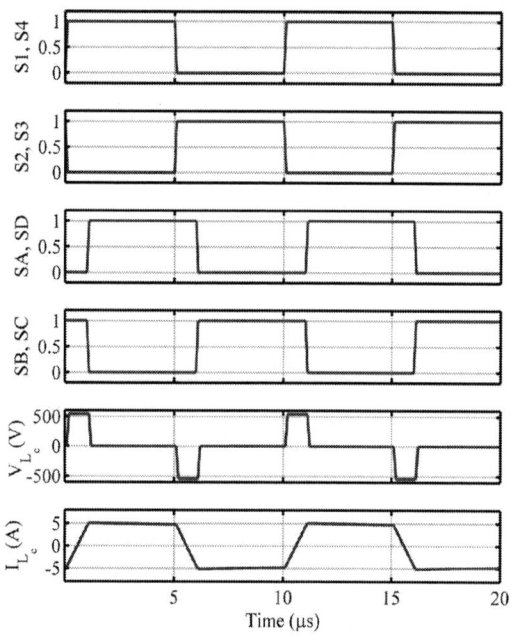

Fig. 8. Representative waveforms of the dual-active bridge converter at 100 kHz.

III. HARDWARE IMPLEMENTATION AND TEST

Table. II summarizes the converter design specifications.

TABLE II
DETAILS OF THE POWER CELL DESIGN

Input Voltage (V_i)	270V
Output Voltage (V_o)	28V
Power Rating (P_o)	1.2kW
Switching Frequency	100kHz
Primary-side switches (SiC MOSFET)	SCT2120AFC
Secondary-side switches (Opti-MOS)	IPA032N06N3 G
Output capacitance (C_o)	720µF
Transformer turns ratio	19:2
Series Inductor (Lc)	45µH

On the low-voltage/high-current secondary side, very large capacitance value is required to enable suitable

voltage smoothing without the need for additional LC filters, which might complicate the circuit dynamics by introducing additional poles. The capacitance is implemented by paralleling a number of ceramic and electrolytic capacitors to achieve an optimum trade-off between volume/cost and performance.

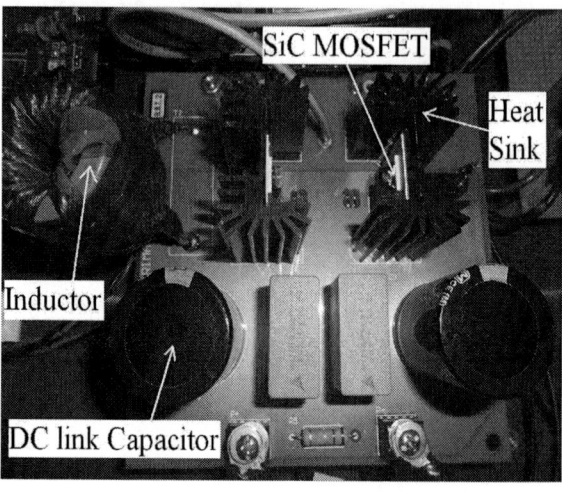

(a)

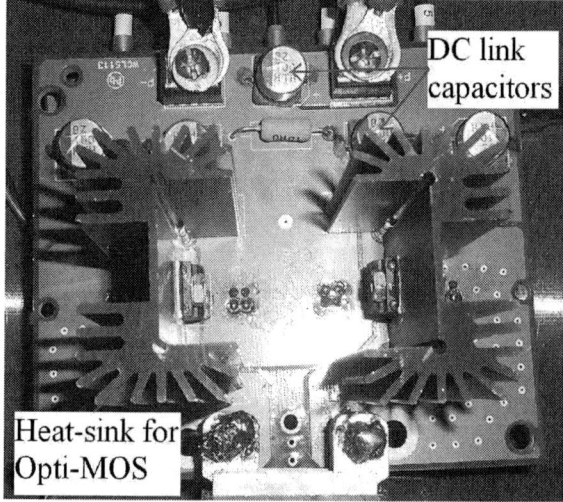

(b)

Fig. 9. Photograph of the experimental set up (power cell): 270V side (a) and 28V side (b).

Fig. 9 shows a photograph of the experimental prototype implementation. The primary and secondary side power cells, respectively, are implemented on separate PCBs and interconnected by the transformer. In both PCBs 2 semiconductor devices shares a common heat sink. The control and protection functions are implemented in the prototype by means of an FPGA/DSP board; for the final converter, we anticipate the use of a fully FPGA based control platform. The gate drivers are fitted from bottom of the power cells. A bespoke design is carried out for the primary side SiC MOSFETs and the secondary side Si MOSFET. Non-symmetrical drive voltages are used for the SiC devices on the primary side (-4 to +20 V), and for the low voltage Si MOSFETs on the secondary side (-4 to +15 V). Insulation at gate-driver level is achieved by means of opto-couplers, in particular in view of greater ease in achieving a non-symmetrical drive voltage waveform as compared to the use of transformers.

In view of the required transformer turns-ratio, to contain overall volume and optimize thermal management, a planar implementation of transformer and inductor was opted for. Indeed, the series inductor could be fabricated as transformer leakage inductance in the forms of integrated magnetics. However, since its value directly affects the converter control and dynamic characteristics, as well as the load level down to which ZVS of the transistors can be achieved, it was decided at this stage to keep the two elements separate, which also enables an optimization of the transformer efficiency and thermal management.

Fig. 10 shows a photograph of the transformer enclosed in a dedicated heat-sink. This component was designed externally upon customary specification. The primary side magnetizing inductance is 2.5 mH, the DC-resistance value is 0.35 mΩ and the 100 kHz value is 45mΩ. The leakage inductance is less than 0.3% of the nominal magnetizing inductance value. The dimensions are 75 x 65 x 31 mm and the total weight, including the heat-sink, also visible in Fig. 10, is less than 300 grams. The estimated loss of efficiency due to the transformer was 2% with a nominal heat-sink temperature of 70 °C and a maximum hot-spot temperature inside the transformer of 115 °C at 1.2 kW power transfer.

Fig. 10. Custom-designed planar transformer.

The inductor was built in-house using toroidal Moly-Permalloy Powder cores to determine the optimum value before moving to a planar fabrication. The planar component is being fabricated externally based on a custom design. Its dimensions are 20 x 20 x 40 mm and the calculated power losses at full load are 10W.

Extensive functional testing of the converter was carried out. Fig. 11 shows the 28V side output voltage.

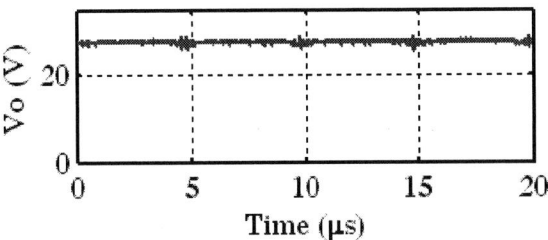

Fig. 11. Experimental waveform of output voltage 28V at full load

More representative experimental waveforms are shown in Fig. 12(a) and Fig. 12(b) for forward and backward power flow directions respectively. The measured efficiencies at various loading are shown in Fig. 13. It can be seen that the converter performance is quite consistent with the direction of power flow. Operation of the switches is fully snubber-less and ZVS is achieved down to about 30% of the maximum load on the primary side and is kept down to about 12% of the maximum laod on the secondary side (Fig. 14).

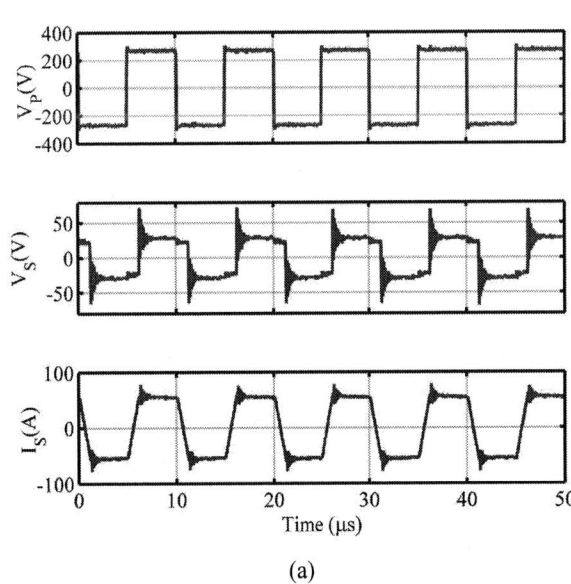

(a)

The 2014 International Power Electronics Conference

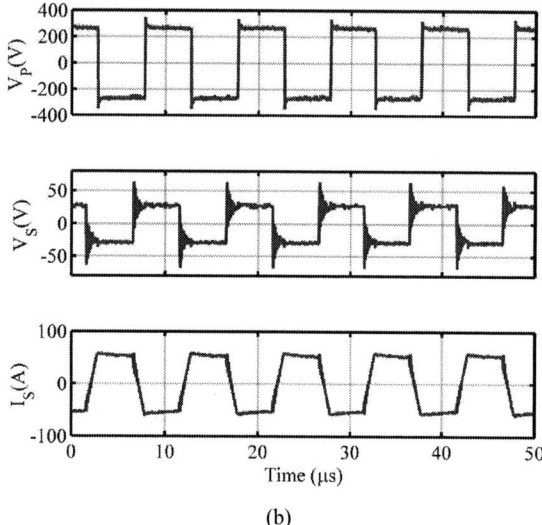

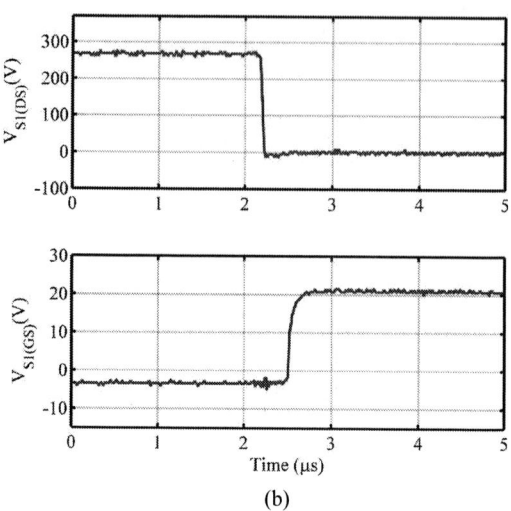

(b)

Fig. 12. Measured results: (a) power flow in forward direction
(b) power flow in reverse direction

(b)

Fig. 14. Soft switching at (a) secondary side and (b) primary
side converter

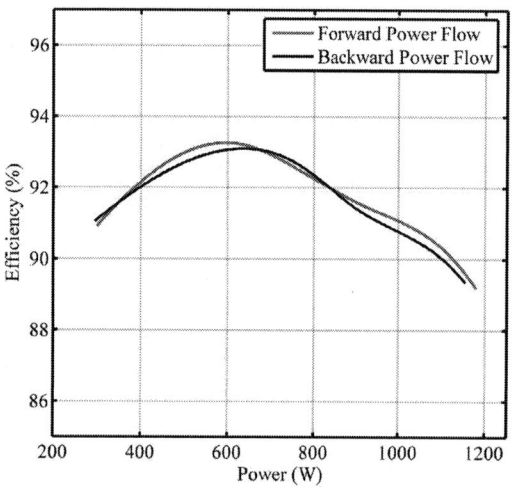

Fig. 13. Measured efficiency of the power converter at forward
and backward direction.

IV. THERMAL CHARACTERISATION AND LOSS DISTRIBUTION ANALYSIS

Finally, infrared thermal characterization of the converter was performed. In particular, the focus was on the thermal performance of the SiC MOSFETs and the planar transformer. Fig. 15 (a) shows temperature distribution for one heat-sink used for cooling two of the primary side bridge SiC MOSFETs (specify which, see Fig. 9): the heat-sink nominal thermal resistance to ambient is 3.7 W/K and as can be seen, with a laboratory ambient temperature of approximately 18 °C and under natural convection conditions, the heat-sink temperature at full load is only 47 °C, indicating the possibility to significantly reduce the heat-sink size and even go fully heat-sink less depending on the operational boundary conditions. Fig. 15 (b) shows the temperature distribution on the transformer heat-sink still at full load (see Fig. 10): in this case, too, there is ample margin for optimization of the thermal management effort depending on the actual operational conditions. Definitely, the planar transformer structure enables easy and efficient cooling. It should be noted here that in Fig. 15 (a) and (b), only the transistors and transformer heat-sink were coated black and had an emissivity $\varepsilon = 0.95$, as indicated in the images. The color scale cannot be transferred to other parts of the image (e.g., the transformer windings), due to very different emissivity values (in other words, the fact that the transformer windings appear to be at a considerably higher temperature is an imaging effect due to different emissivity value).

The loss separation is carried out based on simulation study at different loading condition (for both forward and backward directions). The simulation software requires the datasheet information of the semiconductor devices. Fig. 16 and Fig.17 shows the conduction and switching losses in 270V side (SiC) devices at various loading conditions. Total semiconductor conduction and switching losses are shown in Fig. 18 and Fig. 19

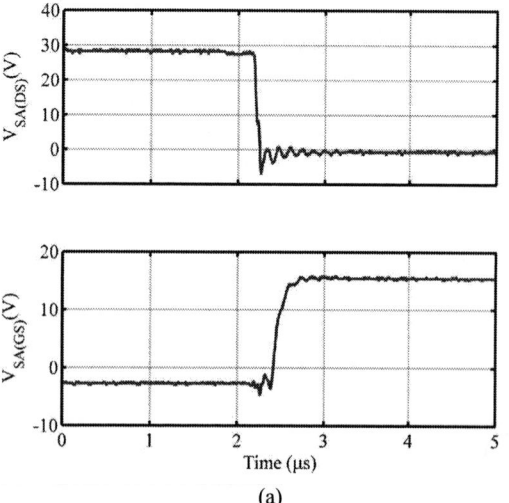

(a)

978-1-4799-2706-7/14 $31.00 © 2014 IEEE   2508

respectively. It can be noticed that the losses in SiC devices are quite small compared to total semiconductor losses. The transformer losses (26 W) and inductor losses (10 W) make the total accounted losses 100W at full load. The remaining losses are taking place in bleeder resistors, capacitors, PCB tracks, connecting wires and joints.

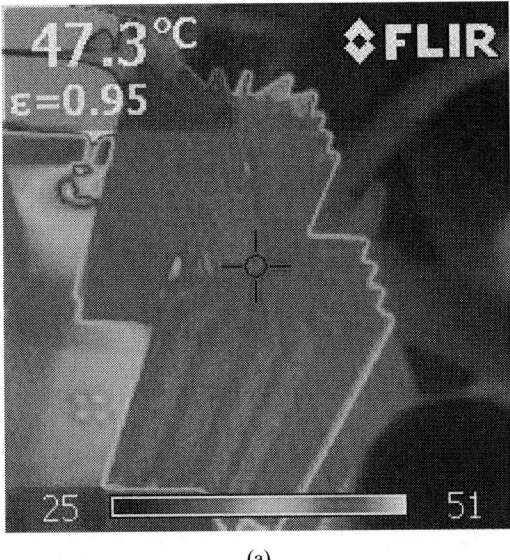

(a)

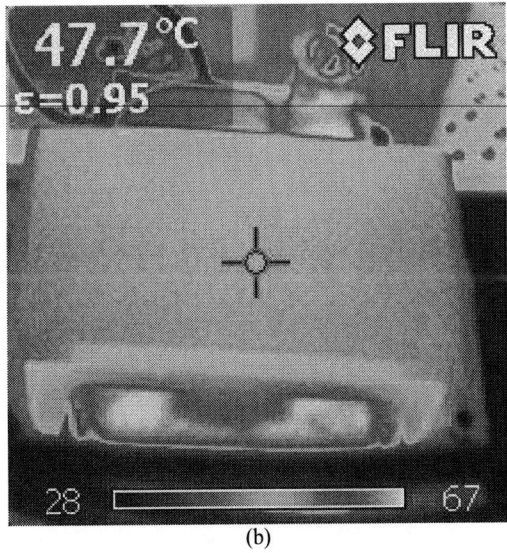

(b)

Fig. 15. Thermal image of (a) Heat sink for SiC devices at primary side (b) high frequency transformer.

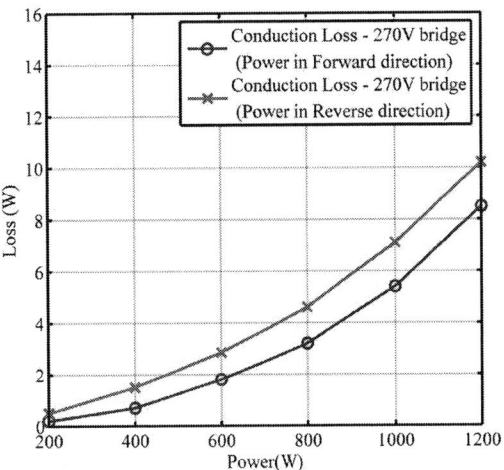

Fig. 16. Conduction loss at 270V SiC MOSFETs from PSIM

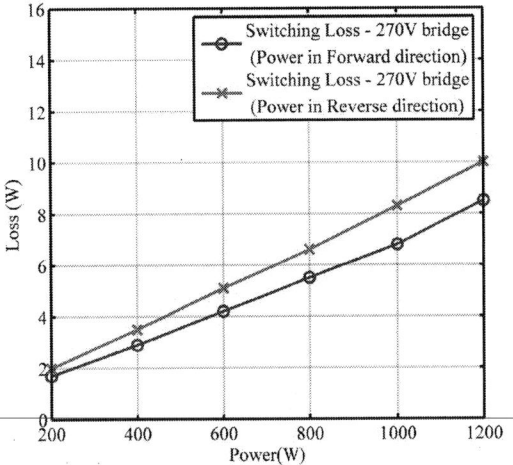

Fig. 17. Switching loss at 270V SiC MOSFETs from PSIM

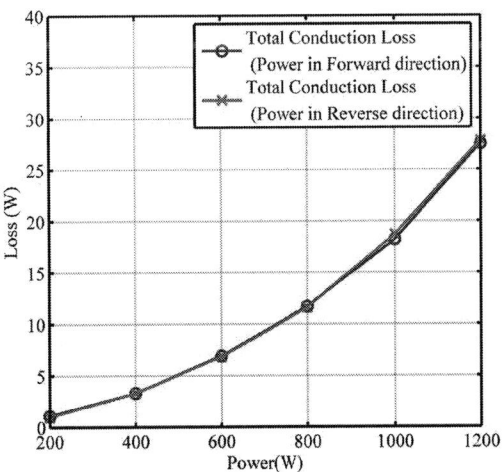

Fig. 18. Total conduction loss of the converter from PSIM

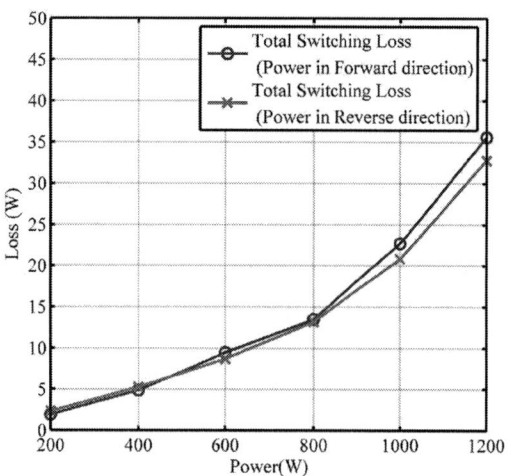

Fig. 19. Total switching loss of the converter from PSIM

V.   CONCLUSION

This paper proposes the design of a 1.2 kW dc-dc converter making use of SiC power MOSFETs and planar magnetics to achieve high switching frequencies without degradation of efficiency, so as to be able to minimize the overall converter size. A comparative study between the isolated high step-down converters reveals the superiority of dual active bridge in the application. The power converter fully bi-directional and its performance are quite consistent with the direction of power flow. Extensive experimental results confirm the validity of the proposed solution. The loss separation and thermal image during continuous full load testing show the effectiveness of SiC MOSFETs and the planer transformer.

ACKNOWLEDGMENT

This work was supported by the EU CleanSky Joint Technology Initiative with the REGENSYS project. The authors wish to acknowledge the appreciated contribution with FPGA programming of Dr. Saul Lopez-Arevalo of the Power Electronics, Machines and Control Group of the University of Nottingham.

REFERENCES

[1] Dong Cao, Shuai Jiang, Fang Z. Peng, and Yuan Li, 'Low Cost Transformer Isolated Boost Half-bridge Micro-inverter for Single-phase Grid-connected Photovoltaic System', *in Proc. IEEE APEC 2012*, pp. 71-78.

[2] Cleber Zanatta, and José Renes Pinheiro, 'A No DC-Gain Error Small-Signal Model for the Zero-Voltage-Switching Phase-Shift Modulated Full-Bridge DC-DC Converter', *in Proc. IEEE IECON 2006*, pp. 1921-1926.

[3] Myoungho Kim, Martin Rosekeit, Seung-Ki Sul, and Rik W. A. A. De Doncker, 'A Dual-Phase-Shift Control Strategy for DualActive-Bridge DC-DC Converter in Wide Voltage Range', *in Proc. IEEE ECCE 2011*, pp. 364-371.

[4] Naayagi, R. T.; Forsyth, A.J.; Shuttleworth, R., "High-Power Bidirectional DC–DC Converter for Aerospace Applications," *Power Electronics, IEEE Transactions on*, vol.27, no.11, pp.4366,4379, Nov. 2012

[5] Xinke Wu, Chen Zhao, Junming Zhang, and Zhaoming Qian, 'A New ZVZCS Full Bridge Converter with an Auxiliary Centre Tapped Rectifier', *in Proc. INTELEC 2004*, pp. 321-327.

[6] A. Castellazzi, R. Kraus, Y. Gerstenmaier, G. Waschutka, *Reliability Analysis and Modeling of Power MOSFETs in the 42-V-PowerNet*, in IEEE Trans. On Power Electronics, Vol. 21, No. 3, May 2006.

[7] V. d'Alessandro, A. Magnani, A. Castellazzi, M. Riccio, G. Breglio, A. Irace, N. Rinaldi, *SPICE Modeling and Dynamic Electrothermal Simulation of SiC Power MOSFETs*, to be published in Proc. ISPSD2014, Hawai, USA, June 2014.

[8] A. S. Babokany, M. Jabbari, G. Shahgholian, M. Mahdavian, "A review of bidirectional dual active bridge converter," *in Proc. ECTI-CON 2012*, pp.1,4, 16-18 May 2012

[9] A. Fayyaz, L. Yang, A. Castellazzi, *Transient robustness testing of silicon carbide (SiC) power MOSFETs*, in Proc. of EPE2013, Lille, France, Sept. 2013.

Analysis of Hysteresis and Eddy-Current Losses for a Medium-Frequency Transformer in an Isolated DC-DC Converter

Mizuki Nakahara and Keiji Wada
Department of Electrical and Electronics Engineering
Tokyo Metropolitan University, 1-1 Minami Osawa, Hachioji, Tokyo, JAPAN
kj-wada@tmu.ac.jp

Abstract—Recently, because of the rapid development of power devices, high-power isolated DC-DC converters using medium-frequency (MF) transformers have been proposed. In order to realize a high-power-density isolated DC-DC converter, it is important to analyze losses of the transformers because the volume of the MF transformer is one of the large parts in the DC-DC converter.

In this paper, a procedure for separating the iron loss into hysteresis and eddy-current losses is presented. The validity of the separation procedure is then considered by comparing the calculation results with the results of the analysis method. Furthermore, the temperature dependencies of the iron loss will be considered by the experiments.

I. INTRODUCTION

Recently, bidirectional power converters for medium-voltage applications have been discussed for a next-generation electrical grid such as a smart grid. A typical candidate for bidirectional circuits is an isolated DC-DC converter with a medium-frequency (MF) transformer, and high-power isolated DC-DC converters have been proposed [1]-[3]. Isolated DC-DC converters for a railway application and a power distribution system have been discussed [4]. Furthermore, a DC-DC converter whose voltage is 2.5 kV and transmitted power is 200 kW have been developed [5].

In order to realize a high-power isolated DC-DC converter, a reduction in size and higher efficiency in an MF transformer are needed. However, a reduction in size results in higher power and a loss in density in the transformer, which may cause serious problems such as a thermal limitation in the core. Because of that, it is necessary to calculate losses, especially the iron loss of the MF transformer. The MF transformer for the DC-DC converters has been discussed in many papers [6], including topics such as the total loss and isolation materials. However, the analysis procedures of the transformer loss for division into the hysteresis and eddy-current losses have never been discussed. In order to design the optimization of the MF transformer, it is necessary to divide the iron loss under actual operation condition of the DC-DC converter. If the characteristics of each iron loss factor are made clear, a dominant content of the iron loss can be clear. For example,

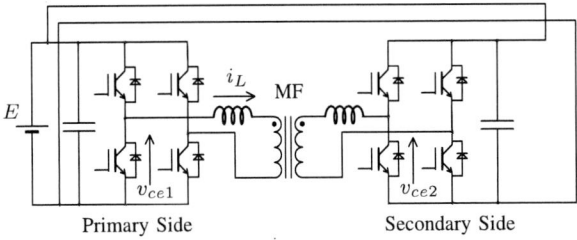

Fig. 1. Experimental configuration

if the eddy-current loss is a dominant component of the iron loss, a better design of a MF transformer which can reduce eddy-current loss can be propose.

In this paper, a simple method for the total iron loss calculation specializing in the isolated DC-DC converter will be shown without using 3D-Finete Element Method (3D-FEM). In order to analyze the iron loss in detail, this paper proposes an iron loss calculation method that can divide the iron loss into hysteresis and eddy-current losses. The validity of the separation procedure will be confirmed by analytical results on the basis of the measurements of the core characteristics. Furthermore, the temperature dependencies of the iron loss for an isolated DC-DC converter are discussed with the experimental results. Three core materials (nanocrystalline, amorphous, and silicon sheet) are used as transformers in the experiment. This is because the three core materials are sold on the market and are suitable for use in an actual high-power DC-DC converter. In order to compare the calculated and measured iron losses, an experimental circuit is designed and implemented.

II. IRON LOSS CALCULATION METHODS FOR ISOLATED DC-DC CONVERTER

A. Circuit Configuration

Fig. 1 shows the experimental circuit of a DC-DC converter. It transmits electrical power from the primary side to the secondary side via the MF transformer, and the transmitted power is regenerated from the secondary side to the primary side. The rated DC voltage is set to 300 V, and the transmitted power is 9 kW. Moreover, the switching frequency f is set

978-1-4799-2706-7/14 $31.00 © 2014 IEEE

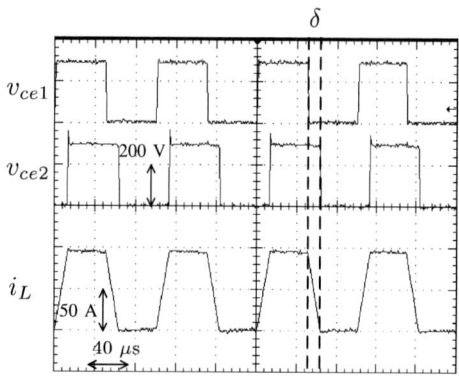

Fig. 2. Waveforms at $E = 300$ V, transmitted power is 9 kW

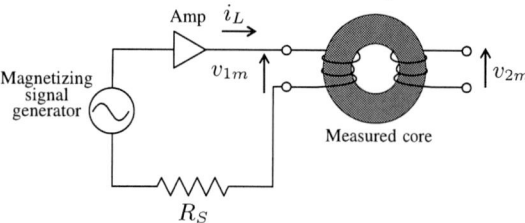

Fig. 3. Measurement circuit in the B-H analyzer

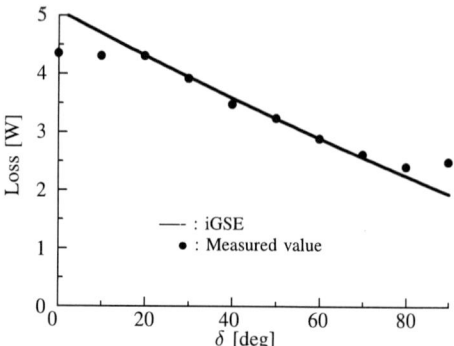

Fig. 4. Iron losses by proposed method and iGSE for amorphous core material

to 10 kHz. Fig. 2 shows the experimental waveforms at the rated power of the voltages and current in the circuit. In this case, the phase difference of the circuit is set to 30°. The output current i_L of the primary side circuit flows 50 A and trapezoidal current waveform.

In this system, the cores of the MF transformer can be changed easily with the same structure of windings. The core shape of CS-125 (121 mm × 63 mm × 35 mm) is used, and it is the laminated core. Table I shows the specifications of the cores. Further, in order to decrease the leakage inductance, a bifilar wound structure is adopted.

B. Calculation Method from Customized iGSE

As an iron loss calculation method under alternative excitation conditions, the improved generalized Steinmetz equation (iGSE) [7] is well known. Applying iGSE to analyze the iron loss of the isolated DC-DC converter yields the following equation:

$$P_{iGSE} = 2^{\alpha+\beta} k_i f^\alpha \left(1 - \frac{\delta}{\pi}\right)^{1-\alpha} B_{max}{}^\beta \ [\text{W/m}^3], \qquad (1)$$

where α, β, and k_i are Steinmetz constants. The δ is the phase difference of the output voltage between the primary and secondary side of the circuit. B_{max} is the maximum flux density in the transformer. From (1), the iron loss P_{iGSE} reaches a maximum when the phase difference is set to zero ($\delta = 0°$).

In order to calculate the Steinmetz constants, a B-H analyzer (SY-8219, IWATSU), which can measure the iron loss at any frequency and flux density, is used. Fig. 3 shows the measurement circuit in the B-H analyzer. In this circuit, i_L and v_{2m} are used for drawing a B-H curve. The Steinmetz constants of the cores as shown in (1) are then calculated with a curve fitting tool in MATLAB and the loss measurement results.

C. Iron Loss Measurement System under DC-DC Converter

In order to measure the iron loss under the DC-DC converter operation, two measurement modules of the power analyzer (PZ4000, YOKOGAWA) are connected to the primary and the secondary side of the MF transformer, respectively. The iron loss P_{iron} and copper loss of windings P_{cop} can be obtained by subtracting the power of the secondary side from that of the primary side. With this system, the iron loss P_{iron} can be separated from the copper loss P_{cop} because P_{cop} can be calculated by measuring the winding resistance beforehand. The litz wire is used as the windings of the transformer, and turn numbers are set to 18 turns in each side. In this case, the winding resistance under DC current is 0.024 Ω.

III. EXPERIMENTAL RESULTS OF THE IRON LOSS

Fig. 4 shows the measurement results and the calculated results of iron loss in the case of the amorphous core. The measurement results correspond to the calculated results in almost all regions. Therefore, the calculation method is confirmed to be valid for the isolated DC-DC converter.

IV. IRON LOSS SEPARATION INTO HYSTERESIS LOSS AND EDDY-CURRENT LOSS

The iron loss separation into hysteresis and other losses has been presented in the case of inductors for switching ripple reduction filters under DC excitation [8]. On the other hand, the ratios of hysteresis and eddy-current losses to the total iron losses of the MF transformer are unclear. If the characteristics of these two factors of the iron loss are made clear, a better core shape or material for a DC-DC converter can be proposed.

Generally, in order to analyze iron losses of a transformer under sinusoidal excitations, a separation equation for hysteresis and eddy-current losses is shown as follows [9]:

$$P_{iron} = \sigma_h f B_{max}{}^2 + \sigma_e f^2 B_{max}{}^2 \ [\text{W/m}^3] \qquad (2)$$

where σ_h and σ_e are the iron loss parameters, which depend on the material and the core shape. These parameters can be calculated with the iron loss data measured with the B-H analyzer under sinusoidal excitation. The first part of the equation handles the hysteresis loss and the second part handles the eddy-current loss.

Under rectangular excitation, the voltage waveform contains significant harmonic component. Therefore, the above

TABLE I.   INFORMATION OF CORE MATERIALS

Core	Sheet thickness	Magnetic path	Cross area	Volume	Weight
Silicon steel	0.1 mm	271 mm	6.18 cm^2	167×10^{-6} m^3	1.250 kg
Amorphous	0.025 mm	292 mm	5.45 cm^2	159×10^{-6} m^3	1.166 kg
Nanocrystalline	0.018 mm	292 mm	5.19 cm^2	152×10^{-6} m^3	1.128 kg

equation cannot be applied under rectangular excitation, and parameters are used for the exponents. The following equation can be applied under rectangular excitation:

$$P_{iron} = \sigma_h f B_{max}{}^{\lambda_1} + \sigma_e f^\gamma B_{max}{}^{\lambda_2} \ [\mathrm{W/m^3}] \qquad (3)$$

where γ, λ_1, and λ_2 are the exponents for the frequency and maximum flux density.

In order to calculate five iron loss parameters, B-H analyzer and a numerical calculation software are used. Loss measurements are done with the B-H analyzer, and the iron loss parameters are obtained by a curve fitting tool in the numerical calculation software, MATLAB.

The calculation procedure is identical to that of the case of Steinmetz constants. In the case of Steinmetz constants, the B-H analyzer is used under sinusoidal excitation. However, in the case of iron loss parameters, the B-H analyzer is used under only the rectangular excitation whose duty ratio is 50%.

A. Correction of Eddy-Current Loss Term

The excitation voltage waveform to the transformer for a DC-DC converter is rectangular. When δ is set to zero, the duty ratio of the excitation waveform is 50%, and the waveform of the flux density is triangular. On the other hand, when δ is not set to zero, the duty ratio of the excitation waveform is not 50%, and the waveform of the flux density is trapezoidal. Therefore, because the waveforms of the flux density are different owing to the operation conditions of the DC-DC converter and (3) is not a function of δ, (3) cannot be applied under $\delta \neq 0°$. Therefore, it should be corrected for the iron loss analysis under $\delta \neq 0°$. The excitation waveform at $\delta = 0°$ is shown in Fig. 5, and that at $\delta \neq 0°$ is shown in Fig. 6. Fig. 7 shows the waveform of the flux density at $\delta \neq 0°$.

In this paper, in order to calculate the iron loss easily, a correction coefficient is added to the eddy-current loss term. This coefficient is $\left(1 - \dfrac{\delta}{\pi}\right)^{1-\gamma}$. Therefore, (3) is rewritten as (4).

$$P_{iron} = \sigma_h f B_{max}{}^{\lambda_1} + \sigma_e \left(1 - \frac{\delta}{\pi}\right)^{1-\gamma} f^\gamma B_{max}{}^{\lambda_2} [\mathrm{W/m^3}] \quad (4)$$

The added coefficient changes its value depending on δ.

B. Concept for the Correction Coefficient

In (4), a correction for the hysteresis loss term is not performed. The hysteresis loss can be recognized as the area of the B-H loop, which is in proportion to the frequency f. Therefore, the hysteresis loss does not depend on the rate of change in the flux density dB/dt. In other words, the correction for the hysteresis loss term is not necessary.

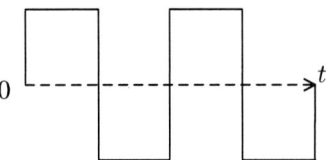

Fig. 5.   Waveform of voltage E ($\delta = 0$)

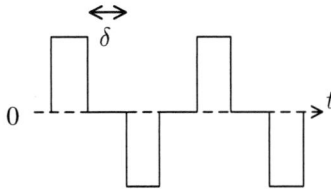

Fig. 6.   Waveform of voltage ($\delta \neq 0$)

On the other hand, the eddy-current loss depends on dB/dt. From Fig. 8, because dB/dt at $\delta \neq 0°$ is different from that at $\delta = 0°$, a correction for dB/dt is needed. The ratio of dB/dt can be calculated as shown in (5).

$$Ratio = \left(1 - \frac{\delta}{\pi}\right)^{1-\gamma} \leq 1, \qquad (5)$$

where the range of δ [rad] is $0 \leq \delta \leq \pi/2$. The correction of dB/dt can be performed by multiplying f by the ratio. However, the frequency of the corrected waveform of B shown as a broken line in Fig. 9 is higher than the actual frequency. Because the eddy-current loss occurs when the flux density B is changing, if the change in B is not corrected, the extra loss will be calculated. Therefore, the correction for the loss should be performed.

The ratio of the loss can be also calculated as $\left(1 - \dfrac{\delta}{\pi}\right)^{1-\gamma}$. Therefore, by dividing the eddy-current loss term by the ratio, the loss can be corrected, as shown in Fig. 10. As a result, the corrected equation (4) is obtained. Equation (4) separates the iron loss into the hysteresis and eddy-current losses and is a specialized equation for the MF transformer used in a DC-DC converter.

C. Calculation Procedure of Iron Loss Parameters

The parameters in (4), including γ, λ_1, λ_2, σ_h, and σ_e, have to be obtained by the experiment because these parameters depend on the core shape and its material. The B-H analyzer can measure the iron loss under arbitrary frequency and flux density. The iron loss data measured with the B-H analyzer are inputted into numerical calculation software (MATLAB), and then the iron loss data are analyzed with a curve-fitting tool in MATLAB. From the above procedure, the parameters γ, λ_1, λ_2, σ_h, and σ_e can be calculated. In this

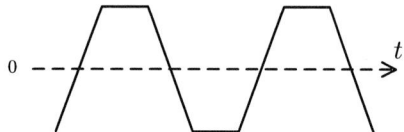

Fig. 7.   Waveform of B at $\delta \neq 0°$

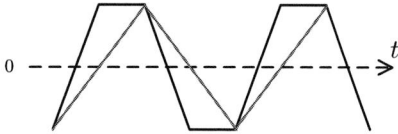

Fig. 8.   Uncorrected waveform of B

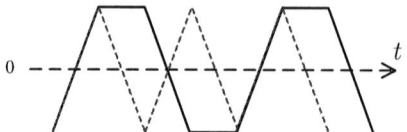

Fig. 9.   Incompletely corrected waveform of B

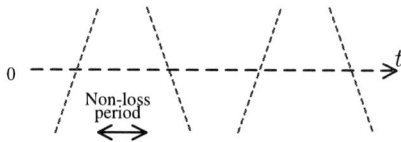

Fig. 10.   Completely corrected waveform of B

paper, the frequency range is 1 to 10 kHz, and the flux density range is 0.06 to 0.30 T. Therefore, there are 130 measurement points in Figs. 11 and 12 for each material, respectively.

The calculated results are shown in Table II. From Table II, γ of the amorphous core is larger than that of the silicon steel core. This shows that the eddy-current loss of the amorphous core has a larger frequency dependency. The σ_e of the silicon steel, which shows the magnitude of the eddy-current loss, is fourteen times larger than that of the amorphous core. In the case of the nanorystalline core, σ_h and σ_e are significantly small. Further, the exponents γ, λ_1 and λ_2 are almost 2. This shows that the nanocrystalline core is hardly affected by harmonic components.

These three cores are laminated sheet structuer. The thickness of the nanocrystalline core is 0.018 mm. And that of the amorphous and the silicon steel core are 0.025 mm and 0.1 mm, respectively. Therefore, the eddy-current loss of the silicon steel is larger than that of the other cores.

D. Separation Results with Proposed Method

The hysteresis and eddy-current loss under the actual operation region of the DC-DC converter can be calculated by using these parameters shown in Tables II and (4). In the case of changing δ of the DC-DC converter, Figs. 13, 14, and 15 show the analysis results of the hysteresis loss and eddy-current loss. In the case of the amorphous core (Fig. 13), the hysteresis loss is almost equivalent to the eddy-current loss in the entire range of δ. On the other hand, in the case of silicon steel (Fig. 14), the eddy-current loss is significantly larger than the hysteresis loss in the entire range of δ. The eddy-current loss of silicon steel is five times larger than that of the amorphous core. This difference is caused by the difference in the thickness of the laminated core and resistivity of its material. In the case of nanocrystalline material, both the iron losses are significantly small in the range of δ.

In order to confirm the validity of the separation procedure, the total iron loss calculated by the proposed method is compared with that by iGSE and the measured iron loss in the same condition for each material. Fig. 16 shows the total calculation results depending on δ. The total iron loss that is calculated by the proposed method corresponds to the loss calculated by iGSE and the measured iron loss.

V.   TEMPERATURE DEPENDENCIES OF IRON LOSS

The temperature dependency of magnetic permeability has been discussed in Ref. [10], that is, the iron loss of the MF transformer also depends on the core temperature. The temperature dependency of the iron loss at $\delta = 0°$ in each material was measured.

A. Experiment Method

In order to experiment a high temperature operation of the core, the core is heated by a boiled water. The core was placed in a heat-resistant container filled with water, and for safety, the core was covered with a heat-resistant bag. For heating water, a hot plate was used. Temperature of the core was measured with a thermocouple placed on the surface of the core.

B. Experimental Results Under $\delta = 0°$ at High Temperature

Fig. 17 shows the experimental results of temperature dependencies. In Fig. 17, in order to control the core temperature, the switching frequency is made constant, and the DC voltage is set to different values (silicon sheet: 150 V, amorphous: 200 V, nanocrystalline: 350 V) among the three materials because the iron losses of each material depend on the flux density of the core.

TABLE III.   IRON LOSS PARAMETERS (SILICON STEEL)

Core temp[°C]	γ	λ_1	λ_2	σ_h	σ_e
25	1.600	1.816	1.866	89.66	1.574
80	1.652	1.416	1.897	30.08	0.9243

TABLE IV.   IRON LOSS PARAMETERS (AMORPHOUS)

Core temp[°C]	γ	λ_1	λ_2	σ_h	σ_e
25	1.715	1.777	2.053	68.64	0.1067
80	1.708	1.786	2.122	49.56	0.1290

TABLE II.   IRON LOSS PARAMETERS

Core material	γ	λ_1	λ_2	σ_h	σ_e
Amorphous	1.715	1.777	2.053	68.64	0.1067
Silicon steel	1.600	1.816	1.866	89.66	1.574
Nanocrystalline	1.847	2.034	1.964	28.38	0.008136

The 2014 International Power Electronics Conference

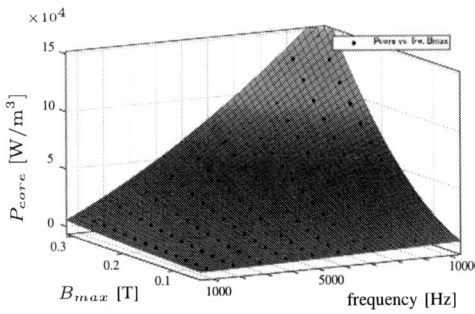

Fig. 11. Fitting result (amorphous)

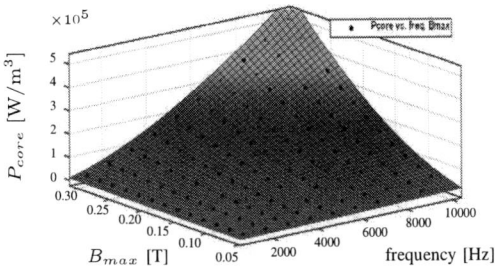

Fig. 12. Fitting result (silicon sheet)

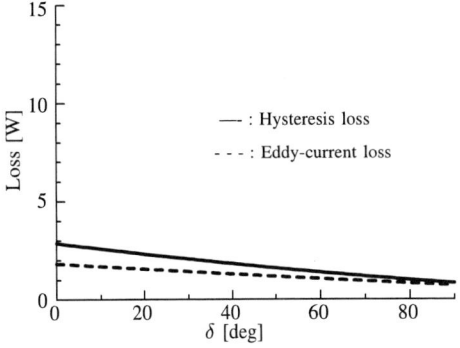

Fig. 13. Separated iron losses into hysteresis and eddy-current losses for amorphous core

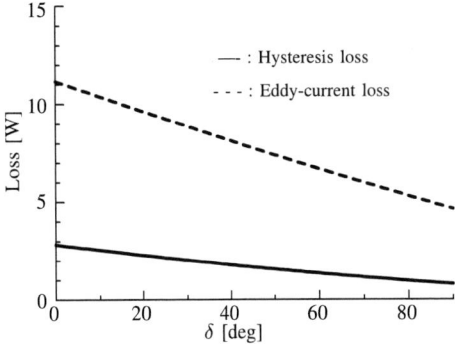

Fig. 14. Separated iron losses into hysteresis and eddy-current loss for silicon steel core

The iron loss of amorphous and silicon sheet materials have strong temperature dependencies. The iron loss of amorphous material at 85 °C reaches a minimum and decreases 14.7% compared to that at room temperature. Moreover, the iron loss of the silicon sheet material at 80 °C reaches a minimum and decreases 19% compared to that at room temperature. In both materials, the iron losses have minimum values near 80 °C. In order to achieve higher efficiency, the MF transformer should be operated at a higher temperature where the iron loss approaches a minimum value. On the other hand, the iron loss of the nanocrystalline material does not have significant temperature dependency.

C. Iron Loss Separation Under High-Temperature Condition

In this section, the iron loss separation into hysteresis and eddy-current losses under a high-temperature condition is shown. In this paper, the iron loss separations are performed in the cases of silicon steel and amorphous core materials under a high-temperature condition of 80 °C. The iron loss separation method is that shown in section IV.

Tables III and IV show the calculated iron loss parameters. In the case of a silicon sheet, γ and λ_2 are almost equivalent to those at 25 °C. This means that the frequency characteristics and the maximum flux density characteristics of the eddy-current loss are almost identical. On the other hand, λ_1 at 80 °C is smaller than that at 25 °C. This means the maximum flux density characteristics of hysteresis loss are changed by temperature. Further, σ_h, which is the magnitude of the hysteresis loss, and σ_e, which is the magnitude of the eddy-current loss, are significantly smaller than those at 25 °C.

In the case of the amorphous core, σ_h at 80 °C is smaller than that at 25 °C. On the other hand, σ_e at 80 °C is larger than that at 25 °C.

D. Iron Loss Separation at High Temperature Under Actual Operation of DC-DC Converter

In this section, iron loss separations at high temperature under actual operation of a DC-DC converter are shown. Figs. 18 and 19 show the δ characteristics of the hysteresis and eddy-current losses under two temperature conditions, 80 °C and 25 °C. In both core materials, the differences in both iron loss components between 80 °C and 25 °C are large in the small-δ region and become smaller as δ increases.

VI. CONCLUSION

This paper discusses an iron loss of the isolated DC-DC converter based on the analytical and experimental results. As a result, the total iron loss at $\delta = 0°$ calculated by iGSE corresponds to that measured by the experimental results, and the validity of the proposed methods are confirmed. It is confirmed that the proposed analysis procedure can divide the iron loss into hysteresis and eddy-current losses.

Moreover, the temperature dependency of the iron loss is also demonstrated by the experimental results. From experiments, the iron loss at high temperature is 19% smaller than that at 25 °C.

REFERENCES

[1] S. Inoue, and H. Akagi, "Operating voltage and loss analysis of a bi-directional isolated dc/dc converter," *IEEJ Transactions on Industry Applications*, vol. 127 (2007), pp. 189-197.

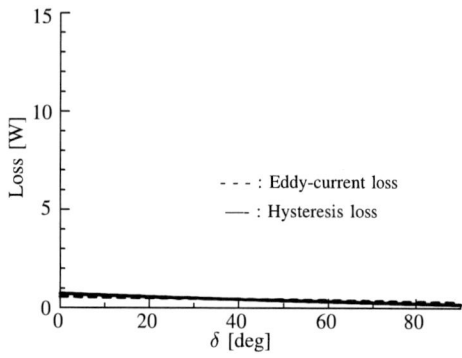

Fig. 15. Separated iron losses into hysteresis and eddy-current loss in nanocrystalline core

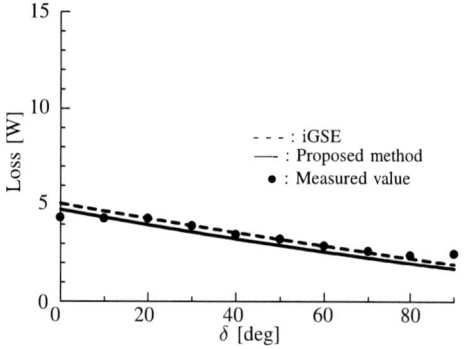

Fig. 16. Iron losses by proposed method, iGSE, and measurement in amorphous core

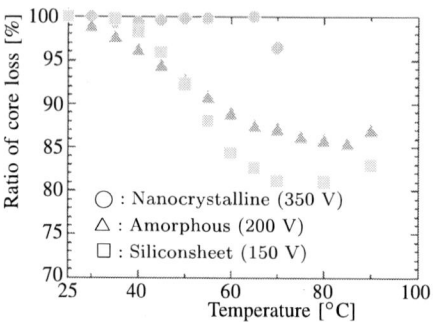

Fig. 17. Temperature dependencies of the iron loss

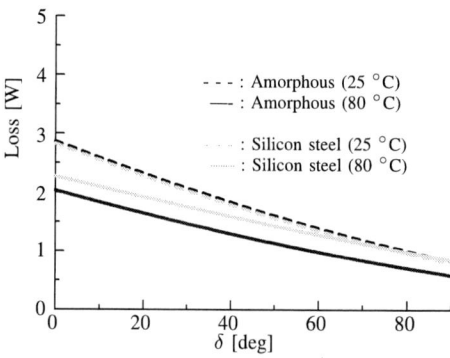

Fig. 18. Temperature dependencies of hysteresis loss in each material

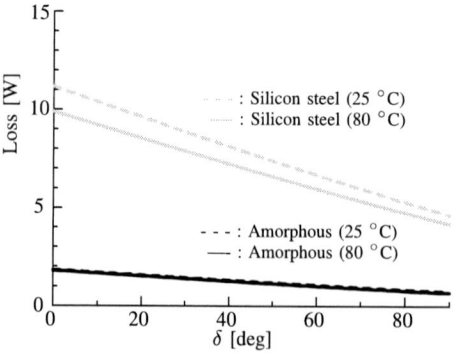

Fig. 19. Temperature dependencies of eddy-current loss in each material

[2] A. Huang, et al. "The future renewable electric energy delivery and management (FREEDM) system: the energy internet." *Proceedings of the IEEE* , vol. 99, No. 1 (2011), pp. 133-148.

[3] T. Yamagishi, H. Akagi, S. Kinouchi, and Y. Miyazaki, "The 750-V, 100-kW, 20-kHz Bidirectional Isolated DC/DC Converter Using SiC-MOSFET/SBD Modules.", *IEEJ Joint Technical Meeting on Semiconductor Power Converter*, SPC-13-116 (2013)

[4] H. Hoffmann, and P. Bernhard, "Medium frequency transformer in resonant switching dc/dc-converters for railway applications.", *Power Electronics and Applications (EPE 2011)* , 2011. pp. 1-8.

[5] Y. Matsuoka, M. Nakahara M. Wada, K. Takao, K. Sung, and S. Nishizawa, "2.5kV, 200kW Bi-Directional Isolated DC/DC Converter for Medium-Voltage Applications," *International Power Electronics Conference (IPEC) ECCE-Asia* (2014)

[6] G. Ortiz, M. Leibl, J. W. Kolar, and O. Appeldorn, "Medium frequency transformers for Solid-State-Transformer applications - design and experimental verification," *IEEE Int. Conf. Power Electronics and Drive Systems (PEDS)* , 2013.

[7] K. Venkatachalam, C. R. Sullivan, T. Abdallah, and H. Tacca,"Accurate prediction of ferrite core loss with nonsinusoidal waveforms using only Steinmetz parameters" *Proc. IEEE Workshop Comput. Power Electron.*, 2002, pp. 36-41.

[8] H. Matsumori, T. Shimizu and A. Moritani, "Discussion on the Iron loss of the silicon-steel for inductors of Electrical Machines", *The 2011 Annual Meeting of IEEJ.*, 2011. no. 4-123

[9] K. Yamazaki, "Loss Calculation of Induction Motors Considering Harmonic Electromagnetic Fields in Stator and Rotor", *IEEJ*, Vol. 123-D, No. 4, pp. 392-400 (2003) (in Japanese).

[10] M. Sippola, and E. S. Raimo, "Accurate prediction of high-frequency power-transformer losses and temperature rise", *IEEE Transactions on Power Electronics*, vol. 17, no. 5 (2002), pp. 835-847.

Experimental Verification of Capacitive Power Transfer using One Pulse Switching Active Capacitor for Practical Use

Tatsuaki Kitabayashi, Hirohito Funato, Hiroya Kobayashi, Katsuya Yamaichi

Department of Electrical and Electronic Engineering

Utsunomiya University

Utsunomiya, Japan

funato@cc.utsunomiya-u.ac.jp

Abstract—The authors have proposed a new capacitive power transfer (CPT) system using one-pulse switching active capacitor (OPSAC). The proposed system improves power transfer efficiency without LC resonance so that it is robust against parameter change. In this paper, a small sized vehicle powered by CPT is investigated to verify the practicability.

Keywords—wireless power transfer, capacitive coupling, capacitive power transfer, active capacitor

I. INTRODUCTION

Recently, wireless power transfer (WPT) is very attractive to provide electric power to mobile equipments, such as mobile phones, electric vehicles and so on. Inductive power transfer (IPT) is currently the most popular way to realize contactless power transfer. However, IPT has some disadvantages, such as power decrease by inexact connection, radiation of unexpected radio wave, unexpected heating of alien metal substance and so on. Electromagnetic resonance coupling is the emerging method suitable for long distance.

Another way of WPT is capacitive power transfer (CPT) with capacitive couplings. There are only few studies about CPT because of small power density compared to another method due to small coupling capacitance. CPT is however investigated recently to utilize the advantages in CPT, such as simple structure of couplings, lightweight, low cost, position flexibility and high frequency applicability [1]~[6]. In [1], a battery charger for soccer robot using CPT was proposed. then it is extended to provide electric power to lights using rotary couplings [2]. The characteristics of the proposed system was analized in [3]. In [4], a battery charging system is also proposed using very small coupling capacitance compared to [1]. In [6], CPT system using resonance of LCL-type is proposed. In these studies, provides very small transfer power less than 30 W. The larger power transfer using CPT was challenged in [5] where 90 W power transfer was succeeded. All of the above mentioned studies, LC resonance is used to enhance transfer power. LC resonance is however fragile for parameter change which may be caused by contact condition of capacitive couplings.

To overcome this problem, one-pulse switching active capacitor (OPSAC) was proposed [7]~[9]. Using OPSAC, coupling capacitance is equivalently increased independent from frequency so that transfer power can be increased for any condition of couplings. OPSAC consists of single phase inverter with dc capacitor at dc bus. The inverter switches only one time per cycle of supply voltage. The biggest advantage of the proposed CPT system with OPSAC is its simple operation with inherent stability. In addition, OPSAC has another advantage which is suitable for cascaded connection. In [7], basic principle was proposed then operational characteristics were discussed. Cascaded connection and new control scheme to enhance transfer power were described in [8]. In [9], the transfer power of CPT system with LC resonance and OPSAC was analyzed for the change of coupling capacitance, and robustness of the CPT with OPSAC was verified by simulations. In this paper, transfer power of CPT system with a small sized vehicle is experimentally analyzed to investigate the practicability and the problem.

II. CPT-SYSTEM USING OPSAC

In this section, the basic principle of CPT system using OPSAC is described.

A. CPT System using OPSAC

Fig. 1 shows the circuit diagram of CPT system using 1st-stage-3level-OPSAC (1S-3L-OPSAC). The dash-dotted area is OPSAC. The left part is power source which provides high frequency square wave voltage v_s. OPSAC outputs three level voltage v_{op}. In Fig. 1, v_{js} is synthesized voltage of v_s and v_{op} which is applied to coupling capacitor C_j. In the load side, choke-input type rectifier is connected to provide dc power to load. D is free wheel diode during non-conduction period of rectifier. OPSAC can increase the conduction period of the coupling capacitor current i_{cj}, which results in enhancement of transfer power. OPSAC is worked as a reactive element so that it is possible to replace dc voltage source by a link-capacitor C_{op}. In this case, initial charge and voltage maintenance are also big problem. In case of OPSAC, the voltage of a dc capacitor is automatically charged up to the same value or double of dc voltage of source inverter E, and it is automatically maintained only

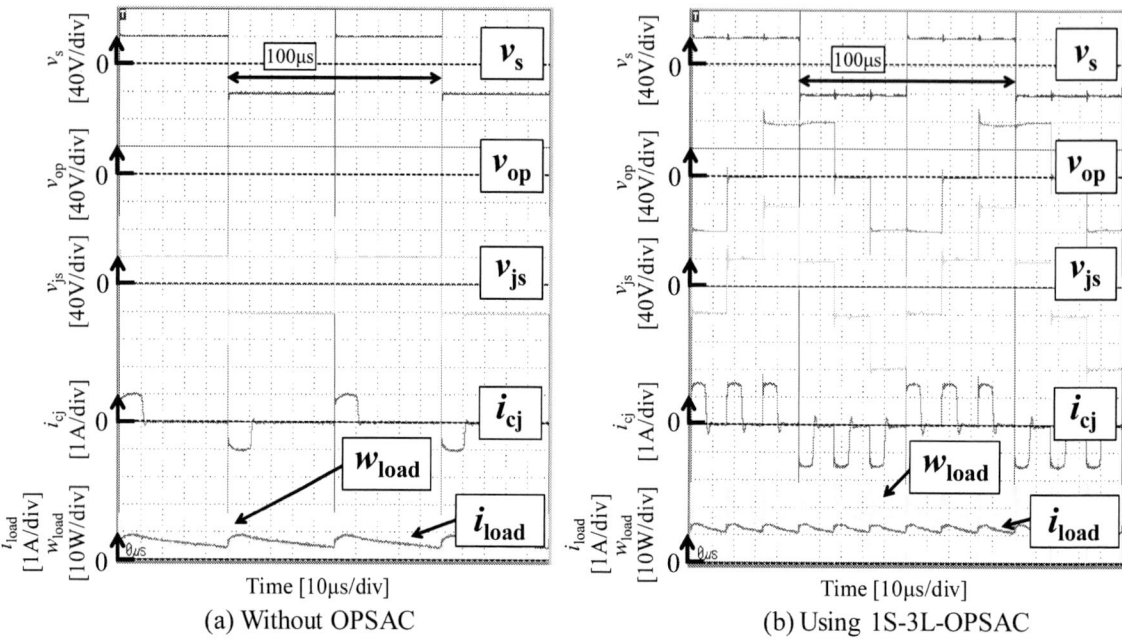

Fig. 2. Experimental waveforms of CPT system with and without 1S-3L-OPSAC (circuit parameters A)

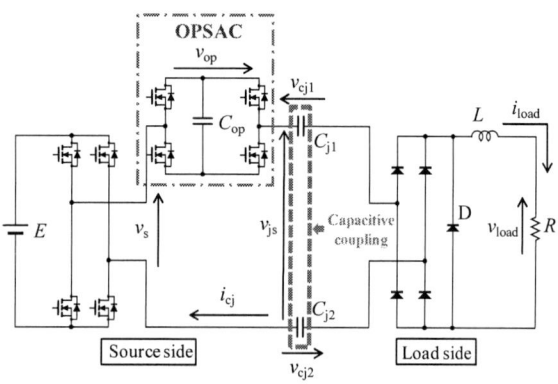

Fig. 1. Circuit configuration of CPT system using 1S-3L-OPSAC

TABLE I. CIRCUIT PARAMETERS A

Circuit parameters A		
Input dc voltage	E	40 V
Coupling capacitance	C_{j1}, C_{j2}	230 nF, 250 nF
Switching frequency	f	10 kHz
Resistive load	R	10 Ω
Inductance	L	1 mH

Fig. 3. Capacitive couplings (230 nF, 250 nF, couplings Y)

driving ordinary switching timings without any feedback loop. This is one of the advantages and detailed analysis was described in [7].

B. Experiments using 1S-3L-OPSAC

Fig. 2 (a) and (b) show the experimental waveforms with and without 1S-3L-OPSAC respectively. The circuit parameters are shown in TABLE I. Fig. 3 shows capacitive couplings. Insulated material of Fig. 3 (Couplings Y) is composed by barium titanate ($BaTiO_3$). In Fig. 2 (b), OPSAC outputs three level voltage in v_{op}. Connecting OPSAC, the number of conduction period of i_{cj} becomes six in one cycle which is triple of that in Fig. 2 (a) so that the transfer power can be enhanced. Transfer power can be further enhanced using three level operation compared with using two level operation and cascade connection. Using OPSAC with two level operation, transfer power

can be theoretically obtained as $(n + 1)$ times of transfer power obtained by CPT system without OPSAC. Using OPSAC with three level operation, transfer power can be obtained as $(2n + 1)$ times. Here, n is the number of cascade connection of OPSAC. The detailed analysis were described in [8].

C. Experiments using 5S-3L-OPSAC

Fig. 4 shows the circuit diagram of CPT system using 5stage-3level-OPSAC (5S-3L-OPSAC). Fig. 5 shows

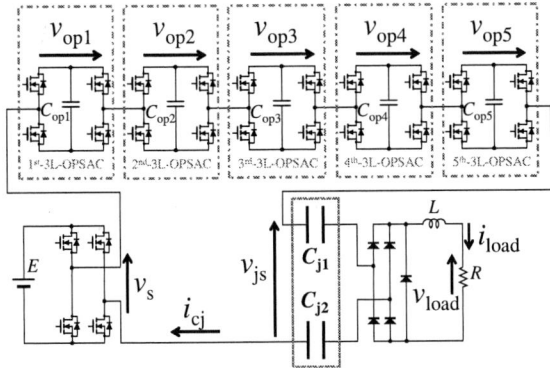

Fig. 4. Circuit configuration of CPT system using 5S-3L-OPSAC

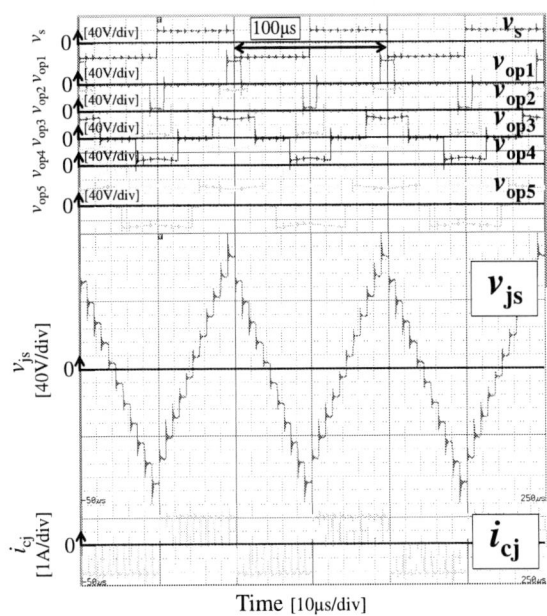

Fig. 5. Experimental results with 5S-3L-OPSAC

the experimental results with 5S-3L-OPSAC. The circuit parameters are the same as shown in TABLE I. From Fig. 5, the number of conduction period of i_{cj} becomes 22, so load current i_{load} is increased. The transfer power is enhanced compared with CPT system using 1S-3L-OPSAC. The transfer power using 3L-OPSAC W_{load} changes by two operational mode, continuous current mode (CCM) and discontinuous current mode (DCM). W_{load} can be calculated as follows;

$$W_{load} = \begin{cases} (2n+1)4C_j\,E^2 f & \text{(DCM)} \\ E^2/R & \text{(CCM)} \end{cases} \quad (1)$$

The boundary condition of DCM and CCM can be calculated as follows;

$$4(2n+1)C_jRf = 1 \quad (2)$$

From Eq. (1), it is verified that n stage-3L-OPSAC enhances transfer power $(2n+1)$ times.

III. SMALL-SIZED VEHICLE APPLIED PROPOSED SYSTEM

In this section, the specification of the produced vehicle is described.

A. Configuration of Small-sized Vehicle

Fig. 6 shows the small-sized vehicle applied CPT system with OPSAC. This vehicle is controlled by line-trace method. Fig. 7 shows capacitive couplings applied the vehicle (Couplings X). The target specification of this vehicle is to drive for 10 minutes with 10 W power consumption. From this specification, the required energy is calculates as 6000 J. Electric double-layer capacitor (EDLC) is used as storage devices of the vehicle. Using the couplings shown in Fig. 7, coupling capacitance C_j is calculated as 2 nF. The coupling capacitance is very small compared with the coupling capacitance used in experiments shown in TABLE I. The reason why the capacitance was very small is that popular plastic wrap is assumed to be used as separator of couplings for easy realization.

TABLE II. CIRCUIT PARAMETERS B

Circuit parameters B		
Input dc voltage	E	40 V
Coupling capacitance	C_{j1}, C_{j2}	16 nF
Switching frequency	f	50 kHz
Capacitance of EDLC	C_{EDLC}	126 F
Inductance	L	1 mH

B. CPT System using OPSAC Applied to Small-sized Vehicle

Fig. 8 shows the proposed circuit diagram of CPT system with 5S-3L-OPSAC and EDLC. In the previous circuit, resistance is only connected as load, but this time, EDLC is connected as storage devices. In the load side, choke-input type rectifier is connected. Because it can be regarded as current source, the system using OPSAC is suitable for wireless charging to storage devices such as EDLC without complicated control. In this system, switch S1 are ON and switch S2 is OFF during power is transferred to capacitance of EDLC C_{EDLC}. At the completion of charging, S1 is turned OFF. Then S2 becomes ON after a while so that power is provided to motors.

C. Simulation Results of Proposed System

Fig. 9 shows simulation results using circuit parameters shown in TABLE II. In this simulation, initial voltage of EDLC is set to 7 V considering actual condition. From Fig. 9, The charging current is observed as 4 A at 7 V. The transfer power is calculated as 28 W which is similar power using resistive load. From this simulation, the proposed CPT is suitable for charging EDLC.

978-1-4799-2706-7/14 $31.00 © 2014 IEEE

The 2014 International Power Electronics Conference

Fig. 6. Small-sized vehicle applied proposed system

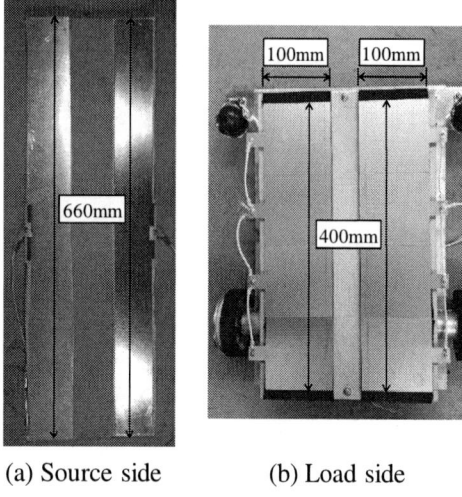

(a) Source side   (b) Load side

Fig. 7. Capacitive couplings of the vehicle (2 nF, couplings X)

IV. EXPERIMENTAL RESULTS OF SMALL-SIZED VEHICLE

In this section, the experimental results of small-sized vehicle equipped with the proposed CPT system are explained.

A. CPT System Equipped on the Small-sized Vehicle

Fig. 10 shows the small-sized vehicle with the proposed CPT system. The circuit of this system is shown in Fig. 8. 5S-3L-OPSAC is connected at source side. In the experiment, the couplings X is used in the CPT system. However, the target specification is not satisfied using couplings X because of very small coupling capacitance. Therefore, couplings Y with higher capacitance are also used for experiments in order to obtain enough transfer power. The circuit parameters are shown in TABLE III. The theoretical transfer power using couplings X and Y are calculated as 3.5 W and 21 W respectively.

B. Experimental Results using Couplings X

Fig. 11 shows the experimental results of the charging using couplings X. In this experiment, the charging cur-

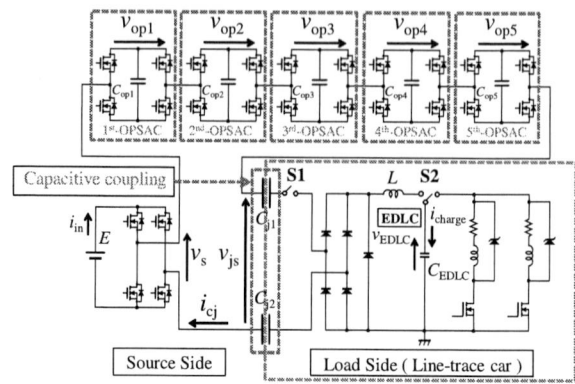

Fig. 8. Proposed circuit configuration of CPT system using 5S-3L-OPSAC

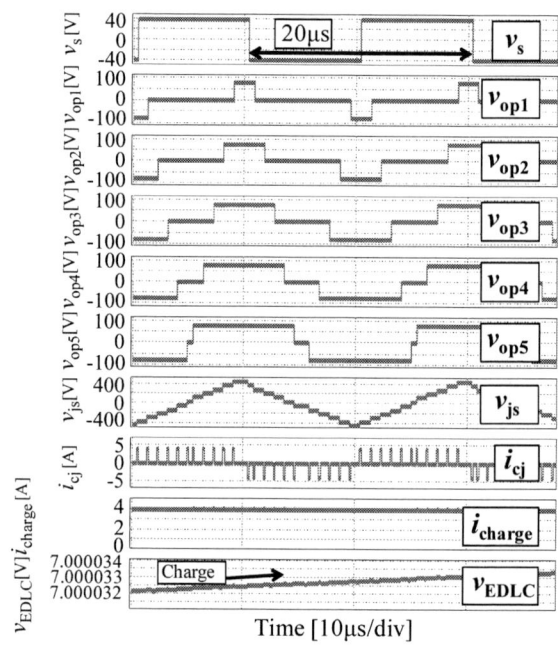

Fig. 9. Simulation results of the proposed system

rent is only 200 mA. The small charging current is due to small coupling capacitance. The transfer power is 2 W because the EDLC voltage is 10 V. The circuit does not operate as theory because of small current.

C. Experimental Results using Couplings Y

Fig. 12 (a) and (b) show the experimental results of charging to EDLC using the CPT without OPSAC and with 5S-3L-OPSAC respectively. In these experiments, the voltage of EDLC v_{EDLC} is set to 7 V. In (a), circuit becomes unstable because transfer power is too small to maintain continuous current of the DC choke. The transfer power is about 1 W. In (b), the transfer power is observed about 9.4 W. Theoretically, the transfer power using 5S-3L-OPSAC becomes 11 times of that without OPSAC. The transfer power enhancement is smaller than theory

978-1-4799-2706-7/14 $31.00 © 2014 IEEE   2520

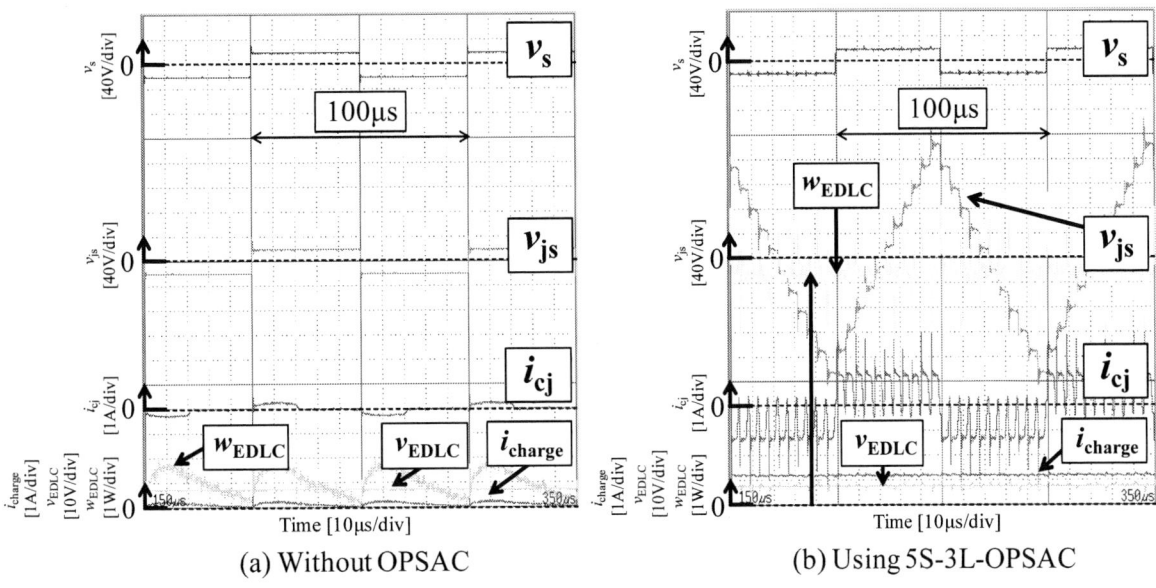

(a) Without OPSAC   (b) Using 5S-3L-OPSAC

Fig. 12. Experimental waveforms with and without 5S-3L-OPSAC using couplings Y ($v_{EDLC} = 7\,$V)

TABLE III.   CIRCUIT PARAMETERS C

Parameters/Coupling type		X	Y
Input dc voltage	E [V]	40	20
Coupling capacitance	C_{j1}, C_{j2} [nF]	2	230, 250
Switching frequency	f [kHz]	50	10
Capacitance of EDLC	C_{EDLC} [F]	133	133
Inductance	L [mH]	2	2
Dead time	t_{dead} [ns]	400	400

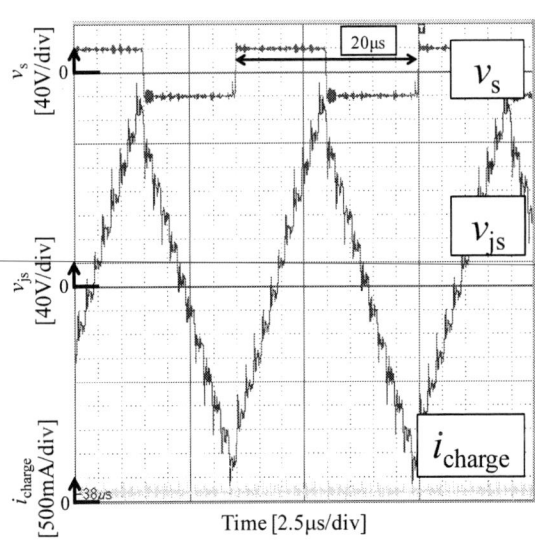

Fig. 11. Experimental results using couplings X ($v_{EDLC} = 10\,$V)

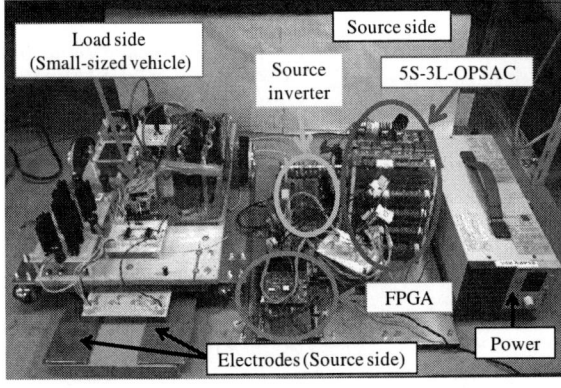

Fig. 10. Experimental equipments of CPT system

due to the loss in the OPSAC. However the transfer power in Fig. 12 (b) is enough for charging to EDLC without any problem.

D. Experimental Results of Charge and Discharge Cycle

Fig. 13 and Fig. 14 are the experimental results of charging and discharging cycle using couplings X and Y respectively. From Fig. 13, it is observed that it takes about four hours to charge EDLC from 6 V to 11.5 V using couplings X. On the contrary, it takes about 32 min-

utes from 6.4 V to 11.5 V using couplings Y from Fig. 14,. In these two experiments, the EDLC voltages sometimes become flat which means zero charging current. This is because fault of controller. The target specification is satisfied using the couplings Y. However the couplings Y is hard and expensive so that the improvement of coupling X is required.

V. CONCLUSIONS

In this paper, the basic principle and characteristics of CPT sytem using one-pulse switching active capacitor (OPSAC) was explained. Using OPSAC, the transfer power of CPT system can be enhanced without LC

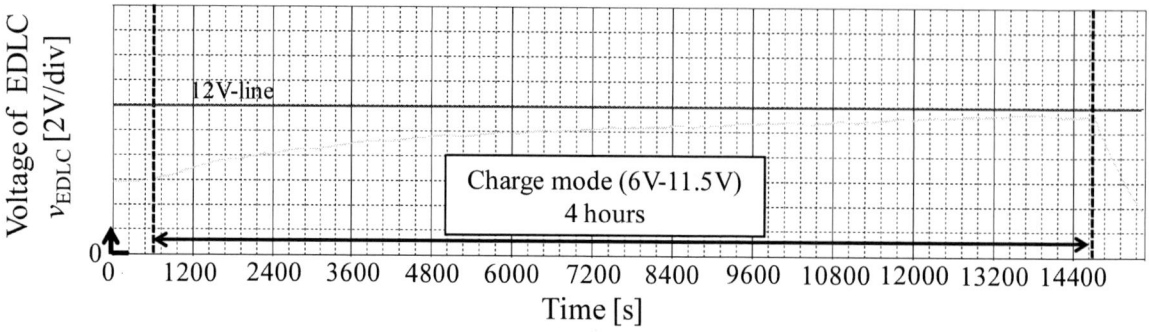

Fig. 13. Experimental results of charge and discharge test using couplings X

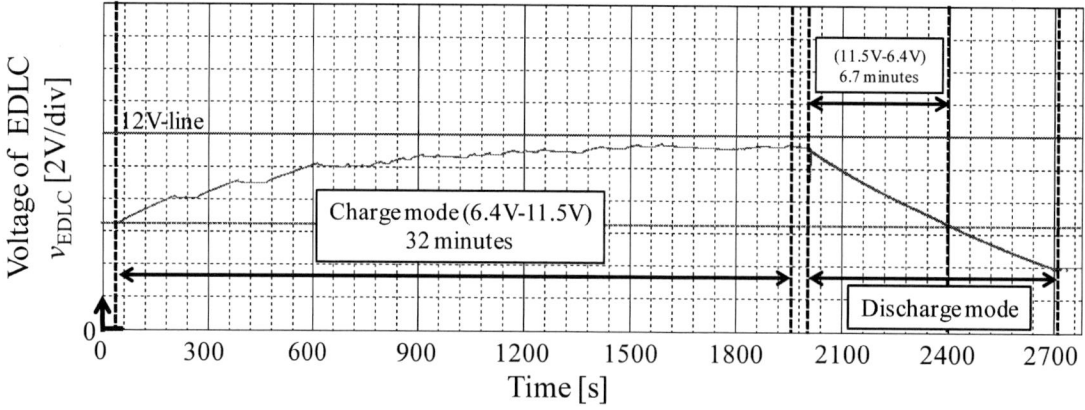

Fig. 14. Experimental results of charge and discharge test using couplings Y

resonance. In this paper, a small-sixed vehicle powered by the proposed CPT was made in order to verify the effectiveness of the proposed CPT in the actual use. The experiments of charging and discharging cycle were done using two different couplings. In the near future, further improvements of couplings will be performed to satisfy the target specifications.

REFERENCES

[1] A.P. Hu, C. Liu, and H.L. Li : "A Novel Contactless Battery Charging System for Soccer Playing Robot", in *Proc. 15th IEEE Int. Conf. Mechatronics and Machine Vision in Practice*, Aucland, New Zealand, Dec.2008, pp.646-650 (2008)

[2] C. Liu, A.P. Hu, and N.-K.C. Nair : "Coupling Study of a Rotary Capacitive Power Transfer System", in *Proc. 2009 IEEE Int. Conf. Industrial Technology*, Melbourne, Australia, pp.1-6 (2009)

[3] C. Liu, A.P. Hu, G.A. Covic, and N.-K.C. Nair : "Comparative Study of CCPT Systems With Two Different Inductor Tuning Positions", *IEEE Transactions on Power Electronics*, Vol.27, No.1, pp.294-306 (2012)

[4] M. Kline, I. Izyumin, B. Boser, and S. Sanders : "Capacitive Power Transfer for Contactless Charging", *2011-26th Annual IEEE, Applied Power Electronics Conference and Exposition* (APEC), pp.1398 - 1404 (2011)

[5] Ken-ichi Harakawa, Kenji Kageyama, and Kazuyuki Miura : "Possibility of Wireless Power Supply by Electric Coupling", *Technology Takenaka technical research report*, No.66 pp.8 (2010) (in Japanese)

[6] Michael P.Theodoridis : "Effective Capacitive Power Transfer", *IEEE Trans. on Power Electronics*, Vol.27, No.12, pp.4906-4913 (2012)

[7] Hirohito Funato, Yuki Chiku, and Ken-ichi Harakawa : "Proposal for Wireless Power Distribution System with Capacitive Coupling Using One-Pulse Switching Active Capacitor", *IEEJ Transactions on Industry Applications*, Vol.132-D, No.1, pp.27-34 (2011) (in Japanese)

[8] Hiroya Kobayashi, Hirohito Funato, and Yuki Chiku : "Enhancement of Transfer Power of Capacitive Power Transfer System using Cascaded One Pulse Switching Active Capacitor (C-OPSAC) with Three-Level Operation", *2012-7th International Power Electronics and Motion Control Conference* (IPEMC), pp.884-888 (2012-6)

[9] Hirohito Funato, Hiroya Kobayashi, and Tatsuaki Kitabayashi : "Analysis of Transfer Power of Capacitive Power Transfer System", *2013-10th IEEE International Conference on Power Electronics and Drive Systems* (PEDS), pp.1015-1020 (2013-4)

A Single-Stage High-PF Driver for Supplying a T8-type LED Lamp

Chun-An Cheng, Chien-Hsuan Chang, Hung-Liang Cheng and Tsung-Yuan Chung

Power Electronics Lab, Department of Electrical Engineering, I-Shou University

Kaohsiung City, Taiwan, R. O. C.

E-mail: cacheng@isu.edu.tw

Abstract—This paper proposes a novel single-stage AC-DC driver for supplying a T8-type light-emitting-diode (LED) lamp with high power-factor (PF). The presented AC-DC driver integrates a dual boost converter with coupled inductors and a half-bridge series-resonant converter into a single-stage power conversion topology. Operational principles and experimental results of one T8-type LED lamp with a rated power of 18 W operating at 50 kHz switching frequency with 110V-rms line input voltage are presented. Satisfactory results demonstrate the feasibilities of the proposed driver. Moreover, the features of the driver are cost-effectiveness, high power factor (PF>0.99), low input current total-harmonic-distortion (THD<4%), high circuit efficiency (>90%), and zero-voltage switching (ZVS) obtained on both two power switches.

Keywords-- Converter, driver, light-emitting-diode (LED)

I. INTRODUCTION

Instead of incandescent lamps, fluorescents lamps, which have features of better lighting efficiency, longer lamp life and less consuming power than incandescent lamps, are commonly used in indoor and household applications [1]. The traditional fluorescent lamps commonly used in indoor lighting circumstances are T8-type ones. The conventional two-stage electronic ballast for T8-type fluorescent lamps is shown in Fig. 1, and it consists of a boost converter (including an inductor L_{PFC}, a power switch S_B, a diode D_B and a DC-linked capacitor C_{DC}) for achieving power-factor-corrections (PFC) in order to meet the input current harmonics requirements of IEC standards and a half-bridge series-resonant parallel-loaded inverter (including a DC-linked capacitor C_{DC}, two switches S_1 and S_2, a resonant inductor L_S, a DC-blocking capacitor C_S, a resonant capacitor C_p and the equivalent lamp resistor R_{lamp}) for converting a DC voltage/power to a high-frequency sinusoidal source in order to supply the lamp [2].

Recently, the LED lamps with high lighting efficiency, energy-savings and long lamp lifetime are going to replace the T8-type fluorescent lamps. Table I shows the comparisons between T8-type fluorescent lamp (China Electric FL40D-EX) and T8-type LED one (EVERLIGHT FBW/T8/857/U/4ft) [3], [4]. According to Table I with almost the same color temperature and color rendering

index, the LED lamp has better lighting efficiency, less consumed power, longer lamp lifetime and no mercury-contained inside the lamp tube than the T8-type counterparts. Therefore, T8-type LED lamps have become increasingly popular lighting sources used for public architectures, offices, classrooms, and so on [5]-[11]. Fig. 2 shows the commercial two-stage driver for supplying a T8-type LED lamp, which is composed of a boost PFC converter as the first stage and a buck converter as the second stage for regulating the voltage/current of the LED lamp [12]. The circuit efficiency is limited due to the two-stage power conversion.

Therefore, this paper presents a novel single-stage high-power-factor (HPF) and high-efficiency AC-DC LED driver for T8-type fluorescent lamps replacements. Theoretical analysis of operating modes, experimental results from a prototype circuit of the presented driver for supplying an 18W-rated T8-type LED lamp are demonstrated in this paper.

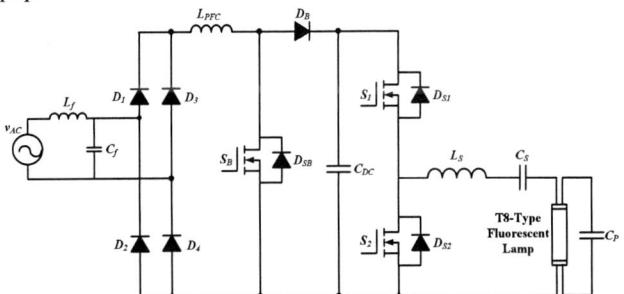

Figure 1. The conventional two-stage high-PF electronic ballast for supplying fluorescent lamps.

TABLE I
COMPARISONS BETWEEN T8-TYPE FLUORESCENT AND LED LAMPS

Items	T8-type Fluorescent Lamp (China Electric FL40D-EX)	T8-type LED Lamp (EVERLIGHT FBW/T8/857/U/4ft)
Consumed Power	40W	18W
Lumen Output	3400 lm	1800 lm
Lighting Efficiency	> 85 lm/W	> 100 lm/W

Color Temperature	6700 K	5700 K
Color Rendering Index R_a	> 80	> 80
Lamp Life	> 10000 hours	> 35000 hours
Lamp Base	G13	G13
Containing Mercury	Yes	No

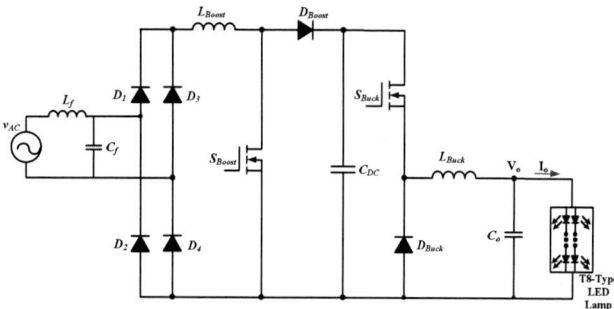

Figure 2.   The commercial two-stage driver for supplying a T8-Type LED lamp.

II. THE DESCRIPTION AND ANALYSIS OF THE PROPOSED LED LAMP DRIVER

Fig. 3 shows the proposed driver, which combines the dual boost converter with coupled inductors (one boost converter including a diode D_{b1}, a coupled inductor L_{PFC1}, a switch S_1, a body-diode of the switch S_2, and a DC-linked capacitor C_{DC}; another boost converter including a diode D_{b2}, a coupled inductor L_{PFC2}, a switch S_2, a body-diode of the switch S_1, and the capacitor C_{DC}) with the half-bridge series-resonant converter (including a DC-linked capacitor C_{DC}, two switches S_1 and S_2, a resonant inductor L_r, a resonant capacitor C_r, a full-bridge rectifier $D_1 \sim D_4$, and a output capacitor C_o) into a single-stage topology for supplying a T8-type LED lamp. In addition, a LC-filter (an inductor L_f and a capacitor C_f) is connected with the input utility-line voltage.

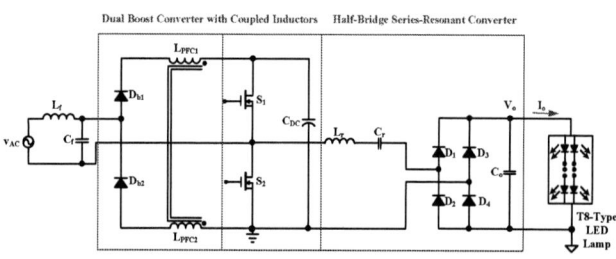

Figure 3.   The proposed single-stage driver for supplying a LED lamp.

In order to analyze the operations of the presented driver for supplying a LED lamp, the following assumptions are made.

(a) Since switching frequencies of two switches S_1 and S_2 are much higher than that of utility-line voltage v_{AC}, the sinusoidal utility-line voltage can be considered as a constant value for each high-frequency switching period.

(b) Power switches are complementarily operated, and their inherent diodes are considered.

(c) For simplifying the analysis, the LC-filter is not shown in the of the operation modes of the driver circuit.

(d) The conducting voltage drops of diodes D_{b1}, D_{b2}, D_1, D_2, D_3 and D_4 are neglected.

(e) Two coupled inductors (L_{PFC1} and L_{PFC2}) are designed to be operated in discontinuous-conduction mode (DCM) for naturally achieving power-factor-correction.

The operating modes and the key waveform of the proposed LED driver operated during the positive half-cycle of input utility-line voltage are shown in Fig. 4 and Fig. 5, respectively, and the analysis of operations are described in detail in the followings.

Mode 1 ($t_0 \leq t < t_1$; in Fig. 4(a)): This mode begins at time t_0 when the body-diode of the switch S_1 is forward biased. The resonant inductor current i_{Lr} provides energy to capacitors C_{DC}, C_r, C_o and the LED lamp through the S_1's body-diode, D_2 and D_3. At time t_1, the drain-source voltage v_{DS1} of power switch S_1 is at zero level resulting in zero-voltage switching (ZVS), and then this mode ends.

Mode 2 ($t_1 \leq t < t_2$; in Fig. 4(b)): This mode begins when the switch S_1 achieves ZVS-turning-on at t_1. The input voltage v_{AC} provides energy to the coupled inductor L_{PFC1} through the diode D_{b1} and the switch S_1. The inductor current i_{LPFC1} linearly increases from zero level and it can be expressed as

$$i_{LPFC1}(t) = \frac{\sqrt{2} v_{AC-rms} \sin(2\pi f_{AC} t)}{L_{PFC1}} (t - t_1) \quad (1)$$

where v_{AC-rms} is rms value of input utility-line voltage, and f_{AC} is the utility-line frequency.

The inductor current i_{Lr} still provides energy to capacitors C_{DC}, C_r, C_o and the LED lamp through the S_1's body-diode, D_2 and D_3. This mode finishes when the current i_{Lr} decrease to be zero at t_2.

Mode 3 ($t_2 \leq t < t_3$; in Fig. 4(c)): The voltage v_{AC} still provides energy to the coupled inductor L_{PFC1} through the diode D_{b1} and the switch S_1. The DC-bus capacitor C_{DC} supplies energy to the inductor L_r, the capacitors C_r and C_o, and the LED lamp through the switch S_1, diodes D_1 and D_4. At t_3, inductor current reaches its peak value, defined as $i_{LPFC1-pk}(t)$, and it is given by

$$i_{LPFC1-pk}(t) = \frac{\sqrt{2} v_{AC-rms} \sin(2\pi f_{AC} t)}{L_{PFC1}} D T_S \quad (2)$$

where D and T_S are the duty cycle and period of the power switch, respectively.

Mode 4 ($t_3 \leq t < t_4$; in Fig. 4(d)): At t_3, the power switch S_1 turns off. The utility-line voltage v_{AC} and the coupled inductor L_{PFC1} supply energy to the drain-source capacitor of S_1 through the diode D_{b1}. The inductor current i_{LPFC1} linearly decreases from the peak level and it can be given by

$$i_{LPFC1}(t) = \frac{\sqrt{2}v_{AC-rms}\sin(2\pi f_{AC}t) - V_{DC}}{L_{PFC1}}(t - t_3) \quad (3)$$

where V_{DC} is the voltage of the DC-bus capacitor C_{DC}.

The drain-source capacitor of S_2 provides energy to the inductor L_r, capacitors C_r and C_o and the LED lamp through diodes D_1 and D_4. When the drain-source voltage v_{DS2} of S_2 is decreased to be zero at t_4, and then this mode ends.

Mode 5 ($t_4 \leq t < t_5$; in Fig. 4(e)): The utility-line voltage v_{AC} and the coupled inductor L_{PFC1} provide energy to the C_{DC} through the diode D_{b1} and the body-diode of the switch S_2. The inductor L_r provides energy to the capacitors C_r and C_o and the LED lamp through diodes D_1 and D_4. At t_5, the inductor current i_{Lr} decreases to be zero, and then this mode ends.

Mode 6 ($t_5 \leq t < t_6$; in Fig. 4(f)): At this mode, voltage v_{AC} and the coupled inductor L_{PFC1} still provide energy to the C_{DC} through the diode D_{b1} and the body-diode of the switch S_2. Additionally, the switch S_2 achieves ZVS-turning-on at t_5. The capacitor C_r provides energy to the inductor L_r, capacitor C_o and the LED lamp through S_2 and diodes D_2 and D_3. At t_6, the inductor current i_{LPFC1} decreases to be zero, and then this mode ends.

Mode 7 ($t_6 \leq t < t_7$; in Fig. 4(g)): During this mode, capacitor C_r provides energy to the inductor L_r, capacitor C_o and the LED lamp through S_2 and diodes D_2 and D_3. The mode ends when the switch S_2 turns off at t_7.

Mode 8 ($t_7 \leq t < t_8$; in Fig. 4(h)): At this mode, resonant inductor L_r and the drain-source capacitor of the switch S_1 provide energy to the capacitor C_r, the drain-source capacitor of S_2, capacitor C_o and the LED lamp through diodes D_2 and D_3. When the drain-source voltage v_{ds1} of S_1 is decreased to be zero at t_8, and this mode ends. Then *Mode* 1 begins for the next high-frequency switching period.

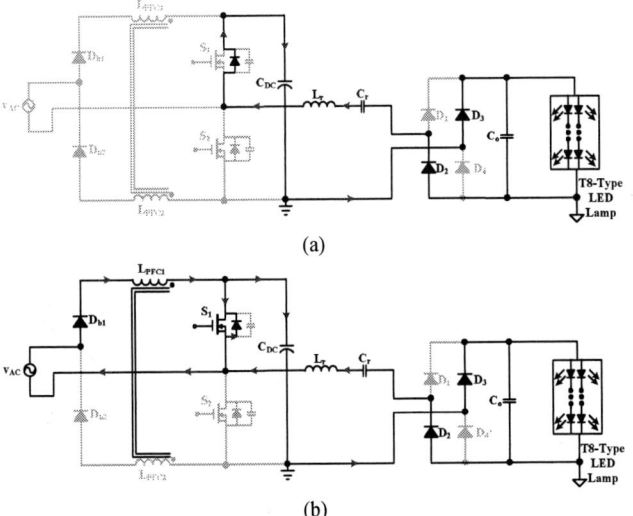

(a)

(b)

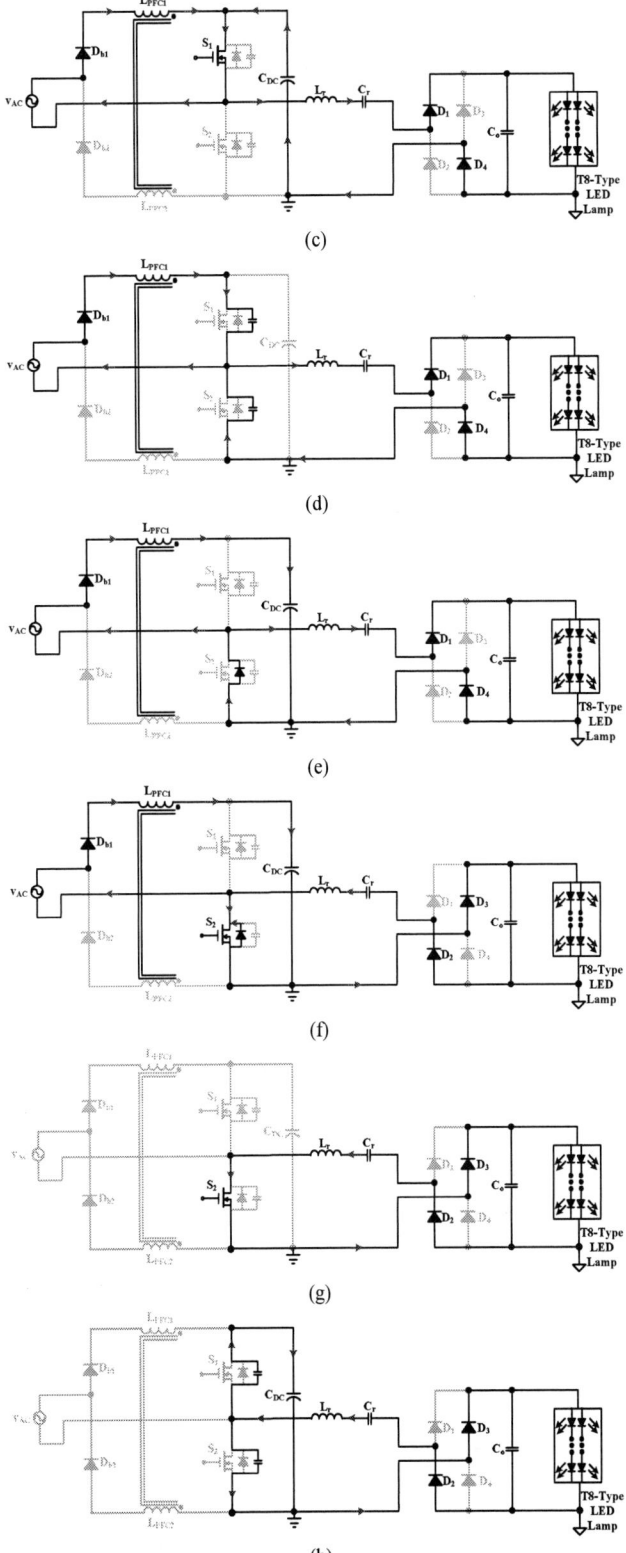

(c)

(d)

(e)

(f)

(g)

(h)

Figure 4. Operation modes of the presented driver during the positive half-cycle of input voltage v_{AC}.

$$f_0 = \frac{1}{2\pi\sqrt{L_r C_r}} \qquad (5)$$

In order to obtain zero-voltage switching of two active switches, the resonant frequency is designed to be smaller than the switching frequency so that the resonant tank can be resembled as an inductive network.

Therefore, the relationship between switching frequency f_S and resonant frequency f_0 is assumed as:

$$f_S = 4f_0 \qquad (6)$$

Combining (5) with (6), the resonant inductor L_r is given by

$$L_r = \frac{4}{\pi^2 f_S^2 C_r} \qquad (7)$$

With an f_S of 50 kHz and a C_r of 82 nF, the resonant inductor L_r is computed as

$$L_r = \frac{4}{\pi^2 f_S^2 C_r} = \frac{4}{\pi^2 \times (50k)^2 \times 82n} = 1.97mH$$

In addition, L_r is selected to be 2 mH.

IV. EXPERIMENTAL RESULTS

A prototype driver has been built and tested for supplying a 18W-rated T8-type LED lamp (EVERLIGHT FBW/T8/857/U/4ft), whose rated voltage and current are 60 V and 0.3 A, respectively. The components utilized in the driver are shown in Table II.

Fig. 6 shows the measured inductor currents i_{LPFC1} and i_{LPFC2}. The measured switch voltage v_{ds2} and inductor current i_{Lr} is shown in Fig. 7, and the series resonant tank can be resembled as an inductive load. Fig. 8 and Fig. 9 present the measured voltages (v_{ds1} and v_{ds2}) and currents (i_{ds1} and i_{ds2}) of the two power switches S_1 and S_2, respectively. Zero-voltage switching (ZVS) are obviously achieved on these power switches in order to boost up the circuit efficiency.

Fig. 10 shows the measured output voltage and current waveforms, and the average values of V_o and I_o are 60 V and 0.3 A, respectively. The input utility-line voltage and current is shown in Fig. 11, and the presented driver offers high power factor (0.99) and low input current THD (3.31%) at an input voltage of 110 V. The measured harmonic components of input current are shown in Fig. 12, and all input current harmonics of the LED lamp driver meets the IEC 61000-3-2 Class C standards under a tested input utility-line voltage of 110 V. In addition, the measured efficiency of the presented LED driver is 90.6%.

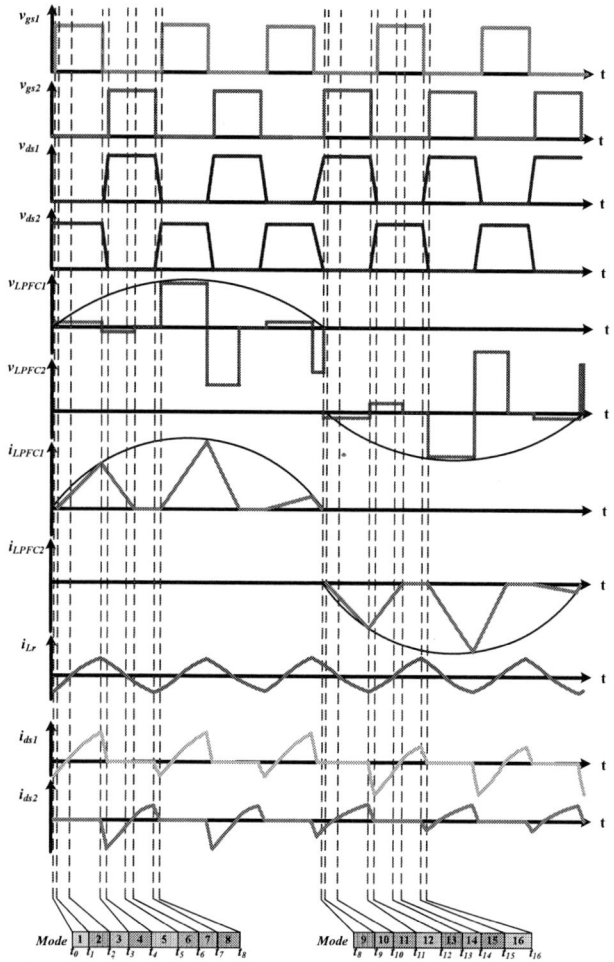

Figure 5. Key waveforms during the positive half-cycle of input voltage v_{AC}.

III. DESIGN EQUATIONS OF KEY COMPONENTS

A. Design of Coupled Inductors L_{PFC1} and L_{PFC2}

The design equation of the coupled inductor L_{PFC1} (L_{PFC2}) is expressed as

$$L_{PFC1} = \frac{\eta v_{AC-rms}^2 D^2}{2 P_{lamp} f_S} = L_{PFC2} \, . \qquad (4)$$

where η is estimated circuit efficiency, P_{lamp} is the rated power of the LED lamp, and f_S is the switching frequency.

With a η of 0.9, a v_{AC-rms} of 110V and a P_{lamp} of 18 W, an f_S of 50 kHz and a D of 0.5, the coupled inductors L_{PFC1} and L_{PFC2} are designed to be 1.5 mH.

B. Design of Series Resonant Tank

Referring to Fig. 3, the series resonant tank is composed of a resonant inductor L_r in series connection with a resonant capacitor C_r, and the resonant frequency f_0 can be expressed by

TABLE II
KEY COMPONENTS USED IN THE PROPOSED LED DRIVER

Component	Value
Filter Inductor (L_f)	3.3 mH
Filter Capacitor (C_f)	0.47 μF/250 V
Power Switches (S_1, S_2)	IRF840
Coupled Inductors (L_{PFC1}, L_{PFC2})	1.5 mH
DC-Linked Capacitor (C_{DC})	220 μF/450 V
Resonant Inductor (L_r)	2 mH
Resonant Capacitor (C_r)	82 nF
Diodes (D_{b1}, D_{b2}, D_1, D_2, D_3, D_4)	MUR460
Output Capacitor (C_o)	1000μF/63V

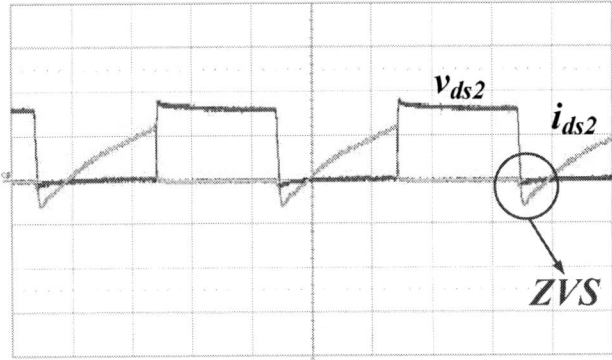

Figure 9.   Measured voltage v_{ds2} (200V/div) and current i_{ds2} (0.5A/div); time scale: 5μs/div.

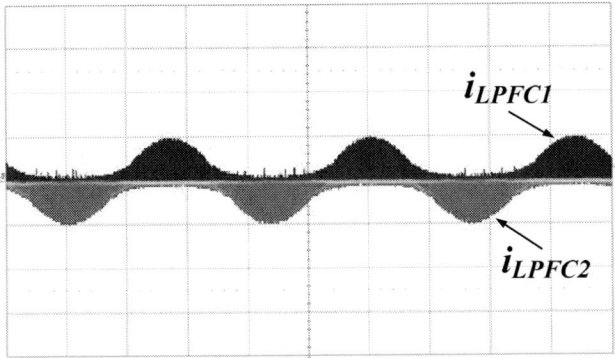

Figure 6.   Measured inductor currents i_{LPFC1} (1A/div) and i_{LPFC2} (1A/div); time scale: 5ms/div.

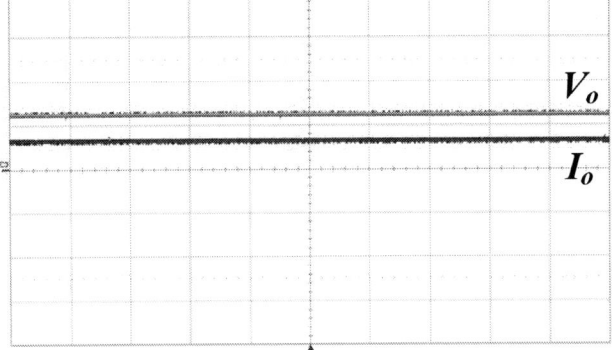

Figure 10. Measured output voltage V_o (50V/div) and current I_o (0.5A/div); time scale: 5ms/div.

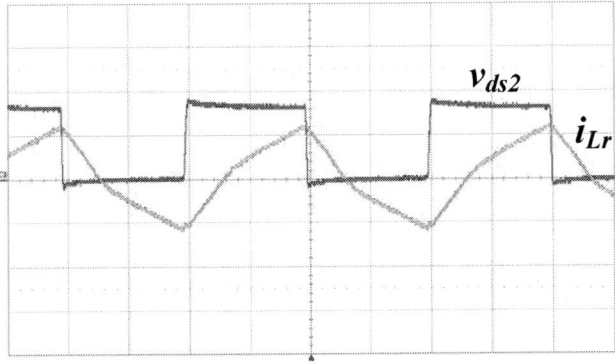

Figure 7.   Measured voltage v_{ds2} (200V/div) and inductor current i_{Lr} (0.5A/div); time scale: 5μs/div.

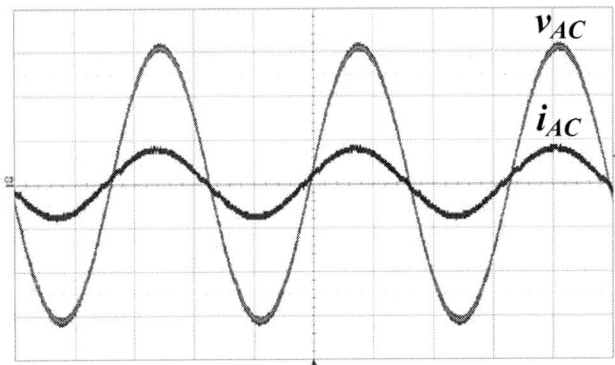

Figure 11. Measured input utility-line voltage v_{AC} (50V/div) and current i_{AC} (0.5A/div); time scale: 5ms/div.

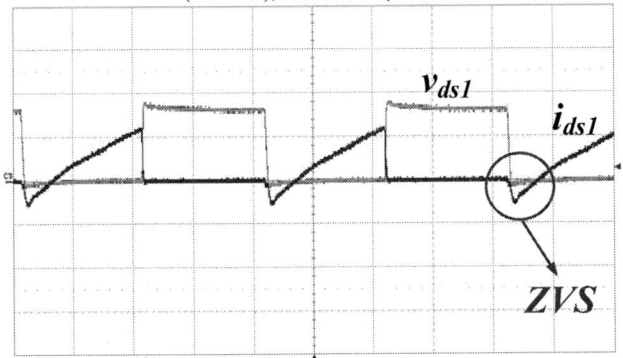

Figure 8.   Measured voltage v_{ds1} (200V/div) and current i_{ds1} (1A/div); time scale: 5μs/div.

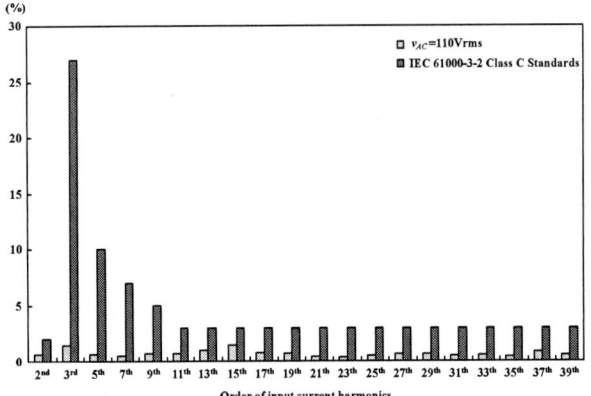

Figure 12. Measured input current harmonics compared with IEC Class C standards.

V. Conclusions

This paper has proposed a single-stage driver, which integrates a dual boost converter with coupled inductors and a half-bridge series-resonant converter, for supplying a T8-type LED lamp with power-factor-corrections. A prototype circuit has been successfully built for supplying an 18W-rated LED lamp from a 110-rms utility-line voltage. The experimental results have demonstrated high-PF (>0.99), low THD ($<4\%$), high efficiency ($>90\%$), so that the functionality of the presented LED driver is verified.

Acknowledgment

This work was supported by the Ministry of Science and Technology in Taiwan, R. O. C., under Grant NSC 102-2221-E-214-024.

References

[1] H. L. Cheng and Y. H. Huang, "Design and implementation of dimmable electronic ballast for fluorescent lamps based on power-dependent lamp model," *IEEE Trans. Plasma Science*, vol. 38, no. 7, pp. 1644-1650, July 2010.

[2] Y. S. Lin, C. A. Cheng, J. F. Chen, T. J. Liang and W. S. Liu, "Design and implementation of electronic ballast for fluorescent lamps with low lighting flicker," in *Proc. of IEEE IPEMC'06, 2006*, pp. 1-5.

[3] Website of China Electric Mfg. corporation: http://www.toa.com.tw/

[4] Website of Everlight Electronics: http://www.everlight.com/

[5] P. S. Almeida, A. L. C. Mello, H. A. C. Braga, M. A. Dalla Costa and J. M. Alonso, "Off-line soft-switched LED driver based on an integrated bridgeless boost-half-bridge converter," in *Proc. of IEEE Industry Applications Society Annual Meeting, 2013*, pp. 1-7.

[6] Y. S. Chen, T. J. Liang, K. H. Chen and J. N. Juang, "Study and implementation of high frequency pulse LED driver with self-oscillating circuit," in *Proc. of IEEE ISCAS'11*, 2011, pp. 498-501.

[7] C. S. Moo, Y. J. Chen and W. C. Yang, "An efficient driver for dimmable LED lighting," *IEEE Trans. Power Electron.*, vol. 27, no. 11, pp. 4613-4618, November 2012.

[8] N. Chen and H.S.-H. Chung, "A driving technology for retrofit LED lamp for fluorescent lighting fixtures with electronic ballasts," *IEEE Trans. Power Electron*, vol. 26, no. 2, pp. 588-601, Feb. 2011.

[9] E. F. Schubert, *Light-emitting diodes*, Cambridge University Press, 2006.

[10] Y. Qin, D. Lin and S. Y. (Ron) Hui, "A simple method for comparative study on the thermal performance of LEDs and Fluorescent lamps," *IEEE Trans. Power Electron*, vol. 24, no. 7, pp. 1811-1818, July 2009.

[11] Nan Chen and H.S.-H. Chung, "A universal driving technology for retrofit LED lamp for fluorescent lighting fixtures," in Proc. of *IEEE APEC'12, 2012*, pp. 980-987.

[12] "LED lighting solutions," ON Semiconductor, pp. 1-48, March 2013. http://www.onsemi.cn/pub_link/Collateral/BRD8034-D.PDF

Elimination of Electrolytic Capacitor in AC-DC System of LED Driver

Rijalul Fahmi Mustapa, Nabil M Hidayat, Rahayu Tukiman
Faculty of Electrical Engineering
Universiti Teknologi MARA
Shah Alam, Selangor, Malaysia
mnabil@salam.uitm.edu.my

Abstract— The objective of this project is elimination of electrolytic capacitor in LED driver in order to enhance the operating time. In this paper, a novel circuit of LED driver is proposed using neutral point type buck-boost converter incorporating valley fill circuit, which can reduce the capacitance of smoothing capacitor. Thus, film capacitor can replace electrolytic capacitor as the smoothing capacitor. 80W prototype of LED driver has been built to confirm the validity of the proposal.

Keywords— Active filters, Converter, Harmonic filters, Inverter, LED driver.

I. INTRODUCTION

T LEDs have several real advantages over conventional lighting. Long life time is the most benefit of LED lights. LEDs have an outstanding operational life time expectation of up to 100 000 hours or around 10 years of continuous operation while it would take 20 years for an average usage of 8 hours per day. LEDs also are different to standard lighting. They do not really burn out and stop working like a conventional lighting; moreover the lighting diodes emit lower output levels over a very long period of time and become less bright.

Unfortunately, due to smoothing capacitor in the AC/DC converter by using electrolytic capacitor, it reduces the life span of LED driver. This is because electrolytic capacitor is constructed in severe condition. So, when LED operates and is applied with heat, the subject tends to evaporate and at certain time will fail due to dryness of electrolytic capacitor. Thus in this research, the main objective of this project is a novel circuit of LED driver that is initially consists of electrolytic capacitor substitute with non-electrolytic capacitor such as polypropylene film capacitor. Beneficially, the LED driver circuit uses a dry type capacitor for smoothing the output of AC to DC rectifier for enhancing LED durability. The circuit can also operate from low DC voltages while its converter turn on and off with negligible voltage across them. This condition is known as soft switching and is desirable to minimize heating of switching.

Literature review shows that elimination of electrolytic capacitor in AC/DC converter is widely done in various methods in order to enhance the life span of circuit. As known, electrolytic capacitor in AC/DC converter is used to balance the instantaneous power difference. Thus, if electrolytic capacitor is eliminated or reduced without considering their capacitance, there will be balancing power issue.

The first approach is to modify the waveform of line input current. Less storage capacitance is needed in order to balance input output power if peak-to-average ratio of input pulsating is reduced. In [1] and [2], injection of third and fifth harmonic signal into input current is done to reduce the peak-to-average ratio of input power. These methods are constructed based on simple main circuits (Boost and Flyback converter) and just need a modification in control circuits. However, these methods have disadvantages which are reduced input power factor.

Another approach is using current with large ripple to drive LEDs. If there is small difference power between input and output, less storage capacitance is needed to balance the energy. In [3] and [4], two types of topologies circuits are proposed for drive the LEDs. These topologies conduct pulsating current at twice the line frequency. A novel passive off-line LED driver is proposed in [5]. This method conduct based on general photo-electro-thermal theory to drive the LED system which is driven with relative large current ripple. Unfortunately, this approach is only suitable for simple system such as public/road lighting system, instead of system compactness.

The third approach is to adopt some energy storage elements to handle power difference [6]-[7]. Energy storage elements can be divided into two types which are inductive and capacitive elements. In [8] and [9], inductors are conducted as energy storage element where output power is larger than input power. When input power is larger than output power, inductor will store the excessive energy and also vice versa. Thus, electrolytic capacitor's capacitance value can be reduced or eliminated.

Recently, several topologies are proposed based on

capacitor parallel structure. In [10] and [11], single stage photovoltaic (PV) power to ac grid module is using current pulsating smoothing parallel active filter. DC bus is connected parallel with bidirectional Buck-Boost circuit.

The target of this project is come out the solution for problem that associated with electrolytic capacitor which is reduced the life span of LED driver. In this paper, neutral point type buck-boost converter incorporating valley fill circuit is chosen to drive a 80W LED system. So that, the output still stable due to low ripple voltage and current harmonic.

During encountered the problem of electrolytic capacitor in LED driver, a new technique was proposed to eliminate electrolytic capacitor without neglect the function of smoothing output current and voltage. Electrolytic capacitor which was used to smoothing capacitor in LED driver is replaced by polypropylene film capacitor in order to maintain the function.

II. PROPOSED CIRCUIT

A. Circuit Topology

The proposed technique of neutral point type buck-boost converter incorporating valley fill circuit is used. This circuit is designing based on two half wave rectifiers, diodes and capacitor [12]. In order to increase the lifetime of this overall system, electrolytic capacitor will be replaced with polypropylene capacitor. To maintain the function of capacitor as smoothing circuit, polypropylene capacitors are arranged as a valley fill circuit. This circuit will consume small current ripple that causes power variation in the LED load. However, this power variation will not effect of luminous variation to human eyes. Besides that, additional diodes in this circuit are used to conduct current flow in one direction between inductor to capacitor during discharging process. There is open loop system between the input sources and capacitor. Consequently, this circuit will reduce the inrush current. This circuit operation depends on the operation of switching elements. Switch S1 will be operated during positive half cycle while switch S2 during negative half cycle. When either switch turn ON, current from input will charges the inductor L while switch turn OFF, counter electromotive force current that stored in the inductor will flow through capacitor to charge it.

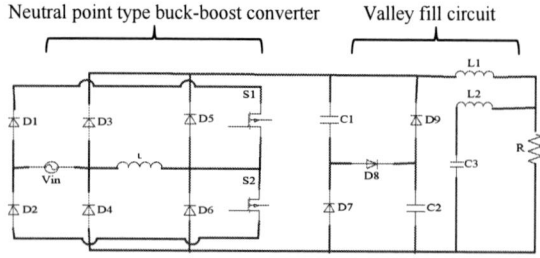

Neutral point type buck-boost converter   Valley fill circuit

Fig. 1. Neutral point type buck-boost converter incorporating valley fill circuit schematic diagram

B. Circuit Operations

In this technique, the circuit operation will follow the 4 assumptions stated below.

1.   The ON-OFF switching frequency of switching elements must be higher that frequency of input voltage (Vin).
2.   Each circuit elements is an ideal circuit element. (No   dissipation energy due to resistance in each circuit   elements).
3.   The output voltage of capacitor will be ideal.
4.   The output load can be expressed as resistor.

Based on the above assumptions, this circuit operation will be explained based on ON-OFF states of switching elements S1 and S2. This circuit can be divided into 6 state operations (Table 1). Depending on the polarity of input voltage source, 3 state operations are during positive half cycle and remains for negative half cycle. State 3 is dead time occurs where at this time both switching elements turn OFF for 1.2us.

TABLE I
SWITCH OPERATION IN EACH STATE

State	1	3	2	4	3	5
Input Voltage	+	+	+	-	-	-
Switch 1	ON	OFF	OFF	OFF	OFF	ON
Switch 2	OFF	OFF	ON	ON	OFF	OFF

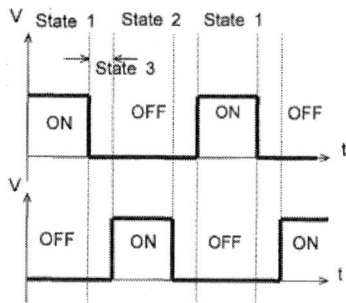

Fig. 2. ON-OFF conditions of switching elements in each state operation

III. EXPERIMENT RESULTS

The neutral point type buck-boost converter incorporated with valley fill circuit was constructed and LED as a load (Fig. 1). A low pass filter (cut-off frequency 2 kHz) was connected in parallel with input voltage for measurement purposes and switching frequency fixed at about 50 kHz. The load is a 15W LED lamp with 40V input voltage. Table 2 shows the experiment component parameter of topology circuit for Fig. 1. The result experiment using electrolytic capacitor and valley fill circuit can be seen in Table 3 and 4.

TABLE 2
EXPERIMENT COMPONENT PARAMETER OF TOPOLOGY CIRCUIT

Part	Specification
FET S1, S2	IRF840L
Diode; D1-D9	5GLZ47
Electrolytic Capacitor;	500V, 20uF
Polypropyline Capacitor,	700V, 20uF
Inductor; L1	1mH
Inductor L2,L3	2.55mH
Half-bridge Driver	IR2153
LED load	15W

TABLE 3
EXPERIMENT RESULTS OF THE LOW POWER LED AS LOAD BY USING
CIRCUIT INCORPORATING ELECTROLYTIC CAPACITOR

Input Voltage	40V	Output Voltage	39.3V
Input Current	0.23A	Output Current	0.161A
Input Power	9.2W	Output Power	6.33W
Power Factor	96.6%	Oscillation Frequency	50kHz
Ripple Voltage	18.58%		

TABLE 4
EXPERIMENT RESULTS OF THE LOW POWER LED AS LOAD BY USING
CIRCUIT INCORPORATING VALLEY FILL CIRCUIT

Input Voltage	40V	Output Voltage	38.2V
Input Current	0.22A	Output Current	0.146A
Input Power	8.8W	Output Power	5.58W
Power Factor	95.1%	Oscillation Frequency	50kHz
Ripple Voltage	18.32%		

A 15W LED system can be developed by using either electrolytic capacitor or valley fill circuit. Fig. 3(a) and (b) show the waveforms of input voltage and input current by using electrolytic capacitor while Fig. 4(a) and (b) show the waveform of input voltage and input current by using valley fill circuit. From the result, it can be seen that the input voltage for both proposed and conventional circuit are smooth and sinusoidal with value of 40V. Besides that, the input current obtained by using valley fill circuit produced less input surge current which was 0.22A, compared to electrolytic capacitor usage which was 0.23A. Furthermore, the measured output voltages were captured in Fig. 3(c) for electrolytic capacitor used and Fig. 4(c) for valley fill circuit used. Based on these results, the waveform using valley fill circuit was fairly smooth and contained small ripple compared to electrolytic capacitor. But, circuit of output ripple cancellation was needed, and implemented by using an inductor and capacitor at the end of valley fill circuit. It was to provide a smooth DC voltage as well as fast regulation. The output voltage for using valley fill was 38.2V and 18.32% of ripple voltage instead of by using electrolytic capacitor which was 39.3V with ripple voltage of 18.58%. As expected, the measurement of using valley fill circuit is nearly to the conventional circuit of using electrolytic capacitor.

The content of line current harmonic for load of low power LED was also measured. Harmonic measurements were made by using Fluke meter analyzer and compared with IEC 6100-3-2 class C standard. Based on Fig. 6, it shows that this proposed circuit has low harmonics content and the value still within the range of IEC 61000-3-2 class C. 3rd harmonic order shows that percentage of harmonic for load of low power LED was below the standard of IEC 61000-3-2 class C. In 5th harmonic order, the percentage of harmonic by using valley fill circuit was lower than using electrolytic capacitor which is 10.9% compared to 11.7%.

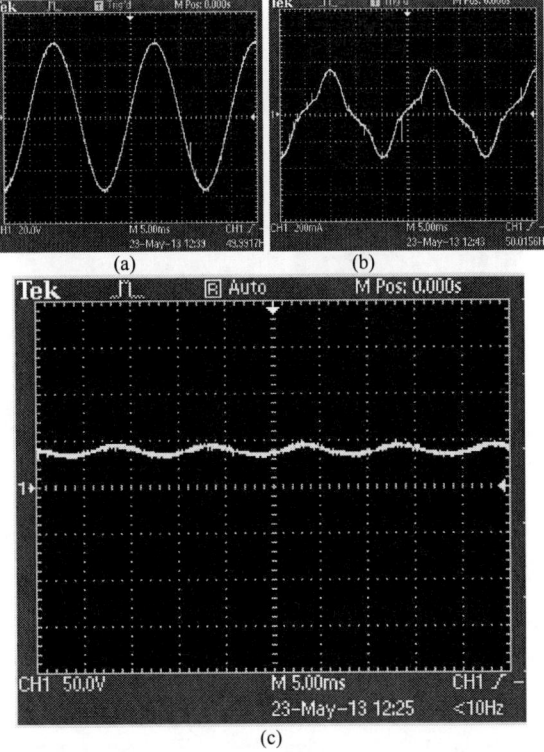

Fig. 4. (a) input voltage and (b) input current and (c) output voltage waveform of the low power LED as a load by using circuit incorporating electrolytic capacitor

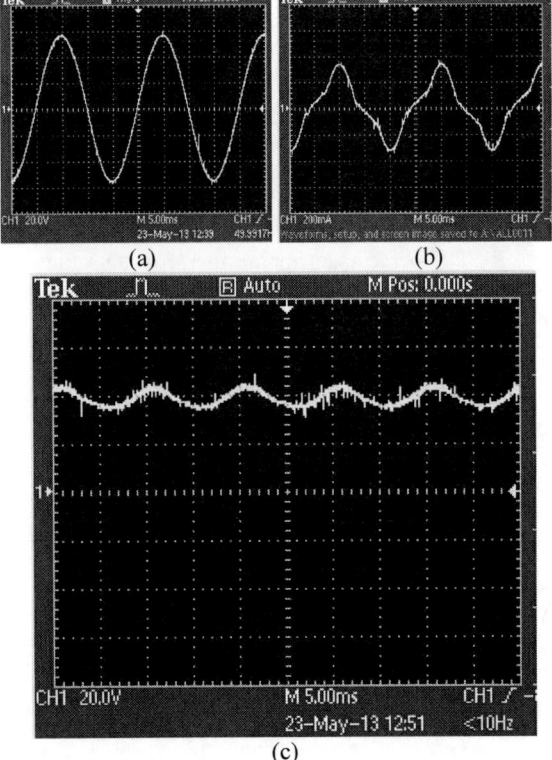

Fig. 5. (a) input voltage and (b) input current and (c) output voltage waveform of the low power LED as a load by using circuit incorporating valley fill circuit

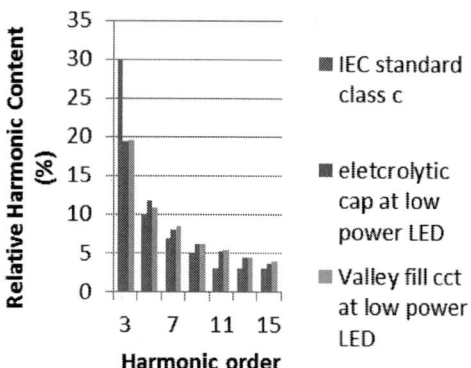

Fig. 6 Comparison rates of harmonics in the input current of proposed system with IEC harmonics limit standard for class C

IV. CONCLUSIONS

The proposed neutral point type buck-boost converter with valley fill circuit was successfully functioned to eliminate electrolytic capacitor which is the main problem of LED driver to expend their life span. This proposed circuit is functioning well without neglecting the conventional function of electrolytic capacitor as smoothing the output voltage and current. Unfortunately due to damage of the high power LED load, experiment only can be tested on bulb and low power LED. But the result from this experiment still followed the objectives of the project. In addition, the output voltage resembled smooth waveform and the value is nearly the input voltage. The percentage of ripple voltage is low and percentage of THD still acceptable. The results have achieved the following objectives:

1. To eliminate the electrolytic capacitor in LED driver by replacing with polypropylene film capacitor.

2. The proposed circuit topology can be implemented as a driver for LED.

3. The percentage ripple voltage using valley fill circuit is low as same as using electrolytic capacitor.

4. The proposed circuit managed to get output value approximately equal to input value.

ACKNOWLEDGMENT

The authors would like thank Research Management Institute of Universiti Teknologi MARA (UiTM) and Kementrian Pelajaran Malaysia for providing financial assistant under grant 600-RMI/RAGS 5/3 (60/2013).

REFERENCES

[1] B. Wang, X. Ruan, K. Yao, and M. Xu, "A method of reducing the peak-to-average ratio of LED current for electrolytic capacitor-less AC–DC drivers," IEEE Trans. Power Electron., vol. 25, no. 3, pp. 592–601, Mar. 2010.
[2] L. Gu, X. Ruan, M. Xu, and K. Yao, "Means of eliminating electrolytic capacitor in AC/DC power supplies for LED lightings," IEEE Trans. Power Electron., vol. 24, no. 5, pp. 1399–1408, May 2009.
[3] G. Spiazzi, S. Buso, and G. Meneghesso, "Analysis of a high-power-factor electronic ballast for high brightness light emitting diodes," inProc. IEEE Power Electron. Spec. Conf. (PESC), 2005, pp. 1494–1499.
[4] E. Mineiro S´ aJr.,C.S.Postiglione,F.L.M.Antunes, andA.J.Perin, "Low cost ZVS PFC driver for power LEDs," inProc. IEEE Ind. Electron. (IECON), 2009, pp. 3551–3556.
[5] S. Y. R. Hui, S. Li, X. Tao, W. Chen, and W. M. Ng, "A novel passive off-line light-emitting diode (LED) driver with long lifetime," inProc. IEEE Appl. Power Electron. Conf. (APEC), 2010, pp. 594–600.
[6] B. Zhang, X. Yang, M. Xu, Q. Chen, and Z. Wang, "Design of Boost-Flyback single-stage PFC converter for LED power supply without elec-trolytic capacitor for energy-storage," inProc. IEEE Int. Power Electron. Motion Control Conf. (IPEMC), 2009, pp. 1668–1671.
[7] K. Yao, M. Xu, X. Ruan, and L. Gu, "Boost-Flyback single-stage PFC converter with large DC bus voltage ripple," inProc. IEEE Appl. Power Electron. Conf. (APEC), 2009, pp. 1867-1871.
[8] T. Shimizu, Y. Jin, and G. Kimura, "DC ripple current reduction on a single-phase PWM voltage-source rectifier," IEEE Trans. Ind. Appl., vol. 36, no. 5, pp. 1419–1429, Sep./Oct. 2000.
[9] S. Li, B. Ozpineci, and L. M. Tolbert, "Evaluation of a current source active power filter to reduce the DC bus capacitor in a hybrid electric vehicle traction drive," in Proc. IEEE Energy Convers. Congr. Expo. (ECCE), 2009, pp. 1185–1190.
[10] A. C. Kyritsis, N. P. Papanicolaou, and E. C. Tatakis, "A novel parallel active filter for current pulsation smoothing on single stage grid-connected AC-PV modules," inProc. Eur. Conf. Power Electron. Appl.(EPE), 2007, pp. 1–10.
[11] A. C. Kyritsis, N. P. Papanicolaou, and E. C. Tatakis, "Enhanced current pulsation smoothing parallel active filter for single stage grid-connected AC-PV modules," Proc. Power Electron. Motion Control Conf. (EPE. PEMC), 2008, pp. 1287–1292.
[12] Nabil M. Hidayat, Yoshito Kato and Yoshio Itoh, "One stage converter method neutral point type voltage free converter and application to an electronic ballast" in Proc. IEEE Power Electron , vol.34,No.1,2010.

A Novel Bridgeless Boost Half-Bridge ZVS-PWM Single-Stage Utility Frequency AC–High Frequency AC Resonant Converter for Domestic Induction Heaters

Tomoakzu Mishima, Yuki Nakagawa
Dept. of Marine Engineering, Grad. Sch. of Maritime Sci.
Kobe University
Higashinada, Kobe, Hyogo 658-0022 Japan

Mutsuo Nakaoka
Elec.Energy Saving Research Center.
Kyungnum University
Wolyoung-Dong, Masan, The Republic of Korea

Abstract—A new prototype of a single-stage utility ac (UFAC) to high-frequency ac (HFAC) resonant power converter for domestic induction heating (IH) applications is presented in this paper. The high frequency resonant ac-ac converter proposed herein can process UFAC to HFAC power conversion without any diode full-bridge rectifier stage, thereby reducing the number of semiconductor power devices and eliminating the relevant conduction power losses. In addition, power factor correction (PFC) can be achieved by the front-end inductor-based ac chopper operation with boosting the UFAC source voltage to a non-smoothed dc link voltage. The operation principle together with a IH load power regulation scheme is described, and the actual performances are demonstrated in an experiment with a 2.8 kW–30 kHz laboratory prototype. Finally, the feasibility of the proposed ac-ac converter is evaluated from a practical point of view.

Keywords—*Induction heating (IH), utility frequency ac (UFAC)–high frequency ac (HFAC) power conversion, diode rectifier-less (bridgeless), boost half-bridge (BHB)*

I. INTRODUCTION

A domestic IH cooker as schematically illustrated in Fig. 1 has been advanced with a wide variety of circuit topologies featuring soft switching technologies in the past decades, and now is getting into a new phase of research and development pursuing for high-efficiency and cost-effective electric power conversion and processing[1]-[3].

A typical IH cooker consists of a three-stage power conversion via the single-phase UFAC–smoothed and boosted dc with PFC function–HFAC by adopting a diode full-bridge rectifier, boost-type dc-dc converter, and HF inverter as depicted in Fig. 2[4][5]. This circuit configuration is supported by the well-established and stable power conversion technologies, and has now been the basic power processing scheme in the domestic and consumer IH appliances. However, the multiple power conversion stages might lead to the efficiency deterioration, especially in a low output power setting, and consequently the efficiency improvement for higher power density will be obstructed.

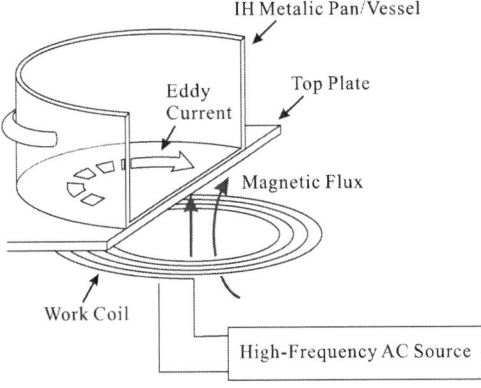

Fig. 1. Principle of induction heating with a HFAC current.

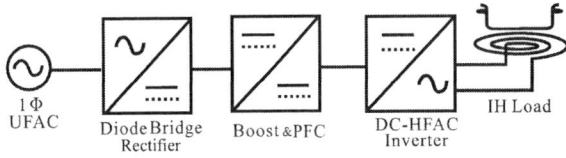

Fig. 2. Power and energy conversion process of the conventional domestic and consumer induction heaters.

In order to improve the efficiency with a cost-effective circuit configuration, the two-stage BHB HF inverter as illustrated in Fig. 3 has been proposed and commercialized[6]. In this two-stage UFAC–HFAC converter, the non-smoothed dc–HFAC power processing can be performed simultaneously in the HF inverter stage, thereby the high efficiency and cost reduction can be attained. However, the diode rectifier connecting the UFAC source with the HF inverter is still demanded, thus the further improve of the total efficiency can not be expected in the two-stage UFAC–HFAC converter.

As a solution for the technical challenge of the two-stage BHB topology as well as the other existing converters for the domestic IH applications, the innovative and cost-effective UFAC–HFAC power converter is proposed in this paper. The proposed ac-ac converter is free of the

978-1-4799-2706-7/14 $31.00 © 2014 IEEE

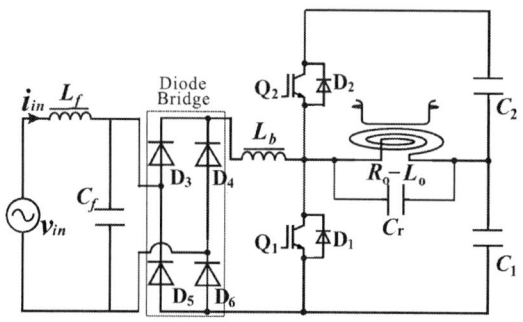

Fig. 3. A two-stage BHB ZVS-PWM UFAC–HFAC converter[6].

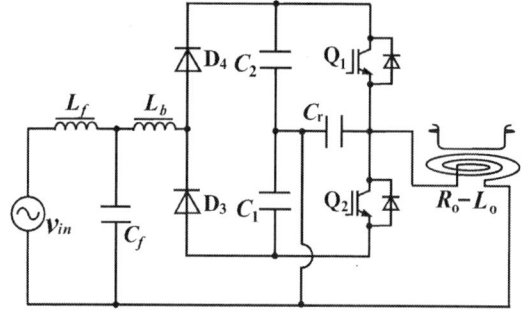

Fig. 4. A single-stage ZVS-PWM UFAC–HFAC converter with a voltage doubler[7][8].

diode rectifier, named as "bridgeless", and the UFAC-HFAC power conversion can be performed in the single-stage process while keeping the PFC and the inductor-assisted voltage boost functions with a reverse conducting insulated gate-bipolar-transistor (RC-IGBT).

The voltage doubler type single-stage ac-ac converters for IH applications have been proposed by the authors [7][8], whereby no actively voltage regulation can be performed. The non-boost type single-stage ac-ac converter has been proposed[9], however the full-bridge diode is necessary and the cost-effectiveness is still in a challenge. Another type of single-stage ac-ac converter has been proposed by using bidirectional switches[10], but the reliability of this new type of power device is not up to par for the practical level.

The rest of this paper is organized as follows. The circuit configuration and operation principle of the proposed ac-ac converter are explained in Section II. The theoretical analysis of the HF resonant current fed to the IH load and the load resonant frequency is described in Section III. Furthermore, the experimental results and evaluations of its laboratory prototype of $2.8\,\mathrm{kW}$–$30\,\mathrm{kHz}$ are demonstrated in Section IV. Finally, the unique and advantageous properties of the proposed ac-ac converter are summarized in Section V.

II. BHB SINGLE-STAGE UFAC–HFAC CONVERTER

A. Circuit Configuration

The main circuit diagram of the proposed ac-ac converter is depicted in Fig. 5[12][13]. Note here that L_f

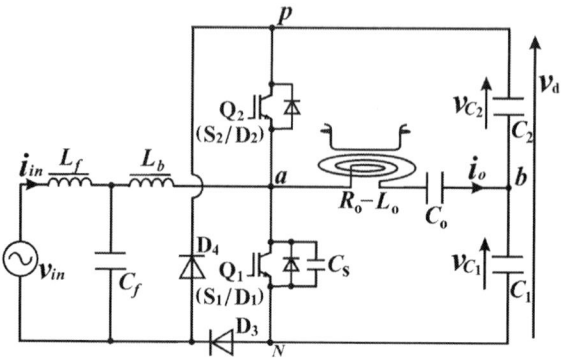

Fig. 5. A proposed single-stage UFAC-HFAC power converter.

and C_f consist of the grid-connected filter for eliminating the switching frequency components from the utility current i_{in}. The active switches Q_1, Q_2 are shared by the BHB[11] and HF resonant inverter operations. The source voltage v_{in} is lifted to the voltages v_{c1} and v_{c2} across the non-smoothed capacitors C_1 and C_2 by the effect of the inductor L_b under the HF switching conditions of Q_1 and Q_2. At the same time, PFC can be achieved in the front-end BHB stage, whereby the low distorted source current can be naturally obtained.

The theoretical voltage and current waveforms for the UFAC cycle of the proposed ac-ac converter are depicted in Fig. 6. The IH load power is controlled by the asymmetrical pulse-width-modulation (A-PWM), where the duty cycle D is defined by referring to Fig. 6 as $D = t_{on}/T_s$. In order to make the single-stage power conversion symmetrically for the positive and negative half-cycles of the power source v_{in}, the gate signal pattern for adjusting D is exchanged in accordance with the polarity of v_{in} as illustrated in Fig. 7. The controller configuration including the A-PWM pulse generator is schematically depicted in Fig. 8.

B. Operation Principle

The voltage and current waveforms of the proposed ac-ac converter for the HFAC cycle is shown in Fig. 9. In addition, the switch-mode transitions and equivalent circuits during the positive and negative half-cycles of v_{in} are depicted in Figs. 10 and 11, respectively. For the sake of paper length restriction, only the positive half-cycle operation is explained below.

[Mode1: single-loop power supply mode] $(t_0 \leq t < t_1)$ The low-side switch Q_1 is on-state, and the UFAC source current i_{in} flows through the network of v_{in}–L_b–S_1–D_3. During this interval, the boost inductor L_b stores the magnetic energy. At the same time, the HFAC resonant current is fed from C_1 to the IH load R_o–L_o until S_1 is turned off at $t = t_1$.

[Mode2: edge resonance mode] $(t_1 \leq t < t_2)$ The low-side active switch Q_1 is turned off at $t = t_1$, and the low-side lossless snubbing capacitor C_s and equivalent effective inductor L_o make the series resonance. Accordingly, C_s is gradually charged and the voltage v_{Q1} across Q_1

The 2014 International Power Electronics Conference

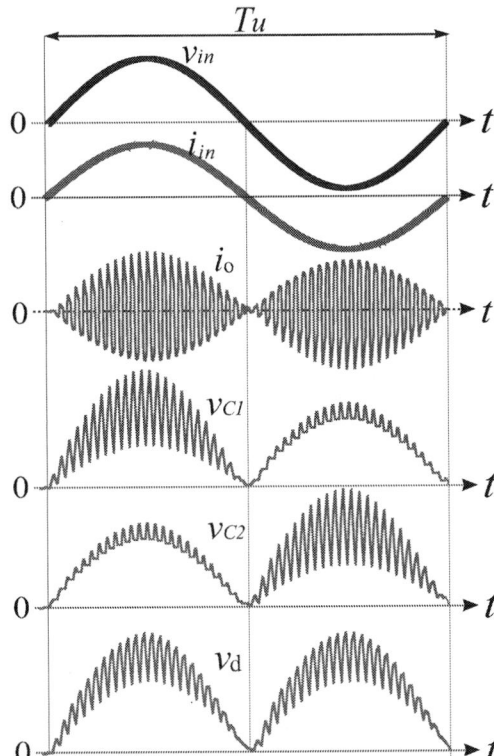

Fig. 6. Relevant voltage and current waveforms for UFAC cycle.

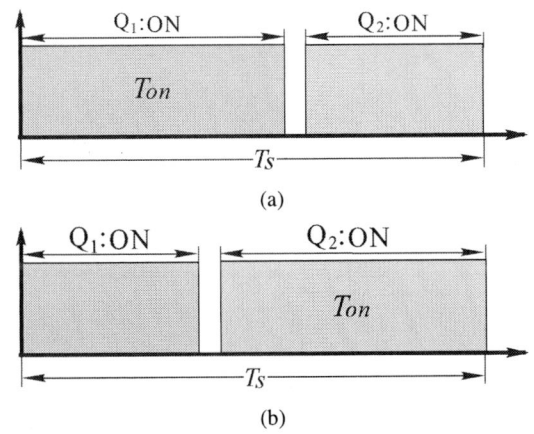

Fig. 7. Asymmetrical PWM gate signal patterns for low- and high-side active switches: (a) $v_{in} > 0$, and (b) $v_{in} < 0$.

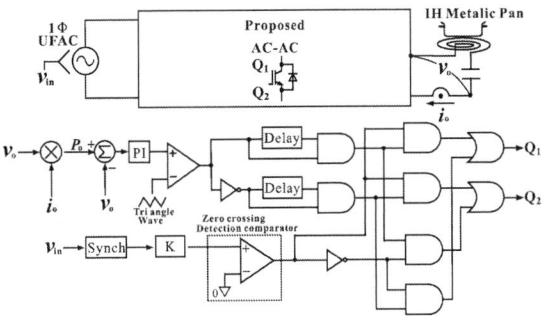

Fig. 8. Schematic diagram of controller logic circuit.

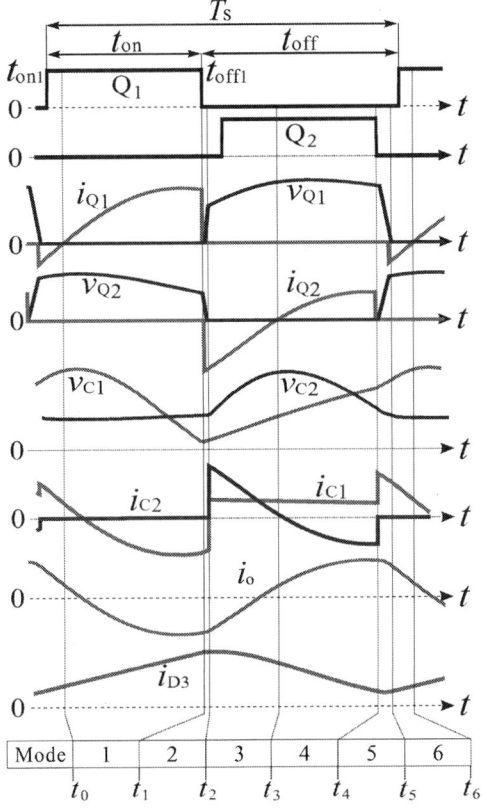

Fig. 9. Relevant voltage and current waveforms for HFAC cycle ($v_{in} > 0$).

rises with a certain slope, while the voltage v_{Q2} across the high-side active switch Q_2 gradually declines. Thus, ZVS turn-off transition starts in Q_1. In order to successfully attain the ZVS operation, the following condition should be satisfied:

$$L_o i_o(t_1)^2 > C_s v_d(t_1)^2 \qquad (1)$$

[Mode3 : high-side free-wheeling mode] ($t_2 \leq t < t_3$) The voltage v_{Q1} reaches the non-smoothed dc-link voltage v_d at $t = t_2$, then ZVS turn-off operation of Q_1 is completed. At the same time, the anti-parallel diode D_2 in Q_2 is naturally forward-biased. During this interval,

the gate-signal is supplied for S_2 in Q_2, whereby zero voltage and zero current soft-switching (ZVZCS) turn-on can be performed in Q_2. The UFAC source current i_{in} is fed to C_1 and C_2 through L_b with releasing the stored magnetic energy into C_1 and C_2, thus the UFAC source voltage v_{in} is boosted.

[Mode4 : dual-loop power supply mode] ($t_3 \leq t < t_4$) The current i_{Q1} through Q_2 commutates from D_2 to S_2 at $t = t_3$. Accordingly, the IH load current i_o reverses its polarity and the load power is supplied from both the UFAC source and the high-side non-smoothed dc-link capacitor C_2.

[Mode5 : edge resonance mode] ($t_4 \leq t < t_5$) The high-side active switch Q_2 is turned off at $t = t_4$, and the

978-1-4799-2706-7/14 $31.00 © 2014 IEEE   2535

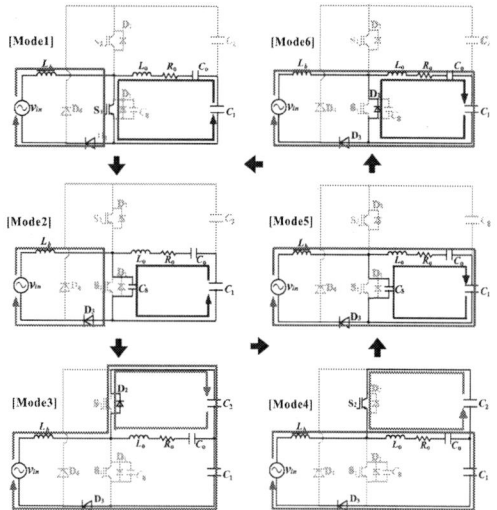

Fig. 10. Mode transitions and equivalent circuits for the positive polarity half-cycle of v_{in}.

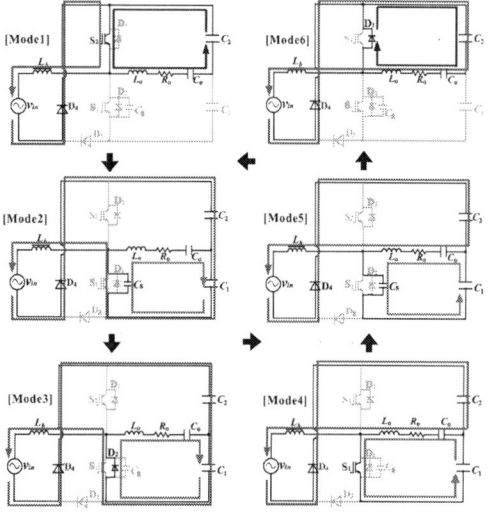

Fig. 11. Mode transitions and equivalent circuits for the negative polarity half-cycle of v_{in}.

lossless snubbing capacitor C_s and equivalent effective inductor L_o produce the series resonate. Accordingly, C_r is gradually charged and v_{Q2} across Q_2 rises with a certain slope, while v_{Q1} across Q_1 gradually decreases. Thus, ZVS turn-off is started in Q_2. In order to successfully attain the ZVS operation, the following condition should be satisfied:

$$L_o i_o(t_4)^2 > C_s v_d(t_4)^2 \qquad (2)$$

[Mode6 : low-side free-wheeling mode] ($t_5 \leq t < t_6$) The voltage v_{Q2} reaches the non-smoothed dc-link voltage v_d at $t = t_5$, then ZVS turn-off operation of Q_2 is completed. At the same time, the anti-parallel diode D_1 in Q_1 is naturally forward-biased. During this interval, the gate-signal is provided for S_1 in Q_1, whereby ZVZCS turn-on is achieved in Q_1. The source current i_{in} is fed to the IH load through L_b with charging C_1. The current

TABLE I. COMPARISON OF CONDUCTING POWER DEVICES IN POSITIVE HALF-CYCLE OF v_{in}.

	Proposed 1-stage BHB	Previous 2-stage BHB
S_1: ON	S_1, D_3	D_3, S_1, D_6
S_1: OFF	D_2, D_3	D_2, D_3, D_6

TABLE II. COMPARISON OF CONDUCTING POWER DEVICES IN NEGATIVE HALF-CYCLE OF v_{in}.

	Proposed 1-stage BHB	Previous 2-stage BHB
S_2: ON	S_2, D_4	S_2, D_4, D_5
S_2: OFF	D_1, D_4	D_1, D_5, D_5

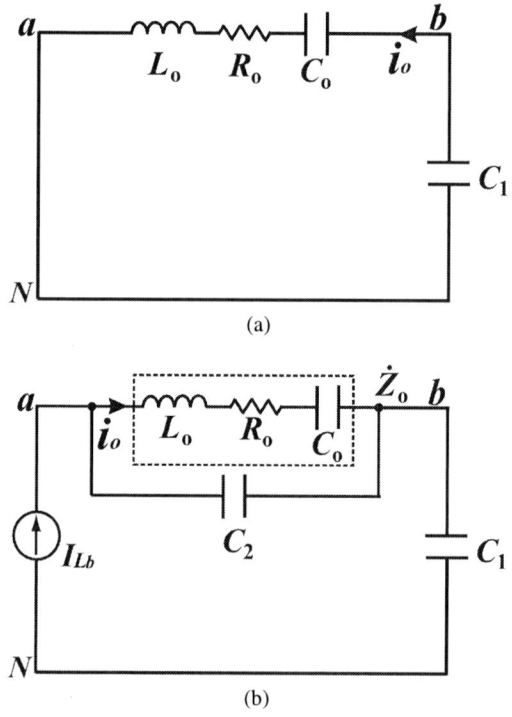

Fig. 12. Simplified equivalent circuits under the condition $v_{in} > 0$: (a) S_1 ON, and (b) S_1 OFF.

i_{Q1} through Q_1 commutates from D_1 to S_1 at $t = t_6$, then the circuit state is initiated into Mode 1 mentioned above.

The power devices conductions are compared between the two-stage and the proposed single-stage ac-ac converters in TABLE I and II. It can be known from the comparison that the number of the conduction power devices can be reduced by the proposed ac-ac converter.

III. TIME-DOMAIN ANALYSIS OF IH LOAD CURRENT WITH SIMPLIFIED EQUIVALENT CIRCUITS

The IH load current and resonant frequency of the proposed ac-ac converter depend on the state of active switches. The six sub-mode transitional equivalent circuits in Fig. 10 for $v_{in} > 0$ can be reconfigured into the two simplified equivalent circuits as depicted in Fig. 12 by neglecting the edge-resonant intervals.

978-1-4799-2706-7/14 $31.00 © 2014 IEEE

In Fig. 10 (a) which corresponds to the Mode 1 (Mode 6) described above, the IH load current i_o can be expressed under the condition of the turn-on timing of S_1, $t = t_{on_1}$ by

$$i_o(t) = e^{-\alpha(t-t_{on_1})}\big\{ I^* \cos \beta(t - t_{on_1}) - \frac{1}{\beta} \sin \beta(t - t_{on_1}) \big\}, \quad (3)$$

$$\alpha = \frac{R_o}{2L_o}, \ \beta = \sqrt{\frac{1}{L_o C_{r1}} - \Big(\frac{R_o}{2L_o}\Big)^2}, \ C_{r1} = \frac{C_o C_1}{C_o + C_1}, \tag{4}$$

where I^* represents the peak value of the IH load current i_o. During this interval, the resonant frequency f_{r1} of the proposed ac-ac converter can be defined under the condition of the turn-off timing of S_1, $t = t_{off_1}$ as

$$f_{r1} = \frac{1}{2\pi\sqrt{L_o C_{r1}}} \tag{5}$$

In Fig. 10 (b) which corresponds to the Mode 3 and Mode 4 described above, the input UFAC side is equivalently expressed by the current source I_{Lb} for the HFAC cycle. Accordingly, the IH load current i_o can be defined by

$$i_o(t) = e^{-\alpha(t-t_{off_1})}\big\{ K_1 \cos \beta'(t - t_{off_1}) + \frac{1}{K_2} \sin \beta'(t - t_{off_1}) + \frac{C_o}{C_1} I_{Lb} \big\}, \tag{6}$$

$$\alpha = \frac{R_o}{2L_o}, \ \beta' = \sqrt{\frac{1}{L_o C_o} - \Big(\frac{R_o}{2L_o}\Big)^2}, \tag{7}$$

$$K_1 = -\Big(I^* + \frac{C_2}{C_1} I_{Lb}\Big), \ K_2 = \frac{\alpha}{\beta'} K_1. \tag{8}$$

During this interval, the resonant frequency f_{r2} of the proposed ac-ac converter can be defined as

$$f_{r2} = \frac{1}{2\pi\sqrt{L_o C_o}} \tag{9}$$

IV. EXPERIMENTAL RESULTS AND EVALUATIONS

A. Experimental Set-Up and Specification

The practical feasibility of the proposed ac-ac converter is investigated in an experiment based on a 2.8 kW-30 kHz laboratory prototype.

The exterior appearance of the prototype is indicated in Fig. 13. The specification of the laboratory prototype and the experimental conditions are indicated in TABLE III. In this experiment, an iron pan is used for the IH load, and the high frequency resonant current is supplied with a LITZ wire. The two active switches $Q_1\,Q_2$ are implemented by the high-speed RC-IGBT module CM100DU-24NFH (1000 V, 100 A), while the reverse-conduction blocking diodes D_1 and D_2 by ultra-fast recovery diode STTH6012 (1200 V, 60 A) .

B. Observed Waveforms and Steady-State Characteristics

The observed switching voltage and currents of Q_1, Q_2 and the IH load current are depicted in Fig. 14 for the two sets of duty cycles $D = 0.4$ and $D = 0.25$

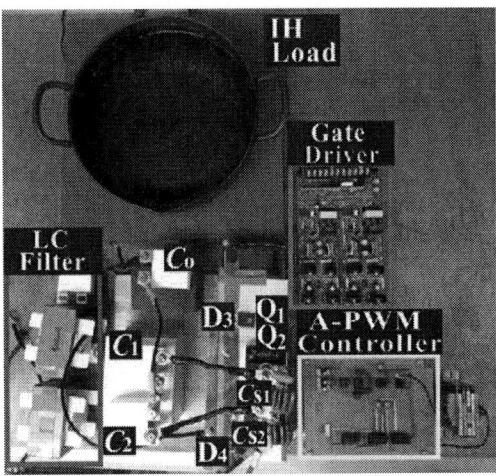

Fig. 13. Experimental set-up of the proposed single-stage UFAC-HFAC converter prototype.

TABLE III.　CIRCUIT PARAMETERS AND SPECIFICATION OF PROTOTYPE FOR SIMULATION AND EXPERIMENT.

Item	Symbol	Value [unit]
Input UFAC Voltage(rms)	v_{in}	200 [V]
Input filtering inductor	L_f	200 [μH]
Input filtering capacitor	C_f	750 [nF]
Boost inductor	L_b	500 [μH]
Switching frequency	f_s	30 [kHz]
Dead time interval	t_d	2 [μs]
Lossless snubbing capacitor	C_s	40 [nF]
Non-smoothed dc-link capacitors	C_1, C_2	2 [μF]
series resonant capacitor	C_o	2 [μF]
IH load equivalent resistance	R_o	3.5 [Ω]
IH load equivalent inductance	L_o	45 [μH]

under the condition of $v_{in} > 0$. It can be proved from the result that the high frequency IH load current can be regulated by changing the control variable D.

The enlarged voltage and current waveforms are depicted in Fig. 15. The complete ZVZCS turn-on and ZVS turn-off operations can be actually observed in the power range 0.5 kW–2.8 kW as depicted in Fig. 15 (a). In the output power range $P_o < 0.5$ kW, the turn-off transitions of Q_1 in the case of $v_{in} > 0$ and Q_2 in the case of $v_{in} < 0$ are out of the complete ZVS behavior due to the inadequate output current for achieving the ZVS operation as depicted in Fig. 15 (b). However, this commutation is far from the hard switching transition, accordingly no significant switching noise occurs in the low output power range.

The voltage and current trajectories relevant to the enlarged waveforms of Fig. 15 are demonstrated in Fig. 16. The overlapped area of voltage and current are well reduced even in the low output power setting, as compared to the hard-switching transitions. Thus, the wide-range ZVS operation is actually proven, and consequently the low switching noises can be expected over the wide range of the output power in the proposed ac-ac converter

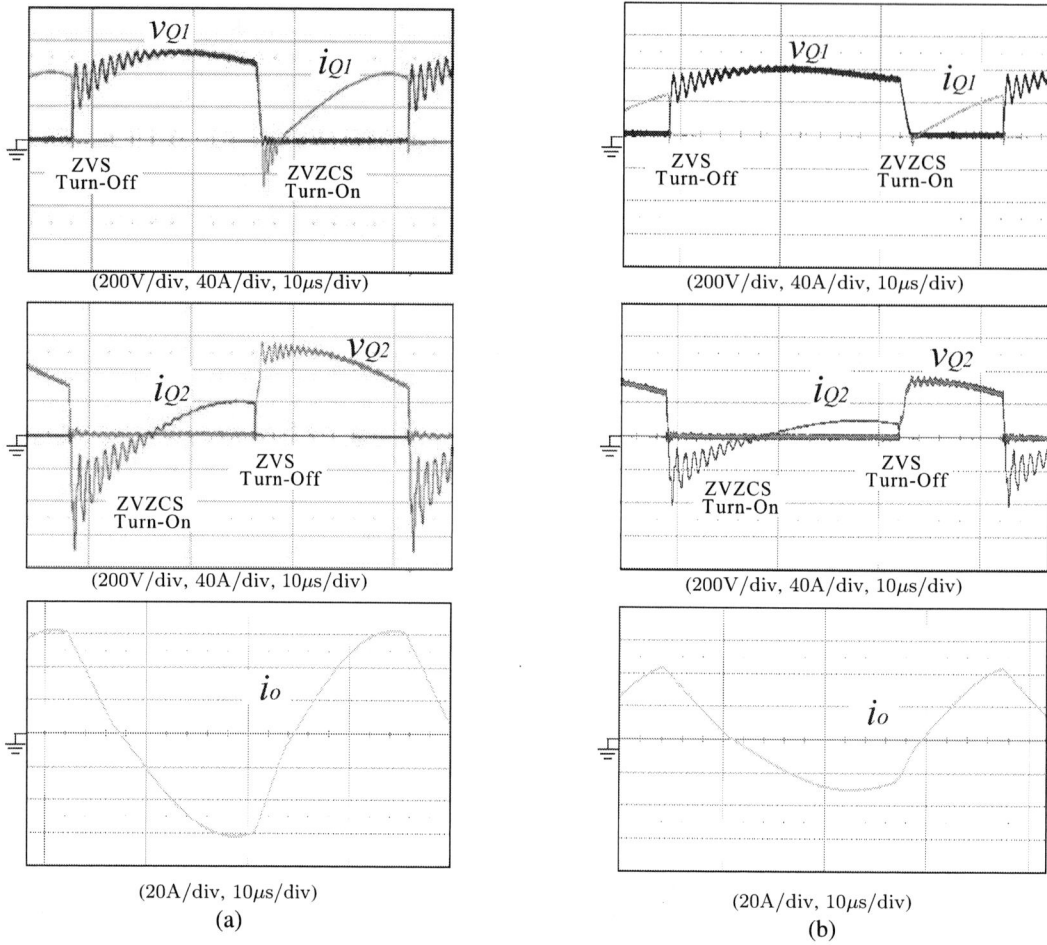

(200V/div, 40A/div, 10μs/div)   (200V/div, 40A/div, 10μs/div)

(200V/div, 40A/div, 10μs/div)   (200V/div, 40A/div, 10μs/div)

(20A/div, 10μs/div)   (20A/div, 10μs/div)

(a)   (b)

Fig. 14. Observed voltage and current waveforms of Q_1 switching and IH load for the switching cycle: (a) $D = 0.4$, and (b) $D = 0.25$.

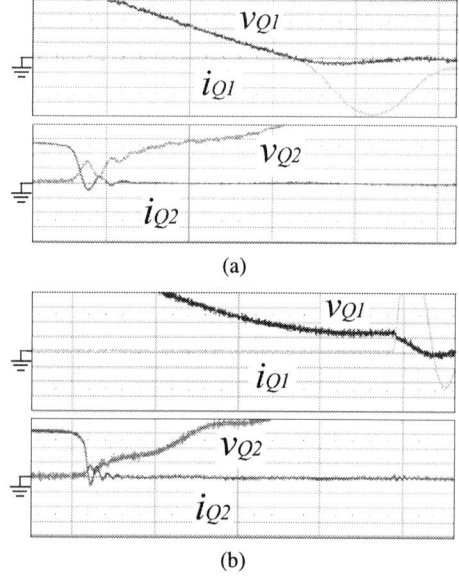

(a)

(b)

Fig. 15. Enlarged switching waveforms at turn-off transitions: (a) $P_o = 2.8\,\mathrm{kW}$ (100V/div, 15A/div, 200ns/div), and (b) $P_o = 500\,\mathrm{W}$ (50V/div, 4A/div, 500ns/div).

prototype.

The non-smoothed dc-link capacitors voltages are demonstrated in Fig. 17 in terms of both the average and peak values. In addition, the measured characteristics of the non-smoothed dc-link voltage v_d versus the duty cycle D are shown in Fig. 17. It can be confirmed from those results that the inductor-assisted boost operation can be attained in accordance with D in the proposed ac-ac converter. Thus, the effectiveness of the BHB circuit topology can be verified.

The output power versus duty cycle curve is presented in Fig. 19, which indicates the validity of A-PWM-based power regulation scheme in the proposed ac-ac converter. The actual power conversion efficiency versus output power characteristics are depicted in Fig. 20. The high and flat curve of efficiency can be maintained in the power range of $2.8\,\mathrm{kW}$ ($100\,\%$) to $0.7\,\mathrm{kW}$ ($40\,\%$) owing to the wide-range ZVS commutation. The maximum efficiency $94.5\,\%$ is measured at the rated output power $P_o = 2.8\,\mathrm{kW}$ in the prototype.

C. Harmonics Analysis of UFAC Current

The observed UFAC voltage and current waveforms are provided in Fig. 21. High power factors 0.98–0.99

978-1-4799-2706-7/14 $31.00 © 2014 IEEE   2538

The 2014 International Power Electronics Conference

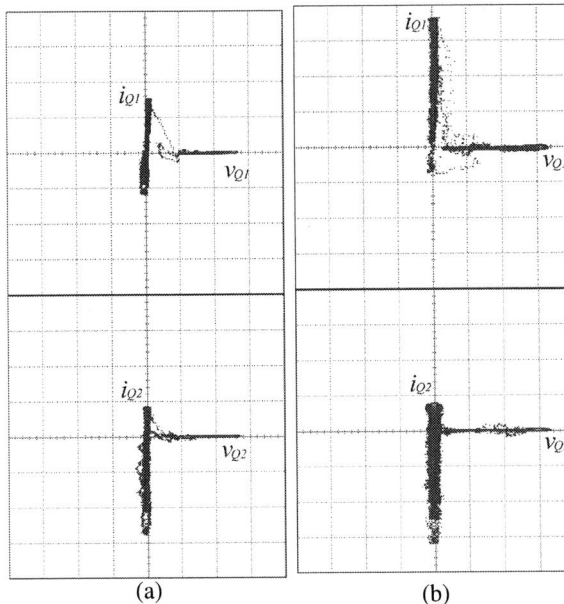

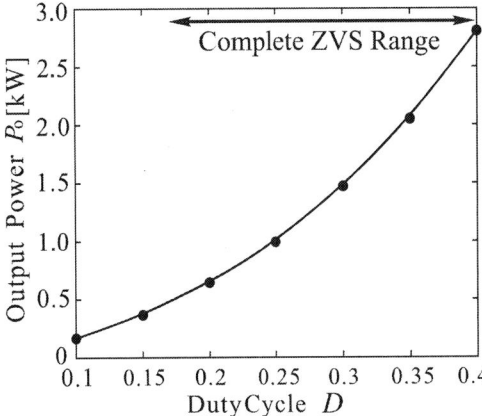

Fig. 19. Experimental characteristics of output power versus duty cycle curve.

Fig. 16. Voltage and current trajectory : (a) $P_o = 2.8\,\text{kW}$ (v_{Q1}, v_{Q2}: 200 V/div, i_{Q1}, i_{Q2}: 50 A/div), and (b) $P_o = 0.5\,\text{kW}$ (v_{Q1}, v_{Q2}: 100 V/div, i_{Q1}: 10 A/div, i_{Q2}: 20 A/div).

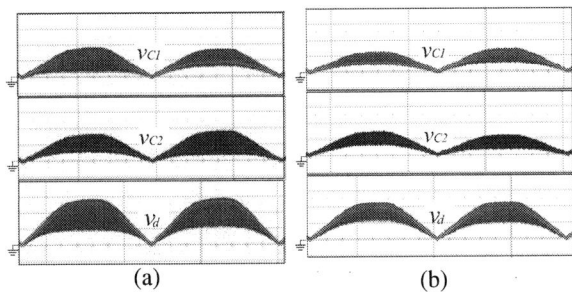

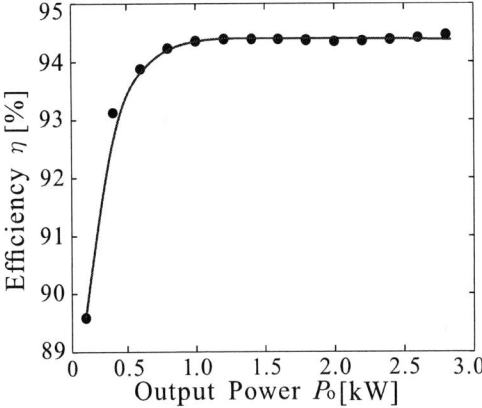

Fig. 20. Actual efficiency versus output power curve.

Fig. 17. Observed non-smoothed dc-link capacitor voltages: (a) $D = 0.4$, and (b) $D = 0.25$ (200V/div, 5.0ms/div).

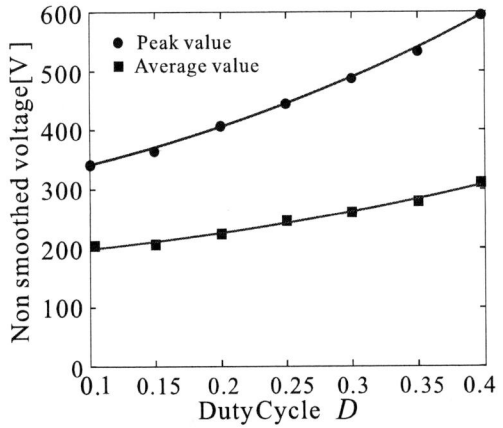

Fig. 18. Measured characteristics of the non-smoothed dc-link capacitor voltage v_d versus duty cycle curve.

can be confirmed from the waveforms. In addition, the corresponding UFAC current harmonics analysis by

Fast Fourier Transformation (FET) is depicted in Fig. 22, where the measured values are compared with the standards of IEC61000-3-2-Class A. Those results demonstrate the effective performances on the PFC operation and the low distortion in the source current.

V. CONCLUSION

The practical effectiveness of the newly proposed single-stage UFAC–HFAC converter has been demonstrated in this paper. The bridgeless boost half-bridge-assisted PFC and the non-smoothed dc–high frequency resonant inverter operation can be attained by the RC-IGBT-based simple and cost-effective circuit topology under the condition of ZVS.

The performances of the ac-ac converter proposed herein have been evaluated in an experiment using its laboratory prototype, and the excellent properties have been actually confirmed as follows:

- A wide range of output power regulation (0.1 kW–2.8 kW) can be achieved by the asymmetrical PWM scheme
- PFC and low harmonics in the UFAC source current can be attained, which clears the relevant

978-1-4799-2706-7/14 $31.00 © 2014 IEEE   2539

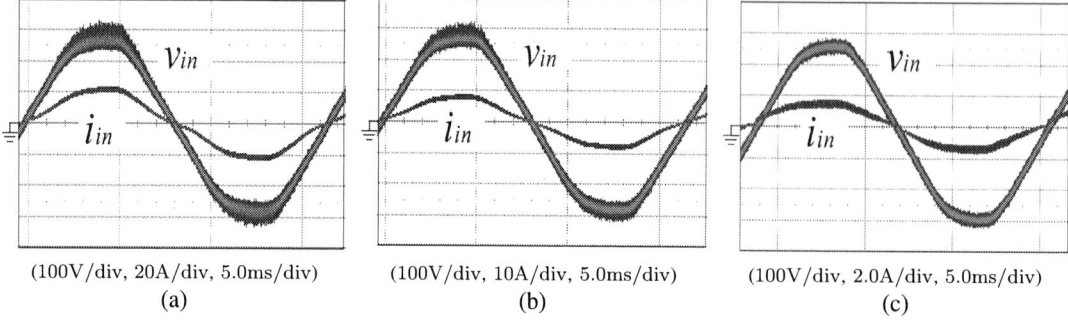

(100V/div, 20A/div, 5.0ms/div)　　(100V/div, 10A/div, 5.0ms/div)　　(100V/div, 2.0A/div, 5.0ms/div)
(a)　　　　　　　　　　　　(b)　　　　　　　　　　　　(c)

Fig. 21. Voltage and current waveforms of the utility power source: (a) $D = 0.4$, (b) $D = 0.25$, and (c) $D = 0.1$.

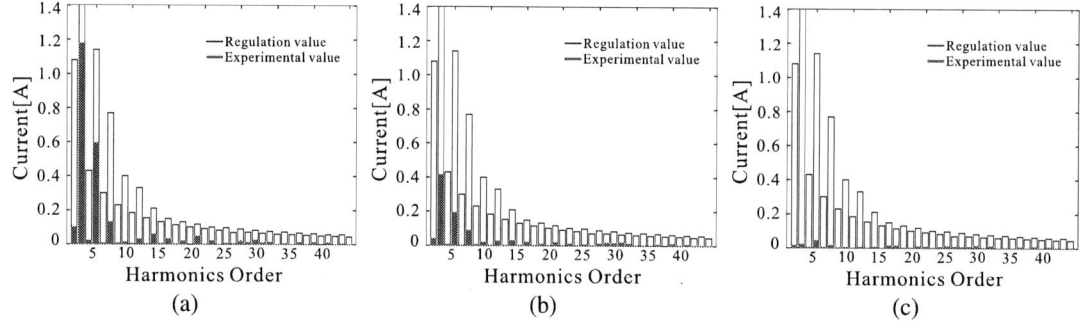

(a)　　　　　　　　　　　　(b)　　　　　　　　　　　　(c)

Fig. 22. FFT analysis of utility-side ac current i_{in}: (a) $D = 0.4$, (b) $D = 0.25$, and (c) $D = 0.1$.

industrial standard

- The UFAC source voltage is effectively lifted in the non-smoothed dc-link capacitor voltage with the aid of the inductor-type boost half-bridge circuit
- The high-efficiency power conversion with the complete ZVS operations can be realized over the wide-rage output power setting (40 %–100 %).

The future challenges for investigating the proposed ac-ac converter topology include the evaluations under the higher switching frequency condition such as 90 kHz–100 kHz.

References

[1] T. Mishima, C. Takami, and M. Nakaoka, "A new current phasor-controlled ZVS twin half-bridge high-frequency resonant inverter for induction heating," *IEEE Trans. Ind. Electron.*, vol.61, no.5, pp.2531–2545, May 2014.

[2] B. Saha and R-Y. Kim, "High power density series resonant inverter using an auxiliary switched capacitor cell for induction heating applications," *IEEE Trans. Power Electron.*, vol.29, no.4, pp.1909–1918, Apr. 2014.

[3] A. Okuno, H. Kawano, J. Sun, M. Kurokawa, A. Kojina, and M. Nakaoka, "Feasible development of soft-switched SIT inverter with load-adaptive frequency-tracking control scheme for induction heating," *IEEE Trans. Ind. Appl.*, vol. 34, no. 4, pp. 713-718, Jul. / Aug. 1998.

[4] H. Shoji, J. Uruno, and M. Isogai, "Induction heating system for all types metal pans employing a method of changing the dc link voltage and the number of turns of the work coil", *IEEJ Trans. IA*, vol.133, no.11, pp.1082–1088 (2013-11) (in Japanese)

[5] Y. Kawaguchi, E. Hiraki, T. Tanaka, M. Nakaoka, H. Sadakata, A. Fujita, and H. Omori, "Feasibility study of a full-bridge inverter for induction heating cooking appliances with discontinuous current mode PFC control," in *Proc. IEEE Power Electron. Sepc. Conf.*, 2008, pp.2948–2953.

[6] B. Saha, S. K Kwon, N.A. Ahmed, H. Omori, and M. Nakaoka, "Commercial frequency ac to high frequency ac converter with boost-active clamp bridge single stage ZVS-PWM inverter," *IEEE Trans. Power Electron.*, vol.23, no.1, pp.412–419, Jan. 2008

[7] H. Sugimura, H. Muraoka, E. Hiraki, I. Hirota, K. Yasui, H. Omori, H.W. Lee, and M. Nakaoka, "A novel soft switching PWM power frequency converter with non dc smoothing filter link for consumer high frequency induction heating", *IEEJ Trans. IA*, vol.125-D, no.11, pp.988–999 (2005) (in Japanese)

[8] N.A. Ahmed, "High-frequency soft-switching ac conversion circuit with dual-mode PWM/PDM control strategy for high-power IH applications," *IEEE Trans. Ind. Electron.*, vol.23, no.34, pp. 1440–1448, Apr. 2011.

[9] H. Sarnago, O. Lucía, A. Mediano, and J.M. Burdío, "Efficient and cost-effective ZCS direct ac-ac resonant converter for induction heating," *IEEE Trans. Ind. Electron.*, vol.61 no.5, pp.2546–2555, May 2014.

[10] M. Salehifar, M. Moreno-Eguilaz, V. Sala, and L. Romeral, "A novel ac-ac converter based SiC for Domestic Induction Cooking Applications," *Proc. 25th IEEE Applied Power Electronics Conference and Exposition* (IEEE-APEC 2013), Mar. 2013, pp.3216-3223.

[11] R. Srinivasan and R. Oruganti, "A unity power factor converter using half-bridge boost topology," *IEEE Trans. Power Electron.*, vol.13 no.3, pp.487–500, May 1998.

[12] Y. Nakagawa, T. Mishima, and M. Nakaoka: "A new bridgeless boost half-bridge ZVS-PWM high frequency resonant ac-ac converter for induction heating", *The Paper Joint Tech. Meeting on Semiconductor Power Conv., Vehicle Tech., and Home and Consumer Appl., IEEJ*, SPC-13-152 / VT-13-35 / HCA-13-57, pp.47–52 (2013-12) (in Japanese)

[13] T. Mishima, Y. Nakagawa, and M. Nakaoka, "Practical evaluation of a new bridgeless boost half-bridge ZVS-PWM single-stage utility frequency ac – high frequency ac resonant converter for induction heating", *The Paper Joint Tech. Meeting on Semiconductor Power Conv. and Motor Drive, IEEJ*, SPC-14-41 / MD-14-41, pp.83–88 (2014-1) (in Japanese)

978-1-4799-2706-7/14 $31.00 © 2014 IEEE

Application of Virtual Validation System for Inverter Heat Pump System

Masaki Kanamori, Koji Noda, Takahisa Endo
Core Technology Center
Toshiba Carrier Corporation
Fuji, Japan

Nobuyuki Suzuki
Corporate Manufacturing Engineering Center
Toshiba Corporation
Yokohama, Japan

Abstract- **Virtual validation system for inverter and motor control of air conditioners and other inverter heat pump systems is developed. This system is composed by hardware-in-the-loop simulation (HILS) with a real-time simulator. The validity and effectiveness of the simulation system is confirmed through application examples: optimization of the motor control parameters and vibration suppression control of compressors.**

Keywords—compressor, heat pump system, hardware-in-the-loop-simulation, motor control

I. INTRODUCTION

In recent years, the use of a combination of an inverter and a compressor driven by a permanent magnet synchronous motor (PMSM) has become common as one measure to increase the energy efficiency of air conditioners and other heat pump systems. In Japan, roughly 100% of the residential and commercial air conditioners install inverters. Inverter and PMSM is not so popular in other countries, but is expected to spread worldwide. Recently, inverters are also beginning to appear in high-capacity heat pump systems. There are, however, a number of problems in the development and design process.

For further improvement in the energy efficiency of heat pump systems, the inverter and PMSM is often redesigned for each system. One inverter is arranged for each PMSM and tuning of control parameters is required for each combination. On the other hand, inverter heat pump systems are required to have various models and lineups in order to respond to various user needs. This complexity of models is causing a delay in product development and provision to the market.

As a solution to this problem, the use of simulations may be an answer. There are many reports on simulating inverter-motor systems [1-5]. However, because of the complexity and specialized nature of the simulations, general simulation methods are not practical. In view of the above, a virtual validation system for an inverter heat pump system is developed as a hardware-in-the-loop simulation (HILS), using a real-time simulator. Features of the system and confirmation of validity through applications are presented in this paper.

II. SYSTEM CONFIGURATION

Fig. 1 shows general equipment for motor drive validation, with an actual inverter and motor installed. Low power motors are relatively easy to use in an experiment. However, high-power systems require correspondingly large power equipment and motor load equipment to reproduce the desired behavior; thus general test methods require extensive preparation time and increase testing costs.

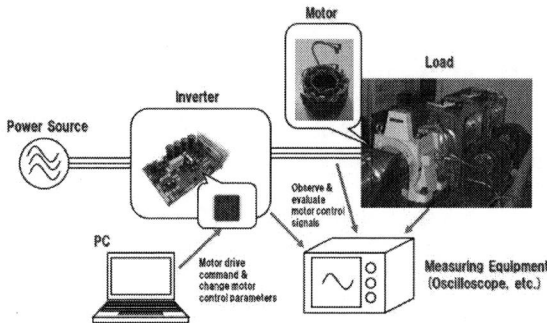

Fig. 1. General test equipment for motor drive validation

In order to simplify the testing procedure and reduce the time taken, a new validation method is developed, using the combined hardware-in-the-loop simulation (HILS), and the configuration of the system is shown in Fig. 2. This system is composed of a virtual motor model, a virtual inverter model, and an actual microcontroller unit (MCU). The MCU may be mounted on a printed-circuit board (PCB), and use actual detection circuits, etc, although, a DC power supply (3.3-5 V) is necessary for the power source for the MCU. The motor design drawing is acquired and imported into the electromagnetic field analyzer, and then the coordinate data from the drawing is converted into a simulation motor model. The results of the electromagnetic field analyzer are exported as data lists of induced voltage, motor current, current phase and rotation angle. These data lists are loaded into the real-time simulator with an inverter model as the virtual inverter and motor model. The motor load model is represented using MATLAB/Simulink. For simplicity, models of the converter circuit and AC power source are replaced by a DC power source. MCU control software, which is

978-1-4799-2706-7/14 $31.00 © 2014 IEEE

installed in a PC, transfers operation commands and parameter changes to the MCU. Then, it imports and displays operational results from the MCU.

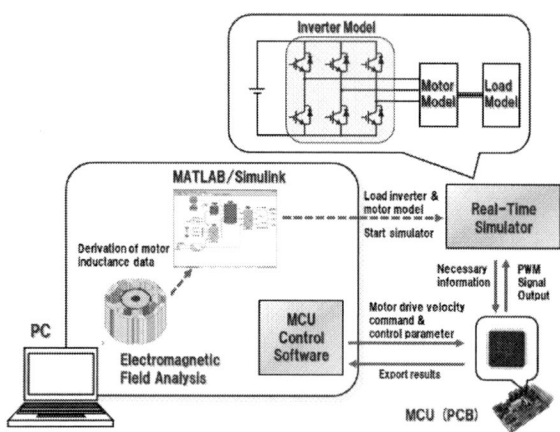

Fig. 2. Configuration of virtual validation system.

III. SYSTEM FEATURES

The greatest advantage of the proposed virtual validation system is that actual MCU, hardware, software, and control methods could be used to simulate the operation of virtual motors and inverters. The inverter and motor simulation model is constructed on a field programmable gate array (FPGA) board in the real-time simulator. Because of this feature, the real-time simulator can output the response of the inverter and motor at a high-speed simulation time step. This enables the MCU to respond and behave as if it were controlling an actual inverter and motor. In addition, the source of error by hardware, such as various detection circuits, and internal arithmetic delays can be considered by using an actual PCB or MCU. Therefore, higher accuracy and consistency could be obtained compared with methods which compose all the elements by simulation.

As mentioned before, optimization of motor control is necessary for each product according to its refrigerating capacity and specification. However, experiments and evaluations of heat pump systems take a relatively long time since various temperature conditions are required. By introducing the virtual validation system, the selection of the MCU or PCB for the subject heat pump system and evaluation of the system becomes simpler. Furthermore, heat pump systems have distinctive operation states and load conditions. The proposed system can reproduce these various phenomena in the product by changing the operating sequence settings and motor load models. Therefore, troubleshooting and software debugging can also be carried out by this virtual validation system without an actual power source or motor. In other words, increases in product development efficiency, quick adaptation to various user needs, and timely product provision are realized by utilizing this virtual validation system.

IV. CONFIRMATION OF SYSTEM VALIDITY BY OPTIMIZATION OF MOTOR CONTROL PARAMETERS

A. Background

In this section, an example application of the proposed system to optimize the motor control parameters considering the distinctive sudden load change is presented, and effectiveness of the system is confirmed.

High robustness is required for motor control whereas heat pump systems essentially have transient states that have to be stabilized. Fig. 3 shows an example of a transient condition of a heat pump system: this is a defrosting operation where the refrigerating cycle is reversed to melt the ice on the heat exchanger. This reverse operation causes a sudden change in the compressor (motor) load, and the inverter and motor must not stall during this period. This makes it necessary to optimize the motor control parameters for every combination of components. However, using the actual inverter and motor with conventional test systems is difficult to reproduce all possible situations. Therefore, the virtual validation system is used to reproduce the sudden load change during the defrost operation and optimization of motor control parameters is carried out to confirm the effectiveness and validity of the system.

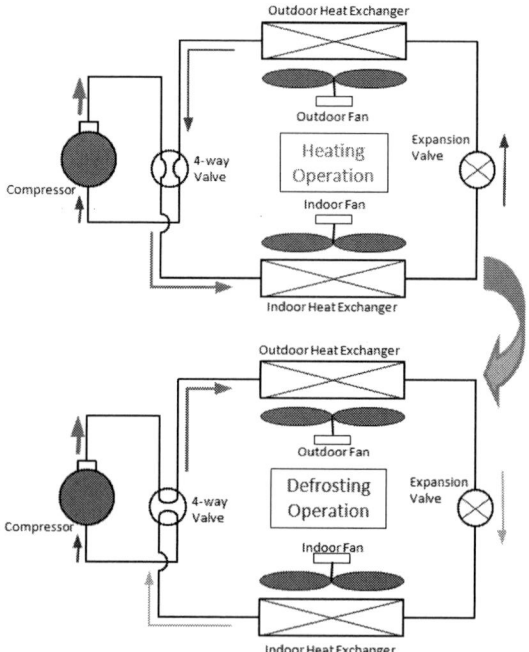

Fig. 3. Example of sudden load change in a heat pump system (from heating to defrosting operation)

B. Confirmation of consistency

First, the consistency of the virtual validation system is evaluated by reproducing the sudden compressor (motor) load changes for the defrosting operation. Fig. 4 shows the evaluation points, Fig. 5(a) shows the waveform results of the actual motor drive and Fig. 5(b) shows the waveforms simulated by the virtual validation system. Position sensor-less vector control is used for the motor drive. The waveforms in Fig. 5 are: the U-phase motor current (Iu: blue line), reference and actual q-axis motor current (Iq: green and pink line), reference and actual d-axis motor current (Id: red and light blue line) and reference and estimated rotation velocity of motor (ω: light green and orange line) which are calculated results of the MCU.

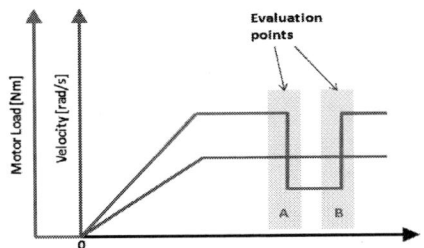

Fig. 4. Evaluation points for sudden load changes.

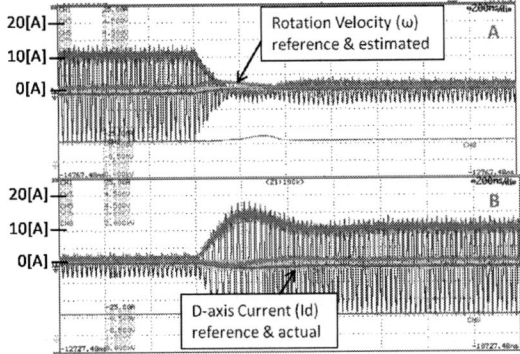

Fig. 5(a) Waveform for actual motor drive

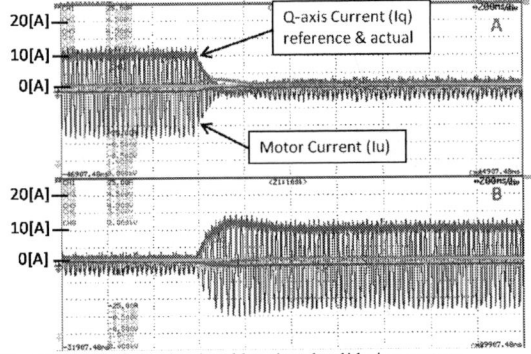

Fig. 5(b) Waveform simulated by virtual validation system

Fig. 5. Comparison between simulation and actual motor drive in a sudden load change condition. States A and B are the results for the corresponding evaluation points shown in Fig. 4.

The evaluation is carried out after the motor has started and when rotating at a stable rate, by varying the load. At the evaluation points, the motor load is suddenly decreased (state A) and then increased (state B) after an interval of 1 second. According to Fig. 5(a) and 5(b), the waveforms of the simulation results are identical to the actual motor waveforms in state A. Although there is some difference in the peak motor current in state B, the simulated results of Iq, Id, and ω behavior agree well. Therefore, it is considered that the proposed system can mostly reproduce the motor control behavior at a sudden load change.

C. Optimization of motor control parameters

Next, the effectiveness of the virtual validation system is confirmed through a process of optimizing the motor control parameters. The PMSM for the compressor of an inverter heat pump system is generally driven by sensor-less vector control. The control block diagram is shown in Fig. 6. When a new product or motor is designed, the following motor control parameters must be adjusted: proportional-integral (PI) control gains for current controller, velocity controller, and velocity/position estimator. These parameters require manual adjustment with the actual motor drive system. Consequently, in applying the virtual validation system, it is combined with general-purpose optimization software for experimental optimization of the PI controller gains.

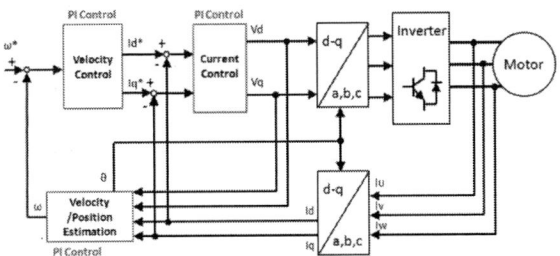

Fig. 6. Block diagram of position sensor-less vector control. (X* stands for the reference of X)

The system configuration is shown in Fig. 7. The optimization software is connected to the MCU control software, and analysis results (reference and actual values of Id, Iq, and ω) from the MCU are collected. Thus, the optimization software learns the past results and selects the subsequent control gains which are most likely to minimize the error between the reference and actual values of Id, Iq, and ω. Differential Evolution (DE), which is one type of genetic algorithm, is used for the search algorithm, in view of its wide applicability and ability to find optimum solutions in a large search space. The new parameter values selected by the optimization software are sent to the MCU, and the analysis is pursued again. This cycle is repeated until all control gain values converge and reach a global solution.

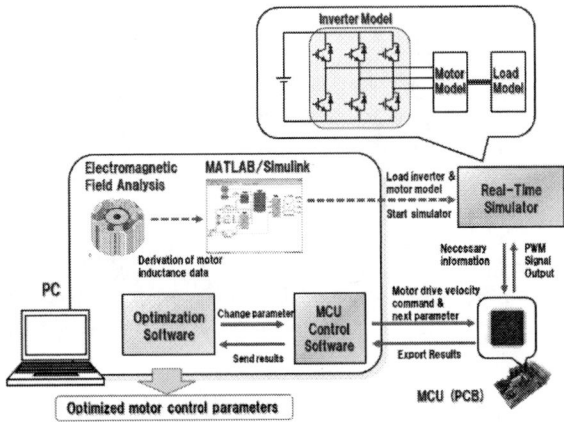

Fig. 7. Configuration of virtual validation system when combining with optimization software.

Fig. 8 shows the progressive optimization process for the current control PI gain. The horizontal axis represents the number of generations, and the vertical axis represents the trial values of PI gain. The black vertical bars represent the parameter search range for the optimization software, and the horizontal thin line shows the optimized result for each generation. With successive iterations, the parameter search range decreases, and it can be seen that the control gain values converge.

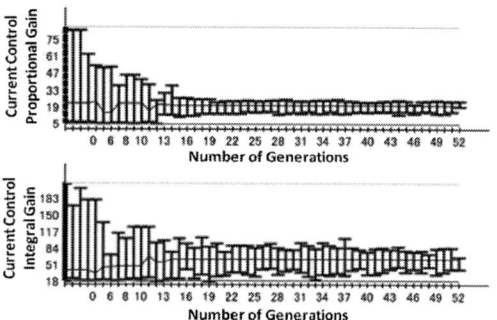

Fig. 8. Optimization progress of current control gains (Top: Current control proportional gain, Bottom: Current control integral gain)

In order to evaluate the validity of the optimized PI control gains obtained above, an actual motor drive is used in the experiment. Fig. 9 shows the waveforms representing the motor behavior when operated with the optimized PI control gains obtained with the virtual validation system and optimization software. As in Fig. 9, reference, actual and estimated value outputs from the MCU are shown for Iq, Id, and ω. The evaluation was carried out with multiple sudden load changes after the motor had reached a stable velocity. In spite of the severe conditions, as seen in Fig. 9, continuous motor control is confirmed without stalling with the optimized PI control gains obtained with the virtual validation system. The actual values of Iq, Id and estimated value of ω closely followed the reference values, and continuous operation in other conditions is also confirmed. Considering the results above, the validity and effectiveness of the virtual validation system for optimizing control parameters is verified.

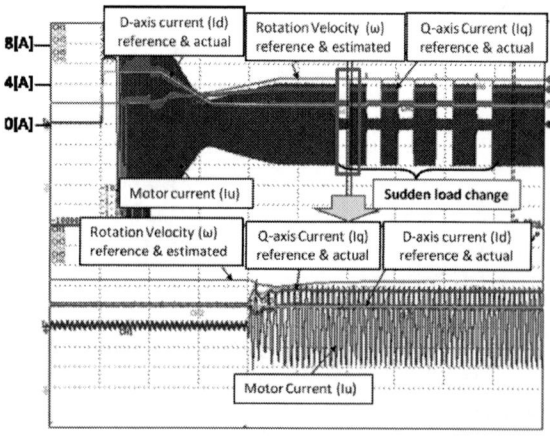

Fig. 9. Motor behavior with actual motor when installing the control PI gains optimized by proposed system.

V. Confirmation of System Validity by Investigation of Vibration Suppression Control

A. Background

In this section, an example application of the proposed system for adjustment of vibration suppression control of rotary compressors is presented and the effectiveness of this system is confirmed.

A rotary compressor is one type of compressor often used in small-to-medium systems, such as residential air conditioners, which is used to compress refrigerant gas. Fig. 10 shows the compression mechanism image of a 1 cylinder rotary compressor. The rolling piston inside the cylinder is connected to the rotation axis eccentrically, which is the shaft connected to the rotor. The refrigerant gas is compressed by the eccentric rotation of the rolling piston as shown in Fig. 10. The vane is forced to move along with the rolling piston by a spring or the backpressure, dividing the cylinder into the suction chamber and the compression chamber. In other words, the suction phase (intake of refrigerant gas through the suction port) and the discharge phase (discharge of compressed refrigerant gas through the discharge port) are carried out simultaneously. The discharge port has a valve that only opens when the pressure inside the chamber is equal or greater than the discharge pressure to prevent the discharge gas from flowing backward. Compressors with only one compression chamber are called 1 cylinder type and compressors with two compression chambers shifted 180 degrees are called 2 cylinder type.

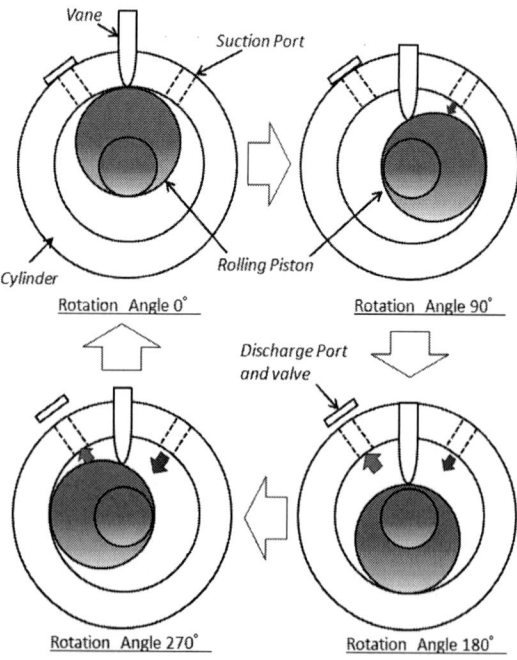

Fig. 10. Compression mechanism image of a 1 cylinder rotary compressor. Refrigerant gas is inhaled into the compression chamber and compressed by the rolling piston.

Because of the distinctive structure of the rotary compressor, the compression load (motor load) becomes a distinctive shape. Fig. 11 shows a compressor load ripple example of a 1 cylinder rotary compressor for one motor rotation. Although, characteristic of the load ripple varies depending on the cylinder size and gas pressure conditions, there is often a large fluctuation in the motor load as the motor rotates. When the motor output torque is controlled at a constant torque, the torque difference between the compressor load and motor output torque causes acceleration in the motor. This fluctuation of motor speed is one of the reasons for the vibration of the compressor. In a heat pump system, refrigerant pipes are connected to the compressor and other refrigerant cycle components. Therefore, when the vibration of the compressor is excessive, it may lead to breakage of the refrigerant pipes.

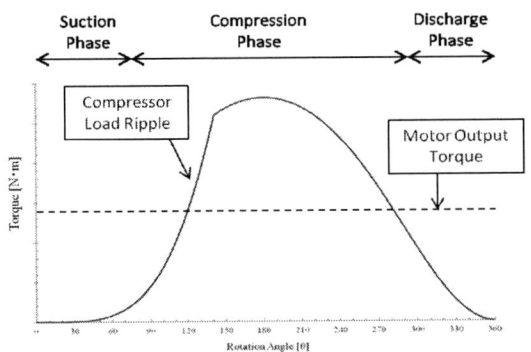

Fig. 11. Compressor load ripple example of a 1 cylinder rotary compressor.

There are mechanical measures such as installing damping materials onto the refrigerant pipes and balancers onto the rotor. On the other hand, there are methods by motor control to suppress the vibration. The vibration suppression control used in this paper is a motor control method to suppress the rotation velocity fluctuation by correcting the Iq. Similar to the motor control parameters, this vibration suppression control tends to require optimization for each motor and heat pump system. In this investigation, 1 cylinder rotary compressor is chosen as the subject since its vibration is generally high and the virtual validation system is used for adjustment of the vibration suppression control.

B. Confirmation of consistency

First, consistency of the proposed system is confirmed by modeling the compressor load ripple of a 1 cylinder rotary compressor as shown in Fig. 11. Fig. 12 shows the motor current (Iu), the estimated rotation angle (θ), the estimated rotation velocity (ω), the rotation velocity reference (ωref), and the vibration displacement of a 1 cylinder rotary compressor operated at 15 rps. Fig. 13 shows the waveforms simulated by the proposed system under the same operating condition. The vibration of the compressor is shown as the amount of displacement in the rotation direction detected by a vibrometer. However, this is not shown in Fig. 13 since the virtual validation

system cannot output an actual vibration. By comparing Figs. 12 and 13, it can be seen that the shape of the motor current is similar. Although the fluctuation of the rotation velocity for the actual measurement is very large and sharp, the amplitude of the rotation velocity and the phase against the rotation angle agree well with the simulated result. In spite of some errors, it is considered that the motor behavior with compressor load ripple is mostly reproduced.

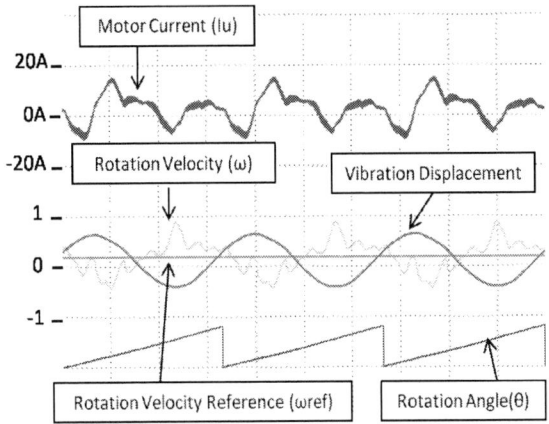

Fig. 12. Motor behavior and vibration displacement of a 1 cylinder rotary compressor.

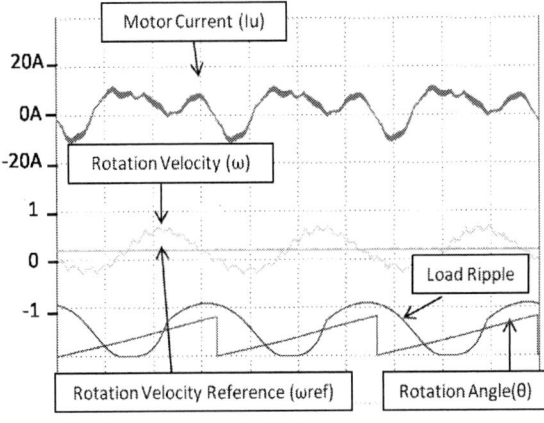

Fig. 13. Reproduction of compressor load ripple and motor behavior by proposed system.

C. Adjustment of vibration suppression control parameters

Next, the parameters of vibration suppression control (correction coefficient of Iq) are adjusted using the proposed system. As shown in Fig. 12, it can be seen that the actual vibration of the compressor and rotation velocity has an interrelation. In the case of using the proposed system, the evaluation of the control method is judged by the rotation velocity fluctuation. The motor behavior after adjusting the vibration suppression control parameters is shown in Fig. 14, where the rotation velocity fluctuation is controlled at its minimum.

Finally, an actual compressor is operated by using the adjusted parameters. The motor behavior and vibration results are shown in Fig. 15. The rotation velocity and vibration displacement is suppressed using the vibration suppression control parameters adjusted by the virtual validation system. Considering these results, the effectiveness and validity of the virtual validation system for investigating motor control methods are verified. By applying this system, various conditions can be simulated and control methods could be verified effectively.

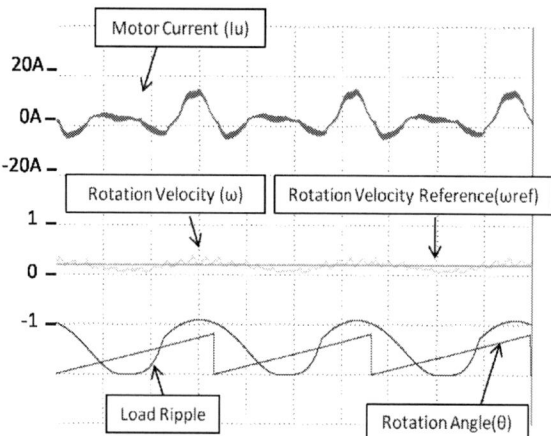

Fig.14. Adjustment results of the vibration suppression control parameters by the proposed system.

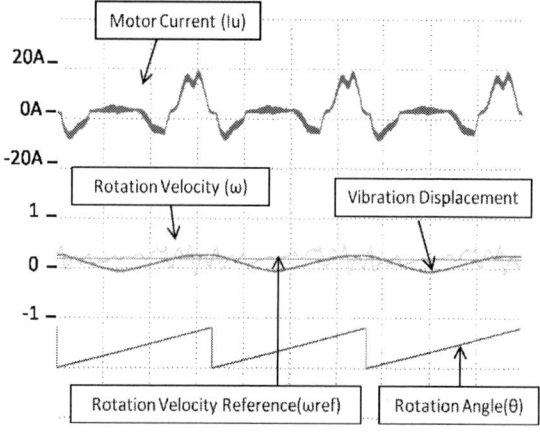

Fig.15. Confirmation of effectiveness of the vibration suppression control tuned by the proposed system.

VI. CONCLUSION

A virtual validation system for inverter heat pump system with actual MCU is proposed, which enables a real time simulation to analyze inverter and motor operation. The validity and effectiveness of the system is confirmed through optimization of the motor control parameters and adjustment of vibration suppression control of compressors. As an application of this system, it has been applied to an automatic optimization system for motor control parameters, which has reduced the time required for the design process. Since this system is widely applicable, further development of the system will be carried out in the future, along with development of energy efficient and high robustness heat pump systems.

REFERENCES

[1] T. Murata, U. Kawatsu, J. Tamura, and T. Tsuchiya, "Modeling and Simulation Technique of Two Quadrant Chopper PWM Inverter-Fed IPMSM Drive System and Its Application to Hybrid Vehicles," *Journal of International Conference on Electrical Machines and Systems*, vol. 1, no. 2, pp. 232-238, 2012.

[2] C. Chakraborty, and Y. Hori, "Fast Efficiency Optimization Techniques for the Indirect Vector-Controlled Induction Motor Drives," *IEEE Trans. on Industry Applications*, vol. 39, no. 4, pp. 1070-1076, 2003.

[3] M. Batool, and A. Ahmad "Mathematical Modeling and Speed Torque Analysis of Three Phase Squirrel Cage Induction Motor Using Matlab Simulink for Electrical Machines Laboratory," *IEEJ International Electrical Engineering Journal*, vol. 4, no. 1, pp. 880-889, 2013.

[4] M. S. Rao and A. Ramakrishma, "Modeling and Simulation of PMSM Using Direct Torque Control Method", International Journal of Emerging trends in Engineering and Development, vol. 5, Issue 2, pp. 87-101, 2012

[5] K. Boby, "Mathematical Modeling of PMSM Vector Control System Based on SVPWM with PI Controller Using MATLAB", International Journal of Advanced Research in Electrical, Electronics and Instrumentation Engineering, vol. 2, Issue 1, pp. 689-695, 2013

Test setup for accelerated test of High Power IGBT modules with online monitoring of V_{ce} and V_f voltage during converter operation

Angel Ruiz de Vega[1], Pramod Ghimire[1], Kristian Bonderup Pedersen[2], Ionut Trintis[1], Szymon Beczckowski[1], Stig Munk-Nielsen[1]

1- Energy Technology 2 –Physics and Nanophysics
Aalborg University
Aalborg Denmark
email: arv@et.aau.dk

Bjørn Rannestad, Paul Thøgersen
KK-electronic a/s
Aalborg, Denmark

Abstract— Several accelerated test methods exist in order to study the failures mechanisms of the high power IGBT modules like temperature cycling test or power cycles based on DC current pulses. The main drawback is that the test conditions do not represent the real performance and stress conditions of the device in real application. The hypothesis is that ageing of power modules closer to real environment including cooling system, full dc-link voltage and continuous PWM operation could lead to more accurate study of failure mechanism. A new type of test setup is proposed, which can create different real load conditions like in the field. Furthermore, collector-emitter voltage (V_{ce}) has been used as indicator of the wear-out of the high power IGBT module. The innovative monitoring system implemented in the test setup is capable of measure the V_{ce} and forward voltage of the antiparallel diode (V_f) during converter operation, which is also demonstrated.

Keywords— *accelerated test setupt, high power IGBT module , online Vce monitoring , wear out .*

I. INTRODUCTION

One of the major factors of failures in wind turbines are related to power converters, which represents approximately 15% of all failures [1]. Within the power converters, high power IGBT modules are the weakest part because of continuous exposition of temperature cycles caused by conduction and switching power losses in the chip [2]. In order to reduce the incidents caused by high power IGBT modules in the power converters, it is necessary to study the failure mechanisms.

This paper presents an improvement and enhancement of a previous test setup [3]-[4], where it is possible to test IGBT power modules in different field application. One example of different field application can be seen in [5], where the test setup has been used as an experimental validation of the efficiency using adaptive control for grid converter (50 Hz). However, in this paper the rotor side of the Doubly Fed Induction Generator has been selected as wind turbine application. A DFIG system is selected as it is the most common used of this generator in the wind turbine industry. Moreover, the rotor side

operating point of converter is within few Hz. This fact causes high stress on the IGBT modules due to the high temperature produced by low frequency operation, and thereby, heavy thermal cycling which stress the different layers of IGBT modules, as they typically have different coefficients of expansion (CTE), and thus, the cycling will cause failure of IGBT module.

In order to study the failure mechanisms a real and continuous PWM power cycle is used in the proposed test setup. Normally, accelerated temperature cycle test or DC current pulses test have been used as a source for wearing out the IGBT devices [6]-[7]. Besides these power cycling methods, a periodic PWM load is shown in [8]-[9], in which PWM is stopped when chip reaches the desired increment of ΔT. This method allows more realistic electrical stress than the other methods.

In addition to the realistic stresses, the proposed method in this paper is directly implementable for continuous sinusoidal applications. Therefore, the test setup uses as ageing component a continuous sinusoidal current which emulates the same working points as the rotor side of DFIG. The advantages of wearing out the IGBT modules applying real working points will give a more accurate study of the failure mechanism in the power modules. However, the test system is not only used for wearing out the IGBT modules, but it is used as a platform to develop and test new devices that may be applied in the field. For example, V_{ce} parameter is well known as indicator of internal failure and it is one of the main methods to evaluate the junction temperature because of V_{ce} is Temperature Sensitive Electrical Parameter (TSEP) [10]. Therefore, an online V_{ce} monitoring system has been developed and implemented in the test setup.

This paper will describe the test system for IGBT power modules including the cooling system. Next the V_{ce} online measurement system will be described and finally some laboratory results of the test setup will be shown.

II. TEST SETUP CONVERTER

In order to realize the accelerated ageing process of high power IGBT modules a test setup has been built which is formed by different parts, such as, the module under test (DUT), the control power modules, protection system, cooling system, and also the online V_{ce} monitoring system.

A. Device Under Test

The high power IGBT modules used in the accelerated test setup are a half bridge 1000A and 1700V. However, it is possible to test any other IGBT module with different current and voltage ratings due to an easy adaptation of the controlled parameters, which allows a great flexibility when testing different IGBT modules and operating points.

Fig. 1. Type of High Power IGBT module used in the test

The only limitation is the package of the IGBT module that can be tested, which must be PrimePack (see Fig. 1). This constrain is due to the liquid cooling interface which is design for these type of modules.

B. Converter

A H-bridge converter is the proposed topology for the converter, which is formed by DC-link, one leg which is Device Under Test (DUT), two parallel legs which are the control side, one main inductor (L_1), and two sharing inductors (L_2 and L_3) as shown in Fig 2.

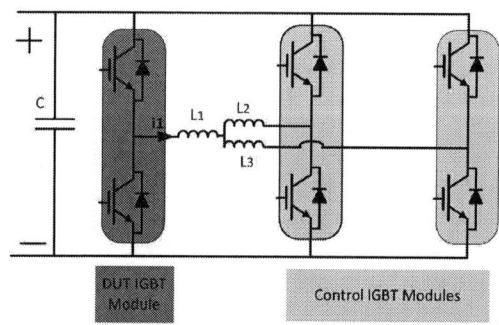

Fig. 2. High power test setup for IGBT modules.

Previous test bench showed in [3], [4], used only one leg in the control side, which has as a consequence that the ageing processes were started simultaneously at both DUT and control IGBT modules. Since the aim is to wear out only the DUT IGBT module and not the control IGBT module, an additional leg and two sharing inductors have been added into the topology, which reduces the total load of the control modules to half of the current. Therefore, the ageing process of the control modules becomes slower in comparison to the DUT ones.

C. Converter mounted on WT power stack

Besides the additional components, the converter was mounted on a power stack which is used in real wind turbines, see Fig 3. This fact provides a realistic test setup and further it offers more safety and reliable test setup than the previous test setup.

Finally, the test setup is designed to make nonstop accelerated test until the desired end point. This means that the system may be running 24/7. Therefore, the test setup must be reliable and safe. Different parameters are under continuous supervision in order to protect the setup such as overcurrent, shoot through, pump-stop, over temperature or even fire. In case that one or more malfunctions could take place the protection system will shut down the power, turns off the PWM and an email will be sent in order to notify a malfunction in the setup.

Fig. 3. A power stack module where the IGBT modules are mounted.

III. COOLING SYSTEM

The cooling system is of paramount importance for the test system as continuous high power is involved in the system. There is a total power loss of 7 kW in nominal operation which must be dissipated from the system.

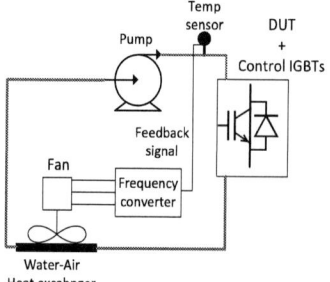

Fig. 4. Cooling system.

The dissipation of this power is done by means of a liquid cooling system as illustrated in Fig4. This is formed by one pump, one expansion vessel, one temperature sensor, one heat exchanger, one fan and one frequency converter which controls the fan speed using a built in PI controller.

The liquid used in the cooling system is a mixture of water and ethylene glycol. This mixture of water-ethylene glycol has the drawback such as reduction thermal conductivity and density increment in comparison to pure water [11]. Nevertheless, the purpose of the test setup is to be as close as possible to real field application where the cooling systems use this mixture (water-glycol) in order to decrease the liquid freezing point. The advantage for the test setup is a slight increment of the boiling point.

Additionally, the power stack interfaces the liquid cooling system with the baseplate of the IGBT modules by means of Shower Power ™ technology which offers lower gradients temperature on the baseplate in comparison to other technologies [12]-[13]. The test system is located in a room where the temperature is controlled and kept constant to 20±2°C. The stability that provides the controlled temperature allows a better control and more stable measurement of the V_{ce}, which initially will cause an easier calibration and estimation of the junction temperature. Calibrations and estimation of the online junction temperature are shown and explained in [14]-[15].

IV. ONLINE V_{CE} MONITORING SYSTEM

A new online V_{ce} monitoring system has been developed [16] and implemented in the present test setup.

Offline V_{ce} monitoring system used in [4] and the new online V_{ce} monitoring could measure either V_{ce} or V_f of the IGBT module. However, the mentioned offline V_{ce} monitoring system is reed relay based as shown in Fig. 5. Using these types of mechanical switches causes a slow operation due to its intrinsic mechanical contraction, which features a minimum switching time of 3 ms. As the minimum pulse of PWM is 100 μs applying test conditions shown in Table I, this V_{ce} monitoring system could only operate during offline operation. Measurement during offline can only offer the wear out status of the IGBT module, but the mayor drawback is that this V_{ce} monitoring system cannot be implemented in the field.

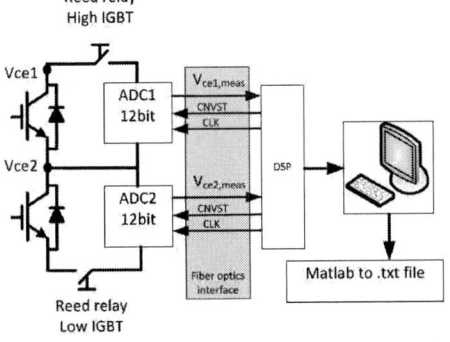

Fig. 5.   Offline V_{ce} monitoring principle.

By contrast, the new online V_{ce} monitoring system is not implemented with reed relays, but it is double diode based as shown in Fig. 6. The principle of the proposed monitoring circuit is based on V_{ce} desaturation measurement which some gate drivers use as protection against short circuits [16]-[17].

The challenge to achieve the online measurement lies in blocking the high voltage when the non-conducting state of the IGBT module in order to protect the ADC circuitry, and being enough accurate in order to measure the V_{ce} and V_f voltage within interval of millivolts when the IGBT module is conducting [18]. The proposed V_{ce} monitoring system is implemented with an ADC of 14 bits which gives a resolution of 0.61mV:

$$V_{ADC\_res} = \frac{V_{ref}}{2^{bits}} \qquad (1)$$

In addition, the V_{ce} monitoring system has been mounted on the top of the IGBT gate driver in order to reduce parasitic components and electromagnetic interferences (EMIs) which may cause inaccuracies during the measurements. Finally, D_1 and D_2 diodes must be identical and must be thermally coupled in order to achieve the adequate measurement.

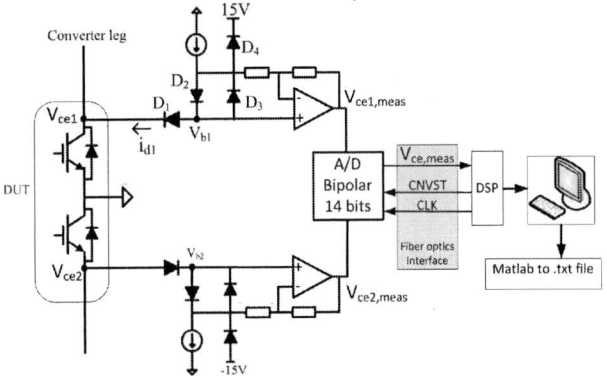

Fig. 6. Online V_{ce} monitoring principle.

V. MEASURING ROUTINE FOR WEAR OUT TEST OF IGBT MODULES

A.   Selected Online and Offline measuring routines

A measurement routine is proposed to use both online and offline operation. Every 5 minutes, the system enters on offline measurement routine. The online measurement routine takes also place every 5 minutes but it is shifted 2.5 minutes in respect to the offline measurement routine, as shown in Fig 7.

Fig. 7. Measuring routine used in the test setup.

Within time interval of 5 minutes, the IGBT modules experience a total of 1.8 kcycles according to the nominal test parameters. The selection of 5 minutes intervals between each measurement has been found optimal. Longer time between measurements would lead to an

978-1-4799-2706-7/14 $31.00 © 2014 IEEE   2549

imprecise accuracy because of the lack of on state voltages variations. Latter fact becomes more critical when the IGBT module is close to the end of its lifetime, where fast and discrete variations of on state voltages take place as shown in Fig. 13. By contrast, shorter time would lead to a massive amount of data which is not necessary to process.

Furthermore, both measurement routines measure the V_{ce}, the V_f and the current through the converter. When the increment of V_{ce} or V_f reaches some predefined limits or wearing out level, the system is stopped and the IGBT module is considered for further study of internal failures.

B. Comparison between old offline routine and new offline routine

The new offline routine is formed by some subroutines where one V_{ce} and one V_f are measured sequentially. The first subroutine is explained in forthcoming paragraphs.

Initially, the control high IGBTs (CTL$_{h1}$ and CTL$_{h2}$) and DUT low IGBT (DUT$_l$) are turned on until the current will rise up to 920A as depicted in next figure.

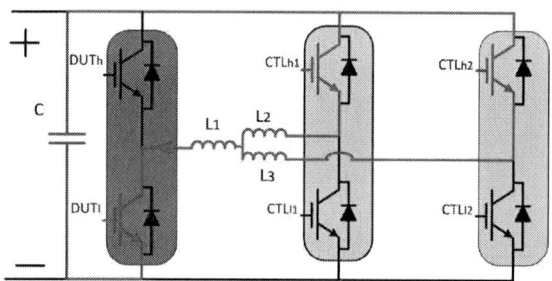

Fig. 8. Initial sequence in first offline subroutine

Afterwards, the IGBTs are turned off and the stored energy in the inductor flows through DUT$_h$ Diode and CTL$_{l1}$ - CTL$_{l2}$ Diodes as shown in next figure.

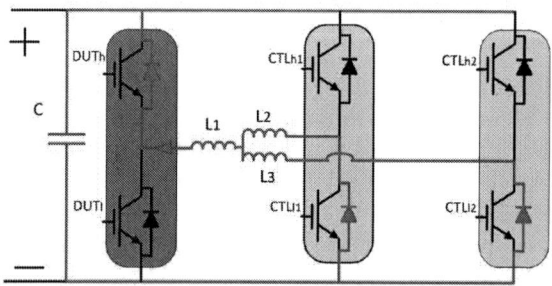

Fig. 9. Final sequence in offline subroutine

In Fig. 10 is illustrated the current through L$_1$ the explained subroutine. It should be noticed that the currents illustrated in the Fig 10. and Fig 11. present a scaling factor of 5000 with respect to the converter current as it was obtained by a hall current sensor.

The segment named D corresponds with the offline subroutine sequence shown in Fig 8. and the segment named E corresponds with offline subroutine sequence

shown in Fig 9. During this subroutine DUT$_l$ IGBT and DUT$_h$ Diode components are measured. To complete the offline routine similar subroutines are used in order to obtain all the values of V_{ce} and V_f from the DUT module.

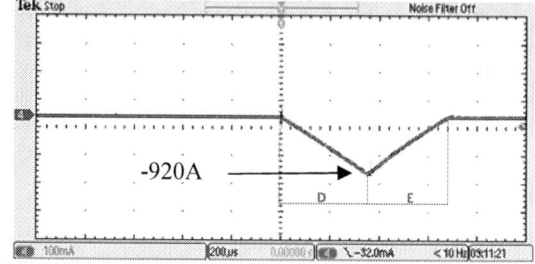

Fig. 10. New offline measuring pulse implemented for the new V_{ce} online monitoring system (double diode)

The offline routine used in [4] had an extra sequence due to the presence of the reed relays as shown in segment B in Fig 11.

The component to be measure (IGBT or diode) must be on state before the reed relay closes and connects the ADC circuitry to measure V_f. If not, the ADC would be connected to the high voltage and the consequence is the destruction of monitoring circuit and module.

Therefore two delays were present previously. The first one is related to the waiting time until the component is fully conducting, and thus, the reed relay can be safely switch on, and the second one is related to minimum switching time of the relay, 3 ms.

The result is lower available current in order to measure V_f at high current level. The additional segment B shows the current flowing through converter and was not used for measurement purposes. The consequence is that the available current was up to 600A and in the new offline routine is up to 920A.

The increment in the current available during measurement is of paramount importance in order to obtain better resolution during calibration process at high current level.

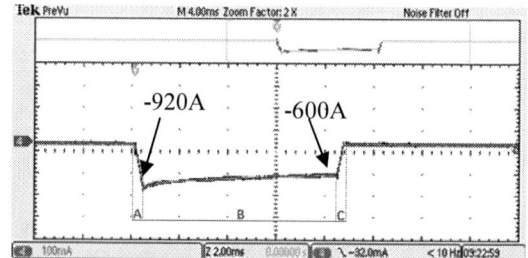

Fig. 11. Old offline measuring pulse implemented in the new V_{ce} online monitoring system.

C. Online measurement routine

As mentioned before, an online measurement of the on state voltage not only offers the status of the power module at any time, but it allows in combination with current measurement and baseplate temperature the possibility of acquire the estimation of the junction temperature at any time during converter operation.

The measuring routine consists of 4 measurements in

total which are the on state voltages (V_{ce} and V_f), the current through converter and liquid temperature of the cooling system. The data is sampled in the middle of PWM when there is no effect of the switching transients of the IGBT module, which is 10 μs for the applied conditions. Moreover, the whole data is obtained for 3 cycles of the current.

In Fig. 12 (upper graph) presents the online measurement of the voltages (V_{ce} and V_f) at high and low side of the IGBT module. Fig. 12 (lower graph) presents the current through the converter.

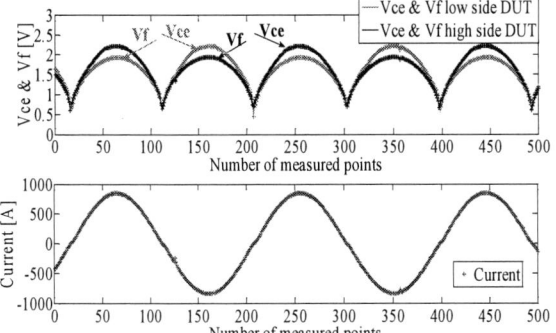

Fig. 12. Online measurement result at one measurement routine

D. Offline vs online measurement results

Offline and online measurement results can show the status of the IGBT modules. However, the results cannot be strictly compared because of the intrinsic measurement principle. During offline measurement the converter and the fun of the cooling system are stopped.

Moreover, the offline routine is run one minute after the stopping of converter. These facts allow more homogenous temperatures along the IGBT module.

However, during offline routine the switching effect can be neglected and only the current flowing through the component causes power dissipation inside the power device. By contrast, during online measurement the power dissipation is affected by converter operation.

Therefore, the thermal behavior will not be the same, and thus, the on state voltages values will differ between offline and online measurements for the same current level.

VI.   ACCELARATED TESTS

A.   Selected test conditions

TABLE I
USED PARAMETERS FOR TESTING IGBT MODULE

Symbol	Meaning	Value
$V_{DC\text{-}LINK}$	Voltage at Dc-Link	1000V
V_{DUT}	Voltage at middle point	253V
I_1	Current through L_1	890A_{peak}
F_{OUT}	Fundamental current frequency	6Hz
F_{SW}	Switching frequency	2.5Khz
T_c	Coolant temperature	80±0.7°C
T_{Lab}	Room temperature	20±2°C

As cited previously, the test setup can be used for testing high power IGBT modules at different scenarios,

where different voltages, current and frequency levels can be applied. However, the results presented in the forthcoming paragraphs have been achieved applying the parameters shown in previous table.

B.   Number of High Power IGBT modules tested

A total of 4 High Power IGBT modules have been tested and presented in following paragraphs.

Initially, one IGBT module was tested until its destruction in order to know approximately its life duration applying the selected test conditions. The total number of cycles until its final destruction was 5100 kcycles.

The other 3 IGBT modules were tested before their destruction and at different wear out levels, such as 2500 kcycles, 3500 kcycles and 4500 kcycles.

When the modules are worn out until predefined level, they are considered for further study of internal degradation or failure. Although detailed results regarding internal degradation are out of the scope of this paper some images will be shown.

C.   Offline measurement result

In Fig. 13 is depicted the offline measurements of Low DUT diode for the 4 tests. The results are shown for the same current level as explained in [3] which was 550A. The 5100 kcycles test led to final failure of the device and presented a total increment of 62 mV.

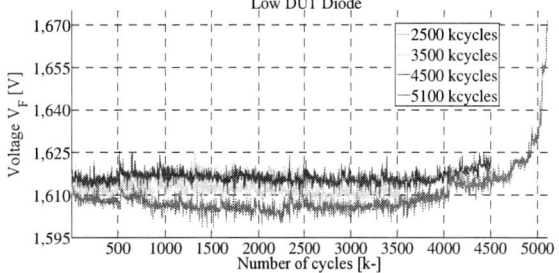

Fig. 13. Offline measurement result

The result for 2500 and 3500 kcycles tests did not show a significant on state voltage increment. The 4500 kcycles test presented a linear increment from 4000 kcycles which may be caused by reconstruction in the metallization surface. However, latter fact must be ratified by physical test of the modules.

Finally, in the case of 5100 kcycles test can be seen how fast and steep are the increments increment of V_f before the failure. These steep steps are caused by bond wire lift off as cited before in [3]-[7].

D.   Online measurement result for 5100 kcycles wear out level module

The online results were achieved as describes in *online measurement routine* paragraph. Furthermore, the temperature in the baseplate is kept constant at 80±0.7°C during the entire testing time.

978-1-4799-2706-7/14 $31.00 © 2014 IEEE

However, the results depicted in forthcoming figures show the on state voltages at 900 A and -900A as depicted in Fig. 14.

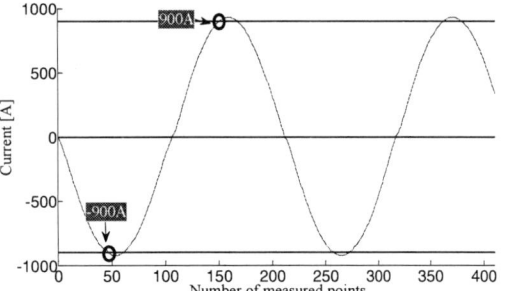

Fig. 14. Current during online measurement

In the next figure can be seen the initial on state voltages of Low side components of DUT module when starting the test and the final on state voltages when the test was stopped.

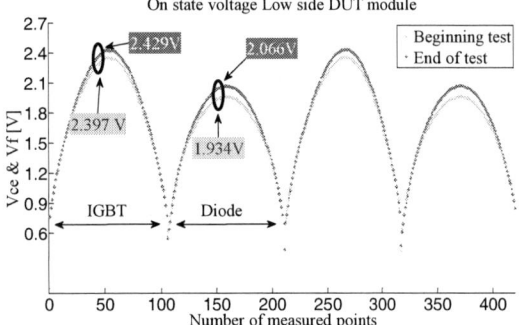

Fig. 15. On state voltage Low side DUT module for 5100 kcycles test. (Green: beginning test- Red: end of test)

The Low side IGBT component varied 32 mV; in turn the Low side Diode component underwent an increment of 132 mV.

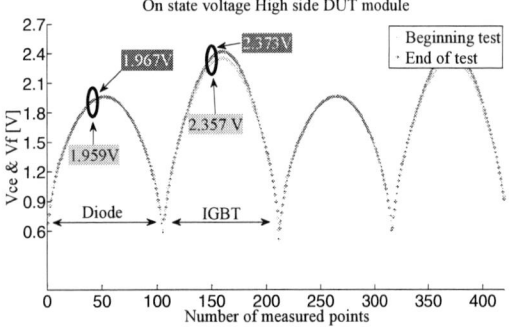

Fig. 16. On state voltage High side DUT module for 5100 kcycles test. (Green: beginning test- Red: end of test)

In the case of the High side components the IBGT increased only 8 mV and the diode varied 16 mV as shown in preceding figure.

From on state voltages results can be derived that the Low side of the IGBT module suffered. In case of Low side Diode the on state voltage presented the greatest increment in comparison to the rest of the components inside the module. In fact, the Low side Diode was the failed component. By contrast, the high side of the module did not show as great increment as low side, especially the High side IGBT which only presented an increment of 8 mV.

VII. BRIEF PHYSICAL ANALYSIS

To understand the variations in the electrical parameters a physical study after wearing out the High Power IGBT modules is of paramount importance. For example, a steep or discrete step in the on state voltage is related to bond wire lift off [7], nonetheless this is not the unique degradation phenomena and more degradation process may be involved as reconstruction of metallization surface, solder fatigue etc. Therefore, a complete and detailed exam must be done to understand the physics of failure mechanism. Although detailed exam of physical failure and analysis is out of the scope of this paper some images will be shown.

A. High Power IGBT module

The IGBT module used as DUT consists of 12 IGBT chips and 12 diode chips. Power terminals, IGBT and diode are interconnected with 10 aluminium bond wires.

Following figure shows an IGBT and a diode chips in one of the sections.

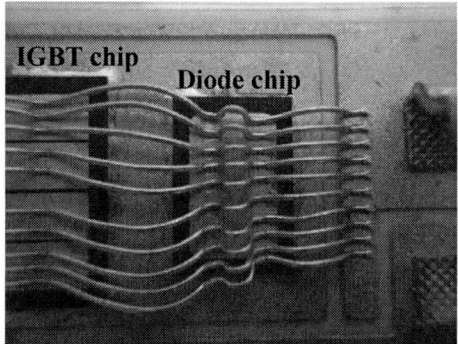

Fig. 17. Half section of High Power IGBT module used in the test setup

A. SEM images for 5100 kcycles test module

The exploded module could not be studied in detail because of the explosion when the device failed. The failure caused high internal destruction of the module as can be seen following image.

Fig. 18. One section of the exploded IGBT module

978-1-4799-2706-7/14 $31.00 © 2014 IEEE

VIII. CONCLUSIONS

A new test setup is presented in this paper to accomplish an accelerated test for high power IGBT modules. The proposed test setup offers more robust, reliable and advanced wear out test system because of the converter structure, controlled room temperature and the online V_{ce} monitoring system. Furthermore, the test setup is flexible to set different loading parameters which are suitable for field applications.

Finally, proposed monitoring technique is a potential method for field application whereas the setup can be used to study the degradation mechanism of power modules in the laboratory realizing a field mission profile.

ACKNOWLEDGMENT

This work is carried out to develop a realistic accelerated power cycle test setup for high power IGBT. The work is conducted under the Intelligent Efficient Power Electronics (IEPE) and Center of Reliable Power Electronics (CORPE) project framework at Aalborg University.

REFERENCES

[1] M. Wilkinson, G. Hassan Partners Ltd, and Reliwind EU project. "Reliability data field study in the reliawind project" *WindTurbine Reliability Workshop Alburquerque, NM USA, 2009.*

[2] Chiappa, M. "Selected failure mechanism of modern power modules. *Microelectronics Reliability, vol. 42, no 4-5 pp. 653-667.*

[3] Nielsen, R.O.; Due, J.; Munk-Nielsen, S. "Innovative measuring system for wear-out indication of high power IGBT modules" *Proceedings of the 3rd IEEE Energy Conversion Congress and Exposition (ECCE 2011).* IEEE Press, s. 1785-1790 10 s

[4] Due, J.; Munk-Nielsen, S.; Nielsen, R. "Lifetime investigation of high power IGBT modules" *Proceedings of the14th European Conference on Power Electronics and Applications (EPE 2011).* IEEE Press, s. 1-8 8s.

[5] Trintis, I.; Munk-Nielsen, S.; Abrahamsen, F.; Thogersen, P.B. "Efficiency and Reliability Improvement in Wind Turbine Converters by Grid converter Adaptive control" *EPE'13 ECCE Europe 15th European Conference on Power Electronics and Applications . 2013.*

[6] IEC 60747-9. "Semiconductor devices- Discrete devices- Part 9: Insulated gate bipolar transistors" 45-48 pp.

[7] Josef Lutz · Heinrich Schlangenotto ·Uwe Scheuermann · Rik De Doncker 2010, "Semiconductor Power Devices: Physics, Characteristics, Reliability" *Springer, 390-392p.*

[8] Smet, V.; Forest, F.; Huselstein, J.-J.; Richardeau, F.; Khatir, Z.; Lefebvre, S.; Berkani, M. "Ageing and Failure Modes of IGBT Modules in High-Temperature Power Cycling" *IEEE Transaction on Industrial Electronics, VOL 58, NO 10, October 2011.*

[9] Smet, V.; Forest, F.; Huselstein, J.; Rashed, A.; Richardeau, F. "Evaluation of Vce as Real-Time Method to Estimate Aging of Bond Wire IGBT Modules Stressed by Power Cycling". *IEEE Transaction on Industrial Electronics, VOL 60, NO 7, July 2013.*

[10] Dupont, L.; Avenas, Y.; Jeannin, P.-O. "Comparison of Junction Temperature Evaluations in a Power IGBT Module Using an IR Camera and Three Thermosensitive Electrical Parameters". *Industry Applications, IEEE Transactions on* (Volume:49, Issue: 4)

[11] Tyfocor ®, "Antifreeze and Anticorrosion Concentrate for Heating and Cooling Circuits. Medium for Ground Source Heat Pump Systems". *Technical information.*

[12] Klaus Olesen, Danfoss Silicon Power GmbH. "Innovative Cooling Concept for Power Modules". *Bodo's power system magazine.*

[13] Klaus Olesen, Dr. Rüdiger Bredtmann, Dr. Prof. Ronald Esele, "ShowerPower: New Cooling concept for Automotive Applications" *APE- June 2006 – Paris.*

[14] P. Ghimire, A. R. de Vega, S. Beczkowski, B. Rannestad, S. Munk-Nielsen, P. Thøgersen, "Improving reliability of power converter using an online monitoring of IGBT modules", IEEE Industrial Electronics Magazine, Accepted.

[15] P. Ghimire, A. R. de Vega, K. B. Pedersen, B. Rannestad, S.Munk-Nielsen, P. Thøgersen "A real time measurement of junction temperature variation in high power IGBT modules for wind power converter application", 8th International Conference on Integrated Power Electronic Systems (CIPS) 2014, Nuremberg, Germany.

[16] Beczkowski, S.; Ghimre, P.; de Vega, A.R.; Munk-Nielsen, S. ; Rannestad, B.; Thøgersen, P. 2013, "Online Vce measurement method for wear-out monitoring of high power IGBT modules" *EPE'13 ECCE Europe 15th European Conference on Power Electronics and Applications.* 2013.

[17] A. Volke, M. Hornkamp 2012. "IGBT Modules: Technologies, Driver and Application*" 2nd Ed. Infineon Technologies AG. 247-254p.*

[18] Bryant, A.; Shaoyong Yang; Mawby, P.; Dawei Xiang; Ran, L.; Tavner, P.; Palmer, P.R." Investigation Into IGBT dV/dt During Turn-Off and Its Temperature Dependence" *Power Electronics, IEEE Transactions on* (Volume:26, Issue:10)

DESIGN OF HIGH-SPEED IGBT-BASED SWITCHING MODULES FOR PULSED POWER APPLICATIONS

Andreas Kluge*, Lutz Goehler‡, Henry Gueldner*, Thomas Trompa§, David Mory§, Karl-Heinz Segsa¶

*TU Dresden

Fakultaet Elektrotechnik und Informationstechnik, Elektrotechnisches Institut, Lehrstuhl Leistungselektronik, Dresden, Germany

Email: andreas.kluge@mailbox.tu-dresden.de

‡HTW Dresden, Fakultaet Elektrotechnik, Professur Leistungselektronik Friedrich-List-Platz 1, 01069 Dresden, Germany

§Lasertechnik Berlin GmbH, Am Studio 2c, 12489 Berlin, Germany

¶Spree Hybrid & Kommunikationstechnik GmbH, Schkopauer Ring 24, 12681 Berlin, Germany

Abstract—**This paper presents the design of IGBT-based switching modules for pulsed power applications. Starting from theoretical and practical aspects for the selection of the IGBTs the required parameters for the design of a single cell are derived. Extended results of the characterization process for different IGBTs are presented. They show that by using a special gate driving method for IGBT devices it is possible to realize the required maximum values of peak current and current slope. From the design of the single cell the conditions for a multiple cell series connection for the target application follow. This includes the development of a high-speed gate drive based on a pulse transformer. The designed cascade is used in a nitrogen gas laser and switches a voltage of $12\,\mathrm{kV}$ and carries a peak current of $500\,\mathrm{A}$ at a maximum current slope of about $28\,\mathrm{A\,ns^{-1}}$.**

I. INTRODUCTION

A. Pulsed Power Systems

Pulsed power systems have the goal to deliver a certain amount of energy to a load in a very short time. The applications range from industrial, biological and military to medical ones. Some typical applications are:

- gas lasers e.g. excimer-, nitrogen lasers [1]
- exhaust treatment, SO_x, NO_x reduction [2]
- food sterilization
- radar systems [3]
- lithotripsy systems

For these systems the current and voltage slopes in the load are of great importance [4]. Therefore special pulsed power switching devices like MCT (MOS Controlled Thyristor), MTO (MOS Turn On) or DSRD (Drift Step Recovery Diode) have been developed. They have limitations in reliability, availability, life time and/or price because they are often provided only by one manufacturer. For these reasons it is desirable to replace these switches by widely commercially available devices like MOSFETs or IGBTs which are not specified for that kind of applications. Their switching speed in typical power electronics applications is at about one order of

magnitude lower than desired for pulsed power applications, especially for the nitrogen laser. Recently, there have been some investigations into the high-speed switching behavior of IGBTs and MOSFETs ([1], [5], [6], [7]) which were promising.

As was shown in [7], single IGBT-chips can reach the desired

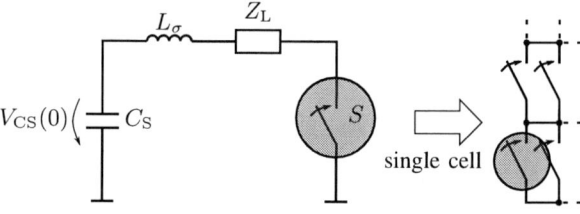

Fig. 1. Pulsed power system with capacitive energy storage

switching speed of some pulsed power applications when using higher gate voltages than recommended in datasheets. It has been found that above a gate voltage level of about $60\,\mathrm{V}$ the switching dynamics are not improving anymore. The switching losses reach a saturation level at this point.

Figure 1 shows the schematic of a pulsed power system with capacitive energy storage. By turning on the switch, the precharged energy storage C_S discharges into the load. Due to different current and voltage demands, the switch is realized by a series-, parallel- or series-parallel connection of single devices with lower ratings. For the module design it is necessary to define and to design a single cell which depends both on the available devices and the required power for the load. Starting from this minimum cell the module can be designed.

The most critical parameter in power modulators is the high required ratio between the average power and the peak power of the switch which is caused by the short pulse lengths and moderate repetition frequencies. This fact leads to the high switching dynamics. A simple calculation according to equation 1 results in the required switching power which must be handled by the switch.

$$P_{\mathrm{S}} = \frac{E_{\mathrm{Load,req}}}{t_{\mathrm{P}}} \qquad (1)$$

For example, the nitrogen gas laser requires an energy amount of approximately $30\,\mathrm{mJ}$ delivered in a time of about $20\,\mathrm{ns}$ which results in a switching power above $1\,\mathrm{MW}$.

B. Nitrogen gas laser

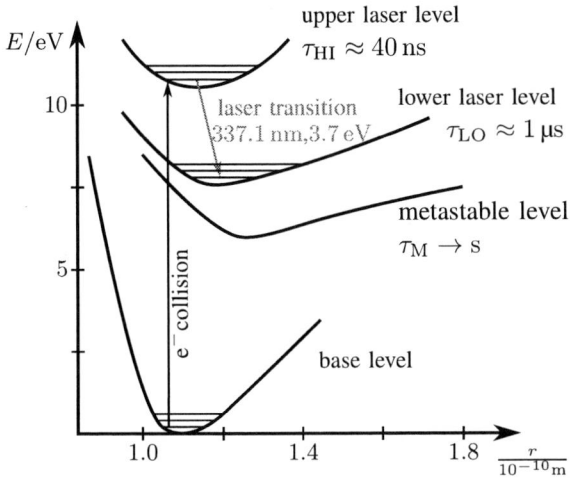

Fig. 2. Energy level diagram of the N_2 laser [8]

The main application for this research is the nitrogen gas laser. Its optical output is realized by a gas filled laser tube at low pressure. By switching of a capacitor, a transverse-electrical discharge (TE) is fired between two electrodes in the tube and thus a laser process starts. The output is realized by two mirrors, one at each end of the tube. The demand for the high-dynamic switching results from the energy level diagram of this laser which is shown in Figure 2. Due to the electron collision caused by the TE, the upper laser level with a lifetime of about 40 ns is filled. The laser transition occurs between the upper and the lower laser level, emitting a photon at an ultraviolet wave length of 337.1 nm. The lower laser level has a lifetime of about 10 µs. To reach the base level, a pass through a metastable state with a lifetime of seconds is necessary. These different lifetimes lead to the fact that the nitrogen gas laser can only be driven in a pulsed mode. Furthermore, the electron collisions have to be provided by the discharge in a time which is in the order of the lifetime of the upper laser level. The switch design is the greatest challenge when developing such a laser. On the other hand, the nitrogen gas laser has an extraordinary high laser gain. Thus, there is no need for very complex mirror systems. To increase the peak power and to decrease the beam divergence a simple high-reflective rear mirror is sufficient.

C. CC-Topology

To provide the TE discharge to the tube, a topology called CC-topology is used. It is already known from thyratron systems and has been described in detail in [7]. Although there are high speed switches like MCT or thyratron, their switching speed is still too low for a nitrogen gas laser tube. The laser tube requires a current slope of up to $400\,\mathrm{A\,ns^{-1}}$. Therefore, the CC-topology sharpens the pulse provided by the switch by a factor of 15 to 20 regarding current slope and factor 3 regarding peak current.

First, the storage capacitor C_{S} is charged by a HV-charger. During this time the peaking capacitor C_{P} is shorted by the impedance Z_{SYM}. The switch turns on and the laser tube is ignited by an energy transfer between C_{S} and C_{P}. This transfer can be influenced by the switching speed. When the laser tube ignites, the high speed transfer of energy to laser tube is mainly provided by the peaking capacitance C_{P}. Thus, the above mentioned gain of peak current and current slope is observed. The switch is protected from that high current slope due to the parasitic inductance L_{σ}. A disadvantage of the CC-topology is the greater amount of energy which has to be charged into C_{S}. This energy is only partially transferred to the laser tube.

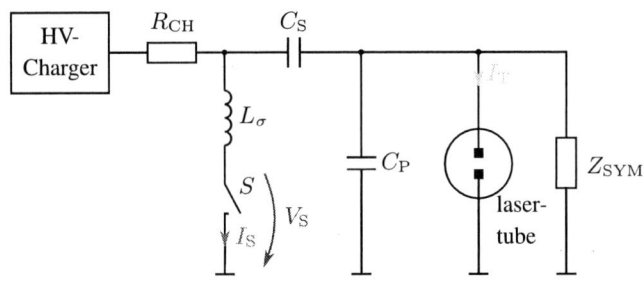

Fig. 3. CC-topology

II. DEVICE SELECTION ASPECTS FOR PULSED POWER APPLICATIONS

A. Theoretical aspects

Applications as described above require fast turn-on of the main power device. Due to the moderate repetition rate, switch-off delay and losses are of lower interest. When an IGBT turns on, the current is determined by the n-channel MOSFET part of the device initially. Its structure is shown in Figure 4. It is clear, that the MOSFET should be turned on as fast as possible and that it should feature a high saturation current which is a function of the doping N_{A} of the $p^+-layer$, the gate-emitter voltage V_{GE} and the aspect ratio $\frac{w_{\mathrm{MOS}}}{L_{\mathrm{MOS}}}$. Since the transit time of an electron in the MOSFET channel of a modern IGBT is far below one nanosecond, channel formation itself is not decisive for MOSFET turn-on delay. The switching delay is instead due to the effective gate-to-emitter capacitance C_{IN} and the effective gate resistance R_{Geff}, which is a well-known fact. In the following, both are assumed to be

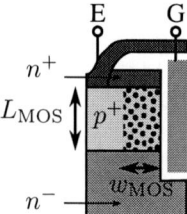

Fig. 4. MOSFET channel structure in a Trench IGBT

approximately constant. The delay can be partly compensated by an elevated driver output voltage. If, for instance, the gate is charged to a voltage V_{GEON} and the driver output voltage is raised from V_{G1} to V_{G2} the gate charge time is reduced by:

$$\Delta t_{\mathrm{c,G}} = \tau \ln\left(\frac{1 - \frac{V_{\mathrm{GEON}}}{V_{\mathrm{G2}}}}{1 - \frac{V_{\mathrm{GEON}}}{V_{\mathrm{G1}}}}\right) \qquad (2)$$

where $\tau = R_{\mathrm{Geff}}C_{\mathrm{IN}}$ at the cost of a higher instantaneous current

$$I_{\mathrm{G}}(0) = \frac{V_{\mathrm{G2}}}{R_{\mathrm{Geff}}} \qquad (3)$$

The latter leads to higher power dissipation. In the above discussion it is assumed, that the driver output voltage is controlled so that gate oxide breakdown is avoided safely. Despite these possible improvements by the gate drive, which will be discussed later in chapter III, the device of choice for a pulsed power application should have minimum parasitic capacitances and a high saturation current of the MOSFET part, as already stated. Trench technology lowers $R_{\mathrm{DS,ON}}$ (i.e. increases the saturation current) at the cost of a higher gate capacitance, so that a higher driver power compared to a planar gate device is required. If possible, it is of advantage not to choose a wide trench pitch device. These devices are designed for a lower short circuit saturation current (or lower MOSFET saturation current). Short circuit ruggedness is no key demand here, too. Figure 5 compares the switching transients of generic trench IGBTs with high and low MOSFET saturation current (low N_{A} and high N_{A}, respectively), as they result from a 2D device simulation with NGSpice [9]. The doping profiles of both structures are compared in Figure 6. To reduce the simulation time, the compared devices have blocking voltages of about 200 V, this way requiring a smaller calculation grid.

When the MOSFET part conducts, it injects carriers into the lightly doped and hence initially high-ohmic n-base. The increasing charge in the base lowers its resistance as time proceeds, which is known as conductivity modulation. To obtain a fast decrease of the n-base resistance the width of that region should be small and the carrier lifetime (or the resulting ambipolar lifetime) high. The latter demand results from the simplified charge control equation:

$$i_{\mathrm{MOSFET}} \approx I_{\mathrm{MOSFET}} = \frac{Q_{\mathrm{n}}}{\tau_{\mathrm{a}}} + \frac{\mathrm{d}Q_{\mathrm{n}}}{\mathrm{d}t} \qquad (4)$$

where τ_{a} stands for the ambipolar lifetime and Q_{n} denotes the

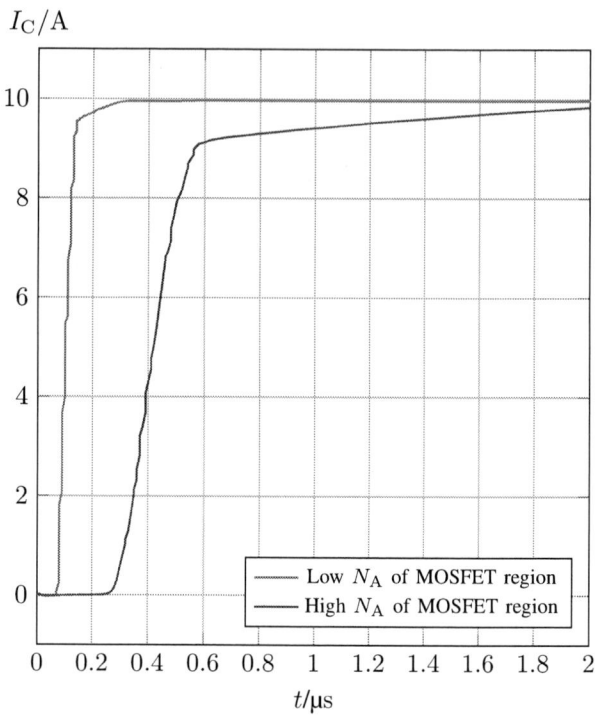

Fig. 5. Comparison of collector current transients during switch-on of a generic trench IGBT (ohmic load) with higher MOSFET saturation current (low N_{A}) and lower MOSFET saturation current (high N_{A}). The gate drive signals are equal in both cases.

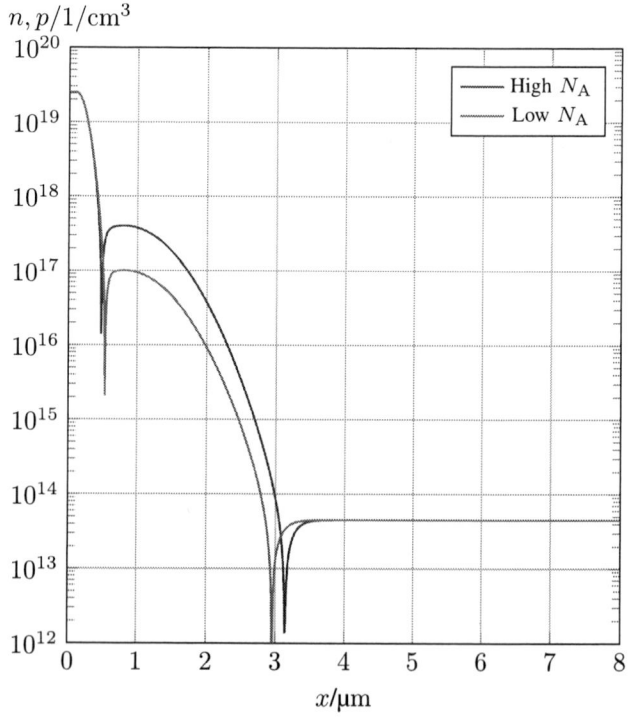

Fig. 6. Doping profiles of the MOSFET part [10] (n-base width w_{n}=100 μm)

978-1-4799-2706-7/14 $31.00 © 2014 IEEE

TABLE I
INVESTIGATED CHIPS

Type	V_{BR}	$I_{\mathrm{C,Peak}}$	$R_{\mathrm{G,int}}$	$R_{\mathrm{G,ext}}$
IGC142T120T6	1200 V	450 A	5 Ω	1 Ω
4 x IGC36T120T6	1200 V	420 A	0 Ω	1 Ω
MCT	1400 V	4000 A	0 Ω	1 Ω

stored charge. It can be seen, that if the MOSFET current is approximately constant and if no electron recombination occurs at the p-emitter side (total recombination of the electrons in the n-base), a given amount of charge is built up faster if the ambipolar lifetime assumes higher values. This is the reason, why IGBT structures with field stop layer and high carrier lifetime are favorable in practical applications.

B. Practical aspects

A design for modulators with blocking voltages above 10 kV requires a series connection of single devices. To limit the effort for driving and symmetrization, only devices with a blocking voltage of at least 1200 V are investigated. In this voltage-class only IGBTs reach sufficient pulse currents for the target application. In comparison, there are just a few available MOSFETs from IXYS with blocking voltages above 1000 V. Due to the unipolar conduction and the strong dependency of $R_{\mathrm{DS,ON}}$ on the blocking voltage (power of 2.3) they are specified for much smaller peak currents than IGBTs. Charge compensated devices could be an alternative but they are only available for blocking voltages below 1000 V. Infineon-IGBTs of the 1200 V- and 1700 V-class reach the highest peak pulse powers among IGBTs, they are specified for peak pulse currents of up to 600 A. Furthermore, they are available as single chip devices which is advantageous for a low inductive assembly.

One possibility is to design a single cell for a blocking voltage of 1200 V and a peak current of 450 A by a single switch. An interesting alternative is a parallel connection of IGBTs with smaller current rating. These devices do not contain an internal gate resistor. Hence, it should be easier to reach high-dynamic switching in comparison to the single-chip cell and the effort for the gate drive is reduced.

Infineon offers 1200 V-IGBTs with a peak current of 105 A without gate resistor. A parallel connection of four such devices could carry nearly the same current as a single chip with a gate resistor. Table I shows the investigated IGBT-chips. For a comparison the state of the art MCT-switch is investigated, too.

III. SINGLE CELL CHARACTERIZATION AND GATE DRIVE STRATEGY

Due to the fact that commercial datasheets do not contain data for pulsed operation, especially the achievable switching dynamics, measurements for characterization is necessary. Two recently published types of operation are possible to characterize a single cell for pulsed power applications:

- Switching with high stored energy, independent $I_{\mathrm{C,Peak}}$ and $\frac{\mathrm{d}I_{\mathrm{C}}}{\mathrm{d}t}$, $I_{\mathrm{C,Peak}}$ controlled by the gate, rectangular current pulse [5]
- Switching based on equivalent energies for a single cell with maximum $I_{\mathrm{C,Peak}}$ and $\frac{\mathrm{d}I_{\mathrm{C}}}{\mathrm{d}t}$, $I_{\mathrm{C,Peak}}$ limited by the load and parasitics [7]

Both feature the gate drive strategy "Gate-Boosting" which delivers the required gate charge faster to the gate of the IGBT as shown in Figure 7. This method is realized by using an elevated gate terminal voltage of up to 80 V. According to equation 2, the gate charging time is then reduced by changing the gate voltage level. Although the gates can withstand the higher voltage, the exact timing of the gate pulse ensures that the gate oxide voltage of the IGBT does not exceed the maximum permissible transient voltage of 30 V.

The usability of the both types of operation mentioned above

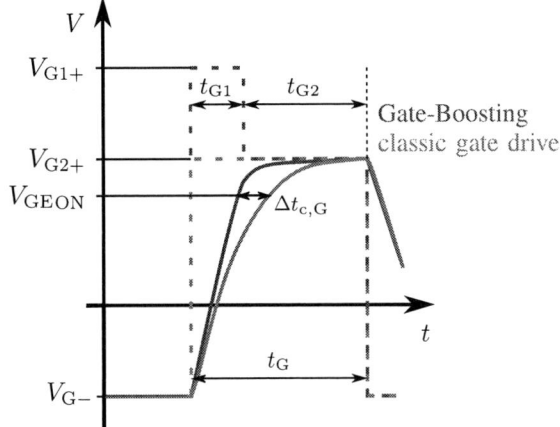

Fig. 7. Principle of Gate Boosting

depends on the load capacitance and the required energy transfer. For the nitrogen gas laser the second type is used because the stored energy of the application is known. A typical characterization circuit for a single cell is shown in Figure 8. It has its origin in a CC-topology to drive the nitrogen gas laser but can be used in the same manner for any application which requires an energy transfer from a storage capacitor to a load. The goal of characterization is to reach similar voltage and current slopes for a single cell as they would occur in the series-parallel connection for the real application. To reach equivalent switching transients for a single device, the storage capacitor should contain an energy amount which scales down from the real application according to equation 5.

$$E_{\mathrm{CS\,Single}} = \frac{E_{\mathrm{CS\,App}}}{m_{\mathrm{P}} \cdot n_{\mathrm{S}}} \tag{5}$$

There, n_{S} is the number of cells in series and m_{P} is the number of cells in parallel. This approach assumes equal voltage and current sharing of all cells which has to be ensured by the design and gate drive of the resulting switch. The strategy leads to comparable voltage and current values and slopes during characterization.

The 1200 V IGBT devices shown in table I have been measured according to the characterization strategy in [7] with the parameters:

- $E_{CS} = 13.4\,\text{mJ}$
- $V_{CS}(0) = 1000\,\text{V}$ (12 stage design)
- $C_S = 26.4\,\text{nF}$
- $C_P = 7.7\,\text{nF}$
- $n_S = 12$, $m_P = 1$

For the MCT switch, which has a higher blocking voltage, the energies have been adapted according to equation 5 for $n_S = 10$. The measurement results for the different devices from table I are shown in Figure 9. As can be seen, the parallel connection of the IGBTs reach a slightly higher peak current and higher dynamics than the single 1200 V IGBT at smaller gate voltages. A maximum di/dt of about 23 A ns^{-1} can be observed for the parallel connection. Furthermore, the turn-on delay is reduced for these chips due to the smaller gate resistor. Both devices could switch a higher peak current. Taking the damping effect of the current viewing resistor $R_{M,S}$ into account, in the target application the maximum switching performance of the IGBTs should be reached. Furthermore, the MCT shows even a slightly better behavior than the parallel connection of four 1200 V IGBTs.

The main result from the single cell characterization is that in the special application IGBTs, especially the parallel connection of 1200 V IGBTs can reach a switching power comparable to the MCT when using the special gate drive strategy without increasing the stored energy compared to the the MCT-switch.   A more general way for characterization

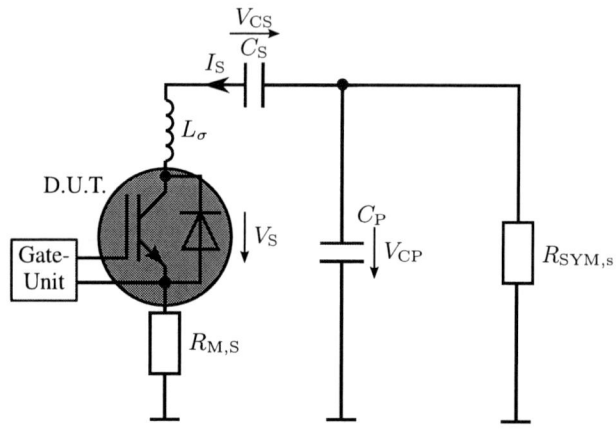

Fig. 8. Characterization circuit

of the maximum pulsed power could be the following. It is assumed that a single cell consists of a single IGBT (or a MOSFET), a parasitic inductance, a damping resistor and a capacitance charged at a certain voltage level. Furthermore, the switch has a finite switching dynamics and a defined peak current. This peak current can be the datasheet value or a modified value depending on planned lifetime. Then, two

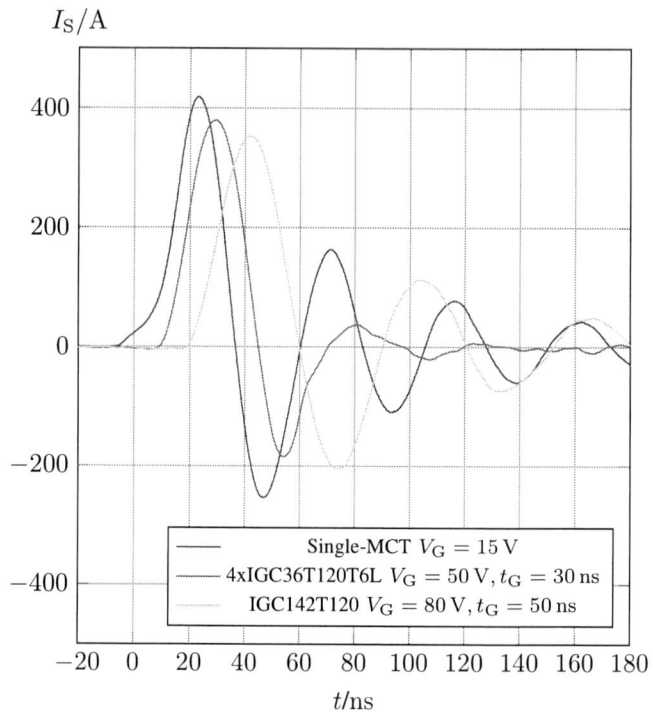

Fig. 9. Comparison of switching currents for different IGBTs

degrees of freedom exist and can be adapted during the design: The switching dynamics controlled by V_{GE} and the amount of energy E_{CS} which is transferred during the switching process. Assuming that the IGBT switches at maximum speed (e.g. maximum V_G) three different cases exist for the relationship between initial energy E_{CS0} and peak current:

- The peak current will be exceeded, decrease of V_{GE} or E_{CS0} necessary to stay within peak current limit.
- The peak current will be reached: Design optimum between switching speed and maximum switching power.
- The peak current will not be reached, further increase of E_{CS0} possible.

Starting from such a characterization of a single cell, the desired voltage and current level for the application can be provided by series, parallel, or series-parallel connection. To reduce the number of cells it is the goal to reach the second case.

In practice, the peak current in the single cell characterization should be about 10 % to 20 % below the maximum vaule in the real application. This considers non scaling conditions for the parasitic inductance and the current viewing resistor. Figure 9 shows, that the IGBTs, especially the IGC36T120T6L, reach nearly the above mentioned design optimum between the switching speed and the maximum switching power. Therefore, the criteria for maximum switching power is the peak current value from the datasheet.

IV. DESIGN OF THE HV-CASCADE

This chapter presents the design of a 12-stage series connection of IGC36T120-based cells, the realized cascade consists of 11 stages. Each cell has four single IGBTs in parallel which share the gate drive and the symmetrization network. The switch design is shown in Figure 10, a picture of the cascade is shown in Figure 11.

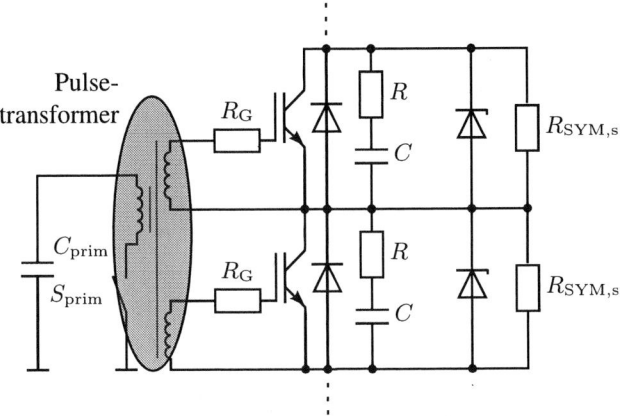

Fig. 10. Structure of HV-cascade

Fig. 11. HV-cascade based on IGC36T120-IGBTs

A. Galvanically isolated gate drive

To control the IGBTs a galvanically isolated Gate-Drive is necessary which takes the special gate drive strategy into account. Due to the fact that gate voltage pulses of about 50 V to 80 V with pulse lengths of about 50 ns are required, the rise time should be below 20 ns. Although a lower gate resistor is advantageous for higher switching dynamics, a minimum gate resistor must be used to avoid voltage oscillations in the gate circuit [11]. These oscillations might lead to a different turn on behavior of the single stages and therefore asynchronous voltage sharing. The solution of the second order differential equation in the gate circuit with the parasitic inductance $L_{\sigma,G\sum}$ and the input capacitance C_{IN} leads to the minimum

gate resistor $R_{G,MIN}$ to reach the critical damped condition:

$$R_{G,MIN} = 2\sqrt{\frac{L_{\sigma,G\sum}}{C_{IN}}} \qquad (6)$$

For typical values of $L_{\sigma,G\sum}$ and C_{IN} for a 1200 V/450 A-IGBT (or four parallel connected 1200 V/105 A-IGBTs) a minimum gate resistor of 1 Ω to 2 Ω should be used. Gate boosting is necessary to obtain the maximum dynamics of the device without voltage oscillations. Different methods for the realization of the gate drive are possible, three basic concepts are:

- Single pulse transformer, switching energy supply at the primary side
- Divided pulse transformer, switching energy supply at the primary side
- Single pulse transformer, switching energy supply at the secondary side

For the nitrogen laser the first option seems to be the best solution. Due to the required compactness and the relatively low number of cells, the turn on energy can be transformed by a single transformer. Furthermore, the synchronization of a single transformer should be better than that of a distributed system and the number of devices required is reduced. For the pulse transformer of the nitrogen laser a toroidal core of nanocrystaline material is used. The turns ratio of the transformer is 4:1. A fast primary circuit switches a voltage of about 800 V at the primary winding with a rise time of 5 ns. The design process of the pulse transformer was assisted by a 3D-FEM transient simulation. For example, Figure 12 shows the flux density in the core 28 ns after applying the voltage to the primary side. The resulting transformer realizes an output slope of up to $6\,\text{V ns}^{-1}$ with a gate equivalent $5\,\Omega$ load resistor and up to $4\,\text{V ns}^{-1}$ in a real cascade. The transformer output voltages with a $5\,\Omega$ resistive load on all output windings is shown in Figure 13. The voltages have been measured simultaneously with four coupled and synchronized oscilloscopes. As can be seen, the rise time from 0 V to 80 V is about 15 ns. The difference between the windings at 80 V level is less than 3 ns. Thus, the requirements regarding dynamics and synchronization of the gate drive for a cascade are fulfilled.

B. Measurement results

Figure 14 shows the measurement result of a IGBT-based cascade with a blocking voltage of about 12 kV. In this mode the laser produces an output energy which is about 20 % above a comparable MCT-based system. The first peak in the switch current I_S determines the quality of the laser output. As planned, it reaches approximately the maximum peak pulse current of the four parallel devices. In comparison to the single cell characterization in chapter III it is increased by about 20 %. This is caused by the reduced damping of the circuit (CVR not scaled) and due to the fact that for the measurement only a 11-stage cascade has been used. Thus, the switching energy of a single cell in the cascade is slightly

978-1-4799-2706-7/14 $31.00 © 2014 IEEE

The 2014 International Power Electronics Conference

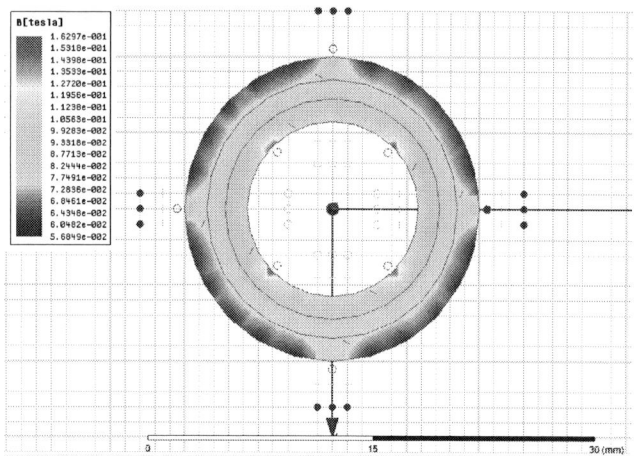

Fig. 12. Flux density of the pulse transformer at $t = 28\,\mathrm{ns}$

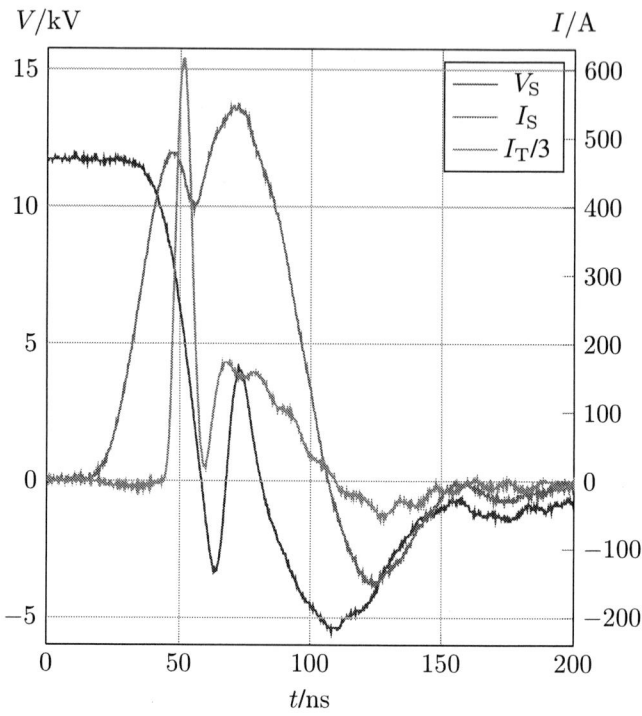

Fig. 14. Switching waveforms of the IGC36T120-based switch and $V_{\mathrm{CS}}(0) = 11.5\,\mathrm{kV}$, $I_{S\,Peak} \approx 550\,\mathrm{A}$, $\left.\frac{dI_S}{dt}\right|_{max} \approx 30\,\mathrm{A\,ns^{-1}}$

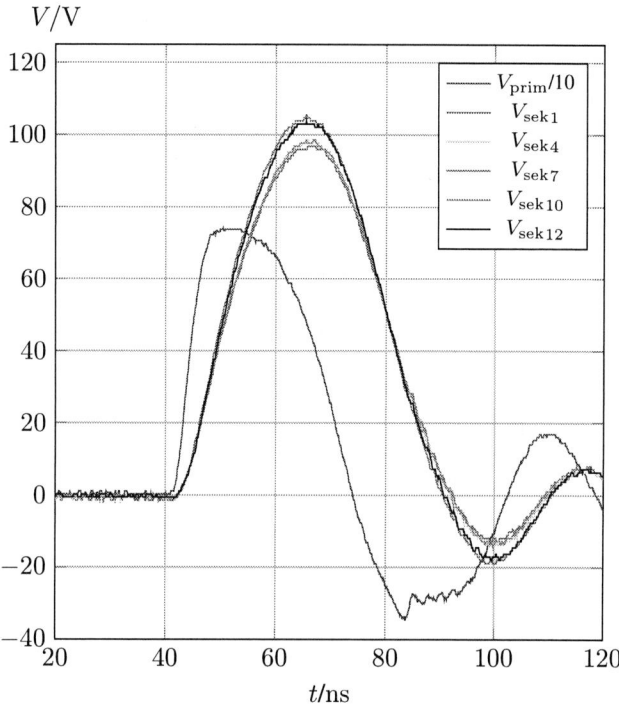

Fig. 13. Transforming a 50 ns pulse to $5\,\Omega$ resistive load on each output winding

increased in comparison to the single cell characterization. Figure 15 compares the different switch currents for the IGBT cascade and the MCT cascade switching a voltage of $11\,\mathrm{kV}$. Before igniting the laser tube, the cascade based on the parallel connection reaches a higher current peak than the MCT. Hence, the energy transferred into the peaking capacitance is increased which leads to the higher laser output energy. Furthermore, the ratio between the first and the second current peak is smaller than in the other variants. This is desired due to a better energy balance of the system. Owing to the fact that the MCT switch current is decreased by about 20 % compared

to the single device characterization, it is assumed that the MCT-based switch is less synchronous than the IGBT-based one.

V. CONCLUSION

Main result of our investigations is that it is possible to use IGBTs in compact pulsed power systems like the nitrogen gas laser which is not a standard up to now. In the target application it has been proved that the IGBT-based systems can reach and beat the performance of the MCT-based switches. A second important achievement is the expansion and the generalization of the results for a single device reported in former publications. Thus, a modular design process for other pulsed power applications is possible. The relevant design steps for the HV-cascade have been presented. Measurements show that the results from the equivalent energy approach can be verified in the cascaded switch. Further investigations have to be done in the IGC142T120-based cascade. Additionally, the usability of 1700 V-IGBTs in the cascades have to be investigated. They behave less dynamic in single device characterization than the 1200 V which will lead to a lower output energy of the laser. The advantage would be the reduced effort due to the less number of required stages. Hence, a comparison of the different cascades is possible and the best option for the target application can be chosen.

The 2014 International Power Electronics Conference

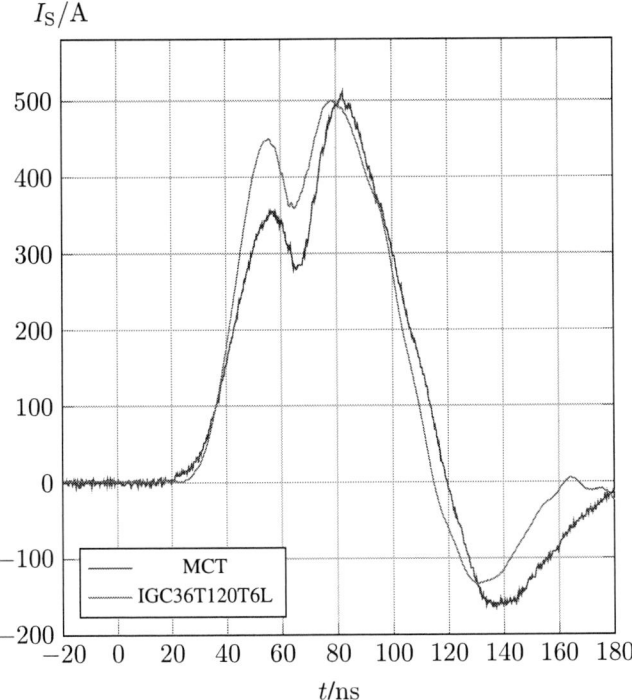

Fig. 15. Comparison of switching currents of IGC36T120 and MCT cascades at 11 kV

REFERENCES

[1] C. Strowitzki, M. Baumann, and P. Zacharias, "A novel solid state pulsed power module for excimer laser," in *27th International Power Modulator Symposium*, 2006, pp. 207–210.

[2] Y.-H. Chung and C.-S. Yang, "All solid-state switched pulser for nox control," in *Conference Record of the 2001 IEEE Industry Applications Conference*, vol. 4, 2001, pp. 2533–2540.

[3] H. B. Knight and L. Herbert, "The development of mercury-vapour thyratrons for radar modulator service," *Journal of the Institution of Electrical Engineers - Part IIIA: Radiolocation*, vol. 93, no. 5, pp. 949–962, 1946.

[4] W. Jiang, K. Yatsui, K. Takayama, M. Akemoto, E. Nakamura, N. Shimizu, A. Tokuchi, S. Rukin, V. Tarasenko, and A. Panchenko, "Compact solid-state switched pulsed power and its applications," *Proceedings of the IEEE*, vol. 92, no. 7, pp. 1180–1196, 2004.

[5] A. Kluge, T. Kramer, and H. Gueldner, "Gate boosting of mos-controlled devices for pulsed power applications," in *Power Electronics South America*, vol. 1, 2012, pp. 1–4.

[6] M. Giesselmann, B. Palmer, A. Neuber, and J. Donlon, "High voltage impulse generator using hv-igbts," in *Pulsed Power Conference, 2005 IEEE*, 2005, pp. 763–766.

[7] A. Kluge, H. Gueldner, and L. Goehler, "Characterization of igbts for high-speed switches for laser applications," in *Pulsed Power Conference (PPC), 2013 19th IEEE*, June 2013, pp. 1–7.

[8] J. Eichler and H.-J. Eichler, *Laser / Bauformen, Strahlfuehrung, Anwendungen*, 7th ed. Berlin und Heidelberg: Springer, 2010.

[9] Ng spice software and documentation. [Online]. Available: http://ngspice.sourceforge.net/

[10] F. Udrea, S. S. M. Chan, J. Thomson, S. Keller, G. Amaratunga, A. D. Millington, P. Waind, and D. E. Crees, "Development of the next generation of insulated gate bipolar tranistors based on trench technology," in *Solid-State Device Research Conference, 1997. Proceeding of the 27th European*, September 1997, pp. 504–507.

[11] M. März, S. Zeltner, and B. Eckardt, "Sind moderne leistungshalbleiter zu schnell? aspekte zum umgang mit schaltungsparasiten," in *Bauelemente der Leistungselektronik und ihre Anwendungen - 6. ETG-Fachtagung*, 2011, pp. 89–95.

978-1-4799-2706-7/14 $31.00 © 2014 IEEE

The 2014 International Power Electronics Conference

Comparative Suitability Evaluation of Reverse-Blocking IGBTs for Current-Source Based Converter

Ankan De, Sudhin Roy, Subhashish Bhattacharya
Electrical and Computer Engineering
North Carolina State University
Raleigh, NC, USA
ade@ncsu.edu, sroy3@ncsu.edu, sbhatta4@ncsu.edu

Abstract— **In this paper, a Reverse Blocking IGBT is compared with various other combinations of switches and diodes keeping Current Source based Converter application in mind. The devices are tested under Reverse Voltage Commutation, Switch Overlap (turn on at non-zero current but zero voltage), Hard Switching, and Zero Current Switching condition. Test Circuits have been constructed and tested at various voltage levels with various combinations of devices. Forward Characteristics have also been compared. The main motivation of the paper is to make a fair judgment on device selection for Partial Resonant Link AC/AC Converter (soft-switched) and Isolated Dynamic Current Converter (hard switched). As for soft switched, majority of the losses are caused due to conduction loss. Therefore, RB-IGBT leads to a huge reduction of losses owing to better forward characteristics (lower conduction loss) as compared to the rest of the set. SiC-MOS and SiC-JBS Diode combination showed significant loss reduction for hard switched converter.**

INTRODUCTION

Series Connection of Switch and Diode has long been used to realize current stiff converters. Though there are several advantages of these topologies, conduction losses owing to increased device count limits the maximum attainable efficiency. In this context, reverse blocking IGBTs (RB-IGBTs) may show better overall performance for current source rectifiers and inverters [1].

Over the last couple of decades a lot of work has been done on converters based on Soft Switching techniques [2-4]. The science behind the typical Zero Voltage Switching (ZVS) characteristics has been presented and an in-depth study has been shown [5-6]. As most available devices are designed for hard switching applications, very little data is available in the literature on device behavior under Current Switch (series connected switch and diode or RB-IGBTs) based soft switching conditions.

The application of current-switch converters such as the High Frequency Link inverter is being actively considered by many manufacturers [7-9]. As the current-switch technology matures, designers push the performance envelope for their circuits until the device once again becomes the limiting

factor. In this paper, an attempt has been made to demonstrate the behavior of several devices working under reverse voltage commutation, hard switched and switch overlap condition at various voltage and current levels. A test circuit has been built and tested with various series connected device combinations, viz. (a) Si-IGBT and Si-Diode, (b) Si-IGBT and SiC-JBS Diode, (c) SiC-MOSFET and SiC-JBS Diode, and (d) RB-IGBTs.

The forward characteristics of the devices were tested using a Curve Tracer. The switching and conduction loss data was then fed to a look up table based circuit simulator (PLECS). The topology under study is the Partial Resonant High Frequency Link Converter [8] and Isolated Dynamic Current Converter (Dyna-C) [9]. All the above mentioned switching characteristics can be witnessed in this converter. The main motivation of the paper is to find the best combination of devices to minimize losses and device stress.

REVERSE RECOVERY LOSS ANALYSIS

The Reverse Voltage Commutation of a current switch occurs when the voltage polarity of a conducting diode either naturally changes or is forcefully changed to negative. This forces the diode to stop conducting. It has specific advantage in current stiff converters as the conducting switch in series with the diode now does not need to be turned off. This is a preferred mode of soft turn off as it reduces loss in the converter.

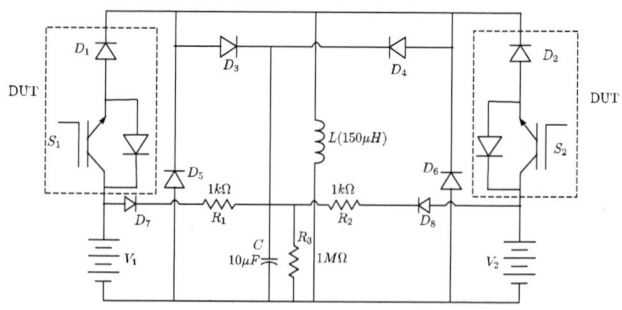

Figure 1: Test Circuit without the leakage inductances

978-1-4799-2706-7/14 $31.00 © 2014 IEEE

Fig. 1 shows the test circuit schematic. The circuit has been so chosen to execute several characteristic behaviors of a typical current switch. Fig. 2 shows possible realization of the current switch by series connection of switch and diode and also by RB-IGBT.

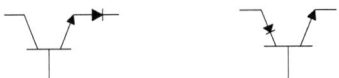

Figure 2: Possible realizations of a Current Switch: (left) realization with series connection of IGBTs and diodes, (right) realization with RB-IGBTs.

Fig. 3 shows the switching sequence of S1 (top) and S2 (bottom). At first the right hand side DC voltage (V2) is set to a higher value than the left hand side DC voltage (V1). The switch S1 is turned on first and the device (S1+D1) current starts to rise linearly. At time t=2µs, S2 is turned on. This makes the diode, D1 reverse biased and it naturally commutes off. This results in soft turn off of Si-IGBT at the low voltage side as the voltage across the switch remains close to zero throughout the operation. Fig. 4 shows the device (RB-IGBT) characteristics of the side with lower voltage.

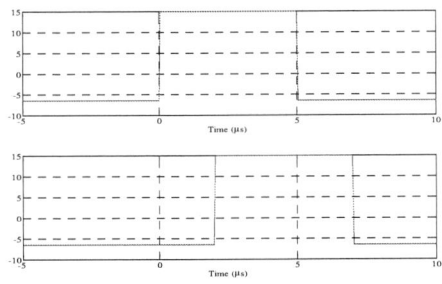

Figure 3: Switching sequence of S1 (top) and S2 (bottom)

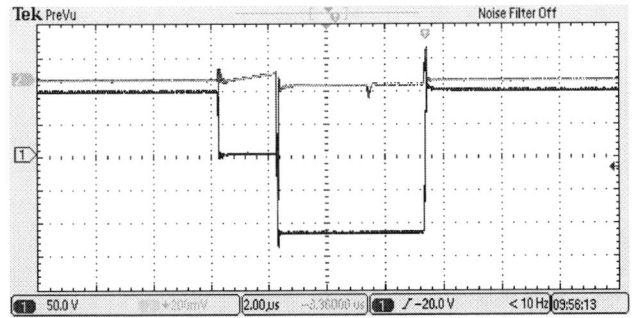

Figure 4: Device characteristics of the side with lower voltage (V1) for RB-IGBT (Test Case). Channel 1: Blocking Voltage of RB-IGBT, Channel2: Conducting Current (10mV/A)

The above experiment is repeated by fixing V2 at 800V and varying V1 from 100V to 700V. This is done to find a relation between voltage and energy loss incurred due to commutation. Fig. 5 shows the variation of reverse recovery current as V1 is varied. Fig. 6 shows the comparison of loss between Si-Diode, SiC-JBS Diode and RB-IGBT (working as a Diode). It shows superior performance of SiC-JBS Diode as

compared to the rest of the set as it has negligible reverse recovery current.

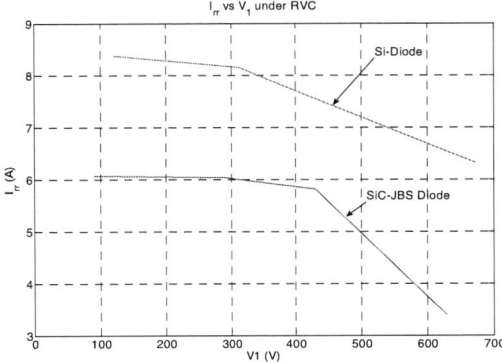

Figure 5: Comparison of Reverse Voltage Commutation of losses

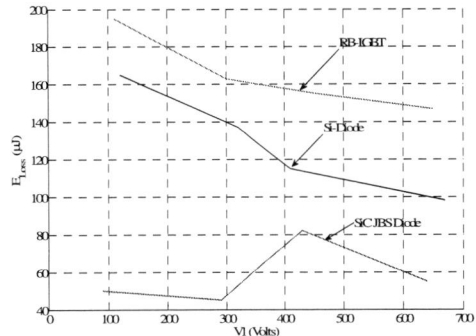

Figure 6: Comparison of Reverse Voltage Commutation of losses

SWITCH OVERLAP LOSS ANALYSIS

A new kind of switching characteristics is discovered with a modified switching scheme. A unique Voltage bump appears at non-zero turn-on current while transferring power from higher voltage to lower voltage. In this case the switch corresponding to higher voltage is turned on first. Consequently, the device current starts rising linearly and after some point of time, the switch corresponding to the lower voltage is turned on. In this case, no reverse voltage appears across the switch corresponding to higher voltage and therefore, it continues conducting. The conducting switch is then gated off and the non-conducting switch starts conducting. This is when a voltage bump appears across the IGBT corresponding to the leg with lower voltage. This system has been tested with Si-IGBT in series with SiC-JBS Diode.

Fig. 7 shows the typical characteristics depicting the voltage bump in Si-IGBT. The RB-IGBT has also been tested under the same condition and an embedded (to the reverse blocking voltage) voltage bump is noticed as shown in Fig. 8 below. Series connection of SiC-MOSFET and SiC-Diode does not witness this voltage bump [10]. Fig. 9 shows a typical behavior. The device has been tested at various voltage levels but this behavior does not appear in any case. This reduces

978-1-4799-2706-7/14 $31.00 © 2014 IEEE   2563

the corresponding loss and also avoids the voltage spike which might cause device failure.

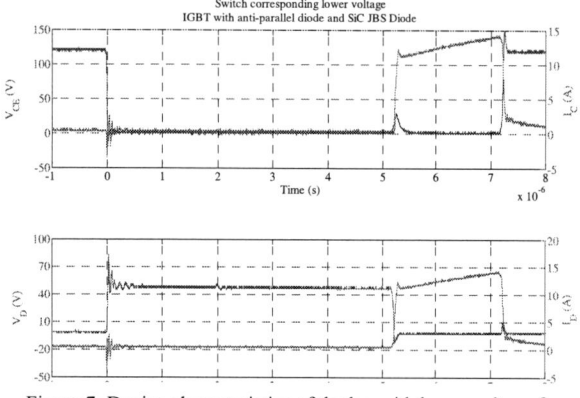

Figure 7: Device characteristics of the leg with lower voltage for Si-IGBT with SiC-Diode

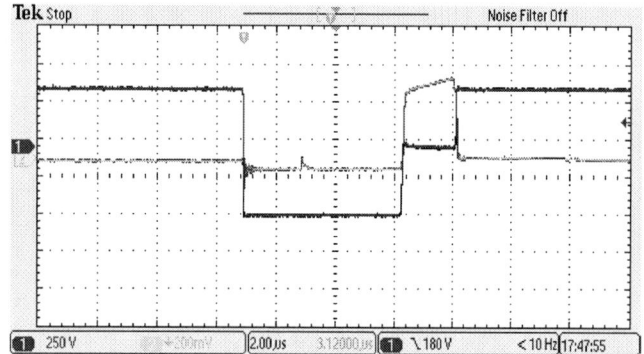

Figure 8: Device characteristics of the leg with lower voltage for RB-IGBT

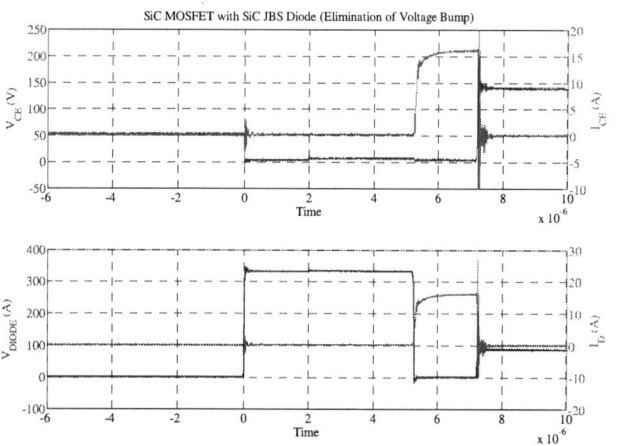

Figure 9: Device characteristics of the leg with lower voltage for SiC-MOS

Fig 10 and 11 show the variation of voltage bump peak and energy loss incurred.

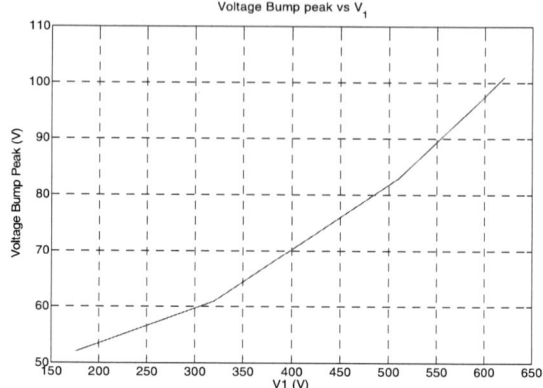

Figure 10: Si-IGBT voltage Bump (peak) under different voltages

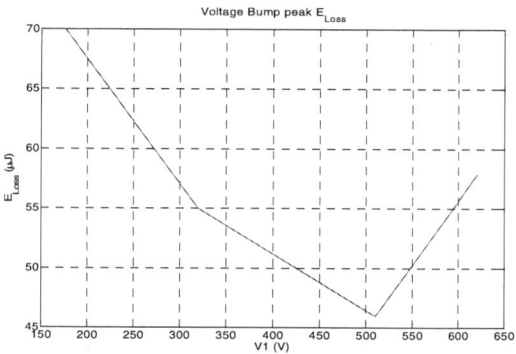

Figure 11: Energy Loss in Si-IGBT due to voltage Bump under different voltages

Fig 11 suggests that there exist minima in the energy loss. Even though the bump peak rises for increasing V1, the settling time reduces with increase in voltage. The overall energy loss depends on both the factors.

HARD SWITCH TEST

As mentioned in previous sections, most of the switching in current stiff converters is based on self/forced commutation. However, some instants may still have hard switch turn-on and off. To address this issue, the circuit has been tested under hard switched conditions. In this test, V2 is set to 800V and V1 is varied from 150 to 700 V. At first S1 is turned on and consequently, the inductor current rises linearly. At some point of time S2 is turned on making D1 reverse biased and as a result the current now starts flowing through the switch combination S2+D2. This results in hard turn on of the switch S2. After some point of time S2 is gated off. It brings hard turn off for S2. Fig. 12 shows the hard turn on and off characteristics of the understudied RB-IGBT.

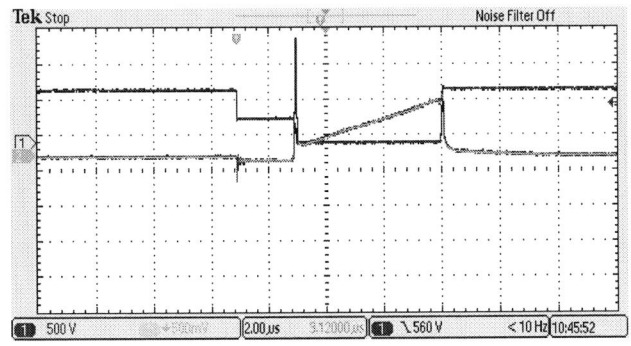

Figure 12: Hard Switch Test; (blue): Voltage across S2; (green): Current through S2. S2 has been realized by RB-IGBT

This test has been repeated for the remaining sets of devices. Fig. 13 and 14 show the comparison of turn on and off losses.

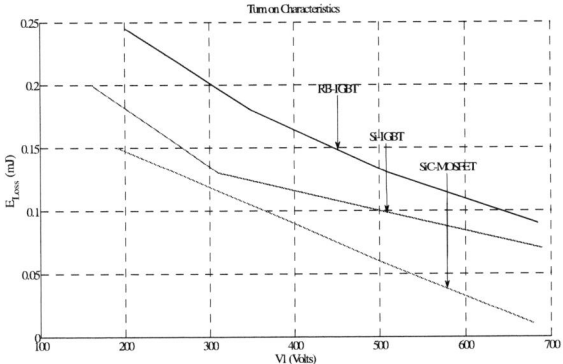

Figure 13: Hard Turn-On (left) Loss Comparison

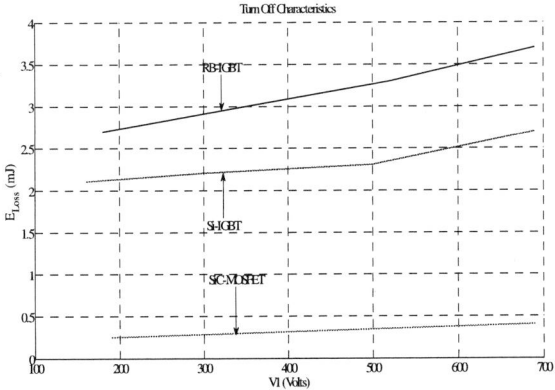

Figure 14: Hard Turn-Off Loss Comparison

The results show the benefit of using SiC-MOSFET as compared to Si-IGBT. It reduces both losses and duration of high device stress.

ZERO CURRENT SWITCHING ANALYSIS

The circuit is further modified to make it work under zero current switching condition. Fig. 15 shows the test circuit diagram.

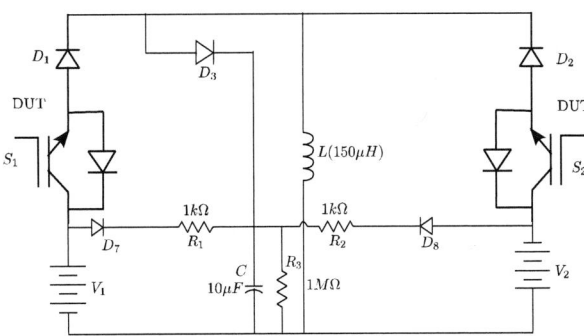

Figure 15: Modified circuit for ZCS Analysis

Here, V1 and V2 (placed in reverse orientation) are both set to a nominal voltage of 100V. The switch S1 is switched on at t=0. Thus, the current through the inductor (L) rises linearly. The switch S2 is turned on after a short interval. As D2 is reverse-biased, the right hand part of the circuit is still not conducting. The moment S1 is turned off, the non-zero current through L starts flowing through V2, S2 and D2. As the voltage across inductor has changed its polarity, now the current through inductor starts decreasing linearly. At one point of time, it goes to zero. Thus, turning off D2 which restricts negative current. S2 is naturally turned off at zero current depicting ZCS behavior. The loss incurred during this turn off is almost negligible. Fig. 16 shows the switch characteristics of D2 under ZCS.

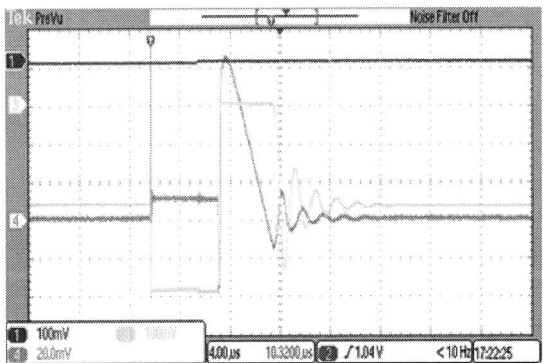

Figure 16: Switch Characteristics under ZCS (green: Voltage across diode ; Pink: Conducting Diode Current)

Figure 17: Hardware setup

It can be seen in Fig. 16, the diode blocking voltage shows some ringing. The ringing vanishes off when another pulse train is applied to the inductor. Thus, even though it exists, it does not destabilize the system. The negative current constitutes a reverse diode current, which actively removes diode stored charge. At some time later, the diode stored charge in the vicinity of the diode junction becomes zero, and the diode junction becomes reverse biased. The inductor current is now negative, and must flow through the diode leakage capacitor. The inductor and the capacitor then form a series resonant circuit, which rings with decaying sinusoidal waveforms as shown. This ringing is eventually damped out by the parasitic loss elements of the circuit, such as the inductor winding resistance, inductor core loss, and capacitor equivalent series resistance. Fig. 17 shows the hardware setup.

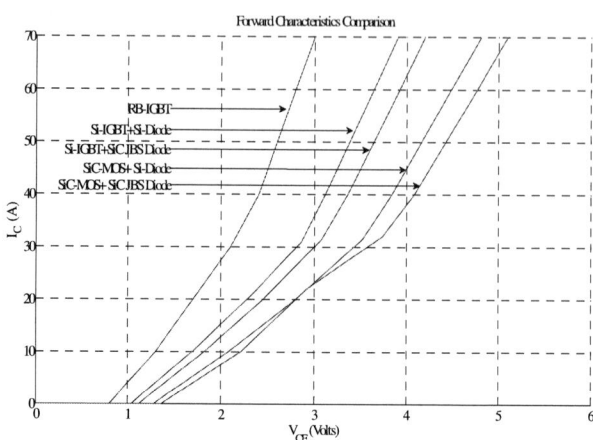

Figure 18: Forward Characteristics Comparison

FORWARD CHARACTERISTICS

In order to make a fair judgment on the overall device performance, it is essential to study the forward characteristics of the individual devices. The forward characteristics govern the overall conduction losses in a circuit. In this section, RB-IGBT is compared with Si-IGBT+Si-Diode, Si-IGBT+SiC JBS Diode, SiC-Mosfet+Si-Diode and SiC-Mosfet+ SiC JBS Diode. Fig. 18 shows the various forward characteristics of the above mentioned devices.

Interestingly it is observed that even though RB-IGBT has worse switching loss characteristics than both Si-IGBT and SiC-MOSFET, it still demonstrates better forward characteristics than all combinations of series connected switch and diodes. This gives it a unique advantage in soft switched converters where conduction loss dominates over the switching loss.

PARTIAL RESONANT LINK CONVERTER

The switching and conduction loss data is collected for each of the individual switches and fed to a look up table based circuit simulator (PLECS). The circuit under study is a simplified Partial-Resonant High Frequency Link Converter. Fig. 19 shows the schematic of the converter topology. Each leg of the converter consists of two series connected IGBT and Diode (or current switch). The converter operates by first charging the inductor from the input supply and then discharging it to the output grid. To draw harmonic-free power from the input, the ratio of the charge drawn from each input phase must be equal to the ratio of the sinusoidal reference input currents. In order to ensure this, the current references for both input and output sides of the converter are needed. Starting from the input side, the link is connected to the input lines having the highest line-line voltage by turning on the appropriate switches to charge it in the positive direction.

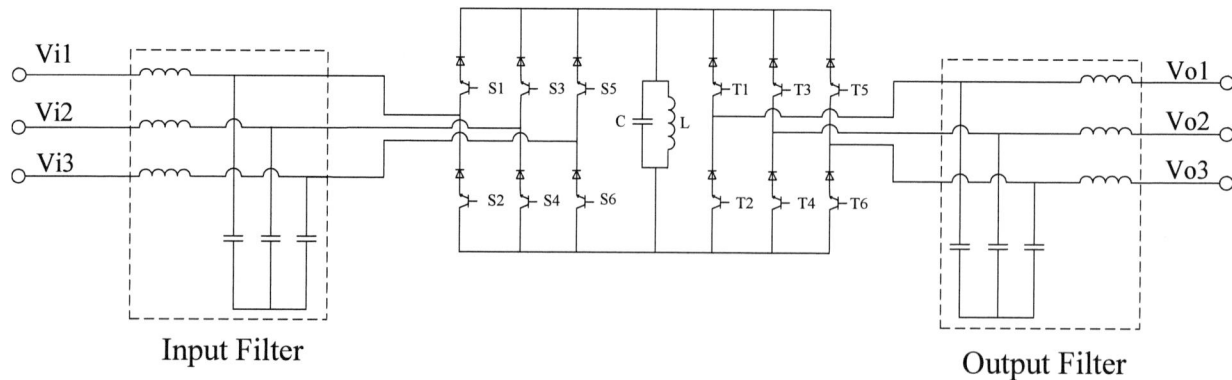

Figure 19: Simplified version of Partial Resonant High Frequency Link Converter

The link is charged till the average value of one of the line currents is equal to the reference set by the controller. The switches are then turned off and the link partially resonates till the link voltage matches with the voltage of the input pair which has the second highest value. At this point, the switches are turned on and the link continues to get charged from the input section. These results in both zero-voltage turn off and turn on, thereby reducing the losses. This is continued till the particular line current is equal to the reference current and then the switches are turned off. The link resonates till the link voltage is equal to the negative of the second highest line voltage of the output grid. The switches are then selected to apply a negative voltage of the link, thereby discharging the stored energy from the link to the output grid. This is continued till the reference current is equal to the output side line current and then the switches are turned off. The link now partially resonates till the link voltage becomes equal to the negative voltage of the output pair which has the highest value. Corresponding switches are turned on and the link continues to discharge. This is continued till the line current is equal to the reference set by the controller. These steps are repeated thereby transferring power from the input supply to the output grid. For galvanic isolation, the link inductor can be replaced with a transformer which can withstand a non-zero average current. A look-up table (LUT) based approach can also be used to estimate average inductor current and appropriate switch-timing using the input and output voltage of the link, thereby reducing the burden on the controller. Fig. 29 shows one such LUT used in PLECS software.

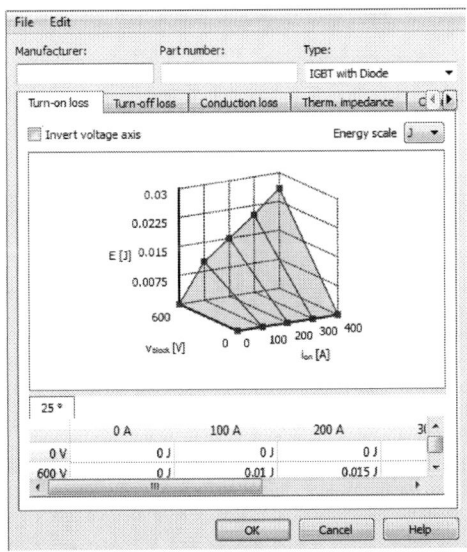

Figure 29: PLECS Look-UP Table for Turn-On, Turn-Off and Conduction Loss Computation.

The system has been simulated for 100kW, 480V, 3 phase system. The switching frequency is set to 20kHz. The estimated peak inductor current is 400A.

The following kind of switching/conduction loss takes place in this converter:-
1) Switch Overlap Loss (in Si-IGBT);
2) Soft Turn off loss (in Si-IGBT);
3) Conduction Loss.

As the peak current requirement is higher than the current rating of the devices, appropriate number of individual devices is connected in parallel to support the current. In order to get realistic values of forward characteristics, each of the devices were tested using a Conduction Loss Curve Tracer. The tracer provided the forward characteristics of each of the devices. The various switching and conduction loss data collected from the switching loss and forward characteristic test has been tabulated and fed to a look-up table in PLECS.

Table 1: Switching and Conduction Loss break-up

Loss	Si-IGBT + Si-Diode	Si-IGBT + SiC JBS Diode	SiC MOS + Si-Diode	SiC MOS + SiC JBS Diode	RB-IGBT
Conduction	1565 W	1718 W	1591 W	1743 W	960 W
Switching	679 W	603 W	267 W	197 W	792 W
Total	2244 W	2321 W	1858 W	1940 W	1752 W

Tab. 1 shows the switching and conduction losses for various sets of devices. As expected, SiC-MOS shows significant reduction of switching loss. Among the chosen devices, RB-IGBT shows better forward characteristics and thus contributed to lower conduction loss. Even though RB-IGBT witnessed considerably high switching losses, it resulted in the best overall performance. This shows the effective selection of RB-IGBT in soft switched converters where conduction loss dominates over the other losses.

ISOLATED DYNAMIC CURRENT ("DYNA-C") CONVERTER

This converter consists of two cascaded full bridge inverters and one high frequency link transformer. The transformer provides electrical isolation between the sources [10]. It also allows power flow between the sources with unequal voltages. As shown in Fig. 30, the converter consists of a clamp circuit which restricts the inverter voltage to rise beyond a specific limit. The two cascaded converters are realized using an IGBT in series with a diode or a RB-IGBT.

In this converter, at first the phases with maximum voltage are connected to the transformer first (but turning on the appropriate switches). The turn on duration is set such that the average current in the phases meet the set reference. Then the phases with second highest voltage are turned on till the phase currents meet the desired reference. After this stage of operation, similar sets of operation are conducted to discharge the transferred power from the transformer to the output grid. Through repetitive cycles of the above mentioned operation, charge can be extracted from the power source and injected to the output.

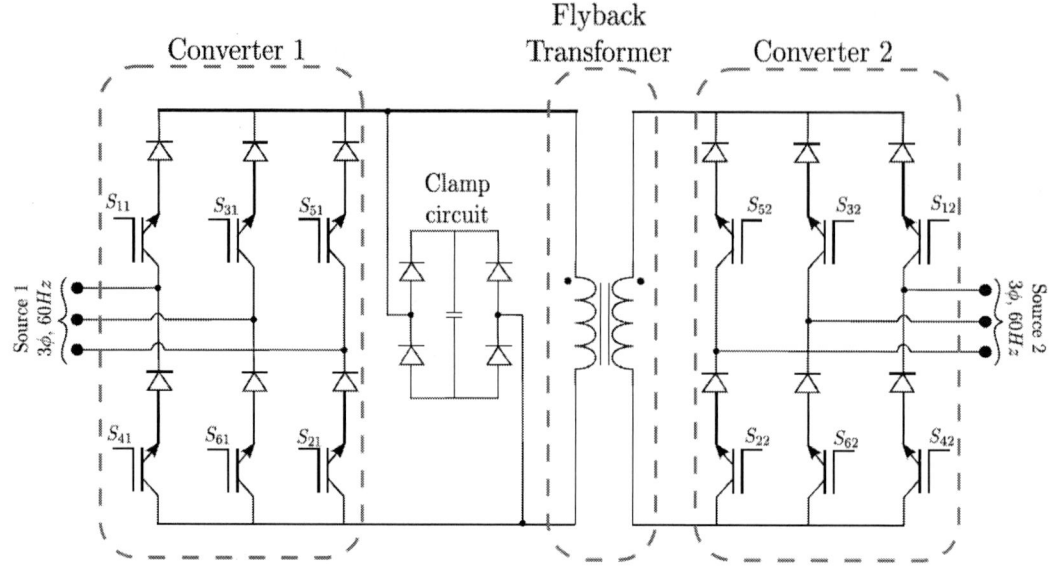

Figure 30: Circuit Schematic of Dyna-C Converter

Identical to the system mentioned in previous section, this converter has been simulated for 100kW, 480V, and 3 phase system. The switching frequency is set to 20 kHz. The estimated peak inductor current is 400A. The following kind of switching/conduction loss takes place in this converter:-

1) Switch Overlap Loss (in Si-IGBT);
2) Reverse Voltage Commutation;
3) Hard Turn on/off;
3) Conduction Loss.

Table 2: Switching and Conduction Loss break-up

Loss	Si-IGBT + Si-Diode	Si-IGBT + SiC JBS Diode	SiC MOS + Si-Diode	SiC MOS + SiC JBS Diode	RB-IGBT
Conduction	1436 W	1668 W	1505 W	1713 W	924 W
Switching	1544 W	1242 W	764 W	496 W	1803 W
Total	2980 W	2910 W	2269 W	2209 W	2727 W

Tab. 2 shows the switching and conduction losses for various sets of devices. As expected, SiC-MOS shows significant reduction of switching loss. This shows the effective selection of SiC-Mos in hard-switched converters where switching loss dominates over the other losses.

CONCLUSION

This paper presents comparison of diode reverse recovery loss of Si-Diode, SiC-JBS diode and RB-IGBT (acting as a diode). SiC-JBS diode shows a considerable reduction of loss. A new form of switching characteristics is reported for Si-IGBT. This switching behavior is not noticed when the same experiment is conducted with SiC-MOS, thus reducing the effective loss. In order to make sense out of the various characterizations, a High Frequency Link converter is simulated with the loss data collected from the various experiments. RB-IGBT though having high switching losses, shows impressive reduction of overall losses. To demonstrate performance of current switch for hard switched converter, SiC-MOS+ SiC-JBS Diode showed exceptional performance for this topology.

ACKNOWLEDGMENT

This work made use of FREEDM ERC shared facilities supported by National Science Foundation under Award Number EEC-0812121.

References

[1] M. Otsuki and Y. Seki, "The authentic reverse blocking igbt (rb-igbt) technologies for bi-directional switching applications," in PCIM Brasil Conference, 2012.

[2] G. L. Skibinski and D. M. Divan, "Characterization of Power Transistors under Zero Voltage Switching", IEEEIAS Annual Conference Records, 1987, pp. 493-503.

[3] G. L. Skibinski and D. M. Divan, "Characterization of GTOs for Soft Switching Applications", IEEE-IAS Annual Conference Records, 1988. pp. 638-645.

[4] G. Venkataramanan, A. Mertens and H-Ch. Skudelny, "Switching Charactersitics of Field Controlled Thyristors", Conference Record of EPE91, pp. 0:220-0:225. [SI Astrid Petterteig and T. Rogne. "IGBT turn-off losses - in Hard Switching and with Capacitive Snubber" Conference Record of EPE-MADEP91, pp. 0:203-0:208.

[5] A.Kurnia, O.H. Stielau; G. Venkataramanan ; D.M. Divan, "Loss mechanisms in IGBTs under zero voltage switching", Power Electronics Specialists Conference, 1992. PESC '92 Record., 23rd Annual IEEE, vol.2, pp. 1011-1017.

[6] Kurnia A., Cherradi H. and Divan D.M., "Impact of IGBT behavior on design optimization of soft switching inverter topologies," IEEE Transactions on Industry Applications, vol. 31 , issue: 2, March-April 1995.

[7] P.K. Sood, T.A. Lipo, and I.G. Hansen, "A versatile power converter for high-frequency link systems," Power Electronics, IEEE Transactions on, vol. 3, no. 4, pp. 383-390, Oct 1988.

[8] A. Balakrishnan, H.A. Toliyat, and W. Alexander, "A Novel High Frequency AC Link Soft-Switched Buck-Boost Converter," IEEE Proc. of APEC '08, Austin, TX, 2008.

[9] US Patent Publication # US20130201733 A1, Aug 8, 2013. "Isolated dynamic current converters", by Deepakraj M. Divan, Anish Prasai, Hao Chen.

[10] De, A.; Roy, S.; Bhattacharya, S.; Divan, D. M., "Performance Analysis and Characterization of Current Switch under Reverse Voltage Commutation, Overlap Voltage Bump and Zero Current Switching", Proceedings of the 28th Applied Power Electronics Conference and Exposition (APEC 2013), Long Beach, California, USA, March 17-21, 2013.

New Reverse-Conducting IGBT (1200V) with Revolutionary Compact Package

K. Takahashi, S. Yoshida, S. Noguchi, H. Kuribayashi, N. Nashida,
Y. Kobayashi, H. Kobayashi, K. Mochizuki, Y. Ikeda and O. Ikawa
Fuji Electric Co. Ltd.
4-18-1 Tsukama, Matsumoto, Nagano, Japan 390-0821
takahashi-kouta@fujielectric.co.jp

Abstract- **Fuji Electric developed a 1200V class RC-IGBT based on our latest thin wafer process. The performance of this RC-IGBT shows the same relationship between conduction loss and switching loss as our 6th generation conventional IGBT and FWD. In addition its trade-off can be optimized for hard switching by lifetime killer. Calculations of the hard switching inverter loss and chip junction temperature (Tj) show that the optimized RC-IGBT can handle 35% larger current density per chip area. In order to utilize the high performance characteristics of the RC-IGBT, we assembled them in our newly developed compact package. This module can handle 58% higher current than conventional 100A modules at a 51% smaller footprint.**

Keywords—RC-IGBT, Reverse Conducting IGBT, hard switching inverter, moulding package

I. INTRODUCTION

Recently, IGBT modules are widely applied in industrial, consumers, hybrid vehicle or renewable energy applications with a main focus on energy savings. In most of the applications customers are in need for more compact IGBT modules to reduce the cost and size of their product. However, while the size of IGBT module becomes smaller the power density and operating temperature of IGBT module are increasing. These trends have negative effects on the reliability and lifetime of IGBT module. In order to enable compact IGBT module with high reliability, chip and/or package innovation are strongly desired. For this purpose, we investigated Reverse Conducting IGBT (RC-IGBT) and applied our newly developed compact package.

In this paper we will explain the functionality and characteristics of our RC-IGBT, which is a type of IGBT having FWD functions at the same time. By using RC-IGBTs FWD chips can be eliminated which leads to more efficient use of available mounting area.

Many researchers have reported RC-IGBT [1]-[4] in the past. Commonly the applications are restricted mainly to soft switching conditions because of large recovery current and switching loss. As a matter of fact most present inverter used hard switching conditions for motion control or power conversion. This is the reason why until now RC-IGBTs had been rarely applied in such applications. In this report, we will show a RC-IGBT designed for hard switching, and demonstrate that RC-IGBT has great potential for motion control and power conversion applications.

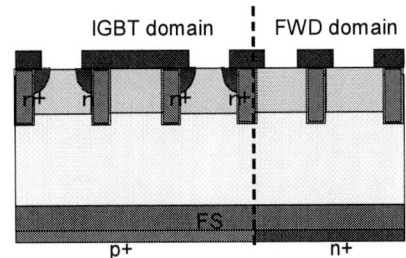

Fig. 1. The cross section diagram of the developed RC-IGBT

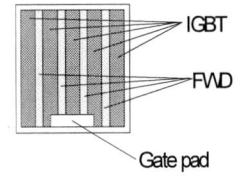

Fig. 2. The top view diagram of the developed RC-IGBT

The cross section diagram of the RC-IGBT is shown in fig. 1, and the top view diagram is shown in fig. 2. As shown in fig. 1 the RC-IGBT is trench-gate field-stop (FS) thin wafer type IGBT. The gate structure and chip thickness of the RC-IGBT are the same as our latest conventional IGBT, 6th generation IGBT (V-Series) [5]. As shown in fig. 2 IGBT- and FWD-domains of the RC-IGBT are arranged in a stripe pattern. The fabrication process is almost the same as our 6th gen. IGBT. Additional processes of the RC-IGBT production are a lifetime control process and a backside photo-etching ion-implantation process for the formation of backside P/N structure.

Finally the RC-IGBT chip was mounted in our newly developed compact package that has low thermal impedance and high reliability [6]-[9].

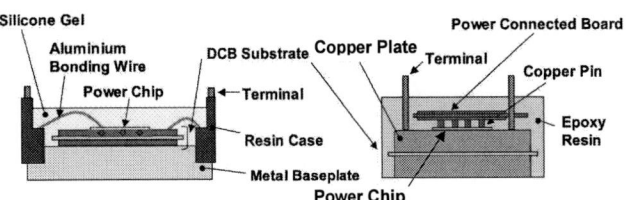

Fig. 3. Comparison between conventional package structure (left) and our new revolutionary package structure (right).

A comparison between conventional package (left) and our new package (right) is shown in fig. 3.

(1) Compact;
Conventional package uses Aluminum wire bonding and a circuit pattern on the DCB substrate for the connection between chips and terminals. These methods need a certain amount of area for the wire bonding and the circuit. In our new package, copper pins are used instead of wire bonding and a power connected board instead of a circuit pattern on the DCB substrate. Copper pins are soldered vertically on the chips and the power connected board is arranged just above the chips. This methodology eliminates space for wire bonding and circuit pattern on the DCB substrate. Consequently the footprint of our new package is less than half of conventional packages as shown in fig. 4.

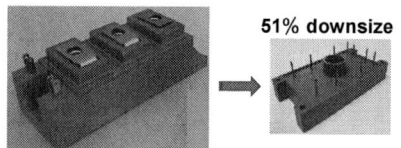

Fig. 4. Footprint comparison of 1200V/100A module

(2) Low thermal impedance;
Our new structure uses Si_3N_4 for insulating material, which thermal conductivity is less than half of conventional materials like Al_2O_3. In addition our new structure uses a thick copper plate on both sides of ceramic substrate. These copper plates effectively spread the heat from the chips to the cooling fin. As a result the thermal impedance of our new structure is less than half compared to conventional structure.

(3) High reliability;

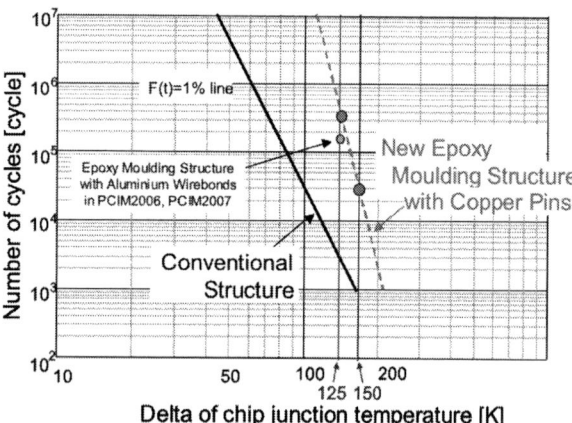

Fig. 5. Comparison of power cycling capability

Regarding the reliability of this IGBT module, power cycling capability has the biggest impact. In the power cycling test, the wire bonding on the chip and the solder layer between chips and the DCB substrate are the weak points of the conventional structure. As explained above our new structure eliminates wire bonding by using copper pins thus the power cycling capability of our new package strongly depends on the solder layer lifetime. Epoxy resin is used as encapsulating material instead of silicone gel in our new package. In the power cycling test, epoxy resin causes the effect of thermal stress dispersion at the solder layer due to its rigidity. Consequently the concentration of thermal stress at the solder layer is greatly reduced which leads to long lifetime of the solder layer. As described we reinforced power cycling capabilities by our new package technology significantly as shown in fig. 5.

II. CHARACTERISTICS OF THE NEW RC-IGBT

The active area and the chip size of the new RC-IGBT are shown in table 1. The reference 6th gen. IGBT and FWD are 1200V/100A rating chips. The active area of the RC-IGBT consists of a sum of IGBT region areas and FWD region areas as shown in Fig. 2. The active area of the RC-IGBT is the same as the 6th gen. IGBT active area and the 6th Gen. FWD active area. Even though the active area is the same, the chip size of the RC-IGBT is 9.4% smaller compared to the 6th Gen. IGBT chip size and the 6th Gen. FWD chip size. This is a result of the fact that the conventional IGBT and FWD need two edge regions for IGBT and FWD, where the RC-IGBT needs only one edge region.

Table 1. Chip size and thermal impedance of the developed RC-IGBT

chip type	6th gen. IGBT/FWD (total)	RC-IGBT
active area (a.u)	0.64 / 0.36 (1.00)	1.00
chip size (a.u.)	0.62 / 0.38 (1.00)	0.91
Thermal resistance Rth(jc) [K/W]	0.24 / 0.40	0.15 / 0.15 (IGBT/FWD)

*Rth(jc) of Conventional package

The thermal resistance between chip and case (Rth(jc)) are shown in table 1 considering that each chip is assembled in a conventional package. The Rth(jc) is 38% lower than that of the conventional IGBT. Nevertheless it is very similar to Rth(jc) of other conventional IGBTs which active area is the same as the RC-IGBT. The reason why the Rth(jc) of the RC-IGBT is so similar to the same size conventional IGBT is clear seeing fig. 2. IGBT regions and FWD regions of the RC-IGBT are arranged alternately and the distance of two adjacent regions is less than 0.3mm. Therefore the temperature of FWD region is nearly equal to that of IGBT region. As a result FWD regions contribute to heat radiation as much as IGBT regions when current flows through IGBT regions. The active area for heat radiation of RC-IGBTs consists of not only IGBT regions but also FWD regions. Consequently Rth(jc) of RC-IGBTs is very similar to same size conventional IGBTs.

The 2014 International Power Electronics Conference

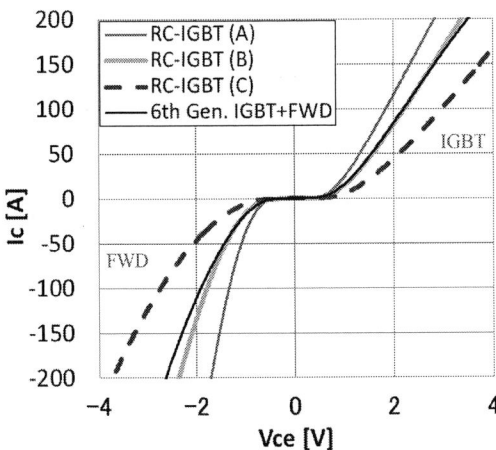

Fig. 6. Output characteristics of the developed RC-IGBT
Condition; Tj=150°C, Vg=15V(Ic>=0), Vge=-15V(Ic<0).

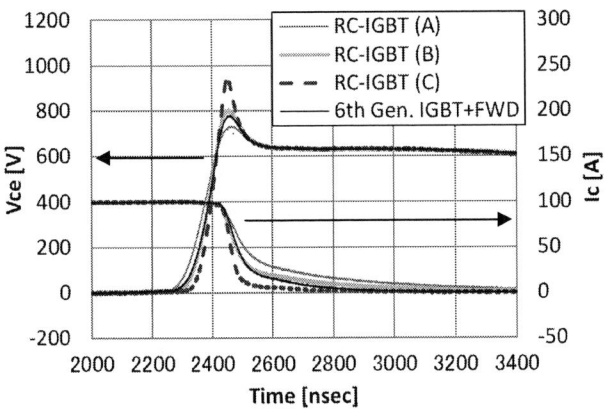

Fig. 7. Turn-off waveform of the developed RC-IGBT
Condition; Tj=150°C, Vcc=600V, Ic=100A,
Vge=+15V/-15V, Rg=9Ω
*Rg includes chip internal resistance.

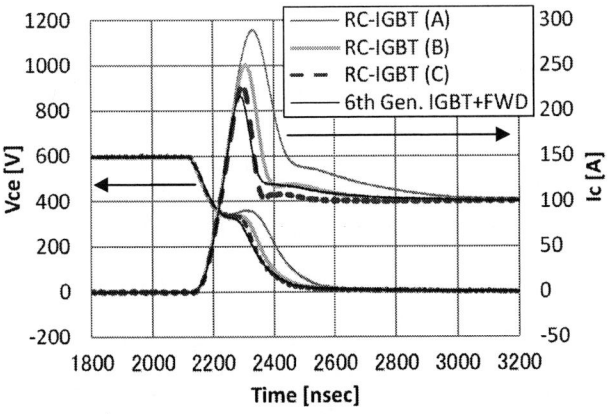

Fig. 8. Turn-on waveform of the developed RC-IGBT
Condition; Tj=150°C, Vcc=600V, Ic=100A,
Vge=+15V/-15V, Rg=9Ω
*Rg includes chip internal resistance.

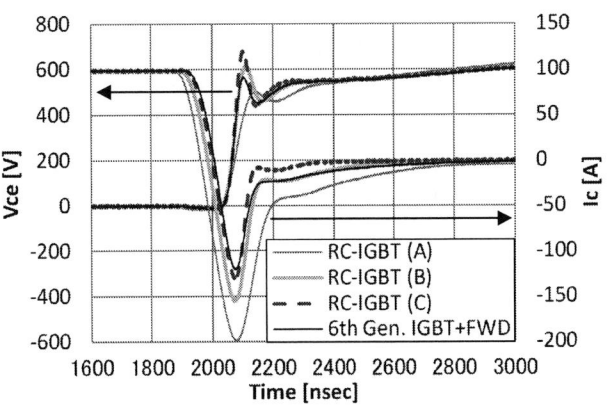

Fig. 9. Reverse recovery waveform of the developed RC-IGBT
Condition; Tj=150°C, Vcc=600V, Ic=100A,
Vge=+15V/-15V, Rg=9Ω
*Rg includes chip internal resistance.

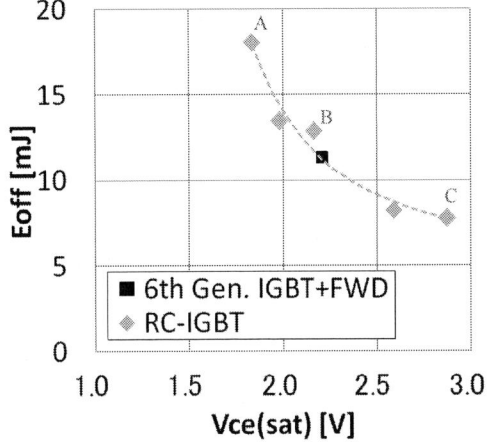

Fig. 10. IGBT trade-off characteristic of the developed RC-IGBT
Condition (Vce(sat)); Tj=150°C, Ic=100A, Vge=15V
Condition (Eoff); Tj=150°C, Vcc=600V, Ic=100A,
Vge=+15V/-15V, Rg=9Ω
*Rg includes chip internal resistance.

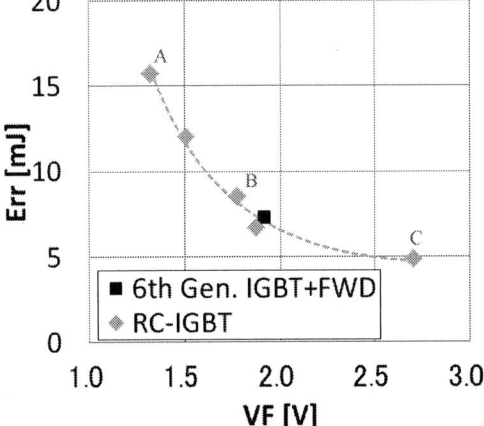

Fig. 11. FWD trade-off characteristic of the developed RC-IGBT
Condition (VF); Tj=150°C, Ic=100A, Vge=-15V
Condition (Err); Tj=150°C, Vcc=600V, Ic=100A,
Vge=+15V/-15V, Rg=9Ω
*Rg includes chip internal resistance.

978-1-4799-2706-7/14 $31.00 © 2014 IEEE

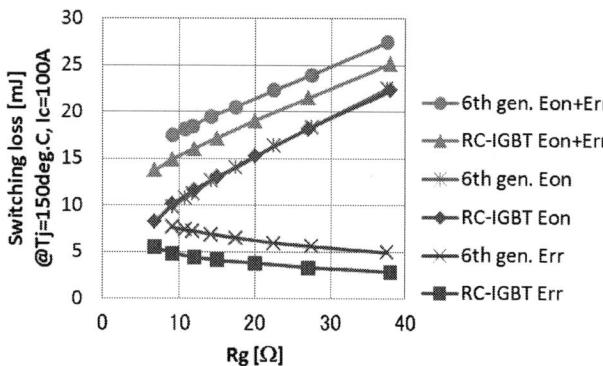

Fig. 12. Rg dependence of switching loss of the developed RC-IGBT
Condition; Tj=150°C, Vcc=600V, Ic=100A, Vge=+15V/-15V
*Rg includes chip internal resistance.

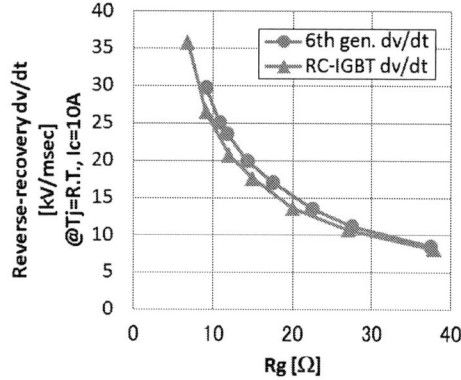

Fig. 13. Rg dependence of dv/dt of the developed RC-IGBT
Condition; Tj=R.T., Vcc=600V, Ic=10A, Vge=+15V/-15V
*Rg includes chip internal resistance.

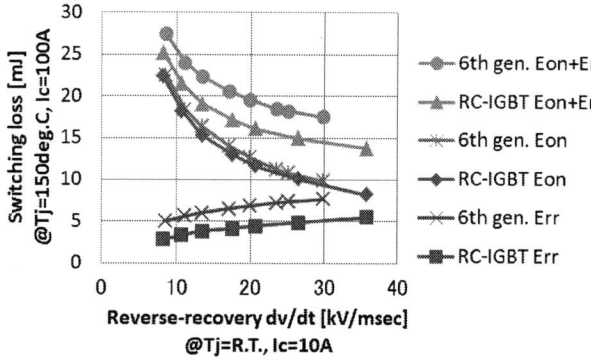

Fig. 14. dv/dt dependence of switching loss of the developed RC-IGBT
Condition; Vcc=600V, Vge=+15V/-15V
*Rg includes chip internal resistance.

The output characteristics of the RC-IGBT and the conventional IGBT and FWD are shown in Fig. 6. The RC-IGBTs have both IGBT and FWD output characteristics at the same time. The Vce(sat) and VF of the RC-IGBT are shifted so widely from sample A to C by lifetime killer.

The turn-off, turn-on and reverse recovery waveforms of the RC-IGBTs are shown in fig. 7, 8 and 9 respectively. The turn-off tail current and the reverse recovery current of the RC-IGBT are changed so widely from sample A to C by lifetime killer. Since the RC-IGBT has the same gate structure as the conventional IGBT, the waveform of sample B is very similar to the conventional IGBT and FWD.

The IGBT and FWD trade-off characteristics of the RC-IGBT are shown in fig. 10, 11 respectively. The IGBT and FWD trade-off characteristics of the RC-IGBT are also as good as those of the conventional IGBT and FWD. No deterioration by the integration of IGBT and FWD can be found. The RC-IGBT trade-off point can be moved by lifetime killer so widely along the trade-off line thus the trade-off characteristics of RC-IGBTs can be optimized for hard switching inverter drive condition, carrier frequency and/or output current.

Furthermore the gate resistance (Rg) dependence of the switching loss and the reverse recovery dv/dt of the RC-IGBT are shown in fig. 12, 13 respectively. It is widely known that the small current switching behavior is strongly affects Electro-Magnetic Interference (EMI) noise. Especially the small current reverse recovery dv/dt has great influence to EMI noise [10]. Accordingly in order to reduce the EMI noise the small current reverse recovery dv/dt needs to be reduced. The switching loss and the dv/dt can be well controlled by adjustment of the Rg value. Generally the bigger the Rg value is, the smaller the dv/dt will be. However at the same time larger Rg values result in larger switching losses. In short, the switching loss and the dv/dt have a very close trade-off relationship and the trade-off between the dv/dt and the switching loss are the most important characteristics of IGBTs. That trade-off characteristic of the RC-IGBT is shown in Fig. 14. This figure shows again that RC-IGBT has the same trade-off between the dv/dt and the switching loss as conventional IGBTs and FWDs. Considering these dv/dt control characteristics, the RC-IGBT is as good as our conventional IGBTs and FWDs.

III. INVERTER LOSS AND CHIP TEMPERATURE OF OUR NEWLY DEVELOPED RC-IGBT

To confirm the performance of the RC-IGBT we calculate the hard switching inverter loss and the IGBT chip temperature (Tj) of the RC-IGBT assuming that the reverse recovery dv/dt is less than 10kV/μsec. Since these are common conditions in motion control inverter drive. The loss characteristics of sample C were used. The calculation result is shown in Fig. 15. The inverter losses of the RC-IGBT are smaller compared to conventional IGBTs and FWDs. Consequently it is demonstrated that the RC-IGBT has a great possibility to replace conventional IGBTs and FWDs in hard switching inverter applications. On the top of that the Tj of the RC-IGBT is also much smaller than that of the conventional IGBTs and FWDs. This is caused a much smaller Rth(jc) than that of the conventional IGBTs. Since the Tj of the RC-IGBT is much smaller, consequently the RC-IGBT can be downsized. Fig. 15 also contains the calculation result of another RC-IGBT which active area is 21% downsized. As a result, the inverter loss and the Tj of the downsized RC-IGBT are very close to that of the conventional IGBTs and FWDs. In other words, the chip size of the downsized RC-IGBT is 74% of that of the conventional IGBT and FWD as shown in Table 2, and therefore the current density of the RC-IGBT is 35% higher than that of the conventional IGBT and FWD. As described, it can be said that the RC-IGBT has great possibility to realize module downsizing.

Table 2. The active area and the chip size of two types of RC-IGBTs

Chip type	6th gen. IGBT/FWD (total)	RC-IGBT	RC-IGBT (downsized)
active area (a.u)	0.64 / 0.36 (1.00)	1.00	0.79
chip size (a.u.)	0.62 / 0.38 (1.00)	0.91	0.74
Thermal resistance Rth(jc) [K/W]	0.24 / 0.40	0.15 / 0.15 (IGBT/FWD)	0.20 / 0.20 (IGBT/FWD)
Current density per chip size (a.u.)	1.00	1.10	1.35

*Rth(jc) of Conventional package

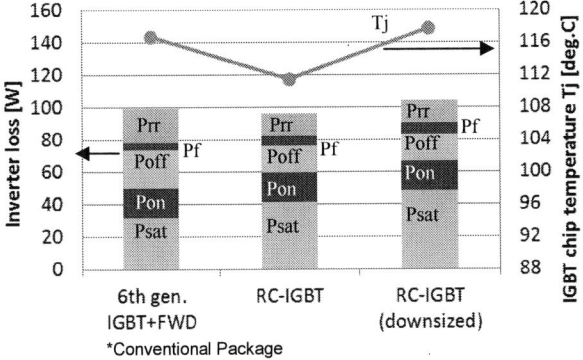

Fig. 15. Calculated Inverter loss and IGBT chip temperature of the developed RC-IGBT
Condition ;Fo=60Hz, Fc=8kHz, Vcc=600V, Iout=50Arms, Cosφ=0.9,
Modulation rate=1.0, Reverse recovery dv/dt <=10kV/µsec
Cooling fin temperature=90°C,
Thermal compound; 2W/mK , thickness=50µm,
Rth(jc); see table 2

IV. THE RC-IGBT WITH OUR NEWLY DEVELOPED COMPACT PACKAGE

Finally we investigated the combination of the RC-IGBT and our revolutionary compact package. Table 3 shows that only a downsized version of our RC-IGBT can be assembled in our new compact package considering the chip size.

The characteristics of the combination of the RC-IGBT and our new package are shown in table 3.

The module footprint of our new package is 51% less than that of the conventional package and the chip size of the downsized RC-IGBT is 26% smaller than that of the conventional IGBT and FWD. The Rth(jc) of the downsized RC-IGBT assembled in our new package is 62% lower than that of the conventional IGBT assembled in conventional package.

Table 3. Characteristics of the combination of the RC-IGBT and our new package

Package type	Conventional package			New Package
Chip type	6th gen. IGBT+FWD	RC-IGBT	RC-IGBT (downsized)	
module footprint [cm2]		31.3		15.5
Ratio		1.00		0.49
chip size (IGBT/FWD)[a.u.]	0.62/0.38 (total 1.00)	0.90	0.74	
Thermal resistance Rth(jc) (IGBT/FWD)[K/W]	0.24/0.40	0.15/0.15	0.20/0.20	0.09/0.09

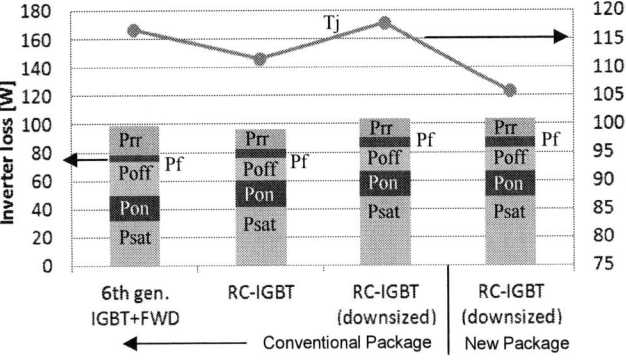

Fig. 16. Calculated Inverter loss and IGBT chip temperature of the developed RC-IGBT using our revolutionary package
Condition ;Fo=60Hz, Fc=8kHz, Vcc=600V, Iout=50Arms, Cosφ=0.9,
Modulation rate=1.0, Reverse recovery dv/dt <=10kV/µsec
Cooling fin temperature=90°C,
Thermal compound; 2W/mK , thickness=50µm,
Rth(jc); see table 3

The calculated inverter loss and the Tj of the downsized RC-IGBT using our revolutionary packaging technology are shown in fig. 16. The inverter loss of this advanced module is close to that of conventional IGBT and FWD in conventional packages with a 100A/1200V rating and the Tj of our innovative module is the lowest in all combinations of fig. 16 and is more than 11°C lower compared to that of the conventional 100A module. Another inverter loss calculation, assuming that the Tj of our advanced module must be less than 116.8°C which is the Tj of the conventional 100A module, our advanced module can handle Io=78.9Arms. This is 58% higher current than the conventional 100A module can handle realized at only half footprint. Besides the power cycling capability of our advanced module is over 100 times larger than that of conventional modules assuming the same ΔTj. As a conclusion it can be said that the combination of the RC-IGBT and our revolutionary package is promising candidate for the downsizing of IGBT module and as a result complete inverter systems.

V. CONCLUSIONS

In this report we described our newly developed 1200V class RC-IGBT which is realized by applying our latest thin wafer process. This RC-IGBT has the same relationship between conduction loss and switching loss as our 6th generation (V-Series) IGBT and FWD. In addition its trade-off can be optimized for hard switching by lifetime killer. On top of all that calculating the hard switching inverter loss and the chip junction temperature (Tj) the calculation result shows that the optimized RC-IGBT can handle 35% larger current density per chip area. Finally the assembly of the RC-IGBT in our revolutionary compact package was described. That module can handle 58% higher current than the conventional 100A module at a 51% smaller footprint.

ACKNOWLEDGMENT

The authors would like to thank Mr. Tamenori, Mr. Yoshimura, Mr. Kajiwara, and Ms. Kodama for their technical supports during device fabrication and their useful advices.

REFERENCES

[1] H. Takahashi et al "1200V Reverse Conducting IGBT", pp. 133-136 Proceeding of ISPSD 2004

[2] K. Satoh et al "A New 3A/600V Transfer Mold IPM with RC (Reverse Conducting)-IGBT", Proceeding of PCIM Europe 2006

[3] H. Rüthing et al "600V Reverse Conducting (RC-)IGBT for Drives Application in Ultra-Thin Wafer Technology", pp. 89-92 Proceeding of ISPSD 2007

[4] S. Voss "Anode Design Variation in 1200-V Trench Field-stop Reverse-conducting IGBTs", pp. 169-172 Proceeding of ISPSD 2008

[5] Y. Onozawa et al "Development of the next generation 1200V trench-gate FS-IGBT featuring lower EMI noise and lower switching loss", pp. 13-16 Proceeding of ISPSD 2007

[6] M. Horio et al "New Power Module Structure with Low Thermal Impedance and High Reliability for SiC Devices", Proceeding of PCIM Europe 2011

[7] Y. Iizuka et al "A Novel SiC Power Module with High Reliability", Proceeding of PCIM Europe 2012

[8] Y. Ikeda et al "Investigation on Wirebond-less Power Module Structure with High-Density Packaging and High Reliability" , pp. 272-275 Proceeding of ISPSD 2011

[9] Y. Ikeda et al "Ultra Compact and High Reliable SiC MOSFET Power Module with 200°C Operating Capability" , pp. 81-84 Proceeding of ISPSD 2012

[10] S.Momota et al "Analysis on the Low Current Turn-On Behavior of IGBT Module", pp. 369-372 Proceeding of ISPSD 2000

An improved modulated carrier control of single-phase CCM boost PFC converter

Hyejin Kim, Bo-Hyung Cho
School of Electrical and Computer Engineering
Seoul National University
Seoul, Korea
gpwls12@snu.ac.kr

Hangseok Choi
Fairchild Semiconductor
8 Commerce drive Suite 3A
Bedford, NH 03110, USA
hangseok.choi@fairchildsemi.com

Abstract-This paper proposes an improved modulated carrier control method for a continuous conduction mode (CCM) power factor correction (PFC) boost converter. The proposed method allows precise sinusoidal line current shaping for a boost PFC converter operating in CCM. This is accomplished by simply comparing the compensated modulated carrier signal with the switch current information, such that the current loop compensation is eliminated and multiplier is not required. Also, the proposed control method overcomes the instability problem of conventional modulated carrier control around the zero crossings of the line voltage at light load. The performance of the proposed control method is experimentally verified on a 300 W PFC converter.

I. INTRODUCTION

As power grid harmonic pollution has attracted growing attention, standards regulating the line current harmonic content and power factor have become more stringent causing power factor correction (PFC) to be indispensable. In low to medium power applications, discontinuous conduction mode (DCM) as well as critical conduction mode (CRM) is widely used due to its low switching loss, low diode reverse recovery current, and inherent current shaping capability. In high power applications above 300 W, continuous conduction mode (CCM) boost converter has been the most preferred topology to achieve high power factor because of lower conduction losses and reduced electromagnetic interference (EMI) filter size. However, the line current shaping needs more complex control, such as average current mode control [1] or charge control [2].

Although the average current mode control is a well-known control technique for CCM PFC due to its good performance, it has relatively complex implementation and also needs the line and output voltage sensing, and the inductor current sensing circuit. In addition, a precision analog multiplier to generate a current reference is a disadvantage because it complicates the control integrated circuit design.

In order to simplify the control scheme, the concept of modulated carrier control method was proposed in [3] and several variations have been proposed in [4]-[10]. These methods do not require the line voltage sensing, the precision analog multiplier, and the current loop compensation. However, [7] is effectively a peak current mode control, hence has limitations to PFC applications.

Also, [8] utilizes the peak value of the inductor or switch current on the assumption that the current ripple is small enough compared to the average value. The disadvantage of the method is a line current distortion because the envelope of the inductor current is controlled to follow a line voltage. In addition, although [9], [10] present average current control with only switch current sensing, they exhibit instability when the converter operates in DCM around the zero crossings of the line voltage at light load.

This paper presents an improved modulated carrier control methods based on the approach presented in [9], [10] for single-phase PFC converters. The proposed method allows precise sinusoidal line current shaping for a CCM PFC without the current loop compensation and the precision analog multiplier. Moreover, it overcomes the instability problem so that the current distortion around the zero crossings of the line voltage can be reduced.

The next section first introduces the operation principle of the proposed control method. A stability analysis of the conventional method is then conducted. In Section III, improvement of the current loop stability is analyzed by adding artificial signals to the sensed switch current and the carrier generator. In Section IV, experimental results are shown to validate the performance of the proposed control method.

II. MODULATED CARRIER CONTROL

A. Principle of Operation

Fig. 1 shows a power circuit of a boost PFC converter. L and C_{dc} are boost inductance and output capacitance. v_{in}, i_{in}, i_L, i_s, v_{gs}, and v_{dc} are line voltage, line current, inductor current, switch current, gate-source voltage and output voltage, respectively. Current sensing resistor R_S is placed in series with MOSFET switch, Q, to obtain the boost inductor current information.

For PFC control, the inductor current is proportional to the rectified line voltage $|v_{in}|$ as

$$i_{L,avg} = \frac{|v_{in}|}{R_e} \qquad (1)$$

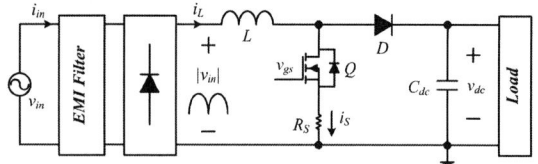

Fig. 1. Power circuit of a boost PFC converter.

where R_e is the emulated resistance determined by the output voltage controller. When the boost PFC converter operates in CCM, the duty ratio d is

$$\frac{v_{dc}}{|v_{in}|} = \frac{1}{1-d} . \qquad (2)$$

Substituting (2) into (1) with a current sensing resistor R_S yields

$$\frac{(1-d)v_{dc}R_S}{R_e} = i_{L,avg}R_S . \qquad (3)$$

Let

$$V_M = \frac{v_{dc}R_S}{R_e} \qquad (4)$$

Then the control key equation for the modulated carrier control method [9], [10] is presented as follow:

$$V_M(1-d) = i_{L,avg}R_S . \qquad (5)$$

Therefore, in each switching cycle if d is controlled to satisfy (5), (1) can be realized.

Figs. 2 and 3 depict the control block diagram and timing waveforms of the modulated carrier control to realize the control key equation (5). The control circuit consists of a PI controller, a clock, a sawtooth carrier signal generator, a comparator, a SR flip-flop (F/F), and an on-time doubler.

As shown in Fig. 2, v_{dc} is sensed and compared with the reference voltage, v_{ref}. This error is processed by a voltage loop compensation to generate a signal V_M which varies with load condition. A carrier signal v_c whose peak-to-peak value is $2V_M$ is generated.

The rising edge of the clock signal sets the SR F/F triggering the turn-on of switch. Once the switch is turned on, i_s increases. When the sensed switch current reaches carrier signal, the output of the comparator resets the SR F/F. An on-time doubler measures the time duration from the rising edge to the falling edge of the SR F/F output, v_{co}, and then outputs a gate signal, v_{gs}, that has twice conduction time of SR F/F output signal.

Since the V_M can be considered constant over a switching cycle, this control circuit realizes

$$V_M(1-\frac{2t}{T_S}) = R_S i_S(t) . \qquad (6)$$

At $t = 0.5dT_S$, (6) can be expressed as

$$V_M(1-d) = R_S i_S(0.5dT_S) = R_S i_{L,avg} . \qquad (7)$$

The duty ratio d of a CCM boost PFC converter is

$$d = 1 - \frac{|v_{in}|}{v_{dc}} . \qquad (8)$$

Therefore, combination of (7) and (8) yields

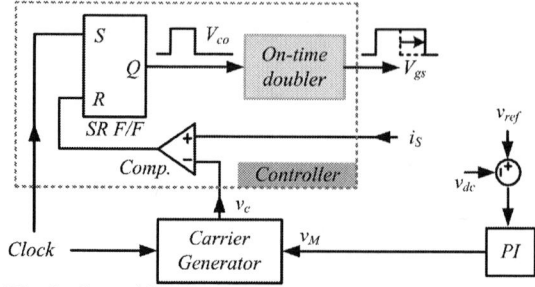

Fig. 2. Control block diagram.

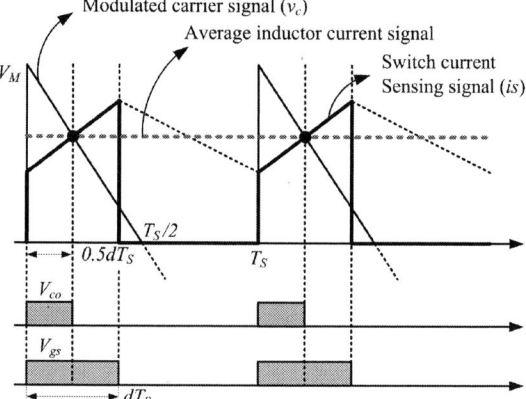

Fig. 3. Modulated carrier control waveforms.

$$V_M \frac{|v_{in}|}{v_{dc}} = R_S i_{L,avg} . \qquad (9)$$

From (9), it can be inferred that average inductor current is proportional to rectified line voltage as long as the boost converter operates in CCM. This allows for a unity power factor with sinusoidal line voltage. Therefore, the resulting duty ratio under this control would perform CCM PFC operation without a line voltage sensing, a precision analog multiplier, and current loop compensation.

B. Stability Characteristics

In this section, a reduced-order sampled-data model for the inductor current is derived on the assumption that variations in the input and output voltage are much slower than the converter system dynamic, so that the input and output voltage can be considered constant [11].

From Fig. 4, when the inductor current, i_L, reaches the carrier signal, v_c, the large-signal relation during one-half of on-time period in the n-th switching cycle is given by:

$$i_n + M_1 \frac{d_n}{2}T_S + M_3 \frac{d_n}{2}T_S = V_M \qquad (10)$$

where the subscript n represents the n-th cycle, $n+1$ represents the $(n+1)$-th cycle, and

$$M_1 = \frac{|v_{in}[n]|}{L}R_S , \qquad (11)$$

$$M_3 = \frac{2V_M}{T_S} \qquad (12)$$

978-1-4799-2706-7/14 $31.00 © 2014 IEEE

where $|v_{in}[n]|$ is the rectified line voltage at the beginning of the n-th switching cycle.

From (10), the duty ratio, d_n, can be determined as follow:

$$d_n = \frac{2}{T_S} \frac{V_M - i_n}{M_1 + M_3} . \quad (13)$$

To determine the value of the inductor current at the end of the n-th switching cycle, denoted as i_{n+1}, note that

$$i_{n+1} = i_n + M_1 d_n T_S - M_2 (1 - d_n) T_S \quad (14)$$

where

$$M_2 = \frac{v_{dc} - |v_{in}[n]|}{L} R_S . \quad (15)$$

Substituting (13) into (14) gives the following sampled-data model for stability analysis:

$$i_{n+1} = \frac{-M_1 - 2M_2 + M_3}{M_1 + M_3} i_n - M_2 T_S + \frac{2(M_1 + M_2) T_S V_M}{M_1 T_S + 2V_M} \quad (16)$$

This first-order model indicates that the current loop is stable if

$$\left| \frac{-M_1 - 2M_2 + M_3}{M_1 + M_3} \right| < 1 . \quad (17)$$

The combination of (11)-(12), (15), and (17) yields the stability criterion as follow:

$$K_s = \frac{2Lf_s V_M}{R_S v_{dc}} > 1 - \frac{|v_{in}|}{v_{dc}} . \quad (18)$$

where K_s is the parameter that represents the stability condition.

It can be noted that (18) shows that the current loop is stable as long as $K_s > 1$, which is the same condition that characterizes the boost PFC operating in CCM over the entire ac line cycle [12]. In conclusion, current loop is fully stable when the converter always operates in CCM at high power condition. However, at medium to light load, the converter operates in DCM around the zero crossings of the line voltage so that the inequality of (18) is no longer satisfied resulting in local instability which appears as bounded oscillation. This phenomenon can be seen from the experimental waveforms shown in Fig. 5. Although the average inductor current follows the line voltage, the line current enters DCM near zero crossing point and the instability is observed.

III. STABILITY IMPROVEMENT

Fig. 6 (a) shows the control block diagram of the proposed method. The stability of the current loop can be improved by addition of an artificial periodic signal to v_c as illustrated in Fig. 6. The artificial signal has slope kM_2. Also, to maintain the control principle (15) in CCM, the second artificial ramp whose slope is kM_1 is added to i_s. Therefore, to implement the improved modulated carrier control, modified carrier signal v_{cm} is compared to i_{sm} and (5) can be reformulated as

$$(V_M + k\frac{M_2}{2f_s})(1 - \frac{2t}{T_S}) = R_s i_s(t) + kM_1 t . \quad (19)$$

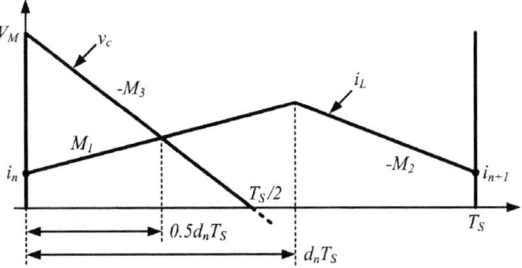

Fig. 4. Steady-state waveforms of the modulated carrier control.

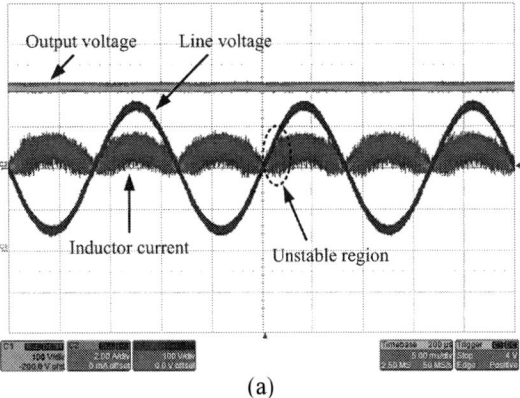

(a)

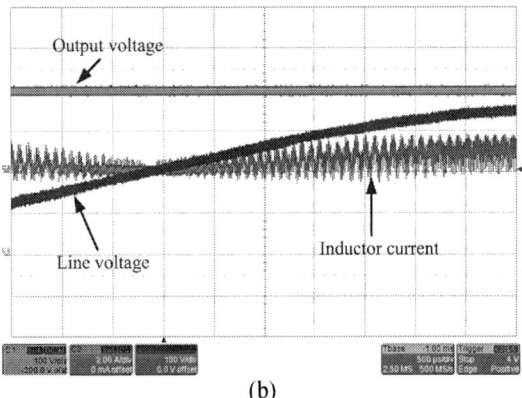

(b)

Fig. 5. (a) Experimental waveforms for light load condition. (b) Zoomed waveforms of the unstable region around the zero crossing of the line voltage at 20% load condition.

When the converter operates in CCM, at $t = 0.5dT_S$,

$$V_M (1-d) + k\frac{M_2}{2f_s}(1-d) = R_s i_{L,avg} + kM_1 \frac{dT_S}{2} . \quad (20)$$

From (20), it can be noted that the second term on the left-hand side and the second term on the right-hand side are cancelled. Thus, in CCM, the basic principle (5) is still satisfied.

To examine how the modified method improves the stability, stability analysis presented in the previous subsection is applied again. It was found that the current loop is stable if

978-1-4799-2706-7/14 $31.00 © 2014 IEEE

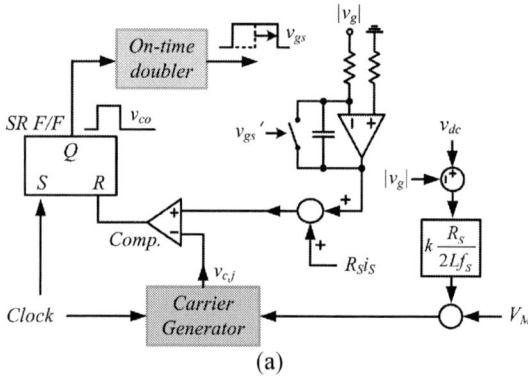

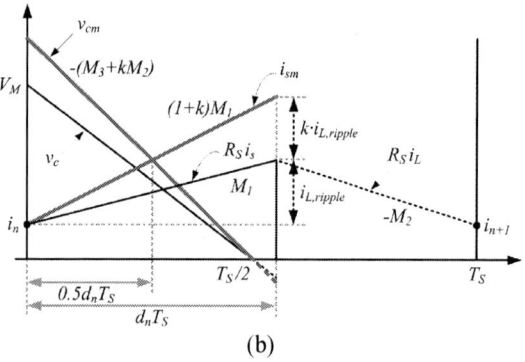

Fig. 6. (a) Control block diagram of the proposed method, (b) Steady-state waveforms of the improved modulated carrier control.

$$\left| \frac{(k-1)M_1 + (k-2)M_2 + M_3}{(k+1)M_1 + kM_2 + M_3} \right| < 1. \qquad (21)$$

Compared to (17), both the numerator and denominator on the left-hand side of (21) contain an extra term, $kM_1 + kM_2$. Therefore, proper value of k can be designed to extend the range of stable operation. If it is necessary to operate the PFC converter stably over the entire line conditions, we can define the design criterion simply. For example, if we choose $k = 1$, (21) is always satisfied.

IV. EXPERIMENTAL RESULTS

The performance of the proposed control method is verified on a 300-W laboratory prototype. The components of the power circuit are listed in Table I. Figs. 7 and 8 show the measured line voltage and line current at different line and load conditions when the current loop is implemented based (19). The output voltage is 390V. Fig. 7 (a) and (b) illustrates the line AC voltage, the line current, and the DC-link voltage at 110-V_{rms} for 100% and 10% load conditions, respectively, whereas Fig. 8 shows them at 220-V_{rms}. As can be seen, at high power condition, in CCM, the line current and the line voltage are almost in phase, thus maintaining almost unity power factor as analyzed in Section II. On the other hand, at light load condition, the converter operates in DCM over

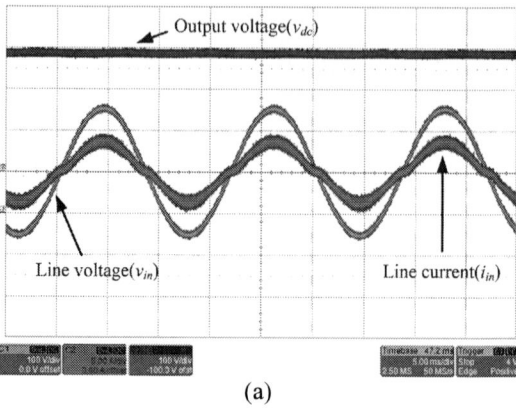

(a)

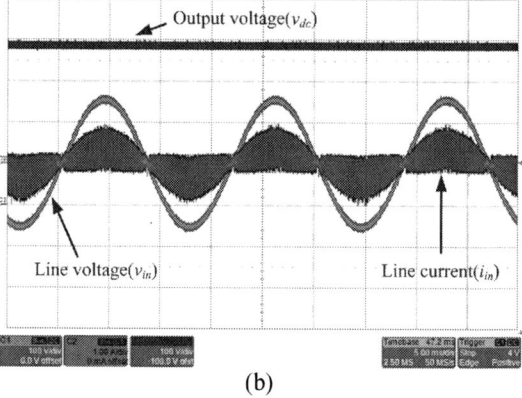

(b)

Fig. 7. Steady-state waveforms of the proposed method: (a) 100% load, $V_{in,rms}$ = 110-V_{rms} (v_{in} : 100V/div, v_{dc} : 100V/div, i_{in} : 5A/div), (b) 10% load, $V_{in,rms}$ = 110-V_{rms} (v_{in} : 100V/div, v_{dc} : 100V/div, i_{in} : 1A/div).

an entire line period. The converter output is still regulated and the line current is reasonably shaped.

Fig. 9 shows the zoomed waveforms around the zero crossing of the line voltage at 20% load which is the same condition as in Fig. 5. Compared to Fig. 5, the implementation of the improved method eliminates the instability phenomenon around the zero crossings of the line voltage as detailed in Section III.

TABLE I
THE COMPONENTS AND CIRCUIT PARAMETERS IN THE PROTOTYPE CIRCUIT

Design parameters	Values
Input voltage	110 V_{rms}/220 V_{rms} 60 Hz
Output power	300 W
Inductance	1.63 mH
Switching frequency	65 kHz

V. CONCLUSION

This paper presented an improved modulated carrier control method for a single-phase boost PFC converter. With the proposed method, the average current controlled boost PFC operation can be realized simply by sensing switch current and greatly simplifies the control circuit due to the elimination of multiplier. The instability problem of the conventional control method was analyzed and the improved method for elimination of the instability was proposed. It enhances the instability around the zero crossings of the line voltage. Experimental results validate the operation of the proposed control scheme.

ACKNOWLEDGMENT

This research was conducted under the industrial infrastructure program for fundamental technologies which is funded by the Ministry of Knowledge Economy (MKE, Korea) and it was supported by the Korea Micro Energy Grid (KMEG) of the Office of Strategic R&D Planning (OSP) grant funded by the Korea government Ministry of Knowledge Economy (No. 2011 TIOO100025).

REFERENCES

[1] L. Dixon, "Average current mode control of switching power supplies," in *Unitrode Power Supply Design Handbook*.

[2] C. Zhou and M. M. Jovanovic, "Design trade-offs in continuous current-mode controlled boost power-factor correction circuit," in *Proc. HFPC'92*, 1992, pp. 209-220.

[3] J. Shires and J. Turner, "Control arrangement for a switch mode power supply," U.S. Patent 4 942 509, Mar. 28, 1989.

[4] Z. Lai and K. M. Smedley, "A family of continuous-conduction-mode power-factor-correction controllers based on the general pulse-width modulator," *IEEE Trans. Power Electron.*, Vol. 13, No. 3, pp. 501-510, May 1998.

[5] J. –H. Chiang, B. –D. Liu, S. –M. Chen, "A simple implementation of nonlinear-carrier control for power factor correction rectifier with variable slope ramp on field-programmable gate array," *IEEE Trans. Ind. Inf.*, Vol. 9, No. 3, pp. 1322-1329, Aug. 2013.

[6] B. A. Mather and D. Maksimovic, "A simple digital power-factor correction rectifier controller," *IEEE Trans. Power Electron.*, Vol. 26, No. 1, pp. 9-19, Jan. 2011.

[7] R. Brown and M. Soldano, "One cycle control IC simplifies PFC designs," in *Proc. IEEE Appl. Power Electron.Conf.*, 2005, pp. 825-829.

[8] D. V. Ghodke, K. Chatterjee, and B. G. Fernandes, "Modified one-cycle controlled bidirectional high-power-factor AC-to-DC converter," *IEEE Trans. Ind. Electron.*, Vol. 56, No. 5, pp. 1499-1510, May 2009.

[9] H. Choi, "Continuous conduction mode power factor correction circuit with reduced sensing requirement," U.S. Patent US8 279 630, Oct. 2, 2012.

[10] H. J. Kim, B. H. Cho, and H. Choi, "Interleaved continuous conduction mode power factor correction boost converter with improved modulated carrier control method," in *Proc. IEEE Appl. Power Electron.Conf.*, 2013, pp. 351-355.

[11] M. Chen, A. Mathew, and J. Sun, "Nonlinear current control of single-phase PFC converters," *IEEE Trans. Power Electron.*, Vol. 22, No. 6, pp. 2187-2194, Nov. 2007.

[12] R. W. Erickson and D. Maksimovic, *Fundamental of Power Electronics*. Springer, 2011.

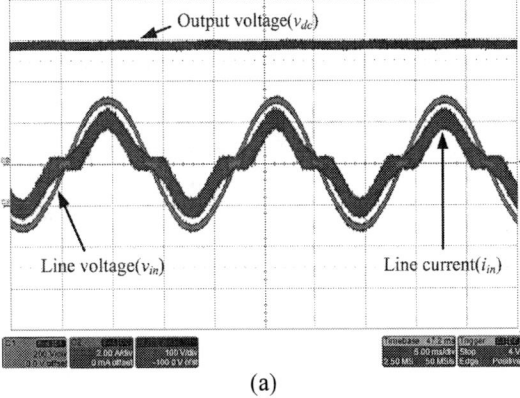

(a)

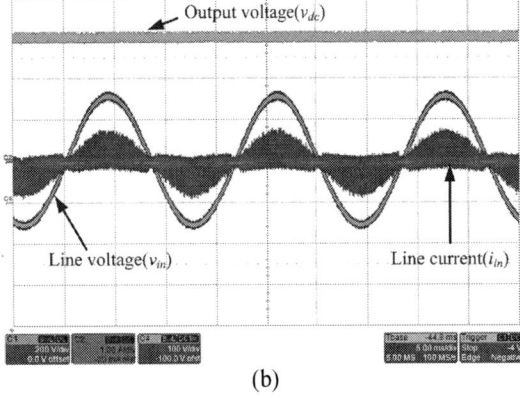

(b)

Fig. 8. Steady-state waveforms of the proposed method: (a) 100% load, $V_{in,rms}$ = 220-V_{rms} (v_{in} : 200V/div, v_{dc} : 100V/div, i_{in} : 2A/div), (b) 10% load, $V_{in,rms}$ = 220-V_{rms} (v_{in} : 200V/div, v_{dc} : 100V/div, i_{in} : 1A/div).

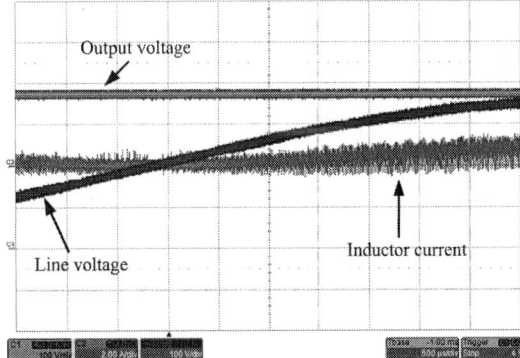

Fig. 9. Zoomed waveforms around the zero crossing of the line voltage at 20% load condition.

Modified Interleaved Current Sensorless Control for Three-Level Boost PFC Converter with Asymmetric Loads

Hung-Chi Chen and Jhen-Yu Liao
Department of Electrical and Computer Engineering (ECE),
National Chiao Tung University (NCTU), Hsinchu, 30010, Taiwan.

Abstract —Three-level boost PFC converter has advantages of lower inductor current ripple and lower switch withstanding voltage than the boost PFC converter. Interleaved current sensorless control (ICSC) had been proposed for multiphase boost PFC converters. In this paper, ICSC is modified for current sensorless control of the three-level boost PFC converter with the asymmetric loads. The modified interleaved current sensorless control (MICSC) is implemented in a field programmable gate array (FPGA) board. Simulation results and experimental results are also provided to validate the proposed MICSC.

Index Terms —Current Sensorless Control, PFC

I. INTRODUCTION

To reduce the power transmission loss and improve the power quality, the power factor correction (PFC) function is required [1] for the commercial products. PFC function not only shapes the input current waveform but also regulates the output voltage. When the converter draws electricity from the main grid, the yielded current harmonics need to comply with the standard to improve the power quality.

Boost converter is a popular topology for the PFC function due to its continuous current [2], but the power switch needs to withstand the output dc voltage. For the high voltage applications, the high withstanding voltage of power switch becomes an issue. Therefore, three-level boost PFC converter was proposed because each power switch needs to withstand half of the dc output voltage [3].

In addition, the three-level boost PFC converter has the advantages of low inductor current ripple and high efficiency [3]. Therefore, the three-level boost PFC converter is often used in the high output voltage [4]-[6], the photovoltaic system [7], and the wind power system [8].

The conventional PFC control for the boost PFC converter needs to sense the input voltage, the output voltage, and the inductor current. Due to the required high resolution and high sampling frequency, the cost of A/D converter for sensing current is usually high.

Instead of sensing the inductor current, a control method sensing the output current was proposed in [9]. The requirement of the A/D converter for sensing the output current is lower than sensing the inductor current because the output current is approximately a constant. In [10] and [11], the new sensing method using only comparators without A/D converter was proposed.

Another way to eliminate current sensor is to develop the current sensorless control method. Many current sensorless control methods had been proposed [12]-[21], and they are summarized in Table I.

[12]-[13] proposed current sensorless control for DC-DC converters such as buck, boost, buck-boost converter, and so on. [14]-[21] proposed current sensorless control for AC-DC converters such as boost converter, full-bridge converter, and the multiphase boost converter.

Table I. Summary of current sensorless control

Topology (DC-DC)	Reference	Topology (AC-DC)	Reference
Buck converter	[12]-[13]	Boost converter	[14], [16]-[20]
Boost converter	[12]-[13]	Full-bridge converter	[15]
Buck-boost converter	[12]-[13]	Multiphase boost converter	[21]
CU'K converter	[13]		
SEPIC converter	[13]		

The interleaved PWM shown in Fig. 1(a) is usually used in the multi-phase boost converter. Recently, the interleaved control is also used for the three-level boost converter in [4]-[5], [7]-[8], [22] as shown in Fig. 1(b). In [21], the interleaved current sensorless control (ICSC) with consideration of both inductor resistance and the conducting voltage was developed for the multi-phase boost converter. In [22], the conventional multiloop control for boost PFC converter is modified to become multiloop interleaved control (MIC) for three-level boost PFC converter.

As shown in Fig. 1(b), the three-level boost PFC converter has two cascaded capacitors. To prevent the capacitor voltages exceed the withstanding voltage of the power switch, the controller needs to balance both capacitor voltages [23].

In this paper, ICSC is modified to the control of three-level boost PFC converter with consideration of the voltage imbalance. Three-level boost PFC converter is able to supply two independent loads. When two loads are asymmetric, the imbalance of capacitor voltages is considered to develop the modified interleaved current sensorless control (MICSC). Simulation results and experimental results are also provided to validate the proposed MICSC.

978-1-4799-2706-7/14 $31.00 © 2014 IEEE

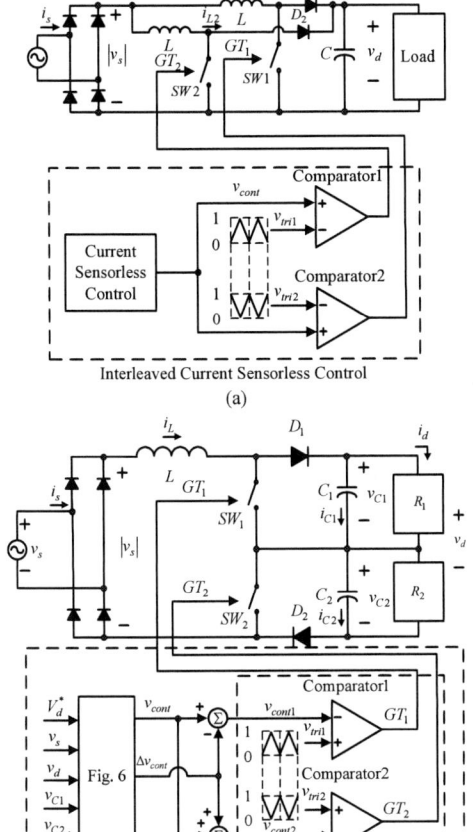

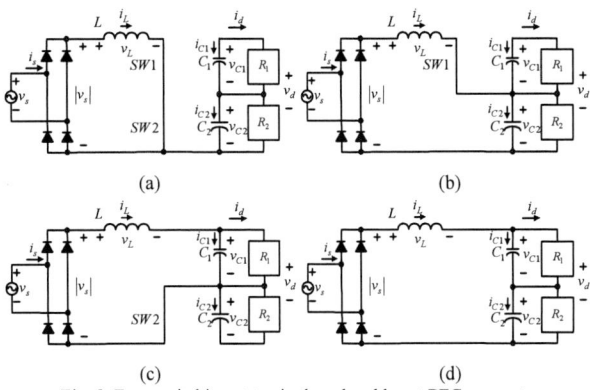

Fig. 2. Four switching states in three-level boost PFC converter: (a) state 1; (b) state 2; (c) state 3; (d) state 4.

Fig. 1. (a) The interleaved current sensorless control for 2-phase boost converter [21]; (b) the proposed modified of interleaved current sensorless control for three-level boost PFC converter.

II. THREE-LEVEL BOOST PFC CONVERTER

As shown in Fig. 1(b), two resistors R_1 and R_2 are connected in parallel to the capacitors C_1 and C_2. For the symmetric loads, two resistors are equivalent to each other $R_1 = R_2$. For the asymmetric loads, two resistors are not equal to each other $R_1 \neq R_2$.

As shown in Fig. 2, there are four switching states in the three-level boost PFC converter. Both switches may conduct at the same time (i.e. state 1), but for another state, both switches may block at the same time (i.e. state 4). In the analysis, the input voltage v_s is assumed to be $v_s = \hat{V}_s \sin(\omega t)$.

Two power switches are turning on in state 1. Both capacitors need to supply the load resistors. The inductor voltage in state 1 can be expressed as

$$v_L = L \frac{di_L(t)}{dt} = \hat{V}_s |\sin(\omega t)| - r_L i_L - V_F \qquad (1)$$

where r_L is the equivalent series resistance (ESR) of the inductor and V_F is the sum of conduction voltage in the current flowing path.

The upper power switch $SW1$ turns on and the other switch $SW1$ turns off in switching state 2. The upper capacitor C_1 supply the load R_1. The inductor current charges the capacitor C_2 charges and supply the load R_2. The inductor voltage in switching state 2 can be expressed as

$$v_L = L \frac{di_L(t)}{dt} = \hat{V}_s |\sin(\omega t)| - v_{C2} - r_L i_L - V_F \qquad (2)$$

Likewise, only the lower power switch $SW2$ turns on in the state 3. The inductor voltage of state 3 can be expressed as

$$v_L = L \frac{di_L(t)}{dt} = \hat{V}_s |\sin(\omega t)| - v_{C1} - r_L i_L - V_F \qquad (3)$$

Both power switches block in state 4. Both capacitors charge by the inductor current. The inductor voltage of state 4 can be expressed as

$$v_L = L \frac{di_L(t)}{dt} = \hat{V}_s |\sin(\omega t)| - v_{C1} - v_{C2} - r_L i_L - V_F \qquad (4)$$

From Fig. 1(b), both gate signals are obtained from the comparison results. The gate signal GT_1 is generated from the comparison of the control signal v_{cont1} and the signal v_{tri1}, and the other gate signal GT_2 is obtained from the comparison of the control signal v_{cont2} and the signal v_{tri2}. Two signals v_{tri1} and v_{tri2} varying between 0 and 1 are interleaved by 180°. It is noted that the control signals v_{cont1} and v_{cont2} are connected to the inverting terminals (-) of the comparators.

Both control signals v_{cont1} and v_{cont2} are generated from the control signal v_{cont} and the balance control signal Δv_{cont}

$$v_{cont1} = v_{cont} - \Delta v_{cont} \qquad (5)$$
$$v_{cont2} = v_{cont} + \Delta v_{cont} \qquad (6)$$

where both signals v_{cont} and Δv_{cont} are generated form the proposed modified interleaved current sensorless control (MICSC) in Fig. 6. From (5) and (6), the sum of the control signals v_{cont1} and v_{cont2} can be expressed as

$$v_{cont1} + v_{cont2} = 2v_{cont} \qquad (7)$$

The analysis of three-level boost PFC converter can be divided into two cases. As shown in Fig. 3, one case is $v_{cont} \geq 0.5$, and the other case is $v_{cont} < 0.5$. For the case $v_{cont} \geq 0.5$ in Fig. 3(a), only switching state 2, state 3 and state

4 can be found. For the other case $v_{cont} < 0.5$, only switching state 1, state 2 and state 3 can be found.

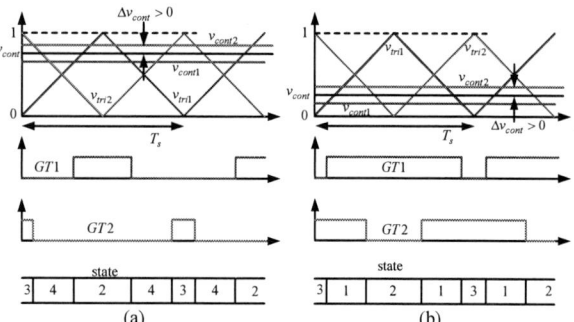

Fig. 3. Two cases of the three-level boost PFC converter
(a) $v_{cont} \geq 0.5$; (b) $v_{cont} < 0.5$.

$v_{cont} \geq 0.5$

With consideration of the voltage imbalance, there are four possible conditions as plotted in Fig. 4.

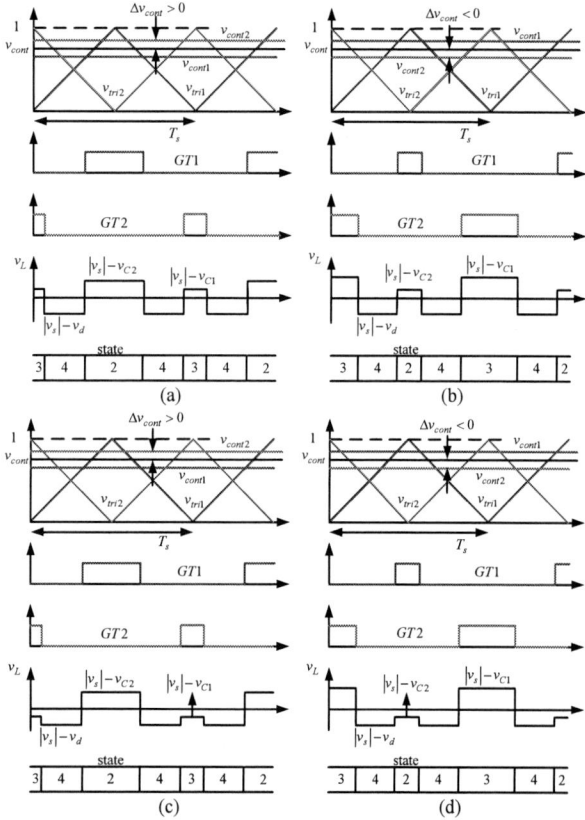

Fig. 4. Illustrated waveforms ($v_{cont} > 0.5$) (a) $|v_s| > v_{C1} > v_{C2}$;
(b) $|v_s| > v_{C2} > v_{C1}$; (c) $v_{C1} > |v_s| > v_{C2}$; (d) $v_{C2} > |v_s| > v_{C1}$.

From Fig. 4, the conducting times of state 2 and state 3 are $(1-v_{cont1})T_s$ and $(1-v_{cont2})T_s$, respectively. For a fixed switching period T_s, the total conducting time for state 4 is $(v_{cont1} + v_{cont2} - 1)T_s$. Therefore, from (2)-(7), the average

inductor voltage $\langle v_L \rangle_{Ts}$ within the switching time T_s can be expressed as

$$\langle v_L \rangle_{T_s} = \hat{V}_s |\sin(\omega t)| - v_{cont} v_d$$
$$+ \Delta v_{cont}(v_{C1} - v_{C2}) - r_L i_L - V_F \qquad (8)$$

$v_{cont} < 0.5$

Likewise, four possible conditions for the case $v_{cont} < 0.5$ are plotted in Fig. 5. From Fig. 5, the conducting times for state 2 and state 3 are $v_{cont2}T_s$ and $v_{cont1}T_s$, respectively. The remaining time for state 1 is $(1 - v_{cont1} - v_{cont2})T_s$. Fortunately, from (1)-(3) and (5)-(6), the average inductor voltage $\langle v_L \rangle_{Ts}$ can be expressed as the same equation (8).

Likewise, four possible conditions for the case $v_{cont} < 0.5$ are plotted in Fig. 5. From Fig. 5, the conducting times for state 2 and state 3 are $v_{cont2}T_s$ and $v_{cont1}T_s$, respectively. The conducting time for the switching state 1 is $(1 - v_{cont1} - v_{cont2})T_s = (1 - 2v_{cont})T_s$. Fortunately, from (1)-(3) and (5)-(6), the average inductor voltage $\langle v_L \rangle_{Ts}$ can be expressed as the same equation (8).

Therefore, the average inductor voltage can be expressed as (8) regardless of the control signal v_{cont}.

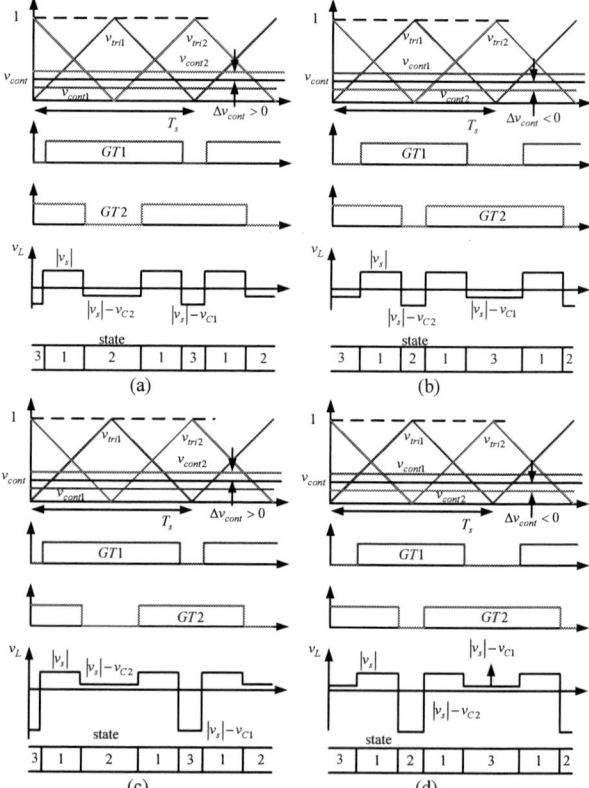

Fig. 5. Illustrated waveforms ($v_{cont} < 0.5$) (a) $v_{C1} > v_{C2} > |v_s|$;
(b) $v_{C2} > v_{C1} > |v_s|$; (c) $v_{C1} > |v_s| > v_{C2}$; (d) $v_{C2} > |v_s| > v_{C1}$.

III. MODIFIED INTERLEAVED CURRENT SENSORLESS CONTROL

Fig. 6 is the modified interleaved current sensorless control (MICSC) for three-level boost PFC converter. In the development of the proposed MICSC, some assumptions are made. i) The output voltage is assumed to be a constant voltage which equals the output voltage command V_d^* .due to the bulk capacitor. ii) The switching period T_s is much smaller than the line period and thus, the control signal is assumed to be constant within the switching period T_s .

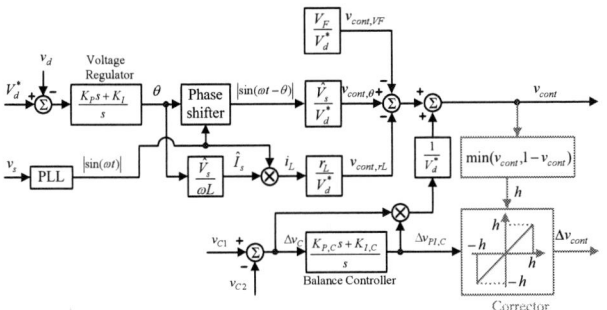

Fig. 6. Modified interleaved current sensorless control.

From Fig. 6, two PI controllers are used. One is voltage regulator, and the other is the balance controller. The output voltage v_d is sensed and the phase shift signal θ is generated through the voltage regulator. After detecting the input voltage v_s to generate the synchronous signal $|\sin(\omega t)|$, the shift signal $|\sin(\omega t - \theta)|$ is obtained.

Both capacitor voltages are sensed. The voltage imbalance signal Δv_C and the control output $\Delta v_{PI,C}$ are generated. Form Fig. 6, the control signal v_{cont} can be expressed as

$$v_{cont} = \frac{\hat{V}_s}{V_d^*}|\sin(\omega t - \theta)| - \theta\frac{\hat{V}_s r_L}{\omega L V_d^*}|\sin(\omega t)| \\ - \frac{V_F}{V_d^*} + \frac{\Delta v_{cont}(v_{C1} - v_{C2})}{V_d^*} \qquad (9)$$

By instituting (9) into (8), the average inductor voltage can be rewritten

$$\langle v_L \rangle_{Ts} = \hat{V}_s|\sin(\omega t)| - \hat{V}_s|\sin(\omega t - \theta)| + \left[\frac{\hat{V}_s \theta}{\omega L}|\sin(\omega t)| - i_L(t)\right]r_L \quad (10)$$

The value of function $\sin(\omega t)$ must approach to θ, and the value of the function $\cos(\theta)$ must approach to 1 as the phase shift signal θ is small. Then, with the trigonometric identity rule $\sin(\omega t - \theta) = \sin(\omega t)\sin\theta + \cos(\omega t)\cos\theta$, (10) can be simplified to become

$$L\frac{di_L(t)}{dt} \cong \hat{V}_s|\sin(\omega t)| - \hat{V}_s|\sin(\omega t) - \theta\cos(\omega t)| \\ + \left[\frac{\hat{V}_s\theta}{\omega L}|\sin(\omega t)| - i_L(t)\right]r_L \qquad (11)$$

By solving (11), the inductor current can be obtained.

$$i_L(t) \cong \frac{\hat{V}_s\theta}{\omega L}|\sin(\omega t)| \qquad (12)$$

Thus, the average power can be expressed as

$$P \cong \frac{\hat{V}_s^2\theta}{2\omega L} \qquad (13)$$

which is directly proportional to the phase shift signal θ. It follows that a PI voltage controller can be included to the proposed MICSC to regulate the DC-side voltage.

To avoid the overmodulation in both comparisons, the corrector block is included. The band of the corrector is varying according to the smaller value h between the signal v_{cont} and $(1 - v_{cont})$. The final signal Δv_{cont} is corrected to

$$\Delta v_{cont} = \begin{cases} \Delta v_{PI,C} & when \ |\Delta v_{PI,C}| < h \\ 0 & when \ |\Delta v_{PI,C}| \geq h \end{cases} \qquad (14)$$

IV. SIMULATION

Some simulation results are provided in this section to evaluate the proposed MICSC. Table II shows the simulation parameters of three-level boost PFC converter.

Table II. Simulated parameters of three-level boost PFC converter

Input voltage	$\hat{V}_s = 155V$, $50Hz$
Output voltage command	$V_d^* = 400V$
Inductance	$4.65mH$
ESR of the inductor	$r_L = 0.6\Omega$
Total conduction voltage	$V_F = 6.85V$
PWM frequency	$40kHz$
Capacitor	$C_1 = C_2 = 1120\mu F$
Balance Controller	$K_{P,C} = 0.001$, $K_{I,C} = 0.05$
Voltage Regulator	$K_P = 0.00001$, $K_I = 1$

Fig. 7 shows the simulation results with symmetric load ($R_1 = R_2 = 200\Omega$). From Fig. 7, the input current is sinusoidal and in phase with the input voltage v_s . The output voltage v_d is well regulated to 400V. Due to the symmetric load, the capacitor voltages are balanced, and thus, the output of the balance controller $\Delta v_{PI,C} = 0$ is near zero. It shows that the output of the corrector block is also near zero, i.e. $\Delta v_{cont} \approx 0$. Thus, The resulting control signals v_{cont1} and v_{cont2} are closed to the control signal v_{cont} .

With output voltage 400V and loads $R_1 = R_2 = 200\Omega$, the total output power is near 400W. From Fig. 7, the value of the steady-state phase shift signal θ is near 0.05rads, which also demonstrates the result in (13).

To evaluate the function of the proposed corrector block, Fig. 8 shows the simulation results with asymmetric load ($R_1 = 100\Omega$ and $R_2 = 200\Omega$) without the corrector block. Because the load power across the capacitor C_1 is larger than that across the capacitor C_2 , the output signal $\Delta v_{PI,C}$ becomes negative. It follows that the control signal v_{cont2} is sometimes smaller than zero, and thus, the overmodulation occurs in the comparison.

978-1-4799-2706-7/14 $31.00 © 2014 IEEE

As shown in Fig. 8, the overmodulation would contribute to the current distortion near the zero-voltage-crossing-points.

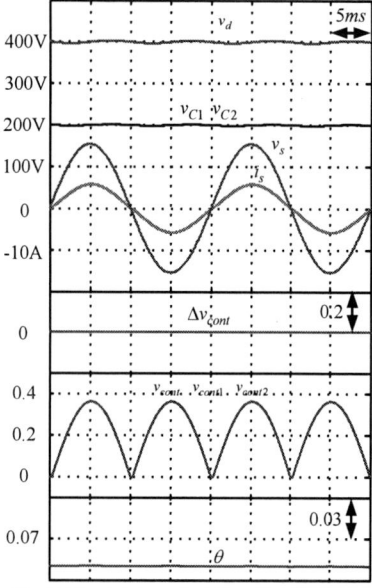

Fig. 7. Simulated results ($R_1 = R_2 = 200\Omega$).

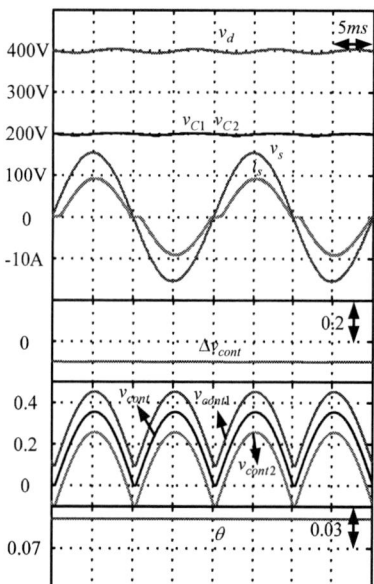

Fig. 8. Simulated results without the corrector block($R_1 = 100\Omega$, $R_2 = 200\Omega$).

Fig. 9 shows the simulation results with the corrector block where the load condition is the same as Fig. 8 ($R_1 = 100\Omega$ and $R_2 = 200\Omega$). Due to the asymmetric load, the balance controller output $\Delta v_{PI,C}$ is negative. Duo to the corrector block, both control signals v_{cont1} and v_{cont2} are corrected to avoid the occurrence of the overmodulation.

From Fig. 9, the yielded input current i_s is sinusoidal waveform in phase with the input voltage v_s . In addition, the phase shift signal θ in Fig. 9 is near 0.75rad because that the total average power now is near 600W.

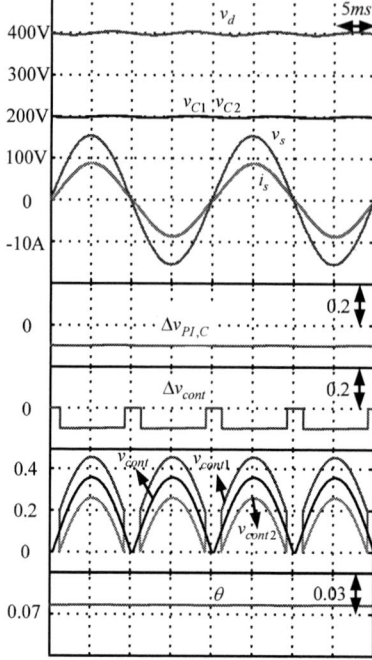

Fig. 9. Simulated results with the corrector block ($R_1 = 100\Omega$, $R_2 = 200\Omega$).

Fig. 10 is simulated waveforms with the corrector block under the change of symmetric load to the asymmetric load. The load R_1 changes from 200Ω to 100Ω , and R_2 is fixed to 200Ω . During the change of load, significant imbalance of capacitor voltages occurs, and the controllers are corrected to avoid overmodulation.

With the increase of average power from 400W to 600W, the phase shift signal θ is also changed from 0.05 to 0.75 to increase the input power from (13).

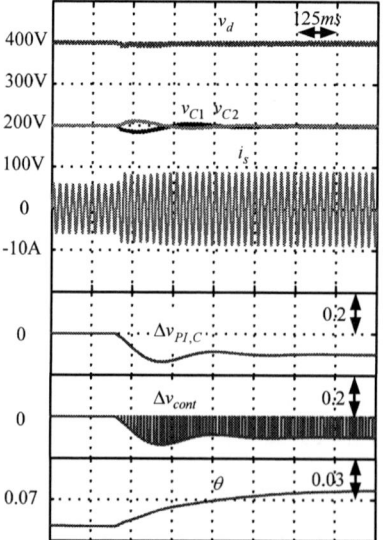

Fig. 10. The simulated results with the corrector block during the change of R_1 from 200Ω to 100Ω and $R_2 = 200\Omega$.

V. EXPERIMENTAL RESULT

The proposed MICSC is implemented by the field programmable gate array (FPGA) board with Xilinx XC3S250. The experimental parameters are the same as the simulated parameters in Table I. The experimental results of symmetric loads are shown in Fig. 11 ($R_1 = R_2 = 200\Omega$). From Fig. 11, the input currents are sinusoidal and in phase with the input voltages. Both capacitor voltages are well kept to half of the output voltage command.

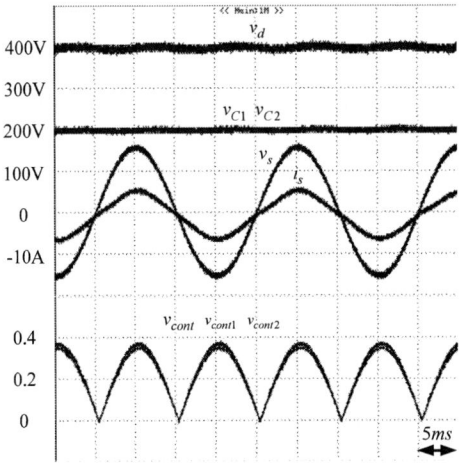

Fig. 11. Experimental results ($R_1 = 200\Omega$, $R_2 = 200\Omega$).

The experimental results of asymmetric loads ($R_1 = 100\Omega$ and $R_2 = 200\Omega$) without the proposed corrector block and with corrector block are plotted in Fig. 12(a) and Fig. 12(b), respectively.

In Fig. 12(a), the output voltage is well regulated and the capacitor voltages v_{C1} and v_{C2} are balanced. Additionally, the control signal v_{cont2} is sometimes negative, and thus, the input current is distorted because of the resulting overmodulation.

Form Fig. 12(b), the proposed corrector block is able to avoid the overmodulation and the current distortion is reduced even the loads are asymmetric.

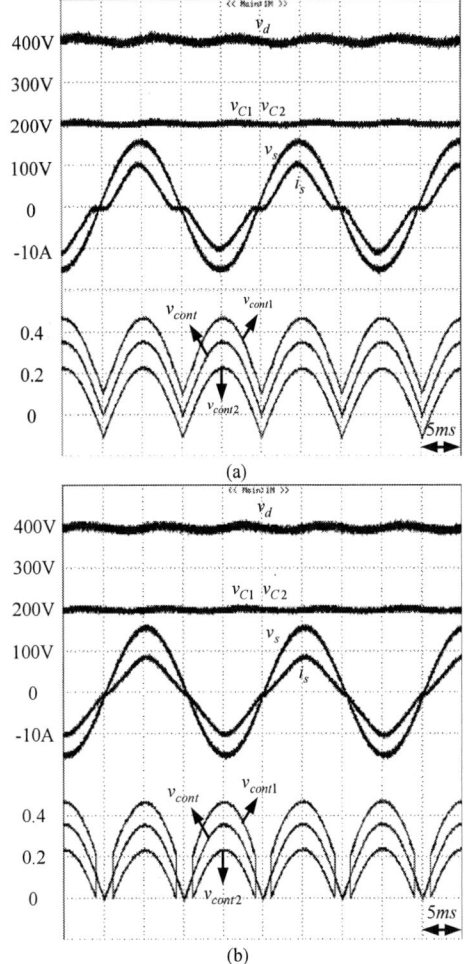

Fig. 12. Experimental results ($R_1 = 100\Omega$, $R_2 = 200\Omega$)
(a) without the proposed corrector block; (b) with the proposed corrector block.

Table III. Harmonics of experimental results

Harmonics	Class A	Class D 400W	Fig. 11	Class D 600W	Fig. 12(a)	Fig. 12(b)
Fundamental	X	X	3.7288A	X	5.9629A	5.7749A
3rd	2.3A	1.36A	0.1855A	2.04A	1.1009A	0.5144A
5th	1.14A	0.76A	0.0639A	1.14A	0.1760A	0.0270A
7th	0.77A	0.4A	0.0336A	0.6A	0.1537A	0.0471A
9th	0.4A	0.2A	0.0238A	0.3A	0.0802A	0.0165A
11th	0.33A	0.14A	0.0194A	0.21A	0.0421A	0.0373A
13th	0.21A	0.118A	0.0172A	0.178A	0.0185A	0.0137A
15th	0.15A	0.103A	0.0146A	0.154A	0.0089A	0.0141A
17th	0.132A	0.091A	0.0153A	0.136A	0.0125A	0.0175A
19th	0.118A	0.081A	0.0135A	0.122A	0.0129A	0.0039A
PF			0.9978		0.9993	1.0000
THD$_i$			7.491%		18.958%	9.438%

To evaluate the performance of current harmonics, Table III shows the standard IEC-61000-3-2 and the current harmonics of experimental results. It shows that all the yielded current harmonics are below the standards.

Fig. 13 shows the experimental results during the change of the load R_1 from 200Ω to 100Ω (R_2 is fixed to 200Ω). At the beginning of the load change, the capacitor voltage v_{C1} drops, and the other capacitor voltages v_{C2} rises in order to regulate the output voltage v_d.

All the provided results show that the proposed MICSC is able to provide PFC function.

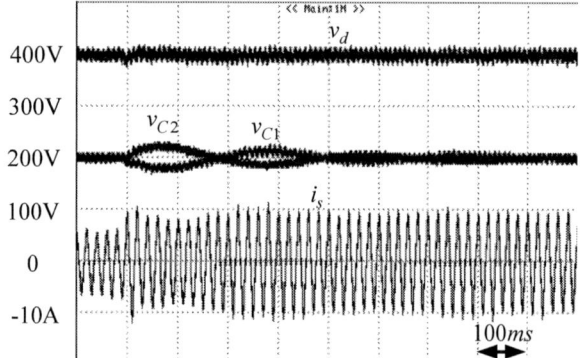

Fig. 13. Experimental results with the corrector block during the change of R_1 from 200Ω to 100Ω and $R_2 = 200\Omega$.

VI. CONCLUSION

In this paper, the interleaved current sensorless control is modified for the three-level boost PFC converter. The corrector block is proposed to improve the current waveforms with asymmetric loads. The simulated results and the experimental results show that the proposed modified interleaved current sensorless control is able to work well and all the current harmonics are below the standard IEC-61000-3-2.

References

[1] B. Singh, B. N. Singh, A. Chandra, K. Al-Haddad, A. Pandey, and D. P. Kothari, "A Review of Single-Phase Improved Power Quality AC-DC Converters," *IEEE Trans. on Industrial Electronics,* vol. 50, no. 5, pp. 962-981, Oct. 2003.

[2] J. C. Crebier, B. Revol, and J. P. Ferrieux, "Boost-Chopper-Derived PFC Rectifiers: Interest and Reality", *IEEE Trans. on Industrial Electronics,* vol. 52, no. 1, pp. 36-45, Feb. 2005.

[3] M. T. Zhang, Y. Jiang, F. C. Lee, and M. M. Jovanovic, "Single-Phase Three-Level Boost Power Factor Correction Converter," in *Proc. IEEE Applied Power Electronics Conference and Exposition (APEC),* Mar. 1995, pp. 434–439.

[4] L. S. Yang, T. J. Liang, H. C. Lee, and J. F. Chen, "Novel High Step-Up DC–DC Converter With Coupled-Inductor and Voltage-Doubler Circuits," *IEEE Trans. on Industrial Electronics,* vol. 58, no. 9, pp. 4196-4206, Sep. 2011.

[5] A. Shahin, M. Hinaje, J. P. Martin, S. Pierfederici, S. Rael, and B. Davat, "High Voltage Ratio DC-DC Converter for Fuel-Cell Applications," *IEEE Trans. on Industrial Electronics,* vol. 57, no. 12, pp. 3944-3955, Dec. 2010.

[6] W. Li and X. He, "Review of Nonisolated High-Step-Up DC/DC Converters in Photovoltaic Grid-Connected Applications", *IEEE Trans. on Industrial Electronics,* vol. 58, no. 4, pp. 1239-1250, April 2011.

[7] J. M. Kwon, B. H. Kwon, and K. H. Nam, "Three-Phase Photovoltaic System with Three-Level Boosting MPPT Control," *IEEE Trans. on Power Electronics,* vol. 23, no. 5, pp. 2319-2327, Sep. 2008.

[8] V. Yaramasu and B. Wu, "Three-Level Boost Converter Based Medium Voltage Megawatt PMSG Wind Energy Conversion Systems," *Energy Conversion Congress and Exposition (ECCE),* pp. 561–567, 2011.

[9] S. Sivakumar, K. Natarajan, and R. Gudelewicz, "Control of power factor correcting boost converter without instantaneous measurement of input current," *IEEE Trans. on Power Electronics,* vol. 10, no. 4, pp. 435–445, Jul. 1995.

[10] M. Rodriguez, V. M. Lopez, F. J. Azcondo, J. Sebastian, and D. Maksimovic, " Average Inductor Current Sensor for Digitally Controlled Switched-Mode Power Supplies," *IEEE Trans. on Power Electronics,* vol. 27, no. 8, pp. 3795-3806, Aug. 2012.

[11] K. I. Hwu, H. W. Chen, and Y. T. Yau, "Average Inductor Current Sensor for Digitally Controlled Switched-Mode Power Supplies," *IEEE Trans. on Power Electronics,* vol. 27, no. 9, pp. 4021-4029, Sep. 2012.

[12] Y. Qiu, X. Chen, and H. Liu, "Digital Average Current-Mode Control Using Current Estimation and Capacitor Charge Balance Principle for DC–DC Converters Operating in DCM," *IEEE Trans. on Power Electronics,* vol. 25, no. 6, pp. 1537-1545, Jun. 2010.

[13] P. Midya, P. T. Krein, and M. F. Greuel, "Sensorless current mode control-an observer-based technique for DC-DC converters," *IEEE Trans. on Power Electronics,* vol. 16, no. 4, pp. 522-526, Jul. 2001.

[14] A. Sanchez, A. d. Castro, V. M. Lopez, F. J. Azcondo, and J. Garrido, "Single ADC Digital PFC Controller Using Precalculated Duty Cycles," *IEEE Trans. on Power Electronics,* vol. 29, no. 2, pp. 996-1005, Feb. 2014.

[15] H.-C. Chen, and J.-Y. Liao, "Bidirectional Current Sensorless Control for the Full-Bridge AC/DC Converter With Considering Both Inductor Resistance and Conduction Voltages," *IEEE Trans. on Power Electronics,* vol. 29, no. 4, pp. 2071-2082, Apr. 2014.

[16] H.-C. Chen, "Single-loop current sensorless control for single-phase boost-type SMR," *IEEE Trans. on Power Electronics,* vol. 24, no. 1, pp. 163–171, Jan. 2009.

[17] H.-C. Chen, C.-C. Lin, and J.-Y. Liao, "Modified Single-Loop Current Sensorless Control for Single-Phase Boost-Type SMR With Distorted Input Voltage," *IEEE Trans. on Power Electronics,* vol. 26, no. 5, pp. 1322–1328, May 2011.

[18] Y.-K. Lo, H.-J. Chiu, and S.-Y. Ou, "Constant-switching-frequency control of switch-mode rectifiers without current sensors," *IEEE Trans. on Industrial Electronics,* vol. 47, no. 5, pp. 1172-1174, Oct. 2000.

[19] M. Pahlevaninezhad, P. Das, G. Moschopoulos, and P. Jain, "Sensorless control of a boost PFC AC/DC converter with a very fast transient response," in *Proc. IEEE Applied Power Electronics Conference and Exposition (APEC),* Mar. 2013, pp. 356–360.

[20] V. M. Lopez-Martin, F. J. Azcondo, and A. d. Castro, "Current error compensation for current-sensorless power factor corrector stage in continuous conduction mode," in *Proc. IEEE Control and Modeling for Power Electronics (COMPEL),* Jun. 2012, pp. 1-8.

[21] H.-C. Chen, "Interleaved Current Sensorless Control for Multiphase Boost-Type Switch-Mode Rectifier With Phase-Shedding Operation," *IEEE Trans. on Industrial Electronics,* vol. 61, no. 2, pp. 766-775, Feb. 2014.

[22] H.-C. Chen and J.-Y. Liao, "Multiloop Interleaved Control for Three-Level Switch-Mode Rectifier in AC/DC Applications," *IEEE Trans. on Industrial Electronics.*

[23] B. R. Lin and H. H. Lu, "A Novel PWM Scheme for Single-Phase Three-Level Power-Factor-Correction Circuit" *IEEE Trans. on Industrial Electronics,* vol. 47, no. 2, pp. 245-252, Apr. 2000.

A Novel Critical-Conduction-Mode Bridgeless Interleaved Boost PFC Rectifier

Guoen Cao and Hee-Jun Kim
Dept. of Electrical System Engineering
Hanyang University
Ansan Gyeonggi-do, 426-791 Korea

Abstract—A novel bridgeless interleaved boost PFC rectifier is proposed for improving power efficiency and system performance in this paper. By combining the conventional bridgeless topology and the interleaved technology, this rectifier is comprised of two interleaved boost branches without the front-end diode bridge. Each branch operates in every half-line cycle, together with the current following through a minimum number of switching devices. By using the interleaved approach, this topology not only decreases the current stress of the switching devices but also reduces the current and voltage ripple. Moreover, as operating in critical-conduction-mode (CrM), all the switches can achieve soft-switching characteristics to reduce the switching loss and evidently raise the conversion efficiency. Using a conventional interleaved controller, simple control scheme can be employed to the proposed converter. The operational principle and theoretical analysis of the proposed converter are presented. Finally, simulations based on actual semiconductor models were carried out to verify the feasibility and exactness of the proposed rectifier.

Keywords—PFC, bridgeless, interleaved, critical conduction mode

I. INTRODUCTION

Active power factor correction techniques (PFC) are popularly employed in many types of electronic equipments to comply with international standards such as IEC-61000-3-2. Most active PFC regulators comprise a bridge rectifier and a high frequency single-ended dc-dc converter such as a boost topology.

Generally, current in a conventional boost PFC converter flows through two rectifier diodes and one power switch during switch on state, and two rectifier diodes and one boost diode during switch off state. At low line input and high power applications, the high condition loss caused by the high forward voltage drop of the bridge diode begins to degrade to overall system efficiency, and the heat generated within the rectifier may destroy individual power devices.

In order to maximize power supply efficiency and optimize system thermal design, significant research efforts have been made on the developments of PFC rectifiers. Two representative techniques are the bridgeless PFC [1]–[4] and the interleaved PFC converters [5], [6].

To reduce conduction losses, the front-end bridge rectifier and the PFC stage have been combined to develop a bridgeless PFC rectifier. Without the input rectifier bridge, this topology may reduce the conduction loss by allowing

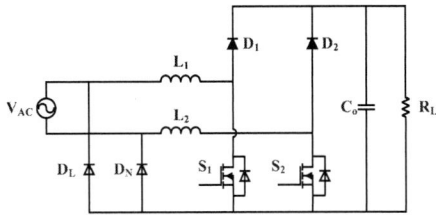

Fig. 1.   Basic bridgeless boost PFC rectifier.

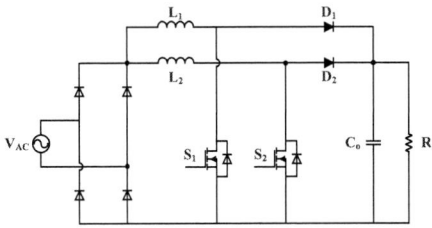

Fig. 2.   Basic interleaved boost PFC rectifier.

the current to follow through a minimum number of switching devices comparing with the conventional PFC rectifiers. In this kind of converter, two boost converters are employed for each half line cycles. As shown in Fig.1, during a positive half line cycle, the boost branch, $L_1 - S_1 - D_1$, is active through D_N, while during a negative half line cycle, the boost branch, $L_2 - S_2 - D_2$, is active through D_L. In addition, in the bridgeless PFC circuit, two inductors compared to a single inductor have better thermal performance. Therefore, this converter is able to increase the system efficiency specially in high power applications.

Another effective topology is interleaved PFC converter which usually combines more than two conventional boost topologies. A two-phase interleaved boost converter is shown in Fig.2. Because of the interleaved operation, the current waveforms exhibit lower ripple and smaller harmonics content than those of the conventional topologies in the same power condition [7]–[9] . Therefore, the size and losses of the energy storage inductors and filtering stages can be reduced and the switching losses can be decreased. In addition, by operating on CrM mode, soft switching condition is achieved for all semiconductor elements without any auxiliary circuit.

This paper proposes a novel boost PFC rectifier

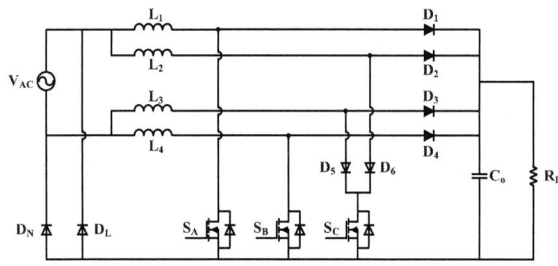

Fig. 3. The proposed bridgeless interleaved boost PFC.

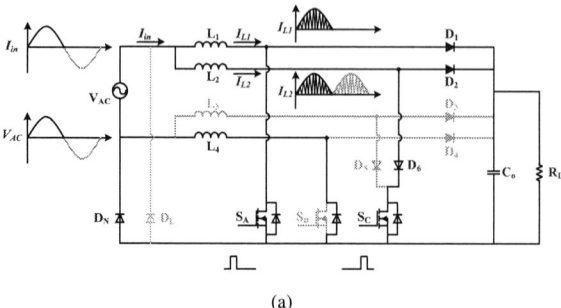

(a)

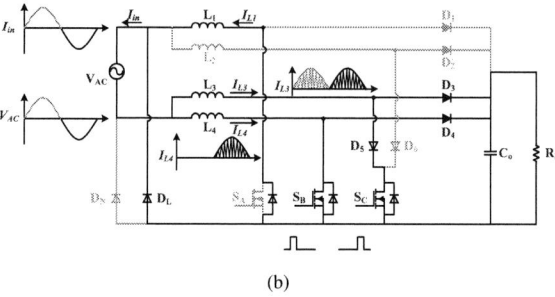

(b)

Fig. 4. Operating stages for the proposed converter in Fig.3. (a) During the positive half-line period. (b) During the negative half-line period of the input voltage.

combining the conventional bridgeless topologies and the interleaved converters. The proposed rectifier is composed of two conventional boost converter, two auxiliary inductors and one auxiliary switch as shown in Fig.3. When operating in CrM mode, the rectifier is able to achieve zero-current-switching (ZCS) during switch turn-on transition, and zero-voltage-switching (ZVS) during turn-off transition to reduce switching losses. The detailed circuit configuration and operation principle are described, along with the comparison study with conventional PFC converters and simulation results.

II. CIRCUIT CONFIGURATION AND OPERATION PRINCIPLE

The main structure of the circuit is formed by combining two interleaved boost branches, branch A and branch B, one for each half-line period of the input voltage, as can be seen in Fig.3. Each of the interleaved branches is comprised of two phase boost converters connected in parallel. The interleaved branch A is consisted of L_1-S_A-D_1 as phase A1, and L_2-D_6-S_C-D_2 as phase A2, while the interleaved branch B is consisted of L_4-S_B-D_4 as phase B1, and L_3-D_5-S_C-D_4 as phase B2, respectively. It should be noted that D_L and D_N are slow-recovery diodes, whereas $D_1 \sim D_4$ are fast-recovery diodes.

As illustrated in Fig.3, the auxiliary switch S_C is shared by the second phases of each interleaved branch, meanwhile, the current from each phase is blocked by employing diodes D_5 and D_6. S_C operates in the whole line cycle of the input voltage while one of S_A and S_B operates as active switch and another one operates as a diode in each half line cycle. As the input line frequency is low enough (50Hz or 60Hz), slow recovery diodes can be used for D_5 and D_6. Moreover, during the reverse recovery interval, enough time could be provided by the slow-recovery diodes to achieve soft switching conditions for S_C.

To guarantee the common-mode EMI performance, D_N conducts in the entire positive half line cycle and D_L conducts in the entire negative half line cycle. During the positive half line cycle, the branch A boost circuit is active through D_N, while during the negative half line cycle, the branch B is active through D_L. Although there are two additional inductors compared to the conventional interleaved topologies, a better thermal performance can be achieved. On the other hand, the two inductors of each

branch in the proposed topologies can be coupled on the same magnetic core, allowing considerable size and cost reduction.

The operation circuit during positive and negative half-line period for the proposed converter are shown in Fig.4(a) and Fig.4(b), respectively. For each branch, there are one slow-recovery diode, one MOSFET switch, and one fast recovery diode in the current flowing path for one phase, while there is one more slow-recovery diode, D_5 or D_6, for another phase, which can reduce the number of conduction devices compared to the conventional interleaved PFC rectifiers. Hence, the conduction losses, as well as the thermal stresses on the semiconductor devices, are further reduced, and the circuit efficiency is improved.

Furthermore, the two branch power switches, S_A and S_B, can be driven by the same control signal while S_C is driven by the interleaved control signal. This operation significantly simplifies the control scheme and can easily implemented by using some industry standard interleaved controller ICs in the market.

Every branch phase can operates in DCM or CCM, but several advantages can be obtained when operating the rectifier in CrM. As a result, soft-switching and low-ripple input current not only increase the converter efficiency but also reduce the conducted EMI noise compared to conventional bridgeless converters. In addition, smaller inductor, reasonable size of capacitor, and less-ideal switches can be used to implement those advantages. The other advantage of the proposed converter is the current stress reduction for the power switch compared to the conventional boost rectifiers because of the interleaved operation.

TABLE I. COMPARISON BETWEEN CONVENTIONAL AND THE PROPOSED PFC RECTIFIER

Item	Bridgeless PFC Rectifier	Interleaved PFC Rectifier	Proposed PFC Rectifier
Boost inductance value	$0.5L$	L	L
Inductance peak current	$2I_{pk}$	I_{pk}	I_{pk}
Boost diode current	$\sqrt{2}I_D$	I_D	I_D
Thermal performance	Moderate	Moderate	High
Current path (switch on state)	1 slow diode, 1 switch	2 slow diodes, 1 switch (Phase 1), 2 slow diodes, 1 switch (Phase 2)	1 slow diodes, 1 switch (Phase 1), 2 slow diodes, 1 switch (Phase 2)
Current path (switch off state)	1 slow diode, 1 fast diode	2 slow diodes, 1 fast diode (Phase 1), 2 slow diodes, 1 fast diode (Phase 2)	1 slow diodes, 1 fast diode (Phase 1), 2 slow diodes, 1 fast diode (Phase 2)

III. DESIGN PROCEDURE AND COMPARISON STUDY

Since the proposed converter is constructed by connecting two converters with each operating as interleaved boost converter in CrM, the switching performance of the converter remains the advantages of the bridgeless topologies as well as the interleaved converters, which results in lower switching losses, condition losses, and current ripple.

As the parallel-operated boost units are identical, design procedure of the proposed converter becomes quite similar to the conventional interleaved boost rectifiers. By operating the converter in CrM mode, according to the relationship between input and output power, the maximum peak inductance current for one phase is derived as:

$$I_{pk} = \frac{\sqrt{2}P_{out}}{\eta V_{inmin}}, \tag{1}$$

where P_{out} is the maximum output power, η is the system efficiency, and V_{inmin} is the minimum dc-link voltage.

If the minimum switching frequency is selected at the minimum dc-link voltage, the inductance value can be obtained as follows:

$$L = \frac{2\eta V_{inmin}^2 (V_{out} - \sqrt{2}V_{inmin})}{V_{out}P_{out}f_s}, \tag{2}$$

where f_s is the maximum switching frequency; V_{out} is the output voltage of the rectifier. A slightly larger inductance value should be selected to ensure CrM operation in the full load range.

Every boost unit contributes half power of the whole circuit. Therefore, the boost diode rms current of the proposed converter can be expressed as:

$$I_D = \sqrt{2}I_{pk}\sqrt{\frac{V_{inmin}}{9\pi V_{out}}}. \tag{3}$$

The difference between the proposed PFC converter and the conventional boost PFC converters are summarized in Table.I. As is described, the input current in the proposed PFC circuit flows through fewer power devices compared with conventional interleaved converters, and the peak inductance current reduces half than conventional bridgeless converters, simultaneously. Although there are more power components, higher system efficiency and better thermal performance can be achieved in the proposed rectifier.

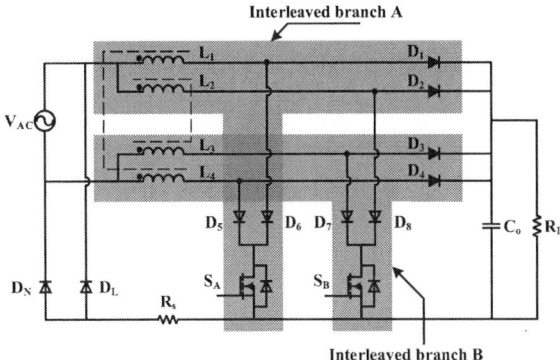

Fig. 5. The proposed bridgeless interleaved boost PFC.

TABLE II. MAJOR COMPONENTS OF THE SIMULATION MODEL AND CIRCUIT PROTOTYPE

Components	Symbol	Part number
Switch MOSFET	$S_A \sim S_C$	IRF840
Low-recovery diode	D_L, D_N, D_5, D_6	GBU6J
Fast-recovery diode	$D_1 \sim D_4$	MURS360T3

For the proposed circuit, the concept illustrated in Section II could be extended to another phase lag. Fig.5 shows an improved bridgeless interleaved boost PFC converter. The implementation only involved adding two auxiliary blocking diodes with one MOSFET less. To improve the utilization of magnetic components, coupled inductors could also be involved to the circuit. The basic operation principles are the same as those of the proposed version in Section. II.

IV. SIMULATION AND EXPERIMENTAL RESULTS

A. Simulation Results

To verify the feasibility of the proposed rectifier, a Pspice simulation model of the circuit has been developed. The model is designed as the power specifications shown in Table.III.

In order to examine the proper performance of the rectifier in practical applications, PSpice actual semiconductor models of the power devices were employed, as listed in Table.II. An interleaving CrM PFC controller from TI, UCC28063, has been used as the system controller. Fig.6 presents the power stage of the simulated model.

978-1-4799-2706-7/14 $31.00 © 2014 IEEE

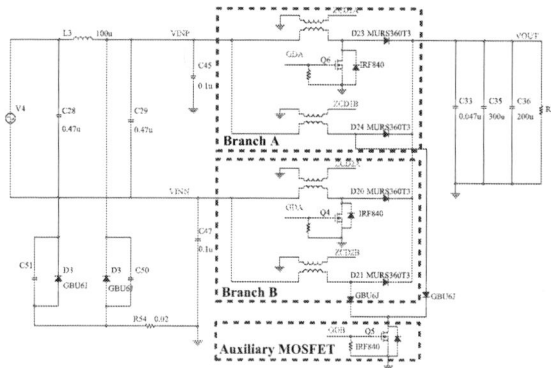

Fig. 6.　Power stage PSpice simulation model of the proposed rectifier

TABLE III.　PARAMETERS OF THE SIMULATION MODEL

Parameters	Symbol	Value
Input voltage	V_{ac}	85VAC~265VAC
Line frequency	f_L	60Hz
Output voltage	V_o	385VDC
Output power	P_{out}	600W
Boost inductance	$L_1{\sim}L_4$	80μH
Output capacitor	C_{out}	500μF

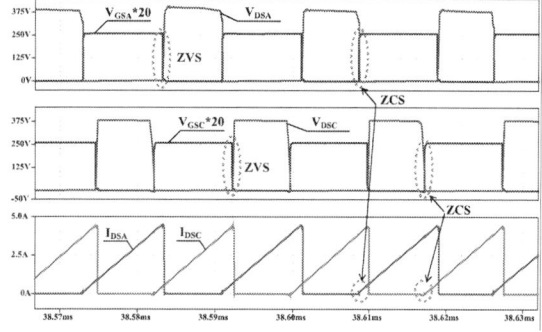

Fig. 7.　Waveforms of driver signal V_{GS}, drain-to-source voltage V_{DS}, and drain-to-source current I_{DS} of S_A and S_C.

Fig.7 illustrates the switching waveforms of branch A. As is obvious, the switches are turned on under ZCS and turned off under ZVS conditions, which could be achieved under CrM operation mode. Therefore, the proposed circuit is characterized with less reverse-recovery noise and significant low switching losses.

The total inductance current and the current of each phase inductor of branch A are shown in Fig.8, along with the gate signals (S_A and S_C). As it can be observed, effective ripple frequency increased twice and peak-to-peak value input ripple current is significantly reduced compared to the two inductance current ripple because of the interleaved operation. Thus, input filter size can be decreased. It can also be indicated that the boost phases operate in CrM.

Fig.9 shows the input current versus input voltage at full load, as well as the output voltage simultaneously. It can be observed that input current is in phase with the

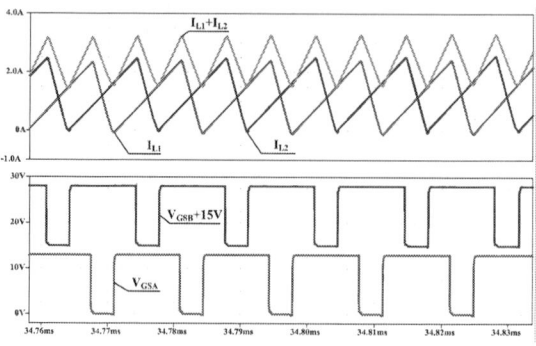

Fig. 8.　Simulation results of the current waveforms of L_1 and L_2 as well as output current.

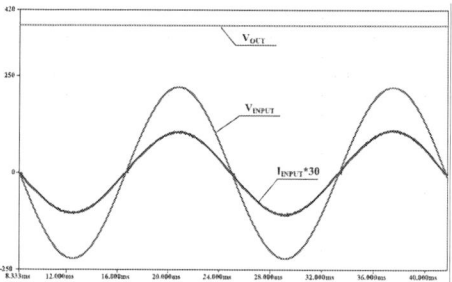

Fig. 9.　Simulation results of the input voltage V_{INPUT}, input current I_{INPUT}, and output voltage V_{OUTPUT}.

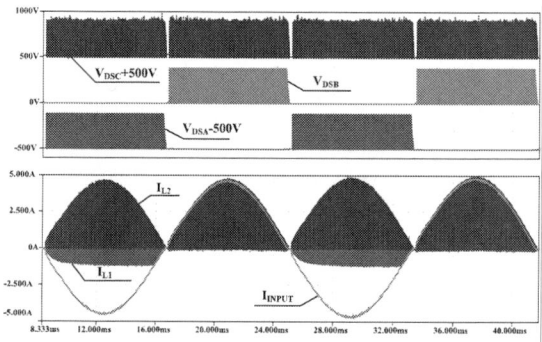

Fig. 10.　Waveforms of drain-to-source voltages of the MOSFET switches, inductance current of branch A, and the input current.

input voltage and is practically sinusoidal with low total harmonic distortion and high power factor.

Fig.10 represents the inductance current waveforms of branch A under 220V_{AC} line voltage and full-load conditions. The input current and drain-to-source voltages (S_A and S_C) are also shown. As expected from the operation analysis, S_A and S_B operate in each half-line cycle, while S_C operates in the whole line cycle.

B. Experimental Results

A 600W experimental prototype circuit, as shown in Fig.11, has been built and tested. The circuit specifications are the same as those of the simulation model described in Table.III.

Fig. 11.   Photograph of the prototype converter.

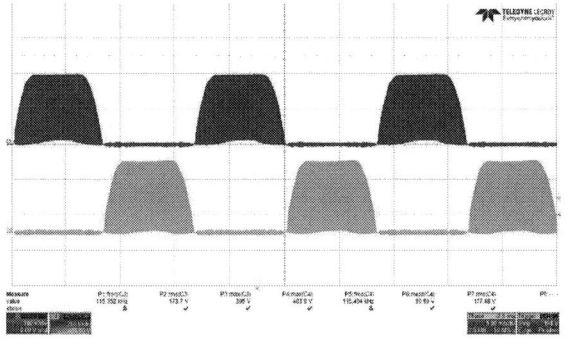

Fig. 12.   Waveforms of drain-to-source voltages of the MOSFET switches (5ms/div). From top to bottom: 200 V/div, 200 V/div.

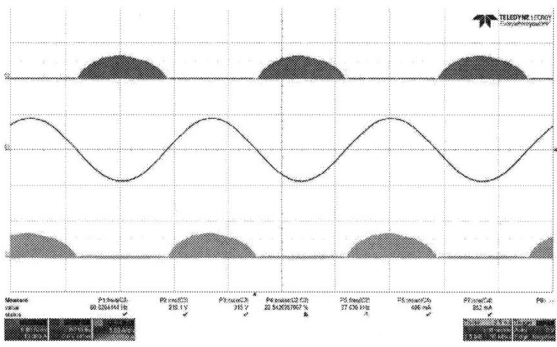

Fig. 13.   Inductance current waveforms of L_1 and L_4, as well as the input voltage (5ms/div). From top to bottom: 5 A/div, 350 V/div, and 5 A/div.

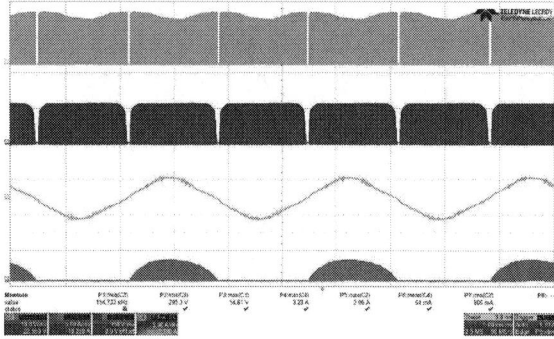

Fig. 14.   Waveforms of gate driver and drain-to-source voltages of S_C, inductance current of L_2, and the input current (5ms/div). From top to bottom: 10 V/div, 350 V/div, 5 A/div, and 5 A/div.

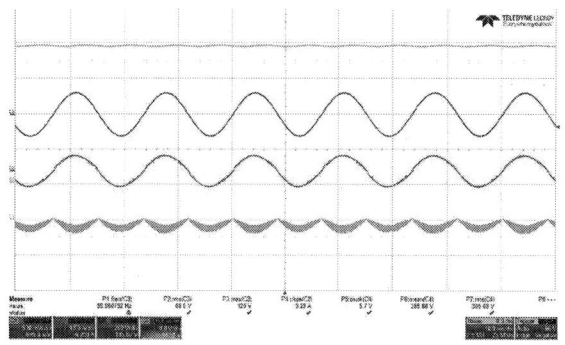

Fig. 15.   Experimental results for the input voltage, input current, output voltage and current sensing signal under full load with 220 V input voltage (10ms/div). From top to bottom: 100 V/div, 200 V/div, 10 A/div and 500 mV/div.

Experiments have been carried out to verify the analysis. The drain-to-source voltages of S_A and S_B are shown in Fig.12. Fig.13 shows the current waveforms of L_1 and L_4. It can be observed that S_A and S_B operate in each half-line cycle, which could improve the system thermal performance. The inductance currents follow the input voltage under CrM operation mode. It should be noted that there is small circulating current between S_A and S_B due to the intrinsic diode of MOSFET. This undesired circulating loop can be blocked by the improved circuit in Fig.5.

The gate driver signal and drain-to-source voltage of S_C, the inductance current of L_2, and input current waveform are shown in Fig.14. It can be observed that S_C operates in the whole-line cycle, while L_2 and L_3 operate in each half-line cycle. During each half-line cycle, two switches (S_A, S_C or S_B, S_C) operate in interleaved mode, which is the same operation with conventional interleaved topology. However, the number of components in current path is reduced by the bridgeless topology.

Fig.15 shows the input current and input voltage at full load, as well as the output voltage simultaneously. The current sensing signal is also shown in the figure. It can be observed that input current is in phase with the input voltage and is practically sinusoidal with low total harmonic distortion and high power factor.

The sum inductance current and the current of each phase inductor of branch A are shown in Fig.16. As it can be seen, effective ripple frequency increased twice and peak-to-peak value of current ripple is significantly reduced compared to the two inductance current ripple, which is benefited from the interleaved operation. As a result, input and output filter size can be decreased. Therefore, reserve-recovery noise could be decreased significantly. It can also be indicated that the boost phases operate in CrM.

Fig.17 shows the efficiency curves of the proposed PFC converter under 85 Vrms and 265 Vrms input voltages. It is illustrated that the efficiency at full load under 85 Vrms is above 95.1% and the maximum efficiency is 97.5%, which is achieved at full load under 265 V. It can

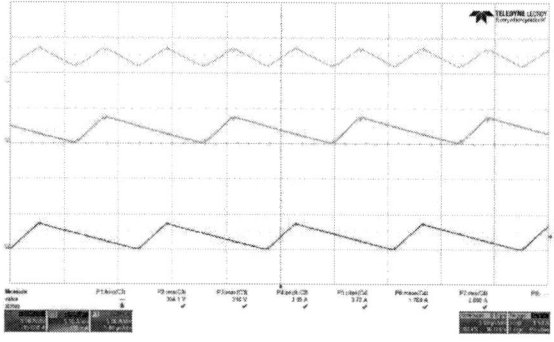

Fig. 16. Waveforms of drain-to-source voltages of the MOSFET switches, inductance current of branch A, and the input current.

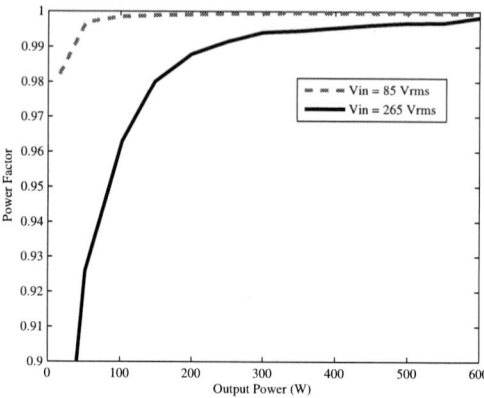

Fig. 18. Waveforms of drain-to-source voltages of the MOSFET switches, inductance current of branch A, and the input current.

ACKNOWLEDGMENT

This work was supported by the Energy Efficiency & Resources Core Technology Program of the Korea Institute of Energy Technology Evaluation and Planning(KETEP) granted financial resource from the Ministry of Trade, Industry & Energy, Republic of Korea. (No.20132020101530)

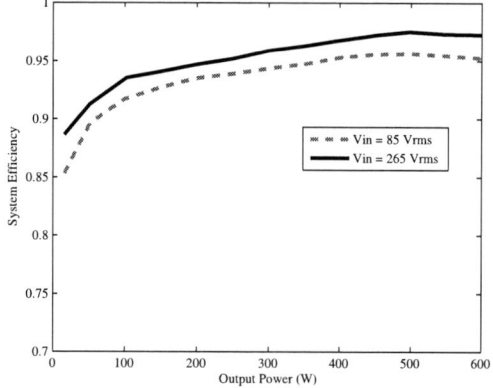

Fig. 17. Waveforms of drain-to-source voltages of the MOSFET switches, inductance current of branch A, and the input current.

be observed that the efficiency is improved at heavy load, which is because the component number in the current flowing path is reduced.

In Fig.18, the power factors of the proposed circuit under 85 Vrms and 265 Vrms input voltages are shown. It is significant in the sense that the proposed rectifier achieves very high power factor under 85 Vrms, which is always high than 98% from 10 % load to full load, while under 265 Vrms input voltage, the power factor is high than 99.6% under the rated load.

V. CONCLUSION

In this paper, a novel bridgeless interleaved boost PFC rectifier is proposed and analysed. The proposed circuit combines the conventional interleaved boost converter with the bridgeless PFC topology. With operating in interleaved and CrM mode, this converter is able to provide higher output power and lower current ripple. To verify the feasibility of the proposed converter, simulation model with actual parameters were developed. A 600 W prototype circuit was designed and tested. Experimental results show that almost unity power-factor and very low THD are achieved. Very high power efficiencies of 95.1% and 97.5% are obtained under full load at 85 V and 265 V input voltages, respectively.

REFERENCES

[1] L. Huber, Y. Jang, and M. Jovanovic, "Performance evaluation of bridgeless pfc boost rectifiers," in *Applied Power Electronics Conference, APEC 2007 - Twenty Second Annual IEEE*, Feb 2007, pp. 165–171.

[2] A. Sabzali, E. Ismail, M. Al-Saffar, and A. Fardoun, "New bridgeless dcm sepic and cuk pfc rectifiers with low conduction and switching losses," *Industry Applications, IEEE Transactions on*, vol. 47, no. 2, pp. 873–881, March 2011.

[3] B. Su, J. Zhang, and Z. Lu, "Single inductor three-level boost bridgeless pfc rectifier with nature voltage clamp," in *Power Electronics Conference (IPEC), 2010 International*, June 2010, pp. 2092–2097.

[4] C. Zheng, H. Ma, B. Gu, R. Chen, E. Faraci, W. Yu, J.-S. Lai, and H.-S. Koh, "An improved bridgeless sepic pfc rectifier with optimized magnetic utilization, minimized circulating losses, and reduced sensing noise," in *Applied Power Electronics Conference and Exposition (APEC), 2013 Twenty-Eighth Annual IEEE*, March 2013, pp. 1906–1911.

[5] G. Yao, A. Chen, and X. He, "Soft switching circuit for interleaved boost converters," *Power Electronics, IEEE Transactions on*, vol. 22, no. 1, pp. 80–86, Jan 2007.

[6] E. Firmansyah, S. Tomioka, S. Abe, M. Shoyama, and T. Ninomiya, "A critical-conduction-mode bridgeless interleaved boost power factor correction," in *Telecommunications Energy Conference, 2009. INTELEC 2009. 31st International*, Oct 2009, pp. 1–5.

[7] M. Pahlevaninezhad, P. Das, J. Drobnik, P. Jain, and A. Bakhshai, "A zvs interleaved boost ac/dc converter used in plug-in electric vehicles," *Power Electronics, IEEE Transactions on*, vol. 27, no. 8, pp. 3513–3529, 2012.

[8] Y. Jang and M. Jovanovic, "A bridgeless pfc boost rectifier with optimized magnetic utilization," *Power Electronics, IEEE Transactions on*, vol. 24, no. 1, pp. 85–93, 2009.

[9] J. Figueiredo, F. Tofoli, and B. Silva, "A review of single-phase pfc topologies based on the boost converter," in *Industry Applications (INDUSCON), 2010 9th IEEE/IAS International Conference on*, 2010, pp. 1–6.

Analysis and Design of a Push-Pull Single-Stage Flyback Power Factor Corrector

Yu-Kang Lo, Huang-Jen Chiu, and Yu-Chen Liu
Department of Electronic Engineering
National Taiwan University of Science and Technology
Taipei City, Taiwan, ROC
Email: yklo@mail.ntust.edu.tw

Chung-Yi Lin, Shih-Jen Cheng, and CS Yang
Flextronics Power
New Taipei City, Taiwan, ROC
Email: keyboard.lin@flextronics.com

Abstract-A novel mixed-mode controlled push-pull single-stage flyback power factor corrector (PFC) composed of two flyback PFCs with a coupled inductor is proposed in this paper. The studied single-stage PFC consists of two power modules which have common secondary winding, output diode, output capacitor, and input filtering capacitor. Besides possessing the capability of sharing the input current and output current equally, integrating two input inductors into one magnetic core makes the operating frequency of the core double the switching frequency. In addition, a cut-in-half duty cycle can reduce the conduction losses of the switches and both the turns and diameters of the inductor windings. Both the continuous conduction mode (CCM) and the transition mode (TM) controls are fulfilled in the proposed single-stage PFC during one line frequency cycle. Under CCM operations, the smaller inductor current ripple reduces both the core loss of the coupled inductor and the conduction losses of the switches, which not only promotes the heavy-load efficiencies, but also alleviates the current stresses of the switches and the output diode. Under TM operations, quasi-resonant (QR) valley witching on the switches and zero-current switching (ZCS) of the output diode can reduce the switching losses. The light-load conversion efficiencies are thus improved. Detailed analysis of the proposed topology is given. The experiments are conducted on a prototype with a universal line voltage, a 19-V output DC voltage, and a 190-W output power to verify its feasibility.

KEYWORDS: Power factor correction, single-stage converter, and flyback converter.

I. INTRODUCTION

A two-stage cascade structure composed of a power factor corrector (PFC) [1-3] and a flyback dc/dc converter is widely used for a single-output ac/dc adaptor in low-power applications, such as laptops and other electronic products of which the input power is higher than 75 W. The front-end PFC stage not only improves the power-factor (PF) value and reduces the total harmonic distortion (THD), but also provides a regulated DC output voltage. The high output voltage of the PFC is fed into the post-stage flyback dc/dc converter and then converted to the required output voltage level. Some advantages such as low output voltage ripple and easy satisfaction of hold-up time requirement can be obtained in this two-stage cascade structure. However, the two-stage cascade structure presents high complexity, large volume, and high cost.

A novel mixed-mode controlled push-pull single-stage flyback PFC composed of two flyback PFCs with a coupled inductor is proposed in this paper. The studied single-stage PFC consists of two power modules which have common secondary winding, output diode, output capacitor, and input filtering capacitor. Besides possessing the capability of sharing the input current and output current equally, integrating two input inductors into one magnetic core makes the operating frequency of the core double the switching frequency. The circuit volume and the current ripple can be thus reduced under the same inductance. In addition, a cut-in-half duty cycle can reduce the conduction losses of the switches and both the turns and diameters of the inductor windings. The proposed single-stage PFC has the voltage step-down and step-up capabilities, which feature a better performance at zero crossings. In addition, the transfer ratio of the output voltage to the input voltage can be adjusted flexibly by the transformer turns ratio.

Both the continuous conduction mode (CCM) and the TM controls are fulfilled in the proposed single-stage PFC during one line frequency cycle. The durations of these two modes depend on both the line voltage and the output load due to the pre-set fixed off-time (FOT) [4, 5]. Under CCM operations, the smaller inductor current ripple reduces both the core loss of the coupled inductor and the conduction losses of the switches, which not only promotes the heavy-load efficiencies, but also alleviates the current stresses of the switches and the output diode. Under TM operations, quasi-resonant (QR) valley switching on the switches and zero-current switching (ZCS) of the output diode can reduce the switching losses. The light-load conversion efficiencies are thus improved.

II. OPERATING PRINCIPLES

Fig. 1 shows the schematics of a push-pull single-stage flyback PFC. Module A consists of the switch S_a, the primary winding N_{Pa}, and the inductor L_a. Module B consists of the switch S_b, the primary winding N_{Pb}, and the inductor L_b. These two modules have common secondary winding N_s, output diode D_o, output capacitor C_o, and input filtering capacitor C_{in}. L_a and L_b are two coupled windings wound on one magnetic core with the same turns and inductances. The proposed PFC under TM is operated with a constant on-time and variable-switching frequency control. The key waveforms are drawn as in Fig. 2(a). The key waveforms of the proposed PFC under CCM are drawn as in Fig. 2(b). The dead times between S_a and S_b are controlled by the pre-set FOT to force the inductor currents i_{La} and i_{Lb} to operate in CCM. To analyze the operating principles, there are some assumptions as listed below.

- The conducting resistances of S_a and S_b are ideally zero. The conduction time interval is DT_s, where D is the duty cycle and T_s is the switching period.
- The forward voltage of D_o is ideally zero.
- The magnetic core for manufacturing L_a and L_b is perfectly coupled without leakage inductance. In addition, the turns of the windings N_{Pa} and N_{Pb} are the same. Therefore, L_a and L_b are also matched.

The operating states of the proposed PFC under TM are analyzed as follows.

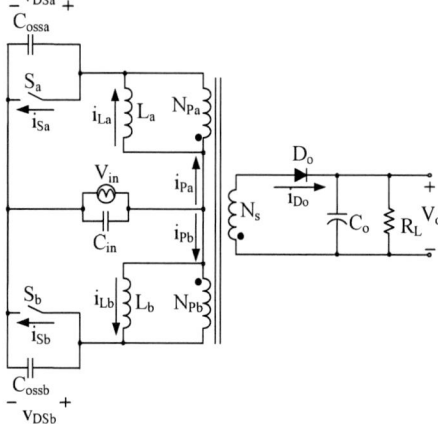

Fig. 1. Power circuit of a push-pull single-stage flyback PFC.

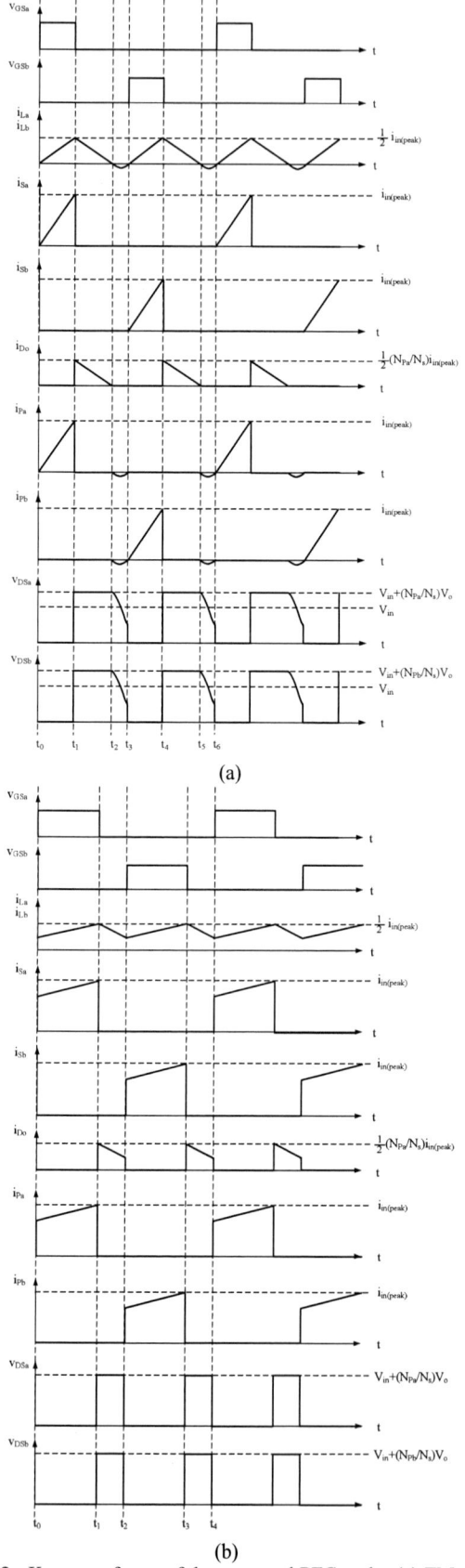

Fig. 2. Key waveforms of the proposed PFC under (a) TM and (b) CCM.

State 1: $t_0 < t < t_1$

As shown in Fig. 3(a), S_a conducts and S_b is off. At the primary side, the voltage across N_{Pa} equals the rectified line voltage V_{in}, which is also coupled to N_{Pb}. Both the windings' dotted terminals are positive. At the secondary side, the reflected voltage across N_s makes D_o reverse-biased. Thus both L_a and L_b store energy, and the inductor currents i_{La} and i_{Lb} increase linearly. In module B, i_{Lb} flows into the non-dotted terminal of N_{Pb}. By the coupling effect, this current flows into the dotted node of N_{Pa}. C_o supplies the energy to the load during this interval.

State 2: $t_1 < t < t_2$

As shown in Fig. 3(b), S_a is turned off and S_b is still off. D_o conducts for i_{La} and i_{Lb} to flow continuously. L_a and L_b release their energies to C_o and the load. Both i_{La} and i_{Lb} are decreasing linearly. For TM, this state ends until L_a and L_b release their energies completely, and i_{La} and i_{Lb} decrease to zero. For CCM, this state is terminated by the pre-set FOT and enters into the next half switching cycle, of which the operating modes are similar to state 1 and state 2.

State 3: $t_2 < t < t_3$

As shown in Fig. 3(c), both S_a and S_b keep turned off. At t_2, D_o is turned off with ZCS since i_{La} and i_{Lb} decrease to zero naturally. In this interval, C_o supplies the energy to the load. At the same time, in module A, the series resonant loop formed by V_{in}, the parallel connection of L_a and L_b, and the output capacitor of the switch S_a, C_{ossa}, starts to resonate. Similarly, in module B, the series resonant loop formed by V_{in}, the parallel connection of L_a and L_b, and the output capacitor of the switch S_b, C_{ossb}, begins to resonate. Therefore, v_{DSa} and v_{DSb} decrease simultaneously. For TM, S_b can be turned on with QR valley switching when the voltage generated by the zero-current-detection (ZCD) circuit decreases to the level pre-set by the controller. Then this state ends and enters into the next half switching cycle, of which the operating modes are similar to the above three states.

The operating states of the proposed topology under CCM are similar to TM, but without state 3.

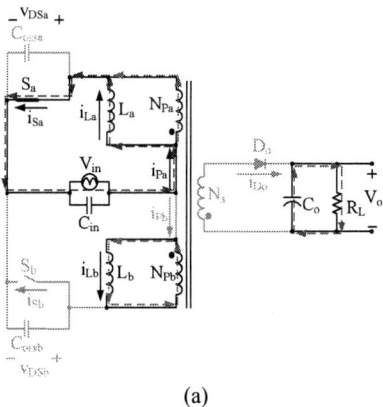

(a)

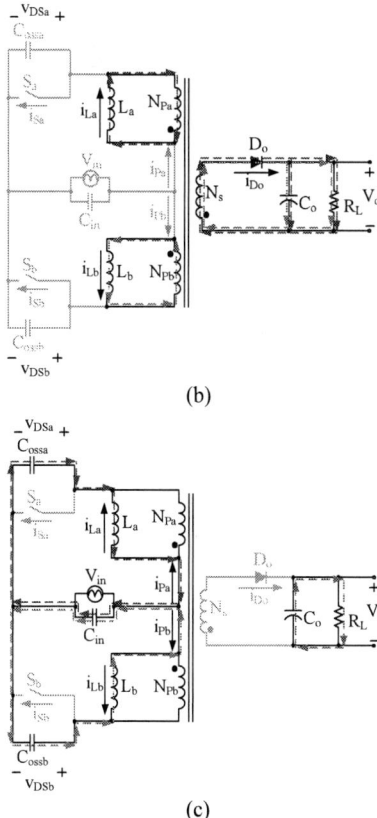

(b)

(c)

Fig. 3. Conduction paths of (a) State 1, (b) State 2, and (c) State 3 for a push-pull single-stage flyback PFC during a half switching period.

The waveform of the rectified line current i_{in}, which equals the sum of i_{La} and i_{Lb} while one of the switches turning on, is shown in Fig. 4. The peak envelope of i_{in}, $i_{in(peak)}\sin(\theta)$, forms into sinusoidal shape by the mixed-mode control composed of CCM and TM. Since the required discharge time for i_{La} or i_{Lb} to drop to zero, T_{FW}, depends on its stored energy, the proposed PFC enters into CCM operation if the pre-set FOT, T_{off}, is shorter than T_{FW}. Otherwise, TM operation is achieved by the ZCD circuit. The transition angle θ_T is defined as the phase angle boundary from TM to CCM. By considering the waveform symmetry, the phase angle ($\pi - \theta_T$) is the boundary from CCM back to TM. Therefore, both TM and CCM operation modes coexist during a line frequency cycle, and each occupied percentages are dependent on both the line voltage and the output load because of the pre-set FOT.

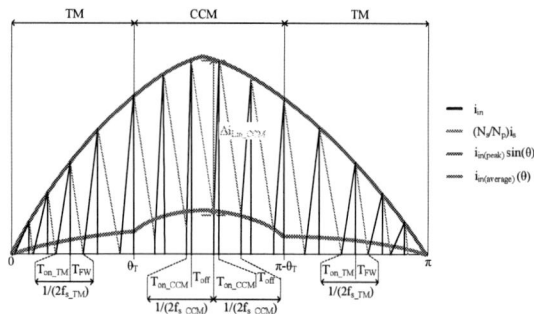

Fig. 4. Waveform of the rectified line current under the mixed-mode controls.

III. EXPERIMENTAL RESULTS

In order to demonstrate the feasibility of the proposed PFC and verify the theorems, some experimental results are shown as follows. The circuit specifications and components are listed in Table 1. The power circuit is designed under CCM. The magnetizing inductance of the coupled inductor is measured as 200 µH from either side of the winding. The inductance of L_a or L_b thus equals to 400 µH due to the same winding turns of N_{Pa} and N_{Pb} which are parallel wound in the same magnetic core.

TABLE 1
THE CIRCUIT SPECIFICATIONS AND COMPONENTS OF THE PFC.

Parameter	Value
Line voltage: V_{in_rms}	90 V_{ac} ~ 264 V_{ac}
Output DC voltage: V_o	19 V_{dc}
Rated output power: $P_{o(rated)}$	190 W
The maximum switching frequency at the peaks of the minimum RMS line voltage: $f_{s\_CCM(max)}$	40 kHz
Maximum duty cycle: D_{max}	0.246
Magnetic core	TDK PQ32/30
Winding turns of module A/B: N_{Pa}/ N_{Pb}/ N_s	26/ 26/ 4 turns

Fig. 5 shows the waveforms of v_{GSa}, v_{GSb}, and v_{DSa}/v_{DSb} under $V_{in\_rms} = 110$ V_{ac} and $P_o = 19$ W. The proposed mixed mode control operates under TM and features valley switching of the switches at turn-on moments. Fig. 6 shows the waveforms of v_{GSa}, v_{GSb}, and v_{DSa}/v_{DSb} under $V_{in\_rms} = 110$ V_{ac} and $P_o = 190$ W. The proposed mixed mode control operates under CCM, and the switches turn on with hard switching.

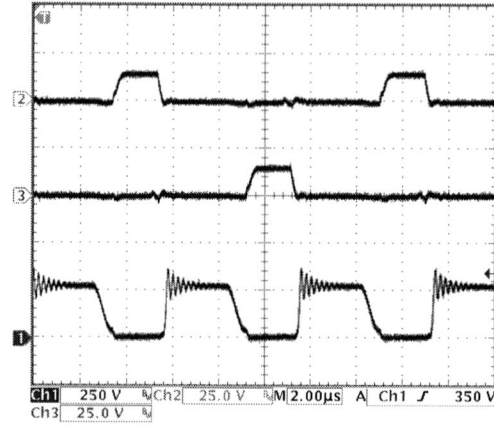

Fig. 5. Measured waveforms of v_{GSa} (Ch2), v_{GSb} (Ch3), and v_{DSa}/v_{DSb} (Ch1) under $V_{in\_rms} = 110$ V_{ac} and $P_o = 19$ W.

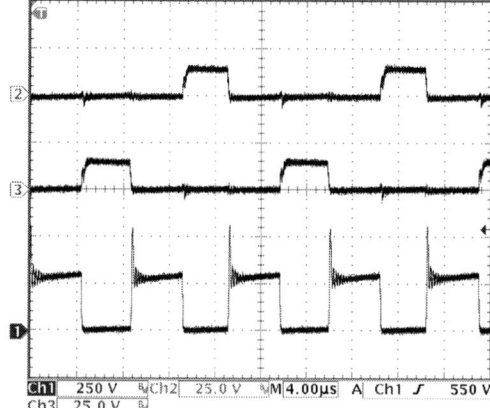

Fig. 6. Measured waveforms of v_{GSa} (Ch2), v_{GSb} (Ch3), and v_{DSa}/v_{DSb} (Ch1) under $V_{in\_rms} = 110$ V_{ac} and $P_o = 190$ W.

Fig. 7 shows the measured efficiencies of the proposed PFC at $V_{in\_rms} = 110$ V_{ac} and $V_{in\_rms} = 220$ V_{ac}, respectively. Fig. 8 shows the measured PF values of the proposed PFC at $V_{in\_rms} = 110$ V_{ac} and $V_{in\_rms} = 220$ V_{ac}, respectively.

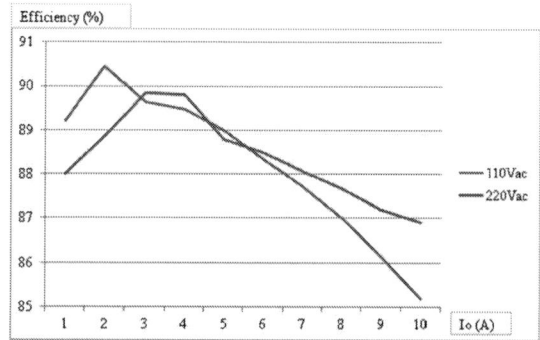

Fig. 7. Measured efficiencies of the proposed PFC.

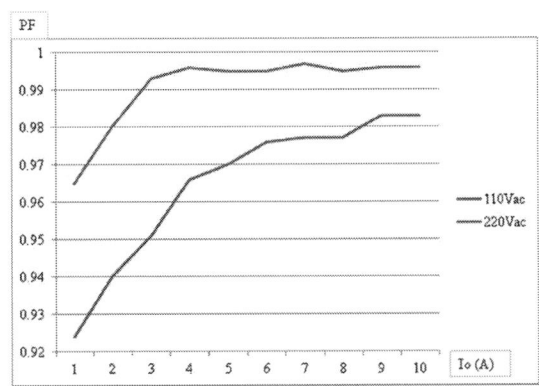

Fig. 8. Measured PF values of the proposed PFC.

IV. CONCLUSIONS

This paper proposes a novel mixed-mode controlled push-pull single-stage flyback PFC composed of two flyback PFCs with a coupled inductor. The features of having common secondary winding, output diode, output capacitor, and input filtering capacitor as well as integrating two input inductors into one magnetic core reduce the complexity, the circuit volume, and the cost. Both the CCM and the TM controls are fulfilled in the proposed single-stage PFC during one line frequency cycle to promote both the heavy-load and light-load efficiencies. A prototype with a universal line voltage, a 19-V output DC voltage, and a 190-W output power demonstrated the theorems mentioned above and verified its feasibility. From the experimental results, the proposed mixed-mode PFC enters into TM when the output power is lower than about 60 W and 90 W at $V_{in\_rms} = 110$ V_{ac} and $V_{in\_rms} = 220$ V_{ac}, respectively. The measured efficiencies are all above 85 % from light load to full load whether at $V_{in\_rms} = 110$ V_{ac} or $V_{in\_rms} = 220$ V_{ac}. The measured PF values are all above 0.96 and 0.92 from light load to full load at $V_{in\_rms} = 110$ V_{ac} and $V_{in\_rms} = 220$ V_{ac}, respectively. The measured load regulation whether at $V_{in\_rms} = 110$ V_{ac} or $V_{in\_rms} = 220$ V_{ac} is within ±1.1%.

REFERENCES

[1] K. Yao, X. Ruan, X. Mao, and Z. Ye, "Variable-duty-cycle control to achieve high input power factor for DCM boost PFC converter," *IEEE Trans. Industrial Electronics.*, vol. 58, no. 5, pp. 1856-1865, Jun. 2011.

[2] J.-M. Kwon, W.-Y. Choi, and B.-H. Kwon, "Single-stage quasi-resonant flyback converter for a cost-effective PDP sustain power module," *IEEE Trans. Industrial Electronics.*, vol. 58, no. 6, pp. 2372-2377, Jun. 2011.

[3] F. Zhang and J. Xu, "A novel PCCM boost PFC converter with fast dynamic response," *IEEE Trans. Industrial Electronics.*, vol. 58, no. 9, pp. 4207-4216, Sep. 2011.

[4] ST Microelectronics, "Design of fixed-off-time-controlled PFC pre-regulators with the L6562," *AN1792, Application Note*, Nov. 2003.

[5] C. Adragna, S. De Simone, and G. Gattavari, "New fixed off-time PWM modulator provides constant frequency operation in boost PFC pre-regulators," *International Symposium on Power Electronics, Electrical Drives, Automation and Motion, 2008*, pp. 656-661, 11-13 Jun. 2008.

Linear Over-Modulation Strategy for Current Control in Photovoltaic Inverter

Yongsoon Park and Seung-Ki Sul
Dept. of Electrical and Computer Engineering
Seoul National University
Seoul, Korea

Ki-Nam Hong
LG U+ Corp.
Seoul, Korea

Abstract— Photovoltaic (PV) inverters autonomously adjust their DC-link voltages to maximize power generation. Around sunrise or sunset, a PV inverter may operate at much lower DC-link voltage than the nominal level due to the low irradiance. The inverter would be under over-modulation if the DC-link voltage is relatively low to the grid voltage at the point of common coupling. In this paper, a series of implementation schemes are proposed to keep the current regulation under over-modulation. After the proposed method is detailed, its fundamental operations are verified by a small-scale prototype inverter and further evaluated by the 250 kW PV inverter installed at a proving ground.

Keywords—current control, over-modulation, photovoltaic inverter.

I. INTRODUCTION

Recently, the number and capacity of the grid-connected photovoltaic (PV) inverters have been enormously increased, and their unit power rating has reached to MW-scale. In general, the grid-connected inverter has a lower limit in the DC-link voltage in order to regulate the grid current into Point of Common Coupling (PCC). The PV panel connected to the inverter may be constructed such that the maximum power point tracking (MPPT) is mostly achieved in a voltage range well above the lower limit. However, in some cases, depending on irradiance and temperature, the DC-link voltage may have to be lower than that limit and PV inverter should be operated to generate available power.

Usually, the minimum DC-link voltage where the PV inverter has to stall is determined by the manufacturer with some margin. If MPPT has to be carried out near the lower limit due to low irradiance, the inverter's stall could occur frequently. It would be inconvenient to disconnect the PV inverter from the grid whenever the DC-link voltage crosses the limit. In addition, these types of disconnections would impede the consistent power transfer from the PV inverter to PCC.

The primary reason to disconnect the inverter at low DC voltages is that voltage outputs cannot be correctly synthesized according to their references. As shown in Fig. 1, a voltage hexagon can be drawn if a DC-link voltage is given [1]. When the voltage reference vector is outside the hexagon, how to synthesize the references is so called as over-modulation (OVM).

One category of OVM is to modify the spatial position of the reference vector in the voltage plane partitioned by the hexagon [1]-[6]. In the other category, the actual fundamental

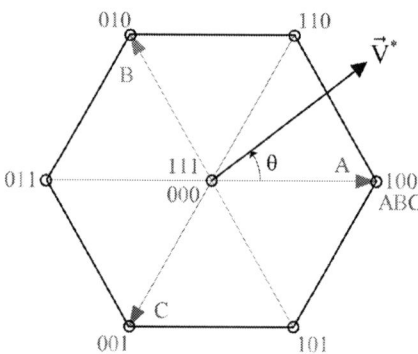

Fig. 1. Voltage reference vector and voltage hexagon.

voltages had been investigated as per pulse width modulation (PWM) methods when the pole voltage magnitudes were simply limited by a half of DC-link voltage under OVM [7]-[9]. In this paper, how to implement OVM is discussed when the switching functions are regarded. In particular, grid-connected inverters should comply with the corresponding regulations on harmonic currents to PCC [10]. In this paper, the appropriate selection of OVM method is discussed to meet those harmonic regulations.

Because the active and reactive powers must be under control in PV applications, the feedback control on output currents should be used under OVM. If the integrator is incorporated in the feedback controller, an anti-windup control block should be considered against the saturation of actuator output [11]. By virtue of the proposed method, the power factor of the fundamental currents could be aptly maintained even if the voltage output is limited under OVM.

The effectiveness and performance of the proposed method are fundamentally confirmed with a laboratory-scale prototype inverter. Then, the feasibility of the proposed method for commercialization is discussed with the results from a 250 kW PV inverter installed in Gochang proving ground.

II. OVER-MODULATION

Over-modulation occurs when a voltage reference vector is outside of the voltage hexagon as shown in Fig. 1. If OVM is applied, the outside vector is replaced with a realizable vector within the hexagon. To minimize this magnitude error, the modified voltage vector should be located at a side of the hexagon.

978-1-4799-2706-7/14 $31.00 © 2014 IEEE

Each switching function, which indicates the on-state of upper switch in a leg, is presented in Fig. 1. For example, the switching function for A-phase is not changed if the reference vector rotates along the sides of the hexagon from $-\pi/3$ to $+\pi/3$. Because the zero vectors are not used for synthesis of the voltage in this rotation, the switching states for A-phase are not changed even during PWM. Based on this observation, the A-phase pole voltage under OVM could be depicted as shown in Fig. 2.

That is, OVM methods can be classified by how to connect the lines between 'a' and 'b' points and between 'c' and 'd' points in Fig. 2. Three OVM methods have been simply induced as shown in Fig. 3. Their common features are straight piecewise lines and odd function with respect to the point x, which can contribute to simple implementation and prevention of even harmonics. Actually, almost the same shapes to OVM1 and OVM2 in Fig. 3 appear respectively when space vector PWM (SVPWM) and Depenbrock's discontinuous PWM are used under OVM [7].

In grid-connected PV inverters, OVM methods should be evaluated in terms of harmonic current into PCC to meet the corresponding regulations [10]. Then, weighted selective harmonic distortion (WSHD) in (1) could be used as a criterion to evaluate each OVM method's harmonic distortion:

$$WSHD = \frac{\sqrt{(V_5/5)^2 + (V_7/7)^2 + (V_{11}/11)^2 + (V_{13}/13)^2}}{V_1} \quad (1)$$

where the subscript numbers are harmonic orders.

In this paper, modulation index (MI) is defined as the ratio of the fundamental voltage to $V_{dc}/2$, which is a half of DC-link voltage. As shown in Fig. 4, the harmonic distortions of each OVM method are differently varying over the entire OVM range, where MI is between 1.1547 and 1.2732. Because the harmonic distortions have to be minimized, one of OVM methods should be selected depending on MI. In the figure, OVM4 reveals the smallest WSHD when MI is between 1.1547 and 1.1971. The implementation of OVM4 can be done as follows.

A. Over-Modulation Mode I

Before introducing OVM4 in detail, it is important to understand the meaning of the limiter at the final stage of pulse width modulator in Fig. 5. Even if any limiter is not explicitly used in the modulator, the actual output synthesized by an inverter shall be saturated as long as the pole voltage is absolutely larger than $V_{dc}/2$. Due to this limitation, the fundamental voltage of actual output deviates from its reference under OVM if any extra treatment is absent.

If pole voltage references for SVPWM are naturally limited, the actual voltage vector can be synthesized with the minimum magnitude error to the reference vector in the voltage plane [3], [12]. This means that the root mean square (RMS) error of voltage vector under OVM can be minimized through the method limiting the pole voltage references from SVPWM. However, as mentioned earlier, due to the limiting action, the actual MI deviated from the

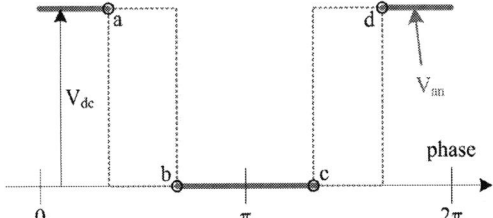

Fig. 2. A-phase pole voltage (V_{an}) under over-modulation.

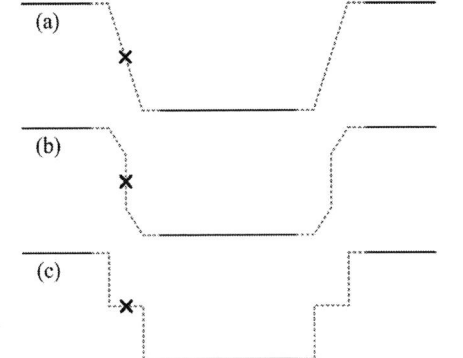

Fig. 3. Over-modulation methods, (a) OVM1, (b) OVM2, (c) OVM3.

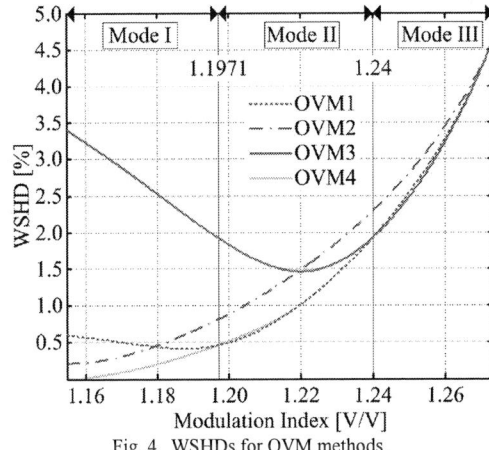

Fig. 4. WSHDs for OVM methods.

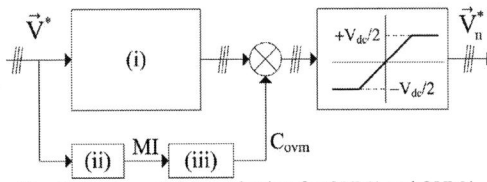

Fig. 5. Implementation mechanism for OVM1 and OVM4.

reference MI as shown in Fig. 6. For instance, in order to achieve the actual MI of 1.20, the reference MI must be given as 1.24. This deviation can be offset by using a sort of compensation gain referred as C_{ovm}. For example, because the actual MI of 1.20 can be achieved with the fictitious MI of 1.24, the necessary C_{ovm} is computed as 1.033 (=1.24/1.2). When considering Fig. 5, the method marked as OVM4 can be described as follows: (i) SVPWM is applied to the original voltage reference $\vec{V}^*$, (ii) the present MI is computed from $\vec{V}^*$, (iii) C_{ovm} is calculated from the present MI, and (iv) the limiter is applied after the output of (i) is multiplied by C_{ovm}.

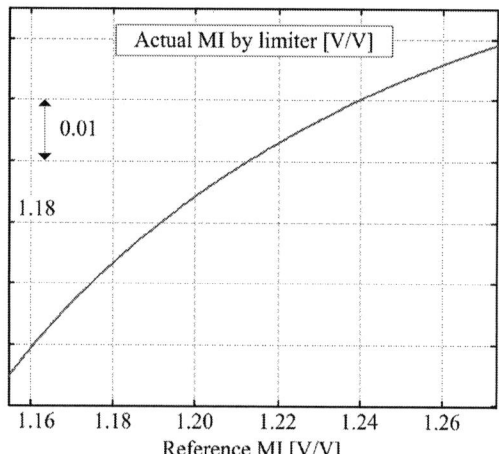

Fig. 6. MI distortion due to V_{dc} limit with SVPWM.

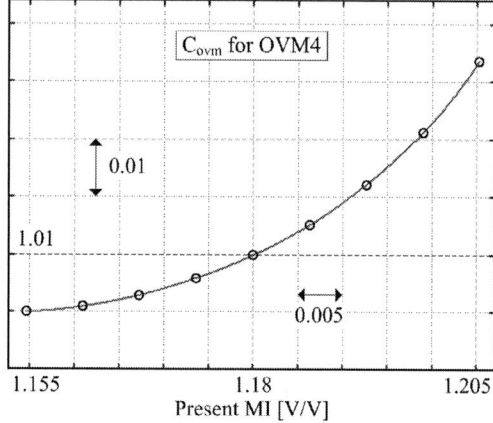

Fig. 7. C_{ovm} table for OVM4.

The MI distortion in Fig. 6 is reversely used to calculate C_{ovm}. Then, a gain table for C_{ovm} can be readily obtained as shown in Fig. 7. For practical implementation, the points indicated by circles in the figure are stored in the table, and the other points are linearly interpolated. As shown in Fig. 8, the pole voltage reference and the actual phase voltage have been captured in a prototype inverter when MI was 1.1812 when V_{dc} is 160 V. In the frequency analysis, the actual fundamental voltage was 93.61 V, which presents -0.94% error to the original reference ($\vec{V}^*$ in Fig. 5).

B. Over-Modulation Mode II

When a required MI is between 1.1971 and 1.24, OVM1 is applied to generate pole voltages. The process to implement OVM1 can be also understood with Fig. 5. In fact, when it comes to pole voltages, OVM4 becomes equal to OVM1 when MI approaches to 1.2732 [7]. However, by utilizing a different table at the block (iii) in Fig. 5, the block (i) can be skipped in OVM1. Namely, the limitation after compensation by C_{ovm} can generate a quasi-trapezoidal voltage as shown in Fig. 9. This scheme has been adopted in that sine-wave presents almost linear variations near zero-crossing.

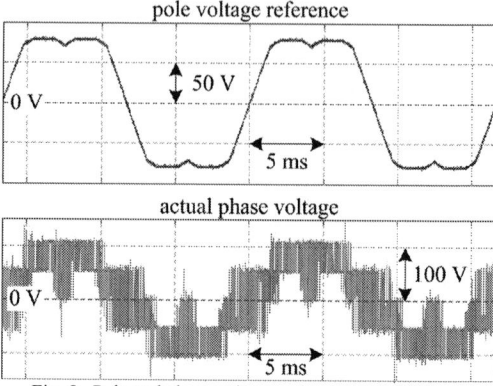

Fig. 8. Pole and phase voltages when MI is 1.1812.

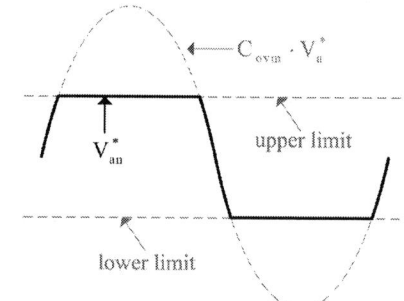

Fig. 9. Pole voltage generation in OVM mode II.

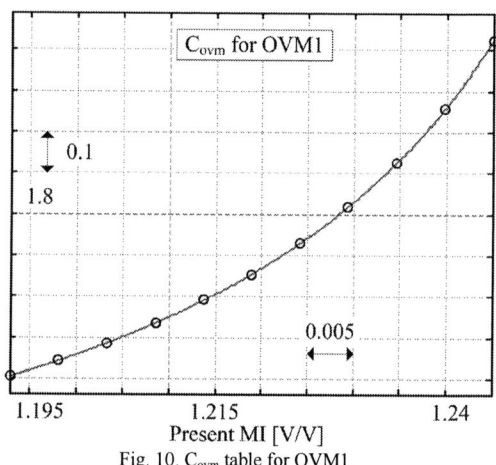

Fig. 10. C_{ovm} table for OVM1.

In the similar way to OVM4, the compensation gain table has been numerically obtained as shown in Fig. 10. The pole voltage reference and the actual phase voltage have been captured as shown in Fig. 11 when MI is 1.225 under the same V_{dc}, 160 V. In the frequency analysis, the actual fundamental voltage was 97.44 V, which means -0.57 % error to the original reference.

C. Over-Modulation Mode III

Both of OVM1 and OVM4 have a demerit when the six-step operation is required. Theoretically, this is because the compensation gain must be infinity to implement the six-step operation. Furthermore, because harmonic currents can be reflected into the voltage references if the current regulation is kept under OVM, the infinite gain can amplify those harmonics as well. In

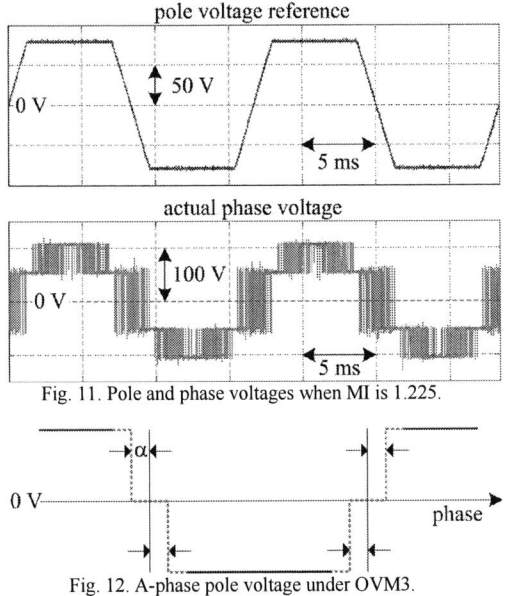

Fig. 11. Pole and phase voltages when MI is 1.225.

Fig. 12. A-phase pole voltage under OVM3.

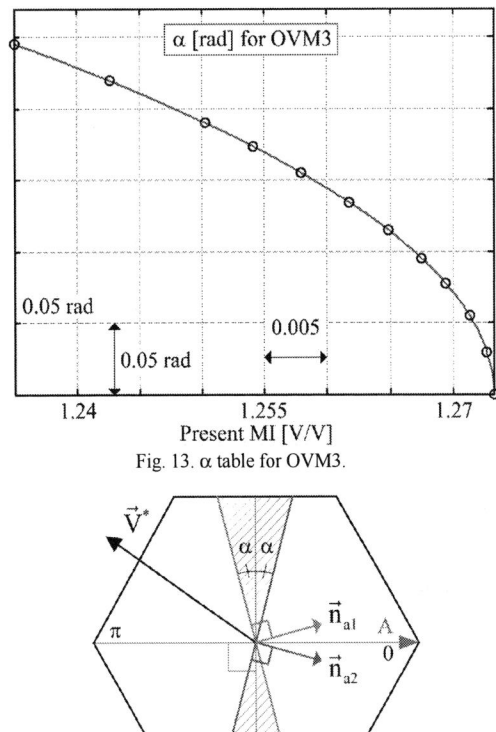

Fig. 13. α table for OVM3.

Fig. 14. α-related vectors when A-phase is considered for OVM3.

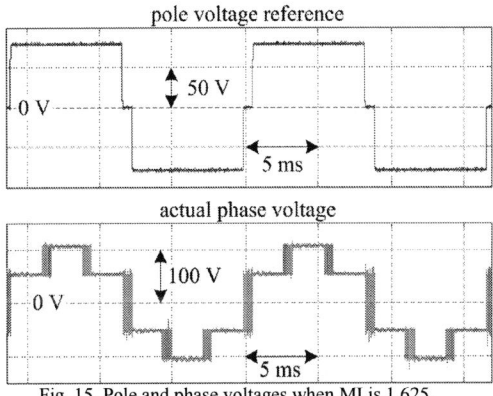

Fig. 15. Pole and phase voltages when MI is 1.625.

addition, OVM3 has a merit in the view point of WSHD as shown in Fig. 4 when MI is greater than 1.24.

The phase angle denoted by α in Fig. 12 is the main factor to adjust harmonic voltages in OVM3. When α decreases, MI increases. In particular, if α becomes zero, the inverter would be under the six-step operation. The required α according to MI has been numerically obtained as shown in Fig. 13.

The intervals where the pole voltage is zero in Fig. 12 can be depicted spatially in the voltage plane. This corresponds to the hatched area in Fig. 14. Two normalized vectors, $\vec{n}_{a1}$ and $\vec{n}_{a2}$, can be considered for implementation, whose phase angle is α and $-\alpha$, respectively. What type of pole voltage has to be synthesized can be determined by utilizing (2) and (3):

$$D_{a1} = \vec{V}^* \cdot \vec{n}_{a1} \qquad (2)$$

$$D_{a2} = \vec{V}^* \cdot \vec{n}_{a2} . \qquad (3)$$

Because D_{a1} is the inner product, D_{a1} is positive if the phase difference of the voltage reference $\vec{V}^*$ to $\vec{n}_{a1}$ is absolutely smaller than $\pi/2$. Then, for instance, if D_{a1} and D_{a2} are all positive, the voltage vector's phase is between $-\pi/2+\alpha$ and $\pi/2-\alpha$. When considering Fig. 1, the switching function for A-phase is 1, which corresponds to the pole voltage of $+V_{dc}/2$. In addition, it can be readily inferred that the zero clamping occurs when the product of D_{a1} and D_{a2} is negative. For the other phases, the pole voltage implementation is similar except using the $2\pi/3$-shifted normalized vectors.

The voltage synthesis under OVM3 has been captured as shown in Fig. 15 when MI is 1.625 under the same V_{dc}, 160 V. In the frequency analysis, the actual fundamental voltage was 100.7 V, which means -0.297 % error to the original reference.

III. CURRENT CONTROL UNDER OVER-MODULATION

The feedback control on output currents at the fundamental frequency should be maintained even under OVM in PV inverters to regulate active and reactive power to PCC [13]. The proposed current control can be understood with Figs. 16-17.

As mentioned earlier, harmonic currents can be reflected into voltage references through the feedback control under OVM. If these harmonics are not aptly mitigated, the voltage outputs would be more distorted than expected. To circumvent this problem, a notch filter in (4) for each harmonic could be used to mitigate the harmonic distortions in the synchronous reference frame. By these notch filters, voltage references can be less distorted under OVM.

$$NF(s) = \frac{s^2 + \omega_{nf}^2}{s^2 + 2\zeta\omega_{nf}s + \omega_{nf}^2} . \qquad (4)$$

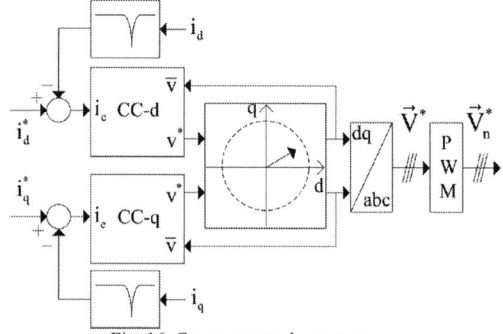

Fig. 16. Current control structure.

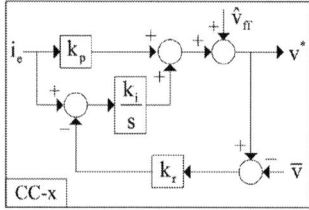

Fig. 17. Current controller.

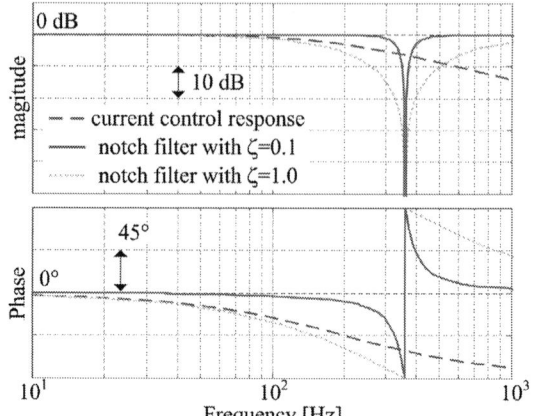

Fig. 18. Bode plots of current control and notch filter.

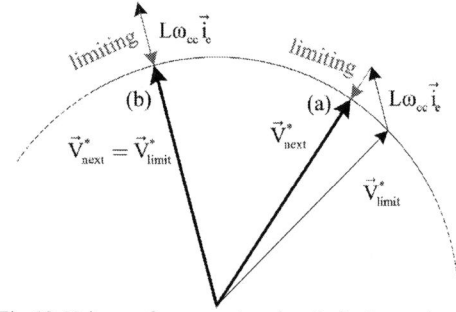

Fig. 19. Voltage reference vector when the limiter works.

The notch frequency, ω_{nf} in (4), is set to $2\pi360$ rad/s to deal with 5th and 7th order harmonics in this paper. In order to specify the damping coefficient ζ in (4), the current control property has been considered together. The current controller in the synchronous reference frame can be designed to present the following closed loop response [14]:

$$CC(s) = \frac{\omega_{cc}}{s + \omega_{cc}} \quad (5)$$

where ω_{cc} is the control bandwidth.

This response is possible only if the pole-zero cancellation is correct with the following PI gains:

$$k_p = L\omega_{cc}, \quad k_i = R\omega_{cc} \quad (6)$$

where L and R are line inductance and resistance between PV inverter and the grid.

If harmonic controllers like resonant controller are separately used [15], the control bandwidth for the fundamental current should be limited to prevent interferences between the fundamental and harmonic controllers. Then, ω_{cc} in (5) has been empirically set below $2\pi200$ rad/s to prevent the interference. Because the notch filters are employed on the feedback loops as presented in Fig. 16, their response should not disturb the original loop property. As shown in Fig. 18, if ζ is set to 0.1 rather than 1, the magnitude and phase of notch filter can be almost 0 dB and 0 ° within the bandwidth of current control. Because 0 dB and 0 ° in Bode plot mean no disturbance in magnitude and phase, ζ was set to 0.1.

As shown in Fig. 16, a voltage reference vector is limited by circle rather than hexagon. This circular limiter can preserve the voltage vector's phase even after its magnitude is reduced. The radius of the limiting circle is determined by MI_{max}:

$$\left|\vec{V}^*\right|_{max} = MI_{max} \cdot \frac{V_{dc}}{2} . \quad (7)$$

where MI_{max} is the maximum available MI. Up to MI_{max}, PWM is expected to be linear in terms of the fundamental

voltage in the proposed method. Because the harmonic distortion monotonically increases with respect to MI as shown in Fig. 4, MI_{max} can be determined by considering the worst degree of harmonic currents.

In addition to the PI gains of k_p and k_i in (6), the feedback gain denoted by k_r, which is an anti-windup gain, must be specified in Fig. 17. This gain prevents integrator's divergence even if the current error i_e does not vanish. From Fig. 17, the following equation can be derived:

$$k_p \cdot i_e + k_i \cdot \frac{\{i_e - k_r(v^* - \bar{v})\}}{s} + \hat{v}_{ff} = v^* \quad (8)$$

where $\bar{v}$ is the processed output by the circular limiter from voltage reference v^*. $\hat{v}_{ff}$ indicates feed-forwarding voltage.

The controller output v^* is derived as (9) from (8):

$$v^* = k_p \cdot i_e + k_i \cdot \frac{(1 - k_r k_p) \cdot i_e + k_r \cdot (\bar{v} - \hat{v}_{ff})}{s + k_r k_i} + \hat{v}_{ff} . \quad (9)$$

Because k_r has no meaning if $\bar{v}$ is equal to v^*, (9) should be regarded when the circular limiter works. Through the midterm in (9), the integrator's response can be explained. If k_r is set to $1/k_p$, distortions caused by i_e can be minimized in integrator's output. In addition, if the PI gains in (6) are used, (10) can be deduced. Under these settings, the converging response to the limiting value $\bar{v}$ depends only on the system parameter of R/L.

$$v^* = L\omega_{cc} \cdot i_e + \frac{R/L}{s + R/L} \cdot \bar{v} + \frac{s}{s + R/L} \cdot \hat{v}_{ff} . \quad (10)$$

When i_e, $\bar{v}$, and $\hat{v}_{ff}$ are assumed to be step-varying, and the final value theorem is applied to (10), v^* finally converges to $L\omega_{cc}i_e + \bar{v}$. This meaning of steady state needs to be extended in vector space as shown in Fig. 19. If the voltage reference vector in the synchronous reference frame is in steady state, its next state is the same with its present state. However, as shown in Fig. 19(a), if the limited voltage vector $\vec{V}^*_{limit}$ is not in phase with the current error vector $\vec{i}_e$, the voltage vector has to be changed after the limiting. That is, as shown in Fig. 19(b), even if the current error vector does not become null, the voltage reference vector is modified by the proposed method so that its limited one is in phase with the current error vector. This would finally enable the inverter to keep the power factor of the fundamental currents even if the circular limiter operates.

IV. EXPERIMENTAL RESULTS

Initially, the proposed method has been fundamentally tested with a laboratory-scale prototype inverter. After the current control performance of the prototype inverter is verified, data logged from a large-scale PV inverter are discussed.

A. Fundamental operation under OVM

The small-scale PV inverter was connected to 110 V_{rms} grid. All proposed algorithms were implemented in a DSP board based on TMS320F28335. The sampling frequency was 15 kHz, and the switching frequency was 7.5 kHz. The phase locked loop (PLL) was based on [16].

To discuss the control dynamics on output currents, the DC-link voltage of the inverter was supplied by a constant voltage source. Initially, when V_{dc} was 179 V, the q-axis current reference was changed in step manner from 5 A to 15 A. As presented in Fig. 20, the normal current control response was captured as reference while MI was always lower than 1.1547.

For comparison, the same step change with Fig. 20 was tested when V_{dc} was 159 V. With this DC-link condition, the inverter operated under OVM after the step change as shown in Fig. 21. Because harmonic currents increased right after entering OVM, the current control dynamics was not exactly the same with that under normal modulation. However, in a series of repeated tests, it has been confirmed that the rising time under OVM at least did not increase when compared to that under normal modulation.

Even though the inverter was under OVM at the test of Fig. 21, the limiter in Fig. 16 did not work at all. This means that the inverter could still synthesize given voltage references at the fundamental frequency. It is important to check the proposed method's performance when the voltage references cannot be synthesized any more.

In order to consider the circular limiter's transient operation, DC input of the inverter was connected to a PV simulator that emulated the power-voltage curve of a solar panel shown in Fig. 22. At the beginning, the inverter regulated its DC-link voltage as 170 V. When MI_{max} was set to 1.26, the DC-link voltage was decreased

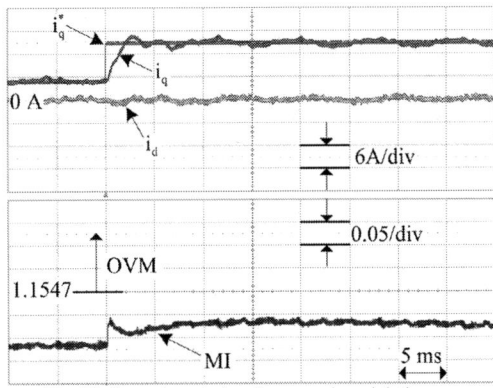

Fig. 20. Current control under normal modulation.

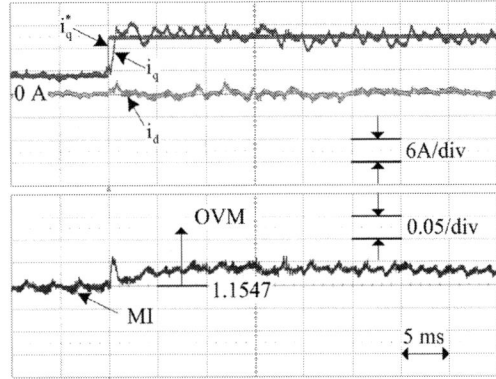

Fig. 21. Current control under over-modulation.

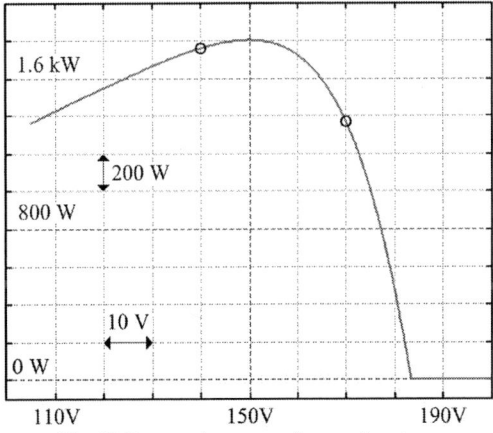

Fig. 22. Power-voltage curve for experiment.

toward 140 V with the rate of 150 V/s as shown in Fig. 23.

The DC-link voltage was settled down to 143 V whereas its reference was 140 V. In result, MI was reached to its preset maximum value of 1.26 in steady state. This indicates that the magnitude of voltage vector was limited by the proposed method. The dashed box in Fig. 23 has been magnified in Fig. 24 to carefully investigate the output currents. Though lots of harmonics were included in the currents, the average of d-axis current was 0.64 A, which was negligible in that its reference was 0 A. This means that the proposed method is capable of maintaining the power factor of fundamental currents as expected.

978-1-4799-2706-7/14 $31.00 © 2014 IEEE

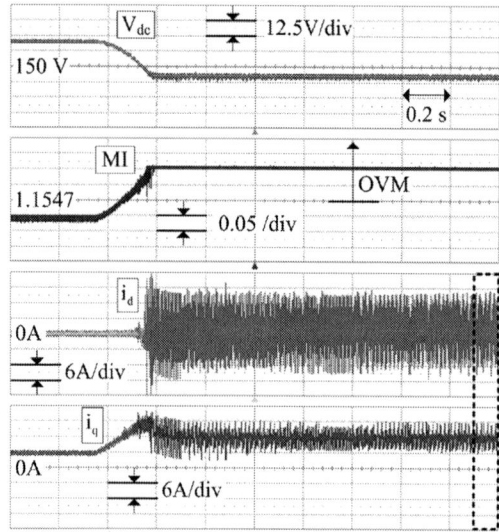

Fig. 23. Circular limiter test.

Fig. 24. Magnified current waveforms of Fig. 23.

B. Discussion on Data from Gochang proving ground

The proposed method under OVM has been tested in a 250 kW PV inverter connected to PV panels in Gochang PV generation proving ground. The inverter and PV panels are shown in Fig. 25. In fact, for higher efficiency, the PV inverter is based on T-type three-level topology shown in Fig. 26. However, if three-level inverter is under OVM, it is hard to use the small vectors [17]. This can cause instability on the neutral potential control. Because the T-type topology is a hybrid one, it can be operated as a two-level inverter by opening the bidirectional switches. Through this two-level operation, the PV inverter under test can operate under OVM.

The test data were obtained at Feb. 28, 2014, and the weather was cloudy. Because the temperature around sunset was about 6 °C, the open circuit voltage of PV panels was relatively high when compared to 280 V_{rms} grid. The OVM mode by the proposed method has been briefly observed on that day. Test data have been logged by analyzing instruments.

As shown in Fig. 27, the inverter continued to transfer power from PV panels to grid even if the DC-link voltage was decreased due to sunset. Because 450 V was officially the lower limit for MPPT in the inverter's catalogue, the power generation after 17:50 (in x-axis of Fig. 27) is related to the proposed system operation. The variation of MI up to the preset MI_{max} of 1.223 can be confirmed with Fig. 28. Considering Figs. 27 and 28, it

Fig. 25. 250 kW PV inverter and Gochang proving ground.

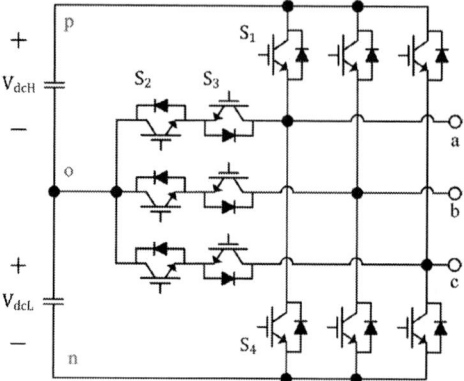

Fig. 26. T-type three-level inverter.

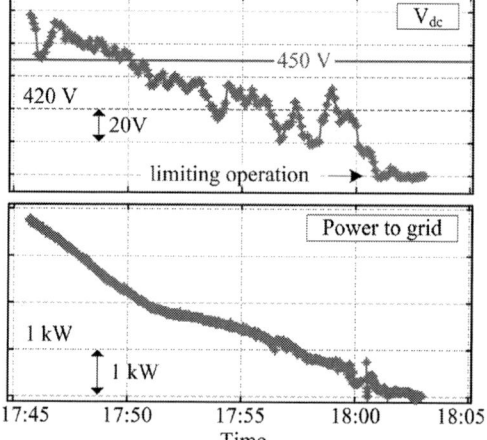

Fig. 27. DC-link voltage and generated power around sunset.

can be inferred that the inverter entered into OVM near 420 V (around 17:55).

It was important to check the harmonic distortions of output currents. Although data associated with harmonic distortions in Fig. 29 were not smoothly obtained, distortion trend could be roughly identified. As shown in Fig. 4, the harmonic distortion increases as MI under OVM increase. Therefore, the inverter would comply with IEEE std. 1547 if MI_{max}, which determine the radius of limiting circle in Fig. 16, were aptly smaller than 1.223.

V. CONCLUSIONS

In this paper, an over-modulation strategy has been proposed for the grid-connected PV inverters. Initially, over-modulation methods have been designed to linearly modulate the fundamental voltage according to its reference. In conjunction with this linear over-modulation,

978-1-4799-2706-7/14 $31.00 © 2014 IEEE

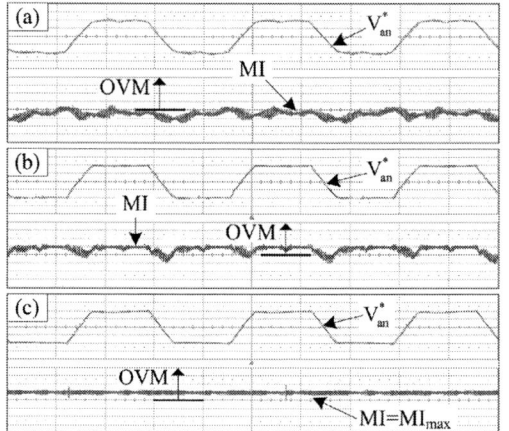

Fig. 28. Pole voltage reference and MI at (a) 17:55, (b) 18:00, (c) 18:02
(V_{an}^* : 100V/div with center at 0V; MI: 0.075/div with center at 1.1547;
time: 5 ms/div).

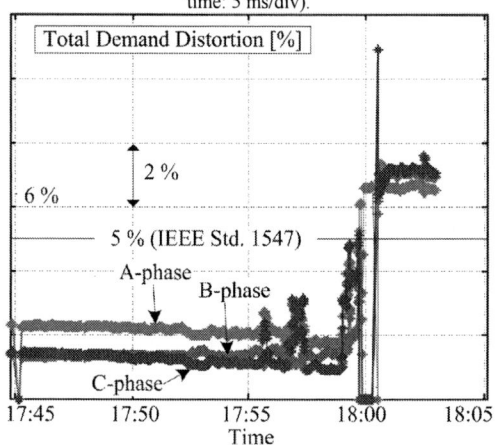

Fig. 29. Total demand distortion for each phase current around sunset.

the circular limiter has been utilized instead of voltage hexagon. The settings for current control have been carefully detailed so that the current control property is kept even under over-modulation. After the proposed method's fundamental operations were tested in a small-scale inverter, its feasibility has been examined with 250kW PV generation system in a proving ground. Although the weather condition was not optimal to extensively test the proposed method, it has been confirmed that the PV inverter can execute MPPT at very low DC-link voltages by the proposed method.

REFERENCES

[1] J. Holtz, W. Lotzkat, and A.M. Khambadkone, "On Continuous Control of PWM Inverters in the Overmodulation range including the six-step mode," *IEEE Trans. Power Electron.*, vol. 8, no. 4, pp. 546-553, Oct. 1993.

[2] A. M. Khambadkone and J. Holtz, "Compensated Synchronous PI Current Controller in Overmodulation Range and Six-Step Operation of Space-Vector-Modulation-Based Vector-Controlled Drives," *IEEE Trans. Ind. Electron.*, vol. 49, no. 3, pp. 574-580, June 2002.

[3] D.R. Seidl, D.A. Kaiser, and R.D. Lorenz, "One-Step Optimal Space Vector PWM Current Regulation Using a Neural Network," *In conf. rec. Ind. Applicat. Society Ann. Meeting 1994*, Oct. 2-6 1994, pp. 867-874.

[4] L. Rossetto, P. Tenti, and A. Zuccato, "Integrated Optimum Control of Quasi-Direct Converters," *IEEE Trans. Power Electron.*, vol. 12, no. 6, pp. 993-999, Nov. 1997.

[5] J.-K. Seok, J.-S. Kim, and S.-K. Sul, "Overmodulation Strategy for High-Performance Torque control," *IEEE Trans. Power Electron.*, vol. 13, no. 4, pp. 786-792, Jul. 1998.

[6] B.-H. Bae and S.-K. Sul, "A Novel Dynamic Overmodulation Strategy for Fast Torque Control of High-Saliency-Ratio AC Motor," *IEEE Trans. Ind. Applicat.*, vol. 41, no. 4, pp. 1013-1019, Jul./Aug. 2005.

[7] A.M. Hava, R.J. Kerkman, T.A. Lipo, "Carrier-Based PWM-VSI Overmodulation Strategies: Analysis, Comparison, and Design," *IEEE Trans. Power Electron.*, vol. 13, no. 4, pp. 674-689, Jul. 1998.

[8] A.M. Hava, S.-K. Sul, R.J. Kerkman, and T.A. Lipo, "Dynamic Overmodulation Characteristics of Triangle Intersection PWM methods," *IEEE Trans. Ind. Applicat.*, vol. 35, no. 4, pp. 896-907, Jul./Aug. 1999.

[9] K. Zhou and D. Wang, "Relationship Between Space-Vector Modulation and Three-Phase Carrier-Based PWM: A Comprehensive Analysis," *IEEE Trans. Ind. Electron.*, vol. 49, no. 1, pp. 186-196, Feb. 2002.

[10] *IEEE Standard for Interconnecting Distributed Resources With Electric Power Systems*, IEEE Std. 1547, 2003.

[11] Y. Peng, d. Vrancic, and R. Hanus, "Anti-Windup, Bumpless, and Conditioned Transfer Techniques for PID Controllers," *IEEE Control Systems*, vol. 16, no. 4, pp. 48-57, Aug. 1996.

[12] D.-W. Chung, "Unified Analysis of PWM Method for Three Phase Voltage Source Inverter Using Offset Voltage," Ph.D. dissertation, Seoul Nat'l Univ., Seoul, Korea, 2000.

[13] *Generating Plants Connected to the Medium-Voltage Network*, BDEW technical guideline, June 2008.

[14] S.-K. Sul, "Design of Regulators for Electric Machines and Power Converters," *in Control of Electric Machine Drive Systems*, Hoboken, NJ:Wiley, 2011, ch. 4, pp. 154-229.

[15] R. Teodorescu, F. Blaabjerg, M. Liserre, and P.C. Loh, "Proportional-Resonant Controllers and Filters for Grid-Connected Voltage-Source Converters," *IEE Proc. Electric Power Applicat.*, vol. 153, no. 5, pp. 750-762, Sept. 2006.

[16] Y. Park, S.-K. Sul, W.-C. Kim, and H.-Y. Lee, "Phase Locked Loop Based on an Observer for Grid Synchronization," *in Proc. IEEE Applied Power Electron. Conf. Exposition (APEC)*, Mar. 17-21, 2013, pp. 308-315.

[17] Y. Park, S.-K. Sul, C.-H. Lim, W.-C. Kim, and S.-H. Lee, "Asymmetric Control of DC-Link Voltages for Separate MPPTs in Three-Level Inverters," *IEEE Power Electron.*, vol. 28, no. 6, pp. 2760-2769, June 2013.

The 2014 International Power Electronics Conference

Design of Decentralized Voltage Control for PV Inverters to Mitigate Voltage Rise in Distribution Power System without Communication

Tzung-Lin Lee
Department of Electrical Engineering
National Sun Yat-sen University
Kaohsiung, TAIWAN.
e-mail : tllee@mail.ee.nsysu.edu.tw

Shih-Sian Yang   Shang-Hung Hu
Department of Electrical Engineering
National Sun Yat-sen University
Kaohsiung, TAIWAN.
e-mail : M003010103@student.nsysu.edu.tw

Abstract- **High penetration of PV generation may result in serious voltage rise in the distribution power system due to reverse power flow. This paper presents a decentralized voltage control for PV inverters to mitigate voltage fluctuations. Instead of operating in MPPT (maximum power point tracking) mode, each PV inverter is able to adjust its real power and reactive power based on resistive and inductive characteristics at its installation location, respectively. In this way, dispersed PV inverters on the feeder are able to cooperatively reduce voltage rise based on individual installation voltage only. Therefore, more PV inverters are allowed to deliver their power to the utility without violating voltage regulation. In this paper, various case studies were conducted to verify effectiveness of the proposed method by using power-flow analysis and time-domain simulation.**

Keywords- PV inverter, voltage rise, decentralized voltage control

I. INTRODUCTION

Due to global concerns on the environmental issues, various renewable energy sources have received much attention. Power electronics-based PV (photovoltaic) generation, which uses an interfaced inverter to deliver output power to the utility, becomes an acceptable renewable energy source due to electricity cost and government policy. However, extensive installation of PV inverters may cause voltage rise and frequency variation due to reverse power flow [1]. Several grid codes, VDE 0126-1-1 [2] and IEC 61727 [3], have defined normal operation of grid voltage as 110%~85%, during which distributed generators should stay at grid-connected mode. In Taiwan, grid voltage is limited in the range of 95%~105% in order to guarantee normal operation by Taiwan Power Company (TAIPOWER) [4].

According to grid codes, the PV inverters must stop their output as grid voltage is not inside the normal operation range. Various methods, such as on-line tap changer (OLTC), line voltage drop compensator (LDC) and switching capacitor/reactor, have been wildly used to cope with voltage variation due to load variation in the conventional power system. However, these methods may not be suitable to mitigate voltage variation due to variable power of PV generation. The PV inverter with

reduced power factor operation has been proposed. In this way, the voltage rise can be alleviated by compensating reactive power from the PV inverter. Some voltage supporting strategies for grid connected PV inverters were also presented [5]-[9]. The compensating reactive power is adjusted according to both voltage and line impedance at the installation point of the PV inverter. However, the improvement is not clear due to large R/X ratio in the low-voltage feeder, in which the voltage magnitude is not sensitive to the reactive power. In this sense, large amount of reactive power is required to provide voltage regulation, resulting in large conducting losses in the feeder.

This paper proposes a decentralized voltage control for PV inverter to alleviate voltage rise. Instead of operating in MPPT (maximum power point tracking) mode, the proposed PV inverter is operated at reduced real power output based on the resistive characteristic at installation location of PV inverter when the voltage rises. Similarly, PV inverter is able to compensate reactive power according to both the inductive characteristic and voltage magnitude. Therefore, entire feeder voltage can be maintained in an acceptable level by cooperative operation of various PV inverters. In this sense, more PV inverters are allowed to deliver their power to the utility. Based on power-flow analysis and time-domain simulation, various case studies are conducted to verify effectiveness of the proposed method.

II. OPERATION PRINCIPLE

Fig.1 shows the proposed control of PV inverter. For simplicity, MPPT part of the PV inverter is replaced by a constant voltage source. The decentralized voltage control is able to curtail real power and compensate reactive power based on voltage magnitude $|V|$ and impedance $Z = R+jX$ at the installation location of PV inverter. Fig. 2 shows the proposed real-power curtailing strategy. When $|V|$ is less than $1+D_P$, PV inverter operates as MPPT mode. The inverter starts to decrease real power command P^* as voltage magnitude $|V|$ is larger than $1+D_P$, and stops real-power output after $|V|$ exceeds V_{OP}. On the other hand, the inverter starts to compensate reactive power Q^* as $|V|$ is larger than $1+D_Q$, and keeps at the

---

This work is funded by the National Science Council of TAIWAN under grant NSC102-3113-P-214-001.

978-1-4799-2706-7/14 $31.00 © 2014 IEEE

maximum reactive power output Q_{MAX} after $|V|$ exceeds V_{OP} as shown in Fig.3. V_{OP} is defined as voltage limitation at which the PV inverter must stop delivering real power.

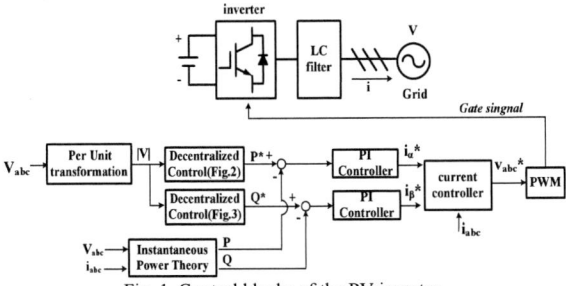

Fig. 1. Control blocks of the PV inverter.

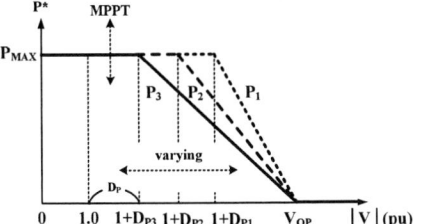

Fig. 2. The proposed active power control.

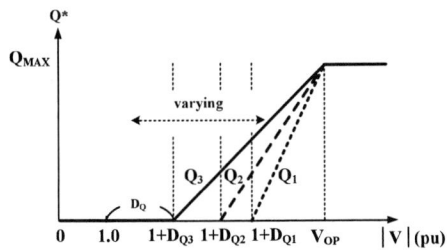

Fig. 3. The proposed reactive power control.

$$
P^* = \begin{cases} P_{MAX} & , \ V < 1 + D_P \\ P_{MAX} \times \dfrac{V_{OP} - V}{V_{OP} - (1 + D_P)} & , \ 1 + D_P \leq V < V_{OP} \\ 0 & , \ V \geq V_{OP} \end{cases}
$$

$$
Q^* = \begin{cases} 0 & , \ V < 1 + D_Q \\ Q_{MAX} \times \dfrac{V - (1 + D_Q)}{V_{OP} - (1 + D_Q)} & , \ 1 + D_Q \leq V < V_{OP} \\ Q_{MAX} & , \ V \geq V_{OP} \end{cases} \tag{1}
$$

The control algorithm is expressed as (1), in which D_P and D_Q are %voltage to start curtailing real power and injecting reactive power of the inverter, respectively. According to sensitivity analysis [8][9], voltage variation is sensitive to real power for resistive feeder, but sensitive to the reactive power in case of inductive feeder. In order to suppress voltage rise effectively, D_P –R droop and D_Q– X droop are proposed as given in (2), which allows both D_P and D_Q to be adjusted according to the resistive part R and the inductive part X of the impedance Z at the installation point of PV inverter, respectively. Fig. 4 shows both droop characteristics. Obviously, larger R is related to smaller D_P, which means PV inverter will start curtailing real power earlier. Similarly, larger X is related to smaller D_Q, which means PV inverter will start

compensating reactive power earlier.

$$
D_P = \begin{cases} D_{MAX} & , \ R_i < R_{MIN} \\ D_{MIN} + \left(\dfrac{D_{MAX} - D_{MIN}}{R_{MAX} - R_{MIN}}\right) \times (R_{MAX} - R_i), & R_{MIN} \leq R_i \leq R_{MAX} \\ D_{MIN} & , \ R_i > R_{MAX} \end{cases}
$$

$$
D_Q = \begin{cases} D_{MAX} & , \ X_i < X_{MIN} \\ D_{MIN} + \left(\dfrac{D_{MAX} - D_{MIN}}{X_{MAX} - X_{MIN}}\right) \times (X_{MAX} - X_i), & X_{MIN} \leq X_i \leq X_{MAX} \\ D_{MIN} & , \ X_i > X_{MAX} \end{cases} \tag{2}
$$

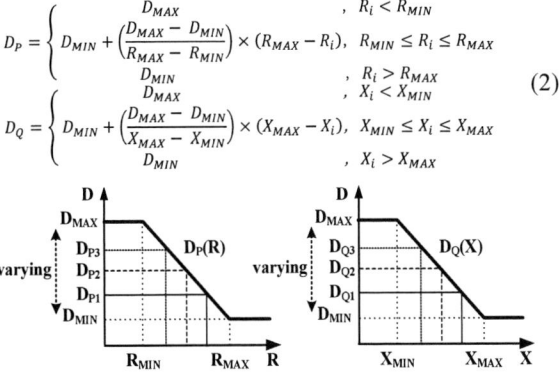

Fig. 4. Droop characteristics of both D_P and D_Q

After determination of current command, PI controllers are applied to generate the current control of the inverter. Based on current command, the current control and PWM are to synthesize gating signals of the inverter to accomplish proper power conversion.

III. TEST RESULTS

In this section, the power-flow analysis is evaluated on MATLAB platform and the corresponding time-domain simulations are conducted by PSCAD software. Fig. 5 and TABLE I show the considered circuit and the associated parameters, where three PV inverters are installed at different locations on a radial feeder.

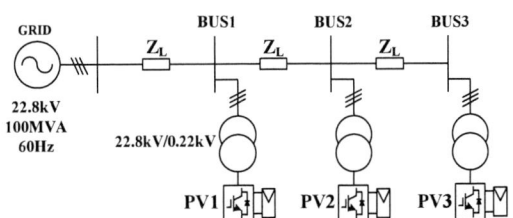

Fig. 5. Simulation circuit.

TABLE I
SYSTEM PARAMETERS

Parameter	Value
System voltage	22.8 kV
System rating	100 MVA
transformer	22.8 kV / 0.22 kV
PV rating	500 kW
Feeder impedance(Z_L)	2.2+j2.856(p.u)

A. Power Flow Analysis

Table II(a) shows bus voltages when all PV inverters are assumed operating at the MPPT mode with the rated real power output. As shown, bus voltages rise and especially voltage at bus 3 exceeds 1.05p.u, which should be disconnected from the grid. Thus, 1.0MW can be supplied by both PV1 and PV2. After applying the proposed control strategy, all bus voltages are reduced under 1.05p.u as shown in Table II(b). That means PV3 can keep connecting to the grid and all PV inverters with the proposed control can supply 1.36MW, which is larger than the supplied power of the previous case.

Inductive and resistive in TABLE III are also considered. In case of inductive feeder, voltage rise is unapparent. Slight reactive power is compensated by the PV 3 with the proposed control as shown in TABLE IV(a) and TABLE IV(b). However, for resistive feeder, voltage is significantly increased as shown in TABLE V(a), for example 1.05pu at BUS 3. In this case, only PV1 is able to deliver its power 0.5MW to the utility. After the proposed control is enabled, TABLE V(b) shows that voltage rise is clearly suppressed. Three PV inverters cooperatively supply real power 0.88MW to the utility without exceeding voltage limitation.

Fig. 6 gives different irradiation of PV generation at MPPT mode in a day. Fig. 7 indicates bus voltages can be well maintained below 1.05pu and power delivery is significantly increased in day time (10:00 to 14:00) as shown in Fig. 8. For example, total power output at 12:00 from 1.44(MW) dropped to 0.96(MW) for MPPT mode, which is less than 1.2542 (MW) for the proposed control.

TABLE II(a)
BUS VOLTAGE AND OUTPUT POWER WITH MPPT CONTROL.

Bus No.	Voltage Mag. (p.u)	Angle Degree.	PV Generation MW	PV Generation MVar
-	1.000000	0.000	-1.430593	0.090279
1	1.029766	2.384	0.500000	0.000000
2	1.045008	3.898	0.500000	0.000000
3	1.060370	4.633	0.500000	0.000000

TABLE II(b)
BUS VOLTAGE AND OUTPUT POWER WITH THE PROPOSED CONTROL.

Bus No.	Voltage Mag. (p.u)	Angle Degree.	PV Generation MW	PV Generation MVar
-	1.000000	0.000	-1.303280	0.321437
1	1.020454	2.379	0.500000	0.000000
2	1.031490	3.915	0.499926	0.047860
3	1.033701	4.635	0.363052	0.196079

TABLE III
DIFFERENT IMPEDANCE OF FEEDER CASE(Z_L)

Feeder Type	Feeder Value(p.u)
inductive feeder	0.866+j3.5
resistive feeder	3.5+0.866j

TABLE IV(a)
INDUCTIVE FEEDER WITH MPPT CONTROL.

Bus No.	Voltage Mag. (p.u)	Angle Degree.	PV Generation MW	PV Generation MVar
-	1.000000	0.000	-1.470859	0.117947
1	1.009964	2.980	0.500000	0.000000
2	1.017582	4.931	0.500000	0.000000
3	1.021676	5.896	0.500000	0.000000

TABLE IV(b)
INDUCTIVE FEEDER WITH THE PROPOSED CONTROL.

Bus No.	Voltage Mag. (p.u)	Angle Degree.	PV Generation MW	PV Generation MVar
-	1.000000	0.000	-1.470703	0.128660
1	1.009604	2.986	0.500000	0.000000
2	1.016870	4.944	0.500000	0.000000
3	1.020615	5.916	0.500000	0.010253

TABLE V(a)
RESISTIVE FEEDER WITH MPPT CONTROL.

Bus No.	Voltage Mag. (p.u)	Angle Degree.	PV Generation MW	PV Generation MVar
-	1.000000	0.000	-1.395693	0.025874
1	1.048683	0.710	0.500000	0.000000
2	1.080786	1.148	0.500000	0.000000
3	1.096735	1.357	0.500000	0.000000

TABLE V(b)
RESISTIVE FEEDER WITH THE PROPOSED CONTROL.

Bus No.	Voltage Mag. (p.u)	Angle Degree.	PV Generation MW	PV Generation MVar
-	1.000000	0.000	-0.845135	0.218525
1	1.027796	0.834	0.499950	0.000024
2	1.038725	1.405	0.241670	0.015295
3	1.041718	1.829	0.138090	0.194650

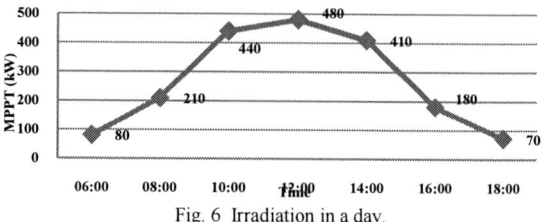

Fig. 6 Irradiation in a day.

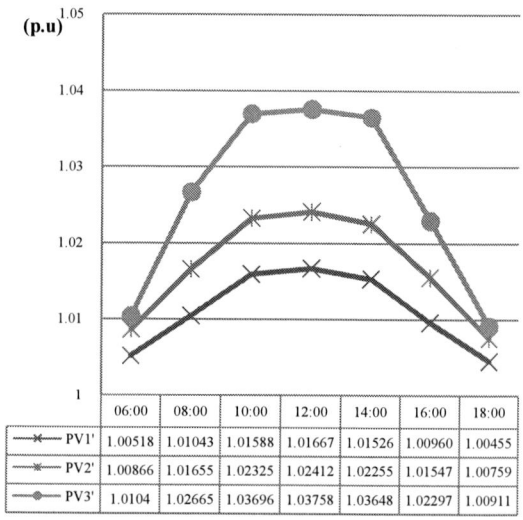

	06:00	08:00	10:00	12:00	14:00	16:00	18:00
PV1'	1.00518	1.01043	1.01588	1.01667	1.01526	1.00960	1.00455
PV2'	1.00866	1.01655	1.02325	1.02412	1.02255	1.01547	1.00759
PV3'	1.0104	1.02665	1.03696	1.03758	1.03648	1.02297	1.00911

Fig. 7 Curve of voltage fluctuation in a day.

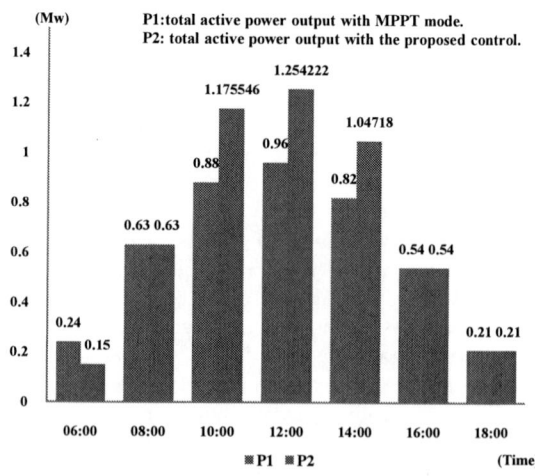

Fig. 8 Total power output of PV inverters in a day.

B. Time-Domain Simulation

Fig. 9 shows time-domain simulations when PV1, PV2, and PV3 are turned on at T_1, T_2, T_3, respectively. As can be seen from Fig. 9(a), bus voltage rises toward the end of the line for PV inverters operated at MPPT mode. Voltage at bus 3 approaches 1.06 pu. After the proposed control method is applied, voltage rise is clearly improved. Fig. 8(b) shows voltage at bus 3 is reduced from 1.06pu to 1.027pu. Fig. 8(c) and Fig. 8(d) illustrate both real power is reduced and reactive power is increased for each PV inverter. Obviously, power-flow analysis is well verified by time-domain simulations.

IV. CONCLUSION

A decentralized voltage control for PV inverter is presented to alleviate voltage rise in the distribution system. Feeder voltage can be maintained at an allowable level by both curtailing real power and compensating reactive power at the same time. Both power flow analysis and time-domain simulation results verify effectiveness of the proposed method.

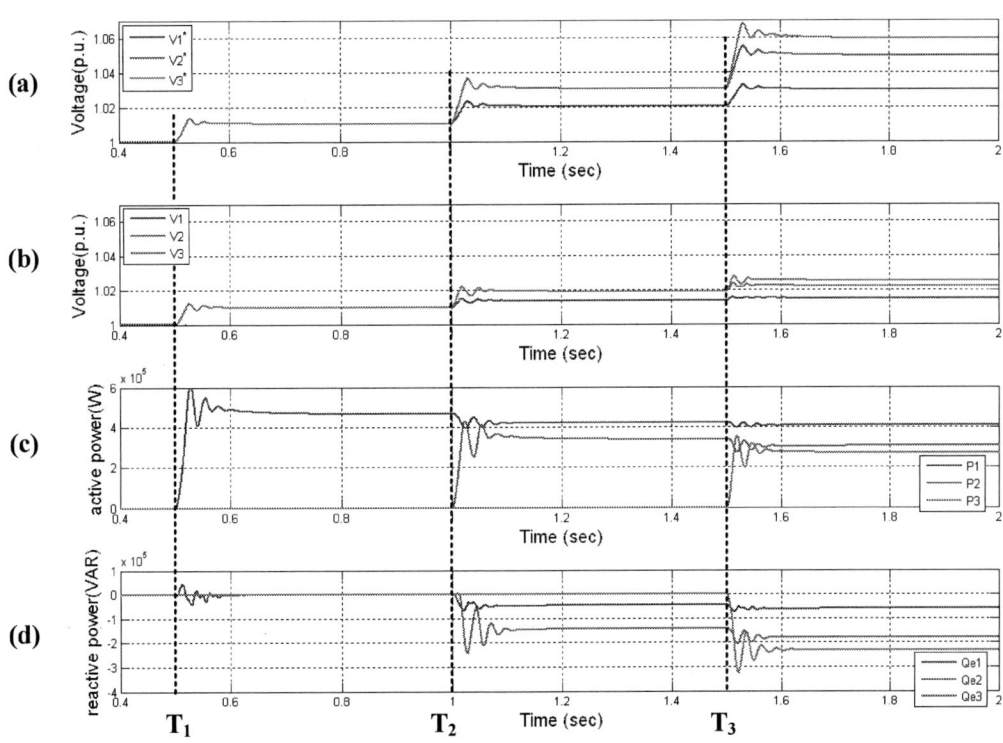

(a)

(b)

(c)

(d)

Fig. 9. Time-domain simulation results. (a) bus voltage with MPPT control (b) bus voltage with the proposed control (c) Real power with the proposed control (d) reactive power with the proposed control.

REFERENCES

[1] E. Demirok, P. C. González, , K. H. B. Frederiksen, D. Sera, P. Rodriguez, and R. Teodorescu, "Local Reactive Power Control Methods for Overvoltage Prevention of Distributed Solar Inverters in Low-Voltage Grids", *IEEE Journal of Photovoltaics,* vol. 1, pp. 174–182, Oct 2011.

[2] Technical requirements for the connection to and parallel operation with low-voltage distribution networks, VDE Verlag, VDE-AR-N4105, 2006.

[3] Photovoltaic (PV) Systems—Characteristics of the Utility Interface, Int. Electrotechn. Comm., Geneva, Switzerland, IEC 61727, Dec. 2004.

[4] Renewable Energy Power System Interconnection Techniques Code, TAIPOWER Inc., TAIWAN, 2009.

[5] M. Braun, T. Stetz, T. Reimann, B. Valov, and G. Arnold, "Optimal reactive power supply in distribution networks—Technological and economic assessment for PV systems," in *the 24th Eur. Photovoltaic Solar Energy Conf.*, Hamburg, Germany, Sep. 2009.

[6] M. Braun, "Reactive power supply by distributed generators," *Power and Energy Society General Meeting -*

Conversion and Delivery of Electrical Energy in the 21st Century, 2008 IEEE , vol., no., pp.1,8, 20-24 July 2008.

[7] J. Backes, C. Schorn, and H. Basse, "Cost-efficient integration of dispersed generation using voltage dependent reactive power control," in *the CIRED Workshop*, Lyon, France, 2010.
E. Demirok, D. Sera, R. Teodorescu, and P. Rodriguez, "Evaluation of the Voltage Support Strategies for the Low Voltage Grid Connected PV", in *IEEE Energy Conversion Congress and Exposition*, pp. 710-717, 2010.

[8] R. Tonkoski and L. A. C. Lopes, "Voltage Regulation in Radial Distribution Feeders with High Penetration of Photovoltaic," in *IEEE Energy 2030 Conference*, vol., no., pp.1,7, 17-18 Nov. 2008.

Stability Analysis and Active Damping for *LLCL*-filter Based Grid-Connected Inverters

Min Huang, Frede Blaabjerg, Poh Chiang Loh
Department of Energy Technology
Aalborg University
Aalborg, Denmark
hmi@et.aau.dk, fbl@et.aau.dk, pcl@et.aau.dk

Weimin Wu
Electrical Engineering
Shanghai Maritime University
Shanghai, China
wmwu@cle.shmtu.edu.cn

Abstract— A higher order passive power filter (*LLCL*-filter) for the grid-tied inverter is becoming attractive for the industrial applications due to the possibility to reduce the cost of the copper and the magnetic material. To avoid the well-known stability problems of the *LLCL*-filter it is requested to use either passive or active damping methods. This paper analyzes the stability when damping is required and when damping is not necessary considering sampling and transport delay. Basic *LLCL* resonance damping properties of different feedback states are also studied. Then an active damping method which is using the capacitor current feedback for *LLCL*-filter is introduced. Based on this method, a design procedure for the control method is given. Last, both simulation and experimental results are provided to validate the theoretical analysis of this paper.

Keywords— *LLCL-filter, grid converter, active damping, current control, resonant frequency, stability.*

I. INTRODUCTION

Recently, due to the energy crisis, distributed generation (DG) systems using clean renewable energy such as solar energy, wind energy, etc., have become an important issue in the technical research. Typically, a simple series inductor L is inserted between a voltage source inverter (VSI) and the grid to attenuate the high-frequency PWM harmonics to a desirable limit. But the high value of L-filter needs to be adopted to reduce the current harmonics around the switching frequency, which would lead to a poor dynamic response of the system and a high power loss. Hence, a low-pass passive power filter, *LCL*-filter, can achieve a high harmonic attenuation performance with less total inductance ($L_1 + L_2$) [1], [2].

In order to further reduce the total inductance, the *LLCL*-filter was proposed [3]. Compared with the *LCL*-filter, the total inductance and volume of the *LLCL*-filter can be reduced a factor of 25% ~40%. So, the *LLCL*-filter for the grid-tied inverter is becoming attractive for industrial applications [4]. The application of *LLCL*-filter for a three-phase three-wire Shunt Active Power Filter (SAPF) [5] and a Large-Scale Wave Power Plant [6] were analyzed.

As a high order filter, the *LLCL*-filter has also a resonant problem. To suppress the possible resonances of an *LCL*-filter or *LLCL*-filter, active damping [7-11] or passive damping [12-15] measures may be adopted. Passive damping is realized by adding additional components in the system but it causes a decrease of the overall system efficiency. For a stiff grid application, a passive damping strategy is more attractive. Due to high efficiency and flexibility, the active damping method might be preferred, although at the risk of higher cost of sensors and more control complexity. Normally, the digital sampling and transport delays caused by the controller and modulation, as well as discretization effects are taken into account. The delay will influence the stability of the system and when resonant frequency varies, active damping is usually required in an *LCL*-filter [8].

Proportional Resonant (PR) compensator and Proportional Integral (PI) compensator are widely used to control the injected grid current in single-phase grid-connected inverters [16]-[18]. The converter control commonly consists of an outer dc-link voltage PI control loop and an inner current control loop. In this paper, all possible feedback states of currents and voltages of *LLCL*-filter capacitors and inductors with different feedback transfer functions are considered and compared in the continuous Laplace domain. The results show how the various feedback signals need to be fed back in order to achieve resonance damping. Based on the available choices of feedback variables, active damping of the capacitor current control strategy is used for *LLCL*-filter.

First, the system is described and its stability is analyzed in Section II. In Section III, a more general analysis of different active resonance damping solutions is carried out. The basic *LLCL* resonance damping properties of different feedback states are studied. In Section IV, the design of current controller and capacitor current feedback damping are described. Last, a 6-kW three-phase grid-connected inverter is built to verify the proposed design method.

II. MODELING AND STABILITY OF *LLCL*-FILTER-BASED GRID-CONNECTED INVERTER

A. Modeling of LLCL-filter-based Grid-connected Inverter

A three-phase voltage source converter connected to the grid via an *LLCL*-filter is studied as shown in Fig. 1. The inverter output voltage and current are represented as u_i (phase voltage) and i_c, and the grid voltage and current are represented as u_g and i_g. L_g is the grid impedance. Grid side current control will be discussed in this paper [1].

978-1-4799-2706-7/14 $31.00 © 2014 IEEE

Neglecting the influence of the grid impedance and Equivalent Series Resistances (ESRs) of the inductors and capacitors, the transfer function i_g (s) / u_i (s) of the *LLCL*-filter can be derived in (1).

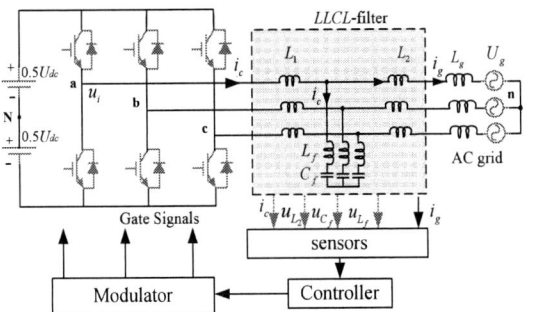

Fig. 1. General structure of three-phase grid-connected inverter with *LLCL*-filter.

$$G_{u_i \to i_g}(s) = \frac{L_f C_f s^2 + 1}{\left(L_1 L_2 C_f + (L_1 + L_2)L_f C_f\right)s^3 + (L_1 + L_2)s} \quad (1)$$

$$\omega_r = \frac{1}{\sqrt{\left(\dfrac{L_1 L_2}{L_1 + L_2} + L_f\right)C_f}} \quad (2)$$

As shown in (2), ω_r is the resonant frequency (in radians per second), f_r is the resonant frequency. If the inductance of L_f is set to zero, then the transfer functions of the *LCL*-filter can also be calculated. Fig. 2 shows bode plots of transfer functions i_g (s)/u_i (s) of *LCL*-filter and *LLCL*-filter when they meet the same harmonic requirement of grid-injected current according to IEEE 519-1992[19].

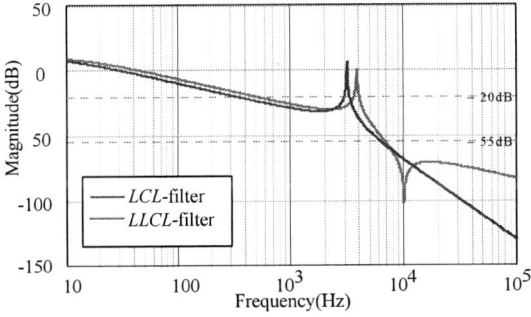

Fig. 2. Bode plots of transfer functions i_g (s) / u_i (s) for different filters.

It can be seen from Fig. 2, the resonant frequency of the *LLCL*-filter is higher than the resonant frequency of the *LCL*-filter. The ratio of the resonant frequency and the sampling frequency is relative to the stability of *LCL*-filter due to the delay. It means that if an *LCL*-filter with a low resonance frequency is chosen for the purpose of high damping of switching harmonics, the design of the active damping gets very difficult and a poor robustness is obtained [11], [20], and [21]. For the *LLCL*-filter, the ratio of the resonant frequency and the sampling frequency also influence the control and stability.

Table I shows the parameters of the selected test system. Table II shows the parameters of the *LLCL*-filter in different resonant frequencies cases according to the design method given in [19].

B. Stability of LLCL-Filter-Based Grid-Connected Inverter with Different Resonant Frequencies

The inverter can be modeled as a linear gain k_{PWM}. The control block diagram of a single loop controller without any damping methods of the three-phase grid-connected inverter with *LLCL*-filter is shown in Fig. 3.

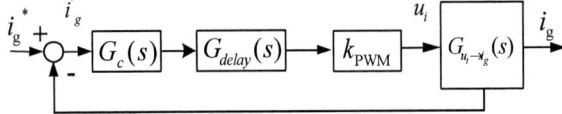

Fig. 3. Block diagram of grid current feedback control.

Other transfer functions in Fig. 3 are shown in (3)-(4) in s-domain. $G_c(s)$ is a PI controller, where k_p and τ are representing its proportional gain, integral time constant respectively. $G_{delay}(s)$ is the delay part in series with the forward path. One sample period delay due to computation and PWM are included. $G(s)$ is the open transfer function of the grid current feedback control.

$$G_c(s) = k_p + \frac{k_p}{s\tau} \quad (3)$$

$$G(s) = G_{delay}(s)k_{PWM}G_c(s)G_{u_i \to i_g}(s) \quad (4)$$

TABLE I
Test System Parameters

DC link voltage U_{dc}	700 V	Grid frequency f_o	50 Hz
Grid phase voltage U_g	220 V	Sample frequency f_d	10kHz
Switching frequency f_s	10kHz	Sampling period T_d	100 us

TABLE II
LLCL-filter Parameters and Resonance Frequency

Case I	Case II	Case III
L_1=2.4mH	L_1=2.4mH	L_1=3mH
L_2=1.2mH	L_2=2mH	L_2=2.4mH
C_f= 2uF	C_f= 8uF	C_f= 8uF
L_f= 128uF	L_f= 32uF	L_f= 32uF
f_r=3.69kHz	f_r=1.68kHz	f_r=1.523kHz
f_r/f_d= 0.369	f_r/f_d=0.168	f_r/f_d=0.153

Fig. 4 shows the bode plot of the forward path transfer function for the single loop.

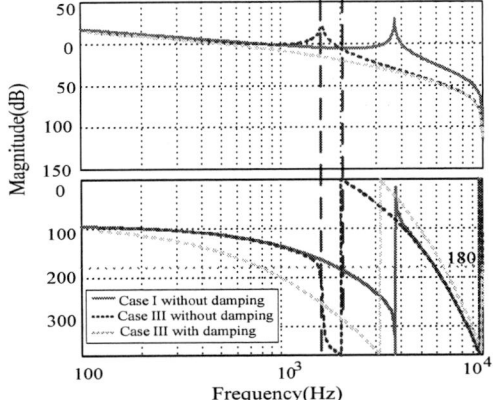

Fig. 4. Bode plot of the forward path transfer function for the single loop shown in Fig. 3.

It can be seen from Fig. 4, that the *LLCL*-filter resonance has no influence on system stability when the resonant frequency is high, since the phase is already well below -180° due to the sampling and transport delay.

This analysis identifies that there is also a resonant frequency critical for *LLCL*-filter, and above it, active damping can be avoided by adjusting the controller gain. When a Zero-Order-Hold (ZOH) is in series of the open loop, discretization of the system introduces delay too, as shown in (5).

$$G_0(s) = H_o(s)G_{u_i \to i_g}(s) = (1 - e^{-T_d s})\frac{G_{u_i \to i_g}(s)}{s} \quad (5)$$

When $\angle G(j\omega_k) = -\pi$, the function is shown in (6). $\omega_k = \pi/3T_d$, it is the root. If a phase angle is already below −180° at this resonant frequency, the system can be stable. It can also deduce that $f_k = f_d/6$, f_k is the critical frequency.

Hence, a single loop is sufficient to be stable when the resonant frequency is above the critical frequency and active damping is necessary when the resonant frequency is below the critical frequency. For the example in Table I, the critical frequency f_k is 1.68 kHz.

$$\angle G(j\omega_k) = \angle \left\{ \begin{array}{c} e^{-j\omega_k T_d} \cdot \dfrac{1 + j\omega_k \tau}{j\omega_k \tau} \cdot \dfrac{1 - e^{-j\omega_k T_d}}{j\omega_k} \\[2mm] \cdot \dfrac{1 - L_f C_f \omega_k^2}{j\left[(L_1 + L_2)\omega_k - (L_1 L_2 C_f + (L_1 + L_2)L_f C_f)\omega_k^3\right]} \end{array} \right\} = -\pi$$

(6)

Fig. 5 shows the closed loop root loci of the three cases in Table II for the single loop grid current feedback.

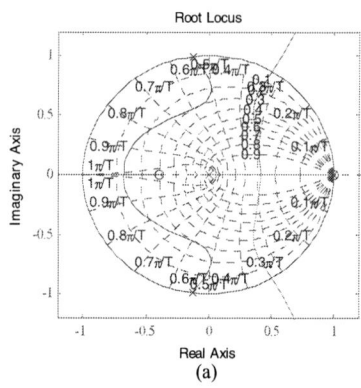

Root Locus

(a)

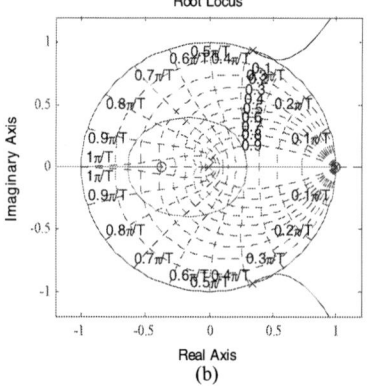

Root Locus

(b)

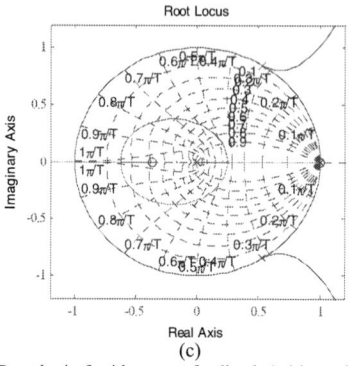

Root Locus

(c)

Fig. 5. Root loci of grid current feedback (without damping) of different cases in Table II. (a) Case I, (b) Case II and (c) Case III.

Fig. 5 (a) depicts the *LLCL* resonant frequency is above the critical frequency. The poles initially track inside the unit circle with the proper proportional gain K_p. Fig. 5(b) shows the *LLCL* resonant frequency at the critical frequency. The system is on the edge to be unstable. When the resonant frequency is less than the critical frequency system will always be unstable regardless of what the proportional gain is without damping, as shown in Fig. 5(c). In this case, a damping method is necessary to be used.

III. STUDY OF ACTIVE DAMPING WITH DIFFERENT FEEDBACK STATES

A. Notch Filter Concept

Active damping methods can be classified into two main classes: multi-loop and filter-based active damping [8]. For passive damping, the resonance frequency of an *LCL*-filter can be damped by connecting a resistor to the filter. The virtual resistance without reducing the efficiency was introduced, which is the equivalent as the passive damping resistors of the filter that resonance damping techniques.

The control structure of inner current loop with a notch concept [22] is shown in Fig. 6.

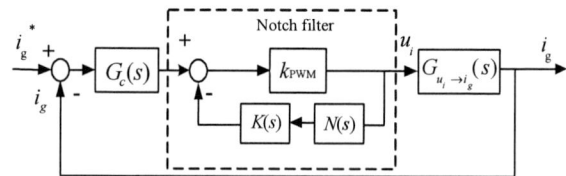

Fig. 6. Active damping based on a notch filter concept.

$$B(s) = \frac{k_{PWM}}{1 + k_{PWM}K(s)N(s)} = \frac{s^2 + \omega_r^2}{s^2 + 2\xi_2\omega_r s + \omega_r^2} \quad (7)$$

In order to eliminate the resonant peak at the frequency ω_r, the notch $B(s)$ should have a negative peak in ω_r, so it can be expressed as (7).

There are different variables can be chosen as control object for *LLCL*- of detection variables, it has different function $K(s)$ with different feedback variables. Table III shows the expressions filter. The structure of $N(s)$ depends on the selection using the filter capacitor voltage u_c, filter capacitor current i_c, filter resonant inductor voltage in grid side u_{L2}, inductor voltage in resonant circuit u_{Lf}. It can be seen from the Table III, the feedback function of grid side inductor voltage is a little complex. The choice of the sensor location depends strongly on the application situations. For different ratios between the resonance frequency and control frequency, the selected approaches behave differently [10].

B. Filter Capacitor Current Feedback

For the LCL-filter, capacitor current feedback and capacitor voltage feedback are often used [10]. When the capacitor current feedback is sensed, $K(s)$ can be easily configured as a proportion link, which is isolated with the system parameters. Fig. 7 shows the capacitor current feedback control scheme according to Fig. 6.

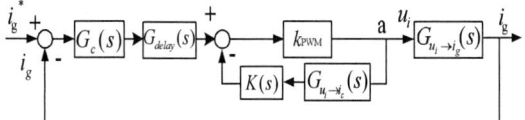

Fig.7. Capacitor current feedback control scheme.

$$N(s) = G_{u_i \to i_c}(s) = \frac{\omega_r^2 L_2 C_f s}{(s^2 + \omega_r^2)(L_1 + L_2)} \quad (8)$$

Then the open loop transfer function of the capacitor current feedback control from i_g to i_g* is shown in (9)

$$G'(s)_{open} = \frac{k_{PWM} \omega_r^2 (L_f C_f s^2 + 1) G_c(s) G_{delay}(s)}{(s^2 + \omega_r^2)(L_1 + L_2) + K_{i_c} G_{delay}(s) k_{PWM} L_2 s \omega_r^2} \frac{1}{s} \quad (9)$$

As shown in Fig. 4, the capacitor current feedback can dampen the resonant peak and make the system stable.

C. Filter Capacitor Voltage Feedback

When the capacitor voltage is sensed, a derivative filter capacitor voltage feedback is required for the resonance damping. Some papers have illustrated this method for an LCL-filter. As shown in Table III, a differential feedback is necessary but it may cause noise problems in the control because it will amplify high-frequency signals [23].

D. Filter Resonant Inductor Voltage Feedback

When the resonant inductor voltage is sensed, the integral feedback can be expected to keep the system stable. Normally, the filter capacitor current is sensed for the LCL-filter but the voltage sensor is cheaper than current sensor. This is a new active damping method for $LLCL$-filter which can reduce the cost of sensor. The resonant inductor voltage feedback can also be transferred in Fig. 7. The feedback coefficient is integral and the structure of $N(s)$ is the transfer function from u_i to u_{Lf}.

The next section will analyze the design of capacitor current feedback only.

TABLE III
TRANSFER FUNCTIONS OF DIFFERENT VARIABLES FEEDBACK

Variable	u_{C_f}	$i_c(s)$	$u_{L_f}(s)$	u_{L_2}
$N(s)$	$\dfrac{\omega_r^2 L_2}{(s^2 + \omega_r^2)(L_1 + L_2)}$	$\dfrac{\omega_r^2 L_2 C_f s}{(s^2 + \omega_r^2)(L_1 + L_2)}$	$\dfrac{\omega_r^2 L_2 L_f C_f s^2}{(s^2 + \omega_r^2)(L_1 + L_2)}$	$\dfrac{\omega_r^2 L_2 (L_f C_f s^2 + 1)}{(s^2 + \omega_r^2)(L_1 + L_2)}$
$K(s)$	$sK_{u_{C_f}}$	K_{i_c}	$K_{u_{L_f}} \Big/ s$	$K_{u_{L2}} \dfrac{C_f s}{L_f C_f s^2 + 1}$

IV. DESIGN OF CURRENT REGULATOR AND CAPACITOR CURRENT FEEDBACK COEFFICIENT

A. PI Controller Gain Design

The maximum possible controller gains for the system can now be analytically determined using the concepts developed in [24] - [26]. The proportional gain is then set to achieve unity gain at the desired crossover frequency f_c / ω_c. The choice of k_p can be decided by the system bandwidth satisfying the desired phase margin Φ_m. For a single loop control, the phase angle at the crossover frequency can be described in (10). In addition, the cross-over frequency ω_c can be determined, when the LCL-filter is approximated to an L-filter.

$$\omega_c = \frac{\pi/2 - \Phi_m}{3T_d/2} \quad (10)$$

Usually, the integral time constant τ can be set as $10/\omega_c$.

The system open-loop gain achieves unity at ω_c. Then the maximum gain can be calculated as

$$|\angle G(j\omega_c)| \approx \left| \frac{k_p \sqrt{(\omega_c \tau)^2 + 1}}{\omega_c \tau} \right| \left| \frac{1 - e^{-j\omega_c T_d}}{j\omega_c} \right| \left| \frac{k_{PWM}}{j(L_1 + L_2)\omega_c} \right|$$

$$k_p \approx \frac{\omega_c (L_1 + L_2)}{k_{PWM}} \quad (11)$$

B. Capacitor current Feedback Coefficient Gain

The denominator of closed loop transfer function of $LLCL$-filter is shown in (12) based on (9). Base on (12) the minimum value of K_{i_c} can be found using the limiting ratio of proportional gain k_p to dampen using Routh's Stability Criterion, as shown in (13).

$$D(s) = \left[L_1 L_2 C_f + (L_1 + L_2) L_f C_f \right] s^4 + K_{i_c} k_{PWM} L_f L_2 C_f s^3$$

$$+ (L_1 + L_2) s^2 + k_p k_{PWM} s + k_p k_{PWM}/\tau \quad (12)$$

$$K_{i_c} \geq \frac{k_p}{(L_1 + L_2)L_2}\left[L_1 L_2 + (L_1 + L_2)L_f\right] \quad (13)$$

GM_1 and GM_2 are defined by the magnitude of (9) at f_r, and $f_d/6$ [27]. Then the limitation of K_{ic} can be obtained from (14) and (15).

$$K_{i_c} = 10^{GM_1/20}\frac{2\pi f_c L_1}{k_{PWM}} \quad (14)$$

$$K_{i_c} = 10^{GM_2/20}(\frac{6f_c}{f_d})^2\frac{2\pi f_c L_1}{k_{PWM}} + \frac{2\pi f_c L_1}{k_{PWM}}\frac{(f_d/6)^2 - (f_r)^2}{f_d/6} \quad (15)$$

K_{ic} should satisfy this region. Hence, the basic design procedures can be addressed as:

1. Determine the specifications of the loop gain. Desired phase margin Φ_m is 40°, GM_2 is 5.1 dB, and GM_1 is −5.8 dB.

2. Obtain the value of k_p and f_c to satisfy all the requirements according to (10), (11) and (13), f_c= 1.1 kHz is chosen to obtain fast dynamic response. Then k_p is calculated according to (11).

3. Taking f_c = 1.1 kHz in (13), (14) and (15), calculate the satisfactory feedback gain in this example.

Fig. 8 shows the loci branches of the system with the capacitor current coefficient increasing. There is a stable range of K_{ic} as shown in Fig. 8. The calculated gain should not pass the limitation.

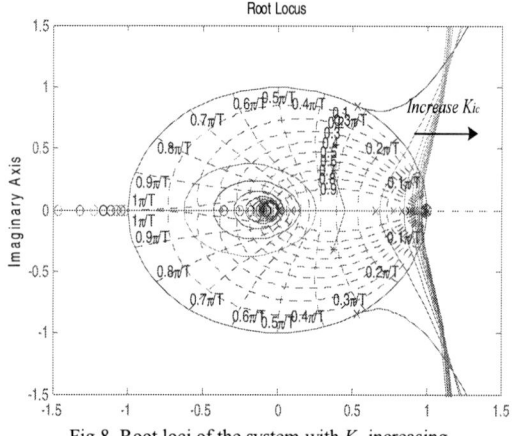

Fig.8. Root loci of the system with K_{ic} increasing.

V. SIMULATION AND EXPERIMENTAL RESULTS

A. Simulation Results

In order to illustrate the stability and verify the active damping method of *LLCL*-filter based grid-connected inverter, a three-phase inverter with 6 kW rated power is simulated using MATLAB. This paper uses a PI controller and SVPWM modulation. The detailed system parameters are listed in Table I and Table II.

First, in order to investigate stability of the system without damping in different resonant frequencies, the *LLCL*-filter is analyzed into three cases with different parameters, one with a high resonance frequency, one with a critical resonance frequency and the other with a

low resonance frequency. According to (11), the controller gain is $k_p = 0.065$ and $\tau = 1.59$ ms.

(1) Case I: high frequency, f_{res}=3.69 kHz.

In the case I, the resonant frequency is high (3.69 kHz) and crossover frequency is set to 1.1 kHz in order to get a fast response and to meet phase margin limitation. It can be seen from Fig. 9 the system is stable.

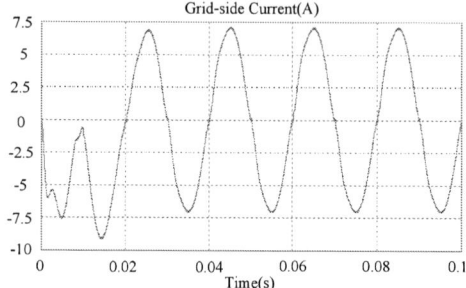

Fig. 9. Grid-side current waveform of Case I (f_{res}=3.69 kHz).

(2) Case II: critical frequency f_{res}=1.68 kHz.

As it mentioned before, there is a critical frequency for *LLCL*-filter. It is calculated as $f_d/6$ based on the function (6). It can be seen from Fig. 10, the system is almost unstable at the critical frequency. When the grid impedance is increased, the resonant frequency will be deduced more.

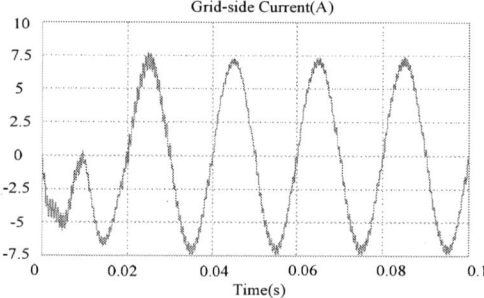

Fig. 10. Grid-side current waveform of Case II (f_{res}=1.68 kHz).

As shown in Fig. 11, when the grid impedance is set to 2.4 mH, the resonant frequency is reduced to 1.6 kHz which is under the critical frequency and the system is unstable.

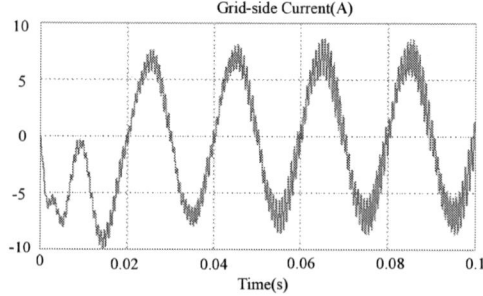

Fig. 11. Grid-side current waveform of Case II when Increase L_2 to 2.4 mH (f_{res}=1.60 kHz).

As shown in Fig. 12, when the grid-side inductance is set to 1.2 mH, the resonant frequency is increased to 1.95 kHz and the system is changed from critical state to stable.

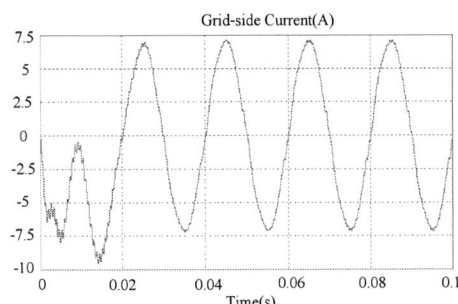

Fig.12. Grid-side current waveform of Case II when increase L_2 to 2.4mH (f_{res}=1.95 kHz).

(3) Case III: Low frequency f_{res}=1.523 kHz

In case III, the resonant frequency is low (1.523 kHz). It can be seen from Fig. 13 that system is completely unstable without damping.

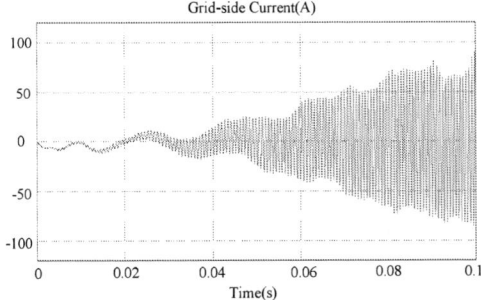

Fig.13. Grid-side current waveform of Case III (f_{res}=1.523 kHz).

So, when designing the parameters it is better to make the resonant frequency higher in order to get a better stability and robustness. When the resonant frequency is lower or nearby the critical frequency, the active damping method is necessary to be used. In this paper, take Case III as an example, the active damping with a capacitor current feedback is used.

Applying the parameters described in Table I and Table II, the crossover frequency is set to 1.1 kHz to get a high response and meet the phase margin limitation.

In case III, the resonant frequency is low (1.53 kHz). In order to get enough phase margin crossover frequency should be set below the resonant frequency. According to (12), the PI controller gains $k_p = 0.043$ and $\tau = 3.18$ ms.

Fig. 14 and Fig. 15 show the dynamic performance of Case I and Case III respectively when the reference current changes from 7 A to 12.9 A at time 0.1s to test the dynamic performance. It can be seen from Fig. 14, there is no oscillation during the transient in the high resonant frequency without damping methods. Fig. 8 shows the same dynamic performance in the low resonant frequency when the capacitor current feedback active damping method is used. It can also be seen that the current ripple in Fig. 15 is smaller than the current ripple in Fig. 14, because the inductance and capacitance in the high resonant frequency case is larger than that in the low resonant frequency case.

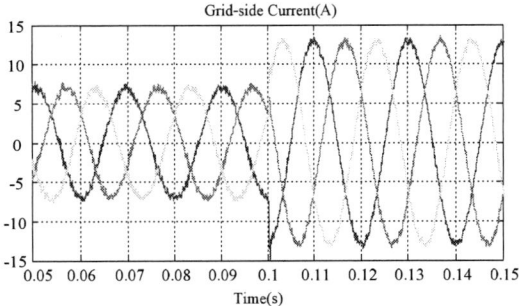

Fig.14. Grid-side currents of Case I (high resonant frequency) without active damping.

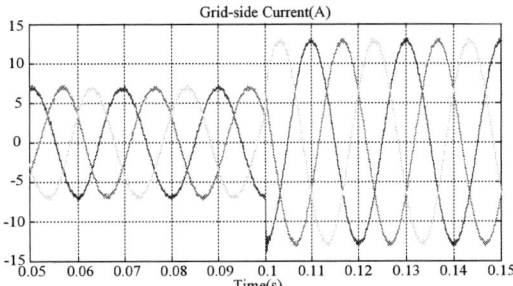

Fig. 15. Grid-side currents of Case III (low resonant frequency) with active damping.

Fig. 16 shows that the system is unstable at the beginning in low resonance frequency case, but it turns out to be stable when active damping is enabled at 0.1 in Case III. Fig.17 shows that the system becomes unstable when the value of the capacitor feedback coefficient is increased. There is a limited stable region for the feedback coefficient.

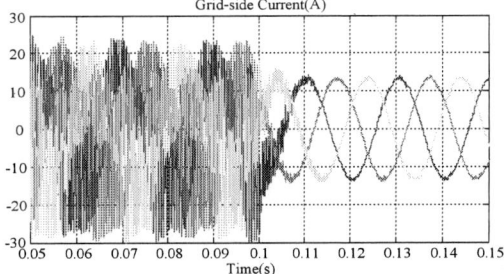

Fig. 16. Grid-side currents of Case III when active damping is enabled at time 0.1s.

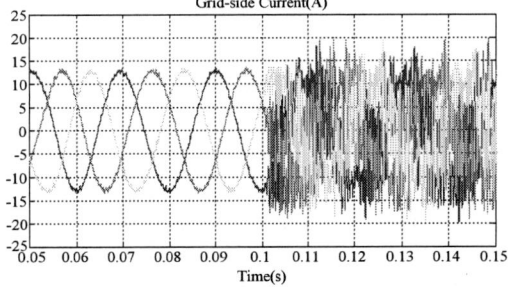

Fig. 17. Grid-side currents of Case III when K_{i_c} is increased at time 0.1s.

B. Experimental Results

The control algorithm is implemented on a dsPACE DS 1006 board. Due to the limitation of the setup, the power in experiment is lower than the simulation. Fig. 18

shows the dynamic transition of grid-side currents in high resonant frequency case when the power is increased without active damping. Fig. 19 shows the grid-side currents when active damping is enabled in the low resonant frequency case.

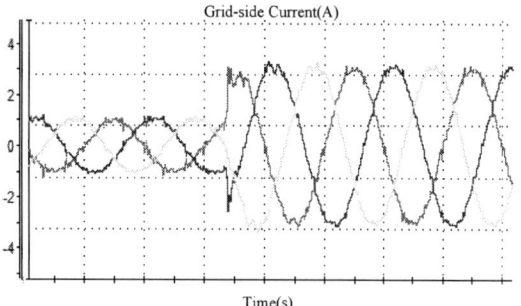

Fig. 18. Experimental grid-side currents in high resonant frequency case without active damping.

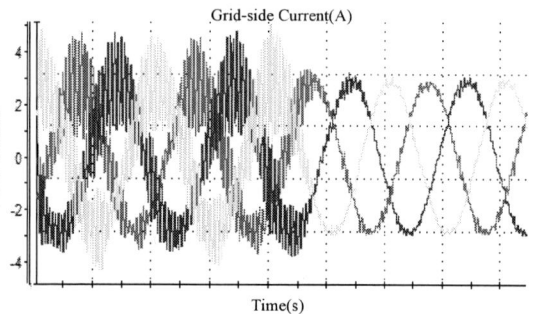

Fig. 19. Experimental grid-side currents in low resonant frequency case when active damping is enabled.

VI. CONCLUSION

Compared with the *LCL*-filter, the resonant frequency of *LLCL*-filter is higher and it is easier to be stable. The work presented in this paper shows when damping method is not required in an *LLCL*-filter as the resonant frequency varies considering different sampling and transport delays. The critical frequency is determined by the delay time and sample frequency.

In the low resonant frequency cases, or critical cases, the resonant frequency is easy to be changed due to the parameter variation and grid impedance variation. Then, damping methods are necessary to be used. Based on the analysis of possible feedback variables, capacitor current feedback active damping control strategy is chosen for the *LLCL*-filter. Simulation and experiment results prove the influence of the resonant frequency and design of capacitor current feedback. The control of *LCL*-filter and *LLCL*-filter are similar and the additional inductor of *LLCL*-filter brings no extra control difficulties.

REFERENCES

[1] M. Liserre, F. Blaabjerg, and S. Hansen, "Design and Control of an LCL-Filter-Based Three-Phase Active Rectifier", *IEEE Trans. Ind. App.*, vol. 41, no. 5, pp. 1281-1291, Sep./Oct. 2005.

[2] V. Salas, and E. Olías, "Overview of the state of technique for PV inverters used in low voltage grid-connected PV systems:

[3] W. Wu, Y. He, and F. Blaabjerg, "An *LLCL* power filter for single-phase grid-tied inverter," *IEEE Trans. Power Electron.*, vol. 27, no. 2, pp. 782-789, Feb. 2012.

[4] J. M. Bloemink, T C. Green, "Reducing Passive Filter Sizes with Tuned Traps for Distribution Level Power Electronics," in *Proc. of IEEE EPE 2011*, pp. 1-9, *Aug.* 2011.

[5] K. Dai, K. Duan, X. Wang, "Yong Kang Application of an *LLCL* Filter on Three-Phase Three-Wire Shunt Active Power Filter," in *Proc. of IEEE INTELEC 2012*, pp. 1-5, Sep. 2012.

[6] A. M. Cantarellas, E. Rakhshani, D. Remon, P. Rodriguez, "Design of the *LCL*+trap filter for the two-level VSC installed in a large-scale wave power plant," in *Proc. IEEE Energy Conversion Congress and Exposition (ECCE)*, pp. 707-712, 2013.

[7] J. Dannehl, M. Liserre and F. Fuchs, "Filter-based active damping of voltage source converters with *LCL* filters," *IEEE Trans. Ind. Electron.*, vol. 58, no. 8, pp. 3623-3633, Oct. 2011.

[8] M. Liserre, A. D. Aquila, and F. Blaabjerg, "Genetic algorithm-based design of the active damping for an *LCL*-filter three-phase active rectifier," *IEEE Trans. Power Electron.*, vol. 19, no. 1, pp. 76–86, Jan. 2004.

[9] R. Teodorescu, F. Blaabjerg, M. Liserre, and A. Dell'Aquila, "A stable three-phase *LCL*-filter based active rectifier without damping," in *Proc. 2003 IEEE Ind. Appl. Soc. Annu. Meeting*, pp. 1552–1557, 2003.

[10] J. Dannehl, F. Fuchs, S. Hansen, "Investigation of active damping approaches for PI-based current control of grid-connected pulse width modulation converters with *LCL* filters," *IEEE Trans. Ind. App.*, vol.46, no. 4, pp.1509-1517, 2010.

[11] S. Parker, B. McGrath , and G Holmes. "Regions of Active Damping Control for *LCL* Filters," in *Proc. IEEE Energy Conversion Congress and Exposition (ECCE)*, Raleigh, NC, pp. 53-60, 2012.

[12] W. Wu, Y. He, and F. Blaabjerg, "A New Design Method for the Passive Damped *LCL*- and *LLCL*-Filter Based Single-Phase Grid-tied Inverter," *IEEE Trans. Ind. Electron.*, vol. 60, no. 10, pp. 4339-4350, Oct. 2013.

[13] W. Wu, M. Huang, Y. Sun, X. Wang, F. Blaabjerg, "A composite passive damping method of the *LLCL*-filter based grid-tied inverter," in *Proc. of PEDG 2012*, Aalborg, Denmark, pp: 759 – 766, 25-28 June 2012.

[14] R. Peña-Alzola, M. Liserre, F. Blaabjerg, R. Sebastián, J. Dannehl, F.W. Fuchs, "Analysis of the Passive Damping Losses in *LCL*-Filter-Based Grid Converters," *IEEE Power Electronics Transactions on*, vol.28, no.6, pp. 2642-2646, June 2013.

[15] P. Channegowda and V. John, "Filter Optimization for Grid Interactive Voltage Source Inverters", *IEEE Trans. on Ind. Electron.*, vol. 57, no. 12, pp. 4106–4114, Dec. 2010.

[16] D. G. Holmes, T. A. Lipo, B. P. McGrath and W. Y. Kong, "Optimized design of stationary frame three phase AC current regulators," *IEEE Trans. Power Electron.*, vol. 24, no. 11, pp. 2417-2425, Nov. 2009.

[17] D. Pan, X. Ruan, C. Bao, W. Li, X. Wang, "Capacitor-Current-Feedback Active Damping With Reduced Computation Delay for Improving Robustness of *LCL*-Type Grid-Connected Inverter Capacitor-Current-Feedback Active Damping With Reduced Computation Delay for Improving Robustness of *LCL*-Type Grid-Connected Inverter," *IEEE Trans. Power Electron.*, vol.29, no. 7, pp.3414-3427, 2014.

[18] D. Ricchiuto, M. Liserre, T. Kerekes, R. Teodorescu, F. Blaabjerg, "Robustness analysis of active damping methods for an inverter connected to the grid with an *LCL*-filter", in *Proc. of Energy Conversion Congress and Exposition (ECCE)*, on page(s): 2028 – 2035, 2011.

[19] M. Huang, W. Wu, F. Blaabjerg, and Y. Yang, "Step by Step Design of a High Order Power Filter for Three-Phase Three-Wire Grid-connected Inverter in Renewable Energy System," in *Proc. of PEDG 2013*, pp.1-8, July 08-11,2013.

[20] S. Yang, Q. Lei, P. F.Z., Z. Qian, "A Robust Control Scheme for Grid-Connected Voltage-Source Inverters," *IEEE Trans. Power Electron.*, vol.58, no. 1, pp.202-212, 2011.

[21] V. Blasko and V. Kaura, "A novel control to actively damp resonance in input *LC* filter of a three-phase voltage source converter," *IEEE Trans. Ind. Appl.*, vol. 33, no. 2, pp. 542–550, Apr. 1997.

[22] C. Liu, X. Zhang, L. Tan and F. Liu "A novel control strategy of LCL-VSC based on notch concept," in *Proc. of PEDG 2010*, pp.343-346, 2010.

[23] P.A. Dahono, "A control method to damp oscillation in the input LC-filter," in *Proc. PESC'02*, vol. 4, pp. 1630–5, 2002.

[24] Y. Tang, P. C. Loh, P. Wang, F. H. Choo and F. Gao, "Exploring inherent damping characteristics of *LCL*-filters for three-phase grid connected voltage source inverters," *IEEE Trans. Power Electron.*, vol. 27, no. 3, pp. 1433-1443, Mar. 2012.

[25] D. G. Holmes, T. A. Lipo, B. P. McGrath and W. Y. Kong, "Optimized design of stationary frame three phase AC current regulators," *IEEE Trans. Power Electron.*, vol. 24, no. 11, pp. 2417-2425, Nov. 2009.

[26] F. Liu, Yan Zhou, S. Duan, J. Yin, B. Liu, F. Liu, "Parameter Design of a Two-Current-Loop Controller Used in a Grid-Connected Inverter System With *LCL* Filter," *IEEE Trans. Ind. Electron.*, vol. 56, no. 11, pp. 4483-4491, 2009.

[27] M. Xue, Y. Zhang, Y. Kang, Y. Yi, S. Li, and F. Liu, "Full feed forward of grid voltage for discrete state feedback controlled grid-connected inverter with *LCL* filter," *IEEE Trans. Power Electron.*, vol. 27, no. 10, pp. 4234–4247, Oct. 2012.

The 2014 International Power Electronics Conference

Integrated Common and Differential Mode Filter Applied to a Single-Phase Transformerless PV Microinverter with Low Leakage Current

Ricardo Souza Figueredo, Kelly Caroline Mingorancia de Carvalho, Lourenço Matakas Junior
Electrical Energy and Automation Department
Polytechnic School of the University of São Paulo
São Paulo, Brazil
ricardoszf@gmail.com, kellymingorancia@gmail.com, matakas@pea.usp.br

Abstract— **This paper proposes an integrated common and differential mode filter with passive damping for single-phase grid-connected transformerless photovoltaic (PV) microinverters, employing the full-bridge voltage source inverter (VSI) topology with continuous unipolar modulation. This strategy achieves low current distortion, low leakage current and high efficiency. A design procedure for the proposed filter is presented including the passive damping components. The effectiveness of the proposed solution is confirmed by simulation and experimental results. Additionally, discussions about safety standards, analysis of leakage current production mechanism and some common mode (CM) filter implementations found in the literature are presented.**

Keywords— *Integrated filter, Leakage current, Passive damping, Transformerless PV microinverter.*

I. INTRODUCTION

Nowadays the majority of residential and commercial PV systems employ string and multi-string inverters; however, according to the report [1], published by the market research firm IHS Electronics & Media, the global market for microinverters will expand by four times its current level by 2017. The main advantages presented by microinverters are: *i)* reduced DC cabling; *ii)* simple and less expensive installation; *iii)* reduced losses due to partial shading and PV module mismatch; *iv)* the installation can be easily expanded. The main drawback is its higher cost compared to string inverters. Therefore, the development of low cost transformerless solutions for microinverters may be helpful to expand its market share [2], [3]. Fig. 1 shows the block diagram of a two-stage transformerless microinverter, which is composed by non-isolated DC/DC and DC/AC converters.

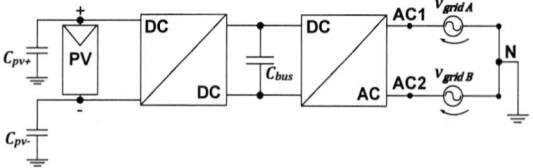

Fig. 1. Two-stage non-isolated grid-connected PV microinverter.

The lack of galvanic isolation between the PV system and the utility grid causes difficulties, such as the possibility of DC current injection and leakage current. However, when transformerless inverters are compared to inverters with high or low frequency transformer they present relevant advantages as higher efficiency, lower weight, volume and cost [4]. The protection against excessive residual currents is an important safety requirement, established by PV inverter standards and grid codes [5]-[7]. The purposes of this requirement are personal protection against electric shock and fire prevention. Due to the PV module constructive structure, parasitic capacitance is formed between the PV cells and the panel structure. The parasitic capacitance (C_{PV}) value is between 50 and 150 nF/kWp for crystalline silicon cells and it can reach 1 µF/kWp for thin-film solar cells [8]. Transformerless inverters can produce time varying common mode voltages, leading to leakage current flow through the parasitic capacitance and ground. The leakage current can cause tripping of residual current protection system [9], increase of the inverter losses, electromagnetic interference (EMI) and harmonic distortion of the current injected in the grid [10].

This paper proposes an integrated differential and common mode filter with passive damping to minimize the leakage current produced by a single-phase grid-connected full-bridge transformerless VSI employing continuous unipolar PWM. The proposed implementation combines, a simple, well known and low cost topology, a modulation strategy that minimizes the differential mode (DM) filtering requirements and a low loss passive damping configuration that simplifies the control strategy, avoiding additional current or voltage transducers and computation time required by active damping solutions.

The following topics are discussed in this paper: Section II discusses some safety standards for PV inverters; Section III shows a simplified CM model; Section IV shows a review of CM filters proposed to minimize the leakage current. Section V presents the modeling and design of an integrated CM and DM filter applied to a transformerless full-bridge PV microinverter; Section VI presents simulation results; Section VII shows experimental results of the proposed filter; Section VIII presents conclusions.

978-1-4799-2706-7/14 $31.00 © 2014 IEEE

II. SAFETY STANDARDS FOR PV INVERTERS

In order to provide protection against electric shock and fire hazard for non-isolated inverters, the standards VDE V 0-126-1-1:2006 and IEC 62109-2:2011, require an additional protection through the application of residual current detectors (RCD's) or by a residual current monitoring unit (RCMU). This additional protection can be integrated or it can be external to the inverter. This requirement can be met by provision of an RCD with a residual current setting of 30 mA between the inverter and the mains [6]. When the protection is provided by an RCMU, the inverter shall disconnect from the mains within 0.3 s when a continuous residual current exceeds 300 mA RMS. When a sudden increase in the RMS residual current is detected, the Table I shall be followed [6], [7].

TABLE I
RESPONSE TIME FOR SUDDEN CHANGE IN RESIDUAL CURRENT

Residual current sudden change	Max disconnection time
30 mA RMS	0.3 s
60 mA RMS	0.15 s
150 mA RMS	0.04 s

III. COMMON MODE EQUIVALENT CIRCUIT

The leakage current magnitude depends on the total common mode voltage ($v_{cm\ total}$) and the common mode impedance of the equivalent circuit. The simplified inverter circuit (Fig. 2), includes the key elements that affect the total common mode voltage and consequently the leakage current. The voltage source v_{bus} represents the PV module and the non-isolated high gain step-up DC/DC converter [3], [11], C_{pv} represents the PV module parasitic capacitances, L_1 and L_2 are the current harmonics filter inductors. The simplified circuit neglects the following parasitic elements: *i)* the parasitic capacitances between the power semiconductors and heat sinks; *ii)* the impedance of the electric grid distribution conductors and *iii)* the ground connection impedance. These simplifications do not have significant influence from the point of view of leakage current analysis [12], [13].

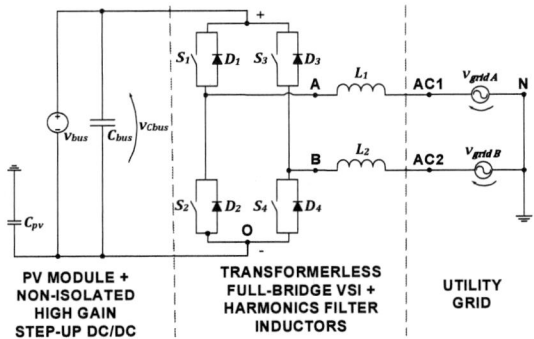

Fig. 2. Simplified circuit of transformerless PV microinverter.

The simplified common mode equivalent circuit obtained from the Fig. 2 circuit is presented in Fig. 3.

The model shown in Fig. 3 is enough to analyze the influence of modulation strategy, harmonics filter configuration (it can use only one inductor, with the total

required inductance or 2 inductors, each one with half the total inductance) and electric distribution system on $v_{cm\ total}$ waveform (open loop voltage between the node O and the ground if the parasitic capacitance is removed). A detailed derivation of the common mode circuit can be found in [12], [13].

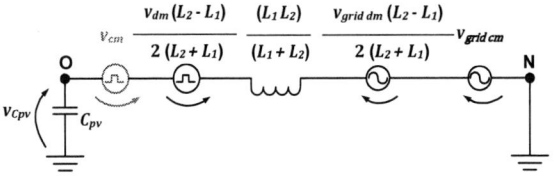

Fig. 3. Simplified common mode equivalent circuit

$$v_{cm\ total} = -v_{cm} - \left[v_{dm} \cdot \frac{L_2 - L_1}{2 \cdot (L_2 + L_1)}\right] + $$
$$v_{grid\ cm} + \left[v_{grid\ dm} \cdot \frac{L_2 - L_1}{2 \cdot (L_2 + L_1)}\right] \quad (1)$$

Where:

$$v_{cm} = (v_{AO} + v_{BO})/2 ; \quad (2)$$
$$v_{dm} = v_{AO} - v_{BO} ; \quad (3)$$
$$v_{grid\ cm} = (v_{gridA} + v_{gridB})/2 ; \quad (4)$$
$$v_{grid\ dm} = v_{gridA} - v_{gridB} ; \quad (5)$$

v_{cm} = Inverter common mode voltage;
v_{dm} = Inverter differential mode voltage;
$v_{grid\ cm}$ = Grid common mode voltage;
$v_{grid\ dm}$ = Grid differential mode voltage;
v_{AO} = Voltage between node A and DC bus (-) terminal;
v_{BO} = Voltage between node B and DC bus (-) terminal;
v_{gridA} = Line A to neutral voltage;
v_{gridB} = Line B to neutral or neutral voltage (depends on the electric distribution system);
v_{Cpv} = voltage across C_{pv}.

Equation (1) is a general expression that defines the total common mode voltage for full-bridge topology with any combination of modulation strategy, harmonics filter configuration and electric distribution system. Among the sources contributing to $v_{cm\ total}$, v_{cm} provides the main contribution to the leakage current, because it presents high frequency PWM waveform when hybrid PWM (discontinuous unipolar modulation) and unipolar PWM (continuous unipolar modulation) strategies are employed [14], [15], [16].

For bipolar modulation, when $L_1 = L_2$, v_{cm} is ideally constant, but due to non-ideal conditions found in practical implementations, high frequency common mode voltage is produced [13], [17], [18]. Bipolar modulation also presents lower efficiency and requires a larger differential mode harmonics filter when compared to unipolar PWM [16], [19]. Therefore, bipolar modulation strategy is not suitable for competitive commercial products. The contribution of v_{dm} and $v_{grid\ dm}$ are minimized employing $L_1 = L_2$. The contribution of $v_{grid\ cm}$, depends on the electric distribution system, nevertheless, considering a grid voltage with low distortion, it presents low frequency sinusoidal waveform, resulting in low leakage current.

IV. LEAKAGE CURRENT MINIMIZATION TECHNIQUES

There are basically two suitable methods that can be applied to limit the leakage current produced by single-phase transformerless PV inverters employing the full-bridge VSI topology using continuous or discontinuous unipolar PWM. The first one uses modified topologies derived from full-bridge with additional switches and the second uses common mode filters [20]. Common mode filters can be passive or active [21], this paper will discuss only passive CM filters.

This section presents two implementations of passive CM filter for leakage current minimization Fig. (4)-(7).

A. Passive CM filter with C_{cm} connected to ground

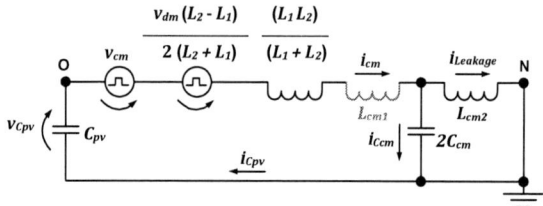

Fig. 4. Passive CM filter with C_{cm} connected to ground

An LCL type EMI filter designed to limit the common mode current (i_{cm}), in a full-bridge transformerless PV inverter is proposed in [22] (Fig. 4). The simplified equivalent CM circuit neglecting the grid voltage influence on i_{cm} is shown in (Fig. 5).

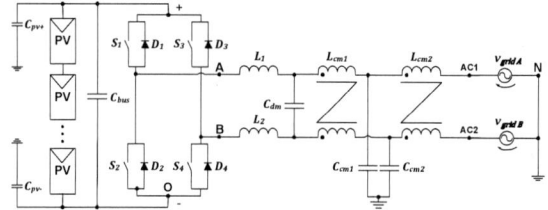

Fig. 5. Equivalent CM circuit proposed in [22]

With this CM filter the current flowing through the parasitic capacitance (i_{Cpv}) is equal to i_{cm}. The common mode current flowing through the grid ($i_{Leakage}$) is strongly reduced, because the common mode capacitor (C_{cm}) provides a low impedance path for high frequency, the current flowing through the common mode capacitor (i_{Ccm}) $\gg i_{Leakage}$.

B. Passive CM filter with C_{cm} connected to DC bus-

A common mode filter cascaded with an LCL type differential mode harmonics filter (Fig. 6) for leakage current minimization in a full-bridge transformerless PV inverter, employing continuous unipolar PWM is proposed in [17]. The CM filter proposed here differs from [22], because the C_{cm} capacitors are connected to the node O (DC bus negative terminal), in this case the major part of i_{cm} circulates inside the converter instead of the ground conductor.

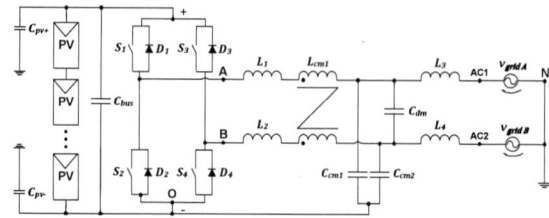

Fig. 6. Passive CM filter with C_{cm} connected to DC bus-

The equivalent CM circuit neglecting the grid voltage influence on i_{cm} is shown in (Fig. 7). The main concept of the filter proposed in [17] can also be found in [20], [23]-[26].

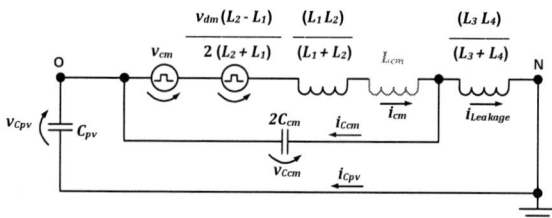

Fig. 7. Equivalent CM circuit proposed in [17]

V. INTEGRATED COMMON AND DIFFERENTIAL MODE FILTER FOR TRANSFORMERLESS PV MICROINVERTER

This paper proposes an integrated CM and DM filter based on [17] and [20] (Fig. 8).

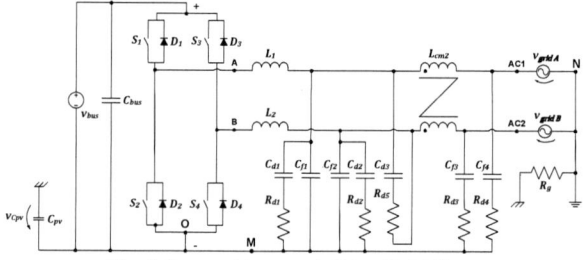

Fig. 8. Proposed integrated CM and DM filter

The proposed filter is applied to a single-phase grid-connected PV microinverter employing continuous unipolar PWM. The inductors L_1 and L_2 performs DM and CM filtering, eliminating a bulky CM inductor (inductors L_{cm1} in Fig. 5 and L_{cm} in Fig. 7), which is needed to minimize the leakage current in the previously proposed CM filter implementations, when the modulation produces time varying v_{cm}.

RC parallel passive damping ($R_{d1} - C_{d1}$), ($R_{d2} - C_{d2}$) and ($R_{d5} - C_{d3}$) are connected across the main filter capacitors ($C_{f1,} C_{f2}$) in order to achieve damping of resonances with low losses, without compromising the attenuation of high frequency harmonics [27]-[32], avoiding additional current or voltage transducers and computation time required by active damping strategies.

The simplified equivalent DM circuit shows an LCL type DM filter (Fig. 9), where $L_{cm2\ Lkg}$ represents the leakage inductance of the common mode inductor L_{cm2} and $L_{grid\ dm}$ represents the inductive part of the grid impedance.

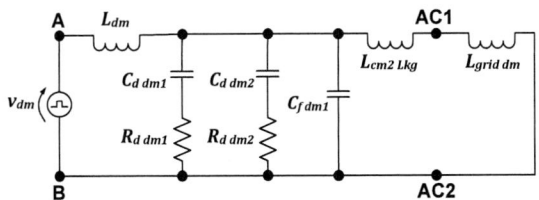

Fig. 9. Simplified equivalent differential mode circuit

The simplified equivalent CM circuit shows a two stage LC type CM filter (Fig. 10). L_{cm2} is a small common mode inductor which blocks the high frequency voltage drop across $C_{f\,cm1}$ (common mode filter capacitor).

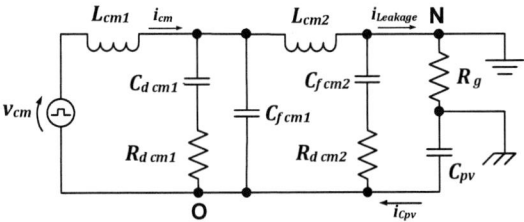

Fig. 10. Simplified equivalent common mode circuit

The main parameters used for the filter design are shown in Table II.

TABLE II
MAIN PARAMETERS OF THE INVERTER

Parameter	Value
f_{grid}	60 Hz
f_{sw}	19.98 kHz
V_{grid}	220 V rms
V_{bus}	385 V dc
P_{max}/S_{max}	300 W/300 VA

From the equivalent CM and DM circuits Fig. (9)-(10) the following parameters can be defined:

$$L_1 = L_2 = (2 \cdot L_{cm1}) = (L_{dm}/2) \tag{6}$$
$$C_{f1} = C_{f2} = (C_{f\,cm1}/2) = (2 \cdot C_{d\,dm1}) \tag{7}$$
$$C_{d1} = C_{d2} = (C_{d\,cm1}/2) = (2 \cdot C_{d\,dm1}) \tag{8}$$
$$C_{f3} = C_{f4} = (C_{f\,cm2}/2) \tag{9}$$
$$C_{d3} = C_{d\,dm2} \tag{10}$$
$$R_{d1} = R_{d2} = (2 \cdot R_{d\,cm1}) = (R_{d\,dm1}/2) \tag{11}$$
$$R_{d3} = R_{d4} = (2 \cdot R_{d\,cm2}) \tag{12}$$
$$R_{d5} = R_{d\,dm2} \tag{13}$$

The DM inductance (L_{dm}) is designed to limit the ripple current and consequently the total harmonic distortion (THD) of the output current, the ripple current is also related to the high frequency ac flux density and inductor core losses [27], [33], [34]. L_{dm} inductance is given by (14) [35]. Choosing the maximum peak to peak value of the inductor ripple current (Δi_{pp}) as 10% of $i_{grid\,peak}$ (r % in equation 15) and substituting the parameters in (15) and (16) results in $L_{dm} \cong 12$ mH.

$$L_{dm} = V_{bus}/(8 \cdot f_{sw} \cdot \Delta i_{pp}) \tag{14}$$
$$\Delta i_{pp} = r\,\% \cdot i_{grid\,peak} \tag{15}$$
$$i_{grid\,peak} = \sqrt{2} \cdot P_{max}/V_{grid} \tag{16}$$

Substituting L_{dm} by 12 mH in (6), results in $L_1 = L_2 = 6$ mH and $L_{cm1} = 3$ mH.

The filter capacitors are designed in order to keep the resonant frequencies $f_{res\,dm}$, $f_{res\,cm1}$ and $f_{res\,cm2}$ between the limits given by (17) [33], (18) and (19) [36], [37].

$$10f_{grid} \leq f_{res\,dm} \leq (f_{sw}/2) \tag{17}$$
$$10f_{grid} \leq f_{res\,cm1} \leq (f_{sw}/10) \tag{18}$$
$$3f_{res\,cm1} \leq f_{res\,cm2} \leq (f_{sw}/2) \tag{19}$$

Simplified DM and CM circuits neglecting the damping resistors, the leakage inductance ($L_{cm2\,Lkg}$) and the ground resistance (R_g) Figures (11)-(12) are considered in order to design the filter capacitors.

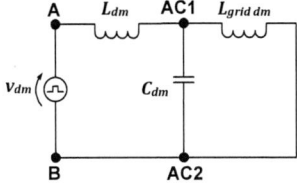

Fig. 11. DM circuit neglecting damping resistors and $L_{cm2\,Lkg}$

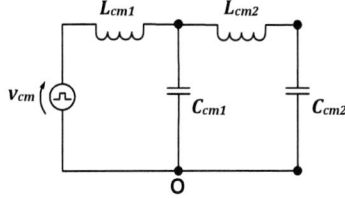

Fig. 12. CM circuit neglecting damping resistors and R_g

From the simplified DM and CM circuits Fig. (11)-(12) the following parameters can be defined:

$$C_{dm} = C_{f\,dm1} + C_{d\,dm1} + C_{d\,dm2} \tag{20}$$
$$C_{cm1} = C_{d\,cm1} + C_{f\,cm1} \tag{21}$$
$$C_{cm2} = C_{f\,cm2} + C_{pv} \tag{22}$$
$$L_{eq\,dm} = (L_{dm} \cdot L_{grid\,dm})/(L_{dm} + L_{grid\,dm}) \tag{23}$$

The DM filter resonant frequency neglecting the damping resistors is given by (24).

$$f_{res\,dm} = \frac{1}{2\pi\sqrt{L_{eq\,dm}C_{dm}}} \tag{24}$$

The resonant frequencies of the two stage LC common mode filter are given by (25) and (26) [36].

$$f_{res\,cm1} \cong \frac{1}{2\pi\sqrt{L_{cm1}\left(C_{cm1} + C_{f\,cm2} + C_{pv}\right)}} \tag{25}$$

$$f_{res\,cm2} \cong \frac{1}{2\pi\sqrt{L_{cm2}\left[\dfrac{C_{cm1}\left(C_{f\,cm2} + C_{pv}\right)}{C_{cm1} + C_{f\,cm2} + C_{pv}}\right]}} \tag{26}$$

The minimum and maximum values of C_{pv} according to [8] are 15 nF and 300 nF respectively. The limits for $f_{res\,cm1}$ (18) are more restrictive than the limits for

$f_{res\ dm}$ (17), therefore $C_{cm1} \cong 6.68\ \mu F$ is calculated considering $C_{pv} = 300\ nF$ (maximum value of C_{pv}), $f_{res\ cm1} = 1.1\ kHz$ and neglecting $C_{f\ cm2}$ in (25).

In order to balance damping performance and power loss $C_{d\ cm1} = C_{f\ cm1}$ is selected [28]-[32]. Combining (21), (7) and (8), results in (27).

$$C_{f1} = C_{f2} = C_{d1} = C_{d2} = (C_{cm1}/4) \qquad (27)$$

Selecting a commercial value $C_{f1} = C_{f2} = C_{d1} = C_{d2} = 1.5\ \mu F$, results in $C_{cm1} = 6\ \mu F$.

Replacing C_{f1}, C_{f2} in (7) and C_{d1}, C_{d2} in (8), gives $C_{d\ cm1} = C_{f\ cm1} = 3\ \mu F$ and $C_{f\ dm1} = C_{d\ dm1} = 750\ nF$.

The CM inductor of the second LC filter stage is designed in order to keep the resonant frequency of this stage between the limits given by (19) [36], [37].

L_{cm2} is calculated considering the lower limit of (19). Replacing $f_{res\ cm2}$ by $(4f_{res\ cm1} = 4.4\ kHz)$, $C_{pv} = 300\ nF$ (maximum value of C_{pv}), $C_{cm1} = 6\ \mu F$ and neglecting $C_{f\ cm2}$ in (26), gives $L_{cm2} \cong 4.6\ mH$.

$C_{f\ cm2}$ is calculated considering the upper limit of (19). Equation (28) is obtained from (26) neglecting C_{cm1}.

$$C_{f\ cm2} \cong [1/(2\pi \cdot f_{res\ cm2})^2 \cdot L_{cm2}] - C_{pv} \qquad (28)$$

Replacing $f_{res\ cm2}$ by $(f_{sw}/2 = 9990\ Hz)$, $C_{pv} = 15\ nF$ (minimum value of C_{pv}) and $L_{cm2} = 4.6\ mH$ in (28), results in $C_{f\ cm2} \cong 42\ nF$. Substituting $C_{f\ cm2}$ in (9) and considering a commercial value, gives $C_{f3} = C_{f4} = 22\ nF$.

The damping resistors R_{d1-5} are designed in order to avoid unwanted oscillations caused by the excitation of resonant frequencies, while keeping the losses as low as possible [27]-[32]. The damping resistors R_{d1} and R_{d2} are designed to optimize the damping performance of the common mode circuit.

$$R_{d\ cm1} = \sqrt{\frac{L_{cm1}}{C_{f\ cm1}}} \cdot \left(\frac{n_{cm1} + 1 \cdot \sqrt{n_{cm1} + 1}}{2 \cdot n_{cm1}} \right) \qquad (29)$$

$$n_{cm1} = C_{d\ cm1}/C_{f\ cm1} \qquad (30)$$

Solving equations (29) and (30) gives $R_{d\ cm1} \cong 53.98\ \Omega$. Substituing $R_{d\ cm1}$ in (11) and selecting a commercial value results in $R_{d1} = R_{d2} = 110\ \Omega$.

Equations (31) [38] and (32) neglecting C_{pv} are employed to design the damping resistors R_{d3} and R_{d4}.

$$R_{d\ cm2} = \frac{1}{3 \cdot \omega_{res\ cm2} \cdot C_{f\ cm2}} \qquad (31)$$

$$\omega_{res\ cm2} \cong \frac{1}{\sqrt{L_{cm2} \cdot (C_{f\ cm2} + C_{pv})}} \qquad (32)$$

Solving equations (31) and (32) gives $R_{d\ cm2} \cong 108\ \Omega$. Substituing $R_{d\ cm2}$ in (12) and selecting a commercial value results in $R_{d3} = R_{d4} = 220\ \Omega$.

The damping resistors R_{d1} and R_{d2} were designed to optimize the damping performance of the common mode circuit, the resulting $R_{d\ dm1} = 220\ \Omega$ (11), does not provide appropriate damping for the differential mode circuit. The damping branch R_{d5}, C_{d3} is designed to improve the damping performance of the differential mode circuit. $C_{d\ dm2} = C_{d3} = 330\ nF$ is selected considering an increase of about 20% for the differential mode capacitance C_{dm} (20) in order to keep reduced reactive power and small capacitor size. Considering the differential mode grid inductance $L_{grid\ dm} = 460\ \mu H$ [39], [40] in (23), gives $L_{eq\ dm} = 443\ \mu H$. Equation (33) and (34) are employed to design the damping resistor R_{d5}.

$$R_{d\ dm2} = \sqrt{\frac{L_{eq\ dm}}{C_{f\ dm1}}} \cdot \left(\frac{n_{dm2} + 1 \cdot \sqrt{n_{dm2} + 1}}{2 \cdot n_{dm2}} \right) \qquad (33)$$

$$n_{dm2} = C_{d\ dm2}/C_{f\ dm1} \qquad (34)$$

Solving equations (33) and (34) gives $R_{d\ dm2} \cong 72.9\ \Omega$, selecting a commercial value results in $R_{d5} = R_{d\ dm2} = 82\ \Omega$.

VI. SIMULATION RESULTS

For simulation and experimental setup the PV module, the non-isolated DC/DC converter and the parasitic capacitance (C_{pv}) were replaced by a DC voltage source with a capacitor connected between the negative node of the DC bus and the grid ground.

Table III summarizes the main parameters used in the simulations performed with PSIM software.

TABLE III
MAIN PARAMETERS AND COMPONENT VALUES USED IN THE SIMULATION

Parameter	Value
$L_1 = L_2$	5.5 mH
L_{cm2}	4.3 mH
$C_{f1} = C_{f2} = C_{d1} = C_{d2}$	1.5 μF
$C_{f3} = C_{f4}$	22 nF
C_{d3}	330 nF
$R_{d1} = R_{d2}$	110 Ω
$R_{d3} = R_{d4}$	220 Ω
R_{d5}	82 Ω
R_g	5 Ω
C_{pv}	150 nF
$L_{grid\ dm}$	460 μH
$R_{grid\ dm}$	0.38 Ω
$i_{grid\ reference}$	1.93 A peak

The modulation strategy is continuous unipolar PWM [15], the switching frequency is 19.98 kHz. The current control is performed by a PI controller with grid voltage feed-forward and single update sampled PWM [41].

Synchronization with the grid voltage is obtained with a phase locked loop (PLL) [42]. The single-phase inverter is connected between two lines of a three-phase wye connected with grounded neutral electric distribution system (220 V_{RMS}). A ground resistance (R_g) of 5 Ω was considered in the simulations. The simulation and experimental measurements without CM filter are performed removing L_{cm2} and opening the connection between nodes O and M (Fig. 8).

Figures (13)-(14) shows simulation waveforms. Table IV summarizes the main results for the inverter operating with and without the proposed filter.

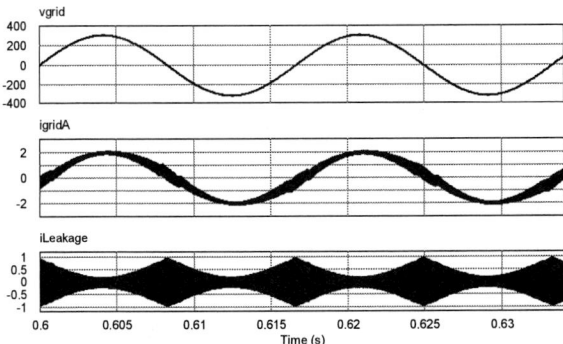

Fig. 13. Without CM filter, C_{pv} = 150 nF.
Waveforms from top to bottom: v_{grid}, i_{grid}, $i_{Leakage}$.

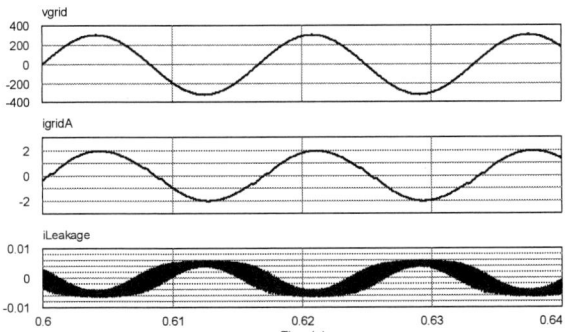

Fig. 14. With CM filter, C_{pv} = 150 nF.
Waveforms from top to bottom: v_{grid}, i_{grid}, $i_{Leakage}$.

TABLE IV
SIMULATION RESULTS

Parameter	CM FILTER OFF	CM FILTER ON
$i_{grid\ RMS}$	1.395 A	1.382 A
$i_{Leakage\ RMS}$	386.103 mA	3.909 mA
P	301.530 W	301.483 W
PF	0.979	0.989
$THD_{v\ (40)}$	0.037 %	0.037 %
$THD_{v\ (671)}$	0.125 %	0.120 %
$THD_{i\ (40)}$	3.127 %	3.159 %
$THD_{i\ (671)}$	14.223 %	3.182 %
$P_{Rd\ total}$	915.3 mW	931.2 mW

The simulation results shows that the leakage current without the proposed filter is 386 mA RMS, which is higher than the limit established by the standards VDE V 0-126-1-1:2006 and IEC 62109-2:2011. When the proposed filter is employed the leakage current is reduced to 3.9 mA RMS, which is lower than the residual current setting of 30 mA, required when the protection against electric shock hazard is provided by an RCD [6]. The total power loss caused by the damping resistors ($P_{Rd\ total}$) represents only 0.31 % of the power injected in the grid.

The simulated and experimental total harmonic distortion (THD_x) of the grid voltage (THD_v) and grid current (THD_i) were calculated according to (35) [43], where (X_1) is the RMS value of the fundamental component of the voltage (V_1) or current (I_1) and ($X_{(h)}$) is the RMS value of each harmonic component of the voltage ($V_{(h)}$) or current ($I_{(h)}$), obtained from FFT analysis performed using Simview software. The maximum order of the harmonics (N) for THD

calculation were selected considering N=40, required by the standard IEEE 1547:2003 [44], N=671 $\left(2f_{sw} + 5f_{grid}\right)$, selected in order to evaluate the filter performance considering the high frequency components up to the first group of the differential mode harmonics, which are located around twice the switching frequency for continuous unipolar PWM [15]-[43].

$$THD_{x\ (N)}(\%) = 100 \cdot \sqrt{\sum_{h=2}^{N}\left(\frac{X_{(h)}}{X_1}\right)^2} \qquad (35)$$

VII. EXPERIMENTAL RESULTS

Table V shows the list of filter components used in the experimental prototype. PWM, PLL and current controller were implemented in a TMS320F28335 DSP.

TABLE V
FILTER COMPONENTS USED IN THE EXPERIMENTAL SETUP

Parameter	Description
L_1, L_2	6mH - 2 stacked powder cores (APH33P60), 222 turns, 20AWG magnetic wire, Size (D=40.0 x H=30.0) mm
L_{cm2}	4.3mH - Ferrite core (ZW42207), 24 turns (each winding), 20AWG magnetic wire, Size (T=12.0 x H=23.0 x W=25.0) mm
$C_{f1}, C_{f2}, C_{d1}, C_{d2}$	1.5µF/305Vac - Metallized polypropylene film capacitor, Size (T=10 x H=18.5 W=26.5) mm
C_{f3}, C_{f4}	22nF/300Vac - Metallized polypropylene film capacitor, Size (T=10.0 x H=18.5 W=26.5) mm
C_{d3}	330nF/275Vac - Metallized polypropylene film capacitor, Size (T=6.0 x H=17.5 W=18.0) mm
R_{d1}, R_{d2}	110Ω/3W - Power metal film resistor PR03
R_{d3}, R_{d4}	220Ω/2W - Power metal film resistor PR02
R_{d5}	82Ω/3W - Power metal film resistor PR03

The leakage current injected in the grid without CM filter was measured with a current probe N2774A (Agilent). A current transducer LEM CT 0.4-P [45] was used to measure the leakage current with the proposed CM filter connected. Figures (15)-(16) shows experimental waveforms.

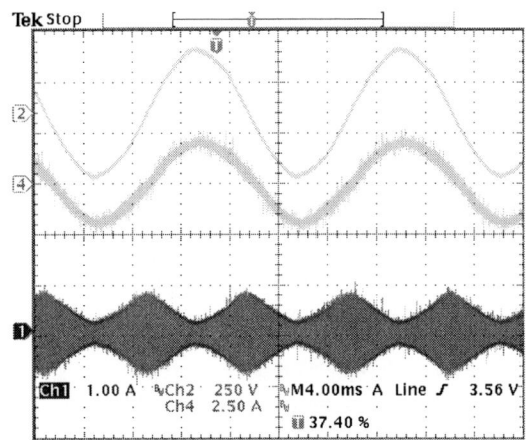

Fig. 15. Without CM filter, C_{pv} = 150nF.
Waveforms from top to bottom: ch2-v_{grid} (250V/div); ch4-i_{grid} (2.5A/div); ch1-$i_{Leakage}$(1A/div).

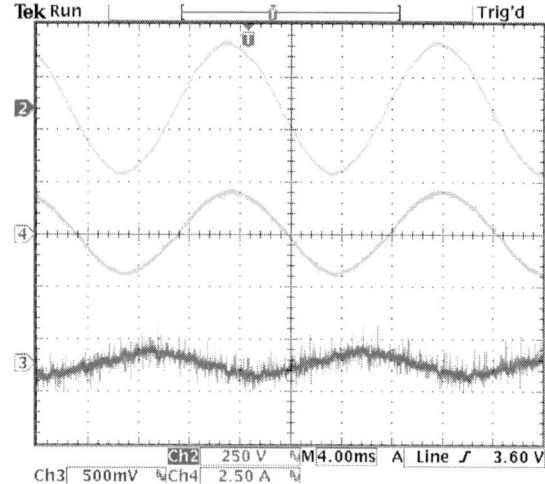

Fig. 16. With CM filter, $C_{pv} = 150$nF.
Waveforms from top to bottom: ch2-v_{grid} (250V/div); ch4-i_{grid} (2.5A/div); ch3-$i_{Leakage}$(40mA/div).

The oscilloscope waveform data was transferred to a PC, the software Simview 9.0 was used to perform FFT analysis of the experimental measurements. The results of FFT analysis and equation (35) were used for THD calculation. Table VI summarizes the main results for the inverter operating with and without the proposed filter.

TABLE VI
EXPERIMENTAL RESULTS

Parameter	CM FILTER OFF	CM FILTER ON
$i_{grid\,RMS}$	1.379 A	1.368 A
$i_{Leakage\,RMS}$	369.526 mA	6.619 mA
P	295.562 W	295.402 W
PF	0.981	0.990
$THD_{v(40)}$	1.6615 %	1.618 %
$THD_{v(671)}$	2.070 %	1.906 %
$THD_{i(40)}$	3.978 %	3.106 %
$THD_{i(671)}$	11.821 %	3.924 %
$P_{Rd\,total}$	-	694.798 mW

The leakage current measured without CM filter was 369 mA RMS, which is higher than the limit established by [5]-[6]. The leakage current measured with the CM filter was reduced to 6.6 mA, which is suitable for operation with an RCD with a residual current setting of 30 mA [6].

The differences between simulation and experimental results are more significant for the case with CM filter; however, the difference of 2.71 mA, represents only 0.6775 % of the nominal measuring range ($I_{PN} = 400$ mA) of the current transducer LEM CT 0.4-P, which is lower than the accuracy limits ($< \pm 1\%$ of I_{PN}) [45].

Despite the differences discussed above, both simulation and experimental results confirmed the effectiveness of the proposed filter showing that without CM filter, the transformerless full-bridge VSI employing continuous unipolar PWM is not able to comply with the established limit of 300 mA RMS for continuous residual current. On the other hand, it was shown that the proposed filter reduces the leakage current to a value below 7 mA RMS, which is suitable even for the most restrictive case, when the protection against electric shock is provided by an RCD with a residual current

setting of 30 mA. The experimental results also confirmed the small power loss caused by the damping resistors, 0.235 % of the power injected in the grid.

VIII. CONCLUSIONS

This paper proposed an integrated common and differential mode filter for the DC/AC converter of a two stage, single-phase grid-connected transformerless PV microinverter, employing the full-bridge VSI topology with continuous unipolar PWM and low loss passive damping strategy. Design of filter and damping components were presented. Simulation and experimental results shows that the proposed filter is able to maintain the leakage current in compliance with the standards VDE 0-126-1-1:2006 and IEC 62109-2:2011. It was confirmed that low power loss and appropriate damping is obtained by employing RC parallel passive damping across the main filter capacitors. The total harmonic distortion of the current injected in the grid, considering up to the first group of differential mode harmonics (40.26 kHz), was maintained below 4% using the proposed filter.

ACKNOWLEDGMENT

The authors would like to thank PHB Eletrônica LTDA., for the material provided for the experimental setup, CAPES, FAPESP and CNPq for financially supporting the co-authors.

REFERENCES

[1] IHS Solar Solutions, Press Releases, "The World Market for PV Microinverters and Power Optimizers-2013 Ed." http://www.imsresearch.com/press-release/solar_microinverter_shipments_to_quadruple_and_reach_2_1_gigawatts_in_2017 (accessed October 1, 2013).

[2] H.A. Sher, K.E. Addoweesh. "Micro-inverters - Promising solutions in solar photovoltaics". *Energy for Sustainable Development*. Vol. 16. 2012. pp. 389-400.

[3] S. V. Araujo, P. Zacharias, B. Sahan, R. P. Torrico-Bascope, and F. L. M. Antunes, "Analysis and proposition of a PV module integrated converter with high voltage gain capability in a non-isolated topology," in Proc. 7th *Int. Conf. Power Electron.*, 2007, pp. 511–517.

[4] M. C. Poliseno, R. Mastromauro and M. Liserre, "Transformer-less photovoltaic (PV) inverters: A critical comparison", in *Energy Conversion Congress and Exposition (ECCE)*, IEEE, 2012.

[5] IEC, "60364-7-712 - Electrical installations of Buildings - Part 7-712: Requirements for special installations or locations. Solar photovoltaic (PV) power supply systems, ed. 2002", 2002.

[6] IEC, "62109-2 ed1.0 - Safety of power converters for use in photovoltaic power systems - Part 2: Particular requirements for inverters", 2011.

[7] DIN, "VDE 0126-1-1 - Automatic disconnection device between a generator and the public low-voltage grid", 2006.

[8] J. M. A. Myrzik and M. Calais, "String and module integrated inverters for single-phase grid connected photovoltaic systems - a review," in *IEEE Power Tech Conference Proceedings*, Bologna, 2003.

[9] SMA, "Leading Leakage Currents - Information on the Design of Transformerless Inverters, Version 2.3," http://files.sma.de/dl/7418/Ableitstrom-TI-en-23.pdf (accessed Feb. 23, 2014).

[10] R. Gonzalez, J. Lopez, P. Sanchis and L. Marroyo, "Transformerless Inverter for Single-Phase Photovoltaic Systems", *IEEE Transactions on Power Electronics*, vol. 22, n. 2, pp. 693-697, 2007.

[11] C. W. Li, X. He, "Review of Non-Isolated High Step-Up DC/DC Converters in Photovoltaic Grid-Connected Applications", *IEEE Transactions on Industrial Electronics*, vol. 58, no. 4, April 2011.

[12] H. Xiao and S. Xie, "Leakage Current Analytical Model and Application in Single-Phase Transformerless Photovoltaic Grid-Connected Inverter", *IEEE Transactions on Electromagnetic Compatibility*, vol. 52, n. 4, pp. 902-913, 2010.

[13] E. Gubia, P. Sanchis, A. Ursua, J. Lopez and L. Marroyo, "Ground currents in single-phase transformerless photovoltaic systems," *Progress in Photovoltaics: Research and Applications*, vol. 15, n. 7, pp. 629-650, 2007.

[14] R.-S. Lai and K. D. T. Ngo, "A PWM method for reduction of switching loss in a full-bridge inverter", in *Applied Power Electronics Conference and Exposition (APEC)*, 1994.

[15] D. G. Holmes and T. A. Lipo, "*Pulse width modulation for power converters: principles and practice,*" IEEE Press Series Power Eng., 2003.

[16] R. Teodorescu, M. Liserre and P. Rodriguez, *Grid Converters for Photovoltaic and Wind Power Systems*, Wiley, 2011.

[17] D. Dong, "Ac-dc Bus-interface Bi-directional Converters in Renewable Energy Systems", Ph.D. dissertation, Virginia Polytechnic Institute and State University, 2012.

[18] K. Zhang, Y. Zhou, Y. Zhang and Y. Kang, "Reduction of Common Mode EMI in a Full-Bridge Converter through Automatic Tuning of Gating Signals", in CES/IEEE 5th *International Power Electronics and Motion Control Conference* (IPEMC), 2006.

[19] I. Patrao, E. Figueres, F. González-Espín, and G. Garcerá, "Transformerless topologies for grid-connected single-phase photovoltaic inverters," *Renewable and Sustainable Energy Reviews*, vol. 15, pp. 3423-3431, 9/ 2011.

[20] Figueredo, R. S. , Carvalho, K. C. M. , Ama, N. R. N. , L. Matakas Jr., "Leakage Current Minimization Techniques for Single-phase Transformerless Grid-connected PV Inverters – An Overview", *Proceedings of the 12th Brazilian Power Electronics Conference* (COBEP), vol., no., pp.517-524, October 27-31, 2013.

[21] Barater, D.; Buticchi, G.; Lorenzani, E.; Concari, C., "Active Common-Mode Filter for Ground Leakage Current Reduction in Grid-Connected PV Converters Operating With Arbitrary Power Factor", *IEEE Transactions on Industrial Electronics*, vol.61, no.8, pp.3940,3950, Aug. 2014.

[22] E. Gubia, P. Sanchis, J. Lopez, A. Ursua and L. Marroyo, "EMI filter inductor size for transformerless PV systems based on the full bridge structure", in 14th *International Power Electronics and Motion Control Conference* (EPE/PEMC), 2010.

[23] D. Dong, F. Luo, D. Boroyevich and P. Mattavelli, "Leakage Current Reduction in a Single-Phase Bidirectional AC-DC Full-Bridge Inverter", *IEEE Transactions on Power Electronics*, vol. 27, n. 10, pp. 4281-4291, 2012.

[24] H. Akagi, H. Hasegawa and T. Doumoto, "Design and performance of a passive EMI filter for use with a voltage-source PWM inverter having sinusoidal output voltage and zero common-mode voltage", *IEEE Transactions on Power Electronics*, vol. 19, n. 4, pp. 1069-1076, 2004.

[25] Hedayati, M.H.; Acharya, A.B.; John, V., "Common-Mode Filter Design for PWM Rectifier-Based Motor Drives," *IEEE Transactions on Power Electronics*, vol.28, no.11, pp.5364,5371, Nov. 2013.

[26] Wu, W.; Sun, Y.; Lin, Z.; He, Y.; Huang, M.; Blaabjerg, F.; Chung, H.S.-h., "A Modified LLCL Filter With the Reduced Conducted EMI Noise," *IEEE Transactions on Power Electronics*, , vol.29, no.7, pp.3393,3402, July 2014.

[27] Wang, T.C.Y.; Zhihong Ye; Gautam Sinha; Xiaoming Yuan, "Output filter design for a grid-interconnected three-phase inverter," *IEEE 34th Annual Power Electronics Specialist Conference, PESC*, vol.2, no., pp.779,784, 15-19 June 2003.

[28] Channegowda, P.; John, V., "Filter Optimization for Grid Interactive Voltage Source Inverters," *IEEE Transactions on Industrial Electronics*, vol.57, no.12, pp.4106,4114, Dec. 2010.

[29] Weimin Wu; Yuanbin He; Tianhao Tang; Blaabjerg, F., "A New Design Method for the Passive Damped LCL and LLCL Filter-Based Single-Phase Grid-Tied Inverter," *IEEE Transactions on Industrial Electronics*, vol.60, no.10, pp.4339,4350, Oct. 2013.

[30] Mukherjee, N.; De, D., "Analysis and improvement of performance in LCL filter-based PWM rectifier/inverter

application using hybrid damping approach," *Power Electronics, IET* , vol.6, no.2, pp.309,325, Feb. 2013.

[31] Weimin Wu; Zhe Lin; Yunjie Sun; Xiongfei Wang; Min Huang; Huai Wang; Chung, H.S.-H., "A hybrid damping method for LLCL-filter based grid-tied inverter with a digital filter and an RC parallel passive damper," *Energy Conversion Congress and Exposition (ECCE)*, vol., no., pp.456,463, 15-19 Sept. 2013

[32] Wu, W.; Sun, Y.; Huang, M.; Wang, X.; Wang, H.; Blaabjerg, F.; Liserre, M.; Chung, H.S.-h., "A Robust Passive Damping Method for LLCL-Filter-Based Grid-Tied Inverters to Minimize the Effect of Grid Harmonic Voltages," *IEEE Transactions on Power Electronics*, vol.29, no.7, pp.3279,3289, July 2014.

[33] Liserre, M.; Blaabjerg, F.; Hansen, S., "Design and control of an LCL-filter-based three-phase active rectifier," *IEEE Transactions on Industry Applications*, vol.41, no.5, pp.1281,1291, Sept.-Oct. 2005.

[34] McLean, J.W., "Inductor design using amorphous metal C-cores," *Circuits and Devices Magazine, IEEE* , vol.12, no.5, pp.26,30, Sep 1996.

[35] H. Kim and K.-H. Kim, "Filter design for grid connected PV inverters," in *Proc. IEEE Int. Conf. Sustainable Energy Technol.*, 2008, pp. 1070–1075.

[36] Ridley, R. B., "Secondary LC-Filter Analysis and Design Techniques for Current-Mode Controlled Converters," *IEEE Transactions on Power Electronics*," Vol. 3, No. 4, October 1988, 499-507.

[37] Timothy M. Ozenbaugh and Richard Lee Pullen, EMI Filter Design, Third Edition, CRC Press, 2011.

[38] Araujo, S.V.; Engler, A.; Sahan, B.; Antunes, F., "LCL filter design for grid-connected NPC inverters in offshore wind turbines," *, 2007. ICPE '07. 7th Internatonal Conference on Power Electronics* , vol., no., pp.1133,1138, 22-26 Oct. 2007.

[39] JIS C 61000-3-2:2011 - Electromagnetic compatibility (EMC) - Part 3-2: Limits - Limits for harmonic current emissions (equipment input current <=20 A per phase), Japanese Standards Association, 2012.

[40] Japanese Guideline for Reduction of Harmonic Emission caused by electrical and electronic equipment for household and general use, IEC SC77A Japanese National Committee, Nov. 2001, http://home.jeita.or.jp/eps/harmonics/guideline/data/guideline_eng .pdf (accessed Feb. 23, 2014).

[41] Martinz, F.O.; Miranda, R.D.; Komatsu, W.; Matakas, L., "Gain limits for current loop controllers of single and three-phase PWM converters", *International Power Electronics Conference (IPEC)*, 2010 , vol., no., pp.201,208, 21-24 June 2010.

[42] Ama, N.; Destro, R.; Komatsu, W.; Kassab, F.; Matakas, L., "PLL performance under frequency fluctuation-compliance with standards for distributed generation connected to the grid," *Innovative Smart Grid Technologies Latin America (ISGT LA), 2013 IEEE PES Conference On* , vol., no., pp.1,6, 15-17 April 2013.

[43] N. Mohan, T. M. Undeland, W. P. Robbins, *Power Electronics: converters, applications, and design*, John Wiley & Sons, 2nd Edition, New York, USA, 1995.

[44] IEEE Standard for Interconnecting Distributed Resources With Electric Power Systems," *IEEE Std. 1547-2003* , 2003.

[45] LEM; "Current transducers CT 0.1. 0.4-P; datasheet; 080909/14".

978-1-4799-2706-7/14 $31.00 © 2014 IEEE

Design and Integration of Interphase Inductors for Interleaved Three Phase Voltage-Source-Inverters in DC-fed Motor Drive Systems

Xuning Zhang, Dushan Boroyevich and Rolando Burgos
Center for Power Electronics Systems (CPES)
The Bradley Dept. of Electrical & Computer Engineering
Virginia Tech, Blacksburg VA 24061
xuning45@vt.edu

Abstract— This paper presents a detail analysis on the design and integration of interphase inductor for interleaved three phase voltage-source-inverters (VSI). The benefits of interleaving on EMI filter weight reduction are presented. Interleaving can cancel certain order of harmonics in the total output current and reduce the size and weight of the EMI filter to meet certain EMI standard. However, it also creates the voltage difference between the two converters that generate circulating current in the system which increase system loss and device stress. In order to limit the circulating current, additional interphase inductors are needed for motor drive system. The penalty of the circulating current and additional interphase inductors are analyzed in detail. Design examples are given for EMI filter and interphase inductor design, which shows that interleaving will increase the total passive weight due to the additional weight of coupled inductors for motor drive systems. In order to reduce the total weight of the passive components of the system, integration design methods for interphase inductor is proposed which indicates that certain value of AC DM inductance can be integrated with interphase inductors with negligible increase on the weight and the maximum integrated inductance is related with interleaving angle. Verifications are carried out through both the simulations and experiments on a 2kW DC-fed motor drive system.

Keywords—interleaving; interphase inductor; motor drive, passive component weight reduction.

I. INTRODUCTION

Adjustable-speed motor drive with a voltage-source pulse-width modulation inverter with long cables brings the EMI issues [1]. Usually EMI filters are parts of the system to suppress EMI noise according to the EMI requirements and take a significant portion of the system weight. To improve system power density and reliability, paralleled and interleaved converter topology is commonly used. By phase-shifting the carrier waveform of individual converters with an appropriate angle, the voltage harmonics and EMI noise can be reduced [2~3] and the line inductor and EMI filter weight can be reduced [3~4]. However, interleaving also create the voltage difference between the two converters and require interphase inductors to limit the circulating current between the two converters [5]. Reference [6] showed the control method to limit the low frequency circulating current and utilized the line inductors to

perform as an interphase inductor for voltage source rectifiers. However, in motor drive applications, there is no such line inductors to perform as interphase inductors since the main load inductor is inside the motor. Thus, additional interphase inductor are necessary which make interleaving topologies less beneficial to motor drive applications. Furthering the past work, this paper studied the design and integration of interphase inductor for a paralleled three-phase voltage-source converters shown as Fig. 1, which contains two VSIs working in an interleaved mode with reverse coupled inductors to limit the circulating current. Since previous design shows that EMI filter weight is dominated by common mode (CM) filters especially the CM choke inductors [7], the benefit of interleaving on CM filter weight reduction is discussed. The results show that interleaving can reduce the weight of EMI filters with different interleaving angles. However, interleaving also creates the voltage difference between the two converters that generate circulating current in the system which create device loss and current stress problems. Thus the design and analysis are also focused on circulating current analysis and coupled inductor design. The results show that interleaving will increase the total passive weight for voltage source inverters due to the additional weight of coupled inductors. In order to reduce the total weight of the passive components of the system, integration design methods for interphase inductors are proposed which indicate that certain value of AC DM inductance can be integrated with interphase inductors with negligible increase on the weight and the maximum integrated inductance is related with interleaving angle. Verifications are carried out through both simulations and experiments on a 2kW DC-fed motor drive system.

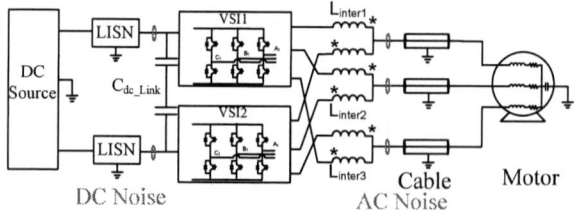

Figure 1: Motor drive system with interleaved VSI

II. EMI Filter Weight Reduction in Interleaving Topology

For the interleaved two-level VSI DC-fed motor drive system shown as Fig.1, previous EMI filter design results showed that EMI filter weight is dominated by CM filters, thus the analysis will focus on the benefit of interleaving on CM filter weight reduction[8]. The equivalent circuit for CM noise analysis of the interleaved system is shown in Fig. 2(a).

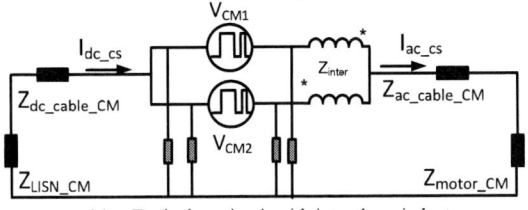

(a) Equivalent circuit with interphase inductor

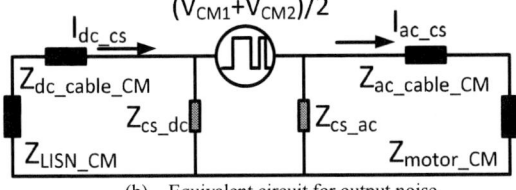

(b) Equivalent circuit for output noise

Figure 2: Equivalent circuit for CM noise analysis

It is clear that the AC-side and DC-side CM noise is coupled and shares the same noise source. Moreover, the noise sources of the two converters are different and the interphase inductor is also included. Since the EMI standard only limits the EMI noise at the output, the equivalent circuit for converter output noise can be simplified as shown in Fig. 2(b) where the reverse coupled inductor is taken out since it do not influence the output current noise. Moreover, the two noise sources are combined and the total noise source in interleaved system can be calculated analytically as shown in Equation (1)

$$C'_{mn\_diff} = C_{mn} \cos(m\kappa/2) \qquad (1)$$

Where C_{mn} represents the magnitude of the harmonics of the CM voltage noise source in non-interleaved VSI, which is related with the modulation scheme and modulation index and can be calculated analytically using double Fourier integral transformation (DFIT) method [9]. C'_{mn_diff} represents the magnitude of the harmonics the CM voltage noise source in interleaved VSI; κ represents the interleaving angle between the two interleaving VSIs.

From (1), it is clear that the impact of interleaving varies for different orders of switching harmonics when the interleaving angles are different, the noise source reduction ratio on different order of switching harmonics with different interleaving angles are shown in Fig.3. The results show that big interleaving angle can cancel certain order of harmonics and small interleaving angle can attenuate noise source in a wide range. Reference [8] shows that different interleaving angles can be selected based on the system operation conditions and noise

propagation path impedance to effectively reduce the weight and size of the EMI filters.

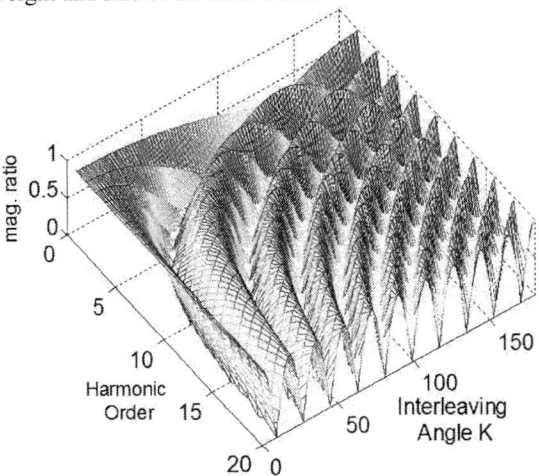

Figure 3: Voltage noise source reduction ratio on different order harmonics of interleaving with different interleaving angle

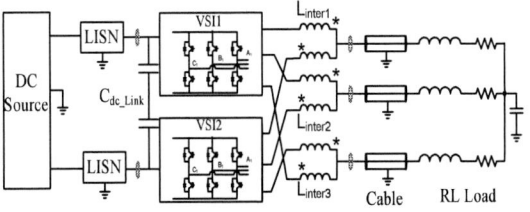

(a) System structure

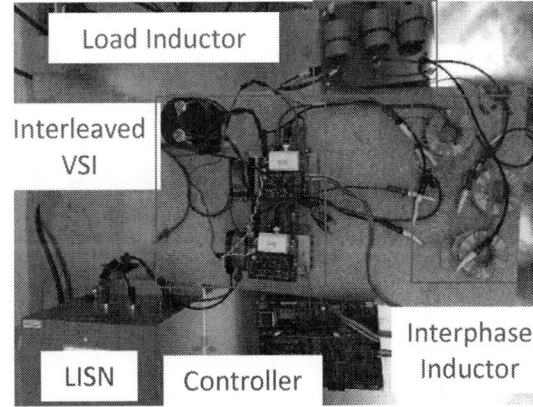

(b) System Setup

Figure 4: 2kW experimental system

To verify the benefit of interleaving on filter weight reduction, a 2kW interleaved 2L VSI prototype is built with the system structure shown as in Fig. 4(a) and system setup is shown in Fig.4(b) where a RL load with grounding capacitors is used instead of motor for easy modification of propagation path impedance. System switching frequency f_s=30 kHz and fundamental frequency f_1=200 Hz.

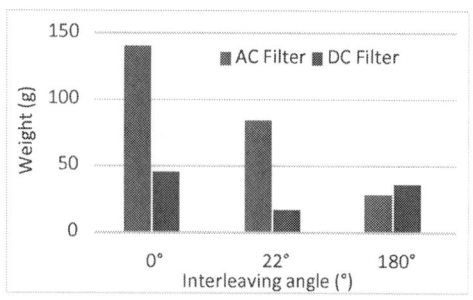

Figure 5: EMI filter weight comparison with different interleaving angle

System EMI noise is measured and EMI filters are designed for both AC and DC sides with different interleaving angles and the results are shown in Fig. 5. It is clear that interleaving can help to reduce the EMI filter weight and the weight reduction ratio of the EMI filters varies with different interleaving angles, the optimized interleaving angle is related with system working condition.

III. CIRCULATING CURRENT ANALYSIS AND COUPLED INDUCTOR DESIGN

In interleaving topology, when the carrier signals of the two converters are phase-shifted, the output voltage of the two converters will also be phase-shifted, then there will be voltage difference between the same phase leg of the two converters which will generate the circulating current between the two converters. In order to limit the circulating current, certain interphase reactors are needed to limit the magnitude of the circulating current [6]. For motor drive applications, since the load inductor is inside the motor, adding reversed coupled inductors as shown in Fig.1 is an effective method to limit the circulating current. Considering the system structure, the equivalent circuit for circulating current is shown as Fig. 6. It is clear that the circulating current is determined by the coupled inductance added into the system and the voltage difference between the two converters that is related with interleaving angle.

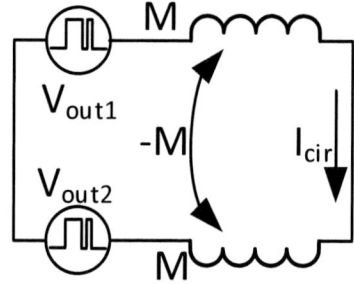

Figure 6: Equivalent circuit for circulating current

Since the circulating current is an additional current generated by interleaving, it will change the current waveform through each devices and change the loss distribution in the converter. Figure 7 shows the device current during switching with interleaving and without interleaving, it shows that the turn-on current is reduced and turn-off current is increased due to the existence of circulating current.

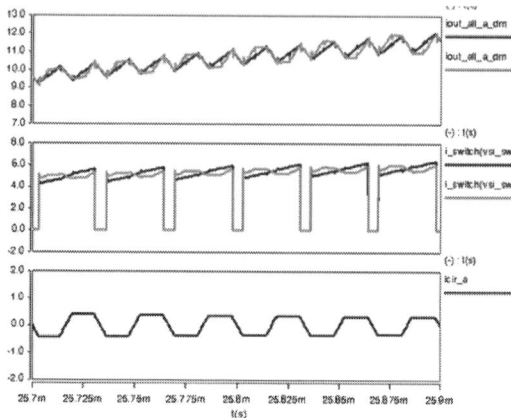

Figure 7 Simulation results of device current during switching

The impact of circulating current on converter loss distribution can be calculated analytically. Assuming output current ripple is small and the carrier ratio (defined as the ratio between switching frequency and fundamental frequency) is high, one can assume that the current through devices during each switching period is constant without interleaving as shown in Fig. 8.

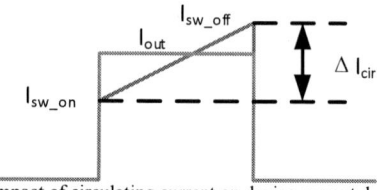

Figure 8: Impact of circulating current on device current during switching

With interleaving, the circulating current will be added to the output current, the changes during each switching period can be calculated from the equivalent circuit. Then the current changes during each switching cycle can be calculated separately. Then conduction loss (P_{cond}), turn-on loss (P_{on}), turn-off loss (P_{off}) and reverse recovery loss (P_{rr}) can be calculated based on the measurement results at rated switching condition which can be got from device datasheets.

In order to show the impact of circulating current on system loss distribution. A loss calculation was conducted on the experimental system with different configurations, the device loss data was obtained from the datasheet of the IPM modules used in the experimental system with different configuration as an example.

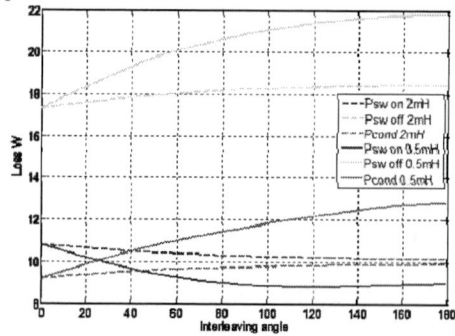

(a) Calculation results of loss change with interleaving angle under different interphase inductance

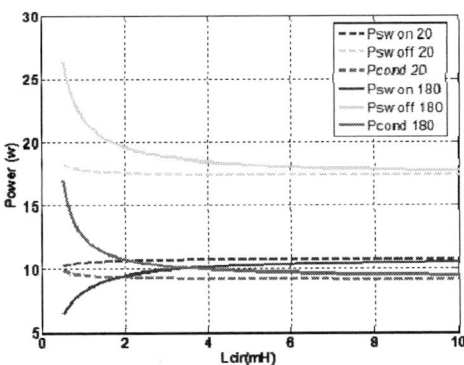

(b) Calculation results of loss change with inductance under different interleaving angles

Figure 9: Calculation results of loss change with inductance and interleaving angles

Figure 9(a) shows the loss changing versus the interleaving angles when different interphase inductors were added into the system. The results indicate that increasing interleaving angle will reduce the turn-on loss but increase the turn-off loss and reverse-recovery loss since it will increase the circulating current in the system. In terms of the impact of circulating current on system total loss, it depends on the characteristics of the devices. In some systems where turn-on loss is dominant, increasing the circulating current may help to reduce the total loss. Figure 9(b) shows the loss change versus the inductance in the circulation loop with different interleaving angles. It shows that increasing the inductance in the circulating loop can reduce the impact of interleaving since it can reduce the circulating current in the system. The results also indicate that when the inductance in the circulating loop is big enough, further increasing the inductance will have less impact since the circulating current is already small enough and the impact is not significant anymore.

The circulating current is an inner design constraint, the limitation should be selected to avoid the voltage overshoot on the power devices during switching and loss limitation of the system (efficiency and thermal design), when the requirement of the circulating current is determined, the additional interphase inductor value can be calculated and physical implantation can be designed based on the value and the working condition of the inductors.

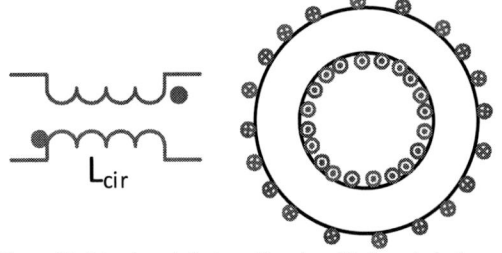

Figure 10: Interphase inductor with reduced leakage inductance

Since coupled inductors are additional components added to the system. It need to be designed carefully to reduce the total weight of the passive components.

Toroidal cores are implemented for the reverse coupled inductor and twisted wires are used to reduce the leakage inductance as shown in Fig. 10. With a relatively small leakage inductance, the volt-second on the interphase inductor is determined by the voltage difference between the two converters that is related with the interleaving angle. To get the minimum weight of the coupled inductor, all of the possible dimensions of the toroidal core are swept under the constraints of physical fit, core saturation and core temperature rise as shown in Fig.11

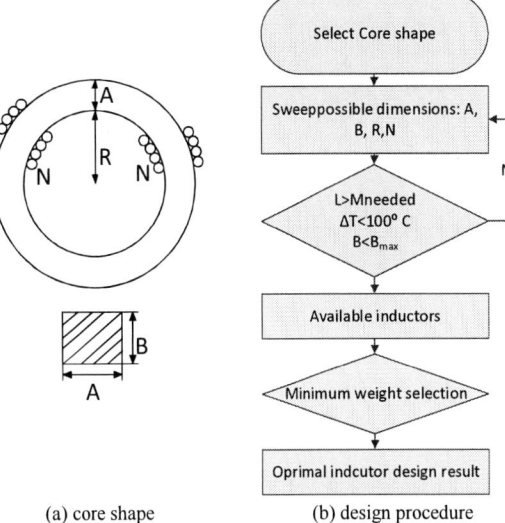

(a) core shape   (b) design procedure

Figure 11: Coupled inductor design method

Fig.12 shows the total passive component weight comparison with different interleaving angles. It is clear that interleaving will increase the passive component weight regardless of the interleaving angle with the interphase inductors. Since the voltage on the interphase inductor is switching frequency excitation, the core loss is relatively high and inductor weight is determined by the temperature rise of the inductor, which make the coupled inductor heavy.

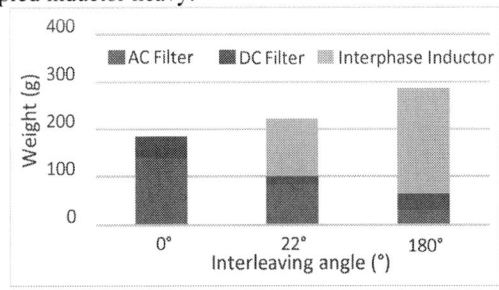

Figure 12: Total passive component weight comparison with different interleaving angle

IV. INTEGRATION OF THE COUPLED INDUCTOR WITH AC DM INDUCTOR

The optimization results of the coupled inductor shows that the weight is determined by the temperature rise of the inductor and the flux density on the core is much smaller than the saturation limit, thus the core is not fully utilized for the implementation of the reverse coupled inductor with twisted winding wire. In order to utilize the core more effectively and reduce the total passive component weight, the interphase inductor can be integrated with the AC DM filter inductor. Instead of twisting the wire of the coupled inductor, the two windings can be separated as shown in Fig. 13 to increase the leakage inductance which can perform as the AC DM filter inductors (L_{DM}). When the L_{DM} is added, the total volt-second will be increased as shown in equation (2).

$$VS_{Total} = L_{DM}I_{out} + VS_{cir} \qquad (2)$$

Where VS_{cir} is the volt-second generated due to the voltage difference between the two converters. $L_{DM}I_{out}$ is the volt-second generated by adding the leakage inductance and it is proportional to output current. Moreover adding L_{DM} will also increase the core loss as shown in equation (3).

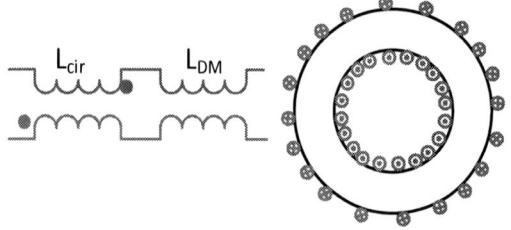

Figure13: Integrated Interphase inductor with coupled inductor and the filter DM inductor

$$P_{fs} = \sum_{n=1..CR} \left(a\left(\frac{VS_n}{NA_c}\right)^c (f_s)^d V \times Ts \right),$$

$$P_{fl} = a\left(\frac{L_{DM}I_{out}}{NA_c}\right)^c (f_1)^d V \times Ts \times f_1, \qquad (3)$$

$$P_{Loss\_core} = P_{fs} + P_{fl}$$

Where P_{fs} is the core loss generated due to the voltage difference of the two converters that is related with switching frequency and P_{fl} is the additional core loss due to the leakage inductance that is relate with fundamental frequency. For motor drive system $f_1 << f_s$, Thus $P_{fl} << P_{fs}$, which means that adding L_{DM} will increase the volt-second on the interphase inductor that is low frequency, thus the additional loss is negligible. Since the coupled inductor weight is determined by temperature rise, adding leakage inductance will not increase the inductor weight if the total volt-second do not exceed the saturation limit. The integrated interphase inductor design result is shown in Fig.14 for different interleaving angles and different leakage inductances. It is clear that when interleaving angle is large, adding leakage inductance will not increase the inductor weight until the volt-second determines inductor weight. Thus, the AC DM inductor is implemented with no weight increase of the interphase inductor. While, for small angle interleaving condition, the maximum allowable DM inductance that can be integrated without increasing the weight of the coupled inductor is smaller.

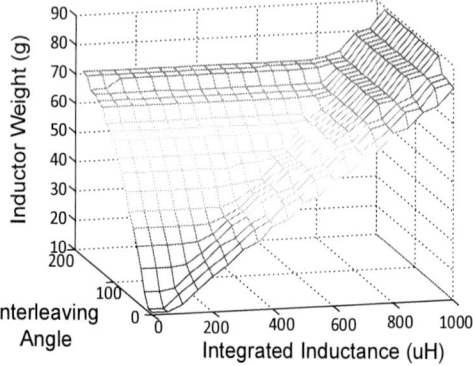

Figure 14: Integrated inductor design results with different interleaving angles

V. EXPERIMENTAL VERIFICATION

To verify the analysis, two kinds of interphase inductor is built using nano-crystaline material with L_{cir}=20mH: one with twisted winding to reduce leakage inductance with L_{DM}=1uH as shown in Fig.15(a); one with separately winding to increase the leakage inductance with L_{DM}=60uH as shown in Fig.15(b). Additional open ended five turns is added to show the flux changing by measuring the voltage (V_{ext}) on the extra winding.

Both inductors are tested in the experimental systems as shown in Fig.4 with 2kW output power and 30° interleaving angle. System operational time domain and frequency domain test results are shown in Fig. 16. Since the interphase inductance is very big, the circulating current is well suppressed, thus the currents through two converters are almost the same, from frequency domain it is clear that interleaving can help to reduce EMI noise and help EMI filter design.

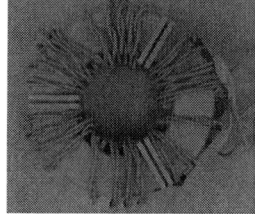 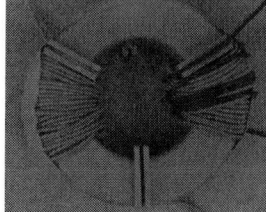

(a) Low Leakage inductor   (b) High Leakage Inductor

Figure 15: Implemented interphase inductor

The test results of interphase inductor are shown in Fig.17. It is clear that the output voltage of the two converters are phase-shifted. With the low leakage inductance, V_{ext} is the same with the voltage difference between the two converters, which shows that the volt-second in the core is determined by the voltage difference. However, for the inductor with high leakage inductance, there are more ringing on V_{ext} due to the leakage inductance added, and the integration of V_{ext} is larger than the results with low leakage inductance, which show the increase of the volt-second. Moreover, since the ringing will hardly change the magnitude of the switching frequency voltage excitation. The increase of core loss will be negligible.

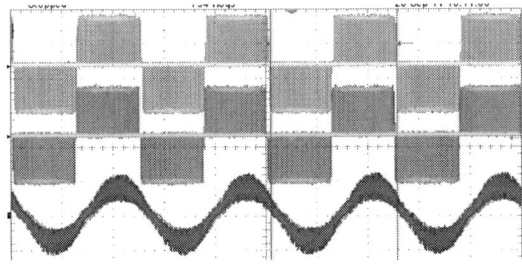

(a) Time domain: V_{out1}: ; V_{out2}:cyan
(200V/div); I_{out1}:pink (10A/div); I_{out2}:green (10A/div)

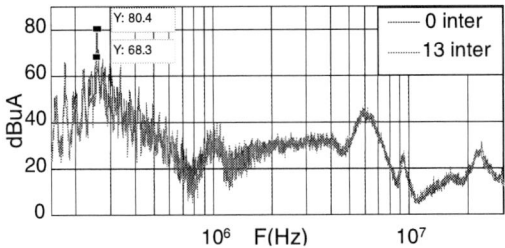

(b) Frequency domain EMI noise measurement results:
0°: blue; 13°: red

Figure 16: Experimental system EMI test results

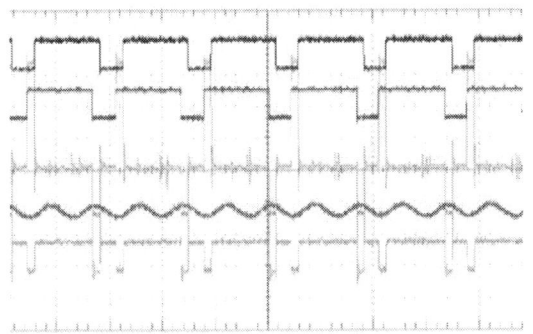

(a) Results of adding interphase inductor with low leakage inductance

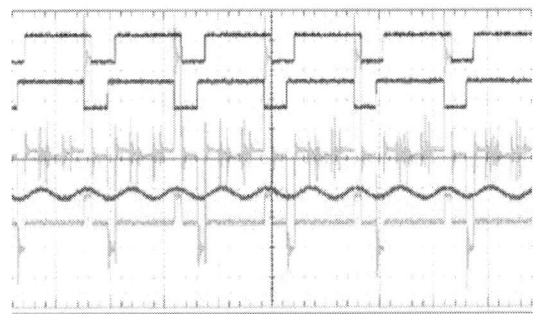

(b) Results of adding interphase inductor with high leakage inductance

Figure 17: Interphase Inductor Test results:
V_{out1}:blue(200V/div); V_{out2}:red(200V/div); V_{ext}: dark green(30V/div);
I_{out}:purple (20A/div); V_{out1}-V_{out2}: light green (200v/div)

VI. CONCLUSIONS AND FUTURE WORK

This paper presents a detailed analysis on the design and integration of interphase inductors for interleaved three phase VSIs for motor drive applications. The benefits and penalties of interleaving on system passive weight reduction are analyzed in detail. The results show that interleaving will increase the total passive weight

due to the additional weight of interphase inductors. To solve this problem, a design method to integrate the interphase inductor with the AC DM inductor is proposed and analyzed in detail, the results show that certain value of AC DM inductance can be integrated with interphase inductor with negligible increase on the weight and the maximum integrated inductance is related with interleaving angle. All the analysis is verified through a 2kW prototype. Future work will focus on inductor core loss analysis to verify the core loss increases with higher leakage inductance.

REFERENCES

[1] H. Akagi, T. Shimizu, "Attenuation of Conducted EMI Emissions From an Inverter-Driven Motor", *IEEE Transactions on Power Electronics*, vol 23, no. 1, pp. 282 - 290, Jan. 2008

[2] S. K. T. Miller, T. Beechner, and J. Sun, "A comprehensive study of harmonic cancellation effects in interleaved three-phase VSCs," in Proc. *IEEE Power Electron. Spec. Conf. (PESC)*, 2007, pp. 29 - 35.

[3] Di Zhang; Wang, F.; Burgos, R.; Rixin Lai; Boroyevich, D. "Impact of Interleaving on AC Passive Components of Paralleled Three-Phase Voltage-Source Converters", in *IEEE Transactions on Industry Application*, vol. 23, no. 1. pp.1042 - 1054

[4] Lei Xing; Jian Sun; "Motor drive system EMI reduction by asymmetric interleaving" in *Proc.COMPEL*, 2010, pp1-7

[5] Z. M. Ye , P. K. Jain and P. C. Sen "Circulating current minimization in high frequency AC power distribution architecture with multiple inverter modules operated in parallel", *IEEE Transactions on Industrial Electronics.*, vol. 54, no. 5, pp.2673 -2687 2007

[6] Di Zhang ; Fei Wang ; Burgos, R. ; Boroyevich, D., "Common-Mode Circulating Current Control of Paralleled Interleaved Three-Phase Two-Level Voltage-Source Converters With Discontinuous Space-Vector Modulation", *IEEE Transactions on Power Electronics*, vol. 26, no.12, pp.3925 - 3935, Dec. 2011

[7] Zhang, Xuning ; Mattavelli, Paolo ; Boroyevich, Dushan ; Wang, Fei. "Impact of interleaving on EMI noise reduction of paralleled three phase voltage source converters", *Applied Power Electronics Conference and Exposition (APEC)*, pp 2487- 2492.Mar. 2013

[8] Xuning Zhang; Dushan Boroyevich ; Rolando Burgos; "Impact of Interleaving on Common-Mode EMI Filter Weight Reduction of Paralleled Three-Phase Voltage-Source Converters", *Energy Conversion Congress and Exposition (ECCE)*, Sep. 2013

[9] Behzad Vafakhah and John Salmon, Member, "Interleaved Discontinuous Space-Vector PWM for a Multilevel PWM VSI Using a Three-Phase Split-Wound Coupled Inductor," in IEEE Trans. Ind. Applicat., vol. 45,no.3, pp. 2015–2023, Sep./Oct. 2010.

[10] Guangyong Zhu, Brent A. McDonald, Kunrong Wang, "Modeling and Analysis of Coupled Inductors in Power Converters", IEEE Transactions on Power Electronics, Volume 26, Issue 5, May 201 1, pp. 1 355-1363.

[11] P.-L.Wong, P.Xu, P.Yang, and F.C.Lee, "Performance Improvements of Interleaving VRMs with Coupling Inductors," IEEE Trans.Power Electron., vol.16, no.4, pp.499-507, 2001

[12] P.Zumel, O.Garcia, J.A.Cobos, and J.Uceda, "Magnetic Integration for Interleaved Converters," in Proc.18th Annual IEEE Applied Power Electronics Conf.and Exposition (APEC), vol.2, 2003, pp.1143-1149

A Novel Transformer Model Using Magnetic Circuit

Fuminori Nakamurame

Mitsubishi Electric Corporation
8-1-1, Tsukaguchi-Honmachi, Amagasaki, Hyogo, Japan

Toshifumi Ise

Division of Electrical, Electronics and Information Eng.,
Graduate School of Eng.
Osaka University
2-1, Yamada-oka, Suita 565-0871, Osaka, Japan

Abstract-We have proposed a novel transformer model using magnetic circuit. Numerical simulations using PSCAD/EMTDC has been performed, and the performance have been examined.

Keywords—Magnetic Circuit, Model, PSCAD/EMTDC, Transformer

I. INTRODUCTION

Recently large capacity STATCOM has been installed and applied for stability enhancement [1]. In order to prevent out of synchronism or suppress over voltage in power systems, STATCOM is required continuous operation under severe condition such as line fault.

On the other hand, voltage distortion causes DC magnetization of voltage source converter transformer [2]. The DC magnetization may cause overcurrent of STATCOM converter. Therefore, the transformer model is necessary for the purpose of developing control scheme of STATCOM. The transformer model for PLECS has been proposed [3], but the model for PSCAD/EMTDC is suitable for the power system simulation.

In this paper, the authors have a proposed novel transformer model using magnetic circuit. The authors also have performed numerical simulations using PSCAD/EMTDC and have examined the performance of the transformer model.

II. PROPOSED MODEL

A. Coupling Interface

Fig. 1. shows the block diagram of the proposed transformer model.

The electric part of the model consists of winding described as a current source. The magnetic part of the model consists of a magnetic circuit described as a magnetic flux source and a magnetic resistance, and the electric part and the magnetic part are coupled with coupling interface which consists of control blocks of PSCAD/EMTDC.

In the electric part, the winding voltage V is obtained. Then, in the coupling interface, the magnetic flux Φ is calculated from equation (1) where R_M is the magnetic resistance in fig. 1.

$$\Phi = \frac{1}{n} \int V dt \qquad (1)$$

In the magnetic circuit, the magnetomotive force F is calculated from equation (2).

$$F = R_M \Phi \qquad (2)$$

In the coupling interface, the winding current I is calculated from equation (3) where n is the number of turns and Int is an integrator.

$$I = \frac{F}{n} \qquad (3)$$

Using the proposed model, the simulation becomes mathematically stable because the coupling interface consists of an integrator.

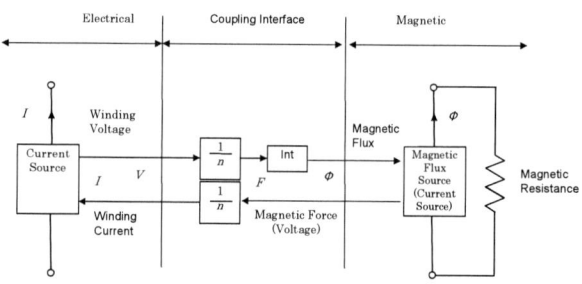

Fig. 1. Proposed transformer model

B. Magnetic Circuit Model

Fig. 2. shows the schematic of a single phase transformer and fig. 3. shows the simulation model of the transformer.

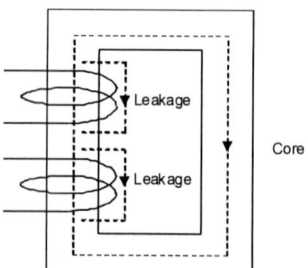

Fig. 2. The schematic of single a phase transformer

978-1-4799-2706-7/14 $31.00 © 2014 IEEE

The primary winding and the secondary winding are coupled with the magnetic circuit through the coupling interface.

Since the major magnetic flux path through primary winding, secondary winding, and core, they are connected in series.

On the other hand, since the leakage magnetic flux path through only each winding and the air, the winding and the leakage impedance is connected in parallel with the magnetic flux source.

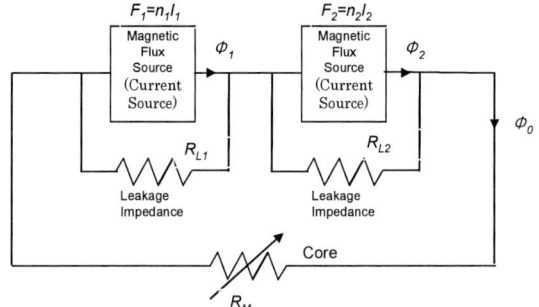

Fig. 3. Magnetic circuit model of a transformer

C. Equivalent Resistance for Leakage Impedance

The leakage inductance L_L is calculated from the voltage difference between the primary voltage V_1 and the secondary voltage V_2 as equation (4).

$$\Delta V = V_1 - \frac{n_1}{n_2} V_2 = L_L \frac{d}{dt} I \qquad (4)$$

The primary voltage V_1 is obtained from equation (5) where R_{L1} is the equivalent resistance for leakage impedance of primary side winding.

$$V_1 = n_1 \frac{d\Phi_1}{dt} = n_1 \frac{d}{dt}\left(\frac{n_1 I_1}{R_{L1}} + \Phi_0\right) \qquad (5)$$

Since the secondary winding current I_2 is described as equation (6), the secondary voltage V_2 is obtained from equation (7), where R_{L2} is the equivalent resistance for leakage impedance of secondary side winding.

$$I_2 = -\frac{n_1}{n_2} I_1 \qquad (6)$$

$$V_2 = n_2 \frac{d\Phi_2}{dt} = n_2 \frac{d}{dt}\left(-\frac{n_1 I_1}{R_{L2}} + \Phi_0\right) \qquad (7)$$

Therefore equation (4) becomes equation (8).

$$\Delta V = n_1^2\left(\frac{1}{R_{L1}} + \frac{1}{R_{L2}}\right)\frac{d}{dt}I_1 = L_L \frac{d}{dt}I \qquad (8)$$

Finally, assuming that the primary and secondary leakage impedance is the same value, the equivalent resistance R_L for leakage impedance is calculated from the leakage inductance L_L as equation (9).

$$R_L = 2\frac{n_1^2}{L_L} \qquad (9)$$

D. Core Model

Fig. 4. shows the core model of the proposal model. The magnetic flux density B is calculated from equation (6), the magnetomotive force F is calculated from equation (7), and magnetic resistance of the core R_M is calculated from equation (8).

$$B = \frac{\Phi}{A} \qquad (6)$$

$$F = LH \qquad (7)$$

$$R_M = \frac{F}{\Phi} \qquad (8)$$

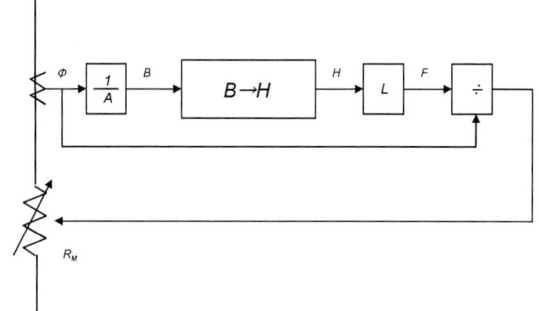

Fig. 4. Core model

In order to calculate the magnetic field H, magnetization characteristic of the core is also necessary.

Fig. 6. shows an example of the magnetization characteristic of the grain-oriented silicon steel core [4][5]. The unit of magnetic field axis is AT/m. Therefore, in order to obtain the magnetic field H, we must use the magnetic path length of the core, 0.97m. Since the magnetic field axis is in logarithm scale, the relational expression of the magnetic flux density B and the magnetic field H can be obtained as equation (9).

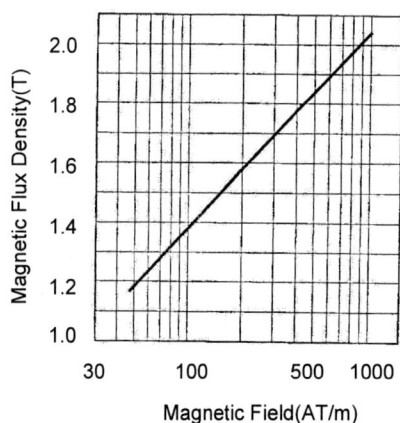

Fig. 5. Magnetization characteristic of the grain-oriented silicon steel core [4]

$$B = 0.66\log_{10} 0.97H + 0.0569 \qquad (9)$$

Hence, the magnetic field H can be calculated from equation (10).

$$H = 10^{\frac{B-0.0569}{0.66}} \qquad (10)$$

III. SIMULATION

A. Transformer Design

In order to confirm the performance of the model, we have performed numerical simulations using PSCAD/EMTDC.

Fig. 6. shows the transformer design and table I indicates the specification of the transformer.

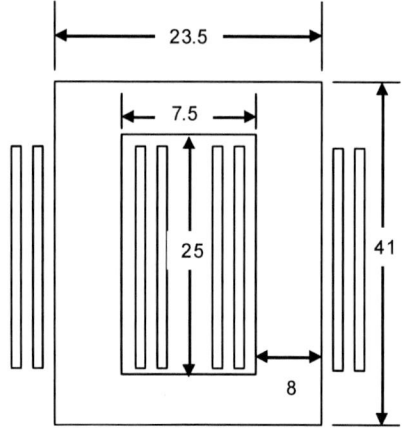

Fig. 6. Transformer design

TABLE I
THE SPEIFICATION OF THE TRANSFORMER

Q	Rated capacity	20kVA
f	Rated frequency	50Hz
V_1	Primary voltage	210V
V_2	Secondary voltage	210V
I_1	Primary current	95.23A
I_2	Secondary current	95.23A
N_1	Primary turn	60
N_2	Secondary turn	60
L_1	Leakage impedance	1.14%
B	Magnetic flux density	1.4T
A	Cross section	$112 \times 10^{-4} m^2$
L	Magnetic path length	0.97m

Fig. 7. indicates the magnetization curve of the core. Since the saturation flux density is 2.0T, we set 1.4T as the rated magnetic flux density.

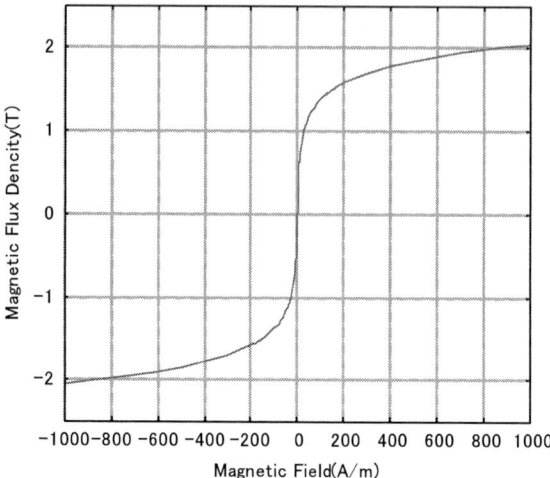

Fig. 7. Magnetization curve

B. Simultain Result When the Primary Voltage is Rated and the Secondary Voltage is 90% of Rated

Fig. 8. shows the simulation model. The primary winding is connected to the AC voltage, and the secondary winding is connected to the voltage source. The voltage source generates the voltage as given voltage reference.

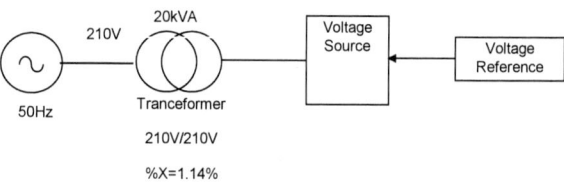

Fig. 8. Simulation model

In order to examine the performance of the leakage impedance model, we have performed the numerical simulation when the primary voltage is rated and the secondary voltage is 98.86% of rated.

The waveforms are indicated in fig. 9. Since the leakage impedance is 1.14%, the primary current is rated 95.23Arms and 90 degree lag behind the primary voltage.

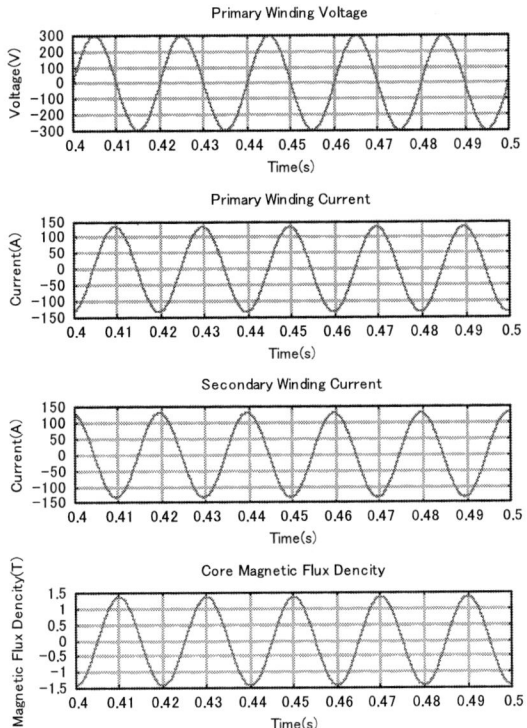

Fig. 9. Waveforms when the primary voltage is rated and the secondary voltage is 90%

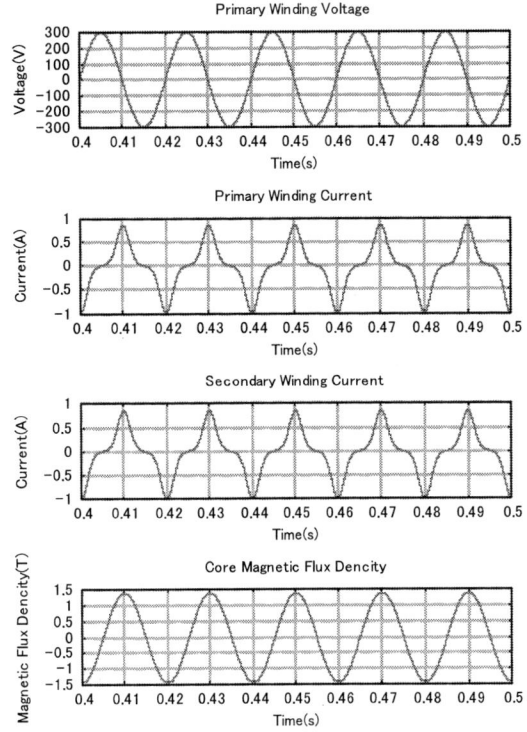

Fig. 10. Waveforms when the primary voltage and the secondary voltage are rated

C. Simuration Result When the Primary Voltage and the Secondary Voltage are Rated

In order to examine the performance of the core model, we have performed the numerical simulation when the primary voltage and the secondary voltage are rated.

The waveforms are indicated in fig. 10.. In this case, the lag or lead current can not be seen. It is because the difference between two winding voltage is zero and the excitation current is negligible small.

Therefore, the primary current and secondary current become rated excitation current and the core magnetic flux density becomes rated.

The rated excitation current is calculated from equation (9) using equation (10), (3), and (7).

$$I = L\frac{10^{\frac{B-0.0569}{0.66}}}{n} = 0.97 \times \frac{10^{\frac{1.4-0.0569}{0.66}}}{60} = 1.80\text{A}$$

(11)

Since the excitation current is supplied from two windings, the primary winding current and secondary winding current become 0.9A respectively.

Compared with fig.9, the primary winding current and the secondary winding current is distorted at the peak. It is because the magnetization curve shown in fig. 7. begins to saturate over 1.0T.

D. Simuration Result When the DC Voltage is Superposed on the Primary Sinusoidal Voltage

In order to examine the performance of the saturated core model, we have performed the numerical simulation when the DC voltage is superposed on the primary sinusoidal voltage. In this case, 5% DC voltage becomes to be applied at 0.2 second.

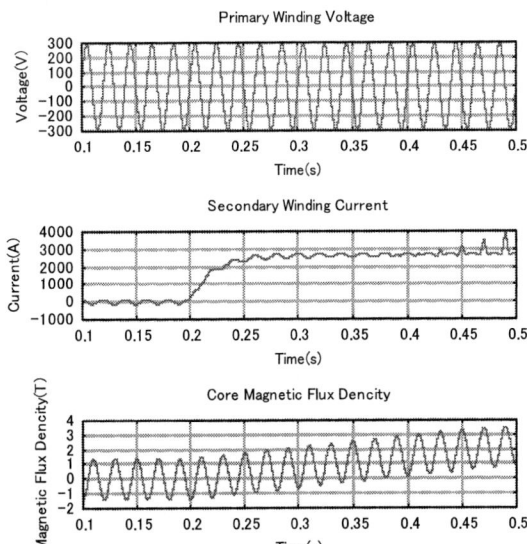

Fig. 11. Waveforms when the DC voltage is superposed

The waveforms are illustrated in Fig. 11. Since DC magnetization occurs, the polarity of the core magnetic flux density becomes positive.

Additionally, the peak of the primary winding current begins to be distorted.

E. Simuration Result When the PWM Converter is Connected to the Secondary Winding

In order to examine the performance of the transformer model when the PWM converter is connected, we have performed the numerical simulation.

Fig. 12. shows the simulation model. In this case, the leakage impedance is changed to 10% as the normal design for the converter transformer. The DC voltage of the converter is 500V.

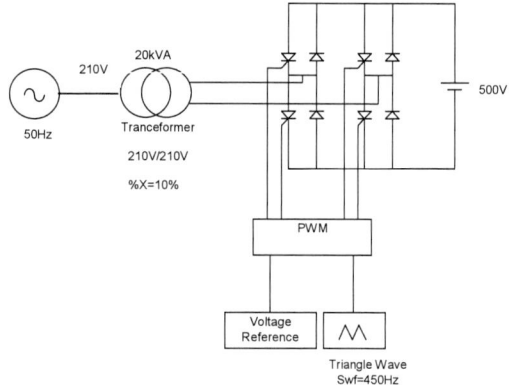

Fig. 12. Simulation model when the PWM converter is connected

Fig. 13 shows the waveforms when the converter output 90% of the rated voltage. The waveforms show that the primary current is rated 95.23Arms and 90 degree lag behind the primary voltage.

Fig. 14 shows the waveforms when the converter output 110% of the rated voltage. The waveforms show that the primary current is rated 95.23Arms and 90 degree lead of the primary voltage.

These results proved that the proposed transformer model can be also applied to the PWM converter simulation.

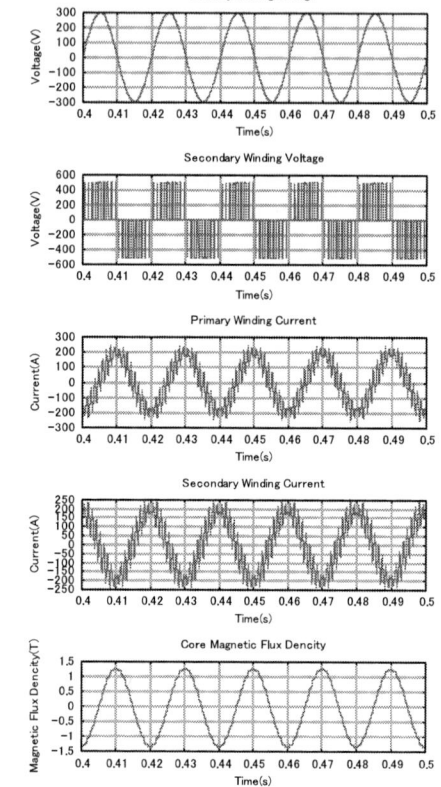

Fig. 13. Waveforms when the converter outputs 90% of rated voltage

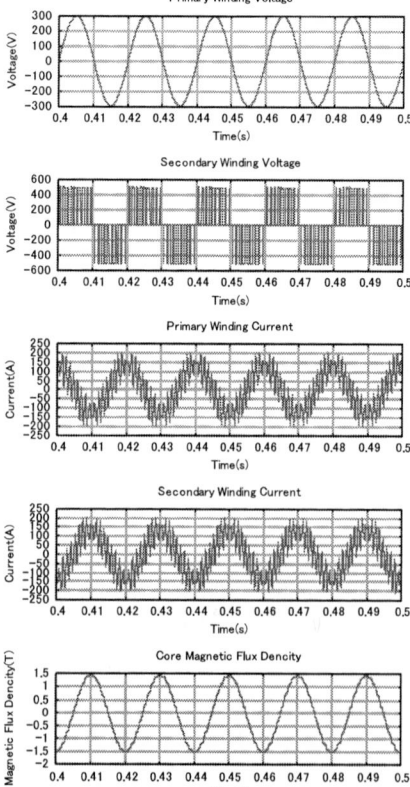

Fig. 14. Waveforms when the converter outputs 110% of rated voltage

IV. CONCLUSIOS

We have proposed a novel transformer model using magnetic circuit.

The model consists of the electrical part, the magnetic part, and the coupling interface. The coupling interface includes the leakage impedance and the core model with the saturation feature.

In order to examine the model, we have performed numerical simulations using PSCAD/EMTDC.

First, we have performed the numerical simulation when the primary voltage is rated and the secondary voltage is 98.86% of rated. Since the leakage impedance is 1.14%, the primary current is rated and 90 degree lag.

We have also performed the numerical simulation when the primary voltage and the secondary voltage are rated. In this case, the excitation current is seen and the amplitude was 0.9A as designed and the distorted winding current is seen since the magnetization curve begins to saturate over 1.0T.

In order to examine the DC magnetization, we have performed the numerical simulation when the DC voltage is superposed on the primary sinusoidal voltage. Since DC magnetization occurs, the polarity of the core magnetic flux density becomes positive and the peak of the primary winding current begins to be distorted.

Finally, we have performed the numerical simulations when the PWM converter is connected. The primary current became rated 95.23Arms and 90 degree lag when the converter output 90% of the rated voltage, and the primary current became rated 95.23Arms and 90 degree lead when the converter outputs 110% of rated voltage.

These results proved that the proposed transformer model can be applied to the numerical simulation using PSCAD/EMTDC and also applied the PWM converter simulation.

The authors are planning to develop a control scheme for DC magnetization suppression using this model.

REFERENCES

[1] T. Akedani, J. Hayashi, K. Temma, and N. Morishima, "450MVA STATCOM installation plan for stability improvement," *Proceedings of 2010 CIGRE*, Paper B4-207, 2010.

[2] T. Nakajima, K. Suzuki, M. Yajima, N. Kawakami, K. Tanomura, and S. Irokawa, "A new control method preventing transformer DC magnetization for voltage source self-commutated converters," IEEE Transactions on Power Delivery, vol. 11, no. 3, pp.1522-1828, 1996.

[3] J. Allmeling, W. Hammer, and J. SchOnberger, "Transient simulation of Magnetic Circuits Using the Permeance-Capacitance Analogy," *Proceedings of COMPEL'12*, Paper PS-01-5, 2012.

[4] JyutaroTakeuchi, *Design engineering of electrical machinery*, Ohmusya, pp.166, 1952.

[5] M. Hagiwara, P. Viet Pham, H. Akagi, "Elucidation of DC Magnetic Deviation in Converter Transformers Used for a Self-Commutated BTB System during Single-Line-to-Ground Faults," *IEEJ Transactions on Industry Applications*, Vol. 127, No. 9, pp. 1013-1022, 2007.

Hardware-In-the-Loop Simulation
of a Machine Model with Real-Time Animation

Xiaojie Zhuang
Research &Development Dept.
Mitsubishi Electric Corporation, Nagoya Works,
Nagoya, Aichi, JAPAN
So.Gyoketsu@df.MitsubishiElectric.co.jp

Shinya Hibino
Inverter System Dept.
Mitsubishi Electric Corporation, Nagoya Works,
Nagoya, Aichi, JAPAN
Hibino.Shinya@dr.MitsubishiElectric.co.jp

Masaya Harakawa
Research &Development Dept.
Mitsubishi Electric Corporation, Nagoya Works,
Nagoya, Aichi, JAPAN
Harakawa.Masaya@cw.MitsubishiElectric.co.jp

Ryosuke Terabe
Research &Development Dept.
Mitsubishi Electric Corporation, Nagoya Works,
Nagoya, Aichi, JAPAN
Terabe.Ryosuke@ds.MitsubishiElectric.co.jp

Takayuki Ozaki
Fundamental Technology Dept.
Mitsubishi Electric Engineering Corporation,
Nagoya Engineering Office
Nagoya, Aichi, JAPAN
Ozaki.Takayuki@ma.mee.co.jp

Tetsuaki Nagano
Research &Development Dept.
Mitsubishi Electric Corporation, Nagoya Works,
Nagoya, Aichi, JAPAN
Nagano.Tetsuaki@ab.MitsubishiElectric.co.jp

Abstract—**This paper deals with a new technology called Hardware-In-the-Loop (HIL) simulation. It is a method by which a simulation is carried out in real time (RT). In this system, a real product (Inverter) is used as a controller instead of a software-based controller; the power circuit, induction motor and machine (crane) are all modeled and loaded into an RT simulator. In this HIL simulation, the motor drive simulation cycle is as short as 10μsec to ensure the accuracy. Furthermore, the movement of the crane is shown as a 3D animation in real time. According to the experiment result, HIL simulation shows its merits of efficiency increase, reliability and safety.**

Keywords—Real-Time, simulation, Hardware-In-the-Loop, HIL, crane, 3D animation, Real-Time animation

I. INTRODUCTION

In the field of Factory Automation (FA), the driving products, such as inverter and servo amplifier, are connected with the customers' mechanical systems to work. Therefore, in developing the driving products, the best way to ensure the performance of the products is to connect them with the mechanical systems of the end users. However, the customers' motors and machines are always not available. For example, the developers can just use the machines by adjusting the time with the customers. That means various running patterns have to be tested in a limited time. Moreover, in the case of large-scale system, such as crane (Fig. 1), there are still some safety risks in the test. In such a situation with the problems of real process, a new developing method of mechanical system seems necessary.

The aim of the present paper is to give a new developing approach for a crane testing system, called Hardware-In-the-Loop (HIL) simulation. The HIL simulation system is processed by a Real-Time (RT) simulator. In this simulation, all of the power circuit, motors and crane are modeled and loaded into the RT simulator instead of the real machines. By connecting the real driving controller, an RT HIL simulation is realized to evaluate the driving product's performance. The simulation cycle of motor drive components can be carried out as short as 10μsec, which ensures the accuracy of the simulation. In addition, by the communication with a mechanical model tool, the movement of the crane can also be checked by a 3D animation.

The remainder of this paper is divided into the following 4 sections. Crane system and HIL system will be described in section II and III. The crane simulation system constructed this time will be explained in section IV, and section V will be a conclusion.

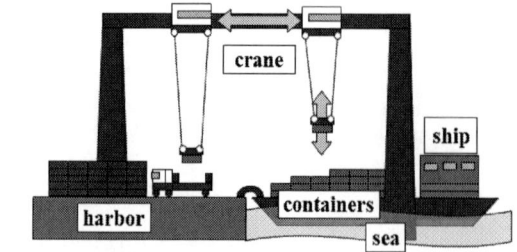

Fig. 1. Image of a container crane

II. CRANE SYSTEM

Crane is a type of machine that can be used both to lift and lower materials and to move them horizontally. It is always used for lifting heavy things and transporting them to other places.

These years, with the developing of industry, the market of crane has increased in demand. Hence, many FA companies started to set foot in the crane market. That led an instance competition. In order to cope with the market competition, the efficiency of developing appeared more important.

In this paper, we modeled a container crane as the controlled object. A container crane, showed as Fig.1, is a type of large dockside gantry crane always found at container terminals for loading and unloading intermodal containers from container ships. To ensure the crane to move safely and efficiently, the sway of the cranes is the very problem, which the developers are always focused on [1]. Due to its large scale, it is hard to test in a real process. It seems impossible to set a same-size crane in a laboratory for testing and going to the real place will cost too much. As a solution to this problem, a simulation approach will be introduced in next section called HIL simulation.

III. HIL SIMULATION

In this section, an overview about HIL simulation is given, and we will illustrate how an RT crane simulation is performed.

According to the previous works, crane could be defined as the followed state equation.

$$\frac{d}{dt}\begin{bmatrix} x \\ v \\ \theta \\ \omega \end{bmatrix} = \begin{bmatrix} 0 & 1 & 0 & 0 \\ 0 & 0 & \dfrac{mg}{M} & 0 \\ 0 & 0 & 0 & 1 \\ 0 & 0 & -\dfrac{(m+M)g}{Ml} & 0 \end{bmatrix}\begin{bmatrix} x \\ v \\ \theta \\ \omega \end{bmatrix} + \begin{bmatrix} 0 \\ \dfrac{1}{M} \\ 0 \\ -\dfrac{1}{Ml} \end{bmatrix} \cdot gr \cdot Tm \quad (1)$$

where x is the trolley travel's distance of the crane, v is the trolley travel's speed, θ is the sway angle, ω is the sway angle speed, m is the spreader payload's mass, M is the trolley's mass, l is the rope length, gr is the gear ratio, and Tm is the motor's torque.

Generally, based on this mathematical model, the off-line simulation could be held as Fig. 2. However, without the real controller, the products could not be really evaluated.

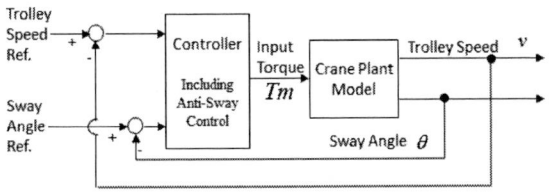

Fig. 2. Crane control system

On the other hand, since an accurate machine model is always complicated, that will lead a long time to simulate it if you adopted a short simulation cycle to gain the simulation's accuracy. Considering about these parts, we raise the HIL simulation which can solve the problems.

HIL simulation, as we have mentioned as Hardware-In-the-Loop simulation, is a complex RT embedded simulation, where some of the control-loop components are real hardware, and some are virtual. It shows its advantages when the real process is not available because of the lack of equipment or the safety problem, or the experiments with the real process are too costly or require too much time to prepare.

In the past few years, there are lots of researches about HIL simulation [2]. The flexibility of the technology showed its merits in the efficiency of expansion and reconfiguration. Several years ago, a complete PMSM (Permanent Magnet Synchronous Motor) drive simulation was carried out by HIL simulation system with an RT simulator called RT-LAB [Opal-RT Technologies inc.] [3][4]. Owing to the high-speed simulation cycle and the precise model, it became possible to simulate the high frequency switching power circuit. It made a remarkable progress in the field of HIL simulation.

In this paper, a crane system was proposed as we mentioned above. Based on the real process, the HIL simulation system was shown as Fig. 3. The motor drive components and a container crane, which was necessary in the real process, were modeled. Instead of a complicated mathematical model, a mechanical modeling tool was used to construct the crane model.

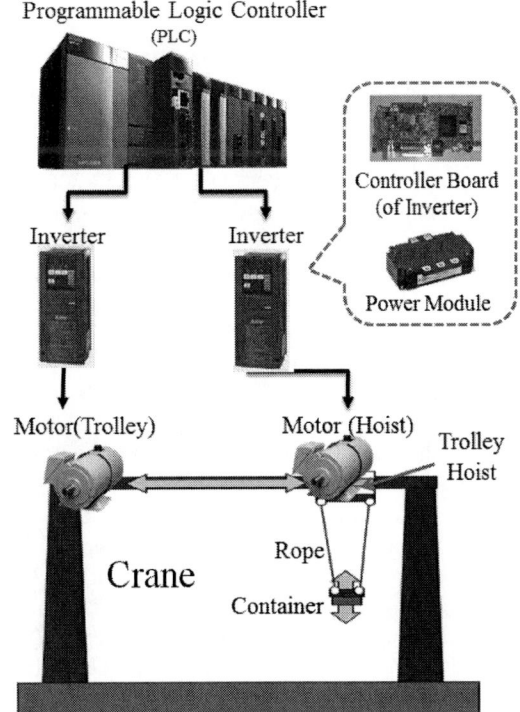

(a) Real Process

978-1-4799-2706-7/14 $31.00 © 2014 IEEE

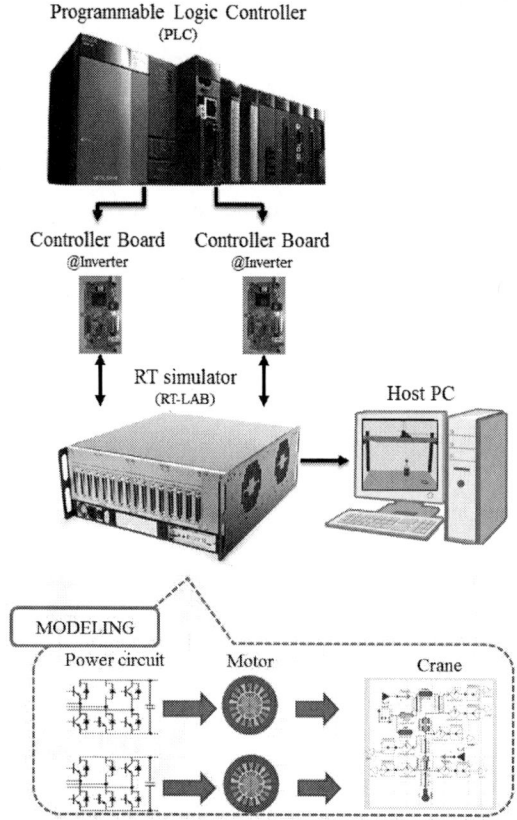

(b) HIL Simulation Process

Fig. 3. Crane testing system by HIL simulation

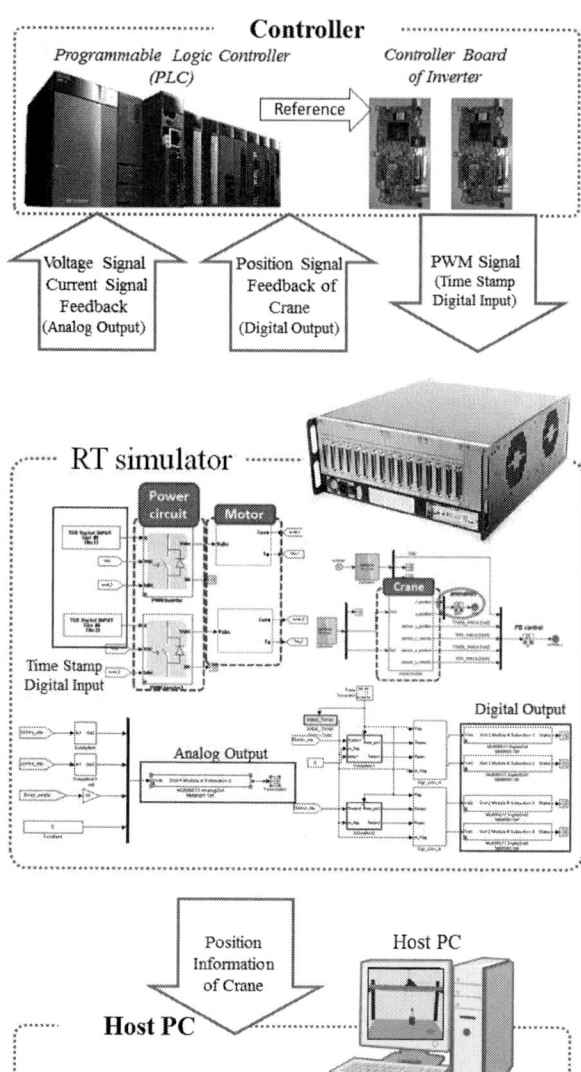

Fig. 4. I/O's overview of crane HIL simulation system

In this way, the model of crane was not only just a block type, which could be calculated in the simulation process, but also a 3D animation type, which could be checked in real time.

All of the models were built based on Simulink in a host PC. The crane model, constructed by the mechanical tool was also compiled into an s-function and quoted as a Simulink block. By the interface of Simulink, the RT model system was compiled and downloaded from the host PC into the RT simulator--RT-LAB, which executed the RT HIL simulation. During the simulation process, the host PC could set the simulation condition and check the result in real time.

Fig. 4 gives I/O's overview of the crane HIL simulation system. A Programmable Logic Controller (PLC) was used as a host controller with the same program as what it performed in the real process. Through the controller board, the commands from the PLC were transformed into PWM (Pulse Width Modulation) signals. The time-stamping digital input, as a unique technology of RT-LAB, could accept the PWM signals as a more accurate digital input. Driven by the PWM signals, a high speed calculation was carried out with all the power circuits, motors and machine (crane) models in the RT simulator. The results of the calculation, including the currents and voltage of motors, as well as the crane's sway angle, were feed backed to the control circuit by analog outputs.

Of course, the system could be evaluated by the numerical data of the crane's movement. Besides that, in this system, the positions' information was also handed to the 3D mechanical model. In this way, with the simulation being implemented, the movement of the crane could also be displayed by a 3D animation in real time. It was possible to evaluate the effects of control algorism more intuitively than by the result of numeral data, especially in multi-axis system.

About the simulation cycle, in this work, although there are two axes, and also, the crane model was added to the system, which made it much heavier than the previous works [4] [5], thanks to the high speed RT simulator—RT-LAB OP5600, the simulation cycle of motor drive components was succeeded to processed by 10μsec, which ensured the accuracy of the simulation. For the crane model, it was set as 50μsec, which was enough for the frequencies of the crane sway about 1Hz.

IV. EXPERIMENTS AND RESULTS

By employing the HIL simulation system, the crane test can be conducted to evaluate the controlling algorism efficiently. Firstly, before applying the HIL simulation, it is important to confirm the accuracy of the system by comparing the results of HIL simulation process and the real process

In this section, real process experiments, off-line simulations and HIL simulations were carried out. The results of the comparison would be explained.

For the real process experiments, a real mini crane was set as Fig. 5 to inspect the reliability of the simulation. A PLC was used as the host controller. By the inverter, two induction motors were driven to control the trolley (13.34kg) and the ropes (1.05m). A weight (5.3kg) was set to be the supposed container. The conditions of the experiments were shown as Table I. The container was commanded to move horizontally for 0.54m. The max velocity was set to be 0.3m/sec. Anti-sway control was set to be on and off. The data of each condition was observed for 20sec.

All the situations of the real process were set same in the off-line simulation and HIL simulation. According to evaluating the effect of anti-sway control, the performance of HIL simulation would be reported.

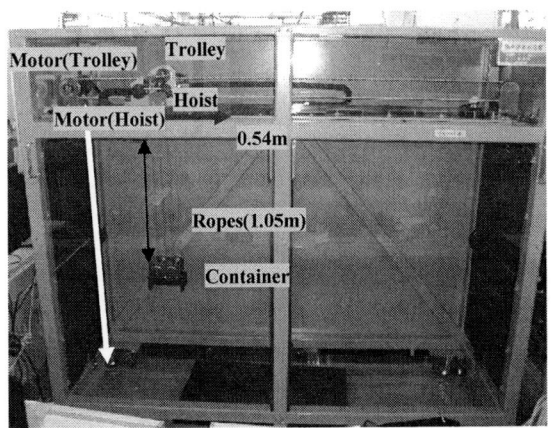

Fig. 5. Mini crane device

TABLE I
EXPERIMENTAL CONDITIONS

Item	Anti-sway Control On	Anti-sway Control Off
Max Velocity	0.3[m/sec]	0.3[m/sec]
Speed-up Time	1[sec]	1[sec]
Speed-down Time	1[sec]	1[sec]
Max Velocity Time	0.8[sec]	0.8[sec]

TABLE II
COMPARISON OF TESTING TIME

Item	Simulation Cycle	Testing Time
Real Process	-	20[sec]
Off-line Simulation	10[μsec]	400[sec]
Off-line Simulation	50[μsec]	20[sec]
HIL Simulation	10[μsec]	20[sec]

A. Result1: Testing time comparison of real process and each simulation

For the 20sec process, the real process obviously cost 20sec. For the HIL simulation, as a RT simulation, also cost 20sec, no matter how long the simulation cycle was. Meanwhile, the off-line simulation cost almost 7 minutes if the simulation cycle was set to be as 10μsec as what was set for HIL simulation in this experiment. Moreover since the PWM signals' generator was hard to be simulate, in the off-line simulation process, just a simple velocity reference was adopted. That means it would cost much more time if a more accurate but heavier modeled was constructed. Actually, in normal developing process, it is hard to wait for such a long time for every simulation. So that the developers always set the simulation cycle longer by losing the accuracy of the simulation. For a comparison, we tried to set a certain simulation cycle, by which, simulation could be restricted to 20sec. The simulation cycle was worked out to be 200μsec. It could be imaged that the 5-time increase of the simulation cycle might cause the falling of the simulation's reliability.

The comparison of the testing was shown as Table II. The merit of HIL simulation could be recognized at the aspect of testing time. In addition, it also showed its advantage at testing the function of the real product such as PWM generating.

B. Result2: Control evaluation comparison of real process experiments, off-line simulation and HIL simulation

Sway angles of ropes, trolley travel distance, trolley travel speed and trolley input torque were observed to evaluate the performance of the anti- sway control. All of the data was under the condition of both with the anti-sway control and without it.

Fig. 6(a) shows the result of the real process, Fig. 6(b) is about the off-line simulation and Fig. 6(c) is the HIL simulation. According to the results, it could be found that each of the experiment showed the effect of the anti-sway control from the decrease of the vibration. However, since the off-line simulation did not use the real controller, the control system was nearly ideal. It is impossible to find the product's problem. Moreover, it cannot test how the real product affected the evaluating results with other facts such as PWM-switch, noise of product and etc.

The 2014 International Power Electronics Conference

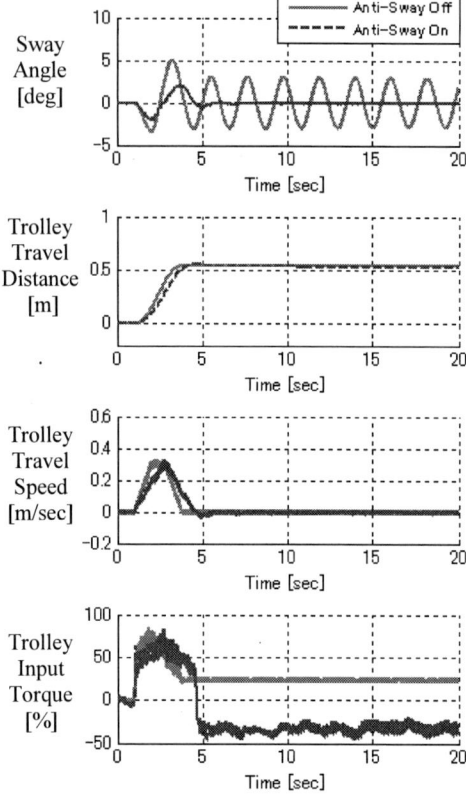

(a) Result of real process experiment

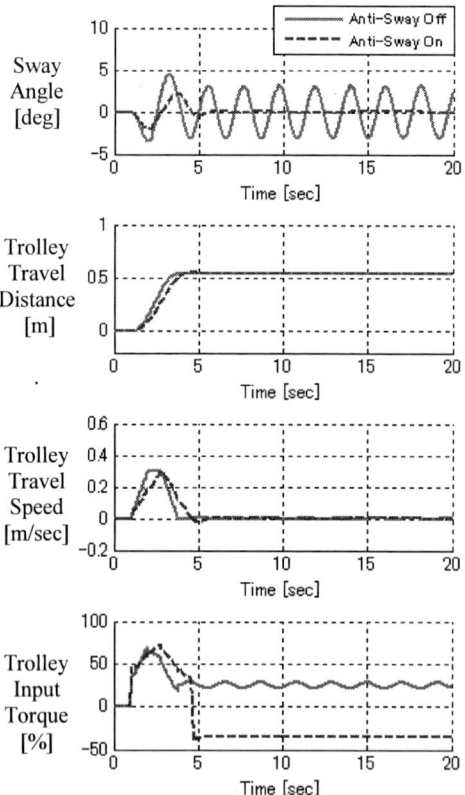

(b) Result of off-line simulation process experiment

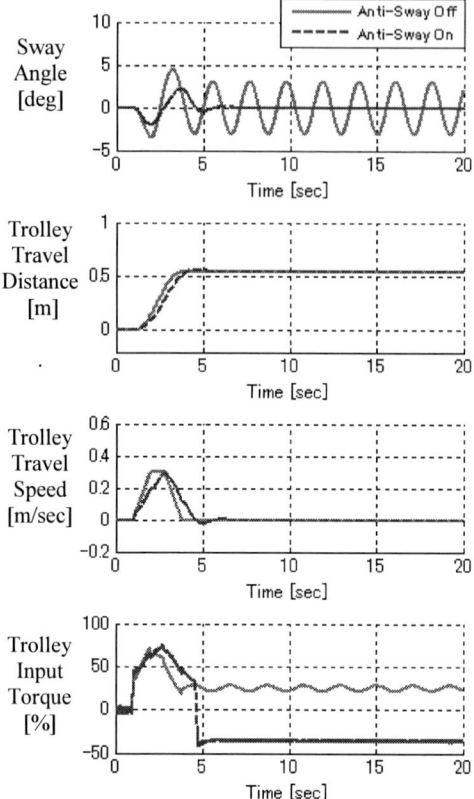

(c) Result of HIL simulation process experiment

Fig. 6. Comparison of Anti-Sway Control Effects

That confirmed the HIL simulation's advantage that the real product's performance could be tested. That meant HIL simulation could be used to examine the products instead of the real process if the machine is accurately modeled. For developers, the flexibility no doubt helped to raise the development efficiency.

C. Result3: 3D animation in real time

The model of the container crane constructed by a mechanical tool was shown as Fig. 7. It was not only compiled to be an s-function but also expressed as a 3D animation. The s-function was used to calculate in HIL simulation system in real time and the 3D animation showed the results of the calculations.

To realize the real-time animation, in the HIL simulation work, we communicated the mechanical tool (in the host PC) with the RT simulator. In RT simulator, the crane model (s-function) calculated the actions by 50μsec and transferred the action information, such as sway angle, trolley position and so on, to the 3D model in the host PC. The transfer cycle was set as 0.01sec (100Hz), since human cannot recognize the refresh rate greater than about 70Hz.

Fig. 7 showed the action of the HIL simulation. The crane was ordered to move to the destination. According to the crane's position information from RT-simulator,

978-1-4799-2706-7/14 $31.00 © 2014 IEEE

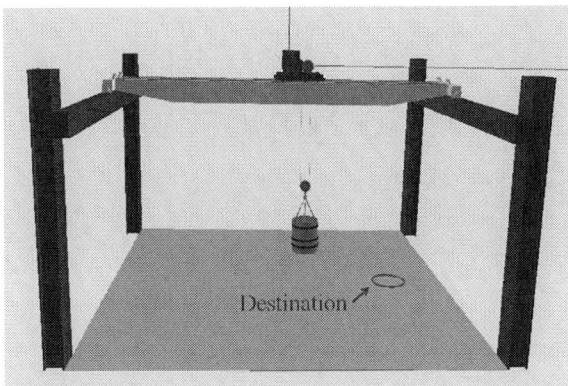

(a) Starting point

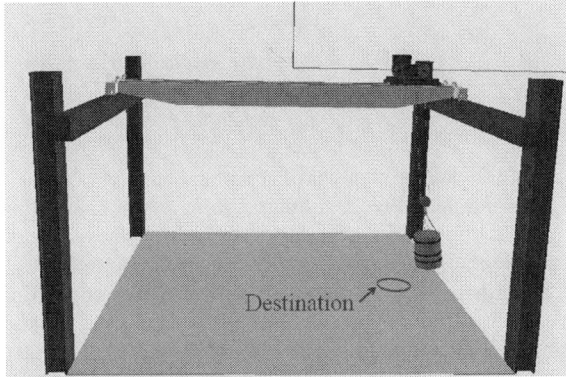

(b) Strong vibration when Anti-Sway off

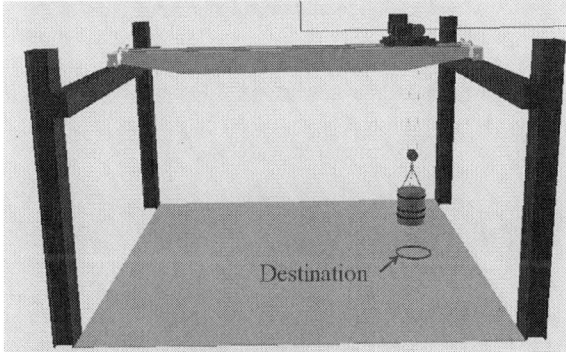

(c) Slight vibration when Anti-Sway on

Fig. 7. Image of real-time animation

the crane 3D model was successfully acting as a real machine in real time. It showed the different behaviors under different situations. The effect of the anti-sway control algorism was able to be judged more intuitively by the animations just during the simulating besides by the data after the experiments.

V. CONCLUSIONS

In this paper, we proposed a new approach to solve the problem of the experiments' difficulties for large scale equipment, such as crane.

An RT simulation, called HIL simulation, was carried out to test the crane anti-sway control system, which was always difficult to perform by a real process. All the power circuits, the motors, and crane in this system were modeled and loaded into an RT simulator. According to the command of the host controller—PLC, the motors drove the crane as a model base and check the effects of the anti-sway control by a numerical data as well as a real-time animation.

As a result, the HIL simulation proved its accuracy by comparing the results with the real process. Through the experiment process, HIL simulation showed its merits as followed.

1. The flexibility of HIL simulation brought the efficiency of expansion and reconfiguration.

2. It became possible to test under the situation of lacking the real equipment by HIL simulation.

3. HIL simulation could be performed safely in spite of large scale system.

4. By HIL simulation, the time and labor cost of preparing and tidying for experiment could be saved.

In addition, by comparing the process of the HIL simulation and off-line simulation, HIL simulation also had its advantages as followed.

5. The testing time of HIL simulation was much shorter than that of an off-line simulation especially for a complex model.

6. HIL simulation made it possible to evaluate the real product's performance because of applying the product's hardware.

All these merits showed that the HIL simulation for a crane could be a reliable as well as an efficient approach in developing period.

Nevertheless, to be a popular simulation approach, there are still some problems which should be solved in future work. For example, customers' machines are always complicated and not easy to model. Since the exactness of the model affects the result of simulation directly, it should be paid attention. On the other hand, with the complexity of the model, the model will become heavier, and the simulation cycle might become longer, which also influence the precision. How to make an efficient and exact model should be an important subject in a HIL simulation process.

REFERENCES

[1] Azdiana M.Y. , "Vibration control of a gantry crane system using dynamic feedback swing controller," *Journal of Mechanical Engineering and Technology,* 1 (1). pp. 63-72. ISSN 2180-1053, 2009

[2] S. Abourida, "Real-time PC-based simulator of electric systems and drives," *Proceedings of the IEEE Applied Power Electronics Conference and Exposition (2002)*

[3] C. Dufour, J. Belanger, "A PC-Based Real-Time Parallel Simulator of Electric Systems and Drives," *Parallel Computing in Electrical Engineering,* 2004. PARELEC 2004, 2004

[4] M. Harakawa, "Real-Time Simulation of a Complete PMSM Drive at 10µs Time Step," *Proceedings of IPEC-Niigata 2005*

[5] S. Abourida, "Real-Time HIL Simulation of a Complete PMSM Drive at 10 µs Time Step," *Proceedings of the 11''' European Conference on Power Electronics and Applcations (EPE 2005),* 2005.

Development of Real Time Digital Simulator for Self-Commutated SVC to Suppress Voltage Flicker

Yutaka Terao, Yasuhiro Shishida, Yoshinori Tsuruma,
Tomotsugu Ishizuka, Fumio Aoyama,
and Teruo Yoshino
Power Electronics Department
TOSHIBA MITSUBISHI-ELECTRIC INDUSTRIAL
SYSTEMS CORPORATION
Fuchu-City, Japan
TERAO.yutaka@tmeic.co.jp

Yutaka Kato
NEAT CO., LTD
Nagoya-City, Japan
Yutaka_kato@neat21.co.jp

Jean Bélanger
OPAL-RT TECHNOLOGIES
Montreal, Canada
Jean.belanger@opal-rt.com

Abstract-This paper deals with real-time simulation of a self-commutated static var compensator (SVC) system for steel plants with electric arc furnaces (EAFs). A voltage flicker often occurs due to the unsteady current and changes in reactive power. Simulation of the flicker suppression system is quite complicated, and new simulation techniques should be developed. We constructed a main circuit and control system models of an SVC for real-time simulation with RT-LAB®. The real-time simulation results were compared with actual operational data of a steel plant system equipped with two furnaces. Finally, we showed that we successfully developed a real-time simulation system for SVC systems.

Keywords— IEGT converter, Real-time digital simulation, SVC, Steel plants

I. Introduction

Steel plants equipped with electrical arc furnaces (EAFs) often cause voltage fluctuations. One of the causes of these fluctuations is irregular currents in the EAFs. These voltage fluctuations, or voltage flicker, affect electrical devices and light bulbs in residential houses connected to the distribution system close to the EAFs. Light bulb flickering sometimes makes people uncomfortable and it should be suppressed. Static var compensator (SVC) systems are one of the solutions for suppressing voltage flicker. These systems are usually installed to compensate for fluctuations in the EAF current. The current factors cause the variations in the instantaneous and effective values of the voltage supplied to the light bulb, such as variations in the reactive component and second or third harmonics.

There are two kinds of SVC for steel plant systems. Line-commutated SVCs are designed with thyristor-controlled reactors (TCRs) [1]. On the other hand, self-commutated SVCs use insulated gate bipolar transistors (IGBTs) or injection enhanced gate transistors (IEGTs) [2, 3]. A self-commutated SVC has faster response speed than that of a line-commutated SVC. Both systems are connected to the load bus for the EAFs and suppress voltage flicker at the point of common coupling (PCC).

SVC simulation models often require a long calculation time, as long as several seconds, because they include many complicated components, such as multiple-stage transformers, IEGT converters, etc. To overcome this issue, we have been developing a real-time digital simulation system for steel plants with SVCs [4]. This simulation system, called RT-LAB® developed by Opal-RT Technologies Inc., is able to carry out real-time simulation for such a plant control system.

In this paper, we describe the real-time simulation of an SVC system for a steel plant and show a comparison with actual site data. We constructed the main circuit and control system models based on an actual steel plant. The simulation data is compared with the real plant data.

II. Characteristics of Self-Commutated SVC

A. Steel Plant System Configuration

Fig. 1 represents the system configuration of the self-commutated SVC for the steel plant, which is equipped with an EAF and a ladle furnace (LF). The self-commutated SVC suppresses voltage flicker and compensates for variations in reactive power like a current source with pulse width modulation (PWM) control. The area "A" surrounded by the dotted line in Fig. 1 is made up of a group of IEGT single-phase bridge units (Fig. 2). Some harmonic filters are also connected to the load bus. These processes are handled by some cubicles shown in Fig. 3. Fig. 4 shows the main digital control board. The main part of the card is the high-performance microprocessor, which operates to improve the current waveform to compensate for the flicker. A power-electronics oriented digital processor PP7 is installed on the card and carries out ultra-high speed calculations.

The control card monitors the EAF current and analyzes the factors causing flicker, including fluctuations in the reactive current component and harmonic current components. The reactive components are further broken

978-1-4799-2706-7/14 $31.00 © 2014 IEEE

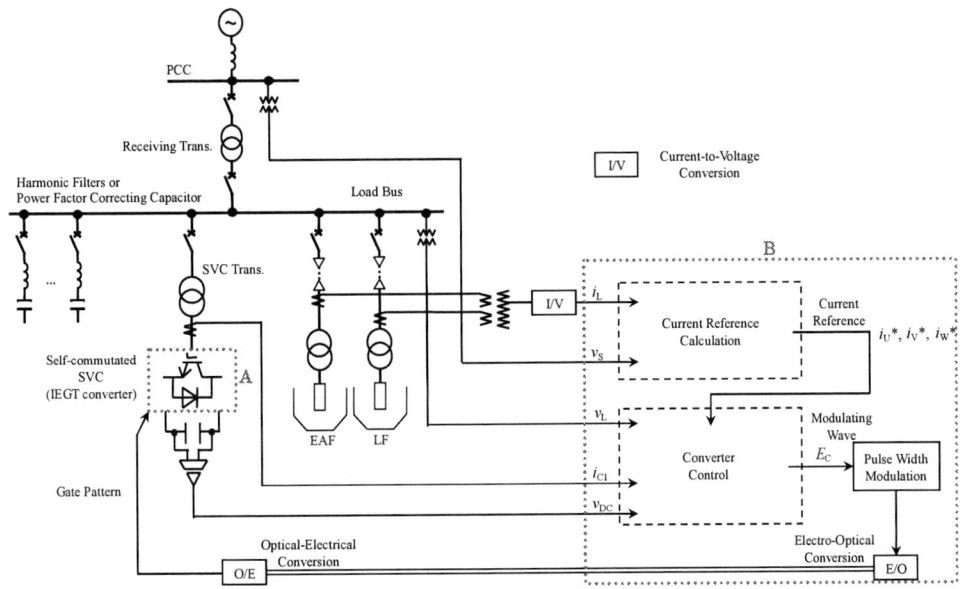

Fig. 1. Self-commutated SVC system for a steel plant.

Fig. 2. IEGT single-phase bridge unit.

Fig. 3. IEGT converter and 10 MVA-class control cubicle.

down into positive and negative sequence components which cause unbalance in the current or voltage.

Fig. 4. Digital control board mounted in SVC system.

To analyze the positive and negative sequence components, a special instantaneous p-q transformation function is employed, in addition to the conventional instantaneous d-q transformation function. Since the card controls the two sequence components, the current reference signals for the SVC output are different from one another for the three phases. Thus, the SVC compensates for rapid and large voltage flickers from the load.

B. Control Block Diagram for Current Reference Calculation

Fig. 5 shows control block diagram for the current reference calculation. Two of the control boards shown in Fig. 4 are required for this calculation. The PCC voltage phase (θ_S) is tracked by the PLL. The features of the calculation are as follows:

(1) Component analyzer: There are two kinds of component analyzers in the control block. The harmonic components (p_H, q_H) are analyzed from the instantaneous power signals (p_L, q_L) analyzed by the component analyzer 1. The component analyzer 2 analyzes the fluctuations of the positive sequence component (q_F) from q_L. Both

The 2014 International Power Electronics Conference

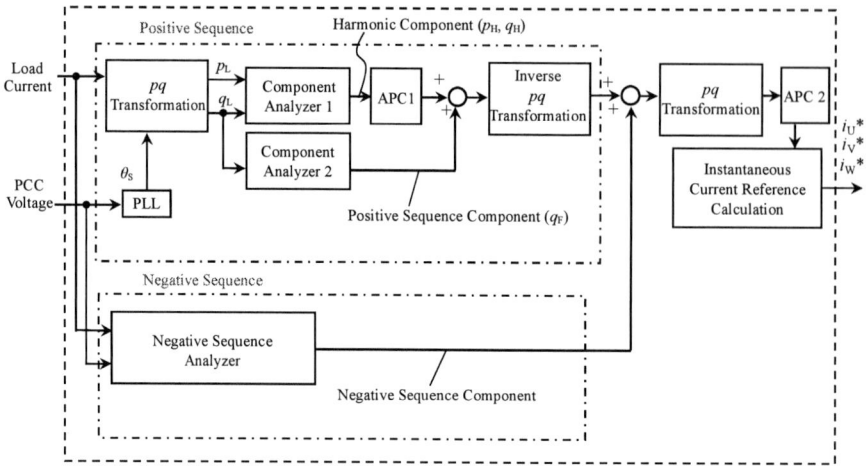

Fig. 5. Control block diagram of a self-commutated SVC for calculating current reference.

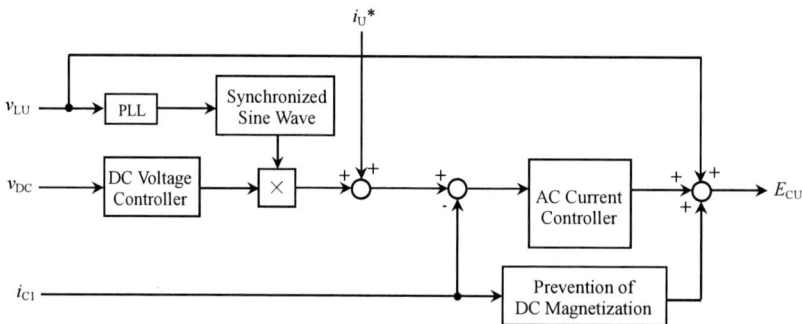

Fig. 6. Control block diagram of a self-commutated SVC for controlling the converter.
This block is shown for the U-phase. The V- and W-phases also have the same control block.

analyzers work as tuned filters for extracting varying components of the target frequencies. The negative sequence block in the second card analyzes the negative sequence component.

(2) APC (Advanced Power Controller): The APC automatically adjusts the gains of the control functions. The APC 1 adjusts the output gain of the harmonic components (p_H, q_H) to limit the SVC output range assigned for harmonic suppression in order to keep the output range for the fundamental frequency reactive power (q_F). After calculating the positive and negative sequence components, they are added and input to the p-q transformation block. The APC 2 reduces the total SVC output gain to keep the SVC output current within the rated capacity when the load reactive power exceeds the SVC rated capacity. In other words, the APC 2 keeps the SVC output currents always in proportion to the current output reference signals by keeping the SVC output current amplitudes within the SVC current ratings.

(3) Instantaneous Current Reference Calculation: The instantaneous current calculator contains an inverse p-q transformation in the block, and the reference

currents (i_U^*, i_V^*, i_W^*) for the three phases are generated in this block.

C. Control Block Diagram of Converter Control

The control block diagram for converter control of the U-phase is shown in Fig. 6. The V-phase and W-phase also have the same block. There are three main functions in this block, as follows:

(1) AC current controller: The AC current controller tunes the SVC output voltage so that the output current follows the reference value, which includes the three current reference values (i_U^*, i_V^*, i_W^*) and the output of the DC voltage controller.

(2) Prevention of DC magnetization: This function prevents DC magnetization of the SVCS transformer.

(3) DC voltage controller: The DC voltage controller regulates the dc voltage to a constant value by compensating for losses in the IEGT converter.

The reference voltage (E_C) is calculated from the AC current control output and the outputs from two other control functions. The PWM controller generates the gate pulse signals for the IEGT converter from the modulating wave (E_C).

III. APPLICATION TO ACTUAL STEEL PLANT SYSTEM

A. Real-Time Analysis Model Configuration

Fig. 7 shows a conceptual diagram and the set-up of the real-time simulation system with RT-LAB®, which has a parallel processing capability. As shown in Fig. 7(a), it is noted that this simulation system is capable of full-digital analysis, from obtaining plant data to simulation with the main circuit models. The overall evaluation system is reduced in size in comparison with a conventional analog simulation system [4]. First, the data from an actual plant is recorded by oscilloscope A. To analyze the voltage flicker of the plant, a large quantity of data is required, for example, on the order of several seconds. However, this amount of data is so large that it is difficult to use it in the MATLAB® system. Therefore, we developed a conversion tool for MATLAB® to solve the issue. The obtained data is converted to a MAT-file (almost 430 MB) with the conversion tool and is then loaded into the

memory of the RT-LAB® system. The system is controlled by a laptop PC, as shown in Fig. 7(b). Simulation with the real-time digital simulator is performed in the main circuit and control blocks. The processing core is optimally separated for parallel processing in this calculation block. Finally, the simulated data is sent to externally connected oscilloscope B and is observed as waveforms. The results can also be displayed on the laptop PC or recorded in real-time on the simulator hard disk.

Fig. 8 shows a conceptual analysis diagram and the conceptual main circuit model of the SVC. The site data is sent to the main circuit components shown in Fig. 8(a). Three sets of waveform data are sent to the main circuit model. They are the data for the system voltage, the EAF, and the LF. Two current sources are used as models for the furnaces. The current source models reproduced the current waveform based on the site data in the form of a load current waveform.

Fig. 8(b) shows the conceptual diagram of the 30 MVA class SVC with IEGTs. The SVC contains multiple-stage transformers and 72 switching devices. In particular, the

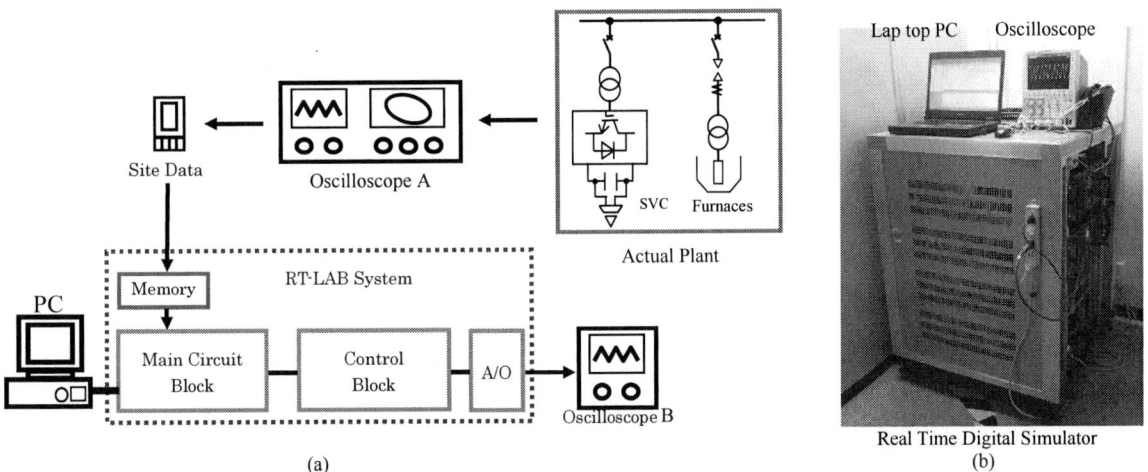

(a)

(b)

Fig. 7. Conceptual system diagram of real-time simulation system with RT-LAB.

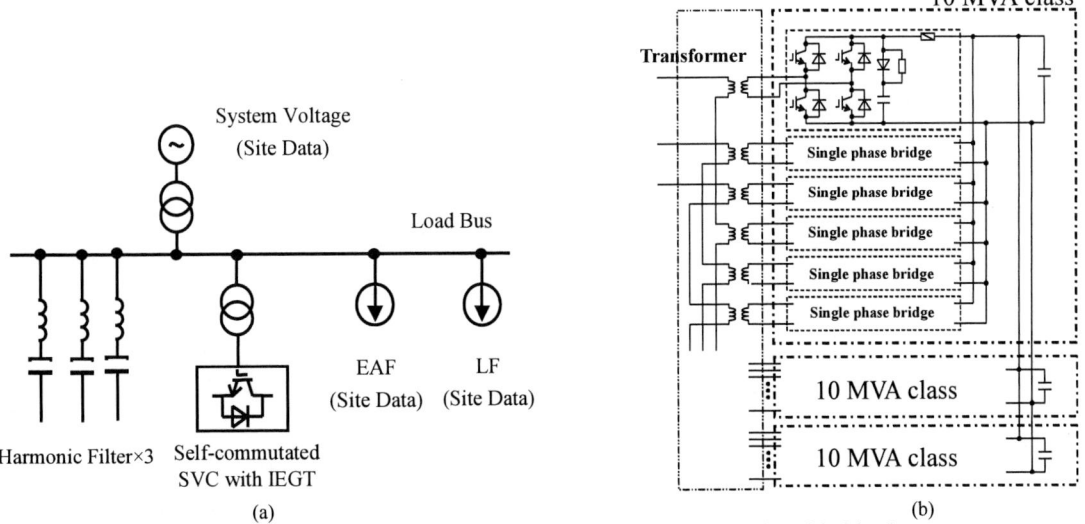

(a)

(b)

Fig. 8. Conceptual diagram of the main circuit block. (a) Simplified circuit model of the plant.
(b) 30 MVA class SVC system model in (a).

978-1-4799-2706-7/14 $31.00 © 2014 IEEE

transformer saturation characteristic is simulated in detail, including core hysteresis. A simulation time step of around 30 microseconds is achieved by using several processor cores of a standard multi-core PC. The converter is simulated with the detailed OPAL-RT converted model with real-time interpolation to achieve high accuracy even when firing pulses occur during the model sampling time [5]. It should be noted that co-development and real-time simulation of such a complicated system was the first challenge for users and OPAL-RT. In the simulation model, we used the rated capacity of the IEGT converter in an actual steel plant, which is around 60 MVA (30 MVA class × 2). This SVC system is activated with 33 kV at 60 Hz.

B. Comparison between actual waveforms and simulated waveforms

Fig. 9 represents the system voltage at the connection point, the load current, and the SVC compensation current. The total length of the actual plant data was 60 seconds, and the time range shown in the three figures is 0.2 s. In addition, Fig. 9 (a), (b), and the yellow line of (c) are the waveforms obtained at an actual plant. On the other hand, two SVC currents are included in Fig. 9(c) to allow comparison of the results of this analysis and the actual plant data. The results show that both waves are almost matched in spite of the complicated waveforms. Also, without the simulation system, it took over 5 minutes to calculate even a 1 second range. On the other hand, real-time calculation was possible with the constructed system. Therefore, we can say that the developed simulation system can reproduce the actual plant system with a self-commutated SVC with high-speed analysis thanks to the real-time simulation.

IV. CONCLUSIONS

We have developed a real-time analysis model of a self-commutated SVC system to suppress voltage flicker in a steel plant. Simulations using RT-LAB® were performed with a detailed converter model simulated with a time step of around 30 microseconds.

The main circuits, including multiple-stage transformers and converters, were modeled and activated in software. The simulation results showed that the real-time analysis can be applied to long-time actual steel plant data. The constructed model has the potential to be applied to many kinds of steel plant analysis for voltage flicker suppression at high speed.

Further evaluations, such as several cases of voltage flicker with the use of actual control devices and a digital flicker evaluation system [4], are required to confirm the effectiveness of the real-time SVC system simulator. Finally, it is important to evaluate the SVC system from various perspectives with real-time simulators.

REFERENCES

[1] A. D. Kolagar, A. Kiyoumarsi, M. Ataei, and R. A. Hooshmand, "Reactive power compensation in a steel industrial plant with several operating electric arc furnaces utilizing open-loop controlled TCR/FC compensators". *European Trans. on Electrical Power*, vol. 21, pp.824-838, 2011

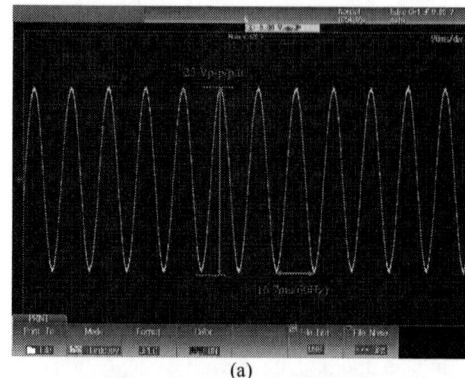

(a)

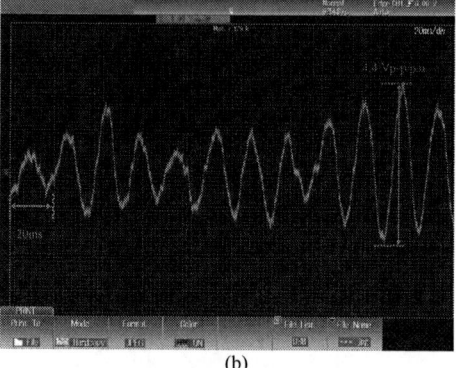

(b)

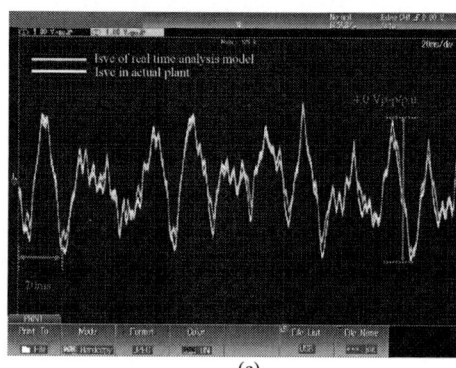

(c)

Fig. 9. Output waveforms of the real-time calculation. (a), (b), and (c) High voltage of power system (U-V phase), load current, and SVC currents, respectively, where the waveforms in (a) and (b) were obtained in an actual plant, and (c) contains both the simulated and actual plant waveforms.

[2] R. Grunbaum, T. Gustafsson, J.-P. Hasler, M. Osada, J. Rasmussen, and K. Thorburn, "STATCOM for voltage flicker suppression from a steel plant connected to a weak 66 kV grid", *Proc. in International Power Electronics Conference 2010 (IPEC 2010)*, pp.1773-1779, 2010.

[3] K. Usuki, F. Aoyama, and M. Hanamatsu, "Development of SVC control for suppressing voltage fluctuations", *Proc. in 2011 IEEE 8th International Conference on Power Electronics and ECCE Asia (ICPE & ECCE)*, pp.2073-2080, 2011.

[4] K. Hidese, T. Ishizuka, F. Aoyama, S. Ota, and K. Ogushi, "Development on 31.5 MVA STATCOM and digital evaluation tool for voltage flicker compensation", *Proc. in 7th Internatonal Conference on Power Electronics 2007 (ICPE'07)*, pp. 1230-1234, 2007

[5] M. Harakawa, H. Yamasaki, T. Nagano, S. Abourida, C. Dufour, and J. Bélanger," Real-Time Simulation of a Complete PMSM Drive at 10 μs Time Step", *Proc. in 2005 International Power Electronics Conference-Niigata (IPEC-Niigata 2005)*, pp.1007-1010, 2005.

Operational aspects and power architecture design for a microgrid to increase the use of renewable energy in wireless communication networks

Alexis Kwasinski
Department of Electrical and Computer Engineering
The University of Texas at Austin
Austin, Texas, USA

Andres Kwasinski
Department of Computer Engineering
Rochester Institute of Technology
Rochester, NY, USA

Abstract-**This extended abstract discusses the design of a power and control architecture for wireless communication networks that enables an increased use of renewable energy sources than in the conventional present approach. In the proposed approach a group of cell sites forms a dc microgrid in which their power consumption is coordinated with the power generated from renewable sources in order to manage local energy resources effectively. That is, load management is coordinated with power generation and use of local energy storage. In addition to environmental benefits, this proposed microgrid enhances power supply availability in cell sites and, thus, may improve wireless networks performance during natural disasters and other extreme events.**

I. INTRODUCTION

This paper discusses operation and power architecture design of a microgrid that allows increased use of renewable energy sources in wireless communication networks. Statistics about wireless networks energy consumption in [1] indicates that these networks account in between 1 % and 2 % of the total annual electric energy consumption in developed countries. These same statistics show that most of this consumption is observed at base stations—note: a base station is the telecommunication equipment that is placed at cell sites in order to establish a link with users' cell phones and process the exchange of data, voice communications and information between the users' cell phones and the wireless communication network. Even when 2 % of electric energy consumption seems to be a small percentage, a key difference from other electrical users is that, electrical power consumption from wireless networks originates in a handful of companies and, thus, improvements are simpler to achieve because of the few parties with responsibility to implement solutions. Moreover, electrical power consumption from wireless networks is increasing much faster than power consumption from other electricity users.

Figure 1 shows a conventional approach to power almost all of the base stations in wireless communication networks. In it, power is normally received from a large power grid which has a relatively low penetration of renewable energy sources. For example, in the U.S. and Japan total electrical power generation from renewable sources account to about 10 % of the total generated electrical power, and only a small number of countries exceed 20 % of electrical power generated from renewable sources, even when hydroelectric power is included. Hence, wireless networks represent a good opportunity to reduce the use of electricity use from non-renewable sources.

However, use of renewable energy sources in wireless communication networks present some challenges. One of these challenges is that renewable energy sources, such as photovoltaic (PV) arrays and wind turbines, have a variable output which require to have them paired with energy storage—e.g. batteries—or some other non-renewable source, such as a main grid or a generator driven by an internal combustion engine in order to meet a power availability of at least 5-nines as required in communication networks best practices. Moreover, renewable energy sources have a relatively large footprint compared with wireless networks loads. For example, a typical base station may consume about 1.5 kW/m²—consumed power over rack or cabinet power consumption—whereas a typical PV module may generate about 250 W/m²—a footprint 6 times larger than that of the load. Although relatively large footprints of PV modules may not be an issue when trying to collocate them at a cell site in a rural environment, large footprints of renewable energy sources is a significant limitation in urban and sub-urban areas where space is limited. Hence, the fact that use of renewable energy sources to power wireless communication networks is limited because of their variable output and their large footprint compared

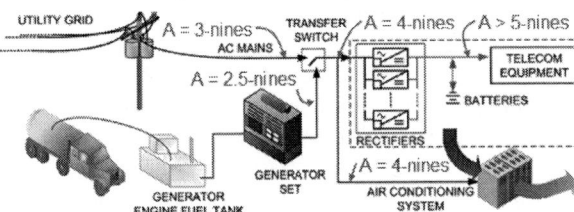

Fig. 1. Basic power architecture for a communication network site including typically observed power availabilities.

with their load provides a hint that use of renewable sources in this application could be increased if loads, energy storage and power output from renewable sources are managed in an integrated way.

This paper discusses a self constrained and independently controlled power system—i.e., a microgrid—that is formed by interconnecting a group of nearby cell sites in a common power distribution architecture. By allowing an increased use of renewable sources, this microgrid creates a *sustainable wireless area* (SWA). The discussion in the next section summarizes the basic power architecture design and integrated operation and management approach in order to increase penetration of renewable energy sources in wireless communication networks.

II. SYSTEM ARCHITECTURE

Figure 2 shows an overview of the proposed architecture that powers a cluster of cellular base stations using a microgrid configuration. The architecture divides the cellular network in clusters of a few base stations (the typical SWA size to be considered is seven base stations). Power for the base stations is obtained primarily from PV modules and wind turbines. As the electrical scheme in Fig. 3 represents, the cell sites are typically equipped with batteries. While in conventional power plants, such as the one in Fig. 1, these batteries are used to power a given cell site during a main power grid outage, in the proposed microgrid the batteries are used to manage the variable power output of the renewable sources. All three components, the PV modules, the wind turbines and the batteries are located where there is sufficient space for them, which may be either located at a centralized location or distributed on the different cell sites, but in all cases they are shared using a microgrid configuration by all the cell sites in the SWA. Hence, stored energy or generated power could be transferred from one cell site to another. Moreover, since power consumption of a base station depends on the traffic and the power of the emitted radio signal through the antenna. Since both of these variables, can be controlled within a reasonable margin without significantly affecting service quality (e.g. power output of one antenna can be reduced so the other surrounding antennas can take calls of the edges of the cell site with the antenna operating at reduced power) the load may partially act as virtual power generation.

In the proposed approach it is assumed that cell sites within an SWA are located nearby. Such assumption of geographic proximity is consistent with having the SWA in an urban or suburban area where the large footprint of renewable energy sources makes it difficult to collocate them in each site. Taking advantage their geographical proximity, base stations within the same SWA are electrically interconnected with dc power lines creating a dc microgrid. The use of dc for the microgrid power distribution architecture provides all the advantages in terms of efficiency, availability and others mentioned in [2]. One of such advantages is that it is possible to use

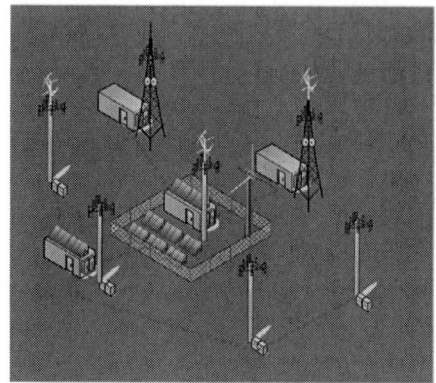

Fig. 2. SWA example. PV arrays are placed where there is sufficient space, which in this example is primarily in a central base station. Microcells and other cell sites with insufficient space for PV arrays, are powered through the microgrid.

dc-fed air conditioning units significantly improving cooling efficiency [1] [3] and avoiding the issues with power availability found in conventional power architectures in which the air conditioning cannot be powered by the batteries [3].

Each SWA may or may not have one or more grid ties. When present, these ties allow the grid to be used as a secondary power source in order to enhance the SWA power availability or to provide more flexibility, for example, to optimize operational costs. Since the power distribution system in the SWA is based on dc, each point of common coupling (PCC) with the main grid is realized with a bidirectional inverter/rectifier. Thus, at times when the microgrid has excess power generation capacity, the extra power can be injected into the power grid so in places where the local utility has net-metering programs, operational cost can be reduced

A preliminary study presented in [4] has shown that the cellular base stations can be powered from renewable sources a significant portion of time. Moreover, work presented in [5] has shown that microgrids can achieve an availability in the order of 5-nines as required in most communication network operators best practices. In order

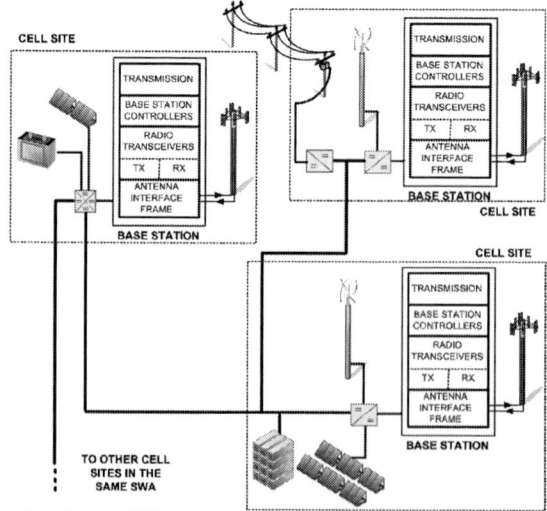

Fig. 3. Simplified electrical schematic of part of an SWA.

to reduce the amount of energy storage necessary to achieve this power availability level, [6] indicated that it is recommended to diversify power sources. Based on the conclusions of this reference, the SWA concept enhances wireless networks cell sites survivability and resiliency to natural disasters [6].

III. OPERATION AND CONTROL

Figure 4 shows the general architecture of the SWA electric controller. As this figure shows, the proposed controller has a hierarchical structure in which the top level is used to optimize SWA operation and to coordinate all power sources, energy storage devices and loads in the SWA. At a bottom level a local autonomous controller is in charge of acting on its corresponding base station traffic and/or local power generation or energy storage devices based on the commands received from the high-level controller. In case of failure in the communication link with the high-level controller or in the communication link to it, the local controllers are still able to manage local resources and load at their corresponding base stations but at a sub-optimal level.

A. High-level Controller

Physically, it is expected that all cell sites would have a module in their operations platform able to perform this top layer function but only one of them in the SWA will be configured to perform this function. In the other cell sites, this function remains "dormant" until it is activated in case it is desirable. In this top level, the optimization algorithm may possible follow different objectives, such as maximize availability or quality of service, or reduce operational costs. Hence, this central top-level controller process general state information data, such as currents and voltage levels, along the SWA and solves the optimization problem that is part of the operational goal and sends to the converters interfacing the power generation sources, such as a PV array converter, the necessary commands—e.g. desired power output—that are the result of solving the optimization problem. This central controller is also in charge of commanding the base stations the necessary traffic limits and antenna power output). Although the optimization portion of the controller is implemented with a centralized controller,

the system is still able to operate sub-optimally without central controller. This approach is still consistent with high availability goals in wireless communication networks as traffic routing is centralized in central offices.

For a given traffic level, base station nominal load $P_{n,BS}$ can be represented based on two components: one base load and one that depends on the traffic. That is, the nominal load at a given traffic level is [7]

$$P_{n,BS}(v) = P_B + vP_T \qquad (1)$$

where P_B is the base load, P_T is a constant term representing the traffic-dependent power consumption of the base station and v is a base station utilization factor that represents how much the communication resources (e.g. channels / spectrum) is used in the base station. That is, the traffic B will directly depend on v which is a factor that varies between 0 and 1, is

$$B \propto v \qquad (2)$$

As the traffic changes, so is v and, in turn, $P_{n,BS}$ changes.

Still, base station load can be further modified by managing its quality of service through traffic shaping. This modification can be explained in the following way: in modern wireless communication networks all information (data and voice signals) are transmitted with digital signals. Such signals will have associated a required number of bits that are transmitted per unit time, i.e., a transmission rate associated to a normal quality of service level (e.g. hear a voice signal with a given clarity level). However, in many instances it is possible by processing those signals to reduce such transmission rate without affecting the quality of level in a significant way (e.g. the pitch of a speaker may become a little less distinct, or some data may take a little longer to download, or some video may lose a little definition). Hence, (1) can be modified by adding a traffic shaping factor σ that takes values between 0 and 1 and represent how much traffic is reduced from the nominal levels. That is, the actual base station load is now

$$P_{BS}(v, \pi) = P_B + \sigma v P_T \qquad (3)$$

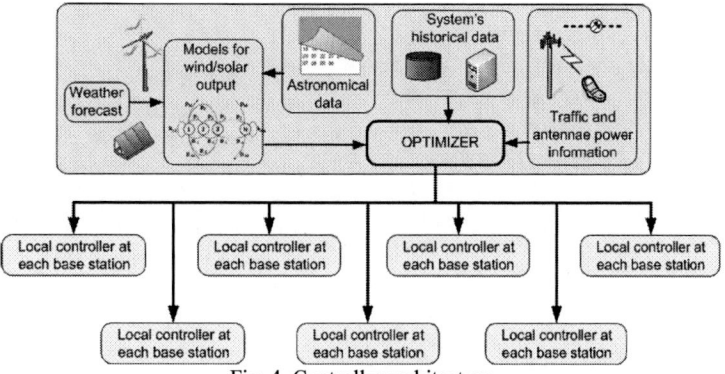

Fig. 4. Controller architecture.

The main function of the high-level controller is to determine an optimal value for σ based on the optimization objective, weather forecast, state of charge of batteries, power generation resources, time of the day, service quality targets, and several other factors. Since v is a random variable that follows the random traffic load at a base station, once σ is determined, each base station is a random load that is related to the value of v.

In addition to manage loads profiles, the high-level controller generates control signals for the desired output of each of the local power generation units, including PV arrays and wind generators. Managing power output of local dispatchable power generators, such as diesel generators, microturbines and fuel cells is relatively simple. However, Since in the case of PV arrays and wind generators, their profile is partially stochastic, one approach for realizing the optimization algorithm that is the basis of the central controller is to use the Markov chain approach presented in [8] in order to model the state of charge of the energy storage components (usually batteries) in the SWA microgrid. A representation of this Markov Chain is shown in Fig. 5. In it, each state represents a given state of charge of an energy storage device—typically, batteries—associated to the partially stochastic generator. The energy difference between adjacent states equals $T_t\Delta$ where T_t is the time between consecutive transitions among states. As the batteries charge and discharge depending on algebraic difference, $m\Delta$, between total generated power and total load, the batteries will transition among states in the Markov Chain in Fig. 5. However, since both the load and the power output of renewable sources is at least partially stochastic, the value of $m\Delta$ is also stochastic. Thus, transition among states are characterized by a given probability. Therefore, the battery state of charge, i.e. the state of the system represented in Fig. 5, is not known with certainty but rather inferred with a certain probability π_i

The algorithm to determine the probabilities π_i is represented in Fig. 6. First, solar and/or wind data is used in combination with astronomical and weather forecast information in order to realize a power generation probability distribution of the expect power output from PV arrays or wind generators. Likewise network historical data and traffic information is used to determine a probability distribution of expected load. The difference between expected power generation and expected load yields a probability distribution of the expected power into or out from the batteries $m\Delta$ in each state transition interval T_t. This distribution is, then, used to determine the transition probabilities p_i among the states of the Markov Chain in Fig. 5 with energy differences $T_t\Delta m$, where m can take integer values in the closed interval [-N, N] where N is the number of states representing the battery state of charge. Once the transition probabilities are known it is possible to calculate the limiting probabilities π_i of the battery being in each of the N state of charge in Fig. 5. This information can be used as exemplified in Fig. 6 and

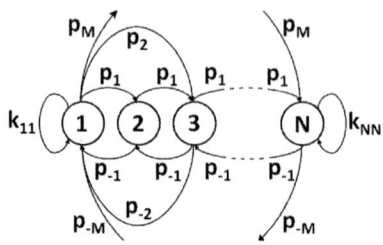

Fig. 5. Markov model for a system combining energy storage and renewable energy sources.

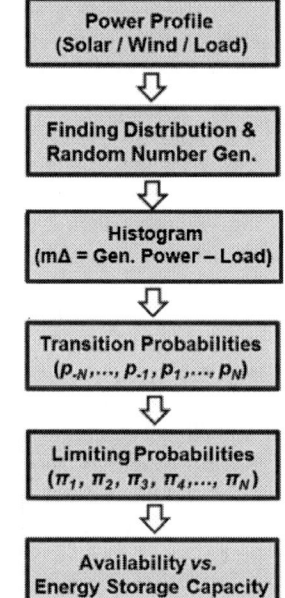

Fig. 6. Algorithm used to determine limiting probabilities in the Markov model in Fig. 5 and its application for planning purposes to determine the optimal size of energy storage in the SWA with respect to availability targets.

described in [8] for planning purposes in order to determine the necessary total battery capacity in order to obtain a total power availability target or it can be used as shown in Fig. 7 based on the application in Fig. 4 in order to shape the traffic through the factor σ in (3) and/or control the various power sources in the SWA (renewables and other sources) to desired optimal values (e.g. if a fuel cell is present and is controlled to its maximum power point). When such optimization is performed, it creates a iterative algorithm in which since the load and power generation profiles are adjusted, then the histogram of the difference between generated power and load is adjusted (because the probability distributions for loads and renewable energy sources are modified). This adjustment yields new transition probabilities values and, in turn, yields new limiting probability π_i values. Iterations are repeated until a given variable used as an optimization goal falls within a given tolerance. Hence, this algorithm realizes an approach for real time control to achieve a given objective based on managing in an integrated way load, generated power and batteries state of charge. In a typical application and since battery autonomy have a significant impact on the SWA power availability [6], the optimization goal could be to obtain probability values for state N or states close to N (i.e. the states representing the battery in full state of charge or close to full charge) as high as possible.

B. Local Controllers

The second level of the controller is implemented by local agents residing in the power electronic interfaces of sources and (in special cases) energy storage devices—because it is unlikely that in a communications network, the batteries will not be directly connected to the main dc bus. Figure 8 shows a representation of the SWA with the local controllers. These agents serve several purposes:

- Control sources or energy storage devices so the optimization commands generated by the top level of the controller are satisfied. In the proposed approach this is accomplished with a droop controller as suggested in [9] which in the case of source interfaces is represented by

$$\hat{E}_{S,i} = \hat{E}^*_{S,i} - m_i (P_i - P_i^*) \qquad (4)$$

where $\hat{E}_{S,i}$ is the internal voltage set point for a given source i, m_i is the slope of the linear droop law, and P_i^* is a reference power corresponding to the reference voltage $\hat{E}^*_{S,i}$. This reference voltage, P_i^*, and m_i are the parameters that define the droop line. In (4) P_i^* is also an input variable equal to the desired dispatched power from source #i [10] [11] as indicated by the optimizer whereas P_i is the actual measured power output of the source. Once $\hat{E}_{S,i}$ has been determined based on (4), the controller adds a virtual voltage drop caused by a virtual resistance $R_{V,i}$ [12] that serves various purposes, including to ensure that the microgrid distribution grid is presented to sources as mostly resistive, so

$$E_{R,i} = \hat{E}_{S,i} - I_{S,i} R_{V,i} \qquad (5)$$

This approach can be used for energy storage interfaces—e.g. if flywheels are used—or when the source interfaces or multiple-input multiple-output bidirectional converters are added in the distribution grid as active power distribution nodes (APDNs) [13].

- In the case of source interfaces and possibly APDNs, the controller may need to regulate the main bus voltage, for example, in order to keep batteries directly connected to the main bus floating. This goal is achieved with a secondary droop loop [12] given by

$$\hat{E}_{R,i} = \delta\hat{E}^* + \hat{E}^*_{S,i} - I_{S,i} R_{V,i} \qquad (6)$$

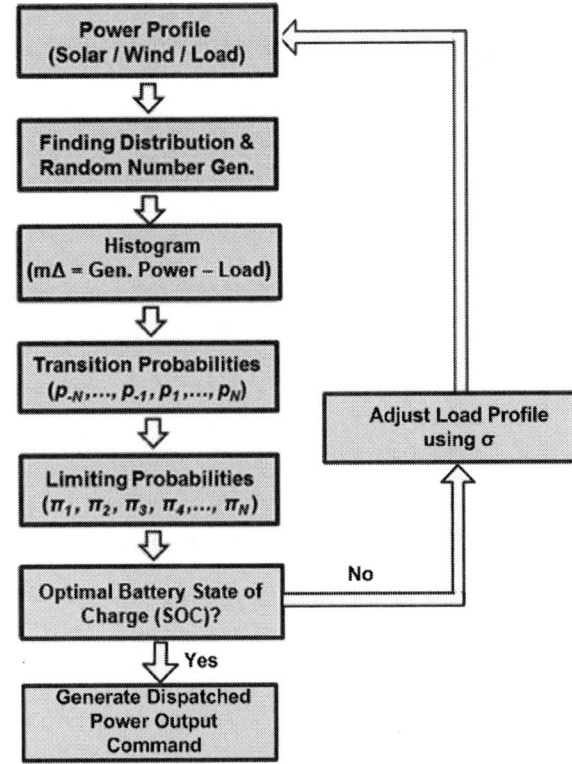

Fig.7. Algorithm used to determine limiting probabilities in the Markov model in Fig. 5 and its application optimizing energy storage resources utilization through real time controls.

where

$$\delta\hat{E}^* = K_i \int (\hat{U}^*_{MG} - v_B) dt \qquad (7)$$

and v_B is the measured main bus voltage, $\hat{U}^*_{MG}$ is the desired main bus voltage and K_i is a gain small enough so that this regulation action is slow compared with the inner control loop dynamics. This secondary droop control could not be necessary if the there is no need to regulate the main bus voltage—e.g., if there is no batteries directly connected to the main bus—or if voltage drops that are the result of the droop controller can be accommodated.

- In the particular case of grid interface, the local controller would regulate current flow to and from the grid as a tertiary controller [12]. Hence, (7) is replaced by

$$\delta\hat{E}^* = K_i \int (\hat{I}^*_G - I_g) dt \qquad (8)$$

where $\hat{I}^*_G$ is the reference absorbed current from the grid and I_g is the measured current on the grid side. This current can also be measured and control on the grid side and add functions, like reactive power control, but this topic is out of the scope of this paper.

- Address issues with constant-power loads (CPLs). It is important to notice that loads in the SWA are expected to show constant-power behavior which leads to stability issues. Approaches to address these issues were presented in [14]. Additionally, the virtual resistance in (5) also contributes to improve stability characteristics and prevent CPL issues as also explained in [14].

IV. CONCLUSIONS

This paper has explained an approach to expand the use of renewable energy sources, such as PVs and wind, in wireless communication networks by integrating power generation and base stations traffic/service control within a microgrid called SWA. Issues found in conventional approaches to integrate renewable sources in wireless communication networks include large footprint of PV arrays and wind generators and the need

for very large energy storage capacity in order to address the variable output profile of wind and solar resources.

Integrated energy management and traffic/service control is achieved with a hierarchical controller which at its top level includes a central controller that optimizes the operation of the SWA in order to enhance the utilization of renewable energy sources. The central controller produces power generation dispatch signals and traffic shaping commands that are send to the individual base stations or power generation centers within the SWA to local controllers. These local controllers have droop control schemes able to manage energy in their domain even when the high-level controller or communication links fail. This autonomous operation mode is, however, suboptimal. This paper has also presented the algorithm used in the central controller and described the droop control laws in the local controllers. These local control laws may also include secondary or tertiary controllers to regulate bus voltages

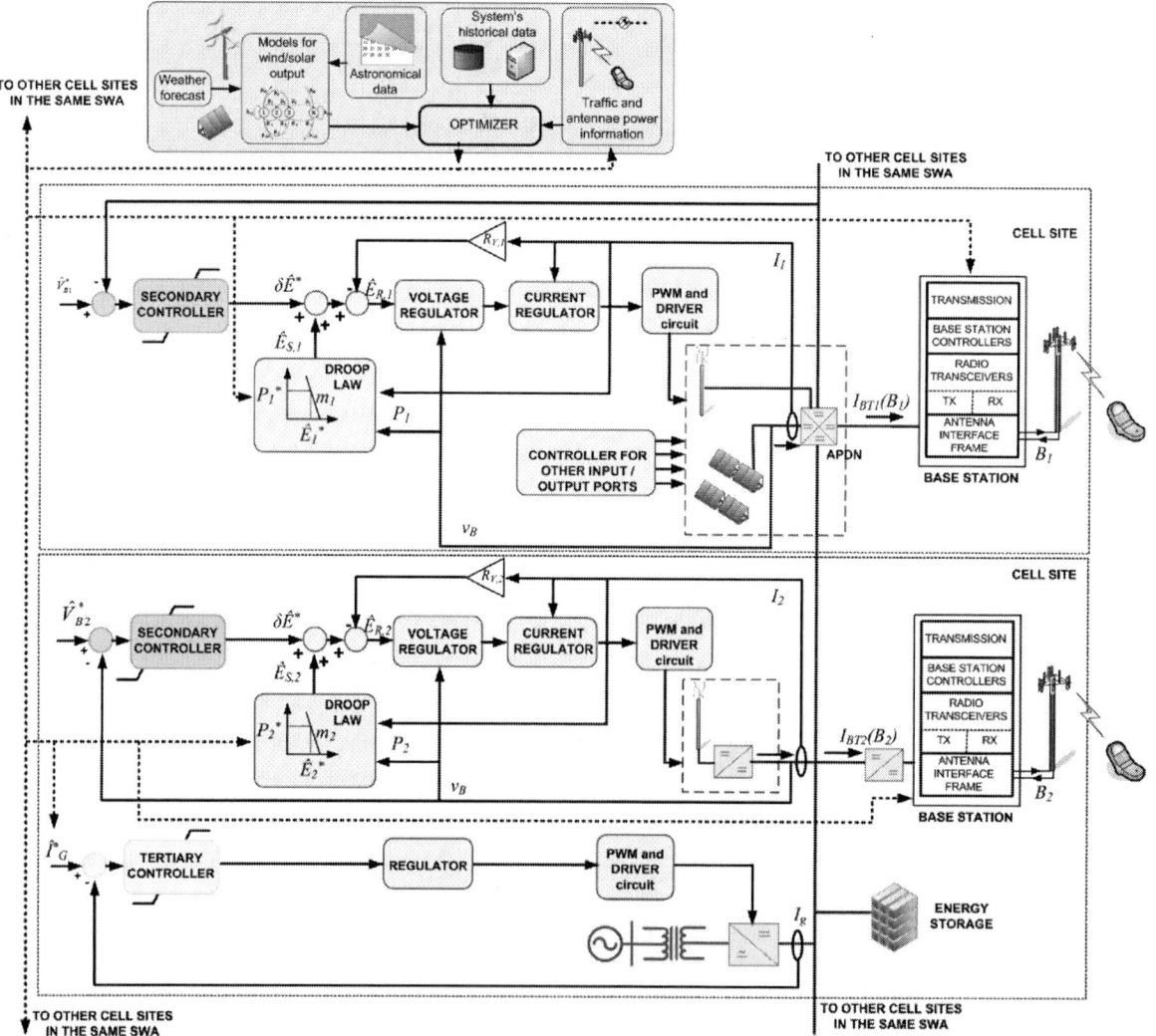

Fig. 8. Local controllers structure integrated with the central controller (optimizer). Notice that the current into a base station I_{BT} is a function of the traffic B.

or manage interactions with a bulk power grid.

V. ACKNOWLEDGMENTS

The work presented in this paper has been supported by US National Science Foundation (NSF) under a CyberSEES award #1331788.

REFERENCES

[1] S. Roy "Energy Logic for Telecommunications" Emerson Network Power White Paper ES-113, September 2008.

[2] A. Kwasinski, "Evaluation of dc Voltage Levels for Integrated Information Technology and Telecom Power Architectures," International Telecommunication Energy Special Conference, 2009, 7 pages.

[3] A. Kwasinski, "Analysis of Electric Power Architectures to Improve Availability and Efficiency of Air Conditioning Systems," 2008 International Telecommunications Energy Conference (INTELEC), vol. 10, no. 2, pp. 1-8.

[4] A. Kwasinski and A. Kwasinski, "Architecture for Green Mobile Network Powered from Renewable Energy in Microgrid Configuration," in Proc. IEEE Wireless Communications and Networking Conference, Shanghai, China, April, 2013, pp. 1525-3511.

[5] A. Kwasinski and P.T. Krein, "Optimal Configuration Analysis of a Microgrid-Based Telecom Power System," in Proc. 2006 International Telecommunications Energy Conference (INTELEC), pp. 602-609.

[6] A. Kwasinski, V. Krishnamurthy, J. Song, and R. Sharma, "Availability Evaluation of Micro-Grids for Resistant Power Supply During Natural Disasters," IEEE Transactions on Smart Grid, vol. 3, no. 4, pp. 2007 - 2018, Dec. 2012.

[7] Auer, G.; Blume, O.; Giannini, V.; Gódor, I.; Imran, M.; et al.; ``Energy Efficiency Analysis of the Reference Systems, Areas of Improvements and Target Breakdown", Energy Aware Radio and neTwork tecHnologies (EARTH) Project deliverable D2.3, document INFSO-ICT-247733 EARTH, (2010).

[8] J. Song, V. Krishnamurthy, A. Kwasinski, and R. Sharma, "Development of a Markov Chain Based Energy Storage Model for Power Supply Availability Assessment of Photovoltaic Generation Plants," IEEE Transactions on Sustainable Energy, vol. 4, issue 2, pp. 491 – 500, April 2013.

[9] T. L. Vandoorn, B. Meersman, J. D. M. De Kooning, and L. Vandevelde, "Analogy Between Conventional Grid Control and Islanded Microgrid Control Based on a Global DC-Link Voltage Droop." IEEE Transactions on Power Delivery, vol. 27, no. 3, pp. 1405 – 1414, July 2012.

[10] J. J. Justo, F. Mwasilu, J. Lee, and J-W. Jung, "AC-microgrids versus DC-microgrids with distributed energy resources: A review." Renewable and Sustainable Energy Reviews, vol. 24, pp. 387–405, Aug. 2013.

[11] J. C. Vasquez, J. M Guerrero, A. Luna, P. Rodriguez, and R. Teodorescu, "Adaptive Droop Control Applied to Voltage-Source Inverters Operating in Grid-Connected and Islanded Modes." IEEE Transactions on Industrial Electronics, vol. 56, no. 10, pp. 4088- 4096, Oct. 2009.

[12] J. M. Guerrero, J. C. Vasquez, J. Matas, L. Garcia de Vicuña, and M. Castilla, "Hierarchical Control of Droop-Controlled AC and DC Microgrids—A General Approach toward Standardization." IEEE Transactions Industrial Electronics, vol. 58, no. 1, pp. 158 – 172, Jan. 2011.

[13] A. Kwasinski, "Advanced Power Electronics Enabled Distribution Architectures: Design, Operation, and Control," in Proc. 2011 IEEE International Conference on Power Electronics - ECCE Asia, pp. 1484-1491

[14] A. Kwasinski and C. N. Onwuchekwa, "Dynamic Behavior and Stabilization of DC Microgrids with Instantaneous Constant-Power Loads," IEEE Transactions on Power Electronics, vol. 26, no. 3, pp. 822-834, March 2011.

P+ Multiple Resonant Control for Output Voltage Regulation of Microgrid with Unbalanced and Nonlinear Loads

Kyungbae Lim, Jaeho Choi
School of Electrical Engineering
Chungbuk National University
Cheongju, Republic of Korea
choi@cbnu.ac.kr

Juyoung Jang, Junghum Lee, Jaesig Kim
Fuel Cell Technology Group
POSCOenergy Co. Ltd.
Pohang, Republic of Korea
Zangju0@poscoenergy.com

Abstract—This paper describes the output voltage regulation for the islanded mode of microgrid based on inverter with LCL filter. When a microgrid is transferred to the islanded mode, it has to supply the full local load demand without supporting of grid and to operate as a voltage source with an appropriate voltage regulation. Here, if the local loads are unbalanced and nonlinear, then it makes the output voltage distorted and so the performance of power quality will be degraded. In this paper, P+ multiple resonant controller which makes the control gain very high on the specific frequency range has been selected instead of PI controller for the compensation of voltage distortion due to the unbalanced and nonlinear load. The effect of voltage regulation method using P+ multiple resonant controller under the nonlinear and unbalanced condition has been verified through the PSIM simulation results. And then it also has been compared with the conventional PI and the advanced PI for harmonic compensation.

Keywords— P+ multiple resonant controller, unbalanced load, nonlinear load, harmonic compensation

I. INTRODUCTION

The 'Microgrid' which is an integration of RES (renewable energy sources) such as the photovoltaic or wind power generator and fuel cell with the engine generator, batteries, and, etc., has been issued due to its flexibility and capability of reliable power supply. Normally, the microgrid is defined as a sub power source connected with the main grid. Hence, grid-connected inverters have to be controlled the output current within the allowable range of harmonic current components. For this, L or LCL filter used be installed at the output side of inverter. But when the microgrid operates as an islanded mode, this output filter works as an internal output impedance of a voltage source and the voltage waveform is distorted by the load current harmonic components under the nonlinear and unbalanced load conditions.

Hence, system operators need the controller which satisfies not only fast and robust system dynamics but also effective harmonic compensation under the above condition. There have been many researches on this issue. The double loop controller with an outer PI voltage loop and an inner P or PI current loop was introduced for the operation of parallel UPS [1] and the other double loop

controller with an outer P+ resonant voltage loop and an inner proportional current control was applied for a three-phase four-wire ac-link inverter [2]. In [3], the inverter parallel operation for sharing the non-linear load demand has been realized by using the droop characteristic and double loop P+ multiple resonant control.

Hence, in this paper, P+ multiple resonant controller [2,4,5] has been selected for the regulation of the voltage control and analyzed the voltage regulation performance to meet the fast dynamics and the harmonics compensation under the nonlinear and unbalanced conditions. And also the advantage of P+ multiple resonant control has been analyzed by comparing with the harmonic compensation method based on a conventional PI or an advanced PI control based on the synchronously rotating reference frame which have more complex implementation than P+ multiple resonant control for harmonic compensation [6].

Aforementioned, an inverter with LCL filter is defined as a current source with sub power sources in grid connected mode and it needs current control for output current regulation. Here, both the output current and the output voltage become distorted due to the presence of non-linear loads when it is connected to the LC filter. In this paper, P+ multiple resonant control which is used for the voltage control in islanded mode has been extended to

The cascaded current-voltage control method in grid connected mode which has an outer current loop and an inner LC filter output voltage loop to solve the distortion of the output current and voltage waveforms when non-linear loads are connected at the output of LC filter [7].

Finally, the performances of P+ multiple resonant voltage control has been verified with the comparison of the performances of PI voltage control and advanced PI voltage control for harmonic compensation. And also the related analyses and results under the cascaded current-voltage control in grid connected mode have been shown as simulation results. All simulations have been done with PSIM simulator.

II. DESIGN OF VOLTAGE CONTROL

The voltage control of three-phase inverter with a conventional PI is implemented under the synchronously rotating dq-reference frame. The basic transfer function of

The 2014 International Power Electronics Conference

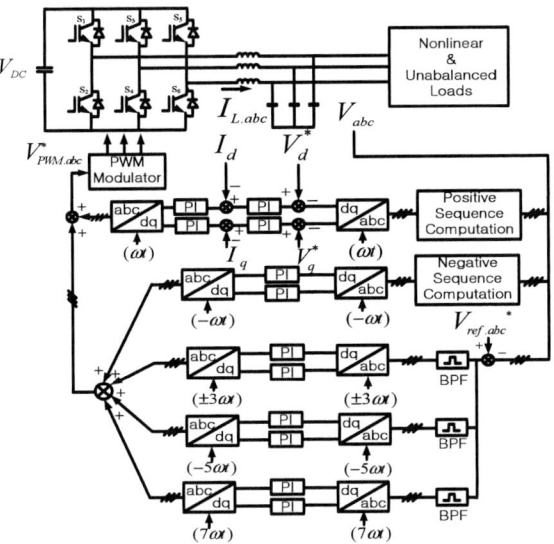

Fig. 1. Control scheme for PI control with harmonic compensation.

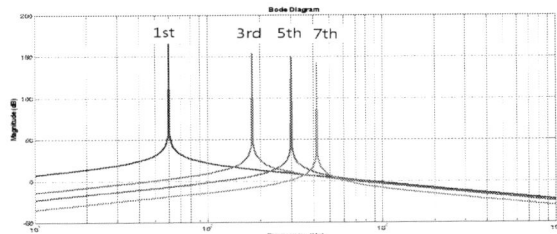

Fig. 2. Bode plot of P+ multiple resonant voltage controller.

conventional PI controller can be written as following:

$$C_{PI} = K_P + \frac{K_i}{s} \qquad (1)$$

This PI controller has been a good solution for regulating a sinusoidal waveform under the balanced three-phase system. But it is not proper when the microgrid is connected with unbalanced or nonlinear loads. Hence, the advanced PI control method was proposed for the harmonic compensation under the unbalanced and nonlinear loads as shown in figure 1 [6]. This method is quite complicate because it needs not only the distributed control for both positive and negative sequence fundamental components but also a string of controllers and digital filters proportional to the specific harmonic components to be compensated. On this, the equation for the compensation of positive and negative sequences in figure 1 which includes digital filters are as following:

$$\begin{bmatrix} v_a^{pn} \\ v_b^{pn} \\ v_c^{pn} \end{bmatrix} = \frac{1}{3}\begin{bmatrix} 1 & -1/2 & -1/2 \\ -1/2 & 1 & -1/2 \\ -1/2 & -1/2 & 1 \end{bmatrix}\begin{bmatrix} v_{ca} \\ v_{cb} \\ v_{cc} \end{bmatrix}$$

$$\mp \frac{1}{j2\sqrt{3}}\begin{bmatrix} 0 & 1 & -1 \\ -1 & 0 & 1 \\ 1 & -1 & 0 \end{bmatrix}\begin{bmatrix} v_{ca} \\ v_{cb} \\ v_{cc} \end{bmatrix} \qquad (2)$$

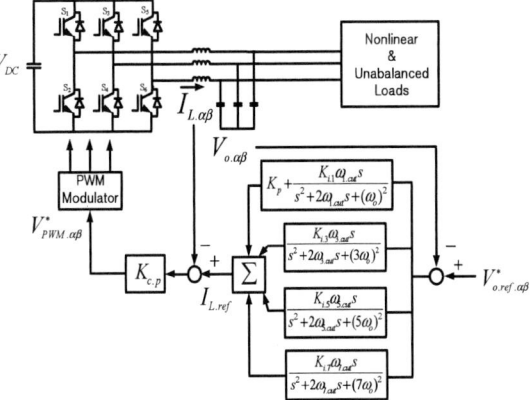

Fig. 3. Control scheme for P+ multiple resonant control in islanded mode.

Moreover the accurate phase information of distorted output voltage due to the unbalanced and nonlinear loads is also necessary for transforming from abc to dq.

Hence, this paper has tried to improve the harmonic compensation capability and to reduce the complexity of implementation by using the P+ multiple resonant control which can be used under the stationary frame [8]. The transfer function of P+ multiple resonant control used in this paper for compensating unbalance and voltage harmonic is as following:

$$C_{PR} = K_P + \frac{K_i \omega_{1.cut} s}{s^2 + 2\omega_{1.cut} s + \omega_o^2}$$

$$+ \sum_{h=3,5,7th} K_{ih} \frac{\omega_{h.cut} s}{s^2 + 2\omega_{h.cut} s + \omega_h^2} \qquad (3)$$

Figure 2 represents the bode diagram of each resonant term in (3). As shown in figure2, the controller has extremely high gain on 1, 3, 5, 7 order harmonic frequency ranges to compensate the harmonics. In other words, the compensation of 1st negative sequence, 3rd zero sequence, 5th and 7th (6n±1) harmonic components can be compensated by (3). (Note that PI control needs each positive and negative sequence controller for the harmonic compensation, but P+R control needs only one controller for regulating both positive and negative sequence components [8].)

Figure 3 represents the overall control scheme of the proposed method. The controller is composed of an outer voltage loop (P+ multiple resonant control) for reducing the steady state error and an inner current loop ($K_{C.P}$) for enhancing system dynamics [9,10]. Figures 4 and 5 are block diagrams of inner current loop and outer voltage loop, respectively. According to figure 5, the voltage loop characteristic equation can be obtained as (4). It is only for the fundamental component voltage loop. As shown in (4), all control gain values and system parameters can affect the system model. Figure 6 is the bode plot of voltage controller with harmonic compensation terms. As shown in figure 6, it has very high gains at the specific harmonic frequency orders of output current and its phase

978-1-4799-2706-7/14 $31.00 © 2014 IEEE

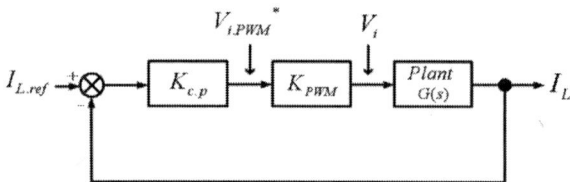

Fig. 4. Block diagram of inner current control loop.

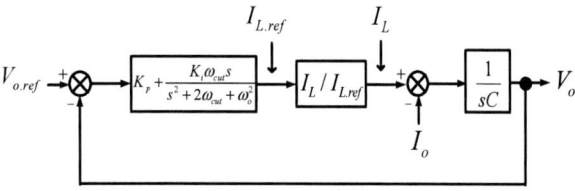

Fig. 5. Block diagram of outer voltage control loop.

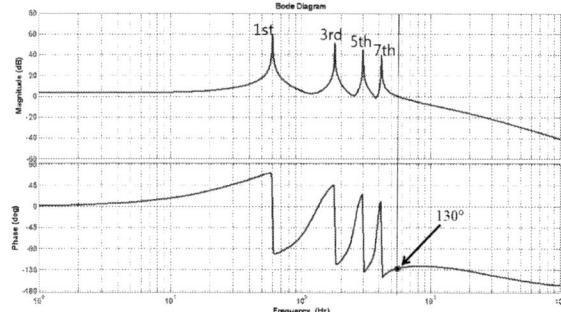

Fig. 6. Bode plot of voltage controller with harmonic compensation

margin has been designed as $50°$ to satisfy the system stability and good control performances.

III. CASCADED CURRENT-VOLTAGE CONTROL FOR GRID CONNECTED MODE OPERATION

Aforementioned in section 1, the microgrid is defined as a current source in the grid-connected mode and a voltage source in the islanded mode, respectively. When non-linear and unbalanced local loads are connected at PCC (output filter capacitor side), then it has to guarantee the low THD of output current and PCC voltage. But it cannot guarantee the low THD of PCC voltage under the limited range due to the absence of voltage control. Considering the seamless transfer between the grid-connected mode and the islanded mode, both harmonic compensations of current and voltage is implemented in this section by using the cascaded current-voltage control. This kind of hybrid control of voltage and current has been used for the distributed generation [7,11]. The inverter is controlled as a current source by one set of a controller in the grid connected mode, while as a voltage source by the other set of controller in the islanded mode. In this paper, P+ multiple resonant controllers are used for the cascaded current-voltage controller as shown in figure 7. When the microgrid operates in the islanded

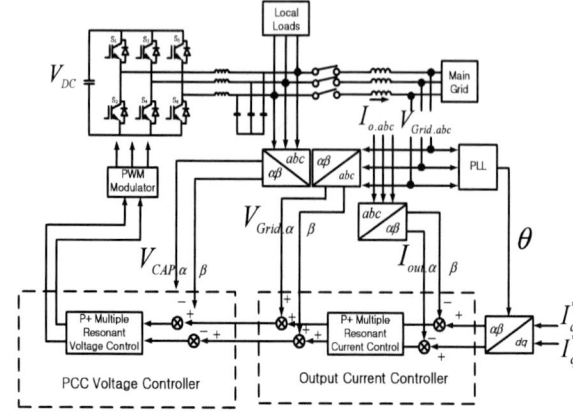

Fig. 7. Control Scheme for P+ multiple resonant control for cascaded current-voltage control method.

mode due to the grid fault, the same controller can be used with zero reference current and opening the static transfer switches. Hence, the seamless transfer between the operation modes is possible by this cascaded current-voltage control. Related analyses and results under the cascaded current-voltage control in both grid connected mode and islanded mode have been shown with the transient response in simulation results.

IV. SIMULATION RESULTS

The overall system configuration and parameters for PI and P+ R controls in figures 1 and 3 are shown in Table 1. The performance of three controllers, conventional PI, advanced PI, and P+ multiple resonant, are evaluated and compared with the simulation results. The switching frequency is 4 kHz and the sampling frequency is 8 kHz in this simulation. The details of the load used for simulation studies are shown in Appendix. Simulations have been performed by using PSIM simulator.

TABLE 1 SIMULATION PARAMETERS

Parameters			Value	Unit
Rated active power			100	kW
DC link voltage			158	Vdc
Rated output voltage			90	Vrms
Switching frequency			4	kHz
Filter capacitor, C_f			2700	uF
Inverter side inductor, L_i			15	uH
PI control gain	1st Controller K_p, K_i	Voltage	1 , 500	Ω^{-1}
		Current	0.08, 16	Ω
	Harmonic compensation		0.5 , 0.01	None
P+R control gain	P gain(K_p), P gain ($K_{c.P}$)		6 , 0.17	Ω^{-1}, Ω
	Resonant gain (1st)		4000	Ω^{-1}
	Resonant gain (3rd)		5000	Ω^{-1}
	Resonant gain (5th)		5000	Ω^{-1}
	Resonant gain (7th)		5000	Ω^{-1}
	Cut-off frequency($\omega_{1,3,5,7.cut}$)		1,3,5,7	rad/s

$$\frac{s^2 K_P K_{C.P} + s(2K_P K_{C.P}\omega_{cut} + K_I K_{C.P}) + K_P K_{C.P}\omega_o^2}{s^4 LC + s^3(K_{C.P}C + 2LC\omega_{cut}) + s^2(2\omega_{cut}K_{C.P}C + K_P K_{C.P} + LC\omega_o^2 + 1) + s(K_{C.P}C\omega_o^2 + K_I K_{C.P} + (K_P K_{C.P} + 1)2\omega_{cut}) + (K_P K_{C.P} + 1)\omega_o^2} \quad (4)$$

$$(K_i.\omega_{cut} \approx K_i, K_{PWM} \approx 1)$$

The 2014 International Power Electronics Conference

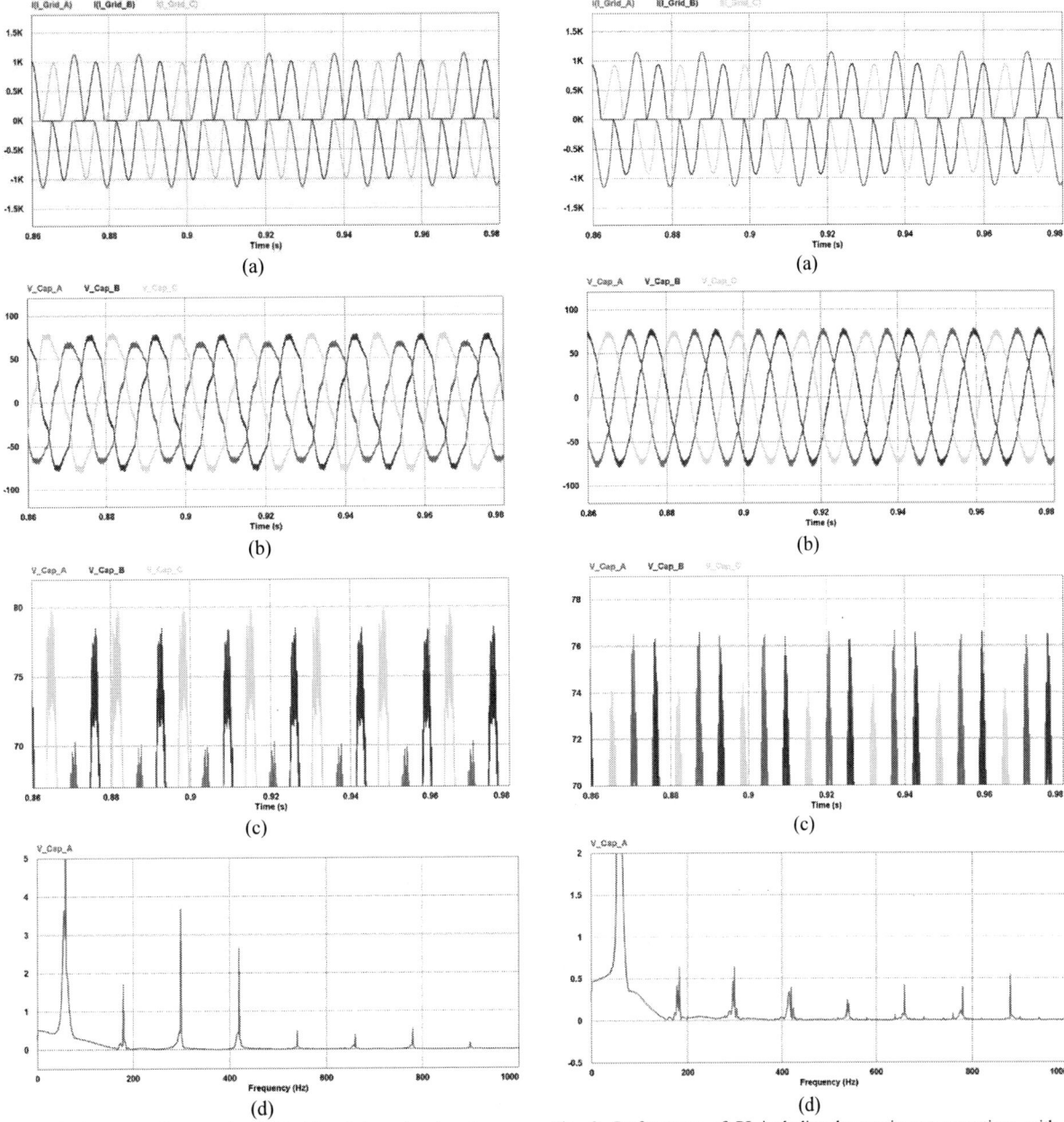

Fig. 8. Performance of conventional PI with unbalanced and nonlinear load:
(a)   Output load current,
(b)   measured output voltage,
(c)   measured output voltage zoomed in,
(d)   harmonic analysis of output voltage on phase A.

Fig. 9. Performance of PI including harmonic compensations with unbalanced and nonlinear load:
(a)   Output load currents,
(b)   measured output voltages,
(c)   measured output voltage zoomed in,
(d)   harmonic analysis of output voltage on phase A.

Figure 8 represents the conventional PI performance with unbalanced and nonlinear loads. Figure 8(a) is the output load current distorted by unbalanced and nonlinear load. Figure 8(b) is the measured output voltage and figure 8(d) is the harmonic analysis of phase A voltage. As shown in figure 8(c), each phase magnitude of output voltage is unbalanced on its peak, because it does not include any harmonic compensation method especially 1st negative sequence and 3rd zero sequence component. As shown in figure 5(d), 3, 5, 7 order harmonic components are very large as like as 9.1% THD.

Figure 9 represents the performances of advanced PI including harmonic compensation method under the same load condition. Figure 9(a) is the output load current distorted. In figure 9(b), each magnitude of output phase voltages is almost balanced compared with that in figure 8(b). Besides, compared with figure 8(d), it is shown that 3, 5, 7 order harmonic components are dramatically reduced as THD 4.2% as in figure 9(d). As a result, the advanced PI has a significant effect for harmonic compensation.(Note that the waveform scales of figures 8 and 9 are different.)

978-1-4799-2706-7/14 $31.00 © 2014 IEEE

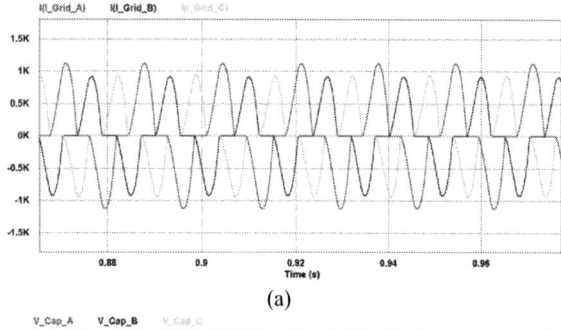

(a)

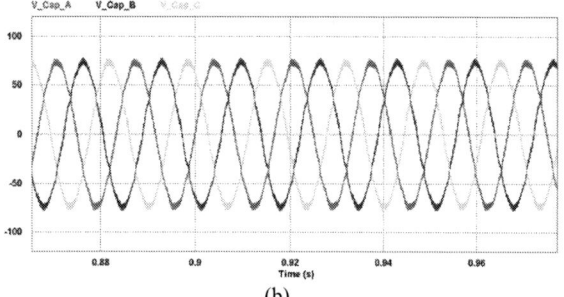

(b)

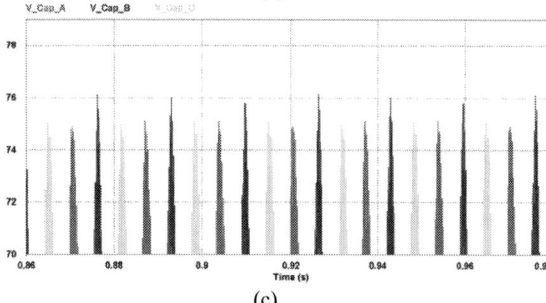

(c)

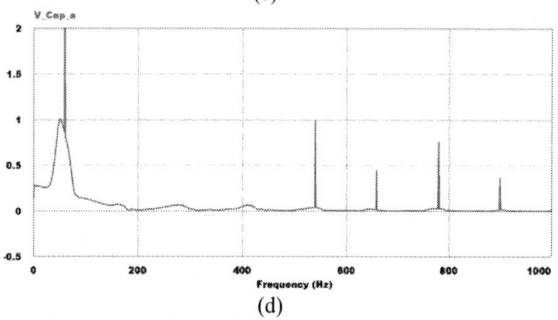

(d)

Fig. 10. Performance of P+ multiple resonant control with unbalanced and nonlinear load:
(a)   Output load current,
(b)   measured output voltage,
(c)   measured output voltage zoomed in,
(d)   harmonic analysis of output voltage on phase A.

Figure 10 represents the performances of P+ multiple resonant control with unbalanced and nonlinear load. Each phase magnitude of output voltage in figure10(b) is almost balanced as in figure 9(b). And also 3, 5, 7 order harmonic components in figure 10(d) is almost zero as THD 3.9%. The balancing accuracy of output voltage also has been upgraded as shown in figure 10(c) which shows the output voltage waveform zoomed in. The little FFT differences between figure 9(d) and 10(d) due to the parameters of harmonic gain can also be adjusted according to the gain of controller.

TABLE 2 PERFORMANCE COMPARISON OF THREE CONTROLS

Parameters	Voltage THD (A phase)	Unbalance (peak to peak)
PI control without harmonic compensation	9.1 %	≈ 11.5 V
PI control with harmonic compensation	4.2 %	≈ 1.8 V
P+ multiple resonant control	3.9 %	≈ 1 V

All results have been put together in Table 2. As shown in this table, the conventional PI method without harmonic compensation cannot guarantee the good output voltage quality when unbalanced and nonlinear loads are connected. Hence, the advanced PI control with harmonic compensation method or the P+ multiple resonant control is necessary on this kind of non-linear and unbalanced load condition.

The next simulation shows the performance of the cascaded current-voltage control method in both grid-connected mode and islanded mode. Parameters of configuration are all same as parameters of previous simulation. Overall controller gains are represented in Table 3 as below.

TABLE 3 PARAMETER OF TWO CONTROLLERS

	Parameters	Value	Unit
Single current control gain	P gain	0.05	None
	Resonant gain (1st)	5	None
	Resonant gain (5th)	2	None
	Resonant gain (7th)	2	None
	Cut-off Freq($\omega_{1,5,7.cut}$)	1,5,7	rad/s
Cascaded current-voltage control gain	Current P gain	0.035	Ω
	Current resonant gain (1st)	0.05	Ω
	Current resonant gain (5th)	0.02	Ω
	Current resonant gain (7th)	0.02	Ω
	Voltage P gain	0.004	None
	Voltage resonant gain (1st)	50	None
	Voltage resonant gain (5th)	10	None
	Voltage resonant gain (7th)	10	None
	Cut-off freq.($\omega_{1,5,7.cut}$)	1,5,7	rad/s

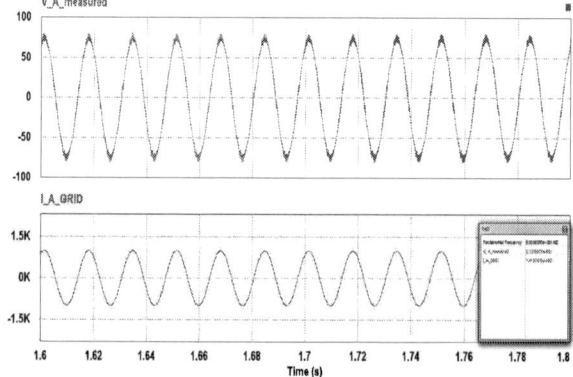

Fig. 11. Single current controller performance with harmonic compensation in grid-connected mode: Output voltage (top), output current (bottom).

The 2014 International Power Electronics Conference

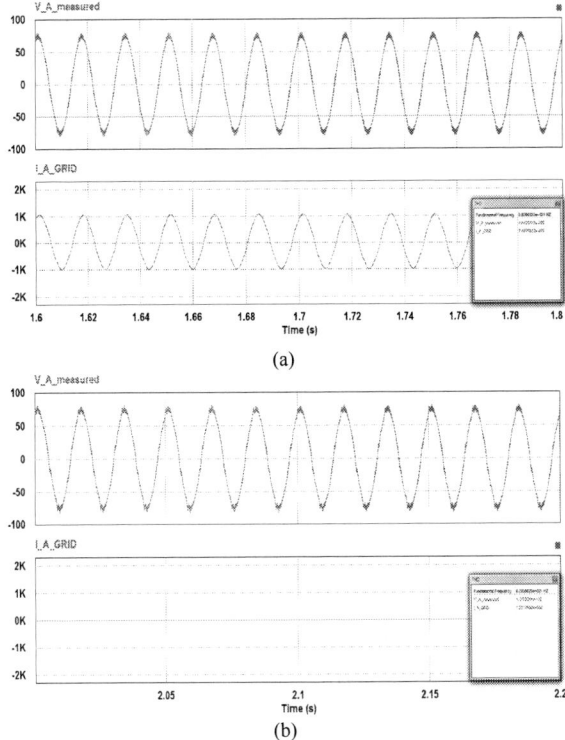

(a)

(b)

Fig. 12. Cascaded current-voltage controller performance without harmonic compensation:
(a) Output voltage and current in grid-connected mode,
(b) output voltage and current in islanded mode.

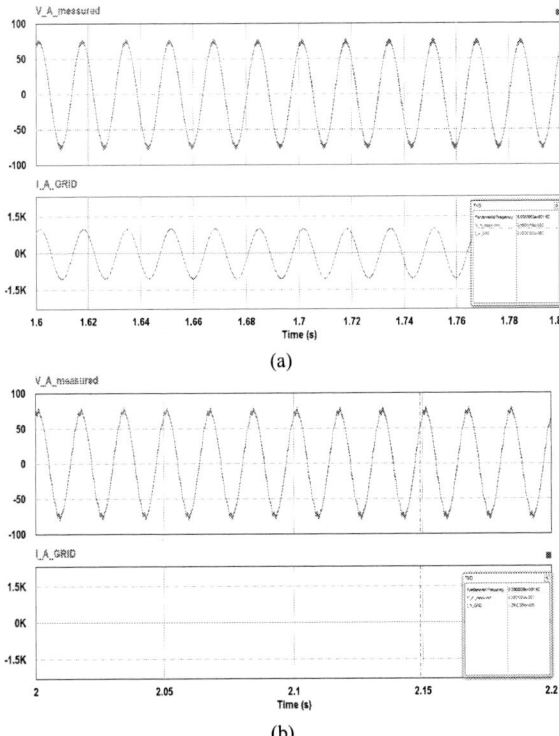

(a)

(b)

Fig.13. Cascaded current-voltage controller performance with harmonic compensation:
(a) Output voltage and current in grid-connected mode,
(b) output voltage and current in islanded mode.

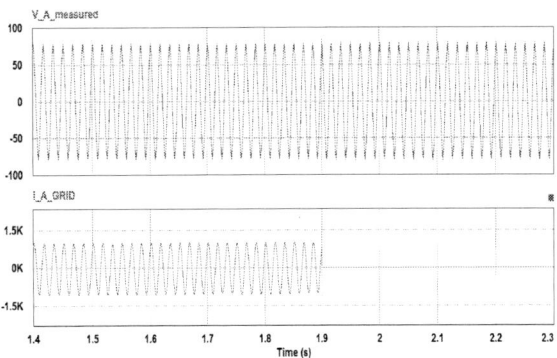

Fig.14. Transient performance of cascaded current-voltage controller with harmonic compensation from grid-connected mode (I_d=1000A) to islanded mode (I_d=0A).

Figure 11 shows the output voltage and output current waveforms, respectively by using a single current control composed of P+ multiple resonant under the condition of non-linear local load connection. In spite of the presence of non-linear local load, 1.81% THD result of output current is satisfied due to the harmonic compensation by P+ multiple resonant control. But THD result of output voltage is quite high as 5.13% because the inverter controls only the output current in this method.

Figure 12 represents the performance of cascaded current-voltage method composed of P+ resonant control just for the fundamental component in both grid-connected and islanded modes. In grid-connected mode operation, THD performances of the output current and voltage are 7.49% and 4.64%, respectively. In this case, the output current THD is higher than the result of figure 11, because the cascaded current-voltage method in figure 12 does not have any harmonic compensation functions. But the output voltage THD is a little bit improved than the result of figure 11 in spite of no harmonic compensation function, because the cascaded current-voltage method has the voltage regulation term. The output voltage THD in islanded mode is about 5.27%.

Figure 13 represents the performances of the cascaded current-voltage method composed of P+ multiple resonant control in both grid-connected and islanded modes. As shown in this figure, THDs of output voltage and current are 3.56% and 2.82%, respectively, which are much more improved than the results of figure 12 in grid connected mode. And, THD of output voltage in islanded mode also has been improved to 4.5%.

Figure 14 represents the transient performance of the output voltage and current when inverter changes the control mode from a grid-connected mode to an islanded mode. As shown in this figure, the seamless mode transfer is realized because a same cascaded current-voltage controller can be used for both grid-connected and islanded mode.

All results are shown in Table 4 for comparison. As shown in table, the cascaded current-voltage control composed of P+ multiple resonant control is effective for the harmonic compensation in comparison with the

978-1-4799-2706-7/14 $31.00 © 2014 IEEE

TABLE 4 PERFORMANCE COMPARISON OF THREE CONTROLLERS

Control method		Voltage THD	Current THD
Single current controller with harmonic compensation	Grid connected mode	5.13%	1.81%
Cascaded current-voltage controller without harmonic compensation	Grid connected mode	4.64%	7.49%
	Islanded mode	5.27%	------
Cascaded current-voltage controller with harmonic compensation	Grid connected mode	3.56%	2.82%
	Islanded mode	4.5%	------

cascaded current-voltage control composed of only P+ resonant controller without harmonic compensation. It has a little THD degradation of the output current than that of the single current controller in grid-connected mode. But its output current THD is satisfactory and also its output voltage THD has been improved within 5% in both the grid connected mode and the islanded mode..

V. CONCLUSION

In this paper, firstly, it has been described and compared the output voltage regulation method by using PI, advanced PI and P+ multiple resonant control in islanded mode of microgrid. Under the unbalanced and nonlinear loads conditions, which the fundamental negative sequence (-1st), zero sequence(3rd), and $(6n \pm 1)$th order harmonic components are generated, the compensation technologies should be considered. Hence, three control methods, the conventional PI control, advanced PI control, and P+ multiple resonant control, are simulated for comparison. As a result, it is verified that the P+ multiple resonant controller can realize the effective unbalance compensation and the small output voltage THD in spite of simpler design than the advanced PI controller with harmonic compensation. Secondly, this paper also has been realized the output voltage and current THD improvement by using the cascaded current-voltage control composed of P+ multiple resonant controller. The proposed method realizes not only the effective THD improvement in both grid-connected and islanded mode but also the seamless mode transfer between the grid-connected mode and the islanded mode.

REFERENCES

[1] Wei Yao, Min Chen, Jose Matas, and Josep M. Guererro. "Design and Analysis of the Droop Control Method for Parallel Inverters Considering the Impact of the Complex Impedance on the Power Sharing," *IEEE Trans. on Ind. Electron.*, vol.58, no.2, pp. 576-588,2011.

[2] Dipankar De and Venkataramanan Ramanarayanan, "A Proportional + Multiresonant Controller for Three-Phase Four-Wire High-Frequency Link Inverter," *IEEE Trans. on Power Electronics*, vol. 25, no. 4. pp. 899-906, 2010.

[3] J. C. Vasquez, J. M. Guerrero, M. Savaghebi, and R.

Teodorescu, "Modeling, Analysis, and Design of Stationary Reference Frame Droop Controlled Parallel Three-phase Voltage Source Inverters," *in Conf. Rec. of ICPE'2011-ECCE Asia,* pp. 272–279, 2011.

[4] Dipankar De and Venkataramanan Ramanarayanan, "Decentralized Parallel Operation of Inverters Sharing Unbalanced and Nonlinear Loads," *IEEE Trans. on Power Electronics*, vol. 25, no. 12. pp. 3015-3025, 2010.

[5] D. N. Zmood, D. G. Holmes, and G. H. Bode, "Stationary Frame Current Regulation of PWM Inverters with Zero Steady-State Error," *IEEE Trans. on Power Electronics*, vol. 18, no. 3, pp. 814–822, 2003.

[6] Kyung-Hwan Kim and Dong-Seok Hyun, "Advanced Synchronous Reference Frame Controller for Three-Phase UPS Powering Unbalanced and Nonlinear Loads," *Trans. of KIPE*, vol. 10, no. 5, pp. 508–517,2005.

[7] Tomas Hornik and Qing-Chang Zhong, "Power Quality in Microgrids," *Ph. D. Thesis,* University of Liverpool, 2010.

[8] F. Blaabjerg, R. Teodorescu, M. Liserre, and A. Timbus, "Overview of Control and Grid Synchronization for Distributed Power Generation System," *IEEE Trans. on Ind. Electron.*, vol. 53, no. 5, pp. 1398-1409. 2006.

[9] Herong Gu, Xiaoqiang Guo, and Weiyang Wu, "Accurate Power Sharing Control for Inverter-Dominated Autonomous Microgrid,"*in Conf. Rec. of IPEMC'2012-ECCE Asia*, pp. 368-372, 2012.

[10] Y. W. Li and Ching Nan Kao, "An Accurate Power Control Strategy for Power-Electronics-Interfaced Distributed Generation Units Operation in a Low Voltage Multibus Microgrid," *IEEE Trans. on Power Electronics*, vol. 24, pp.2977-2988, 2009.

[11] Zeng Liu, Jinjun Liu, and Yalin Zhao, "A Unified Control Strategy for Three-Phase Inverter in Distributed Generation," *IEEE Trans. on Power Electronics*, vol. 29, no. 3, pp. 1176-1190, 2014.

APPENDIX

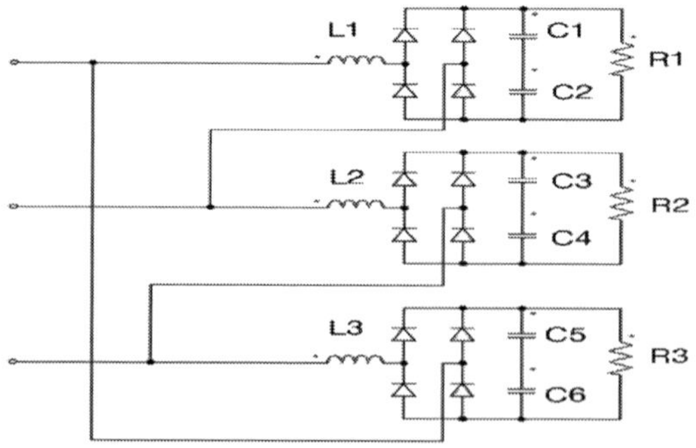

Fig. A.1. Details of nonlinear and unbalanced loads:
$L_1=L_2=L_3=85\mu H$, $C_1=C_2=C_3=C_4=C_5=C_6=130mF$, $R_1=0.2\Omega$, and $R_2=R_3=0.3\Omega$.

130MVA-STATCOM
for Transient Stability Improvement

Takao Imanishi*, Yoshinobu Nagatomo*, Shinya Iwasaki*, Kenji Masaki[†], Toshiyuki Fujii[†] and Jun Ieda[‡]

*The Kansai Electric Power Co., Inc.,
3–6–16, Nakanoshima, Kita-ku, Osaka, Japan
[†]Mitsubishi Electric Corp.,
1–1–2, Wadasakicho, Hyogo-ku, Kobe, Japan
[‡]Toshiba Mitsubishi-Electric Industrial Systems Corp.,
1–1–2, Wadasakicho, Hyogo-ku, Kobe, Japan

Abstract—The authors developed two novel control schemes for STATCOM (STATic synchronous COMpensator) to improve transient stability. First scheme is "Coordination Control" which coordinates the AVR (Automatic Voltage Regulator) control and the PSS (Power System Stabilizer) control depending on the status of the power system to maximize the effect of transient stability improvement. Second scheme is "Double Bundle control" which operates parallel STATCOMs as one lump STATCOM equivalently and realizes lower harmonics and high-speed current control. We performed verification test using real-time power system simulator in order to confirm the effect of the control schemes. We proved that these novel schemes improve transient stability under various severe faults.

Keywords—STATCOM, Transient Stability, Gate Commutated Turn-off thyristor, Steady-State Stability

I. INTRODUCTION

There may be some problems such as system stability and voltage fluctuation depending on a power system condition. FACTS (Flexible AC Transmission Systems) device has advantage of inexpensive price and is one of effective countermeasures to solve these problems. Therefore, it is applied to a power system which has these problems.

World's first STATCOM was installed in Inuyama Switching Station in 1991. One of the functions of the STATCOM is improvement of steady-state stability. Furthermore, recently a STATCOM is required to improve transient stability from the point of view of grid operation [1].

II. 130MVA-STATCOM

A. Purpose

Fig.1 shows an example of steady-state stability problem and transient stability problem of a power system. Power flow from the group of generators is supplied to C-S/S through A-SW/S and B-S/S via 154kV long-distance transmission line. The distance from the group of generators to C-S/S is about 250km. Therefore, it makes the power system unstable under heavy load condition, and makes full-power operation of all generators impossible (about 300MW). Moreover, system faults outside of the 154kV line in Fig.1 may cause loss of synchronism of some generators. Then it is necessary to install STATCOM in A-SW/S which is the electrical median point of the system and improve steady-state stability and transient stability by the STATCOM.

B. Transient Stability improvement Control

The synchronizing power and the damping force of the system shown in Fig.1 are weak especially under heavy load flow because of its long-distance transmission line. Its steady-state stability is also weak. Increase of these forces is required in order to improve steady-state stability. The STATCOM

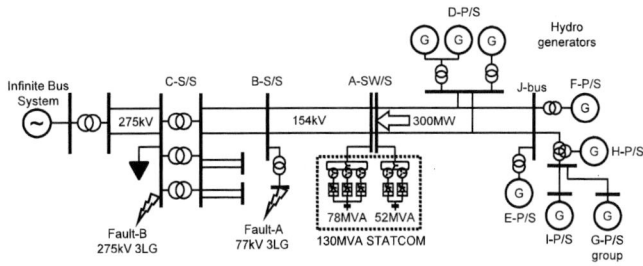

Fig. 1. Power System

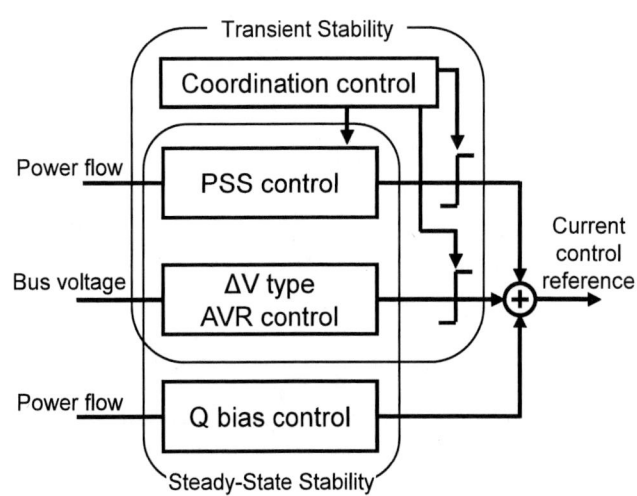

Fig. 2. STATCOM control

control scheme is configured by AVR control, PSS control and Q bias control as shown in Fig.2. The AVR control improves the synchronizing power. The PSS control improves the damping force. In addition, the Q bias control reduces transmission loss and improves its steady-state stability. The Q bias control outputs preset reactive power under heavy load conditions.

Transient stability is also improved by the improvement of the synchronizing power and the damping force. In order to improve transient stability much more, STATCOM is operated in three stages in consideration of the system phenomenon at the system fault. In the first stage, STATCOM detects the system fault from which the STATCOM must prevent the loss of synchronism of some generators and PSS control restricts its output until the fault is cleared. That is, the AVR control outputs reactive power to support system voltage against voltage drop until clearance of the fault. Improvement of the synchronizing power is achieved and first swing step-out is prevented by STATCOM. In the second stage, AVR control and PSS control change these limiters depending on the status of the power system after the system fault is cleared. Especially when the power flow passing through A-SW/S is heavy, the limiters are removed and the gain of PSS is raised in order to improve the damping force much more. STATCOM is able to output a reactive power which attenuates power swing effectively. Improvement of the damping force is achieved and STATCOM prevents N-wave step-out. In the last stage, after STATCOM prevents the loss of synchronism of generators and detects convergence of power swing, the limiters and the gain are returned to normal condition. Such control scheme which is called coordination control maximizes the effect of improving transient stability among fault-on and post-fault. Therefore, STATCOM prevents the loss of synchronism.

C. STATCOM Capacity

In order to determine the capacity of STATCOM, we performed transient stability analyses. The following Fault-A and B are the severest fault points outside of the 154kV line in Fig.1. STATCOM must prevent all generators from losing synchronism against Fault-A or B.

Fault-A: B-S/S 77kV Bus Feeder Line 3LG fault
      (Fault period : 233ms)
Fault-B: C-S/S 275kV Bus 3LG fault
      (Fault period : 100ms)

Fig.3 shows a result of transient stability analysis of Fault-A. Fig.4 shows one of Fault-B. In both cases, transient stability is most severe because of heavy power flow which all generators output at rated oparation. As shown Fig.3 and Fig.4, generators lose synchronism at first wave when no STATCOM exists. At least 130MVA-STATCOM is required to prevent the loss of synchronism using the control scheme described above against the severest Fault-A and B.

We performed eigenvalue analyses to evaluate steady-state stability enhancing effects by the 130MVA-STATCOM. Fig.5 shows the eigenvalue of the power system. In this case, all

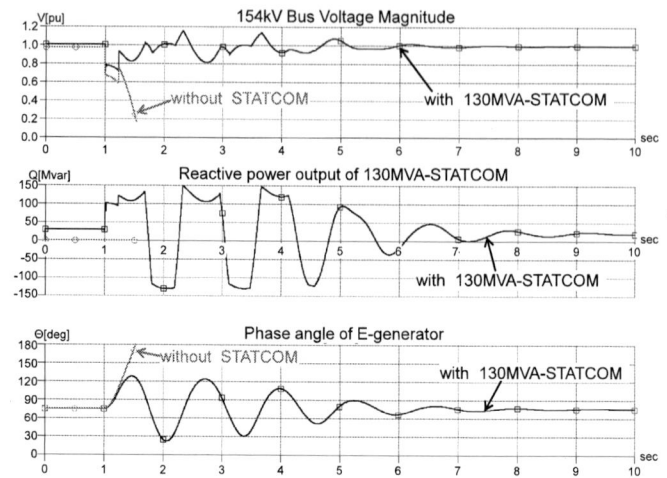

Fig. 3.   System study for transient stability at Fault-A by simulation.

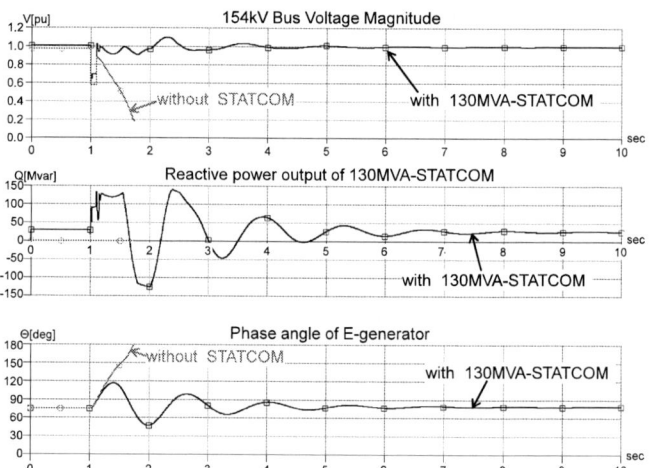

Fig. 4.   System study for transient stability at Fault-B by simulation.

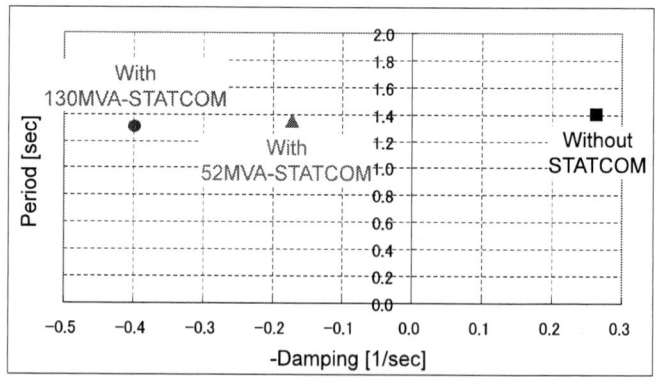

Fig. 5.   System study for steady-state stability by eigenvalue analysis.

generators run at rated power and the steady-state stability is the severest. As shown in Fig.5, real part of the eigenvalue is positive when no STATCOM exists, which means that the power system is unstable. However, real part of the eigenvalue is negative when the 130MVA STATCOM exists. Moreover, the eigenvalue is negative when capacity of STATCOM is 52MVA. It is found that even 52MVA-STATCOM ensures steady-state stability.

D. STATCOM Configuration

Fig.6 shows the system configuration of the 130MVA-STATCOM. Table.I shows the specification of 130MVA-STATCOM. The STATCOM is a parallel system of 3-stage-converter (the left-side : 78MVA-STATCOM) and 2-stage-converter (the right-side : 52MVA-STATCOM). Each converter is connected to 154kV bus through multi-stage transformers. The STATCOM system is redundant against steady-state stability, meaning that each STATCOM can ensure steady-state stability alone. Moreover, by both STATCOMs running, the STATCOM system which is equivalent of 130MVA-STATCOM ensures transient stability. The large capacity 6kV-6kA GCT (Gate Commutated Turn-off) thyristor in Fig.7, which achieves high reliability, downsizing, and low loss, is employed for the 3-level converter unit.

In order to improve transient stability by STATCOM, it is required to output reactive power by STATCOM in the event of a system fault. To do this, it is necessary to enhance Fault Ride-Through (FRT) capability of STATCOM. In general, it is necessary to increase the pulse number in order to enhance FRT capability. However, in order to realize an efficient system, it is desirable to reduce the pulse number as much as possible. PWM pulse number was selected 5 pulses with consideration for FRT capability and efficiency.

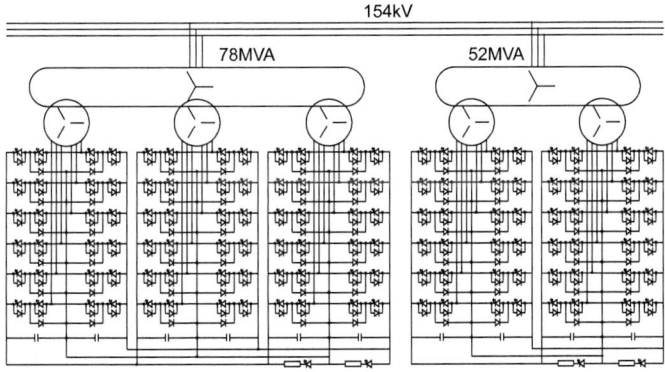

Fig. 6.   130MVA-STATCOM configuration.

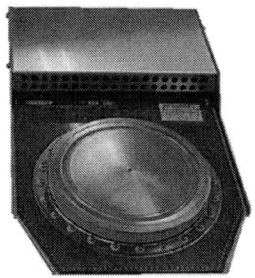

Fig. 7.   6kV-6kA GCT thyristor.

TABLE I.   SPECIFICATION OF 130MVA-STATCOM FOR TRANSIENT STABILITY

Specification of STATCOM System	
Capacity	±130MVA
AC grid voltage	154kV
System configuration	2-parallel
Capacity of each STATCOM	±78MVA, ±52MVA
Specification of78MVA-STATCOM	
Capacity	±78MVA
Power device	6kV-6kA GCT, 1S1P
Inverter circuit	3-level full-bridge × 3phase
Multiplex transformer	3-stage multiplex
Capacity / stage of inverter	26MVA
Voltage, Current / stage of inverter	3846V, 2253A
Specification of52MVA-STATCOM	
Capacity	±52MVA
Power device	6kV-6kA GCT, 1S1P
Inverter circuit	3-level full-bridge × 3phase
Multiplex transformer	2-stage multiplex
Capacity / stage of inverter	26MVA
Voltage, Current / stage of inverter	3846V, 2253A

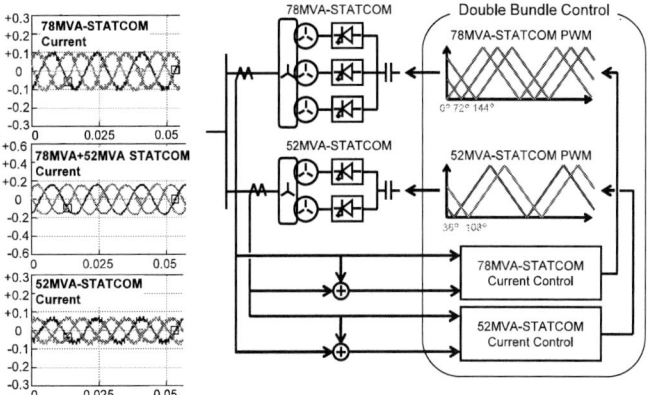

Fig. 8.   Double bundle control scheme.

E. Double Bundle Control Scheme

As shown in Fig.8, a double bundle control scheme by which a parallel system is operated in one lump system to offer technical advantages is used. It is to operate two parallel STATCOMs as one lump STATCOM equivalently. The equivalent STATCOM is operated as 5-stage-converter. The double bundle control scheme adjusts the PWM pulse timing of each stage. Composite voltage is equivalent to the voltage of 5-stage serial connection STATCOM and its harmonics are reduced. That is, it is possible that a parallel system applies more sinusoidal current to a power system. Furthermore, each STATCOM detects composite current and controls it individually. High-speed current control equivalent to that of 5-stage serial connection STATCOM is realized. This control scheme suppresses overcurrent by voltage fluctuation at the system fault and reduces the frequency of gate block (GB). Moreover, even if this STATCOM performs GB, STATCOM is possible to perform gate de-block immediately at the high-speed.

In order to check FRT capability which satisfies required specification, we modeled the power system as shown in Fig.1, STATCOM main circuit, and STATCOM control scheme in detail by electromagnetic transient analysis. We performed electromagnetic transient analyses of the case of Fault-A and B. Fig.9 shows result of Fault-B. The Fault-B is a severe condition for the FRT capability of the STATCOM. Immediately after bus 3LG fault, transient phenomenon with voltage distortion occurs. It is found that overcurrent which reaches GB level does not occur in the STATCOM at the Fault-B and STATCOM operated satisfactorily. From this result, it has checked that FRT capability required for the improvement in transient stability can be expected at the time of the system fault. The double bundle control scheme improved both harmonics in steady-state and FRT capability in transient-state.

III. REAL-TIME SIMULATOR TEST

We performed analog type real-time simulator test to confirm the performance of 130MVA-STATCOM. Fig.10 shows the configuration of real-time simulator test. The system shown in Fig.1 is modeled in detail by the real-time simulator. Bulk power system was simulated as infinite bus and the short-circuit impedance. Transmission lines, transformers, loads and phase modifying equipment were simulated as combination of resistor, inductor and capacitor. Transmission line models were simulated as PI sections with mutual induction. Transformer models were equipped with saturation characteristic. Generators were simulated as detailed models which include synchronous machine (Park model), AVR, PSS and Governor by digital controller. The detailed generator model is almost same as the transient stability analysis model.

Fig.11 shows the frequency-impedance characteristics (f-z characteristics) of the power system. The upper figure is f-z characteristic of simulator model. The lower figure is one of electromagnetic transient analysis model. Both characteristics are very well matched. It is confirmed that the power system is simulated with great accuracy by the real-time simulator.

We connected the simulator and STATCOM miniature model which is composed of the same configuration as the

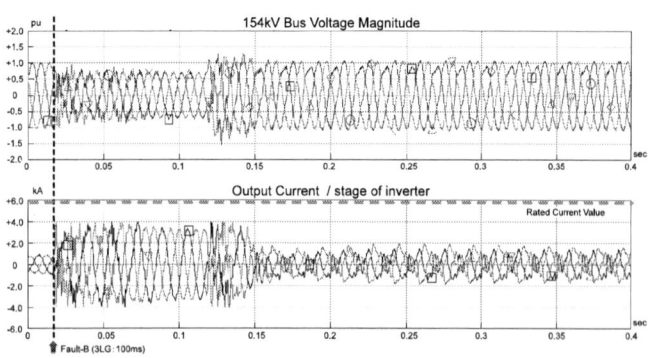

Fig. 9. System study for FRT capability by electromagnetic transient analysis.

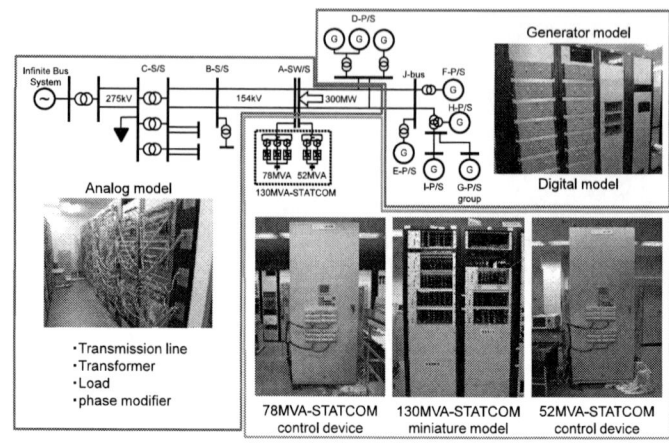

Fig. 10. Configuration of real-time simulator test.

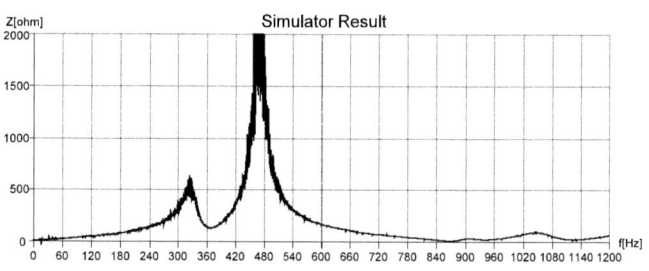

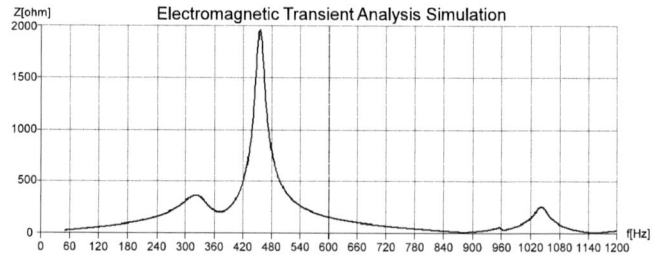

Fig. 11. Frequency-Impedance characteristics.

actual and verified operation of the STATCOM under various system faults.

Fig.12 shows simulator test result of steady-state stability. In this case, all generators run at rated power and the steady-state stability is the severest. Smaller 52MVA-STATCOM operates solely (78MVA-STATCOM stopped) to check the redundancy. In this case, small disturbance was applied (J bus 3LG fault (fault period: 10ms)). We checked the stabilizing effect of the STATCOM for small disturbance. Fig.12 shows STATCOM connection bus Voltage V_a, V_b, V_c, output current of 52MVA-STATCOM I_a, I_b, I_c, output of PSS control, output of AVR control, reactive current reference value I_{dref}, output of reactive current I_d, power flow P and positive phase sequence voltage V_1.

The AVR control and the PSS control of the STATCOM respond depending on voltage variation and power swing, and get the STATCOM to output reactive power. As shown in Fig.12, the power swing is converging in about 20 seconds

978-1-4799-2706-7/14 $31.00 © 2014 IEEE

after disturbance occurs. It is found that steady-state stability is improved by the STATCOM.

Fig.13 shows simulator test result of transient stability. Fig.13 is a case of Fault-A. Transient stability of this case is most severe because all generators run at rated power. In this case, both of 78MVA-STATCOM and 52MVA-STATCOM are operated by the double bundle control scheme. As shown in Fig.13, the each STATCOM outputs preset reactive power by Q bias control. After the Fault-A occurs, the coordination control scheme restricts the PSS control to prevent first swing step-out and STATCOM improves the synchronizing power. Output of AVR control is dominant in the output of reactive power by the STATCOM among fault-on. The STATCOM outputs full leading reactive current depending on the voltage drop. The STATCOM is running without GB even when the bus voltage falls to about 0.7pu. After the Fault-A is cleared, coordination control scheme removes the limiters of AVR control and PSS control and raises the gain of PSS to improve the damping force. The STATCOM prevents N-wave step-out.

IV. CONCLUSIONS

We proposed these novel STATCOM control schemes, which are the coordination control and the double bundle control, to improve transient stability. The coordination control adjusts AVR control and PSS control so as to maximize the effect of improving transient stability. The double bundle control enhances the FRT capability and reduces its harmonics by operating parallel STATCOMs as one lump STATCOM. These proposed control methods were applied to the STATCOM control and the required capacity to prevent generators from losing synchronism was evaluated. Main circuit configuration was designed in view of the redundancy for the steady-state stability. The STATCOM system was a parallel system of 3-stage 78MVA-STATCOM and 2-stage 52MVA-STATCOM. The transient stability improvement effect by the STATCOM using the novel schemes was verified by the real-time simulator.

REFERENCES

[1] S. Mori, "Initiatives perspectives by the power industry towards a low carbon emission society," Keynote address of Paris Session CIGRE 2010.

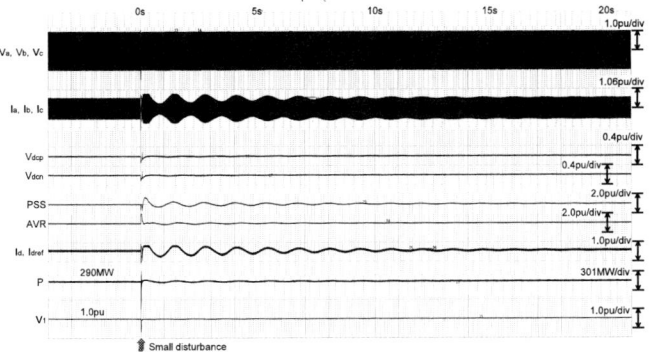

Fig. 12.   Simulator test result of steady-state stability.

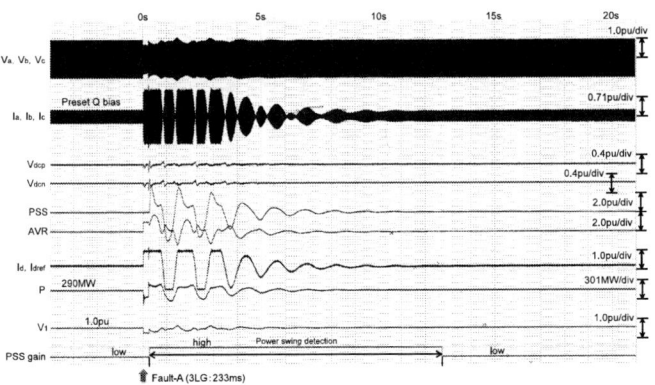

(a) Full waveform.

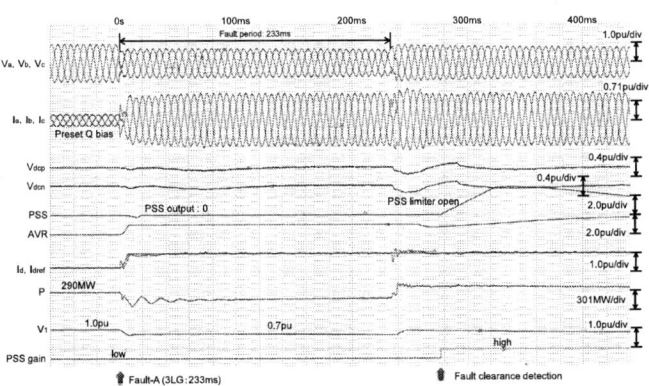

(b) Enlarged view.

Fig. 13.   Simulator test result of Fault-A.

Improved Droop Controller for Microgrid Inverter Considering the Line Impedance Mismatching

Du Yan   Liuchen Chang   Meiqin Mao   Jianhui Su   Ning Liu
Department of Electrical Engineering
Hefei University of Technology
Heifei, China

Abstract—In islanding microgrid, the impact of line impedance on the proportional power sharing cannot be neglected if multi-inverters are operated to form a microgrid. In this paper, an improved droop controller is proposed to reduce the unexpected dynamic and steady power sharing errors caused by the mismatching of line impedance. With this proposed combined compensation, the power quality, the dynamic and steady performances are benefited. The design principle of the parameters in the proposed controller is also presented. The simulation and experimental results are provided to validate the proposed controller and the parameters design methodology.

Keywords— Microgrid, droop controller, line impedance, load distribution

I. INTRODUCTION

As a smart integration mechanism of distributed energy sources (DER), microgrid is a promising direction of DER applications because of its outstanding performance on eliminating the negative impact caused by the large scale penetration of DERs into the distribution network[1-3]. In a microgrid, the voltage source inverter equipped with energy storage unit (microgrid inverter, MGI) is a critical element that guarantees reliable and stable operation of microgrid[4-5].

Nowadays, droop control is the most used controller for MGIs in multi-MGI-formed microgrid applications since it has been successfully applied for controlling parallel uninterruptible power supply[6]. However, the conventional droop controller has an inherent limitation, that is, the line impedance of parallel-operated MGIs should follow a strict restriction when the load sharing is achieved[7].

In microgrid application, the feeder impedance among MGIs is not negligible due to the different locations of MGIs. As a result, the load sharing performance of droop control especially the accuracy of reactive power sharing can be affected by both the output impedance and the feeder impedance.

The proportional load sharing of MGIs considering the mismatched line impedance has been intensively investigated. In order to match the line impedance, the algorithms of virtual impedance are proposed in [8][9] to design the output impedance. Because the local voltage and current are used, these algorithms may still be affected by the unequal feeder impedances. In [10], a compensation of feeder impedance is proposed to improve accuracy of reactive power sharing. However, this strategy is sensitive to computational errors, parameter drifts and component mismatches. In [11][12], the Q-U droop controller is improved by adding an extra integral action. However, this method cannot eliminate the power sharing error cause by the feeder impedance if the local power is adopted in droop control. A Q–$\dot{U}$ droop control method was proposed in [13] to improve the accuracy of reactive power sharing but the embedded $\dot{U}$ restoration mechanism costs the accuracy of the reactive power sharing.

In this paper, a detailed analysis of line impedance's constraint to share the load accurately is presented when the different measured power is adopted in the conventional droop controller. After that, an improved droop controller is proposed to reduce the unexpected dynamic and steady power sharing errors caused by the mismatching of line impedance. The simulation and experimental results are also presented to validate the proposed controller.

II. CONSTRAINT ON LINE IMPEDANCE

Conventional P–ω and Q–U droop controller has been successfully adopted in the microgrid, and its function is written as:

$$\omega = \omega^* + m(P^* - P) \qquad (1)$$

$$U = U^* + n(Q^* - Q) \qquad (2)$$

Where m is the P-ω droop coefficient, n is the Q-U droop coefficient, ω^* and U^* are the nominal frequency and the nominal voltage magnitude, P^* and Q^* are the references of active power and reactive power; P and Q are the measured active power and reactive power.

Usually, P and Q are measured locally so that communication line is not needed. However, the extra active power and reactive power consuming on the line impedance may result in the unexpected power sharing error.

A simplified circuit in Fig. 1 is used to explain the restriction on the line impedance to share the power accurately if the local P and Q are used.

This work is supported by Guangdong Innovative Research Team Program (GIRTP).

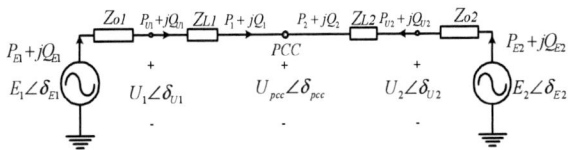

Fig. 1. Simplified circuit of the system

In Fig. 1, the common load is at the PCC($U_{pcc}\angle\delta_{pcc}$), the ith(i=1,2) MGI(MGI$_i$) can be modeled as a voltage source ($E_i\angle\delta_{Ei}$) with an output impedance ($Z_{oi}=R_{oi}+jX_{oi}$), thus the output voltage of MGI$_i$ is $U_i\angle\delta_{Ui}$. And MGI$_i$ is connected to the common load through the feeder impedance ($Z_{li}=R_{li}+jX_{li}$). So the line impedance ($Z_i=R_i+jX_i$) includes the output impedance(Z_{oi}) and the feeder impedance(Z_{li}). P_i+jQ_i is the power of MGI$_i$ injected into PCC, $P_{Ui}+jQ_{Ui}$ is the output power of MGI$_i$, $P_{Ei}+jQ_{Ei}$ is the virtual power generated by the voltage source ($E_i\angle\delta_{Ei}$).

The goal of the proportional power sharing is to achieve MGI's output current in proportion to its power rating, so the following equation should be satisfied:

$$\frac{i_1}{i_2}=\frac{S_1}{S_2}=k,\frac{m_1}{m_2}=\frac{n_1}{n_2}=k \tag{3}$$

Where i_i(i=1,2) and S_i are the output current and the power rating of MGI$_i$, m_i and n_i are the droop coefficients of MGI$_i$.

From(3), the power of MGI$_i$ injected into PCC should follows:

$$P_1=kP_2,Q_1=kQ_2 \tag{4}$$

Therefore, (4) is the necessary condition for the proportional power sharing. The following part will discuss the constraint on line impedance when the different power is used to control the frequency and the voltage's magnitude of MGI$_i$.

Case1: P_i+jQ_i is used in the droop controller
The power of MGI$_i$ injected into PCC is shown as:

$$P_i=\frac{X_iE_iU_{pcc}}{R_i^2+X_i^2}\sin\delta_i+\frac{R_iU_{pcc}(E_i\cos\delta_i-U_{pcc})}{R_i^2+X_i^2} \tag{5}$$

$$Q_i=-\frac{R_iE_iU_{pcc}}{R_i^2+X_i^2}\sin\delta_i+\frac{X_iU_{pcc}(E_i\cos\delta_i-U_{pcc})}{R_i^2+X_i^2} \tag{6}$$

Where, $\delta_i=\delta_{Ei}-\delta_{pcc}$ is the power angle between E_i and U_{pcc}.

It is obvious that an accurate real power sharing can be achieved as long as $m_1=km_2$ and $\omega_1^*=\omega_2^*=\omega^*$, because all MGIs share the same frequency at the steady state.

Combining(2),(3),(5),(6) and considering $P_1=kP_2$, $n_1=kn_2$, the following formula should be satisfied to eliminate the reactive power sharing error:

$$(U_{pcc}E_1-U_{pcc}^2)(\frac{1}{X_1}-\frac{k}{X_2})+P_1(\frac{R_2}{X_2}-\frac{R_1}{X_1})=0 \tag{7}$$

Eq.(7) leads to the constraint on line impedance shown as:

$$X_2=kX_1,R_1=kR_2\ \&\&\ E_1=E_2 \tag{8}$$

Case2: $P_{Ei}+jQ_{Ei}$ is used for droop controller
The relations between $P_{Ei}+jQ_{Ei}$ and P_i+jQ_i are shown as:

$$P_{Ei}\approx P_i-\frac{R_i(E_i-U_{pcc})^2}{R_i^2+X_i^2} \tag{9}$$

$$Q_{Ei}\approx Q_i-\frac{X_i(E_i-U_{pcc})^2}{R_i^2+X_i^2} \tag{10}$$

Similarly, $P_{E1}=kP_{E2}$ can be obtain. Thus, the following equations should be followed to obtain $P_1=kP_2$ and $Q_1=kQ_2$

$$\frac{R_1(E_1-U_{pcc})^2}{R_1^2+X_1^2}=k\frac{R_2(E_2-U_{pcc})^2}{R_2^2+X_2^2} \tag{11}$$

$$-\frac{E_1}{n_1}+\frac{X_1(E_2-U_{pcc})^2}{R_1^2+X_1^2}=-\frac{E_2}{n_1}+k\frac{X_2(E_2-U_{pcc})^2}{R_2^2+X_2^2} \tag{12}$$

From (11) and(12), a constraint on line impedance is obtained as follows:

$$X_2=kX_1,R_1=kR_2\ \&\&\ E_1=E_2 \tag{13}$$

Case3: $P_{Ui}+jQ_{Ui}$ is used for droop controller
The relations between $P_{Ui}+jQ_{Ui}$ and P_i+jQ_i are shown as:

$$P_{Ui}=P_i+P_{z_{li}}=P_i+\frac{P_i^2+Q_i^2}{U_{pcc}^2}R_{Li} \tag{14}$$

$$Q_{Ui}=Q_i+Q_{z_{li}}=Q_i+\frac{P_i^2+Q_i^2}{U_{pcc}^2}X_{Li} \tag{15}$$

To eliminate the real and reactive power sharing errors the following formulas should be satisfied:

$$\frac{(E_1^2+U_{pcc}^2-2E_1U_{pcc})R_{L1}}{R_1^2+X_1^2}=k\frac{(E_2^2+U_{pcc}^2-2E_2U_{pcc})R_{L2}}{R_2^2+X_2^2} \tag{16}$$

$$\frac{(E_1^2+U_{pcc}^2-2E_1U_{pcc})X_{L1}}{R_1^2+X_1^2}-k\frac{(E_2^2+U_{pcc}^2-2E_2U_{pcc})X_{L2}}{R_2^2+X_2^2}+\frac{E_2-E_1}{n_1}=0 \tag{17}$$

Assuming (8) is satisfied in (16) and (17), (16) and (17) can be rewritten as

$$(E_1^2+U_{pcc}^2-2E_1U_{pcc})kR_{L1}=(E_2^2+U_{pcc}^2-2E_2U_{pcc})R_{L2} \tag{18}$$

$$kX_{L1}(E_1^2+U_{pcc}^2-2E_1U_{pcc})-$$
$$X_{L2}(E_2^2+U_{pcc}^2-2E_2U_{pcc})+(R_1^2+X_1^2)\frac{E_2-E_1}{n_1}=0 \tag{19}$$

The conditions of (18) and (19) are satisfied if

$$X_{l2}=kX_{l1},R_{l1}=kR_{l2}\ \&\&X_2=kX_1,R_1=kR_2\ \&\&\ E_1=E_2 \tag{20}$$

In the summary, *Case1* has the weakest constraint on the line impedance and only the reactive power sharing is affected by the line impedance. But *Case1* needs the high-speed communication line to measure the voltage of PCC so that the approach deteriorates the flexibility and stability of microgrid. Meanwhile, the local power is used in both *Case2* and *Case3*. In *Case2* and *Case3*, both active and reactive power sharing are dependent on the line impedance. Compared to P_{Ei} and Q_{Ei}, P_{Ui} and Q_{Ui} are the most used variables in the conventional droop controller because it is convenient to measure. However, this convenience is at the expenses of stronger constraint on the line impedance. Nevertheless, notice that those constraints shown as(8),(13) and (20) are sufficient but not necessary, so it opens up opportunities to improve the droop controller to overcome the impact caused by the line impedance.

978-1-4799-2706-7/14 $31.00 © 2014 IEEE

III. IMPROVED DROOP CONTROLLER TO ACHIEVE THE ACCURATE LOAD SHARING

To overcome the impact of line impedance, an improved droop controller is proposed in this paper, shown as:

$$\omega = \omega^* + m(P^* - P) - m_d \frac{dP}{dt} \tag{21}$$

$$E = E^* + \Delta E + f(Q)(Q^* - Q) - n_d \frac{dQ}{dt} \tag{22}$$

Where ΔE is a compensation of the voltage drop caused by the line impedance, m and $f(Q)$ are the steady droop coefficients of P–ω and Q–U droop controller, m_d and n_d are the dynamic droop coefficients of P and Q. The following part will introduce the function of those proposed parts.

1) ΔE

In Fig. 1, assuming $\delta_{pcc}=0$, δ_{Ui} is small enough and (2) is adopted to control the output voltage, the relation between the reference of voltage E^* and the voltage of PCC U_{pcc} is shown as

$$U_{pcc} = (E^* - nQ) - (\frac{R_{Li}P_{Ui}}{U_i} + \frac{X_{Li}Q_{Ui}}{U_i}) \tag{23}$$

The first part on the right side of (23) takes the forms of Q-U droop equation; the second part represents the voltage drop caused by feeder impedance.

Therefore, the accuracy of reactive power sharing can be improved if the following equation is satisfied

$$\Delta E = \frac{R_{Li}P_{Ui}}{U_i} + \frac{X_{Li}Q_{Ui}}{U_i} \tag{24}$$

However, the unexpected power sharing can be eliminated only if the information of line impedance is correct. Therefore, this approach is sensitive to the measurement error of line impedance.

2) Steady droop coefficient- $f(Q)$

To compensate the power sharing error caused by impedance measurement, a variable coefficient of Q-U is proposed. This proposed droop coefficient is a function of absolute value of Q to guarantee its effectiveness in both inductive and capacitive load condition, and it is shown as

$$f(Q) = n^* + K_q |Q| \tag{25}$$

Where, n^* is the nominated droop coefficient, K_q is the coefficient of the steady compensation.

To limit the voltage droop and better power sharing, the following restriction should be satisfied when optimizing the parameter in (25)

$$\begin{cases} \Delta U = Q_{max} f_1(Q) \le 2\% \Delta U_{max} \\ \dfrac{K_{q1} \cdot |Q_1|}{Z_1} = \dfrac{K_{q2} \cdot |Q_2|}{Z_2} = ... = \dfrac{K_{qn} \cdot |Q_n|}{Z_n} = C \end{cases} \tag{26}$$

3) Dynamic droop coefficient- m_d and n_d

Because the conventional droop controller only has one tunable parameter m and n, the mismatching of line impedance deteriorates the dynamic power. In order to improve the dynamic performance, a dynamic droop coefficient is proposed presented as

$$m_d = K_{p1}P^2 + K_{p2} \tag{27}$$

$$n_d = K_{q2} - K_{q3}Q \tag{28}$$

Considering the line impedance is predominantly inductive due to the transformers or the use of LCL filters, a small signal model of the system at ($\delta_u^0, \delta_{pcc}^0, U^0, U_{pcc}^0$) is shown as

$$\Delta P_U = \frac{A_0}{(A_0 m_d + 1)s + A_0 m}(\Delta \omega^* - \Delta \omega_{pcc}) \tag{29}$$

$$\Delta Q_U = -\frac{A_1}{C}\Delta U_{pcc} + \frac{B_1}{C}\Delta E^* \tag{30}$$

Where

$$A_0 = \frac{U^0 U_{pcc}^0 \cos(\delta_u^0 - \delta_{pcc}^0)}{sX_L}$$

$$A_1 = -\frac{U^0 \cos(\delta_u^0 - \delta_{pcc}^0)}{X_L},$$

$$B_1 = \frac{2U^0 - U_{pcc}^0 \cos(\delta_u^0 - \delta_{pcc}^0)}{X_L},$$

$$C = B_1 n_d s + B_1 n + B_1 K_q (2|Q^0| - Q^*) + 1$$

Thus the dominated poles of active power and reactive power are

$$\lambda_P = -\frac{mA_0}{1 + m_d A_0} \tag{31}$$

$$\lambda_Q = -\frac{1 + B_1(n + K_q(2|Q^0| - Q^*))}{n_d B_1} \tag{32}$$

The location of λ_P and λ_Q can be designed by adjusting the value of m_d and n_d, according to the pole-zero placement Notice that A_0 and B_1 representing the operational point also have impact on the location of λ_P and λ_Q. Therefore, the values of m_d and n_d are determined by the location of the desired λ_P, λ_Q and the output active and reactive power.

Assuming m=0.5e-4, n=2e-4, k_q=1e-8, Q^*=0, $\lambda_P=\lambda_Q$=-100rad/s, Fig. 2 shows how the active power affects the desired m_d

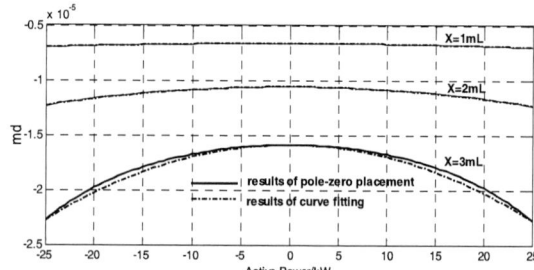

Fig. 2 Relationship between md and the output power

This function of m_d and P can be fitted by a quadratic function presented as(27) and the parameters in (27) can be expressed as

$$K_{p2} = md_0, K_{p1} = \frac{md_{max} - md_0}{P_{max}^2} \tag{33}$$

Where, m_{d0} is value of m_d at no load, m_{dmax} is the value of m_d at the full load P_{max}.

The comparisons of the results of pole-zero placement and curve fitting are shown In Fig. 2.

Fig. 3 presents the relationship between n_d and the reactive power. In Fig. 3, when Q>0, n_d decreases with the increasing of reactive power. When Q<0, n_d increases with the increasing of absolute value of reactive power. Fig. 3 also presents the trend of n_d with k_q=0,1e^{-6}and1e^{-5}. The results show the larger value K_q has, the larger value n_d has especially in the case of heavy load. Thanks to the restriction shown as (26), K_q 's impact on n_d is negligible.

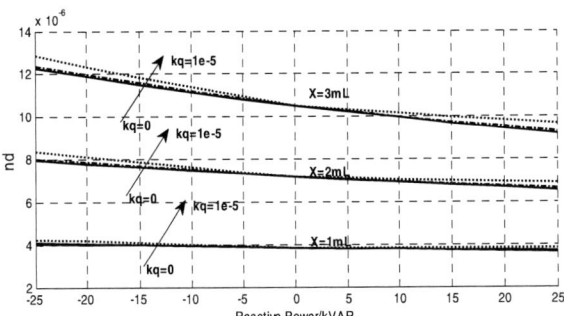

Fig. 3 Relationship between n_d and the reactive power

Taking account of dominated inductive load, Q>0 has the priority of the curve fitting. Fig. 4 shows the results of curve fitting. The function of n_d and reactive power can be fitted by (28) with the parameters selected as

$$K_{q2} = nd_0, K_{p1} = \frac{md_{max} - md_0}{Q_{max}} \quad (34)$$

Where, n_{d0} is value of n_d at no load, n_{dmax} is the value of n_d at the full inductive load Q_{max}.

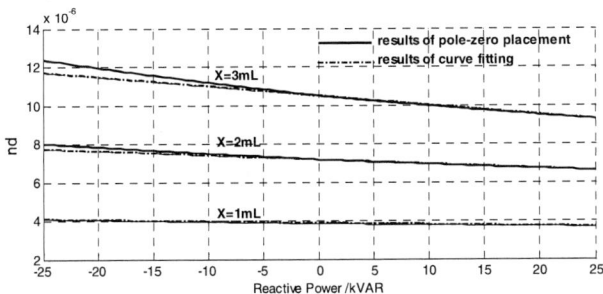

Fig. 4. The results of curve fitting

IV. SIMULATION AND EXPERIMENTAL RESULTS

The proposed droop controller has been verified in Matlab/Simulink simulations and experimentally. In the simulations and experiments, a microgrid with three MGI systems, as shown in Fig.5, is employed.

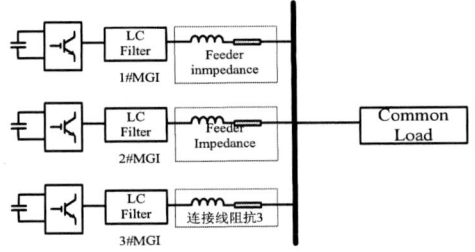

Fig. 5. Three parallel MGIs formed microgrid

In the simulation, three MGIs are served by the proposed droop controller to share the common load. The phase voltage is 220V, the power rating of 1#,2# and3#MGI are 5kW , 5kW and10kW and their line impedances are 0.2+j0.56,0.1+j0.94and0.2+j0.56.The simulation parameters are shown in Tab.1.

TABLE I
SIMULATION PARAMETERS

	m	n	K_q	$kp1$	$kp2$	$kp1$	$kq2$
1#	1e^{-5}	2e^{-4}	2.5e^{-8}	-1.434e^{-12}	-1.318e^{-5}	-7.541e^{-10}	3.005e-6
2#	1e^{-5}	2e^{-4}	1.3e^{-8}	-6.696e^{-12}	-1.982e-5	-1.734e^{-10}	1.553e-6
3#	0.5e^{-5}	1e^{-4}	1.3e^{-8}	-5.513e^{-12}	-9.913e-6	-4.210e^{-10}	2.213e-6

The initial load is 4+j3.14, and an extra load 20+j6.28 is added at 0.4s. Fig. 6 gives the power sharing performance of three MGIs. Fig. 7 gives the corresponding value of m_d and n_d during the load step.

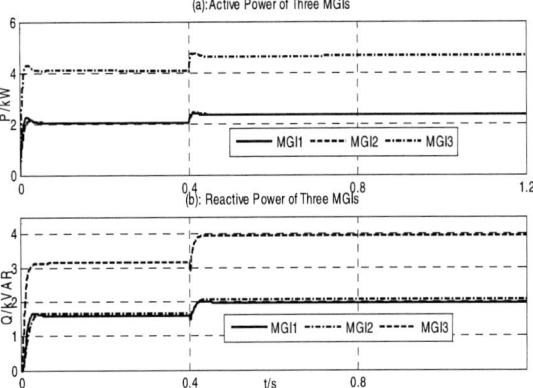

Fig. 6. Power sharing performance of three MGIs

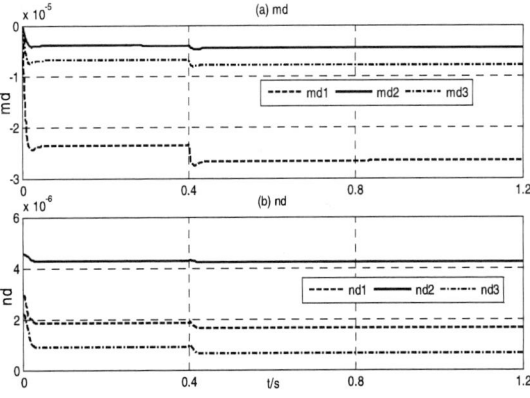

Fig. 7. Values of md and nd during the load changing

In the physical experiment, 1# and 2# MGI are governed by the proposed controller with control parameters shown as Tab.1 while 3# MGI serves as reactive load by injecting the inductive or capacitive current. An additional 2mL inductor is added at the terminal of 1#MGI ac output to introduce the effects of unequal impedance.

Fig. 8(a) gives reactive power sharing with the traditional droop control. Fig. 8(b) gives the improved performance of reactive power sharing with the proposed controller.

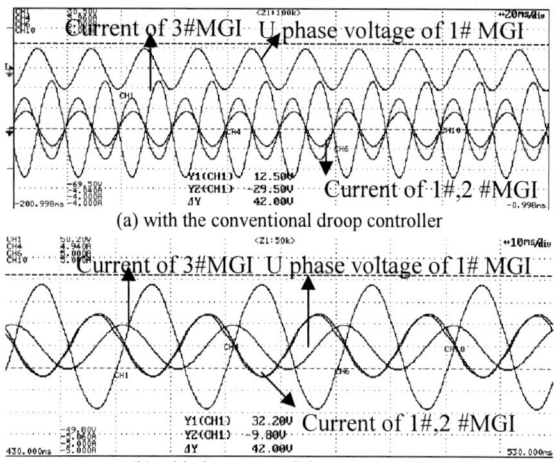

(a) with the conventional droop controller

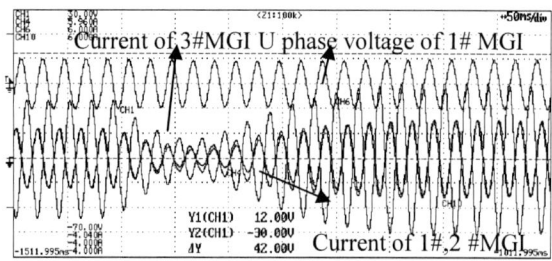

(b) with the proposed droop controller

Fig. 8. Reactive power sharing of the proposed controller

Fig. 9 gives the dynamic performance of 1# and 2# MGI. Initially, 1#and 2# shared the capacitive load, and then the output of 3#MGI switched from capacitive to inductive. Fig. 9 shows that the proposed controller is effective in both inductive and capacitive load.

Fig. 9. Reactive power sharing of the proposed controller

V. CONCLUSIONS

In this paper, the constraint on line impedance is analyzed when different power is used to drop the MGI's voltage and frequency. To overcome the impact of line impedance on accurate power sharing, an improved droop control strategy is proposed for MGI in microgrid. The proposed control strategy contains a compensation of line impedance voltage drop and the variable steady and dynamic droop coefficient. Both simulation and experimental results are provided to verify the effectiveness of the proposed control strategy.

REFERENCES

[1] Cavlovic M, "Challenges of optimizing the integration of distributed generation into the distribution network," *2011 8th International Conference on the European, Zagreb,* 2011.

[2] Chris Marnay,F Javier Rubio,Afzal S Siddiqui. "Shape of the microgrid," *Power Engineering Society Winter Meeting ,* 2001.Columbus,OH,2001.

[3] Lasseter R.H. "Smart Distribution: Coupled Microgrids," *Proceedings of the IEEE,*vol.99,no.6,pp: 1074-1082,2011.

[4] Joan Rocabert,Alvaro Luna,Frede Blaabjerg, "Control of Power Converters in AC Microgrids," *IEEE Trans On Power Electronics,* vol. 27,no. 11,pp: 4734-4748,2012.

[5] Chun-Xia Dou, Bin Liu. "Multi-Agent Based Hierarchical Hybrid Control for Smart Microgrid." *IEEE Trans on Smart Grid,* vol. 4, no. 2, pp: 771 – 778,2013.

[6] J. M. Guerrero,J. C. Vasquez,J. Matas,M. Castilla,L. G. de Vicuna. "Control Strategy for Flexible Microgrid Based on Parallel Line-Interactive UPS Systems." *IEEE Trans On Industrial. Electronics,* vol. 56,no. 3,pp: 726–736, 2009.

[7] Jinwei He, YunWeiLi, Josep M. Guerrero, Frede Blaabjerg. "An Islanding Microgrid Power Sharing Approach Using Enhanced Virtual Impedance Control Scheme." *IEEE Trans On Power Electronics,* vol.28, no.11, pp: 5272-5282, 2013.

[8] Wei Yao, Min Chen, Matas J, Guerrero J.M. "Design And Analysis Of The Droop Control Method For Parallel Inverters Considering The Impact Of The Complex Impedance On The Power Sharing." *IEEE Trans on Industrial Electronics,*vol.58, no.2, pp:576–588, 2011.

[9] Juan C. Vasquez, Josep M. Guerrero, Mehdi Savaghebi, etc. "Modeling, Analysis, and Design of Stationary Reference Frame Droop Controlled Parallel Threephase Voltage Source Inverters." *ECCE Asia 2011 IEEE 8th International Conference on,* Jeju, Korea,2011.

[10] YunWei Li, Ching-Nan Kao. "An Accurate Power Control Strategy for Power-Electronics-Interfaced Distributed Generation Units Operating in a Low-Voltage Multibus Microgrid." *IEEE Trans On Power Electronics,* vol. 24, no. 12, pp: 2977-2988, 2009.

[11] Charles K. Sao, Peter W. Lehn. "Autonomous Load Sharing of Voltage Source Converters" *IEEE Trans On Power Delivery,* vol. 20, no. 2, pp:1009-1016, 2005.

[12] Qing-Chang Zhong. "Robust Droop Controller for Accurate Proportional Load Sharing among Inverters Operated in Parallel." *IEEE Trans On Industrial. Electronics,* vol. 60,no. 4,pp: 1281–1290,2013.

[13] Chia-Tse Lee, Chia-Chi Chu, Po-Tai Cheng. "A New Droop Control Method for the Autonomous Operation of Distributed Energy Resource Interface Converters." *IEEE Trans On Power Electronics,* vol.28, no.4, pp: 1980-1993,2013.

978-1-4799-2706-7/14 $31.00 © 2014 IEEE

Suppression Control Method for Iron Loss of MATRIX Motor under Flux Weakening Utilizing Individual Winding Current Control

Hiroki Hijikata
Student member, IEEJ
Shibaura Institute of
Technology
Tokyo, Japan
nb13509@shibaura-
it.ac.jp

Kan Akatsu
Member, IEEJ
Shibaura Institute of
Technology
Tokyo, Japan
akatsu@sic.shibaura-
it.ac.jp

Yoshihiro Miyama
Member, IEEJ
Mitsubishi Electric
Corporation
Hyogo, Japan
Miyama.Yoshihiro@bc.
mitsubishielectric.co.jp

Hideaki Arita
Member, IEEJ
Mitsubishi Electric
Corporation
Hyogo, Japan
Arita.Hideaki@ab.mitsu
bishielectric.co.jp

Akihiro Daikoku
Member, IEEJ
Mitsubishi Electric
Corporation
Hyogo, Japan
Daikoku.Akihiro@ab.mi
tsubishielectric.co.jp

Abstract— An interior permanent magnet synchronous motor experiences large harmonic iron loss in the stator teeth caused by the harmonic magnetomotive forces. In particular, the eddy current loss remarkably increases at high rotational speeds under the flux weakening control. In this paper, a novel suppression control method for the stator iron loss is proposed by utilizing the individual winding current control. Although the proposed method requires multi-phase winding and multiple full-bridge inverters, the individual winding current control can observe and regulate the flux density in each tooth. This paper reveals that the proposed control method can achieve the extensive reduction of the total iron loss.

Keywords— *Harmonic iron loss, Individual winding current control, Interior permanent magnet synchronous motor, MATRIX motor.*

I. INTRODUCTION

An interior permanent magnet synchronous motor (IPMSM) is widely used in high performance applications due to lots of benefits. One of the main advantages is to extend the operating area facilitated by the flux weakening control [1]. This control can reduce the back-EMF and can generate the reluctance torque. On the other hand, the iron loss of IPMSM is popularly known for increasing at high rotational speeds under the flux weakening control [2]-[4]. In particular, the harmonic eddy current loss remarkably increases in the stator teeth of the distributed wound IPMSM [5]. The cause of increasing iron loss is the harmonic magnetomotive forces by the spatial harmonics due to the permanent magnet and armature reaction. The flux weakening control can only cancel the fundamental component of the magnet flux linkage. Some literatures research the optimized design methods [6]-[9]. It can be considered that the harmonic magnetomotive forces can be reduced by optimizing shapes of the rotor or stator core. These methods, however, often reduce the output torque due to decreasing magnet flux linkage and q-axis

inductance. As a result, these methods take difficult to exactly obtain high efficiency.

In this paper, a novel suppression control method for the stator iron loss is proposed by utilizing the individual winding current control. The authors have previously proposed the MATRIX motor as expecting to offer additional advantage by using lots of power electronics devices [10]. The proposed method can be achieved by multi-phase winding and decoupling d_1-q_1-d_3-q_3-d_5-q_5 sequence current control with multiple inverters [11]-[12]. Although the proposed method requires the multi-phase winding and multiple full-bridge inverters, the individual winding current control can regulate the flux linkage of each tooth. Furthermore, the harmonic flux density in each tooth is also controlled by injecting harmonic current into each winding. The purpose of this study is to suppress the total iron loss by the proposed method even though the copper loss increases. As a result, the proposed method is expected to improve the motor efficiency at high speed area. This paper reveals that the proposed method can achieve the extensive reduction of the total iron loss by the 2-D finite element analysis (FEA) and some experiments.

II. PROPOSED MOTOR DRIVE SYSTEMS

A. Motor and Inverter Configuration

Fig. 1(a) shows the tested FEA model created by JMAG (electromagnetic field analysis software). The model is the distributed wound IPMSM and includes the open-winding either wye- or delta-connection [13]-[15]. Although the open-winding motor requires more number of power electronics devices than the number of the conventional wye-connection uses, the configuration is expected to offer additional advantages, for example, adopting the zero-sequence, the high utilization ratio of DC bus voltage, and its fault tolerance with an individual drive unit. The armature windings connect the 12 full-bridge inverters shown in Fig. 1(b) because the motor includes 12-phase armature windings for flux control in

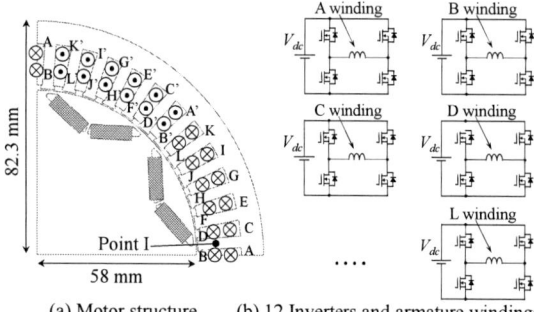

(a) Motor structure (b) 12 Inverters and armature windings

Fig. 1. Proposed motor drive system component.

TABLE I
ANALYTICAL MOTOR SPECIFICATIONS

Parameters	Value	Unit
Number of poles	8	[-]
Number of slots	48	[-]
Stack length	66.6	[mm]
Turn number per slot	3	[turn/slot]
Winding resistance	5	[mΩ/winding]

each tooth of 8-pole and 48-slot IPMSM. The motor specifications are given in Table 1.

B. Decoupling Transformation Matrix

The proposed motor is applied the 6-phase current despite the motor has 12 armature windings because each slot includes 2-layer winding. Decoupling transformation matrix for 6-phase to control the harmonics is given by (1) at the bottom of the page [11]-[12]. First and second rows in (1) define the fundamental component of the machine variables and the k th order harmonics with $k = 12m \pm 1$ ($m = 1, 2, 3 ...$), such as 11th, 13th, 23rd, 25th ... harmonics, which are transformed into the d_1-q_1 subspace or d_1-q_1 plane. Third and fourth rows in (1) show harmonics with $k = m - 3$ ($m = 1, 3, 5 ...$), such as 3rd, 9th, 15th, 21st ... harmonics, which mapped into the d_3-q_3 subspace or d_3-q_3 plane. Last two rows in (1) indicate harmonics with $k = 6m \pm 1$ ($m = 1, 3, 5 ...$), such as 5th, 7th, 17th, 19th ... harmonics, which mapped into the d_5-q_5 subspace or d_5-q_5 plane. The fundamental component, d_3-q_3 subspace, and d_5-q_5 subspace are individually controlled since these subspaces are orthogonal each other.

C. Iron Loss Calculation

In this paper, the FEA calculates the iron loss by following equation [3]

$$W_i = W_h + W_e = K_h B_m^2 f + K_e B_m^2 f^2 \qquad (2)$$

where, W_i, W_h, W_e, K_h, K_e, B_m, and f indicate the iron loss, hysteresis loss, eddy current loss, coefficient of the hysteresis loss, coefficient of the eddy current loss, maximum flux density, and frequency of the flux density, respectively. In (2), it is clear that the eddy current iron loss depends on the flux density and its frequency.

D. Distorted Flux Density under Flux Weakening Control

It is generally known that the flux weakening control distorts the flux density waveform in stator teeth [3]. Fig. 2 shows the radial component of the flux density distribution waveforms at point I in Fig. 1 and their FFT results. The applying phase current is 37.5 A_{rms} and current phase shifts from 0 to 80 deg. As shown in these graphs, it is found that the 3rd, 5th, 7th, and 9th harmonics component of the flux density remarkably increase with increasing current phase even though the fundamental component decreases.

Fig. 3 shows the FEA results of the classified iron loss in case of driving at the low and high speed area. The conditions of the phase current and rotating speed are 37.5 A_{rms} and 100 or 6000 rpm, respectively. These values are normalized by each total iron loss at 0 deg. As shown in this graph, the total iron losses decrease with increasing current angle in each rotating speed. In case of the low speed conditions, it seems that the total iron loss depends on the hysteresis loss in the stator core due to low frequency in (2). On the other hand, it can be found that the eddy current losses in the high speed conditions, especially in stator core, are much larger than those in the low speed conditions. In particular, these losses cannot be decreased with increasing current angle because the harmonic eddy current loss caused by the distorted flux density in the stator teeth remarkably increases.

E. Individual Control for Iron Loss Suppression

The flux linkage is estimated by following equation [16]

$$T = \begin{bmatrix}
\cos\theta & \cos\left(\theta - \dfrac{\pi}{6}\right) & \cos\left(\theta - \dfrac{2\pi}{3}\right) & \cos\left(\theta - \dfrac{5\pi}{6}\right) & \cos\left(\theta - \dfrac{4\pi}{3}\right) & \cos\left(\theta - \dfrac{3\pi}{2}\right) \\
-\sin\theta & -\sin\left(\theta - \dfrac{\pi}{6}\right) & -\sin\left(\theta - \dfrac{2\pi}{3}\right) & -\sin\left(\theta - \dfrac{5\pi}{6}\right) & -\sin\left(\theta - \dfrac{4\pi}{3}\right) & -\sin\left(\theta - \dfrac{3\pi}{2}\right) \\
\cos3\theta & \cos3\left(\theta - \dfrac{\pi}{6}\right) & \cos3\left(\theta - \dfrac{2\pi}{3}\right) & \cos3\left(\theta - \dfrac{5\pi}{6}\right) & \cos3\left(\theta - \dfrac{4\pi}{3}\right) & \cos3\left(\theta - \dfrac{3\pi}{2}\right) \\
-\sin3\theta & -\sin3\left(\theta - \dfrac{\pi}{6}\right) & -\sin3\left(\theta - \dfrac{2\pi}{3}\right) & -\sin3\left(\theta - \dfrac{5\pi}{6}\right) & -\sin3\left(\theta - \dfrac{4\pi}{3}\right) & -\sin3\left(\theta - \dfrac{3\pi}{2}\right) \\
\cos5\theta & \cos5\left(\theta - \dfrac{\pi}{6}\right) & \cos5\left(\theta - \dfrac{2\pi}{3}\right) & \cos5\left(\theta - \dfrac{5\pi}{6}\right) & \cos5\left(\theta - \dfrac{4\pi}{3}\right) & \cos5\left(\theta - \dfrac{3\pi}{2}\right) \\
-\sin5\theta & -\sin5\left(\theta - \dfrac{\pi}{6}\right) & -\sin5\left(\theta - \dfrac{2\pi}{3}\right) & -\sin5\left(\theta - \dfrac{5\pi}{6}\right) & -\sin5\left(\theta - \dfrac{4\pi}{3}\right) & -\sin5\left(\theta - \dfrac{3\pi}{2}\right)
\end{bmatrix} \qquad (1)$$

978-1-4799-2706-7/14 $31.00 © 2014 IEEE

The 2014 International Power Electronics Conference

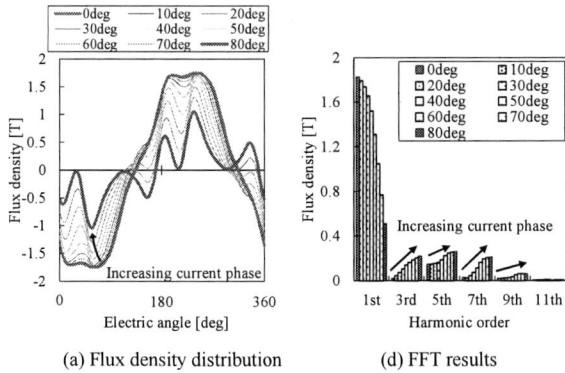

(a) Flux density distribution   (d) FFT results

Fig. 2. Radial component of the flux density distribution and FFT results at point I. The applying phase current is 37.5 Arms and current phase shifts from 0 to 80 deg.

TABLE II
SIMULATION CONDITIONS

Conditions	Value	Unit
Phase current of the fundamental component	37.5	[A$_{rms}$]
Current phase	80	[deg]
Rotating speed	6000	[rpm]

$$\lambda = \int (v - Ri)\, dt + \lambda_0 \qquad (3)$$

where, λ, v, R, i, and λ_0 indicate the total flux linkage, armature voltage, armature winding resistance, armature current, and initial value of the total flux linkage, respectively. Although the system requires 12 voltage sensors and 12 current sensors, it can obtain the flux linkage in each tooth by

$$\lambda_{AC} = \lambda_A - \lambda_C \qquad (4)$$

where, λ_{AC}, λ_A, and λ_C indicate the total flux linkage in tooth between A winding and C winding, flux linkage of A winding, and C winding, respectively. As calculating the tooth between A winding and C winding, (3) and (4) estimate flux linkage in 6 teeth among 12 windings. Using d_1-q_1-d_3-q_3-d_5-q_5 transformation, the distorted flux linkage is expressed by the rotational orthogonal frame. The difference between the distorted flux linkage and flux reference make the current reference fluctuate in order to cancel the distorted flux linkage. The proposed

method allows the flux linkage in each tooth deformed to substantially sinusoidal waveform. Furthermore, the above method also makes the flux density distribution deform to substantially sinusoidal waveform. Thus, the harmonic iron loss can be suppressed.

III. SIMULATION RESULTS AND DISCCUSSION

In this section, the effectiveness of the suppression method for the iron loss is verified by the FEA. The applying harmonic current is used by the sequence of points which are calculated to suppress the harmonic flux linkage in each tooth.

A. Comparison of Conventional Flux Weakening Control and Proposed Control

This section describes a comparison between the conventional flux weakening control and proposed suppression control for the iron loss. The simulation conditions are summarized in Table II. Fig. 4 shows the waveforms with or without compensation and Fig. 5 shows their FFT results. As shown in Figs. 4(a) and 5(a), the flux density at point I is much distorted without compensation, especially 3rd, 5th, 7th, 9th harmonics, whereas the flux density with compensation has few harmonic component. The voltage waveform also include few harmonic component with compensation compared with the results compensation in Figs. 4(c) and 5(c). On the other hand, the current waveform with compensation includes the 3rd, 5th, 7th, and 9th harmonics in Figs. 4(b) and 5(b) to cancel the harmonic flux density. As shown in Figs. 4(d) and 5(d), the proposed control method generates the large torque pulsation caused by harmonic injection. However, the torque ripples less effect at high-speed revolution when the flux weakening control is performed.

This part compares the iron loss with or without compensation between the 6-phase and 3-phase motor. Fig. 6 shows the winding structure of the comparative 3-phase motor which is popularly employed for the traction motor. Although the structure of the rotor and stator core is exactly same as Fig. 1(a) which is the 6-phase motor, the motor includes the 3-phase distributed lap-winding. The 3-phase motor is applied additional current in order to produce same output torque of the 6-phase motor which includes full-pitch winding. The simulation

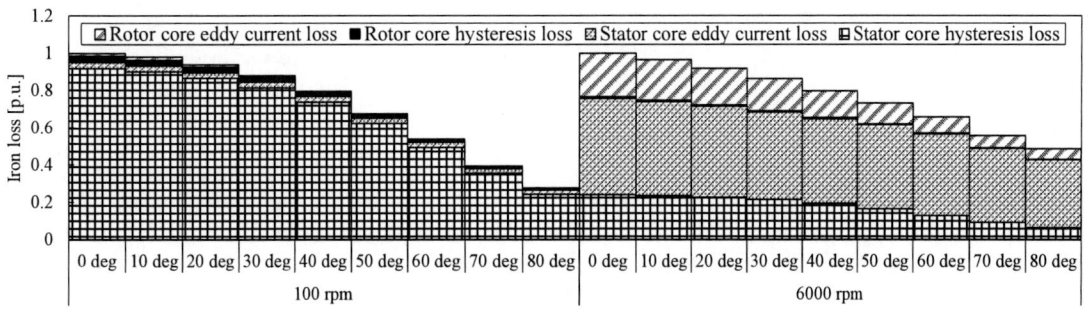

Fig. 3. Classified iron loss in case of the low and high speed operation. The applying phase current is 37.5 A$_{rms}$ and current phase shifts from 0 to 80 deg. These values are normalized by each total iron loss at 0 deg.

978-1-4799-2706-7/14 $31.00 © 2014 IEEE   2675

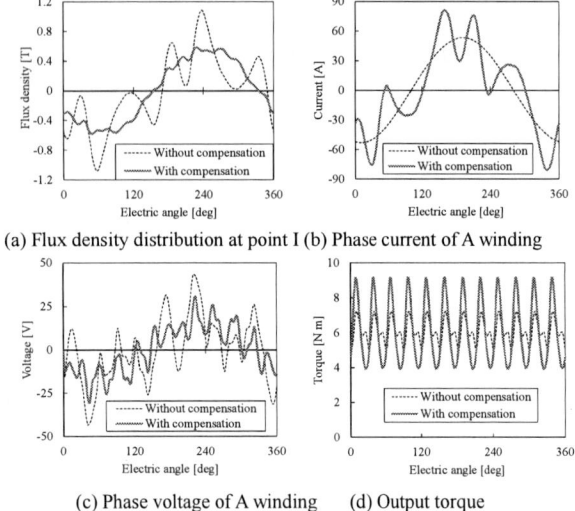

(a) Flux density distribution at point I (b) Phase current of A winding

(c) Phase voltage of A winding   (d) Output torque

Fig. 4.   FEA results of each waveform with or without compensation.

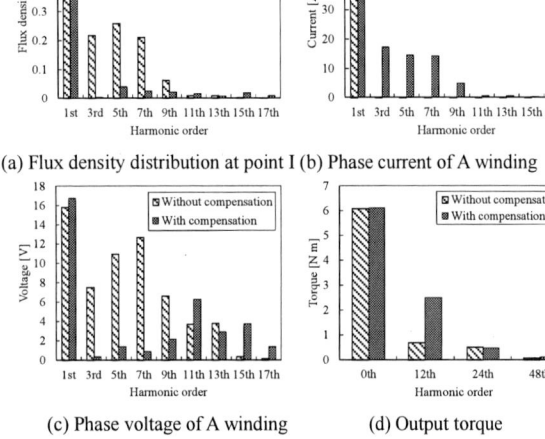

(a) Flux density distribution at point I (b) Phase current of A winding

(c) Phase voltage of A winding   (d) Output torque

Fig. 5.   FFT results of each waveform.

conditions are given in TABLE II. Fig. 7 shows the iron loss density distribution with or without compensation. As shown in Fig. 7(a) and (b), the iron loss in the stator teeth is clearly reduced with compensation in the 6-phase drive. It can be said that the proposed method can remarkably suppress the iron loss because the flux density waveform in each tooth becomes sinusoidal waveform as shown in Fig. 5(a). Thus, the proposed control method can largely inhibit the expression of the harmonic iron loss. The 3-phase motor without compensation obtains similar iron loss density distribution of 6-phase motor without compensation. On the other hand, it is found that the proposed control method increase the stator iron loss in case of the 3-phase motor. The reason is the proposed method can only control the flux density of the teeth between phases. Therefore, it is said that the proposed control method of

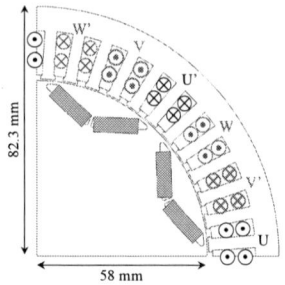

Fig. 6. Proposed motor with 3-phase armature winding configuration.

the 8-pole 48-slot motor is suitable for 6-phase drive.

Figs. 8 and 9 show the results of the classified iron losses due to their origins and the output power and loss analysis, respectively. It can be seen that the stator core eddy current loss of the 6-phase motor with compensation is remarkably decreased in Fig. 8. Furthermore, other losses are also decreased such as the stator core hysteresis loss, rotor core hysteresis loss, and rotor core eddy current loss. The total iron loss decreases approximately 70 % whereas the copper loss increases about 30 % due to applying additional harmonic current in Fig. 9. Therefore, it is expected that the proposed method makes the efficiency improve.

B.   Speed and Loss Characteristics

This part discusses about a useable speed range of the proposed iron loss reduction control in 6-phase motor. These simulation results are summarized in Fig. 10 which indicates amount of compensated loss by black color. The compensated loss is defined that the amount of increased copper loss is subtracted from amount of decreased iron loss. As shown in this graph, it is proved that the iron loss is reduced with increase of the current phase and rotating speed. It is also found that the boundary line between the amount of increased copper loss and amount of decreased iron loss is drawn in approximately 2500 to 4100 rpm. As a result, it is said that the efficiency is improved in more than 2500 rpm region because the proposed control method obtains the effectiveness in these area.

IV.   EXPERIMENTAL RESULTS AND DISCUSSION

This section experimentally confirms the effect of the proposed method. The reference current to cancel the distorted flux density is decided by the trial-and-error method. The picture of the experimental equipment and experimental conditions are shown in Fig. 11 and TABLE III.

The waveforms of the measured current and observed flux linkage and their FFT results are shown in Figs. 12 and 13, respectively. As shown in Figs. 12(a) and 13(a), the conventional flux weakening control, without compensation, applies the sinusoidal current, whereas the proposed control method, with compensation, applies distorted current. In case of linkage flux in Figs. 12(b) and 13(b), it is found that the observed flux linkage waveform in stator teeth without compensation includes

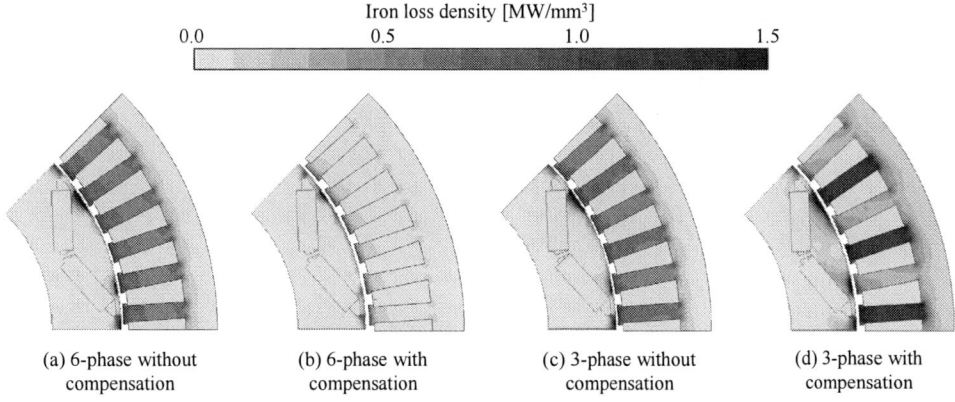

(a) 6-phase without compensation (b) 6-phase with compensation (c) 3-phase without compensation (d) 3-phase with compensation

Fig. 7. Iron loss density distribution in the rotor and stator core.

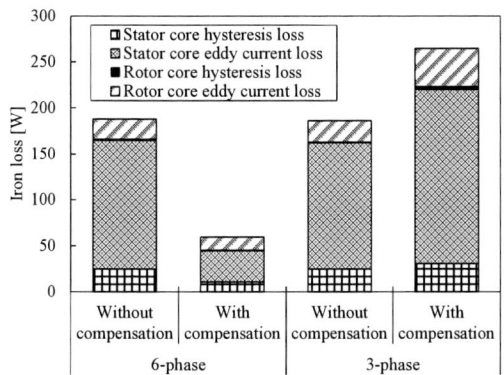

Fig. 8. Classified iron losses with or without compensation. The model in Figs. 1 or 6 are employed for the 6-phase or 3-phase drive, respectively.

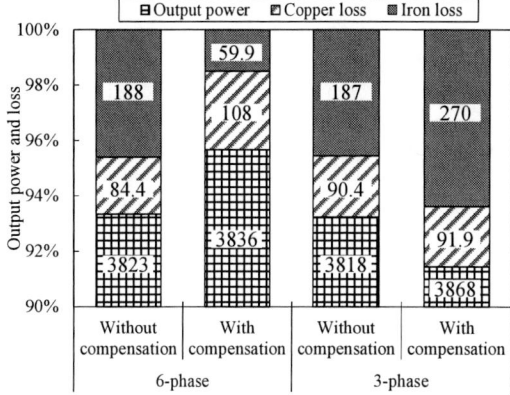

Fig. 9. Output power and loss abalysis. Each value in the bar is expressed in watt (W)

much harmonic component as with the FEA results. On the other hand, it indicates that the harmonic component with compensation became very small, and the waveform of the flux linkage became nearly sinusoidal. In particular, the 3rd, 5th, and 7th harmonic components are decreased with compensation. This effect is owing to the proposed control method. In the future work, the detailed experimental data of the efficiency measurement will be shown. It is expected to improve the motor efficiency by the proposed method.

V. CONCLUSIONS

This paper described the suppression control method for the harmonic iron loss under the flux weakening control. The results of this study reached the following conclusions.

- The proposed method can regulate each tooth flux utilizing the individual current control. As a result, the harmonic iron losses are reduced at the high rotational speed by the harmonic current injection.

- Some simulation results reveal that the proposed control method obtains the effectiveness at more than approximately 2500 rpm. In particular, the iron loss can be reduced up to about 70 % even though the copper loss increases about 30 %.

- The proposed control method generates the large torque pulsation. However, the torque ripples less effect at high-speed revolution when the flux weakening control is performed.

- Some experimental results revealed that the proposed method can reduce the harmonic flux linkage in the stator teeth.

In the future work, the experimentally effectiveness of the proposed method in the iron loss will be shown.

REFERENCES

[1] T.M. Jahns, G.B. Kliman, and T.W. Newmann, "Interior Permanent-Magnet Synchronous Motor for Adjustable-Speed Drives," *IEEE Tran. on Industry Applications*, Vol. IA-22, No. 4, pp. 738-747, Jul./Aug. 1986.

[2] R. Schiferl and T.A. Lipo, "Core Loss in Buried Magnet Permanent Magnet Synchronous Motors," *IEEE Trans. on Energy Conversion*, Vol. 4, No. 2, pp. 279-284. Jun. 1989.

[3] K. Yamazaki and Y. Seto, "Iron Loss Analysis of Interior Permanent-Magnet Synchronous Motors — Variation of Main Loss Factors Due to Driving Condition," *IEEE Trans. on Industry Applications*, Vol. 42, No. 4, pp. 1045-1052. Jun./Aug. 2006.

[4] V. Zivotic-Kukolj, W.L. Soong, and N. Ertugrul, "Iron Loss Reduction in an Interior PM Automotive Alternator," *IEEE Trans. on Industry Applications*, Vol. 42, No. 6, pp. 1478-486, Nov./Dec. 2006.

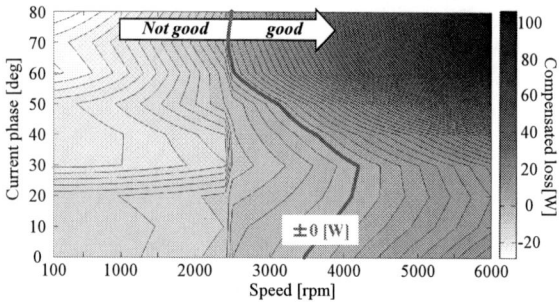

Fig. 10. Loss different characterisitics.

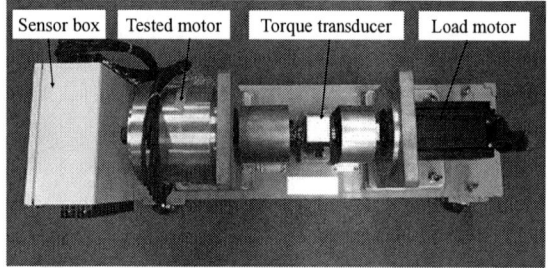

Fig. 11. Picture of experimental equipment.

TABLE III
EXPERIMENTAL CONDITIONS

Conditions	Value	Unit
Phase current of the fundamental component	37.5	[A_{rms}]
Current phase	80	[deg]
Rotating speed	500	[rpm]

[5] K. Yamazaki, Y. Fukushima, and M. Sato, "Loss Analysis of Permanent Magnet Motors With Concentrated Windings — Variation of Magnet Eddy-Current Loss Due to Stator and Rotor Shapes," *IEEE Trans. on Industry Applications*, Vol. 45, No. 4, pp. 1334-1342, Jul./Aug. 2009.

[6] K. Yamazaki and H. Ishigami, "Rotor-Shape Optimization of Interior-Permanent-Magnet Motors to Reduce Harmonic Iron Losses," *IEEE Trans. on Industrial Electronics*, Vol. 57, No. 1, pp. 61-69, Jan. 2010.

[7] M. Barcaro, N. Bianchi, and F. Magnussen, "Rotor Flux-Barrier Geometry Design to Reduce Stator Iron Losses in Synchronous IPM Motors Under FW Operations," *IEEE Trans. on Industry Applications*, Vol. 46, No. 5, pp. 1950-1958, Sep./Oct. 2010.

[8] S.H. Han, W.L. Soong, T.M. Jahns, M.K. Guven, and M.S. Illindala, "Reducing Harmonic Eddy-Current Losses in the Stator Teeth of Interior Permanent Magnet Synchronous Machines During Flux Weakening," *IEEE Trans. on Energy Conversion*, Vol. 25, No. 2, pp. 441-449, Jun. 2010.

[9] S.H. Han, T.M. Jahns, and Z.Q. Zhu, "Analysis of Rotor Core Eddy-Current Losses in Interior Permanent-Magnet Synchronous Machines," *IEEE Trans. on Industry Applications*, Vol. 46, No. 1, pp. 196-205, Jan./Feb. 2010.

[10] H. Hijikata and K. Akatsu, "Principle and Basic Characteristic of MATRIX Motor that Achieves Variable Parameters by Arbitrary Winding Connections," *IEEJ Trans. Industry Applications*, Vol.2, No.6, pp.283-291, Nov. 2013.

[11] M.A. Abbas, R. Christen, and T.M. Jahns, "Six-Phase Voltage Source Inverter Driven Induction Motor," *IEEE*

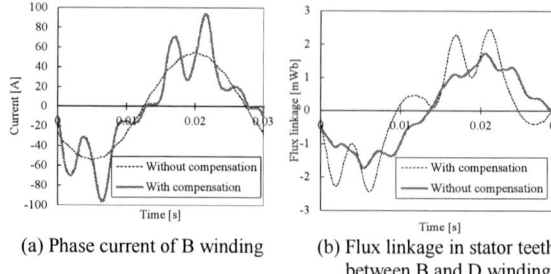

(a) Phase current of B winding (b) Flux linkage in stator teeth between B and D winding

Fig. 12. Experimental results of each waveform with or without compensation.

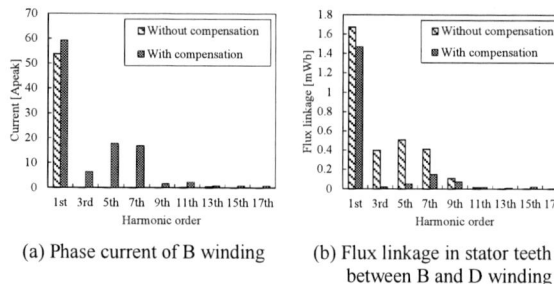

(a) Phase current of B winding (b) Flux linkage in stator teeth between B and D winding

Fig. 13. FFT results of each experimental waveform with or without compensation.

Trans. on Industry Applications, Vol.IA-20, No.5, pp.1251-1259, Sep./Oct. 1984.

[12] Y. Zhao and T.A. Lipo, "Space Vector PWM Control of Dual Three-Phase Induction Machine Using Space Vector Decomposition," *IEEE Trans. on Industry Applications*, Vol.31, No.5, pp.1100-1109, Sep./Oct. 1995.

[13] Y. Kawabata, M. Nasu, T. Nomoto, E.C. Ejiogu and T. Kawabata, "High-Efficiency and Low Acoustic Noise Drive System Using Open-Winding AC Motor and Two Space-Vector-Modulated Inverters," *IEEE Trans. on Industrial Electronics*, Vol. 49, No. 4, pp. 783-789, Aug. 2002.

[14] B.A. Welchko, T.A. Lipo, T.M. Jahns, and S.E. Schulz, "Fault Tolerant Three-Phase AC Motor Drive Topologies: A Comparison of Features, Cost, and Limitations," *IEEE Trans. on Industry Applications*, Vol. 19, No. 4, pp. 1108-1116, Jul. 2004.

[15] R.U. Haque, M.S. Toulabi, A.M. Knight, and J. Salmon, "Wide Speed operation of PMSM using an Open Winding and a Dual Inverter Drive with a Floating Bridge," in *Proc. of IEEE Energy Conversion Congress and Exposition (ECCE) 2013*, pp.3784-3791, Sep. 2013.

[16] I. Takahashi, T. Noguchi, "A New Quick-Response and High-Efficiency Control Strategy of an Induction Motor," *IEEE Trans. on Industry Applications*, Vol. 22, No. 5, pp. 820-827, Sep./Oct. 1986.

Performance analysis of a new concentrated-winding interior permanent magnet synchronous machine under Field Oriented Control

D. Nguyen, R. Dutta, J. Fletcher, F. Rahman
School of Electrical Eng. and Telecommunications
University of New South Wales
Sydney, Australia
Email: Dai.nguyen@unsw.edu.au

Howard Lovatt
CSIRO, Lindfield Laboratories
Material Science and Engineering
Sydney, Australia

Abstract— **This paper analyzes the performance of a new fractional-slot, concentrated-winding (FSCW) Interior Permanent Magnet Synchronous Machines (IPMSM) under Field Oriented Control (FOC). The FSCW IPMSM is being developed vigorously in recent years because of its high-power density, high efficiency, wide field-weakening capability, and high fault tolerance compared to the distributed-wound IPMSMs. The major disadvantage of concentrated winding is often cited to be its non-sinusoidal stator MMF. However, using appropriate combination of slots and poles, nearly sinusoidal EMF and very low cogging torque can be achieved. Although steady-state performance such as efficiency and constant power speed range of this type of PM machines have recently been published, dynamic performances have yet not been reported. The FOC scheme which rely on the machine model, were applied to a prototype a 14-pole/ 18-slot, double layer concentrated-wound IPM machine. The *dq* model for distributed wound IPMSM was used for the control. This paper investigates performances of the test machine in terms of torque and current ripple, transient responses for the step variations of speed and load under maximum torque per ampere and field weakening control regimes. The experimental results on the concentrated winding IPM machine indicate that the distributed windng *dq* model is not sufficiently accurate to harness the full capability of the new machine.**

Keywords— *Field oriented control, fractional-slot concentrated winding, flux weakening, interior permanent magnet synchronous machine.*

I. INTRODUCTION

The interior permanent magnet synchronous machine (IPMSM) in which magnet poles are buried inside the rotors has been an attractive choice for many high performance applications due to its high power and torque density and efficiency. In the conventional ac machines, distributed winding (DW) which result in nearly sinusoidal back EMF and MMF waveforms are desired for achieving high efficiency and low torque ripple [1]. However, there are several disadvantages such as difficulty in winding automation, long end-winding in the stator and higher copper losses compared to the concentrated winding (CW). In contrast to DW, it is well-known that the CW machines generate non-sinusoidal

MMF which contains harmonics and sub-harmonics that do not rotate at synchronous speed. These additional harmonics are the main cause of increased frequency related losses.

Recently, several IPM machines with fractional-slot, concentrated-windings (FSCW) have been reported in the literature [2-7]. These machines have very short end-windings, which are more compact and thus increases the power density of these machines beyond the DW IPMSMs. The windings, being localized for each stator slot, offer better reliability, compactness and high direct-axis inductance which assists in achieving wider field weakening speed range than the DW IPMSM. However, apart from the non-sinusoidal MMF low winding factor and higher torque ripples are also commonly cited short-comings of the CW. Recent studies [2, 8] have shown that a winding factor close to unity and very low cogging torque can be achieved in FSCW machine if an appropriate combination of slots and poles are chosen.

This has opened the path for the FSCW PM machine to many applications, in particular, to automotive drives where high field weakening and compact size are important. It has been noted that FSCW machines have low mutual inductance between two phases. Consequently, the saliency ratio of the CW IPMSM is often much lower than a DW IPMSM [3]. In spite of the low saliency ratio, FSCW machines can offer wide flux-weakening range [9]. The 14-pole/18-slot, double layer FSCW IPMSM machine of reference [3] demonstrates field weakening range in excess of 8:1. Figure 1 shows the rotor and stator views of this prototype machine.

The Field Oriented Control (FOC) is a high performance control scheme and is widely used in many drive systems when a high-resolution, ripple-free shaft position sensor is available. Under this scheme, the IPMSM drives can be controlled according to the maximum torque per ampere (MTPA), maximum torque per voltage (MTPV), field weakening (FW) and loss minimization trajectories [10-13]. One of the main short-comings of the FOC scheme, being model based, is its reliance on machine parameters. Several control strategies

978-1-4799-2706-7/14 $31.00 © 2014 IEEE

which can compensate for the parameter variations and enhance the performance of the FOC are available for the DW IPMSM [14-17]. Another major drawback of FOC is the requirement of rotor position feedback, although several sensorless version of FOC have also been reported in the recent years again mostly for DW IPMSM [16, 17]. Availability of this sensor, however, guarantees high performance down to zero speed.

Figure 1 Rotor and stator of the FSCW IPMSM

This paper reports, for the first time, an evaluation of the steady-state and dynamic performance of a FSCW IPMSM under FOC scheme. The machine model used for FSCW IPMSM is based on the conventional rotor dq frame. The purpose of this experimental study is, firstly, to understand the control characteristics of the FSCW machine compared to a DW machine, and secondly, to investigate the appropriateness of application of the rotor dq model of the DW machine to the FSCW machine, in order to harness its full power-speed capability.

The paper is organized as follows:

The field oriented control principles of the IPMSM in rotating reference frame are discussed in section II. The experimental results and the analysis of the model based controllers for the FSCW machine are presented in section III. The section IV discusses some of the issues encountered in high flux weakening range while implementing the FOC in the prototype CW IPMSM. Section V presents concluding remarks of this study.

II. FIELD ORIENTED CONTROL PRINCIPLES

A. Mathematical model of the IPMSM in rotor reference frame

In the d-q coordinate that rotate synchronously with rotor, the flux linkage and voltage equations of the IPMSM are expressed as follow

$$\begin{bmatrix} \lambda_d \\ \lambda_q \end{bmatrix} = \begin{bmatrix} L_d & 0 \\ 0 & L_q \end{bmatrix} \begin{bmatrix} i_d \\ i_q \end{bmatrix} + \begin{bmatrix} \lambda_{pm} \\ 0 \end{bmatrix} \quad (1)$$

$$\begin{bmatrix} v_d \\ v_q \end{bmatrix} = \begin{bmatrix} R_s + pL_d & -\omega L_q \\ \omega L_d & R_s + pL_q \end{bmatrix} \begin{bmatrix} i_d \\ i_q \end{bmatrix} + \begin{bmatrix} 0 \\ \omega \lambda_{pm} \end{bmatrix} \quad (2)$$

where,

R_s	stator resistance
L_d, L_q	d- and q- axis inductances
λ_{pm}	permanent magnet flux linkage
i_d, i_q	d- and q- axis currents
v_d, v_q	d- and q- axis voltages
ω	rotor angular velocity in electrical rad/s

It should be noted that in this paper, the three-phase voltages, currents and flux linkages are transformed into d-q coordinate by using the well-known Park's transformation.

$$\begin{bmatrix} d \\ q \end{bmatrix} = \frac{2}{3} \begin{bmatrix} \cos\theta & \cos(\theta - \frac{2\pi}{3}) & \cos(\theta + \frac{2\pi}{3}) \\ -\sin\theta & -\sin(\theta - \frac{2\pi}{3}) & -\sin(\theta + \frac{2\pi}{3}) \end{bmatrix} \begin{bmatrix} a \\ b \\ c \end{bmatrix} \quad (3)$$

where θ is the rotor position in electrical radian.

The torque expression for a permanent magnet machine in the d-q coordinate is given by

$$T = \frac{3}{2} P_p \left[\lambda_{pm} i_q + (L_d - L_q) i_d i_q \right] \quad (4)$$

The phasor diagram for IPMSM at steady-state is presented in the Figure 2

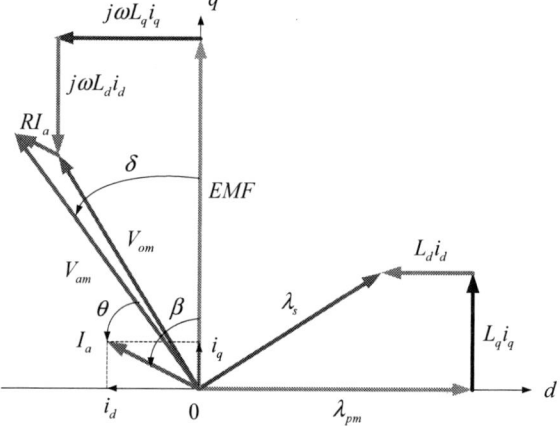

Figure 2 Phasor diagram for IPMSM in rotating reference frame

B. Trajectories control of IPMSM

1) Maximum torque per ampere (MTPA) trajectory

The MTPA condition can be achieved by seeking the maximum output torque for particular current amplitude. Based on equation (4), the relation between trajectories for maximum torque per ampere and voltage and current limits can be found as follow

$$i_d = \frac{\lambda_{pm}}{2(L_q - L_d)} - \sqrt{\frac{\lambda_{pm}^2}{4(L_q - L_d)^2} + i_q^2} \quad (5)$$

2) Flux weakening (FW) control

The flux linkage of the IPMSM is delivered from

$$\lambda_s = \sqrt{(L_d i_d + \lambda_{pm})^2 + (L_q i_q)^2} = \frac{V_{om}}{\omega} \quad (6)$$

where $V_{om} = V_{max} - R_s I_{am}$, V_{max} is the maximum phase voltage of the machine.

In the flux weakening region, when the voltage reaches the rated value, the flux linkage can be reduced by adjusting negative d-axis current. The relation between d-axis and q-axis current for flux weakening control algorithm can be derived from (6) by replacing V_o with V_{om} so that resistive drop can be accounted for.

$$i_d = -\frac{\lambda_{pm}}{L_d} \pm \frac{1}{L_d}\sqrt{(\frac{V_{om}}{\omega})^2 - (L_q i_q)^2} \quad (7)$$

3) Voltage and current constraint

Considering the constraints of voltage, the terminal voltage is limited as follow

$$V_o = \omega\sqrt{(L_d i_d + \lambda_{pm})^2 + (L_q i_q)^2} \le V_{om} \quad (8)$$

The stator current is limited by the rated current of the machine and current limit is given as,

$$I_a = \sqrt{i_d^2 + i_q^2} \le I_{am} \quad (9)$$

where I_{am} is the maximum current of the machine

The Figure 3 shows the current limit, voltage limit, MTPA trajectory and movement of the operating point during flux weakening for a typical IPMSM under FOC.

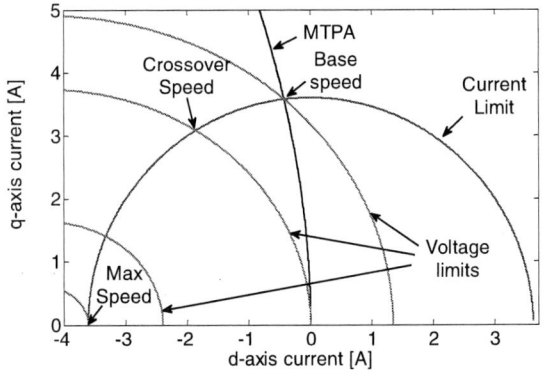

Figure 3 Calculated control trajectories in $i_d - i_q$ plane for the FSCW IPMSM

C. Transition of control mode

1) Below the base speed

The IPMSM operates under MTPA (5) until the speed at which voltage reaches the maximum limit voltage. The d- and q-axis currents of the operating point at base speed, under MTPA is given by

$$\begin{cases} i_{dA} = \frac{\lambda_{pm}}{4(L_d - L_q)} - \sqrt{\frac{\lambda_{pm}^2}{16(L_d - L_q)^2} + \frac{I_{om}^2}{2}} \\ i_{qA} = \sqrt{I_{om}^2 - i_{dA}^2} \end{cases} \quad (10)$$

The base speed is derived from (8) as follow

$$\omega_b = \frac{V_{om}}{\sqrt{(\lambda_{pm} + L_d i_d)^2 + (L_q i_q)^2}} \quad (11)$$

2) The crossover speed

Crossover speed is the speed at which the back-EMF

$\omega\lambda_{pm}$ equals to V_{om}. Hence, the crossover speed is

$$\omega_{cr} = \frac{V_{om}}{\lambda_{pm}} \quad (12)$$

Between the base speed and crossover speed, the machine can be operated either at MTPA or FW depending on load and speed. The control mode can be determined by comparing the flux linkage derived from voltage limit

$$\lambda_{sFW} = \frac{V_{om}}{\omega} \quad (13)$$

and from MTPA

$$\lambda_{sMTPA} = \sqrt{(\lambda_{pm} + L_d i_{dMPTA})^2 + (L_q i_{qMTPA})^2} \quad (14)$$

If $\lambda_{sMTPA} \ge \lambda_{sFW}$ then the MTPA is selected, otherwise the FW is selected.

3) Above the crossover speed

Above the crossover speed, there is no common point between the MTPA and the voltage limit ellipses, thus the FOC is controlled by the flux weakening control only. The maximum torque in flux weakening can be obtained by seeking an intersecting point of the voltage ellipses of operating speeds and current limit circle. The maximum d- and q-axis currents can be determined from using (6) and (9) as

$$\begin{cases} i_d = \dfrac{-\lambda_{pm}L_d + \sqrt{\lambda_{pm}^2 L_d^2 - (L_d^2 - L_q^2)(I_{om}^2 L_q^2 + \lambda_{pm}^2 - \dfrac{V_{om}^2}{\omega^2})}}{L_d^2 - L_q^2} \\ i_q = \sqrt{I_{om}^2 - i_d^2} \end{cases} \quad (15)$$

shows the flow chart of selection of control modes for FOC.

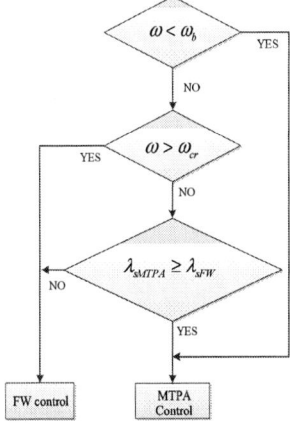

Figure 4 Selection of control modes

4) Maximum speed

For machine in which the characteristic current $I_{ch} = -\dfrac{\lambda_{pm}}{L_d}$ is higher than the rated maximum current,

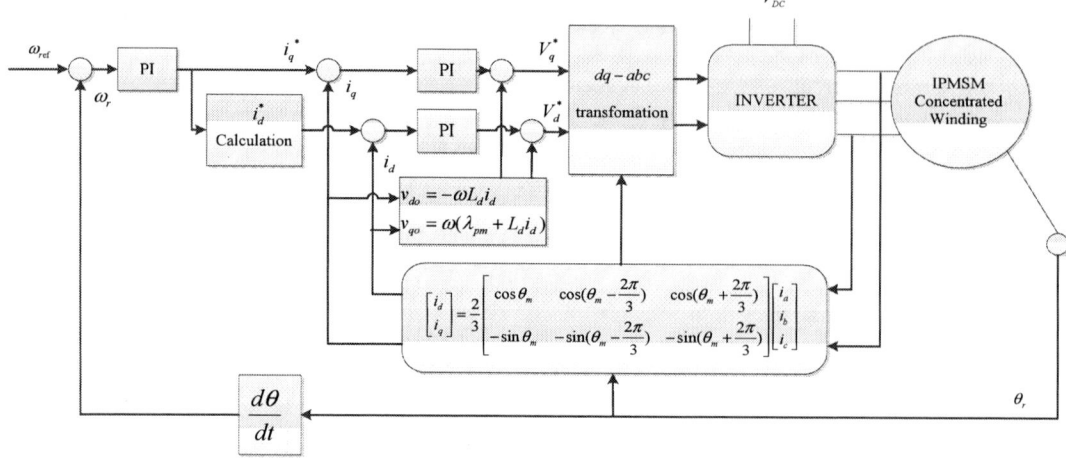

Figure 5 The block diagram of FOC for an IPMSM with CW

center of the concentric ellipses $(0, -I_{ch})$ stays outside of the current limit circle. In such machines, there is a finite maximum rotor speed at which the motor torque drops to zero. This maximum speed can be easily determined by

$$\omega_{max} = \frac{V_{om}}{\lambda_{pm} - I_{am} L_d} \quad (16)$$

III. EXPERIMENTAL RESULTS

TABLE I

Symbol	Meaning	Value
P_p	Number of pole pairs	7
R_s	Stator resistance	9.6 Ω
λ_{pm}	Magnet flux linkage	0.4 Wb
L_d	d-axis inductance	0.08644 H
L_q	q-axis induction	0.09757 H
V_{rated}	Rated Line voltage	320 V
I_a	Rated current	2.55 A

A. System Configuration

The block diagram of FOC is shown in . The control algorithms were implemented using DS1104 controller board. The 3- phase Mitsubishi intelligent power modules with IGBT switches were used in the voltage source inverter. A rotary encoder is used for speed and rotor position feedback.

B. Dynamic Performance

Figure 6 shows the speed and torque responses for a step speed change from 200 rpm to 700 rpm. Below the base speed of 450 rpm, the FOC was controlled along the MTPA trajectory of Figure 3. The motor is accelerated by maximum available torque. The flux weakening started when speed increases above base speed of 450 rpm.

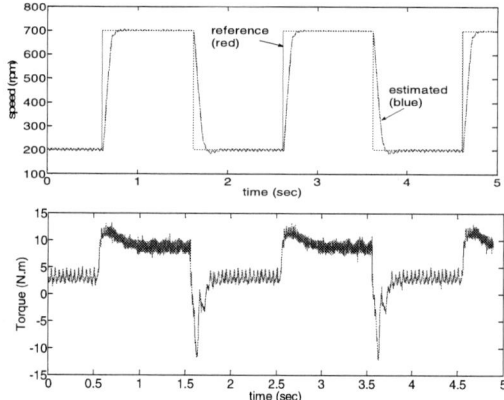

Figure 6 Speed and torque response of the FOC drive for speed change

The $i_d - i_q$ current responses of the FOC drive for a load step change is illustrated in Figure 7. The machine initially was run at 450 rpm under no-load condition, then the full load was applied to the shaft of the motor. The load torque was then abruptly removed and the machine came back to the no-load condition. The peak ripples in the $i_d - i_q$ current response were approximately 0.5 and 0.3 ampere respectively. These ripples were also reflected on the torque response of Figure 6.

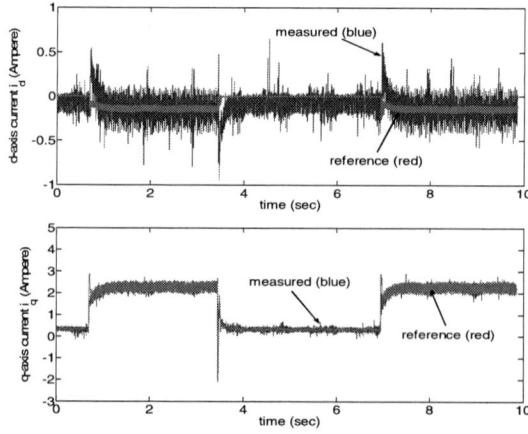

Figure 7 $i_d - i_q$ current responses of the FOC for load disturbances

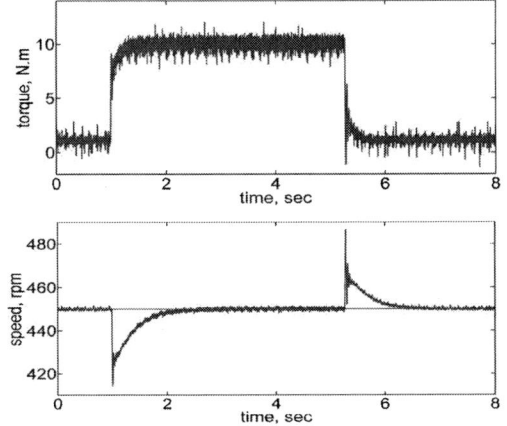

Figure 8 Speed and torque response at 450 rpm with load disturbances

Figure 8 illustrates the speed and torque response under the above mentioned load disturbances. From the figure, it can be observed that the speed dropped approximately 5% when the step load was applied to the shaft of the motor. The speed and torque regulation times are 0.8 and 0.2 sec respectively.

IV. PERFORMANCE ISSUES IN HIGH FLUX WEAKENING RANGE

A. Effect of Rotor position error at extremely high speed

Due to the limitation of the equipment, the system operates at the PWM switching frequency of 10kHz. Thus, the input signals including rotor position from encoder is updated by using DSpace1104 controller board for every 100µs. This reading delay corresponds to some rotor position error. The rotor position error in term of electrical rotor position is small and can be ignored at low speed. However, at high speed in flux weakening region, this delay will be significant and needs to be compensated. The delay in rotor position can be easily found as follow

$$\theta_{delay} = \omega T_s \qquad (17)$$

Figure 9 shows the back EMF waveform and rotor position read from encoder at 4,500 rpm under no-load test. At this speed, the delay of 100µs corresponds to the rotor position error of 25.7 electrical degree.

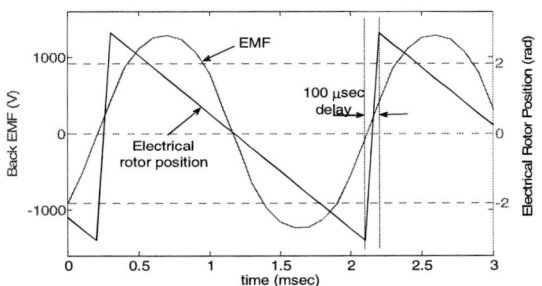

Figure 9 The back EMF and rotor position with no-load test at 4,500rpm

To demonstrate the effect from rotor position error, the torque-speed characteristics computed with (4) using measured id and iq current from the drive system with and without the delay are shown in Figure 10. As can be seen from this figure, there is an error which increases with speed due the increased delay. At the maximum speed of 4500 rpm, the error is about 6 Nm.

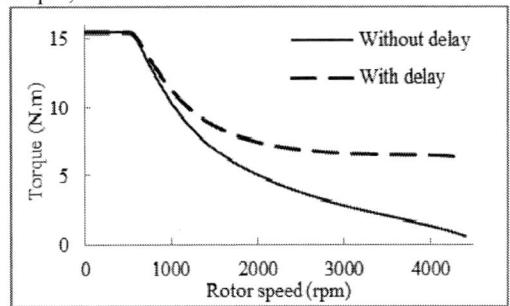

Figure 10 Calculated torque in an IPMSM with and without rotor position error

The rotor position error also affects to the d-axis and q-axis current reference values because of which without the delay compensation, the machine is not able to follow the trajectories mentioned in section II in deep flux weakening. By using a simple closed-loop speed-feedback compensation using (17), the rotor position error due to delays from the system was compensated. The Figure 11 shows the measured power vs speed of the machine with and without rotor position compensation.

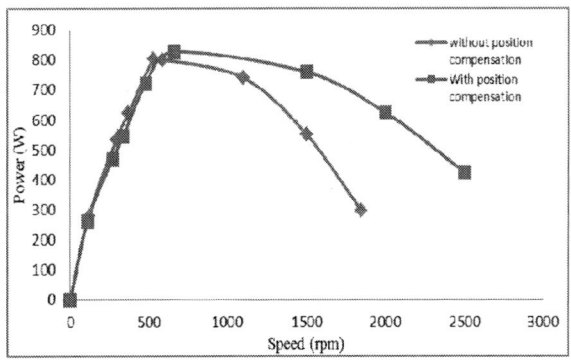

Figure 11 Power Speed characteristic with and without rotor position compensation

B. Effect from Concentrated Winding

The FOC scheme is based on the mathematical model of the machine which assumes sinusoidal MMF. However, the FSCW produces MMF which is rich in harmonics and sub-harmonics. It is useful to compare the flux linkage density in the air-gap by using the finite element analysis method. The Figure 12 shows magnetic flux distribution of a CW IPMSM.

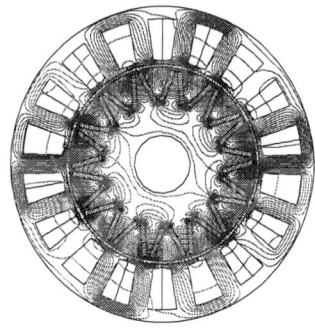

Figure 12 Flux distribution of concentrated winding 14-pole/18-slot IPMSM

Figure 13 shows the flux density distribution due to the stator current only for a concentrated winding and distributed winding.

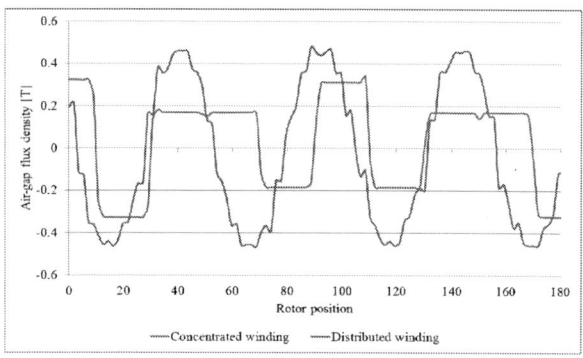

Figure 13 Air gap flux density (T) distribution due to stator current only

Since the concentrated winding produces MMF which is rich in harmonics and sub-harmonics, the conventional mathematical model of the IPM machine with DW is not accurate for the IPM machine with CW. The inaccuracy increases under flux-weakening condition. It is well known that harmonics in the flux-linkage increases under flux-weakening condition for the IPMSM with DW. Same is also true for the IPMSM with CW because of which mathematical model of the machine deteriorates further under flux-weakening. Although the FOC were applied successfully to the prototype IPMSM with CW, a noticeable steady-state error was observed between the measured and estimated flux-linkage of the machine, especially in the flux-weakening region. Figure 14(a) illustrates the estimated flux linkage of the machine by using the rotating (dq) reference frame and stationary ($\alpha\beta$) reference frame at 10 rpm which is well below the base speed. Figure 14(b) presents the same flux linkages at 700 rpm which is at 1.4 time base speed.

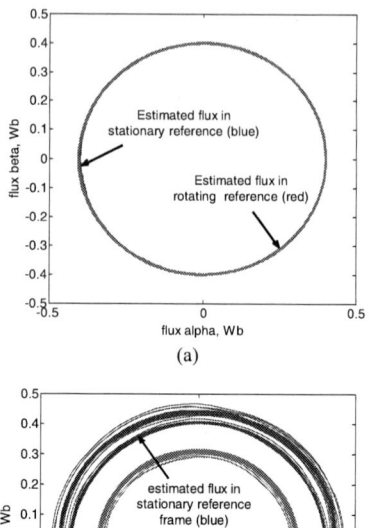

(a)

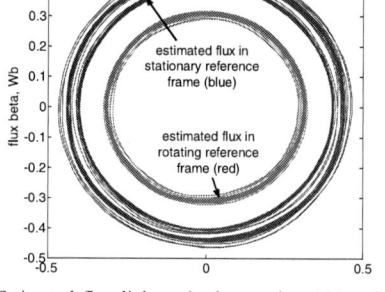

Figure 14 Estimated flux linkage in the rotating (dq) reference frame and stationary ($\alpha\beta$) reference frame at (a) 10 rpm, (b) 700rpm.

Figure 14 (a) shows that the estimated flux in the stationary reference frame ($\alpha\beta$) follows very closely to the estimated flux of the rotating reference frame (dq) when the speed was 10 rpm. However, the same fluxes measured at 700rpm presents a different picture. In this case, the estimated flux of the (dq) reference frame is no longer same as the estimated flux of ($\alpha\beta$) reference frame as it can be seen in Figure 14(b). This observation of estimated flux indicates that mathematical model of the FSCW IPMSM deteriorates as the speed increases.

Theoretically, the stationary reference frame can be found from the relationship between actual rotor position from sensor and the back EMF of the machine which assumes sinusoidal flux density in the air-gap. However the stationary reference frame of the CW IPM motors can be shifted by harmonic and sub-harmonic components in the air gap flux density of concentrated winding as can be seen in Figure 13. In order to improve the performance of the concentrated winding IPMSM, the phase shift between fundamental components between concentrated winding and distributed winding was studied by using finite element analysis. For this specific CW IPMSM, the phase difference between CW and DW was found to be 28 electrical degrees.

Figure 15 shows the power-speed characteristics of the CW IPMSM with phase shift angles to the reference frame under FOC.

The 2014 International Power Electronics Conference

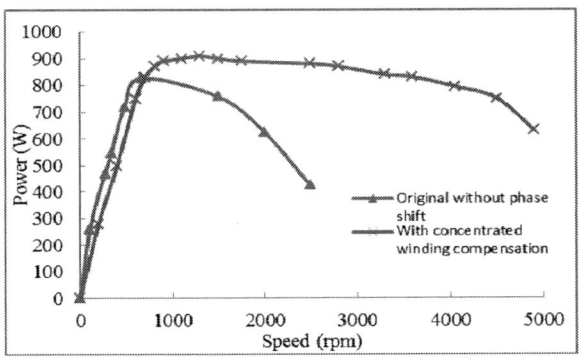

Figure 15 Comparison of the power-speed characteristics of the CW IPMSM at different stationary reference frame

As can be seen from this figure, by adding 28 degree of phase shift from the original axis, the maximum wide constant power speed range 9:1 can be achieved, around twice power speed range compared to the unmodified axis. It is also noted that theoretically the maximum speed of the 14-pole/18-slot concentrated winding IPMSM is calculated only around 4,500 rpm according to , compared to the experimental maximum speed of 4,900 rpm achieved as shown in

Figure 15. The experimental results confirmed the effectiveness of the proposed adjustment method according to the investigation of difference between concentrated and distributed windings of IPMSM.

V. CONCLUSION

This paper analyzed the operation of the field oriented control of a new fractional-slot, non-overlapping, concentrated winding IPMSM. Because the implemented FOC is based on the mathematical model of the IPMSM with distributed winding, it was found to be inadequate for the fractional-slot, concentrated winding in accessing its full power-speed range. The experimental result indicates that mathematical model starts to deteriorate in the field-weakening region resulting in a steady-state error between calculated flux linkage from rotating reference frame and flux linkage from stationary reference frame. The comparison between the flux density of the CW and DW obtained from the finite element model is studied to investigate the sources of this error which causes deterioration of the mathematical model. A suitable position adjustment of the stationary reference frame which includes characteristics of the concentrated winding was found by repeated experimentation which confirmed the effectiveness of the proposed adjustment. Further research is being carried out to determine the reasons for the adjustment and to find its soundness in terms of mathematical representation of the dynamics of the CW IPMSM.

REFERENCES

[1]   R. Dutta and M. F. Rahman, "Design and Analysis of an

Interior Permanent Magnet (IPM) Machine With Very Wide Constant Power Operation Range," *Energy Conversion, IEEE Transactions on,* vol. 23, pp. 25-33, 2008.

[2]   N. Bianchi, S. Bolognani, M. D. Pre, and G. Grezzani, "Design considerations for fractional-slot winding configurations of synchronous machines," *Industry Applications, IEEE Transactions on,* vol. 42, pp. 997-1006, 2006.

[3]   R. Dutta, L. Chong, and M. F. Rahman, "Design and Experimental Verification of an 18-Slot/14-pole Fractional-Slot Concentrated Winding Interior Permanent Magnet Machine," *Energy Conversion, IEEE Transactions on,* vol. 28, pp. 181-190, 2013.

[4]   J. Cros and P. Viarouge, "Synthesis of high performance PM motors with concentrated windings," *Energy Conversion, IEEE Transactions on,* vol. 17, pp. 248-253, 2002.

[5]   J. J. Germishuizen and M. J. Kamper, "IPM Traction Machine With Single Layer Non-Overlapping Concentrated Windings," *Industry Applications, IEEE Transactions on,* vol. 45, pp. 1387-1394, 2009.

[6]   L. Hyung-Woo, L. Ki-Doek, K. Won-Ho, J. Ik-Sang, K. Mi-Jung, L. Jae-Jun, and L. Ju, "Parameter Design of IPMSM With Concentrated Winding Considering Partial Magnetic Saturation," *Magnetics, IEEE Transactions on,* vol. 47, pp. 3653-3656, 2011.

[7]   A. M. El-Refaie, T. M. Jahns, P. J. McCleer, and J. W. McKeever, "Experimental verification of optimal flux weakening in surface PM Machines using concentrated windings," *Industry Applications, IEEE Transactions on,* vol. 42, pp. 443-453, 2006.

[8]   A. M. El-Refaie, "Fractional-Slot Concentrated-Windings Synchronous Permanent Magnet Machines: Opportunities and Challenges," *Industrial Electronics, IEEE Transactions on,* vol. 57, pp. 107-121, 2010.

[9]   L. Chong, R. Dutta, and M. F. Rahman, "Field weakening performance of a concentrated wound PM machine with rotor and magnet geometry variation," in *Power and Energy Society General Meeting, 2010 IEEE,* 2010, pp. 1-4.

[10]   S. Morimoto, Y. Tong, Y. Takeda, and T. Hirasa, "Loss minimization control of permanent magnet synchronous motor drives," *Industrial Electronics, IEEE Transactions on,* vol. 41, pp. 511-517, 1994.

[11]   S. Morimoto, Y. Takeda, T. Hirasa, and K. Taniguchi, "Expansion of operating limits for permanent magnet motor by current vector control considering inverter capacity," *Industry Applications, IEEE Transactions on,* vol. 26, pp. 866-871, 1990.

[12]   T. M. Jahns, "Flux-Weakening Regime Operation of an Interior Permanent-Magnet Synchronous Motor Drive," *Industry Applications, IEEE Transactions on,* vol. IA-23, pp. 681-689, 1987.

[13]   T. M. Jahns, G. B. Kliman, and T. W. Neumann, "Interior Permanent-Magnet Synchronous Motors for Adjustable-Speed Drives," *Industry Applications, IEEE Transactions on,* vol. IA-22, pp. 738-747, 1986.

[14]   L. Kyu-Wang, J. Doo-Hee, and H. In-Joong, "An online identification method for both stator resistance and back-EMF coefficient of PMSMs without rotational transducers," *Industrial Electronics, IEEE Transactions on,* vol. 51, pp. 507-510, 2004.

[15]   M. N. Uddin and M. M. I. Chy, "Online Parameter-Estimation-Based Speed Control of PM AC Motor Drive in Flux-Weakening Region," *Industry Applications, IEEE Transactions on,* vol. 44, pp. 1486-1494, 2008.

[16]   S. Morimoto, M. Sanada, and Y. Takeda, "Mechanical Sensorless Drives of IPMSM With Online Parameter Identification," *Industry Applications, IEEE Transactions on,* vol. 42, pp. 1241-1248, 2006.

[17]   B. Nahid-Mobarakeh, F. Meibody-Tabar, and F. M. Sargos, "Mechanical sensorless control of PMSM with online estimation of stator resistance," *Industry Applications, IEEE Transactions on,* vol. 40, pp. 457-471, 2004

The 2014 International Power Electronics Conference

Online Particle Swarm Optimization for Sensorless IPMSM Drives Considering Parameter Variation

Z. Q. Song

School of Electrical and Automotive Engineering
Yangzhou Polytechnic College
Jiangsu, China
zhengqiangsong@yahoo.com.cn

D. Xiao, and M. F. Rahman

School of Electrical Engineering and Telecommunications
The University of New South Wales
Sydney, Australia
d.xiao@unsw.edu.au, f.rahman@unsw.edu.au

Abstract— **In this paper, a novel online particle swarm optimization method is proposed to design speed and current controllers of sensorless vector controlled interior permanent magnet synchronous motor drives. The sliding mode observer is used for joint stator flux and rotor speed estimation. The stator resistance variation is compensated with a speed correction term which is derived from the estimation error of d-axis current. The speed and current controller gains are optimized with particle swarm optimization online, and the fitness function is changed according to the system dynamic and steady states. The proposed optimization algorithm is compared with conventional PI controller design method in the condition of step speed change and load disturbance. The offline and real-time simulation results are shown to confirm the effectiveness of the proposed method, with which the sensorless IPMSM drive exhibits better robustness and dynamic characteristics compared with the conventional PI controller design method.**

Keywords— *particle swarm optimization, permanent magnet synchronous motor, space vector modulation*

I. INTRODUCTION

Proportional-integral (PI) control technique has been widely used in high performance field orientation controlled interior permanent magnet synchronous motor (IPMSM) drives. However, fixed PI gains have to be designed based on a precise mathematical model of the drive for guaranteeing certain control performance. It is unavoidable that the uncertainties are caused by parametric variations such as flux linkage or stator resistance and unstructured dynamics in a practical IPMSM drives [1, 2].

Thus, to obtain a high dynamic performance for an IPMSM drive, current controllers should be optimized together with speed controller at the same time, because the current controller influences directly the drive dynamics.

There are numerous researches on the applications of computational intelligence techniques to controller parameters design for PMSM [3-9]. Among these, Particle Swarm Optimization [4], first introduced by Kennedy and Eberhart in 1995, is one of the modern heuristic algorithms. Because of its simplicity and computational efficiency, PSO has been widely used to solve a broad range of optimization problems, such as adaptive tuning of controller gains and parameters

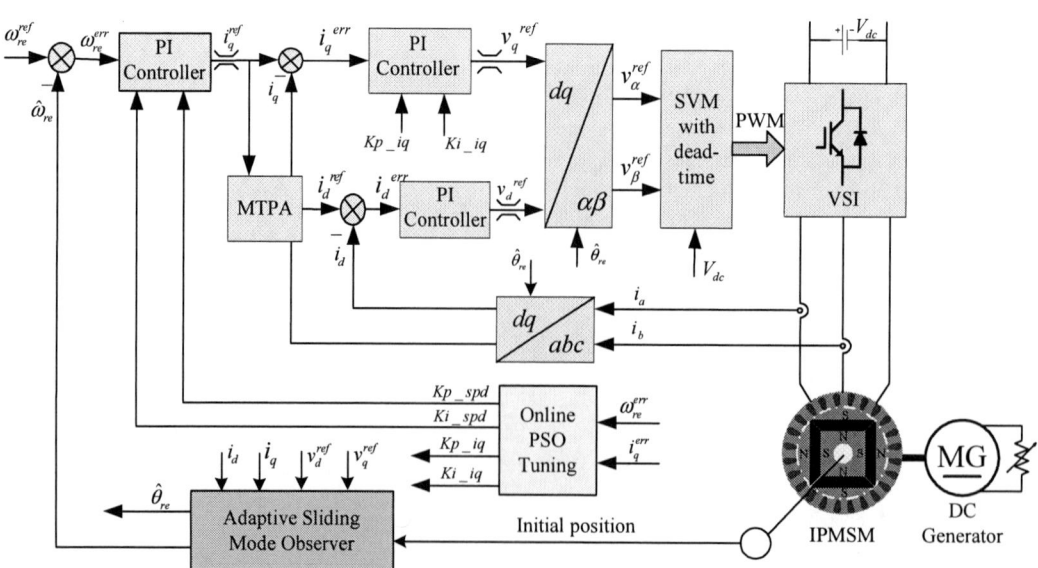

Fig. 1. Block diagram of sensorless FOC IPMSM drive based on PSO.

978-1-4799-2706-7/14 $31.00 © 2014 IEEE

identification. However, there still exist some problems/limitations with this method on the optimization of controller gains. Firstly, the PSO optimization applications in designing controller gains for PMSM rely on offline precise calculations of responses of PMSM using mathematical model [3]. It makes optimization effectiveness rely on the fixed PMSM model excessively. Secondly, the online PSO for PMSM controller optimization, however only applied to the speed controller gains without considering current controller optimization. It is difficult to achieve high dynamics and robustness for PMSM drive system. Thirdly, much effort has been made on real-time PSO application in parameters identification, and results show effectiveness because parameters of PMSM are changed slowly [3]. This should be incorporated with the optimization process online.

In this paper, an online PSO method is applied to adjust speed and q-axis current controller gains for the FOC IPMSM drives to achieve good performance in both dynamic and steady states. Different fitness function has been adopted based on the speed and currents of the motor. The reference flux value was selected according to the maximum torque per ampere (MTPA) trajectory to increase the efficiency of the overall drive system. An adaptive flux observer is proposed and implemented in the rotating (d-q) reference frame to estimate the rotor speed and position. The influence of stator resistance variation is compensated by adding in the adaptive model a speed correction term derived from the current estimation error. The proposed online controller optimization scheme has been integrated in a sensorless FOC IPMSM drive, as shown in Fig. 1. Both Simulink and real-time simulations have been carried out, confirming the performance improvement of the proposed scheme in comparison with the traditional controller design method.

II. ADAPTIVE FLUX OBSERVER

The adaptive observer is implemented in the estimated rotor (d-q) reference frame with the d-axis oriented along the permanent magnet flux [10]. The structure of the adaptive sliding mode observer can be expressed as

$$\begin{pmatrix} \dot{\hat{\lambda}}_d \\ \dot{\hat{\lambda}}_q \end{pmatrix} = \begin{pmatrix} -\dfrac{\hat{R}_s}{L_d} & \hat{\omega}_{re} \\ -\hat{\omega}_{re} & -\dfrac{\hat{R}_s}{L_q} \end{pmatrix} \begin{pmatrix} \hat{\lambda}_d \\ \hat{\lambda}_q \end{pmatrix} + \begin{pmatrix} v_d \\ v_q \end{pmatrix} + \begin{pmatrix} \dfrac{\hat{R}_s}{L_d}\lambda_f \\ 0 \end{pmatrix} + \mathbf{KS} \quad (1)$$

$$\begin{pmatrix} \hat{i}_d \\ \hat{i}_q \end{pmatrix} = \begin{pmatrix} 1/L_d & 0 \\ 0 & 1/L_q \end{pmatrix} \begin{pmatrix} \hat{\lambda}_d \\ \hat{\lambda}_q \end{pmatrix} - \begin{pmatrix} \lambda_f/L_d \\ 0 \end{pmatrix} \quad (2)$$

where the superscript ^ denotes estimated values and $\mathbf{K}$ is feedback gains of the observer. The sliding hyperplane $\mathbf{S}$ is defined upon the stator current errors.

$$\mathbf{S} = \begin{pmatrix} S_1 \\ S_2 \end{pmatrix} = \begin{pmatrix} i_d - \hat{i}_d \\ i_q - \hat{i}_q \end{pmatrix} = \begin{pmatrix} 1/L_d & 0 \\ 0 & 1/L_q \end{pmatrix} \begin{pmatrix} \lambda_d - \hat{\lambda}_d \\ \lambda_q - \hat{\lambda}_q \end{pmatrix} \quad (3)$$

The flux estimation error dynamics is given by

$$\begin{pmatrix} \dot{\tilde{\lambda}}_d \\ \dot{\tilde{\lambda}}_q \end{pmatrix} = (\mathbf{A}-\mathbf{KC}) \begin{pmatrix} \tilde{\lambda}_d \\ \tilde{\lambda}_q \end{pmatrix} + \tilde{\omega}_{re} \begin{pmatrix} \hat{\lambda}_q \\ -\hat{\lambda}_d \end{pmatrix} - \tilde{R}_s \begin{pmatrix} \dfrac{\hat{\lambda}_d - \lambda_f}{L_d} \\ \dfrac{1}{L_q}\hat{\lambda}_q \end{pmatrix}$$

$$\mathbf{A} = \begin{pmatrix} -R_s/L_d & \omega_{re} \\ -\omega_{re} & -R_s/L_q \end{pmatrix}, \mathbf{C} = \begin{pmatrix} 1/L_d & 0 \\ 0 & 1/L_q \end{pmatrix} \quad (4)$$

where the superscript ~ stands for the estimation errors.

The adaptation mechanism is derived from the Lyapunov stability analysis and the output of the adaptation mechanism is the rotor speed, used as the main correction in the adaptive model.

$$\dot{\hat{\omega}}_{re} = (\tilde{\lambda}_d\hat{\lambda}_q - \hat{\lambda}_d\tilde{\lambda}_q)/\gamma_1 \quad (5)$$

In order to improve the dynamic behavior of the speed estimation, a proportional term is added. It becomes a PI estimator.

$$\hat{\omega}_{re} = (K_P + K_I/s)*(\tilde{\lambda}_d\hat{\lambda}_q - \hat{\lambda}_d\tilde{\lambda}_q) \quad (6)$$

The observer gain $\mathbf{K} = k_1\mathbf{I} + k_2\mathbf{J}$ can be obtained by designing the observer poles with identical imaginary parts of the motor poles, but shifted to the left by $k(k>0)$ in the complex plane.

$$k_1 = 2k(L_d+L_q)/L_dL_q \quad (7)$$

$$k_2 = \frac{1}{2}\Big\{-\omega_{re}(L_d+L_q) + $$
$$sign(\omega_{re})\sqrt{\omega_{re}^2(L_d+L_q)^2 + 4k[L_dL_qk+R_s(L_d+L_q)]}\Big\} \quad (8)$$

The effect of the variation of the stator resistance on a sensorless IPMSM drive is to degrade the speed dynamic performance and lead to steady-state errors between the estimated and actual values of current and position. It will result in the instability of the sensorless control. In order to compensate the effects of stator resistance variation on the system performance, a speed correction is added to the estimated speed in the adaptive model. The correction term, derived from the d-axis current estimation error due to the stator resistance mismatch, is used for augmenting the observer by correcting the estimated flux direction.

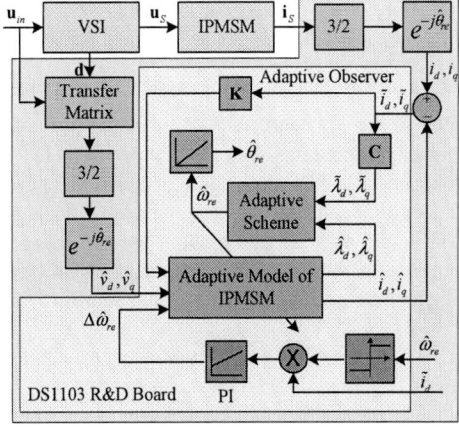

Fig. 2. Structure of the proposed adaptive observer.

$$\begin{pmatrix} \dot{\hat{\lambda}}_d \\ \dot{\hat{\lambda}}_q \end{pmatrix} = \begin{pmatrix} -R_s/L_d & \hat{\omega}_{re} - \Delta\hat{\omega}_{re} \\ -\hat{\omega}_{re} + \Delta\hat{\omega}_{re} & -R_s/L_q \end{pmatrix} \begin{pmatrix} \hat{\lambda}_d \\ \hat{\lambda}_q \end{pmatrix} + \begin{pmatrix} v_d \\ v_q \end{pmatrix}$$
$$+ \begin{pmatrix} R_s/L_d \cdot \lambda_f \\ 0 \end{pmatrix} + \mathbf{KS} \qquad (9)$$

$$\Delta\hat{\omega}_{re} = \text{sgn}(\hat{\omega}_{re}) \cdot (K_{P1}\tilde{i}_d + K_{I1}\int\tilde{i}_d dt) \qquad (10)$$

where K_{P1}, and K_{I1} are the PI controller gains for generating the speed correction due to the stator resistance variation. The structure of the proposed adaptive observer is shown in Fig. 2.

III. ONLINE PSO FOR CONTROLLER GAINS TUNING

A. Basic principle of the PSO algorithm

Particle swarm optimization algorithm is an evolutionary computation technique developed by Kennedy and Eberhart in 1995 [4]. It finds global optimum solution in search space through the interactions of individuals in a swarm of particles. Similar to other evolutionary algorithms, the PSO algorithm firstly produces initial swarm of particles in search space. Each particle represents a candidate solution to the problem and it has its own position X and velocity V. Each row in the position matrix X shows each particle's position, through which we can acquire the evaluation value of the particle. At each iteration, each particle memorizes and follows the tracks of its personal best (*Pbest*) and the global best position (*Gbest*) vectors to update the velocity matrix V.

Known these two best positions, particles can change velocities and positions using the following rules:

$$v_{j,g}^{(t+1)} = w \cdot v_{j,g}^{(t)} + c_1 r_1 \cdot (pbest_{j,g}^{(t)} - x_{j,g}^{(t)}) + c_2 r_2 \cdot (gbest_g^{(t)} - x_{j,g}^{(t)}) \quad (11)$$

$$X^{(t+1)} = X^{(t)} + V^{(t)} \qquad (12)$$

where $j=1, 2, \ldots m$; $g=1, 2, \ldots n$. The superscripts t and $t+1$ denote the time index of the current and the next iterations respectively. The parameters c_1 and c_2 are called acceleration constant which adjust the maximum step of the particle flight towards *Pbest* and *Gbest* position. Usually, parameters c_1 and c_2 are equal to 2, r_1 and r_2 are uniformly distributed random numbers in the interval (0, 1). Parameter w is inertia weight factor that usually decreases linearly from 0.9 to 0.4 in according to (13) over the course of the run. In this way, the algorithm can easily escape from local optimal solution in the early iteration stage as well as speed the convergence in the later iteration stage, and increase the reliability of finding the global optimal solution.

$$w^{(t)} = w_{max} - t \cdot (w_{max} - w_{min})/iter_{max} \qquad (13)$$

where w_{max} and w_{min} are the maximum and minimum values of w, and $iter_{max}$ is the maximum iteration times.

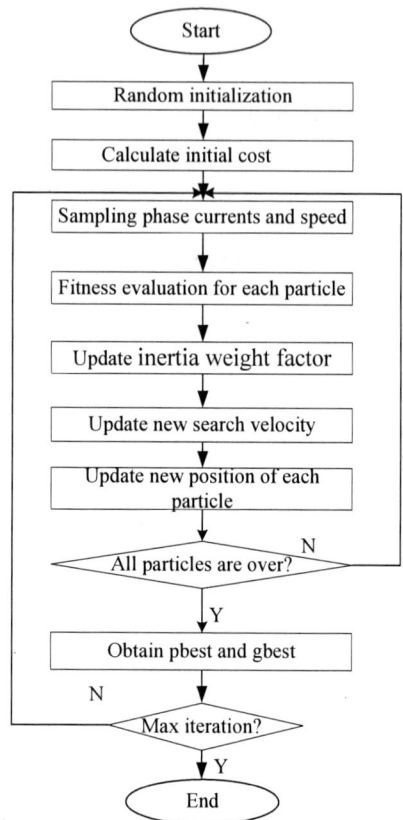

Fig. 3. Online PSO flowchart within one sampling time.

In order to reduce the likelihood of particles leaving the search space, the value of each dimension of the velocity $v_{j,g}^{(t)}$ is clamped to the interval $[-v_g^{max}, v_g^{max}]$. The value of v_g^{max} is usually chosen to be

$$v_g^{max} = k \cdot x_g^{max}, \ 0.1 \le k \le 0.5 \qquad (14)$$

where x_g^{max} is the upper bound of search region in the g-th dimension.

B. Implementation of online PSO

Most of existing PSO based gains tuning or parameter identifications were only implemented offline [3]. During the optimization process, to evaluate a candidate solution, such as PI gains were kept constant under the whole system simulation, at the same time output errors such as speed errors were added up to evaluate the candidate solution. Usually, PSO algorithm requires a number of iterations to obtain a satisfactory solution. For iteration, the system model has to be simulated once. Then, the model needs to be simulated a number of times to search the best solution. In fact, we cannot make IPMSM drive system repeat starting continually. That means such simulation results cannot be acquired in real system. However, it is difficult for each particle of candidate PI parameters to be evaluated within one sampling time, which is usually from 10-100 μsec. So, in this paper, we adopted a new method for online PSO update

978-1-4799-2706-7/14 $31.00 © 2014 IEEE

TABLE I: PARAMETERS OF IPMSM

Number of pole pairs	Pp	2
Stator resistance	R	5.95 Ω
Magnet flux linkage	λ_f	0.533 Wb
d-axis inductance	L_d	0.0448 H
q-axis inductance	L_q	0.1024 H
Phase voltage (rms)	V	230 V
Phase current (rms)	I	3 A
Rated torque	T_b	6 Nm

TABLE III: PARAMETERS OF CONTROL SYSTEM

Sampling frequency	10 kHz
Speed controller (Kp, Ki)	4.63, 0.2
d-axis current controller	150, 20
q-axis current controller	340, 50
Speed estimator	161, 273700
Maximum current for MTPA	6 A
Observer gain, k	100
Dead time	2 µs

TABLE II: CONTROLLER PARAMETERS

Iteration	k_{p_s}	k_{i_s}	Fitness
1	2.39635	0.515588	0.135628
10	7.372698	9.754793	0.114823
20	6.321931	0.50782	0.093713
30	2.973025	6.254318	0.05207

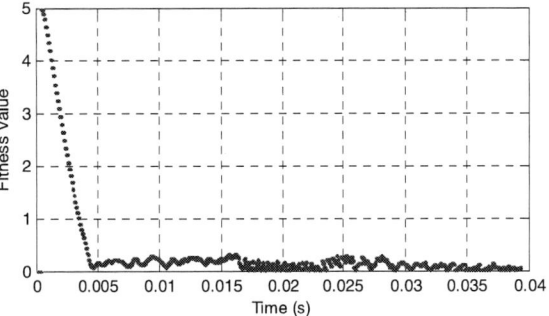

Fig. 4. Fitness value of PSO for IPMSM drives.

calculations of particle positions X, velocity V, $Pbest$ and $Gbest$ of the PSO algorithm. We measured speed and current as sampling values, while updating present particle information instead of updating whole swarm information. For example, there is a particle swarm with 30 particles for optimization. That means if the sampling of IPMSM drive system is 10µsec, each particle's information such as position and velocity updates once within 100 µs, however, $Pbest$ and $Gbest$ would be updated completely within 30*10 µsec. Fig. 3 shows the flowchart of online PSO implemented in this paper.

C. Definition of the Evaluation Function

To optimize the overall response of the motor drive, the fitness function is a weighted sum of several performance index based on measurements of the speed and currents outputs. It is shown in (15):

$$F(k_{p1}, k_{p2}, k_{i1}, k_{i2}) = \sum_{i=1}^{3}(a_i \cdot f_i) \tag{15}$$

$$f_1 = (\omega_{re}(k) - \omega_{re}(k-1))^2 \tag{16}$$

$$f_2 = (\omega_{re}(k) - \omega_{re}^{ref})^2 \tag{17}$$

$$f_3 = ((i_a(k) - i_a^{ref}(k))^2 + (i_b(k) - i_b^{ref}(k))^2 \\ +((i_c(k) - \hat{i}_c(k))^2] \tag{18}$$

where a_i represent positive weights, and f_i are three performance indices defined in the following:
1) f_1, speed transient response index;

2) f_2, speed steady index;

3) f_3, current reference oscillations for constant speed reference.

ω_{re}, i_a, i_b, i_c are speed and phase currents of the motor, respectively, from IPMSM output; ω_{re}^{ref}, i_a^{ref}, i_b^{ref}, i_c^{ref} are the speed reference and the current references in the three-phase reference frame; During the

transient state, $a_i = [0.6, 0.1, 0.3]$. When IPMSM nearly reaches steady state, $a_i = [0.1, 0.6, 0.3]$.

IV. OFFLINE SIMULATION

The offline simulation has been carried out in MATLAB/SIMULINK to examine the performance of the online PSO and adaptive observer for the sensorless FOC IPMSM drive. The parameters of IPMSM are given in Table I. The controller sampling time is chosen as 100 µsec, which is the same as the controller sampling time in the real-time experiment.

Table II shows the values of controller parameters and fitness functions when PSO iterations are 1, 10, 20 and 30 respectively. Initial swarm size of particles is 30, and each particle has two variables (k_{p_s}, k_{i_s}) representing its position vectors in search space. After 30 iterations, the results of $Gbest$ would be updated as optimized outputs, and the Fitness of speed errors is smaller step by step.

Fig. 4 shows the change of fitness value of PSO for IPMSM drives and its decline rate is very fast.

In order to test the proposed method in this paper, two different controllers were applied on IPMSM drive, and comparison results between conventional PI controllers and online PSO-PI controllers for IPMSM drives are presented in Fig. 5 and Fig. 6, where (a) is i_d current response curve; (b) is i_q current response curve; (c) is speed response curve; (d) is torque response curve; (e) are three phase current curves. It can be seen that the performance of step response and load disturbance of the drive system with PSO tuned controllers is improved compared with that of traditional controllers with fixed PI gains.

V. REAL-TIME SIMULATION

The effectiveness of the proposed sensorless drive scheme was tested with a real-time simulation platform,

The 2014 International Power Electronics Conference

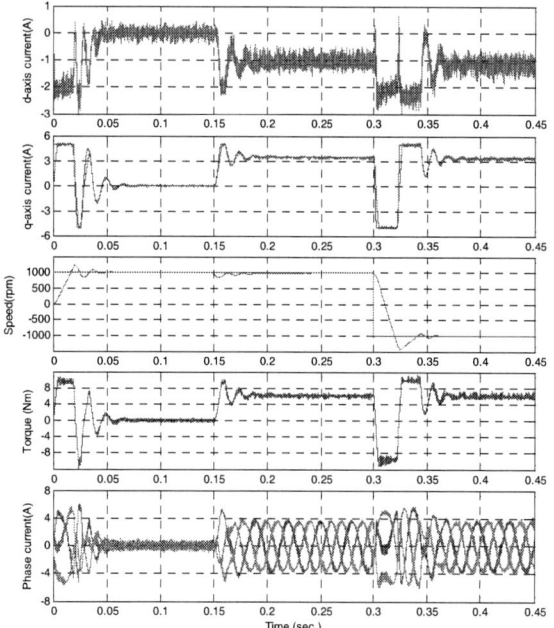

Fig. 5. Performance of IPMSM drive with conventional PI controllers.

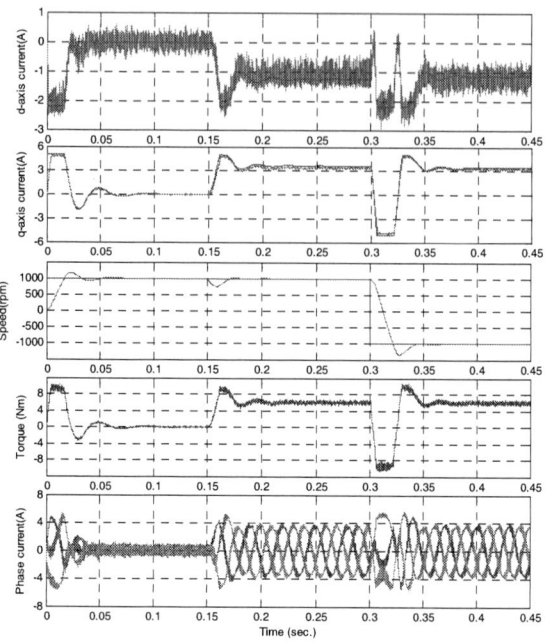

Fig. 6. Performance of IPMSM drive with online PSO

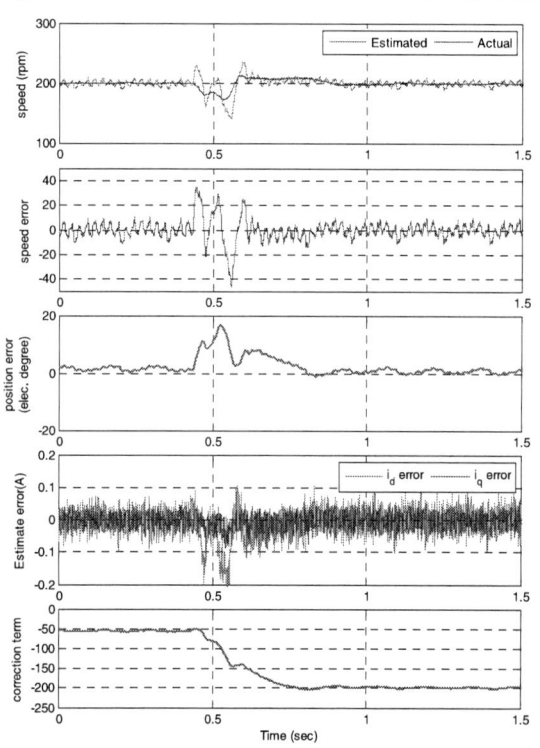

Fig. 7. Performance of stator resistance step change to 158% rated value at 200rpm with 63% rated load (RT).

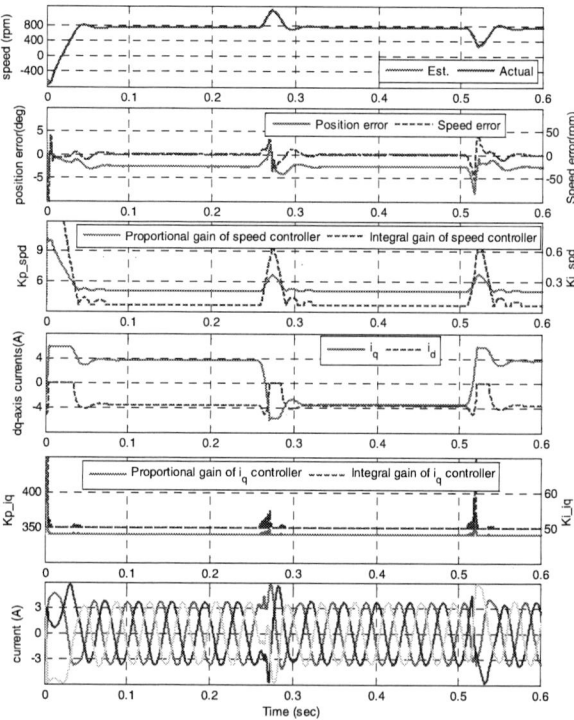

Fig. 8. Performance of the sensorless drive with PSO at 750rpm and ±5N·m square-wave load torque. (real-time).

dSPACE DS1103. The sensorless drive control and space vector modulation strategy are implemented within the PWM synchronization interrupt service routine. Three-phase PWM signals with predefined dead-band are generated by DS1103. A three-phase voltage source inverter is emulated by measuring the three-phase PWM signals and DC-bus voltage which are provided by the controller via timing I/Os and DAC channels. The

mathematical models of permanent magnet dc machine and the target IPMSM are implemented with Timer0 interrupt. The armature current of PMDC generator is separately regulated to emulate the load. The IPM machine used in this experiment has the same parameters as the one used for the simulation, as given Table I. The pameters of the control system are given in Table III. The switching frequency of the VSI was set to 10 kHz and

978-1-4799-2706-7/14 $31.00 © 2014 IEEE

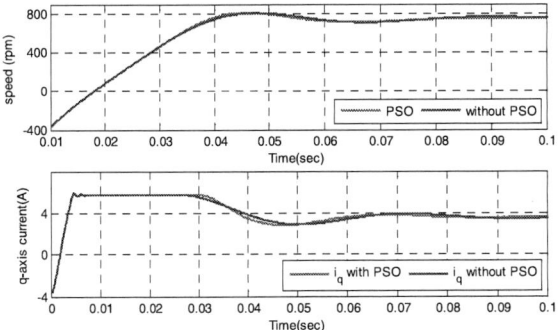

Fig. 9 (a). Performance comparison of the speed step responses of the sensorless drive with PSO and conventional PI controller at 750rpm and 5N·m (real-time).

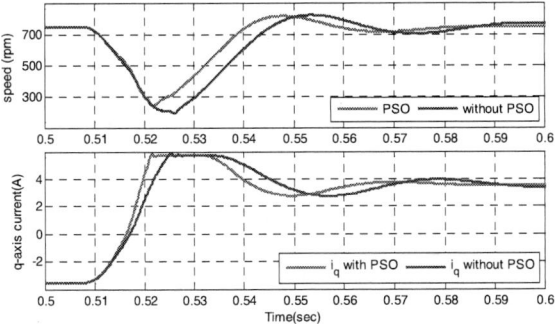

Fig. 9 (b) Performance comparison of the load disturbance with 5N·m load abruptly applied at 750rpm (real-time).

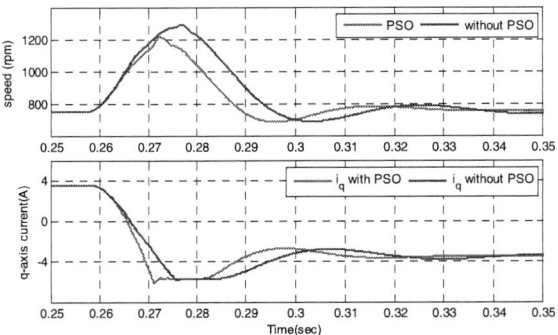

Fig. 9 (c). Performance comparison of the load disturbance with - 5N·m load abruptly applied at 750rpm (real-time).

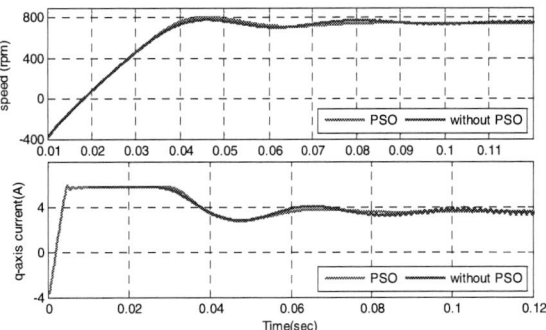

Fig. 10 (a). Performance comparison of the speed step responses of the sensorless drive with PSO and conventional PI controller at 750rpm and 5N·m (real-time).

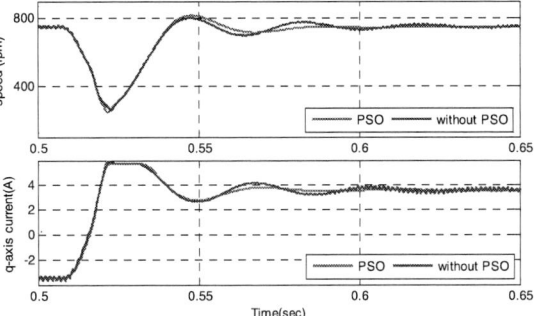

Fig. 10 (b) Performance comparison of the load disturbance with 5N·m load abruptly applied at 750rpm (real-time).

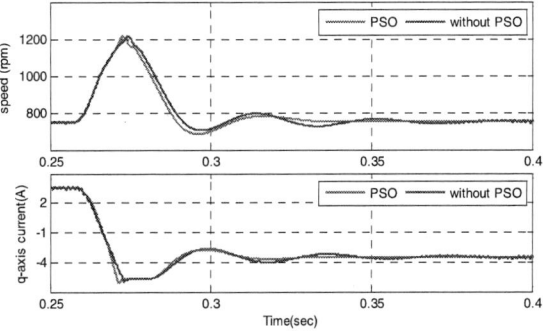

Fig. 10 (c). Performance comparison of the load disturbance with - 5N·m load abruptly applied at 750rpm (real-time).

DC-bus voltage is 340V.

Fig. 7 shows the compensation of the stator resistance variation by the correction term in the observer. It has been tested by abruptly adding 3.4Ω three-phase external resistance to the stator windings when the machine was running at 200rpm with 63% rated load. Speed, position and estimation errors are shown in Fig. 7. The effects of the step change in stator resistance can be compensated by the correction term $\Delta\hat{\omega}_{re}$ in the observer within 0.5s. The current and speed estimation errors reduce to zero, exhibiting the effectiveness of both the modified sliding mode observer and speed estimation scheme.

Fig. 8 shows the performance of proposed sensorless control scheme under load disturbance and speed step response. Speed estimation error is bounded by ±50rpm during transients and converges to zero in steady-state. A sudden nominal load reversal gives 9 electrical degrees transient error and 2 electrical degree steady state error to the position estimation. The gains of the speed and q-axis current controllers are tuned online by the PSO algorithm, as shown in Fig. 8. In addition, you can see the effect of the MTPA algorithm on the dq-axis current control. The d-axis current reference is generated based on the q-axis current and maximum current. During the transients, q-axis current reaches the maximum value while d-axis current is kept at zero. During the steady states, d-axis goes to negative in order to produce the desired torque to overcome the load torque with minimum possible current magnitude.

As shown in Fig. 9 (a)-(c), the performance of setpoint tracking and load disturbance rejection of the sensorless drive with PSO tuning and fixed PI gains are compared with ±750rpm square-wave speed reference and abruptly adding ±5Nm to the PMDC generator after the speed is stable. The conventional PI controllers are tuned offline in order to achieve the best setpoint regulation. It can be seen that the fixed PI and PSO tuned PI controllers have the similar performance at the speed and current step responses. However, the PSO tuned controllers exhibit better performance during the load disturbance test, which have lower overshoot, faster response and less oscillations, as show in Fig. 9 (b) and (c). The trade-offs between setpoint tracking and disturbance rejection, which often exists during the traditional PI controllers design, can be alleviated with PSO tuning the PI gains online.

The comparison between the PSO and conventional PI controller with a higher band-width is shown in Fig. 10 (a)-(c). In order to increase the bandwidth of the controller, the new gains (Kp_spd=6, Ki_spd=0.27,) are used for the speed controller. It can be seen that the overshoot of the responses of the conventional PI controller is smaller than before. However, it creates more oscillations and ripples on the current and speed waveforms in the steady states, which results in torque pulsation, and hence audible noise and additional power losses in the machine.

VI. CONCLUSIONS

In this paper, a novel online particle swarm optimization method is proposed to design speed and current PI controllers of sensorless FOC IPMSM drives, taking into account stator resistance variation. The speed and current controllers are optimized and updated online using PSO in each sampling cycle. The dynamic and steady-state performance of the drive system has been improved by optimization of speed and current controller compared with the traditionally tuned PI controllers. In addition, the variation of the stator resistance is compensated with a speed correction term introduced into the adaptive observer.

REFERENCES

[1] S. H. Chang and P. Y. Chen, "Self-tuning gains of PI controllers for current control in a PMSM," *IEEE 5th Conference on Industrial Electronics and Applications*, pp. 1282–1286, Jun. 2010.

[2] M. F. Rahman, M. E. Haque, L. Tang and L. Zhong, "Problems associated with the direct torque control of an Interior permanent-magnet synchronous motor drive and their remedies," *IEEE Trans. on industrial electronics*, vol. 51, no. 4, pp. 799–808, Aug. 2004.

[3] W. Liu, L. Liu, I. Y. Chung, and D. A. Cartes, "Real-time particle swarm optimization based parameter identification applied to permanent magnet synchronous machine," *Elsevier, Applied Soft Computing*, vol. 11, no. 2, pp. 2556–2564, Mar. 2011.

[4] J. Kennedy, and R. Eberhart, "Particle swarm optimization," in *Proceedings of International Conference on Neural Network (ICNN)*, vol. 4, pp. 1942–1948 Perth, Australia 1995.

[5] S. Yan, D. Xu, X. Gui, M. Yang and B. Li, "Online particle swarm optimization design of speed controller considering anti-windup for PMSM Drive System," *IEEE Annual Power Electronics Specialists Conference*, pp. 2273–2277, Jun. 2007.

[6] F. J. Lin, and C. H. Lin, "On-line gain-tuning IP controller using RFNN" *IEEE Trans. on Aerospace and Electronic Systems*, vol. 37, no. 2, pp. 655–670, Apr. 2001.

[7] Y.A.-R.I. Mohamed, "Adaptive self-tuning speed control for permanent-magnet synchronous motor drive with dead time," *IEEE Trans. on energy conversion*, vol. 21, no. 4, pp. 855–862, Dec. 2006.

[8] A. Lidozzi, L. Solero, F. Crescimbini, and A. Di Napoli, "Direct tuning strategy for speed controlled PMSM drives," *IEEE International Symposium on Industrial Electronics*, pp.1265–1270, Jul. 2010.

[9] S. B. Lee, "Closed-loop estimation of permanent magnet synchronous motor parameters by PI controller gain tuning," *IEEE Trans. on energy conversion*, vol. 21, no. 4, pp. 863–869, Dec. 2006.

[10] D. Xiao and M. F. Rahman, "Sensorless direct torque and flux controlled IPM synchronous machine fed by matrix converter over a wide speed range," *IEEE Trans. on Ind. Informatics*, vol. 9 no. 4, pp. 1855-1867, Nov. 2013.

A DTC-PWM Control Scheme of PMSM based on 12-Sectors Division and Speed Information

Yunchang Kwak, Jin-Woo Ahn, Dong-Hee Lee

Dept. of Mechatronics Engineering
Kyungsung University
Busan, Korea
kyc9281@ks.ac.kr, jwahn@ks.ac.kr, leedh@ks.ac.kr

Abstract— **This paper presents a modified DTC-PWM scheme which uses 12-side sectors and 12 voltage vectors to reduce torque and flux ripple. In the proposed control scheme, three voltage vectors can be selected in a sector with the torque and flux error conditions. The 3 voltage vectors are set as big-small, medium-medium and small-big pairs of torque and flux component according to the sign of the torque and flux error. So, the suitable voltage vector can be determined by the error magnitudes of the torque and flux. And the duty ratio to determine the switching time of the selected voltage vector is calculated by the torque, flux error and speed information. Unlike the conventional approaches, the properly selected voltage vector among three voltage vectors can increase the dynamic response of torque and flux control. And the properly calculated duty ratio based on the flux, torque error and speed information, can decrease the flux and torque ripple with the fixed switching frequency.**

The proposed DTC-PWM based on the 12-side sectors and duty ratio calculation method is verified by the computer simulation and experimental comparisons of PMSM.

Keywords— Direct Torque Control, PWM(Pulse Width Modulation), Duty ratio, 12 sides sector division

I. INTRODUCTION

The PMSMs(Permanent Magnet Synchronous Motor) are the most popular drives for the precise industrial control applications and robotic systems. For the high performance of the practical systems which use PMSMs, the higher torque control performance is essential [1-2]. The most popular torque control schemes are FOC(Field Oriented Control) and DTC(Direct Torque Control).

The FOC which is called as Vector Control can control the torque and flux independently using d-q axis transformation. So, the flux and torque of the PMSM can be easily separated. However, the practical implementation of the torque control is done by the PWM(Pulse Width Modulation) such as SPWM(Sinusoidal Pulse Width Modulation) and SVPWM(Space Vector Pulse Width Modulation) method. When the d-q axis voltage commands to control the flux and torque components are calculated, PI(Proportional and Integral) controller or fuzzy controller can be used. Consequently, the torque and flux

cannot be directly controlled. And the integral term of the controller can make saturation problem with parameter dependency [3-4].

The DTC is very easy to be implemented, but the torque and flux ripple are dependent on the sampling frequency, and the switching frequency is not constant according to the torque and flux error. In order to improve the control performance and overcome the problem of the conventional DTC, the DTC-SVM(Direct Torque Control-Space Vector Modulation) and DTC-PWM(Direct Torque Control-Pulse Width Modulation) with 6-side sectors and 12-side sectors are investigated .

DTC-SVM can produce low torque ripple with advance performance, but the calculation is very complex and it uses PI controller to generate voltage commands. Consequently, it may have same problem such as saturation effects of integral term and gain dependency of the control performance.

DTC-PWM is very useful idea to reduce torque ripple with the fixed switching frequency. However, the previously researched DTC-PWMs use 6-side sectors, so the selected voltage vector cannot supply enough voltage to produce flux around the edges of the sector. And speed information is not used to the calculation of the duty ratio. And the duty ratio which is determined by the torque and flux error is not proper according to the speed variation.

This paper presents a modified DTC-PWM scheme which uses 12-side sectors and voltage vectors to reduce torque and flux ripple. The proposed method is based on the voltage analysis of the conventional 6-side sectors and 12-side sectors methods. The selected voltage in every sector can be divided as torque and flux voltage components. And the voltage characteristics by the selected voltage vectors can be analysed. In the proposed control scheme, three voltage vectors can be selected in a sector with the torque and flux error conditions. The 3-voltage vectors are set as big-small, medium-medium and small-big pairs of torque and flux component according to the sign of the torque and flux error. So, the suitable voltage vector can be determined by the error magnitudes of the torque and flux. And the duty ratio to determine the switching time of the selected voltage vector is calculated by the torque, flux error and speed information. Unlike the conventional approaches, the properly selected voltage vector among three voltage vectors can increase the dynamic response of torque and flux control. And the properly calculated duty ratio based on the flux, torque error and speed information, can decrease the flux and torque ripple with the fixed switching frequency [5-6].

The proposed DTC-PWM based on the 12-side sectors and duty ratio calculation method is verified by the experimental comparisons of PMSM.

II. ANALYSIS OF CONVENTIONAL DTC(CDTC)

Fig. 1 shows the 6 effective voltage vectors of the VSI(Voltage Source Inverter) and sector division in the stationary coordinate. The sectors are divided as 6 regions as electrical 60 degree according to the rotor position. In each sector, the switching voltage vector of the next sampling step is determined by the torque and flux error shown in Table I.

In Fig. 1, the rotor position is defined as θ_{re} and the rotor angle based on the sector reference is defined as θ_s. The d-q axis defines the flux and torque axis of PMSM according to the rotor position. As well known, the positive d-axis means positive flux axis. And the positive q-axis is the axis for positive torque. According to the rotor position, space vectors can be divided as 4 parts : positive torque and flux region, positive torque and negative flux region, negative torque and positive flux region and negative torque and flux region. And, the Table I shows the switching voltage vector according to the rotor position and torque and flux error sign. In the Table I, s_φ and s_τ are the sign of flux and torque error.

TABLE I

SELECTED VOLTAGE VECTORS IN CDTC

s_φ	s_τ	Rotor position(sector number)					
		S1	S2	S3	S4	S5	S6
1	1	V2	V3	V4	V5	V6	V1
	-1	V6	V1	V2	V3	V4	V5
-1	1	V3	V4	V5	V6	V1	V2
	-1	V5	V6	V1	V2	V3	V4

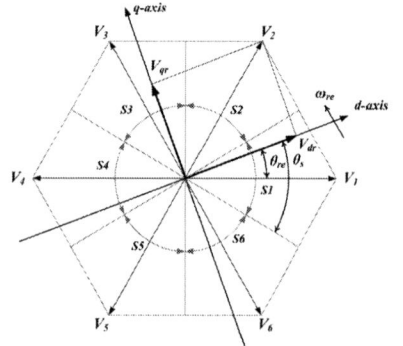

Fig. 1. Voltage vectors and sectors of the conventional CDTC

In the S1 sector, voltage vector V2 is selected for the positive torque and flux generation. As shown in Fig. 1, the torque angle between S1 sector and V2 is changed from 90 to 30 degree during S1 sector. Similarly, the torque angle between S1 sector and V6 for positive flux with negative torque is changed from 30 to 90 degree. And the selected voltage vector can be transferred to the d-q axis voltages to produce flux and torque current according to the rotor position and sector number.

The voltage vector can be analyzed as d-q voltage component with coordinate transform using the rotor position θ_{re} and sector position.

Where, V_{ak}, V_{bk} and V_{ck} are the phase voltages of the selected voltage vector V_k, where the k is voltage vector number.

Fig. 2 shows the d-q axis voltage analysis results according to the selected voltage vector for the desired flux and torque. As shown in Fig. 2, the output voltages in d-q axis are varied according to the torque angle variation. The d-axis voltage V_{dr} for the flux component is changed from 0 to 70.7[%] of DC-link voltage at every sector and the error of the flux-torque condition. And the q-axis voltage V_{qr} for the flux component is changed from 40.8 to 81.6[%] of DC-link voltage. Consequently, the d-q axis current can be changed by the supplied voltage vector. Furthermore, the d-axis voltage is very low around the edge of the sector area. For this reason the selected voltage vector can supply proper sign torque and flux voltages in every sector, the supplied voltages are not enough in some region. Especially, the d-axis voltage is almost zero around sector changing region, and the q-axis voltage variation is much serious in this region. These sudden voltage variation and zero voltages can make high torque and flux ripple in the conventional DTC method.

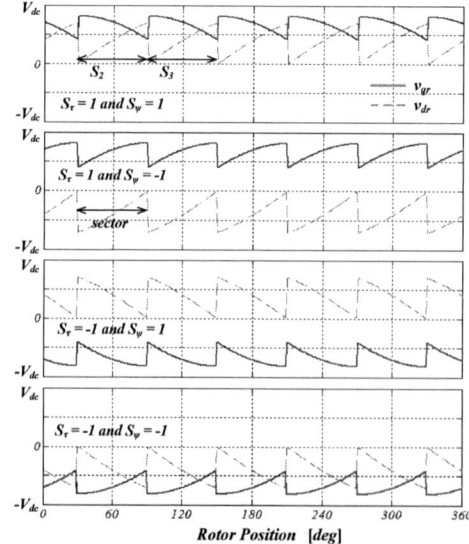

Fig. 2. The d-q axis voltage analysis according to the selected voltage

978-1-4799-2706-7/14 $31.00 © 2014 IEEE

III. ANALYSIS OF 12 SECTORS DTC VOLTAGES

In order to reduce the torque and flux ripple in CDTC, 12 sectors DTC method is introduced . In this approach, the sector division is divided as 12 parts according to the rotor position, and the results can decrease the torque ripple due to the reduction of torque angle. However, the previous researches, the selected switching voltage vectors are six effective voltage vectors of voltage source inverter same as the CDTC (Conventional DTC). Consequently, the selected voltage vector is not enough to reduce torque and flux ripple. Some researches, additional voltage vectors between the six effective voltage vectors are used to produce proper torque and flux.

Fig. 3 shows the 12 sided sectors and the 6 additional voltage vectors V12, V23, V34, V45 and V61. Each sector is divided as 30 electrical degrees respectively. And the additional voltage vectors are produced by the combinations of the adjacent two voltage vectors in the voltage source inverter. Unlike the CDTC, the switching voltage vector is not one in the 12 sided sectors method. When the rotor position is on S1 sector in Fig. 3, V12, V2 and V23 can be selected for the positive torque and flux generation. In S1 sector, voltage vector 23 is the positive big torque vector, and voltage vector V12 is for the positive big flux vector. And the voltage vector V2 is for the positive medium torque and positive medium flux vector.

For the easy comprehension, the selected voltage vectors can be described as PB(Positive Big), PM(Positive Medium), PS(Positive Small), NS(Negative Small), NM(Negative Medium) and NB(Negative Big) according to the torque and flux production.

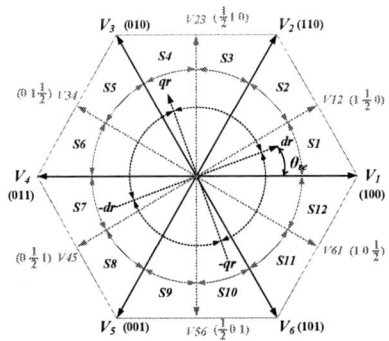

Fig. 3. 12 sided sectors and additional 6 voltage vectors

The three voltage vectors which can be selected in every sector have different torque and flux characteristics. When the torque and flux are required in positive direction, the one voltage vector has PB torque voltage with PS flux voltage. The second voltage vector has PM torque voltage with PM flux voltage. The last has PS torque voltage with PB flux voltage. Fig. 4 shows the analyzed voltage components from the selected voltage vectors in every sector. Fig. 4(a) shows the analyzed voltages in case of positive torque and positive flux, and the negative torque and negative flux are shown in Fig. 4(b). As shown in Fig. 4(a) and Fig. 4(b), three voltage vectors can be selected in every sector, and the selected voltage vectors have different torque and flux effects. For the higher torque or flux voltage, flux and torque voltage are zero and are not enough around sector changing region. And the zero voltage effect is same as the CDTC method. Some researches uses medium torque and medium flux voltage vector selection in 12 sided sectors approach. The selected voltage vectors in every sector can produce non-zero torque and flux voltage, and the torque and flux voltage ripples are smaller than the conventional DTC. This method can reduce torque and flux ripple. However, the torque component of the selected voltage vector is lower than the CDTC and Big torque voltage vector of 12-side sectors method. The dynamics of the torque control and the maximum output torque are lower than a CDTC.

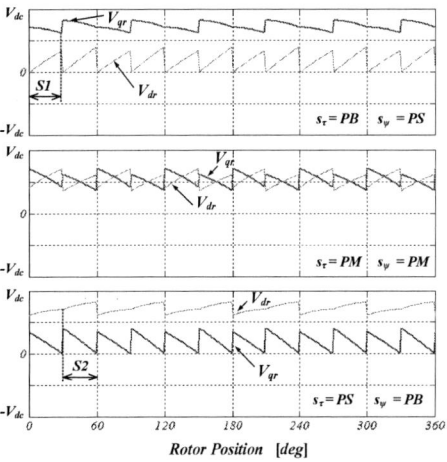

(a) In case of 12 sided sectors $s_\tau > 0$ and $s_\varphi > 0$

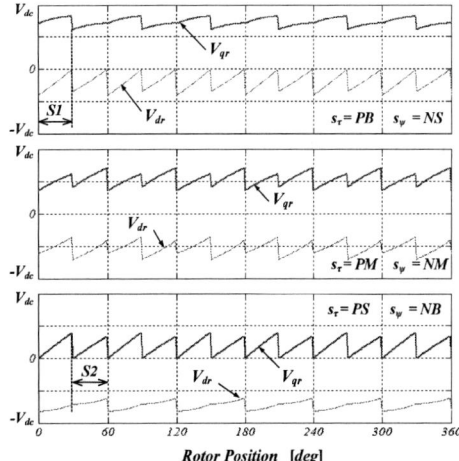

(b)In case of 12 sided sectors $s_\tau > 0$ and $s_\varphi < 0$

Fig. 4. d-q axis voltage analysis with selected voltage vectors

Consequently, the 12-side sectors method can reduce the torque and flux ripple due to the enough d-q axis voltage, but the maximum torque and control dynamics are lower than a CDTC.[7-8]

In this paper, switching voltage vector is selected by the torque and flux condition, and the switching time of the selected voltage vector is determined by the torque, flux error and the speed condition to consider the back E.M.F effect on d-q axis voltage.

IV. THE PROPOSED DTC-PWM METHOD

In order to reduce torque and flux ripple, the voltage characteristics of the PMSM are considered in this paper. The voltage equation at d-q axis can be described as follows.

$$v_{dr} = R_s i_{dr} + L_s \frac{d}{dt} i_{dr} - L_s \omega_{re} i_{qr} \tag{1}$$

$$v_{qr} = R_s i_{qr} + L_s \frac{d}{dt} i_{qr} + L_s \omega_{re} i_{dr} + K_e \omega_{re} \tag{2}$$

Where, v_{dr}, v_{qr}, i_{dr} and i_{qr} are the d-q axis voltages and currents respectively. R_s and L_s are the winding resistance and inductance of the motor. K_e is the back E.M.F constant which has V/rad/s unit. And, ω_{re} is the electrical speed of PMSM. In the steady state, current and voltage relationship can be expressed as follows.

$$i_{dr(k)} = \frac{v_{dr(k)} + L_s \cdot \omega_{re} \cdot i_{qr(k-1)}}{R_s} \tag{3}$$

$$i_{qr(k)} = \frac{v_{qr(k)} - L_s \cdot \omega_{re} \cdot i_{dr(k-1)} - K_e \cdot \omega_{re}}{R_s} \tag{4}$$

From, (3) and (4), the d-axis current i_{dr} to generate additional flux is increased by the speed and q-axis current in the same d-axis voltage. On the other side, the q-axis current i_{qr} to produce the torque is decreased by the speed increasing in the same q-axis voltage supply from the selected voltage vector. From these reason, the flux and torque are not constant with same duty ratio in a wide speed range. In order to overcome this problem, the calculation of torque ratio has to be considered the speed information [9-10].

Unlike the previous 12 sided sector approaches, the proposed DTC uses three voltage vectors according to the torque, flux and speed conditions. Table II shows the switching voltage vectors according to the torque and flux error. In Table II, e_τ and e_φ are the torque and flux error level. They are divided as 6 levels according to the error sign and magnitude. The PB, PM, PS, NS, NM and NB mean positive big, positive medium, positive small, negative small, negative medium and negative big, respectively. As shown in Table II, the selected voltage vector does not have BIG torque and BIG flux component.

TABLE II

(a) SELECTED VOLTAGE VECTORS IN THE PROPOSED DTC(S1~S6)

s_τ	s_φ	Rotor Position(Sector Number)					
		S1	S2	S3	S4	S5	S6
PB(3)	PS	V23	V3	V34	V4	V45	V5
	NS	V3	V34	V4	V45	V5	V56
PM(2)	PM	V2	V23	V3	V34	V4	V45
	NM	V34	V4	V45	V5	V56	V6
PS(1)	PB	V12	V2	V23	V3	V34	V4
	NB	V4	V45	V5	V56	V6	V61
NS(-1)	PB	V1	V12	V2	V23	V3	V34
	NB	V45	V5	V56	V6	V61	V1
NM(-2)	PM	V61	V1	V12	V2	V23	V3
	NM	V5	V56	V6	V61	V1	V12
NB(-3)	PS	V6	V61	V1	V12	V2	V23
	NS	V56	V6	V61	V1	V12	V2

(b) SELECTED VOLTAGE VECTORS IN THE PROPOSED DTC(S7~S12)

s_τ	s_φ	Rotor Position(Sector Number)					
		S1	S2	S3	S4	S5	S6
PB(3)	PS	V56	V6	V61	V1	V12	V2
	NS	V6	V61	V1	V12	V2	V23
PM(2)	PM	V5	V56	V6	V61	V1	V12
	NM	V61	V1	V12	V2	V23	V3
PS(1)	PB	V45	V5	V56	V6	V61	V1
	NB	V1	V12	V2	V23	V3	V34
NS(-1)	PB	V4	V45	V5	V56	V6	V61
	NB	V12	V2	V23	V3	V34	V4
NM(-2)	PM	V34	V4	V45	V5	V56	V6
	NM	V2	V23	V3	V34	V4	V56
NB(-3)	PS	V3	V34	V4	V45	V5	V56
	NS	V56	V6	V61	V1	V12	V2

In this paper, the torque and flux error are defined as 6 levels such as PB(+3), PM(+2), PS(+1), NS(-1), NM(-2) and NB(-3). Fig. 5 shows the selection of the torque error level s_τ. And the flux error level s_ψ is determined according to the flux error value.

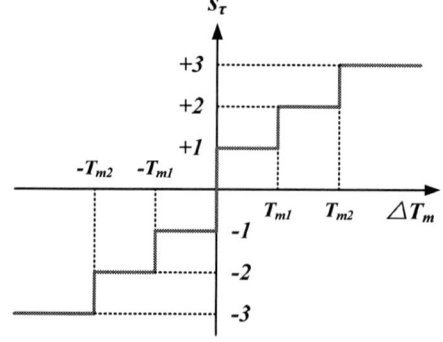

Fig. 5. Selection of torque error level s_τ

The torque and flux error can be defined as follows.

$$\Delta T_m = T_m^* - T_m \tag{5}$$

$$\Delta \Psi_m = \Psi_m^* - \Psi_m \tag{6}$$

Where, T_m^* and Ψ_m^* are the torque and flux command. The T_m and Ψ_m are the actual torque and flux which are calculated by the torque and flux estimator.

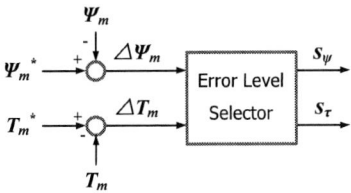

Fig. 6. Selection of torque error level s_τ

Fig. 6 shows the error level sector of the torque and flux and its error calculation. For example, the voltage vector V2 shown in Fig. 1, can be selected in the positive s_τ and the positive s_ψ in the conventional DTC method during the sector 1 region. However, the voltage vector V23, V2 and V12 shown in Fig. 3 can be selected according to the s_τ and s_ψ which is determined by the error level selector. The selected voltage vector can supply the proper voltage components according to the required flux and torque. In the proposed control scheme, the torque and flux error level are used to determine the voltage vector shown in Table II. And the torque and flux error are used to calculate the duty ratio of the selected voltage vector based on the speed information as follows.

$$d = C_\Psi |\Delta\Psi_m| + C_\tau |\Delta T_m| + C_\omega |\omega_{re}| \qquad (7)$$

d is the calculated duty ratio of the proposed DTC-PWM method. C_Ψ and C_τ are the constant gains of the flux and torque error. And C_ω is the constant gain of the speed. This term from speed can decrease torque and flux ripple by the speed variation.

Fig. 7 shows the proposed DTC-PWM scheme using 12-side sectors and the duty ratio calculation with speed information. As shown in Fig. 7, the error level sector and duty ratio control block are the novel block compare than the conventional method. And the vector table is used to complex 12-side sector[11-13].

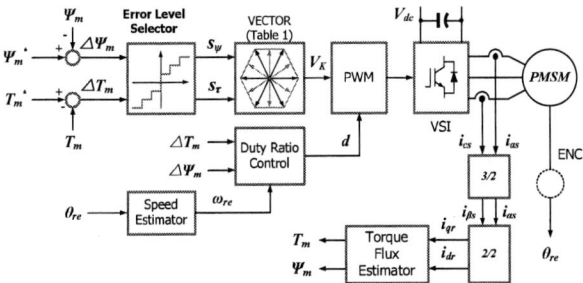

Fig. 7. The block diagram of the proposed DTC-PWM scheme

V. EXPERIMENTAL RESULTS

In order to verify the proposed DTC-PWM scheme, practical experiments are done in this paper. Table III shows the specifications of the proto-type PMSM which is used to the experiment.

The controller for PMSM is designed by the TMS320F28335 which has embedded PWM and ADC module. The PS21997 IPM(Intelligent Power Module) by Mitsubishi is used to the inverter power devices. It has 20[kHz] maximum switching frequency. From this switching frequency, the sampling time of the controller is set as 50[µs]. The speed and position of the motor are measured by the 17bit serial communicated encoder. The position data of the encoder are transferred as serial data. The minimum communication time is 65[µs]. So, the rotor position is measured once in every twice sampling period. Phase currents of the motor are measured by the chip type current sensors and feedback by the ADC7865AS ADC module which has 14bit resolution. And the results are transferred to the DSP. The DC-link voltage is measured by the isolated voltage sensor and 14bit ADC module can measure the link voltage.

TABLE III

SPECIFICATIONS OF THE PROTO-TYPE MOTOR

Parameter	value	Parameter	Value
Pole number	8	Rated Power	750 [W]
Resistance	0.475 [Ω]	Rated Speed	3,000
Inductance	2.7 [mH]	Rated Torque	2.4 [Nm]

For the easy comparison and advanced performance of the proposed control scheme, CDTC with fixed duty ratio in 6 sector method, CDTC-PWM in 6 sectors and the proposed method are tested in this paper.

Fig. 8 shows the compared experimental results of the CDTC, CDTC-PWM in 6 sectors and the proposed method. As shown in Fig. 8(a), the torque current i_{qr} and flux current i_{dr} have large ripple.

The flux and torque current ripple can be reduced by the DTC-PWM method shown in Fig. 8(b). However, the flux and torque current ripples are much reduced in the proposed method shown in Fig. 8(c).

In the conventional 6-sector divisions, the output voltages of the selected voltage vector have insufficient voltages to satisfy the torque and flux error. And the voltage variations are much higher than the proposed method. The less variation of the voltages can reduce the flux and torque current.

The 2014 International Power Electronics Conference

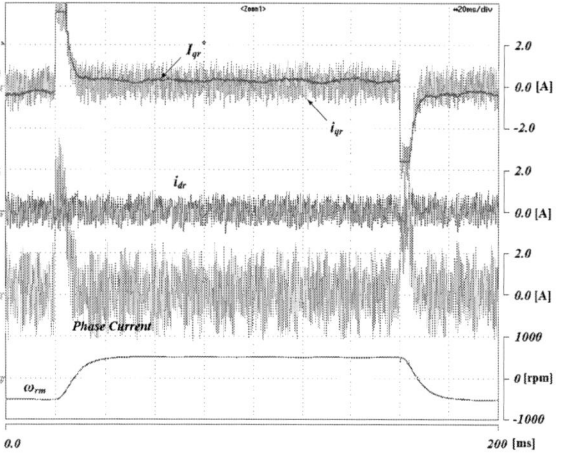

(a)   Experimental result of CDTC with fixed duty ratio in 6 sectors

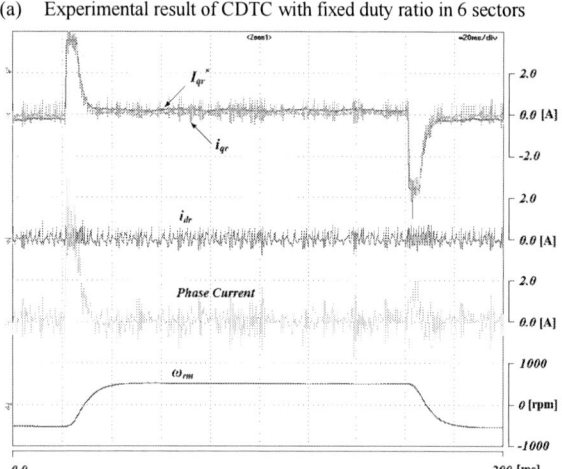

(b)   Experimental result of CDTC-PWM in 6 sectors

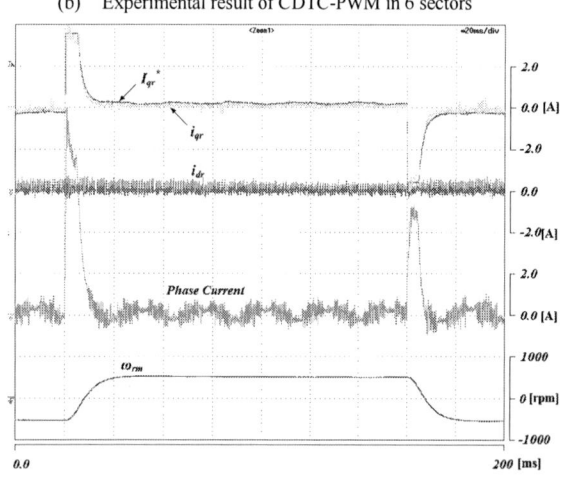

(c)   Experimental result of the proposed method
Fig. 8. Compared experimental results at ±500[rpm]

Fig. 9 shows the experimental results at 1,000[rpm]. Similar to the results of Fig. 8, the proposed method can reduce the current ripples due to the error level selector and selected voltage vector with PWM duty ratio.

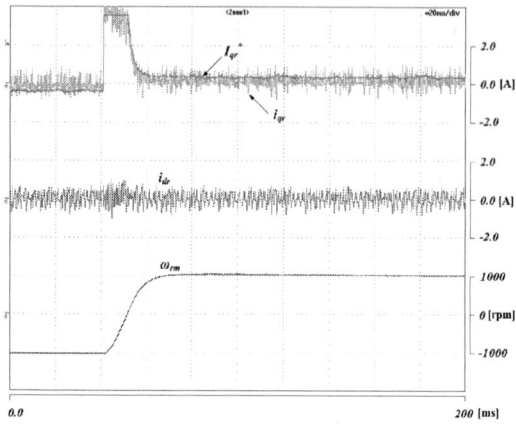

(a)   Experimental result of CDTC-PWM in 6 sectors

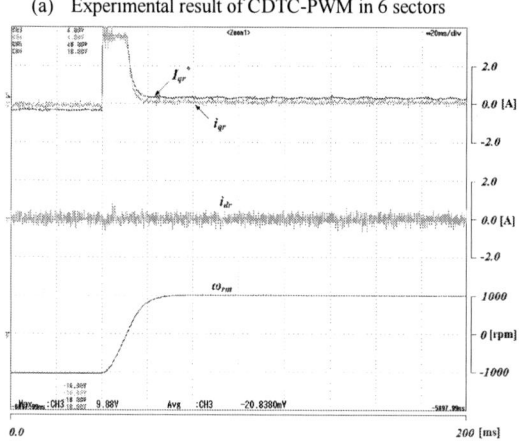

(b)   Experimental result of the proposed method
Fig. 9. Compared experimental results at ±1000[rpm]

Fig. 10 shows the variations of the duty ratio in the proposed method. As shown in Fig. 10, the duty ratio of the proposed method is changed according to the torque and flux error with speed information. During the variation region, the duty ratio is declined according to the speed decreasing, and the duty ratio is inclined according to the speed increasing as shown in Fig. 10.

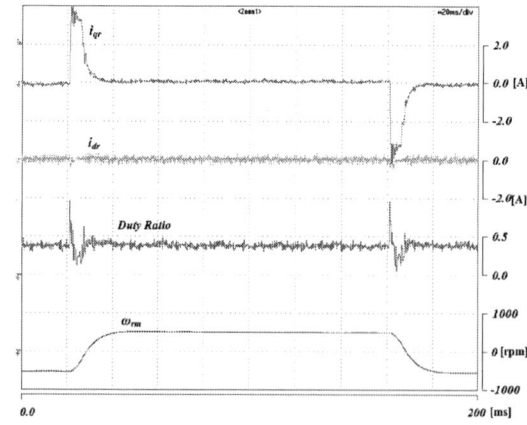

Fig. 10. PWM duty ratio of the proposed method

978-1-4799-2706-7/14 $31.00 © 2014 IEEE

Fig. 11 and Fig. 12 show the selected voltage vector according to the sector number and error levels of the torque and flux in the proposed method. As shown in Fig. 11 and Fig. 12, the error levels of the torque and flux are changed according to the torque and flux state between PB(+3) and NB(-3). And the proper voltage vector is selected by the torque and flux error level at every sector number.

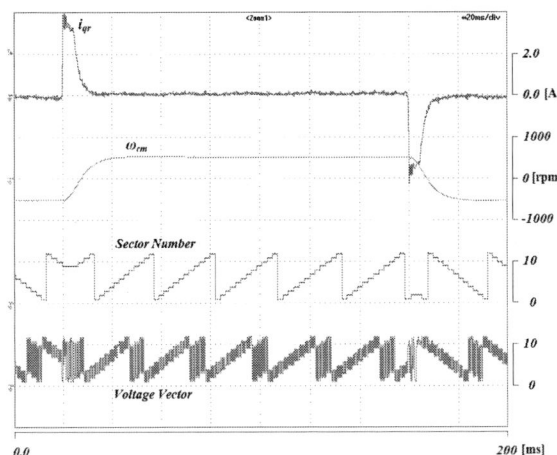

Fig. 11. Sector and selected voltage vectors

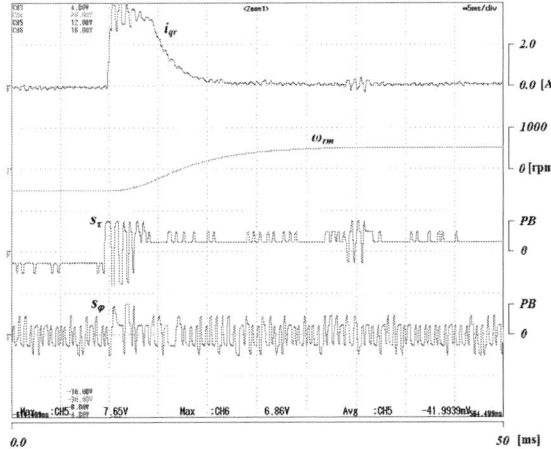

Fig. 12. Torque and flux error level

VI. CONCLUSINS

This paper presents an advanced DTC-PWM method using error level sector and speed information of the motor. The proposed error level selector is used determine the proper voltage vector among 12 voltage vectors for the torque and flux error at every sector position. The selected voltage vector has 12 type characteristics and they can supply enough and less d-q axis voltage ripples during a same sector. In the calculation of the PWM duty ratio, the speed is included.

The speed information which is used to calculate the PWM duty ratio can compensate the back E.M.F. of the motor. So, the fast and stable torque and flux control can be implemented by the compensation method.

In the experimental results, the proposed method shows an advanced performance compare than the conventional DTC and DTC-PWM method.

ACKNOWLEDGMENT

This work (Grants No.2013791) was supported by Business for Cooperative R&D between Industry, Academy, and Research Institute funded Korea Small and Medium Business Administration in 2014 and BK 21+.

REFERENCES

[1] M. Depenbrock, "Direct self-control of inverter-fed machine," *IEEE Trans. Power Electron*, vol. 3, pp. 420-429, Oct. 1988.

[2] I. Takahashi and T. Naguchi, "A new quick-response and highefficiency control strategy of an induction motor," *IEEE Trans.Ind. Applicat*, vol. IA-22, pp. 820-827, Sept./Oct. 1986.

[3] C. French and P. Acarnley, "Direct torque control of permanent magnet drives," *IEEE Trans. Ind. Applicat.*, vol. IA-32, pp. 1080-1088, Sept./Oct. 1996.

[4] L. Zhong, M. F. Rahman, W. Y. Hu, and K. W. Lim, "Analysis of direct torque control in permanent magnet synchronous motor drives," *IEEE Trans. Power Electron*, vol. 12, pp. 528-536, May 1997.

[5] M. F. Rahman, L. Zhong, and K. W. Lim, "A direct torque controlled interior permanent magnet synchronous motor drive incorporating field weakening," *IEEE Trans. Ind. Applicant.*, vol.34, pp. 1246-1253, Nov./Dec. 1998.

[6] I. Takahashi and T. Noguchi, "Take a look back upon the past decade of direct torque control," in Proc. *IEEE-IECON'97 23rd Int. Conf*, vol. 2, 1997, pp. 546-551.

[7] D. Casadei, G. Serra, and A. Tani, "Implementation of a direct torque control algorithm for induction motors based on discrete space vector modulation," *IEEE Trans. Power Electron*, vol. 15, pp. 769-777, July, 2000.

[8] C. G. Mei, S. K. Panda, J. X. Xu, and K. W. Lim, "Direct torque control of induction motor-variable switching sectors," *in Proc. IEEE-PEDS Conf*, Hong Kong, July 1999, pp. 80-85.

[9] A. Tripathi, A. M. Khambadkone, and S. K. Panda, "Space-vector based, constant frequency, direct torque control and dead beat stator flux control of AC machines," *in Proc. IEEE-IECON'01 Conf*, Nov. 2001, pp. 1219-1224.

[10] Yongchang Zhang, Jianguo Zhu, "Direct Torque Control of Permanent Magnet Synchronous Motor With Reduced Torque Ripple and Commutation Frequency", *Power Electronics, IEEE Transactions*, Jan. 2011 ,vol.26 ,pp. 235-248.

[11] Ke Wang, Yaohua Li, Liming Shi, Qiongxuan Ge, "A Novel Switching Scheme for Direct Thrust Control of LIM with Reduction of Thrust Ripple", *Electrical Machines and Systems (ICEMS), 2010 International Conference*, Oct. 2010 , 1491-1494

[12] Chintan Patel, Rijil Ramchand, Rajeevan.P.P,K Sivakumar, Anandarup Das, K.Gopakumar, Marian P.Kazmierkowski, "Direct Torque Control Scheme of IM Drive with 12-sided Polygonal Voltage Space Vectors", *Power Electronics and Applications (EPE 2011), Proceedings of the 2011-14th European Conference* , Sept 2011, 1-11.

[13] Xing Shaobang, Luo Yinsheng, "PMSM twelve sectors DTC system with minimum iron loss", 2011 *Electric Information and Control Engineering (ICEICE)* , 15-17 April 2011

Control of Power Flow between the Wind Generator and Network

Péter Stumpf[*,**], István Nagy[*,**], István Vajk[*,**]

*Department of Automation and Applied Informatics Budapest University of Technology and Economics
Budapest, Magyar tudósok krt. 2. QB108.
**MTA-BME Control Engineering Research Group, nagy@get.bme.hu

Abstract-The paper is concerned with the active and reactive power flow between the doubly fed induction generator harvesting wind power and the grid controlled by the rotor side converter. After deriving the basic equations in Appendices it builds up a block diagram for studying the dynamic processes and it applies them for investigating the slip – rotor current characteristics where rotor voltage components are the parameters. A deeper insight can be gained into the power control process by the evaluation of the characteristics.

Keywords-Wind power, Control of Doubly fed Induction Generator, current-slip characteristics

I. INTRODUCTION

Many trends are becoming apparent in electric distribution driven from both the demand side where better reliability and efficiency are desired, and from the supply side where the integration of generation from renewables and peak shaving have to be accommodated. One main trends is the application of Microgrid Renewable (MR, Green) systems. MR systems are cluster of microsources, storages and loads, many cases controllable loads.

They can be controlled as a united system. Within the MR system on the one hand the implementation of advanced metering infrastructure facilitates the real time pricing to the end users and reduces feeder peak demand and even minimizing feeder losses by controlling plug-in electric vehicles and low priority loads, on the other hand, excellent opportunities are offered to host various renewable energy sources.

One benefit of MR is that it can be designed and operated to compensate for intermittency associated with renewable sources like wind power transporting the intermittency management inside the Microgrid this way contributing to the increased penetration of renewable sources. Microgrid Renewable Distribution System connected to a local utility power grid can incorporate various energy sources such as wind, solar energy, fuel cell, internal combustion engines, micro turbines, energy from CHP, hydro, biomass, tidal and wave energies etc. to supply local loads and interchange energy with smart grid. Wind energy is one of our most abundant resources and one of the fastest growing renewable energy technologies resulting in dropping its generation cost from 37 cent US / kWh in 1980 down to 4 cents US/ kWh in 2008 and almost 300 GWs of global installed wind generation capacity at the end of year 2012.

One of the key elements of MR systems is the cyber control of the whole systems and within it the intelligent control of its units. MR strives for optimized operation not only when connected to the grid but also in a stand-alone mode. There are three main aggregate control features connected with MR.

The first one coordinates the controllable entities of MR to make them at the point of interconnection (POI) to the grid resemble one single dispatchable entity in steady-state and transient. It incorporates power limit control, power frequency control and ramp rate limit control.

The second control features are in operation in stand-alone mode when the distribution of active and reactive power among the sources and loads, the stable operation have to be taken care of. Now neither the voltage nor the frequency is forced on MR. Usually the voltage regulation is performed by voltage droop and the power sharing by frequency droop.

Finally to perform either the first or the second control tasks the intelligent and flexible controls of all units of MR, the generation, energy storage units and controllable loads have to be appropriately connected to the MR system bus. Most cases the connection is done via power converters when required bidirectional ones providing the highly desired individual control.

Interconnection, step-up down function and isolation can be done without heavy-bulky network transformer and in many applications without thermo-mechanical switchgears. As at the end of the line almost all of the control function mentioned is carried out by the power converters, their controls and performances are key features of the operation of MR and in wind power source as well. The focus point of our paper is the control of power converters applied in doubly fed induction generator (DFIG) incorporated in wind power production.

We are interested in the paper in particular in the utilization of wind energy among others in MR environment. The paper is organized as follows: First a short description of the DFIG configuration in Chapter II and that of its control in Chapter III are given. Next in Chapter IV the basic equations for stator flux oriented vector control derived in Appendices are summarized. The slip – rotor current characteristics and their evaluation are included in Chapter V. Chapter VI deals with the dynamics of the stator flux oriented vector control of Rotor Side Converter (RSC). It develops the block diagram of the control of RSC and includes the simulation results of the rotor current loops. The last Chapter is the Conclusions.

II. WIND ENERGY CONVERSION SYSTEM (WECS) APPLYING DFIG

The paper is concerned only with WECs applying Doubly Fed Induction Generator (DFIG) since it is the most popular configuration (Fig.1.) Here the stator is directly connected to the grid while the rotor side converter (RSC) and the grid side converter (GSC) with DC link between them are placed between the wound rotor terminals and the grid. The converters offer excellent control possibilities, partly decoupling the turbine and grid. The rotor speed can be changed by $\pm 30\%$ around synchronous speed (SS) or in some cases around 80-90% of SS. It allows harvesting maximum wind power with varying velocity, and it reduces the investment costs and losses of the converters. The variable speed permits fatigue reduction of mechanical components. The drawback of DFIG is the brushes. They need regular inspection and replacement.

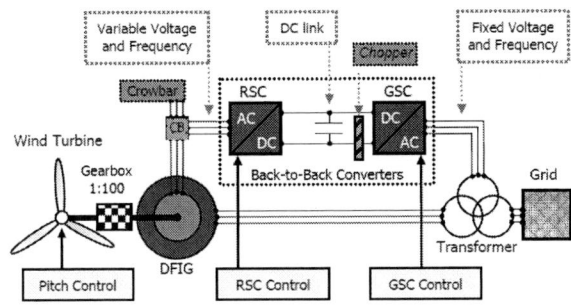

Fig. 1 Wind energy conversion system (WECS) with DFIG [17]

III. CONTROLS IN WECs APPLYING DOUBLY FED INDUCTION GENERATOR (DFIG)

Variable speed wind turbines are capable to harvest most of the wind power therefore they are considered to be the right solution in the wind energy industry. The short overview of the controls in WECS with DFIG can be divided into two parts: Turbine side controls and Generator side controls. The controls in the turbine side consist of pitch control and yaw control. The pitch mechanism performs the rotation of the three blades of the turbine on their longitudinal axis. The angle of attack of the blades is modified by the rotation and it affects the captured wind power and the power conversion efficiency. They can be optimized by pitch control. It even offers protection of the turbine structure against damage caused by strong wind.

The other turbine side control is the yaw control which keeps the area swept by the blades facing always into the directions of the wind maximizing the captured wind power.

The generator side controls are performed by the RSC and the GSC. The RSC control can regulate the stator active P_s and reactive Q_s power independently by rotor current component I_{qr} and I_{dr}, respectively. On the other hand, the output voltage of RSC will eventually

determine I_{dr} and I_{qr}. Therefore the stator active and reactive power can be modified by the two components of rotor supply voltage: V_{dr} and V_{qr} but with them the independent control possibility is lost due to cross-coupling. Both V_{dr} and V_{qr} can change the active and the reactive power but the quantity or size of changes are quite different (See details later).

Finally the GSC control regulates the DC link voltage in the rotor circuit and the rotor reactive power flow Q_r between GSC and the grid by the grid side rotor current component I_d and I_q. (I_d and I_q are different from I_{dr} and I_{qr}). Regulating the DC link voltage modifies the P_r active power of the rotor.

The operation modes are classified in four modes in the wind speed range In the "Parking Mode1" below the so called Cut-in wind speed the turbine generator set is turned-off, it is in stall. The captured power P_{turb} would be less than the loss of the system. Above the Cut-out wind speed in "Parking Mode2" the wind turbine is stopped again ($\omega_{turb} = 0$) to prevent the mechanical damage resulting in very high wind. The "Generator Speed Control section" starts from the Cut-in wind speed to the rated turbine speed or $\omega_{turb} = 1$ p.u. The Maximum Power Point Tracking (MPPT) control sets ω_{turb} at the peak power point (MPP) of each wind speed. Along the trajectory of MPP curve P_{turb} is proportional to ω^3_{turb}. Finally in the "Pitch Control section" the harvested power is kept at $P_{turb} = 1$ p.u. to avoid higher mechanical and electrical stress than their rated values. As it has been mentioned the stator and the rotor active and reactive power can be controlled by RSC. The paper treats in next sections in some detail the RSC control.

IV. STATOR FLUX ORIENTED VECTOR CONTROL. BASIC EQUATIONS

The stator flux oriented vector control (SFOVC) is the most popular control strategy in the RSC. In SFOVC the basic equations are written in Rotating Reference Frame (RRF) revolving with synchronous angular speed ω_s determined by the stator frequency f_s. The SFOVC means that the d axis of RRF is fixed to space vector of the stator linkage flux Ψ_s (Fig. A1.1). Therefore $\Psi_{qs} = 0$. All vectors are space vectors in the paper. In this study the stator copper, iron and ventilation losses are neglected ($R_s=0$).

Considering steady-state, the grid voltage V_s and its angular frequency ω_s are assumed constant: $V_s = j\omega_s \Psi_{ds} = jV_{qs}$. Both $\Psi_s = \Psi_{ds}$ and V_{qs} are constant and $V_{ds} = 0$.

The basic equations are derived in Appendix 1 and 2. Substituting V_{ds} from (A1.3) and V_{qs} from (A1.4) into relation (A1.14) the stator active power in steady-state is

$$P_s = \frac{3}{2}\left(\Psi_{ds}I_{qs} - \Psi_{qs}I_{ds}\right)\omega_s = T_e\omega_s \tag{1}$$

where T_e is the electric torque. Applying $\Psi_{qs} = 0$ and $V_{qs} = \omega_s \Psi_{ds}$, the stator active power from (A1.10) and (1) is

$$P_s = \frac{3}{2}V_{qs}I_{qs} = \frac{3}{2}\frac{L_m}{L_s}V_{qs}I_{qr} = K_pI_{qr} \tag{2}$$

and the stator reactive power from (A.1.15) and (A.1.9)

$$Q_s = \frac{3}{2} V_{qs} I_{ds} = \frac{3}{2} V_{qs} \frac{1}{L_s} (\Psi_{ds} + L_m I_{dr}) = K_{Q1} + K_{Q2} I_{dr} \quad (3)$$

Applying (1) and (A1.10), the electric torque is

$$T_e = \frac{3}{2} \Psi_{ds} I_{qs} = \frac{3}{2} \frac{L_m}{L_s} \Psi_{ds} I_{qr} \quad (4)$$

Equation $\Psi_{qs} = 0$ and $I_{qs} = (L_m/L_s) I_{qr}$ is taken into account in (2) and (4) from (A1.10). Equation $I_{ds} = (1/L_s)$ $(\Psi_{ds} + L_m I_{dr})$ from (A1.9) is used in (3). The main message from (2) and (3) is that both the stator active power P_s and the reactive power Q_s can be changed independently by the rotor current component I_{qr} and I_{dr}, respectively. Furthermore the torque together with P_s can be changed by I_{qr}.

V. Control by Voltage Component V_{dr} and V_{qr}

The stator active and reactive power can be controlled independently by rotor current component I_{qr} and I_{dr}, respectively ((2) and (3)). On the other hand, the RSC changes directly the rotor supply voltage, its two components V_{dr} and V_{qr}. Due to the cross-coupling manifested by voltage component V_{drq} and V_{qrd} (see (A1.16)…(A1.21)) the reference voltage V^*_{dr} depends on both current components I_{dr} and I_{qr} and similar statement holds for reference voltage V^*_{qr}. Conversely, both rotor voltage components vary the two rotor current components, that is, it is not possible to keep the independent control of stator active and reactive power by the two rotor voltage components. On the basis of equations derived in Appendix 1 and 2 the slip(I_{dr}) and the slip(I_{qr}) curves are depicted in Fig.2 and Fig.3 when $V_{qr}=0$ and V_{dr} is the parameter and they are shown in Fig.4 and Fig.5 when $V_{dr}=0$ and V_{qr} is the parameter.

Various conclusions can be drawn from the figures. At zero slip the stator active power P_s can be changed only by component V_{qr} (Fig.5) and the stator reactive power can be changed only by component V_{dr} (Fig.2). On the other hand, when the slip is high ($s=\pm 0.2$ or 0.3) larger stator active power change can be achieved by voltage component V_{dr} (Fig.3) than by V_{qr} (Fig.5). Similarly for high slip values larger stator reactive power change can be achieved by voltage component V_{qr} (Fig.4) than by V_{dr} (Fig.2.).

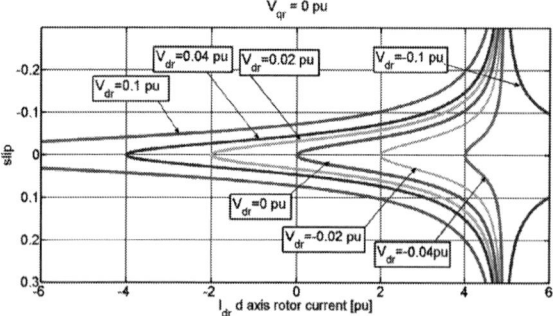

Fig. 2 Slip-d axis rotor current characteristics. Parameter V_{dr}. Voltage component $V_{qr}=0$

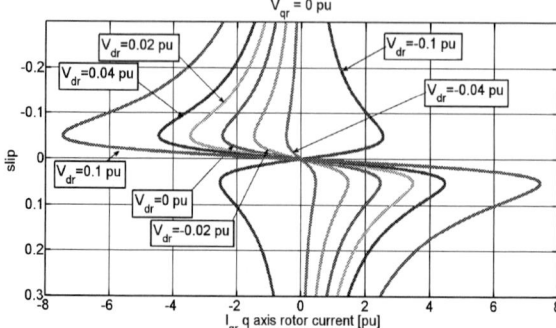

Fig. 3 Slip-q axis rotor current characteristics. Parameter V_{dr}. Voltage component $V_{qr}=0$

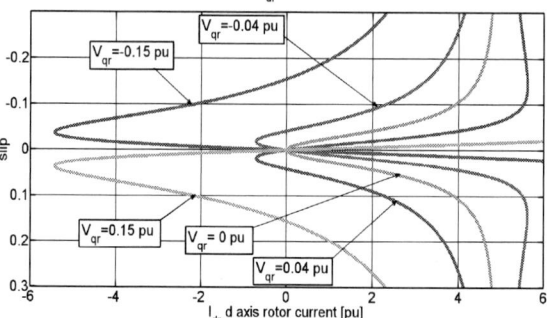

Fig. 4 Slip-d axis rotor current characteristics. Parameter V_{qr}. Voltage component $V_{dr}=0$

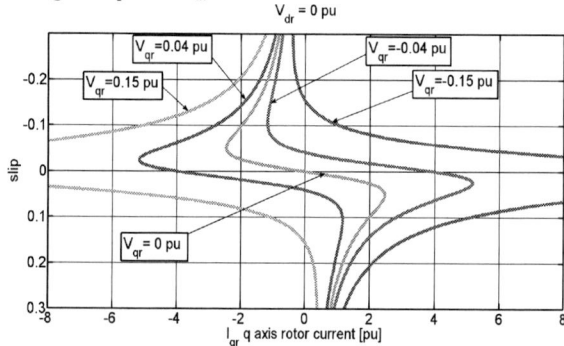

Fig. 5 Slip-q axis rotor current characteristics. Parameter V_{qr}. Voltage component $V_{dr}=0$

It is obvious from the last four figures that the slip can be varied by v_{qr} (or by i_{qr}) rather than v_{dr} (or by i_{dr}). Furthermore the operation point can be on the unstable region of the characteristic slip(i_{qr}) at high slip values. In addition to the current control loops an outer speed control loop is needed for stable operation.

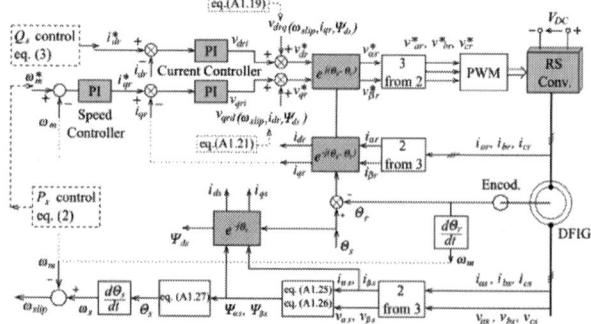

Fig. 6 Stator flux oriented vector control of RSC ($\Psi_{qs}=0$)

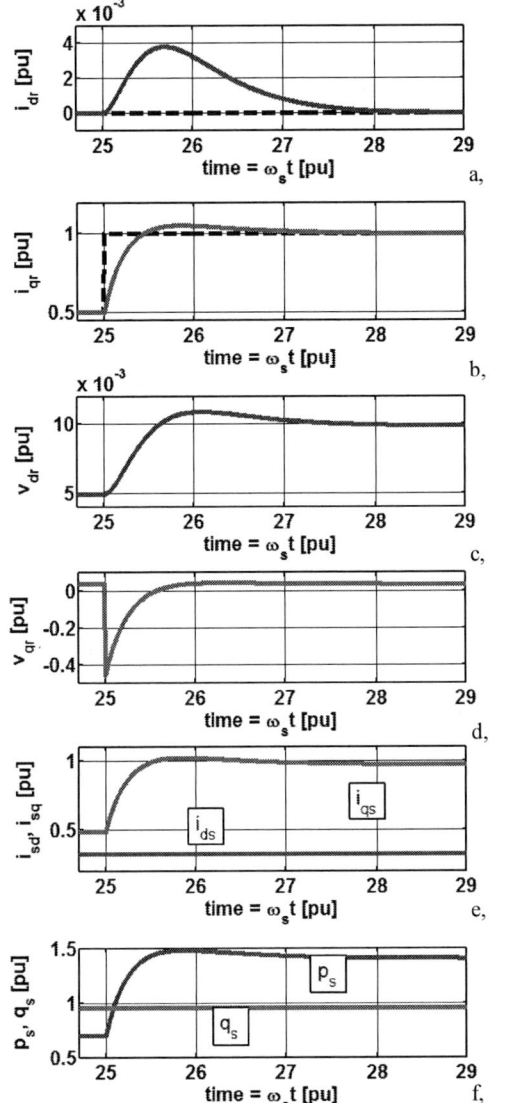

Fig. 7 Dynamic behavior of current loops due to step-wise change in reference current $i*_{qr}$ ($\Delta i*_{qr}$=-0.5 pu), ω_{slip}=-0.05 pu

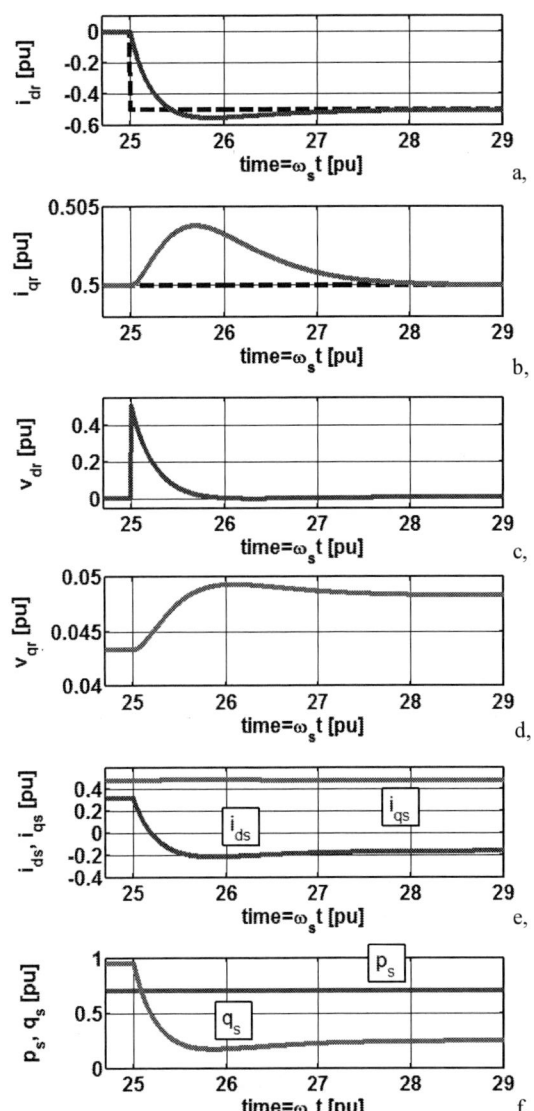

Fig. 8 Dynamic behavior of current loops due to step-wise change in reference current $i*_{dr}$ ($\Delta i*_{dr}$=-0.5 pu), =-0.05 pu

VI. BLOCK DIAGRAM OF THE RSC CONTROL

The RSC control based on the principle of stator flux oriented vector control is shown in block diagram form in Fig. 6. The equations needed to build up the block diagram is derived in Appendix 1.

The stator voltages and currents as well as the rotor currents of DFIG are measured and first they are converted into α-β components (see (A1.23) and (A1.24)), The α-β components of the stator flux linkage $\Psi_{\alpha s}$ and $\Psi_{\beta s}$ are calculated by (A1.25) and (A1.26). Using these two components, the angle θ_s of axis d of RRF can be determined by (A1.27) and the stator angular frequency ω_s by (A1.27) and the stator angular frequency ω_s by (A1.28). Now the absolute value of Ψ_s is calculated from (A1.29) and its space vector is $\Psi_s = \Psi_s e^{j\theta s}$.

Measuring the rotor angle θ_r and its angular frequency ω_m by encoder the angular slip frequency $\omega_{slip}=\omega_s -\omega_m$ as

well as $\Psi_{ds} = \Psi_s e^{-j\theta s}$ are known. Knowing $\theta_s - \theta_r$, we can transform space vector i_r from rotor reference frame into synchronous RRF and determine its two components i_{dr} and i_{qr}. The block diagram has outer loops determining the P_s active and Q_s reactive power, while the inner loop controls the rotor currents in the d and q axis. Their control is made through the d and q components of the rotor voltage, $v*_{dr}$ and $v*_{qr}$ considering them as reference values. First they are transformed to α-β components and finally to three phase values $v*_{ar}$, $v*_{br}$ and $v*_{cr}$ and applied to the PWM block of the RSC.

Fig. 7 and Fig.8 presents the simulation results of the two rotor current control loops. The simulations were carried out on the basis of block diagram (Fig.6) in Matlab environment. Assuming that the electric transients are much faster than the mechanical one, the slip value was kept constant. Step-wise $\Delta i*_{qr}$ and $\Delta i*_{dr}$ reference current change was applied in Fig.7b and Fig.8a, respectively. The time functions of the rotor voltage components v_{dr}, v_{qr} (Fig. c and d) and those of the stator current components i_{ds}, i_{qs} (Fig. e) and finally those of the stator active and reactive power (Fig. f) are also presented in Fig.7 and 8. The cross couplings between d and q directions are visible in Fig.7a and Fig.8b. The numerical values of the parameters used in the simulation are included in Appendix 2.

As a conclusions it is demonstrated that Δi^{*}_{qr} changes only the stator active power without varying q_s while Δi_{dr} changes only the stator reactive power without modifying p_s.

VII. Conclusions

The main concern of the paper is to understand the effect of the two rotor current and voltage components, V_{dr} and V_{qr} to the stator active and reactive power flow in the doubly fed induction generator. Changing reference i^{*}_{dr} or i^{*}_{qr}, the possibility for changing the stator active and reactive power independently on each other is maintained. The dynamic behavior of the two rotor current control loops are shown to step-wise reference current change and their independent effect to the stator active and reactive power flow is demonstrated.

VIII. Acknowledgement

The authors wish to thank the Hungarian Research Fund (OTKA K100275) and the Control Research Group of the Hungarian Academy of Sciences (HAS). This work is connected to the scientific program of the "Development of quality-oriented and harmonized R+D+I strategy and functional model at BME" project. This project is supported by the New Széchenyi Plan (Project ID: TAMOP-4.2.1/B- 09/1/KMR-2010- 0002).

IX. Appendix 1 Basic Relations

The aim is to derive a few basic relations for doubly fed induction machine (DFIM). Space vectors and Rotating Reference Frame (RRF) fixed to the stator linkage flux space vector Ψ_s are used (Fig. 4). RRF is fixed to vector Ψ_s revolving with angular frequency ω_s.
The stator and rotor voltage equations in space vector form are

$$v_s = R_s i_s + p\Psi_s + j\omega_s \Psi_s \tag{A1.1}$$

$$v_r = -R_r i_r + p\Psi_r + j(\omega_s - \omega_m)\Psi_r \tag{A1.2}$$

where $\omega_m = P\omega_{mech}$, P = Nr. of pole pairs, $p = d/dt$. v_s, v_r and i_s, i_r are the stator and rotor voltage and current space vectors, respectively. The resistance of one phase is R_s and R_r.
Here ω_{mech} is the rotor mechanical angular frequency. The d and q coordinate equations of eq. (A1.1) and (A1.2), respectively

$$v_{ds} = R_s i_{ds} + p\Psi_{ds} - \omega_s \Psi_{qs} \tag{A1.3}$$

$$v_{qs} = R_s i_{qs} + p\Psi_{qs} + \omega_s \Psi_{ds} \tag{A1.4}$$

$$v_{dr} = -R_r i_{dr} + p\Psi_{dr} - \omega_{slip}\Psi_{qr} \tag{A1.5}$$

$$v_{qr} = -R_r i_{qr} + p\Psi_{qr} + \omega_{slip}\Psi_{dr} \tag{A1.6}$$

The stator and rotor linkage flux equation in space vector form are

$$\Psi_s = L_s i_s - L_m i_r \tag{A1.7}$$

$$\Psi_r = L_m i_s - L_r i_r \tag{A1.8}$$

where L_s and L_r are the stator and rotor inductances, L_m the mutual inductance. The d and q coordinate equations of (A1.7) and (A1.8), respectively

$$\Psi_{ds} = L_s i_{ds} - L_m i_{dr} \tag{A1.9}$$

$$\Psi_{qs} = L_s i_{qs} - L_m i_{qr} \tag{A1.10}$$

$$\Psi_{dr} = L_m i_{ds} - L_r i_{dr} \tag{A1.11}$$

$$\Psi_{qr} = L_m i_{qs} - L_r i_{qr} \tag{A1.12}$$

The instantaneous stator power

$$s_s = p_s + jq_s = \frac{3}{2} v_s i_s^* = \frac{3}{2}\left\{\operatorname{Re}\left[v_s i_s^*\right] + j\operatorname{Im}\left[v_s i_s^*\right]\right\} \tag{A1.13}$$

where i_s^* is the complex conjugate of space vector i_s. 3 stands for 3 phases, 2 stands for peak values of v_s and i_s.
Taking into account $v_s = v_{ds} + jv_{qs}$ and $i_s = i_{ds} + ji_{qs}$ the active power p_s and reactive power q_s are

$$p_s = \frac{3}{2}\left(v_{ds} i_{ds} + v_{qs} i_{qs}\right) \tag{A1.14}$$

$$q_s = \frac{3}{2}\left(v_{qs} i_{ds} - v_{ds} i_{qs}\right) \tag{A1.15}$$

Considering that $\Psi_s = \Psi_{ds}$, that is, $\Psi_{qs} = 0$, and substituting the linkage flux equations (A1.11), (A1.12) into the rotor voltage equations (A1.5), (A1.6) the results are

$$v_{dr} = v_{dri}(i_{dr}) + v_{drq}(i_{qr}, \Psi_{ds}) \tag{A1.16}$$

$$v_{qr} = v_{qri}(i_{qr}) + v_{qrd}(i_{dr}; \Psi_{ds}) \tag{A1.17}$$

where

$$v_{dri} = -(R_r + \sigma L_r p)i_{dr} \tag{A1.18}$$

$$v_{drq} = \omega_{slip}\sigma L_r i_{qr} + \frac{L_m}{L_s} p\Psi_{ds} \tag{A1.19}$$

and

$$v_{qri} = -(R_r + \sigma L_r p)i_{qr} \tag{A1.20}$$

$$v_{qrd} = -\omega_{slip}\left[\sigma L_r i_{dr} - \frac{L_m}{L_s}\Psi_{ds}\right] \tag{A1.21}$$

The two stator quantities i_{ds} and i_{qs} had been eliminated from (A1.16)…(A1.21) by using the stator flux linkage relation (A1.9), (A1.10) and $\Psi_{qs} = 0$. The value σ is

$$\sigma = 1 - \frac{L_m^2}{L_s L_r} \tag{A1.22}$$

Note, that the term v_{drq} in (A1.16) and v_{qrd} in (A1.17) are expressing cross-coupling between the d and q components as v_{dr} depends not only on i_{dr} but on i_{qr} as well and similarly v_{qr} depends not only on i_{qr} but on i_{dr} as well.

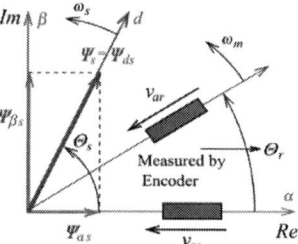

Fig. A.1.1 Stationary α-β and synchronous d-q reference frame. Determination of Θ_s and ω_s

The determination of angle Θ_s is needed for the calculation of dynamics and for the control of DFIG (Fig.A1.1 and A1.2). In Stationary Reference Frame (SRF) α, β

$$v_s = v_{\alpha s} + jv_{\beta s} \tag{A1.23}$$

$$i_s = i_{\alpha s} + ji_{\beta s} \tag{A1.24}$$

The calculation of the α and β component of the stator flux linkage space vector Ψ_s can be done by

$$\Psi_{\alpha s} = \int\left(v_{\alpha s} - R_s i_{\alpha s}\right)dt \tag{A1.25}$$

$$\Psi_{\beta s} = \int\left(v_{\beta s} - R_s i_{\beta s}\right)dt \tag{A1.26}$$

The instantaneous value of angle Θ_s is (Fig. A1.2)

$$\Theta_s = \tan^{-1}\left(\Psi_{\beta s} / \Psi_{\alpha s}\right) \tag{A1.27}$$

and the angular frequency

$$\omega_s = \frac{d\Theta_s}{dt} \tag{A1.28}$$

The magnitude of Ψ_s is

$$\Psi_s = \Psi_{ds} = \sqrt{\Psi_{\alpha s}^2 + \Psi_{\beta s}^2} \tag{A1.29}$$

X. APPENDIX 2. TORQUE, CURRENT AND ROTOR POWER RELATIONS

The aim is the derivation of the relations in steady-state as follows: d and q components of stator and rotor currents, stator and rotor active and reactive power and the electric torque. The independent variables in all relations have to be the rotor voltage components V_{dr} and V_{qr} as well as the rotor angular frequency ω_{slip}, what is the same, the slip s. In term of physical background the generator input variables are the stator or grid voltage V_s, the rotor voltage (V_{dr}, V_{qr}) and the mechanical power. The last one will be represented in the relations by the slip and it is assumed that V_s = const. Applying stator flux oriented vector control, the flux component $\Psi_{qs} = 0$.

Neglecting the stator resistance, from (A1.1)

$$\mathbf{V}_s = j\omega \mathbf{\Psi}_s = j\omega \Psi_{ds} \tag{A2.1}$$

From (A1.9) and A(1.10)

$$I_{ds} = \frac{1}{L_s}\left(\Psi_{ds} + L_m I_{dr}\right) = \frac{V_s}{X_s} + \frac{X_m}{X_s} I_{dr} \tag{A2.2}$$

$$I_{qs} = \frac{L_m}{L_s} I_{qr} \tag{A2.3}$$

Let us express V_{dr} and V_{qr} [see (A1.5), (A1.6)] as a function of I_{dr} and I_{qr} by substituting Ψ_{dr} and Ψ_{qr} from (A1.11) and (A1.12), respectively

$$V_{dr} = -R_r I_{dr} + \omega_{slip}\sigma L_r I_{qr} \tag{A2.4}$$

$$V_{qr} = -R_r I_{qr} - \omega_{slip}\left(\sigma L_r I_{dr} - \frac{L_m}{L_s}\Psi_{ds}\right) \tag{A2.5}$$

Here eq. (A2.3) and (A1.9) were used as well. From the last two equations

$$I_{dr} = \frac{1}{D(s)}\left[-\frac{R_r}{X_r}V_{dr} - \sigma s V_{qr} + \sigma s^2 \frac{X_m}{X_s}V_s\right] \tag{A2.6}$$

$$I_{qr} = \frac{R_r/X_r}{D(s)}\left\{V_{qr} + s\left[(X_r/R_r)\sigma V_{dr} + (X_m/X_s)V_s\right]\right\} \tag{A2.7}$$

where

$$D(s) = X_r\left[\left(\frac{R_r}{X_r}\right)^2 + (\sigma s)^2\right] \tag{A2.8}$$

Here $V_s = \omega_s\Psi_{ds}$ the grid voltage, $X_r = \omega_s L_r$, $X_s = \omega_s L_s$ and $s = \omega_{slip}/\omega_s$ the slip. Substituting I_{qr} from (A2.7) into the torque equation (5) the torque

$$T_e(s) = \frac{3}{2}\frac{X_m}{X_s}V_s I_{qr} = \frac{3}{2}\frac{X_m}{X_s}\frac{R_r/X_r}{D(s)} \cdot$$
$$\cdot \left\{V_{qr}V_s + s\left[(X_r/R_r)\sigma V_{dr}V_s + (X_m/X_s)V_s^2\right]\right\} \tag{A2.9}$$

The active and reactive power of stator and rotor from (A1.14) and (A1.15), respectively.

$$P_s = \frac{3}{2}V_s I_{qs} = \frac{3}{2}\frac{X_m}{X_r}V_s I_{qr} \tag{A2.10}$$

$$Q_s = \frac{3}{2}\left(\frac{V_s^2}{X_s} + \frac{X_m}{X_s}V_s I_{dr}\right) \tag{A2.11}$$

Here (A2.2) and (A2.3) were used.
The instantaneous rotor power likewise to stator power (A1.13)

$$S_r = P_r + jQ_r = \frac{3}{2}\mathbf{V}_r \mathbf{I}_r^* = \frac{3}{2}\left\{R_e\left[\mathbf{V}_r, \mathbf{I}_r^*\right] + j\text{Im}\left[\mathbf{V}_r, \mathbf{I}_r^*\right]\right\} \tag{A2.12}$$

Substituting here $\mathbf{V}_r = V_{dr} + jV_{qr}$ and $\mathbf{I}_r^* = I_{dr} - jI_{qr}$

$$P_r = \frac{3}{2}\left(V_{dr} I_{dr} + V_{qr} I_{qr}\right) \tag{A2.13}$$

$$Q_r = \frac{3}{2}\left(V_{qr} I_{dr} - V_{dr} I_{qr}\right) \tag{A2.14}$$

	Notation	Value p.u.
Grid voltage	V_s	1
Grid angular frequency	ω_s	1
Stator linkage flux	$\Psi_s = \Psi_{ds}$	1
Mutual inductance	$X_m = L_m$	3
Stator inductance	$X_s = L_s$	3.1
Rotor inductance	$X_r = L_r$	3.1
Stator resistance	R_s	0.01
Rotor resistance	R_r	0.01
Current controller gain	K_{pid}/K_{piq}	1
Current controller time constant	T_{iid}/T_{iiq}	1

$$\sigma = 1 - \frac{L_m^2}{L_s L_r} = 0.0634$$

REFERENCES

[1] Maurício B. C. Salles , Kay Hameyer , José R. Cardoso , Ahda. P. Grilo and Claudia Rahmann "Crowbar System in Doubly Fed Induction Wind Generators" Energies 2010, 3, 738-753

[2] J. Wang, J. Zhang, Y Zhong, „Study on a Super Capacitor Energy Storage System for Improving the Operating Stability of Distributed Generation System" Conference on Electric Utility Deregulation and Restructuring and Power Technologies, 2008, DRPT 2008, Nanjing, China, 6-9 April 2008, pp. 2702 – 2706

[3] J. Leuchter, P. Bauer, V. Rerucha, V. Hajek, „Dynamic Behavior Modeling and Verification of Advanced Electrical-Generator Set Concept" IEEE Transactions on Industrial Electronics, January, 2009, vol. 56, nr:1, pp. 266-279

[4] C. Sourkounis, J. Wenske, „Cascaded State Control for Dynamic Power Conditioning in Wind Parks", Proceeding of the International Conference on Electrical Power Quality and Utilisation, EPQU 2011, Lisbon, 17 October 2011 through 19 Ocotber, 2011

[5] A. El Aroudi, B. Robert, A. Cid-Pastor, L. Martinez-Salamero, „Modeling and Design Rules of a Two-Cell Buck Converter Under a Digital PWM Contoller" IEEE Transactions on Power Electronics, 2008, Vol 23, Issue 2, pp. 859-870

[6] C. Sourkounis, J. Wenske, "Electronic synchronous machine for dynamic power conditioning in wind parks", Compability and Power Electronics (CPE 2011), Tallin, Estonia, 1-3 June, 2011, pp. 56-61, *ISBN: *978-1-4244-8805-6

[7] T. Ghennam, EM. Berkouk, B. Francois, K. Aliouane, "A New Space-Vector Based Hysteresis Current Control Applied on Three-Level Inverter to Control Active and Reactive Powers of Wind Generator" International Conference on Power Engineering, Energy and Electrical Drives, 2007. POWERENG 2007, pp. 636-641

[8] Li Hui Yang, Zhao Xu, Østergaard, J, Zhao Yang Dong, Xi Kui Ma, "Hopf bifurcation and eigenvalue sensitivity analysis of doubly fed induction generator wind turbine system", Power and Energy Society General Meeting, 2010 IEEE, Minneapolis, MN, pp. 1-6, 25-29 July, 2010, ISBN:978-1-4244-6549-1

[9] Zhou Y, Bauer P, Ferreira JA, et al.:" Control of DFIG under unsymmetrical voltage dip" 38th IEEE Power Electronic Specialists Conference, JUN 17-21, 2007 Orlando, FL, pp. 933-938

[10] Zhou Y, Bauer P, Ferreira JA, et al.:"Operation of Grid-Connected DFIG Under Unbalanced Grid Voltage Condition", IEEE TRANSACTIONS ON ENERGY CONVERSION Volume: 24 Issue: 1, pp: 240-246, 2009 March

[11] R.T. Pinto, S.Rodrigues, P.Bauer, J.Pierik: Description and Comparison of DC voltage Control Strategies for Offshore MTDC Networks: Steady-State and Foult Analysis; European Power Electronics and Drives Journal (EPE Journal) ISSN 0939-8368, March 2013

[12] Peroutka, Z., Glasberger, T.: "Algorithms of Space Vector PWM in Overmodulation Area". Advances in Electrical and Electronic Engineering (AEEE) Journal. Vol. 5, No. 1-2, 2006. pp. 90-93. ISSN 1336-1376.

[13] Lihui Yang, Zhao Xu, J, Østergaard, Zhao Yang Dong, Kit Po Wong, Xikui Ma: "Oscillatory Stability and Eigenvalue Sensitivity Analysis of A DFIG Wind Turbine System", IEEE Transactions on Energy Conversion, Vol.26, No.1, March, 2011

[14] C. Sourkounis, J. Wenske, "Cascaded State Control for Dynamic Power Conditioning in Wind Parks" Proceeding of the International Conference on Electrical Power Quality and Utilisation, EPQU 2011, Lisbon, 17 October 2011through 19 Ocotber,2011

[15] P.M. Lacko M. Olejar, J. Dudrik,"DC-DC Push&Pull Converter with Turn-Off Snubber for Renewable   Energy Sources" EDPE 2009, Dubrovnik, Croatia, 12-14 Oct, 2009, CD Rom ISBN: 953-6037-56-8

[16] Sergey Ryvkin, "Elimination of the Voltage Oscillation Influence in the 3-Level VSI Drive Using Sliding Mode Control Technique" AUTOMATIKA, 51(2), 2010, pp. 138-148

[17] Salles, M.B.C.; Grilo, A.P.; Cardoso, J.R.; Lessa, L.L., "The influence of the applied rotor voltage on ride-through capability of doubly fed induction generator," International Conference and Utility Exhibition on Power and Energy Systems: Issues & Prospects for Asia (ICUE), 2011., pp.1,4, 28-30 Sept. 2011

[18] Jiabing Hu; Hailiang Xu; Yikang He, "Coordinated Control of DFIG's RSC and GSC Under Generalized Unbalanced and Distorted Grid Voltage Conditions," IEEE Transactions on Industrial Electronics, vol.60, no.7, pp.2808-2819, July 2013

[19] Marwali, Mohammad N., Jin-Woo Jung, and Ali Keyhani. "Control of distributed generation systems-Part II: Load sharing control.", IEEE Transactions on Power Electronics Vol.19 No.6 2004, pp1551-1561.

[20] Ye, Zhongming, Praveen K. Jain, and Paresh C. Sen. "Phasor-domain modeling of resonant inverters for high-frequency AC power distribution systems." IEEE Transactions on Power Electronics, Vol. 24. No.4, 2009, pp 911-923.

[21] Komrska, Tomáš, et al. "Current reference generator for 50-Hz and 16.7-Hz shunt active power filters." International Journal of Electronics Vol.97 No.1 2010, pp63-81.

[22] Huber, J.E.; Korn, A.J., "Optimized Pulse Pattern modulation for Modular Multilevel Converter high-speed drive," 15th International Power Electronics and Motion Control Conference (EPE/PEMC), 2012, pp.LS1a-1.4-1,LS1a-1.4-7, 4-6 Sept. 2012

[23] Tekwani, P. N., R. S. Kanchan, and K. Gopakumar. "Current-error space-vector-based hysteresis PWM controller for three-level voltage source inverter fed drives." Electric Power Applications, IEE Proceedings-. Vol. 152. No. 5. IET, 2005.

[24] Kouro, S.; Malinowski, M.; Gopakumar, K.; Pou, J.; Franquelo, L.G.; Bin Wu; Rodriguez, J.; Perez, M.A.; Leon, J.I., "Recent Advances and Industrial Applications of Multilevel Converters," IEEE Transactions on Industrial Electronics,, vol.57, no.8, pp.2553,2580, Aug. 2010

Advances in Nanogrid Technology and Its Integration into Rural Electrification in India

Santanu Mishra, *Senior Member, IEEE* and Olive Ray, *Student Member, IEEE*
Indian Institute of Technology Kanpur, India.
Email: santanum@iitk.ac.in, Phone: +91-512-259-6249

Abstract- Nanogrids are small residential power systems with renewable sources, storage, and domestic loads. It may or may not have connectivity to utility grid. This paper discusses the role of advanced converter technology for Nanogrids in solving acute power shortage problem in Rural India. The paper is divided into two sections. In the first section, the advances in Nanogrid converter technology are discussed. The implementations and advantages of multi-output converters are delineated. In the second section, some proposals are presented to incorporate these new technologies for easy adaption in rural electrification. A prototype is built using off-the-shelf components and commercial loads to prove the usability of the proposed concepts.

I.   Introduction

An Indian village would be declared as electrified [1] by the 'Ministry of Power' if the following definitions are met.

i. *Basic infrastructure such as Distribution Transformer and Distribution lines are provided in the inhabited locality as well as the Dalit Basti/ hamlet where it exists. (For electrification through Non Conventional Energy Sources a Distribution transformer may not be necessary).*

ii. *Electricity is provided to public places like Schools, Panchayat Office, Health Centers, Dispensaries, Community centers, etc.*

iii. *The number of households electrified should be at least 10% of the total number of households in the village.*

There are two major observations about the above definition: (a) The definition doesn't say anything about the number of hours in a day the electricity should be available, (b) If 10 % of the village is supplied with basic electricity infrastructure, the complete village is deemed electrified. Even with these liberal definitions of electrification, by 2009, about 20 % of the villages in India didn't meet the above standard and are un-electrified. [2]

The solutions to electrify the Indian villages are easier said than done. There are three major challenges to rural electrification: (a) Grid infrastructure is neither adequate nor economical in rural areas due to sheer geographical area of the country and sparsity of villages; (b) There is a 30-40 % unaccounted transmission and distribution loss in the grid infrastructure [3]; (c) There is a huge shortage of electric energy each year. Fig. 1a shows the energy shortage in the last five years (in Million Units (MU)). The shortage invariably affects the rural population as the ever increasing industrial centers are mostly in the urban or semi-urban areas.

Renewable power is seen as a viable solution to this acute rural power shortage. Various government agencies like Ministry of New and Renewable Energy (MNRE), Ministry of Power (MoP), Rural Electrification Corporation Limited (RECL), Scientific and Engineering Research Board (SERB) through its thrust on R &D effort and commercialization, have stepped up their endeavor in this direction. Apart from these, there is also a concerted effort by the Government of India to establish policy and guidelines for easy accommodation of renewable energy in current distribution system. [4]

There are two ways of using renewable power for rural electrification: (a) Install large renewable farms at higher power level and feed this energy to the transmission grid to augment the existing energy; (b) Use renewable power at individual house or community level and consume the energy at the point-of-generation and exchange the surplus energy with the distribution grid. In the second choice, these small sources are referred to as distributed generators (DGs) [5-6]. It comes under the umbrella of 'Smart-grid' activity. The article concentrates on this second strategy as

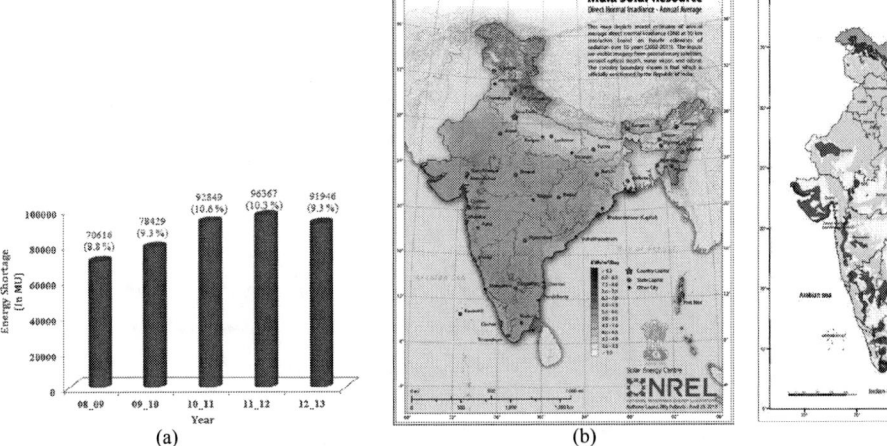

Fig. 1. (a) Energy shortage in India over the last five years (in **Million Units**), (b) Solar Resource in India, and (c) Wind resource in India.

978-1-4799-2706-7/14 $31.00 © 2014 IEEE

a viable alternative for rural electrification.

Fig. 1b shows the map of a solar resource in India [7]. If the renewable energy infrastructure is studied closely, solar comes out to be a clear winner as far as rural electrification is concerned. This is due to the fact that it is available all across the country for most of the days in a year, which essentially avoids costly transmission infrastructure to geographically dispersed rural localities.

Just to show the availability of alternate renewable resources, Fig. 1c shows the wind resources in India. It can be seen that wind energy is available in pockets across some costal and high lying areas. It can only be used in rural areas if the energy is fed into the transmission grid infrastructure, and therefore it is not a good option under present scenario.

This article discusses the advances in converter technology for small residential renewable systems, also called Nanogrids, and how it can aid in rural electrification in India. The article is structured in two sections. The first part of the article discusses the advances in Power electronics converter technology suitable for Nanogrids. The second part discusses the challenges and suggested procedures to ease these new technologies into rural electrification.

II. Advances in Nanogrid Technology

a. *What is a Nanogrid?*

Fig. 2 shows a domestic renewable power system consisting of solar panels as sources, battery storage, loads, and distribution grid interconnection (if available). In many ways, it is a miniature power grid of less than 5 kW capacity, and is referred to as a Nanogrid. [8] All the sources and loads are interfaced using power electronics (PE) converters. The supply cables are represented by red lines and the distribution cables inside the house are represented by green lines in this figure. The distribution bus (green lines) inside the house can be DC or AC. In case of a DC distribution, it is called a DC Nanogrid and with an AC distribution it is called an AC Nanogrid. [9-10] As mentioned, a Nanogrid can either be connected to a utility distribution grid or can be operated in a standalone fashion.

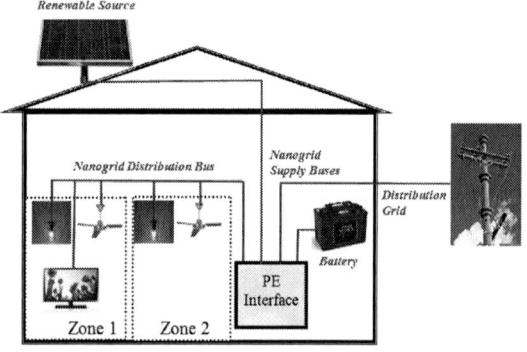

Fig. 2. A typical Nanogrid with renewable source, battery, loads, and distribution grid back-up.

Fig. 3 shows the implementation of an AC Nanogrid. It has a charge controller which interfaces the solar panel and the battery. A DC to AC inverter converts the charge

controller output into an AC. A step-up power transformer is typically required as the Solar PV panels produce an output at a relatively lower voltage level (12 V to 48 V) and utility is at 230 V AC. [11] Another approach can be to incorporate the transformer isolation inside the conversion stage by using isolated converters. The second approach reduces the size of the transformer as high frequency isolation can now be used. Both these designs are common in conventional implementations of an AC Nanogrid.

As the power generation by a renewable source is in form of a DC, a DC Nanogrid is also explored as an option. This will reduce the number of conversion stages from source to load and result in lower cost as well as high overall efficiency [12]. However, for smooth transition to a DC Nanogrid, commercial loads which work with a DC input (at appropriate level) has be available. In the second section of the paper, the author explores the commonly available appliances in India and how they fit into both AC as well as DC Nanogrids.

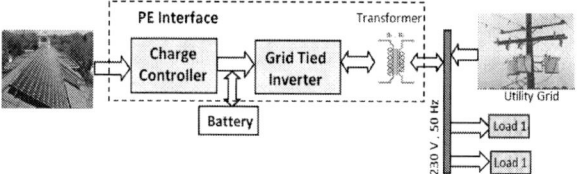

Fig. 3. Conventional implementation of PE interfaces.

b. *Nanogrid as a part of a Smart-grid*

It is envisioned that a smart grid will consist of numerous distributed sources (houses, Electric vehicles, etc.) which will also work as load. This essentially implies that the current power distribution system will require having bidirectional power flow characteristics. This is a significant deviation from its operation as currently it is only used to carry power from the substation to customers through many distribution transformers. What makes it even more challenging is that most of the Nanogrids are in fact small sources of energy.

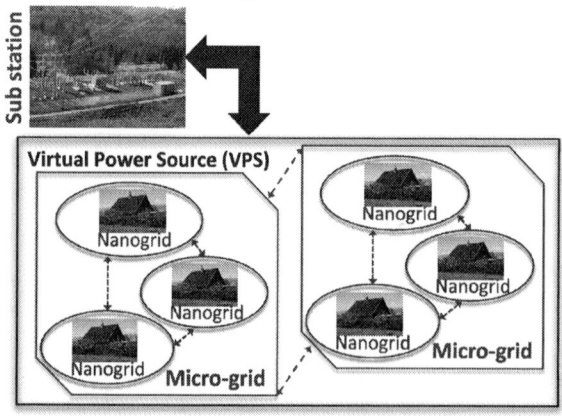

Fig. 4. Incorporating a Nanogrid into a Smart-grid.

Connecting numerous Nanogrids to the utility distribution grid is not a good option as energy management will be quite difficult. One proposal to solve this issue is to first interconnect many Nanogrid to create a cluster called Microgrid which is of about 10-50 kW rating. Many Microgrids can be interconnected to form a virtual power source (VPS). [13] This virtual power source can be connected to the sub-station. All these blocks can exchange power between them as indicated by the arrows, as shown in Fig. 4, thus making the system truly smart.

The above approach has good fault isolation capability as a fault can be monitored and isolated at a Microgrid level. It is similar to the presently used AC distribution system where the fault is isolated by tripping the 11kV/440 V distribution transformer to a specific area where fault has occurred. This localizes the impact of the fault as far as overall power system is concerned.

c. Conventional Nanogrid Technology

Typical converter architectures to implement the Nanogrid (in Fig. 3) are shown in Fig. 6. It consists of an input boost stage which acts like a charge controller and extracts maximum power from the panels, followed by a voltage source inverter. The advantages of this structure are: (a) the input current is continuous (b) Even with a smaller panel voltage; a higher DC link voltage can be obtained (about four times). The storage can be interfaced to the DC link either directly or through a converter. If the DC link voltage is higher (e.g., more than 48 V or so), a (bi-directional) synchronous buck converter can be interfaced at its output to feed it to a lower voltage battery as shown in Fig. 6b.

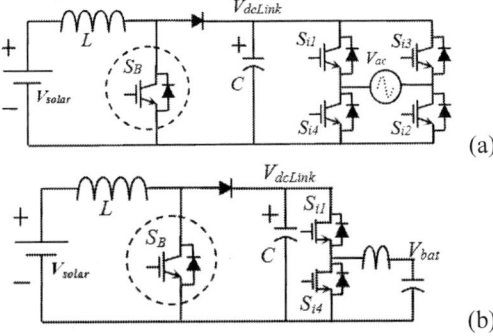

Fig. 6. (a) Conventional charger and grid-tied inverter (b) Conventional topology to interface DC link to a low voltage battery.

Buck based topologies, like a Voltage source inverter (VSI) or a synchronous buck converter, have two major

draw backs: (a) their output is always smaller than the input, (b) there needs to be a shoot-through protection between two switches of a leg. The second drawback results in distortion of output and makes the design immune to EMI noise and spurious turn-on of switches.

d. Advances in Nanogrid Converter Technology

As shown in Fig. 6, the conventional technology uses multiple converters coupled using a capacitive link (V_{dcLink}). This necessitates the need to have shoot-through protection for the subsequent converter stage.

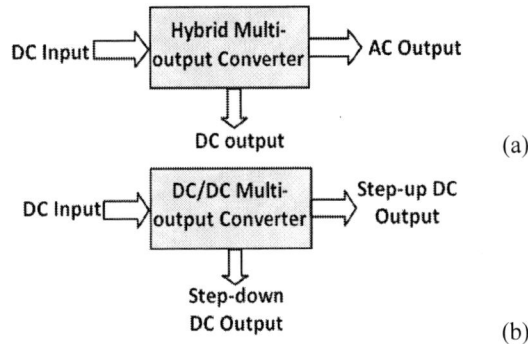

Fig. 7. Block diagram of Multi-output converters (MOC) (a) Hybrid MOC, (b) DC-MOC.

What if the operation of charge controller and grid-tied inverter are combined by making them share some of the switches? The fundamental idea is to reuse the switches of a grid tied inverter for charge controller application. The resulting architecture leads to reduced number of switches and allows shoot-through of the inverter or buck converter leg switches. For low power applications, this strategy will result in improved reliability and better power density. These new converters are called multi-output converters (MOCs).

Fig. 7a shows the block diagram of a hybrid architecture which can generate a step-up DC and an AC output, simultaneously, from a single DC input. It is referred to as Hybrid-MOC as the outputs can be AC as well as DC. The converter is realized by replacing the boost switch (S_B) in Fig. 6a with the inverter legs S_{i1}-S_{i4}. Therefore, boost action (turning on of S_B) is realized by

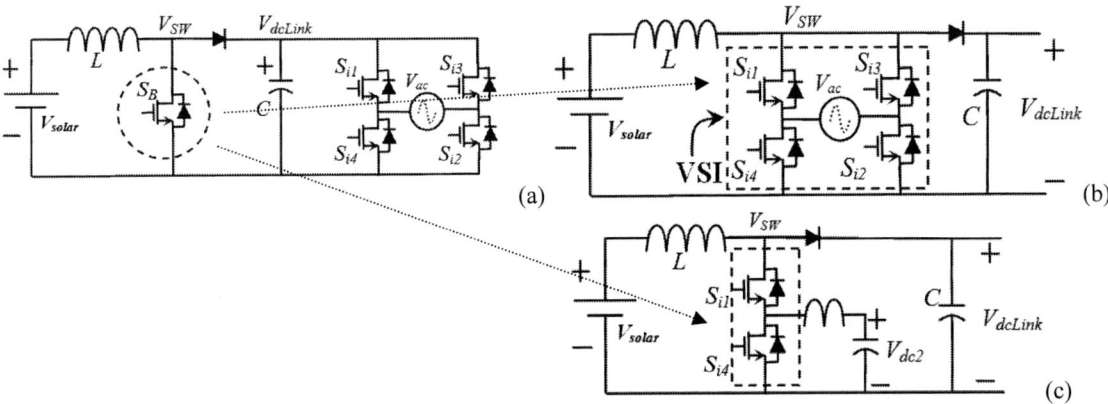

Fig. 5. Development of Multi-output converters (MOC) from a conventional Boost-VSI structure. (a) Boost-VSI Schematic (b) Boost based Hybrid-MOC schematic (c) Boost based DC-MOC schematic.

turning on any inverter leg switches simultaneously. For example, Si1 and Si4 can be turned on simultaneously to realize S_B turn-on interval [14-15]. The inverter is controlled in the rest of the interval in a switching cycle.

This idea can be extended to a DC-DC converter as well. Fig. 7b shows the block diagram of a DC-MOC where instead of the inverter legs, a buck converter is connected in place of S_B. The converter can produce two DC outputs, one higher (V_{dcLink}) and one lower (V_{dc2}) than the input source voltage (V_{solar}).

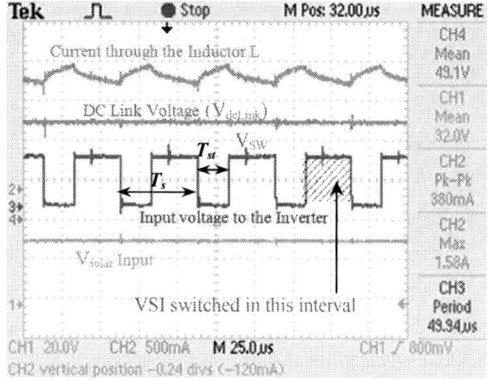

Fig. 8. Operational waveform of a boost based hybrid-MOC. (D_{st}=T_{st}/T_s)

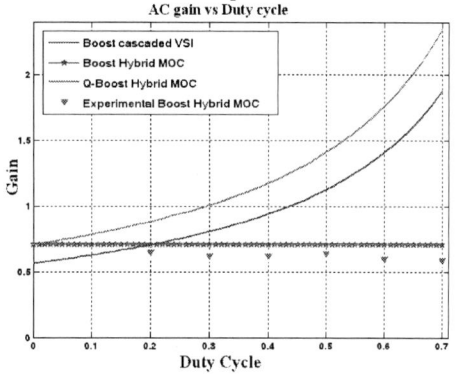

Fig. 9. AC gain of a hybrid-MOC. (Modulation Index (M_a) =1-D_{st})

Operationally, hybrid-boost MOC (Fig. 6b) is different from a conventional VSI circuit (in Fig. 6). In this case, the input to the inverter or synchronous buck is not fixed, and is switching between 0 and V_{dcLink} (which is higher than V_{solar}). The input to the inverter is a uni-polar square wave as shown in Fig. 8. Thus, the power interval of the inverter can be achieved when the inverter input is high. In order words, when the legs (S_{i1}/S_{i4} or S_{i2}/S_{i3}) are turned on at the same time, inductor is charged. In the rest of the interval, the inverter legs switch similar to a VSI. During these intervals, the inductor current divides between the load at the boost output (V_{dcLink}) and inverter output (V_{ac}). A similar explanation holds good for the multi-output DC-DC converter in Fig. 6c.

The maximum gain of the hybrid-MOC limited as both boost interval as well as the inverter PWM interval are accommodated within a switching cycle (T_s). The inverter is operated only during (T_s-T_{st}) interval (Fig. 8). The gain plot of the AC output as a function of duty cycle (D_{st}) is shown in Fig. 9. It is seen that irrespective of the

value of duty (D_{st}), the gain between the output AC and input DC (V_{solar}) equals 0.707. Higher gain can be achieved by using quadratic boost structure as shown in Fig. 10 With a Quadratic boost ([14], [16]) based Hybrid-MOC the rms value of AC output voltage is higher compared to a boost based Hybrid MOC for the same duty cycle as shown in Fig. 9.

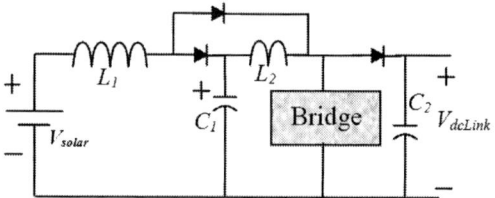

Fig. 10. Schematic of a Quadratic Boost based Hybrid-MOC.

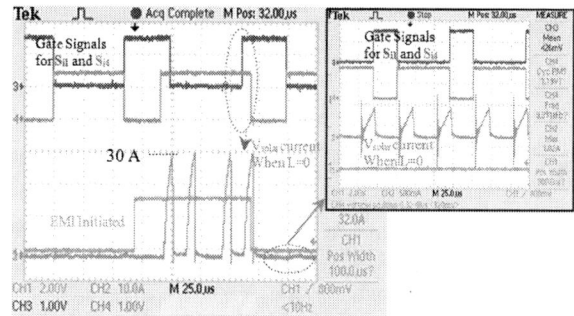

(a)

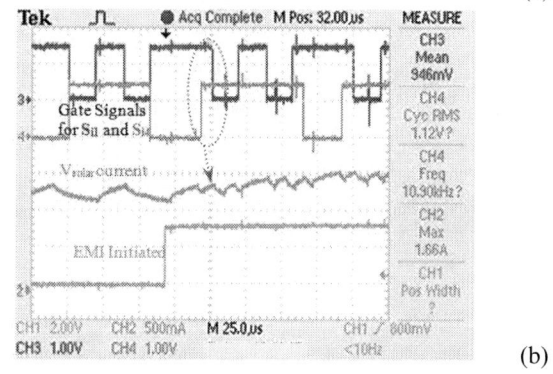

(b)

Fig. 11. (a) VSI performance with similar EMI (b) Superior EMI performance of the inverter.

There are five primary advantages of the aforementioned MOCs (a) A single stage generates multiple outputs (DC & AC or DC &DC) (b) It has less number of switches compared to a conventional structure, (c) the DC Link voltage is higher than the solar input (V_{solar}), (d) the VSI and synchronous buck stage is not affected by shorting of legs thus providing it higher EMI noise immunity, and (e) In a properly designed circuit the input current is continuous. In fact, shoot-though is a valid state for this type of converter. Because of reuse of switches in a MOC structure, it is better suited for low power application.

In order to visualize the fourth advantage, a test is performed where a small overlap period is inserted into the switching signals of a particular leg of a VSI. In this test

978-1-4799-2706-7/14 $31.00 © 2014 IEEE

after the EMI initiation signal is high in every cycle a small overlap time is introduced in the switching legs. This leads to a high short circuit current of almost 60 times the nominal and it is only limited by the PCB trace resistance and a current limiting resistor of 0.5 Ω (*500 mA Vs 30 A peaks*) as shown in Fig. 11 (a). In case of the hybrid MOC in Fig. 11(b), a dead short during switching overlap has no impact on the source current as shown in Fig. 11 (b). Thus, a dead short doesn't affect the DC link improving the converter's reliability.

III. Rural Electrification

a. *Issues in a Practical Nanogrid*

While using a solar PV based solution to electrify the Indian villages, it is paramount to think of the investment that will be affordable for a rural house hold. Keeping this in mind, it is easier to realize that as the investment is limited, the generated power will be limited. Thus, only is the critical low wattage loads can be supported for prolonged use.

For example, it is impractical to believe investment of multi-kilowatt rating solar PV for an individual house-hold that can power air conditioners or room heaters in a typical rural scenario. Nevertheless, commercially either AC or DC based loads are available as per Table 1, and they can be readily incorporated in AC or DC Nanogrid implementation.

Another issue that is encountered when a practical Nanogrid is implemented is choosing the appropriate voltage levels of distribution. In an AC Nanogrid, this voltage is 230 V at 50 Hz or 110 V at 60 Hz. However, what is the appropriate distribution voltage for DC Nanogrids?

In general, the selection of voltage level is dependent on the maximum power handled by the power converters [18]. Higher the power level, higher should be the distribution voltage. In [11], a 380 V distribution bus is selected for a Nanogrid of about 10 kW. A secondary distribution bus voltage of 48 V is used for low power appliances. In this article, the selection of voltage level is based on commercially available products, which will ultimately result in easy integration for a rural scenario.

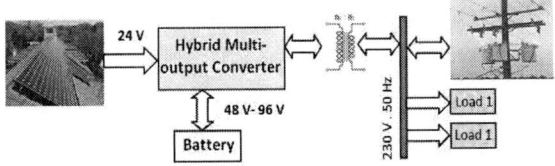

Fig. 12.MOC integration into a conventional AC Nanogrid.

b. *Commonly Available Appliances*

From a designer's prospective, domestic loads are typical as (a) Individual power consumption of each load is small (b) the loads are redundant (e.g., a house can have many fans and many lights). A list of typical Indian rural loads and their power consumptions are given in Table 1.

Table 1: Commercially Available Products in Rural India

Electrical	Operational	Power	Technology

Fixture	Input Voltage Level	Consumption	
CFL Light	6/12 VDC 230 VAC	5-25 W	CFL or LED Lightings
Fans	12 V DC 120 V (DC) 230 VAC	15 W 32 W 45 W	DC series (universal) BLDC Fans, Single phase AC
Cell Phone Chargers	12 V DC 230VAC	3.3 W 7.5 W	Standard Linear Chargers or SMPS based chargers
Pumps	230 V (DC or AC)	120 W-270 W	1-Φ Induction Motor or Universal Motor Based
TV	230 VAC	50-100 W	LED or LCD
Refrigeration	230 VAC	150-400 W	1-Φ Induction Motor Based

Most modern loads don't use the utility directly and need a front-end local converter. For example, a LED TV converts the utility supply to a DC to activate LED action where as a CFL lamp converts the utility input into a high frequency AC using a ballast circuit. This is why it is not particularly advantageous to have an AC Nanogrid, other than the fact that utility supplied loads are readily available in the market. In fact, wide spread power outages has slowly given way to other more reliable load alternatives which are battery powered. Fig. 13 (a) shows a commonly used CFL lighting system that is battery operated and can be charged when utility supply is available. It is common in most rural areas including local vegetable markets as shown in Fig. 13 (b). Battery powered DC based cell phone chargers and fans are also common. At an international level, Emerge Alliance, is developing products and standards to streamline DC based system. [17]

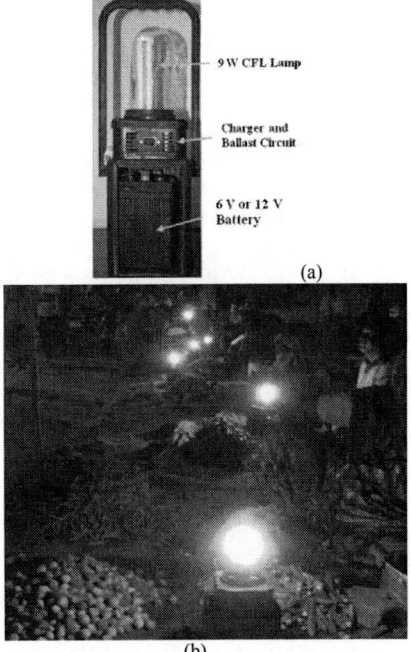

Fig. 13. (a) A battery powered CFL lamp popularly known as 'Emergency.' (b) 'Emergency' in action in a local Vegetable Market.

c. System Implementation

As pointed out earlier, the loads in a Nanogrid are of smaller rating. Therefore, a MOC based structure can be used to realize a Nanogrid. This will not only reduce the number of conversion stages, but also improve the reliability and power density of the overall solution due to reduced number of switching devices.

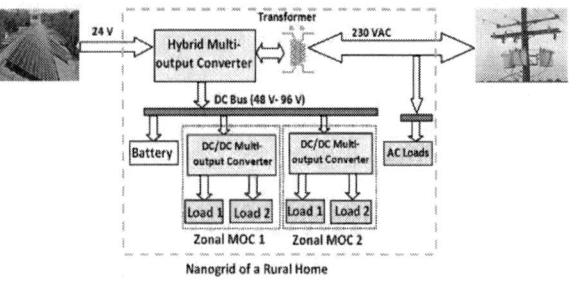

Fig. 14. A DC Nanogrid with Hybrid-MOC as well as DC-MOC.

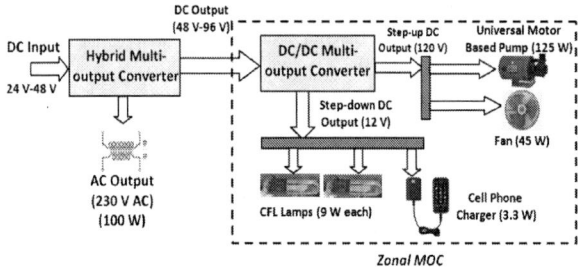

Fig. 15. Model Nanogrid using MOCs.

A hybrid-MOC (Fig. 6b) can be used to implement an AC Nanogrid architecture. The charge controller and the VSI are replaced with a boost based Hybrid-MOC and the DC terminal is interfaced to the battery where as the AC terminal is interfaced to a step up transformer for 230 V AC interfacing, similar to conventional architecture. Fig. 12 shows an example of this implementation where a 24 V solar panel is used to generate a 48 V battery voltage and a 230 V AC at its transformer interfaced terminal.

Fig. 14 shows an example where the 48 V bus is used as a distribution bus inside the house. In this case the MOC will be of more advantage because one MOC can power multiple domestic loads. This is particularly an advantage for rural house hold because, the loads are repetitive and one or multiple of these MOC can power multiple loads. However, higher power load are still supplied from the grid and only operational when grid is available. This strategy will optimize battery usage. This concept of using one MOC to power multiple domestic loads is called 'Zonal MOC.' In a house, each room can equipped with one Zonal MOC to supply the lighting, Fan, and charging needs of the room.

d. Model Rural Nanogrid Implementation

A MOC based Nanogrid has been readily built using off-the-shelf components and commercially available loads in local market to show its ready usability in a rural electrification application. Fig. 15 shows the block diagram of the Nanogrid along with some typical rural loads. It consists of a 24 V DC input to emulate a solar panel which

is used by the hybrid-MOC to generate a 48 VDC and a 230 VAC transformer interfaced output. The 48 VDC bus is used to supply a DC-MOC to supply multiple commercial loads. The loads are indicated in the figure.

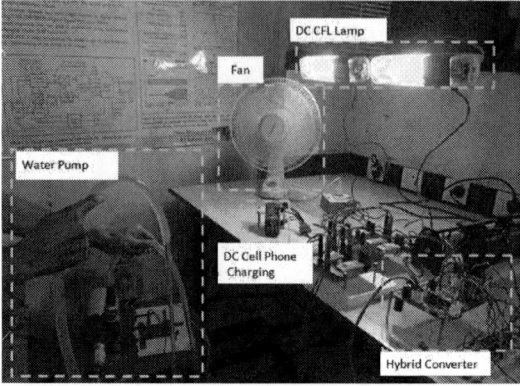

Fig. 16. A rural Nanogrid with typical loads in operation.

A picture of the Nanogrid in operation is shown in Fig. 16. The performance of the hybrid MOC is shown in Fig. 17a. As multiple loads are present, cross regulation is an important issue that needs attention. The cross regulation of the converter is tested as follows.

While all the other loads are kept on, a CFL lamp is turned off and its impact on the output voltages are observed. The result is shown in Fig. 17b. It can be seen that there is minimal impact on the rest of the system indicating excellent cross-regulation property of the converter based on MOC architecture.

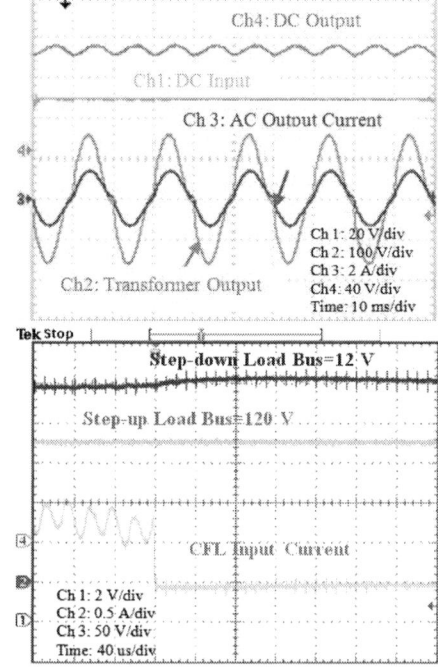

Fig. 17. (a) Operation of the Hybrid-MOC. (b) Dynamic behavior of the DC-MOC when CFL lamp turned-off with all other load present.

IV. Conclusions

Advances in Nanogrid technology is discussed in the paper. Operation and advantages of a multi-output converter are delineated. Multi-output converters are single stage converters which can simultaneously produce a DC as well as AC/DC outputs. These converters have superior EMI immunity leading to better reliability. The paper discusses its adaptation into electrification of Rural India. It proposes different systems that will help easy adaption of the proposed technology in a practical implementation.

Acknowledgement

This work was supported by Science and Engineering Research Board (SERB), Govt. of India, under grant no. SR/S3/EECE/0187/2012.

References

[1] Ministry of Power, http://powermin.nic.in/rural_electrification/definition_village_electrification.htm.

[2] Bharat Nirman, Ministry of Power, http://www.bharatnirman.gov.in/download.pdf .

[3] Ministry of Power, PPT presentation on T & D Loss in the country. http://www.powermin.nic.in/distribution/apdrp/projects/pdf/Presentation_on_AT&C_Losses.ppt

[4] S. Mukhopadhyay and Bhim Singh, "Distributed Generation - Basic Policy, Perspective Planning, and Achievement so far in India," in PES 2009, pp. 1-7

[5] R. C. Dugan and T. E. McDermott, "Distributed generation," in *IEEE Industry Applications Magazine*, pp. 19-25, 2002

[6] M. B. Nissen, High performance development as distributed generation, *IEEE Potentials*, pp. 25-31, Nov. 2009.

[7] India Solar Resource Maps, NERL, http://www.nrel.gov/international/ra_india.html.

[8] Dushan Boroyevich, Igor Cvetković, Dong Dong, Rolando Burgos, Fei Wang, Fred Lee, "Future Electronic Power Distribution Systems – A contemplative view –," in *IEEE OPTIM*, pp. 1369-1380, 2010.

[9] R. Adda, O. Ray, S. Mishra, and A. Joshi, "Synchronous Reference Frame Based Control of Switched Boost Inverter for Standalone DC Nanogrid Applications," *IEEE Trans. on PE*, Volume: 28, Issue: 3, 2013, pp. 1219 - 1233.

[10] Ander Goikoetxea, Ander Goikoetxea, Roberto, and Sanchez Pablo Zumeta, "DC versus AC in residential buildings: Efficiency comparison," in *EUROCON*, pp. 1-5, July 2013.

[11] D. Dong, Igor Cvetkovic, Dushan Boroyevich, Wei Zhang, Ruxi Wang, Paolo Mattavelli, "Grid-Interface Bidirectional Converter for Residential DC Distribution Systems—Part One: High-Density Two-Stage Topology," in *IEEE Trans. On PE*, pp. 1655-1666, Apr. 2013.

[12] Gab-Su Seo, Jongbok Baek, Kyusik Choi, Hyunsu Bae, Bohyung Cho, "Modeling and analysis of DC distribution system," in *ECCE-Asia*, Korea, pp. 223-227, May 2011.

[13] B. Kroposki, R. Lasseter, T. Ise, S. Morozumi, S. Papatlianassiou, N. Hatziargyriou, "Making microgrids work," in *Power and Energy Magazine*, Vol. 6, Issue 3, pp. 40-53, 2008.

[14] Olive Ray and Santanu Mishra, "Boost-Derived Hybrid Converter with simultaneous DC and AC outputs," in *IEEE Trans. On IAS*, Vol. 99, pp. 1, April 2014.

[15] Olive Ray, A. Prasad, and Santanu Mishra, "A multi-port DC-DC converter topology with simultaneous buck and boost outputs," in *IEEE ISIE*, pp. 1-6, 2013. *DOI: 10.1109/TIA.2013.2271874*

[16] B.-R Lin, J.-J Chen and F.-Y Hsieh, "Analysis and implementation of a bidirectional converter with high conversion ratio", *IEEE ICIT'08*, pp. 1-6, 2008.

[17] Emerge Alliance Webpage, http://www.emergealliance.org/

[18] S. Anand and B. Fernandes, "Optimal voltage level for DC microgrids," in *IEEE-IECON 2010*, pp. 3034-3039.

Study and Implementation of Seven-Level Inverter Using Coupled Inductor and Switched-Capacitor

Yi-Chun Lin
Electrical Engineering
National Cheng Kung
University
Tainan, Taiwan

Jiann-Fuh Chen
Electrical Engineering
National Cheng Kung
University
Tainan, Taiwan

Wen-Chien Hsu
Electrical Engineering
National Cheng Kung
University
Tainan, Taiwan

Sheng-Kai Kao
Electrical Engineering
National Cheng Kung
University
Tainan, Taiwan

Abstract –This paper proposes a novel single-phase seven-level inverter. The topology of the proposed structure is composed of a DC source, a switched-capacitor circuit, and a coupled inductor. Compared to conventional seven-level inverter structure, numbers of switches and capacitors are reduced. The voltages of capacitors are self-balanced by using coupled inductor without complex control method. Therefore, the output voltage total harmonic distortion can be reduced. Finally, the simulation and experimental result show the 350-V input voltage, 380-V$_{ac}$ output voltage, and 0.52% output total harmonic distortion under 3kW output power condition to verify the feasibility of the proposed multilevel inverter.

Index Terms –coupled inductor, multilevel inverter, switched-capacitor, total harmonic distortion

I. INTRODUCTION

Multilevel inverter has been widely used for power conversion in medium-power and high-power system applications such as utility and large renewable energy systems. Renewable energy systems can be divided into wind [1], photovoltaic, and fuel cells, etc. Multilevel inverter is originally used for ac motor, and it is also the medium stage to transmit power to many electrical devices such as uninterruptible power supply, induction motor, electric vehicle, and photovoltaic system [2-4]. Compared to conventional H-bridge inverter, the voltage stress of switch in multilevel inverters and power loss is low. Besides, the filter size of multi-level inverter is small and total harmonic distortion (THD) of output voltage is reduced.

In recent years, many circuit structures of multi-level inverters have been proposed, such as diode-clamped multilevel inverter [5-8], flying capacitor inverter [9-11], cascaded H-bridges inverter [12-16], modified H-bridge inverter [17-19], and multilevel inverter with coupled inductor [20-23]. In general, control methods of switches usually use sinusoidal pulse width modulation (SPWM) [24] and space vector pulse width modulation (SVPWM) [25-26].

When the number of output level increases, more input sources, switches, and capacitors are needed. For example, conventional seven-level inverter structures use three capacitors to divide input voltage source [27] or use more than one input voltage source [28] to make seven-level output. Traditional seven-level inverters use three capacitors to parallel input voltage source, and it will unfortunately make the voltage of capacitor unbalanced in one cycle and

results in poor sine waveform. Furthermore, the unbalanced capacitors may cause high total harmonic distortion, and the voltage stress of power switches and diodes are increased. Recently, many researchers have proposed other control methods of switches to make the capacitor voltage balanced [29-30], and others proposed voltage-balance circuit to the input of multilevel inverter [31-33]. In a word, these methods listed above have to use more power devices to solve the problem.

The design consideration of multi-level inverter is the number of capacitors, switches, and diodes. In addition to the consideration previously described, others include total harmonic distortion of output voltage and voltage-balancing of capacitors. In this paper, a novel seven-level inverter without using any extra voltage-balancing circuit is proposed. At the input side, two voltage levels are built by switched-capacitor circuit, while at the output side two different voltage levels is built by coupled inductor, thus output voltage is seven-level. When the number of input source increases, the level of output voltage increased. Moreover, the total harmonic distortion of the proposed inverter is lower than those of conventional seven-level inverters.

In this paper a new structure of seven-level inverter is proposed. The novel seven-level inverter uses only one DC source, two capacitors, ten switches, two diodes, and a coupled inductor to achieve seven-level output without voltage balancing-circuit. In Section II, the proposed seven-level inverter is presented and discussed in detail. In Section III, the control method for seven-level inverter will be introduced. In Section IV, the simulation results and experimental results of the proposed circuit are shown to verify the theoretical analysis. Finally, the conclusion of this paper is given in Section V.

II. PROPOSED SINGLE PHASE SEVEN-LEVEL INVERTER

Fig. 1 shows the circuit topology of the proposed single-phase seven-level inverter. It consists of a switched-capacitor circuit and coupled inductor technology. The proposed multilevel is composed of a coupled inductor, ten switches S_A-S_J, two diodes D_1 and D_2, and capacitors C_1 and C_2. A switched-capacitor circuit and coupled inductor technology is used to make seven-level output voltage. The novel topology is simpler than other seven-level inverters.

978-1-4799-2706-7/14 $31.00 © 2014 IEEE

In addition, the advantage of the proposed multi-level inverter is fewer capacitor, switches, and capacitors. In section A, the operation of coupled inductor will be introduced. In section B, the operation of proposed inverter is presented and discussed in detail. Finally, the design rule of coupled inductor is shown in section C.

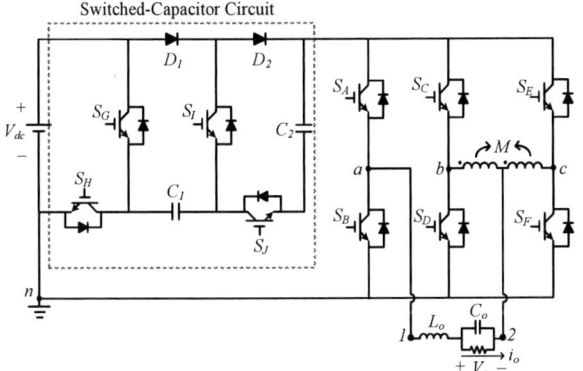

Fig. 1. Proposed single-phase seven-level inverter.

A. The Operation of coupled inductor

The coupled inductor diagram is shown in Fig. 2. The mutual inductance and voltage of coupled inductor can be expressed as

$$M = k \cdot \sqrt{L_1 \cdot L_2} \tag{1}$$

$$V_{b2} = V_{bn} - V_{2n} = L_1 \cdot \frac{di_b}{dt} - M \cdot \frac{di_c}{dt} = (L_1 + M) \cdot \frac{di_b}{dt} - M \cdot \frac{di_L}{dt} \tag{2}$$

$$V_{c2} = V_{cn} - V_{2n} = L_2 \cdot \frac{di_c}{dt} - M \cdot \frac{di_b}{dt} = (L_2 + M) \cdot \frac{di_c}{dt} - M \cdot \frac{di_L}{dt} \tag{3}$$

where n is neutral point, k is coupled coefficient, and M is mutual inductance. Assume coupled inductor L_1 and L_2 are equal and the coupled coefficient k is 1. Hence, the mutual inductance M is equal to L_1 and L_2.

According to Kirchhoff's current law, load current can be expressed as

$$i_L = i_b + i_c \tag{4}$$

From equation (2) to (4), the voltage V_{2n} can be derived in (5) with voltage V_{bn} and voltage V_{cn}.

$$V_{2n} = \frac{V_{bn} + V_{cn}}{2} \tag{5}$$

According to (5), the output voltage V_{12} and V_{bc} of the proposed seven-level inverter can be derived as

$$V_{12} = V_{1n} - V_{2n} = V_{an} - \frac{V_{bn} + V_{cn}}{2} \tag{6}$$

$$V_{bc} = 2M \cdot (\frac{di_b}{dt} - \frac{di_c}{dt}) \tag{7}$$

Equation (6) shows the relationship between output voltage V_{12}, voltages V_{bn}, and voltage V_{cn}.

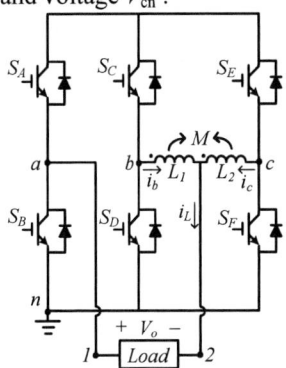

Fig. 2. Coupled inductor diagram.

B. The Operation of Proposed Seven-level Inverter

In this circuit, proposed seven-level inverter is combined with switched-capacitor circuit and coupled-inductor. In order to simplify the analysis, it is assumed that the capacitors C_1 and C_2 are large enough to be considered as constant voltage $V_{dc}/2$. In addition, switches S_A-S_J and diodes D_1 and D_2 are ideal, and turn-on voltage drop is zero.

Based on the aforementioned assumptions, the operation of proposed seven-level inverter can be divided into twelve operational modes as shown in Fig. 3(a)-(l). Furthermore, Table I shows the switching states that produced seven levels of output voltage ($3V_{dc}/2$, V_{dc}, $V_{dc}/2$, 0, -$V_{dc}/2$, -V_{dc}, -$3V_{dc}/2$).

1) Mode 1: During this time interval, output voltage level V_o is 0. Switches S_A, S_C, S_E, S_H, and S_J are turned on, and diodes D_1 and D_2 are turned on during this mode. Thus, V_{an} = V_{dc}, V_{bn} = V_{dc}, and V_{cn} = V_{dc}. According to equation (6), voltage V_{12} is 0. Energy of capacitors is provided by input source. Fig. 3(a) shows the current path of this mode.

2) Mode 2: During this time interval, output voltage level V_o is $V_{dc}/2$. Switches S_A, S_C, S_F, S_H, and S_J are turned on, and diodes D_1 and D_2 are turned on during this mode. Thus, V_{an} = V_{dc}, V_{bn} = V_{dc}, and V_{cn} = 0. According to equation (6), voltage V_{12} is $V_{dc}/2$. Energy of output and capacitors is provided by input source. Fig. 3(b) shows the current path of this mode.

3) Mode 3: During this time interval, output voltage level V_o is $V_{dc}/2$. Switches S_A, S_D, S_E, S_H, and S_J are turned on, and diodes D_1 and D_2 are turned on during this mode. Thus, V_{an} = V_{dc}, V_{bn} = 0, and V_{cn} = V_{dc}. According to equation (6), voltage V_{12} is $V_{dc}/2$. Energy of output and capacitors is provided by input source. Fig. 3(c) shows the current path of this mode.

4) Mode 4: During this time interval, output voltage level V_o is V_{dc}. Switches S_A, S_D, S_F, S_H, and S_J are turned on, and diodes D_1 and D_2 are turned on during this mode. Thus, V_{an} = V_{dc}, V_{bn} = 0, and V_{cn} = 0. According to equation (6), voltage V_{12} is V_{dc}. Energy of output and capacitors is

provided by input source. Fig. 3(d) shows the current path of this mode.

5) Mode 5: During this time interval, output voltage level V_o is $3V_{dc}/2$. Switches S_A, S_D, S_F, S_G and S_I are turned on, and diode D_2 is turned on during this mode. Thus, $V_{an} = 3V_{dc}/2$, $V_{bn} = 0$, and $V_{cn} = 0$. According to equation (6), voltage V_{12} is $3V_{dc}/2$. Energy of output is provided by input source and capacitor C_1. Fig. 3(e) shows the current path of this mode.

6) Mode 6: During this time interval, output voltage level V_o is $3V_{dc}/2$. Switches S_A, S_D, S_F, S_I, and S_J are turned on, and diode D_1 is turned on during this mode. Thus, $V_{an} = 3V_{dc}/2$, $V_{bn} = 0$, and $V_{cn} = 0$. According to equation (6), voltage V_{12} is $3V_{dc}/2$. Energy of output is provided by input source and capacitor C_2. Fig. 3(f) shows the current path of this mode.

7) Mode 7: During this time interval, output voltage level V_o is 0. Switches S_B, S_D, S_F, S_H, and S_J are turned on, and diodes D_1 and D_2 are turned on during this mode. Thus, $V_{an} = 0$, $V_{bn} = 0$, and $V_{cn} = 0$. According to equation (6), voltage V_{12} is 0. Energy of capacitors is provided by input source. Fig. 3(g) shows the current path of this mode.

8) Mode 8: During this time interval, output voltage level V_o is $-V_{dc}/2$. Switches S_B, S_C, S_F, S_H, and S_J are turned on, and diodes D_1 and D_2 are turned on during this mode. Thus, $V_{an} = 0$, $V_{bn} = V_{dc}$, and $V_{cn} = 0$. According to equation (6), voltage V_{12} is $-V_{dc}/2$. Energy of output and capacitors is provided by input source. Fig. 3(h) shows the current path of this mode.

9) Mode 9: During this time interval, output voltage level V_o is $-V_{dc}/2$. Switches S_B, S_D, S_E, S_H, and S_J are turned on, and diodes D_1 and D_2 are turned on during this mode. Thus, $V_{an} = 0$, $V_{bn} = 0$, and $V_{cn} = V_{dc}$. According to equation (6), voltage V_{12} is $-V_{dc}/2$. Energy of output and capacitors is provided by input source. Fig. 3(i) shows the current path of this mode.

10) Mode 10: During this time interval, output voltage level V_o is $-V_{dc}$. Switches S_B, S_C, S_E, S_H, and S_J are turned on, and diodes D_1 and D_2 are turned on during this mode. Thus, $V_{an} = 0$, $V_{bn} = V_{dc}$, and $V_{cn} = V_{dc}$. According to equation (6), voltage V_{12} is $-V_{dc}$. Energy of output and capacitors is provided by input source. Fig. 3(j) shows the current path of this mode.

11) Mode 11: During this time interval, output voltage level V_o is $-3V_{dc}/2$. Switches S_B, S_C, S_E, S_G, and S_I are turned on, and diode D_2 is turned on during this mode. Thus, $V_{an} = 3V_{dc}/2$, $V_{bn} = 0$, and $V_{cn} = 0$. According to equation (6), voltage V_{12} is $-3V_{dc}/2$. Energy of output is provided by input source and capacitor C_1. Fig. 3(k) shows the current path of this mode.

12) Mode 12: During this time interval, output voltage level V_o is $-3V_{dc}/2$. Switches S_B, S_C, S_E, S_I, and S_J are turned on, and diode D_1 is turned on during this mode. Thus, $V_{an} = 3V_{dc}/2$, $V_{bn} = 0$, and $V_{cn} = 0$. According to equation (6), voltage V_{12} is $-3V_{dc}/2$. Energy of output is provided by input

source and capacitor C_2. Fig. 3(l) shows the current path of this mode.

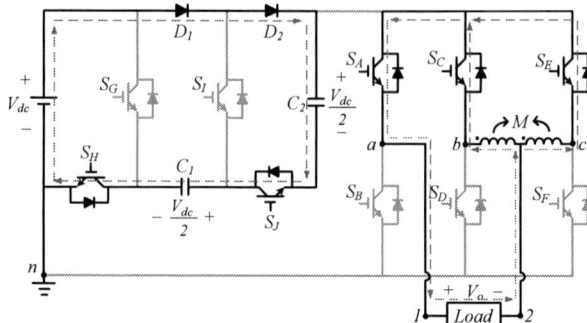

Fig. 3(a) Mode 1: $V_{12} = 0$

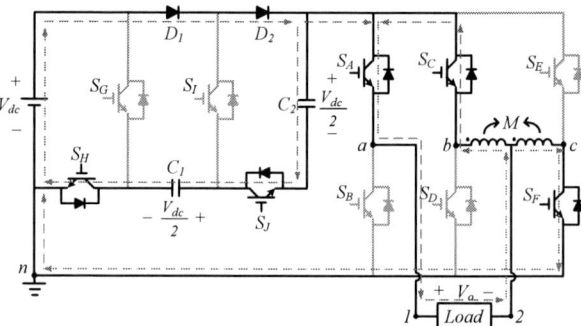

Fig. 3(b) Mode 2: $V_{12} = V_{dc}/2$

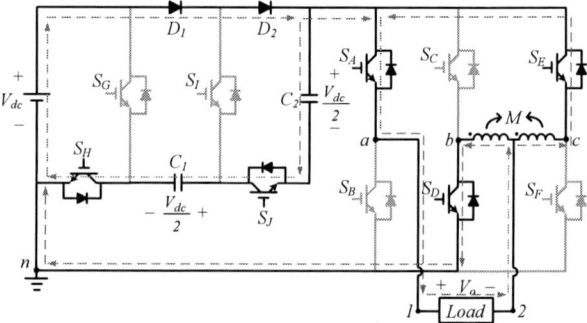

Fig. 3(c) Mode 3: $V_{12} = V_{dc}/2$

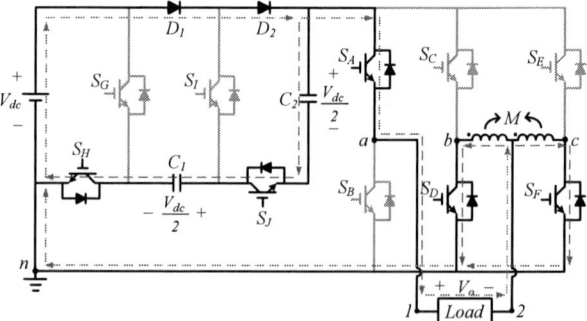

Fig. 3(d) Mode 4: $V_{12} = V_{dc}$

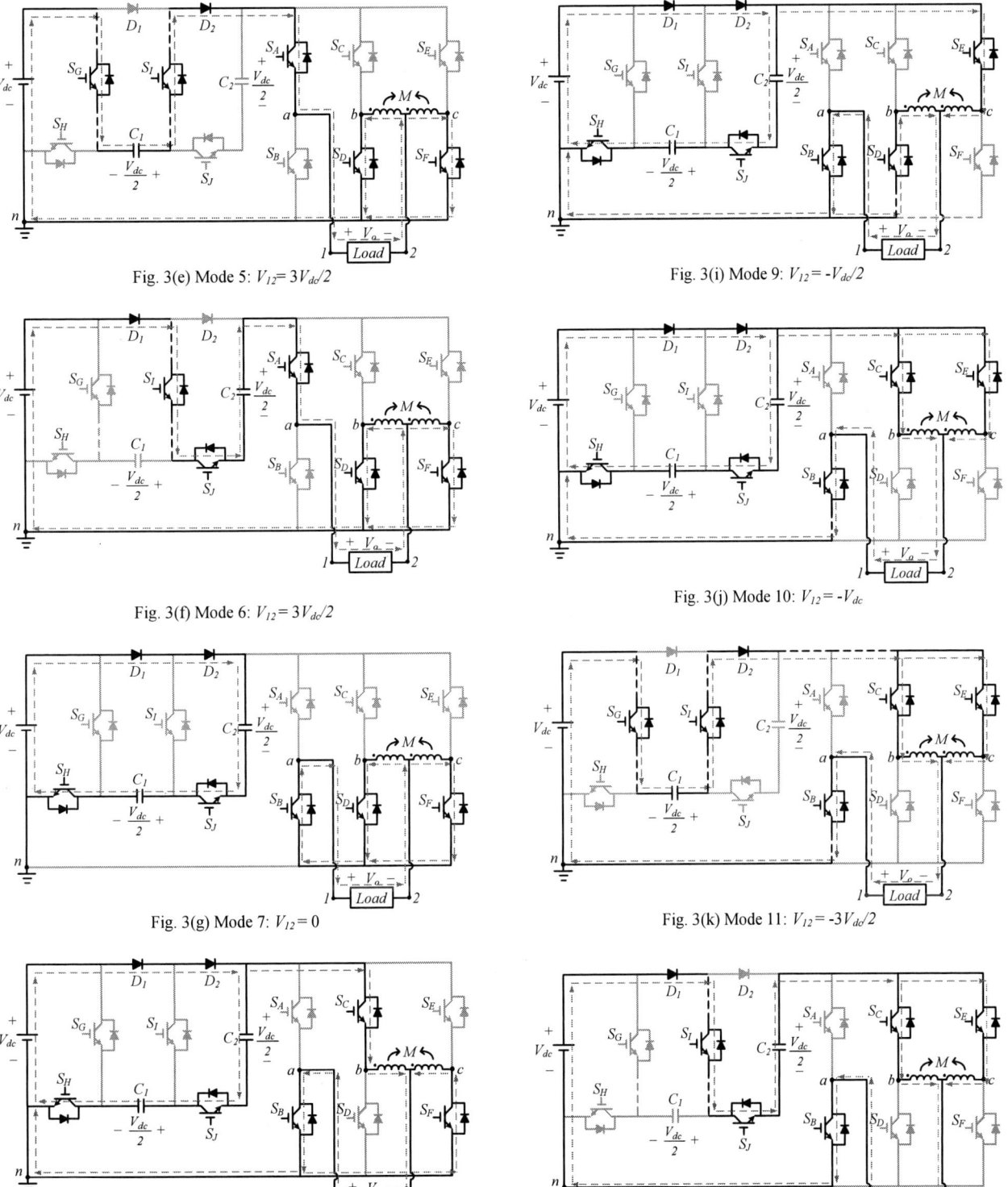

Fig. 3(e) Mode 5: $V_{12} = 3V_{dc}/2$

Fig. 3(i) Mode 9: $V_{12} = -V_{dc}/2$

Fig. 3(f) Mode 6: $V_{12} = 3V_{dc}/2$

Fig. 3(j) Mode 10: $V_{12} = -V_{dc}$

Fig. 3(g) Mode 7: $V_{12} = 0$

Fig. 3(k) Mode 11: $V_{12} = -3V_{dc}/2$

Fig. 3(h) Mode 8: $V_{12} = -V_{dc}/2$

Fig. 3(l) Mode 12: $V_{12} = -3V_{dc}/2$

Fig. 3 Current flow of the proposed seven-level inverter (a) Mode 1. (b) Mode 2. (c) Mode 3. (d) Modes 4. (e) Mode 5. (f) Modes 6. (g) Modes 7. (h) Mode 8. (i) Mode 9. (j) Mode 10. (k) Mode 11. (l) Mode 12.

Table I. Switching state of the proposed seven-level inverter

Mode	S_A	S_B	S_C	S_D	S_E	S_F	S_G	S_H	S_I	S_J
1	on	off	on	off	on	off	off	on	off	on
2	on	off	on	off	off	on	off	on	off	on
3	on	off	off	on	on	off	off	on	off	on
4	on	off	off	on	off	on	off	on	off	on
5	on	off	off	on	off	on	on	off	on	off
6	on	off	off	on	off	on	off	off	on	on
7	off	on	off	on	off	on	off	on	off	on
8	off	on	on	off	off	on	off	on	off	on
9	off	on	off	on	on	off	off	on	off	on
10	off	on	on	off	on	off	off	on	off	on
11	off	on	on	off	on	off	on	off	on	off
12	off	on	on	off	on	off	off	off	on	on

C. Design of The Coupled Inductors

The coupled inductor diagram is shown in Fig. 2. From equation (7), the ripple current of coupled inductor can be derived as

$$i_{ripple} = \frac{1}{2M}\int V_{bc}\,dt \qquad (8)$$

In one period, the maximum output voltage is equal to $3V_{dc}/2$. The maximum ripple current of coupled inductors can be expressed as

$$i_{Ripple(\max)} = \frac{1}{2M}\times\int \frac{3}{2}V_{dc}\,dt = \frac{3T\cdot V_{dc}}{4M} \qquad (9)$$

Then, the inductance L_1 and L_2 could be determined by the following equation:

$$M > \frac{3T\cdot V_{dc}}{4\cdot i_{Ripple(\max)}} \qquad (10)$$

III. Control Method of Proposed Topology

In this paper, the control method of switches S_A-S_J is Phase Disposition PWM. According to carrier position, the control methods can be divided into six kinds of PWM, such as Hybrid PWM (H PWM), Phase Disposition PWM (PD PWM), Phase Opposition Disposition PWM (POD PWM), Alternate Phase Opposition Disposition PWM (APOD PWM), Phase Shift PWM (PS PWM), and Super Imposed carrier PWM. The control method of Phase Disposition PWM is simple and has low harmonic distortion. The number of high-frequency carriers is determined by output voltage level. There are six high-frequency carriers ($v_{\mathrm{tri},1}$-$v_{\mathrm{tri},6}$) and a low-frequency sine wave (v_{control}), as shown in Fig. 4.

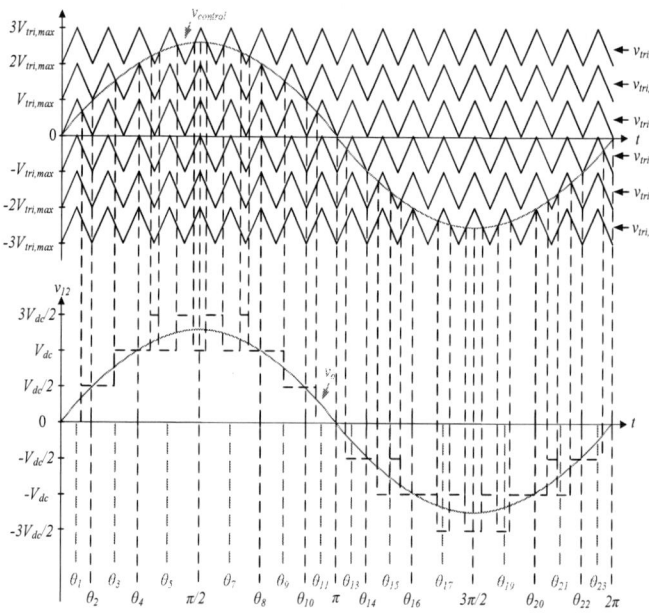

Fig. 4 Phase disposition PWM

The high-frequency carrier signals have the same frequency (f) and amplitude ($V_{\mathrm{tri,max}}$). There are twelve states of proposed seven-level inverter control in one cycle, and operational principles are explained in detail.

1) State 1 ($0 < \omega t < \theta_1$ or $\theta_{10} < \omega t < \theta_{11}$): In this mode, v_{control} operates between 0 and $V_{\mathrm{tri,max}}$. When v_{control} is greater than $v_{\mathrm{tri},3}$, the circuit operates in mode 2, and output voltage is $V_{dc}/2$. On the contrary, when v_{control} is smaller than $v_{\mathrm{tri},3}$, the circuit operates in mode 1, and output voltage is 0.

2) State 2 ($\theta_1 < \omega t < \theta_2$ or $\theta_{11} < \omega t < \pi$): In this mode, v_{control} operates between 0 and $V_{\mathrm{tri,max}}$. When v_{control} is greater than $v_{\mathrm{tri},3}$, the circuit operates in mode 3, and output voltage is $V_{dc}/2$. On the contrary, when v_{control} is smaller than $v_{\mathrm{tri},3}$, the circuit operates in mode 1, and output voltage is 0.

3) State 3 ($\theta_2 < \omega t < \theta_3$ or $\theta_8 < \omega t < \theta_9$): In this mode, v_{control} operates between $V_{\mathrm{tri,max}}$ and $2V_{\mathrm{tri,max}}$. When v_{control} is greater than $v_{\mathrm{tri},2}$, the circuit operates in mode 4, and output voltage is V_{dc}. On the contrary, when v_{control} is smaller than $v_{\mathrm{tri},2}$, the circuit operates in mode 2, and output voltage is $V_{dc}/2$.

4) State 4 ($\theta_3 < \omega t < \theta_4$ or $\theta_9 < \omega t < \theta_{10}$): In this mode, v_{control} operates between $V_{\mathrm{tri,max}}$ and $2V_{\mathrm{tri,max}}$. When v_{control} is greater than $v_{\mathrm{tri},2}$, the circuit operates in mode 4, and output voltage is V_{dc}. On the contrary, when v_{control} is smaller than

$v_{tri,2}$, the circuit operates in mode 3, and output voltage is $V_{dc}/2$.

5) State 5 ($\theta_4 < \omega t < \theta_5$ or $\pi/2 < \omega t < \theta_7$): In this mode, $v_{control}$ operates between $2V_{tri,max}$ and $3V_{tri,max}$. When $v_{control}$ is greater than $v_{tri,3}$, the circuit operates in mode 5, and output voltage is $3V_{dc}/2$. On the contrary, when $v_{control}$ is smaller than $v_{tri,3}$, the circuit operates in mode 4, and output voltage is V_{dc}.

6) State 6 ($\theta_5 < \omega t < \pi/2$ or $\theta_7 < \omega t < \theta_8$): In this mode, $v_{control}$ operates between $2V_{tri,max}$ and $3V_{tri,max}$. When $v_{control}$ is greater than $v_{tri,1}$, the circuit operates in mode 6, and output voltage is $3V_{dc}/2$. On the contrary, when $v_{control}$ is smaller than $v_{tri,1}$, the circuit operates in mode 4, and output voltage is V_{dc}.

7) State 7 ($\pi < \omega t < \theta_{13}$ or $\theta_{22} < \omega t < \theta_{23}$): In this mode, $v_{control}$ operates between $-V_{tri,max}$ and 0. When $v_{control}$ is greater than $v_{tri,4}$, the circuit operates in mode 7, and output voltage is 0. On the contrary, when $v_{control}$ is smaller than $v_{tri,4}$, the circuit operates in mode 8, and output voltage is $-V_{dc}/2$.

8) State 8 ($\theta_{13} < \omega t < \theta_{14}$ or $\theta_{23} < \omega t < 2\pi$): In this mode, $v_{control}$ operates between $-V_{tri,max}$ and 0. When $v_{control}$ is greater than $v_{tri,4}$, the circuit operates in mode 7, and output voltage is 0. On the contrary, when $v_{control}$ is smaller than $v_{tri,4}$, the circuit operates in mode 9, and output voltage is $-V_{dc}/2$.

9) State 9 ($\theta_{14} < \omega t < \theta_{15}$ or $\theta_{20} < \omega t < \theta_{21}$): In this mode, $v_{control}$ operates between $-2V_{tri,max}$ and $-V_{tri,max}$. When $v_{control}$ is greater than $v_{tri,5}$, the circuit operates in mode 8, and output voltage is $-V_{dc}/2$. On the contrary, when $v_{control}$ is smaller than $v_{tri,5}$, the circuit operates in mode 10, and output voltage is $-V_{dc}$.

10) State 10 ($\theta_{15} < \omega t < \theta_{16}$ or $\theta_{21} < \omega t < \theta_{22}$): In this mode, $v_{control}$ operates between $-2V_{tri,max}$ and $-V_{tri,max}$. When $v_{control}$ is greater than $v_{tri,5}$, the circuit operates in mode 9, and output voltage is $-V_{dc}/2$. On the contrary, when $v_{control}$ is smaller than $v_{tri,5}$, the circuit operates in mode 10, and output voltage is $-V_{dc}$.

11) State 11 ($\theta_{16} < \omega t < \theta_{17}$ or $3\pi/2 < \omega t < \theta_{19}$): In this mode, $v_{control}$ operates between $-3V_{tri,max}$ and $-2V_{tri,max}$. When $v_{control}$ is greater than $v_{tri,6}$, the circuit operates in mode 10, and output voltage is $-V_{dc}$. On the contrary, when $v_{control}$ is smaller than $v_{tri,6}$, the circuit operates in mode 11, and output voltage is $-3V_{dc}/2$.

12) State 12 ($\theta_{17} < \omega t < 3\pi/2$ or $\theta_{19} < \omega t < \theta_{20}$): In this mode, $v_{control}$ operates between $-3V_{tri,max}$ and $-2V_{tri,max}$. When $v_{control}$ is greater than $v_{tri,6}$, the circuit operates in mode 10, and output voltage is $-V_{dc}$. On the contrary, when $v_{control}$ is smaller than $v_{tri,6}$, the circuit operates in mode 12, and output voltage is $-3V_{dc}/2$.

The parameter values of angles are defined by (6)-(11), and the amplitude modulation index is defined as (12).

$$\theta_0 = 0 \tag{10}$$

$$\theta_6 = \frac{\pi}{2} \tag{11}$$

$$\theta_{12} = \pi \tag{12}$$

$$\theta_{18} = \frac{3\pi}{2} \tag{13}$$

$$\theta_{24} = 2\pi \tag{14}$$

$$\theta_{m+1} = \frac{\theta_m + \theta_{m+2}}{2}, \text{ m=0,2,4} \cdots \text{20,22} \tag{15}$$

$$m_a = \frac{V_{control,max}}{3V_{tri,max}} \tag{16}$$

where $V_{control,max}$ is the maximum value of sine waveform, and $V_{tri,max}$ is the amplitude of triangle waveform. The phase angle displacement can be expressed as

$$\theta_2 = \sin^{-1}\left(\frac{V_{tri,max}}{V_{control.max}}\right) \tag{17}$$

$$\theta_4 = \sin^{-1}\left(\frac{2V_{tri,max}}{V_{control.max}}\right) \tag{18}$$

$$\theta_8 = \pi - \theta_4 \tag{19}$$

$$\theta_{10} = \pi - \theta_2 \tag{20}$$

$$\theta_{14} = \pi + \theta_2 \tag{21}$$

$$\theta_{16} = \pi + \theta_4 \tag{22}$$

$$\theta_{20} = 2\pi - \theta_4 \tag{23}$$

$$\theta_{22} = 2\pi - \theta_2 \tag{24}$$

IV. SIMULATION AND EXPERIMENTAL RESULTS

In this paper, the proposed structure is shown in Fig.1. In order to verify the proposed structure, the laboratory prototype is implemented with the following circuit specification and component parameter in Table II- III.

Simulation software PSIM® is used to generate the simulation circuit of the proposed seven-level inverter. To verify the proposed structure, the simulation of seven-level inverter uses a 100 Ω load at the output. The simulation results of the proposed single-phase seven-level inverter in steady state when the m_a is 1 is shown in Fig. 5, and the feasibility of the proposed circuit is verified.

Digital signal processor (DSP- TMDSDOCK28035) is used to generate the control signals of switches, and verify the simulation results. Figs. 6 shows the experimental results of the proposed inverter output voltage under 3000W condition with $m_a = 1$. The output voltage total harmonic distortion (THD) under 3000W condition is shown in Fig. 7. The THD is 0.52% under 3000W. The experimental results show that the THD is reduced and the output voltage has seven levels. Furthermore, the voltage of capacitors waveform shows the capacitor voltages are uniform, and output voltage of sine waveform is balanced. Efficiency curve is shown in Fig. 12. The efficiency under 1259W is 96.8%, and 3000W is 95.7%. Therefore, the proposed structure is a useful topology for renewable energy without using voltage-balancing circuit.

Table II. Circuit specification of proposed seven-level inverter

Input Voltage : V_{dc}	DC 350 V
Output Voltage : V_o	AC 380 V_{rms}, 60 Hz
Output Power : P_o	3000W
Switching Frequency ($S_A \sim S_J$)	18 kHz

Table III. Component parameter of proposed seven-level inverter

Capacitors : C_1, C_2	3.6 mF
Switches : $S_A \sim S_J$	IXGH 32N60A
Diodes : D_1, D_2	DSEP 30-06A
Output inductor : L_o	1 mH
Output Capacitor : C_o	1.36 µF

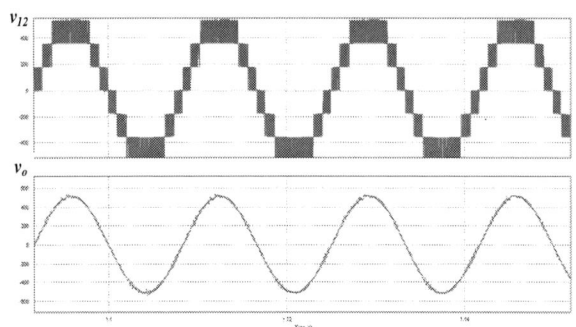

(V_{12}: 200V/div ; V_o: 200V/div ; Time: 20msec/div)

Fig. 5. Simulation result of proposed seven-level inverter

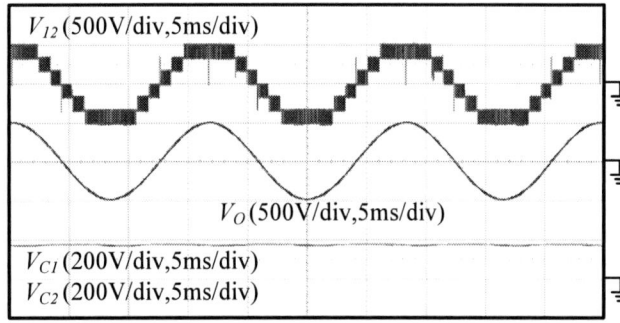

Fig. 6. Experimental result of output voltage and voltages of capacitor C_1 and C_2 under 2930W

Fig. 7. Output voltage total harmonic distortion of the proposed seven-level inverter under 2930W

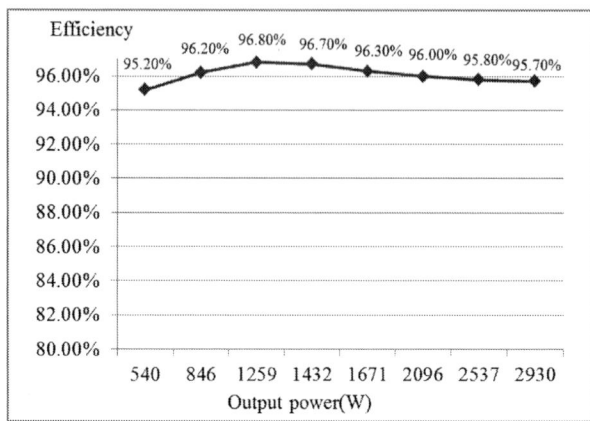

Fig. 8. Efficiency curve of the proposed seven-level inverter

V. CONCLUSION

In this paper, a novel single-phase seven-level inverter is proposed. Compared with conventional multilevel inverter structure, the benefits of proposed seven-level inverter are fewer components, self-voltage balance, and low output voltage total harmonic distortion (THD). Moreover, the proposed structure does not need any extra capacitor voltage-balancing circuit, and the control methods of switches are simple.

Furthermore, the other advantage of the proposed inverter is low cost. Operational principles and coupled inductor design method for proposed inverter are analyzed in detail. The laboratory prototype of the presented seven-level inverter has been realized. Finally, the simulation and experimental waveform are shown to verify the feasibility of the proposed circuit.

ACKNOWLEDGE

The authors greatly appreciate Ministry of Science and Technology, Taiwan for supporting funds so that the experiment could be done smoothly. The number of the project is MOST 101-2622-E-006-021-CC2.

REFERENCES

[1] Paulson Samuel, Rajesh Gupta, and Dinesh Chandra, "Grid Interface of Wind Power With Large Split-Winding Alternator Using Cascaded Multilevel Inverter," *IEEE Transactions on Energy Conversion.*, vol. 26, no. 1, pp. 299–309, Mar. 2011.

[2] Nasrudin A. Rahim, and Jeyraj Selvaraj, "Multistring Five-Level Inverter With Novel PWM Control Scheme for PV Application," *IEEE Trans. Ind. Electron.*, vol. 57, no. 6, pp. 2111–2123, Jun. 2010.

[3] V. T. Somasekhar, K. Gopakumar, M. R. Baiju, Krishna K. Mohapatra, and L. Umanand, "A Multilevel Inverter System for an Induction Motor With Open-End Windings," *IEEE Trans. Ind. Electron.*, vol. 52, no. 3, pp. 824–836, Jun. 2005.

[4] Javier Pereda, and Juan Dixon, "23-Level Inverter for Electric Vehicles Using a Single Battery Pack and Series Active Filters," IEEE Transactions on Vehicular Technology, vol. 61, no. 3, pp. 1043–1051, Mar. 2012.

[5] Mohan M. Renge and Hiralal M. Suryawanshi, "Five-Level Diode Clamped Inverter to Eliminate Common Mode Voltage and Reduce dv/dt in Medium Voltage Rating Induction Motor Drives," *IEEE Trans. on Power Electronics*, vol. 23, no.4, pp. 1598-1607, Jul. 2008.

[6] Grain P. Adam, Stephen J. Finney, Ahmed M. Massoud, and Barry W. Williams, "Capacitor Balance Issues of the Diode-Clamped Multilevel Inverter Operated in a Quasi Two-State Mode," *IEEE Trans. Ind. Electron.*, vol.55, no. 8, pp. 3088–3099, Aug. 2008.

[7] M. Marchesoni and P. Tensa, "Diode-Clamped Multilevel Converters: A Practicable Way to Balance DC-Link Voltages," *IEEE Trans. Ind. Electron.*, vol. 49, no. 4, pp. 752 - 765, Aug. 2002.

[8] O. Bouhali, B. Francois, E. M. Berkouk, and C. Saudemont, "DC Link Capacitor Voltage Balancing in a Three-Phase Diode Clamped Inverter Controlled by a Direct Space Vector of Line-to-Line Voltages," *IEEE Trans. on Power Electronics*, vol. 22, no.5, pp. 1636-1647, Sep. 2007.

[9] Dae-Wook Kang, Byoung-Kuk Lee, Jae-Hyun Jeon, Tae-Jin Kim, and Dong-Seok Hyun, "A Symmetric Carrier Technique of CRPWM for Voltage Balance Method of Flying-Capacitor Multilevel Inverter," *IEEE Trans. Ind. Electron.*, vol. 52, no. 3, pp. 879 - 888, Jun. 2005.

[10] Jing Huang, and Keith A. Corzine "Extended Operation of Flying Capacitor Multilevel Inverters," *IEEE Trans. on Power Electronics*, vol. 21, no.1, pp. 140-147, Jan. 2006.

[11] Anshuman Shukla, Arindam Ghosh, and Avinash Joshi "Improved Multilevel Hysteresis Current Regulation and Capacitor Voltage Balancing Schemes for Flying Capacitor Multilevel Inverter," *IEEE Trans. on Power Electronics*, vol. 23, no.2, pp. 518-529, Mar. 2008.

[12] Zhong Du, Burak Ozpineci, Leon M. Tolbert, and John N. Chiasson, "DC–AC Cascaded H-Bridge Multilevel Boost Inverter With No Inductors for Electric/Hybrid Electric Vehicle Applications," *IEEE Trans. Ind. Appl.*, vol.45, no. 3, pp. 963–970, May/Jun 2009.

[13] R. Gupta, A. Ghosh and A. Joshi, "Switching Characterization of Cascaded Multilevel-Inverter-Controlled Systems," *IEEE Trans. Ind. Electron.*, vol. 55, no. 3, pp. 1047–1058, Mar. 2008.

[14] Rajesh Gupta, Arindam Ghosh, and Avinash Joshi "Multiband Hysteresis Modulation and Switching Characterization for Sliding-Mode-Controlled Cascaded Multilevel Inverter," *IEEE Trans. on Industrial Electronics*, vol. 57, no. 7, pp. 2344-2353, Jul. 2010.

[15] Zhong Du, Leon M. Tolbert, Burak Ozpineci, and John N. Chiasson, "S Fundamental Frequency Switching Strategies of a Seven-Level Hybrid Cascaded H-Bridge Multilevel Inverter," *IEEE Trans. on Power Electronics*, vol. 24, no.1, pp. 25-33, Jan. 2009.

[16] Patricio Cortés, Alan Wilson, Samir Kouro, Jose Rodriguez, and Haitham Abu-Rub, "Model Predictive Control of Multilevel Cascaded H-Bridge Inverters," *IEEE Trans. Ind. Electron.*, vol. 57, no. 8, pp. 2691–2699, Aug. 2010.

[17] C. M. Wang, C. H. Su, M. C. Jiang and Y. C. Lin, "A ZVS-PWM Single-Phase Inverter Using a Simple ZVS-PWM Commutation Cell," *IEEE Trans. Ind. Electron.*, vol. 55, no. 2, pp. 758 - 766, Feb. 2008.

[18] E. Babaei, S. H. Hosseini, G. B. Gharehpetian, M. Tarafdar Haque and M. Sabahi, "Reduction of DC Voltage Sources and Switches in Asymmetrical Multilevel Converters Using a Novel Topology," Elsevier Journal of Electric Power Systems Research, 2007, no. 77, pp. 1073-1085.

[19] F. S. Kang, S. J. Park, S. E. Cho, C. U. Kim and T. Ise, "Multilevel PWM Inverters Suitable for the Use of Stand-Alone Photovoltaic Power Systems," *IEEE Trans. Energy Convers.*, vol. 20, no. 4, pp. 906–915, Dec. 2005.

[20] A. M. Knight, J. Ewanchuk, and J. C. Salmon, "Coupled three-phase inductors for interleaved inverter switching," IEEE Trans. Magn., vol. 44, no. 11, pp. 4199–4122, Nov. 2008.

[21] J. Salmon, A. Knight, and J. Ewanchuk, "Single phase multi-level PWM inverter topologies using coupled inductors," in Proc. IEEE Power Electron. Spec. Conf. (PESC), 2008, pp. 802–808.

[22] D. Floricau, E. Floricau, and G. Gateau, "A Novel Single-Phase Five-Level Inverter With Coupled Inductors," *IEEE Trans. Ind. Electron.*, vol. 58, no. 12, pp. 5344–5351, Jul. 2011.

[23] J. Salmon. J. Ewanchuk, A. Knight, "PWM inverters using split-wound coupled inductors," *IEEE Trans. Ind. Appl.*, vol. 45, no. 6, pp. 2001–2009, Nov. 2009.

[24] Zheng Zhao, Jih-Sheng Lai, and Younghoon Cho, "Dual-Mode Double-Carrier-Based Sinusoidal Pulse Width Modulation Inverter With Adaptive Smooth Transition Control Between Modes," *IEEE Trans. Ind. Electron*, vol. 60, no. 5, pp. 2094-2103, May. 2013.

[25] Wenxi Yao, Haibing Hu, and Zhengyu Lu, "Comparisons of Space-Vector Modulation and Carrier-Based Modulation of Multilevel Inverter," *IEEE Trans. on Power Electronics*, vol. 23, no.1, pp. 45-51, Jan. 2008.

[26] Subrata K. Mondal, Bimal K. Bose, Valentin Oleschuk, and Joao O. P. Pinto, "Space Vector Pulse Width Modulation of Three-Level Inverter Extending Operation Into Overmodulation Region," *IEEE Trans. on Power Electronics*, vol. 18, no.2, pp. 604-611, Mar. 2003.

[27] Nasrudin A. Rahim, Krismadinata Chaniago, and Jeyraj Selvaraj, "Single-Phase Seven-Level Grid-Connected Inverter for Photovoltaic System," *IEEE Trans. Ind. Electron*, vol. 58, no.6, pp. 2435-2443, Jun. 2011.

[28] Ehsan Najafi and Abdul Halim Mohamed Yatim, "Design and Implementation of a New Multilevel Inverter Topology," *IEEE Trans. on Industrial Electronics*, vol. 59, no. 11, pp. 4148-4154, Nov. 2012.

[29] Javier Chavarría, Domingo Biel, Francesc Guinjoan, Carlos Meza, and Juan J. Negroni, "Energy-Balance Control of PV Cascaded Multilevel Grid-Connected Inverters Under Level-Shifted and Phase-Shifted PWMs," *IEEE Trans. Ind. Electron*, vol. 60, no.1, pp. 98-111, Jun. 2013.

[30] Dae-Wook Kang, Byoung-Kuk Lee, Jae-Hyun Jeon,Tae-Jin Kim, and Dong-Seok Hyun, "A Symmetric Carrier Technique of CRPWM for Voltage Balance Method of Flying-Capacitor Multilevel Inverter," *IEEE Trans. Ind. Electron*, vol. 52, no.3, pp. 879-888, Jun. 2005.

[31] Kenichiro Sano, and Hideaki Fujita, "A New Control Method of a Resonant Switched-Capacitor Converter and its Application to Balancing of the Split DC Voltages in a Multilevel Inverter," IEEE ,2007.

[32] Anshuman Shukla, Arindam Ghosh, and Avinash Joshi, "Flying Capacitor-Based Chopper Circuit for DC Capacitor Voltage Balancing in Diode-Clamped Multilevel Inverter," *IEEE Trans. Ind. Electron*, vol. 57, no. 7, Jul. 2010.

[33] Zeliang Shu, Xiaoqiong He, Zhiyong Wang, Daqiang Qiu, and Yongzi Jing"Voltage Balancing Approaches for Diode-Clamped Multilevel Converters Using Auxiliary Capacitor-Based Circuits," *IEEE Trans. on Power Electronics*, Vol. 28, no. 5, May 2013.

Cascaded Multilevel Converter based Bidirectional Inductive Power Transfer (BIPT) System

Bac Xuan Nguyen, D. M. Vilathgamuwa,

Gilbert Foo, Andrew Ong, Prasad K. Sampath

School of Electrical and Electronics Engineering

Nanyang Technological University, Singapore

Email:{xuanbac001, cong008, prasadku001}@e.ntu.edu.sg;
{emahinda, gilbert.foo}@ntu.edu.sg;

Udaya K. Madawala

Department of Electrical & Computer Engineering

University of Auckland, New Zealand

Email: u.madawala@auckland.ac.nz

Abstract— **A typical low power IPT system employs an H-Bridge converter with a simple control strategy to generate a high frequency current from DC power supply. This paper proposes a cascaded multilevel converter for bidirectional IPT (BIPT) systems, which is suitable for low to medium power applications as well as for situations such as PV cells where several individual DC sources are to be utilized. A novel modulation strategy is proposed for the multilevel converter with the aim of minimizing switching losses. Series – Series (SS) compensation circuit is adopted for the IPT system and a mathematical model is presented to minimize the coil losses of the system under varying output power. Theoretical results presented in comparison to the simulations to demonstrate the applicability of the proposed concept and the validity of the developed model. The experimental results show the feasibility of the proposed phase shift modulation.**

Keywords — *Multilevel Converter, Bidirectional Inductive Power Transfer (BIPT), phase shift modulation, efficiency.*

I. INTRODUCTION

Inductive Power Transfer systems are gaining popularity as they offer numerous advantages over conventional wired power transfer systems with regards to convenience, safety, isolation, operation in hostile conditions and flexibility. Consequently, there is an increased demand for IPT powered applications from low power applications, such as mobile charging systems and implant systems, to medium power systems such as electric vehicle charging systems, lighting, material handling, grid-tied PV converters [1-5].

A number of different power converters have been proposed for IPT systems [6-8]. As there are some limitations in power switch characteristics in terms of voltage, current, operating temperature, etc., this paper proposes a cascaded multilevel converter topology for IPT front end converter with a method to calculate the optimal phase shift angle of each H-Bridge module at which the switching losses are kept minimal.

Multilevel converter is known to be one of the most popular converters, suitable for medium to high power applications. The development of multilevel converter topologies and the control strategies can be found in [9-10]. The most significant benefit of the cascaded multilevel converter is its inherent ability to connect multiple low power modules to increase the overall power rating. This is useful in the case of multiple low power sources such as PV cells.

In addition, an optimal control algorithm for maximizing the efficiency of the given system in Fig. 1 (a) is also proposed. The proposed control algorithm aims to minimize the conduction losses in the inductive coils. For the conventional control method, output power is controlled by adjusting the phase shift angle at the secondary side converter, while the phase shift angle of the primary side converter is employed to maintain the input current amplitude [11]-[13]. In this work, the output power is controlled by the phase shift angles of the converters from both sides so that the efficiency is optimized for a wide range of output power.

II. TOPOLOGY DESCRIPTION AND ANALYSIS

A. Topology description

Fig.1 shows the proposed topology with cascaded multilevel converter. The SS compensation circuit is employed for both resonant sides of the IPT system.

By using H-Bridge converter in the pickup side, the power flow can be bidirectional. There are many types of multilevel converter topologies as well as control strategies proposed [8-10]. Most of industrial applications, the output voltage should have low total harmonic distortion (THD). Therefore, modulation strategies focus on eliminating the high order harmonic components of the output voltage. However, THD plays a less important role in the IPT system. Therefore, the phase shift modulation is employed

The 2014 International Power Electronics Conference

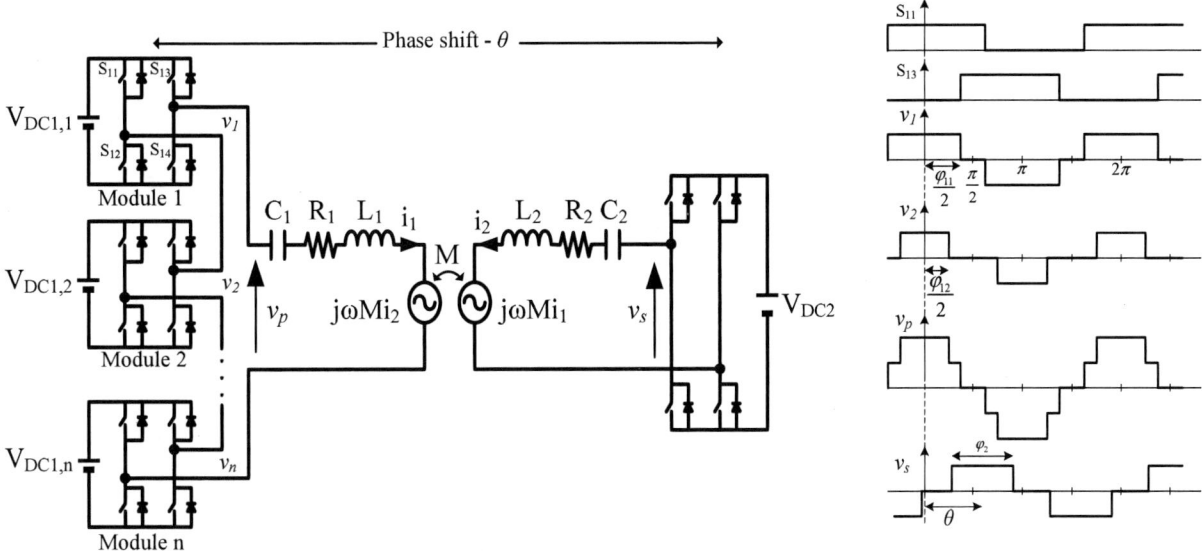

a) Topology　　　　　　　*b) Phase modulated voltages generated by both side converters*

Fig. 1. Cascaded multilevel converter based Bidirectional IPT system.

in most of the studies on IPT systems to adjust the amplitude of the fundamental component of output voltage [11-14]. Fig. 1 (b) shows the output voltage waveform v_p with two modules of H-Bridge converters in the cascaded multilevel converter with phase shift modulation.

One of the most important considerations of applying the multilevel converter to IPT system is its power loss that consists of converter switching loss and conduction loss.

This paper proposes a phase shift modulation strategy to drive the proposed cascaded multilevel converter with minimized switching loss in the switching devices. Depending on the number of cascaded modules, the conduction loss of multilevel converter is larger than the conventional H-Bridge converter if the number of switching devices employed is identical. However, in the situation where input DC voltages of all multilevel converter modules are low, the OptiMOS power MOSFET with extremely low on – state resistance could be applied. Thus, the conduction loss of multilevel converter can be kept even lower than that of the conventional converter. To illustrate this assumption, we compare 12 modules of multilevel converter with conventional converter. Multilevel converter and conventional converter employ the OptiMOS MOSFET (IPB025N10N: 100V, 180A, 2.5 mΩ) and the high speed SiC MOSFET (C2M0080120D: 1200V, 32A, 80 mΩ), respectively. From this example, one can see that the conduction loss of multilevel converter is lower than that of conventional converter for the same system voltage rating.

B. Derivation of proposed phase shift modulation strategy

Firstly, let us consider the switching loss of the H-Bridge converter with phase shift control method.

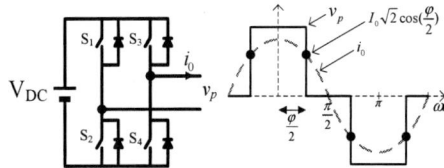

Fig. 2. H – Bridge converter output waveforms.

Assume that the output voltage and current of the H-Bridge converter are in phase as shown in Fig. 2. The instantaneous switching current of H-Bridge converter is $I_0\sqrt{2}\cos(\varphi/2)$. Thus the switching loss is zero when phase shift angle between converter legs is equal to π. Furthermore, to achieve zero switching loss when the phase shift angle is zero, (S_1, S_3) and (S_2, S_4) should be turned on and off for the whole duration required, respectively. This is the idea to derive the modulation strategy with minimum switching loss described in the following section.

C. The proposed phase shift modulation strategy

Let's consider the multilevel converter that consists of n – modules of H – Bridge converters as shown in Fig. 1 (a). The output voltage of each H – Bridge converter and the output voltage of multilevel converter are shown in Fig. 1 (b).

The phase shift angle is calculated so that the switching loss is minimized.

The output voltage of each module is given as follows,

$$v_i(t) = \frac{4V_{DC1,i}}{\pi} \sum_{k=1,3,5...}^{+\infty} \left[\frac{1}{k}\cos(k\omega_I t)\sin(\frac{k\varphi_{1,i}}{2}) \right] \quad (1)$$

And the output voltage of multilevel converter is given as follows,

$$v_p(t) = \sum_{i=1}^{n} v_i(t) \qquad (2)$$

The amplitude of the fundamental component of output voltage is as follows,

$$V_{pm} = \frac{4}{\pi} \sum_{i=1}^{n} V_{DC1,i} \sin(\frac{\varphi_{1,i}}{2}) \qquad (3)$$

Let's define $V_{pm,ref}$ as the desired amplitude voltage of multilevel converter which is obtained from the closed loop controller. Assume that the input DC voltages are sorted as follows,

$$0 = V_{DC1,0} < V_{DC1,1} \le V_{DC1,2} \le \dots \le V_{DC1,n} \qquad (4)$$

The phase shift angle of each converter is defined using the following rules,

1. If $\dfrac{4}{\pi} V_{DC1,j} < V_{pm,ref} \le \dfrac{4}{\pi} V_{DC1,j+1}$

$$\Rightarrow \begin{cases} \sin\dfrac{\varphi_{1,j+1}}{2} = \dfrac{V_{pm,ref}}{\dfrac{4}{\pi} V_{DC1,j+1}} \\ \varphi_{1,j} = 0 \ \text{for } j \ne i+1 \\ i = 0,1, \dots , (n-1) \\ j = 1,2, \dots , n \end{cases} \qquad (5)$$

2. If $\dfrac{4}{\pi} \sum_{i=k}^{n} V_{DC1,i} < V_{pm,ref} \le \dfrac{4}{\pi} \sum_{i=k-1}^{n} V_{DC1,i}$

$$\Rightarrow \begin{cases} \sin\dfrac{\varphi_{1,k-1}}{2} = \dfrac{V_{pm,ref} - \dfrac{4}{\pi} \sum_{i=k}^{n} V_{DC1,i}}{\dfrac{4}{\pi} V_{DC1,k-1}} \\ \varphi_{1,i} = \pi \ \text{for } k \le i \le n \\ \varphi_{1,i} = 0 \ \text{for } i < (k-1) \\ k = 1,2, \dots , n \end{cases} \qquad (6)$$

The first rule describes the case where normalized reference voltage $\frac{\pi}{4} V_{pm,ref}$ is less than the largest input dc voltage while the second rule describes the case where normalized voltage is greater than all the input dc voltages. Fig. 3 illustrates these rules in greater clarity.

With the proposed modulation scheme, at any point in time, only one module is controlled with phase shift angle: $0 < \varphi_{1,i} < \pi$, while phase shift angles of other modules are either zero or π. Therefore, as mentioned in Section II – B, the switching loss of these modules is zero.

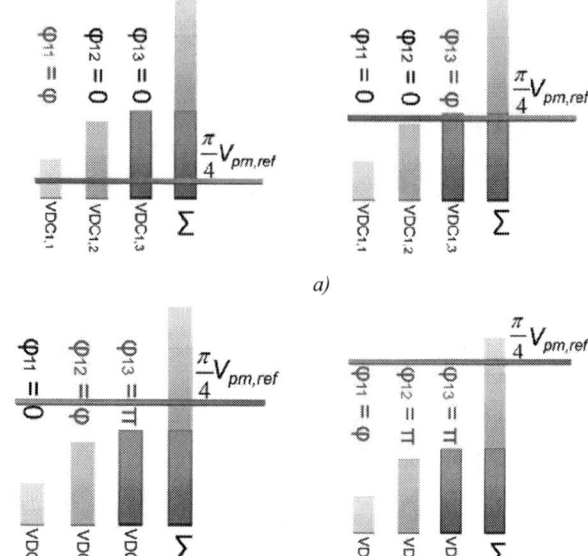

a)

b)

Fig. 3. Phase shift modulation strategies illustration.
    a)   For the rules in (5)
    b)   For the rules in (6)

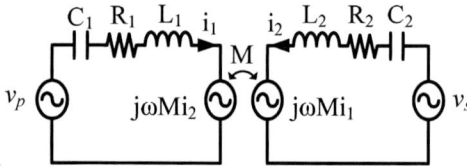

Fig. 4. The equivalent steady state model of the topology in Fig. 1.

III. AN ANALYSIS FOR MINIMIZING THE COIL LOSS

In recent literature [11]-[13], the phase shift angle of the primary side converter in a BIPT system is adjusted to control the input current while the output power is controlled by adjusting the phase shift angle of the pickup side converter. This study presents a novel algorithm in obtaining the phase shift angle with minimized coil loss. With the above phase shift modulation strategy, the fundamental components of the input and output side voltages can be given as follows,

$$v_p(t) = \frac{4}{\pi}\left(\sum_{i=1}^{k} V_{DC1,i} + V_{DC1,k+1} \sin(\frac{\varphi_{1,k+1}}{2}) \right)\cos(\omega_T t) \quad (7)$$

$$v_s(t) = \frac{4}{\pi} V_{DC2} \sin(\frac{\varphi_2}{2})\cos(\omega_T t + \theta) \qquad (8)$$

where θ is the phase shift angle between the primary side and the secondary side converters.

The input and output side currents (Fig. 4.) are,

$$i_1 = \frac{v_p - j\omega M i_2}{R_1 + j\omega L_1 + \dfrac{1}{j\omega C_1}} \tag{9}$$

$$i_2 = \frac{-j\omega M i_1 + v_s}{R_2 + j\omega L_2 + \dfrac{1}{j\omega C_2}} \tag{10}$$

IPT operates at the resonance angular frequency of ω_T, $\omega_T^2 = 1/L_1 C_1 = 1/L_2 C_2$. Substituting (7) and (8) into (9) and (10) gives,

$$i_1 = \frac{R_2 V_{pm} - j\omega M V_{sm} \angle \theta}{\omega^2 M^2 + R_1 R_2} \tag{11}$$

$$i_2 = \frac{j\omega M V_{pm} + R_1 V_{sm} \angle \theta}{\omega^2 M^2 + R_1 R_2} \tag{12}$$

The input and output active and reactive powers can be calculated by,

$$
\begin{aligned}
P_{in} &= \frac{1}{2} \mathrm{Re}\{v_p.i_1^*\} \\
&= \frac{1}{2} \frac{V_{pm}(R_2 V_{pm} + \omega M V_{sm}\sin\theta)}{\omega^2 M^2 + R_1 R_2}
\end{aligned} \tag{13}
$$

$$
\begin{aligned}
Q_{in} &= \frac{1}{2} \mathrm{Im}\{v_p.i_1^*\} \\
&= \frac{1}{2} \frac{\omega M V_{pm} V_{sm}\cos\theta}{\omega^2 M^2 + R_1 R_2}
\end{aligned} \tag{14}
$$

$$
\begin{aligned}
P_{out} &= \frac{1}{2} \mathrm{Re}\{v_s.i_2^*\} \\
&= \frac{1}{2} \frac{V_{sm}(R_1 V_{sm} - \omega M V_{pm}\sin\theta)}{\omega^2 M^2 + R_1 R_2}
\end{aligned} \tag{15}
$$

$$
\begin{aligned}
Q_{out} &= \frac{1}{2} \mathrm{Im}\{v_s.i_2^*\} \\
&= -\frac{1}{2} \frac{\omega M V_{pm} V_{sm}\cos\theta}{\omega^2 M^2 + R_1 R_2}
\end{aligned} \tag{16}
$$

The reactive power components in both sides of the system can be minimized by keeping the phase shift angle between the primary and the secondary side converters to be either $+90^0$ or -90^0. When $\theta = +90^0$, power will be transferred from the primary side to the secondary side, while power flow is reversed when $\theta = -90^0$.

When $\theta = +90^0$, the efficiency is given as follows,

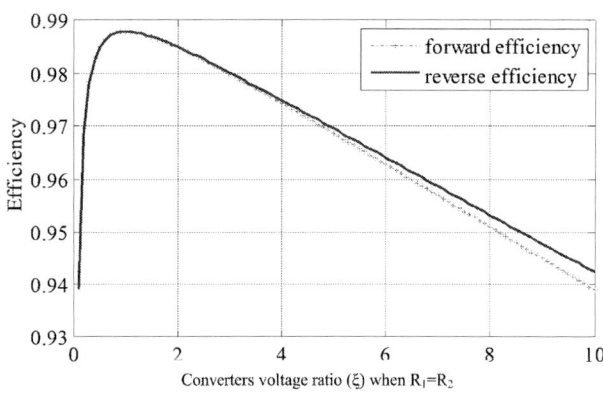

Fig. 5. The dependency of the efficiency on the converter voltages.

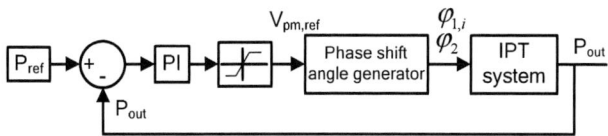

Fig. 6. A PI controller for the proposed IPT system.

$$
\begin{aligned}
\eta_{for} &= \frac{|P_{out}|}{P_{in}} = \frac{V_{sm}(\omega M V_{pm} - R_1 V_{sm})}{V_{pm}(\omega M V_{sm} + R_2 V_{pm})} \\
&= \frac{\xi(\omega M - R_1 \xi)}{\omega M \xi + R_2}
\end{aligned} \tag{17}
$$

where $\xi = V_{sm}/V_{pm}$ is the ratio of secondary and primary output voltage of the converters.

Similarly, when $\theta = -90^0$, we get the efficiency in the reverse direction as follows,

$$
\begin{aligned}
\eta_{rev} &= \frac{|P_{in}|}{P_{out}} = \frac{V_{pm}(\omega M V_{sm} - R_2 V_{pm})}{V_{sm}(\omega M V_{pm} + R_1 V_{sm})} \\
&= \frac{\omega M \xi - R_2}{\xi(\omega M + R_1 \xi)}
\end{aligned} \tag{18}
$$

It's obvious from (17) and (18) that the efficiency of the resonant sides of IPT system depends on the ratio of both converter output voltages. Fig. 5 shows the dependency of efficiency on the voltage ratio. It is obvious that efficiencies can reach a maximum value at a proper voltage ratio. The maximum forward and reverse efficiency takes place when the following optimized voltage ratios are used respectively,

$$\xi_{opt,for} = \frac{-R_1 R_2 + \sqrt{R_1^2 R_2^2 + R_1 R_2 \omega^2 M^2}}{R_1 \omega M} \tag{19}$$

$$\xi_{opt,rev} = \frac{R_1 R_2 + \sqrt{R_1^2 R_2^2 + R_1 R_2 \omega^2 M^2}}{R_1 \omega M} \tag{20}$$

Assuming that $R_1 \& R_2 \ll \omega M$, (19) and (20) become,

$$\xi_{opt,for} \approx \xi_{opt,rev} \approx \sqrt{R_2/R_1} \tag{21}$$

From (8), the calculation of the phase shift angle used in the pickup side converter is as follows,

$$\sin\left(\frac{\varphi_2}{2}\right) = \frac{\pi}{4}\frac{V_{sm,ref}}{V_{DC2}} = \frac{\pi}{4}\frac{\xi_{opt}V_{pm,ref}}{V_{DC2}} \tag{22}$$

A closed loop controller can be applied for the given system to minimize coil losses by choosing an appropriate voltage ratio between the primary side and secondary side converters. To ensure that the phase shift angle of the primary and secondary side converters is feasible, a saturation block should be added after the controller (Fig. 6). In that case, $V_{pm,ref}$ must satisfy the following condition,

$$V_{pm,ref} \le \min\left(\frac{4V_{DC1}}{\pi}, \frac{4V_{DC2}}{\pi\xi_{opt}}\right) \tag{23}$$

Fig. 6 presents a PI controller with the phase shift generator block using the set of equations (5), (6) and (22) to calculate the phase shift angle of both side converters.

IV. SIMULATION AND EXPERIMENT RESULTS

The IPT system in Fig. 1 is simulated using the controller in Fig. 6 with the parameters in Table I.

Fig. 7 and Fig. 8 show the simulation results of input and output voltages and currents of the given system during the control process. Fig. 9 shows the power response of the controller with zero steady state error and zero over shooting. The efficiencies of the coupling coils are maintained at around 99% for a wide range of output power as shown in Fig. 10.

Fig. 11 shows the hardware prototype with single H-Bridge converter at both sides of the BIPT system. The experiment is set up to verify the efficiencies of the proposed phase shift modulation when the system operates for a wide range of output power. A complete experiment set up of the proposed system will be presented in future works.

Due to the power loss through the power converters and filters, the overall efficiency of the proposed system is around 92% and 90% when delivering powers of 450 W and 300 W respectively.

Due to the difference of switching losses in two cases, the efficiencies between two experiments are different. The dependency of efficiency on the phase shift angle has been reported in [15].

TABLE I
SIMULATION AND EXPERIMENTATION PARAMETERS

Parameter	Symbol	Simulation	Experimental setup
DC input voltage (V)	$V_{DC1,1}$	150	100
	$V_{DC1,2}$	140	
	$V_{DC1,3}$	160	
DC output voltage (V)	V_{DC2}	500	115
Coils inductance (µH)	$L_i = L_1 = L_2 = L_o$	183	183.2
Equivalent AC resistances (Ω)	$R_i = R_1 = R_2 = R_o$	0.1	0.14
Compensator capacitances (Ω)	$C_1 = C_2$	0.06	0.059
Switching frequency (kHz)	f_T	48	48
Coupling coefficient	k	0.38	0.38
Airgap (mm)			60

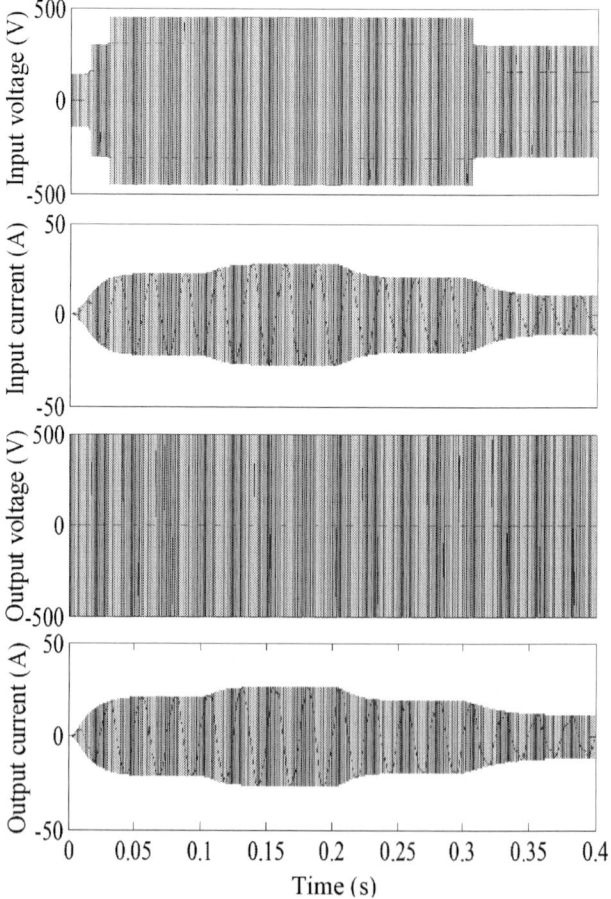

Fig. 7. The voltage and current waveforms of converters.

The 2014 International Power Electronics Conference

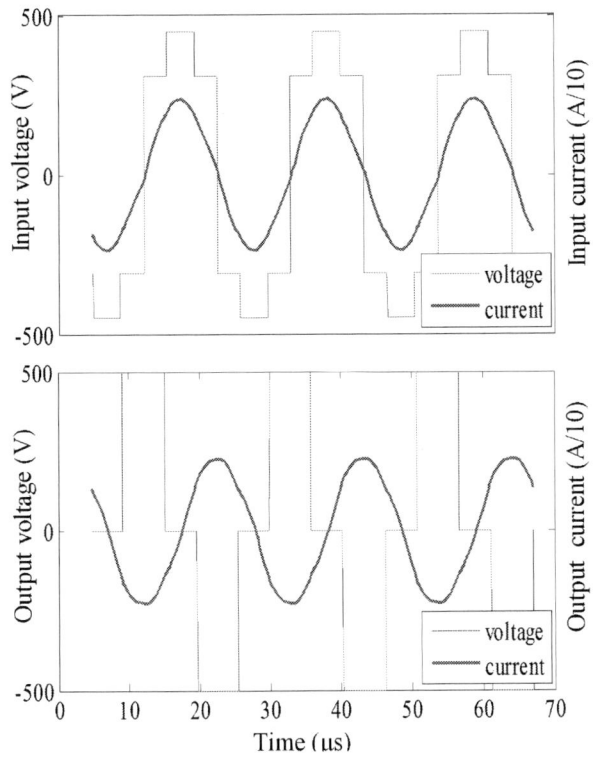

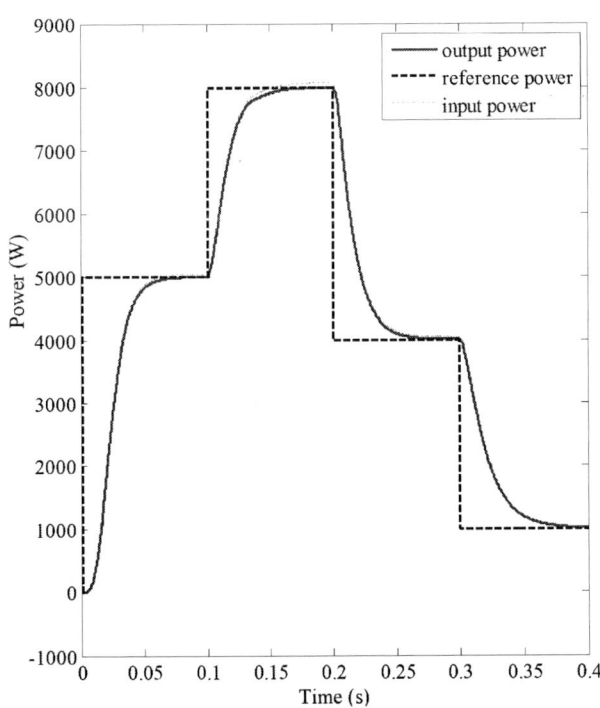

Fig. 8. Instantaneous voltages and currents of converters.

Fig. 9. Power response of the PI controller.

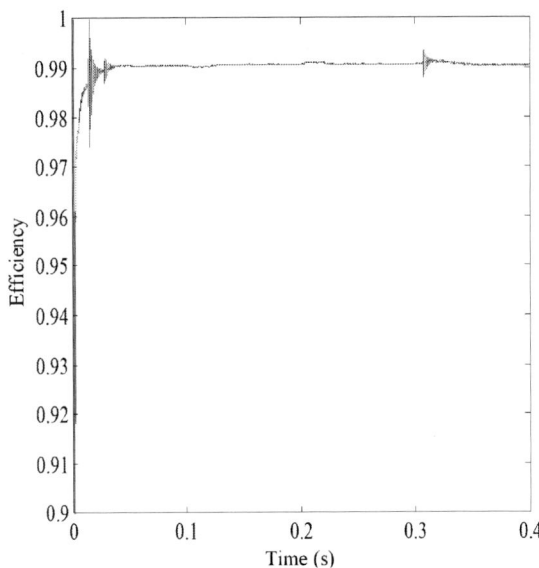

Fig. 10. Efficiency of the proposed system.

Fig. 11. Experimental set up.

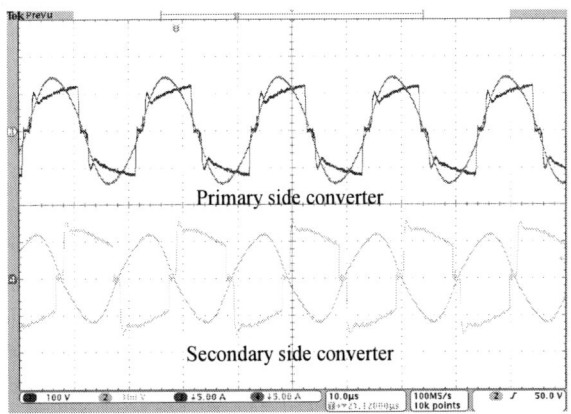

Fig. 12. Experimental results: Voltage and current waveforms when delivering 450 W with 92% efficiency.

978-1-4799-2706-7/14 $31.00 © 2014 IEEE

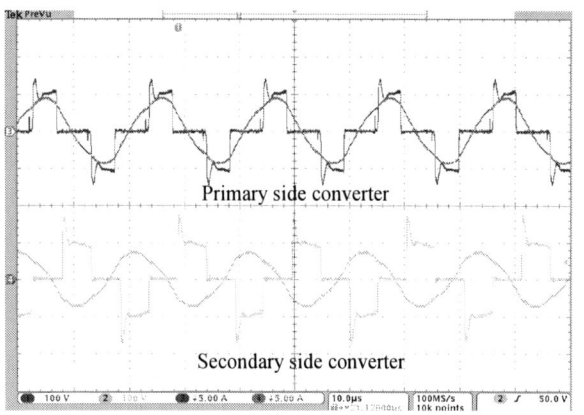

Fig. 13. Experimental results: Voltage and current waveforms when delivering 300 W with 90% efficiency.

V. CONCLUSIONS

A cascaded multilevel converter based IPT system has been proposed to improve the scalability of low to medium power wireless transmission for a new generation of IPT systems. This paper presents the algorithms to optimize the efficiency of the whole system by,

- proposing the multilevel converter modulation strategy to minimize the switching loss in the converters.
- proposing the phase shift modulation algorithm to minimize coil loss.

A PI controller is proposed to regulate the output power with fast response and zero steady state error. The simulations show that the system is highly efficient for a wide range of output power.

The experimental results show the feasibility of the propsed phase shift modulation.

REFERENCES

[1] G. A. Covic, and J. T. Boys, "Modern Trends in Inductive Power Transfer for Transportation Applications," *IEEE Journal of Emerging and Selected Topics in Power Electronics*, vol. 1, no. 1, pp. 28 – 41, 2013.

[2] H. H. Wu, J. Boys, G. Covic and D. Robertson, "A practical 1.2kW Inductive Power Transfer lighting system using AC processing controllers," in 6th *IEEE Conf. on Industrial Electronics and Applications (ICIEA)*, pp. 345-350, 2011.

[3] P. Si, A. P. Hu, S. Malpas and D. Budgett, "A frequency control method for regulating wireless power to implantable devices," *IEEE Trans. Biomed. Circuits Syst.*, vol. 2, no. 1, pp.22 -29, 2008.

[4] S. Y. R. Hui and W. W. C. Ho, "A new generation of universal contactless battery charging platform for portable consumer electronic equipment," *IEEE Trans. Power Electron.*, vol. 20, no. 3, pp. 620–627, 2005.

[5] U. K. Madawala and D. J. Thrimawithana, "Modular-based inductive power transfer system for high-power applications," *IET Trans. Power Electro.*, vol. 5, pp. 1119–1126, 2013.

[6] H. L. Li, A.P. Hu, G. A. Covic, "A Direct AC–AC Converter for Inductive Power-Transfer Systems," *IEEE Trans. Power Electron.* vol. 27, no. 2, pp. 661 – 668, 2012.

[7] H. Hao, G. A. Covic, J. Boys, "A parallel topology for Inductive Power Transfer power supplies," *IEEE Trans. Power Elect.* vol. 29, no. 3, pp. 1832 – 1837, 2013.

[8] B. S. Riar, U. K. Madawala, "A novel Modular Multi-level Converter topology with Voltage Correcting Modules (M2LC-VCMs)," in *IEEE International Conference on Industrial Technology (ICIT)*, pp. 451 – 456, 2013.

[9] J. Rodriguez, Jih-Sheng Lai, Z. P. Fang, "Multilevel inverters: a survey of topologies, controls, and applications," *IEEE Trans. Ind. Elect.* vol. 49, no. 4, pp.724 – 738, 2002.

[10] M. Malinowski, K. Gopakumar, J. Rodriguez, M. A. Pérez, "A Survey on Cascaded Multilevel Inverters". *IEEE Trans. Ind. Elect.* vol. 57, no. 7, pp. 2197 – 2206, 2010.

[11] D. J. Thrimawithana, U. K. Madawala, "A Generalized Steady-State Model for Bidirectional IPT Systems," *IEEE Trans. Power Elect.*, vol. 28, no. 10, pp. 4681 – 4689, 2013.

[12] D. J. Thrimawithana, U. K. Madawala, Michael Neath, "A Synchronization Technique for Bidirectional IPT Systems," *IEEE Trans. Ind. Elect.*, vol. 60, no. 1, pp. 301 – 309, 2013.

[13] D. J. Thrimawithana, U. K. Madawala, Michael Neath, "A steady-state analysis of bi-directional inductive power transfer systems," in *IEEE International Conference on Industrial Technology (ICIT)*, pp. 1618 – 1623, 2013.

[14] N. X. Bac, D. M. Vilathgamuwa, U. K. Madawala, "A matrix converter based Inductive Power Transfer system," in *IEEE Conf. on Power & Energy (IPEC)*, pp. 509 – 514, 2012.

[15] N. X. Bac, D. M. Vilathgamuwa, U. K. Madawala, "A SiC based Matrix Converter Topology for Inductive Power Transfer System," *IEEE Trans. Power Elect.* 2013 (Early Access).

The 2014 International Power Electronics Conference

Undersampling Control of a Bidirectional Cascaded Buck+Boost Dc-Dc Converter

Martin Rosekeit*, Philipp Joebges*, Markus Lelie*†, Dirk Uwe Sauer*† and Rik W. De Doncker*

*Institute for Powerelectronics and Electrical Drives
RWTH Aachen University
Jägerstr. 17-19, 52066 Aachen, Germany
Email: post@isea.rwth-aachen.de

†Jülich Aachen Research Alliance
JARA-Energy
Germany

Abstract—This paper describes the design of a current control for a bidirectional cascaded buck+boost dc-dc converter that is used to stabilize a 48 V dc-grid using a battery. While the hardware is designed for a switching frequency of 100 kHz, the control frequency is limited to 10 kHz. A model based predictor of the inductor current is introduced to compensate the misalignment between update of the PWM and the current sampling. Despite the restriction of the control frequency a dead-beat behavior of the control of a continuous conducting current was reached. The effect of an undersampled control to the sensitivity of measurement noise and limitations due to quantized PWM are analyzed. To protect the battery of negative currents a switching strategy for discontinuous inductor current is presented and the transition between both strategies is discussed.

Index Terms—bidirectional cascaded buck+boost converter, dc-dc power converters, dead-beat control, digital control design

I. INTRODUCTION

Uninterruptible power supplies (UPS) are used to increase the reliability of the electric grid [1]. Small UPS systems with a power of a few kilo watts often use a battery as energy source. As a result the active operation time of the system is limited to a few hours determined by the battery capacity. However, the hurricanes Sandy and Isaac have shown that this is insufficient for infrastructure installations [2, 3]. Possible solutions are fuel cell systems with reformer powered by natural gas or propane gas. The gas needed to supply the loads for days can be stored in gas bottles. In case of a longer black out, gas bottles can be exchanged easily.

The UPS system described in this paper is shown in Fig. 1. It has a nominal power of 1 kW at a dc-link voltage of 48 V. The target application requires full operation at environmental temperatures from −40 °C to 50 °C. A lithium iron phosphate (LFP) battery is used to buffer load fluctuations and therefore increases the life time of the fuel-cell system [4].

During stand-by operation the battery and power electronics are not heated to reduce the overall energy consumption. In case of active operation and negative temperature of the battery at system start, the battery needs time to heat up. During the heat-up phase the battery can supply the load, but it will be damaged if it is charged [5]. To control the dc-link voltage even at negative battery temperatures a chopper has been integrated into the system. The chopper transforms the

excess energy from the fuel-cell system into heat in an external resistor.

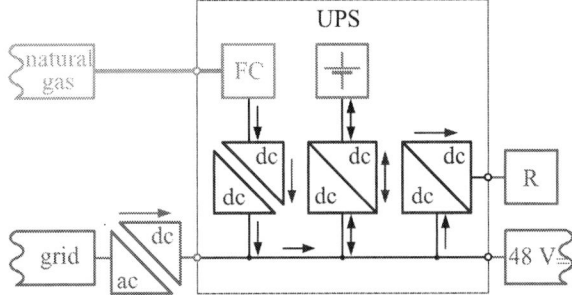

Fig. 1. Hybrid UPS system with fuel cell as primary and battery as dynamic energy supply

The battery voltage is depending on the state of charge (SOC) and the battery current. The voltage range from 70 to 105 % of the nominal voltage. To be able to control the battery current and hence the SOC, the battery is connected to the dc-link via a dc-dc converter. The nominal voltage of the battery is close to the nominal voltage of the dc-link. A bidirectional cascaded buck+boost converter, as shown in Fig. 2, was selected to control the power flow.

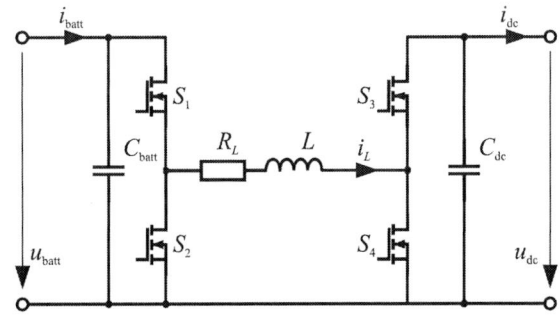

Fig. 2. Topology of bidirectional cascaded buck+boost converter

The bidirectional cascaded buck+boost converter is a widely used topology [6–8]. It features a compact inductor, high utilization of the semiconductor devices, and a potentially high efficiency [6]. Due to the two independent half-bridges the

978-1-4799-2706-7/14 $31.00 © 2014 IEEE

topology has an additional degree of freedom. Therefore different switching strategies have been presented: The majority of the literature focuses on control with pulse width modulation (PWM) [8]. Under PWM control hard switching events occur therefore soft switching strategies were developed. In [9] the control strategy uses the so-called critical conduction mode to ensure soft switching, but it requires a dynamic switching frequency and results in a high ripple of the inductor current at nominal load. The switching strategy shown in [10] generates a trapezoidal current waveform. This strategy ensures soft switching with a reduced ripple of the inductor current, but it needs solving of complex equations and precise switching.

The hardware design shown in this paper supports a maximum voltage of 60 V. The used 100 V-class MOSFETs with fast switching behavior and low reverse-recovery charges enable a hard-switched designed with a switching frequency of $f_s = 1/T_s = 100\,\text{kHz}$. Therefore PWM with a continuous inductor current is used. The duty cycles of the top switches are maximized to minimize the conduction losses, but the duty cycles are also limited due to control stability. Furthermore a simplified triangular/trapezoidal modulation strategy based on [10] is used to ensure positive battery currents by hardware. The battery voltage u_{batt}, dc-grid voltage u_{dc}, battery current i_{batt}, current i_L of the inductor L, and the output current i_{dc} are measured synchronously with the switching events. The control frequency is $f_c = 1/T_c = 10\,\text{kHz}$. In Fig. 3 the used hardware is shown with the two converters combined on one heat sink.

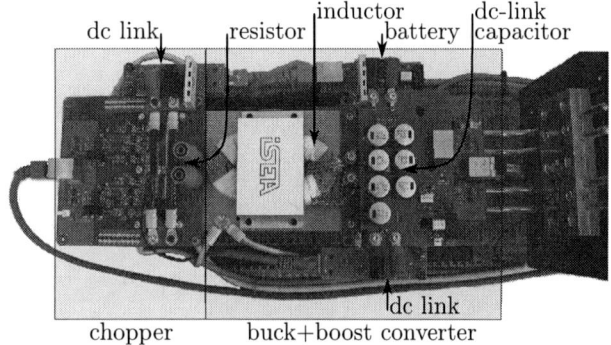

Fig. 3. Photo of dc-dc converters

II. CONTROL

The dc-link voltage is controlled by three cascaded stages. In Fig. 4 the control structure with focus on the current control is shown.

The outer loop consists of a PI controller that generates the reference value for the output current $\bar{i}^*_{\text{dc}}$. The following stage transforms the output current into the reference value for the transfer power $\bar{p}^*$ and the reference value for the inductor current $\bar{i}^*_L$. This stage also ensures that the battery current does not exceed the safety limits. The inner loop consist of two parallel controls. One control is a process-based dead-beat (DB) controller that controls a continuous

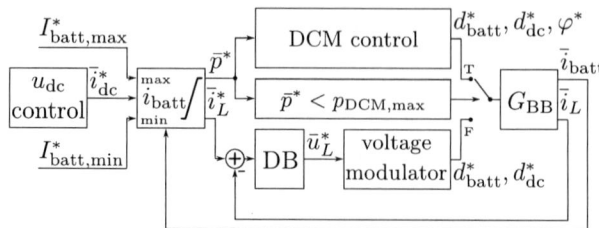

Fig. 4. Control structure

inductor current and gives a reference value of the inductor voltage $\bar{u}^*_L$. The DB controller passes the computed value to a voltage modulator that calculates the duty cycles of both half bridges d^*_{batt} and d^*_{dc}. If charging is prohibited, S_3 is constantly turned off. This results at low transfered power in a discontinuous inductor current. The open-loop discontinuous conduction mode (DCM) control uses a triangular/trapezoidal current waveform and calculates the duty-cycles d^*_{batt} and d^*_{dc} and the phase shift between the half-bridges φ^* from $\bar{p}^*$. G_{BB} represents the transfer function of the buck+boost converter.

In the following all controlled quantities are averaged over one switching cycle:

$$\bar{x} = \frac{1}{T_s} \cdot \int_{T_s} x(t)\, \text{d}t \qquad (1)$$

The average voltages at the switches S_2 and S_4 are defined as $\bar{u}_{S,2} = d_{\text{batt}} \cdot u_{\text{batt}}$ and $\bar{u}_{S,4} = d_{\text{dc}} \cdot u_{\text{dc}}$.

A. Predictive Inductor-Current Measurement

In power-electronic systems operated with synchronous sampling a delay of one sample occurs from the measurement of the controlled current to the PWM update. However, in case of this specific application the control frequency is smaller than the switching frequency. Thus a deeper analysis of the voltage sampling and PWM update is needed in order to obtain the correct system states for the control.

In Fig. 5 a timing diagram of the different events during a control cycle is shown. With mid-cycle sampling the mean inductor-current $\bar{i}_L$ is measured synchronously to the PWM by an analog digital converter (ADC) at the time $t = t_n$. When the ADC finishes the conversion an interrupt function in the digital signal processor (DSP) is called and the next duty cycles are calculated, which takes multiple switching periods. The new duty cycles are set at the beginning of the next interrupt at the time $t = t'_{n+1}$ to ensure a constant control delay.

The time delay between sampling and changing the PWM is $T_d = 12.5\,T_s$ as indicated in Fig. 5. With a discrete controlled system it is difficult to represent a system with a fractionized delay time. To compensate the fractionized delay time, a model based prediction of the inductor current is used, Thereby, the inductor current at the update of the PWM $(t = t'_n)$ is determined based on the sampled current obtained at $t = t_n$ (Fig. 5). By neglecting the series resistance of the inductance R_L and assuming constant mean voltages $\bar{u}_{S,2}$ and $\bar{u}_{S,4}$ between two updates of the PWM unit, a linear change

The 2014 International Power Electronics Conference

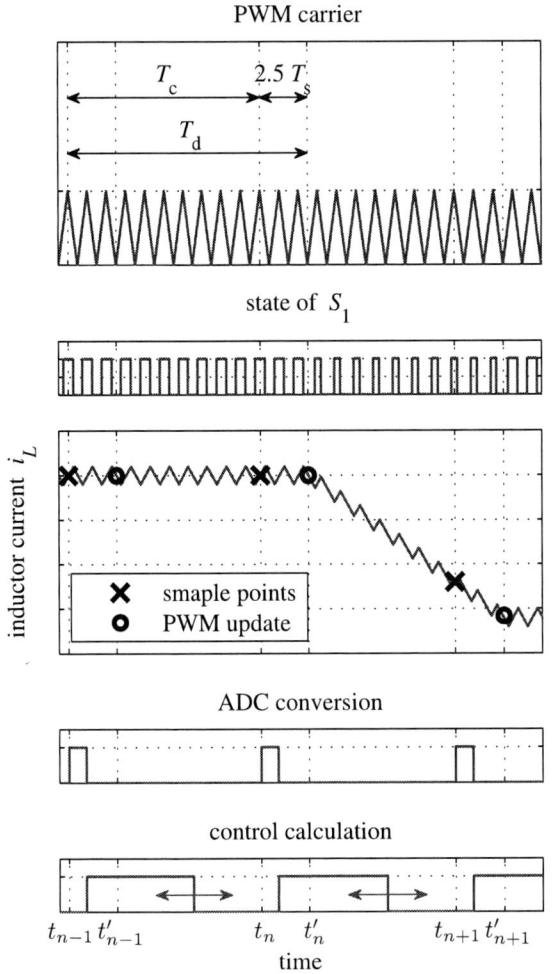

Fig. 5. Timing diagram of current sampling and PWM update

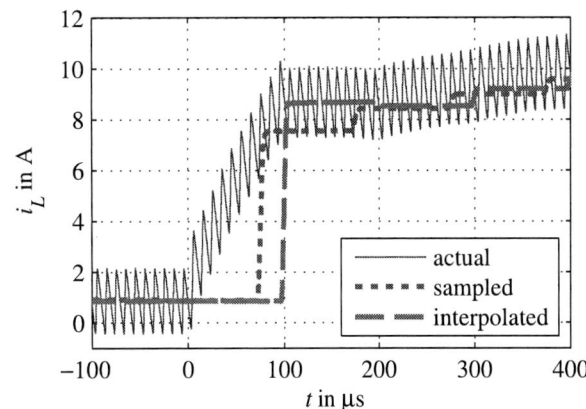

Fig. 6. Comparison of i_L with sampled and interpolated current

of the mean inductor current $\bar{i}_L$ is assumed. The current value at update of the PWM unit is then interpolated to:

$$\bar{i}_L[n] = i_L(t'_n) \tag{2}$$

$$= i_L(t_n + 0.25\,T_c) \tag{3}$$

$$= i_L(t_n) + (i_L(t_n) - \bar{i}_L[n-1]) \cdot 0.25 \tag{4}$$

where $i_L(t_n)$ is the measured current at the time t_n, $\bar{i}_L[n-1]$ is the previous interpolated value, and $\bar{i}_L[n]$ is the new interpolated value.

The effect of the interpolation is tested with a current step as shown in Fig. 6. The interpolated inductor current shows the mean current value of the last switching period before the new PWM values are set.

B. Closed-Loop Control in Continuous-Conduction Mode

The CCM control consists of a process-based second-order dead-beat controller that calculates the reference voltage for the inductance u^*_{L+R} from the error of the inductor current.

The DB controller is followed by a voltage modulator that decouples variation of the input and output voltages and calculates the duty cycles.

1) Process Model: The relation between voltage and current in the inductor is given by $\bar{u}_L = L \cdot \mathrm{d}\bar{i}_L/\mathrm{d}t$ or after transformation into the Laplace domain: $\bar{i}_L = \frac{\bar{u}_L}{s \cdot L} + i_L(t_0)$, where $i_L(t_0)$ is the inductor current at t_0. It follows with $\bar{u}_{R,L} = R_L \cdot \bar{i}_L$ and $\bar{u}_L = \bar{u}_{S,2} - \bar{u}_{S,4} - \bar{u}_{R,L}$:

$$\bar{i}_L = (\bar{u}_{S,2} - \bar{u}_{S,4} - R_L \cdot \bar{i}_L) \cdot \frac{1}{s \cdot L} + i_L(t_0) \tag{5}$$

The continuous transfer function is given by:

$$G_{\mathrm{P}}(s) := \frac{\bar{i}_L}{\bar{u}_{S,2} - \bar{u}_{S,4}} = K \cdot \frac{1}{s + \alpha} \tag{6}$$

with $K = 1/L$, $\alpha = R_L/L$, and assuming $i_L(t_0) = 0$.

By representing the PWM unit with a zero order hold (ZOH) element and using tables from literature [12], the continuous transfer function (6) is transformed into the discrete time domain to:

$$G_{\mathrm{P},0}(z) := K/\alpha \cdot \left(1 - e^{-\alpha T_0}\right) \cdot \frac{z^{-1}}{1 - e^{-\alpha T_0} \cdot z^{-1}} \tag{7}$$

with $T_0 = T_c$.

The delay time of the control calculation is added by multiplying $G_{\mathrm{P},0}(z)$ with z^{-a}, where $a \in \mathbb{N}$ is the number of additional dead times T_0. The delay time of this control is one due to the prediction of the inductor current.

The overall transfer function of the control path results in:

$$G_{\mathrm{P},1}(z) := K/\alpha \cdot \left(1 - e^{-\alpha T_0}\right) \cdot \frac{z^{-2}}{1 - e^{-\alpha \cdot T_0} z^{-1}} \tag{8}$$

2) Control Parameters: A dead-beat control on process models was presented in [12]. For a control path that can be represented by

$$G_{\mathrm{S},m}(z) = \frac{b_1 z^{-1} + \ldots + b_m z^{-m}}{1 + a_1 z^{-1} + \ldots + a_m z^{-m}} \tag{9}$$

978-1-4799-2706-7/14 $31.00 © 2014 IEEE

the transfer function of a dead-beat control is given by

$$G_{C,m}(z) = \frac{q_0 + q_1 z^{-1} + \ldots + q_m z^{-(m)}}{1 - p_1 z^{-1} - \ldots - p_{m+1} z^{-(m)}} \quad (10)$$

with

$$q_0 = \frac{1}{b_1 + b_2 + \ldots + b_m}$$

$$q_1 = a_1 q_0 \qquad\qquad p_1 = b_1 q_0$$

$$\vdots \qquad\qquad\qquad \vdots$$

$$q_m = a_m q_0 \qquad\qquad p_m = b_m q_0$$

For the control path with the order $m = 2$, the transfer function is given by

$$G_{C,2}(z) = \frac{q_0 + q_1 z^{-1} + q_2 z^{-2}}{1 - p_1 z^{-1} - p_2 z^{-2}} \quad (11)$$

For (8) the dead-beat control transfer-function is given by the coefficients:

$$q_0 = \frac{\alpha}{K \cdot (1 - e^{-\alpha T_0})} \quad (12)$$

$$q_1 = -e^{-\alpha \cdot T_0} \cdot q_0 \qquad p_1 = 0 \quad (13)$$

$$q_2 = 0 \qquad\qquad p_2 = 1 \quad (14)$$

The resulting controller can be interpreted as two parts. The first part sets a voltage on the inductor $\bar{u}_L^*$ for one control step. After the step the error should be removed. The second part compensates the conduction losses by integrating the given errors to gain the inductor current and calculating a compensation voltage $\bar{u}_R^*$. The sum of both components is the output $\bar{u}_{L+R}^*$ that is fed into the voltage modulator to calculate the duty cycles.

3) Fine Tuning: Using R_L and L of the inductor to calculate q_0 and q_1 will result in a stable control as shown in Fig. 7. However the parameters depend on production tolerances and parasitic effects (e.g. skin effect or semiconductor losses). To improve the dynamic control the control parameters can be tuned by analyzing the result of the step responses of the system.

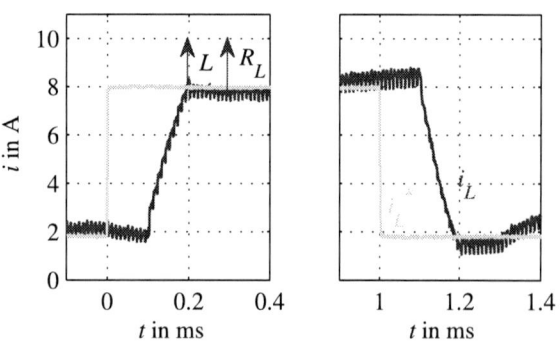

Fig. 7. Step-up and step-down response of CCM current control

A self tuning algorithm for first-order dead-beat control was presented in [13]. For a second-order control, the algorithm

can be adapted. The error after the first control step correlates essentially to the parameter L and the error after the second step correlates to the parameter R_L as indicated in Fig. 7.

4) Voltage Modulation: The voltage modulation in CCM calculates the duty cycle for both half-bridges from a given voltage $\bar{u}_{L+R}^*$ and the measured input and output voltages u_{batt} and u_{dc}. Losses of the semiconductor devices and the resistance of the inductor are neglected and will be compensated by the closed-loop control. A derivation of the duty cycles can be found in [14].

The selected switching strategy maximizes the turn-on time of the two top switches S_1 and S_3. This strategy minimizes the mean inductor current per transfered power and therefore reduces the conduction losses. Three operation modes are used: The buck mode is used when the battery voltage is higher than the dc-grid voltage. The half bridge consisting of S_1 and S_2 is PWM controlled and S_3 is continuously on. The topology behaves like a synchronous buck converter. The boost mode is used when the battery voltage is lower than the dc-grid voltage. S_1 is continuously on and the half bridge of S_3 and S_4 is PWM controlled. The topology behaves like a synchronous boost converter. The equations for the duty cycles in both modes are given in TABLE I.

MOSFETs have a finite switching speed and therefore the behavior of the topology at duty cycles close to one is strongly nonlinear. To avoid critical operation of the MOSFETs a third operation mode is used at nearly equal battery and dc-link voltages. In this mode both half bridges are PWM controlled with one of both half bridges at a maximum duty cycle of $d_{max} = 95\%$. The limitation of the duty cycle ensures linear behavior of the topology. Two cases are possible. In the first case the half-bridge of the battery side has a variable duty cycle and $d_{S,3} = d_{max}$. In the second case the duty cycle of the dc-link half-bridge is variable and $d_{S,1} = d_{max}$. The duty cycles over the complete voltage range are given in TABLE I and depicted in Fig. 8.

TABLE I
VOLTAGE MODULATION IN CCM

Mode	$d_{S,1}$	$d_{S,3}$
boost	1	$\frac{u_{batt} - \bar{u}_{L+R}^*}{u_{dc}}$
boost+buck	0.95	$\frac{0.95 \cdot u_{batt} - \bar{u}_{L+R}^*}{u_{dc}}$
buck+boost	$\frac{0.95 \cdot u_{dc} + \bar{u}_{L+R}^*}{u_{batt}}$	0.95
buck	$\frac{u_{dc} + \bar{u}_{L+R}^*}{u_{batt}}$	1

The decoupling of the input and output voltages has the advantage that the dead-beat controller is independent of the voltages. However the direct feed through of the measured voltages makes the system sensitive to errors in the voltage measurements. The undersampled control of the current has a leverage effect on the sensitivity of the system. Errors in scale and offset of sensors are compensated by the integral component of the DB controller and therefore only have an effect on the dynamic behavior. Noise on the measured values

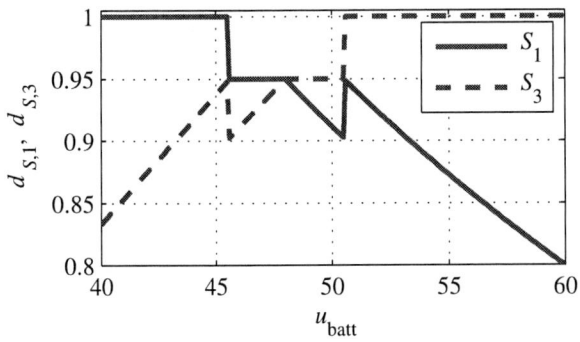

Fig. 8. Duty cycle of both half bridges ($u_{dc} = 48\,$V, $\bar{u}_{L+R}^* = 0\,$V)

cannot be compensated by the controller and directly leads to an error in the current

$$\epsilon_{i,L,\mathrm{nois}} = (\epsilon_{u,\mathrm{batt}} + \epsilon_{u,\mathrm{dc}}) \cdot {}^1\!/_L \cdot T_c \qquad (15)$$

with $\epsilon_{u,\mathrm{batt}}$ and $\epsilon_{u,\mathrm{dc}}$ being the error values of the voltage measurements due to noise.

5) Quantized PWM and Limit Cycling: The effect of an undersampled control is not limited to the dynamic response of the system, but it also effects the quality of the steady state.

The quantization of the PWM resolution leads to a quantized voltage $\bar{u}_{L+R}$ and therefore the current $\bar{i}_L$ has limit cycling as shown in Fig. 9. For a controller with ideal parameters, the error in steady state current $\epsilon_{i,L}$ is given by

$$\epsilon_{i,L,\mathrm{lc}} = {}^1\!/_2 \cdot \max(u_{\mathrm{batt}}, u_{\mathrm{dc}}) \cdot {}^1\!/_L \cdot {}^1\!/_{N_{\mathrm{PWM}}} \cdot a \cdot T_c \quad (16)$$

where N_{PWM} is the resolution of PWM unit per switching period T_s.

One solution to reduce the limit cycling is to increase the effective PWM resolution by dithering the duty cycle in each switching cycle as presented in [15]. However, this solution could not be realized on the given hardware. Therefore the limit cycling of the investigated system is $\epsilon_{i,L} = 325\,$mA.

C. Open Loop Control for Discontinuous Conduction Mode

S_3 is constantly turned off at negative battery temperatures to ensure a positive battery current i_{batt}. At low currents a discontinuous current is flowing through the inductor and the above presented control method fails due to the nonlinear behavior of the system. Thus for DCM-operation an adaption of the trapezoidal modulation strategy presented in [10] is used.

The current I_0 was introduced in [10], which represent the negative current value at the moment S_3 is turned of. With S_3 constantly turned off I_0 is forced to 0 A. Furthermore the turn-on moment of S_4 is moved to t_0 of the following cycle and t_3 is limited to $95\,\% \cdot T_p$. These modifications ensure that the inductor current falls back to zero and no unwanted CCM operation can occur. The resulting modulation strategy is shown in Fig. 10.

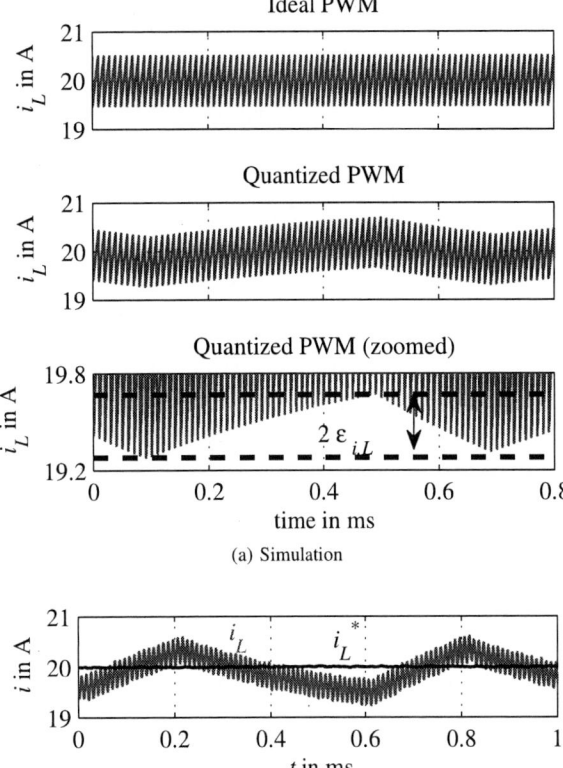

(a) Simulation

(b) Meassurement

Fig. 9. Comparison of ideal PWM and quantized PWM in steady state

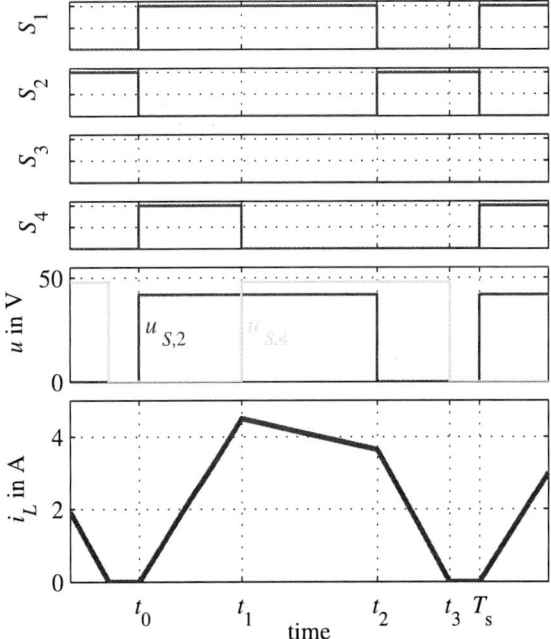

Fig. 10. Modulation strategy for DCM

The switching times t_1 and t_2 can be calculated for a transferred power $\bar{p}^*$ by solving the equations given in [10] considering the restriction for I_0 and t_3. The number of equations can be reduced by mirroring the battery and dc-link side respecting the direction of the power flow:

$$u_1 = \begin{cases} u_{\text{batt}}, & \text{if } \bar{p}^* > 0 \\ u_{\text{dc}}, & \text{otherwise} \end{cases} \tag{17}$$

$$u_2 = \begin{cases} u_{\text{dc}}, & \text{if } \bar{p}^* > 0 \\ u_{\text{batt}}, & \text{otherwise} \end{cases} \tag{18}$$

$$Q^* = 2 \cdot L \cdot |\bar{p}^*| \cdot T_{\text{s}} \tag{19}$$

$$q = 95\,\% \tag{20}$$

For a low power $\bar{p}^*$, a triangular current waveform occurs. Depending on the ratio between the voltage on both sides the following sets of equations has to be selected.

$\mathbf{u_2 > u_1}$ $\qquad\qquad$ $\mathbf{u_2 < u_1}$

$$t_1 = \sqrt{\frac{Q^* (u_2 - u_1)}{u_2\, u_1^2}} \qquad t_1 = 0 \tag{21}$$

$$t_2 = t_1 \left(1 + \frac{u_1}{u_2 - u_1}\right) \qquad t_2 = \sqrt{\frac{Q^*}{u_1^2 - u_1 u_2}} \tag{22}$$

$$t_3 = t_2 \qquad\qquad t_3 = t_2 \cdot \frac{u_1}{u_2} \tag{23}$$

The following set of equations has to be used if the triangular current does not fall back to zero during the switching period ($t_3 > q\,T_{\text{s}}$) or for equal voltages on both sides. In these cases, the trapezoidal current waveform has to be used.

$$t_3 = q\,T_{\text{s}} \tag{24}$$

$$t_1 = t_3 - \frac{u_1}{u_2 \left(\sqrt{u_1}\, u_2^2 + u_1^{\frac{3}{2}}\, u_2 + u_1^{\frac{5}{2}}\right)} \cdot$$

$$\left(\sqrt{u_2 \left(q^2\, T_{\text{s}}^2\, u_1^2\, u_2^2 - Q^* \left(u_1^2 + u_2^2 + u_1\, u_2\right)\right)}\right.$$

$$\left. + q\,T_{\text{s}} \sqrt{u_1}\, u_2^2 + q\,T_{\text{s}}\, u_1^{\frac{3}{2}}\, u_2\right) \tag{25}$$

$$t_2 = (t_3 - t_1) \cdot u_2/u_1 \tag{26}$$

With t_1 and t_2 the duty cycles and the phase shift between the two half bridges are given by

$\mathbf{p^* > 0}$ $\qquad\qquad$ $\mathbf{p^* < 0}$

$$d_{\text{batt}} = t_2 \cdot f_{\text{s}} \qquad d_{\text{batt}} = 1 - t_1 \cdot f_{\text{s}} \tag{27}$$

$$d_{\text{dc}} = 1 - t_1 \cdot f_{\text{s}} \qquad d_{\text{dc}} = t_2 \cdot f_{\text{s}} \tag{28}$$

$$\varphi = t_1 \cdot f_{\text{s}} \qquad \varphi = 1 - t_1 \cdot f_{\text{s}} \tag{29}$$

The resulting waveform for the trapezoidal current waveform is shown in Fig. 11(a).

The drop of the current i_L to negative value I_0 occurs due to the discharge of the parasitic drain-source capacitors of S_3 and S_4. The stored energy is used during the dead-time between S_2 and S_1 to charge the drain-source capacitors of S_1 and S_2. Therefore zero-voltage switching is achieved at all switches and the negative current has no impact on the

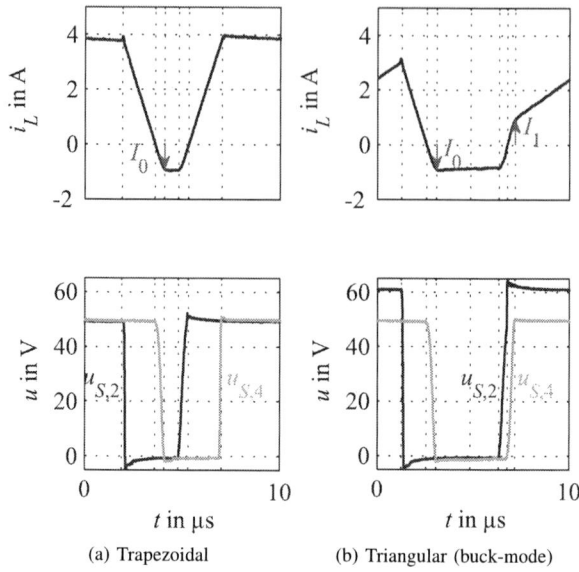

(a) Trapezoidal $\qquad$ (b) Triangular (buck-mode)

Fig. 11. Measurement of inductor current waveform in trapezoidal DCM operation

transferred power. During the period where negative current is flowing either S_2 is closed or the current is charging the drain-source capacitors of S_1 and S_2. Therefore the negative current does not flow through the battery.

The DCM control is used as open loop control. In Fig. 12 the error between the reference power p^* and the transferred power on the battery side is shown. The losses increases the transfered power slightly above the desired value. However, the transfered power strongly differ from the desired value in triangular buck mode as the dark area in Fig. 12 indicates. The measured current waveform of this operation point is shown in Fig. 11(b). The current I_1 indicates a boost of the current i_L during the charging of the drain-source capacitors of the dc-link half-bridge ($u_{S,4}$). The boosted current lead to a higher mean current over the switching period while the calculation in (21) assumes a immediate change of $u_{S,4}$ and $I_1 = 0$.

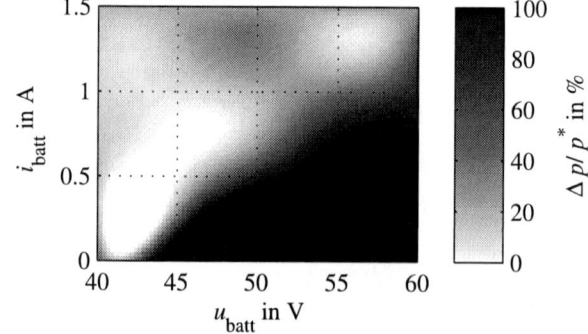

Fig. 12. Error between reverence power p^* and battery discharge power

D. Transition between CCM and DCM Operation

The advantage of the trapezoidal current waveform is that the same power can be transfered using DCM and CCM. This gives a comfortable range to do the transition and also gives a margin for a hysteresis for switching between the two control structures. The boundaries are indicated in Fig. 13 for a fixed dc-link voltage of 48 V.

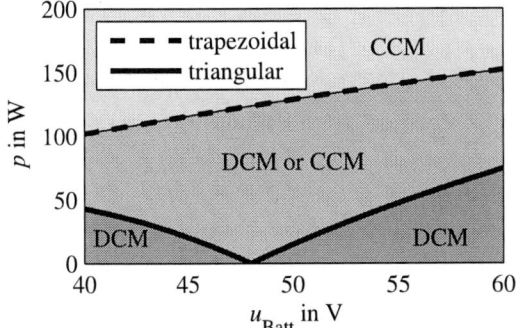

Fig. 13. Operation boarders of triangular and trapezoidal DCM and triangular CCM current waveform for $U_{dc} = 48$ V

The switching patterns of the CCM and DCM operation are different and the behavior of the PWM unit has to be well understood if a direct transition is done. Fortunately, joining or leaving the DCM operation means that the current at the beginning or end of the switching period should be zero and the transition is done at rather low power. Therefore, one switching period is used where all four switches are turned off, before the new switching vector is set in the next switching cycle. The transition from CCM to DCM is shown in Fig. 14. The transition from DCM to CCM is equivalent.

In Fig. 15 a measured step responses for switching between DCM and CCM are shown. The difference between reverence and actual current at 200 µs results from the difference between the measured current and the current at the beginning of the slope. The measured inductor current at the sample points at 0 s and 100 µs are approximately 1 A and the DB controller sees and therefore compensates an error of 5 A, while the actual error is 6 A.

E. Control of the Battery Current

The battery management system (BMS) provides current limits namely the maximum discharge current $I^*_{\text{Batt,max}}$ and the maximum charge current $I^*_{\text{Batt,min}}$ (negative value) to protect the battery from critical states. Small transient violations of the boundaries are permitted due to the high time constant of the battery, but in steady state the boundaries have to be met.

The battery-current control is implemented as a second loop around the inductor-current control as indicated in Fig. 4. The battery-current control consist of one PI controller for the lower and one for the upper boundary. The two PI controllers are limited by the boundaries of the battery and 0 A as shown

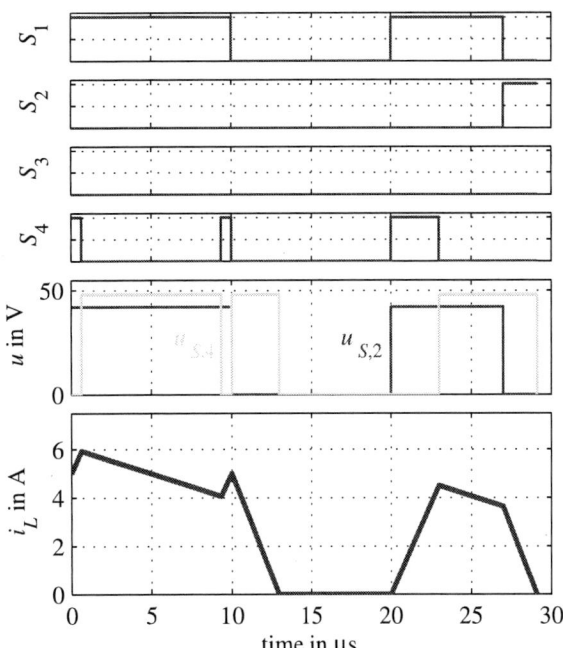

Fig. 14. Transition from CCM to DCM

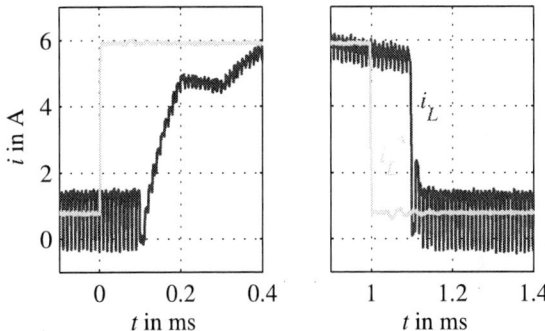

Fig. 15. Step-up and step-down response between CCM and DCM

in Fig. 16. While the battery current is inside the boundaries one integrators is limited by the upper and the other one by the lower boundary of the battery current. If the battery current exceeds one boundary, the corresponding PI controller will limit the reference inductor current $\bar{i}^*_L$.

The effect of this control is that a change of $\bar{i}^*_{dc}$ is at first limited by the idealized relation between $\bar{i}_L$ and $\bar{i}_{\text{Batt}}$, given by the voltage ratio u_{batt}/u_{dc} and the duty cycles d_{Batt} and d_{dc}. In the second step the losses are compensated.

The power difference between the command from the voltage control and the reference value for the transfered power is forwarded to the chopper in case of negative power. This enable a voltage control even with a fully charged battery.

The reaction of the current limitation is shown in Fig. 17. In this measurement the fuel cell supplies a constant power to the dc-link. After the reduction of the load current i_{load},

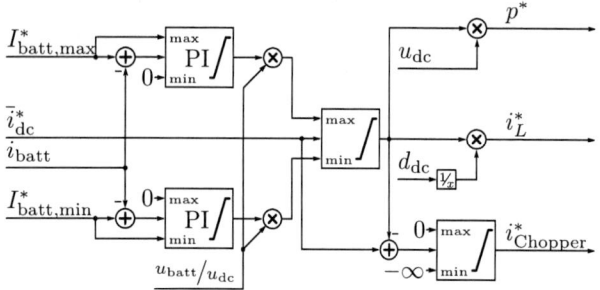

Fig. 16. Apply current limits of battery

the dc-link voltage u_{dc} increases and thus the voltage control increases the charging current of the battery. When the battery current i_{batt} crosses the boundary $I^*_{batt,min}$, the chopper is activated. After 1 ms the battery current is controlled to its given limit.

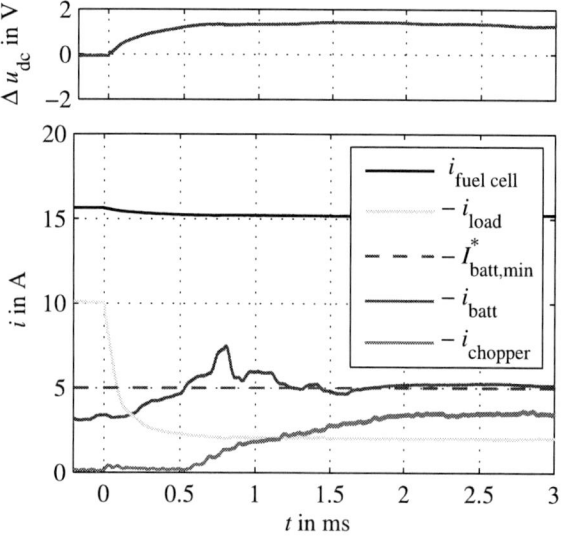

Fig. 17. Limitation of battery current after load step

III. CONCLUSION

In this paper the design of a current control for a bidirectional cascaded buck+boost converter has been shown. The control frequency is ten times slower than the switching frequency. A predictive method has been used to correct the sampled current to the corresponding value at the end of the control period. The design and fine tuning of a second-order dead-beat controller for the CCM has been shown as well as the voltage modulation to compensate the input and output voltages. The result of the current control has been presented with the step-response of the inductor current. Furthermore the limits of the steady state error of the control due to the limited PWM resolution have been discussed. At low temperatures the

battery has to be protected from charging currents, therefore a DCM control has been introduced, to ensure positive currents by hardware. The transition between between DCM and CCM control has been discussed and shown by measurements.

ACKNOWLEDGMENT

This work was kindly financed by the German Federal Ministry of Education and Research within the project EURHOPE as part of the CLIENT program.

REFERENCES

[1] Peter M. Curtis. "UPS Systems: Applications and Maintenance With an Overview of Green Technologies". In: *Maintaining Mission Critical Systems in a 24/7 Environment*. John Wiley & Sons, Inc., 2011, 223–264. ISBN: 9781118041642. URL: http://dx.doi.org/10.1002/9781118041642.ch10.

[2] Alexis Kwasinski. "Effects of Hurricanes Isaac and Sandy on Data and Communications Power Infrastructure". English. In: Hamburg: VDE, Oct. 2013.

[3] Scott Spink and Sandra Saathoff. "Superstorm Sandy: Fuel Cell Design for Disaster vs. Backup Power". English. In: Hamburg: VDE, Oct. 2013.

[4] L. Schindele, M. Braun, and H. Spath. "The influence of power electronic dynamics on PEM fuel cell-system". In: *Power Electronics and Applications, 2005 European Conference on*. 2005, 9 pp. –P.9. DOI: 10.1109/EPE.2005.219540.

[5] Markus Lelie et al. "Design of a Battery System for a Fuel Cell Powered UPS ApplicationWith Extreme Temperature Conditions". In: *Telecommunications Energy Conference 'Smart Power and Efficiency' (INTELEC), Proceedings of 2013 35th International*. Oct. 2013, pp. 1–6.

[6] R.M. Schupbach and J.C. Balda. "Comparing DC-DC converters for power management in hybrid electric vehicles". In: *Electric Machines and Drives Conference, 2003. IEMDC'03. IEEE International*. Vol. 3. 2003, 1369–1374 vol.3. DOI: 10.1109/IEMDC.2003.1210630.

[7] S. Hamasaki, R. Mukai, and M. Tsuji. "Control of power leveling unit with super capacitor using bidirectional buck/boost DC/DC converter". In: *Renewable Energy Research and Applications (ICRERA), 2012 International Conference on*. 2012, pp. 1–6. DOI: 10.1109/ICRERA.2012.6477316.

[8] I. Aharon, A. Kuperman, and D. Shmilovitz. "Analysis of bi-directional buck-boost converter for energy storage applications". In: *Industrial Electronics Society, IECON 2013 - 39th Annual Conference of the IEEE*. Nov. 2013, pp. 858–863. DOI: 10.1109/IECON.2013.6699246.

[9] Na Su et al. "Study of bi-directional buck-boost converter with different control methods". In: *Vehicle Power and Propulsion Conference, 2008. VPPC '08. IEEE*. 2008, pp. 1–5. DOI: 10.1109/VPPC.2008.4677528.

[10] S. Waffler and J.W. Kolar. "A Novel Low-Loss Modulation Strategy for High-Power Bidirectional Buck+Boost Converters". In: *IEEE Transactions on Power Electronics* 24.6 (June 2009), pp. 1589 –1599. ISSN: 0885-8993. DOI: 10.1109/TPEL.2009.2015881.

[11] Eliahu I. Jury. "Analysis and synthesis of sampled-data control systems". In: *American Institute of Electrical Engineers, Part I: Communication and Electronics, Transactions of the* 73.4 (Sept. 1954), pp. 332–346. ISSN: 0097-2452. DOI: 10.1109/TCE.1954.6372164.

[12] Rolf Isermann. *Digitale Regelsysteme*. German. 2nd ed. Vol. 1. Berlin: Springer Verlag, 1988. ISBN: 3-540-16596-7.

[13] W. Stefanutti et al. "Digital Deadbeat Control Tuning for dc-dc Converters Using Error Correlation". In: *Power Electronics Specialists Conference, 2006. PESC '06. 37th IEEE*. 2006, pp. 1–6. DOI: 10.1109/PESC.2006.1712130.

[14] Xiaoyong Ren et al. "Three-Mode Dual-Frequency Two-Edge Modulation Scheme for Four-Switch Buck-Boost Converter". In: *Power Electronics, IEEE Transactions on* 24.2 (Feb. 2009), pp. 499–509. ISSN: 0885-8993. DOI: 10.1109/TPEL.2008.2005578.

[15] A.V. Peterchev and S.R. Sanders. "Quantization resolution and limit cycling in digitally controlled PWM converters". In: *Power Electronics, IEEE Transactions on* 18.1 (Jan. 2003), pp. 301–308. ISSN: 0885-8993. DOI: 10.1109/TPEL.2002.807092.

Sub-microsecond Response Digital Controller for POL

Hirotaka Nonaka /Nagasaki University
Graduate School of Engineering
Nagasaki Univ.
Nagasaki,Japan
bb35210144@cc.nagasaki-u.ac.jp

Yoichi Ishizuka /Nagasaki University
Graduate School of Engineering
Nagasaki Univ.
Nagasaki,Japan
isy2@nagasaki-u.ac.jp

Kenji Mii/Nagasaki University
Graduate School of Engineering
Nagasaki Univ.
Nagasaki,Japan

Fumiaki Takenami
Graduate School of Engineering
Nagasaki Univ.
Nagasaki,Japan

Daisuke Kanemoto/University of Yamanashi
Faculty of Engineering
The Univ. of Yamanashi
Yamanashi,Japan
dkanemoto@yamanashi.ac.jp

Abstract—**This paper will discuss about the proposed hard-ware logic type digital controller for on-board SMPS which has a very small time-delay in control loop. Some experimental has been done including estimation of the load current change experiment and the frequency characteristic of open loop transfer function. These result reveal the proposed circuit could be suppressed the time delay to sub-microsecond order. To use the multi-phase system reduces the output ripple.**

Keywords—DPWM,Point of Load(POL),Digital Control ,Multi-Fhase

I. INTRODUCTION

Digital electronic products have been spreading quickly by the advancement of the integrations technologies. ICs, DSPs and FPGAs require a high performance and a high speed due to the trend. Along with the situations, the power consumption is increasing. To suppress the power consumption, the power supply voltage is getting lower toward to sub 2V. Figure 1 shows the relation between size of LSI and margin of V_{DD}. The trend of future size of LSI and margin of V_{DD} are going to become lower and more severe. Because of the severe voltage margin by the lower power supply voltages, special SMPS, point of load (POL) is disposed very near to the load. The requirements of the control circuit of POL are high accurate, high speed, adaptive and low cost. For the control purpose, pulse width modulation (PWM) control is a one of appropriate technique. Digital control or DPWM can accomplish robust and flexible power control with soft-tuned parameters and will become popular control technique. Although, there are some disadvantages in cost and speed, against analog control circuit. Proposed hardware-logic based digital PWM control circuit is effective to such requirements.

In this paper, trend and problems of DPWM controller for POLs is introduced at first. At the next, the proposed DPWM control method's principles of operation and circuit configurations are described. At last, the effectiveness of the proposed technique is confirmed with some experiments and comparison between prior works and prototype proposed circuit is introduced.

II. DPWM CONTROL METHOD FOR POL

A. Current State of DPWM Controller for POL

Before describing the proposed digital control system for POL, trend of digital controllers for POLs are summarized. The trend is categorized with some keywords, which are treated in these papers [2]-[21] that treat DPWM control, in Fig. 2.

We have found that most of the paper treats the transient response performance, small size and power consumption. In contrast, the cost has not been discussed so much. But this point cannot be ignored because the cost is also important in the real electronic products.

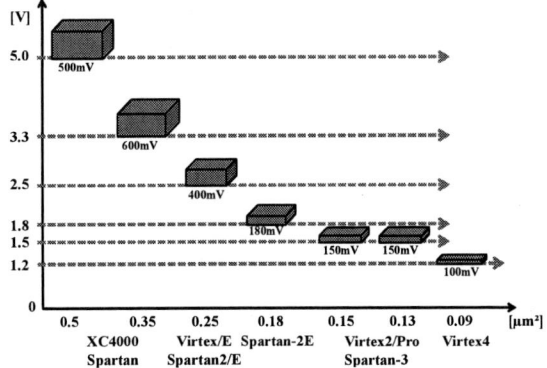

Fig. 1 The relation between size of LSI and margin of V_{DD}

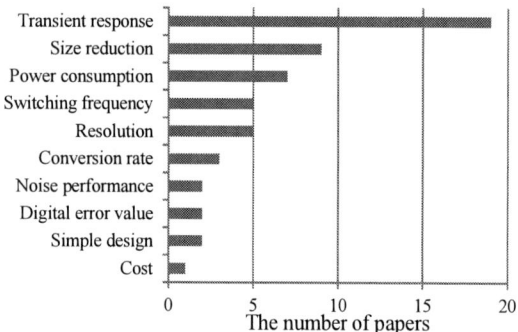

Fig.2 Key point of DPWM controller and number of papers

Therefore, designing good DPWM controller, is taking balance of all of the factors including the cost. .

In the next section, the problems about the restrict factor of high-speed response of DPWM control are described.

B. The Circuit Configuration of General DPWM Controller for POL

The circuit configuration of general DPWM controlled POL is shown in Fig.3(a). This topology has two major time-delay problem. First, time-delay occurs at A/D converter with the conversion-delay. And, the calculation time of digital controller is another problem. Both of the time-delay directly effects on the response speed of the control circuit and influences stability of the control. Total of the delay time will be described as the discrete delay factor e^{-Ls} in the control loop shown in Fig.3(b). In general, total of the delay time is 600nsec at least excepting the delay time of driver circuit.

C. The Circuit Configratin of Proposed DPWM Controller for POL

Main POL circuit is a quite ordinary non-isolated buck converter. The control circuit is composed with D/A converter, analog comparator, digital controller and drive circuit.

The analog timing converter (ATC) block shown in Fig.4(a) is composed with D/A converter and analog comparator. In Fig.4(b), the control block diagrams are shown. ATC block, PID control calculation block and up-counter block for gate pulse creation, are all in parallel and synchronized with the system clock f_{CLK}.

D. The control circuit configuration of the proposed DPWM-POL

Figure 5 shows the precise control circuit configuration, and Fig. 6 shows control signal flow of the proposed DPWM-POL. Let's take a look at the signal flow.

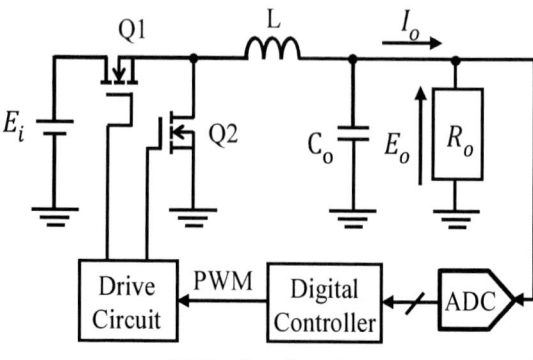

(a) Circuit configuration

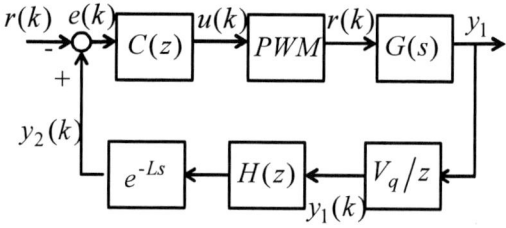

(b) Control block

Fig.3 General DPWM-POL

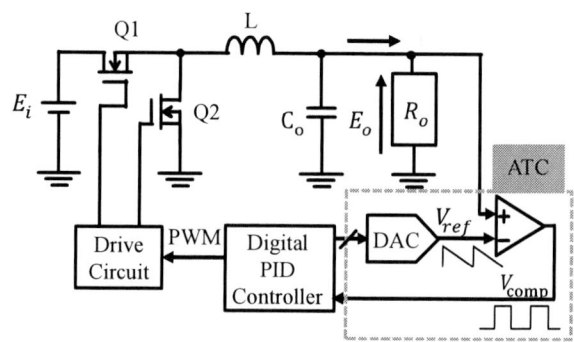

(a) Proposed DPWM-POL

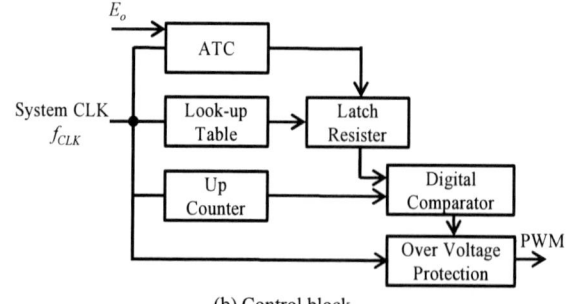

(b) Control block

Fig.4 Circuit configuration of proposed DPWM-POL

978-1-4799-2706-7/14 $31.00 © 2014 IEEE

As mentioned above, all blocks are synchronized with the system clock f_{CLK}. Memory 1 is used the look-up table method, and can store waveform values not only triangle or saw tooth but any waveforms. In this paper, the step-down saw tooth wave form is employed. V_{ref}^{+} is the maximum output voltage of DAC. The output voltage E_o of POL is compared with the output voltage of D/A converter in the ATC block, successively. The comparator's output is read out to the latch signal to each D-ff at the timing that E_o was sensed. Also, the look-up table method is used for the duty ratio calculation with memory 2, 3 and 4. Especially, the duty ratio data which is pre-calculated with the value of E_o, are stored in memory2.

The duty ratio data is read out from memory2 according to f_{CLK}. At the timing of E_o sensed, one of the duty ratio data is chosen and transferred to $u(k)$ in D-ff4, where k is the number of switching term.

Because of the small delay of this control technique, the sensed E_o data can reflect on-term of the same switching term.

At the digital comparator, $u(k)$ is compared with up-counter data, and converted to real-time analog PWM waveforms. On-term $T_{on}(k)$ of DPWM signal of switching term k is decided by $u(k)$. In parallel with the processing of ATC block, the $u(k)$ is called with system clock and latched by ATC output as trigger. At the last of the switching term, $u(k)$ is preset maximum value by PR signal generator for preparing next term.

The delay of the proposed control circuit is mostly dominated with the calling and the loading time of memory 2.

E. PID control with Look-up Table

$u(k)$ which is stored in memories is pre-calculated by general PID digital control laws as

$$u(k)=u_{ref}+K_P e(k)+K_I n_I(k)+K_D(e(k)-e(k-1)) \quad (1)$$

where u_{ref} is a reference value of $u(k)$, $e(k)$ is an digitalized error value between r which is digitalized reference voltage V_{ref} and, $y_1(k)$ is output data of up-counter in switching term k as

$$e(k) = y_1(k) - r \quad (2)$$

K_P, K_I and K_D are a proportional gain, an integral gain and an derivative gain, respectively, $n_I(k)$ is integral factor, that is

$$n_I(k)=n_I(k-1)+e(k) \quad (3)$$

At the timing of the latch signal is becoming high, $y_1(k-1)$ is latched to $y_2(k-1)$ as

$$y_2(k-1)=y_1(k-1) \quad (4)$$

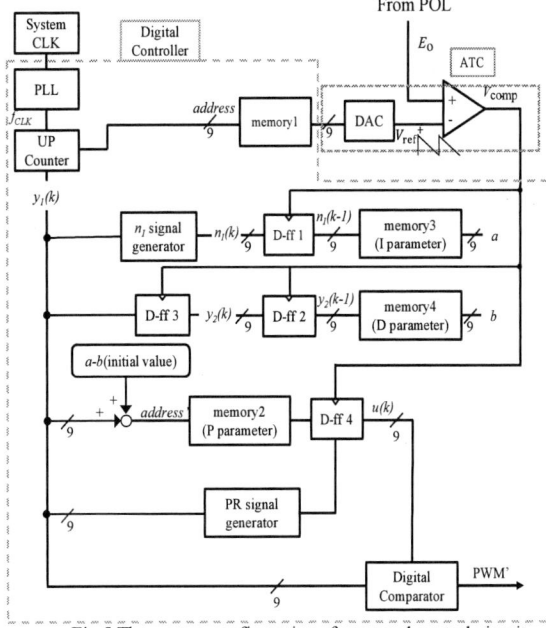

Fig.5 The system configuration of proposed control circuit

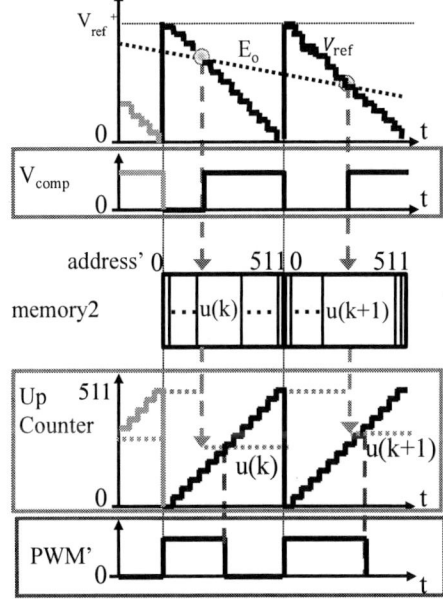

Fig.6 Control signal flow

TABLE I Address and data of memory2 at $K_P = 5, K_I = 0$

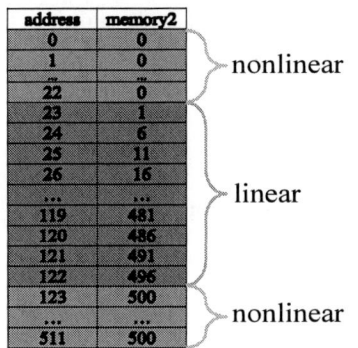

address	memory2	
0	0	nonlinear
1	0	
...	...	
22	0	
23	1	
24	6	
25	11	
26	16	
...	...	linear
119	481	
120	486	
121	491	
122	496	
123	500	
...	...	nonlinear
511	500	

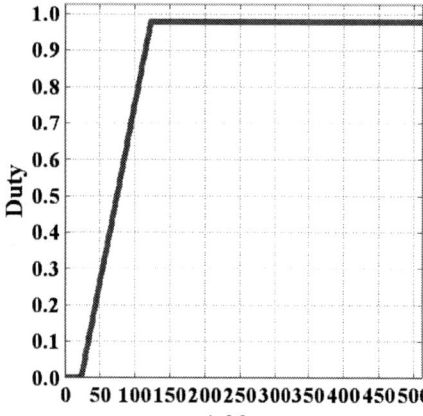

Fig.7 Address vs. duty ratio at Table I

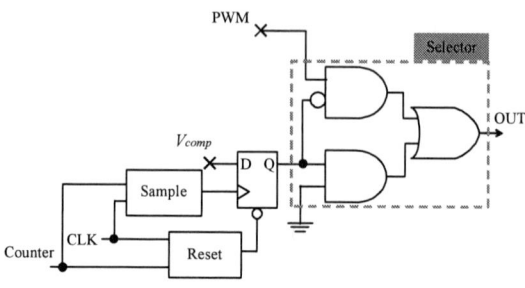

Fig.8 The system configuration of overvoltage protection logic circuit

From above equations, (1) can be transformed to

$$u(k) = u_{ref} - (K_P + K_I)r + A\{y_1(k) + \frac{K_I}{A}n_I(k-1) - \frac{K_D}{A}y_2(k-1)\} \quad (5)$$

Where

$$A = K_P + K_I + K_D \qquad (6)$$

$$a = \frac{K_I}{A}n_I(k-1) \qquad (7)$$

$$b = \frac{K_D}{A}y_2(k-1) \qquad (8)$$

Memory 3 and memory 4 store *a* and *b*, respectively.

In (5), $a - b$ in the term *k* is pre-calculated in the term *k-1* and the obtained value becomes the initial value of programmable counter of the term *k*. And, *address'* which indicates address of memory 2 is incremented with system clock and *u(k)* is called from memory 2, simultaneously.

$$address' = y_1(k) + a - b \qquad (9)$$

From (5) and (9),

$$u(k) = u_{ref} - (K_P + K_I)r + A\{address'\} \qquad (10)$$

Therefore, *u(k)* is determined as soon as E_o is sensed.

Table I shows the address and data of memory2 at $K_P = 5, K_I = 0$, u_{ref} =86, r= 40 in (10). Figure 6 shows address and duty ratio at Table I. You can see, Table I data have linear and nonlinear domain. And, if K_P becomes more higher, linear domain becomes shorter.

F. Overvoltage protection logic circuit

If trigger does not occur, proposed circuit outputs overvoltage. Therefore, we use overvoltage protection logic circuit.

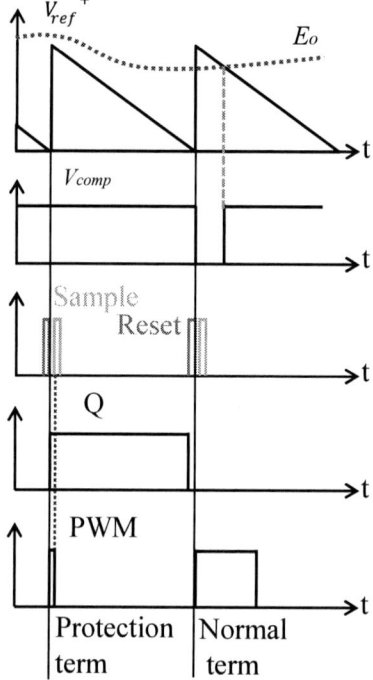

Fig.9 Protection signal flow

Table II   Truth table of selector of protection logic circuit

Q	OUT
0	PWM
1	Gnd

Figure 8 shows overvoltage protection circuit for proposed circuit. It is composed with selector, D-ff, sample pulse generator and reset pulse generator for D-ff.

Figure 9 shows protection signal flow, and Table II shows truth table of the selector. First, sample pulse generator occurs pulse, second, value of Q is decided by sample pulse and V_{comp}, at last OUT is decided by Table II. When sensed output voltage $E_o > V_{ref}^+$, PWM signal becoming off by force.

978-1-4799-2706-7/14 $31.00 © 2014 IEEE   2740

G. The proposed circuit with Multi-phase method

Figure 10 shows the main and the proposed control circuit configuration using a multi-phase method. The circuit is composed of five-phase.

There are some benefits to use multi-phase converter. For example, multi-phase converter is able to supply stable output current for increasing of load current because of reducing of current per phase than single-phase and reduction of ripple current. By some factors mentioned above, transient response is improved. Therefore, it has become an effective method in recently. However, there are some disadvantages that increasing of cost and circuit scale, difficulty of control by increasing of the switching element than single-phase. Therefore, appropriate design is needed to compensate above disadvantages. In proposed control system, multi-phase converter is designed easily.

Figure.11 shows how to adopt the multi-phase method to the proposed circuit. The signal flow of multi-phase system block is shown in Fig.12. These figures show how to recognize system block of second-phase.

Timing of the falling edge of PWM2 signal is determined by that of PWM signal which is first-phase. First, the timing of the rising edge of the PWM2 signal is determined by the block of Start in Fig.11. Duty ratio of PWM2 is created by adding two values of counter which are the timing of PWM falling edge and PWM2 rising edge. Block of Const creates constant value which is same value of PWM2 rising. In this time, it is 102. Q of D-ff I and output of block of Const are added by Adder block. At block of Phase, output of Adder and value of counter are compared. And the just timing of the value of counter becomes equal to output of Adder, 0 is output form Phase block. Output of Phase block can reset output of D-ff II and determine the falling edge of PWM2 signal.

III. EXPERIMENTAL RESULTS

A. Specifications

The proposed control system with prototype circuit is shown in Fig.13. The digital controller part is designed with FPGA Altera Cyclone IV. Texas Instruments DAC900 is used as DAC. Linear Technology LT1719 is used as analog comparator. FDMF6705V is used as MOSFEET and driver. The DC-DC converter topology is basically same as buck converter in Fig.4(a). The buck converter with proposed controller was verified with the experimental conditions are shown in Table III. The experimental conditions with multi-phase are shown in Table IV.

B. Experimental set-up

Some experiments are performed to verify the proposed controller. Figure.14 shows experimental set-up of load current change dynamic response. We measured output voltage E_o, V_{comp}, PWM signal and output current I_o with analog probe, V_{comp} and $u(k)$ in Fig.5 with digital probe. Current change slew rate is 50A/μs.

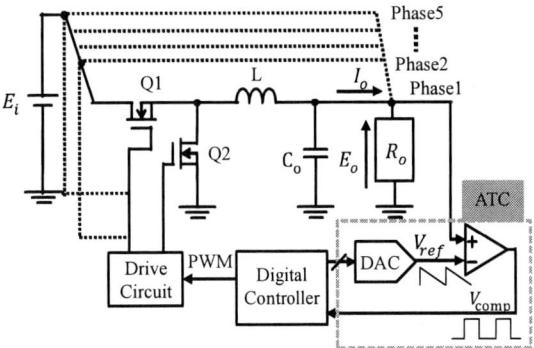

Fig.10   Proposed circuit with multi-phase

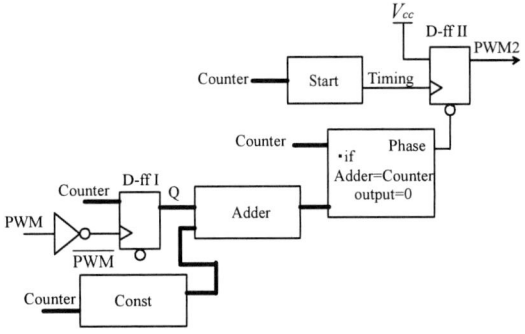

Fig.11 The system configuration of logic circuit with multi-phase method

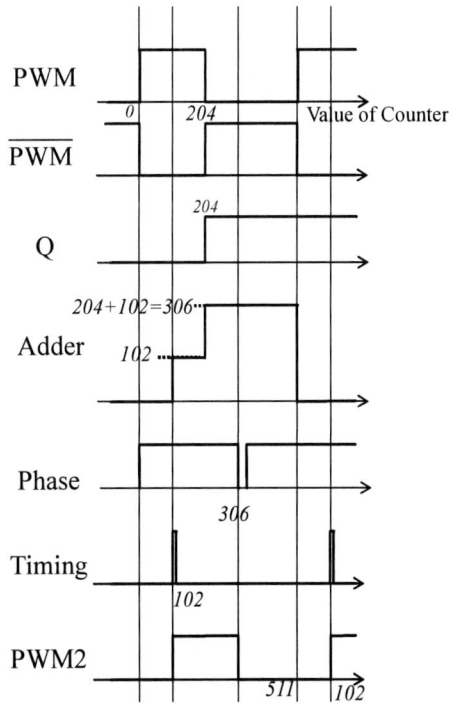

Fig.12   Signal flow of proposed circuit with multi-phase

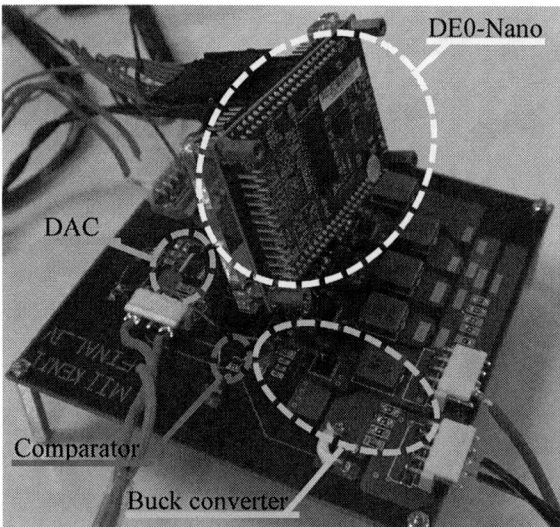

Fig.13 Prototype circuit

TABLE III

Parameters / Components	Value / name
Input voltage E_i	12V
Output target voltage	1.5V
Output current I_o per phase	0.3A ~ 0.9A
Switching frequency f_s	1MHz
Choke inductor L	3.3μH
Output capacitor C_o	10μF(single)/57μF(five)
Proportional gain K_P	10
Integral gain K_I	0
Differential gain K_D	0
V_{ref}^+	1.6
f_{CLK}	500MHz
Current change slew rate	50A/μs
Digital PWM resolution	9bit
Dr. MOS (MOSFET and driver)	Fairchild FDMF6705V
Analog comparator	Linear Technology LT1719
D/A converter	Texas Instruments DAC900
FPGA	Terasic DE0-Nano (Cyclone IV)

C. Measurement Results

1) Static characteristics

Figure 15 shows static characteristic at KP=10. It shows when load current with single-phase is 0.3~0.9, output voltage is 3% of target voltage.

2) Load current change transient response

Figure 15 and 16 show experimental waveforms of load current change transient response. Both figures show E_o: 100mV/div(CH1), PWM: 3V/div(CH2), V_{comp}: 3V/div(CH3), I_o: 1A/div (CH4), V_{comp} (digital):D9, $u(k)$:D0-D8, respectively. Bottom figure is an enlargement of a part of top figure and it shows reflection time. Figure 16 and 17 show 0.3~0.9A load transient response at K_P=10. Figure 16 shows light to heavy load transient response and output voltage settled in 10.4μs, under shoot is 160mV. Figure 17 shows heavy to light load transient response and output voltage settled in 3.2μs, over shoot is 58mV. From these figures, the reflection time from sensed V_{comp} to $u(k)$ change is 11.0ns even in the worst case.

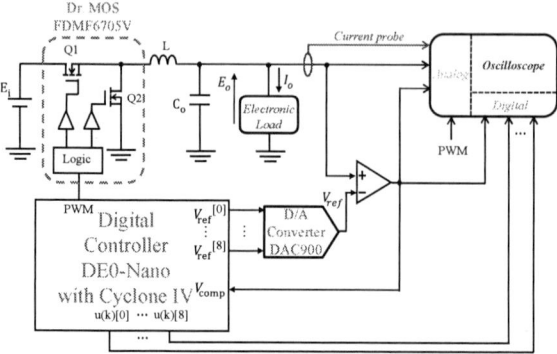

Fig. 14 Experimental set-up ststic characteristic and load current change transient response

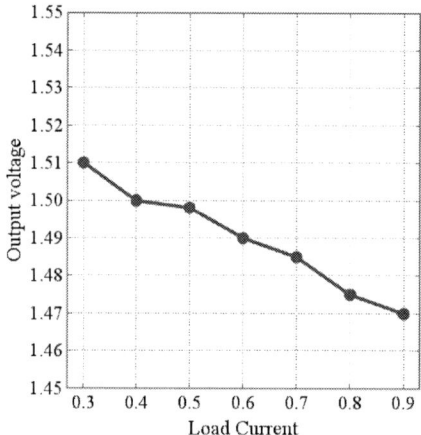

Fig. 15 Static characteristic of converter at K_P=10

Table IV is comparing the settling time of the output voltage from the proposed digitally controlled switching converter with prior work. It shows that the proposed digital controller achieves faster output settling time than existing digital controller.

1) Load current change transient response with multi-phase method

Fig18 and 19 show experimental waveforms of load current change transient response using multi-phase method. Measurement range used in this experiment is the same as single-phase. They are 0.3~0.9A load transient response at K_P=10. Figure 18 shows light to heavy load transient response and output voltage settled in 15.6μs, under shoot is 190mV. Figure 19 shows heavy to light load transient response and output voltage settled in 9.0μs, over shoot is 170mV.

978-1-4799-2706-7/14 $31.00 © 2014 IEEE

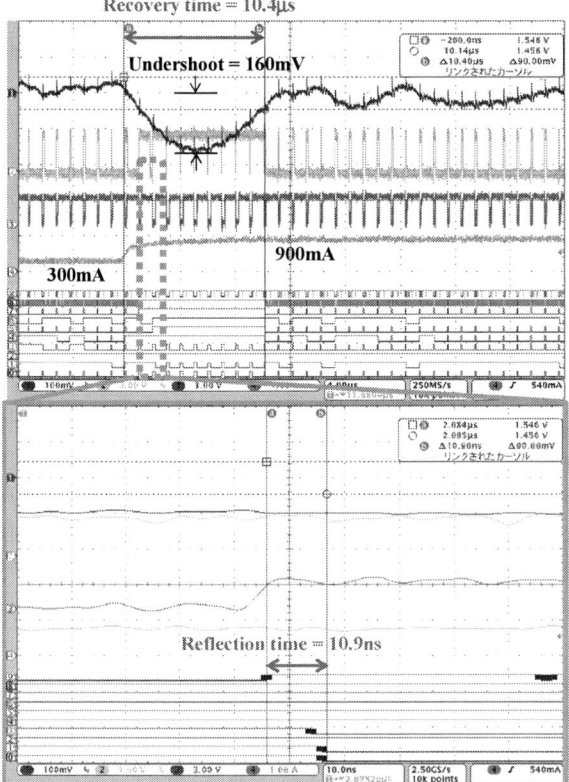

Fig. 16 Load change from 0.3 to 0.9A (K_P=10)

TABLE IV COMPARISON WITH PRIOR WORKS.

	[18]	[19]	[20]	[21]	This work
L	10µH	4.7µH	4.7µH	4.7µH	3.3µH
C	10µH	4.7µH	4.7µH	22µF	10µF
Switching frequency	1M	4M	1M	780k	1M
Load Current step	0.45A	0.16A	0.6A	0.59A	0.6A
Input voltage	1.8 -3.8V	3.3V	3V	3.6V	12V
Output target voltage	1.2V	1.5V	1.8V	1.2V	1.5V
Settling time	3.5µs	18µs	15.5µs	60µs	10.4µs

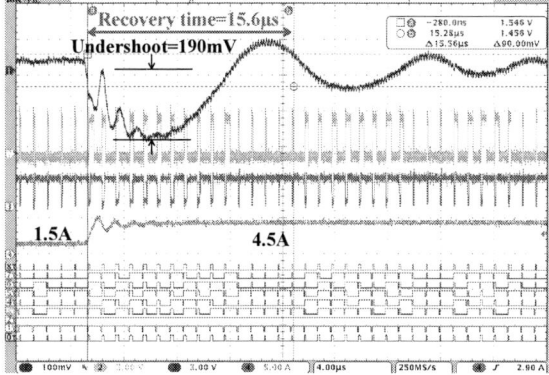

Fig18 Load change with multi-phase 1.5A to 4.5A (K_P=10)

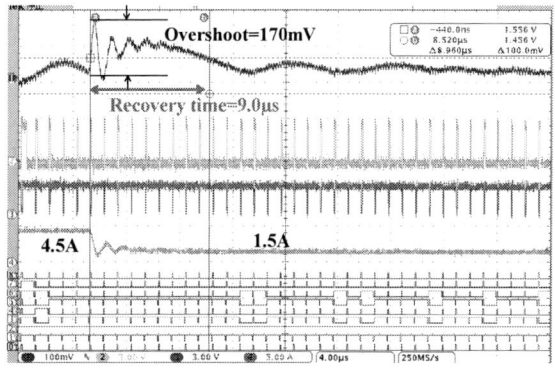

Fig.19 Load change with multi-phase 4.5A to 1.5A (K_P=10)

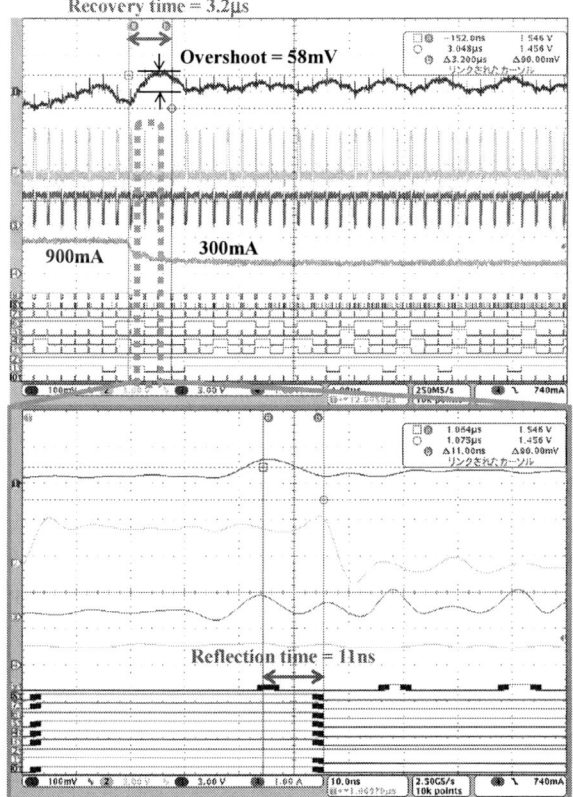

Fig. 17 Load change from 0.9 to 0.3A (K_P=10)

IV. CONCLUSIONS

In this paper, the hardware logic type digital controller for on-board SMPS, which has a very small time-delay in control loop, is confirmed with some experiments including frequency characteristic. In single-phase, settling time of proposed prototype circuit is 10.4 µs, and undershoot is 160 mV. This result with the specifications of table III is superior in DPWM-POLs. And the reflection time from sensed V_{comp} to $u(k)$ change is 11.0ns even in the worst case. The total reflection time of control system except driver circuit is in 50n sec. Therefore, it is confirmed that the proposed circuit is able to respond sub-microsecond. In addition, multi-phase proposed circuit achieved reduction of the output ripple. Settling time is 15.6μs and undershoot is 190mV. From these results, it is confirmed that the proposed circuit can increase output current stably by using multi-phase. As the future work, to further enhance the PID control added to the integral control and to devise new control system, such as can be achieved faster.

REFERENCES

[1] Masatugu Tasaki. The Intranet Architecture:. "Importance of the power supply circuit in the latest FPGA board," Bellnix Corp., Japan. [Online]. Available: http://japan.xilinx.com/xcell/xl51/jp-xcell51_13.pdf

[2] Radic, A., Lukic, Z., Prodic, A. and de Nie, R.H., "Minimum-Deviation Digital Controller IC for DC–DC Switch-Mode Power Supplies," *IEEE Transactions on*, vol.28, Issue.9, Sept 2013.

[3] Yue Wen, and Trescases, O., "DC-DC converter with digital adaptive slope control in auxiliary phase to achieve optimal transient response," *Applied Power Electronics Conference and Exposition, IEEE*, April 2014.

[4] Corradini, L., Bjeletic, A., Zane, R. and Maksimovic, D., "Fully digital hysteretic modulator for DC-DC switching converters," *Energy Conversion Congress and Exposition* IEEE,Sept 2009

[5] Wangxin Huang, Abu Qahouq, J. A. "Tuning of a digital proportional-integral compensator for DC-DC power converter," *Applied Power Electronics Conference and Exposition IEEE,*March 2013.

[6] Huerta,S.C, Alou, P., Garcia, O., and Oliver, J.A., "Hysteretic Mixed-Signal Controller for High-Frequency DC-DC Converters Operating at Constant Switching Frequency," *Power Electronics, IEEE Transactions on* Vol.27, Issue. 6 June 2012.

[7] Yingyi Yan, Lee, F.C., and Mattavelli, P., "Comparison of Small Signal Characteristics in Current Mode Control Schemes for Point-of-Load Buck Converter Applications," *Power Electronics, IEEE Transactions on* Vol.28 , Issue, 7 July 2013

[8] Ahmad, H., and Bakkaloglu, B., "A digitally controlled DC-DC buck converter using frequency domain ADCs," *Applied Power Electronics Conference and Exposition, IEEE* Feb. 2010

[9] Barai, M., SenGupta, S., and Biswas, J., "Dual-Mode Multiple-Band Digital Controller for High-Frequency DC−DC Converter," *Power Electronics, IEEE Transactions on* , Vol24, Issue3 March 2009

[10] Babazadeh, A., Corradini, L., and Maksimovic, D., "Near time-optimal transient response in DC-DC buck converters taking into account the inductor current limit," *Energy Conversion Congress and Exposition, IEEE* Sept. 2009

[11] Ahmad, H., and Bakkaloglu, B., "A DC-DC digitally controlled buck regulator utilizing multi-bit $\sum\Delta$. frequency discriminators," *Applied Power Electronics Conference and Exposition, IEEE,* Feb. 2008

[12] Jakobsen, L.T., Schneider, H., and Andersen, M.A.E., "Comparison of state-of-the-art digital control and analogue control for high bandwidth point of load converters," *Applied Power Electronics Conference and Exposition, IEEE* Feb. 2008

[13] Jing Wang, Prodic, A., and Wai Tung Ng, "Flyback transformer based transient suppression method for digitally controlled buck converters," Energy Conversion Congress and *Exposition,IEEE,* Sept. 2011

[14] Yingyi Yan, Lee, F.C., and Mattavelli, P., "Dynamic performance comparison of current mode control schemes for Point-of-Load Buck converter application," *Applied Power Electronics Conference and Exposition, IEEE,* Feb. 2012

[15] Shi-Quan Fan, Li Geng, and Sheng-Lei Wang, "An algorithmic ADC applied in digital controlled switched DC-DC converters," *Solid-State and Integrated Circuit Technology, IEEE International Conference on,* Nov. 2010

[16] Ng, W.T., Wang, J., Ng, K., Prodić, A., Kawashima, T., Sasaki, M., and Nishio, H., "Digitally controlled integrated DC-DC converters with fast transient response," *Radio-Frequency Integration Technology,IEEE International Symposium on,* Jan. 9 2009-Dec. 11 2009

[17] Abbas, G., Sturtzer, E., and Abouchi, N., "Design and implementation of a PWM-based digital controller for a high-frequency dc-dc buck converter working in CCM using classical control techniques," *NEWCAS Conference, IEEE International on* , June 2010

[18] Huey Chain Foong, Yuanijin Zheng, Yen Kheng Tan, and Meng Tong Tan "Fast-Transient Integrated Digital DC-DC Converter With Predictive and Feedforward Control," *IEEE Transactions on* Vol.59,Issue,7,July 2012

[19] Shuibao Guo, Yanxia Gao, Yanping Xu, and Xuefang "Digital PWM controller for high-frequency low-power DC-DC switching mode power supply," *Power Electronics and Motion Control Conference, 2009. IPEMC '09. IEEE 6th International*, May 2009

[20] Lee A.T.L, and Chan,P.C.H, "Adaptive prediction in digitally controlled buck converter with fast load transient response," *IEEE 13th Workshop on* June 2012

[21] Chien-Hung Tsai, Chun-Hung Yang, Jiunn-Hung Shiau, and Bo-Ting Yeh, " Digitally Controlled Switching Converter With Automatic Multimode Switching," *Power Electronics, IEEE Transactions on.*Vol.29,Issue.4,April 2014

[22] Yoichi Ishizuka, Masao Ueno, Ichiro Nishikawa, Akira Ichinose and Hirofumi Matsuo, "A Low-Delay Digital PWM Control Circuit for DC-DC Converters", in Proc. 2008 IEEE Applied Power Electronics Conference (APEC), CA, USA, 2008, pp.579–586 .

[23] Y. Ishizuka, F. Hirose, Y. Yamada, H. Matsuo, "A time-delay suppression technique for DPWM control circuit", in Proc. 2009 IEEE Telecommunications Energy Conference (INTELEC), pp.1–6.

[24] Y. Ishizuka, Y. Yamada, F. Hirose, M. Nishi, H. Matsuo, "A design for small time-delay control circuit for DPWM- POL" in Proc. 2010 IEEE Applied Power Electronics Conference and Exposition (APEC), pp. 1879–1884.

[25] Y. Ishizuka, F. Hirose, Y. Yamada, H. Matsuo, "A fast transient response technique for DPWM DC-DC converters", in Proc. 2010 IEEE Telecommunications Energy Conference (INTELEC), pp. 1–6.

[26] Yoichi Ishizuka, Kenji Mii, Daisuke Kanemoto, Tamotsu Ninomiya, "Frequency Response Analysis of Proposed Digital Control System for DPWM-POL", in Proc. 2013 IEEE Power Electronics and Applications European Conference (EPE), pp. 1–10.

The 2014 International Power Electronics Conference

Gain controlled high efficiency power factor correction circuit

Yu Yonezawa, Hiroshi Nakao, Tomotake Sasaki, Yoshinobu Matsui,Yoshiyasu Nakashima, Junji Kaneko
Hiroshi Shimamori*, Yukio Yoshino*
Hosoyama Hisato**, Manabe Atsushi**
Shun Motizuki***, Shigeharu Yamashita***

Fujitsu Laboratories Ltd., * Fujitsu Ltd., **Fujitsu Advanced Technologies Ltd., *** Fujitsu Telecom Networks Ltd.,
Kawasaki, Japan
E-mail: yonezawa.yu@jp.fujitsu.com

Abstract— **The purpose of this study is to improve the efficiency of a boost-type power factor correction (PFC) circuit with a digital controller. One of the biggest losses in PFC circuits is the switching loss. Switching loss can be reduced by low boost ratio operation, but these results in a greater voltage drop at transient response. The voltage drop can be suppressed by using a high gain setting for the voltage control loop, but this lowers the power factor, which creates a harmonic distortion of the AC power line. To improve this trade-off, we developed a gain control method that adjusts the gain to high only at the moment of transient response. This method improves both the efficiency and the power factor of PFC circuits. We implemented the proposed method using a digital controller and demonstrated a power factor improvement of 0.08 points and an efficiency improvement of 0.4 points for a 2.5 kW PFC circuit.**

Keywords—PFC, High efficiency, High power factor, Gain control

I. INTRODUCTION

As the energy consumption of data centers continues to increase, the development of a more efficient energy supply has become an increasingly important issue [1], [2]. In addition, a high-power supply unit (PSU) is required for servers as the servers grow more powerful. We are currently trying to make a high power (over 2 kW) and high efficiency (97%) PSU for servers [3], [4], [5], [6] and are also attempting to improve the PSU efficiency.

In this paper, we propose a new gain control method for improving boost-type power factor correction (PFC) circuit efficiency. One of the biggest losses in PFC boost converters is switching loss. To minimize the switching loss, the boost ratio should be low. PFC efficiency at various output voltages is shown in Fig. 1. A lower output voltage results in a lower switching loss and higher efficiency. However, lower boost ratio operation causes a larger voltage drop at transient response. Figure 2 shows an example of the transient response of a PFC circuit when the output voltage is 375 V, which results in a minimum voltage at the transient response lower than 350 V. The typical minimum input voltage of a back-end DC-DC converter is 350 V or higher because lower input

voltage degrades the efficiency of the converter. In other words, this voltage drop (Fig. 2) is larger than acceptable for our back-end converter. The transient response of the PFC is dependent on the voltage control loop gain, and a higher control gain improves the transient response, as shown in Fig. 3. However, the higher gain results in a lower power factor because the high voltage gain suppresses the current shaping ability of the PFC [7]. Further, the high gain control does not satisfy the 80 Plus Titanium certification, which requires a power factor of more than 0.95 for more than 20% of the load, as shown in Fig. **4**. These issues illustrate the trade-off relationship between transient response and power factor.

In order to solve these problems, we propose a gain control method that adjusts the gain to high only at the moment of transient response. This method is implemented with a digital controller and improves both the efficiency and the power factor of PFC circuits. We demonstrated the effectiveness of the proposed method through experiments and established a design procedure for the gain controller.

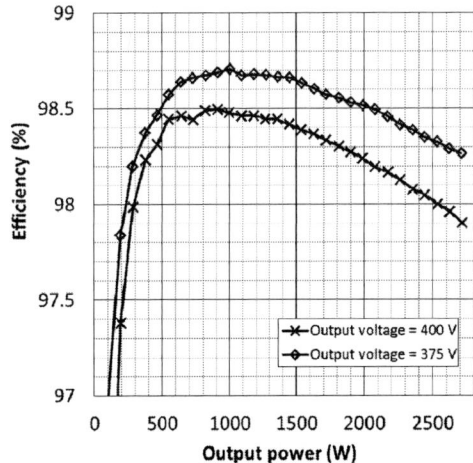

Fig. 1 PFC efficiency at output voltages of 375 and 400. A lower output voltage results in higher efficiency.

978-1-4799-2706-7/14 $31.00 © 2014 IEEE

The 2014 International Power Electronics Conference

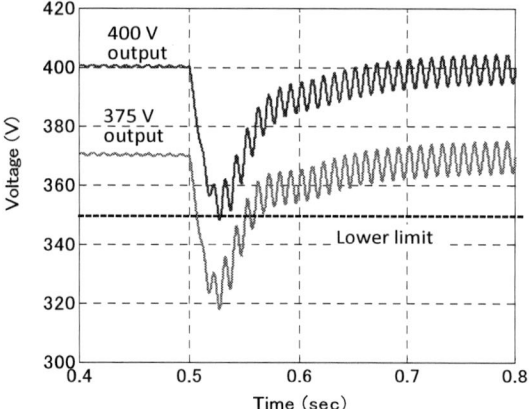

Fig. 2 Transient response for output voltages of 375 and 400. In the case of an output voltage of 365, the minimum output becomes lower than 320 V and is not able to meet the requirements of our DC-DC converter.

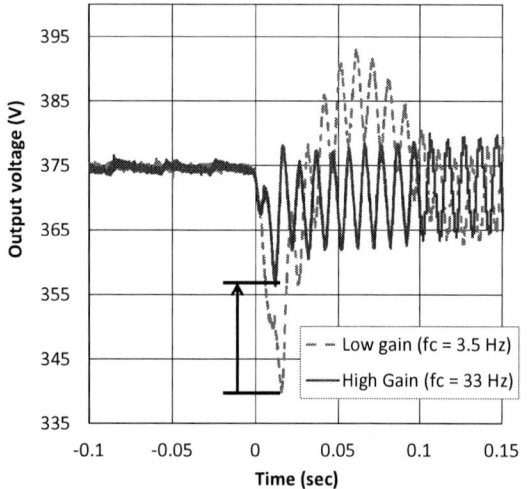

Fig. 3 Transient response for two voltage gains. Higher voltage gain improves transient response.

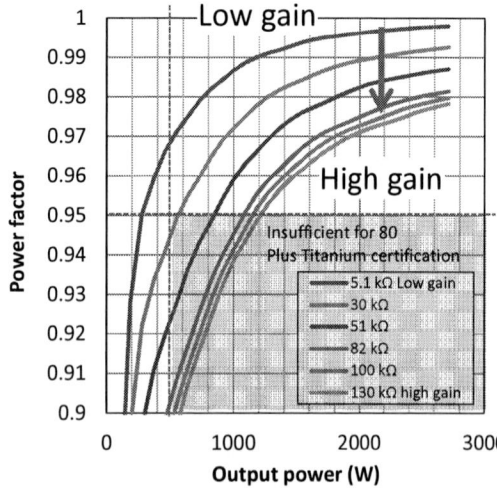

Fig. 4 Power factor for various voltage gains. Higher voltage gain lowers the power factor.

II. PROPOSED GAIN CONTROL METHOD

The control block diagram of the proposed PFC is shown in Fig. **5**. A digital signal processor (DSP) was used for the control. The DSP acquired the output current of a DC-DC converter and controlled the voltage and gain of the PFC. The control sequence diagram is shown in Fig. **6**. The control has three states selected from the values of |di/dt| and of current states. These three states are defined as follows.

a. State I: |di/dt| < threshold value

PFC gain is set to low for high power factor operation and PFC voltage is set to low for high efficiency operation.

b. State II: |di/dt| > threshold value

PFC voltage gain is set to high for suppression of voltage drop. After a certain period (100 ms), the voltage gain is held to high.

c. State III: One-second period after State II

PFC gain is set to low for high power factor operation. PFC voltage is set to high for suppression of voltage drop caused by transient load current. State III continues for one second. In this state, the PFC characteristics are the same as conventional PFCs.

The threshold value of |di/dt| is chosen from the acceptable voltage drop for an output current transient and the output voltage drop of PFC is estimated by the circuit simulation. A control flow chart is shown in Fig. 7. First, the absolute value of the differential of the output current |di/dt| is measured. If |di/dt| is more than the threshold value, the gain of the voltage compensator is set to high. This high gain suppresses the output voltage drop. After a certain period (100 ms), the voltage gain is set to low and the reference voltage is set to high. The high reference voltage continues for a period of 10 times longer than the high gain period (1 s).

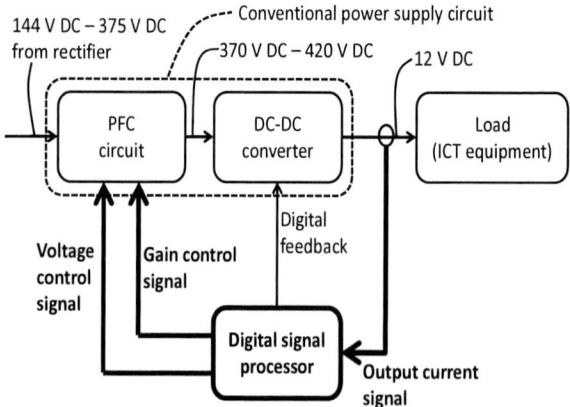

Fig. 5 Control block diagram of proposed PFC.

978-1-4799-2706-7/14 $31.00 © 2014 IEEE

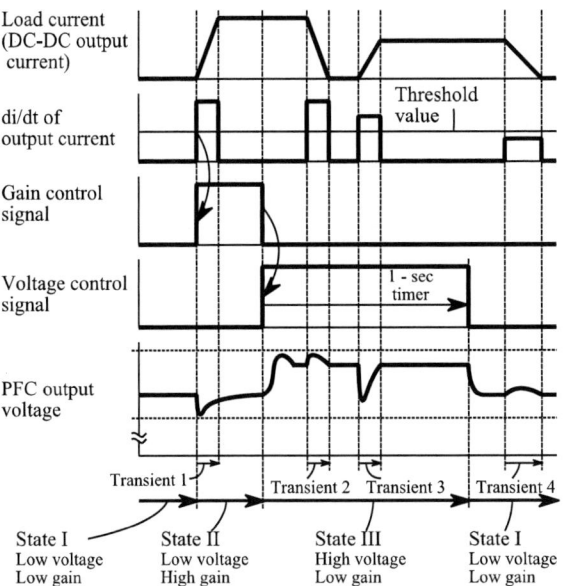

Fig. 6 Control sequence diagram of proposed PFC.

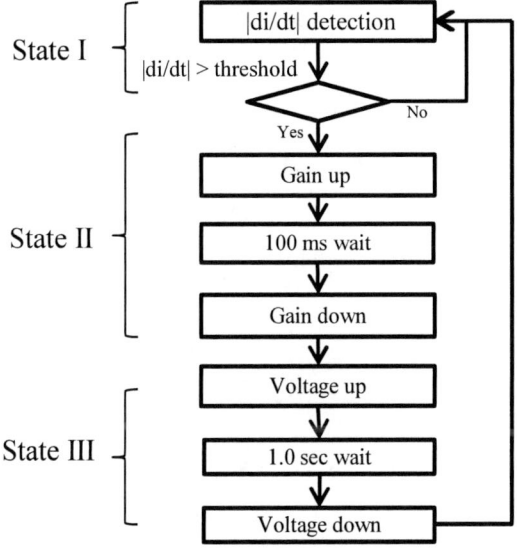

Fig. 7 Control flow chart of gain control method.

III. CONTROL DESIGN

The voltage loop of the boost converter for PFC must have sufficient gain to suppress the voltage drop during the high gain period. A control block diagram of the PFC is shown in Fig. 8. The topology of this circuit is that of a typical current mode PFC. Figure 9 shows the circuit of the voltage compensator. Gain is controlled by a switch connected in parallel to the $R_{ZV}2$. Gain is high at switch open. The relation between the gain crossover frequency and the voltage drop of the PFC, estimated from the circuit simulation, is shown in Fig. 10. Equation (1) is the approximated relation of gain crossover frequency f0 and voltage drop, indicated by the dashed line in Fig. 10.

$$f_0 = \left(\Delta v / 114.2\right)^{-\frac{1}{0.53}} \tag{1}$$

Acceptable voltage drop Δvlim is calculated from the minimum input voltage Vmin of the backend DC-DC converter.

$$\Delta v_{\lim} = V_{OUT} - V_{MIN} - V_{margin} \tag{2}$$
$$= 375 - 350 - 5 = 20\,\mathrm{v}$$

Gain crossover frequency is 27 Hz from Δvlim and equation (1) and the gain characteristics of the PFC boost converter are calculated from the transfer function. Gain block for the voltage loop is defined as shown in Table 1. Table 2 shows the gain of each component for the PFC converter in Fig. 8. Transfer functions of the boost converter are derived by a state space averaging method [8], [9], [10], [11]. Loop gain is expressed as

$$LoopGain = G_{v\det ect} \cdot G_{Vcomp} \cdot G_{multi} \cdot GI_{COMP} \cdot G_{PWM} \cdot$$
$$GV_{BOOST} \cdot \frac{1}{1 + G_{PWM} \cdot GI_{COMP} \cdot GI_{DETECT} \cdot GI_{BOOST}} \tag{3}$$

Rz1 and Rz2 are adjusted to 30 kΩ and 100 kΩ, respectively, from the simulation. C_{ZV} and C_{PV} are designed using a conventional design method [12], [13], [14].

Fig. 11 shows the simulated loop gain, where 10 Hz and 30 Hz gain crossover frequency for low and high gain control are exhibited.

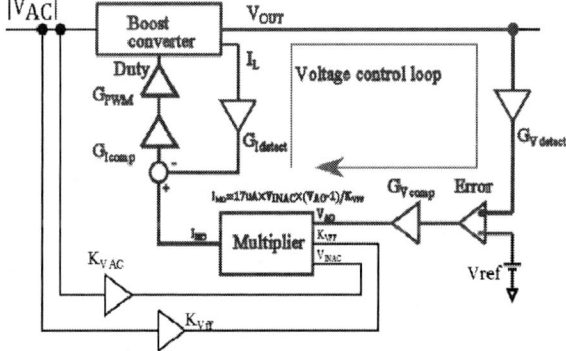

Fig. 8 Control block diagram of PFC.

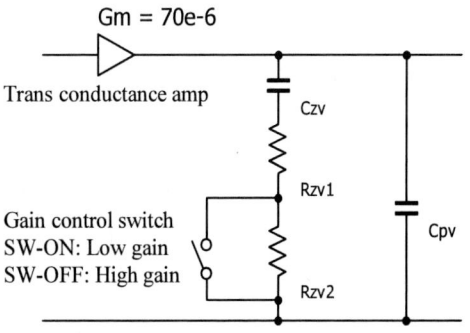

Fig. 9 Voltage compensator circuit.

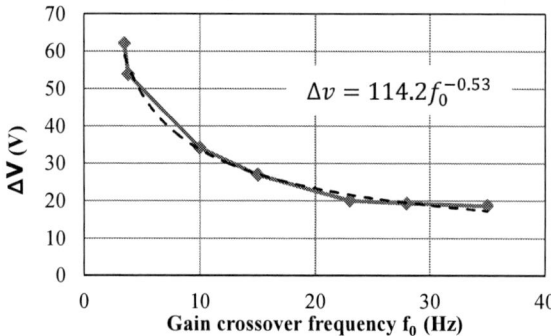

$$\Delta v = 114.2 f_0^{-0.53}$$

Fig. 10 Relation between gain crossover frequency and voltage drop of the PFC.

Table 1 Definition of gain block.

Block	Definition
GV_BOOST	Voltage gain of boost converter for a duty to output voltage.
GI_BOOOST	Current gain of boost converter for a duty to output current.
G_vdetect	Voltage gain of voltage detector.
G_Vcomp	Voltage gain of voltage compensator.
G_MULTI	Voltage gain of multiplier.
G_ICOMP	Voltage gain of current compensator.
G_PWM	Modulation gain of compensator.

Table 2 Transfer function of gain block.

Block	Transfer function
GV_BOOST	$\dfrac{\left(-\dfrac{1}{1-D}\dfrac{1}{CR}s + \dfrac{1}{LC}(1-D) - \dfrac{1}{1-D}\dfrac{r_L}{LCR}\right)V_{ref}}{s^2 + \left(\dfrac{r_L}{L} + \dfrac{1}{CR}\right)s + \dfrac{1}{LCR} + \dfrac{1}{LC}(1-D)^2}$
GI_BOOOST	$\dfrac{\left(\dfrac{1}{L}s + \dfrac{2}{LCR}\right)V_{ref}}{s^2 + \left(\dfrac{r_L}{L} + \dfrac{1}{CR}\right)s + \dfrac{r_L}{LCR} + \dfrac{1}{LC}(1-D)^2}$
G_vdetect	$3/400$
G_Vcomp	$g_{mv} \cdot \dfrac{s(R_{zv1}+R_{zv2})C_{zv} + 1}{s^2(R_{zv1}+R_{zv2})C_{zv}C_{pz} + s(C_{zv} + C_{pz})}$
G_MULTI	$R_{mo} \times 17 \times 10^{-6}\dfrac{V_{INAC}}{k_{VFF}}$
G_ICOMP	$g_m \cdot \dfrac{sR_{zc}C_{zc} + 1}{s^2 R_{zc}C_{zc}C_{pc} + s(C_{zc} + C_{pc})}$
G_PWM	$1/4$

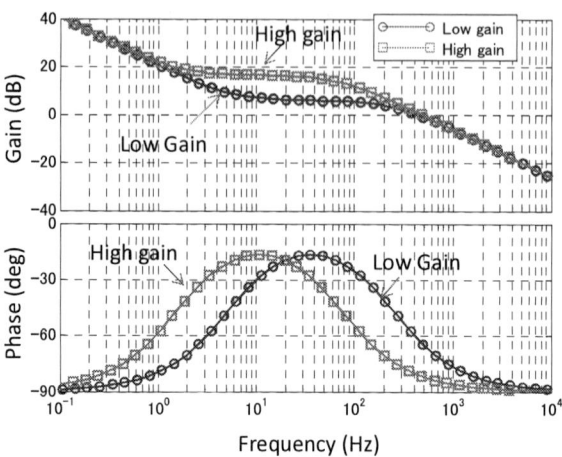

Fig. 11 Gain of voltage loop for high-low gain control operation.

IV. RESULTS AND DISCUSSION

We implemented and examined the designed compensator gain resistance. Transient response with and without controls is shown in Fig. 12. The load current step was 0 to 3.3 A and the time step was 100 μs. By applying this gain-control method, we were able suppress the voltage drop and keep the minimum voltage above 350 V. Voltage up occurred 100 ms after the gain up operation and the output voltage reached more than 400 V. The circuit efficiency was improved 0.4 points at maximum load, as shown in Fig. 13. The power factor was improved 0.08 points at 20 % load, as shown in Fig. 14, which satisfies the 80 Plus Titanium certification.

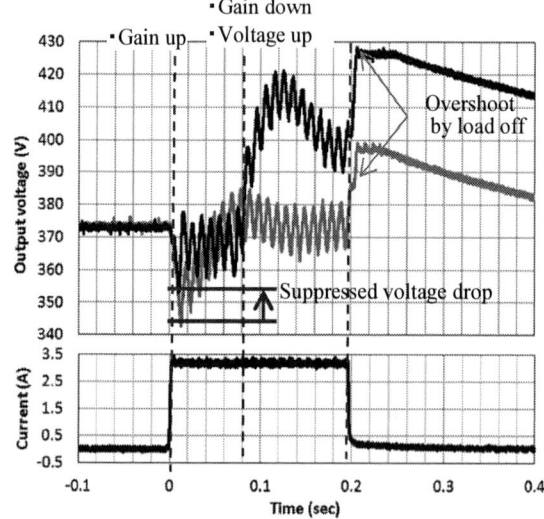

Fig. 12 Transient response of proposed method.

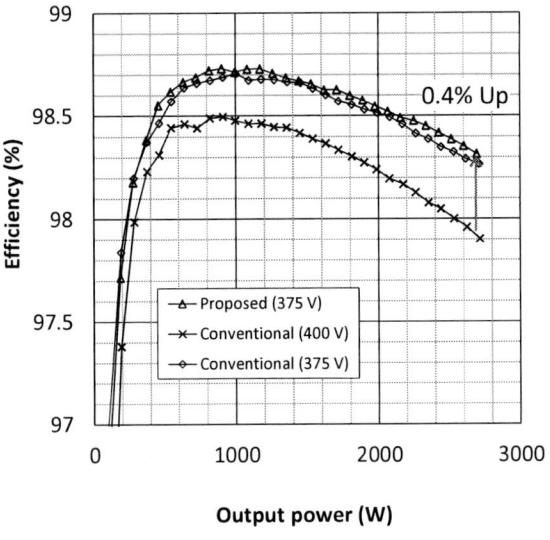

Fig. 13 Efficiency improvement of proposed method.

Fig. 14 Power factor improvement of proposed method.

REFERENCES

[1] U.S. Department of Energy, "Data Center Energy Consumption Trends," [Online]. Available: http:/www1.eere.energy.gov/femp/program/dc_energy_consumption.html.

[2] 80 PLUS titanium, http://www.plugloadsolutions.com/80PlusPowerSupplies.aspx

[3] H. Nakao, et al., "Soft switching for power factor correction for server power supply unit .", Power Electronics, Machines and Drives (PEMD 2012), 2012

[4] H. Nakao, et al., "2.5-kW power supply unit with semi-bridgeless PFC designed for GaN-HEMT.", Applied Power Electronics Conference and Exposition (APEC), 2013

[5] Y. Yonezawa, et al., "Digital dead-time control for two phase double-ended forward converter.", Power Electronics and Drive Systems (PEDS), 2013

[6] Hosoyama, et al., "Dead-Time Controlled Digital-PSU for Server and Automatic Optimization Method of its Dead-Time Table Suitable for Mass-Production", IEICE Technical Report, EE2013-34 (2014-1), In Japanese.

[7] L.H.Dixon, "High Power Factor Preregulators for Off-Line Power Supplies", Texas Instruments application note slup087

[8] F. A. Huliehel, et al., "Small-signal modeling of the single-phase boost high power factor converter with constant frequency control", In IEEE Power Electronics

[9] R. D. Middlebrook, et al., "A general unified approach to modeling switching-converter power stages", Int. J. ELECTRONICS, 1977, VOL.42, NO. 6, 521-550

[10] Wei Tang, et al., "Small-Signal modeling of Average Current-Mode Control" , Applied Power Electronics Conference and Exposition, 1992. APEC '92. Conference Proceedings 1992.

[11] Sasaki S., Watanabe .H, "Analysis of multiple operating points for dynamical control of switching power converters", Technical Report of IEICE, EE2004-71, in Japanese.

[12] L.H.Dixon, "Optimizing the Design of a High Power Factor Switching Preregulator"

[13] "UC3854 Controlled Power Factor Correction Circuit Design", Philip C. TODD, Texas Instruments application note U-134,1999

[14] "UCC28070 Interleaving Continuous Conduction Moe PFC Controller", Texas Instruments application note slus794E, 2007

V. CONCLUSION AND FUTURE WORK

We propose a gain control method that improves both the efficiency and the power factor of a PFC circuit by adjusting the gain to high only at the moment of transient response. The design method for the transfer function of the controller was considered theoretically. We implemented the proposed method using a digital controller and confirmed a power factor improvement of 0.08 points and an efficiency improvement of 0.4 points for a 2.5 kW PFC circuit. The power factor of the circuit also exhibited a sufficient value for the 80 Plus Titanium requirements. Our future work is to verify the reliability of this method for mass production.

The 2014 International Power Electronics Conference

Design of Quasi-Resonant Flyback Converter Control IC with DCM and CCM Operation

Kai-Hui Chen, *Student Member, IEEE*, Tsorng-Juu Liang, *Senior Member, IEEE*

Green Energy Electronics Research Center (GREERC)/ Advanced Optoelectronic Technology Center (AOTC)
Department of Electrical Engineering, National Cheng Kung University, Tainan, Taiwan
Email: tjliang@mail.ncku.edu.tw

Abstract- **A Quasi-resonant (QR) flyback converter control IC is design and implemented. Flyback converter is applied in small to medium power rating applications mostly, due to low cost and simple circuit topology. The conventional fix frequency operated flyback converter is the most adapted power converter, but the switching losses cause lower conversion efficiency. The flyback converter can be operated in BCM and DCM by quasi-resonant control. The quasi-resonant control utilizes the resonant action between magnetize inductor of transformer and parasitic capacitor of power MOSFET to achieve valley voltage turn-on and reduce the turn on losses increasing the conversion efficiency. The switching frequency increases with reducing load, and the switching losses will cause poor efficiency. Flyback converter operated in BCM and DCM in heavy load condition will lead to higher conduction losses. To improve the conversion efficiency at light load and heavy load condition, the proposed control circuit is integrated with frequency clamp and maximum off-time limit functions.**

Keywords: Quasi-resonant, flyback converter, power ICs

I. INTRODUCTION

In recent years, the progress of technology increase the usage of electrical products such as laptop computer, LCD monitor, and cell phone which demand higher efficient power supplies to reduce energy consumption and worse environmental issues. California Energy Commission (CEC)and Energy Star set up the mandatory regulation of average power converter efficiency at 25%, 50%, 75% and 100% load should be greater than 87% with output power above 49 W, and the standby power consumptions at no load should be less than 0.5 W which the output power from 50 to 250 W [1]

The flyback converter is the most applied converter topology in the power rating bellow 250W. Conventional fix frequency controlled flyback converter has the benefit of easy design the simple control, but it is suffer from the switching losses of high voltage stress and hard switching of MOSFET.

Fig. 1 shows the typical waveform of QR flyback converter. The switching frequency increases with reducing load. Switching losses in light load condition is the significant drawback to the efficiency performance. To overcome this issue the highest switching frequency clamping control must be used, and force the converter operated in DCM. In heavy load condition, the peak currents of the power MOSFET and major current path are very high, and it will cause higher conduction losses and lead to poor efficiency. To solve this problem, the flyback converter is operated in CCM with the lowest switching frequency limit control. [2][3]

This paper is organized as follows. In Section II, the operating principle of the QR flyback converter and the converter behavior of the proposed frequency limit control is shown. Section III describe the frequency limitation control behavior and control circuit. Section IV shows the simulation results of the proposed control to verify the feasibility. Section V makes the conclusion.

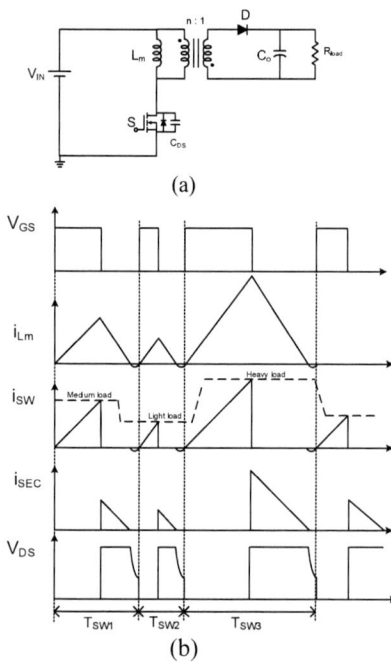

Fig. 1 (a) Flyback converter (b) Switching periods versus different load condition

II. OPERATING PRINCIPLE OF QR FLYBACK CONVERTER

There are three operating mods in one switching period of the QR flyback converter. Fig. 2 shows the current path of the converter during each modes. Fig. 3 shows the key waveforms of the flyback converter with QR control.

To simplify the analysis of the operating principle, some assumptions are made.

1) All components are ideal. Only the magnetizing inductor, L_m, anti-connected parallel diode of S, and the output capacitance of S are take into consideration, because they are the key components of QR flyback

2) The capacitance of output capacitor C_o is large enough, and the voltage across C_o is treated as constant voltage.

3) The leakage inductance is simplified to make the analysis simple.

978-1-4799-2706-7/14 $31.00 © 2014 IEEE

Mode 1: The power switch, S, is turned on. The output diode, D, is reverse biased with voltage and is turned off. The voltage across the output diode as shown in (1)

$$V_D = \frac{1}{n}V_{in} + V_o \qquad (1)$$

The output capacitance C_o provides the energy to load, R_{load}. The magnetizing inductor L_m is charged by Vin, so the current i_{Lm} is increasing linearly in this mode. When power switch, S, is turned off, next mode starts.

Mode 2: The power switch, S, is turned off. The output diode is turned on. The output voltage is reflected to the primary side of the transformer, and the V_{DS} of the power switch is shown in (2)

$$\boldsymbol{V_{DS} = nV_o + V_{in}} \qquad \textbf{(2)}$$

The energy stored in the magnetizing inductance, L_m, is released to the output capacitor, C_o, and load. When the magnetizing current of the transformer, i_{Lm}, reaches zero, this mode ends and goes to mode 3.

Mode 3: The power switch, S, and output diode, D, are turned off in mode 3. As the magnetizing inductor L_m is demagnetized with the output. The energy stored in the output capacitance, C_{DS}, of the power switch, S, start resonating with the magnetizing inductance, L_m. The energy required by the load is provided by the C_o. When V_{DS} reaches the valley voltage, as shown in (3), the power switch, S, is turned on, and next switching cycle starts.

$$V_{valley} = V_{in} - nV_o \qquad (3)$$

As the operating principle revealed above, it is clear to get the relationship between the load condition and switching period, as shown in Fig. 1.

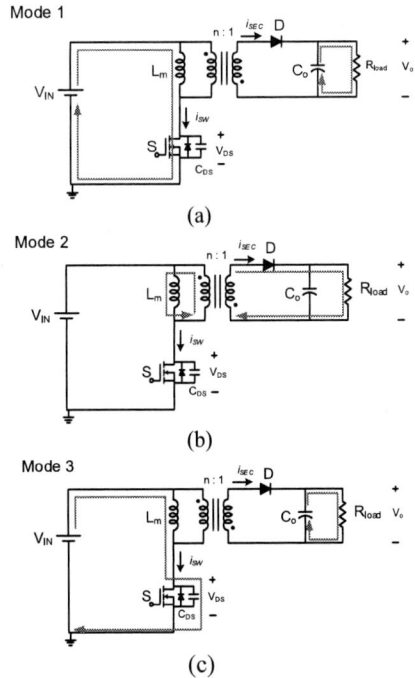

(a)

(b)

(c)

Fig. 2 Current path of the QR flyback converter
(a) mode 1 (b) mode2 (c) mode 3

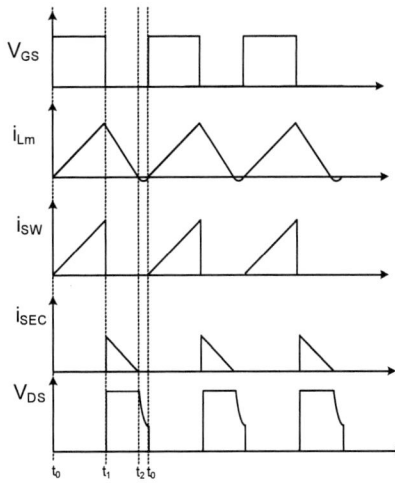

Fig. 3 Key waveforms of QR flyback converter

III. PROPOSED CONTROL CI PROPOSED CONTROL CIRCUIT

The key waveforms of the proposed control are shown in Fig. 4. The converter is operated in BCM, that the power switch turn on at first v_{DS} valley, when that output power is in the middle range of the rated power. The switching frequency increases as the power demanded is reduced. The high switching frequency will cause the lower conversion efficiency during light load operation. To overcome the growing switching frequency, the maximum switching frequency must be clamped. The converter will be operated in DCM and valley turned on in light load condition.

When the output power demand increased, the BCM control with valley switch will cause the current peak goes to high and cause higher conduction losses in heavy load condition. To reduce the peak current of the switch and output diode, the off-time of the power switch is limited. The proposed QR control allows the converter operated in CCM during heavy load.

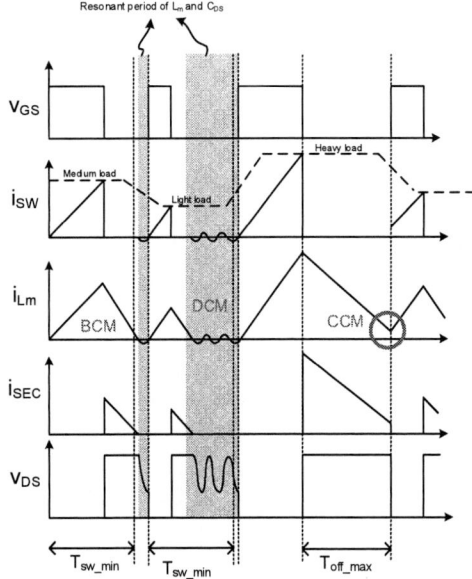

Fig. 4 Key waveforms of the proposed control

978-1-4799-2706-7/14 $31.00 © 2014 IEEE

The proposed control circuit is based on the conventional current-mode control as shown in Fig. 5. The Tigger-on signal is generated from the valley detect circuit, switching frequency clamping circuit, and maximum off-time limit circuit with proposed design, as shown in Fig. 6. The waveforms of the proposed controller in light load and heavy load is described in Fig. 7. The key waveforms of the proposed controller is shown in Fig. 8.

The capacitor C_{FC} in frequency clamping circuit is charged by the 100 uA current source when the power switch is turned on. If the voltage of C_{FC} is less than 3 V, the valley_detect signal will be blind and does not trigger the power switch keeping the power switch off. The minimum switching period can be designed by the equation (4). For example, to get a 10µS T_{s_min} needs a 330pF capacitor.

$$T_{s\_min} = C_{FC} \cdot 2.8 \times 10^4 \qquad (4)$$

The capacitor C_{Toff} is used to count the off-time of the power switch. When the power switch is off ($\overline{Gate}$ is high) capacitor C_{Toff} is charged. As the voltage reached 3 V, the power switch will be forced turned on regardless the valley_detect signal from valley detect circuit. As the power switch turned on, one valley_detect pulse will be generated to reset the capacitor in frequency clamping circuit, C_{FC}. The maximum off-time can be derived from equation (5). For example, to get a 15µS off-time limitation of the power switch needs a 500pF capacitor.

$$T_{s\_min} = C_{FC} \cdot 3 \times 10^4 \qquad (5)$$

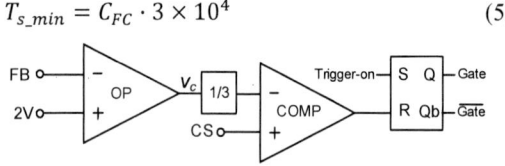

Fig. 5 Conventional current mode control

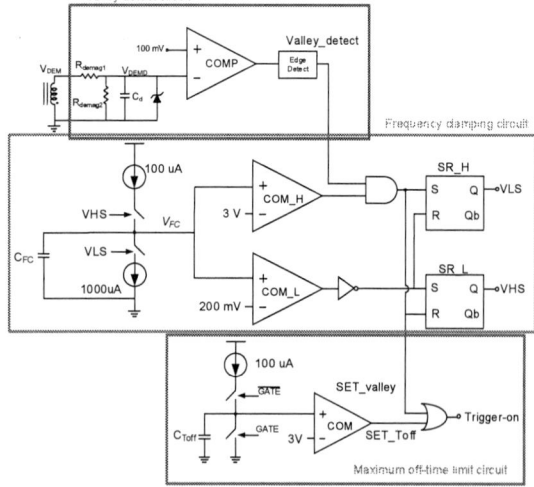

Fig. 6 Proposed frequency clamping and maximum off-time limit circuits

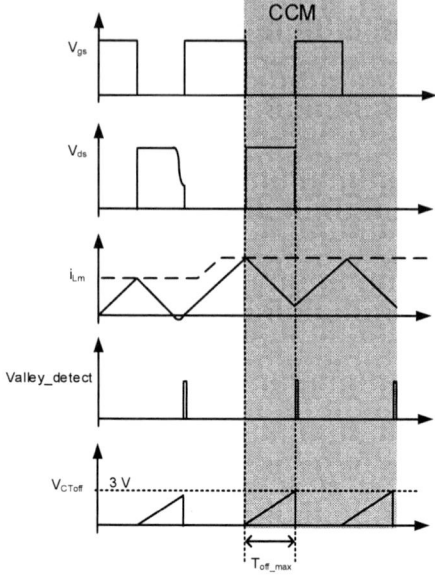

(a)

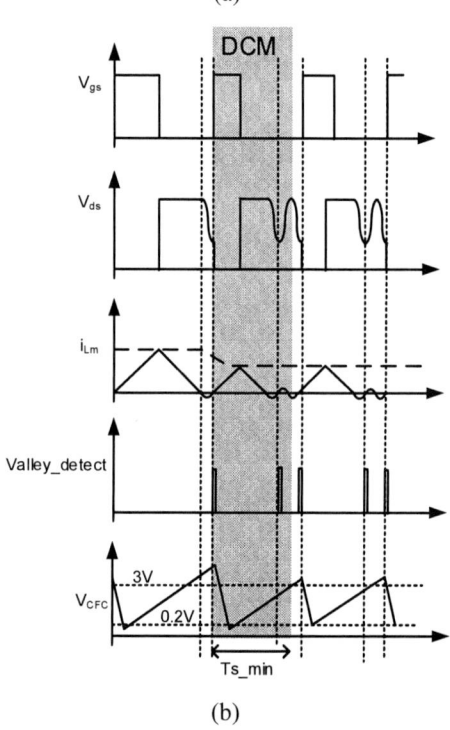

(b)

Fig. 7 Key waveforms of the proposed control(a) Maximum off-time limit circuit(b) Frequency clamp circuit

The valley detect circuit is very simple and easy to implement. To turn on the power switch at valley voltage exactly, the delay circuit composed of R_{demag2} and C_d should be deigned to get a delay time as 1/4 resonant period of L_m and C_{DS}.

IV. SIMULATION RESULTS AND EXPERIMENTAL RESULTS

The proposed controller IC is designed with TSMC T25BCD 60V process. The circuit simulations is made to verify the feasibility. Fig. 8 shows the IC layout. This IC area is about 0.81 mm^2. Fig. 9 shows the waveforms of frequency clamping circuit. The first valley trigger signal is blinded by the proposed frequency circuit, and the power switch is turned on at the second valley. Fig. 10 shows the waveforms of the maximum off-time limit circuit. The converter is operated in CCM during heavy load as the off-time exceed the maximum off-time limitation. This circuit can reduce the conduction losses effectively. Fig. 11 shows the frequency clamp circuit blind the valley voltage and prevent high switching frequency leading to high switching losses during light load condition.

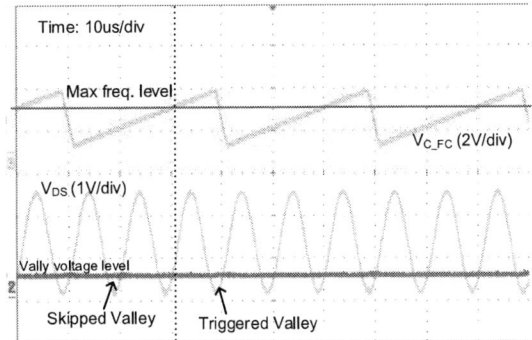

Fig. 11 Frequency clamp circuit measured waveform

V. CONCLUSIONS

In this paper, a QR flyback converter control IC is designed with practical process. The proposed control method will reduce the switching losses by clamping the switching frequency at light load. When the converter is operated in heavy load, it is allowed to be operated in CCM to reduce the conduction losses caused by the high peak current. Compared to the conventional QR flyback control, using the proposed control IC can get a better average efficiency.

ACKNOWLEDGEMENT

The authors gratefully acknowledge financial support from the Ministry of Science and Technology, Taiwan under project No. 102-2221-E-006-097-MY2, and chip implementation support from National Chip Implementation Center (CIC), Taiwan.

Reference

[1] "Energy Star Program Requirements for External Power Supplies"version 1.1, Energy Star, March 2006
[2] Philips Semiconductors, "GreenChip^{TM}II SMPS control IC," TEA1532 datasheet, Feb. 2005.
[3] On Semiconductor "PWM Current-Mode Controller for Free-RunningQuasi-Resonant Operation," NCP1377 datasheet, Oct. 2004.
[4] R. B. Ridley, A. Lotfi, V. Vorperian, and F. C. Lee, "Design and control of a full-wave quasi-resonant flyback converter," in Proc. IEEE APEC'88, pp. 41-49
[5] F. C. Lee, "High-frequency quasi-resonant converter technologies," IEEE Proc., vol. 76, pp. 377-390, April 1988.
[6] D.M. Sable, and R.B. Ridley,"Comparison of performance single-loop and current-injection control for PWM converters that operate in both continuous and discontinuous modes of operation"*Power electronics, IEEE Transactions* on Volume 7, Issue 1, Jan. 1992 Page(s):136-142
[7] J. Lai and D. Chen, "Design considerations for power factor correct ion boost converter operating at the boundary of continuous conduction mode and discontinuous conduction mode," *IEEE Applied Power Electronics Conference*, pp. 267-273, Mar. 1993

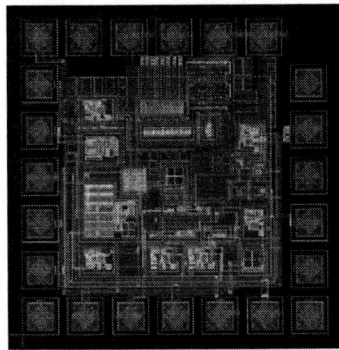

Fig. 8 Proposed control IC layout

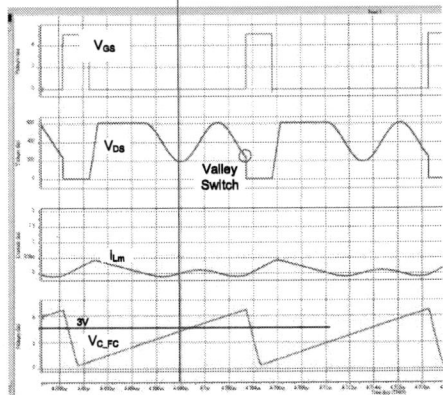

Fig. 9 Waveforms of frequency clamping circuit

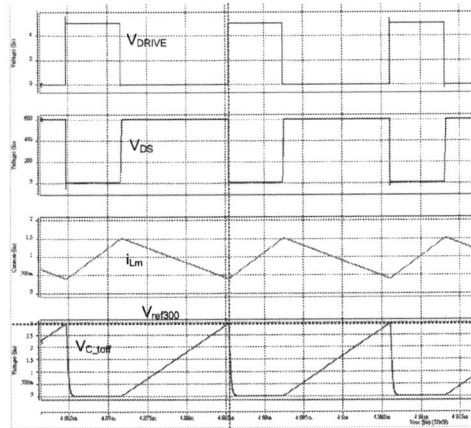

Fig. 10 Waveforms of the maximum off-time limit circuit

Load Transient Response Improvement Based on PID Control

Y. T. Yau[1], *Member, IEEE*, and K. I. Hwu[2], *Member, IEEE*
Department of Electrical Engineering, National Taipei University of Technology, Taipei, Taiwan
E-mail[1]: tsmc35@yahoo.com.tw, E-mail[2]: eaglehwu@ntut.edu.tw

Abstract– In this study, a PID controller with the proposed control strategy is presented, which is used to improve the load transient response. During the load transient period, the parameters of the proposed controller are on-line adjusted and make the controller design easy. As compared with the traditional PID control, the proposed control strategy does not need additional hardware circuits. In addition, the corresponding calculation flow chart is varied as minimum as possible, such that the parameters of the PID controller is of easiness and generality.

Index Terms—Load transient response, PID controller.

I. INTRODUCTION

The traditional analog control technique applied to the switching power supply possesses many merits, such as simplicity, easiness in design, convenience in implementation, etc. But, in some applications requiring high performances of load responses, the analog control has been gradually replaced by the digital proportional-integral-derivative (PID) control. However, in some applications, such as voltage-regulated module (VRM), communication power supply, digital control, etc., the traditional PID controller can not meet some requirements to some extent. Consequently, there are many researches on control technology, providing high control performances for the switching power supply. As generally acknowledged, the merits of the digital control contain high elasticity, programmability, low tolerance, and no aging problem. Therefore, the digital control is very suitable for nonlinearity control and sequence control. However, a key demerit in digital control is that the analog-to-digital converter (ADC) is required to sample the analog signal, and the faster the speed is or the higher the resolution is, the more the system clock and the more the cost. Accordingly, a good controller must take into account low cost, simple structure, easy design and easy implementation, as well as high elasticity and programmability. In the following, some existing researches are provided for analysis convenience.

In order to obtain high performances of load transient responses, the fuzzy control is first presented to control the switching power supply [1][2]. However, many hardware circuits are required to do fuzzy rule calculation, and load transient responses and steady-state performances seem not good enough. Some people present nonlinear control [3]-[7], which via variations in the feedback control loops or improvements in PID controllers, the load transient responses are improved, but the steady-state performances are only for some special operating points and are deteriorated as compared with the traditional PID controller. In the literatures [8]-[12], the adaptive control technique or the auto-tuning technique is adopted, but the number of current sensing components and

ADCs is to be increased, thereby leading to relatively high cost. The literature [13] has the same problem as the literatures [8]-[12] except that the former has to do try-and-error to obtain the parameters of the PID controller for each quiescent operating point, and accordingly, a good control result can be attained by the refined divisions, thus implying that more development time is required. Besides, in the literature [13], there are no researches and experiments under different input voltage levels. In such a case, if various input voltages are taken into consideration, the corresponding size of the parameter table would be increased, thereby causing the parameter memory space and the development time to be increased.

In the literatures [14]-[16], the parameters of the PID controllers are dynamically tuned, so as to upgrade the load transient responses. According to various feedback errors, the parameters of the PID controllers are altered. Hence, if the output voltage is quite different from the voltage reference, then the control effort is relatively large, thus shortening the response time. In [14][16], on the one hand, a prescribed minimum value of integral gain K_I can be used to reduce the steady-state error, and on the other hand, the maximum feedback error during the transient period is used as a base, and as the output voltage is close to the voltage reference gradually, the values of the proportional gain K_P and integral gain K_I, based on this base, are proportionally calculated. However, in this case, under different quiescent operating points and various load transient requirements, the controllers use the same maximum values of K_P and K_I, thereby leading to overshoot or undershoot. In [15], there is a given error reference, and accordingly, during the load transient period, if the output feedback voltage error is larger than this given value, then the parameters of the PID controller are fixed; otherwise, the parameters K_P and K_I are set to zero, thereby causing the steady-state error to appear in the output voltage. Besides, in each of the literatures [14]-[16], only a single quiescent operating point is taken on experiments, without considering different values of voltages and currents.

According to the mentioned above, the proposed control method, based on the traditional PID controller structure, is used to optimize the load transient response, and this method possesses good performances for quiescent operating points. Also, during the load transient period, the parameters of the proposed controller are on-line adjusted and make the controller design easy. As compared with the traditional PID control, the proposed method does not require additional hardware circuits, such as ADCs, current sensing components, etc. Furthermore, the corresponding calculation flow chart is reduced as minimum as possible, so that the design of the parameters of the PID controller is of easiness and generality.

II. OVERALL SYSTEM WITH PROPOSED CONTROL STRATEGY

Fig. 1 shows the overall system with the proposed control strategy, in which the control parameters, containing the proportional gain K_P, the integral gain K_I and the derivative gain K_D, are on-line tuned. In Fig. 1, the main power stage is constructed by a synchronously-rectified buck converter in which both the switches are driven by a half-bridge gate driver. The sensed output voltage is subtracted from the voltage reference V_{ref}, and the resulting error, v_e, is sent to a 9-bit ADC. After this, the resulting digital error, E_{rr}, is sent to the proposed tuning rule and the PID controller. In the tuning rule, the parameters of K_P, K_I and K_D are adjusted according to the proposed method to create a digital control force so as to generate 10-bit DPWM gate driving signals to drive two MOSFETs after a half-bridge gate driver.

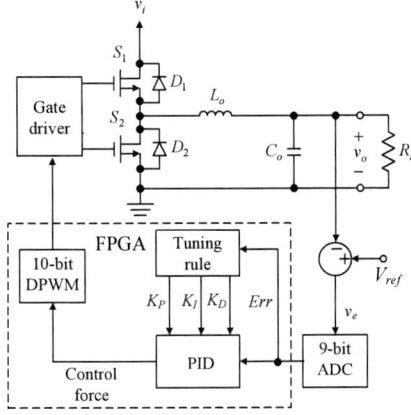

Fig. 1. Overall system with the proposed control strategy.

Fig. 2 shows the concept of the proposed control strategy, which is a combination of two methods mentioned below. Fig. 2(a) illustrates a load transient response due to the traditional PID controller. In order to obtain a first-order response, the parameters K_P and K_I are relatively small and the parameter K_D is relatively large, so as to obtain a stable response. This means that the rising or falling speed of the output voltage during the transient period is slow. Fig. 2(b) shows the transient waveforms created by the radicalized parameters of the PID controller. Via design of high values of K_P and K_I, and low value of K_D, a low damping response can be obtained, but the corresponding stable time is relatively long and the corresponding overshoot/undershoot is relatively large. Besides, the merit of this method is that the first response speed is forced to be fast during the transient period. Therefore, in this study, as shown in Fig. 2(c), combining these two methods mentioned above is used to achieve improvement of the load transient response and to reduce oscillation.

Fig. 3 shows two board lines to separate the high steady-state region and the fast transient regions. As generally acknowledged, the feedback error v_e is equal to V_{ref} minus v_o. Therefore, in Fig. 3, if v_o is larger than V_{ref}, then the upper

bound is $-v_{e\_th}$, whereas if v_o is smaller than V_{ref}, the lower bound is v_{e_th}. And hence, the region over the lower bound or upper bound belongs to the fast transient region, whereas the region between the lower bound and upper bound belongs to the high steady-state region.

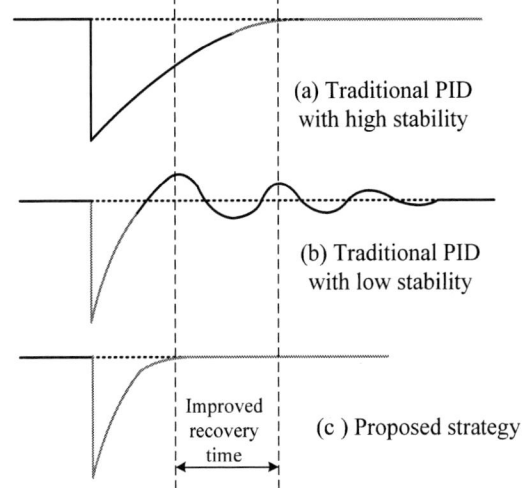

Fig. 2. Concept of the proposed control strategy.

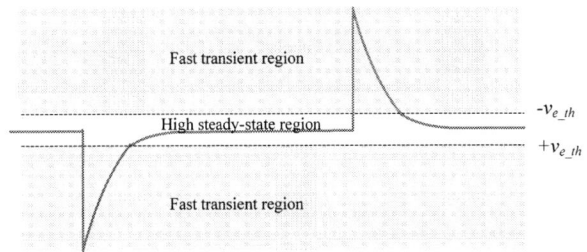

Fig. 3. Anticipated load transient responses.

III. DESIGN OF PID CONTROLLER PARAMETERS

In the following, the design of the PID controller parameters is considered. First, the parameters of the PID controller for the high steady-state region are obtained. Since at this moment, the output voltage is close to the setting voltage reference V_{ref}, the load transient response is not taken into account. Hence, a relatively small value of K_I and a relatively large value of K_D are used so as to upgrade system damping, and hence to reduce system oscillation and steady-state error. At the same time, a suitable value of K_P is selected so as to make the system stable. After this, the parameters of the PID controller in the fast transient region are to be determined. Unlike other research methods, the proposed control strategy does not need many try-and-errors or many parameters to be tuned. Since there are three parameters in the PID controller, there are many possible combinations [1][5][13]-[15]. In order to reduce variables for design convenience, only the parameter K_I is adjusted during the fast transient region, and at the same time, K_D is set to zero. That is, in Fig. 4, as soon as $|v_e| > v_{e\_th}$, the operating region goes into the fast transient region, the value of K_P is kept constant and the value of K_I is increased proportional to $|v_e|$ so as to upgrade the increasing velocity of the control force. At the same time, in order to speed up the velocity of v_o close to V_{ref}, let the value of K_D be zero. By doing so, the fast response shown in Fig. 2(c) can be obtained. As v_o is gradually close to V_{ref}, K_I is set to $\alpha \cdot |v_e|$, where α is a slope with a positive value of smaller than one, so as to reduce the accumulation velocity of the control effort, such that as $|v_e|$ is smaller than v_{e_th}, the operating region will go to the high steady-state region with smooth transfer, so as to avoid creating overshoot or undershoot. Fig. 5 shows the flow chart of the PID controller parameter adjustment. There are some differences in calculation between the proposed and traditional PID controllers. The main reason is that for entering into the fast transient region of the proposed control strategy, one additional adjustment block diagram with one product operator is needed, i.e., $K_I = K_{i\_trans} = \alpha \cdot |v_e|$, which is quite simple. Practically, the parameters of the proposed PID controller are defined according to Fig. 4. Due to the limitation of digital bits, the maximum value of $|v_e|$ is 256, and hence, the maximum value of K_I is 256×0.0625, equal to 16.

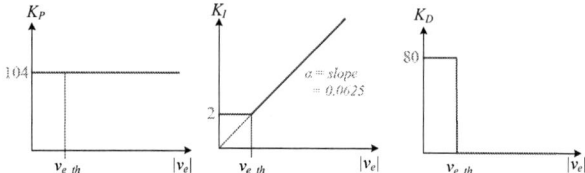

Fig. 4. Design of the parameters of the proposed PID controller.

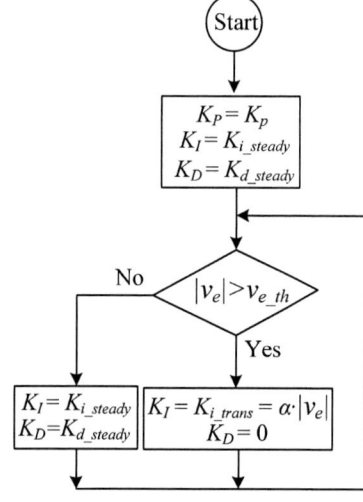

Fig. 5. Flow chart for the adjustment of PID controller parameters.

To speak lucidly, the values of K_P=104, K_I=2 and K_D=80 are firstly set by try-and-error so as to make the converter operated with good performance within the high steady-state region to some extent, under the rated input voltage all over the load range. Afterwards, during the fast transient region, the values of K_P and K_D are set to 104 and zero, respectively, but the value of K_I is varied. This is because tuning v_o based on K_I is faster than tuning v_o based on K_P and K_D. Therefore, in this study, the values of K_P and K_D are kept constant so as to reduce operating resources and to simplify parameter adjustment.

IV. SYSTEM AND COMPONENT SPECIFICATIONS

The following show the system and component specifications: (i) the input voltage range is from 5V to 15V; (ii) the output voltage is 3.3V; (iii) the rated output current is 9A; (iv) the switching frequency is 240kHz; (v) the product name of S_1 and S_2 is PHD96NQ03LT; (vi) the product name of the gate driver is ISL2100A; (vii) one 10µF MLCC capacitor is chosen for the output capacitor; (viii) the value of the output inductor L_o is 2µH; (ix) the product name of FPGA is Altera EP3C5 with a system clock of 240MHz; (x) the bit number of ADC is 9 bits; and (xi) the bit number of DPWM is 10 bits.

V. EXPERIMENTAL RESULTS

Under the rated input voltage of 10V, for the traditional PID controller with parameters fixed at K_P=104, K_I=2 and K_D=80, Fig. 6 shows load transient responses, containing the load enable signal, $LOAD\_EN$, the output voltage, v_o, and the output inductor, i_{Lo}, due to load change from 0% to 100% of the rated load, whereas Fig. 7 shows load transient responses, containing the load enable signal, $LOAD\_EN$, the output voltage, v_o, and the output inductor, i_{Lo}, due to load change from 100% to 0% of the rated load.

On the other hand, under the rated input voltage of 10V, for the proposed PID controller with parameters set according to Fig. 4, Fig. 8 shows load transient responses, containing the load enable signal, $LOAD\_EN$, the output voltage, v_o, and the output inductor, i_{Lo}, due to load change from 0% to 100% of the rated load under the input voltage of 10V, whereas Fig. 9 shows load transient responses, containing the load enable signal, $LOAD\_EN$, the output voltage, v_o, and the output inductor, i_{Lo}, due to load change from 100% to 0% of the rated load.

From Figs. 6 and 8, it can be seen that both the undershoots are the same but the former has the recovery time of about 50µs and the latter has the recovery time of about 25µs; From Figs. 7 and 9, it can be seen that both the overshoots are the same but the former has the recovery time of about 60µs and the latter has the recovery time of about 40µs. From Figs. 6 to 9, it can be seen that the proposed method has better performance of load transient responses than the traditional method.

978-1-4799-2706-7/14 $31.00 © 2014 IEEE

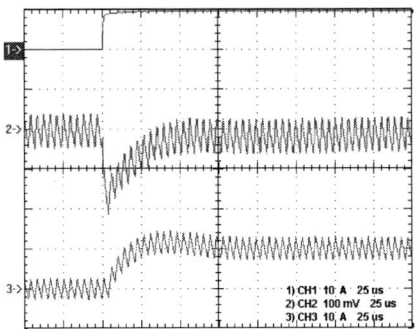

Fig. 6. Traditional load transient response due to load change from 0% to 100% under input voltage of 10V: (1) *LOAD_EN*; (2) v_o; (3) i_{Lo}.

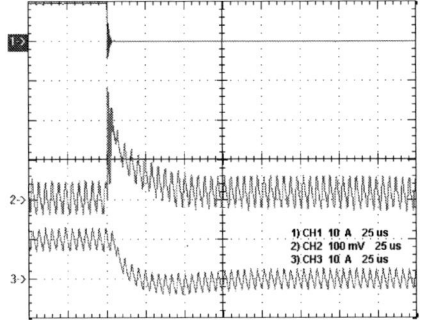

Fig. 7. Traditional load transient response due to load change from 100% to 0% under input voltage of 10V: (1) *LOAD_EN*; (2) v_o; (3) i_{Lo}.

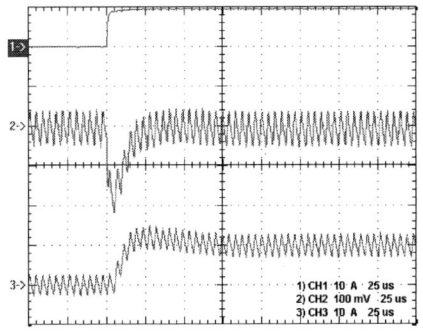

Fig. 8. Proposed load transient response due to load change from 0% to 100% under input voltage of 10V: (1) *LOAD_EN*; (2) v_o; (3) i_{Lo}.

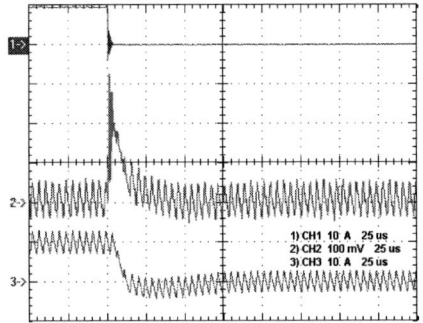

Fig. 9. Proposed load transient response due to load change from 100% to 0% under input voltage of 10V: (1) *LOAD_EN;* (2) v_o; (3) i_{Lo}.

VI. CONCLUSION

A PID controller with the proposed control strategy is presented herein. During the fast transient region, only the integral gain of the PID controller is on-line adjusted, thereby rendering this controller design easy. From experimental results, it can be seen that as compared with the traditional PID controller, the proposed controller can enhance load transient responses significantly.

REFERENCES

[1] A. G. Perry, Guang Feng, Yan-Fei Liu and P. C. Sen, "A design method for PI-like fuzzy logic controllers for DC-DC converter," *IEEE Transactions on Industrial Electronics*, vol. 54, no. 5, pp. 2688-2696, 2007.

[2] Liping Guo, J. Y. Hung and R. M. Nelms,. "Evaluation of DSP-based PID and fuzzy controllers for DC-DC converters," *IEEE Transactions on Industrial Electronics*, vol. 56, no. 5, pp. 2237-2248, 2009.

[3] Haitoa Hu, V. Yousefzadeh and D. Maksimovic, "Nonlinear control for improved dynamic response of digitally controlled DC-DC converters," *IEEE PESC '06*, pp. 1-7, 2006.

[4] Haitao Hu, V. Yousefzadeh and D. Maksimovic, "Nonuniform A/D quantization for improved dynamic responses of digitally controlled DC-DC converters," *IEEE Transactions on Power Electronics*, vol. 23, no. 4, pp. 1998-2005, 2008.

[5] V. Yousefzadeh and S. Choudhury, "Nonlinear digital PID controller for DC-DC converters," *IEEE APEC'08*, pp. 1704-1709, 2008.

[6] Mingzhi He and Jianping Xu, "Nonlinear PID in digital controlled buck converters," *IEEE APEC'07*, pp. 1461-1465, 2007.

[7] L. Corradini, A. Costabeber, P. Mattavelli and S. Saggini, "Time optimal, parameters-insensitive digital controller for VRM applications with adaptive voltage positioning," *IEEE COMPEL'08*, pp. 1-8, 2008.

[8] Guang Feng, E. Meyer and Yan-Fei Liu, "A new digital control algorithm to achieve optimal dynamic performance in DC-to-DC converters," *IEEE Transactions on Power Electronics*, vol. 22, no. 4, pp. 1489-1498, 2007.

[9] G. Eirea and S. R. Sanders, "Adaptive output current feedforward control in VR applications," *IEEE Transactions on Power Electronics*, vol. 23, no. 4, pp. 1880-1887, 2008.

[10] Chun-Yu Hsieh and Ke-Horng Chen, "Adaptive pole-zero position (APZP) technique of regulated power supply for improving SNR," *IEEE Transactions on Power Electronics*, vol. 23, no. 6, pp. 2949-2963, 2008.

[11] Yu-Huei Lee, Shih-Jung Wang, Chun-Yu Hsieh and Ke-Horng Chen, "Current mode DC-DC buck converters with optimal fast-transient control," *IEEE ISCAS' 08*, pp. 3045-3048, 2008.

[12] J. Morroni, R. Zane and D. Maksimovic, "Design and implementation of an adaptive tuning system based on desired phase margin for digitally controlled DC-DC converters," *IEEE Transactions on Power Electronics*, vol. 24, no. 2, pp. 559-564, 2009.

[13] Jen-Ta Su, Chih-Wen Liu and De-Min Liu, "Adaptive control scheme for interleaved DC/DC power converters," *IEEE IPEC'10*, pp. 3105-3111. 2010.

[14] V. Arikatla and J. A. Abu Qahouq, "An adaptive digital PID controller scheme for power converters," *IEEE ECCE'10*, pp. 223-227, 2010.

[15] V. Arikatla and J. A. A. Qahouq, "DC-DC power converter with digital PID controller," *IEEE APEC'11*, pp. 327-330, 2011.

[16] V. P. Arikatla and J. A. Abu Qahouq, "Adaptive digital proportional-integral-derivative controller for power converters," *IET Power Electronics*, vol. 5, no. 3, pp. 341-348, 2012.

An Active-Clamping Forward Converter with Non-linear Step-down Conversion

Jing-Yuan Lin
Dept. of Electrical Engineering

National Taitung College

Taitung City, Taiwan R.O.C
jylin@ntc.edu.tw

Yu-Kang Lo
Dept. of Electronic Engineering

National Taiwan University of
Science and Technology
Taipei City, Taiwan R.O.C
yklo@mail.ntust.edu.tw

Huang-Jen Chiu
Dept. of Electronic Engineering

National Taiwan University of
Science and Technology
Taipei City, Taiwan R.O.C
hjchiu@mail.ntust.edu.tw

Chao-Fu Wang
Dept. of Electronic
Engineering
National Taiwan University of
Science and Technology
Taipei City, Taiwan R.O.C
D9902204@mail.ntust.edu.tw

Chien-Yu Lin
Dept. of Electronic
Engineering
National Taiwan University of
Science and Technology
Taipei City, Taiwan R.O.C
D9902209@mail.ntust.edu.tw

Abstract- **This paper propose a forward converter with non-linear step-down conversion ratio, comparison with traditional forward converter, it has more better duty utilization ratio, besides, two output inductors can share the load current naturally, the current stress of inductors and output diodes can be reduced. Thus, the proposed converter is suitable for high step-down application. At primary side, active-clamp control can cancel the voltage spike of main switch, it also achieve zero voltage switching operation of main switch and auxiliary switch, so that the switching loss can be reduced. Detailed analysis and design of this proposed converter are described. Experimental results are recorded for a prototype converter with an AC input voltage ranging from 90 to 264 V, an output voltage of 12 V and an output power of 240 W.**

I. INTRODUCTION

The isolated active clamp forward converter (ACFC) is widely using in the switching mode power supply design to provide regulated output voltages for middle or high power applications. The active-clamping technique has been proposed to improve the voltage spike of switches [1][2]. Also active-clamping helps switches to operate in ZVS for reduce the switching loss. The output inductor of forward is used to storage and it can reduce rms current of output rectifiers. This paper proposed a novel forward converter is composed of two identical inductors and two diodes. Compare with traditional forward, two output inductors are parallel operate to provide output load, it can share load current naturally and both core size can be reduced [4][5]. Otherwise, the proposed forward has a higher step-down ratio and duty utilization ratio than conventional forward.

Detailed analysis of the circuit operations, steady-state behaviors, and DC gain of the proposed converter are given in the following sections. Theoretical discussions are validated with experimental results on a prototype converter that delivers a 12-V/20-A output from an AC input ranging from 90 to 260 V.

II. Circuit description and principle operation of proposed converter

A. Circuit Description

Fig. 1 shows the schematic of the proposed novel rectifier using in active clamp forward converter. This converter consists of the clamping capacitor C_1, the active switch S_1 and S_2, external resonant inductor L_r, the output rectifier diode composed of forward diode D_1, flywheel diode D_2 and D_3, the output inductor L_1 and L_2, the output capacitor Co, Cr is equal to the combination of output capacitor and parasitic capacitor of the transformer primary winding. L_m is the magnetizing inductor of transformer T.

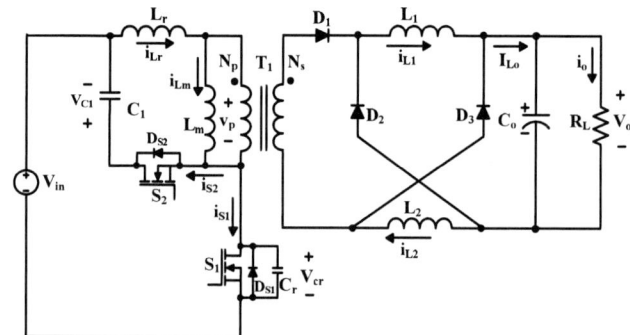

Fig. 1 schematic of the proposed active-clamping converter

To analyze the proposed active-clamping forward converter, the following assumptions are made.

- The clamping capacitor C1 is larger than resonant capacitor Cr, moreover, the steady-state voltage of clamping capacitor V_{C1} equal to constant.
- Both output inductors are very large, the current of each inductor can be assume to constant.
- The turns ratio of the transformer is n = Np/Ns
- The conduction times of S_1 and S_2 are DT_s and $(1-D)T_s$, respectively, where D is the duty cycle of S_1, and Ts is the switching period.

When S_1 is turned on, the voltage across the N_p (or the magnetizing inductor L_m) approximates the input voltage V_{in}. On the other hand, when S_2 turns on, the voltage across the L_m is about $-V_{C1}$. From the flux balance of L_m under the steady state, V_{C1} can be determined as

$$V_{C1} = \frac{D}{1-D} V_{in} \qquad (1)$$

Similarly, the voltage transfer ratio of Vin to Vo can be found by the flux balance of L1 or L2. When S1 is turned on, the voltage across the L1 and L1 as following

$$V_{L1} = V_{L2} = \frac{V_{in} \dfrac{N_s}{N_P} - V_o}{2} \qquad (2)$$

When S_1 turns off, both the voltage across L1 and L1 are −Vo. Then from (2) and the inductor flux balance, the voltage transfer ratio of V_{in} to Vo can be found to be

$$\frac{V_{in}\dfrac{N_s}{N_p}-V_o}{2}DT_s = V_o(1-D)T_s \tag{3}$$

$$\frac{V_o}{V_{in}} = \frac{1}{n}\frac{D}{2-D} \tag{4}$$

Fig. 2 shows the key waveforms of the proposed active-clamping forward converter. There are eight states in a complete switching cycle. The conduction paths for each operating state are illustrated in Fig. 3.

B. Principals of Operation

State 1 ($t_1 < t < t_2$):

As shown in Fig. 3(a), S_1 is turned off at t_1. In this interval, C_r is charged from 0 to $(V_{in} + V_{C1})$. The diode D_1 still conducting and V_{in} deliver power to output load, the current I_{L1} mapping to primary side with turn ratio n. Because of C_r is very small and charged quickly. This interval ends at t_2 when v_{Cr} is charged to $(V_{in} + V_{C1})$.

State 2 ($t_2 < t < t_3$):

As shown in Fig. 3(b), in this interval, C_r is charged to $(V_{in} + V_{C1})$. The voltage of primary side $v_p(t)$ decrease to zero and D_1 turned off. The primary side circuit is composed of C_r, L_r and L_m. Because of C_r is very small and charged quickly. State 2 is ended when D_{S2} conducts at t_2.

State 3 ($t_3 < t < t_4$):

As shown in Fig. 3(c), at t_2, v_{Cr} equals $(V_{in} + V_{C1})$ and the body diode across S_2 conducts, in this state, S_2 can achieve ZVS turn on, the voltage of primary side v_p approximate equal to $V_{C1}(L_m + L_r)/L_m$. The current of secondary diode D_1 decreasing, both D_2 and D_3 current are increasing. Since the clamping capacitance C_1 is much larger than the resonant capacitance C_r, almost all the energy stored in the inductors L_m and L_r are released to charge C_1,

State 4 ($t_4 < t < t_5$):

As shown in Fig. 3(d), at t_3, S_2 achieved ZVS turn on, the secondary diode D_1 turned off. Both D_2 and D_3 are conduct to share load current. The resonant inductor L_r and the clamping capacitor C_1 begin to resonant. During this state, i_{Lr} will change its polarity and flows through S_2. In order to ensure the ZVS operation of S_2, it should be turned on before i_{Lr} goes negative.

State 5 ($t_5 < t < t_6$):

As shown in Fig. 3(e), at t_5, S_2 is turned off. In this state, the output diode D_2 and D_3 are still conducting. The resonant inductor current i_{Lr} starts to discharge the energy of resonant capacitor C_r from $(V_{in}+V_{C1})$ to V_{in}. This time interval ends when the voltage v_{Cr} equals to V_{in} at t_6.

State 6 ($t_6 < t < t_7$):

As shown in Fig. 3(f), at t_6, v_p equals to zero and the secondary side diode D_1 conduct. In this state, the output diode D_2 and D_3 are still conducting and the current of each one decreasing. The resonant inductor current i_{Lr}

discharging the energy of resonant capacitor C_r to zero. To ensure the ZVS turn-on of S_1, the energy stored in L_r must be greater than the energy stored in C_r. That is

$$L_r > \frac{C_r V_{IN}^2}{i_{Lr}^2(t_6)} \tag{5}$$

State 6 is ended when v_{Cr} decreases to zero.

State 7 ($t_7 < t < t_8$):

As shown in Fig. 3(g), in this state, C_r is discharged to zero and the body diode D_{S1} conducts. The voltage across the magnetizing inductor L_m equals V_{in} and i_{Lm} is linearly increasing. The current of secondary diode D_1 linear increasing until $i_{D1}=I_{Lo}$. Before the resonant inductor current i_{Lr} goes positive, S_1 must turn on to ensure a ZVS operation. This state ends when S_1 is turned on.

State 8 ($t_8 < t < t_9$):

As shown in Fig. 3(h), at t_8, The diodes D_2 and D_3 are truned off after the current of each one decrease to zero. The voltage v_p equals to V_{in} and the current i_{Lr} increases linearly. The secondary diode D1 turned on and V_{in} delivers power to output load.

When State 8 ends, the operating state returns to State 1 to begin the next switching cycle.

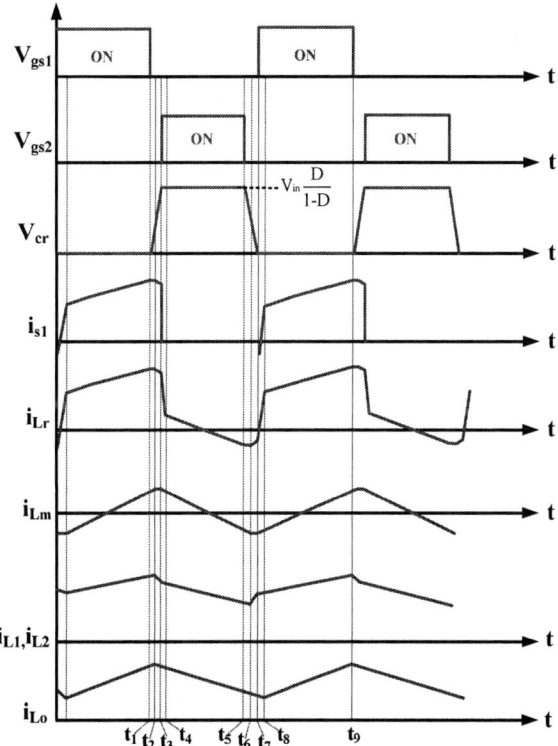

Fig. 2 The key waveform of the proposed active-clamping forward converter

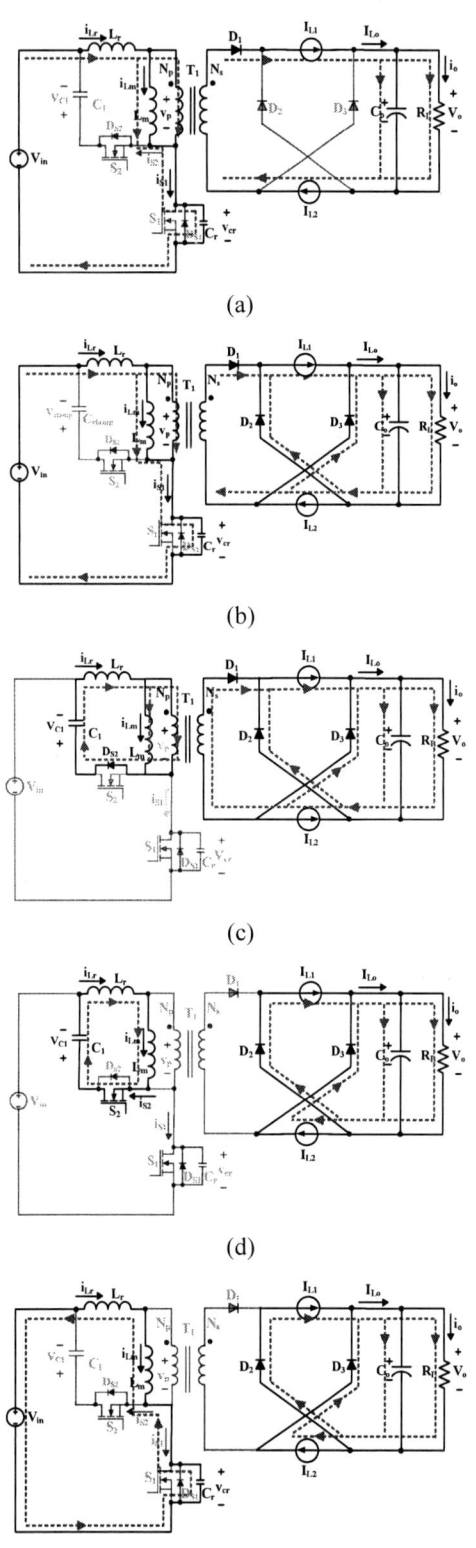

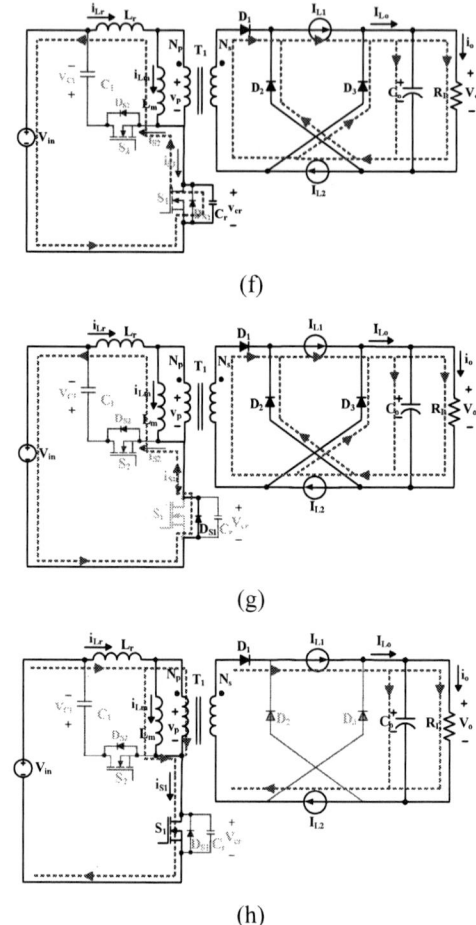

Fig. 3 Conduction paths of (a) State 1, (b) State 2, (c) State 3, (d) State 4, (e) State 5, (f) State 6, (g) State 7, and (h) State 8 for the proposed forward converter during one switching period

C. Circuit Parameters Analysis

According to application of this proposed converter, it should operate in CCM. A comparison between the proposed converter and active clamp forward, which contains the discussed equations, is in Table I.

Table I.
Comparison between the proposed converter and active clamp forward

	ACFC	The Proposed Converter
V_o/V_{in}	D	$\dfrac{D}{2-D}$
The maximum voltage stress of S_1 and S_2	$V_{in}+V_{C1}=\dfrac{1}{1-D}\cdot V_{in}$	$V_{in}+V_{C1}=\dfrac{1}{1-D}\cdot V_{in}$
The maximum current stress of S_1	$\dfrac{1}{n}\sqrt{D}\cdot I_O$	$\dfrac{1}{n}\dfrac{\sqrt{D}}{2-D}I_O$
The current stress of forward diode	$D\cdot I_O$	$\dfrac{D}{2-D}I_O$

The current stress of flywheel diodes	$(1-D) \cdot I_O$	$\dfrac{1-D}{2-D} I_O$
The voltage stress of forward diode	$\dfrac{V_{in}}{n} + V_O$	$\dfrac{V_{in}}{n} + V_O$
The voltage stress of flywheel diodes	$\dfrac{V_{in}}{n} + V_O$	$\dfrac{1}{2}(\dfrac{V_{in}}{n} + V_O)$

Following above analysis and Table I, the conclusions can be drawn [6][7]:

- At the same input voltage and D as ACFC, the output voltage of the proposed converter is reduced by (2-D) times more, or the same output voltage is obtained for a higher value of D. It means that for a low conversion ratio, the duty cycle will not be an extreme value, and it also suitable for wide input voltage range.

- Smaller rms current values of the proposed converter give reduced conduction losses in output rectifiers for this converter.

- Considering the same operation conditions (the same input and output voltages and equal output power), the average transistor currents are the same

III. Experimental Results

In order to verify the theoretical analysis, a 240-W prototype converter is built and tested in the laboratory. The implementation of the proposed converter with non-linear step-down conversion ratio is shown in Fig. 5. The experimental results are obtained with the following parameters.

Input AC voltage range: 90 ~ 264 V_{rms}

Output voltage: $V_O = 12$ V

Rated output current: $I_O = 20$ A

Switching frequency: $f_s = 100$ kHz

From these specifications, choosing $D_{max} = 0.8$ then the parameter of following can be found :

Turns ratio: $n = N_p/N_s = 54 / 4$

Clamping capacitances: $C_1 = 33$ nF

Output inductors : $L_1 = 50 \mu$ H, $L_2 = 50 \mu$ H

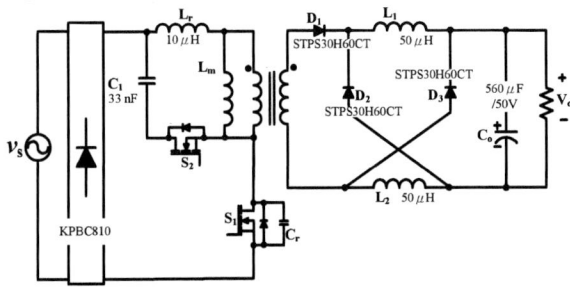

Fig. 5 Schematic of the proposed converter.

Fig. 6 shows the measured waveform of drain-to-source voltages and drain-to-source current of S_1 and S_2, at 20% and 100% load condition with 230-V_{rms} input voltage. It illustrate ZVS feature. Fig. 8 shows the measured efficiencies of the proposed active-clamping forward converter with Non-linear Step-down Conversion at a

115-V_{rms} and 230-V_{rms} input voltage for different output powers. The peak efficiency of the system is above 91%.

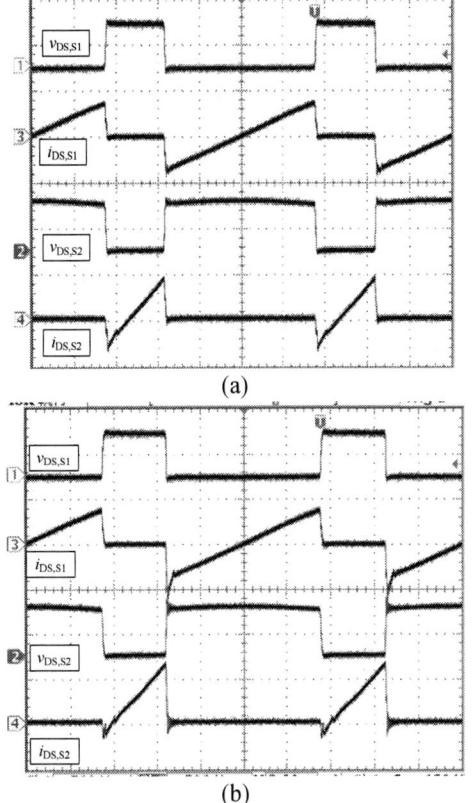

(a)

(b)

Fig. 7 Waveforms of vDS and iDS of S1 and S2 at (a) 20% load (b) 100% load (vDS,S1/ vDS,S2: 500 V/div., iDS,S1/ iDS,S2: 5 A/div., Time: 4μs/div.)

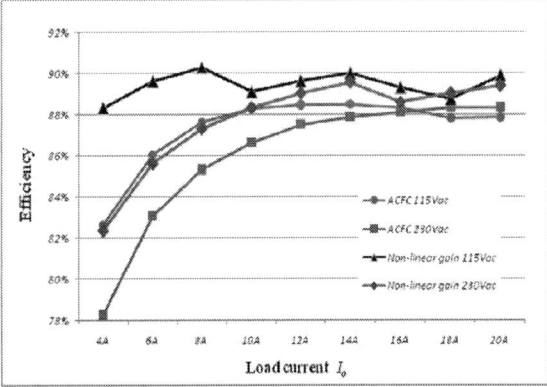

Fig. 10 Efficiencies of the proposed converter

IV. Conclusion

This paper presents a active-clamping forward converter with non-linear step-down conversion to facilitate a low-profile implementation while gaining a high conversion efficiency via ZVS of the active switches. The load current is equally shared between the two inductors and two flywheel diodes to reduce the conduction loss. Comparison with traditional ACFC, the current stress of the proposed converter is smaller than ACFC, it has more better efficiency performance. Moreover, the non-linear gain achieve high voltage transfer ratio, it also can be applied in high step-down application. The experimental results on a 240-W prototype are recorded to verify the theoretical analysis.

REFERENCES

[1] C. S. Leu, G. Hua, F. C. Lee, and C. Zhou, "Analysis and design of RCD clamp forward converter," in *Proc. 7th High Frequency Power Conversion Conf.*, 1992, pp. 198–208.

[2] H. Huang, "Design guidelines on the effect of resonant transitions of forward converter on efficiency with active clamp," 23rd Annual IEEE Applied Power Electronics Conference and Exposition, 2008, pp. 600-606.K. Harada and H. Sakamoto, "Switched snubber for high frequency switching," in Proc. Power Electron. Spec. Con ., 1990, pp. 181-187.

[3] I. Rezaei and M. Akhbari, "Transformerless hybrid buck converter with wide conversion ratio," Power Electronics, Drive Systems and Technologies Conference, 2011, pp. 599-603.

[4] Axelrod, B.; Berkovich, Y.; Ioinovici, A., "Switched-Capacitor/Switched-Inductor Structures for Getting Transformerless Hybrid DC–DC PWM Converters," *Circuits and Systems I: Regular Papers, IEEE Transactions on* , vol.55, no.2, pp.687,696, March 2008

[5] B. Axelrod, Y. Berkovich and A. Ioinovici, "Transformerless dc-dc converters with a very high dc line-to-load voltage ratio," in Proc. of the 2003 International Symposium on Circuits and Systems, 2003, pp. 435-438.

[6] A. Ioinovici, "Switched-capacitor power electronics circuits," IEEE Circuits and Systems Magazine, vol. 1, no. 3, pp. 37-42, Jan. 2001.

Switching Loss Minimization of 3-phase Interleaved Bidirectional DC-DC Converter

Eui-Cheol Nho, Jae-Hun Jung,
Hak-Soo Kim, In-Dong Kim
Dept. of Electrical Engineering
Pukyong National University
Busan, Korea
nhoec@pknu.ac.kr

Heung-Geun Kim
Dept. of Electrical Engineering
Kyungpook National University
Daegu, Korea

Tae-Won Chun
Dept. of Electrical Engineering
University of Ulsan
Ulsan, Korea

Abstract— A new switching scheme for 3-phase interleaved bidirectional DC-DC converter is proposed. The proposed scheme provides soft switching condition for all the switches in both charging and discharging modes regardless of the output power level. Analysis and simulation are carried out for 3kW bidirectional DC-DC converter to charge and discharge a battery of an electric vehicle. Simulation results show the usefulness of the proposed scheme.

Keywords— *bidirectional DC-DC converter, Soft switching.*

I. INTRODUCTION

Bidirectional DC-DC converters are widely used in the field of battery charging and discharging, supercapacitor energy storage, and power converters for electric vehicles. Electric vehicle, especially, requires small size and light weight DC-DC converter. In order to reduce the size and weight switching frequency of the converter should be increased, however, that results in increased switching loss and heat sink volume. To obtain high switching frequency with low switching loss various soft switching methods have been reported [1-5].

Among the soft switching DC-DC converters ZVT and ZCT DC-DC converters have good feature of constant switching frequency, however, they require additional auxiliary circuit that results in complex circuit topology and cost increase. [6] suggested 3-phase interleaved soft switching DC-DC converter with simple structure and constant switching frequency. The circuit operation of [6] is very good in full load condition, however, conduction loss become large as the load power decreases.

This paper proposes new switching scheme to obtain not only low switching loss but also reduced conduction loss even in low power condition. Circuit operation is analysed in charging and discharging mode with 3kW bidirectional DC-DC converter. Simulation results show the usefulness of the proposed scheme.

II. OPERATING PRINCIPLE OF THE PROPOSED SCHEME

Fig. 1 shows the conventional 3-phase interleaved bidirectional DC-DC converter. V_B and V_{DC} mean battery voltage and dc-link voltage of a PWM inverter,

respectively. S1 (Sa1, Sb1, and Sc1) and S2 (Sa2, Sb2, and Sc2) operate complimentary, and power is controlled with duty ratio. In charging and discharging mode each pole of the converter operates as buck and boost mode, respectively.

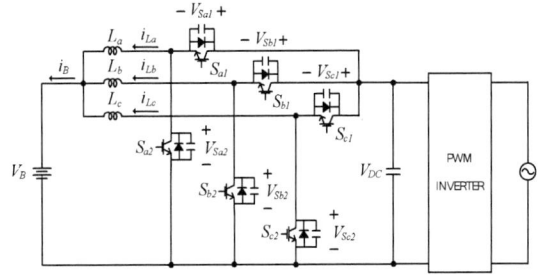

Fig. 1. Conventional bidirectional DC-DC converter

Fig. 2 shows proposed 3-phase interleaved bidirectional DC-DC converter. It is shown that the capacitors connected to the upper side switches, Sa1, Sb1, and Sc1 in parallel are omitted. Besides, the reduction of the number of capacitors the switching scheme is newly developed to get low conduction loss as well as switching loss.

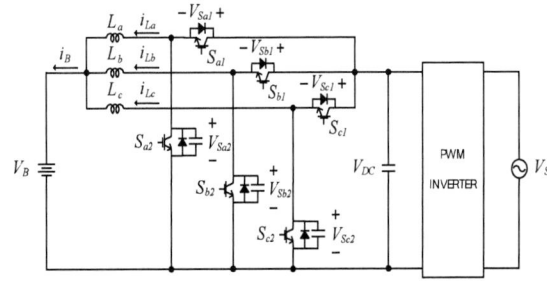

Fig. 2. Proposed bidirectional DC-DC converter

A. Full load Condition

First, the circuit operation is analysed in full load condition. Fig. 3 shows switching signals, voltage, current, and current path in charging mode for 1-pole of the converter. It is shown that there are 4 operating modes.

978-1-4799-2706-7/14 $31.00 © 2014 IEEE

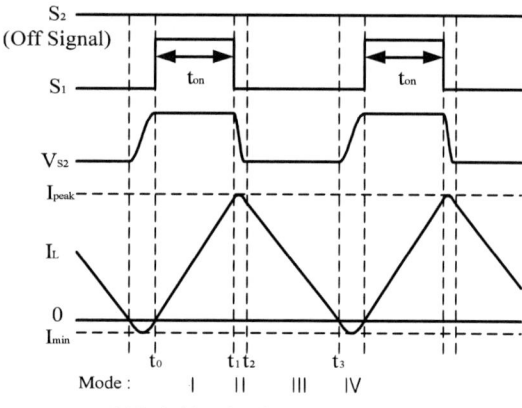

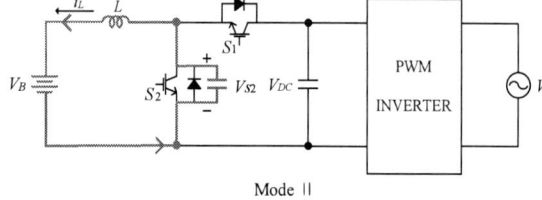

Mode I

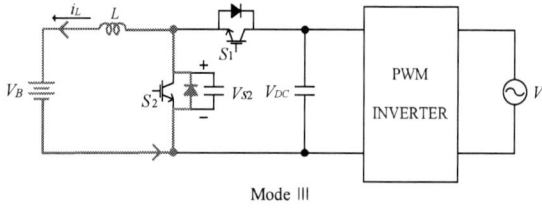

Mode II

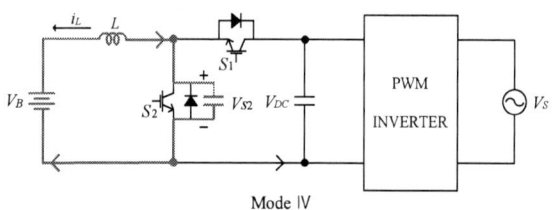

Mode III

Mode IV

(b) Current path in each mode

Fig. 3. Waveforms and current path in charging mode

· Mode – I $(t_0 \le t < t_1)$

The switch S_1 is turned on under the condition of $V_{S2} = V_{DC}$. Furthermore, the inductor current i_L begins to start from zero. Therefore, ZVS and ZCS conditions are provided at the instant of turn-on of the switch S_1. The current increases as follows.

$$i_L = \frac{1}{L} \int_{t_0} (V_{DC} - V_B) dt \qquad (1)$$

· Mode – II $(t_1 \le t < t_2)$

As soon as the switch S_1 is turned-off at t_1 L-C resonant circuit begins to operate, and the resonant current and voltage are as follows.

$$i_L = i(t_1) \cos \omega_o (t - t_1) + \frac{V_B - V_{S2}(t_1)}{Z_o} \sin \omega_o (t - t_1) \qquad (2)$$

$$v_{S2} = V_B - (V_B - V_{DC}) \cos \omega_o (t - t_1) + Z_o i_L (t_1) \sin \omega_o (t - t_1) \qquad (3)$$

where, $\omega_o = 1/\sqrt{LC}$ and $f_o = \sqrt{L/C}$.

Therefore, the switch S_1 can be turned off under zero voltage switching condition. This mode ends when V_{S2} decreases to zero.

· Mode – III $(t_2 \le t < t_3)$

After the voltage V_{S2} decreases to zero the anti-parallel connected diode starts to conduct to provide the freewheeling of the inductor current. During this mode the inductor current decreases as follows.

$$i_L = i_L (t_2) - \frac{1}{L} \int_{t_2} V_B dt \qquad (4)$$

· Mode – IV $(t_3 \le t < t_4)$

As soon as the inductor current decreases to zero another resonant operation occurs. The source voltage for the resonance is V_B, therefore, the current and switch voltage V_{S2} are as follows.

$$i_L = \frac{V_B}{Z_o} \sin \omega_o (t - t_3) \qquad (5)$$

$$v_{S2} = V_B \{1 - \cos \omega_o (t - t_3)\} \qquad (6)$$

This mode ends when the voltage V_{S2} reaches V_{DC}.

Fig. 4 shows each waveform and current path in each operating mode of 1-pole of the bidirectional DC-DC converter in discharging mode. In this mode the converter operates in boost mode.

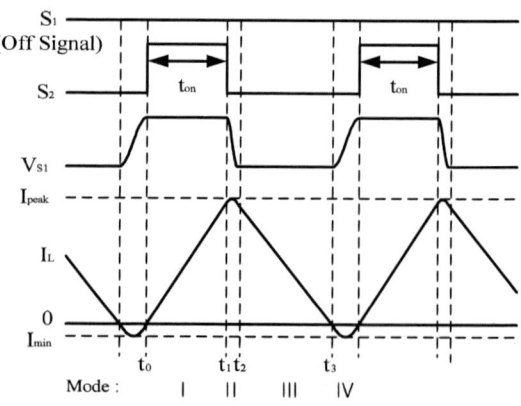

(a) Switching signal, current, and voltage

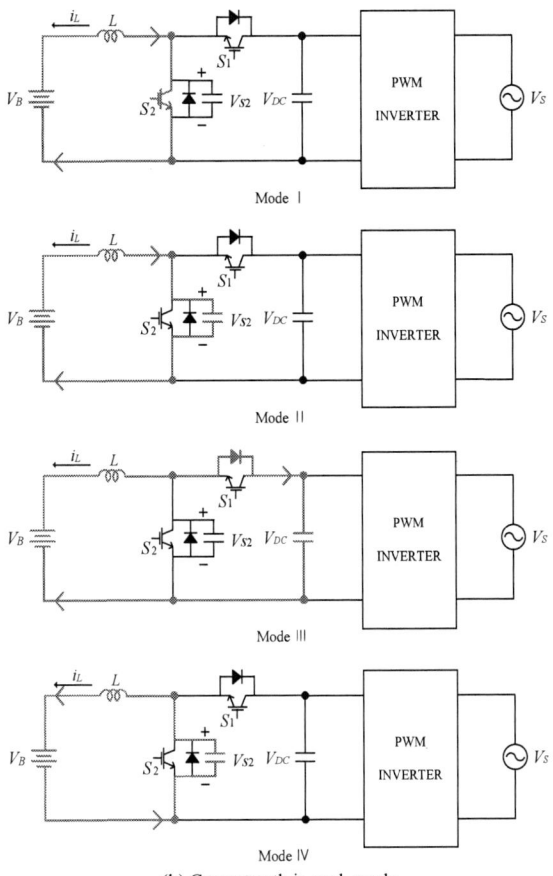

(b) Current path in each mode

Fig. 4. Waveforms and current path in discharging mode

· Mode – I $(t_0 \leq t < t_1)$

This mode starts by turn-on of the switch S_2 at t_0 under zero current switching condition, and the current begins to increase linearly as follows.

$$i_L = \frac{1}{L} \int_{t_0}^{t} V_B dt \qquad (7)$$

· Mode – II $(t_1 \leq t < t_2)$

The switch S_2 is turned off at t_1 under zero voltage condition, and L-C resonant operation starts with the following current and voltage.

$$i_L = i_L(t_1) \cos \omega_o (t - t_1) + \frac{V_B}{Z_o} \sin \omega_o (t - t_1) \qquad (8)$$

$$v_{s2} = V_B \{1 - \cos \omega_o (t - t_1)\} + Z_o i_L(t_1) \sin \omega_o (t - t_1) \qquad (9)$$

As soon as the voltage V_{S2} reaches the DC-link voltage, this mode ends.

· Mode – III $(t_2 \leq t < t_3)$

The stored energy in inductor L begins to discharge, and the current decreases as follows

$$i_L = i_L(t_2) + \frac{1}{L} \int_{t_2}^{t} (V_B - V_{DC}) dt \qquad (10)$$

When the current decreases to zero this mode ends.

· Mode – IV $(t_3 \leq t < t_4)$

As soon as the current becomes zero the capacitor begins to discharge making L-C resonant circuit, and the resonant current and voltage are as follows.

$$i_L = \frac{V_B - V_{DC}}{Z_o} \sin \omega_o (t - t_3) \qquad (11)$$

$$v_{s2} = V_B \{1 - \cos \omega_o (t - t_3)\} + Z_o i_L(t_3) \sin \omega_o (t - t_3) \qquad (12)$$

B. Below Full Load Condition

Fig. 5 shows voltage and current waveforms in case of the load power is below full Load in charging mode.

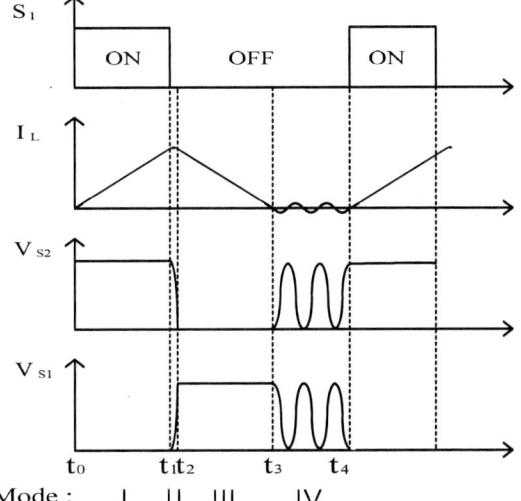

(a) Current and voltage waveforms

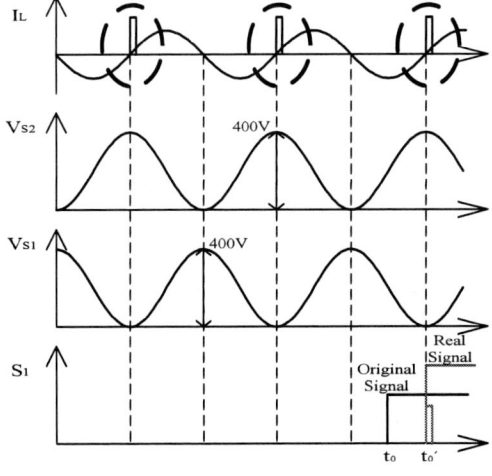

(b) Detailed waveforms

Fig. 5. Waveforms below full load condition in discharging mode

It is found that there is oscillation during Mode-IV($t_3 \leq t < t_4$) as shown in Fig. 5(a). Fig. 5(b) shows detailed waveforms for Mode–IV. Basically the converter switching frequency is constant, therefore, the time instant of t_0 in Fig. 5(b) corresponds to a determined switching instant of S_1.

However, the voltage of V_{S1} is not zero at t_0, therefore, ZVS and ZCS switching can not be obtained in this case.

To provide the soft switching condition, the switching time should be delayed until the voltage V_{S1} reduce to zero. Therefore, the actual turn-on time of S_1 is $t_0`$ instead of t_0. In case of discharging mode, the soft switching condition can be obtained by the similar method.

(c) SIMULATION RESULTS

Simulation parameters are shown in table I.

TABLE I
SIMULATION PARAMETERS

Parameter	Value
V_{DC}	400 V
V_B	200 ~ 280 V
L_a, L_b, L_c	200 μH
C_{a2}, C_{b2}, C_{c2}	2 nF
f_S	50 kHz
Max. output power	3 kW

Fig. 6 shows each waveform in charging mode with full load condition. It is found that S_1 turns on with ZVS and ZCS conditions, and turns off with ZVS condition.

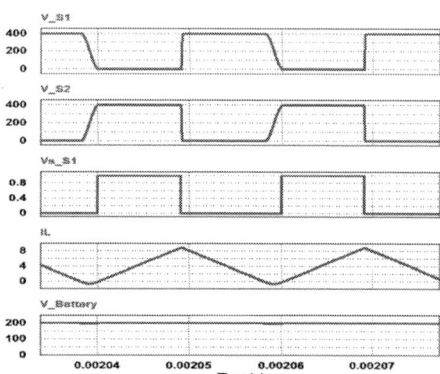

Fig. 6. Waveforms in charging mode with full load

Fig. 7 shows waveforms in charging mode with half load condition. Operating Mode I–IV are the same with those in case of full load, but Mode–IV operation is different. It is found that the switch can be turned on and off under soft switching condition. The simulation results in discharging mode are almost similar to those in charging mode.

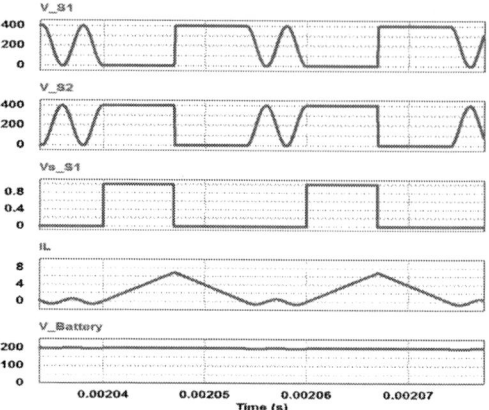

Fig. 7. Waveforms in charging mode with half load

Fig. 8 and Fig. 9 show current and voltage waveforms in 3-phase interleaved bidirectional DC-DC converter with full load and half load condition, respectively, in charging mode.

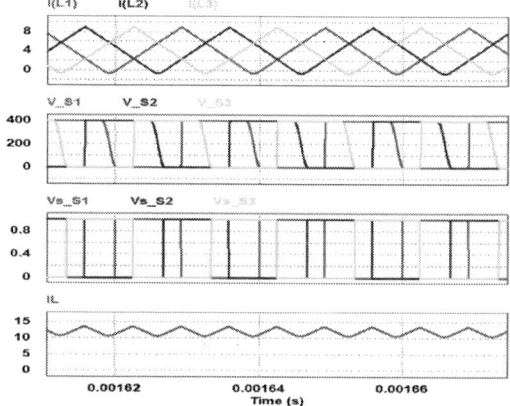

Fig. 8. Waveforms of 3-phase interleaved converter in charging mode with full load

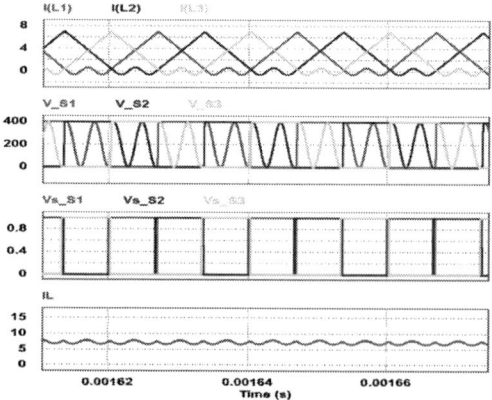

Fig. 9. Waveforms of 3-phase interleaved converter in charging mode with half load

It is found that the battery current ripple component is reduced considerably. The waveforms in discharging mode are similar to those in charging mode.

(d) Conclusions

A new switching scheme for 3-phase interleaved bidirectional DC-DC converter is proposed. The proposed scheme provides not only reduced switching loss but also low conduction loss. The major feature of the proposed scheme is as follows.

- The number of capacitors for ZVS is reduced to 3 that is half of the conventional converter.
- ZVS and ZCS conditions are guaranteed at the instant of turn-on of the switches.
- ZVS condition can be provided at the instant of turn-off of the switches.
- The soft switching conditions can be obtained regardless of the output power level.

The validity of the proposed scheme is shown through simulations and experimental set up is under construction. It is expected that the proposed scheme can be used in the field of bidirectional DC-DC converters for battery, supercapacitor, electric vehicle, etc. with small size, light weight, and high efficiency.

Acknowledgment

This work was supported by the Power Generation & Electricity Delivery of the Korea Institute of Energy Technology Evaluation and Planning (KETEP) grand founded by the Korea government Ministry of Knowledge Economy. (No. 20111020400260)

References

[1] P.J. Grbović, P. Delarue, P. Le Moigne, P. Bartholomeus, "A bidirectional Three-level DC-DC converter for the ultracapacitor applications," *IEEE Trans. on Industrial Electronics*, vol. 57, no. 10, pp. 3415-3430, 2010.

[2] R. Ayyanar, N. Mohan, "A novel full-bridge DC-DC converter for battery charging using secondary-side control combines soft switching over the full load range and low magnetics requirement," *IEEE Trans. on Industry Applications*, vol. 37, no. 2, pp. 559-565, 2001.

[3] Zhan Wang, Hui Li, "A soft switching three-phase current-fed bidirectional dc-dc converter with high efficiency over a wide input voltage range," *IEEE Trans. on Power Electronics*, vol. 27, no. 2, pp. 669-684, 2012.

[4] G. Eason, B. Noble, and I.N. Sneddon, "On certain integrals of Lipschitz-Hankel type involving products of Bessel functions," *Phil. Trans. Roy. Soc. London*, vol. A247, pp. 529-551, April 1955.

[5] J. Dudrik, N.-D. Trip, "Soft-switching PS-PWM DC-DC converter for full-load range applications," *IEEE Trans. on Industrial Electronics*, vol. 57, no. 8, pp. 2807-2814, 2010.

[6] Junhong Zhang, Jih-Sheng Lai, Rae-young Kim, Wensong Yu, "High-power density design of a soft-switching high-power bidirectional DC-DC converter," *IEEE Trans. on Power Electronics*, vol. 22, no. 4, pp. 1145-1153, 2007.

Modified Three-Phase Three-Level DC-DC Converter -Adopting Asymmetrical Duty Cycle Control

Yue Chen*, Xuling Chen**, Fuxin Liu*, Xinbo Ruan*

*: Jiangsu Key Laboratory of New Energy Generation and Power Conversion	**: College of Mechanical and Electrical Engineering
Nanjing University of Aeronautics and Astronautics	Nanjing University of Aeronautics and Astronautics
Nanjing, China	Nanjing, China

Abstract—**Three-phase three-level (TPTL) dc/dc converter has the advantages of lower voltage and current stress on switches, which is suitable for high power and high input voltage applications. Adopting a symmetrical control strategy, the ripple frequency of input and output current can be increased significantly, resulting in a reduced filter requirement. However, all the switches are hard-switching, leading to a considerable switching loss. In this paper, an asymmetrical duty cycle control strategy was proposed to the TPTL dc/dc converter. The converter can keep all the advantages compared with the original control strategy, meanwhile, soft-switching can be achieved using the energy stored in output filter inductance and leakage inductances of transformers (or resonant inductances). In this paper, the operation principle with asymmetrical duty cycle control was analyzed, along with the characteristics and soft-switching condition of the converter. Experimental results from a 540~660V input and 48V/20A output are presented to verify the theoretical analysis and the performance of the proposed converter.**

Keywords— Asymmetrical duty cycle control, three-phase three-level, zero-voltage-switching, dc/dc converter

I. INTRODUCTION

Full-bridge dc/dc converters have been widely used in the medium-to-large power applications for the simple structure, constant frequency control strategy and soft-switching characteristics. To further reduce the power rating on switches[1-3], a prominent three-phase full-bridge dc/dc converter was put forward[1]. It has the superior features including lower current rating of switches, increased input and output current pulse frequency allowing small-size filter requirement. However, the switches still sustain the whole input voltage due to the fact it is derived by full bridge converters.

To reduce the voltage stress on switches, three-phase three-level (TPTL) dc/dc converters were proposed[4-6]. The converter in [6] has the advantages of simple structure, low voltage and current stress on switches, which is suitable in high voltage and high power applications. Adopting a symmetrical duty cycle control, the converter

can increase the frequency of input and output current ripple greatly to reduce the size and weight of the filter. However, it also has the disadvantages of hard-switching, high switching loss and low efficiency.

In order to realize soft-switching, an asymmetrical control strategy is proposed in this paper, which keeps the advantages of original symmetrical control strategy; meanwhile, the output filter inductance and leakage inductance of transformer can be used to achieve zero-voltage-switching (ZVS) for switches. In this paper, the operation principle with asymmetrical control strategy is introduced briefly. The characteristics of the converter and conditions to realize soft-switching are discussed. The experimental verification from a prototype of 540~660V input and 48V/20A output is carried out.

II. THREE-PHASE THREE-LEVEL DC/DC CONVERTER

Fig. 1 shows the main topology of TPTL dc/dc converter, in which the three-phase transformer is Δ-Y connection. C_{d1} and C_{d2} are large enough and they share evenly the input voltage which can be equivalent to constant voltage. Three switching bridges of the converter are composed of Q_1 and Q_4, Q_2 and Q_5, Q_3 and Q_6 respectively. D_{f1} and D_{f2} are freewheeling diodes. C_{ss} is the flying capacitor, which is in favor of decoupling the switching transition of Q_1, Q_3, Q_4 and Q_6. L_{lka}, L_{lkb}, L_{lkc} are leakage inductance of three-phase transformer, and L_{ra}, L_{rb}, L_{rc} are resonant inductances of each phase. D_{R1}~D_{R6} are rectifier diodes. The output filter is composed of L_f and C_f, and R_{Ld} is the load.

III. ASYMMETRICAL DUTY CYCLE CONTROL STRATEGY FOR TPTL CONVERTER

Fig. 2 shows the switching sequences of the original control strategy and the modified control strategy, as shown in Fig. 2(a), Q_1~Q_6 are switched on in turn with interval of one-sixth switching period, the duty cycles of all switches are equal, and each switch has a maximum conduction period of 120°. The required range for the duty cycle of any switch is from 0.167 to 0.33. Obviously, the two interleaved switches have a simultaneous turn-off interval, during which, the intrinsic capacitors of two switches will resonate with the leakage inductances of transformers for several periods. As the duty cycle of the

---

Sponsored by grants from the Power Electronics Science and Education Development Program of Delta Environmental & Educational Foundation, NO. DREG2013005, the Fundamental Research Funds for the Central Universities, China, NO. NS2014025, and Jiangsu province university outstanding science and technology innovation team project.

978-1-4799-2706-7/14 $31.00 © 2014 IEEE

The 2014 International Power Electronics Conference

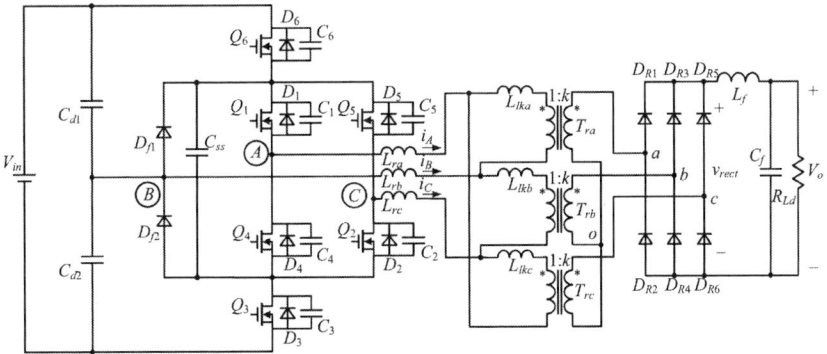

Fig. 1. Three-phase three-level DC/DC converter

switches varies with the input voltage and the load, the incoming switch cannot be ensured to turn on exactly when its drain-to-source voltage resonates to zero within the operation range, therefore, the switches suffer hard-switching and a considerable switching loss occurs. To realize the soft-switching for switches, the original interleaved switches should be designed in a complementary manner, and a short delay time is necessary to be introduced between the two complementary switches to provide an interval for the ZVS commutation, which is similar to the control strategy of asymmetrical half-bridge converter. Accordingly, the duty cycles of Q_1, Q_3 and Q_5 are served to regulate the output voltage, while the drive signals of Q_4, Q_6 and Q_2 are complementary to that of the Q_1, Q_3 and Q_5, respectively. The obtained control strategy is illustrated in Fig. 2(b).

when the duty cycle varies between (0, 1/3), (1/3, D_r) and (D_r, 1/2), where D_r is a critical duty cycle that depends on the load current and the parameters of converter. The related waveforms in different operation modes are referred to Fig. 3(a)-(c).

As seen, the rectified voltage v_{rect} has two voltage levels, $0.75kV_{in}$ and kV_{in} respectively, where k represents secondary-to-primary turns ratios of the transformer. In addition, the output voltage can be modulated by adjusting the duty cycle under the medium and small duty cycle modes. However, as shown in Fig. 3(c), the time intervals for two voltage levels of the rectified voltage are constant during a switching period under the large duty cycle mode. As a result, the output voltage is uncontrollable in this mode, which should be avoided in the practical design.

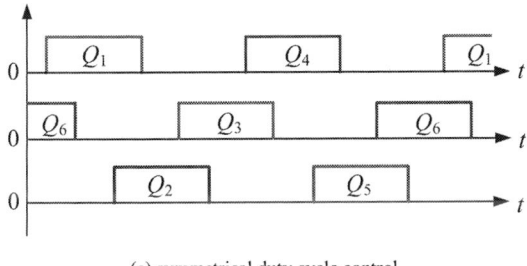

(a) symmetrical duty cycle control

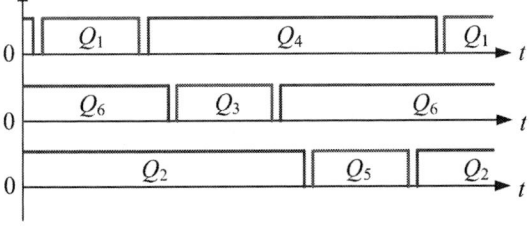

(b) asymmetrical duty cycle control
Fig. 2. Two kinds of control strategies for TPTL converter

Fig. 3 shows the key waveforms of the TPTL converter with asymmetrical duty cycle control, as seen, the operation of the TPTL converter can be classified by different modes, according to the duty cycle range and the load current. The corresponding operation modes are defined as the small duty cycle mode, the medium duty cycle mode, and the large duty cycle mode, respectively,

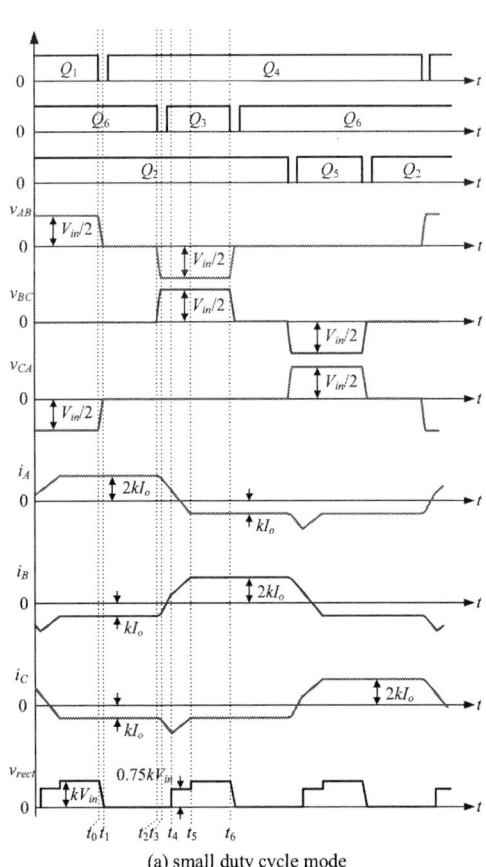

(a) small duty cycle mode

978-1-4799-2706-7/14 $31.00 © 2014 IEEE   2769

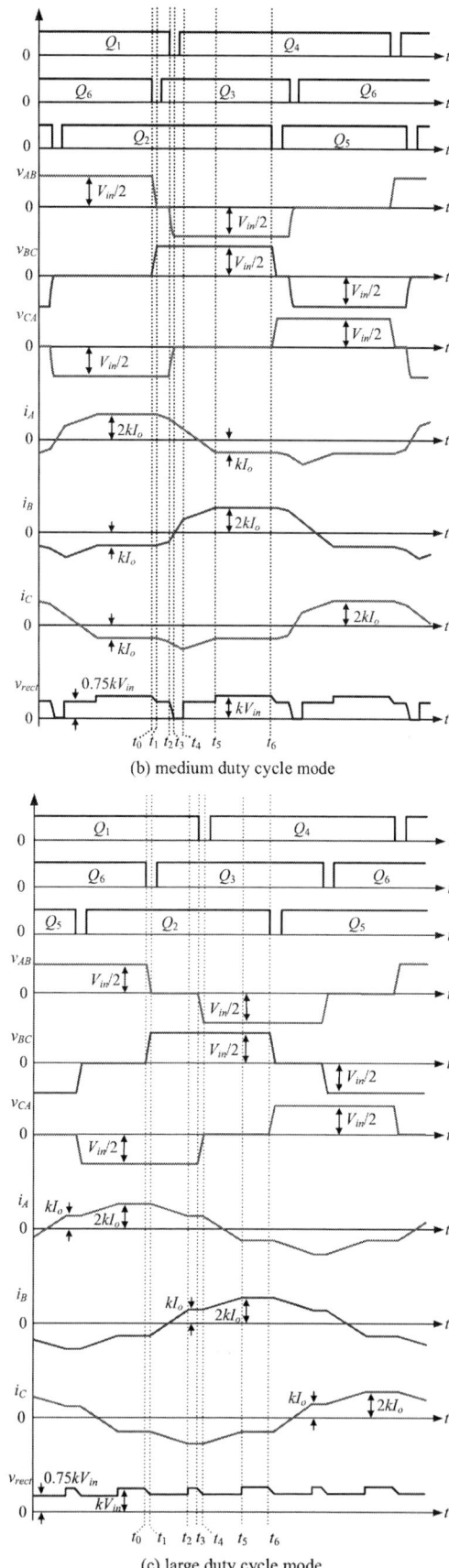

(b) medium duty cycle mode

(c) large duty cycle mode

Fig. 3. Key waveforms under different operation modes

IV. CHARACTERISTICS OF THE CONVERTER

A. Relationship between the Input and Output Voltage

The average value of the rectified voltage is the output voltage. From Fig. 3, the expressions of the output voltage under different operation modes are as follows,

$$
V_o = \begin{cases}
\dfrac{0.75k \cdot V_{in} \cdot t_{45} + k \cdot V_{in} \cdot t_{56}}{T_s/3} & \left(0 < D < \dfrac{1}{3}\right) \\[3mm]
\dfrac{0.75k \cdot V_{in} \cdot (t_{12} + t_{45}) + k \cdot V_{in} \cdot t_{56}}{T_s/3} & \left(\dfrac{1}{3} < D < D_r\right) \\[3mm]
\dfrac{0.75k \cdot V_{in} \cdot (t_{12} + t_{45}) + k \cdot V_{in} \cdot (t_{23} + t_{56})}{T_s/3} & \left(D_r < D < \dfrac{1}{2}\right)
\end{cases} \quad (1)
$$

One can derive the relationship between input and output voltage by substituting the expressions of different intervals under each mode into (1),

$$
V_o = \begin{cases}
3k \cdot V_{in} \cdot D - 9k^2 \cdot I_o \cdot L_p \cdot f_s & \left(0 < D < \dfrac{1}{3}\right) \\[3mm]
\dfrac{3k \cdot V_{in}}{4} \cdot (1+D) - 9k^2 \cdot I_o \cdot L_p \cdot f_s & \left(\dfrac{1}{3} < D < D_r\right) \\[3mm]
k \cdot V_{in} - 6k^2 \cdot I_o \cdot L_p \cdot f_s & \left(D_r < D < \dfrac{1}{2}\right)
\end{cases} \quad (2)
$$

where $L_p = L_r + 3L_{lk}$, f_s is switching frequency.

As seen in (2), under the large duty cycle mode, the output voltage is independent of duty cycle. As a result, the output voltage cannot be adjusted.

B. Input capacitor balancing analysis

It has been known that by using PWM to control the converter, the input capacitor balancing is a function of the duty cycle and the charging/discharging current. If the symmetrical duty cycle control is utilized, each input capacitor presents one-half of the input voltage. While using an asymmetrical duty cycle control, an analysis of the input capacitor energy must be made. Fig. 4 shows the ideal charging/discharging waveforms for the input capacitor under different operation modes, in which the influence of the leakage inductance and the resonant inductance are omitted without detriment to the analysis. Table I shows energy balance of input capacitors under different operation modes during a switching period, where "↑" represents the capacitor is discharging and "↓" represents charging.

From Fig. 4 and Table I, it can be seen that the capacitor C_{d1} is discharged in the interval ΔT_1 and is charged in the interval ΔT_2, here, $\Delta T_2 = 2\Delta T_1$ and the charging current is one-half of the discharging current. The opposite operation occurs in C_{d2}. As a result, the total energy variations of input capacitors are equal to zero, considering that the amount of energy variations is equal in two intervals. Therefore, during a switching period the total voltage variation in each input capacitor is equal to zero, and all input capacitors voltages remain equal to one-half of the input voltage.

C. Output Filter Inductance

For three-phase configuration is adopted, the frequency of rectified voltage is 1.5 times of that in single-phase converters, which can reduce the current ripple of output filter inductance and filter requirement. Fig. 5 and Fig. 6

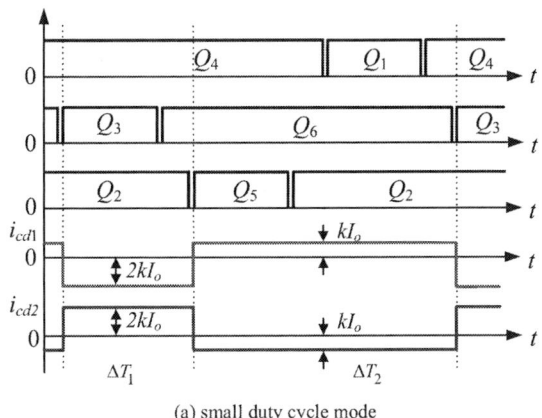

(a) small duty cycle mode

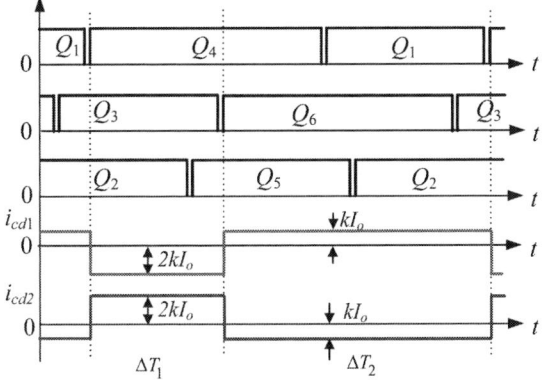

(b) medium duty cycle mode

Fig. 4. Charge and discharge waveform of C_{d1}、C_{d2}

TABLE I
ENERGY BALANCE OF INPUT CAPACITORS UNDER DIFFERENT
OPERATION MODES

Small duty cycle mode			Medium duty cycle mode		
Time interval	ΔT_1	ΔT_2	Time interval	ΔT_1	ΔT_2
C_{d1}	↑	↓	C_{d1}	↑	↓
C_{d2}	↓	↑	C_{d2}	↓	↑
Total energy	0	0	Total energy	0	0

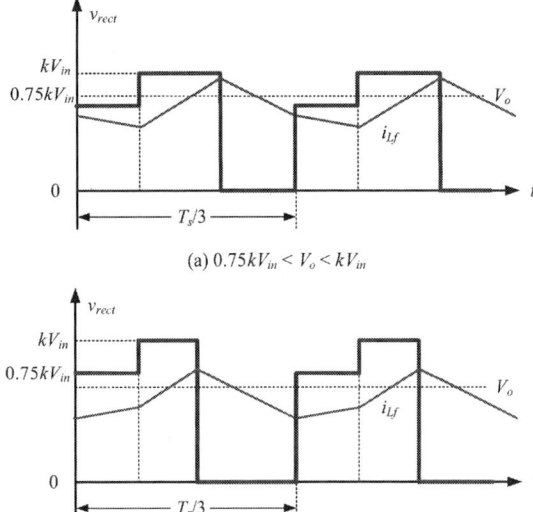

(a) $0.75kV_{in} < V_o < kV_{in}$

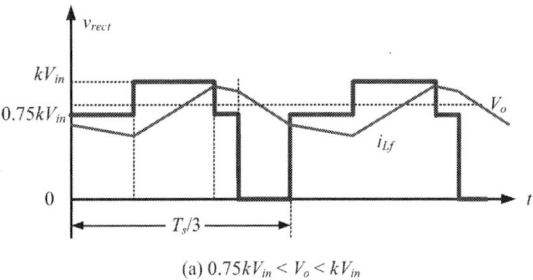

(b) $V_o < 0.75kV_{in}$

Fig. 5. Waveforms of the rectified voltage and output filter inductance current under small duty cycle mode

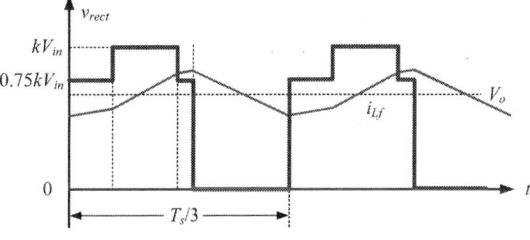

(a) $0.75kV_{in} < V_o < kV_{in}$

(b) $V_o < 0.75kV_{in}$

Fig. 6. Waveforms of the rectified voltage and output filter inductance current under medium duty cycle mode

shows the rectified voltage v_{rect} and output filter inductance current i_{Lf} under small and medium duty cycle modes, respectively. The output filter inductance under different operation modes can be derived as follows,

$$
L_{f\_TP} = \begin{cases}
\frac{k \cdot V_{in} - V_o}{\Delta i_{Lf} \cdot f_s} \cdot \left(\frac{V_o}{k \cdot V_{in}} - \frac{3k \cdot I_o \cdot L_p \cdot f_s}{V_{in}} \right) & \left(0.75kV_{in} < V_o < kV_{in}, 0 < D < \frac{1}{3} \right) \\
\frac{V_o}{3\Delta i_{Lf} \cdot f_s} \cdot \left(1 - \frac{V_o}{k \cdot V_{in}} - \frac{3k \cdot I_o \cdot L_p \cdot f_s}{V_{in}} \right) & \left(V_o < 0.75kV_{in}, 0 < D < \frac{1}{3} \right) \\
\frac{k \cdot V_{in} - V_o}{\Delta i_{Lf} \cdot f_s} \cdot \left(1 - \frac{2V_o}{3k \cdot V_{in}} - \frac{12k \cdot I_o \cdot L_p \cdot f_s}{V_{in}} \right) & \left(0.75kV_{in} < V_o < kV_{in}, \frac{1}{3} < D < D_r \right) \\
\frac{2V_o}{3\Delta i_{Lf} \cdot f_s} \cdot \left(1 - \frac{V_o}{k \cdot V_{in}} - \frac{6k \cdot I_o \cdot L_p \cdot f_s}{V_{in}} \right) & \left(V_o < 0.75kV_{in}, \frac{1}{3} < D < D_r \right)
\end{cases}
\tag{3}
$$

where Δi_{Lf} is the output filter inductance current ripple.

In order to illustrate the superior characteristics of the TPTL converter, the half-bridge TL converter is adopted to make the comparison. The output filter inductance is

$$
L_{f\_HB} = \frac{V_o \cdot \left(1 - \frac{2V_o}{k_{HB} \cdot V_{in}} \right)}{2\Delta i_{Lf} \cdot f_s}
\tag{4}
$$

where k_{HB} is the turns ratio of the transformer in the half-bridge TL converter. Fig. 7 shows the comparison of output filter inductance between TPTL converter and half-bridge TL converter, where V_{in}=540~660V, V_o=48V, f_s=50kHz, I_o=20A, Δi_{Lf}=4A. As shown, the TPTL converter which adopted asymmetrical control strategy can save the output filter inductance significantly, which is reduce by a factor of about 52% compared with the half-bridge TL converter.

D. Current Stress and Voltage Stress on Switches

Ignoring the affection of leakage inductance and output filter inductance, ideal current waveforms flowing through switches under small and medium duty cycle mode are shown in Fig. 8. The rms currents through the switches Q_1 and Q_4 are given by

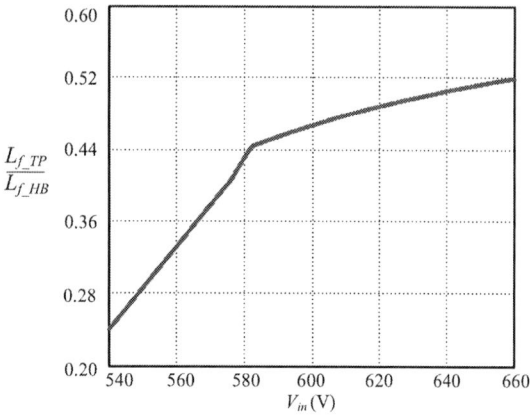

Fig. 7. Comparison of L_f between TPTL converter and half-bridge TL converter

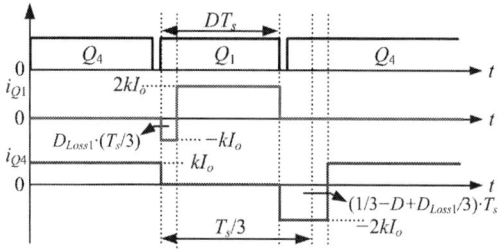

(a) small duty cycle mode

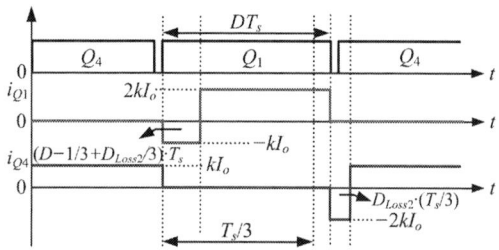

(b) medium duty cycle mode

Fig. 8. Ideal current waveforms flowing through switches

$$
I_{rms\_Q1} = \begin{cases} k \cdot I_o \cdot \sqrt{4D - D_{loss1}} & \left(0 < D < \dfrac{1}{3}\right) \\[3mm] k \cdot I_o \cdot \sqrt{1 + D - D_{loss2}} & \left(\dfrac{1}{3} < D < D_r\right) \end{cases} \quad (5)
$$

$$
I_{rms\_Q4} = \begin{cases} k \cdot I_o \cdot \sqrt{2 - 4D + D_{loss1}} & \left(0 < D < \dfrac{1}{3}\right) \\[3mm] k \cdot I_o \cdot \sqrt{1 - D + D_{loss2}} & \left(\dfrac{1}{3} < D < D_r\right) \end{cases} \quad (6)
$$

where D_{loss1} and D_{loss2} are duty cycle loss under small and medium duty cycle modes respectively, which are expressed as follows,

$$
D_{loss1} = \frac{6k \cdot I_o \cdot L_p}{V_{in} \cdot T_s} \quad (7)
$$

$$
D_{loss2} = 2 - \frac{2V_o}{k \cdot V_{in}} - \frac{12k \cdot I_o \cdot L_p}{V_{in} \cdot T_s} \quad (8)
$$

Fig. 9 shows the comparison between rms values of current flowing through switches in the TPTL converter

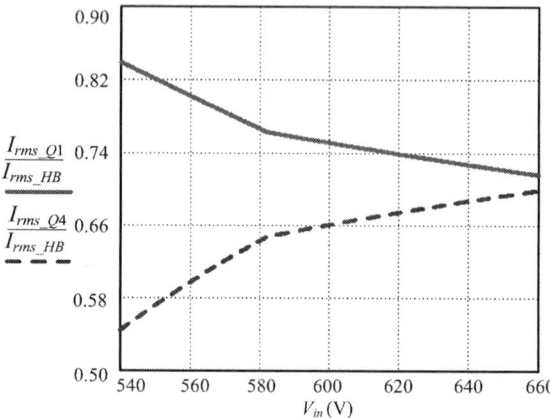

Fig. 9. Comparison between rms values of current flowing through switches

with asymmetrical control and half-bridge TL converter with phase-shifted control. As seen, the rms value of current in TPTL converter can also be reduced compared with half-bridge TL converter. However, the current stresses on different switches are unequal for the asymmetrical duty cycle.

As for the voltage rating on switches, thanking for the TL configuration, the voltage stress on switches will be limited at half of the input voltage.

E. Soft-switching Conditions

When switches are switched off, the intrinsic capacitors limit the rising rate of voltage, which make it possible for all switches to realize zero-voltage turn-off. The conditions of zero-voltage turn-on are discussed as follows.

● Small Duty Cycle Mode

As Fig. 3(a) shows, before Q_2, Q_4 and Q_6 are switched on, the line current remains constant and voltage of intrinsic capacitors varies linearly. In order to realize zero-voltage turn-on, the line current needs to discharge the intrinsic capacitor of the incoming switch fully during the delay time. The minimum load current is given by

$$
I_{o\_min\_Q_2(Q_4,Q_6)} = \frac{V_{in} \cdot C_p}{2t_d \cdot k} \quad (9)
$$

where t_d is the delay time between complementary switches, C_p is the intrinsic capacitor of switches.

Before Q_1, Q_3 and Q_5 turn on, the resonance among the intrinsic capacitors, the leakage inductances and resonant inductances occurs, while the energy to realize ZVS is provided by the leakage inductances and resonant inductances. The minimum load current needed is given by

$$
I_{o\_min\_Q_1(Q_3,Q_5)} = \frac{V_{in}}{Z_{r1} \cdot k} \quad (10)
$$

where $Z_{r1} = \sqrt{L_p / C_p}$.

From (9) and (10), Fig. 10(a) shows the minimum load current to realize ZVS under small duty cycle mode. From Fig. 10(a), it can be known that the minimum load current to realize ZVS increases with the input voltage. In

978-1-4799-2706-7/14 $31.00 © 2014 IEEE

addition, it is more difficult for Q_1, Q_3 and Q_5 to realize ZVS compared with Q_2, Q_4 and Q_6.

● *Medium Duty Cycle Mode*

Likewise, the conditions to realize ZVS for switches under this mode can be obtained by from Fig. 3(b),

$$I_{o\_min\_Q_1(Q_3,Q_5)} = \frac{V_{in} \cdot C_p}{t_d \cdot k} \quad (11)$$

$$I_{o\_min\_Q_2(Q_4,Q_6)} = \frac{\sqrt{3}V_{in}}{4Z_{r2} \cdot k} \quad (12)$$

where $Z_{r2} = \sqrt{L_p / 4C_p}$.

It should be noted that the critical point between medium and small duty cycle modes is related to the duty cycle and load current from (2). Substituting $D=1/3$ into the second expression in (2), the minimum load current to ensure the converter into medium duty cycle mode can be obtained,

$$I_{o\_min} = \frac{k \cdot V_{in} - V_o}{9k^2 \cdot L_p \cdot f_s} \quad (13)$$

According to (11)-(13), Fig. 10(b) shows the relationship between the minimum load current and input voltage. As seen, within the input voltage range, the minimum load current which makes the converter operate under medium duty cycle mode is more than the minimum load current to realize ZVS for switches. As a result, all switches can realize ZVS under medium duty cycle mode.

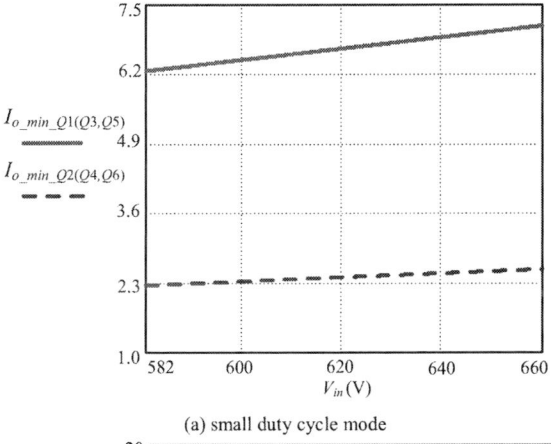

(a) small duty cycle mode

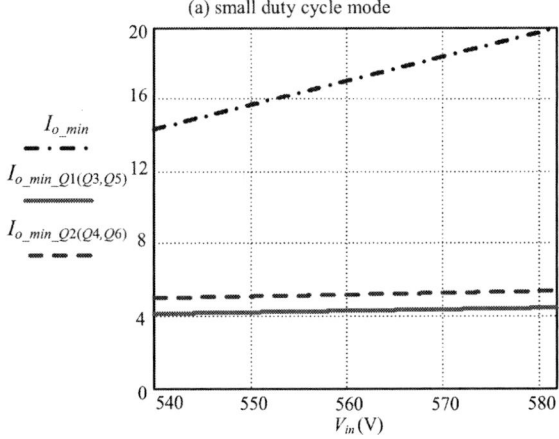

(b) medium duty cycle mode

Fig. 10. Minimum load current to realize ZVS for switches

V. EXPERIMENTAL VERIFICATIONS

In order to verify the operation principle of TPTL converter with asymmetrical control strategy, a prototype about 1kW was built in the laboratory. The specifications are given as follows: V_{in}=540~660V, V_o=48V, I_o=20A, f_s=50kHz, k=1:9, L_r=44μH, L_{lk}=18μH, L_f=22μH, Q_1~Q_6: IPW65R080CFD, D_{f1}~D_{f2}: DSEI30-06A, D_{R1}~D_{R6}: DSEP30-03A. The asymmetrical duty cycle control is implemented with TMS320F2812.

Fig. 11 shows the experimental waveforms at full load with V_{in}=540V and V_{in}=660V, under medium and small duty cycle modes respectively. As shown, the frequency of the rectified voltage is three times the switching frequency and two voltage levels appear in the rectified voltage, which are well in agreement with the theoretical analysis.

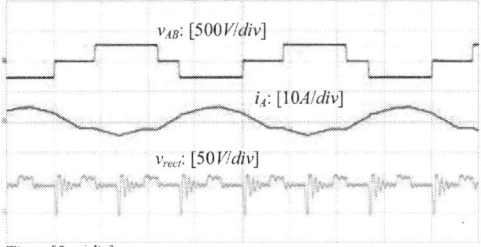

(a) medium duty cycle mode

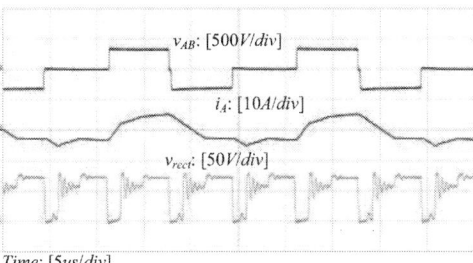

(b) small duty cycle mode

Fig. 11. Experimental waveforms

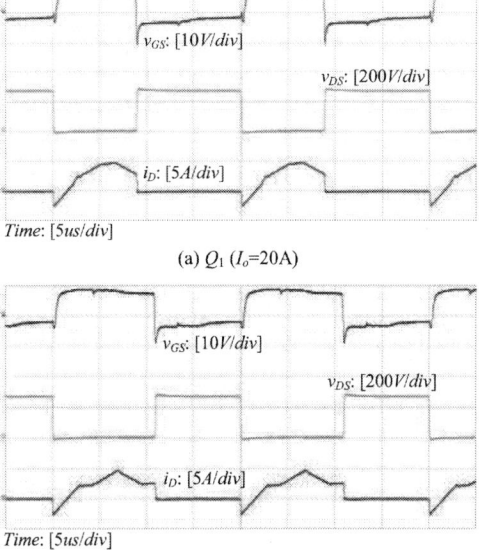

(a) Q_1 (I_o=20A)

(b) Q_4 (I_o=20A)

Fig. 12. ZVS waveforms of under medium duty cycle mode

Fig. 12 shows ZVS waveforms of V_{in}=540V at full load while Fig. 13 shows that of V_{in}=660V at light load. As shown, the voltage stress on switches is half of the input voltage. Furthermore, before each switch turns on, v_{DS} reduces to zero and is clamped by their body diodes, so they are all zero-voltage turn-on. In addition, the minimum load current for switches to realize ZVS in Fig. 13 is in agreement with the theoretical value in Fig. 10(a).

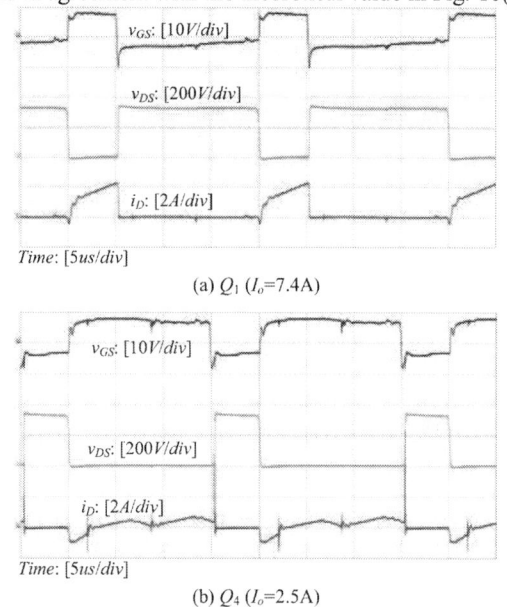

(a) Q_1 (I_o=7.4A)

(b) Q_4 (I_o=2.5A)

Fig. 13. ZVS waveforms of under small duty cycle mode

Fig. 14 shows comparison of efficiency with different control strategies. It is can be observed that the converter with asymmetrical duty cycle control has higher efficiency because it can realize ZVS for switches.

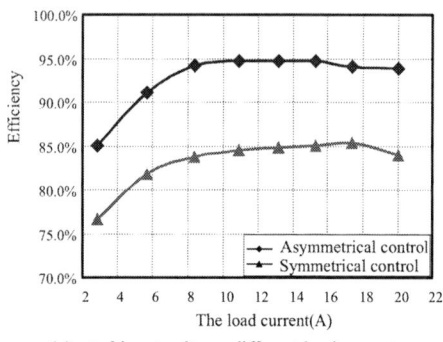

(a) rated input voltage, different load current

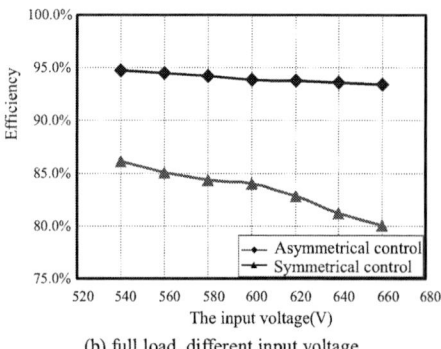

(b) full load, different input voltage

Fig. 14. Comparison of efficiency under different control strategies

VI. CONCLUSIONS

This paper proposed an asymmetrical duty cycle control strategy for the TPTL converter, which has the followings characteristics. 1) The converter can reduce the current rating of switches. All power switches sustain only half of the input voltage; meanwhile, the output rectified inductance is reduced. 2) Under different duty cycles and load, the converter has three operation modes, and the output voltage cannot be adjusted under large duty cycle mode. 3) All switches can realize ZVS due to the energy of output filter inductance and leakage inductances of the transformer, so that the converter has high efficiency. 4) Complementary switches have different current stress.

ACKNOWLEDGEMENT

Sponsored by grants from the Power Electronics Science and Education Development Program of Delta Environmental & Educational Foundation, NO. DREG2013005, the Fundamental Research Funds for the Central Universities, China, NO. NS2014025, and Jiangsu province university outstanding science and technology innovation team project.

REFERENCES

[1] A. R. Prasad, P. D. Ziogas, and S. Manias. "Analysis and design of a three-phase off-line dc/dc converter with high-frequency isolation," *IEEE Trans. on Ind. Appl.*, vol. 28, no. 4, pp. 824–832, 1992.

[2] H. Cha and P. Enjeti, "A novel three-phase high power current-fed DC/DC converter with active clamp for fuel cells," *IEEE PESC*, pp. 2485-2489, 2007.

[3] H. Kim, C. Yoon, and S. Choi, "A three-phase zero-voltage and zero current switching dc–dc converter for fuel cell applications," *IEEE Trans. on Power Electronics*, vol. 25, no. 2, pp. 391-398, 2010.

[4] E. Agostini and I. Barbi, "Three-phase three-level PWM DC-DC converter," *IEEE Trans. on Power Electronics*, vol. 26, no. 7, pp. 1847-1856, 2011.

[5] D. V. Ghodke, K. Chatterjee, and B. G. Fernandes, "Three-phase three level, soft switched, phase shifted PWM dc–dc converter for high power applications," *IEEE Trans. on Power Electronics*, vol. 23, no. 3, pp. 1214-1227, 2008.

[6] F. Liu, G. Hu and X. Ruan, "Three-phase three-level DC/DC converter for high input voltage and high-power applications adopting symmetrical duty cycle control," *IEEE Trans. on Power Electronics*, vol. 29, no. 1, pp. 56-65, 2014.

The 2014 International Power Electronics Conference

Deadbeat Control of Power Leveling Unit with Bidirectional Buck/boost DC/DC Converter

Shin-ichi Hamasaki, Ryosuke Mukai, Yoshihiro Yano, Mineo Tsuji
Division of Electrical Engineering and Computer Science,
Nagasaki University,
Nagasaki, Japan
hama-s@nagasaki-u.ac.jp

Abstract— **As a distributed generation system increases, a stable power supply becomes difficult. Thus control of power leveling (PL) unit is required to maintain the balance of power flow for irregular power generation. The unit is required to respond to change of voltage and bidirectional power flow. So the bidirectional buck/boost DC/DC converter is applied for the control of PL unit in this research. The PL unit with Electric double-layer capacitor (EDLC) is able to absorb change of power, and it is examined whether the stable power supply is possible. The output current of PL unit is controlled so as to keep power balance and DC bus voltage. The effectiveness of the deadbeat control for power leveling unit is proved in simulation and experiment.**

Keywords— *power leveling, Bidirectional DC/DC converter, EDLC, deadbeat control*

I. INTRODUCTION

In recent years, problems of exhaustion of fossil fuel and global warming by CO_2 emission being focused, a distributed generator of renewable energy sources such as the photovoltaic (PV) and the wind power generation is attracting attention. New power supply systems, such as a smart grid using distributed generators are expanding[1]-[4]. Since the output of renewable power sources has fluctuation and instability, however, problems of reverse power flow and voltage optimization occur. Especially, as the renewable power sources increase, change of the power supply becomes intense. Therefore, power fluctuation should be absorbed and the system to equalize the power supply is required as a power leveling. An electric double-layer capacitor (EDLC) has advantages of low inner resistance, large capacity, and long life compared with the secondary battery. EDLC is suitable for absorbing frequent change of power.

In this research, the control system of power leveling (PL) unit combined EDLC with the bidirectional buck/boost DC/DC converter[2] is investigated. The bidirectional buck/boost DC/DC converter can work four-quadrant operation, which is positive and negative voltages and currents. Therefore, the current is able to flow for power charge to EDLC or power supply from EDLC. Even if EDLC is maintaining high voltage or voltage becomes small, it is possible for the PL unit to work adequately. DC bus voltage must be maintained

constant, thus the PL unit should operate the leveling quickly for instantaneous fluctuation of power generation. Buck/boost DC/DC Converter applying the deadbeat control[5] based on linearization is proposed to obtain quick response. Effectiveness of the proposed system is verified by simulation and experiment.

II. CIRCUIT STRUCTURE AND CONTROL METHOD

A. Circuit structure

Fig.1 shows a circuit structure of the system in this research. In Fig.1, PL unit consists of Buck/boost DC/DC converter and EDLC. A grid connected inverter is installed between AC bus and DC bus. LC filters are connected to the output terminal of the PL unit and the inverter in order to suppress the switching ripple. The boost chopper is connected to the PV cell for regulation of PV power.

The control is performed by combining two controls, for active and reactive power in AC side by the inverter, and for input and output power control in DC side by the PL unit. The rate of the output power supplied to the load in AC side can be controlled by the inverter. A current reference I_{ed}^{*} of the PL unit is calculated from the active component current I_{Gd} and the output current I_{pv} of a PV cell. Further, in order to keep the DC bus voltage constant, feedback control of the V_{dc} is carried out.

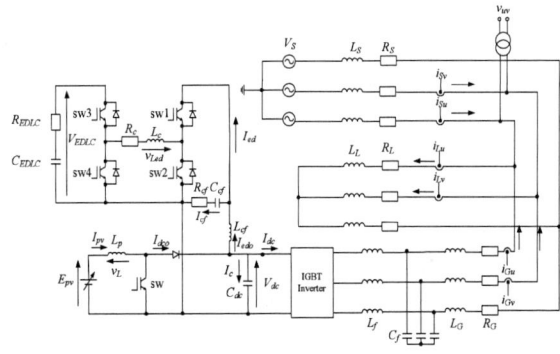

Fig. 1. Circuit structure

978-1-4799-2706-7/14 $31.00 © 2014 IEEE

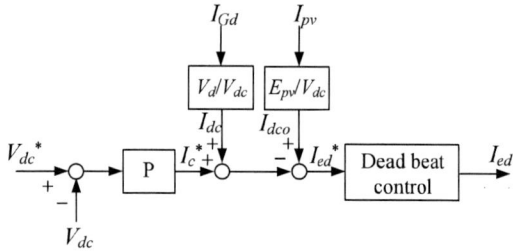

Fig. 2. Block diagram of power leveling unit control

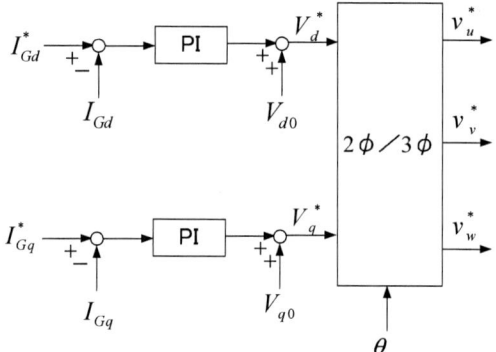

Fig. 3. Block diagram of inverter control.

B. Control of power leveling and inverter

Fig.2 shows a block diagram of the PL unit control and Fig.3 shows a block diagram of the inverter control respectively. Fig.2 is a block diagram in case that the DC/DC converter performs in a buck operation as seen from EDLC. Deadbeat control for output current regulator is performed. I_{edo} is obtained through LPF of I_{ed}.

The following equations are derived from Fig.2.

$$V_{dc}I_{dc} = V_d I_{Gd} \tag{1}$$

$$E_{pv}I_{pv} = V_{dc}I_{dco} \tag{2}$$

$$I_{ed} \simeq I_{dco} - I_c - I_{dc} \tag{3}$$

$$I_c^* = K_{pc}\left(V_{dc}^* - V_{dc}\right) \tag{4}$$

(1)-(3) express a balance of active power and current. LPF is neglected due to few losses in (3). DC bus voltage V_{dc} is regulated by the P control in (4).

The inverter control is performed by a block diagram in Fig.3. The dq-component converting from the output voltage and currents of the distributed generation system determine the active component and the reactive component. Using them, power flow can be controlled by PI control in (5) and (6). Then V_{d0} and V_{q0} are reference voltage calculated from line-to-line voltages on AC bus.

$$V_d^* = V_{d0} + K_{pd}\left(I_{Gd}^* - I_{Gd}\right) + K_{id}\int_0^t \left(I_{Gd}^* - I_{Gd}\right)dt \tag{5}$$

$$V_q^* = V_{q0} + K_{pq}\left(I_{Gq}^* - I_{Gq}\right) + K_{iq}\int_0^t \left(I_{Gq}^* - I_{Gq}\right)dt \tag{6}$$

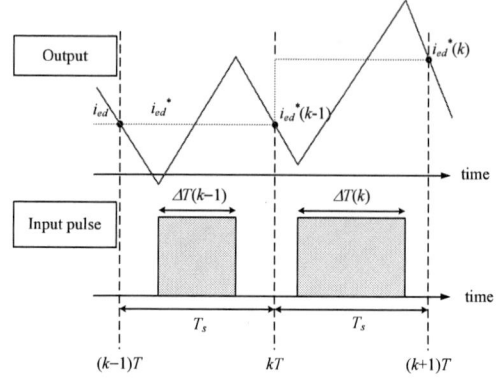

Fig. 4. Principle of deadbeat control

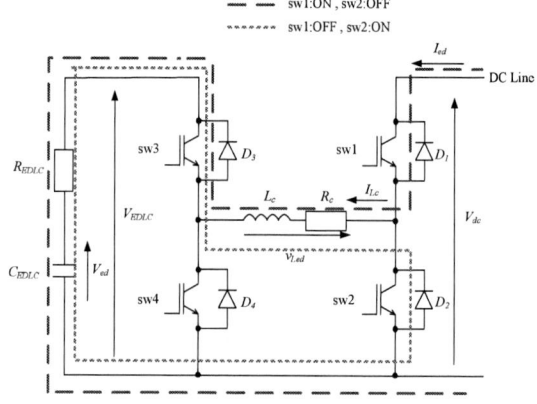

Fig. 5. The operation of boost chopper

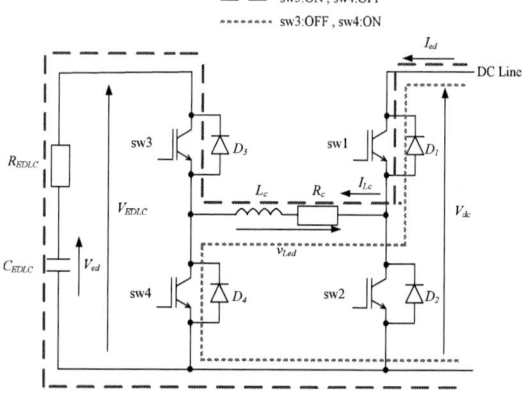

Fig. 6. The operation of buck chopper

C. Deadbeat control of DC/DC converter

In order to operate the PL units at high speed, the deadbeat control is applied. Fig.4 illustrates a principle of the deadbeat control. In this control, it calculates the optimum duty ratio of the PWM at every one period.

Fig.5 shows the operation of boost chopper. The duty ratio of switching is calculated from I_{Lc}^* and the current I_{Lc} of L_c when the bidirectional buck/boost DC/DC converter performed a boost operation as seen from EDLC.

Sw1 and sw2 are alternatively switched according to the calculated duty ratio. Then State equations with respect to on and off operations of sw1 are as follows.

sw1 : ON (sw2 : OFF)

$$\frac{d}{dt}\begin{bmatrix} i_{Lc} \\ v_{ed} \end{bmatrix} = \begin{bmatrix} -\dfrac{R_c + R_{EDLC}}{L_c} & -\dfrac{1}{L_c} \\ \dfrac{1}{C_{EDLC}} & 0 \end{bmatrix} \begin{bmatrix} i_{Lc} \\ v_{ed} \end{bmatrix} + \begin{bmatrix} \dfrac{1}{L_c} \\ 0 \end{bmatrix} V_{dc} \quad (7)$$

sw1 : OFF (sw2 : ON)

$$\frac{d}{dt}\begin{bmatrix} i_{Lc} \\ v_{ed} \end{bmatrix} = \begin{bmatrix} -\dfrac{R_{EDLC} + R_c}{L_c} & -\dfrac{1}{L_c} \\ \dfrac{1}{C_{EDLC}} & 0 \end{bmatrix} \begin{bmatrix} i_{Lc} \\ v_{ed} \end{bmatrix} + \begin{bmatrix} 0 \\ 0 \end{bmatrix} V_{dc} \quad (8)$$

The expression of the operation is derived by the state-space averaging method. d_1 is a duty ratio on the basis of sw1.

$$\frac{d}{dt}\begin{bmatrix} \overline{i_{Lc}} \\ \overline{v_{ed}} \end{bmatrix} = \begin{bmatrix} -\dfrac{R_{EDLC} + R_c}{L_c} & -\dfrac{1}{L_c} \\ \dfrac{1}{C_{EDLC}} & 0 \end{bmatrix} \begin{bmatrix} \overline{i_{Lc}} \\ \overline{v_{ed}} \end{bmatrix} + \begin{bmatrix} \dfrac{d_1}{L_c} \\ 0 \end{bmatrix} \overline{v_{dc}} \quad (9)$$

Linearization of the equation is performed using minute variations defined by the following equations:
$\overline{v}_{dc} \equiv V_{dc} + \Delta v_{dc}$, $\overline{v}_{ed} \equiv V_{ed} + \Delta v_{ed}$, $\overline{i}_{Lc} \equiv I_{Lc} + \Delta i_{Lc}$, $d_1 \equiv D + \Delta d_1$.

$$\frac{d}{dt}\begin{bmatrix} \Delta i_{Lc} \\ \Delta v_{ed} \end{bmatrix} = \begin{bmatrix} -\dfrac{R_{EDLC}}{L_c} & -\dfrac{1}{L_c} \\ \dfrac{1}{C_{EDLC}} & 0 \end{bmatrix} \begin{bmatrix} \Delta i_{Lc} \\ \Delta v_{ed} \end{bmatrix}$$
$$+ \begin{bmatrix} \dfrac{V_{dc}}{L_c} \\ 0 \end{bmatrix} \Delta d_1 + \begin{bmatrix} \dfrac{D_1}{L_c} \\ 0 \end{bmatrix} \Delta v_{dc} \quad (10)$$

Replacing $\Delta i_{LC} = \Delta i_{Lc}{}^*$, $\Delta d_1(k)$ is obtained by backward difference method of z-transform from (10).

$$\Delta d_1(k) = \frac{\Delta i_{Lc}^*(k) - k_2 \Delta i_{Lc}(k-1) - k_3 \Delta v_{ed}(k-1) - k_4 \Delta v_{dc}(k)}{k_1} \quad (11)$$

,where k_1, k_2, k_3 and k_4 are constants to be determined by circuit parameters. The duty ratio $D_1(k)$ at sampling period kT is obtained by (12).

$$D_1(k) = D_1(k-1) + \Delta d_1(k) \quad (12)$$

It is possible to determine the optimal PWM duty ratio $D_1(k)$ by (12). And the output current I_{Lc} can follow the reference value $I_{Lc}{}^*$ with one sample delay.

Fig.6 shows the operation of buck chopper as seen from EDLC. Sw3 and sw4 are alternatively switched according to the calculated duty ratio. Then State equations with respect to on and off operations of sw3 are as follows.

sw3 : ON (sw4 : OFF)

$$\frac{d}{dt}\begin{bmatrix} i_{Lc} \\ v_{ed} \end{bmatrix} = \begin{bmatrix} -\dfrac{R_c + R_{EDLC}}{L_c} & -\dfrac{1}{L_c} \\ \dfrac{1}{C_{EDLC}} & 0 \end{bmatrix} \begin{bmatrix} i_{Lc} \\ v_{ed} \end{bmatrix} + \begin{bmatrix} \dfrac{1}{L_c} \\ 0 \end{bmatrix} V_{dc} \quad (13)$$

sw3 : OFF (sw4 : ON)

$$\frac{d}{dt}\begin{bmatrix} i_{Lc} \\ v_{ed} \end{bmatrix} = \begin{bmatrix} -\dfrac{R_c}{L_c} & 0 \\ 0 & 0 \end{bmatrix} \begin{bmatrix} i_{Lc} \\ v_{ed} \end{bmatrix} + \begin{bmatrix} \dfrac{1}{L_c} \\ 0 \end{bmatrix} V_{dc} \quad (14)$$

The expression of the operation is derived by the state-space averaging method. d_2 is a duty ratio on the basis of sw3.

$$\frac{d}{dt}\begin{bmatrix} \overline{i_{Lc}} \\ \overline{v_{ed}} \end{bmatrix} = \begin{bmatrix} -\dfrac{d_2 R_{EDLC} + R_c}{L_c} & -\dfrac{d_2}{L_c} \\ \dfrac{d_2}{C_{EDLC}} & 0 \end{bmatrix} \begin{bmatrix} \overline{i_{Lc}} \\ \overline{v_{ed}} \end{bmatrix} + \begin{bmatrix} \dfrac{1}{L_c} \\ 0 \end{bmatrix} \overline{v_{dc}} \quad (15)$$

Then the equation is performed by the linearization using minute variations the same as the boost operation in (9) and (10).

$$\frac{d}{dt}\begin{bmatrix} \Delta i_{Lc} \\ \Delta v_{ed} \end{bmatrix} = \begin{bmatrix} -\dfrac{R_c + D_2 R_{EDLC}}{L_c} & -\dfrac{D_2}{L_c} \\ \dfrac{D_2}{C_{EDLC}} & 0 \end{bmatrix} \begin{bmatrix} \Delta i_{Lc} \\ \Delta v_{ed} \end{bmatrix}$$
$$+ \begin{bmatrix} \dfrac{R_{EDLC} I_{Lc} - V_{ed}}{L_c} \\ \dfrac{I_{ed}}{C_{EDLC}} \end{bmatrix} \Delta d_2 + \begin{bmatrix} \dfrac{1}{L_c} \\ 0 \end{bmatrix} \Delta v_{dc} \quad (16)$$

$\Delta d_2(k)$ is calculated by backward difference method of z-transform from (16).

$$\Delta d(k) = \frac{\Delta i_{Lc}^*(k) - m_2 \Delta i_{Lc}(k-1) - m_3 \Delta v_{ed}(k-1) - m_4 \Delta v_{dc}(k)}{m_1} \quad (17)$$

,where m_1, m_2, m_3 and m_4 are constants to be determined by circuit parameters. The Duty ratio $D_2(k)$ is obtained by (18).

$$D_2(k) = D_2(k-1) + \Delta d_2(k) \quad (18)$$

III. SIMULATION

Simulation is performed to verify the proposed control method by using the circuit in Fig.1. Table I shows the parameters of the circuit.

Figs.7 and 8 show simulation results in the case of boost and buck operation respectively. The PV gives practical output with fluctuation using random function.

The 2014 International Power Electronics Conference

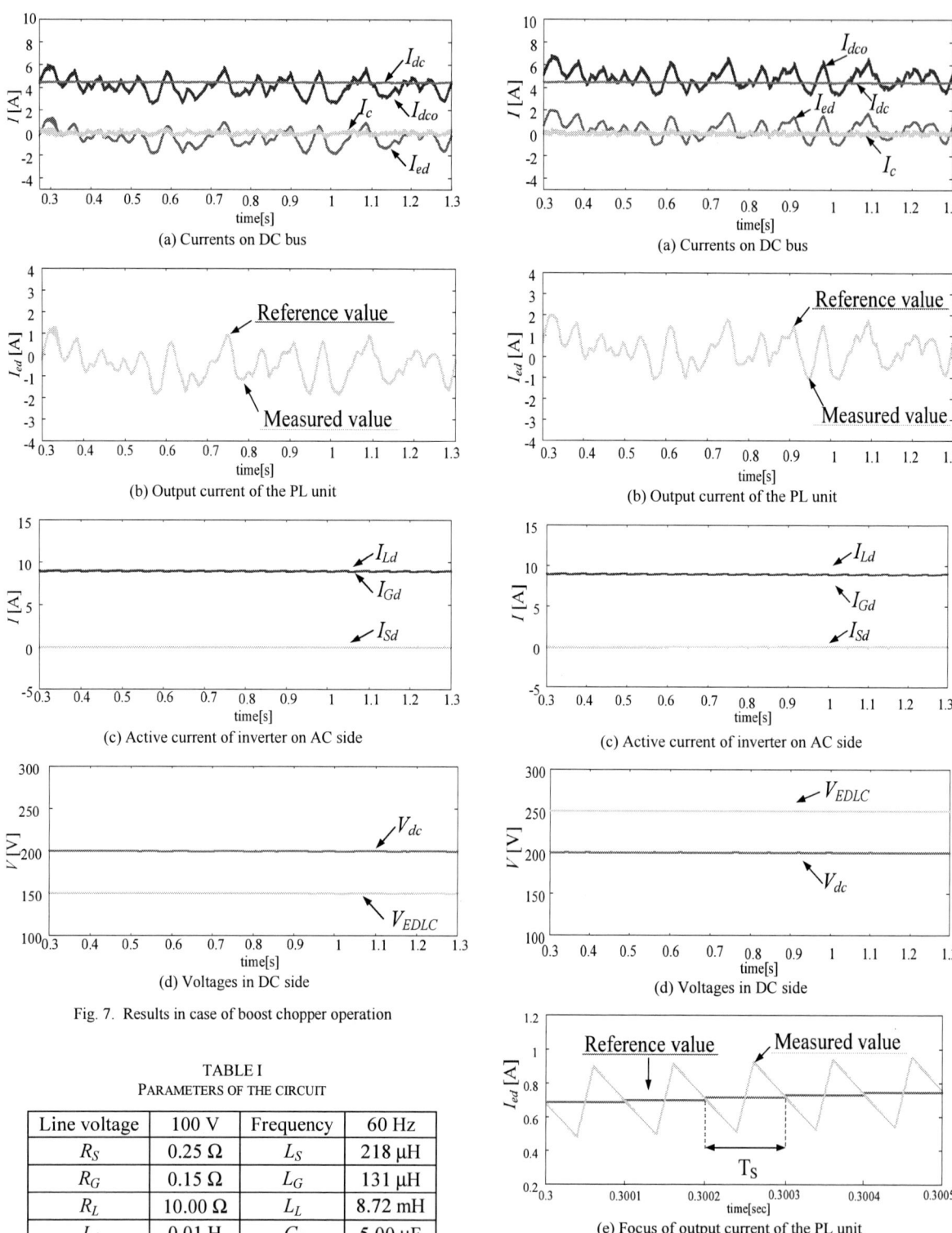

(a) Currents on DC bus

(b) Output current of the PL unit

(c) Active current of inverter on AC side

(d) Voltages in DC side

Fig. 7. Results in case of boost chopper operation

(a) Currents on DC bus

(b) Output current of the PL unit

(c) Active current of inverter on AC side

(d) Voltages in DC side

(e) Focus of output current of the PL unit

Fig. 8. Results in case of buck chopper operation

TABLE I
PARAMETERS OF THE CIRCUIT

Line voltage	100 V	Frequency	60 Hz
R_S	0.25 Ω	L_S	218 μH
R_G	0.15 Ω	L_G	131 μH
R_L	10.00 Ω	L_L	8.72 mH
L_f	0.01 H	C_f	5.00 μF
V_{dc}	200 V	C_{dc}	5000 μF
L_c	10.0 mH	R_c	1.5 mΩ
R_{EDLC}	0.027 Ω	C_{EDLC}	10.0 F
L_p	10.0 mH	E_{pv}	80.0 V
C_{cf}	30.0 μF	L_{cf}	1.5 mH

Fig.7 shows results in case that the bidirectional chopper works as a boost chopper ($V_{EDLC}<V_{dc}$). It is controlled in order to keep balance of the charge and discharge of EDLC against the change of PV output in Fig.7(a) and (b). In Fig.7(b), the positive current means charge to

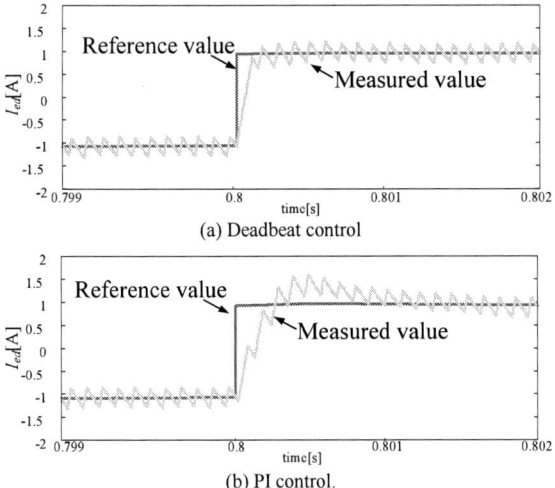

(a) Deadbeat control

(b) PI control.

Fig. 9. Results of comparison in case of buck chopper operation

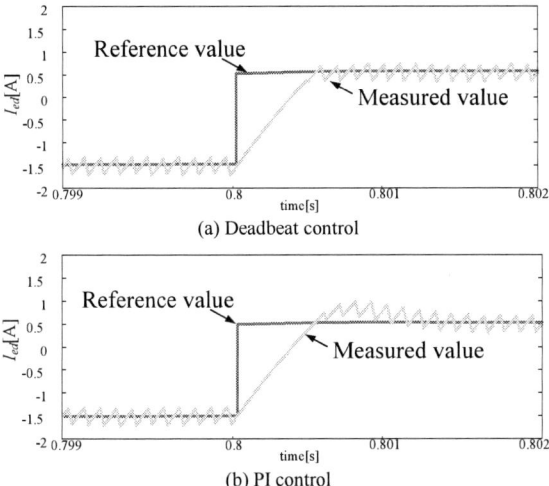

(a) Deadbeat control

(b) PI control

Fig. 10. Results of comparison in case of boost chopper operation

EDLC and the negative one means power supply to the DC bus, respectively. And output current can follow to the reference by the deadbeat control. Fig.7(c) shows results with the power control based on the main AC bus. I_{Sd} becomes almost zero. I_{Gd} and I_{Ld} are the same value. This means that all the power is supplied from only distributed generation system and the main system does not send out the power. In Fig.7(d), DC bus voltage keeps constant, and the bidirectional buck/boost DC/DC converter is operating properly as a boost chopper.

Fig.8 shows results in case that the bidirectional chopper works as a buck chopper ($V_{EDLC} > V_{dc}$). It is controlled in order to keep balance of the charge and power supply of EDLC against change of PV output in Fig.8 (a) and (b). In Fig.8(b), the output current follows to the reference and can compensate the fluctuation of PV output properly. The power supplied to AC side keeps constant in Fig.8 (c). In Fig.8(d), DC bus voltage keeps constant, and the bidirectional buck/boost DC/DC converter is operating properly as a boost chopper. Fig.8(e) shows an enlarged

figure of Fig.8(b). We can see that the measured value follows to the reference value every one sample and is able to confirm quick response for the operation of the PL unit.

Fig.9 shows results of comparison with the deadbeat control and conventional PU control in case of buck chopper operation. The gain design of PI control is selected for fast response and small overshoot. In Fig.9(b), settling time is almost 0.002s and overshoot is 0.5A. On the other hand, settling time of deadbeat control is shorter than the PI control and overshoot is nothing in Fig.9(a). Fig.10 shows results of comparison in case of boost chopper operation. Settling time of the deadbeat control is shorter than the PI control and overshoot is nothing the same as Fig.9.

IV. EXPERIMENT

In order to verify the proposed PL unit operation, Experimental system for DC bus is constructed as shown in Fig.11. The PL unit consists of IGBT (Mitsubishi Elec. : PM50CL1A060) and EDLC (Shizuki Elec. : FML-3A). The PV is simulated by a power AMP. An electric load is used for a constant current consumption.

All the controls are executed by the DSP (TI : TMS320C33-150MHz). PWM switching frequency is 10 kHz and sampling period of DSP is 100μs in the experiment.

Figs.12-14 show the experimental result in three situations.

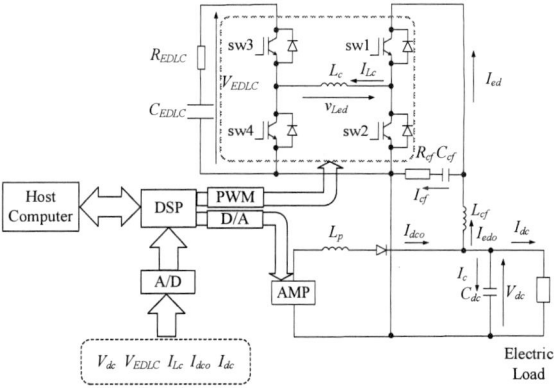

Fig. 11. Configuration of experimental system

TABLE II
PARAMETERS OF EXPERIMENTAL CIRCUIT

L_c	5.0 mH	C_{dc}	10.0 mF
C_{EDLC}	60.0 F	R_{EDLC}	0.027 Ω
C_{cf}	30.0 μF	L_{cf}	1.5 mH
L_p	0.2 mH	$V_{dc}{}^*$	25.0 V

978-1-4799-2706-7/14 $31.00 © 2014 IEEE

The 2014 International Power Electronics Conference

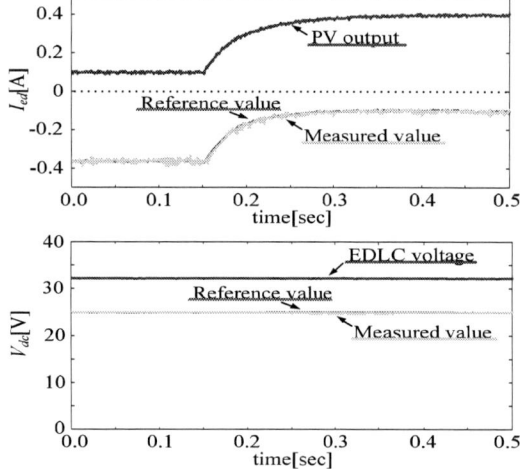

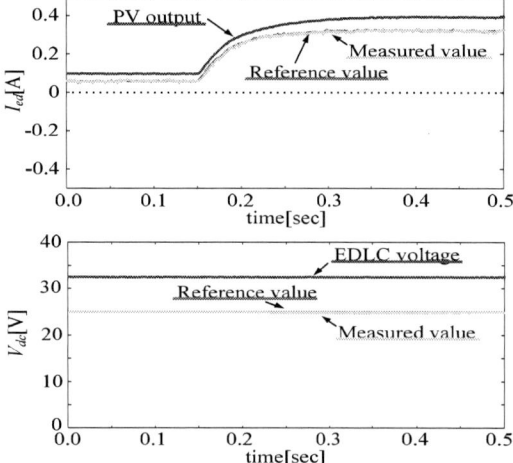

Fig. 12. Experimental result (operation of buck and supply)

Fig. 13. Experimental result (operation of buck and charge)

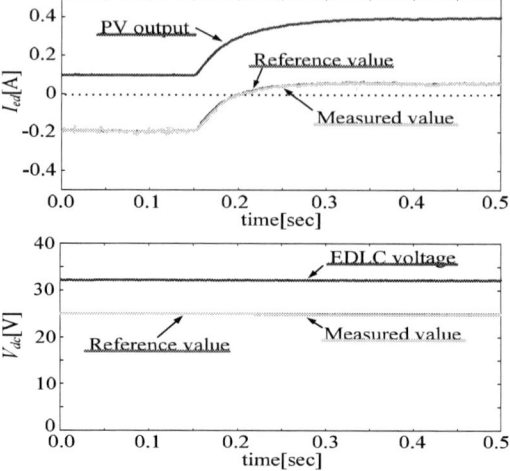

Fig. 14. Experimental result (change supply to charge)

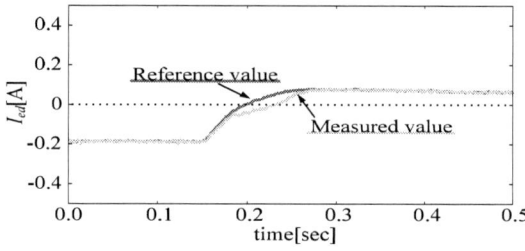

Fig. 15. Experimental result by PI control

EDLC, PV power fluctuation and so on. The output current followed the reference value every one sample and is able to confirm the quick response without error by the proposed method.

Figs.12 and 13 are cases of buck operation ($V_{EDLC} > V_{dc}$) for power supply from EDLC ($I_{ed} < 0$) and charge to EDLC ($I_{ed} > 0$), respectively. Voltage of EDLC is 32V at the start. The simulated PV output increases at t = 0.15s. Output current of PL unit precisely follows to the reference by the deadbeat control. As a result, DC bus voltage keeps constant by the voltage control correctly.

Fig.14 is a case that output current changes from discharge to charge. In this case, operation of PL unit is good the same as Figs.12 and 13. Fig.15 shows an example of current control by conventional PI control for comparison. An error occurs at around 0A due to discontinuous of current in a reactor. However, the proposed method is able to control the current without error.

V. CONCLUSIONS

In this study, the deadbeat control of PL unit with EDLC using the bidirectional buck/boost DC/DC converter is proposed. The effectiveness of the deadbeat control was confirmed by the simulation and the experiment in various conditions such as charge or discharge of

REFERENCES

[1] Y. Ito, Z. Yang, and H. Akagi, "A Control Method of a Small-Scale DC Power System Including Distributed Generators", *IEEJ Trans. IA*, Vol.126, No.9, 2006

[2] S. Funabiki, M. Yamamoto : "Estimation of Bidirectional Buck/boost DC/DC Converters with Electric Double-Layer Capacitors for Energy Storage Systems", *IEEJ Trans. IA*, Vol.129, No.6, pp.658-663, 2009

[3] H. Kakigano, T. Ise, et al. : "DC Voltage Control of the DC Micro-Grid for Super High Quality Electric Power Distribution" (*in Japanese*), *JIEE Trans. on Industrial Application*, vol. 127, No. 8, pp.890-897, 2007

[4] K. Yukita, Y. Shimizu, Y. Goto, et al., "Study of AC/DC Power Supply System with DGs using Parallel Processing Method", *The 2010 International Power Electronics Conference (IPEC)*, pp.722-725, 2010

[5] S. Hamasaki and A. Kawamura, "Improvement of Current Regulation of Line-Current-Detection-type Active Filter based on Deadbeat Control", *IEEE Trans. on Industrial Application*, Vol.39, No.2, pp.536-541, 2003

[6] S. Hamasaki, M. Tsuji, E. Yamada : "A Study on Power Flow Control for Distributed Generator with EDLC", *SYMPOSIUM ON POWER ELECTRONICS, ELECTRICAL DRIVES, AUTOMATION AND MOTION (SPEEDAM)*, Vol.1, pp.1502-1507, 2010

[7] S. Hamasaki, R. Mukai, M. Tsuji : "Control of Power Leveling Unit with Super Capacitor using Bidirectional Buck/boost DC/DC Converter", *ICRERA2013*, Vol.1, 2013

978-1-4799-2706-7/14 $31.00 © 2014 IEEE 2780

Design of Optimized On-off Control to Improve Efficiency of Paralleled Converter System

Teruhiko Kohama, Yuki Sogawa, and Satoshi Tsuji
Dept. of Electrical Engineering
Fukuoka University
Fukuoka 814-0180, JAPAN

Abstract— This paper describes design guideline on optimized on-off switching control to improve overall power conversion efficiency of paralleled converter system. The optimized on-off control changes the number of active modules according to load current. Key point of the control is to determine the on-off points of converter modules. Design guideline on determining optimized switching points is shown by introducing simple point-estimating manner. Effectiveness of the estimating method is confirmed by experiment.

Keywords— *DC-DC converter, improvement of efficiency, optimized on-off control, paralleled converter system*

I. INTRODUCTION

Paralleled converter system is widely used for low-voltage and high-current power application such as CPUs and memories in recent personal computers and mainframe computers. In the paralleled system, identical converter modules are connected in parallel to share load current equally. However, serious current imbalance occurs due to slight difference between output-voltages of actual modules. The imbalance causes several problems of thermal stress concentration, reverse current flow, and degradation of power conversion efficiency. In a practical use, an active current sharing control[1-3] is employed to solve those problems. However, the current-sharing control degrades power conversion efficiency in the paralleled system at light-load conditions. This is because module currents become relatively low by the current-sharing control. Generally, power conversion efficiency in converter module has very low efficiency in light-load range and a maximum efficiency point in middle-load range. Recently, energy saving techniques and improving over-all power efficiency in electrical equipment have been key issues for prevention of global warming in earth. Latest loads change rapidly and frequently from heavy to light to save the power consumption. As a result, period of light-load is increasing compared with heavy-load time. Therefore it is important to improve the efficiency not only at heavy-load but also at light or middle-load condition. In order to improve the over-all efficiency, we propose optimized on-off controlling method. It changes the number of active modules according to the load current. Key point of the control is to determine the on-off points for converter modules to improve overall efficiency of paralleled converter system. Optimized switching points are easily determined by introducing simple point-estimating manner. Effectiveness of the method is confirmed by experiment.

II. EFFICIENCY ESTIMATION OF PARALLELED CONVERTER SYSTEM

Power conversion efficiency of paralleled converter system is estimated in this section. At first, power conversion efficiency of a buck type converter with synchronous rectifier is approximately expressed as following equation [4].

$$P_{loss} = aI_o{}^2 + bI_o + c \tag{1}$$

where,
a: power loss factor proportional to square value of output current I_o
b: power loss factor proportional to I_o
c: constant power loss.

Figure1 shows two-paralleled converter system consuming power P_{loss} in Eq.(1) as an internal loss of converter module. Two-modules are assumed to be the same. Module currents I_{o1} and I_{o2} are represented by load current I_L with a current sharing ratio α as follows:

$$I_{o1} = \alpha I_L, \ I_{o2} = (1 - \alpha)I_L \tag{2}$$

where, $0 < \alpha < 1$.
The loss of each module is represented by the following equations.

$$P_{loss1} = a(\alpha I_L)^2 + b(\alpha I_L) + c \tag{3}$$

$$P_{loss2} = a\{(1 - \alpha)I_L\}^2 + b(1 - \alpha)I_L + c \tag{4}$$

Entire power loss P_{total} in the two-paralleled system is given by the sum of P_{loss1} and P_{loss2}.

$$P_{total} = a(2\alpha^2 - 2\alpha + 1)I_L{}^2 + bI_L + 2c \tag{5}$$

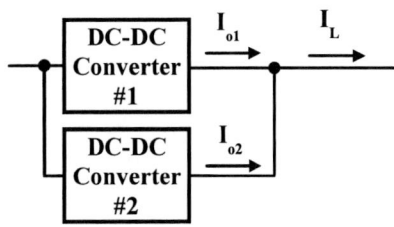

Fig.1 Two-paralleled converter system.

978-1-4799-2706-7/14 $31.00 © 2014 IEEE

From this equation, P_{total} is strongly affected by coefficient $(2\alpha^2\text{-}2\alpha+1)$ of $I_L{}^2$. It is noted that $(2\alpha^2\text{-}2\alpha+1)$ becomes minimal in the case of $\alpha=0.5$ which obtains maximum efficiency of the paralleled system. As a result, maximum efficiency is achieved by sharing load current equally. Figure 2 shows measured efficiency of two-paralleled converter system with a parameter of α. The result shows that the efficiency for $\alpha=0.5$ remains maximum at any load condition.

Maximum efficiency for N-paralleled converter system as shown in Fig.3 is discussed for the next step. From Eq.(5) power loss of factor a is dependent on current sharing ratio α, while losses due to factor b and c are completely independent on α. The same principle is easily obtained for N-paralleled converter system. Therefore only the loss of factor a should be discussed to minimize total power loss in N-paralleled system. From Fig.3 relationship between module currents and load current is derived as follows:

$$I_{o1} + I_{o2} + I_{o3} \ldots + I_{oN} = I_L. \qquad (6)$$

From Eq.(1) power loss P_{square} due to factor a is defined by

$$P_{square} = a(I_{o1})^2 + a(I_{o2})^2 + \cdots + a(I_{oN})^2$$
$$= a\{(I_{o1})^2 + (I_{o2})^2 + \cdots + (I_{oN})^2\} \qquad (7)$$

On the condition that load current I_L is constant in Eq.(6), it is evident that P_{square} is minimized when $I_{o1} = I_{o2} = \cdots = I_{oN}$. Therefore current sharing control is effective to improve overall efficiency of paralleled system.

III.   EFFICIENCY IMPROVEMENT BY ON-OFF CONTROL

In a practical use, an active current-sharing control is employed to achieve maximum power conversion efficiency. However, the current-sharing control degrades the efficiency at light-load conditions. This is because module currents become relatively low by the current-sharing control. Generally, characteristic of the power conversion efficiency of converter module has very low efficiency in light-load range and a maximum efficiency point in middle-load range. Figure 4 shows an example of efficiency. Figure 5 shows a system of two parallel

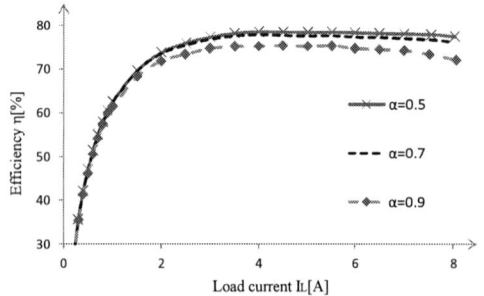

Fig.2 Efficiency of two-paralleled converter system with parameter of current sharing ratio α.

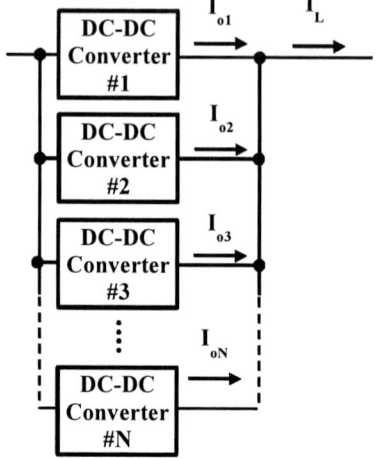

Fig.3 N-paralleled converter system.

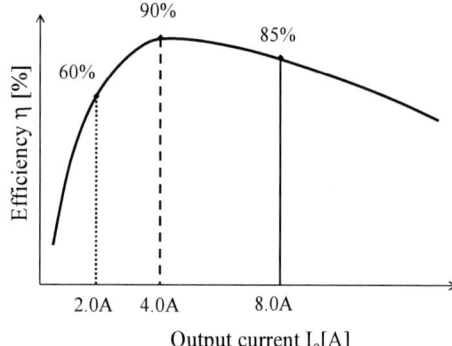

Fig.4 Efficiency of converter module

modules with the same efficiency curve as shown in Fig.4. In the case of $I_L =16.0[A]$ each module supplies output current of 8.0[A] which achieves efficiency of 85%. In the case of $I_L= 4.0[A]$, however, efficiency is seriously reduced to 60% since each module only supplies 2[A]. From the view point of efficiency improvement, module #2 should be turn off to increase output current in module #1 as shown in Fig.6. As a result, efficiency is improved from 60% to 90%. This on-off controlling method is the best way of improvement of power conversion efficiency in paralleled converter system at any load conditions. The main concern of proposed on-off control is to determine the switching point of the module.

IV.   DESING OF OPTIMIZED SWITCHING POINTS BY SIMPLE ESTIMATING METHOD

Figure 7 shows examples of efficiencies of one-module and two-paralleled converter system. It is noted that there exists cross point Psw of efficiency curves. In the range of $I_L>I_{LP}$ two-paralleled system is more efficient than that of one-module operation. On the other hand one-module operation is more efficient than two-paralleled system in the range of $I_L<I_{LP}$. As a result, two-

paralleled system should halt one module to improve efficiency for light load of $I_L<I_{LP}$.

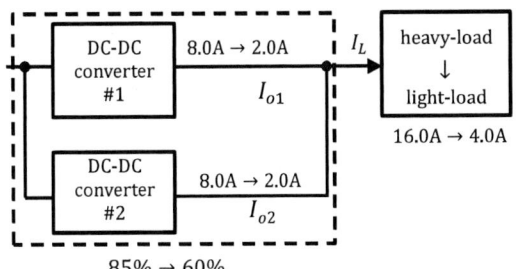

Fig.5 Two-paralleled converter system.

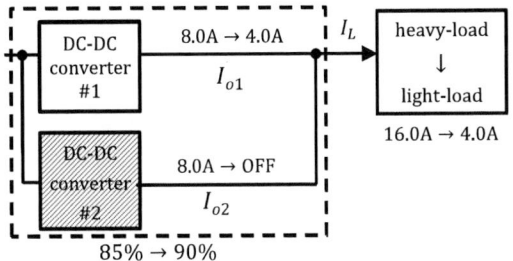

Fig.6 Two-paralleled converter system with on-off control.

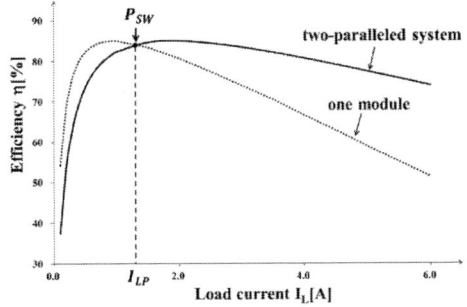

Fig.7 Switching point of two-parallel systems.

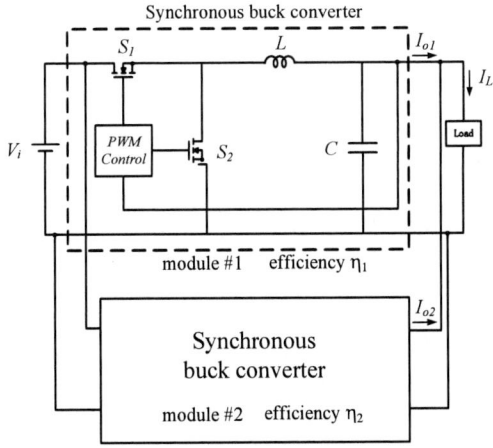

η_{Total} :efficiency of paralleled system

Fig.8 Two-paralleled converter system with synchronous rectifier.

In order to obtain the switching point P_{SW} and current threshold I_{LP}, measurement of efficiency curves is required. P_{SW} and I_{LP} is derived by measuring two efficiency curves. Figure 8 shows two-paralleled system to be tested here. Synchronous buck converter is used as a module. In a practical application the measurement to obtain P_{SW} should be simplified. As long as same modules are used in paralleled converter system, efficiency curve for two-paralleled system is easily estimated by following steps.

Step1. Measure an efficiency curve for single module. The data is stored in memories of computer or measuring instrument.

Step2. Draw the efficiency curve with the data in Step1 as shown in Fig.9.

Step3. Multiply output current data for horizontal axis in Step1 by 2.

Step4. Draw curve with the data in Step3. The curve is the estimation of efficiency for two-paralleled system as shown in Fig.9.

The estimation is performed under the assumption that two modules share load current equally, that is

$$I_{o1} = I_{o2} = I_o \qquad (8)$$

where

$$I_o = I_L/2. \qquad (9)$$

In Fig.9 efficiency for two-paralleled system at $I_L=2I_o$ is considered to be the same efficiency of single-module at $I_L=I_o$ since module efficiencies η_1, η_2 for module #1 and #2 are also the same, that is

$$\eta_1 = \eta_2 = \eta_{total}. \qquad (10)$$

This idea is easily applied to N-paralleled converter system as follows:

$$I_{o1} = I_{o2} = \cdots = I_{on} = I_o \qquad (11)$$

$$\eta_1 = \eta_2 = \cdots = \eta_n = \eta = \eta_{total}. \qquad (12)$$

Overall efficiency η_{total} for N-paralleled system at $I_L = N \times I_o$ is the same as efficiency of single-module efficiency η at $I_o = I_L/N$. From the above discussion efficiency curve is easily obtained by measuring efficiency of single-module and multiplying output current by N.

V. EXPERIMENTAL RESULTS

Figure 10 shows comparison between estimated efficiency and measured efficiency of two-paralleled converter system as shown in Fig.8. Estimated data fit measured data perfectly, which confirms the effectiveness of the proposed estimating method in section IV. Therefore estimated switching point P_{SW} is also reliable and is used for design of optimized on-off control. Figure 11 shows measured efficiencies of two-paralleled system with and without on-off control. By controlling the number of active module with current threshold I_{LP}, overall efficiency is improved especially in light load range.

978-1-4799-2706-7/14 $31.00 © 2014 IEEE

Circuit simulation is performed for three-paralleled converter system with optimized on-off control as shown in Fig.12. Load current is detected and used to turn on and turn off module #2 and #3. All efficiencies for single-module, two-module, and three-module operation are drawn in Fig.13 by using the estimation in section IV. Switching points P_{SW1} and P_{SW2} in Fig.13 divide load range into three parts, that is $I_L<I_{LP1}$, $I_{LP1}<I_L<I_{LP2}$, and $I_{LP2}<I_{LP}$. Only module #1 operates at light load of $I_L<I_{LP1}$. Module #1 and #2 are on-state for middle load range of $I_{LP1}<I_L<I_{LP2}$. All three modules are on-state when $I_{LP2}<I_{LP}$. From Fig.13 optimized on-off control achieves maximum efficiency for all load ranges.

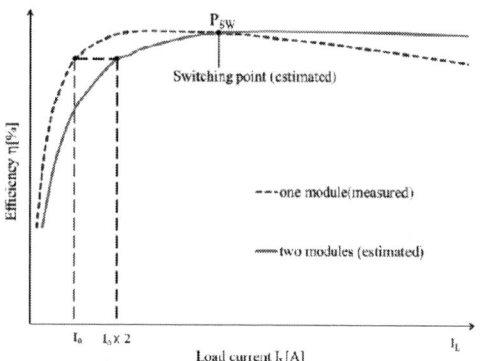

Fig.9 Efficiency estimation of two-paralleled converter system.

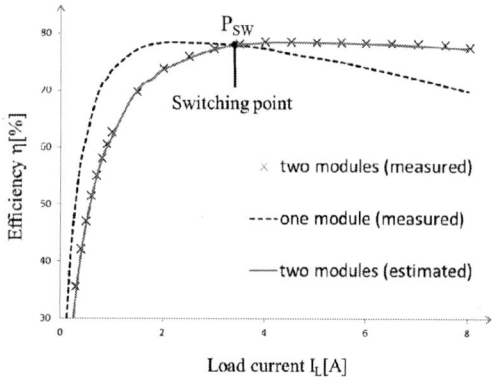

Fig.10 Comparison between measured efficiency and estimated efficiency.

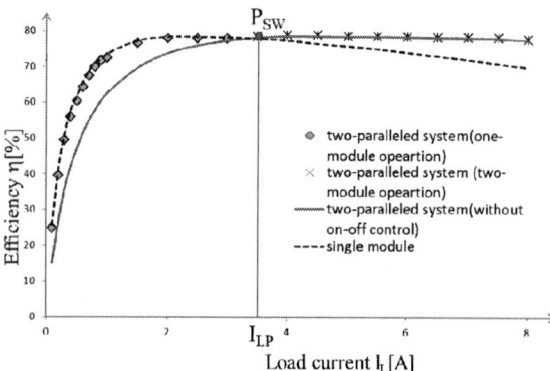

Fig.11 Measured efficiencies of two-paralleled system.

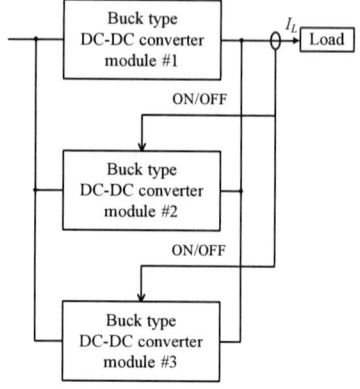

Fig.12 Three-paralleled converter system(circuit simulation).

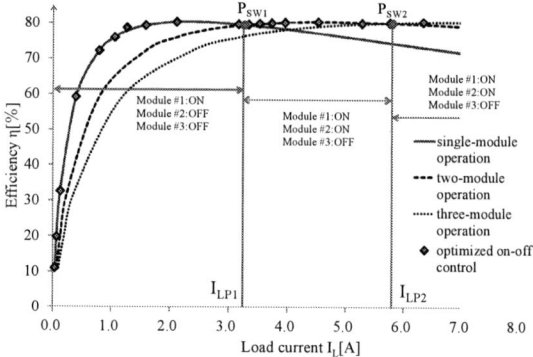

Fig.13 Efficiency characteristics of three-paralleled system by optimized on-off control.

VI. FURTHER FFICIENCY IMPROVEMENT BY USING DIFFERENT CAPACITY MODULE

In Section II to V, we have discussed efficiency of paralleled converter system where identical converter modules are used. The proposed on-off controlling method improves over-all efficiency of paralleled converter system. However, degradation of the efficiency is still observed in light load range in Fig.13. This decrease is improved by using an additional different capacity module.

A. Power loss in two-paralleled converter system with different capacity modules.

Entire power loss in two-paralleled system shown in Fig.1 with different capacity modules is discussed in this section. Considering that power loss for each module is different from the other, following equations are derived from Eq.(1).

$$P_{loss1} = a_1 I_{o1}{}^2 + b_1 I_{o1} + c_1 \qquad (13)$$

$$P_{loss2} = a_2 I_{o2}{}^2 + b_2 I_{o2} + c_2 \qquad (14)$$

where,

P_{loss1}: power loss in module #1
a_1, b_1, c_1: power loss factors in module #1
P_{loss2}: power loss in module #2

978-1-4799-2706-7/14 $31.00 © 2014 IEEE

a_2, b_2, c_2: power loss factors in module #2.

Substituting Eq.(2) into (13) and (14), the loss of each module is represented by the following equations.

$$P_{loss1} = a_1(\alpha I_L)^2 + b_1(\alpha I_L) + c_1 \tag{15}$$

$$P_{loss2} = a_2\{(1-\alpha)I_L\}^2 + b_2(1-\alpha)I_L + c_2 \tag{16}$$

Entire power loss P_{total} in the two-paralleled system is given by the sum of P_{loss1} and P_{loss2} as follows:

$$P_{total} = A\alpha^2 + B\alpha + C \tag{17}$$

where,

$$A = (a_1 + a_2)I_L^2 \tag{18}$$

$$B = (b_1 - b_2)I_L - 2a_2 I_L^2 \tag{19}$$

$$C = c_1 + a_2 I_L^2 + b_2 I_L + c_2 . \tag{20}$$

B. Optimized current sharing ratio to achieve maximum efficiency.

Eq.(17) shows that P_{total} depends on current sharing ratio α. In order to achieve maximum efficiency, P_{total} must be minimized. Therefore the differential equation of Eq.(17) should be equal to zero.

$$2A\alpha + B = 0 \tag{21}$$

As a result, optimized current sharing ratio α_{op} is given by following equation.

$$\alpha_{op} = -\frac{B}{2A}$$
$$= \frac{a_2}{a_1 + a_2} - \frac{b_1 - b_2}{2(a_1 + a_2)I_L} \tag{22}$$

C. Design of optimized on-off control for different capacity modules.

Maximum efficiency is achieved by using the ratio α_{op}. Figure 14 shows efficiency of each converter module in Fig.1. Figure 15 shows efficiencies of two-paralleled converter system with a parameter of α, where two modules in Fig.14 are connected in parallel. From Fig.15 the efficiency with optimized current sharing ratio α_{op} remains maximum at any load condition. Therefore the efficiency curve for $\alpha = \alpha_{op}$ is applied and superimposed on Fig.14. Then, finally we obtain Fig.16 which shows optimized switching points for two-paralleled system with different capacity modules. From Fig.16, only module #1 is on-state in range 1. In range 2, only module #2 is on-state. For range 3, two modules are on-state. Comparing to efficiency curve in Fig.13, efficiency in the range of 0 to 1.5[A] can be improved.

The results in section VI are only theoretical. Further experimental study will be done in the near future.

VII. CONCLUSIONS

Improvement of overall efficiency of paralleled system is achieved by using proposed on-off controlling method with optimized switching points. Practical design of switching points are easily obtained by simple estimating method. The validity of the estimating method is confirmed by experiment and circuit simulation. The guide line on determining optimized switching points with different capacity modules is also shown to achieve further improvement at light load.

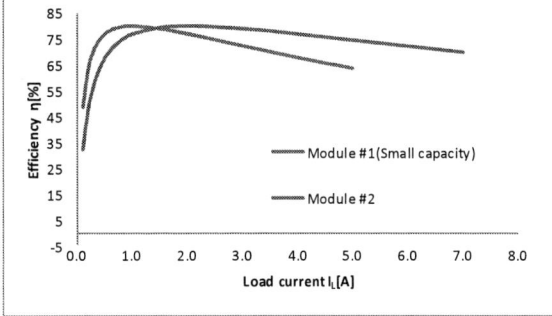

Fig.14 Efficiency curves of each module in two-paralleled system.

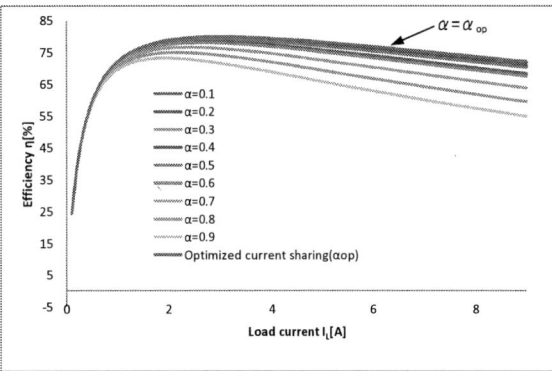

Fig.15 Efficiency curves of two-paralleled converter system for various current sharing rate.

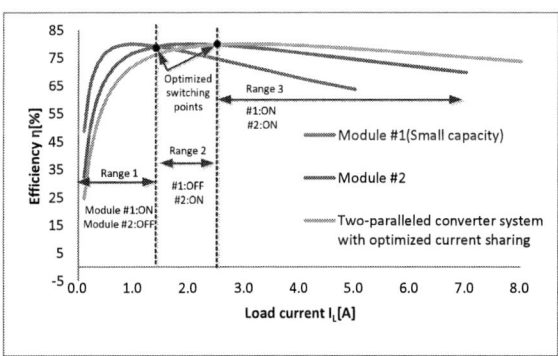

Fig.16 Efficiency improvement of two-paralleled converter system with optimized on-off control.

REFERENCES

[1] R.H.Wu, T.Kohama, Y.Kodera, T.Ninomiya, F.Ihara ,"Load-Current-Sharing Control for Parallel Operation of DC-to-DC Converters", *IEEE 24th Power Electronics Specialists Conference Record*, pp.101-107, 1993.

[2] T.Kohama, T.Ninomiya, M.Shoyama, and F.Ihara, "Dynamic Analysis of a Parallel-Module Converter System with Current Balance Controllers," *Proceedings of IEEE 16th International Telecommunications Energy Conference*, pp.190-195, 1994.

[3] D.K.W.Cheng, X.C.Liu, and Y.S.Lee,"Parallel operation of DC-DC converters with synchronous rectifiers," *Proceedings of Power Electronics Specialists Conference Record*, pp.1225-1229, 2011.

[4] T.Kohama and T.Tahara,"Estimation of Power Conversion Efficiency for Low-Voltage Buck-Converter," *Fukuoka University Review of Technological Science*, No.87,pp.17-23, 2011.

[5] H.Dejima and T.Kohama," *Improvement of Overall Efficiency of Paralleled Converter System*," 2011 Annual Conference of I.E.E. of Japan, Industry Applications Society, No.Y-36, 2011.

Efficiency improvements in a Single Active Bridge modular DC-DC converter with snubber capacitance optimisation

Yeh Ting, Sjoerd de Haan, Jan A. Ferreira

Electrical Power Processing Group
Delft University of Technology
Delft, The Netherlands
y.ting@tudelft.nl, s.w.h.dehaan@tudelft.nl, j.a.ferreira@tudelft.nl

Abstract— **Modular DC-DC converters are commonly used to increase the power capacity of a DC-DC converter for high power applications. To obtain the required power, multiple modules are connected in the input-parallel and output-parallel (IPOP) configuration. In addition, phase-shedding are used to increase efficiency of such modular DC-DC converters at light-loads. With Single Active Bridge (SAB) modules, efficiency of the system can be further improved by increasing the snubber capacitance which results in Partial Resonant Single Active Bridge (PR-SAB) modules. As the load range of the PR-SAB load range is reduced, certain loads within the entire load range are therefore not attainable when they are connected in the IPOP configuration. This paper describes how snubber capacitance in each module is optimised such that the entire existing load range is maintained while improving system efficiency. In addition, by utilising unequal module power distribution, the overall IPOP converter efficiency is increased over a wide load range to above 94 %. Peak efficiency is also increased to about 96 %. Finally, experimental results for the IPOP converter are obtained with the optimised snubber capacitances in the SAB modules.**

Keywords— Modular IPOP converter, Single Active Bridge, snubber capacitance, module power distribution

I. INTRODUCTION

Modular DC-DC converters are made up of multiple units of DC-DC converters modules which are primarily used to achieve higher current [1] or voltage capability [2]. In addition, modular DC-DC converters also enhance availability and reliability of the system with the use of redundant modules. Also modular DC-DC converters allow devices with lower voltage and current ratings that have better figures-of-merit (FoM) to be used for higher efficiency. As DC-DC converters are easily scalable to the required power output with additional identical modules, modular DC-DC converters are commonly used to increase power capacity of DC-DC converters. In modular DC-DC converters where input voltage, U_{in}, and output voltage U_{out}, do not exceed the maximum module voltage ratings, series connection is not required. Hence, the increase in power capacity is achieved with input and output parallel connections of the modules resulting in the input parallel and output parallel (IPOP) modular DC-DC converter shown in Fig. 1. An IPOP modular

converter with $n = 3$ modules is depicted here. The total input current, I_{in}, and output current, I_{out}, are shared among the modules. Hence I_{in} and I_{out} are the sum of the input currents, $I_{in,m}$, and output currents, $I_{out,m}$, from each of the module m. In addition, the module m input voltage, $U_{in,m}$, and output voltage, $U_{out,m}$, have the same values as the input voltage, U_{in}, and the output voltage, U_{out}, of the IPOP converter respectively.

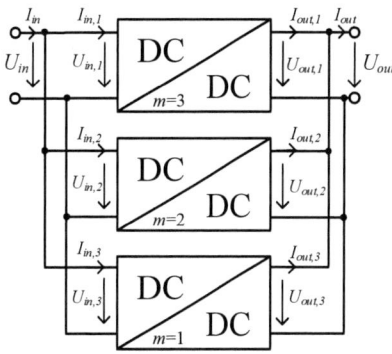

Fig. 1. An input parallel and output parallel (IPOP) modular DC-DC converter consisting of $n = 3$ modules ($m = 1$, 2 and 3).

With the increase in output power, the power at which efficiency starts to decrease also increases. As a result, the efficiency of a single DC-DC converter module is higher than the IPOP converter with multiple modules at the same light-load. In order to mitigate this drawback, phase-shedding [4] is used to improve light-load efficiency. Selected modules are shut down at light-load such that the remaining modules operate at heavier loads with higher efficiencies. This concept is illustrated in Fig. 2 with the plot of the efficiency, η_n, against the output power, P_{out}, of the IPOP converter for $n = 3$ modules. It compares the efficiency of the IPOP converter with and without phase-shedding.

With phase-shedding, the IPOP converter is able to operate at $n < 3$, i.e. $n = 2$ and $n = 1$, by shutting down one and two modules respectively. This achieved efficiency improvement of the $n = 3$ IPOP converter at light-load is highlighted in Fig. 2. From Fig. 2, it can be seen that the threshold power levels, $P_{th,m}$, where the module m shutdown occurs are the intersects of the

978-1-4799-2706-7/14 $31.00 © 2014 IEEE

functions $\eta_n = f(P_{out})$ and $\eta_{n-1} = f(P_{out})$. In addition, the IPOP converter is also able to operate at a lower minimum power, $P_{min,1}$, which is the minimum power of a single converter module as depicted in Fig. 2.

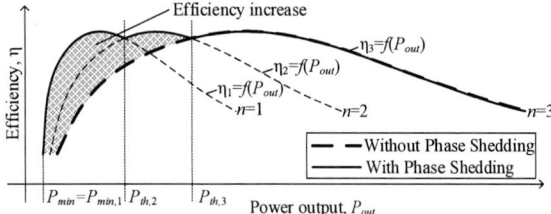

Fig. 2. The efficiency as a function of power output for an $n = 3$ IPOP modular converter showing efficiency improvements at light-load with phase-shedding.

Although it is possible to effectively increase light-load efficiency of the IPOP modular DC-DC converter with the shutdown of the excess modules, the maximum efficiency is still limited by the peak efficiency of the individual modules. Despite this, there are two possibilities to further the efficiency of the IPOP converter. One of them is to increase the efficiency of the individual modules which is normally not easily achievable without drastic changes in the topology. The other method is to deal with the way power is controlled among the modules. This paper investigates how these methods can be applied to increase the light-load as well as peak efficiency of an IPOP converter for modules based on the Single Active Bridge topology.

II. IPOP MODULAR DC-DC CONVERTER WITH SINGLE ACTIVE BRIDGE

There are many topologies that can be used for the modules in the IPOP modular DC-DC converter. The Single Active Bridge (SAB) [3] shown in Fig. 3 is selected as modules in the IPOP modular DC-DC converter as it offers high power density, galvanic isolation and requires minimal components.

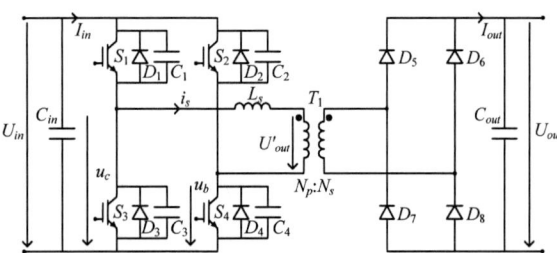

Fig. 3. Topology of the Single Active Bridge used in the input-parallel and output-parallel (IPOP) modular converter.

However, one of the disadvantages of the SAB is that it suffers from large switching losses at light-loads. This is due to hard switching of S_2 and S_4 as described in [6]. To enable more efficient operation over a wider load range, two methods are described in [6] where efficiency of the SAB at light-load is increased by reducing its

switching losses when operating in discontinuous conduction mode (DCM).

With these measures in place, multiple SAB modules are connected in parallel to form an IPOP modular converter with high power output and efficiency across a wide load range. The specifications of each SAB converter module are given in TABLE I. The resulting efficiency of an $n = 4$ IPOP modular converter with SAB modules is shown in Fig. 4 over the entire load range from $P_{out} = 74$ W to 17.8 kW. Furthermore, light-load efficiency of the IPOP modular converter is further enhanced with phase-shedding. The efficiency improvement with phase-shedding at $P_{out} = 300$ W is also indicated where the efficiency is about 8 % higher.

TABLE I
SAB CONVERTER MODULE SPECIFICATIONS

Parameter	Value
Input voltage, U_{in}	600 V
Output voltage, U_{out}	660 V
Minimum power output, P_{min}	74 W
Maximum power output, P_{max}	4.45 kW
Snubber capacitors, C_1 & C_3	5 nF
Snubber capacitors, C_2 & C_4	5 nF

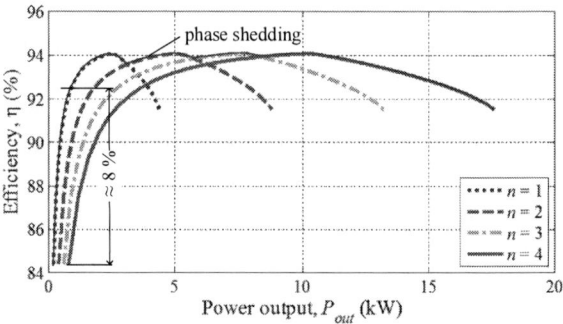

Fig. 4. Efficiency as a function of power output for the $n = 4$ SAB IPOP modular DC-DC converter showing improved efficiency at light-load with phase-shedding from $n = 4$ to $n = 1$.

III. EFFICIENCY IMPROVEMENT WITH THE PARTIAL RESONANT SINGLE ACTIVE BRIDGE MODULES

In [5], conduction losses in the SAB are significantly reduced by increasing its snubber capacitance. As described, the resulting topology known as the partial resonant SAB (PR-SAB) has higher efficiency than the SAB. As explained in [5], the conduction losses in the capacitors due to its equivalent series resistance are lower than that of IGBTs at 25 kHz. With large snubber capacitance, IGBT conduction duty-cycles and RMS current are significant reduced while those of the snubber capacitors increases. Conduction losses in the IGBT are reduced more than their increase from the capacitors. In addition, as snubber capacitance is increased, the turn-off losses of the IGBTs S_1 and S_3 are also reduced. Total conduction and switching losses of the converter are therefore reduced. The relationship of the module peak efficiency, η_{peak}, and the snubber capacitance C_1 and C_3 in the SAB module in Fig. 3 is illustrated in Fig. 5.

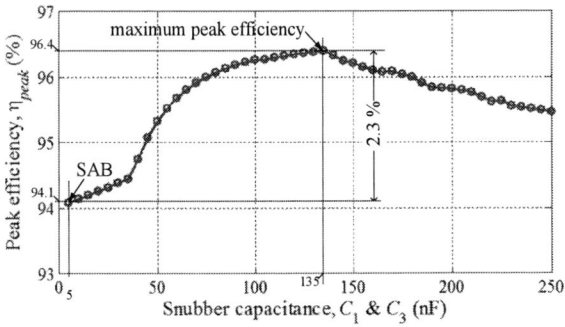

Fig. 5. The relationship is shown between module peak efficiency, η_{peak}, and snubber capacitance values, C_1 & C_3, in the PR-SAB module.

The indicated η_{peak} at C_1 and $C_3 = 5$ nF corresponds to η_{peak} of the SAB with the specifications in TABLE I. As shown, as C_1 and C_3 increases, η_{peak} increases to a maximum of 135 nF. Further increase of C_1 and C_3 of more than 135 nF lowers η_{peak} instead. Hence by increasing C_1 and C_3 to 135 nF, the PR-SAB module is able to operate at its highest possible efficiency at 96.4 % as compared to the SAB at 94.1 %.

IV. IPOP MODULAR DC-DC CONVERTER WITH PARTIAL RESONANT SINGLE ACTIVE BRIDGE

From Fig. 5, it is hence possible to increase the SAB IPOP modular converter's peak efficiency by replacing C_1 and C_3 of the SAB modules from 5 nF with 135 nF. However as shown in Fig. 6, the minimum output power, P_{min}, of the SAB module significantly increases with increasing C_1 and C_3 while the maximum output power, P_{max}, decreases slightly. Since the power range of the converter is defined by difference between P_{max} and P_{min} as indicated in

Fig. 6, the load range is effectively reduced with increasing C_1 and C_3. In fact, the module load range with C_1 and C_3 at 135 nF is reduced by more than half as compared to C_1 and C_3 at 5 nF. The specifications of the PR-SAB module with C_1 and C_3 at 135 nF are given in TABEL II. The significant load range reduction can be seen by comparing P_{min} and P_{max} in TABLE II with that in TABLE I.

TABLE II
PR-SAB CONVERTER MODULE SPECIFICATIONS

Parameter	Value
Input voltage, U_{in}	600 V
Output voltage, U_{out}	660 V
Minimum power output, P_{min}	2.34 kW
Maximum power output, P_{max}	4.38 kW
Snubber capacitors, C_1 & C_3	135 nF
Snubber capacitors, C_2 & C_4	5 nF

When these PR-SAB modules with C_1 and C_3 at 135 nF are used in the IPOP converter, the resulting power output and efficiencies of the $n = 1$ to $n = 4$ IPOP converter are illustrated in Fig. 7. It shows the efficiency and load range of the $n = 4$ PR-SAB IPOP modular converter with phase-shedding, similar to what is achieved with the SAB modules in Fig. 4. As shown, the η_{peak} of the $n = 4$ modules IPOP converter is about 2.3 % higher than that of the SAB, i.e. at about 96.4 % instead of 94.1 %. This is due to the peak efficiency increase with C_1 and C_3 at 135 nF as shown in Fig. 5.

In the PR-SAB, η_{peak} occurs at the minimum power as shown in the efficiency curves in Fig. 7. This means that the threshold power levels to shut down a PR-SAB module during phase-shedding occurs at the minimum power. A decrease in power output when operating with n modules at minimum power will trigger the shutdown of the n-th module. However with the reduced load range of each PR-SAB module as indicated in Fig. 6, operating in certain load range is not possible. This occurs when the IPOP converter operating with n modules has a minimum voltage higher than the maximum voltage when it operates with $n - 1$ modules. In this case, if the required power is between the maximum of $n - 1$ modules and minimum of n modules, the shutdown of the n-th module will be triggered. This causes the remaining $n - 1$ modules to operate at its maximum power which is below the required power. This again results in the activation of the n-th module. Hence the PR-SAB IPOP modular converter oscillates between the maximum power of $n - 1$ modules and minimum power of n modules without being able to attain the required power. The power ranges that are not within reach of the IPOP converter with PR-SAB modules are defined as the forbidden load zones here.

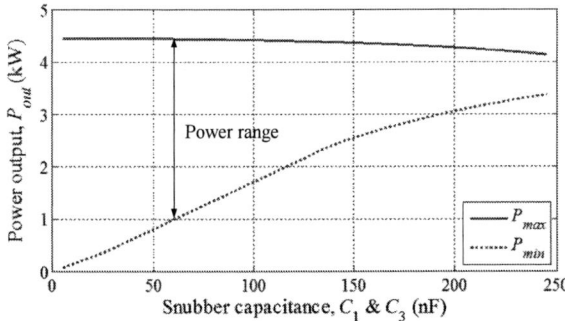

Fig. 6. The minimum and maximum power output, P_{min} and P_{max}, with increasing C_1 & C_3 in the SAB module.

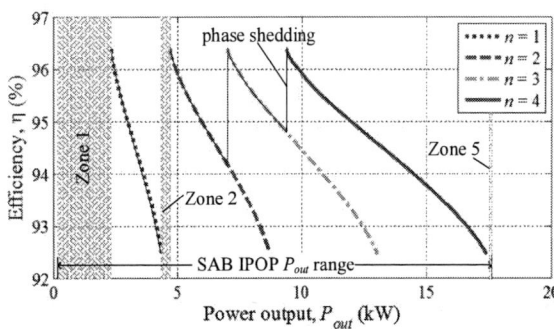

Fig. 7. Efficiencies for a PR-SAB IPOP modular converter for $n = 1$ to $n = 4$ indicating the forbidden load zones with phase-shedding for $n = 4$.

In the PR-SAB IPOP converter, there can be a maximum of $n + 1$ forbidden load zones for the PR-SAB IPOP converter with n modules. The increase in minimum power, P_{min}, results in the first forbidden zone whereas the decrease in maximum power, P_{max}, results in the $(n + 1)$-th or last forbidden zone. The remaining $n - 1$ forbidden zones result from the power gaps between the modules. Considering the entire load range of the former $n = 4$ module SAB IPOP converter from 74 W to 17.8 kW, there are three forbidden load zones when C_1 and C_3 equal 135 nF. These forbidden load zones are indicated in Fig. 7 and their respective load range is given in TABLE III. Zone 5 is neglected here as the percentage decrease in maximum power is minimal.

TABLE III
FORBIDDEN LOAD ZONES IN THE PR-SAB IPOP CONVERTER

Forbidden load zone	Range
Zone 1	74 W – 2.34 kW
Zone 2	4.35 kW – 4.68 kW
Zone 5	17.52 kW – 17.80 kW

Although the resulting PR-SAB IPOP converter has higher peak efficiency, the forbidden zones restricts the phase-shedding operation over the required load range if the entire load range of the SAB IPOP converter is to be maintained. Hence a way to optimise the snubber capacitance C_1 and C_3 has to be established in order to effectively increase the efficiency of the SAB IPOP converter without sacrificing its full power output range.

V. OPTIMISATION OF SNUBBER CAPACITANCE IN THE SINGLE ACTIVE BRIDGE MODULES

In most IPOP converters, current and also P_{out} are controlled such that they are distributed equally among the n modules. As a result, when n modules are in operation, its minimum power, $P_{min,n}$, and maximum power, $P_{max,n}$, are limited by the highest minimum and lowest maximum power of the module with the largest C_1 & C_3. $P_{min,n}$ and $P_{max,n}$ of the IPOP converter from $n = 1$ to $n = 4$ with increasing C_1 & C_3 are shown in Fig. 8. In each diagram, the $P_{min,n}$ and $P_{max,n}$ are shown together with $P_{min,n-1}$ and $P_{max,n-1}$. As C_3 is equal to C_1, $P_{min,n}$ and $P_{max,n}$ are expressed as polynomials in terms of C_1 in (1) and (2) with coefficients obtained from curve fitting of the functions in Fig. 8. The obtained coefficients from $n = 1$ to $n = 4$ for (1) and (2) are listed in TABLE IV and TABLE V respectively.

$$P_{min,n} = a_{n4}C_1^4 + a_{n3}C_1^3 + a_{n2}C_1^2 + a_{n1}C_1 + a_{n0} \quad (1)$$

$$P_{max,n} = b_{n3}C_1^3 + b_{n2}C_1^2 + b_{n1}C_1 + b_{n0} \quad (2)$$

TABLE IV
COEFFICIENTS OF $P_{min,n}$ FOR THE IPOP CONVERTER FROM $n = 1$ TO $n = 4$

n	a_{n4}	a_{n3}	a_{n2}	a_{n1}	a_{n0}
1	$1 \cdot 10^{-9}$	$-8 \cdot 10^{-7}$	$1 \cdot 10^{-4}$	$1.08 \cdot 10^{-2}$	$2.73 \cdot 10^{-2}$
2	$2 \cdot 10^{-6}$	$-1.5 \cdot 10^{-3}$	0.2538	21.559	54.546
3	$4 \cdot 10^{-9}$	$-2 \cdot 10^{-6}$	$4 \cdot 10^{-4}$	$3.23 \cdot 10^{-2}$	$8.18 \cdot 10^{-2}$
4	$5 \cdot 10^{-9}$	$-3 \cdot 10^{-6}$	$5 \cdot 10^{-4}$	$4.31 \cdot 10^{-2}$	0.1091

TABLE V
COEFFICIENTS OF $P_{max,n}$ FOR THE IPOP CONVERTER FROM $n = 1$ TO $n = 4$

n	b_{n3}	b_{n2}	b_{n1}	b_{n0}
1	$-2 \cdot 10^{-8}$	$1 \cdot 10^{-6}$	$-3 \cdot 10^{-4}$	4.4605
2	$-4 \cdot 10^{-5}$	$2.5 \cdot 10^{-3}$	-0.5705	8921
3	$-7 \cdot 10^{-6}$	$4 \cdot 10^{-6}$	$-9 \cdot 10^{-4}$	13.382
4	$-9 \cdot 10^{-8}$	$5 \cdot 10^{-6}$	$-1.1 \cdot 10^{-3}$	17.842

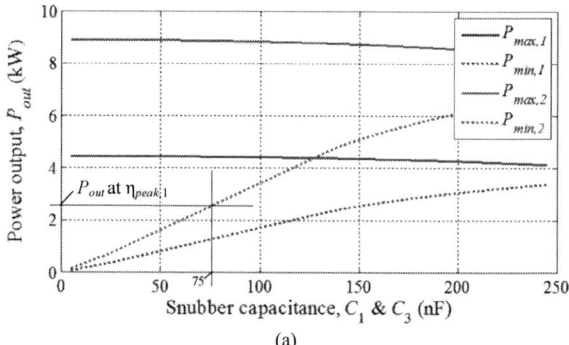

(a)

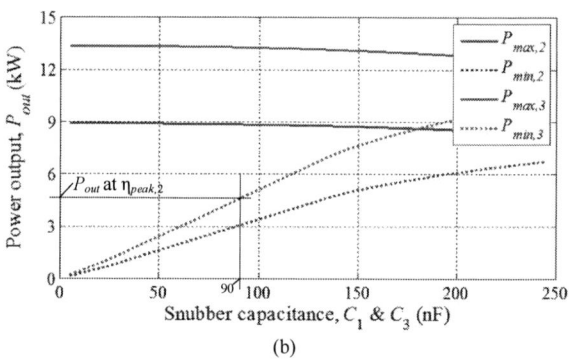

(b)

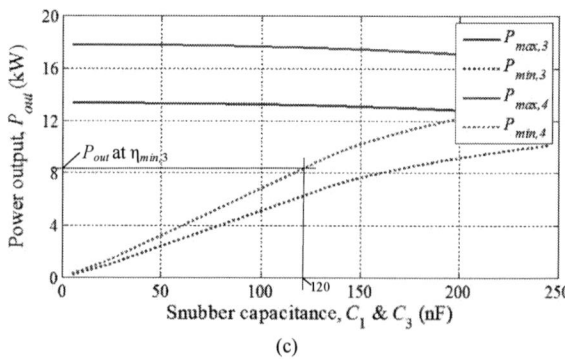

(c)

Fig. 8. The minimum and maximum power output, $P_{min,n}$ and $P_{max,n}$, of the IPOP converter for: (a) $n = 1$ and $n = 2$; (b) $n = 2$ and $n = 3$; and (c) $n = 3$ and $n = 4$ shown as functions of C_1 & C_3.

To eliminate the forbidden zones of the respective modules in Fig. 7, the values of C_1 and C_3 in each SAB module will have to be lowered and optimised accordingly using Fig. 8. C_1 and C_3 shall be lowered such that the condition $P_{min,n} \leq P_{max,n-1}$ is fulfilled. There are two methods to determine $P_{min,n}$ for $n > 1$. The first one uses the intersect of $P_{min,n}$ and $P_{max,n-1}$ which gives the highest overall peak efficiency at higher load at the expense of lower efficiency at light load. The other method makes use of the peak efficiency of $n - 1$ modules, $\eta_{peak,n-1}$, to determine $P_{min,n}$ of n modules. Although its peak efficiency is not the highest, the latter method produces the highest average efficiency with least variation across the entire load range and is discussed in this paper.

For P_{min} to be maintained at 74 W in the non-optimised SAB IPOP converter, $C_1 = 5$ nF is chosen for module $m = 1$. Larger C_1 can be used for higher efficiency if a higher P_{min} is acceptable. $\eta_{peak,1}$, is obtained with the function $\eta_1 = f(P_{out})$ which results in $P_{out} = P_{min,2} = 2.6$ kW at $\eta_{peak,1}$ as shown in Fig. 8(a). The corresponding C_1 for module $m = 2$ is then obtained from (1) by solving for its roots which results in 75 nF as indicated in Fig. 8(a). This is repeated for module $m = 3$ with $P_{out} = P_{min,3} = 4.8$ kW at $\eta_{peak,2}$ giving $C_1 = 90$ nF as shown in Fig. 8(b). However as $\eta_3 = f(P_{out})$ decreases monotonically, $\eta_{peak,3}$ occurs at $P_{min,3}$. Hence $\eta_{peak,3}$ can no longer be used to size C_1 in $m = 4$ module. The lowest required efficiency for $n = 3$, $\eta_{min,3}$ is used here instead. The lowest required efficiency here is selected to be 94.5 % which results in $C_1 = 120$ nF as shown in Fig. 8(c). The obtained optimised C_1 & C_3 values for modules $m = 1$ to $m = 4$ and their respective P^m_{min} and P^m_{max} are summarised in TABEL VI.

TABLE VI
OPTIMISED SNUBBER CAPACITANCE FOR THE SAB MODULES

Module	C_1 & C_3	P^m_{min}	P^m_{max}
$m = 1$	5 nF	74 W	4.45 kW
$m = 2$	75 nF	1.26 kW	4.43 kW
$m = 3$	90 nF	1.53 kW	4.42 kW
$m = 4$	120 nF	2.07 kW	4.40 kW

The overall efficiency, $\eta_{IPOP,n}$, for an arbitrary n modules IPOP converter can be expressed with (3) where P^m_{out} and η^m are the power output and efficiency for each module m respectively. Since power is equally distributed among the modules, $P^m_{out} = P_{out}/n$. Using (3), the resulting efficiency of the IPOP converter with the optimised C_1 and C_3 is calculated and shown in Fig. 9. The load thresholds for phase-shedding are at $P_{min,n}$ for $n = 2$ to $n = 4$ as seen in Fig. 9. The improvement in efficiency as compared to the SAB IPOP converter without snubber optimisation from Fig. 4 is indicated.

$$\eta_{IPOP,n} = \frac{P_{out}}{\displaystyle\sum_{m=1}^{n} \frac{P^m_{out}}{\eta^m}} \qquad (3)$$

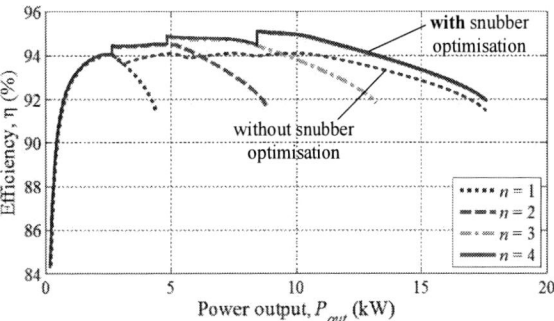

Fig. 9. A comparison of the SAB IPOP converter efficiencies with and without optimised C_1 & C_3, both with phase-shedding.

VI. UNEQUAL POWER DISTRIBUTION IN MODULES

Using the optimised C_1 & C_3, further increase in IPOP converter efficiency at light-loads can be achieved if power distribution among the individual modules is controlled. In [7], unequal power distribution for an n-module IPOP converter is introduced in an attempt to attain highest possible efficiency across the entire load range. To achieve this, the phase-shedding scheme described earlier and shown in Fig. 2 is modified. All modules except one operate at their highest efficiencies. When this is applied to the SAB IPOP converter with equal and non-optimised C_1 & C_3, efficiency gain is in fact less than if equal power distribution is used [7]. However with optimised C_1 & C_3 where efficiencies in each of the modules m, η^m_{peak}, are different, the advantage of unequal module power distribution becomes obvious. Efficiency gain is significant when the most efficient module is given the priority to operate at its highest efficiency. Similar to [7], power in every module need not be the same. Power in each individual module is controlled to obtain maximum overall efficiency. The power distribution across the modules that gives the highest efficiency is calculated with an algorithm that maximises overall efficiency which is calculated with (3). The algorithm takes into account the efficiencies of all the possible module power combinations at a given P_{out} and selects the combination that gives the maximum possible efficiency. The obtained power distribution for all the modules across the entire load range is shown in Fig. 10.

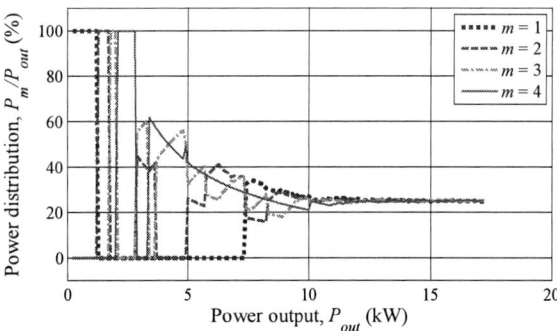

Fig. 10. The power distribution among the modules from $m = 1$ to $m = 4$ when they are optimised for the highest overall efficiency.

In Fig. 11, it is shown that with optimised C_1 & C_3, the efficiency increase for unequal power distribution at light-loads is more significant than for equal power distribution as compared to without snubber C_1 optimisation. A comparison of the peak efficiencies, $\eta_{peak,4}$, and the efficiency range for $\eta_{IPOP,4} > 94$ % are shown in the three types of IPOP modular converters with SAB and PR-SAB modules is tabulated in TABLE VII. Hence the IPOP converter with both snubber optimisation and unequal power distribution will yield the highest peak efficiency as well as the widest load range in which high efficiency can be maintained.

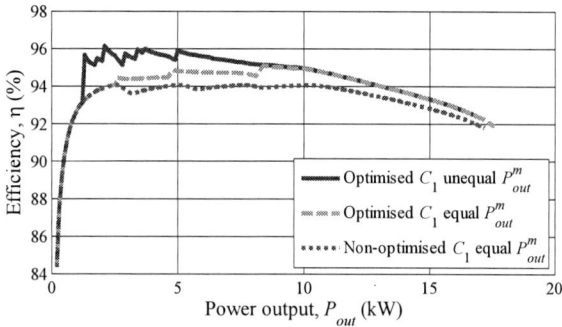

Fig. 11. The efficiencies of the IPOP converter are shown for unequal and equal module power distribution with C_1 & C_3 optimisation; and equal module power distribution without C_1 & C_3 optimisation.

TABLE VII

A COMPARISON OF THE PEAK EFFICIENCIES AND RANGE OF HIGH EFFICIENCIES IN THE DIFFERENT IPOP MODULAR DC-DC CONVERTERS

Modules	$\eta_{peak,4}$	$\eta_{IPOP,4} > 94$ %
SAB non-optimised C_1	94.1 %	2.3 kW – 2.7 kW 4.6 kW – 5.4 kW 6.9 kW – 8.1 kW 9.2 kW – 10.8 kW
SAB and PR-SAB optimised C_1	95.1 %	2.3 kW – 13.2 kW
SAB and PR-SAB optimised C_1 & unequal power distribution	96.2 %	1.3 kW – 13.2 kW

VII. EXPERIMENTAL RESULTS

In Fig. 12, the experimental setup with $n = 4$ DC-DC converter modules is shown connected in the IPOP configuration. Due to the optimisation shown earlier, the setup comprises one SAB module and three PR-SAB modules with larger C_1 & C_3. The measured values of C_1, C_3 and L_s, in the modules from $m = 1$ to $m = 4$ in the experimental setup are listed in TABLE VI. Synchronisation of the modules is not implemented in the hardware. The L_s current waveforms, $i_{s,m}$, for the respective modules are shown in Fig. 13 during equal power distribution at $P_{out} = 10$ kW. Here, the power in all modules are equal and hence $P^m_{out} = 2.5$ kW from $m = 1$ to $m = 4$. U_{in} and U_{out} are at 600 V and 660 V respectively. Although the module output power are equal, $i_{s,m}$ waveforms are different due to different values of C_1 & C_3.

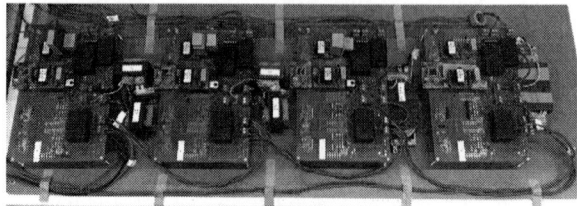

Fig. 12. The experimental setup is shown with $n = 4$ modules connected as an IPOP modular converter.

TABLE VIII

MEASURED SNUBBER CAPACITANCE AND INDUCTANCE FOR THE SAB AND PR-SAB MODULES IN THE EXPERIMENTAL SETUP

Module	C_1 & C_3	C_2 & C_4	L_s
$m = 1$	5 nF	5 nF	107.3 μH
$m = 2$	75.5 nF	5 nF	104.9 μH
$m = 3$	90.5 nF	5 nF	111.4 μH
$m = 4$	122.5 nF	5 nF	112.6 μH

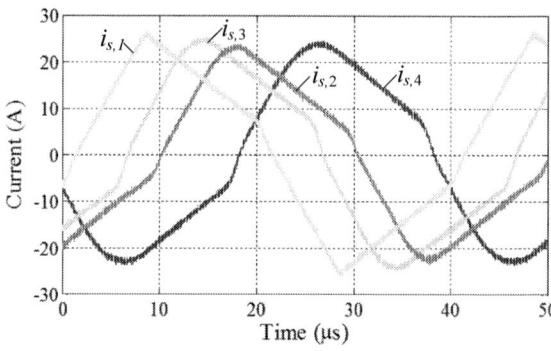

Fig. 13. The L_s current measurements, $i_{s,m}$, are shown from $m = 1$ to $m = 4$ with equal power distribution for the PR-SAB IPOP modular converter at $P_{out} = 10$ kW.

Fig. 14 shows the respective L_s current waveforms, $i_{s,m}$, for the case of unequal power distribution at $P_{out} = 4.6$ kW. Phase-shedding is also implemented to obtain high light-load efficiency as mentioned. Here, modules $m = 1$ and $m = 2$ are shut down while modules $m = 3$ and $m = 4$ are operating at different power outputs at $P^3_{out} = 2.5$ kW and $P^4_{out} = 2.1$ kW. As described earlier, the modules are loaded differently to obtain the highest efficiency in the IPOP converter.

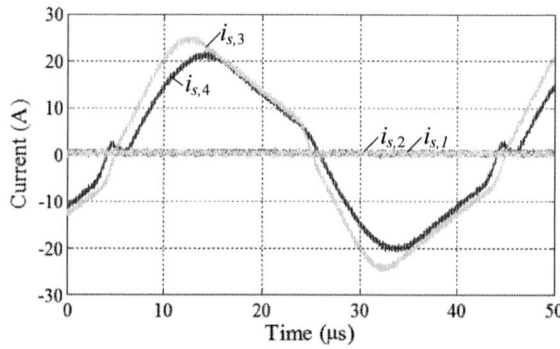

Fig. 14. The L_s current measurements, $i_{s,m}$, are shown from $m = 1$ to $m = 4$ with unequal power distribution for the PR-SAB IPOP modular converter at $P_{out} = 4.6$ kW. Here P^1_{out} and $P^2_{out} = 0$ kW; $P^3_{out} = 2.5$ kW and $P^4_{out} = 2.1$ kW.

For both equal and unequal power distribution, efficiencies are measured across the entire load range. These values are shown in Fig. 15 in comparison with the calculated values. The efficiency improvements of unequal over equal module power distribution especially can be observed. At light-loads, the measured efficiency is up to about 2.5 % higher than with optimised snubber capacitance. This shows the effectiveness of having unequal power distribution in the snubber capacitance optimised SAB IPOP modular DC-DC converter.

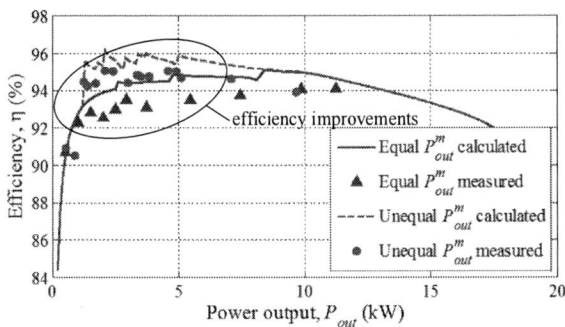

Fig. 15. Measurement results obtained from the experimental setup for optimised C_1 with equal and unequal module P^m_{out} distribution.

VIII. CONCLUSIONS

By utilising the advantage of lower losses in the PR-SAB as compared to the traditional SAB, a method to optimise the modular SAB IPOP system for high efficiency over a wide load range is described. It uses different snubber capacitor values in each of the SAB modules of the IPOP modular converter to maximise its efficiency. With snubber capacitance optimisation in each SAB module, both light-load and peak efficiency are enhanced without compromising the load range. More importantly, a much wider operating range with high efficiency can be achieved as compared to the SAB IPOP converter with non-optimised snubbers. Finally, with unequal power distribution in the modules, these efficiency improvements are demonstrated to be even more significant.

ACKNOWLEDGMENT

The authors would like to acknowledge DSO National Laboratories in Singapore for funding the research of Yeh Ting research its scholarship program. Also to acknowledge is Kewei Huang for his help in the hardware implementation.

REFERENCES

[1] J. Shi, L. Zhou, and X. He, "Common-Duty-Ratio Control of Input-Parallel Output-Parallel (IPOP) Connected DC–DC Converter Modules With Automatic Sharing of Currents," *IEEE Trans. on Power Electronics*, vol. 27, no. 7, pp. 3277-3291, July 2012.

[2] S. N. Manias, and G. Kostakis, "Modular DC-DC convertor for high-output voltage applications," *IEE Proceedings B*, vol. 140, no. 2, pp. 97-102, March 1993.

[3] R. W. De Doncker, D. M. Divan, and M. H. Kheraluwala, "A Three-phase Soft-Switched High-Power-Density DC-DC Converter for High-Power Applications", *IEEE Transactions On Industry Applications*, vol. 27, no. 1, pp. 63-73, Jan/Feb 1991.

[4] S. Waffler, and J. W. Kolar, "Efficiency Optimization of an Automotive Multi-Phase Bi-directional DC-DC Converter," *Proceedings of Power Electronics and Motion Control Conference, 2009, IPEMC '09, IEEE 6th International*, pp. 566-572, 2009.

[5] Y. Ting, S. de Haan, and J. A. Ferreira, "The partial-resonant single active bridge DC-DC converter for conduction losses reduction in the single active bridge," *Proceedings of ECCE Asia Downunder (ECCE Asia), 2013 IEEE*, pp. 987-993, 2013.

[6] Y. Ting, S. de Haan, and J. A. Ferreira, "Elimination of switching losses in the single active bridge over a wide voltage and load range at constant frequency," *Proceedings of ECCE Europe (EPE), 2013 IEEE*, 2013.

[7] P. Bartal, J. Hamar and I. Nagy, "Efficiency Improvement in Modular DC/DC Converters Using Unequal Current Sharing," *Proceedings of Power Electronics, Electrical Drives, Automation and Motion (SPEEDAM), 2012 International Symposium on*, pp. 275-280, 2012.

A Wireless Power Transfer System Optimized for High Efficiency and High Power Applications

Mohammad BANI SHAMSEH
Department of Electrical
& Computer Enginneering
Yokohama National University
Yokohama, Japan
Email: bani-husam-bk@ynu.jp

Atsuo KAWAMURA
Department of Electrical
& Computer Enginneering
Yokohama National University
Yokohama, Japan
Email: kawamura@ynu.ac.jp

Itsuo YUZURIHARA, Atsushi TAKAYANAGI
Kyosan Electric Manufacturing Co., Ltd.
Yokohama, Japan

Abstract—This paper summarizes the research progress in the field of Wireless Power Transmission (WPT) using Magnetic Resonance Coupling. To make the technique of WPT feasible for industrial and commercial applications it is vital to achieve a stable operation of power transmission from the source to the load under different conditions such as load and distance variations and antennas misalignment. To achieve a highly efficient transmission there are two parts that need to be improved. The AC/DC converter and the antenna system. We will show that a rectifier operating at 13.56 MHz with efficiency of 94% at power level of 1kW or more is feasible. On the other hand, the theoretical approach for optimizing the antenna system for optimum operation is derived along with its equivalent circuit and simulation. An overall efficiency of 89.11% at 1.5kW power and 12cm distance for the complete system from the RF source to the DC load was estimated by simulation and confirmed experimentally. A 3.7kW maximum power could be transferred in the experiment.

I. INTRODUCTION

Wireless power transfer using the non-radiative energy generated by two strongly coupled objects is discussed in[2] . This method is based on the fact that two objects which have the same resonant frequency can exchange energy efficiently in the medium-range distances. It was reported in [1] that 60W power was transferred over a 2 and 1 meters distances with 40% and 90% efficiencies, respectively. This highly-efficient method can be used in many applications to power devices ranging from small devices rated at a few watts up to high power applications such as charging the batteries of Electric Vehicles without the need to connect wires.

However, for applications that require a constant DC voltage, this high frequency voltage must be converted into a DC voltage by means of a rectifier. At such frequency, the abnormal behaviour of the diode and the parasitic inductance and capacitance lead to high distortion in the voltage and current waveforms and to a low power factor[3], which causes reduction in the efficiency of rectification. Several papers discussed the efficiency of rectifiers in WPT systems [3], [4] in which a total maximum efficiency of 75.2% and 60% from the RF transmitter to the DC load was reported respectively, with a SiC diode used in the rectifier circuit. The main cause of low efficiency of the system is the losses that occur in the rectifier stage. This paper is mainly divided into two parts.

The first part concerns the rectifier stage in which we show that the transition of the output current from the Discontinuous Mode to the Continuous Mode can increase the efficiency of the rectifier at high power levels and an efficiency of 94% at 1kW power or more is feasible.

In the second part, an equivalent circuit which accurately describes the operation of the helical antenna is proposed and its parameters are measured experimentally using the network analyser. Using the equivalent circuit of the antenna, its efficiency under different conditions such as variation in the distance between the coils and load variation is calculated by simulation. Finally, the system was tested experimentally and the results confirm the theory.

A. Electromagnetic Resonance Coupling

A WPT system is divided into two parts: the transmitter side and the receiver side. The transmitter consists of a function generator which generates a high frequency AC voltage. This voltage is fed into a transmitting antenna which is a helical type antenna[4]. The receiving antenna has a similar structure as the transmitting one. Fig.1 shows the block diagram of the wireless power transfer system. The impedance matching system is used to match the load impedance to the characteristic impedance so that the reflected power is reduced.

Unlike electromagnetic induction method which utilizes frequencies in the 50KHz range and can be used in the range of only a few millimetres due to the low quality factor of the antennas, in the resonance enhanced electromagnetic coupling,

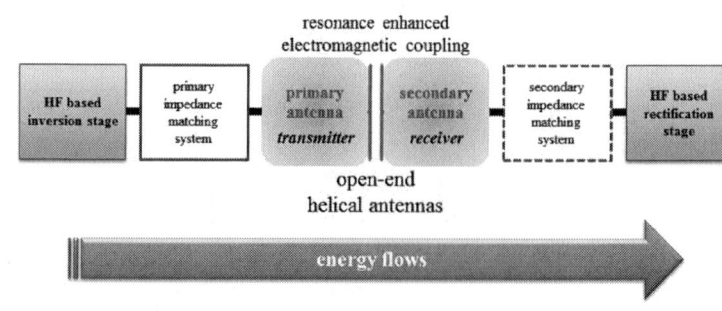

Fig. 1. Block diagram of the WPT system

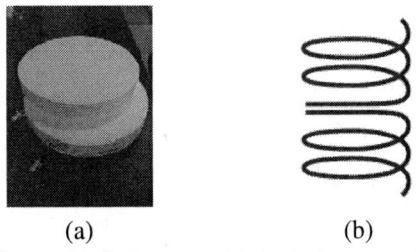
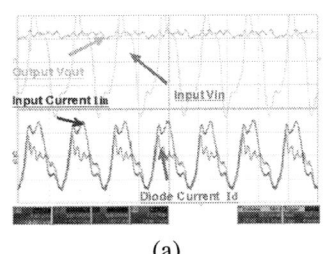
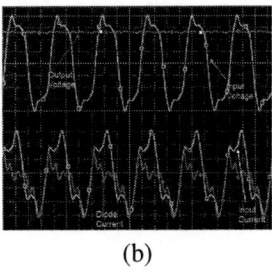

(a)   (b)

Fig. 2. Open-end helical antenna; (a) physical structure, (b) symbol

(a)   (b)

Fig. 3. Input voltage, input current, output voltage and diode current measured from; (a) experiment, (b) simulation

although the coupling coefficient $k(1)$ between the antennas is low and drops drastically with the increase of distance between the antennas, but the quality factor Q of each antenna is increased by the resonance between the inductance and distributed capacitance of the windings which allows high efficient energy transfer with less interaction with nearby, off-resonant objects[6].

$$k = \frac{L_m}{\sqrt{L_1 L_2}} \qquad (1)$$

where L_m : magnetizing (mutual) inductance
          $L_{1,2}$: inductances of transmitter and receiver coils.

B. Helical Antenna

There are two types of helical antennas: Open end helical antenna[7], in which the feeding point is in the center, and short end helical antenna in which the feeding point is at the edge. Open end helical antenna with 6 turns is used in the experiment. Fig.2 shows the structure of open-end helical antenna.

II. HIGH FREQUENCY AC/DC RECTIFIER

To analyse the performance of the rectifier, a high frequency supply is used without the wireless part. The supply is connected to the rectifier through SWR meter to measure the forward and reflected power and an impedance matcher to reduce the reflected power from the load. A coaxial cable with 50 ohms characteristic impedance is used. The components used in the rectifier circuit are summarized in Table 1.

TABLE I
COMPONENTS USED IN THE RECTIFIER

Parameter	Value	Description
Input frequency	14MHz	100W rated power
Load	100~ 240 Ω	Metal resistor
Smoothing Capacitor	3x10nF	Ceramic Capacitor
Diode	600V/2A	SiC Schottky Diode

Four diodes of SiC type are used in the rectifier circuit. SiC diodes are used since they have higher switching speed and lower forward drop voltage compared to their silicon counterparts[3].

A. Parasitic Elements

Parasitic inductances and capacitances in the circuit play an important role at such a high frequency, causing distortion in the waveforms of the voltages and currents. Measurements of

the voltages and currents were done and the results in Fig.3(a) show the input voltage, input current, output voltage and diode current in the rectifier taken from the experiment. The power supply used in this experiment has a low rated power. From the figure we observe that the input voltage and input current are distorted from their normal sinusoidal shape, also high ripples appear in the diode current.

Using the PSpice simulator we built the circuit in an attempt to get similar results as those obtained from the experiment. It was observed that the reason of distortion and ripples in the voltage and current waves is the existence of parasitic elements in the circuit. Fig.4 shows the circuit used in the simulation of the circuit. A precise equivalent model for the SiC diode is used in the PSpice simulation which accurately reflects the behaviour of the diode in the experiment. The parameters in the circuit simulation are defined as follows:

R_s : source input impedance(50Ω)
L_{cable} : inductance of the coaxial cable($0.1uH$)
C_{cable} : capacitance of the coaxial cable($10pF$)
L_{para} : parasitic inductance of the wire at the output side($0.05uH$)
C_{smooth} : smoothing capacitor($10nF x3$)
R_{load} : load($100 \sim 240\Omega$)
R_s : series resistance of the smoothing capacitor(0.5Ω)
L_s : series inductance of the smoothing capacitor ($0.02uH$)

Fig.3(b) shows the voltages and currents waves taken from the simulation. Comparing the two cases and noticing the similarity between the waves we conclude that the reason of distortion and appearance of higher harmonics in the waves is the existence of these parasitic elements in the circuit.

B. Output Current Conduction Modes

The conduction modes of the output current can be divided into two modes depending on whether the current goes to

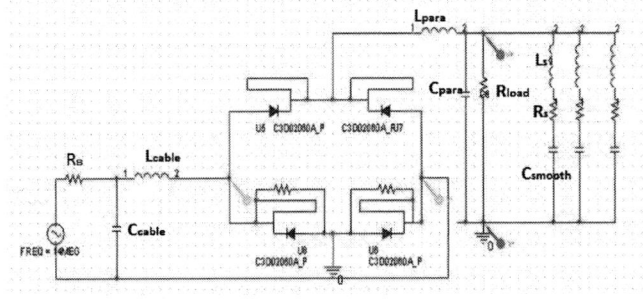

Fig. 4. Circuit diagram of the rectifier including parasitic elements

978-1-4799-2706-7/14 $31.00 © 2014 IEEE

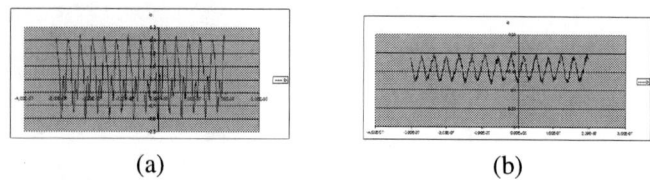

| (a) | (b) |

Fig. 5. Output current in the discontinuous and continuous modes; (a) discontinuous, (b) continuous

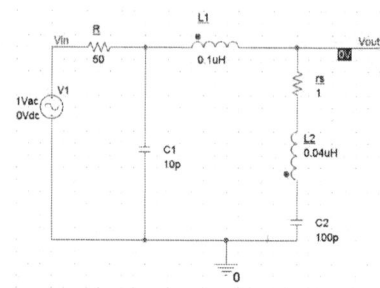

Fig. 7. Input of the rectifier after adding the capacitor

zero or not: discontinuous and continuous current modes[5]. In discontinuous mode the output current decreases to zero and maintains this value until the beginning of the next cycle. While in the continuous mode the current never falls to zero and is always continuous. However, in the case of 13.56 MHz frequency, the current not only is discontinuous, but it also sweeps to the negative side for some periods of time. Which means that the smoothing capacitor at the output discharges and the current flaws in the negative direction to the ground through the diodes. Fig.5(a) demonstrates this. The energy stored in an inductor that is added at the output of the rectifier as shown in Fig.6 will keep the current flowing in the same direction until commutation is done and the next cycle starts. To filter out the high frequency odd harmonics a capacitor is added in parallel to the input side of the rectifier as shown in the circuit diagram of Fig.6.

Consider the input side of the rectifier, the equivalent circuit after adding the capacitor the circuit is shown in Fig.7. This is a capacitor-input low pass filter with a transfer function expressed in (2), where L_1 and C_1 are the inductance and capacitance of the cable, L_2 is the equivalent series inductance of the capacitor and r_s is its equivalent series resistance which can be measured experimentally using the network analyser. The bode diagram for different capacitors is shown in Fig.8. To reduce the harmonics it is essential to design the filter such that the cut off frequency is higher than the fundamental frequency (14MHz) and less than the 3rd harmonic frequency (41MHz).

$$H(j\omega) = \frac{AS^2 + BS + 1}{CS^3 + DS^2 + ES + 1} \quad (2)$$

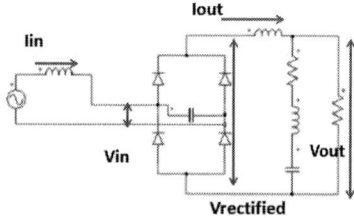

Fig. 6. New circuit with a commutation capacitor at the input and an inductor at the output

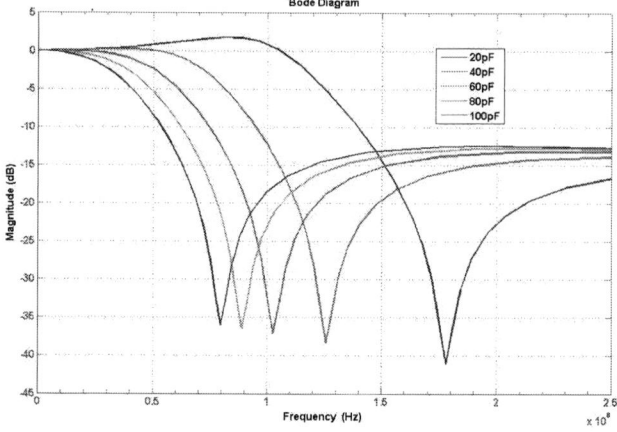

Fig. 8. Bode diagram of the capacitor-input filter for different capacitors

Where:

$$
\begin{aligned}
A &= L_2 C_2 \\
B &= C_2 r_s \\
C &= C_1 C_2 R_s (L_1 + L_2) \\
D &= C_2 (L_1 + L_2 + R_s r_s C_1) \\
E &= C_2 R_s + C_1 R_s + C_2 r_s
\end{aligned}
\quad (3)
$$

The output current now is in the continuous mode as shown in Fig.5(b), and the input voltage and current are now less distorted and include less harmonics than the original case. Fig.9(b) shows the input voltage and input current obtained by the experiment after modifying the circuit.

C. Efficiency Calculation

Reflected power at high frequency is a serious problem that causes sever degradation in the efficiency[8]. The reflected power depends on the value of reflection coefficient Γ. The reflection coefficient depends on the mismatch between the characteristic impedance Z_o (50 ohm in our case) and the load impedance Z_L[3].

An experimental set-up was used to measure the efficiency of rectification with a power supply with higher rated power, and with a higher quality impedance matcher that reduces the reflected power to almost zero. In Fig.10 the relation between

978-1-4799-2706-7/14 $31.00 © 2014 IEEE

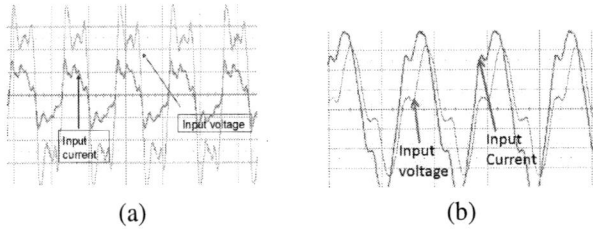

(a)   (b)

Fig. 9. Input voltage and current before and after circuit modification; (a) before, (b) after

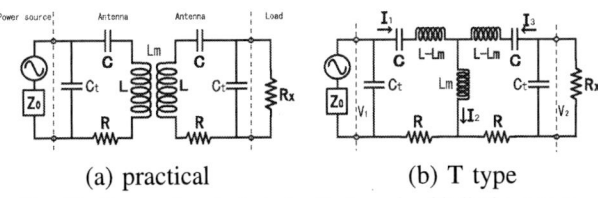

(a) practical   (b) T type

Fig. 11. Proposed equivalent circuit of a pair of helical antennas

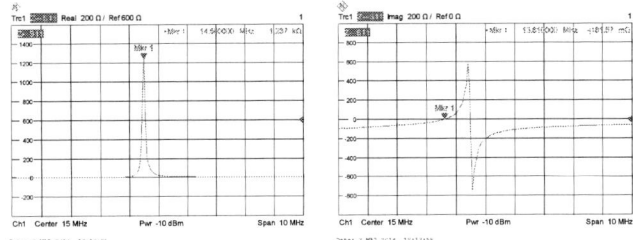

(a) real impedance   (b) imaginary impedance

Fig. 12. Real and imaginary parts of the antenna taken from the network analyser

the power and the efficiency is shown. At 1kW input power or more, the efficiency is 94%. The diodes are SiC with 20A rated current.

III. WIRELESS POWER TRANSFER SYSTEM OPTIMIZATION

A. Antenna equivalent circuit

The first step to analyse the antenna system is to introduce an accurate model of the antenna. In Fig.11 an accurate model for the open-end helical antenna is introduced[6]. Where R, L, and C are the resistance, distributed capacitance and self inductance of the coil, respectively. And C_t is the terminal parallel capacitance of the antenna. Those parameters can be calculated by solving the equations of real and imaginary parts of the input impedance at resonance and anti-resonance frequencies.

If we connect one antenna to the network analyser we can identify the real and imaginary parts of its input impedance as a function of frequency. As an example to show how to calculate the antenna parameters, consider an antenna with input impedance response as taken from the network analyser as shown in Fig.12. The resonance frequency for this antenna is $13.61 * 10^6 Hz$ (imaginary part is zero). And the anti-resonance frequency is $14.5 * 10^6 Hz$ (real impedance is high). To calculate the parameters of the antenna we need four equations. The real and imaginary parts of the input impedance of the antenna can be derived and are given by equations(4) and (5). The four equations to solve the system are given by (6), (7), (8), and (9).

$$Re[Z_{11}] = \frac{R}{\{1 - \omega C_t(\omega L - \frac{1}{\omega C})\}^2 + \omega^2 R^2 C_t^2} \quad (4)$$

$$Im[Z_{11}] = \frac{j(\omega L - \frac{1}{\omega C})\{1 - \omega C_t(\omega L - \frac{1}{\omega C})\} - j\omega R^2 C_t}{\{1 - \omega C_t(\omega L - \frac{1}{\omega C})\}^2 + \omega^2 R^2 C_t^2} \quad (5)$$

$$\omega_0 = \frac{1}{\sqrt{LC}} = 2\pi * 13.61 * 10^6 \quad (6)$$

$$Re\{Z'_{11}\} = \frac{R}{1 + \frac{1}{LC}R^2 C_t^2} = 3.5\Omega \quad (7)$$

$$Re\{Z''_{11}\} = \frac{1}{\frac{1}{L}(\frac{1}{C_t} + \frac{1}{C})RC_t^2} = 1237\Omega \quad (8)$$

$$\omega_a = \sqrt{\frac{1}{L}(\frac{1}{C_t} + \frac{1}{C})} = 2\pi * 14.5 * 10^6 \quad (9)$$

Where ω_0 and ω_a are the resonance and anti-resonance radian frequencies respectively, Z'_{11} and Z''_{11} are the input impedances of the antenna at resonance and anti-resonance, respectively.

Solving equations (6), (7), (8), and (9) yields to the solution for the four parameters as: $R = 3.5\Omega$, $C = 29.7pF$, $L = 4.6uH$, and $C_t = 220pf$. Substituting these values into equations (4) and (5) then plotting the real and imaginary parts as a function of frequency yields to the results shown in Fig.13.

The resistance of the antenna is high which limits the maximum efficiency of the system. In an attempt to maximize the efficiency of the antenna we replaced the wire with one that has lower resistance and is Teflon-isolated. The higher isolation will increase the maximum power that can be trans-

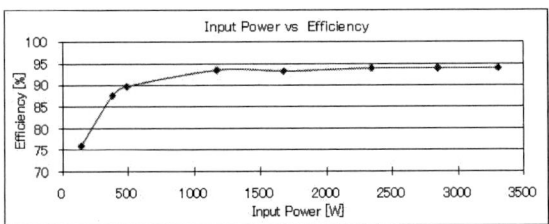

Fig. 10. Input power vs. efficiency for high input power

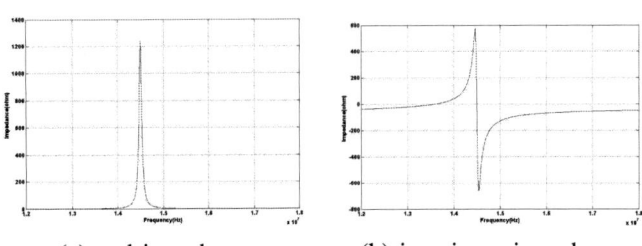

(a) real impedance   (b) imaginary impedance

Fig. 13. Real and imaginary parts of the antenna-simulation

TABLE II
PARAMETERS OF THE HELICAL ANTENNA USED IN THE EXPERIMENT

parameter	unit	value
L	uH	4.7
C	pF	27.3
C_t	pF	203.4
R	Ω	1.2

ferred by the antenna. Furthermore, at the edges of the wire the electric field is high and might exceed the disruptive potential gradient which is $30kV/cm$ causing corona discharge, especially when the distance is increased. Corona caps are used in the experiment in order to reduce the potential gradient.
Table II shows the values of the parameters for the antenna used in the experiment.

B. S-Parameters optimization

The S-parameters of a helical antenna can be described in terms of the antenna parameters and the characteristic impedance. S_{21} is the power transfer coefficient and S_{11} is the power reflection coefficient. Equations for both coefficients are shown in (10) and (11)[6].

$$S_{11} = \frac{CLL_m{}^2 - C^2L^2Z_0{}^2 + L_m{}^2C_t{}^2Z_0{}^2}{CLL_m{}^2 + C^2L^2Z_0{}^2 - L_m{}^2C_t{}^2Z_0{}^2 + 2jL_m{}^2C_tZ_0\sqrt{LC}} \tag{10}$$

$$S_{21} = \frac{2jL_mZ_0LC\sqrt{LC}}{CLL_m{}^2 + C^2L^2Z_0{}^2 - L_m{}^2C_t{}^2Z_0{}^2 + 2jL_m{}^2C_tZ_0\sqrt{LC}} \tag{11}$$

Where L_m is the mutual inductance between the coils. To maximize the power transfer efficiency the reflection coefficient should be zero. Equating equation(10) to zero we get an expression of the mutual inductance for maximum power transfer as in(12). For the antenna used in the experiment the mutual inductance can be calculated as 0.42 uH.

$$L_m = \frac{CLZ_0}{\sqrt{C_t{}^2Z_0{}^2 + CL}} \tag{12}$$

Fig.14 shows the relation between the mutual inductance and the magnitude of the power transfer coefficient $|S21|$ using equation (11). It is confirmed from the figure that the maximum power transfer is achieved at 0.42 uH.

C. Optimization of power transfer using the equivalent circuit of the antenna

Since the two coils are magnetically coupled, we can combine their equivalent circuits into a one T-type circuit by using the mutual inductance to link the two parts. Fig.15 shows the T-type circuit of the antennas connected to the source power and the load. Where Zs is the impedance of the source which is equal to the characteristic impedance 50 Ohms.
If the frequency of the source equals the resonance frequency of the antenna then the system resonates. At this particular frequency the impedance of L equals the impedance of C and as a result they cancel each other out. Consequently, the series combination of the impedance of C and L-Lm will be -jXm. We are particularly interested in the T shape part and how does the impedance from the source side look like. As shown in

Fig.16 if a resistance is connected at the output, the equivalent input impedance Z_{in} can be calculated as in equation(13).

$$Z_{in} = \frac{X_m^2}{R_L} \tag{13}$$

The load impedance is inverted and multiplied by X_m^2 factor when viewed from the source side. In other words, if no load is connected at the load side the source will "see" a short circuit, on the other hand, if the output is shorted the equivalent impedance to the source is an open circuit with no load attached. Moreover, series connected impedances would look as a parallel connected impedances and parallel connected impedances would equivalently be series connected. Applying this rule to the WPT system would result in the new circuit in Fig.17 as viewed from the source side.

IV. MAXIMUM POWER AND MAXIMUM EFFICIENCY

We would like to investigate the conditions for maximum power and maximum efficiency. For maximum power, applying the maximum power transfer theorem so that the load impedance is equal to the Thevenin equivalent impedance and assuming a pure resistive load we get equations (14) and (15).

$$R_L = Z_s \tag{14}$$

$$L_m = \frac{1}{\omega}\frac{Z_sX_{ct}}{\sqrt{Z_s^2 + X_{ct}^2}} \tag{15}$$

Where ω is the angular frequency and X_{ct} is the impedance of the terminal capacitor of the antenna C_t. Equations (14) and (15) give the conditions for maximum power transfer. There are two restrictions: the first one is on the load value which should be equal to the characteristic impedance (50 Ohm), the second one is on the distance between the antennas. If we substitute the values in the equation of the mutual inductance we get a 0.424 uH which is the same as the one obtained by

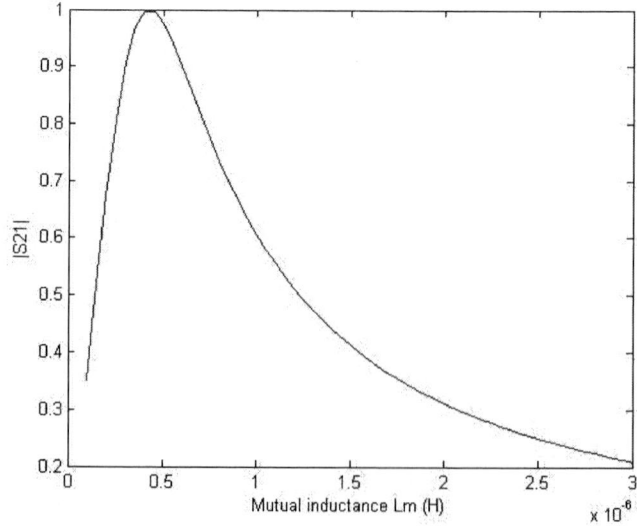

Fig. 14. Mutual inductance vs. transmission coefficient

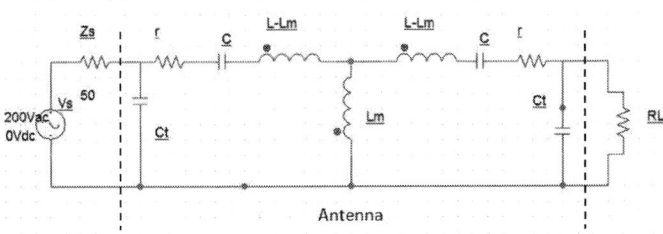

Fig. 15. Complete WPT system including the source, antenna and load

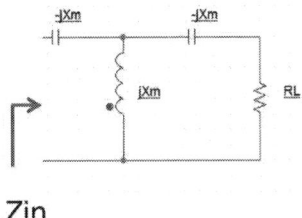

Zin

Fig. 16. Impedance inverter

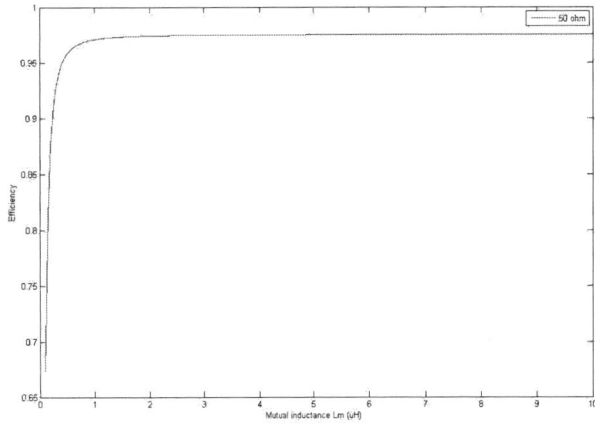

Fig. 18. Mutual inductance vs. antenna efficiency derived from the antenna equivalent circuit

the S-parameters analysis previously.

Using the equivalent circuit of the system shown in Fig.17 we can calculate the efficiency of the antenna as a function of L_m. The result is shown in Fig.18. We conclude that the higher L_m (the shorter the distance is), the higher the efficiency is, up to a certain point at which the maximum efficiency will not be exceeded. However, this result is not complete since it does not take into consideration the transmission and reflection coefficients of the coils(S_{11} and S_{21}). In Fig.14 we already showed that maximum transmission can be achieved at one point (0.42 uH). Consequently, the efficiency in Fig.18 must be multiplied by S_{21} and the efficiency for different loads is shown in Fig.19.

Fig.19 confirms two things: 1− maximum efficiency is transferred at 50 Ohms load, and 2− this point is at a certain distance which is equivalent to a mutual inductance which was calculated previously as 0.42 uH for this antenna. Simulation shows that the maximum efficiency of the antenna is 96%. This means that if the rectifier is connected to the antenna and based on the results in Fig,10 the efficiency of the rectifier

is 94% at 1000W input power, based on that we expect a total efficiency of the system of $96\% * 94\% = 90.24\%$. A very important thing to notice is that unlike electromagnetic induction in which a high coupling coefficient means a higher efficiency, in magnetic resonance coupling a high coupling coefficient does not necessarily result in higher efficiency. On the contrary, a high coupling coefficient will limit the ability to transfer power.

In Fig.20 L_m vs. source current is shown for different loads. Increasing the distance between the antennas beyond the optimum point (small L_m) will cause high currents to flow in the sending coil. This is logical since if the two antennas are placed far away from each other the linkage flux is very small which means the load effect on the source is almost negligible and the source will "see" a short circuit.

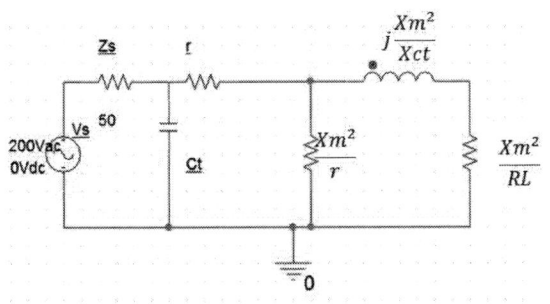

Fig. 17. Circuit viewed from the source after considering the impedance inverter

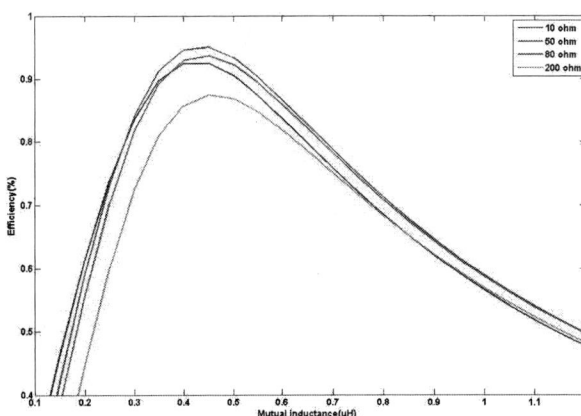

Fig. 19. Mutual inductance vs. efficiency of antenna for different loads-simulation

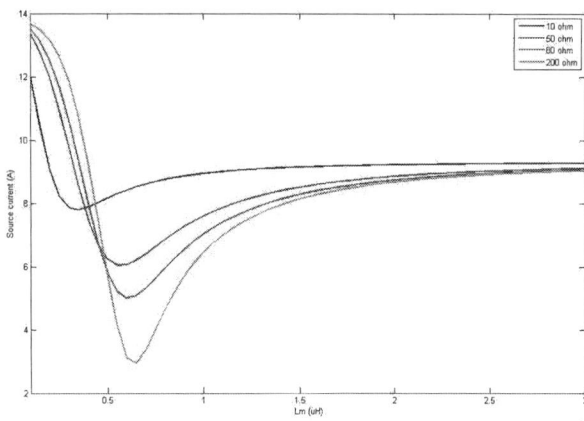

Fig. 20. Mutual inductance vs. source current for different loads-simulation

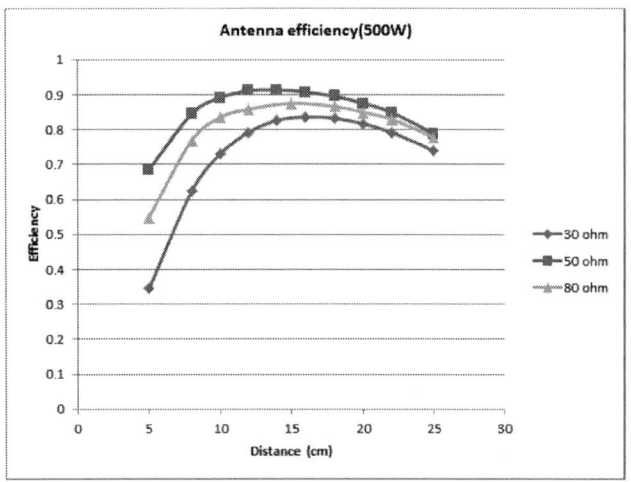

Fig. 21. Distance vs. antenna efficiency-experiment

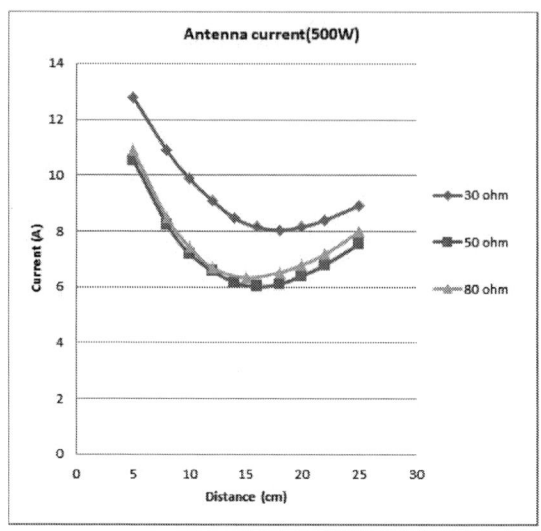

Fig. 22. Distance vs. source current-experiment

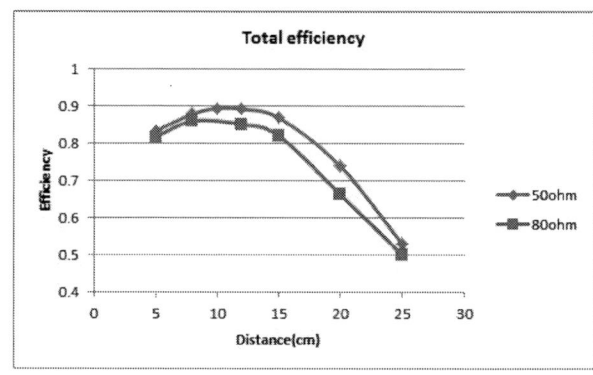

Fig. 23. Distance vs. efficiency of complete system-experiment

V. EXPERIMENT

A. Antenna characteristics

In this part RF source with frequency 13.56MHz is connected to the antenna through an impedance matcher. A load is connected at the receiving coil. We tested the system under three different load conditions: 30Ω, 50Ω, and 80Ω. Fig.21 shows the results of the experiment. An important thing to point out is that maximum efficiency (92%) is achieved at 50Ω load and 15cm distance. Deviation from this optimum point causes reduction in efficiency which was predicted by simulation in Fig.19. The source power in this case is 500W. Next, the source current is measured and the results are shown in Fig.22. Again the experimental result agree with the simulation in that increasing the distance beyond a certain point (small L_m) causes increase in the source current.

B. WPT system and rectifier

In this part the complete system is tested including the antenna and the rectifier as shown in the block diagram in Fig.1. The system was tested under two load conditions: 50Ω

and 80Ω, results are shown for input power ranging from 800W to 1500W. The data is summarised in TableIII for 50Ω load and the results are shown in Fig.23. Where Pr is the reflected power.

Maximum total efficiency of the system is 89.11% at 1500W and 12cm distance.

Next, the experiment was carried out at a constant output DC voltage of 300V by changing the load and the input power. Results are shown in TableIV for input power 3.5kW and different distances. Fig.24 shows the efficiency against input power at 300V constant output voltage.

TABLE III
EXPERIMENTAL RESULTS FOR DC LOAD 50Ω

D(cm)	Pin(W)	Pr(W)	$V_o(Vdc)$	$I_o(Adc)$	Efficiency(%)
5	1512.9	47.6	248	4.92	83.27
8	1500	0.4	257.1	5.12	87.78
10	1502.4	0.1	259.5	5.17	89.3
12	1499.9	0.2	259	5.16	89.11
15	1500.7	0.4	255.9	5.09	86.818
20	1017	1.4	194.1	3.88	74.154
25	803.9	2.2	146	2.92	53.177

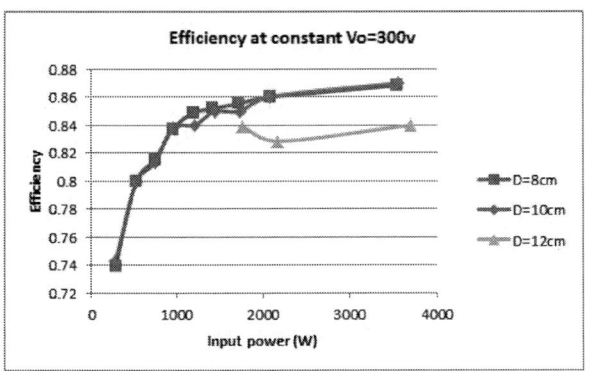

Fig. 24. Input power vs. efficiency (output voltage is 300Vdc constant)-experiment

VI. CONCLUSION

We made an accurate model for the high frequency AC/DC rectifier by simulation, then made changes on the circuit to make the output current continuous. The efficiency of the rectifier is 94% for input power more than 1kW.

The theory for calculating the efficiency of the antenna system was established and the conditions for maximum efficiency was depicted by theory and simulation and confirmed by experiment.

Maximum efficiency can be achieved at 50Ω load and 0.42uH mutual inductance between the antennas. Maximum efficiency of 90.24% was estimated by simulation and 89.11% efficiency at 1.5kW input power and 12cm distance was achieved experimentally.

Unlike electromagnetic induction in which the efficiency of transmission can be increased by reducing the distance between the two coils, in magnetic resonance coupling there is a point at which this condition is achieved. Smaller distance does not mean a higher power transfer capability in magnetic resonance.

At small distance the current is high and it will decrease as distance is increased up to a certain point. Any further increase of the distance will result in increase in the current again. Long distance results in high currents in the source side because of the effect of inverted impedance that is viewed by the source which is discussed earlier. At long distance the load effect is very small on the source and appears like a low impedance.

REFERENCES

[1] A. Kurs, A. Karalis, R. Moffatt, J. D. Joannopoulos, P. Fisher, and M. Soljacic, "Wireless Power Transfer via Strongly Coupled Magnetic Resonances," *Science,* Vol. 317, No. 5834, pp. 83-86, 2007.

[2] A. Karalis, J. D. Joannopoulos, and M. Soljacic, "Efficient Wireless nonradiative mid-range energy transfer," *Annuals of Physics,* Vol. 323, No. 1, pp. 34-48, 2008.

[3] K. Kusaka, and J. Itoh "Experimental verification of rectifiers with SiC/GaN for wireless power transfer using a magnetic resonance coupling," *Proceedings of IEEE International Conference on Power Electronics and Drive Systems,* pp. 1094-1099, 2011.

[4] Y. Moriwaki, T. Imura, and Y. Hori, "Basic study on reduction of reflected power using DC/DC converters in wireless power transfer system via magnetic resonant coupling," *Proceedings of IEEE International Telecommunications Energy Conference,* pp. 1-5, 2011.

[5] S.B. Dewan, "Optimum Input and Output Filters for a Single-Phase Rectifier Power Supply," *IEEE Transactions on Industry Applications,* Vol. IA-17, No. 3, pp. 282-288, 1981.

[6] T. Anazawa, A. Kawamura, and T.W. Kim, "Parameter Optimization of Wireless Power Transmission Based on Electrical Equivalent Circuit and Efficiency Analysis of HF Based Rectifier," *Proceedings of Japan-Korea Joint Technical Workshop on Semiconductor Power Converter,* pp. 93-98, 2011.

[7] T. Imura, H. Okabe, T. Uchida, and Y. Hori, "Study on Open and Short End Helical Antennas with Capacitor in Series of Wireless Power Transfer using Magnetic Resonant Couplings," *Proceedings of Annual Conference of IEEE Industrial Electronics,* pp. 3848-3853, 2009.

[8] K. Kusaka, and J. Itoh, "Input impedance matched AC-DC converter in wireless power transfer for EV charger," *Proceedings of International Conference on Electrical Machines and System,* pp. 1-6, 2012.

TABLE IV
EXPERIMENTAL RESULTS FOR CONSTANT OUTPUT VOLTAGE 300V

D(cm)	Pin(W)	Pr(W)	$I_o(Adc)$	Efficiency(%)
8	3542	0.2	10.25	86.82
10	3552	0.075	10.3	86.995
12	3692	30.1	10.25	83.97

978-1-4799-2706-7/14 $31.00 © 2014 IEEE

Non-iterative LCL Filter Design for Three-phase Two-level Voltage-source PWM Converters

Byung-Geuk Cho

Electrical and Computer Engineering
Seoul National University
Seoul, Korea
bk8089@eepel.snu.ac.kr

Seung-Ki Sul

Electrical and Computer Engineering
Seoul National University
Seoul, Korea
sulsk@plaza.snu.ac.kr

Abstract— **This paper describe a design method of an LCL filter for the three-phase two-level voltage-source PWM converters. The grid current harmonics regulation is the primary condition for the operation of the grid-connected converters to satisfy. If the grid voltage is assumed to be ideally sinusoidal, PWM switching is the only factor to produce harmonics on the grid current, especially around the bands of the switching frequency. However, every single harmonic cannot be evaluated for the filter design. In this paper, it is found that there exists the most significant harmonic frequency at which the harmonic current is dominant. Theoretical description of the PWM waveform and the transfer function of the filter's impedance are needed for the derivation of the most significant frequency. Once the most significant frequency is recognized, the total inductance to satisfy the grid current harmonic regulations is determined by selecting proper resonant frequency. Along with the information of the designed total inductance and preset resonant frequency, additional conditions such as the filter's total energy, converter current ripple and converter capacity are utilized to specify the values of each component of the filter. The effectiveness of the proposed method is verified with simulation and experimental results.**

Keywords— LCL filter, Non-iterative design, PWM converters.

I.   INTRODUCTION

Three-phase PWM converters are increasingly employed for grid connection of regenerative energy sources such as wind and solar power generation [1]. In the operation of the grid-connected PWM converters, high frequency harmonics are intrinsically produced by the PWM switching and therefore current harmonics are subject to the regulation such as IEEE 519-1992 or IEC 61000-3-2 [2].

To confine the current harmonics, single inductor L filters are simply inserted or LCL filters are employed between the converter and the grid [3][4]. However, as higher value of inductance is required for L filters than the case for LCL filters, LCL filters are more popularly used in various applications.

Unfortunately, however, there is no fixed reference or guide line for the LCL filter components selection. Numerous approaches have been reported before but most of the methods are based on the trial-and-error methods or iterative processes [5][6].

In this paper, a non-iterative way of designing the LCL components for three-phase two-level voltage source converters is proposed. As in [2], the proposed method is based on the identification for the most significant harmonic frequency of the grid current with the theoretical derivation of the PWM output voltage but space vector PWM (SVPWM) is implemented instead of SPWM, which achieves higher modulation index. After obtaining the most significant harmonic frequency, the total inductance can be calculated to satisfy the grid harmonic regulation at the corresponding frequency by taking the switching frequency, the resonant frequency and the DC link voltage as inputs. Along with the calculated total inductance, several cost functions such as the filter's total energy, converter current ripple and converter capacity for the optimal selection of the each component are investigated and the design process ends.

To validate the proposed design method, simulation results are included and experiments are carried out.

II.   SYSTEM CONFIGURATION

The grid-connected converter system considered in this paper is depicted in Fig. 1 and the specification of the variables is shown in TABLE I. To match the experimental results with the simulation ones, the power level is set as low as 10kW but the switching frequency (f_{sw}) is determined to be 2.5kHz to address that the proposed filter design method is also applicable to higher power systems in which available switching frequency is limited within a few kHz. It is assumed that the grid side inductance (L_g) involves the filter inductance, the grid impedance and the leakage inductance of the transformer if required for the system configuration. In addition, unintentional resistances including any equivalent series resistances (ESRs) of the components and any stray resistances existing in the circuit are ignored to simplify the design process.

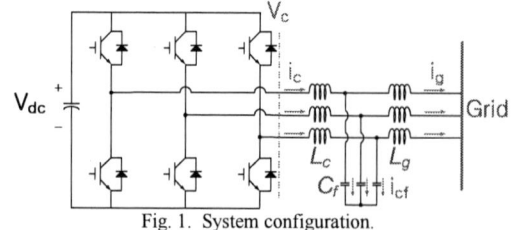

Fig. 1. System configuration.

TABLE I
SYSTEM SPECIFICATION

Rated power(P_b)	10[kW]	Switching frequency(f_{sw})	2.5[kHz]
Rated grid voltage(V_b)	220[V_{rms}]	Grid frequency(f_g)	60[Hz]
Rated current(I_b)	26.24[A_{rms}]	Base inductance(L_b)	12.8[mH]
DC link voltage(V_{dc})	380[V]	Base capacitance(C_b)	548[μF]

III. LCL FILTER DESIGN

A. LCL Filter Model

Mathematical description of the LCL filter is important for the filter design process and two primary transfer functions are given in (1) ~ (2), where V_c, i_c, i_g stand for the converter output voltage, converter current and grid current, respectively. L_c represents the converter side filter inductance while L_g implies the grid side inductance as mentioned in II. C_f means the filter capacitance. As can be seen from the equations, the LCL filters suffer from the resonance.

$$\frac{i_g}{V_c} = \frac{1}{s^3 C_f L_c L_g + s\left(L_c + L_g\right)} \quad (1)$$

$$\frac{i_c}{V_c} = \frac{s^2 C_f L_g + 1}{s^3 C_f L_c L_g + s\left(L_c + L_g\right)} \quad (2)$$

B. Harmonic Regulation

The reason for the filter to be inserted between the converter and the grid is to suppress the injected current harmonics into the grid under certain levels. IEEE 519-1992 addresses specific numbers for each harmonic and it is stated as in TABLE II. The converter system with generation capability belongs to the shaded area and therefore those numbers in the first row must be satisfied.

TABLE II
Current Distortion Limits of IEEE 519-1992
(120V through 69000V)

Maximum Harmonic Current Distortion in Percent of I_L						
Individual Harmonic Order (Odd Harmonics)						
I_{sc}/I_L	< 11	11< h <17	17< h <23	23< h <35	35≤ h	TDD
< 20*	4.0	2.0	1.5	0.6	0.3	5.0
20<50	7.0	3.5	2.5	1.0	0.5	8.0
50<100	10.0	4.5	4.0	1.5	0.7	12.0
100<1000	12.0	5.5	5.0	2.0	1.0	15.0
>1000	15.0	7.0	6.0	2.5	1.4	20.0

Even harmonics are limited to 25% of the odd harmonic limits above.

Current distortions that result in a dc offset, e.g., half-wave converters, are not allowed.

*All power generation equipment is limited to these values of current distortion, regardless of actual I_{sc}/I_L.

Where
I_{sc} = maximum short-circuit current at PCC.
I_L = maximum demand load current (fundamental frequency component) at PCC

C. Identification of the Most signficant Frequency

To ideally design the LCL filter, the grid current must be examined to see if the regulation is met at every individual harmonic frequency. However this makes the design process unnecessarily exhausting and complex. Therefore, in this paper, the most significant frequency from the LCL filter design point of view is investigated hereinafter. Once it is identified, only the current harmonic at the corresponding frequency needs to be

checked for the design of the LCL filter.

The grid current is determined by the equation (1) and it can be modified for each harmonic as in (3), where $i_{g,har}$, $V_{c,har}$, $Z_{LCL,har}$ stand for the grid current harmonics, the converter output voltage harmonics and the impedance of the LCL filter at harmonic frequencies, respectively. In addition, ω_{har}, ω_{res} and L_T represent the frequency of the harmonics, the resonant frequency and the total inductance ($=L_g+L_c$), respectively.

$$\left|i_{g,har}\right| = \frac{\left|V_{c,har}\right|}{\left|Z_{LCL,har}\right|} = \frac{\omega_{res}{}^2 \left|V_{c,har}\right|}{\omega_{har}\left(\omega_{har}{}^2 - \omega_{res}{}^2\right)L_T} \quad (3)$$

From (3), it can be recognized that the grid current harmonics are determined by the magnitude of the converter output voltage harmonics, the harmonic frequency, the resonant frequency and the total inductance of the filter. However, in this sector, as identification of the most significant frequency is the main concern for a given LCL filter, ω_{res} and L_T can be assumed to be fixed and therefore simply characteristics of $V_{c,har}$ and ω_{har} are investigated.

1) Converter Output Voltage Harmonics($V_{c,har}$)

According to [7], the theoretical expression for the converter output voltage in the case of asymmetrical regular sampled SVPWM is given as (4), where m is the carrier index variable and n is the baseband index variable. M is the modulation index with reference to $0.5V_{dc}$ while J is the Bessel function and q is defined as $m+nf_g/f_{sw}$. From (4), it is known that the magnitude of the converter output voltage varies depending on the values of M, m, n, V_{dc} and f_{sw}. Because V_{dc} and f_{sw} are fixed as in TABLE I, M, m and n are the factors which actually have effects on the magnitude of the converter output voltage.

$$V_{c,har} = \frac{4V_{dc}}{m\pi^2}\left\{\begin{array}{l}\left[\frac{\pi}{6}\sin\left(\left[m+n\right]\frac{\pi}{2}\right)\left\{J_n\left(q\frac{3\pi}{4}M\right)+2\cos\frac{n\pi}{6}J_n\left(q\frac{\sqrt{3}\pi}{4}M\right)\right\}\right. \\ \left.+\frac{1}{n}\sin\frac{m\pi}{2}\cos\frac{n\pi}{2}\sin\frac{n\pi}{6}\left\{J_0\left(q\frac{3\pi}{4}M\right)-J_0\left(q\frac{\sqrt{3}\pi}{4}M\right)\right\}\right]_{n\neq0} \\ +\sum_{\substack{k=1 \\ (k\neq-n)}}\left[\begin{array}{l}\frac{1}{[n+k]}\sin\left([m+k]\frac{\pi}{2}\right)\cos\left([n+k]\frac{\pi}{2}\right)\sin\left([n+k]\frac{\pi}{6}\right) \\ \times\left\{J_k\left(q\frac{3\pi}{4}M\right)+2\cos\left([2n+3k]\frac{\pi}{6}\right)J_k\left(q\frac{\sqrt{3}\pi}{4}M\right)\right\}\end{array}\right] \\ +\sum_{\substack{k=1 \\ (k\neq n)}}\left[\begin{array}{l}\frac{1}{[n-k]}\sin\left([m+k]\frac{\pi}{2}\right)\cos\left([n-k]\frac{\pi}{2}\right)\sin\left([n-k]\frac{\pi}{6}\right) \\ \times\left\{J_k\left(q\frac{3\pi}{4}M\right)+2\cos\left([2n-3k]\frac{\pi}{6}\right)J_k\left(q\frac{\sqrt{3}\pi}{4}M\right)\right\}\end{array}\right]\end{array}\right\} \quad (4)$$

2) Harmonic Frequency(ω_{har})

To include the effect of the harmonics frequencies on the identification of the most significant frequency, 8 candidates of $m=1$, $m=2$ and $n=\pm1$, ±2, ±3, ±4 are considered. As the converter output voltages at those harmonic frequencies are dominant among sideband harmonics of the switching frequency, the grid currents at those harmonics frequencies are suspicious in terms of the most significant harmonics.

3) Grid Current Harmonics in LCL Filter

By exploring the 8 candidates, the frequency which gives the most significant effect on the grid current harmonics can be identified. To do so, modification of the equation (1) needs to be made. If the resonant frequency is sufficiently lower than the switching frequency, then the transfer function of (1) can be approximated as in (5). To demonstrate the validity of the approximation, magnitude plots of (1) and (5) are presented in Fig. 2 and it is noted that the approximation is effective for the frequency ranges of the 8 candidates.

$$\frac{i_g}{V_c} = \frac{1}{s^3 C_f L_c L_g + s\left(L_c + L_g\right)} \approx \frac{1}{s^3 C_f L_c L_g} = \frac{\omega_{res}^2}{s^3 L_T} \quad (5)$$

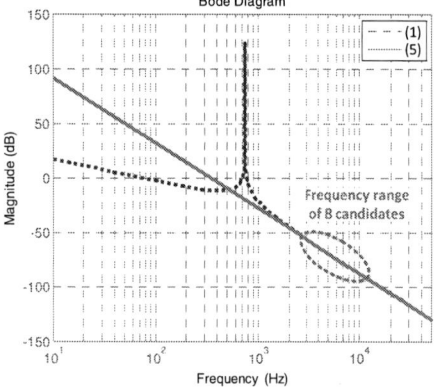

Fig. 2. Comparison of transfer functions between (1) and (5).

Then, the grid current harmonics around the 8 candidates are calculated as in (6), which shows that the current harmonics are proportional to output voltage harmonics divided by the cubic of corresponding harmonic frequency. This indicates that the volt per cubic Hertz (V/Hz^3) for the 8 candidates has to be observed to identify the most significant frequency.

$$\left|i_{g,har}\right| \approx \frac{\left|V_{c,har}\right|\omega_{res}^2}{\omega_{har}^3 L_T} \propto \frac{\left|V_{c,har}\right|}{\omega_{har}^3} \quad (6)$$

In Fig. 3, the results of the volt per cubic Hertz (V/Hz^3) for the 8 candidates are shown. As the values of the volt per cubic Hertz for each candidate may be different with different M and f_{sw}, the volt per cubic Hertz is inspected by varying M and f_{sw}. Furthermore, to notice the result clearly, ratios of the volt per cubic Hertz for $m=1$, $n=-2$ on the basis of the other 7 candidates are depicted. As it demonstrates, the ratios are always larger than 1 and this concludes that the most significant frequency is $f_{sw}-2f_g$. Therefore, the current harmonic at $f_{sw}-2f_g$ is only to be checked if the current harmonic regulation is satisfied for the LCL filter design.

However, if the 8 significant frequencies are located in different sectors of TABLE II, the regulation values for the harmonics become different and thus the conclusion derived above may be changed. For example, if the frequency of $m=1$ and $n=-2$ is in the area of 23<h<35 and the frequency of $m=1$ and $n=2$ is in the area of 35<h, the harmonic regulation values are two times different. In Fig. 3(b) and Fig. 3(c), then the ratio is less than 2 and it should be concluded that the most significant frequency is the case of $m=1$ and $n=2$ or $m=1$ and $n=4$. Thus if those frequencies are placed in other areas, additional investigation must be made. In this paper, the switching frequency is set as 2.5[kHz] and this guarantees that all the sidebands of the switching harmonics are located in the same sector. More specifically, if the lowest

frequency of $m=1$ and $n=-4$ is higher than 35th harmonics, which means that the switching frequency is higher than 2.34[kHz], the design process in this paper can effectively be applied without any further consideration.

D. Total Inductance Design

In this sector, the total inductance ($L_T = L_g + L_c$) can be initially designed with the proposed non-iterative method. As proved from the previous sectors, if the current harmonic at the most significant frequency (f_{sig}[Hz] or ω_{sig}[rad/s]) satisfies the regulation, then it can be said that all the harmonics meet the regulation. Hence only the harmonic at ω_{sig} needs to be examined and the equation (3) can be replaced by (7)

$$\left|i_{g,har}\right|_{m=1,n=-2} = \frac{\omega_{res}^2\left|V_{c,har}\right|_{m=1,n=-2}}{\omega_{sig}\left(\omega_{sig}^2 - \omega_{res}^2\right)L_T} \quad (7)$$

Then the current harmonic at ω_{sig} is determined by the variables such as ω_{sig}, ω_{res}, L_T and $V_{c,har}$ for $m=1$ and $n=-2$. As $V_{c,har}$ is dependent on M according to (4) and the most significant frequency, ω_{sig}, is known when the switching frequency is fixed, the final factors to determine the magnitude of $i_{g,har}$ are M, ω_{res} and L_T.

The modulation index, M, is based on the magnitude of the converter voltage reference at the fundamental grid frequency. If the resonant frequency is sufficiently larger than the fundamental grid frequency, the LCL filter can be assumed as a filter of pure inductance with the summation of L_c and L_g. Accordingly, when the modulation index is considered, only the total inductance can be included.

Then, to calculate the modulation index, the converter voltage reference is calculated for the unity power factor and it is shown in the nominator of (8) and conclusively the modulation index is determined by L_T with the predetermined rated current, the DC link voltage and the nominal grid frequency.

$$M = \frac{\sqrt{V_b^2 + \left(2\pi f_g L_T i_b\right)^2}}{0.5 V_{dc}} \quad (8)$$

In conclusion, the final factors to determine the magnitude of $i_{g,har}$ are modified from M, ω_{res} and L_T to ω_{res} and L_T. Based on this derivation, if the resonant frequency is determined, the total inductance is calculated with the non-iterative method in this paper.

If the resonant frequency is selected to be 800Hz for an example, then the final equation for the grid current harmonic at the most significant frequency is derived by combining (4), (7) and (8). As expected, the equation is solely dependent on L_T because other variables are already fixed. Then if the regulation of TABLE II is applied, an inequality is made as in (9).

$$\left|i_{g,har}\right|_{m=1,n=-2} = \frac{\omega_{res}^2\left|V_{c,har}\right|_{m=1,n=-2}}{\omega_{sig}\left(\omega_{sig}^2 - \omega_{res}^2\right)L_T} \leq 0.3[\%] \times i_b \quad (9)$$

As there is only single variable of L_T in (9), the solution can be numerically calculated. It is found that the total inductance satisfying the condition (9) needs to be larger than 0. 2035[p.u.].

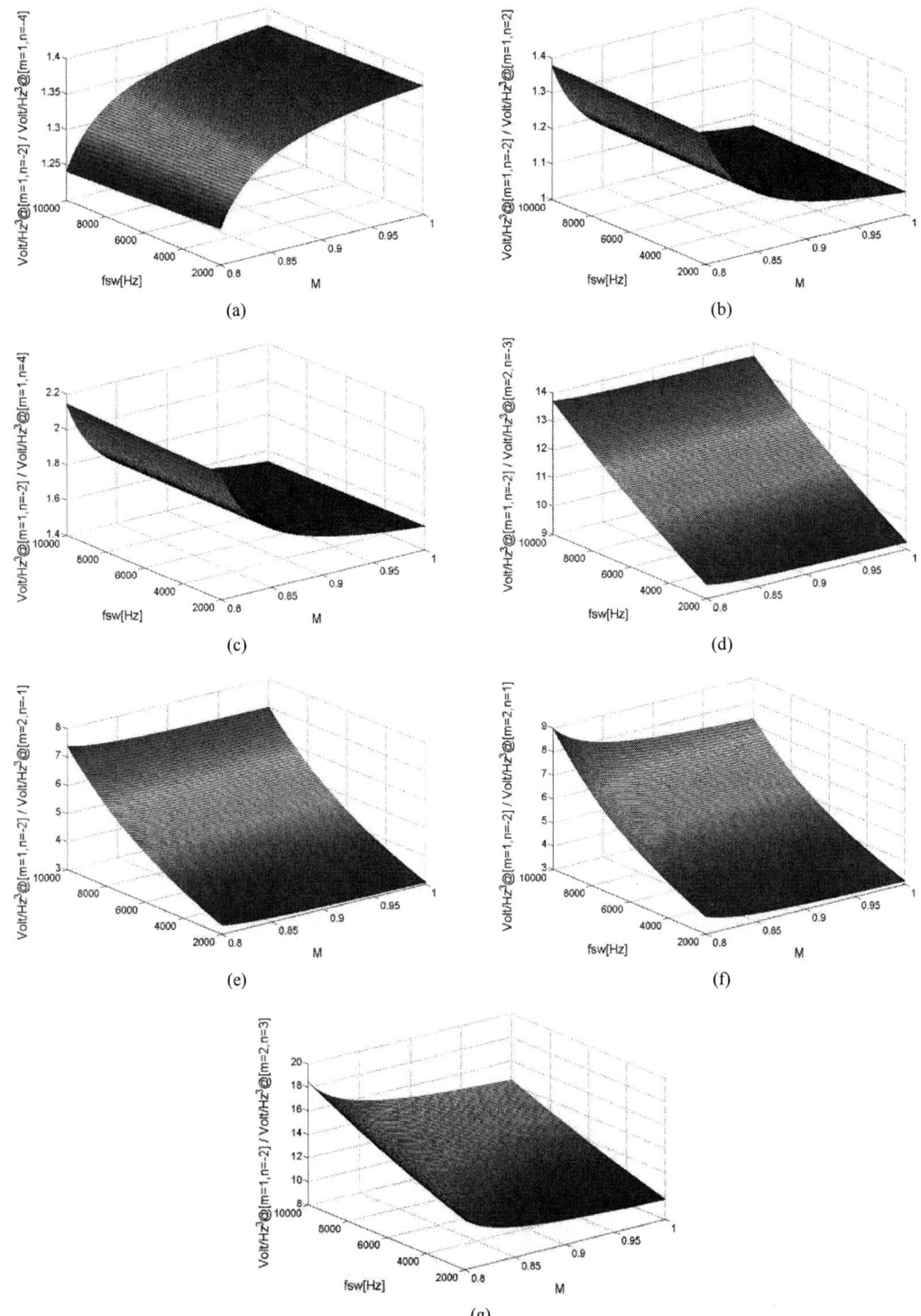

Fig. 3. Ratios of volt per cubic Hrtz at significant frequencies.

Similarly, if another resonant frequency is applied, the threshold value for the total inductance satisfying the harmonic regulation can be easily computed.

As for the selection of the resonant frequency, it is advantageous if the resonant frequency becomes lower at the cost of the inductor size because more attenuation can be achievable. However, it is more likely for the resonance to appear when the resonant frequency is closer to the grid frequency due to the low order harmonics of the grid voltage, which threatens the stability of the system.

In this paper, 800Hz is taken as an example for the development of the proposed method because the total inductance of 0.2035[p.u.] is considered to be appropriate and allowable in terms of the size while the resonant frequency is sufficiently large to guarantee acceptable bandwidth of the current controller.

E. Final Selection of Filter Component

Even though the resonant frequency and the total inductance can be designed to satisfy the grid current regulations, the two designed values are insufficient to designate the three filter components. Therefore additional conditions need to be introduced. In this paper, the converter current ripple, the total energy of the filter and the converter capacity are considered as the cost functions for the final selection of the LCL filter.

1) Converter Current Ripple

Converter current ripple has a deep impact on the cost and the size of the converter side inductor and the power semiconductors in the converter. If the converter side inductance is inappropriately chosen, the filter implementation is not feasible even though the grid current harmonic regulation is satisfactory. Therefore the converter current ripple needs to be analyzed.

The maximum converter current ripple is also investigated at the most significant frequency as for the grid current. The converter current is described with respect to the converter output voltage as in (2). At the frequencies around the sideband of the switching frequency, (2) can be modified to (10) and this approximation is commonly used for the converter side inductance selection in numerous literatures.

$$\frac{i_c}{V_c} = \frac{s^2 C_f L_g + 1}{s^3 C_f L_c L_g + s\left(L_c + L_g\right)} \approx \frac{s^2 C_f L_g}{s^3 C_f L_c L_g} = \frac{1}{sL_c} \quad (10)$$

According (10), to derive the most significant frequency for the converter current, volt per Hertz (V/Hz) need to be examined as in the design of a pure inductor filter and this is conducted in [2], which proves that the maximum converter current ripple occurs at the same significant frequency as for the grid current. Therefore the harmonic in the converter current at the most significant frequency simply needs to be analyzed.

The converter current is related to the grid current as in (11). Thus the converter current harmonic at ω_{sig} is variable depending on the values of the filter capacitance and the grid side inductance. As the total inductance is fixed as 0.2035[p.u.] for the given resonant frequency of 800Hz in this design example, the maximum converter current ripple can be inspected by changing the ratio between L_c and L_g. The maximum converter ripple is depicted as the ratio varies in Fig. 4. It indicates that as the converter side inductance becomes larger the ripple is expectedly reduced but the rate of the reduction decreases as the ratio increases. Generally, the converter current ripple of 20% or less is said to be acceptable [8].

$$\frac{i_c}{i_g} = s^2 C_f L_g + 1 \quad (11)$$

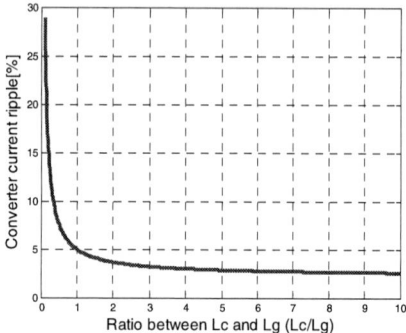

Fig. 4. Converter current harmonic at the most significant frequency with respect to L_c/L_g.

2) Filter's Total Energy

The filter size is also a critical design factor because it has an intensive effect on the system's volume, cost and etc. Unfortunately, however, there is no specific index for the size. In this paper, the total energy of the filter is assumed to be an index for the filter size.

The total energy is calculated for the unity power factor at the grid side as in (12). To analyze the effect of the filter parameters combination on the total energy, the calculated total energy is shown in Fig. 5 as the ratio of the inductances varies. The result indicates that it is preferable if the ratio is close to 1 for the minimization of the energy.

$$E = \frac{1}{2} L_g i_g^2 + \frac{1}{2} L_c i_c^2 + \frac{1}{2} C_f V_{cf}^2 \quad (12)$$

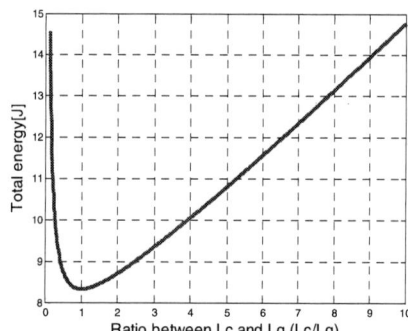

Fig. 5. Total energy of filter with respect to L_c/L_g.

3) Converter Capacity

The converter capacity also depends on the values of the filter parameters while supplying the same power to the grid. In terms of the size, the converter should be minimized to save the cost and the space of the system. Therefore the apparent power of the converter, described as in (13) where '*' denotes the conjugate and Vc represents the rms value of the converter output voltage whereas Ic stands for the phasor value of converter current, is investigated by varying the ratio of L_c/L_g and it is depicted in Fig. 7. Even though the converter capacity is the maximum when the ratio is almost 1, the variation of the capacity is less than 1.2% for the range of the ratio shown in the figure.

$$S = 3\left| V_c I_c^* \right| \quad (13)$$

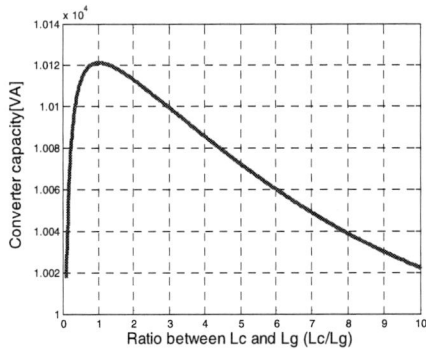

Fig. 6. Converter capacity with respect to L_c/L_g.

IV. SIMULATION

Simulation has been conducted to validate the effectiveness of the proposed LCL filter design method with the computer simulating tool, PSIM. The system specification such as the switching frequency and the DC link voltage is the same as in the TABLE I. For the simulation and the experiments following next, closed-loop current control with the filter-based active damping in synchronous frame is applied. The grid voltages and the grid currents are measured to supply the intended power to the grid and notch filters are inserted in series to the current controller as a way of the active damping [9]. This paper only addresses the issues concerning the filter design and thus control aspects are not discussed any further.

TABLE III specifies the final values for the designed filter components. The total inductance of 0.2035[p.u.] is chosen so that the grid current harmonic at the most significant frequency is expected to be 0.3%. In the design process, ESRs of each component and any stray resistances in the circuit were ignored. However, in the simulation, 0.005[p.u.] resistances are added in series to each reactance component because otherwise the simulation results have offsets or become unstable due to the lack of the damping factors. In the actual system, more resistances exist as will be shown in the experimental results

The harmonics characteristics of the grid current and the converter current need to be checked to see if the results are identical to the designed values. In Fig. 8(a), the frequency spectrum of the grid current is shown. In terms of the most significant frequency, the component with m=1 and n=-2 is the largest among components around the multiples of the switching frequency. In addition, the magnitude of the corresponding harmonic is the same as the designed value, 0.3%. The reason that higher harmonic exist at 800Hz is that the resonance is excited due to the lack of sufficient passive damping components in the system even though the active damping is implemented [10]. The low order harmonics are generated due to the intrinsic feature of SVPWM.

In Fig. 8(b), the frequency spectrum of the converter current is depicted. As expected, the magnitude of the harmonic at the most significant frequency is almost 5% when L_c and L_g are the same and thus the result is corresponding to the designed value.

Through the simulation results, the feasibility of the proposed design method is shown.

TABLE III
Filter Parameters for Simulation

L_c	L_g	C_f	f_{res}
1.3[mH], 0.10175[p.u.]	1.3[mH], 0.10175[p.u.]	60.6[μF], 0.1106[p.u.]	800[Hz]

(a)

(b)

Fig. 8. Simulated waveforms of frequency spectrum for (a) grid current and (b) converter current.

V. EXPERIMENTAL RESULTS

To verify the effectiveness of the proposed method, experiments in the laboratory scale have been conducted with the system specification of TABLE I. The specification of the filter used for the experiments is intended to be as in TABLE III but due to the manufacturing tolerance, the actual filter parameters become slightly different as in TABLE IV. Unfortunately, this difference can make noticeable deviation of the switching harmonic magnitude from the theoretically calculated value. So, through the comparison between the simulation results with the actual filter values for the experiment and the experimental results, the validity of the proposed design method is scrutinized.

TABLE IV
Filter Parameters for Experiment

L_c	L_g	C_f	f_{res}
1.42[mH], 0.1106[p.u.]	1.25[mH], 0.0974[p.u.]	57[μF], 0.104[p.u.]	817[Hz]

Before identifying the performance of the filter, the frequency characteristics of the grid voltage is recognized to see if the resonance could be excited by the grid voltage harmonics. The measured grid line voltage and its frequency spectrum are shown in Fig. 9. It is noticed that the 7th harmonic (420Hz) dominantly reaches almost 1% of the fundamental component and the 13th (780Hz) harmonic is around 0.5%.

In Fig. 10 the experimental results for the grid current

in the time domain and the frequency domain are displayed. There exists resonant harmonic around 800Hz and it is likely that the 13th harmonic in the grid voltage may excite the resonance because the converter output voltage reference must include the harmonic to inject pure sinusoidal current into the grid. The 7th harmonic in the grid voltage also makes noticeable peak in Fig. 10(b). The performance of the proposed LCL filter design method can be proved by analyzing the switching frequency harmonics. As can be seen in the small window in Fig. 10(b), the most significant component is corresponding to the expectation in this paper and the value is slightly larger than the designed value of 0.3[%], which is because of the inductor manufacturing tolerance.

In Fig. 11, simulation results are depicted with the same filter parameters. The waveforms in the time domain and its frequency spectrum are nearly the same as in the experiment.

To enhance the effectiveness of the proposed design method, the converter current is also observed in Fig. 12 and Fig. 13. Because the ratio between L_c and L_g is 1.1355, the converter current ripple at the most significant frequency is expected to be a bit less than 5% according to Fig. 4 and it agrees with the experimental result and the simulation result.

VI. CONCLUSIONS

This paper proposed a non-iterative LCL filter design method. The system parameters such as the switching frequency and the DC link voltage are assumed to be determined before commencing the filter design. Then, the resonant frequency is selected and once the resonant frequency is chosen, the total inductance can be calculated by using one single inequality equation without any iterative process for satisfying the grid current harmonics regulation at the most significant frequency.

These two conditions, the resonant frequency and the total inductance, are the minimal requirements for the mandatory current harmonics regulation of the grid. Afterwards, various additional conditions can be adopted into the design process. In this paper the converter current ripple, the total filter energy and the converter capacity are considered. If the ratio between L_c and L_g is chosen to be 1, the converter current ripple is almost 5% and the total energy is minimized. The converter capacity becomes 10.12[kVA] while providing 10[kW] to the grid.

The proposed method does not suffer from any iterative process which appears variously in the existing research. The proposed method is verified by simulation and experimental results.

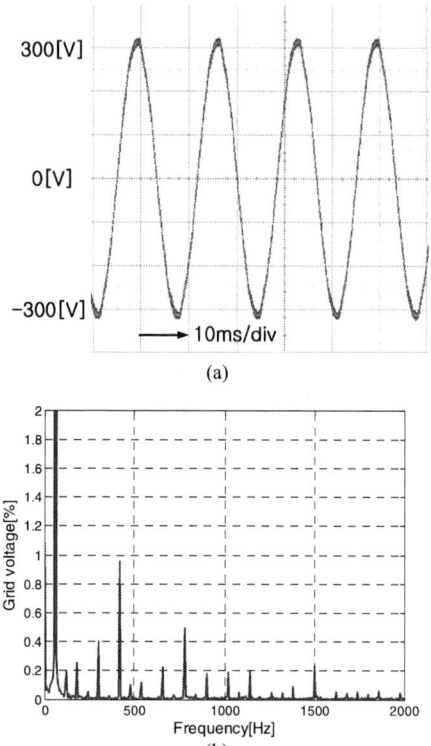

Fig. 9. Measured grid voltage in (a) time domain and (b) frequency domain.

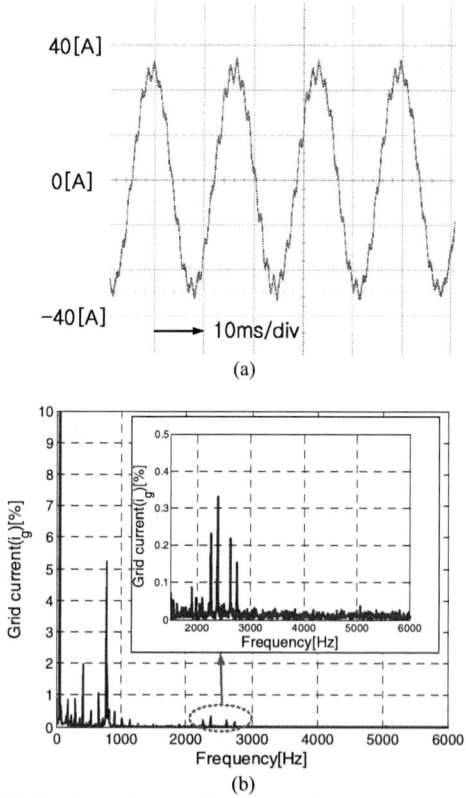

Fig. 10. Experimental results of grid current in (a) time domain and (b) frequency domain..

The 2014 International Power Electronics Conference

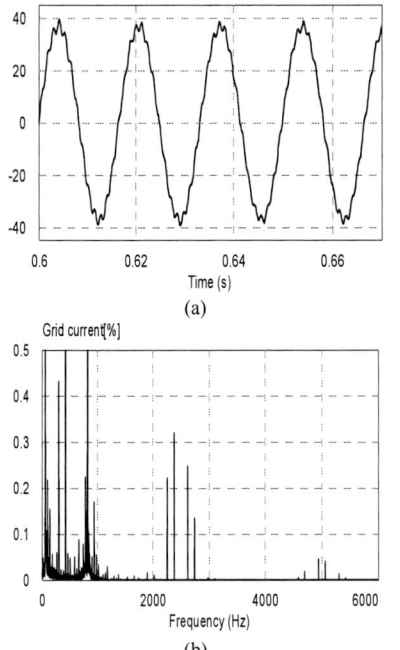

(a)

(b)

Fig. 11. Simulated results of grid current in (a) time domain and (b) frequency domain.

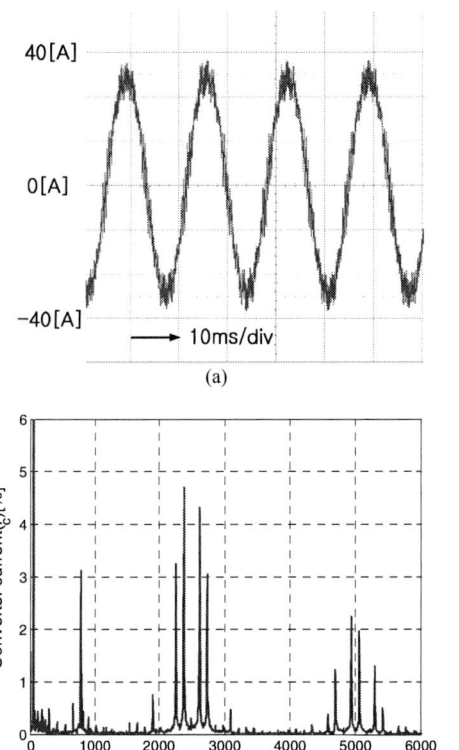

(a)

(b)

Fig. 12. Experimental results of converter current in (a) time domain and (b) frequency domain.

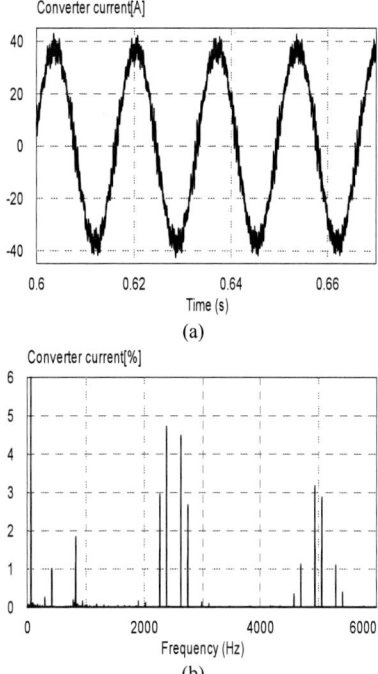

(a)

(b)

Fig. 13. Simulated results of converter current in (a) time domain and (b) frequency domain..

REFERENCES

[1] R. Teodorescu, M. Liserre and P. Rodriguez, *Grid Converters for Photovoltaic and Wind Power Systems*, West Sussex:John Wiley & Sons, 2011

[2] K. Jalili and S. Bernet, "Design of LCL Filters of Active-Front-End Two-Level Voltage-Source Converters," *IEEE Trans. Industrial Electronics*, vol. 56, pp. 1674-1689, May 2009

[3] M. Malinowski and S. Bernet, "A Simple Voltage Sensorless Active Damping Scheme for Three-Phase PWM Converters With an LCL Filter," *IEEE Trans. Industrial Electronics*, vol. 55, pp. 1876-1880, April 2008

[4] J. Dannehl, F. W. Fuchs, S. Hansen and P. B. Thogersen, "Investigation of Active Damping Approaches for PI-Based Current Control of Grid-Connected Pulse Width Modulation Converters With LCL Filters," *IEEE Trans. Industry Applications*, vol. 46, pp. 1509-1517, July/August 2010

[5] M. Liserre, F. Blaabjerg and S. Hansen, "Design and Control of an LCL-Filter-Based Three-Phase Active Rectifier," *IEEE Trans. Industry Applications*, vol. 41, pp. 1281-1291, September/October 2005

[6] Y. Lang, D. Xu, Hadianamrei S.R. and H. Ma, "A Novel Design Method of LCL Type Utility Interface for Three-Phase Voltage Source Rectifier," *IEEE Power Electronics Specialists Conf.*, 2005, pp. 313-317

[7] D. G. Holmes and T. A. Lipo, *Pulse Width Modulation for Power Converters:Principles and Practice*, New Jersey:John Wiley & Sons, 2003

[8] A. A. Rockhill, M. Liserre, R. Teodorescu and P. Rodriguez, "Grid-Filter Design for a Multimegawatt Medium-Voltage Voltage-Source Inverter," *IEEE Trans. Industrial Electronics*, vol. 58, pp. 1205-1217, April 2011

[9] J. Dannehl, M. Liserre and F. W. Fuchs, "Filter-Based Active Damping of Voltage Source Converters With LCL Filter," *IEEE Trans. Industrial Electronics*, vol. 58, pp. 3623-3633, August 2011

[10] Byung-Geuk Cho and Seung-Ki Sul, "LCL Filter Design for Grid-connected Voltage-source Converters in High Power Systems," in *Proc. IEEE Energy Conversion Congress and Exposition Conf.*, 2012

DSP-Based Interleaved Buck Power Factor Corrector

Yu-Chen Liu
Dept. of Electrical Engineering
National Taiwan University of
Science and Technology
Taipei City, Taiwan R.O.C
D9802204@mail.ntust.edu.tw

Tsan Chen
Dept. of Electronic Engineering
National Taiwan University of
Science and Technology
Taipei City, Taiwan R.O.C
M10002232@mail.ntust.edu.tw

Po-Jung Tseng
Dept. of Electronic Engineering
National Taiwan University of
Science and Technology
Taipei City, Taiwan R.O.C
D9702201@mail.ntust.edu.tw

Yu-Kang Lo
Dept. of Electronic Engineering
National Taiwan University of
Science and Technology
Taipei City, Taiwan R.O.C
yklo@mail.ntust.edu.tw

Huang-Jen Chiu
Dept. of Electronic Engineering
National Taiwan University of
Science and Technology
Taipei City, Taiwan R.O.C
hjchiu@mail.ntust.edu.tw

Abstract-An interleaved Buck PFC with digital and clamping current control is proposed in this paper. Inductor current can be operated in CCM or DCM, which depends on different input line voltages and load conditions. The interleaved topology is composed of two Buck converters, which can have higher power density, lower input and output ripple currents and improve input current harmonics. Microprocessor TMS320X228035 is adopted to be the controller to implement clamping current mode control and raise the PF value at universal input. And the phase management can be adopted to lift light-load efficiency. A 300-W/80V-output digital interleaved Buck PFC with universal input and light-load efficiency all greater than 95.5% is implemented. When input voltage is over 115Vrms, the mid-to-full load PF values are all greater than 0.94.

I. INTRODUCTION

Because the AC power system is the mainstream in the world, the AC-DC power conversion is indispensable to lots of electric equipments. However the phase shift or distortion will be occurred in the input current to affect the converter performance and power quality during the power conversion, utilizing the PFC to solve these issues is needed. PFC can keep input current sinusoidal and in phase with source voltage to increase the utility of real power and reduce surge current and input current harmonics. Some regulations related to power quality have been proposed to improve power conversion efficiency to save energy. So the conversion efficiency and PF value should be considered in the process of circuit design.

Many circuit topologies of PFC adopt the boost-type in the applications. And the PWM method is applied to control input current to be sinusoidal and in phase with source voltage to gain higher PF value and lower input current harmonics.

In this paper, the clamping current mode control is adopted in the digital interleaved Buck PFC to fulfill power factor correction [1]. Moreover, the interleaved gate-driven signals are used to lift the circuit efficiency and power density.

Clamping current mode control is derived from the peak current mode control. Its operation adopts constant frequency and limited maximum duty cycle. Besides the above, external ramp slope compensation is taken as well. So the Buck PFC can meet the requirement of high PF value at universal input.

Compared with other common control modes, clamping current mode control is simple and low-cost. Because the control is interleaved between the switches of two phases, there are some merits such as lower differential-mode noise, cancelled ripple current, decreased current harmonics. Since the power is shared out, furthermore, reduced output inductance and capacitance, higher power density and lower heat production can be obtained as well. Besides, phase management can be applied to raise system efficiency. By the DSP and detecting circuits, the clamping current mode control and interleaved gate-driven signals can be gained to implement the digital interleaved Buck PFC.

As the conventional clamping current mode control is applied to the interleaved Buck PFC, there will be a drawback [2]. It is that PF values can't keep high enough at universal input as the slope compensation value is fixed. However, the problem can be solved by digital control. The slope compensation value will be varied with input voltage to keep high PF at universal input. Although the price of DSP is still high, the usage of DSP will be a trend. By using a few of simple external analog circuits and writing DSP code, complex control mechanism can be fulfilled. When the mechanism is needed to be adjusted, it can be done by modifying DSP code. Due to the reduction of hardware complexity of control circuit, the circuit size and system power consumption can be mitigated as well.

II. INTERLEAVED BUCK POWER FACTOR CORRECTOR

A. Topology

As Fig1, interleaved Buck PFC is the studied topology in this paper [3-5]. It is composed of two paralleled Buck converters. In order to simplify the gate drivers, the power switches are placed in the bottom of legs. Some components are used to form this circuit such as input bridge diode, power switches Q1and Q2, inductors L1and L2, output capacitor Co and freewheeling diodes D1and D2.

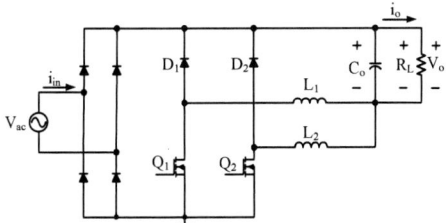

Fig 1. Interleaved Buck PFC converter

Interleave control method means the gate-driven signals are interleaved with 180˚. Compared with single-phase operation, it can handle higher power conversion and mitigate the ripple currents of input and output. Besides the inductance can be reduced, the input filtering and output capacitances can be decreased as well. Incidentally, the speed of transient response and power density can be raised at the same rated power. Because of the phase-angle difference between gate-driven signals, duty cycle can be divided into two cases D>0.5 and D<0.5. For D>0.5 and D<0.5, the operation timings of inductor currents at CCM are depicted in Fig.2(a) and Fig.2(b).

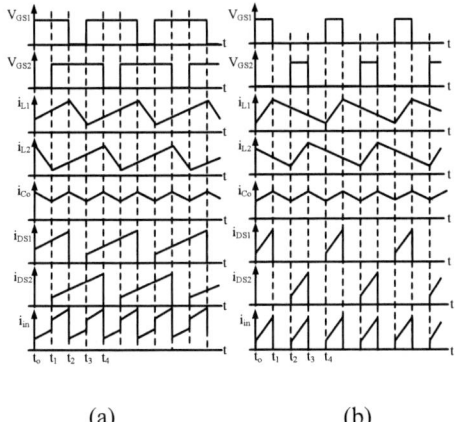

(a)   (b)

Fig 2. Interleaved Buck converter operation in CCM :
(a) D > 0.5、 (b) D < 0.5

B. Clamped-current Mode

Clamping current mode control is similar to peak current mode control [6,7]. Inductor current Isense is compared with reference signal during each switching period. But the maximum duty cycle (Dmax) of clamping current mode control is limited. Therefore, clamping current mode control is mainly applied to the mid-to-low power application. It features simple control, higher PF and low cost.

Fig.3 shows the diagram of clamping current mode control applied to single-phase Buck PFC. It can be seen that feedback error signal Ve is sent to the PWM generator to be compared with the combined signal (Isense+IR).

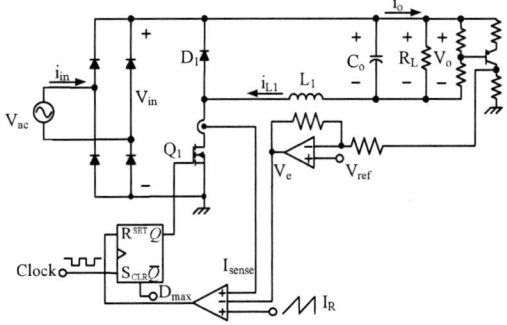

Fig 3. Clamped-current mode applied to single-phase inverse buck PFC

III. DIGITAL CONTROL SYSTEM DESIGN

A. Digital System

In this paper, the series of DSP TMS320C/F28x is adopted [8]. It has complete functions and is popular in the field of power electronics. Fig.4 presents the diagram of digital clamping current mode control applied to the interleaved Buck PFC. In which five main signals are sensed such as switch currents of two phases, input and output voltages and output current [9]. After the processes of sampling and amplifying, these signals will be sent to DSP to be computed. Finally, gate-driven signals will be generated by the ePWM module. Through the gate drivers, gate-driven signals will be amplified and sent to power switches.

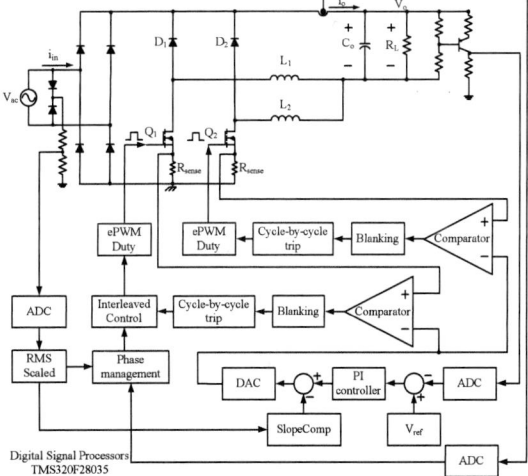

Fig 4. DSP applied to interleaved Buck PFC

The circuit also contains functions of slope adjustment and phase management. Different input voltage will correspond to different slope compensation value to keep higher PF at universal input. The function of phase management is to lift system efficiency. However, the light-load efficiency will become lower for interleaved operation when the load power is within the power level of single-phase operation. So the starting timing of interleaved operation depends on load power if it is greater than the power level of single-phase operation.

B. Digitized Clamped-current Mode

According to the Fig.5, the needed interleaved control mode is implemented by the comparator of DSP TMS320F28035. The switch current Isense which is an external signal and will be processed by sampling circuit and ADC module is sent to plus terminal (+) of comparators. The difference between the amplified error of DAC module and ramp compensation value is sent to minus terminal (-) of comparators. This difference is computed by Ramp Generation module and it will be continuously sent to minus terminal (-) of comparators.

Fig.9 shows the waveforms of clamping current mode control. As Fig.9(a), comparator will change the status of output register COMPSTS when inductor current is greater than the difference generated from the DAC module and D< Dmax. The register COMPSTS is used to provide triggering signal to ePWM module. When it receives this signal, ePWM module will shut down the gate driver to control the duty modulation. As Fig.9(b), if ePWM module receives the triggering signal when D Dmax, the signal will be ignored.

When the ramp generation module receives the synchronous signal comes from the ePWM module. The next switching period will be operated. Then, gate-driven signal VGS will be sent out and the difference between the error amplifier signal and slope compensation value will be reset. And the ramp generation module will be reset as well.

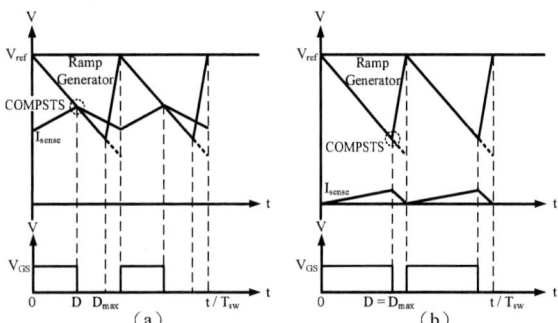

Fig 5. Clamped-current mode waveform : (a) D < Dmax 、 (b) D Dmax

C. System Control Flow Chart

The whole system control flow chart can be divided to three blocks.They are the initialization, service interrupt and background loop. The system code is wrote with language C. The system control flow chart is shown as Fig.6.

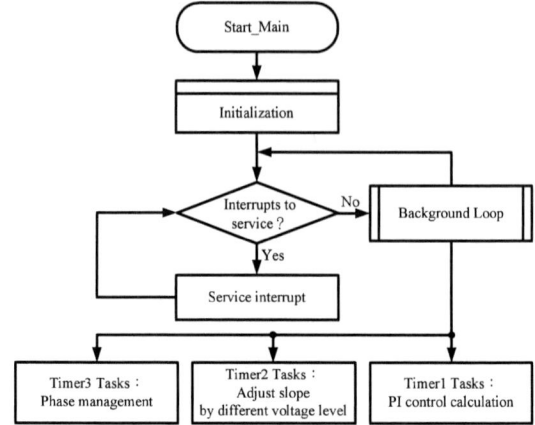

Fig 6. System main code control flow chart

The flow chart of system background loop is shown as Fig.7. After the system initialization, the loop will be executed. The loop contains the functions of PI computation, slope adjustment and phase management.

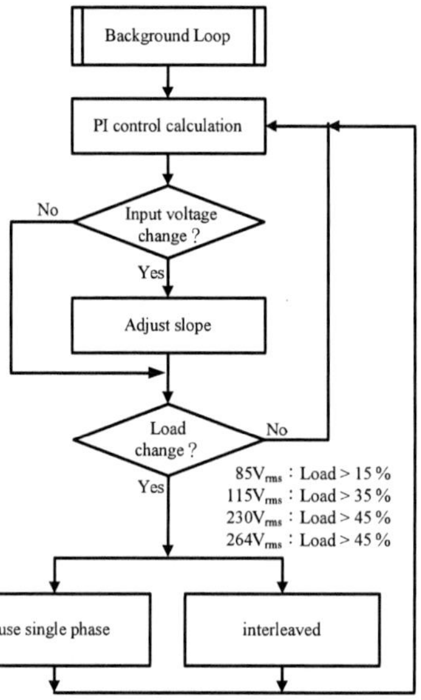

Fig 7. Background loop flow chart

IV. EXPERIMENTAL RESULTS

TABLE I
KEY DESIGN PARAMETERS OF A PROTOTYPE CONVERTER

Design Parameters	Values
Input voltage range	85 Vrms ~264 Vrms
Output voltage	80V
Output power	300W
Line frequency	60Hz
Switching frequency	65kHz
Bridge diode	LL15XB60(600V/15A)
Switching MOSFETs	IPB60R099CP(600V/31A)
Freewheeling diodes	C3D06060(600V/6A)
Output inductor	90μH (PQ3230)
Output capacitor	1800 μF (100V)

A 300W/ 80V laboratory prototype with circuit parameters listed in Table I was built and tested to verify the feasibility of the proposed converter. Fig.8 shows the curves between efficiency and output power. Different input voltage will correspond to different efficiency curve. This circuit has the function of phase management. The efficiency with phase management is drawn by a solid line. It can be seen that phase-management point will varied with input voltage. And the light-load efficiency is better than the one without phase management at universal input. Each full-load efficiency is all greater than 95% at universal input.

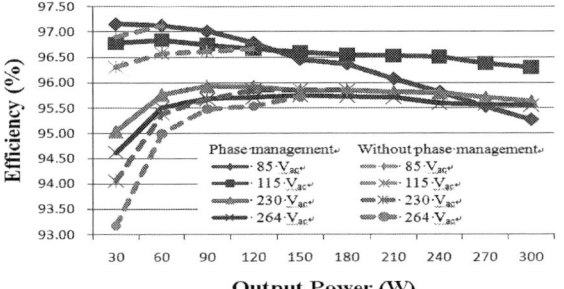

Fig 8. Efficiency versus output power

Fig.9 presents the curves between PF and output power. It is known that PF is an important index for power conversion performance and power quality. It can be seen that the PF value is considerably small at lower input voltage between 85-Vac and 115-Vac inputs. The reason is that input current conduction angle is smaller when input voltage is lower. And it makes input current too distorted to follow the source voltage. Then, a poor power factor correction will be produced.

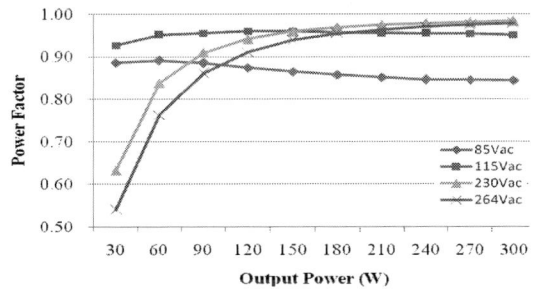

Fig 9. PF value versus output power

Fig.10 shows the input and output voltage and input current of the digital interleaved Buck PFC at 264-Vrms input and full-load condition. It can be observed that the input current follows the input voltage to achieve the function of power factor correction within the conduction angle of input current.

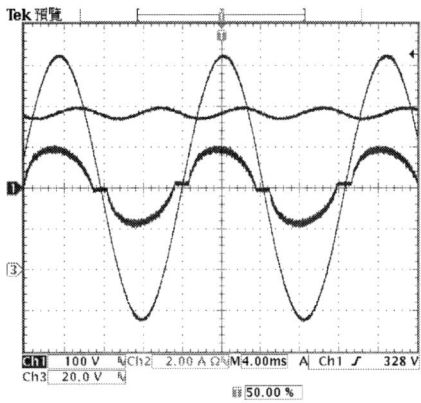

Fig 10. Line voltage, line current and output voltage waveform (Pout = 300W, Vin = 264 Vrms)(Ch1 = vin: 100 V/div、 Ch2 = iin 2 A/div、 Ch3 = Vo: 20 V/div、 Time: 4 ms/div)

Fig.11 shows the inductor currents iL1、 iL2 and output inductor current iLo of the interleaved Buck PFC at 264-Vrms input and full-load condition.

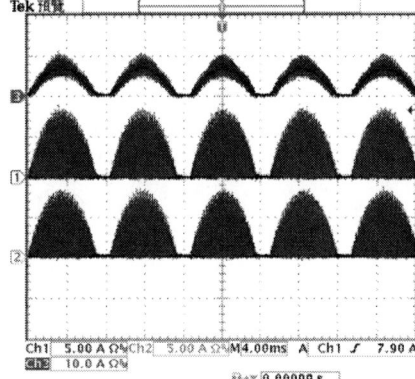

Fig 11. Phase A、 B inductor current and output inductor waveform(Pout = 300W、 Vin = 264 Vrms)(Ch1/Ch2 = iL1/iL2: 5 A/div、 Ch3 = iLo: 10 A/div、 Time: 4 ms/div)

V. CONCLUSION

In this paper, a DSP TMS320F28035 is selected to be the controller. A 300-W/80V-output digital interleaved Buck PFC with universal input is implemented. For design specification, its maximum efficiency is 97.16% and light-load efficiency are all greater than 95.51% within all load conditions. The PF values are all greater than 94% at mid-to-full load condition for 115Vrms-to-264Vrms input.

REFERENCES

[1] L. Huber and M.M. Jovanović, "Design-Oriented Analysis and Performance Evaluation of Clamped-Current-Boost Input-Current Shaper for Universal-Input-Voltage Range," IEEE Transactions on Power Electronics, vol. 13, no. 3, pp. 528-537, May 1998.

[2] L. Huber, L. Gang, and M. M. Jovanovic, "Design-Oriented Analysis and Performance Evaluation of Buck PFC Front-End," IEEE Transactions on Industrial Electronics, vol. 25, no. 1, pp. 85-94, May 2010.

[3] M. Orabi and T. Ninomiya, "A Unified Design of Single-Stage and Two-Stage PFC Converter," IEEE Power

Electronics Specialists Conference, vol. 4, pp. 1720-1725, Jun. 2003.

[4] X. Wu, J. Yang, J. Zhang, and Z. Qian, "Variable On-Time (VOT)-Controlled Critical Conduction Mode Buck PFC Converter for High-Input AC/DC HB-LED Lighting Applications," IEEE Transactions on power Electronics, vol. 27, no. 11, pp. 4530-4539, Nov. 2012.

[5] Texas Instruments, "UCC29910A Buck PFC Controller," Datasheet, 2011.

[6] D. Maksimovic, "Design of the Clamped-Current High-Power-Factor Boost Rectifier," IEEE Transactions on Industrial Electronics, vol. 31, no. 5, pp. 986-992, Oct. 1995.

[7] M. Ilic and D. Maksimovic, "Interleaved Zero-Current-Transition Buck Converter," IEEE Transactions on Industrial Electronics, vol. 43, no. 6, pp. 1619-1627, Dec. 2007.

[8] Texas Instruments, "TMS320F28035 Piccolo Microcontrollers," Datasheet, 2011.

[9] Texas Instruments, "Digital Peak Current Mode Control with Slope Compensation Using the TMS320F2803x," Application Report, 2011.

The Average Model of a Three-Phase Three-Stage Power Electronic Transformer

Shaodi Ouyang, Jinjun Liu, Xinyu Wang, Xiaojian Wang, Fei Meng, Javid Riffat

School of Electrical Engineering, Xi'an Jiaotong University

Xi'an, China

oysd1989@stu.xjtu.edu.cn

Abstract—**Power Electronic Transformer (PET) has been the focus of interest for many years. The PET opens up the possibility to replace bulky low-frequency transformers, and has more functions and better performance. In this paper, the model and the control strategy of a three-phase three-stage PET are analyzed, and its average model is derived to increase the simulation speed. However, the average model would miss a special current distortion as it neglects the voltage drop of the switches, which limits the effective range of the average model.**

Keywords—average model, current distortion, power electronic transformer .

I. INTRODUCTION

The idea of replacing the bulky traditional power transformer with power electronic transformer (PET) was proposed in the 1970s [1]. As compared with the traditional transformer, PET offers many advantages, such as smaller size and weight, input current and output voltage regulation, etc.

There are various types of PET. Depending on the topology and application, PET can be divided into three types: direct AC-AC type, AC-DC-DC-AC type and AC-DC-DC type.

The AC-AC type consists of an input converter and an output converter. The input line-frequency waveform is modulated by the input converter to a high frequency square waveform and passed through a high frequency transformer, and then be demodulated back into a line-frequency waveform by the output converter [2]-[9]. As the magnetic material is working in high frequency mode, the size and weight would be significantly reduced. The AC-AC type PETs have advantages of high efficiency, simple in structure and less number of switches. However, they suffer from some difficulties: need for bi-directional semiconductor switches, less control flexibility, more harmonics and lack of ability to improve the power quality.

In order to deal with the defects of AC-AC PETs, AC-DC-DC-AC PETs are proposed. The AC-DC-DC-AC PETs have three stages: an AC-DC stage, a DC-DC stage and a DC-AC stage [10]-[12]. Compared with the AC-AC technique, the AC-DC, DC-DC and DC-AC conversion techniques are more mature, and the conventional IGBT/MOSFETs with reverse diodes can be used. The AC-DC-DC-AC type PETs have better control performance, and they are more capable of output adjustment and power quality improvement. However, as more stages are used, they will suffer from lower efficiency and stability challenges. This type of PET is mainly assumed to be applied in the power distribution system.

The AC-DC-DC type PETs are used to connect an AC system to a DC system, and their application on PV system and railway traction can be found [13][14].

The author's research group has proposed a three-phase AC-DC-DC-AC PET topology in [15], as shown in Fig.1. In this paper, the average model of this PET is derived, and the simulation comparison between the average model and the switching model is made. The waveforms of the average model simulation matches well with those of the switching model simulation, and the simulation speed of the average model is much faster than that of the switching model. However, as the average model neglects the IGBT's voltage drop, the simulation results would miss a special current distortion, which limits its effective range.

This paper is organized as follows: section II discusses the configuration, control and the average model of each stage; section III gives the simulation results; section IV shows that special current distortion and gives a brief analyze; section V gives the conclusions.

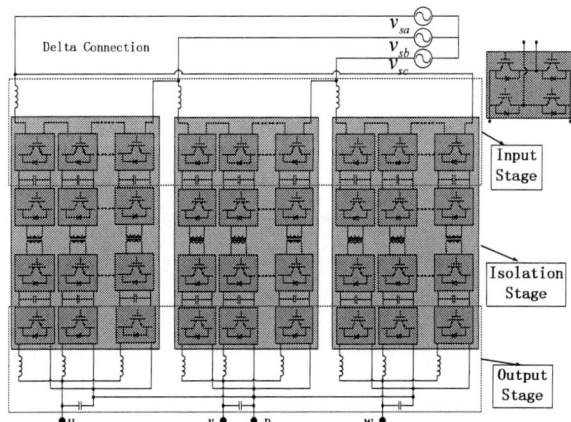

Fig. 1. Three-stage, Three-phase power electronic transformer

II. MODEL AND CONTROL OF PET

A. Model and control of the rectifier stage

The input rectifier stage is a delta-connected cascade H bridge rectifier. Each cluster of the rectifier consists of three cascaded H bridges, and the three clusters are connected head-to-tail. Each H bridge module supports its own load through a DC-DC converter. The voltage and current equations of the AC-DC stage are as (1).

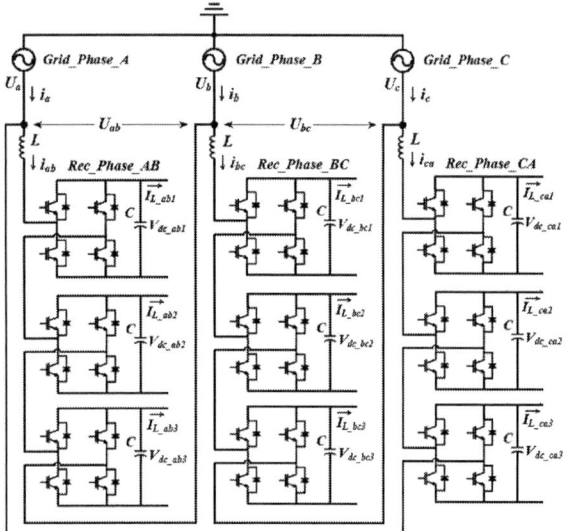

Fig. 2. The AC-DC stage of PET

$$\begin{cases} L\dfrac{di_{ab}}{dt}=U_{ab}-(d_{ab1}\cdot V_{dc\_ab1}+d_{ab2}\cdot V_{dc\_ab2}+d_{ab3}\cdot V_{dc\_ab3}) \\[2mm] L\dfrac{di_{bc}}{dt}=U_{bc}-(d_{bc1}\cdot V_{dc\_bc1}+d_{bc2}\cdot V_{dc\_bc2}+d_{ab3}\cdot V_{dc\_bc3}) \\[2mm] L\dfrac{di_{ca}}{dt}=U_{ca}-(d_{ca1}\cdot V_{dc\_ca1}+d_{ca2}\cdot V_{dc\_ca2}+d_{ca3}\cdot V_{dc\_ca3}) \\[2mm] C\dfrac{dV_{dc\_abn}}{dt}=d_{abn}\cdot i_{ab}-i_{Load\_abn}\to n=1,2,3 \\[2mm] C\dfrac{dV_{dc\_bcn}}{dt}=d_{bcn}\cdot i_{bc}-i_{Load\_bcn}\to n=1,2,3 \\[2mm] C\dfrac{dV_{dc\_can}}{dt}=d_{can}\cdot i_{ca}-i_{Load\_can}\to n=1,2,3 \end{cases} \quad (1)$$

where U_{ab}, U_{bc} and U_{ca} are AC line voltages; i_{ab}, i_{bc} and i_{ca} are rectifier input currents flowing inside the delta; V_{dc_abn}, V_{dc_bcn} and V_{dc_can} ($n=1,2,3$) are DC link voltages the H bridge modules; d_{abn}, d_{bcn}, and d_{can} ($n=1,2,3$) are duty cycles the H bridge modules; i_{Load_abn}, i_{Load_bcn} and i_{Load_can} are the load currents.

As have illustrated in [15], the AC-DC stage has three goals: unity power factor, voltage balancing and zero sequence current injection. To achieve these goals, the control strategy of the AC-DC stage consists of three layers. The first layer is to adjust the overall DC link voltage by a conventional d-q decoupled dual loop control shown in Fig.3. The i_q is set to zero to achieve unity power factor. The second layer is set to redistribute power among three clusters by injecting a zero sequence current, thus the unbalanced cluster loads would be compensated and the grid current would be balanced. The control block is shown in Fig.4 and the calculation is as (2)[16]. The third layer is to balance the DC voltages within one cluster, shown in Fig.5. Each DC link voltage is compared with the average value of its belonging cluster, and then the regulators would adjust the amplitudes of the modulation waves.

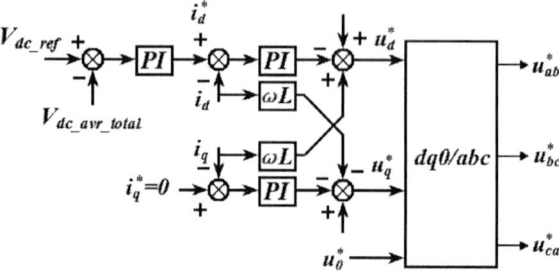

Fig. 3. First layer of the rectifier control

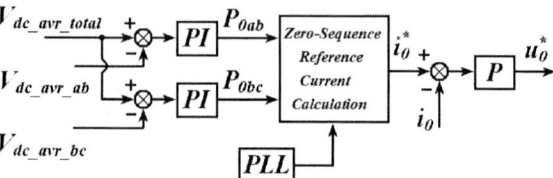

Fig. 4. Second layer of the rectifier control

$$\begin{cases} I_o=\dfrac{2}{V_S}\sqrt{(p_{0ab})^2+\dfrac{1}{3}(p_{0ab}+2\cdot p_{0bc})^2} \\[3mm] \varphi_o=\tan^{-1}[-\dfrac{1}{\sqrt{3}}(1+\dfrac{2\cdot p_{0bc}}{p_{0ab}})] \end{cases} \quad (2)$$

where I_o and φ_o are the magnitude and phase angle of the injected zero sequence current, respectively; V_S is the magnitude of line-to-line grid voltage; T_S is the line period.

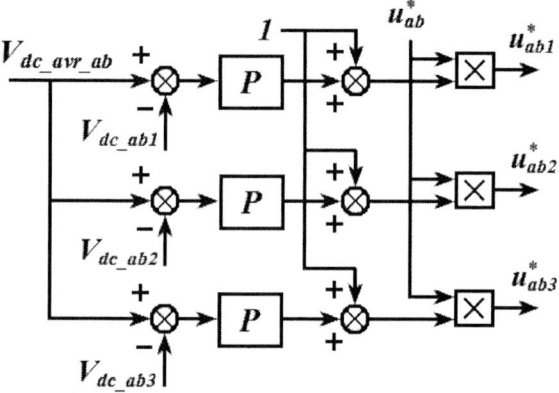

Fig. 5. The third layer of the rectifier control strategy

The average model developed for each phase is shown in Fig.6.

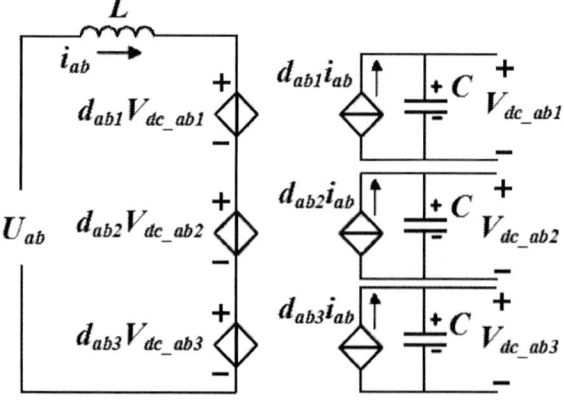

Fig. 6. Average model for one phase of the rectifier stage

B. Model and control of the dc-dc stage

The converter used in DC-DC stage is dual active bridge (DAB). The DAB topology is shown in Fig.6. As the topology is asymmetrical, the power can flow either side to another, which is suitable for the PET application.

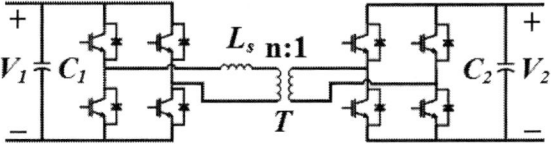

Fig. 7. DAB configuration

For the DAB converter, both the duty cycle and phase shift angle can be adjusted to control the power flow. In this paper, the single phase shift modulation (PSM) is applied to the DAB stage. The voltage and current waveform of PSM is shown in Fig.8, and the power equation is (3).

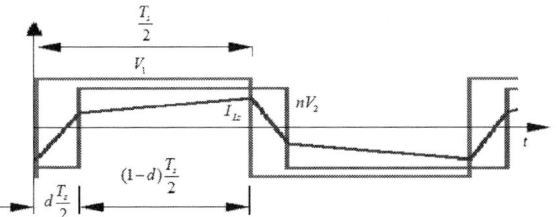

Fig. 8. Phase shift modulation of DAB

$$P = \frac{nV_1V_2}{2f_sL_s}d(1-d) \qquad (3)$$

where V_1 is the input side DC voltage; V_2 is the output side DC voltage; L_s is the leakage inductance; n is the transfer ratio of the transformer; f_s is the switching frequency; d is defined as ϕ / π as shown in the figure.

A single output voltage loop is used to control the DAB stage. An average model which neglects the dynamics of inductor current is introduced in [16]. By replacing the H bridges, leakage inductor and transformer with two controlled current sources, we obtain the average model of DAB, as shown in Fig.9.

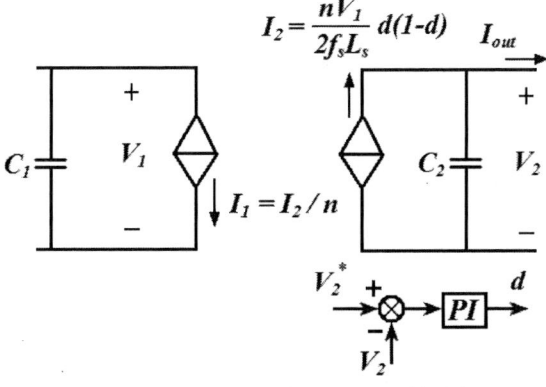

Fig. 9. Control strategy and the average model of DAB

C. Model and control of the inverter stage

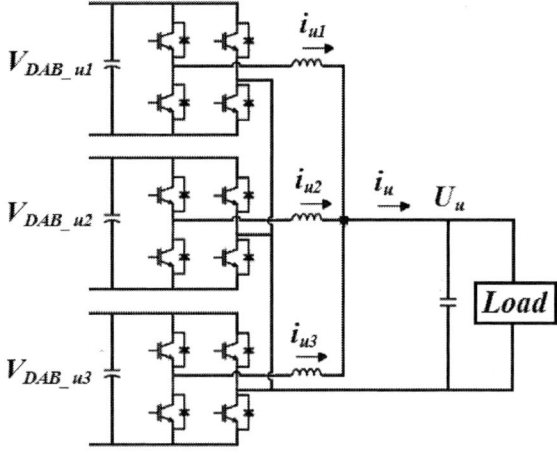

Fig. 10. DC-AC stage of PET

The DC-AC stage of the PET is configured as three single phase inverters. The inverter phase u, v and w are connected to rectifier phase ab, bc and ca through the DC-DC converters, individually. Each inverter phase consists of three H-bridge inverter modules in parallel, as shown in Fig.10.

To achieve current sharing between modules, three current regulators share the same reference, which is given by the voltage regulator. Paralleled modules are interleaved so as to increase the equivalent switching frequency, and Quasi-PR controllers are used to improve the quality of output waveforms. The control strategy is shown in Fig.10, and the average model of the DC-AC stage is shown in Fig.11.

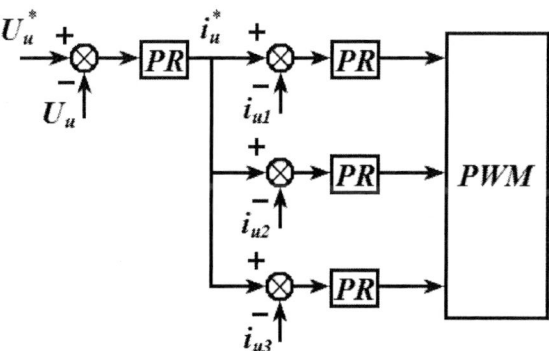

Fig. 11. Control strategy of the inverter stage

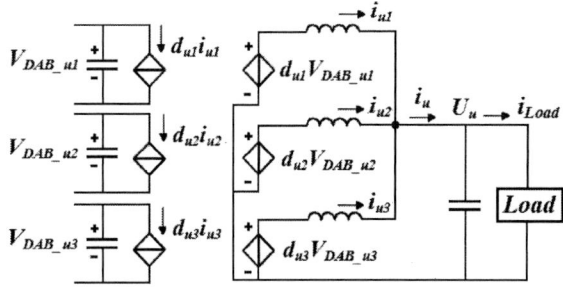

Fig. 12. Average model of the inverter stage

III. SIMULATION RESULTS

In order to verify the control strategy and the average model developed in this paper, MATLAB/SIMULINK was used. Simulation of switching model and average model were carried out individually. Parameters of the PET are shown in Table I.

TABLE I
PARAMETERS OF PET

Grid Line-Line RMS Voltage	380V
Rectifier input Inductor	2mH
Primary DC link capacitor	4700uF
Primary DC link voltage	200V
DAB Leakage Inductor	0.076mH
HF Transformer Ratio	0.5:1
Secondary DC link capacitor	560uF
Secondary DC link voltage	400V
Inverter Output Inductor	1mH
Inverter Output Capacitor	70uF
Output Phase-Ground RMS Voltage	220V
Load Impedance of Phase u	14+3.768j
Load Impedance of Phase v	14+3.768j
Load Impedance of Phase w	11.2+3.014j

The load power of inverter phase w is 25% higher than that of phase u and v. During the simulation, a symmetrical 10kW resistive load was added to the inverter stage at 0.06s, and then was cut off from the system at 0.12s, thus to check the dynamic response. The waveforms of the switching model and those of the average model are put together to make a comparison.

Fig.13 shows the rectifier input currents, and Fig.14 shows the grid currents. The waveforms of the average model simulation are very close to those of the switching model simulation in both steady state and during dynamic process, and the only difference is a small magnitude difference, which is due to the power loss on the IGBTs and diodes.

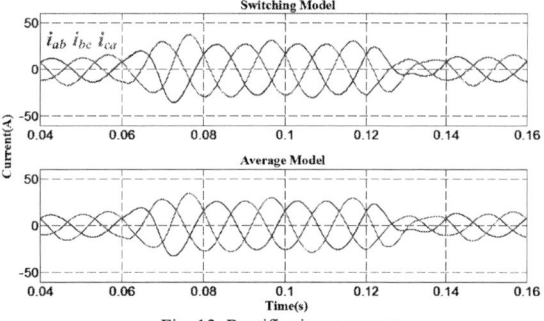

Fig. 13. Rectifier input current

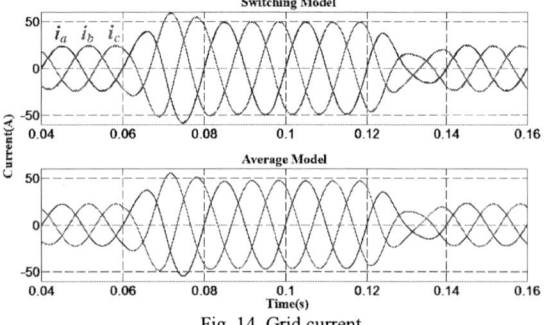

Fig. 14. Grid current

It can also be seen that, although the load power is unbalanced, the grid current is still balanced, verifying the second layer control of the rectifier stage.

Fig.15 shows the DAB primary side DC voltage of phase u, and Fig.16 shows the DAB secondary side DC voltage of phase u. It can be seen that, both in the steady stage and during the dynamic process, the average model for DAB is effective. Additionally, the primary side DC voltages are well balanced, thus the third layer control of the rectifier stage is effective.

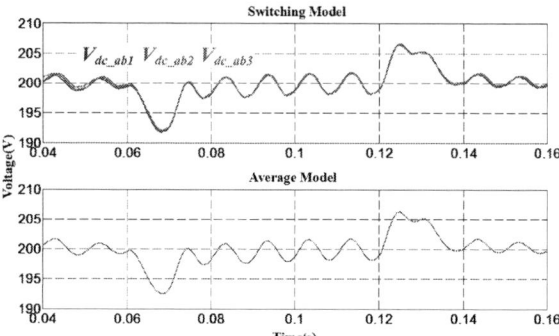

Fig. 15. DAB primary side DC voltage

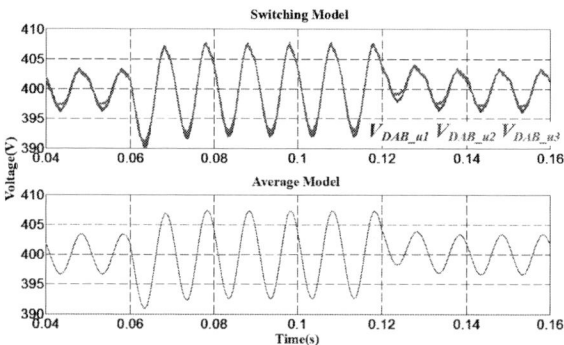

Fig. 16. DAB secondary side DC voltage

Fig.17 shows the inverter output current. The waveforms of average model match well with those of the switching model.

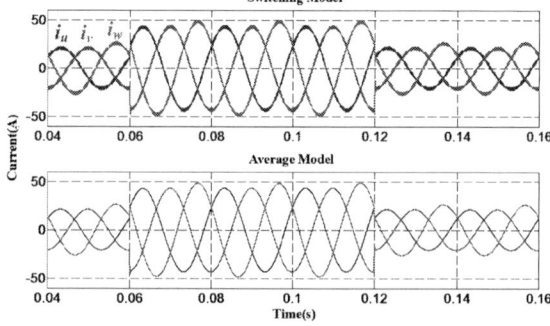

Fig. 17. Inverter output current

The waveforms above show that the average models are effective and they are suitable for both steady state and dynamic process simulation. The speed of average model simulation is 10 times faster than that of the switching model simulation.

IV. A SPECIAL CURRENT DISTORTION

The Average models ignore the IGBT's forward voltage drop and the reverse diode voltage drop, thus the simulation results would miss the distortions caused by these non-ideal characteristics. In this application, there would be a special grid current distortion in the AC-DC stage, especially when the input AC voltage is low.

According to the rectifier's control strategy above, by injecting a zero sequence current inside the delta, the grid current should be balanced even when the load is unbalanced. However, this is correct only in average model or ideal switch model simulation. In the non-ideal switching model simulation, when the load is unbalanced, the rectifier currents i_{ab}, i_{bc} and i_{ca} would have different magnitudes, and their zero-crossing distortion levels are different, too. Due to these "unbalanced" distortion parts of the rectifier currents, the grid current would not be strictly balanced and a special distortion occurs.

To illustrate this distortion, simulation under the condition shown in TABLE II is carried out. This is the extreme case of unbalanced load condition, as the load power of inverter phase w is zero. The input AC voltage is low so that the zero-crossing distortion level is high. The rectifier currents and the grid currents are shown in Fig.18 and Fig.19, individually.

TABLE II
SIMULATION CONDITIONS WITH LOW AC VOLTAGE

Grid Line-Line RMS Voltage	85V
Primary DC link voltage	80V
IGBT Forward Voltage Drop	1.6V
Reverse Diode Voltage Drop	1.2V
Load Power of Inverter Phase u	400W
Load Power of Inverter Phase v	400W
Load Power of Inverter Phase w	0W

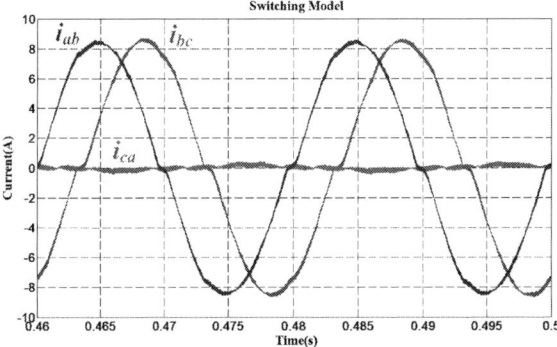

Fig. 18. Rectifier input currents at low AC voltage

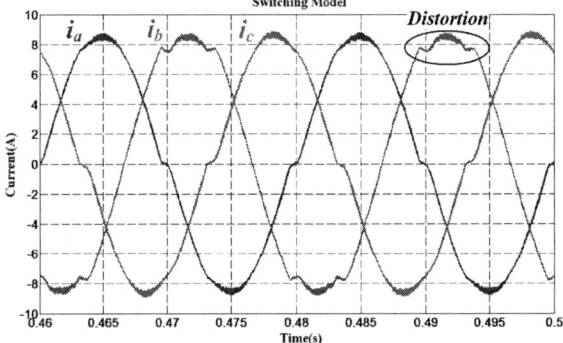

Fig. 19. Grid currents at low AC voltage

As can be seen that, when the load power of phase w is zero, the rectifier input current i_{ca} is approximately equal to zero. Therefore, according to Kirchhoff's current law, we have the relationships:

$$\begin{cases} i_a \approx i_{ab} \\ i_b = i_{bc} - i_{ab} \\ i_c \approx -i_{bc} \end{cases} \quad (4)$$

The grid current i_b reveals a special current distortion, which is a mixture of the distortions from both i_{bc} and i_{ab}, as marked in Fig.19. As i_a and i_c do not have such a kind of distortion, the grid currents are not strictly balanced. The THD and RMS value of i_a, i_b and i_c are shown in TABLE III. It can be seen that, the distortion not only makes the THD of i_b significantly higher than that of i_a and i_c, but also influences the RMS value, for the RMS value of i_b is 6% larger than that of i_a and i_c.

TABLE III
SIMULATION RESULTS

Grid Current	THD	RMS Value
i_a	5.18%	6.01A
i_b	7.00%	6.39A
i_c	5.71%	6.01A

Therefore, when the input AC voltage is low, the zero sequence injection control is not able to strictly balance the grid currents due to the distortions of the rectifier input currents. This phenomenon is missing in the average model simulation, as shown in Fig. 20.

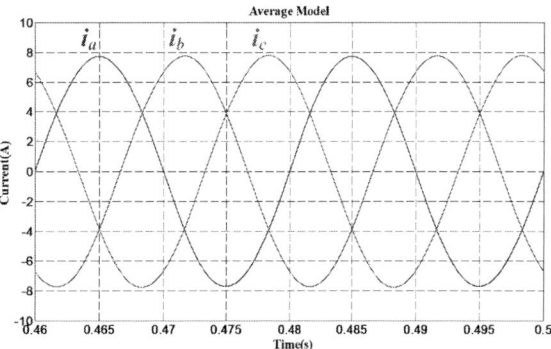

Fig. 20. Grid currents of average model simulation

This current distortion is noticeable when the input AC voltage is low, but negligible when the AC voltage is much higher than the IGBT's forward voltage drop. Under that condition the zero-crossing distortion level of the rectifier currents is low, thus wouldn't affect the grid currents much. To verify this, simulation under conditions shown in TABLE IV is done, where the input AC voltage is high enough.

TABLE IV
SIMULATION CONDITIONS WITH HIGH AC VOLTAGE

Grid Line-Line RMS Voltage	380V
Primary DC link voltage	200V
IGBT Forward Voltage Drop	1.6V
Reverse Diode Voltage Drop	1.2V
Load Power of Inverter Phase u	2400W
Load Power of Inverter Phase v	2400W
Load Power of Inverter Phase w	0W

The grid currents are shown in Fig.21 and their THD/RMS values are shown in TABLE V. It can be seen that the distortion is negligible and the magnitudes of the grid currents are well balanced. Therefore, the average model would be effective under this condition.

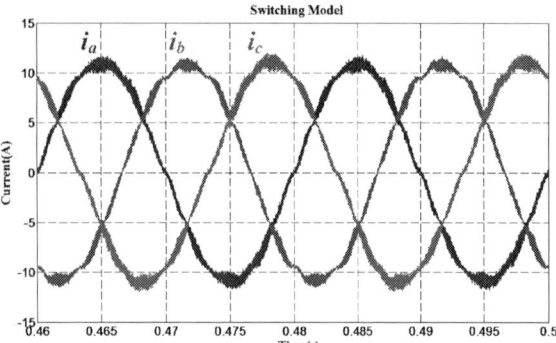

Fig. 21. Grid currents at high input AC voltage

TABLE V
SIMULATION RESULTS

Grid Current	THD	RMS Value
i_a	5.34%	7.78A
i_b	3.89%	7.81A
i_c	5.81%	7.78A

V. CONCLUSIONS

In this paper, the model and control strategy of a 20KW, 380V-380V PET is discussed, and the average model of each stage is derived. The simulation results validate the average model under both steady state and transient state. The simulation with average models is significantly faster than that with switching models.

As the average model neglects the voltage drop of the switches, the simulation results would miss some distortions. In this application, there would be a special grid current distortion under unbalanced load, and this distortion is obvious when the input AC voltage is low. The distortion would make the grid currents unbalanced, thus the second layer control of the rectifier stage would not be very effective, and the average model would not be effective, too. Only when the input AC voltage is high enough, this distortion would be negligible and the average model would be effective.

REFERENCES

[1] W. McMurray, "Power converters circuits having a high-frequency link," *U.S. Patent, 3517300, 1970.*

[2] M. Kang, P.N. Enjeti, I.J. Pitel, "Analysis and design of electronic transformers for electric power distribution system," *IEEE Trans. on power electronics, vol.14, no.6, pp. 1133-1141, 1999.*

[3] M.D. Manjrekar, R. Kieferndorf, G. Venkataramanan, "Power electronic transformers for utility applications," *Proceedings of the 2000 IEEE-IAS Annual Meeting, pp. 2496-2502, 2000.*

[4] H. Qin, and J. W. Kimball, "AC-AC dual active bridge converter for solid state transformer," *in Proc. IEEE ECCE, 2009, pp. 3039-3044.*

[5] X. Wang, J. Liu, T. Xu, et al, "Research of Different Modulation Methods for Single-Phase Single-Stage AC/AC Converter," *in Proc. IEEE ECCE Asia, 2013, pp. 613-619.*

[6] S. Krishnamurthy, "Half bridge AC-AC electronic transformer," *in Proc. IEEE APEC, 2012, pp. 1414-1417.*

[7] E.R. Ronan, S.D. Sudhoff, S.F. Glover, et al, "A power electronic-based distribution transformer," *IEEE Trans. on Power Delivery, vol. 17, no. 2, pp. 537-543, 2002.*

[8] M. Sabahi, S.H. Hosseini, M.B.B. Sharifian, M.B.B, et al, "A Three-Phase Dimmable Lighting System Using a Bidirectional Power Electronic Transformer," *IEEE Trans. on power electronics, vol. 24, no. 3, pp. 830-837, 2009.*

[9] K. Basu, N. Mohan, "A power electronic transformer for three phase PWM AC/AC drive with loss less commutation and common-mode voltage suppression," *in Proc. IEEE IECON, 2010, pp. 315-320.*

[10] E.R. Ronan, S.D. Sudhoff, S.F. Glover, et al, "A power electronic-based distribution transformer," *IEEE Trans. on power delivery, vol. 17, no. 2, pp. 537-543, 2002.*

[11] D. Wang, C. Mao, J. Lu, "Coordinated Control of EPT and Generator Excitation System for MultiDouble-circuit Transmission Lines System," *IEEE Trans. On Power Delivery, vol.23, no.1, pp. 371-379, 2008.*

[12] T. Zhao, G. Wang, S. Bhattacharya, et al, "Voltage and Power Balance Control for a Cascaded H-Bridge Converter-Based Solid-State Transformer," *IEEE Trans. on power electronics, vol. 28, no. 4, pp. 1523-1532, 2013.*

[13] G. Brando, A. Dannier, R. Rizzo, "Power Electronic Transformer application to grid connected photovoltaic systems," *in Proc. IEEE ICCEP, 2009, pp. 685-690.*

[14] D. Dujic, A. Mester, T. Chaudhuri, et al, "Laboratory scale prototype of a power electronic transformer for traction applications," *in Proc. EPE 14th European Conference on, 2011, pp. 1-10.*

[15] X. Wang, J. Liu, T. Xu, et al, "Control of a Three-Stage Three-Phase Cascaded Modular Power Electronic Transformer," *in proc. IEEE Applied Power Electronics Conference and Exposition,2013, pp. 1309-1315.*

[16] S. Du, J. Liu, J. Lin, et al, "A Novel DC Voltage Control Method for STATCOM Based on Hybrid Multilevel H-Bridge Converter," *IEEE Trans. on Power Electronics, vol. 28, no. 1, pp. 101-111, 2013.*

[17] H.K. Krishnamurthy, R. Ayyanar, "Building Block Converter Module for Universal (AC-DC, DC-AC, DC-DC) Fully Modular Power Conversion Architecture," *in Proc. IEEE PESC, 2007, pp. 483-489.*

A Multi-carrier PWM for AC-DC-AC Converter without DC Link Electrolytic Capacitor

Chung-Chuan Hou
Dept. of Electrical Engineering
Chung Hua University
Hsinchu, Taiwan, R.O.C.

Hsin-Ping Su
Dept. of Electrical Engineering
Chung Hua University
Hsinchu, Taiwan, R.O.C.

Abstract—**This study proposes a multi-carrier pulse width modulation (PWM) for AC-DC-AC converter without DC link electrolytic capacitor. The AC-DC-AC converter consists of an AC-DC active front-end converter, a DC-AC voltage source inverter, and a 10 uF ceramic capacitor in dc link. The DC bus voltage controller with load compensator is utilized to maintain the voltage of the DC link capacitor at designed value. Furthermore, the multi-carrier PWM is utilized to reduce the common-mode voltage (CMV) of AC-DC active front-end converter and DC-AC inverter. The simulation and test results are presented to validate the performances of the proposed scheme.**

Keywords—*AC-DC-AC converter; active front-end converter; inverter; multi-carrier PWM .*

I. INTRODUCTION

In variable speed drives systems, diode rectifiers are used as the AC-DC front-end. The advantages of the diode rectifiers include low cost, simplicity and high reliability. However, these rectifiers draw significant harmonic current from the utility grid and lack the regeneration capability. Moreover, the electrolytic capacitor is used in dc link with low cost and high density energy storage. The disadvantages of the electrolytic capacitor are short life time, low reliability, and slow down system response.

This study proposes an AC-DC-AC converter system as shown in figure 1 [1, 2]. The AC-DC-AC converter consists of an AC-DC active front-end (AFE) converter, a DC-AC voltage source inverter (VSI), and a 10 uF ceramic capacitor in dc link. The advantages of the AFE converter are unity power factor with low current distortion and regeneration capability [3, 4]. The load of the AFE converter is a VSI supplying a permanent magnet synchronous motor (PMSM) running at constant speed ω_r. Instead of the electrolytic capacitor, ceramic capacitor is utilized for DC link energy storage to improve the reliability and system response.

II. DYNAMIC MODEL FOR AC-DC-AC CONVERTER

A. AFE converter model

Figure 2 shows the equivalent circuit of the AC-DC-AC converter system where the switches are replaced by dependent voltage and current sources representing the

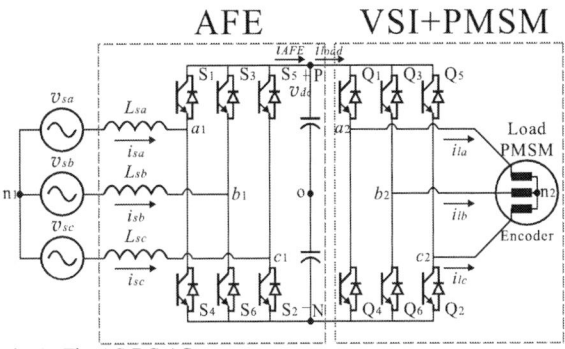

Fig. 1. The AC-DC-AC converter system.

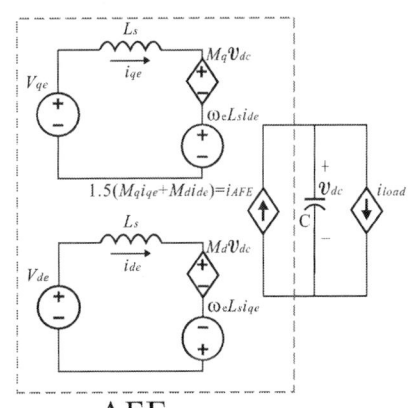

(a) Equivalent circuit of the AFE in a synchronous reference frame (ω_e)

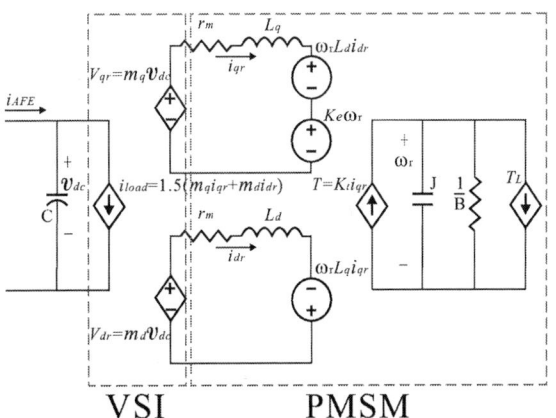

(b) Equivalent circuit of the VSI and PMSM in a rotor reference frame (ω_r)

Fig. 2. Equivalent circuit of the AC-DC-AC converter system.

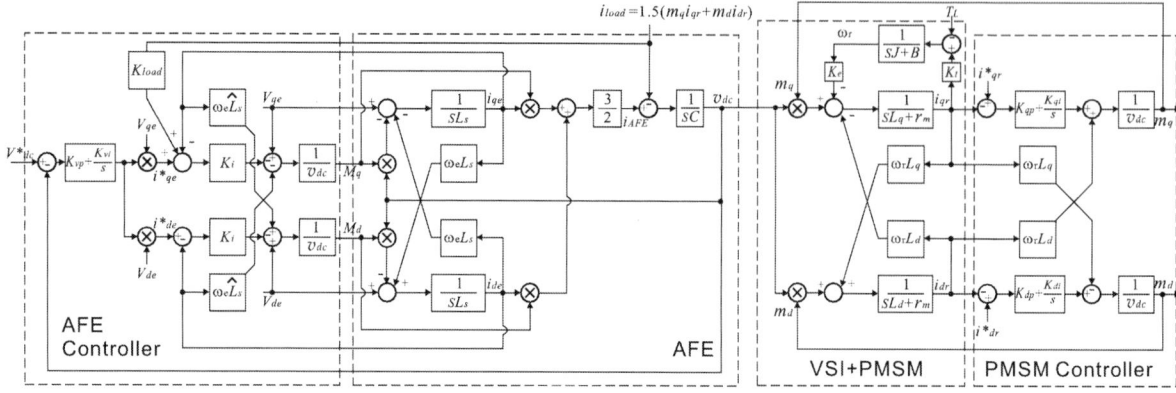

Fig. 3. The modeling and control of the AC-DC-AC converter system in the synchronous reference frame.

switching behavior [5]. The phase voltages v_{sa}, v_{sb}, and v_{sc} are transformed into the synchronous reference frame (V_{qe} and V_{de}) as shown in figure 2(a). The frequency ω_e is synchronized to the utility by a software phase lock loop. The phase currents i_{sa}, i_{sb}, and i_{sc} are transformed into the synchronous reference frame (i_{qe} and i_{de}). The following assumptions are made for the AFE converter: 1) line inductance $L_s = L_{sa} = L_{sb} = L_{sc}$; 2) the VSI-PMSM load is represented as a dependent current source i_{load}.

The duty cycle of the upper switches for phases a, b, and c is M_a, M_b, and M_c, and then is transformed into the synchronous reference frame (M_q and M_d). The equivalent DC bus capacitance is C. Therefore, the equivalent equations of the AFE converter in the synchronous reference frame are given in (1).

$$\frac{d}{dt} i_{qe} = \frac{1}{L_s}(V_{qe} - M_q v_{dc} - \omega_e L_s i_{de})$$

$$\frac{d}{dt} i_{de} = \frac{1}{L_s}(V_{de} - M_d v_{dc} + \omega_e L_s i_{qe})$$

$$\frac{d}{dt} v_{dc} = \frac{1}{C}[\frac{3}{2}(M_q i_{qe} + M_d i_{de}) - i_{load}]$$

$$= \frac{1}{C}[i_{AFE} - i_{load}]$$

(1)

As indicated in (1), the model of the AFE converter in the synchronous reference frame is as shown in figure 3.

B. Controller design of AFE converter

The error between the DC bus voltage command V^*_{dc} and the actual DC bus voltage v_{dc} is fed into a proportional-integral (PI) regulator ($K_{vp}+K_{vi}/s$) to establish the close-loop control of the dc bus voltage. In order to draw power from the ac side at unity power factor to maintain the dc bus voltage, the output of the PI regulator is then multiplied by the phase voltages V_{qe} and V_{de} in the synchronous reference frame to generate the current commands i^*_{qe} and i^*_{de} that are exactly in-phase with the phase voltages. The load current i_{load} is multiplied by K_{load} and is utilized as a feed-forward term of the current command. The close loop current control

adopts the proportional control gain K_i to determine the modulation index M_q and M_d of the pulse width modulation (PWM) operation as in (2).

$$M_q = [V_{qe} - \omega_e \hat{L}_s i_{de} - K_i(i^*_{qe} - i_{qe})]/v_{dc}$$

$$M_d = [V_{de} + \omega_e \hat{L}_s i_{qe} - K_i(i^*_{de} - i_{de})]/v_{dc}$$

(2)

Assuming the estimated value $\hat{L}_s$ approximates to the line inductance L_s. Applying equation (2) to (1) and solving the state equation, the transfer function can be obtained as in (3).

$$v_{dc}(s) = \frac{1.5M_q}{(sC) \cdot (1 + s/\omega_i)} \cdot i^*_{qe}(s) - \frac{1}{sC} i_{load}(s)$$

$$+ \frac{1.5M_d}{(sC) \cdot (1 + s/\omega_i)} \cdot i^*_{de}(s)$$

$$G_q(s) = \frac{v_{dc}(s)}{i^*_{qe}(s)} = \frac{1.5M_q}{(sC) \cdot (1 + s/\omega_i)}$$

(3)

$$G_{iq}(s) = \frac{i_{qe}(s)}{i^*_{qe}(s)} = \frac{1}{1 + s/\omega_i}$$

$$G_{id}(s) = \frac{i_{de}(s)}{i^*_{de}(s)} = \frac{1}{1 + s/\omega_i}$$

where
$\omega_i(=K_i/L_s)$ is the pole related to the current control loop;
$G_{id}(s)$ is the transfer function between i_{de} and i^*_{de};
$G_{iq}(s)$ is the transfer function between i_{qe} and i^*_{qe};
$G_q(s)$ is the transfer function between v_{dc} and i^*_{qe}.

The transfer functions have two poles as given in (3). The pole on the origin is related to the capacitor value for the DC bus voltage control, and the pole ω_i is related to the current regulator for the current control. The control block diagram of the AFE converter system in synchronous reference frame is as shown in figure 4. The DC bus voltage is affected by the load and the current commands of q-axis and d-axis.

978-1-4799-2706-7/14 $31.00 © 2014 IEEE

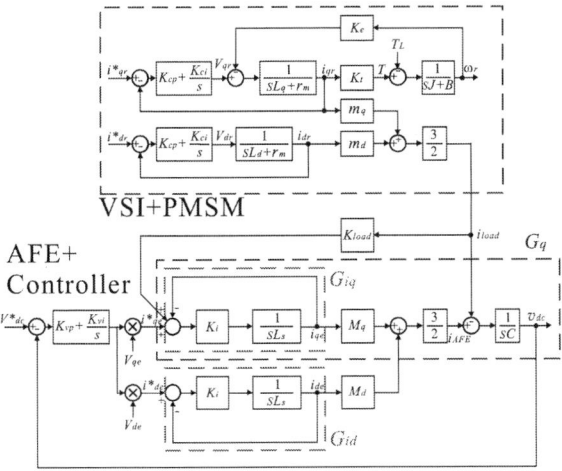

VSI+PMSM

AFE+
Controller

Fig. 4. Control block diagram of the AC-DC-AC converter system.

C. VSI-PMSM load

The VSI-PMSM load is represented as a dependent current source i_{load} on the previous discussion. Figure 2(b) shows the equivalent circuit of the VSI and PMSM in a rotor reference frame (ω_r) [6]. The quantities V_{qr} and V_{dr} represent the phase voltages in the rotor reference frame. The quantities i_{qr} and i_{dr} represent the phase currents in the rotor reference frame. The resistance of the armature winding is r_m. The quantities L_q and L_d represent the inductance in the rotor reference frame. The cross coupling terms ($\omega_r L_d i_{dr}$ and $-\omega_r L_q i_{qr}$) result from the rotor reference frame transformation. The induced voltage $K_e \omega_r$ is proportional to the rotor speed of the PMSM. The induced torque T is proportional to the current i_{qr}. The equivalent equations of the VSI-PMSM load are given in (4)

$$V_{qr} = m_q \upsilon_{dc} = r_m i_{qr} + L_q \frac{di_{qr}}{dt} + \omega_r L_d i_{dr} + K_e \omega_r$$

$$V_{dr} = m_d \upsilon_{dc} = r_m i_{dr} + L_d \frac{di_{dr}}{dt} - \omega_r L_q i_{qr} \qquad (4)$$

$$T = K_t i_{qr} = T_L + J \frac{d\omega_r}{dt} + B\omega_r$$

As indicated in (4), the model of the VSI-PMSM load in the rotor reference frame is as shown in figure 3.

D. Controller design of VSI-PMSM

The current commands of the PMSM in rotor reference frame are i^*_{qr} and i^*_{dr}. The current errors ($i^*_{qr} - i_{qr}$) and ($i^*_{dr} - i_{dr}$) are fed into the PI regulators ($K_{qp}+K_{qi}/s$) and ($K_{dp}+K_{di}/s$) to establish the close loop control of the PMSM current. The modulation index m_q and m_d are the duty cycle in the rotor reference frame as designed with the decoupling terms ($\omega_r L_d i_{dr}$ and $-\omega_r L_q i_{qr}$) in (5) as follow:

$$m_q = [(K_{qp} + K_{qi}/s)(i^*_{qr} - i_{qr}) + \omega_r L_d i_{dr}]/\upsilon_{dc} \qquad (5)$$

$$m_d = [(K_{dp} + K_{di}/s)(i^*_{dr} - i_{dr}) - \omega_r L_q i_{qr}]/\upsilon_{dc}$$

As indicated in (5), the controller of the VSI-PMSM load in the rotor reference frame is as shown in figure 3.

The following assumptions are made in developing the control block diagram for the VSI-PMSM: 1) voltage $V_{dr} = 0$; 2) the cross coupling terms ($\omega_r L_d i_{dr}$ and $-\omega_r L_q i_{qr}$) are perfectly decoupling by the current controller of the PMSM; 3) proportional gain $K_{qp} = K_{dp} = K_{cp}$ and integral gain $K_{qi} = K_{di} = K_{ci}$. Therefore, the control block diagram of the VSI-PMSM is as shown in figure 4.

III. DC BUS VOLTAGE CONTROL

Figure 4 shows the control block diagram of the AC-DC-AC converter system. There are two major control loops in the AFE converter system, one is the current control loop and the other is the DC bus voltage control loop. The PI regulator ($K_{vp} + K_{vi}/s$) is utilized for the compensation of the DC bus voltage. The function of the current control loop is to maintain the line currents tracking the current commands with predicted current control gain K_i. As indicated in (3), the DC bus voltage is affected by load disturbance. Therefore, the load compensation gain K_{load} is utilized to suppress the disturbance of load.

The d-axis current command i^*_{de} is controlled to be of zero value for unity power factor. The pole of current regulator ω_i is ten times large than K_{vi}/K_{vp} and $1.5M_qK_{vp}/C$. Therefore, the close loop transfer function of the DC bus voltage controller is given in (6).

$$\frac{\upsilon_{dc}(s)}{V^*_{dc}(s)} = \frac{(K_{vp} + \dfrac{K_{vi}}{s}) \cdot \dfrac{1.5M_q}{s^2(C/\omega_i)}}{1 + \dfrac{\omega_i}{s} + (K_{vp} + \dfrac{K_{vi}}{s}) \cdot \dfrac{1.5M_q}{s^2(C/\omega_i)}}$$

$$\approx \frac{(K_{vp} + \dfrac{K_{vi}}{s}) \cdot \dfrac{1.5M_q}{s^2(C/\omega_i)}}{(1 + \dfrac{\omega_i}{s})(1 + \dfrac{\omega_{vp}}{s})(1 + \dfrac{\omega_{vi}}{s})}$$

$$(6)$$

where poles ω_{vp} and ω_{vi} (near the origin) are defined as

$$\omega_{vp} = (1 + \sqrt{1 - \frac{4C}{1.5M_q} \cdot \frac{K_{vi}}{K_{vp}^2}}) \cdot \frac{1.5M_qK_{vp}}{2C}$$

$$\omega_{vi} = (1 - \sqrt{1 - \frac{4C}{1.5M_q} \cdot \frac{K_{vi}}{K_{vp}^2}}) \cdot \frac{1.5M_qK_{vp}}{2C}$$

underdamped : $\quad K_{vi}/K_{vp}^2 > 1.5M_q/4C$

criticallydamped : $K_{vi}/K_{vp}^2 = 1.5M_q/4C$

overdamped : $\quad K_{vi}/K_{vp}^2 < 1.5M_q/4C$

The performance of the AC-DC-AC converter system is affected by load disturbance. Therefore, the DC bus voltage disturbance rejection is defined as the magnitude of i_{load} needed to affect a unit deviation in the DC bus voltage. The transfer function $i_{load}(s) / \upsilon_{dc}(s)$ is given in (7) according to figure 4.

978-1-4799-2706-7/14 $31.00 © 2014 IEEE

$$\frac{i_{load}(s)}{v_{dc}(s)} = \frac{1 + \dfrac{\omega_i}{s} + (K_{vp} + \dfrac{K_{vi}}{s}) \cdot \dfrac{1.5 M_q}{s^2 (C / \omega_i)}}{\dfrac{-1}{sC}(1 + \dfrac{\omega_i}{s}) + K_{load} \cdot \dfrac{1.5 M_q}{s^2 (C / \omega_i)}} \quad (7)$$

$$\approx \frac{(1 + \dfrac{\omega_i}{s})(1 + \dfrac{\omega_{vp}}{s})(1 + \dfrac{\omega_{vi}}{s})}{\dfrac{-1}{sC}(1 + \dfrac{\omega_i}{s}) + K_{load} \cdot \dfrac{1.5 M_q}{s^2 (C / \omega_i)}}$$

The VSI-PMSM load is represented as i_{load} ($= 1.5 m_q i_{qr} + 1.5 m_d i_{dr}$). Then, the current loop transfer function of VSI-PMSM is given in (8) according to figure 4.

$$\frac{i_{qr}(s)}{i^*_{qr}(s)} = \frac{s K_{cp} + K_{ci}}{s^2 L_q + s(r_m + K_{cp}) + K_{ci}}$$

$$\approx \frac{1 + \dfrac{s}{\omega_{cp}}}{(1 + \dfrac{s}{\omega_{cp}})(1 + \dfrac{s}{\omega_{ci}})} \quad (8)$$

The pole ω_{cp} ($= K_{ci}/K_{cp}$) is related to the PI regulator. The pole ω_{ci} ($= K_{cp}/L_q$) is related to the current control loop. The transfer function i_{qr}/i^*_{qr} is equivalent a low pass filter to keep current i_{qr} follows the current command i^*_{qr}. The d-axis current command i^*_{dr} is controlled to be of zero value for voltage $V_{dr} = 0$ and current $i_{dr} = 0$, and then i_{load} is equivalent to $1.5 m_q i_{qr}$. Therefore, the DC bus voltage is affected by q-axis current command as in (9).

$$\frac{i^*_{qr}(s)}{v_{dc}(s)} = \frac{i^*_{qr}(s)}{i_{qr}(s)} \cdot \frac{i_{qr}(s)}{v_{dc}(s)} = \frac{i^*_{qr}(s)}{i_{qr}(s)} \cdot \frac{2}{3 m_q} \cdot \frac{i_{load}(s)}{v_{dc}(s)} \quad (9)$$

IV. MULTI-CARRIER PWM

In the space vector PWM (SVPWM) scheme, the AFE reference voltages v^*_{a1}, v^*_{b1}, and v^*_{c1} are utilized to calculate a zero sequence signal v_{01} and the VSI-PMSM reference voltages v^*_{a2}, v^*_{b2}, and v^*_{c2} are utilized to calculate a zero sequence signal v_{02}.

$$v_{01} = -\frac{1}{2}[\max(v^*_{a1}, v^*_{b1}, v^*_{c1}) + \min(v^*_{a1}, v^*_{b1}, v^*_{c1})] \quad (10)$$

$$v_{02} = -\frac{1}{2}[\max(v^*_{a2}, v^*_{b2}, v^*_{c2}) + \min(v^*_{a2}, v^*_{b2}, v^*_{c2})]$$

Adding v_{01} to the AFE reference voltages, the modified AFE reference voltages v^{**}_{a1}, v^{**}_{b1}, and v^{**}_{c1} are obtained and then utilized to generate gating pulse for PWM. Adding v_{02} to the VSI-PMSM reference voltages, the modified reference voltages v^{**}_{a2}, v^{**}_{b2}, and v^{**}_{c2} are obtained also. The CMV of the AC-DC-AC converter is defined as the neutral points of the voltage source and PMSM to the center of the DC bus in (11).

$$v_{n1o} = (v_{a1o} + v_{b1o} + v_{c1o})/3$$
$$v_{n2o} = (v_{a2o} + v_{b2o} + v_{c2o})/3 \quad (11)$$

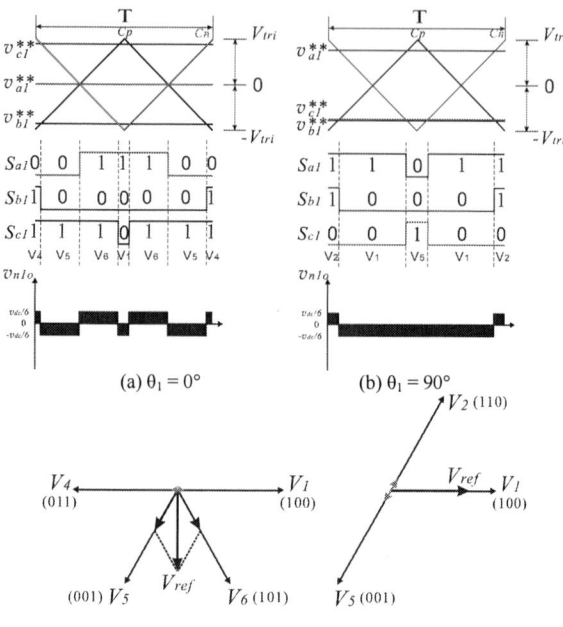

(a) $\theta_1 = 0°$ (b) $\theta_1 = 90°$

(c) vector ($\theta_1 = 0°$)   (d) vector ($\theta_1 = 90°$)

Fig. 5. Switching vector of one bi-carrier SVPWM cycle.

Figure 5 shows the switching vector of one bi-carrier SVPWM cycle [7]. As shown in figure 5(a) ($\theta_1 = 0°$), among the modified reference voltages, the minimum magnitude one (v^{**}_{a1}) compares with the triangle wave Cn, the others (v^{**}_{b1} and v^{**}_{c1}) compare with the triangle wave Cp respectively, and then generate the active vectors ($V_1 = [1,0,0]$, $V_4 = [0,1,1]$, $V_5 = [0,0,1]$, and $V_6 = [1,0,1]$). The zero vectors ($V_0 = [0,0,0]$ and $V_7 = [1,1,1]$) are replaced by the opposite active vectors ($V_1 = [1,0,0]$ and $V_4 = [0,1,1]$). The CMV of the AFE converter is contained within $\pm v_{dc}/6$. Figure 5(b) shows the switching vector of one bi-carrier SVPWM cycle ($\theta_1 = 90°$). Among the modified reference voltages, the minimum magnitude one (v^{**}_{c1}) compares with the triangle wave Cn, the others (v^{**}_{a1} and v^{**}_{b1}) compare with the triangle wave Cp respectively, and then generate the active vectors ($V_1 = [1,0,0]$, $V_2 = [1,1,0]$, and $V_5 = [0,0,1]$). The zero vectors ($V_0 = [0,0,0]$ and $V_7 = [1,1,1]$) are replaced by the opposite active vectors ($V_2 = [1,1,0]$ and $V_5 = [0,0,1]$). The CMV of the AFE converter is contained within $\pm v_{dc}/6$. Figure 5(c) and 5(d) show the reference voltage vector composition of the bi-carrier SVPWM. This bi-carrier SVPWM can synthesize the desired output voltage by using opposite active vectors instead of using zero vectors. The choice of the triangle wave to be compared with the modified reference voltages is given in table I and table II.

TABLE I
AFE MODIFIED REFERENCE VOLTAGES AND CARRIER

θ_1	v^{**}_{a1}	v^{**}_{b1}	v^{**}_{c1}
$-30° \sim 30°$	$Cn\ (v_{mid})$	$Cp\ (v_{min})$	$Cp\ (v_{max})$
$30° \sim 90°$	$Cp\ (v_{max})$	$Cp\ (v_{min})$	$Cn\ (v_{mid})$
$90° \sim 150°$	$Cp\ (v_{max})$	$Cn\ (v_{mid})$	$Cp\ (v_{min})$
$150° \sim 210°$	$Cn\ (v_{mid})$	$Cp\ (v_{max})$	$Cp\ (v_{min})$
$210° \sim 270°$	$Cp\ (v_{min})$	$Cp\ (v_{max})$	$Cn\ (v_{mid})$
$270° \sim 330°$	$Cp\ (v_{min})$	$Cn\ (v_{mid})$	$Cp\ (v_{max})$

TABLE II
VSI-PMSM REFERENCE VOLTAGES AND CARRIER

θ_2	v^{**}_{a2}	v^{**}_{b2}	v^{**}_{c2}
$-30° \sim 30°$	$Cn\ (v_{mid})$	$Cp\ (v_{min})$	$Cp\ (v_{max})$
$30° \sim 90°$	$Cp\ (v_{max})$	$Cp\ (v_{min})$	$Cn\ (v_{mid})$
$90° \sim 150°$	$Cp\ (v_{max})$	$Cn\ (v_{mid})$	$Cp\ (v_{min})$
$150° \sim 210°$	$Cn\ (v_{mid})$	$Cp\ (v_{max})$	$Cp\ (v_{min})$
$210° \sim 270°$	$Cp\ (v_{min})$	$Cp\ (v_{max})$	$Cn\ (v_{mid})$
$270° \sim 330°$	$Cp\ (v_{min})$	$Cn\ (v_{mid})$	$Cp\ (v_{max})$

V. SIMULATION AND TEST RESULTS

The parameters of the simulation and test bench are given as follows:
1) Utility: three-phase 220 Vrms/60Hz; L_s = 2.0 mH.
2) AC-DC-AC converter: switching frequency is f_{sw} = 10 kHz, sampling frequency is 20 kHz.
3) DC link: υ_{dc} = 370V; 10 uF ceramic capacitor.
4) Controller: DSP is TMS320C28335, 150 MHz operating speed; CPLD is LCMXO640C, 50 MHz.

Figure 6 shows the AC-DC-AC converter operated under load changed at time 0.8 second. As shown in figure 6(a), line currents are 60 Hz sinusoidal wave and are increased under load changed at time 0.8 second. As shown in figure 6(b), load currents are 50 Hz sinusoidal wave. The DC bus voltage is maintained within 20V deviation from the operated value 370V with 10 uF ceramic capacitor.

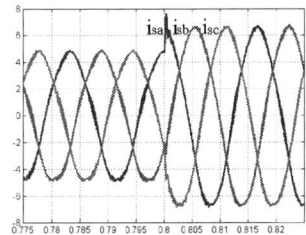

(a) Increased line currents at time 0.8 second.

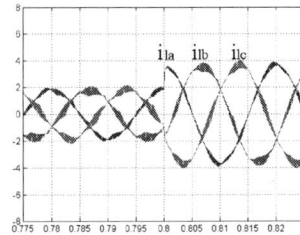

(b) Increased load currents at time 0.8 second.

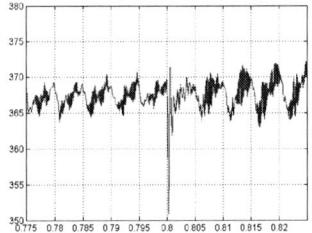

(c) DC bus voltage disturbance by load changed.
Fig. 6. AC-DC-AC converter operates under load changed.

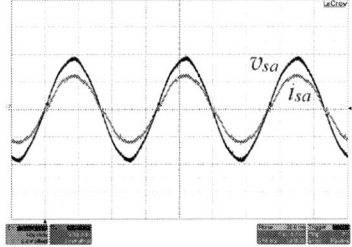

Fig. 7. Phase voltage and line current of AC-DC AFE converter (υ_{sa} : 100V/div; i_{sa} : 5A/div; time base: 5ms/div).

Figure 7 shows the phase voltage and line current of AC-DC AFE converter with constant load. The phase voltage and line current are in phase.

VI. CONCLUSIONS

This study proposes a multi-carrier PWM for AC-DC-AC converter without DC link electrolytic capacitor. The dynamic model of the AC-DC-AC converter is presented. According to the dynamic model, the controllers of the AC-DC AFE converter and DC-AC VSI-PMSM are designed. The DC bus voltage control and load disturbance rejection are analyzed to improve the performances of the AC-DC-AC converter. Furthermore, a multi-carrier PWM is utilized to reduce the CMV of the AC-DC-AC converter. Simulation and test results are presented to validate the performances of the proposed AC-DC-AC converter.

REFERENCES

[1] Thomas G. Habetler, "A space vector-based rectifier regulator for AC/DC/AC converters," *IEEE Trans. on Power Electronics*, vol. 8, no. 1, pp. 30-36, January 1993.

[2] Anno Yoo, Seung-Ki Sul, Hyeseung Kim, and Kyung-Seo Kim, "Flux-weakening strategy of an induction machine driven by an electrolytic-capacitor-less inverter," *IEEE Trans. on Industry Applications*, vol. 47, no. 3, pp. 1328-1336, May/June 2011.

[3] Y. Suh and T. A. Lipo, "Control scheme in hybrid synchronous stationary frame for pwm ac/dc converter under generalized unbalanced operating conditions," *IEEE Trans. on Industry Applications*, vol. 42, no. 3, pp. 825-835, May/June 2006.

[4] Johann W. Kolar, Thomas Friedli, Jose Rodriguez, and Patrick Wheeler, "Review of three-phase PWM AC-AC converter topologies," *IEEE Trans. on Industrial Electronics*, vol. 58, no. 11, pp. 4988-5006, November 2011.

[5] Chung-Chuan Hou and Po-Tai Cheng, "Experimental verification of the active front-end converters dynamic model and control designs," *IEEE Trans. on Power Electronics*, vol. 26, no. 4, pp. 1112-1118, April 2011.

[6] F. Morel, J.-M. Retif, X. Lin-Shi, and C. Valentin, "Permanent magnet synchronous machine hybrid torque control," *IEEE Trans. on Industrial Electronics*, vol. 55, no. 2, pp. 501-511, February 2008.

[7] Chung-Chuan Hou, "A multicarrier PWM for parallel three-phase active front-end converters," *IEEE Trans. on Power Electronics*, vol. 28, no. 6, pp. 2753-2759, June 2013.

A Decoupling Offset-Based PWM Control for a Multilevel Inverter Under DC Voltage Unbalance

Nho Van Nguyen   Tam Khanh Tu Nguyen
Dept. of Electrical and Electronics Engineering
Hochiminh City University of Technology
Hochiminh City, Vietnam

Hong-Hee Lee
School of Electrical Engineering
University of Ulsan
Ulsan, Korea

Abstract— **In this paper, a novel offset-based two-level pulse width modulation (PWM) scheme for a multilevel inverter with unbalanced dc sources is presented. The offset voltage is introduced as an indispensable variable in the control model for a multilevel voltage source inverter. An expression for the inverter parameters in** *abc* **coordinates can enable the determination of the operating ranges of related voltages for the unbalanced dc source condition. The PWM performance is controlled through the design of two offset components in a subsequence. One offset defines the operating points of the three-phase leg voltages among the dc levels, while the second offset restricts its effect on the quality of PWM control in related dc levels. A nominal switching time diagram and voltage vector diagram for a virtual two-level inverter reveal the typical characteristics of offset-based two-level modulation. With regards to the actual application of the scheme, the voltage outputs remain balanced even with abnormal dc sources. An improving of the PWM quality as reduction of the switching loss in Discontinuous PWM mode can be obtained by setting the local offset related to the load currents. The validity of the proposed algorithm was verified by experimental results.**

Keywords— PWM technique, Decoupling offset, Unbalance dc sources, Multilevel inverter.

I. Introduction

Multilevel inverters play an important role in current high performance applications. Two topologies have become especially popular: the neutral point clamped multilevel inverter and the cascaded multilevel inverter. A circuit diagram of a three-level NPC inverter is shown in Fig. 1. Well-known PWM methods include carrier-based (CPWM) and space vector (SVPWM) techniques [1]-[8]. Carrier-based PWM techniques are commonly used in practical applications because of their simple implementation. In a recent study, a novel carrier PWM method offered flexible control of vector redundancies in an offset generator [4]. The researchers showed that the algorithms of a space vector PWM scheme with the nearest three voltage vectors (NTV) can be completely realized by a corresponding carrier PWM method.

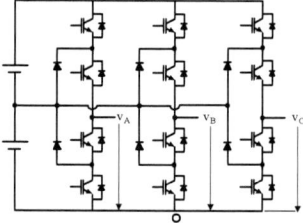

Fig. 1. Circuit diagram for a three-level NPC inverter.

The offset feature in a two-level inverter has been extensively investigated in previous studies [9]-[11]. Offset control is also available in space vector modulation by redistributing the switching time duties of the two redundant states [10]-[13]. PWM control is more sophisticated in a multilevel inverter. Space vector modulation requires the determination of a small vector triangle in the hexagonal diagram. It then selects the switching states and calculates their corresponding time duty ratios. In order to apply the NTV principle, a lookup table has often been used to determine the area and related vectors [14], [15]. After determining the three pivot vectors, up to four active switching states can be selected. The algorithm may be difficult if the offset is used as a variable to control the dc neutral currents and thus, balance the dc voltages and reduce the impact of the common mode voltage (CMV) on the electrical drives [16], [17]. In a previous study pertaining to Neutral point clamped (NPC) multilevel inverters, a carrier PWM algorithm for balancing the ac output with unbalanced dc sources was theoretically described [18]. Its feed-forward compensating algorithm modified either the modulating signals or the carrier waves. In a subsequent study, similar work was performed for a cascaded multilevel inverter [19]. Nho et al. theoretically extended the aforementioned compensating idea for a three-phase inverter to a multiphase multilevel inverter with unbalanced dc sources [20]. More results can be found in other works where the PWM algorithm for balanced dc sources [21], [22] and the feed-forward PWM algorithm for unbalanced dc sources [23] were separately solved. These studies were mainly reduced to an algorithm for computing the switching time duties for a previously defined reference. In later works [19], [21]-[23], a sinusoidal reference was assumed and the offset voltage was fixed as a constant. The modulation index was also limited to a value below 0.866 for balanced dc sources. Since there was no offset voltage in the inverter model, the researchers could not take further PWM improvements into consideration. Therefore, in comparison with the conventional sinusoidal PWM technique, these methods were not convincing for their contribution to improving the PWM quality. Because of relying on the use of the redundant states, the loss of the vector redundancy under the dc voltage imbalance prevents this algorithm from the application. The offset voltage has been proven to have a large impact on converter performance. Proper offset voltage selection can reduce the number of switching and the harmonic distortion factor in different PWM techniques [4], [24]; its regulation with load currents may help to minimize switching loss [25]. Offset adjusting can generate different PWM modes [3], [26]. Therefore, the offset voltage is practically indispensable and an important variable in the control model of a multilevel voltage source inverter.

Several offset functions have been introduced in previous research [1], [3], [26]-[31]. However, each PWM method had its own algorithm and was available for some particular demand. In order to flexibly utilize the advantages of the offset PWM technique, it is necessary to establish a universal offset-based model for a multilevel inverter where the offset becomes a control variable in the converter system.

In this paper, a novel decoupling offset-based two-level PWM scheme in *abc* coordinates is proposed. Its main contributions are clarified in the following points:

(a) Offset based PWM quality control: Unlike many methods that taking into account sinusoidal reference input without any modification [19],[21]-[23],[32], in the proposed PWM scheme, the offset voltage is defined as a control variable. The offset regulation enables controlling common mode voltage [33] and PWM quality of the converter.

(b) General dc source condition: Unlike recent work conducted under the assumption of balanced dc sources [22],[25], the problems addressed here will be solved under the condition of variable dc voltage sources. The influence of the dc voltage imbalance will be corrected by producing proper nominal modulating signals in the proposed PWM scheme.

(c) Flexible offset controllability: A so-called decoupling offset algorithm helps to control the PWM performance for an entire fundamental period as well as in each sampling time period. The decoupling offset technique is concerned with two offset components. The first, named the global offset, provides an initial CMV function. The second, named the local offset, provides an additional CMV function. The underlying principle of the technique is shown in Fig. 2. The global offset together with the fundamental voltages establish an "operating point" for the multilevel inverter (i.e., v_{LX} ; $X = A, B, C$) that their values will determine a working virtual two-

level inverter. In addition to extending the maximum voltage range, the global offset can be utilized to reduce the CMV, lower the voltage imbalance among dc-cell levels [30], and optimize the switching loss in the fundamental period [24],[25]. The local offset allows the leg voltages to be adjusted within two closet dc-levels and may help to increase the PWM quality in the working virtual two-level inverter. Thus, the main contribution of the local offset is an improvement in the PWM performance in each sampling period [31].

(d) Simplicity of implementation: The use of virtual two-level inverter, its switching sequences and corresponding nominal switching time diagrams enables users to attain PWM algorithm in several simple math operators. The coordinate transformation can be done with simple calculation. The nominal switching time diagram can be flexibly modified for different applications. Its particular case is effective reducing CMV [33].

(e) Universal and complete PWM scheme: two offset controllers, N-to-two level inverter transformation and PWM generator with a feed-forward dc unbalanced voltage correction establish a main structure of a universal PWM algorithm in multilevel inverters.

(f) General applicability: The proposed method is applicable to any number of levels. Its principle may be extended to a multiphase power converter [20]. The offset scheme in Fig. 2 is unchanged irrespective of PWM patterns. In this paper, a Phase Disposition PWM (PD PWM) technique is considered, but applying different PWM patterns to the offset based PWM scheme is available.

The proposed algorithm will be analyzed for use with an NPC inverter. However, with a proper modification the algorithm can be used for other topologies. The validity of the method will ultimately be demonstrated with simulated and experimental results.

II. Proposed Decoupling Offset PWM Scheme

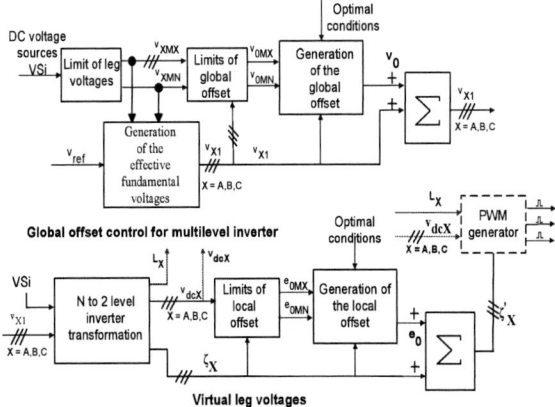

Fig. 2. Decoupling offset-based PWM scheme for a multilevel inverter under input voltage unbalance.

A. Decoupling Offset Principle

The global offset $\mathbf{v}_0$ first defines the operating points by setting the leg voltages and the corresponding virtual two-level inverter. In the virtual inverter, the local offset will be adjusted so that the switching state sequence and the corresponding switching time duties will satisfy demanding outputs. The advantage of the decoupling offset PWM technique is the subsequent control of the two offset components.

The decoupling offset-based PWM method may be expressed as:

$$\mathbf{v}_X' = \mathbf{v}_{X1} + \mathbf{v}_0 + \mathbf{e}_0 \; ; X = A, B, C \quad (1)$$

The reference leg voltage $\mathbf{v}_X'$ is defined as the sum of the fundamental voltage $\mathbf{v}_{X1}$ and two control variables: the global offset $\mathbf{v}_0$ and the local offset $\mathbf{e}_0$, as shown in Fig. 3(b).

B. The Global Offset Control

The global offset controller helps produce the primitive leg voltages $\mathbf{v}_X$, which consist of the fundamental voltage $\mathbf{v}_{X1}$ and the CMV $\mathbf{v}_0$ as follows:

$$\mathbf{v}_X = \mathbf{v}_{X0} = \mathbf{v}_{X1} + \mathbf{v}_0 \; ; X=A, B, C. \quad (2)$$

The limit of the fundamental load voltages VX1

The limits of the leg voltages allow the maximum output voltage range in linear modulation to be defined. They also confine the offset voltage range. Suppose VSA, VSB, and VSC are three groups of voltage sources that supply the converter and SA, SB, and SC are the corresponding groups of switching devices (see Fig. 3(a)). Let us also define "O" as the virtual zero voltage point, v_A, v_B, v_C as the three-phase leg voltages (see Fig. 3(b)), and the maximum and minimum phase-leg voltage levels as $v_{AMX}, v_{BMX}, v_{CMX}$ and $v_{AMN}, v_{BMN}, v_{CMN}$, respectively. From the previous limits, any instantaneous value of the leg voltage has to be set in the following range:

$$\mathbf{v}_{XMN} \le \mathbf{v}_X \le \mathbf{v}_{XMX} \; ; X = A, B, C \quad (3)$$

The limit of the fundamental load voltages VX1

An instantaneous line-to-line voltage can be expressed as:

$$\mathbf{v}_{XY} = \mathbf{v}_X - \mathbf{v}_Y \; ; X, Y = A, B, C \; ; X \ne Y \quad (4)$$

The condition for the reference line-line voltages can also be expressed as:

$$V_{LMN} \le \mathbf{v}_{XY} \le V_{LMX} \quad (5)$$

where the two extreme functions define a maximum and minimum value from all variables as:

$$V_{LMX} = \text{Min}(v_{AMX} - v_{BMN}, v_{BMX} - v_{CMN}, v_{CMX} - v_{AMN}) \quad (6)$$

$$V_{LMN} = \text{Max}(v_{AMN} - v_{BMX}, v_{BMN} - v_{CMX}, v_{CMN} - v_{AMX}) \quad (7)$$

In order to produce symmetrical and sinusoidal three-phase outputs, the maximum amplitude of the line-line voltages can be determined as the minimum absolute value from the instantaneous three-phase line-line voltages during a fundamental period as follows:

$$V_{LM} = \text{Min}(V_{LMX}, |V_{LMN}|) \quad (8)$$

As a result, the maximum amplitude of the fundamental phase voltage can be determined as:

$$V_{1m} = V_{LM}/\sqrt{3} \quad (9)$$

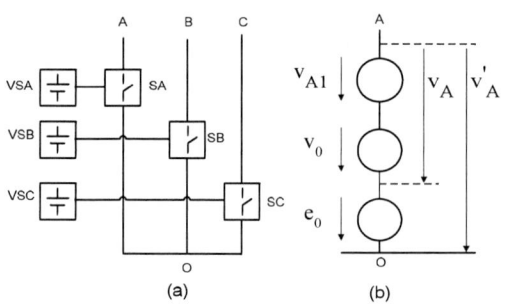

Fig. 3. (a) Block circuit diagram of the three-phase power converter, (b) an analysis of the A-phase leg voltage.

The limits of the offset voltages v_{0MX} and v_{0MN}

The offset voltage can be varied in the following range:

$$v_{0MN} \le \mathbf{v}_0 \le v_{0MX} \quad (10)$$

$$v_{0MX} = \text{Min}(v_{AMX} - v_{A1}, v_{BMX} - v_{B1}, v_{CMX} - v_{C1}) \quad (11)$$

$$v_{0MN} = -\text{Min}(v_{A1} - v_{AMN}, v_{B1} - v_{BMN}, v_{C1} - v_{CMN}) \quad (12)$$

The limits of the offset controller of the multilevel inverter are based on the differences between the fundamental voltages and the limits of the leg voltages given by (10)-(12).

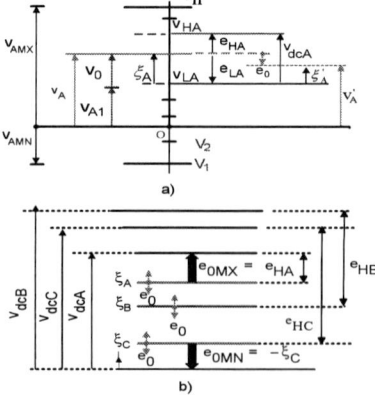

Fig. 4. (a) A detailed analysis of the A-phase leg voltage in the active dc-level and (b) a detailed analysis of the voltage components in a virtual two-level inverter.

Generation of the global offset v_0 in the optimization block

The function of the global offset for a given modulation index can be selected through a consideration of the output behavior for the entire fundamental period. From (10)-(12), the global offset can be expressed as a function of variable η_1 as:

$$v_0 = (1 - \eta_1)v_{0MN} + \eta_1 v_{0MX}; \quad 0 \le \eta_1 \le 1 \quad (13)$$

Medium common mode voltage [9]-[12] can be selected as the global offset and is defined as the average of two extreme CMV values by setting $\eta_1 = 0.5$ in (13) as follows:

$$v_0 = (v_{0MX} + v_{0MN})/2 . \quad (14)$$

Minimum common mode voltage: Suppose that the ground G has a potential of v_{GO} with respect to the virtual point "O". The offset function for attaining a minimum CMV can be defined as :

$$v_0 = \begin{cases} v_{0MX} & \text{if} & v_{0MN} \le v_{0MX} \le v_{G0} \\ v_{0MN} & \text{if} & v_{G0} \le v_{0MN} \le v_{0MX} \\ v_{GO} & \text{if} & \text{else} \end{cases} \quad (15)$$

The parameter η_1 obtains a value among 0, 1 and η_0, $\eta_0 = \dfrac{v_{GO} - v_{0MN}}{v_{0MX} - v_{0MN}}$. The resulting sinusoidal PWM method is a particular case of the PWM technique with a minimum CMV for a modulation index range lower than 0.866.

C. N-to- 2 level inverter transformation, virtual 2-level inverter

Virtual 2-level inverter

A useful characteristic of n-to-2 level transformation is that it determines three actual dc voltages of the virtual two-level inverter and the related leg voltages as a virtual reference. For the X-phase $X = A, B, C$, suppose the two-leg voltages v_{HX} and v_{LX} are the active higher and active lower levels, respectively. With regard to the dc voltage limits, the relationships between v_{HX}, v_{LX}, and the reference leg voltages are given by:

$$\begin{cases} v_{XMN} \le v_{LX} \le v_X < v_{HX} \\ v_{LX} < v_X \le v_{HX} \le v_{XMX} \end{cases} \quad (16)$$

The parameters of the indexes H_x and L_x correspond, respectively, to the lower and higher voltage levels closest to the reference leg voltages v_X

$$H_x = L_x + 1; X = A, B, C \quad (17)$$

A virtual two-level inverter is characterized by its centered vector $\vec{V}_L = [V_{LA}, V_{LB}, V_{LC}]^T$; its

corresponding vector diagram is determined by three dc cells v_{dcX} defined by (18):

$$v_{dcX} = v_{HX} - v_{LX}; X=A,B,C \quad (18)$$

Eight switching states establish a voltage vector diagram of a virtual 2-level inverter, as shown in Fig. 5. Two redundant states disappear if the virtual inverter is supplied with three different dc voltage values (see Fig. 5(b)). Different two-level PWMs can be established for this virtual inverter through the use of the offset controller and PWM pattern selection.

Virtual leg voltages

The leg voltages ξ_X are considered to be the references of the virtual two-level inverter and are defined as the difference between the reference leg voltage and its lower dc level v_{LX}, X=A,B,C as follows:

$$\xi_X = v_X - v_{LX} ; X=A, B, C \quad (19)$$

or in vector form:

$$\vec{\xi} = \vec{v} - \vec{v}_L \quad (20)$$

The PWM behavior of the virtual inverter is important for attaining the output voltage quality of the multilevel inverter.

D. The offset control in a virtual two-level inverter

The limits of the local offset

From (1), the local offset e_0 becomes the only controllable variable of the 3-phase virtual leg voltages for given global voltages v_X . The local offset controller produces modified reference leg voltages of the virtual inverter $\vec{\xi}$, as shown in Fig. 5:

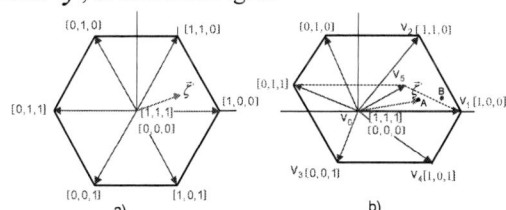

Fig. 5. Vector diagram of a virtual two-level inverter for (a) balanced dc sources and (b) unbalanced dc sources $v_{dcA} > v_{dcB} > v_{dcC}$.

$$\xi'_X = \xi_X + e_0 ; X = A, B, C \quad (21)$$

or in vector form, $\vec{I} = [1,1,1]^T$:

$$\vec{\xi}' = \vec{\xi} + e_0 \vec{I} \quad (22)$$

The limits of the local offset voltage e_{0MN} and e_{0MX}

The local offset e_0 range is dependent on the reference local leg voltage and the three active dc-cells. It can be varied in the following range (see Fig. 4a):

$$e_{0MN} \le e_0 \le e_{0MX} \quad (23)$$

The resulting reference leg voltages of the virtual two-level inverter can be expressed as follows:

$$e_{0MX} = \text{Min}(e_{HA}, e_{HB}, e_{HC}) \tag{24}$$

$$e_{0MN} = -\text{Min}(\xi_A, \xi_B, \xi_C) \tag{25}$$

$$e_{HX} = v_{HX} - v_X = v_{dcX} - \xi_X \tag{26}$$

The offset controller for the virtual two-level inverter is designed based on the virtual leg voltages $\vec{\xi}$ and the active dc sources. From (23)-(25), the local offset can be expressed as a function of variable η_2, as follows:

$$e_0 = (1 - \eta_2)e_{0MN} + \eta_2 e_{0MX} \; ; 0 \leq \eta_2 \leq 1 \tag{27}$$

Equation (27) shows that with a proper selection of the variable η_2, a new local offset e_0, and thus a new controlling mode for the multilevel inverter can be achieved. For example, a discontinuous PWM method can be obtained if the value of the parameter η_2 is set equal to 1 or 0, as follows:

$$e_0 = \begin{cases} e_{0MX}, & \text{Condition} \quad 1 \\ e_{0MN}, & \text{Condition} \quad 2 \end{cases} \tag{28}$$

The setting can be applied on a current comparison basis to obtain the minimum switching loss [25]. With no current sensors, the offset may be defined by considering the leg voltage. Several variants of the global offset may be considered when determining an optimized switching loss DPWM method.

The SVPWM method with the same switching time duties of two zero redundant states in a two-level inverter can also be approximately extended to a multilevel inverter by setting the parameter $\eta_2 = 0.5$ in (27) as follows:

$$e_0 = \frac{e_{0MX} + e_{0MN}}{2} \tag{29}$$

For a small level of unbalanced dc sources, (29) would satisfy for a low current ripple in the entire modulation index.

E. PWM generator- Space vector and carrier PWM implementation variants

For a decoupling offset two-level PWM scheme, both the space vector PWM and carrier PWM variants can be equivalently deduced as shown in Figs. 6 and 7. The main algorithm of the offset-based PWM scheme defines a virtual two-level inverter with reference virtual leg voltages ξ'_X. These voltages can be utilized by different space vector PWM and carrier PWM algorithms. In Figs. 6 and 7, the virtual leg voltages will first be rescaled (normalized) to produce nominal modulating signals $\xi^{*'}_X$. The simplicity of the carrier-based PWM approach lies in the use of comparators. To deduce the switching time duties in the space vector approach, the nominal signals

can be rearranged in a "max-mid-min" descending order, as shown in Fig. 8.

The normalization compensates for the influence of the input dc voltage imbalance on the ac output load voltages by producing proper nominal modulating signals. When compared with 2-D space vector PWM [35] and 3-D space PWM methods [23], the offset PWM scheme depicted in a nominal switching time diagram can simplify modifications and explanations of the commutations in the unbalanced input voltage condition. The change in the switching states and the corresponding time duties caused by the offset adjusting can also be deduced from the nominal switching time diagram.

For the Phase Disposition PWM method, the virtual leg voltage vector is normalized as:

$$\vec{\xi}^{'*} = [\xi_A^{*'}, \xi_B^{*'}, \xi_C^{*'}]^T \tag{30}$$

where each component nominal modulating signal $\xi_X^{*'}$, $X=A,B,C$ is defined as the ratio of the virtual leg voltage from (22) to the corresponding dc cell voltage from (18) as follows:

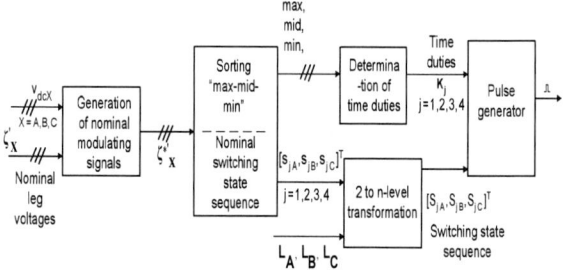

Fig. 6. PWM generator- Space vector PWM implementation.

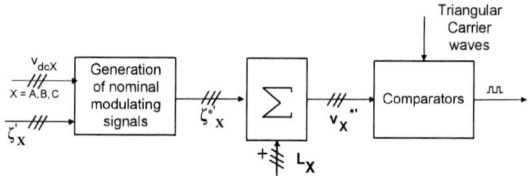

Fig. 7. PWM generator - Carrier PWM implementation.

$$\xi_X^{'*} = \frac{v_X^{'} - v_{LX}}{v_{HX} - v_{LX}} = \frac{\xi_X^{'}}{v_{dcX}} \; ; \; 0 \leq \xi_X^{'*} \leq 1 \tag{31}$$

For multi-carrier implementation in Fig.7, the normalized modulating signals of the NPC multilevel inverter can be expressed as follows:

$$\vec{v}^{'*} = \vec{L} + \vec{\xi}^{'*} \tag{32}$$

where $\quad \vec{L} = [L_A, L_B, L_C]^T \tag{33}$

Nominal switching time diagram

For a better view, the nominal modulating signals can be rearranged in "max-mid-min" order. This arrangement enables the switching time duties of the corresponding four switching states involved in a sequence to be deduced as follows:

$$K_1 = 1 - \max; K_2 = \max - mid \qquad (34)$$
$$K_3 = mid - \min; K_4 = \min$$

where:

$$\max = Max(\xi_A^{*}, \xi_B^{*}, \xi_C^{*})$$
$$mid = Mid(\xi_A^{*}, \xi_B^{*}, \xi_C^{*}) \qquad (35)$$
$$\min = Min(\xi_A^{*}, \xi_B^{*}, \xi_C^{*})$$

The nominal switching states $[s_{jA}, s_{jB}, s_{jC}]^{T}$

$[s_{jA}, s_{jB}, s_{jC}]^{T}$; $j = 1, 2, 3, 4$ can be derived by comparing the nominal modulating signals of $\xi_A^{*}, \xi_B^{*}, \xi_C^{*}$. A simple procedure for determining the nominal switching states and the switching time duties can be found in Refs. [20] and [24]. The nominal switching states can also be deduced by using the mapping function Table as shown in Table 1.

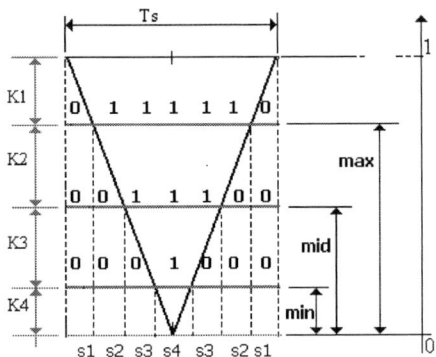

Fig. 8. Nominal switching time diagram in the PD PWM.

TABLE I
NOMINAL SWITCHING STATE SEQUENCE IN PD PWM
TECHNIQUE

CONDITIONS			Nominal switching state sequence in half triangle ($s_A s_B s_C$)
ξ_A^{*}	ξ_B^{*}	ξ_C^{*}	$\vec{s}_1 \to \vec{s}_2 \to \vec{s}_3 \to \vec{s}_4$
max	mid	min	$000 \to 100 \to 110 \to 111$
max	min	mid	$000 \to 100 \to 101 \to 111$
mid	max	min	$000 \to 010 \to 110 \to 111$
mid	min	max	$000 \to 001 \to 101 \to 111$
min	max	mid	$000 \to 010 \to 011 \to 111$
min	mid	max	$000 \to 001 \to 011 \to 111$
PWM Pattern in "max-mid-min" coordinates			$000 \to 100 \to 110 \to 111$

III. Experimental results

Experimental hardware was built for the 3-level NPC inverter so as to validate the proposed theory. The algorithm is implemented using the eZdsp TMS320F2812 control kit. Two DC voltages were measured with LEM LV25 NP Hall sensors and filtered through a low pass filter with a cut-off frequency of 1 kHz.

The current sensors Hall LEM LA25NP were used to measured phase load currents. The frequency of the triangle carrier waveform was set at 5 kHz. The experimental parameters were $v_{C1} = 50V$, $v_{C2} = 70V$, $f_0 = 50Hz$, $R = 10\Omega$ and $L = 60mH$.

The PWM algorithms without compensation were gradually realized with m = 0.3 and 0.8. The phase load voltage, currents, and FFT diagrams of the current without compensation for m = 0.3 are shown in Fig. 11(a), respectively. The low-order harmonics, particularly the second-order harmonic, were rejected from the output current after the compensating algorithm was applied. The load voltage, currents, and FFT diagrams with compensation for m = 0.3 are shown in Fig.11(b), respectively. Further results for m = 0.8 without compensation are displayed in Fig. 12(a), respectively. The second-order harmonic was reduced by compensation to a negligible value of around 0.3% in the diagram of the FFT of the load current, as shown in Fig. 12(b). The algorithm was ultimately verified with two different dc capacitors with capacitances of $C_1 = 4700$ μF and $C_2 = 220$ μF. Because of the low capacitance, the dc voltage on this capacitor had a considerably high ripple and its dc voltage varied between 48V and 64V, while the dc voltage on the second capacitor was held approximately smooth at 70V. The different capacitors produced different dc sources, which in turn caused asymmetrical output voltages/currents. The FFT diagram in Fig. 13(a) revealed a significant increase in the second-order harmonics. After applying the compensating algorithm, the spectrum of the load current was improved, as shown in Fig. 13(b). Under the same conditions, the FFT diagram of the load current with the compensating algorithm revealed that the second harmonic content of 1.26% for the uncompensated algorithm had been reduced to a negligible value for the compensating algorithm. As a result, the outputs regained their balance.

In other experiments with compensated outputs, the dc voltage sources were set as $v_{C1} = 100$ V, $v_{C2} = 80$ V, the RL load was $R = 10\Omega$, L= 30mH and the switching frequency was 10kHz. The influence of the unbalanced dc sources on the voltage waveform for different CMV setting was demonstrated in Fig.14. For instance, the offset for m = 0.866, in minimum CMV PWM had to be set to an extreme value and the control was turned into a discontinuous PWM mode, while in the medium CMV PWM, the voltage diagram was always in the discontinuous PWM mode irrespective of the change of the dc sources. The effect of the local offset controller was demonstrated in medium CMV Discontinuous PWM mode as shown Figs.15 (m = 0.3) and 16 (m = 0.866). The local offset was added to gain a DPWM mode depending on the phase load current values. The inverter leg appeared without commutation during the intervals that its corresponding load current attained a maximum or medium absolute value in compared with the two remaining phases. As a result of that current controlled DPWM mode, the switching loss can be significantly reduced, similar to the case of input voltage balance [25].

The 2014 International Power Electronics Conference

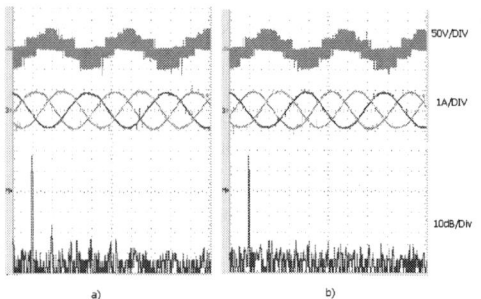

Fig. 11. Diagram of the phase load voltage, the load currents and the FFT of the load current (a) without compensation and (b) with compensation for m = 0.3.

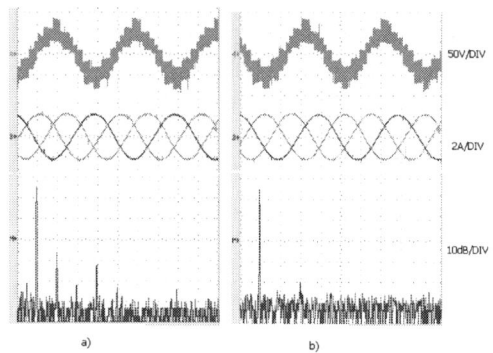

Fig. 12. Diagram of the phase load voltage, the load currents and the FFT of the load current (a) without compensation and (b) with compensation for m = 0.8.

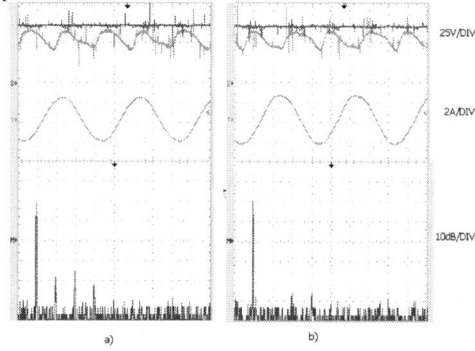

Fig. 13. Diagram of the dc sources and the output current for the algorithm (a) without compensation and (b) with compensation under a variable dc voltage: C_1 = 220 μF , C_2 = 4700 μF , R = 10 Ω , L = 60 mH, V_{c2} = 70 V, and V_{c1} = variable (48-64 V).

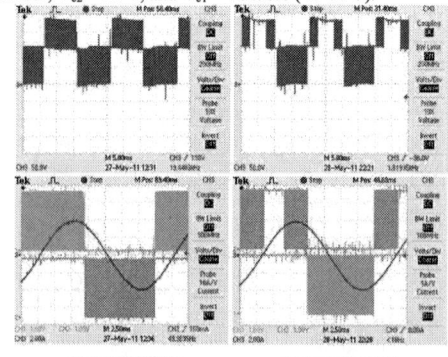

Medium CMV PWM   Minimum CMV PWM

Fig.14. Diagrams of the phase to neutral point voltage [50V/div, 5ms/div], and trigger pulse versus load current [2A/div, 2.5ms/div] of a phase for m = 0.866 in the medium CMV PWM and the minimum CMV PWM methods.

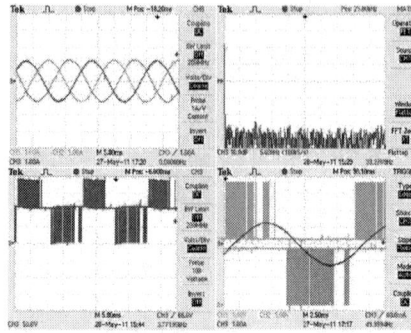

Fig.15: Diagrams of the load currents [1A/div, 5ms/div] and corresponding FFT [10dB/div, 5kHz/div], a phase to neutral point voltage [50V/div, 5ms/div] and trigger pulse versus load current [1A/div, 2.5ms/div] in the current based Discontinuous PWM for m = 0.3, and medium CMV.

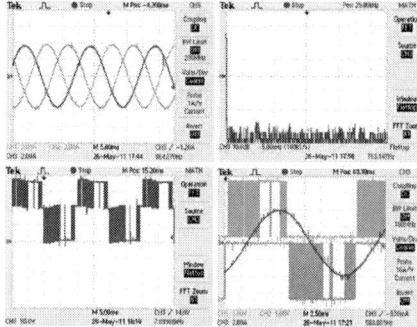

Fig.16: Diagrams of the load currents [2A/div, 5ms/div] and corresponding FFT [10dB/div, 5kHz/div], a phase to neutral point voltage [50V/div, 5ms/div] and trigger pulse versus load current [2A/div, 2.5ms/div] in the current based Discontinuous PWM for m=0.866 and medium CMV.

IV. Conclusion

In this paper, a novel offset-based PWM scheme for unbalanced dc sources was proposed. The offset voltage is an important variable of the controlling model for a multilevel inverter. The global offset allows the entire output voltage range to be obtained and reduces both the CMV and switching loss in the fundamental period. The local offset can be flexibly modified in order to establish different PWM modes. It can also reduce both switching loss in discontinuous PWM mode and the current ripple with a local offset. The global and local offset can be properly designed for attaining other behaviors related to dc-link balance and eliminating CMV. The decoupling offset PWM scheme is valid for each PWM method under any dc source condition. The N-to-two level inverter transformation and the characteristics of the virtual two-level inverter both play a major role in the offset-based two-level modulation scheme. The PWM performance of the multilevel inverter can be actually regulated in the decoupling offset based PWM scheme. The decoupling offset-based scheme can be applied to a higher number of phases. The experimental results for the feed-forward compensating output voltages and for optimizing the switching loss in a DPWM mode under a dc imbalance confirm the validity of the proposed PWM method.

978-1-4799-2706-7/14 $31.00 © 2014 IEEE

Acknowledgement

This research is funded by Vietnam National University HoChiMinh City (VNU-HCM) under grant number C2013-20-10

References

[1] G. Carrara, S.Gardella, M. Marchesoni, R. Salutari, and G.Sciutto, "A new multilevel PWM method- A theoretical analysis," IEEE Trans. Power Electronics, Vol.7, pp.497-505 1992

[2] J.Rodríguez, J.S.Lai, and F. Z. Peng,"Multilevel Inverters: A Survey of Topologies, Controls, and Applications", IEEE Transactions on Industrial Electronics, Vol. 49, No. 4, August 2002

[3] McGrath, B.P.; Holmes, D.G.; Lipo, T., "Optimized space vector switching sequences for multilevel inverters", IEEE Transactions on Power Electronics, Vol.18, No.6, Nov.2003, pp.1293-1301

[4] N.V.Nho, M.J.Youn,"Comprehensive study on Space Vector PWM and carrier based PWM correlation in multilevel invertors " , IEE Proceedings Electric Power Applications , Vol.153, No.1, Jan 2006, pp.149-158

[5] S.B-Monge, J.Bordonau, D. Boroyevich, and S.Somavilla, " The Nearest Three Virtual Space Vector PWM—A Modulation for the Comprehensive Neutral-Point Balancing in the Three-Level NPC Inverter", IEEE Power Electronics Letters, Vol. 2, No. 1, March 2004 , pp.11-15

[6] N. Celanovic, and D. Boroyevich, "A fast space vector modulation algorithm for multilevel three phase converters," *IEEE Trans on Industry Applications,* Vol.37, No.2, 2001, pp. 637-641.

[7] J. Rodriguez, P. Correa, and L. Moran, "A vector control technique for medium voltage multilevel inverters," *Applied Power Electronics Conference and Exposition(APEC),* Vol.1, 2001, pp.173-178.

[8] Sanmin Wei and Bin Wu Fahai Li and Congwei Liu, "A General Space Vector PWM Control Algorithm for Multilevel Inverters", Proceedings of the Applied Power Electronics Conference and Exposition, 2003., APEC '03.

[9] Ahmet M. Hava, Russel J. Kerkman, and Thomas A. Lipo ,"A High-Performance Generalized Discontinuous PWM Algorithm", IEEE Transactions On Industry Applications, Vol. 34, No. 5, September/October 1998, pp. 1059-1071

[10] Blasko, V.: 'A hybrid PWM strategy combining modified space vector and triangle comparison methods'. Proc. IEEE PESC Conf., 1996, pp.1872–1878

[11] C.B. Jacobina, A.M..N Lima , E.R.C. da Silva, R.N.C. Alves, and P.F. Seixas, -"Digital Scalar Pulse-Width Modulation: A Simple Approach to Introduce Non-Sinusoidal Modulating Waveforms", IEEE Transactions On Power Electronics, Vol. 16, No. 3, May 2001, pp. 351-359

[12] H. Van Der Broeck, H. Skudelny, and G. Stanke, "Analysis and realization of a pulse width modulator based on voltage space vectors," in IEEE-IAS Conf. Rec., 1986, pp. 244–251.

[13] Keliang Zhou and Danwei Wang , "Relationship Between Space-Vector Modulation and Three-Phase Carrier-Based PWM: A Comprehensive Analysis", IEEE Transactions On Industrial Electronics, Vol. 49, No. 1, February 2002, pp.186-196

[14] J. Pou,,D. Boroyevich, and R. Pindado, "Effects of Imbalances and Nonlinear Loads on the Voltage Balance of a Neutral-Point-Clamped Inverter", EEE Transactions on Power Electronics, Vol. 20, No. 1, January 2005, pp. 123-131

[15] Josep Pou, Rafael Pindado, and Dushan Boroyevich," Voltage-Balance Limits in Four-LevelDiode-Clamped Converters With Passive Front Ends", IEEE Transactions On Industrial Electronics, Vol. 52, No. 1, February 2005, pp.190-196.

[16] K.Celanovic, D.Boroyevich," A Comprehensive Study of Neutral-point voltage balancing problem in three-level voltage source PWM Inverters", Transactions on Power Electronics, Vol.15, No.2, 2000, pp. .242-249

[17] P.C. Loh, D.G. Holmes, Y. Fukuta, and T.A. Lipo , "Reduced Common-Mode Modulation Strategies for Cascaded Multilevel Inverters", IEEE Transactions On Industry Applications, Vo 39, No. 5, September/October 2003, pp. 1386 -1395

[18] N.V.Nho, H.H. Lee,"Generalized Carrier PWM Algorithms For Multilevel Inverters With Unbalanced DC Voltages", *Proceeding of the 37th IEEE Power Electronics Specialists Conference* PESC 18-22nd June 2006, Jeju , Korea.

[19] Kouro, S.; Lezana, P.; Angulo, M.; Rodriguez, J.; , "Multicarrier PWM With DC-Link Ripple Feedforward Compensation for Multilevel Inverters," Power Electronics, IEEE Transactions on , vol.23, no.1, pp.52-59, Jan. 2008

[20] N.V.Nho, H.H.Lee,"Carrier PWM Algorithm For Multi-leg Multilevel Inverters", EPE 2007 - 12th European Conference on Power Electronics and Applications 2 - 5 September 2007, Aalborg, Denmark

[21] O. Lopez, J. Alvarez, J.I. Leon, J. Doval-Gandoy, F.D. Freijedo, "Multilevel Multiphase space vector PWM algorithm", IEEE Transactions on Industrial Electronics, May 2008 Vol. 55, No. 5, pp.1933-1942

[22] J.I. Leon, S. Vazquez, J.A. Sanchez, R. Portillo, L.G. Franquelo, J.M. Carrasco, and E.Dominguez," Conventional Space-Vector Modulation Techniques Versus the Single-Phase Modulator for Multilevel Converters", IEEE Transactions on Industrial Electronics, July 2010, Vol. 57, No. 7, pp. 2473 – 2482

[23] J.I.Leon, Sergio Vazquez, Ramon Portillo, Leopoldo G. Franquelo, Juan M. Carrasco, Patrick W. Wheeler, and Alan J., "Three- Dimensional Feedforward Space Vector Modulation Applied to Multilevel Diode-Clamped Converters", Industrial Electronics, IEEE Transactions on, Jan. 2009, Vol. 56, No. 1, pp.101-109

[24] N.V. Nho, Q.T. Hai and Hong-Hee Lee, " Carrier based Single-state PWM Technique Of Minimized Vector Error In Multilevel Inverter ", Journal of Power Electronics, Vol.10, No.4, July, 2010

[25] N.V. Nho, N.X. Bac and H-H. Lee, "An Optimized Discontinuous PWM Method to Minimize Switching Loss for Multilevel Inverters", IEEE Transactions on Industrial Electronics, 2011

[26] T. Brückner, and D.G. Holmes, "Optimal Pulse-Width Modulation for Three-Level Inverters", IEEE Transactions On Power Electronics, Vol. 20, No. 1, January 2005, pp 82-89.

[27] K.A.Corzine, and J.R. Baker, "Multilevel Voltage-Source Duty-Cycle Modulation: Analysis and Implementation" IEEE Transactions On Industrial Electronics, Vol.. 49, No. 5, October 2002, pp.1009-1016

[28] F. Wang, "Sine-triangle versus space-vector modulation for three-level PWM voltage-source inverters," IEEE Transactions On Industry Applications, Vol. 38, No. 2, Mar./Apr. 2002, pp. 500–506

[29] B.P. McGrath, D.G. Holmes, and T. Meynard, "Reduced PWM Harmonic Distortion for Multilevel Inverters Operating Over a Wide Modulation Range", IEEE Transactions On Power Electronics, Vol. 21, No. 4, July 2006, pp. 941-949

[30] J. Pou, J. Zaragoza, S. Ceballos, M. Saeedifard, D. Boroyovich, "A Carrier based PWM strategy With Zero sequence Voltage Injection for a three-level Neutral Point Clamped Converter", IEEE Transactions on Power Electronics, available online 2010.

[31] W. Chenchen, L. Yongdong, "Analysis and Calculation of Zero sequence voltage considering Neutral-point potential Balancing in Three-level NPC Converters", IEEE Transactions On Industrial Electronics, Vol.57, No.7, July 2010, pp. 2262 - 2271

[32] N.Y.Dai, M.C. Wong, Y.H. Chen, and Y.D.Han, "A 3-D Generalized Direct PWM Algorithm for Multilevel Converters", IEEE Power Electronics Letters, Vol. 3, No..3, September 2005, pp.85-88

[33] N.V.Nho,H.H.Lee,"Analysis of Carrier PWM Method for Common Mode Elimination in Multilevel inverter", EPE 2007 - 12th European Conference on Power Electronics and Applications 2 - 5 September 2007, Aalborg , Denmark

[34] N.V.Nho and H.H.Lee,"Linear Overmodulation Control in Multiphase multilevel inverters For Unbalance DC voltages", The 7th IEEE Int'l Conference on Power Electronics and Drive Systems PEDS 2007, 27-30 November, Bangkok, Thailand

[35] J.Pou, D.Boroyevich, and R. Pindado,"New Feedforward Space-Vector PWM Method to Obtain Balanced AC Output Voltages in a Three-Level Neutral-Point-Clamped Converter", IEEE Transactions on Industrial Electronics, Vol. 49, No. 5, October 2002, pp.1026-1034

978-1-4799-2706-7/14 $31.00 © 2014 IEEE

The 2014 International Power Electronics Conference

η-ρ Pareto Optimization of 3-Phase 3-Level T-type AC–DC–AC Converter Comprising Si and SiC Hybrid Power Stage

Hirofumi Uemura*, Florian Krismer*, Yasuhiro Okuma[†] and Johann W. Kolar*

*Power Electronic Systems Laboratory, ETH Zurich, Physikstrasse 3, 8092 Zurich, Switzerland

Email: uemura@lem.ee.ethz.ch, www.pes.ee.ethz.ch

[†]Fuji Electric Co., Ltd, 1, Fuji-machi, Hino-city, Tokyo 191-8502, Japan

Abstract—**In this paper, the η-ρ (efficiency‐power density) Pareto front of a 20 kVA Uninterruptible Power Supply (UPS), comprising a Si and SiC hybrid 3-level T-type rectifier and inverter stage, and input/output side filters, is determind. A multi-objective optimization procedure is detailed, which employs an electrical converter model and coupled electro-thermal component models. Each component model determines the corresponding losses, volumes, and temperature rises. A detailed description of the inductor model is given which considers high frequency winding losses as well as core losses, and calculates the temperature rises inside the inductor, e.g. the hot-spot winding temperature. Based on the determined η-ρ Pareto front, the most suitable realization of the proposed UPS system is determined, which is based on the resulting sensitivities of the design parameters (switching frequency, current ripples, magnetic materials, etc.) on the performance of power converter system (efficiency and power density). As a result, a switching frequency of 16 kHz, maximum relative input and output side current ripples of 20 %, *amorphous* core materials for DM inductors, and *nanocrystalline* core materials for CM inductors are selected in order to achieve a converter efficiency of 96.6 % at a power density of 2.3 kVA/dm^3.**

I. INTRODUCTION

The optimization of power converter systems is a multi-domain procedure [1], which considers thermal effects and limitations besides electric and/or magnetic characteristics of active and passive power components. Early implementations of power converter optimizations are limited to the optimization with respect to a single performance index, e.g. power density ρ or efficiency η [2]–[4]. Simple-objective converter optimizations, however, often yield unsatisfying remaining converter characteristics, i.e. a converter optimized for high power density may generate high losses, due to increasing losses of high power density magnetic components. Thus, the multi-objective optimization of a PFC rectifier based on the η-ρ Pareto front is proposed in [5]. This Pareto front identifies the highest efficiency for a given power density (and vice versa) and can be used as basis for initial decisions concerning the converter design parameters. Details on the calculation of the η-ρ Pareto front and the extension to a third variable (e.g. cost) are presented in [6].

Examples of η-ρ Pareto optimizations of power converters include the realizations of an ultra-high efficiency DC–DC converter ($\eta = 99\%$ at $\rho = 2.3\,\text{kW}/\text{dm}^3$) [7] and an ultra-high efficiency PFC rectifier ($\eta = 99.2\%$ at $\rho = 1.1\,\text{kW}/\text{dm}^3$) [8]. Recent publications related to η-ρ Pareto optimizations further

TABLE I.	CONVERTER SPECIFICATIONS	
Input phase voltage	V_{in}	230 V RMS
Input frequency	f_{in}	50 Hz
Max. input phase voltage ripple	$\gamma_{\text{v,in}}$	7.6 %
EMI requirement		CISPR class A
Output phase voltage	V_{out}	230 V RMS
Output frequency	f_{out}	50 Hz
Output power	S_{out}	20 kVA
Max. output voltage ripple	$\gamma_{\text{v,out}}$	1.0 %
DC voltage	V_{dc}	720 V
Ambient temperature	T_{amb}	55 °C
Max. junction temperature	$\vartheta_{\text{j,max}}$	150 °C
Max. inductor winding temperature	$\vartheta_{\text{whs,max}}$	150 °C
Max. inductor core temperature	$\vartheta_{\text{c,max}}$	Depends on material.

confirm that this approach can be advantageously used for designing power converters, e.g. for the design of a 50 kW bidirectional resonant converter in [9], and the selection of a suitable five-level inverter topology in [10].

This paper determines the η-ρ Pareto front for a complete 20 kVA UPS system consisting of a rectifier and an inverter stage and corresponding input and output filters in order to enable the converter design with respect to power density at a given efficiency. **Section II** presents the converter topology and specifications, **Section III** outlines the optimization procedure and the employed loss and volume models, and **Section IV** details the inductor models (loss and thermal models), which are verified based on experimental results. **Section V** discusses the obtained η-ρ Pareto front and the selection of optimum design parameters. Based on the results, a power density of 2.3 kVA/dm^3 is calculated for a converter efficiency of 96.6%.

II. THREE-PHASE THREE-LEVEL T-TYPE CONVERTER

Fig. 1 depicts the rectifier and inverter stages of the considered UPS system. It consists of three-phase three-level T-type inverter and rectifier circuits, a mains-side EMI filter, and a two stage output filter. Both power stages employ hybrid Si and SiC power semiconductor configurations and reverse-blocking IGBTs in order to achieve minimum semiconductor losses, which is detailed in [11].

The input and output filters are both realized with two-stage differential mode (DM) filters and include damping networks in order to avoid instabilities due to resonances. The input filter, in addition, contains a fifth-order common mode (CM) filter with three CM inductors in order to fulfil the EMI requirements. **Tab. I** summarizes the converter requirements and specifications.

978-1-4799-2706-7/14 $31.00 © 2014 IEEE

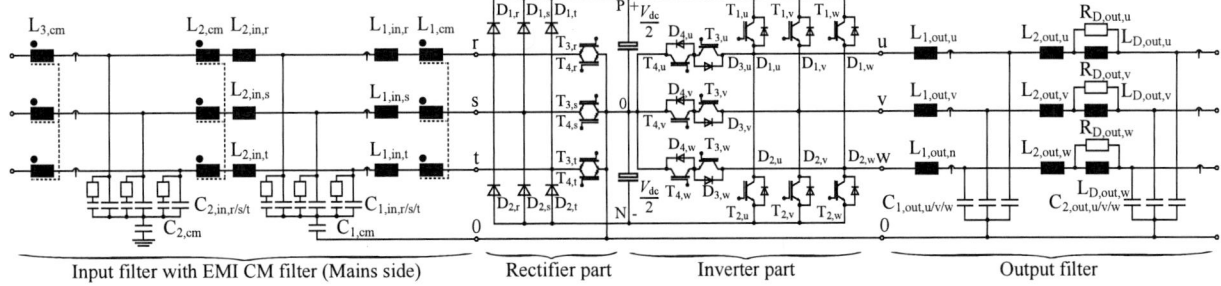

Fig. 1. Schematic of the 3-phase 3-level T-type AC–DC–AC converter comprising a Si and SiC hybrid rectifier and inverter stage.

TABLE II. DESIGN VARIABLES FOR OPTIMIZATION

Switching frequency	$f_{\mathrm{sw},i,n}$	8 kHz . . . 40 kHz
Max. input phase current ripple	$\gamma_{i,\mathrm{in},n}$	5 % . . . 40 %
Max. output current ripple	$\gamma_{i,\mathrm{out},n}$	5 % . . . 40 %

III. MULTI-OBJECTIVE OPTIMIZATION PROCEDURE

The multi-objective optimization approach presented in this paper considers the converter and component models listed below:

- Electrical converter model: voltages and currents, predictions of CM and DM noise levels and filter component values to fulfil the design specifications; cf. [12], [13].

- Inductor model: winding and core losses, thermal model, boxed volume.

- Capacitor model: only boxed volume, capacitor losses are neglected; cf. [2].

- Power semiconductors: conduction and switching losses, thermal model; the volume of the semiconductors is not considered; cf. [11].

- Heat sink: thermal model for forced air cooling, power demand of the fan, boxed volume (heat sink plus fan); cf. [14].

Each component model facilitates the calculation of the respective losses, the temperature rises, and the component volumes. The losses and the temperature rises of the components are coupled, e.g. the inductors' winding and core losses depend on the respective winding and core temperatures (according to the material properties) and vice versa (according to the properties of the inductor's thermal network which includes the heat flux to the ambient). For this reason, not only accurate loss models, but also accurate thermal models are required. According to the list given above, a large number of publications related to component models is readily available. This is particularly true for the semiconductors [11] and the heat sink [14]. However, no detailed discussion of coupled electro-thermal models for inductors is available. Therefore, in this paper a coupled electro-thermal inductor model is developed, implemented, and verified.

Fig. 2 depicts the flow chart of the optimization procedure. It conducts fully automatic converter designs and calculates

and stores the respective components' losses, temperatures, and volumes in a pool of design results. The models employed in this automatic design procedure are highly non-linear and, therefore, a high number of converter designs is carried out in order to find the global optimum. Thus, in a first step, the design space is created for the considered power converter, which, for the AC–DC–AC part, is based on Tab. II:[1]

- switching frequencies $f_{\mathrm{sw},n}$: from 8 kHz to 40 kHz with a step size of 1 kHz;

- rel. input and output inductor current ripples $\gamma_{i,\mathrm{in},n}$ and $\gamma_{i,\mathrm{out},n}$: from 5% to 40% with a step size of 5%; the ratio of the maximum peak-to-peak inductor current ripple, $\Delta I_{\mathrm{L},n}$, to the peak fundamental current, $\hat{I}$, defines the relative current ripple:

$$\gamma_{i,\mathrm{in},n} = \Delta I_{\mathrm{L}_1,\mathrm{in},n}/\hat{I}_{\mathrm{in}},$$
$$\gamma_{i,\mathrm{out},n} = \Delta I_{\mathrm{L}_1,\mathrm{out},n}/\hat{I}_{\mathrm{out}};$$

If required, the optimization procedure could also consider further design variables, e.g. the numbers of semiconductors operated in parallel for each switch or diode (here: constant; the actual numbers are given in [11]), the base plate temperature of the heat sink (here, a base plate temperature of $\vartheta = 100\,^\circ\mathrm{C}$ is assumed), and the DC link voltage (here: 720 V).

In a second step the design procedure selects the n-th set of design variables from the design space and conducts a single converter design. It calculates the semiconductor losses, $P_{\mathrm{SC},n}$, and the respective junction temperatures for the assumed heat sink base plate temperature. With known semiconductor losses the heat sink can be optimized according to [14], which yields the total boxed volume of the cooling system, $V_{\mathrm{CS},n}$, and the power demand of the fans, $P_{\mathrm{CS},n}$. The values of the filter components are calculated according to [13] in order to fulfil the design specifications and design variables. The total losses and the boxed volumes of the DM and CM inductors ($P_{\mathrm{LDM},n}$, $P_{\mathrm{LCM},n}$, $V_{\mathrm{LDM},n}$, $V_{\mathrm{LCM},n}$) are calculated with the inductor optimization procedure detailed in Section IV, which employs an inductor model that considers electro-thermal and magneto-thermal couplings. Furthermore, the total volume of all DM

---

[1]This paper is confined to the optimization of the AC–DC–AC part of the UPS, since the optimization of the DC–DC converter (not shown in Fig. 1) which is employed for interfacing the battery storage to the UPS DC link, can be implemented in an analogous manner.

The 2014 International Power Electronics Conference

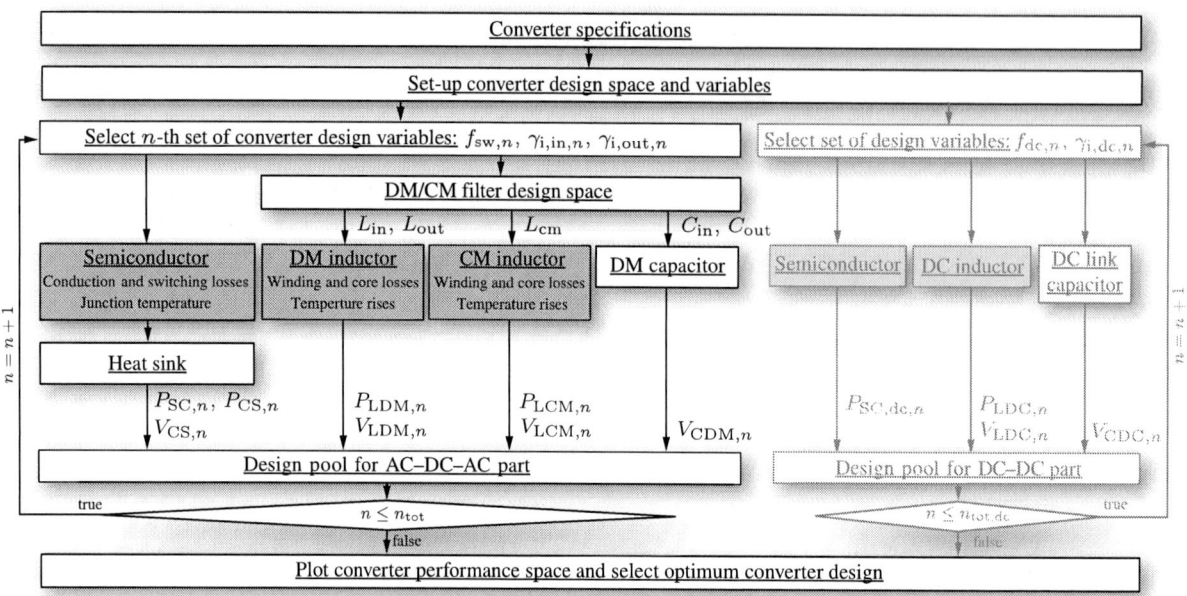

Fig. 2. Block diagram of the optimization procedure used to determine the η-ρ Pareto front. $P_{SC,n}$, $P_{CS,n}$, $P_{LDM,n}$ and $P_{LCM,n}$ denote the total losses of the semiconductors with the n-th set of design variables, the fans of the cooling system, the DM inductors and the CM inductors, respectively. $V_{CS,n}$, $V_{LDM,n}$, $V_{LCM,n}$, and $V_{CDM,n}$ denote the total boxed volume of the cooling system (heat sink with fans), the DM inductors, the CM inductors, and the DM capacitors, respectively. The optimization of the DC–DC converter part which interfaces the UPS battery storage to the DC link can be perform in analogous manner.

capacitors, $V_{CDM,n}$, is determined for each converter design (the total volume of all CM capacitors is found to be relatively small, i.e. negligible in a first step).

Based on all power densities and efficiencies calculated for all converter designs the η-ρ Pareto front can be generated which, finally, facilitates the identification of the optimum converter design, cf. Section V.

IV.   INDUCTOR MODEL

A. Procedure overview and loss model

The calculation of the inductor losses is implemented according to the flow chart depicted in **Fig. 3**. The proposed procedure requires the specifications listed below:

- The inductance value L.

- The fundamental peak current $\hat{I}$ and mains frequency f_{main}.

- The maximum peak-to-peak inductor current ripple ΔI_L and the switching frequency f_{sw}.

- The maximum allowed winding hot-spot and core temperatures $T_{whs,max}$ and $T_{c,max}$ and the ambient air temperature T_{amb}.[2]

The implemented automatic inductor design procedure is based on a high number of different inductor designs and the automatic evaluation of the resulting ρ-η-Pareto front similar to the converter optimization procedure discussed in Section III. In a

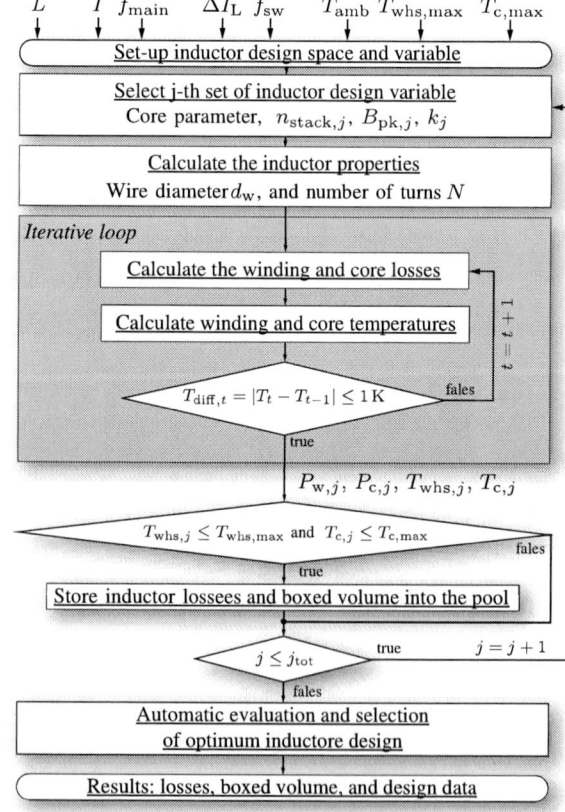

Fig. 3.   Flow chart of the inductor design procedure.

---

[2]T_{amb} is the absolute ambient air temperature, e.g. $T_{amb} = \vartheta_{amb} + 273\,\mathrm{K}$. The same representation manner is applied for the other temperatures.

first step, the inductor design space is created. The considered inductor design variables are:

- core material and geometry (selected from a predefined list of available magnetic cores),

- number of stacked cores n_{stack} ($n = 1$, 2, or 3),

- peak flux density B_{pk} (50% B_{sat} to 100% B_{sat} with a step size of 10%),

- filling factor k (50% to 100% with a step size of 10%).

The procedure starts with selecting the first set of design parameters from the inductor design space and calculates the inductor properties that can be directly calculated, such as the wire diameter, the number of turns, and the air gap length. In a next step, the procedure calculates the winding and core losses. The winding losses are separated into low (mains) frequency and high (switching) frequency losses. The low frequency losses are calculated with the DC resistance and the calculation of the high frequency losses is based on the mirroring method detailed in [12]. The inductor design considers two types of conductors, i.e. solid copper wire and high frequency litz wire. The core losses are calculated based on the improved Generalized Steinmetz Equations (iGSE) [15]. The winding and core losses, however, depend on the winding and core temperatures, which are determined using the thermal model discussed in the next Subsection IV-B. The winding and core losses and temperatures are calculated in an iterative loop according to Fig. 3. This loop is repeated until all differences between previously and currently calculated temperatures fall below 1 K. The losses and volumes resulting for both types of wires are evaluated with respect to the maximum allowable winding hot-spot and core temperatures (T_{whs} and T_{c}, respectively). The design results of every inductor design, which features hot-spot temperatures less than the specified maximum values is stored in the pool of inductor design results. Thereafter, the design procedure selects the next set of design parameters and conducts the next inductor design. After processing the complete inductor design space, the results available in the pool of inductor design results is analysed in order to determine the optimal inductor according to the following procedure:

1. The inductor with minimum boxed volume, $V_{\mathrm{L,min}} = \min(V_{\mathrm{L},j})$, is identified in the pool.

2. The inductor designs with $V_{\mathrm{L},j} > 1.2 \times V_{\mathrm{L,min}}$ are removed from the pool.

3. The inductor design with minimum losses in the remaining pool of inductor design results denotes the considered optimal choice.

The losses of the inductors contribute relatively little to the total losses, therefore, a major optimization criterion is maximum power density. Very compact inductor designs, however, yield comparably high losses. A considerable loss reduction is achieved if a somewhat higher boxed volume is allowable, which is taken into account by the procedure given above.

TABLE III.   MAGNETIC MATERIALS FOR OPTIMIZATION

Name	μ_{r}	B_{sat}	$\vartheta_{\mathrm{c,max}}$
EPCOS N87 (*Ferrite*)	2200	0.25 T	120 °C
Magnetics Kool-Mu (*Iron powder*)	14...125	1.0 T	100 °C
Metglas 2605SA1 (*Amorphous*)	45000	1.5 T	150 °C
Finemet FT-3M (*Nanocrystalline*)	70000	1.23 T	150 °C

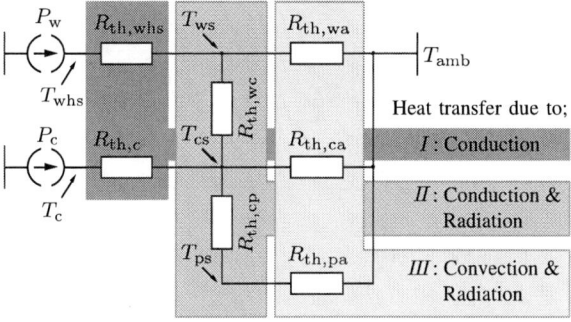

Fig. 4.   Extended thermal network of the inductor. P_{w} and P_{c} denote the total winding losses and the core losses, respectively. T_{whs} and T_{c} denote the hot-spot winding temperature and the temperature inside the core. T_{ws}, T_{cs} and T_{ps} are the surface temperatures of the winding, the core and the metal base plate, respectively. The base plate is considered only for the evaluation of the inductor model and it is neglected in the converter optimization in order to keep some safety margin for the inductor design.

For DM inductors the inductor design procedure considers E-core shapes and four different magnetic materials (*Nanocrystalline, amorphous, ferrite,* and *iron powder,* cf. **Tab. III**) with different magnetic properties, in particular different saturation flux densities (B_{sat}), operating core temperatures ($\vartheta_{\mathrm{c,max}}$), and core loss characteristics. For CM inductors, toroidal cores made of *nanocrystalline material* are considered, due to the high permeability and the low core losses required. N.B.: Inductors using tape wound cores made of *amorphous* or *nanocrystalline* magnetic materials are subject to increased core losses in presence of orthogonal components of the magnetic flux density, i.e. if the direction of the magnetic flux vector is not aligned to the hard magnetic direction of the magnetic material, which particularly happens close to the inductor's air gap [16]–[18]. Currently, there is no loss model known to the authors, which accurately takes this effect into account. In order to still consider the expected increase of the core losses due to this effect, the core losses are multiplied with a correction factor $k_{\mathrm{Pc}} = 2$, which has been determined based on the results of [16].

B. Thermal model

The thermal model is implemented on the basis of [19] and estimates the hot-spot temperature inside the winding (T_{whs}), the temperature inside the core (T_{c}), the surface temperature of the winding (T_{ws}), and the surface temperature of the core (T_{cs}). It takes three different heat transfer mechanisms into account, i.e. *conduction*, *convection* and *radiation*. The thermal resistance network of [19] is extended by three additional thermal resistances in order to determine T_{c} and to consider the heat transfer from the inductor to the metal base plate the inductor is mounted on. **Fig. 4** depicts the resulting thermal model.

978-1-4799-2706-7/14 $31.00 © 2014 IEEE

TABLE IV.	THERMAL PROPERTIES OF THE CORE MATERIALS	
Material	$\lambda_{c,xy}$	$\lambda_{c,z}$
Ferrite	4.18 W/(mK)	4.18 W/(mK)
Iron powder	8 W/(mK)	8 W/(mK)
Amorphous	7.65 W/(mK)	9 W/(mK)
Nanocrystalline	7.65 W/(mK)	9 W/(mK)

TABLE V.	TEST INDUCTOR DESIGN PARAMETERS
Core material	Metglas 265SA1 *Amorphous*
Core size	AMCC06R3 (2 sets × 2 stacked)
Air gap width	2 × 0.5 mm
Number of turns	20 turns (5 turns × 4 layers)
R_{dc} of wire	≈ 7.8 mΩ (at 25 °C)

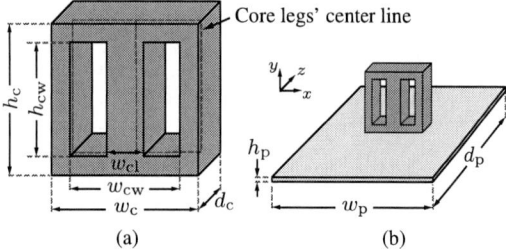

Fig. 5. Simplified core dimensions (a) and base plate dimensions (b).

The thermal resistances $R_{th,whs}$, $R_{th,wc}$, $R_{th,wa}$, and $R_{th,ca}$ are calculated according to [19].

The calculation of the temperature inside the core, T_c, assumes a constant temperature along the core legs' center lines (cf. **Fig. 5(a)**). Thus, the additional resistance $R_{th,c}$ represents the thermal resistance between the core legs' center lines and the core surface and is expressed as:

$$\frac{1}{R_{th,c}} = \frac{1}{R_{th,c,xy}} + \frac{1}{R_{th,c,z}}, \tag{1}$$

$$R_{th,c,xy} = \frac{l_c}{\lambda_{c,xy} A_{c,xy}}, \tag{2}$$

$$R_{th,c,z} = \frac{d_c/2}{\lambda_{c,z} A_{c,z}}, \tag{3}$$

where l_c is an effective thermal distance, $\lambda_{c,xy}$ and $\lambda_{c,z}$ are direction dependent effective thermal conductivities of the core (cf. Fig. 5(a) regarding x-, y-, and z-directions), $A_{c,xy}$ denotes the total surface area of the core in x- and y-directions (top, bottom, and both sides; $A_{c,xy}$ also includes the surface areas inside the core windows), and $A_{c,z}$ is the surface area of the core in z-direction (front and back). These parameters can be calculated using the core dimensions shown in **Fig. 5(a)**:

$$l_c = \frac{w_c - w_{cw}}{4}, \tag{4}$$

$$A_{c,xy} = 2\{(h_c + w_c) + (h_{cw} + w_{cw} - w_{cl})\} d_c, \tag{5}$$

$$A_{c,z} = 2\{h_c w_c - h_{cw}(w_{cw} - w_{cl})\}. \tag{6}$$

The thermal conductivities of the core materials are given in **Tab. IV**.

The second additional resistance, $R_{th,cp}$, represents the thermal resistance between the core and the base plate due to the combination of *conduction* and *radiation*. It is expressed as

$$\frac{1}{R_{th,cp}} = \frac{1}{R_{th,cond,cp}} + \frac{1}{R_{th,rad,cp}}, \tag{7}$$

$$R_{th,cond,cp} = \frac{l_{eg}}{\lambda_{air} A_{cp}}, \tag{8}$$

$$R_{th,rad,cp} = \frac{1}{h_{rad,cp} A_{cp}}$$

$$= \frac{T_{cs} - T_{ps}}{\epsilon_{cs} \sigma \left(T_{cs}^4 - T_{ps}^4\right) A_{cp}}. \tag{9}$$

An equivalent air gap length of $l_{eg} = 0.2$ mm between the core surface and the base plate top surface is found to yield reasonably accurate and reproducible results for all inductors measured with the test set-up detailed in Subsection IV-C.[3] Moreover, $\lambda_{air} = 0.03$ W/(mK) is the thermal conductivity of air at 80 °C, $A_{cp} = w_c d_c$ is the contact area of the core and the base plate, $\epsilon_c = 0.9$ is the assumed emissivity of the core surface, and σ is the Stefan-Boltzmann constant [= 5.67×10^{-8} W/(m²K⁴)] [19].

The last additional resistance, $R_{th,pa}$, represents the thermal resistance between the surface of the base plate and the ambient air and considers heat flux due to *convection* and *radiation*. The dimensions of the base plate are depicted in **Fig. 5(b)** and the expressions for calculating $R_{th,pa}$ are given in [19]. Further parameters are: the assumed emissivity of the base plate surface, $\epsilon_p = 0.04$ (aluminum, polished), the total open surface area of the base plate, $A_{pa} = 2\{w_p d_p + (w_p + d_p)\} - A_{cp}$, and the total distance passed by the air that cools the base plate, $L = d_p + h_p$.

C. Evaluation of Inductor Model

In order to evaluate the accuracy of the inductor model, i.e. the accuracies of the calculated losses and temperatures, the inductor's total losses and core and winding temperatures have been measured. The inductor test set-up and a test inductor are built as shown in **Fig. 6**. The inductor test set-up contains an aluminum base plate (dimensions: 220 mm × 300 mm × 5 mm), which supports a duct made of acrylic glass; the data logger used to process the measurement results is located below the base plate of the duct. The back side of the duct shown in Fig. 6 is terminated with a fan that can be used to control the air flow inside the duct. The test set-up is equipped with K-type thermocouples and air speed sensors to measure ϑ_{amb} and the air speeds at different locations inside the duct, respectively. The test inductor, shown in **Tab. V**, employs *Amorphous* core material and solid copper wire and is equipped with thermocouples, located inside the winding, at the surface of the winding, and at the surface of the core, in order to measure ϑ_{whs}, ϑ_{ws}, and ϑ_{cs} (cf. **Fig. 6(b)**).

The inductor losses are simultaneously measured with a calorimeter and a power analyzer for a low frequency (LF) sinusoidal current with a superimposed triangular high frequency (HF) AC current. The test conditions are specified in **Tab. VI**.

Fig. 7 compares the obtained measured and calculated inductor losses and temperatures. A good agreement between

---

[3]Experimental results confirm the existence of a non-zero value of $R_{th,cp}$, however, the exact value of the equivalent air gap length, l_{eg}, is difficult to determine and may vary depending on the inductor core and the set-up.

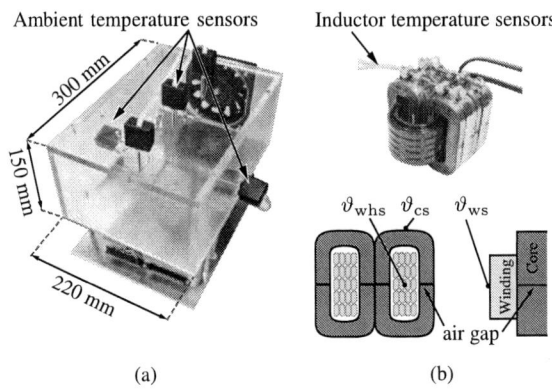

(a)   (b)

Fig. 6. (a) Inductor test set-up with ambient air temperature sensors and air-speed sensors; (b) test inductor.

TABLE VI. TEST CURRENT CONDITIONS

HF component	$\Delta I = 9.5\,\mathrm{A}$
(peak-to-peak)	$f_{\mathrm{sw}} = 16\,\mathrm{kHz}$
LF component	$\hat{I} = 0, 20, 48\,\mathrm{A}$
	$f_{\mathrm{main}} = 50\,\mathrm{Hz}$

measured and calculated results for all considered operating points (losses and temperatures) is obtained if the core loss correction factor is reduced to $k_{\mathrm{Pc}} = 1.65$ (instead of $k_{\mathrm{Pc}} = 2$, cf. Subsection IV-A):[4] The relative error for all considered total losses and winding hot-spot temperatures is less than 10 %.

V. OPTIMIZATION RESULTS AND DISCUSSION

Fig. 8(a) shows the power densities and efficiencies calculated for all considered converter design points. Besides different switching frequencies and current ripples (cf. Section II) also different core materials are considered for the DM inductors. According to these results, highest efficiencies and highest power densities are achieved with *nanocrystalline materials*, due to the high saturation flux densities and the low core losses of *nanocrystalline materials* even including increased core losses with air gap. The high saturation flux densities of *amorphous materials* allows high power densities

---

[4]The value of k_{Pc} is expected to depend on the air gap length and is, therefore, not expected to be constant. For this reason, the optimization results presented in Section V are still based on $k_{\mathrm{Pc}} = 2$ to avoid unrealistic inductor designs. Thus, further research related to a more accurate calculation of the core losses of tape-wound cores with air gaps is required.

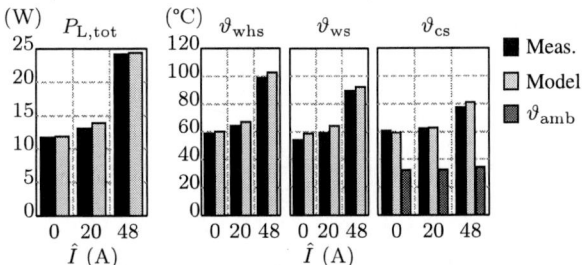

Fig. 7. Comparison of measured and calculated inductor losses and temperatures. Measurement results are shown by black bars (meas.) and calculation results are shown by light gray bars (model). Ambient temperatures are shown by dark gray bars ($\vartheta_{\mathrm{amb}} \approx 33\,°\mathrm{C}$ inside the closed box of the calorimeter).

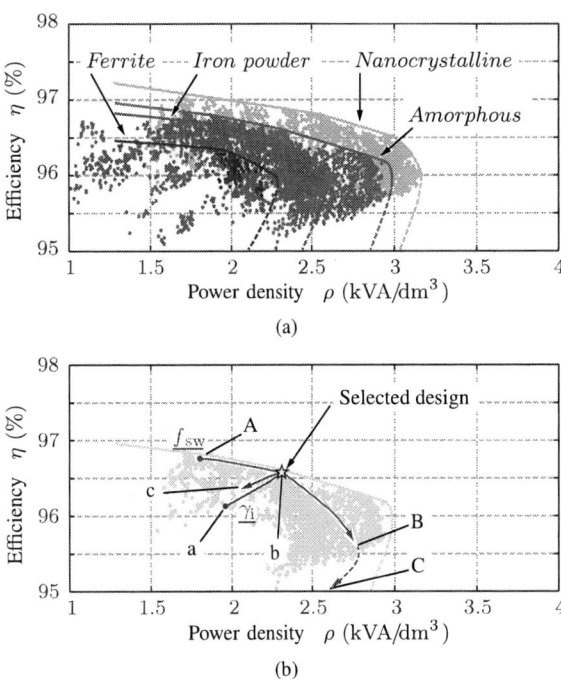

Fig. 8. η-ρ Pareto front of the proposed UPS system with four different magnetic materials (a) and Pareto front with *amorphous material* (b) used for everything the mapping of the design space into the system performance.

too, however, due to higher core losses, the achievable converter efficiencies are less than the efficiencies achievable for *nanocrystalline materials*. Also *iron powder materials* can be used to realize the inductors, however, the permeabilities of these materials are less than those of the other materials. In addition, the permeabilities drop with increasing flux densities, i.e. the inductances of inductors using *iron powder cores* show a non-linear dependency on the inductor current. As a consequence, the number of turns needs to be increased in order to maintain a given inductance value at an elevated current, which increases the winding losses. Due to the increased winding losses the inductor volume has to be increased in order to maintain the winding hot-spot temperature limit, which decreases the power density. Thus, for the given application, inductors using *iron powder cores* are found to be less suitable. For the given application also *ferrite cores* are found to give comparably low power density and efficiency values, which is due to the low saturation flux density of *ferrite materials* and the relatively low switching frequencies used by reason of the employed semiconductors. Due to the high cost of *nanocrystalline materials* (23 €/kg) [20] and the comparably small difference in efficiency and power density, the *amorphous material* (16 €/kg) is selected for realizing the inductors of the converter.

Fig. 8(b) shows, how the design space is projected into the performance space:

• **between points A and B on the led line in Fig. 8(b)**, the *increase of the switching frequency* at constant current ripple of 20 % causes the power density to increase and the efficiency to drop. If the switching frequency exceeds a certain

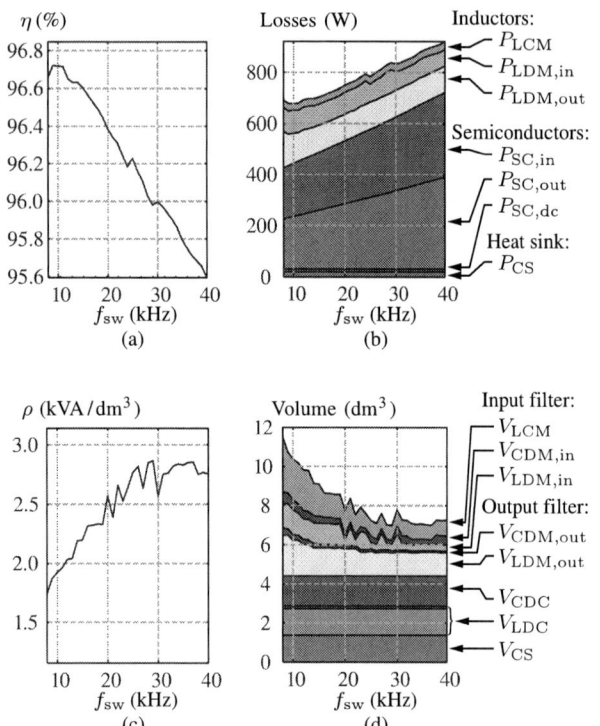

Fig. 9. Influence of the design parameter switching frequency f_{sw} on the efficiency (a), the loss distribution (b), the power density (c), and the volume distribution (d). In this figure, the current ripple is kept constant ($\gamma_i = 20\,\%$).

Fig. 10. Influence of the design parameter current ripple γ_i on the efficiency (a), the loss distribution (b), the power density (c), and the volume distribution (d). In this figure, the switching frequency is kept constant ($f_{sw} = 16\,$kHz).

value, however, the power density drops (cf. **B–C in Fig. 8(b)**), due to the additional converter volume needed to dissipate the heat (increased heat sink volume, additional volume of passive components due to thermal limitation).

• **between points a and b on the blue line in Fig. 8(b)** the *increase of the current ripple* at constant switching frequency of 16 kHz as the selected design causes the power density and the efficiency to increase, due to the reduction of the inductances, which helps to reduce the inductor winding losses and volumes; **between points b and c** increasing current ripple start to deteriorate the overall efficiency and power density due to increase in the inductor losses and the capacitor volumes. With higher current ripple at same switching frequency, the inductor core losses is increased. In addition, in order to dissipate the increased heat, the inductor volumes can not be reduced (thermal limit). Moreover, the volume of the EMI filter (CM inductor and DM capacitor) increases, since the required filter attenuation increases with increasing current ripple.

Efficiencies, loss distributions, power densities, and volume distributions calculated for different switching frequencies, constant relative input and output side current ripples of 20%, and inductors made of *amorphous cores* are shown in **Fig. 9**. According to Fig. 9(a), the efficiency drops with an increase in switching frequency, which is mainly due to the switching losses of the power semiconductors. The power density improves with an increase in switching frequency due to the volume reduction in the passive components. However, it shows a maximum for a switching frequency of approximately

29 kHz, and no further improvement is seen with increasing switching frequency, due to an increase in HF core losses of the DM inductors and the associated overall component surfaces needed to maintain the thermal limits. A switching frequency of 16 kHz is selected in order to achieve a high efficiency and to avoid audible noise. At $f_{sw} = 16\,$kHz, the contribution of the total semiconductor losses on the total converter losses is 65% and the contribution of the total passive components' losses is 27%. The contribution of the passive components (DM inductors, capacitors, and CM inductors) on the total converter volume is 49%.

Fig. 10 depicts the results calculated for different current ripples and constant switching frequency, $f_{sw} = 16\,$kHz. In this case the maxima of efficiency and power density occur for similar current ripples (15% for maximum efficiency and 20% for maximum power density) on the input and output sides. According to Fig. 10(b), the DM inductor losses decrease significantly between 5...15%, since a reduction of the inductance value significantly reduces the LF winding losses in the DM inductors. This helps to reduce the volume of the DM inductor and improves the overall power density (cf. Fig. 10(c) and (d)). However, the DM inductor losses slightly increase for current ripples between 15%...40%, due to increasing HF winding losses and core losses. Since the DM inductor losses are not decreasing, the DM inductor volume is also not decreasing in that range of the current ripple. However, the DM capacitor volume increases, since the voltage ripple is kept constant and the current ripple increases. Therefore, the power density drops if the relative current ripples exceed 20%. In order to achieve a high power density, a current ripple of 20% is selected.

VI. CONCLUSION

The optimization procedure presented in this paper employs a multi-domain approach which is based on component models that consider electric, magnetic, and thermal aspects. The paper especially details a coupled electro-thermal and magneto-thermal model of the inductors that features accurate calculation of losses, volume, and hot-spot temperature inside the winding. This presented results reveal that the global converter optimization based on the η-ρ Pareto front enables a high converter efficiency of 96.6% at a maximum power density of $2.3\,\mathrm{kVA/dm^3}$ for the investigated $20\,\mathrm{kVA}$ UPS system including the EMI input filter and the output filter. The relationship between the η-ρ performance space and the f_{sw}-γ_{i} design space is discussed and suitable design parameters of the converter system are determined by projecting selected performance points back into the design space. As a result, a switching frequency of $16\,\mathrm{kHz}$, a current ripple of $20\,\%$, and dedicated inductor realizations (DM inductors: *amorphous cores* and solid copper wires; CM inductors: *nanocrystalline cores* and solid copper wires) are selected.

Future research may focus on further improvements related to the inductor model, e.g. modelling of the increased core losses in the tape wound core in presence of an air gap, and including forced air cooling in the thermal model of the inductor. In summary, the multi-objective optimization approach gives a clear picture of the achievable overall converter performances (efficiency, power density, etc.) as coordinates in the performance space including all power components. It, therefore, provides a very good basis for decision making and helps to shorten the development time for an optimized power electronics system.

REFERENCES

[1] H. Ohashi, "Research Activities of the Power Electronics Research Centre with Special Focus on Wide Band Gap Materials," in *Proc. of the 4th International Conference on Integrated Power Systems (CIPS)*, pp. 1–4, 2006.

[2] J. W. Kolar, U. Drofenik, J. Biela, M. L. Heldwein, H. Ertl, T. Friedli, and S. D. Round, "PWM Converter Power Density Barriers," in *Proc. of the Power Conversion Conference (PCC)*, pp. 9–29, 2007.

[3] U. Badstuebner, J. Biela, and J. W. Kolar, "Power Density and Efficiency Optimization of Resonant and Phase-Shift Telecom DC–DC Converters," in *Proc. of the 23rd Annu. IEEE Applied Power Electronics Conference and Exposition (APEC)*, pp. 311–317, 2008.

[4] U. Badstuebner, J. Biela, B. Faessler, D. Hoesli, and J. W. Kolar, "An Optimized 5 kW, 147 W/in³ Telecom Phase-Shift DC–DC Converter with Magnetically Integrated Current Doubler," in *Proc. of the 24th Annu. IEEE Applied Power Electronics Conference and Exposition (APEC)*, pp. 21–27, 2009.

[5] J. Biela and J. W. Kolar, "Pareto Optimal Design and Performance Mapping of Telecom Rectifier Module Concepts," in *Proc. of the Power Conversion and Intelligent Motion (PCIM)*, 2010.

[6] J. W. Kolar, J. Biela, S. Waffler, T. Friedli, and U. Badstuebner, "Performance Trends and Limitations of Power Electronic Systems," in *Proc. of the 6th International Conference on Integrated Power Electronics Systems (CIPS)*, 2010.

[7] U. Badstuebner, J. Biela, and J. W. Kolar, "An Optimized, 99 % Efficient, 5 kW, Phase-Shift PWM DC–DC Converter for Data Centers and Telecom Applications," in *Proc. of the International Power Electronics Conference (IPEC)*, pp. 626–634, 2010.

[8] U. Badstuebner, J. Miniboeck, and J. W. Kolar, "Experimental Verification of the Efficiency / Power-Density (η-ρ) Pareto Front of Single-Phase Double-Boost and TCM PFC Rectifier Systems," in *Proc. of the 28th Applied Power Electronics Conference and Exposition (APEC)*, 2013.

[9] J. Huber, G. Ortiz, F. Krismer, N. Widmer, and J. W. Kolar, "η-ρ Pareto Optimization of Bidirectional Half-Cycle Discontinuous-Conduction-Mode Series-Resonant DC/DC Converter with Fixed Voltage Tranfer Ratio," in *Proc. of the 28th Applied Power Electronics Conference and Exposition (APEC)*, 2013.

[10] Y. Kashihara and J. Itoh, "Parformance Evaluation among Four Types of Five-level Topologies using Pareto Front Curves," in *Proc. of the IEEE Energy Conversion Congress and Exposition (ECCE USA)*, 2013.

[11] H. Uemura, F. Krismer, and J. W. Kolar, "Comparative Evaluation of T-Type Topologies Comprising Standard and Reverse-Blocking IGBTs," in *Proc. of the IEEE Energy Conversion Congress and Exposition (ECCE USA)*, 2013.

[12] J. Mühlethaler, H. Uemura, and J. W. Kolar, "Optimal Design of EMI Filters for Single-Phase Boost PFC Circuits," in *Proc. of the 38th Annual Conference of the IEEE Industrial Electronics Society (IECON)*, 2012.

[13] D. O. Boillat, T. Friedli, J. Mühlethaler, J. W. Kolar, and W. Hribernik, "Analysis of the Design Space of Single-Stage and Two-Stage LC Output Filters of Switched-Mode AC Power Sources," in *Proc. of the IEEE Power and Energy Conference at Illinois (PECI)*, 2012.

[14] U. Drofenik, A. Stupar, and J. W. Kolar, "Analysis of Theoretical Limits of Forced-Air Cooling Using Advanced Composite Materials with High Thermal Conductivities," *IEEE Trans. on Computers, Packaging, and Manufacturing Technology*, vol. 1, pp. 528–535, 2011.

[15] K. Venkatachalam, C. R. Sullivan, T. Abdallah, and H. Tacca, "Accurate Prediction of Ferrite Core Loss with Nonsinusoidal Waveforms using only Steinmetz Parameters," in *Proc. of the IEEE Workshop Computers in Power Electronics*, pp. 36–41, 2002.

[16] H. Fukunaga, T. Eguchi, Y. Ohta, and H. Kakehashi, "Core Loss in Amorphous Cut Cores with Air Gaps," *IEEE Trans. on Magnetics*, vol. 25, pp. 2694–2698, 1989.

[17] B. Cougo, A. Tüysüz, J. Mühlethaler, and J. W. Kolar, "Increase of Tape Wound Core Losses Due to Interlamination Short Circuits and Orthogonal Flux Components," in *Proc. of the 37th Annu. Conference of the IEEE Industrial Electronics Society (IECON)*, 2011.

[18] C. Marxgut, J. Mühlethaler, F. Krismer, and J. W. Kolar, "Multiobjective Optimization of Ultraflat Magnetic Components with PCB-Integrated Core," *IEEE Trans. on Power Electronics*, vol. 28, pp. 3591–3602, 2013.

[19] A. van den Bossche and V. C. Valchev, *Inductors and transformers for power electronics*. Taylor & Francis, 2005.

[20] R. Burkart and J. W. Kolar, "Component Cost Models for Multi-Objective Optimizations of Switched-Mode Power Converters," in *Proc. of the IEEE Energy Conversion Congress and Exposition (ECCE USA)*, 2013.

978-1-4799-2706-7/14 $31.00 © 2014 IEEE

Practical investigation of the gate bias effect on the reverse recovery behavior of the body diode in power MOSFETs

Kristian Lindberg-Poulsen, Lars Press Petersen, Ziwei Ouyang, Michael A. E. Andersen
Department of Electrical Engineering
Technical University of Denmark
Email: krili@elektro.dtu.dk, lpet@elektro.dtu.dk, zo@elektro.dtu.dk, maea@elektro.dtu.dk

Abstract—**This work considers an alternative method of reducing the body diode reverse recovery by taking advantage of the MOSFET body effect, and applying a bias voltage to the gate before reverse recovery. A test method is presented, allowing the accurate measurement of voltage and current waveforms during reverse recovery at high di/dt. Different bias voltages and dead times are combined, giving a loss map which makes it possible to evaluate the practical efficacy of gate bias on reducing the MOSFET body diode reverse recovery, while comparing it to the well known methods of dead time optimization. A selection of 60V devices for synchronous rectification are compared for their suitability for gate bias, while a selection of 600V devices are compared for the efficacy of gate bias for the zero voltage transition converter application. The results show that many of the tested devices benefit from greatly reduced reverse recovery after the application of gate bias.**

Keywords—*body diode, body effect, MOSFET, reverse recovery*

I. INTRODUCTION

MOSFET body diode reverse recovery is a familiar issue associated with synchronous rectification. In order to avoid shoot-through, it is necessary to have a small dead-time where both MOSFETs in a switching pair are off, and the inductor current freewheels through the synchronous MOSFET body diode. The body diode turn-on at the beginning of the dead-time is associated with an accumulation of charge carriers. When the main switch is turned on at the end of the dead-time, this charge must be removed before the drain-source voltage of the synchronous switch can rise, which results in the characteristic reverse recovery peak and losses in the main switch.

If the dead time is shorter than the turn-on time of the body diode, less charge will be stored in the intrinsic region. Consequently, a lower charge will have to be removed during reverse recovery, resulting in lower reverse recovery current peak, and lower losses. [1] This fact motivates the designer to choose as low a dead-time as possible, even if the diode conduction loss during the dead time is negligible for converter efficiency. However, component tolerance and temperature dependent delays, for example gate driver delay and MOSFET V_{gs} threshold variation, necessitate a safety margin in the dead time in order to guarantee the avoidance of shoot-through.

This has led to the development of several active dead time optimization schemes, where a control loop is used to continually adjust the dead time in order to obtain the lowest reverse recovery peak while avoiding shoot-through, thus maximizing converter efficiency at any operating point. [2]–[4]

A second application is that of zero voltage transition converters, where current flows in the body diode after each resonant transition. Normally the ZVT converters do not experience hard reverse recovery, but multiple failure modes have been reported where the hard reverse recovery can take place. [5], [6] For this reason, the market for 600V to 900V devices, such as the Infineon CoolMOS range, include "fast body diode" options, which are specifically recommended for ZVT converters, such as LLC topology and phase shifted full bridge.

Whereas the previous work has focused on simulation and modelling, the aim of this work is to experimentally verify the gate bias effect, as well as compare its efficacy to that of the well known methods of dead time optimization.

II. BODY EFFECT AND GATE BIAS THEORY

The body effect causes a change in the threshold voltage of a MOSFET when there is a voltage difference between the source and the body, according to (1).

$$V_{th} = V_{th,0} + \gamma \left(\sqrt{V_{SB} + |2\phi_p|} - \sqrt{|2\phi_p|} \right) \quad (1)$$

where

$$\gamma = \frac{\sqrt{2qN_A\epsilon_s}}{C_{ox}} \quad (2)$$

$V_{th,0}$ is the threshold voltage at zero V_{SB}, ϕ_p is the surface potential and γ is the body effect parameter. N_A is the effective channel doping, ϵ_s is the silicon permittivity, q is the charge of an electron and C_{ox} is the oxide capacitance. [7] The effect is well known by integrated circuit designers, but is ignored in power MOSFETs due to the fact that body and source are shorted by the source metallization. However, when a power MOSFET is conducting from source to drain, for example when used for synchronous rectification, the drain terminal is the source of charge carriers flowing through the device. The mechanical drain terminal thus behaves electrically as the source, and the channel resistance depends on the voltage drop between the gate and mechanical drain, rather than the mechanical source terminal.

Figure 1 shows a simple diagram of a MOSFET conducting current in the reverse direction, where the electrical gate-source voltage is denoted by $v_{gs}*$. If the gate voltage is clamped to ground, $v_{gs}* = V_f$, and if a bias is inserted, then $v_{gs}* = V_f + V_{bias}$. Additionally, there is now a voltage difference between the body and the electrical source equal to the body diode forward voltage drop, such that $V_{SB} = -V_f$. According to (1), this will cause a decrease of the threshold voltage. If the body effect lowers the threshold voltage sufficiently, current will partially commutate from the body diode to the channel. The lowered amount of current flowing in the body diode results in a lower charge accumulation, resulting in reduced reverse recovery effect. [8] By applying a small voltage bias to the MOSFET gate, the body diode reverse recovery can be reduced further, at the cost of increased shoot-through due to $C_{gd}dv/dt$ current. [8]

III. MEASUREMENT SETUP

The device parameters γ and ϕ_F, which influence the body effect, are not commonly available for commercial power MOSFETs, hence it is necessary to experimentally evaluate the efficacy of gate bias for reducing reverse recovery.

A double pulse tester, as shown in figure 2, can be used to measure the reverse recovery behavior. The operation of the circuit is indicated by the timing diagram on figure 3. The reverse recovery behavior can then be compared for any combination of dead time (DT) and gate bias V_{bias}.

In order to provide the required gate voltage bias, the bias circuit shown in figure 4 is used. The figure shows the device under test, S_{SR}, including parasitic capacitances, C_{gd}, C_{gs}, C_{ds}, the common source inductance L_{cs} and the internal gate resistance R_{Gint}. After S_{HS} is turned on at t_4, and the reverse recovery charge has

been removed, the drain source voltage v_{dsSR} increases quickly, causing a current to flow through C_{gd}. The total charge flowing through C_{gd} is denoted Q_{gd}, and this charge will increase v_{gsSR}, potentially causing shoot through if the voltage reaches threshold. When adding a gate bias, less additional charge will be required to charge v_{gsSR} to the threshold voltage, so it is important to have a low impedance local turn off capable of quickly extracting charge through R_{Gint}. To maximize the voltage drop over R_{Gint} and thus the current, a small MOSFET is used rather than a BJT. To control this MOSFET, a complementary output gate driver is used. The non-inverting output turns on S_{SR}, while the inverting output turns on the local turnoff MOSFET, which then turns off S_{SR} by discharging v_{gsSR} to the bias voltage set by the external voltage source, V_{bias}.
The completed test circuits are shown in figure 5.

A. Current measurement

At di/dt over 1A/ns, high care must be taken to obtain an accurate current measurement. The sense resistor and return path form a current loop, giving rise to a an inductance. In order to capture only the resistive voltage drop, the measurement loop must include the sense resistor, without enclosing any of the changing flux induced by the current loop. A coaxial shunt resistor can be used to obtain very low inductive coupling to the measurement loop, but this will insert a stray inductance in the current loop of several nH [9], [10], reducing the maximum switching speed and increasing circuit oscillations. An alternative using a wide row of parallel mounted SMD resistors, forming a so-called "circuit integrated coaxial shunt", has been reported [11]. The key difference to a normal surface mounted sense resistor is that the measurement loop is placed on top of the PCB rather than in the layers, thus reducing the coupling. The bandwidth achieved with this type of current sensor has been shown to exceed that of the 500Mhz oscilloscope probes [12], so it is deemed sufficient for the required measurements.

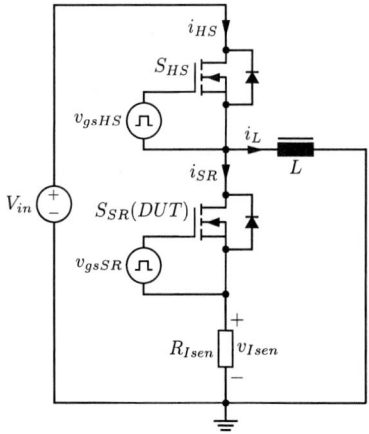

Fig. 2: Double pulse tester circuit including shunt resistor, R_{Isen}

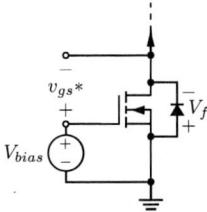

Fig. 1: Power MOSFET conducting current through the body diode. The charge carriers are entering the drain terminal, hence the drain terminal is behaving electrically as the source of the MOSFET. The electrical gate-source voltage is denoted by $v_{gs}*$. A gate bias of V_{bias} is applied.

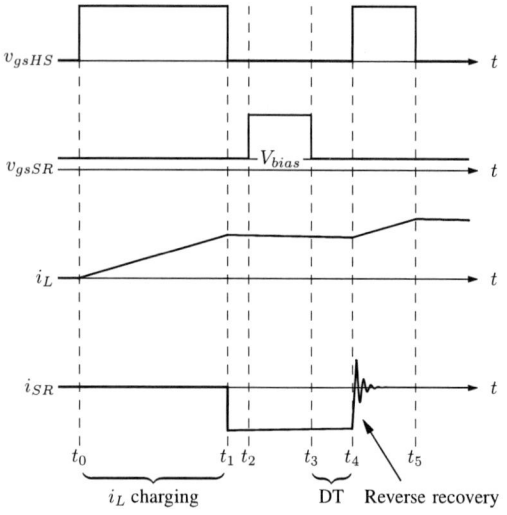

Fig. 3: Timing diagram of the double pulse tester.

B. Voltage measurement

The drain to source and gate to source voltages of both switches are measured differentially, by measuring drain, gate and source voltages with ground referenced probes,

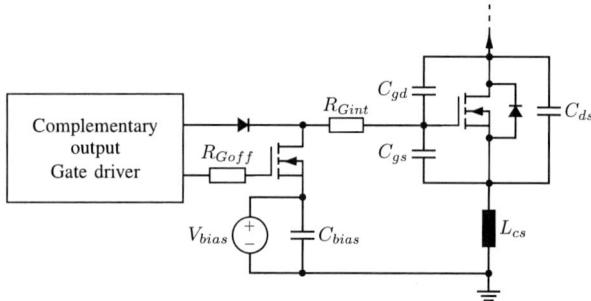

Fig. 4: MOSFET gate driver, local turn off and bias circuit, including parasitic capacitances, C_{gd}, C_{gs}, C_{ds}, the common source inductance L_{cs} and the internal gate resistance R_{Gint}.

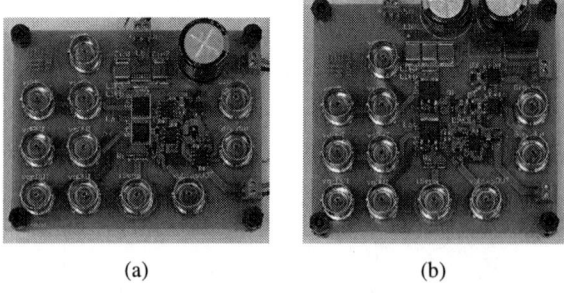

Fig. 5: Double pulse test circuit for measuring body diode reverse recovery. a) shows the test circuit for 60V devices in Power SO8 package; b) shows the test circuit for 600V devices in DPAK package.

and subtracting to obtain the differential voltages. The high di/dt of the main current loop may induce an error voltage in the voltage measurements. This is avoided by ensuring that the measurement loops are orthogonal to the current loop and by having a separate ground plane for the measurement loops. In order to achieve this, a 4 layer PCB is necessary. Coaxial connectors were used to ensure high low inductive coupling and high repeatability. For the 60V devices, the Power SO8 package was selected, due to low parasitics and wide adoption. For the 600V devices, the DPAK package was used. Lower parasitic packages are available, such as the ThinPAK package, but not from all the relevant manufacturers.

C. 60V devices

Table I shows an overview of the selected 60V devices with datasheet "typical" values at room temperature, with R_{dsON} at 10V V_{gs}. These devices were chosen for their similar on resistance, and their availability in the Power SO8 package. For the given level of on resistance, the best devices available were chosen from each manufacturer, corresponding to the devices with the lowest figure of merit ($Q_{gtot} R_{dsON}$).

The table shows the on resistance, total gate charge Q_{gtot} and gate drain "miller" charge Q_{gd}. The Q_{gd} value is especially interesting for the application of gate bias, as a higher Q_{gd} generally corresponds to a higher suscep-tibility to gate bounce. The datasheet values for reverse recovery charge is not included, as this value is measured at different diode forward current for each device, and is thus not comparable. Additionally, the reverse recovery datasheet values typically specify a di/dt of 100A/μs, which is far from the practical circuit implementation, where the switching typically occurs as fast as possible, with di/dt over 1A/ns. As will be shown in the measure-ment results, a switching speed of 4A/ns was achieved.

TABLE I: 60V devices selected for measurement

Device	R_{dsON}[mΩ]	Q_{gtot}[nC]	Q_{gd}[nC]
Infineon BSC028N06NS	2.5	37	7
Infineon BSC031N06NS3G	2.5	98	8
IRF IRFH5006TRPBF	3.5	69	20
NXP PSMN5R5-60YS	3.6	56	11.2
Fairchild FDMS86540	3.4	65	12

D. 600V devices

Table II shows an overview of the selected 600V de-vices with datasheet "typical" values at room temperature, with R_{dsON} at 15V V_{gs}. These devices were chosen for their similar on resistance, and their availability in the DPAK package. For the given level of on resistance, the best devices available were chosen from each manufac-turer, corresponding to the devices with the lowest figure of merit ($Q_{gtot} R_{dsON}$). Additionally, "fast body diode" devices from Infineon and ST were chosen; the Infineon IPD65R420CFD and the ST STD13NM60ND.

TABLE II: 600V devices selected for measurement

Device	$R_{dsON}[m\Omega]$	$Q_{gtot}[nC]$	$Q_{gd}[nC]$
Infineon IPD60R385CP	350	17	6
Infineon IPD65R225C7	199	20	6
Infineon IPD65R420CFD	378	31.5	18.6
ST STD13N60M2	350	17	9
ST STD18N65M5	198	31	14
ST STD13NM60ND	320	24.5	17
Fairchild FCD380N60E	320	34	13

IV. EXPERIMENTAL RESULTS

The switching waveforms during reverse recovery were measured at a number of combinations of V_{bias} and dead time. Figure 6 shows an example of the captured waveforms for both switches. Figure 6a shows the drain to source voltages and Figure 6b shows the drain current. The data points for voltages and currents are multiplied to give the instantaneous power, which is then integrated to give the cumulative energy, shown in figure 6c, where it is noted that E_{SR} includes diode conduction energy loss, as well as the energy that is stored in C_{oss} at turn off, while E_{HS} does not include the C_{oss} energy, due to the fact that this is dissipated in the channel when the high side MOSFET turns on.

In order to compare the devices fairly, the same high side switch is used for all cases of 60V and 600V devices, respectively. This ensures that the di/dt is the same for all measurements.

For the 60V devices, the Infineon BSC031N06NS3G is used as high side switch. The di/dt used is 4A/ns, the switching voltage is 40V, and the forward current before reverse recovery is 20A.

For the 600V devices, the Infineon IPD60R385CP is used as high side switch. The di/dt used is 750A/μs, the switching voltage is 300V, and the forward current before reverse recovery is 1A.

Figure 6d shows the measured gate to source voltages, which are used to log the dead time and V_{bias} values for each measurement, such that the effect of these can be compared. The waveforms are transferred to Matlab, where the dead time is automatically measured as the time between the crossing of 4V for the two v_{gs} measurements.

In this example, the deadtime is measured to be 190ns. As can be seen in figure 6c, there is a constant increase in E_{SR} during the deadtime. This is caused by the fact that the current is flowing through the body diode, with a forward voltage drop of 0.6V, and partially through the channel as caused by the body effect and gate bias. The loss during deadtime can be approximated by (3):

$$E_{DT} = \int_0^{T_{DT}} v_{dsSR} \, i_{dsSR} \, dt \leftrightarrow$$
$$E_{DT} = V_{dsSR} I_{dsSR} T_{DT} = (-0.6V) \cdot (-20A) \cdot 190ns \leftrightarrow$$
$$E_{DT} = 2.3\mu J$$

$$(3)$$

This corresponds well with the value of E_{SR} at the end of the deadtime, as seen in figure 6c.

Figure 7 shows the reverse recovery at three interesting conditions:

- At zero gate bias and maximum (190ns) deadtime.

- At optimal gate bias and maximum (190ns) deadtime.

- At zero gate bias and optimal deadtime.

The reverse recovery is most severe at high deadtime and zero gate bias, where the voltage overshoot exceeds the voltage rating of the device, potentially causing avalanche. The results for optimal bias marked by purple, and optimal deadtime marked by green, both decrease the reverse recovery significantly. In fact, the result from having optimal deadtime seems to be only slightly better than when using gate bias. However, the gate bias method suffers from the fact that there is still a long period of body diode and partial channel conduction, giving a higher total loss than the case of optimal deadtime. This can be seen in figure 8, showing that the total loss is only

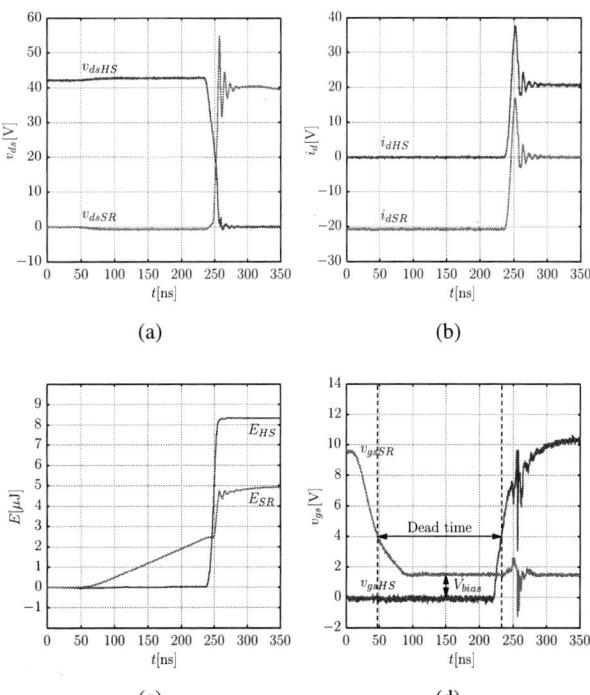

Fig. 6: Example measurements of reverse recovery of Infineon BSC028N06NS. The v_{ds} measurements in a) are multiplied with the current measurements in b) giving instantaneous power, which is integrated over time to produce the total energy loss shown in c). The gate voltage measurements in plot d) are used to automatically log the dead time and gate voltage bias for later analysis.

978-1-4799-2706-7/14 $31.00 © 2014 IEEE

slightly reduced by using gate bias, whereas the total loss is minimized by using optimal dead time.

A. Reverse recovery measurement results for 60V devices

In order to further compare the efficacy of gate bias versus deadtime optimization - as well as their combination - a sweep of measurements were made of

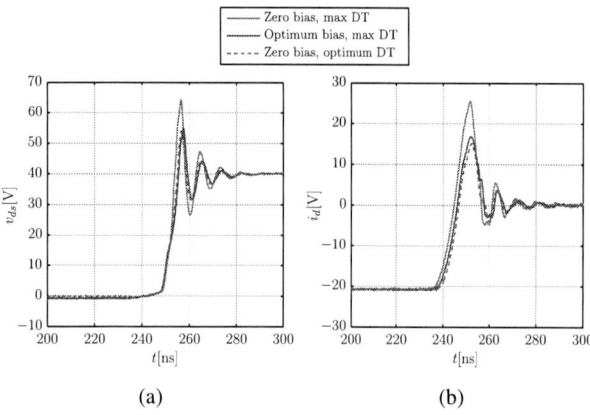

(a)   (b)

Fig. 7: Reverse recovery of Infineon BSC028N06NS, comparing measurements at zerobias, max DT (red); optimum bias, max DT (purple); zero bias, optimum DT (green). a) shows v_{dsSR} and b) shows i_{dSR}.

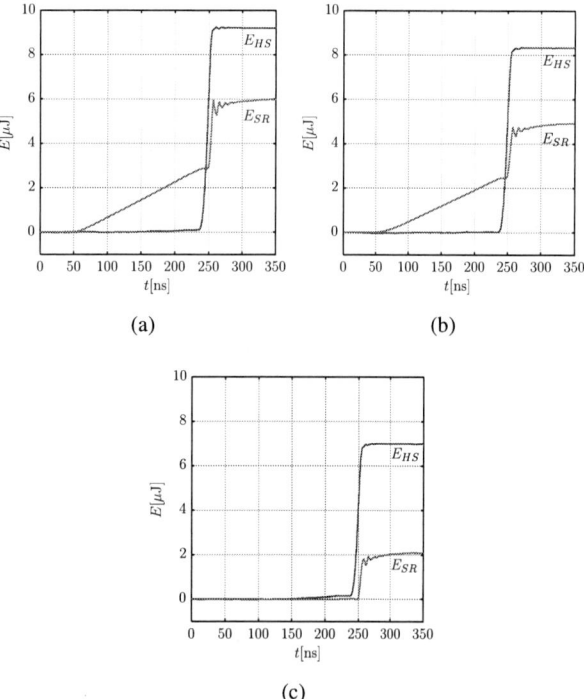

(a)   (b)

(c)

Fig. 8: Reverse recovery loss of Infineon BSC028N06NS, comparing measurements at zerobias, max DT (a); optimum bias, max DT (b); zero bias, optimum DT (c).

combinations of different values of deadtime and gate bias. A sweep of measurements was completed for each of the selected devices. The results are presented as contour plots, where the relative benefit of gate bias and dead-time optimization is shown, both with regards to reverse recovery loss and current peak. Finally, an overview of all the devices is presented, where the absolute loss values and current peak values can be easily compared.

The contour plots show percentage values of loss and reverse recovery peak compared to the measurement result at maximum deadtime and zero bias. These results are presented as relative values in order to make it easy to compare the efficacy of dead time optimization and gate bias for each individual device. The axis scaling and color coding is consistent between the different devices, allowing comparison of the relative effects between all devices, making it possible to identify any similarities or differences.

Figure 9 shows the reverse recovery loss and reverse recovery current peak maps for Infineon BSC028N06NS. This is the same device used for the previous measurement examples in figures 6, 7 and 8. The three measurement points shown in figures 7 and 8 can be easily seen on the contour maps. The max deadtime, zero bias measurement corresponds to the lower right measurement point, which acts as the 100% reference point for both loss and reverse recovery current peak. The max deadtime, optimal bias point is located at the local minimum on the right edge of the plots. The optimal deadtime, zero bias point is located at the left on the lower edge of each plot, in the local minimum there.

As noted in figure 8, the deadtime optimization method achieves the lowest loss. However, the reduction in reverse recovery peak is nearly as good when using bias as when operating with optimal deadtime. The reverse recovery current peak contour in figure 9b shows an unexpected local maximum at 50ns deadtime and 1V to 2V gate bias. This is caused by a small error in the measured gate bias value, due to the fact that there is a small undershoot of v_{gsSR}, as seen in figure 6d.

Figures 9, 10 and 11, show the results for the Infineon BSC028N06NS, IRFH5006TRPBF and NXP PSMN5R5-60YS, respectively.

B. Discussion of measurement results for 60V devices

Figure 12 shows an overview of the absolute reverse recovery loss and current peak measurements. The Infineon devices performed similarly, with similar relative benefit of optimized deadtime and gate bias. However, the newest device, BSC028N06NS, has the lowest absolute losses and current peak of the two, which correlates with its lower Q_{gd}, as shown in table I.

The contour maps of International Rectifier IRFH5006TRPBF in figure 10 are drastically different, showing that any application of gate bias only increases the loss. As seen in table I, this device has the highest Q_{gd} of all the selected 60V devices, making it highly

The 2014 International Power Electronics Conference

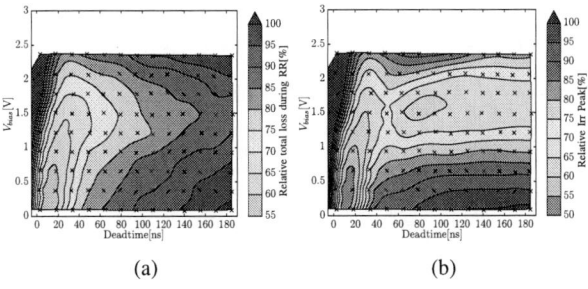

(a)   (b)

Fig. 9: Reverse recovery contours for Infineon BSC028N06NS. a) shows the total reverse recovery loss, while b) shows the value of the reverse recovery current peak. Both are shown as relative percentage values of the measurement at maximum deadtime and zero gate bias. Measurement points are marked by x.

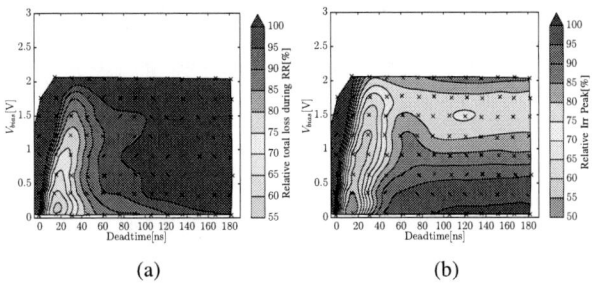

(a)   (b)

Fig. 10: Reverse recovery contours for International Rectifier IRFH5006TRPBF. a) shows the total reverse recovery loss, while b) shows the value of the reverse recovery current peak.

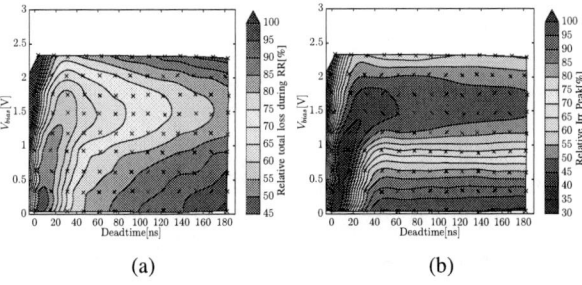

(a)   (b)

Fig. 11: Reverse recovery contours for NXP PSMN5R5-60YS. a) shows the total reverse recovery loss, while b) shows the value of the reverse recovery current peak.

susceptible to gate bounce, which corresponds with the observation that any application of gate bias causes increased losses. However, there is still a moderate reduction in reverse recovery peak current, although this comes at a cost of higher loss.

The results for the NXP PSMN5R5-60YS in figure 11 show that this is the 60V device that gets the greatest relative benefit from gate bias. The reverse recovery current peak is reduced to 45% with optimal bias and maximum deadtime, compared to 35% for zero bias and optimal dead time.

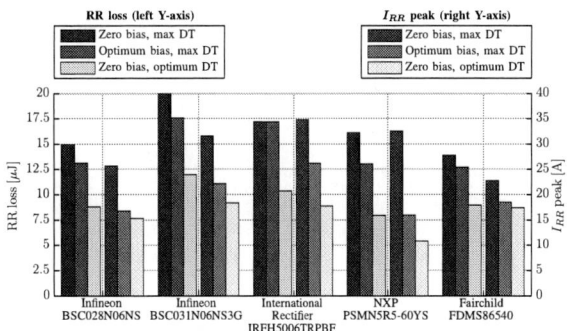

Fig. 12: Comparison of measurement results for the selected 60V devices. For each device is shown the measurement results at zero bias, max deadtime; optimum bias max deadtime; and zero bias, optimum deadtime. The blueish bars show the absolute loss measurement using the left Y-axis, and the reddish bars show the reverse recovery peak values.

Comparing the absolute values of current peaks in figure 12, the Infineon BSC028N06NS and NXP PSMN5R5-60YS achieved the lowest reverse recovery current peaks with optimum gate bias and maximum dead time. However, when judging the devices as a whole, it must be noted that the Infineon devices both have 2.5mΩ on resistance compared to 3.4 to 3.6mΩ for the other manufacturers, as shown in table I.

Overall, the results support the argument that a low Q_{gd} leads to a higher possible reduction of reverse recovery current peak by application of gate bias, leading to a reduction in EMI. The application of gate bias is not as beneficial for the reduction of losses, due to the fact that it will not remove the diode conduction loss during dead time. However, it may be beneficial to choose a fixed moderately low deadtime combined with a small gate bias, providing a compromise of very low EMI and moderately low loss, as an alternative to the costly and complicated implementation of active dead time optimization. The presented contour maps of the reverse recovery can thus be used to choose a suitable combination of gate bias and deadtime.

C. Reverse recovery measurement results for 600V devices

The reverse recovery measurements for 600V devices were conducted using the Infineon IPD60R385CP as high side device for all cases, using an external gate resistance of 100Ω, giving $di/dt = 750\text{A}/\mu\text{s}$. The switching voltage was limited to 300V, allowing normal 400V oscilloscope probes to be used for measurement, and the forward current before reverse recovery was 1A.

Figure 7 shows the reverse recovery at the following conditions:

- At zero gate bias and maximum (250ns) deadtime.

978-1-4799-2706-7/14 $31.00 © 2014 IEEE   2847

- At optimal gate bias and maximum (250ns) dead-time.

- At zero gate bias and optimal deadtime.

At figure 13b, the zero bias, maximum deadtime case shows an initial reverse recovery current peak of 20A, followed by a snappy reduction in current during charging of the MOSFET output capacitance, followed by a second, larger peak reaching 38A. The second peak is a brief shoot through current caused by self turn on of the device, because of the high dv/dt and di/dt. The high dv/dt causes a high current to flow through C_{gd}, thus increasing v_{gsSR}, while the high negative di/dt after the first peak causes a negative voltage drop over the common source inductance, further increasing v_{gsSR}.

The benefit of gate bias is clear in figure 13b, giving a reverse recovery behavior very close to the optimal deadtime case. Examples of the resulting contour maps for 600V devices are shown in figures 14 and 15.

D. Discussion of measurement results for 600V devices

Figure 12 shows an overview of the absolute reverse recovery loss and current peak measurements. It is clear from these results that, at zero bias and maximum deadtime, the "fast body diode" devices, the Infineon IPD65R420CFD and the ST STD13NM60ND, have the lowest reverse recovery loss. However, they are also the two devices that benefit the least from gate bias; the Infineon device showed no reduction in loss while the ST device showed only a 20% reduction.

The Infineon devices not marketed as "fast body diode", the IPD60R385CP and the IPD65R225C7, showed the greatest gain from the application of bias, which corresponds to the fact that they have the lowest Q_{gd} of the selected devices, as shown in table II.

The ST devices not marketed as "fast body diode", the

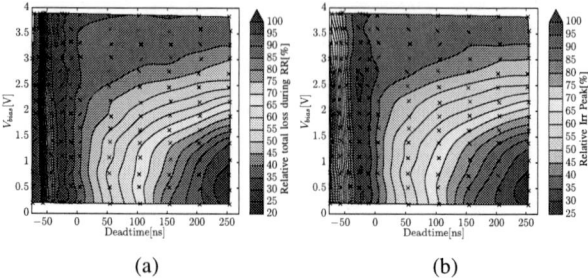

(a)   (b)

Fig. 14: Reverse recovery contours for Infineon IPD60R385CP. a) shows the total reverse recovery loss, while b) shows the value of the reverse recovery current peak.

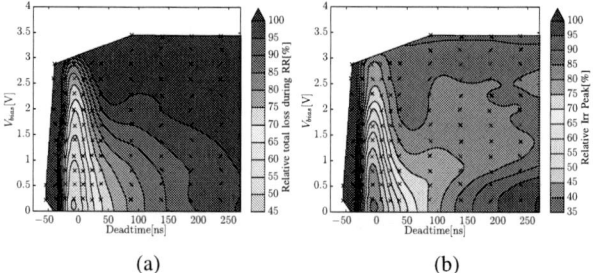

(a)   (b)

Fig. 15: Reverse recovery contours for Infineon IPD65R420CFD. a) shows the total reverse recovery loss, while b) shows the value of the reverse recovery current peak.

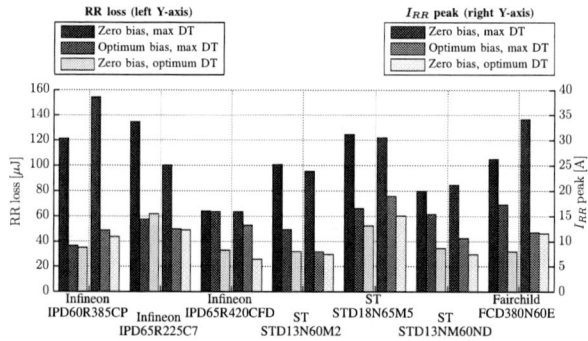

Fig. 16: Comparison of measurement results for the selected 600V devices. For each device is shown the measurement results at zero bias, max deadtime; optimum bias max deadtime; and zero bias, optimum deadtime. The blueish bars show the absolute loss measurement using the left Y-axis, and the reddish bars show the reverse recovery peak values.

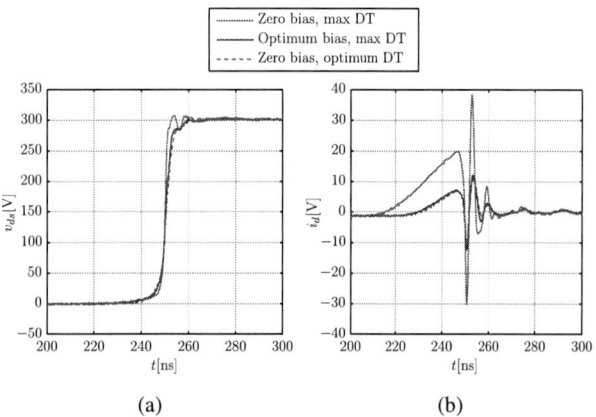

(a)   (b)

Fig. 13: Reverse recovery of Infineon IPD60R385CP, comparing measurements at zerobias, max DT (red); optimum bias, max DT (purple); zero bias, optimum DT (green). a) shows v_{dsSR} and b) shows i_{dSR}.

STD13N60M2 and STD18N65M5, benefit less from gate bias than the Infineon devices, corresponding to their higher Q_{gd}, while the Fairchild device performs slightly worse.

The most interesting conclusion from the measurements is that the normal devices with gate bias perform better than the "fast body diode" devices. It has been reported that the fabrication of these fast diode devices generally requires compromises in the overall MOSFET performance, [6] corresponding well with the relatively higher Q_{gd} of these devices, as seen in table II. Considering that the fast diode devices are generally more expensive due to the extra manufacturing steps required [6], the use of gate bias may provide a strong alternative. This allows the designer to choose the best performing device in terms of on resistance and device capacitances, while keeping a reverse recovery performance that is better than the fast body diode alternatives. The drawback is the requirement of the external biasing circuit, but optimally this would be included in the gate driver.

V. FUTURE RESEARCH

There are many aspects of the effect of gate bias on MOSFET body diode reverse recovery, which require further investigation in order to fully understand and successfully utilize the effect. A key aspect is the negative temperature coefficient of the threshold voltage and body diode. It is possible that an optimal gate bias voltage found at room temperature will cause shoot through at higher temperatures, due to the lowering of the threshold voltage. In that case, it may be necessary to implement a bias circuit which also has a negative temperature coefficient, for example by the use of diodes.

VI. CONCLUSION

This paper provides the first practical investigation into the efficacy of using gate bias for reducing MOSFET body reverse recovery. The measurement methology has been described, allowing the results to be replicated or extended to other devices. A comprehensive set of measurements have been carried out, comparing multiple commercially available MOSFETs. Two sets of devices were chosen for comparison: 60V devices for synchronous rectification application, and 600V devices for zero voltage transition converter application.

The results for the 60V devices show that a low Q_{gd} leads to a higher possible reduction of reverse recovery current peak by application of gate bias, leading to a reduction in EMI. Compared to having optimal deadtime, the application of gate bias is not as beneficial for the reduction of losses, due to the fact that it will not remove the diode conduction loss during dead time. However, it may be beneficial to choose a fixed moderately low deadtime combined with a small gate bias, providing a compromise of very low EMI and moderately low loss, as an alternative to the costly and complicated implementation of active dead time optimization.

The results for the 600V devices show a much better reduction in losses from applying gate bias, due to the fact that the diode conduction loss during the deadtime is not significantly greater than the channel conduction loss. An especially interesting conclusion from the 600V

device measurements is that the normal devices with gate bias perform better than the "fast body diode" devices, with or without bias. A number of performance sacrifices are made to produce the "fast body diode" devices, including added manufacturing costs. By using a normal device with gate bias instead of the fast recovery devices, the designer may choose the best performing device in terms of on resistance and device capacitances, while keeping a reverse recovery performance that is better than the fast body diode alternatives.

REFERENCES

[1] D. Polenov, T. Reiter, R. Baburske, H. Probstle, and J. Lutz, "The influence of turn-off dead time on the reverse-recovery behaviour of synchronous rectifiers in automotive DC/DC-converters," in *Power Electronics and Applications, 2009. EPE '09. 13th European Conference on*, 2009, pp. 1–8.

[2] V. Yousefzadeh and D. Maksimovic, "Sensorless optimization of dead times in DC-DC converters with synchronous rectifiers," *Power Electronics, IEEE Transactions on*, vol. 21, no. 4, pp. 994–1002, 2006.

[3] T. Reiter, D. Polenov, H. Probstle, and H. G. Herzog, "Observer based PWM dead time optimization in automotive DC/DC-converters with synchronous rectifiers," in *Power Electronics Specialists Conference, 2008. PESC 2008. IEEE*, 2008, pp. 3451–3456.

[4] A. Zhao, A. Fomani, and J. Ng, "One-step digital dead-time correction for DC-DC converters," in *Applied Power Electronics Conference and Exposition (APEC), 2010 Twenty-Fifth Annual IEEE*, 2010, pp. 132–137.

[5] A. Fiel and T. Wu, "Mosfet failure modes in the zero-voltage-switched full-bridge switching mode power supply applications," in *Applied Power Electronics Conference and Exposition, 2001. APEC 2001. Sixteenth Annual IEEE*, vol. 2, 2001, pp. 1247–1252 vol.2.

[6] W. suk Choi, S. mo Young, and D. wook Kim, "Analysis of mosfet failure modes in llc resonant converter," in *Telecommunications Energy Conference, 2009. INTELEC 2009. 31st International*, 2009, pp. 1–6.

[7] G. Dolny, S. Sapp, A. Elbanhaway, and C. Wheatley, "The influence of body effect and threshold voltage reduction on trench MOSFET body diode characteristics," in *Power Semiconductor Devices and ICs, 2004. Proceedings. ISPSD '04. The 16th International Symposium on*, 2004, pp. 217–220.

[8] R. Elferich and T. Lopez, "Impact of gate voltage bias on reverse recovery losses of power MOSFETs," in *Applied Power Electronics Conference and Exposition, 2006. APEC '06. Twenty-First Annual IEEE*, 2006, pp. 6 pp.–.

[9] Z. Chen and D. Boroyevich, "Modeling and simulation of SiC MOSFET fast switching behavior under circuit parasitics," in *Proceedings of the 2010 Conference on Grand Challenges in Modeling & Simulation*, ser. GCMS '10. Vista, CA: Society for Modeling & Simulation International, 2010, pp. 352–359. [Online]. Available: http://dl.acm.org/citation.cfm?id=2020619.2020668

[10] X. Huang, Q. Li, Z. Liu, and F. Lee, "Analytical loss model of high voltage GaN HEMT in cascode configuration," *Power Electronics, IEEE Transactions on*, 2013.

[11] J. Ferreira, W. Cronje, and W. A. Relihan, "Integration of high frequency current shunts in power electronic circuits," in *Power Electronics Specialists Conference, 1992. PESC '92 Record., 23rd Annual IEEE*, 1992, pp. 1284–1290vol.2.

[12] J. C. H. Botella, "Ultrafast switching super junction mosfets for single phase pfc applications," in *Applied Power Electronics Conference and Exposition, 2014. APEC 2014. IEEE*, vol. TBD, 2014, p. TBD.

978-1-4799-2706-7/14 $31.00 © 2014 IEEE

An online V_{ce} measurement and temperature estimation method for high power IGBT module in normal PWM operation

Pramod Ghimire, Angel Ruiz de Vega,
Szymon Beczkowski, Stig Munk-Nielsen
Department of Energy Technology
Aalborg University
Aalborg, Denmark
e-mail: pgh@et.aau.dk

Bjørn Rannested, Paul Bach Thøgersen
kk-electronic a/s
Denmark

Abstract—An on-state collector-emitter voltage (V_{ce}) measurement and thereby an estimation of average temperature in space for high power IGBT module is presented while power converter is in operation. The proposed measurement circuit is able to measure both high and low side IGBT and anti parallel diode voltages for a half bridge module which are also used to monitor the electrical degradation of the module. The V_{ce} load current is proposed to estimate the variation of average temperature in space at every fundamental cycle of sinusoidal loading current. Initially, the calibration of voltage and junction temperature for load current level is presented and a trend of change in calibration factor for the IGBT is presented. Finally, the variation in temperature for sinusoidal variation of current is presented at initial stage and after an ageing of the IGBT. The measurement technique is simple and easy to implement into a gate driver for field applications.

Keywords—IGBT power module, junction temperature, real time monitoring, reliability.

I. INTRODUCTION

A real time monitoring of electrical and thermal characteristics during conduction and switching of power electronic devices are still a challenge in order to improve reliability of power converters. A physical architecture of standard multichip power IGBT consists of multilayers of materials with mismatched coefficient of thermal expansion (CTE), which is a major drawback for the device because the operating temperature and thermal cycling severely affects its performance [1]. Specially for traction and wind power converter applications, large temperature oscillation occurs at low modulating frequencies even in normal operation [2]. In normal operation, the semiconductor power devices shows ageing due to both electrical and thermal degradation after certain number of cycles of operation. An electrical degradation changes the electrical resistance which is monitored by observing the on-state V_{ce} drop or on-state resistances. However the thermal degradations changes the thermal resistance which can be monitored by observing the junction temperature or thermal impedance of the device. For example, in the electrical degradation the rise in on-state electrical resistance can be the effect of wire bonding degradations or aluminum reconstruction which increases

sheet resistance of the aluminum layer. On the other hand the thermal degradation increases the thermal resistance by increasing the temperature inside the power module. Hence, a real time estimation of junction temperature during normal PWM operation of power module is in priority to improve the overall reliability of power converters [3] [4] [5]. In practice, the junction temperature T_j of an IGBT can be measured using optical methods, physical contacting methods and electrical methods. In case of a real time application, the optical methods have limitations to implement physically in field applications. Similarly, the physical contacting methods have slow response in the measurement [6] [7]. Therefore, the temperature sensitive electrical parameter (TSEP) such as an on-state V_{ce} is preferable to measure the T_j and ageing of IGBT power module in real time operation [8]. It has been proven that the IGBT chip itself can be used as a temperature sensor [9] [8].

IGBT has negative temperature coefficient (NTC) at lower current level whereas positive temperature coefficient (PTC) at higher current level. The estimation of T_j using V_{ce} requires an accurate calibration of $V_{ce}(T_j)$. Therefore, to avoid a temperature rise on chip due to self-heating, the $V_{ce}(T_j)$ calibration is conducted at small current mainly at 100mA [9], where IGBT posses the NTC characteristics. On the other hand, a very accurate calibration of $V_{ce}(T_j)$ is required at load current calibration method [10], where IGBT shows both NTC and PTC characteristics. The T_j estimation method is proposed using offline characterization method of the module [11]. The on-state V_{ce} is influenced by the collector current, coolant temperature and the gate voltage. Excluding the collector current, every other parameters are kept constant during the characterization at load current method in this paper.

This paper presents a measurement of real time on-state V_{ce} and thereby an estimation of average T_j in space at V_{ce} load current while converter is in operation. In this method, the $V_{ce}(T_j)$ calibration is conducted at load current level for both IGBT and free-wheeling diode. A calibration factor (K-factor) [12] is obtained at different current levels by increasing the load current up to the nominal rated value in a very short period of time. Finally,

978-1-4799-2706-7/14 $31.00 © 2014 IEEE

the average temperature rise after 5.1 million cycles of operation is presented for both sigh side and low side of the IGBT. However, this paper only deal with the temperature estimation for IGBT. A measurement set up and the (V_{ce}) measurement methods are also briefly described. The rise in V_{ce} due to electrical degradation is also presented.

II. THE CONVERTER SETUP

An H-bridge topology is used as a test converter [13] as shown in Fig. 1, where half bridge modules are used on each half leg as a device under test (DUT) and control side of the converter. Two control IGBT modules are used in order to ensure the slower wear-out of the control side. In Fig. 1,

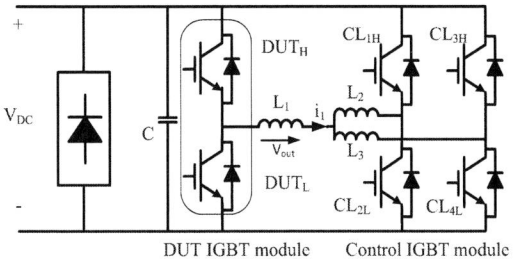

Fig. 1. The power converter set up for testing IGBT modules

DUT_H: High side IGBT device under test
DUT_L: Low side IGBT device under test
L_1, L_2 and L_3: Load inductors

A separate DC supply is used to charge DC link capacitor bank in the converter. Eventually, a $890A_{peak}$ peak current is circulated through the inductors during normal operation of the converter, where the major power loss is dissipated through the power modules such as DUT. The converter operating parameters are given in Table I.

TABLE I. THE CONVERTER OPERATING PARAMETER

Symbol	Meaning	Value
V_{DC}	DC link voltage	$1000V$
V_{DUT}	Forward voltage reference	$253V_{rms}$
I_L	Load current	$890A_{peak}$
F	Fundamental frequency	$6Hz$
F_{SW}	Switching frequency	$2.5kHz$
C	DC link capacitance	$4mF$

III. VOLTAGE MEASUREMENT METHOD

The on-state voltage of the IGBT module is measured using double diode circuit as shown in Fig. 2. The circuit is able to measure the voltages for both IGBT and free wheeling diode of a half-bridge module with

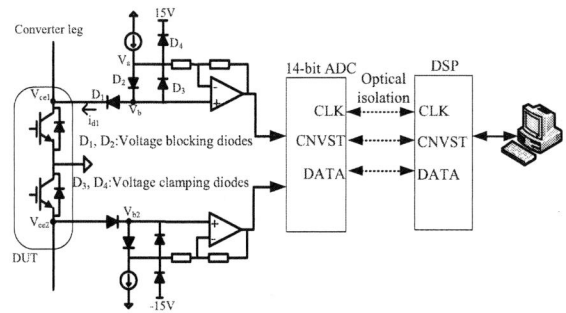

Fig. 2. The power converter test set up

1mV accuracy [14]. Equation (1) shows V_{ce} measurement formulation for high side IGBT as shown in Fig. 2.

$$V_{ce1} = V_b - V_{D2} = V_b - (V_a - V_{b)} = 2V_b - V_a \qquad (1)$$

The measurement circuit is connected to kelvin terminals of the device and is directly connected on top of a IGBT gate driver. The measurement technique has no influence in the switching and operation of the converter during operation [15].

IV. VOLTAGE-TEMPERATURE CALIBRATION

In order to transfer knowledge about current and V_{ce} into temperature an initial calibration is needed. Therefore, the output characteristics of the DUT is calibrated in the same converter and the measurement set up as shown in Fig. 1 and Fig. 2. Each calibration is completed for all

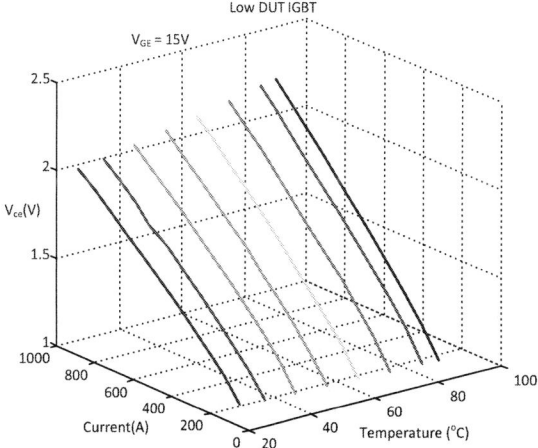

Fig. 3. The IV characteristics for the low side IGBT

four IGBTs and free-wheeling diode keeping the steady baseplate temperature and ramping up the current until $890A_{peak}$ through the load inductors within $340\mu S$. The calibration is started at room temperature normally 22^oC and the converter is then operated to increase the coolant temperature using self-heating until 85^oC. As the coolant content a mixture of water and glycol, the calibration process stopped at 85^oC due to safety issues. Prior to the calibration, all PWM signals are disabled for a minute

The 2014 International Power Electronics Conference

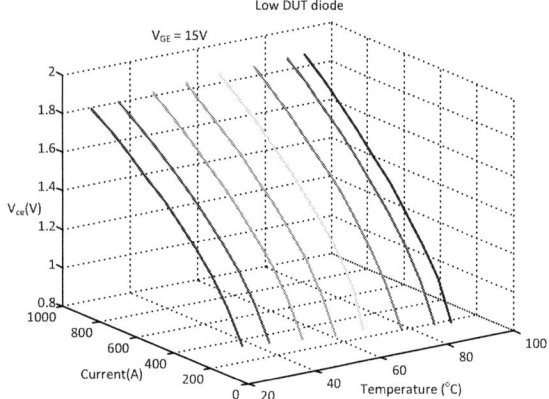

Fig. 4. The IV characteristics for the low side free wheeling diode

to maintain the homogeneous temperature distribution across the baseplate and thereby in the chips. Then after, the calibration is conducted for each temperature by switching the DUT and the control side IGBT in a controlled fashion.

The $V_{ce}(T_j)$ calibration at high current is shown in Fig. 3 and Fig. 5 for low side IGBT and the corresponding free-wheeling diode respectively. The calibration is completed in $680\mu S$ to minimize the error due to self heating for both sides of the half bridge module. Just before the calibration the surface of the chip temperature inside power module is maintained at the same level by running the cooling water continuously without applying the load for a minute. The homogeneous distribution of temperature is also confirmed by the finite element modelling as described in paper [16].

Although the calibration is conducted very fast, because

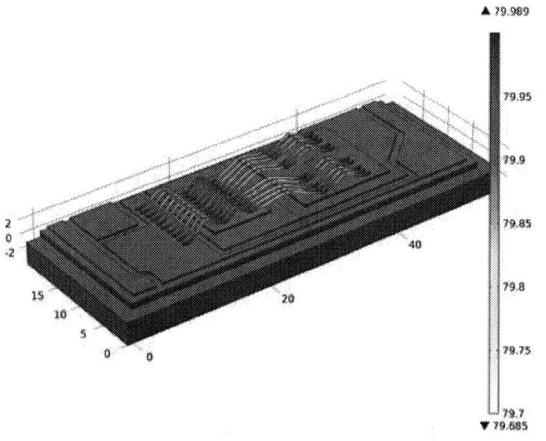

Fig. 5. The temperature distribution on surface of the chip before applying load current

of the high load current surface of the chip experiences small energy loss which will increase the surface temperature above than the cooling temperature. Therefore, the calibration factor is corrected by using thermal impedance

to estimate the chip temperature on the surface of the module [5].

The calibration factor is calculated at different current level as given in Equation (2).

$$K = |\frac{T_{J2} - T_{J1}}{V_{CE2} - V_{CE1}}|^{\circ}C/mV \tag{2}$$

The gain of the calibration factor at crossover point from

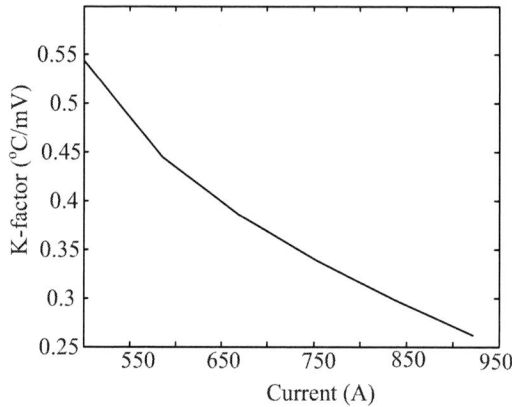

Fig. 6. The calibration factor

NTC to PTC is infinite which cannot be used to calculate the average T_j of the device. In fact, the major interest of study is also at higher current level, hence the K-factor for the higher current level is considered for the temperature estimation. The IGBT and diode has NTC to PTC crossover point at 152A and 419A respectively for the DUT. A 1700V/1000A P3 IGBT module is used in the DUT, which consists of six identical sections of half bridge structure. Fig. 6 shows the corrected calibration factor which is used to estimate the average junction temperature of the IGBT.

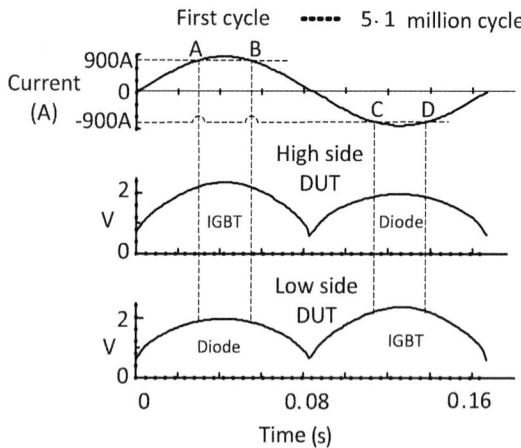

Fig. 7. The on-state voltage measurement

V. EXPERIMENTAL RESULTS

A. Online voltage monitoring

An online measurement is conducted at every five minutes in normal operation to limit the amount of data. In every single measurement, slightly above 2 fundamental cycles of collector current and corresponding conductive voltage drops are recorded as exhibited in Fig. 7 [5]. As given in Table I, the cooling temperature is kept steady with $80 \pm 0.5^{o}C$ throughout the test as exhibited in Fig. 8. At 6Hz sinusoidal current loading

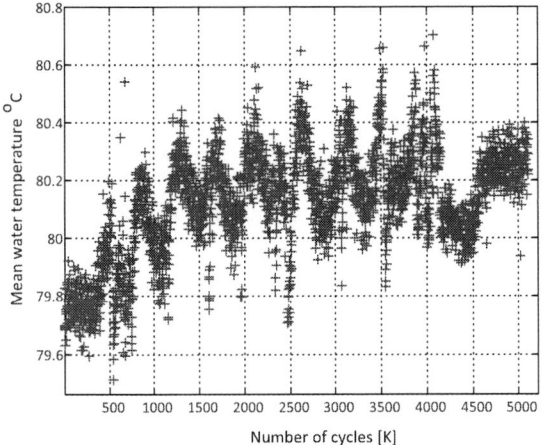

Fig. 8. The variation in water cooling temperature during the test

the power module is failed after 5.1 million cycles of operation due to ageing. The real time voltage measurement shows that the trend of on-state voltage increment is different for IGBTs and free-wheeling diodes. The IGBT has nearly linear increment in on-state V_{ce} drop until the module fails, on the other hand the corresponding free wheeling -diode has minimal change in on-state V_{FD} drop until certain cycles of operation. For instance, in this case only above around 3.5 million

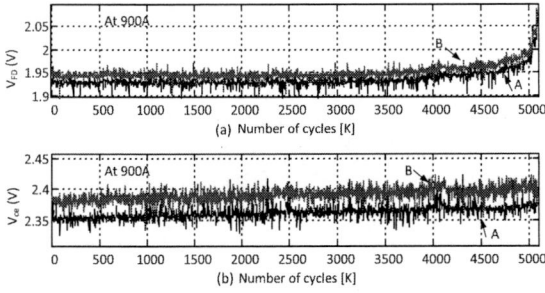

Fig. 9. The on state voltage variation at 900A (a)The on state V_{FD} drop for low side diode at point A and B from Fig. 7 and (b)The on state V_{ce} drop for high side IGBT at point A and B as shown in Fig. 7

cycles the rate of rise of V_{FD} is increasing until the module fails. After the degradation has been started

the rate of rise of voltage is also increasing. In the measurement, the step increment of nearly 7mV in the on-state V_{FD} is also witnessed due to bond wire lift-off. The on-state V_{FD} of low side diode is increased in total by 132mV from the beginning before it fails at 900A as shown in Fig. 9(a) and Fig. 10. However, the on state V_{ce} of high side IGBT is only increase by 22mV after the 5.1 million cycles.

The fig. 9 exhibits the trend of change in on-state V_{ce} and V_{FD} for low side diode and high side IGBT at 900A. The on state voltage drops at the rising edge and falling edge of the sinusoidal current are compared at 900A. Although the electrical parameters are similar at the same current level, the on-state voltage drop is higher at the falling edge of the sinusoidal current because the temperature rises slowly in comparison to the load current due to thermal impedance of the chip and the module. Assuming similar electrical degradation for a half cycle of the loading current, the voltage gradient is mainly due to the rise of the surface temperature.

Fig. 10 exhibited an extended plot of the rate of increase

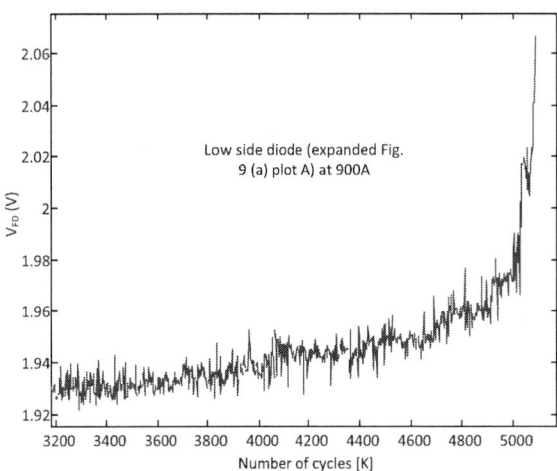

Fig. 10. An extended figure of on state V_{FD} for low side diode from 9 at point

in on-state V_{FD} with respect to number of cycles after for above 3.3 million cycles of operation. As mentioned above, the on state V_{ce} and V_{FD} gradient at 900A are calculated as follows from fig. 7;

High side IGBT V_{ce} at rising side point A = $A(V_{ceRise})$
High side IGBT V_{ce} at falling side point B = $B(V_{ceFall})$
High side IGBT V_{ce} gradient (ΔV_{ceHIgh})= $B(V_{ceFall}) - A(V_{ceRise})$
Low side IGBT V_{ce} at rising side point C = $C(V_{ceRise})$
Low side IGBT V_{ce} at falling side point D= $D(V_{ceFall})$
Low side IGBT V_{ce} gradient (ΔV_{ceLow})= $D(V_{ceFall}) - C(V_{ceRise})$

Similarly for corresponding free-wheeling diode;
High side diode V_{FD} at rising side point C = $C(V_{FDRise})$
High side diode V_{FD} at falling side point D = $D(V_{FDFall})$
High side diode V_{FD} gradient (ΔV_{FDHIgh})=

978-1-4799-2706-7/14 $31.00 © 2014 IEEE

$D(V_{FDFall}) - C(V_{FDRise})$
Low side diode V_{FD} at rising side point A = $A(V_{FDRise})$
Low side diode V_{FD} at falling side point B= $B(V_{FDFall})$
Low side diode V_{FD} gradient $(\Delta V_{FDLow})=$
$B(V_{FDFall}) - A(V_{FDRise})$

In fig. 11,

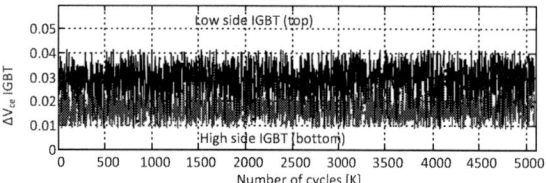

Fig. 11. The rise in V_{FD} gradient at $900A$ for diode

Low side IGBT (top)= $D(V_{ceFall}) - C(V_{ceRise})$
High side IGBT (bottom) = $B(V_{ceFall}) - A(V_{ceRise})$

In fig. 12,

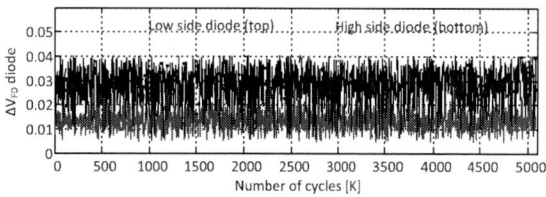

Fig. 12. The rise in V_{ce} gradient at 900A bentween for IGBT

Low side diode (top)= $B(V_{FDFall}) - A(V_{FDRise})$
High side diode (bottom) = $D(V_{FDFall}) - C(V_{FDRise})$

Fig. 11 and fig. 12 shows that the change of rise in gradient voltage drop between falling and rising edge of sinusoidal voltage at $900A$ for IGBT and diode in all cycles. In both IGBT and diode, the trend of change in gradient voltage is nearly constant throughout the cycles, even though the rate of rise of on state voltage drops are different at different number of cycles. Hence, it can be predict that the voltage increment in Fig. 9 is mainly due to electrical degradation.

B. Temperature estimation

As shown in Fig. 6, the voltage sensitivity is in between 2mV/°C to 3mV/°C for the current range 550A to 922A. Hence, for an accurate estimation of average T_j in space, the measurement circuitry should have mV accuracy in the real time operation. Similarly, the current measurement should have higher accuracy. The voltage measurement circuitry requires some settling time which is fulfilled by measuring at the switching pulse. In addition this makes the implementation into control simple. Actually, the measurement could be beneficailly made at the end of the switching period, but this complicates the implementation [5].

As mentioned in previous section, because of the thermal impedance of the chip the rise in temperature takes longer than the electrical signals. The average T_j is calculated based on the real time measurement and $V_{ce}(T_j)$ calibration as given in equation (3).

$$T_j = T_{ref} + K \times (V_{cemeas} - V_{ceref}) \qquad (3)$$

Where,
T_{ref}: reference temperature taken during calibration process
T_{cemeas}: real time on-state V_{ce} measurement
V_{ceref}: on-state Vce measured at given reference temperature which is at 30^oC in this paper

Fig. 13 demonstrates the T_j variation at different

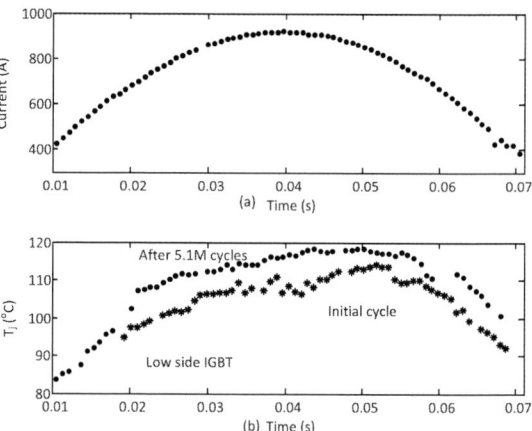

Fig. 13. The average T_j estimation (a)The collector current flowing through the IGBT (b) The Tj variation on low side IGBT

current during half fundamental cycle at initial and after 5.1 million cycles of operation. In both cases the collector currents are same. The initial calibration factor measured for the low side IGBT is used to obtain the T_j in both cases for the comparison. But for the final cycle the new V_{ceref} which is measured at 5 million cycles is used in equation (3). The peak temperature is raised by close to 5^oC from the beginning of the cycle cycle until module fails due to ageing process. Fig. 14 demonstrates the calculated T_j variation of the high side IGBT during half fundamental cycle of the current in the beginning and after 5.1 million cycles of operation where the peak of calculated T_j is increased nearly by 9^oC. The calibration factor measured at the beginning of the high side IGBT is used to compare the calculated T_j for both initial cycle and final cycle, but for this side also the new new V_{ceref} which is measured at 5 million cycles is used in equation (3). At the initial cycle the peak of calculated T_j is higher by close to 10^oC in the low side than the high side. The cooling temperature is maintained at 80^oC as shown in fig. 8 through out the test period. As exhibited in the figures 13 and 14, high peak temperature is observed in low side IGBT than the high side IGBT at the beginning and at the end of the

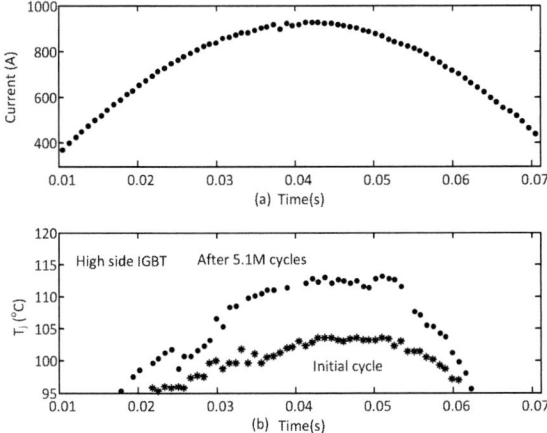

Fig. 14. The average T_j estimation (a)The collector current flowing through the IGBT (b) The Tj variation on high side IGBT

cycles but the temperature rise is higher in high side IGBT in comparison to the initial state of both sides. On the voltage measurement, the trend of rise in V_{FD} is higher in the low side diode leading to the failure of the DUT.

VI. CONCLUSIONS

This paper describes a potential method for real time measurement of on state collector emitter and forward voltage drop in continuous sinusoidal loading of the current. The proposed V_{ce} load current method which is used to estimate the average junction temperature in space shows that a good agreement between voltage and temperature measurement. The proposed method is able to detect the mean junction temperature variation for the low side and high side IGBT separately in every fundamental cycle of the current. As observed in the measurement result, the lower side of the tested IGBT module is failed earlier during the test. The verification of the accuracy of temperature estimation by this method is in progress using direct measurement technique. The proposed technique could be a potential method for the field application such as for wind power converter, automotive application etc.

ACKNOWLEDGEMENT
This work is under progress with in Center of Reliable Power Electronic (CORPE) and Intelligent and Efficient Power Electronic (IEPE)project framework at Department of Energy Technology, Aalborg University, Denmark.

REFERENCES

[1] M. Ciappa, "Selected failure mechanisms of modern power modules", Microelectronics Reliability, 42 (2002)pp. 653-657.

[2] I. Bahun, N. Cobanov, Z. Jakopovic, "Real-time measurement of IGBT's operating temperature", AUTOMATIKA 52, 4, pp 295-305, 2011.

[3] P. Ghimire, S. Beczkowski, S. Munk-Nielsen, B. Rannestad, P. B. Thgersen, A review on real time physical measurement techniques and their attempt to predict wear-out status of IGBT in Proc. EPE13 ECCE Europe 15th European Conference on Power Electronics and Applications, Lille, France. Sep. 2013.

[4] I. Masayasu, K. Tsungu, A simple approach for dynamic junction temperature estimation of IGBTs on PWM operating conductions, in Proc. Power Electronics Specialist Conference, 2007, PESC2007, IEEE, vol. no.916.920, pp 17-21, June 2007.

[5] P. Ghimire, A. R. de Vega, S. Beczkowski, B. Rannestad, S. Munk-Nielsen, P. Thgersen, "Improving reliability of power converter using an online monitoring of IGBT modules", IEEE Industrial Electronics Magazine, Accepted.

[6] L. Dupont, Y. Avenas, and P. Jeannin, Comparison of junction temperature evaluations in a power IGBT module using an IR camera and three thermo-sensitive electrical parameters in Proc. IEEE 2012 Applied Power Electronics Conference and Exposition (APEC), 2012 Twenty-Seventh Annual IEEE, pp. 182-189 Feb. 2012.

[7] D-L. Blackburn, Temperature measurement of semiconductor decices A Review,in Proc. 20th Annual Semiconductor Thermal Measurement and Measurement Symposium, pp. 70-80, 2004.

[8] V. Smet, F. Forest, J.J. Huselstein, A. Rashed, and F. Richardeau, Evaluation of Vce monitoring as a real-time method to estimate aging of bond wire-IGBT modules stressed by power cycling, in Proc. IEEE Transaction on Industrial Electronics, vol. 60, No- 7, July 2013.

[9] R. Schmidt and U. Scheuermann, "Using the chip as a temperature sensor The influence of steep lateral temperature gradients on the Vce(T)-measurement," in Proc. Power Electronics and Applications, 2009. EPE '09. 13th European Conference on, pp. 1-9, 2009.

[10] X. Perpina, J. F. Serviere, J. Saiz, D. Barlini, M. Mermet-Guynner, J. Millan, Temperature measurement on series resistance and devices in power packs based on on-state voltage drop monitoring at high current, Microelectronics Reliability, Vol. 46, Issuses 9-11, pp 1834-1839, Sep.-Nov. 2006.

[11] Y.S. Kim and S.K. Sul, On-line estimation of IGBT junction temperature using on-state voltage drop, in Proc. Industrial Applications Conference, 1998. Thirty-Third IAS Annual Meeting. The 1998 IEEE, vol. 2, pp 853-859 Oct. 1998.

[12] JEDEC Standard: Thermal Impedance Measurement for Insulated gate Bipolar Transistors: JESD24-12: JEDEC Solid State Technology Association 2004.

[13] R.O. Nielsen, J. Due and S. Munk-Nielsen, "Innovative measuring system for wear-out indication of high power IGBT modules," in Proc. Energy Conversion Congress and Exposition (ECCE), 2011 IEEE, pp. 1785-1790, 2011.

[14] S. Beczkowski, P. Ghimire, S. Munk-Nielsen, P. B. Thgersen, B. Rannested Online Vce measurement method for wear out monitoring of high power IGBT modules,in Proc. EPE13 ECCE Europe 15th European Conference on Power Electronics and Applications, 2013.

[15] P. Ghimire, A. R. de Vega, S. Munk-Nielsen, B. Rannestad, P. Thgersen, "A Real Time Vce Measurement Issues for High Power IGBT Module in Converter Operation", IFEEC 2013, Nov. 3-6, 2013.

[16] P. Ghimire, A. R. de Vega, K. B. Pedersen, B. Rannestad, S. Munk-Nielsen, P. Thgersen, "A real time measurement of junction temperature variation in high power IGBT modules for wind power converter application", 8th International Conference on Integrated Power Electronic Systems (CIPS) 2014, Nuremberg, Germany.

Evaluation on Iron Loss Characteristics in Series Connection and Parallel Connection of Loads with Inverter Excitation

Shunya Odawara (Toyota Technological Institute)
Department of engineering
2-12-1 Hisaskata, Tempaku
Nagoya, Aichi, Japan
sodawara@toyota-ti.ac.jp

Keisuke Fujisaki (Toyota Technological Institute)
Department of engineering
2-12-1 Hisaskata, Tempaku
Nagoya, Aichi, Japan
fujisaki@toyota-ti.ac.jp

Abstract— **It is revealed that plural loads should be driven connected in series to reduce an iron loss. In this research, three ring specimens as load are connected in series or in parallel and these rings are excited by a single phase PWM inverter. An iron loss on series connection becomes smaller than parallel connection because a minor loop on series connection which makes iron loss increase is smaller than parallel connection. The differential of minor loop size is resulting from the value of on-voltage on power semiconductor. This on-voltage becomes small in the series connection because the on-voltage is divided into each load.**

Keywords— *Iron loss, on-voltage, minor loop, inverter excitation.*

I. INTRODUCTION

Demand of electrical instrument is increasing due to progress of the power electronics and the human society. Therefore, in order to use electrical energy efficiently, the driving system for electrical instrument which can be driven efficiently is necessary. In some cases, plural electrical machines are excited by an electrical source. Especially, plural induction motors are driven by using an inverter in train [1]. The connection method of motor is decided from convenience of motor control, and researches on the control of plural motors are performed [2]-[4]. However, it seems that the relationship between the iron loss and the connecting method of plural loads is hardly evaluated in conventional research [5].

Therefore, the connection method with which iron loss can be made small is examined on measurement from the view point of magnetic material characteristics in this research. Three ring specimens as load connected in series and in parallel are excited by using single PWM inverter so that the flux density becomes the same on each condition. In inverter, there are power semiconductors as the switching device and the free wheel device, and the on-voltage characteristics of these semiconductors affect to the making minor loop on BH curve and the iron loss [6]-[7]. This on-voltage affecting one ring specimen is changed by connection method and it is expected that the iron loss is changed by connection method because size of minor loop which makes iron loss increase is resulting from on-voltage value.

II. METHOD FOR MEASUREMENT

A. Evaluation System

The overview of measurement system for evaluation of iron loss characteristics is shown in Fig. 1. The ring part in Fig. 1 is shown in Fig. 2. Three rings which are connected in series and in parallel are excited by using single phase PWM inverter. For reference, the evaluation by using just one ring without connection is also performed. These rings are created so that the specifications of each ring specimen become the same. The specifications of rings are shown in Table II. The used inverter (MWINV-9R122A) is produced by Myway Plus Corporation.

B. Exciting Conditions

In single phase PWM inverter excitation, the fundamental frequency f_o, the carrier frequency f_c and the modulation factor m are set to 50 Hz, 10 kHz and 0.6, respectively. The applied DC voltage V_{dc} in Fig. 1 is decided so that the maximum flux density B_{max} in each ring becomes 1 T.

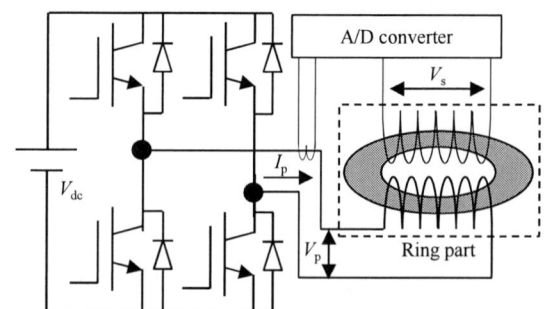

Fig. 1. Evaluation system.

TABLE I. SPECIFICATIONS OF RING SPECIMEN.

Material	35H300
Outside diameter	127 mm
Inside diameter	102 mm
Height	7 mm
Primary coil winding number	254 turn
Secondary coil winding number	254 turn

978-1-4799-2706-7/14 $31.00 © 2014 IEEE

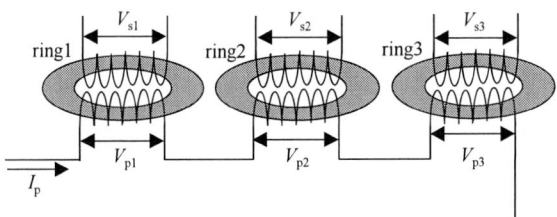

(a) Connected in series

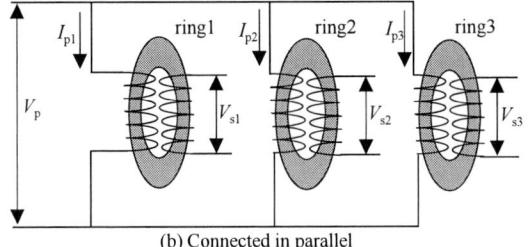

(b) Connected in parallel

Fig. 2. Connection of three rings in ring part.

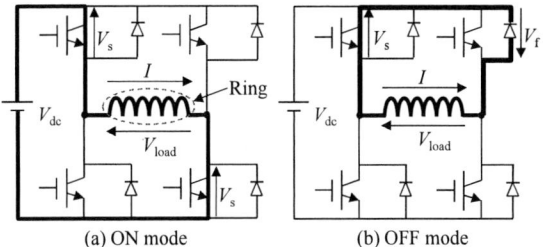

(a) ON mode   (b) OFF mode

Fig. 3. ON/OFF mode of PWM inverter.

III. ON-VOLTAGE OF POWER SEMICONDUCTOR DEVICE IN PWM INVERTER

A. PWM Voltage Waveform Including On-voltage

When a power semiconductor device as a switching device becomes conduction state, the resistance value of semiconductor should be ideally zero. However, there is a small resistance value in semiconductor and the voltage depression occurs due to current which is flowed in this small resistance. This voltage is on-voltage. In PWM inverter, several semiconductors as switching device and free wheel device are used, so the influence of on-voltage is included in output PWM voltage.

Fig. 3 shows the ON and OFF mode of PWM inverter excitation. At ON mode shown in Fig. 3(a), the voltages can be expressed as following equation (1) due to Kirchhoff's laws. In the same way, the voltages at OFF mode can be expressed as following equation (2).

$$V_{load} = V_{dc} - 2V_s \tag{1}$$

$$V_{load} = -V_s - V_f \tag{2}$$

where V_{load}, V_s and V_f are the voltage applied to rings, the on-voltage of switching device and the on-voltage of free wheel device, respectively. Since V_s and V_f are enough smaller than V_{dc}, V_{load} becomes almost V_{dc} at ON mode. On the other hand, at OFF mode, V_{load} does not become zero due to existence of on-voltage V_s and V_f.

Fig. 4 shows the current-voltage waveform of one-cycle obtained from PWM inverter excitation on measurement. The enlargement of voltage waveform near maximum current is also shown. In Fig. 4, it is confirmed that the voltage at OFF mode does not become zero due to on-voltage. The phenomenon in which the voltage does not become zero by on-voltage of semiconductor affects making of minor loop on BH curve.

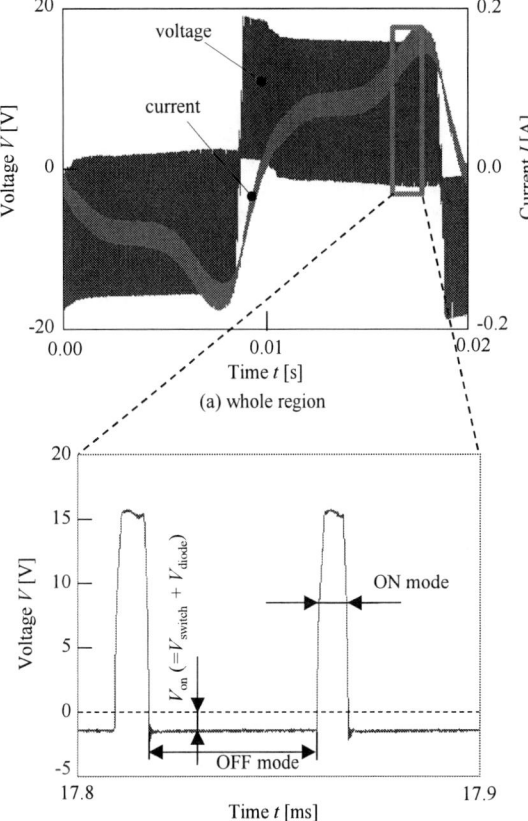

(a) whole region

(b) enlargement of voltage near maximum current

Fig. 4. Current-voltage waveform with inverter excitation (f_o = 50 Hz, f_c = 10 kHz, m = 0.6, without connection).

B. Relationship between On-voltage and Minor Loop

Magnetic flux density B is obtained from the time integration of secondary voltage V_s and Magnetic field intensity H are obtained from the primary current I_p as following equations (3) and (4), respectively.

$$B = \frac{1}{N_2 S} \int V_s dt \ [\text{T}] \tag{3}$$

$$H = \frac{N_1 I_p}{L} \ [\text{A/m}] \tag{4}$$

where N_1, N_2, S and L are the number of primary coil windings, the number of secondary coil windings, the cross-section area of ring and the length of flux path of ring, respectively.

When magnetic flux densities at one on-pulse and one off-pulse are defined as ΔB_{on} and ΔB_{off}, ΔB_{on} and ΔB_{off} is expressed as following equation (5), (6).

$$\Delta B_{on} \propto \int_{on\,mode} V_{dc}\,dt \tag{5}$$

$$\Delta B_{off} \propto \int_{off\,mode} (V_s + V_f)\,dt \tag{6}$$

In equation (5), V_s is neglected because V_s is enough smaller than V_{dc}. In equation (6), when $(V_s + V_f)$ becomes negative as shown in Fig. 4(b), ΔB_{off} also becomes negative.

This phenomenon makes minor loop on BH curve because the magnetic flux density B rises at ON mode and the B falls at OFF mode as shown in Fig. 5. Therefore, it is expected that the minor loop becomes large when the on-voltage is large. Generally, since minor loop makes iron loss increase, the small on-voltage is better in order to reduce the iron loss.

C. On-voltage and Connectionection Method

The on-voltage is changed by connection method. Fig. 6 shows the relationship between on-voltage and connection method.

When on-voltage V_{on} and primary current I_p in single coil without connection are defined as shown in Fig. 6(a), the each on-voltage of three rings connected in series becomes $(V_{on} / 3)$ by voltage dividing. On the other hand, that connected in parallel becomes V_{on}' which is not V_{on} because the current flowing in power semiconductor is three times the I_p and the on-voltage characteristics are changed as shown in Fig. 7. It is confirmed by Fig. 7 that V_{on}' becomes somewhat larger than V_{on}. Therefore, it is expected that the iron loss become smallest in series connection because the on-voltage is small and the minor loop will become small.

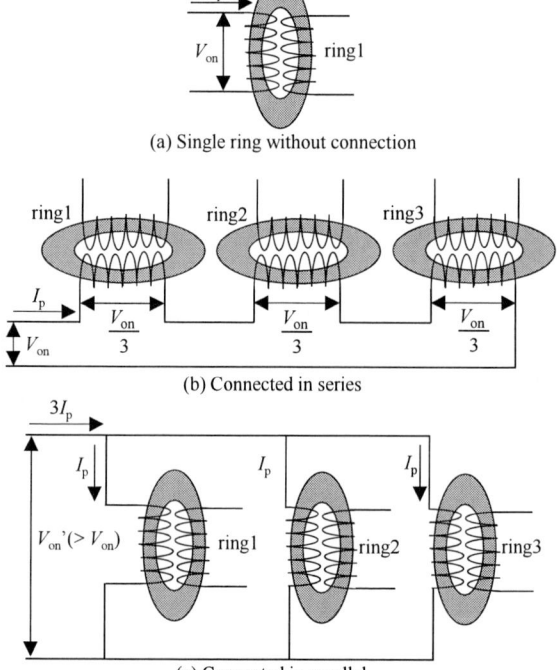

(a) Single ring without connection

(b) Connected in series

(c) Connected in parallel

Fig. 6. Relationship between on-voltage and connection method.

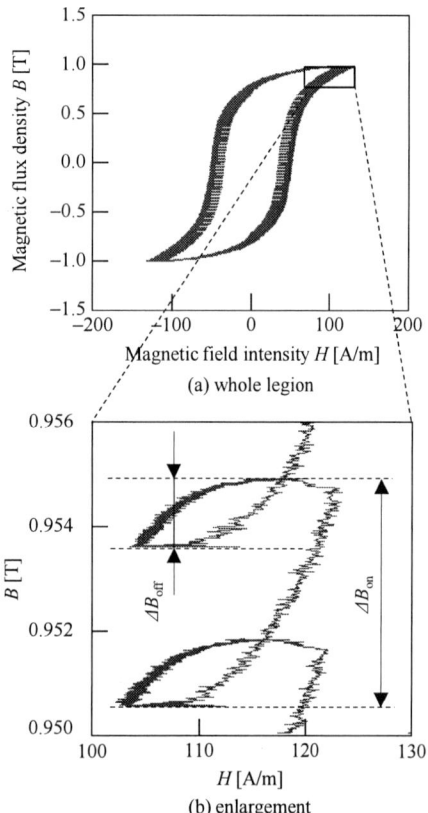

(a) whole legion

(b) enlargement

Fig. 5. Minor loop on BH curve
(f_o = 50 Hz, f_c = 10 kHz, m = 0.6, without connection)

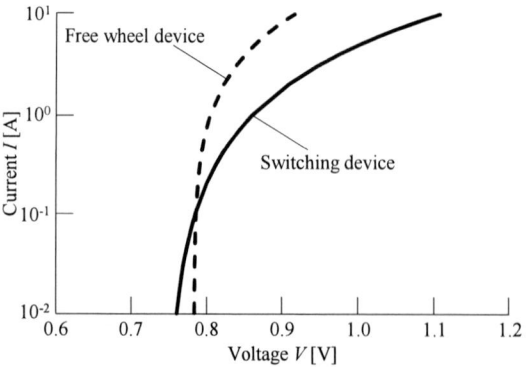

Fig. 7. I-V characteristics of power semiconductor devices.

IV. RESULT OF MEASUREMENT

A. Iron Losses on Each Connection

Iron loss is calculated by the magnetic flux density B and the magnetic field intensity H obtained from each connection as following equation (7). In measurement, since the maximum flux density B_{max} does not become just 1 T, the normalization of iron loss on just $B_{max} = 1$ T is performed by using following equation (8) because iron loss is proportional to the square of magnetic flux density.

$$W = \frac{f_o}{\rho} \int H dB \text{ [W/kg]} \tag{7}$$

$$W_{fe}^* = W_{fe} \times 1^2 / B_{max}^2 \tag{8}$$

where ρ is density of electrical steel sheet (35H300), B_{max} is measured value, and super script (*) means the normalized value.

Table II shows the iron losses obtained from each connection. Although the iron loss of 35H300 on datasheet obtained from Epstein method due to sinusoidal excitation is 1.05 W/kg, the measured iron loss is increased by inverter excitation [8]-[11]. In Table II, the iron losses of series connection are smaller than that of parallel connection. Moreover, the total iron loss of parallel connection is larger about 3.8 % than that of series connection. This result is caused by difference of on-voltage, and the difference of on-voltage due to connecting method is shown in detail in next session.

TABLE II. IRON LOSSES OBTAINED FROM EACH CONNECTION.

W_{fe}^* [W/kg]	Single coil	Series connection	Parallel connection
Ring 1	1.55	1.49	1.54
Ring 2	1.51	1.48	1.51
Ring 3	1.51	1.48	1.51
Total	4.57	4.45	4.56

B. On-voltages Obtained from Each Connection

Fig. 8 shows the waveforms of voltage V_{load} and current I_p near the maximum current. First, the on-voltage in series connection becomes almost one-third of that in single. Second, the on-voltage in parallel connection is larger than that in single, since the current in parallel connection is almost three times of that in single ring. These values follow the theory mentioned above. Next, in order to confirm the relationship between the on-voltage and the I-V characteristics of power semiconductor is verified by using the obtained current and the datasheet

of semiconductors in Fig. 7. Figs. 9 and 10 show the operating point on switching device and that on free wheel device. The switching device is IGBT (PM75RSA060) and the free wheel device is diode (RM30TB-H). In addition, since the I-V characteristics below 1 A cannot read from the datasheet, the approximation by a quadratic expression is carried out by using data of above 1 A on datasheet. Table III shows the on-voltage obtained from measurement and datasheet. In Table III, the values of on-voltage $V_{on}^{(meas.)}$ obtained from measurement correspond well with the values of on-voltage $V_{on}^{(ideal)}$ obtained from datasheet.

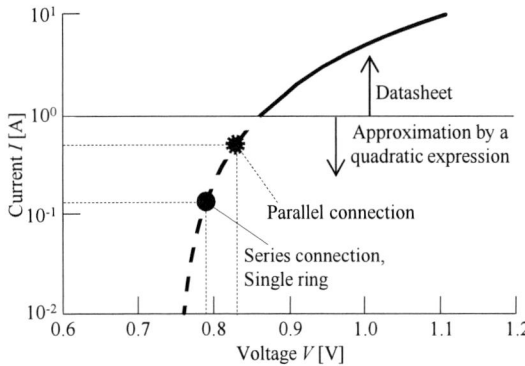

Fig. 9. Operating point on switching device (IGBT: PM75RSA060).

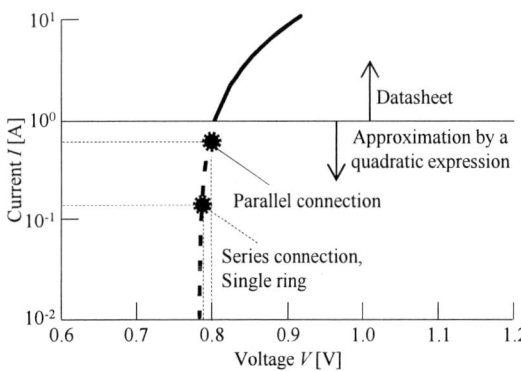

Fig. 10. Operating point on free wheel device (diode: RM30TB-H).

TABLE III. ON-VOLTAGE.

Ring	I_C, I_F [A]	V_C [V]	V_F [V]	$V_{on}^{(ideal)}$ [V] $(V_C + V_F)$	$V_{on}^{(meas.)}$ [V]
Single	0.15	0.79	0.79	1.58	1.42
Series	0.15	0.79 / 3	0.79 / 3	0.53	0.54
Parallel	0.50	0.83	0.80	1.63	1.71

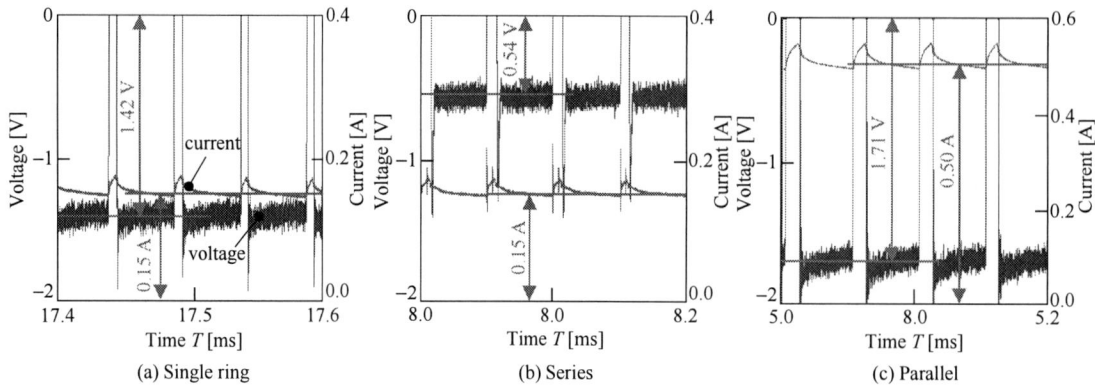

Fig. 8. I-V characteristics of power semiconductor devices.

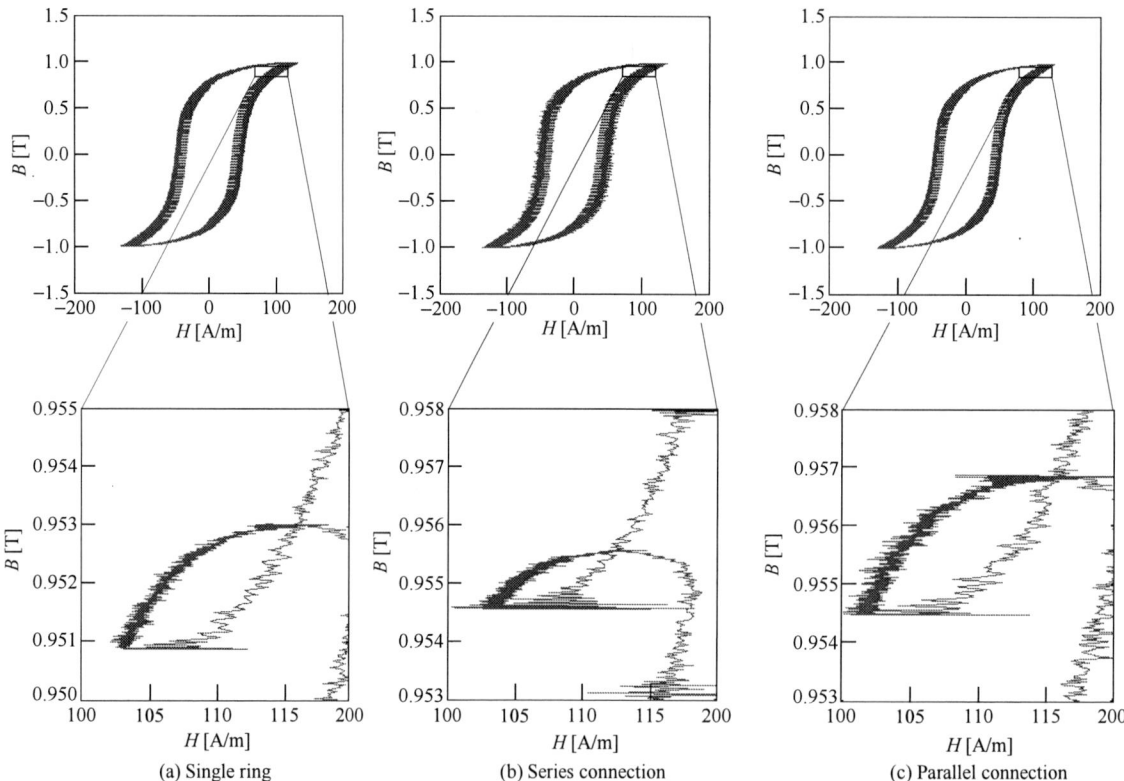

Fig. 11. Minor loops on each connection.

C. Minor Loops on BH Curves

Fig. 11 shows the BH curve and the enlargement of minor loop near the maximum flux density. Although the major loops of each connection are almost the same, the size of minor loops is different. The minor loop of series connection is smallest and that of parallel connection is largest. This relationship is the same as the on-voltage. Since this minor loop makes iron loss increase, the iron loss on series connection becomes smallest and that on parallel connection becomes largest as shown in Table II. Therefore, in order to reduce the iron loss, the load should be connected in series.

D. Investigation on Other Carrier Frequency

The investigation mentioned above is carried out in carrier frequency $f_c = 10$ kHz. Therefore, on other carrier frequency, it is investigated whether the same result is obtained.

Fig. 12 shows the comparison between total iron loss on series connection and on parallel connection. The iron loss on series connection is smaller than that on parallel connection on each carrier frequency. Therefore, the conclusion that iron loss in series connection becomes low is corroborated from the result in other career frequency.

978-1-4799-2706-7/14 $31.00 © 2014 IEEE

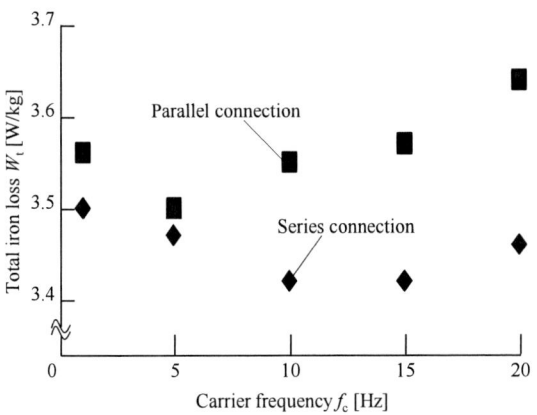

Fig. 12. Total iron losses by changing carrier frequency.

V. CONCLUTION

In this research, the connection method of plural loads with which iron loss can be made small is examined from the view point of magnetic material characteristics. The conclusion obtained from measurement is as follows.

1) On-voltage of each ring on series connection becomes smaller than that on parallel connection, because the voltage dividing of on-voltage is caused by series connection.

2) Since on-voltage on series connection is small, the minor loop which is made by influence of on-voltage becomes smaller than parallel connection.

3) Minor loop of series connection which makes iron loss increase is small, so iron loss on series connection is smaller than parallel connection.

4) Therefore, in order to reduce the iron loss, plural loads should be connected in series when these are driven.

REFERENCES

[1] T. Hasebe, and H. Yamamoto: "Next-Generation Shinkansen Drive Systems", Toshiba review, vol. 61, no. 9, 2006 (in Japanese).

[2] M. Hua, H. Hu, Y. Xing, and Z. He: "Distributed Control for AC Motor Drive Inverters in Parallel Operation," *IEEE Trans. on.* vol. 58, no. 12, pp. 5361-5370, 2011.

[3] T. Nagano, Y. Nakajima, Y. Noge, and J. Itoh: "A Method for a Parallel Operation System of Multiple Units Permanent Magnet Synchronous Motor", *IEEJ Industry Applications Society Conference*, L2177B, 2012 (in Japanese).

[4] K. Matsuse, H. Kawai, Y. Kouno, and J. Oikawa: "Characteristics of Speed Sensorless Vector Controlled Dual Induction Motor Drive Connected in Parallel Fed by a Single Inverter," *IEEE Trans. on.* vol. 40, no. 1, pp. 153-161, 2004.

[5] N. Urasaki, T. Senjyu, and K. Uezato: "Relationship of Parallel Model and Series Model for Permanent Magnet Synchronous Motors Taking Iron Loss Into Account," *IEEE Trans. on.* vol. 19, no. 2, pp. 265-270, 2004.

[6] D. Kayamori, and K. Fujisaki: "Influence of Power Semiconductor On-Voltage on Iron Loss of Inverter-fed", *IEEE Power Electronics and Drive Systems (PEDS)*, 9034, 2013.

[7] K. Fujisaki, and S. Liu : "Magnetic Hysteresis Curve Influenced by Power-Semiconductor Characteristics in PWM Inverter", *Journal of Applied Physics*, 115, 17A321, 2014.

[8] A. Boglietti, A. Cavagnino, and M. Lazzari: "Fast Method for the Iron Loss Prediction in Inverter-Fed Induction Motors," *IEEE Trans. on.* vol. 46, no. 2, pp. 806-811, 2010.

[9] M. Kawabe, T. Nomiyama, A. Shiosaki, H. Kaihara, N. Takahashi, and M. Nakano: "Behavior of Minor Loop and Iron Loss Under Constant Voltage Type PWM Inverter Excitation," *IEEE Trans. on, vol. 48, no. 11*, pp. 3458-3461, 2012.

[10] A. Matsusita, T. Kusakabe, K. Yun, and K. Fujisaki: "Comparison of Hysterisis Curve by Sine Wave Excitation and Inverter Excitation", *IEEJ Industry Applications Society Conference*, 1-44, 2012 (in Japanese).

[11] K. Fujisaki, R. Yamada, and T. Kusakabe: "Difference in Iron Loss and Magnetic Characteristics for Magnetic Excitation by PWM Inverter and Linear Amplifier", *IEEJ transactions on industry applications*, 133(1), 69-76, 2013 (in Japanese).

Loss and Thermal Model for Power Semiconductors Including Device Rating Information

K. Ma, A. S. Bahman, S. M. Beczkowski, F. Blaabjerg

Department of Energy Technology
Aalborg University, Aalborg 9220, Denmark
kema@et.aau.dk, asb@et.aau.dk, sbe@et.aau.dk, fbl@et.aau.dk

Abstract – The electrical loading and device rating are both important factors that determine the loss and thermal behaviors of power semiconductor devices. In the existing loss and thermal models, only the electrical loadings are focused and treated as design variables, while the device rating is normally pre-defined by experience with poor design flexibility. Consequently a more complete loss and thermal model is proposed in this paper, which takes into account not only the electrical loading but also the device rating as input variables. The quantified correlation between the power loss, thermal impedance and silicon area of Insulated Gate Bipolar Transistor (IGBT) is mathematically established. By this new modeling approach, all factors that have impacts to the loss and thermal profiles of power devices can be accurately mapped, enabling more design freedom to optimize the efficiency and thermal loading of power converter. The proposed model can be further improved by experimental tests, and it is well agreed by both circuit and Finite Element Method (FEM) simulation results.

I. INTRODUCTION

In many emerging and important energy conversion applications like renewable energy, traction, aerospace, and electric vehicles, etc. the power electronics are essential parts which need to process relatively large amount of power even up to MW level [1]-[4]. Due to the limited space and high cost of failures, the power density and reliability performances are especially focused in these applications. Consequently, the power loss and thermal loading of power semiconductor device are becoming more important considerations which are required to be accurately designed and optimized [5]-[7], because they are sensitive parameters that are related to the efficiency, cost, and reliability performances of the whole converter system [8]-[14].

There are mainly two design freedoms related to the loss and thermal behaviors of power devices. It is well known that the electrical loading of converter is one of the determine factors (e.g. switching frequency, modulation, DC bus voltage, power factor, load current, etc.). Intensive works have been done to translate and optimize the electrical profiles of converter into the corresponding loss/temperature profiles of power semiconductor devices [15]-[23]. On the other hand, the device rating is another important design freedom which was less focused and utilized: If devices with larger current rating are applied, the power handling ability of converter can be enhanced, thereby the junction temperature will be relieved. It is expected that for a given converter specifications, the loss and themral loading can be also optimized by tuning the rating of devices. However, this mathmatical corelation between thermal loading and device rating has not been comprehensively established before.

With the existing models and design approaches, there are some limits to achieve fully designable and optimal power loss and thermal loading of power semiconductor devices. Normally in these models the power devices have to be first selected as a pre-known factor according to the applied current/voltage stresses [24]-[26]. Afterwards the electrical profiles of converter treated as input variables are translated to the loss/temperature profiles of the selected devices [15]-[19]. Loss and thermal optimizations are normally restrained by tuning the electrical parameters in a limited range [27]-[32]. If the loss and temperature can still not be satisfied, the power device and cooling system have to be re-selected with several try-and-error iterations until the requirements are satisfied. It can be seen that in this design process, the loss and temperature information is unexpected before the power devices are decided, thereby the design freedom of device rating is not well utilized with poor flexibility for loss/thermal optimization.

As a result, a more complete loss and thermal model is proposed in this paper, which takes into account not only the electrical loading but also the device rating as input variables. The quantified relationship between the power loss, thermal impedance and silicon area of power device Insulated Gate Bipolar Transistor (IGBT) is mathematically established. It is concluded that by this new model, all factors that have impacts to the loss and thermal profiles of power devices can be accurately mapped, enabling full design freedom to optimize the efficiency and thermal loading of converter by tuning the device rating. The proposed model is improved by experimental tests and is well agreed by both circuit and FEM simulations.

II. BASIC IDEA AND CONDITIONS FOR MODELLING

In order to include the device rating into the loss and thermal models of power device, the proper rating definition has to be first clarified. Normally the manufacturers use the current rating to quantify power handling ability of devices, it is defined as an RMS current at which the junction temperature achieves 125 °C at a given case temperature without switching [24], [26]. It can be seen that this current rating is an ambiguous and general definition which may vary a lot depending on the actual operating conditions of converter, normally some margins of current rating have to be reserved by experience to ensure a safe temperature in the device [24]. As a result the current rating provided by manufacturer is not a suitable parameter to quantify the rating of power devices for the loss and thermal modeling.

In most of the cases if the converter specifications are decided, the voltage level of converter will be also settled as well as the voltage rating of applicable power devices. But different amount of transistor and diode dies/chips, which are connected in parallel by means of bond wire and copper layer, may be used to achieve different current handling ability, as demonstrated in Fig. 1. Therefore the silicon area or paralleled chip number have more direct physical meaning to scale the power handling capability of power devices. Moreover the silicon area or paralleled chip number is a fixed parameter which does not change with the electrical loading of converter, making it more suitable to be modeled and quantified.

The flow of loss and thermal modeling used in this paper are shown in Fig. 2, where the corresponding input/output variables are highlighted. The electrical loading of converter is first translated to the power loss on devices, then the thermal impedance model is applied to generate the junction temperature, which is then feedback to the loss model in order to enable the temperature dependency of power losses [2], [15]. It is noted that besides the electrical loading of converter, the device rating quantified by paralleled chip numbers is also included as input variable.

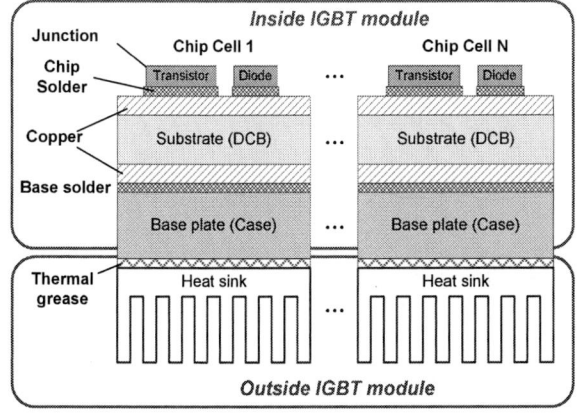

Fig. 1. IGBT physical structure with multi-cells.

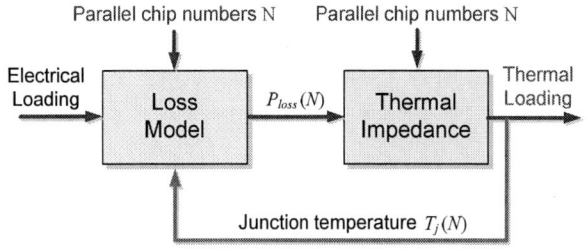

Fig. 2. Flow and idea for loss/thermal modelling including device rating information.

A popular two-level voltage source converter (2LVSC) used in the wind power application are chosen as a study case in this paper, as shwon in Fig. 3. The detail converter specifications and used transitor/diode chips are listed in Table I, which are typical values for a grid-side inverter at low-voltage level. It is noted that other operating conditions and converter solutions can be also applied sharing the similar modelling idea and process.

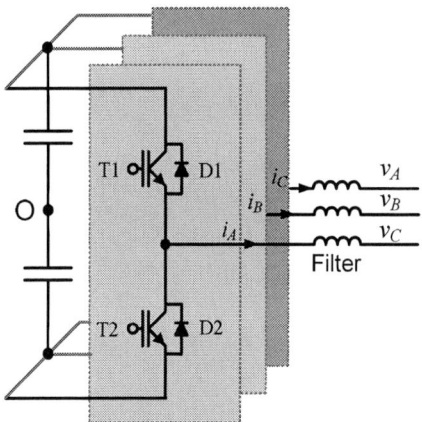

Fig. 3. Two-level voltage source DC-AC converter for case study (2L-VSC).

Table I. Converter parameters in Fig. 3.

Rated output active power P_o	250kW
Output power factor PF	1.0
DC bus voltage V_{dc}	1050 VDC
*Rated primary side voltage V_p	690 V rms
Rated load current I_{load}	209 A rms
Fundamental frequency f_o	50 Hz
Switching frequency f_c	2 kHz
Filter inductance L_f	1.2 mH (0.2 p.u.)
IGBT chip IGC186T170R3	1700V/150A
Diode chip SIDC85D170H	1700V/150A

* Line-to-line voltage in the primary windings of transformer.

III. LOSS MODEL INCLUDING DEVICE RATING INFORMATION

As has been well investigated in [15]-[23], the power loss for the power semiconductor devices is composed of two parts: conduction loss (or steady-state loss) and switching loss (or dynamical loss), the modeling process will be detailed in the following:

A. Conduction loss

The total conduction loss of a IGBT module is composed of the conduction loss on individual transistor/diode chips, therefore the instantaneous conduction loss of IGBT module with N chips paralleled ($p_{condT/D}$) is a function of N and time t, and can be expressed as the sum of conduction loss on each chip p_{condT/D_chipx}:

$$p_{condT/D}(N,t) = \sum_{x=1}^{N} p_{condT/D\_chipx}(t) \qquad (1)$$

For simplicity, assuming that the loss is evenly distributed on each transistor/diode chip, then the conduction loss of the whole IGBT module can be calculated as:

$$\begin{aligned} p_{condT/D}(N,t) &\approx N \cdot p_{condT/D\_chip}(t) \\ &= N \cdot v_{ce/f\_chip}(i_{chip}(t)) \cdot i_{chip}(t) \cdot D_{T/D}(t) \end{aligned} \qquad (2)$$

where v_{ce/f_chip} is the conduction voltage of transistor/diode chip, $i_{chip}(t)$ is the current flowing in each chip, $D_{T/D}(t)$ is the duty ratio for transistor or diode. If assuming the current is evenly distributed among paralleled chips, then:

$$i_{chip}(t) \approx \frac{|i_{load}(t)|}{N} \qquad (3)$$

The average conduction loss of IGBT Module P_{condT/D_Ave} can be expressed as the integral of instantaneous loss $p_{condT/D}$ within a fundamental cycle $1/f_o$:

$$\begin{aligned} P_{condT/D\_Ave}(N) &= f_o \int_0^{1/f_o} p_{condT/D}(N,t)\, dt \\ &= N \cdot f_o \int_0^{1/f_o} \left[(v_{ce/f\_chip}(i_{chip}(t)) \cdot i_{chip}(t) \cdot D_{T/D}(t) \right] dt \end{aligned} \qquad (4)$$

As can be seen in (2) and (4), the only unknown parameter is the conduction voltage for transistor/diode chip. Unfortunately, the conduction characteristic of the individual chip v_{ce/f_chip} is not always accessible information. If the current deviation among the chips is not significant when characterizing the IGBT modules, the v_{ce/f_chip} can be acquired by normalizing the conduction characteristic of IGBT module with function:

$$V_{ce/f\_chip}(I_{chip}) = V_{ce/f}(\frac{I_{ce/F}}{N}) \qquad (5)$$

Where $V_{ce/f}$ is the conduction voltage of the IGBT module and $I_{ce/F}$ is the corresponding load current, their relationship can be easily acquired from the device

datasheets[33], or by experimental characterization [16], [19].

As an example, the nominal conduction characteristic for a set of transistor chips are shown in Fig. 4, where the datasheet values from IGBT modules with different current ratings are used. It can be seen that the conduction characteristic of transistor or diode chip is quite consistent among different IGBT modules. Thereby the doted curve shown in Fig. 4 is used as the conduction characteristic of chips, which can be expressed as a function of chip current i_{chip} and two fitting constants V_{ce/f_chip0} and $B_{ce/f}$:

$$v_{ce/f\_chip}(i_{chip}) = V_{ce/f\_chip0} + (i_{chip})^{B_{ce/f}} \qquad (6)$$

B. Switching loss

Similarly, the total switching loss of a whole IGBT module is composed of the switching loss on individual chip. The instantaneous switching loss of IGBT module with N chips can be calculated as:

$$\begin{aligned} p_{sw/rr}(N,t) &= N \cdot p_{sw/rr\_chip}(t) \\ &= f_s \cdot N \cdot E_{sw/rr\_chip}(i_{chip}(t)) \end{aligned} \qquad (7)$$

Where E_{sw/rr_chip} is the switching energy for a transistor or diode chip, which is a function of chip current. f_s is the switching frequency of converter. The average switching loss of IGBT Module P_{swT/D_Ave} can be expressed as the integral of instantaneous loss $p_{swT/D}$ within a fundamental cycle $1/f_o$ as:

$$\begin{aligned} P_{swT/D\_Ave}(N) &= f_o \int_0^{1/f_o} p_{sw/rr\_chip}(N,t)\, dt \\ &= N \cdot f_s \cdot f_o \int_0^{1/f_o} E_{sw/rr\_chip}(i_{chip}(t))\, dt \end{aligned} \qquad (8)$$

If assuming the current is evenly distributed among each paralleled chip, then function (3) can be applied. As can be seen in (7) and (8), the only unknown parameter is the switching energy of the individual transistor/diode chip. Similarly, E_{sw/rr_chip} can be acquired by normalizing the characteristic of IGBT modules by:

$$E_{sw/rr\_chip}(I_{chip}) = \frac{1}{N} E_{sw/rr}(\frac{I_{ce/F}}{N}) \qquad (9)$$

where $E_{sw/rr}$ is the switching energy of the IGBT module and $I_{ce/F}$ is the corresponding load current, their relationship can be easily acquired from the device datasheet or by experimental tests.

As an example, the nominal switching characteristic for a set of transistor/diode chips are shown in Fig. 5, where the turn-on energy from IGBT modules with different current ratings are indicated. It can be seen that unlike the conduction characteristic in Fig. 4, the switching characteristic per chip in different modules distribute in a larger range. This is because the switching characteristic of power devices depends on many factors, such as the drive

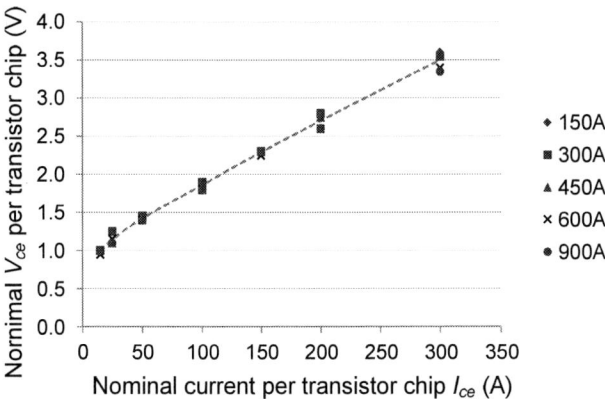

Fig. 4. Nominal conduction voltage for single transistor/diode chip in 1700V IGBT modules having different current ratings (transistor, datasheet values, recommended testing condition, Tj=125°C).

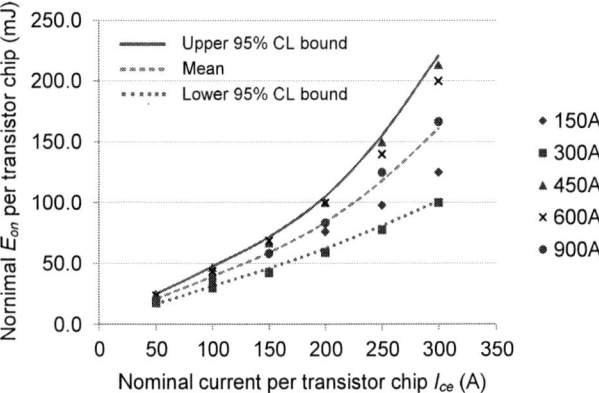

Fig. 5. Nominal switching loss for single IGBT/Diode chip in a series of modules having different current ratings (turn-on energy of transistor, datasheet values, recommended testing condition, Tj=125°C).

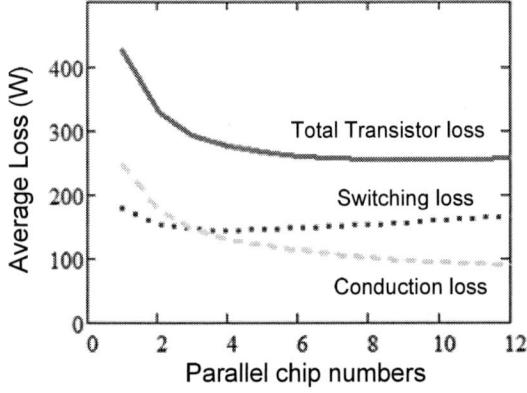

Fig. 6. Average power loss of transistor/diode in IGBT module with relation to paralleled chip numbers (Tranistor, converter condition in Table I, Tj=125°C).

resistance, *di/dt*, stray inductance, etc. These factors are sensitive to the circuit configurations and can be easily deviated when different numbers of chips are packaged.

In order to cope with this distribution of switching characteristic, a statistics description is used to cover the upper and lower 95% confidence level as well as the mean value of the nominal switching energy [34]. The three curves can be chosen depending on the required margin and confidence, in this paper the mean value is used. All of the three curves in Fig. 5 can be expressed as a function of current flowing in chip i_{chip} and three fitting constants $S_{T/D1}$, $S_{T/D2}$, $S_{T/D3}$,

$$E_{sw/rr\_chip}(i_{chip}) = S_{T/D1} \cdot (i_{chip})^2 + S_{T/D2} \cdot i_{chip} + S_{T/D3} \quad (10)$$

With the loss characteristics of individual transistor/diode chip in Fig. 4 and Fig. 5, the relationship between the IGBT power loss and corresponding chip numbers in parallel can be established. The results are shown in Fig. 6, where the converter is operating under the conditions shown in Table I, and the average conduction loss, switching loss and total loss are indicated respectively at the junction temperature of 125 °C.

It can be discovered that the increasing of paralleled chip numbers (device rating) only benefits to the conduction loss reduction of IGBT module, while the switching loss cannot be improved by increasing the chip numbers or device rating: In transistor the switching loss start to saturate and dominate after the device rating increase to a certain level.

IV. THERMAL IMPEDANCE MODEL INCLUDING DEVICE RATING INFORMATION

As can be seen from Fig. 1, the thermal impedance of IGBT module is composed of two parts: the part for device internal packaging structure Z_{jc} (junction to case) and the part for the external cooling structure Z_{ca} (case to ambient) [35]-[28], the modelling process will be detailed in the following:

A. Thermal impedance inside IGBT module

Practically in order to limit the stray inductance, the chips of transistor and its freewheeling diode are packaged closely to form a "chip pair" unit, as illustrated in Fig. 1. In order to facilitate the packaging, the electrical and thermal performance of a chip pair unit is optimized, and various current rating for the IGBT modules is achieved by scaling the standard chip pair units in parallel. In a good packaging of IGBT module, the distance between chip pair units is optimized to minimize the thermal coupling effects. Therefore from the point view of thermal modeling, the internal structure of IGBT module can be virtually divided into many chip cells/units, as also illustrated in Fig. 1.

Normally the thermal impedance for power semiconductors is modelled by a series of RC lumps which are composed of a thermal resistance R and a thermal capacitance C. Typically multi RC lumps are used to represent different layers of the materials from the chip to the case of the IGBT module [35]-[28]. According to the

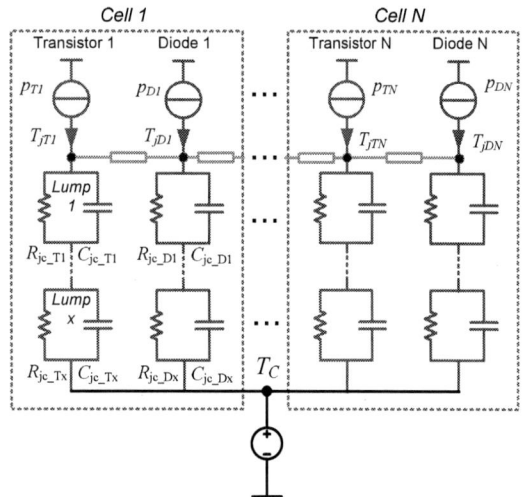

Fig. 7. Internal network of thermal impedance for IGBT module based on Fig. 1.

cell-based structure shown in Fig. 1, the network of thermal impedance for the internal structure of an IGBT module can be drawn in Fig. 7, in which the RC lumps is divided into multi-cells (cell 1-cell N) and multi-layers (lump 1–lump x), and the transistor/diode chips are represented by "thermal sources" governed by their instantaneous power losses.

Assuming the multi-layers structure from chip to case is evenly distributed for each of the divided cell, and then the thermal resistance for the xth layer RC lump of IGBT module $R_{jcT/Dx}$ is equivalent to the corresponding thermal resistance for each cell R_{jcT/Dx_chip} with N in parallel:

$$R_{jcT/Dx}(N) = \frac{R_{jcT/Dx\_chip}}{N} = \frac{d_x}{\lambda_x \cdot A_{T/Dx\_chip}} N^{-1} \quad (11)$$

where N is the number of paralleled chips, A_{T/Dx_chip} is the equivalent heat transfer area for a transistor/diode chip at x^{th} layer, d_x is the thickness of material at x^{th} layer, and λ_x is the thermal conductivity of material at x^{th} layer (W/m*K). By knowing the physical geometry and material of the IGBT structure, each of the thermal resistance in Fig. 7 can be calculated with the help of FEM simulation [39]. Unfortunately, the module structure/geometry and FEM simulation may not be always available. Another easier way used in this paper is to fit the function (11) with the thermal resistances from datasheets.

As shown in Fig. 8 (a), in which the overall thermal resistances for the transistor in IGBT modules with different current ratings are plotted, the values are acquired from datasheets and four layers of RC lumps (R_{jc1}-R_{jc4}) are used to represent the thermal behavior from chip to case. As can be observed, each layer of the thermal resistance in different IGBT modules can be fitted with a function that is inverse proportional to the chip number N, this relation is consistent with the physical model shown in (11), in which the parameters R_{jcT/Dx_chip} can be fitted with R_{jcx} shown in Fig. 8 (a). As a result, the relationship between thermal

resistance and corresponding device rating (paralleled chip numbers) are established.

In respect to the thermal capacitance, the modeling approach is similar. The thermal capacitance for the xth layer RC lump of IGBT module $C_{jcT/Dx}$ is equivalent to the corresponding thermal capacitance for each cell R_{jcT/Dx_chip} with N in parallel:

$$C_{jcT/Dx}(N) = N \cdot C_{jcT/Dx\_chip} = c \cdot \rho_x \cdot d_x \cdot A_{T/Dx\_chip} \cdot N \quad (12)$$

where the c is the specific heat factor proportional to heat in (Ws/g*K), ρ_x is the density of materials at x^{th} layer (g/cm^3). Similarly, the thermal capacitances in Fig. 7 can be acquired either by calculations or by fitting datasheet values. It is noted that, instead of thermal capacitance, the time constant of the RC lump $\tau_{jcT/Dx}$ is more commonly used by datasheets, it can be calculated as:

$$\tau_{jcT/Dx} = R_{jcT/Dx} \cdot C_{jcT/Dx} = \frac{c \cdot \rho}{\lambda} \cdot d_x^2 \quad (13)$$

In the function (9) it can be seen that there is no item for the chip numbers N as well as A_{T/Dx_chip}. That means the time constant of RC lump is only related to the physical prosperity and thickness of the xth layer material. Thereby with the same multi-layer structure and material, $\tau_{jcT/Dx}$ should be the same either between transistor and diode or between IGBT modules having different paralleled chip numbers.

The overall time constant for the transistor in IGBT modules with different current ratings are plotted in Fig. 8 (b), the values are acquired from datasheets and four layers of RC lump are used [37]. It can be seen that the time constants of the RC lump inside the IGBT modules keep constant under different device ratings, this characteristic is consistent with the physical model described in (13), in which the parameters $\tau_{jcT/Dx}$ can be fitted with t_{jcx} shown in Fig. 8 (b). As a result the relationship between thermal capacitance/time constant and corresponding device rating (paralleled chip numbers) are established.

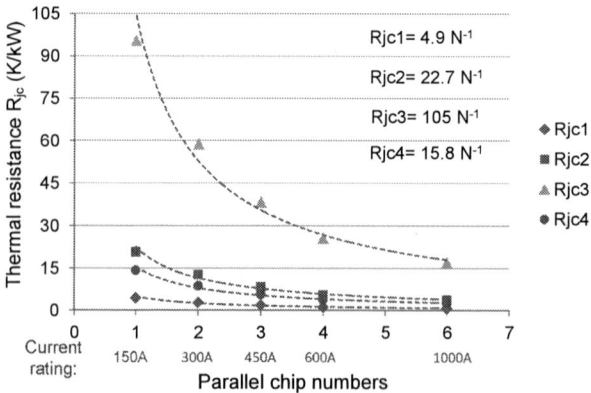

(a) Thermal resistance vs. chip numbers

The 2014 International Power Electronics Conference

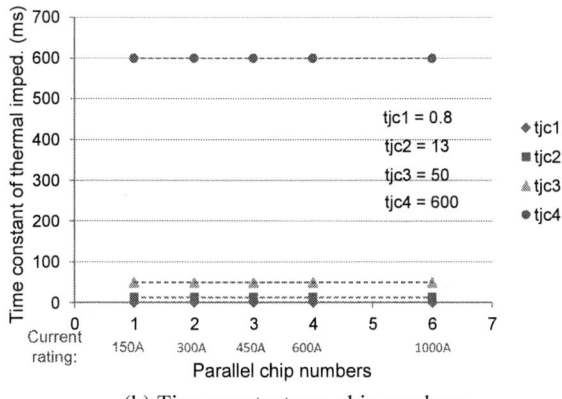

(b) Time constants vs. chip numbers

Fig. 8. Thermal impedance of RC lumps for transistor inside the IGBT modules (junction to case) with relation to paralleled chip numbers.

In a summary the time domain thermal impedance of IGBT module including the rating information (paralleled chip numbers) can be written as:

$$Z_{jcT/D}(N,t) = N^{-1} \cdot \sum_{x=1}^{4} R_{jcT/Dx\_chip} \cdot (1 - e^{-t/\tau_{jcT/Dx}}) \qquad (14)$$

where R_{jcTx} and τ_{jcTx} can be acquired from Fig. 8 (a) and Fig. 8 (b) respectively.

B. Thermal impedance outside IGBT module

The thermal impedance outside IGBT module mainly consists of two parts: the thermal grease part from case to heat sink, and the heat sink part from grease to the ambient, as indicated in Fig. 1. Unlike the thermal impedance inside the IGBT module, thermal modelling of this part normally involves a lot of uncertainties which are hard to predict. Due to the variations of applied grease type, thicknesses, mounting force, heat sink type and ambient temperature, the thermal impedance outside IGBT module may significantly deviate. Therefore the relatively accurate way is to seek the help of experimental testing and FEM simulation case by case.

In this paper, it is assumed that the heat sink is selected in a way to be able to maintain its temperature at 60 °C, which is a typical design target for the heat sink system [26]. The thermal grease is treated as a thermal resistance from case to heat sink R_{ch}, which is calculated based on thermal conductivity of 1 W/m*K, and a thickness of 150 μm. It is noted that R_{ch} is related to the base plate area of IGBT module, and it is no longer inverse proportional to the paralleled chip numbers N.

V. ANALYTICAL SOLUTION FOR JUNCTION TEMPERATURE

With the loss model in (4), (8) and thermal resistance model in (11), it is possible to acquire the information of the steady-state junction temperature of power device, and the temperature dependency of power losses can be also considered.

A. Analytical solution for power loss considering temperature dependency

It is well known that the loss characteristics of power semiconductor are temperature dependent [19] [24]-[26]. With the thermal resistance information, it is possible to include the junction temperature information into the power loss models. Considering the average power loss is linearly distributed with the junction temperature, then the updated average loss for IGBT module considering temperature dependency can be analytically solved by:

$$P_{LossT/D}(N) = \frac{P_{T/D\_Ave@Tr2}(N) \cdot (T_{r1} - T_{r2}) + \Delta P_{T/D\_Ave}(N) \cdot (T_h - T_{r2})}{(T_{r1} - T_{r2}) - [R_{jcT/D}(N) + R_{chT/D}(N)] \cdot \Delta P_{T/D\_Ave}(N)}$$
$$(15)$$

where

$$\Delta P_{T/D\_Ave}(N) = P_{T/D\_Ave@Tr1}(N) - P_{T/D\_Ave@Tr2}(N) \qquad (16)$$

$$P_{T/D\_Ave@Tr1/Tr2}(N) = P_{swT/D\_Ave} @ T_{r1} / T_{r2} \\ + P_{condT/D\_Ave} @ T_{r1} / T_{r2} \qquad (17)$$

The T_{r1} and T_{r2} are two reference temperatures which are normally chosen at 125°C and 25 °C respectively. The updated power loss of IGBT with relation to the device rating (paralleled chip numbers) is shown in Fig. 9, where the temperature dependency of losses is considered. The average transistor loss, diode loss, and total loss are indicated respectively, it can be see that with the given condition in Table I, there is an optimal point where the total loss of IGBT can be minimum when 6 chips are paralleled.

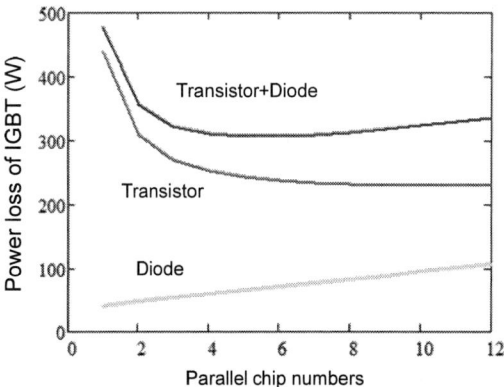

Fig. 9. Average power loss of transistor/diode vs. Paralleled chip numbers considering temperature dependency (converter condition in Table I).

B. Analytical solution for junction temperature

With the thermal impedance model (14) and instantaneous power loss model (2) and (7), it is possible to

978-1-4799-2706-7/14 $31.00 © 2014 IEEE

calculate the instantaneous junction temperature of power device by convoluting the loss and thermal impedance:

$$T_j(N,t) = T_h + \int_0^t \left[\frac{d}{dz} P_{\text{LossT/D}}(N,z) \right] \cdot Z_{jh}(N,t-z) dz \quad (18)$$

However, this calculation is too complicated and is not suitable for the design purpose. Actually only the mean junction temperature T_m and junction temperature fluctuation ΔT_j caused by the current alternating are interested parameters for the thermal design and optimization [20]-[23]. As a result, simplified solutions which can directly extract the information of T_m and ΔT_j are developed in this paper.

According to [36], the mean value of steady state junction temperature T_m can be written as follows, where the information of paralleled chip numbers is included:

$$T_{jm\_T/D}(N) = T_h + P_{\text{LossT/D}}(N) \cdot [R_{jcT/D}(N) + R_{chT/D}(N)] \quad (19)$$

In DC-AC converter, the alternating of the load current at fundamental frequency will result in quasi-sinusoidal distribution of instantaneous power loss on the transistor and diode. As a result corresponding temperature swing will be observed on the chips [40] and it also needs to be taken into account in the thermal design. Because the exact time when the junction temperature achieves its maximum/minimum value is hard to be derived by (12), approximation by loss pulses which have the same loss-time area as the original loss are used to calculate the ΔT_j. In this case the time when the junction temperature achieves its maximum value can be easily determined. In Fig. 10 loss approximation of two steps loss pulses which share the same loss-time area are applied to the same thermal impedance. It can be seen that the loss approximation can achieve good accuracy in respect to the temperature fluctuation amplitude.

As a result, the fluctuation of junction temperature ΔT_j with relation to the chip numbers of IGBT module can be analytically solved by:

$$\Delta T_{j\_T/D}(N) = P_{\text{LossT/D}}(N) \cdot Z_{jcT/D}\left(N, \frac{3}{8f_o}\right)$$
$$+ 2P_{\text{LossT/D}}(N) \cdot Z_{jcT/D}\left(N, \frac{1}{4f_o}\right) \quad (20)$$

And the maximum junction temperature of transistor or diode can be calculated as:

$$T_{jmax\_T/D}(N) = T_{jm\_T/D}(N) + \Delta T_{j\_T/D}(N) \quad (21)$$

In Fig. 11 the maximum junction temperature of transistor with relation to the device rating (parallel chip numbers) is shown, in which the conditions of converter in Table I with several variants of output powers and switching frequencies are indicated. It can be seen that different from the traditional thermal models, the device rating are also treated as variables. As a result, plenty of design freedoms to control and optimize the maximum

temperature of power device can be achieved by tuning both the device rating and the electrical loading.

VI. CONCLUSION

The loss and thermal loading of power devices is becoming critical design considerations which are closely related to the cost, efficiency and reliability performances of converter. In the existing loss and thermal models, only the electrical loadings are focused and treated as design variables, while the device rating is normally pre-defined by experience with poor design flexibility.

A more complete loss and thermal model is proposed in this paper, which takes into account not only the electrical loading but also the device rating as input variables. The quantified correlation between the power loss, thermal impedance and silicon area of Insulated Gate Bipolar Transistor (IGBT) is mathematically established. By this

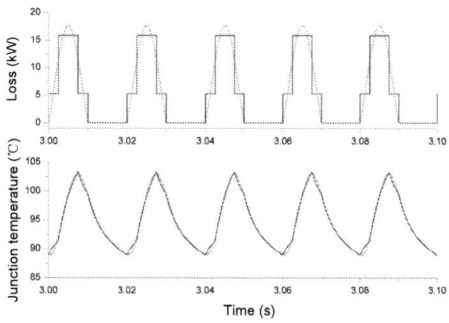

Fig. 10. Loss approximation to get analytical solution of junction temperature fluctuation. (green: original loss and corresponding junction temperature, red: Approximate loss pulses and corresponding junction temperature)

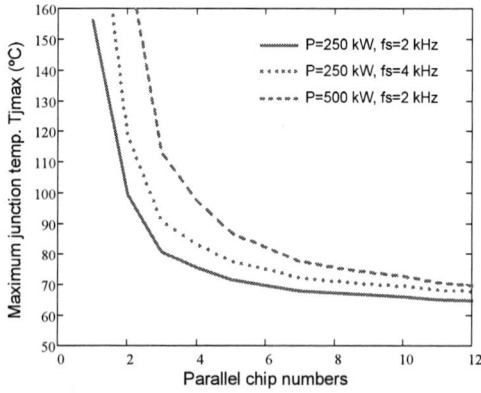

Fig. 11. Maximum junction temperature Tj_{max} vs. parallel chip numbers under different operating conditions of converter (only transistor is shown).

new modeling approach, all factors that have impacts to the loss and thermal profiles of power devices can be accurately mapped, enabling more design freedom to optimize the efficiency and thermal loading of power converter. The proposed model can be further improved by experimental tests, and it is well agreed by both circuit and Finite Element Method (FEM) simulation results.

REFERENCE

[1] F. Blaabjerg, K. Ma, "Future on Power Electronics for Wind Turbine Systems," *IEEE Journal of Emerging and Selected Topics in Power Electronics*, vol. 1, no. 3, pp. 139-152, 2013.

[2] D. Krug, S. Bernet, S.S. Fazel, K. Jalili, M. Malinowski, "Comparison of 2.3-kV Medium-Voltage Multilevel Converters for Industrial Medium-Voltage Drives," *IEEE Trans. on Industrial Electronics*, vol. 54, no. 6, pp. 2979-2992, 2007.

[3] S. Kouro, M. Malinowski, K. Gopakumar, J. Pou, L. G. Franquelo, B. Wu, J. Rodriguez, M. A. Perez, J. I. Leon, "Recent Advances and Industrial Applications of Multilevel Converters," *IEEE Transactions on Power Electronics*, vol. 57, no. 8, pp. 2553 – 2580, 2010.

[4] O. Apeldoorn, B. Odegard, P. Steimer, S. Bernet, "A 16 MVA ANPC-PEBB with 6 kA IGCTs," in *Proc. IAS'05*, vol.2, pp. 818-824, 2005.

[5] H. Wang, K. Ma, F. Blaabjerg, "Design for reliability of power electronic systems," *Proc. of IECON' 2012*, pp. 33-44, 2012.

[6] K. Ma, F. Blaabjerg, M. Liserre, "Operation and Thermal Loading of Three-level Neutral-Point-Clamped Wind Power Converter under Various Grid Faults," *Proc. of ECCE' 2012*, pp.1880-1887, 2012.

[7] S. Bernet, S. Ponnaluri, R. Teichmann, "Design and loss comparison of matrix converters, and voltage-source converters for modern AC drives," *IEEE Trans. on Industrial Electronics*, vol. 49, no. 2, pp. 304-314, Apr 2002.

[8] C. Busca, R. Teodorescu, F. Blaabjerg, S. Munk-Nielsen, L. Helle, T. Abeyasekera, P. Rodriguez, "An overview of the reliability prediction related aspects of high power IGBTs in wind power applications," *Microelectronics Reliability*, Vol. 51, no. 9-11, pp. 1903-1907, 2011.

[9] E. Wolfgang, "Examples for failures in power electronics systems," presented at *ECPE Tutorial on Reliability of Power Electronic Systems*, Nuremberg, Germany, April 2007.

[10] S. Yang, A. T. Bryant, P. A. Mawby, D. Xiang, L. Ran, and P. Tavner, "An industry-based survey of reliability in power electronic converters," *IEEE Trans. on Ind. Appl.*, vol. 47, no. 3, pp. 1441-1451, May/Jun. 2011.

[11] J. Due, S. Munk-Nielsen, Rasmus Nielsen, "Lifetime investigation of high power IGBT modules", in *Proc. of EPE'2011 – Birmingham*, 2011.

[12] J. Berner, "Load-cycling capability of HiPak IGBT modules," *ABB Application Note 5SYA 2043-02*, 2012.

[13] U. Scheuermann, "Reliability challenges of automotive power electronics," *Microelectronics Reliability*, vol. 49, no. 9-11, pp. 1319-1325, 2009.

[14] D. Hirschmann, D. Tissen, S. Schroder, R.W. De Doncker, "Reliability Prediction for Inverters in Hybrid Electrical Vehicles," *IEEE Trans. on Power Electronics*, vol. 22, no. 6, pp. 2511-2517, Nov. 2007.

[15] S. Dieckerhoff, S. Bernet, D. Krug, "Power loss-oriented evaluation of high voltage IGBTs and multilevel converters in transformerless traction applications," *IEEE Trans. on Power Electronics*, vol. 20, no. 6, pp. 1328-1336, Nov. 2005.

[16] F. Blaabjerg, U. Jaeger, S. Munk-Nielsen and J. Pedersen, "Power Losses in PWM-VSI Inverter Using NPT or PT IGBT Devices," *IEEE Trans. on Power Electronics*, vol. 10, no. 3, pp. 358–367, May 1995.

[17] M.H. Bierhoff, F.W. Fuchs, "Semiconductor losses in voltage source and current source IGBT converters based on analytical derivation," in *Proc. of PESC' 2004*, pp.2836-2842, 2004.

[18] A. M. Bazzi, P.T. Krein, J.W. Kimball, K. Kepley, "IGBT and Diode Loss Estimation Under Hysteresis Switching," *IEEE Trans. on Power Electronics*, vol. 27, no. 3, pp. 1044-1048, 2012.

[19] Jun Wang, Tiefu Zhao, Jun Li, A. Q. Huang, R. Callanan, F. Husna, A. Agarwal, "Characterization, Modeling, and Application of 10-kV SiC MOSFET," *IEEE Trans. on Electron Devices*, vol. 55, no. 8, pp. 1798-1806, Aug. 2008.

[20] O. S. Senturk, L. Helle, S. Munk-Nielsen, P. Rodriguez, R. Teodorescu, "Power Capability Investigation Based on Electrothermal Models of Press-Pack IGBT Three-Level NPC and ANPC VSCs for Multimegawatt Wind Turbines," *IEEE Trans. on Power Electronics*, vol. 27, no. 7, pp. 3195-3206, 2012.

[21] I. Swan, A. Bryant, P. A. Mawby, T. Ueta, T. Nishijima, K. Hamada, "A Fast Loss and Temperature Simulation Method for Power Converters, Part II: 3-D Thermal Model of Power Module," *IEEE Trans. on Power Electronics*, vol. 27, no. 1, pp. 258-268, 2012.

[22] Z. Zhou, M.S. Kanniche, S.G. Butcup, P. Igic, " High-speed electro-thermal simulation model of inverter power modules for hybrid vehicles," *IET Electric Power Applications*, vol. 5, no. 8, pp. 636-643, 2011.

[23] A. Ammous, S. Ghedira, B. Allard, H. Morel, "Choosing a thermal model for electrothermal simulation of power semiconductor devices," *IEEE Transactions on Power Electronics*, vol. 14, no. 2, pp. 300-307, Mar. 1999.

[24] FUJI Electric Co.,Ltd., "FUJI IGBT Moudules Application Manual," May 2011.

[25] ABB Application Note: Applying IGBT and diode dies, March 2010.

[26] A. Wintrich, U. Nicolai, T. Reimann, "Semikron Application Manual," 2011.

[27] M.Z. Sujod, I. Erlich, S. Engelhardt, "Improving the Reactive Power Capability of the DFIG-Based Wind Turbine During Operation Around the Synchronous Speed," *IEEE Transactions on Energy Conversion*, vol. 28, no. 3, pp. 736-745, Aug. 2013.

[28] F. Blaabjerg, J.K. Pedersen, "Optimized design of a complete three-phase PWM-VS inverter," *IEEE Transactions on Power Electronics*, vol. 12, no. 3, pp. 567-577, 1997.

[29] A. Isidori, F.M. Rossi, F. Blaabjerg, K. Ma, "Thermal Loading and Reliability of 10 MW Multilevel Wind Power Converter at Different Wind Roughness Classes," *IEEE Trans. on Industry Applications*, 2013.

[30] K. Ma, F. Blaabjerg, "Modulation Methods for Neutral-Point-Clamped Wind Power Converter Achieving Loss and Thermal Redistribution Under Low-Voltage Ride-Through.", *IEEE Trans. on Industrial Electronics*, vol. 61, no. 2, pp. 835-845, Feb 2014.

[31] T. Bruckner, S. Bernet, "Estimation and Measurement of Junction Temperatures in a Three-Level Voltage Source Converter," *IEEE Transactions on Power Electronics*, vol. 22, no. 1, pp. 3-12, 2007.

[32] K. Ma, M. Liserre, F. Blaabjerg, "Reactive Power Influence on the Thermal Cycling of Multi-MW Wind Power Inverter," *IEEE Trans. on Industry Applications*, vol. 49, no. 2, pp. 922-930, 2013.

[33] Website of Infineon Power Semiconductor, (Avaiable: https://www.infineon.com/cms/en/product/igbt/igbt-module/channel.html?channel=ff80808112ab681d0112ab69e66f036 2)

[34] Wikipedia "Confidence interval," Mar 2014. (Avaiable: http://en.wikipedia.org/wiki/Confidence_interval).

[35] M. Marz, P. Nance, "Thermal modeling of Power electronic System," Infineon Application Note.

[36] Infineon Application Note: "Thermal Resistance Theory and Practice," Jan 2000.

[37] Infineon Application Note AN2008-03: "Thermal equivalent circuit models", June 2008.

[38] ABB Application Note 5SYA 2093-00: "Thermal design and temperature ratings of IGBT modules", 2012.

[39] Z. Luo, A. Hyungkeun, M.A.E.Nokali, "A thermal model for insulated gate bipolar transistor module," *IEEE Trans. on Power Electronics*, vol. 19, no. 4, pp. 902-907, 2004.

[40] K. Ma, F. Blaabjerg, "Lifetime Estimation for the Power Semiconductors Considering Mission Profiles in Wind Power Converter," *IEEE Trans. on Power Electronics*, 2014.

The 2014 International Power Electronics Conference

Improving Reliability of IGBT Surface Electrode for 200 °C Operation

Tomohiro Nishimura, Yoshinari Ikeda, Hiroaki Hokazono, Eiji Mochizuki, Yoshikazu Takahashi

Fuji Electric Co., Ltd., 4-18-1 Tsukama, Matsumoto, Nagano, 390-0821 Japan

E-mail : nishimura-tomohiro-m@fujielectric.co.jp

Abstract— The surface barrier effect due to nickel (Ni) film on aluminum (Al) surface electrode of insulated gate bipolar transistor (IGBT) via power cycling (P/C) test at the maximal junction temperature (Tjmax) of 200 °C and thermal cycling (T/C) test in the -55 °C to 200 °C range has been carefully investigated. The difference of coefficient of thermal expansion (CTE) between Al and Ni and the stiffness of Ni played a key role to prevent mass transfer phenomena such as a migration of Al grain boundaries. We show that long P/C and T/C lifetime under high temperature operating condition can be achieved with the conventional IGBT module using our technique.

Keywords— *surface electrode, barrier layer, 200 °C operation, reliability*

I. INTRODUCTION

Rapid progress has been recently achieved in the research and development of power modules for extreme temperature operation leading to high efficient driving, downsizing, and cost cutting of power electronics equipment, in which novel technologies such as an interconnection and soldering are well discussed [1-3]. The degradation of surface electrode of IGBT devices is a cause to decrease the reliability of power modules at high Tj operation. Frequently it is considered that the degradation of surface electrode with increasing Tj leads to the increase of thermal resistance, the interfacial fatigue between bonding wire and electrode, and the failure of electrical contacting point of power devices. Although the application of high melting point materials to surface electrode and/or bonding wire in place of Al is an effective technique for suppression of electrode degradation, the reliability involving failure physics and the production challenge have not clarified.

We proposed the Ni plating technique on Al surface electrode as an improved Al based electrode to eliminate the topside failure of power devices during high Tj operation [4]. Consequently a failure part in power modules was limited to Al bonding wire with a combination of Ni plated Al surface electrode and high reliable die-attachment, the electrical breaking in case of a failure of power modules has been mechanically obtained. Thus we have established a reliable packaging technique for safety and maintainability of power electronics. In this study we expanded Ni plating technique on IGBT surface electrode to 200 °C operation of power modules, reported the improving reliability of IGBT surface electrode based on detailed experiments.

II. EXPERIMENTAL

A. Preparation of Surface Barrier Layer on IGBT Devices

The barrier layer of Ni on Al surface electrode of IGBT devices was prepared with a plating process. Darkened surface of Ni plated electrode was confirmed in comparison to bare Al electrode as shown in Fig. 1.

Fig. 1. Photograph of bare (a) and Ni plated (b) IGBT device.

B. Fablication of IGBT Modules

Power modules including IGBT devices with bare Al or Ni plated electrode for reliability tests were assembled with conventional components as shown in Fig. 2. Static characteristics for assembled power modules was carefully checked, no problem in all power modules was confirmed.

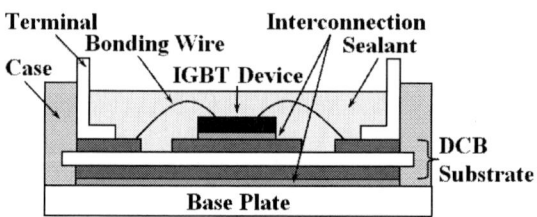

Fig. 2. Schematic of conventional IGBT module.

C. Reliablity Tests and Evaluation

All P/C tests were performed at Tjmax = 200 °C due to Tc = 25 °C and dTj = 175 °C. The holding temperature in all T/C tests was alternately repeated from -55 °C to 200 °C.

Structural observations for surface electrodes of IGBT devices after reliability tests were carried out with a scanning electron microscope (SEM) and a focused ion beam - scanning ion microscope (FIB-SIM).

978-1-4799-2706-7/14 $31.00 © 2014 IEEE

III. RESULTS AND DISSCUSSION

A. Power Cycling Reliability

P/C reliability is an important issue as a simulation of switching numbers involving a heat generation for power modules. Thus an improvement and an increase of P/C capability advantageously affect to the long term reliability of power electronics equipment.

A SEM image of bare Al surface electrode of IGBT device prior to P/C operation was shown in Fig. 3 (a). Remarkable structures in Al grains or grain boundaries were not recognized in acquiring SEM image. However, the drastic structural change indicating degradation of bare Al surface electrode after P/C operation was confirmed in Fig. 3 (b). Al electrode degradation connected to the stress migration (SM) phenomenon as in Fig. 3(b) has been well investigated [5]. Detailed mechanism has been actively discussed, was originated to the tensile behavior in Al grains and/or grain boundaries due to thermal stress [6]. Although the application of copper added AlSi has been proposed as a candidate for SM suppressing material, the improvement has been slightly exhibited [7].

On the other hand, remarkable degradation suppression effect for P/C reliability due to Ni plated Al surface electrode was appeared. SEM images of Ni plated surface electrode of IGBT devices prior to and after P/C operation were shown in Fig. 3 (c) and (d). Comparison of Fig. 3 (c) and (d) illustrates no clear differences, indicates that Ni layer on Al surface electrode has effectively behaved as the surface barrier metal. It can be considered that sandwiched Al electrode with a power die and Ni layer has a difficulty for SM and/or other mass transfer phenomena due to the stiffness of Ni. Additionally, the difference of CTE between Al and Ni played a key role to suppress mass transfers of Al.

Surface barrier effect of Al electrode due to Ni layer was evidently demonstrated from SEM observations. However, the degradation such as voids inside an electrode is an important issue for a reliability of power modules, has not been characterized. Thus we have rigorously investigated the depth directional degradation of surface electrode via P/C operations with FIB-SIM. FIB-SIM image of degraded bare Al electrode via P/C operation was shown in Fig. 4 (a). It can be seen that Al grains and grain boundaries were heavily deformed in comparison to initial Al electrode with the bamboo structure. Complex of the bamboo grain boundary SM model [8] and the stone walling grain boundary SM model [9] has been required for explanation of present result. Understanding of SM of electrode on power devices will be under discussion. Comparison of Fig. 4 (a) and a typical cross section FIB-SIM image of Ni plated Al electrode, Fig. 4 (b), obviously exhibited the suppression effect of degradation of Al surface electrode. Ni layer indicating with arrows in Fig. 4 (b) behaved as a barrier film for SM expansion of Al, prevented electrode degradation. However, a crack on Ni layer has infrequently occurred via P/C operation, behaved as a trigger of SM expansion. Fig. 4 (c) shows an irregularly observed cross section FIB-SIM image of Ni plated Al electrode with cracked Ni indicating arrowed gap. It was demonstrated that the stress relaxation region added due to Ni crack induced the SM of Al basis. The present results do not

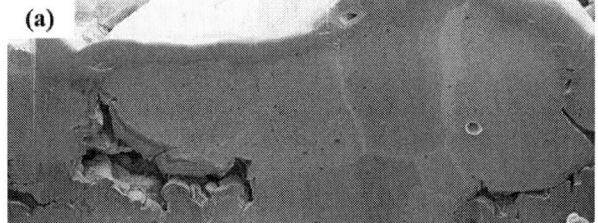

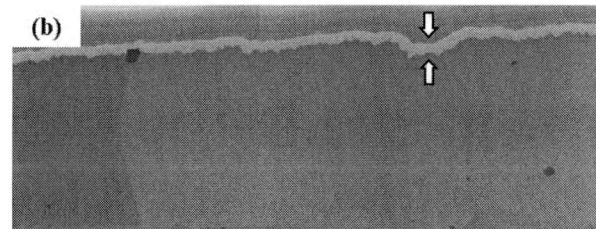

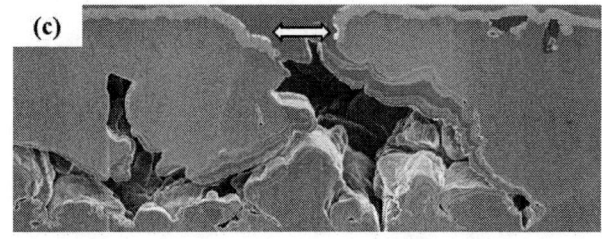

Fig. 4. Cross section FIB-SIM images of IGBT surface electrodes after P/C operation.
(a) Bare Al surface electrode.
(b) Typical and (c) irregularly cracked Ni plated electrode.

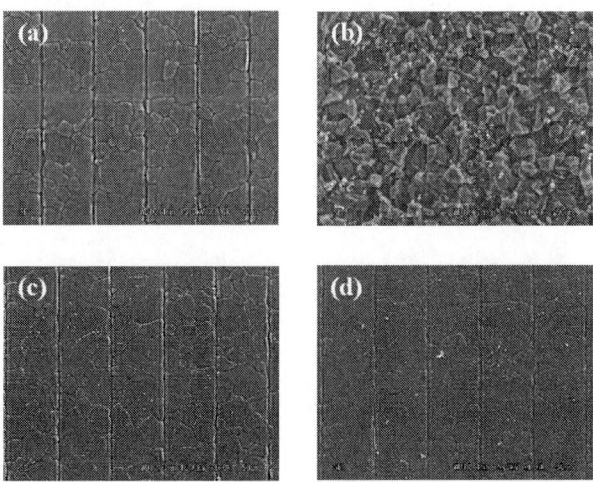

Fig. 3. SEM images of IGBT surface electrodes.
Al electrode prior to (a) and after (b) P/C operation.
Ni plated electrode prior to (c) and after (d) P/C test.

discount the surface barrier effect with Ni plating, but these results highlights the need for precise design of electrodes.

Continuous P/C test at Tjmax = 200 °C was performed with conventional type IGBT power modules as shown in Fig. 2 assembled with improved Ni plated Al surface electrode and high reliable components. Consequently over 100,000 P/C lifetimes at F(t) = 1 % marked with red line in Fig. 5 was taken, 50 times P/C lifetimes was achieved in comparison to the P/C lifetime of typical power modules marked with blue line in Fig. 5.

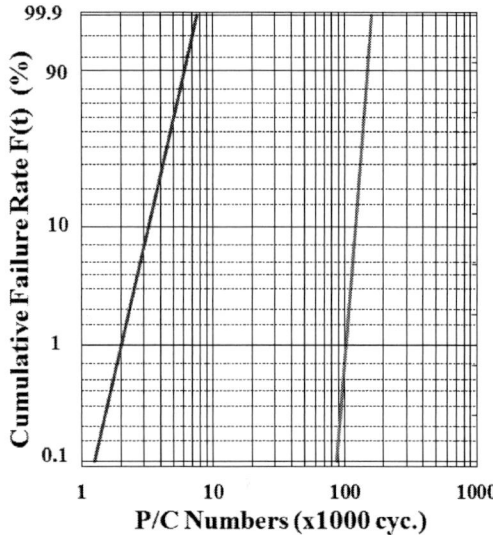

Fig. 5. Weibull plot of continuous P/C operation.
Bule line indicates reliability of typical power modules.
Red line indicates reliability of developed one.

B. Thermal Cycling Reliability

T/C reliability is a key issue as a simulation of holding temperature for power modules involving surface electrode of power dies. Reliability of surface electrode of IGBT devices via T/C operation was investigated.

A SEM image of bare Al surface electrode after T/C operation was shown in Fig. 5 (a). The structural degradation of surface electrode was slightly appeared. Mechanism of electrode degradation has been also considered in SM phenomenon. However, a significant difference in magnitude of electrode degradation via T/C operation was recognized in

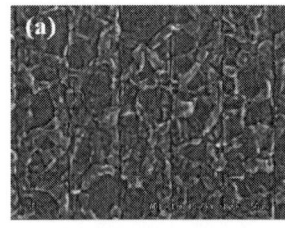 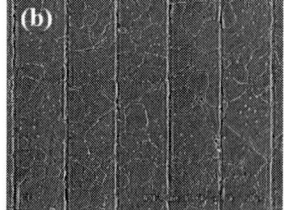

Fig. 6. SEM images of IGBT surface electrodes after T/C test.
Bare Al electrode (a) and Ni plated Al electrode (b).

contrast to that via P/C operation as shown in Fig. 3, can be explained in the difference of thermal stress distribution. Although thermal stress in case of P/C operation localizes around center of power devices, that in case of T/C operation entirely distributes in power modules. SM has been generally accepted as a mass transfer phenomenon along the stress gradient, Al transfers during T/C operation was suppressed in comparison to that of P/C operation. While Fig. 6 (a) indicates the electrode degradation during T/C operation is slight, it also points to an advantage of Ni layer behaved as a surface barrier layer on Al surface electrode as shown in Fig. 6 (b). It can be considered that Ni layer plays a key role to suppress the delamination of bonding wires due to Al structural changes from surface electrode.

Additionally the depth directional electrode degradation was investigated with FIB-SIM. Fig. 7 (a) and (b) show cross section FIB-SIM images of bare Al electrode and Ni plated Al electrode after T/C operation. Although the mass transfer of Al grains and/or grain boundaries partially occurred, significant changes inside Al electrode did not precipitated.

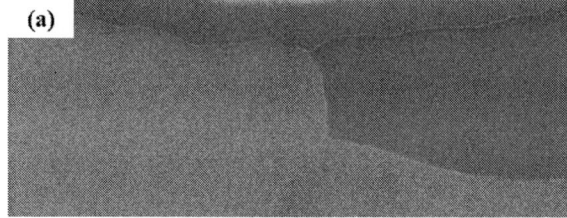

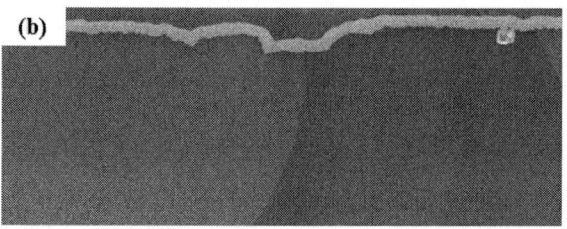

Fig. 7. Cross section FIB-SIM images of IGBT surface electrodes after T/C operation.
(a) Bare Al electrode and (b) Ni plated Al electrode

IV. CONCLUSION

Ni layered Al surface electrode on IGBT devices leading to the improvement and increase of reliability for power modules was proposed. Remarkable advantages were revealed via detailed studies at high temperature based on P/C operation, T/C operation, and other results which will be presented in the meeting. Additionally, a combination of our developed Ni plated Al surface electrode and high reliable components provides long term P/C lifetime in comparison to typical modules.

However, we confirmed challenges of this technique. A challenge is the crack of Ni layer causes of degradation inside Al electrode during P/C operation. Instability of Ni layer discounted the reliability of surface electrode, indicates the imperfection with conventional modules. Another is a challenge of power module as electrical circuit. While improved our technique provides the increase of reliability of power modules, a reduction of inductance and a downsizing are limited due to a technique based on conventional packaging technologies. The requirement of premium reliability for power modules has been recognized for the increase of power electronics quality, it has been established with new concept power module [10]. We have expected the developed technique in this study will bridge the conventional power modules and future developments.

ACKNOWLEDGMENT

We would like to thank Ms. Yohko Kakiki for her help in our experiments and valuable suggestions for results.

REFERENCES

[1] R.A. Amro, "Packaging and Interconnection Technologies of Power Devices, Challenges and Future Trends", *Proc. of World Academy of Science, Enginnering and Technology*, Vol. 25, pp. 691-694, 2009.

[2] K. Guth, D. Siepe, J. Görlich, H. Torwesten, R. Roth, F. Hille and F. Umbach, "New assembly and interconnects beyond sintering methods", *Poc. of PCIM 2010*, pp. 232-237, 2010.

[3] T. Saito, Y. Nishimura, K. Kido, F. Momose, E. Mochizuki and Y. Takahashi, "New assembly technologies for Tjmax=175°C continuous operation guaranty of IGBT module", *Proc. of PCIM 2013*, pp. 455-461, 2013.

[4] Y. Ikeda, H. Hokazono, S. Sakai, T. Nishimura and Y. Takahashi, "A study of the bonding-wire reliability on the chip surface electrode in IGBT", *Proc. of the 22nd ISPSD*, pp. 289-292, 2010.

[5] J. Curry, G. Fitzgibbon, Y. Guan and R. Muollo, "New Failure Mechanisms in Sputtered Aluminum-Silicon Films", *Proc. Int. Reliability Physics Symp.*, pp. 6-8, 1984.

[6] K. Hinode, I. Asano, T. Ishiba and Y. Homma, "A study on stress-induced migration in aluminum metallization based on direct stress measurements", *J. Vac. Sci. & Tchnol. B*, Vol. 8, No.3, pp. 495-498, 1990.

[7] S. Mayumi, T. Umemoto, M. Shishino, H. Nanatsue, S. Ueda and M. Inoue, "The Effect of Cu Addition to Al-Si Interconnects on Stress Induced Open-Circuit Failures", *Proc. Int. Reliability Physics Symp.*, pp. 15-21, 1987.

[8] F.G. Yost, D.E. Amos and A.D. Romig, "Stress-driven diffusive voiding of aluminum conductor lines", *Proc. Int. Reliability Physics Symp.*, pp. 193-201, 1989.

[9] F.G. Yost, "Voiding due to thermal stress in narrow conductor lines", *Scr. Metall.*, Vol. 23, No. 8, pp. 1323-1328, 1989.

[10] Y. Ikeda, Y. Iizuka, Y. Hinata, M. Horio, M. Hori and Y. Takahashi, "Investigation on Wirebond-less Power Module Structure with High-Density Packaging and High Reliability", *Proc. of the 23rd ISPSD*, pp. 272-275, 2011.

Influence of Carrier Frequency on Iron Loss Taking Account of Dead Time Effect

Ryosuke Kogi
(Toyota Technological Institute)
Dept. of advanced of engineering
2-12-1 Hisaskata, Tempaku
Nagoya, Aichi, Japan
sd08047@toyota-ti.ac.jp

Shunya Odawara
(Toyota Technological Institute)
Department of engineering
2-12-1 Hisaskata, Tempaku
Nagoya, Aichi, Japan
sodawara@toyota-ti.ac.jp

Keisuke Fujisaki
(Toyota Technological Institute)
Department of engineering
2-12-1 Hisaskata, Tempaku
Nagoya, Aichi, Japan
fujisaki@toyota-ti.ac.jp

Abstract— In order to confirm the iron loss characteristics of electrical steel sheet magnetized by PWM inverter excitation, its carrier frequency characteristics are evaluated by taking account of dead time effect. Usually the iron loss decreases when the carrier frequency becomes large in less than 10 kHz or so. However, the iron loss increase tendency is measured in the higher carrier frequency as 40 kHz. The reason is considered to be that DC voltage increases and it is attributable to dead time in PWM signal phase in this paper.

Keywords—PWM Inverter, Iron Loss, Carrier Frequency, Dead Time

I. INTRODUCTION

Electrical equipment prevails widely in the world by progress of human society, and the demand of electrical energy has been expanded. In recent years, however, there is an urgent need to establish the more high-efficient electrical equipment due to the problem of nuclear power plants and the worldwide environmental issues [1].

To use an electrical power effectively, a power electronics technology is also progressing. Especially, by using new power semiconductor device, high-speed switching becomes possible, and the loss reduction on power conversion circuit and the miniaturization of circuit become possible [2]-[4]. Although the conventional switching frequency, which is called carrier frequency in this paper, was about 10 kHz, it is possible to use until MHz order or so due to the realization of new material devices as SiC or GaN. Since previous research shows that the iron loss increases by inverter excitation [5]-[8], we suggest that it is necessary to evaluate not only the inverter circuit loss but also the iron loss on inverter excitation. Moreover, the iron loss characteristics at high carrier frequency are considered to be an important item to be evaluated.

In this paper, the iron loss characteristics on more than carrier frequency of 10 kHz used conventionally are evaluated by the measured data and the obtained iron loss characteristics are discussed from view point of the applied DC voltage and the dead time on inverter.

II. METHOD FOR MEASUREMENT

A. Measurment System

In this research, the single phase PWM (Pulse-Width-Modulation) inverter which is made in person is used, because the carrier frequency of 20 kHz or more cannot be treated in the commercial inverter owned at the laboratory. Fig.1 shows the measurement system. The ring specimen composed of electrical steel sheet is excited by inverter. This ring specimen is for evaluating the iron loss seen with magnetic material characteristics. The electrical steel sheet is processed by wire-cut as shown in Fig. 2(a). It is laminated and placed in acrylic case like Fig. 2(b), and the primary coil and the secondary coil are winded. The specifications of ring are shown in Table I.

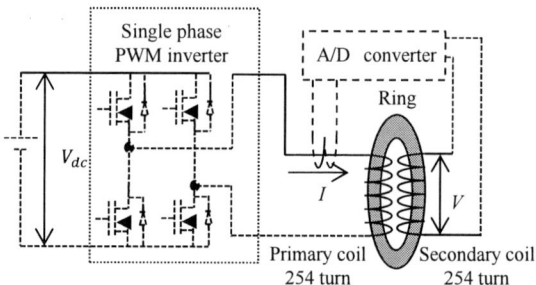

Fig.1. Overview of measurement system

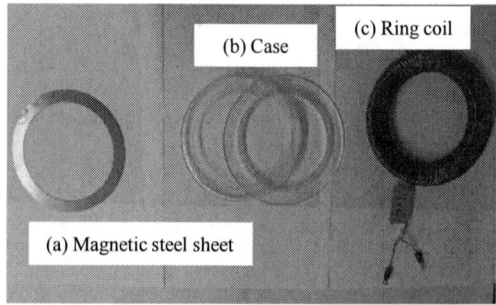

Fig.2. Overview of ring specimen.

TABLE I. SPECIFICATIONS OF RING SPECIMEN.

Material	35H300
Outside diameter	127 mm
Inside diameter	102 mm
Height	7 mm
Primary coil winding number	254 turn
Secondary coil winding number	254 turn

B. Excitation Conditions

The ring specimen is excited so that the maximum magnetic flux density B_{max} in ring becomes 1 T. The fundamental frequency f_0 is 50 Hz, the modulation factor m is 0.8. The carrier frequency f_c is changed from 1 kHz to 40 kHz.

C. Calculation of Iron Loss

The iron loss W of ring is obtained from integration of the magnetic field intensity H calculated from primary current I and the magnetic flux density B calculated from secondary voltage V, as follows:

$$H = \frac{N_1 I}{L} \ [A/m] \tag{1}$$

$$B = \frac{1}{N_2 S} \int V dt \ [T] \tag{2}$$

$$W = \frac{f_0}{\rho} \int H dB \ [W/kg] \tag{3}$$

where N_1, N_2, L, S and ρ in equation the primary windings, the secondary windings, the magnetic path length of ring, the cross section area of ring, and the density of electrical steel sheet, respectively.

III. EXPERIMENTAL RESULT

Fig. 3 shows the iron losses on each carrier frequency f_c. In our conventional research used carrier frequency of 10 kHz or less, the iron loss is decreased when the carrier frequency becomes large. Therefore, it is expected that the iron loss is decreased or saturated when the carrier frequency becomes larger. However, although this tendency is also obtained until f_c = 10 kHz in Fig. 3, the iron loss begins to increase when f_c is set to more than 20 kHz. The results of more than f_c = 50 kHz cannot obtain because the applied DC voltage V_{dc} becomes large and the used voltage source cannot apply such voltage value. In following chapter, the cause which the iron loss is increased on more than f_c = 10 kHz is described.

IV. DISCUSSION

A. Comparison of B-H Curve

Fig. 4 shows the B-H curves of 10 kHz and 40 kHz and Fig. 5 shows the enlargement near B = 0 T. In Fig. 5, the pulsating component ΔH of magnetic field intensity on 40 kHz becomes larger than that of 10 kHz. Therefore, BH curve of 40 kHz is expanded to H component by pulsation ΔH than the BH

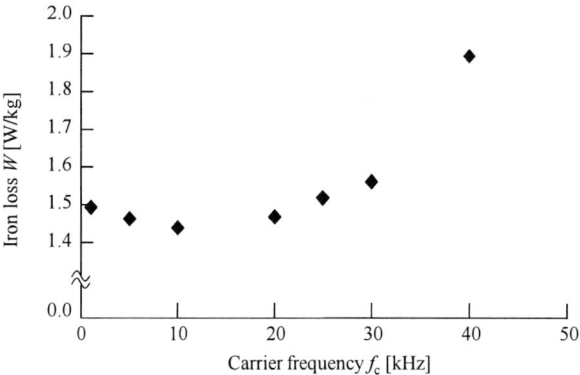

Fig.3. Relationship between iron loss and carrier frequency (f_0 = 50 Hz, m = 0.8).

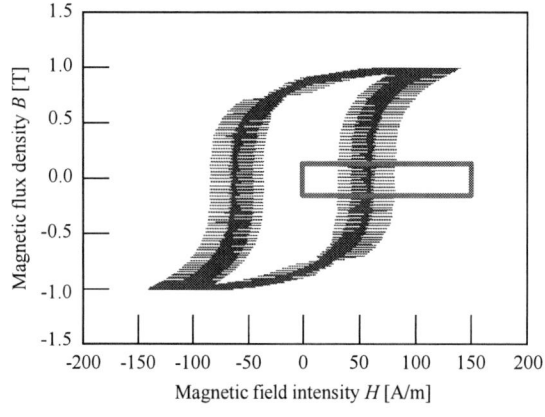

(a) f_c = 10 kHz

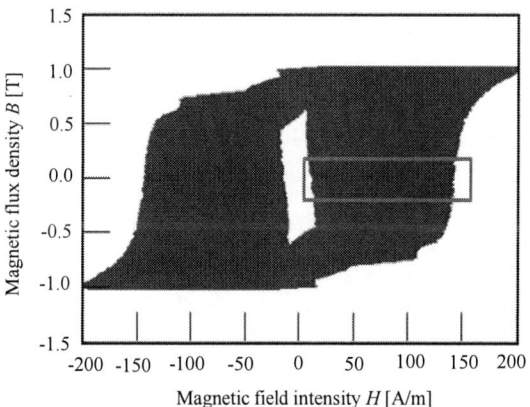

(b) f_c = 40kHz

Fig. 4. B-H curves (f_0 = 50 Hz, m = 0.8).

curve of 10 kHz. Since the iron loss obtained from equation (3) is proportional to the inside area of BH curve, the iron loss becomes large in f_c = 40 kHz in which the area of BH curve is larger than f_c = 10 kHz. In next section, the reason of increasing ΔH is described.

978-1-4799-2706-7/14 $31.00 © 2014 IEEE

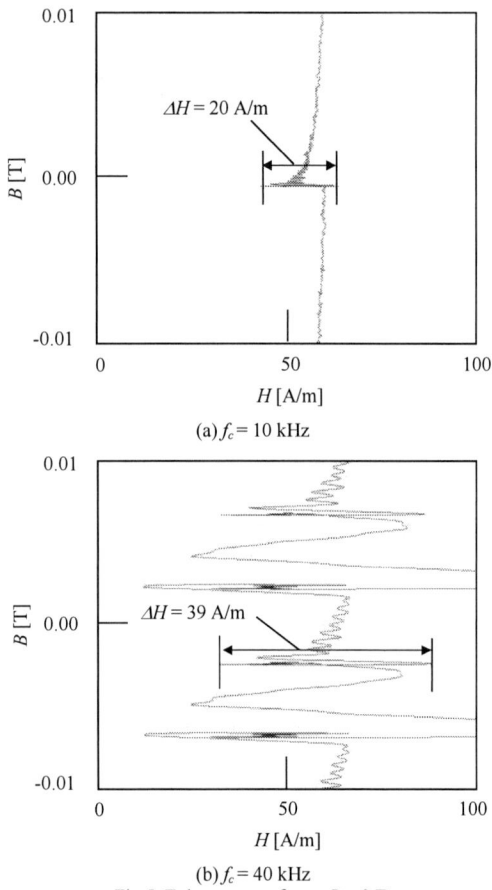

(a) f_c = 10 kHz

(b) f_c = 40 kHz
Fig.5. Enlargement of near B = 0 T.

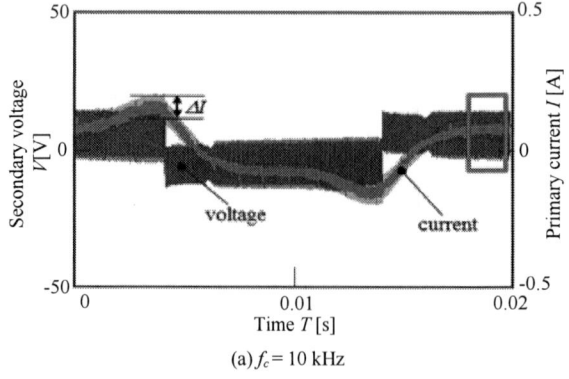

(a) f_c = 10 kHz

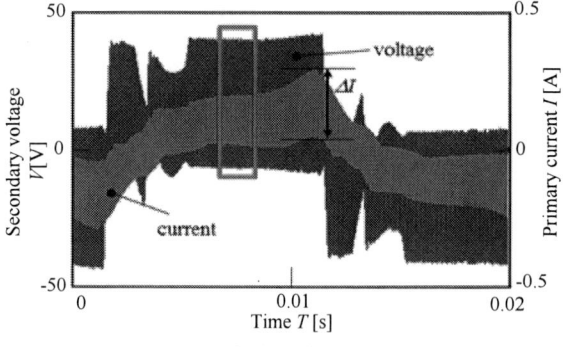

(b) f_c = 40 kHz
Fig.6. Voltage and current waveforms (f_0 = 50 Hz, m = 0.8).

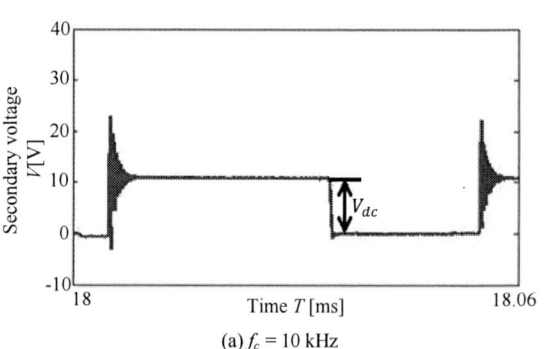

(a) f_c = 10 kHz

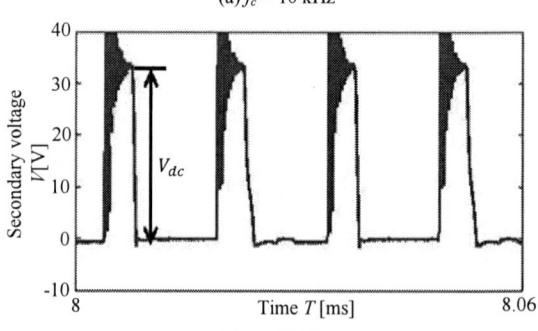

(b) f_c = 40 kHz
Fig.7. Enlargement of voltage near B = 0 T.

B. Cause of Increasing ΔH

Fig. 6 shows the primary current waveform and the secondary voltage waveform. The enlargement of voltage near B = 0 T is shown in Fig. 7. In Fig. 6, the secondary voltage is increased when the f_c becomes large. In this case, although the secondary voltage differs from applied DC voltage V_{dc} strictly, primary coil windings and secondary coil windings of used ring specimen are the same, so the primary voltage corresponds to the secondary voltage one-to-one. Therefore, since V_{dc} becomes large due to increasing f_c, the current ripple ΔI of primary current I as shown in Fig. 6 is increased. Magnetic field intensity H is obtained from the primary current as equation (1), so the current ripple ΔI affect ΔH directly.

Fig. 7 shows the relationship between f_c and V_{dc}. The V_{dc} is increased when the f_c becomes large. This phenomenon is caused by dead time T_{dt} [9]. The dead time, in which the OFF-signal is given, is included in gate-signal in PWM control so that the circuit cannot be short. In next chapter, the relationship between T_{dt} and V_{dc} is described analytically.

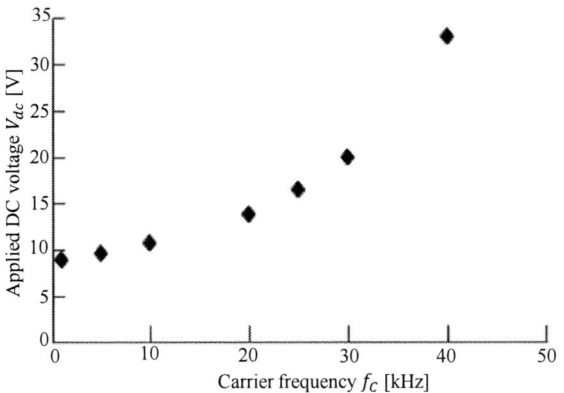

Fig. 8. Relationship between applied DC voltage and carrier frequency ($f_0 = 50$ Hz, $m = 0.8$).

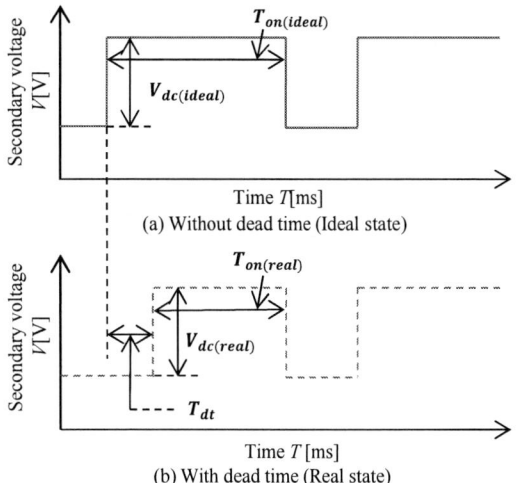

(a) Without dead time (Ideal state)

(b) With dead time (Real state)

Fig. 9. Voltage waveforms of ideal state and real state.

V. ANALYTICAL INVESTIGATION

A. Theoretical Analysis

It is expressed in previous chapter that the T_{dt} has the factors increasing V_{dc}. In this chapter, the cause is explained and the influence rate of dead time is investigated. Although the equation (2) is an expression for obtaining the magnetic flux density, this equation (2) can be said to be a summation of the product of applied DC voltage and ON-time of PWM. Thus, equation (2) can be expressed as the total area of pulse width of the secondary voltage waveform. When the maximum magnetic flux density B_{max} is constant, the total area becomes constant if the carrier frequency is changed. The number of ON-pulse N is expressed as follow:

$$N = \frac{2 f_c}{f_0} \qquad (4)$$

The ideal state in which the dead time is not included and the real state in which the dead time is included are defined. In the Fig. 9, the ON-time of real state is narrower than that of ideal state because the pulse width of ON-time of real state becomes as following equation (5):

$$\sum_{i=1}^{N} T_{on(real)i} = \sum_{i=1}^{N} T_{on(ideal)i} - N \cdot T_{dt} . \qquad (5)$$

Moreover, in this study, the maximum magnetic flux density to be applied to the ring is constant, so the total area of pulse width is constant. It can be described both of the real state and the ideal state. Thus, following equation (6) is true.

$$\sum_{i=1}^{N} T_{on(real)i} \cdot V_{dc(real)} = \sum_{i=1}^{N} T_{on(ideal)i} \cdot V_{dc(ideal)} \qquad (6)$$

By using equation (4), (5), (6), following equation (7) can be obtained.

$$V_{dc(real)} = V_{dc(ideal)} + \frac{\dfrac{2 f_c}{f_0} T_{dt}}{\displaystyle\sum_{i=1}^{N} T_{on(real)i}} V_{dc(ideal)} \qquad (7)$$

The applied voltage V_{dc} is increased by the second term of equation (7) due to the influence of dead time.

B. Theoretical Modelling

In order to investigate the influence of dead time affecting to applied voltage V_{dc}, the numerical simulation is carried out. Table II shows the conditions of simulation. The enlargement of secondary voltage waveform obtained from simulation of each carrier frequency is shown in Fig. 10.

In Fig. 10, when the applied voltage V_{dc} of ideal state is assumed to be 1, V_{dc} in the real state of each carrier frequency is increased. Fig. 11 shows the increase rates of V_{dc} on numerical simulation and on actual measurement when the carrier frequency is increased. In Fig. 11, the increase rates of V_{dc} of simulation and measurement are increased when the carrier frequency becomes large. However, there is difference between the increase rates in simulation and measurement. Finally, voltage waveform obtained from simulation is compared with that obtained from measurement.

TABLE II. CONDITIONS FOR SIMULATION

Fundamental frequency f_0	50 Hz
Modulation factor m	0.8
Maximum magnetic flux density B_{max}	Constant
Dead time T_{dt}	4000 ns
Carrier frequency f_c	10~40 kHz
The point of enlargement	Near the $B_{max} = 0$ T

978-1-4799-2706-7/14 $31.00 © 2014 IEEE   2877

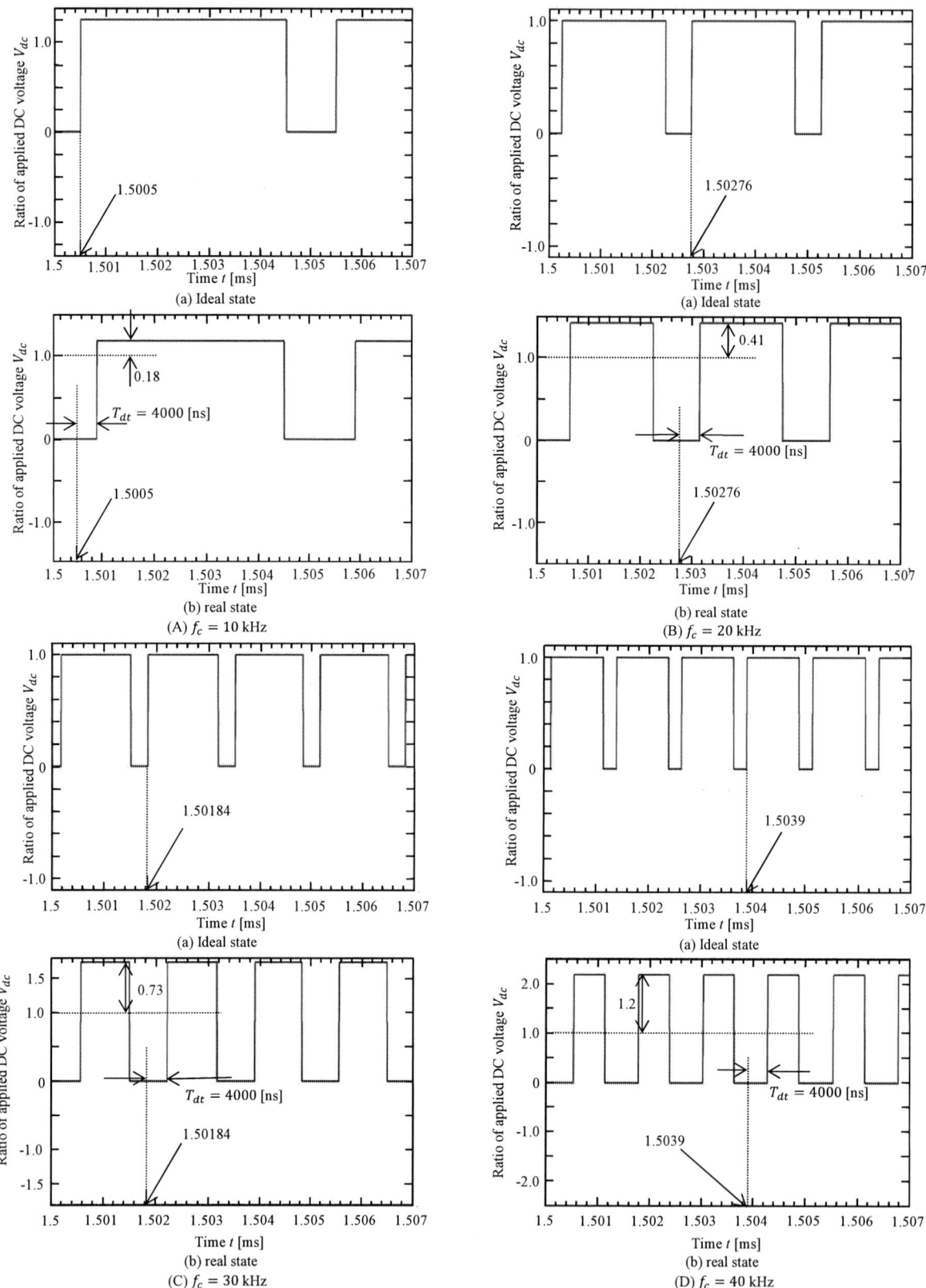

Fig.10. Simulation result.

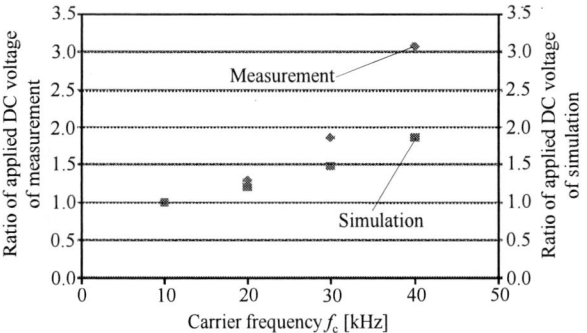

Fig.11. Increase rate of V_{dc} on each f_c.

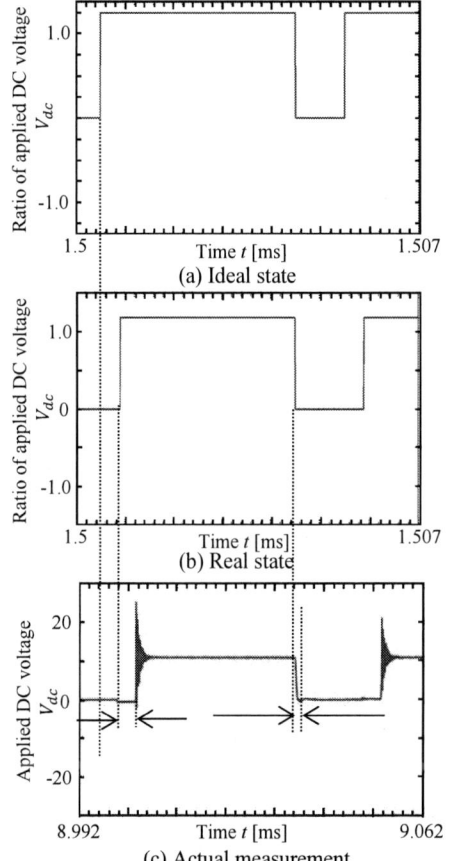

Fig.12. Comparison of simulation and measurement.

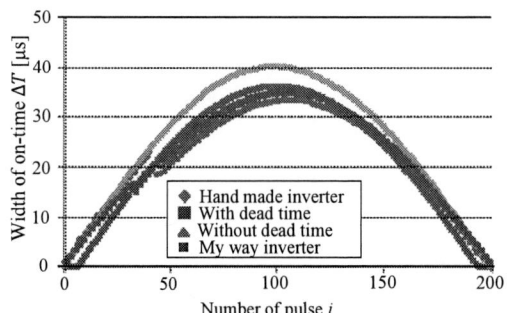

Fig.13. Difference of ON-time on each result.

C. Comparison with Simulation and Measumant

Fig. 12 shows the enlargements of voltage waveform near $B = 0$ T of ideal state, real state and actual measurement on f_c = 10 kHz. In Fig. 12, the pulse width of actual measurement is narrower than that of simulation including dead time. As one of this reason, it is caused due to the influence of rise time and of fall time on the power semiconductor device, because the rise time and the fall time are neglected in simulations.

Fig. 13 shows the width of pulse 200 pieces of the secondary voltage waveform of a half of one-cycle on the fundamental wave. In Fig.12, the horizontal axis is the series number of the pulse, and the vertical axis is its pulse width. As the measured data, the results by using the handmade inverter and the commercial inverter (MWINV-9R122A) produced by Myway Plus Corporation are shown in Fig. 12. In Fig. 12, the pulse widths obtained from each method are different. This difference is caused due to difference in characteristics rise-time and fall-time of the semiconductor device.

VI. CONCLUSION

In this research, the iron loss characteristics on higher carrier frequency than 10 kHz or so are evaluated and discussed from a view point of applied DC voltage and dead time of PWM inverter. The conclusion obtained from this research is as follows.

1. On higher carrier frequency than the conventional one, iron loss increases.

2. The cause of the iron loss increase is that the current ripple ΔI and the magnetic field intensity ripple increase by the applied DC voltage increase.

3. The cause of the applied DC voltage increase is an effect of dead time in PWM inverter.

REFERENCES

[1] Agency for Natural Resources for and Energy (Japan),"Annual Energy Report",2010

[2] Y. Nakabayashi, S. Fujisaki, K. Terazono, H. Hara, and K. Ideno, "Development of High Power Density AC-AC converter", *IEEJ Industry Applications Society Conference*,1-O1-2, 2012

[3] Y. Matsumoto, Y. Kondo, Y. Kobayashi, and H. Kimura, "Industrial Motor Drive Inverter Adopting SiC Power Device", IEEJ Industry Applications Society Conference,1-O1-3, 2012

[4] H. Kaihara, et al,"Effect of Carrier Frequency and Circuit Resistance on Iron Loss of Electrical Steel Sheet Under Single-Phase Full-Bridge PWM Inverter Excitation", *IEEE Trans. on,* vol. 48, no.11, pp.3454-3457,2012

[5] D. Kayamori, K. Fujisaki, "Comparison with SiC-MOSFET and IGBT of Iron Loss on Electrical Steel Sheet in Single Phase Inverter", *IEEJ Technical Meeting on Magnetics*, MAG-12-156, 2012

[6] M. Sakai, M. Takiguchi, T. Koyama, K. Kodachi, K. Ogura, "A evaluation of SiC 15kVA Invarter", *IEEJ Industry Applications Society Conference*,1-O1-4, 2012

[7] K. Fujisaki, S. Liu, "Magnetic Hysteresis Curve Influenced by Power-Semiconductor Characteristics in PWM Inverter", Journal of Applied Physics, vol.115, 17A321, 2014

[8] R. Kogi, S. Odawara, K. Fujisaki, "Iron Loss Increase due to Increase of Carrier Frequency on Inverter", IEEJ Technical Meeting on Magnetics, MAG-13-150, 2013

978-1-4799-2706-7/14 $31.00 © 2014 IEEE

[9] S. M. Dehgham and K. Fujisaki, "Impact of Dead-Time on Iron Loss in Inverter-Fed Magnetic Materials", *IEEE Energy Conversion Congress & Exposition (ECCE)*, pp.3166-3171, 2013

Decrease of SiC-BJT driver losses by one-step commutation

Henry Barth, Wilfried Hofmann
Department of Electrical Machines and Drives
TU Dresden
Dresden, Germany
Henry.Barth@tu-dresden.de

Abstract— The silicon carbide bipolar junction transistor (SiC-BJT) is a promising power semiconductor device for high efficient motor drive inverters. Especially with the very low on-state voltage drop and high switching speed it challenges the state of the art device – the silicon IGBT. The main disadvantage is the high driver loss in on-state compared to its voltage driven competitors, though. With the one-step commutation in voltage source inverters (VSI) a new approach on decreasing on-state losses is presented. An "all-SiC" inverter with DC-link consisting of SiC-BJTs and SiC-diodes has been designed and build-up. First measurements indicate that by using one-step commutation instead of the conventional one driver losses can be cut in half. This leads to an increase of efficiency of SiC-BJT voltage source inverters.

Keywords— *SiC-BJT, silicon carbide motor drive inverter, decrease of driver losses, one-step commutation*

I. INTRODUCTION

After decades of research into wide band gap semiconductor devices silicon carbide diodes and transistors are commercially available today [1]. These devices are strong competitors for conventional IGBTs in electric motor drives, if high temperature and high efficiency operation with high power density are required. Especially the silicon carbide bipolar junction transistor is an interesting alternative due to the lowest specific on-state resistance of its class ($R_{ON} = 53$ mΩ @ 15 A / 25°C). Furthermore its on-state resistance has the weakest temperature dependency of all SiC transistors. At the same time the total switching energy losses equal basically the ones of a Cree SiC MOSFET [1]. Thus, possible applications for SiC-BJT inverters are traction drives for electric vehicles or electric power transmission [2].

The main drawback however is its control mechanism. During on-state there is DC-current of several hundred milliamperes flowing into the base. This causes driver and base-emitter diode losses. To minimize those, several driving concepts have been proposed [3], [4]. Instead of designing a new driver, the aim of this work is to decrease the driver and base-emitter diode losses by one-step commutation algorithm [5] that is well known in connection with VSIs to eliminate the dead time effect [6].

II. REDUCING ON-STATE LOSSES BY ONE-STEP-COMMUTATION

In VSIs a short circuit of the DC-link must be prevented in any case, i.e. switches S1 and S2 of the phase-leg in Fig. 1 a must not be in on-state at the same time. The conventional commutation algorithm (see Fig. 1 b) switches alternating between upper and lower device of one phase-leg.

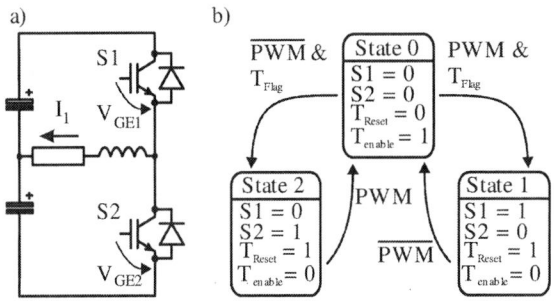

Fig. 1. Half-Bridge setup (a) and finite state machine (FSM) of conventional commutation (b)

In order to prevent a shoot through, a blanking time greater than the longest expected transistor turn-off time is inserted between turn-off of one transistor and turn-on of the other. In the finite state machine (FSM) of Fig. 1 b the blanking time equals the time from setting T_{enable} until the timer flag T_{Flag} is set. By setting T_{Reset} the timer is stopped and T_{Flag} reset.

This algorithm is simple to implement but causes current waveform distortion and fundamental output voltage loss [6] as Fig. 3 a illustrates. One way to eliminate the dead-time effect is the one-step commutation [5], [6]. The idea of this type of commutation shows the FSM in Fig. 2. With additional information about the polarity of the load current only one of the two phase-leg transistors is turned on and off while the other one is in off-state. In case of positive current of I_1 S1 is switched on and off while S2 stays switched-off. Thus, there is no need to insert any blanking time as long as the current polarity of I_1 is positive. Accordingly, S2 is switched on and off while S1 stays switched-off at negative I_1.

978-1-4799-2706-7/14 $31.00 © 2014 IEEE

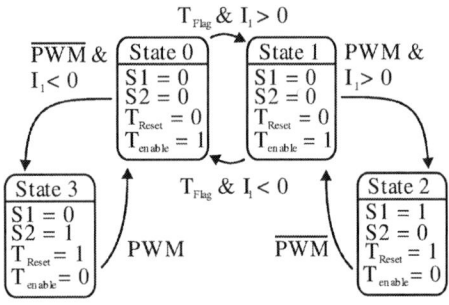

Fig. 2. FSM of one-step commutation

Only when I_1 is close to a value of zero ampere the conventional commutation algorithm is used. The result is a sinusoidal fundamental without noteworthy harmonics and a higher current amplitude (see Fig. 3 b).

The information about current-polarity is determined by measuring the phase currents with commercially available current sensors [7]. These noisy signals are smoothed by a Kalman-Filter using simplified models of motor and inverter.

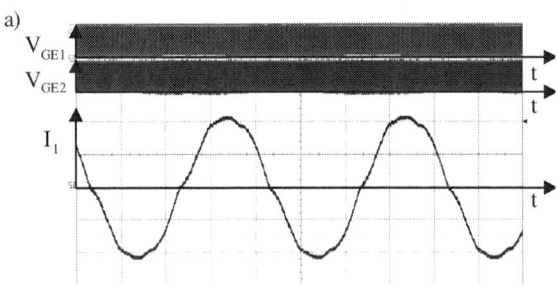

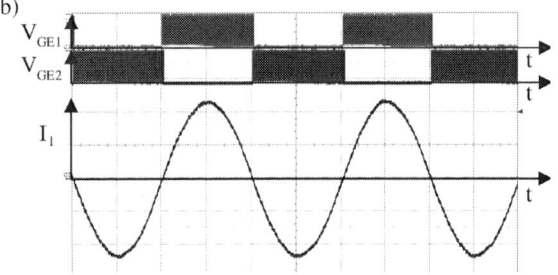

Fig. 3. Measurements on laboratory IGBT inverter ($\hat{I}_l$=4A)
a) conventional commutation
b) one-step commutation

For a phase-leg consisting of SiC-BJTs each base driver needs energy for only half a current period. Thus, compared to conventional commutation, the one-step commutation algorithm eliminates unnecessary driver and base-emitter diode losses.

III. ON–STATE LOSS CALCULATION

The on-state losses of a SiC-BJT [8] and a state of the art Si-IGBT [9] have been calculated numerically. In case of a switched-on BJT on-state loss energy consists of energywhile conducting the current E_{con}, driver- and base- emitter-loss energies (E_{drv}, E_{BE}). The instantaneous

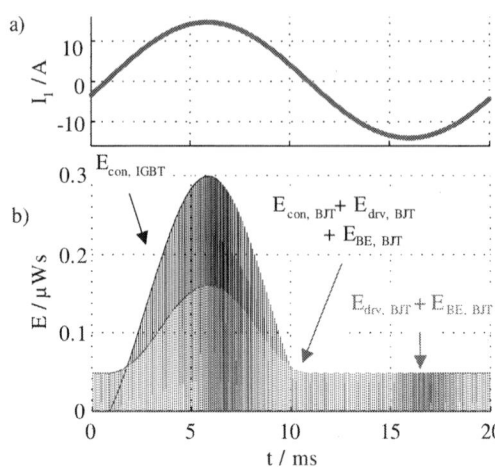

Fig. 4. Numerical calculation of on-state losses at $\hat{I}_l = 15$ A

on-state loss energy calculations shown in Fig. 4 b are based on I_C-V_{CE}-datasheet diagrams and a phase current with an amplitude of 15 A (see Fig. 4 a) generated by space vector modulation. Furthermore a junction temperature of 25°C, a switching frequency of $f_S = 10$ kHz and a base current of 400 mA have been assumed. The actual on-state losses have been calculated with (1).

$$P_{con} = \frac{\sum_{k=1}^{100}\left(E_{con}(k)+E_{BE}(k)+E_{drv}(k)\right)+\overbrace{\sum_{k=101}^{200}\left(E_{BE}(k)+E_{drv}(k)\right)}^{\to\,0\text{ for one-step commutation}}}{T} \quad (1)$$

for $f_S = 10$ kHz, $f_0 = 50$ Hz

The proposed one-step-commutation eliminates the red marked area of driver and base-emitter diode losses in Fig. 4 b. Accordingly, the overall on-state losses P_{con} of this SiC-BJT can be reduced by 25% at load current with peak value of 15 A compared to conventional commutation (see Fig. 5). In comparison to the IGBT the on-state losses of the SiC-BJT with one-step commutation are only 60%.

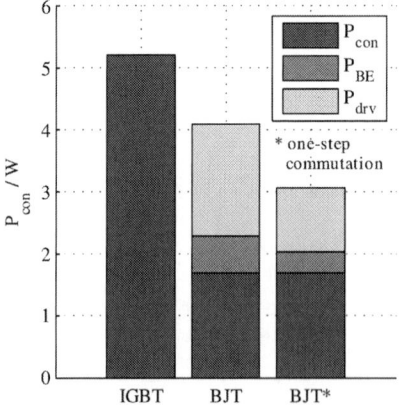

Fig. 5. Comparison of on-state losses between IGBT [7] and SiC-BJT [8] with conventional- and one-step-commutation (*) at $\hat{I}_l = 15$ A

At lower modulation indices an even higher driver loss reduction is expected, due to increasing impact of driver and base-emitter diode losses on total losses (see Table I). Indeed there is a reduction of on-state losses up to 40 % at 5 A, but those losses are still higher than the ones of the IGBT at 25°C. Due to the weak temperature-dependency of the SiC-BJT's on-resistance this might even change at higher temperatures.

TABLE I
CALCULATED ON-STATE LOSSES

$\hat{I}$	P_{con}(IGBT)	P_{con}(BJT)	P_{con}(BJT*)	P_{con}(*)-decrease
5 A	1.2 W	2.6 W	1.5 W	> 40 %
10 A	3.1 W	3.2 W	2.1 W	≈ 35 %
15 A	5.2 W	4.1 W	3.1 W	≈ 25 %

IV. SiC–BJT INVERTER

An "all-SiC" inverter of a VSI has been designed and assembled. SiC-BJTs [8] and SiC – Schottky Barrier Diodes (SiC-SBD) [10] with 15 A / 1200 V rating have been applied. The inverter is shown in Fig. 6.

Fig. 6. SiC-BJT inverter

Two foil-capacitors with a capacity of 80 µF each that can withstand voltages far above 600 VDC even at higher temperatures have been installed as DC-link capacitors. Both low- and high-side drivers are electrically isolated. IGBT-drivers [11] decouple galvanically the control signals. Additionally features like desaturation protection and "gate-clamping" come in handy. The latter allows driving the SiC-BJTs with a unipolar driver. As mentioned in the introduction, the effort of providing galvanic isolated power supply to drive SiC-BJTs is much higher than for IGBT- or MOSFET-inverters. Since during on-state there is a continuous base current required. This leads to an increase of power and size of the applied DC/DC converters. The converters of this laboratory inverter can provide the required base current permanently. This means the three high side drivers are equipped each with a 5 W device [12] and the three low-side drivers use one 15 W converter [13].

V. TEST SET-UP

A. For Switching behavior of SiC-BJT

A base driver using a push-pull stage has been designed to control one SiC-BJT [8]. Additionally to the IGBT driver [11] a low side MOSFET driver [14] has been installed to be able to provide the base current continuously. A SiC-SBD [10] was used as freewheeling diode. The driver and the devices under test are used later on in the laboratory inverter described above. In buck converter setting (see Fig. 7) a double-pulse was applied to the device under test. Besides the base resistor R_B a clamp resistor R_{clamp} is used to ensure the off-state of the BJT when it is required. This feature allows using a unipolar driver with an output voltage between 0 V and 12 V.

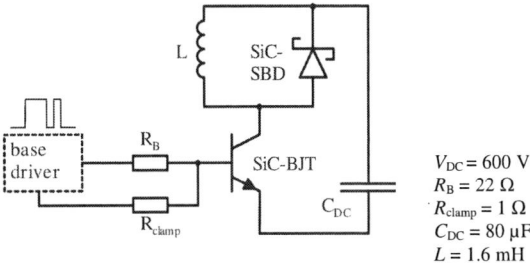

$V_{DC} = 600$ V
$R_B = 22$ Ω
$R_{clamp} = 1$ Ω
$C_{DC} = 80$ µF
$L = 1.6$ mH

Fig. 7. Double pulse test set-up

The measuring instruments used for recording the voltage and current waveforms are specified in Table II.

TABLE II
MEASURING INSTRUMENTS – SWITCHING LOSSES

Dimension	Scale	Bandwidth	Device/Probe
V_{CE}	1:100	300 MHz	TESTEC TT-HV150
I_C	1:10	50 MHz	FLUKE i50s
		600 MHz	LeCroy Waverunner XI

B. For inverter operation

Besides measuring equipment and the control unit the test set-up consists of a transformer, a laboratory voltage source inverter and an electrical machine set (see Fig. 8).

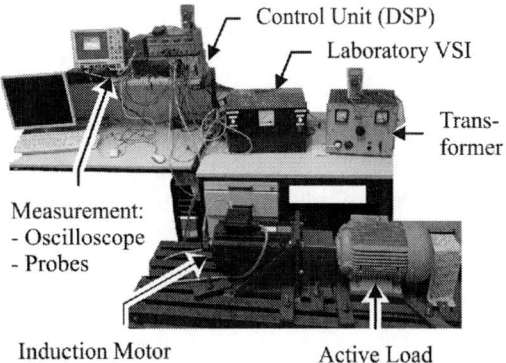

Fig. 8. Test set-up

The one-phase transformer shown in the picture has been replaced by a voltage-adjustable three phase one in order to stabilize the DC-link voltage.

First studies have been done with an inverter applying discrete silicon IGBTs with internal free-wheeling diodes [9]. Three current sensors [7] and one voltage sensor provide the inverter controlling DSP with the phase current and DC-link voltage values. Linear V/f control is used to drive the induction motor. Space vector modulation is used to generate the pulse patterns for all six drivers. Both, conventional and one-step commutation algorithms are implemented on the DSP. After successful commissioning of the test set-up and comprehensive measurements, the IGBT inverter has been replaced by the SiC-BJT one.

The load consists of two induction machines that are rigidly coupled. The one connected to the inverter has a rated power of 3.6 kW and rated voltage of 330 V. Due to this voltage rating and to a conservative safety margin concerning the isolation of the winding when using fast switching semiconductor devices the DC-link voltage for the measurements is set to 200 V.

The measurement set-up to determine the total driver losses accordingly to the commutation algorithm is illustrated in Fig. 9.

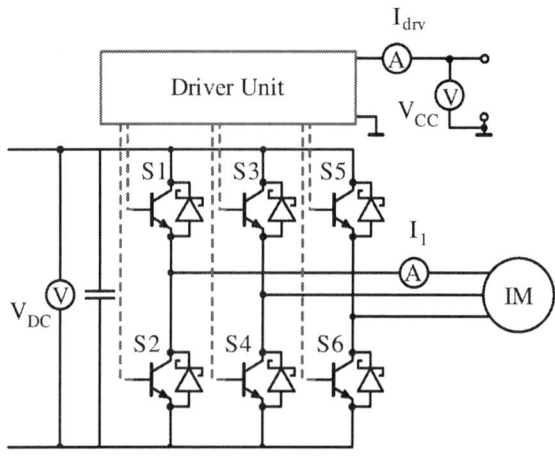

Fig. 9. Measurement of driver losses

Besides measuring the supply voltage V_{CC} and current I_{drv} of the driver unit, the DC-link voltage V_{DC} and one phase-current I_1 are recorded. Additionally the control voltage V_{S1} is monitored in order to detect the commutation algorithm that is used.

The driver unit includes all six BJT - driver units and the DC-chopper driver. In Table III the measuring equipment is listed.

TABLE III
MEASURING INSTRUMENTS – DRIVER LOSSES

Dimen-sion.	Osc.-Channel	Scale	Bandwidth	Device/Probe
V_{S1}	C1	1:10	400 MHz	LeCroy PP008
I_{DRV}	C2	1:1	50 MHz	FLUKE i50s
V_{CC}	---	1:1	---	FLUKE 175 Multimeter
I_1	C3	1:10	50 MHz	FLUKE i50s
V_{DC}	C4	1:500	25 MHz	Tektronix P5200
	---		600 MHz	LeCroy Waverunner XI

VI. EXPERIMENTAL RESULTS

A. Switching behavior of SiC-BJT

The switching behavior and switching energy losses during turn-on (E_{ON}) and turn-off (E_{OFF}) at rated current of the SiC-BJT are shown in Fig. 10 and Fig. 11.

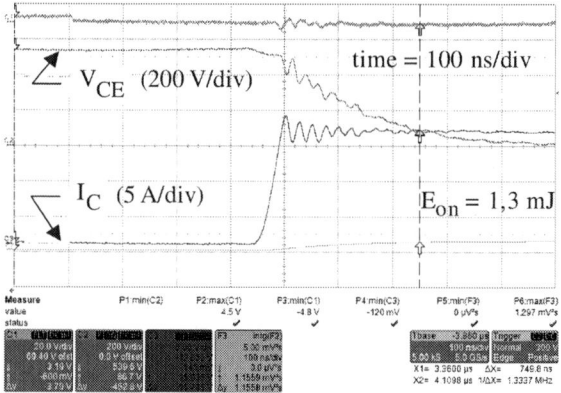

Fig. 10. SiC-BJT [8] - Switch-ON; 15A/600V

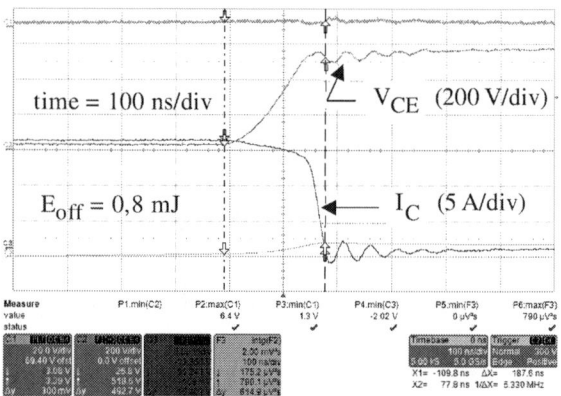

Fig. 11. SiC-BJT [8] - Switch-OFF; 15A/600V

The measured switching energy losses of the SiC-BJT and the datasheet-values of the Si-IGBT are compared in Table IV. Due to lower E_{OFF} the total switching energy loss (E_{TS}) of the BJT is only slightly lower than the one of the IGBT.

TABLE IV
SWITCHING ENERGY LOSSES FOR
V_{CE} = 600 V, I_C = 15 A, R_G = 56 Ω,

	IGBT in mJ	BJT in mJ
E_{ON}	1.3	1.3
E_{OFF}	1.4	0.8
E_{TS}	2.7	2.1

In order to take full advantage of the possible low switching energy losses of the SiC-BJT, e.g. a speed-up capacitor [3] needs to be implemented. Since the focus of this work is on the decrease of total driver losses, switching energy losses comparable to the ones of an IGBT are tolerable. At the same time voltage peaks and ringing due to very fast switching actions are minimized.

This leads to safer inverter operation because of better electromagnetic compatibility and less stress for motor winding isolation.

B. SiC-BJT inverter operation

The SiC-BJT inverter is running idle with a fundamental current frequency of 5 Hz and a pulse frequency of 10 kHz. As described in Chapter II it is characteristic of the conventional commutation algorithm that it switches alternating upper and lower device of one phase-leg. This can be seen in Fig. 12 by means of channel C1. The control signal for BJT S1 alternates between 0 V and 5 V.

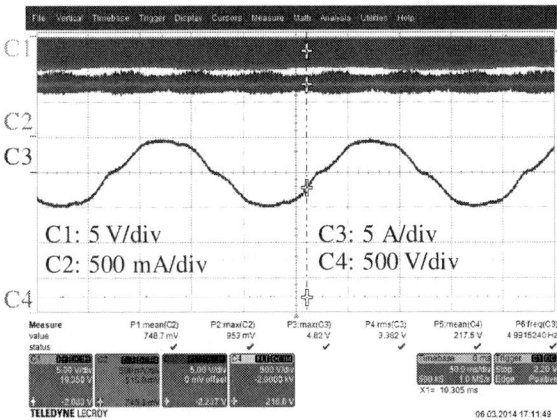

Fig. 12. Phase current waveform and driver current when using conventional commutation algorithm

This is not the case when using one step communication (see channel C1 in Fig. 13). There is a pulse pattern for BJT S1 as long as the current is positive. During negative current half wave the transistor S1 stays switched-off.

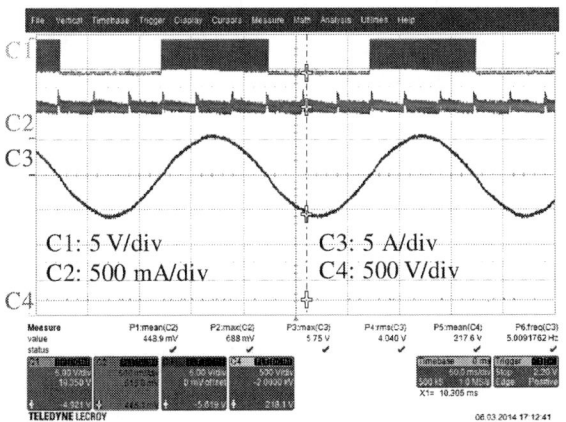

Fig. 13. Phase current waveform and driver current when using one-step commutation algorithm

This has at low modulation indices a remarkable effect on the current waveform as can be seen when looking at channel C3 or Table V.

However, more interesting for this work is the average driver unit current I_{drv}. In idle state the current I_{drv}(idle) equals 119 mA for both commutation algorithms. During

inverter operation the driver unit current I_{drv} varies with the commutation algorithm. When I_{drv}(idle) is subtracted from I_{drv} the average driver unit current that corresponds to the particular commutation algorithm I_{drv}^* is calculated. With I_{drv}^*(one-step) = 330 mA the current is approximately only half of the value than I_{drv}^*(conv.) = 630 mA.

In terms of driver losses P_{drv}^*, that are total driver losses corrected by idle state losses, this means that by using the one-step commutation algorithm P_{drv}^* can be cut in half.

In comparison to driver unit losses of IGBT inverters with similar rating this is still a lot, because there isn't any remarkable difference of driver unit losses between idle state and operation at IGBT driver units for VSIs.

TABLE V
COMPARISON OF COMMUTATION ALGORITHMS
$f_0 = 5$ Hz, $f_P = 10$ kHz, $V_{DC} = 200$ V, $V_{CC} = 24$ V

Commutation Algorithm	I_{drv} in mA	I_{drv}(idle) in mA	I_{drv}^* in mA	P_{drv}^* in W	$\hat{I}_1$ in A	I_{RMS} in A
conventional	749	119	630	15	4,8	3,4
one-step	**449**	**119**	**330**	**7,9**	**5,75**	**4,0**

VII. CONCLUSIONS

A new approach on decreasing the driver losses of SiC-BJTs in VSIs has been presented. Instead of a new driver design the loss reduction is achieved by one-step commutation algorithm, i.e. any unnecessary switching actions are eliminated. First measurements indicate that by using one-step commutation instead of the conventional one driver losses can be cut in half.

Compared to IGBT driver unit losses this is still a lot, since when controlling IGBTs the losses in idle state almost equal the ones in inverter operation.

Future work will focus on efficiency measurements in order to analyze what impact the low on-resistance of the SiC-BJT and the by one-step commutation decreased base emitter losses have on the efficiency of the inverter. Another key aspect is to minimize the losses of each driver.

VIII. ACKNOWLEDGMENTS

The authors would like to thank the German Science Foundation (Deutsche Forschungsgemeinschaft) for funding this project.

Furthermore we would like thank Infineon for providing us with free samples of the SiC Schottky Diodes and IGBT drivers.

REFERENCES

[1] C. DiMarino, Z. Chen, D. Boroyevich, Burgos R. and P. Mattavelli, "Characterization and comparison of 1.2 kV SiC power semiconductor devices", *EPE'13 ECCE Europe*, 2013

[2] Q. Zhang, A.K. Agarwal,. "Design and technology considerations for SiC bipolar devices", *Silicon Carbide – Volume2: Power Devices and Sensors*, WILEY-VCH Weinheim, 2009, S. 389

[3] J. Rabkowski, G. Tolstoy, D. Peftitsis, H. Nee, "Low-Loss High-Performance Base-Drive Unit for SiC BJTs" *IEEE Transactions on Power Electronics* vol.27 no.5 pp.2633-2643 May 2012

[4] L. Wang, H. Baengtsson, "How to Control SiC BJT with High Efficiency?" *7th International Conference on Integrated Power Electronics Systems (CIPS)* 2012

[5] S. Krauss, W. Hofmann, "One-step commutation in voltage source inverters using pulse wire sensor based current sign detection" *AFRICON 2007* Sept. 2007

[6] L. Chen, F. Z. Peng, "Dead-Time Elimination for Voltage Source Inverters" *IEEE Transactions on Power Electronics* vol.23 no.2 pp.574-580 March 2008

[7] LEM, "Current Transducer LAH 25-NP I_{PN}=8-12-25 A", LAH 25-NP datasheet, Version 18, Nov. 2012

[8] Fairschild Semiconductor, "57mΩ Silicon Carbide NPN Power Transistor", datasheet FSICBH57A120 (1200V/15A), Rev. 0.0.7, 2012

[9] Infineon, "IGBT in Trenchstop and Fieldstop technology with soft, fast recovery anti-parallel EmCon HE diode", IKW15T120, 1200V/15A datasheet, Rev. 2.3, Sep. 2008

[10] Infineon,"thinQ!™ SiC Schottky Diode", IDH15S120 1200V/15A datasheet, Rev. 2.0, April 2010

[11] Infineon, "EiceDRIVER™ Single IGBT Driver IC", 1ED020I12FA datasheet, Rev. 3.0, May 2013

[12] Traco Electronic, "DC/DC Converters, 5W", TEL 5 Series datasheet, Rev. April 2013

[13] Traco Electronic, "DC/DC Converters, 15W", TEL 15 Series datasheet, Rev. Feb. 2014

[14] CLARE, "9-Ampere Low-Side Ultrafast MOSFET Drivers", IXD_609 datasheet, Aug. 2010

Power profile based selection and operation optimization of parallel-connected power converter combinations

T. Vogt, A. Peters, N. Fröhleke, J. Böcker
Power Electronics and Electrical Drives
University of Paderborn
D-33095 Paderborn, Germany
vogt@lea.upb.de

S. Kempen
AEG Power Solutions GmbH
Emil-Siepmann-Str. 32
D-59581 Warstein-Belecke, Germany

Abstract—**Many electrical sources and loads are operating at different power levels, partly with full and partly with only a small percentage of their rated power (e.g. photovoltaic systems, elevators or electric heaters). For such devices the connection to the grid can be more efficient with several parallel-connected converters instead of a larger single one. A system of parallel-connected converters facilitates the potential to increase the overall efficiency of the system, either by the selection of different topologies and switching devices of the particular converters or by different operating strategies to split the power between the single converters. This paper presents a method to optimize the operating strategy of such systems with different parallel-operating converters, including the optimal choice of an arrangement based on a specific power demand profile. These two topics sound different but the optimal converter combination can only be selected if the most efficient operation strategy is available. Therefore these topics are interrelated.**

I. Introduction

Developing power electronic systems the converter topology selection is one of the big challenges to reach highest efficiencies. One of the results on the run after high conversion efficiency is, that many different converter topologies are available. The efficiency of most converters depends significantly on the operation point. As shown in [1], it is possible to raise the efficiency for particular operation points by using several equal converters in parallel instead of a large single one. Such an arrangement is conceivable for both, loads like elevators or electrical heaters and sources like photovoltaic or wind generators. For larger influence on the shape of the efficiency curve, obviously different converter types can also be used. The basis of this paper is an arrangement as shown in general in Figure 1. It consists of several different converters operating in parallel connecting an electrical sink to an electrical source.

For applications of such parallel converter stages two questions appear: How to split up the power between the parallel-connected converters for an efficiency-optimal operation strategy? Which arrangement of parallel-operated converters provides highest efficiency for a specific application? The answer obviously depends on the efficiency curves of the considered converters and the behaviour of the regarded application. The consideration begins with the different efficiency curves. The

differences are caused by several influences like the used topology, the kind of switching devices, the mode of operation and more.

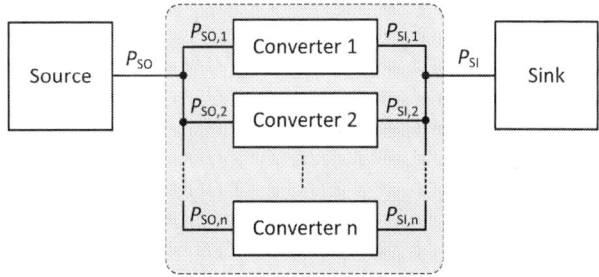

Fig. 1. Principle of the treated arrangement in this paper.

For the method presented in this paper, the reasons for the different efficiency profiles are subsidiary. Therefore it is renounced to give an explanation, as done in [2]. Figure 2 shows exemplary shapes of efficiency curves; the differences between the converters are simply indicated by "configuration".

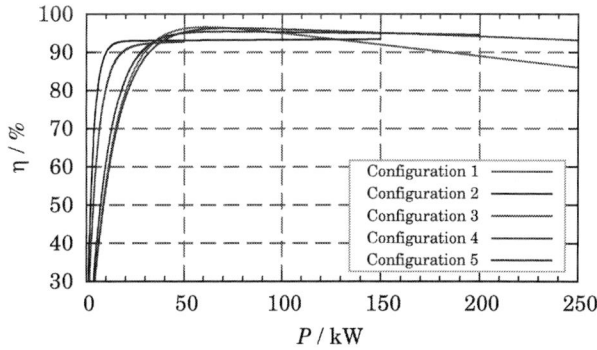

Fig. 2. Exemplary efficiency curves of different configurations.

The common way for parallel connection of converters is to use several converters of the same type and rated power and to split up the power symmetrically between the converters. In the low power ratings commonly some converters are just

The 2014 International Power Electronics Conference

switched off. This yields a small potential to shape and enhance the resulting efficiency curve. The parallel connection of converters with different power ranges and topologies provides more degrees of freedom to influence the overall efficiency curve. Especially for this purpose the optimal split of the power differs from a symmetrical split and therefore optimization algorithms have to be applied.

If an efficiency optimized split of the power is executed, an overall efficiency curve for any arrangement of converters results. The selection of the most efficient converter arrangement out of a set of considered converter configurations is the second step of optimization, as shown in Figure 3. The individual power requirement of the connected device (sink or source) can be investigated using power demand profiles. The power profiles show the power ranges which are frequently required/provided by the device.

With knowledge about the power profile the best combination of different converters can be selected. For this purpose, a weighting function can be calculated, which is based on the power profile and also the information of the absolute power losses at each operation point. This weighting function is multiplied with the overall efficiency curve, the integral of the resultant curve represents the cumulated weighted overall efficiency for the investigated arrangement.

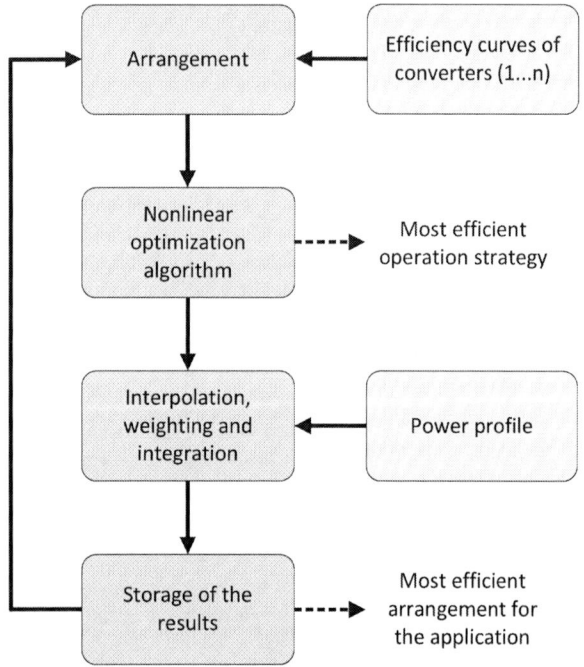

Fig. 3.   Fundamental procedure of the presented method.

This procedure is executed for all considered options, the most efficient arrangement is identified and selected. Accordingly, this method provides both, a power profile dependent recommendation for the most efficient converter combination and the optimal operating strategy for this arrangement.

The method presented in this paper deals with static efficiency curves. The calculations for the selection of the converter combination have to be executed offline. The optimization of the operation strategy can also be executed offline for static efficiency curves. The results can be used at runtime using a lookup table. As prospective, it seems feasible to also take care of temperature influences and other parameters, which change the efficiency curves during runtime and optimize the power scheduling of individual converters online.

II.   PROCEDURE OF OPTIMIZATION

As mentioned, the comparison of different converter arrangements is only significant with consideration of the operation strategy aiming on highest efficiency. For this the overall efficiency curve can be used as an objective function. Neglecting energy storage elements, the equation of the overall efficiency curve is

$$\eta_S(P_{SO}, P_{SI}) = \frac{P_{SI}}{P_{SO}} \qquad (1)$$

where η_S is the efficiency, P_{SO} is the power of the source and P_{SI} is the power of the sink of the overall system. That means that P_{SO} is the summmarized power on the source side of all parallel connected converters and P_{SI} is the summmarized power on the sink side of all parallel-connected converters, as shown in Figure 1. This results in

$$\eta_S(P_{SO}, P_{SI}) = \frac{P_{SI}}{P_{SO}} = \frac{\sum\limits_{i=1}^{n} P_{SI,i}}{\sum\limits_{i=1}^{n} P_{SO,i}} \qquad (2)$$

Either the power on the sink or on the source side determines the setpoint, depending on the application[1]. Therefore two equation are utilized for the optimization process. For applications with fixed power on the sink side results

$$\eta_S(P_{SI}) = \left(\sum_{i=1}^{n} P_{SI,i} \right) \left(\sum_{i=1}^{n} \frac{P_{SI,i}}{\eta_i(P_{SI,i})} \right)^{-1} \qquad (3)$$

and for applications with fixed power on the source side accordingly

$$\eta_S(P_{SO}) = \left(\sum_{i=1}^{n} P_{SO,i}\ \eta_i(P_{SO,i}) \right) \left(\sum_{i=1}^{n} P_{SO,i} \right)^{-1} \qquad (4)$$

In order to determine the optimal operating strategy for the sharing of power in all operating points, these equations can be used as an objective functions and maximized for each operating point. The objective function is in general nonlinear

---

[1]For a better understanding two examples for this: a) If a PV system is connected to the grid, the PV system is the source and the grid is the sink. The source power is set by the application. b) If a heater is conneced to the grid the grid is the source and the heater is the sink. The sink power is set by the application.

978-1-4799-2706-7/14 $31.00 © 2014 IEEE

and can be optimized with several different optimization methods for nonlinear optimization. The optimization problem can be formulated for sinks as

$$\underset{P_{\text{SI},1},\cdots,P_{\text{SI},n}}{\text{minimize}} \quad -\eta_{\text{S}}(P_{\text{SI},1},\cdots,P_{\text{SI},n}) \tag{5}$$

$$\text{subject to} \quad \sum_{i=1}^{n} P_{\text{SI},i} = P_{\text{SI}} \tag{6}$$

$$P_{\text{SI}} \in [0, P_{\text{R}}] \tag{7}$$

where P_{R} is the rated power of the considered overall system. For sources the same structure applies, using (4) instead of (3). One possible solution of the optimization problem is to calculate all possible combinations to split up the power between the converters with a useful resolution of operating points $(0,\cdots,P_{\text{R}})$ and to choose the most efficient one. This may require much calculation time, depending on the number of considered converters. However, this method has the advantage of always calculating the global maximum.

To give a more detailed explanation and illustration of the optimization problem, an example with two parallel-connected converters is shown in the following. In this example a device with a sink characteristic is being considered, so that P_{SI} is the reference power. The efficiency curves of the converters in this example are largely differing to highlight the effect of the method. In reality less differences are expected, but the principle of the problem becomes clear with this example. Figure 4 shows the efficiency curves of two converters, which have different shapes but the same rated power of $P_{\text{R},1} = P_{\text{R},2} = 100$ kW. The total rated power of the device in this example is $P_{\text{R}} = 200$ kW.

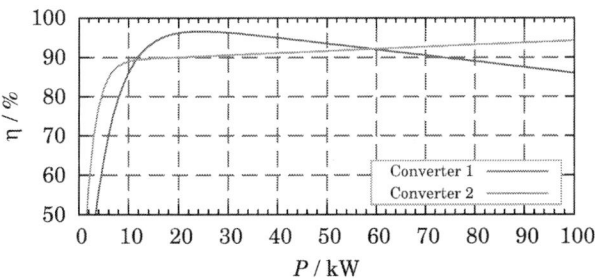

Fig. 4. Different efficiency characteristics of the exemplary case.

The objective function to optimize the efficiency of the parallel operation follows according to (3) as

$$\eta_{\text{S}}(P_{\text{SI},1}, P_{\text{SI},2}) = \frac{P_{\text{SI},1} + P_{\text{SI},2}}{\dfrac{P_{\text{SI},1}}{\eta_1(P_{\text{SI},1})} + \dfrac{P_{\text{SI},2}}{\eta_2(P_{\text{SI},2})}} \tag{8}$$

and can be represented for each combination of $P_{\text{SI},1}$ and $P_{\text{SI},2}$ as shown in Figure 5. The sum of $P_{\text{SI},1}$ and $P_{\text{SI},2}$ results to the required power of the sink P_{SI} in this specific operating point. The optimization result is a trajectory of the most efficient operation points for $0 \le P_{SI} \le 200$ kW (resolution of 1kW). The projection of the trajectory onto the horizontal bottom

plane shows the power split of $P_{\text{SI},1}$ and $P_{\text{SI},2}$ at the most efficient operating points. The resolution of 1kW and the rated power of 200kW leads to 200 operating points. If no enhanced optimization method is used, 100^2 calculations of the objective function are necessary to find the most efficient power split. However, in this case it is recommended to follow this approach, because it avoids solutions in local maxima without any additional efforts and within few seconds.

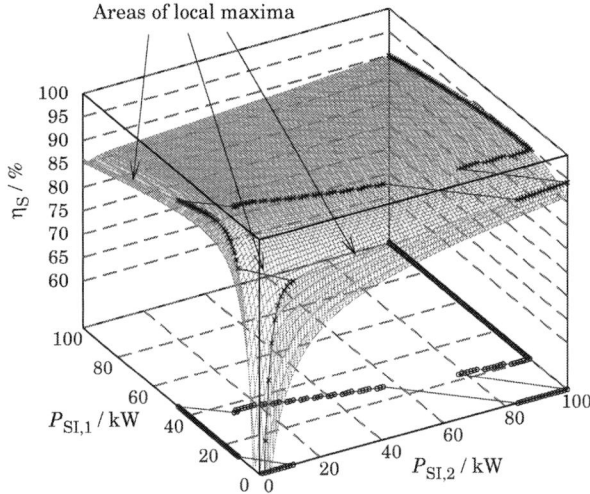

Fig. 5. Three dimensional representation of overall efficiency with two parallel-connected converters. The most efficient operation points are indicated by the trajectory and its projection onto the horizontal bottom plane.

As also indicated in Figure 5 several local maxima of the objective function arise at different operating points, in addition to the searched global maximum. This is clearly illustrated for the treated two dimensional problem by analyzing different operating points. Figure 6 shows the system efficiency for three different operating points ($P_{\text{SI}} = 30\text{kW}, 60\text{kW}$ and 90kW) with varied power of the first converter $P_{\text{SI},1}$.

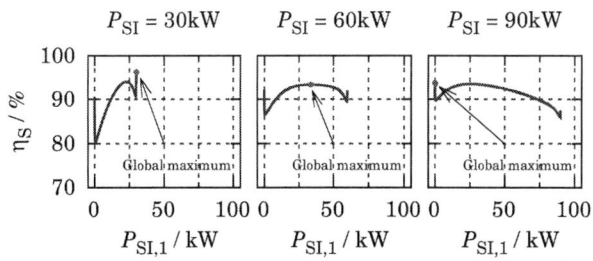

Fig. 6. System efficiency depending on the first converter's power $P_{\text{SI},1}$ at different operating points (thus $P_{\text{SI},2} = P_{\text{SI}} - P_{\text{SI},1}$).

For better clarity the optimization results are presented in two 2D diagrams in Figure 7. The first diagram shows the efficiency curves of the inverter and the resulting systems efficiency curve. The second diagram shows the most efficient power split at each operating point. These can directly be used in a lookup table to operate the system.

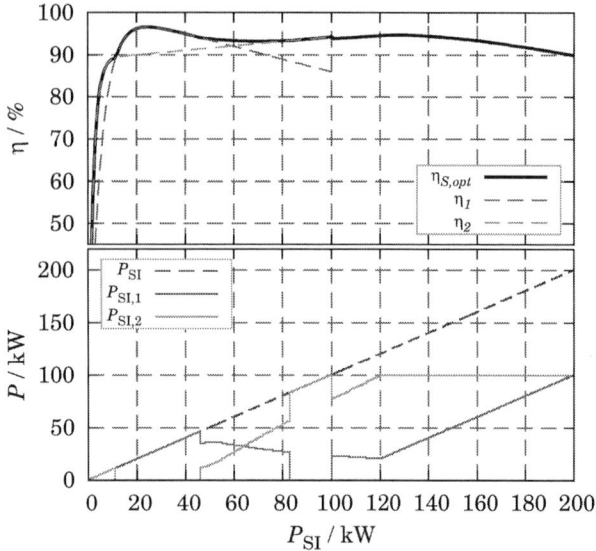

Fig. 7. Two dimensional result of system efficiency and power split between the converters.

The described approach is well suited for arrangements of up to 4 or 5 converters depending on the amount of operating points and resolution of calculation. However, even for the described simple example the use of an enhanced optimization method for arrangements with at least six converters is recommended, since the calculation time and data storage requirements becomes high. The result of an optimization with six converters is shown in Figure 8.

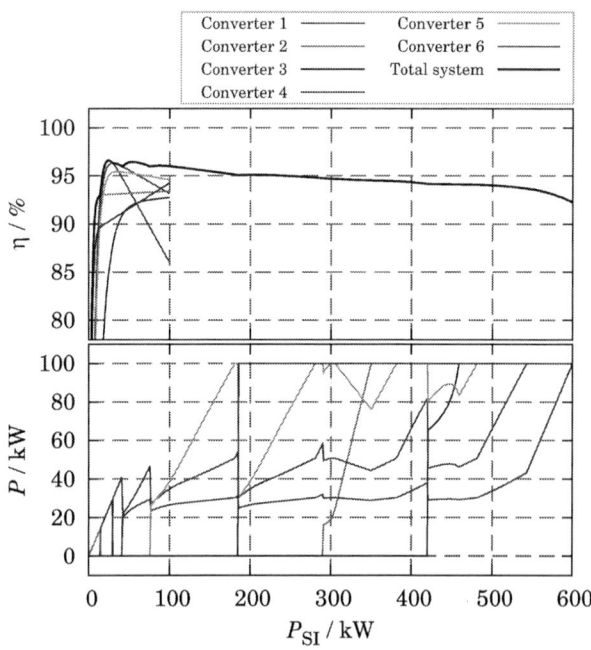

Fig. 8. Exemplary result of the optimization for an arrangement with six different converters.

To reach this result in acceptable time and acceptable storage ressources a global optimization method is used. Without going into details of optimization methods, it turned out that the gradient-based successive approximation method "sequential quadratic programming, (SQP)", in combination with a global multiple starting point approach is beneficial for this optimization problem. Beside of stochastic based starting points also starting points, for which local or global solutions are expected, are used. This applies for points, where the complete power is supplied by a single converter. Thus it is feasible to achieve a significant optimization speedup compared with the approach to calculate all possible combinations. Also the resolution of the result is increased. Of course this approch is just a proposal for this optimization problem and many other approaches are conceivable. More information about SQP are given in [3]. The global multiple starting point approach is described in [4] and [5].

III. Comparison of solutions

The common way to use converters is to use one single converter for the whole power range of an application or to split up the power to several small converters of the same kind. If the power is split up to equal converters, one way is to split up the power symmetrical between the converters, which results in an efficiency curve of the same shape as one single converter with a higher rated power. Another way is to stepwise activate the converters. For this example four converters with a power range of 25kW each are considered, which leads to an overall power rating of 100kW.

For a set of four converters of the same type, the resulting efficiency curve of stepwise activate operation and symmetrical power split is shown in Figure 9.

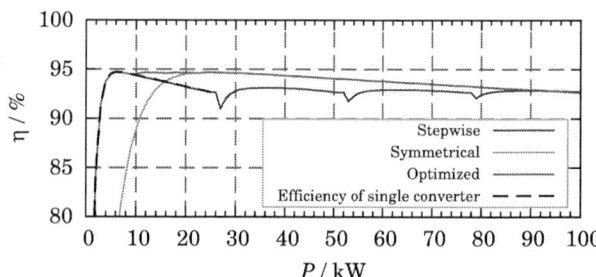

Fig. 9. Total efficiency curves of a set of converter of same kind.

Obviously stepwise activating operation has the advantage of much higher efficiency for small power (Figure 9, blue line), but less efficiency in the middle power range, compared to symmetrical power split (Figure 9, green line). The red line in Figure 9 shows the efficiency curve of an optimal power split strategy between the four converters. Here the advantage of an optimal power split becomes very clear. The resulting efficiency curve enables better efficiency than the other two methods in a wide range of operation.

Figure 10 shows the switching strategy to split the power among the converters for symmetrical power split (upper plot), stepwise power split (middle plot) and for one optimal strategy (lower plot).

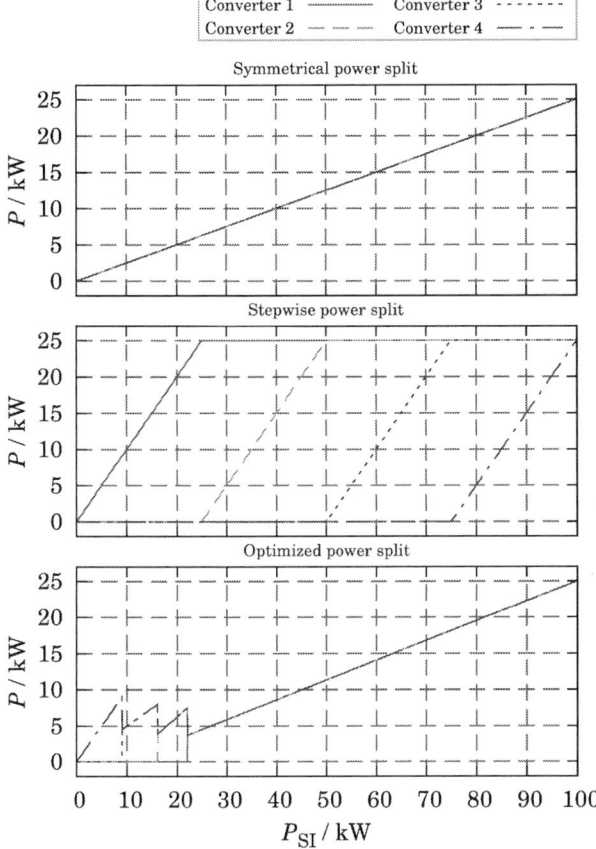

Fig. 10.   Optimal power split strategy between 4 different converter.

Using different converter types allows to influence the shape of the resulting efficiency curve by a higher degree and facilitates finding an optimal set of converters for every application, like it is shown in chapter II.

IV.   WEIGHTING WITH POWER PROFILE

One initial question remains: Which combination of converters yields to highest total efficiency of a specific application? Therefore detailed information about the power requirement of the application is necessary. Because of the power being a continuous variable, a probability density function seems to be an appropriate way to indicate the power ranges which are frequently required by the application. The approximation of a probability density based on measurement results is however combined with several problems, such as the measurement is discrete and the amount of measurements is limited. A useful alternative to determine the probability density is to determine the frequency distribution. That is implemented by quantization of the power and counting the measurement

results in the resulting power ranges. Figure 11 shows an exemplary power profile as a histogram which is determined as described. The profile shows the power of a sink (e.g. a three step heater). For further processing, this power profile is interpolated and normalized with the respective appropriate power in each point, which leads to a weighting function as shown in Figure 11.

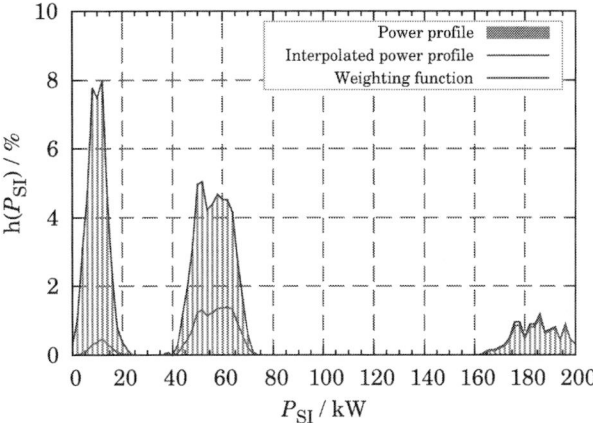

Fig. 11.   Sink power profile and the resulting weighting function.

The same procedure can also be used to extract a power profile for sources. An example for this is shown in Figure 12, which is taken from a photovoltaic power station.

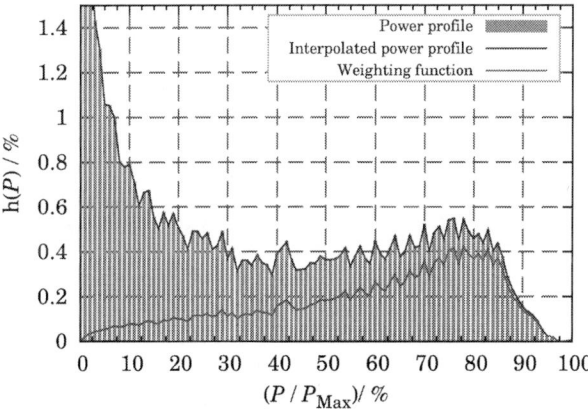

Fig. 12.   Power profile of a PV system and the resulting weighting function.

The weighting function indicates the power ranges including the information of the frequency of use and the absolute power. If this function is multiplied by the overall efficiency curve of a specific arrangement of converters, a weighted efficiency curve for the investigated application results. To compare different arrangements, this weighted efficiency curve can be integrated and normalised to the rated power of the application. For sink power profile it follows

$$J_k = \frac{1}{P_R} \int_0^{P_R} \eta_{S,\text{opt},k}(P_{SI})\ f_w(P_{SI})\ dP_{SI} \quad k = 1 \cdots m \quad (9)$$

where variable J is the weighted accumulated efficiency for the arrangement k and m is the number of considered arrangements. This approach is similar to the weighted efficiency for PV converters (e.g. η_{EUR} or η_{CEC} explained in [6]) but explicit for the considered application. The higher the value of J, the lower the losses of the respective arrangement k for the considered power profile. The arrangement k with the highest value of J should be selected to achieve highest efficiency. If the price of different arrangements should also be included in the comparison, the values of J can be used to create a cost-benefit ratio function.

V. EXEMPLARY RESULT

The shown power profiles of the section before are used for an exemplary result of the described method, using 19 different arrangements. For each arrangement an efficiency optimized power split is calculated and weighted with both power profiles. The resulting values for $J_{k,r}$ with $k = [1, 2, \cdots, 19]$ and $r = [A, B]$ are shown in Figure 13.

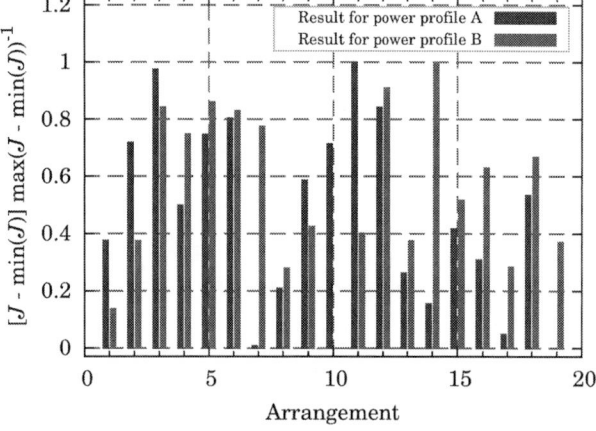

Fig. 13. Results of an optimization process for 19 different arrangements.

In order to enhance comprehension of comparison of the different arrangements, the representation of J is normalized by $[J - \min(J)]\max(J - \min(J))^{-1}$. The results can directly be used to choose the efficiency optimal arrangement based on the power profile. It becomes clear that the power profile has significant influence on the choice, e.g. the converter arrangement 14 is suited for the best application with power profile B but not with power profile A. The same applies for arrangement 11 the other way round.

VI. CONCLUSION

The presented method facilitates the choice of an arrangement of converters based on efficiency considerations for a specific application. It becomes clear, that the efficiency optimal operating strategy and choice of an arrangement are coupled. The result of a simple example shows that, even if there are only two converters working in parallel, optimized operating strategies are not trivial. The approach proposes the variable J,

which is suitable to summarize the influences and information and can be easily used to compare different arrangements for a specific application. This is completely independent from the kind of application, which can be sources like PV and wind power but also sinks like heaters or elevators.

VII. SYMBOLS AND INDICES

TABLE I. EXPLANATION OF USED SYMBOLS AND INDICES

Symbol	Meaning
f_{w}	Weighting function
J_k	weighted accumulated efficiency for arrangement k and power profile normalized to the rated power
η_i	Efficiency of converter i
η_{S}	Efficiency of the converter arrangement
$\eta_{\mathrm{S,opt}}$	Optimized efficiency of the converter arrangement
n	Number of converters in arrangement
m	Number of arrangements
r	Index of power profile
k	Index of arrangement
P_{R}	Rated power of the converter arrangement
P_{SI}	Power on sink side
P_{SO}	Power on source side

VIII. ACKNOWLEDGEMENT

This research and development project is funded by the German Federal Ministry of Education and Research (BMBF) within the Leading-Edge Cluster "Intelligent Technical Systems OstWestfalenLippe" (it's OWL) and managed by the Project Management Agency Karlsruhe (PTKA). The author is responsible for the contents of this publication.

REFERENCES

[1] H. Wetzel, N. Fröhleke, J. Böcker, P. Ide, "High Efficient 3kW Three-Stage Power Supply", *Applied Power Electronics Conference (APEC) and Exposition. 2006. APEC '06. Twenty-First Annual IEEE*, pp. 1361-1367, 2006.

[2] B. Burger, D. Kranzer, "Extreme High Efficiency PV-Power Converters", *Power Electronics and Applications, 2009. EPE '09. 13th European Conference on*, 2009.

[3] J. Nocedal, S. Wright, "Numerical Optimization", *Springer Science+Business Media, New York, 2nd Edition*, 2006.

[4] L. Liberti, N. Maculan, "Global Optimization: From Theory to Implementation", *Springer Science+Business Media, New York*, 2006.

[5] J. Pintér, "Global Optimization in Action, Continuous and Lipschitz Optimization", *Kluwer Academic Publishers*, 1996.

[6] European Standard, "EN 50530:2010 - Overall efficiency of grid connected photovoltaic inverters", 2010

A Novel Power Loss Calculation Method for IGBTs in Power Converters via Chaotic SPWM Control.

Boyu Wang, Hong Li, Xiaojie You, Trillion Zheng
School of Electrical Engineering
Beijing Jiaotong University
Beijing, China
hli@bjtu.edu.cn

Abstract—**Chaotic sinusoidal pulse width modulation (SPWM) has been proposed and applied into suppressing the electromagnetic interference (EMI) generated by power converters. However, the power loss of the power switches under chaotic SPWM control have not been mentioned and analyzed yet; the power loss is an very important factor, since it can partly decide the dynamic performance and reliability of power converters. Furthermore, the existing power loss calculation method is no longer suit for the situation with chaotic SPWM control. In this paper, a novel power loss calculation method is proposed, and tested by the IGBTs in an AC-DC converter under Chaotic SPWM control. According to the calculation results, different power loss of IGBTs under SPWM and chaotic SPWM control are observed and analyzed. Finally, the experimental results are given to prove the effectiveness of the power loss calculation method proposed in this paper.**

Keywords— Chaotic SPWM, Power loss, Calculation, Temperature.

I. INTRODUCTION

Chaotic SPWM control can be used to spread the harmonic spectra and reduce the EMI of power converters efficiently [1-5]. In [1], chaotic SPWM control has been proposed to reduce EMI in grid-connected inverters and it has the advantages in design flexibilities and low cost [1]. While in electric vehicles (EVs), chaotic SPWM control can be designed to improve the electromagnetic compatibility (EMC) and the mechanical performance for EV motor drives [2]. In [3], a new chaotic PWM control scheme was used to approve the harmonic peaks can be reduced greatly by chaotic SPWM control since a chaotic switching frequency was adopted. Furthermore chaotic SPWM control is as well as an effective way to reduce current ripple and torque in induction motor drives [4]. And chaotic SPWM control is easy to be realized and implemented by digital and analog ways [5]. However the power loss and the temperature feature of power switches in converters under chaotic SPWM control have not been analyzed yet. The working temperature is a key factor to the power switches and will affect the whole system's reliability and efficiency, especially in the middle and high power converters. The high power density subject power switches to high thermal stress, especially in the harsh

electrical environments [6]. Besides, the power loss can further influence the power switches' dynamic performance, reliability and life time [7]. Therefore, the power loss calculation of the power switches under chaotic SPWM control is very important for practical application.

Because the switching frequency under chaotic SPWM control is time-varying, the power loss calculation methods for power switches, such as IGBT, MOSFET and etc., are no longer suit for the calculation of the power loss under chaotic SPWM control. In this paper, a novel power loss calculation method is proposed for the power switches under chaotic SPWM control. The power loss of IGBT under chaotic SPWM control are calculated and compared to that under SPWM control. Finally, a serious of experiments is also designed to verify the accuracy of the proposed calculation method.

This paper is organized as follows. In section II, the chaotic SPWM control principle is presented. A calculation method for IGBT in power converters under chaotic SPWM control is proposed in section III. The comparison of the IGBT's loss under chaotic SPWM control and under SPWM control is analyzed and calculated in section IV. In section V, the temperature comparison experiments are designed based on an AC-DC converter. Finally, the conclusions are given.

II. CHAOTIC CONTROL PRINCIPLE

SPWM control has been widely used in the control of power converters due to its high power factor and good output waveform. The problem in SPWM control is driving signal has large harmonic content [8] .Chaotic PWM control is an effective method to reduce harmonic content, which has plenty of advantages. The control principles of chaotic SPWM control and SPWM control is similar, therefore chaotic PWM control can be implemented in almost all the circuits in which SPWM control can be used.

The difference between chaotic SPWM control and SPWM control is lie in the carrier frequency f_c, as shown in Fig.1, the chaotic carrier includes a reference switching frequency and an additional chaotic frequency, namely, $f_c = f_s + \Delta f$. The chaotically switching frequency can spread the energy over the total frequency band and suppress the peaks on the spectrum and then reduce the EMI of power converters [9].

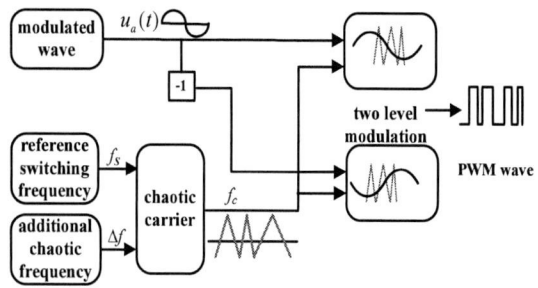

Figure 1: The diagram of chaotic SPWM control.

However, the calculation of power loss of power switches in converters differs from the traditional calculation method due to the chaotifying switching frequency. So a calculation method for power loss of power switches under chaotic PWM control is needed to analyze the power loss and temperature performance of power switches.

III. POWER LOSS CALCULATION METHOD OF IGBTs UNDER CHAOTIC SPWM CONTROL

IGBT has been widely used in power converters as it was characterized by a high rated voltage, current, high switching speed and easy to drive [10]. In this paper, we propose a calculation method based on a single phase AC-DC converter, and analyze the difference of power loss of IGBT.

The power loss of IGBTs can be divided into conduction loss and switching loss (turn-on and turn-off loss) [11].The calculation method is bound up with the practical working condition of IGBTs under chaotic SPWM control and SPWM control in the AC-DC converter, as well as the electric parameters and information in the IGBT datasheet [12].

TABLE I
FUNDAMENTAL PHYSICAL CONSTANTS

Symbol	Meaning
D	Duty cycle
T	Cycle time
I	Collector current
E_{on}	Turn-on energy
E_{off}	Turn-off energy
P_{con}	Conduction loss
P_{sw}	Switching loss
U_S	IGBT actual working voltage
U_N	IGBT rated voltage
M	Modulation ratio

Since the switching frequency changed chaotically in chaotic SPWM control, as shown in Fig.2, the duty cycle, conduction loss and switching loss under chaotic SPWM control cannot be analyzed by (1) - (3)., which are for the power loss calculation under SPWM control[13]

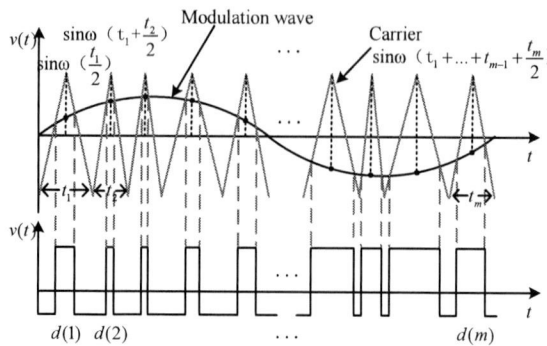

Figure 2: The diagram of Chaotic SPWM

$$D(t) = \frac{1 - M \cdot \sin(\omega t + \varphi)}{2} \tag{1}$$

$$P_{SS} = \frac{1}{T} \int_0^T U_{CE}(t) \cdot I_C(t) \cdot D(t) dt \tag{2}$$

$$P_{SW} = \frac{U_S}{U_N} \cdot \frac{F}{2} \cdot (E_{on} + E_{off}) \tag{3}$$

According to Fig.2, it is obvious that the duty cycle of each period can be calculated from the modulation process and then calculate the power loss by discretized voltage and current of IGBT.

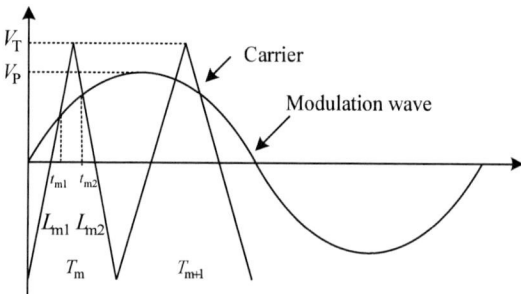

Figure 3: The modulation process under chaotic SPWM control.

In Fig.3, the two lines of mth triangular carrier L_{m1} and L_{m2} can be expressed by (4) and (5), respectively. The modulation wave can be expressed by (6).And we can derive the duty cycle and the current i_C of IGBT per switching cycle from the equations (4)-(6), which are shown in (7):

$$y = \frac{4V_T}{t_m} \left\{ x - \left(t_1 + \dots + t_m - \frac{t_m}{2} \right) \right\} + V_T \tag{4}$$

$$y = -\frac{4V_T}{t_m} \left\{ x - \left(t_1 + \dots + t_m - \frac{t_m}{2} \right) \right\} + V_T \tag{5}$$

$$v(t) = V_P \cdot \sin(\omega \cdot x) \tag{6}$$

And, the duty cycle is given in (7):

$$d(\mathrm{m}) = \frac{t_2 - t_1}{T}$$

$$= \frac{1 - \dfrac{1}{2}M(\sin \omega \cdot t_{2m-1} + \sin \omega \cdot t_{2m})}{2} \quad (7)$$

Where, $M = \dfrac{V_P}{V_T}$.

Since the carrier's frequency is much higher than modulation wave's frequency, the approximation in (8) can be made:

$$\sin \omega t_{2m-1} = \sin \omega t_{2m} = \sin \omega (\ t_1 + ... t_{m-1} + \frac{t_m}{2}) \quad (8)$$

So the duty cycle can be following expressed as:

$$d(m) = \frac{1}{2} - \frac{M}{2} \cdot \sin\omega (\ t_1 + t_2 + ... + t_{m-1} + \frac{t_m}{2}) \quad (9)$$

In the modulation process, I_c can also be expressed, as shown in (10), and I_c in m^{th} carrier period can be expressed by the same approximation method, as seen in (11).

$$i_C(t) = I_0 \cdot \sin(\omega \cdot x + \varphi) \quad (10)$$

$$i_C(m) = I_0 \cdot \sin\left\{ \omega \cdot (t_1 + t_2 + ... + t_{m-1} + \frac{t_m}{2}) + \varphi \right\} \quad (11)$$

Because in the single phase AC-DC converter the current wave of each IGBT is shown in Fig.4, when $I_c < 0$, the IGBT's power loss of that cycle is 0,so the conduction and switching power loss of mth cycle can be expressed as (12) and (13):

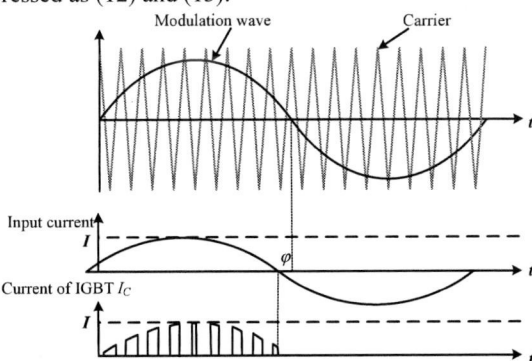

Figure 4: The modulation process of single phase AC-DC converter.

$$\begin{cases} P_{con}(m) = v_{CE}(m) \cdot i_C(m) \cdot d(m), i(m) > 0 \\ P_{con}(m) = 0, i(m) \le 0 \end{cases} \quad (12)$$

$$\begin{cases} P_{sw}(m) = \dfrac{U_S}{U_N} \cdot (E_{on} + E_{off}), i(m) > 0 \\ P_{sw}(m) = 0, i(m) \le 0 \end{cases} \quad (13)$$

To simplify the calculation in (12) and (13), the relationship curves from the IGBT's datasheet can be used. Fig.5 and Fig. 6 shows the relationship between

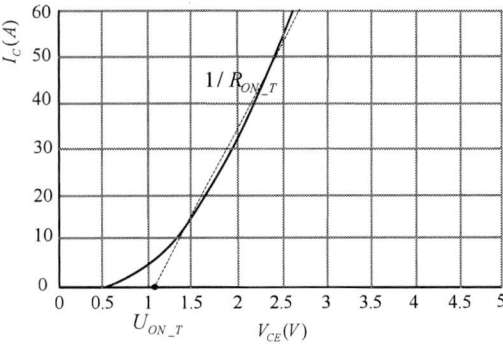

Figure 5: The relation curve between I_C and U_{CE} from IGBT datasheet.

From Fig.5, the relationship between I_C and V_{CE} can be expressed as follow:

$$V_{CE} = U_{ON\_T} + R_{ON\_T} \cdot I_C \quad (14)$$

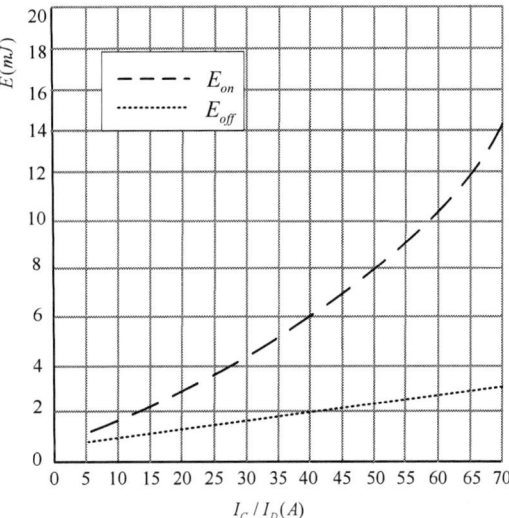

Figure 6: The relation curve between current and switching energy from IGBT datasheet.

From Fig.6, the relationship between current and switching energy can be expressed as follow:

$$\begin{cases} E_{on} = K_1 \cdot I_C \\ E_{off} = K_2 \cdot I_C \end{cases} \quad (15)$$

To calculate the power loss of IGBT under chaotic SPWM control, we discretize the voltage and current of IGBT .Then the calculation of conduction loss and switching loss can be expressed as (16) and (17) which derived from the traditional calculation formulas (1)-(3).

$$P_{con} = \lim_{m \to +\infty} \frac{\sum\limits_{i=1}^{m} P_{con}(i) \cdot t(i)}{t} \quad (16)$$

$$P_{sw} = \lim_{m \to +\infty} \frac{\sum_{i=1}^{m} P_{sw}(i) \cdot t(i)}{t} \qquad (17)$$

In (16) and (17), the $t(i)$ is bounded by Logistic chaotic mapping $f(x)$, which probability distribution density is shown in Fig.7. The relation can be expressed in (18):

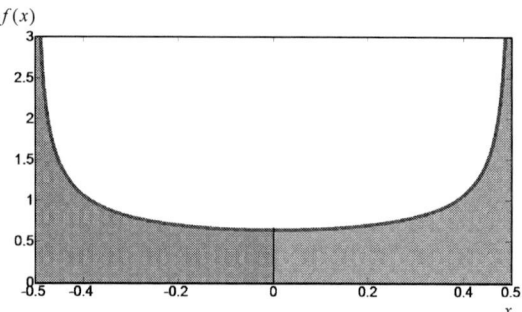

Figure 7: The probability distribution density of logistic sequence.

$$t(i) = \frac{1}{f(i)} = \frac{1}{F + f(x) \cdot F_0} \qquad (18)$$

Where, F_0 is the reference frequency we defined.

IV. COMPARISON OF LOSS BETWEEN SPWM CONTROL AND CHAOTIC SPWM CONTROL

Based on the proposed calculation method, the power loss of IGBTs under chaotic SPWM control can be calculated, but the difference of power loss under chaotic SPWM control and under SPWM control have not been analyzed, which could supply the merits and demerits of chaotic SPWM control in thermal design.

From the calculating method proposed in Section III, the duty cycles D_0 and the current I_{C0} by SPWM control can also be expressed as follows by the discretized method:

$$d_0(m) = \frac{1}{2} - \frac{M}{2} \cdot \sin\omega(2m - 1) \cdot T \qquad (19)$$

$$i_{C0}(m) = I_0 \cdot \sin\omega\{2m - 1) \cdot T + \varphi\} \qquad (20)$$

So the difference of switching loss of IGBT by chaotic SPWM control and SPWM control can be expressed as:

$$\Delta_1 = P_{sw} - P_{sw}$$

$$= \lim_{t \to \infty} \frac{\sum_{i=1}^{m} P_{sw}(i) \cdot t(i)}{t} - \frac{\sum_{i=1}^{n} P_{sw0}(i) \cdot T}{t} \qquad (21)$$

Fig.7 shows the probability distribution density of logistic sequence. And taking account of the feature of Logistic mapping as shown in Fig.7, we can

know $\lim_{m \to \infty} \frac{\sum_{i=1}^{m} f(i)}{m} = F$. According to (22), the relationship between the carrier's period under chaotic SPWM control and SPWM control can be derived as (23).

$$\frac{m}{\sum_{i=1}^{m} \frac{1}{f(i)}} \le \frac{\sum_{i=1}^{m} f(i)}{m} = F \qquad (22)$$

$$\frac{\sum_{i=1}^{m} t(i)}{m} \ge T \qquad (23)$$

Namely, the average switching period of IGBTs under chaotic SPWM control is higher than that under SPWM control. So the switching losses under SPWM control are higher than that under chaotic SPWM control.

The difference of conduction loss $\Delta 2$ by chaotic SPWM control and SPWM control can be expressed as:

$$\Delta_2 = \lim_{t \to \infty} \frac{\sum_{i=1}^{m} P_{sw}(i) \cdot t(i)}{t} - \frac{\sum_{i=1}^{n} P_{sw0}(i) \cdot T}{t}$$

$$= \lim_{m \to \infty} \frac{\sum_{i=1}^{m} a \cdot \left(|\sin(A)| - |\sin(B)|\right)}{t} \qquad (24)$$
$$\frac{+ b \cdot \left(|\sin(A)^2| - |\sin(B)^2|\right)}{+ c \cdot \left(|\sin(A)^3| - |\sin(B)^3|\right)}{t}$$

In (24), a, b and c are positive constants. A and B are shown in (25)

$$\begin{cases} A = \omega(t_1 + t_2 + \ldots + t_{m-1} + \dfrac{t_m}{2}) \\ B = \omega(2m-1) \cdot T \end{cases} \qquad (25)$$

From the chaotic sequence's feature we can know that $\lim_{m \to \infty} \sum_{i=1}^{m} |\sin(A)| - |\sin(B)| = 0$. Therefore, we can safely come to the conclusion that the conduction loss under chaotic SPWM control and that under SPWM control are approximately equal.

The calculation results obtained by Matlab are shown in Fig.8-9, the red curves represent IGBT loss under chaotic SPWM control and the blue curves are that under SPWM control. Table I shows the calculation and experimental parameters.

TABLE I
PARAMETERS OF CALCULATION AND EXPERIMENTS

Symbol	Quantity
IGBT	H20R1202
I_0	5A
F_0	256Hz
F(Frequency)	1KHz、10KHz

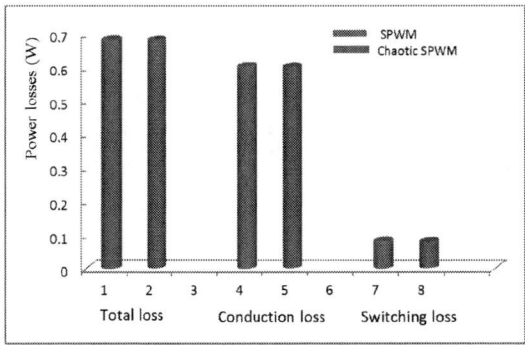

Figure 8: Power loss of IGBT under 1KHz.

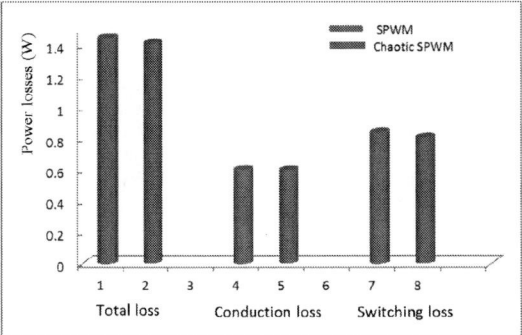

Figure 9: Power loss of IGBT under 10KHz.

Based on the analysis and calculation of the difference of power loss under chaotic SPWM control and SPWM control, we can get the conclusions that:

(1) The IGBT's switching loss under chaotic SPWM control is lower than that under SPWM control.

(2) The conduction loss under chaotic SPWM control and that under SPWM control are approximately equal.

(3) When the switching frequency is lower than 5K, the total loss under chaotic SPWM control and that under SPWM control are approximately equal, as the conduction loss plays the main role in the total loss.

V.   SIMULATION AND EXPERIMENT

A series of experiments had been designed to verify the accuracy of the calculation method we proposed. The experiments parameters are also shown in Table I.

A. FFT analysis experiments

Fig9-10 are the FFT analysis of driving waves under SPWM control and chaotic SPWM control, respectively.

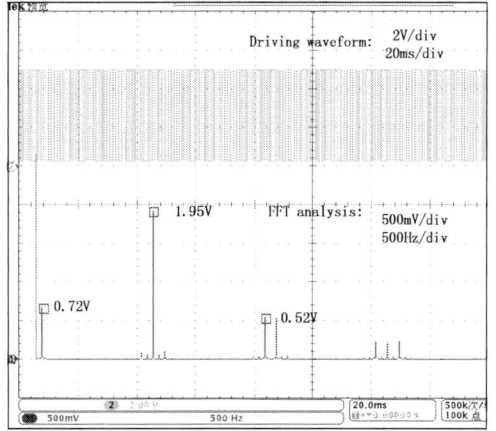

Figure 10: The FFT analysis of driving wave under SPWM control.

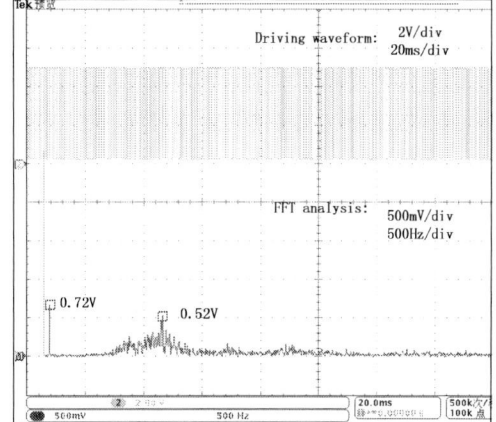

Figure 11: The FFT analysis of driving wave under chaotic SPWM control.

The effective of EMI suppressing are shown from the comparison in Fig10-11 apparently, as a result of that, chaotic SPWM control has got widespread attention in EMI suppressing area.

B. Temperature experiments

To verify the correctness of the proposed power loss calculation method for chaotic SPWM control, temperature measurement experiments for IGBTs in the AC-DC converter are carried out in this paper, as shows in Fig.12.

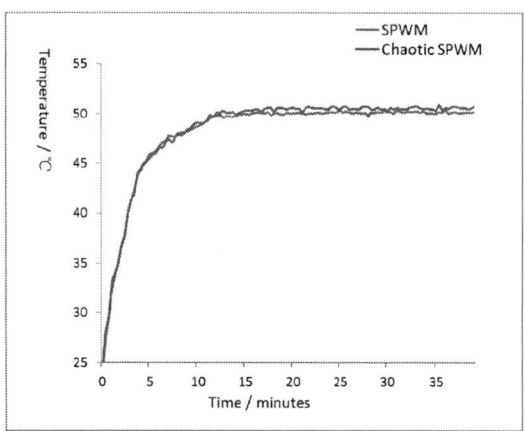

Figure 12: Case temperature of IGBT.

Fig.12 shows the experimental results, in which the red curve represents the IGBT's case temperature under chaotic SPWM control and the blue one represents that under SPWM control. The IGBT's temperature approximately equal under 1KHz, therefore, the experimental results are coincident with the calculation results, which prove that the power loss calculation method for power switches under chaotic SPWM control is correct and accurate.

VI. CONCLUSION

This paper has proposed an accurate method to calculate the power loss of IGBTs under chaotic SPWM control and the calculation method is further applied into an AC-DC converter. Based on the proposed method, the comparison and analysis show the power loss under chaotic SPWM control and SPWM control are approximately equal due to the features of logistic mapping with low switching frequency(<5KHz). Finally, the calculation and experimental results prove the correctness and accuracy of the proposed calculation method.

ACKNOWLEDGMENT

This work was supported by the National Natural Science Foundation of China under Grants 51007004 and 50937001; by the Fundamental Research Funds for the Central Universities under Grant 2012JBM096; by Beijing Natural Science Foundation under Grant 3142015; by Beijing Higher Education Young Elite Teacher Project under Grant YETP0569.

REFERENCES

[1] H. Li, T. Q. Zheng, Z. Li and F. L. Wang, "EMI suppression for single-phase grid-connected inverter based on Chaotic SPWM control," *Electromagnetic Compatibility (APEMC)*, 2012 Asia-Pacific Symposium on, May21-24, 2012.

[2] Z. Wang and K. T. Chau, "Design and analysis of a Chaotic PWM inverter for electric vehicles," *Industry Applications Conference*, Sept23-27, 2007.

[3] D. C. Hamill and J. H. B. Deane, "Improvement of power supply EMC by chaos," *Electron.Lett*, vol. 32, no. 12, pp. 1045, Jun. 1996.

[4] A. M. Trzynadlowski, C. Stancu, J. M. Nagashima and M. H. Zelechowski, "Comparative investigation of PWM techniques for a new drive for electric vehicles," *IEEE Trans.Ind.Appl.*, vol. 39, No. 5, 2003, pp. 1396-1403.

[5] A. M. Trzynadlowski, J. K. Pedersen, R. L. Kirlin and S. Legowski, "Random Pulse Width Modulation Techniques for Converter-Fed Drive Systems-A Review," *IEEE Trans. Ind. Appl.*, vol. 30, No. 5, 1994, pp. 1166-1175.

[6] Z. G. Pan, X. J. Jiang, H. W. Lu and L. P. Huang, "Junction temperature analysis of IGBT devices," *Power Electronics and Motion Control Conference*, vol.3, pp. 1068-1073, 2000.

[7] M. M. R. Ahmed, G. A. Putrus and L. Ran, "Predicting IGBT junction temperature under transient condition," *Industrial Electronics*, vol 3,pp874-877, 2002.

[8] J. F. Gao, K. Zhao and M. M. Huang Jones, "Suppressing Boost conberter EMI with chaotic control," *Power electronics*, vol. 38(3), pp. 82-85, 2004.

[9] M. S. Khanniche, M. Towers,P. A. Mawby,P. Igic,S. T. Kong and Z. Zhou " A Fast Power Loss Calculation Method for Long Real Time Thermal," *EPE2005-Dresden*, Sep11-14, 2005.

[10] D. W. Berning and A. R. Hefner, "IGBT model validation for soft-switching applications," *IEEE Trans.Ind.Appl.*, vol. 37, pp. 650-660, 2001

[11] T. J Kim, D. W. Kang, Y. H. Lee and D. S. Hyun, "The Analysis of Conduction and Switching Losses in Multi-Level Inverter System," *in Proc. of the 32th Power Electronics Specialists. Conference and Applications*, pp. 1363-1368, 2001.

[12] M. S. Khanniche, M. Towers, P. A. Mawby, P. Igic,S. T. Kong and Z. Zhou, "A Fast Power Loss Calculation Method for Long Real Time Thermal," *EPE2005-Dresden*, Sep11-14, 2005.

[13] T. J Kim, D. W. Kang, Y. H. Lee and D. S. Hyun, "The Analysis of Conduction and Switching Losses in Multi-Level Inverter System," *in Proc. of the 32th Power Electronics Specialists. Conference and Applications*, pp. 1363-1368, 2001.

Loss Analysis and Soft-Switching Characteristics of Flyback-Forward High Gain DC/DC Converter with GaN FET

Zhang Yajing, Trillion Q. Zheng, Li Yan
School of Electrical Engineering
Beijing Jiaotong University
Beijing, China
E-mail: zyj04291298@163.com

Abstract— Compared with Si MOSFET, the GaN FET devices have advantages in the electrical characteristics, thermal properties and mechanical properties. This paper compares electrical properties of the GaN FET and Si MOSFET. Evaluation of the GaN FET based on flyback-forward high gain DC/DC converter at soft-switching condition are presented in detail. In addition, the power loss analysis of GaN FET based flyback-forward high gain DC/DC converter is discussed in detail. Finally a 200W GaN FET based flyback-forward high gain DC/DC converter is established, experiment results verified that the GaN FET is superior to the silicon MOSFET in switching characteristic and efficiency.

Keywords— *GaN FET, flyback-forward, high gain, loss analysis*

I. INTRODUCTION

Wide bandgap semiconductor materials such as silicon carbide (SiC) and gallium nitride (GaN) have many advantages including wide band gap, high saturation drift velocity, high critical breakdown field, etc, so wide bandgap semiconductor device is more suitable for high-frequency, high-temperature, high power density applications [1-4].Currently, a series of research achievements on SiC devices have been made, however, the research and application of GaN devices is still limited [5-8].

Virginia Polytechnic University and American Electric Power Electronic Systems Center studied the application of GaN devices on MHz Buck converter and the LLC resonant converter, and discussed the impact of GaN device layout, magnetic components and distribution parameters on the circuit [9-12].

This paper compares electrical properties of the GaN FET and Si MOSFET. Evaluation of the GaN FET based on flyback-forward high gain DC/DC converter at soft-switching condition is presented in detail. The power loss analysis of GaN FET based flyback-forward high gain DC/DC converter is discussed in detail. Finally a 200W GaN FET based flyback-forward high gain DC/DC converter is established, experiment results verified that the GaN FET is superior to the silicon MOSFET in switching characteristic and efficiency.

II. STRUCTURE AND CHARACTERISTICS OF GaN DEVICE

A. Structure and Characteristics of GaN FET

Fig. 1 is a structure of GaN FET. Si semiconductor material is used as the substrate in the GaN FET, and GaN crystal layer with high resistance is grown on the basis of the Si substrate. Aluminum nitride (AlN) insulating layer is added in between the GaN layer and the Si substrate layer isolate the device and the substrate. AlGaN layer exists between the GaN layer and the gate (G), the source (S) and drain (D) electrodes, and two-dimensional electron gas (2DEG) with high electron mobility, low resistance can be generated between the AlGaN layer and the GaN layer.

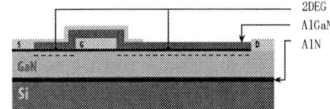

Fig. 1 Structure of GaN FET

The device is voltage-controlled device. When the positive gate-source voltage is greater than the threshold voltage, the gate is enabled, 2DEG form and the transistor is turned on; contrarily, the transistor is turned off.

The GaN FET is a lateral structure device, as shown in Fig. 1. Different from Si MOSFET, GaN FET has no parasitic body diode. There is no P-type parasitic bipolar region connected to the source electrode under the gate electrode of GaN FET. This structure makes GaN FET has a symmetrical transfer characteristic, which is GaN FET can be driven either by positive gate-to-source voltage (V_{gs}) or positive gate-to-drain voltage (V_{gd}).

B. Characteristics of GaN Schottky diode

Compared with Si Schottky diode and SiC diodes, GaN Schottky diode has lower forward voltage, as shown in Fig. 2. GaN Schottky diode has lower on-state losses. In addition, GaN Schottky diode no minority carriers, which will greatly reduce the transient voltage spikes.

978-1-4799-2706-7/14 $31.00 © 2014 IEEE

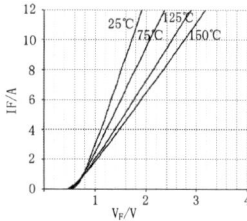

Fig. 2 Forward Characteristics of TPS3410PK 600V/6A

III. TOPOLOGY AND SIMULATION OF FLYBACK-FORWARD HIGH GAIN DC/DC CONVERTER

A. Topology of Flyback-Forward DC/DC Converter

The flyback-forward high gain DC/DC converter is shown in Fig. 3. The main switches S_1 and S_2 work in the interleaved mode, and their control signals have a 180 degree phase shift. The active-clamp circuits are mainly composed of the auxiliary switches S_{c1} and S_{c2} and the clamp capacitors C_{c1} and C_{c2}. The clamp switches S_{c1} and S_{c2} are driven complementarily with the main switches S_1 and S_2, which can recycle the leakage energy, suppress the turn-off voltage spikes on the main switches, and realize ZVS for all the primary devices. Besides, there are two coupled inductors in the converter named L_1 and L_2, in which the primary inductors L_{1a} and L_{2a} are coupled with their secondary inductors L_{1b} and L_{2b} respectively. L_{lk} is the total leakage inductance, which is equivalent to the secondary side.

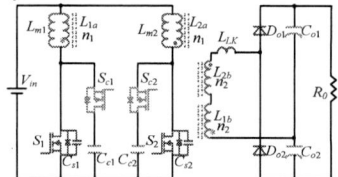

Fig. 3 Topology of flyback-forward high gain DC/DC converter

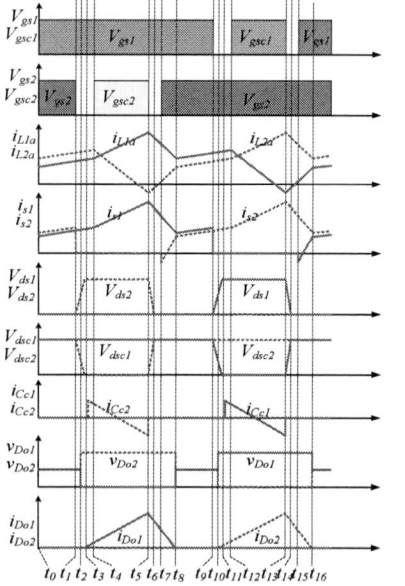

Fig. 4 Waveforms of flyback-forward high gain DC/DC converter

B. Simulation of Flyback-Forward DC/DC Converter

PSIM is utilized to verify the principle of the circuit. The simulation parameters are list as following : V_{in}=25V, f_s=100kHz, V_o=380V, resistive load R_o=722Ω.

The waveforms of primary side current I_{L1a}, I_{L2a} of coupled inductor, and secondary diode current I_{Do1}, I_{Do2} and drive signal V_{gs1}, V_{gs2} is shown in fig. 5 (a). The waveform of the drain-source voltage V_{ds} of the main switches S_1, S_2 is shown in fig. 5 (b). The voltage and current of S_1 without active-clamp circuit and with active-clamp circuit are shown, respectively in Fig. 5 (c) and Fig. 5 (d). Compared with the simulation result without active-clamp, the circuit with active-clamp achieves ZVS on and ZCS off, eliminating voltage spikes of S_1 and S_2.

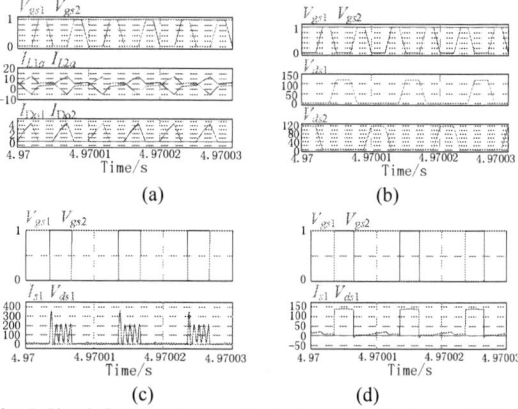

Fig. 5 Simulation waveforms of flyback-forward high gain DC/DC converter(a) IL1a, IL2a IDo1, IDo2, (b) Vds of S1and S2, (c) results without active-clamp, (d) results with active-clamp

IV. DEVICE SELECTION AND LOSS ANALYSIS OF GAN FET BASED FLYBACK-FORWARD HIGH GAIN DC/DC CONVERTER

The design specifications of GaN FET based flyback-forward high gain DC/DC converter are shown in Table I.

TABLE I
DESIGN SPECIFICATION OF FLYBACK-FORWARD HIGH GAIN DC/DC CONVERTER

Parameter	Range
Input voltage	18-40V
Output voltage	380V
Power	200W
Switch frequency	100kHz

A. Device Selection of GaN FET Based Flyback-Forward High Gain DC/DC Converter

Main switch is selected by the voltage levels and current levels [13]. The voltage gain is

$$\frac{V_{out}}{V_{in}} = \frac{2N}{1-D} \qquad (1)$$

For duty ratio D>0.5, when V_{in}=40V, $N < 2.375$ .We select N=2. The main switch voltage will be

$$V_{ds\_S} = V_{ds\_Sc} = V_{in} + \frac{V_{out}}{N} = \approx 140V \qquad (2)$$

The main switch current amplitude is

$$I_{S1} = \frac{P}{V_{in.min}} = \frac{100}{18} \approx 5.6A \qquad (3)$$

Therefore, two pieces of EPC2010 in parallel are employed as the main switches. Both of main and clamp switches have the same voltage stress. So the parameters of clamp switches should be the same as main switches. We also select EPC2010 as the clamp switches.

When the primary-side switch is turned off, the maximum voltage drop of rectified diodes is about 350V. The peak current of the diodes is

$$I_{Do\_peak} = ((N \cdot V_{Cc1} - V_{dc}/2) \cdot (1-D)/f_S)/L_{lk} \qquad (4)$$

We choose GaN schottky diode TPS3410PK 600V/6A produced by Transphorm Company as rectified diodes. The key parameters are shown in detail as following: V_F=1.3V, I_R= 25uA, Q_c=54nC, C =81pF.

When coupled inductor turns ratio N=2, the leakage inductance need

$$L_{Lk} \leq \frac{2 \cdot R_0 N^2 - M \cdot N \cdot R_0 (1-D)}{2 \cdot M^2 \cdot f_s} \qquad (5)$$

The magnetizing inductor L_m can be determined by setting an acceptable current ripple, which is given by

$$L_m = \frac{V_{in} \cdot D}{0.2 \cdot I_{Lm} \cdot f_S} \quad f_s = 100kHz \qquad (6)$$

The magnetizing inductor is chosen as $L_m = 80\mu H$, $f_s = 100kHz$ the leakage inductance of the primary side L_{Lk_P}=1μH .

B. Loss Analysis of GaN FET Based Flyback-Forward High Gain DC/DC Converter

According to the symmetry of the circuit, we take switch S_1 as example for loss analysis. P_o=200W, V_{in}=25V, turn-on time of main switch $T_1 = D*T_S$, the clamp switches conduction time is $T_2 = (1-D)*T_S$, dead time is 1% of the switching period Ts.

Active-clamp circuit leads to ZVS of S_1, whose loss is mainly composed of conduction loss and switching loss. According to Fig. 4, during one switching period, the current of S_1 in every stage is:

[t_0-t_3]: The current in one which is following the primary of coupled inductor, defined by

$$i_{L1a}(t) = I_{Lm1} = \frac{P_o}{2*V_{in}} \qquad (7)$$

[t_3-t_5]: Because of the effect of active clamp circuit, the leakage current of the secondary I_{Lk} increases linearly. Converting it to the primary, the main switch current is

$$i_{L1a}(t) = I_{Lm1} + Ni_{Lk}(t)$$
$$= I_{Lm1} + N\frac{(NV_{Cc2} - V_o/2)}{L_{Lk}}(t-t_3) \quad (t_3 < t < t_5) \quad (8)$$

Where $\Delta V = NV_{Cc2} - V_o/2$=2V , L_{Lk}=5μH , $t_5 - t_3$=T_2

[t_5-t_8]:
$$i_{L1a}(t) = I_{L1a}(t_5) - Ni_{Lk}(t)$$
$$= I_{L1a}(t_5) - N\frac{V_o}{2L_{Lk}}t \quad (t_5 < t < t_8) \quad (9)$$

Where $t_5 - t_8 = \frac{(2*NV_{Cc2} - V_o)}{V_o}*T_2$

[t_8-t_9]: During this stage, the main switches S_1 and S_2 are in the turn-on state. The current flowing through S_1 is the one of coupled inductor, According to the analyses above, the RMS current of S_1 is given by

$$I^2_{L1a\_RMS} = \frac{\int_0^{T_s} i_{L1a}^2(t)dt}{T_s} \qquad (10)$$

For the R_{DS} of EPC2010 is $25m\Omega$, the switch S_1 conduction loss is

$$P_{S_1\_on} = I_{L1a\_RMS}^2 * R_{DS}/2 \qquad (11)$$

Turn-off loss is

$$P_{S_1\_of} = t_f f_s V_{peak\_S_1} I_{peak\_S_1}/6 \qquad (12)$$

In the above formula, t_f is the overlap time of I_{DS} and V_{DS}, V_{peak_S1} is the peak voltage of drain-source voltage.

Since the clamp switch is ZVS turn on, the current flowing through the clamp switch can be expressed as

$$i_{S_{c1}}(t) = I_{Lm1} - Ni_{Lk}(t)$$
$$= I_{Lm1} - N\frac{(NV_{Cc2} - V_o/2)}{L_{Lk}}(t-t_3) \quad (t_{11} < t < t_{13}) \quad (13)$$

The RMS current of S_{c1} is

$$I^2_{S_{c1}\_RMS} = \frac{\int_0^{T_s} i_{S_{c1}}^2(t)dt}{T_s} \qquad (14)$$

Active clamp switches have the same type with main switches, so the conduction loss of S_{c1} is

$$P_{S_{c1}\_on} = I^2_{S_{c1}\_RMS} * R_{DS} \qquad (15)$$

Turn-off loss is of S_{c1} is

$$P_{S_{c1}\_sw} = t_f f_s V_{peak\_S_{c1}} I_{peak\_S_{c1}}/6 \qquad (16)$$

The current of secondary side diode is in discontinuous current mode, so the loss is mainly conduction loss.
[t_3-t_5]: The current increases linearly,

$$i_{D_{o1}}(t) = \frac{NV_{Cc2} - V_o/2}{L_{Lk}}(t-t_3) \quad (t_3 < t < t_5) \quad (17)$$

[t_5-t_8]: The secondary-side current decreases linearly,

$$i_{D_{o1}}(t) = I_{D_{o1}\_peak} - \frac{V_o/2}{L_{Lk}}(t-t_5) \qquad (18)$$

The average current of D_{o1} is

$$I_{D_{o1}\_avg} = \frac{\int_0^{T_s} i_{D_{o1}}(t)dt}{T_s} \qquad (19)$$

Conduction loss is

$$P_{D_{o1}\_on} = I_{D_{o1}\_avg} V_F \qquad (20)$$

Where V_F is the forward voltage drop of GaN schottky dioe, which is 1.3V.

In this design, we choose Kool toroid core. For f_s=100 kHz, we take B_{pk}=0.6. The power loss of A125-330 is P_0=500mW/cm^3. The magnetic circuit volume of the core is V_e=4.150cm^3, so magnetic core loss is P_{core}=2.74W.

The losses of the GAN FET based flyback-forward high gain DC/DC converter is shown in Table II. The circuit total loss is about 7.54W, theoretical efficiency can reach 96.2%.

TABLE II
POWER LOSS OF GAN FET BASED FLYBACK-FORWARD HIGH GAIN DC/DC CONVERTER

Device	Power losses
Conduction loss of S1	0.18W
Turn off loss of S1	0.35W
Conduction loss of S_{c1}	0.03W
Turn off loss of S_{c1}	0.23W
Conduction loss of D01	0.24W
Loss of a coupled inductor	2.74W

V. EXPERIMENTAL RESULTS OF GAN FET BASED FLYBACK-FORWARD HIGH GAIN DC/DC CONVERTER

We apply EPC2010 into flyback-forward high gain DC/DC converter, the prototype is shown in Fig. 6. As the unique LGA package of EPC2010 products, parasitic parameters in the main power circuit and driver circuit can be significantly reduced, improving the driving stability, while reducing voltage stress of the main switches.

Taking test under the condition that the input voltage is 25V, the power is 200W and the ratio of possession and void is 0.74, we obtain the experimental waveform shown in Fig. 7 (a), where channel 1 is output voltage (100V/div), channel 3 is V_{ds} of main switches S_1 (50V/div), channel 4 is primary current of coupled inductors I_{L1a} (5A/div). We tested the temperature of the devices with 30°C room temperature, as shown in the Fig. 7 (b).

Fig. 8 shows the experimental results when the main switch is selected as Si MOSFET IPB107N20N3G 200V/88A in the same operating condition. Taking test under the condition that the input voltage is 25V and the output voltage is 380V. We only change the main switches and the driver.

In Fig .8 (a), Channel 1 and 2 are V_{ds} of main switch S_1 and S_2 (50V/div), channel 4 is primary current of coupled inductors I_{L1a} (10A/div), channel 3 is the output voltage (100V/div). We tested the temperature of the devices with 30°C room temperature, as shown in the Fig. 8 (b).

From the Fig. 7 (a), we can conclude that no voltage spikes generate when the main switch S_1 is turned off, and switching loss is small. The output DC voltage of the circuit can be stabilized at 380V. Contrarily, visible voltage spikes generate of Si MOSFET based flyback-forward high gain DC/DC Converter，which will reduce the efficiency. The Si MOSFET has higher Temperature in the same operating condition.

The highest efficiency is 95.8% at 200W power point of the GaN FET based flyback-forward high gain DC/DC Converter. While, it's only 94% at 200W power point of the Si MOSFET based Converter. From the experimental results, we can see that the application of GaN device can reduce the voltage stress of the switch tube and then enhance the circuit efficiency.

Fig. 6 PCB board of GaN FET based flyback-forward high gain DC/DC Converter

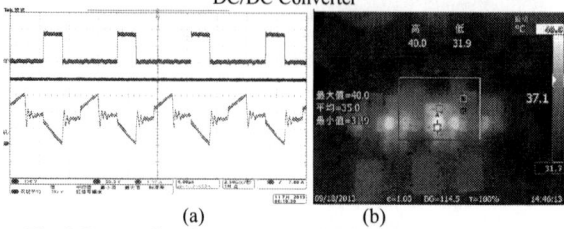

(a)   (b)

Fig. 7 Test results of GaN FET based flyback-forward high gain DC/DC Converter (a) experiment waveforms, (b) temperature test results

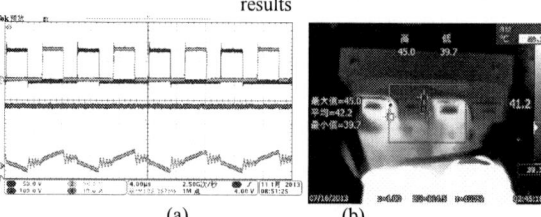

(a)   (b)

Fig. 8 Test results of Si MOSFET flyback-forward high gain DC/DC Converter (a) experiment waveforms, (b) temperature test results

VI. CONCUSIONS

This paper describes the development of GaN FET and presents its structure and electrical properties. We apply EPC2010 in flyback-forward high gain DC/DC converter. Loss analysis is discussed in detail for GaN FET based flyback-forward high gain DC-DC Converter. Application of GaN FET can significantly reduce the switch off voltage spike, and reduce switching losses. Finally a 200W GaN FET based flyback-forward high gain DC/DC converter is established, experiment results verified that the GaN FET is superior to the silicon MOSFET.

REFERENCES

[1] A. Lidow, J. Strydom, M. de Rooij, and Y. Ma, "GaN transistorsfor efficient power conversion,"*Power Conversion Publications,* 2012.

[2] K. Boutros, R. Chu, B. Hughes, "Gan power electronics for automotive application," *Energytech*, pp.1-4, 2012.

[3] T. Morita, S. Tamura, Y. Anda, M. Ishida, Y. Uemoto, T. Ueda,T. Tanaka, and D. Ueda. "99.3 gan-based gate injection transistors," *Applied Power Electronics Conference and Exposition (APEC)*, pp.481-484 ,2011.

[4] S. Tamura, Y. Anda, M. Ishida, Y. Uemoto, T. Ueda, T. Tanaka, D. Ueda, "Recent advances in gan power switching devices," *Compound Semiconductor Integrated Circuit Symposium (CSICS)*,pp.1-4, 2010.

[5] M.J., Jinzhu Li, Jin Wang, "Applications of Gallium Nitride in power electronics," *Power and Energy Conference at Illinois (PECI)*,pp. 1-7, 2013.

[6] Scott, M.J, Ke Zou, Inoa, E, Duarte, R., Yi Huang, Jin Wang, "A Gallium Nitride switched-capacitor power inverter for photovoltaic applications," *Applied Power Electronics Conference and Exposition (APEC)*, pp.46-52, 2012.

[7] Scott, M.J., Ke Zou, Jin Wang, Chingchi Chen, Ming Su, Lihua Chen, "A Gallium Nitride Switched-Capacitor Circuit Using Synchronous Rectification," *IEEE Transactions on Industry Applications*, vol. 49, no. 3,pp.1383-1391, 2013.

[8] Delaine, Johan, Jeannin, Pierre-Olivier, Frey, David, Guepratte, Kevin, "Improvement of GaN transistors working conditions to increase efficiency of A 100W DC-DC converter," *Applied Power Electronics Conference and Exposition (APEC)* 3, pp.656-66, 2013.

[9] Xiaoyong Ren, Reusch, D., Shu Ji, Zhiliang Zhang, Mingkai Mu, Lee, F.C, "Three-level driving method for GaN power transistor in synchronous buck converter," *Energy Conversion Congress and Exposition (ECCE)*, pp.2949-2953, 2012.

[10] Reusch, D., Lee, F.C., Gilham, D., Yipeng Su, "Optimization of a high density gallium nitride based non-isolated point of load module," *Energy Conversion Congress and Exposition (ECCE)*, pp.2914-292, 20120.

[11] Shu Ji, Reusch, D., Lee, F.C, "High-Frequency High Power Density 3-D Integrated Gallium-Nitride-Based Point of Load Module Design," *IEEE Transactions on Power Electronics*, vol. 28, no. 9, pp.4216-4226, 2013.

[12] Lee, F.C., Qiang Li, "High-Frequency Integrated Point-of-Load Converters: Overview," *IEEE Transactions on Power Electronics*, vol. 28, no. 9, pp.4127-4136, 2013.

[13] Jung-Min Kwon, Kim Eung-Ho, Bong-Hwan Kwon, Kwang-Hee Nam, "High-efficiency fuel cell power conditioning system with input current ripple reduction," *IEEE Transactions on Industrial Electronics*, vol. 56, no. 3, pp.826-834, 2009.

The 2014 International Power Electronics Conference

Insulated Metal Substrate for Power Modules using Anodic Oxide Film of Aluminum

Takeshi Tokuyama, Jumpei Kusukawa, Kinya Nakatsu
Department of Power Electronics Systems Research
Hitachi Research Laboratory, Hitachi, Ltd.
Hitachi, Japan
takeshi.tokuyama.xs@hitachi.com

Abstract— **To improve the insulation reliability of insulated metal substrates for power modules, we propose an insulated metal substrate that has a porous alumina anodic oxide film along the interface between an aluminum base and resin insulating layer. By having two anodic oxide films, i.e., inorganic and organic acids, compared with conventional substrates, the proposed substrate improves the adhesiveness with the resin insulating layer, resulting in 160% improvement in insulation resistance after 2400 h under atmospheric condition in which the temperature was 85°C and relative humidity was 85%, and 30% improvement in breakdown voltage compared with conventional substrates.**

Keywords—Insulated Metal Substrate, Power Module, Anodic Oxide Film of Aluminum, Migration.

I. INTRODUCTION

To conserve energy towards mitigating global warming, efficient electric-energy-conversion technology, which uses power electronics, is essential. The inverter, which is a typical power electronics apparatus that transforms DC electric power into AC electric power, high-power density and high reliability are required. An essential part in the inverter is the power module, which requires low thermal resistance and high reliability [1][2]. Insulated substrates are important components in the power module, which determine cooling capability and insulating performance. From the viewpoint of reliability, improvements in "insulation reliability" and "structure reliability" are required for an insulated substrate.

An insulated substrate can be mainly divided into a ceramic substrate and insulated metal substrate. Ceramics have the advantage of high thermal conductivity (20-140 W/mK). However, they have a low coefficient of linear expansion (4.0-7.0 ppm/K). Therefore, thermal stress is generated due to the difference in the coefficient of linear expansion between the ceramic and metal components as a heat spreading base. This stress generates cracks in the ceramics, which degrade insulation reliability. Thus, it is necessary to fully study structure reliability [3]-[5]. On the other hand, an insulated metal substrate consists of a wiring conductor, resin insulating layer, and metal base. The wiring conductor is adhered to the metal base with the resin insulating layer. Because of the low thermal conductivity (3.0-5.0 W/mK) of the layer, it is used as a thin film (0.1-0.2 mm) to lower thermal resistance [6].

This substrate is widely used for application with several hundred volts. However, ion migration to the substrate occurs. Ion migration is a phenomenon in which the metal material is ionized by moisture absorption and electric field supply and moves in the resin insulating layer. Ion migration develops over time and reduces insulation resistance. Thus, migration durability needs to be sufficiently studied to achieve high insulation reliability [7][8].

To apply an insulated metal substrate to high-voltage application, our goal was to improve insulation reliability. To achieve this, insulation resistance and breakdown voltage must be improved. We propose a substrate that has a heat-spreading aluminum base coated with a two-layer anodic oxide film. We conducted an element production trial and confirmed improvement in insulation resistance and breakdown voltage.

II. TECHNOLOGICAL PROBLEMS WITH INSULATED METAL SUBSTRATE FOR POWER MODULES

Figure 1 shows a sectional drawing of an insulated metal substrate for power modules and the associated general problems. This substrate consists of a lead frame, resin insulating layer, and heat dissipation aluminum base. The lead frame is adhered to the base with the resin insulating layer, and the power device is soldered onto the lead frame. The resin insulating layer insulates the power device from the base and transfers the heat that the power device generates to the aluminum base. In the insulated metal substrate, the problems with insulation reliability are (1) ion migration and (2) partial discharge. Ion migration is a phenomenon in which a metal material moves in the resin insulating layer. The metal is ionized by moisture absorption, and the electric field supply mat-

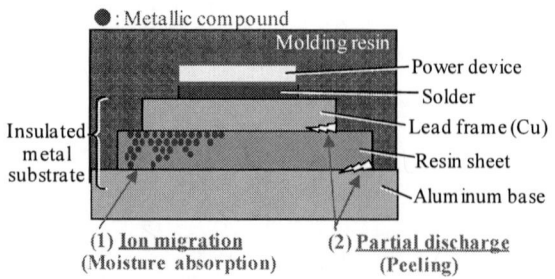

Fig. 1. Technological problems with insulated metal substrate

978-1-4799-2706-7/14 $31.00 © 2014 IEEE   2904

erial is in the lead frame. Ion migration develops with time and reduces the insulation resistance. Ion migration durability is improved by reducing the ionic ingredients or increasing heat resistance, moisture resistance, and insulation resistance. On the other hand, partial discharge is the phenomenon of reducing the breakdown voltage over time. The main cause of this phenomenon is cavities formed due to interfacial peeling, which occurs during the thermal cycle. Hence, partial discharge is improved due to the increase in adhesiveness of the resin insulating layer with the lead frame and aluminum base.

The proposed substrate uses an anodic oxide film of aluminum to solve the above-mentioned problems of the insulated metal substrate for power modules.

III. PROPOSED SUBSTRATE

The structure of our substrate improves adhesiveness and insulation by using a two-layer anodic oxide film of aluminum.

First, we explain the features of the anodic oxide film of aluminum. This film is a porous Al_2O_3 film formed on the surface of the aluminum by anodization. Figure 2 shows a structural drawing of the film. This drawing is the Keller model, which is a general model of anodic oxide film [9]. This film is porous and consists of cells in the shape of hexagonal columns that have a pore in their centers. The pore size and wall thickness of the cell change depending on the electrolyte and electrolytic conditions. This results in the formation of a film with large pores or thick walls. These characteristics are used for several applications that require corrosion resistance, adhesiveness, electric insulation, and wear resistance. Anodization can also result in a

multilayer film with two or more functions [10]. Our proposed substrate uses this multilayer film formation. Figure 3 shows a cross-sectional drawing of the proposed substrate. This substrate has a two-layer anodic oxide film along the interface between the resin insulating layer and aluminum base. The first layer is an adhesive film for improving adhesion strength, which is placed at the adhesion interface with the resin insulating layer. The second layer is an insulation film for improving insulation resistance and breakdown voltage, which is placed at the lower layer of that film. With this two layered film, adhesiveness and insulation improve. Regarding migration, the insulation resistance gradually decreases with the development of ion migration in conventional substrates, as shown in Fig. 4(a) [11]. In the proposed substrates, however, the development of ion migration is prevented by the anodic oxide film, as shown in Fig. 4(b), because this film consists of alumina of ceramics, which is an inorganic material. Hence, it is believed that our proposed substrate improves migration durability.

We first evaluated the anodic oxide film of aluminum and examined its adhesiveness and insulation. Next, we conducted a trial production of the proposed substrate and verified its effectiveness.

IV. EVALUATION OF ANODIC OXIDE FILM

A. Evaluation of anodic oxide film for adhesion

The anodic oxide film for adhesion needs to be in a shape that provides an anchoring effect. The resin insulating layer permeates the pores of the film, which achieves this effect. Thus, it is believed that film with a large pore diameter provides strong adhesiveness. The pore diameter changes depending on the type of electrolyte acid. Table 1 shows cross-sectional scanning electron microscope (SEM) images of the films we evaluated. The electrolytes of these films are inorganic acid 1, inorganic acid 2 and an organic acid. The pore diameters are as follows: inorganic acid 1, 0.08~0.13 μm; inorganic acid 2, 0.4~1.2 μm; organic acid, 0.4~0.6 μm. Each pore value is different, and the smallest one is inorganic acid 1 and the largest is inorganic acid 2.

Their adhesive strength measurement results are discussed below. The measurement conditions were as follows. The resin insulating layer was a 120-μm epoxy resin, the glass transition temperature was 124°C, and the

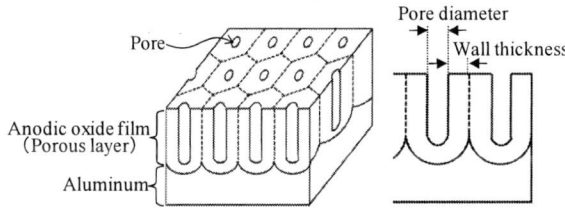

Fig. 2. Structural drawing of anodic oxide film

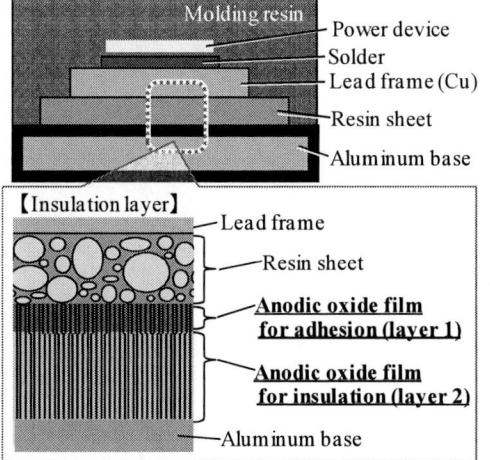

Fig. 3. Cross-sectional drawing of proposed substrate

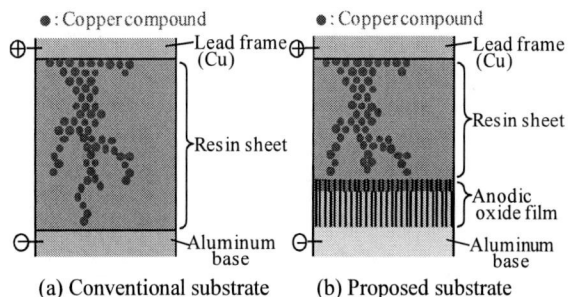

Fig. 4. Cross-sectional drawing of migration prevention effect of proposed substrate

Table. I. Cross-sectional SEM image of each Anodic Oxide film

Acid for electrolyte	Inorganic acid 1	Inorganic acid 2	Organic acid
Cross-sectional SEM image	Pore Anodic oxide film ×30.0k 1.00um	Pore Anodic oxide film ×30.0k 1.00um	Pore Anodic oxide film ×30.0k 1.00um
Pore diameter (mm)	0.08～0.13	0.4～1.2	0.4～0.6

aluminum plate of the adherend was A6063, which was anodized. The measurement method of adhesive strength was based on a shearing test method of JISK6850, and the temperature was -20°C.

Figure 5 shows the measurement results. The inorganic acid 2 film was the highest with respect to adhesive strength, which was proportional to the pore diameter size. This result suggests that film with large pores results in a high anchoring effect and that the inorganic acid 2 film is suitable as an adhesion film for our proposed substrate.

B. Evaluation of anodic oxide film for insulation

We evaluated the insulation resistance and breakdown voltage and selected the film thickness that exhibits high insulation resistance and high breakdown voltage. For application to power modules, the target insulation resistance should be more than 10^8 Ω and that of breakdown voltage should be more than 600 V_{rms}.

Figure 6 shows a test piece and the test equipment. The test piece was an aluminum plate, which was anodized, and the material type of the aluminum plate was A1050-H24. We used an organic acid film as the insulation film because it has a thick wall and is strong. In the test equipment, the test piece was pressed with an electrode, and the voltage was applied between the test piece and electrode. The film thickness of the test piece was measured with an eddy-current film thickness tester ISOSCOPE manufactured by Phisher. The film thickness of the test piece was the mean value in the set of measurements.

Figure 7 shows the results of insulation resistance. We calculated the insulation resistance from the leakage current of the test piece. The leakage current was measured by applying DC 300 V for 1 min with a digital pico-ammeter R8340 manufactured by ADVANTEST in an atmospheric condition in which the temperature was

27°C and relative humidity was 40%. A columnar electrode of 20 mmφ was used for the test equipment. According to Fig. 7, the thicker the film is, the higher the insulation resistance becomes. First, the mean value of the insulation resistance of the 25-μm-thick film was 4.0×10^8 Ω. Next, the mean value of insulation resistance of 50-μm-thick film was 1.4×10^9 Ω. Furthermore, the mean value of insulation resistance of the 66-μm-thick film was 2.5×10^9 Ω.

Figure 8 shows the measurement results of breakdown voltage. The entire test equipment was immersed in insulating oil to prevent partial discharge in the electrode circumference. We define the breakdown voltage as a voltage when a 3-mA short circuit current flows. A rectangular electrode of 45×30 mm^2 was used for the test equipment and insulating oil FC-3283 manufactured by 3M. The thicknesses of the films in Figs. 8 were also a mean value. The thicker the film is, the higher the breakdown voltage becomes. First, the mean value of the breakdown voltage of the 25-μm-thick film was 520 V_{rms}. Next, the mean value of the insulation resistance of the 50-μm-thick film was 920 V_{rms}. Furthermore, the mean

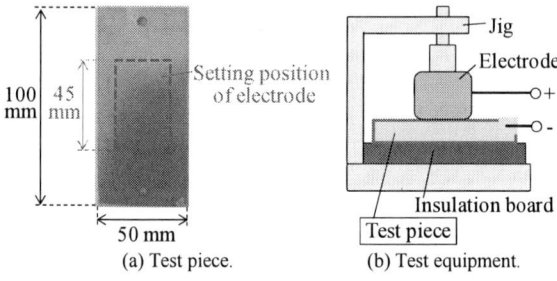

Fig. 6. Schematic of test piece and test equipment

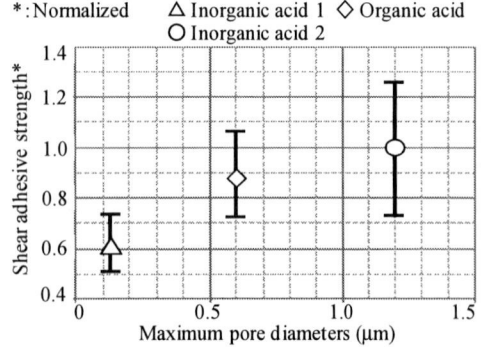

Fig. 5. Measurement results of adhesive strength

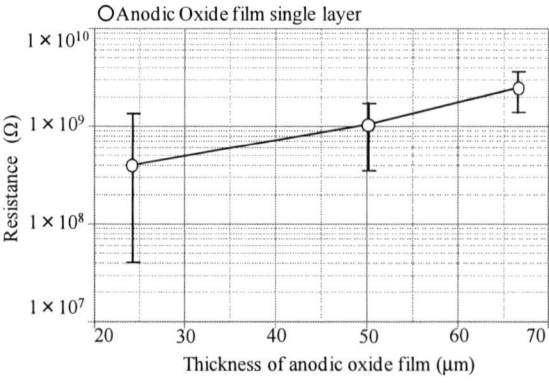

Fig. 7. Measurement results of insulation resistance test

978-1-4799-2706-7/14 $31.00 © 2014 IEEE

value of the insulation resistance of the 65-μm-thick film was 1163 V$_{rms}$.

Figure 9 shows the temperature dependency of insulation resistance. The atmospheric temperature was 27, 85, and 120°C. The anodic oxide film of aluminum was the 50-μm-thick organic acid film. The resin insulating layer was the above-mentioned 120-μm epoxy resin. In the resin insulating layer, the higher the atmospheric temperature is, the lower the insulation resistance becomes. On the other hand, in the anodic oxide film of aluminum the higher the atmospheric temperature is, the higher the insulation resistance becomes. This is caused by humidity around these films. The results indicate that relative humidity around the anodic oxide film of aluminum decreases due to an increase in the atmospheric temperature [12][13]. The relative humidity values are as follows: 27°C, 40%; 85°C, 2.5%; 120°C, 0.7%. Hence, the insulation resistance of the anodic oxide film of aluminum increased as the atmospheric temperature increased. It appears that the proposed substrate exhibits higher migration durability than a conventional substrate because the proposed substrate exhibits inorganic material ceramic and high insulation resistance in high atmospheric temperature.

The above results show that the target insulation resistance of 10^8 Ω and breakdown voltage of 600 V$_{rms}$ was obtained with a film thickness of 50 μm or more. Because a thick film requires a long process time, we selected 50 μm, which is the minimum film thickness that can achieve the targets.

V. RESULTS FROM TRIAL PRODUCTION OF PROPOSED SUBSTRATE

A. Structure of proposed substrate sample

We fabricated and evaluated a sample of the proposed substrate using the inorganic acid 2 film for adhesion and organic acid film for insulation. Figure 10 shows the structure of the proposed substrate sample. The thickness of the inorganic acid 2 film was about 1.0 μm and that of the organic acid film was 50 μm. The resin insulating layer was the above-mentioned epoxy resin. The material of the heat dissipation base was A1050-H24 and that of the lead frame was C1020.

The trial manufacture process is as follows: (1) the anodic oxide film for adhesion is formed on the aluminum base by anodization in the electrolyte of inorganic acid 2, (2) the anodic oxide film for insulation is formed by anodization in the electrolyte of organic acid, and (3) the resin insulating layer is adhered on the aluminum base, which has been anodized before.

B. Results and discussion

Figure 11 shows the cross-sectional SEM image of the proposed substrate sample. The inorganic acid 2 film was formed on the interface, and the organic acid film was formed in the lower layer. The boundary of these films is the position where the size of the pore diameter changes.

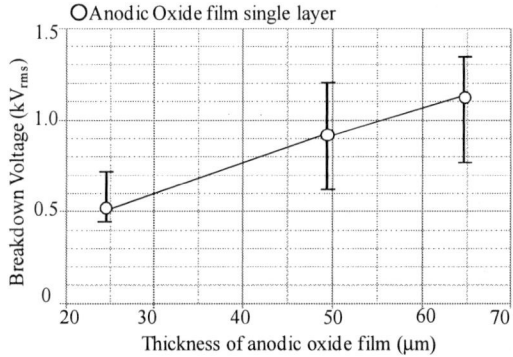

Fig. 8. Measurement results of breakdown voltage

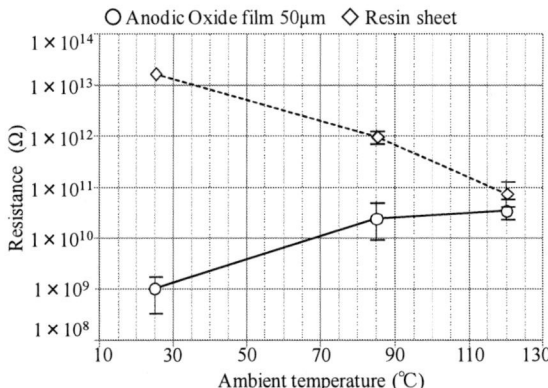

Fig. 9. Measurement results of temperature dependability of insulation resistance

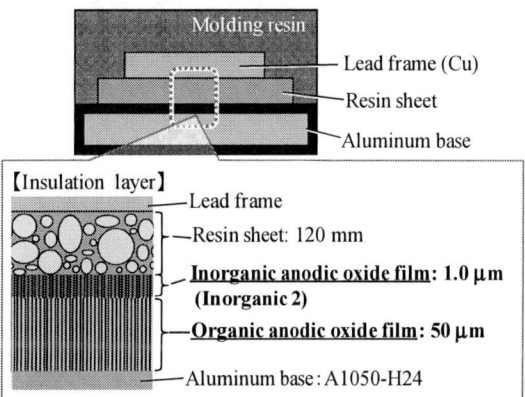

Fig. 10. Cross-sectional drawing of proposed substrate sample

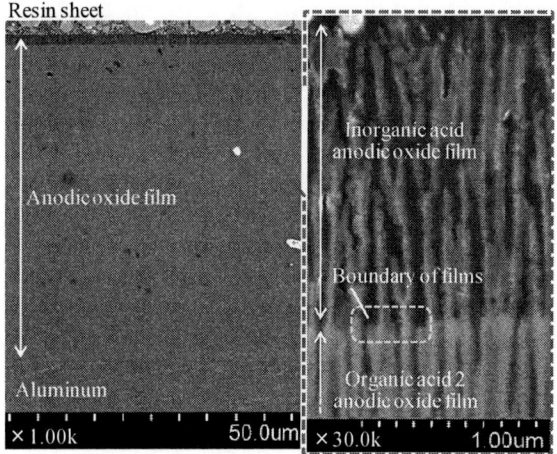

Fig. 11. Cross-sectional SEM image of proposed substrate sample

Figure 13 shows the measurement results of migration durability. The time to reach 1.0×10^9 Ω with the proposed substrate was the longest; 900 h. After 2400 h, the insulation resistance of the proposed substrate was 6.4×10^8 Ω. The insulation resistance of a conventional substrate with 120-μm-thick resin insulating layers decreased to 2.4×10^8 Ω. The insulation resistance of a conventional substrate with 160-μm-thick resin insulating layers decreased to 4.4×10^8 Ω.

Figure 14 shows the cross-sectional image of the trial samples. Figure 14(a) shows the development of ion migration in the insulation layer of the proposed substrate. The image suggests that the anodic oxide film prevents the development of ion migration. On the other hand, Figure 14(b)(c) shows that the ion migration of the conventional substrates developed at the bottom of the resin insulation layer. As a result, it appears that the proposed substrate maintained the highest insulation resistance value in these samples by preventing the development of migration.

Figure 15 shows the measurement results of breakdown voltage. The straight line shows the relation between the resin insulating layer thickness and break-down voltage. The proposed substrate exhibited about a 1.0-k V_{rms} improvement from the 120-μm conventional substrate. Compared with the same thick insulation layer of 170 μm, the proposed substrate exhibited about a 0.5-kV_{rms} improvement.

From the above mentioned results, the proposed substrate sample exhibited improvements in insulation resistance and breakdown voltage compared to the conventional substrates.

VI. CONCLUSION

To improve the insulation reliability of the insulated metal substrate for power modules, we proposed an insulated metal substrate that has an anodic oxide film along the interface between an aluminum base and a resin insulating layer. From the evaluation of an anodic oxide film of aluminum, the inorganic acid 2 film was found to be suitable as an adhesive film, and an organic acid film with a thickness of 50 μm achieved the target insulation resistance of 10^8 Ω and breakdown voltage of 600 V_{rms}. With these films, the proposed substrate improved adhesiveness, exhibited 160% improvement in insulation resistance after 2400 h under an atmospheric condition in which the temperature was 85°C and relative humidity was 85%, and 30% improvement in breakdown voltage compared to the 120-μm conventional substrate.

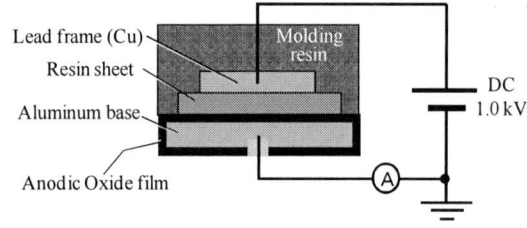

Fig. 12. Structure of sample for migration durability

— Resin sheet 120 μm +anodic oxide film 50 mm (Proposed)
-·- Resin sheet substrate: 120 μm sheet (Conventional)
···· Resin sheet substrate: 160 μm sheet (Conventional)

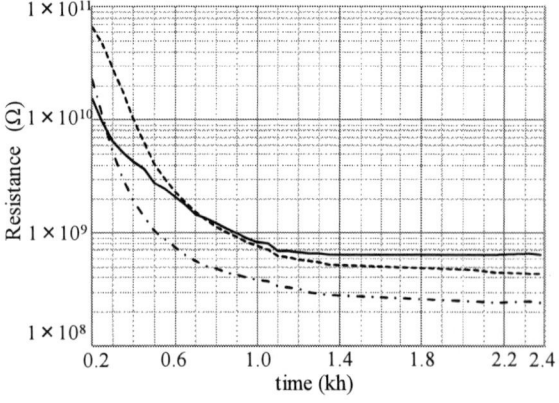

Fig. 13. Measurement results of migration durability

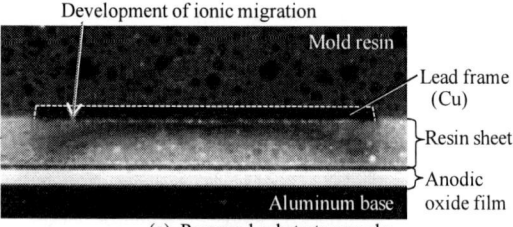

(a) Proposed substrate sample

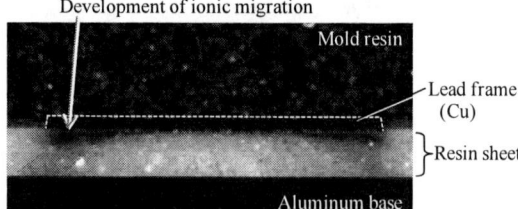

(b) Conventional substrate sample: resin sheet: 120 μm

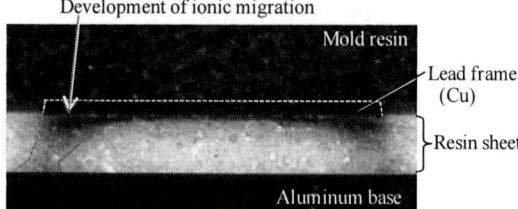

(c) Conventional substrate sample: resin sheet: 160 μm.

Fig. 14. Cross-sectional image of each sample

● Resin sheet 120μm + Anodic oxide film 50μm (Proposed)
◇ Resin sheet (Conventional)

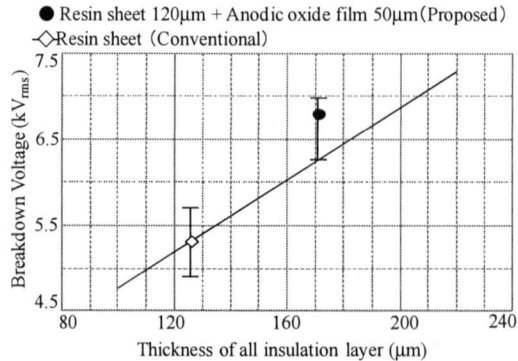

Fig. 15. Measurement results of breakdown voltage

978-1-4799-2706-7/14 $31.00 © 2014 IEEE

ACKNOWLEDGEMENTS

We would like to thank Iwatani Materials Corporation and Shimajiri Technical Center for cooperating with trial production.

REFERENCES

[1] K. Nakatsu, H. Suzuki, A. Nishihara, K. Sasaki, "Next generation Inverter Technology Supporting Environm-entally Conscious Vehicle," *Hitachi Review*, Vol. 94, No.04, pp. 34-37, 2012

[2] N. Hirano, Y. Sakamoto, "Looking back on power electronics products for automobile, and mounting technologies of double-side cooling power module," *Journal of the Japan Welding Society*, Vol. 80, No. 4, pp. 294-298, 2011

[3] T. Nishimura, K. Mimura, S. Hiramatsu, H. Shiota, T. Ueda, "High Performance Transfer-molded Power Module," *MITSUBISHI ELECTRIC ADVANCE Magazine*, Vol. 84, No. 4, pp. 219-223, 2010

[4] K. Yamaguchi, H. Masubuchi, T. Okamoto, "Quality and Reliability Integration Technology in Automotive Semiconductor Products", *Fuji Electric Journal*, Vol. 84, No. 2, pp. 17-21, 2011

[5] Y. Nishimura, A. Morozumi, E. Mochizuki, and Y. Takahashi, "Investigations of all lead free IGBT module structure with low thermal resistance and high reliability," *IEEE ISPSD*, pp. 1-4, 2006

[6] M. Tsuji, T. Kawahira, Y. Takase, K. Fukushima, Y. Takezawa, "The High Thermal Conductive Laminates of the Next Generation," *Shin-Kobe Technical Report*, No. 19, pp. 43-48, 2009

[7] T. Tsukui, "Insulation Deterioration and the Prevention Method by Electrochemical Migration of Electric Equipment (Pt.1)", *Journal of Japan Institute of Electronics Packaging*, Vol. 8, No. 4, pp. 339-345, 2005

[8] T. Tsukui, "Insulation Deterioration and the Prevention Method by Electrochemical Migration of Electric Equipment (Pt.2)", *Journal of Japan Institute of Electronics Packaging*, Vol. 8, No. 6, pp. 523-530, 2005

[9] S. Asahara, S. Ueda, M. Sato, H. Nagasaka, M. Matsunaga, M. Mukai, "Anodic Oxidation Metal Finishing", *J Soc. of Japan*, pp. 1-6, Asakura Publishing Co., Ltd

[10] T. Nakayama, "Surface Treatment of aluminum", pp. 148-150, Nikkan Kogyo Shimbun Ltd.

[11] K. Okamoto, T. Maeda, K. Haga "Study of Copper Ion Migration in Metal Base Printed Wiring Boards", *The journal of Japan Institute for Interconnecting and Packaging Electronic Circuits*, Vol. 10, No. 2, pp. 108-112, 1995

[12] M. Maejima, K. Saruwatari, M. Hirata, H. Itoh, K. Teramoto, Y. Kaneko "Insulation Properties of Oxide Film of Aluminum", *Fujikura Technical Journal*, Vol. 1, No. 96, pp. 49-53, 1999

[13] M. Maejima, M. Takaya "Dielectric Properties of Anodic Oxide Film of Aluminum –Introduction on Rikenn Iho and Dr. Miyata's Writings-", *ALUTOPIA*, Vol. 37, No. 2, pp. 37-44, 2007

The 2014 International Power Electronics Conference

A Fast-Transient-Response Buck Converter with Split-Type III Compensation and Charge-Pump Circuit Technique

Jiann-Jong Chen, Wei-Ting Hsu, Jih-Hua Yu, Yuh-Shyan Hwang and Cheng-Chieh Yu

Department of Electronic Engineering, National Taipei University of Technology, Taipei 106, Taiwan, R.O.C

Abstract— In order to achieve fast transient response and high efficiency on a CMOS DC-DC buck converter, Split-Type III (ST3) compensation technique and a charge-pump circuit would be proposed and presented in this paper. The proposed buck converter has been fabricated with a TSMC 0.35-μm CMOS 2P4M process. This compensation design tries to add a charge pump circuit into the compensator for improving its transient response time. The operating frequency of the proposed buck converter is 1MHz for input voltage 3.3V and output voltage range 1.0V-2.5V applications. The proposed structure using charge pump circuit technique can achieve fast transient response when load current changes between light load and heavy load, and vice versa. Compared with conventional Type III compensator, ST3 can reduce unnecessary area and power consumption of passive components. Measured result shows that the settling time of the converter output is less than 2 μs for a load current step of 200 mA. Peak power efficiency of 90.8% is obtained at 100 mA load current. The whole chip area is 1.49 mm×1.46 mm(include PADs).

Keyword- fast transient response, high efficiency, charge-pump circuit, Type III compensation

I. INTRODUCTION

DC-DC converters are required in many applications, such as cell phones and tablet PC, especially it with a fast transient response get more and more essential nowadays [2-3]. So how to define a good products not only power efficiency but also transient response. Because electronic devices need to save power and life time of the battery, having efficient charging are very important. On the other hand, chip size reducing also is a challenge for power management integrated circuits. The design of system compensation needs to consider recovery time and stability of buck converters. Type III compensation used to extend the crossover frequency and promote the dc gain for the system stability, the frequency response of the conventional Type III compensator which can include above mention considerations for buck converter. For the reason that a Split-Type III compensation technique in Fig. 1 be proposed which imitates [1] would be presented. Another thing that Split-Type III compensation apply currents be PWM inputs for reduce passive component, it has some different between PT3 and ST3. Instead of a large area of conventional Type III compensator which is implemented on-chip, the proposed

compensator circuit can reduce the on-chip passive component size and power consumption [4]. X(s) is designed as an operational transconductance amplifier (OTA) with wide input range and Y(s) is a low-power bandpass filter. Because of the line and load variations may be large on portable products, requirements of line and load regulations must be important in changing period. In order to enhance the transient response of buck converter, it adds a charge-bump circuit to the compensation scheme. For one input of pulse width modulation (PWM), the proposed converter using adder to sum two current signals that ramp generator through Voltage-to-Current converter (V-to-I) and band-pass filter. The mentioned band-pass filter is a low-power circuit and meanwhile is a part of ST3 compensation; another is an operational transconductance amplifier.

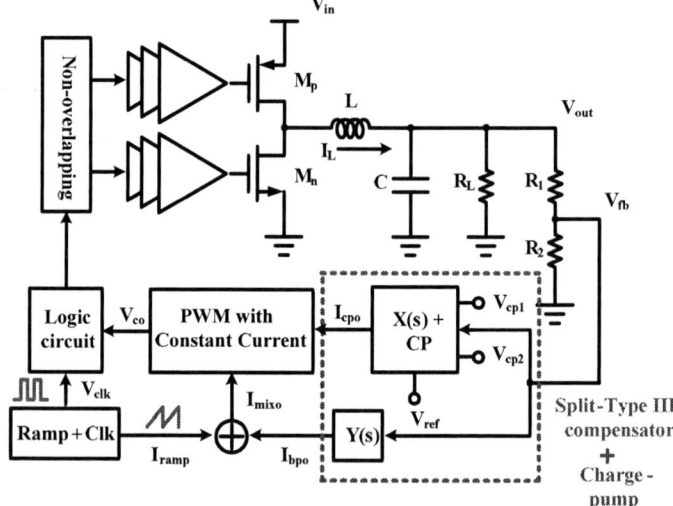

Fig. 1. Block diagram of Split-Type III and charge-pump circuit buck converter.

II. CIRCUIT DESCRIPTIONS

There are disadvantages of conventional Type III compensator are described in section II. The input signals of Split-Type III compensator which is split to paths for slow and fast portions, including parts X(s) and parts Y(s), its' input by the voltage divider of V_{out}, also is for charge-pump circuit. A pole be generated by parts X(s) for whole transform function. The parts Y(s), bandpass filter can makes two poles, and the compensated locations of two

978-1-4799-2706-7/14 $31.00 © 2014 IEEE

zeros can be generated by the sums of parts X(s) and parts Y(s). The charge-pump circuit is added to be combined with parts X(s) for makes the transient response faster through its control circuits with compared equipment. The small-signal model of the proposed Split-Type III compensation buck converter is illustrated in Fig. 2, and loop gain T(s) would be derived as

$$T(s) = \frac{k}{V_m} G_{LC}(s)[X(s) + Y(s)] \qquad (1)$$

where k is the factor of voltage divider $R_2/(R_1+R_2)$, V_m is the amplitude of ramp signal, and $G_{LC}(s)$ is transfer function of LC filter, X(s) and Y(s) are transfer function of operational transconductance amplifier and bandpass filter, respective [8]. Finally, X(s) and Y(s) can be derived as

$$X(s) = \frac{G_{OTA}}{1 + \dfrac{s}{\omega_{p0}}} \qquad Y(s) = \frac{G_{BPF}\left(1 + \dfrac{s}{\omega_{zbpf}}\right)}{\left(1 + \dfrac{s}{\omega_{p1}}\right)\left(1 + \dfrac{s}{\omega_{p2}}\right)} \qquad (2)$$

where G_{OTA} and G_{BPF} are the dc gain of operational transconductance amplifier and bandpass filter, respective. There are two zeros and three poles if sums X(s) and Y(s) to Z(s).

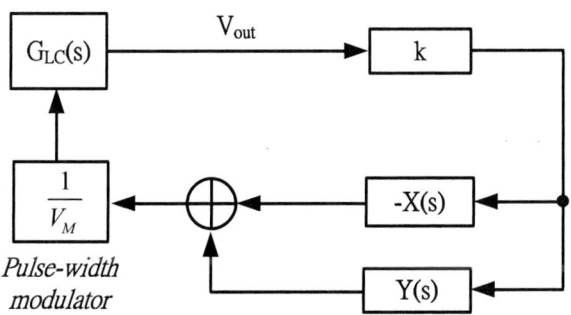

Fig. 2. Small-signal block diagram of the converter with ST3 compensation.

Additionally, the charge-pump and its control circuit can be used to enhance the transient response. The ramp signal generated by using current to charge or discharge a capacitor, and then it would through a V-to-I converter to provide a current with ramp signal. Another function of the ramp generator is produces clock cycle signals by discharging the capacitance immediately.

The pulse width modulation (PWM) circuit consists of a current comparator which input signals are output currents of parts X(s) with charge-bump circuit, another is sum of two paths signals by a band-pass filter combines with a ramp generator. Finally, the non-overlapping circuit and buffer are designed to prevent those power MOSFET transistors from conducting suddenly at the same time, and also to drive the power MOSFET transistors.

Fig. 3(a) shows the two stage OTA, which used to

improve the linearity if input voltage beyond the range of active in common [9-10], and OTA has to convert voltage to current for one input of current comparator. Moreover, it can generate a pole for whole circuit.

The band-pass filter and a voltage-to-current converter are shown in Fig. 3(b) [1]. In order to offset the poles or zeros of the output LC filter, there are two pole and one zero would be generate by the amplifier, M_{bp}, C_{bp}, R_{bp2}, and R_{bp3}. The amplifier, M_{bp} and R_{bp1} convert the input voltage V_{fb} to current I_{bpfo} that it mixed with ramp signal which PWM comparator needs.

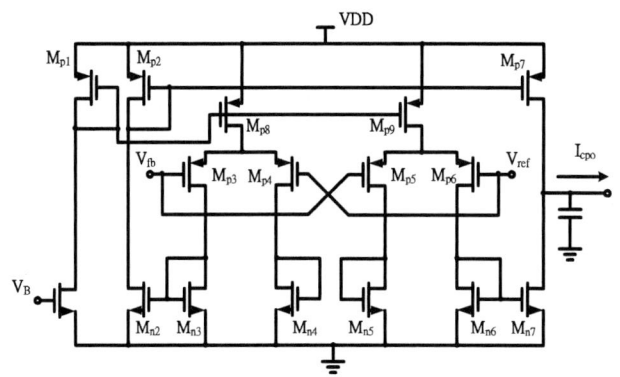

(a)

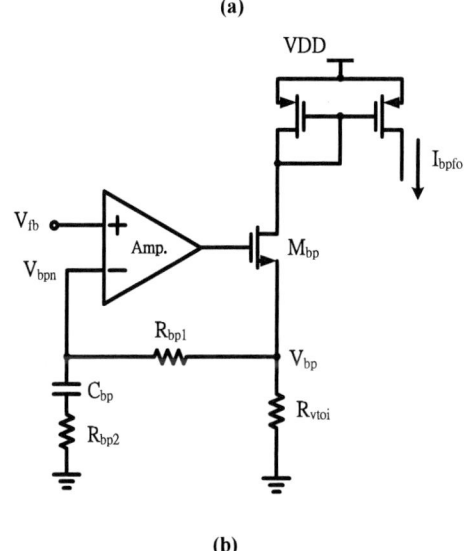

(b)

Fig. 3. ST3 compensator. (a) Schematic of charge-pump circuit and OTA. (b) Schematic of bandpass filter.

Fig. 4 shows the charge-pump of the proposed buck converter, it provides two reference voltages V_{cp1} and V_{cp2} to detect the output current variance and decide how output currents of the charge-bump flow (sourcing or sinking). V_{cp1} is set higher than V_{cp2}, so a region between V_{cp1} and V_{cp2}, and V_{fb} in the region. When V_{fb} is between the two referenced voltages, comparators would not work, so the additional charge-bump idle; it also can't affect the whole circuit. If $V_{fb} > V_{cp1}$, this charge-bump would offer some currents to charge the short current of the PWM comparator so that output currents can be stable much faster than conventional compensator [11]. For the other case where V_{fb}

978-1-4799-2706-7/14 $31.00 © 2014 IEEE

$< V_{cp2}$, it has the opposite situation.

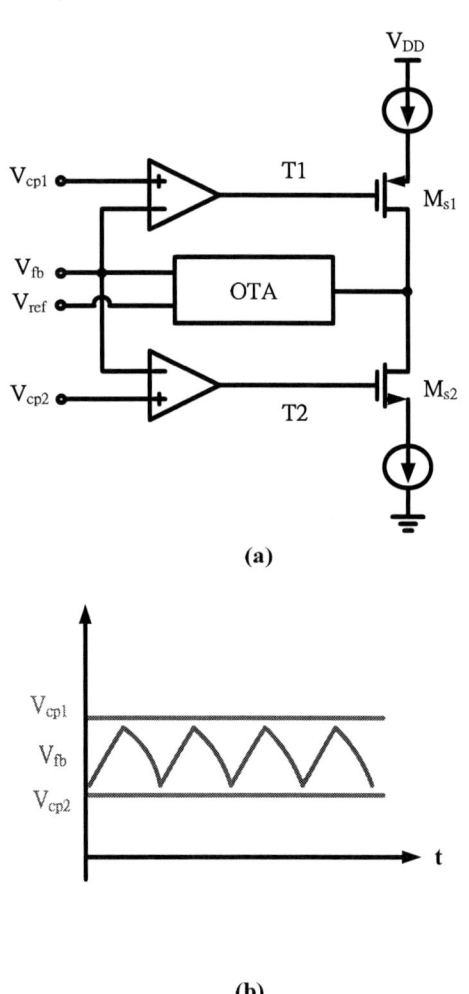

(a)

(b)

Fig. 4. Charge-pump circuit (a) Schematic of charge-pump (b) V_{fb} compare with V_{cp1} and V_{cp2}

III. EXPERIMENTAL RESULTS

The fast transient response buck converter is proposed and implemented with TSMC 0.35-μm 2P4M CMOS processes. With PWM-Controlled-Mode DC-DC buck converter has been used 3.3V supply voltage. The maximum efficiency of the buck converter up to 90.8% is obtained when output voltage is 2.5V and load current is 100 mA. Therefore, the output voltage range of the proposed converter is designed from 1.0V to 2.5V and maximum load current is 400 mA in this work. The transient responses of proposed buck converter are shown in Figs. 5 and 6 when the output voltage is 2.5V. Fig. 5 shows the recovery time is 2μs when the load current I_{Load} changes from 0 mA to 200 mA. Fig. 6 shows the response time is 2μs when the load current I_{Load} changes from 200 mA to 0 mA. The output voltage V_{OUT} is always stable whether load current variation. Both of heavy/light load changing, this buck converter offer a very fast recovery times is about 2μs.

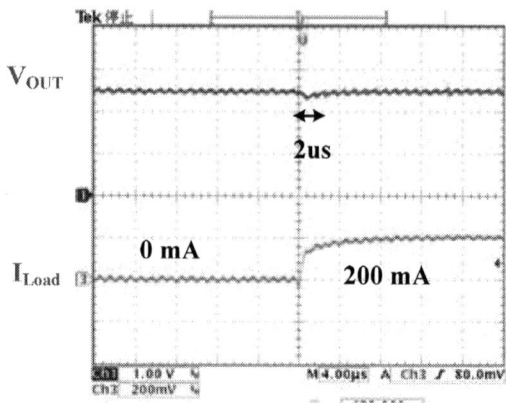

Fig. 5. The measurement result under input/output voltage of 3.3V/2.5V. Load current I_{Load} from 0 mA to 200 mA.

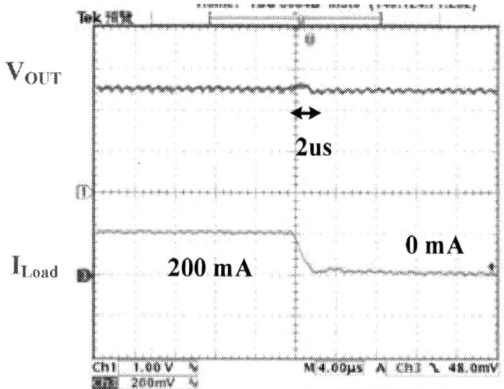

Fig. 6. The measurement result under input/output voltage of 3.3V/2.5V. Load current I_{Load} from 200 mA to 0 mA..

Fig. 7 shows the chip micrograph and composition of the proposed buck converter. The total chip area is about 1.46×1.49 mm^2 (include PADs). Fig. 8 shows the power efficiency under voltage is 1.2V, 1.8V, 2.5V. The peak efficiency of this chip is 90.8% at output voltage is 2.5V with load current is 100 mA.

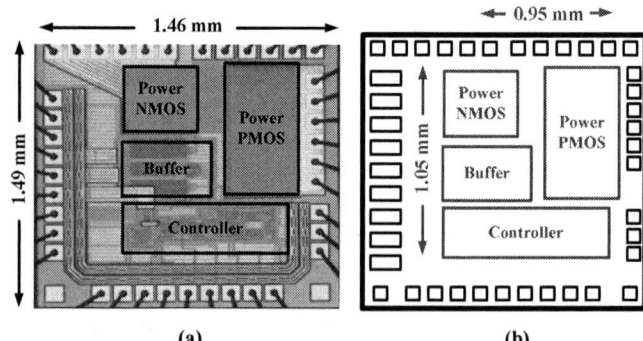

(a)   **(b)**

Fig. 7. Chip micrograph and configuration of the proposed buck converter. (a) Chip micrograph. (b) Chip composition.

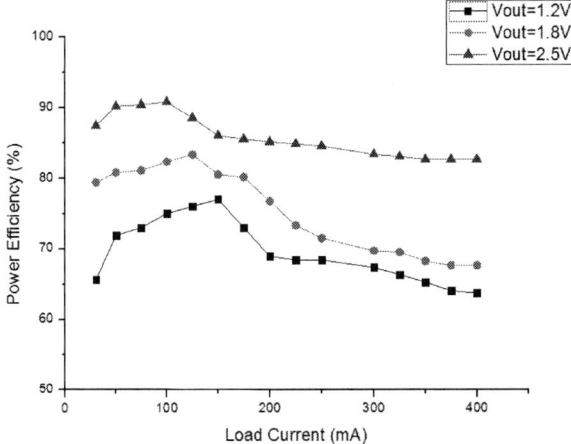

Fig. 8. Power efficiency of the proposed buck converter.

In this design, the input voltage is 3.3V and the output voltage range of the buck converter is designed from 1.0V to 2.5V, and the maximum load current is 400 mA. The transient response is very fast about 2μs at input/output voltage of 3.3V/2.5V by changing load current. The switching frequency is 1 MHz.

IV. CONCLUSIONS

According to conventional Type III compensator has high gain and enough phase margin at low frequency or high frequency, so that compensator can help loop of circuit stabilizing, it also has disadvantage which large chip area with passive components by R-C pairs. This paper presents a fast transient response buck converter with Split-Type III compensation and charge-pump circuit technique. It was designed with TSMC 0.35-μm 2P4M CMOS process. The buck converter with ST3 compensator and the ramp generator using Schmitt-trigger circuit really can reduce chip size for power consumption and circuit complexity improving. Finally, the efficiency of the buck converter can reach above 80% in load is 50 to 400mA, which peak efficiency achieve 90.8% with 3.3V supply voltage and 2.5V output voltage. There are some important advantages of the proposed converter and those are very important features on the SOC applications.

ACKNOWLEDGMENT

The authors would like to thank National Science Council (NSC) for project supporting and Chip Implementation Center (CIC) for chip fabrication.

REFERENCES

[1] P. Y. Wu, S. Y.S. Tsui, and P. K. T. Mok, "Area- and power-efficient monolithic buck converters with pseudo-type III compensation," *IEEE J. Solid-State Circuits*, vol. 45, no. 8, pp. 1446–1455, Aug. 2010.

[2] R. Miftakhutdinov, "Analysis and optimization of synchronous buck converter at high slew-rate current transients," in *IEEE Power Electronics Specialists Conf.*, 2000, pp. 714-720.

[3] M. T. Zhang, M. M. Jovanovic, and F. C. Lee, "Design considerations for low-voltage on-board dc/dc modules for next generations of data processing circuits," *IEEE Transactions on Power Electronics*, vol. 11, no. 2, pp. 328-337, Mar. 1996.

[4] S. S. Kelkar and F. C. Lee, "A fast time domain digital simulation technique for converters: Application to a buck converter with feed-forward compensation," *IEEE Trans. Aerosp. Electron. Syst.*, vol. 1, no. 1, pp. 21–31, Jan. 1986.

[5] A. Pressman, *"Switching Power Supply Design,"* McGraw-Hill Inc., 1991.

[6] Vishay Siliconix. "Design a high performance buck or boost converter with Si9165,"[Online Document], http://www.ieechina.com/Upload/Tech/200405101503130 781.pdf

[7] L. Guo, J.Y. Hung, and R.M. Nelms, "PID controller modifications to improve steady-state performance of digital controllers for buck and boost conveners," *IEEE APEC Proc.*, 2002, pp. 381-388.

[8] Y. Wu and P. K. T. Mok, "Comparative studies of common control schemes for reference tracking and application of end-point prediction," in *IEEE Custom Integrated Circuits. Conf.*, 2007, pp. 559-562.

[9] A. Demosthenous and M. Panovic, "Low-voltage MOS linear transconductor/squarer and four-quadrant multiplier for analog VLSI," *IEEE Trans. Circuits Syst. I, Reg. Papers*, vol. 52, no. 9, pp. 1721–1731, Sep. 2005.

[10] S. Chatterjee, Y. Tsividis, and P. Kinget, "0.5-V analog circuit techniques and their application in OTA and filter design," *IEEE J. Solid-State Circuits*, vol. 40, no. 12, pp. 2373–2387, Dec. 2005.

[11] Kichang Jang, Jungsoo Choi, Chulkyu Park, and Joongho Choi, "A Voltage-Mode DC-DC Converter with Enhanced Transient Responses," in *Proc. IEEE International Symposium on Circuits and Systems (ISCAS)*, 2012, pp. 974–977.

[12] S. L. Chen and M. D. Ker, "A new schmitt trigger circuit in a 0.13-μm 1/2.5-V CMOS process to receive 3.3-V input signals," *IEEE Trans. Circuits Syst. II: Express Briefs*, vol. 52, no. 7, pp. 361-365, Jul. 2005.

[13] Y. S. Hwang, S. C. Wang, F. C. Yang, and J. J. Chen, "New Compact CMOS Li-Ion Battery Charger Using Charge-Pump Technique for Portable Applications" *IEEE Transactions on Circuit and Systems I: Regular Papers*, Vol. 54, No. 14, April 2007, pp. 705-712.

[14] W. Qiu, S. Mercer, Z. Liang, and G. Miller, "Driver deadtime control and its impact on system stability of synchronous buck voltage regulator," *IEEE Trans. on Power Electronics*, vol. 23, no.1, pp. 163-171, Jan. 2008.

[15] Y. H. Lee, S. J. Wang, K. H. Chen, "Quadratic Differential and Integration Technique in V^2 Control Buck Converter with Small ESR Capacitor", *IEEE Transactions on Power Electronics*, vol. 25, no. 4, pp 829-838, April 2010.

[16] W. Yan, W. Li, and R. Liu, "A noise-shaped buck DC-DC converter with improved light-load efficiency and fast transient response," *IEEE Trans. Power Electron.*, vol. 26, no. 12, pp. 3908–3924, Dec. 2011.

[17] Y. H. Lee, S. C. Huang, S. W. Wang, and K. H. Chen, "Fast Transient (FT) Technique With Adaptive Phase Margin (APM) for Current Mode DC-DC Buck Converters," *IEEE Trans. Very Large Scale Integration (VLSI) Syst.*, vol. 20, no. 10, pp. 1781–1793, Oct. 2012.

Advantages of Low Parasitic Inductance Packages of Power MOSFET for Server Power Applications

Wonsuk Choi, Dongkook Son and Dongwook Kim
Fairchild Semiconductor
HV PCIA, PSS Team
Bucheon-si, Republic of Korea
wonsuk.choi@fairchildsemi.com

Abstract— Super -junction MOSFETs and shielded gate trench MOSFETs are widely used especially for PFC and DC-DC converters for networking and computing power supply systems that require high efficiency and power density. With smaller parasitic capacitances, the latest SJ MOSFETs have extremely fast switching characteristics and therefore reduced switching losses. But, naturally this switching behavior leads to device stress include voltage and current spikes and greater dv/dt and di/dt by self-inflicted voltage transients during switching transition time due to the stray parasites in devices and printed circuit board. In this paper, the impact of parasitic inductances in Power MOSFET and PCB layout during switching transient by analytical Pspice simulation and experimental results The benefits of low inductance packages, called Power88 and Power 56, for high voltage and medium voltage MOSFETs are shown in server power system.

Keywords— *Sper-junction MOSFET, Shielded gate trench MOSFET, Server Power, Power56/Power88 pacakge*

I. INTRODUCTION

A growing trend in the power conversion world is saving energy. Power factor (the ratio of real power to apparent power in an ac power system) is a widely used measure of the quality of the input power of AC/DC converters. Power factors much less than one lead to inefficiency and instability of a power distribution network. Power factor correction (PFC) circuits are used to ensure that the AC/DC converter power factor is close to one. To comply with power quality and line harmonics standards (e.g. IEC 61000-4-3) Switch-mode power supplies are increasingly being designed with an active power-factor correction (PFC) at the input stage to meet international regulations for harmonics. Power supplies not only convert energy but they consume it. Improving efficiency is an important subject in modern power supplies. Many soft-switching techniques have been developed for higher efficiency and high power density by reducing power losses during the on/off transition of power switches. For secondary side of DC-DC converter, a synchronous rectifier is an essential building block since it enables to improve efficiency for these conversion stages with both lower conduction loss and switching losses as shown in Fig. 1. Simultaneously, power MOSFET technology has been developed towards

higher cell density for lower on-resistance. For high voltage MOSFETs($500 \sim 650V_{DSS}$), the super-junction device utilizing charge balance theory is introduced to semiconductor industry ten years back, and it set a new benchmark in high voltage power MOSFET market [1][2]. The super-junction MOSFETs enable higher power conversion efficiency. However, the extremely fast switching performance of super-junction MOSFETs creates unwanted side effects, like high voltage or current spikes or poor EMI performance. Based on recent system trends, improving efficiency is critical goal, and going with slow switching device just for EMI is not an optimized solution. For low and medium voltage MOSFETs ($40 \sim 200V_{DSS}$), it has been rapidly developed to accomplish significant performance improvements by trench gate technology after introduction of the planar technology in early 1970s. The trench gate MOSFETs are the most preferred power devices for medium to low voltage power application. These trench MOSFETs implement a gate structure embedded into a trench region that is carefully etched into the device structure. The specific on-resistance improved about 30% with this new technology thanks to the ability to increase channel density as well as eliminating the JFET resistance component. Power losses in synchronous rectification can be lowered when the product of MOSFET's on resistance and drain current is less than the diode forward voltage drop.

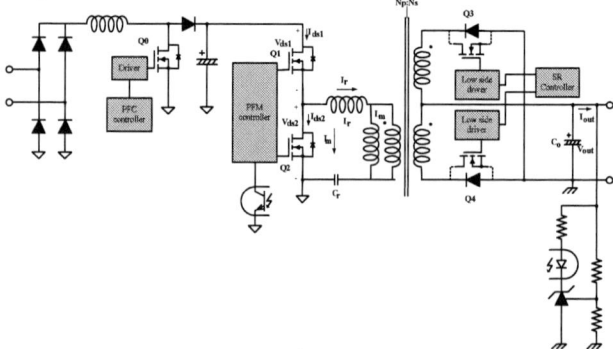

Fig. 1. Server Power supply using PFC and LLC resonant converter with synchronous rectification

978-1-4799-2706-7/14 $31.00 © 2014 IEEE

However, low on-resistance is not the only requirement for the power switches in terms of synchronous rectification. They should have small gate charge to reduce driving losses especially for light load condition. Voltage spikes are critical issue in PFC and synchronous rectification. General guidelines to minimize the undesirable voltage spikes include short and thick board copper trace patterns that will minimize current loops. However, it is not easy to apply all of them due to size and cost limitations. Sometimes designers need to consider mechanical structures, like heat sinks and fans, and sometimes are forced to use single-sided printed circuit board due to cost constraints. For a practical alternative, snubbers can be used to manage voltage spikes within the maximum drain-source voltage ratings. Additional power losses are inevitable in this case. In addition, the power losses due to the snubbers are not negligible at light load. Besides the circuit board parameters, device characteristics also affect to the voltage spike level. In this paper, the effect of parasitic source inductance in PFC and synchronous rectification and benefits of the new SMD packages are verified switching performance including losses, gate oscillation and current and voltage spikes through experimental results.

II. EFFECT OF PARASITIC INDUCTANCE IN PFC AND SYNCHRONOUS RECTIFICATION

A. Parasitic Inductance effects in PFC

Parasitic inductance can strongly influence MOSFET switching characteristics, usually causing increased switching losses and deviations from expected performance. Parasitic inductance arising from both component packaging and circuit layout is the reality of any circuit. From a general perspective, there are several oscillation circuits that affect the switching behavior of the MOSFET, including internal and external oscillation circuits [3]. Fig. 2 shows a simplified schematic of a PFC circuit with both internal parasitics of the power MOSFET and external oscillation circuits given by the external coupling capacitance, $C_{gd\_ext.}$, and parasitic inductance of drain, source, and gate of the board layout. Parasitic components in the devices and boards are involving switching characteristics more as the switching speed is getting faster. In Fig. 2, L, C_o, and D_{boost} are the inductor, output capacitor, and boost diode in PFC circuit. C_{gs}, C_{gd_int}, and C_{ds} are parasitic capacitances of the power MOSFET. L_{d1}, L_{s1}, and L_{g1} are the parasitic lead inductances of drain, source, and gate; including wire bonding of the power MOSFET. R_{g_int} and R_{g_ext} are the internal gate resistors of the power MOSFET and the external gate driving resistor of circuit. C_{gd_ext} is the parasitic gate-drain capacitance of the circuit. L_D, L_S, and L_G are the drain, source, and the gate copper trace stray inductances of the printed circuit board. Gate parasitic oscillation occurs in a resonant circuit formed by gate-drain capacitance, C_{gd}, and gate lead inductance, L_{g1} when MOSFET is turned on and off. Oscillation voltage

in drain-source of the MOSFET passes through gate-drain capacitance, C_{gd}, due to load wiring inductance L_D when MOSFET switching is getting fast, and particularly when it is turned off. A resonant circuit with gate lead inductance, L_{g1}, is formed. As gate resistor (R_{g_int} and R_{g_ext}) is extremely small, the oscillation circuit, Q, becomes rather large. When the resonance condition occurs, a large oscillation voltage is generated between that point and C_{gd} or L_{g1}. The voltage drop across L_S and L_{s1}, which can be represented by (1), is cause by negative drain current in turn-off transient.

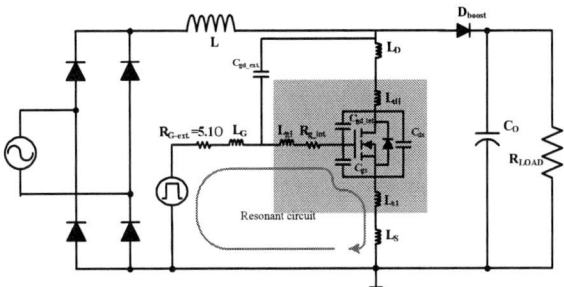

Fig. 2. Simplified schematic of PFC circuit with internal and external parasitics of Power MOSFET

This voltage drop across stray source inductances, L_S and L_{s1}, generates oscillation in gate-source voltage. The parasitic oscillation can cause gate-source breakdown, bad EMI, large switching losses, losing gate control, and can lead to MOSFET failure.

$$\Delta V_{GS} = (L_S + L_{s1}) \cdot \frac{di_d(t)}{dt} \quad (1)$$

To explain oscillation mechanism, transient switching behavior is simulated using analytical PSPICE simulation, focusing on the impact of parasitic inductances, L_G and L_S, in a power MOSFET and printed circuit board layout during turn-off transient. 0 shows the PSPICE simulation waveforms of the gate-to-source voltage, V_{GS}; the internal gate-to-source voltage, V_{GS_int}; the drain-to-source voltage, V_{DS}; the current channel of MOSFET, $I_{channel}$; and the drain current, I_D, in clamped inductive load switching circuit. To explain the gate oscillation of the power MOSFET with the effect of parasitic inductances, the turn-off transient is divided into two intervals ($t_1 \sim t_2$). Fig 4 shows the MOSFET equivalent circuit including parasitic capacitances and inductances.

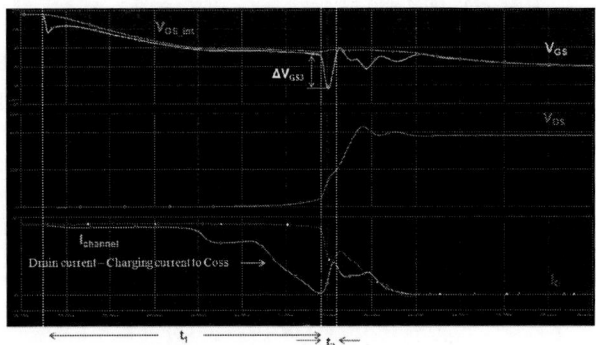

Fig. 3. Simulation Waveforms at Turn-Off Transient

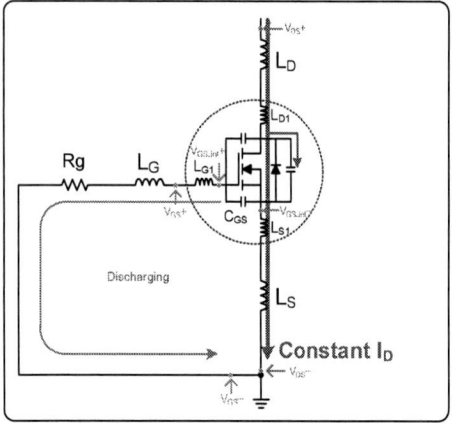

(a) MOSFET Operation during t_1

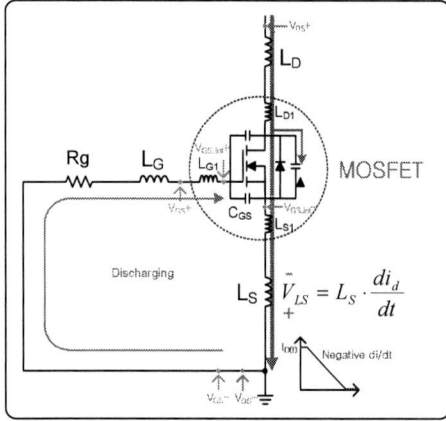

(b) MOSFET Operation during t_2

Fig.4. MOSFET Equivalent, including Parasitic Inductances

1) During Time Interval t_1

The voltage V_{GS} decreases exponentially due to discharging of input capacitance, which the gate-to-source capacitance, C_{GS}, and the gate-to-drain capacitance, C_{GD}, via the gate resistance, R_g, as shown in 0(a). When the gate voltage reaches the gate plateau voltage, the channel current in the MOSFET is reduced due to output characteristics in MOSFET, which is a characteristic curve between the gate voltage and drain current. At the same time, the output capacitance is charged up slowly.

2) During Time Interval t_2

The C_{oss} of the SJ MOSFET becomes strongly non-linear, that is, C_{oss} decreases very rapidly around 30~50 V drain-source voltage in the SJ MOSFET. These effects give an extremely fast dv/dt and di/dt, as shown in 0. At the same time, the voltage drop across the common-source inductance L_S is caused by negative drain current slope (-di_D/dt). This voltage drop leads to negative gate voltage. Due to this effect, the discharged current flows in the opposite direction, as shown in 0 4(b).

B. Parasitic Inductance effects in Synchronous Rectification

Body diode performance is depends on not only silicon design but also package inductance in synchronous rectification. The parasitic inductance can strongly influence MOSFET switching characteristics, usually causing increased switching losses and deviations from the expected performance. Parasitic inductance arising from both component packaging and circuit layout is a reality of any circuit. 0 shows a simplified schematic for a power MOSFET with all parasitic components in synchronous rectification. Low $R_{DS(on)}$ and low inductance packages are a must for low- and medium-voltage MOSFETs to achieve best switching performance and reduced conduction losses for highest efficiency. The length of the package lead introduces quite a bit of the stray source inductance. Industry standard through-hole type TO-220 package has 7 nH of typical lead inductance, while typical lead inductance of PQFN56 SMD package is only 1 nH. Other important parasitic components are layout parasitic inductance and capacitance. In circuit board layout, 1 cm of trace pitch has an inductance of 6-10 nH. These parasitic inductances directly affect to body diode reverse-recovery characteristics and peak voltage spikes.

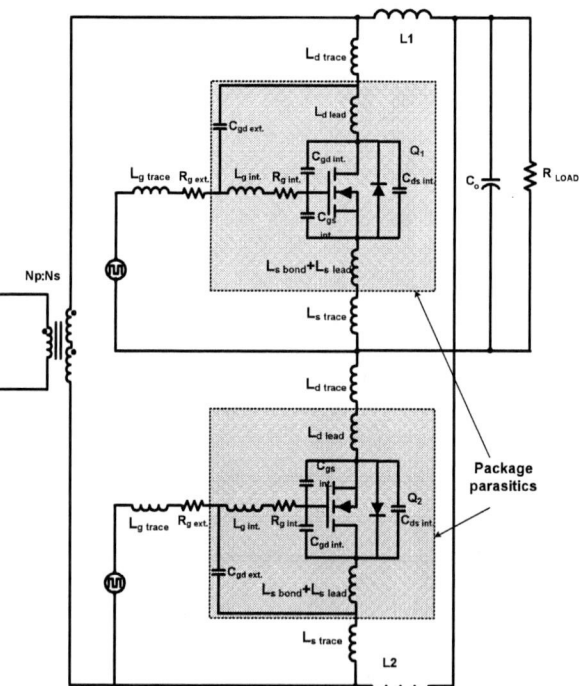

Fig.5. Power MOSFET with Internal and External Parasitic Components in Synchronous Rectification

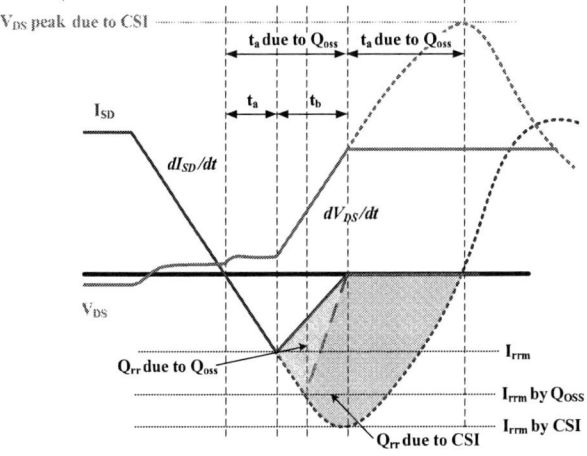

Fig.6. Body Diode Reverse Recovery Waveforms Considering Common Source Inductance (CSI)

The body diode recovery charge on a datasheet is generally the sum of C_{OSS} displacement current, the recovered minority carrier current, and the reactive currents arising from common-source inductance of the test circuit, as shown 0 6. The addition of common-source inductance effectively reduces system efficiency by increasing Q_{rr} and inducing a voltage across the MOSFET.

III. ADVANTAGES OF LOW PARASITIC INDUCTANCE PACKAGES OF POWER MOSFET

1) Power88 Package of HV MOSFETs

To drive fast switching by using the super-junction MOSFETs in different applications, it is also necessary to understand the influence of parasitic components in MOSFET package. Low $R_{DS(on)}$ and low parasitic component are important factors to achieve reduced conduction losses and best switching performance for highest efficiency in low voltage packages. The super-junction MOSFETs are mainly used in the voltage range of 500-600V. In these voltage ratings, the clearance and creepage distance requirements have to be considered. Power Quad Flat No-lead package, which is widely used in low voltage applications, is a surface mount package to achieve power density and small form factor requirements of many applications. Power88 package is a new leadless package for high voltage super-junction MOSFET.

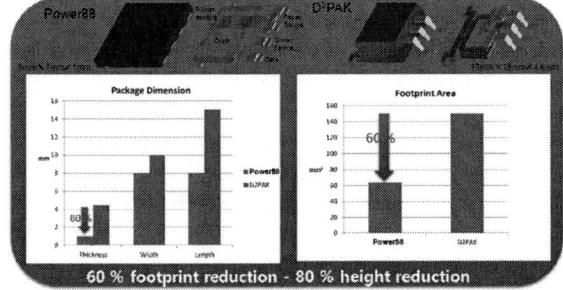

Fig. 7. Comparisons of Package Dimension and Footprint area between standard SMD package, D²PAK and New Leadless Package, Power88

As shown in Fig. 7, this new package has very low profile with 1mm ultra slim thickness, 8mm width and 8mm length and smaller foot print of 64mm² compared to D²PAK which is industrial standard SMD package for high voltage MOSFET. As shown in Fig 8, standard gate drive circuitry includes common source inductance but power88 package separate power and driver source to minimize common source inductance influence during switching transient. As shown in Table I, power88 package provides very lower parasitic inductance to achieve excellent switching performance compared to D²PAK. The approximately common source inductance value of Power88 package is 3nH but one of D²PAK is 7nH, Power88 package can reduce both source and gate inductances compared to D²PAK

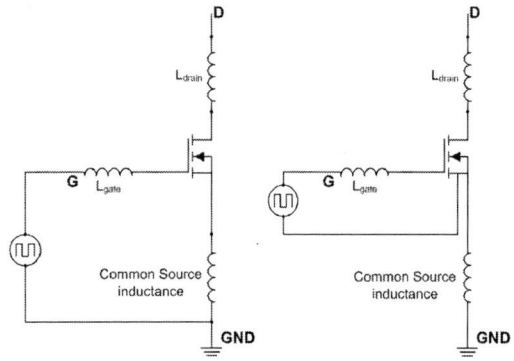

(a) Standard gate drive schematic (D²PAK)

(b) Kelvin source gate drive schematic (Power88)

Fig. 8. Comparison of Gate drive schematic with parasitic source inductance between standard SMD package, D²PAK and new leadless package, Power88 (Kelvin Source Configuration)

TABLE I
COMPARISONS OF APPROXIMATE PARASITIC INDUCTANCES FOR EACH PACKAGE

DUTs	L_S [nH]	L_D [nH]	L_G [nH]
Power88 Package	2	1	4
D²PAK package	7	1	7

2) Power56 Package of MV MOSFETs

For high-voltage MOSFETs, the resistance of packages has not been a concern. $R_{DS(ON)}$ can be achieved at 1~2 mΩ in a TO-220 standard package, depending on the voltage rating, by using modern medium voltage MOSFETs technology. Unlike high-voltage MOSFETs, the package itself contributes a significant portion of the total resistance for medium-voltage MOSFETs due to wire bonding, lead, and source metal. For example, up to around 33% of the $R_{DS(ON)}$ is accounted for by the package resistance in a 75 V/2.3 MOSFET, as shown in **Error! Reference source not found.**. SO-8 packages were popular before upgraded power package Power56. Total on resistance of medium-voltage MOSFET can be dramatically reduced by using an SMD package, such as Power56. It can also reduce package inductance that

978-1-4799-2706-7/14 $31.00 © 2014 IEEE

causes undesirable voltage spikes. It enables use of lower $R_{DS(ON)}$ MOSFETs by replacing lower voltage rating MOSFETs.

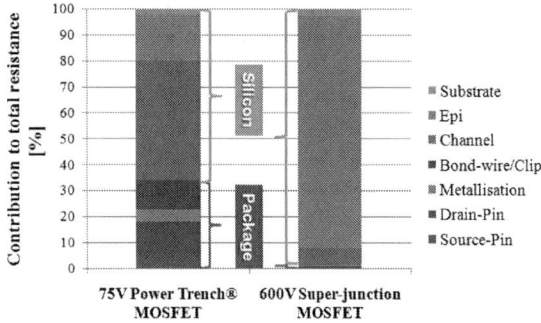

Fig. 9. Relative Contribution to $R_{DS(ON)}$ with 75V MOSFET and 600V MOSFET

IV. EXPERIMENTAL RESULTS

Switching waveforms of the DUTs under PFC is shown in Fig. 10. This figure shows the MOSFET turn-on and turn-off transients of DUTs. 600V / 6A SiC Schottky diode was used for PFC diode. The measurements are at V_{DD}=400V, I_D=16A, R_g=10Ω. Fig 10 shows the waveforms comparisons of 600V/190mΩ SuperFET® II MOSFET in both Power88 (color) and D₂PAK (white). Turn-on loss of SuperFET® II MOSFET in Power88 is 51.22% (41.67uJ) less than one of D₂PAK (85.43uJ) as shown in Fig. 10(a). The turn-off loss is 26.51% (52.25uJ) less for SuperFET® II MOSFET in Power88 compared to one of D₂PAK (71.10uJ) as shown in Fig 10(b). The cross-over area of the drain-source voltage and drain current with SuperFET® II MOSFET in Power88 is less than one of D²PAK at turn-on and turn-off transient. Furthermore, the gate oscillation is highly reduced due to the optimized switching performance of SuperFET® II MOSFET with lower parasitic lead inductances in Power88 package.

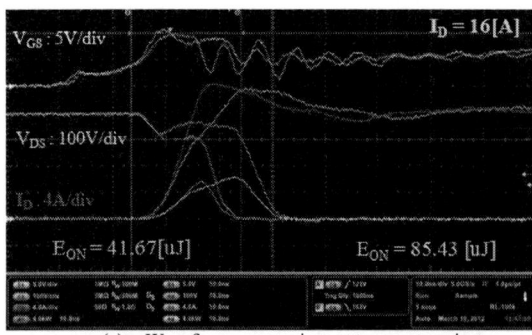

(a)   Waveforms comparison at turn-on transient

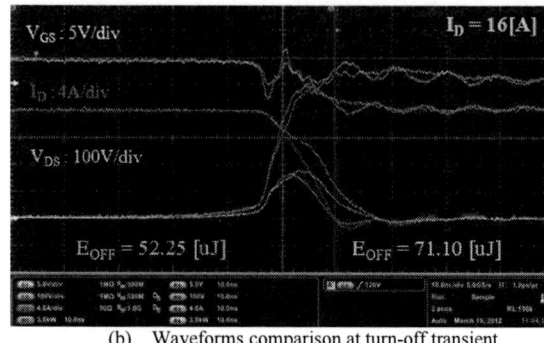

(b)   Waveforms comparison at turn-off transient

Fig. 10. Comparisons of switching performance between SuperFET® II MOSFET in Power88 (Color) and SuperFET® II MOSFET in D²PAK(White) under V_{DD}=400V, I_D=16A R_g=10Ω

Fig. 11 shows the summary of the switching losses with variable gate resistor under V_{DD}=400V, I_D=8A. As shown in the switching loss analysis, turn-on loss of SuperFET® II MOSFET in Power88 is greatly reduce because the di/dt slew rate of SuperFET® II MOSFET is higher while minimizing voltage and current ringing thanks to lower common source inductance of Power88 package with separated power and driver GND.

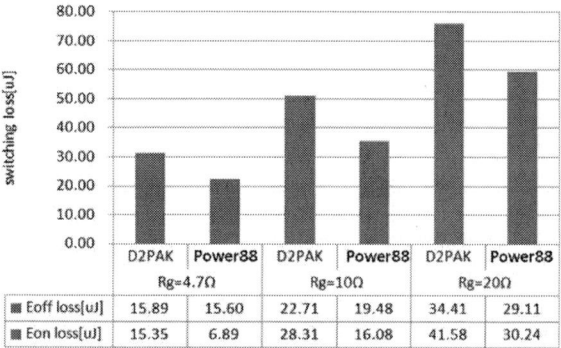

	D2PAK	Power88	D2PAK	Power88	D2PAK	Power88
	Rg=4.7Ω		Rg=10Ω		Rg=20Ω	
Eoff loss[uJ]	15.89	15.60	22.71	19.48	34.41	29.11
Eon loss[uJ]	15.35	6.89	28.31	16.08	41.58	30.24

Fig. 11. Comparisons of switching performance summary between SuperFET® II MOSFET in Power88 between SuperFET® II MOSFET in D²PAK under under V_{DD}=400V, I_D=8A

0 shows the simulation and the experimental waveforms of body diode reverse recovery with the same MOSFET in TO-220 and Power56 packages. It is clear that higher inductance can cause larger Q_{rr} and higher peak voltage. Body diode performance of the same device in TO-220 and Power56 package is also evaluated.

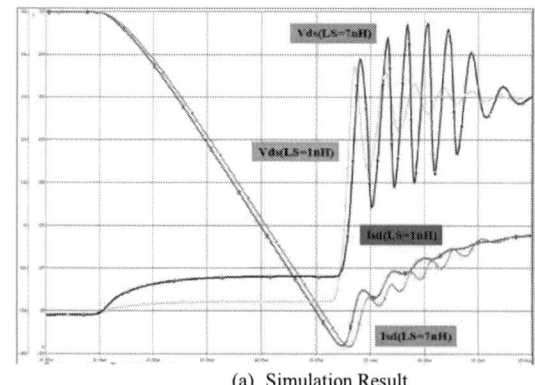

(a)   Simulation Result

978-1-4799-2706-7/14 $31.00 © 2014 IEEE

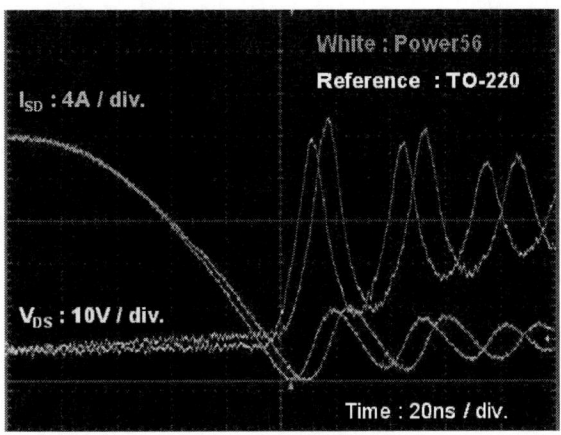

(b) Experimental Result

Fig.12. Body Diode Reverse Recovery Waveforms Comparisons According to Source Inductance

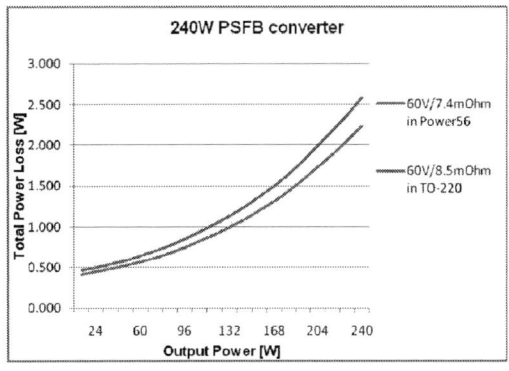

Fig. 13. Comparison of power losses in 240W PSFB converter

TABLE II
BODY DIODE PERFORMANCE COMPARISON OF THROUGH HOLE AND SMD PACKAGE IN SAME SILICON DIE

DUTs	V_{ds}(peak)	I_{rrm}	Q_{rr}
60V/7.4mΩ in Power56	49.6 [V]	3.04 [A]	36.21 [nC]
60V/8.5mΩ in TO-220	54.45 [V]	3.20 [A]	41.23 [nC]

As shown in Table II, $R_{DS(ON)}$ is reduced 0.9mohm, Qrr is reduced 12% and the peak voltage is reduced from 49.6V to 54.45V by using Power56 package instead of TO-220 package. Minimizing common source inductance is therefore critical to system efficiency. Table III shows the simulation result of 60V/7.4mOhm in Power56 and 60V/8.5mOhm in TO-220 in the 240W(12V/ 20A) PSFB DC-DC converter with 300kHz switching frequency. In terms of the conduction loss, 60V/7.4mOhm device is 14.7% lower than other. And also, this device has 12.1% low switching loss compared to TO-220 device. As a result, 60V/7.4mOhm in Power56 has 13.8% power loss saving compared to 60V/8.5mOhm in TO-220. Fig. 13 shows the comparison of power losses according to output load condition. MOSFET's parameters have the most impact on the efficiency of synchronous rectification.

V. CONCLUSION

As power conversion efficiency becomes more and more critical and technology of discrete devices advances every day. Extremely fast switching of super-junction MOSFET and trench gate MOSFETs is essential choice for higher efficiency but it is not easy to control than previous generation MOSFETs. For high voltage SJ MOSFETs, the new optimized leadless Power88 maximize switching performance. For medium voltage trench gate MOSFETs, Power56 enable lower $R_{DS(ON)}$, smaller Q_{RR} and lower peak voltage spikes in synchronous rectification thanks to minimized parasitic. There optimized leadless packages enable achieving high efficiency and low profile in power supply systems.

REFERENCES

[1] Lorenz L., Deboy G., Knapp A.,Maerz. M, "CoolMOS – a new milestone in high voltage Power MOSFET", *ISPSD* 1999.
[2] G. Deboy, M. März, J.-P. Stengl, H. Strack, J. Tihanyi, and H. Weber, "A new generation of high voltage MOSFET's breaks the limit line of silicon," in *IEDM Tech. Dig.*, 1998, pp. 683–685.
[3] T. Fujihira, T. Yamada, Y Minoya; "New Oscillation Circuits Discovered in Switching-Mode Power Supplies." *Proceedings of Power Semiconductor Devices and ICs, 2008. ISPSD '08*, pp. 193-196.

TABLE III
DISTRIBUTION OF POWER LOSSES AT FULL LOAD CONDITION

DUTs	$P_{conduction}$	P_{drive}	$P_{switching}$	P_{total}
60V/7.4mΩ in Power56	1.67[W]	0.07[W]	0.49[W]	2.23[W]
60V/8.5mΩ in TO-220	1.96[W]	0.07[W]	0.55[W]	2.58[W]

Modular Integration of a Matrix Converter

Adane Kassa Solomon, Robert Skuriat, Alberto Castellazzi, Pat Wheeler

Power Electronics, Machines and Control Group
University of Nottingham
Nottingham, UK
alberto.castellazzi@nottingham.ac.uk

Abstract— **The future development of high performance power electronics will rely increasingly on system level integration, where semiconductor devices are co-packaged with other active and passive components (e.g., gate-drivers, filter capacitors and inductors) into a power module. In view of the widespread electrification of pivotal elements of the energy generation and distribution infrastructure (e.g., smart-grids, electric aircraft, electric vehicles), *modularity* is also increasingly gaining importance as a means of enhancing overall system performance and reducing long-term maintenance costs.**

This paper focuses on the development of a highly integrated 3-to-1 phase matrix converter for avionic applications. It proposes an integration approach which enhances the volumetric and gravimetric power handling capability, with enhanced electro-magnetic and electro-thermal performance as compared with established solutions. Maintenance is also simplified by the modular assembly approach.

Keywords— *Integrated power converter/switches, double-sided cooling, matrix converter.*

I. INTRODUCTION

AC-to-AC power conversion requires switches which enable bi-directional current flow between the power source and the load which are capable of blocking voltage irrespective of its polarity. Steady advances in semiconductor technology keep enhancing the electro-thermal performance of solid state power switches and packaging is presently very often the limiting factor of performance and reliability. In order to overcome the limitations of standard bond-wire packaging technology, novel approaches have been proposed over the last decades based on the replacement of bond-wires with solid bumps. These enable a dramatic improvement of thermal and electromagnetic performance as well as allowing for optimum space exploitation and advanced integration schemes targeting application-related switch performance optimization [1-7].

Fig. 1 shows the circuit schematic of a bi-directional matrix-converter type switch (in the following, BD switch). In the case of IGBTs being used as the active devices, a series diode is necessary to ensure reverse blocking capability (i.e., between emitter and collector terminals). During operation, current conduction is either through the device pair T1-D1 or through T2-D2. Fig. 2 shows the cross-section and the actual structure of a vertical IGBT and a vertical diode chip (in this case latest generation 70μm thin vertical IGBTs and diodes, rated at

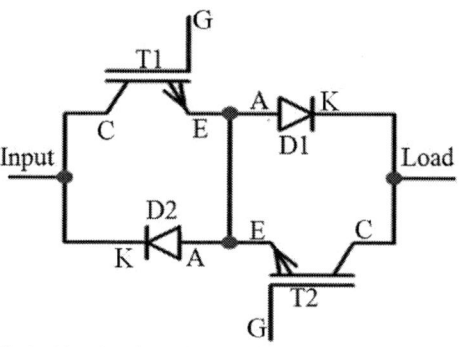

Fig 1. Circuit schematic of a bi-directional power switch

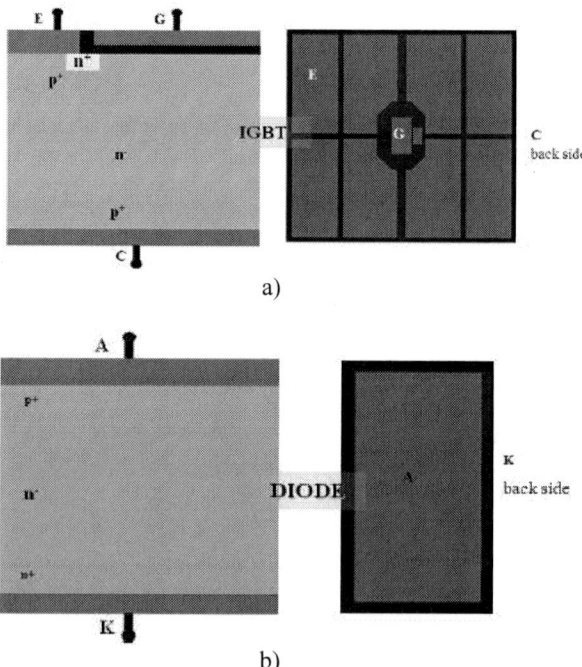

Fig 2. Schematic cross-section and actual structure of IGBT, a), and diode, b), chips, where the top metallization is treated with AlSiCu finish to be solderable.

200A-600V) with an indication of their electrical terminals. The most effective interconnection solution between IGBT emitter (E) and diode anode (A) terminals is achieved by flipping, for instance, the diode upside down and contacting its top surface to the top surface of the IGBT. The interconnection can be achieved by means of surface power bumps instead than bond-wires, a technological feature already demonstrated which enables to keep the back-side of each device in contact with a substrate for optimum thermal management (see [1, 3, 5], for example).

II. POWER SWITCH INTEGRATION

The mounting scheme is as illustrated in Fig. 3: the upper devices are flipped upside down and the interconnection between the two layers of chips is implemented by means of bumps soldered directly between the two device surfaces: this way, the switch parasitic inductance can be greatly reduced by removing bond-wires and by ensuring that the flow of current through the switch is entirely vertical; thermal management is also improved since all devices have their backside in contact with a principal cooling plane while partial heat removal also takes place via the surface, through the interconnection posts and the stacked device. The design of the basic switch and the choice of specific technology features, such as the shape, material and size of the bumps, for instance, relies on a *built-in reliability* design approach, consisting of extensive structural analysis (e.g., finite elements) of the electro-thermal and thermo-mechanical stress both during the assembly process and under real operational mission profiles [5, 6]. So, for instance, here the bumps are implemented by means of stacked Copper-Molybdenum layers to minimize the creep strain accumulation in the solder joints as compared to solid copper bumps.

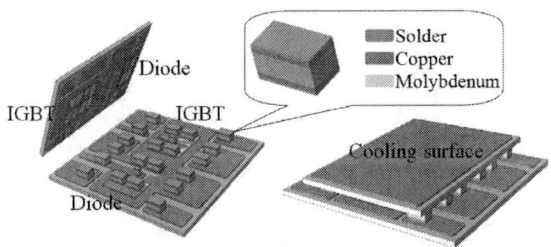

Fig. 3 Stacked assembly of substrate-chip-bump-chip-substrate for a bi-directional switch

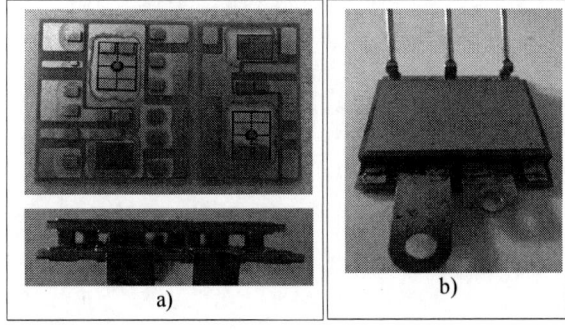

Fig. 4 Bi-directional switch prototype: in a), switch components view and side view of assembled switch; in b), fully terminated and insulated switch.

The switches are then assembled and can be gel-filled for insulation to deliver hardware prototype parts as shown in Fig. 4: here, all interconnects are soldered with Sn-3.5Ag solder alloy. An important feature of these switches is the ability to separate the transistor driving path (gate and emitter terminals) from the power terminals.

For the switches of Fig. 4, preliminary reliability tests for technology validation, summarized in Fig. 5, have delivered an initial lifetime estimate of around 240 cycles (ca. 430 hours), based on thermal cycling between -60°C and 150°C (in these tests, the variation of the IGBT on-state voltage for a fixed on-state current and temperature was used as the monitor of interconnect degradation [7]).

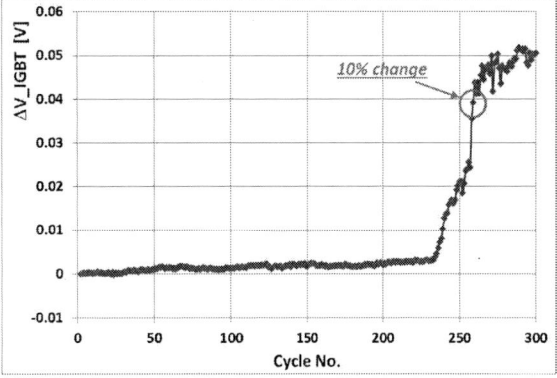

Fig. 5 Preliminary lifetime estimation results for the integrated bi-directional switches.

III. POWER MODULE INTEGRATION

At least three switches as per Fig. 1 are required to build, a 3-to-1 phase matrix converter, as shown in the circuit schematic of Fig. 6.

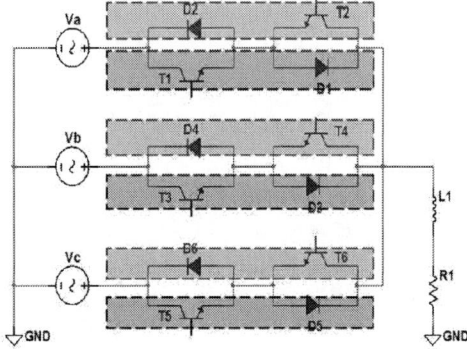

Fig 6. Circuit schematic of a 3-to-1 Matrix converter.

In such topology, current commutations take place not between device pairs in the same BD switch, but within device pairs in different phases and switches, conducting current in the same direction. So, for instance, in the case of Fig. 6, current commutations would be between the pairs T1-D1 and T3-D3, T1-D1 and T5-D5, T3-D3 and T5-D5 during the positive half wave of the input voltage; between the pairs T2-D2 and T4-D4, T4-D4 and T6-F6,

T2-D2 and T6-D6 during the negative half-wave. So, for minimising the parasitic inductance associated with layout and interconnection, it is better to refer to positive and negative cells in each bi-directional switch: it is between such cells, identified with different colours in Fig. 6, that it is important to minimize the electromagnetic path for enhanced performance in the application.

Bespoke matrix-converter module integration in standard packaging technology is intrinsically limiting in the achievable system performance (e.g., maximum switching frequency), due to the well-known shortcomings of bond-wire technology and 2D layout For instance, Fig. 7 shows the photograph of a failed integrated 3-to-3 phases BD power module. Current commutations, as discussed above, would necessarily imply a relatively high inductance and asymmetric loops (i.e., some switch pairs are closer between them than others), even with optimized external bus-bar design; power and temperature distribution are not easy to keep uniform over the module during transient operation; unification of source drive-terminal and source power-terminal prevents very high speed transistor commutation. Moreover, as is evident from Fig. 7, failure of a single-chip implies the need to replace and dispose of the whole

module, with a major disproportion between cause and effect (i.e., cost of a single chip as compared to the cost of the module) resulting in non-negligible long-term running costs of the power system. Clearly, the impact of a single chip failure is even more significant in the case of passives, gate-drivers, sensors and logic circuits being co-packaged within the same module.

Fig 7. Convectional wire-bond Matrix converter power module.

To overcome such limitations and drawbacks, here, an alternative module integration approach is pursued.

a) b) c)

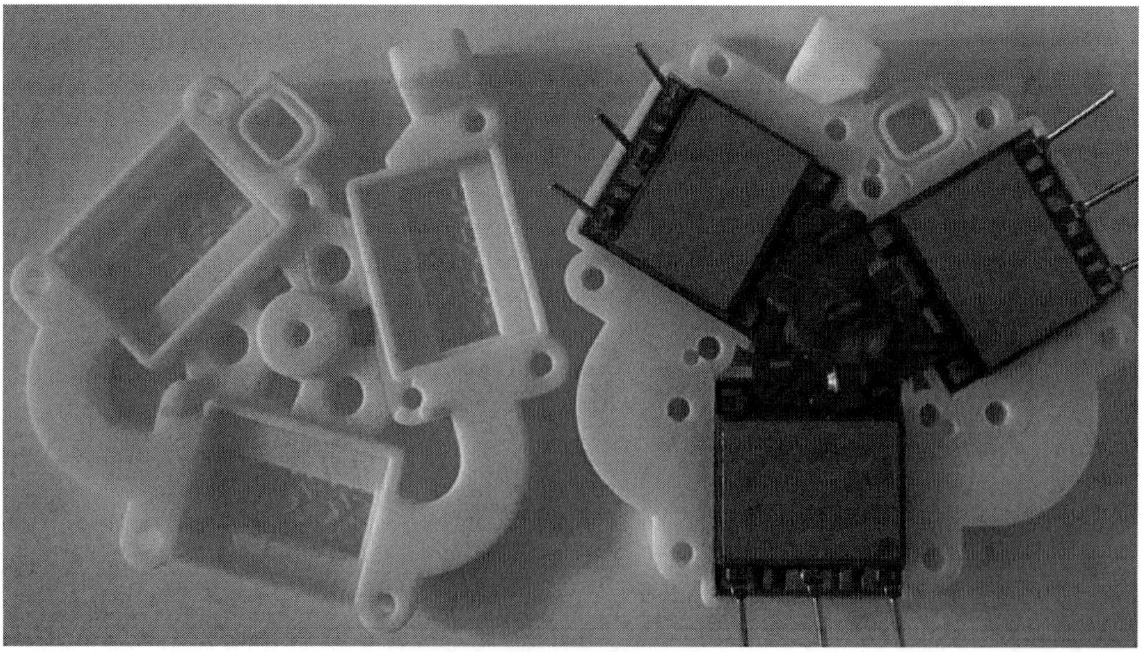

Fig. 8 Internal view of the assembly and the interconnection setup.

Referring to the schematic diagram of Fig 6, the module is constructed using three independent bi-directional switches. The switches are enclosed within a forced liquid based cooler that can cool top and bottom side of the switch. A common output terminal (load side) and three input terminals (input side) are designed to be connected only by means of bolts and screws (no soldering) as shown in Fig 8. Three rectangular blocks of copper are placed on each phase to allow gap between the load and phases (Fig 8(b)). As from Fig 8(c), properly shaped interconnect enables to enclose phase-to-phase high-frequency filtering capacitors (here, a total of 2.2 µF ceramic capacitance is introduced between the phases; the value can be easily increased by stacking the capacitors one on top of the other. The module design is fully symmetrical, so that each switch sees identical electro-magnetic and electro-thermal conditions. Fig. 9 shows the closed module: the power and gate signals are completely separated and the top-side of the enclosure (the cooler) can be used to mount gate drivers and additional filter elements and control to deliver a self-contained unit.

conditions for the switches, 3D simulations carried out with established industrial computational fluid dynamic design tools, delivered an indication delivered an indication that the temperature gradient would be low enough to make it a viable easier solution for initial testing. In the simulations, the liquid flow rate is defined as inlet boundary condition with initial liquid temperature set at 300 K. At the outlet, the external gauge pressure is set to zero. Thermal properties of the materials involved in the model have been assigned to evaluate the heat conduction through all solid bodies due to the fluid flow. For this simulation, a heat transfer coefficient of $15 W/m^2 K$ was used for each wall which is exposed to the air considered as a free convection. An evenly distributed heat flux of $10,000 W/m^2$ is used for top surface of each three silicon block. This is equivalent to the typical value for using water cooler in power electronics [8]. A velocity of 0.08m/s (a flow rate of 0.5litre/minute) is used to the inlet while the outlet is fixed at zero pressure. This flow rate corresponds to the realistic heat transfer value between 500 - 2000 $W/m^2 k$. Fig. 10 shows the resulting temperature distribution. Although the cooler design is not fully symmetrical, this initial study ensured that the asymmetry is negligible under realistic operational conditions and the cooler design can easily be optimized.

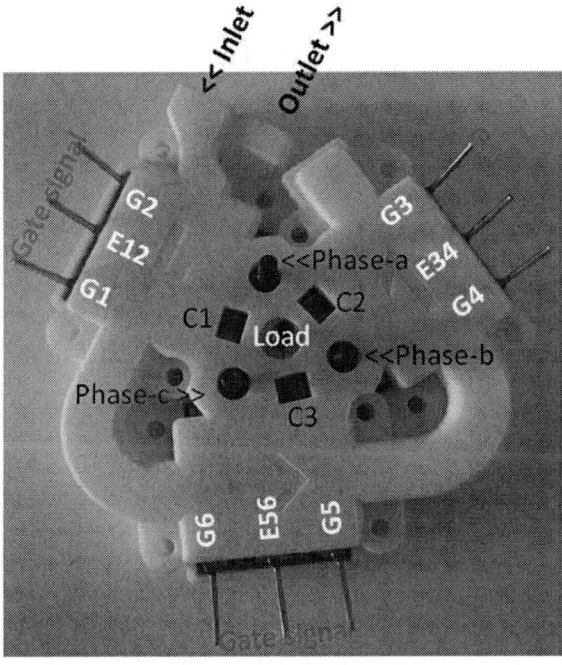

Fig. 9 Three phase to single phase Matrix converter assembly with water cooler; where G1-G6 are six IGBT gate connections and E12, E34 and E56 are common emitter connections for each bi-directional switch

The water cooling system flow path inside the power module is designed to directly target to the hot spot (light blue area shown in Fig 8(top) where the power devices are located. The flow of liquid will start from the inlet and goes to the top surface of each three switches; and once it passes the third switch it turns down and flows via the bottom side of each switch to get to the outlet. Although this initial cooler design, with a single inlet and outlet, cannot guarantee exactly symmetrical thermal

Fig 10. Heat transfer temperature contour (in K) using ANSYS Fluent of 0.01m/s water flow.

IV. FUNCTIONAL TESTS

The fully assembled module was subjected to some preliminary functional tests. Fig. 11 shows the assembled hardware and test setup, corresponding to the circuit schematic of Fig. 6. In the initial set-up, a dedicated test board was built, which accommodates passives and interface with the control platform; also, the gate drivers were connected to the switch terminals still using additional connectors to enable more easily a thorough characterization of the converter performance. In the final implementation, the PCBs can be mounted directly onto the switch terminals (see Fig. 9), and a number of passives can be allocated directly on the sides of the cooler for higher integration levels and more compact overall design.

978-1-4799-2706-7/14 $31.00 © 2014 IEEE

The 2014 International Power Electronics Conference

Fig 11. Hardware and test set-up for preliminary functional testing.

Initially, the test conditions were very conservative to avoid unnecessary failures. The cooling liquid temperature was regulated at 20 °C. Fig. 12 shows typical matrix converter waveforms: from bottom to top, gate-drive signals (C2), load output current (C3) and load output voltage. In these tests, the output voltage was set to be DC, but a purely resistive-inductive RL load was used, without any filter capacitors, which explains the load voltage and current waveforms. The switching frequency is 10 kHz.

modular matrix converter.

Fig. 13 proposes a zoomed-in view of the waveforms, with the addition of the voltage drop measured across one of the bi-directional switches (C4): as can be seen the switching waveform is free from overshoots of any sort, indicating a significantly contained value of the overall switch parasitic inductance. Finite-element electro-magnetic analysis indicated a value of just a few nH for each current conduction path in a switch at 10 kHz.

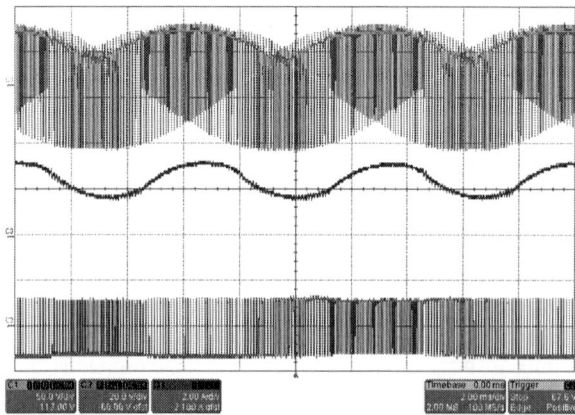

Fig 12. Representative preliminary test waveforms for the

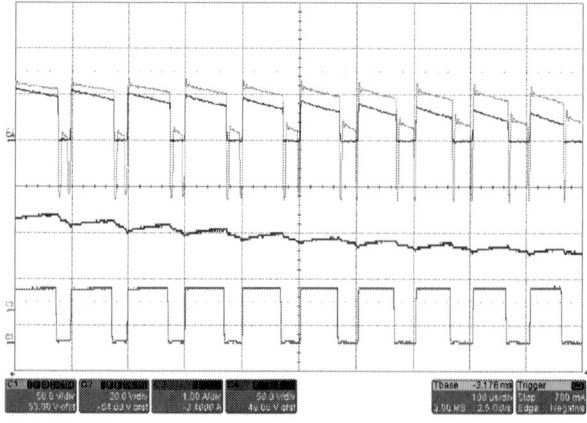

Fig 13. Detailed view of test waveforms, including the voltage across a single bi-directional switch.

978-1-4799-2706-7/14 $31.00 © 2014 IEEE

V. CONCLUSIONS

This paper presents the modular integration of a 3-to-1 phase matrix converter. It aims to progress beyond the state of the art in power system assembly by proposing a solution which significantly improves the electro-magnetic and electro-thermal performance of the semiconductor switches, as a result of both an original switch design and assembly process and system-level integration of the switches in the converter. In particular, fully bond-wire-less, double-sided cooling and layout symmetry are key aspects. The proposed approach is transferable to a number of topologies and has the additional important benefit of limiting the impact of single device/switch failure on the overall system availability. The solution can be of interest to all applications in which weight and volume reduction are highly favored, such as aerospace, automotive, traction.

ACKNOWLEDGMENT

The authors gratefully acknowledge the appreciated contribution of Dr. Jianfeng Li, Dr. Andrew Trentin, Mr. Safari Saeed and Mr. Tianxiang Dai of the Power Electronics, Machines and Control Group of the University of Nottingham.

REFERENCES

[1] M. Mermet-Guyennet, *New structure of power integrated module*, in Proc. of the 4th International Conference on Integrated Power Systems (CIPS 2006), Naples, Italy, June 2006.

[2] C. M. Johnson, C. Buttay, S. J. Rashid, F. Udrea, G. A. J. Amaratunga, P. Ireland, R. K. Malhan, *Compact Double-Side Liquid-Impingement-Cooled Integrated Power Electronic Module*, in Proc. of the 19th International Symposium on Power Semiconductor Devices & ICs (ISPSD2007), Jeju, Korea, May 2007.

[3] P. Solomala, A. Castellazzi, M. Mermet-Guyennet, M. Johnson, *New Technology and Tool for Enhanced Packaging of Semiconductor Power Devices*, in Proc. of IEEE International Symposium on Industrial Electronics, 2009 (ISIE 2009), Seoul, South Korea, July 2009.

[4] A. Castellazzi, A. Solomon, P. Agyakwa, J. Li, A. Trentin, C.M. Johnson, *High power density, low stray inductance, double sided cooled matrix-converter type switch*, in Proc. IPEC2010, Sapporo, Japan, June 2010.

[5] J.F. Li, A. Castellazzi, A. Solomon, C.M. Johnson, *Reliable Integration of Double-Sided Cooled Stacked Power Switches based on 70 μm Thin IGBTs and Diodes*, in proc. of CIPS2012, March 2012, Nuremberg, Germany.

[6] A. Castellazzi, T. Dai, J.F. Li, A. Solomon, A. Trentin, P. Wheeler, *Integrated Matrix Converter Switch*, in Proc. of the 10th International Conference on Power Electronics and Drive Systems (PEDS2013), Kitakyushu, Japan, April 2013.

[7] T. Dai, J.F. Li, M. Corfield, A. Castellazzi, P. Wheeler, *Real-time degradation monitoring and lifetime estimation of 3D integrated bond-wire-less double-sided cooled power switch technologies*, in Proc. EPE'13 ECCE Europe.

[8] M. J. Rizvi, R. Skuriat, T. Tilford, C. Bailey, C. M. Johnson, H. Lu, Modelling of Jet-Impingement Cooling for Power Electronics, 10th. Int. Conf. on Thermal, Mechanical and Multiphysics Simulation and Experiments in Micro-Electronics and Micro-Systems, EuroSimE 2009

A Modular Nanosecond Pulse Generation System for Plasma-assisted Ignition

Peng Gao
School of Electrical Engineering and Telecommunications
Sydney, Australia
p.gao@student.unsw.edu.au

John Fletcher
School of Electrical Engineering and Telecommunications
Sydney, Australia
john.fletcher@unsw.edu.au

Sean O'Byrne
School of Engineering and IT
University of NSW, Canberra, Australia
s.obyrne@adfa.edu.au

Abstract—**Plasma-assisted ignition technology has been proposed to boost the combustion efficiency of scramjets during high speed flight. One technique utilizes high-voltage nanosecond-duration pulses, which can generate free radicals thereby initiating ignition earlier in the combustion chamber and improving fuel efficiency. A high-voltage nanosecond pulse generator is an integral part of the system.**

In this paper, a modular nanosecond pulse generation system, utilizing multiple high-speed, high-voltage MOSFETs, is developed and tested. The modular system can generate width-adjustable pulses (from 20 ns to 50 ns) with fast rise time (<6 ns), fast fall time (< 6 ns) and variable amplitude using multiple switch cells. Employing the inductive voltage adder, the system is configured in two different ways: two switch cells coupled in parallel and two switch cells coupled in series.

The parallel-coupled two-switch configuration increases the peak current capability of the system for a given MOSFET current rating. The increased distributed capacitance is a dominant factor, which leads to mismatch of the drain-to-source voltage at turn-off and increases the output pulse width. The series-coupled two switch configuration doubles the peak voltage of the output pulse. However, the increased leakage inductance is a major contributor to increased rise and fall time of the output pulse and this is demonstrated experimentally.

Keywords—pulse transformer; nanosecond pulse generation; inductive voltage adder;

I. INTRODUCTION

The supersonic combustion ramjet (scramjet) requires the fuel and air to mix rapidly with subsequent ignition. The time required to ignite the fuel/air mix has an important impact on the overall efficiency of the scramjet [1], because the ignition delay time is of the same order of magnitude as the residence time in the combustor. Reducing the ignition delay time improves combustion efficiency, as more of the heat release chemistry, which produces thrust, occurs within the combustion chamber. Millisecond improvements in the time to ignite the fuel/air mix enhances the output power considerably because of the short-time the gas remains in the chamber before the hypersonic flow sweeps the gas out of the chamber [2]. Several methodologies have been proposed

to reduce the ignition delay time; one is plasma-assisted hypersonic ignition, which generates reactive species (atomic oxygen, atomic hydrogen or hydroxyl radical) to the fuel during the combustion process [3, 4]. These species have the effect of accelerating the reaction rate and thus reduce the ignition delay time. The plasma-assisted hypersonic ignition technology has the benefit of shortening the ignition delay time especially in high speed combustion.

It is also important to ensure that these radicals are produced without significantly heating the flow. The thrust produced by the pressure rise due to combustion is strongly related to the temperature difference induced in the flow by the combustion reactions. Thus, a method for generating radicals in a non-thermal plasma is highly desirable. One such method is the repetitively pulsed nanosecond-duration discharge. Several successful methods have been developed, including aerodynamic stabilization [5], electron-beam-controlled discharges [6-8] and repetitively pulsed nanosecond-duration discharge [9, 10]. From a practical point of view, external ionization by nanosecond-duration pulses offers key advantages compared with other methods [3]. This is because short duration pulse generators are easy to operate and maintain.

The plasma is generated by high-voltage nanosecond-duration pulses. The high voltage will form electrons with energies of 10-15 eV, elevating the probability of breaking down more molecules in either air or fuel [11]. In other words, more reactive chemicals are generated. The plasma-assisted hypersonic ignition technology therefore has the potential of reducing ignition delay, increasing peak pressure and raising the heat release ratio in comparison with conventional ignition technologies [12].

There are several popular methods of generating a high-voltage pulse, Marx generator, inductive voltage adder, Blumlein line pulse forming network and magnetic pulse compression [13]. The Inductive Voltage Adder (IVA) is the most suitable high voltage pulse generation topology in this application because of its minor pulse delay, variable pulse duration and peak amplitude. Most importantly, it provides easy scalability through modularization for both current and voltage rating.

This paper investigates a power-electronics-based technology for improvement to hypersonic ignition technology. The overall objective is to develop a high-voltage, modular nanosecond pulse generation system. It should be capable of delivering a nearly square-shaped voltage pulse with the following parameters, fast rise time (< 5 ns), fast fall time (< 5 ns), variable pulse duration ranging from 20 ns to 50 ns and modular architecture such that the system is scalable in voltage and current. In addition, this paper also focuses on examining the effect of series-and parallel-coupling of the scalable system.

II. SYSTEM DESCRIPTON

The overall functional block diagram of the modular nanosecond pulse generation system employing inductive voltage adder is shown in Fig. 1. It utilizes several switch cells (in this case, 2) which are supplied by a common DC voltage supply, V_{dc}. Each switch cell includes a high-speed gate driver, a set of DC capacitors, a HV MOSFET switch and ancillary components.

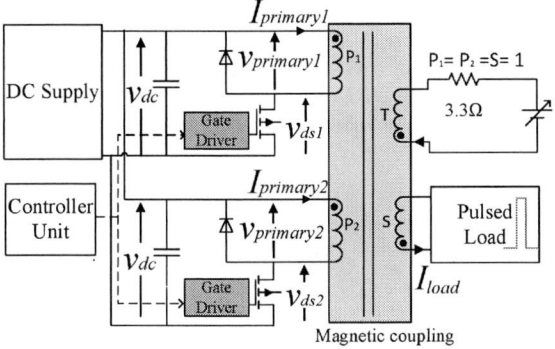

Fig. 1. Diagram of the proposed prototype design showing two switch cells series coupled and core reset circuit.

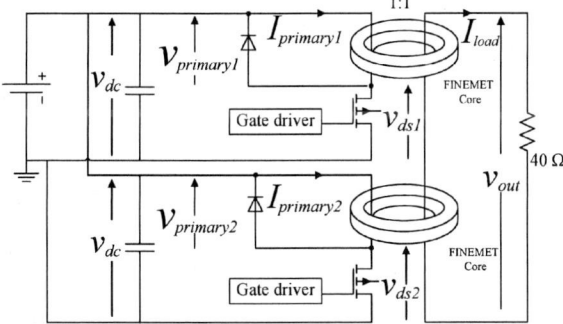

Fig. 2. Arrangement of two switch cells coupled in series.

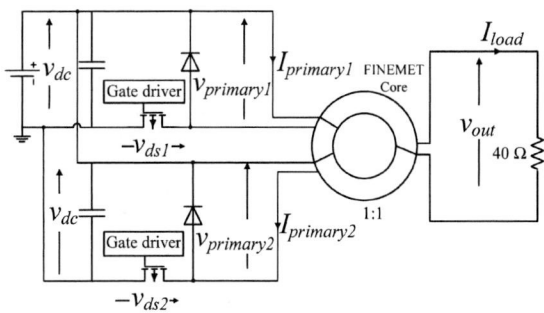

Fig. 3. Arrangement of two switch cells coupled in parallel.

The gate driver (IXYS IXRFD630) receives the variable pulse-width logic signal from the controller unit so as to control rapidly the state of the MOSFET (IXYS DE475-102N21A). Once the MOSFET turns on, a high-voltage output will be induced at the secondary windings, because of the rapid change of magnetic flux in the primary winding. Three different capacitors are used to bypass V_{dc}, 0.01 µF (surface mount), 1000 pF (surface mount) and 0.15 µF (leaded polypropylene). An ultra-fast recovery diode (Fairchild RHRG30) is in parallel with the primary winding so as to freewheel the magnetizing and leakage current when the switch turns off. To achieve a minimum leakage inductance, the primary and secondary windings have only one turn.

The transformer reset circuit, including the variable DC voltage source and 3.3 Ω load resistor, is vital for demagnetizing (resetting) the transformer cores prior to the next pulse. This allows the maximum available swing in the magnetic flux density for the core material, thus maximizing core utilization. Reset is realized by magnetizing the core in the opposite direction to the main pulse.

Using the inductive voltage adder provides a solution to easily integrate low power switch cells to expand the output power capability. Fig. 2 shows the transformer arrangement used to series-couple output voltages from the two cells using toroidal cores with a common secondary conductor. The cells can also be parallel coupled where each cell drives its own single-turn primary winding on a single toroidal core with one secondary turn, as shown in Fig. 3. The FINEMET nano-scale crystalline core (Hitachi Metals Ltd. FINEMET F1AH0898) has good magnetic properties such as a high saturation magnetic flux density (B_s) of 1.23 T, high permeability and low core loss, the material responds quickly enough to be used in the pulsed power generation system.

III. EXPERIMENTAL RESULTS

The following experiments demonstrate the performance of the nanosecond pulse generation system configured in two different ways □ two switch cells coupled in parallel, and two switch cells coupled in series. Each configuration is examined under two load conditions: the transformer is unloaded (open circuit) and loaded with a 40 Ω resistor. A 20 ns gate signal is fed into the gate driver in both configurations.

The 2014 International Power Electronics Conference

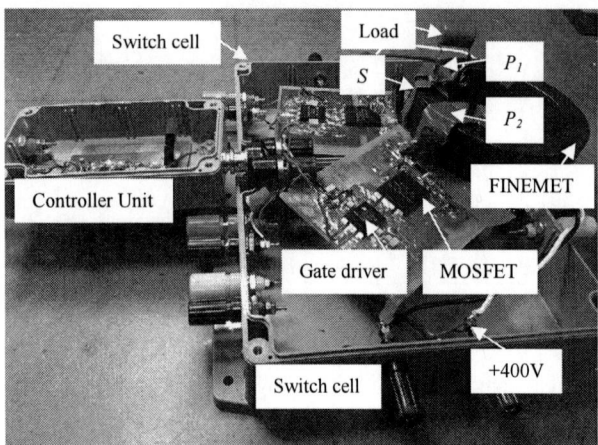

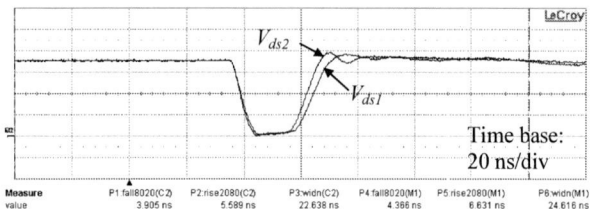

Fig.4. The prototype of two switch cells coupled in parallel.

Measure value	P1:fall8020(C2) 3.905 ns	P2:rise2080(C2) 5.589 ns	P3:widn(C2) 22.638 ns	P4:fall8020(M1) 4.366 ns	P5:rise2080(M1) 6.631 ns	P6:widn(M1) 24.616 ns

Fig. 5. The measured waveforms under the loaded condition (40 Ω resistor) showing V_{ds1} (M1, 200 V/div) with t_{rise} = 6.6 ns, t_{fall}= 4.4 ns, and V_{ds2} (C2, 200 V/div) with t_{rise} = 5.6 ns and t_{fall} = 3.9 ns.

A. Two switch cells coupled in parallel

Fig. 4 shows the prototype of the nanosecond pulse generation system with two switch cells, magnetically coupled in parallel and only one high-voltage DC supply, V_{dc}, (700 V) is employed to power both switch cells. The secondary winding, S, is sandwiched between two primary windings P_1 and P_2 in order to improve the magnetic coupling and reduce leakage inductance. The main feature of the configuration is that the peak current capability of the system is increased because each primary winding shares half the load current. The equivalent leakage inductance seen by the load (40 Ω) is reduced by half. However, the distributed capacitance accounting for the inter coil capacitance of two primary windings increases.

The loaded condition with a 40 Ω resistor is examined first. Fig. 5 shows both MOSFETs in the switch cells are triggered simultaneously. At turn on, the two voltage trajectories V_{ds1} and V_{ds2} are synchronized and well matched. When the switches are turned off, the turn-off voltage transients are no longer synchronized and are mis-matched. As a result, the pulse width of V_{ds1} is stretched to 24.6 ns. The second primary winding affects the distributed capacitance between the windings, leading to the mismatch between V_{ds1} and V_{ds2}.

Fig. 6(a) and 6(b) show the waveforms of $I_{primary1}$ and $I_{primary2}$, as well as V_{ds1} and V_{ds2}. The two primary windings share equal current with maximum amplitude of 28 A. They are composed of I_m and a half of I_{load}, indicating I_m is ~20 A.

The I_{load} and V_{out} are illustrated in Fig. 6(c). It is clear that V_{out} experiences an overshoot and reaches 650 V with a width of 23.4 ns and a small overshoot. Both the fall time (5.6 ns) and rise time (3.5 ns) are smaller than 6 ns. Clearly, the shape of I_{load} suffers from distortion; that is, longer rise time and fall time in comparison with V_{out}. Part of the reason for this is the limited bandwidth (100 MHz) of the current probe. The peak load current is 17 A.

Fig. 7 shows the experimental results under the no-load condition. The pulse widths of V_{ds1} and V_{ds2} are also affected by the distributed capacitance and extended to 24.1 ns and 24.9 ns, respectively. The rise time of V_{ds1} and V_{ds2} are 2 ns longer than that in the loaded condition because of the reduced primary current (≈ 18 A). A reduced primary current leads to slower MOSFET turn-off as C_{ds} is charged slower. Fig. 7(c) shows V_{out} with a width of 26.1 ns has a fall time of 5.6 ns and a rise time of 2.9 ns, which is shortest transient time in all the experiments. The peak in V_{out} is almost 700 V.

The experimental results prove that the two switch cells coupled in parallel can generate a 23.4 ns voltage pulse, which satisfies the requirement of rise time and fall time. The parallel coupling relieves the current burden on the MOSFET. It increases the peak current capability of the system for a given MOSFET current rating. The increased equivalent distributed capacitance in the parallel configuration is considered as the dominant factor shaping the output pulse response.

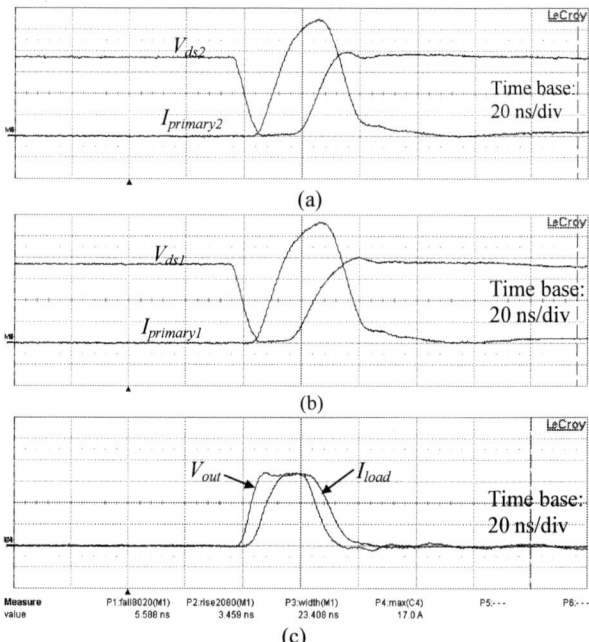

Measure value	P1:fall8020(M1) 5.588 ns	P2:rise2080(M1) 3.459 ns	P3:width(M1) 23.408 ns	P4:max(C4) 17.0 A	P5:---	P6:---

Fig. 6. Experimental results of the two switch cells coupled in parallel under the loaded condition (40 Ω resistor) showing (a) V_{ds2} (M4, 200 V/div) and $I_{primary2}$ (M3, 5 A/div) (b) V_{ds1} (M4, 200 V/div) and $I_{primary1}$ (M2, 5 A/div), and (c) V_{out} (M1, 200 V/div) and I_{load} (C4, 5 A/div).

978-1-4799-2706-7/14 $31.00 © 2014 IEEE

The 2014 International Power Electronics Conference

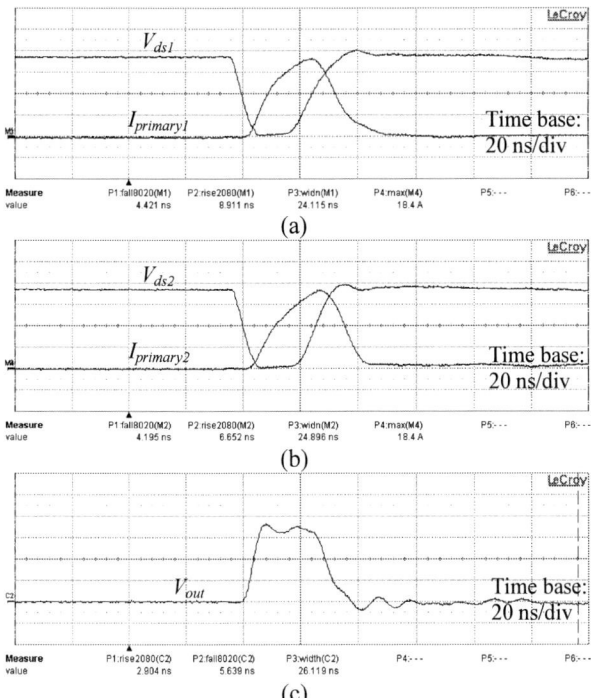

(a)

(b)

(c)

Fig. 7. Experimental results of the two switch cells coupled in parallel under no load condition showing (a) V_{ds1} (M1, 200 V/div) with t_{rise} = 8.9 ns and t_{fall} = 4.4 ns and $I_{primary1}$ (M3, 5 A/div), (b) V_{ds2} (M2, 200 V/div) with t_{rise} = 6.7 ns and t_{fall} = 4.2 ns and $I_{primary2}$ (M4, 5 A/div), and (c) V_{out} (C2, 200 V/div).

B. Two switch cells coupled in series

The experimental set up of two switch cells coupled in series is shown in Fig. 8. The same V_{dc} (700 V) and load resistor (40 Ω) are employed. Each FINEMET core associated with a switch cell is stacked to inductively add the cell output voltages. Therefore, V_{out} can be made a multiple of V_{dc} (in this case, 2). However, the leakage inductance is increased due to the increased length of foil in the practical set-up and the equivalent leakage inductance seen by the load resistor is doubled due to series coupling.

Fig. 9 shows the drain to source voltage when the system is tested under the no load condition. V_{ds1} and V_{ds2} are synchronized and well matched during both the turn-on and turn-off, unlike the parallel-coupled two switch cell configuration. This means the influence of distributed capacitance is not as significant as in the parallel configuration. The pulse width is around 18 ns.

The output load voltage, V_{out} displays distortion, as shown in Fig. 10(a), has a rise time of 7.7 ns and fall time of 6.3 ns, which are longer than that in parallel configuration. $\hat{V}_{out}$ reaches 1044 V, demonstrating the doubling effect of series coupled configuration. $\hat{I}_{load}$ is approximately 22 A, as displayed in Fig. 10(b). Also, $\hat{i}_{primary1}$ (36.2 A) and $\hat{i}_{primary2}$ (37.6 A) in Fig. 11 are close to expectation (40 A) during the 18 ns pulse. The increase in rise time is due to the doubled equivalent leakage inductance seen by load resistor, which also contributes to a doubled load current time constant.

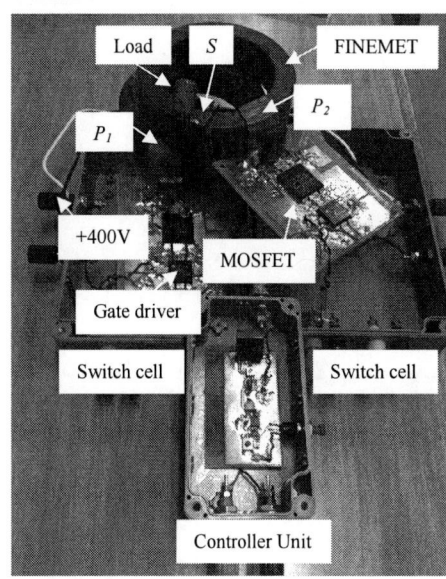

Fig. 8. The prototype of two switch cells coupled in series.

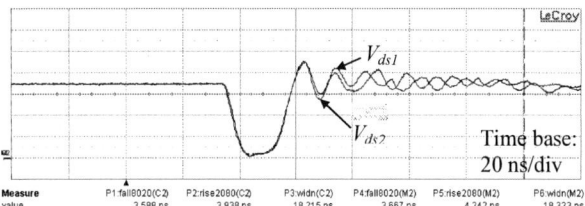

Fig. 9. The measured waveforms under loaded condition showing V_{ds1} (C2, 200 V/div) with t_{rise} = 3.9 ns, t_{fall} = 3.6 ns, width = 18.2 ns, and V_{ds2} (M2, 200 V/div) with t_{rise} = 4.2 ns, t_{fall} = 3.7 ns, width = 18.2 ns.

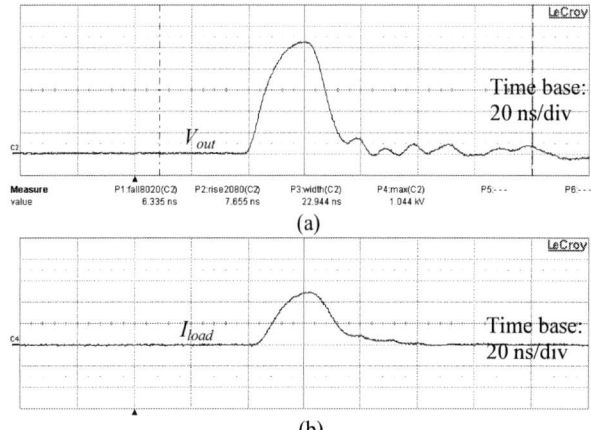

(a)

(b)

Fig.10. Experimental results of the two switch cells series coupled under loaded condition (40 Ω resistor) showing (a) V_{out} (C2, 200 V/div) and (b) I_{load} (C4, 10 A/div).

978-1-4799-2706-7/14 $31.00 © 2014 IEEE

The 2014 International Power Electronics Conference

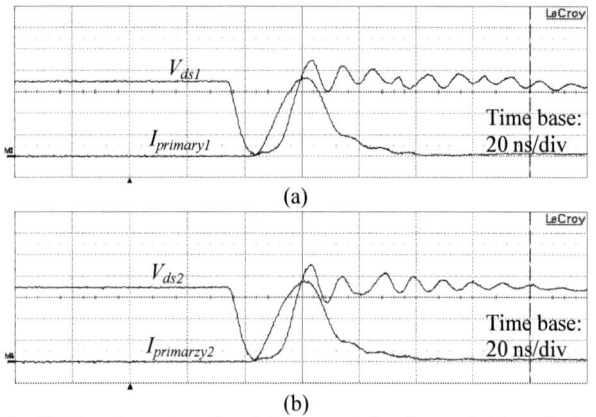

(a)

(b)

Fig. 11. Experimental results of the two switch cells coupled in series in loaded condition (40 Ω resistor) showing (a) V_{ds1} (M1, 200 V/div) and $I_{primary1}$ (C4, 10 A/div), and (b) V_{ds2} (M4, 200 V/div) and $I_{primary2}$ (M3, 10 A/div).

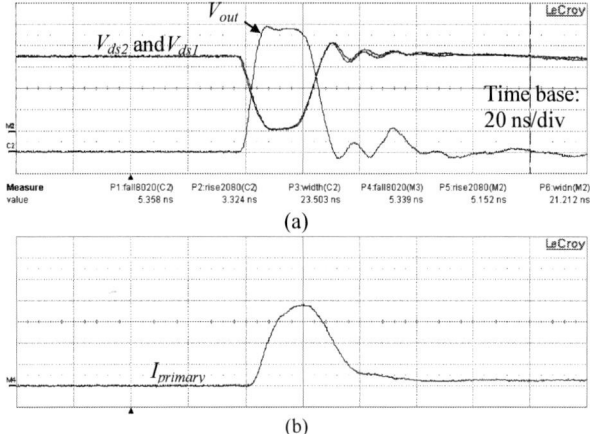

(a)

(b)

Fig. 12. The measured waveforms in no load condition, showing (a) V_{ds} (M1 and M2, 200 V/div), V_{out} (C2, 200 V/div) and (b) $I_{primary}$ (M4, 5 A/div).

Fig. 12 shows the measured waveforms when the load is disconnected. V_{out} reaches 1200 V in 3.3 ns at turn-on and 0 V in 5.4 ns at turn-off. The MOSFET is turned on and off in 5.3 ns and 5.2 ns, respectively. The magnetizing current, I_{m1}, which equals to $I_{primary}$, is still 19 A. It is demonstrated that, when the transformer is disconnected from the load, the series coupling configuration can deliver a narrow pulse (18.2 ns) with fast rise time (<6 ns) and fall time (<6 ns), which meets the design criteria.

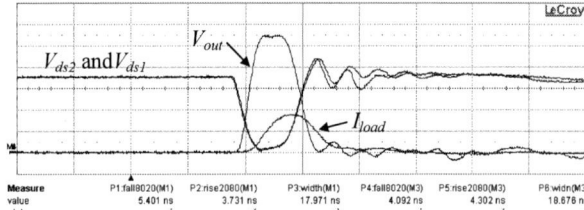

Fig. 13. The demonstration of generating V_{out}, which has improved rise time and fall time with R = 120 Ω. V_{out} (M3, 200 V/div), V_{ds} (M1 and M2, 200 V/div) and I_{load} (M4, 5 A/div) are displayed.

When the load is connected, the series coupling can still generate a voltage pulse as requested but only when the R_{load} is increased high enough to cancel out the impact of increased leakage inductance. This is shown in Fig. 13. A 1100 V/18 ns pulse with rise time of 4.3 ns fall time of 4.1 ns is produced when R_{load} = 120 Ω. This confirms that the increased leakage inductance is a major factor affecting the shape of V_{out} and I_{load}.

C. Discussion

Table I and Table II compares V_{ds} and V_{out} in terms of t_{rise}, t_{fall} and the pulse width (t_{on}) measured in parallel and series coupled configurations. It is clear that fall time of V_{ds}, ranging from 3.6 ns to 5.3 ns, is not affected by the load condition and the configuration. The gate current generated by the gate driver is the determining factor. On the contrary, t_{rise} is strongly related to $I_{primary}$, which changes with load condition. $I_{primary}$ charges the C_{gd} and C_{ds} capacitance of the MOSFET switch in order to switch off the MOSFET. That is the reason t_{rise} in the no load condition is always longer than that in the loaded condition in each configuration.

The load time constant, which is dominated by the ratio of equivalent leakage inductance and load resistance, determines the rise time and fall time of V_{out}. Due to the influence of the increased equivalent leakage inductance in series-coupled two switch cell configuration, the rise time (7.7 ns) of V_{out} are more than doubled, compared with that in parallel configuration (3.5 ns). The fall time (6.3 ns), however, is slightly longer than (5.6 ns). This is due to the increased peak load current, which also has an impact on t_{fall} of V_{out}.

V_{out} in the parallel-coupled switch cell configuration has shorter rise time (3.5 ns) due to reduced equivalent leakage inductance; however, t_{on} of V_{out} and V_{ds} are longer because of increased distributed capacitance, compared with the series-coupled configuration.

TABLE I
COMPARISON OF V_{DS} AND V_{OUT} WITH PARALLEL COUPLING

Parameter	Parallel Coupling					
	Without load			With load (40 Ω)		
	V_{ds1}	V_{ds2}	V_{out}	V_{ds1}	V_{ds2}	V_{out}
t_{rise} (ns)	6.7	8.9	2.9	6.6	5.6	3.5
t_{fall} (ns)	4.2	4.4	5.6	4.4	3.9	5.6
t_{on} (ns)	24.9	24.1	26.3	24.6	22.6	23.4

TABLE II
COMPARISON OF V_{DS} AND V_{OUT} WITH SERIES COUPLING

Parameter	Series Coupling					
	Without load			With load(40 Ω)		
	V_{ds1}	V_{ds2}	V_{out}	V_{ds1}	V_{ds2}	V_{out}
t_{rise} (ns)	5.2	5.2	3.3	3.9	4.2	7.7
t_{fall} (ns)	5.3	5.3	5.4	3.6	3.7	6.3
t_{on} (ns)	21.2	21.2	23.5	18.2	18.2	22.9

978-1-4799-2706-7/14 $31.00 © 2014 IEEE

The upper voltage for the single switch cell is ~800 V due to voltage overshoot and the maximum voltage rating of the DE475-102N21A MOFET is 1000 V. The scramjet application will require voltage of tens of kV. Therefore, a multiple number of switch cells are required to scale the system up to 10 kV using series coupling. The equivalent load resistance of the system, which is a contributor to the voltage rise time in the scramjet application, is unknown. If the load resistance is large, it is possible to deliver output pulse with nanosecond rise time and fall time, even with the expected increase in leakage inductance. If the load resistance is small, the rise time and fall time will be increased and much attention during the design phase will be required to reduce as far as possible the leakage inductance. Nonetheless, the combination of series and parallel coupling may provide a sufficiently fast rise time. More research investigating the equivalent load of the ignition system is required to better predict the load and optimize the system size and design.

The experimental results show that the parallel coupling configuration can be used to reduce the time constant inferred by the rising edge of V_{out}. The series coupling configuration increases the time constant because of increased leakage inductance. Reducing the leakage inductance and distributed capacitance plays a key role in delivering pulses with fast rise and fall times, and demonstrates that when using series coupling it may well be a requirement to also parallel-couple cells to limit the consequent increase in equivalent leakage inductance.

IV. CONCLUSION

A MOSFET-based nanosecond pulse generation system utilizing an inductive voltage adder is proposed, developed and tested. It satisfies the design criteria successfully. Employing the inductive voltage adder topology, the switch cell can be coupled in series or in parallel to form systems with expanded power capability.

The series-coupled two cell configuration has the feature of amplifying the input supply by the number of switch cells. The increased leakage inductance is responsible for the increased load time constant, which in turn lengthens the rise time and fall time of the output pulse.

The parallel-coupled two switch cell configuration has the advantage of increasing the peak current capability of the system for a given MOSFET current rating. The increased equivalent distributed capacitance leads to the mismatch of the drain-to-source voltages at turn-off and distortion of the output pulse. However, the load time constant is decreased.

REFERENCES

[1] S. O. Macheret, M. N. Shneider, R. C. Murray, and R. B. Miles, "Ionization in strong electric fields and dynamics of nanosecond-pulse plasmas," *Physics of Plasmas,* vol. 13, pp. 23502-23502, 2006.

[2] E. Mintusov, M. Nishihara, N. Jiang, I. Choi, M. Uddi, A. Dutta, *et al.,* "Nanosecond Pulse Burst Ignition of Ethylene and Acetylene by Uniform Low-Temperature Plasmas," in *39th AIAA Plasmadynamics and Lasers Conference,* 2008.

[3] I. V. Adamovich, W. R. Lempert, J. W. Rich, Y. G. Utkin, and M. Nishihara, "Repetitively pulsed nonequilibrium plasmas for magnetohydrodynamic flow control and plasma-assisted combustion," *Journal of Propulsion and Power,* vol. 24, pp. 1198-1215, 2008.

[4] N. L. Aleksandrov, S. V. Kindysheva, I. N. Kosarev, S. M. Starikovskaia, and A. Y. Starikovskii, "Mechanism of ignition by non-equilibrium plasma," *Proceedings of the Combustion Institute,* vol. 32, pp. 205-212, 2009.

[5] R. McLeary and W. E. K. Gibbs, "CW CO_2 at atmospheric Pressure," *Quantum Electronics, IEEE Journal of,* vol. 9, pp. 828-833, 1973.

[6] N. Basov, I. Babaev, V. Danilychev, M. Mikhailov, V. Orlov, V. Savelev, *et al.,* "Closed cycle CW CO2 laser employing an electron beam ionizer," *Soviet Journal of Quantum Electronics,* vol. 6, pp. 772-781, 1979.

[7] A. Kovalev, E. Murtov, A. Ozerenkeo, A. Rakhimov, and N. Suetin, "Structure of a beam-driven RF discharge in gas flow," *Fizika Plazmy,* vol. 11, pp. 882-888, 1985.

[8] W. Witteman and V. N. Ochkin, *Gas lasers: recent developments and future prospects* vol. 10: Kluwer Academic Pub, 1996.

[9] N. Generalov, V. Zimakov, V. Kosynkin, Y. P. Raizer, and D. Roitenburg, "Method for significantly increasing the stability limit of the discharge in fast-flow large-volume lasers," *Sov. Tech. Phys. Lett.(Engl. Transl.);(United States),* vol. 1, 1975.

[10] N. Generalov, V. Zimakov, V. Kosynkin, Y. P. Raizer, and D. Roitenburg, "Method for significantly increasing the stability limit of the discharge in fast-flow large-volume lasers," *Sov. Tech. Phys. Lett.(Engl. Transl.);(United States),* vol. 1, 1975.

[11] A. Dutta, I. Choi, M. Uddi, E. Mintusov, A. Erofeev, Z. Yin, *et al.,* "Cavity Flow Ignition and Flameholding in Ethylene-Air by a Repetitively Pulsed Nanosecond Discharge," *AIAA Pap,* vol. 2009, p. 821, 2009.

[12] E. Mintusov, A. Serdyuchenko, I. Choi, W. Lempert, and I. Adamovich, "Mechanism of plasma assisted oxidation and ignition of ethylene–air flows by a repetitively pulsed nanosecond discharge," *Proceedings of the Combustion Institute,* vol. 32, pp. 3181-3188, 2009.

[13] J. Mankowski and M. Kristiansen, "A review of short pulse generator technology," *Plasma Science, IEEE Transactions on,* vol. 28, pp. 102-108, 2000.

Development of a Single Switch Cell for Modular Nanosecond Pulse Generation Systems

Peng Gao
School of Electrical Engineering and Telecommunications
Sydney, Australia
p.gao@student.unsw.edu.au

John Fletcher
School of Electrical Engineering and Telecommunications
Sydney, Australia
john.fletcher@unsw.edu.au

Sean O'Byrne
School of Engineering and IT
University of NSW, Canberra, Australia
s.obyrne@adfa.edu.au

Abstract-The development of a single switch cell for a modular nanosecond pulse generation system is described with the proposed use in plasma-assisted ignition for scramjets. Using the inductive voltage adder topology and high voltage MOSFETs, the proposed system can generate nanosecond pulses with variable duration from 20 ns to 50 ns, fast rise time (< 6 ns), fast fall time (< 6 ns) and variable amplitude. The minimum pulse width (17 ns) is achieved.

A simulation model for the modular system with single switch cell is developed. The outputs from the simulation model agree with the experimental results. The model predicts the characteristics of the single switch configuration accurately, such as the propagation delay of V_{out} compared with V_{ds} and the load time constant. It is also concluded that the ratio of leakage inductance and the load resistance is mainly responsible for the rise time and fall time of output pulse; the propagation delay of V_{out} mainly depends on the inductive elements, $L_{leakage}$, $L_{parasitic}$ and L_{load}.

Keywords—pulse transformer; nanosecond pulse generation; inductive voltage adder;

I. Introduction

There are several popular schemes employed in pulsed power systems to develop high-voltage pulses, such as the inductive voltage adder [1-5], the magnetic pulse compressor [6-8] and the Blumlein line pulse forming network [9-12]. The generated pulse can differ widely in peak power and average power; currently the highest energy achieved is around 100 MJ delivered in a pulse of 5 ms duration [13]. The range of voltage and current amplitudes lies between 10 kV and 50 MV and between 1 kA and 10 MA, respectively. The pulse is also specified by rise and fall time, duration, repetition rate and shape.

The shape of the electrical pulse is critically related to the requirement of the pulsed power application. The applications involving generation of ozone and water treatment, for instance, require repetitive operation and long lifetime [14]. High energy physics, particle accelerators, and commercial semiconductor and metal treatment processes require pulses with high power, variable duration, flat top and high voltage. Ion implantation applications prefer the generated pulse with fast fall and rise time. In a radar transmitter, the top of the pulse should be flat to reduce the phase noise. This paper focuses on the application of plasma-assisted ignition technology, which requires pulses as close as possible to ideal square shape in order to boost the combustion efficiency of scramjets during high speed flight [15].

The overall objective of this paper is to develop a single switch cell that can be magnetically coupled in parallel and series in a modular nanosecond pulse generation system. The system should deliver a nearly square-shaped voltage pulse with the following specification: fast rise time (<6 ns), fast fall time (<6 ns), variable pulse duration ranging from 20 ns to 50 ns and variable amplitude. A simulation model is developed for the single switch cell, which predicts the behavior of the system and therewith the influence on the pulse shape.

II. Design

Fig. 1 shows a functional block diagram of the modular nanosecond pulse generation system with a single switch cell using the inductive voltage adder topology. The inductive voltage adder is chosen as the candidate topology in the scramjet application because the system is easily scaled through modularisation.

The proposed design includes the controller unit, the switch cell and the transformer reset circuit. The switch cell drives its own single-turn primary winding on a toroidal core with one secondary turn. The single turn on P_1 and S has the advantage of minimizing the leakage inductance.

The reset circuit restores the magnetic flux density of the core to its initial state. This allows the maximum available swing of the magnetic flux density, thus maximizing core utilization. Reset is realized by magnetizing the core in the opposite direction prior to the next pulse.

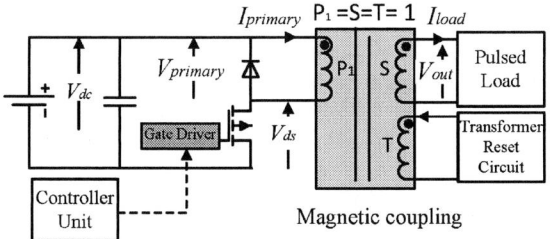

Fig. 1. The illustration of the modular nanosecond pulse generation system with single switch cell.

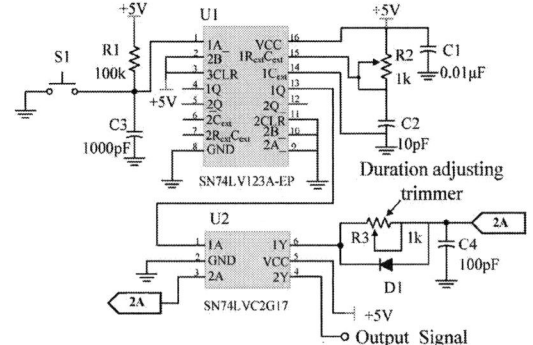

Fig. 2. The schematic diagram of the controller unit showing the pulse generation unit and pulse adjustment circuit.

A. The controller unit

The controller unit is designed to generate a debounced and adjustable-width signal out. Fig. 2 is a schematic diagram of the controller unit, which is constructed using a SN74LV123A-EP (dual retriggerable monostable multivibrator) and a SN74LVC2G17 (Dual Schmitt Trigger Buffer).

The retriggerable monostable multivibrator (U1) is mainly responsible for producing a debounced signal every time the bush-button switch S1 is pushed. The produced signal depends upon the value of timing capacitor (C_2) and resistor (R_2). The internal Schmitt triggers at $\overline{A}$ input have enough hysteresis (2.5 V) to cope with slow transition input signals and the switch bounce problem. This feature is important as the fast switching transient created by the switching stage may generate significant EMI. The 10 nF surface mount ceramic capacitor C_1 is selected to bypass V_{cc} and is located as close as possible to U1 to filter noise on the supply line.

The dual Schmitt Trigger buffer (U2) and its associated circuitry are used to reduce the duration of the pulse to 20 ns. The Schmitt Trigger has enough hysteresis (maximum 2.5 V) to speed up the edge of the input transition, hence generates jitter-free output. The variable resistor R_3 is used to change the transition rate of the input signal, which means the output pulse duration can be altered by varying R_3.

B. The switch cell

Fig. 3 shows the schematic diagram of the switch cell, the transformer and the R_{load} (40 Ω). The switch cell utilizes a gate driver to control a HV power MOSFET with a short duration pulse. Magnetic series or parallel

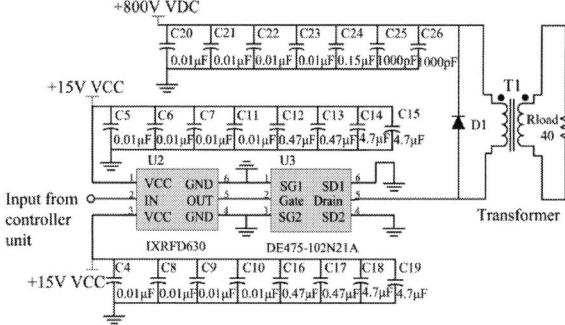

Fig. 3. The schematic diagram of the gate drive and HV MOSFET circuit board.

coupling of multiple switch cells can expand the output power capability in voltage or current rating, respectively.

An IXYS nMOS DE475-102N21A (U3) rated at 1000 V, 21 A and 30 MHz is selected because of its nanosecond switching speed (typically 5ns), low turn on resistance R_g (0.41 Ω), low stray inductance at the drain terminal and the source terminal, L_G (1 nH) and L_S (0.5 nH).

The MOSFET is powered by a +800 V V_{dc} supply in this case. The source is connected to the circuit ground and the drain is connected to the primary winding of the transformer (T1). V_{DC} is bypassed by several capacitors comprised of C_{20} through C_{26}. All the capacitors are surface mount types with C0G dielectric except C_{25}, which is a polypropylene capacitor. These capacitors have the advantage of high voltage and high pulse current rating, making them good choice for pulse applications. An ultra-fast recovery diode D1 (RHRG30120) is in parallel with the primary winding so as to freewheel the magnetizing and leakage current when the switch turns off. The anode of the diode should also be placed as close as possible to the drain of the MOSFET, which reduces the overshoot due to commutation current flowing through the stray inductance.

The performance of the MOSFET switch in a pulsed-power generation system is significantly affected by the gate drive circuitry, especially when it comes to achieving the nanosecond rise time and fall time. A CMOS high speed high current gate driver IXYS IXRFD630 (U2) is used because it can source and sink sufficient peak current, 30 A, to fully charge the MOSFET C_{gs} within 4 ns. Also, the minimum pulse width (8 ns), generated by the gate driver, is shorter than the design requirement (20 ns). The IXRFD630 driver is a surface-mount device and incorporates patented RF layout techniques and a similar low inductance design to the MOSFET package.

The gate driver is powered by a +15 V supply, V_{cc}. V_{cc} is bypassed via a group of surface mount capacitors (C_4 through C_{19}), which are symmetrically located on the two sides of U2. The surface mount types use much less space on the printed circuit board and allow for greater packing densities and lower interconnected inductance.

978-1-4799-2706-7/14 $31.00 © 2014 IEEE

These surface mount capacitors can provide the 30A gate drive current in 2-6 ns and are low inductive, low ESR (equivalent series resistance) and high pulse current service capacitors. They also help strengthen the V_{cc} supply and avoid sags or ringing on the gate signal. Three different capacitance values are selected, 0.01 μF, 0.47 μF and 4.7 μF.

The surface mount tantalum capacitors (TPSB475K025R1500) are the choice for the 4.7 μF capacitor because of their superior performance in pulse applications. They also have the merit of inherently low ESR (1.5 Ω). The 0.01 μF and 0.47 μF capacitors use surface mount ceramic C0G types capacitors. Compared with Class-I dielectric, typically X7R/X5R materials, C0G dielectrics (belonging to Class-II) exhibit low dielectric loss.

III. PCB LAYOUT DESCRIPTION

In order to verify the design, single sided printed circuit boards (PCBs) were fabricated. The PCB layout for the controller unit and the gate drive circuitry are shown in Fig. 4 and Fig. 5, respectively.

In Fig. 4, the SN74LV123A-EP (U1) monostable multivibrator is a hybrid digital and analog IC, which means it is more sensitive to noise, especially the pins connected to timing functions. Therefore, two important issues with respect to circuit design were considered. The loop area encircled by the DC supply (+5 V) to ground path is reduced to minimize the stray inductance. Then, the timing components (R_t and C_t) are placed as close as possible to the chip but also away from noise-producing sources and tracks carrying currents with a high rate of change.

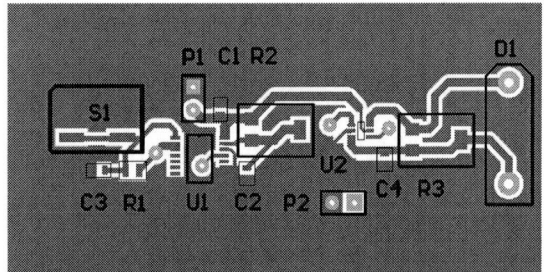

Fig. 4. The top layer of the controller unit.

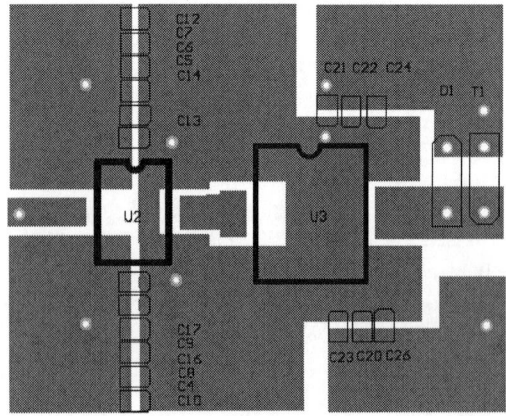

Fig. 5. The top layer of the gate drive circuitry.

To achieve 5 ns switching performance, care should be taken in the placement of components and the track routing to reduce the stray inductance, as shown in Fig. 5.

The priority is given to the power components, including the MOSFET, bypass capacitors (C_{20} through C_{26}) and the fast switching diode. This is because they carry high current and high rate of change of current. The placement of those power components should be arranged in a way that the path lengths are minimized in order to minimise the stray inductance. The anode of the diode should be located as close as possible to the drain of the MOSFET, which reduces the overshoot due to commutation current owing through the stray inductance. All the bypass capacitors should also be close to the MOSFET, especially the polypropylene capacitor, which is placed on the bottom side of the PCB to facilitate this.

The output pin of IXRFD630 gate driver should be as close to the MOSFET as possible by trimming the tabs of both gate driver and MOSFET. It should be noted that only the ground return path of the gate signal is routed at the bottom layer to minimize the noise propagation, and the vias with large area for interconnection are used. This assists in keeping the gate signal noise free. The most important trace in this section is between bypass capacitors (C_4 through C_{19}), IXRFD630 and MOSFET. Every effort was made to keep this loop as short as possible with wide tracks.

The ground tabs of gate driver and MOSFET have a balanced symmetrical layout to achieve low loop inductance target; this technique should be extended to the external circuit as well. The whole PCB layout should have a symmetrical configuration where the ground tracks for the gate signal and the drain signal lies on either of devices, forming a distributed coplanar transmission line. Also, large copper islands for V_{DC} and V_{CC} sit on either side of the signal tracks. It is also important that the signal ground and power ground should interconnect at only one point.

IV. SIMULATION

The transformer uses nano-scale crystalline toroidal core (FINEMET F1AH0898) with o.d. 104 mm i.d. 76 mm and height 25 mm. It has good magnetic properties such as a high saturation magnetic flux density (B_s) of 1.23 T, high permeability and low core loss. One layer of PVC insulation tape is wrapped around the primary and secondary foils fabricated from copper strip of thickness 0.4 mm and width 22 mm. The following relations are used to estimate the magnetizing inductance (L_m), the leakage inductance ($L_{leakage}$) [16], the interwinding capacitance for the single layer winding (C_{ds}) [17] and the stray capacitance (C_s) between the foil winding and the core [18]. The calculated values of C_d, C_s, $L_{leakage}$ and L_m are given in Table I.

The magnetizing inductance is estimated using

$$L_m = \frac{N^2 t \mu_r \mu_0}{2\pi} \ln \frac{b}{a} \qquad (1)$$

Where μ_r is the relative permeability of the magnetic material, μ_0 is the permeability of free space, b is the outer diameter of the toroidal core in meter, a is the inner diameter of the toroidal core in meter, t is the thickness of the toroidal core in meter and N is the number of the primary winding.

The leakage inductance is estimated using

$$L_{leakage} = \frac{2\pi MLTN^2\left(c + \dfrac{d_1 + d_2}{3}\right)}{L_c} \times 10^{-8}\,H \qquad (2)$$

Where a is the height of the winding in meter, d_1 is the thickness of the secondary winding, d_2 is the thickness of primary winding, c is the distance between the primary and secondary windings, L_c is the thickness of the core and MLT is the mean length of a turn.

The interwinding capacitance is

$$C_d = \frac{\varepsilon \varepsilon_0 S}{2h_1 + h_2} F \qquad (3)$$

Where h_1, and h_2 are the insulation thickness of PVC and air respectively, ε and ε_0 are the relative and free space permeability respectively, S is the surface area of the edges of the secondary foil (m^2).

The stray capacitance is [18]

$$C_s = \frac{\varepsilon \varepsilon_0 S}{h_1 + h_2} F \qquad (4)$$

Where S is the surface area of the foil adjacent to the core.

The modular nanosecond pulse generation system with single switch cell is simulated using the LTspice software. Fig. 6 illustrates the schematic diagram. The LTspice models of IXRFD630 and DE475-102N21A are included to closely represent the physical processes of turn-on and turn-off. It is clear that the switch cell with a single primary turn is magnetically coupled to a single secondary winding. The switch cell comprises C_1, S_1 (MOSFET), D_1 and the interconnection parasitic inductance $L_{parasitic}$. L_{load} accounts for the parasitic inductance of R_{load}.

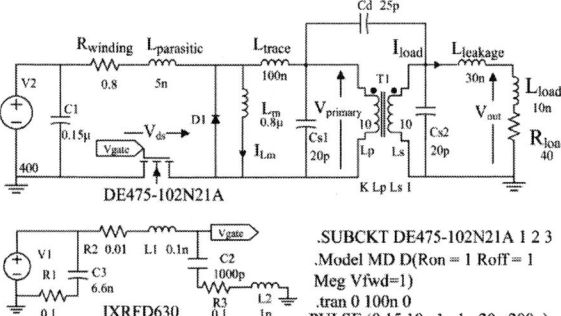

Fig. 6. The LTspice model of the nanosecond pulse generation system with single switch cell

TABLE I
SPECIFICATION OF PARAMETER VALUES USED IN SIMULATION

Parameter	Meaning	Value
V_{dc}	High voltage supply	800 V
R_{load}	Load resistance	40 Ω
$L_{parasitic}$	Parasitic inductance	5 nH
$L_{leakage}$	Leakage inductance	30 nH
L_{trace}	Trace inductance	100 nH
$R_{winding}$	Winding Resistance	0.8 Ω
L_m	Magnetizing inductance	0.8 µH
C_d	inter winding capacitance	25 pF
C_s	Stray capacitance	20 pF
t_{on}	Pulse width	20 ns

The equivalent circuit of the transformer is considered as a lumped circuit model to simplify the analysis of the circuit [19]. This is because the single layer secondary is wound near a one-layer primary and the distance between them is small. The transformer model includes the winding resistance ($R_{winding}$), the primary inductances L_p, the secondary inductance L_s, the magnetizing inductance L_m, the leakage inductance ($L_{leakage}$) and the interwinding capacitance C_d. C_d models the interwinding capacitance of P and S windings. The winding resistance $R_{winding}$ is measured at DC in the laboratory. Table I specifies these circuit elements, as well as other parameter values used in the simulation.

Eqns. (4) and (5) give formulas for the turn-on and turn-off periods of the circuit based on Kirchhoff's Voltage Law.

During rise time:

$$I_{load} = \frac{V_{dc}}{R_{load}}\left(1 - e^{-\frac{t}{\tau}}\right) \qquad (4)$$

During fall time:

$$I_{load} = \frac{V_{dc}}{R_{load}}\left(e^{-\frac{t}{\tau}}\right) \qquad (5)$$

Where τ is the load time constant and expressed as the ratio of the inductive elements ($L_{leakage}$ and L_{trace}) and the load resistance (R_{load}). The calculated load time constant (5 ns) determines the rise time and fall time of V_{out}.

The simulation result is shown in Fig. 7. Referring to Fig. 7(a), the MOSFET turns on and off in 3.9 ns and 4.2 ns, respectively. V_{out} has a propagation delay of 3 ns, at turn on, compared with V_{ds}. The delay is mainly contributed from $L_{parasitic}$, $L_{leakage}$, L_{trace} and L_{load}, which delay the load current from increasing during turn on. There is an overshoot of 100 V caused by $L_{parasitic}$ at the end of turn-off transient.

V_{out} rises to 63% of its peak value (700 V) in 3.2 ns, which is close to the calculated time constant (2.9 ns). The 100 V voltage drop at peak results from stray elements ($L_{leakage}$, L_{trace} and $L_{parasitic}$) and on state resistance R_d of the MOSFET.

In Fig. 7(b), the load current, I_{load}, is displayed. It reaches 18 A and starts to decrease when the MOSFET turns off. The maximum value of the magnetizing current, I_m, as shown in Fig. 7(c), is ~19 A.

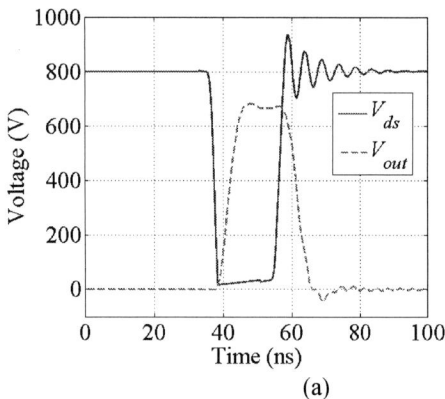

(a)

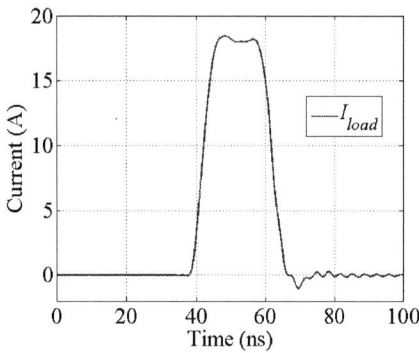

(b)

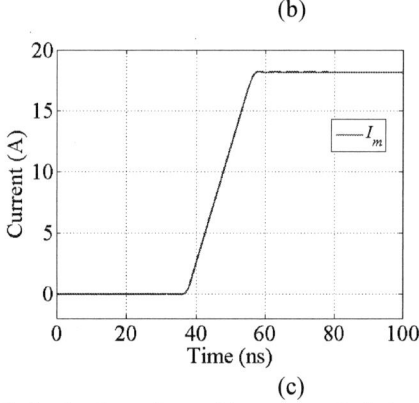

(c)

Fig. 7. Simulated waveforms of the system with single switch cell: (a) V_{ds} and V_{out} with rise time of 3.4 ns and fall time of 4.2 ns. (b) The load current I_{load} and, (c) the magnetizing current I_m.

V. EXPERIMENTAL RESULT

The prototype of the modular nanosecond pulse generation system with a single switch cell, shown in Fig. 8, was built and tested to prove the feasibility of the design concept. The DC voltage supply (800 V) is employed to power the switch cell; the transformer is loaded with a 40 Ω resistor.

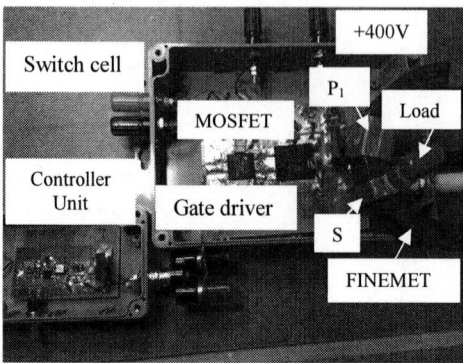

Fig. 8. The prototype of the nanosecond pulse generation system with a single switch cell.

Fig. 9(a) displays the measured waveforms, V_{dc}, $V_{primary}$ and V_{ds}, when the transformer is loaded. The V_{dc} (800 V) is clean and stable although there is a voltage drop of approximately 9 V when the MOSFET turns on. A V_{ds} fall time of ~4.1 ns and a V_{ds} rise time of ~5.5 ns are achieved. V_{ds} overshoots to 900 V at turn-off but is within the limitation of the MOSFET voltage rating. V_{out} also has propagation delay (4 ns) in comparison to V_{ds}, similarly to the simulation result. At turn off, $V_{primary}$ has a rise time and fall time of 4.2 ns and 5.8 ns, respectively.

Referring to Fig. 9(b), V_{out} rises and falls quickly within 3.4 ns and 4.2 ns, which meets the requirement of rise time (6 ns) and fall time (6 ns). V_{out} drops nearly 100 V to 700 V peak and $\hat{I}_{load}$ is ~17.5 A.

The load current I_{load} with a peak amplitude of 9 A is shown in Fig. 9(c). Since the difference between $I_{primary}$ and I_{load} is I_m, it is concluded that $\hat{I}_m$ is ~17 A.

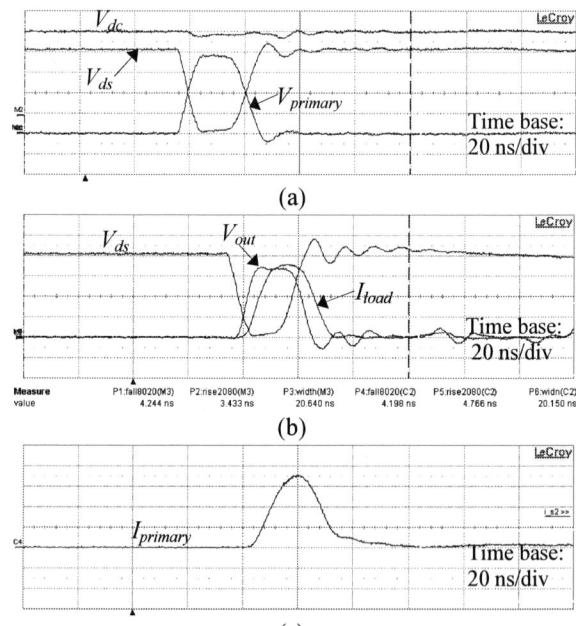

Fig. 9. Experimental result of single switch cell with 40 Ω load resistor: (a) $V_{primary}$ (Math, M1 and M2), DC supply voltage V_{dc} (M2, 200 V/div) and V_{ds} (M1, 200 V/div) with 5.5 ns rise time, 4.1 ns fall time and 21.7 ns pulse width (b) V_{out} with 3.4 ns rise time and 4.2 ns fall time (M3, 100 V/div), V_{ds} (C2, 200 V/div) and I_{load} (M4, 5 A/div) (c) $I_{primary}$(C4, 10 A/div).

978-1-4799-2706-7/14 $31.00 © 2014 IEEE

The 2014 International Power Electronics Conference

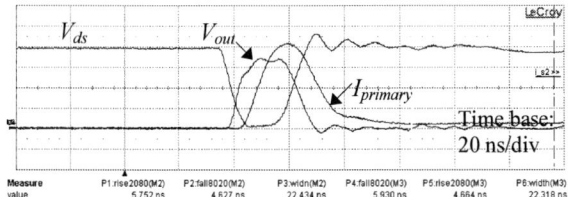

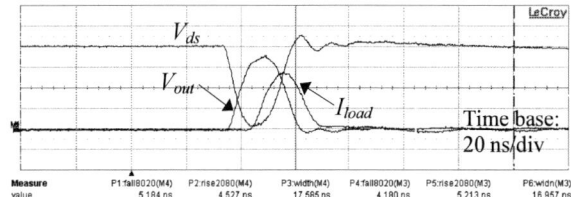

Fig. 10. Waveforms of V_{ds} (M3, 200 V/div), $I_{primary}$ (C4, 5 A/div) and output voltage V_{out} (M2, 200V/div) observed on oscilloscope in the no-load condition.

Fig. 12. Waveforms of V_{out} (M3, 200 V/div) with minimum duration (17 ns), I_{load} (M2, 5 A/div), and V_{ds} (M4, 200 V/div) under the loaded condition (40 Ω).

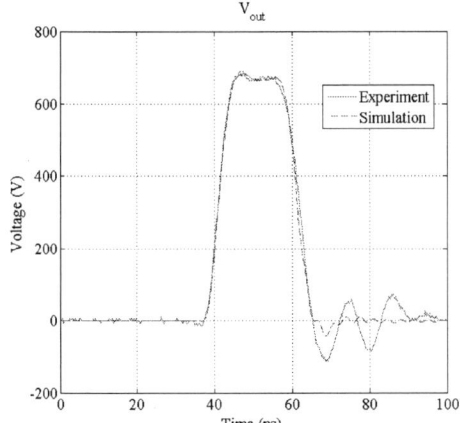

Fig. 11 Comparison of experimental and the simulation result.

Fig. 10 shows the measured waveforms of V_{ds}, $I_{primary}$ and V_{out} when the load resistor 40 Ω is removed. V_{out} reaches 700 V. The rise time is 5.8 ns, longer than 3.4 ns on the loaded condition. The fall time is extended to 4.6 ns. V_{out} experiences a minor oscillation of 50 V. The waveform of $I_{primary}$ (equal to I_m) peaks at 20 A, as expected.

The comparison between the simulation and the experiment result in terms of the key timings and magnitudes is provided in Table II. Simulated result agrees very well with the experimental result with only a small error, which is acceptable. The simulated $\hat{I}_{load}$ (9 A) is the same as the experimental result.

Fig. 11 shows the measured and simulated pulse response of V_{out}. Clearly, they agree with each other during the turn on and turn off transient.

The simulation model also successfully estimates the characteristics of the modular system, the propagation delay and the load time constant of V_{out}. V_{out} in both the simulation and experimental result has a propagation delay (<5 ns), caused by $L_{leakage}$ and L_{trace}. The load time constant (τ) inferred from the simulation (3.2 ns) and the experiment (3.4 ns). It is concluded that reducing $L_{leakage}$ and L_{trace} could deliver better pulse response.

It is usually recommended that an RC snubber is employed to suppress the voltage spike at V_{ds} while the switch turns off. However, for the prototype, it will increase the rise time of V_{ds}. To turn on the MOSFET, the snubber capacitor has to be fully discharged as well as the gate drain capacitance (C_{GD}). This introduces extra delay at turn on.

The experimental results meet all the design criteria. V_{out} has a rise time (< 6 ns) and fall time (< 6 ns) and the minimum pulse width is 17ns, as shown in Fig. 12.

VI. CONCLUSION

A modular nanosecond pulse generation system has been successfully modelled and built. It is able to generate a minimum 17 ns pulse and meet the requirement of rise time and fall time.

The simulation result is in good agreement with the experimental result, especially the transient time of V_{ds} and V_{out}. The simulation model also successfully predicts the propagation delay and the load time constant of the output pulse.

REFERENCES

[1] J. Biela, D. Bortis, and J. Kolar, "Modeling of pulse transformers with parallel-and non-parallel-plate windings for power modulators," *Dielectrics and Electrical Insulation, IEEE Transactions on,* vol. 14, pp. 1016-1024, 2007.

[2] D. Bortis, J. Biela, and J. W. Kolar, "Transient Behavior of Solid-State Modulators With Matrix Transformers," *Plasma Science, IEEE Transactions on,* vol. 38, pp. 2785-2792, 2010.

[3] C. P. Burkhart and J. R. Bayless, "Modulator for generating high voltage pulses," ed: Google Patents, 2000.

[4] S. Kohno, Y. Teramoto, I. V. Lisitsyn, S. Katsuki, and H. Akiyama, "High-current pulsed power generator ASO-X using inductive voltage adder and inductive energy storage system," *Japanese Journal of Applied Physics,* vol. 39, p. 2829, 2000.

[5] K. Yatsui, K. Shimiya, K. Masugata, M. Shigeta, and K. Shibata, "Characteristics of pulsed power generator by versatile inductive voltage adder," *Laser and Particle Beams,* vol. 23, p. 573, 2005.

[6] D. Barrett, "Parameters which influence the performance of practical magnetic switches," in *Pulsed Power Conference, 1995. Digest of Technical Papers., Tenth IEEE International,* 1995, pp. 1154-1159.

[7] H. Deguchi, T. Hatakeyama, E. Murata, Y. Izawa, and C. Yamanaka, "Efficient design of multistage magnetic pulse compression," *Quantum Electronics, IEEE Journal of,* vol. 30, pp. 2934-2938, 1994.

[8] R. Narsetti, R. D. Curry, K. F. McDonald, T. E. Clevenger, and L. M. Nichols, "Microbial inactivation in water using pulsed electric fields and magnetic pulse compressor technology," *Plasma Science, IEEE Transactions on,* vol. 34, pp. 1386-1393, 2006.

[9] J. Crouch and W. Risk, "A Compact High Speed Low Impedance Blumlein Line for High Voltage Pulse Shaping," *Review of Scientific Instruments,* vol. 43, pp. 632-637, 1972.

[10] A. de Angelis, J. F. Kolb, L. Zeni, and K. H. Schoenbach, "Kilovolt Blumlein pulse generator with variable pulse duration and polarity," *Review of Scientific Instruments,* vol. 79, pp. 044301-044301-4, 2008.

[11] M. Joler, C. G. Christodoulou, and E. Schamiloglu, "Limitations to compacting a parallel‐plate Blumlein pulse‐forming line," *International Journal of RF and Microwave Computer‐Aided Engineering,* vol. 18, pp. 176-186, 2008.

[12] J. F. Kolb, S. Kono, and K. H. Schoenbach, "Nanosecond pulsed electric field generators for the study of subcellular effects," *Bioelectromagnetics,* vol. 27, pp. 172-187, 2006.

[13] T. Martin, M. Williams, and M. Kristiansen, *JC Martin on pulsed power* vol. 3: Springer, 1996.

978-1-4799-2706-7/14 $31.00 © 2014 IEEE

[14] H. Akiyama, T. Sakugawa, T. Namihira, K. Takaki, Y. Minamitani, and N. Shimomura, "Industrial applications of pulsed power technology," *Dielectrics and Electrical Insulation, IEEE Transactions on,* vol. 14, pp. 1051-1064, 2007.

[15] I. V. Adamovich, W. R. Lempert, J. W. Rich, Y. G. Utkin, and M. Nishihara, "Repetitively pulsed nonequilibrium plasmas for magnetohydrodynamic flow control and plasma-assisted combustion," *Journal of Propulsion and Power,* vol. 24, pp. 1198-1215, 2008.

[17] A. Naderian-Jahromi, J. Faiz, and H. Mohseni, "Calculation of distribution transformer leakage reactance using energy technique," in *Australasian Universities Power Engineering Conference, AUPEC,* 2002.

[18] H. Taskar, M. Dorlikar, M. Patil, H. Mangalvedekar, and D. Chakravarthy, "Numerical validation of amorphous core pulse transformer," in *Power Modulator and High Voltage Conference (IPMHVC), 2010 IEEE International,* 2010, pp. 271-273.

[19] M. Akemoto, S. Gold, A. Krasnykh, and R. Koontz, "Pulse transformer R&D for NLC klystron pulse modulator," in *Pulsed Power Conference, 1997. Digest of Technical Papers. 1997 11th IEEE International,* 1997, pp. 724-729.

Advantage of super junction MOSFET for power supply application

K. Tabira, S. Watanabe, T. Shimatou, T. Watashima, S. Takenoiri

Fuji Electric Co., Ltd.
4-18-1Tsukama, Matsumoto, Nagano, 390-0821Japan
tabira-keisuke@fujielectric.co.jp

Abstract— Characteristics of fast recovery diode (FRED) type super junction MOSFET (SJ-MOS) are reported. The reverse recovery ruggedness (-di/dt ruggedness) of the SJ-MOS FRED is dramatically improved compared to that of non-FRED type, which is almost 16 times better.

Keywords — Super junction MOSFET, Fast recovery diode type, reverse recovery ruggedness, power supply efficiency

I. INTRODUCTION

Rising concerns over the environmental issues, electric equipments have been attracting attention due to their superior function of energy savings and resource conservation. In order to reduce the power loss of electric equipments, more efficient power converters would be needed. Highly-efficient - low conduction loss and low switching loss - power MOSFETs are able to fulfill these demands.

Super junction MOSFET (hereafter SJ-MOS) is one of the promising candidates for the purpose of use in efficient power supply applications due to their excellent tradeoff characteristics between specific on-resistance (RonA) and breakdown voltage (BV). And it is also cost effective compared to the current wide-band-gap switching devices, such as those made of SiC or GaN.

Super junction structures are comprised of alternating n/p-pillars in drift region and can achieve low on resistance by increasing the drift region doping concentrations, narrowing the n/p pillar widths [1, 2].

Commercially available SJ-MOS has been launched since the middle of 2000, but still has not become a major part of the MOSFETs' market because of the heavy demand of the conventional double diffused planar MOSFET (hereafter DMOSFET). The major reason of this would be as follows; there is a concern that SJ-MOS which has relatively weak reverse recovery capacity compared to that of a DMOSFET could be damaged when it is used in a DC-AC inverter circuit or an LLC converter circuit. Reverse recovery in diode mode could be occurred in case of startup, abrupt shift in a load or startup with shorted load when operating frequency tends to decrease and be afield from a resonant frequency. Because of this situation, application of SJ-MOS is considered to be limited.

To avoid failures of SJ-MOS, a range of measures such as a schottky barrier diode being placed in parallel with it or optimized drive conditions being used so that body diode of it may not work have been implemented. However, reverse recovery capacity of SJ-MOS itself must be improved to contribute to better efficiency, higher energy-saving rate and downsizing for power supply units. Characteristics of the reverse recovery capacity improved Fast Recovery Diode (hereafter FRED) type SJ-MOS is reported in this paper.

II. BASIC DESIGN OF THE SJ-MOS DEVICE

Basically, a super junction structure makes it possible to reduce the RonA dramatically results in a significant decrease in the parasitic capacitance due to the smaller die size compared to that of a conventional planar MOSFET for the similar current rating. Therefore, surface MOS as well as super junction structure should be carefully designed so that both low switching loss and moderate switching speed could be achieved simultaneously. In order to improve the Eoff -turn-off dV/dt tradeoff, it is necessary to suppress high turn-off dV/dt. Regarding the turn-off dV/dt, it is expressed as eq (1), if a gate-to-source voltage (Vgs) is assumed to be constant in Miller-period during a turn-off [3].

$$\frac{dV}{dt} = \frac{\dfrac{I_d}{g_{fs}} + V_{th}}{C_{gd} \cdot V_{ds} \cdot R_g} \tag{1}$$

where Id is a drain current, gfs is a transconductance, and Vds is a drain-to-source voltage.

The eq. (1) shows that larger gate-to drain capacitance (Cgd) and lower Vth can reduce the turn-off dV/dt when Rg, Id and Vds are fixed. A schematic cross section of a SJ-MOS is shown in Fig.1. Cgd is determined by the p-base spacing that is related to gate length (Lg). Vth is determined by the impurity concentration of the p-base. In order to increase Cgd and decrease Vth, the expansion of Lg and the reduction of the impurity concentration of the p-base are necessary, respectively. By increasing Cgd and decreasing Vth, the turn-off dV/dt is reduced and the controllability is improved. [4]

978-1-4799-2706-7/14 $31.00 © 2014 IEEE

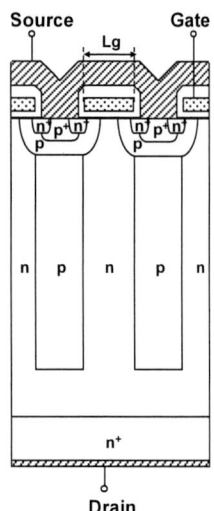

Fig. 1. Schematic cross-section of SJ-MOSFET

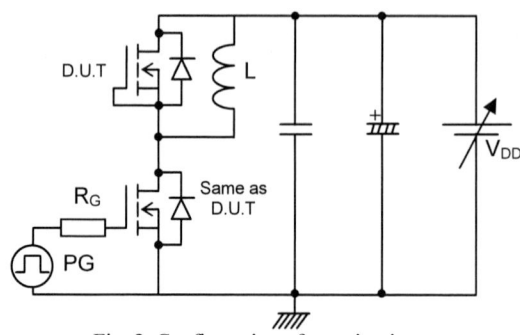

Fig. 2. Configuration of test circuit.

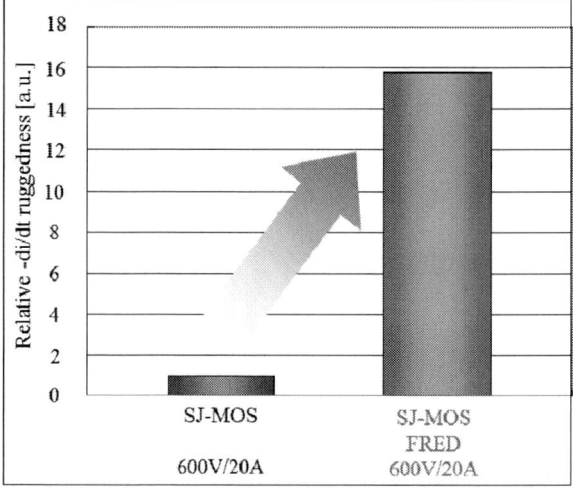

Fig. 3. Comparison of -di/dt ruggedness
between SJ-MOS and SJ-MOS FRED.

III. FRED TYPE SJ-MOS

The reverse recovery ruggedness (-di/dt ruggedness) of SJ-MOS FRED was compared with the SJ-MOS in equivalent current rating (Table.1). We evaluated these -di/dt ruggedness using a step down chopper circuit (Fig.2) under the following conditions; power supply (VDD) = 400V, body diode forward current (IF) = 20A, temperature = 25°C. Fig.3 shows the comparison of -di/dt ruggedness between the SJ-MOS and SJ-MOS FRED. The -di/dt ruggedness of the SJ-MOS FRED is dramatically improved compared to that of SJ-MOS non-FRED type, which is almost 16 times better. It is the reason that the reverse recovery time (Trr), the reverse recovery charge (Qrr) and the reverse recovery current (Irp) of SJ-MOS FRED are smaller than that of SJ-MOS as shown in table2 and Fig4 under the following conditions; power supply (VDD) = 400V, body diode forward current (IF) = 20A, (-di/dt)=100A/us, temperature = 25°C. Judging from these results, it becomes possible to apply SJ-MOSFET to a resonant circuit in which conventional DMOSFETs are used.

TABLE 1
Rating of SJ-MOS and SJ-MOS FRED.

Series	Device	BV$_{DSS}$ [V]	I$_D$ [A]
SJ-MOS	FMP20N60S1	600	20
SJ-MOS FRED	FMP20N60S1FD	600	20

TABLE 2
Comparison of Reverse recovery characteristics
between SJ-MOS and SJ-MOS FRED.

Series	Device	Trr [ns]	Qrr [uC]	Irp [A]
SJ-MOS	FMP20N60S1	406	6.09	30
SJ-MOS FRED	FMP20N60S1FD	160	1.04	13

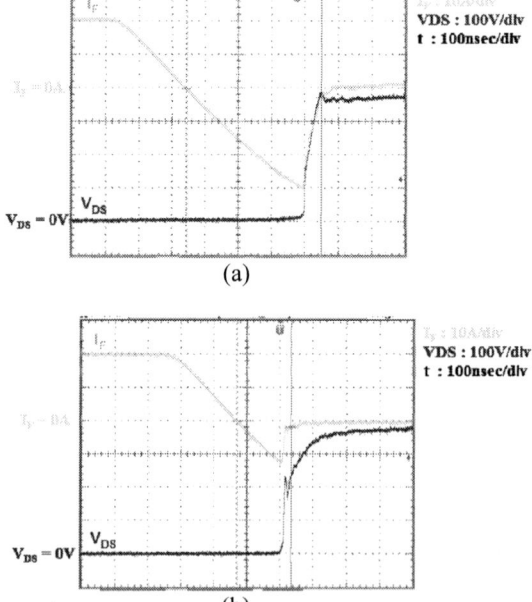

Fig.4 Reverse recovery waveforms of
(a)SJ-MOS, (b)SJ-MOS FRED under same –di/dt(=100A/us).

However, the power supply efficiency may fall because the drain leakage current (IDSS) of SJ-MOS FRED is larger than that of SJ-MOS. Table.3 shows the comparison of IDSS between the SJ-MOS and SJ-MOS FRED under the following conditions; drain voltage (VDS) = 500V, temperature = 25°C.

TABLE 3
Comparison of drain leakage current between SJ-MOS and SJ-MOS FRED.

Series	Device	I_{DSS} [nA]
SJ-MOS	FMP20N60S1	2.2
SJ-MOS FRED	FMP20N60S1FD	800

Then, in order to evaluate the performance of the SJ-MOS FRED, the device is set to the LLC converter circuit in the 1000W-class power supply as indicated in Fig.6, and Fig.5 shows the comparison of Efficiency vs. Load in the power supply between the SJ-MOS FRED and the SJ-MOS. The input voltage is 230Vac, the switching frequency is 200 to 250kHz, and the output voltage is 53V.The power average efficiency in the SJ-MOS FRED is 91.6%. It is almost equivalent to the SJ-MOS.

In order to confirm the effect of power efficiency due to the difference of device, we compared the SJ-MOS FRED loss and SJ-MOS loss based on the device characteristics and operation waveform. Here, we estimate the loss of MOSFET Q1 and MOSFET Q2 to be substantially the same, and have calculated the loss from the waveform of MOSFET Q1. Fig.7 shows the waveform of the current flowing to the transformer and gate-to-source voltage (Vgs) of MOSFET Q1, drain-to-source voltage (Vds) of MOSFET Q1 (as shown in Fig.6). Moreover Fig.8 shows the switching waveform of the SJ-MOS and SJ-MOS FRED under same condition (Load =50%). The waveform of the SJ-MOS FRED is almost equivalent to the SJ-MOS. In the period A, Vgs of MOSFET Q2 is turned off and the current flows in the forward direction of the body diode of MOSFET Q1. Drain-source capacitance of MOSFET Q2 is charged, as a result, Vds of MOSFET Q2 is decreased. Vgs of MOSFET Q1 is turned on during this period (as shown in Fig.9). Therefore, for the sake of convenience, in this paper, the loss of period A (hereafter diode loss) are calculated as the loss due to the body diode forward voltage (VF) and current. In the period B, Vgs of MOSFET Q1 is turned on, and the current is flowing through the channel of the MOSFET Q1.

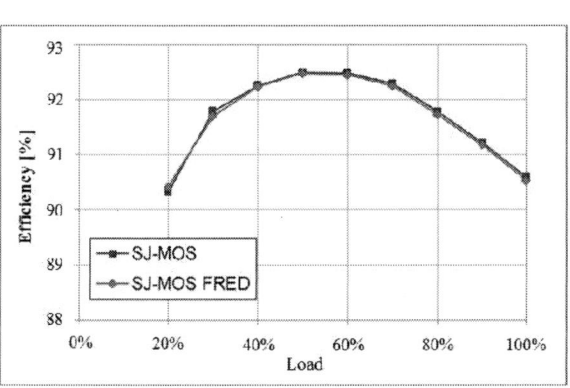

Fig.5 Comparison of Efficiency vs. Load.

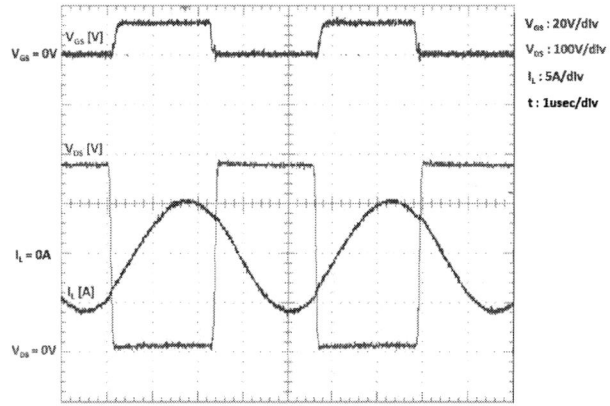

Fig.7 waveform of LLC converter circuit.

Fig.6 Configuration of test circuit diagram

978-1-4799-2706-7/14 $31.00 © 2014 IEEE

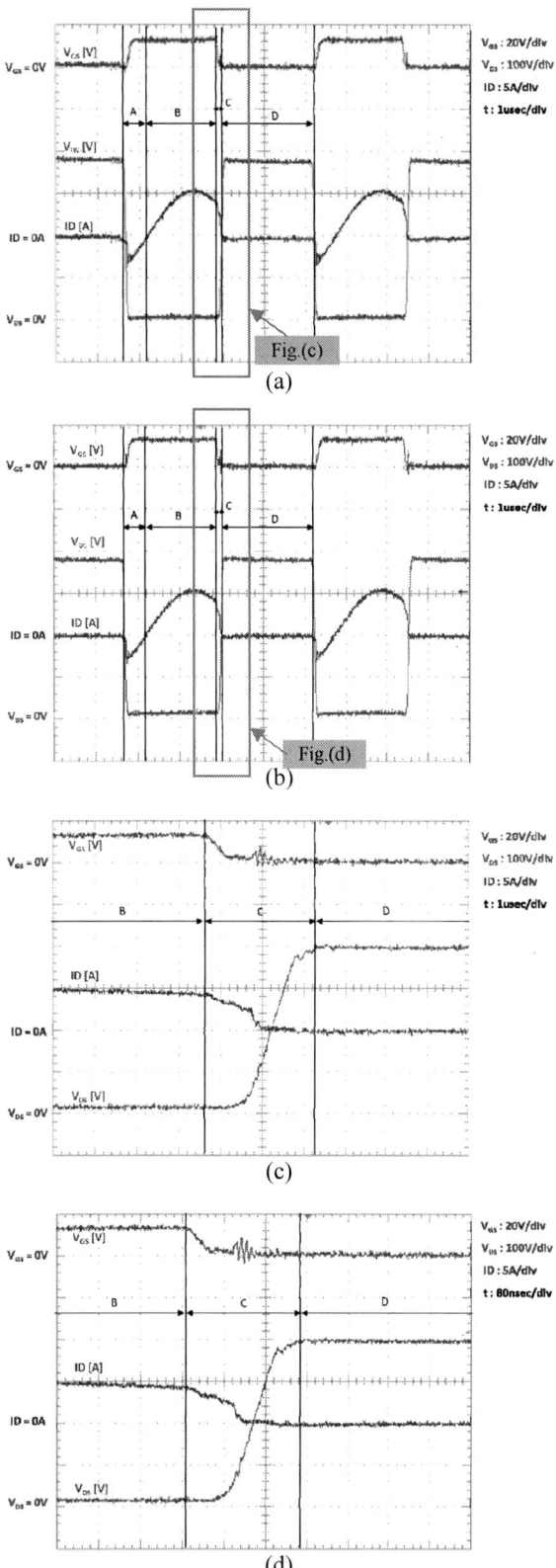

The loss of period B (on-state loss) depends on the current and on-resistance (Ron) of the MOSFET Q1. In the period C, Vgs of MOSFET Q1 is turned off and and the current flows in the forward direction of the body diode of MOSFET Q2. Drain-source capacitance of MOSFET Q1 is charged, as a result, the ID of MOSFET Q1 is reduced and the Vds of MOSFET Q1 is increased. Therefore, the loss of period C (turn-off loss) depends on the drain voltage and current, switching speed of the device. In the period D, MOSFET Q1 is off-state, and IDSS flows through the MOSFET. The loss of period D (off-state loss) depends on the drain voltage and the IDSS of the MOSFET.

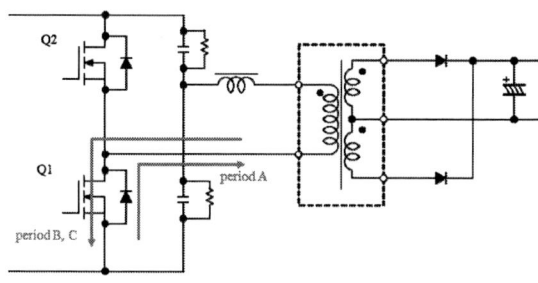

Fig.9 Image view of current direction.

Fig.10 shows the MOSFET loss of each period, which is calculated from the device characteristics and operation waveform. Fig.11 shows the total loss of the whole circuit and MOSFET. As shown in Fig.10, the loss of SJ-MOS is 2.12W, and that of SJ-MOS FRED is 2.25W. As a result, there is no significant differences between the two devices. As shown in Fig.11, the total loss ratio of the MOSFET is around 10% among losses of the whole circuit. In addition, the off-state loss ratio of the MOSFET is around 0.01% among total losses of MOSFET. Therefore, even if IDSS of the SJ-MOS FRED is large than SJ-MOS (as shown in Table 3), effect on the efficiency of the power supply is negligibly small.

Fig.8 (a)(c)SJ-MOS, (b)(d)SJ-MOS FRED waveform of LLC converter circuit under same Load(=50%).

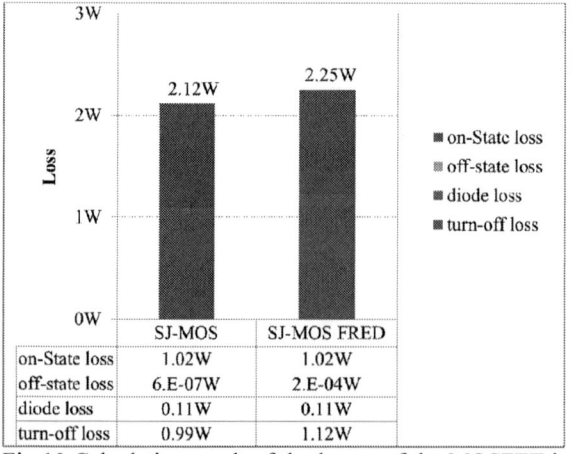

	SJ-MOS	SJ-MOS FRED
on-State loss	1.02W	1.02W
off-state loss	6.E-07W	2.E-04W
diode loss	0.11W	0.11W
turn-off loss	0.99W	1.12W

Fig.10 Calculation result of the losses of the MOSFET in LLC converter circuit under same Load(=50%).

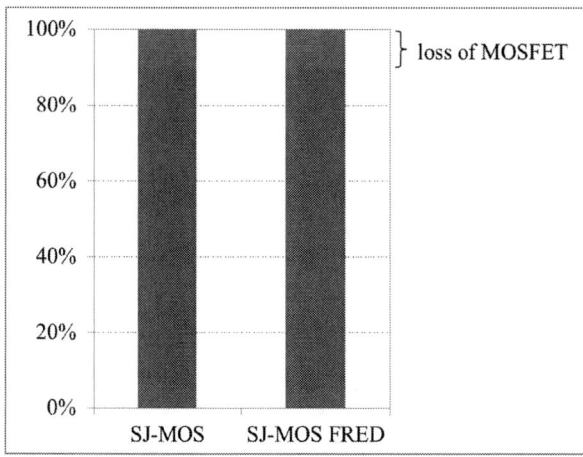

Fig.11 Calculation result of the losses of LLC converter circuit under same Load(=50%).

IV. CONCLUSION

Since the RonA in SJ-MOS is dramatically reduced by about 75% compared to that of conventional DMOSFET, efficiency of power supply will be improved by application of SJ-MOS. Moreover, it is possible to lower the possibility of the reverse recovery destruction when operating frequency tends to decrease and be afield from a resonant frequency because -di/dt ruggedness of SJ-MOS FRED was dramatically improved compared to that of SJ-MOS.

Experimental results shows that it is very effective to apply SJ-MOS FRED to the resonant circuit from a viewpoint of efficiency and safety.

REFERENCES

[1] T. Fujihira : Theory of Semiconductor Superjunction Devices, Jpn. J. Appl. Phys.

[2] G. Deboy et al. : A New Generation of High Voltage MOSFETs Breaks the Limit Line of Silicon, Proc. IEDM, pp. 683-685, 1998.Line of Silicon, Proc. IEDM, pp. 683-685, 1998.J. Clerk

[3] B. J. Baliga : Modern Power Devices, John Wiley & Sons, Inc., pp. 305-314, 1987.1 US Patent applied for (2008)

[4] T. Tamura et al.: Reduction of Turn-off Loss in 600V-class Superjunction MOSFET by Surface Design, PCIM Asia 2011, pp. 102-107, 2011

Study on an Accurate Calculation of the Conducted EMI Noise of the Power Converters

Shinpei Omata and Toshihisa Shimizu
Dept. of Electrical Engineering
Tokyo Metropolitan University
Tokyo, Japan
shimizut@tmu.ac.jp

Abstract: **The efficient reduction of EMI noise generated by power converters has been an indispensable technology. In order to meet the requirements of CISPR22, EMI filters have to be equipped with power converters. It is widely known that an EMI filter can be designed more optimally if the designer could know its inherent EMI noise in advance. This paper presents an accurate simulation method for the conducted EMI noise in power converters and verifies its effectiveness through experiments. By using the results, it is possible to reduce the design time for the power converter as well as the volume of the converter. In addition, a novel common-mode current reduction method in which only two small capacitors are added to the inverter is proposed.**

I. INTRODUCTION

Power converters that are used in various products (home appliances, electric cars, etc.) are required to provide higher specifications such as a high power density, high efficiency, and low cost. High-frequency switching with advanced power devices, such as SiC and GaN, is one effective method used to satisfy the above requirements. On the other hand, electromagnetic interference (EMI) noise emitted from the power converter tends to increase due to a high-frequency switching operation accompanied by high dv/dt or di/dt transients of the SiC and GaN devices. In order to meet EMI regulations, such as CISPR22 and IEC61000, we need to use EMI filters at the AC port or DC port of the power converters. However, it is not easy to optimize the EMI filter design because the inherent EMI noise generated by the power converters cannot be evaluated in advance of the EMI filter design [1]–[3]. Hence, a reiterative design process based on trial and error has to be implemented during the design stage. As a consequence, some problems result such as long-duration design and test processes, which increase cost.

In order to eliminate the vague process and to achieve an optimum EMI filter design, accurate calculation methods for the conducted EMI noise have been studied [4]–[9]. Based on the results, the problems in designing the EMI filter have been improved. However, some problems still remain in realizing accurate calculations. For instance, very complicated equivalent circuits for the passive components, the power devices, and the wiring pattern of the printed circuit board have to be determined in advance of the EMI calculation. These processes make the design process complicated. By taking these problems into account, the authors propose an electromagnetic compatibility (EMC) design method for the power converters, as shown below.

This paper first presents the modeling method for the components used in power converters, such as the passive device, the active device, and the copper conductor on the printed circuit board. Then, the conducted noise that appears on the line impedance stabilization network (LISN) calculated by the circuit simulator is shown. Also, the calculated results are compared with the measured results from the experimental setup, and the effectiveness of the proposed calculation method is verified. In addition, based on the investigation of the current path and the mechanism of the common-mode noise, a novel common-mode current reduction method in which only two small capacitors are added to the inverter is proposed. This is valuable for reducing the volume of the common-mode noise filter.

II. MODELING METHODS

In order to perform an accurate calculation of the conducted EMI noise using the circuit simulator, it is necessary to provide an equivalent circuit model that reflects the parasitic parameters of the component. The authors developed a simple equivalent circuit model that helps to realize both a short calculation time and high calculation accuracy.

A) Passive Elements

With regard to capacitors and inductors, it is necessary to use equivalent circuit models that satisfy the impedance characteristics of the regulatory frequency band of the conducted EMI noise (150 kHz–30 MHz). Fig. 1 shows the equivalent circuits and the impedance and phase characteristics of the passive components. The circuit parameters are determined by the least-square method to match the measured value [8].

B) Switching Devices

There are two kinds of parasitic components in the switching devices shown in Fig.2. One is a parasitic capacitance C_{DS}, and the other is a parasitic inductor L_S; both of them exist inside the package of the power

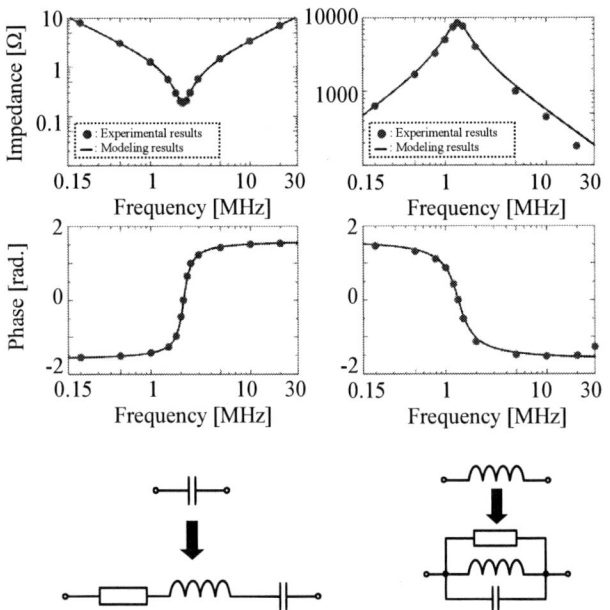

Fig.1. Impedance characteristics and equivalent circuits of the passive elements.

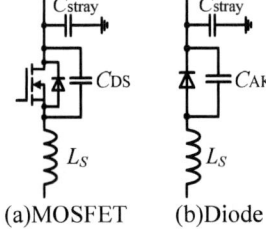

(a)MOSFET   (b)Diode

Fig.2. Equivalent circuits of the switching devices.

Fig.3. Equivalent circuits of the wiring pattern.

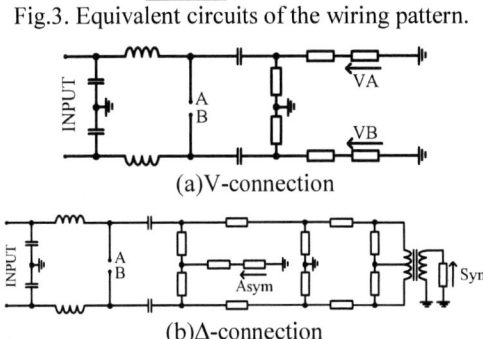

(a)V-connection

(b)Δ-connection

Fig.4. Equivalent circuits of the LISN (KNW-403D).

device. The parasitic capacitance C_{DS} is usually shown in the supplier's data sheet, and the parasitic inductance L_S can be modeled based on Reference [10]. The other one is a stray capacitance C_{stray}, which exists between the heat sink and the power device. This stray capacitance has structural dependence, and it strongly influences the common-mode noise [11]. In order to clarify the influence of the capacitance and to perform accurate EMI noise calculations, the authors use ceramic capacitors with the same capacitance as the actual stray capacitances.

C) Wiring Pattern

The wiring conductors contain wiring resistance and parasitic inductance. The authors use a model that does not take into account the skin effect, as shown in Fig. 3. The wiring resistance is calculated using Equation (1), and the parasitic inductance is calculated using the circular division method proposed by K. Wada [12].

$$R = \rho \frac{l}{S} \qquad (1)$$

where, ρ is resistivity, l is length, and S is cross-section area.

D) LISN

The authors used a line impedance stabilization network (LISN), which is defined by CISPR16. In this study, KNW-403D (Kyoritsu Co. Ltd.), which meets the EMI regulation, is used as the LISN. The LISN has two functions: V-connection and Δ-connection. The sum of the differential-mode noise and common-mode noise is measured in the V-connection, and the differential-mode noise and common-mode noise are measured separately in the Δ-connection. It is necessary to use the V-connection mode in order to

evaluate the conducted noise level, which is defined by the CISPR22 and IEC61000 regulations. However, the noise value measured only by the V-connection LISN is not sufficient to optimally design the noise filter. Usually, we need to provide individual designs for both the attenuation characteristics of the common-mode noise filter and the differential-mode noise filter based on the potential noise of the converter. Hence, the authors also calculate both of the EMI noises measured by the Δ-connection mode. The calibration value and the circuit configuration are shown in Fig. 4 by referring to the data sheet.

III. EVALUATION SYSTEMS

The authors calculate the conducted EMI noise of both the buck chopper circuit and the single-phase inverter circuit, as show in Figs. 5 and 7, respectively.

A) Buck Chopper

Fig. 5 shows the equivalent circuit of the buck chopper. In order to perform an accurate calculation, the parasitic components given by the modeling method presented in the previous section are used.

The operating conditions are shown in Table 1. The main noise sources of this circuit are as follows: (a) the current flowing into the DC capacitor, which mainly influences the differential-mode noise, (b) the current flowing into the snubber capacitor, which mainly influences the differential-mode noise, and (c) a common-mode voltage, which appears between the GND and the neutral point of the stray capacitor (when the connection between those two terminals are opened).

In order to calculate the common-mode voltage

978-1-4799-2706-7/14 $31.00 © 2014 IEEE

The 2014 International Power Electronics Conference

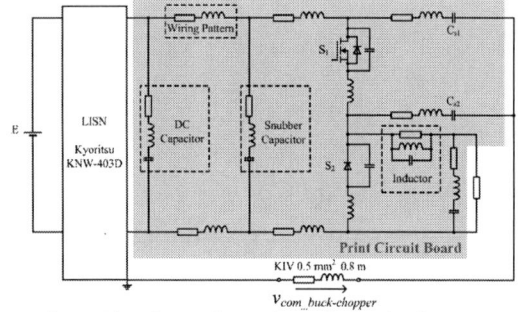

Fig.5. Circuit configuration of a buck-chopper.

TABLE1. Operating conditions of a buck-chopper.

Input voltage	Input current	Switching frequency	Duty ration
100 V	2.5 A	10 kHz	50 %

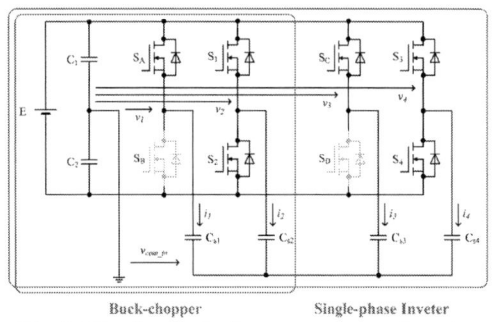

Fig.6. Circuit to calculate common-mode voltage.

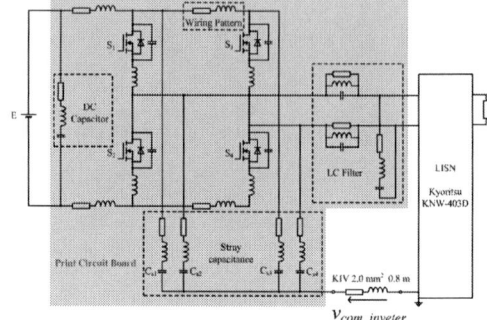

Fig.7. Circuit configuration of a single-phase inverter.

TABLE2. Operating conditions of a single-phase inverter.

Input voltage	Input current	Switching frequency	Control
100 V	2.5 A	1 kHz	180°

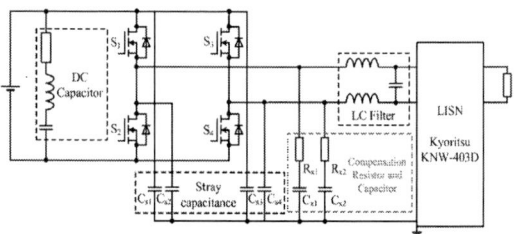

Fig.8. Single-phase inverter with compensation capacitor.

Fig.9. Experimental setup.

appearing at the neutral point of the stray capacitors, a modified method of calculating the common-mode voltage proposed by Ref. [13] is used. Fig. 6 shows the equivalent circuit for the calculation of the common-mode voltage, where S_A and S_B represent the assistant switches. S_A is always ON, and S_B is always OFF. Also, S_1 and S_2 represent the virtual switches. The stray capacitances C_{S1} and C_{S2} are connected as shown in Fig. 6. So, we can calculate the common-mode voltage in the buck chopper as the output voltage of the single-phase inverter for the special conditions. For this condition, the following equations are satisfied:

$$v_1 = \frac{1}{C_{s1}} \int i_1 \, dt + v_{com\_in} \qquad (2)$$

$$v_2 = \frac{1}{C_{s2}} \int i_2 dt + v_{com\_in} \qquad (3)$$

$$i_1 + i_2 = 0 \qquad (4)$$

Therefore, the common-mode voltage in the buck chopper is calculated using Equation (5).

$$v_{com\_buck-chopper} = \frac{C_{s1}v_1 + C_{s2}v_2}{C_{s1} + C_{s2}} \qquad (5)$$

where, $v_1 = \frac{E}{2}$ and $v_2 = \pm\frac{E}{2}$.

After all, the common-mode voltage changes half of the input voltage in the switching. These changes in the common-mode voltage and the impedance cause the common-mode noise.

B) Single-phase Inverter

Next, the conducted EMI noise in the single-phase inverter circuit shown in Fig. 7 is studied. The LISN is connected to the AC output terminal of the inverter because we are considering the utility interactive inverter. The operating conditions are shown in Table 2. The main noise sources of this circuit are as follows: (a) the current flowing through each inverter leg, which mainly influences the differential-mode noise, (b) the current flowing into each inverter leg from the DC capacitor, which mainly influences the differential-

978-1-4799-2706-7/14 $31.00 © 2014 IEEE   2946

The 2014 International Power Electronics Conference

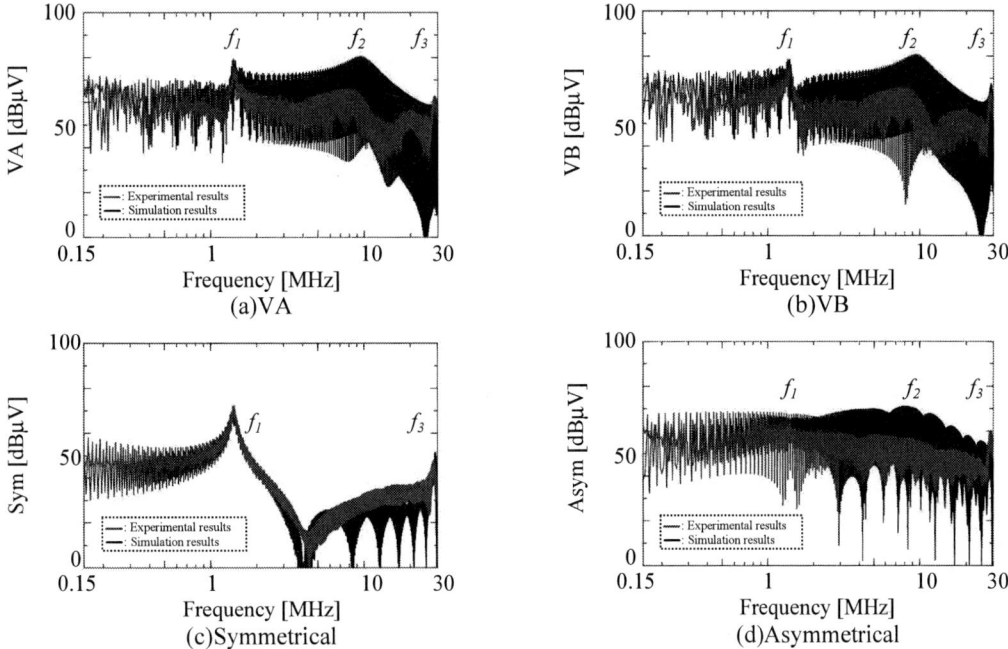

Fig.10. Comparisons of simulated and experimental results of the EMI noise (buck-chopper).

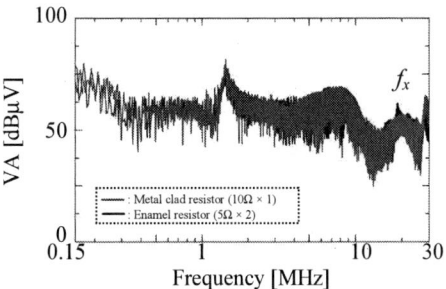

Fig.11. Change in conducted emission for different loads.

(a)Metal clad resistor   (b)Enamel resistor

Fig.12. Resistors used in the experiments.

mode noise, and (c) the common-mode voltage, which appears between the GND and the neutral point of the stray capacitance (when the connection between those two terminals is opened).

S_A, S_B, S_C, and S_D represent the assistant switches. S_A and S_C are always ON, and S_B and S_D are always OFF. S_1, S_2, S_3, and S_4 represent the virtual switches. The stray capacitances C_{S1}, C_{S2}, C_{S3}, and C_{S4} are connected as shown in Fig. 6. So, we can calculate the common-mode voltage of the single-phase inverter as the output voltage of the fourth-phase inverter for the special conditions. In this condition, the following equations are satisfied as well as the buck chopper circuit:

$$v_1 = \frac{1}{C_{s1}} \int i_1 dt + v_{com\_in} \qquad (6)$$

$$v_2 = \frac{1}{C_{s2}} \int i_2 dt + v_{com\_in} \qquad (7)$$

$$v_3 = \frac{1}{C_{s3}} \int i_3 dt + v_{com\_in} \qquad (8)$$

$$v_4 = \frac{1}{C_{s4}} \int i_4 dt + v_{com\_in} \qquad (9)$$

$$i_1 + i_2 + i_3 + i_4 = 0 \qquad (10)$$

Therefore, the common-mode voltage in the buck chopper is calculated using Equation (11).

$$v_{com\_in} = \frac{C_{s1}v_1 + C_{s2}v_2 + C_{s3}v_3 + C_{s4}v_4}{C_{s1} + C_{s2} + C_{s3} + C_{s4}} \qquad (11)$$

where, $v_1 = v_3 = \frac{E}{2}$ and $v_2 = v_4 = \pm \frac{E}{2}$.

In the interconnection inverter shown in Fig. 7, the common-mode voltage is calculated using Equations (12)–(15) applying Equation (11).

978-1-4799-2706-7/14 $31.00 © 2014 IEEE

Mode1(S1,S4; ON)

$$v_{com\_inverter} = \frac{E}{2} - \frac{C_{s4}\,E}{C_{s1} + C_{s2} + C_{s3} + C_{s4}} \quad (12)$$

Mode2(S1,S3; ON)

$$v_{com\_inverter} = 0 \quad (13)$$

Mode3(S2,S3; ON)

$$v_{com\_inverter} = \frac{E}{2} - \frac{C_{s2}\,E}{C_{s1} + C_{s2} + C_{s3} + C_{s4}} \quad (14)$$

Mode4(S2,S4; ON)

$$v_{com\_inverter} = E - \frac{(C_{s2} + C_{s4})\,E}{C_{s1} + C_{s2} + C_{s3} + C_{s4}} \quad (15)$$

The changes in the common-mode noise voltage shown in Equations (12)–(15) and the impedance cause the common-mode noise. No changes in the common-mode noise voltage can prevent the common-mode noise in theory. If the stray capacitances C_{S2} and C_{S4} are larger than C_{S1} and C_{S3}, the common-mode voltage is nearly 0 in all modes based on Equations (12)–(15).

Therefore, the authors propose a way to reduce the common-mode noise with compensation capacitors C_{x1} and C_{x2} that is larger than C_{S1} and C_{S3}, as shown in Fig. 8. This compensation capacitor is connected to the output of the single-phase inverter and makes the changes in the common-mode voltage nearly 0. The resistances R_{x1} and R_{x2} connected in series with the compensation capacitor reduce the overcurrent in the compensation capacitor. The compensation capacitor has a small capacity, so the loss of resistance is small. Also, it is possible to meet the standard with a minimum volume of the common-mode noise filter by using this method.

IV. SIMULATED AND EXPERIMENTAL RESULTS

The simulated and experimental results for the conducted EMI noise in each converter are discussed. The experiment was performed in a shielded room.

A) Buck Chopper

Figs. 10(a) and (b) shows the results of the conducted EMI noise for VA and VB measured in the V-connection mode of the LISN. The simulated values and the experimental values coincide very well for both VA and VB. It is confirmed that the circuit model used in this study provides an accurate calculation of the conducted EMI noise.

Figs. 10(c) and (d) show the results of the conducted EMI noise measured in the Δ-connection mode. The frequencies f_1 and f_3 are the frequency components that appeared either in the differential-mode noise or the asymmetrical common-mode noise. The frequency component f_2 is the common-mode noise. The simulated values and the experimental values also coincide very well for both symmetrical and asymmetrical noise. So, it can be confirmed whether the frequency components are differential-mode noise

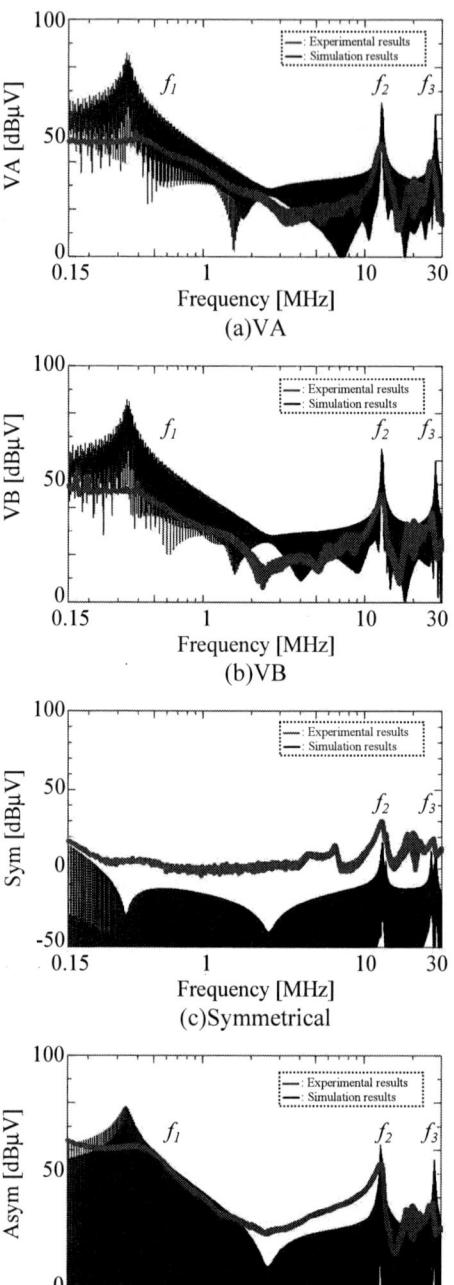

Fig.13. Comparisons of simulated and experimental results of the EMI noise (Single-phase Inverter).

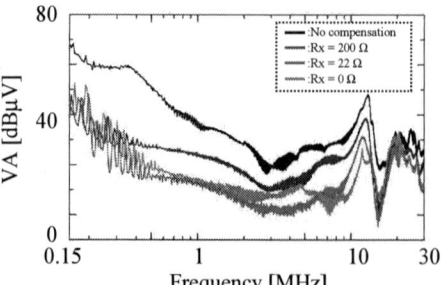

Fig.14. Experimental results with compensation capacitors.

or common-mode noise. Hence, the proposed method provides useful information for properly designing the common-mode noise filter and the differential-mode noise filter.

Fig. 11 shows the experimental results for the conducted EMI noise in the VA mode when two kinds of load resistors are used. One resistor is a metal clad resistor (10 Ω × 1), and the other is an enamel resistor (5 Ω × 2 series), and exterior views of these resistors are shown in Fig.12. In the case where the enamel resistor is used, a higher noise level is observed in f_x, but it is not observed in the metal clad resistor. This phenomenon is expected because the radiated noise from the resistor and the connecting wire may have some interaction with the wiring conductors on the main circuit and can affect the conducted EMI noise.

This phenomenon cannot be calculated in the circuit analysis used in this study. Therefore, we need to use analysis such as electromagnetic field analysis if we need a more accurate value.

B) Single-phase Inverter

Figs. 13(a) and (b) shows the results for the conducted EMI noise for VA and VB measured in the V-connection mode of the LISN. The trends of the simulated and experimental results coincide very well except for the frequency f_1. The cause of this difference is the fact that the model of the wiring pattern does not take into account the skin effect.

Figs. 13(c) and (d) show the results for the symmetrical noise and asymmetrical noise in the Δ-connection mode. The frequencies f_2 and f_3 are the frequency components that appeared either in the differential-mode noise or the asymmetrical common-mode noise. The frequency f_1 appeared in the common-mode noise. The trends of the simulated and experimental results coincide very well, but the magnitude has some discrepancy in the symmetrical noise. The reason for this discrepancy is the fact that the calculated symmetrical noise level is very low, and it is difficult to measure a value of 0 dBμV or less in the spectrum analyzer used in this study. In any case, the proposed method provides useful information for properly designing the common-mode noise filter and the differential-mode noise filter as well as the buck chopper. Therefore, the proposed method is also useful in a real-scale system.

Next, Fig. 14 shows the experimental result used for the common-mode noise suppression method proposed in Fig. 8. We confirm the effect of the compensation capacitor in Fig. 14, and we observe the different effects in the resistor series of the compensation capacitor. For a small resistance, the noise level trends downward. This phenomenon is a result of the difference in discharge time between the compensation capacitor and the stray capacitance. For a large resistance, the common-mode voltage has a high pulse voltage in the switching transients because the discharge time of the compensation capacitor is longer than that of the stray capacitance. This high pulse voltage makes the noise level increase. However, the

noise increases in a specific spectrum due to the new differential-mode noise when the compensation resistor is small. Therefore, we have to design considering both cases. In this circuit, the optimum resistance is 22 Ω.

V. CONCLUSION

This paper presented a calculation method for the conducted EMI noise and a reduction method for the common-mode noise in an interconnection inverter. From this study, the following conclusions were made:
(1) A simple method for calculating the conducted EMI noise was proposed, and its effectiveness was verified.
(2) In the high-frequency region, the radiated noise influences the conducted EMI noise. Hence, proper modeling of the physical structure may be needed in order to consider the influence of the radiation noise.
(3) A reduction method for common-mode noise that suppresses the changes in the common-mode noise voltage was demonstrated, and it successfully decreases the noise level.

Applications of the proposed method for the optimum design of the EMI filter will be a topic for future research.

REFERENCES

[1] M.L. Heldwein, T. Nussbaumer, and J.W. Kolar, "Differential Mode EMC Input Filter Design for Three-Phase AC-DC-AC Sparse Matrix PWM Converters", *Proceedings of the 35th IEEE Power Electronics Specialists Conference*, pp.284–291(2004)

[2] T. Nussbaumer, M.L. Heldwein, and J.W. Kolar, "Common Mode EMC Input Filter Design for a Three-Phase Buck-Type PWM Rectifier System", *Proceedings of the 21st Annual IEEE Applied Power Electronics Conference and Exposition*, Vol.3, pp.1617–1623 (2006)

[3] T. Chida, N. Kusuno, A. Mishima, M. Kurita, and S. Ibori, "Filter Design Technique for Inverter Complied with European EMC standard", *IPEC Sapporo*, pp.218-221(2010)

[4] B. Revol, J. Roudet, J.L. Schanen, P. Loizelet, "Fast EMI Prediction method for three phase inverter based on Laplace Transforms", *Proceedings of 34th IEEE Annual Power Electron Specialists Conference*, pp.1133–1138(2003)

[5] M. Sato, S. Doki, M. Ishida, "High-Frequency Equivalent Circuit of PM Synchronous Motor Driven by PWM Inverter", *IEEJ*, Vol.124-D, No.5, pp.464-470 (2004) (in Japanese)

[6] A. Mishima, "Switching analysis methods using Power Device Models and Magnetic Field Coupling System", *IPEC Nigata*, pp.2063-2068 (2005)

[7] J. Lai, X. Huang, E. pepa, S. Chen, T.W.Nehl, "Inverter EMI Modeling and Simulation Methodologies", *IEEE Transactions on Industrial Electronics*, vol.53, No.3, pp.736-744 (2006)

[8] M. Tamate, T. Sasaki, A. Toda, "Quantitative Estimation of Conducted Emission from an Inverter System", *IEEJ*, Vol.128-D, No.3, pp.193-200 (2008) (in Japanese)

[9] S. Maekawa, J. Tsuda, A. Kuzumaki, S. Matsumoto, H. Mochikawa, H. Kubota, "Quantitative analysis of analytical accuracy and precision in the analysis model of the inverter EMI", *IEEJ Industry Applications Society Conference*, p.191–196(2013) (in Japanese)

[10] S. Hashino, T. Simizu, "Separation Measurement of Parasitic Inductance and Stray Capacitance using the TDR Method", *IEEJ*, Vol.133-D, No.4, pp.443-449 (2013) (in Japanese)

[11] T. Shimizu, G. Kimura, J. Hirose, "High Frequency Leakage Current Caused by the Transistor Module ant Its Suppression Technique", *IEEJ*, Vol.116-D, No.7, pp.758-766 (1996) (in Japanese)

[12] M. Ando, K. Wada, "High-Speed Analysis of Bus Bar Inductance for a Laminated Structure", *IEEJ Journal of Industry Applications*, Vol.2, No.4, pp.189-194 (2013)

[13] S. Ogasawara, H. Fujita, H. Akagi, "Modeling and Analysis of High-Frequency Leakage Currents Caused by Voltage-Source PWM Inverters", *IEEJ*, Vol.115-D, No.1, pp.77-83 (1995) (in Japanese)

978-1-4799-2706-7/14 $31.00 © 2014 IEEE

An Exact Discrete-Time Model Considering Dead-Time Nonlinearity for an H-Bridge Grid-Connected Inverter

Ruiliang Xie, Xiang Hao, Xu Yang, Wenjie Chen, Lang Huang and Chao Wang
State Key Laboratory of Electrical Insulation and Power Equipment
Xi'an Jiaotong University , Shaanxi, Xi'an, China
xie19902009@163.com

Abstract-**With the development of the renewable energy, grid-connected H-bridge inverter plays a more important role than ever. However, the dead-time and the zero crossing clamping (ZCC) effect caused by dead-time can cause THD problem. As a consequence, more and more researches have focused on this issue. Most of the researches use the harmonics to replace the effect of dead-time, which ignores ZCC effect. Some other researches use the average current to distinguish the ZCC effect, which is not accurate enough. This paper analyzes the whole changing process of inductor current ripples during power cycle and a complete discrete-time model of the H-bridge grid-connected inverter with dead-time is established. On the basis of the model, the analysis of dynamic behavior is carried out, the result of which shows that the length of dead-time can cause nonlinear phenomenon and reduces the stability region of the controller parameters. Besides, this model is also useful to the nonlinearity compensation of the system.**

I. INTRODUCTION

With the increasing concerns about the energy crisis, the renewable energy is gaining more and more attention. As a bridge between the renewable energy and the grid, the grid-connected voltage source inverter (VSI) plays an important role in the grid-connected system [1]-[5]. To satisfy the strict harmonic limits of current power quality standards, the Total Harmonic Distortion (THD) of the system needs to be very low. However, the dead-time and the zero crossing clamping (ZCC) effect caused by dead time can cause serious THD problem. As a consequence, more and more researches are carried out to reduce the impact of dead-time. Some researchers are devoted to replace the dead-time effect by harmonics [6], but the ZCC effect has not been considered, which makes the modeling of the system not accurate enough. Some other researchers consider the effect of ZCC and distinguish it by using the average value of inductor current during one switching cycle [7]. As there are some approximations in the calculation of average current and critical condition [8]-[10], the modeling using the average current can also influence the accuracy of the modeling of the system.

In this paper the proportional controlled first order single phase grid-connected inverter is studied. After analyzing the whole changing process of the current

ripples around zero crossing points, the current ripples can be divided into several cases. Then the criteria and the discrete-time model of each case are established. As a result a complete model of VSI system with dead-time is established, which is accurate enough by using the discrete-time model. Based on this model, the studies of dynamic behavior and the stability of the system are carried out, the results of which show that the length of the dead-time can cause period-doubling bifurcation around zero crossing points and can reduce the stability region of the controller parameter. Besides, this model can also be used to do the nonlinearity compensation for the system [9]-[10], which will be our work in the future.

II. CIRCUIT OPERATION AND INTRODUCTION OF DEAD-TIME

The schematic circuit of a proportional controlled single-phase grid-connected VSI system is shown in Fig.1.

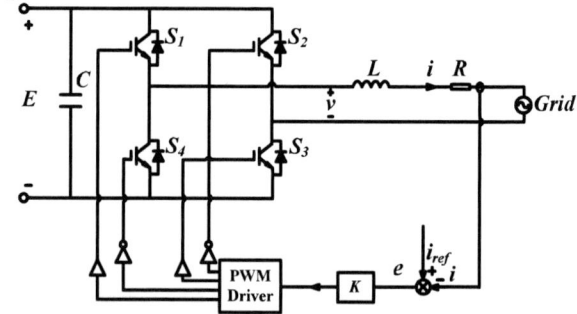

Fig. 1. Topology of single-phase grid-connected VSI with a proportional controller

Where, E is the DC bus voltage; L is inductor; R is the sum of the parasitic-resistance of the inductor and the resistance of wire; v is the output voltage; i is inductor current; e is the error of the inductor current; k is the parameter of proportional controller.

Without dead-time, the VSI system has two working states: in case of state 1, S1 and S3 on, S2 and S4 off; in case of state 2, S1 and S3 off, S2 and S4 on. The state equation of the system can be expressed as:

This work was supported by the National Natural Science Foundation of China under Project 51177129

$$\text{State 1:} \quad \frac{d}{dt}\mathbf{i} = -\frac{R}{L}\mathbf{i} + \frac{E}{L} - \frac{A\sin(\omega t)}{L} \qquad (1)$$

$$\text{State 2:} \quad \frac{d}{dt}\mathbf{i} = -\frac{R}{L}\mathbf{i} - \frac{E}{L} - \frac{A\sin(\omega t)}{L} \qquad (2)$$

where A is the amplitude of grid voltage, ω is the line angular frequency.

As is shown in Fig.2, during dead times both switches of the same leg are off, and the output current i flows through one of the freewheeling diodes. When $i>0$, the output voltage is $-E$ and the current will decrease during the dead time. When $i<0$, the output voltage is E and the current will increase. As a result, the current waveforms will be changed with the import of dead-time.

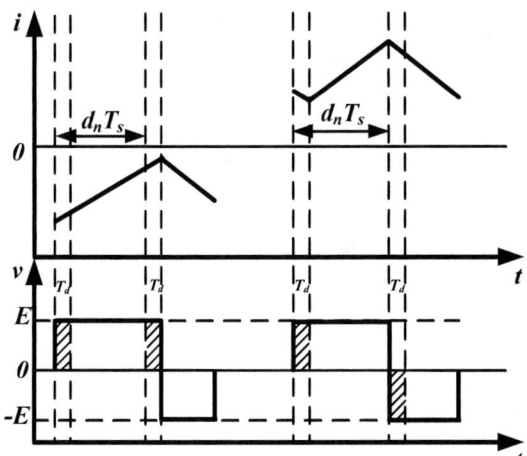

Fig. 2. Inverter output waveforms for dead-time disturbance analysis without ZCC phenomenon

Regardless of ZCC effect, the discrete-time iterative model can be expressed as:

Case $i<0$:

$$\mathbf{i}(n+1) = \mathbf{i}(n) \cdot e^{\alpha T_s} + \beta[e^{\alpha T_s} - 2e^{\alpha((1-d_n)T_s - T_d)} + 1]$$
$$- \frac{A\sin(\omega n T_s)}{R}(1 - e^{\alpha T_s}) \qquad (3)$$

Case $i>0$:

$$\mathbf{i}(n+1) = \mathbf{i}(n) \cdot e^{\alpha T_s} - \frac{A\sin(\omega n T_s)}{R}(1 - e^{\alpha T_s})$$
$$+ \beta[-e^{\alpha T_s} - 2e^{\alpha(1-d_n)T_s} + 2e^{\alpha(T_s - T_d)} + 1] \qquad (4)$$

where $\alpha=-R/L$, $\beta=-E/R$, d_n is the duty ratio of the nth switching cycle, which can be expressed as:

$$d_n = \frac{k[\mathbf{i}_{ref}(n) - \mathbf{i}(n)] + 1}{2} \qquad (5)$$

III. THE COMPLETE DISCRETE-TIME MODEL CONSIDERING ZCC PHENOMENON

When the current is very close to zero, the current will be clamped to zero when the current flows through the freewheeling diode reaches zero during dead-time, which is called the ZCC phenomenon. In most cases the ZCC phenomenon can cause more serious harmonic distortion. As a result, the exact discrete-time model is necessary,

with the help of which the nonlinearity compensation for the system can be realized.

As is shown in Fig.3, there will be seven steps during whole changing process of inductor current including the process of ZCC phenomenon (Step B to Step F). During its rising section, the order of the steps is from A to G. And it is in the opposite order, during the falling section.

The discrete-time iterative model of these steps can be expressed as:

Step A: same as (3).

Step B and Step C:

$$\mathbf{i}(n+1) = -\frac{E + A\sin(\omega n T_s)}{R}(1 - e^{\alpha((1-d_n)T_s - T_d)}) \qquad (6)$$

Step D:

$$\mathbf{i}(n+1) = \mathbf{i}(n) \cdot e^{\alpha T_s} + \beta[e^{\alpha T_s} - 2e^{\alpha(1-d_n)T_s} + 1]$$
$$- \frac{A\sin(\omega n T_s)}{R}(1 - e^{\alpha T_s}) \qquad (7)$$

Step E and F:

$$\mathbf{i}(n+1) = \beta(1 - 2e^{\alpha(1-d_n)T_s} + e^{\alpha(T_s - T_d)})$$
$$- \frac{A\sin(\omega n T_s)}{R}(1 - e^{\alpha(T_s - T_d)}) \qquad (8)$$

Step G: same as (4).

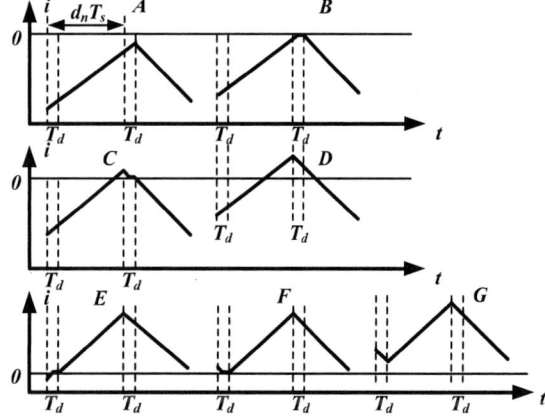

Fig. 3. Seven steps during the whole changing process of inductor current

On the basis of the discrete-time models of each step, the criteria that can distinguish different steps are needed for the build of the complete model. As is shown in Fig.4, the current of the dead-time can be divided into two cases:

Case 1: $i>0$ at the beginning of the dead-time.

Case 2: $i<0$ at the beginning of the dead-time.

The diagonal dashed lines present the critical situations for two cases. Assuming that the current at the beginning of a dead-time equals to i_0, the criteria that can distinguish whether there exists ZCC phenomenon are as follows:

If $i_0<i_1$, ZCC phenomenon will not happen and the current will keep increasing during the dead-time.

If $i_0>i_2$, ZCC phenomenon will not happen and the current will keep decreasing during the dead-time.

If $i_1<i_0<i_2$, ZCC phenomenon will happen and the current will become zero when the dead-time finishes.

Where
$$\mathbf{i}_1 = \frac{[E - A\sin(\omega n T_s)](e^{\alpha T_d} - 1)}{Re^{\alpha T_d}} \quad (9)$$

$$\mathbf{i}_2 = \frac{[-E - A\sin(\omega n T_s)](e^{\alpha T_d} - 1)}{Re^{\alpha T_d}} \quad (10)$$

As i_1 and i_2 are deduced on the basis of discrete-time model of the current during dead-time, (9) and (10) are accurate enough to detect the appearance of ZCC.

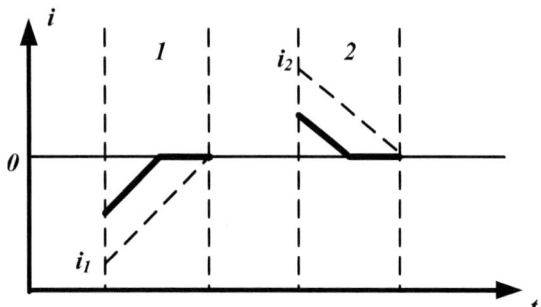

Fig. 4. Schematic of the current during the dead-time when ZCC phenomenon happens

Based on the criteria of ZCC, a complete discrete-time model of the system can be obtained by detecting whether there are ZCC phenomena in the first and second dead-time section of each switching cycle during an entire power frequency cycle. And the complete model can be expressed as follow with an iterative number of $2\pi/(\omega \cdot T_s)$.

$$
\begin{cases}
\mathbf{i}(n) < \mathbf{i}_1 : \begin{cases} \mathbf{i}(nT_s + d_nT_s) < \mathbf{i}_1 : CaseA \\ \mathbf{i}_1 < \mathbf{i}(nT_s + d_nT_s) < \mathbf{i}_2 : CaseB \& C \\ \mathbf{i}(nT_s + d_nT_s) > \mathbf{i}_2 : CaseD \end{cases} \\
\\
\mathbf{i}_1 < \mathbf{i}(n) < \mathbf{i}_2 : CaseE \& F \qquad (11) \\
\\
\mathbf{i}(n) > \mathbf{i}_2 : CaseG
\end{cases}
$$

IV. THE ANALYSIS OF DYNAMIC BEHAVIOR AND STABILITY

As it is shown in Fig 5, the simulation is carried out and the results show that the period-doubling bifurcation will happen and the system will be instable with dead-time increasing from $0.03T_s$ to $0.08T_s$ in case of the same controller parameter. The system parameters are shown in Table 1.

TABLE I
PARAMETER VALUES OF THE INVERTER SYSTEM

Symbol	Meaning	Value
E	Input voltage	500 V
R	Parasitic resistance	0.8 Ω
L	Inductance	1 mH
ω	Line angular frequency	314 rad/s
f_s	Switching frequency	30 kHz
f	Power frequency	50 Hz
I_{ref}	The reference of current	50sin(ωt) A

In most cases, the dynamic behavior and the stability are distinguished by using the eigenvalues of Jacobian matrix. However, the nonlinearity imported by ZCC phenomenon could add the complexity of this method. After considering the dead-time effect, the duty ratio will keep its monotonicity under stable operation and will lose its monotonicity if the period-doubling bifurcation occurs, as is shown in Fig.6. As a result, a simplified judgment method which uses the monotonicity of the duty ratio can be proposed.

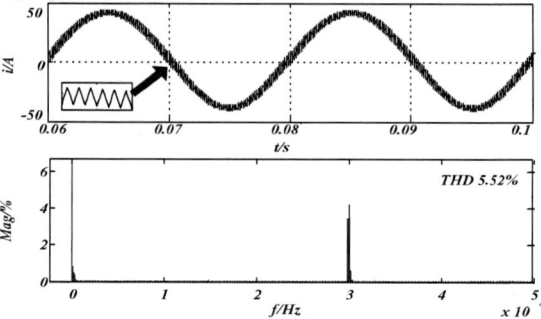

(a) Time domain waveform and FFT result with dead-time nonlinearity when $k=0.08$, $T_d=0.03T_s$

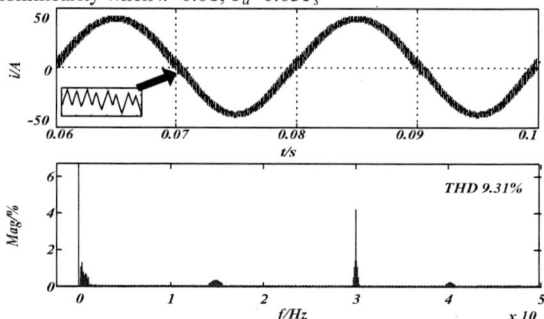

(b) Time domain waveform and FFT result with dead-time nonlinearity when $k=0.08$, $T_d=0.08T_s$

Fig. 5. Simulation result with dead-time effect

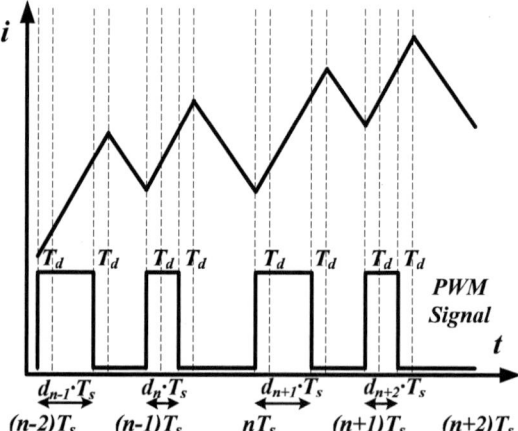

Fig. 6. Schematic of the current ripple and PWM signal under instable operation

By using M switching cycles (M is large enough to include all the zero crossing points) in the decreasing section of current, the judgement method can be expressed as:

$$P = \sum_{N}^{N+M-1} \frac{[\mathbf{i}_{ref}(n-1)-\mathbf{i}_{ref}(n)]-[\mathbf{i}(n-1)-\mathbf{i}(n)]}{\left|[\mathbf{i}_{ref}(n-1)-\mathbf{i}_{ref}(n)]-[\mathbf{i}(n-1)-\mathbf{i}(n)]\right|} \quad (12)$$

where both N and $N+M-1$ are in the decreasing section of current. If $P=M$ the system is stable, otherwise the system is instable. On the basis of the judgement method, the stability region of the system is shown in Fig.7 (T_d represents the proportion of dead-time and k represents the proportional controller parameter).

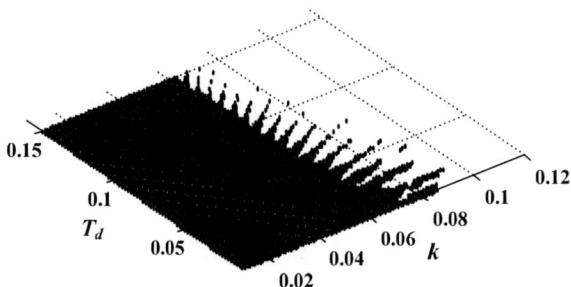

Fig. 7. The stability region of the system

Besides, the nonlinearity compensation of dead-time and the ZCC phenomenon caused by dead-time can be carried out by using the discrete-time model proposed in this paper, which will be our work in the future.

V. CONCLUSION

With the consideration of the ZCC effect during the dead-time in the VSI grid-connected system, the whole changing process of the current ripples is presented, which can be divided into seven cases. And the discrete-time models and the criteria of each case are established. As a consequence, an exact discrete-time iterative model considering dead-time nonlinearity is presented in this paper, on the basis of which the stability and dynamic behavior of the system are analyzed. And the analysis result shows that with the increase of the length of dead-time, period-doubling bifurcation occurs in the system and the stability region of the controller parameters is reduced. By using the proposed judgement method of the instable phenomenon, the stability region of the controller parameters is obtained, which can be used to design the dead-time length and the controller parameter. Besides, the proposed model can also be used to do the nonlinear compensation of the dead-time, which will be our future work.

REFERENCES

[1] L. Dominic, L. Lai, C. Chen, T. LaBella, and B. Chen, "High Reliability and Efficiency Single-Phase Transformerless Inverter for Grid-Connected Photovoltaic Systems," *IEEE Trans. Power Electronics*, vol.28, no. 5, pp. 2235-2245, May 2013.

[2] X. Hao, X. Yang, T. Liu, L. Huang and W. Chen, "A Sliding-Mode Controller With Multi-Resonant Sliding Surface for Single-Phase Grid-Connected VSI With an LCL Filter," *IEEE Trans. Power Electronics*, vol.28, no. 5, pp. 2259-2268, May 2013.

[3] X. Hao, X. Yang, R. Xie, T. Liu and L. Huang, "A Fixed Switching Frequency Integral Resonant Sliding Mode Controller for Three-Phase Grid-Connected Photovoltaic Inverter With LCL-Filter," in *Proc. IEEE IPEMC*, vol. 1, pp.793-798, Jun. 2013.

[4] J. Matas, L. G. de Vicuna, L. Miret, J. M. Guerrero, and M. Castilla, "Feedback linearization of a single-phase active power filter via sliding mode control," *IEEE Trans. Power Electron.*, vol. 23, no. I, pp. 116-125, Jan. 2008.

[5] R. L. Wai, and W. H. Wang, "Grid-connected photovoltaic generation system," *IEEE Trans. Circuits Syst. I, Reg. Papers*, vol. 55, no. 3, pp.953-964, Apr. 2008.

[6] X. Hao, T. Liu, X. Yang, and L. Huang, "A Discrete-Time Integral Sliding-Mode Controller with Nonlinearity Compensation for Three-Phase Grid-connected Photovoltaic Inverter," in *Proc. IEEE IPEMC*, vol. 2, pp. 831-835, Jun. 2012.

[7] M. Herran, J. Fischer, S. Gonzalez, M. Judewicz, and D. Carrica, "Adaptive Dead-Time Compensation for Grid-Connected PWM Inverters of Single-Stage PV Systems," *IEEE Trans. Power Electronics*, vol.28, no. 6, pp. 2816-2825, June 2013.

[8] T. J. Summers and R. E. Betz, "Dead-Time Issues in Predictive Current Control," *IEEE Trans. Industry Applications*, vol. 40, no. 3, May/June 2004.

[9] A. Cichowski, and J. Nieznanski, "Self-Tuning Dead-Time Compensation Method for Voltage-Source Inverters," *IEEE Power Electronics Letters*, vol.3, no. 2, June 2005.

[10] L. Ben-Brahim, "On the Compensation of Dead Time and Zero-Current Crossing for a PWM-Inverter-Controlled AC Servo Drive," *IEEE Trans. Industry Electronics*, vol. 51, no. 5, October 2004.

The 2014 International Power Electronics Conference

Theoretical Analysis of the Duality Principle Applied to Interleaved Topologies

M.L.A. Caris, H. Huisman, and J.L. Duarte
Eindhoven University of Technology
Electromechanics and Power Electronics group

Abstract—**The neutral-point clamped, flying-capacitor and cascaded H-bridge converter topologies are examples of multilevel topologies in which switch devices are connected in series. One of the advantages of this series connection is that the supply voltage is distributed over the switch devices, reducing its stress. Although the topologies are creating a multilevel waveform at the output side, at the dc input-voltage side, however, these series-type converter topologies are still creating a two-level rectangular-wave input current producing harmonic distortion which is undesired in applications where the converter is directly connected to a dc grid. Interleaved topologies, on the other hand, are characterized by switches connected in parallel, resulting in the property that current is equally distributed over the switch devices. Additionally, the interleaved topology has the advantage of creating multilevel output voltage as well as creating multilevel input current. Therefore, a circuit which has this property of the interleaved converter and is characterized by switch devices in series will result in a desirable topology and may be useful in many applications. This paper performs a theoretical analysis of the duality principle and applies it to the interleaved topology, resulting in the interleaved-dual topology, a series topology with the same property as the interleaved topology. Simulation results are presented to show the dual multilevel behavior.**

I. INTRODUCTION

The principle of duality has been used many times in the past in lumped-parameter modeling to provide electrical engineers more insight into electronic circuits, and it is, therefore, a tool in basic circuit analysis [1]. This principle can also be applied to power-electronics networks. One can imagine finding a relationship between different types of converters [2] and using the technique to extend, for instance, modulation or control strategies to other converter topologies [3], [4]. As an example, in 1979 it was shown how four basic switching DC-DC converters - the boost, the buck, the buck-boost and the Ćuk converter - can be related by using the duality principle [5]. Coincidentally, the concept of the canonical switching cell [6] was introduced with the same intention, that is, to show the relationship between different switching topologies.

These examples show that there is a need for generalization of topologies in power electronics theory, especially as the hybrid multilevel structures become more complex. Two approaches for the generalization process in the literature can be identified: First, the introduction of basic switching structures [7] or generalized topologies [8], and secondly, the search for dual and symmetric relations between topologies [9], [10]. It is assumed that one important constraint to find the dual counterpart of a certain topology is planarity. In other words, it is not possible to find a dual of a topology which cannot

be drawn as a planar graph (i.e. cannot be drawn such that it has no crossing branches). The buck converter with more than three cells is an example of such a non-planar topology.

However, it is shown below that this assumption is incorrect, and that the duality principle can be applied to non-planar circuits, such as the interleaved buck converter with $N \geq 3$ mentioned above and multi-phase voltage source converters [11], [3]. To be able to find a non-planar dual transformation the concept of electrical duality as opposed to topological duality has to be introduced. A crucial step in obtaining electrical duality is the transformation of the original circuit to a functionally equivalent planar realization, i.e. which has the has the same properties imposed by Kirchhoff's mesh and junction laws as the original circuit.

Two methods can be used for creating an equivalent planar circuit. The first method is to add ideal transformers, segregating the non-planar circuit into two planar circuits and thereby enforce constraints identical to Kirchhoff's laws to the new equivalent circuit [12], [11]. The second method is to evaluate each switching state of the original converter, assuming each state results in a planar circuit [3]. After obtaining a functionally equivalent planar circuit by either method, the classical method of graph transformation and interchange of elements is applied to find the topological dual of the planar equivalent circuit(s). In this paper the first method is applied to a generalized interleaved topology, which is shown in Fig. 1.

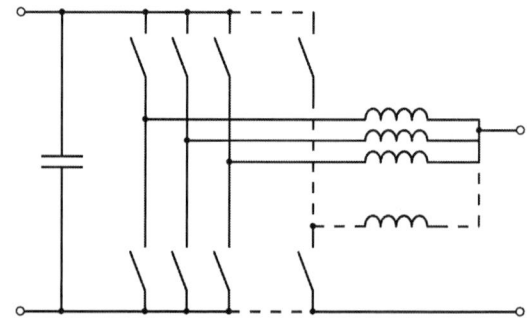

Fig. 1. Basic N-cell interleaved converter structure

This topology is basically a parallel connection of buck-stages, or canonical switching cells. The cell inductors are star-connected, where the star point acts as the output of the entire topology. Therefore, the interleaved buck could be categorized as a parallel converter. This topology has some interesting

978-1-4799-2706-7/14 $31.00 © 2014 IEEE

features, such as the unique characteristic that it is creating multilevel signals on both the input current as well as on the output voltage of the converter in contrast to the classical series multilevel converters, i.e. neutral-point clamped, flying-capacitor or cascaded topologies. In this paper the dual transformation is applied to this interleaved topology and the resulting dual behavior is verified by means of simulations.

II. METHODOLOGY

A power-electronics system may be represented as an interconnection of various one- and two-port elements. Fig. 2 shows an example of such a block-diagram representation of a conventional power-electronics topology. Consider the situation in which the input is a voltage source, and the output is represented by a current source (simplified representation of an inductive load). When the filters are neglected for the sake of argument, then, the hard-switched two-level converter would produce a rectangular-wave current signal as well as a rectangular-wave voltage signal. These signals result in high distortion on the input as well as on the output of the converter. This is shown in Fig. 2(a). As a result, as distortion levels as shown in Fig. 2(a) are unacceptable in almost any application, an input filter and output filter are required to reduce the higher-order harmonic components and to ensure that the system will satisfy the requirements on electromagnetic interference (EMI) at the input and maximum allowed distortion on the output waveform.

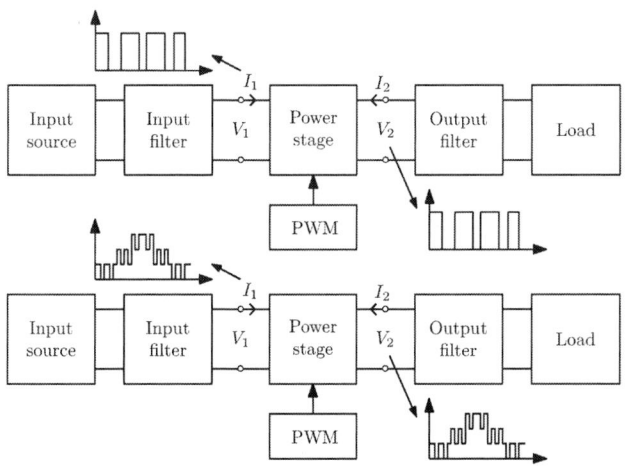

Fig. 2. Conventional power amplifier block diagram

To enhance the converter topology such that the waveforms at both input and output are less distorted, a multilevel approach could be used. This is shown in Fig. 2(b). The multilevel circuit produces waveforms with reduced and more high-frequent content. Subsequently, less effort has to be spent in the filter design. Most of the multilevel topologies that have been proposed in the last decades produce multilevel output waveforms but do not produce multilevel waveforms at the input side. This is a result of the series connection of switches. On the other hand, the interleaved converter, as shown in Fig. 1, has the exact characteristic as shown in Fig. 2(b). This

topology has, therefore, the advantage of creating a multilevel output-voltage as well as creating a multilevel input-current. In contrast to conventional multilevel topologies, interleaved topologies are characterized by switches connected in parallel, resulting in the property that current is equally distributed over the switching cells, instead of the property in which voltage is equally distributed over the switch devices. A topology which has the property of a series connection of switches and the advantage of the interleaved converter will result in a desirable topology which may be useful in various applications.

Finding this topology could be accomplished by searching for the dual topology as shown in Fig. 3. In this figure, the dual of a converter is shown and the converter is treated as a single element. One drawback in relating power amplifiers by means of duality are the restriction in handling the input and output sources and loads. If, for example, we choose our power amplifier to be in a configuration in which a voltage source is connected at the input and a current source at the output, then, consequently, a dual converter topology will be found in which a current source is connected at the input and a voltage source at the output. This is a result which is based on the fact that a dual transformation of a cascaded connection of two-port elements results in a cascaded connection of the individual two-port elements [13]. For the dual transformation, here, the standard convention is used as defined in [14], which means that the current direction for all elements (also the voltage source) is always defined from the positive to the negative terminal. Care should be taken with the current/voltage directions. As can be seen, in the original circuit the elements have been placed such that the positive voltage is defined at the top nodes of the two-port elements. However, in the dual circuit this is not the case anymore and the elements are oriented such that the current is circulating through the cascaded connection. The dual topology can now be inverted, such that the topology is again a voltage-supplied current source configuration. However, it would be more convenient if the sources are uncoupled from the topology itself, because the primary interest is the converter dual. Afterwards, it is determined how sources should be connected, and it is defined what the input and output of the topology will be.

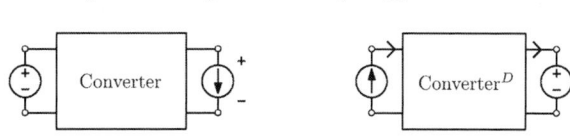

Fig. 3. Converter shown as a two port element, and the converter dual

III. THE DUALITY PRINCIPLE

A dual transformation of an electric circuit results in a circuit which has equivalent behavior, but currents and voltages are exchanged. The duality principle is not a concept which is limited to circuit theory. In many other fields such as mathematics and physics, the concept of duality is also used. However, there is no proper definition of duality and there does not exist a mathematical description which generalizes

978-1-4799-2706-7/14 $31.00 © 2014 IEEE

all types of dualities in all fields. The fact that the exact same transformation applied twice to something leads to exactly the original again could be seen as the main characteristic of duality between two objects, equations, etc. Originals and their dual objects have a unique symmetrical relationship.

Graph theory is an example of a field within mathematics in which duality plays a major role, and it concerns all fundamental concepts related to graphs. Graphs are used to represent networks in various scientific fields; i.e. computer science, communication theory, electrical engineering etc. An electric circuit, built up from lumped elements, is an example of a network and can, therefore, be drawn as a graph. The duality principle is one of the tools in graph theory and it can also be applied to electric circuits. The dual circuit can be constructed by, first, drawing the electrical circuit as a planar graph, denoted as $\mathcal{G}$. After that, a transformation needs to be applied to the graph to find the dual graph $\mathcal{G}^{\mathcal{D}}$. By interchanging all the elements by their counterparts in the new graph, the dual circuit is found. This dual circuit has the property that the voltage over all the elements in $\mathcal{G}$ is transformed into a current through the dual counterpart elements in $\mathcal{G}^{\mathcal{D}}$. Throughout this paper the superscript letter $\mathcal{D}$ will be used to assign all terms which pertain to the dual graph. If the dual transformation is applied again to the dual circuit, then the original circuit will be found again.

The duality principle in graph theory is restricted only to planar graphs. That is because graph theory only concerns the geometrical outline of a graph and not the laws which are imposed on the network. These laws describe the behavior of the circuit, and, when taken into consideration, these laws will give extra freedom. As already introduced in [12], the concept of topological duality as opposed to electrical duality needs to be defined. Topological duality refers to conventional duality as defined within graph theory, and this type of duality can only be applied to planar networks. Topological duality can be seen as exact duality between two circuits, because not only are currents and voltages dual, also the layout itself is dual. Electrical duality applies to non-planar electric circuits, as certain tools from electric circuit theory can be used to convert non-planar electric circuits into a subset of planar circuits. In other words, electrical duality allows the conversion of a circuit in an equivalent circuit.

A. The duality principle applied to planar circuits

A dual transformation of an electrical circuit is explained in many textbooks about circuit theory [1] or graph theory, and a short summary and an example are given in this section. First, the concept of topological duality is defined:

Definition 1. *Two electric circuit networks $\mathcal{G}$ and $\mathcal{G}^{\mathcal{D}}$ are each other's topological dual if*

(a) There is a one-to-one correspondence between all elements of $\mathcal{G}$ and their counterparts in $\mathcal{G}^{\mathcal{D}}$ in such a way that voltages across elements in one circuit are represented as currents through their counterpart element in the other circuit.

(b) There is a one-to-one correspondence between the meshes of $\mathcal{G}$ and the nodes of $\mathcal{G}^{\mathcal{D}}$.

The transformation consists of two steps. The first step is a so-called graph transformation. This is done by drawing the circuit as a simplified interconnection of nodes with branches, called a graph. Every branch represents an element and the element itself is not drawn. This is slightly different than the conventional way of drawing a circuit, where also nodes are sometimes drawn as an interconnection of nodes and branches. The dual graph can be found by putting nodes in every mesh, including the exterior mesh, of the original graph $\mathcal{G}$ and interconnecting the nodes such that every branch is crossed once, which is shown in Fig. 4(b).

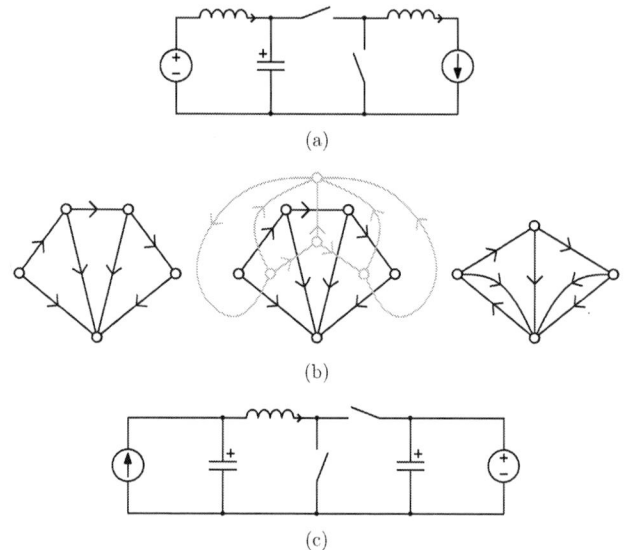

Fig. 4. Example of a dual transformation applied to a power electronics circuit: a) The example circuit b) The graph transformation c) The dual circuit

Note that because often circuits have most of their elements connected to ground, a good starting point is to put the ground as the lowest node. The node which is placed in the exterior mesh will be the ground connection in the dual graph $\mathcal{G}^{\mathcal{D}}$. The second step involves the replacement of the original elements by their dual counterparts [14]. Further note that arrows have been placed in the graphs. These arrows indicate the current direction. When performing the graph transformation, the arrows should be placed counter-clockwise with respect to the original arrows. This leads to a correct current direction in the dual graph.

B. The duality principle applied to non-planar circuitss

The application of the duality principle in graph theory only applies to planar circuits. When applied to non-planar circuits, problems occur because the graph transformation will not work properly. To verify whether a particular graph is planar or not, two specific non-planar graphs are useful to recognize. These two non-planar graphs are known as Kuratowski's graphs and are shown in Fig. 5. K_5 is the non-planar graph with the smallest number of vertices, and $K_{3,3}$ is the non-planar graph with the smallest number of edges. The theorem of Kuratowski states that $\mathcal{G}$ is planar iff $\mathcal{G}$ contains no sub-graph $K_{3,3}$ or K_5.

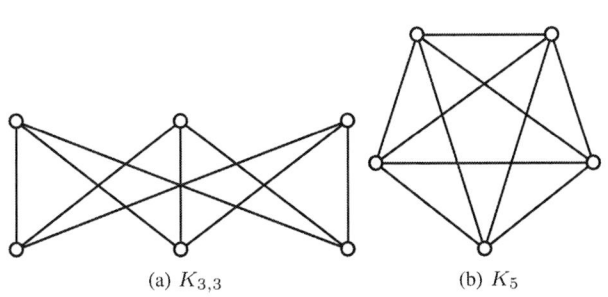

(a) $K_{3,3}$ (b) K_5

Fig. 5. Kuratowski's graphs: $\mathcal{G}$ is planar iff $\mathcal{G}$ contains no sub-graph $K_{3,3}$ or K_5.

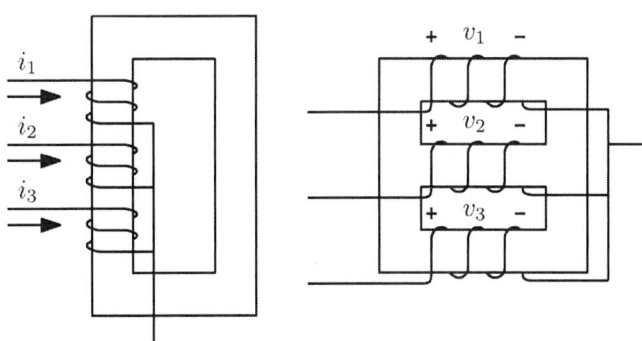

Fig. 6. Example of (a) Junction transformer (b) Mesh transformer

Methods in which duality is applied to non-planar circuits rely on the idea that the non-planar circuit first needs to be converted into a planar circuit. One can think of, for instance, the addition of elements and making certain branches unnecessary, or alternatively, by only considering the switching states. The resulting dual circuit will not be a topological dual circuit, because the graphs are not dual. However, the behavior of the circuits is still dual. Electrical duality is, therefore, defined as:

Definition 2. *Two electric circuit networks $\mathcal{G}$ and $\mathcal{G}^{\mathcal{D}}$ are each other's electrical dual if the electric circuit $\mathcal{G}$ is equivalent to an equivalent circuit $\mathcal{G}^{eq}$ and $\mathcal{G}^{eq}$ is topologically dual with $\mathcal{G}^{\mathcal{D}}$. By equivalence, here, is meant that $\mathcal{G}$ and $\mathcal{G}^{eq}$ have identical behavior.*

Basically, this definition means that the duality principle is not applicable for all the elements in the entire circuit. Only certain voltages or currents, which are of interest, are considered. For a power electronics circuit this means that we are, for instance, only interested in a converter topology which has dual input and output behavior. In other words, the dual of the topology as a whole - defined as topological duality - is for these circumstances not of our interest.

C. Method of addition of transformers

One method of converting a non-planar electrical circuit into a planar circuit is the addition of transformers. Two types of transformers can be used, and depending on the non-planar circuit's layout a certain choice is more practical. The first type of ideal transformer is a junction transformer and it can replace a junction in a circuit. By putting one winding in each of the branches leading towards the node of interest, the junction transformer is actually enforcing Kirchhoff's current law (KCL) in that particular node. In other words, the junction transformer enforces

$$\sum i_k = 0, \qquad (1)$$

where i_k is the kth current flowing in the node and k is the number of branches connected to the node.

The second type of transformer is the mesh transformer, and it is, obviously, the dual of the junction transformer. It enforces the Kirchhoff's voltage law on a certain mesh in the network, hence

$$\sum v_k = 0, \qquad (2)$$

where v_k is the kth voltage in a mesh and k is the number of branches enclosing the entire mesh.

IV. RESULTS

The most basic interleaved structure is shown in Fig. 1. When N is defined as the number of cells which are connected in parallel, then $N = 1$ results in the canonical switching cell, which is self-dual. A dual transformation will, in this case, result in the exact same circuit again. The case $N = 2$ results in a planar graph, and a dual transformation can easily be applied. $N = 3$ results in Kuratowski's graph $K_{3,3}$, and, by definition, it is non-planar. In general, every interleaved converter with $N > 3$ has $K_{3,3}$ as a sub-graph and is non-planar too.

The dual of these non-planar interleaved converters can be found by adding a junction transformer in the star-connected output node, as discussed in section III-C. After that, the node is connected to ground. By adding this ideal transformer the total circuit's behavior will not change and the resulting circuit is, therefore, an equivalent circuit of the interleaved converter. This equivalent circuit is shown in Fig. 7a, and this equivalent circuit is actually a combination of two planar circuits. As a result, these planar circuits can be transformed by using the graph transformation. The electrical dual circuit is shown in Fig. 7b.

A. Simulation results

Both the interleaved topology of Fig. 1 and the dual topology of Fig. 7b have been simulated in Matlab Simulink [15] by using the tool PLECS [16]. This is done by connecting a voltage source to the input of the topology and a current source to the output as shown in Fig. 3. The voltage source is set to 100 V and the current source to 10 A. In both topologies phase-shifted carrier pulse-width modulation has been used. The reference is a sine wave of 50 Hz, and the switching frequency is 9 times higher than the reference frequency. A high cell-inductor inductance has been chosen to show near ripple-free voltage and current levels and to show the multilevel behavior. The result of this simulation is shown in Fig. 8a. It can be seen

978-1-4799-2706-7/14 $31.00 © 2014 IEEE

The 2014 International Power Electronics Conference

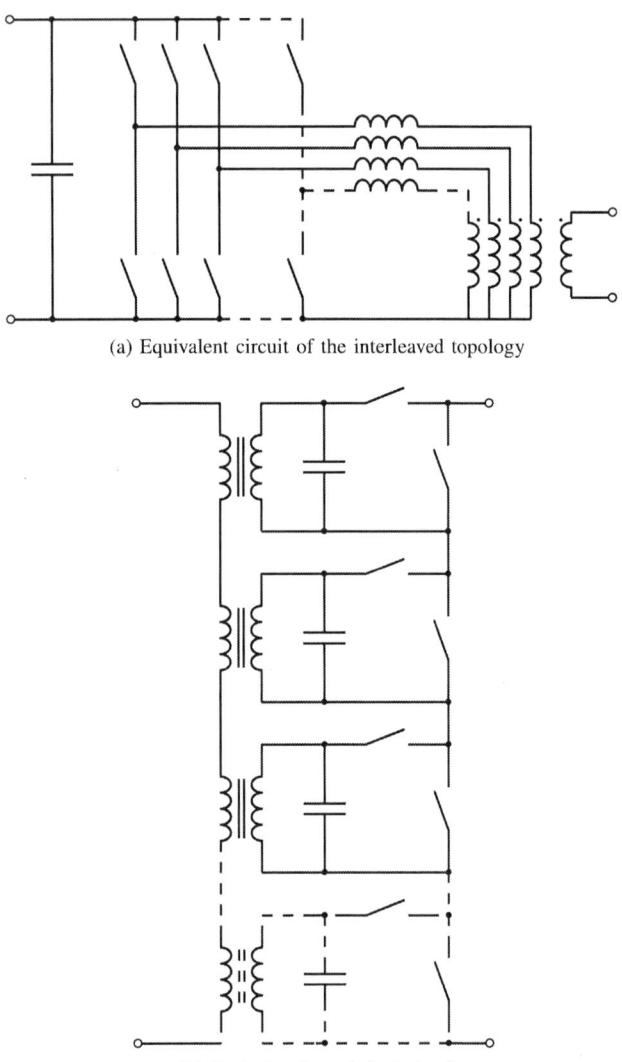

(a) Equivalent circuit of the interleaved topology

(b) Basic interleaved dual circuit

Fig. 7.   Dual transformation of interleaved topology

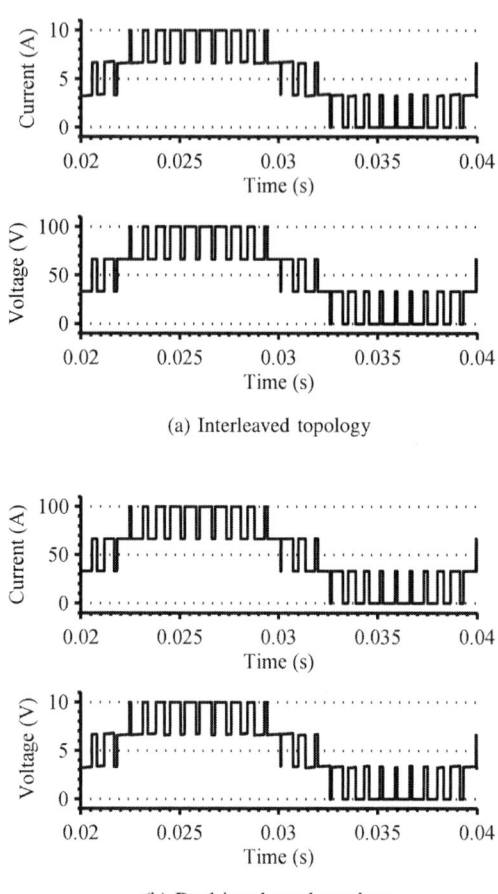

(a) Interleaved topology

(b) Dual interleaved topology

Fig. 8.   Simulation of dual behavior between the two topologies

that the converter is creating both multiple input and output levels.

Similarly, the dual topology has been simulated and the results are shown in Fig. 8b. Clearly, this simulation shows that the waveforms are identical to the waveforms of the simulation of the interleaved and, hence, this simulation shows that the topologies are each others electrical dual. Also the waveform of the currents through the inductors in the interleaved topology are identical to the waveform of the voltages across the capacitors in the interleaved-dual topology.

Unfortunately, the dual topology in Fig. 7 has no direct practical use. The transformers which have been added are assumed to be ideal, whereas real transformers are not able of transferring both dc currents and voltages. It is not possible to remove the transformers in the dual circuit. Evaluation of both circuits reveals that the star-connection of the switching cells in interleaved converters results in a common voltage for all the switching cells at the output. This property has to be transformed in the dual circuit into a common current for the series cells and this can only be obtained when the switching cells are isolated from each other. The series connection of the ideal transformers ensures that all the switching cells will have the exact same current.

As a solution, the transformers could be replaced by isolated DC-DC converters. This is shown in Fig. 9. As it was allowed to add transformers, the addition of such ideal converters does not influence the multilevel input/output behavior. Furthermore, up till now half-bridge interleaved converters have been used as example. However, if the topology of Fig. 1 would have been drawn with full-bridge buck stages, then, the resulting dual circuit would also have a full-bridge configuration. The resulting circuit is one of the basic multilevel converters, the cascaded H-bridge multilevel converter with isolated dc-sources. However, in these type of multilevel converters usually the DC-DC converters are connected in parallel at the input side. The input-series connection of the isolated DC-DC converters (dual-active bridge) is also employed in practice, although the output is then connected directly either in series [17] or in parallel [18]. Here, each output consists additionally of a full-bridge inverter.

978-1-4799-2706-7/14 $31.00 © 2014 IEEE   2958

The 2014 International Power Electronics Conference

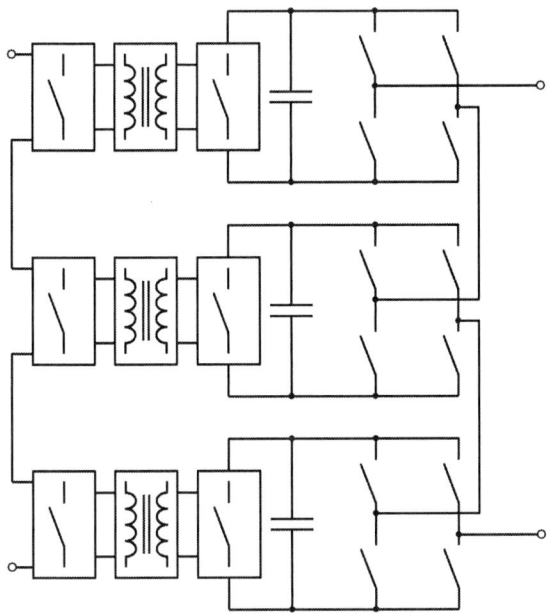

Fig. 9.   Transformers replaced by isolated DC-DC converters

[10] Z. Bai and Z. Zhang, "Conformation of Multilevel Current Source Converter Topologies Using the Duality Principle," *Power Electronics, IEEE Transactions on*, vol. 23, no. 5, pp. 2260 –2267, Sept. 2008.

[11] P. Wolfs, G. Ledwich, and K. Kwong, "The application of the duality principle to nonplanar circuits," *Power Electronics, IEEE Transactions on*, vol. 8, no. 2, pp. 104 –111, Apr 1993.

[12] A. Bloch, "On methods for the construction of networks dual to nonplanar networks," *Proc. Phys. Soc.*, vol. 58, pp. 677–694, 1946.

[13] D. Shmilovitz, "Application of duality for derivation of current converter topologies," *HAIT Journal of Science and Engineering B*, vol. 2, Issues 3-4, pp. pp. 529–557, 2005.

[14] S. Freeland, "Techniques for the practical application of duality to power circuits," *Power Electronics, IEEE Transactions on*, vol. 7, no. 2, pp. 374 –384, Apr 1992.

[15] MATLAB, *8.1.0.604 (R2013a)*. Natick, Massachusetts: The MathWorks Inc., 2013.

[16] PLECS, *version 3.5.3*. Zurich, Switzerland: Plexim GmbH, 2013.

[17] R. Giri, R. Ayyanar, and E. Ledezma, "Input-series and output-series connected modular DC-DC converters with active input voltage and output voltage sharing," in *Applied Power Electronics Conference and Exposition, 2004. APEC '04. Nineteenth Annual IEEE*, vol. 3, 2004, pp. 1751–1756 Vol.3.

[18] D. Dujic, C. Zhao, A. Mester, J. Steinke, M. Weiss, S. Lewdeni-Schmid, T. Chaudhuri, and P. Stefanutti, "Power electronic traction transformer-low voltage prototype," *Power Electronics, IEEE Transactions on*, vol. 28, no. 12, pp. 5522–5534, Dec 2013.

V. DISCUSSION

In this paper, the duality principle has been applied to interleaved converter topologies, and the resulting electrical dual circuit has been deduced. This resulting interleaved dual topology is shown in Fig. 7b. Simulations have been presented to show that the interleaved dual topology has, indeed, dual input/output behavior. As an aside, this paper also discusses the definition of the duality principle itself, and shows how other tools from circuit theory can be used to expand the duality principle.

REFERENCES

[1] C. A. Desoer and E. S. Kuh, *Basic Circuit Theory*. McGraw-Hill, 1969.

[2] A. Bhat and F. Tan, "A unified approach to characterization of PWM and quasi-PWM switching converters: topological constraints, classification, and synthesis," *Power Electronics, IEEE Transactions on*, vol. 6, no. 4, pp. 719 –726, Oct 1991.

[3] J. Kolar, H. Ertl, and F. Zach, "Quasi-dual modulation of three-phase PWM converters," *Industry Applications, IEEE Transactions on*, vol. 29, no. 2, pp. 313 –319, Mar/Apr 1993.

[4] B. McGrath and D. Holmes, "Natural Current Balancing of Multicell Current Source Converters," *Power Electronics, IEEE Transactions on*, vol. 23, no. 3, pp. 1239 –1246, May 2008.

[5] S. Cuk, "General Topological Properties of Switching Structures," in *IEEE Power Electronics Specialists Conference*, 1979, pp. 109–130.

[6] E. Landsman, "A unifying derivation of switching DC-DC converter topologies," in *IEEE Power Electronics Specialists Conference*, 1979, pp. 239–243.

[7] L. Tolbert, F. Z. Peng, F. Khan, and S. Li, "Switching cells and their implications for power electronic circuits," in *Power Electronics and Motion Control Conference, 2009. IPEMC '09. IEEE 6th International*, May 2009, pp. 773 –779.

[8] F. Z. Peng, "A generalized multilevel inverter topology with self voltage balancing," *Industry Applications, IEEE Transactions on*, vol. 37, no. 2, pp. 611 –618, Mar/Apr 2001.

[9] T. Meynard, B. Cougo, F. Forest, and E. Laboure and, "Parallel multicell converters for high current: Design of intercell transformers," in *Industrial Technology (ICIT), 2010 IEEE International Conference on*, Mar 2010, pp. 1359 –1364.

978-1-4799-2706-7/14 $31.00 © 2014 IEEE   2959

A New Impedance Measurement Method Based on High Frequency Compensation

Xiaolong Yue, Fang Zhuo, Hao Yi*
School of Electrical Engineering
Xian Jiaotong University
Xian, China
Email: xiaolong_yue@163.com

Abstract—Impedance is very important in Micro-grid and other distributed power system based on power electronics because of the close relationship between impedance and stability. This paper presents a new current injection method for impedance measurement. Sinusoidal amplitude modulation is a method to shift signals to different frequency bands. A signal with large frequency bandwidth and small amplitude attenuation can be obtained by superposing square pulse signals after sinusoidal amplitude modulation. When the superposed signal is injected to the system as a current source, impedance information in the frequency range of interest can be calculated with short time cost and high accuracy. The simulation results demonstrate the effectiveness of this method.

Index Terms—impedance measurement,high Frequency Compensation, perturbation current injection, sinusoidal amplitude modulation, superposition

I. INTRODUCTION

Impedance is important for power electronic based systems because of the close relationship between stability and impedance [1-3]. For power system, impedance is the basis for many other parameters calculation [4]. Because it is difficult to obtain the impedance by modeling, measurement becomes an effective way.

Much work has been done on the impedance measurement methods and two of them are the most widely used. The first one is injecting a narrow current spike into the network by active power filter [5] or capacitor switching induced transients [6]. It is fast because one time injection is enough, but cannot measure the impedance in high frequency accurately because of the amplitude attenuation. The second is injecting a series of small sinusoidal signals over the frequency range of interest by network analysis, chopper circuit, etc. [7-10]. It can measure high frequency accurately but take a lot of test time for so many times injection. This paper proposes a method based on high frequency compensation with modulated periodic square pulse, combining advantages of the two methods mentioned above. The proposed method can measure impedance with shorter test time cost and higher accuracy in both low and high frequency.

---

*Corresponding author. This work was supported by 2011 Natural Science Foundation of China (No.51177130).

II. PERIODIC SQUARE PULSE AND ITS SPECTRUM

The pulse signal has large frequency bandwidth. When the duty ratio of a periodic square wave towards to zero, the signal will approximate to a pulse. The expression is written in (1), and the waveform is shown in Fig. 1.

$$x(t) = \begin{cases} 1, & |t| < T_1 \\ 0, & T_1 < |t| < T/2 \end{cases} \quad (1)$$

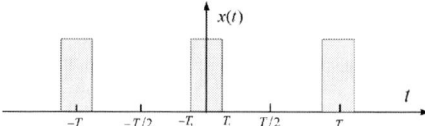

Fig. 1.   Waveform of the periodic square wave.

The Fourier series is:

$$a_0 = \frac{1}{T} \int_{-T_1}^{T_1} dt = \frac{2T_1}{T} = d \quad (2)$$

$$a_k = \frac{\sin(k\omega_0 T_1)}{k\pi}, k \neq 0 \quad (3)$$

where $\omega_0 = 2\pi/T$ is the fundamental frequency, d is the duty ratio. In practical, the signal we usually used has the follow form, which is shown in Fig2.

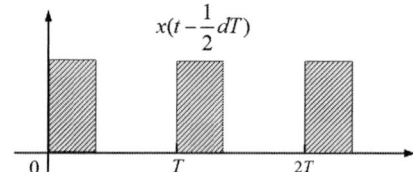

Fig. 2.   Periodic square wave after time shifting.

This periodic square pulse can be rewritten as:

$$x_0(t) = x(t - \frac{1}{2}dT) \quad (4)$$

According to (2), (3), (4), the Fourier series of the signal defined in (4) has the form

$$\tilde{a}_0 = \frac{1}{T} \int_0^{dT} dt = \frac{dT}{T} = d \quad (5)$$

978-1-4799-2706-7/14 $31.00 © 2014 IEEE

$$\tilde{a}_k = \frac{\sin(k\omega_0 T_1)}{k\pi} e^{jk\omega_0 T_1} = \frac{\sin(k\pi d)}{k\pi} e^{jk\pi d}, k \neq 0 \quad (6)$$

When $\tilde{a}_k$ is defined as:

$$\tilde{a}_k = A_k e^{j\theta_k} \quad (7)$$

Eq. (4) can be represented as:

$$x_0(t) = d + 2 \sum_{k=1}^{\infty} A_k \cos(k\omega_0 t + \theta_k) \quad (8)$$

From (6) and (7), we know that:

$$A_k = \left| \frac{\sin(k\pi d)}{k\pi} \right| \leq \frac{1}{k\pi} \quad (9)$$

When $d = 0.5$, A_k will get the maximum value, and the signal has the highest signal-to-noise ratio. Therefore, we will choose $d = 0.5$ in the following discussion. Under this condition, $x_0(t)$ will take the form

$$x_0(t) = \frac{1}{2} + \frac{2}{\pi} \sum_{k=0}^{\infty} \frac{1}{2k+1} \sin(2k+1)\omega_0 t \quad (10)$$

III. MODULATION AND SUPERPOSITION OF PERIODIC SQUARE PULSE

A. Sinusoidal amplitude modulation for Signal Square Pulse

Sinusoidal amplitude modulation can shift signal spectrum to a different frequency band. In frequency domain, This process can be described in Fig.3.

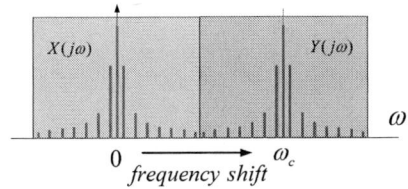

Fig. 3. The modulation process in frequency domain.

In time domain, the process can be expressed as follows:

$$y(t) = x(t) \cdot c(t) \quad (11)$$

where $x(t)$ is the modulating signal, $c(t)$ is the carrier signal and has the form

$$c(t) = \sin(\omega_c t + \theta_c) \quad (12)$$

When $x(t) = x_0(t)$, $c(t) = \sin(\omega_c t + \theta_c)$, (11) can be rewritten as

$$
\begin{aligned}
y(t) =& d\cos(\omega_c t + \theta_c - \frac{\pi}{2}) + \\
& \frac{1}{\pi} \cdot \sum_{k=0}^{\infty} \frac{1}{2k+1} \cos\left\{ [\omega_c + (2k+1)\omega_0] t + (\theta_c - \pi) \right\} + \\
& \frac{1}{\pi} \cdot \sum_{k=0}^{\infty} \frac{1}{2k+1} \cos\left\{ [\omega_c - (2k+1)\omega_0] t + \theta_c \right\}
\end{aligned}
$$
$$(13)$$

When $\omega_c < (2k+1)\omega_0$, the frequency will be minus, as

$$
\begin{aligned}
\cos\left\{ [\omega_c - (2k+1)\omega_0] t + \theta_c \right\} = \\
\cos\left\{ [(2k+1)\omega_0 - \omega_c] t - \theta_c \right\} \quad (14)
\end{aligned}
$$

the frequency point at $(2k+1)\omega_0 - \omega_c$ is actually the same with that at $\omega_c - (2k'+1)\omega_0$, where $\omega_c > (2k'+1)\omega_0$, k and k' are determined by the following equation:

$$(2k+1)\omega_0 - \omega_c = \omega_c - (2k'+1)\omega_0 \quad (15)$$

To make the sum of these two parts be the maximum, they should have the same phase angle, from (14), $\theta_c = -\theta_c \pm 2k\pi (k = 1, 2, \cdots)$. Then we have

$$\theta_c = \pm k\pi (k = 0, 1, 2, \cdots) \quad (16)$$

To avoid truncation effect by FFT, the phase angle should be the same with that shown in (16). From (13), the magnitude and phase characteristic has the form

$$
|Y_1(\omega)| = \begin{cases}
\frac{1}{\pi} \left(\dfrac{\omega_0}{\omega_c - \omega} + \dfrac{\omega_0}{\omega_c + \omega} \right) \\
\quad , 0 < \omega < \omega_c \ \& \ \omega = \omega_c - (2k+1)\omega_0 \\
d \ , \omega = \omega_c \\
\frac{1}{\pi} \left(\dfrac{\omega_0}{\omega - \omega_c} - \dfrac{\omega_0}{\omega_c + \omega} \right) \\
\quad , \omega > \omega_c \ \& \ \omega = \omega_c + (2k+1)\omega_0 \\
0, the\ other\ \omega
\end{cases}
\quad (17)
$$

$$
\angle Y_1(\omega) = \begin{cases}
\theta_c, 0 < \omega < \omega_c \ \& \ \omega = \omega_c - (2k+1)\omega_0 \\
\theta_c - \dfrac{\pi}{2}, \omega = \omega_c \\
\theta_c - \pi, \omega > \omega_c \ \& \ \omega = \omega_c + (2k+1)\omega_0 \\
0, the\ other\ \omega
\end{cases}
\quad (18)
$$

When $T = 0.1s$, $\omega_c = 1600\pi rad/s$, the spectrum of $y(t)$ is shown in Fig.4.

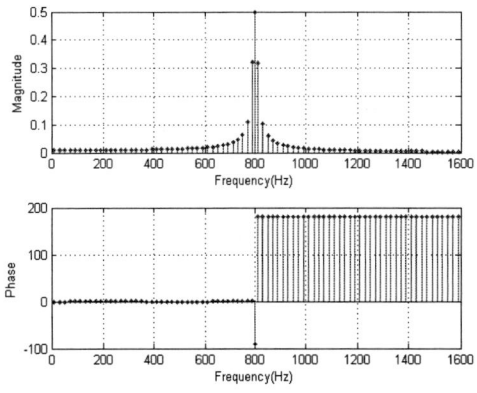

Fig. 4. The spectrum of $y(t)$ when $T = 0.1s$, $\omega_c = 1600\pi rad/s$.

It can be seen from the figure that when the frequency is lower than 800Hz, the phase angle is zero and the magnitude is almost the same but a little higher than those frequency beyond 800Hz, whose phase angle is -180 degree. The magnitude of 800Hz is the highest and the phase angle is -90 degree. All these are coincide with (17) and (18).

B. Superposition of two modulated square pulse

From what has been discussed above, we can obtain a signal with large frequency bandwidth and small amplitude attenuation by superposing sinusoidal amplitude modulated signals with different frequencies. Considering two different carrier signals $c_1(t) = \sin(\omega_1 t + \theta_1)$ and $c_2(t) = \sin(\omega_2 t + \theta_2)$, where $\omega_1 < \omega_2$, then

$$y_1(t) = \begin{cases} \sin(\omega_1 t + \theta_1), \ 0 \leq t \leq 0.5T \\ 0, \ 0.5T < t < T \end{cases} \tag{19}$$

$$y_2(t) = \begin{cases} \sin(\omega_2 t + \theta_2), \ 0 \leq t \leq 0.5T \\ 0, \ 0.5T < t < T \end{cases} \tag{20}$$

Suppose that $Y_1(\omega)$ and $Y_2(\omega)$ are the Fourier transform of signal $y_1(t)$ and $y_2(t)$. If these two signals are superposed directly with $\theta_1 = \theta_2 = 0$, then we have

$$|Y_1(\omega) + Y_2(\omega)| = \begin{cases} |Y_1(\omega)| + |Y_2(\omega)|, \omega < \omega_1 \\ ||Y_1(\omega)| - |Y_2(\omega)||, \omega_1 < \omega < \omega_2 \\ |Y_1(\omega)| + |Y_2(\omega)|, \omega > \omega_2 \end{cases} \tag{21}$$

The magnitude of $|Y_1(\omega) + Y_2(\omega)|$ in the frequency range around $\omega = (\omega_1 + \omega_2)/2$ will approach to zero, which will be adverse to impedance measurement because of the low signal-to-noise ratio.

In fact, the magnitudes of both $Y_1(\omega)$ and $Y_2(\omega)$ in the frequency range $\omega_1 < \omega < \omega_2$ are higher than other ranges. Actually, on the frequency band $\omega_1 < \omega < \omega_2$, $||Y_1(\omega)| + |Y_2(\omega)||$ is better than $||Y_1(\omega)| - |Y_2(\omega)||$ for the measurement, according to (18), if we choose $\theta_2 = 2k\pi + (\theta_1 - \pi)$ $(k = 0, 1, 2, \cdots)$, then

$$\theta_2 = \pm(2k - 1)\pi + \theta_1 \ (k = 0, 1, 2, \cdots) \tag{22}$$

With the condition in (22), we have

$$|Y_1(\omega) + Y_2(\omega)| = \begin{cases} |Y_1(\omega)| - |Y_2(\omega)|, \omega < \omega_1 \\ ||Y_1(\omega)| + |Y_2(\omega)||, \omega_1 < \omega < \omega_2 \\ |Y_2(\omega)| - |Y_1(\omega)|, \omega > \omega_2 \end{cases} \tag{23}$$

When $\theta_1 = 0$, $\theta_2 = \pi$, $\omega_1 = 1600\pi rad/s$, $\omega_2 = 3200\pi rad/s$ and $T = 0.1s$, the spectrum of (19) and (20) are shown in Fig.4 and Fig.5. When superposing the two modulated signals, the spectrum is shown in Fig.6. It can be seen from the figure that the phase angle is equal between 800Hz and 1600Hz.

C. Superposition of multiple Modulated Square pulse

Part B talks about the superposition of two modulated signals with different frequency. When the frequency bandwidth needed to measure is large, more modulated signals should

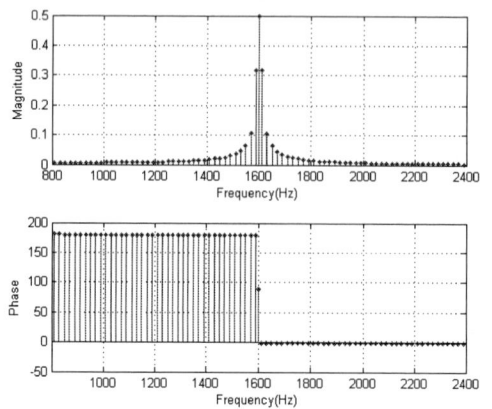

Fig. 5. The spectrum of $y(t)$ when $T = 0.1s$, $\omega_c = 3200\pi rad/s$.

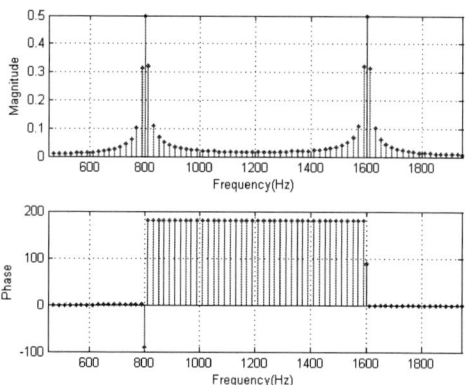

Fig. 6. The superposition of two modulated signal in Fig.4 and Fig.5.

be superposed to obtain enough bandwidth. Suppose that the carrier signals are:

$$c_i(t) = \sin(\omega_i t + \theta_i) \ (i = 1, 2, \cdots, n) \tag{24}$$

where $\omega_{i+1} = \omega_i + k\omega_0$, k is an integer, $\theta_i = (i - 1)\pi$.

If they are superposed directly, then in time domain

$$y(t) = \sum_{i=1}^{n} y_i(t) = \sum_{i=1}^{n} x_0(t) \cdot c_i(t), 0 \leq t \leq T \tag{25}$$

However, the signal shown in (25) has negative effect on the system, when $n = 5$, $\omega_i = 200(2i - 1)\pi rad/s, i = 1, 2 \cdots, 5$, the waveform of in time domain is shown in Fig.7.

It can be seen from the above figure that at some time points, the amplitude is very high, which may cause strong impact in the system. To solve this problem, the different modulated signals are separated when they are injected and sampled, but superposed together when doing calculations. This process can be shown in Fig.8.

978-1-4799-2706-7/14 $31.00 © 2014 IEEE

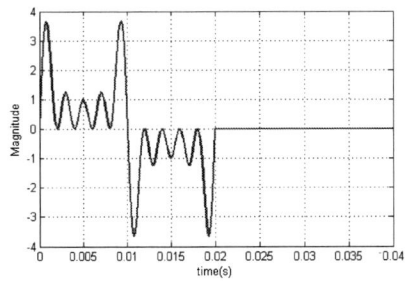

Fig. 7. Superposed signal in time domain

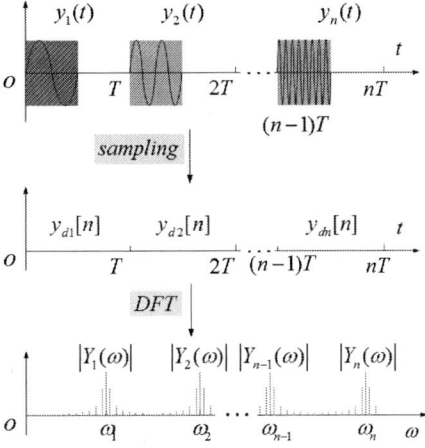

Fig. 8. Injection and calculation process

Then the magnitude in frequency domain will be

$$|Y(\omega)| = \begin{cases} \displaystyle\sum_{k=1}^{n} (-1)^{k+1} |Y_k(\omega)|, 0 < \omega < \omega_1 \\ \displaystyle\sum_{k=1}^{i} (-1)^{i-k} |Y_k(\omega)| + \sum_{k=i+1}^{n} (-1)^{k-i-1} |Y_k(\omega)| \\ \quad, \omega_i < \omega < \omega_{i+1}, i = 1, 2, \cdots, n-1 \\ \displaystyle\sum_{k=1}^{n} (-1)^{n-k} |Y_k(\omega)|, \omega > \omega_n \end{cases}$$

(26)

IV. MATHEMATICAL ANALYSIS AND USE INTRODUCTION

A. Mathematical Analysis for the proposed method

In fact, according to the process shown in Fig.8, the digital signal obtained by sampling is independent and can be processed separately. By using band pass digital filtering with the form

$$Y_i(\omega) = \begin{cases} Y_i(\omega), \omega_{i-1} < \omega < \omega_{i+1} \\ 0, the \ other \ \omega \end{cases}$$

(27)

then (26) can be simplified. Detailed analysis will be talked for each interval one by one.

1) $0 < \omega < \omega_1$: In the interval $0 < \omega < \omega_1$, beside $Y_1(\omega)$, the amplitude of the other $Y_k(\omega)(k = 2, 3, \cdots, n)$ are all zero, so we have

$$|Y(\omega)| = \sum_{k=1}^{n} (-1)^{k+1} |Y_k(\omega)| = |Y_1(\omega)|$$

(28)

When $0 < \omega < \omega_1$ & $\omega = \omega_1 - (2k+1)\omega_0, k = 1, 2, \cdots$

$$|Y_1(\omega)| = \frac{1}{\pi} \left(\frac{\omega_0}{\omega_1 - \omega} + \frac{\omega_0}{\omega_1 + \omega} \right)$$

(29)

$$|Y_1(\omega)|_{\min} = |Y_1(\omega_0)| \approx \frac{2}{\pi} \cdot \frac{\omega_0}{\omega_1}$$

(30)

2) $\omega_i < \omega < \omega_{i+1}, i = 1, 2, \cdots, n-1$: In the interval $\omega_i < \omega < \omega_{i+1}, i = 1, 2, \cdots, n-1$, beside $Y_i(\omega)$ and $Y_{i+1}(\omega)$, the magnitude of the other $Y_k(\omega)(k = 1, 2, \cdots, n)$ are all zero, so we have

$$|Y(\omega)| = \sum_{k=1}^{i} (-1)^{i-k} |Y_k(\omega)| + \sum_{k=i+1}^{n} (-1)^{k-i-1} |Y_k(\omega)|$$
$$= |Y_i(\omega)| + |Y_{i+1}(\omega)|$$

(31)

Because $\omega_i/\omega_0 \gg 1, \omega_{i+1}/\omega_0 \gg 1$. in this interval, according to (17), when $\omega > \omega_i$ & $\omega = \omega_i + (2k+1)\omega_0, k = 1, 2, \cdots$

$$|Y_i(\omega)| = \frac{1}{\pi} \left(\frac{\omega_0}{\omega - \omega_i} - \frac{\omega_0}{\omega_i + \omega} \right) \approx \frac{1}{\pi} \cdot \frac{\omega_0}{\omega - \omega_i}$$

(32)

when $\omega < \omega_{i+1}$ & $\omega = \omega_{i+1} - (2k+1)\omega_0, k = 1, 2, \cdots$

$$|Y_{i+1}(\omega)| \approx \frac{1}{\pi} \cdot \frac{\omega_0}{\omega_{i+1} - \omega}$$

(33)

According to (18), $Y_i(\omega)$ and $Y_{i+1}(\omega)$ has the same phase angle, so (31) can be rewritten as

$$|Y(\omega)| \approx \frac{1}{\pi} \cdot \frac{\omega_0}{\omega - \omega_i} + \frac{1}{\pi} \cdot \frac{\omega_0}{\omega_{i+1} - \omega}$$

(34)

If we define

$$\omega_{i+1} - \omega_i = h\omega_0$$

(35)

where h is an integer and represents the frequency distance between two carrier, then

$$|Y(\omega)|_{\min} = \left| Y(\frac{\omega_i + \omega_{i+1}}{2}) \right| = \frac{4}{\pi} \cdot \frac{\omega_0}{\omega_{i+1} - \omega_i} = \frac{4}{h\pi}$$

(36)

3) $\omega > \omega_n$: In the interval $\omega > \omega_n$, beside $Y_n(\omega)$, the magnitude of the other $Y_k(\omega)(k = 1, 2, \cdots, n-1)$ are all zero, so we have

$$|Y(\omega)| = \sum_{k=1}^{n} (-1)^{n-k} |Y_k(\omega)| = |Y_n(\omega)|$$

(37)

Because $\omega_n/\omega_0 \gg 1$, when $\omega = \omega_n + (2k+1)\omega_0, k = 1, 2, \cdots$

$$|Y_n(\omega)| = \frac{1}{\pi} \left(\frac{\omega_0}{\omega - \omega_n} - \frac{\omega_0}{\omega_n + \omega} \right) \approx \frac{1}{\pi} \cdot \frac{\omega_0}{\omega - \omega_n}$$

(38)

B. Use introduction for the proposed method

To make the measured result accurate, the minimum value of the injection signal should be large enough to guarantee enough signal-to-noise ratio. If the result shown in (36) is chosen as the minimum magnitude value and defined

$$\delta = \frac{4}{h\pi} \tag{39}$$

From (30) and (38), we know that

$$\delta = \frac{4}{h\pi} = \frac{2}{\pi} \cdot \frac{\omega_0}{\omega_1} = \frac{1}{\pi} \cdot \frac{\omega_0}{\omega_u - \omega_n}$$

where $\omega_u = \omega_n + (2k+1)\omega_0, k = 1, 2, \cdots$, then

$$\omega_1 = \frac{h}{2}\omega_0 \tag{40}$$

$$\omega_u = \omega_n + \frac{h}{4}\omega_0 \tag{41}$$

From (24) and (35), we have

$$\omega_n = \omega_1 + (n-1)h\omega_0$$

then the measuring range is

$$\omega_0 \leq \omega \leq (n - \frac{1}{4})h\omega_0 \tag{42}$$

When using the method proposed in this paper to measure impedance of $0 < \omega \leq \omega_b$, from (42), we have

$$\omega_b = (n - \frac{1}{4})h\omega_0 \tag{43}$$

From what had been discussed above, the spectral resolution is $2\omega_0$ except the point of carrier frequency. To achieve the proposed method, the first step is to determine the spectral resolution ω_0 of the measured result; the second step is to choose the signal-to-noise ratio by (39) to determine h; the third step is to calculate the number of the carrieries n by (43), if n is not an integer, then round up. In this example,

$$n = \frac{\omega_b}{h\omega_0} + \frac{1}{4} \tag{44}$$

The last step is to choose carrier frequencies, ω_1 is determined by (40), for the others,

$$\omega_i = \omega_1 + (i-1)h\omega_0 \tag{45}$$

where $i = 2, 3, \cdots, n$, n is determined by (42). From (39) and (43), the measuring bandwidth is

$$\omega_{band} \approx nh\omega_0 = \frac{4}{\pi\delta}n\omega_0$$

When h is a small integer, then n and δ will be large, this method will approach to the method by injecting a series of sinusoidal signals. When h is a large integer, then n and δ will be small, this method will approach to the method by injecting a narrow spike. That is to say, this method is a combination of the two methods.

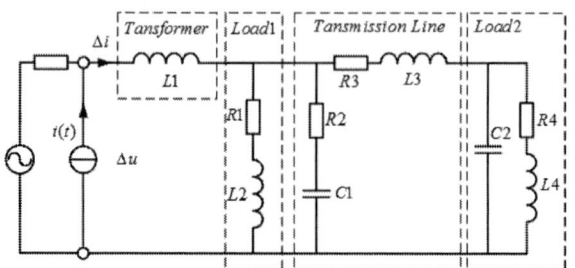

Fig. 9.   Load system model.

V. SIMULATION RESULTS

In order to validate the proposed method, a detailed power system model shown in Fig.9 is simulated by using MAT-LAB SIMULINK. The model includes components such as power transformers, transmission lines, feeder load etc [12]. A transformer is represented by a series inductive branch L1 while an R-C shunt branch R2, C1 and an R-L series branch R3, L3 represent a transmission line. In addition, two R-L loads are represented in the model as R-L branches: one in front of transmission line is R1 and L2, and the other is R4 and L4. The capacitor C2, which placed in front of R4 and L4, represented the reactive power compensator. Parameters of the above components are shown as follows: L1=0.06mH, R1=1.033Ω, L2=1.6mH, R2=1mΩ, C1=1.2mF, R3=0.036Ω, L3=0.1mH, C2=1.5mF, R4=0.775Ω, L4=1.2mH. The power supply is a sinusoidal voltage source of 240V RMS, 50Hz. The harmonic impedance of this network will be calculated by the injected current disturbance Δi and voltage disturbance Δu caused by the injection current.

In this simulation, the waveform of the injection current is the same with that had been discussed in Fig.8. The sampling frequency is 100 kHz, and the frequency band we need to measure is 0-4kHz.

From what had been discussed in section IV, firstly, we choose the spectral resolution, then we can calculate the period of the periodic square wave $T = 0.1s$. Secondly, we choose $h = 40$ to grantee the minimum magnitude of the injection signal has enough signal-to-noise ratio. The third step, calculating the number of carrier by (43)

$$n = \frac{\omega_b}{h\omega_0} + \frac{1}{4} = \frac{4000}{40 \times 10} + \frac{1}{4} \approx 10$$

Then we can determine the carrier frequencies

$$\omega_1 = \frac{h}{2}\omega_0 = 400\pi rad/s$$

$$\omega_i = \omega_1 + (i-1)h\omega_0 = 400(2i-1)rad/s (i = 2, 3, \cdots, 10)$$

The following injection and calculation process is the same as what has been shown in Fig.8. The total measurement time will be $t_m = nT = 1s$, which is much shorter than that by using a series of sinusoidal signals. The spectrum of the injected current signal is shown in Fig.10. It can be seen from the figure that the magnitude of whole frequency band is relatively high, that is to say, the injected signal has sufficient ability to resist noise.

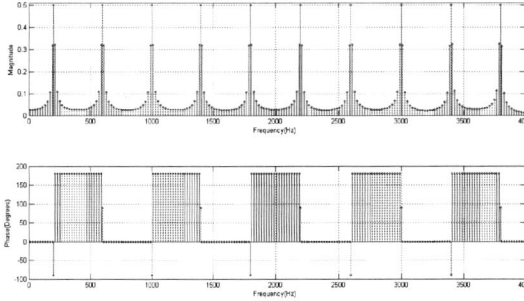

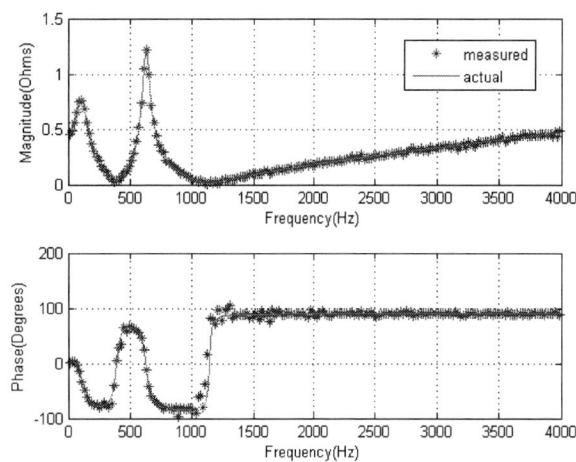

Fig. 10. The spectrum of the injection current signal.

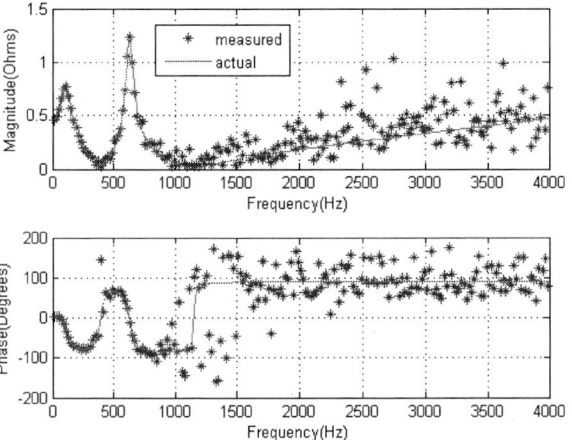

Fig. 12. Impedance in simulation by using proposed method with noise.

magnitude attenuation in high frequency leads to low signal-to-noise ratio. In this paper, a new injection form obtained by superposing modulated periodic square waves is proposed. With this method, impedance can be measured accurately both in low and high frequency with only several times injection. The new method is a combination of sinusoidal signals injection method and impulse signal injection method. The simulation results show that it is an efficient and accurate method when measuring impedance.

Fig. 11. Impedance in simulation by using single pulse with noise.

To make the simulation model closer to the practical system, some random noise are added to the system. Fig.11 shows the measured impedance of the network when the injection current is a pulse signal as defined in (10). Compared with the theoretical value, the measured results are accurate in low frequency, but have a large error in high frequency. The reason is that pulse signal has high magnitude attenuation in high frequency, that is to say, the influence of the noise grows, which leads to the observed error.

When the proposed new method is used, the measured impedance is shown in Fig.12. It can been seen from the figure that the measured results coincide well with the theoretical value range from 10Hz-4kHz, only the phase characteristic of several points are not good. The reason is that magnitude of this points are very small, which lead to the phase characteristic easy to be affected by noise.

VI. Conclusion

Current injection method is widely used in impedance measurement. When the injection current is a series of sinusoidal signals, we need to inject many times to obtain the impedance information in interested frequency band. The impulse signal only needs to be injected once, but the results in high frequency bands are not accurate, because the high

References

[1] S.D.Sudhoff, S.F.Glover, P.T.Lamm, D.H.Schmucker, and D.E.Delisle, "Admittance space stability analysis of power electronic systems," *IEEE Trans Aerosp Electron Syst*, vol.36, no.3, pp.965-973, Jul.2000.

[2] X. Feng, J. Liu, F.C. Lee, "Impedance specifications for stable DC distributed power systems" *IEEE Trans. Power Electron.*, vol. 17, no. 2, pp. 157-162, Mar. 2002.

[3] J. J. Liu, X. G. Feng, F. C. Lee, "Stability margin monitoring for DC distributed power systems via perturbation approaches" *IEEE Trans. Power Electron.*, vol. 18, no. 6, pp. 1254-1261, Nov. 2003.

[4] Rhode, Jason P., A. W. Kelley, and M. E. Baran. "Line impedance measurement: a nondisruptive wideband technique." *Industry Applications Conference*, Vol. 3. 1995.

[5] Palethorpe.B, Sumner.M, Thomas.D.W.P. "System impedance measurement for use with active filter control" *Power Electronics and Variable Speed Drives*, 2000. Eighth International Conference on (IEE Conf. Publ. No. 475).

[6] A.A. Girgis, R. B. McManis, "Frequency domain techniques for modeling distribution or transmission networks using capacitor switching induced transients," *IEEE Power Engineering Review*, pp. 74, July, 1989.

[7] Jing Huang, KeithA.Corzine, and Mohamed Belkhayat, "Small-signal impedance measurement of powerelectronicsbased ac power systems using line-to-line current injection," *IEEE trans on power electronics*, vol.24, no.2, Feb.2009, pp.445-455

[8] Y. L. Familiant, K. A. Corzine, J. Huang, and M. Belkhayat, "AC impedance measurement techniques," in *Proc. IEEE Electric Mach. Drive Conf.*, May 2005, pp. 1850-1857.

[9] J. Sun and K. J. Karimi, "Input impedance modeling of line-frequency rectifiers for aircraft power system stability analysis," *IEEE Trans. Aerosp. Electron. Syst.*, vol. 44, no. 1, pp. 217-226, Jan. 2008.

[10] P. Xiao, G. K. Venayagamoorthy, K. A. Corzine, and J. Huang, "Recurrent Neural Networks Based Impedance Measurement Technique for Power Electronic Systems," *IEEE Trans. Power Electron.*, vol.10, Feb.2010, pp.382-389.

[11] Oppenheim, Alan V., Alan S. Willsky, and Syed Hamid Nawab. *Signals and systems*. Vol. 2. Englewood Cliffs, NJ: Prentice-Hall, 1983.

[12] X. L. Yue, F. Zhuo, Z. H. Zhang, H. Shi, X. Bao, Z. Yang. "A New Current Injection Method for Impedance Measurement Using Superposed Modulated Square Pulse," in *Proc. IEEE Appl. Power Electron. Conf.*, Mar. 2013, pp. 1452-1456.

Numerical and Experimental Investigation of parasitic edge capacitance for Photovoltaic Panel

Wenjie Chen, Xiaomei Song, Hao Huang and Xu Yang
State Key Laboratory of Electrical Insulation and Power Equipment
Xi'an Jiaotong University , Shaanxi, Xi'an, China
cwj@mail.xjtu.edu.cn

Abstract- The occurrence of leakage current (also called ground current) that can occur in photovoltaic (PV) system depends strongly on the value of parasitic capacitance between PV cell and its metal frame, usually earth connected. This paper presents a straightforward approach to calculate the involved panel parasitic capacitance in order to predict the likeliness of such leakage current. A novel 2-D parasitic edge capacitance model is developed to accurately calculate the grounding capacitance of PV panel. Experimental results are obtained on five different PV panels of mono-crystalline, polycrystalline and thin-film type with rated power 10W, 50W and 175W. It is demonstrated that the proposed approach combines easy of applications and satisfying accurateness.

I. INTRODUCTION

Transformerless photovoltaic (PV) grid-connection converters may become a trend for future PV system because of their higher efficiency, greater flexibility, easier installation and lower cost. However, the key problem, which arises in absence of galvanic isolation, is the presence of a common mode current (i.e. ground leakage current) on both the inverter DC side and AC side. The amplitude and spectrum of this current depends on many factors such as, the inverter topology, the controlling strategy, the filter and the parasitic capacitance between the panel and the ground.

Numerous papers have investigated the influence of these factors on the leakage current[1-3]. Nevertheless, there is a lack of manuscript that give detailed discussion about the capacitance coupling between the PV panel and the ground. As we know, due to the large area of PV array as well as different configuration types of the array, the generalized used single parasitic capacitance model as shown in Fig. 1 is not accurate enough.

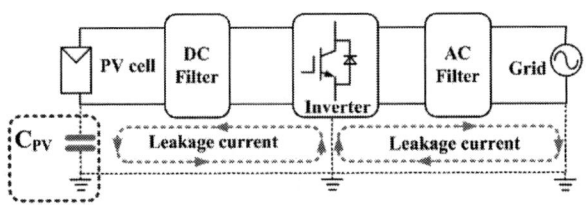

Fig. 1. Simplified common mode leakage current coupling model in photovoltaic system

In this paper, an analytical model for parasitic capacitances between the PV panel and the ground is developed. A straightforward approach to calculate the values of these capacitances is proposed in this paper. It combines ease of application and satisfying accurateness. Experimental results are obtained on five different PV panels of mono-crystalline, polycrystalline and thin-film type with rated power 10W, 50W and 175W.

II. PARASITIC CAPACITANCE MODELING

A 3-D schematic view of a typical PV panel with a simplified mounting structure is shown in Fig. 2. The silicon cell is packaged in a sandwich structure with such encapsulated material as glass, Ethylene-vinyl acetate (EVA), back sheet (Tedlar) and aluminum frame. From the capacitance calculation point of view, the silicon cell acts as one conductor while the panel frame, the mounting rack or the ground constructs the other conductor of the capacitor.

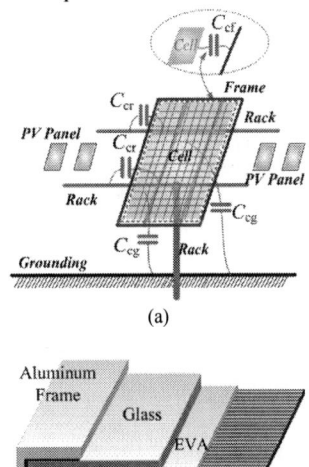

(a)

(b)

Fig. 2. (a) Distributed parasitic capacitance model for a mounted PV panel. (b) Three-dimensional schematic diagram of a laminating PV panel.

Therefore, the parasitic capacitances in PV panel can be split into three parts. 1) Cell-to-frame capacitance C_{cf}, 2) Cell- to-rack capacitance C_{cr} and 3) Cell- to-ground capacitance C_{cg}. Cell-to-ground capacitance C_{cg} is generally considered to be a main parasitic parameter,

---

This work was supported by the National Natural Science Foundation of China under Project 51277145

978-1-4799-2706-7/14 $31.00 © 2014 IEEE

while it can be calculated through a generalized method easily[4]. Cell- to-rack capacitance C_{cr} changes with the types of different mounting rack structures. Due to the length limitation of the manuscript, only the Cell-to-frame capacitance C_{cf} is study in detail in this paper. As we know, although the effective surface area of this capacitor is not very large, the distance between the capacitor electrodes (silicon cell and its aluminum frame) is very small. However, this factor is omitted by most of the published papers.

III. ANALYSIS OF THE CELL-TO-FRAME PARASITIC CAPACITANCE C_{CF}.

A. Two-dimensional computation model.

Electromagnetic field simulation is employed firstly to investigate the electric field line distribution responsible for the capacitive coupling between any two conducting electrodes. The 2-D simulated field line contour is shown in Fig. 3. The most important feature demonstrated by the simulation is that the capacitive coupling not only through the space between the silicon cell and the top of the panel frame but also through the space between every two adjacent sidewall of frame and cell.

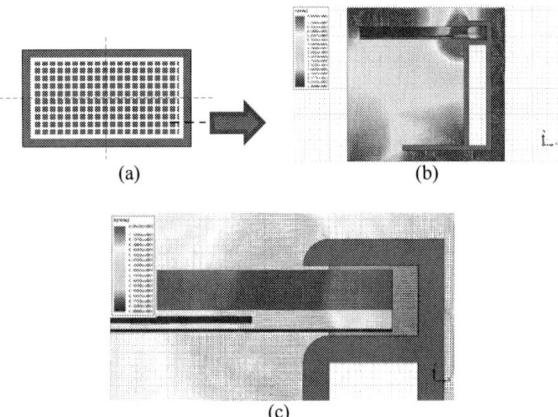

(a)

(c)

Fig. 3. Calculated electric field line contour. (a) Plan view of a PV module with aluminum frame. (b) 2-D computation domain of the laminated structure and the aluminum frame. (c) Zoom in on the top side of the frame.

Therefore, by considering the main coupling factors, the computation domain are divided into five subdomains, as proposed in Fig. 4, to account for five components of capacitance. And the cell-to-frame capacitance C_{cf} is decoupled into five uncorrelated capacitances.

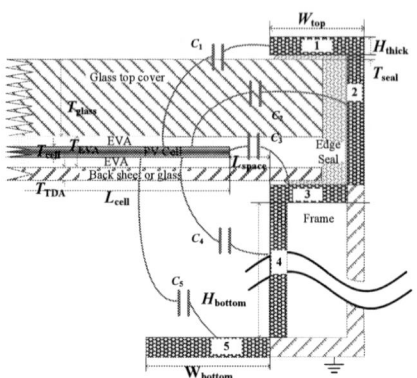

Fig. 4. Field line contours are partitioned into five regions with capacitance components defined as $C_1 \sim C_5$.

B. Theoretical analysis of $C_1 \sim C_5$.

For the computation region separated by $C_1 \sim C_5$, each component can be calculated through a fringe capacitance model, in which two conducting plates are not identical to each other. As shown in Fig. 5, the edge capacitance (per unit length) is composed of four components, generally, C_{top}, C_{side}, C_{bottom} and C_{in}. By using Schwarz–Christoffel mapping[5-6], these capacitance are derived as

$$C_{top} = \frac{\varepsilon}{\pi} \ln(1 + \frac{L}{a+b})$$

and

$$C_{in} = \frac{\varepsilon_0 \varepsilon_r L}{a} \tag{1}$$

$$C_{bottom} = \frac{(2 - \ln 4)\varepsilon}{\pi} \tag{2}$$

where

$$C_{side} = \frac{\varepsilon}{\pi} \ln[2k\sqrt{k^2 - 1} + 2k - 1] \tag{3}$$

with

$$k = 1 + \frac{b}{a} \tag{4}$$

$$C_{side1} = \frac{2\varepsilon}{\pi} \ln\left[\frac{a + \eta b + \sqrt{c^2 + (\eta b)^2 + 2a\eta b}}{c + a} \right] \tag{5}$$

$$+ \frac{M\varepsilon}{\pi} \ln\frac{\pi}{\sqrt{c^2 + a^2}} \exp\left(-\frac{c-a}{c+a} \right)$$

$$\eta = \exp\left(\frac{d + c - \sqrt{c^2 + b^2 + 2ab}}{\tau d} \right) \tag{6}$$

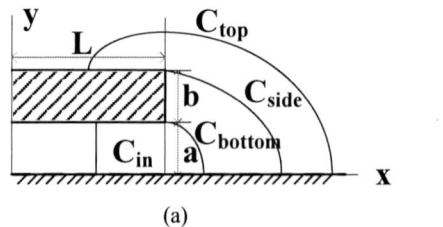

(a)

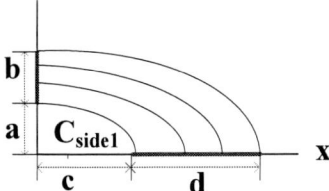

Fig. 5. Schematic diagram of electrode structure with certain thickness. (a) The electrode is right above the ground plate. (b) The ground plate is not right below the electrode.

Based on that analysis, the coupling capacitance between the cell and the #1 frame piece can be taken as a sum of C_{top}, C_{side} and C_{bottom}. Similar conclusion can be drawn on the other capacitance. So, the decoupled five capacitances can be derived as

$$C_1 = C_{top} + C_{side1} + C_{bottom} \approx C_{top} + C_{side1}$$
$$C_2 = C_{side1} + C_{in} \approx C_{side1}$$
$$C_3 = C_{bottom-side1} + C_{top-side1} \qquad (7)$$
$$C_4 = C_{side1}$$
$$C_5 = C_{top} + C_{side} + C_{in} \approx C_{in}$$

The calculated results of these five parasitic capacitances are shown in Fig. 6(a)-(f). The calculation is based on a typical PV panel frame parameter as shown in Table I. For this parasitic capacitance, an effective fringe computation range is set to be 50mm (L_{cell}) within the edge of the cell. Based on that, a wide range of variations around the typical size has been conducted to investigate the scaling effect.

TABLE I
TYPICAL PV PANEL FRAME PARAMETERS USED IN THE CALCULATION

Parameters	Value (mm)
Length of #1 frame, W_{top}	10
Thickness of all the frame, H_{thick}	2
Horizontal distance from #3 frame to the cell, L_{space}	10
Height of #4 frame, H_{bottom}	35
Effective length of #5 frame, W_{bottom}	20
Length of the PV panel, L_{PV}	1620
Width of the PV panel, W_{PV}	920

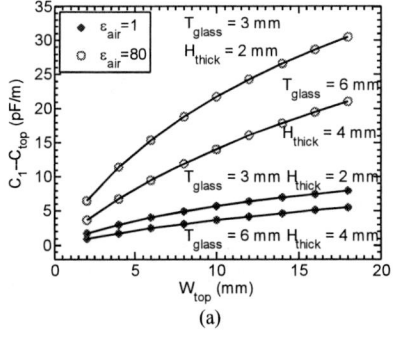

(a)

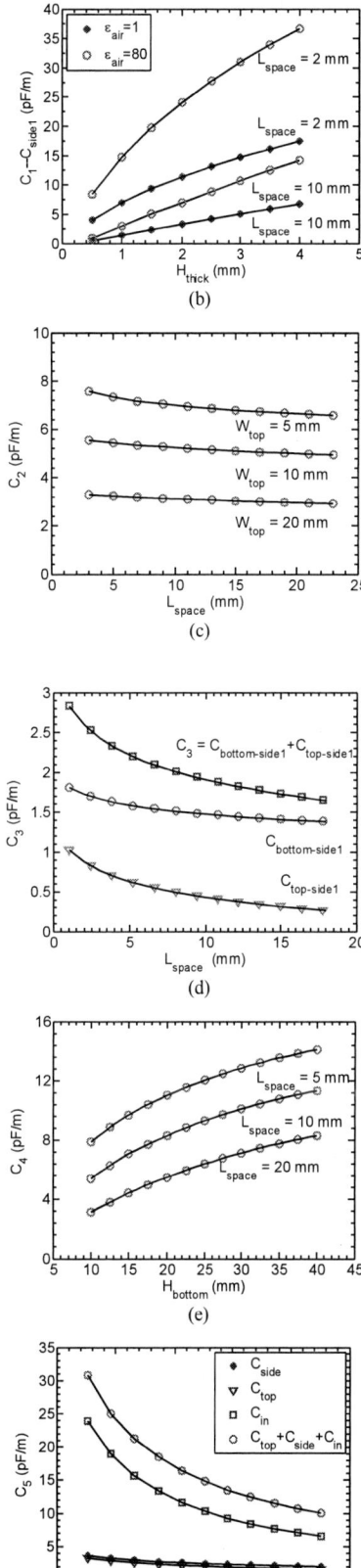

Fig. 6. Parasitic capacitance Ccf calculated through the two-dimensional capacitor model as a function of various frame geometry parameters. (a) C_1- C_{top} as a function of W_{top}. (b) C_1- C_{side1} as a function of H_{thick}. (c) C_2 as a function of L_{space}. (d) C_3 as a function of L_{space}. (e) C_4 as a function of H_{bottom}. (f) C_5 as a function of H_{bottom}.

IV. RESULTS AND DISCUSSIONS

Measurements were performed on five selected commercial PV panels with rated power from 10W to 175W. Detailed parameters of these panels are shown in Table II. The parameters were measured through Agilent E5061B network analyzer. A copper covered plane is used to model the ground. By omitting the influence of metal mounting rack, the PV panel is put on the equivalent ground with insulated stand. And the distance between the panel and the ground is maintained as a fixed value (5mm) in all the measurements. So, in this experimental multi-conductor test system, we can have cell-to-frame capacitance, cell-to-ground capacitance and frame-to-ground capacitance. The tested impedance and phase curve results are transferred to capacitance value as shown in Fig. 7.

TABLE II
SIZE AND PROPERTIES OF EACH PV PANEL.

Type	Power (W)	Length (mm)	Width (mm)	Cell to frame (mm)
1.Polycrystalline	10	36	30	15
2.Polycrystalline	50	67	54	15
3.Thin film	10	69	32	3
4.Monocrystalline	50	67	54	15
5.Monocrystalline	175	158	88	25

It can be seen that for frequency lower than 1MHz, the impedance between PV cell and PV frame, PV cell and ground, PV frame and ground are all capacitance. That demonstrated the correctness of the proposed 2-D model. Note that in the real PV system, the metal frame is always grounded. So C_{fg} will not exit in the real PV system. It is noticed the C_{cg} of No.1 panel is smaller than that of No.5 panel, because the value of C_{cg} is proportional to the panel area. Also, the value of C_{cf} is proportional to the total frame length of a panel. That is why the C_{cf} of No.5 is the largest among these five panels.

Fig.7 (b) and Fig.7 (d) has nearly the same parasitic capacitances C_{cf} and C_{cg}. It is because they have the same panel size. That means cell-to-frame capacitance has nothing to do with the solar cell material. For the same cell material with different distance away from the ground (as shown in Fig.7 (c) and Fig.7 (f) with thin film solar cell), the cell-to-frame capacitance remains unchanged while the cell-to-ground capacitance decreased quickly. It is because the value of C_{cg} is inversely proportional to the cell to ground distance while the value of C_{cf} is not impact by that distance. The C_{cf} of No.3 thin film panel is bigger than the C_{cf} of No.2 or No.4. It is because the cell to frame distance of No.3 is much closer than that of No.2 or No.4. Therefore, cell to frame distance might influence the value of cell-to-frame capacitance to a great extent.

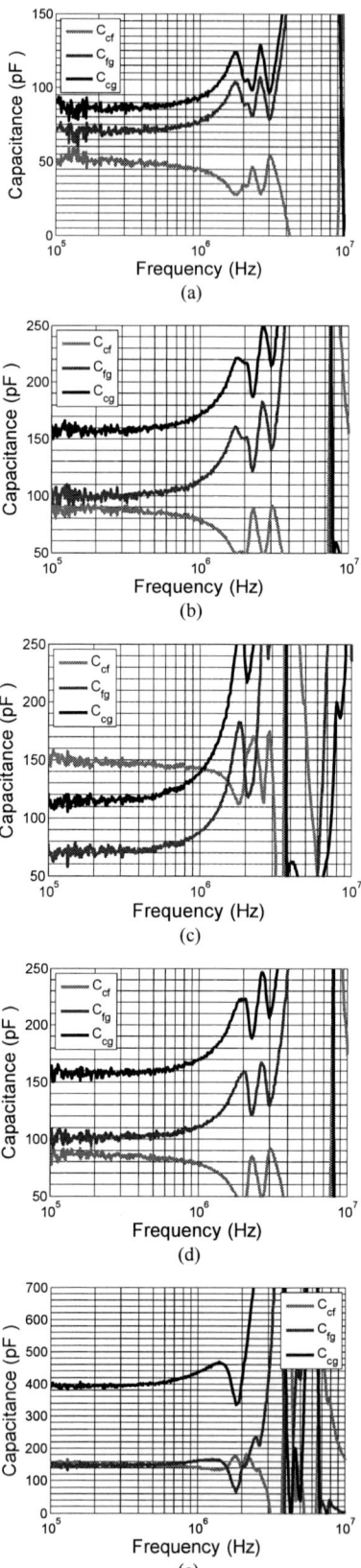

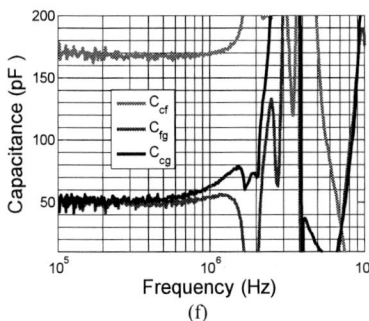

(f)

Fig. 7. Parasitic capacitance of different PV panels measured through impedance analyzer. (a) 10W polycrystalline panel. (b) 50W polycrystalline panel. (c) 10W thin film panel. (d) 50W mono-crystalline panel. (e) 175W mono-crystalline panel. (f) 10W thin film panel with 300mm distance away from the ground.

VI Conclusion

A new 2-D parasitic ground capacitance model, as well as a straightforward calculation approach for PV panel has been presented. The accuracy has been proven by an excellent match with experimental measurements on five different PV panels. This model predicts that PV panel parasitic ground capacitances are mainly composed of cell-to-frame capacitance and cell-to-ground capacitance. Cell-to-frame capacitance is an intrinsic parameter since it is mainly determined by panel design parameters. And Cell-to-frame capacitance should not be omitted in the future study. Cell-to-ground capacitance will decrease quickly with the distance between PV cell and the ground. It is not a dominant parameter especially when the panel is mounted far away from the ground.

REFERENCES

[1] G. Buticchi, D. Barater, E. Lorenzani and G. Franceschini, "Digital Control of Actual Grid-Connected Converters for Ground Leakage Current Reduction in PV Transformerless Systems," IEEE Trans. Industrial Informatics, vol. 8, no. 3, pp. 563–572, Aug. 2012.

[2] T. Kerekes, R. Teodorescu, P. Rodríguez, G. Vázquez and E. Aldabas, "A New High-Efficiency Single-Phase Transformerless PV Inverter Topology," IEEE Trans. Industrial Electronics, vol. 58, no. 1, pp. 184–191, Jan. 2011.

[3] B. Yang, W. Li, Y. Gu, W. Cui and X. He, "Improved Transformerless Inverter With Common-Mode Leakage Current Elimination for a Photovoltaic Grid-Connected Power System," IEEE Trans. Power Electronics, vol. 27, no. 2, pp. 752–762, Feb. 2012.

[4] M. Zahn, Electromagnetic Field Theory: A Problem Solving Approach. New York: Wiley, 1979.

[5] L. Wei, J. Deng, and H.-S P. Wong, "Modeling and performance comparison of 1-D and 2-D devices including parasitic gate capacitance and screening effect," IEEE Trans. Nanotechnol., vol. 7, no. 6, pp. 720–727, Nov. 2008.

[6] S. Oh. and H.-S. Philip Wong, " Physics-Based Compact Model for III–V Digital Logic FETs Including Gate Tunneling Leakage and Parasitic Capacitance," IEEE Trans. Electron Devices, vol. 58, no. 4, pp. 1068–1075, April. 2011.

Vehicle Interior Noise Control of Ultra-Compact Electric Vehicle (Fundamental Consideration Using Rectangular Enclosure)

Taro Kato
Tokai University
Undergraduate, Dep. of Prime Mover Eng.
4-1-1, Kitakaname, Hiratsuka, Kanagawa, Japan

Ryosuke Suzuki
Tokai University
Graduate Student, Course of Mech. Eng.
4-1-1, Kitakaname, Hiratsuka, Kanagawa, Japan

Hideaki Kato
Tokai University
Graduate Student, Course of Science & Tech.
4-1-1, Kitakaname, Hiratsuka, Kanagawa, Japan

Shinya Hasegawa
Tokai University
Dep. of Prime Mover Eng.
4-1-1, Kitakaname, Hiratsuka, Kanagawa, Japan

Yasuo Oshinoya
Tokai University
Dep. of Prime Mover Eng.
4-1-1, Kitakaname, Hiratsuka, Kanagawa, Japan
ossy@keyaki.cc.u-tokai.ac.jp

Abstract—The demand for noise reduction inside a vehicle has been recently increasing in the pursuit of improved ride comfort. In consideration of requirement of quietness improvement and reduction of weight, an active noise control (ANC) system has been attracting much attention. This method is effective for noise with low frequency. In this study, we propose an ANC system by using the vibration of giant magnetostrictive actuator and we aim to reduce the noise level in an interior of a vehicle using this proposed ANC system. The giant magnetostrictive actuator has a high output that is sufficiently controllable inside a vehicle space and at the same time has high space efficiency. In this paper, the giant magnetostrictive actuator is installed at windshield of the enclosure that act as a vehicle, and the noise reduction effect to ANC were examined as the noise applied to the box from outside.

Keywords— Active noise control, Giant magnetostrictive actuator, Boundary surface vibration

I. INTRODUCTION

In recent years, the automobile industry has high interest in manufacturing electric vehicles which have low noise level compared to engine vehicles. Therefore, the demand for noise reduction control inside vehicle has been increasing in the pursuit of improved ride comfort. As a countermeasure, although passive noise reductions occurred by using sound insulation and sound-absorbing material, there are issues in the reduction of vehicle interior space efficiency and increasing in weight. Active noise control (ANC) system has becoming quite much of attention due to the trade-off between the improvement of quietness and reduction of weight as well as the requirement for highly advanced technology.

Preciously, group carried out ANC test by using a small vehicle and a giant magnetostrictive actuator as a sound output source (Fig. 1). By having high space efficiency even with its small size and lightweight, this giant magnetostrictive actuator acquires a high output which is sufficiently controllable from the inside of a vehicle space. The giant magnetostrictive actuator adopts a feedback algorithm. Our next study in the future, we are planning to conduct an experiment on how to control the complexity of noise that changes depending on the surrounding environment [1]-[6].

In this study, we utilized an enclosure as an ultra-compact electric vehicle. We installed acrylic plate into the enclosure as the electric compact vehicle window and assumed the noise is coming from outside of the vehicle. We also used a giant magnetostrictive actuator to manage boundary surface vibration.

II. PRINCIPLE OF SYSTEM

A. Classification of sound

People in vehicles are surrounded by various sounds.

978-1-4799-2706-7/14 $31.00 © 2014 IEEE

These sounds include 1) information required to drive the vehicle, 2) comfortable sounds such as music, and 3) noise such as the sound of driving on the road (road noise) and wind noise generating during high-speed driving. These three types of sound are mutually related psychological factors. The purpose of this study is to reduce only the that how a person perceives these sounds is influenced by sensory and noise such as road noise and wind noise, and to make other sounds, such as information and comfortable sounds, clearer to provide a comfortable vehicle environment. In the actual driving of vehicles, noise in the frequency range of 200-300 Hz is most frequently heard [7].

Fig. 1. Ultra-compact electric vehicle.

B. Giant magmetostrictive actuator

A magnetic material is divided into domains called magnetic domains. Each magnetic domain shows spontaneous magnetization, and the crystal lattice of a magnetic material is spontaneously distorted along with the generation of spontaneous magnetization. Upon the application of external magnetic field, the directions of the spontaneous magnetization of the magnetic domains align. Along with the rotation of the spontaneous magnetization, the distortions of the spontaneous magnetization also align, leading to a deformation in the external shape of the magnetic material. A coil is wound around a giant magnetostrictive material composed of terbium, dysprosium, and iron, to form a giant magnetostrictive actuator. Control sound waves with a high sound pressure are expected to be produced by generating a large excitation force using the magnetostrictive effect exceeding 1000 ppm generated upon the application of external magnetic field [8].

The giant magnetostrictive actuator is driven utilizing the deformation of the magnetic material. The shape of the magnetic material changes in an ultrashort period of microsecond order, and the force generated in association with the deformation is very large.

In this study, the actuator was brought into contact with other materials such as an acrylic plate to produce the sound used in ANC. Figure 2 shows the giant magnetostrictive actuator used in the experiment [9].

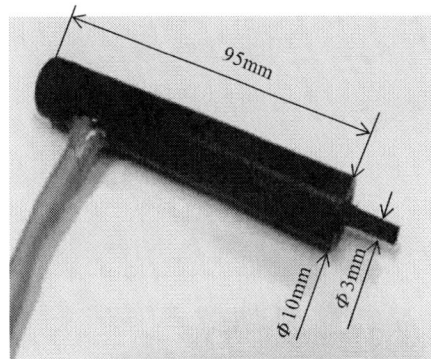

Fig. 2. Magnetostrictive actuator.

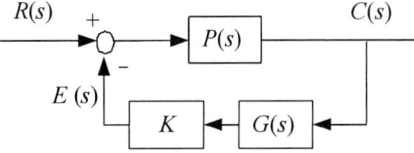

Fig. 3. Block diagram of phase-compensated feedback control system.

C. Control algorithm

Conventionally, in the adaptive signal processing which is abundantly used in ANC, optimal signal-processing characteristic is improved by repeating the correction of the signal-processing characteristics using noise and error information over a control target. However, operating the ANC system using adaptive signal processing in the acoustic field in closed space where resonance occur will result in problems such as too many corrections in the optimal signal-processing characteristic as well as the operations are time consuming [10].

In this report, due to the cavernous resonance inside a fixed tire by the frequency of noise source, changing the gain is inessential, and thus we considered that by only a phase is compensated using all pass filter, sound

cancellation effect is improvable.

All-pass filter has unity gain and can provide a phase shift ranging from 0 to 180 deg. if noninverting, or 180 ~360 deg. if inverting.

$$G(s) = \pm\left(\frac{s+z}{s-p}\right) \qquad (1)$$

Where, z is the zero location of the all-pass filter and p is the pole location, where the pole and zero must be equal, $(z = p)$.

If $0 < \theta < 180$, then a noninverting all-pass is used and the pole/zero location are calculated using:

$$p = z = \frac{\omega}{cot\left(\dfrac{\pi\theta}{360}\right)} \qquad (2)$$

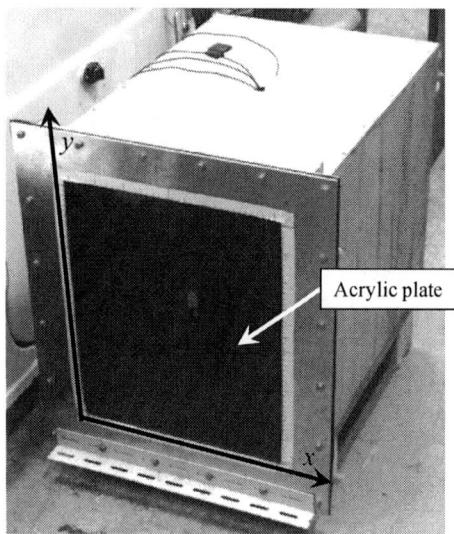

Fig. 4. Reduction experimental enclosure.

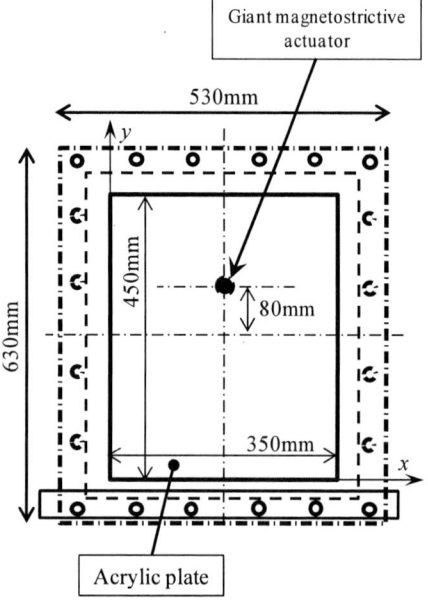

Fig. 5. Schematic illustration of reduction experimental enclosure. (front view)

However, if $180 < \theta < 360$, than an inverting all-pass is used and the pole/zero location are calculated using:

$$p = z = -\frac{\omega}{\tan\left(\dfrac{\pi\theta}{360}\right)} \tag{3}$$

Where, θ is phase difference and ω is target frequency [11].

Also, as shown in Fig. 3, feedback control is done that compensate for the delay of the control voltage to the input wave. Where, $R(s)$ is control target, $P(s)$ is error path, $C(s)$ is error signal $G(s)$ is phase compensation, K is

feedback gain and $E(s)$ is control output. The gain K was determined experimentally to be 80 by trial and error.

III. NOISE REDUCTION EXPERIMENT

A. Active noise control

Conventionally, a technique called passive noise control has been used to reduce noise. In passive noise control, sound-absorbing materials and reflectors are used to reduce the noise through absorption and reflection. High-frequency noise is absorbed by the absorbing material and reflected by the reflectors owing to its linearity; as a result, the noise intensity decreases.

However, the effect of the absorbing material is limited for low-frequency noise, which also weaves through the gaps between reflectors. Therefore, the reduction of such noise is difficult. The size of a system for reducing noise with long wavelength and low frequency inevitably becomes large, which is a significant demerit of passive noise control.

In the ANC system, pseudo noise with the same amplitude and with a phase opposite to phase opposite to the target noise is artificially produced from a control sound source to cancel the target noise by means of interference between the pseudo noise and the target noise. Because the frequency and amplitude of the noise constantly change, those of the pseudo noise used to cancel the target noise should also change accordingly. To achieve this, the signal obtained near the noise source and the noise at the control point should be monitored continuously to calculate the frequency and amplitude and output the sound that can cancel the target noise. This method is particularly effective for noise with long wavelength and low frequency. In the conventional passive noise control using a sound-absorbing material, a large space is required to control noise with low frequency.

In contrast, in the ANC system, no large space is required and only equipment to produce sound with a phase opposite to that of the target noise is necessary. In general, a frequency range where the ANC system is effective is considered to be 500 Hz or lower. By bringing the control sound source closer to the control point, the frequency range that is controllable with the ANC system can be expanded [1], [6].

B. Experimental equipment and method

In this study, as shown in Fig. 4, we used 350 mm × 450 mm with a 15 mm thick plate as a window and set it on the enclosure which is 630 mm long, 430 mm wide and 530 mm high. The window was installed on the enclosure within a frame which is 610 mm × 510 mm, with the thickness of 20 mm thick (Fig. 5). The inside of the enclosure is covered with the acoustic material of urethane sponge with 30 mm thick.

As shown in Fig. 6, it is sine wave was emitted toward the acrylic plate from a speaker used as the noise source. The sound wave was measured using another

978-1-4799-2706-7/14 $31.00 © 2014 IEEE   2974

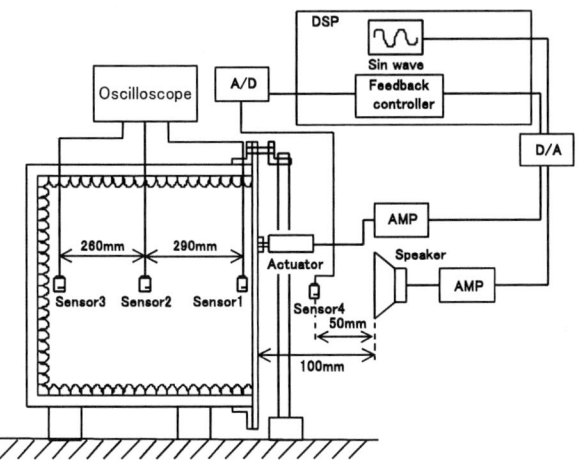

Fig. 6. Active noise control system

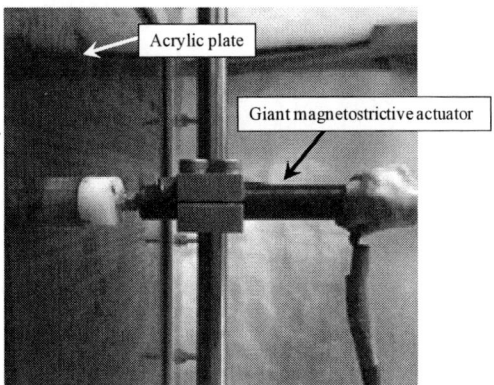

Fig. 7. Installation of giant magnetostrictive actuator.

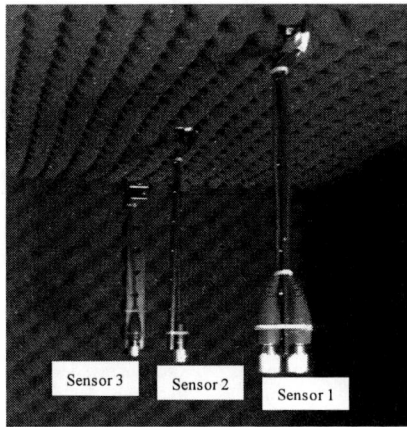

Fig. 8. Position of sensors.

sensor (sensor 4) outside the enclosure. We input noise data into a computer via a digital signal processor (DSP). We control the noise by emitting wave with certain phase in opposite direction. The noise reduction effect is measured by sensor 1, sensor 2, and sensor 3. The sensor 1 is 290 mm away from sensor 2, and sensor 3 is 260 mm from the sensor 2.

In this experiment, the output from the speaker recorded was load noise and frequency of acrylic plate based on the tire cavernous resonance of sign wave of 200 Hz. The giant magnetostrictive actuator was installed on the plate at a distance of 80 mm above the center (Fig. 5, Fig. 7). There are three sensors sensor 1 is placed just behind the plate, sensor 2 is placed in the center of the box, and sensor 3 is placed behind the sensor 2 (Fig. 8).

In this experiment, speaker with a sine wave of 200Hz as noise which apply toward the acrylic plate with enclosure is set as an output. As sensor 4 receive noise from the speaker, a control sound wave is created using the date acquired by the gain and phase compensation which set forth in section C of II. Therefore, we cancel the noise from the speaker of the vibration of the acrylic plate by giant magnetostrictive actuator.

In this study, we aim to suppress the noise inside enclosure by cancelling out the noise from the speaker on the outside.

C. Results

Figures 9~14 show the results of sound pressure levels both when the noise was controlled and uncontrolled. Noise reductions of 6.5 dB were recorded in the sensor 1 (Fig. 9, Fig. 10), 16.7 dB in the sensor 2 (Fig. 11, Fig. 12), and 14.4 dB in the sensor 3 (Fig. 13, Fig. 14). As shown in Fig. 10, there were sound pressure level differences in different positions within the enclosure. Also, we confirmed the highest noise reduction effect occurred at sensor 2 in the enclosure inside.

In this experiment, we set an output using giant magnetostrictive actuator which control sound wave that created based on the noise. As a result, it was found that it is possible to suppress the noise inside enclosure by cancelling the noise between acrylic plate and the speaker.

IV. CONCLUSION

In this research study, we performed an experiment of the ANC system by installing a giant magnetostrictive actuator into the ultra-compact vehicle, and analyzed noise reduction effect inside the enclosure. As a result, compared with when the ANC system was deactivated, the highest noise reduction effect recorded was at sensor 2 inside the enclosure. The type of output of the sound wave from the giant actuator depends on the shape and material of the acrylic plate vibratedby the actuator. Therefore, we also concluded that a better performance of the ANC system could be accomplished if we could design an acrylic plate that enables giant magnetostrictive actuator to produce sine wave without distortion. Therefore, it is necessary to consider the material and shape of the plate installed with the giantmagnetostrictive actuator in order to develop an effective sound source control for the ANC system.

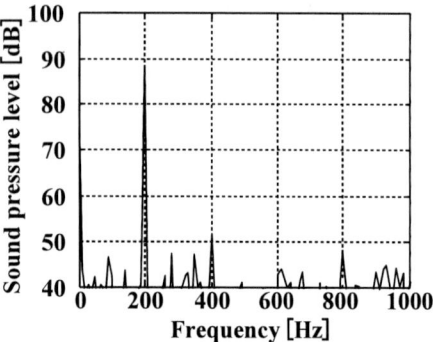

Fig. 9. Experimental result for sensor 1.
(without control, noise frequently : 200Hz)

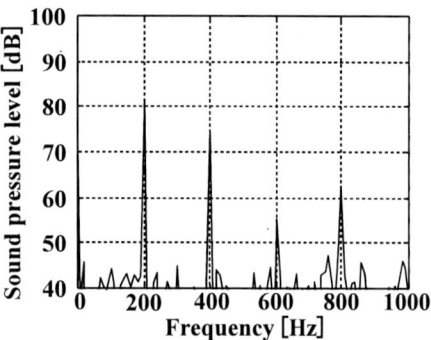

Fig. 10. Experimental result for sensor 1.
(control, noise frequently : 200Hz)

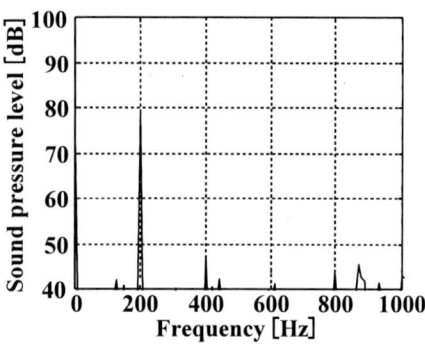

Fig. 11. Experimental result for sensor 2.
(without control, noise frequently : 200Hz)

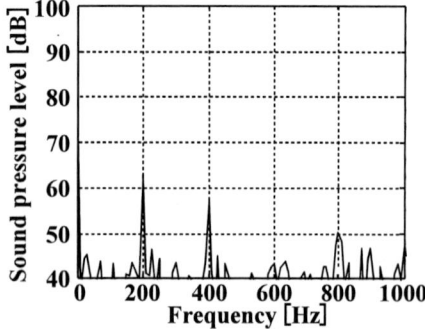

Fig. 12. Experimental result for sensor 2.
(control, noise frequently : 200Hz)

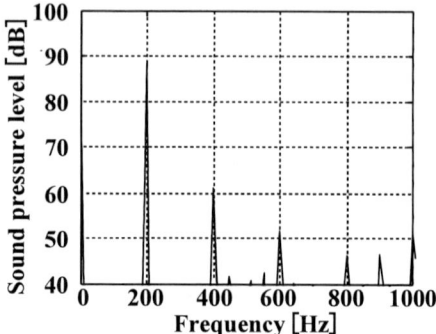

Fig. 13. Experimental result for sensor 3.
(without control, noise frequently : 200Hz)

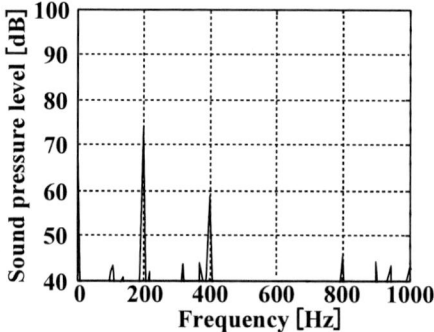

Fig. 14. Experimental results for sensor 3.
(control, noise frequently : 200Hz)

REFERENCES

[1] H. Kato, Y. Oshinoya, S. Hasegawa, T. Morishita, H. Kasuya: "Basic Study on Reduction Technology for Small Vehicle with Giant Magnetorictive Actuator," *Proc. Schl. Eng. Tokai Univ., Ser. E*, pp. 43-47, 2011, (in Japanese).

[2] T. Tabata, S. Hasegawa, H. Hyodo, Y. Nakaji, T. Shikata: "The Development of the Nissan Active Noise Control System," *Nissan technical review*, 30, pp. 43-47, 1991, (in Japanese).

[3] S. Hasegawa: "Active Noise Control System (ANC) for Cars," *Techno marine bulletin of the Society of Naval Architects of Japan*, 761, pp. 839-842, 1993, (in Japanese).

[4] M. Nishimura, T. Usagawa, S. Ise "*Active Noise Control*," *The sound technology* series 9, Corona Publishing Co., Ltd, pp. 94-99, 2006, (in Japanese).

[5] T. Inoue, H. Sano, H Matsuoka, "Active Noise and Vibration Control Technology for New V6 Engine with Variable Cylinder Management system," *IEICE Technical Report* , pp. 16-20, pp. 37-42, 2006, (in Japanese).

[6] R. Takahasi, S. Hasegawa, Y. Oshinoya, "Basic Study of Active Noise Control using Giant Magnetostrictive Actuator," *Academic lecturers of JSAE Kanto of The Japan Society of Mechanical Engineers*, pp. 619-620, 2011, (in Japanese).

[7] K. Terai, H. Sano, "Active Control System for Low-Road Noise," *The journal of the acoustical society of Japanese*, vol. 126, no. 10, pp. 424-425, 2003, (in Japanese).

[8] Research committee on new generation electric and magnetic actuators, "New Generation Electric and Magnetic Actuators," *Technical Report of IEE Japan*, vol. 1169, pp. 34-41, 2009, (in Japanese).

[9] Shounan Metaltec Co.,:Giant magnetostrictive material, [Online] . Available http://www.shonan-metaltec.com/ , (in Japanese).

[10] S. Tsujii "Adaptive Signal Processing," *Shoukoudou*, pp. 9-11, 1993, (in Japanese).

[11] Jesse B. Bisnette, Adam K. Smith, Jeffrey S. Vipperman, Daniel D. Buduy, "Active Noise Control Using Phase Compensated, Damped Resonant Filter," *Journal of Vibration and Acoustics*, vol.128, issue 2, pp. 148-155, 2005.

Consideration for the Propagation Path of Conductive Noise in Air Conditioners

Tsuyoshi Tokiwa, Masaki Kanamori, Takahisa Endo
Core Technology Center
Toshiba Carrier Corporation
Fuji, Japan

Mikiya Iida
Corporate Manufacturing Engineering Center
Toshiba Corporation
Yokohama, Japan

Satoshi Ogasawara, Yizhanyi Tang
Graduate School of Information Science & Technology
Hokkaido University
Sapporo, Japan

Abstract—The disturbance voltage of air conditioners may be degraded, depending on the combination of the outdoor and indoor units. This paper describes a technique to simulate the disturbance voltage of an air conditioner and examines the propagation path of conductive noise when two or more indoor units are connected to the outdoor unit.

Keywords— Air Conditioner, Disturbance Voltage, Electromagnetic Compatibility (EMC), Simulation.

I. INTRODUCTION

More than 30 years have passed since an inverter was applied to an air conditioner for residential use for the first time in the world in 1981. Ever since, while inverters have made a great contribution to improving comfort and the energy conservation capability of air conditioners, electromagnetic noise caused by the switching of power devices has posed a great problem as more and more inverter air conditioners have been used. This problem is expected to stand out more as next-generation devices, such as SiC (silicon carbide) devices, spread in the future. In addition, an outdoor unit of an air conditioner may connect to more than one indoor unit in some cases, which may sometimes increase the electromagnetic noise depending on the combination of those units. Against this background, some examples of noise simulation intended to improve countermeasures against noise were reported [1, 2], but there is no report on simulations where two or more indoor units are connected and on the principle of noise in that case. This paper describes a system that simulates conductive electromagnetic noise in an air conditioner and examines the influences of connecting two or more indoor units on the disturbance voltage.

II. STRUCTURE OF AIR CONDITIONER

Fig. 1 shows the structure of a general air conditioner for commercial use. The outdoor unit and the indoor unit are connected to each other with a power cable and a copper pipe for circulating refrigerant. The outdoor unit has a fan motor for heat exchange, a compressor motor, and an inverter unit that drives each of these permanent magnet (PM) motors. The indoor unit also has a PM motor as a fan motor for heat exchange. Therefore, an air conditioner has more than one inverter circuit and multiple conductors intricately connected.

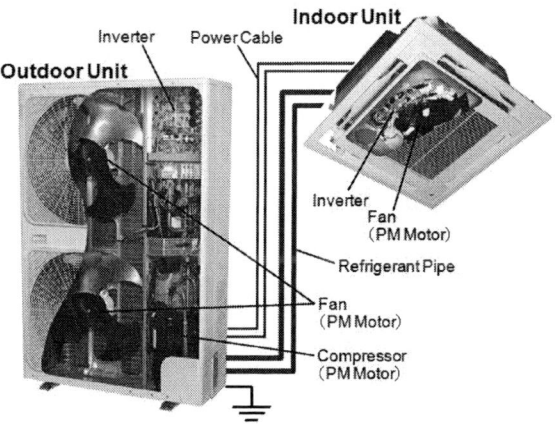

Fig. 1. Structure of air conditioner for commercial use

III. REGULATION FOR DISTURBANCE VOLTAGE OF AIR CONDITIONERS

Disturbance voltage on power terminals is regulated in Chapter 5 of the Electrical Appliance and Material Safety Law (ordinance article 1) conforming to international standard CISPR14-1 as shown in Table 1 in Japan. Fig. 2 is a scene of measuring the disturbance voltage of an air conditioner. The schematic diagram in Fig. 3 shows the connection. The indoor unit and outdoor unit are connected by a refrigerant pipe 5 m long. The outdoor

978-1-4799-2706-7/14 $31.00 © 2014 IEEE

unit is placed at a height of 10 cm from the grounding plate on the floor. The measurement is carried out with a line impedance stabilization network (LISN) placed 80 cm away from the outdoor unit and a spectrum analyzer.

TABLE I
DISTURBANCE VOLTAGE LIMIT VALUE IN CHAPTER 5 OF ELECTRICAL
APPLIANCE AND MATERIAL SAFETY LAW

Frequency	Quasi-Peak Detection
0.5265 to 5.0 MHz	56 dBuV
5.0 to 30 MHz	60 dBuV

Fig. 2. Measuring disturbance voltage of air conditioner

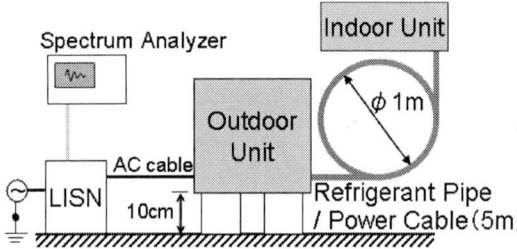

Fig. 3. Schematic diagram for measurement of disturbance voltage of air conditioner

IV. SIMULATION OF DISTURBANCE VOLTAGE

A. Simulation procedure

Fig. 4 shows the flow of simulation. In the first step, a model of noise propagation path is created. Equivalent circuit models are created for the compressor and filter parts based on measured impedance characteristics, and the equivalent circuit model of the PCB wiring is created through electromagnetic field analysis. In the second step, a step waveform indicating changes in the voltage of the power device that generates noise is input to a circuit connected to the equivalent circuit model created in step 1. The voltage waveform at the LISN where the disturbance voltage is observed is simulated (on the time axis). Then the simulated voltage waveform is transformed to the frequency domain waveform by using a similar method used in spectrum analyzers.

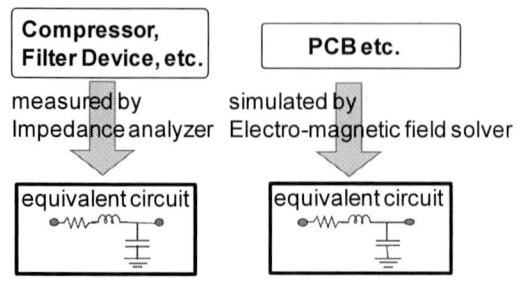

(a) Step 1. Modeling noise propagation path

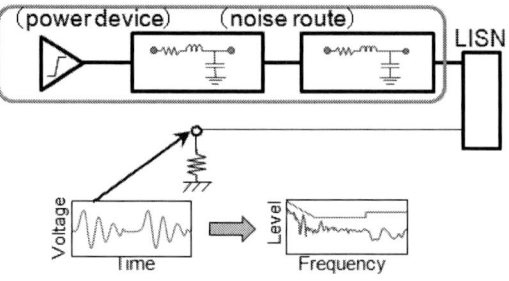

(b) Step 2. Circuit analysis and Fourier transformation
Fig. 4. Simulation flow

B. Modeling the noise propagation path

Modeling is carried out for compositions that are suspected to be the propagation path of common-mode noise caused by the power device, which is expected to be the noise generation source. The subjects of the modeling were common-mode countermeasure parts (common-mode choke, Y capacitor, the ferrite core attached to the compressor cable, etc.), compressor cable, compressor, and the conductors of the outdoor unit frame. As an example, modeling method of an equivalent circuit model of the compressor is explained (Fig. 5). First, the compressor cable is removed from the compressor and impedance characteristics from the input terminal to the compressor frame are measured using the impedance analyzer. Next, the circuit topology and circuit constant are adjusted to create an equivalent circuit model that fits with the measurement results, as shown in Fig. 4(b).

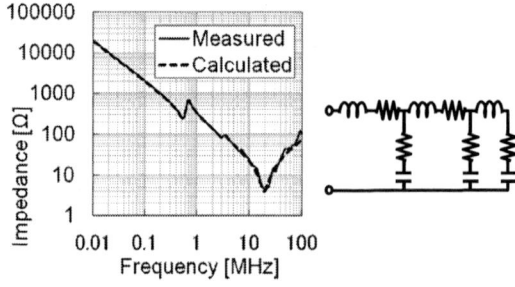

(a) Impedance characteristics   (b) Equivalent circuit model
Fig. 5. Modeling method of the compressor

C. Analysis of the Combined Circuit Model

Disturbance voltage at LISN is simulated when a step waveform is input at the position of the power device of the inverter with the created equivalent circuit model connected as shown in Fig. 6. Fig. 7 shows the result of converting the simulated voltage fluctuation into a value on a frequency axis and the result of actual measurement. As can be seen, the simulation result indicates that the noise reduction effect of the value of Y capacitor is close to the actual measurement result.

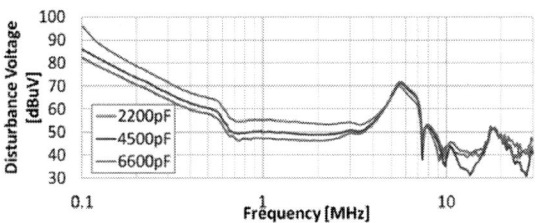

(a) Simulation result

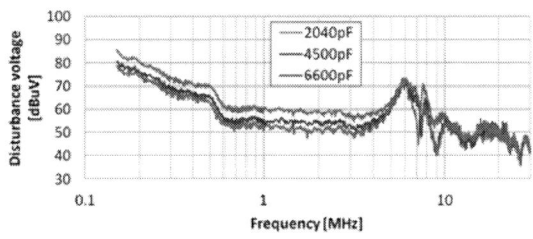

(b) Measurement result

Fig. 7. Influences of value of Y capacitor on disturbance voltage

V.   ANALYSIS OF INFLUENCES OF AIR CONDITIONER STRUCTURE

This chapter considers the influences of the indoor unit and the refrigerant pipe on the disturbance voltage. Fig. 9(a) shows the result of measuring the disturbance voltage with an insulated hose inserted into the connection point between the outdoor unit and refrigerant pipe as shown in Fig. 8. Incidentally, Fig. 9(a) shows a case where the insulated hose is inserted into the refrigerant pipe as "Insulated (1)" and a case where a ferrite core is attached to the power cable of the indoor unit, in addition to "Insulated (1)," as "Insulated (2)." Fig. 9(b) shows the result of the simulation conducted under the same conditions as the measurement. For this simulation, only the compressor and fan inside the outdoor unit are considered noise sources, and the fan inside the indoor unit is not considered a noise source. It can be seen from Fig. 9 that the simulation result shows a tendency similar to the measurement. Since the disturbance voltage shows no change at 1 MHz or lower, it can be said that the disturbance voltage in this frequency band is determined by the outdoor unit and LISN. It can also be said that the noise peak of the disturbance voltage that peaks at 6 MHz disappears when the refrigerant pipe and the power cable of the indoor unit are electrically opened-circuited. It can be estimated from this that the noise peak in this frequency band is generated because of LC resonance among the refrigerant pipe, indoor unit, and power cable of the indoor unit.

Next, the cause of an increase in the disturbance voltage at around 6 MHz is considered. Fig. 10(a) is a simplified block diagram that shows an air conditioner with an outdoor unit and an indoor unit connected on a one-to-one basis. The block in Fig. 10(a) consists of only the main common-mode noise route. A power cable, a refrigerant pipe, and the indoor unit are connected to the outdoor unit that consists of a compressor, a common-mode choke, and a Y capacitor. Since the impedance of the LISN in each phase under the frequency band of the disturbance voltage is about 50 Ω, the common-mode impedance of the LISN in the case of three-phase input is

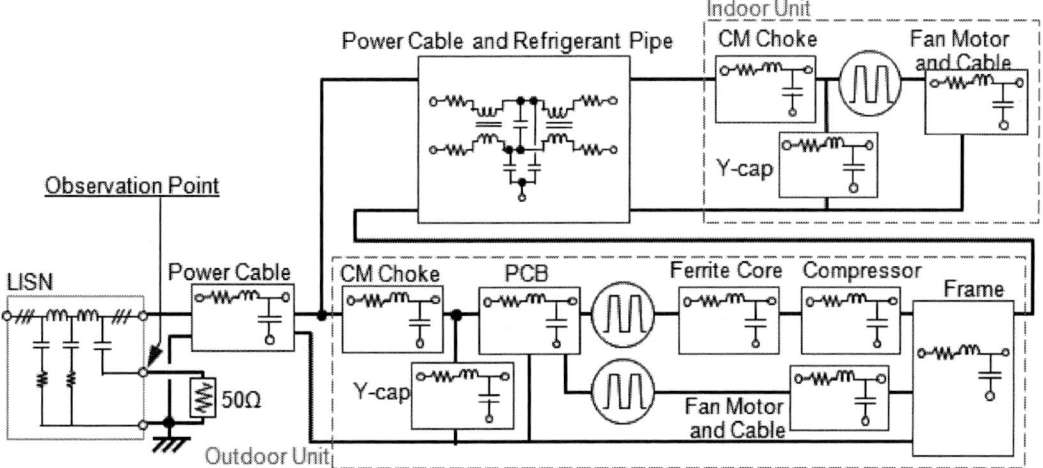

Fig. 6. Structure of model circuit for simulating disturbance noise

$50\ \Omega/3 \fallingdotseq 17\ \Omega$ [1]. Points b and c in Fig. 10(a) are at the same potential. Therefore, Fig. 10(a) can be rearranged to Fig. 10(b). Voltage applied to the LISN is proportional to voltage V_{ab} across points a and b. At the frequency of which the circuit connecting Path1 and Path2 in parallel in Fig. 10(b) makes parallel resonance, the impedance of the parallel circuit of Path1 and Path2 becomes maximum, raising V_{ab} to the maximum. The voltage applied to the LISN also reaches the maximum. This is considered the cause that the disturbance voltage increases to the maximum at around 6 MHz. Fig. 11 shows the result of simulating the impedance characteristics observed between points a and b in Fig. 10. Since the impedance of the common-mode choke is high, these impedance characteristics are considered to be almost equal to the impedance of the parallel circuit of Path1 and Path2. From Fig. 11, it can be confirmed that the frequency at which the impedance becomes maximum almost coincides with the frequency at which the disturbance voltage becomes maximum.

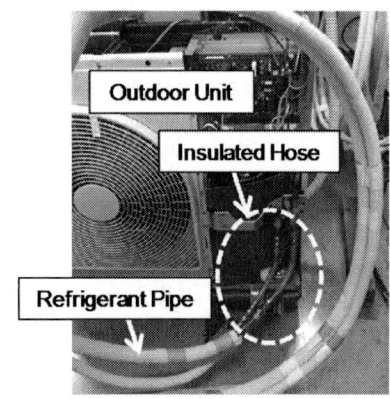

Fig. 8. Inserting insulated hose into refrigerant pipe

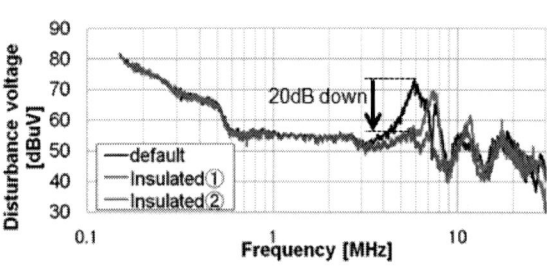

(a) Measurement result

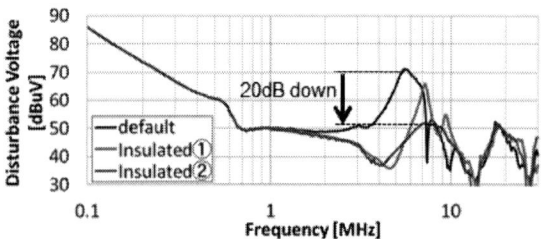

(b) Simulation result

Fig. 9. Influences of insulation of refrigerant pipe on disturbance voltage

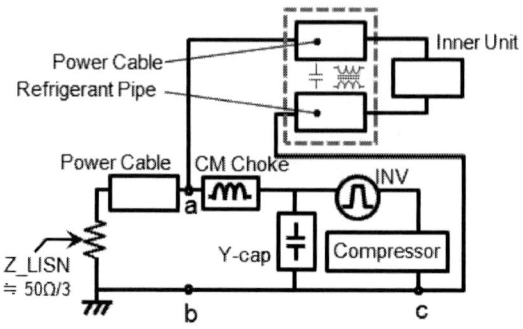

(a) Before change of indication

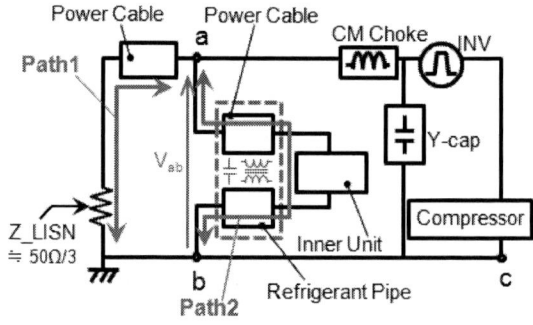

(b) After change of indication

Fig. 10. Simplified block diagram of air conditioner

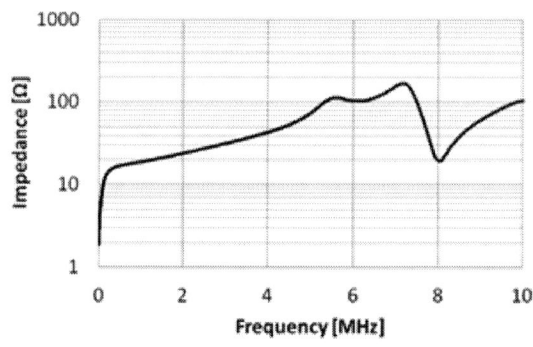

Fig. 11. Result of simulating impedance characteristics of parallel circuit of Path1 (LISN and power cable) and Path2 (refrigerant pipe and power cable for indoor unit)

VI. Disturbance Voltage when Two Indoor Units are Connected

Disturbance voltage when two indoor units are connected to one outdoor unit is evaluated. Fig. 12 shows two types of positional relations between the refrigerant pipe and the power cable for the indoor units. In this paper, these two types of positions of the power cable are called Twin1 and Twin2, respectively. For both connections, both the refrigerant pipe and the power cable are 5 m long. The refrigerant pipes of the Indoor Unit 1 and Indoor Unit 2 are connected in parallel from the outdoor unit. The power cable for Indoor Unit 1 is routed completely in parallel with the refrigerant pipe for Indoor Unit 1. For Twin1, the power cable for Indoor Unit 2 is routed in parallel with the refrigerant pipe for Indoor Unit 2 as much as possible. For Twin 2, the power cable for Indoor Unit 2 is routed in parallel with the refrigerant pipe for Indoor Unit 1 as much as possible. Fig. 13 shows the result of measuring the disturbance voltage. This figure shows a case where only one indoor unit is connected as Single and a case where two indoor units are connected as Twin1 and Twin2. Fig. 14 shows the result of simulating the disturbance voltage in the same conditions as the measurement. As shown in Fig. 15, the electromagnetic coupling between the refrigerant pipe and power cable of Indoor Unit 2 is partially deleted. It can be seen that the simulation result shows a tendency similar to that of the experiment. Even though two indoor units are connected, the disturbance voltage at 1 MHz or below showed no change because the outdoor unit is not changed. When the number of indoor units is increased from one to two, a new noise peak appeared at around 4 MHz. The power cable for Indoor Unit 2 is connected in series with the power cable for Indoor Unit 1. Consequently, the inductance of the power cable and the stray capacitance between the Outdoor Unit and Indoor Unit 2 are greater than those between the Outdoor Unit and Indoor Unit 1. Since the resonant frequency is in reverse proportion to the square root of the product of the inductance and capacitance, when two indoor units are connected, resonance caused by the power cable and refrigerant pipe for the second indoor unit (Indoor Unit 2) generates at a frequency lower than that of resonance caused by the power cable and refrigerant pipe for the first indoor unit (Indoor Unit 1). On the other hand, resonance caused by the power cable and refrigerant pipe between Indoor Unit 1 and the outdoor unit is almost equal to that when only one indoor unit is connected because the positional relation between the power cable and refrigerant pipe of the first indoor unit (Indoor Unit 1) is almost equal to connecting only one indoor unit. Therefore, a rise in the disturbance voltage that emerges at 6 MHz when two indoor units are connected is considered to result from the resonance between the power cable and refrigerant pipe for Indoor Unit 1, and an increase in disturbance voltage at 4 MHz is considered to result from by resonance between power cable and refrigerant pipe for Indoor Unit 2. The difference in the frequency at which the disturbance voltage becomes maximum at 4 MHz between Twin1 and Twin2 is considered to originate from the difference in the routing of the power cable for Indoor Unit 2 between the two connections.

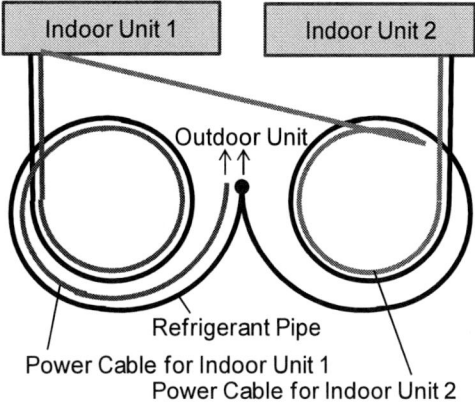

(a) Twin1

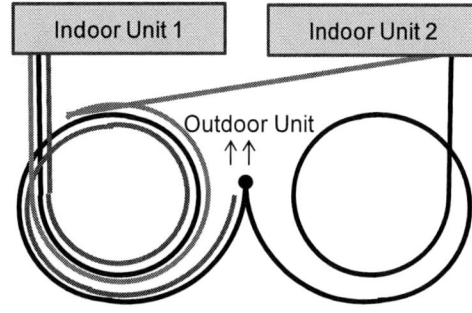

(b) Twin2

Fig. 12. Positional relations of refrigerant pipe and power cable when two indoor units are connected to one outdoor unit

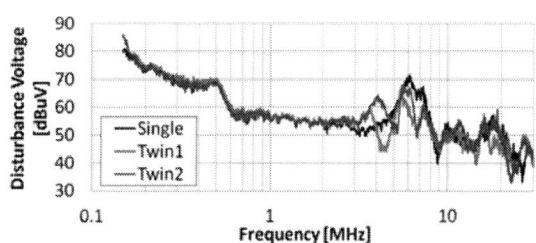

Fig. 13. Influences of number of indoor units on disturbance voltage (measurement)

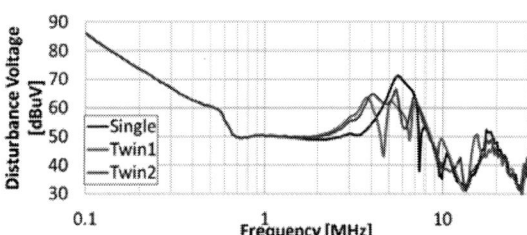

Fig. 14. Influences of number of indoor units on disturbance voltage (simulation)

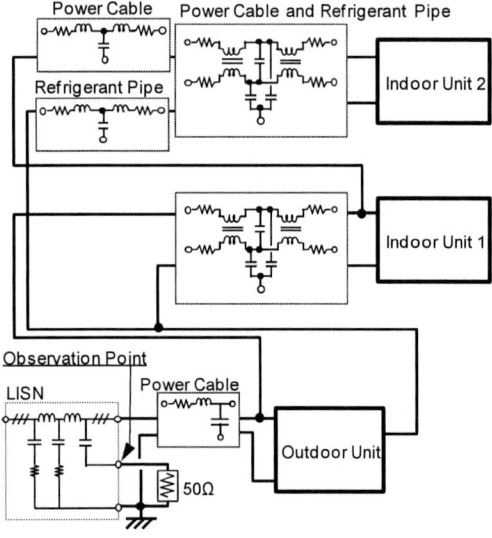

(a) Twin1

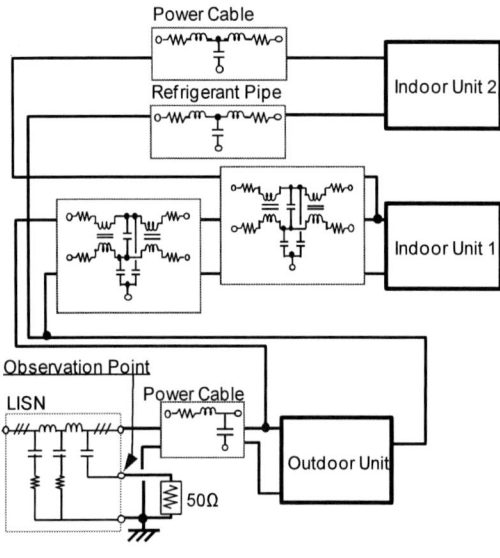

(b) Twin2

Fig. 15. Simulation model when two indoor units are connected to one outdoor unit

VII. CONCLUSION

This report described a technique for simulating the disturbance voltage of an air conditioner and examined the propagation path of conductive noise as a step to address the electromagnetic compatibility (EMC) issue of an air conditioner. As a heat pump solution company, we will continue focusing on development of noise suppression technologies using noise simulation for inverter heat pump systems.

REFERENCES

[1] S. Ogasawara, H. Akagi, "Modeling and damping of high-frequency leakage currents in PWM inverter-fed AC motor drive systems," IEEE. Trans. on Industry Applications, Vol.32, no. 5, pp.1105-1114, 1996.

[2] Y. Koyama, M. Tanaka, H. Akagi, "Modeling and analysis for simulation of common-mode noises produced by an inverter-driven air conditioner," IEEE Trans. on Industry Applications, vol. 47, no. 5, pp. 2166-2174, 2011.

[3] M. Tanaka, T. Endo, "EMC of air conditioning equipment", Institute of Electric Engineers of Japan, Industrial Application Department Convention (2010)

The 2014 International Power Electronics Conference

Iron loss evaluation of iron powder core suitable for inductor used in power converters

Tomohiro Mori, Kazunori Igarashi, Kinji
Kanagawa, Nobuyuki Yamashita
Central Research Institute
MITSUBISHI MATERIALS CORPORATION
Saitama, Japan

Toshihisa Shimizu, Yosio Bizen
Department of Electrical and Electronic Engineering
Tokyo Metropolitan University
Tokyo, Japan

Abstract— **A study evaluating the iron loss of a developed iron powder core under DC bias conditions with sinusoidal and square magnetizing voltage waveforms was conducted. It was found that the iron loss of the powder core decreases when the DC bias magnetic flux density increases. Also, the variation in the iron loss of the core material caused by the difference in the magnetizing voltage waveform is smaller than that of a conventional steel core. Hence, the developed core material is suitable for applications such as inductors used in power converters.**

Keywords— Iron Loss, Powder Core, DC bias, Square waveforms.

I. INTRODUCTION

Many kinds of power converters that provide high power density conversion have been used in industrial applications as well as consumer electronics. In particular, the market volume of power converters used in hybrid electric vehicles and photovoltaic power generation systems has grown. In these systems, inductors are used in the chopper circuit or the PWM inverter circuit, and a lower loss and smaller volume for the inductor is demanded. However, the inductors are operated under the DC bias condition with a squared magnetizing voltage waveform, and the loss characteristics are sometimes different from those of the sinusoidal magnetizing waveform conditions.

The authors have developed a specially made iron powder core (hereinafter called the Powder Core) suitable for inductors used in chopper circuits or PWM inverter circuits. Since such inductors are usually magnetized by square voltage waveforms under DC bias conditions, the iron loss characteristics are different from those of the sinusoidal voltage magnetizing waveforms without the DC bias condition. Hence, the authors studied the difference in iron loss between the Powder Core and a conventional core and verified that the developed core has an advantage with regard to lower iron loss.

The specifications and basic magnetizing characteristics of the Powder Core are explained in Section II. In Section III, the measurement method for the iron loss is proposed. The measured results for the proposed core and the conventional core are shown in

Section IV, and the differences in their loss characteristics are discussed. Finally, the conclusions are summarized in Section V.

II. SPECIFICATIONS OF MEASURED CORE

It is assumed that the Powder Core is used for the inductor in the chopper circuit with a switching frequency of a few tens of kHz. Fig. 1 shows a schematic diagram of the fine structure of the Powder Core.

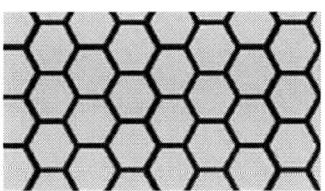

Fig. 1. Schematic diagram of the fine structure of the Powder Core; the gray portion indicates the magnetic material, and the black lines indicate the insulating boundary.

The Powder Core is composed of iron powder, and an MgO coating with high heat resistance is used as the insulating material for the iron powder. In addition, in order to reduce hysteresis losses, an annealing treatment at up to 650 °C is performed. Because the grain size of the iron powder is a few micrometers and each powder grain is insulated, eddy current loss is lower than that of a core using electrical steel sheet. Since the magnetic permeability of the magnetic powder in the Powder Core is isotropic, a three-dimensional magnetic circuit design is possible, enabling the formation of a complex shape because the shape is easily given by the powder compacting process.

III. MEASUREMENT METHOD FOR IRON LOSS

A. Core material used in the study

A Powder Core material with a toroidal core shape is used, and for comparison, two kinds of cores made from electrical steel sheet are also used. Table I shows the specifications of each core.

978-1-4799-2706-7/14 $31.00 © 2014 IEEE

TABLE I
MEASURED CORE SPECIFICATIONS

Name	Constitution	Ring size φ1: Outer diameter φ2: Inside diameter	Weight
Powder Core	Developmental iron powder core	φ1=35 mm φ2=25 mm h=5 mm	17.3 g
Core A	Si6.5% electrical steel, Board thickness = 0.1 mm	φ1=42 mm φ2=25 mm h=5 mm	30.8 g
Core B	Si3.5% electrical steel, Board thickness =0.2 mm	φ1=42 mm φ2=25 mm h=5 mm	31.7 g

The measuring frequency is 10 kHz. One of the cores (hereinafter called Core A) is a non-oriented electrical steel sheet that is made from 6.5% Si steel, and it exhibits a low iron loss at the measuring frequency. The second core is a typical core (hereinafter called Core B) made from non-oriented electrical steel sheet with 3.5% Si.

Core A and Core B are fabricated into a toroidal shape by wire electro-discharge machining, and these steel sheets were bonded to create a laminated structure. In the manufacturing process, the steel sheets are handled carefully in order to avoid damage caused by mechanical stress and are made keeping their original magnetizing characteristics.

B. DC magnetization curve

Fig. 2 shows the DC magnetization characteristics of the measured core. We used a typical DC B-H curve tracer for the measurement.

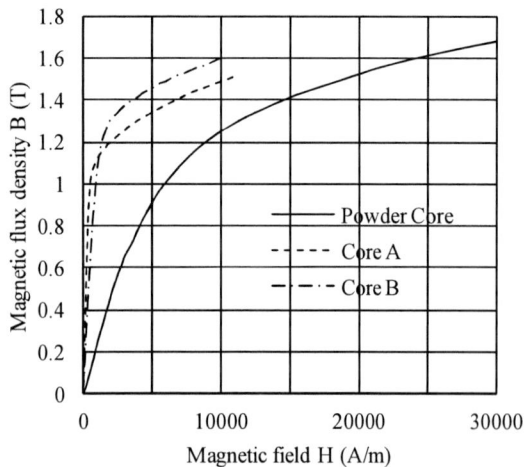

Fig. 2. DC magnetization curve.

Because the Powder Core is a compressed powder core having a structure with many gaps existing inside of the magnetic material, the magnetic flux density B increases slowly with an increase in the induced magnetic field strength H. On the other hand, the flux densities of Core A and Core B increase rapidly with an increase in the magnetic field strength.

C. Iron loss measurement method

Fig. 3 shows the measurement circuit used in this study, which is essentially the same as that adopted by Gmyrek [3]. A primary-side coil and a secondary-side coil are wound into a toroidal core, and AC current is then applied to the primary side. In order to measure the iron loss, a B-H analyzer (SY-8218: IWATSU Corp.) is used. In this system, the magnetizing current at the primary winding and the magnetizing voltage at the secondary winding are used.

A third-side coil is used to inject the DC bias current. In order to apply the DC bias current, a canceling coil is provided by preparing another core with the same size and same material as those of the measured core in order to prevent the AC current applied to the primary side from leaking to the DC power supply. The third-side coil is wound in a state with the measured core and the cancelling coil core overlapped and connected as shown in Fig. 3 to cancel the electromotive force induced in the third-side coil. In this system, it is confirmed that the iron loss value measured without applying a current to the third-side coil becomes nearly equal to the value of iron loss measured without the canceling coil.

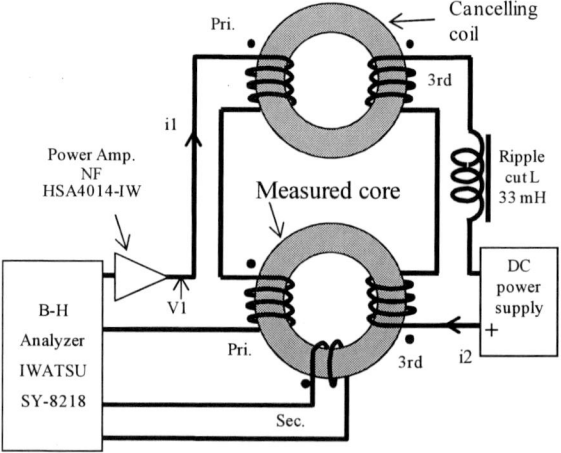

Fig. 3. Circuit for measuring iron loss.

The AC magnetizing voltage waveform of the primary side is set to a sinusoidal wave and a square wave with a 50% duty factor in order to compare the iron loss values due to the difference in the waveform. Fig. 4 shows the waveforms for that condition. ΔB represents the AC magnetic flux density generated by the AC current. B_{DC} represents the DC bias magnetic flux density generated by the DC bias current.

The 2014 International Power Electronics Conference

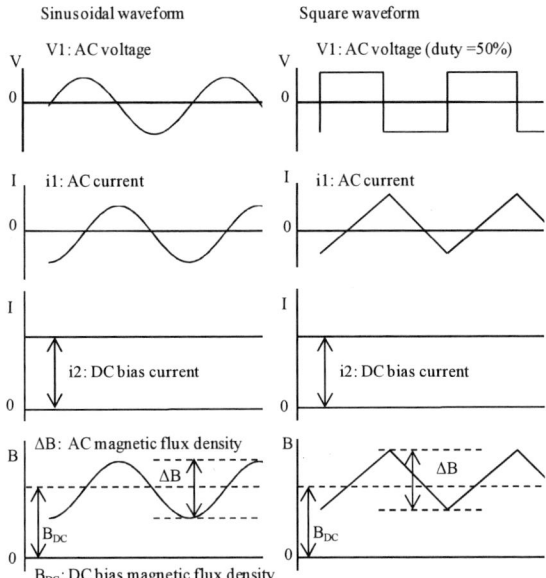

Fig. 4. Sine wave and square wave AC magnetizing voltage waveforms.

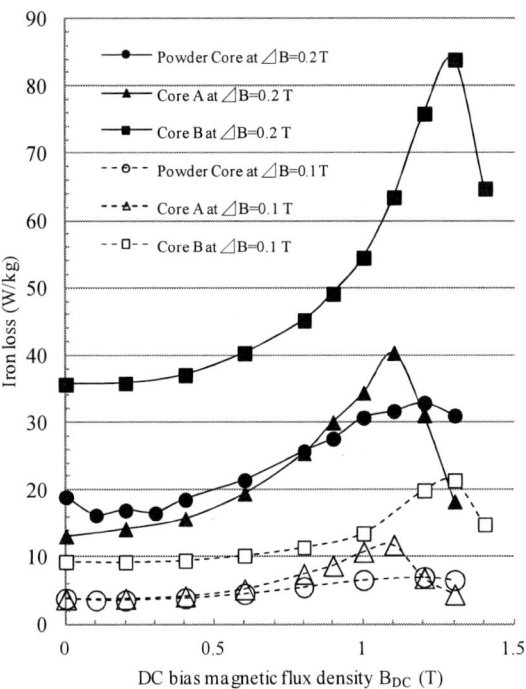

Fig. 5. Iron loss values of measured cores (ΔB is 0.1 T and 0.2 T, f=10 kHz, and AC magnetizing voltage waveform is a sinusoidal wave).

IV. MEASURED RESULTS AND DISCUSSION

A. Evaluation of iron loss under DC bias conditions

Fig. 5 shows the iron loss values of the cores for the condition where the AC magnetizing voltage waveform is a sinusoidal wave, the frequency is f=10 kHz, and ΔB is 0.1 T and 0.2 T. The horizontal axis in Fig. 5 represents the DC bias magnetic flux density B_{DC} generated by the DC bias current. It can be seen that the iron loss for each core increases with an increase in ΔB because the iron loss depends on the area of the dynamic minor loop, which increases as ΔB increases. In addition, the iron loss changes depending on the value of the DC bias magnetic flux density, and those are characterized as follows:

· from B_{DC}=0 T to 0.8 T, the core losses of the three materials are Core A ≤ Powder Core < Core B

· from B_{DC}=0.8 T to 1.1 T, the core losses of the three materials are Powder Core < Core A < Core B

· from B_{DC}=1.1 T to 1.3 T, the iron loss of Core A shows a peak value at 1.1 T and then decreases. On the other hand, the iron loss of the Powder Core decreases gradually.

In the small DC bias magnetic flux density condition, the iron loss of the Powder Core is almost the same as that of Core A. However, when the DC bias magnetic flux density increases to a range of B=0.8 T to 1.1 T, Core A shows a larger iron loss than the Powder Core. Iron loss peaks like those of Core A and Core B do not appear even for B=1.1 T or greater. These characteristics of the Powder Core are considered suitable for an inductor operating under a DC bias condition in the chopper circuit.

These different iron loss behaviors found in the electrical steel sheet and Powder Core can be interpreted as follows. It is well known that the magnetization process consists of two stages in homogeneous magnetic materials such as electrical steel sheet: the first is the domain wall displacement, followed by magnetization rotation with an increasing magnetic field. In the transition region between these two stages, which corresponds to the bending point on the magnetization curves shown in Fig. 2, it is thought that the irreversibility of the magnetization change becomes larger because it is associated with the disappearance and nucleation of reverse magnetic domains. This irreversibility possibly enlarges the minor hysteresis loop and increases the iron loss.

In practice, the peaks of the iron loss for Core A and Core B found in Fig. 5 are very sharp, which is closely related to the distinct bending points on the magnetization curves. Furthermore, the peak positions of the iron loss for Core A and Core B are coincident with the magnetic flux densities at the bending points on the magnetization curves for each core.

On the other hand, the Powder Core has an inhomogeneous microstructure due to the broad distribution of the powder particle size and shape as well as the consequent vacancies. For this reason, the effective magnetic field applied to each powder particle is different because of the different demagnetizing field. This means that the Powder Core exhibits no apparent transition region where the iron loss shows a distinct peak.

978-1-4799-2706-7/14 $31.00 © 2014 IEEE   2985

B. Evaluation of iron loss due to the difference in AC magnetizing voltage waveform

In the evaluation of iron loss in the previous chapter, although the iron loss measurement uses a sinusoidal wave for the AC magnetizing voltage waveform, the AC magnetizing voltage waveform applied to the inductor for the buck and step-down choppers is a square wave. Hence, the iron loss of the Powder Core is measured using a square wave for the AC magnetizing voltage waveform, and it is compared with the measured value for the iron loss using a sinusoidal wave.

Fig. 6 shows the iron losses at the AC magnetizing voltage waveform set to a square wave and a sinusoidal wave. The measurement conditions are a frequency of f=10 kHz and ΔB=0.1 T. For comparing the properties of the core materials, Core A and Core B are evaluated with a similar measurement.

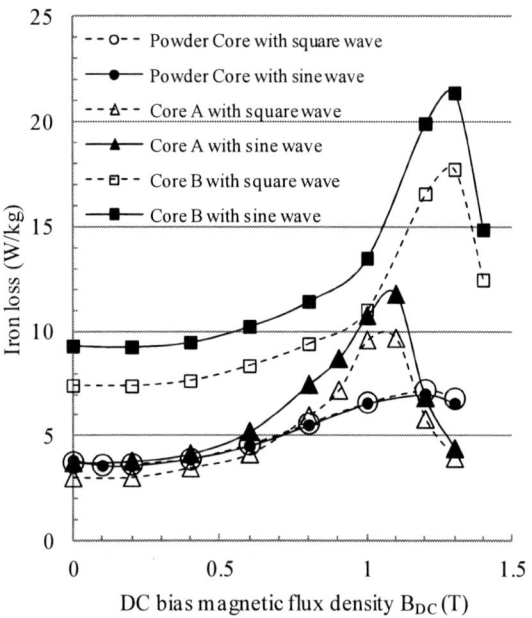

Fig. 6. Iron loss values when the AC magnetizing voltage waveform is a sinusoidal wave and a square wave (ΔB=0.1 T, f=10 kHz).

From Fig. 6, the difference between iron losses due to the difference in the AC magnetizing voltage waveform hardly appeared in the Powder Core. In contrast, it is confirmed that the iron loss with a sinusoidal wave is larger by about 1 to 3 (W/kg) compared with that of the square wave in Core A and Core B.

Here, the influence of the magnetizing waveform on both the hysteresis loss and eddy current loss is studied. In general, the hysteresis loss is proportional to the frequency, and the eddy current loss is proportional to the square of the frequency. Hence, based on the relationship between the frequency and the iron loss, the iron loss can be separated into the hysteresis loss and the eddy current loss. The loss separation is conducted based on the iron loss measured at frequencies of 2 kHz, 5 kHz, 10 kHz, and 20 kHz. Figs. 7 and 8 show the measured results for the hysteresis loss and the eddy current loss, respectively.

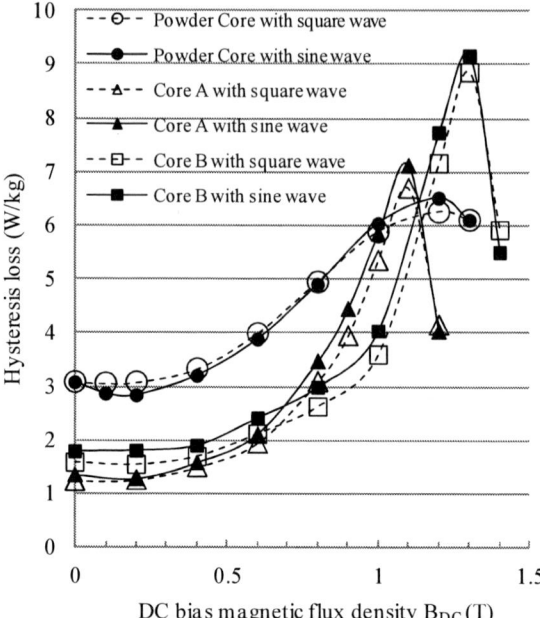

Fig. 7. Hysteresis loss (calculated value) when the AC magnetizing voltage waveform is a sinusoidal wave and a square wave (B=0.1 T, f=10 kHz).

The hysteresis loss of each core material has almost the same value, even when the magnetizing waveform is different, as shown in Fig. 7. On the other hand, the eddy current loss of the Powder Core has a constant value, even when the DC bias magnetic flux density B is increased, as shown in Fig. 8. Also, the value itself is smaller than those for the other materials.

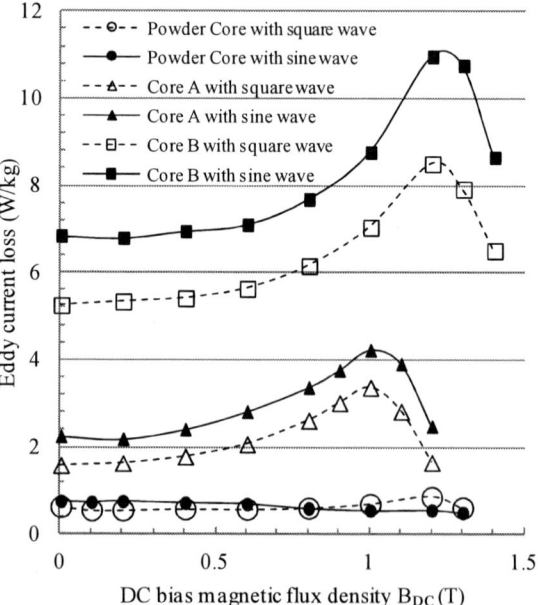

Fig. 8. Eddy current loss (calculated value) when the AC magnetizing voltage waveform is a sinusoidal wave and a square wave (ΔB=0.1 T, f=10 kHz).

By contrast, the eddy current losses of Core A and Core B are larger for a sinusoidal wave irrespective of whether the bias magnetic flux density is large or small. It is considered that this is caused by the fact that the gradient change rate of the sinusoidal magnetizing waveform is larger than that of the triangular magnetizing waveform. A more detailed analysis will be performed in future work.

V. CONCLUSION

In this paper, the iron loss characteristics of the newly developed iron powder core and conventional cores were discussed under the condition of DC bias magnetization and different magnetizing waveforms, i.e., a sinusoidal waveform and a square waveform. The Powder Core shows a lower iron loss compared with that of the core made from electrical steel sheet under the DC bias condition. It is also shown that the difference in iron loss caused by the magnetizing waveform (i.e., a sinusoidal waveform and a square waveform) is caused by the eddy current loss. Also, the Powder Core exhibits a smaller iron loss regardless of the magnetizing waveforms compared with the other magnetic materials. It is concluded that the proposed Powder Core has desirable loss characteristics for an inductor used in a chopper circuit or a PWM inverter.

REFERENCES

[1] K. Kakazu, T. Shimizu, A. Moritani, K. Takano, H. Ishii, "Evaluation of Iron Loss of Filter Inductor on a PWM Inverter," IEEJ SPC-11-37, pp.71–76, (2011)

[2] K. Emori, T. Shimizu, Y. Bizen, "Discussion on Design Optimization of Inductors Focused on Copper and Iron Loss," IEEJ SPC-12-131, pp.1–6, (2012)

[3] M. Albach, T. Dürbaum, A. Brockmeyer, Calculating Core Losses in Transformer for Arbitrary Magnetizing Currents: A Comparison of Different Approaches," Proc. 27th Annual IEEE Applications of Power Electronics Conference, Vol.2, pp.1463–1468, (1996)

[4] Z. Gmyrek, A. Boglietti, A. Cavagnino, "Iron Loss Prediction with PWM Supply using Low- and High-Frequency Measurements: Analysis and Results Comparison," IEEE Transactions on Industrial Electronics, Vol.55, No.4, pp.1722–1728, (2008)

[5] T. Shimizu, K. Kakazu, K. Takano, H. Ishii, "Verification of Iron Loss Calculation Method using a High-Precision Iron Loss Analyzer," IEEJ Transactions on Industry Applications, Vol.133, No.1, pp.84–93, (2012)

[6] A. Brockmeyer, "Experimental Evaluation of the Influence of DC-Premagnetization on the Properties of Power Electronic Ferrites," APEC IEEE Conference Proceedings, pp.454–460, (1996)

[7] S. Imamori, H. Ohguchi, A. Toba, "DC-biased Magnetic Properties of Soft Magnetic Composite," 2010 National Convention Record IEEJ, No.2, p.146

[8] T. Kusakabe, R. Yamada, K. Fujisaki, "Iron Loss Increase by the Inverter Excitation of the Electromagnetic Steel Sheet Under the High Magnetic Flux Density," 2012 National Convention Record IEEJ, No.4, p.120

The 2014 International Power Electronics Conference

Optimized Tuning Method of Stationary Frame Proportional Resonant Current Controllers

Fernando Ortiz Martinz, Kelly Caroline Mingorancia de Carvalho, Naji Rajai Nasri Ama, Wilson Komatsu, Lourenço Matakas Junior

Electrical Energy and Automation Department
Polytechnic School of the University of São Paulo
São Paulo, Brazil
fernando_martinz@yahoo.com, matakas@pea.usp.br, wilsonk@usp.br

Abstract— **This paper presents an optimized tuning method of Proportional Resonant (PR) current controllers implemented in the stationary frame. The approach consists of defining three operation regions based on behavior of the closed loop poles, and of determining simplified tracking and disturbance transfer functions for each of these regions. Charts of proportional and resonant normalized gains versus settling time and overshoot are then built, allowing the designer to easily visualize the influence of these gains on the transient response. It is shown that the disturbance response performance is more significant than the tracking one, and that minimum settling time and overshoot for this case are obtained if the resonant gain is set equal to the controller resonant frequency, providing a simple and efficient rule-of-thumb for the designer. Simulation and experimental results are presented to validate the method.**

Keywords— PWM Converters, Current Control, Proportional Resonant Controller, Design Optimization.

I. INTRODUCTION

Proportional Resonant (PR) current controllers are adopted in several power electronics applications of grid connected converters [1]. The reasons for using these controllers are: they ideally provide zero steady state error at the resonant frequency ω_o [1],[2]; they are equivalent to the synchronous frame Proportional Integral (PI) controllers [2] and; only one controller is able to compensate for both positive and negative sequences [2].

Since the steady state error is null, transient response performance must be the main concern when evaluating a PR current control loop. However, few works in literature investigate the PR transient response performance, and many times a clear or simple tuning method is not provided. For some papers in which this is done, the design is based on the choice of the crossover frequency ω_{CR} [2], but the most common procedure consists in choosing the control system phase margin (PM) [3]. However, as stated in [4], for closed loop systems which cannot be reduced to the standard second order form (i.e., underdamped systems without zeros), or for systems without two dominant complex poles, the phase margin cannot be used as a performance criteria. That is the case of the closed loop systems analyzed in [1],[2],[3], and also in this paper: neglecting the computation and actuation delay, the continuous time transfer functions

have three poles and two zeros. Moreover, as shown in [1], depending on the choice of the resonant gain k_R, there will be no complex poles. That is, for the system under analysis the phase margin cannot be the only parameter to be considered for obtaining an optimized transient response. This is the reason why the authors of [5] state that the design proposed in [3] is not optimized for the transient performance. In this way, [1] presents an optimized design for a stationary PR controller, based on the analysis of the closed loop poles. However, the results are only valid for the power converter described in [1], i.e., analytical expressions of the optimized k_R were not presented. On the other hand, [5] derives expressions of optimized gains, but the results are obtained for a positive sequence synchronous PI controller, which is different from the stationary PR controller proposed in this paper.

Thus, considering these previous aspects of PR design found in literature, the contributions of this paper are:

1) A different approach from that presented in [1],[2],[3] is developed in Sections II and III to design the PR controllers. The tracking and disturbance transfer functions are individually investigated, and then simplified expressions based on the analysis of the behavior of the closed loop poles are derived. These expressions allow to analytically obtain optimized gains, settling time and overshoot when designing PR current controllers, extending the results to other power converters;

2) Low, medium and high resonant gain regions are defined in Section III for the closed loop system operation and it is shown that the transient performance highly depends on which region the resonant gain lies. Moreover, this approach mathematically explains the pole behavior observed, but not detailed in [1];

3) Charts of normalized proportional and resonant gains *versus* overshoot and settling time parameters are built in Section IV, as it is done in [6], allowing to clearly visualize the effect of the variation of gains on the transient response. Since these gains are normalized in terms of grid and converter parameters, the results can be extended for any grid connected power converter with an inductor filter;

978-1-4799-2706-7/14 $31.00 © 2014 IEEE

4) It is shown in Section IV that the disturbance response is preponderant over the tracking response on transient analysis, and that the settling time is the critical parameter to be considered;

5) An optimized resonant gain is derived in Section IV from the transfer functions and from the charts of overshoot and settling time. It is shown that minimum settling time is achieved if the resonant gain is set equal to the controller resonant frequency. Thus, a simple and efficient rule-of-thumb is provided for the designer.

Simulation and experimental results are presented to validate the proposed method.

II. System Modeling

The control system is composed by a Pulse Width Modulation (PWM) power converter, with an inductor filter of resistance R and inductance L, whose output is connected to the mains voltage $V(s)$. References [2] and [6] have demonstrated that the power converter plus the PWM block may be expressed by an averaged model, which consists of a voltage controlled source $V_C(s)$ with unitary gain, since $V_{CREF}(s)$ is divided by the DC link voltage. Thus, the per-phase closed loop system is represented by Fig. 1, where $I(s)$ is the output current. The main objective of this loop is to achieve good tracking performance of the reference current $I_{REF}(s)$, i.e., to minimize the current error $ERR(s)$. Another important task of the current loop is to reject the disturbance represented by the output voltage $V(s)$.

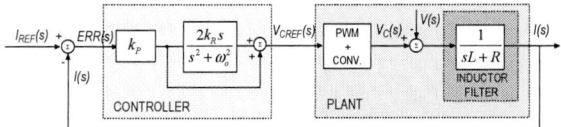

Fig. 1. Closed current loop model with PR controller.

The current controller is a Proportional Resonant type and it is implemented in stationary coordinates, as it is done in [3], i.e.,

$$G_C(s) = k_P\left(1 + \frac{2k_R s}{s^2 + \omega_o^2}\right), \tag{1}$$

Assuming $k_P \gg R$, the closed loop tracking and disturbance transfer functions are, respectively, given by

$$G_T(s)\big|_{V=0} = \frac{(k_P/L)(s^2 + 2k_R s + \omega_o^2)}{s^3 + s^2\left(\dfrac{k_P}{L}\right) + s\left(\omega_o^2 + \dfrac{2k_P k_R}{L}\right) + \omega_o^2\left(\dfrac{k_P}{L}\right)} \tag{2}$$

$$G_D(s)\big|_{iref=0} = \frac{-(s^2 + \omega_o^2)/L}{s^3 + s^2\left(\dfrac{k_P}{L}\right) + s\left(\omega_o^2 + \dfrac{2k_P k_R}{L}\right) + \omega_o^2\left(\dfrac{k_P}{L}\right)}. \tag{3}$$

If $s = j\omega_o$ is substituted in (2) and (3), the magnitudes of tracking and disturbances transfer functions are respectively equal to unity and zero, and the phase angles are respectively zero and ninety degrees. This well

known result [2] shows that, in steady state, perfect tracking and disturbance responses for the PR controller are achieved at the resonant frequency, for any k_P and k_R. Consequently, the PR closed loop performance shall be investigated by means of the transient response, as it is done in next section, where the following assumptions are made:

- the analysis is made in the continuous time, i.e., the actuation and computation time delay is neglected, and then validated for the discrete time case by means of experimental results;

- the resonant frequency ω_o is at least one decade below the open loop (0 dB) crossover frequency ω_{CR};

- the controller parameters are expressed in terms of grid and inductor filter values, with the purpose of extending the results of this paper to other power converter ratings, i.e.,

$$k_P = k_P^{pu} \cdot \omega L, \tag{4}$$

$$\omega_o = k\omega, \tag{5}$$

$$k_R = k_R^{pu} \cdot \omega, \tag{6}$$

where k is the relation between the resonant frequency and the base (or fundamental) frequency ω, k_P^{pu} is the normalized proportional gain defined in [6], and k_R^{pu} is the normalized resonant gain.

III. Analysis of Closed Loop Poles and Zeros

The closed loop poles are the roots of the denominators of (2) and (3), which are cubic equations that can be solved by the Tartaglia's method. This method defines a discriminant d given by

$$d = \frac{\omega^6}{27}\left[k^2 + 2k_P^{pu}k_R^{pu} - \frac{\left(k_P^{pu}\right)^2}{3}\right]^3 +$$
$$+ \left(k_P^{pu}\right)^2 \omega^6 \left[\frac{k^2 + 2k_P^{pu}k_R^{pu}}{6} - \frac{\left(k_P^{pu}\right)^2}{27} - \frac{k^2}{2}\right]^2. \tag{7}$$

Considering that the coefficients of the denominators (2) and (3) are real, the following conditions determine three regions for the resonant gain: (i) if $d > 0$, one pole is real and two poles are conjugate complex; (ii) if $d = 0$, all poles are real, and at least two of them are equal to each other, and (iii) if $d < 0$, all three poles are real and different from each other. These conditions mathematically explain the pole behavior observed in [1].

However, solutions of the third order denominator of (2) and (3) are not trivial, since they depend on k_P, k_R and ω_o, and thus, poles shall be obtained through numerical simulation and inspection. In this way, Fig. 2 shows the behavior of the closed loop poles (represented by 'X') and zeros (represented by 'O'), for the *tracking transfer function* and three different normalized resonant gains k_R^{pu}. The simulation parameters are $L=10mH$, $\omega=377\ rad/s$, $k=1$ and $k_P^{pu}=63.66$. The normalized proportional gain k_P^{pu} is chosen to be the maximum permitted value for a two-level PWM and a single-phase converter, for a switching frequency of 12 kHz, according to [6]. The first reason to choose high values for k_P^{pu} is that it is important to assure that the three resonant gain regions will always occur, and simulations of (7) show that, for given values of ω and k, this is true only if $k_P^{pu} \geq 5k$. The second reason, to be demonstrated in Section IV, is that as the proportional gain is increased, the closed loop tracking performance is improved.

Initially a low resonant gain region, where $d > 0$, is defined and shown for $k_R^{pu} = 0.16$ at the top of Fig. 2. The zeros of the tracking transfer function (2) are

$$z_{T1,2} = -k_R \pm \sqrt{k_R^2 - \omega_o^2} \ . \tag{8}$$

When $k_R^{pu} = 0$, two poles are equal to $\pm j\omega_o$ and are cancelled by the two zeros of (8), and thus a distant third pole at $-k_P/L$ determines the dynamic response. For a very low range of $0 < k_R^{pu} \leq 0.011 k_P^{pu}$ (obtained by simulation), this pole-zero cancellation persists, and the system behaves like a first order system, expressed by

$$G_{TVLOW}(s) \cong \frac{\left(k_P/L\right)\left(s^2 + 2k_R s + \omega_o^2\right)}{\left(s + k_P/L\right)\left(s^2 + 2k_R s + \omega_o^2\right)} \cong \frac{\left(k_P/L\right)}{\left(s + k_P/L\right)} . \tag{9}$$

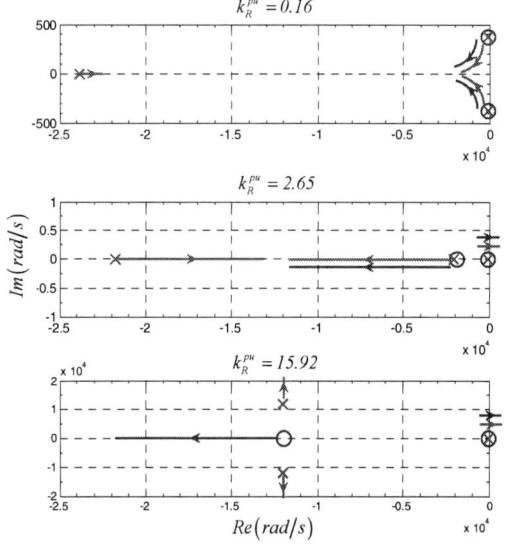

Fig. 2. Closed loop poles and zeros behavior for the tracking transfer function, when $k_R^{pu} = 0.16$ (top), $k_R^{pu} = 2.65$ (center) and $k_R^{pu} = 15.92$ (bottom).

By increasing k_R in the low gain region, the system behaves like a second order one, not modeled in this paper, with two complex poles and two complex zeros, since the distant real pole $-k_P/L$ may be neglected. In this case, the rising of k_R forces these two poles and zeros to move to the left, and they tend to become real, while the distant third real pole slowly moves to the

origin. The first region limit occurs when $d = 0$, two poles are real and equal to each other and the two zeros are equal to $-\omega_o$. In this case, $k_R \cong \omega_o$ (for the simulation of Fig. 2, $k_R^{pu} \cong k = 1$), which is equivalent to the optimized tuning for the PR of [1] for the disturbance response, to be discussed in Section IV. Despite the dynamic response being slow and given by these two poles and zeros close to the origin, tuning at this low resonant gain region, which is restricted to $k > k_R^{pu} \geq 0$, is commonly adopted in the literature [2], since the optimized gain for tracking response lies in this region.

A medium gain region is determined by the condition $d < 0$. Now there are three real poles, as it is shown at the center of Fig. 2 for $k_R^{pu} = 2.65$. Two poles move in opposite directions as k_R^{pu} increases, till they become coincident at the left side of Fig. 2. A third pole moves to the origin, and tends to be cancelled with one real zero at $z_{T1} \cong 0$. The tracking transfer function is approximately a second order system expressed by (10), with a real zero $z_{T2} \cong -2k_R$ and two real poles of (11), that is,

$$G_{TMED-HIGH}(s) \cong \frac{\left(k_P/L\right)\left(s + 2k_R\right)}{s^2 + s\left(k_P/L\right) + \left(2k_R k_P/L\right)}, \tag{10}$$

$$p_{MED1,2} \cong \left(\frac{k_P}{2L}\right)\left(1 \pm \sqrt{1 - \frac{8Lk_R}{k_P}}\right). \tag{11}$$

The second occurrence of $d = 0$ takes place for $k_R = k_P/8L$ (or $k_R^{pu} = k_P^{pu}/8$), when there are two real and equal poles, and a third pole that cancels a zero at the origin. That is, the medium gain region is restricted to $k_P^{pu}/8 > k_R^{pu} > k$.

Finally, a high gain region is defined, where again $d > 0$, two poles are conjugate complex, and a third pole is real, as shown at the bottom of Fig. 2 for $k_R^{pu} = 15.92$. As k_R^{pu} increases, only the imaginary parts of the dominant complex poles are affected, one real zero moves away from the origin, and a pole-zero cancellation occurs close to the origin. The closed loop system is a second order type, whose transfer function is also given by (10), however with two conjugate complex poles, i.e.

$$p_{HIGH1,2} \cong \left(\frac{k_P}{2L}\right)\left(1 \pm j\sqrt{\frac{8Lk_R}{k_P} - 1}\right). \tag{12}$$

Regarding the *disturbance transfer function*, since the zeros are constant and equal to $\pm j\omega_o$, there is no pole-zero cancellation. Thus, for the low gain region, the dynamic response is given by the two conjugate complex poles at $p_{1,2} \cong -k_R \pm j\sqrt{k_R^2 - \omega_o^2}$. Neglecting the distant pole

$$G_D(s) \cong \frac{-\left(s^2 + \omega_o^2\right)/k_P}{\left(s^2 + 2k_R s + \omega_o^2\right)} . \tag{13}$$

For the medium and high gain region, the closed loop poles are given by (11) and (12), respectively. Since there is no pole-zero cancellation, by inspection, the dominant pole for these regions is $p_3 \cong -k_R + \sqrt{k_R^2 - \omega_o^2}$.

IV. TRANSIENT RESPONSE EVALUATION

Simulations in the Appendix show that for tracking and disturbance transfer functions, the 'settling time' for a sinusoidal input is approximately equal to that obtained with a step input, when the phase of the reference current is 90° and when the phase of the grid voltage is 0°. Furthermore, the critical overshoot for a sinusoidal input occurs when the phase angle is 90°, and it is also approximately equal to that obtained with a step input. Consequently, this paper adopts the step response to derive expressions of settling time and overshoot, and to build the charts of transient parameters *versus* controller gains.

Figs. 3 and 4 show simulations in *Matlab* of the 2% settling time charts, normalized in terms of the fundamental period, for the tracking and disturbance transfer functions, respectively, while Figs. 5 and 6 show the overshoot charts for these two transfer functions, respectively. A unit step input is applied to the tracking (2) and disturbance (3) transfer functions, for $L= 10mH$ and $\omega=377rad/s$. Selecting $k=1$, the three k_R^{pu} regions are highlighted in these figures.

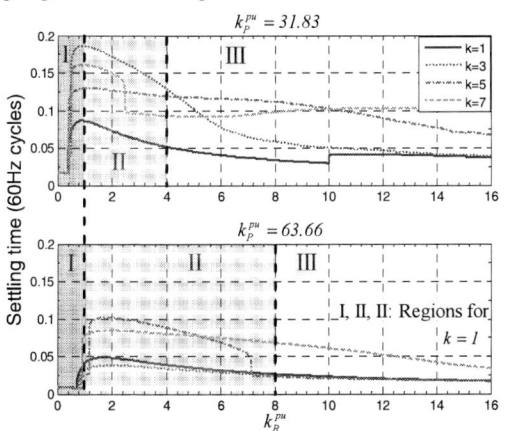

Fig. 3. Settling time of the tracking transfer function for a unit step input, when $k_P^{pu} = 31.83$ (top) and $k_P^{pu} = 63.66$ (bottom).

Figs. 3 to 6 show that:
- the disturbance settling time (Fig. 4) is more significant than the tracking one (Fig. 3), for all regions;
- the disturbance overshoot at the low gain region reaches its maximum value (Fig. 6). On the other hand, the tracking overshoot is very large in medium and high gain regions (Fig. 5);
- at the high gain region, the disturbance settling time (Fig. 4) is prohibitively long if $k=1$ (in other words, $\omega_o = \omega$);
- the tracking settling time (Fig. 3) and overshoot (Fig. 5) are highly dependent on the proportional gain (see also Table I). If the proportional gain is increased, the tracking performance is improved;
- the disturbance settling time (Fig. 4) and overshoot (Fig. 6) are independent from the proportional gain (see

also Table I): the charts for $k_P^{pu} = 31.83$ and $k_P^{pu} = 63.66$ are superimposed in Figs. 4 and 6.

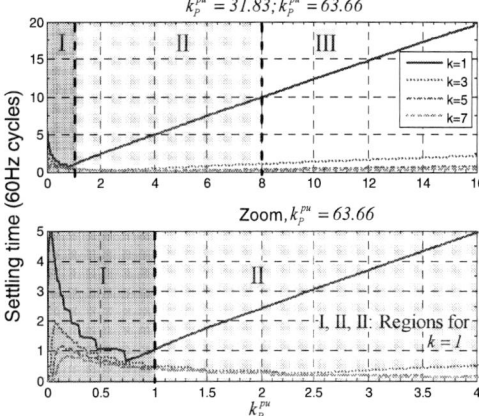

Fig. 4. Settling time of the disturbance transfer function for a unit step input, when $k_P^{pu} = 31.83$ and $k_P^{pu} = 63.66$ (top, superimposed). Zoom of the first region when $k_P^{pu} = 63.66$ (bottom).

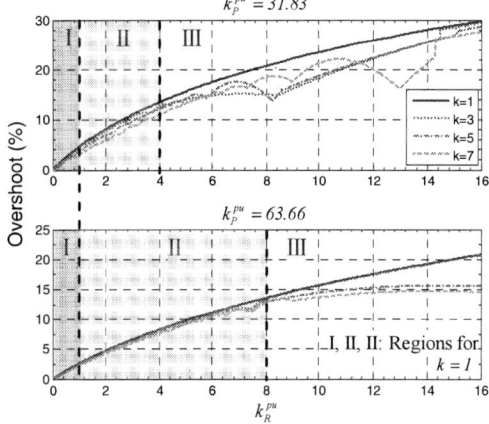

Fig. 5. Overshoot of the tracking transfer function for a unit step input, when $k_P^{pu} = 31.83$ (top) and $k_P^{pu} = 63.66$ (bottom).

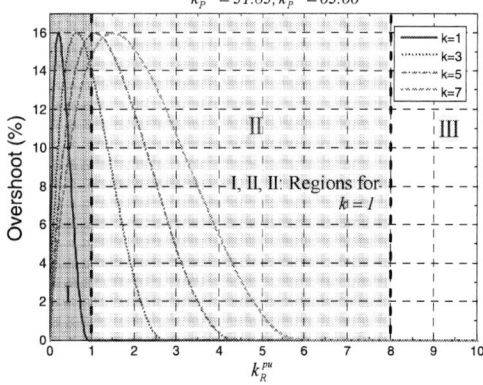

Fig. 6. Overshoot of the disturbance transfer function for a unit step input, when $k_P^{pu} = 31.83$ and $k_P^{pu} = 63.66$, superimposed.

That is, there are optimized gains for the tracking and disturbance responses, as observed by the authors of [1]. Results of Figs. 3 to 6 have shown that the optimized gain for the tracking response is limited to the validity of

The 2014 International Power Electronics Conference

TABLE I – OVERSHOOT AND SETTLING TIME EXPRESSIONS

k_R	Tracking transfer function		Disturbance transfer function	
	Overshoot	Settling time (2%)	Overshoot	Settling time
Very low	Negligible	$4L/k_P$	$-\dfrac{200k_R}{\sqrt{\omega_o^2-k_R^2}}\exp\left(-\dfrac{3k_R\pi}{2\sqrt{\omega_o^2-k_R^2}}\right)$	$\dfrac{-\ln\left(0,01\omega_o/k_R\right)}{k_R}$
Medium	$\dfrac{50}{x}\left[(x+1)^{-1/x}(x-1)^{1+1/x}+(x-1)^{1/x}(x+1)^{1-1/x}\right]$ $x=\sqrt{1-8k_RL/k_P}$	$\dfrac{2L}{k_P(1-x)}\cdot\left[\ln\left(1-\dfrac{1}{x}\right)+3,22\right]$ $x=\sqrt{1-8k_RL/k_P}$	Null	$\dfrac{4}{k_R-\sqrt{k_R^2-\omega_o^2}}$
High	$100\exp\left\{-\left[\left[\pi-\tan^{-1}\left(\dfrac{\sqrt{k_P(8Lk_R-k_P)}}{4Lk_R-k_P}\right)\right]\cdot\left(\dfrac{k_P}{8Lk_R-k_P}\right)^{1/2}\right]\right\}$	$8L/k_P$		

the first order model, i.e., for very low gains at the first region in the range $0 < k_R^{pu} \le 0.011k_P^{pu}$. Furthermore, the optimized design for the disturbance response is achieved if $k_R^{pu} \cong k$ (in other words, if $k_R \cong \omega_o$), when the disturbance settling time is close to its minimum value (Fig. 4) and the overshoot for the disturbance is null (Fig. 6). In this case $d = 0$, two poles are real and equal to each other and the two zeros are $-\omega_o$. Since the disturbance settling time is critical in PR transient response, it is recommended to set $k_R^{pu} = k$.

Table I shows expressions of overshoot and settling times, derived for the step response of (9), (10) and (13), according to method presented in [7]. As can be seen, the minimum settling time for the disturbance response, when the disturbance optimized gain is chosen, is $T_{SMIN} \cong 4/k_R = 4/\omega_o$.

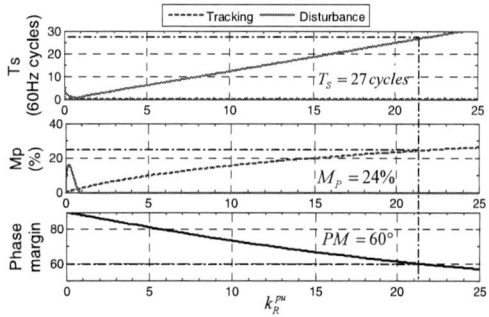

Fig. 7. Settling time (top), overshoot (center) and phase margin (bottom) of the tracking and disturbance transfer functions, when $k_P^{pu} = 63.66$ and $k = 1$.

As it was highlighted in the Introduction, for the system under analysis the phase margin cannot be the only parameter to be considered when designing PR current controllers (as it is done in [3]), especially if an optimized response is desired. This statement can be confirmed in simulated charts of Fig. 7, in which the phase margin is set to a normally recommended value of 60° for $k = 1$, resulting in $k_R^{pu} \cong 21$, an excessive 27 cycles disturbance settling time and 24% tracking overshoot. In this way, Fig. 7 reinforces the conclusions of [4], which affirm that for this type of

system, phase margin indicates robustness of the system, but does not assure good transient performance.

In summary, *the rule-of-thumb for design PR current controllers* in continuous time, and in terms of normalized parameters is:

(1) Set the proportional gain equal to the maximum limit k_{PMAX}^{pu} and be sure to satisfy the condition $k_P^{pu} \ge 5k$. For a two-level converter, [6] states that $k_{PMAX} = 2f_{TRI} \cdot L$, where f_{TRI} is the switching frequency;
(2) If the tracking response must be optimized, the resonant gain must be set in the range $0 < k_R^{pu} \le 0.011k_P^{pu}$;
(3) If the disturbance response must be optimized, set the resonant gain equal to $k_R^{pu} = k$.

V. EXPERIMENTAL RESULTS

Experimental tests are made in the three phase setup of Fig. 8 (only phase A represented), for the parameters of Table II. The control loop is implemented in a Texas TMS320F28335 Digital Signal Processor (DSP). The PR discretization method is Tustin with pre-warping [1], and the discrete proportional gain is set according to [6], considering only the proportional action and a damping factor of 0.55. Only two current controllers are implemented, and the reference voltage of phase C is the negative sum of the references voltages of phases A and B, as it is done in [3],[8]. This strategy eliminates the coupling between phases and thus the controllers are adjusted with the same gains of the single phase controller modeled in previous sections.

The three phase PLL is based on a multiplier phase detector with a moving average filter, and a Proportional Integral controller whose gains are adjusted as described in [9], resulting in $k_{PPLL}=139.426$ and $k_{IPLL}=0.574$.

Voltage sags are generated by short circuiting the grid side for a few cycles, as shown in Fig. 8, and the DC-link is composed by capacitors connected to three

The 2014 International Power Electronics Conference

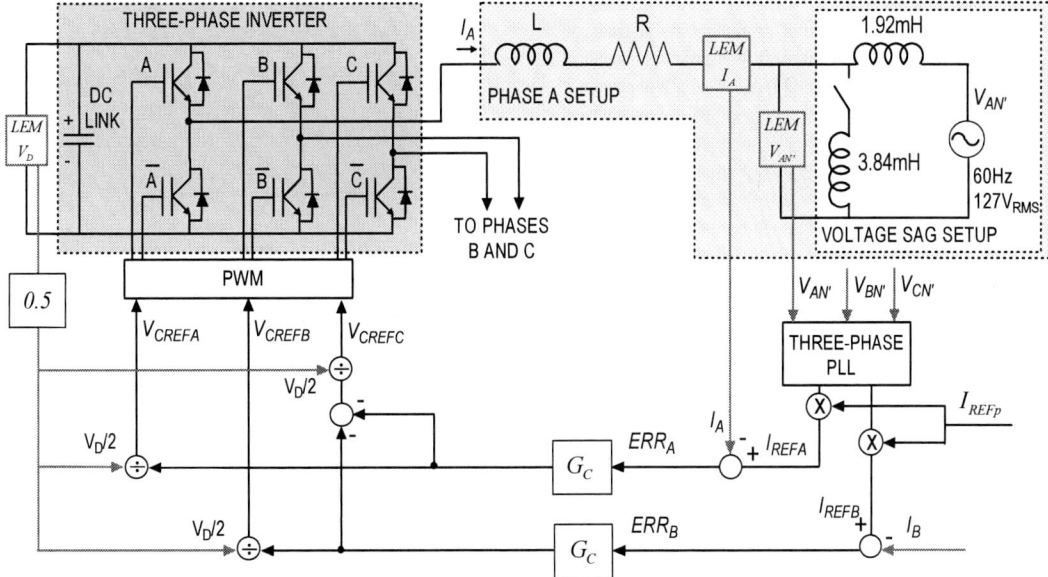

Fig. 8. Experimental setup and control loop

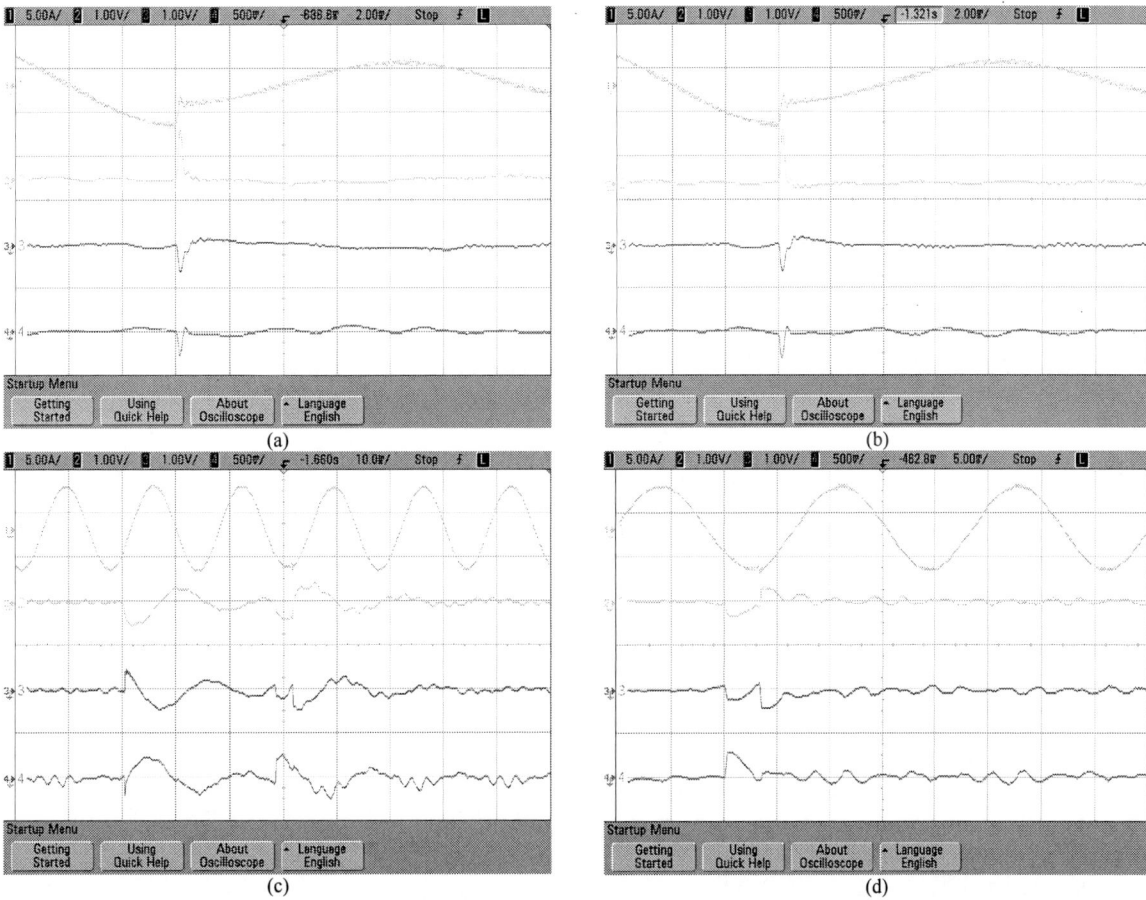

Fig. 9. Experimental results of the discrete time PR. Reference current step for (a) $k_R = 60$ and (b) $k_R = 377$ (optimized). Grid voltage step for (c) $k_R = 60$ and (d) $k_R = 377$ (optimized). CH1 (5A/div): phase A output current. CH2 (1A/div): phase A current error. CH3 (1A/div): phase B current error. CH4 (1A/div): phase C current error. Time scale: 2ms/div for (a) and (b), 5ms/div for (c) and 10ms for (d).

non-controlled rectifier fed by variable autotransformers. All voltages and currents are measured by means of Hall effect sensors (LEM in Fig. 8). Experimental waveforms are: the phase A output current, measured with an Agilent N2775 current probe, and discrete time current errors. These errors are internal variables that are measured at general purpose PWM outputs, used in the setup as digital to analog converters (DAC). The switching frequency of these PWM signals is 120kHz, and they are filtered by means of two cascaded RC filters, each one with $R=560\Omega$ and $C=100nF$. All signals are acquired with a DSO6014 Agilent oscilloscope.

TABLE II - EXPERIMENTAL SETUP AND CONTROL LOOP PARAMETERS

DC-link voltage	370V_{DC}
Grid line to line voltage/frequency	220V_{RMS} /60Hz
L-Filter inductances	10mH
L-Filter resistances	0.72Ω
Switching frequency	12kHz
Sampling frequency	24kHz
Reference (output) current	5A_{PEAK}, +90°, 60Hz
Resonant frequency ω_0	377rad/s
Current controller proportional gain	120 V/A

Fig. 9 shows experimental results for a reference current variation from 1 to 0.5 per unit, and for a step at the disturbance (i.e., a grid voltage sag to 66%), for the disturbance optimized design ($k_R^{pu}=1$ or $k_R=377rad/s$) and for one possible tracking optimized design ($k_R^{pu}=0.16$ or $k_R=60rad/s$) at very low gain region. The following restrictions must be observed when analyzing these results:
- the disturbance and tracking transfer function effects are not separated in the experimental setup, and thus the values of settling time and overshoot will not be exactly the parameters of Table II and Figs. 3 to 7;
- since the steady state errors are null for the fundamental component, the definition of settling time is not valid. In this way, settling times for sinusoidal inputs are estimated by the time the current errors stays in the range of ±2% of its maximum fundamental peak value;
- grid voltages are distorted (V_5=1.38%; V_7=0.4%, V_{11}=0.34%, V_{13}=0.42%), and thus the current errors for harmonic orders different from the fundamental component will not be null;
- the instant in which the voltage sag occurs cannot be finely controlled, in the experimental setup, to the disturbance response worst case, which is 90° phase angle (see Appendix);
- Figs. 3 to 7 show simulations for tracking and disturbance transfer functions, but experimental results show the waveforms of discrete time current errors $E_{RR}(z)$. Appendix demonstrates that for the settling time, the values of Figs. 3 and 4 can be directly compared to experimental results of current errors in Fig. 9.

The settling times for the experimental setup are presented in Table III, demonstrating that results are very compatible with the simulations of Section IV,

confirming the optimized design method for both tracking and disturbance responses. The effect of the RC filter of the DACs on settling times is negligible.

TABLE III – SIMULATION AND EXPERIMENTAL SETTLING TIMES

Case	k_R (k_R^{pu})	Simulations		Experimental	
		T_S	Fig.	T_S	Fig.
Reference current step	60 (0.16)	0.33ms (0.02 cycles)	3 (top)	1.08ms (0.07 cycles)	9a
	377 (1.00)	1.50ms (0.09 cycles)		2.22ms (0.13cycles)	9b
Grid voltage step	60 (0.16)	46.00 ms (2.76 cycles)	4	55.00ms (3.30 cycles)	9c
	377 (1.00)	10.61ms (0.64 cycles)		14.15ms (0.85 cycles)	9d

Overshoots may be obtained from phase A output current in Fig. 9, according to the definition for sinusoidal inputs presented at the Appendix. However, due to switching frequency components and to the small values of overshoot observed, it is hard to measure this parameter from the output current. Another possibility to evaluate overshoot is by means of the peak of the current error, but in the experiments of this paper that could not be precisely made, since the digital to analog conversion based on RC filtering of DSP PWM outputs attenuates the current errors amplitude. Thus, only a qualitative analysis of overshoot is done for the presented results, based on the evaluation of phase A current error peak value. In this way, Fig. 9 shows that the overshoot of the disturbance, especially for the optimized gain, is lower than that obtained for the tracking response, as predicted by Figs. 5 and 6. Moreover, for the disturbance, the overshoot for $k_R=60rad/s$ (see Fig. 9c) is higher than that achieved for $k_R=377rad/s$ (see Fig. 9d), as expected from Fig. 6.

Finally, it must be noted that if a feedforward of the grid voltage is implemented, as it is done in [3], the disturbance is cancelled in the continuous time model. Thus, settling times are drastically reduced, and the transient response is improved. In this case, the optimized design for the tracking response seems to be a better option, to be investigated in future paper.

VI. CONCLUSIONS

This paper has presented an optimized tuning method of PR current controllers implemented at the stationary frame. The analysis of the behavior of closed loop poles and zeros in Section III led to the definition of three operation regions, resulting in approximated tracking and disturbance transfer functions.

Charts of proportional and resonant normalized gains versus settling time and overshoot were built in Section IV, allowing the designer to visualize these three regions and the influence of gains on the transient response. It was also proved that, when analyzing the transient response of PR controller, T_S and M_P must be always considered, instead of taking solely the phase margin into account.

Analytical expressions for tracking and disturbance optimized gains, as well for overshoot and settling time, not previously found in literature, were derived in Section IV. It was concluded that, concerning settling time, the disturbance response is more significant that the tracking one. For this case, this paper has demonstrated that minimum settling time and overshoot are obtained if the resonant gain is set equal to the controller resonant frequency, providing a simple and efficient rule-of-thumb for the designer. Simulation and experimental results have validated the proposed method.

APPENDIX I – RELATIONS BETWEEN SINUSOIDAL AND STEP INPUTS

Simulations of (2) and (3) for the parameters of Section III were made in *Matlab* to compare the behavior of settling time and overshoot, when a unit step and a 1 p.u. sinusoidal input signals are applied at the reference current and at the grid voltage. Simulation parameters are $\omega = 377\,rad/s$, $k = 1$, $k_P^{pu} = 63.66$, and $k_L = 0.1$. Figs. 10 and 11 show the results of the settling time for the tracking and disturbances transfer functions, respectively.

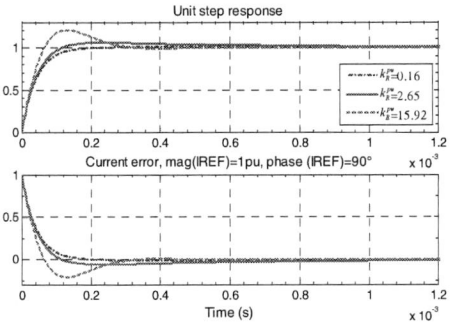

Fig. 10. Settling time for unit step (top) and sinusoidal (bottom) inputs for the tracking transfer function.

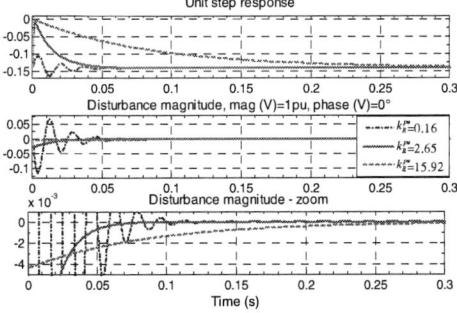

Fig. 11. Settling time for unit step (top) and sinusoidal (center, bottom) inputs for the disturbance transfer function.

The resonant gains of Section III are chosen, and the current error for the sinusoidal input is compared to the step response. It can be concluded that the settling time of the step input is approximately equal to the time that the current error stays in the range of ±2% of its maximum fundamental peak value for the sinusoidal input. Critical cases occur when the phase of the reference current is 90° and when the phase of the grid voltage is 0°.

Since the overshoot is not defined for sinusoidal inputs, this parameter is evaluated by means of the

relation between the maximum peak values of the output current for a sinusoidal input I_{pSIN} and for a step input I_{pSTEP}. Thus, Fig. 12 shows the results when the angles of the reference current and of the grid voltage are varied from 0° to 180°, for $k_P^{pu} = 63.66$ and $k = 1$. It can be concluded that for both tracking and disturbance transfer functions, the maximum peak values are identical when the phases of the reference current and of the grid voltage are equal to 90°.

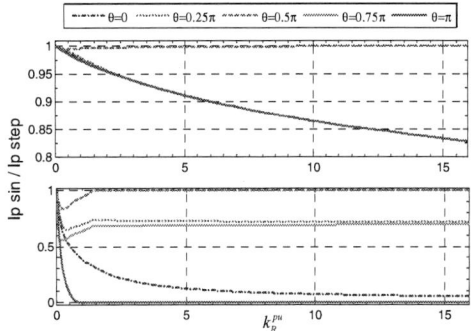

Fig. 12. Relation between the maximum peak values for step and sinusoidal inputs. Tracking (top) and disturbance (bottom) transfer functions.

ACKNOWLEDGMENT

The authors are grateful to Texas Instruments Academic Program, for donating the Evaluation Kit and software, and to FAPESP, CAPES, CNPq and FDTE for financial support.

REFERENCES

[1] A. Vidal, F.D. Freijedo, A.G. Yepes, P. Fernandez-Comesana, J. Malvar, O. Lopez, J. Doval-Gandoy, "Assessment and Optimization of the Transient Response of Proportional-Resonant Current Controllers for Distributed Power Generation Systems," IEEE Transactions on Industrial Electronics, vol. 60, no. 4, pp. 1367-1383, 2013.

[2] S. Buso, and P. Mattavelli, "Digital Control in Power Electronics," 1st ed., Morgan & Claypool, 2006.

[3] D.G. Holmes, T.A. Lipo, B.P. McGrath, and W.Y. Kong, "Optimized Design of Stationary Frame Three Phase AC Current Regulators," IEEE Transactions on Power Electronics, vol. 24, no. 11, pp. 2417-2426, 2009.

[4] K. Ogata, "Modern Control Engineering," 3rd ed., Prentice Hall, 2000.

[5] A.G. Yepes, A. Vidal, J. Malvar, O. Lopez, and J. Doval-Gandoy, "Tuning Method Aimed at Optimized Settling Time and Overshoot for Synchronous Proportional-Integral Current Control in Electric Machines," IEEE Transactions on Power Electronics, vol. 29, no. 6, pp. 3041-3054, 2014.

[6] F.O. Martinz, R.D. Miranda, W. Komatsu, and L. Matakas Jr., "Gain limits for current loop controllers of single and three-phase PWM converters," Proceedings of 2010 IPEC, Sapporo, vol.1, pp. 201-208, 2010.

[7] J.C.P. Jones and D.P. Atherton, "Root locus diagrams and the effect of zeros on system response", International Journal on Electrical Engineering Education, v. 34, p. 48-69, 1997.

[8] T. E. Nuñez-Zuniga, and Pomilio, J. A., "Shunt active power filter synthesizing resistive loads", IEEE Transactions on Power Electronics, vol. 17, no. 2, pp. 273-278, 2002.

[9] N.R.N. Ama, W. Komatsu, F. Kassab Jr., and L.Matakas Jr., "Adaptive single phase moving average filter PLLs: analysis, design, performance evaluation and comparison", *Przeglad Elektrotechniczny*, Issue 3, 2014.

Instantaneous Power Theory applied to Power Conditioning under Distorted Mains Voltages: a MATLAB/Simulink Approach

Petre-Marian Nicolae
Lucian-Dinuț Popa
Marian-Ștefan Nicolae
Dept. of Electrical, Energetic and Aero-Spatial Engineering
University of Craiova
Craiova, Romania
pnicolae@elth.ucv.ro, lpopa@ elth.ucv.ro

Ileana-Diana Nicolae
Department of Computers and Information Technology
University of Craiova
Craiova, Romania
nicolae_ileana@software.ucv.ro

Abstract— The paper presents some problems and solutions for instantaneous power theory applied to active power conditioning under distorted mains voltages in a MATLAB/Simulink approach. Simulation of a real system is made, namely a transformer which feeds a controlled rectifier bridge in an excitation system from a power system generator group, based on measurements acquired with a dedicated data acquisition system. Replacing the old auxiliary excitation generator eliminates the disadvantage of using electrical machines with the mass in motion inertia and wear over time, but with the price of facing two important problems: the line voltage notching due to thyristor switching and also the distorting of line current. The MATLAB/Simulink model allows for testing the instantaneous power theory implementation and dealing with problems arising with distortion of mains voltages, respectively observing its effectiveness based on the results.

Keywords— *Active filters, instantaneous power theory, power quality, static power converters, voltage distorsion.*

I. INTRODUCTION

Synchronous generators can have two types of excitation systems, namely with auxiliary generator and static excitation respectively. Auxiliary generator excitation system can be upgraded by replacing it with a static one. This eliminates the disadvantage of using electrical machines with the mass in motion inertia, maintenance and wear over time, but with the price of facing some problems related to harmonic distortion.

The instantaneous active and reactive power theory (IRP theory) or the *p-q* theory is widely used to design controllers for active filters. Some problems could arrive in calculation of reference current for the active filter, mainly due to distorted voltages at the point of common connection (PCC), leading to unsatisfactory performance [1].

In Section II one presents the operational context submitted to analysis and details on the results of the preliminary tests.

Section III is dedicated to matters related to reducing the voltage notches, whilst the 4-th section is dedicated to the simulation of the shunt active power filter and of the line reactors.

Section V deals with the control of APF using the Instantaneous Reactive Powers (IRP) theory, a final section being dedicated to conclusions.

II. POWER QUALITY MEASUREMENTS FOR STATIC EXCITATION SYSTEM WITHOUT COMPENSATION

The electric equivalent scheme for the main power source with static excitation system is depicted by Fig. 1, where the energy sources of 330 MW, 24kV three-phase synchronous generator.

Within this internal network, the excitation system of the main synchronous generator is supplied by a thyristors based rectifier (TBR). The excitation system is supplied from the main terminals of the synchronous generator trough an excitation transformer 24/0.65 kV (Fig. 1).

Monitoring system's own group current measuring transformers (CMT) and voltage measuring transformers (VMT) are connected between the excitation transformer and TBR. Waveforms recordings by data acquisition system are also performed for these transformers.

The data acquiring and processing was imposed by abnormally frequent operations of transformers' replacement imposed by their deterioration. A strong distorting regime was suspected from the beginning.

The recordings were made at different generator loads, for almost all energetic group quantities, but are focused mainly on the excitation transformer's secondary winding. Some recorded data are presented in this paper, waveforms corresponding to three-phase line currents and phase voltages being depicted by Fig. 2.

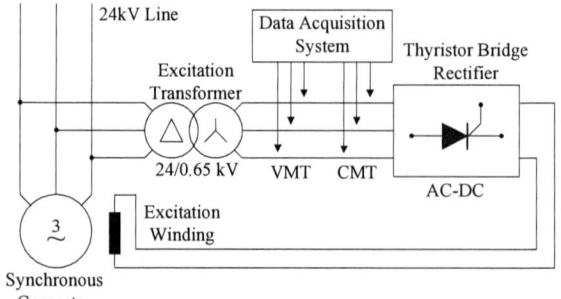

Fig. 1. Electrical equivalent scheme of the analyzed system.

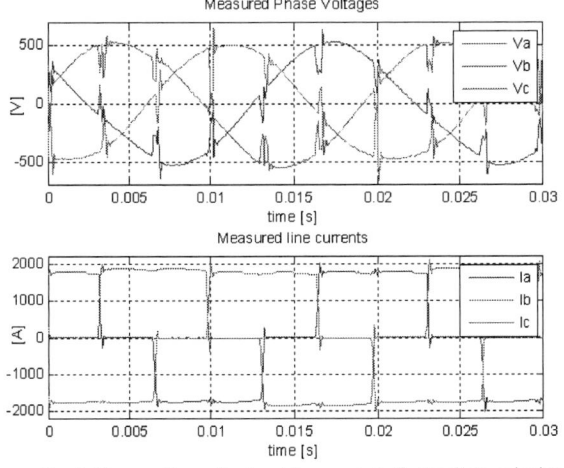

Fig. 2. Phase voltages (top) and line currents (bottom) at excitation transformer's secondary winding.

Original processing software packages were used for numerical processing, based on Fast Fourier Transform (FFT).

One can notice in these waveforms the effects of thyristors' switching under the form of small peaks at currents and notching at voltages.

The thyristor bridge is composed of six power switching devices operated in pairs to convert three phase AC to DC by switching the excitation winding equivalent RL load among the various thyristor pairs six times per cycle. During this process, six notches are produced for each complete cycle of the three phase voltages. When the thyristor Th_1 conducts, the current in the converter is flowing from the phase A through the RL excitation winding's equivalent load (Fig. 3). When the thyristor Th_3 fires, the current begins to transfer from the phase A to the phase B. Transformer's secondary winding inductance prevents instantaneous transfer, thus the required switching time becomes the notch width of the corresponding overlap angle μ. A short-circuit takes place for a short period when both thyristors Th_1 and Th_3 are conducting simultaneously, producing two deep notches in the line voltages, the other four notches reflecting the action of the thyristor on the other phases [2], [3].

The results yielded by FFT based decompositions of the voltages and currents from Fig. 2 are gathered by Table 1.

As suspected, the excitation system's distorting regime is significant, with a maximum total harmonic distortion factor for voltage (THDV) equal to 19.6 %, and respectively for current (THDI) equal to 29%. The significant odd harmonics and respectively small harmonics of order 2 and 3 reveal the unbalance.

The excitation transformer harmonic currents are flowing through the transformer's impedance and generator stator winding. Therefore certain harmonic voltage drops occur along the impedance, but their amplitude is proportional to a certain limit, which indicates that the harmonic source behaves more like as a current source. Here, the voltage harmonics are mainly caused by line notches.

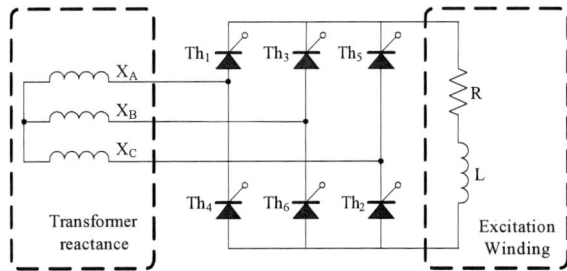

Fig. 3. Three phase equivalent circuit of converter.

TABLE I
SIGNIFICANT PHASE VOLTAGES AND CURRENTS HARMONICS AT THE EXCITATION TRANSFORMER'S SECONDARY WINDING

Harm. order	Phase Voltages [% of fundamental]			Currents [% of fundamental]		
	VA	VB	VC	IA	IB	IC
5	3.99	4.21	4.82	20.72	20.12	20.17
7	4.80	4.18	4.93	13.44	13.83	13.85
11	4.25	4.36	4.53	9.22	8.67	8.72
13	4.38	3.87	4.95	7.11	7.45	7.48
17	4.18	4.25	4.30	5.91	5.34	5.43
19	4.38	3.90	4.55	4.73	5.05	5.16
23	3.99	3.99	3.96	4.29	3.75	3.82
25	4.04	3.72	4.52	3.44	3.71	3.85

The distorting residue's maximum value equals 443 A. For a full harmonic compensation this value should be the RMS value of the current through the corrective shunt active power filter.

For the generator's rated operation, the current's RMS is 1480 A, obtained as an average over the three phases. Similarly the voltage's RMS is 370 V.

The effects of voltage and current harmonics are power loss, disturbances, measurement errors, and control malfunctions.

In the distorted voltage waveform, the notches are the causes for the most significant problems because they are associated to sudden variations and fallings to zero, creating extra zero crossing points.

The first measure to be taken is to reduce the depth of voltage notches, because any device that uses zero crossing reference or is based on sinusoidal voltage will not operate properly and/or will heat up. On the other hand, the power conditioning equipment might operate inefficiently.

The notch depth and width depend on the firing delay angle. When the DC output voltage is lowered by delaying the thyristor firing angle α, the notch width is getting smaller whilst the notch depth is getting bigger [2]. The relationship between line notching and total harmonic distortion is based on (1) [3]:

$$V_h = \sum_{h=5}^{\infty} V^2{}_h = \sqrt{\left(2V^2{}_N t_N + 4\left(V_N/2\right)^2 t_n\right) \cdot f_1} \quad (1)$$

where: t_N is the notch time period,
 f is the fundamental frequency.

The constant "2" addresses two deep notches while the constant "4" corresponds to the 4 small notches.

978-1-4799-2706-7/14 $31.00 © 2014 IEEE

III. REDUCING LINE NOTCHING

Line notching can be attenuated by installing line reactors. Any additional inductance will reduce the notch depth but will slightly increase the notch width, according with (2) [2]:

$$\mu = \cos^{-1}\left(\cos(\alpha) - \frac{2\omega L I_{dc}}{V_{phase}}\right) - \alpha \qquad (2)$$

where: μ is the overlap angle; ω is the throb; L represents the equivalent line inductance; I_{DC} is the DC current; V_{phase} is the phase voltage.

Also, the overlap angle μ of line voltage notches increases as the firing angle moves closer to 0° and as the DC current becomes larger.

Considering these aspects, if the inductor is selected such that the voltage drop across it should reach about 3-4% of the phase voltage at rated current, the notches at the transformer's secondary winding's terminals should be limited to an acceptable level.

IV. USING A SHUNT ACTIVE POWER FILTER

When trying to derive proper corrective measures, several solutions have been considered. The most effective is represented by the active filtering.

As revealed by Table 1, the thyristor bridge introduces harmonic currents of orders 5, 7, 11, 13, 17, 19, 23 and 25 with significant weights. It means that the solution through passive filtering is not efficient (too many filtering cells would have been involved).

The solution with shunt active power filter (APF) is chosen because, in conjunction with 3-4% line reactors, it is able to reduce significantly the line notching. The

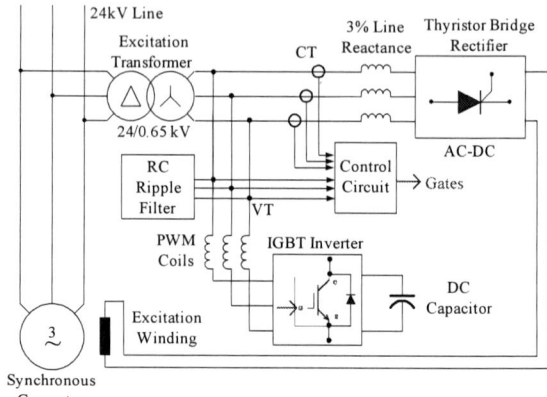

Fig. 4. Compensation system schematic.

power factor compensation and load balancing are also possible when using APF. Fig. 4 depicts a possible scheme for a compensation system. It is mandatory to install the APF before the line reactors, in order to avoid the zero crossing points corresponding to voltages and to improve performances. In order to design the APF's components one should firstly analyze the acquired waveforms. For the beginning, the load currents from Fig. 2 were imported into MATLAB workspace to obtain the compensation currents.

The fundamental active current is calculated and subtracted from the load current. Afterward the result is inverted to obtain the reference current for the APF [4].

The compensation system is firstly simulated in MATLAB/Simulink software module to test its effectiveness and weaknesses. The system's general scheme is given in Fig. 5.

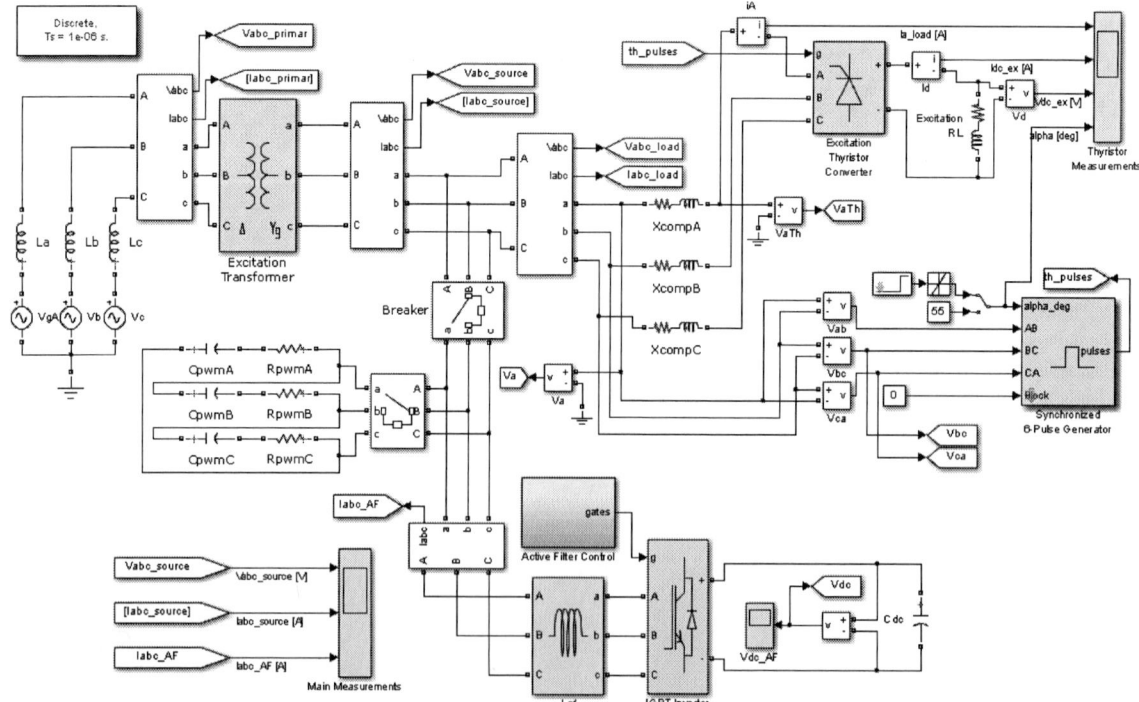

Fig. 5. The MATLAB/Simulink general scheme of the system.

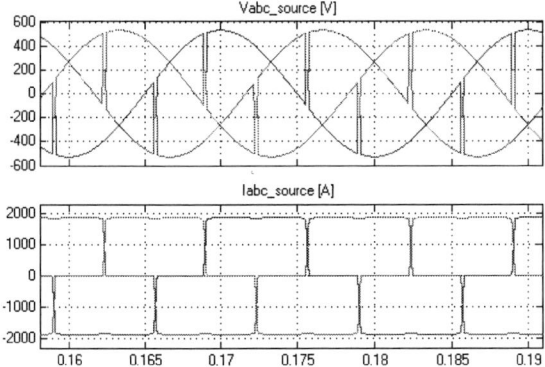

Fig. 6. Phase voltages (top) and currents (bottom) simulated waveforms.

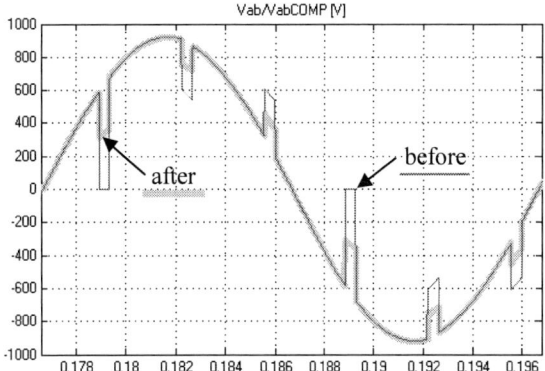

Fig. 7. V_{AB} line voltage notch after line reactors installation.

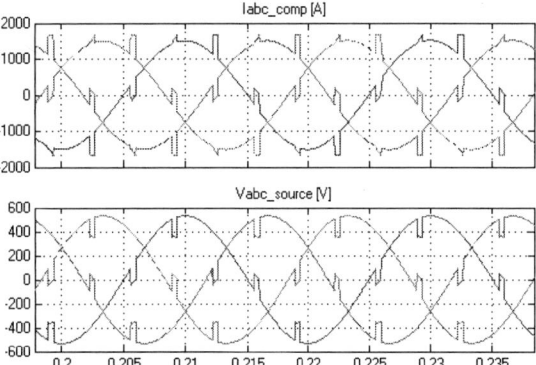

Fig. 8. Distorted theoretical source currents (top) and phase voltages (bottom).

The APF's switching frequency is chosen considering the highest harmonic order to be compensated. Theoretically it is possible to compensate harmonics up to half of the switching frequency. When choosing the switching frequency, the RC filter switching ripple cutoff frequency must be taken into account, which is:

$$f_{RC\,cutoff} = 1/(2\pi\, RC) \qquad (3)$$

The MATLAB/Simulink scheme comprises the equivalent synchronous generator stator winding, the excitation transformer, the TBR that provides the excitation and the distorting regime compensation system composed of line reactors along with APF. The active filter contains the low-power switching ripple RC filter, the high-power IGBT inverter section and the control system based on the IRP theory.

The simulated waveforms for the phase voltages and currents in the absence of corrective measures are depicted by Fig. 6.

The excitation voltage and current are varied through the thyristors firing angle control. This way the compensation system effectiveness is evaluated in every situation, not only at rated operation.

V. CONTROL OF APF USING IRP THEORY

Obtaining IGBT inverter currents is based on IRP Theory, widely used as a basis for the control algorithm of active filters. It provides very accurate results if the voltages are balanced and sinusoidal. If the voltages used in the calculation of the compensation currents are not sinusoidal and balanced, the current waveforms in the source are not purely sinusoidal. This is not a problem of the theory, only a misinterpretation [5], [6].

Obtaining the compensation currents is based on the fact that harmonic currents and voltages produce oscillations in instantaneous active and reactive powers.

To estimate the current and voltage distortion levels, the FFT Analysis tool from PowerGUI interface was used. The values yielded by the harmonic decomposition are in good agreement with those acquired from system with respect to the significant harmonics orders and values of THDs.

In Fig. 7 one can observe the depth reduction in phase voltage notch after line reactors installation. So THD was reduced up to 10.45%.

Fig. 8 reveals that theoretical source currents obtained by adding compensation currents to load currents are not sinusoidal, containing ripples that increase the peak values and are able to produce voltage harmonics due to high di/dt flowing through the source impedance. In the absence of 3% line reactances, the resulting source currents would be worst. The definition of IRP theory shows that for distorted and/or imbalanced voltages it is impossible to obtain sinusoidal source currents and respectively draw a constant instantaneous active power.

The problem caused by the voltage distortion can be solved using a phase locked loop (PLL) circuit, which is used to detect the fundamental positive-sequence component of the voltage at the PCC. The control scheme using PLL circuit is presented in Fig. 9 and the resulting theoretical source currents in Fig. 10. The current waveforms are sinusoidal even if the voltage waveforms are not [6], [7].

In order to obtain the oscillating instantaneous power components $\tilde{p}$ and $\tilde{q}$, two Butterworth 5th-order low-pass filters with 100Hz cutoff frequency are used and then the oscillating powers are subtracted from the total powers (p and q).

This method is insensible to the phase shift introduced by the high pass filters. The reason why the entire value of q is not compensated is that the power factor's correction is desired. This involves the division of the constant

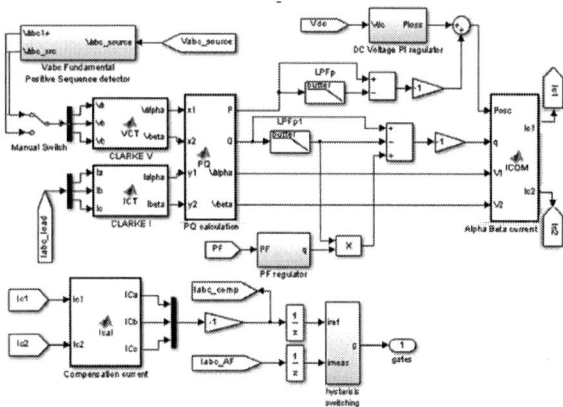

Fig. 9. The control scheme using positive sequence fundamental voltage detector PLL circuit, IRP theory and hysteresis switching

instantaneous reactive power $\overline{q}$ in order to achieve the target power factor.

The APF dynamically compensates the power factor to the Romanian neutral value of 0.92 and, because the power factor drops, the RMS current through APF is maintained constant by means of an original power factor regulator function:

$$k_{\overline{q}} = \frac{\sin(\text{acos}(PF))}{(PF - 0.18)} + \sqrt{2} \cdot (0.93 - PF) \qquad (4)$$

In this expression, $k_{\overline{q}}$ is the gain for the constant IRP part and PF is the instantaneous measured power factor.

The 100 Hz cutoff frequency of Butterworth filters is chosen such as to prevent interferences with power oscillations due to the firing angle variation. Also the lowest order harmonic to compensate (5th) had to be considered.

The d.c. voltage regulator is added to the control strategy, because a small amount of real power must be continuously drawn from the supply in order to compensate for switching and ohmic losses in the PWM IGBT converter. Otherwise, this energy would be supplied by the d.c. capacitor, which would discharge. The IGBT converter of the shunt active filter is a boost–type converter. Therefore, in order to guarantee the controllability and performance of the active filter, the d.c. voltage must be kept higher than the peak value of the ac bus voltage, in this case 1125 V. Because mains voltages are distorted, a higher value of the d.c. capacitor voltage and/or higher capacity value is needed for obtaining good harmonic compensation for all thyristor firing angles. Otherwise current peaks may appear [5].

As for the dynamic response of APF, Fig. 11 depicts waveforms corresponding to the excitation transformer secondary winding phase voltages and currents, and respectively the APF compensation current.

The firing angle is changed from 40° to 55° while the PF value is kept at the neutral value 0.92.

The d.c. current flowing through the excitation winding has a smooth transition, being safe and making

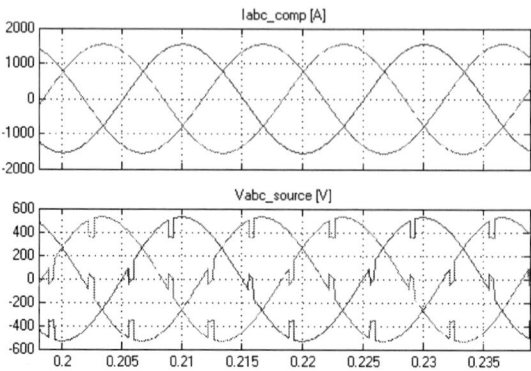

Fig. 10. Sinusoidal theoretical source currents (top) and phase voltages (bottom) with PLL positive sequence voltages detector.

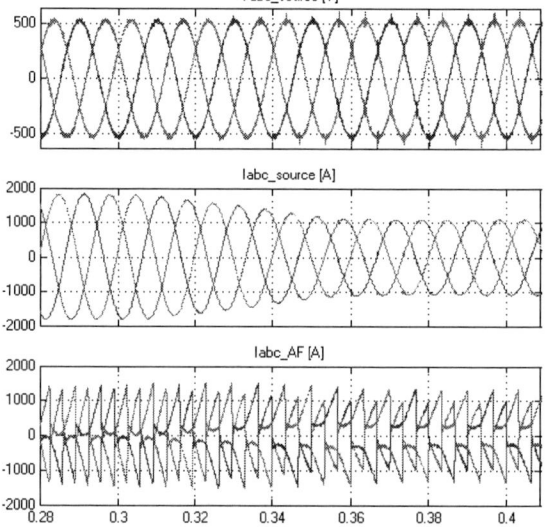

Fig. 11. Phase voltages (top), currents absorbed from source (middle) and currents through APF (bottom) during compensation – dynamic response.

the job easier for the voltage regulators (Fig. 12). In this way the small electromechanical oscillations can be attenuated. Proper protection systems and additional measures are required for large oscillations. High attenuation for phase voltage notches can be observed when both line reactors and APF are installed, especially at 40°.

For an overview of the performance of active filter, waveform distortion data were gathered in Table II.

It can be seen that when the firing thyristor switching angle increases, the compensation system efficiency is reduced by means of voltage notch depth increase.

Since the synchronous generator operates most of the time in rated regime, the firing thyristor switching angle is kept low, harmonic distortion being maintained implicitly at low levels.

Three cases were simulated for the power factor (PF) compensation. For each of them the RMS active filter current was evaluated. For harmonic compensation only, active filter RMS current measured value is 430 A (leaving the PF unchanged to the value 0.75). For a constant PF target equal to 0.92, the RMS current is 650 A.

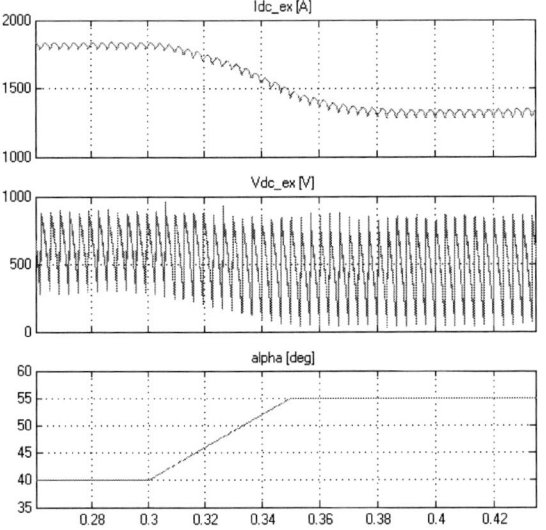

Fig. 12. Excitation winding current (top) and excitation winding voltage (middle) vs. firing angle alpha (bottom) during compensation – dynamic response.

TABLE II
CURRENT AND VOLTAGE TOTAL HARMONIC DISTORTIONS FOR DIFFERENT FIRING ANGLES

Thyristor firing angle	Voltage THD [% of fundamental]	Current THD [% of fundamental]
35	1.37	0.3
40	1.41	0.32
45	1.44	0.35
50	1.89	0.62
55	2.12	0.93
65	2.67	1.3

For power factor of almost 1, the current through filter was 1080 A. The higher the target PF it is- the higher the costs due to higher rated IGBT modules are.

VI. CONCLUSIONS

Measurements of power quality indices for the static excitation system revealed severe distorting regime.

The simulation of the real system along with the power conditioning system consisting of line reactors and power active filter, using MATLAB/Simulink, made possible the efficiency testing for the proposed compensation solution.

For the studied case, instantaneous reactive power theory based control of active filter currents revealed some problems when used in the absence of positive sequence voltage detector and line reactors for reducing voltage notches.

In the ideal case, the compensation system is able to reduce the total harmonic distortions. At rated operation, for current the reduction was from 29% up to 0.3% and for voltage it was from 19.6% to 1.37%. Moreover, the notches were properly mitigated.

The recordings performed before simulations were acquired from a real power group's excitation system of 330 MW. A real model of the simulated APF is under implementation in order to reduce the corresponding distorting regime. Considering the significant powers that APF will have to handle, firstly validation tests concerning its control will be performed using dSPACE.

REFERENCES

[1] E. H. Watanabe, M. Aredes, H. Akagi "The p-q theory for active filter control: some problems and solutions" *Sba Controle & Automação*, vol. 15, no. 1, pp. 78-84, 2004.

[2] N. Mohan, T. M. Undeland, W. P. Robbins, *Power Electronics: Converters, Applications, and Design, 2nd Edition*, Wiley, pp. 150, 1995.

[3] IEEE Recommended Practices and Requirements for Harmonic Control in Electric Power Systems, IEEE Std. 519-1992.

[4] H. Akagi, E. H. Watanabe, M. Aredes, *Instantaneous power theory and applications to power conditioning*, Wiley-Interscience, pp. 109-145, 2007.

[5] D. L. Popa, P.M. Nicolae, "Analysis and Simulation of Line Notching Attenuation for a Static Excitation System in a Power System", *4-th International Youth Conference on Energy IYCE2013*, Hungary, 2013.

[6] R. Pregitzer, T. A. Sousa, J. L. Afonso, "Comparison of fundamental positive-sequence detectors for highly distorted and unbalanced systems" *IEEE 1st International Conference on Electrical Engineering CEE'05*, pp. 78-84, 2005.

[7] F. M. Serra, D. G. Forchetti, C. H. De Angelo, "Comparison of positive sequence detectors for shunt active filter control", *IEEE International Conference on Industry Applications INDUSCON 2010*, pp. 1-6, 2010.

The 2014 International Power Electronics Conference

The Research on Reliability and Real-time of the Scheme of Process Layer GOOSE Network in Smart Substation Based on Artificial Cobweb Topology Structure

Xiaosheng Liu, Honglin Zhu, Dianguo Xu, Yanxiang Li
Institute of Power Electronics and Electrical Driving
Harbin Institute of Technology
Harbin China
Liuxsh2004@126.com, zhuhonglinhit@sina.com

Abstract—According to GOOSE defined in the IEC61850 in the smart substation, it explicitly stipulates the requirements for the transmission in reliability and real-time of the process layer network scheme. The cobweb structure is an artificial communication network topology based on cobweb as it occurs in nature. On the basis of the cobweb in the nature, a novel cobweb topology structure, based on process layer GOOSE network in smart substation is proposed in this paper. With the method of fault tree analysis, this paper demonstrates the reliability of the network scheme in one certain smart substation. Industrial Ethernet transmission analysis on delay and OPNET Modeler are also employed to proof the cobweb topology performs well in reliability and real-time. The results of the theoretical analysis and simulation indicate that the cobweb topology structure exhibits excellent reliability and real-time in smart substation GOOSE communication network.

Keywords— Goose, Cobweb, Reliability, Real-time, OPNET Simulation

I. INTRODUCTION

Generic Object Oriented Substation Event (GOOSE) is defined in the IEC61850, which is mainly used in the field of substation relay protection, in order to transmit simple binary Boolean values, such as real-time signals packet of tripping or closing [1]. IEC61850 explicitly stipulates the requirements for the transmission delay of fast message, where the delay time of GOOSE message signal must be less than 3ms[2]. So the reliability and real-time of the process layer network plays an important role on GOOSE.

The topology structure of process layer network can be mainly divided into star network, ring network and bus network [3]. The analysis of bus topology, single star topology and ring topology in references [4-8] proved that there are serious problems such as low reliability, long network delay, high redundancy etc. On the basis of the three kinds of structure above, seamless ring network structure with high reliability is derived into the network topology, while it belongs to the ring network redundancy node structure where each node has two network adapters so that the message can be sent to the network ring on both directions at the same time. It is in essence bidirectional ring network with low resistance to destruction.

The cobweb structure is an artificial communication network topology based on cobweb as it occurs in nature. The structure of cobweb can ensure the reliability and rapidity of prey information transmitting. Cobweb is not only similar to popular communication network topology in structure, but also in the behavior of building the communication network. This paper organized the process layer network with one certain smart substation, to construct the GOOSE web, calculating, analyzing and verifying the real time and reliability of network mode according to the cobweb network topology.

II. BASED ON THE REDUNDANT COBWEB GOOSE PROCESS OF SMART SUBSTATION LAYER NETWORK TOPOLOGY

A. Artificial Cobweb Network Topology

The central region of cobweb is similar to the network core switch in communication network. Accordingly single-layer artificial cobweb model is shown in figure 1. Compared to star or ring topology, which has been widely used in industry, not only can the data be exchanged through radial path, but also through the ring structure switch. So the transmission link redundancy is very high.

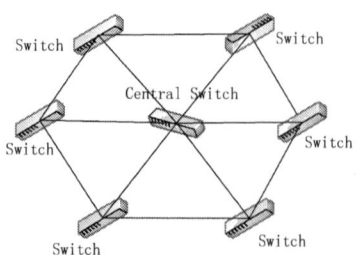

Fig.1. Artificial cobweb network topology model

B. Process Layer GOOSE Network Topology in Substation Based on Cobweb

Process layer GOOSE communication network in substation consists of switches, routers, IED (Intelligent Electronic Device) network interface devices, servers and some hardware. Illustrated by the example of one smart substation, there are 30 IED devices realizing all kinds of protection. In the substation, every 6 groups of IED devices share a switch, and all the 5 switches connect to the central switch in the form of a single cobweb. According to figure 1, the process of layer GOOSE single web topology is constructed and shown in figure 2. According to the requirements of IEC61850, a double layer of redundancy design is needed in the process of substation communication network [9]. Fig3 is cobweb redundant topology realized by using double central switches and fig4 is cobweb redundant topology realized by using smart dual port IED.

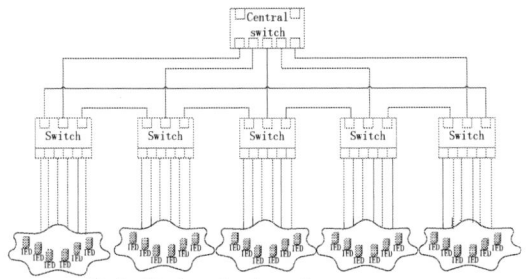

Fig.2. Singler cobweb topology structure

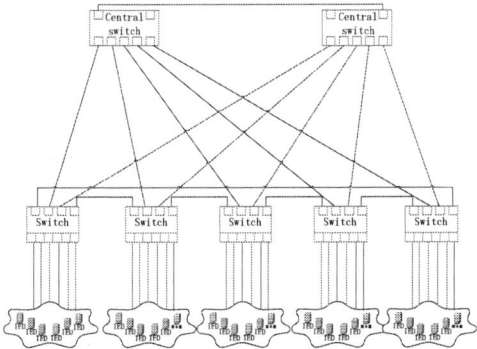

Fig.3. Dual cobweb redundant topology structure

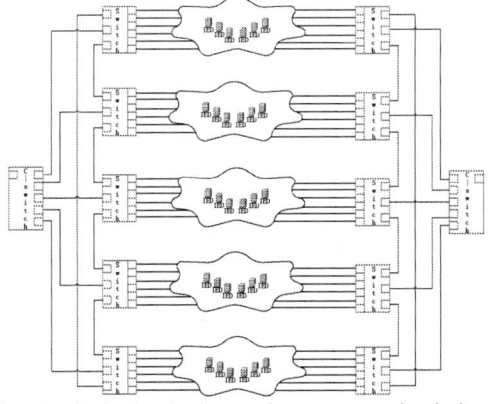

Fig.4. Dual cobweb redundant topology structure using dual-port IED

III. BASED ON FAULT TREE ANALYSIS GOOSE WEB COBWEB TOPOLOGY STRUCTURE RELIABILITY ANALYSIS

A. The Fault Tree Analysis Method

Fault tree analysis, as a powerful tool in analyzing failure systems and predicting the probability of system failure, can be applied to the reliability of the substation communication system analysis[10]. The basic steps of fault tree analysis are as follows:

1) Define a system and confirm system failures with the "top event".

2) Build the fault tree.

3) Analysis qualitatively and quantitatively

According to the fault tree analysis, the reliability of smart substation layer GOOSE topology network can be analyzed.

B. The Fault Tree Analysis of Cobweb Topological Structure in GOOSE Web

Take the same smart substation with fault tree analysis for example, "top event" is defined as "the failure of the process layer network communication ". The reason for process layer network communication system failure includes:

1) The IED equipment failure - I1

2) The IED equipment network interface failure - R1.

3) The central switch failure - I2

4) The central switch network interface failure – R2

5) The underlying switch failure - I3

6) The underlying switches connecting the network interface with IED failure - R3

7) The underlying network connection between switches interfaces failure- R4

8) The underlying switches connecting to the central switches network interface failure -- R5

According to the hardware reliability index [11], we can obtain the failure degree shown in Tab1 by calculating the failure degree of the equipments and device interfaces above.

TABLE I
THE FAILURE RATE OF DEVICES AND PORTS

Device and port	Switch	IED device	IED device network interface	Switch network interface
failure degree $(\times 10^{-6})$	477	55	285	200
Symbol of device and port	I_2 ; I_3	I_1	R_1	R2 ; R3 ; R4 ; R5

Illustrated by the example of the communication between an IED selected in figure 2 and the central switch as shown in figure 5, the multipath redundant transmission links with a serial number are set up. As shown in table2, the components of transmission links fault in table3.

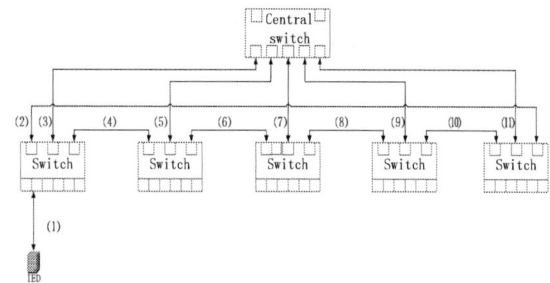

Fig.5. Transmission links in single cobweb topology

TABLE II
POSSIBLE TRANSMISSION LINKS IN SINGLE COBWEB
TOPOLOGY

Number of switches	Possible transmission link	Link number
2	(1)←→(3)	1
3	(1)←→(2)←→(11)	2
	(1)←→(4)←→(5)	3
4	(1)←→(2)←→(10)←→(9)	4
	(1)←→(4)←→(6)←→(7)	5
5	(1)←→(2)←→(10)←→(8)←→(7)	6
	(1)←→(4)←→(6)←→(8)←→(9)	7
6	(1)←→(2)←→(10)←→(8)←→(6)←→(5)	8
	(1) ←→(4)←→(6)←→(8)←→(10)←→(11)	9

TABLE III
COMPOSITE OF TRANSMISSION LINKS FAULT

Transmission link	Components of transmission links fault				
Link 1	R_1	R_3		R_5	R_2
Link 2	R_1	R_3	$2 \times R_4$	R_5	R_2
Link 3	R_1	R_3	$2 \times R_4$	R_5	R_2
Link 4	R_1	R_3	$4 \times R_4$	R_5	R_2
Link 5	R_1	R_3	$4 \times R_4$	R_5	R_2
Link 6	R_1	R_3	$6 \times R_4$	R_5	R_2
Link 7	R_1	R_3	$6 \times R_4$	R_5	R_2
Link 8	R_1	R_3	$8 \times R_4$	R_5	R_2
Link 9	R_1	R_3	$8 \times R_4$	R_5	R_2

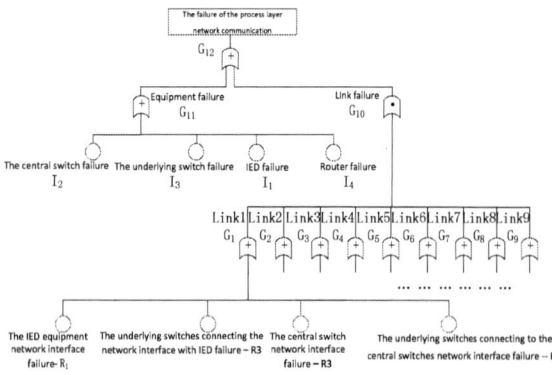

Fig.6 The fault tree based on single cobweb topology

According to the analysis with the method of minimum cut sets above, the establishment of fault tree is shown in figure 6., that is, from the bottom level, using basic events to represent events in this level, so that the events of basic and the corresponding level can represent events at the next level in turn until the highest level.

The followings can be acquired according to the principle:

$$G_1 = R_1 + R_3 + R_5 + R_2 \tag{1}$$

$$G_2 = R_1 + R_3 + 2 \times R_4 + R_5 + R_2 \tag{2}$$

$$G_3 = R_1 + R_3 + 2 \times R_4 + R_5 + R_2 \tag{3}$$

$$G_4 = R_1 + R_3 + 4 \times R_4 + R_5 + R_2 \tag{4}$$

$$G_5 = R_1 + R_3 + 4 \times R_4 + R_5 + R_2 \tag{5}$$

$$G_6 = R_1 + R_3 + 6 \times R_4 + R_5 + R_2 \tag{6}$$

$$G_7 = R_1 + R_3 + 6 \times R_4 + R_5 + R_2 \tag{7}$$

$$G_8 = R_1 + R_3 + 8 \times R_4 + R_5 + R_2 \tag{8}$$

$$G_9 = R_1 + R_3 + 8 \times R_4 + R_5 + R_2 \tag{9}$$

$$G_{10} = 30 \times (G_1 \times G_2 \times G_3 \times G_4 \times G_5 \times G_6 \times G_7 \times G_8 \times G_9) \tag{10}$$

$$G_{11} = I_1 + I_2 + I_3 \tag{11}$$

$$G_{12} = G_{11} + G_{10} \tag{12}$$

The number of 30 is for IED equipments, and each IED equipment has 9 same transmission links.

According to the analysis above, equipment failure degree of process layer GOOSE network in cobweb topology structure G11is 0.4512%, the total order of magnitude of link failures G10 is 10^{-18}. Compared with G11, it can be treat as 0. So the total failure degree of the process layer GOOSE network in cobweb topology structure system G12 is 0.4512%, which means the total effectiveness of the communication is 1-0.4512% = 99.5488%.

Seen from the analysis above, the link failure degree G10 is very low, because of the multi-link redundant structure of cobweb, and the total failure degree of the system G12 is completely determined by the equipment failure rate of the total failure degree G11. So for a single cobweb structure of GOOSE network, the link reliability is very high.

Star topology structure of substation is widely used at the present stage. In condition of the same equipment, the equipment failure degree G11 remains the same by the method of fault tree,. However, due to its singularity, the failure rate of the link G10 is 2.6550%. The system failure degree in all G12 = 3.1062%. The system effectiveness is 96.8938%, which means the total effectiveness of the cobweb topology structure system is very high.

Hardware failure can be solved by double cobweb structure. The total failure degree of double cobweb structure is quadratic, compared to single network failure degree, G12 x G12 = 0.002 0%. So the total effectiveness of double cobweb structure process layer GOOSE network communication is 1-0.0020% = 99.9980%.

IV. REAL-TIME ANALYSIS OF PROCESS LAYER GOOSE NETWORK REDUNDANT LINK BASED ON COBWEB STRUCTURE AND OPNET SIMULATION

A. Real-time analysis of process layer GOOSE network redundant link based on cobweb structure

As shown in figure 7, IEC61850-5 defines the message transmission delay. The delay of sending and receiving message Ta, Tc is associated with the speed of hardware communication devices. The real-time of transmission lines mainly depends on the message transmission delay Tb. While the delay in message transmission is related to the size of the transmission of message, the number of switches that transmission links go through and other factors. In this section, we calculate and analyze the delay time of process layer GOOSE network transmission links on the basis of cobweb structure, and validate the real-time of topology structure.

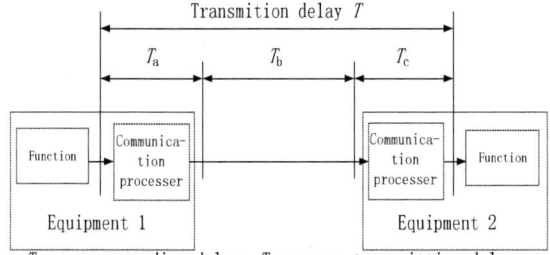

T_a=message sending delay; T_b=message transmitting delay; T_c=message receiving delay

Fig.7.Definition of total transmission time

Network transmission delay is formed by the following delay [12]:

1) Store-and-forward delay of switches T_{SF}. Modern switches are based on the principle of store and forward. As a result, store-and-forward delay of a single switch is equal to the frame length divides the transmission rate. In smart substation, for instance, the speed of optical fiber communication is 100 Mbit/s. But in practical application, relay protection devices usually only transfer a small amount of Boolean values. GOOSE message is generally less than 300b, combined with synchronous frame header 8b. So the longest switches store-and-forward delay is 24.64 μ s.

2) Switch exchange delay T_{SW}. Switch exchange delay is a fixed value, depending on the processing speed of switch chip. Generally, industrial Ethernet switch delay is within 10 μ s.

3) Optical fiber transmission delay T_{WL}. Fiber optic cable transmission delay is cable length divided by the speed of light. For the process layer network in the substation, the total length of fiber could not be very long. If length of fiber is 1 km, the optical transmission delay is about 3.3 μ s.

4) Switch frame queuing time T_Q. Use queuing order delivery when switch frame conflict occurs. If there are K ports in one switch, then the most crowded moment is that other K-1 ports send message to another port at the same time. The longest frame queuing delay is $(K-1)T_{SF}$, the shortest is 0, and the average queuing delay for T_{SF} (K-1) / 2. Based on the analysis above, the most crowded

moment can be estimated, the delay time of each kind of transmission link T_{ALL} for the longest packet network transmission is

$$T_{ALL}=N(T_{SF}+T_{SW}+T_Q)+T_{WL} \qquad (13)$$

Among them, N is the number of switches that transmission link passed.

According to the transmission links shown in table 2, the table 4 shows the delay of each transmission link calculated by formula (13).

TABLE IV
THE TIME-DELAY OF TRANSMISSION LINKS

Number of switches	Possible transmission link	Time (ms)
2	(1)←→(3)	0.82
3	(1)←→(2)←→(11) (1)←→(4)←→(5)	1.23
4	(1)←→(2)←→(10)←→(9) (1)←→(4)←→(6)←→(7)	1.64
5	(1)←→(2)←→(10)←→(8)←→(7) (1)←→(4)←→(6)←→(8)←→(9)	2.05
6	(1)←→(2)←→(10)←→(8)←→(6)←→(5) (1)←→(4)←→(6)←→(8)←→(10)←→(11)	2.46

Seen from table 4, the delay of cobweb structure in each link is less than 3ms where it is stipulated in process layer network GOOSE of smart substation.

B. OPNET simulation of cobweb structure process layer network GOOSE redundant link

According to the analysis above, the simulation is carried out from link 1 in table 2 as a typical link. Data transmission model is constructed by two sets of IED equipment consist of two switches, then simulate it by the OPNET. The simulation results are shown in figure 8.

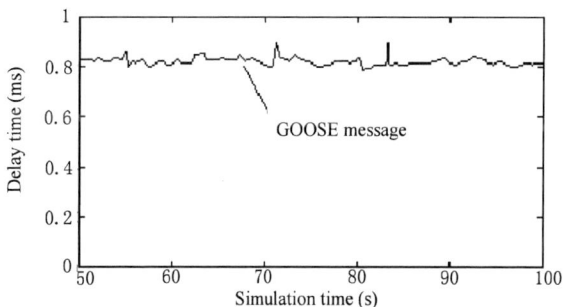

Fig.8.Simulation result of transmission link real-time analysis based on OPNET

The simulation results in process layer GOOSE network communication of real-time are consistent with theoretical analysis in figure 8, and the link delay meets requirements of GOOSE message delay in smart substation.

The reason for the time delay fluctuation in simulation results is related to the different waiting time for message queuing at different moment and the processing speed of hardware not the redundancy.

V. Conclusions

Based on the reliability in structure and real-time in redundancy link, the process layer network GOOSE cobweb topology structure is analyzed, calculated and simulated. The results of the theoretical analysis and simulation indicate that the cobweb structure possess excellent reliability, redundancy, reliability and certainty. It can totally satisfy the requirements for the reliability and real-time in the process layer GOOSE network in smart substation.

References

[1] IEC. IEC61850 Communication networks and systems in substations[S]. Geneva, Switzerland: IEC, 2005.

[2] T. Schaeffler, H. Bauer, W. Fischer, D. Gebhardt, J. Glock, C. Hoga, R. Kutzner, U. Nolte, W. Steingraeber, F. Steinhauser, T. Stirl, K. Viereck, "Process communication in switchgear according to IEC 61850-architectures and application examples," *in Proc. 2008 International Council on Large Electric systems (CIGRE) Conf.*, pp. B5-106.

[3] T. Fencl, P. Burget, J. Bilek, "Network topology design," *Control Engineering Practice*, vol. 19, pp. 1287-1296, July, 2011.

[4] Iqbal Ali, Mini S. Thomas. Substation communication networks architecture[C]. *Power System Technology and IEEE Power India Conference*. New Delhi, India: IEEE,:pp. 1-8, 2008.

[5] C. Fan, Y. M. Ni, A. G. Zhao, S. M. Xu, G. F. Huang, "The research about the scheme of process layer network in smart substation of China," *in Proc. 2011 Asia-Pacific Power and Energy Engineering Conf.*, pp. 1-4.

[6] Chen Fan, Ni Yimin, Zhao Anguo, et al. The research about the scheme of process layer network in smart substation of China[C]. *Power and Energy Engineering Conference*. Wuhan, China: IEEE, pp 1-4, 2011.

[7] S. Zschokke, "Form and function of the orb-web." *in Proc. 2000 19th European Colloquium of Arachnology Conf.*, pp. 99-106.

[8] K.Watcharasitthiwat;P.Wardkein, "Reliability optimization of topology communication network design using an improved ant colony optimization," *Computers and Electrical Engineering*, vol. 35, pp. 730-747, Mar.2009.

[9] X. L. Zhu, J. P. Sang, L. L. Wang, S. Y. Huang, X. W. Zou, "Structure properties and synchronizability of cobweb-like networks," *Physica A*, vol. 387, pp. 6646-6656, Aug., 2008.

[10] B. D. Opell, J. E. Bond, "Capture thread extensibility of orb-weaving spiders: testing punctuated and associative explanations of character evolution." *Biological Journal of the Linnean Society*, vol. 70, pp. 107-120, 2000

[11] K. Tsilipanos, I. Neokosmidis, D. Varoutas, "A system of systems framework for the reliability assessment of telecommunication networks." *IEEE Systems Journal*, vol. 7, pp. 114-124, Mar., 2013.

[12] T. S. Sidhu, Y. J. Yin, "Modelling and simulation for performance evaluation of IEC61850-based substation communication systems," *IEEE Trans. on Power Delivery*, vol. 22, pp. 1482-1489, July, 2007.

Efficiency Improvement of a Self-Start Type Permanent Magnet Synchronous Motor

H. Saikusa, S. Arikawa, T. Higuchi, Y. Yokoi and T. Abe
Graduate School of Engineering
Nagasaki University
1-14 Bunkyo, Nagasaki 852-8521, Japan

Abstract- Recently, a permanent magnet type synchronous motor (PMSM) has been widely used. The PMSM cannot start without an inverter. The authors have proposed in a previous paper a novel self-start type PMSM which moves as induction motor at asynchronous speed region and moves as synchronous motor at synchronous speed. It was surface type PMSM and had relatively large area squirrel-cage region. This paper shows design method for improving efficiency analytically.

Keywords— Permanent magnet motor, surface type, squirrel cage, line start

I. INTRODUCTION

The authors has proposed a novel self-start type surface permanent magnet synchronous motor (Self-start type PMSM) [1]. It had both function of a squirrel-cage induction motor and a permanent magnet synchronous motor. The rotor had unique structure; it had just two magnets for 4-pole motor. The magnets had the same polarity. The motor produced larger torque at induction motor motion and same torque at synchronous motion compared with the self-start type interior permanent magnet synchronous motor.

This paper shows design method for improving efficiency analytically.

II. SELF -START TYPE PMSM

A Motor Construction and Principle

The stator of the proposed self-start type PMSM is the same structure as an usual induction motor. By applying three-phase alternating current to the three-phase stator windings, a revolving magnetic flux is generated. The magnetic flux induces the secondary current on the secondary conductor and torque is generated by an interaction between the magnetic flux and the secondary current. This is the torque production principle for the induction motor motion. For the synchronous motor motion, the torque is produced between the revolving magnetic flux and the permanent magnets.

The rotor structure is shown in Fig. 1. We call it model 1. Two same polarity permanent magnets magnetized in radial direction are mounted on the surface for a 4-pole machine [1, 2]. Since it has large secondary conductor region, the torque is higher than the self-start type IPMSM on the induction motor motion. It also maintains high efficiency and high torque at synchronous operation. Even if the motor loses synchronization, the motor can continue to drive as a synchronous motor with damper windings.

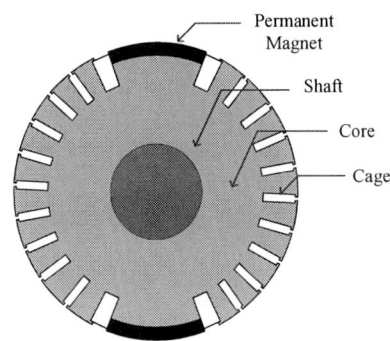

Fig. 1.   Rotor of self-start type PMSM (model 1)

B Torque Characteristics

At starting process, the revolving magnetic field produced by the armature current of the stator windings generates torque at the squirrel-cage windings of the rotor. So the motor operates from start to around synchronous speed as an induction motor. The torque at asynchronous operation is given by

$$T = \frac{P}{4\pi f} 3V_1^2 \frac{r_2' / s}{(r_1 + r_2' / s)^2 + (x_1 + x_2')^2} \tag{1}$$

P: Number of poles V_1: Input voltage
r_1: Primary resistance
r_2': Secondary resistance seen from the primary side
x_1: Primary leakage reactance
x_2': Secondary leakage reactance seen from the primary side

At around the synchronous speed, the rotor is synchronized by the pull-in torque and the motor operates as a synchronous motor [3]. In this state, the torque given by the equation (1) is zero, and the torque is given by the following equation (2). The first term is magnet torque and second term is reluctance torque.

$$T = \frac{P}{4\pi f} \left\{ \frac{3V_1 E_0}{x_d} \sin\delta + \frac{3V_1^2 (x_d - x_q)}{2x_d x_q} \sin 2\delta \right\} \tag{2}$$

Where,

P: Number of poles V_1: Input voltage
E_0: Induced voltage
x_d: d-axis synchronous reactance
x_q: q-axis synchronous reactance
δ: Power angle

III. TORQUE AND EFFICIENCY IMPROVEMENT

A Model 2

Fig. 2 shows the rotor structure of the proposed model 2 to increase the efficiency. Two small magnets are added to converge the magnetic flux there. The squirrel-cage winding is taken away in the middle of winding region between two main magnets of the model 1. Two spacers, flux barriers, are provided at the both ends of the small permanent magnet to reduce leakage magnetic flux.

B Model 3

Fig. 3 shows the rotor structure of the proposed model 3. It has also small magnets, but the magnets are attached on surface of the rotor teeth to prevent the leakage magnetic flux.

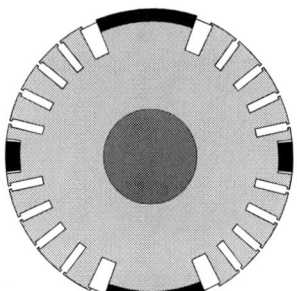

Fig. 2.   Rotor of model 2

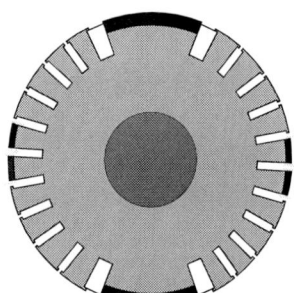

Fig. 3.   Rotor of model 3

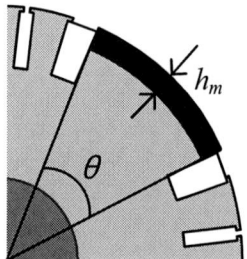

Fig.4.   Magnet angle and height

IV. PERFORMANCE ESTIMATION

A Design Parameter

Table I shows design parameters of the model 1. The rated output power is 2.2 kW, the rated voltage 200 V, and the rated speed 1800 rpm. Fig. 4 shows dimension of the magnet. The angle θ of the main magnet is 40 degree and the height h_m is 4 mm. Fig. 5, Fig. 6 and Fig. 7 show the FEM initial model of the model 1, 2 and 3, respectively.

TABLE I   DESIGN PARAMETER

Output power	2.2 kW
Number of poles	4
Voltage	200 V
Frequency	60 Hz
Rated speed	1800 rpm
Stator diameter	157 mm
Rotor diameter	99.4 mm
Shaft diameter	32 mm
Motor length	90 mm
Air gap	0.3 mm
Number of slots / phase / pole	3
Number of slots	36
Coil pitch	8/9
Connection	Y

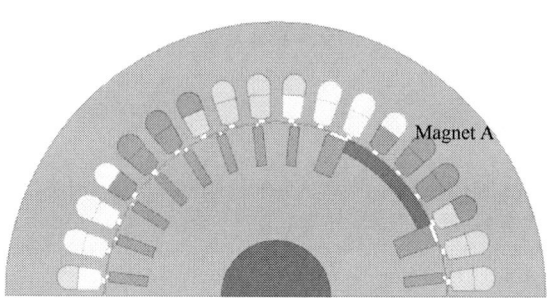

Fig. 5.   2-dimensional model of model 1

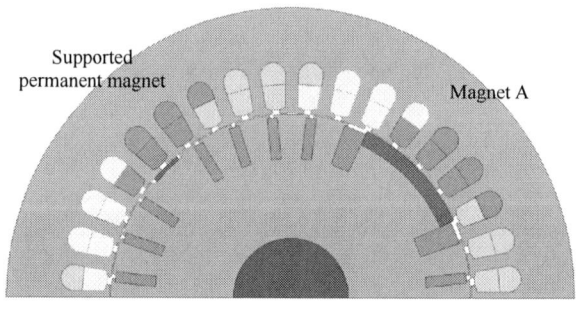

Fig. 6.   2-dimensional model of model 2

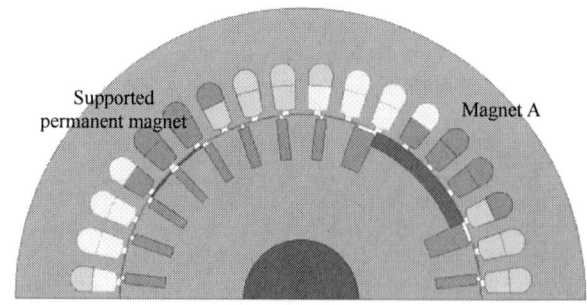

Fig. 7.   2-dimensional model of model 3

B Design of Model 2

B-1 Spacer for additional small magnet

The magnetic flux distributions near the permanent magnet of model 2 is shown in Fig. 8. The leakage magnetic flux is detected at the both ends of the permanent magnet. So, the spacer is provided to reduce the leakage flux. Here, spacer width is 1.15 mm. The flux distributions is shown in Fig. 9.

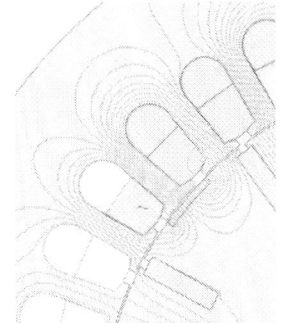

Fig. 8.   Flux distribution of model 2 without the spacer

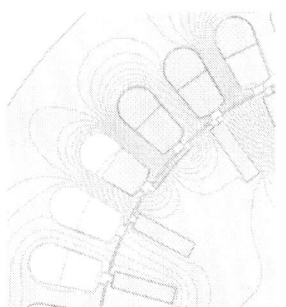

Fig. 9.   Flux distribution of model 2 with the spacer

B-2 Design of additional small magnet

The effect of varying the small magnet angle θ on the torque and efficiency is analyzed. The magnet height h_m is fixed at 1 mm. Fig. 10 shows the torque-slip characteristics under 200 V. The torque of an induction motor of the same construction is shown too. Table II shows the torque, current and efficiency at synchronous speed. Table III shows the efficiency at the rated torque and speed.   The efficiency is given by the following equatoin.

$$\eta = \frac{P}{P + P_c + P_i} \times 100 \ [\%] \qquad (3)$$

P: Output power P_c: Copper loss P_i: Iron loss

It is shown that the torque increases and efficiency increases slightly with increasing θ.

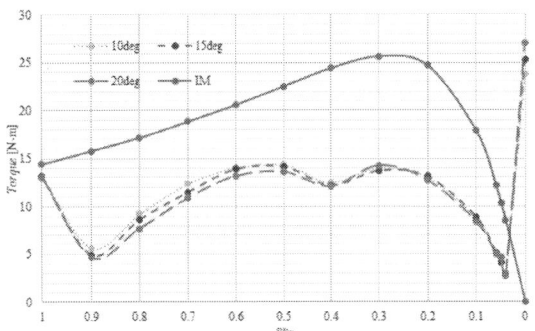

Fig. 10.   Torque-slip characteristics

TABLE II   CHARACTERISTICS AT SYNCHRONOUS SPEED

θ	Torque [N-m]	Current [A]	Efficiency [%]
10 deg	23.79	22.40	78.80
15 deg	25.28	23.36	78.84
20 deg	27.06	24.27	79.06

TABLE III   CHARACTERISTICS AT SYNCHRONOUS SPEED

θ	Torque [N-m]	Current [A]	Efficiency [%]
10 deg	11.74	8.20	87.87
15 deg	11.72	7.89	88.13
20 deg	11.68	7.37	88.60

B-2 Influence by the permanent magnet

The influence of the permanent magnet on the torque is investigated. Fig. 11 shows the torque produced by the permanent magnet when the secondary current is ignored. It is shown that the torque produced by the permanent magnet is negative torque and it has the maximum value at slip of 0.9.

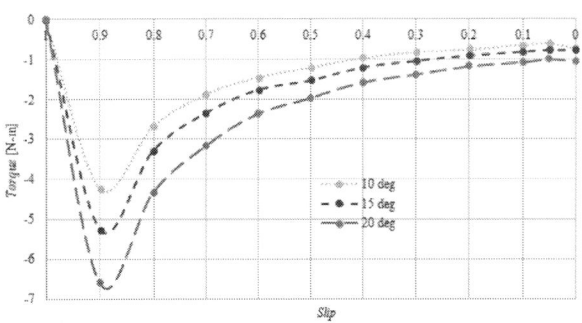

Fig. 11.   Torque by the permanent magnet

B-3 Influence of magnet materials

The performance is examined when a ferrite magnet is used for the additional small magnet. The characteristics are shown in Table. IV. The relative permeability of the ferrite magnet is 1.2, coercive force is 250 A/m, residual flux density is 0.38 T.

The torque of the ferrite magnet is lower than that of the neodymium magnet of the Table II (relative permeability 1.05, coercive force 1000000 A/m, residual flux density 1.05 T). However, efficiency is little higher.

TABLE IV
CHARACTERISTICS OF THE FERRITE MAGNET AT SYNCHRONOUS SPEED

θ	Torque [N-m]	Current [A]	Efficiency [%]
10 deg	21.19	21.54	78.78
15 deg	22.78	21.83	79.08
20 deg	23.47	22.13	79.27

IIV. COMPARISON OF EACH MODEL PERFORMANCE

The three type models are compared on torque and efficiency performances. Their permanent magnet is neodymium. The model 1 is shown in Fig. 5. The magnet angle θ is 40 degree and the height h_m is 4 mm. The model 2 has additional small magnet (θ = 10 degree and h_m = 1 mm) as shown in Fig. 6.

The model 3 has additional small magnet (θ = 10 x 2 = 20 degree and h_m = 0.5 mm) on the rotor teeth of the model 1 as shown in Fig. 7.

A. Magnetic flux distributions

The magnetic flux distributions of the three models are shown in Fig. 12, Fig. 13 and Fig. 14, respectively. Flux lines of the Fig. 13 and Fig. 14 are concentrated at small magnet region compared with Fig. 12.

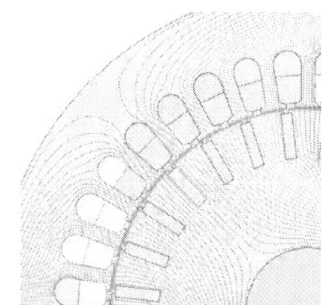

Fig. 12.   Flux distribution of model 1

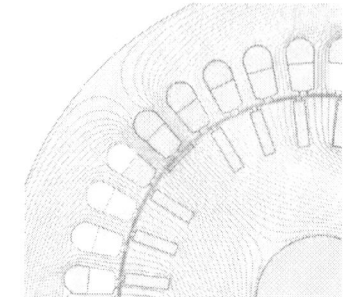

Fig. 13.   Flux distribution of model 2

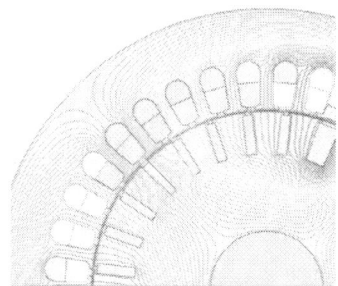

Fig. 14.   Flux distribution of model 3

B. Torque and efficiency characteristics

Static characteristics are compared in Fig. 15. The torque characteristics of the same sized induction motor are shown in this figure. Input voltage is 200 V. The torques of the model 1 to 3 are smaller than induction motor, but not so worse. The torque of the model 1 is little higher at asynchronous speed region and little lower at synchronous speed than the model 2 and model 3.

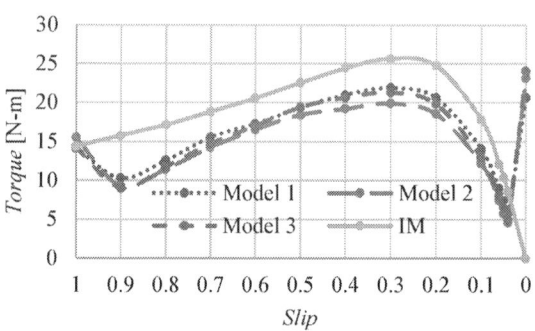

Fig. 15.   Torque slip characteristics

TABLE V   CHARACTERISTICS AT THE SYNCHRONOUS SPEED

	Torque [N-m]	Current [A]	Efficiency [%]
Model 1	20.65	22.09	77.31
Model 2	23.79	22.40	78.80
Model 3	24.03	21.83	79.95

TABLE VI CHARACTERISTICS AT THE SYNCHRONOUS SPEED

	Torque [N-m]	Current [A]	Efficiency [%]
Model 1	11.77	8.67	87.62
Model 2	11.74	8.20	87.87
Model 3	11.50	7.64	88.29

Table V shows their torque, current and efficiency at synchronous speed under 200 V. The torques and efficiencies of the model 2 and 3 are higher than the model 1 and the efficiency of the model 3 is the highest.

Table VI shows the efficiency at the rated torque. The efficiency of the model 3 has the highest value of 88.3 % and is higher than the efficiency of induction motor of 86 %.

C. Starting characteristic

Dynamic characteristics are simulated from start to synchronous speed. Equation of motion of the drive system is shown as follow;

$$\left(J_M + J_L\right)\beta + \lambda\beta = T_{em} + T_{load} \qquad (4)$$

J_M: Moment of inertia of the motor
J_L: Moment of inertia of the load
β: Angular acceleration
λ: Attenuation coefficient
ω: Angular speed
T_M: Torque of the motor
T_L: Torque of the load

Here, the value of the $J_M + J_L$ is 0.05 kg-m^2, and the T_L is 10 N-m. Fig. 16 shows speed-time characteristics. The speed characteristics of the proposed two models are not as different as the self-start type PMSM.

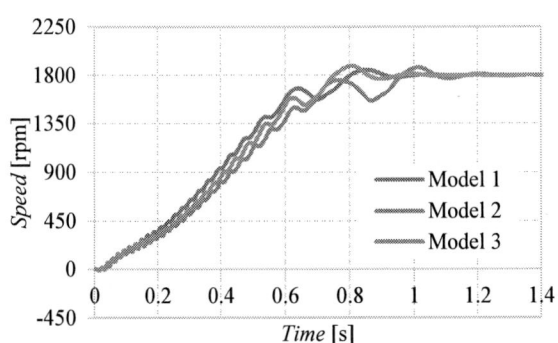

Fig. 16.   Speed characteristic

V. CONCLUSIONS

Two models, model 2 and 3 were proposed to increase in torque and efficiency from the basic self-start type PMSM model (model 1). The model 3 had the highest in efficiency. The torque of the model 3 was smaller than the

other model slightly, but the starting characteristic was not different from the others. It was confirmed that all self-start type models had good starting characteristics.

REFERENCES

[1] T. Higuchi, T. Abe, Y. Miyamoto, M. Ohto and T. Egawa, "Characteristics analysis of a novel self-start type permanent magnet synchronous motor," *2011 IEEJ The Paper of Technical Meeting on Rotating Machinery*, RM-11-104 No. 55.

[2] T. Higuchi, T. Abe, H. Shibuta and T. Egawa, "Fundamental characteristics of a novel self-start type permanent magnet synchronous motor," *2011 IEEJ Industry Applications Society conference*, YPC No. Y-114.

[3] T. Higuchi, T. Abe, H. Shibuta and S. Arikawa, "Characteristics of a Novel Self-Start Type Permanent Magnet Synchronous Motor" *2012 IEEJ Industry Applications Society conference*, YPC No. Y-78.

Consideration of Optimal Number of Poles and Frequency for High-efficiency Permanent Magnet Motor

Daisuke Misu, Makoto Matsushita,
Katsutoku Takeuchi

Toshiba Corporation
1-Toshiba-cho, Fuchu-city, Tokyo, Japan

Koji Oishi, Mitsuhiro Kawamura

Toshiba Mitsubishi-Electric Industrial systems Corporation
6-14 Maruo-cho, Nagasaki-city, Nagasaki, Japan

Abstract— This paper describes an approach to the optimal design of high efficiency permanent magnet reluctance motors (PRM) based on the Taguchi method coupled with finite element analysis (FEA) because of the requirements of the impending IE4 Super Premium Efficiency classification. The objectives of the optimal design are to increase the efficiency and to decrease an amount of the permanent magnet of the motor. We present multi-objective optimization design results in number of poles and frequency for PRM using Taguchi Method.

Keywords— *IE4, Permanent magnet reluctance motor, Pole number, Frequency, Taguchi method*

I. INTRODUCTION

In recent years, country rules for improving efficiency of industrial motors have been set on the basis of new standards [1]. The demand of high-efficiency motors is expected to increase because of enforcing MEPS (Minimum Energy Performance Standard) in the principal nations. The standards also reserve an IE4 class (Super Premium Efficiency) in the near future and the efficiency regulations in accordance with IE4 class become effective in each country. Thus, the development of motors designed with IEC's IE4 requirements is being carried out.

The permanent-magnet synchronous motors (PMSMs) using NdFeB magnets are one solution to meet the IE4-class efficiency levels. It is important to reduce the amount of NdFeB magnet used in PMSMs under the influence of supply risk of rare earth materials and a jump in permanent magnet price.

On the basis of these situations, this paper describes a study of an optimal pole numbers to achieve the IE4-class efficiency levels and to decrease an amount of permanent magnet in case of adopting the permanent magnet reluctance motors (PRM) using finite element analysis (FEA).

In optimizing pole numbers in the PRM, the modified Taguchi method [2] is used. The Taguchi method is useful to find optimal conditions and its appropriateness decreases the experiment times. Therefore, the Taguchi method, successfully applied in the motor optimization design, is normally used for the single-objective purpose optimization such as a reduction of cogging torque to a minimum or an increase of average torque to a maximum. This means that it is insufficient or impossible in analyzing multi-optimization problems. To improve the weakness of the Taguchi method, the modified Taguchi method is applied to analyze the multi-objective optimization problems by an allocation of weight value to each objective function such as an amount of permanent magnet and the efficiency of the PRM.

II. DESIGN CONSIDERATION OF PRM

The authors have been developing the PRM to resolve those defects of the IPM type motors by largely employing the reluctance torque by changing the magnet position and magnetic circuit design. Increase of reluctance torque leads to decrease of an amount of permanent magnet and smaller back EMF. They allow a large variable speed range over 1:5, smaller flux weakening current and higher efficiency.

A. Motor Specification

The cross sections of the PRM are shown in Fig. 1 to Fig. 3 and Table I shows the specification of the PRM. The rating of the motors is 90kW-400V-1500min[-1]. Three models of the PRM, which pole numbers are 4 poles, 6 poles and 8 poles, are analyzed. The coil turn number in those motors is 3turn-2Y in all models and the core length is adjusted to become the same back EMF value. The magnets are placed in V-shape so as to utilize reluctance torque effectively. The V-shape angle of the magnets is 160 degree in 4poles, 140 degree in 6 poles and 120 degree in 8 poles, respectively. The magnet width in each motor is set to the maximum length to be capable of locating magnets in the design of the magnet configuration. The magnet thickness is 10mm and the width of the outer and the center bridge is 2mm in all models.

978-1-4799-2706-7/14 $31.00 © 2014 IEEE

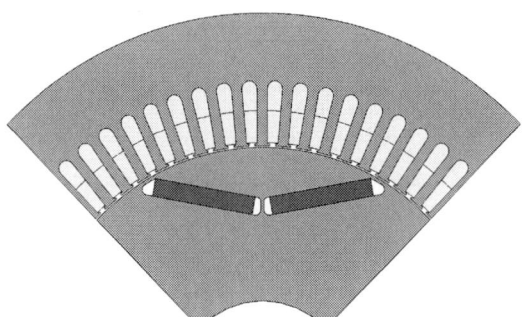

Fig. 1. Cross section of 4 poles motor

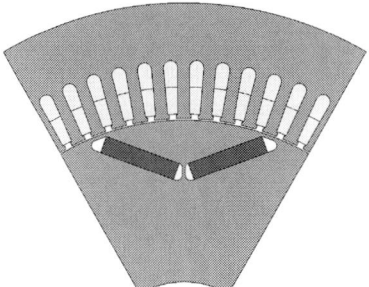

Fig. 2. Cross section of 6 poles motor

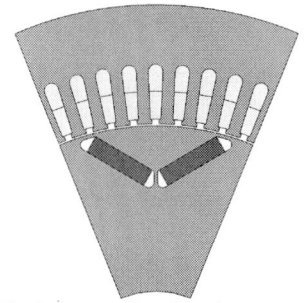

Fig. 3. Cross section of 8 pole motor

Table I Specification of Motors

Item	Unit	4 Pole	6 Pole	8 Pole
Output power	kW	90	90	90
Speed	min⁻¹	1500	1500	1500
Frequency	Hz	50	75	100
Outside diameter of stator	mm	430.0	430.0	430.0
Number of slots	-	72	72	72
Number of turns	turn	3	3	3
Number of parallel circuits	-	2	2	2
Connection	-	Y	Y	Y

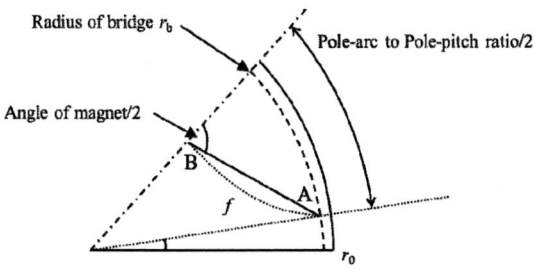

Fig. 4. Magnet configuration of analysis model

B. Design of magnet angle

The design method of the V-shape PM angle is shown in Fig. 4. The magnet geometry to maximize the saliency ratio of the PRM is the geometry to pass the magnetic flux very well in the q-axis direction and not to pass the magnetic flux in the d-axis direction. It is good to be the magnet geometry which follows the magnetic flux line calculated on the assumption that the rotor is a solid iron. The V-shape PM angle should be adopted to pass the magnetic flux very well in the q-axis direction.

In this paper, the V-shape PM angle in each models is calculated on the condition that the pole-arc to the pole-pitch ratio rate of magnet is 60% as follows.

The magnetic potential ϕ of the rotor surface is expressed as follows [3].

$$\phi = \phi_0 \cos(p\theta) \tag{1}$$

where p is a pole pair, θ is a mechanical angle and is a peak value of the magnetic potential.

The magnetic flux line of the rotor core is calculated as follows [3].

$$f = (r/r_0)^p \sin(p\theta) \tag{2}$$

where is the radius of the rotor.

The magnetic flux line is expressed as a contour line of the f.

C. Design variables

The control factors and their levels are shown in Table II and the design variables are shown in Fig. 5. There are five control factors, the magnet width (l_m), the magnet thickness (t_m), the core length (l_c), the pole-arc to pole pitch ratio (θ_p), and the pole number (N). The magnet thickness (t_m) has two levels and other control factors have three levels.

Table II Control factors and levels

Control factor		Level 1	Level 2	Level 3
Magnet thickness	t_m	10mm	8mm	-
Magnet width	l_m	100%	87%	75%
Core length	l_c	100%	105%	110%
Pole-Arc to Pole-Pitch ratio	θ_p	60%	70%	80%
Number of Poles	N	4	6	8

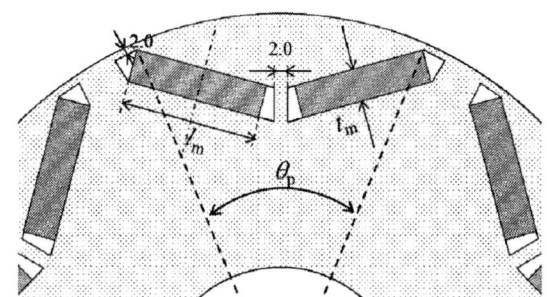

Fig. 5. Control factors

III. Conduct the Experiment

A standard Taguchi's L-18 orthogonal array used for numerical experiments is shown in Table III. We can find optimum settings of many control factors, based on orthogonal array experiments which are used for gathering dependable information about control factors with a small number of experiments. The effects of the control factors are calculated and the results are analyzed to select an optimum setting of the control factors. The L18 orthogonal array allowing no scope for estimation of interactions is used in this paper.

There are eighteen experiments required for us to determine an optimum combination of the levels of these factors as shown in Table III. The performance of the motor in the matrix experiments are computed using the FEA software. The L18 orthogonal array has six factors. Four factors are used as control factors as mentioned above and two factors are used as noise factors.

The calculation results are shown in Fig. 4. Those are Copper loss(W_c), Core loss(W_i), Copper loss + Core loss(W), Efficiency(η), the ratio of reluctance torque to motor torque(τ_r), the amount of magnet(M_m). The value of the current and the phase of the current are adjusted to generate the rated torque (573Nm) at the rating (90kW-1500min^{-1}). The efficiency is calculated as follows.

$$\text{Efficiency} = \text{output} / (\text{output} + \text{copper loss} + \text{core loss} + \text{mechanical loss} + \text{stray loss}) \quad (3)$$

IV. Analysis of Results

After conducting these 18 numerical experiments in Table III and obtaining all the experimental data from FEM simulation, the value of its signal-to-noise (S/N) ratio which represents the quality of the design from motor performance is calculated.

The S/N ratio represents the mean quality characteristics and the noise represents the variance.

A. Calculation formulae of S/N ratios

There are three calculation formulae of S/N ratios to obtain the best settings of the control factors depending on the following strategy.

For "the nominal the best" cases, the S/N ratio can be calculated below.

$$\text{S/N ratio} = 10 \times \log(m^2/\sigma^2) \quad (4)$$

where m represents an average value and σ represents a standard deviation.

For "the smaller the better" case, the S/N ratio can be calculated below.

$$\text{S/N ratio} = -10 \times \log(\sigma^2) = 10 \log(1/\sigma^2) \quad (5)$$

$$\sigma^2 = \frac{1}{n} \times \left(y_1^2 + y_2^2 + \cdots + y_n^2 \right) \quad (6)$$

where y represents the quality characteristics.

For "the larger the better" case, the S/N ratio can be calculated as follows:

Table III L18 orthogonal array

	A t_m	B l_m	C l_c	D θ_p	E N	F e
No.1	1	1	1	1	1	1
No.2	1	1	2	2	2	2
No.3	1	1	3	3	3	3
No.4	1	2	1	1	2	2
No.5	1	2	2	2	3	3
No.6	1	2	3	3	1	1
No.7	1	3	1	2	1	3
No.8	1	3	2	3	2	1
No.9	1	3	3	1	3	2
No.10	2	1	1	3	3	2
No.11	2	1	2	1	1	3
No.12	2	1	3	2	2	1
No.13	2	2	1	2	3	1
No.14	2	2	2	3	1	2
No.15	2	2	3	1	2	3
No.16	2	3	1	3	2	3
No.17	2	3	2	1	3	1
No.18	2	3	3	2	1	2

Table IV Calculation results

	Copper loss W_c [W]	Iron loss W_i [W]	Total loss W [W]	Efficiency η [%]	Reluctance torque τ_r [%]	Magnet weight M_m [kg]
No.1	2150.0	992.4	3142.4	95.6	53.2	6.30
No.2	1443.1	1106.5	2549.6	96.3	38.1	6.96
No.3	1111.7	1294.9	2406.6	96.4	20.2	8.83
No.4	1656.9	1080.2	2737.1	96.1	47.7	5.76
No.5	1241.0	1272.8	2513.8	96.3	29.7	7.31
No.6	2161.1	927.3	3088.4	95.7	55.7	6.02
No.7	2671.1	835.9	3507.0	95.3	64.1	4.73
No.8	1968.7	1090.3	3059.0	95.7	51.0	5.30
No.9	1225.4	1328.3	2553.7	96.3	38.1	6.95
No.10	1342.2	1255.6	2597.9	96.2	27.7	6.72
No.11	2088.0	999.0	3087.0	95.7	54.5	5.30
No.12	1665.2	1112.0	2777.3	96.0	44.7	5.84
No.13	1421.0	1216.7	2637.8	96.2	36.9	5.85
No.14	2657.3	924.3	3581.6	95.2	62.1	4.59
No.15	1770.3	1073.2	2843.5	96.0	54.2	5.09
No.16	2218.2	1099.3	3317.5	95.5	58.7	4.03
No.17	1240.2	1287.9	2528.1	96.0	41.5	5.30
No.18	2423.5	899.7	3323.2	95.5	65.5	4.17

$$\text{S/N ratio} = -10 \times \log(\sigma^2) = 10 \log(1/\sigma^2) \quad (7)$$

$$\sigma^2 = \frac{1}{n} \times \left(1/y_1^2 + 1/y_2^2 + \cdots + 1/y_n^2 \right) \quad (8)$$

B. Calculation of S/N ratios

Quality characteristics are calculated by the S/N ratio. The experimental data are obtained by FEM analysis, including the S/N ratio of the copper loss, the core loss, the copper loss + core loss, the efficiency, the ratio of reluctance torque to motor torque and the amount of magnet. The factor effect diagrams for each control factors are shown in Fig. 6 to Fig. 11.

Compared with the factor effect diagrams of the copper loss and the core loss (those control factors are "the smaller the better" case), the characteristics of these factor effect diagrams are opposed, that is, are the trade-off relation. On the other hands, the characteristic of factor effect diagram of the total loss is the same as the characteristic of the copper loss. This shows the copper loss is dominant in these designed models. The magnet width and the pole number are very effective to increase the efficiency from Fig.9. This shows that the efficiency

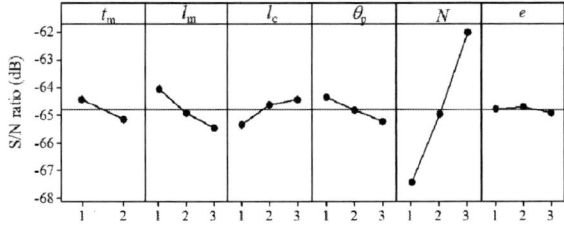

Fig. 6. Factor effect diagram of copper loss

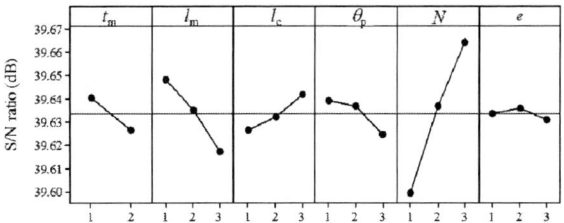

Fig. 9. Factor effect diagram of efficiency

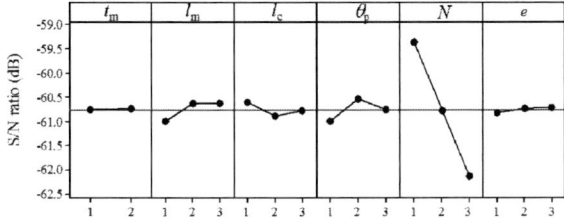

Fig. 7. Factor effect diagram of core loss

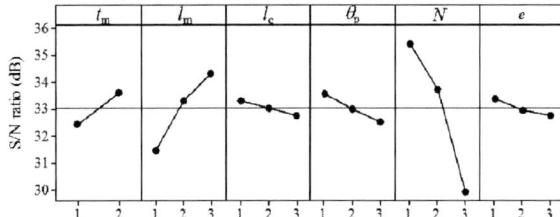

Fig. 10. Factor effect diagram of reluctance torque

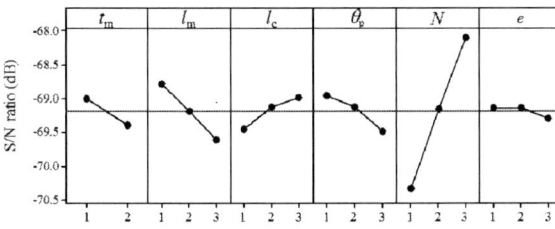

Fig. 8. Factor effect diagram of total loss

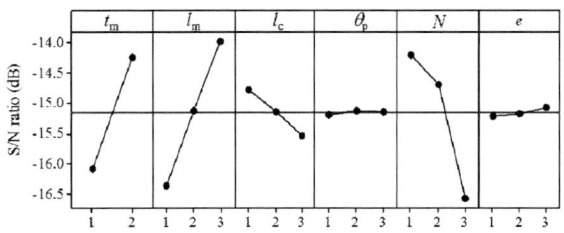

Fig. 11. Factor effect diagram of magnet weight

becomes higher when the magnet width is wide and the pole number is high. Compared with the factor effect diagrams of the ratio of reluctance torque to motor torque and the amount of magnet, the characteristics of these factor effect diagrams are the same. This shows that the ratio of reluctance torque to motor torque is larger when the amount of magnet is smaller.

C. Optimize the multi objective quality characteristics [4] [5] [6]

The above figures present robust optimizations of the copper loss, the core loss, the copper loss + core loss, the efficiency, the ratio of reluctance torque to motor torque and the amount of magnet. These optimization values are single quality characteristics of the motor performance. In order to consider multi objective quality characteristics, such as efficiency and amount of magnet, the procedures for optimizing multi objective function are described below.
(1) Calculate the S/N ratio of the experimental data as defined by the original Taguchi method.
(2) Normalize the S/N ratio of each object using the formula defined as follows:

$$ SN_p^* = \frac{SN_p - \overline{SN_p}}{s.d(SN_p)} \tag{9} $$

where SN_p is the pth S/N ratio, $\overline{SN_p}$ is an average of S/N ratio and s.d(SN_p) is the standard deviation of S/N ratio.

The normalized S/N ratio of the efficiency and the amount of magnet are shown in Table V.
(3) Calculate the total S/N ratio (TS) for the multiple characteristic S/N ratio of each experimental S/N ratio by the following definition:

$$ TS = \sum_i^p w_i SN_i^* \tag{10} $$

where w_i is the weight factor of the ith characteristics that represents the relative importance of the ith characteristics. The sum of the weight value is one.
The multiple characteristics problem is transformed into a single characteristics problem with *TS* being the quality index to be maximized.

The weights of S/N ratio are shown in Table VI. The TS ratio is calculated in three ways that is (a) attaching greater importance to the efficiency, (b) weighting equality and (c) attaching greater importance to the amount of the magnet.

The three factor effect diagrams of the total S/N ratio (TS) are shown in Fig. 12. This figure shows that the pole number is very effective to the efficiency. Eight poles is a good choice when efficiency is important. Six poles is a good choice when the amount of magnets is important or when both of them are important .The optimized settings of control factors are shown in Table VII.

The analysis results calculated using the optimized settings in Table VII are shown in Table VIII.

Table V Normalized S/N ratio

	Normalized SN ratio [dB]	
	Efficiency η	Magnet weight M_m
No.1	-0.69	-0.47
No.2	0.98	-0.96
No.3	1.44	-2.13
No.4	0.41	-0.03
No.5	1.09	-1.20
No.6	-0.55	-0.25
No.7	-1.57	0.94
No.8	-0.48	0.38
No.9	0.96	-0.96
No.10	0.83	-0.79
No.11	-0.55	0.38
No.12	0.29	-0.10
No.13	0.70	-0.11
No.14	-1.74	1.09
No.15	0.10	0.58
No.16	-1.12	1.73
No.17	1.04	0.38
No.18	-1.14	1.56

Table VI Weight of SN ratio

	Weight	
	Efficiency η	Magnet weight M_m
(a)	0.70	0.30
(b)	0.50	0.50
(c)	0.30	0.70

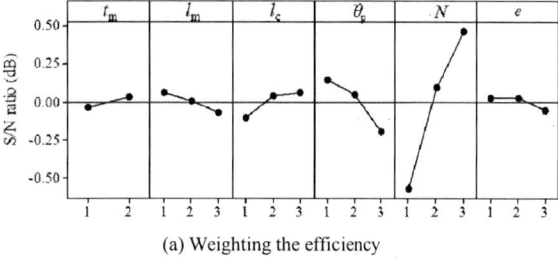

(a) Weighting the efficiency

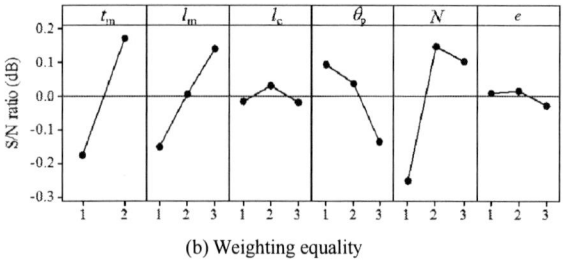

(b) Weighting equality

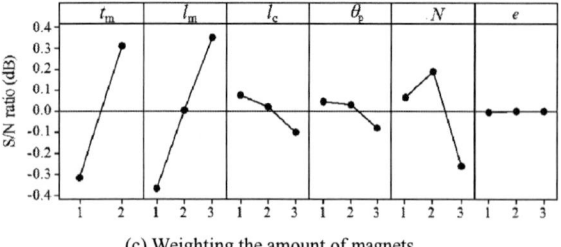

(c) Weighting the amount of magnets

Fig. 12. Factor effect diagrams of total S/N ratio
Table VII Optimization conditions

	Magnet thickness t_m (mm)	Magnet width l_m (%)	Core length l_c (%)	Pole-Arc to Pole-Pitch ratio θ (%)	Number of Poles N Poles
(a)	10mm	100%	110%	60%	8
(b)	8mm	75%	105%	60%	6
(c)	8mm	75%	100%	60%	6

Table VIII Analysis results of optimization conditions

	Copper loss W_c [W]	Iron loss W_i [W]	Total loss W [W]	Efficiency η [%]	Reluctance torque τ_r [%]	Magnet weight M_m [kg]
(a)	1239.0	1380.2	2619.3	96.2	36.5	5.93
(b)	1560.9	1114.8	2675.8	96.1	40.8	4.19
(c)	1710.7	1089.1	2799.8	96.0	41.2	3.98

The efficiency is 96.2% and the amount of magnet is 5.93kg in case of (a), attaching greater importance to the efficiency. The efficiency is 96.0% and the amount of magnet is 3.98kg in case of (c), attaching greater importance to amount of magnet. The IE4 nominal efficiency limit is 96.2% for 90kW-1500min⁻¹ motor. It is confirmed not to achieve the IE4 nominal efficiency limit in case of optimized settings (b) and (c). And it is confirmed that the efficiency becomes higher when copper loss and core loss is almost same in case of (a).

V. COMPARISON OF ANALYSIS RESULTS AND MEASUREMENT RESULTS

The optimized settings are defined using 2D FEM analysis. It is important to carry out the verification of FEM analysis accuracy, compared with FEM analysis results and the measurement results. We performed the verification of FEM analysis accuracy with the PRM designed before. The specification of the motor is shown in Fig. 9. The motor is 8P-90kw-1500min⁻¹.

The comparison of the analysis results and the measurement results of the motor is shown in Fig. 10. The FEM analysis is carried out on the same condition of the measurement current. The induced voltage, the voltage and the torque of FEM analysis are almost same as the measurement results. The total loss is calculated as the sum of copper loss, core loss, mechanical loss and stray loss. The total loss of the analysis results is a little larger than the measurement results. It is confirmed that the efficiency obtained by FEM analysis is slightly underestimated. Therefore, the analysis results of Table VIII calculated with the optimized settings in Table VII is possible to achieve the IE4 nominal efficiency limit.

Table IX Specifications of the comparison motor

Item	Unit	value
Output power	kW	90
Number of poles	pole	8
Speed	min⁻¹	1500
Frequency	Hz	100
Outside diameter of Stator	mm	430
Number of slots	-	72

Table X Results of measurement and analysis

Item	Unit	Experiment	Analysis
Induced voltage	V	355	374
Voltage	V	350	354
Current	A	156	156
Phase angle	deg	35	35
Torque	Nm	577	602
Power factor	%	99.3	99.0
Efficiency	%	96.7	96.4
Copper loss	W	-	422
Iron loss	W	-	2171
Total loss	W	3136	3553

VI. CONCLUSIONS

This paper proposes the optimal pole number and frequency for 90kW-1500min^{-1} motor to be compatible with higher efficiency and smaller amount of magnet. To optimize the design of the motor, we make use of the modified Taguchi method that considered multi objective quality characteristics. As results of the factor effect diagrams of each S/N ratio, the pole number and the magnet width are great influence on increasing the efficiency. The pole number is very effective in improving the efficiency and decreasing the amount of magnet from the multiple objective quality characteristics using normalized S/N ratio. It is confirmed that eight poles is an optimal pole number when efficiency is important and six poles is an optimal pole number when the amount of magnets is important or when both of them are important. The Taguchi method is effectively employed which considered multiple objective quality characteristics as a suitable optimization method in electric machine applications.

REFERENCES

[1] International Electrotechnical Commission "Rotating electrical machines-Part 30: Efficiency classes of single-speed, three-phase, cage-induction motors(IE-code)", IEC 60034-30 (2008)

[2] International Electrotechnical Commission "Rotating electrical machines-Part 31: Selection of energy-efficient motors including variable speed applications - Application guide", IEC/TS 60034-31 (2010)

[3] Minoru Kondo, Keiichiro Kondo, Yasushi Fujishima, Shinji Wakao : "Rotor Dwsign of Permanent Magnet Synchronous Motor for Railway Vehicle", IEEJ Trans. IA, Vol.124-D, No.1 pp.124-130 (2004) (in Japanese)

[4] Yoshinori Matsuda, Ikuro Morita : "Multi-objective Optimum Design for IPMSM with Concentrated Winding", RM-07-55(2007) (in Japanese)

[5] Naoki Sakamaki, Ikuro Morita : "Study on Multi-objective Optimum Design for IPMSM with Concentrated Winding", RM-08-37(2008) (in Japanese)

[6] Hyun Kag Park, Byoung Yull Yang, Sang Bong Rhee and Byung Il Kwon : "Novel Design of Flux Barrier in IPM type BLDC motor using the Taguchi Method", Ninth International Conference on Electrical

Basic Study on the Suitable Structure of a Permanent Magnet Synchronous Motor with a Powder Magnetic Core

Shizuka Hashimoto, Masayuki Sanada, Shigeo Morimoto, and Yukinori Inoue

Osaka Prefecture University
1-1 Gakuen-cho, Naka-ku, Sakai, Osaka, Japan

Abstract— **A high-efficiency motor is required in order to address recent environmental issues. Permanent magnet synchronous motors (PMSMs) are widely used because of their high efficiency. However, even higher efficiency is required in order to reduce the power consumption of motors. The structures of PMSMs using a powder magnetic core are considered. Since the powder magnetic core has very little eddy current loss, the core loss is expected to be reduced. Therefore, multi-pole PMSMs with a powder magnetic core are examined based on the torque, the losses, and the efficiency characteristics. As a result, the efficiency of 16-pole models improved in the high-speed region and a 50% reduction in weight was achieved by using a powder magnetic core. In addition, a torque equal to that generated by 4-pole models was achieved by expanding the motor diameter of the lightweight model. However, the model using a powder magnetic core was not able to achieve an efficiency equal to that of 4-pole models.**

Keywords— *High efficiency, Multi-pole SPMSMs, Powder magnetic core, Surface permanent magnet synchronous motor (SPMSM)*

I. INTRODUCTION

At present, motors are used in a wide range of new technical fields, such as machines for industrial machines and electric vehicles [1]. Among these motors, the permanent magnet synchronous motor (PMSM) is able to operate with high efficiency [2]-[4]. Therefore, the PMSM has become more popular in recent years with the increased need for energy savings. However, even higher efficiency is required because motors are responsible for more than half of the power consumption in Japan. The efficiency depends on the copper loss and core loss. The use of a powder magnetic core is expected to decrease the core loss [4], [5] because the powder magnetic core has very little eddy current loss. The use of multi-pole PMSMs with a powder magnetic core is expected to improve the PMSM. Thus, suitable motor structures that can make use of the characteristics of the powder magnetic core to raise efficiency are examined. In addition, if the weight of the PMSM is reduced while maintaining high efficiency, the amount of materials used in the PMSM will be reduced, thereby decreasing the environmental impact. Therefore, this weight reduction technique becomes useful in the present when the resource problem is paid attention.

In this paper, multi-pole PMSMs with a powder magnetic core are examined in order to determine whether the efficiency and torque can be improved by this structure [6], [7]. Therefore, the PMSM using a powder magnetic core is examined by comparing the characteristics of the PMSM with those of conventional silicon steel sheets.

We herein investigate the characteristics of lightweight PMSMs that use a powder magnetic core. Here, the number of poles and slots of the PMSM are increased without changing the magnet volume. The influence of the weight reduction by trimming the magnetic core, which the magnetic flux density is low, is examined. Furthermore, in order to improve the efficiency, the characteristics of the model of a lightweight extended-diameter motor are investigated.

II. ANALYSIS MODELS

Fig. 1 shows basic analysis models. These models are surface permanent magnet synchronous motor (SPMSM). Outer diameter of the rotor of these models is 60 mm. Figs. 1(a) and 1(b) show the spm4 and spm16-1 models, where a 'P' following the model designation indicates that the model has a powder magnetic core. Table I lists the common specifications of the analysis models, and Table II lists the specifications of the 4-pole and 16-pole stators. The stator winding is a concentrated winding. Fig. 2 shows the *B-H* characteristics of the core material. As shown in Fig. 2, the difference in the core material characteristics can be confirmed. The silicon steel sheet is 35H300, and that of powder magnetic core is soft

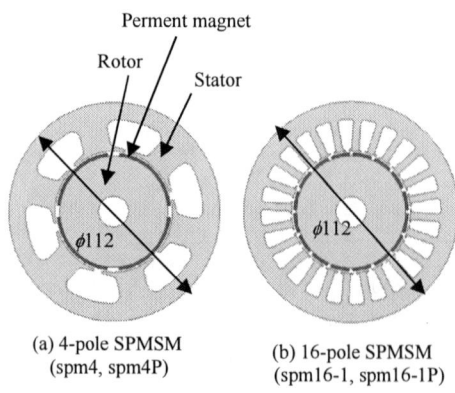

(a) 4-pole SPMSM
(spm4, spm4P)

(b) 16-pole SPMSM
(spm16-1, spm16-1P)

Fig. 1. Analysis models of the SPMSM (unit : (mm)).

978-1-4799-2706-7/14 $31.00 © 2014 IEEE

TABLE I
COMMON SPECIFICATIONS OF ANALYSIS MODELS

Item (Unit)	Value
outer diameter of stator (mm)	112 (100, only spm16-2P)
outer diameter of rotor (mm)	60 (79.4, only spm16-3P)
Stack length (mm)	40
Air gap length (mm)	1.0
Number of turn (turn/phase)	232
Rated current (A)	3
Magnet coercive force (kA/m)	915
Amount of magnet (cm^3)	8.80
Core material	silicon steel sheet(35H300) Powder magnetic core (SMC)

TABLE II
SPECIFICATIONS OF 4-POLE AND 16-POLE STATORS

Item (Unit)	4-pole	16-pole
Number of poles	4	16
Number of slots	6	24
Slot fill factor (%)	50	46.1
Winding resistance (Ω)	0.828	0.497

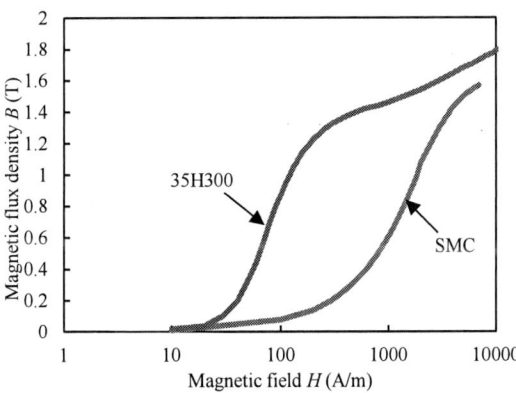

Fig. 2. B-H characteristics of the core materials.

TABLE III
MATERIAL CONSTANTS

Item Unit	SMC	35H300
Hysteresis loss coefficient k_h (mWs/kg)	44	19.09
eddy current loss coefficient k_e (mWs^2/kg)	0.024	0.064

magnetic composite (SMC). Table III lists the material constants. As shown in Table III, the hysteresis loss coefficient of the powder magnetic core is approximately twice that of silicon steel sheet, but the eddy current loss

coefficient of the powder magnetic core is less than half that of silicon steel sheet. Therefore, by using a powder magnetic core, the core loss in the high-speed region is expected to decrease because the eddy current loss is small.

III. ANALYSIS RESULTS

Fig. 3 shows the maximum torque of the 4-pole and 16-pole SPMSMs. As shown in Fig. 3, the maximum torque of spm4P is 0.17 Nm smaller than that of spm4. The maximum torque of spm16-1P is 0.2 Nm smaller than that of spm16-1. Therefore, the characteristics of PMSMs with a powder magnetic core are slightly deteriorated because the *B-H* characteristics of the powder magnetic core are inferior to those of the silicon steel sheet. The maximum torque of spm4 is 2.42 Nm, and the maximum torque of spm16-1 is decreased by approximately 0.03 Nm. In addition, the maximum torque of spm4P is 2.25 Nm, and the maximum torque of spm16-1P is decreased by 0.06 Nm. Multipolarization has little influence on maximum torque because the maximum torque is relatively constant. Fig. 4 shows the output power vs. rotational speed characteristics of the 4-pole and 16-pole SPMSMs. This output power is calculated by considering the core loss. The expression of the output power P (W) is as follows:

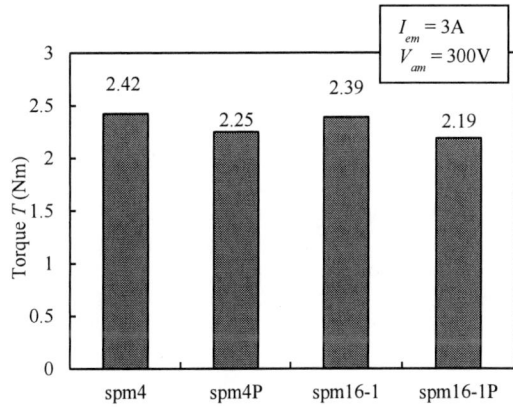

Fig. 3. Maximum torque of the 4-pole and 16-pole SPMSMs.

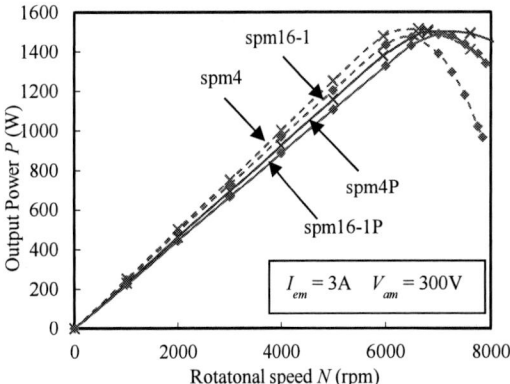

Fig. 4. Output power vs. rotational speed characteristics.

$$P = \frac{2\pi NT}{60} - W_i \tag{1}$$

where W_i is the core loss (W).

Fig. 5 shows the core loss vs. rotational speed characteristics of the 4-pole and 16-pole SPMSMs. As shown in Fig. 5, the core loss is increased by increasing the number of poles. The core loss is the sum of the hysteresis loss and the eddy current loss, as shown in the following equations [8], [9]:

$$W_i = W_h + W_e \tag{2}$$

$$W_h = \sum_{i=1}^{e} \left[m_i \sum_{k=1}^{n} \left\{ k_h(kf) B_{rk-i}^{\;2} + k_h(kf) B_{\theta k-i}^{\;2} \right\} \right] \tag{3}$$

$$W_e = \sum_{i=1}^{e} \left[m_i \sum_{k=1}^{n} \left\{ k_e(kf)^2 B_{rk-i}^{\;2} + k_e(kf)^2 B_{\theta k-i}^{\;2} \right\} \right] \tag{4}$$

where W_h is the hysteresis loss (W), W_e is the eddy current loss (W), k is the harmonic order, n is the maximum harmonic order considered herein, f is the basis frequency (Hz), i is the element number in the FEM, e is the number of elements in the region, m_i is the mass of the i-th element (kg), B_{rk-i} is the amplitude of the k-th harmonic order of the magnetic flux density of the radial direction of the i-th element (T), and $B_{\theta k-i}$ is the amplitude of the k-th harmonic order of the magnetic flux density of tangential direction of the i-th element (T).

Equations (3) and (4) reveal that the core loss increases as the frequency increases. Therefore, the output is reduced as compared to the torque by being multipolarized when the core loss is considered. In addition, the output power is decreased by using a powder magnetic core.

Fig. 6 shows the efficiency vs. rotational speed characteristics of the 4-pole and 16-pole SPMSMs. Table IV shows the efficiencies at various rotational speeds for each model. As shown in Fig. 6 and Table IV, in the low-speed range, the efficiency of spm16-1 is higher than that of spm4. In the high-speed range, the efficiency of spm16-1 decreases compared to that of spm4. The characteristics of the efficiency of the multipolarized model using a powder magnetic core have a similar tendency to those using silicon steel sheet. Fig. 7 shows the copper loss characteristics; as shown, the copper loss is decreased by increasing the number of poles because the winding resistance per phase becomes small. The efficiency decreases because the increment of the core loss is larger than the decrement of the copper loss. As a result, the efficiency is decreased by the multipolarization. In addition, the efficiency of the 16-pole model in the low-speed region is decreased by using a powder magnetic core, but increase in the high-speed region. As shown in Fig. 5, in the low-speed region, the core loss of spm16-1P is larger than that of spm16-1. However, in the high-speed region, the core loss of spm16-1P is lower

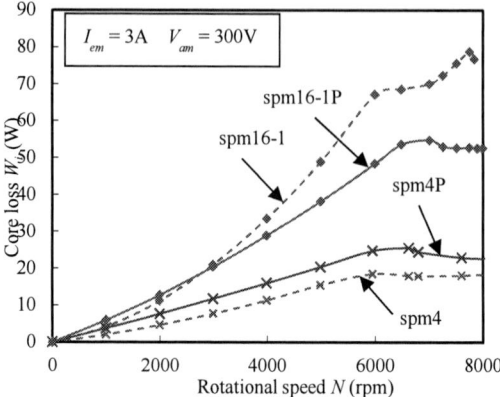

Fig. 5. Core loss vs. rotational speed characteristics.

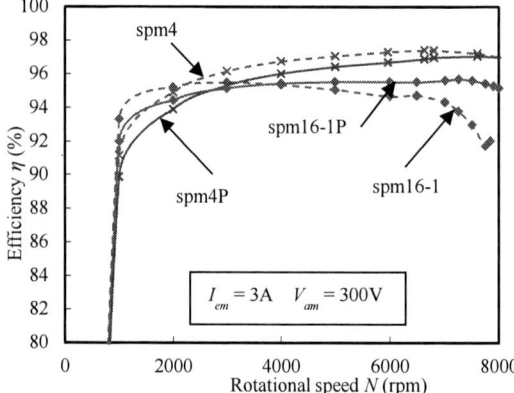

Fig. 6. Efficiency vs. rotational speed characteristics.

TABLE IV

EFFICIENCIES FOR VARIOUS ROTATIONAL SPEEDS FOR EACH MODEL   (UNIT : (%))

	Rotational speed N (rpm)	
	2000	5000
spm4	94.90	97.06
spm4P	93.91	96.44
spm16-1	95.20	95.07
spm16-1P	94.43	95.54

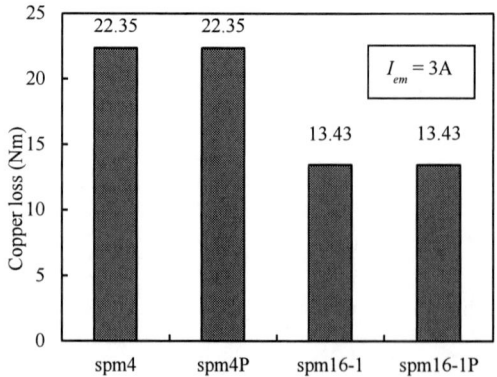

Fig. 7. Copper loss characteristics.

than that of spm16-1. Therefore, the influence of the hysteresis loss is large because the hysteresis loss coefficient of the SMC is large. As shown in Fig. 7, the copper losses of spm16-1 and spm16-1P are the same. Based on these considerations, the efficiency is highly influenced by the increase of the core loss. Thus, the use of a powder magnetic core in 16-pole models in a high-speed region is advantageous but that was inferior to the efficiency of the 4-pole model as the criterion.

IV. WEIGHT REDUCTION OF 16-POLE SPMSM

The weight reduction is expected to improve the torque/volume ratio. Fig. 8 shows the distribution of the magnetic flux density. As shown in Fig. 8, the magnetic flux density is low at several areas of spm16-1P. Here, the magnetic flux density of spm4P is approximately 0.4 T in at the outer edges of the stator and that of spm16-1P is less than 0.3 T. Therefore, the weight of the motor is reduced by replacing the magnetic core to air that magnetic flux density is less than 0.3 T, because the influence on the characteristics is assumed to be small. Fig. 9 the shows analysis model for the lightweight SPMSM. The outer diameter of the rotor of spm16-2P is maintained at 60 mm. The core is a powder magnetic core. Table V lists the weight of each model. The mass considering the core, the coil, and the permanent magnet is 3.03 kg for the spm4P, 2.65 kg for the spm16-1P, and 1.50 kg for the spm16-2P. Thus, the mass of the SPMSM is decreased by 50%.

TABLE V

MODEL WEIGHTS

	spm4P	spm16-1P	spm16-2P
weight of core (kg)	2.12	2.07	0.92
weight of copper wire (kg)	0.84	0.51	0.51
weight of permanent magnet (kg)	0.067		
total weight (kg)	3.03	2.65	1.50
Weight reduction (%)	0	-12.5	-50.5

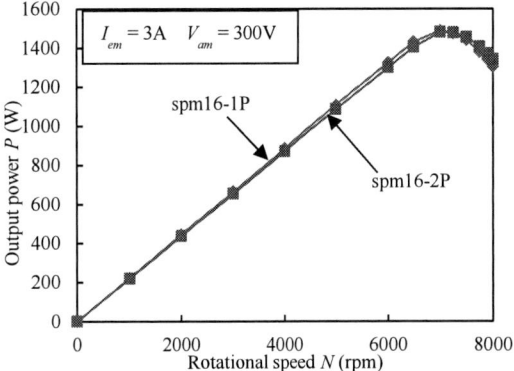

Fig. 10. Output power vs. rotational speed characteristics.

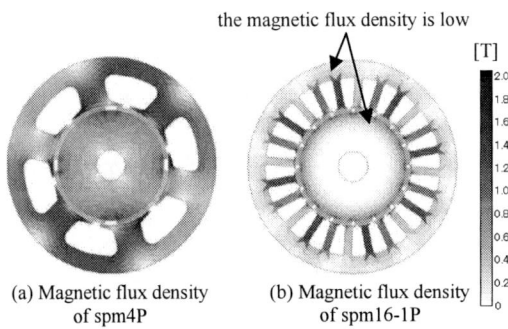

(a) Magnetic flux density of spm4P　(b) Magnetic flux density of spm16-1P

Fig. 8. Magnetic flux density distribution.

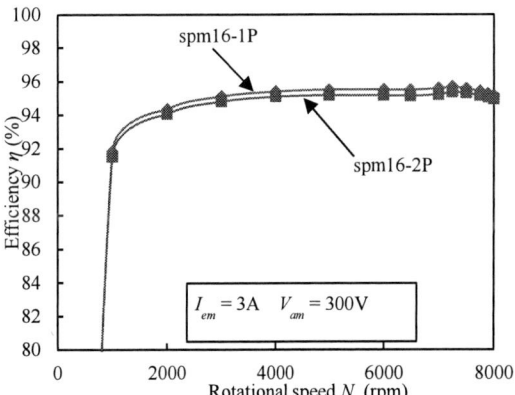

Fig. 11. Efficiency vs. rotational speed characteristics.

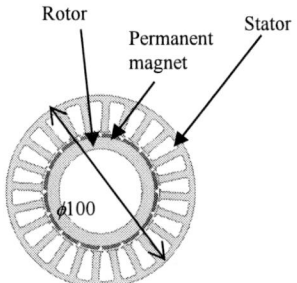

Fig. 9. Analysis model of the light-weight SPMSM (spm16-2P) (unit : (mm)).

Based on the results of the analysis, the maximum torque of spm16-1P is 2.19 Nm and that of spm16-2P is 2.15 Nm for a rated current of 3 A. Fig. 10 shows the output power characteristics, which are approximately equal. Therefore, the weight reduction influences the maximum torque and output power only slightly.

Fig. 11 shows the efficiency vs. rotational speed characteristics. Fig. 12 shows the core loss vs. rotational speed characteristics. The efficiency of spm16-2P is lower than that of spm16-1P because the core loss increases according to the increase in magnetic flux density associated with the weight reduction in the high-

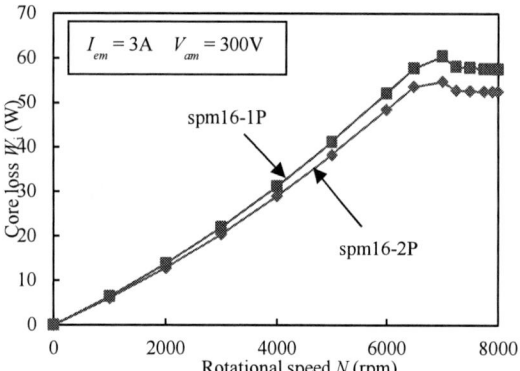

Fig. 12. Core loss vs. rotational speed characteristics.

speed region. However, the efficiency in spm16-2P is decreased slightly compared to spm16-1P, although the mass is decreased significantly. Therefore, spm16-2P is expected to be useful.

V. EFFECT ON THE CHARACTERISTICS BY EXPANDING MOTOR DIAMETER

The model using a powder magnetic core was not able to achieve a torque equivalent to that of spm4 even in the high-speed region, as mentioned above. Therefore the characteristics when the motor diameter is increased to that of spm16-1 are examined.

Fig. 13 shows the lightweight analysis model, for which the motor diameter is changed. As shown in Table I, the outer diameter of the rotor of spm16-3P is 79.4 mm

Fig. 14 shows maximum torque of the 4-pole and 16-pole SPMSMs. As shown in Fig. 14, the maximum torque of spm16-3P is 2.42 Nm, which is equal to that of spm4. In addition, the maximum torque increased approximately 0.3 Nm by extending the motor diameter in comparison with spm16-2P. Fig. 15 shows the output power vs. rotational speed characteristics. The output power characteristics of spm16-3P are approximately equivalent to those of spm4 and are higher than those of spm16-2P at 6,000 rpm or less. This is because the radius of the rotor increases, and the action length, which is related to the torque, increases. Therefore, the torque and the power are confirmed to have increased by extending the motor diameter. However, the critical operating speed of spm16-3P decreases. Table VI shows the motor constant of each model. The minimum d-axis armature

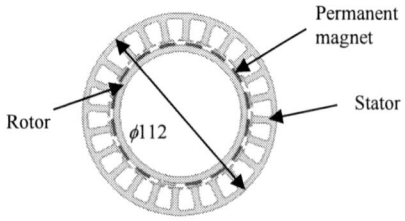

Fig. 13. Light-weight analysis model with a variable motor diameter (spm16-3P) (unit : (mm)).

flux linkage $\Psi_{d\min}$ was obtained from equation [10].

$$\Psi_{d\min} = \Psi_a - L_d i_{am} \qquad (5)$$

The critical operating speed was obtained from Table VI and the following equation:

$$\omega_c = \frac{V_{om}}{\Psi_{d\min}} \qquad (6)$$

As shown in Table VI, the minimum d-axis armature flux linkage of spm16-3P is large. Therefore, the critical operating speed of spm16-3P decreased.

Fig. 16 shows the efficiency vs. rotational speed characteristics, and Fig. 17 shows the core loss vs. rotational speed characteristics. As shown in Fig. 16, the

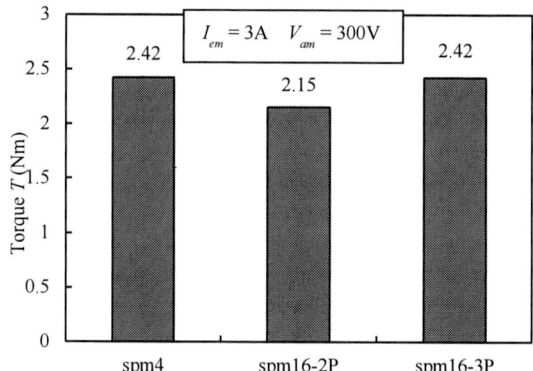

Fig. 14. Maximum torque of the 4-pole and 16-Pole SPMSMs

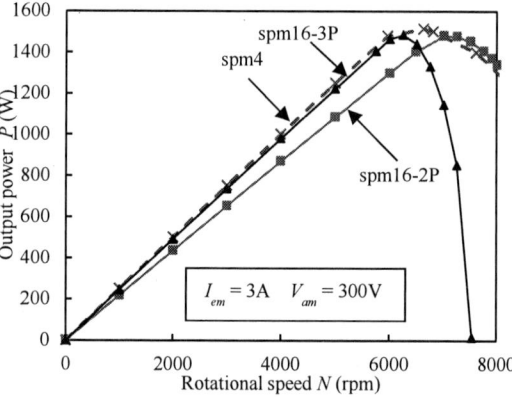

Fig. 15. Output power vs. rotational speed characteristics.

TABLE VI

MOTOR CONSTANT OF EACH MODEL

	Ψ_a (Wb)	L_d (H)	$\Psi_{d\min}$ (Wb)
spm16-1	0.058	2.97	0.042
spm16-1P	0.053	2.77	0.038
spm16-2P	0.052	2.72	0.038
spm16-3P	0.062	2.08	0.051

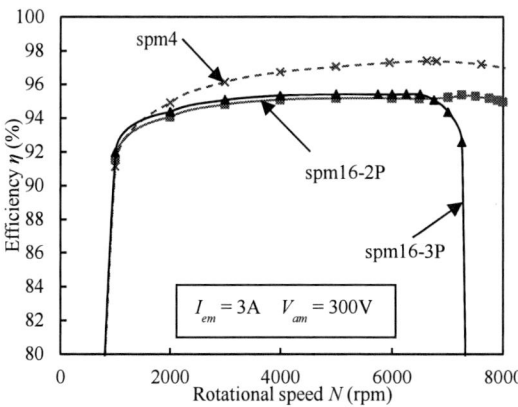

Fig. 16. Efficiency vs. rotational speed characteristics.

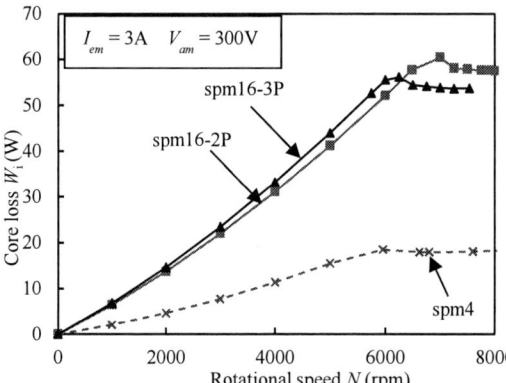

Fig. 17. Core loss vs. rotational speed characteristics.

efficiency improved slightly by increasing the motor diameter. As shown in Fig. 17, the core loss increases by extending the motor diameter. Here, the reason of efficiency is improved that increase in the output power is greater than the increase in the core loss though the core loss increase. However, the efficiency of spm16-3P was lower than that of spm4 because the hysteresis loss coefficient of the powder magnetic core is much larger than that of silicon steel sheet.

VI. CONCLUSIONS

In this paper, the structures of multi-pole SPMSMs using a powder magnetic core were discussed. The results of the present study revealed that the characteristics of 4-pole models using a powder magnetic core were not as good as those of the model using silicon steel sheet. The efficiency of the 16-pole model with a powder magnetic core increased in the high-speed region. However, efficiency of 16-pole model with a powder magnetic core is inferior to the efficiency of the 4-pole models.

In addition, a method by which to reduce the weight of the PMSM without reducing the efficiency was investigated. The number of slots and poles of the PMSM were increased without changing the magnet volume. As a result, the weight of the PMSM could be reduced by approximately 50% while decreasing the efficiency and maximum torque only slightly.

Moreover, a lightweight analysis model for which the

outer diameter of the stator was increased was investigated. The maximum torque of this model is equal to spm4, but the efficiency is inferior to that of the 4-pole models because the hysteresis loss coefficient of the powder magnetic core is much greater than that of silicon steel sheet.

In this paper, pure sine wave current was used in the FEM analysis. The core loss must also consider the PWM carrier loss, which depends on the eddy current loss. In the future, a structure that can reduce the influence of the hysteresis loss should be inbestigated.

REFERENCES

[1] S. Morimoto, "Trend of permanent magnet synchronous machines," *IEEE Transactions on Electrical and electronic enginerring*, Vol. 2, pp. 101-108, 2007.

[2] C. S. Jin, D. S. Jung, K. C. Kim, Y. D. Chun, H. W. Lee, and J. Lee. "A study on improvement magnetic torque characteristics of IPMSM for direct drive washing machine," *IEEE Transactions on Magnetics*, Vol. 44, No. 6, pp. 2811-2814, 2009..

[3] L. W. Song, Y. Dai, and S. M. Cui.; "Development of PMSM Dives for Hybrid Electric Car Applications," *IEEE Transactions on Magnetics*, Vol. 43, No. 1, pp. 434-437, 2007.

[4] K. T. Kim, K. S. Kim, S. M. Hwang, T. J. Kim, and Y. W. Jung, "Comparison of magnetic force for IPM and SPM motor with rotor eccentricity," *IEEE Transactions on Magnetics*, Vol. 37, No. 8, pp. 3448-3451, 2001.

[5] J. H. Seo, D. K. Woo, T. K. Chung, and H. K. Jung, "A study on loss characteristic of IPMSM for FCEV Considering the Rotating Field," *IEEE Transactions on Magnetics*, Vol. 46, No. 8, pp. 3213-3216, 2010.

[6] P. W. Jang, and B. H. Lee, "Effects of resistivity on eddy current loss of compressed powder cores studied by FEM," *IEEE Transactions on Magnetics*, Vol. 44, No. 6, pp. 2781-2783, 2009.

[7] S. M. A. Sharkh, and M. T. N. Mohammad, "Axial field permanent magnet DC motor with powder iron armature," *IEEE Transactions on Energy Conversion*, Vol. 22, No. 3, pp. 608-613, 2007.

[8] J. H. Seo, S. Y. Kwak, S. Y. Jung, C. G. Lee, T. K. Chung, and H. K. Jung, "A research on iron loss of IPMSM with a fractional number of slot per pole," *IEEE Transactions on Magnetics*, Vol. 45, No. 3, pp. 1824-1827, 2009.

[9] A. Boglietti, A. Cavagnino, D. M. Ionel, M. Popescu, and D. A. Staton, "A general model to predict the iron losses in PWM inverter-fed induction motors," *IEEE Transactions on Industry Applications*, Vol. 46, No. 5, pp. 1882-1890, 2010.

[10] W. H. Kim, K. s. Kim, S. J. Kim, D. W. Kang, S. C. Go, Y. D. Chun, and J. Lee, "Optimal PM Design of PMA-SynRM for Wide Constant-Power Operation and Torque ripple Reduction," *IEEE Transactions on Magnetics*, Vol. 45, No. 10, pp. 4660-4663, 2009.

Characteristics of a Half-Wave Rectified Brushless Synchronous Generator

Yuki Hirakawa, Tsuyoshi Higuchi, Yuichi Yokoi and Takashi Abe
Graduate School of Engineering
Nagasaki University
1-14 Bunkyo, Nagasaki 852-8521, Japan
thiguchi@nagasaki-u.ac.jp

Abstract— **The paper proposes the half-wave rectified brushless synchronous generator and analyzes the basic characteristics using the finite element method. It is based on the half-wave rectified excitation theory and doesn't need brush and slip ring system or permanent magnets for field excitation.**

Keywords— *Synchronous generator, half-wave rectified excitation theory, wind power generation, finite element analysis*

I Introduction

Previously we developed the half-wave rectified brushless synchronous motor [1] [2], as an AC servo motor. In this paper , we propose a novel synchronous generator for wind power generation using the half-wave rectified excitation theory. The structure is the same as conventional salient pole type synchronous generator, but whose rotor windings are short circuited with a diode. The field current in the rotor field windings is induced from the stator excitation current and produces a field magnetic flux. The field flux is controllable by varying the amplitude of the excitation current. The controllability makes it possible to easily perform the field weakening operation at high speed region of the vertical axis wind turbine type generation system. It doesn't need any permanent magnet for field excitation and so cut-in wind speed is smaller than permanent magnet type generator. The basic characteristics are calculated using the finite element method (FEM) analysis.

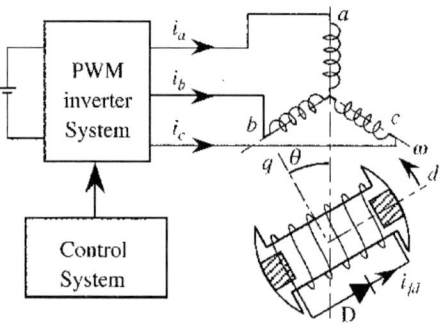

Fig. 1 Motor system configuration.

II Half-Wave Rectified Brushless Synchronous Motor

Figure 1 shows the system configuration of the half-wave rectified brushless synchronous motor[1] [2]. The machine structure is the same as conventional salient pole type synchronous motor, but whose rotor winding is short circuited through a diode. On the rotor, permanent magnets are attached for additional torque generation [3], but they are not necessary for torque generation fundamentally. The control system produces 3-phase current commands necessary for the motor drive. The PWM inverter is implemented to generate 3-phase currents according to the commands.

Fig. 2 shows the dq-axis model of the motor. The dq-axis voltage equations are,

$$e_d = (d/dt)\lambda_d - \omega\lambda_q + r_a$$
$$e_q = (d/dt)\lambda_q + \omega\lambda_d + r_a i_q \qquad (1)$$
$$e_{fd} = (d/dt)\lambda_{fd} + r_{fd}i_{fd}$$

where, e_d is the d-axis voltage, e_q is the q-axis voltage, e_{fd} is the excitation voltage, i_d is the d-axis current, i_q is the q-axis current, i_{fd} is the excitation current, r_a is the stator winding resistance, r_{fd} is the field winding resistance, λ_d is the d-axis flux linkage with the stator winding, λ_q is the q-axis flux linkage with the stator winding, λ_{fd} is the flux linkage with the field winding.

The flux linkages are expressed in terms of self-inductance L and mutual-inductance M, as follow;

$$\lambda_d = L_d i_d + M_{fd}i_{fd} + (M_{fd}/L_{fd})\lambda_{PM}$$
$$\lambda_q = L_q i_q \qquad (2)$$
$$\lambda_{fd} = M_{fd}i_d + L_{fd}i_{fd} + \lambda_{PM}$$

Fig. 3 illustrates the principle of the brushless excitation and torque generation. The following 3-phase currents are supplied to the 3-phase stator windings in Fig. 1;

$$i_a = A_f(t)\sin\theta + \sqrt{2}I_t\cos\theta$$
$$i_b = A_f(t)\sin(\theta - 2\pi/3) + \sqrt{2}I_t\cos(\theta - 2\pi/3) \qquad (3)$$
$$i_c = A_f(t)\sin(\theta - 4\pi/3) + \sqrt{2}I_t\cos(\theta - 4\pi/3)$$

The first term on the right-hand side of (3) is excitation current, which varies with sin of the rotor position θ and whose amplitude is a modulation function $A_f(t)$. $A_f(t)$ is a triangular wave function with the effective value of I_f and whose frequency is bias frequency ω_b. The second term of the equation is torque current component.

The dq-axis currents become;

$$i_d = \sqrt{3/2}A_f(t)$$
$$i_q = \sqrt{3}I_t \qquad (4)$$

We can obtain such rotating field as if the single phase current i_d and the single phase DC current i_q are supplied to the dq-axis windings which rotate synchronously with the rotor. As long as the flux linkage λ_{fd} is increasing, electro motive force in the field winding biases the diode negatively and the diode turns off. When the flux linkage begins to decrease, the diode turns on and the field current starts to flow and compensates the flux decrease. If the time constant is large enough, the flux is almost constant and is kept its maximum value by the diode.

The torque is obtained from the following equation;

$$\tau = \lambda_d i_q - \lambda_q i_d \qquad (5)$$

Though a pulsating torque exists in this motor as shown in Fig. 3, it is not serious problem for practical usage, by choosing the bias angular frequency much greater than the mechanical resonance frequency.

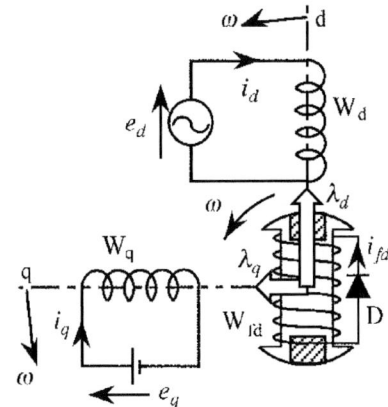

Fig. 2 Motor principle on dq-axis.

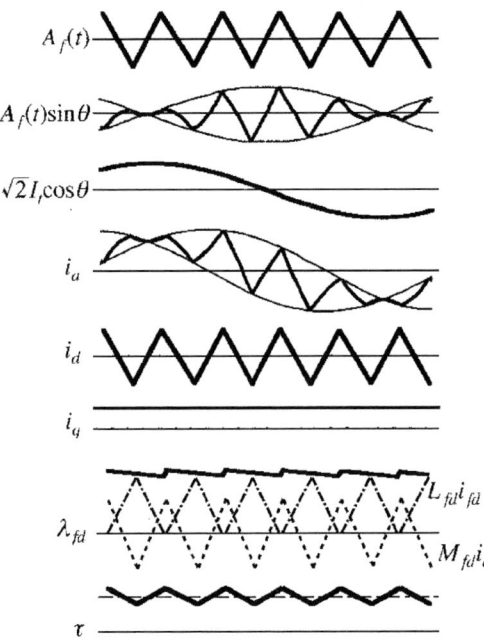

Fig. 3 Waveform of current, flux and torque.

III PRINCIPLE OF HALF-WAVE RECTIFIED BRUSHLESS SYNCHRONOUS GENERATOR

Fig. 4 shows 2 kW analytical generator model and its dimensions. We call it model 1. It has an excitation winding and the armature winding within the same stator slot. They are short pitch distributed winding and their short pitch factor is 7/9. The generator has a simple and robust brushless structure and is maintenance free, in which the field winding is short-circuited with a diode. It has no permanent magnets.

Equation (6) is the excitation current, which varies with sine of the mover position and whose amplitude is modulated by a function $A_f(t)$. Where, $A_f(t)$ is a triangular wave.

$$\begin{cases} i_a = A_f(t)\sin\theta \\ i_b = A_f(t)\sin(\theta - 2\pi/3) \\ i_c = A_f(t)\sin(\theta - 4\pi/3) \end{cases} \qquad (6)$$

Then, the d-axis current becomes equation (7).

$$i_d = \sqrt{3/2}\,A_f(t) \tag{7}$$

If three-phase alternating current is given by equation (6) in the excitation winding, on the d-axis, magneto motive force that alternates bias frequency is generated by the current i_d as shown in Fig. 5. The magneto motive force alternating with the bias frequency ω_b is generated on the d-axis of the rotor. For the increase of the flux linkage with the field winding, the diode turns off. When the flux linkage decreases, the diode turns on. In other words, the field current i_{fd} flows through the field winding to keep the flux linkage constant. The field current is easily controllable by varying the effective value of $A_f(t)$.

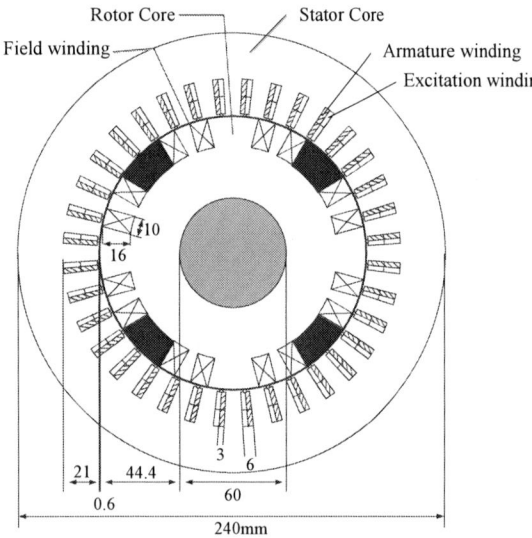

Fig. 4 Analytical model (model 1).

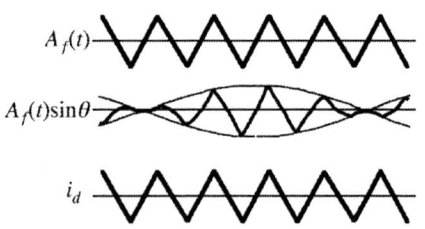

Fig. 5 Wave form of d-axis current.

IV ANALYTICAL RESULTS

A. Induced Voltage

Two-dimensional FEM analysis is carried out with varying the load resistance. Two functions are considered for the modulation function $A_f(t)$; triangular function with a peak value of 12.9 A and sine wave function with the same RMS current value. The no-load induced voltage at the armature winding and flux linkage under triangular function modulation are shown in Fig. 6. Fig. 7 shows the characteristics for 10deg skew model as shown in Fig. 8. Those of sine wave function modulation are shown in Fig. 9 and Fig. 10.

There are many pulse components in the induced voltage under triangular function modulation. The pulse components are produced at turn on and off of the diode. Under sine wave function modulation, the pulse components are decreased.

B. Output Power and Efficiency

Load characteristics are simulated using the circuit of Fig. 11. L and R loads are connected to the armature winding. Their values can be varied.

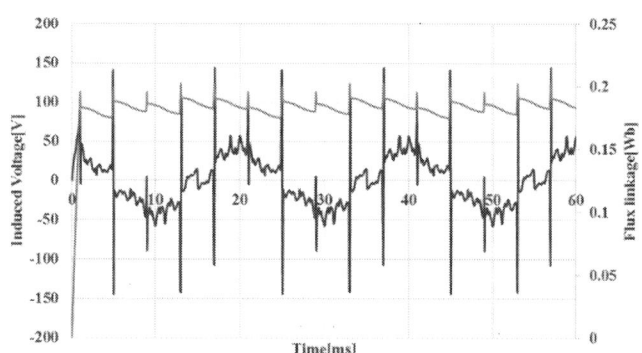

Fig. 6 Induced voltage and flux linkage (triangular wave).

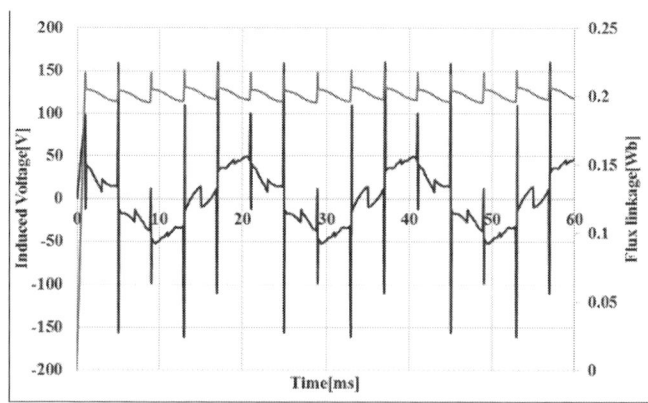

Fig. 7 Induced voltage and flux linkage (triangular wave). (Skew)

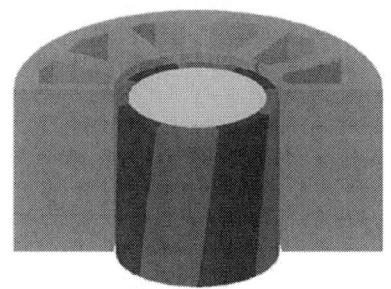

Fig. 8 Skew model

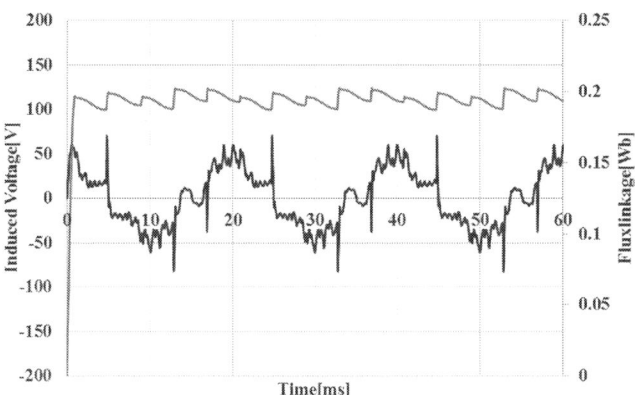

Fig. 9 Induced voltage and flux linkage (sin wave).

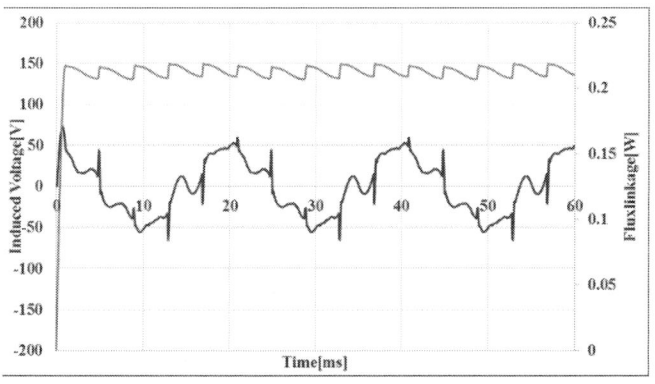

Fig. 10 Induced voltage and flux linkage (sin wave) (Skew)

Fig. 12 shows the output power characteristics for varying R. L is constant, 0.01 mH. The following three models are simulated; (1) the number of turns of the excitation winding is 16 and number of turns of the armature winding is 12, (2) 14 and 14, (3) 12 and 12, respectively. The model (1) produces the largest output and is 2 kW.

The efficiency is calculated using the following formula.

$$\eta = \frac{P_o}{P_o + P_c + P_i} \times 100 \qquad (8)$$

P_o : Output power P_i : Iron loss P_c : Copper loss of the excitation and armature windings

Fig. 13 shows the maximum output power and efficiency for varying the number of turns of the armature winding at 12 turns of excitation winding. Fig. 14 shows the same simulation at 16 turn's excitation winding. Table 1 shows the maximum values of output power and efficiency. The space factor is 49.3 and 61.6%, respectively. It is shown that the output power is 2 kW and the efficiency is 85 % at the combination of 16 turns excitation winding and 13 turns armature winding.

C. Performance Improvement

Fig. 15 shows a new model (model 2). The rotor field winding is changed to concentrated windings. The short pitch factor β of the stator fractional pitch winding is 7/9. Fig. 16 shows the model 3 of β =8/9. The rotor is the same as model 2. The output power and efficiency are shown in table 2.

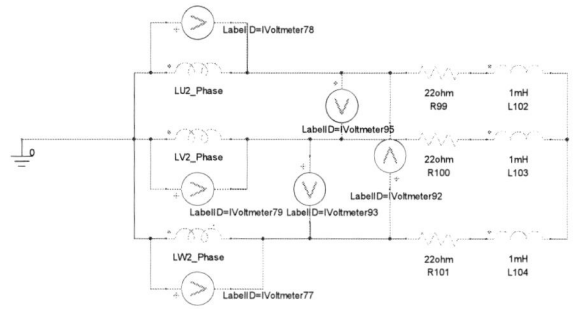

Fig. 11 External circuit.

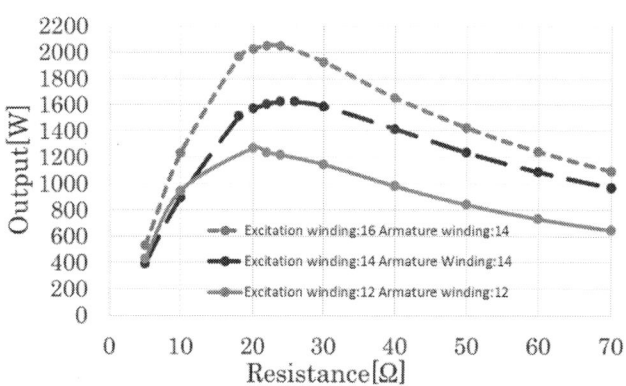

Fig. 12 Output power characteristics (L=0.01mH).

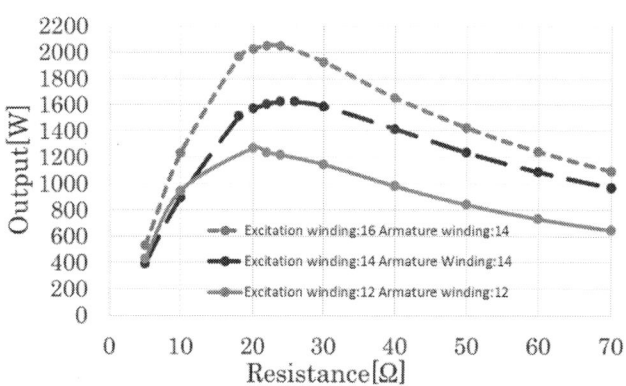

Fig. 13 Output power and efficiency characteristics. (Excitation winding: 12)

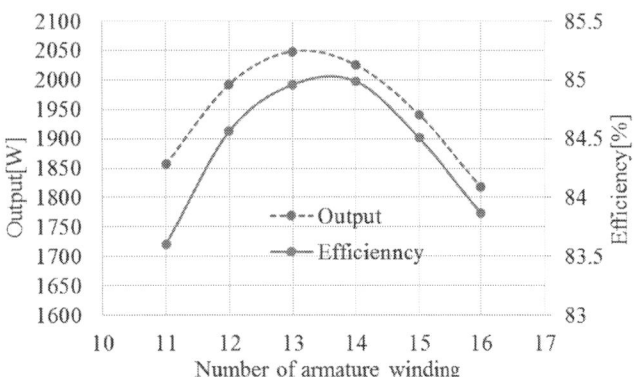

Fig. 14 Output power, efficiency characteristics.
(Excitation winding: 16)

Table. 1 Output power characteristic and space factor

Excitiation winding	Armature winding	Output Power[W]	Efficiency [%]
12	12	1268.7	85.0
16	14	2048.2	85.0

Both output power and efficiency of the model 2 increase compared with model 1. Because the model 1 is designed the field magnetic the flux to be sinusoidal wave form and the fundamental component is smaller than the model 2. It is shown that output power and efficiency of model 3 are improved compared with the model 2.

Their induced electromotive force wave forms are shown in Fig. 17 and Fig. 8. It is difficult to detect the difference of the wave forms. We can confirm the harmonic content of them in Fig. 19.

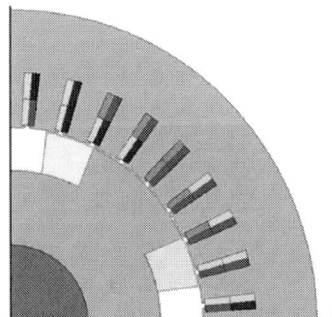 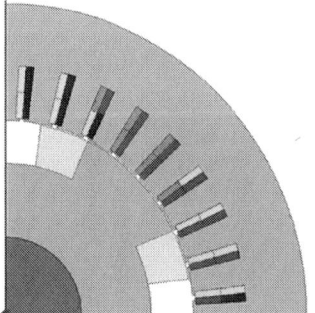

Fig. 15 Model 2 (β=7/9)   Fig. 16 Model 3 (β=8/9)

Table. 2 Output power and efficiency

Model (β)	Output power [W]	Efficiency [%]
Model 2 (7/9)	2234.7	88.4
Model 3 (8/9)	2487.7	88.7

V   Conclusions

The half-wave rectified brushless synchronous generator was proposed and the output characteristics were analyzed.

Harmonic contents of the induced voltage were decreased by using sine wave modulation function. The output power and efficiency was improved by design of windings and rotor construction. The efficiency was 88.7 % at output power of 2.5 kW.

REFERENCES

[1] J. Oyama, S. Toba, T. Higuchi and E. Yamada, "The Characteristics of Half-Wave Rectified Brushless Synchronous Motor", *in Proc. of Beijing International Conf. on Electrical Machines*, pp.654-657, 1987.

[2] J. Oyama, S. Toba, T. Higuchi and E. Yamada, "The Principle and fundamental Characteristics of Half-Wave Rectified Brushless Synchronous Motor", *Trans. IEE of Japan*, vol.107-D, pp.1257-1264, 1987.

[3] J. Oyama, T. Higuchi, T. Abe and E. Yamada, "Analysis of Half-Wave Rectified Brushless Synchronous Motor with Permanent Magnets", *Conf. Rec. IEEE IAS Annu. Meeting*, pp.781-786, 1990.

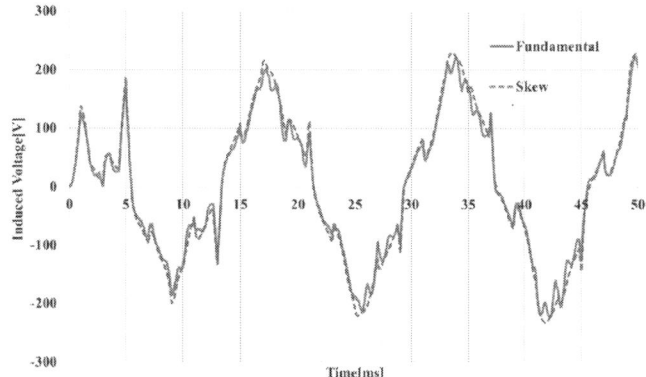

Fig. 17 Induced voltage (β=7/9)

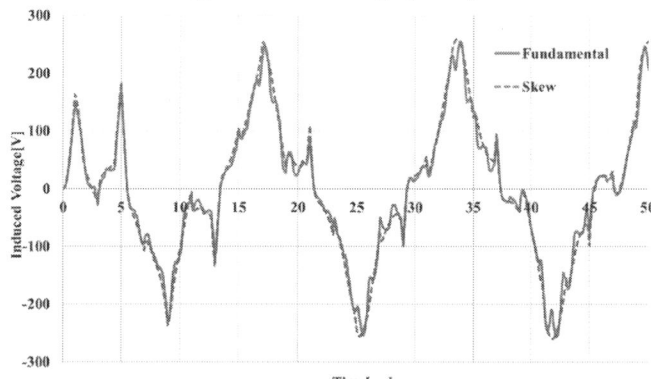

Fig. 18 Induced voltage (β=8/9)

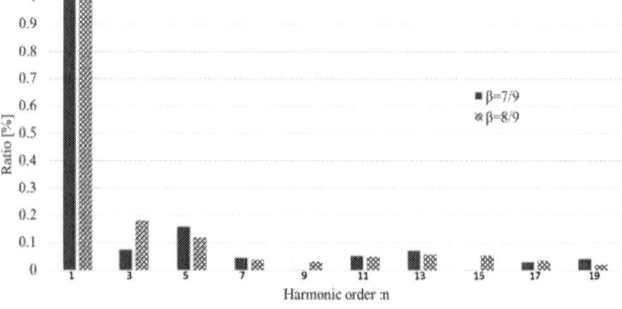

Fig. 19 Number of harmonic of induced voltage

The 2014 International Power Electronics Conference

Modeling of Wound Rotor Synchronous Machines considering Harmonics, Geometric Saliencies and Saturation induced Saliencies

Alexander Rambetius, Sven Luthardt and Bernhard Piepenbreier, *Senior Member, IEEE*

Chair of Electrical Drives and Machines, University Erlangen-Nuremberg
Cauerstrasse 9, 91058 Erlangen, Germany
alexander.rambetius@fau.de

Abstract—**Advanced control techniques for wound rotor synchronous machines need precise machine models. Often a simple fundamental wave model is used to design the control algorithms. Using such a model, it is not possible to account for non-ideal effects that distort the control algorithm. Therefore a general machine model is derived by extending classical winding function theory. This analytical machine model considers higher harmonics as well as geometric saliencies and saturation induced secondary saliencies and leads to new insights for wound rotor synchronous machines. Firstly it is shown that, under certain conditions, the stator slots induce higher harmonics in the rotor winding. Moreover cross saturation effects are described. Last but not least the machine model provides information about the origin of certain harmonics and about their phase depending on saturation. The suggested machine model is verified experimentally by measurement results.**

Keywords—harmonics, saturation, winding function theory, wound rotor synchronous machine, synchronous motor, EESM.

I. INTRODUCTION

Wound rotor synchronous machines (WRSM) feature certain characteristics which are advantageous in automotive traction drives [1–4]. In this field high torque density is required and consequently saturation cannot be neglected. Fig. 1 depicts the electromotive force (EMF) and the rotor current of a highly utilized WRSM which is designed for automotive traction. It can be seen that both, the EMF and the rotor current, contain significant harmonic components. This machine will be used to verify the model proposed in this paper. If advanced control techniques are applied to such a machine, a precise machine model is an essential prerequisite.

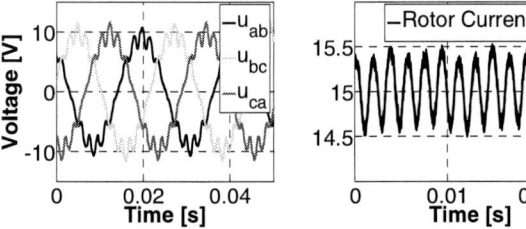

Fig. 1. Armature voltages (left) and rotor current (right) of the investigated WRSM. The machine is operated at constant speed and constant rotor voltage with open-circuited stator windings

Possible applications for the proposed model include the compensation of non-ideal effects in sensorless drives [4–8] or the model based compensation of harmonic components [8–10].

Fundamental wave models of permanent magnet synchronous machines (PMSM) which consider a salient rotor are well-known and can be derived using different techniques. [11] for instance uses winding function theory while [12] suggests a stator-oriented magnetic circuit approach. The method used in [12] is extended for WRSMs in [13]. All these approaches consider saturation but neglect harmonics and stator slots. Harmonics in the EMF of a machine can be caused by different effects. In [9] the harmonics produced by the permanent magnets of a PMSM are taken into account. [10] and [14] additionally consider saturation induced harmonics. These models however are not directly applicable to WRSMs and neglect stator slots.

An analytical evaluation of the magnetic field is carried out for PMSMs in [15] and for WRSMs [16–17]. These models however are more suitable for machine design than for model based control. In [18–19] a WRSM is modeled based on a magnetic equivalent circuit (MEC). The drawback of this approach is that modeling harmonics is complex. Other methods make use of FE-simulations [20] or experimental data [21] and consequently no general machine model is obtained.

The approach suggested in this paper derives an analytical machine model by extending classical winding function theory. The resulting model considers harmonic components as well as geometric saliencies and saturation induced saliencies. Furthermore the origin of harmonics and of cross saturation becomes clear. The cross section which will be used to derive the analytical machine model is depicted in Fig. 2. A stator fixed air gap variation (stator slots) as well as a rotor fixed air gap variation (salient rotor) is considered. Moreover saturation is taken into account by increasing the magnetic reluctance. The maximum of this increase is aligned with the total flux vector Ψ_G. The paper is organized as follows: First a fundamental wave model is derived in Section II. This model is extended by taking harmonics (Section III) and saturation (Section IV) into account. Finally the model is verified experimentally in Section V.

978-1-4799-2706-7/14 $31.00 © 2014 IEEE

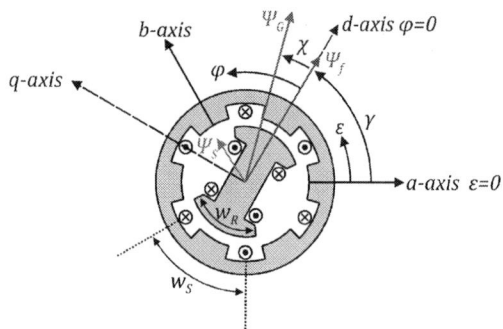

Fig. 2. Cross section of a WRSM considering a stator fixed air gap variation (stator slots) and a rotor fixed air gap variation (rotor geometry)

II. FUNDAMENTAL WAVE MODEL

In this section a fundamental wave model for a salient pole WRSM is derived based on Fig. 2 considering one pole pair. The three stator phases are represented by the coils a, b and c. The rotor features a geometric saliency and carries the field winding, which is represented by the coil f. The general voltage equations of this system are given by:

$$
\begin{pmatrix} u_a \\ u_b \\ u_c \\ u_f \end{pmatrix} = \underline{R} \begin{pmatrix} i_a \\ i_b \\ i_c \\ i_f \end{pmatrix} + \frac{d}{dt} \begin{pmatrix} \Psi_a \\ \Psi_b \\ \Psi_c \\ \Psi_f \end{pmatrix}
\tag{1}
$$

$\underline{R}$ denotes the diagonal matrix of the ohmic resistors. The magnetic flux linkages in (1) are composed of self- and mutual flux linkages and can be expressed as:

$$
\begin{pmatrix} \Psi_a \\ \Psi_b \\ \Psi_c \\ \Psi_f \end{pmatrix} = \begin{pmatrix} \Psi_{aa} + \Psi_{ab} + \Psi_{ac} + \Psi_{af} \\ \Psi_{ba} + \Psi_{bb} + \Psi_{bc} + \Psi_{bf} \\ \Psi_{ca} + \Psi_{cb} + \Psi_{cc} + \Psi_{cf} \\ \Psi_{fa} + \Psi_{fb} + \Psi_{fc} + \Psi_{ff} \end{pmatrix}
\tag{2}
$$

In this section only the fundamental wave is considered. Moreover it is assumed that the magnetic conductivity of the iron is ideal. This means that only the air gap represents a magnetic reluctance. The stator slots are neglected for now. The salient rotor is taken into account by assuming a rotor fixed air gap variation. The relation between rotor fixed φ-coordinates and stator fixed ε-coordinates is given by (3).

$$
\varphi = \varepsilon - \gamma
\tag{3}
$$

γ denotes the rotor position in stator fixed ε-coordinates. Assuming the aforementioned simplifications, the magnetic permeance function in stator fixed ε-coordinates is composed of the constant term Λ_0 and the position dependent second order term $\Lambda_{R,2}$ (2 salient rotor poles per pole pair) [11–15]:

$$
\Lambda_\delta(\varepsilon, \gamma) = \Lambda_0 + \Lambda_{R,2} \cos(2(\varepsilon - \gamma))
\tag{4}
$$

Furthermore the fundamental wave of magnetomotive force (MMF) of the stator coil i (with $i \epsilon \{a, b, c\}$) is given by:

$$
\Theta_i(\varepsilon) = \frac{2}{\pi} \xi_{s_1} N_s i_i \cos(\varepsilon - \varepsilon_i)
\tag{5}
$$

ξ_{s_1} denotes the winding factor of the fundamental wave of the stator winding, N_s the number of turns of the stator coils, i_i the current flowing in the coil i and ε_i the position of the coil center of the coil i ($\varepsilon_a = 0, \varepsilon_b = 2\pi/3, \varepsilon_c = -2\pi/3$). The MMF created by the field winding can be described in rotor fixed φ-coordinates:

$$
\Theta_f(\varphi) = \frac{2}{\pi} \xi_{f_1} N_f i_f \cos(\varphi)
\tag{6}
$$

ξ_{f_1} denotes the winding factor of the fundamental wave of the rotor, N_f the number of turns of the field winding and i_f the current flowing in the field winding. The air gap flux density created by a coil is given by the product of the MMF of the respective coil and the permeance function in (4) [15]:

$$
B_{\delta i}(\varepsilon, \gamma) = \Theta_i(\varepsilon) \Lambda_\delta(\varepsilon, \gamma)
\tag{7}
$$

$$
B_{\delta f}(\varphi) = \Theta_f(\varphi) \Lambda_\delta(\varphi)
\tag{8}
$$

The magnetic flux linkages in (2) can now be calculated by integrating the flux densities in (7) and (8) over the area of the respective coil and multiplying the result by the number of turns of the respective coil [20]. With $i, j \epsilon \{a, b, c\} \wedge i \neq j$, one yields:

$$
\Psi_{ii} = N_s l_{Fe} \int_{-\frac{w_s}{2} + \varepsilon_i}^{+\frac{w_s}{2} + \varepsilon_i} B_{\delta i}(\varepsilon, \gamma) \frac{\tau_p}{\pi} d\varepsilon
\tag{9}
$$

$$
\Psi_{ji} = \Psi_{ij} = N_s l_{Fe} \int_{-\frac{w_s}{2} + \varepsilon_j}^{+\frac{w_s}{2} + \varepsilon_j} B_{\delta i}(\varepsilon, \gamma) \frac{\tau_p}{\pi} d\varepsilon
\tag{10}
$$

$$
\Psi_{ff} = N_f l_{Fe} \int_{-\frac{w_R}{2}}^{+\frac{w_R}{2}} B_{\delta f}(\varphi) \frac{\tau_p}{\pi} d\varphi
\tag{11}
$$

$$
\Psi_{fi} = \Psi_{if} = N_s l_{Fe} \int_{-\frac{w_s}{2} + \varepsilon_i}^{+\frac{w_s}{2} + \varepsilon_i} B_{\delta f}(\varphi) \frac{\tau_p}{\pi} d\varepsilon
\tag{12}
$$

l_{Fe} denotes the active iron length, τ_p the pole pitch, w_s the coil width of the stator coils and w_R the coil width of the rotor coil (See Fig. 2). To simplify the resulting expressions the constants K_s (Stator), K_f (Rotor) and K_{sf} (Coupling) are introduced.

$$
K_s = 2N_S^2 l_{Fe} \tau_p / \pi^2
\tag{13}
$$

$$
K_f = 2N_f^2 l_{Fe} \tau_p / \pi^2
\tag{14}
$$

$$
K_{sf} = 2N_S N_f l_{Fe} \tau_p / \pi^2
\tag{15}
$$

Dividing the magnetic flux linkages in (9)-(12) by the current flowing in the respective coil, the machine inductances in stator fixed coordinates can be calculated. The self- and mutual inductances of the stator are given by:

$$
L_{ii}(\gamma) = \frac{\Psi_{ii}}{i_i} = L_m + \Delta L \cos(2\gamma + \varepsilon_i)
\tag{16}
$$

$$
M_{ij}(\gamma) = \frac{\Psi_{ij}}{i_j} = -\frac{L_m}{2} + \Delta L \cos(2\gamma - \varepsilon_i - \varepsilon_j)
\tag{17}
$$

(16) and (17) show that both, the self- and the mutual

inductances, are composed of the constant term

$$L_m = 2K_s \Lambda_0 \xi_{s_1}^2 \tag{18}$$

and of the position dependent term

$$\Delta L_{R,2} = K_s \Lambda_{R,2} \xi_{s_1}^2. \tag{19}$$

The self-inductance of the rotor and mutual inductances between the rotor and the stator are given by:

$$L_{ff} = \frac{\Psi_{ff}}{i_f} = K_f \big(2\Lambda_0 + \Lambda_{R,2}\big) \xi_{f_1}^2 \tag{20}$$

$$M_{if}(\gamma) = \frac{\Psi_{if}}{i_f} = K_{sf} \big(2\Lambda_0 + \Lambda_{R,2}\big) \xi_{f_1} \xi_{s_1} \cos\,(\gamma - \varepsilon_i) \tag{21}$$

The magnetic flux linkages obtained in (9)-(12) are now inserted into (1)-(2) and are then transformed to rotor fixed coordinates using the well-known amplitude invariant transformation matrices which are given in [13]. The resulting voltage equations in the dq-reference frame can be written as:

$$\begin{pmatrix} u_d \\ u_q \\ u_f \end{pmatrix} = \big(\underline{R} + \omega \underline{L}_{EMF}\big) \begin{pmatrix} i_d \\ i_q \\ i_f \end{pmatrix} + \underline{l}_{diff} \frac{d}{dt} \begin{pmatrix} i_d \\ i_q \\ i_f \end{pmatrix} \tag{22}$$

The matrix $\underline{l}_{diff}$ describes the transformer EMF:

$$\underline{l}_{diff} = \begin{pmatrix} L_{dd} & 0 & M_{df} \\ 0 & L_{qq} & 0 \\ \frac{3}{2}M_{df} & 0 & L_{ff} \end{pmatrix} \tag{23}$$

The new rotor fixed inductances in (23) are given by:

$$L_{dd} = \frac{3}{2}(L_m + \Delta L_{R,2}) \tag{24}$$

$$L_{qq} = \frac{3}{2}(L_m - \Delta L_{R,2}) \tag{25}$$

$$M_{df} = K_{sf} \big(2\Lambda_0 + \Lambda_{R,2}\big) \xi_{f_1} \xi_{s_1} \tag{26}$$

The fact that (23) is no symmetrical matrix is caused by the amplitude invariant transformation from the 3-phase stator model to the equivalent 2-phase stator model. The matrix $\underline{L}_{EMF}$, which describes the motional EMF, can be written as:

$$\underline{L}_{EMF} = \begin{pmatrix} 0 & -L_{qq} & 0 \\ L_{dd} & 0 & M_{df} \\ 0 & 0 & 0 \end{pmatrix} \tag{27}$$

Since the last line in (27) contains zeros only, it can be concluded that there is no voltage in the field winding induced by motion.

III. MACHINE MODEL CONSIDERING HARMONICS

A. General machine model

The iron is still assumed to be ideally conductive. Besides the mean permeance Λ_0 a rotor fixed (rotor saliency) and a stator fixed permeance variation (stator slots) is considered. Consequently the permeance function in stator fixed ε - coordinates is given by:

$$\Lambda_\delta(\varepsilon, \gamma) = \Lambda_0 + \Lambda_R(\varepsilon, \gamma) + \Lambda_s(\varepsilon) \tag{28}$$

The two additional permeance terms $\Lambda_R(\varepsilon, \gamma)$ (Rotor) and $\Lambda_s(\varepsilon)$ (Stator) are Fourier expansions, which take their fundamental wave and the respective higher harmonics into account. In the case of the rotor fixed permeance variation the fundamental wave is a second order harmonic (2 salient rotor poles per pole pair) and consequently the permeance term $\Lambda_R(\varepsilon, \gamma)$ is defined as follows:

$$\Lambda_R(\varepsilon, \gamma) = \sum_{l=1}^{\infty} \Lambda_{R,2l} \cos(2l(\varepsilon - \gamma)) \tag{29}$$

In the case of the stator fixed permeance variation the fundamental wave depends on the number of stator slots per pole pair and consequently $\Lambda_s(\varepsilon)$ is given by:

$$\Lambda_s(\varepsilon) = \sum_{n=1}^{\infty} \Lambda_{s,\frac{Q_s}{p}n} \cos\left(\frac{Q_s}{p}n\varepsilon\right) \tag{30}$$

Q_s denotes the number of stator slots and p the number of pole pairs. Besides the permeance function the MMFs have to be modified as well if harmonics are to be considered. This is done by expressing the MMFs in Fourier series [20]:

$$\Theta_i(\varepsilon) = \frac{2}{\pi} N_s i_i \sum_{k=1,3,5,\ldots}^{\infty} \xi_{s_k} \cos\big(k(\varepsilon - \varepsilon_i)\big) \tag{31}$$

$$\Theta_f(\varphi) = \frac{2}{\pi} N_f i_f \sum_{k=1,3,5,\ldots}^{\infty} \xi_{f_k} \cos\,(k\varphi) \tag{32}$$

ξ_{s_k} denotes the winding factor of the kth harmonic of the stator winding, ξ_{f_k} the winding factor of the kth harmonic of the rotor winding. The stator fixed inductances are found by using the technique suggested in Section II and replacing the permeance function (4) with (28). Furthermore the MMFs (5)-(6) are replaced by (31)-(32). The resulting inductances can be written as:

$$L_{ii}^H = L_{ii}^{\Lambda_0} + L_{ii}^{Slot} + L_{ii}^{Rotor}(\gamma) \tag{33}$$

$$M_{ij}^H = M_{ij}^{\Lambda_0} + M_{ij}^{Slot} + M_{ij}^{Rotor}(\gamma) \tag{34}$$

$$L_{ff}^H = L_{ff}^{\Lambda_0} + L_{ff}^{Slot}(\gamma) + L_{ff}^{Rotor} \tag{35}$$

$$M_{if}^H = M_{if}^{\Lambda_0}(\gamma) + M_{if}^{Slot}(\gamma) + M_{if}^{Rotor}(\gamma) \tag{36}$$

Analyzing (33)-(36), it can be seen that all the stator fixed inductances are composed of three addends. The addends with the superscript Λ_0 denote the mean inductances if harmonics are considered, but no geometric saliencies are present (cylindrical WRSM disregarding stator slots). These mean inductances are given by:

$$L_{ii}^{\Lambda_0} = K_s \sum_{k=1,3,5,\ldots}^{\infty} 2\Lambda_0 \xi_{s_k}^2 \tag{37}$$

$$M_{ij}^{\Lambda_0} = K_s \sum_{k=1,3,5,\ldots}^{\infty} 2\Lambda_0 \xi_{s_k}^2 \cos\big(k(\varepsilon_j - \varepsilon_i)\big) \tag{38}$$

978-1-4799-2706-7/14 $31.00 © 2014 IEEE

$$L_{ff}^{\Lambda_0} = K_f \sum_{k=1,3,5,\ldots}^{\infty} 2\Lambda_0 \xi_{fk}^2 \tag{39}$$

$$M_{if}^{\Lambda_0} = K_{sf} \sum_{k=1,3,5,\ldots}^{\infty} 2\Lambda_0 \xi_{sk} \xi_{fk} cos\big(k(\varepsilon_i - \gamma)\big) \tag{40}$$

The addends with the superscript *Slot* in (33)-(36) describe the effect of the stator slotting and can be found in the appendix (A1)-(A4). It can be seen that the stator slots only chance the mean value of the stator self- and mutual inductances. The reason is that the stator slots represent a stator fixed air gap variation and consequently no position dependent terms are introduced. This effect is usually taken into account by using the Carter coefficient [15]. From the rotor winding point of view the stator slotting represents a position dependent air gap variation and consequently a position dependency is introduced in (35). The impact of the rotor saliency in (33)-(36) is included in the addends with the superscript *Rotor*. Since the rotor saliency represents a rotor fixed air gap variation, it introduces position dependent terms in the stator self- and mutual inductances (see Appendix (A5)-(A6)), but not in the rotor self-inductance (see Appendix (A7)).

B. Reduced model of the investigated machine

To verify the suggested approach the machine model is transformed to the dq-reference frame considering the major effects of the investigated WRSM. These effects are:

- A significant 11th and 13th harmonic (in stator fixed coordinates) is produced by the rotor. Furthermore the stator winding reacts strongly to these harmonics. This means that in (32) $k = 1,11,13$ are taken into account.

- The tested machine features a salient pole rotor. Therefore $l = 1$ is taken into account in (29).

- The fundamental wave of the air gap variation due to the stator slots is taken into account ($n = 1$ in (30)).The tested machine has got 12 stator slots per pole pair.

Assuming these effects as the major ones, the stator fixed general machine model from Subsection A is transformed into the rotor fixed dq-reference frame. In the following the interaction between the rotor saliency and the higher harmonics of the MMFs is neglected. Due to the limited amount of space only the inductance matrix describing the motional EMF will be investigated. This matrix is composed of a fundamental part and of a part describing higher harmonics and is given in (48) at the bottom of this page.

1) Fundamental wave of the motional EMF
The position independent fundamental wave inductances in (48) chance their value due to the stator slotting. The mean

value L_m of the stator self-inductances in (24) and (25) is replaced by:

$$L_{m,Slot} = \underbrace{2K_s \Lambda_0 \xi_{s_1}^2 + K_s \Lambda_{s,12} \xi_{s_1}\big(\xi_{s_{11}} + \xi_{s_{13}}\big)}_{L_m} \tag{41}$$

The saliency $\Delta L_{R,2}$ on the other hand does not chance due to the stator slotting.

2) Harmonics in the motional EMF of the stator
The stator fixed 11th and 13th harmonic, which are produced by the rotor, appear as a 12th harmonic in the dq-reference frame. The amplitude of this harmonic is determined by:

$$M_{af,11} = K_{sf} \xi_{f_{11}}\big(2\xi_{s_{11}}\Lambda_0 + \xi_{s_1}\Lambda_{s,12}\big) \tag{42}$$

$$M_{af,13} = K_{sf} \xi_{f_{13}}\big(2\xi_{s_{13}}\Lambda_0 + \xi_{s_1}\Lambda_{s,12}\big) \tag{43}$$

As expected, $M_{af,11}$ is unequal to zero if the rotor produces the 11th harmonic and if the stator can receive the 11th harmonic ($\xi_{f_{11}}\xi_{s_{11}} \neq 0$). Interestingly, even if the stator cannot receiver the 11th harmonic, $M_{af,11}$ might be unequal to zero. This is the case if the rotor produces the 11th harmonic and at the same time the number of stator slots equals 12. The same line of thought can be carried out for $M_{af,13}$. Generally speaking, the qth harmonic in the rotor fixed dq-reference frame may appear if the following condition is fulfilled:

$$\xi_{f_{q\pm1}}\xi_{s_{q\pm1}} \neq 0 \quad \vee \quad \xi_{f_{q\pm1}}\Lambda_{s,q} \neq 0 \tag{44}$$

This part of the machine model can be used for model-based compensation of harmonics in the stator currents like in [8–9].

1) Harmonics in the rotor winding
The MMFs of the stator winding produce harmonics in the rotor winding (First two elements of the last line in the harmonic part of (48)), which are determined by:

$$M_{df,12} = M_{af,11} + M_{af,13} \tag{45}$$

$$M_{qf,12} = -M_{af,11} + M_{af,13} \tag{46}$$

Additionally, harmonics appear in the field winding even if the stator is open-circuited. The reason is that for the self-inductance of the field winding the stator slots are a position dependent reluctance variation and therefore they create a motional EMF in the field winding. This EMF is determined by:

$$L_{ff,12} = K_f \Lambda_{s,12} \xi_{f_1}\big(\xi_{f_{11}} + \xi_{f_{13}}\big) \tag{47}$$

From (47) it can be concluded that the stator slotting induces a motional EMF in the field winding if the field winding produces harmonics of the order $Q_s/p \pm 1$. This effect is used in [8] to estimate the speed of a sensorless WRSM drive system.

$$\underline{L}_{EMF} = \underbrace{\begin{pmatrix} 0 & -L_{qq} & 0 \\ L_{dd} & 0 & M_{df} \\ 0 & 0 & 0 \end{pmatrix}}_{Fundamental\ wave} + \underbrace{\begin{pmatrix} 0 & 0 & -\big(11M_{af,11} + 13M_{af,13}\big)\sin(12\gamma) \\ 0 & 0 & \big(-11M_{af,11} + 13M_{af,13}\big)\cos(12\gamma) \\ -\frac{3}{2}12M_{df,12}\sin(12\gamma) & \frac{3}{2}12M_{qf,12}\cos(12\gamma) & 12L_{ff,12}\sin(12\gamma) \end{pmatrix}}_{Harmonic\ components} \tag{48}$$

IV. Machine Model considering Saturation

A. General machine model

In this section the machine model presented in Section III is extend in order to account for saturation. Saturation reduces the magnetic permeance in the saturated areas of the machine. The maximum of this reduction is aligned with the total flux vector Ψ_G in Fig. 2 and is shifted by the load angle χ with respect to the d-axis. This saturation induced permeance variation is considered by adding it to (28). The additional permeance term is expressed in a Fourier series and is given by [10]:

$$\Lambda_{Sat}(\varphi,\chi) = \sum_{u=1}^{\infty} \Lambda_{Sat,2u} \cos(2u(\varphi - \chi)) \tag{49}$$

The amplitude of this permeance variation is a function of the absolute value of the total flux vector. In the following the permeance variation in (49) is split up into a sine and a cosine shaped part (in rotor fixed coordinates):

$$\Lambda_{Sat,2u}\cos\big(2u(\varphi - \chi)\big) =$$
$$= \Lambda_{Sat,2u}^{Re}(\chi)\cos(2u\varphi) + \Lambda_{Sat,2u}^{Im}(\chi)\sin(2u\varphi) \tag{50}$$

The amplitudes of these two components are given by:

$$\Lambda_{Sat,2u}^{Re}(\chi) = \Lambda_{Sat,2u}\cos(2u\chi) \tag{51}$$

$$\Lambda_{Sat,2u}^{Im}(\chi) = \Lambda_{Sat,2u}\sin(2u\chi) \tag{52}$$

Form (51) and (52) it can be concluded that if there is no flux in the q-axis ($\chi = 0$) the sine shaped part of the permeance reduction due to saturation is zero. Since the inductances are now current dependent, the transformer EMF in stator fixed coordinates is now longer determined by the absolute inductances, but by the differential inductances [11–13]. These differential inductances can be calculated by partial derivation of the magnetic flux linkages and are denoted by lowercase letters from here on. With $i, j \epsilon \{a, b, c, f\} \wedge i \neq j$ they are given by:

$$l_{ii} = \frac{\partial \Psi_i}{\partial i_i} \qquad m_{ij} = \frac{\partial \Psi_i}{\partial i_j} \tag{53}$$

The approach to find the machine inductances in stator fixed coordinates is the same like presented in Section III. Considering the additional permeance variation in (49), the machine inductances can be written as:

$$L_{ii}^{Sat} = L_{ii}^{H}(\gamma) + L_{ii}^{Sat,Re}(\gamma) + L_{ii}^{Sat,Im}(\gamma) \tag{54}$$

$$M_{ij}^{Sat} = M_{ij}^{H}(\gamma) + M_{ij}^{Sat,Re}(\gamma) + M_{ij}^{Sat,Im}(\gamma) \tag{55}$$

$$L_{ff}^{Sat} = L_{ff}^{H}(\gamma) + L_{ff}^{Sat,Re} \tag{56}$$

$$M_{if}^{Sat} = M_{if}^{H}(\gamma) + M_{if}^{Sat,Re}(\gamma) + M_{if}^{Sat,Im}(\gamma) \tag{57}$$

Analyzing (54)-(57), it can be seen that apart from the addends already present in Section III (see (33)-(36)) new addends appear. The additional addends with the superscript Sat, Re describe the effect of d-axis saturation and can be found in the

Appendix (A9)-(A12). The additional addends with the superscript Sat, Im describe the effect of q-axis saturation and can be found in the Appendix (A13)-(A15). The differential inductances have got the same structure like the absolute inductances [12–13] and are therefore not mentioned separately.

B. Fundamental wave model considering saturation

The general machine model derived in subsection A is now transformed to the dq-reference frame considering only the fundamental wave ($u = 1$ in (49)). Due to the limited amount of space only the inductance matrix describing the transformer EMF is analyzed. This matrix is given by:

$$\underline{l}_{diff} = \begin{pmatrix} l_{dd} & m_{dq} & m_{df} \\ m_{dq} & l_{qq} & m_{qf} \\ \frac{3}{2}m_{df} & \frac{3}{2}m_{qf} & l_{ff} \end{pmatrix} \tag{58}$$

Comparing $\underline{l}_{diff}$ in (23) with (58), it can be seen that the transformer EMF is no longer determined by absolute inductances, but by differential inductances [12–13]. Furthermore the additional mutual inductances m_{dq} and m_{qf} appear. They describe the non-ideal coupling between the d-axis and the q-axis and between the q-axis and the field winding. This effect is often referred to as cross saturation and is of major importance in carrier signal based sensorless control ([5–7] and [11–14]). The absolute cross saturation inductance between the q-axis and the field winding is given by:

$$M_{qf} = K_{sf}\xi_{f_1}\xi_{s_1}\Lambda_{Sat,2}^{Im} = K_{sf}\xi_{f_1}\xi_{s_1}\Lambda_{Sat,2}\sin(2\chi) \tag{59}$$

The differential cross saturation inductance has got the same structure. From (59) it can be concluded that cross saturation only appears if there is a flux component aligned with the q-axis ($\chi \neq 0$). Moreover a statement about the sign of the cross saturation inductances can be made. The term $\Lambda_{Sat,2}$ describes a permeance reduction. Consequently the sign of the mutual cross saturation inductance is determined by the load angle:

$$m_{qf} = \begin{cases} < 0 & for \ \chi > 0 \\ > 0 & for \ \chi < 0 \end{cases} \tag{60}$$

The sign of the load angle on the other hand depends on the sign of the magnetic flux linkages in the d- and q-axis. In addition to this statement about symmetry conditions can be made. Since the amplitude of the permeance variation $\Lambda_{Sat,2}$ depends on the absolute value of total flux vector, the following conditions hold:

$$m_{qf}(\Psi_d, \Psi_q, \Psi_f) = m_{qf}(-\Psi_d, -\Psi_q, -\Psi_f) \tag{61}$$

$$m_{qf}(\Psi_d, \Psi_q, \Psi_f) = -m_{qf}(\Psi_d, -\Psi_q, \Psi_f) \tag{62}$$

C. Machine model considering saturation induced harmonics

Since the stator reacts strongly to the 11th and 13th harmonic (12th harmonic in the dq-reference frame), we consider a saturation induced 12th harmonic ($u = 6$ in (49)) in this subsection. Furthermore the dominant effects of the investigated WRSM, which are mentioned in Section III.B, are taken into account.

The resulting machine model in rotor fixed coordinates consists of very complex terms and is therefore not mentioned in this paper. Instead we analyze the magnetic flux linkage of the d-axis for a d-current equal to zero as an example. The magnetic flux linkage of the d-axis produces a motional EMF in the q-axis. Making the aforementioned assumptions, the 12th harmonic of this motional EMF is given by:

$$u_{q_{EMF,12}} = \omega \Psi_{d,12} = \omega(\tilde{M}_{df,12} i_f + \tilde{M}_{dq,12} i_q) \tag{63}$$

Since it is assumed that the current in the d-axis is equal to zero, only the field current and the q-current produce harmonics. The harmonic flux linkage produced by the field current is determined by:

$$\tilde{M}_{df,12} = M_{df,12}^{Re} \cos(12\gamma) - M_{df,Sat,12}^{Im} \sin(12\gamma) \tag{64}$$

The amplitude of the cosine shaped part in (64) is given by:

$$M_{df,12}^{Re} = -11 M_{af,11} + 13 M_{af,13} \tag{65}$$

$M_{af,11}$ and $M_{af,13}$ are produced by the MMF of the field winding and are independent of saturation (see (42)-(43)). The amplitude of the sine shaped part in (64) is given by:

$$M_{df,Sat,12}^{Im} = -11 M_{af,11,Sat} + 13 M_{af,13,Sat} \tag{66}$$

The saturation induced harmonic mutual inductances in (66) can be written as:

$$M_{af,11,Sat} = K_{sf} \xi_{f_1} \xi_{s_{11}} \Lambda_{Sat,12}^{Im} \tag{67}$$

$$M_{af,13,Sat} = K_{sf} \xi_{f_1} \xi_{s_{13}} \Lambda_{Sat,12}^{Im} \tag{68}$$

These terms only appear for a q-current unequal to zero ($\chi \neq 0$). Furthermore we can deduce from (67)-(68) that the qth saturation induced harmonic in rotor fixed coordinates only appears if the winding factor of the order q±1 is unequal to zero. Apart from the field current, the q-current introduces harmonics in (63) as well. The amplitude of this harmonic flux linkage is determined by:

$$\tilde{M}_{dq,12} = M_{dq,Sat,12}^{Re} \cos(12\gamma) + M_{dq,Sat,12}^{Im} \sin(12\gamma) \tag{69}$$

The amplitude of the sine and cosine shaped parts in (69) can be expressed as:

$$M_{dq,Sat,12}^{Re} = -K_s \xi_{s_1} \left(12\frac{3}{2} \Lambda_{Sat,12}^{Im} (\xi_{s_{11}} + \xi_{s_{13}}) + 2\sqrt{3} \Lambda_{Sat,12}^{Re} (-\xi_{s_{11}} + \xi_{s_{13}}) \right) \tag{70}$$

$$M_{dq,Sat,12}^{Im} = -K_s \xi_{s_1} \left(12\frac{3}{2} \Lambda_{Sat,12}^{Re} (\xi_{s_{11}} + \xi_{s_{13}}) + 2\sqrt{3} \Lambda_{Sat,12}^{Im} (-\xi_{s_{11}} + \xi_{s_{13}}) \right) \tag{71}$$

From (70)-(71) we can deduce that the harmonic components in (69) are purely saturation induced. Generally speaking, it can be said that the phase of harmonic components depends on the load angle χ. For the example considered in this subsection it can be stated, that the sine shaped part of the harmonic flux linkage in (63) is produced by saturation only. Consequently the sine shaped part (imaginary part) in (63)

vanishes for a q-current equal to zero:

$$Im\{\Psi_{d,12}(i_q = 0)\} = 0 \tag{72}$$

Furthermore it can be stated that the sign of the sine shaped harmonic flux in (63) is determined by the sign of the load angle χ and therefore by the sign of the q-current:

$$Im\{\Psi_{d,12}(i_q)\} = -Im\{\Psi_{d,12}(-i_q)\} \tag{73}$$

This part of the machine model can be used for model-based feed-forward control of harmonics in the stator currents [10].

V. EXPERIMENTAL MODEL VERIFICATION

The WRSM summarized in Table I will be used to verify some important parts of the machine model.

A. Effect of stator slots in WRSM

Since the stator slots represent a stator fixed reluctance variation they produce a motional EMF in the rotor winding which is determined by (47). To verify this effect the investigated machine is driven with constant speed. The stator is open-circuited and field winding is supplied with DC-voltage. The measured rotor current is depicted in Fig. 1 and contains a significant harmonic component. Fig. 3 shows a Fourier analysis of the field current in Fig. 1. Since the stator of the investigated machine has got 12 stator slots per pole pair the dominant harmonic is the 12th.

B. Cross saturation

To verify the deduced statements about cross saturation the differential mutual inductance m_{qf} is identified by applying the approach presented in [7]. The identification results are depicted in Fig. 4 and prove the statements in (60) and (62): Cross saturation always acts flux weakening and consequently m_{qf} is negative for positive q-current and positive for negative q-current.

TABLE I
PARAMETERS OF THE INVESTIGATED WRSM

Rated power/rated speed	$P_N = 12\ kW\ /\ n_N = 3325\ rpm$
Rated phase voltage/current (RMS)	$U_N = 18.3\ V\ /\ I_{sN} = 250\ A$
Rated field current	$I_{fN} = 16\ A$
Number of stator slots per pole pair	$Q_s/p = 36/3 = 12$
Number of turns per pole stator/rotor	$N_s = 2\ /\ N_f = 43$
Winding factors stator (1st,11th,13th)	$0.933\ ,\ -\frac{1}{11}0.933\ ,\ \frac{1}{13}0.933$
Winding factors rotor (1st,11th,13th)	$0.844\ ,\ \frac{1}{11}0.991\ ,\ -\frac{1}{13}0.481$

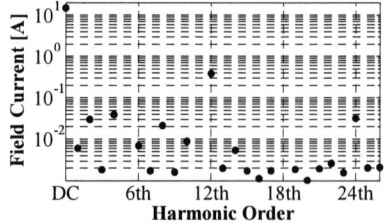

Fig. 3. Fourier analysis of the field current for constant speed, open-circuited stator and constant field voltage

C. Phase Angle of Harmonic Components

The harmonic flux of the q-axis is identified using the flux identification technique presented in [22]. Additionally resonant controllers are used to compensate harmonics like described in [8]. The outputs of these resonant controllers are then used to extract the harmonic fluxes. The amplitude of the sine shaped part (imaginary part) of the harmonic flux linkage in (63) is depicted in Fig. 5. The identification results prove, that the sine shaped part is saturation induced only and therefore vanishes for a q-current equal to zero ($\chi = 0$). Moreover the symmetry condition in (73) is approved by Fig. 5.

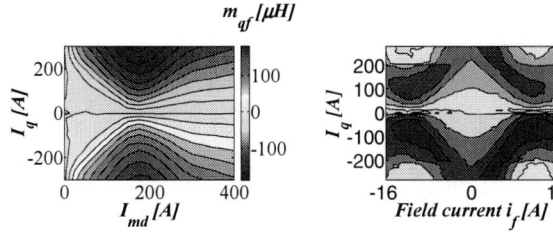

Fig. 4. Differential cross saturation inductance between the q-axis and the field winding

Fig. 5. Amplitude of the sine shaped component of the harmonic flux in the d-axis

VI. CONCLUSION

This paper deduced an analytical machine model of a wound rotor synchronous machine by extending winding function theory. The resulting model considers harmonics as well as geometric saliencies and saturation induced saliencies. Using the suggested technique, it is possible to determine the origin of harmonics and of saturation effects. This leads to new insights for WRSMs. Firstly it can be stated, that the stator slots induce a motional EMF in the rotor winding if certain conditions are fulfilled. Moreover cross saturation effects can be explained. Last but not least the model shows that the phase of harmonics is not constant, but varies depending on the direction of the total flux vector.

APPENDIX

For the sake of completeness this appendix shows the components of the machine inductances of the general machine model in stator fixed coordinates. (A1)-(A8) represent the components of the machine inductances in (33)-(36) if harmonics and geometric saliencies are considered but saturation is neglected. (A9)-(A15) are the additional components of the machine inductances in (54)-(57) if saturation is taken into account.

$$L_{ii}^{Slot} = K_s \sum_{k=1,3,5,\dots}^{\infty} \xi_{sk} \left\{ \sum_{n=1}^{\infty} \Lambda_{s,\frac{Q_s}{p}n} \left[\left(\xi_{s_{\frac{Q_s}{p}n-k}} + \xi_{s_{\frac{Q_s}{p}n+k}} \right) \cos\left(\frac{Q_s}{p}n\varepsilon_i \right) \right] \right\} \tag{A1}$$

$$M_{ij}^{Slot} = K_s \sum_{k=1,3,5,\dots}^{\infty} \xi_{sk} \left\{ \sum_{n=1}^{\infty} \Lambda_{s,\frac{Q_s}{p}n} \left[\xi_{s_{\frac{Q_s}{p}n-k}} \cos\left(k(\varepsilon_i - \varepsilon_j) + \frac{Q_s}{p}n\varepsilon_j \right) + \xi_{s_{\frac{Q_s}{p}n+k}} \cos\left(k(\varepsilon_j - \varepsilon_i) + \frac{Q_s}{p}n\varepsilon_j \right) \right] \right\} \tag{A2}$$

$$L_{ff}^{Slot}(\gamma) = K_f \sum_{k=1,3,5,\dots}^{\infty} \xi_{fk} \left\{ \sum_{n=1}^{\infty} \Lambda_{s,\frac{Q_s}{p}n} \left[\left(\xi_{f_{\frac{Q_s}{p}n-k}} + \xi_{f_{\frac{Q_s}{p}n+k}} \right) \cos\left(\frac{Q_s}{p}n\gamma \right) \right] \right\} \tag{A3}$$

$$M_{if}^{Slot}(\gamma) = K_{sf} \sum_{k=1,3,5,\dots}^{\infty} \xi_{fk} \left\{ \sum_{n=1}^{\infty} \Lambda_{s,\frac{Q_s}{p}n} \left[\xi_{s_{\frac{Q_s}{p}n-k}} \cos\left(k(\varepsilon_i - \gamma) - \varepsilon_i \frac{Q_s}{p}n \right) + \xi_{s_{\frac{Q_s}{p}n+k}} \cos\left(k(\varepsilon_i - \gamma) + \varepsilon_i \frac{Q_s}{p}n \right) \right] \right\} \tag{A4}$$

$$L_{ii}^{Rotor}(\gamma) = K_s \sum_{k=1,3,5,\dots}^{\infty} \xi_{sk} \left\{ \sum_{l=1}^{\infty} \Lambda_{R,2l} \left[(\xi_{s_{2l-k}} + \xi_{s_{2l+k}}) \cos(2l(\varepsilon_i - \gamma)) \right] \right\} \tag{A5}$$

$$M_{ij}^{Rotor}(\gamma) = K_s \sum_{k=1,3,5,\dots}^{\infty} \xi_{sk} \left\{ \sum_{l=1}^{\infty} \Lambda_{R,2l} \left[\xi_{s_{2l-k}} \cos\left(k(\varepsilon_i - \varepsilon_j) + 2l(\varepsilon_j - \gamma) \right) + \xi_{s_{2l+k}} \cos\left(k(\varepsilon_j - \varepsilon_i) + 2l(\varepsilon_j - \gamma) \right) \right] \right\} \tag{A6}$$

$$L_{ff}^{Rotor} = K_f \sum_{k=1,3,5,\dots}^{\infty} \xi_{fk} \left\{ \sum_{l=1}^{\infty} \Lambda_{R,2l} \left[\xi_{s_{2l-k}} + \xi_{s_{2l+k}} \right] \right\} \tag{A7}$$

$$M_{if}^{Rotor}(\gamma) = K_{sf} \sum_{k=1,3,5,\dots}^{\infty} \xi_{fk} \left\{ \sum_{l=1}^{\infty} \Lambda_{R,2l} \left[\xi_{s_{2l-k}} \cos((2l-k)(\varepsilon_i - \gamma)) + \xi_{s_{2l+k}} \cos((2l+k)(\varepsilon_i - \gamma)) \right] \right\} \tag{A8}$$

$$L_{ii}^{Sat,Re}(\gamma) = K_s \sum_{k=1,3,5,\dots}^{\infty} \xi_{sk} \left\{ \sum_{u=1}^{\infty} \Lambda_{Sat,2u}^{Re} \left[(\xi_{s_{2u-k}} + \xi_{s_{2u+k}}) \cos(2u(\varepsilon_i - \gamma)) \right] \right\} \tag{A9}$$

$$M_{ij}^{Sat,Re}(\gamma) = K_s \sum_{k=1}^{\infty} \xi_{s_k} \left\{ \sum_{u=1}^{\infty} \Lambda_{Sat,2u}^{Re} \left[\xi_{s_{2u-k}} \cos\left(k(\varepsilon_i - \varepsilon_j) + 2u(\varepsilon_j - \gamma)\right) + \xi_{s_{2u+k}} \cos\left(k(\varepsilon_j - \varepsilon_i) + 2u(\varepsilon_j - \gamma)\right) \right] \right\} \quad \text{(A10)}$$

$$L_{ff}^{Sat,Re} = K_f \sum_{k=1,3,5,\dots}^{\infty} \xi_{f_k} \left\{ \sum_{u=1}^{\infty} \Lambda_{Sat,2u}^{Re} \left[\xi_{s_{2u-k}} + \xi_{s_{2u+k}} \right] \right\} \quad \text{(A11)}$$

$$M_{if}^{Sat,Re}(\gamma) = K_{sf} \sum_{k=1,3,5,\dots}^{\infty} \xi_{f_k} \left\{ \sum_{u=1}^{\infty} \Lambda_{Sat,2u}^{Re} \left[\xi_{s_{2u-k}} \cos\left((2u-k)(\varepsilon_i - \gamma)\right) + \xi_{s_{2u+k}} \cos\left((2u+k)(\varepsilon_i - \gamma)\right) \right] \right\} \quad \text{(A12)}$$

$$L_{ii}^{Sat,Im}(\gamma) = K_s \sum_{k=1,3,5,\dots}^{\infty} \xi_{s_k} \left\{ \sum_{u=1}^{\infty} \Lambda_{Sat,2u}^{Im} \left[\left(\xi_{s_{2u+k}} + \xi_{s_{2u-k}}\right) \sin\left(2u(\varepsilon_i - \gamma)\right) \right] \right\} \quad \text{(A13)}$$

$$M_{ij}^{Sat,Im}(\gamma) = K_s \sum_{k=1,\dots}^{\infty} \xi_{s_k} \left\{ \sum_{u=1}^{\infty} \Lambda_{Sat,2u}^{Im} \left[\xi_{s_{2u-k}} \sin\left(k(\varepsilon_i - \varepsilon_j) + 2u(\varepsilon_j - \gamma)\right) + \xi_{s_{2u+k}} \sin\left(k(\varepsilon_j - \varepsilon_i) + 2u(\varepsilon_j - \gamma)\right) \right] \right\} \quad \text{(A14)}$$

$$M_{if}^{Sat,Im}(\gamma) = K_{sf} \sum_{k=1,3,5,\dots}^{\infty} \xi_{f_k} \left\{ \sum_{u=1}^{\infty} \Lambda_{Sat,2u}^{Im} \left[\xi_{s_{2u-k}} \sin\left((2u-k)(\varepsilon_i - \gamma)\right) + \xi_{s_{2u+k}} \sin\left((2u+k)(\varepsilon_i - \gamma)\right) \right] \right\} \quad \text{(A15)}$$

REFERENCES

[1] D. G. Dorrell, "Are wound-rotor synchronous motors suitable for use in high efficiency torque-dense automotive drives?," in *IECON 2012 - 38th Annual Conference of the IEEE Industrial Electronics Society*, 2012, pp. 4880–4885.

[2] M. Märgner and W. Hackmann, "Control challenges of an externally excited synchronous machine in an automotive traction drive application," in *Emobility - Electrical Power Train, 2010*, 2010, pp. 1–6.

[3] C. Rossi, D. Casadei, A. Pilati, and M. Marano, "Wound Rotor Salient Pole Synchronous Machine Drive for Electric Traction," in *Industry Applications Conference, Conference Record of the 2006 IEEE 41st IAS Annual Meeting*, 2006, pp. 1235–1241.

[4] I. Boldea, G.-D. Andreescu, C. Rossi, A. Pilati, and D. Casadei, "Active flux based motion-sensorless vector control of DC-excited synchronous machines," in *IEEE Energy Conversion Congress and Exposition 2009, ECCE 2009*, 2009, pp. 2496–2503.

[5] A. Rambetius, S. Ebersberger, M. Seilmeier, and B. Piepenbreier, "Carrier signal based sensorless control of electrically excited synchronous machines at standstill and low speed using the rotor winding as a receiver," in *15th European Conference on Power Electronics and Applications (EPE)*, 2013, pp. 1–10.

[6] A. Rambetius and B. Piepenbreier, "Sensorless control of wound rotor synchronous machines using the switching of the rotor chopper as a carrier signal," in *2013 IEEE International Symposium on Sensorless Control for Electrical Drives and Predictive Control of Electrical Drives and Power Electronics (SLED/PRECEDE)*, 2013, pp. 1–8.

[7] A. Rambetius and B. Piepenbreier, "Effectiveness of carrier signal based sensorless control of wound rotor synchronous machines," in *2014 International Symposium on Power Electronics Electrical Drives Automation and Motion (SPEEDAM)*, 2014

[8] A. Rambetius, S. Luthardt, and B. Piepenbreier, "Speed Estimation and Compensation for Harmonics in Self-Sensing Wound Rotor Synchronous Machines," in *IEEE International Symposium on Sensorless Control for Electrical Drives 2014*, 2014, pp. 1–8.

[9] M. Seilmeier, S. Arenz, B. Piepenbreier, and I. Hahn, "Model based closed loop control scheme for compensation of harmonic currents in PM-synchronous machines," in *2010 International Symposium on Power Electronics Electrical Drives Automation and Motion (SPEEDAM)*, 2010, pp. 1–6.

[10] T. Orlik, M. Lux, and W. Schumacher, "Saturation induced harmonics in permanent magnet synchronous motors," in *Proceedings of the 2011 14th European Conference on Power Electronics and Applications (EPE 2011)*, 2011, pp. 1–10.

[11] I. Hahn, "Differential magnetic anisotropy - prerequisite for rotor position detection of PM-synchronous machines with signal injection methods," in *2010 First Symposium on Sensorless Control for Electrical Drives (SLED)*, 2010, pp. 40–49.

[12] M. Seilmeier and B. Piepenbreier, "Modeling of PMSM with multiple saliencies using a stator-oriented magnetic circuit approach," in *IEEE International Electric Machines & Drives Conference (IEMDC)*, 2011, pp. 131–136.

[13] M. Seilmeier, "Modelling of electrically excited synchronous machine (EESM) considering nonlinear material characteristics and multiple saliencies," in *Proceedings of the 2011 14th European Conference on Power Electronics and Applications (EPE 2011)*, 2011, pp. 1–10.

[14] M. Seilmeier, S. Ebersberger, and B. Piepenbreier, "PMSM model for sensorless control considering saturation induced secondary saliencies," in *2013 IEEE International Symposium on Sensorless Control for Electrical Drives and Predictive Control of Electrical Drives and Power Electronics (SLED/PRECEDE)*, 2013, pp. 1–8.

[15] G. Dajaku and D. Gerling, "Stator Slotting Effect on the Magnetic Field Distribution of Salient Pole Synchronous Permanent-Magnet Machines," *IEEE Transactions on Magnetics*, vol. 46, no. 9, pp. 3676–3683, 2010.

[16] O. Laldin, S.D. Sudhoff, and S.D. Pekarek, "An analytical design model for wound rotor synchronous machines," *2013 IEEE Electric Ship Technologies Symposium (ESTS)*, pp.228,236, 22-24 April 2013

[17] H. Bali, Y. Amara, G. Barakat, R. Ibtiouen, and M. Gabsi, "Analytical Modeling of Open Circuit Magnetic Field in Wound Field and Series Double Excitation Synchronous Machines," *Magnetics, IEEE Transactions on* , vol.46, no.10, pp.3802,3815, Oct. 2010

[18] M. L. Bash and S. D. Pekarek, "Modeling of Salient-Pole Wound-Rotor Synchronous Machines for Population-Based Design," *IEEE Transactions on Energy Conversion*, vol. 26, no. 2, pp. 381–392, 2011.

[19] R. Wang, M. Bash, S. Pekarek, A. Larson, and R. van Maaren, "A voltage input-based magnetic equivalent circuit model for wound rotor synchronous machines," in *2013 IEEE International Electric Machines & Drives Conference (IEMDC)*, 2013, pp. 586–593.

[20] A. Tessarolo, "Accurate Computation of Multiphase Synchronous Machine Inductances Based on Winding Function Theory," *IEEE Transactions on Energy Conversion*, vol. 27, no. 4, pp. 895–904, 2012.

[21] Ying Yan, Jianguo Zhu, Haiwei Lu, Youguang Guo, and Shuhong Wang, "A PMSM model incorporating structural and saturation saliencies," in *Proceedings of the Eighth International Conference on Electrical Machines and Systems, ICEMS 2005*, 2005, pp. 194–199

[22] M. Seilmeier and B. Piepenbreier, "Identification of steady-state inductances of PMSM using polynomial representations of the flux surfaces," in *39th Annual Conference of the IEEE Industrial Electronics Society, IECON 2013*, 2013, pp. 2899–2904.

The 2014 International Power Electronics Conference

Design and Comparison of High Frequency Transformers using Foil and Round Windings

Topic number: 14

Kartik V Iyer, William P Robbins and Ned Mohan

Department of Electrical and Computer Engineering

University of Minnesota

Email: iyerx070@umn.edu, robbins@umn.edu, mohan@umn.edu

Abstract—**High frequency transformers are widely used in Switched-mode power supplies and now are being proposed to be used with power electronic converters to replace line-frequency transformers. This paper presents a winding design procedure for minimizing the power losses using foils and solid round wires under sinusoidal excitation to limit the temperature rise. This paper derives the range from which the thickness of the layers can be chosen to obtain the minimum power loss. This thickness range is a function of the number of layers and does not include the "optimum" based on the previous literature. Using this design procedure, it is shown that interleaving is not necessary in foil -wound transformers to obtain the minimum loss. A comparison of winding losses between foil windings and round conductors is also given. The analytical results are verified by designing six different winding configurations for the same specifications using 2-D Ansys Maxwell finite element design package.**

I. INTRODUCTION

With better availability of fast switching semiconductor switches there is an increase in trend to employ switch-mode power converters at high frequencies. Magnetic components like transformers are an integral part of Switch mode power supplies (SMPS). High frequency transformers determine about 25 % overall volume and more than 30 % of the overall weight of SMPS [1]. Hence, in order to reduce the footprint of SMPS the reduction in transformer volume is imperative. An increase in frequency allows for the volume reduction of transformers but at the cost of increased losses [2].

The losses occur in the core and in the windings of the transformer. The transformer is designed to be operated at low flux densities to avoid core saturation resulting in lower core losses. The winding losses in the transformer depend on the number of turns and the winding dimensions. At high frequencies due to eddy currents the effective conduction area reduces, which increases the losses due to skin effect. The increase in losses due to skin and proximity effects depend on the frequency and the winding type. Foil, solid-round wire, litz wires are the different conductors used in transformer winding [3].

A proper estimation of the winding losses is crucial for the design of high power density transformer. For the design of low/medium power high frequency transformers, area product is the most common method for core selection. With solid-round, foil and litz wires being the different winding types, a lot of different winding design schemes are available, which might fit the given window specifications. Hence, a comparison of the different winding types is terms of winding losses is

inevitable to determine the winding design to be used for a given specification.

In [4] winding losses are computed by considering a 1-D model of a transformer with foil conductors. The optimum thickness of foil conductors obtained by the 1-D analysis, as a function of number of layers, were used to design the foil conductors. However, the designs at optimal thicknesses resulted in ac losses more than the dc losses. Losses can be reduced if the winding can be designed such that the ac losses match the dc losses. A comparison between foil and solid-round conductors for a particular design was shown in [5] which showed that the foil conductors are less lossy than round conductors but still are more than dc losses. In [1] it was shown that the foil conductors should be extremely thin to have low losses. This led to the use of extremely thin strips of conductor at high frequencies known as litz wires, which have ac losses same as the dc losses. The optimal design and the cost analysis of litz wires were done in [6] and [7] respectively. But the litz wires are expensive and the window utilization factor is low. The choice between foil and litz wire is difficult and it depends on various design trade offs as shown in [8]. In literature, the comparison between different winding types are more for specific cases than a generalized comparison [5], [8].

The winding losses for both foil and solid-round conductor are analyzed. For foil conductors, for a given number of layers, instead of an optimal thickness as stated in literature previously, there exists a range of thickness, designing at which leads to an ac-to-dc resistance ratio less than 1.35-1.4. The paper shows that it is possible to design foil, so that the ac to dc resistance is close to 1 thereby giving an alternative to the expensive litz wires. The paper demonstrates a design procedure with foil conductors for ac-to-dc resistance ratio equal to 1.05 for low power high frequency transformers. It is also shown in the paper that if designed effectively the interleaving to minimize winding losses can be avoided.

The paper is organized in the following way: Section II and III show the conventional and the proposed way of winding design using foil conductors respectively. Section IV describes the step-by-step winding design procedure for foil and solid round conductor. Section V gives a comparison between the foil and solid-round winding in terms of power loss. Section VI shows six different winding configuration for the same specifications and power loss for each is computed and validated using 2-D FEM. Finally section VII gives the conclusion.

978-1-4799-2706-7/14 $31.00 © 2014 IEEE

II. Foil Winding Design Based on Conventional Approach

The ac to dc resistance ratio F_R for foil conductors and sinusoidal current is [4],

$$F_R = \Delta \left[\frac{\sinh(2\Delta) + \sin(2\Delta)}{\cosh(2\Delta) - \cos(2\Delta)} + \frac{2}{3}(p^2 - 1)\frac{\sinh(\Delta) - \sin(\Delta)}{\cosh(\Delta) + \cos(\Delta)} \right] \tag{1}$$

where Δ is the ratio of the layer thickness, d and the skin depth, δ at the operating frequency.

$$\Delta = \frac{d}{\delta} \tag{2}$$

Fig. 1 shows a transformer EE-core with primary and sec-

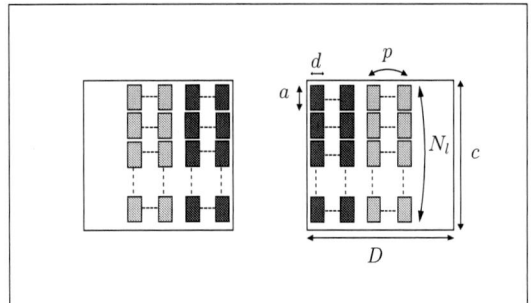

Fig. 1: Multi-turn Multi-layer foil winding of a transformer.

ondary windings.

Fig.2a shows a plot of F_R/Δ versus Δ for different layers. The conventional foil winding design procedure is:

1. Decide the number of layers, p individually for primary and secondary.
2. Locate Δ_{opt} corresponding to the number of layers chosen from Fig. 2a.
3. Compute d_{opt}, the foil conductor thickness from Δ_{opt} as the skin depth is known.
4. Compute a, the width of the conductor from d_{opt} and J, the current density, using $a = \frac{I}{Jd_{opt}}$.

$$N_l \times a/\eta_1 < c \tag{3}$$

5. Check if (3) is satisfied, where N_l is the number of turns per layer, η_1 is the layer porosity factor, generally $0.8 - 0.9$ and c is the height of winding.
   a) If yes, then compute the power loss for the chosen number of layers, p and thickness, d.
   b) If no, then increase $d > d_{opt}$ to fit in required N_l or change the number of layers, p.

The disadvantage of the above procedure is that it is iterative and there is no fixed way to determine the number of layers to be chosen.

III. Foil Winding Design Based on Proposed Method

The ac losses of a single winding of a transformer is given by (4),

$$P_{ac} = I^2 \times R_{dc}|_{d=\delta} \times \frac{F_R}{\Delta} \tag{4}$$

where, $R_{dc}|_{d=\delta} = \frac{N\rho(MLT)}{a\delta}$. N is the total number of turns, ρ is the resistivity of copper. Since, $a = \frac{I}{Jd}$, P_{ac} is,

$$P_{ac} = \frac{IJN(MLT)}{\sigma} \times (F_R) \tag{5}$$

Fig. 2a shows the plot of F_R/Δ versus Δ for different layers. The Table III shows, in the conventional design, the optimal value of Δ as a function of number of layers. The F_R value varies in a range of 1.36-1.4 for all layers. Hence, if designed at Δ_{opt}, using the conventional method, the foil conductors for any number of layers will give approximately the same loss, which obviates the need of interleaving to minimize losses, provided (3) is satisfied. If not, then the thickness of the conductor has to be increased to fit in the required N_l, thereby compensating for increased losses. The reason being, with increasing number of layers the thickness of conductor reduces which increases the width of the foil conductor, and this increase in width counters the increase in losses by the proximity effect.

As seen in Fig. 2b, by designing the foil conductors in a specific range of Δ, the F_R value can be reduced to close to 1. Fig. 3 shows that, using $\Delta_{opt} = 0.43$, which is obtained from Fig. 2a for $p = 10$ will give $F_R = 1.379$, but by designing at $\Delta < \Delta_{opt}$ will lead to $F_R < 1.379$ and hence lower losses. By designing at Δ such that F_R is close to 1, the losses can be reduced by 37% compared to the conventional design. Even in the proposed method, by designing the foil conductors such that F_R is close to 1, for all layers, the need of interleaving to minimize the losses can be avoided.

F_R can be approximated as [9] for $\Delta < 1$,

$$F_R = 1 + \frac{5p^2 - 1}{45}\Delta^4 \tag{6}$$

It is not possible to take $F_R = 1$, but it is definitely possible to design with foil thickness, such that F_R is close to 1. For a particular F_R value, Δ as a function of number of layers, can be computed using (7). Here, for the designs shown in section VI, $F_R = 1.05$ is considered. Table III based on the proposed method provides the value of $p\Delta$ product for 1-20 layers for which $F_R = 1.05$.

$$\Delta_{proposed} = \sqrt[0.25]{\frac{45(F_R - 1)}{5p^2 - 1}} \tag{7}$$

IV. Winding Design Procedure for Foil winding and Solid-Round Conductors

The computation of the minimum winding thickness for a sinusoidal waveform was given by [4] and extended for non-sinusoidal waveform by [9]. But in these methods it is assumed that the number of layers are already known.

The flowchart shown in Fig. 4 is a detailed winding design procedure for both foil and solid round conductor. From the given specifications: Voltage, V, Current, I, frequency, f, Current Density, J, window fill factor, k_w using area product method compute the number of turns, N and the window dimensions, c and D from the chosen core. The proposed winding design procedure is:

978-1-4799-2706-7/14 $31.00 © 2014 IEEE

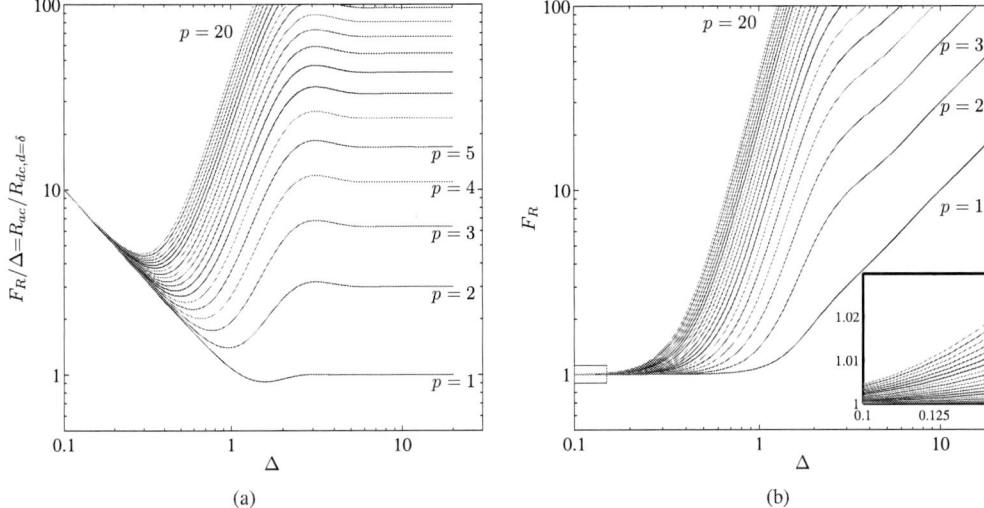

Fig. 2: (a) Plot of F_R/Δ versus Δ for different layers. (b) Plot of F_R versus Δ for different layers.

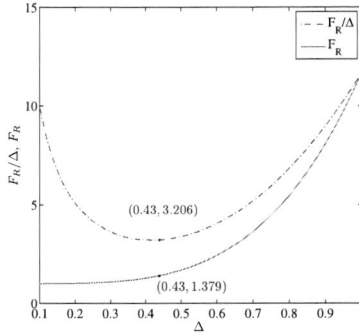

Fig. 3: Plot of $F_R/\Delta, F_R$ versus Δ for 10 layers.

1   For foil conductors,

$$J = \frac{I}{ad} \qquad (8)$$

$$\eta_1 = \frac{N_l a}{c} \qquad (9)$$

$$N_l = \frac{N}{p} \qquad (10)$$

$$p\Delta = \frac{NI}{J\eta_1\delta c} \qquad (11)$$

Rearranging, (8), (9) , (10) and (2), we get, (11). All the quantities on the right hand side of (11), are known.

2   From the look up Table III the number of layers, p and corresponding Δ can be chosen.

3   For the number of layers chosen, check if N_l is an integer, in order to fit integral number of turns in each layer.

    a)   If yes, then compute d, the foil conductor thickness from Δ as δ, the skin depth, is

known and then compute a, the width of the conductor using (8) as d and J are already known.

    b)   If N_l is not an integer, and if (3) is satisfied for a specific number of layers greater than p obtained from Table III, such that N_l is an integer, then the winding can be designed using that many layers. If not, then consider the number of layers less than the layers obtained using $p\Delta$ product, to satisfy (3), thereby increasing the winding losses.

4   Compute the winding ac losses using (5)

For solid round wires, the F_R equation as shown in [3] is modified. In case of round wire, the conductor area depends only on one variable, the conductor diameter. Fig. 4 shows a detailed design procedure of solid-round conductor as well.

1   For solid-round wire, the current, I and current density, J are fixed for a specific design,hence the diameter of round conductor is fixed, which can be computed using, (12)

$$d_R = \sqrt{\frac{4J}{\pi I}} \qquad (12)$$

$$N_{l_R} = \frac{2\eta_1 c}{\sqrt{\pi}d_R} \qquad (13)$$

$$\frac{2\sqrt{\pi}d_R p_R}{\eta_2} < D \qquad (14)$$

$$\Delta_R = {}^{0.75}\sqrt{\frac{\pi}{4}}\frac{d_R}{\delta}\sqrt{\eta_1} \qquad (15)$$

where, η_2 is the distance between two consecutive layers and η_1 is the layer porosity factor as defined in [3].

2   The N_{l_R} can be determined from window height, c using (13)

The 2014 International Power Electronics Conference

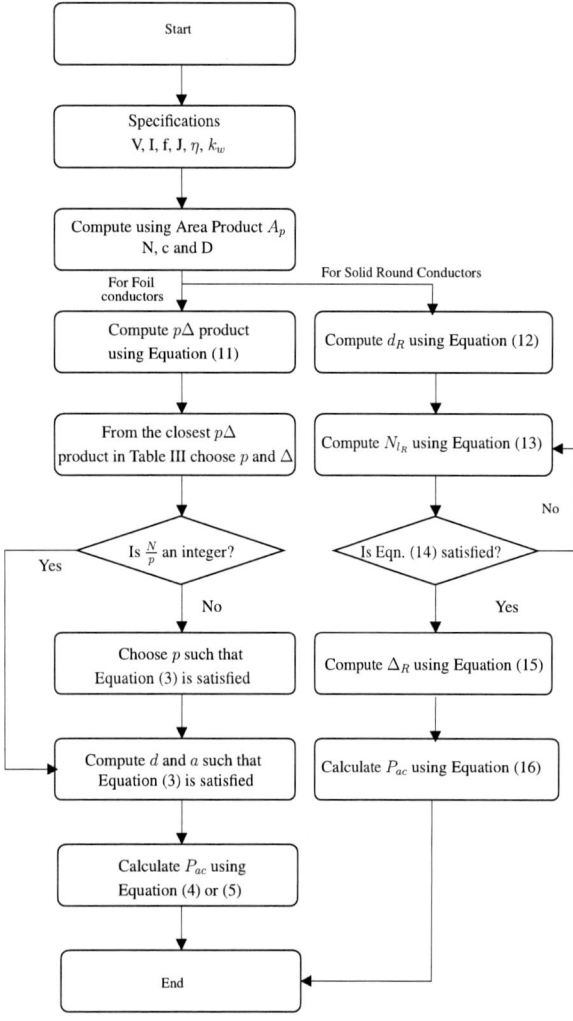

Fig. 4: Winding design procedure for foil and solid-round conductor.

V. Comparison of Foil and Solid-Round Winding

The power loss equation for Foil and round windings is given by (5) and (16) respectively. Equating the two expression for comparing gives,

$$I\eta_1 \sqrt[1.5]{\frac{\pi}{4}} \frac{1}{\frac{\pi\delta^2}{4}} \times \frac{F_{R_{\Delta_R}}}{\Delta_R^2} = J F_{R_{\Delta_F}} \tag{17}$$

Substituting, $\dfrac{I}{J} = A_c = \dfrac{\pi d^2}{4}$, gives,

$$\eta_1 \sqrt[1.5]{\frac{\pi}{4}} \frac{d^2}{\delta^2} \times \frac{F_{R_{\Delta_R}}}{\Delta_R^2} = F_{R_{\Delta_F}} \tag{18}$$

Now, $\eta_1 \sqrt[1.5]{\dfrac{\pi}{4}} \dfrac{d^2}{\delta^2} = \Delta_R{}^2$, substituting which gives,

$$F_{R_{\Delta_R}} = F_{R_{\Delta_F}} \tag{19}$$

which shows that the the round wire should be used if and only if, $F_{R_{\Delta_R}} < F_{R_{\Delta_F}}$.

VI. Foil winding design using $p\Delta$ product and Validation of Results with 2-D FEM.

The example which follows is for the design of a high frequency transformer with turns ratio 1:1 and with the specifications given in Table I.

TABLE I: Specifications

Parameter	Value
Power	$1\ kW$
Voltage	$200\ V$
Frequency, f	$20\ kHz$
Current Density, J	$5\ A/mm^2$
B_{max}	$0.3\ T$
k_w	0.4

Based on the given specifications as in Table I, using area-product method the core is determined. The chosen core is OP44721EC, ferrite material, and from core datasheet: $c = 24.2$ mm and $D = 7.78$ mm, where c and D are window height and window width respectively, as shown in Fig. 1. The number of turns, $N = 36$, can be computed as core area is known. $\eta = 0.9$ is assumed for all computations.

A 2-D Finite element analysis is done to verify the proposed design methodology using ANSYS MAXWELL 16.0. A double EE-core with dimensions of OP44721EC is used. The same core with 6 different winding configurations is analyzed. Four cases are of the foil conductors with/without interleaving for the conventional and proposed method and two cases are for round conductors with/without interleaving.

A detailed description about the computation of analytical winding loss for all six cases is presented.

3 From the number of layers, N_{l_R} compute p_R and check if (14) is satisfied.

   a) If yes, then compute Δ_R using (15).

   b) If no, then re-compute, N_{l_R} by changing η_1.

4 Compute the winding losses using, (16),

$$P_{ac} = I^2 \eta_1 \sqrt[1.5]{\frac{\pi}{4}} \times R_{dc}|_{d=\delta} \times \frac{F_R}{\Delta_R^2} \tag{16}$$

where, $R_{dc}|_{d=\delta} = \dfrac{N\rho(MLT)}{\frac{\pi\delta^2}{4}}$.

The section VI will go through the detailed winding design procedure which will take into account the interleaving as well.

978-1-4799-2706-7/14 $31.00 © 2014 IEEE

A. Case-A: Non-interleaved Foil winding designed using the conventional method

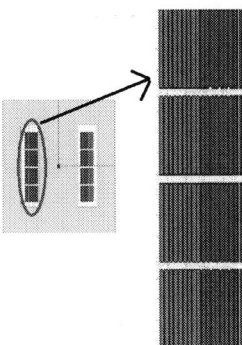

Fig. 5: Non-interleaved Foil winding with 9 layers designed using conventional method: (red)-primary winding, (blue)-secondary winding

The number of layers are chosen to be 9 for both primary and secondary, as done in the conventional method. Fig. 5 shows the winding configuration designed using the existing method with 9 layers. As shown in the Fig. 5, it is a non-interleaved winding structure with 9 layers of primary having 4 turns/ layer followed by 9 layers of secondary. From Fig. 2a or the Table I the value of Δ can be computed as 0.45. Now, a, which is the height of foil conductor can be computed from 20,

$$a = \frac{I}{J\delta\Delta} \tag{20}$$

which gives $a = 4.75mm$. As (3) is satisfied, the ac power loss is computed using (4) which gives $42.393W/m$. The ac power loss can also be computed using (5) which gives the same result and it shows that $F_R = 1.366$. Even if, $p = 18$ would have been chosen then still the losses would have been around the same value as in that case the width of the conductor would be increased such that the F_R is around $1.36 - 1.4$.

B. Case-B: Interleaved Foil winding designed using the conventional method

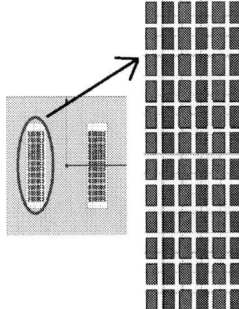

Fig. 6: Interleaved Foil winding with 1 layer designed using conventional method: (red)-primary winding, (blue)-secondary winding

If interleaving of primary and secondary winding is considered, then the best possible way is to have 1-layer of primary and 1-layer of secondary. Fig. 6 shows the above described winding configuration. For such a configuration, from Fig. 2a or the Table I the value of Δ can be computed as 1.58. Using, (3), N_l can be chosen to be 12 to fit the conductors inside the window height. The winding arrangement will be 3 sets of 12 turns of primary followed by 12 turns of secondary. In this case, $a = 1.35mm$ using (20) and the ac power loss is $44.96W/m$.

The analytical ac loss is around the same as that of the case with non-interleaved winding. This is because, the width of the conductor is less in case of single layer as compared to the non-interleaved case. If the design is such that the thickness is the same and the number of layers are increased then the losses for single layer would be less. But in general, with the increase in the number of layers, the thickness is reduced as seen by "conventional method" column in Table III, as proposed in previous literature, hence the losses are almost the same.

C. Case-C: Non-interleaved Foil winding designed using the proposed method

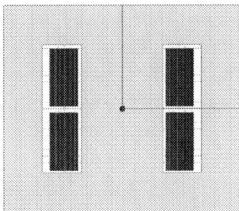

Fig. 7: Non-interleaved Foil winding with 18 layers designed using proposed method: (red)-primary winding, (blue)-secondary winding

Fig. 7 shows the winding configuration designed using the proposed method with 18 layers. As shown in the Fig. 7, it is a non-interleaved winding structure with 18 layers of primary having 2 turns/ layer followed by 18 layers of secondary. If instead of using Dowell's curve, the curve as shown in Fig. 2b is used, then the Δ can be selected such that F_R is close to 1, provided the computed width of the conductor satisfies (3). Here, in this analysis Δ is chosen such that $F_R = 1.05$. The " proposed method" column in Table III is constructed considering $F_R = 1.05$. But the Table can be designed for any other value of F_R using (7). Depending on the chosen value of F_R the analytical winding losses can be reduced compared to the conventional method.

For the example considered, the $p\Delta = 3.537$. From the "proposed method" column in Table III, the number of layers, p, for which $p\Delta$ is close to 3.537 is $p = 18$ which also makes $N_l = 2$, which is an integer. From Table III the value of Δ can be computed, which is 0.19. Now, a, which is the height of foil conductor can be computed from 20, which gives $a = 11.2629mm$. This value of a will not satisfy (3). Hence, Δ, should be chosen so as to allow 2 turns/layer. Considering, $\Delta = 0.1965$ which is obtained by substituting $p = 18$ in $p\Delta = 3.537$ will allow 2 turns/layers but, $F_R = 1.0536 > 1.05$ but still less than 1.37. In this case, $a = 10.89mm$. The ac power loss is computed using (4) which gives $32.698W/m$. The ac power loss can also be computed using (5) which gives the same result and it shows that $F_R = 1.0536$.

D. Case-D: Interleaved Foil winding using the proposed method

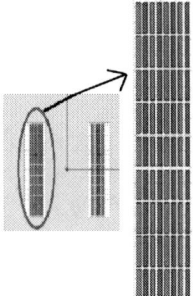

Fig. 8: Interleaved Foil winding with 1 layer designed using proposed method: (red)-primary winding, (blue)-secondary winding

Fig. 8 shows the winding configuration designed using the proposed method with interleaving of primary and secondary winding. As shown in the Fig. 8, it is an inter-leaved winding structure with 4 sets of 9 turns of primary followed by 9 turns of secondary. For such a configuration, number of turns in 1-layer of primary winding can be selected from the "proposed method" column in Table I using $p\Delta = 0.88$ in (21) .

$$N = \frac{p\Delta J\delta\eta_1 c}{I} \qquad (21)$$

Using, the above data, number of turns in 1-layer of primary winding can be selected from, (21) which gives $N = 8.9$. In order to fit in, 9 turns per layer, Δ is chosen to be 0.9. Using, $\Delta = 0.9$ gives $F_R = 1.058$ slightly more than 1.05. In this case, the ac power loss is $32.586W/m$.

E. Case-E: Non-interleaved Round conductors

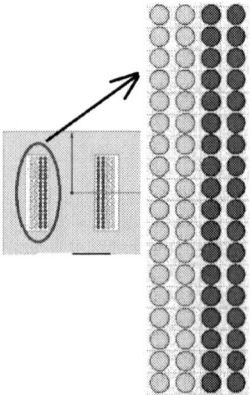

Fig. 9: Non-interleaved Round conductors with 2 layers: (green)-primary winding, (red)-secondary winding

Fig. 9 shows the winding configuration designed using the round conductors without interleaving the primary and secondary winding. The flowchart as shown in Fig. 4 is followed for round conductors. As the given case is a non-interleaved structure, both layers of primary are wound together. The diameter d_R can be computed using (12), which is, $d_R = 1.128$ mm. AWG-17 is the nearest solid-round wire. The number of turns per layer, N_{l_R} can be computed using (13) and as the windings satisfy (14), Δ_R and hence, the losses can be computed.

F. Case-F: Interleaved Round conductors

Fig. 10 shows the winding configuration designed using the round conductors by interleaving the primary and secondary winding. The computation of d_R and N_{l_R} is the same as Case-E. But there is interleaving between primary and secondary windings as a result of which, the losses are reduced in comparison to the non-interleaved case.

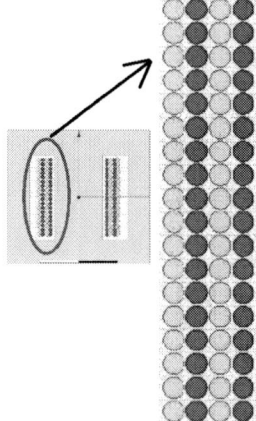

Fig. 10: Interleaved Round conductors with 1 layer: (green)-primary winding, (red)-secondary winding

The Table II compares the losses for all cases. The analytical and simulated losses are computed per unit length. The average mean length of turns can be chosen as 0.125 m according to [10]. The winding thickness using proposed method due to window dimension restrictions for 1 layer was 0.9 instead of 0.88 and 0.1965 instead of 0.19 for 18 layers. The results match closely with the analytical calculation for foil conductors but are different for round winding due to 2-D effects. Also, the losses for solid-round conductor for both interleaving and non-interleaving cases are more than the foil conductors. In case of foil winding, interleaving was not a concern but for solid round conductors as diameter is fixed, Δ is fixed and hence, interleaving has to be done to reduce losses.

VII. CONCLUSION

The paper demonstrates that, if the winding is designed at a thickness lower than the "optimal" thickness, as given in literature, for a specific layer then the losses can be minimized. Theoretically, power loss can be reduced around 36-40%, depending on the number of layers. The ac-to-dc resistance

978-1-4799-2706-7/14 $31.00 © 2014 IEEE

TABLE II: Validation of results with 2-D FEM.

Case	$p\Delta$	p	N_l	Δ	a	Analytical P	Simulated P	F_R
Existing Method Non-interleaving	3.537	9	4	0.45	4.75 mm	5.3 W	5.34 W	1.366
Existing Method Interleaving	1.58	1	12	1.58	1.35 mm	5.62 W	4.85 W	1.449
Proposed Method Non-Interleaving	3.537	18	2	0.1965	10.89 mm	4.087 W	4.23 W	1.0536
Proposed Method Interleaving	0.9	1	9	0.9	2.378 mm	4.073 W	4.044 W	1.058
Round winding Non-Interleaving	-	2	18	1.9488	-	18.22 W	15.93 W	-
Round winding Interleaving	-	1	18	1.9488	-	6.86 W	6.3625 W	-

ratio close to 1 can be achieved using foil conductors provided the winding can fit inside the window. Hence, it can replace the expensive litz wires which also have low window utilization factor. The paper also demonstrates that a proper design can help to avoid interleaving in case of foil conductors for minimizing winding loss. The paper also presents a comparison between foil and round conductors in terms of winding losses. Six different winding configurations with/without interleaving, foil/round conductors for the same specification and on the same EE-core are analyzed using 2-D FEM simulations. For all six cases losses are computed and compared with analytical results. The results show that interleaving of winding is a must in case of solid- round conductors whereas interleaving can be avoided in case of foil winding if designed properly.

TABLE III: Parameters to compute $p\Delta$ with the conventional/proposed method .

p	conventional method			proposed method		
-	Δ_{opt}	$p\Delta_{opt}$	F_r	Δ	$p\Delta$	F_r
1	1.58	1.58	1.449	0.88	0.88	1.052
2	0.97	1.94	1.361	0.59	1.18	1.051
3	0.78	2.34	1.356	0.48	1.44	1.052
4	0.67	2.68	1.351	0.41	1.64	1.05
5	0.6	3	1.355	0.37	1.85	1.052
6	0.55	3.3	1.362	0.33	1.98	1.047
7	0.51	3.57	1.365	0.31	2.17	1.05
8	0.48	3.84	1.374	0.29	2.32	1.05
9	0.45	4.05	1.366	0.27	2.43	1.048
10	0.43	4.3	1.376	0.26	2.6	1.05
11	0.41	4.51	1.377	0.25	2.75	1.052
12	0.39	4.68	1.368	0.24	2.88	1.053
13	0.38	4.94	1.387	0.23	2.99	1.052
14	0.36	5.04	1.364	0.22	3.08	1.051
15	0.35	5.25	1.373	0.21	3.15	1.049
16	0.34	5.44	1.377	0.21	3.36	1.055
17	0.33	5.61	1.378	0.2	3.4	1.051
18	0.32	5.76	1.375	0.19	3.42	1.047
19	0.31	5.89	1.368	0.19	3.61	1.052
20	0.3	6	1.359	0.18	3.6	1.046

REFERENCES

[1] R. Petkov, "Optimum design of a high-power, high-frequency transformer," *Power Electronics, IEEE Transactions on*, vol. 11, no. 1, pp. 33–42, 1996.

[2] W.-J. Gu and R. Liu, "A study of volume and weight vs. frequency for high-frequency transformers," in *Power Electronics Specialists Conference, 1993. PESC '93 Record., 24th Annual IEEE*, 1993, pp. 1123–1129.

[3] R. Wojda and M. Kazimierczuk, "Analytical optimization of solid x2013;round-wire windings," *Industrial Electronics, IEEE Transactions on*, vol. 60, no. 3, pp. 1033–1041, 2013.

[4] P. Dowell, "Effects of eddy currents in transformer windings," *Electrical Engineers, Proceedings of the Institution of*, vol. 113, no. 8, pp. 1387 –1394, august 1966.

[5] W. Hurley, W. Wolfle, and J. Breslin, "Optimized transformer design: inclusive of high-frequency effects," *Power Electronics, IEEE Transactions on*, vol. 13, no. 4, pp. 651–659, 1998.

[6] C. R. Sullivan, "Optimal choice for number of strands in a litz-wire transformer winding," in *Power Electronics Specialists Conference, 1997. PESC'97 Record., 28th Annual IEEE*, vol. 1. IEEE, 1997, pp. 28–35.

[7] C. Sullivan, "Cost-constrained selection of strand diameter and number in a litz-wire transformer winding," *Power Electronics, IEEE Transactions on*, vol. 16, no. 2, pp. 281–288, 2001.

[8] M. Kheraluwala, D. Novotny, and D. Divan, "Design considerations for high power high frequency transformers," in *Power Electronics Specialists Conference, 1990. PESC '90 Record., 21st Annual IEEE*, 1990, pp. 734–742.

[9] W. Hurley, E. Gath, and J. Breslin, "Optimizing the ac resistance of multilayer transformer windings with arbitrary current waveforms," *Power Electronics, IEEE Transactions on*, vol. 15, no. 2, pp. 369 –376, mar 2000.

[10] N. Mohan, T. M. Undeland, and W. P. Robbins, "Power electronics: converters, applications, and design," 1989.

A Method to Calculate the Performance of Linear Induction Motors Using Simple Two-Phase Model

Hideaki Hirahara
Tokyo Metropolitan University
Graduate School of
Science and Engineering
1-1 Minami-Osawa,
Hachioji-shi, Tokyo, Japan
hirahara@uitec.ac.jp

Shu Yamamoto
Polytechnic University
Electrical System
Engineering
2-32-1 Ogawa-Nishimachi,
Kodaira-shi, Tokyo, Japan
yamamoto@uitec.ac.jp

Takahiro Ara
Polytechnic University
Electrical System
Engineering
2-32-1 Ogawa-Nishimachi,
Kodaira-shi, Tokyo, Japan
ara@uitec.ac.jp

Toshihisa Shimizu
Tokyo Metropolitan University
Graduate School of
Science and Engineering
1-1 Minami-Osawa,
Hachioji-shi,Tokyo, Japan
shimizut@tmu.ac.jp

Abstract— **A new two-phase mathematical model of Linear Induction Motor (LIM) is discussed. This model is simpler than conventional model, but can correctly calculate both asymmetric primary current and thrust ripple performances. In addition, an improved DC testing method is proposed to determine all circuit parameters in the model. This test is completed by only one simple standstill test. Hence, both no-load steady-state operation and lock tests which are difficult to carry out on built-in LIMs are unnecessary. The proposed calculation method is implemented on a 3-phase, 4-pole and 12-slot single-sided LIM, and is validated by experimental and simulation results.**

Keywords— Asymmetry, DC testing method, Linear induction motor, Two-phase model

I. Introduction

Small-size and low-speed linear induction motors (LIMs) feature strong-build, maintenance-free in straight-line drives, and are widely applied to conveying machines in various factories. However, critical issues remain.

The first issue is asymmetry of magnetic circuit of LIM, which causes unbalanced current and thrust ripple. Since they cannot be evaluated by general mathematical models used for rotational-type symmetric induction machines, some models for LIMs with the asymmetric nature have been studied. In [1], a model with voltage and thrust equations in the two-axis reference frame was proposed. This model can consider both asymmetric primary current and thrust ripple performances. However, its expression is complicated. In [2], three-phase equivalent circuit is proposed. However, the transient characteristic cannot be evaluated.

The second issue is that there is no standardized method for determining circuit model parameters of LIM. In [1] and [2], methods based on lock test and equivalent no-load test were presented. However, a time-consuming work, removing a secondary-side conductor, is necessary

to carry out the no-load test. In addition, these methods are applicable only to LIM in which neutral point is pulled out. In [3], the authors proposed a method to calculate the performance of LIMs by a simple standstill testing method using a small-capacity DC power supply (tentatively called the DC decay testing method in this paper). This method is suited in performing the parameter measurement test of LIM. This is because it is unnecessary to both lock a mover and dismantle the secondary-side conductor. However, determining the parameters of asymmetric models has not been discussed. In [4], an examination equipment using disk type secondary-side conductor was reported. If this equipment is used, it is possible to perform actual steady-state operation of LIM. However, installing the equipment is considerably difficult.

To address this situation, the authors derive a new two-phase model of LIM. This model is simpler than conventional model [1], but can correctly calculate both asymmetric primary current and thrust ripple performances. In addition, the authors propose an improved dc decay testing method suitable for determining the parameters of the asymmetric model. When the proposed method is employed, the proposed parameter measurement test is completed by only one simple standstill test. Moreover, troublesome tasks to pull out the neutral point, to lock a mover and to dismantle the secondary-side conductor are all unnecessary.

The validity of the proposed method is verified with experimental and simulation results on a 3-phase, 4-pole and 12-slot single-sided LIM.

II. Profile of Test Machine

Table I, Fig. 1 and Fig. 2 show specifications, photo and structure of the tested LIM respectively.

From Fig. 2, it can be seen that the coupling coefficient for a- and b-phase windings is the same as that for a- and c-phase windings, but they are larger than

TABLE I
SPECIFICATIONS OF THE TESTED LIM

Ratings and Dimensions		Mover (primary winding)	
Voltage	200 (V)	No. of phases	3
Current	16.1 (A)	No. of poles	3
frequency	50 (Hz)	No. of slots	12
Stator core size	29.5×222 (mm)	No. of turns	102 (T)*¹ 204 (T)*²
Stator core length	50 (mm)	Coil pitch	#1 to #4
		Connection	Star
Pole pitch	54 (mm)	Armature resistance	1.8 (Ω/phase)
Gap length	0.5 (mm)	Coil space factor	50 (%)

*¹ : No. of turns in the slot #1, #2, #3, #10, #11 and #12 is 102(T).
*² : No. of turns in the slot #4, #5, ..., #9 is 204(T).

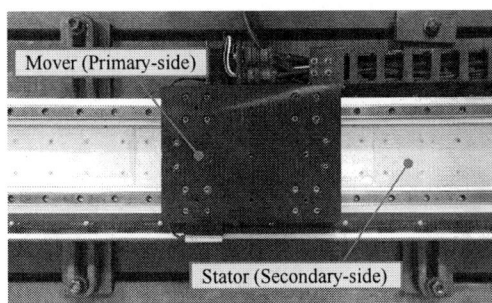

(a) Overview

(b) Primary-side

(c) Secondary-side

Fig. 1. Photo of the tested LIM.

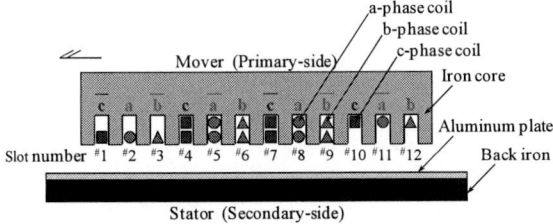

Fig. 2. Sectional view of the tested LIM.

the coupling coefficient for b- and c-phase windings. This is because the ends of the mover core exist in case of the LIM.

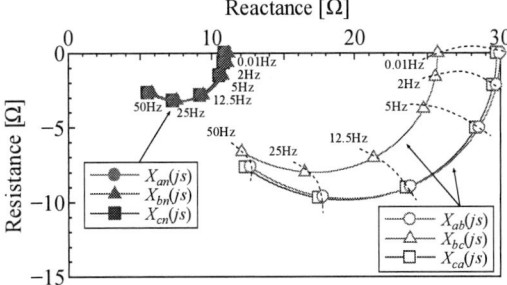

Fig. 3. Operational impedance loci of the tested LIM.

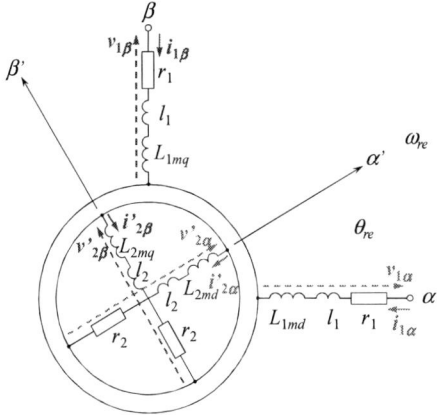

Fig. 4. Two-phase IM model.

Fig. 3 shows operational impedance loci obtained by the DC decay testing method for a-n, b-n, c-n, a-b, b-c nd c-a winding terminals. The details of the DC decay testing method are shown in [3]. From the figure, it is con-firmed that $X_{an}(js) \approx X_{bn}(js) \approx X_{cn}(js)$ and $X_{bc}(js) < X_{ab}(js) \approx X_{ca}(js)$. This result agrees with the magnitude relation of mutual reactances guessed from the structure of Fig. 2.

III. Simple Two-Phase Model

Fig. 4 shows a circuit model of a two phase rotational-type induction motor. Here, r_1 and r_2 are primary and secondary resistances. l_1 and l_2 are primary and secondary leakage inductances. L_{1md} and L_{2md} are primary and secondary d-axis inductances based on the main magnetic flux. L_{1mq} and L_{2mq} are primary and secondary q-axis inductances based on the main magnetic flux. θ_{re} is the rotor electrical angle, and ω_{re} is the angular velocity.

In this paper, LIM is regarded as a two phase induction motor with the following conditions.

1) Winding resistances between winding terminal and neutral point are equal. Namely, $r_{1a} = r_{1b} = r_{1c} = r_1$, and $r_{2a} = r_{2b} = r_{2c} = r_2$.
2) Permeances of leakage magnetic paths are balanced. As a result, $l_{1a} = l_{1b} = l_{1c} = l_1$, and $l_{2a} = l_{2b} = l_{2c} = l_2$.
3) In $\alpha\beta$ stational reference frame shown in Fig.4, the permeance of the magnetic path of α-axis main magnetic flux differs from that of β-axis main

magnetic flux. As a result, α-axis mutual inductance m_d is not equal to β-axis one m_q.

The reasons of 1) and 2) can be explained from the symmetric nature of the operational impedances of a-phase, b-phase and c-phase ($X_{an}(js) \approx X_{bn}(js) \approx X_{cn}(js)$). The reason of 3) can be done from the asymmetric nature of the operational impedances between a-b, b-c and c-a terminals ($X_{bc}(js) < X_{ab}(js) \approx X_{ca}(js)$).

The voltage equation of the circuit model of Fig. 4 is as follows [5]:

$$[\boldsymbol{v_{\alpha\beta}}'] = [\boldsymbol{Z}'][\boldsymbol{i_{\alpha\beta}}']$$
$$= [\boldsymbol{R}'][\boldsymbol{i_{\alpha\beta}}'] + p([\boldsymbol{L}'][\boldsymbol{i_{\alpha\beta}}']) + p([\boldsymbol{M}'][\boldsymbol{i_{\alpha\beta}}']) \tag{1}$$

where p is a differential operator ($p = d/dt$),

$$[\boldsymbol{v_{\alpha\beta}}'] = [v_{1\alpha} \quad v_{1\beta} \quad v_{2\alpha}' \quad v_{2\beta}']^T \tag{2}$$

$$[\boldsymbol{i_{\alpha\beta}}'] = [i_{1\alpha} \quad i_{1\beta} \quad i_{2\alpha}' \quad i_{2\beta}']^T \tag{3}$$

$$[\boldsymbol{R}'] = \mathrm{diag}[r_1 \quad r_1 \quad r_2 \quad r_2] \tag{4}$$

$$[\boldsymbol{L}'] = \mathrm{diag}\left[\frac{N_1^2}{R_{1l}} + \frac{N_1^2}{R_d} \quad \frac{N_1^2}{R_{1l}} + \frac{N_1^2}{R_q} \quad \frac{N_2^2}{R_{2l}} \quad \frac{N_2^2}{R_{2l}}\right] \tag{5}$$

$$[\boldsymbol{M}'] = \begin{bmatrix} 0 & 0 \\ 0 & 0 \\ \dfrac{N_1 N_2}{R_d}\cos\theta_{re} & \dfrac{N_1 N_2}{R_q}\sin\theta_{re} \\ -\dfrac{N_1 N_2}{R_d}\sin\theta_{re} & \dfrac{N_1 N_2}{R_q}\cos\theta_{re} \end{bmatrix} \quad *$$

$$* \quad \begin{bmatrix} \dfrac{N_1 N_2}{R_d}\cos\theta_{re} & -\dfrac{N_1 N_2}{R_d}\sin\theta_{re} \\ \dfrac{N_1 N_2}{R_q}\sin\theta_{re} & \dfrac{N_1 N_2}{R_q}\cos\theta_{re} \\ \dfrac{N_2^2}{R_d}\cos^2\theta_{re} + \dfrac{N_2^2}{R_q}\sin^2\theta_{re} & -\dfrac{1}{2}\left(\dfrac{N_2^2}{R_d} - \dfrac{N_2^2}{R_q}\right)\sin 2\theta_{re} \\ -\dfrac{1}{2}\left(\dfrac{N_2^2}{R_d} - \dfrac{N_2^2}{R_q}\right)\sin 2\theta_{re} & \dfrac{N_2^2}{R_d}\sin^2\theta_{re} + \dfrac{N_2^2}{R_q}\cos^2\theta_{re} \end{bmatrix}. \tag{6}$$

In these equations, N_1 and N_2 are numbers of turns of primary and secondary windings. R_{1l} and R_{2l} are the magnetic resistances of the leak magnetic path of primary and secondary. R_d and R_q are d-axis and q-axis magnetic resistances.

Here, if it is considered that $N_1 = N_2 \equiv N$, $l_1 = N^2/R_{1l}$, $l_2 = N^2/R_{2l}$, $m_d = N^2/R_d$ and $m_q = N^2/R_q$, (5) and (6) can be rewritten as (7) and (8) respectively :

$$[\boldsymbol{L}'] = \mathrm{diag}[l_1 + m_d \quad l_1 + m_q \quad l_2 \quad l_2] \tag{7}$$

$$[\boldsymbol{M}'] = \begin{bmatrix} 0 & 0 \\ 0 & 0 \\ m_d\cos\theta_{re} & m_q\sin\theta_{re} \\ -m_d\sin\theta_{re} & m_q\cos\theta_{re} \end{bmatrix} \quad *$$

$$* \quad \begin{bmatrix} m_d\cos\theta_{re} & -m_d\sin\theta_{re} \\ m_q\sin\theta_{re} & m_q\cos\theta_{re} \\ m_d\cos^2\theta_{re} + m_q\sin^2\theta_{re} & -\dfrac{1}{2}(m_d - m_q)\sin 2\theta_{re} \\ -\dfrac{1}{2}(m_d - m_q)\sin 2\theta_{re} & m_d\sin^2\theta_{re} + m_q\cos^2\theta_{re} \end{bmatrix}. \tag{8}$$

If the following coordinate transformation,

$$[\boldsymbol{C}] = \begin{bmatrix} 1 & 0 & 0 & 0 \\ 0 & 1 & 0 & 0 \\ 0 & 0 & \cos\theta_{re} & -\sin\theta_{re} \\ 0 & 0 & \sin\theta_{re} & \cos\theta_{re} \end{bmatrix} \tag{9}$$

is applied to (1), (1) can be changed as follows:

$$[\boldsymbol{C}][\boldsymbol{v_{\alpha\beta}}'] = [\boldsymbol{C}][\boldsymbol{Z}'][\boldsymbol{i_{\alpha\beta}}']$$
$$[\boldsymbol{C}][\boldsymbol{v_{\alpha\beta}}'] = [\boldsymbol{C}][\boldsymbol{Z}'][\boldsymbol{C}]^{-1}\cdot[\boldsymbol{C}][\boldsymbol{i_{\alpha\beta}}']$$
$$[\boldsymbol{v_{\alpha\beta}}] = [\boldsymbol{Z}][\boldsymbol{i_{\alpha\beta}}] \tag{10}$$

where,

$$[\boldsymbol{v_{\alpha\beta}}] = [v_{1\alpha} \quad v_{1\beta} \quad 0 \quad 0]^T \tag{11}$$

$$[\boldsymbol{i_{\alpha\beta}}] = [i_{1\alpha} \quad i_{1\beta} \quad i_{2\alpha} \quad i_{2\beta}]^T \tag{12}$$

$$[\boldsymbol{Z}] = \begin{bmatrix} r_1 + p(l_1 + m_d) & 0 \\ 0 & r_1 + p(l_1 + m_q) \\ pm_d & \omega_{re}m_q \\ -\omega_{re}m_d & pm_q \end{bmatrix} \quad *$$

$$* \quad \begin{bmatrix} pm_d & 0 \\ 0 & pm_q \\ r_2 + p(l_2 + m_d) & \omega_{re}(l_2 + m_q) \\ -\omega_{re}(l_2 + m_d) & r_2 + p(l_2 + m_q) \end{bmatrix}. \tag{13}$$

Therefore, with the asymmetric nature, an equivalent circuit model of LIM in the stational $\alpha\beta$ frame is expressed as Fig. 5.

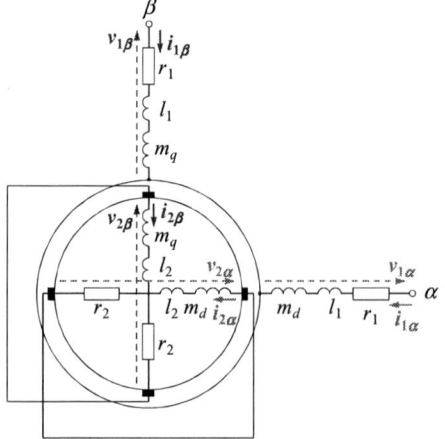

Fig. 5. The equivalent circuit model of LIM in the stational $\alpha\beta$ frame.

Next, the authors drive a thrust equation. The input electric power P_i is written as follows:

$$
\begin{aligned}
P_i &= [i_{\alpha\beta}{}']^T [v_{\alpha\beta}{}'] \\
&= [i_{\alpha\beta}{}']^T [Z'][i_{\alpha\beta}{}'] \\
&= [i_{\alpha\beta}{}']^T [R'][i_{\alpha\beta}{}'] + [i_{\alpha\beta}{}']^T p([L'][i_{\alpha\beta}{}']) + [i_{\alpha\beta}{}']^T p([M'][i_{\alpha\beta}{}']) .
\end{aligned}
$$

(14)

The first term of (14) is copper loss. This is expressed as follows:

$$
[i_{\alpha\beta}{}']^T [R'][i_{\alpha\beta}{}'] = r_1(i_{1\alpha})^2 + r_1(i_{1\beta})^2 + r_2(i_{2\alpha}')^2 + r_2(i_{2\beta}')^2 . \tag{15}
$$

The second term of (14) is a time differentiation of the energy of inductance, and is rewritten as follows:

$$
[i_{\alpha\beta}{}']^T p([L'][i_{\alpha\beta}{}']) = p\left(\frac{1}{2}[i_{\alpha\beta}{}']^T [L'][i_{\alpha\beta}{}'] \right) . \tag{16}
$$

The third term of (14) is expressed as follows:

$$
[i_{\alpha\beta}{}']^T p([M'][i_{\alpha\beta}{}']) = p\left(\frac{1}{2}[i_{\alpha\beta}{}']^T [M'][i_{\alpha\beta}{}'] \right) + \frac{1}{2}[i_{\alpha\beta}{}']^T p[M'][i_{\alpha\beta}{}']
$$

(17)

Since the first term of the right side of (17) is a time differentiation of the energy of mutual inductance, the second term serves as a mechanical output. Moreover, since $[M']$ is a function of θ_{re}, an equation of mechanical output P_m can be obtained as follows:

$$
P_m = \frac{1}{2}[i_{\alpha\beta}{}']^T p[M'][i_{\alpha\beta}{}'] = \frac{1}{2}[i_{\alpha\beta}{}']^T \theta_{re} \frac{\partial}{\partial \theta_{re}}([M'])[i_{\alpha\beta}{}'] . \tag{18}
$$

The speed v_m on the secondary side of LIM is denoted by the following equation:

$$
v_m = \frac{\tau}{\pi} \theta_{re} \tag{19}
$$

where τ is pole pitch.

Therefore, a thrust equation of LIM can be derived as following equation:

$$
\begin{aligned}
F_m &= \frac{P_m}{v_m} = \frac{\pi}{\tau} \left\{ \frac{1}{2}[i_{\alpha\beta}{}']^T \frac{\partial}{\partial \theta_{re}}([M'])[i_{\alpha\beta}{}'] \right\} \\
&= \frac{\pi}{\tau} \left\{ \frac{1}{2}([C]^T [i_{\alpha\beta}])^T \frac{\partial}{\partial \theta_{re}}([M'])([C]^T [i_{\alpha\beta}]) \right\} \\
&= \frac{\pi}{\tau} \left\{ (i_{1\beta} m_q i_{2\alpha} - i_{1\alpha} m_d i_{2\beta}) - (m_d - m_q) i_{2\alpha} i_{2\beta} \right\} .
\end{aligned}
$$

(20)

IV. Determination of Circuit Parameters

Fig. 6 shows the circuit configuration used for the DC decay testing method. The procedure of the DC decay testing method is as follows. First, a DC voltage V_{DC} is applied to primary winding terminals (ex. between the a and b terminals) so that a DC current I_{DC} flows between these terminals. Next, these terminals are short-circuited by a relay switch, denoted by MC in Fig. 6. IGBT used to

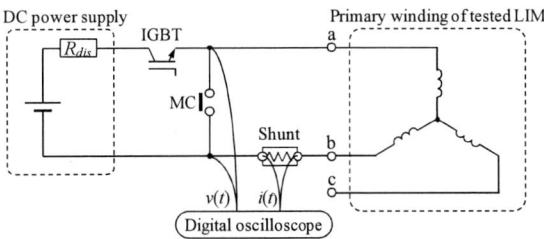

Fig. 6. System configuration of the DC decay testing.

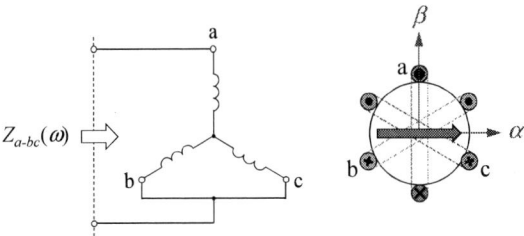

(a) connection   (b) Magnetomotive force

Fig. 7. Wire connection of a DC decay testing to α-axis.

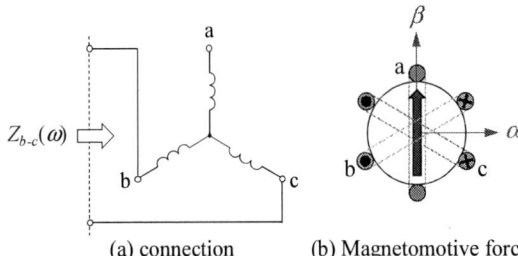

(a) connection   (b) Magnetomotive force

Fig. 8. Wire connection of a DC decay testing to β-axis.

break I_{DC} when the MC is turned on. V_{DC} and I_{DC} before short-circuiting, and the DC decay current $i(t)$ after short-circuiting are measured with a digital oscilloscope. The impedance for each frequency $Z_{a-b}(\omega)$ can be obtained by substituting values of V_{DC}, I_{DC} and $i(t)$ into

$$
Z_{a-b}(\omega) = \frac{1}{\dfrac{\omega}{jV_{DC}} \displaystyle\int_0^{+\infty} i(t)\varepsilon^{-j\omega t} dt + \dfrac{I_{DC}}{V_{DC}}} . \tag{21}
$$

To obtain the circuit parameters (l_1, l_2, m_d, m_q, r_2) of (13), the authors show an improved DC decay testing method. The test procedure is as follows. First, a DC decay testing method is carried out under the connection of Fig. 7(a), then $Z_{a-bc}(\omega)$ is calculated. Since magnetomotive force shown in Fig. 7(b) is alined with α-axis, α-axis operational impedance $X_\alpha(js)$ is determined as follows:

$$
X_\alpha(js) = \frac{2}{3} \frac{Z_{a-bc}(\omega) - r_{a-bc}}{js} \tag{22}
$$

where s is the slip ($= \omega/\omega_0$), ω_0 is the angular frequency of the power source (rad/s), and r_{a-bc} is the winding resistance between a and bc terminals.

Next, after changing the connection as shown in Fig. 8 (a), a DC decay testing is carried out, then $Z_{b-c}(\omega)$ is calculated. In this case, magnetomotive force occurs as

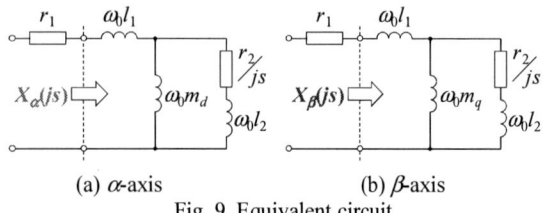

(a) α-axis   (b) β-axis

Fig. 9. Equivalent circuit.

Fig. 10. Operational impedance loci of α-axis and β-axis.

TABLE II
THE EQUIVALENT CIRCUIT PARAMETERS IDENTIFICATION RESULT

Symbol	Value	Unit
r_1	1.80	Ω
l_1	14.4	mH
m_d	34.5	mH
m_q	26.8	mH
l_2	0	mH
r_2	4.42	Ω

shown in Fig. 8(b). So, β-axis operational impedance $X_\beta(js)$ is obtained as follows:

$$X_\beta(js) = \frac{1}{2}\frac{Z_{b-c}(\omega) - r_{b-c}}{js} \qquad (23)$$

where r_{b-c} is the winding resistance between b and c terminals.

Using the results mentioned above, the equivalent circuit parameters l_1, m_d, and r_2 of Fig. 9 (a) are identified by a least-squares method [6][7]. The secondary side of the tested LIM consists of a secondary conductor of an aluminum plate with an iron plate. Therefore, one can determine as $l_2 = 0$ from its physical image. A similar technique can be applied to identify the parameters l_1, m_q, and r_2 of Fig. 9 (b). Finally, l_1 and r_2 are determined as average values of two obtained values. Performance calculation of LIM can be carried out by applying the identified parameters to (10) and (20).

V. Experimental Results

The DC decay testing was implemented on the test machine, and $X_\alpha(js)$ and $X_\beta(js)$ were calculated. Fig. 10 shows the results. Using them, the authors determined the equivalent circuit parameter by the method described in Chapter IV. The result is shown in Table II.

In order to verify the validity of the proposed method, the authors carried out some experiments.

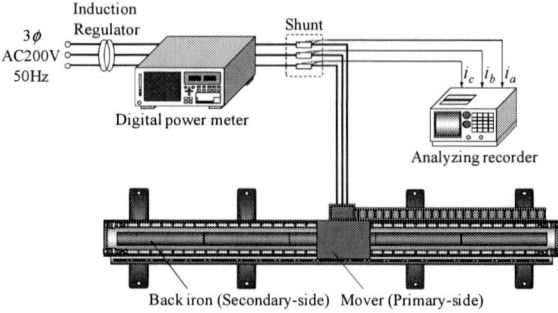

Fig. 11. System configuration of equivalent no-load test.

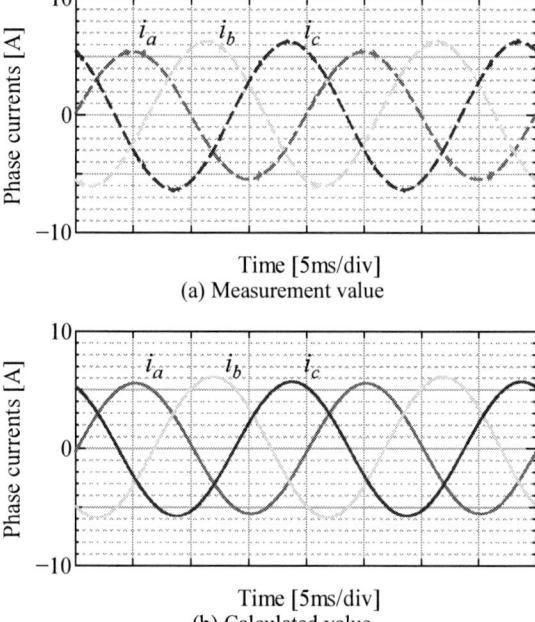

Fig. 12. Phase currents waveform at no-load state
(100V/50Hz) .

First, in order to change into an equivalent no-load state, the secondary conductor (aluminum plate) was removed. While a three-phase voltage (100V/50Hz) was applied in this state, phase voltages and currents were measured by the analyzing recorder (Fig. 11). Fig. 12 (a) shows the result. On the other hand, Fig. 12 (b) shows simulation results of phase currents by (10) using the value of Table II. From Fig. 12, it can be confirmed that the simulation results agree well with the measurement results.

Next, after returning the secondary conductor, the authors locked the mover with a load-sell and measured steady-state thrust at standstill using the test circuit shown in Fig. 13. Fig. 14 (a) shows the simulated thrust obtained by (10) and (20) using the values of Table II. From Fig.14, it can be seen that the simulated thrust ripple components (amplitude and phase) agree well with the measured result. Thus, the results of Fig. 13 and Fig. 14 suggest the validity of the proposed two-phase model and its parameter measurement method. The obtained

978-1-4799-2706-7/14 $31.00 © 2014 IEEE   3048

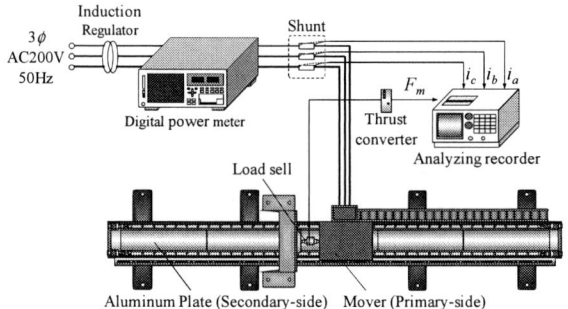

Fig. 13. System configuration of lock test.

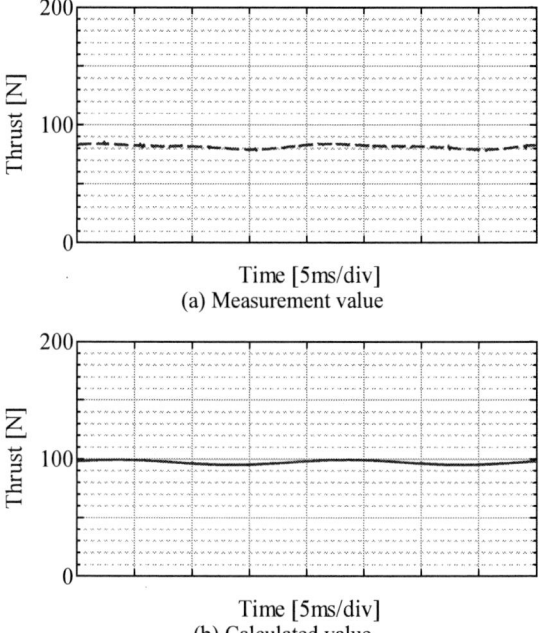

Time [5ms/div]
(a) Measurement value

Time [5ms/div]
(b) Calculated value

Fig. 14. Thrust waveform when the mover is locked (100V/50Hz).

average thrust was 81.1N. On the other hand, measured the calculated one was 96.8N. It is considered that the difference is caused by neglecting iron loss effect in (10) and (20).

VI. Conclusions

The results are summarized as follows.

1) A new simple two-phase circuit model of LIM was proposed. This model is simpler than conventional model, but can correctly calculate both asymmetric primary current and thrust ripple performances.
2) The improved DC decay testing method suitable for determining the parameters of the asymmetric model of LIM was proposed. This method is applicable to LIM in which the neutral point is not pulled out.
3) The validity of 1) and 2) were demonstrated with simulation and experimental results on a 3-phase, 4-pole and 12-slot single-sided LIM.

Since the proposed method requires neither the lock of a mover, nor the disassembly of apparatus for testing, it is possible to adapt to built-in LIMs.

References

[1] H. Sugimoto, M. Tomoe, S. Matsumura and H. Ishii, "A Method of Calculating Asymmetrical Constants Based on Lock Test for Single-Sided Linear Induction," (in Japanese), *IEEJ Trans. on Industry Applications*, vol. 113-D, no. 2, pp. 247-255, 1993.

[2] T. Utumi and I. Yamaguchi, "Parameter-Value Estimation of 3-Phase Equivalent Circuit of Linear Induction Motor under Starting Condition," (in Japanese), *IEEJ Trans. on Industry Applications*, vol. 120-D, no. 11, pp. 1283-1288, 2000.

[3] S. Yamamoto, H. Hirahara, T. Ara, "Comparison of the Characteristics of Linear Induction Motors with Various Secondary-Side Conductors on the Basis of the Simple Calculation Method Using DC Decay Test," (in Japanese), *IEEJ Trans. on Industry Applications*, vol. 121-D, no. 11, pp. 1117-1125, 2001.

[4] T. Morizane, and N. Kimura, "Measurement of Drive Characteristics of Linear Induction Motor with Experimental Equipment Implemented Disc-shaped Secondary Side," (in Japanese), *IEEJ Trans. on Industry Applications*, vol. 131-D, no. 10, pp. 1256-1257, 2011.

[5] A. Nabae, T. Kin, I. Takahashi, S. Nakamura, H.Yamada, "Basic electrical machinery," (in Japanese), pp. 76-87, 1985.

[6] *IEEE Standard Procedures for Obtaining Synchronous Machine Parameters by Standstill Frequency Response Testing*, IEEE Std. 115A, 1987.

[7] S. Yamamoto, T. Ara, S. Oda, K. Matsuse, "Prediction of Starting Performance of PM Motors by DC Decay Testing Method," *IEEE Trans. on Industry Applications*, vol. 36, no. 4, pp. 1053-1060, 2000.

The 2014 International Power Electronics Conference

An ESP downhole parameters monitoring system based on current loop transmission method

Jin Miaoxin
School of Electrical Engineering
Harbin Institute of Technology
Harbin, China
E-mail: jinmiaoxin1986@163.com

Gao Qiang
School of Electrical Engineering
Harbin Institute of Technology
Harbin, China
E-mail: gq_hit@163.com

Zhang Wei
School of Electrical Engineering
Harbin Institute of Technology
Harbin, China
E-mail: zhangwei_0231@sina.cn

Xu Dianguo
School of Electrical Engineering
Harbin Institute of Technology
Harbin, China
E-mail: xudiang@hit.edu.cn

Abstract-**In order to achieve online monitoring the downhole information related to ESP working state, an ESP downhole parameters monitoring system based on current loop data transmission method was developed. Compared with other methods, the current loop data transmission method is relatively insensitive to interference in harsh environment. By means of ESP motor's winding and symmetrical reactance located on the ground, the loop for current transmission could be built. The principle of current loop transmission and involved circuit were introduced. An experiment simulating the hostile downhole environment was conducted to test the transmission effect, by which the reliability and accuracy of current loop data transmission was demonstrated.**

Keywords-ESP; Current loop; Downhole parameter; Monitoring

I.   INTRODUCTION

As high-performance equipment for oil exploitation, electric submersible pump (ESP) features high efficiency and powerful capability of artificial lift. ESP enjoys a wide application area ranging from onshore oil field to intricate offshore platform [1]. With the development of oil industry, ESP makes up an increasingly large portion of oil production equipment. Though ESP's technologies involving oil exploitation has been gradually mature, methods surrounding how to monitor the information of oil reservoir under wells and ESP equipment's running status in the downhole to meet the demand for high-performance oil exploitation as well as prolong ESP's life still need to go further.

Fig.1 shows a typical system of electric submersible pump. The uphole part is mainly composed of control cabinet, junction box and transformer. Inside the wellbore casing, oil drain device, check value, pump, gas separator, protector, submersible motor, sensor and coadjutor constitute the downhole part in series. Besides, an armor-protected power cable connecting surface power with submersible motor supplies power for the motor over a long distance [2]. As is

shown in Fig.1, the narrow space between wellbore casing and downhole part's periphery combined with the relatively deep position where the sensor is located have restricted the way sensor is powered as well as the signal's transmission. In addition, the hostile downhole environment, such as high temperature, high pressure, strong corrosive fluid and high-frequency noise, imposes huge physical and chemical stress on downhole monitoring instrument [3]. Therefore, conventional monitoring methods widely used in other industrial fields couldn't be transplanted directly and thus developing monitoring technique catering for this unique structure is needed.

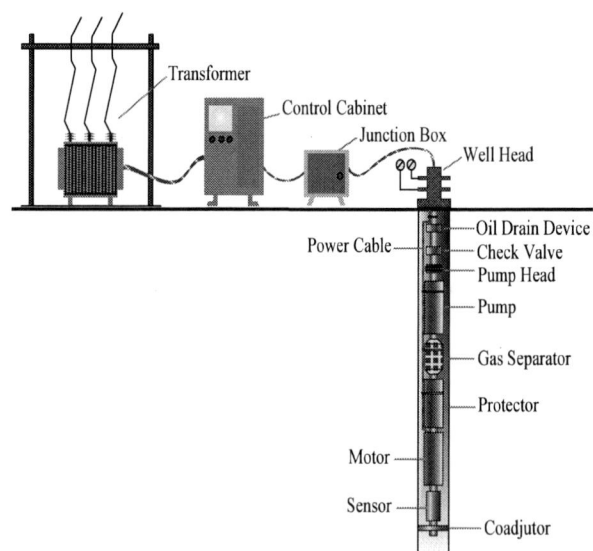

Fig.1 the structure diagram of ESP system

Traditionally, the off-line measurement was a primary method to gain the above-mentioned information, but its complex and time-consuming operation and weakness in real-

978-1-4799-2706-7/14 $31.00 © 2014 IEEE

time performance set a barrier to its further application. Moreover, the price of deferred production is costly [4]. Therefore, developing a real-time monitoring system may be a more feasible and appealing solution. Power line carrier communication and GPRS wireless communication are two essential methods capable of transmitting real-time data in harsh environments, but the former needs an extra armor-protected cable for transmission that is expensive and complicated for operation. In addition, they are both suitable for data transmission under the condition of power frequency voltage, but most of ESP equipments are driven with variable frequency voltage. Therefore, the accuracy of data transmission may suffer from strong interference from inverters and electromagnetic wave in the downhole.

In this paper, a downhole multi-parameter monitoring system based on current loop transmission method was presented. This system includes uphole unit and downhole unit. By means of three-phase reactance placed on the ground and ESP's power cable, the downhole unit could be powered by uphole unit without an extra power cable. Meanwhile, a loop for current signal transmission was established, eliminating the need for an extra communication cable. Moreover, the carrier of signal is current instead of voltage, the influence of voltage drop and electromagnetic interference on the quality of transmission could be largely crippled, which guarantees accuracy and reliability of signal's transmission

II. THE PRINCIPLE OF POWERING DOWNHOLE CIRCUIT

Fig 2 shows the principle of powering the downhole circuit. A three-phase symmetrical star connection reactance in parallel with ESP motor's three-phase winding is placed on the ground and its parameters is approximately equal to ESP motor's three-phase winding.

According to three-phase circuit's basic principle, we have equations as below:

$$\mathbf{V}_a = \mathbf{V}_n + Z_a \cdot \mathbf{i}_a \qquad (1)$$

$$\mathbf{V}_b = \mathbf{V}_n + Z_b \cdot \mathbf{i}_b \qquad (2)$$

$$\mathbf{V}_c = \mathbf{V}_n + Z_c \cdot \mathbf{i}_c \qquad (3)$$

here $\mathbf{V}_a$, $\mathbf{V}_b$, $\mathbf{V}_c$: three-phase supply voltage

$\quad \mathbf{i}_a$, $\mathbf{i}_b$, $\mathbf{i}_c$: three-phase current through ground reactance

$\quad Z_a$, Z_b, Z_c: three-phase impedance of ground reactance

$\quad \mathbf{V}_n$: the voltage in neutral point of ground reactance

After transformation, we could get:

$$\mathbf{V}_n = \frac{\mathbf{V}_a + \mathbf{V}_b + \mathbf{V}_c - (Z_a \cdot \mathbf{i}_a + Z_b \cdot \mathbf{i}_b + Z_c \cdot \mathbf{i}_c)}{3} \quad (4)$$

Since the three-phase windings of submersible motor share approximately equal reactance value, we could substitute equation 5 for equation 4.

$$\mathbf{V}_n = \frac{\mathbf{V}_a + \mathbf{V}_b + \mathbf{V}_c}{3} \qquad (5)$$

Similarly,

$$\mathbf{V}_m = \frac{\mathbf{V}_a + \mathbf{V}_b + \mathbf{V}_c}{3} \qquad (6)$$

here $\mathbf{V}_m$: the voltage in neutral point of submersible motor

Since the ESP motor's neutral point shares the identical voltage with its counterpart in three-phase symmetrical reactance, the downhole circuit could be powered with equivalent voltage value when the uphole circuit exerts the DC voltage on the neutral point of the three-phase symmetrical reactance.

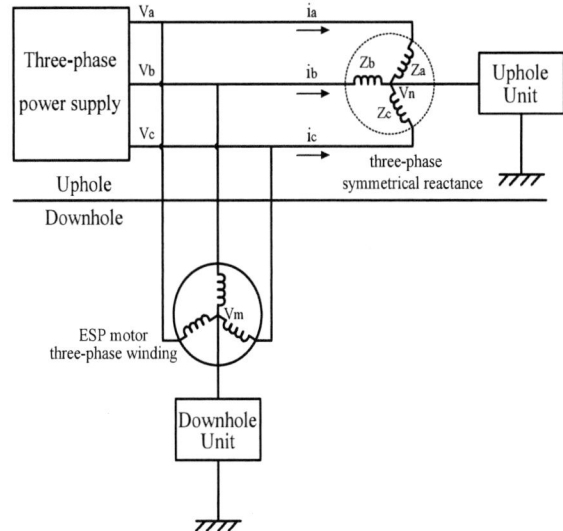

Fig.2 the schematic diagram of powering downhole circuit

III. THE PRINCIPLE OF CURRENT LOOP

Compared with voltage, it's more suitable for current to serve as the signal's carrier because it could avoid the signal's inaccurate transmission resulting from voltage drop and it's more insensitive to noise.

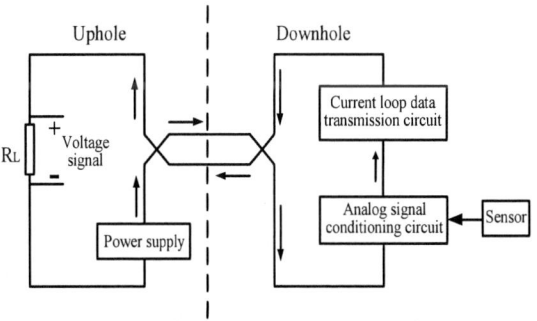

Fig.3 the principle of current loop

As is shown in Fig 2, the uphole unit, three-phase symmetrical reactance, ESP motor's three-phase winding, the downhole circuit and the ground have combined to provide a closed loop for current.

Fig.3 shows the principle of current loop. After uphole unit exerts a DC voltage on the neutral point of the three-phase symmetrical reactance to supply power for downhole unit, a DC current flows through downhole unit. Then, the amplitude of the current is regulated according to the magnified voltage signal originating from sensor. When the current flows through the sampling resistance located in uphole circuit, the amplitude of current is converted into corresponding voltage signal.

IV. THE CIRCUIT OF CURRENT LOOP DATA TRANSMISSION

In this system, there are six parameters needing to be monitored but only one loop for current transmission. Hence, some part of the downhole circuit should be capable of achieving a mechanism of switching the signals in proper order. In the downhole circuit, MCU with eight A/D channels is utilized to fulfill this demand.

Fig.4 shows the MCU time-sharing switching circuit. In this part, six parameters are connected to MCU's six A/D pins respectively. When signals come, MCU conducts sampling and A/D conversion. Then, MCU outputs PWM pulse to approach the signal's real value. A second order Butterworth low pass filter is used to eliminate harmonic wave with high frequency in PWM pulse. In addition, the filter's output terminal is connected to another A/D pin to achieve closed-loop control.

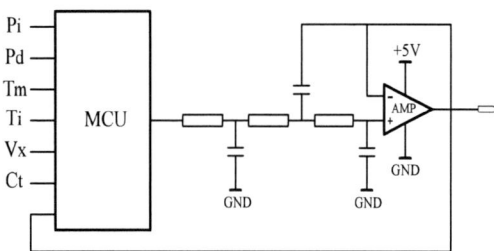

Fig.4 the circuit of MCU time-sharing switching

The main part of current loop data transmission circuit is shown in Fig.5.

R1 and R4 are two resistances of 100 with high precision. For the two resistances, R1 is located in the downhole circuit for current's regulation, while R4 is utilized as data receiver in the uphole circuit.

When the amplified signal comes, a discrepant value appears between Point 1 and Point 2. Due to the integrator formed by R2, C1, C2 and amplifier, the voltage between grid electrode and source electrode of MOSFET is adjusted to make I1 gradually changes. As a result, the voltage in Point 1 follows the current's change to narrow the gap with Point 2. When Point 1 shares the identical voltage value with the signal, the output voltage of integrator reaches a steady state. In this steady state, the amplitude of I1 is in proportion to the signal's amplitude. Consequently, when I1 flows through R4 in uphole, the signal could be extracted in the form of voltage.

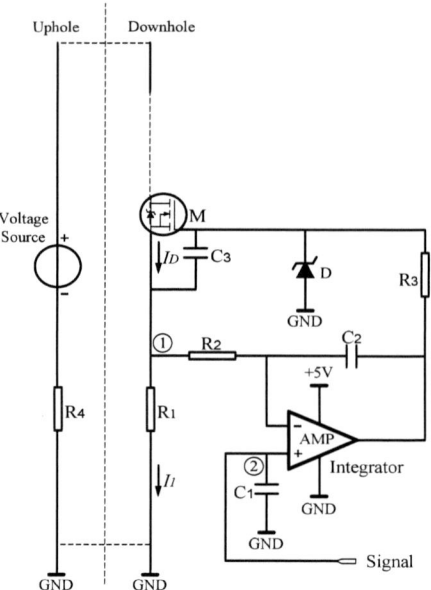

Fig.5 the circuit of the current loop data transmission

V. ESP UPHOLE DATA PROCESSING SYSTEM

As is mentioned above, the uphole circuit undertakes double tasks of dealing with data as well as supplying power for the downhole circuit.

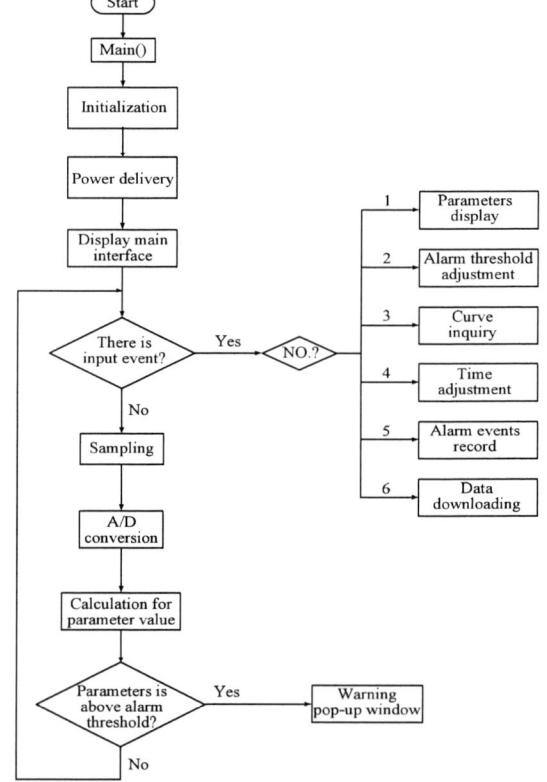

Fig.6 uphole unit flow chart

Fig.6 shows the software flowchart of the uphole unit. In the initial phase, the functions of serial port communication and SD card storage are initialized. Then, the main interface appears and power delivery is conducted. In the main interface, six icons are used to represent six operations, including parameters display, alarm threshold adjustment, historical record inquiry in the form of curve, time adjustment, alarm events record and data downloading. Selecting any icon will trigger corresponding subroutine. In main program, sampling and A/D conversion are made to get the signal's value. Then, a conversion from the signal's value to the parameter's value is conducted. After the parameter's value is obtained, a comparison is made to judge whether the value is beyond the alarm threshold. If it is above the threshold, a warning pop-up window would be triggered and the alarm record would be updated.

VI. EXPERIMENTAL RESULTS

To test the reliability and accuracy of the system, an experiment simulating the downhole data transmission was conducted. In this experiment, six parameters were measured, including pump intake pressure, discharge pressure, intake temperature, motor temperature, pump vibration and leakage current.

Since the environment in downhole is so harsh, some auxiliary instruments are necessary for simulating the hostile condition of high temperature, high pressure and strong vibration in laboratory. Fig.7 shows the oven, piston digital gauge and vibration test platform from left to right.

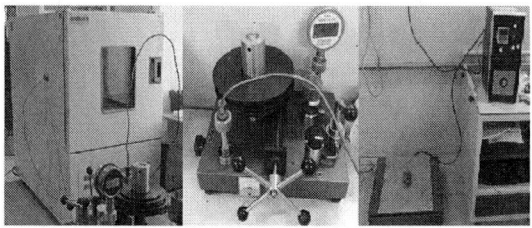

Fig.7 oven, piston digital gauge and vibration test platform

The waveform of experiment is shown in Fig.8. In this experiment, the pressure of 5MPa was exerted on the intake pressure sensor; the oven's temperature was set to 80℃; the vibration amplitude of vibration test platform was adjusted to 2g.

On the oscilloscope's screen, the top curve shows the waveform of the received signal in the uphole, while the bottom one displays the waveform of the transmitted signal in the downhole. It could be seen that they share the similar shape and amplitude. The transmission cycle was 18s, within which the voltage of 1V and 2V indicating the start of signal and six parameters mentioned above respectively taking up 2s. Besides, there was still 2s left for the uphole unit to process data.

The resolutions of pressure sensor, intake temperature sensor and vibration sensor are 30mV/MPa, 10mV/℃ and 24.3mV/g. Moreover, all signals are overlaid with reference voltage of 1V. Thus, the voltages' amplitudes in downhole

signal corresponding to intake pressure, intake temperature and pump vibration approximated 1.15V, 1.8V and 1.048V, respectively. In addition, PT1000 serves as motor temperature sensor and the current through it is 0.5mA constantly, so the voltages' amplitudes in downhole signal corresponding to motor temperature approximates 1.65V.

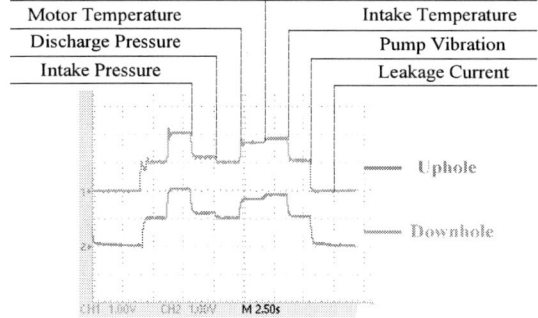

Fig.8 the parameters' waveform in experiment

In the uphole circuit, each part of the signal corresponding to a certain parameter was sampled and AD conversion for hundreds of times within 2s, which could guarantee sufficient extraction accuracy alone with digital filtering.

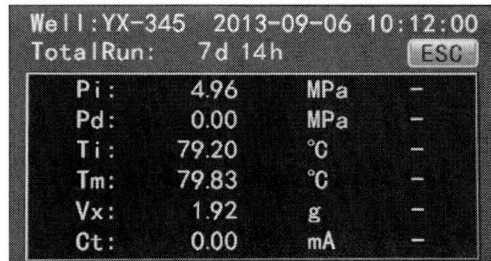

Fig.9 the parameter display interface of system

After sampling and filtering, the extracted voltage amplitudes corresponding to intake pressure, motor temperature, intake temperature and pump vibration are 1.1489V, 1.6525V, 1.7983V and 1.0467V. By reverse conversion, intake temperature was 4.96MPa; intake temperature was 79.20℃; motor temperature was 79.83℃; pump vibration was 1.92g. The error ratios of above four parameters are 0.8%, 1%, 0.21% and 4%, respectively. The parameter was showed on the screen of the uphole instrument as Fig.9, which could fulfill the requirement for accuracy appropriately.

Fig.10 the system real objects

978-1-4799-2706-7/14 $31.00 © 2014 IEEE 3053

Fig.10 shows the real object of the system. The downhole circuit is exhibited on the left, while the uphole circuit board is shown on the right.

VII. CONCLUSION

In this paper, an ESP downhole parameters monitoring system was introduced. A current loop data transmission technology was utilized in this system. Without extra cable, the current loop is able to achieve reliable and accurate data transmission. In addition, because the carrier of signal is current instead of voltage, the influence of voltage drop and electromagnetic interference on the quality of transmission could be largely crippled. To test the effect of current loop data transmission, oven, piston digital gauge and vibration test platform were used to simulate the complicated downhole environment. By comparing the simulated circumstance parameters with the experiment's result, the reliability and accuracy of transmission was demonstrated.

REFERENCES

[1] Ritchie Pragale, David D. Shipp, "Investigation of premature ESP failures and oil field harmonic analysis", Petroleum and Chemical Industry Technical Conference (PCIC), 2012 Record of Conference Papers Industry Applications Society 59th Annual IEEE, Chicago, IL, 24-26 September 2012, PP. 1-8.

[2] Gary Skibinski, Stephen Breit, "Line and Load Friendly Solutions for Long Length Cable Applications in Electrical Submersible Pump Applications", Proceedings of IEEE 51 Petroleum and Chemical Industry Technical Conference (PCIC), 13-15 September 2004, PP. 269-278.

[3] F. Quintaes, A. O. Salazar, A. L. Maitelli, F. Fontes, M. A. A. Vieira, and T.Eslley,"Magnetic sensor used to detect contamination of inulating oil in motors applied to electrical submersible pump", IEEE Trans. Magn., vol. 47, no. 10, pp. 3756–3759, Oct. 2011.

[4] Clark D. Shaver, Sean A. Cain, "Partial discharge and partial discharge testing for ESP motors", PCIC Europe 2010 Conference Record, Oslo, 15-17 June 2010, PP. 1-7.

Bending Magnetic Levitation Control for Thin Steel Plate

(Experimental Consideration Using Sliding Mode Control)

Hikaru Yonezawa

Tokai University
Undergraduate, Dep. of Prime Mover Eng.
4-1-1, Kitakaname,Hiratsuka-shi,Kanagawa,Japan

Takayoshi Narita

Tokai University
Graduate Student, Course of Sience & Tech.
4-1-1, Kitakaname,Hiratsuka-shi,Kanagawa,Japan

Yasuo Oshinoya

Tokai University
Dep. of Prime Mover Eng.
4-1-1, Kitakaname,Hiratsuka-shi,Kanagawa,Japan
ossy@keyaki.cc.u-tokai.ac.jp

Hiroki Marumori

Tokai University
Graduate Student, Course of Mech. Eng.
4-1-1, Kitakaname,Hiratsuka-shi,Kanagawa,Japan

Shinya Hasegawa

Tokai University
Dep. of Prime Mover Eng.
4-1-1, Kitakaname,Hiratsuka-shi,Kanagawa,Japan

Abstract— **In conveyance manufacturing process of thin steel plate, since the plate always in contact with the roller, there is problem that the surface quality deteriorates over time. To solve this problem, electromagnetic levitation technologies have been studied. However, when a flexible thin steel plate with a thickness of less than 0.3 mm is to be levitated, levitation control becomes difficult because the thin plate's flexure increased. We propose a levitation of an ultrathin steel plate that bent to an extent that does not induce plastic deformation. It has been confirmed that vibrations are suppressed and levitation performance improved. In this study, in order to examine the levitation stability and to compare the levitation performance, bending levitation experiments were carried out on the basis of the optimal control and sliding mode control theory using thin steel plate with a thickness of 0.27 mm.**

Keywords— *Bending levitation control, Electromagnetic levitation system, Sliding mode control, Thin steel plate*

I. INTRODUCTION

Nowadays, thin steel plates are widely used in various products, such as automobiles and electrical appliances. It because possible to manufacture an ultra-thin steel plate in recent years. The quality of those surface is required with increasing in demand of various fields. However, in conveyance manufacturing process, thin steel plate always in contact with the roller resulting the surface quality deteriorates over time. To overcome the problem, studies of electromagnetic levitation technology have been carried out [1]-[5]. However, as the steel plate becomes thinner, the vibration caused by minute unpredictable factors, including the nonlinearity of the attractive force of the electromagnet and the change in resistance due to heat generation by the electromagnet, makes it difficult to maintain the levitation state. Furthermore, when an ultrathin steel plate with a thickness of less than 0.3 mm is targeted for levitation, levitation control becomes difficult because the thin plate undergoes increased in flexure. We propose the levitation of an ultrathin steel plate that bent to an extent that does not induce plastic deformation [6], [7]. There is no case which conducted the experiment which uses thin steel plate, or either bended thin steel plate in the past. It has been confirmed that vibrations with mainly low frequencies are generated when a steel plate is bent and levitated [8]. However, a model when steel plate is bent and levitated is not established and levitation performance is not always sufficient because of a parameter variation and modeling error. Accordingly, we aim to improve the levitation performance using sliding mode control theory which has robustness[9].

In this study, we are targetting the rectangular steel plate of the quiescent state and examined the levitation stability by conducting bending levitation experiments on the basis of the optimal control theory (OPT) and sliding mode control theory (SMC) using thin steel plates with a thickness of 0.27mm.

978-1-4799-2706-7/14 $31.00 © 2014 IEEE

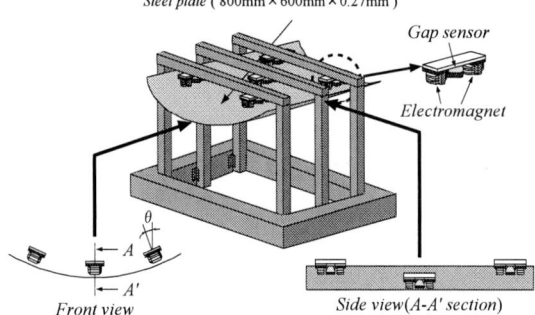

Fig. 1. Schematic illustration of experimental apparatus.

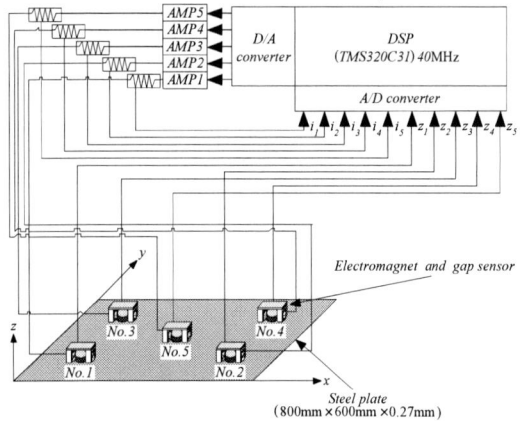

Fig. 2. Electromagnetic levitation control system.

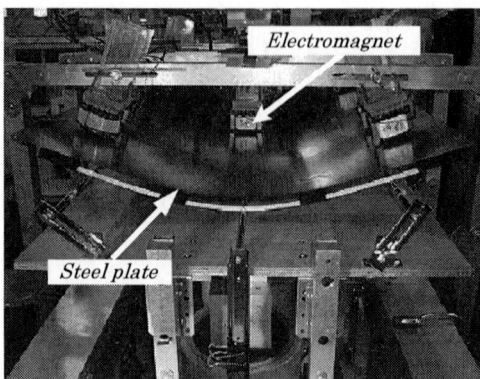

Fig. 3. Photograph of experimental apparatus.

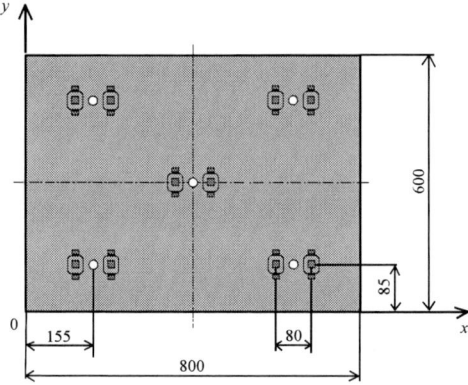

Fig. 4. Arrangement of electromagnets.

II. SYSTEM FOR CONTROL EXPERIMENT

Figure 1 and Figure 2 show an experimental apparatus and outline of the control system. Figure 3 shows a photograph of the experimental apparatus. The object of electromagnetic levitation is a rectangular zinc-coated steel plate (SS400) with length a = 800 mm, width b = 600 mm. To accomplish noncontact support of a rectangular thin steel plate using 5 pairs of electromagnets (Nos. 1~5) as if the plate was hoisted by strings, the displacement of the steel plate is measured using five eddy-current gap sensors. Here, the electric circuits of paired electromagnets are connected in series, while an eddy-current gap sensor is positioned between the two magnets of each pair. The detected displacement is converted to velocity using digital differentiation. In addition, the current in the coil of the electromagnets is calculated from the measured external resistance, and a total of 15 measured values are input into the digital signal processor (DSP) via an A/D converter to calculate the control law. A control voltage is output from the D/A converter into a current-supply amplifier to control the attractive force of the 5 pairs of electromagnets in order that the steel plate is levitated below the surface of the electromagnets by 5 mm.

Among the 5 pairs of electromagnets, the 4 pairs at the corners can be inclined, as shown in the front view in Fig.3. In addition, a central electromagnet can be moved

in vertical direction. Thus, by moving 5 electromagnets, bended magnetic levitation of the steel plate is carried out.

Figure 4 shows the arrangement of the electromagnets. Regarding electromagnets arrangement, it determined as follows. Electromagnets arrangement of the long direction (x-axis) are positioned where the attractive force which supports the steel plate is the same when setting the electromagnets tilt angle as 0 degree. Also, electromagnets arrangements of the short direction (y-axis) are positioned where the whole deflection of steel plate becomes the smallest when steel plate is bended and levitated. The positions of the outside electromagnets (Nos. 1~4) are controlled so that the attractive force is applied to the fixed positions on the thin steel plate even when the thin steel plate is bent (Fig. 5). As a tilt angle of electromagnets becomes large, restorative force of the steel plate increases. Therefore, the attractive forces of outside electromagnets become large and central electromagnet force becomes small with the increase in tilt angle. Finally, steel plate is supported with 4 pairs of outside electromagnets. In this situation, the bending angle of a supporting point is defined as a natural deflection angle. The natural deflection angle θ_N is the angle at the two supporting points when the distributed load due to gravitational force is applied to the thin steel plate under the assumption that a thin steel plate that is not supported at the center can be regarded as both-ends-free beam. The natural deflection angle θ_N is obtained as eq. (1) [10].

$$\theta_N = \frac{\rho g b}{2Eh^2}\left(b^2 - 6d^2\right) \qquad (1)$$

Where E: Young's modulus of the thin steel plate [N/m²], h: plate thickness [m], ρ: plate density [kg/m³], g: acceleration due to gravity [m/s²], b: plate width [m] and d: distance from edge of steel plate to support point [m].

From eq. (1), the natural deflection angle of a 0.27 mm steel plate is 8.5°. The attractive forces applied to each electromagnet when bending a steel plate are calculated by considering a steel plate to be a beam and computed from the relationship between the bending angle of a supporting point and the reactive force.

III. MODELING

In this model, independent control is carried out, in which information on detected values of displacement, velocity and coil current of the electromagnets under study at one position are fed back only to the same electromagnet. Accordingly, as shown in Fig. 6, the steel plate is divided into 5 hypothetical masses and each part is modeled as a lumped constant system (local model).

In an equilibrium levitation state, magnetic forces are determined so as to balance with gravity. The equation of small vertical motion around the equilibrium state of the steel plate subjected to magnetic forces is expressed as

$$m_z \ddot{z} = 2f_z \qquad (2)$$

Where $m_z = m/5$ [kg], z: vertical displacement [m], and f_z: dynamic magnetic force [N].

Figure 7 shows a schematic illustration of the electromagnet. The number of turns of the electromagnet coil are 1005 (wire diameter is 0.5 mm), and the sectional area passing the magnetic flux of the E-type core, which was made from ferrite, is 225 mm². The characteristics of the electromagnets are estimated on the basis of the following assumptions:

The permeability of the core is infinite, the eddy current inside the core is negligible, and the inductance of the electromagnetic coil is expressed as the sum of the component inversely proportional to the gap between the steel plate, magnet and the component of leakage inductance.

If deviation around the static equilibrium state is too small, the characteristic equations of the electromagnet are linearized as

$$f_z = \frac{2F_z}{Z_0}z + \frac{2F_z}{I_z}i_z \qquad (3)$$

$$\frac{d}{dt}i_z = -\frac{L_{eff}}{L_z}\cdot\frac{I_z}{Z_0^2}\dot{z} - \frac{R_z}{2L_z}i_z + \frac{1}{2L_z}v_z \qquad (4)$$

$$L_z = \frac{L_{eff}}{Z_0} + L_{lea} \qquad (5)$$

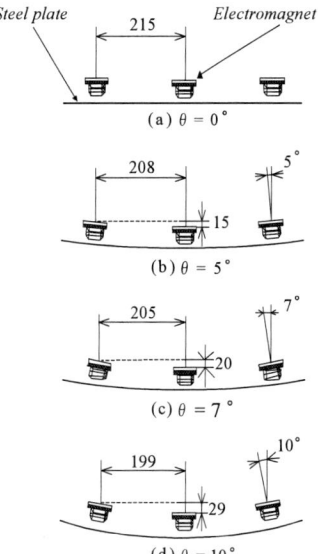

Fig. 5. Relationship between tilt angle of electromagnet and shape of steel plate.

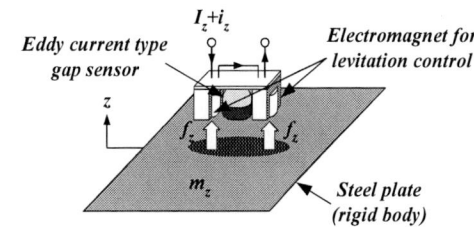

Fig. 6. Theoretical model of levitation control of the steel plate.

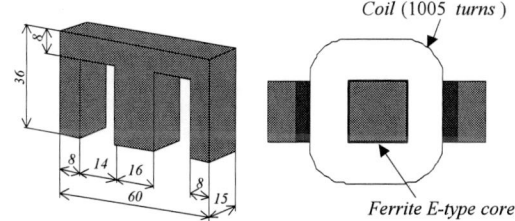

Fig. 7. Configuration of electromagnet.

Using the state vector, the eqs. (2) - (5) are written as the following state equations:

$$\dot{z} = A_z z + B_z v_z \qquad (6)$$

$$z = \begin{bmatrix} z & \dot{z} & i_z \end{bmatrix}^{\mathrm{T}} \qquad (7)$$

$$A_z = \begin{bmatrix} 0 & 1 & 0 \\ \dfrac{2F_z}{m_z Z_0} & 0 & \dfrac{2F_z}{m_z I_z} \\ 0 & -\dfrac{L_{eff}}{L_z}\cdot\dfrac{I_z}{Z_0^2} & -\dfrac{R_z}{2L_z} \end{bmatrix} \qquad (8)$$

$$B_z = \begin{bmatrix} 0 & 0 & \dfrac{1}{2L_z} \end{bmatrix}^{\mathrm{T}} \tag{9}$$

Where F_z: magnetic force of the coupled magnets in equilibrium state [N], Z_0: gap between the steel plate and electromagnet in equilibrium state [m], I_z: current of the coupled magnets in equilibrium state [A], i_z: dynamic current of the coupled magnets [A], L_z: inductance of one magnet coil in equilibrium state [H], R_z: resistance of the coupled magnet coils [Ω], v_z: dynamic voltage of the coupled magnets [V], and L_{lea}: leakage inductance of the one magnet coil [H].

IV. CONTROL THEORY

A. Optimal Control in the Discrete Time System

In this study, a control system is constructed using a discrete time system; therefore, the evaluation function of a continuous system is digitized, and the control law is obtained based on the OPT of the discrete time system. Here, the following discrete time system is considered.

$$z_d(i+1) = \Phi z_d(i) + \Gamma v_d(i) \tag{10}$$

$$\Phi = exp(A_z T_s) \tag{11}$$

$$\Gamma = \int_0^{T_s} [exp(A_z \tau)] d\tau B_z \tag{12}$$

Here, the evaluation function of the discrete time system is expressed as follows.

$$J_d = \sum_{i=0}^{\infty} \left[z_d(i)^{\mathrm{T}} Q_d z_d(i) + v_d(i)^{\mathrm{T}} r_d v_d(i) \right] \tag{13}$$

$$\begin{aligned} M = &\Phi^{\mathrm{T}} M \Phi + Q_d \\ &- \Phi^{\mathrm{T}} M \Gamma \left(r_d + \Gamma^{\mathrm{T}} M \Gamma \right)^{-1} \Gamma^{\mathrm{T}} M \Phi \end{aligned} \tag{14}$$

$$v_d^{opt} = -F_d z_d \tag{15}$$

$$F_d = \left(r_d + \Gamma^{\mathrm{T}} M \Gamma \right)^{-1} \Gamma^{\mathrm{T}} M \Phi \tag{16}$$

Where $Q_d{}^{opt}$ and $r_d{}^{opt}$ are weighting coefficients, M^{opt} is the solution of the algebraic matrix, the Riccati equation, and T_s is a sampling interval. MATLAB command " lqrd " was used to solve eq. (14) and the digital controller was designed by using SIMULINK in the DSP.

B. Sliding Mode Control in the Discrete Time System

SMC is known to be one of the most effective nonlinear control techniques, due to its invariance property to any matched uncertainties on the sliding mode surface. It has been studied extensively and widely used in practical applications because of its simplicity and robustness against parameter variations and

TABLE I
SYMBOLS AND VALUES.

Symbol	Value
m	9.72×10^{-1}kg
Z_0	5.00×10^{-3}m
$R_1 \sim R_5$	21.0Ω
ρ	7.50×10^{3}kg/m³
d	0.43m
E	206GPa
ν	0.30
L_z	1.26×10^{-1}H
L_{lea}	8.97×10^{-2}H
T_s	1.00×10^{-3}s

disturbances.

The discrete time system of eq. (10) is considered. Here we designate the switching hyperplane as S_d, and express the switching function of input as

$$\sigma(i) = S_d z_d(i) \tag{17}$$

The equivalent control input is given as

$$v_{eq}^{smc}(i) = -\left(S_d \Gamma\right)^{-1} S_d \left(\Phi - I\right) z_d(i) \tag{18}$$

Substituting eq. (18) into eq. (10), the equivalent control system can be expressed as

$$z_d(i+1) = \left\{ \Phi - \Gamma \left(S_d \Gamma\right)^{-1} S_d \left(\Phi - I\right) \right\} z_d(i) \tag{19}$$

Here, S_d should be selected such that the system represented by eq. (19) becomes stable. In this study, we used a method of utilizing the zero point of the system and applied the OPT theory of the discrete time system to obtain S_d.

$$S_d = \left(r_d^{smc} + \Gamma^{\mathrm{T}} M^{smc} \Gamma \right)^{-1} \Gamma^{\mathrm{T}} M^{smc} \Phi \tag{20}$$

Where $r_d{}^{smc}$ is weighting coefficients for control input. M^{smc} is a solution of eq. (13).

Next, a control input which converges the state into a hyperplane and generates the sliding mode is considered. In this study, we design a SMC for the discrete time system, wherein chattering is suppressed. The SMC law to satisfy this condition is described below.

$$\left. \begin{aligned} v_d^{smc}(i) &= v_{eq}^{smc}(i) + v_{nl}^{smc}(i) \\ v_{eq}^{smc}(i) &= -\left(S_d \Gamma\right)^{-1} S_d \left(\Phi - I\right) z_d(i) \\ v_{nl}^{smc}(i) &= -\left\{ \alpha(i) + \beta(i) \right\} sgn\{\sigma(i)\} \end{aligned} \right\} \tag{21}$$

$$\alpha(i) = \eta \frac{\|\sigma(i)\|}{\|S_d \Gamma\|} \tag{22}$$

$$\left(0 < \eta < 2, \ \beta(i) \ge F_{max} \right)$$

F_{max} : Maximum value of disturbance.

V. Experiment Of Bending Levitation

A. Condition of experiment

Table I shows the specifications of the system. OPT and SMC are applied for levitation control of the thin steel plate to compare the results under different electromagnet tilt angles. In the bending levitation experiment, the electromagnet tilt angle θ is increased at intervals of 5° from 0°.

In this study, the standard deviation of displacement and levitation probability are measured. The standard deviation of displacement is measured 10 times for each electromagnet tilt angle, and the mean is used as the experimental result. To avoid the effect of the transient state of the thin steel plate, the measurement is started approximately 10 s after the start of levitation. Levitation is considered successful when it continues for at least 30 s and the levitation performance is calculated as a percentage of successful levitations among 50 trials. Under these circumstances stated above, the parameters of each control system are set such that the standard deviation of displacement at sensor No.1 is within an error range of ±10% for 4.0×10^{-2} m during the application of any control method.

The weighting matrices of OPT (eq. (13)) are

$$\boldsymbol{Q}_d^{opt} = \begin{bmatrix} 1.5 \times 10^5 & 0 & 0 \\ 0 & 0.5 \times 10^{-1} & 0 \\ 0 & 0 & 5.5 \times 10^1 \end{bmatrix} \tag{23}$$

$$r_d^{opt} = 1 \times 10^{-1} \tag{24}$$

The weighting matrices (eq. (13)) used for determining the switching function $\boldsymbol{S}_d$ in SMC, and parameters in the nonlinear input terms in eq. (22) are

$$\boldsymbol{Q}_d^{smc} = \begin{bmatrix} 2.0 \times 10^5 & 0 & 0 \\ 0 & 10^{-4} & 0 \\ 0 & 0 & 2.0 \times 10^6 \end{bmatrix} \tag{25}$$

$$r_d^{smc} = 20 \times 10^{-1} \tag{26}$$

$$\eta = 0.05 \tag{27}$$

$$\beta = 1.09 \tag{28}$$

B. Levitation performance applied to OPT

Table II shows the standard deviation of displacement and the levitation probability for the tilt angle of electromagnets under OPT. Although the standard deviation of displacement increases once at tilt angle of 5°, after that it decreases and the levitation probability increases with increasing electromagnet tilt angles. At a tilt angle of 7°, the standard deviation of displacement is the smallest and levitation probability is the highest.

TABLE II
RELATIONSHIPS BETWEEN TILT ANGLE OF ELECTROMAGNETS, STANDARD DEVIATION OF DISPLACEMENT AT SENSOR No.1 AND LEVITATION PROBABILITY USING OPT.

Tilt angle of electromagnets θ [°]	Standard deviation of displacement [mm]	Levitation probability [%]
0	0.041	92
5	0.044	96
7	0.039	100
10	0.062	20

TABLE III
RELATIONSHIPS BETWEEN TILT ANGLE OF ELECTROMAGNETS, STANDARD DEVIATION OF DISPLACEMENT AT SENSOR No.1 AND LEVITATION PROBABILITY USING SMC.

Tilt angle of electromagnets θ [°]	Standard deviation of displacement [mm]	Levitation probability [%]
0	0.042	100
5	0.041	100
7	0.042	100
10	0.042	56

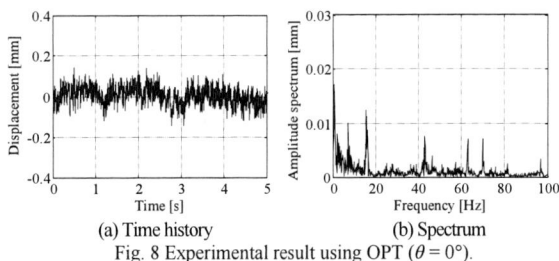

(a) Time history   (b) Spectrum
Fig. 8 Experimental result using OPT ($\theta = 0°$).

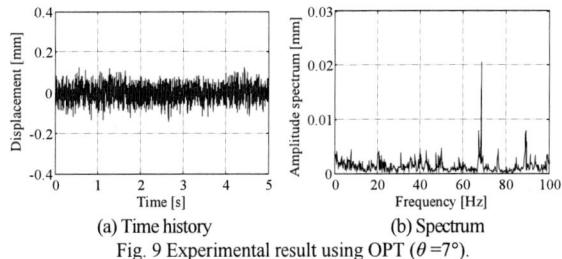

(a) Time history   (b) Spectrum
Fig. 9 Experimental result using OPT ($\theta = 7°$).

Figures 8 and 9 show the experimental results obtained when tilt angles of the electromagnets are 0° and 7°. In these figures, (a) shows the displacement of the steel plate over time, (b) shows its amplitude spectrum obtained using OPT. By bending steel plate, the plate vibrates with low frequency. The above results show that stable levitation of thin steel plate with a thickness of 0.27 mm can be realized by bending the plate under OPT. At a tilt angle of 10°, the standard deviation of displacement increases and the levitation probability decreases. The levitation probability becomes 20% at a tilt angle of 10°. The reason behind the levitation probability of 0% at a tilt angle of 10° is that a tilt angle of 10° markedly exceeds the natural deflection angle (8.5°) of the steel plate with a thickness of 0.27 mm, leading to difficulty in levitation because of a large restoring force applied to the thin steel plate. Therefore, it is effective for thin steel plate with thickness of 0.27 mm to be bent to an extent that does not exceed natural deflection angle.

Table III shows the standard deviation of displacement and the levitation probability for the tilt angle of electromagnets under SMC.

Using SMC, the standard deviations of displacement and levitation probabilities are constant regardless of tilt angle of electromagnets. When steel plate is levitated at tilt angle of 0°, the levitation probability is 100% unlike the result of the OPT. When tilt angle of electromagnets exceeds the natural deflection angle, levitation probability decreases, but standard deviation of displacement still maintains 0.04mm approximately, and shows a good levitation performance. Figures 10 and 11 show the comparison results of OPT and SMC obtained when the tilt angles of the electromagnets are 10°. From above results, the vibrations of steel plate are obviously more suppressed than OPT. In amplitude spectrum, when a steel plate is bent, while high frequency vibration near 80 Hz arises using OPT, the frequency vibration is controlled in SMC. As a result, applied to SMC, the vibration of steel plate is suppressed and the standard deviation of displacement become small rather than OPT. From Tables II and III, compared OPT with SMC at tilt angle of 7°, the standard deviations of displacement and levitation probabilities show the respectively almost same value. Thereby, when suitable tilt angle of electromagnets is known, by bending steel plate at the angle, levitation performance is improved. Additionally, applied to robust control like SMC, the outstanding levitation performance is shown irrespective of bending angle. Therefore, it is confirmed that bending steel plate is effective for levitation system and applied to SMC, levitation performance of steel plate is robust not affected with tilt angle of electromagnets. In conclusion, we proved the validity of applying robust control to a bended magnetic levitation system. Since we are not conducting the experiment on the conditions which inflicted the disturbance right now, we are going to try an experiment in the condition of having inflicted the disturbance.

VI. Conclusions

In this study, we applied bending levitation control system to optimal control and sliding mode control and examined the validity of the robust control. As a result, bending steel plate is effective for levitation system by using sliding mode control. Moreover, its levitation performance is constant regardless of tilt angle of electromagnet and higher compared to optimal control. Therefore, we confirmed the utility of sliding mode control. Since the levitation experiment which conveys a steel plate is not conducted, we are going to try transport-levitation experiment.

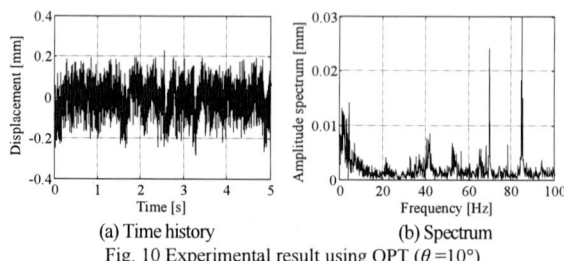

(a) Time history   (b) Spectrum

Fig. 10 Experimental result using OPT (θ =10°).

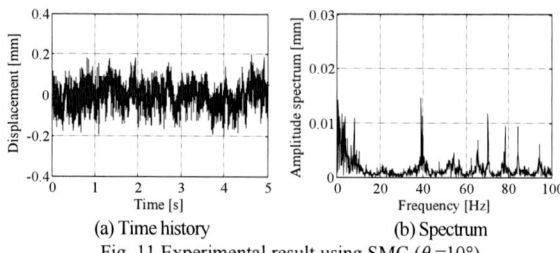

(a) Time history   (b) Spectrum

Fig. 11 Experimental result using SMC (θ =10°).

References

[1] T. Nakagawa, M. Hama, T. Furukawa, "Study of Magnetic Levitation Technique Applied to Steel Plate Production Line," *IEEE Transactions on Magnetics*, vol. 36, no. 5, pp. 3686-3689, 2000.

[2] M. Sase, S. Torii, "Magnetic levitation control with real-time vibration analysis using finite element method," *International Journal of Applied Electromagnetics and Mechanics*, vol. 13, no. 1-4, pp. 129-136, 2001/2002, (in Japanese).

[3] T. Sakurada, T. Mizuno, M. Takasaki, Y. Ishino, "Multiple Magnetic Suspension Systems (2nd Report: Realization of Double Parallel Magnetic Suspension)," *Transactions of the Japan Society of Engineers*, Series C, vol. 77, no. 779, pp. 2684-2694, 2011, (in Japanese).

[4] H. Aburano, H. Miyazaki, T. Ohji, K. Amei, M. Sakui, "Motion Tests on a Levitated Body Using a Magnetic Levitation System for Three-Dimensional Motion with Partial Zero Power Control," *Journal of the Magnetics Society of Japan*, vol. 35, no. 2, pp. 123-127, 2011, (in Japanese).

[5] F. Kubota, S. Matsumoto, Y. Arai, T. Nakagawa, "Control Techniques of Levitation and Guidance for Processing Carrying Very Thin Steel Plates," *IECON 2013-39th Annual Conference of the IEEE*, pp. 3439-3444, 2013.

[6] Y. Oshinoya, N. Nakamura, S. Hasegawa, K. Ishibashi, H. Kasuya, "Electromagnetic Levitation Control for Ultra-Thin Steel Plate (Fundamental Considerations on Stable Levitation of Bended Steel Plate)," *The 15th MAGDA Conference*, pp. 304-305, 2006, (in Japanese).

[7] Y. Oshinoya, N. Nakamura, S. Hasegawa, K. Ishibashi, H. Kasuya, "Bended Levitation Control for Ultra-Thin Steel Plate (Experimental Consideration on Levitation Performance)," *The 23rd Symposium on Electromagnetics and Dynamics*, pp. 37-38, 2007, (in Japanese).

[8] T. Masaki, K. Urakawa, T. Narita, Y. Oshinoya, S. Hasegawa, K. Ishibashi, H. Kasuya, "Bended Levitation Control for Flexible Steel Plate (Experimental Consideration on Electromagnet Position)," *The 16th Conference on Kanto Branch of The Japan Society of Mechanical Engineers*, pp. 177-178, 2010, (in Japanese).

[9] H. Marumori, T. Narita, S. Hasegawa, Y. Oshinoya, "Bending Levitation Control for Flexible Steel Plate (Experimental Consideration on Robustness)," *International Conference of the Asian Union of Magnetic Societies*, pp. 407, 2012.

[10] K. Suzuki, K. Mikata, "Material Mechanics," *Shokodo*, pp. 66-79, 1971, (in Japanese).

The 2014 International Power Electronics Conference

Transformer Winding Losses with Round Conductors for Duty-Cycle Regulated Square Waves

Kartik V Iyer, William P Robbins, Kaushik Basu and Ned Mohan
Department of Electrical and Computer Engineering
University of Minnesota
Email: iyerx070@umn.edu, robbins@umn.edu, basux017@umn.edu, mohan@umn.edu

Abstract—**One of the limiting factor in the course of reducing the size of high frequency transformer is the temperature rise. The knowledge of transformer power loss is important to make an estimate of the temperature rise. The transformer winding loss due to a duty-cycle regulated square current waveform can be estimated by summing the copper loss due to each harmonic using Dowell's formula. The paper shows that a large number of harmonics have to be considered for the loss computation. It is shown that for solid-round wire conductors the losses decrease with increasing diameter and there is no optimal diameter for which the losses are minimum. This paper presents a closed form approximate expression for power loss for a particular range of diameters of round wire that does not require a large series summation. Results from this closed form expression are shown to have reasonable accuracy in comparison to the fourier series method and also validated using 2-D finite element method.**

I. INTRODUCTION

Transformers designed with solid round-wire winding are widely used in resonant power converters, switch mode dc-dc converters, ac-ac converters [1]. High frequency transformers are compact in size and hence there is a need to correctly estimate the losses to limit the temperature rise. The losses in a transformer occur primarily in the winding and the core. The winding losses in high frequency transformer increase with frequency due to skin and proximity effects. The winding losses for foil conductors are computed in [2] by assuming a 1-D model of a transformer. The 2-D effects can increase the losses but occur only if the conductors are spaced significantly apart as shown in [3], [4]. The above mentioned applications allow the use of closely packed conductors and hence 1-D analysis can be used.

The winding loss analysis was extended to solid-round conductor by approximating the round conductors as foils [2], [5], [6]. But an orthogonality existed between skin and proximity effects which resulted in a more generalized analytical approach for loss computation of solid round conductors [7]. The method was further improved in [8]. At high frequencies both Dowell's [2] and Ferreira's [7] methods resulted in substantial errors of 60 % in eddy current loss computation for round conductors [9]. Analytical expressions using a look up table to compute the winding losses more accurately for round conductors were developed in [10]. If the layer porosity factor is high, Dowell's 1-D expression with some modifications can be used to compute the losses with reasonable accuracy [10], [11], [12]. Dowell's 1-D expression for round conductors was modified further in [13]. The modified winding loss expression was used to compute losses for a solid-round wire inductor for sinusoidal excitation [14].

Switching circuits in power electronics lead to non-sinusoidal current waveforms through transformer windings [15], [16]. The losses in the transformer windings will be more in case of non-sinusoidal currents because of harmonics. In [17], for a non-sinusoidal waveform, the winding loss at each harmonic frequency is evaluated and then summed up to give the total winding losses. In [17] winding losses are calculated for a unipolar rectangular waveform as encountered in forward converter topology. In [18], the winding losses are computed for non-sinusoidal currents in solid-round conductors. The winding losses in case of round conductors for non-sinusoidal waveforms will always be less as compared to the analytical losses computed using Dowell's approach [19], [20]. For a duty-cycle regulated square waveform the winding losses depend on a large number of harmonics [21]. In [21] an approximate expression to compute the power loss independent of harmonics for foil windings was shown. This paper computes the winding loss for duty-cycle regulated square waveform in a transformer with solid round-wire windings.

Section II shows the AC power loss expression for duty-cycle regulated square waveform and also shows that there is no optimal diameter of the conductor for which the AC power loss is minimum. A generalized relation between current and normalized diameter (independent of frequency) is also shown. Section III presents a range of normalized diameter which should be avoided. It also shows approximate power loss expressions for other ranges of normalized diameter. The approximate expression is verified using numerical computation. Section IV outlines a procedure to design the windings using round conductors for given specifications and hence compute the losses. Section V validates the proposed method with the analytical expression and FEM results. Finally, Section VI presents the conclusion.

II. WINDING POWER LOSS FOR ROUND WIRE WINDING

The ac to dc resistance ratio F_R for solid round-wire windings and sinusoidal current is [2], [14]

$$F_R = \Delta \left[\frac{\sinh(2\Delta) + \sin(2\Delta)}{\cosh(2\Delta) - \cos(2\Delta)} + \frac{2}{3}(p^2 - 1)\frac{\sinh(\Delta) - \sin(\Delta)}{\cosh(\Delta) + \cos(\Delta)} \right] \tag{1}$$

where Δ is,

$$\Delta = {}^{0.75}\sqrt{\frac{\pi}{4}}\frac{d}{\delta}\sqrt{\eta} \tag{2}$$

The diameter of solid round-wire winding is d. δ is the skin depth at the fundamental frequency and η is d/l where, l is the distance between centers of adjacent round conductors as

978-1-4799-2706-7/14 $31.00 © 2014 IEEE   3061

The 2014 International Power Electronics Conference

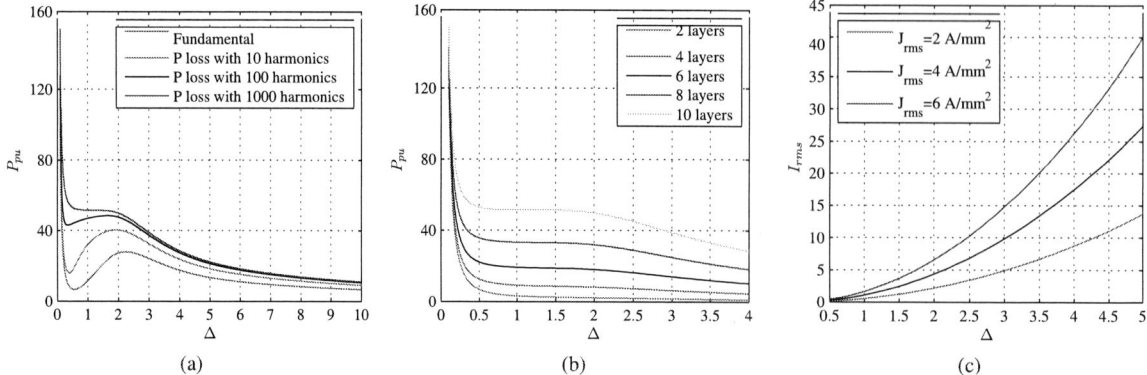

(a)   (b)   (c)

Fig. 3: (a) Plot of P_{pu} using (5) versus Δ for p=10 and D=1 for different numbers of harmonics (b) Plot of P_{pu} versus Δ for different layers for D=1 and using 1000 harmonics (c) Plot of I_{rms} versus Δ for different J_{rms}, for $f = 20kHz$ and $\eta = 0.9$

shown in Fig. 1. A bipolar duty-cycle regulated square current waveform is considered as shown in Fig. 2. This current can be represented by the following Fourier series:

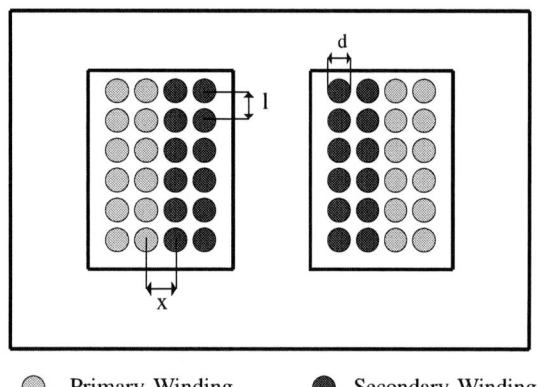

Primary Winding   Secondary Winding

Fig. 1: Solid-round wire winding of a transformer.

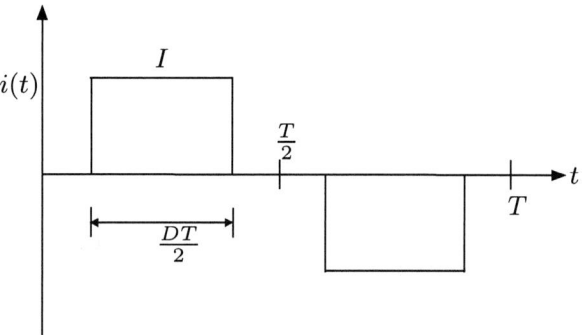

Fig. 2: Duty-cycle regulated current waveform.

$$i(t) = I_0 + \sum_{k=1}^{\infty} \sqrt{2}I_k \sin(k\omega t) \quad \text{where,} \quad I_k = \frac{2\sqrt{2}I \sin(\frac{k\pi D}{2})}{k\pi} \tag{3}$$

and as the waveform is symmetric, $I_0 = 0$. The power loss expression for solid round-wire conductors for one winding can

be written as [2],

$$P = \sum_{k=1}^{\infty} \frac{8I^2 \sin^2(\frac{k\pi D}{2})}{k^{\frac{3}{2}}\pi^2} \frac{\rho(MLT)N\eta}{\delta^2 \Delta} \sqrt{\frac{\pi}{4}}$$
$$\times \left[\frac{\sinh(2\sqrt{k}\Delta) + \sin(2\sqrt{k}\Delta)}{\cosh(2\sqrt{k}\Delta) - \cos(2\sqrt{k}\Delta)} \right.$$
$$\left. + \frac{2}{3}(p^2 - 1)\frac{\sinh(\sqrt{k}\Delta) - \sin(\sqrt{k}\Delta)}{\cosh(\sqrt{k}\Delta) + \cos(\sqrt{k}\Delta)} \right] \tag{4}$$

where, P is the AC power loss, D is the Duty ratio of square waveform, p is the number of layers, k is the harmonic number, ρ is the resistivity of copper, I is the peak value of square current, N_l is the number of turns per layer, N is the total number of turns ($= N_l p$) and MLT is the Mean Length of Turns. Here P_{base} is assumed to be, $P_{base} = \frac{I^2 \eta R_{dc}|_{d=\delta}}{\sqrt{\pi}}$ and the derivation is shown in APPENDIX. Hence,

$$\frac{P}{P_{base}} = P_{pu} = \sum_{k=1}^{\infty} \frac{\sin^2(\frac{k\pi D}{2})}{\Delta k^{\frac{3}{2}}}$$
$$\times \left[\frac{\sinh(2\sqrt{k}\Delta) + \sin(2\sqrt{k}\Delta)}{\cosh(2\sqrt{k}\Delta) - \cos(2\sqrt{k}\Delta)} \right.$$
$$\left. + \frac{2}{3}(p^2 - 1)\frac{\sinh(\sqrt{k}\Delta) - \sin(\sqrt{k}\Delta)}{\cosh(\sqrt{k}\Delta) + \cos(\sqrt{k}\Delta)} \right] \tag{5}$$

Fig. 3a, shows a plot of P_{pu} as the number of harmonics considered is increased. From Fig. 3a, it can be seen that for a sinusoidal waveform as shown in [14], there exists a valley diameter, hill diameter and a critical diameter whereas for duty-cycle regulated square waveform there is no such minimum. Fig. 3b shows a plot of P_{pu} for different layers by considering 1000 harmonics. The reduction in loss is not appreciable from Δ=0.5 to Δ=4. The implication of the results shown in Fig. 3b is that, any value of $\Delta > 0.5$ is acceptable if no other factors are considered. However, choosing $\Delta = 0.5$ so as to minimize the amount of copper is not possible in most cases because of

978-1-4799-2706-7/14 $31.00 © 2014 IEEE   3062

thermal considerations. The current density, can be written as,

$$J_{rms} = \frac{I_{rms}}{A} = \frac{I_{rms}}{\frac{\pi d^2}{4}} \tag{6}$$

Using, (2) and (6), yields,

$$\frac{I_{rms}}{J_{rms}} = \frac{2\Delta^2}{^{1.5}\sqrt{\pi}f\mu\sigma\eta} \tag{7}$$

Simple thermal considerations will yield a maximum allowable J_{rms} for a given core size. Thus for a given J_{rms} and I_{rms} in (7), there will be an associated value of Δ which may be greater than 0.5. Examples of I_{rms} versus Δ are shown in Fig. 3c for different values of J_{rms}.

III. Approximate Power Loss Expression for Different Ranges of Δ.

As shown in the previous section, a large number of harmonics have to be considered to compute the AC power loss accurately for a duty-cycle regulated square waveform. As shown in Fig. 3b there are three ranges of Δ for which P_{pu} has distinct behavior. This allows for considerable simplification in the estimation of P_{pu}.

A. Case A : $\Delta < 0.5$

Fig. 3b, shows a plot of P_{pu} for different layers. From Fig. 3b, it can be seen that the P_{pu} curve for all layers is very steep for $\Delta < 0.5$. For a fixed current I, as J is increased, Δ will reduce, resulting in less copper. Hence, operating at a value of J allowed by thermal consideration should be appropriate. But in this case, even if the current density value is permissible for $\Delta < 0.5$, the use of that diameter should be avoided to prevent significant losses. The designer can use a bigger diameter to reduce the losses drastically at the cost of increased copper or use either of litz wires or foil conductors. Solving (7), by considering that the maximum value of current density is $6A/mm^2$, the limiting value of $\Delta = 0.5$ and $\eta = 0.9$:

$$I_{rms} = 8.213/f \tag{8}$$

where, f is in kHz. As an example, for $f = 10kHz$, $I_{rms} > 0.82A$ for a current density of $J_{rms} = 6A/mm^2$. If for a particular application $I_{rms} < 0.82A$, a smaller value of current density should be chosen to avoid the value of $\Delta < 0.5$.

B. Case B: $0.5 < \Delta < 2.5$

The power loss computation for $0.5 < \Delta < 2.5$ depends on the harmonics and hence is computationally difficult. For $\Delta > 2.5$, both underlined functions in (9) become 1 as shown in [22] and hence, the power loss expression can be split as a finite sum and an infinite sum as shown in (9) below,

$$P_{pu} = \sum_{k=1}^{N_\Delta} \frac{\sin^2(\frac{k\pi D}{2})}{k^{\frac{3}{2}}\Delta} \\
\times \left[\underbrace{\frac{\sinh(2\sqrt{k}\Delta) + \sin(2\sqrt{k}\Delta)}{\cosh(2\sqrt{k}\Delta) - \cos(2\sqrt{k}\Delta)}}_{} \right. \\
\left. + \frac{2}{3}(p^2-1)\underbrace{\frac{\sinh(\sqrt{k}\Delta) - \sin(\sqrt{k}\Delta)}{\cosh(\sqrt{k}\Delta) + \cos(\sqrt{k}\Delta)}}_{} \right] \\
+ \left[\left(\frac{2p^2+1}{3\Delta}\right) \sum_{k=N_\Delta+1}^{\infty} \frac{\sin^2\left(\frac{k\pi D}{2}\right)}{k^{\frac{3}{2}}} \right] \tag{9}$$

where, N_Δ is the harmonic number beyond which the underlined functions become independent of Δ. N_Δ can be found out using, $N_\Delta = \left(\frac{2.5}{\Delta}\right)^2$. The maximum value of N_Δ is 25, corresponding to $\Delta = 0.5$. The infinite sum is a convergent function and hence the power loss can be written as,

$$P_{pu} = \sum_{k=1}^{N_\Delta} \frac{\sin^2(\frac{k\pi D}{2})}{k^{\frac{3}{2}}\Delta} \times \\
\left[\frac{\sinh(2\sqrt{k}\Delta) + \sin(2\sqrt{k}\Delta)}{\cosh(2\sqrt{k}\Delta) - \cos(2\sqrt{k}\Delta)} \right. \\
\left. + \frac{2}{3}(p^2-1)\frac{\sinh(\sqrt{k}\Delta) - \sin(\sqrt{k}\Delta)}{\cosh(\sqrt{k}\Delta) + \cos(\sqrt{k}\Delta)} - \frac{2p^2+1}{3} \right] \\
+ \left[\left(\frac{2p^2+1}{3\Delta}\right) \times S \right] \tag{10}$$

where,

$$S = \sum_{k=1}^{\infty} \frac{\sin^2\left(\frac{k\pi D}{2}\right)}{k^{\frac{3}{2}}} \tag{11}$$

TABLE I: Infinite series summation value for different D

Duty cycle	$D = 0.25$	$D = 0.5$	$D = 0.6$	$D = 0.75$	$D = 0.8$	$D = 1$
S	1.0785	1.4413	1.5336	1.6293	1.6508	1.6886

The value of the infinite sum depends on duty-ratio D and its value is known as shown in Table I. Hence only a series summation of 25 harmonics is to be considered instead of an infinite series to compute the AC losses.

C. Case C: $\Delta > 2.5$

As shown in the previous section, the P_{pu} for $\Delta > 2.5$ is simplified as under,

$$P_{pu} = \left[\left(\frac{2p^2+1}{3\Delta}\right) \sum_{k=1}^{\infty} \frac{\sin^2\left(\frac{k\pi D}{2}\right)}{k^{\frac{3}{2}}} \right] \tag{12}$$

Solving (7), by considering that the maximum value of J is $6A/mm^2$, the limiting value of $\Delta = 2.5$ and $\eta = 0.9$:

$$I_{rms} = 205/f \tag{13}$$

978-1-4799-2706-7/14 $31.00 © 2014 IEEE

where, f is in kHz. If the rms current value is greater than the specified value, then $\Delta > 2.5$. As an example, if $I_{rms} > 4.1A$ then $\Delta > 2.5$ for $f = 50kHz$, $J_{rms} = 6A/mm^2$ or less . For $\Delta > 2.5$, the computation of AC losses is extremely simple and the P_{pu}, can be expressed as under,

$$P_{pu} = S \times \left(\frac{2p^2 + 1}{3\Delta} \right) \tag{14}$$

S can be obtained from Table I. Fig. 5b shows the percentage error in power loss computation by considering 2000 harmonics using (5) and, using (14) which is independent of harmonics, for a range of Δ from $2.5 - 10$ and for $1 - 10$ layers.

IV. PROCEDURE TO COMPUTE LOSSES IN A TRANSFORMER WITH ROUND CONDUCTORS

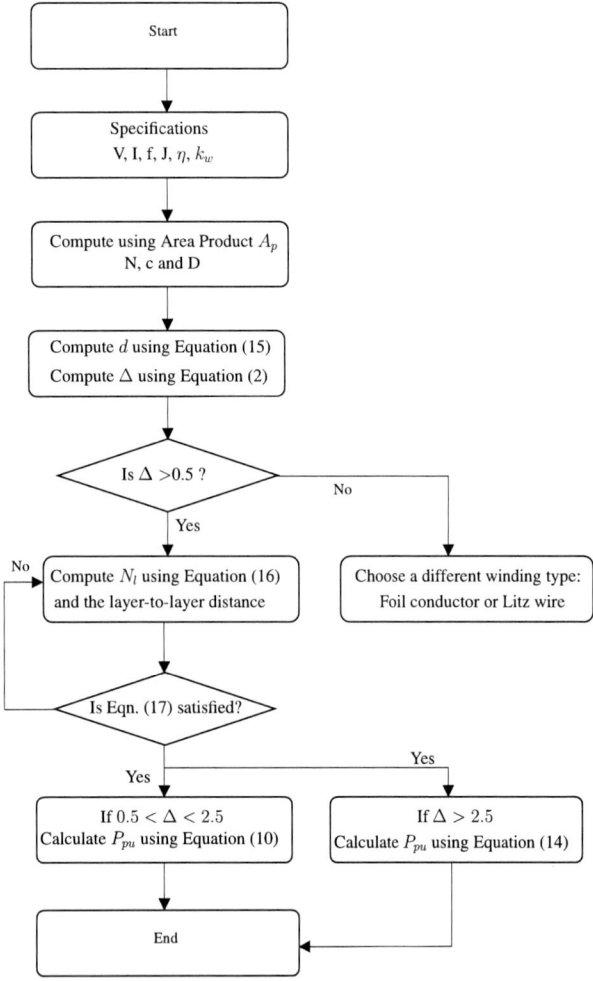

Fig. 4: A Step-by step procedure to compute winding losses in a high frequency transformer with round conductors.

For round conductors, the conductor area depends only on diameter, d. Fig. 4 shows a detailed design procedure to compute the losses in a high frequency transformer designed with solid-round conductors due to a duty-cycle regulated square waveform .

1 From the given specifications: Voltage, V, Current, I, frequency, f, Current Density, J, and window fill factor, k_w the core dimensions and the Number of turns, N can be computed using area product method.

2 For solid-round wire, the current, I and current density, J are fixed for a specific design,hence the diameter of round conductor is fixed, which can be computed using, (15). Compute Δ using (2).

$$d = \sqrt{\frac{4J}{\pi I}} \tag{15}$$

$$N_l = \frac{2\eta_1 c}{\sqrt{\pi}d} \tag{16}$$

$$\frac{2\sqrt{\pi}dp}{\eta_2} < D \tag{17}$$

where, η_2 is $\dfrac{d}{x}$, where, x is the distance between centers of two consecutive layers. For a 1:1 transformer the distance between primary and secondary layer can also be considered to be x, but for a transformer with a different transformation ratio the above simplification does not hold. For a different transformation ratio, the insulation distance between the primary and secondary layer will be more as compared to layer-to-layer insulation. η_1 is the layer porosity factor as defined in [14].

3 If $\Delta < 0.5$, then as explained in Case-A, it is better to use Litz wire or Foil conductors.

3 If not, then N_l can be determined from window height, c using (16)

4 From the number of layers, N_l compute p. For solid-round conductors the interleaving between primary and secondary windings will lead to lower losses. But if the transformation ratio is different, then the distance between primary and secondary winding will be more than x and hence it is important to check if (17) is satisfied.

   a) If yes, then go to step 5.
   b) If the inequality in (17) is not satisfied, then two or more primary layers have to be stacked together, resulting in increased losses.

5 If $0.5 < \Delta < 2.5$, compute the power loss using (10) but if $\Delta > 2.5$, use (14).

V. VALIDATION OF RESULTS

A 2-D Finite element analysis is done to verify the power loss for square waveform of duty ratio $D = 1$, with a rise time of 0.001% using ANSYS MAXWELL 16.0. Simulations are done at $10kHz$, for 2 and 4 layers and for per unit current excitation. Δ depends on the current and the frequency. The validation is done for a range of Δ from $(0.4 - 4.0)$. This allows to validate the results for a wide range of current $(0.05 - 100)$A and frequency $(5 - 100)$kHz and hence a wide range of design specifications. For different design specifications the core sizes will be different. Hence, the window and core sizes are scaled according to Δ to fit the required number of turns/layer. 60 turns for primary and secondary each and 4 values of Δ ranging from $0.4 - 4$ are considered. Fig. 5a shows the current density distribution on one side of window of an EE-core for both primary and secondary round conductors containing 4 layers

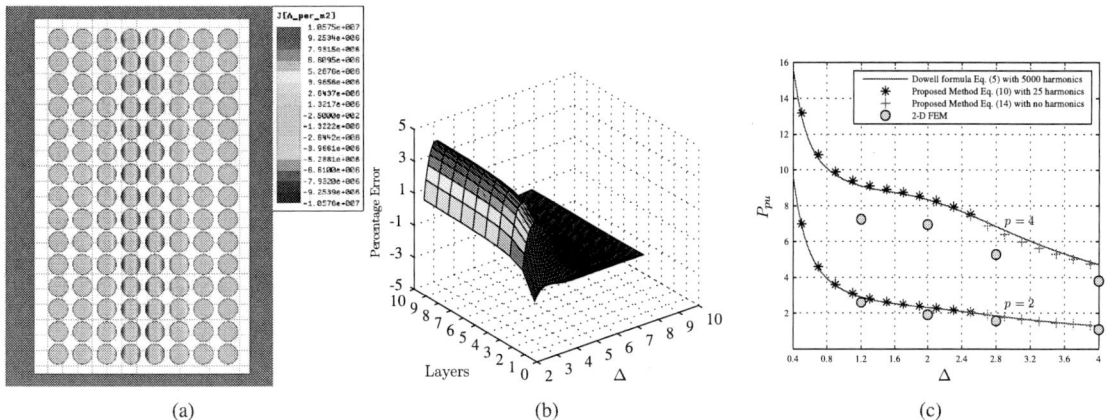

(a)   (b)   (c)

Fig. 5: Validation of results : (a) Current density distribution by 2-D FEM for 4 layers (b) Plot of percentage error in P_{pu} for 1-10 layers and $\Delta > 2.5$ by solving (5) considering 2000 harmonics and solving (14) (c) Plot of P_{pu} comparison between analytical and 2-D FEM simulation for 2 and 4 layers for $D = 1$ square waveform

TABLE II: Specifications

Parameter	Power	Voltage	Frequency, f	Current Density, J	B_{max}	k_w
Design I	5 kW	500 V	50 kHz	4 A/mm^2	0.25 T	0.4
Design II	1 kW	200 V	20 kHz	5 A/mm^2	0.3 T	0.4

each for a unit excitation of current. $\eta = 0.9$ is assumed for all simulations.

P_{pu} is computed analytically using (5) considering 5000 harmonics. P_{pu} is also computed using (10) considering 25 harmonics for $0.5 < \Delta < 2.5$ and using (14) for $\Delta > 2.5$ for 2 and 4 layers. The AC losses are also computed using 2-D FEM and converted into per unit for a square waveform excitation of $1.11072A$ corresponding to $1A$ rms fundamental. Fig. 5c show a comparison of P_{pu} computed using different methods for 2 and 4 layers respectively. As shown in Fig. 5c, the approximate expressions are quite close to the analytical values with negligible errors.

Fig. 5c validates the proposed power loss formula for a general case.

The design flowchart as given in Fig. 4 is validated for two different specifications and the winding losses are computed.

A. Design I and Design II

Fig. 6a and 6b show the 2-D model of a transformer winding designed using the round conductors. The specifications for Design I and Design II are outlined in Table II. Table III shows the core chosen for the particular design using the area product method. The number of turns in each layer and the number of layers for both primary and secondary winding are determined using Fig. 4. For Design I Δ=4.898 and as $\Delta >$2.5, P_{pu} can be computed using (14) which is 125.4 W. The losses can also be computed using (5), which gives 125 W close to the proposed method. The losses computed using 2-D FEM are 99.8 W.

For Design II as Δ=1.9488, P_{pu} is computed using (10) which is 33.2 W. The losses can also be computed using (5), which gives 33.1 W and using 2-D FEM gives 27.7 W.

The losses computed using the proposed method match closely with the analytical results but are more when compared

TABLE III: Computed Parameters for Loss Estimation

Parameter	Core	No. of Turns, N	Δ	MLT
Design I	OP45528EC	28	4.898	0.1376 m
Design II	OP44721EC	36	1.9488	0.125 m

to 2-D FEM as shown in [20] and further validated by [19]. But due to the consideration of large number of harmonics the losses computed using the proposed method are much closer to 2-D FEM results.

VI. CONCLUSION

In this paper, an approximate expression for computing AC power losses of solid round-wire transformer windings for duty-cycle regulated square waveform has been presented. The AC losses keep decreasing with increasing diameter. Hence, there is no optimum diameter of solid round-wire where the AC losses are minimum. It has been shown that for a given current density, depending on the frequency, there is a maximum current value below which the AC losses will be higher. For currents below the maximum, it is advisable to operate at lower current densities to achieve reduced losses at the cost of increased copper. To compute power loss for currents above the maximum current value, the infinite sum is represented as two finite summations. In this paper, it is also shown that for a given current density, depending on the frequency, there is a minimum current value above which the AC power loss can be represented by a simplified expression independent of harmonics. The accuracy of the expression is verified using numerical computation as well as with 2-D FEM.

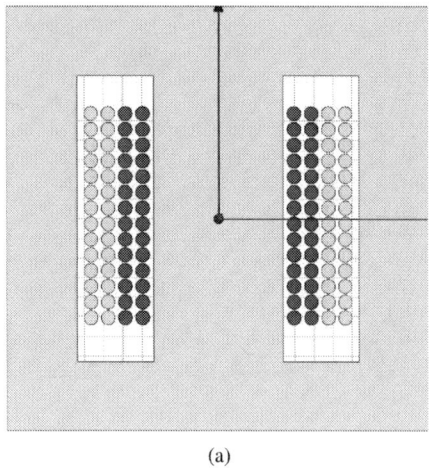 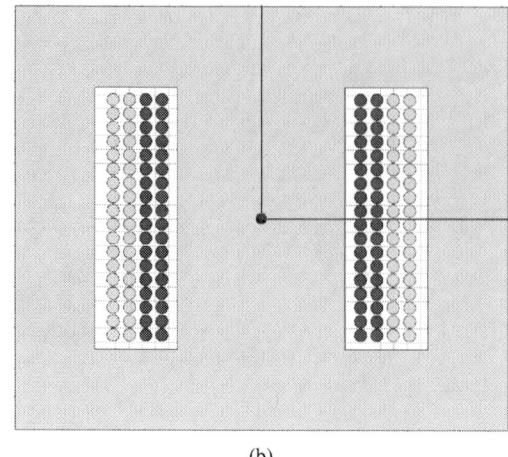

(a)   (b)

Fig. 6: 2-D Transformer model with round conductors: (red)-primary winding, (green)-secondary winding: (a) Design I (c) Design II

APPENDIX

The dc resistance of a transformer containing N multiple turns at a diameter equal to the skin depth of the conductor is given in (A.1). Now P_{base} given by (A.2) can be written in terms of $R_{dc}|_{d=\delta}$ as shown in (A.3).

$$R_{dc}|_{d=\delta} = \frac{N\rho(MLT)}{\frac{\pi}{4}\delta^2} \tag{A.1}$$

$$P_{base} = \frac{8I^2}{\pi^2} \frac{N\rho(MLT)\eta}{\delta^2} \sqrt{\frac{\pi}{4}} \tag{A.2}$$

$$P_{base} = \frac{I^2 \eta R_{dc}|_{d=\delta}}{\sqrt{\pi}} \tag{A.3}$$

REFERENCES

[1] N. Mohan and T. M. Undeland, *Power electronics: converters, applications, and design.* Wiley. com, 2007.

[2] P. Dowell, "Effects of eddy currents in transformer windings," *Electrical Engineers, Proceedings of the Institution of*, vol. 113, no. 8, pp. 1387 –1394, august 1966.

[3] F. Robert, J. Sprooten, P. Mathys, J.-P. Schauwers, and B. Velaerts, "Eddy current losses in smps transformers round wire windings: a semi-analytical closed-form formula," in *Power Electronics and Applications, 2005 European Conference on*, 2005, pp. 9 pp.–P.9.

[4] M. Dale and C. Sullivan, "General comparison of power loss in single-layer and multi-layer windings," in *Power Electronics Specialists Conference, 2005. PESC '05. IEEE 36th*, june 2005, pp. 582 –589.

[5] J.-P. Vandelac and P. Ziogas, "A novel approach for minimizing high-frequency transformer copper losses," *Power Electronics, IEEE Transactions on*, vol. 3, no. 3, pp. 266–277, 1988.

[6] M. Perry, "Multiple layer series connected winding design for minimum losses," *Power Apparatus and Systems, IEEE Transactions on*, vol. PAS-98, no. 1, pp. 116–123, 1979.

[7] J. Ferreira, "Aring;nalytical computation of ac resistance of round and rect-angular litz wire windings," *Electric Power Applications, IEE Proceedings B*, vol. 139, no. 1, pp. 21–25, 1992.

[8] Ferreira, "Improved analytical modeling of conductive losses in magnetic components," *Power Electronics, IEEE Transactions on*, vol. 9, no. 1, pp. 127–131, 1994.

[9] X. Nan and C. Sullivan, "Simplified high-accuracy calculation of eddy-current loss in round-wire windings," in *Power Electronics Specialists Conference, 2004. PESC 04. 2004 IEEE 35th Annual*, vol. 2, 2004, pp. 873–879 Vol.2.

[10] Nan and C. Sullivan, "An improved calculation of proximity-effect loss in high-frequency windings of round conductors," in *Power Electronics Specialist Conference, 2003. PESC '03. 2003 IEEE 34th Annual*, vol. 2, 2003, pp. 853–860 vol.2.

[11] F. Robert, "A theoretical discussion about the layer copper factor used in winding losses calculation," *Magnetics, IEEE Transactions on*, vol. 38, no. 5, pp. 3177–3179, 2002.

[12] E. C. Snelling, *Soft Ferrites: Properties and Applications.* U.K.: Illife Books, 1969.

[13] M. K. Kazimierczuk, *High-frequency magnetic components.* Wiley. com, 2009.

[14] R. Wojda and M. Kazimierczuk, "Analytical optimization of solid x2013;round-wire windings," *Industrial Electronics, IEEE Transactions on*, vol. 60, no. 3, pp. 1033–1041, 2013.

[15] C. Nan and R. Ayyanar, "Dual active bridge converter with pwm control for solid state transformer application," in *Energy Conversion Congress and Exposition (ECCE), 2013 IEEE*, 2013, pp. 4747–4753.

[16] K. D. Hoang and J. Wang, "Design optimization of high frequency transformer for dual active bridge dc-dc converter," in *Electrical Machines (ICEM), 2012 XXth International Conference on*, 2012, pp. 2311–2317.

[17] P. Venkatraman, "Winding eddy current losses in switch mode power trans-formers due to rectangular wave currents," in *Proceedings of Powercon*, vol. 11, 1984, pp. 1–11.

[18] R. Wojda and M. Kazimierczuk, "Analytical optimisation of solid-round-wire windings conducting dc and ac non-sinusoidal periodic currents," *Power Electronics, IET*, vol. 6, no. 7, pp. 1462–1474, 2013.

[19] F. Robert, P. Mathys, and J.-P. Schauwers, "Ohmic losses calculation in smps transformers: numerical study of dowell's approach accuracy," *Magnetics, IEEE Transactions on*, vol. 34, no. 4, pp. 1255–1257, 1998.

[20] S. Crepaz, "Eddy-current losses in rectifier transformers," *Power Apparatus and Systems, IEEE Transactions on*, vol. PAS-89, no. 7, pp. 1651 –1656, sept. 1970.

[21] K. Iyer, K. Basu, W. Robbins, and N. Mohan, "Determination of the optimal thickness for a multi-layer transformer winding," in *Energy Conversion Congress and Exposition (ECCE), 2013 IEEE*, 2013.

[22] W. Hurley, E. Gath, and J. Breslin, "Optimizing the ac resistance of multilayer transformer windings with arbitrary current waveforms," *Power Electronics, IEEE Transactions on*, vol. 15, no. 2, pp. 369 –376, mar 2000.

978-1-4799-2706-7/14 $31.00 © 2014 IEEE

The 2014 International Power Electronics Conference

Simulation of Resin Molded Type Sensor in Pole Switch for Power Delibery Systems

Tatsuya Furukawa, Shoichiro Muta, Hisao Fukumoto, Hideaki Itoh
Graduate School of Science and Engineering
Saga University
Saga, JAPAN

Masashi Ohchi
Faculty of Engineering
Chiba Institute of Technology
Chiba, JAPAN

Abstract—**The current sensor of the voltage–current sensor system proposed by us is two air–core sensing coils connected in series to each other with an opposite polarity. By doing so, it can measure the harmonics and power factor while removing the influence from power lines of other phases. When the sensors are surrounded by an iron case, emf may be influenced. However, it has been confirmed that a large change is not observed with regard to measurement of harmonics, and the proposed sensors can be used without problem. Therefore it was found out that the proposed sensor is able to perform power measurements in a state of being incorporated in the pole switch.**

I. Introduction

In recent years, with the development of smart grids and electric power selling by consumers, observations of the power state in the power delivery system are strongly required. Therefore a pole switch equipped with a voltage–current sensor has been proposed. However, it has not been possible to measure the harmonics since those current sensor have adopted iron cores [1]. In contrast, the sensor proposed in our laboratory is two air–core sensing coils connected in series to each other with an opposite polarity [2],[3]. By doing so, it can measure the harmonics and power factor while removing the influence from power lines of other phases [4]. However, the resin molded type voltage–current sensor is not suitable to permanent use because of its physical dimensions. Also, it takes time to set up the sensor each time of measurement. In this study, the resin molded type voltage–current sensor is incorporated into a pole switch in order to grasp the power states. In this paper, we will evaluate the effectiveness of the proposed sensor by performing a simulation in which we put it in an iron case. The three dimensional FEM simulation is conducted taking into account three–phase alternating current.

II. Concept of Phase Shift Compensation

Figure 1 shows the resin molded type of voltage–current sensor proposed by us. In the present sensing system, a power line will be located between two search coils, which are connected to each other with both electrically and magnetically opposite polarities as shown in Fig. 2.

Figure 2 (a) shows the flux linkage around the u–phase current sensor and its equivalent circuit in open circuit condition when only the u–phase power line is active and the sensor coils are assumed to be concentric, that is, dimensionless. Also

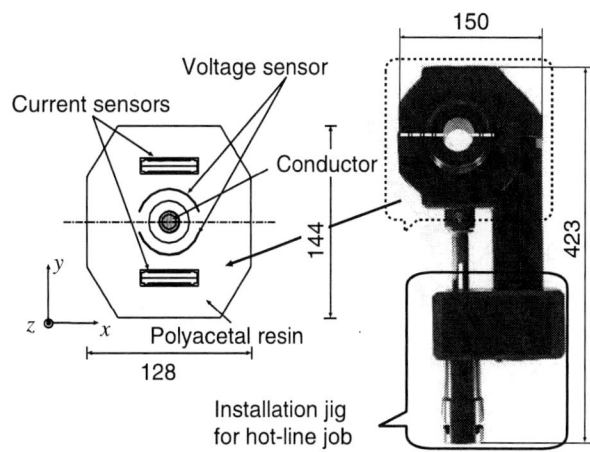

Fig. 1. Preproduction prototype of resin molded type of voltage–current sensor.

Fig. 2 (b) shows the flux linkage due to only the u–phase current around the v–phase current sensor and its equivalent circuit in the same condition in Fig. 2 (a). The arrows in Fig. 2 indicate the flux density and the inducted emf approximately proportional to the practical magnitudes. The emf induced by the flux due to the self phase current will be added in–phase but the induced emfs by the flux due to other phase currents will be offset owing to connecting with opposite polarities.

From Faraday's electromagnetic induction law, emf is proportional to the derivative of the magnetic flux linkage Φ.

$$e = -\frac{d\Phi}{dt} \tag{1}$$

From the relationship in (1), the current sensor outputs a voltage value that is obtained by differentiating the power line current. Therefore, it is necessary to use an integrating circuit when measuring the actual waveform of the current using the sensor.

Figure 3 shows a cross–sectional view of the power line and the current sensor. There is an airspace in the center of the bobbin. The number of winding is 2,000 turns. Thickness of the trunking that covers the distribution line is 2.5 [mm].

978-1-4799-2706-7/14 $31.00 © 2014 IEEE

The 2014 International Power Electronics Conference

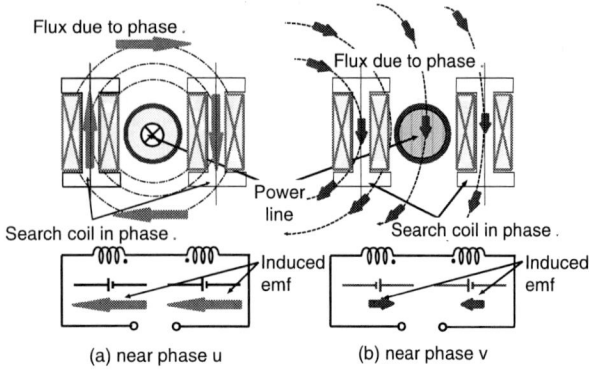

Fig. 2. Schematic flux linkage in u–phase and v–phase sensors.

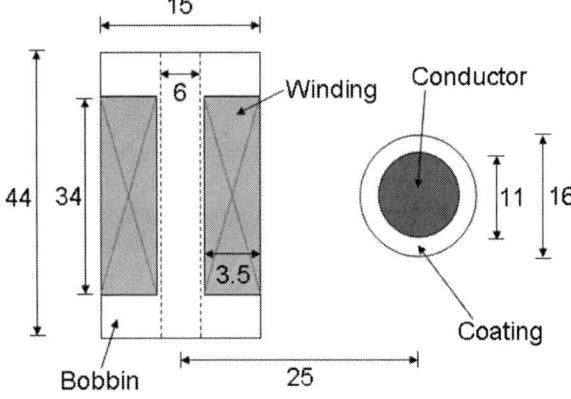

Fig. 3. Size of model.

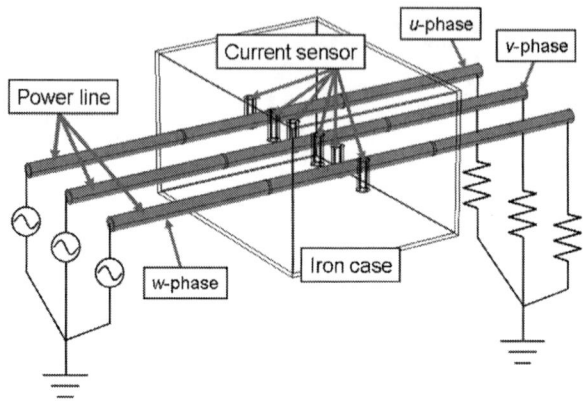

Fig. 4. Analysis model (vertical).

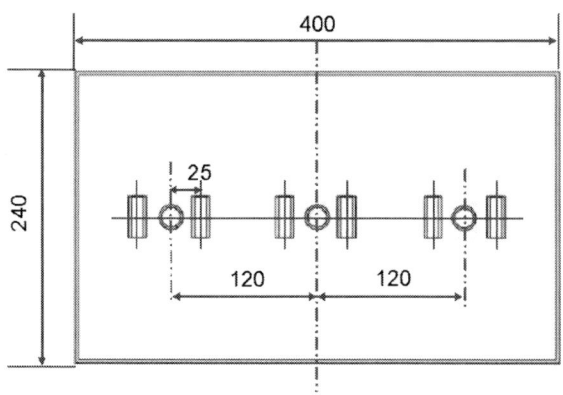

Fig. 5. Front view of the iron case.

III. PATTERN TO BE INSTALLED TO THE LEFT AND RIGHT SENSOR

A. Analysis Model

Figure 4 shows a model for analysis. The long cylindrical three objects represent power lines, and the short cylindrical six objects represent the current sensors. The sensors are placed on the left and right sides of the power lines in Fig. 4. We will call this model "vertical". The interior of the pole switch has the air region. Each sensor seems to be independent in Fig. 4, in fact, the sensors adjacent to each other are connected as shown in Fig. 2, that is, the three–phase circuits have been solved simultaneously with FEM analysis.

Three–phase voltages are applied to the power lines. The frequency is 60 [Hz] and the line voltage is 6.6 [kV]. It is about 3.8 [kV] in terms of the phase voltage. In addition, we connected resistors of 12.7 [Ω] to the power lines so as to turn on the electricity of 300 [A]. Factors such as capacitance to ground are ignored because it is thought that they do not have a significant effect on this magnetic simulation.

Figure 5 shows the front view of the iron case. Figure 6 shows the top view of the model. The distance between the power lines is 120 [mm]. The cuboid surrounding the sensor represents the simulated pole switch whose length, width and

height are 300 [mm], 400 [mm] and 240 [mm] respectively. The material of the pole switch is iron whose thickness is set to 3 [mm].

Figure 7 shows the mesh of the model. We divided the model into volume elements and analyzed using a three dimensional finite element method. The number of tetrahedral elements are about 254,700.

To avoid the remeshing of the whole analysis domain, the existence of the switch case will be modeled by the specific magnetic permeability, whose value takes 500 or 1. The trunkings of the distribution lines are assumed to be made of polyethylene whose specific magnetic permeability was set to 2.3.

B. Analysis Result

First, we have investigated the effect of the presence or absence of the iron case. We will show only the v–phase case below because other phases are similar. Second, we have investigated the effect of the presence or absence of the harmonics.

1) Presence or absence of the iron case: Figure 8 shows the results of analyzing the relationship between the induced

The 2014 International Power Electronics Conference

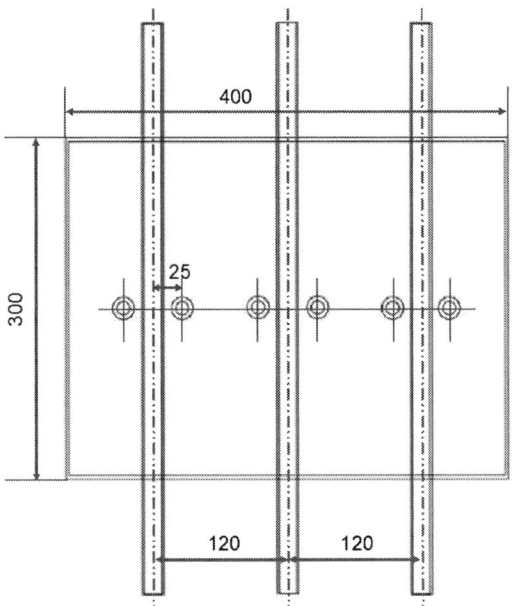

Fig. 6.   Top view of the iron case.

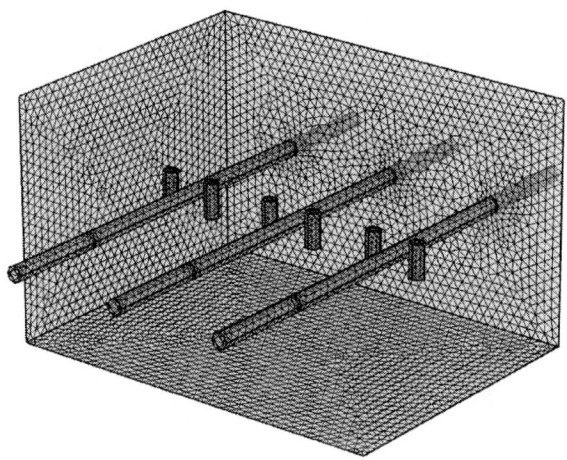

Fig. 7.   Mesh of the model (vertical).

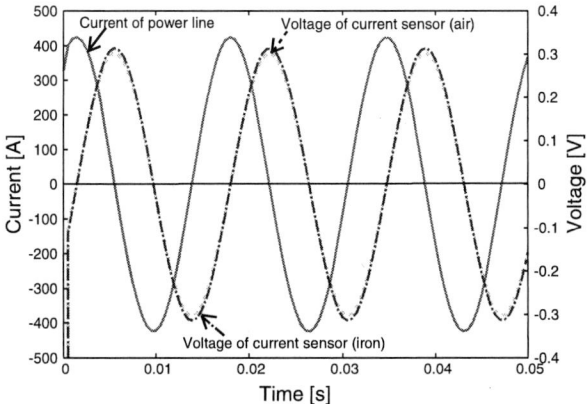

Fig. 8.   Output voltage of current sensor (vertical).

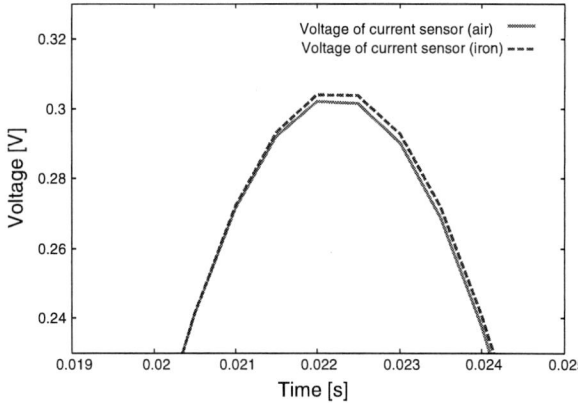

Fig. 9.   The graph of the expanded Fig. 8.

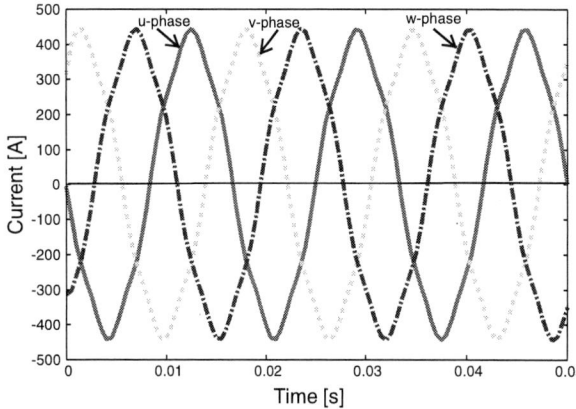

Fig. 10.   Three–phase alternating current, including 5% of the fifth harmonics.

emf generated by the current sensors and the current flowing through the power lines. In Fig. 8, the solid line indicates the current flowing through the v–phase power line, the broken line indicates the emf of sensor when not using the iron case, and the dashed line indicates the emf of sensor when using the iron case. For the phase of the emf, it can be seen that it is delayed by 90 degrees compared to the current in power line regardless of the presence or absence of the iron case. Figure 9 is an expanded figure of Fig. 8. For the magnitude of emf, the maximum value was increased by approximately 0.66% when using the iron case.

2) Presence or absence of the harmonics: Figure 10 shows the applied currents. This currents contain the 5% 5th harmonics. Figure 11 shows the emf of sensors when not using the iron case. Figure 12 shows the emf of sensors when using the iron case.

The solid line indicates the u–phase power line, the broken line indicates the v–phase power line, and the dashed line indicates the w–phase power line.

For the phase of emf, as if it does not contain harmonics, it is delayed by 90 degrees compared to the current regardless of the iron case. For the magnitude of emf, the maximum

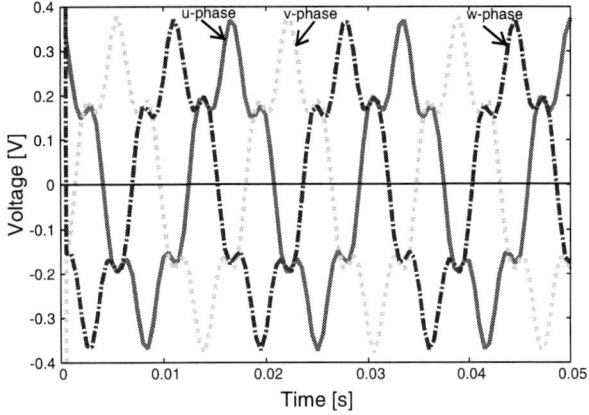

Fig. 11. Comparison of *emf* and harmonic current not using the iron case (vertical).

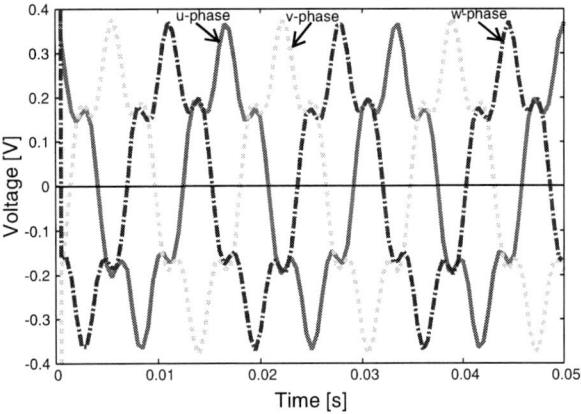

Fig. 12. Comparison of *emf* and harmonic current using the iron case (vertical).

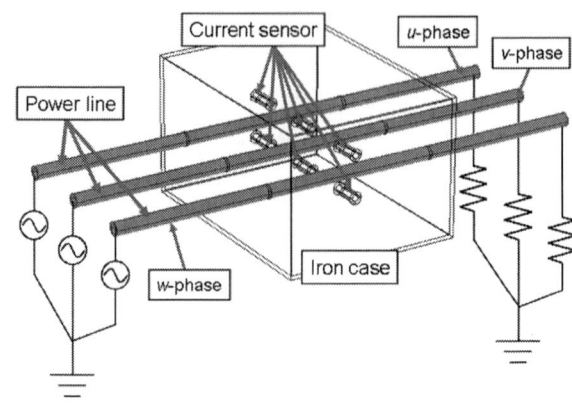

Fig. 13. Analysis model (horizontal).

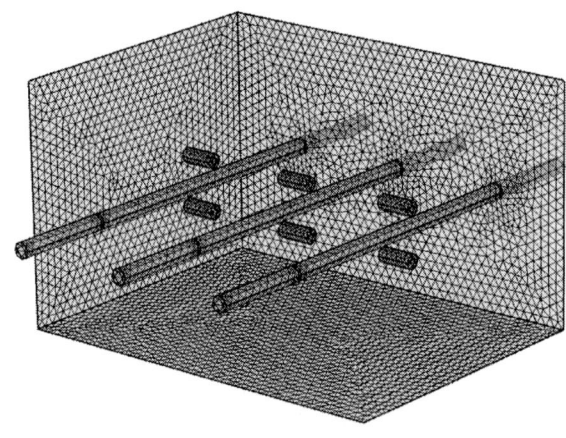

Fig. 14. Mesh of the model (horizontal).

value was increased by approximately 1.3% when using the iron case. Further, regardless of the iron case, the waveform of the *emf* is distorted more as compared to the current. It can be said that it is a phenomenon which is obtained by differentiating the wave including harmonics. Thus, the harmonics of the current can be made remarkable by differentiating.

IV. PATTERN TO BE INSTALLED UP AND DOWN THE SENSOR

A. Analysis Model

Figure 13 shows a model used in the analysis. The sensors are placed up and down of the power lines in Fig. 11, not like in Fig. 4. We will call this model "horizontall". Figure 14 shows the mesh of the model. The number of tetrahedral elements are about 255,300.

B. Analysis Result

As before, first we investigated the effect of the presence or absence of the iron case on the *v*–phase. Second, we

investigated the effect of the presence or absence of the harmonics.

1) Presence or absence of the iron case: Figure 15 shows the results of analyzing the relationship between the induced *emf* generated by the current sensors and the current flowing through the power lines. In Fig. 15, the solid line indicates the current flowing through the *v*–phase power line, the broken line indicates the *emf* of sensors when not using the iron case, and the dashed line indicates the *emf* of sensors when using the iron case. For the phase of *emf*, the same result as that the case in "vertical" model was obtained. Figure 16 is an expanded figure of Fig. 15. For the magnitude of *emf*, the maximum value was decreased by approximately 0.91% when using the iron case.

2) Presence or absence of the harmonics: Figure 17 shows the *emf* of sensors when not using the iron case. Figure 18 shows the *emf* of sensors when using the iron case. Assume that same current as shown in Fig. 8 is applied. The solid line indicates the *u*–phase power line, the broken line indicates the *v*–phase power line, and the dashed line indicates the *w*–phase power line.

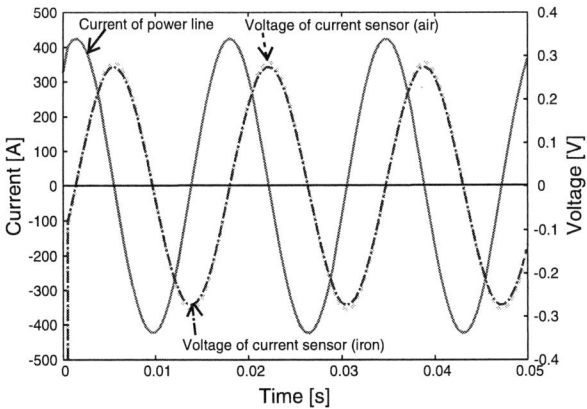

Fig. 15. Output voltage of current sensor (horizontal).

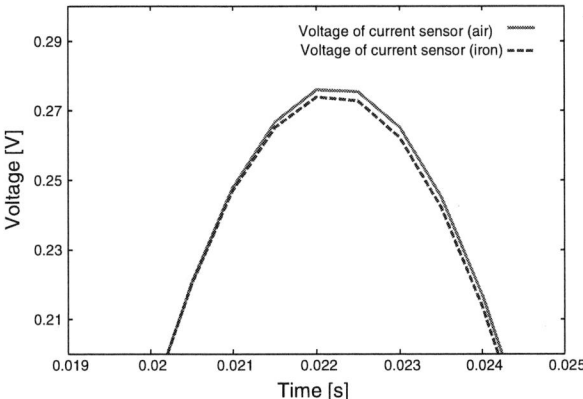

Fig. 16. The graph of the expanded Fig. 15.

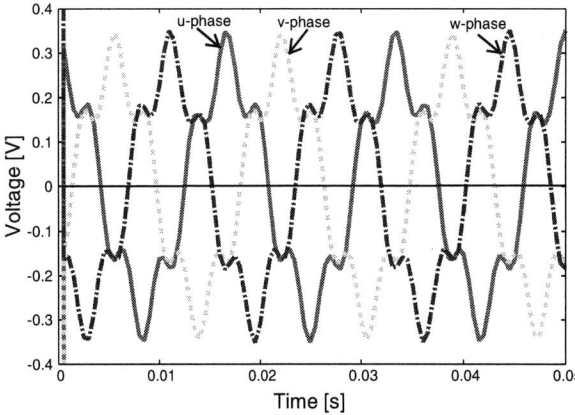

Fig. 17. Comparison of *emf* and harmonic current not using the iron case (horizontal).

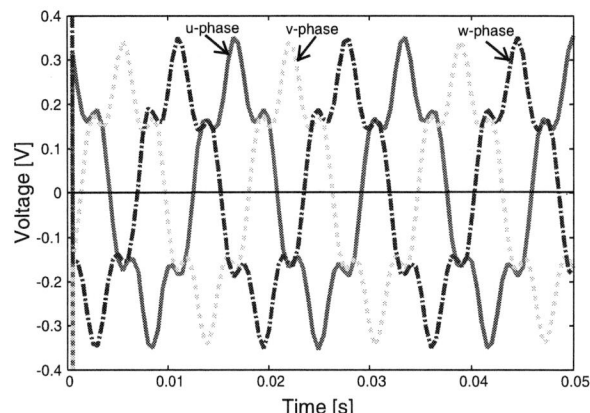

Fig. 18. Comparison of *emf* and harmonic current using the iron case (horizontal).

For the phase of emf, one can view the ideal result that it was delayed by 90 degrees. For the magnitude of emf, the maximum value was decreased by approximately 2.5% when using the iron case.

V. CONCLUSION

In this study, we have modeled the simplest 6.6 [kV] three–phase power lines, and simulated the proposed current sensor with and without the iron case. When the sensors are surrounded by the iron case, emf will vary slightly [5]. However, it is considered to be able to reconstruct the current waveform without problems because that change is almost negligible. Further, it is considered that a large change was not observed with regard to measurement of harmonics, and the proposed sensors could be used without problem. Furthermore, it was found that results are not affected by the geometry of the sensors. In the future, we will plan to perform a simultaneous analysis of the voltage sensor and current sensor in all three phases.

ACKNOWLEDGMENT

Thanks to Mr. M. Hashiguchi, Keisoku Engineering System Co. Ltd. for useful discussion of the FEM analysis.

REFERENCES

[1] http://www.ineltek.com/de/files/Current_Transform ers&Sensors.pdf

[2] T. Furukawa, M. Ashikawa and M. Ohchi, "Steady State Analysis of Resin Molded Type Voltage–Current Sensor for Real–Time Observation of Power Factor in Power Distribution System", *IEEJ Trans. PE*, Vol. 128, No. 6, pp. 811–819, 2008.

[3] T. Furukawa, M. Hirakawa, H. Fukumoto, S. Yoshino and M. Ohchi, "Experimental Proof of Voltage–Current Waveform Sensor Output of Resin Molded Type for Measurement of Power Factor and Harmonics in General Power Distribution System", *IEEJ Trans. FM*, Vol. 129, No. 4, pp. 197–204, 2009 (In Japanese).

[4] T.Kubo, T. Furukawa, H. Itoh, H. Fukumoto, H. Wakuya and M. Ohchi, "Numerical Electric Field Analysis of Power Status Sensor Observing Power Distributing System Taking into Account of Voltage Divider Measurement", *IEEJ Trans. FM*, Vol. 131, No. 3, pp. 171–177, 2011 (In Japanese).

[5] S. Muta, T. Furukawa, H. Fukumoto, H. Itoh, M. Ohch, D. Hirabayashi, "Evaluation Simulation of Resin Molded Type Current Sensor in Iron Case", *The 66th Joint Conference of Electrical and Electronics Engineers in Kyushu*, 05–1P–03, CD–ROM, 2013 (In Japanese).

Robust Startup Control of Sensorless PMSM Drives with Self-Commissioning

Chiao-Chien Lin
Institute of Electrical and Control Engineering
National Chiao Tung Univ.
Hsinchu, Taiwan
straightflush808@hotmail.com

Ying-Yu Tzou
Institute of Electrical and Control Engineering
National Chiao Tung Univ.
Hsinchu, Taiwan
yytzou@mail.nctu.edu.tw

Abstract—This paper presents a robust sensorless startup control method with self-commissioning for permanent magnet motors for smooth ramping speed up without reversing oscillations. Low cost sinusoidal PMSM motors used for fan and blower applications may have large parameter variations and static friction uncertainties. A reliable startup control with smooth operation thus becomes an indispensable requirement for practical applications. A self-commissioning strategy is developed to identify the fan motor parameters when a new application is initiated. These identified parameters are stored in an EEPROM and are used for the field oriented sensorless control of the of the PMSM motor drive. Based on the identified motor parameters, a start-up scheme will apply a series of voltage pulses with proper time durations to each stator winding and the corresponding peak currents are measured for the detection of initial commutation state of the rotor. The proposed scheme will not cause visible rotor vibrations and audio noises during the detection process. The sensorless startup with self-commissioning scheme has been implemented based on a single-chip MCU controller (STM32F0). Experimental results reveal that the proposed scheme can achieve reliable and smooth startup with robust performances.

Index Terms—*Initial rotor position detection, permanent magnet motors, self-commissioning, sensorless start-up control.*

I. INTRODUCTION

Permanent magnet synchronous motors (PMSMs) are widely used in household and industrial applications on account of high efficiency, high power density, and maintenance free. However, conventional PMSM drive systems require Hall sensors or encoders to provide necessary information of the rotor position. These Hall sensors are sensitive to temperature variations, and need to be installed with proper arrangement. Sensorless techniques can be used to eliminate the installation of rotor position sensors.

One major concerned issue for the practical applications of sensorless control of PMSMs is the startup control when the rotor is at standstill or in rotation state due to external forces. Most sensorless control methods are based on back-EMF estimation with measured phase currents and output phase voltages. However, these signals are either unmeasurable or too

small for reliable position estimation when the rotor is at standstill or under low speed operation. A simple solution is to use piecewise linear ramp up V/Hz control with a forced alignment of the rotor. The motor is running under open loop control during this ramp up process, the control is then switched to sensorless control when certain predefined conditions are satisfied. However, this startup control scheme is highly sensitive to motor parameters and load conditions, and therefore, require time consuming tuning in practical applications. Another shortcoming of this startup control scheme is that during the alignment process there may sometimes causes a reversing rotation or temporary vibrations, and this is unacceptable in some applications. Therefore, the initial position information of the permanent magnet at standstill is essential for a sensorless startup control without reversing rotation.

During the past several years, lots of researches have been carried out for the initial position detection of a PMSM with salient or non-salient rotor. Most initial rotor estimation methods are based on the saturation effect of stator iron core of the non-salient PMSMs [1]-[4]. In reference [5] the initial position is identified by the phase angle of short-circuit current vector and rotor frequency by the derivative of each phase angle of current vectors among the three short-circuit durations. However, this method suffers from environment noises and can not provide reliable operation for noise corrupted environment. Another initial position detection method [6] detects the initial position by the time periods of discharge of stator windings, which are excited before discharge. The procedure in [7] combines an iterative sequence of voltage pulses with a fuzzy logic processing of the current responses and phase currents derivation based on the DC-link current measurements. These methods, however, are relatively complicated and sensitive to the indirect measured quantities.

This paper presents a robust sensorless startup control method with self-commissioning for permanent magnet motors for smooth ramping speed up without reversing oscillations. The proposed scheme can detect the initial position with a resolution of 60 electrical degrees at standstill. In order to cope with wide application ranges, a self-commissioning strategy is developed to combine with the startup control scheme to achieve robust

The 2014 International Power Electronics Conference

performance [8], [9]. During the self-commissioning process, all tests and parameter estimation procedures are executed before its normal operation. In contrast, on-line parameter identification method [10] has the advantage that the motor parameters are identified recursively in real time during its normal operation. However, the on-line method requires heavy computation and is not suitable for motor drive with consistent operating condition.

This paper describes the implementation issues for the startup control of sensorless PMSM drive with uncertain parameters. Fig. 1 shows the block diagram of the proposed robust startup method with self-commissioning procedure for the sensorless FOC control of a PMSM. The organization of this paper is as follows. Sec. II introduces the self-commissioning strategy, Sec. III describes the initial rotor position estimation, Sec. IV is the starting procedure, Sec. V gives the experimental results, and Sec. VI is the conclusion.

II. SELF-COMMISSIONING STRATEGY

In order to meet high performance and requirements for universal drives, a self-commissioning strategy has been developed. Fig. 2 shows the flowchart of the proposed self-commissioning procedure for the sensorless drive. The electrical drive is started with a power on initialization and then executes a self-diagnostic routine. If the self diagnosis is passed, the system will check whether it is required to initiate the self-commissioning process for a new application.

The self-commissioning process includes initial setting of necessary motor parameters, identification of the unknown motor parameters and determination of initial control parameters. The self-commissioning is activated when a new application is applied to same type PMSMs with similar load characteristics. The number of poles (N) can not be identified without the rotor mechanical position information. However, the number of poles can be easily measured with an off-line method and can be set as an initial parameter.

The resistance and inductance of stator windings are on-line identified with periodical PWM waveforms. Estimated average value and variance of the measured phase currents are used to calculate the resistance and inductance. The back EMF constant can be identified with a free running rotation by measuring the terminal voltages. Once the sensorless drive completes its self-commissioning process, it will stay in drive ready mode of the background running task and wait for the motion command.

III. INITIAL ROTOR POSITION ESTIMATION

A. Basic Principle

For a PMSM with saliency, the inductances of the d-q axes are different and it is easier to measure the inductance variations by detecting the peak value of phase current with a series of proper arranged voltage pulses. However, for PMSMs without saliency, the inductances of the d-q axes are almost the same, and it

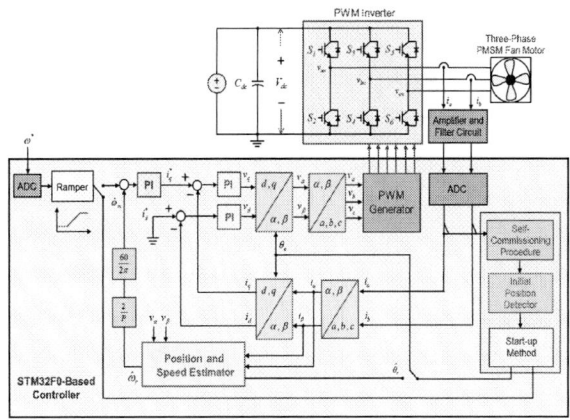

Fig. 1. Block diagram of proposed sensorless drives system.

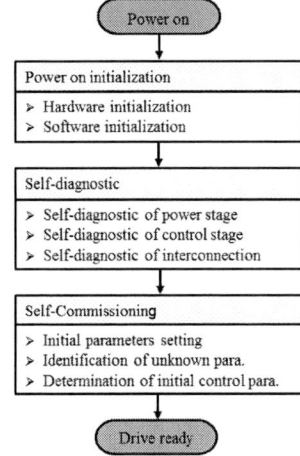

Fig. 2. Self-commissioning procedure for an electrical drive.

becomes more difficult for the detection rotor position by measuring the inductance variations. The magnetic field of the rotor will produce extra fluxlinkage across the stator iron core, this property can be explored by driving the stator winding into soft saturation region, and the inductance variations become prominent. If the rotor magnetic pole is aligned with a pair of excited stator windings, the corresponding winding inductance become smallest duo to the saturation effect of stator iron core.

The detection of the stator inductance variations can be explained with reference to the schematic of a PMSM drive shown in Fig. 3. A series of voltage vectors $\mathbf{V}_n$ with defined time intervals are applied to the stator windings and the corresponding current responses can be measured for the detection of inductance variations due to the rotor position. Fig. 4(a) shows the inductance of a line-to-line stator windings with its third line in floating state is a function of the rotor electrical angular position due to the variation of rotor fluxlinkage. An external stator magnetic field with fixed magnitude and position is applied to the motor, such as $\mathbf{\Psi}_{ab}$ which driving the stator winding into soft saturation region, the corresponding winding inductance and current will be varied due to the coupled fluxlinkage with two windings as shown in Fig. 4(b), when the rotor north pole is aligned with stator flux,

978-1-4799-2706-7/14 $31.00 © 2014 IEEE   3073

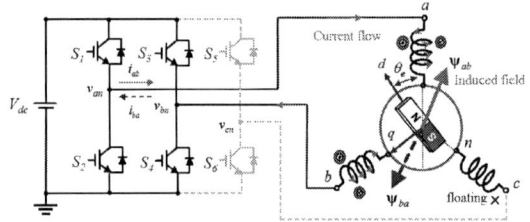

Fig. 3. Two current paths in oppose direction to excite two phases of three-phase windings and corresponding induced field.

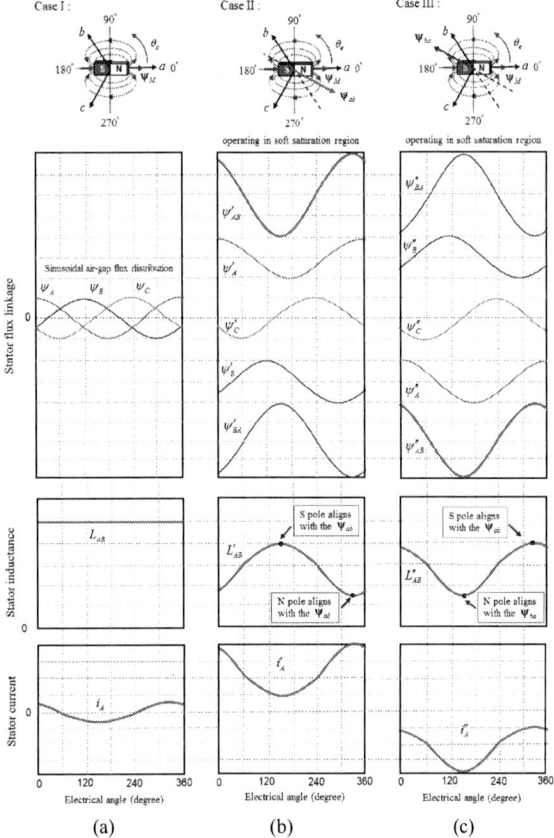

(a)   (b)   (c)

Fig. 4. Stator winding inductance as a function of rotor flux and stator flux. (a) Case I. (b) Case II. (c) Case III.

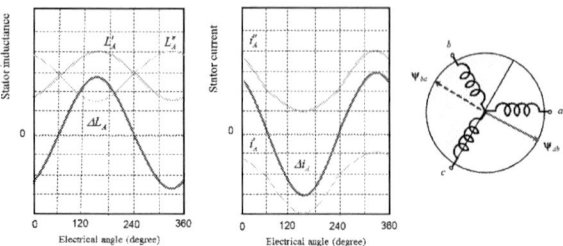

Fig. 5. Variations of stator inductance as a function of rotor position.

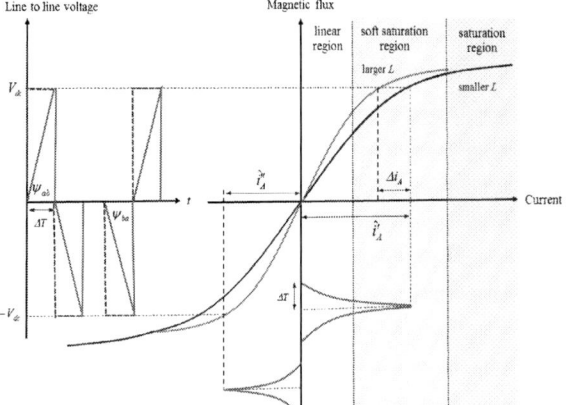

Fig. 6. Nonlinear magnetization characteristics of the stator core.

the external magnetic field let the total magnetic fluxlinkage reach its maximum and inductance comes to its minimum, this can be observed at 330° electrical angle of the rotor flux. On the other hand, if the south pole is aligned with the stator flux, the stator inductance comes to its maximum at 150° electrical angle. When an inversed stator flux vector Ψ_{ba} is applied to the motor and the stator winding is operated in soft saturation region, due to the saturation effect and the total fluxlinkage is negative, the corresponding winding inductance and current will be varied as shown in Fig. 4(c).

By using a pair of applied stator flux vectors with inversed polarity, we can estimate the position of rotor pole within 180 electrical degrees by measuring its inductance variations, as shown in Fig. 5. In order to know the varying of the stator inductance, it needs to use the nonlinear magnetization characteristics of the stator core as shown in Fig. 6. If the magnetic flux is increased to close to its saturation region, the corresponding inductance becomes smaller and will result a larger current spike. The peak current difference between the complementary voltage pulses is a measure of their closeness to the saturation region. If the rotor flux vector is closer to the applied voltage vector, the corresponding current spike becomes larger. The measured inductance can be expressed as

$$L = \frac{\Delta\Psi}{\Delta i_L} = \frac{\Delta v_L \cdot \Delta T}{\Delta i_L} = \frac{(V_{dc} - i_L \cdot R)\Delta T}{\Delta i_L} \approx \frac{V_{dc}\Delta T}{\Delta i_L} \quad (1)$$

where ΔT is time duration of the voltage pulse. In practical conditions, the voltage drop across the resistor can be neglected. For non-salient motor, the stator winding can be driven into its saturation region so that the peak values of the positive and negative currents can be more significant for the detection of the rotor polarity.

To avoid rotor oscillations during the signal injection process, we can apply a series of symmetric complementary pulses to the selected stator windings as shown in Fig. 7(a) and the corresponding current responses is shown in Fig. 7(b). The shape of the current responses is similar with an isosceles triangular wave due to the duration of the voltage pulse ΔT is much smaller than the electrical time constant of the stator windings. The elapsed time between the complementary voltage pulses should be longer than ΔT to let the current decayed to zero. In ideal conditions, variations of the peak current are symptoms for the degree of alignment of the rotor to the composite vector of the excited stator

978-1-4799-2706-7/14 $31.00 © 2014 IEEE

The 2014 International Power Electronics Conference

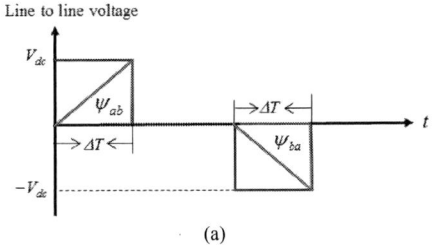

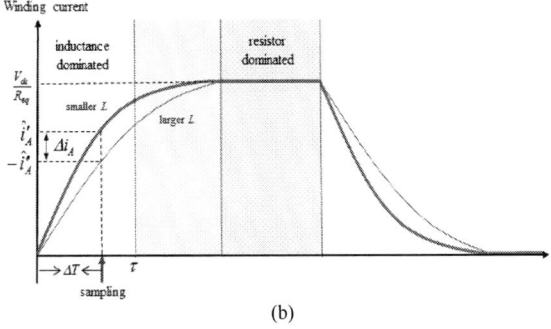

(b)

Fig. 7 Excitation voltage and measured current. (a) Excitation signals. (b) Current response with different inductance.

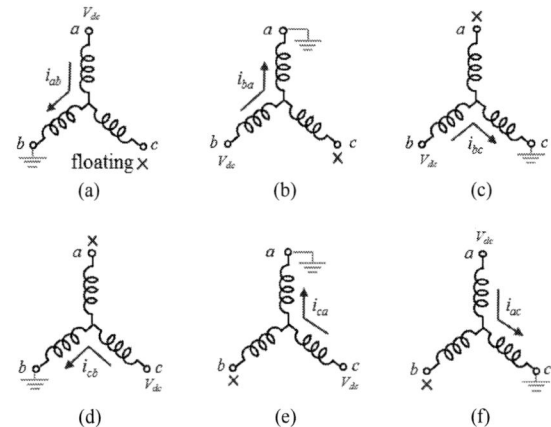

Fig. 8. Six excitation configurations.

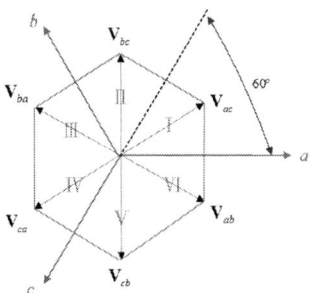

Fig. 9. Six voltage sectors for the determination of rotor flux vector.

windings. For the given example, if the current difference is positive, then the stator flux in the forward direction is in the same direction as the magnetic flux of the rotor.

The complementary voltages pulses are applied to each two-phase windings of the three-phase stator windings with the third phase left unconnected. The circuit configurations are shown in Fig. 8 and the applied voltage pulses are controlled by the connected PWM inverter.

B. Initial Rotor Position Estimation Method

The voltage pulses applied to the stator windings are equivalent to voltage vectors applied to the stator windings with specified directions. The corresponding current responses can be used to determine the rotor flux vector is belong to which sector within the six voltage sectors as shown in Fig. 9. The two excited phase windings of a three-phase PMSM motor at standstill can be modeled as an R-L circuit. The current response can be expressed as

$$i(t) = \frac{V_{dc}}{R_{eq}}(1 - e^{-\frac{R}{L}t}) \qquad (2)$$

where R_{eq} is the equivalent resistance and t is the turn-on time of the applied voltage. If the time duration of ΔT is much smaller than its time constant, the corresponding peak current is proportional ΔT.

Three pairs of the complementary voltage vectors with suitable time durations and elapsed time intervals are applied to the two-phase windings of the Y-connected three-phase windings. Six voltage vectors are applied to the stator as shown in Fig. 9 and the corresponding current peak becomes larger if the rotor flux vector is residing in that sector. Therefore, the initial rotor position can be detected within 60 electrical degrees with enough accuracy for proper commutation during motor start-up

without reversing rotation. Fig. 10 shows the detailed flowchart for the initial rotor position detection of a PMSM at standstill.

There exits possibility that the rotor is around the boundary of two sectors, and results uncertainties in determination of its belonged sector. However, this makes no difference for the startup control of the motor with its assigned rotation direction. A proper commutation voltage vector can be selected in leading direction of the rotor flux vector.

IV. STARTING PROCEDURE

Once the initial position of the rotor flux vector has been detected, the d-axis of the stator current vector is aligned with the selected voltage vector. For example, if the rotor flux vector is belong to sector I as shown in Fig. 11(a), then the d-axis is aligned with the voltage vector $\mathbf{V}_{ac}$ as shown in Fig. 9. The actual position of the rotor flux vector may have a maximum error of 30° electrical degrees. However, the q-axis is aligned with a leading phase angle of 90° for a CCW rotation, as shown in Fig. 11(a). On the other hand, for a CW rotation, the q-axis is aligned with a lagging phase angle of 90°.

It should also be noted, a proper commutation voltage vector sequences with suitable magnitudes should be applied to the stator windings during the startup process. For CCW rotation, the sequences of the voltage vector are scheduled with the same direction as shown in Fig. 11, with an order from I, II ... to VI. For CW rotation, the sequence of the voltage vectors is reversed.

978-1-4799-2706-7/14 $31.00 © 2014 IEEE

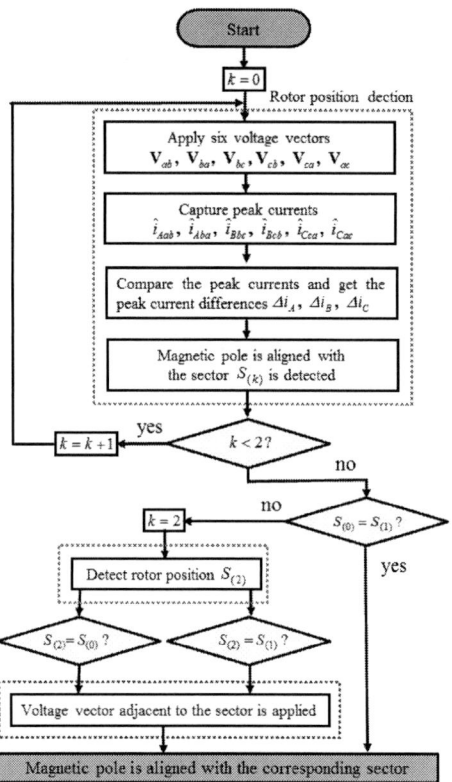

Fig. 10. Flowchart of the initial rotor position detection procedure.

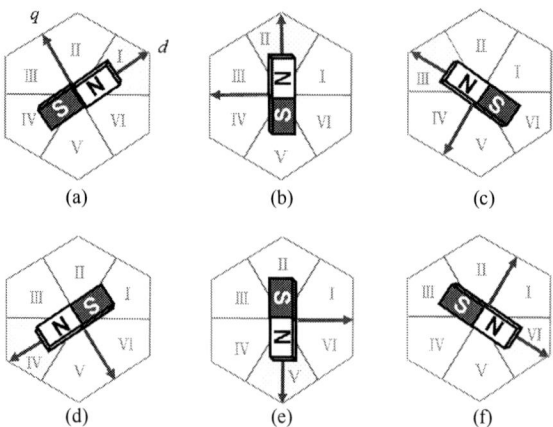

Fig. 11. Possible initial positions of the rotor flux vector within six voltage vector sectors.

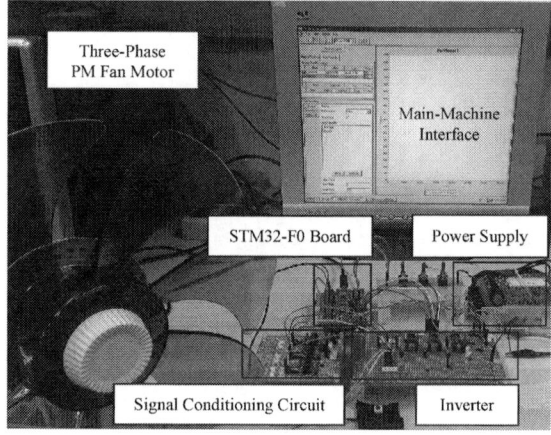

Fig. 12. A view of the implemented experimental system.

During the startup process, the motor is running in open-loop V/Hz control mode with a specified acceleration profile. The static characteristics of the V/Hz for the sensorless drive can be defined with a piecewise linear curve based on the motor and load characteristics. The acceleration rate of sensorless drive depends on the current limit of the PWM inverter and the motor-load characteristics. In practical applications, the linear acceleration profile can be characterized with a tuning time constant.

Once the motor is startup, it is linear accelerated to a minimum target speed for sensorless closed-loop control. A criterion is used to check and switch the drive from open-loop V/Hz mode to sensorless FOC mode. A wide range of sensorless speed control is essential for reliable and smooth startup control for various applications.

V. EXPERIMENTAL RESULTS

The controller was realized with a low-cost single-chip 32-bit MCU controller, the STM32F0, from ST Microelectronics. The STM32F0 is an ARM Cortex-M0-based 32-bit microcontroller particularly designed for cost-sensitive applications. STM32 F0 MCUs combine real-time performance, low-power operation, advanced architecture and peripherals associated with digital motor control [11]. Fig. 12 shows the experimental system for the performance verification of the proposed control scheme. The DC-link voltage of the inverter is 12 V and the tested motor is a PM motor with a connected fan load. The parameters are listed in Table I.

TABLE I
PARAMETERS OF THE TESTED MOTOR

Symbol	Meaning	Value
V_{DC}	Rated voltage	12 V
P	Poles	8 poles
R	Stator resistance	1.5 Ω
L	Stator inductance	1.4 mH
τ_e	Elec. time constant	0.93 msec
K_e	Back-EMF constant	4.27 mV/rpm
J	Rotor inertia	$0.39\cdot10^{-3}$ kg·m^2
B	Viscous coefficient	$1.4\cdot10^{-4}$N·m·s/rad

All the control functions are realized with software, the control program can be down loaded to the on-chip flash memory via an USB interface with PC. The switching frequency of the PWM inverter is 20 kHz, and the sampling frequency of the digital controller is a synchronized 10 kHz using the internal programmable timers. Self-commissioning, startup control, and communication programs are running in the initialization phase and background control loop. Control parameters and variables can be tuned and monitored via the USB interface.

Fig. 13 shows the differences of the peak current when the complementation voltage pulses are applied to each pair of the stator windings. For each pair of the excited windings, the change of their corresponding peak

978-1-4799-2706-7/14 $31.00 © 2014 IEEE

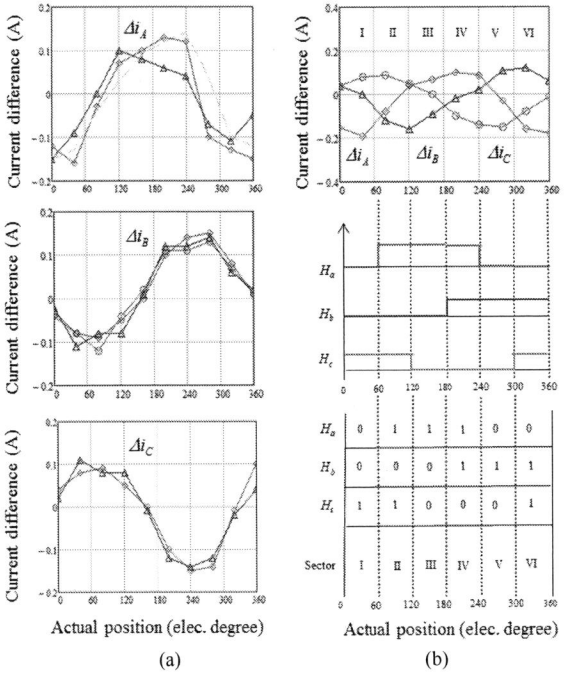

Fig. 13. (a) Three phase peak current differences. (b) The relationship of the current differences and the Hall signals.

currents can be recorded with the rotor randomly placed. The recorded current changes relative their corresponding rotor position exhibits a nearly sinusoidal distribution. This reveals the fact that the rotor fluxlinkage coupled to the stator winding also exhibits sinusoidal distribution for a sinusoidal PMSM. The maximum positive change of the peak currents reveals the rotor flux vector is fixed at the corresponding voltage vector section.

Fig. 14 shows the measured peak currents when the rotor is in the sector I. The measured current differences are about 12.5% of its peak value. Fig. 15 shows the distribution of the estimated rotor position versus the actual position. The estimated position is within ±30° electrical degrees of the applied voltage vector.

The time spent for the initial position detection procedure is about 60~120 ms as shown in Fig. 16. From the measured Hall signals, it can be seen the proposed scheme can startup without reversing oscillations. Fig. 17 shows the phase current with the typical closed-loop sensorless control from standstill to 1000 rpm. It can be observed that the proposed sensorless startup control scheme can rotate from standstill to rated speed with smooth operation.

VI. CONCLUSION

For conventional sensorless PMSM drives, a reliable and smooth startup operation is a most difficult and challenging task, it requires knowledge of the system integration for specific applications and time consuming parameters tuning of the available sensorless dives. This paper has presented a robust sensorless startup control method with self-commissioning for permanent magnet motors for smooth ramping speed up without reversing oscillations with minimum information of the motor and

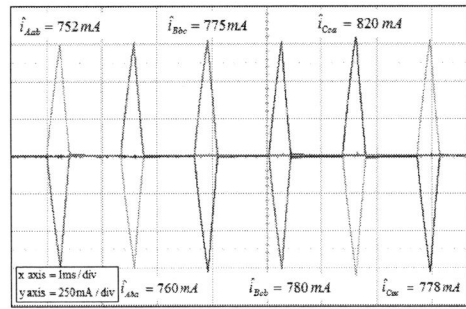

Fig. 14. The measured peak currents when the rotor is in the sector I.

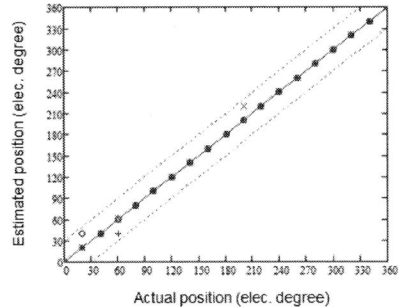

Fig. 15. Estimated rotor position versus the actual position.

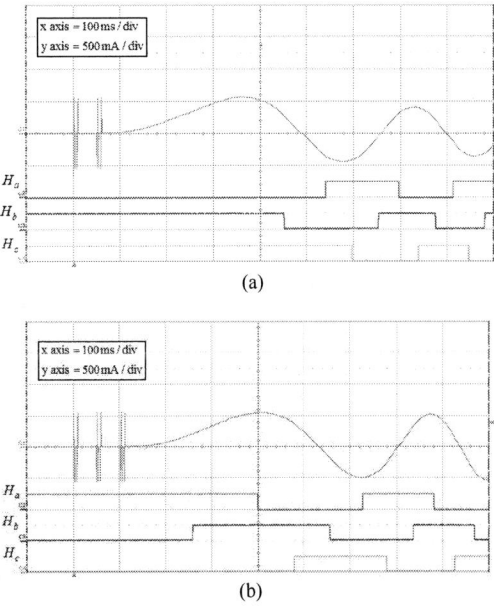

Fig. 16. The startup phase current and Hall signals. (a) Rotor position detection twice. (b) Rotor position detection three times.

load. The proposed scheme will not cause visible rotor vibrations and audio noises during the detection process. The sensorless startup with self-commissioning scheme has been implemented based on a low-cost single-chip 32-bit MCU controller (STM32F0). Experimental results reveal that the proposed scheme can achieve reliable and smooth startup with robust performances. The proposed startup control scheme with self-commissioning strategy is expected to be used for adjustable speed PMSM dries with centrifugal loads.

978-1-4799-2706-7/14 $31.00 © 2014 IEEE

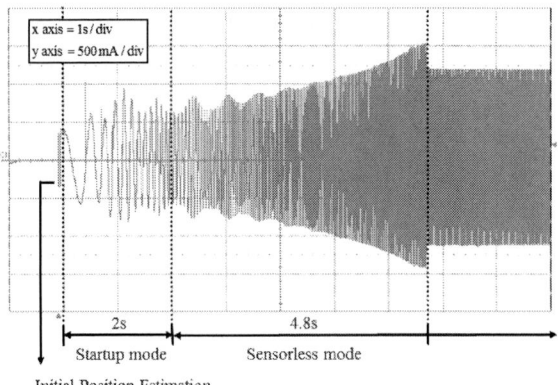

Fig. 17. Sensorless speed control from standstill to 1000 rpm with the proposed start-up method.

REFERENCES

[1] P. B. Schmidt, M. L. Gasperi, R. Glen, and A. H. Wijenayake, "Initial rotor angle detection of a nonsalient pole permanent magnet synchronous machine," *IEEE IAS Annual Meeting*, vol. 1, pp. 459-463, 1997.

[2] Y.-C. Chang, Y.-Y. Tzou, "A new sensorless starting method for brushless DC motors without reversing rotation," *IEEE IAS Annual Meeting*, pp. 619-624, 2007.

[3] G.-M. Kim, B.-G. Park, I.-S. Jung, and D.-S. Hyun, "An improved back-EMF based initial rotor position estimation for IPMSM," *IEEE 8th International Conference on Power Electronics and ECCE Asia*, pp. 1244-1249, 2011.

[4] N. Chen, Z.-H. Wang, S.-Y. Yu, W.-H. Gui, and Y.-Q. Guo, "A new starting method of sensorless PMSM motors based on STM32F103B," *29th Chinese Control Conference*, pp. 4964-4968, 2010.

[5] S. Taniguchi, S. Mochiduki, T. Yamakawa, S. Wakao, K. Kondo, and T. Yoneyama, "Starting Procedure of Rotational Sensorless PMSM in the Rotating Condition," *IEEE Trans. on Industry Applications*, vol. 45, pp. 194-202, 2009.

[6] Y.-S. Lai, F.-S. Shyu, and S.-S. Tseng, "New initial position detection technique for three-phase brushless DC motor without position and current sensors," *IEEE Trans. on Industry Applications*, vol. 39, pp. 485-491, 2003.

[7] M. Tursini, R. Petrella, and F. Parasiliti, "Initial rotor position estimation method for PM motors," *IEEE Trans. on Industry Applications*, vol. 39, no. 6, pp. 1630-1640, 2003.

[8] J. Kania, T. H. Panchal, V. Patel, and K. Patel, "Self commissioning: a unique feature of inverter-fed induction motor Drives," *Nirma University International Conference on Engineering*, pp. 1-6, 2011.

[9] N. Urasaki, T. Senjyu, and K. Uezato, "Self-commissioning for surface-mounted permanent magnet synchronous motors," *Journal of Power Electronics*, vol. 3, no. 1, 2003.

[10] M. Khov, J. Regnier, and J. Faucher, "On-line parameter estimation of PMSM in open loop and closed loop," *IEEE International Conference on Industrial Technology*, pp. 1-6, 2009.

[11] STM32F0 Series ARM-based 32-bit MCU with 16 to 64-KB Flash, timers, ADC, communication interfaces, 2.4-3.6 V operation, datasheet, STMicroelectronics, July, 2013.

Position Sensorless control of PMSM with a low-frequency signal injection

Tomohiro Nimura, Shinji Doki
: Dept. of Electrical Engineering and Computer Science
Nagoya University
Furo-cho, Chikusa-ku, Nagoya, Aichi 464-8603, Japan
E-mail: nimura@nagoya-u.jp, doki@nagoya-u.jp

Masami Fujitsuna
DENSO CORPORATION
1-1, Showa-cho, Kariya 448-8661
E-mail: MASAMI_FUJITSUNA@denso.co.jp

Abstract—In this paper, a new position sensorless control of PMSM for low speed and standstill by using a low-frequency signal injection is proposed. Conventional sensorless control method for a low speed range uses injection signal of a frequency around kHz. The problem with this approach is harsh acoustic noise by the injection signal. Low-frequency signal injection solve the problem of noise but it makes new problems as follows. One is that the filter for signal separation degrades the current control characteristics. The other, PLL(Phase Locked Loop) design for position estimation would be limited. To overcome these problems, we proposed a new position sensorless control methods for a low-frequency signal injection. The proposed method performs subtraction by using the specific frequency component estimated from known information such as motor model and injected signal. It achieves stable control by using a low frequency overlapping with the current control band.

Keywords—acoustic noise, high frequency voltage injection, PMSM, sensorless

I. Introduction

PMSM has excellent performance, it is used as a drive motor such as a train or an electric vehicle. Vector control is mainly used as a control method for it. Vector control requires a magnetic pole position information of the rotor. Therefore, it's basic to use a position sensor. However, in order to realize the cost-saving, space-saving and high reliability, sensorless control is expected. It is already in practical use home appliances, industrial equipment, etc., extensive studies have been carried out aimed at expanding range of applications further.

In general, the position sensorless control method for low speed and standstill is a method of injecting a high frequency signal to the voltage command. Voltage typical shape is rectangular shape which is synchronized to the carrier frequency[1]~[3] or sinusoidal shape with a frequency close to the current control band[4]~[6]. So, range of the frequency of the injection signal is from a current control band to carrier frequency. And, it corresponds to frequency from several hundred to several kilo hertz.

Authors have been focusing on the sine shape injection voltage that has a single frequency component of several hundred Hz. In this approach, depending on the injected signal frequency and current control bandwidth, it is necessary to pay close attention to the design of such

frequency discrimination filter, BPF(Band Pass Filter) and BSF(Band Stop Filter) for signal separation. The reason is that when the frequency of motor drive signal, such as transients across the wide band component in proximity to the frequency of the injected signal may adversely affect the signal separation by the filter[5]~[10]. Therefore, as the frequency of the injected signal, a frequency sufficiently higher than the current control band is usually chosen.

However, a high level acoustic noise can be perceive by human hearing because sensitivity region of human hearing is around the frequency. In automotive applications, such acoustic noise makes it unsuitable. The advantage of the low-frequency signal injection is the reduction of loss and harsh acoustic noise. However, Frequency discrimination filter is disadvantageous in the separation of two components that overlap the band. Thus, it could lead to instability and degradation of the current control and position estimation performance by selecting the low frequency.

In this paper, for low-frequency(300Hz) injection whose purpose reducing the loss and harsh acoustic noise, we propose a new current control system effective in conditions the injection signal frequency and current control band is close or overlapping.

II. Current control system with signal injection

Figure 1 shows the general configuration of the current control system in the case where the position sensorless control by the high-frequency voltage injection. Subscript f means fundamental frequency component for drive and h means injected frequency one for position detection.

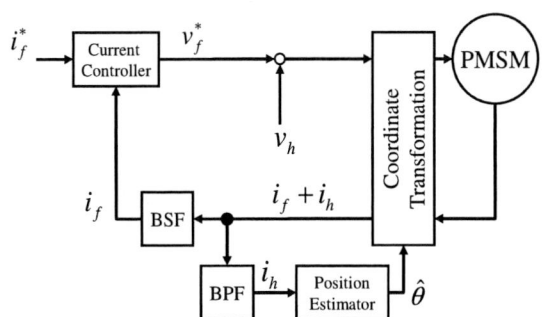

Fig. 1. Conventional current control system

Signal injected to the voltage command is generally a sine wave with frequency of around kHz. By injection of the high-frequency voltage v_h, injection frequency component i_h appears in the current response. i_h is a signal including information of the position. This frequency component is extracted by the BPF, and is inputted to the position estimator. Meanwhile, since it is a disturbance for the current control system, it is removed from the feedback current using the BSF such as NF(Notch Filter).

Some type of signal have been proposed as injection signal v_h[6]. In this paper, the sinusoidal signal as shown in equation (1) with constant amplitude V_h and frequency ω_h is selected as injection signal v_h. As shown in Figure 2, this signal is injected in the rotating coordinate $\gamma\delta$ axis having a phase difference $\Delta\theta$ between the dq axis.

$$v_h = V_h \begin{bmatrix} \cos\omega_h t \\ \sin\omega_h t \end{bmatrix} \quad (1)$$

In this case, current i_h with same frequency of injection voltage v_h is given by equation (2).

$$
\begin{aligned}
i_h &= \begin{bmatrix} i_{\gamma h} \\ i_{\delta h} \end{bmatrix} \\
&= \frac{V_h}{L_d L_q \omega_h} [-L_1 \mathbf{R}(2\Delta\theta)\mathbf{u}_n(\omega_h t) + L_0 \mathbf{u}_p(\omega_h t)] (2)
\end{aligned}
$$

$\mathbf{R}(2\Delta\theta)$ is a vector rotator. $\mathbf{u}_p(\omega_h t)$ is a unit vector of positive phase, and $\mathbf{u}_{jp}(\omega_h t)$ is a vector with 90 degrees phase lead to $\mathbf{u}_p(\omega_h t)$. $\mathbf{u}_n(\omega_h t)$ is a unit vector of negative phase, and $\mathbf{u}_{jn}(\omega_h t)$ is a vector with 90 degrees phase lead to $\mathbf{u}_n(\omega_h t)$. Figure 3 shows these relationships.

$$\mathbf{R}(2\Delta\theta) = \begin{bmatrix} \cos 2\Delta\theta & -\sin 2\Delta\theta \\ \sin 2\Delta\theta & \cos 2\Delta\theta \end{bmatrix} \quad (3)$$

$$\mathbf{u}_p(\omega_h t) = \begin{bmatrix} \sin\omega_h t \\ -\cos\omega_h t \end{bmatrix} \quad (4)$$

$$\mathbf{u}_n(\omega_h t) = \begin{bmatrix} \sin\omega_h t \\ \cos\omega_h t \end{bmatrix} \quad (5)$$

$$\mathbf{u}_{jp}(\omega_h t) = \mathbf{J}\mathbf{u}_p(\omega_h t) = \begin{bmatrix} \cos\omega_h t \\ \sin\omega_h t \end{bmatrix} \quad (6)$$

$$\mathbf{u}_{jn}(\omega_h t) = \mathbf{J}\mathbf{u}_n(\omega_h t) = \begin{bmatrix} -\cos\omega_h t \\ \sin\omega_h t \end{bmatrix} \quad (7)$$

Inductance L_0 and L_1 have a relationship represented by equation (8) for L_d and L_q.

$$\begin{bmatrix} L_0 \\ L_1 \end{bmatrix} = \frac{1}{2} \begin{bmatrix} 1 & 1 \\ 1 & -1 \end{bmatrix} \begin{bmatrix} L_d \\ L_q \end{bmatrix} \quad (8)$$

Figure 4 shows the configuration of the position estimator. Position estimator is composed of signal processing by the heterodyne method and error convergence systems based on PLL.

Heterodyne method is shown in Equation (9). It is possible to produce a u_{PLL} by taking the inner product of $\mathbf{u}_{jn}(\omega_h t)$ and i_h. $\Delta\theta$ and u_{PLL} have positive correlation. The operation of the PLL block use this property. By outputting the estimated position $\hat{\theta}$ according to the u_{PLL}, $\Delta\theta$ converges to zero. As a side note, bandwidth of the PLL block should be designed to be sufficiently attenuated the component of two times the injection frequency($2\omega_h$) in the u_{PLL}.

$$
\begin{aligned}
u_{PLL} &= \mathbf{u}_{jn}(\omega_h t) \cdot i_h \\
&= (-\cos\omega_h t)i_{\gamma h} + (\sin\omega_h t)i_{\delta h} \\
&= \frac{-L_1 V_h}{L_d L_q \omega_h} \sin 2\Delta\theta + \frac{-L_0 V_h}{L_d L_q \omega_h} \sin 2\omega_h t \quad (9)
\end{aligned}
$$

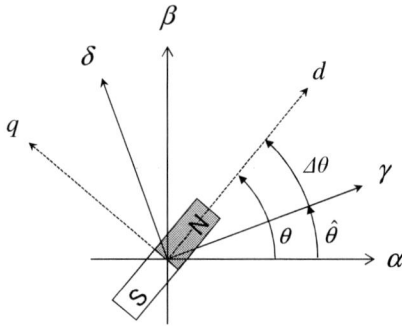

Fig. 2. Relationship of the magnetic pole position and coordinate system

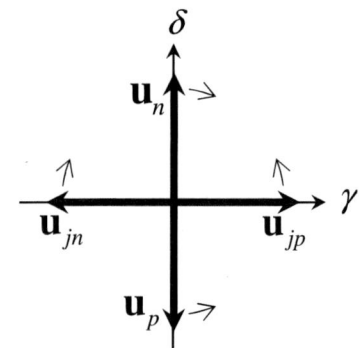

Fig. 3. positive and negative phase vectors

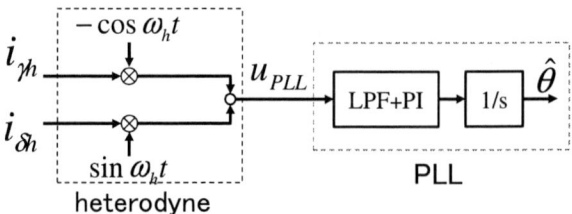

Fig. 4. Configuration of position estimator

III. PROBLEMS FOR LOW-FREQUENCY SIGNAL INJECTION

In this section, problems with low-frequency signal injection is discussed. In the case of lowering the injection frequency, it should be noted the two point. One is that the frequency discrimination filter for signal separation degrades the current control characteristics. The other, PLL design for position estimation would be limited.

A. Filters for signal separation

Lowering the injection frequency is equivalent to lowering the center frequency of the filter, such as BSF. It means that filter bandwidth is close to the current control bandwidth. In a steady state, the driving current component i_f can be regarded as DC component in a rotating frame. So, the frequency discrimination filter can extract or remove only the injection frequency component.

On the other hand, in the transient state, driving-current could contain the component which is close to the injection frequency. In this case, the frequency discrimination filter can not distinguish between these two components. As the results, the Filters for low-frequency signal injection deteriorate current control performance and stability.[10]

B. Residual disturbance involved to the bandwidth of the PLL

The component of $2\omega_h$ frequency in u_{PLL} is disturbance for the position estimator. This disturbance component should be eliminated by frequency characteristic of PLL. But, the lower the injection frequency, the more disturbance component is close to the PLL bandwidth.

As the results, accuracy of position estimation is degraded because the disturbance component remains largely in $\hat{\theta}$. Then, low-frequency signal injection method makes it difficult to design a PLL to achieved both performance and stability of position estimation.

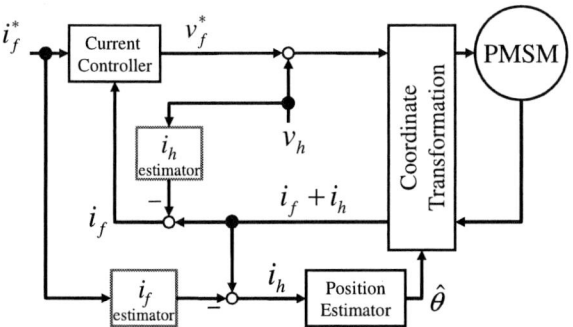

Fig. 5. Proposed current control system

IV. PROPOSED CURRENT CONTROL SYSTEM CONFIGURATION

Figure 5 shows the configuration of the proposed current control system.

A. signal separation method

"i_h estimator" in the figure is in place of the BSF. It estimates a pseudo injection-frequency current component $\hat{i}_h$, which is a estimated value of a injection-frequency current component i_h at $\Delta\theta = 0$, from input signal v_h. Figure 6 shows the block diagram and Equation (10) shows the equation of state.

$$\begin{bmatrix} \dot{\hat{i}}_{\gamma h} \\ \dot{\hat{i}}_{\delta h} \end{bmatrix} = \begin{bmatrix} -\dfrac{R}{L_d} & 0 \\ 0 & -\dfrac{R}{L_q} \end{bmatrix} \begin{bmatrix} \hat{i}_{\gamma h} \\ \hat{i}_{\delta h} \end{bmatrix} + \begin{bmatrix} \dfrac{1}{L_d} & 0 \\ 0 & \dfrac{1}{L_q} \end{bmatrix} \begin{bmatrix} v_{\gamma h} \\ v_{\delta h} \end{bmatrix} \quad (10)$$

In this configuration, the feedback current is obtained by subtracting the estimated value $\hat{i}_h$ from the detected current.

"i_f estimator" in the figure is in place of the BPF. It is for outputting the estimated value $\hat{i}_f$ of the driving-current component, which is nominal response of current control system from current command i_f^*. Figure 7 shows the block diagram and Equation (11) shows the equation of state.

$$\begin{bmatrix} \dot{\hat{i}}_\gamma \\ \dot{\hat{i}}_\delta \\ \dot{\hat{\varepsilon}}_\gamma \\ \dot{\hat{\varepsilon}}_\delta \end{bmatrix} = \begin{bmatrix} -\dfrac{R+K_{Pd}}{L_d} & 0 & \dfrac{K_{Id}}{L_d} & 0 \\ 0 & -\dfrac{R+K_{Pq}}{L_q} & 0 & \dfrac{K_{Iq}}{L_q} \\ -1 & 0 & 0 & 0 \\ 0 & -1 & 0 & 0 \end{bmatrix} \begin{bmatrix} \hat{i}_\gamma \\ \hat{i}_\delta \\ \hat{\varepsilon}_\gamma \\ \hat{\varepsilon}_\delta \end{bmatrix}$$
$$+ \begin{bmatrix} \dfrac{K_{Pd}}{L_d} & 0 \\ 0 & \dfrac{K_{Pq}}{L_q} \\ 1 & 0 \\ 0 & 1 \end{bmatrix} \begin{bmatrix} i_\gamma^* \\ i_\delta^* \end{bmatrix} \quad (11)$$

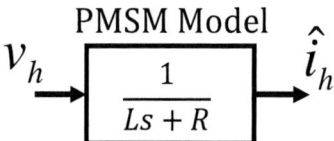

Fig. 6. i_h estimator

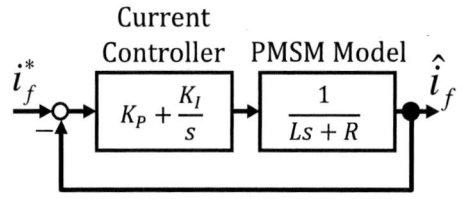

Fig. 7. i_f estimator

In this configuration, a signal obtained by subtracting the estimated value $\hat{i}_f$ from the detection current is input to the position estimator.

The proposed method performs signal separation by the filter by model based estimation instead of the frequency discrimination filter, BPF and BSF. By using our proposed system, each signal can be well separated even in transient response.

B. Position estimator

Figure 8 shows the proposed position estimator. From the second term of equation (9), the component of two times the injection frequency contained in u_{PLL} can be calculated as the estimated value using the motor parameters and the injected signal. By subtracting the estimated component from u_{PLL}, it is possible to reduce the pulsation remaining in $\hat{\theta}$ without relying on design of the PLL bandwidth. As a result, without deteriorating the accuracy of position estimation, it is available in the PLL for low order and stable wideband. u'_{PLL} in the figure indicates a value after subtraction.

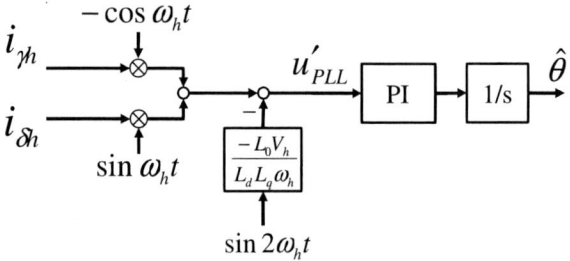

Fig. 8. Proposed position estimator configuration

V. EXPERIMENT

A. signal separation in the current control system

We have some experiments to evaluate our proposed system. Under the condition that load motor is controlled at 100rpm, current command for drive motors is changed suddenly. Table I shows the parameters of the test motor. PLL bandwidth is approximately 150 rad/s.

Figure 9 shows the results obtained when giving step command 100% and the injection frequency 900Hz. Conventional filters are used to signal separation. From top to bottom, it shows γ axis current command I_γ^* and its response I_γ, δ axis current command I_δ^* and its response I_δ, true position θ, estimated position $\hat{\theta}$ and position error $\Delta\theta$.

Figure 10 shows the results obtained when giving step command 50% and the injection frequency 300Hz. Other conditions is the same as Figure 9. When given a step command to more than 60%, the estimated position diverges. Thus, injected with low frequency such as to overlap the current control band, the estimated position is divergent to step immediately.

Figure 11 shows the results obtained when giving step command 100% and the injection frequency 300Hz. In

this case, the proposed method of signal separation is used. It can be seen that the control without degrading the responsiveness and stability.

B. Residual disturbance and bandwidth of the PLL

Next, we also show the results obtained when the widening PLL bandwidth. Under the condition that load motor is controlled at standstill, current command for drive motor is changed suddenly.

Figure 12 shows the results obtained when giving step command 80% and the injection frequency 900Hz. And, PLL bandwidth is approximately 450 rad/s. Input signal u_{PLL} to the PLL is shown in the third row. Figure 13 is the case of using the injection frequency 300Hz under the same conditions. For low-frequency injection, remaining in the estimated position a size that can not be ignored $2\omega_h$ frequency component even in the band setting no problem conventionally.

Figure 14 shows a waveform obtained by applying the proposed method under the same conditions as in Figure 13. We can see that the pulsation of the estimated position is reduced by $2\omega_h$ frequency component is reduced by the subtraction.

TABLE I.   MOTOR PARAMETERS

Symbol	Meaning	Value
P	Number of Pole Pairs	3
R	Stator Resistance	0.17Ω
L_d	d-axis Inductance	1.8mH
L_q	q-axis Inductance	2.7mH
K_E	Back-EMF Constant	0.055V $\cdot$ s/rad
ω_{cc}	Current Cntl. Bandwidth	$2\pi * 300$rad/s

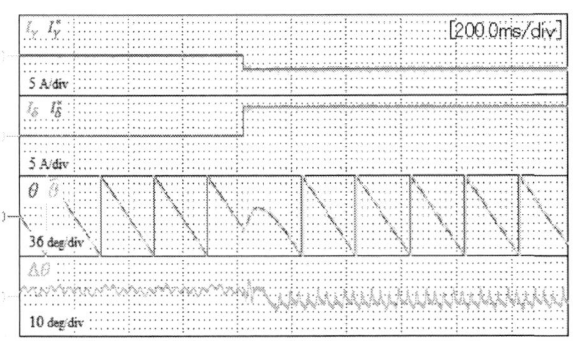

Fig. 9.   injection frequency: 900Hz, signal separation method: filter

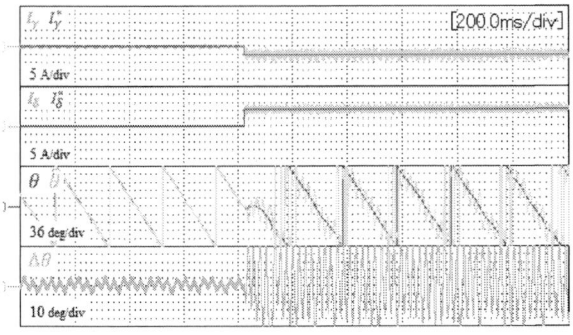

Fig. 10.   injection frequency: 300Hz, signal separation method: filter

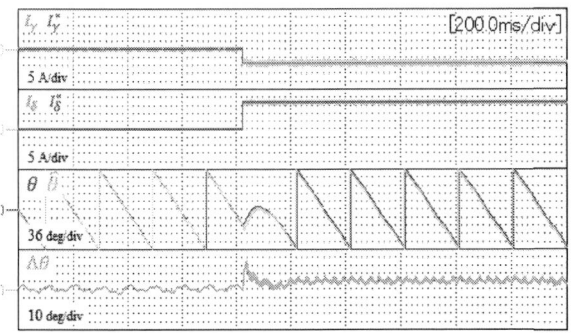

Fig. 11.   injection frequency: 300Hz, signal separation method: proposed method

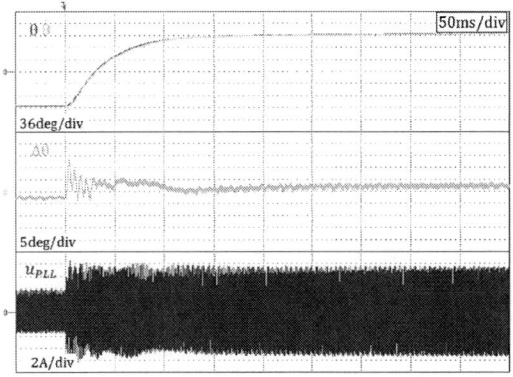

Fig. 12.   injection frequency: 900Hz, the PLL pre-processing: None

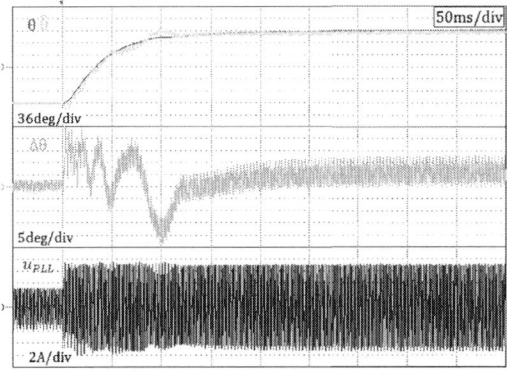

Fig. 13.   injection frequency: 300Hz, the PLL pre-processing: None

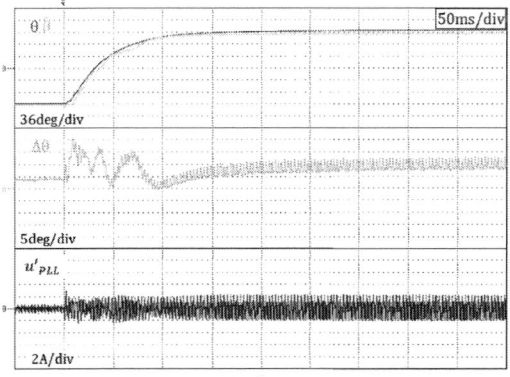

Fig. 14.   injection frequency: 300Hz, PLL pretreatment: the proposed method

C. Parameter error

Proposed method uses motor parameters to estimate some signals. Therefore, attention must be paid to the impact of parameter variations. According to the results that have been discussed in previous papers, most dominant parameter is the inductance in a current estimation, which is sensitive to load[10][11]. Therefore tables of the inductances L_d, L_q as a function of currents i_d, i_q have been prepared for these experiments.

VI. IN CONCLUSION

In this paper a new position sensorless control of PMSM for a low speed and standstill by using a low-frequency signal injection was proposed. In the proposed method, a sensorless control with a lower-frequency signal injection became possible by a model-based signal separation, by using information of motor model and injection signal, instead of a frequency separation filter. To evaluate our proposed method, experimental results of sensorless control showed that proposed methods with injection signal of 300Hz has same performance with conventional methods with injection signal of 900Hz.

REFERENCES

[1] S.M.Kim,J.I.Ha,S.K.Sul. "PWM Switching Frequency Signal Injection Sensorless Method in IPMSM" Proceeding of IEEE-ECCE,pp. 3021-3028,2011.

[2] Y.D.Yoon,S.K.Sul,S.Morimoto,K.Ide. "High Bandwidth Sensorless Algorithm for AC Machines Based on Square-wave Type Voltage Injection" Proceeding of IEEE-ECCE,pp. 2123-2130,2009.

[3] R.Masaki,S.Kaneko,M.Hombu,T.Sawada. "Development of a Position Sensorless Control System on an Electric Vehicle Driven by a Permanent Magnet Synchronous Motor" Proceeding of PCC,pp. 571-576,2002.

[4] J.H.Jang,J.I.Ha,M.Ohto,K.Ide,S.K.Sul. "Analysis of Permanent-Magnet Machine for Sensorless Control Based on High-Frequency Signal Injection" IEEE Transactions on Industry Applications,vol. 40,no. 6,pp. 1595-1604,2004.

[5] F.Briz,A.Diez,M.W.Degner. "Dynamic Operation of Carrier-Signal-Injection-Based Sensorless Direct Field-Oriented AC Drives" IEEE Transactions on Industry Applications,vol. 36,no 5,pp. 1360-1368,2000.

[6] Shinji Shinnaka. "A Generalized Heterodyne Method Incorporating a High-Frequency Integral-Type PLL for Sensorless Drives of PMSMs" IEEJ Transactions on Industry Applications Vol. 130 (2010) No. 8 P 973-986

[7] R.Leidhold,P.Mutschler. "Interaction between the current controller and the injection of alternating carriers in sensorless drives" Proceeding of SPEEDAM,pp. 262-267,2008.

[8] F.Briz,M.W.Degner,P.Garcia,A.B.Diez. "Transient Operation of Carrier Signal Injection Based Sensorless Techniques" Proceeding of IEEE-IECON,pp. 1466-1471,2003.

[9] A.Consoli,A.Gaeta,G.Scarcella,G.Scelba,A.Testa. "Optimization of Transient Operations in Sensorless Control Techniques Based on Carrier Signal Injection" Proceeding of IEEE-ECCE,pp. 2115-2122,2009.

[10] S.H.Jung, S.Doki, S.Okuma, Masami Fujitsuna. "Adjusted Current Controller for Signal-Injection Based Control Algorithms" Proceeding of IEEE-IECON,pp. 1120-1125,2011.

[11] S.Lerdudomsak,S.Doki,S.Okuma. "Current Control System for PMSM in Overmodulation Range of Inverter - Analysis for Robustness to Parameter Variations" Proceeding of IEEE-IECON,pp. 1216-1221,2008.

A comparison of different sensorless position acquisition methods at low speeds for a permanent magnet synchronous machine in vehicle applications

Oliver Lehmann, Matthias Zehelein, Johannes Schuster and Jörg Roth-Stielow
Institute of Power Electronics and Electrical Drives
University of Stuttgart
Stuttgart, Germany
{lehmann, zehelein, schuster, roth-stielow}@ilea.uni-stuttgart.de

Abstract— In this paper three sensorless position acquisition methods at standstill and low speeds for a permanent magnet synchronous machine based on the evaluation of the phase current step responses are introduced and compared in terms of usability for electric vehicles. Besides the well-known INFORM method a new method is described which uses switching commands as test impulses and therefore does not require any additional test signals which yields to an increase of the system efficiency. All three methods were implemented on a test system and measurements were done to compare their performance.

Keywords— *Comparison, permanent magnet synchronous machine, sensorless control, standsill and low speeds*

I. INTRODUCTION

Nowadays more and more electric and hybrid electric vehicles reach the state of series-production. Most of them are equipped with permanent magnet synchronous machines (PMSM) due to high efficiency and high torque-per-volume-ratio. With a special arrangement of the permanent-magnets and a special rotor geometry, investigated in [1], it is possible to reach a wide range of constant power, which is demanded in vehicular applications. This leads to an anisotropy in the magnetic circuit which allows an increased efficiency with maximum torque per ampere techniques (MTPA) [2] and furthermore enables the use of sensorless control methods at low speeds and standstill. The common control method of machines in vehicular applications is a torque control using a current control in a rotor fixed reference frame. Therefore, the actual rotor position is needed to transform the measured phase currents into this reference frame. This kind of control is known as the field oriented control (FOC) [3]. The actual rotor position is typically measured with a shaft encoder, which causes additional costs and decreases the system reliability. To avoid these disadvantages, efforts were made to adapt sensorless control techniques known from industrial applications to vehicular applications. Even if it is not possible to realize a control system without a shaft sensor, e.g. for safety

reasons, sensorless control still gives the opportunity to use simplified and cheaper shaft sensors. Furthermore sensorless control methods can be used to realize a redundancy without using an additional mechanical encoder.

Sensorless control methods can be divided into three main tasks: sensorless position acquisition, position processing and speed calculation. For a PMSM the sensorless position acquisition can be classified according to the physical effects used for the position estimation. One is the detection of the induced voltage or EMF [4]. With a state observer or the calculation of the flux linkage the rotor position φ_{el} can be obtained. Sensorless control methods based on this effect are well investigated and on duty in various applications. Unfortunately these methods need a certain minimum speed to detect the induced voltages. At standstill and low speeds these methods cannot be used. Therefore another concept is needed to obtain a sensorless position signal at low speeds and standstill. These concepts are based on the evaluation of saturation effects or saliency. Saturation or saliency can be detected by measuring the machine inductance, which will show a position dependency, especially under the assumption that the machine shows a measureable magnetic anisotropy, which is always given in this application. There are two common ways to measure an inductance, either by measuring the impedance via high-frequency signal injection [5], [6] or by evaluation of the differential equation of the inductance via voltage test pulses and detection of the current transient [7]. Sensorless control methods used in vehicular application must fulfill strict requirements in terms of safety and efficiency. This paper compares three sensorless position acquisition methods, based on the evaluation of voltage test pulses concerning the use in vehicular applications.

978-1-4799-2706-7/14 $31.00 © 2014 IEEE

II. SENSORLESS CONTROL METHODS FOR LOW SPEEDS

A. The INFORM-method (INFORM 3)

As a well-known method for low speed sensorless control, the INFORM method must be taken into account for this comparison. The basic idea of the INFORM method was described in [7], where it is suggested to use a test vector in each phase direction. Therefore the superimposed control and the PWM is paused while the test impulses are applied. The current response to each test impulse is measured and after the test impulse sequence the control and the PWM are restarted. This procedure repeats with the period T_A, until a maximum speed is reached, where an insufficient amount of measurements per revolution leads to an inaccurate position information. The principle of an INFORM measurement is depicted in Fig. 1.

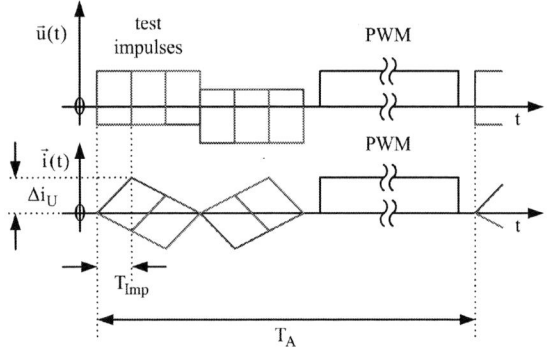

Fig. 1.   Typical INFORM measurement sequence

When regarding only a positive current derivative, the resulting curves of the three current step responses are depicted in Fig. 2.

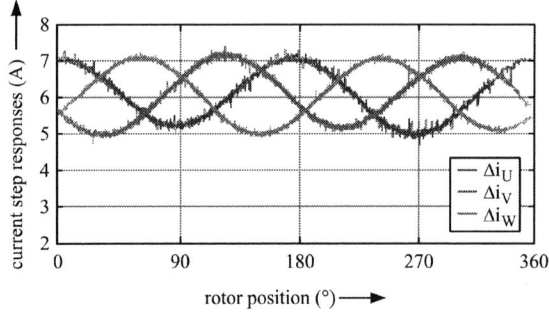

Fig. 2.   Current step responses using INFORM 3

The position can be calculated according to (1) with a remaining 180 degree uncertainty. This formula is only valid if the quadrature inductance L_q is larger than the direct inductance L_d.

$$2\varphi_{el} = \pi - \arg\left[\begin{array}{c} \left(-\Delta i_U + \frac{1}{2}\Delta i_V + \frac{1}{2}\Delta i_W\right) \\ + j \cdot \left(\frac{\sqrt{3}}{2}\Delta i_W - \frac{\sqrt{3}}{2}\Delta i_V\right) \end{array}\right] \quad (1)$$

The main advantage of this method is the parameter independency. The rotor position φ_{el} is only a function of the current responses to the voltage test impulses.

$$2\varphi_{el} = f\left(\Delta i_U, \Delta i_V, \Delta i_W\right) \quad (2)$$

The sequential application of the test impulses allows even a one shunt current measurement [8].

The disadvantages, besides the additional losses generated through the test impulses are the interruptions of the PWM pattern to apply the test impulses and the long duration of the test sequence.

B. The modified INFORM method using only one test impulse (INFORM 1)

To optimize the measurement sequence in terms of efficiency, there is the possibility to use only one voltage test vector instead of three and its inverted. Therefore the responses of all phase currents are measured for the position calculation. This should decrease the losses of the position acquisition. Fig. 3 shows the resulting curves of the phase current step responses.

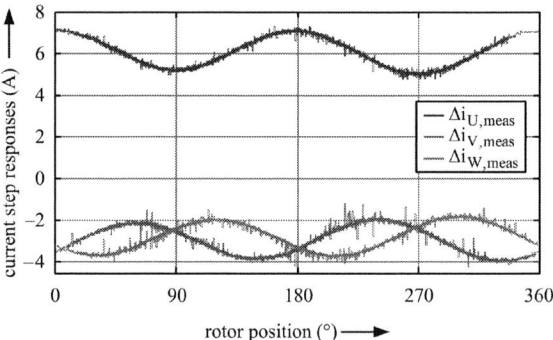

Fig. 3.   Current step responses using the modified INFORM 1

The mean values of the three curves are different and therefore some adjustment is needed before calculating the position. One possibility to adjust the current step responses is shown in (3).

$$\Delta i_U = \Delta i_{U,meas}$$
$$\Delta i_V = \Delta i_{V,meas} + V_{DC} \cdot \frac{T_{Imp}}{L_0} \quad (3)$$
$$\Delta i_W = \Delta i_{W,meas} + V_{DC} \cdot \frac{T_{Imp}}{L_0}$$

With these adjustments a similar equation to (1) can be used to calculate the rotor position. Now the amount of test impulses is reduced but the parameter independency of the standard INFORM 3 method is lost. To adjust the measured current step responses the knowledge of the DC-link voltage V_{DC} as well as the constant part of the inductance L_0 must be known. Using the INFORM 1 method the calculation of the rotor position is a function of the current responses, the DC-link voltage, the constant

978-1-4799-2706-7/14 $31.00 © 2014 IEEE

part of the line inductance L_0 and the duration of the test impulse T_{Imp}.

$$2\varphi_{el} = f\left(\Delta i_{U,meas}, \Delta i_{V,meas}, \Delta i_{W,meas}, V_{DC}, L_0, T_{Imp}\right) \quad (4)$$

Since the responses of all three phase currents to one test impulse are needed, a one shunt measurement, cannot be used with the INFORM 1 method.

The accuracy of both INFORM methods can be improved by taking the stator resistance R_s into account which leads to a further dependency in the position calculation.

C. A new method without additional impulses by evaluating the switching commands

1) Basic idea

Instead of generating additional test impulses, the best way in terms of efficiency is to consider the switching pattern of the inverter output as test impulses. Besides that no additional losses are generated, the measuring period T_A is shortened to a minimum. In literature some approaches have already been investigated [9], [10]. There are several aspects which must be considered if the switching commands should be analyzed. The durations of the switching commands are not fixed and varies with the controller output. This means that the standard INFORM equation cannot be used to calculate the rotor position. In [9] some equations are described, but in this paper a noval set of equations is used, based on the phase equations of a PMSM with saliency as described in [11].

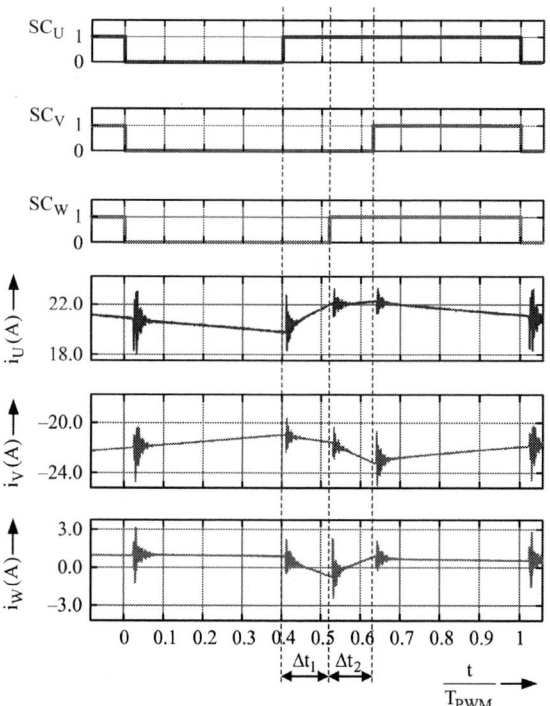

Fig. 4. Basic idea of the switching command evaluation

Doing so, the calculation of the position is a function of the machine's line resistance R_s and the machine inductances L_d and L_q as well as the measured phase voltages, phase currents and the derivatives of the phase currents.

$$2\varphi_{el} = f\left(v_U, v_V, v_W, i_U, i_V, i_W, R_s, L_d, L_q, \frac{di_U}{dt}, \frac{di_V}{dt}, \frac{di_W}{dt}\right) (5)$$

2) Calculation of the rotor position

According to [11] the voltage phase equations for a PMSM with saliency at standstill and low speeds $(\omega_{el} \approx 0)$ can be written as in (6).

$$\begin{pmatrix} v_U \\ v_V \\ v_W \end{pmatrix} = R_s \cdot \begin{pmatrix} i_U \\ i_V \\ i_W \end{pmatrix} + \underbrace{\begin{pmatrix} L_{UU} & L_{UV} & L_{UW} \\ L_{VU} & L_{VV} & L_{VW} \\ L_{WU} & L_{WV} & L_{WW} \end{pmatrix}}_{L} \cdot \begin{pmatrix} \dfrac{di_U}{dt} \\ \dfrac{di_V}{dt} \\ \dfrac{di_W}{dt} \end{pmatrix} (6)$$

The elements of the inductance matrix L can be calculated in dependency of L_d, L_q and the rotor position φ_{el}.

$$L_{UU} = \frac{L_d + L_q}{3} + \cos(2\varphi_{el}) \cdot \frac{L_d - L_q}{3}$$

$$L_{UV} = -\frac{L_d + L_q}{6} - \cos(2\varphi_{el}) \cdot \frac{L_d - L_q}{6} + \sqrt{3}\sin(2\varphi_{el}) \cdot \frac{L_d - L_q}{6}$$

$$L_{UW} = -\frac{L_d + L_q}{6} - \cos(2\varphi_{el}) \cdot \frac{L_d - L_q}{6} - \sqrt{3}\sin(2\varphi_{el}) \cdot \frac{L_d - L_q}{6}$$

$$L_{VU} = L_{UV}$$

$$L_{VV} = \frac{L_d + L_q}{3} - \cos(2\varphi_{el}) \cdot \frac{L_d - L_q}{6} - \sqrt{3}\sin(2\varphi_{el}) \cdot \frac{L_d - L_q}{6} \quad (7)$$

$$L_{VW} = -\frac{L_d + L_q}{6} + \cos(2\varphi_{el}) \cdot \frac{L_d - L_q}{3}$$

$$L_{WU} = L_{UW}$$

$$L_{WV} = L_{VW}$$

$$L_{WW} = \frac{L_d + L_q}{3} - \cos(2\varphi_{el}) \cdot \frac{L_d - L_q}{6} + \sqrt{3}\sin(2\varphi_{el}) \cdot \frac{L_d - L_q}{6}$$

To simplify the following derivation the resistive voltage drop is subtracted in equation (8) resulting in the voltage vector $\vec{v}_L$ applied to the position depending inductance.

$$\vec{v}_L^{UVW} = \begin{pmatrix} v_{LU} \\ v_{LV} \\ v_{LW} \end{pmatrix} = \vec{v}^{UVW} - R_s\vec{i}^{UVW} = L \cdot \frac{d\vec{i}^{UVW}}{dt} \quad (8)$$

Taking into account that the PMSM is star connected and therefore the system of equations (8) is overdetermined, this system can be converted according to (9).

$$
\left(\begin{array}{c} \dfrac{v_{LU} - \dfrac{1}{2}(L_d + L_q)\dfrac{di_U}{dt}}{\dfrac{1}{2}(L_d - L_q)} \\[2ex] \dfrac{v_{LV} - \dfrac{1}{2}(L_d + L_q)\dfrac{di_V}{dt}}{\dfrac{1}{2}(L_d - L_q)} \end{array}\right) = \boldsymbol{A} \cdot \begin{pmatrix} \cos(2\varphi_{el}) \\ \sin(2\varphi_{el}) \end{pmatrix}
\tag{9}
$$

$$
\text{with } \boldsymbol{A} = \left(\begin{array}{cc} \left[\dfrac{di_U}{dt}\right] & \left[\dfrac{1}{\sqrt{3}}\dfrac{di_U}{dt} + \dfrac{2}{\sqrt{3}}\dfrac{di_V}{dt}\right] \\[2ex] \left[-\dfrac{di_U}{dt} - \dfrac{di_V}{dt}\right] & \left[\dfrac{1}{\sqrt{3}}\dfrac{di_U}{dt} - \dfrac{1}{\sqrt{3}}\dfrac{di_V}{dt}\right] \end{array}\right)
$$

The desired vector containing the position information can be calculated with the inverse of the Matrix **A**.

$$
\begin{pmatrix} \cos(2\varphi_{el}) \\ \sin(2\varphi_{el}) \end{pmatrix} = \boldsymbol{A}^{-1} \left(\begin{array}{c} \dfrac{v_{LU} - \dfrac{1}{2}(L_d + L_q)\dfrac{di_U}{dt}}{\dfrac{1}{2}(L_d - L_q)} \\[2ex] \dfrac{v_{LV} - \dfrac{1}{2}(L_d + L_q)\dfrac{di_V}{dt}}{\dfrac{1}{2}(L_d - L_q)} \end{array}\right)
\tag{10}
$$

Now the best way to obtain the rotor position is to define a complex number $\underline{c}$ consisting of the sin and cos terms whose argument corresponds to the rotor position:

$$
\begin{aligned}
\underline{c} &= \cos(2\varphi_{el}) + \mathrm{j}\cdot\sin(2\varphi_{el}) \\
\arg(\underline{c}) &= 2\varphi_{el}
\end{aligned}
\tag{11}
$$

3) Acquisition of the required measures

The proposed calculation of the rotor position without any additional test impulses still needs some measurement information. The voltage across the position depending inductance, and the according current derivative di/dt. Usually a PMSM for traction applications has no starpoint to measure the phase voltages. Therefore the phase voltages are estimated using the switching state of the inverter and the DC-link voltage V_{DC}. In doing so, a symmetric behavior of the three phases of the PMSM is assumed.

$$
\begin{pmatrix} v_U \\ v_V \\ v_W \end{pmatrix} = \frac{1}{3}V_{DC}\begin{pmatrix} 2 & -1 & -1 \\ -1 & 2 & -1 \\ -1 & -1 & 2 \end{pmatrix}\begin{pmatrix} SC_U \\ SC_V \\ SC_W \end{pmatrix}
\tag{12}
$$

The actual current values $\vec{i}$ needed to calculate the resistive voltage drop can be measured with common current sensors which are usually already integrated in the system to obtain the feedback for the current controller.

To capture the current derivatives two different possibilities have been investigated. The first is the

calculation of the derivatives by using two current measurements and detect the switching command duration Δt as shown in Fig. 4.

Doing so the current derivative of one phase can be calculated according to (13), neglecting the resistive part of the windings.

$$
\frac{di}{dt} = \frac{\Delta i}{\Delta t}
\tag{13}
$$

The other possibility is to measure the current derivative directly, by using adequate sensors, in this study Rogowski coils were used [12]. The output voltages v_{Rog} of these coils are proportional with the coupling inductance M to the derivatives of the measured currents.

$$
v_{Rog} = M \cdot \frac{di}{dt}
\tag{14}
$$

III. IMPLEMENTATION

A. The test system

To compare the three sensorless position acquisition methods a test system, as depicted in Fig. 5, was set up. A field orientated current controller with a superimposed speed controller was realized on a dSpace DS1103 rapid prototyping system. The three sensorless position acquisition methods were implemented on a FPGA Cyclone III industrial board extended by high speed ADCs. This enables a precise current measurement, a switching command analysis and a fast computation of the algorithms. A modular setup allows a quick change of the different current sensors. All values processed by the FPGA can be sent to the dS1103 system via a 32 Bit parallel bus for analysis. The PMSM was coupled to a load machine to compare the performance of the sensorless control algorithms under constant burden and load transients. Within the regarded torque range, which was limited by the load machine, no saturation effects were observed, allowing the use of the described methods without any additional adaptions.

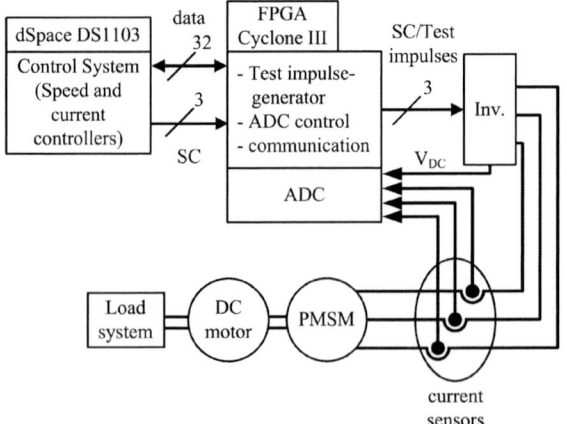

Fig. 5. System overview

B. The current sensors

To capture the currents, different current sensors are compared like a standard current sensor (LEM 505) and a precise laboratory current measurement device (ILA SMZ 200). To capture the current derivative directly, high accuracy Rogowski coils, assembled in own production were used.

C. Encounterd problems and solutions

The measurement of the current gradient during one PWM period shows high frequency disturbances during the switching process. Unfortunately these disturbances need some time to decay. In this test setup they last about 8 µs, as depicted in Fig. 4. In case of the sensorless control method based on switching command analysis, this means that a single switching command needs to last at least 8 µs before a useable current signal can be obtained. At standstill and very low speeds, below 10 rpm, the duration of the switching commands even under load are shorter than 8 µs. To expand the duration two approaches were implemented. First, an asymmetric, sawtooth PWM was used to double the length of the switching commands as depicted in Fig. 4. Second, a speed-variable DC-link voltage was used, to reduce the voltage at low speeds.

With all three introduced methods the rotor position can only be estimated with a 180 degree uncertainty, based on the rotor anisotropy. To get a 360 degree resolution a sector detection was implemented. The actual position is compared to the previous position. If the difference $\Delta\varphi_{obsv}$ crosses a certain limit the sector has changed and an offset of either zero or π must be added to the actual position, according to (15). The starting information $\varphi_{offset}(0)$ can be obtained with a method described in [13].

When using the switching command evaluation the obtained position signals have some high frequency noise. Therefore a Kalman filter was designed to suppress these disturbances. In addition the Kalman filter also calculates the actual speed $\omega_{mech,obsv}$ and load values $M_{L,obsv}$.

Fig. 6 shows the resulting control system used for this investigation, consisting of the sensorless position acquisition methods, the Kalman filter, the 360 degree

dissolver and the control system. For the evaluation a signal from a shaft encoder $\varphi_{el,g}$ is fed back.

$$\Delta\varphi_{obsv} = \varphi_{el,obsv}(n) - \varphi_{el,obsv}(n-1)$$

$$\left(\Delta\varphi_{obsv} > \frac{\pi}{2} \vee \Delta\varphi_{obsv} < -\frac{\pi}{2}\right)$$

$$\Rightarrow \varphi_{offset}(n) = \mathrm{mod}\left(\varphi_{offset}(n-1) + \pi, 2\pi\right) \quad (15)$$

$$\left(\Delta\varphi_{obsv} > \frac{\pi}{2} \vee \Delta\varphi_{obsv} < -\frac{\pi}{2}\right)$$

$$\Rightarrow \varphi_{offset}(n) = \varphi_{offset}(n-1)$$

$$\varphi_{el,obsv}(n) = \varphi_{el,obsv}(n) + \varphi_{offset}(n)$$

D. Measurement parameters

The current control unit was implemented as a PI-controller and the speed control unit is a state regulator with an additional disturbance stabilizer. The duration of the INFORM test impulses T_{Imp} was varied and the period of INFORM measurements T_A was set according to values found in literature [13], [14], [15]. Measurements under different speed and load characteristics were done. To measure the efficiency a power measuring device (ZES Zimmer LMG500) was used and the time-averaged DC input power over one minute was measured at fixed operating points. Table I gives an overview of the most important parameters.

TABLE I
MEASUREMENT PARAMETERS

Symbol	Meaning	Value
T_A	Period of INFORM measurement	1 ms
T_{Imp}	Duration of a test impulses	10 µs…50 µs
n	Machine speed	10 rpm…100rpm
M_L	Load torque	0 Nm…50 Nm
f_{PWM}	PWM-frequency	8 kHz
V_{DC}	DC-link voltage	50 V

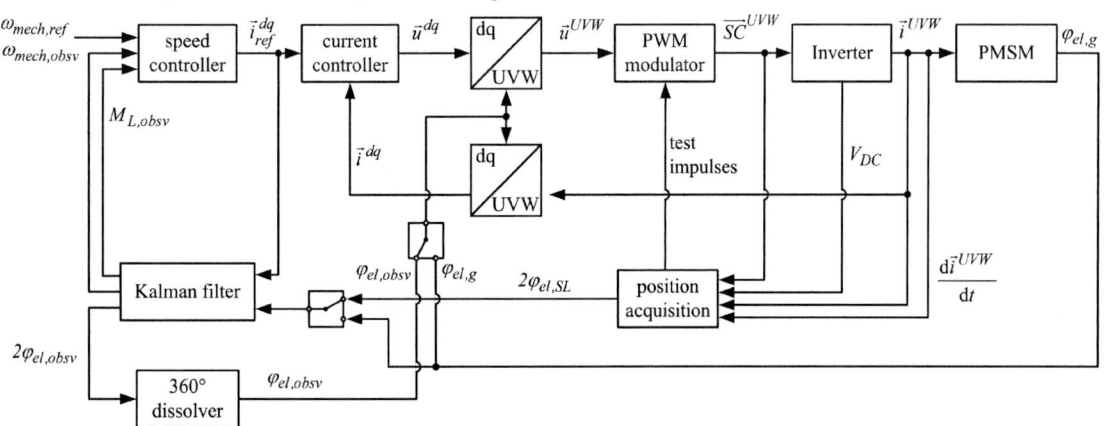

Fig. 6. The control system

IV. MEASUREMENT RESULTS

The following measurements are divided into results concerning the INFORM methods, results concerning the switching command evaluation (SCE) and results concerning the comparison of all methods.

A. Results concerning the INFORM methods

In a first measurement the duration of the INFORM test impulses was varied and the standard deviation to the sensor signal was analysed. It can be seen that the accuracy of the INFORM methods increases with the duration of the test impulse T_{Imp} as depicted in Fig. 7 and Fig. 8.

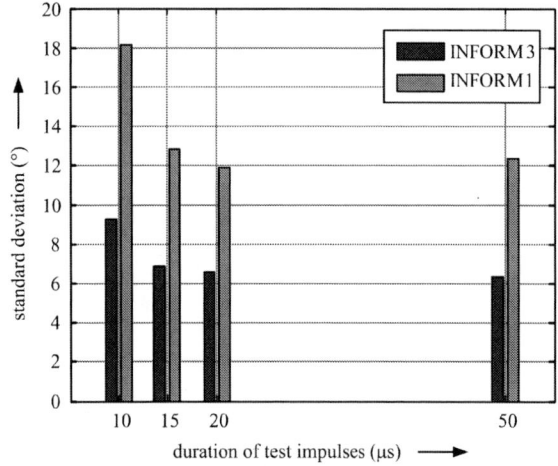

Fig. 7. Comparision of the INFORM methods by variing the duration of the test impulses using SMZ at zero torque and 10 rpm

The INFORM 3 method produces a more accurate position information than the INFORM 1 method. The smaller current rise and the parameter dependent adjustment when using INFORM 1 are the cause. In Fig. 8 the difference between a precise laboratory current sensor (SMZ) and an industrial current sensor (LEM) can be seen.

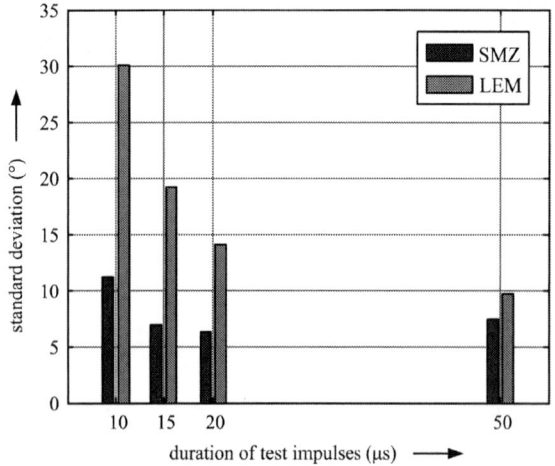

Fig. 8. Comparison of the current sensors using INFORM 3 under 50 Nm load at 10 rpm

Again longer test impulses result in higher currents and therefore to a higher accuracy. On the other hand the longer the test impulses are, the more losses are produced.

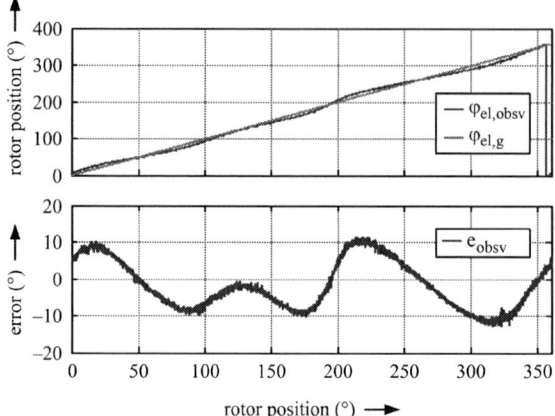

Fig. 9. Sensorless control at 10 rpm using INFORM 3 and 50 µs test impulses under zero torque

Above an impulse duration of about 20 µs no further accuracy is gained. Fig. 9 shows that the error between the sensorless determined rotor position and the encoder signal is possessing a significant course. The reasons are additional effects in the magnetic circuit not covered by the INFORM equations yet, which gives room for further optimization.

B. Results concerning the switching command evaluation (SCE)

Fig. 10 shows a comparison between the possibilities of capturing the current derivation at zero torque and under load.

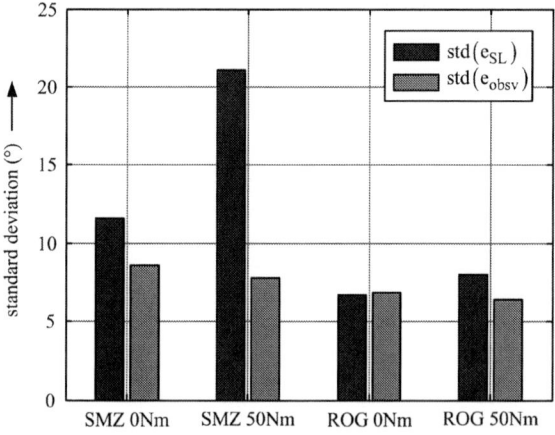

Fig. 10. Comparison of the different current sensores using SCE at 10 rpm

The results, done with the Rogowski coils, tend to lead to less deviation. Furthermore by using the Rogowski coils there is no need for a filtering effort. With the standard current sensor (LEM) no reasonable sensorless operation could be achieved due to the short switching command durations. Again the characteristic of the difference between encoder and sensorless determined rotor position shows some significant deviations as depicted in Fig. 11. This course is independent of the used current derivative capturing method.

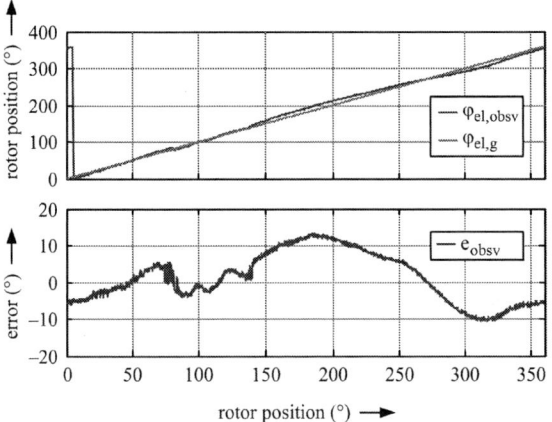

Fig. 11. Sensorless control at 10 rpm using SCE

In summary the proposed method of evaluating the switching commands operates very well and an operation in vehicular application is possible.

C. Speed characteristics

In Fig. 12 an acceleration operation under load done with all presented sensorless methods is depicted.

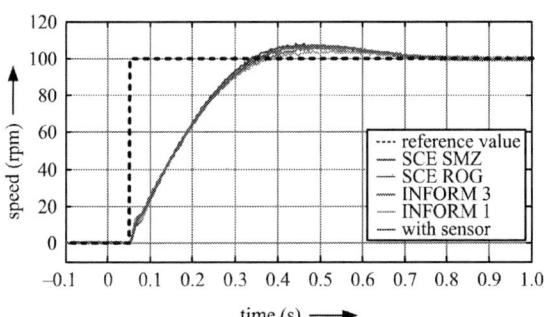

Fig. 12. Sensorless acceleration process under 50 Nm load

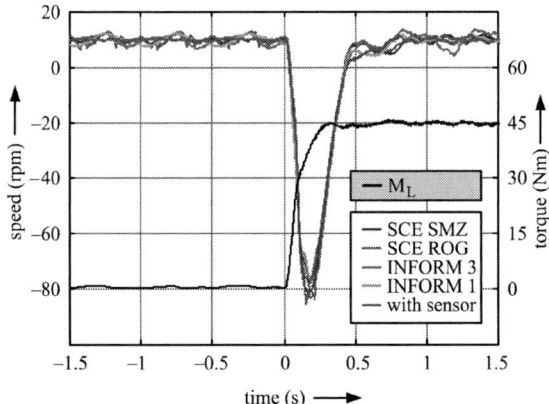

Fig. 13. Response to a load transient

The results show that all methods have quite the same transient response. The same conclusion can be done concerning Fig. 13, which shows the behaviour of the speed controller to a load transient.

In this test setup the control's disturbance reaction needs some improvement, but the behaviour of the control system with position sensor is quite the same as the sensorless control systems

D. Efficency

In Fig. 14 the influence of the test impulses on the measured input power at 10 rpm and zero torque is depicted. This measurement shows that the losses increase with longer test impulses. The INFORM 1 method does not reduce the losses that much.

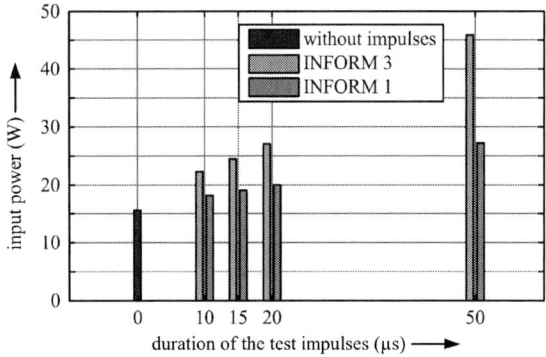

Fig. 14. Influence of the test impulses on the input power at 10 rpm and zero torque

To compare the efficiency of the presented sensorless control methods the efficiency factor at different speeds with a constant load of 50 Nm was measured and compared to the efficiency factor of the system with position sensor. The differences between the efficiency factors are depicted in Fig. 15.

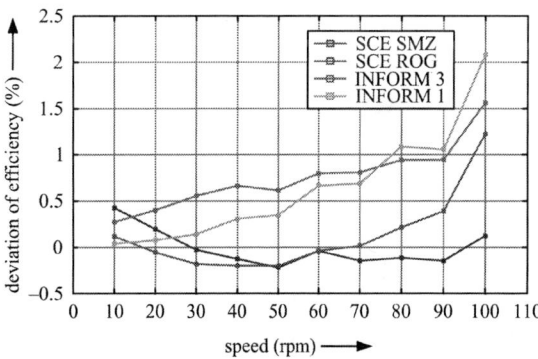

Fig. 15. Deviation of the efficency factors from the control system with encoder

Fewer test impulses result in a better efficiency. Although the difference between INFORM 1 and INFORM 3 is not as obvious as expected. The efficiency of the control system using the switching command evaluation is nearly as good as the system with position sensor, since no additional test impulses are needed.

V. CONCLUSION

The implemented test system is able to control the machine with all three presented sensorless control methods. The achieved accuracy of all methods, using only the basic equations, is quite good. When using an industrial current sensor the duration of test impulses must be increased to obtain an usable signal for the sensorless control methods. The capability of the presented new method of sensorless position acquisition using the evaluation of the switching commands was shown. This new method is able to measure the actual rotor position each PWM period which is an improvement compared to the methods based on test impulses. In addition, by omitting additional test impulses the efficiency of the whole traction system is increased.

Disturbances in the phase currents caused by the semiconductors during switching demand a minimum switching command duration. Unfortunately by using a PWM pattern this minimum duration is not guaranteed in all operation points. The interaction of the SCE with a modulator which guarantees a minimum switching command duration should be investigated in future works.

APPENDIX

All investigations are done on the traction machine shown in Fig. 16 with parameters given in table II.

TABLE II
MACHINE PARAMETERS

Symbol	Meaning	Value
R_s	Stator resistance	$0.051\ \Omega$
L_d	Direct inductance	$180\ \mu H$
L_q	Quadrature inductance	$245\ \mu H$
Ψ_{PM}	Flux linkage	$0.0365\ Vs$
pp	Pole pairs	12
J	Inertia	$0.211\ kgm^2$
I_{max}	Maximum current	$300\ A$
n_{max}	Maximum speed	$6000\ rpm$

Fig. 16. Photograph of the test bench

REFERENCES

[1] Y. Honda, T. Nakamura, T. Higaki, Y. Takeda, "Motor Design Considerations and Test Results of an Interior Permanent Magnet Synchronous Motor for Electric Vehicles" *IEEE Industry Applications Conference 1997*, vol. 1, pp. 75-82, 1997.

[2] D. Schroeder, "Elektrische Antriebe – Regelung von Antriebssystemen 3. Auflage" *Springer-Verlag*, Berlin Heidelberg, pp. 870-875, 2009.

[3] F. Blaschke, "Das Prinzip der Feldorientierung, Die Grundlage für die TRANSVEKTOR-Regelung von Asynchronmaschinen" *Siemens Zeitschrift*, vol. 45 pp. 757-760, 1971.

[4] P.P. Acarnley, J.F. Watson, "Review of Position-Sensorless Operation of Brushless Permanent-Magnet Machines" *IEEE Transactions on Industrial Electronics*, vol. 53, no. 2, pp. 352-362, 2006.

[5] M. Linke, R. Kennel, J. Holtz, "Sensorless speed and position control of synchronous machines using alternating carrier injection" *IEEE Int. Electric Machines and Drives Conference 2003*, pp. 1211-1217, 2003.

[6] M. Corley, R.D. Lorenz, "Rotor position and velocity estimation for a salient-pole permanent magnet synchronous machine at standstill and high speeds" *IEEE Transactions on Industry Applications*, vol. 34, no. 4, pp. 784-789, 1998.

[7] M. Schroedl, "Sensorless Control of AC Machines at Low Speed and Standstill Based on the "INFORM" Method" *IEEE Industry Applications Conference 1996*, vol. 1, pp. 270-277, 1996.

[8] U.H. Rieder, M. Schroedl, A. Ebner, "Sensorless Control of an External Rotor PMSM in the Whole Speed Range including Standstill using DC-link Measurements only" *IEEE Power Electronics Specialists Conference 2004*, vol. 2, pp. 1280-1285, 2004.

[9] S. Bolognani, S. Calligaro, R. Petrella, M. Sterpellone, "Sensorless Control for IPMSM using PWM Excitation: Analytical Developments and Implementation Issues" *Symposium on Sensorless Control for Electrical Drives*, pp. 64-73, 2011.

[10] M. Vogelsberger, S. Grubic, T. Habetler, T. Wolbank, "Using PWM-Induced Transient Excitation and Advanced Signal Processing for Zero-Speed Sensorless Control of AC Machines" *IEEE Transactions on Industrial Electronics*, vol. 57, no. 1, pp. 365-374, 2010.

[11] P. Vas, "Electrical Machines and Drives – A Space-Vector Theory Approach" *Oxford University Press*, New York, pp. 80-115, 1992.

[12] W.F. Ray, C.R. Hewson, "High Performance Rogowski Current Transducers" *IEEE Industry Applications Conference*, vol. 5, pp. 3083-3090, 2000.

[13] M. Braun, O. Lehmann, J. Roth-Stielow, "Sensorless rotor position estimation at standstill of high speed PMSM drive with LC inverter output filter" *IEEE ICIT 2010*, pp. 410-415, 2010.

[14] E. Robeischl, M. Schroedl, "Optimized INFORM-measurement sequence for sensorless PM synchronous motor drives with respect to minimum current distortion" *IEEE Transactions on Industry Applications*, vol. 40, issue 2, pp. 591-598, 2004.

[15] F. Demmelmayr, M. Troyer, M. Schroedl, "Advantages of PM-machines Compared to Induction Machines in Terms of Efficiency and Sensorless Control in Traction Applications" *IECON 2011*, pp. 2762-2768, 2011.

Stability Comparison of IPMSM Sensorless Vector Control Systems Using Extended EMF

Mineo Tsuji, Hiroshi Mizusaki, Sin-ichi Hamasaki

Nagasaki University, Graduate School of Engineering

1-14 Bunkyo-machi, Nagasaki, 852-8521, Japan

Abstract—Sensorless vector control of an interior permanent magnet synchronous motor (IPMSM) has been studied in many papers. The extended electromotive force (EMF) based method is one of the representative systems. We proposed a simplified method which estimates the rotor position using the controlled voltage without using observer. In this paper we have compared these methods through trajectories of poles and experimental transient responses for stability study.

Keywords— interior permanent magnet synchronous motor, sensorless vector control, extended EMF, stability analysis

I. INTRODUCTION

Because of high efficiency and small size, PMSM is used in many application. Especially, IPMSM is possible to produce high torque by using reluctance torque.

To realize a vector control of IPMSM, an encoder for detecting rotor position is used because it is necessary to control currents in accordance with the position of the rotor. However, there are problems; the cost of the position sensor, size of the apparatus and wiring of the signal line. Therefore, many sensorless control methods without using the position sensor are proposed [1]-[6]. The method using the extended EMF is one of the representative sensorless control methods. The extended EMF is estimated by a disturbance observer, and the positional information is obtained by using it [1][2]. On the other hand, we have proposed a simple sensorless method based on the extended EMF[7]. The rotor speed is estimated by the output voltage of the *d*-axis PI current controller with the non-interference control without using the disturbance observer. Although similar simplification is proposed in earlier papers[5][6] for a non-salient pole PMSM, our proposed method can be applied to IPMSM by using the extended EMF.

In this paper, we discuss the stability by comparing the results of the proposed method with those of the conventional extended EMF sensorless one. The root loci of linear models, transient responses obtained by non-linear models and experiments are demonstrated.

II. SENSORLESS SYSTEMS

A. Extended EMF Model of IPMSM

Voltage equation of IPMSM on *d-q* rotating reference frame synchronized with the magnetic pole position is expressed as follows[1]:

$$\begin{bmatrix} v_d \\ v_q \end{bmatrix} = \begin{bmatrix} R_s + pL_d & -\omega_r L_q \\ \omega_r L_q & R_s + pL_d \end{bmatrix} \begin{bmatrix} i_d \\ i_q \end{bmatrix} + \begin{bmatrix} 0 \\ E_{ex} \end{bmatrix} \quad (1)$$

where,

$$E_{ex} = \omega_r \left\{ \left(L_d - L_q \right) i_d + \psi \right\} - \left(L_d - L_q \right) p i_q \quad (2)$$

By converting the reference frame which rotates at an angular speed $\hat{\omega}$ as shown in Fig.1, the following equation can be obtained from (1):

$$\begin{bmatrix} v_\gamma \\ v_\delta \end{bmatrix} = \begin{bmatrix} R_s + pL_d & -\omega_r L_q \\ \omega_r L_q & R_s + pL_d \end{bmatrix} \begin{bmatrix} i_\gamma \\ i_\delta \end{bmatrix} + \begin{bmatrix} e_\gamma \\ e_\delta \end{bmatrix} \quad (3)$$

The second term of the right side is called the extended EMF and expressed as

$$\begin{bmatrix} e_\gamma \\ e_\delta \end{bmatrix} = E_{ex} \begin{bmatrix} \sin \theta_e \\ \cos \theta_e \end{bmatrix} + \left(\hat{\omega} - \omega_r \right) L_d \begin{bmatrix} -i_\delta \\ i_\gamma \end{bmatrix} \quad (4)$$

B. Conventional Method

Conventional sensorless vector control system is shown in Fig.2[1][2]. The extended EMF is estimated by using disturbance observer as follows;

$$\hat{e}_\gamma = \frac{g}{s+g} \left\{ v_\gamma^* + \hat{\omega} L_q i_\delta - \left(R_s + sL_d \right) i_\gamma \right\} \quad (5)$$

$$\hat{e}_\delta = \frac{g}{s+g} \left\{ v_\delta^* - \hat{\omega} L_q i_\gamma - \left(R_s + sL_d \right) i_\delta \right\} \quad (6)$$

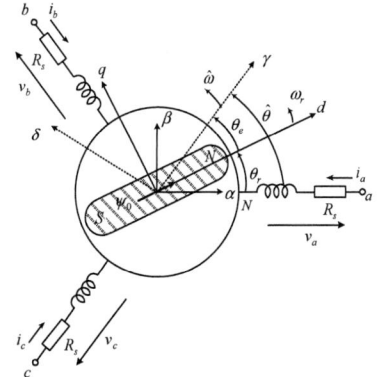

Fig. 1. Model of IPMSM.

978-1-4799-2706-7/14 $31.00 © 2014 IEEE

From (4), the magnetic pole position error $\hat{\theta}_e$ is computed by the following equation:

$$\hat{\theta}_e = \tan^{-1}\left(\frac{\hat{e}_\gamma}{\hat{e}_\delta}\right) \tag{7}$$

Using (7), the rotor angular speed is estimated by

$$\hat{\omega} = -\left(K_{ep} + \frac{K_{ei}}{s}\right)\hat{\theta}_e \tag{8}$$

By using damping coefficient ζ and natural angular frequency ω_n, the PI speed estimation gains of Fig.3 are designed as follows[2]:

$$K_{ep} = 2\zeta\omega_n, \quad K_{ei} = \omega_n^2 \tag{9}$$

To reduce the influence of noise, the rotor angular speed is estimated using a low-pass filter as follows:

$$\hat{\omega}_r = \frac{\omega_c}{s + \omega_c}\hat{\omega} \tag{10}$$

The pole position is computed by the following equation.

$$\hat{\theta} = \frac{1}{s}\hat{\omega} \tag{11}$$

By considering the difference between the γ-δ axis and d-q axis, the actual position estimation error θ_e can be expressed as follows from Fig.1:

$$\theta_e = \hat{\theta} - \theta_r \tag{12}$$

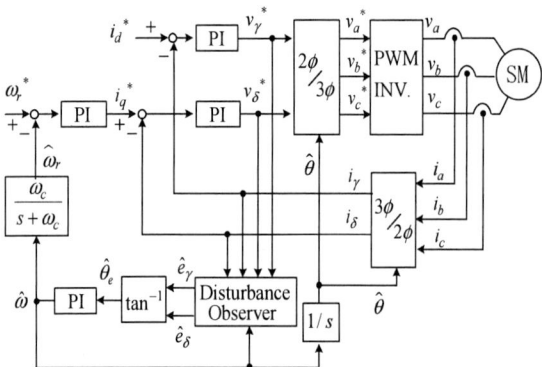

Fig.2. Conventional Sensorless Vector Control System.

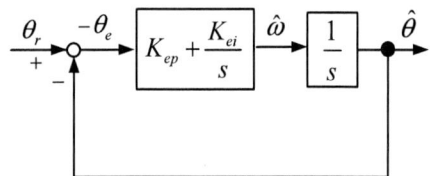

Fig.3. Approximated block diagram for speed estimation.

C. Proposed Method

Proposed sensorless vector control system is shown in Fig.4.[7].

From (4), when the error of the magnetic pole position is small, the following equation is obtained by approximating $\sin\theta_e \simeq \theta_e$ and $\cos\theta_e \simeq 1$.

$$\begin{bmatrix} e_\gamma \\ e_\delta \end{bmatrix} \simeq E_{ex} \begin{bmatrix} \sin\theta_e \\ \cos\theta_e \end{bmatrix} \simeq E_{ex} \begin{bmatrix} \theta_e \\ 1 \end{bmatrix} \tag{13}$$

From (13), the position error θ_e is approximated as

$$\theta_e \simeq \frac{e_\gamma}{E_{ex}} \tag{14}$$

Therefore, it is possible to estimate the position error by using e_γ.

In Fig.4, the d-axis PI current control is given by the following equation.

$$e_\gamma^* = \left(K_{pd} + \frac{K_{id}}{s}\right)(i_d^* - i_\gamma) \tag{15}$$

The γ- axis voltage reference v_γ^* is calculated by the following equation ignoring the differential term of (3).

$$v_\gamma^* = R_s^* i_d^* - L_q^* \hat{\omega}_r i_\delta + e_\gamma^* \tag{16}$$

The rotor angular speed $\hat{\omega}$ is estimated by the following equation using the output voltage e_γ^* from (14).

$$\hat{\omega} = -\left(K_{ep} + \frac{K_{ei}}{s}\right)\frac{e_\gamma^*}{E_{ex}^*} \tag{17}$$

E_{ex}^* is changed as a function of speed by neglecting differential term of (2).

$$E_{ex}^* = \hat{\omega}_r \left\{ (L_d - L_q)i_d^* + \psi \right\} \tag{18}$$

By using damping coefficient ζ and natural angular frequency ω_n, the PI control gains of (17) are designed as same as (9). Figure 3 shows also an equivalent block diagram for proposed speed estimation shown in Fig.4.

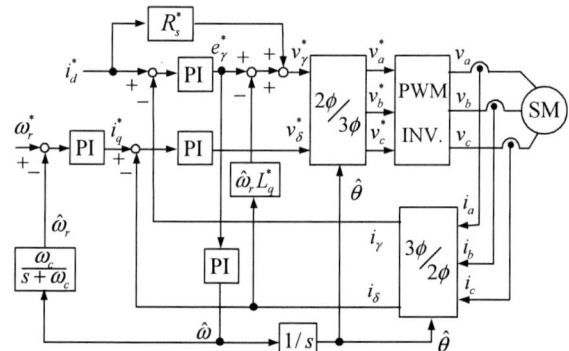

Fig.4. Proposed sensorless vector control system.

III. STABILITY ANALYSIS

A. Nonlinear Model

In order to analyze the system shown in Fig.2, we choose the γ–δ axis of Fig.1. By assuming ideal voltage control of PWM inverter, we have

$$v_a = v_a^* , \quad v_b = v_b^* , \quad v_c = v_c^* \tag{19}$$

Since $\gamma\delta$ transformation of controller is the same as the analysis, the following relation is obtained from (19).

$$v_\gamma = v_\gamma^* , \quad v_\delta = v_\delta^* \tag{20}$$

By using the actual error angle θ_e, the d-q variables are expressed by the following co-ordinate transformation.

$$\begin{bmatrix} v_d \\ v_q \end{bmatrix} = \begin{bmatrix} \cos\theta_e & -\sin\theta_e \\ \sin\theta_e & \cos\theta_e \end{bmatrix} \begin{bmatrix} v_\gamma^* \\ v_\delta^* \end{bmatrix} \qquad (21)$$

$$\begin{bmatrix} i_d \\ i_q \end{bmatrix} = \begin{bmatrix} \cos\theta_e & -\sin\theta_e \\ \sin\theta_e & \cos\theta_e \end{bmatrix} \begin{bmatrix} i_\gamma \\ i_\delta \end{bmatrix} \qquad (22)$$

The *d-q* state equations of IPMSM are obtained by Park's equation as follows:

$$pi_d = \frac{1}{L_d}(v_d - R_s i_d + \omega_r L_q i_q) \qquad (23)$$

$$pi_q = \frac{1}{L_q}(v_q - R_s i_q - \omega_r L_d i_d - \omega_r \psi) \qquad (24)$$

$$p\omega_r = \frac{P^2}{4J}\{\psi i_q + (L_d - L_q)i_d i_q\} - \frac{P}{2J}T_L \qquad (25)$$

where, P is number of poles, J is moment of inertia, and T_L is load torque.

The disturbance observer is described as follows:

$$\hat{e}_\gamma = gz_1 - gL_d i_\gamma \qquad (26)$$

where, $z_1 = \dfrac{1}{s+g}(v_\gamma^* + \hat{\omega}L_q i_\delta - R_s i_\gamma + gL_d i_\gamma) \qquad (27)$

$$\hat{e}_\delta = gz_2 - gL_d i_\delta \qquad (28)$$

where, $z_2 = \dfrac{1}{s+g}(v_\delta^* - \hat{\omega}L_q i_\gamma - R_s i_\delta + gL_d i_\delta) \qquad (29)$

The PI controllers and low pass filter are expressed by the following equations.

PI speed controller:

$$pw_1 = \omega_r^* - \hat{\omega}_r \qquad (30)$$

$$i_q^* = K_{ps}(\omega_r^* - \hat{\omega}_r) + K_{is}w_1 \qquad (31)$$

D axis PI controller:

$$pw_2 = i_d^* - i_\gamma \qquad (32)$$

$$v_\gamma^* = K_{pd}(i_d^* - i_\gamma) + K_{id}w_2 \qquad (33)$$

Q axis PI controller:

$$pw_3 = i_q^* - i_\delta \qquad (34)$$

$$v_\delta^* = K_{pq}(i_q^* - i_\delta) + K_{iq}w_3 \qquad (35)$$

PI speed estimator:

$$pz_3 = -\hat{\theta}_e \qquad (36)$$

$$\hat{\omega} = -K_{ep}\hat{\theta}_e + K_{ei}z_3 \qquad (37)$$

Low pass filter of speed estimation:

$$p\hat{\omega}_r = -\omega_c\hat{\omega}_r + \omega_c\hat{\omega} \qquad (38)$$

By taking the derivative of (12), we have

$$p\theta_e = \hat{\omega} - \omega_r \qquad (39)$$

A nonlinear model of the sensorless system shown in Fig.2 has been obtained. By using this model, we can compute transient responses.

The steady-state values are obtained by setting $p = 0$ or $s = 0$ in the nonlinear model except for (11). From (36), we have

$$\hat{\theta}_{e0} = 0 \qquad (40)$$

Then, $\hat{e}_r$ is obtained from (7) as

$$\hat{e}_{\gamma 0} = 0 \qquad (41)$$

When the motor resistance and inductance are equal to those actual values, the following equation is obtained.

$$\theta_{e0} = 0 \qquad (42)$$

In this case, the $\gamma - \delta$ axis coincides with the *d-q* axis.

B. Linear Model

By considering a small variation in the vicinity of the equilibrium point of nonlinear differential equations, a linear model is obtained.

A linear model of the IPMSM on *d-q* rotating reference frame can be expressed as follows[7]:

$$p\Delta x_s = A_s \Delta x_s + B\Delta u_s + B_T \Delta T_L \qquad (43)$$

where, $\Delta x_s = \begin{bmatrix} \Delta i_d & \Delta i_q & \Delta \omega_r \end{bmatrix}^T$, $\Delta u_s = \begin{bmatrix} \Delta v_d & \Delta v_q \end{bmatrix}^T$

By taking small perturbation of (21) and (22), we have

$$\Delta v_d = \Delta v_\gamma^* - v_{q0}\Delta\theta_e \qquad (44)$$

$$\Delta v_q = \Delta v_\delta^* + v_{d0}\Delta\theta_e \qquad (45)$$

$$\Delta i_\gamma = \Delta i_d + i_{q0}\Delta\theta_e \qquad (46)$$

$$\Delta i_\delta = \Delta i_q - i_{d0}\Delta\theta_e \qquad (47)$$

From (7), the following equation is obtained.

$$\Delta\hat{\theta}_e = \frac{\hat{e}_{\delta 0}}{\hat{e}_{\delta 0}^2 + \hat{e}_{\gamma 0}^2}\Delta\hat{e}_\gamma - \frac{\hat{e}_{\gamma 0}}{\hat{e}_{\delta 0}^2 + \hat{e}_{\gamma 0}^2}\Delta\hat{e}_\delta \qquad (48)$$

By using (41), we have

$$\Delta\hat{\theta}_e = \frac{1}{\hat{e}_{\delta 0}}\Delta\hat{e}_\gamma \qquad (49)$$

A linear model of the controller of Fig.2 is expressed as follows(refer to Appendix):

$$p\Delta w = A_w \Delta w + A_x \Delta x_s + B_r \Delta r \qquad (50)$$

where ,

$$\Delta w = \begin{bmatrix} \Delta z_1 & \Delta z_2 & \Delta z_3 & \Delta\hat{\omega}_r & \Delta w_1 & \Delta w_2 & \Delta w_3 & \Delta\theta_e \end{bmatrix}^T,$$

$$\Delta r = \begin{bmatrix} \Delta i_d^* & \Delta\omega_r^* \end{bmatrix}^T$$

The Δw is state vector. Δz_1 and Δz_2 for disturbance observer, Δz_3 for speed estimation, Δw_1 for speed PI control, Δw_2 and Δw_3 for current PI control are necessary.

The relationship between the controller and the motor input is expressed as follows (see Appendix):

$$\Delta u_s = F_w \Delta w + F_x \Delta x_s + F_r \Delta r \qquad (51)$$

From (43) ,(50) and (51), the linear model of overall system is derived as follows:

$$p\begin{bmatrix} \Delta x_s \\ \Delta w \end{bmatrix} = \begin{bmatrix} A_s + B_s F_x & B_s F_w \\ A_x & A_w \end{bmatrix}\begin{bmatrix} \Delta x_s \\ \Delta w \end{bmatrix}$$
$$+ \begin{bmatrix} B_s F_r \\ B_r \end{bmatrix}\Delta r + \begin{bmatrix} B_T \\ 0 \end{bmatrix}\Delta T_L \qquad (52)$$

As for the proposed method shown in Fig.4, a linear model is derived in the same manner[7].

Using the linear models described above, the stability analysis is performed by computing the eigenvalues of system matrix. The tested IPMSM has the following rated and nominal values: rated output 800W , rated speed 3000rpm , $P = 8$, $R_s = 0.4\Omega$, $L_d = 3.42\text{mH}$, $L_q = 3.82\text{mH}$, $\psi = 0.0845\text{Wb}$ and $J = 0.0048\text{kgm}^2$. We discuss the effect of the variable parameters which are the damping coefficient ζ, the natural angular frequency ω_n of the PI speed estimator and the cross angular frequency ω_{sc} of the PI speed controller. In this case, the speed reference $N_r^* = 500\text{rpm}$, the cut off frequency of LPF $\omega_c = 300\text{rad/s}$, the load torque $T_L = 0.6\text{Nm}$. The observer gain of the conventional system is set as $g = 600\text{rad/s}$. The cut-off frequencies of all PI current controllers are designed 1000rad/s.

Fig.5 shows the trajectories of poles for changes of ω_{sc} and ω_n. The root loci of the conventional method and the proposed method are shown in (a) and (b) respectively. From Fig.5, it is observed that the system becomes oscillating when ω_n is small and ω_{sc} is large. By comparing the root loci of the proposed method and those of the conventional method, it is found that the dominant root loci are very close. Fig.6 shows the trajectories of poles for changes of ω_{sc} and ζ. From Fig.6, the system becomes unstable when ζ is small and ω_{sc} is large. The dominant root loci are also very close for the conventional method and the proposed one.

Fig.7 shows the root loci when the pole position is estimated by the output of the LPF without using (11) in the conventional method as

$$\hat{\theta} = \hat{\omega}_r / s \qquad (53)$$

By comparing the root loci of Fig.5(a) and those of Fig.7, the difference occurs when ω_n is large. In Fig.7, the system becomes unstable when ω_n is large because of LPF. In a real system, a delay occurs in $\hat{\theta}$ for per- forming PWM control by a DSP. Consequently, it is considered

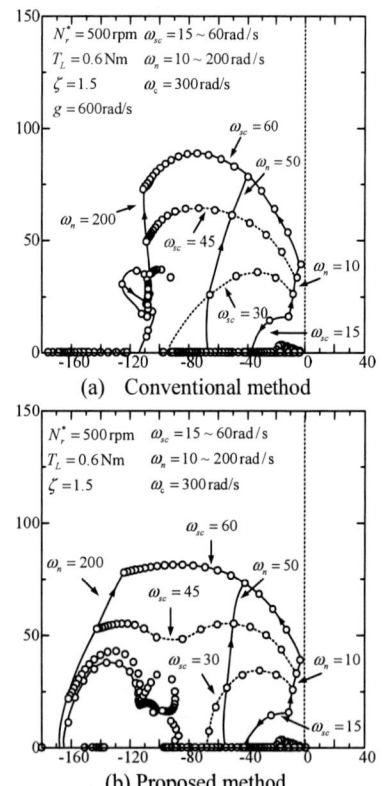

(a) Conventional method

(b) Proposed method

Fig.5. Trajectories of poles for changes of ω_{sc} and ω_n.

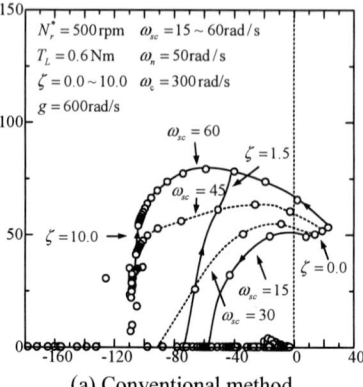

(a) Conventional method

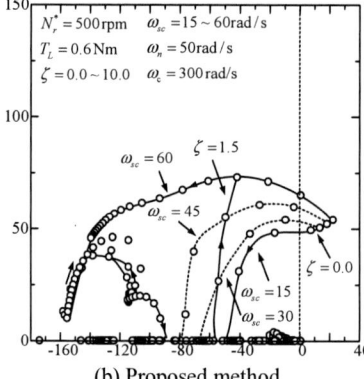

(b) Proposed method

Fig.6. Trajectories of poles for changes of ω_{sc} and ζ.

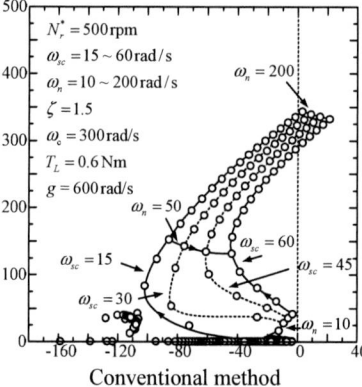

Conventional method

Fig.7. Trajectories of poles for changes of ω_{sc} and ω_n when $\hat{\theta}$ is computed by $\hat{\omega}_r$.

that there is a limit of ω_n as shown in Fig.7 in practical system even if we use (11).

IV. SIMULATION AND EXPERIMENTAL RESULTS

With the control parameters that we used in the stability analysis, we simulated the transient responses for the step change of speed reference from 500rpm to 550rpm by the non-linear models. In addition, we performed the experiment at the same conditions.

Simulation results are shown in Figs.8, 10, 12, 14 and 16. Corresponding experimental results are shown in Figs.9, 11, 13, 15 and 17. The parameters ω_{sc} is fixed and ω_n and ζ are changed. N_r is the actual speed, $\hat{N}_r$ is the estimated speed, θ_e is the magnetic pole position error.

In Fig.8, the oscillation of low frequency is observed in both methods and the position error θ_e is large. In Fig.9, the experimental results are similar to the simulation results. When $\omega_n = 12$, the dominant pole is close to the imaginary axis from Fig.5. Therefore, it is considered that the oscillation is observed. If ω_n in less than 12, the system becomes unstable experimentally. When $\omega_n = 50$, the root is stabilized in Fig.5, the oscillation such as Fig.6 disappears in Fig.10. The experimental results of Fig.11 are in good agreement with the simulation results except for the high frequency ripples. Fig.12 is the case where ω_n is further increased. Speed estimation gain is increased, the position error becomes considerably small. In the experimental results of Fig.13, the ripples of high frequencies are observed in $\hat{N}_r$ and θ_e. However, these values are processed only in the DSP and high frequency ripple of the actual speed N_r which is important for the application is small. From Figs. (14) – (17) show the results for the change of ζ. It is confirmed that the system becomes oscillating or unstable when ζ is small.

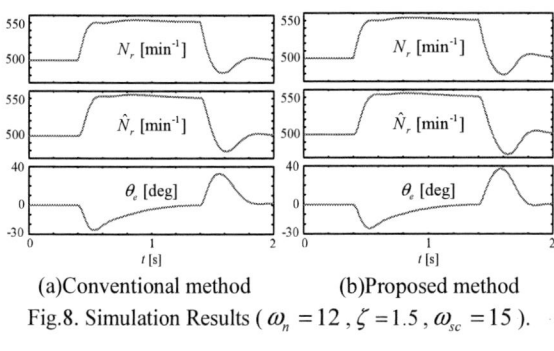

(a)Conventional method   (b)Proposed method

Fig.8. Simulation Results ($\omega_n = 12$, $\zeta = 1.5$, $\omega_{sc} = 15$).

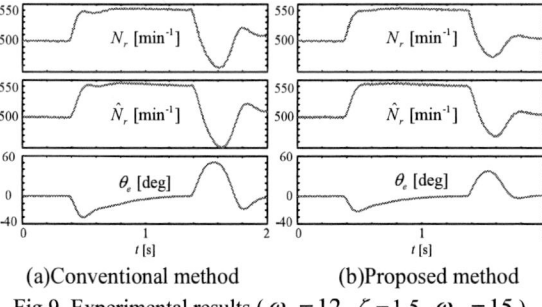

(a)Conventional method   (b)Proposed method

Fig.9. Experimental results ($\omega_n = 12$, $\zeta = 1.5$, $\omega_{sc} = 15$).

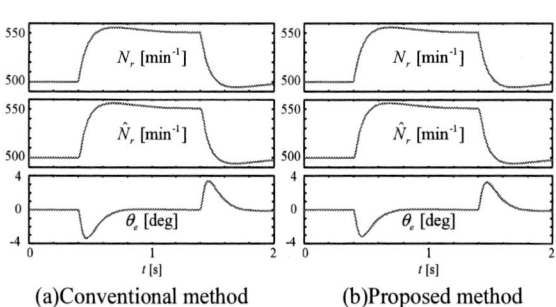

(a)Conventional method   (b)Proposed method

Fig.10. Simulation results ($\omega_n = 50$, $\zeta = 1.5$, $\omega_{sc} = 15$).

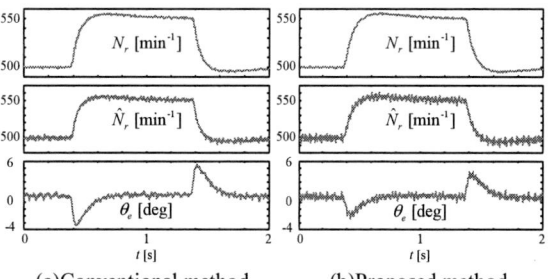

(a)Conventional method   (b)Proposed method

Fig.11. Experimental reults ($\omega_n = 50$, $\zeta = 1.5$, $\omega_{sc} = 15$).

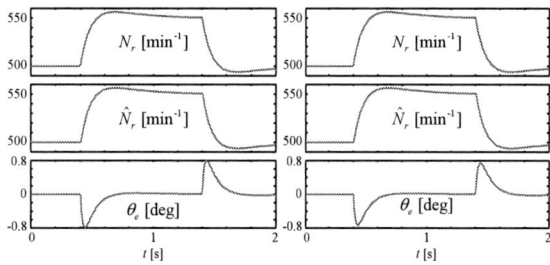

(a)Conventional method   (b)Proposed method

Fig.12. Simulation results ($\omega_n = 120$, $\zeta = 1.5$, $\omega_{sc} = 15$).

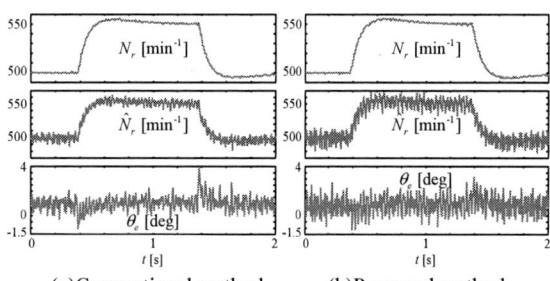

(a)Conventional method   (b)Proposed method

Fig.13. Experimental results ($\omega_n = 120$, $\zeta = 1.5$, $\omega_{sc} = 15$).

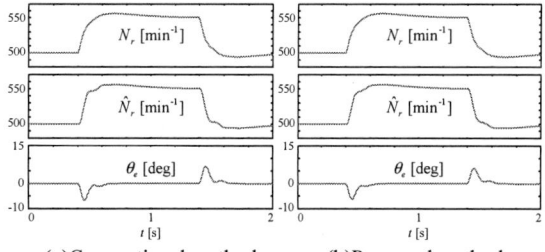

(a)Conventional method   (b)Proposed method

Fig.14. Simulation results ($\omega_n = 50$, $\zeta = 0.5$, $\omega_{sc} = 15$).

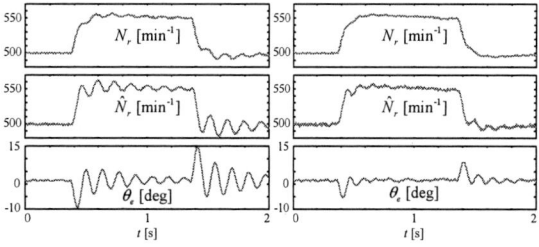

(a)Conventional method   (b)Proposed method

Fig.15. Experimental results ($\omega_n = 50$, $\zeta = 0.5$, $\omega_{sc} = 15$).

The 2014 International Power Electronics Conference

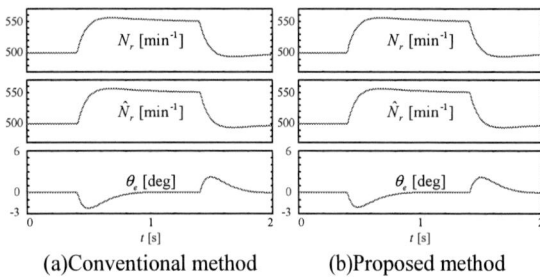

(a)Conventional method   (b)Proposed method

Fig.16. Simulation results ($\omega_n = 50$, $\zeta = 3.0$, $\omega_{sc} = 15$).

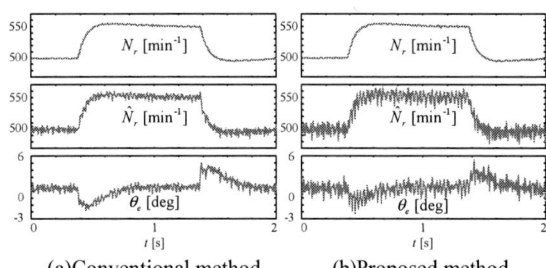

(a)Conventional method   (b)Proposed method

Fig.17. Experimental results ($\omega_n = 50$, $\zeta = 3.0$, $\omega_{sc} = 15$).

Conventional method improves high frequency ripples of estimated speed by the help of disturbance observer. However, the ripples of actual speed which is important for the applications are almost same. The cut-off frequencies of all PI current controllers are designed 1000rad/s. The influence of this frequency for proposed method is not sensitive.

V. CONCLUSIONS

For IPMSM sensorless vector control using the extended EMF, we have compared the proposed method with the conventional method by the root loci, non-linear simulation and experiment. Concerning these points, the results of the proposed method is almost the same as those of the conventional method. Therefore, it is concluded that the proposed method is useful because of its simpler structure.

REFERENCES

[1] Z. Chen, M. Tomita, S. Doki, and S. Okuma, "An Extended Electromotive Force Model for Sensorless Control of Interior Permanent-Magnet Synchronous Motors," *IEEE Trans. Ind. Electron*, vol.50, no.2, pp.288-295, 2003.

[2] S. Morimoto, K. Kawamoto, M. Sanada, and Y. Takeda, "Sensorless Control Strategy for Salient-Pole PMSM Based on Extended EMF in Rotating Reference Frame," *IEEE Trans. Ind. Applicat*, vol.38, no4, pp.1054-1061, 2002.

[3] K. Sakamoto, Y. Iwaji, T. Endo, T. Taniguchi, T. Niki, M. Kawamata, and A. Kawamura, "Position Sensorless Vector Control of Permanent Magnet Synchronous Motors for Electrical Household Appliances," Proc. of PCC-Nagoya 2007, LS4-3-3, pp.1119-1125, 2007.

[4] K. Yamanaka, T. Ohnishi, and M. Hojo, "A Novel Position Sensorless Vector Control of Permanent-Magnet Synchronous Motors," Proc. of PCC-Nagoya 2007, DS8-3-1, pp.290-295, 2007.

[5] B. H. Bae, S. K. Sul, J. H. Kwon and J.S. Byeon, "Implementation of Sensorless Vector Control for Super-High-Speed PMSM of Turbo-Compressor," *IEEE Trans. Ind. Appl.*, vol.39, no3, pp.811-818, 2003.

[6] J. K. Seok, J. K. Lee, and D.C. Lee, "Sensorless Speed Control of Nonsalient Permanent-Magnet Synchronous Motor Using Rotor-Position-Ttracking PI Controller," *IEEE Trans. Ind. Electron*, vol.53, no2, pp.399-405, 2006.

[7] M. Tuji, K. Kojima, G. Mangindaan, D. Akafuji, S. Hamasaki, "Stability Study of a Permanent Magnet Synchronous Motor Sensorless Vector Control System Based on Extended EMF Model," *IEEJ Journal of Industry Applications*, vol.1 no.3 pp.148-154, 2012.

APPENDIX

$$A_w = \begin{bmatrix} -g - k_1 i_{\delta 0} & 0 & L_q i_{\delta 0} K_{ei} & 0 & 0 \\ k_1 i_{\gamma 0} & -g & -L_q i_{\gamma 0} K_{ei} & -K_{pq}K_{ps} & K_{pq}K_{is} \\ -\dfrac{g}{\hat{e}_{\delta 0}} & 0 & 0 & 0 & 0 \\ -\dfrac{g\omega_c K_{ep}}{\hat{e}_{\delta 0}} & 0 & \omega_c K_{ei} & -\omega_c & 0 \\ 0 & 0 & 0 & -1 & 0 \\ 0 & 0 & 0 & 0 & 0 \\ 0 & 0 & 0 & -K_{ps} & K_{is} \\ -\dfrac{K_{ep}g}{\hat{e}_{\delta 0}} & 0 & K_{ei} & 0 & 0 \end{bmatrix}$$

$$* \begin{bmatrix} K_{id} & 0 & k_3 i_{q0} i_{\delta 0} - \hat{\omega}_0 L_q i_{d0} + k_4 i_{q0} \\ 0 & K_{iq} & -k_3 i_{q0} i_{\gamma 0} - \hat{\omega}_0 L_q i_{q0} - k_5 i_{d0} \\ 0 & 0 & \dfrac{gL_d i_{q0}}{\hat{e}_{\delta 0}} \\ 0 & 0 & k_2 \omega_c i_{q0} \\ 0 & 0 & 0 \\ 0 & 0 & -i_{q0} \\ 0 & 0 & i_{d0} \\ 0 & 0 & k_2 i_{q0} \end{bmatrix}$$

where, $k_1 = \dfrac{L_q K_{ep} g}{\hat{e}_{\delta 0}}$, $k_2 = \dfrac{L_d K_{ep} g}{\hat{e}_{\delta 0}}$, $k_3 = \dfrac{L_d L_q K_{ep} g}{\hat{e}_{\delta 0}}$

$k_4 = gL_d - R_s - K_{pd}$, $k_5 = gL_d - R_s - K_{pq}$

$$A_x = \begin{bmatrix} k_3 i_{\delta 0} + k_4 & \hat{\omega}_0 L_q & 0 \\ -k_3 i_{\gamma 0} - \hat{\omega}_0 L_q & k_5 & 0 \\ gL_d/\hat{e}_{\delta 0} & 0 & 0 \\ k_2 \omega_c & 0 & 0 \\ 0 & 0 & 0 \\ -1 & 0 & 0 \\ 0 & -1 & 0 \\ k_2 & 0 & -1 \end{bmatrix} , \quad B_r = \begin{bmatrix} K_{pd} & 0 \\ 0 & K_{pq}K_{ps} \\ 0 & 0 \\ 0 & 0 \\ 0 & 1 \\ 1 & 0 \\ 0 & K_{ps} \\ 0 & 0 \end{bmatrix}$$

$$F_w = \begin{bmatrix} 0 & 0 & 0 & 0 & 0 & K_{id} & 0 & -v_{q0} - K_{pd}i_{q0} \\ 0 & 0 & 0 & -K_{pq}K_{ps} & K_{pq}K_{is} & 0 & K_{iq} & v_{d0} + K_{pq}i_{d0} \end{bmatrix}$$

$$F_x = \begin{bmatrix} -K_{pd} & 0 & 0 \\ 0 & -K_{pq} & 0 \end{bmatrix}$$

$$F_r = \begin{bmatrix} K_{pd} & 0 \\ 0 & K_{pq}K_{ps} \end{bmatrix}$$

Induction Machine based Flywheel Speed Estimation at stand-by mode

Rongqiang Liu
Department of Electrical and Computer Engineering
Ryerson University
Toronto ON, Canada M5B 2K3
Email:rongqiang.liu@ryerson.ca

David Xu, *Member, IEEE*
Department of Electrical and Computer Engineering
Ryerson University
Toronto ON, Canada M5B 2K3
Email:dxu@ee.ryerson.ca

Abstract—**This paper presents an novel approach for stand-by mode sensorless speed estimation scheme for flywheel energy storage system (FESS). The proposed scheme eliminates the sensors required for flywheel speed monitoring, and estimates rotor speed in very short duration without affecting FESS in standby mode. This proposed algorithm only needs minimal change to the existing sensorless-based field orientation control(FOC) scheme of the flywheel. The algorithm is evaluated by simulation and experiment with short duration of excitation to the flywheel. The results also shows this method can estimate the speed without affecting the energy stored in the flywheel.**

Index Terms—**Flywheel energy storage system,sensor-less speed estimation, stand-by mode, hybrid flux estimation,**

I. INTRODUCTION

Flywheel Energy Storage System(FESS) works in three modes: charge/discharge and standby mode, and is characterized as fast-response, high-efficiency energy storage devices with excellent cycling capabilities and high power density. FESS is ideally suited to renewable micro-grids because of the frequent power fluctuations. Induction motor (IM) is a good candidate in flywheel applications for high power long term storage system. Being robust and cheap, the IM has better performance over the permanent magnetic synchronous motor (PMSM) since it has no iron loss at standby mode, which is very essential for long term storage in power systems.

The total kinetic energy stored in the FESS is given by Eqn. [1]

$$K_E = \frac{1}{2}J\omega^2 \qquad (1)$$

where J is the moment of inertia and ω is angular speed of the flywheel/motor.

Speed information of the flywheel is required to estimate the energy stored in the flywheel. Therefore, a speed sensor is usually required to be mounted on the shaft to measure the motor speed. However, a speed sensor increases the cost and requires a connection line between the control system and the motor. Moreover, the reliability of the high speed sensor is low when the whole flywheel (including motor) is floating by magnetic bearing in a vacuum tank. Such a sensor can be the most expensive and fragile component in the whole electric drive. To avoid the problems caused by speed sensors, several sensorless

methodologies have been proposed and developed in recent years: slip frequency calculation method; [2] [3]; speed estimation using state equation; [4] [5];estimation based on slot space harmonic voltages; [6]; Kalman filtering techniques; [7]; neural network based or Fuzzy-logic based sensorless control; [8]; and etc. However, these approaches are mainly based on on-line estimation strategies when the flywheel is continuously energized, and can be computationally intensive depending on the accuracy of the model of the motor. The performance of these methods varies in terms of their sensitiveity to parameter variations and their speed tracking capability.

The main objective of this study is to provide an approach that can achieve sensorless speed estimation of FESS at standby mode. This scheme focuses on the standby mode of FESS, and can be easily extended into charge/discharge modes. If a FESS keeps energized at standby mode, the speed will slow down gradually due to its iron loss. In order to further reduce the power loss, the FESS enters standby mode with a disconnection from its power converter by disabling all its gating signals. As the result, the iron loss of the IM reduced to zero at standby mode. When speed information of FESS is required, the induction motor of FESS will only be energized for a very short duration. An excitation current is supplied from the power converter to the stator side of the IM to create a stationary magnetizing flux. The revolving rotor will generate rotating flux field at the same frequency as the rotor speed, which reversely produces induced stator voltage and current at the same frequency in the stator side. By measuring the stator voltage and current, the rotor speed information can be extracted. Since the speed of a FESS will not change dramatically at standby mode due to its large inertia, the speed can be monitored periodically on a regular basis without deterioration of its performance. In order to achieve the decoupled control of the rotor flux and electrical torque for the FESS, a modified field-oriented control is presented to make the algorithms suitable for standby mode speed estimation. In addition, this approach can be easily applied to any IM based flywheel system, and this design is independent of the feedback controller from the control system.

978-1-4799-2706-7/14 $31.00 © 2014 IEEE

II. FLYWHEEL AND CONTROL SCHEME

A. *System Configuration*

A typical FESS application is shown in Fig.1. The FESS is coupled into the grid via a back-to-back converter and LCL filter. The voltage source converter 1 (VSC1) is connected to grid through a filter to obtain a high performance for the grid side PQ regulation. VSC2 is used to drive the squirrel cage induction motor. VSC1 and VSC2 are coupled through DC link and the DC bus voltage is usually regulated by VSC2 [9]. In the charging mode, power grid injects energy into the flywheel through the bi-directional converter when the DC link voltage is higher than the set value. In discharging mode, the grid extracts energy from DC link and the DC link voltage starts to drop. Then FESS provides energy to the dc link by regulating the DC link voltage. If there is no charging or discharging command from grid, the flywheel will enter the standby mode after a certain time. The FESS enters into standby mode by disabling all the gating signals from VSC2.

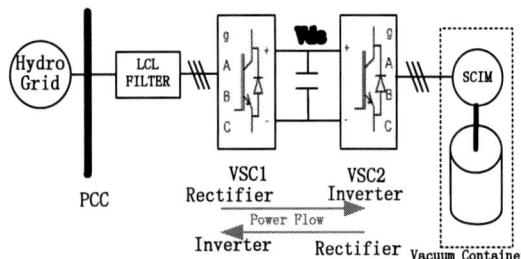

Fig. 1: Basic Circuit Diagram of FESS

Field-orientated control(FOC) scheme is achieved by aligning the rotor flux on the d-axis of the orthogonal frame which results in the mathmatical constrain $\lambda_{qr} = 0$ and $\lambda_{dr} = \lambda_r$. The FOC is used to control the torque and flux independently as shown in Fig. 2. In order to estimate the speed during standby mode, the motor need to be excited for a short period of time. A square wave enable signal is used to switch the FOC from normal operation to standby mode for speed estimation. In the charge/discharge mode, i_{ds}^* and i_{qs}^* are controlled independently as a conventional FOC scheme, while in the standby mode The flux-producing current i_{ex} is used to excite the induction motor. In order to minimize the impact to the energy stored in FESS, the torque-producing current component i_{qs}^* is switched to zero during the speed estimation period in order to achieve zero torque since the electrical torque and the torque-producing current component i_{qs}^* has a linear relation under the decoupled control scheme [10]. As the speed of flywheel will not change dramatically within a short period at standby mode, a zero order holder(ZOH) is used to maintain the speed information during the interval when the excitation is removed. The modified FOC control scheme is shown in Fig. 2

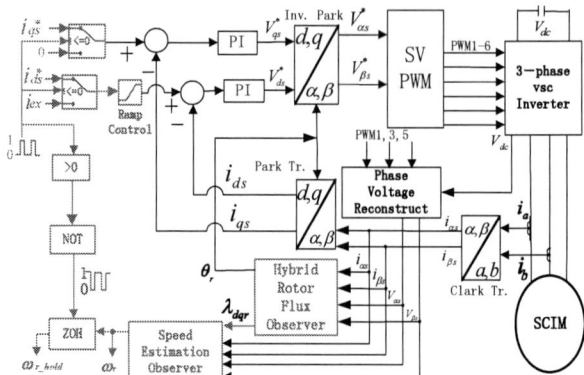

Fig. 2: Modified Field-Oriented Control Scheme

B. *Induction Motor Model*

To validate the proposed approach for standby-mode speed estimation, a squirrel cage induction machine (SCIM) is used for simulation and experiment. Squirrel cage induction machine has many advantages when compared with DC machines and permanent magnetic machines. First of all, it is less expensive. Next, it has a very compact structure and insensitive to environment. Furthermore, it does not require periodic maintenance for brush replacement like DC motors, or iron loss like permanent magnetic machines. However, because of its highly non-linear and coupled dynamic structure, a SCIM requires more complex control schemes than DC motors. A SCIM based FESS model can be described in following equations: [10]

$$v_{dqs}^e = R_s i_{dqs}^e + j\omega_e \lambda_{dqs}^e + p\lambda_{dqs}^e \tag{2}$$

$$0 = R_r i_{dqr}^{'e} + j(\omega_e - \omega_r)\lambda_{dqr}^{'e} + p\lambda_{dqr}^{'e} \tag{3}$$

$$\lambda_{dqs}^e = L_s i_{dqs}^e + L_m i_{dqr}^{'e} \tag{4}$$

$$\lambda_{dqr}^{'e} = L_s i_{dqr}^{'e} + L_m i_{dqs}^e \tag{5}$$

$$\lambda_{dqs}^e = \int (v_{dqs}^e - i_{dqs}^e R_s)dt \tag{6}$$

$$T_e = \frac{3PL_m}{2L_r}(\lambda_{dr} i_{qs}^e - \lambda_{qr} i_{ds}^e) \tag{7}$$

where
^{e}: the synchronous reference frame; v_{dqs}^e: stator d,q axis voltage; i_{dqs}^e:stator d,q axis current; $i_{dqr}^{'e}$: rotor d,q axis current refer to stator side; R_s: stator winding resistance; R_r: rotor winding resistance; L_s: stator self-inductance; L_r: rotor self-inductance; L_m: mutual linkage flux; ω_e: synchronous reference frame; λ_{dqs}^e:stator d,q axis linkage flux; $\lambda_{dqr}^{'e}$:rotor d,q axis linkage flux refer to stator side;P: number of pole pairs

Take $\lambda_{qr}, \lambda_{dr}, i_{qs}, i_{ds}, \omega_r$ as state variables, and reorganize the above equations, a SCIM model can be derived:

Substitute i'^s_{dqr} from Eqn. (4) in the stationary reference frame($\omega_e = 0$) into Eqn. (3),

$$\lambda^s_{dqr} = \left(\frac{L_m R_r}{L_r} i^s_{dqs} + j\omega_r \lambda^s_{dqr} \right) \frac{1}{s + \frac{R_r}{L_r}} \qquad (8)$$

Substitute i'^s_{dqr} from Eqn. (4) in the stationary reference frame($\omega_e = 0$) into Eqn. (2),

$$i_{dqs} = \left(v_{dqs} - j\frac{L_m}{L_r}\omega_r \lambda_{dqr} + \frac{L_m R_r}{L_r^2}\lambda_{dqr} \right) \times \ldots$$
$$\frac{1}{\left(L_s - \frac{L_m^2}{L_r} \right)s + \left(R_s + \frac{L_m^2 R_r}{L_r^2} \right)} \qquad (9)$$

By applying the mathematical constrain $\lambda_{qr} = 0$ from FOC to the Eqn. (7), the linear control of the electrical torque is achieved:

$$T_e = \frac{3PL_m}{2L_r}\lambda_{dr}i^e_{qs} \qquad (10)$$

Therefore, the derived SCIM model is list below: [11]

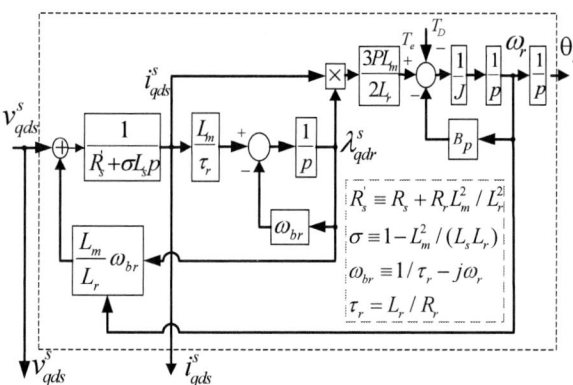

Fig. 3: Squirrel Cage Induction Motor and Load Model

C. Speed estimation

Sensorless speed estimation schemes proposed in the past few years can be achieved over a fairly large speed range with very good dynamic performance. In FESS applications, the operational speed of the FESS usually is limited to high speed range since the very low speed of flywheel should be avoided due to its low stored energy. This speed estimation scheme is derived from the SCIM in stationary reference frame, and the derived instantaneous speed is updated for each sampling period.
From Eqn.(4) and Eqn.(6)

$$i'_{dqr} = \frac{1}{L_m p}(v_{dqs} - R_s i_{dqs}) - \frac{L_s}{L_m}i_{dqs} \qquad (11)$$

From d, q component of Eqn.(3) , eliminating R_r

$$\omega_r = \frac{p\lambda_{qr}i_{dr} - p\lambda_{dr}i_{qr}}{\lambda_{dr}i_{dr} + \lambda_{qr}i_{qr}} \qquad (12)$$

substitute Eqn.(11) into Eqn.(12)

$$\omega_r = \frac{(\lambda_{ds} - L_s i_{ds})p\lambda_{qr} - (\lambda_{qs} - L_s i_{qs})p\lambda_{dr}}{(\lambda_{ds} - L_s i_{ds})\lambda_{dr} - (\lambda_{qs} - L_s i_{qs})\lambda_{qr}} \qquad (13)$$

Similarly, the synchronous electrical speed and slip speed can also be derived

$$\omega_e = \frac{p\lambda_{qs}\lambda_{ds} - p\lambda_{ds}\lambda_{qs}}{\lambda_{ds}^2 + \lambda_{qs}^2} \qquad (14)$$

$$\omega_{sl} = \frac{L_m}{\tau_r}\frac{(\lambda_{ds}i_{qs} - \lambda_{qs}i_{ds})}{\lambda_{ds}^2 + \lambda_{qs}^2} \qquad (15)$$

Therefore

$$\omega_{sl} = \omega_e - \omega_r \qquad (16)$$

D. Rotor Flux Observer

In order to estimate the speed, a rotor flux observer should be developed to compute the instantaneous rotor flux. The ability to obtain accurate flux data is crucial to implement the speed observer since the performance is seriously influenced by the estimated flux accuracy. In most FOC based IM drives, the rotor flux is estimated by measuring stator currents and control reference voltages. The rotor flux observers can be categorized as following based on different models:

1) Voltage model rotor flux observer: The voltage model rotor flux observer is based on the detection of stator voltage and stator current to compute the rotor flux. A pure integration of the stator-induced voltage is used to obtain the stator flux. Although pure integration is simple in concept, it has the following significant challenges in providing adequate real-time performance, especially low speed. 1) A small drift or dc offset inherently present in current measurement channels can cause an integrator to saturate. 2) Variations in the stator resistance result in a magnitude and phase error of the estimated flux at low speeds. 3) An initial condition error produces a constant output dc offset in the estimated flux. [12] [13].
From Eqn. (4) and Eqn.(5)

$$\lambda_{dqr} = \frac{L_r}{L_m}(\lambda_{dqs} - \sigma L_s i_{dqs}) \qquad (17)$$

substitute Eqn.(6) into Eqn. (17)

$$\lambda_{dqr} = \frac{L_r}{L_m}(\int (v_{dqs} - R_s i_{dqs})dt - \sigma L_s i_{dqs}) \qquad (18)$$

where $\sigma = 1 - \frac{L_m^2}{L_s L_r}$

Since the voltage model flux observer adopts the integral, it causes DC bias and the initial values problems. Practically, the problem is solved by a high pass first order filter:

$$\lambda_{dqr} = \frac{L_r}{L_m}\left(\frac{T_c}{1 + T_c s}(v_{dqs} - R_s i_{dqs} - \sigma L_s s i_{dqs}) \right) \qquad (19)$$

The replacement can largely alleviate the drift and the initial value problem while the magnitude and phase angle error are significantly introduced when the stator frequency reaches down to a few hertz.

2) Current model rotor flux observer: The current model rotor flux observer is based on the detection of stator current and rotor speed. It works well at low speed, however, the accuracy of the estimated rotor flux can be easily affected by the rotor parameters, and these parameters will vary with rotor temperature.

From Eqn.5

$$i_{dqr} = \frac{1}{L_r}(\lambda_{dqr} - L_m i_{dqs}) \qquad (20)$$

substitute i_{dqr} into equation(5)

$$\lambda_{dqr} = \frac{1}{\tau_r s + 1}(L_m i_{dqs} - \tau_r \omega_r \lambda_{qdr}) \qquad (21)$$

where $\quad \tau_r = \frac{L_r}{R_r}$

3) Hybrid rotor flux observer: The current model performs best at zero and low speed, and the other performs best at high speeds. Both voltage model and current model have its own pros and cons. Therefore, it is better to use voltage model at high speed, and current model at low speed, and this leads to the hybrid rotor flux observer.

From Eqn.(18)

$$e_{dqr} = s\lambda_{dqr} = \frac{L_r}{L_m}(v_{dqs} - R_s i_{dqs} - \sigma L_s s i_{dqs}) \qquad (22)$$

therefore,

$$\begin{aligned}\lambda_{dqr} &= \left(\frac{1 + T_c s}{1 + T_c s}\right)\lambda_{dqr} \\ &= \left(\frac{T_c}{1 + T_c s}\right)e_{dqr} + \left(\frac{1}{1 + T_c s}\right)\lambda_{dqr}\end{aligned} \qquad (23)$$

The first part of Eqn.(23) is the observed flux value from voltage model, and the second part is the error between the real value and the observed flux value which is compensated by current model. The selection of T_c should make the error as smaller as possible which means a large T_c value is desired; however, a large T_c will lead to a slow dynamic response of the flux observer. Therefore, the magnitude of the T_c value is selected based on the full consideration of requirements to reach its best performance.

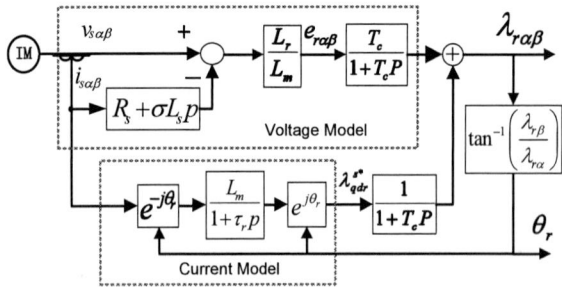

Fig. 4: Hybrid Rotor Flux Observer

The formulation of a hybrid flux observer is shown in Fig.4. It allows better performance to be achieved by

TABLE I: Induction Machine Parameters

Parameters	IM1	IM2
Power	$1/2hp$	$200hp$
Voltages	$220/440V$	$575V$
Frequency	$60Hz$	$400Hz$
Rated Speed	$1750rpm$	$11900rpm$
R_s	0.6405Ω	0.02475Ω
L_s	$0.0452H$	$0.014534H$
R_r	0.3625Ω	0.0133Ω
L_r	$0.0452H$	$0.014534H$
L_m	$0.0418H$	$0.01425H$
Pole pairs	2	2

seamlessly incorporating two models for high-performance flux regulation as well as flux orientation. The combined approach compromises the advantages of both models in performance and parameter insensitivity. The transition between models is governed by the bandwidth of the hybrid observer which is selected by Tc. For this simulation, Tc=0.2, which gives a cut-off frequency at 5Hz for current model rotor flux observer.

III. SIMULATION AND EXPERIMENT RESULTS

For this study, two simulations are presented with asynchronous SCIM models(IM1 and IM2 in Table. 1) at rated speed to simulate a rotating FESS at standby mode. IM1 is an actual induction motor used for the experiment to evaluate the proposed approach. IM2 is a high speed induction motor model used to simulate a high speed flywheel system. The simulation is conducted on both IM1 and IM2 while the experiment is only conducted on IM1. The parameters of the two SCIMs are listed in table.1. For the experiment, The IM1 is connected as a SCIM and coupled mechanically with a DC motor as prime mover to bring the SCIM to high speed. The parameter of IM1 is extracted by no-load test and blocked rotor test during the experiment.

A. Simulation results

In order to simulate the working conditions of the FESS, the running time of IM1 is divided into 3 segments:

$0s \sim 0.1s$: The SCIM model is set to mechanical speed input at its rated condition to simulate a rotating high-speed flywheel. No excitation current is supplied to the SCIM during this period. This segment simulates that the flywheel is working at the standby mode when no flywheel speed information is required.

$0.1s \sim 0.4s$: The gating signals of the VSC2 are enabled (Fig. 5A). The flux-producing current i_{ex} is supplied through the modified FOC to the SCIM to excite the machine for speed measurement. The speed information is extracted from induced terminal parameters and updated at every sampling period.

$0.4s \sim 0.5s$: The speed measurement process is completed, and the excitation current i_{ex} is removed by setting it to zero. The FESS enters into standby mode again. The flux change from the hybrid rotor flux observer is shown in Fig. 5B.

The excitation level is selected based on the consid-

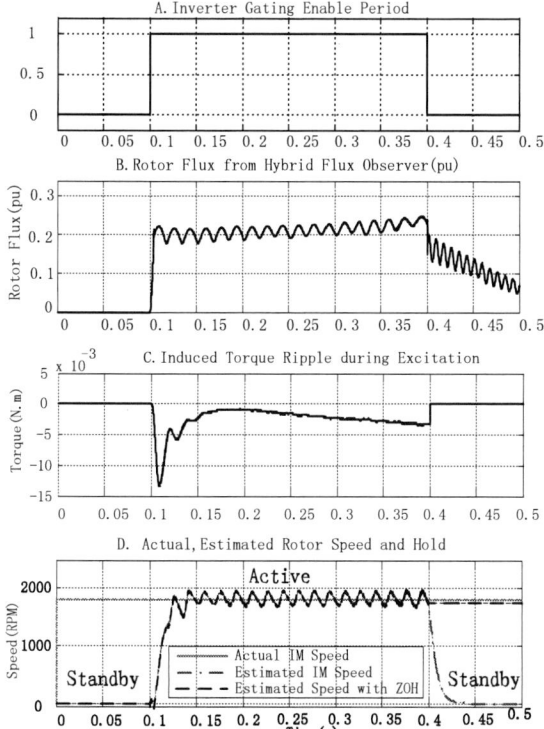

Fig. 5: Excitation, Torque and speed results from simulation of IM1

eration of the impact of the measurement to the exiting FESS. A higher excitation level will produce a more accurate result for the speed estimation; however, the higher the excitation, the larger the transient torque will be generated, therefore, more disturbance to the FESS. For this simulation, the excitation current reference is set to 0.2 pu.

In order to minimize the impact from the process of the speed estimation to the FESS, the torque from the modified FOC scheme is maintained to zero by setting $i_{qs}^*=0$ along with mathematical constrain $\lambda_{qr}=0$ from the modified FOC. However, due to the sudden change of the excitation current i_{ds}^*, a torque ripple is generated as shown in (Fig. 5C) during the transient measurement period. This generated short-period torque is very small in magnitude and can be negligiable, especially when the measurement duration is short enough.

After the excitation is removed from the induction motor, the speed estimation is disabled and the result from last estimation is maintained by a Zero-Order holder(ZOH) to next excitation (Fig. 5D). In order to minimize the torque impact caused by the flux change due to the excitation, the excitation period should be as short as possible. However, the duration of the excitation time need to be larger enough than the settling time of the estimation. The settling time of the estimated speed takes about 60 ms to reach steady state. This value can be further improved by adjusting the cut-off frequency of the speed filter in the speed estimation

observer for a trade-off of higher percentage of overshoot caused by the flux change. The trade-off of settling time of the estimated speed and accuracy should be carefully investigated for different machines.

In order to evaluate the proposed approach for a high speed FESS at standby mode, a simulation is conducted on IM2. The running time of the simulation is divided into 4 segments:

$0s \sim 0.1s$: The SCIM model is set to mechanical speed input at its rated condition to simulate a rotating high-speed flywheel. No excitation current is supplied to the SCIM during this period. This segment simulates that the flywheel is working at the standby mode with no flywheel speed information required.

$0.1s \sim 0.4s$: A flux-producing current i_{ex} is supplied through the modified FOC to the IM2 to excite the machine for speed measurement. The speed information is extracted from induced terminal parameters.

$0.4s \sim 0.5s$: The speed measurement process is completed, and the excitation current is removed. The FESS enters into standby mode again.

$0.5s \sim 1s$: This period shows another speed estimation cycle to simulate the repeating pattern of speed estimation process.

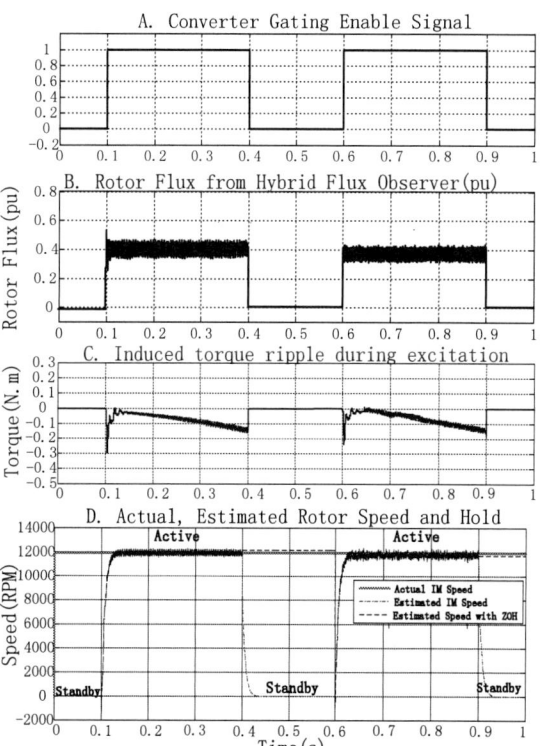

Fig. 6: Excitation, Torque and speed results from simulation of IM2

B. Experimental results

The experiment is conducted with the induction motor(IM1), connected as a SCIM and coupled with a DC

978-1-4799-2706-7/14 $31.00 © 2014 IEEE

motor as prime mover to bring the SCIM to high speed. The SCIM is fed through a 300 V, 30A, MOSFET inverter with Pulse Width Modulation(PWM). The DC link voltage of the inverter is set to 100V for a common DC panel connection. Field oriented Control with pulse width modulation, is implemented on a Digital Signal Processor (DSP) using the Texas Instrument (TI) F2812 microprocessor based converter. To avoid sampling distortion, the sampling frequency is set to 10kHz, which is much higher than the rated frequency of the stator voltage and current at 60Hz. For the experiment, the SCIM stator voltage and current are measured from the Analog/Digital converter on the DSP. For this experiments, the actual measured speed from a tachometer reading is about 832 rpm. To minimize the torque ripple with least settling time, the practical excitation level is selected between $0.2 \sim 0.45$. For this experiment, the excitation current reference is set to 0.4pu.

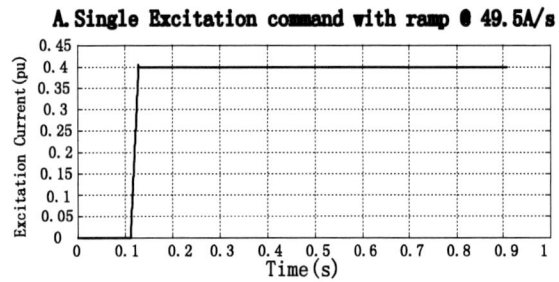

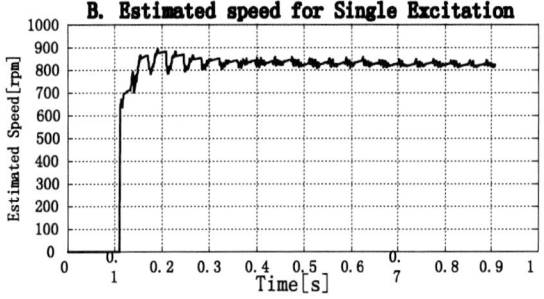

Fig. 7: Single Excitation with ramp @ 49.5A/s and Estimated Speed

From Fig. 7A, a single excitation current command that goes through the ramp control is supplied to the stator side of the SCIM while it is mechanically driven by a DC motor. The resulted estimated speed is shown in Fig. 7B, with an speed magnitude ranging from 800 to 850rpm, which is consistant with the measured speed at 832 rpm.

In order to further test the repeating measurement of the estimated speed, two excitation pulses are supplied to the SCIM in Fig. 8A. The estimated speed is shown in Fig. 8B.

C. Investigation on Settling time vs. ramp rate control

For a lower excitation level, the accuracy of the estimated speed can be easily distorted by the small magnitude of the induced voltage and current; while a large excitation level will cause a large induced current at the instant when

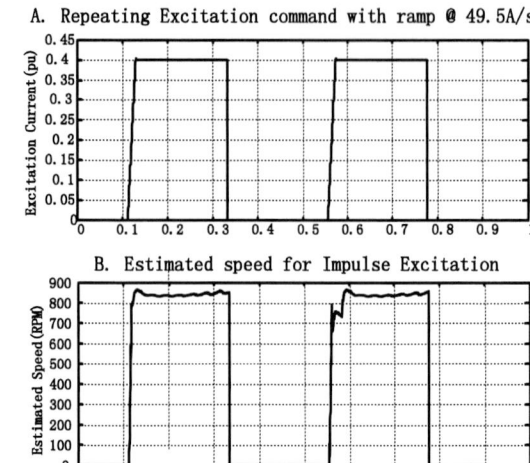

Fig. 8: Repeating Excitation with ramp @ 49.5A/s and Estimated Speed

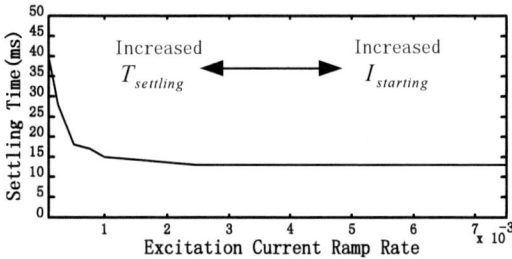

Fig. 9: Settling time of estimated speed at various flux excitation ramp rate

an excitation command is given, which will trigger the over current protection on the converter in the experiment, or large overshoot and oscillation problems which will increase the settling time of the estimated speed. In order to overcome the problems caused by sudden change of the applied excitation command, the excitation current reference is applied with a ramp control(Fig. 2). The relationship between the settling time vs. ramp rate is investigated in Fig. 9 for IM1. Based on the Fig.9, the selected i_{ds}^* ramp rate is $0.005(0.011A/T_{sampling}$ or 49.5A/s). The settling time for this measurement is defined at 90% of the average of steady state speed.

IV. CONCLUSIONS

In this paper, a novel approach for standby mode sensorless speed estimation for FESS is presented with minimal time cost and least disturbance to the existing FESS at standby mode. A hybrid flux observer is used to estimate the instantaneous rotor flux during the speed estimation period. The speed information is extracted from terminal parameters in every sampling period.

The speed information is held to and updated by next

978-1-4799-2706-7/14 $31.00 © 2014 IEEE

estimation cycle. Moreover, since the FESS does not need to be continuously energized for speed estimation, the power loss of the FESS can be further reduced while the speed information is maintained.

The optimal excitation level is investigated in this paper to minimize the impact to the energy stored in the flywheel. The simulation and experimental results validate the proposed algorithm for sensorless speed estimation at standby mode.

REFERENCES

[1] K. Veszpremi and I. Schmidt, "Different flywheel energy storage drives for renewables x2014; limits and optimization," in *Power Electronics and Motion Control Conference (EPE/PEMC), 2010 14th International*, 2010, pp. T12–131–T12–137.

[2] Y. Fan, W. Qu, H. Lu, X. Cheng, X. Zhang, L. Wu, and S. Jiang, "A slip frequency correction method applied to induction machine indirect vector control system," in *Electrical Machines and Systems, 2008. ICEMS 2008. International Conference on*, 2008, pp. 1122–1125.

[3] T.-W. Kim and A. Kawamura, "Slip frequency estimation for sensorless low speed control of induction motor," in *Advanced Motion Control, 1996. AMC '96-MIE. Proceedings., 1996 4th International Workshop on*, vol. 1, 1996, pp. 156–161 vol.1.

[4] K. Zhao and X. You, "Speed estimation of induction motor using modified voltage model flux estimation," in *Power Electronics and Motion Control Conference, 2009. IPEMC '09. IEEE 6th International*, 2009, pp. 1979–1982.

[5] T. Nag, A. Sen, and D. Chatterjee, "An observer based on-line rotor speed estimation technique for a vector controlled induction machine," in *Industrial Electronics, 2008. ISIE 2008. IEEE International Symposium on*, 2008, pp. 630–633.

[6] A. Ferrah, P. Hogben-Laing, K. Bradley, G. Asher, and M. Woolfson, "The effect of rotor design on sensorless speed estimation using rotor slot harmonics identified by adaptive digital filtering using the maximum likelihood approach," in *Industry Applications Conference, 1997. Thirty-Second IAS Annual Meeting, IAS '97., Conference Record of the 1997 IEEE*, vol. 1, 1997, pp. 128–135 vol.1.

[7] H.-W. Kim and S.-K. Sul, "A new motor speed estimator using kalman filter in low-speed range," *Industrial Electronics, IEEE Transactions on*, vol. 43, no. 4, pp. 498–504, 1996.

[8] Y. Xiaoting, Z. Qingchun, and Z. Tao, "Speed estimation of induction motor based on neural network," in *Intelligent Control and Information Processing (ICICIP), 2011 2nd International Conference on*, vol. 2, 2011, pp. 619–623.

[9] L. Zhou and Z. ping Qi, "Modeling and control of a flywheel energy storage system for uninterruptible power supply," in *Sustainable Power Generation and Supply, 2009. SUPERGEN '09. International Conference on*, 2009, pp. 1–6.

[10] N. Z. S. K. Bin Wu, O.Yongqiang Lang, *Power Conversio and Control of Wind Energy Systems.* WILEY: IEEE Press, 2010.

[11] B. K. Bose, *Power Electronics and Variable Frequency Drives,* J. B. Anderson, Ed. IEEE: IEEE Press, 1997.

[12] G. Pellegrino, P. Guglielmi, E. Armando, and R. Bojoi, "Self-commissioning algorithm for inverter nonlinearity compensation in sensorless induction motor drives," *Industry Applications, IEEE Transactions on*, vol. 46, no. 4, pp. 1416–1424, 2010.

[13] K.-R. Cho and J.-K. Seok, "Pure-integration-based flux acquisition with drift and residual error compensation at a low stator frequency," *Industry Applications, IEEE Transactions on*, vol. 45, no. 4, pp. 1276–1285, 2009.

978-1-4799-2706-7/14 $31.00 © 2014 IEEE

Symmetrical signaling system for sensor-less SRM drive

Kenji Yamamoto and Hisashi Takahashi
Department of Electrical and Electronic Engineering
Shizuoka Institute of Science and Technology
2200-2 Toyosawa, Fukuroi, Shizuoka,
437-8555 Japan
yamaken@ee.sist.ac.jp

Nobumasa Ushiro and Koki Shirasawa
SINFONIA TECHNOLOGY CO., LTD
150 Aza-Motoyashiki, Mitsuya, Toyohashi, Aichi,
441-3195, Japan

Abstract— **SRMs, switched reluctance motors, are being expected as an excellent candidate for power sources in harsh environments due to the simple structure. In spite of many methods proposed to eliminate the rotor position sensor, which is needed to commutate the current correctly but not compatible with harsh environments, no successful sensor-less method has been seen so far. The method proposed in this paper enables one to eliminate the sensor by introducing a signal onto windings of a phase in a common mode manner and detecting the induced voltages on the other phases. The method has proved to be practical through simulations and experiments on a test setup. The SRM on the test setup rotated at revolving speed of 100,000 min^{-1} and more with the system.**

Keywords— *Rotary encoder, Sensor-less, SRM, Symmetrical Signaling*

I. INTRODUCTION

SRMs, switched reluctance motors, are being expected to be an excellent candidate for power sources in harsh environments, for example where high revolution rates are required at high ambient temperatures, since they have the simplest rotor structure among the motor designs and do not need any permanent magnet[1][2].

In those harsh environments, the rotary encoders which is available at affordable costs for industrial applications to get the rotor position information are not enough robust to work properly. On the other hand, an SRM, when it is used as a variable speed power source,

still needs some kind of rotor position sensor(rotary encoder) since the current flowing into the motor needs to be properly commutated between the windings to continue its revolution. Those are the reasons why many research and development efforts have been carried out to eliminate the sensor in controlling an SRM.

In the efforts to get the rotor position information without a rotary encoder, included are the passive methods where phase currents and voltages are measured and used to estimate the rotor position based on analytical, observer-based, or artificial intelligence inference[3]-[24] and the active sensing, where the system injects a signal, which can be a pulse or single tone current, into a winding and detects it with one of the other windings or other windings[25]-[32]. Fig. 1 shows a classified tree of sensor-less rotor position estimations for SRMs, summarizing the classification explained above.

The sensor-less scheme proposed in this paper falls into the active category in Fig. 1, is unique in the way the system injects the signal and detects it, and have not been proposed so far. In the system, the connecting point(center tap) of two windings in each phase is wired out of the SRM and connected to a network(CMN) along with the other terminals of the phase windings. Where a phase means a group of windings which are driven simultaneously for torque production. A high frequency signal is injected through CMN to one of the phases currently excited for torque production and detected at the other two phases adjacent to the excited phase at a time.

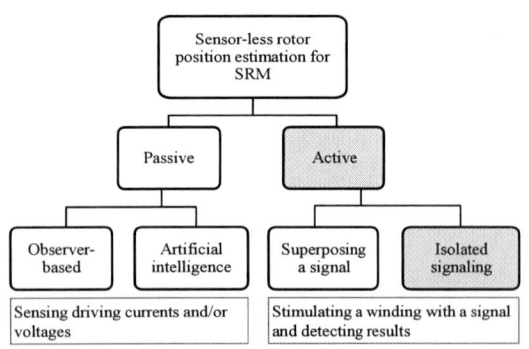

Fig. 1. Researches for SRM sensor-less control

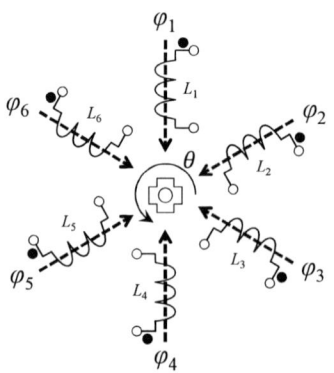

Fig. 2. The disposition of windings and the definition of flux linkage of a four-pole/six-slot SRM

978-1-4799-2706-7/14 $31.00 © 2014 IEEE

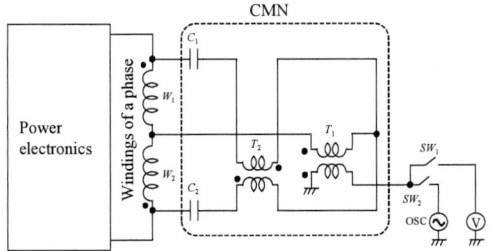

Fig. 3. Common mode network(CMN)

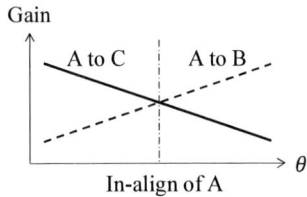

Fig. 4. Detecting in-align position of phase A

II. THEORY

A. Detecting in-align position

Shown in Fig.2 is an outline of the structure of a four-pole/six-slot SRM. The cross section of rotor is drawn in the center of the figure, which has four salient poles. Inductances L_1 to L_6 are individual windings wound on the salient poles on the stator and each pair of them which are opposite to each other forms a phase called A, B, or C. Assuming L_1 and L_4 form phase A, they are connected in series so that the flux which they induce be in the same direction. Phase B and C are formed in the same manner, phase B with L_2 and L_5, phase C with L_3 and L_6. The dots(.) denote that the flux $\varphi_i(i = 1, 2, 3, 4, 5, 6)$ is generated toward the rotor when a current is fed into the terminal.

Illustrated in Fig. 3 is an example of CMN, with which a high frequency signal is injected and detected on each phase. In the figure, W_1 and W_2 are the windings of a phase.

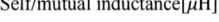

Self/mutual inductance[μH]

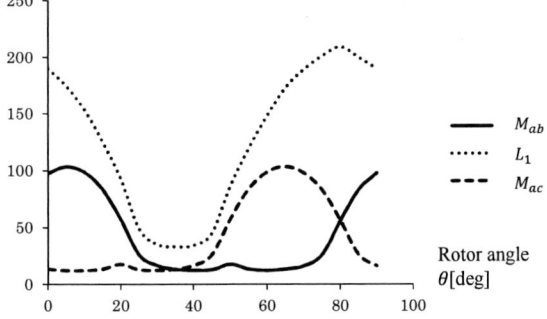

Fig. 5. An example of self and mutual inductance variations along with rotor position change.

The high frequency signal is injected into a phase and signals reproduced over the mutual inductances between the windings are detected on the other phases with the CMNs which connected with them. With a CMN, the interference from the power electronics circuit is effectively blocked and the high frequency signal is also greatly attenuated between the power and signal circuit due to the symmetrical characteristic of mutual inductances in an SRM.

In driving an SRM, one need to know when the excited phase reaches in-align position since the drive circuit need to stop providing the phase with power so that a torque contrary to the rotating direction will not be generated.

Fig. 4 illustrates how to determine the in-align position. When phase A is excited to generate torque the phase is also excited with a high frequency signal. The gains of the high frequency signal from A to C or B is plotted on the graph. When the gains become the same, phase A windings are in align with one pair of the salient poles of the rotor and consequently the drive circuit can commutate the motor current to the next phase. In actual SRM drives, some amount of time from the commutating point to the aligning timing(lead time) is needed for making sure that there is enough time for the commutated current to decay.

An offset voltage can be added to the comparator circuit to introduce the lead time, which shall be an easy electronic circuit design.

B. Actual inductances

Fig.5 shows an example of measured variations of self and mutual inductances along with the rotor position change of an SRM. In the figure, M_{ab} is the mutual inductance between Phase A and Phase B, and M_{ac} is that of A and C. Self-inductance L_1 is shown along with those mutual inductances in the figure. In-align position of Phase A is identified as the peak of L_1 and at that rotor angle the mutual inductances M_{ab} and M_{ac} cross each other. The point of that crossing point of M_{ab} and M_{ac} corresponds to the in-align position shown in Fig.4 since the same mutual inductances cause the same gains of A to C and A to B.

III. SIMULATION

The simulation of the scheme is performed on LTspice[33] with models of voltage controlled self-inductance and mutual inductance, which have been developed for this project. Fig. 6 shows the simulated circuit. In the figure, the circuit elements denoted as VCL are the voltage controlled inductances, with which each individual winding is simulated. The voltage controlled mutual inductances are not indicated in the figure but incorporated in the simulation. L/M controller generates rotor angle-dependent voltage patterns of the self and mutual inductances. The voltage controlled self and mutual inductances are controlled by the voltages. The output voltage of oscillator, OSC, is connected to the CMN of currently torque-excited phase. The CMNs on

978-1-4799-2706-7/14 $31.00 © 2014 IEEE

The 2014 International Power Electronics Conference

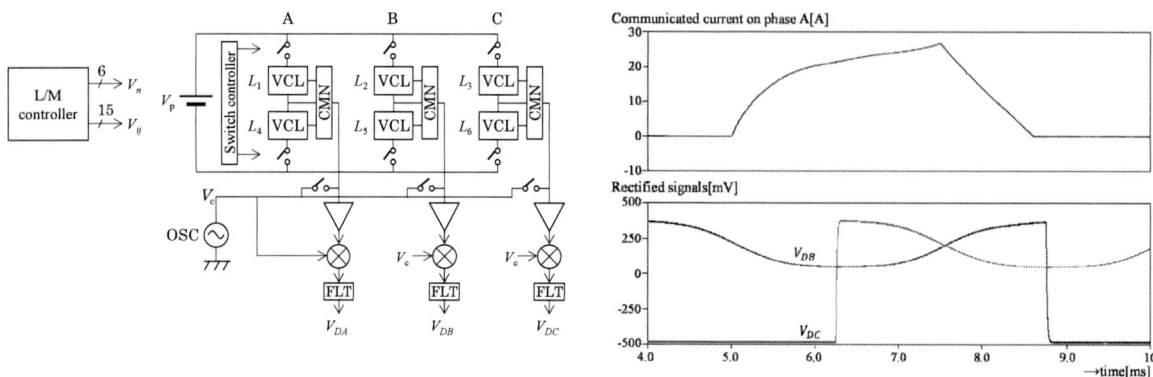

Fig. 6. The block diagram of simulated circuit

Fig. 7. Simulated commutated current(upper) and detected/rectified signals(lower)

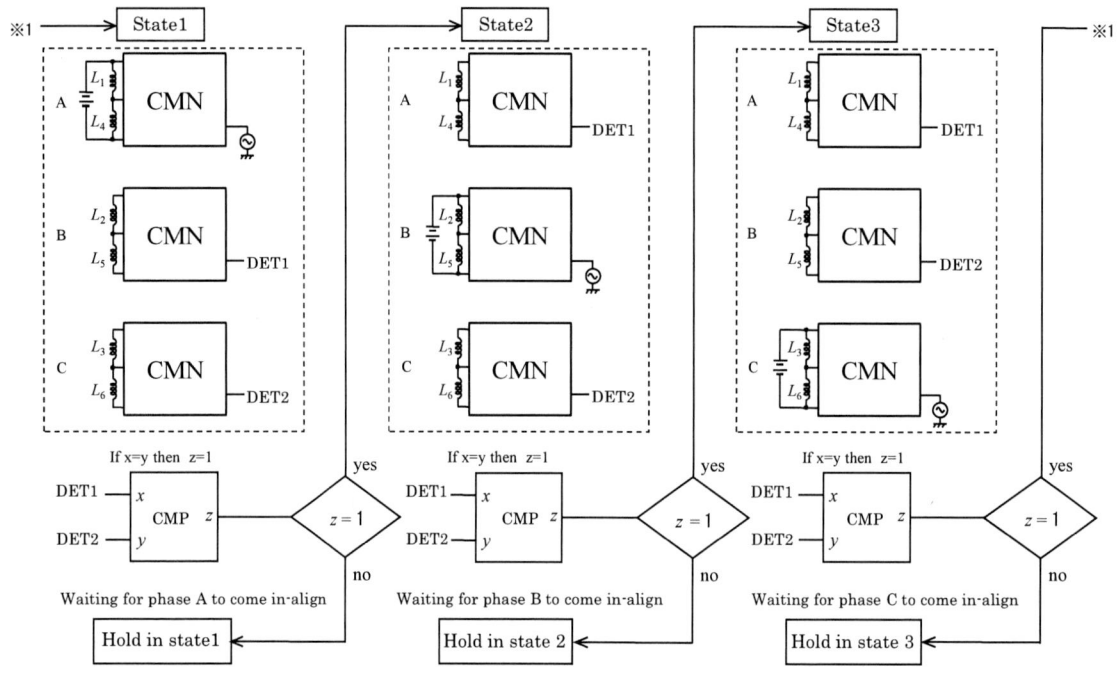

(a) State change diagram of the test setup.

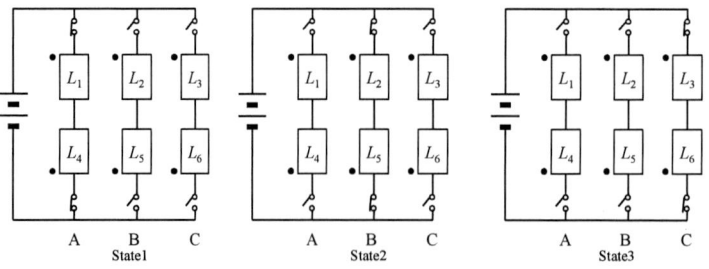

(b) Commutating chart of the test setup.

Fig. 8. Operation of the experimental setup of the SRM drive.

978-1-4799-2706-7/14 $31.00 © 2014 IEEE   3108

the other phases detect voltages induced over the mutual inductances. Each detected voltage is rectified with an envelope detection circuit, resulting voltages V_{DB} and V_{DC}.

The results of the simulation is shown in Fig. 7. In the figure the upper waveform is the commutated current for torque production and the lower is the detected signals, V_{DB} and V_{DC}. The frequency of the OSC used in the simulation is one megahertz. The in-align point is distinguished as the peak point of the commutated current waveform. As shown in the graph, one can detect the timing of the in-align of a phase if the crossing point of the envelopes is detected. In an actual SRM drive, the current shall be commutated earlier than the case of the figure to allow the circuit a time to decay the current for avoiding reverse torque generation. Even in those actual cases with the current for torque production, the scheme will not have any problem since the waveform of the detected signal is independent of the torque-exciting current due to the symmetry of mutual inductances of SRM and the function of CMN as shown in the simulation results.

IV. TEST SETUP

Fig. 8 illustrates the state change and switching diagram of the power electronics circuit designed in the test setup to rotate an SRM with the signaling system. In each state the system connects the signal source to the phase which is excited for torque production and compares the rectified signals from the other phases. When the compared signals cross over, the system switches the state to the next, where the excited and signal-injected phase change to the next phase and the signals to be compared are also changed.

Shown in Fig. 9 are waveforms of detected signals for the input signal on Phase A of $0.4V_{pp}$. The frequency of the signal is one megahertz. The upper graph (a) is for the rotor angle of 80 degrees and corresponds to the in-align position of Phase A. The lower graph (b) is for the rotor

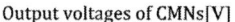
Output voltages of CMNs[V]

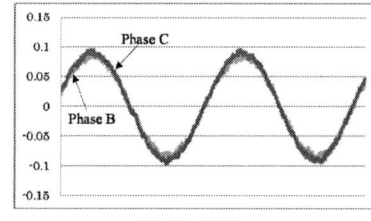

(a) $\theta = 80°$

Output voltages of CMNs[V]

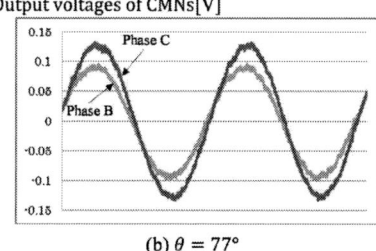

(b) $\theta = 77°$

Fig. 9 Output voltages of CMNs of Phase B and C

Fig. 10. The picture of the SRM
From left to right, the SRM, a torque sensor, and a hysteresis break

angle of 77 degrees. As shown in the graphs, when Phase A is in-align position, the voltage swing of the reproduced signal of Phase B and C are the same but the rotor moves out of in-align position, the voltages differ from each other.

Fig. 10 is the picture of the SRM used in the test setup. From left to right, the SRM, a torque sensor, and a hysteresis break. The SRM is rotated with no loads in the test.

The system has an offset voltages to the comparators, which are denoted as CMP. The offset voltages have been adjusted to have an appropriate lead time of commutation prior to the in-align position. As a result, it drives an SRM at revolving speeds over $100,000 \ \text{min}^{-1}$.

V. CONCLUSIONS

The method is unique in the way it signals over mutual inductances in an SRM to detect the rotor position. No complex calculations nor inferences to estimate the in-align position are needed with the system. Further, the more precise rotor position can be calculated if the rotor position dependencies of the mutual inductances are known, which is practically measurable. Also the signaling system can detect out the rotor position even when the rotor stands still by observing the positions of the detected envelope voltages.

It can be said that the system will be able to be practically designed in any SRM drive system to eliminate the rotor position encoder.

ACKNOWLEDGMENT

We would like to express our very great appreciation to Mr.Hirotoshi Kawamura and Mr.Hirohiko Murata of SINFONIA TECHNOLOGY CO., LTD. for their valuable and constructive comments. We would also like to thank the other staff of the company for their help in performing the experiments, which enabled us to collect the valuable data.

REFERENCES

[1] R. Krishnan, "Switched Reluctance Motor Drives : Modeling, Simulation, Analysis, Design, and Applications", CRC Press, 2001

[2] T. Kenjo, "SR Motor", THE NIKKAN KOGYO SHIMBUN, LTD., 2012

[3] M. Divandari, R. Brazamini, A. Dadpour and M. Jazaeri, "A Novel Dynamic Observer and Torque Ripple Minimization via Fuzzy Logic for SRM Drives", IEEE International Symposium on Industrial Electronics, Seoul Olympic Parktel, Seoul, Korea July 5-8, 2009

[4] M. Divandari, A. Koochaki, M. Jazaeri and H. Rastegar, "A Novel Sensorless SRM Drive via Hybrid Observer of Current Sliding Mode and Flux linkage", Electric Machines & Drives Conference, 3-5 May 2007, IEEE International, vol.1, pp.3-5, 2007

[5] Mohammad S. Islam and M. N. Anwar, "A sensorless wide-speed range SRM drive with optimally designed critical rotor angles", Industry Applications Conference, 08-12 Oct 2000, Conference Record of the 2000 IEEE, vol.3, p.1730 – 1737, 2000

[6] Bin-Yen Ma, Tian-Hua Liu, Ching-Guo Chen, Tsen-Jui Shen, and Wu-Shiung Feng, "Design and implementation of a sensorless switched reluctance drive system", Aerospace and Electronic Systems, IEEE Transactions, vol.34, issue:4, p.1193 – 1207, 1998

[7] Mohammad S. Islam, Iqbal Husain, Robert J. Veillette and Celal Batur, "Design and Performance Analysis of Sliding-Mode Observers for Sensorless Operation of Switched Reluctance Motors", IEEE Transactions on Control Systems Technology, vol.11, no.3, pp.383-289, 2003

[8] Michael T. DiRenzo, "Developing an SRM Drive System Using the TMS320F240", APPLICATION REPORT: SPRA420 ,1998

[9] Tian-Hua Liu, Ming-Tsan Lin, Ching-Guo Chen and Chih-An Tai, "Implementation of a Position Control System of a Sensorless Synchronous Reluctance Drive", IEEE 2002 28th Annual Conference of the Industrial Electronics Society,vol.1, pp.514-519, 2002

[10] Hongwei Gao, Farzad Rajaei Salmasi and Mehrdad Ehsani, "Inductance Model-Based Senseless Control of the Switched Reluctance Motor Drive at Low Speed", IEEE Transactions on Power Electronics, vol.19, issue 6, pp.1568 - 1573, 2004

[11] G. Bhuvaneswari, Thakurta Sarit Guha, P. Srinivasa Rao and S.S. Murthy, "Modeling of a switched reluctance motor in sensorless and "with sensor" modes", J Power Electron, vol.6 , no.4, pp. 315–321, 2006

[12] Husain, and Ehsani, M., "Rotor position sensing in switched reluctance motor drives by measuring mutually induced voltages", IEEE Transactions on Industry Applications, vol.30, issue 3, pp.665 - 672, 1994

[13] Islam, M.S. and Husain, I., "Self-tuning of sensorless switched reluctance motor drives with online parameter identification", Conference Record of the 2000 IEEE Industry Applications Conference, vol.3, pp.1738 - 1744, 2000

[14] Estanislao Echenique, Juan Dixon, Roberto Cárdenas and Ruben Peña, "Sensorless Control for a Switched Reluctance Wind Generator, Based on Current Slopes and Neural Networks", IEEE Transactions on Industrial Electronics, vol.56, issue 3, pp. 817 - 825, 2009

[15] Hyong-Yeol Yang, Duck-Shick Shin and Young-Cheol Lim, "Sensorless Control of a Single-Phase Switched Reluctance Motor Using Residual Flux", Journal of Power Electronics, vol. 9, no. 6, pp.911-918, 2009

[16] X. D. XUE, K. W. E. CHENG and S. L. HO, "Sensorless Control Scheme for Continuously Estimating Rotor Position and Speed of Switched Reluctance Motor Drives Based on Two-Dimensional Least Squares", 2004 1st International Conference on Power Electronics Systems and Applications Proceedings, pp.183-187, 2004)

[17] Geon-Il Kang, Hyong-Yeol Yang, Seung-Hak Yang and Young-Cheol Lim, "Sensorless Drive Method SRMs in Hysteresis Current Control", The International Conference on Electrical Engineering, 2008)

[18] P. Laurent, B. Multon, E. Hoang, and M. Gabsi, "Sensorless position measurement based on PWM eddy current variation for switched reluctant machine", 1995 European Conference on Power Electronics and Applications, pp.264-257, 1995)

[19] K. Akitomo and M. Ichiro, "A Position Sensorless Control Method for SRM Based on Variation of Phase Inductance", The transactions of the Institute of Electrical Engineers of Japan D, A publication of Industry Applications Society vol.127, no.9, pp.1023-1029, 2007

[20] V. Vasan Prabhu and V. Rajini, "Design of GT-FLC Speed Controller and Position Sensorless Control using ANN for 8/6 SRM", International Journal of Electrical Engineering, vol.5, no.4 , pp. 475-488, 2012)

[21] Luis Oscar A. P. Hendriques, Luis G. B. Rolim, Walter I. Suemitsu and P. J. Costa Branco, "Development and implementation of a neuro-fuzzy technique for position sensor elimination in a SRM", 2004 IEEE International Symposium on Industrial Electronics, 4-7 May 2004, vol.1, p.465 – 470, 2004)

[22] N. Prabhu Ram and L.Sheela, "Sensorless Control of Switched Reluctance Machine Based On ANFIS at Wide Speed", International Conference on Computing and Control Engineering (ICCCE 2012), 2012)

[23] Ooi, H.S. and Green, T.C., "Sensorless switched reluctance motor drive with torque ripple minimization", 2000 Power Electronics Specialists Conference 2000, vol.3, pp.1538-1543, 2000)

[24] T. Kosaka, Y. Nabeya, K. Ohyama and N. Matui, "Position Sensorless Drive of SRM Mounted on Hydraulic Pump Unit", The transactions of the Institute of Electrical Engineers of Japan. D, A publication of Industry Applications Society, vol.123, no. 2, pp.105-111, 2003

[25] H. Y. Yang, J. S. Song, J. G. Kim, Y. C. Lim and Y. G. Jung, "A new position sensing method for witched reluctance motor drives using search coils", Proceedings of the ICEE, Kowloon, Hong Kong, July 6-10, 2003, pp.A125, 2003

[26] H. J. Guo, M. Takahashi, T. Watanabe, and O. Ichinokura, "A New Sensorless Drive Method of Switched Reluctance Motors Based on Motor's Magnetic Characteristics", IEEE Transactions on magnetics, vol. 37, no. 4, pp.2831-2833, 2001

[27] E. Afjei, O. Hashemipour, M. M. Nezamabadi and M. A. Saati, "A self-tunable sensorless method for rotor position detection in switched reluctance motor drives", Iranian Journal of Science & Technology, Transaction B, Engineering, vol.31, no.B3, pp.317-328, 2007

[28] A. Brosse and G. Henneberger, "Different models of the SRM in state space format for the sensorless control using a Kalman filter", 1998. Seventh International Conference on Power Electronics and Variable Speed Drives, pp.269-274, 1998

[29] Chi, H.-P., Liang, T.J, Chu, C.L., Chen, J.F., Chang, M.T., "Improved Mutual Voltage Technique of Indirect Rotor Position Sensing in Switched Reluctance Motor", Power Electronics Specialists Conference 2002, IEEE 33rd Annual, vol.1, pp.271 – 275, 2002

[30] Ehsani, M., Husain, I, Mahajan and Ramani, K.R., "New modulation encoding techniques for indirect rotor position sensing in switched reluctance motors", IEEE Transactions on Industry Applications, vol.30, issue 1, pp.85 – 91, 1994)

[31] Laurent, P., Gabsi, M. and Multon, B., "Sensorless rotor position analysis using resonant method for switched reluctance motor", 1993 Industry Applications Society Annual Meeting, vol.1, pp.687 – 694, 1993

[32] K. Akitomo, B. Tatsunori and M. Ichiro ; "Estimation of Rotor Position in a 3-Phase SRM at Standstill and Low speeds", The transactions of the Institute of Electrical Engineers of Japan. D, A publication of Industry Applications Society, vol.129, no.3, pp.311-318, 2009

[33] Linear Technology Corporation, "LTSPICE IV", http://www.linear.com/designtools/software/#LTspice

Digital integrators for condition monitoring: a DC and multitone signal analysis

L. Peretti

Department of Power Technologies
ABB Corporate Research
Västerås, Sweden
luca.peretti@se.abb.com

Abstract—**The digital implementation of the integral operator is a delicate part in electrical drives. Many solutions are available in the literature, which avoid the well-known output drift of ideal digital integrators when a spurious DC offset is present at their input. Such solutions try to extend the integration accuracy toward the zero frequency as well, aiming to broaden their usability range for real applications (low-speed region of electrical machines, for example). So far, very few investigations have been performed as regards to the performances of drift-free integrators for cases where the integration is used within estimation and conditioning monitoring algorithms. This paper aims to investigate different known digital offset-free integrals, in terms of their output accuracy with respect to DC and multitone input signals, for a specific application where they could be exploited for condition monitoring. Simulation and experimental results are presented as well.**

I. INTRODUCTION

The frequency analysis of variables measured in an electrical machine is an aspect of rising interest in the drive control scenario, especially when considering condition monitoring and diagnostics. In these applications, when a deviation from the normal behaviour is detected, the monitoring algorithm could trigger warning messages to the operator or initiate counterbalancing control actions [1].

In a typical case, signals such as the phase currents could be used to determine the presence of a failure in the machine, as in [2]. The phase currents are not the only signals available in a conventional drive: phase voltages could be either measured or estimated through the compensation of inverter non-linearities [3], and may bring information on harmonics when closed-loop controls are employed. The torque could also be interpreted as an indicator of malfunctioning [4]. Vibration signals could also be properly placed on the equipment under test as part of a condition monitoring tool [5].

The frequency analysis is not necessarily performed on raw signals coming from the drive sensors. A typical example is represented by the electromagnetic torque, which could be analysed and estimated through the cross-product between the currents and the flux linkages.

In general, when dealing with raw signals, a prerequisite for a successful harmonic analysis is the selection of a suitable sensor which does not distort the signal in the bandwidth of interest. When the analysed variables are a combination of raw and estimated signals, as the electromechanical torque case

where the currents are measured and the flux linkages must be estimated, it is also essential that the models used for the estimation preserve the original harmonic information without amplitude and/or phase distortion.

Extensive literature has been produced as regards to flux linkage estimation, a major distinction being closed-loop observers of open-loop estimators. In the first case, the electrical machine parameters are used to implement closed-loop observers. These observers are usually exploited for more purposes rather than the flux linkage estimation only. Examples are found in [6], [7], [8], [9], [10], [11], [12], [13], [14].

Open-loop estimators, instead, calculate the flux linkages as the integral of the back-electromotive force. In this case, the integrators are not embedded in closed-loop observers and do not require, in principle, any tuning. The only required machine parameter for the calculation of the back-electromotive force is the stator resistance, while other parameters as the machine inductances do not affect the result of the estimation, as it happens in closed-loop estimators. However, the biggest issue of open-loop estimators is their output drift in presence of spurious DC signals at their input, which are unavoidable in modern digital electrical drives. Extensive research have been performed in the field of drift-free open-loop estimators, as reported in [1], [15], [16], [17], [18], [19], [20], [21].

When considering a frequency analysis for condition monitoring purposes, drift-free open-loop integrators might be considered for their potential lower parameter sensitivity. However, very few investigation has been made as regards to the evaluation of the integrators in terms of DC signal rejection (transient required to obtain an offset-free signal in output) and the capability of performing a correct integration in presence of two (or more) frequencies at once.

This paper investigates the behaviour of different drift-free open-loop integrators, based on existing solutions found in the literature and briefly described in Sect. II. By setting some requirements for the application of interest in Sect. III, the integrators will be tuned according to the DC-input rejection test described in Sect. IV. Then, each integrator performance to multitone input signals, described in Sect. V-A, will be evaluated in Sect. V-B and Sect. V-C. Experimental evidence is included in Sect. VI, showing the existence of integration mismatches that may impact condition monitoring and diagnostics applications.

978-1-4799-2706-7/14 $31.00 © 2014 IEEE

II. OVERVIEW OF DIGITAL INTEGRATORS

This section describes the different algorithms for offset-free integration of the back-electromotive force, for flux linkage estimation and thus torque estimation in electrical drives, listed in chronological order as they have been found in the literature. Each of the solution is briefly analysed.

A. Cascade of low-pass filters with varying time constant

This algorithm, one of the first in the field, was proposed in [15]. The integrator is decomposed in a cascade of low-pass filters, the number of which can be varied but once selected, it will be fixed during real-time operation. The low-pass filters are equal to each other, and their time constant is the inverse of the actual electrical speed. In overall, the cascade of low-pass filters must have a magnitude and phase shift equal to that of an ideal integrator only for the measured electrical speed. To this purpose, [15] establishes two equations that must be fulfilled for any change of electrical speed, resulting in an on-line update of a correction gain and of the filters time constant.

B. Offset cancellation by means of scalar product

The solution in [16] considers that the input and the output signals of an integrator should always be shifted by $\pi/2$ radians with respect to each other, and therefore the scalar product between them should always be equal to zero.

As a first step, the integrator $1/s$ is divided into two transfer functions $1/(s + \omega_c)$ and $\omega_c/(s + \omega_c)$ rearranged in a sum and a loop, where ω_c is an angular frequency selected by the designer. As a second step, a cartesian-to-polar transformation calculates the amplitude and the phase of the flux linkage vector, which is the output of the integrator. The amplitude of the flux linkage is used to normalize an error signal, which is obtained as the scalar product between the flux linkage components and the back-electromotive force components in the $\alpha\beta$ reference frame. A proportional-integral regulator tries to force the normalised error to zero, its output being transformed back in cartesian coordinates using the flux linkage angle. This component component is then processed as input of the transfer function $\omega_c/(s+\omega_c)$, closing the loop.

C. Low-pass filter with changing time constant

The solution in [17] is not very different from the one in [15] in its concept: it is a low-pass filter (only one, in this case), whose time constant is varied with respect to the value of the electrical speed. A gain compensator and a phase adjustment block are present as well. The main difference is that the time constant is calculated as a multiple (by a factor of k) of the inverse of the electrical speed, while in [15] the time constant of the filter is exactly the inverse of the electrical speed. The electrical speed is calculated and fed back from the output of the integrators.

Another difference with respect to [15] is the vectorial nature of the solution. This has no real impact on the integration itself, except in the phase compensation block. In [15], the cascade of the filters was designed in such a way that the overall phase shift of the output would be equal to $-\pi/2$ radians. In [17], the required phase shift is calculated and used for a vectorial rotation, which shifts the output of the required amount to obtain the overall $-\pi/2$ radians lag.

D. Vector-based low-pass filter with changing time constant

The solution presented in [18] gives to the integration operation a complete vectorial perspective - for many aspects, it can be considered a generalisation of the solution in [17], with an easier implementation. A pure integrator is preceded by an high-pass filter which removes the DC components of the incoming signals. A vectorial approach is used. As in [17], the solution in [18] also calculates the electrical speed from the output signal of the integrator, feeding back the result to the blocks that require it.

E. Offset calculator based on vector distortion in $\alpha\beta$

The solution in [19] starts by describing a condition in which a pure integration is performed, and the flux linkage in either the α or β axis drifts to infinite (positive or negative values).

Then, a limitation of the flux linkage is introduced, justified by considering that in electrical machines there will always be a maximum absolute value of the flux linkage. Any time the drift will incur in the limitation, there will be a difference between the maximum peak and the minimum peak of the saturated flux linkage calculated over one electrical period in the $\alpha\beta$ stationary frame.

By computing the average between the maximum peak and the minimum peak for both the α and β axis, and indication of the offset vector direction is given. This direction is filtered and fed back to the input of the pure integrator, in addition to the normal input signal. The overall effect is the offset elimination from the back-electromotive force.

F. High-pass filtering instantaneous compensation

The solution proposed in [20] combines high-pass filters and pure integrators. The high-pass filters are first applied to the $\alpha\beta$ input components. The input and the output of the high-pass filters are routed to a compensation block that calculates the magnitude and the phase shift of the filters transfer function at each time instant. The calculation is vectorial.

The calculated magnitude is used to correct the output of the pure integration which follows the high-pass filter. The phase shift is used for a rotation of the vector obtained from the pure integration, in order to re-establish the $-\pi/2$ radians shift condition.

G. Low-pass filter with changing time constant and PLL for angle estimation

The solution presented in [21], which has also been mentioned and used in [1], is a modification of the one in [17]. The main integration block remains the same, being composed by a low-pass filter with a time constant that is function of the electrical speed. In this case, however, the electrical speed is obtained with a phase-locked loop (PLL) acting on the integrator output signals, rather than directly from the components of the output vector as in [17].

III. REQUIREMENTS OF THE APPLICATION

In the following sections, the integrators described in Sect. II are evaluated with respect to their ability of rejecting a DC input component and performing a correct flux linkage estimation in presence of multitone signals. Since the performances of the integrators vary depending on their tuning, a set of requirements is necessary before starting the evaluation.

Rather than the focus on the correct estimation for frequencies very close to zero, which is a typical requirement in model-based sensorless control algorithms [3], here the interest is towards applications where the presence of multitone signals is quite common. The selected application is the control of wind turbine generators, either synchronous or asynchronous ones, by means of electrical drives.

In this case, the generator is part of a system where many different components contribute to the presence of harmonics in measurable signals. Apart from the generator itself, harmonics could be caused by effects like wind shear and tower shadow [22], tower oscillations [23], gearbox vibrations [5], misalignments of the drive train shaft [4], and many others. Overall, the consideration that the possible harmonics in the measurable signals are found at the fundamental electrical frequency of rotation or its multiples is a simplification of the problem. An example is represented by the resonance effects of a wind turbine tower, as documented in [24]. In this case, the wind gusts hitting the tower lead to temporary oscillations with a main frequency equal to the dominating mechanical resonance of the tower structure. Such oscillations are visible in the torque exerted by the wind on the wind blades, because of the relative movement of the tower with respect to the wind pattern, and therefore in the rotating speed as well [24]. This resonant frequency is completely independent from the rotating electrical frequency.

Within this context, the work in [25] has been used as a main reference to determine the lower limits for the integrators. In [25], different direct-drive permanent-magnet synchronous generators designs are evaluated in terms of vibrations caused by the cogging torque/torque ripple on the mechanical drive train of the turbine. A lower limitation of 4 rpm is used as a cut-in speed for the torque (and thus power) production. Based on the generator parameters reported in [25], the cut-in frequency corresponds to 4 Hz. In order to slightly release this limitation, and thus make the evaluation process easier for the integrators, a limit of 6 Hz is set as the minimum frequency of the flux linkage estimation that is expected to be correctly performed by the drive.

IV. THE DC-INPUT REJECTION TEST

Based on the considerations of Sect. III, the seven integrators described in Sect. II were tuned by considering the effects of a test input signal with the following characteristics:

- it was composed by a pure sinusoid with unity amplitude and frequency of 6 Hz;
- a DC signal of small amplitude (0.01 V) was superimposed to the sinusoid.

The transient required to stabilise the output of each integrator was analysed and post-processed to extract it length and the residual DC offset present at the output. A subsequent tuning process was performed for each integrator, such that their transient was concluded within one second after the superimposition of the DC component.

The imposition of such condition is explained as follows. The work in [19] reports that the offset drift is mainly a thermal effect that changes the DC offset very slowly, therefore the response time of the rejection method is not critical. Although true, there are other considerations that may lean towards a fast rejection of the DC component. In particular, non-perfect DC offset compensation in current sensors at drive commissioning stage may lead to a drift of the flux linkage estimation from the very beginning of the drive operation. At the same time, all the solutions analysed in this work are not able of completely rejecting the DC offset, as shown later in Fig. 2, acting similarly to a low-pass filter instead of a pure integrator. For this reason, it becomes very important to stabilize the output of the integrators around a spurious DC level as fast as possible: although empirically, it has been observed that the longer the transient, the larger is the spurious DC component that accumulates at the output of the integrators. This may only worse the behaviour of the remaining control structure.

An example of an integrator transient after the tuning process is shown in Fig. 1, taken from the output of the integrator described in [16]. The transient is well concluded within one second.

Fig. 2 reports the amount of DC offset of the integrators output that have been calculated after each integrator was tuned according to the aforementioned considerations. With the exception of two cases (the integrators of [15] and [19]) that are going to be explained separately, the other integrators behave in a similar manner as regards to the spurious DC offset at their output.

As regards to [15], the nature of its solution is such that the low-pass filters have a time constant equal to the inverse of the actual electrical speed. The main consequence is that the DC rejection transient time is not selectable, but rather dependant on the value of the fundamental harmonic under integration. For the case of this test, the 6-Hz main harmonic returns a time constant of 0.0265 s for each filter, corresponding to a transient time of approximately 0.1 s. Tihs ten-fold faster transient time has a beneficial effect on the amount of spurious DC effect in output, as shown in Fig. 2. The price to pay, however, will be clear in Sect. V, where multitone signals will be employed for the validation of the integrator Bode diagram.

As regards to [19], Fig. 2 shows a relatively high spurious DC component at the end of the transient. It is worth to note that the nature of the solution relies on the determination of the period of the fundamental harmonic for the calculation of the maximum and minimum flux within one period. In other words, this integrator solution waits the fundamental harmonic to conclude its period before updating a new value of the offset estimation, and several fundamental harmonic periods are required to conclude the transient. As a matter of fact, the transient length for this solution was brought to 25 s in this test, with no means of further reduction. As a consequence, the accumulated spurious DC offset value was increased, as shown in Fig. 2.

Tab. I reports the value of the parameters for all integrators tuned with a DC rejection transient of one second, with the exceptions mentioned above. The name of the parameters is the same reported in the corresponding papers, with the same meaning.

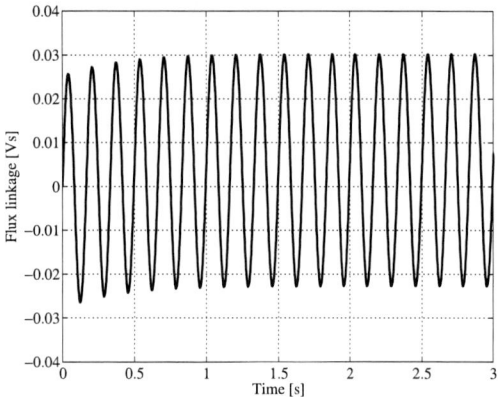

Fig. 1. Example of transient for DC offset rejection concluded within one second, obtained with the integrator [16].

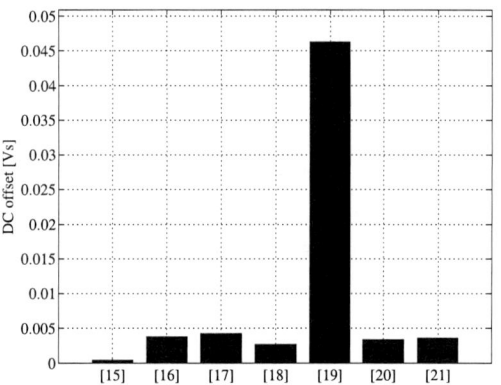

Fig. 2. Residual DC error after the transients for the integrators under evaluation.

TAB. I
PARAMETERS OF THE INTEGRATORS.

Integrator solution	Parameter(s)	Value(s)
[15]	n	3
[16]	ω_c, k_p, k_i	$2\pi \cdot 0.5, 0.01, 0.0$
[17]	k	15
[18]	λ	0.25
[19]	λ_s^*, τ_1	0.0398, 0.0159
[20]	ω_c	$2\pi \cdot 0.5$
[21]	$k, k_{p,PLL}, k_{i,PLL}$	0.08, 1000, 200

V. THE AC-INPUT INTEGRATION TEST

A. The multitone signals as input

The evaluation with AC signals was required in order to understand whether the integrators were able to correctly detect multiple harmonics in the input signal. This mode of operation might recall the case where the measured signals are composed by the fundamental harmonic (related to the electrical speed) and other harmonics that carry information on malfunctioning or failures in the electrical machine or the mechanical load.

Signals carrying more than one harmonics at once are known as multitone signals [26]. The construction of a multitone signal is not a difficult task, however one must keep in mind that the sum of many harmonics could lead to signals with high crest factor (ratio between the peak and the RMS value of the signal). As a matter of fact, it is not a good idea to apply high-crest factor signals for system testing, mainly because this could lead to poor signal-to-noise ratio in the processing of the output results. In the worst case, the output signal might not be post-processed by algorithms that try to reconstruct a transfer function. This problem occurs more frequently in experimental tests, but it is a good practice to prevent this situation even in simulation environments, in order to avoid possible numerical errors.

Good multitone signals are created with a special phase relationship between the tones. One of the better solutions is the Newman phase relation, whose mathematical description can be found in [26]. In brief, the Newman phase for multitone signals requires each tone phase to follow the expression:

$$\varphi_k = \frac{\pi(k-1)^2}{N} \qquad (1)$$

where φ_k is the phase of the kth-harmonic (tone), N is the number of generated harmonics (tones) and k is the counter for the k-th element. An example of multitone signal with Newman phases containing frequencies from 1 to 20 Hz (with steps of 1 Hz) is shown in Fig. 3a (time domain) and Fig. 3b (corresponding FFT). Note that the signal amplitude gets closer to the maximum peak for most of its oscillations, showing a low crest factor. It is easy to prove that the very same multitone signal without the Newman phases has a larger crest factor, with clear consequences on the signal-to-noise ratio of the post-processing results.

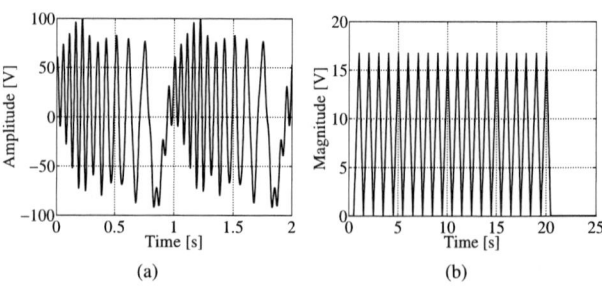

Fig. 3. Multitone signal with Newman phase: (a) time domain, (b) FFT.

B. Description of the test

An input signal resembling a back-electromotive force with many frequencies was created. According to the considerations of Sect. III, the minimum frequency of the input signal was set to 6 Hz, and the maximum was shifted to 1 kHz. Steps of 1 Hz were used in order to cover a good spectrum interval. As for the magnitude of the tones, a more realistic situation was reproduced, where the tones from 7 Hz onwards had a magnitude 100 times or 10 times lower than then tone at 6 Hz. The results have shown no major differences between the two cases: in this work, only the results with tones 10 times lower than the main harmonic are shown. Note that the input signal has no DC component, since the interest was focused on the integrator performances in presence of multitone signals and the DC rejection test was already performed as described in Sect. IV.

The input signal was fed to all the investigated integrators at the same time, including an ideal digital integrator based on the trapezoidal (Tustin) approximation. Since no DC component was present in the input signal, the Tustin integrator performed a perfect integration that has been used as main reference for the results of all the other integrators.

The simulation was performed for 60 seconds, of which the last 54 seconds have been used for the evaluation of the transfer function of each integrator. Nine windows of six seconds each were extracted from the last 54 seconds, calculating the FFT for each window and each integrator output. The average FFT of the nine windows was taken, and the cross power spectrum between the input and output FFTs was obtained. From the cross power spectrum, an estimation of the transfer function of the integrators was calculated. The sample time of the digital integration was set to 100 μs. In addition to the transfer functions, the standard deviation error was calculated for each integrator with respect to the transfer function of the ideal integrator, for both the magnitude and the phase.

C. Results and comments

The results of the magnitude and phase analysis of the integrators are reported in Fig. 4 and Fig. 6. Fig. 5 and Fig. 7 report the magnitude and phase standard deviation errors for the five best solutions, excluding the worst ones ([15] and [20]) that would make the plots non-readable. Some considerations follow for each integrator under analysis.

Solution [15]: this integrator is with no doubt one of the easiest to implement, but its performances are lacking with respect to the other solutions. Due to the fact that the digital integrator is replicated with low pass filters tuned to fulfil the magnitude/phase requirements for only the main fundamental frequency, all the other harmonics are not correctly reproduced in amplitude and distorted in phase. On the other side, as mentioned in Sect. IV, it is the best solution in terms of DC component rejection. Thus, this solution might be an option only in drives where no spurious harmonics are expected rather than just the main one.

Solution [16]: although showing an interesting linearity in the magnitude Bode diagram (see Fig. 4c), there was a detected phase deviation in the lower region of the spectrum (see Fig. 6c). This behaviour was explained considering the non-linear approach of the solution itself. Specific tests on the same integrator using lower-tone signals, not shown in this work, have shown a tendency of the magnitude to deviate from the ideal integrator towards zero, in a similar way as the phase. Nevertheless, the overall behaviour of the integrator for the application of interest was not bad, although not perfect either.

Solution [17]: there was still a visible sort of linearity in the magnitude Bode diagram (see Fig. 4d), although noisier than [16]. The phase Bode diagram in Fig. 6d was quite noisy. An explanation of this effect might be the automatic calculation of the main electrical frequency from the output of the integrator, required to set the constant in the filter of this solution. A better calculation is probably sought, with the clear consequence of a complication of the scheme.

Solution [18]: it was found to be noisier than [16] and [17], as visible in both Fig. 4e and Fig. 6e. Again, the problem might rely in the self-calculation of the fundamental harmonic of the output signal. By looking at Fig. 5, one might argue that the standard deviation error for the magnitude is not large, being approximately 1.3 dB for this solution. However, such a value means approximately a 16% estimation error of the harmonic amplitude. Although probably good enough for closed-loop control applications, such result might be not good enough for condition monitoring solutions, where a precise magnitude estimation is required for a correct prevention of faults.

Solution [19]: it was the best performing one in this test. However, more considerations must be made. On one side, the results in Fig. 4f and Fig. 6f are not surprising, as the solution does not modify the nature of the integrator itself, rather than changing the input signal in order to make it offset-free. On the other side, this solution has inherent drawbacks as the slowness of the DC rejection transient (see Sect. IV). It was also found that the performances of the algorithm were dependent on a convenient saturation level for the estimated flux linkage. Not last, the algorithm needs to know the fundamental harmonic period in order to estimate the maximum and minimum flux within one period. This is difficult in presence of other harmonics not multiple of the fundamental one. In this work, the knowledge of the fundamental period was given in advance to the algorithm.

Solution [20]: it was found to have a relatively noisy behaviour for higher frequencies (Fig. 4g and Fig. 6g). The reason might reside in the instantaneous compensation of the magnitude and phase difference of a high-pass filter with respect of a normal integrator: this might work for a single harmonic, but probably not as well in presence of multiple harmonics.

Solution [21]: as expected, the performances are very similar to [17], and smoothed out because of the presence of the PLL for the estimation of the main fundamental harmonic. The magnitude behaviour is better than the phase, as documented in Fig. 4h and Fig. 6h.

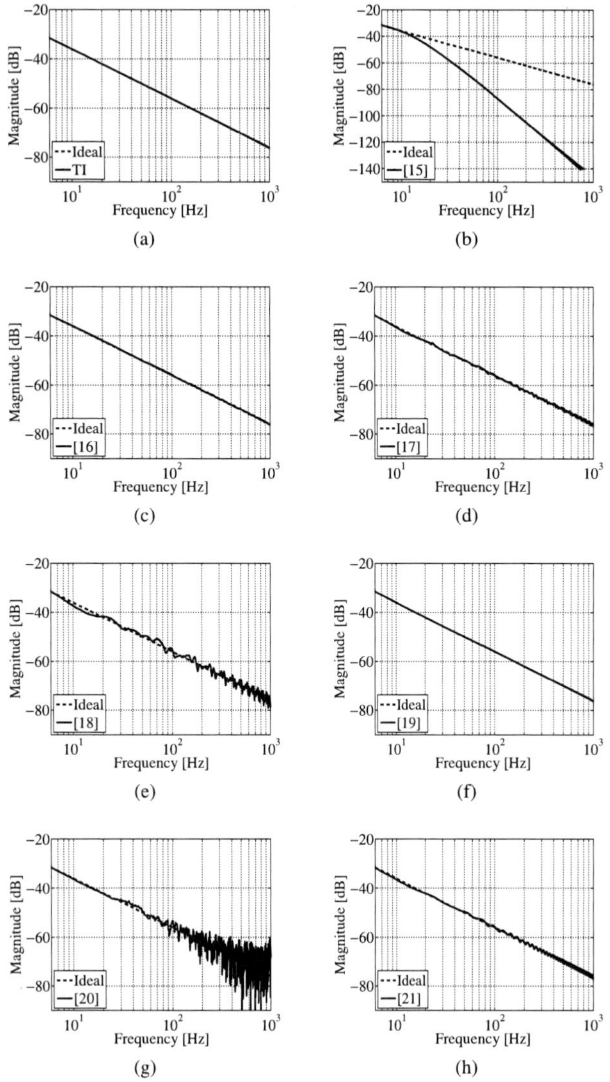

The 2014 International Power Electronics Conference

Fig. 4. Simulation results, magnitude: (a) Tustin integrator, (b) [15], (c) [16], (d) [17], (e) [18], (f) [19], (g) [20], (h) [21].

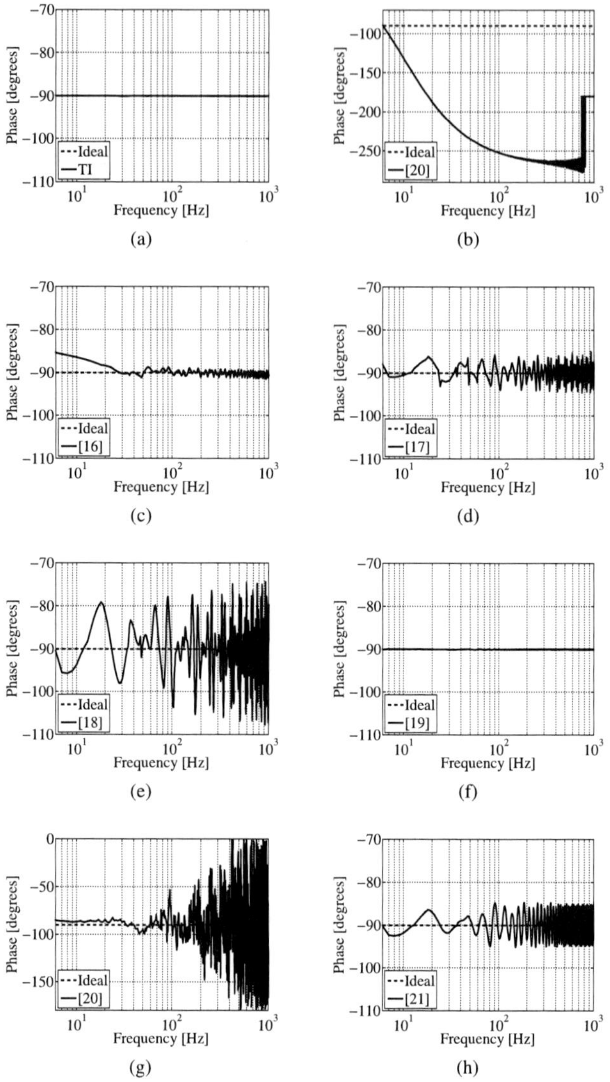

Fig. 6. TSimulation results, phase: (a) Tustin integrator, (b) [15], (c) [16], (d) [17], (e) [18], (f) [19], (g) [20], (h) [21].

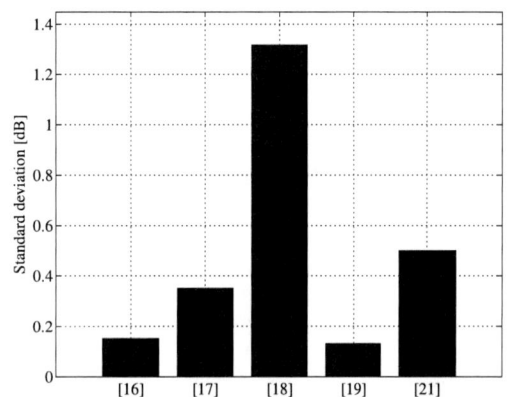

Fig. 5. Magnitude standard deviation error for the best five integrators.

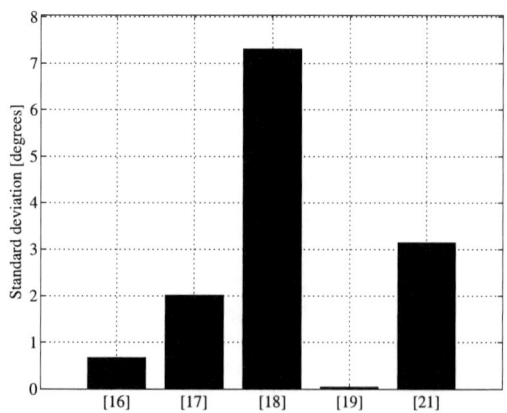

Fig. 7. Phase standard deviation error for the best five integrators.

978-1-4799-2706-7/14 $31.00 © 2014 IEEE

VI. EXPERIMENTAL RESULTS

A set of experimental tests were performed in laboratory premises to validate the simulations of Sect. V. The test bench was composed by two 2-kW synchronous machines (surface permanent magnets) in a back-to-back connection, each controlled with a variable speed drive (ABB ACSM1). One drive was controlled through a custom control board mounting a Xilinx Virtex 6 FPGA, where the algorithms have been developed. A picture of the setup is shown in Fig. 11.

In order to reproduce some frequencies of the spectrum analysed in Sect. V, one of the two machines was controlled with a simple open-loop algorithm where the main voltage vector was rotating at 6 Hz, as the minimum frequency requested in Sect. III. On the top of the reference voltage vector, voltage harmonics were injected at the following frequencies: 15 Hz, 40 Hz, 70 Hz, 168 Hz, 503 Hz and 794 Hz.

The back-electromotive profile in time and FFT domain, computed from the recorded voltages and currents at a sampling rate of 100 μs, are shown in Fig. 10, reporting the peaks corresponding to the injected frequencies.

The integrators analysed in Sect. V were implemented and subjected to the back-electromotive force input as shown in Fig. 10. Their output was recorded and post-processed in a similar manner as the simulation results of Sect. V. The outcome of the experimental analysis, overlapped to the simulation results of Sect. V, is reported in Fig. 8 for the magnitude and Fig. 9 for the phase.

Provided that the measurements are slightly affected by the unavoidable measurement noise, the behaviour is still the same as found in the simulations. For example, the largest deviations from the ideal integrators are visible for the solutions [15], [20] (for the magnitude) and [16], [18], [20] (for the phase).

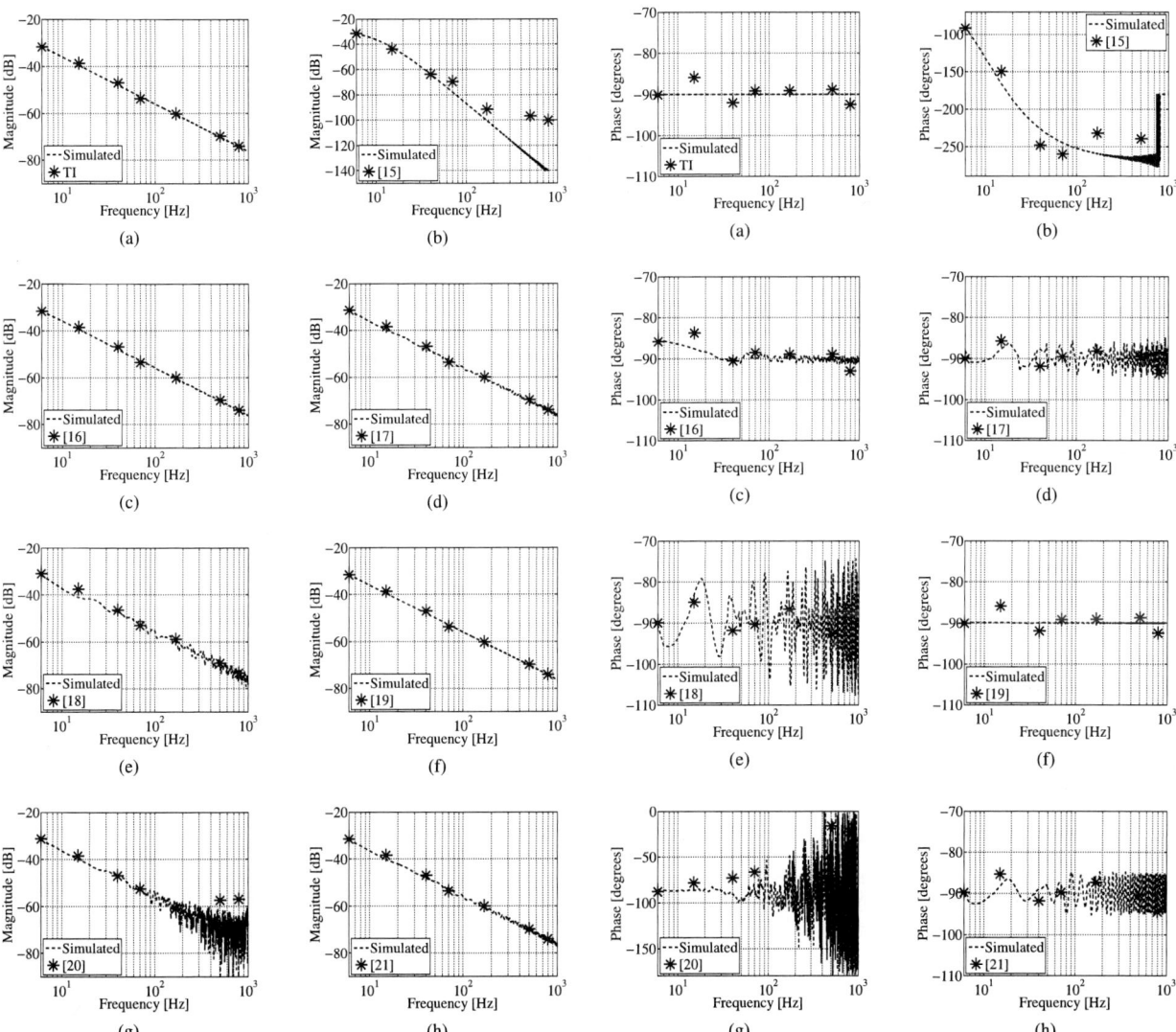

Fig. 8. Experimental results, magnitude: (a) Tustin integrator, (b) [15], (c) [16], (d) [17], (e) [18], (f) [19], (g) [20], (h) [21].

Fig. 9. Experimental results, phase: (a) Tustin integrator, (b) [15], (c) [16], (d) [17], (e) [18], (f) [19], (g) [20], (h) [21].

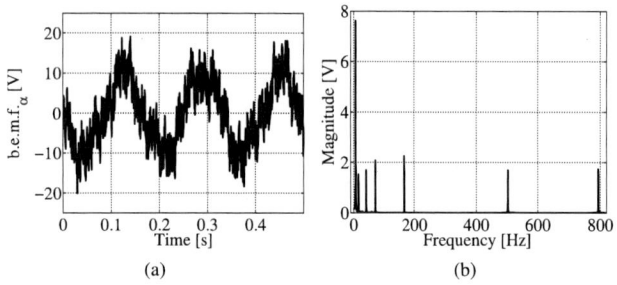

Fig. 10. Back-electromotive force: (a) time domain, (b) FFT.

Fig. 11. The experimental test bench.

VII. CONCLUSIONS

Different algorithms for drift-free open-loop digital integration of the back-electromotive force in electrical machines were analysed. Their performances were evaluated in presence of DC components and multitone signals at their input, recording their deviation with respect to the ideal integrator case. The simulation and experimental analysis showed that some integrator solutions may deviate from the ideal behaviour. This result must be taken into account when performing harmonic analysis of estimated flux linkages or torque, if the focus is on condition monitoring and diagnostics applications.

REFERENCES

[1] S. H. Kia, H. Henao, and G.-A. Capolino, "Torsional vibration assessment using induction machine electromagnetic torque estimation," *IEEE Trans. Ind. Electron.*, vol. 57, no. 1, pp. 209–219, Jan. 2010.

[2] W. L. Roux, R. G. Harley, and T. G. Habetler, "Detecting rotor faults in low power permanent magnet synchronous machines," *IEEE Trans. Power Electron.*, vol. 22, no. 1, pp. 322–328, Jan. 2007.

[3] J. Holtz and J. Quan, "Sensorless vector control of induction motors at very low speed using a nonlinear inverter model and parameter identification," *IEEE Trans. Ind. Appl.*, vol. 38, no. 4, pp. 1087–1095, Jul./Aug. 2002.

[4] B. M. Ebrahimi and J. Faiz, "Diagnosis and performance analysis of three-phase permanent magnet synchronous motors with static, dynamic and mixed eccentricity," *IET Electr. Power Appl.*, vol. 4, no. 1, pp. 53–66, Jan. 2010.

[5] K. Wang, "Phase information at tooth mesh frequency for gear crack diagnosis," in *Proceedings of the 2nd IEEE Conference on Industrial Electronics and Applications (ICIEA 2007)*, Harbin, China, May 23-25 2007, pp. 2712–2717.

[6] C. P. Bottura, J. L. Silvino, and P. de Resende, "A flux observer for induction machines based on a time-variant discrete model," *IEEE Trans. Ind. Appl.*, vol. 29, no. 2, pp. 349–354, Mar./Apr. 1993.

[7] P. L. Jansen and R. D. Lorenz, "A physically insightful approach to the design and accuracy assessment of flux observers for field oriented induction machine drives," *IEEE Trans. Ind. Appl.*, vol. 30, no. 1, pp. 101–110, Jan./Feb. 1994.

[8] J. Stephan, M. Bodson, and J. Chiasson, "Real-time estimation of the parameters and fluxes of induction motors," *IEEE Trans. Ind. Appl.*, vol. 30, no. 3, pp. 746–759, May/Jun. 1994.

[9] D.-W. Chung and S.-K. Sul, "Analysis and compensation of current measurement error in vector-controlled AC motor drives," *IEEE Trans. Ind. Appl.*, vol. 34, no. 2, pp. 340–345, Mar./Apr. 1998.

[10] T. G. Habetler, F. Profumo, G. Griva, M. Pastorelli, and A. Bettini, "Stator resistance tuning in a stator-flux field-oriented drive using an instantaneous hybrid flux estimator," *IEEE Trans. Power Electron.*, vol. 13, no. 1, pp. 125–133, Jan. 1998.

[11] H. Kubota, Y. Kataoka, H. Ohta, and K. Matsuse, "Sensorless vector controlled induction machine drives with fast stator voltage offset compensation," in *Conference Record of the 34th IEEE Industrial Application Society Annual Meeting*, vol. 4, Phoenix, Arizona, USA, Oct. 3-7 1999, pp. 2321–2324.

[12] M. Rodič and K. Jezernik, "An analysis of speed sensorless torque and flux controller for induction motor," in *Proceedings of the Power Eelectronics Specialist Conference (PESC)*, vol. 2, Galway, Ireland, Jun. 18-23 2000, pp. 867–872.

[13] H.-S. Jung, S.-H. Hwang, J.-M. Kim, C.-U. Kim, and C. Choi, "Diminution of current-measurement error for vector-controlled AC motor drives," *IEEE Trans. Ind. Appl.*, vol. 42, no. 5, pp. 1249–1256, Sep./Oct. 2006.

[14] R. Bojoi, P. Guglielmi, and G.-M. Pellegrino, "Sensorless direct field-oriented control of three-phase induction motor drives for low-cost applications," *IEEE Trans. Ind. Appl.*, vol. 44, no. 2, pp. 475–481, Mar./Apr. 2008.

[15] B. K. Bose and N. R. Patel, "A programmable cascaded low-pass filter-based flux synthesis for a stator flux-oriented vector-controlled induction motor drive," *IEEE Trans. Ind. Electron.*, vol. 44, no. 1, pp. 140–143, Feb. 1997.

[16] J. Hu and B. Wu, "New integration algorithms for estimating motor flux over a wide speed range," *IEEE Trans. Power Electron.*, vol. 13, no. 5, pp. 969–977, Sep. 1998.

[17] M.-H. Shin, D.-S. Hyun, S.-B. Cho, and S. Choe, "An improved stator flux estimation for speed sensorless stator flux orientation control of induction motors," *IEEE Trans. Power Electron.*, vol. 15, no. 2, pp. 312–318, Mar. 2000.

[18] M. Hinkkanen and J. Luomi, "Modified integrator for voltage model flux estimation of induction motors," *IEEE Trans. Ind. Electron.*, vol. 50, no. 4, pp. 818–820, Aug. 2003.

[19] J. Holtz and J. Quan, "Drift- and parameter-compensated flux estimator for persistent zero-stator-frequency operation of sensorless-controlled induction motors," *IEEE Trans. Ind. Appl.*, vol. 39, no. 4, pp. 1052–1060, Jul./Aug. 2003.

[20] M. Zerbo, P. Sicard, and A. Ba-Razzouk, "Accurate adaptive integration algorithms for induction machine drive over a wide speed range," in *Proceedings of the IEEE International Conference on Electric Machines and Drives (IEMDC)*, San Antonio, TX, USA, May 15-18 2005, pp. 1082–1088.

[21] M. Comanescu and L. Xu, "An improved flux observer based on PLL frequency estimator for sensorless vector control of induction motors," *IEEE Trans. Ind. Electron.*, vol. 53, no. 1, pp. 50–56, Feb. 2006.

[22] D. S. L. Dolan and P. W. Lehn, "Simulation model of wind turbine 3p torque oscillations due to wind shear and tower shadow," *IEEE Trans. Energy Convers.*, vol. 21, no. 3, pp. 717–724, Sep. 2006.

[23] J. D. Grunnet, M. Soltani, T. Knudsen, M. Kragelund, and T. Bak, "Aeolus toolbox for dynamic wind farm model, simulation and control," in *Proceedings of the European Wind Energy Conference and Exhibition (EWEC) 2010*, Warsaw, Poland, Apr. 20-23 2005.

[24] T. Thiringer and J.-Å. Dahlberg, "Periodic pulsations from a three-bladed wind turbine," *IEEE Trans. Energy Convers.*, vol. 16, no. 2, pp. 128–133, Jun. 2001.

[25] J. Sopanena, V. Ruuskanen, J. Nerg, and J. Pyrhönen, "Dynamic torque analysis of a wind turbine drive train including a direct-driven permanent-magnet generator," *IEEE Trans. Ind. Electron.*, vol. 58, no. 9, pp. 3859–3867, Sep. 2011.

[26] S. Boyd, "Multitone signals with low crest factor," *IEEE Trans. Circuits Syst.*, vol. CAS-33, no. 10, pp. 1018–1022, Oct. 1986.

Audible Noise Reduction Method in IPMSM Position Sensorless Control based on High-Frequency Current Injection

Yuki Tauchi, Hisao Kubota

Graduate School of Electrical Engineering
Meiji University
Kawasaki, Japan

Abstract— **Methods for sensorless control involving the injection of a high-frequency signal in the low-speed range of the Interior Permanent Magnet Synchronous Motor (IPMSM) have been proposed. However, audible noise is a concern in these methods. We propose a method to reduce the audible noise due to a high-frequency current by minimizing the injection signal amplitude suitable for IPMSM driving states.**

Keywords— *Audible Noise, High-Frequency, IPMSM, Position Sensorless Control*

I. INTRODUCTION

IPMSMs are often used in elevators, air conditioners, and HEVs because they are more compact and efficient as compared to induction motors. To drive an IPMSM, the pole position information is necessary. Hence, position sensors such as rotary encoders or hole-effect sensors are used. Because position sensors have disadvantages such as reduced reliability and increased costs, many sensorless control methods have been proposed.

In this study, we investigated the sensorless control system based on high-frequency current injection described in [1], with a focus on the low-speed range drive of the IPMSM. Audible noise is a concern when a high-frequency signal is injected. To solve the problem, a method to determine the high-frequency amplitude from experimental results has been proposed [2].

In this paper, to reduce the audible noise due to the high-frequency current, we propose a method for minimizing the injection signal level suitable IPMSM for driving states without providing the lower limit setting of the high-frequency current amplitude command value. The proposed method uses the speed error of the IPMSM drive.

II. HIGH-FREQUENCY SIGNAL INJECTION METHODS

In the rotating dq coordinate frame, the voltage equation of the IPMSM is represented by Eq. (1).

$$\begin{bmatrix} v_d \\ v_q \end{bmatrix} = \begin{bmatrix} R_a + pL_d & -\omega L_q \\ \omega L_d & R_a + pL_q \end{bmatrix} \begin{bmatrix} i_d \\ i_q \end{bmatrix} + \begin{bmatrix} 0 \\ \omega \psi_a \end{bmatrix} \quad (1)$$

where p is a differential operator.

The estimated axes are represented by the γ-δ coordinate system. A high-frequency current that is injected into the γ-δ axes is expressed as in Eq. (2).

$$\begin{bmatrix} i_{\gamma h} \\ i_{\delta h} \end{bmatrix} = I_h \begin{bmatrix} \sin \theta_h \\ 0 \end{bmatrix} \quad (2)$$

where I_h is the amplitude of the high frequency current. $\theta_h = \omega_h t$ is the phase angle of the high-frequency current.

As a result, a high-frequency voltage represented by Eq. (3) in the γ-δ axes is generated.

$$\begin{aligned} \begin{bmatrix} v_{\gamma h} \\ v_{\delta h} \end{bmatrix} = & R_a I_h \begin{bmatrix} \sin \theta_h \\ 0 \end{bmatrix} \\ & + \frac{L_0 I_h}{2} \left[(\omega_h + \omega_{est}) \begin{bmatrix} \cos \theta_h \\ \sin \theta_h \end{bmatrix} + (\omega_h - \omega_{est}) \begin{bmatrix} \cos \theta_h \\ \sin \theta_h \end{bmatrix} \right] \\ & + \frac{L_1 I_h}{2} \left[(\omega_h - 2\omega + \omega_{est}) \begin{bmatrix} \cos(-\theta_h - 2\Delta\theta) \\ \sin(-\theta_h - 2\Delta\theta) \end{bmatrix} \right. \\ & \left. \qquad + (\omega_h + 2\omega - \omega_{est}) \begin{bmatrix} \cos(\theta_h - 2\Delta\theta) \\ \sin(\theta_h - 2\Delta\theta) \end{bmatrix} \right] \\ & + \omega\psi \begin{bmatrix} \sin(\Delta\theta) \\ \cos(\Delta\theta) \end{bmatrix} \end{aligned} \quad (3)$$

where $L_0 = (L_d + L_q)/2$, $L_1 = (L_d - L_q)/2$, and $\Delta\theta$ is the magnetic pole position error.

Assuming that $\Delta\theta$ is small enough, $\omega\psi\cos\Delta\theta$ of Eq. (3) can be considered a DC component that is removed by using a BPF with a center frequency of ω_h. Then, $v_{\delta h}$ multiplied by $\cos \theta_h$ is given by Eq. (4):

$$\begin{aligned} v_{\delta h} \times \cos \theta_h = & \frac{L_0 \omega_{est} I_h}{2} \sin(2\theta_h) \\ & - \frac{L_1 I_h (\omega_h - 2\omega + \omega_{est})}{4} \{ \sin(2\Delta\theta) \\ & \qquad\qquad\qquad + \sin(2\theta_h + 2\Delta\theta) \} \\ & + \frac{L_1 I_h (\omega_h + 2\omega - \omega_{est})}{4} \{ \sin(2\theta_h - 2\Delta\theta) \\ & \qquad\qquad\qquad - \sin(2\Delta\theta) \} \end{aligned} \quad (4)$$

Eq. (4) includes the position error $\Delta\theta$. The magnetic pole position can be estimated by the average of one cycle of Eq. (4).

III. AUDIBLE NOISE BY HIGH-FREQUENCY CURRENT

During high-frequency signal injection, the problem of audible noise arises. We measured the audible noise at the actual machine steady state. The measured values of audible noise are listed in Table I. The ratings and motor parameters of the IPMSM are listed in Table II.

Table I shows the relationship between audible noise and high-frequency amplitude. It can be confirmed that the audible noise is reduced by decreasing the amplitude. However, with decreasing amplitude, the precision of position estimation deteriorates. When the precision of position estimation deteriorates, the audible noise is increased. In addition, when the amplitude is decreased, it is impossible to control the IPMSM in the transient state. The optimal current amplitude depends on conditions such as rotational speed.

Therefore, it is necessary to control the high-frequency current amplitude command value in to reduce the audible noise. This control is exercised when the high-frequency current amplitude is small at the steady state as well as when the amplitude is high at the transient state.

TABLE I
RELATIONSHIP BETWEEN HIGH-FREQUENCY CURRENT AMPLITUDE AND AUDIBLE NOISE

Speed		audible noise(dB)	
		90 min⁻¹	30 min⁻¹
With sensor, No current injection		45.8–46.6	46.0–46.3
Without sensor, Current amplitude command value	0.4A	Loss of control	Loss of control
	0.5A	46.4–49.0	
	0.7A	46.9–47.7	
	1.0A	47.7–48.4	48.1–50.0
	1.2A	48.7–49.4	47.6–48.6
	1.5A	49.2–50.1	48.4–49.4
	2.5A	53.4–55.8	53.9–55.0

TABLE II
RATINGS AND PARAMETERS OF MOTOR

Parameter	Value
Rated power W	1500
Rated voltage V	180
Rated current A	6.1
Rated speed min⁻¹	1800
Rated torque T/N · m	7.9
Number of pole pairs	6
Stator resistance R Ω	0.783
d-axis inductance L_d mH	9.77
q-axis inductance L_q mH	22.4
Rotor magnet flux Ψ_m mWb	0.2606
Rotor inertia J Kg·m2	0.00144

IV. AUDIBLE NOISE REDUCTION METHOD BASED ON OPTIMIZATION OF HIGH-FREQUENCY CURRENT AMPLITUDE

As described in the previous section, it is necessary to control the high-frequency current amplitude command value in order to reduce the audible noise. In general, the high-frequency current amplitude is maintained at a required constant value in the transient state. This general approach results in a large audible noise.

Here, to reduce the audible noise due to the high-frequency current, we propose a method for minimizing the injection signal to a level suitable for the driving state without providing the lower limit setting of the high-frequency current amplitude command value. We show the proposed method in Fig.1.

We now proceed to explain the principle of the proposed method. Speed error ω_{err} can be expressed as in Eq. (5).

$$\omega_{err} = \omega_{ref} - \omega_{est} \qquad (5)$$

where ω_{ref} is the directive speed, and ω_{est} is the estimated speed.

The value of ω_{err} is low in the steady state, and at the transient state it is high. When the high-frequency current amplitude is reduced, the value of ω_{err} is increased because the estimation accuracy deteriorates. Therefore, the value of ω_{err} can distinguish the state. Our proposed method uses the threshold values of ω_{err}, vis. C1, C3, and C5 to distinguish the steady and transient states. The high-frequency current amplitude command value is then determined. The threshold values C1, C3, and C5 are determined by experiments explained in Section V.

In the proposed method, the average value of the absolute value of ω_{err} is used for determining the optimal current amplitude command value because ω_{err} fluctuates.

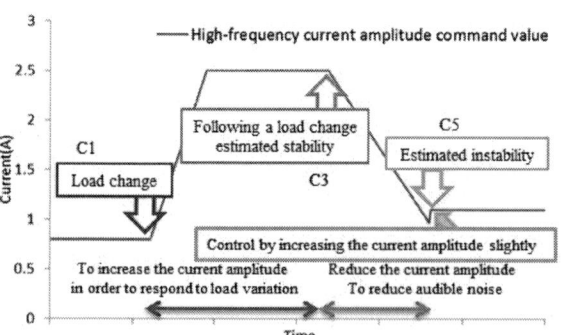

Fig. 1. Method proposed for minimizing the injection signal level suitable for IPMSM driving states.

V. METHOD FOR DETERMINING THE THRESHOLD

In this section, we explain how to determine the three threshold values C1, C3, and C5. It is necessary to carry out two tests.

A. Method for deciding C5

When the high-frequency current amplitude command value was reduced as shown in Fig. 2, we measured the average value of the absolute value of ω_{err}. The experiment was performed at 90min^{-1} (5% of the rated speed) under no-load condition. (Fig. 3)

Fig.3 shows that the average of the absolute value of the speed error is increased, when the high-frequency current amplitude command value is reduced. From the value of the average of the absolute value of the speed error, we can distinguish the position estimate stability from instability.

The threshold value C5 is determined by the permissible range of the position error. C5 is the average value of the absolute value of the speed error when the position error exceeds the permissible range the first time. This time, the permissible range of the position was set at 20°. Then, the value of C5 is decided at 2.0.

B. Method for deciding C1 and C3

We changed the load at a constant speed to decide C1 and C3. We measured the absolute value of the speed error at a constant speed (90min^{-1}) and constant high-frequency current amplitude command (1.5 A). The result of this experiment is shown in Fig. 4.

From Fig.4, it can be confirmed that the speed error is affected when load is applied. From the absolute value of the speed error, we can distinguish the transient state. The threshold values C1 and C3 are determined by the results of this experiment. C1 should be smaller than the absolute value of the speed error at the condition after the change of the load. C3 should be as small as possible, but C3 should be larger than the average of the absolute value of the speed error at the steady state. C1 is set at 8.0 and C3 is set at 0.6.

We determined C1, C3, and C5 by the experiment. The high-frequency current amplitude command value is determined as shown in Fig. 5.

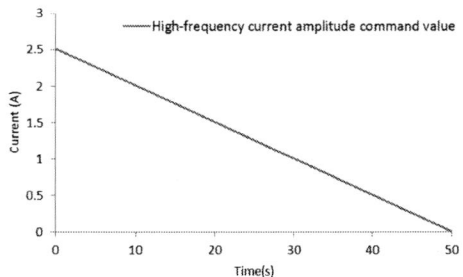

Fig. 2. High-frequency current amplitude command value (I_h)

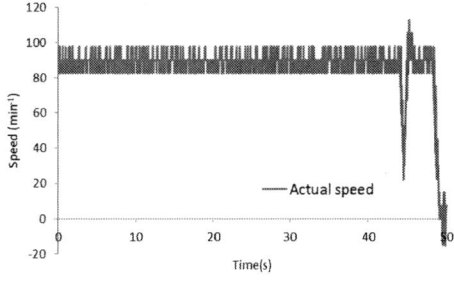

(a) Actual speed

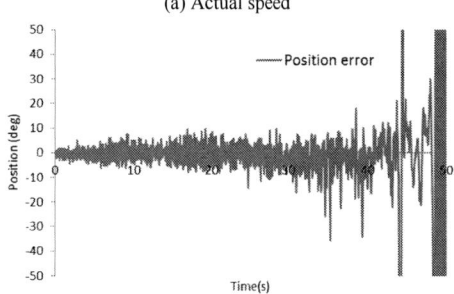

(b) Position error

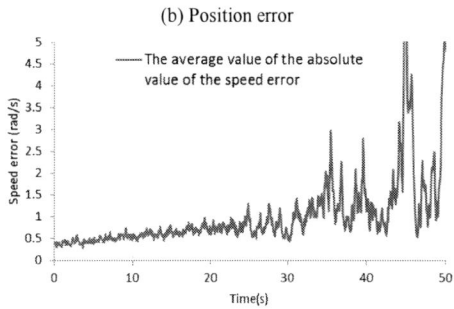

(c) Average value of the absolute value of the speed error
Fig. 3. Speed error and speed (90 min^{-1}, no-load)

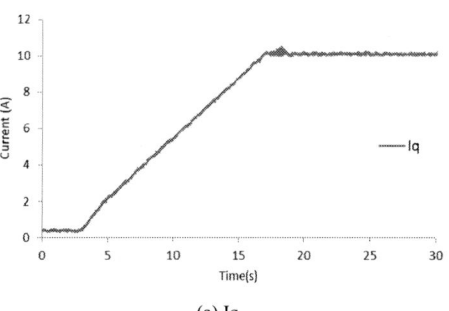

(a) Iq

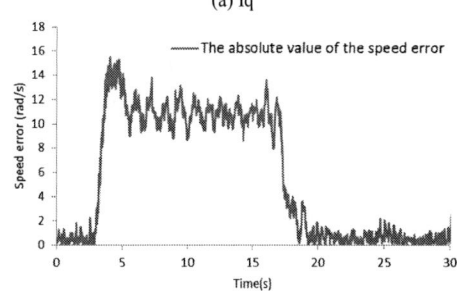

(b) Absolute value of the speed error
Fig. 4. Speed error
(90 min^{-1}, Load torque changes slowly from 0 Nm to 7.9 Nm.)

C1: *Load distinction threshold C2: Increase current Gain*
C3: *Stability criterion threshold C4: Decrease current Gain*
C5: *Instability distinction threshold C6: Increase current Gain*

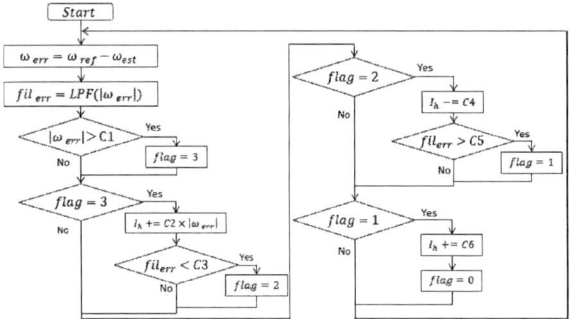

Fig. 5. Flowchart for high-frequency current amplitude determination

VI. EXPERIMENTAL RESULTS

In this section, we demonstrate the effectiveness of the proposed method by explaining experimental results. We conducted the experiment under the condition with the motor speed at 90min^{-1} with no-load. The optimum high-frequency current amplitude command value was decided by our proposed method. After the current amplitude was decided, we changed the motor speed to 30min^{-1} (Fig. 6). To verify the effectiveness of the proposed method at other speeds, we carried out the experiment between 60min^{-1} and 180min^{-1} (Fig.7). Fig. 6 and Fig. 7 show that the high-frequency current amplitude command value is decided over the entire permissible range. Fig. 6 and Table 1 show that audible noise is greatly reduced by this approach. The results of the experiments in which the load is changed slowly and quickly are presented in Fig. 8 and Fig. 9, respectively. These figures show that the proposed method is effective under the rated load condition. Fig. 10 shows experimental results, when the permissible range of the position is set at $10°$. We carried out the experiment between 30 min^{-1} and 90 min^{-1} with the value of C5 set at 0.95.

VII. CONCLUSION

In this paper, we propose an audible noise reduction method using the speed error for sensorless IPMSM drives based on high-frequency current injection. The experimental results demonstrated the effectiveness of the proposed method. This method was successful in the rated load condition. The threshold value C5 is determined by the permissible range of the position error. By determining the threshold value of the speed error, the high-frequency current amplitude command value is minimized depending on operating conditions.

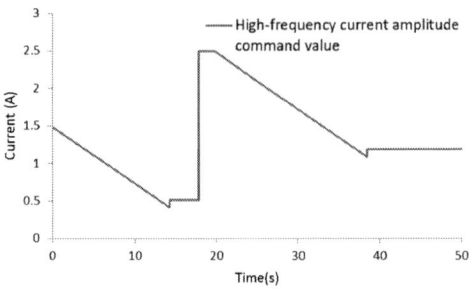

(a) High-frequency current amplitude command value

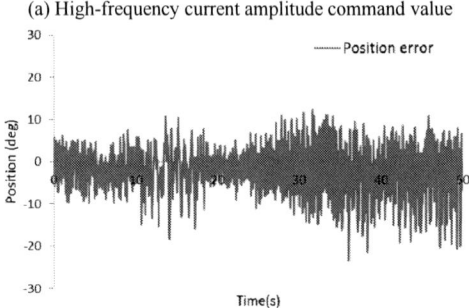

(b) Position error

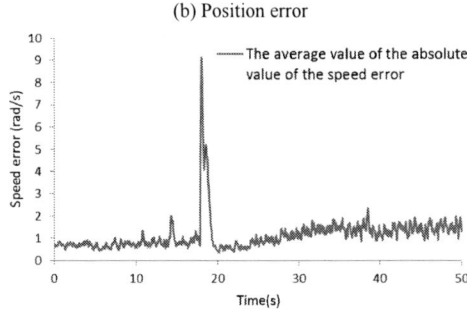

(c) Average value of the absolute value of the speed error

Fig. 6. Experimental results of the proposed method
(Speed varies from 90 min^{-1} to 30 min^{-1})

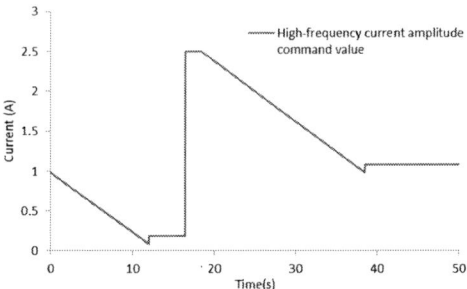

(a) High-frequency current amplitude command value

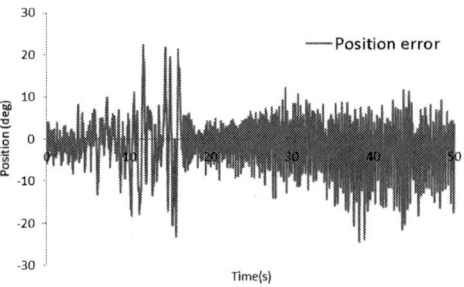

(b) Position error

Fig. 7. Experimental results of the proposed method
(Speed varies from 180 min^{-1} to 60 min^{-1})

The 2014 International Power Electronics Conference

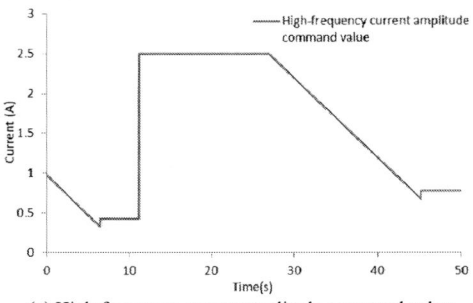

(a) High-frequency current amplitude command value

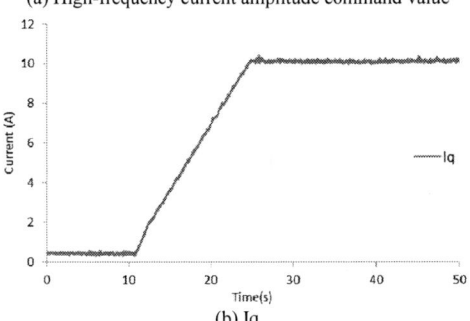

(b) Iq

Fig. 8. Experimental results of the proposed method
(Load torque changes slowly from 0 Nm to 7.9 Nm.)

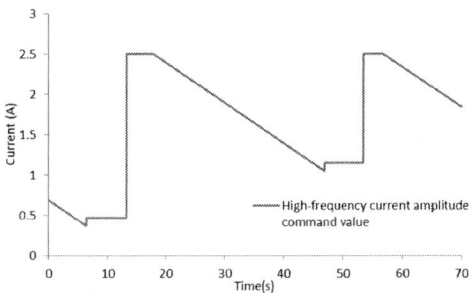

(a) High-frequency current amplitude command value

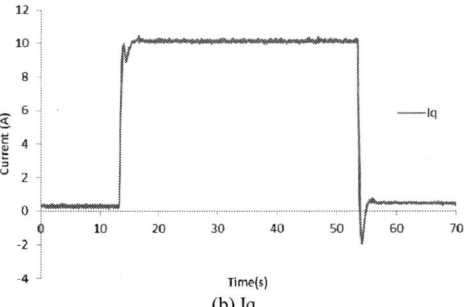

(b) Iq

Fig. 9. Experimental results of the proposed method
(Load torque changes quickly from 0 Nm to 7.9 Nm.)

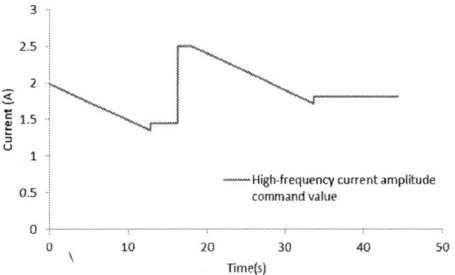

(a) High-frequency current amplitude command value

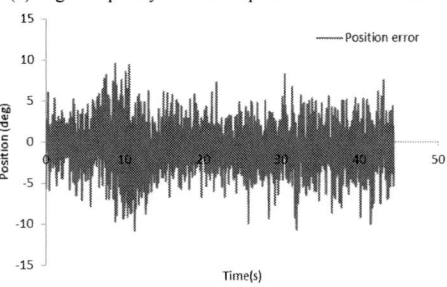

(b) Position error

Fig. 10. Experimental results of the proposed method
(Speed variation from 90 min^{-1} to 30 min^{-1}, C5=0.95.)

REFERENCES

[1] Y. Furukawa, S. Kondo, T. Noguchi, "Sensorless Rotor Position Estimation using Only One-phase High Frequency Test Current Injected to d-axis for Salient PM Motor," National Convention of IEEJ, no. 60, 1998.

[2] S. Taniguchi, T. Homma, S. Wakao, K. Kondo, T Yoneyama, "Control Method for Harmonic Voltage Injection to Achieve Noise Reduction in Position-Sensorless Control of Permanent-Magnet Synchronous Motors at Low Speeds,"IEEJ Transactions on Industry Applications , vol. 129, no. 4, pp. 382-388, 2009.

A Novel Design for Induction Motor Flux Estimation Using Impulsive Observer

Peng Wang[1,3], Yan Li[2], Jianwen Zhang[1], Xu Cai[1], Zhengzhi Han[3]

1: Wind Power Research Center, Department of Electrical Engineering,
Shanghai Jiao Tong University, China
2: Renewable Energy Department, China Electric Power Research Institute, China
3: Electrical & Electronic Experimental Center, Department of Electrical Engineering,
Shanghai Jiao Tong University, China

Abstract- **This paper considers the estimation of flux of induction motor. A novel observer named impulsive observer is designed firstly for induction motor, which is treated as a kind of linear system in this paper. The condition for existence of the impulsive observer is given by a set of linear matrix inequalities, then the asymptotic convergence of the error system is analyzed by the theory of Lyapunov. Finally, simulation is provided to show the effectiveness of the proposed observer.**

I. INTRODUCTION

In the last decade, the field-oriented vector control method is widely used for induction motor (IM) drives. In these applications, flux estimation is of importance in implementing high-performance AC IM drivers. There exist some methods for flux estimation traditionally; these methods are based on the different models, such as current model and voltage model [1-4]. These methods are suitable to be applied in practice because of the simple open-loop structures, but the robustness is poor. In order to improve the robustness of the observer, the closed-loop flux observer for the flux was proposed in 1990s and it then became a hot topic in the field of motor control. [5-7] presented the full-order flux observer, especially, an adaptive observer is used to estimate speed simultaneously [5, 7]. For further improvement, [8] gave systematic design approach for the reduced-order flux observer. More recently, extended Kalman filter [9, 10], sliding-mode flux observer has been exploited for flux estimation to deal with the nonlinear problem in IM control [11, 12]. It should be noted that the mentioned observers all belong to Luenberger-type.

Recently, a new synchronization technique, named impulsive observer, has been reported in [13-17]. The impulsive observer is different from the Luenberger observers. Generally, it consists of two parts: continuous part and discrete part. The error is only added to the discrete part, and nothing need be fed back to the continuous one. Thus, the impulsive observer can reduce the cost of computation obviously. In [13-17], the authors mainly focused on the synchronization of a kind of chaos

The work is sponsored by The National High Technology Research and Development of China (863 Program) (0032012AA050203) and Shanghai Science Foundation project (11dz1200204).

system. To the best knowledge of the authors', the application of impulsive observer to the IM control system has not been paid much attention.

Motivated by the above discussion, we propose a novel flux estimator implemented in stator coordinates for field orientation IM control by using impulsive observer. The rest of paper is organized as follows: In Section II, the IM electrical dynamics in stator coordinates is given. In Section III, the design of impulsive observer for IM flux estimation is presented. Section IV deal with the vector control for IM by using impulsive flux observer. In Section V, a simulation is provided to testify the validity of the designed observers. Finally, conclusion is drawn in Section VI.

II. MODEL OF IM SYSTEM

The IM system can be described by the following state equation in the stationary $\alpha\beta$ reference frame.

$$\begin{cases} \dot{x}(t) = Ax(t) + Bu(t) \\ \quad y(t) = Cx(t) \end{cases} \quad (1)$$

where

$$A = \begin{bmatrix} -\dfrac{R_s L_r^2 + R_r L_m^2}{\sigma L_s L_r^2} & 0 & \dfrac{L_m}{\sigma L_s L_r T_r} & \dfrac{L_m}{\sigma L_s L_r}\omega \\ 0 & -\dfrac{R_s L_r^2 + R_r L_m^2}{\sigma L_s L_r^2} & -\dfrac{L_m}{\sigma L_s L_r}\omega & \dfrac{L_m}{\sigma L_s L_r T_r} \\ \dfrac{L_m}{T_r} & 0 & -\dfrac{1}{T_r} & -\omega \\ 0 & \dfrac{L_m}{T_r} & \omega & -\dfrac{1}{T_r} \end{bmatrix}$$

$$,B = \begin{bmatrix} 1/(\sigma L_s) & 0 & 0 & 0 \\ 0 & 1/(\sigma L_s) & 0 & 0 \end{bmatrix}^{\mathrm{T}}, C = \begin{bmatrix} 1 & 0 & 0 & 0 \\ 0 & 1 & 0 & 0 \end{bmatrix},$$

$$x = \begin{bmatrix} i_{s\alpha} & i_{s\beta} & \psi_{s\alpha} & \psi_{s\beta} \end{bmatrix}^{\mathrm{T}}, u = \begin{bmatrix} u_{s\alpha} & u_{s\beta} \end{bmatrix}^{\mathrm{T}}.$$

In system (1), R_s and R_r are the stator and rotor resistances, respectively. L_s and L_r are the stator and rotor self-inductances, respectively. L_m is the mutual inductance, T_r is the rotor time constant, $i_{s\alpha,s\beta}$, $u_{s\alpha,s\beta}$, $\psi_{r\alpha,r\beta}$

are the stator currents, stator voltages, and rotor fluxes in the stationary αβ reference frame, respectively. ω is the rotor electrical angular velocity, and σ is the motor leakage coefficient. For simplicity, the velocity can be treated as a constant during the switching time.

III. IMPULSIVE OBSERVER DESIGN FOR IM SYSTEM

In this section, we design the impulsive observer for the IM system. Firstly, we introduce some notations which will be used later. Denote that $x(t_k^-) = \lim_{h \to 0+} x(t - h)$, $x(t_k^+) = \lim_{h \to 0+} x(t + h)$, I always represents the identity matrix with an appropriate dimension and R^+ means the set of positive real numbers.

We now design the impulsive observer for system (1) as follows

$$\begin{cases} \dot{\hat{x}}(t) = A\hat{x}(t) + Bu(t), & t \in [t_{k-1}, t_k) \\ \Delta \hat{x}(t) = F(y(t) - C\hat{x}(t)), & t = t_k \end{cases} \quad (2)$$

where the gain $F \in R^{4 \times 2}$, which will be designed later.

Define error $e(t) = x(t) - \hat{x}(t)$, then subtracting eq. (2) from (1) yields

$$\begin{cases} \dot{e} = Ae, & t \in [t_{k-1}, t_k) \\ e(t_k^+) = (I - FC)e(t_k^-), & t = t_k \end{cases} \quad (3)$$

In the sequel, we always assume that the length of interval $[t_k, t_{k+1})$ is a constant, i.e., there is a constant $\eta > 0$ such that $t_k - t_{k-1} = \eta$ for every k. We are now ready to state the following theorem.

Theorem Let $t_k - t_{k-1} = \eta$. Consider the impulsive error system (3), if there exists positive definite matrix $P \in R^{4 \times 4}$, constant $0 < \gamma_3 < 1$, and pre-specified constants $\gamma_1 \neq 0$, $\gamma_2 > 0$, such that

$$PA + A^T P - \gamma_1 P < 0 \quad (4)$$

$$ln(\gamma_2 / \gamma_3) + \gamma_1 \eta < 0 \quad (5)$$

$$\begin{bmatrix} -\gamma_2 P & P - C^T \bar{F}^T \\ P - \bar{F}C & -P \end{bmatrix} < 0 \quad (6)$$

where $\bar{F} = PF$. Then system (3) is asymptotically stable, i.e., $\lim_{t \to \infty} e(t) = 0$.

Proof. Using the Schur complement, eq. (6) is equivalent to

$$(I - FC)^T P(I - FC) < \gamma_2 P \quad (7)$$

Consider the following Lyapunov function candidate

$$V(e(t)) = e^T(t)Pe(t) \quad (8)$$

In the following, we argue it in two cases.
1. When $t \in [t_{k-1}, t_k)$, evaluating the derivative of V along the trajectory of the system (3), we have

$$\dot{V}(e(t)) = 2e^T(t)P\dot{e}(t) = e^T(t)(PA + A^T P)e(t), t \in [t_{k-1}, t_k)$$

$$(9)$$

Substituting eq. (4) into (9) yields

$$\dot{V}(e(t)) < \gamma_1 e^T(t)Pe(t) = \gamma_1 V(e(t)), t \in [t_{k-1}, t_k) \quad (10)$$

Then eq. (10) implies that

$$V(e(t)) < \exp(\gamma_1(t - t_0))V(e(t_{k-1}^+)), t \in [t_{k-1}, t_k) \quad (11)$$

2. When $t = t_k$, from system (3), we can obtain

$$V(e(t_k^+)) = e^T(t_k^+)Pe(t_k^+) = e^T(t_k^-)(I - FC)^T P(I - FC)e(t_k^-)$$

$$(12)$$

Combining eq. (12) with (7), we have

$$V(e(t_k^+)) < \gamma_2 V(e(t_k^-)) \quad (13)$$

In the sequel, eq. (11) and (13) are essential to our further proof.

Firstly, let us consider $t \in [t_0, t_1)$, by eq. (11), the following holds

$$V(e(t_1^-)) < \exp(\gamma_1 \eta)V(e(t_0^+)) \quad (14)$$

Using eq. (5), (13) and (14), we can get

$$V(e(t_1^+)) < \gamma_3 V(e(t_0^+)) \quad (15)$$

Secondly, for $t \in [t_1, t_2)$, using the same treatment as that used for $t \in [t_0, t_1)$, we have

$$V(e(t_2^-)) < \gamma_3 \exp(\gamma_1 \eta)V(e(t_0^+)) \quad (16)$$

$$V(e(t_2^+)) < \gamma_3^2 V(e(t_0^+)) \quad (17)$$

Repeating the above derivation, for $t \in [t_{k-1}, t_k)$, we have

$$V(e(t_{k-1}^+)) < \gamma_3^{k-1}V(e(t_0^+)) \quad (18)$$

Then, for $t \in [t_{k-1}, t_k)$,

$$V(e(t)) < \exp(\gamma_1(t - t_{k-1}))\gamma_3^{k-1}V(e(t_0^+)) \quad (19)$$

eq. (19) means that

$$V(e(t)) < \begin{cases} \exp(\gamma_1 \eta)\gamma_3^{k-1}V(e(t_0^+)) & \gamma_1 > 0 \\ \gamma_3^{k-1}V(e(t_0^+)) & \gamma_1 < 0 \end{cases} \quad (20)$$

Since $t \to \infty$, $k \to \infty$, consider the fact that $0 < \gamma_3 < 1$, we can conclude that $\lim_{t \to \infty} V(e(t)) = 0$, i,e, $\lim_{t \to \infty} e(t) = 0$. Thus, we have completed the proof.

Remark 1. It is obvious that eq. (4) - (6) are a set of linear matrix inequalities (LMIs) provided that $\gamma_1 \neq 0$, $\gamma_2 > 0$ are pre-specified constants. We can use the LMI toolbox in Matlab to solve the feasible solution of eq. (4) - (6), which is shown in the simulation part.

Remark 2. Different from the Luenberger type observer, the form of which is as follows:

$$\dot{\hat{x}}(t) = A\hat{x}(t) + Bu(t) + L(y - C\hat{x}(t)), \quad t \in R^+ \quad (21)$$

where L is the observer gain. It is also the full-order observer in [5]. Comparing with eq. (21), we need not design the gain L in the continuous time interval in eq. (2).

We just need to add the gain F to the observer on the discrete time point t_k Instead. Thus, the advantage of the impulsive observer is obvious.

Here, we employ the impulsive observer designed to estimate the rotor flux, and the estimated flux is used for the control of IM.

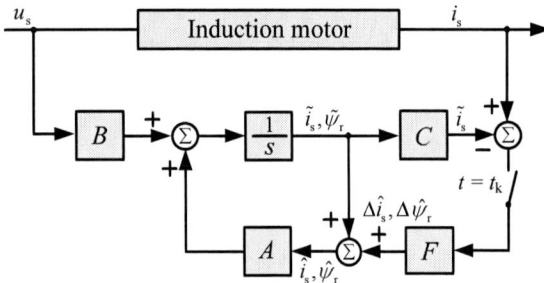

Fig. 1. Block diagram of impulsive flux observer

Fig.1 shows the block diagram of the proposed flux estimation by impulsive observer, where A, B, C are all constant matrices as given in system (1) and F is a compensation matrix working at a pulse moment which is dependent on t_k, u_s is stator voltage as the input of system which could be reconstructed by using the inverter switching states S_{abc} and the dc voltage V_{dc}[18], i_s is stator current as the output of system. According to the result obtained in this Section, the flux estimation is composed of two parts, the continuous estimated values $\tilde{i}_s$, $\tilde{\psi}_r$, and the compensated value $\Delta\hat{i}_s$, $\Delta\hat{\psi}_r$ are calculated on the discrete time point t_k which is determined by the time interval η. It should be noted that the performance of this type of observer is also influenced by the dc offset of input signals and integral saturation. And we could use the technique in [19] to deal with these problems.

IV. ANALYSIS AND RESULTS

For testifying the effectiveness of the present method, the impulsive flux observer is developed in motor control system simulated by Matlab/Simulink/Powersystem. The control block diagram of rotor flux orientation control by using impulsive observer is shown in Fig.2. An outer speed PI controller and two inner current PI controllers are used to complete the system control in the rotating dq reference frame. The outer PI controller is used to make the motor speed ω follow the track of speed given value ω^*, and the inner PI controllers can realize the decoupling control of flux and torque components. In Fig.2, the inputs of flux impulsive observer are stator currents, electrical angular frequency of IM and the inverter switching states; and the outputs are the rotor flux estimation $\hat{\psi}_r$ and its electrical angle θ in synchronously rotating reference frame. The rotor flux can be estimated by the results in Section III.

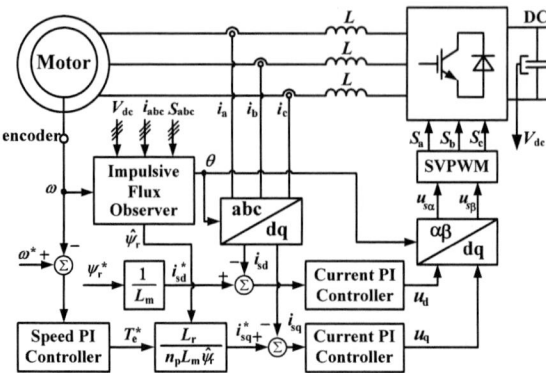

Fig. 2. Block diagram of vector control using impulsive flux observer

In the simulation, system parameters are given in Tab.1. The initial mechanical angular speed of motor is about 72.9 rad/s and the electrical angle of the synchronously rotating reference frame is 120°. The speed reference which determines the reference value of torque current i_{sq} is given as 90 rad/s and changed to 60 rad/s at 10 s, the flux reference which determines the reference value of flux current i_{sd} is given as a constant 1.79 Wb. The simulation results include steady state performance and transient state performance with load torque changing from 5000 N·m to 10000 N·m at 15 s.

TABLE I
SYSTEM PARAMETERS

symbol	quantity	value
P_e	Nominal power of IM (kVA)	3450
V_n	Nominal line-line voltage of IM (V)	690
f_n	Nominal frequency of IM (Hz)	50
L_r	Rotor inductance (mH)	3.331e-2
L_s	Stator inductance (mH)	3.694e-2
L_m	Mutual inductance (mH)	1.187
R_r	Rotor resistance (Ω)	6.928e-4
R_s	Stator resistance (Ω)	8.487e-4
J	Rotational inertia (kg·m^2)	2300
P_n	Pole pairs of IM	3
V_{dc}	Rated DC link voltage (V)	1100
f_s	Switching frequency (kHz)	3
T_c	Calculation period (μs)	20

We design the impulsive observer by using (4) - (6) according to the speed values in real time. For example, during the start-up period, let $\eta = 0.01$, $\gamma_1 = -2$, $\gamma_2 = 1.2$. Solving the LMIs, the gain F could be obtained.

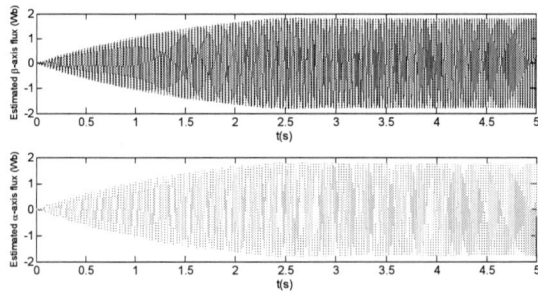

Fig. 3. The estimations of rotor flux in stationary αβ reference frame

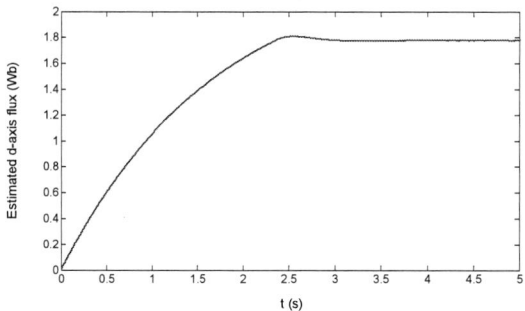

Fig. 4. The estimation of d-axis rotor flux in rotating dq reference frame

In Fig.3, the estimated values of rotor flux in stationary αβ reference frame are provided, and Fig.4 shows the estimated d axis flux in rotating reference frame, it can be noted that the d axis flux follows the track of given flux value 1.79 Wb and reaches the stable value around 3 s. The speed response of impulsive observer vector control system has been shown in Fig.5. It is observed that the speed of IM denoted by solid line can completely track the speed reference value signed by dashed line in finite time. The tracking process depends on the estimated performance of IM states and the rotational inertia of IM. Fig.6 is the torque dynamic response during the speed tracking. The electromagnetic torque increases gradually between 0 - 3 s during the process of flux established. Subsequently, the torque limit value (15000 N·m) is outputted until the speed reached the reference value. Then, the electromagnetic torque is the same as load torque at 5000 N·m. At 10 s, the outputted electromagnetic torque reached the negative limit (-15000 N·m) until 12.3 s because the reference value changes from 90 rad/s to 60 rad/s. And by the change of load torque at 15 s, we can see that the electromagnetic torque is well-tracked. In the end, the errors of stator current which are obtained by comparison of the estimated values by impulsive observer and the measured values are shown in Fig.7. It should be pointed that the current errors are tracking 0 and limited in 100 A after 0.5 s, where the peak current of IM is about 4000 A, as well as the errors of observer are increased at the electromagnetic torque transient. The errors change can rapidly converge in a short time. From the simulation results, we can conclude that the vector control of IM by using impulsive observer for rotor flux estimation has good performances.

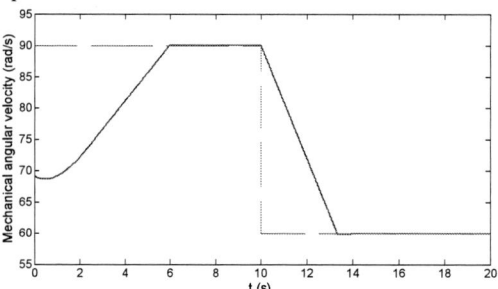

Fig.5. The speed response of impulsive observer vector control system

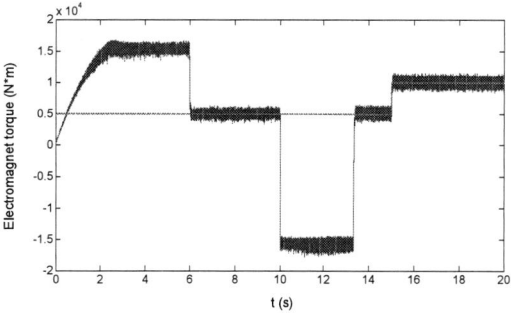

Fig.6. The torque response of impulsive observer vector control system

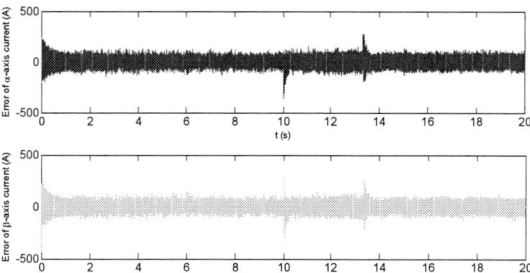

Fig.7. The errors of stator current in stationary αβ reference frame

V. CONCLUSION

In this paper, the impulsive observer for IM has been presented. The impulsive observer is used for the rotor flux estimation. The advantage of this kind observer lies in that it can not only guarantee the stability of system but also reduce the cost. Finally, the simulation has been shown that the proposed observer is effective.

References

[1] T. Ohtani, N. Takada, K. Tanaka, "Vector control of induction motor without shaft encoder," *IEEE Trans. on Industry Applications*, vol. 28, no. 1, pp. 157-164, 1992.

[2] T. Bhattacharya, L. Umanand, "Improved flux estimation and stator-resistance adaptation scheme for sensorless control of induction motor," *Electric Power Applications,* vol. 153, no. 6, pp. 911-920, 2006.

[3] H. K. Khalil, E. G. Strangas, S. Jurkovic, "Speed observer and reduced nonlinear model for sensorless control of induction motors," *IEEE Trans. on Control Systems Technology*, vol. 17, no. 2, pp. 327-339, 2009.

[4] L. Harnefors, "Design and analysis of general rotor-flux-oriented vector control systems," *IEEE Trans. on Industrial Electronics*, vol. 48, no. 2, pp. 383-390, 2001.

[5] H. Kubota, K. Matsuse, T. Nakano, "DSP-based speed adaptive flux observer of induction motor," *IEEE Trans. on Industry Applications*, vol. 29, no. 2, pp. 344-348, 1993.

[6] P. L. Jansen, R. D. Lorenz, "A physically insightful approach to the design and accuracy assessment of flux observers for field oriented induction machine drives," *IEEE Trans. on Industry Applications*, vol. 30, no. 1, pp. 101-110, 1994.

[7] H. Kubota, I. Sato, Y. Tamura, K. Matsuse, H. Ohta, Y. Hori, "Regenerating-mode low-speed operation of sensorless induction motor drive with adaptive observer," *IEEE Trans. on Industry Applications*, vol. 38, no. 4, pp. 1081-1086, 2002.

[8] M. Montanari, S. M. Peresada, C. Rossi, A. Tilli, "Speed sensorless control of induction motors based on a reduced-order adaptive observer," *IEEE Trans. on Control Systems Technology*, vol. 15, no. 6, pp. 1049-1064, 2007.

[9] Y. R. Kim, S. K. Sul, M. H. Park, "Speed sensorless vector control of induction motor using extended Kalman filter," *IEEE Trans. on Industry Applications*, vol. 30, no. 5, pp. 1225-1233, 1994.

[10] M. Barut, S. Bogosyan, M. Gokasan, "Experimental evaluation of braided EKF for sensorless control of induction motors," *IEEE Trans. on Industrial Electronics*, vol. 55, no. 2, pp. 620-632, 2008.

[11] A. B. Proca, A. Keyhani, "Sliding-mode flux observer with online rotor parameter estimation for induction motors," *IEEE Trans. on Industrial Electronics*, vol. 54, no. 2, pp. 716-723, 2007.

[12] M. Tursini, R. Petrella, F. Parasiliti, "Adaptive sliding mode observer for speed sensorless control of induction motors," *IEEE Trans. on Industry Applications*, vol. 36, no. 5, pp. 1380-1387, 2000.

[13] T. Raff, F. Allgower, "An impulsive observer that estimates the exact state of a linear continuous-time system in predetermined finite time," *Mediterranean Conference on Control Automation,* pp. 1-3, 2007.

[14] A. Mahmoudi, A. Momeni, A.G. Aghdam, P. Gohari, "On observer design for a class of impulsive switched systems," *American Control Conference*, pp. 4633-4639, 2008.

[15] E. A. Medina, D. A Lawrence, "State estimation for linear impulsive systems," *American Control Conference*, pp. 1183-1188, 2009.

[16] M. Ayati, H. Khaloozadeh, "Novel Adaptive Impulsive Observer for Synchronization of Uncertain Chaotic Systems," *International Conference on Signals and Electronic Systems*, pp. 331-334, 2010.

[17] M. Ayati, H. Khaloozadeh, L. Xinzhi, "Synchronizing chaotic systems with parametric uncertainty via a novel adaptive impulsive observer," *Asian journal of Control*, vol. 13, no.6, pp. 809-817, 2011.

[18] S. Myoung-Ho, H. Dong-Seok, C. Soon-Bong, C. Song-Yul, "An improved stator flux estimation for speed sensorless stator flux orientation control of induction motors," *IEEE Trans. on Power Electronics*, vol. 15, no. 2, pp. 312-318, 2000.

[19] J. Hu and B. Wu, "New integration algorithms for estimating motor flux over a wide speed range," *IEEE Trans. on Power Electronics*, vol. 13, pp. 969-977, 1998.

Load Torque and Inertia Simulation Based on Double-stator Permanent-magnet Synchronous Motor

Zhe Wang, Mingyan Wang, Ben Guo, Chai Feng
Dept. of Electrical Engineering
Harbin Institute of Technology
Harbin, China
wangz040230112@163.com

Abstract— **A novel structure using dual-stator motor is proposed and the corresponding controller is designed. Two stators execute independently based on the method of torque decoupling control, in which one stator is used to simulate the load torque in expectation; the other stator is used to provide inertia moment. In this paper, an adaptive disturbance observer is presented to acquire acceleration information and derives inertia moment to be simulated, in which a simple adaptive scheme is derived using Popov's hyperstability theory. The control performance of the system is analyzed. Simulation and experiment results verify the validity of the theoretical analysis and the feasibility of the method.** [1]

Keywords— *electric dynamic load simulator; double-stator permanent-magnet synchronous motor; electric inertia simulation; acceleration observer.*

I. INTRODUCTION

Position servo system needs to test its transmission performance, and the test content generally falls into two main aspects: load test and mechanical inertia test. Electric dynamic load simulator (EDLS), which is a typical application of the passive torque servo system in engineering practice, is used to reproduce desired load torque acting on equipment to test its performance in hardware-in-the-loop simulation [1]. EDLS is required of high rapid response and good accuracy, because it must exert load torque on the equipment under test while tracking down its motion[2]. The control strategies, including the intelligent control method and the traditional control theory were introduced to EDLS, [3] ‒ [7]. Usually EDLS is used to simulate load torque while mechanical test is executed by mechanical inertia block, because the structure and control strategies of single stator make it difficult to achieve two different functions and evaluate the system performance. This paper proposes a new hardware structure in which a parallel double stator permanent magnet synchronous motor (PDSPMSM), which has two independent electrical ports and two control freedoms, is used as the execution device of the EDLS[8]. Two stators are coupled with each other

though the same shaft, without electrical and magnetic coupling. Loading dynamic torque is divided into two parts: load torque and mechanical inertia moment, which are exerted by two stators respectively, based on adaptive disturbance observer and torque decoupling control[9]. Linear mathematical model of new EDLS is built to analyze the operating performance under this method. Simulation and experiment verifies the validity of the theoretical analysis and the feasibility of the method.

II. ELECTRIC SIMULATION OF LOAD TORQUE AND MECHANICAL INERTIA

The electric dynamic load simulator uses the motor as the loading actuator, which shall be linked with the target position servo system in the co-axial connection to form a double-servo system that is coupled with each other. In the traditional test, the inertia is usually simulated by the inertial flywheel that is mounted on the target system output shaft and has the equal inertia. See Fig. 1a). The mechanical inertia simulation method, however, has many disadvantages such as inconvenient assembly /disassembly, unable to smooth adjustment of the flywheel size etc. When the motor is used to simulate the target system inertia, an extra inertial torque shall be applied to the original simulation load torque. The traditional single-stator load simulator, however, will result in mixing of two torque signals. Accordingly, the double-stator motor shall be used as the load simulator actuator for independent control. See Fig. 1b).

The double-stator motor is designed with two electrical interfaces, i.e., it is designed with two control redundancies so that the two stators can be designated with different tasks to apply the load torque and simulate the rotational inertia, respectively, realizing control decoupling. The inertia simulation is based on the inertial torque simulation. The expected inertial torque is proportional to the acceleration applied to the loaded side. See (1) where T_{J*} refers to the expected inertial torque on the shaft, J^* refers to the desired rotational inertia, J refers to the EDLS's rotational inertia.

$$T_{j^*}(t) = -(J^* - J)a(t) \qquad (1)$$

This work was supported in part by the National Natural Science Foundation of China under grants (51077025)

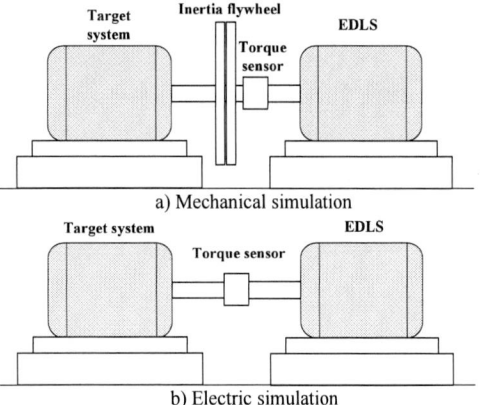

a) Mechanical simulation

b) Electric simulation

Fig. 1. Mechanical/Electric simulation systems.

To accurately output the expected inertial torque, an acceleration observer of accuracy and fast response is a must to acquire the acceleration signal of the loaded system in an accurate manner.

A. Adaptive Acceleration Observer

The observer shall be connected to the control system to acquire the acceleration signal of the loaded system, according to (1). The acceleration observer uses the directly measurable position signal of the loaded system as the feedback to estimate the acceleration with the help of the motion equation of the target system. The load torque on the shaft of the load simulator can be obtained via the torque sensor. The electromagnetic torque of the loaded system, however, is generally unable to be obtained. Moreover, the interference of gap and other nonlinear torques will also significantly affect the observation results of the observer. As a result, the simple adaptive law is added to the observer to form an adaptive disturbance observer, which can observe the output electromagnetic torque and the interference torque of the loaded system so as to observe the acceleration. The output electromagnetic torque and the interference torque will be called as disturbance torque for convenience. Below is the equation of motion state of the loaded system:

$$\begin{cases} \dot{x} = A_1 x + B_1 T_{EB} - B_2 T_L \\ y = C_1 x \end{cases} \tag{2}$$

Where, $A_1 = \begin{bmatrix} 0 & 1 \\ 0 & -B_B/J_B \end{bmatrix}$; $B_1 = B_2 = \begin{bmatrix} 0 \\ 1/J_B \end{bmatrix}$;

$C_1 = \begin{bmatrix} 1 \\ 0 \end{bmatrix}^T$; $x = \begin{bmatrix} \theta_B \\ \omega_B \end{bmatrix}$.

θ_B, ω_B, T_L, T_{EB}, J_B, B_B refers to the shaft angle (rad), speed (rad/s), load torque (N·m), disturbance torque (N·m), equivalent rotational inertia converted to the drive shaft (kg·m^2), the shaft friction coefficient (N·s/m). The following observer can be designed for the output electromagnetic torque of the loaded system according to the equation of state.

$$\hat{x} = A_1 \hat{x} + B_1 \hat{T}_{EB} - B_2 T_L + G(C_1 x - C_1 \hat{x}) \tag{3}$$

Where, ^ is the state variable to be observed and the output electromagnetic torque of the loaded system is taken as the disturbance variable to be observed. Work out the difference between the (2) and (3), and you can get the error equation of the observer as follows:

$$\dot{e} = (A_1 - GC_1)e + B_1(T_{EB} - \hat{T}_{EB}) = A'e + W_1 \tag{4}$$

Where, $G = [g_1 \quad g_2]^T$; $W = -W_1 = B_1(\hat{T}_{EB} - T_{EB})$; $A' = A_1 - GC_1$; $e = x - \hat{x}$.

According to Popov's hyperstability theory, Inequality (6) shall be established if the condition to make the error vector equation convergence is to make (5) positive definite:

$$H = (sI - A')^{-1} \tag{5}$$

$$\int_0^t (y - \hat{y})W dt \ge -\gamma^2 \tag{6}$$

$$\hat{T}_{EL} = \left(K_P + \frac{K_I}{s}\right)(y - \hat{y})B_1 \tag{7}$$

A reasonable gain matrix G can be designed to ensure that (5) is positive definite. The PI adaptive algorithm shall be used. It can be proved that the adaptive algorithm can make (6) established[9]. (7) shows the observed output electromagnetic torque of the loaded motor. To get the acceleration observer, we can rewrite (3) as follows:

$$\hat{a}_B = s\hat{\omega}_B = \frac{s}{J_B s + B_B}\left[\left(K_P' + \frac{K_I'}{s}\right)(\theta_B - \hat{\theta}_B) - T_L\right] \tag{8}$$

Where, $K_P' = K_P/J_B + J_B g_2$; $K_I' = K_I/J_B$.

Fig. 2 shows the structure of the acceleration observer. Based on the position of the loaded motor and the load torque acquired via the torque sensor, the acceleration of the loaded system can be worked out. See (9) for the transfer function of the observer. If the appropriate parameters are set, we can obtain a reasonable observation bandwidth while suppressing the higher-order noise.

$$G_a = \frac{\hat{a}_B}{a_B} = \frac{K_P' s + K_I'}{J_B s^3 + (J_B g_1 + B_B)s^2 + (K_P' + B_B g_1)s + K_I'} \tag{9}$$

Fig. 2. Adaptive acceleration observer.

B. Load Simulator of Double-stator Motor

The double-stator motors refer to those designed with two stators, which can be divided into two types: the concentric type and the parallel type according to the relative position of the two stators. In the paper, the non-salient parallel double-stator permanent-magnet synchronous motors shall serve as the loading actuator where the non-linear factors will be neglected. Take Stator 1 as an example. The stator armature equation is as follows at the synchronous rotational coordinate system:

$$T_{e1} = \frac{3}{2}P[\psi_{f1}i_{q1} + (L_{d1} - L_{q1})i_{d1}i_{q1}] = K_{T1}i_{q1} \tag{10}$$

$$\begin{cases} u_{d1} = R_{s1}i_{d1} + L_{d1}\dfrac{di_{d1}}{dt} - P\omega L_{q1}i_{q1} \\ u_{q1} = R_{s1}i_{q1} + L_{q1}\dfrac{di_{q1}}{dt} + P\omega(L_{d1}i_{d1} + \psi_{f1}) \end{cases} \tag{11}$$

Where, u_{d1}, u_{q1}, i_{d1}, i_{q1} refers to the motor direct/quadrature-axis voltage, current, respectively; ψ_{f1}, R_{s1}, L_{d1}, L_{q1} refers to the permanent magnet flux linkage, the stator resistance and the equivalent direct/quadrature-axis inductance, respectively, and $L_{d1}=L_{q1}$; ω_e, ω refers to the electric/mechanical angle speed of the load simulator, respectively; P is the number of poles.

Similarly, the armature equation of Stator 2 can be obtained. The motion equation and sensor model of the double-stator motor are given below:

$$T_E = T_{e1} + T_{e2} = T_L + B\omega + J\frac{d\omega}{dt} \tag{12}$$

$$T_L = K_\theta(\theta - \theta_B) \tag{13}$$

Where, T_E, J, B refers to the electromagnetic torque, the rotational inertia, the shaft friction coefficient of the load simulator, respectively.

Divide the two stators, and we can achieve independent control where Stator 1 can be used to simulate the load torque and Stator 2 to conduct the electric simulation of the mechanical inertia. In order to

achieve rapid loading of the inertial torque during control operation, Stator 2 can be designed with closed-loop control of electromagnetic torque. PI control will be used. According to (11) and (10), the inner loop of the electromagnetic torque can be seen as a first-order inertial link, see (14). The friction coefficient is neglected here for easy analysis.The transfer function of θ_B and T_{e2} can be worked out as follows with combination of (9):

$$\begin{cases} T_{e2}^* = -(J^* - J)\hat{a}_B = -(J^* - J)G_a\ddot{\theta}_B \\ G_{TE2} = \dfrac{T_{e2}}{T_{e2}^*} = \dfrac{K}{s + K} \end{cases} \tag{14}$$

But electromagnetic torque is not real load torque on shaft. According to (12) and (13), taking no account of stator 1, below is the transfer function of the inertial torque on shaft that Stator 2 simulates:

$$T_L = -\frac{(J^* - J)K_\theta G_a G_{TE2}}{Js^2 + K_\theta}\ddot{\theta}_B \tag{15}$$

Resonance is present in the system, which is caused by the system mechanical structure. The resonance frequency is $\sqrt{K_\theta / J}$. A trap can be connected in series to the control circuit of Stator 2 to suppress the resonance. Similar to the traditional single-stator load simulator, Stator 1 is designed with three-loop controls−the load torque, the speed and the electromagnetic torque. Fig. 3 shows the overall control block diagram of the double-stator motor as load simulator, where torque decoupling means effect on speed and load torque of stator 1 by stator 2 is compensated by feedforward control signal which derived from electromagnetic torque of stator 2 with motion model.

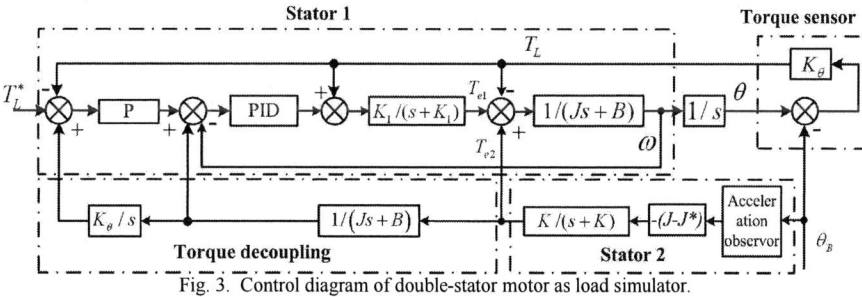

Fig. 3. Control diagram of double-stator motor as load simulator.

C. Simulation and experimental result

TABLE I
PARAMETERS OF SYSTEM

Parameter	Target motor	Loading motor
Rated power	1500W	1500W
Rated torque	7.2 N·m	7.14 N·m
Rated current	3.6A	5A
Rated speed	2000r/min	2000r/min
Number of poles	4	4
Rotational inertia	7.2×10^{-4} kg·m^2	5.64×10^{-4} kg·m^2
Resolution ratio of coded	2500	2500
Rigidity coefficient of flexible shaft	1350 N·m/rad	
Inertia of block	0.031 kg·m^2	

In the environment of MATLAB/Simulink, a system model is established to verify the feasibility of the double-stator motor used as the load simulator to conduct the electric simulation of the load torque and the mechanical inertia. With reference to the data described in the associated manuals (See Table 1), the appropriate parameters of the inner loop are chosen, and a simulation model based on the double-stator permanent-magnet synchronous motor is built. The experiment system is connected in the co-axial method via JN338 static torque sensor to achieve finite position motion and loading. The NC chip (TMS320F2812) is connected with the upstream IPC via the bus to realize real-time data exchange and human-machine interaction based on the VB interface.

$$T_L^* = K_T \theta_B \tag{16}$$

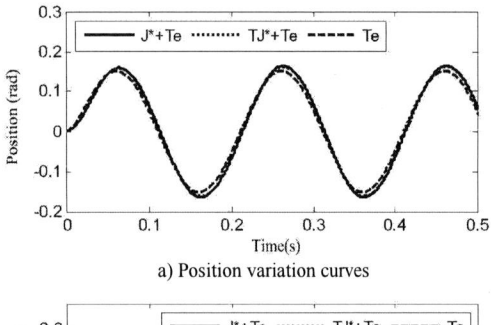

a) Position variation curves

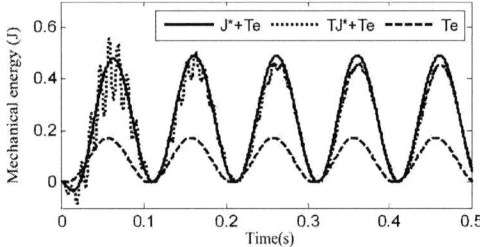

b) Energy variation curves

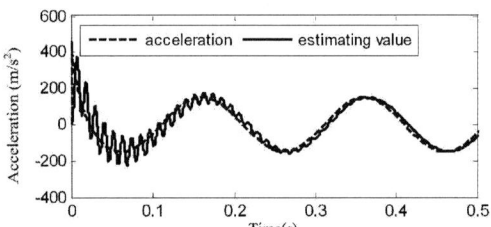

c) Acceleration observation waveform

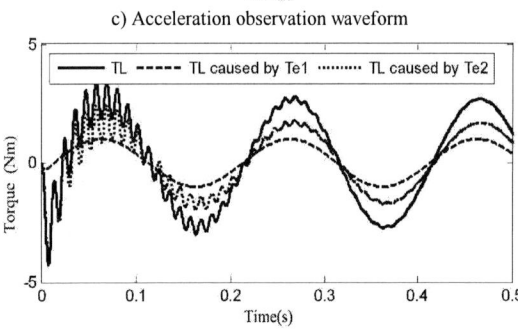

d) Load torque produced by the two stators
Fig. 4. Simulation results.

To verify the correctness of the analysis, a comparison of the simulation and the experiment is conducted where the double-stator load simulator is used to apply the expected load to the target system, and the rotational inertia is simultaneously simulated, and it is compared with the method of applying the mechanical simulation. The given magnitude of the loaded system position is 0.15 *Hz*. The load torque is worked out by the gradient (16), where the gradient coefficient K_T is 6 *N·m/rad*. For the mechanical inertial block, which is rigidly fixed to the loaded system shaft by a keyed jointed, the rotational inertia is $J^*=0.031$ *kg·m²*. Fig. 4a) shows the position waveform of the loaded motor where T_e means load torque; J^* means mechanical

simulation; T_{J*} means electric simulation of inertia moment. The position waveform, which can serve as the evaluation index of inertia simulation performance, shows the position responses of the mechanical inertia and electric simulation system are basically the same, and they are significantly different from that without inertia simulation. The output mechanical energy waveform of the target system, which can serve as a complement to the position evaluation index, is showed in Fig. 4b). It is clear that the mechanical inertia simulation, the electric simulation, and the mechanical energy output by the target system are basically the same. Fig. 4a) and Fig. 4b) prove that the inertial forces applied by the mechanical inertia and the electric inertia to the shaft are identical. Fig. 4c) shows the comparison between the observed values on the adaptive acceleration observer and the actual values. Fig. 4d) shows the load torque on the shaft during electric simulation, whose value equals to the result under the co-action of Stator 1 load torque loading and Stator 2 inertial torque simulation. Fig. 4d) shows that the two controls are completely decoupled from each other to realize their function independently. Accordingly, it will exert little influence over the original loading that the electric simulation is conducted with the double-stator load simulator.

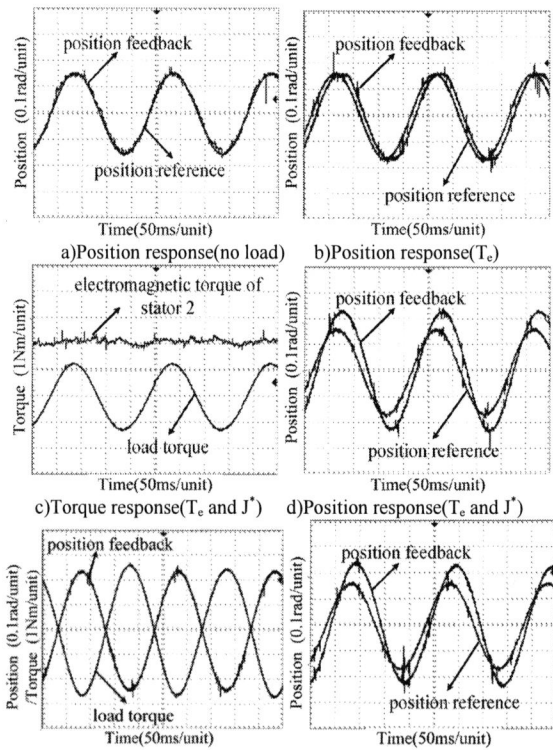

a)Position response(no load)   b)Position response(T_e)

c)Torque response(T_e and J^*)   d)Position response(T_e and J^*)

e)Torque response(T_e and T_{J*})   f) Position response(T_e and T_{J*})
Fig. 5. Experiment results.

Fig. 5a) shows the waveform of the loaded system with no load. Fig. 5b) shows the waveform where the load torque is applied but the inertia simulation is not conducted. Fig. 5d) shows the position response of the loaded system after the load torque and mechanical inertial block are applied. Fig 5c) shows the load torque on the shaft corresponding to Fig. 5d) and the electromagnetic torque of Stator 2. Fig. 5f) shows the position response of the loaded system after the load torque and the electric simulation are applied. Fig. 5e) shows the response relations of shaft torque and loaded system position with Fig. 5f).

III. CONCLUSIONS

The technical principle of electric simulation is analyzed for the mechanical inertia in the load simulator. To conduct the electric simulation of the mechanical inertia, the load torque simulated by the load simulator shall contain a compensation force equivalent to the simulation inertia.

The load simulator in the structure of single-stator motor has technical disadvantages in terms of performing simulation of load torque simultaneously with the electric simulation of mechanical inertia due to structural constraints. To conduct the electric simulation of mechanical inertia, the load simulator in the structure of double-stator motor is proposed.

The load simulator model in the structure of double-stator motor is built and the control method is designed. The simulation experiment proves that the load simulator in the structure of double-stator motor can simulate the load torque while effectively conducting the electric simulation of the mechanical inertia.

REFERENCES

[1] Y. NAM, "QFT force loop design for the aerodynamic load simulator," *IEEE Transaction on aerospace and electronic systems*, vol. 4, no. 37, pp. 1384-1392, 2001.

[2] Q. Fang, Y. Yao, L. Fan, "Modeling and validating of electric torque load simulator," *Proceedings of the First Asia International Symposium on Mechatronics*, pp. 83-88, 2004.

[3] B. Guo, M. Wang, J. Zhang, "A dynamic fuzzy neural networks controller for dynamic load simulator," *Proceedings of the Fifth International Conference on Machine Learning and Cybernetics*, pp. 375-379, 2006.

[4] M. Wang, B. Guo, Y. Guan, H. Zhang, "Design of electric dynamic load simulator based on recurrent neural networks," *IEEE International Electric Machines and Drives Conference*, pp. 207-210, 2003.

[5] Xia Liu, Rui FengYang, JianFang Jia, "The analysis and research of electric dynamic load simulator," *Applied Mechanics and Materials*, pp. 849-852, 2012.

[6] Rong Zhang, Jian Chen, "Repetitive control algorithms for a real-time dynamic electronic load simulator," *IEEE Applied Power Electronics Conference and Exposition*, pp. 919-922, 2006.

[7] Xin Wang, "A study on dynamics of electric load simulator using spring beam and feedforward control technique," *Control and Decision Conference*, pp. 301-306, 2009.

[8] Wang Zhe, Wang Mingyan, Wang Guoqiang, "Simulation analysis of a new electric dynamic load simulator based on double-stator permanent-magnet synchronous motor," *IEEE 7th International Power Electronics and Motion Control Conference*, pp. 2617-2620, 2012.

[9] Dianguo Xu, Yang Gao, "An approach to torque ripple compensation for high performance PMSM servo system," *IEEE Power Electronics Specialist Conference*, pp. 3256-3259, 2004.

Independent Speed and Position Control of Two Permanent Magnet Synchronous Motors Fed by a Four-Leg Inverter

Yuji Kubo
Non Member, IEEE
Meiji University
Higasi-mita, Tama-ku
Kawasaki, Japan
ce31035@meiji.ac.jp

Takayuki Moroi
Non Member, IEEE
Meiji University
Higasi-mita, Tama-ku
Kawasaki, Japan
ce21075@meiji.ac.jpKouki

Matsuse Kouki
Fellow, IEEE
Meiji University
Higasi-mita, Tama-ku
Kawasaki, Japan
matsuse@isc.ac.jp

Hisao Kubota
Menber, IEEE
Meiji University
Higasi-mita, Tama-ku
Kawasaki, Japan
kubota@isc.ac.jp

Kaushik Rajashekara
Fellow, IEEE
The University of Texas at Dallas
800W. Campbell Rd. EC33,
Richardson, TX 75080
K.Raja@utdallas.ed

Abstract - **This paper presents the experimental results of the independent speed and position control of two permanent magnet synchronous motors (PMSMs) fed by a four-leg inverter (4LI) with vector control method. The inverter consists of four legs and two capacitors connected in a series. 4LI have several disadvantages. First, it is not possible to modulate w phase of the PMSMs. Second, the neutral point potential of two capacitors fluctuates. This paper shows an expanded two-arm modulation and compensation method to solve them. Next, the independent speed and position controls of two PMSMs fed by the 4LI with the vector control system is demonstrated by the experimental results. Moreover, it shows the effectiveness of those methods.**

Index terms- four-leg inverter (4LI), neutral point potential compensation, permanent magnet synchronous motor (PMSM), two-arm modulation

I. INTODUCTION

Recently, a four switch inverter [1-4] and a four-leg inverter [5-7], having a lower number of switches, have been studied. The 4LI consists of four legs and two capacitors connected in a series. One phase of both motors is shared and connected to the neutral point of two capacitors in common. 4LI requires eight-switching devices. It can decrease four-switches compared with dual three-phase voltage source inverter systems. Reduce the number of switches is attractive in terms of reduction of switching loss and low cost.

However, independent speed and position controls of two PMSMs fed by a 4LI with vector control have several disadvantages. First, it is not possible to modulate w phase of the motors. Second the neutral point potential of two capacitors fluctuates. First of all, this paper shows the main circuit architecture of the 4LI. Next, this paper shows the expended two-arm modulation and the method of achieving a balanced three-phase current automatically using the vector control method. The effective compensation method for two capacitor voltages

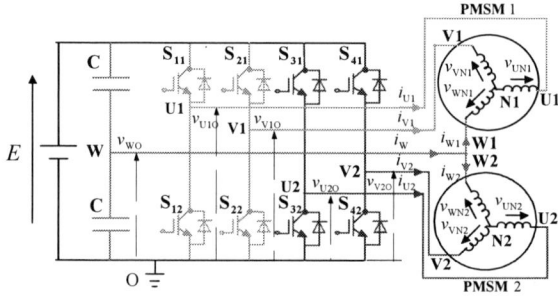

Figure 1. Main circuit of a 4LI

fluctuation is also shown. Moreover, the validity of the compensation method and independent speed and position control of two PMSMs fed by the 4LI with vector control were experimentally demonstrated and the results are presented.

II. MAIN CIRCUIT ARCHITECTURE OF FOUR-LEG INVERTER

Figure. 1 shows the 4LI to supply two three-phase AC motors. The 4LI consists of four legs and two capacitors connected in a series. Inverter U1 and V1 phases are connected to U and V phases of PMSM1 respectively. Inverter U2 and V2 phases are connected to U and V phases of PMSM2 respectively, whereas W phase of both motors is shared and connected to the neutral point of two-split capacitors in common. v_{UNi}, v_{VNi} and v_{WNi} are phase voltages in the PMSMi (i=1,2). v_{xO} (x=U1, V1, U2, V2, W) is inverter x phase voltage. v_{WO} indicates the neutral point potential of two-split capacitors. i_{Ui}, i_{Vi} and i_{Wi} are phase currents in the PMSMi and i_W is inverter phase current. E expresses magnitude of the DC-bus voltage. C is capacitance of two-split capacitors. In this work, a based-point is chosen to the negative side of DC-bus for simplicity of analysis.

III. EXPANDED TWO-ARM MODULATION

Since inverter W phase is connected to the neutral point of two-split capacitors, the modulation of the W phase is impossible. The 4LI must be modulated only two phase of U and V. Therefore, the PWM technique in three-phase VSI is not directly applicable for the 4LI. To obtain balanced three-phase AC voltage in 4LI, the phase difference between U-W line voltage and V-W line voltage need to be controlled at π/3. Then, this paper shows the expanded two-arm modulation. Figure. 2 shows the block diagram of expanded two-arm modulation in 4LI. Reference signal of U (V) phase voltage in the PMSMi compared with the carrier signal is obtained by subtracting W phase voltage command from U (V) phase voltage command.

The reference signal of U (V) phase voltage is as follows.

$$\begin{cases} v_{Ui}{}^* = v_{UNi}{}^* - v_{WNi}{}^* \\ v_{Vi}{}^* = v_{VNi}{}^* - v_{WNi}{}^* \end{cases} \qquad (1)$$

Where, $v_{kNi}{}^*$ is actual k phase voltage command in the motor i. "*" is command value. $v_{kNi}{}^*$ can be defined as follows.

$$\begin{cases} v_{UNi}{}^* = \dfrac{1}{2}M_i{}^*E\sin(\omega_i{}^*t - \varphi_i{}^*) \\ v_{VNi}{}^* = \dfrac{1}{2}M_i{}^*E\sin(\omega_i{}^*t - \dfrac{2\pi}{3} - \varphi_i{}^*) \\ v_{WNi}{}^* = \dfrac{1}{2}M_i{}^*E\sin(\omega_i{}^*t - \dfrac{4\pi}{3} - \varphi_i{}^*) \end{cases} \qquad (2)$$

Where, $M_i{}^*$ and $\omega_i{}^*$ are modulation index and fundamental angular frequency in the PMSMi respectively. $\varphi_i{}^*$ is initial phase angular of phase voltage in the PMSMi.

IV. UNBALANCED CURRENT COMPENSATION

The neutral point potential of two-split capacitors v_{WO} is given by following equation.

$$\begin{aligned} v_{wo} &= \frac{1}{2}E - \frac{1}{2C}\int (i_{W1} + i_{W2})dt \\ &= \frac{1}{2}E + \Delta v_{wo} \end{aligned} \qquad (3)$$

Where, Δv_{wo} is a fluctuated component of v_{WO}.

From equation (3), v_{WO} changes around *E/2*. The reason v_{WO} fluctuates is that W phase currents of each motors flow through capacitors. The fluctuation of the v_{WO} affects motor phase currents and makes it unbalance. The fluctuated component v_{WO} depends on the fundamental wave frequency and peak value of the both motor currents. In other words, decreasing the fluctuation of the v_{WO} is possible if the motors are driven at lighter load and higher speed condition. Using the capacitor with larger capacitance can also decrease the fluctuation. However, the too large capacitor is undesirable because

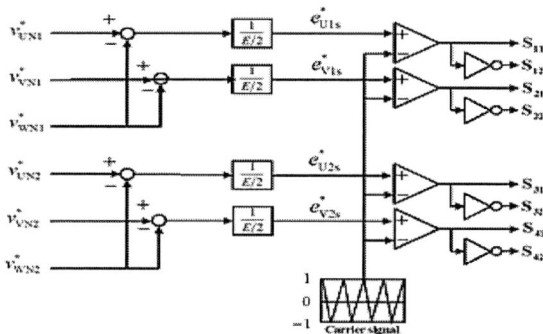

Figure 2. Block diagram of carrier-based PWM

one of the strong points of the 4LI is saving space. Therefore, it needs to consider the compensation method that keeps three-phase current balancing. Since two capacitor voltages fluctuate shown in equation (3), the

balanced three-phase current cannot be obtained under open-loop control such as V/f control system. In order to obtain the balanced three-phase current, capacitor voltage fluctuation Δv_{wo} must be added to the U, V phase terminal voltage[3], v_{UiO}, v_{ViO} respectively such that

$$\begin{cases} v_{UiO} = v_{UNi}{}^* - v_{WNi}{}^* + \Delta v_{wo} \\ v_{ViO} = v_{VNi}{}^* - v_{WNi}{}^* + \Delta v_{wo} \end{cases} \qquad (4)$$

However, under vector control, command values to be balanced three-phase current are automatically obtained because vector control method is usually taken place d, q axis control to get three-phase current. Therefore, under vector control. U, V phase terminal voltage are automatically obtained(4).

V. NEUTRAL POINT POTENTIAL OF TWO-SPLIT CAPACITOR COMPENSATION

The drift phenomenon of the neutral point potential of two capacitors v_{WO} is observed in the starting time of motors. As the result, since both capacitor voltages are not equally separate in E/2, the drift causes the reduction of inverter DC-bus voltage utility factor. In conclusion, v_{WO} must be maintained to E/2 with restraining the drift.

Jaehong Kim, Jinseok Hong, and Kwaghee Nam analyzed about the drift phenomenon of the four switch inverter by utilizing space vector [4]. Also, they proposed compensation method of the drift and demonstrated effectiveness of it by experimental results.

We conducted circuit analysis of each switching mode of the 4LI to calculate compensation method of neutral point potential voltage. Two motors in the 4LI can be regarded as independent. Therefore, in order to analyze easily, we conducted circuit analysis about one motor.

Figure. 3 shows a vector control block diagram with drift compensation.

Figure. 4 shows the current passes of each switching mode; $P_{i1}(0,0)$, $P_{i2}(0,1)$, $P_{i3}(1,0)$, $P_{i4}(1,1)$. It is defined as "1" in the parenthesis when the upper switch of each phase is on, and as 0 when the upper switch is off. In mode $P_{i1}(0,0)$, equation (5) obtained from Figure. 4 (a) as

The 2014 International Power Electronics Conference

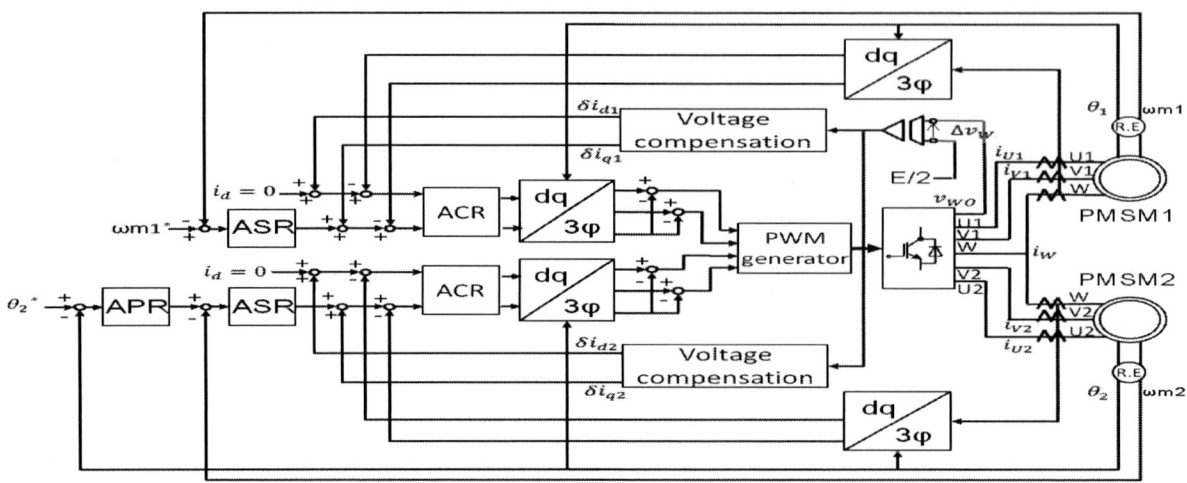

Figure 3. The block diagram of the vector control with compensation

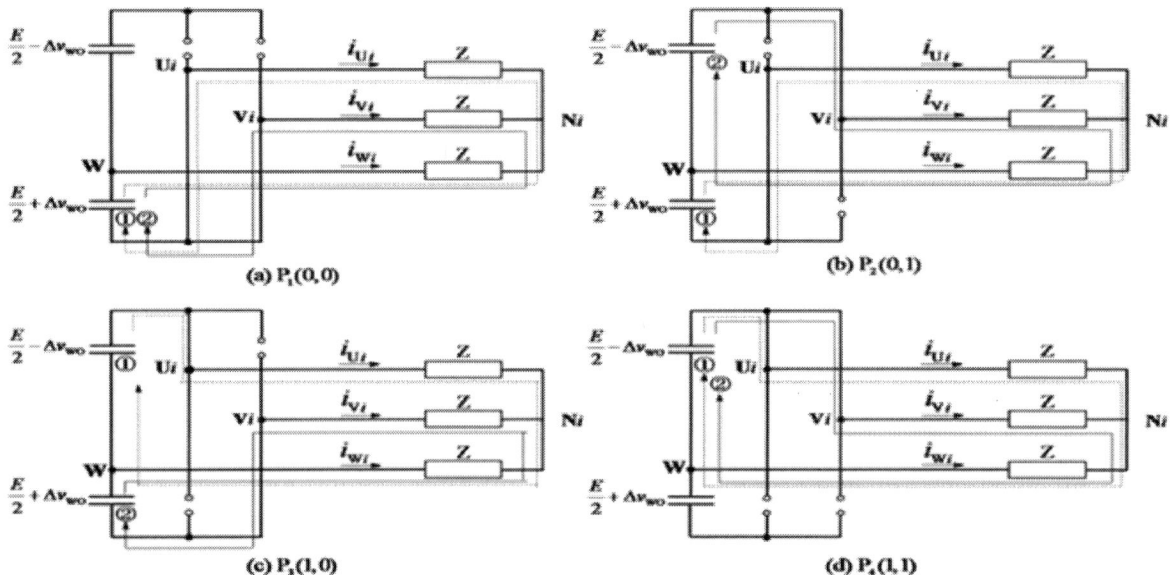

Figure 4. Current pass of each switching mode

$$\begin{cases} Zi_{Wi} - Zi_{Ui} = \dfrac{1}{2}E + \Delta v_{wo} \\[2mm] Zi_{Wi} - Zi_{Vi} = \dfrac{1}{2}E + \Delta v_{wo} \\[2mm] i_{Ui} + i_{Vi} + i_{Wi} = 0 \end{cases} \tag{5}$$

equation (5) can be rewritten as

$$\begin{cases} Zi_{Ui} = -\dfrac{1}{6}E - \dfrac{1}{3}\Delta v_{wo} \\[2mm] Zi_{Vi} = -\dfrac{1}{6}E - \dfrac{1}{3}\Delta v_{wo} \\[2mm] Zi_{Wi} = \dfrac{1}{3}E + \dfrac{2}{3}\Delta v_{wo} \end{cases} \tag{6}$$

$$\begin{cases} Zi_{Ui} = -\dfrac{1}{2}E - \dfrac{1}{3}\Delta v_{wo} \\[2mm] Zi_{Vi} = \dfrac{1}{2}E - \dfrac{1}{3}\Delta v_{wo} \\[2mm] Zi_{Wi} = \dfrac{2}{3}\Delta v_{wo} \end{cases} \tag{7}$$

$$\begin{cases} Zi_{Ui} = \dfrac{1}{2}E - \dfrac{1}{3}\Delta v_{wo} \\[2mm] Zi_{Vi} = \dfrac{1}{2}E - \dfrac{1}{3}\Delta v_{wo} \\[2mm] Zi_{Wi} = \dfrac{2}{3}\Delta v_{wo} \end{cases} \tag{8}$$

Also, we analyzed (b) $P_{i2}(0,1)$, (c)$P_{i3}(1,0)$, (d)$P_{i4}(1,1)$ and obtained next equations.

978-1-4799-2706-7/14 $31.00 © 2014 IEEE

$$\begin{cases} Zi_{Ui} = \dfrac{1}{6}E - \dfrac{1}{3}\Delta v_{wo} \\[2mm] Zi_{Vi} = \dfrac{1}{6}E - \dfrac{1}{3}\Delta v_{wo} \\[2mm] Zi_{Wi} = -\dfrac{1}{3}E + \dfrac{2}{3}\Delta v_{wo} \end{cases} \qquad (9)$$

The equations (6)-(9) show relation of each phase voltage and ΔV_{wo}. From these equations, we analyzed the relation of the inverter phase voltage and drift phenomenon of the neutral point potential by using space vector.

Each Voltage vectors of $P_{i1}(0,0)$, $P_{i2}(0,1)$, $P_{i3}(1,0)$, $P_{i4}(1,1)$ are expressed $\mathbf{V}_{i1}$, $\mathbf{V}_{i2}$, $\mathbf{V}_{i3}$, $\mathbf{V}_{i4}$.

$\mathbf{V}_{i1}$ as follows

$$V_{i1} = \frac{2}{3}\left(Zi_{Ui} + Zi_{Vi}e^{j\left(\frac{2}{3}\pi\right)} + Zi_{Wi}e^{j\left(\frac{4}{3}\pi\right)}\right) \qquad (10)$$

Substituting equation (6) into (10) and then

$$V_{i1} = \left(-\frac{1}{6} - j\frac{1}{2\sqrt{3}}\right)E + \frac{2}{3}\Delta v_{wo}e^{j\left(\frac{4}{3}\pi\right)}$$

$$= \left(-\frac{1}{6} - j\frac{1}{2\sqrt{3}}\right)E + \Delta V_{wo} \qquad (11)$$

where

$$\Delta V_{wo} = \frac{2}{3}\Delta v_{wo}e^{j\left(\frac{4}{3}\pi\right)}$$

$\mathbf{V}_{i2}$, $\mathbf{V}_{i3}$, $\mathbf{V}_{i4}$ are also

$$V_{i2} = \left(-\frac{1}{2} + j\frac{1}{2\sqrt{3}}\right)E + \Delta V_{wo} \qquad (12)$$

$$V_{i3} = \left(\frac{1}{2} - j\frac{1}{2\sqrt{3}}\right)E + \Delta V_{wo} \qquad (13)$$

$$V_{i4} = \left(\frac{1}{6} + j\frac{1}{2\sqrt{3}}\right)E + \Delta V_{wo} \qquad (14)$$

From these equations, ΔV_{wo} influences all switching modes.

Figure. 5 shows vector diagram of ΔV_{wo}, V_{Ui} and V_{Vi}. ΔV_{wo} is compensated by adding ΔV_{wo} which divided U and V vector $\Delta \mathbf{V}_{wo}$ is showed like

$$\Delta \mathbf{V}_{wo} = \frac{2\Delta V_{wo}}{3\sqrt{3}}e^{j\frac{\pi}{6}} + \frac{2\Delta V_{wo}}{3\sqrt{3}}e^{j\frac{\pi}{2}} \qquad (15)$$

From equation (15), compensation terms which added U or V phase voltage command are next equation.

$$\begin{cases} \delta v_{Ui} = -\dfrac{2}{3\sqrt{3}}\Delta V_{wo} \\[2mm] \delta v_{Vi} = -\dfrac{2}{3\sqrt{3}}\Delta V_{wo} \end{cases} \qquad (16)$$

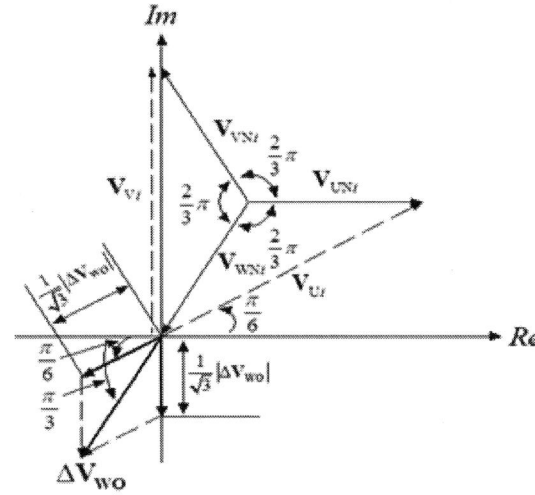

Figure 5. Space voltage vectors

Under vector control method, even if the compensation terms are added to the reference signals of U and V phases just before compared with the carrier signal, it does not improve the performance because the response of Automatic Current Regulator (ACR) in vector control system is very fast. Therefore, in order to use the compensation terms (16) effectively under vector control method, it must be added before ACR, not after ACR. In conclusion, it is necessary to transform (16) into d and q axis coordinate, and compensate as the compensation term in d and q axis currents.

In the 4LI, transformation from d and q reference voltages to U and V phase sinusoidal reference voltage signals are

$$\begin{bmatrix} v_{UiO} \\ v_{ViO} \end{bmatrix}$$

$$= \sqrt{\frac{2}{3}}\begin{bmatrix} 1 & 0 & -1 \\ 0 & 1 & -1 \end{bmatrix}\begin{bmatrix} \cos\theta & -\sin\theta \\ \cos\left(\theta - \dfrac{2}{3}\pi\right) & -\sin\left(\theta - \dfrac{2}{3}\pi\right) \\ \cos\left(\theta - \dfrac{4}{3}\pi\right) & -\sin\left(\theta - \dfrac{4}{3}\pi\right) \end{bmatrix}\begin{bmatrix} v_d \\ v_q \end{bmatrix}$$

$$= \sqrt{2}\begin{bmatrix} \cos\left(\theta - \dfrac{\pi}{6}\right) & -\sin\left(\theta - \dfrac{\pi}{6}\right) \\ \sin\theta & \cos\theta \end{bmatrix}\begin{bmatrix} v_d \\ v_q \end{bmatrix} \qquad (17)$$

Also, inverse transformation of (17) is obtained as,

$$\begin{bmatrix} v_d \\ v_q \end{bmatrix} = \sqrt{\frac{2}{3}}\begin{bmatrix} \cos\theta & \sin\left(\theta - \dfrac{\pi}{6}\right) \\ -\sin\theta & \cos\left(\theta - \dfrac{\pi}{6}\right) \end{bmatrix}\begin{bmatrix} v_{UiO} \\ v_{ViO} \end{bmatrix} \qquad (18)$$

Substituting δv_{UiO}, δv_{ViO} of equation (16) for v_{UiO}, v_{ViO}, it follows that

$$\begin{bmatrix} \delta v_{di} \\ \delta v_{qi} \end{bmatrix} = \sqrt{\frac{2}{3}}\begin{bmatrix} \cos\theta & \sin\left(\theta - \dfrac{\pi}{6}\right) \\ -\sin\theta & \cos\left(\theta - \dfrac{\pi}{6}\right) \end{bmatrix}\begin{bmatrix} -\dfrac{2}{3\sqrt{3}}\Delta v_{wo} \\ -\dfrac{2}{3\sqrt{3}}\Delta v_{wo} \end{bmatrix} \quad (19)$$

Then, (19) divided by the impedance of stator winding, $R+Ls$, d and q axis current compensation terms δi_{di}, δi_{qi} can be obtained.

The d and q axis current commands in the 4LI are obtained by adding the compensation current δi_{di}, δi_{qi} to the standard d and q axis current commands of vector control. By controlling the difference between the commands and the actual current values with Proportional and Integral (PI) controller, the drift in neutral point voltage fluctuations can be restrained.

VI. EXPERIENTAL RESULTS

A) Experimental system

In order to demonstrate the independent drives of two PMSMs, and the usefulness of the compensation method for the drift phenomenon, a 4LI experimental prototype to supply two three-phase PMSMS has been implemented. The system configuration of the prototype is shown in Figure. 6. The main circuit consists of four IGBT-modules and two-split capacitors. PE-Expert3 is used as the control system. The PE-Expert3 is the digital control system equipping a DSP, which also has an Analog-Digital (AD) Converter, digital input-output and PWM function. A program written by user is implemented with transmitting the program from a host computer to the DSP on the PE-Expert3. Figure. 7 depicts a photo of experimental prototype. Table 1 shows the ratings and parameters of tested PMSMs. The ratings and parameters of both PMSMs are identical. Both PMSMs are driven by vector control. DC bus voltage is 282[V]. A slider is used to boost-up three-phase input voltage. The inverter DC-bus voltage can be obtained by adjusting the input voltage of a three-phase diode –rectifier with the slider. Capacitance of capacitor is 9900[μF]. Carrier frequency is 10[kHz].

B) speed and position control experimental result

Figure. 8 show the rotor speed of the PMSM1. The speed commands give -400[rpm] (0-2[sec]). Figure. 9 show the rotor angle of the PMSM2. The rotor position commands give 6π[rad] to PMSM2. Figure. 10 shows the neutral point potential with compensation. Figure. 11 and 12 show the three-phase current waveforms of the PMSMs.

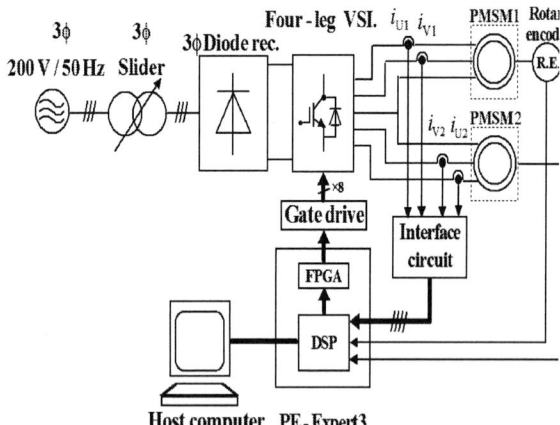

Figure 6. The system configuration of the prototype

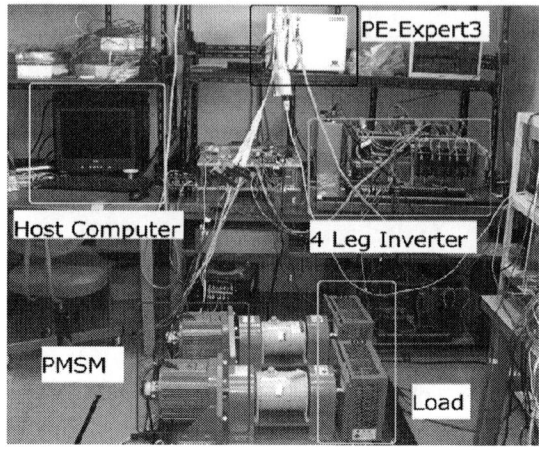

Figure 7. Experimental prototype of 4LI

TABLE I   Rating and parameters of tested permanent magnet synchronous motors

Rated Output	0.75 kw
The Number of Poles	12
Rated Voltage	116 Vrms
Rated Current	4.5 A
Rated Frequency	120 Hz
Rated Speed	1200 rpm
Stator Resistance	0.36 Ω
d-axis Inductance	2.76 mH
q-axis Inductance	2.87 mH
Inertia(PMSM + Load)	12.8 $mkgm^2$

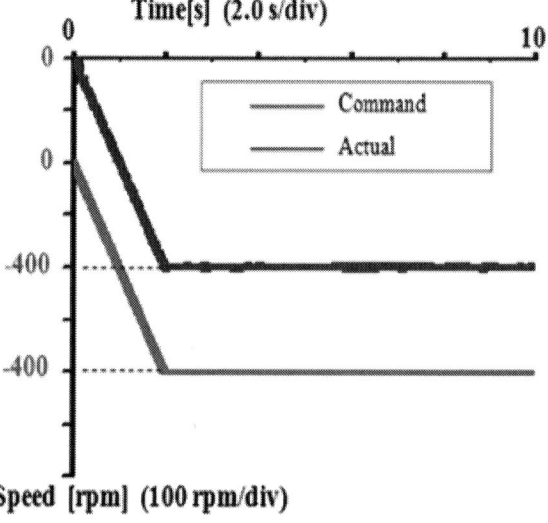

Figure 8. Speed of the PMSM1

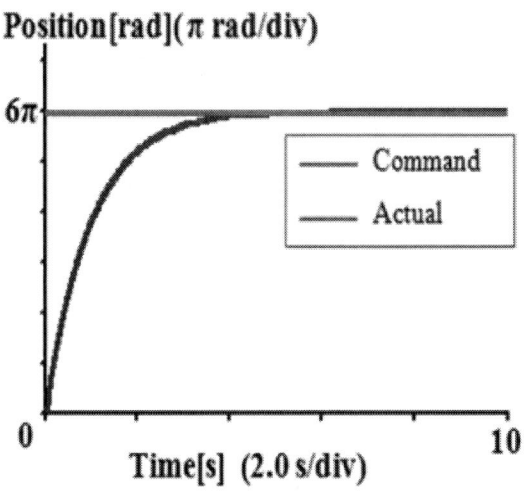

Figure 9. Rotor angle of the PMSM2

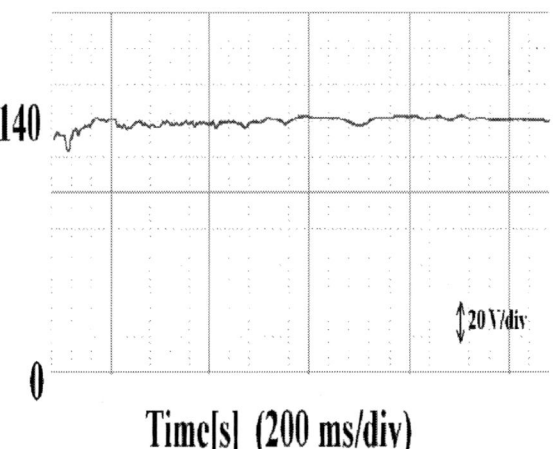

Figure 10. The neutral point potential of two capacitors

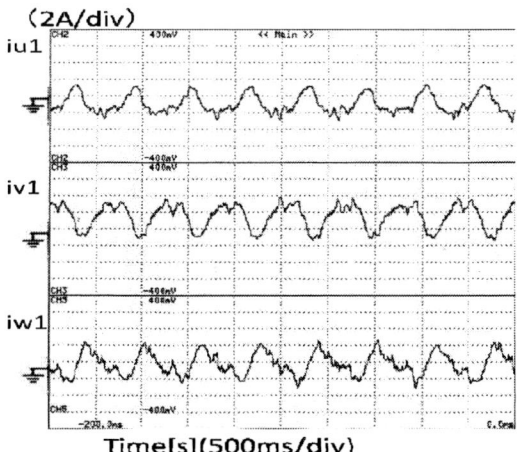

Figure 11. The three-phase current waveforms of the PMSM 1

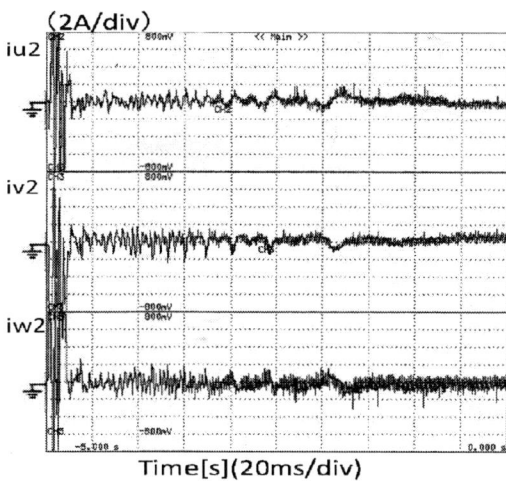

Figure 12. The three-phase current waveforms of the PMSM2

VII. CONCLUSION

This paper presents the experimental results of the independent speed and position controls of two PMSMs fed by a 4LI with vector control method. The effectiveness of the compensation method and the independent speed and position control of two PMSMs fed by a 4LI was supported by the experimental results.

REFERENCES

[1] M. B. de R. correa, C.B. jacobina, E. R. C. da Silva and A. M. N. Lima ,"A General PWM Strategy for Four-Switch Three-Phase Inverters", *IEEE Trans. Power Electon* vol. 21, no. 6, pp. 1618-1627, November, 2006.

[2] J. Kim, J. Hong and K, Nam, "A Current Distortion Compensation Scheme for Four-Switch Inverters", *IEEE Trans. Power Electon* vol. 24, no. 4, pp. 1032-1040, April, 2009.

[3] T. D. Nguyen, H. M. Nguyen and H. H. Lee, "An Adaptive Carrier-Based PWM Method for Four-Switch Three- Phase Inverter", IEEE International symposium on Industrial Electronics, pp. 1552-1557, July 5-8, 2009.

[4] Jaehong Kim, Jinseok Hong, and Kwanghee Nam : "A Current Distortion Compensation Scheme for Four-Switch Inverters", *IEEE TRANSACTIONS ON POWER ERECTRONICS, VOL. 24, NO. 4, APRIL* 2009

[5] K.Oka and K.Matsuse, "A Performance Analysis of a Four-Leg Inverter in Two AC Motor Drives with Independent Vector Control", *TRANSACTIONS ON ELECTRONICAL AND ELECTRONIC ENGINEERING IEEJ Trans* 104-107, 2006, 1

[6] Y.Katagiri, N.Kezuka, H.tanaka, K.Matsuse: "Performance of Independent two Induction Motor Drives Fed by a Four-Leg Inverter with vector control method", IEEE IAS Annual Meeting, 2011-IACC-160,2011

[7] K.Matsuse N.Kezuka K.Oka: "Characteristics of Independent Two Induction Motor Drives Fed by a Four-Leg Inverter", IEEE Trans. Industry Applications, Vol.47, NO.5, pp.2125-2134, Sep./Oct. 2011

Minimization of Stator Currents for Mono Inverter Dual Parallel PMSM Drive System

Yongjae Lee and Jung-Ik Ha
Dept. of Electrical and computer engineering
Seoul National University
Seoul, Korea
yongjaelee@snu.ac.kr, jungikha@snu.ac.kr

Abstract— **This paper proposes maximum torque per ampere (MTPA) control and the active damping control of mono inverter dual parallel (MIDP) permanent magnet synchronous motor (PMSM) drive system. MIDP motor drive system is one of the configurations for reducing the cost of the power conversion circuits. However, it is hard to apply the MIDP motor drive system to PMSMs due to the instability unlike the induction motors. In this paper, optimal current control strategy to minimize the current and the loss is proposed. The MTPA control is executed with not look-up table but real time calculation from the current informations. By minimizing the current, system efficiency can be enhanced. The control methods for MTPA and the active damping are proposed. The proposed control algorithm and strategy are verified through the experiments with 1kW motors.**

Keywords— *Current minimization, Mono inverter dual parallel motor drive system, Permanent magnet synchronous motor.*

I. Introduction

Since the development of IGBTs and MOSFETs, the price of the switching devices has been steadily decreased and the features such as current rating, break down voltage, on voltage drop, and switching loss have been bettered [1]. Although the price of the power electronic devices has been lowered enough, it still takes a large portion in the energy conversion system cost. Especially the fields where the cost is critically concerned such as home appliances or other small motor drives require the circuit reduction studies including position or speed sensor eliminations [2, 3], shunt resistor current sensing [4], and dual/multi motor drives [5–11]. Among the various challenges, dual motor drive technique failed to arouse the interests of researchers and manufacturers due to its limited applications and characteristic compared to other techniques. The dual motor drive system, however, has studied for the specific applications because the dual motor drive system can save lots of the system cost.

Mono inverter dual parallel (MIDP) motor drive system has been widely researched for the induction motors thanks to the slip which makes the parallel system stable [5-7]. While various control strategies have been researched for the MIDP induction motors, control method for permanent magnet synchronous motors

(PMSMs) attracts less attention due to its inherent instability risk. There are several previous studies about the MIDP PMSM drive system by several researchers [8-11]. Analysis of multi PMSM drive and the control strategy based on higher torque motor was proposed in [8]. Controlling both motors using optimal voltage vector prediction was proposed in [9]. The average current control method was proposed in [10] and sum current control method with current prediction was proposed in [11]. Although these researches succeed to drive the dual parallel PMSM with single inverter, these researches tend to rely on mechanical damping or model based control. The optimal current driving strategy was also proposed in [9, 11], however, [9] used the fully model based control and [11] used the complex look-up table.

This paper presents simple and robust controller for MIDP PMSM drive system and maximum torque per ampere (MTPA) control method. The MTPA conditions in the steady state are analyzed to minimize the currents. The novel control method based on the MTPA strategy is also proposed. The proposed controller is not only more stable but also more irrelevant to the mechanical damping of the system.

II. MTPA Strategy

Fig. 1 shows the structure of the MIDP PMSM drive system. The system consists of one inverter and two PMSMs connected in parallel. Voltage equations of both motors in steady state can be written as

$$\begin{bmatrix} v_{dk} \\ v_{qk} \end{bmatrix} = \begin{bmatrix} R_s & -\omega_r L_s \\ \omega_r L_s & R_s \end{bmatrix} \begin{bmatrix} i_{dk} \\ i_{qk} \end{bmatrix} + \begin{bmatrix} 0 \\ \omega_r \lambda_f \end{bmatrix}, \qquad (1)$$

where both motors are surface mounted PMSMs (SPMs), R_s is stator resistance, L_s is stator inductance, λ_f is flux from permanent magnets, ω_r is electrical speed of motor, v_{dk}, v_{qk}, i_{dk}, i_{qk} are voltages and currents of the motors in synchronous reference frame where the subscript "k" indicates the respective motor, 1 for the master motor and 2 for the slave motor. If the motors are in transient state, ω_r have to be differentiated by ω_{r1} and ω_{r2}. In this analysis, however, the steady state is assumed.

Because two motors are tied to the single voltage source inverter (VSI), voltages supplied to the motors are identical. Thus, amplitudes of voltage vector for the two

978-1-4799-2706-7/14 $31.00 © 2014 IEEE   3140

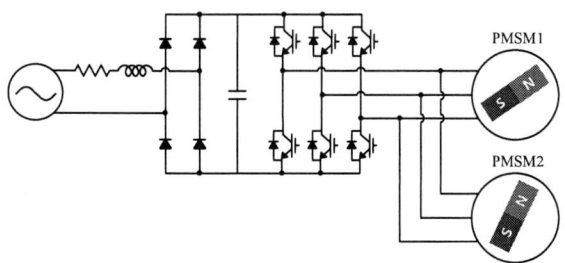

Fig. 1. Configuration of mono inverter dual parallel PMSM drive system.

motors are also same regardless of the angle difference between the motors and can be descripted as

$$
\begin{aligned}
v_o^2 &= v_{d1}^2 + v_{q1}^2 = v_{d2}^2 + v_{q2}^2 \\
&= (R_s i_{d1} - \omega_r L_s i_{q1})^2 \\
&\quad + \left[R_s i_{q1} + \omega_r (\lambda_f + L_s i_{d1}) \right]^2 \\
&= (R_s i_{d2} - \omega_r L_s i_{q2})^2 \\
&\quad + \left[R_s i_{q2} + \omega_r (\lambda_f + L_s i_{d2}) \right]^2.
\end{aligned}
\tag{2}
$$

Because i_{q1} and i_{q2} are decided by load torque in steady state, (2) can be rewritten as (3) by substituting complex coefficients.

$$
\left(i_{d1} + \frac{\beta}{2\alpha} \right)^2 - \left(i_{d2} + \frac{\beta}{2\alpha} \right)^2 = \frac{\gamma_2 - \gamma_1}{\alpha},
\tag{3}
$$

where

$$
\begin{aligned}
\alpha &= R_s^2 + \omega_r^2 L_s^2; \\
\beta &= 2\omega_r^2 \lambda_f L_s; \\
\gamma_1 &= \left(R_s^2 + \omega_r^2 L_s^2 \right) i_{q1}^2 + 2 R_s \omega_r \lambda_f i_{q1} + \omega_r^2 \lambda_f^2; \\
\gamma_2 &= \left(R_s^2 + \omega_r^2 L_s^2 \right) i_{q2}^2 + 2 R_s \omega_r \lambda_f i_{q2} + \omega_r^2 \lambda_f^2.
\end{aligned}
\tag{4}
$$

Every point of (3) indicates the steady state operating point and it can be drawn as the curves which has positive gradient in Fig. 2 according to the variation of load torque of the one motor where another motor is under unit load torque with parameters listed in Table 1. The goal of the MTPA control is minimization of the motor current and loss which is represented as

$$
P_{copper} = \frac{3}{2} R_s \left(i_{d1}^2 + i_{q1}^2 + i_{d2}^2 + i_{q2}^2 \right).
\tag{5}
$$

Because the q axis currents are decided by the load condition, the (5) can be simplified into

$$
g(i_{d1}, i_{d2}) = i_{d1}^2 + i_{d2}^2.
\tag{6}
$$

With the *Lagrange* multiplier, objective function can be written as

$$
\begin{aligned}
L(i_{d1}, i_{d2}, \lambda) &= g(i_{d1}, i_{d2}) \\
&+ \lambda \left(\left(i_{d1} + \frac{\beta}{2\alpha} \right)^2 - \left(i_{d2} + \frac{\beta}{2\alpha} \right)^2 - \frac{\gamma_2 - \gamma_1}{\alpha} \right),
\end{aligned}
\tag{7}
$$

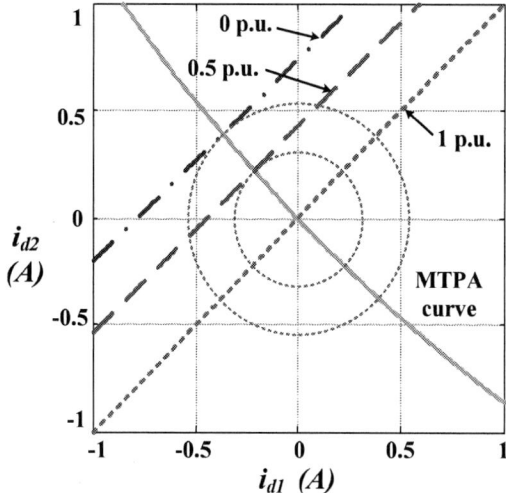

Fig. 2. Steady state operating condition curves and MTPA curve.

where λ is *Lagrangian* coefficient. The partial derivatives of the (7) give the conditions for minimizing copper loss. The partial derivatives of the (7) can be written as

$$
\begin{aligned}
\frac{\partial L(i_{d1}, i_{d2}, \lambda)}{\partial i_{d1}} &= 2 i_{d1} + 2\lambda \left(i_{d1} + \frac{\beta}{2\alpha} \right), \\
\frac{\partial L(i_{d1}, i_{d2}, \lambda)}{\partial i_{d2}} &= 2 i_{d2} - 2\lambda \left(i_{d2} + \frac{\beta}{2\alpha} \right), \\
\frac{\partial L(i_{d1}, i_{d2}, \lambda)}{\partial \lambda} &= \left(i_{d1} + \frac{\beta}{2\alpha} \right)^2 - \left(i_{d2} + \frac{\beta}{2\alpha} \right)^2 \\
&\quad - \frac{\gamma_2 - \gamma_1}{\alpha}.
\end{aligned}
\tag{8}
$$

Minimum copper loss condition is satisfied if and only if all the partial derivatives are zero. By rearranging the first to partial derivatives of (8), the simplified equations can be achieved as

$$
\begin{aligned}
i_{d1} &= -\frac{\lambda}{1+\lambda} \frac{\beta}{2\alpha}, \\
i_{d2} &= -\frac{\lambda}{1-\lambda} \frac{\beta}{2\alpha}.
\end{aligned}
\tag{9}
$$

By merging the two equations of (9), the MTPA condition for the MIDP PMSM drive system can be achieved as

$$
\frac{1}{i_{d1}} + \frac{1}{i_{d2}} = -\frac{4\alpha}{\beta}.
\tag{10}
$$

It is also simply calculated with condition of the tangent line from (3) and (6). The MTPA condition can be drawn as cyan curve which has negative gradient in Fig. 2.

Because (10) is independent to the torque and only dependent to the speed, MTPA curves are identical for certain speed. It is obvious that the cross points of (3) and (10) become optimal and stable operating points for the given load condition, respectively. Because the exact point is hard to solve analytically, look-up table methods are normally used for MTPA control. For the proposed

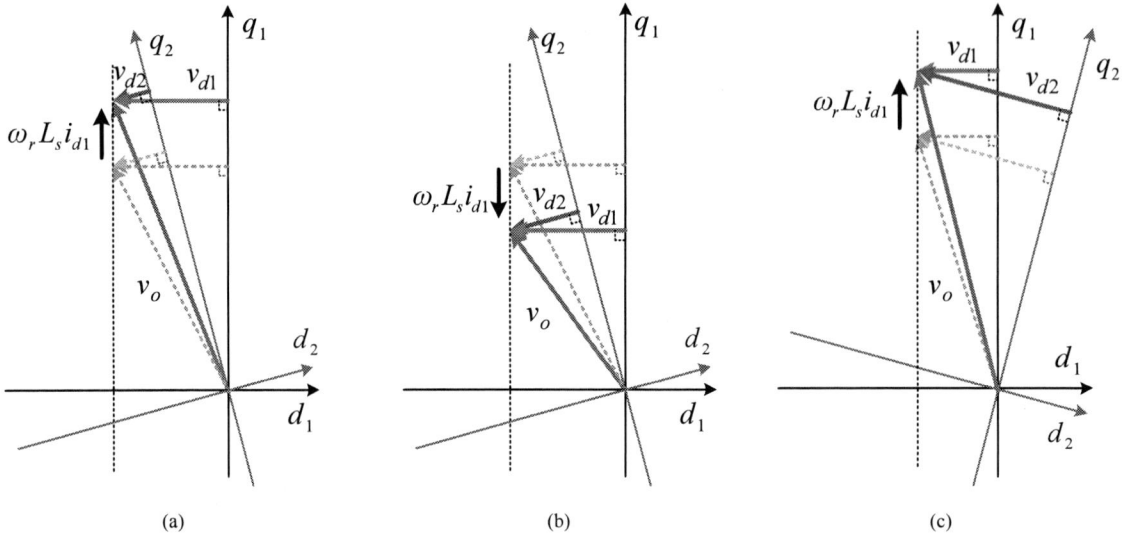

Fig. 3. Vector diagrams of MIDP PMSM drive system. (a) Slave motor leading case with positive i_{d1}. (b) Slave motor leading case with negative i_{d1}. (c) Master motor leading case with positive i_{d1}.

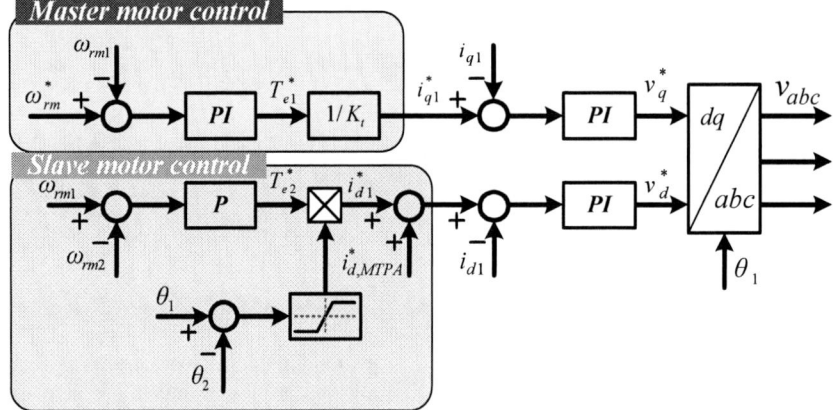

Fig. 4. Block diagram of controller for the proposed MIDP PMSM drive system.

method, the optimal point can be tracked in real time with the numerical method such as bisection method and *newton-raphson* method [12]. Thus cumbersome table making is not required.

TABLE I
PARAMETERS OF THE MOTOR

Meaning	Unit	Value
Number of poles	-	8
Phase resistance	Ω	3.25
Phase inductance	mH	28
Back-EMF constant	$V \cdot sec$	0.2
Rated speed	r/min	1200
Rated torque	$N \cdot m$	5

III. CONTROL STRATEGY

For the proposed control method, d and q axis currents of the one motor, master motor, is treated as control objects. Another motor, called slave motor, is indirectly controlled with the d axis current of the master motor. Because the proposed method actively controls speed of the slave motor with d axis current, it does not require the master selection according to the angle difference and it can remove the transition shocks which was caused when

changing the control object. (11) represents the generated torque by the motor when the resistance voltage drop is neglected.

$$T_e = \frac{3}{2}\frac{P}{2}\lambda_f i_q = -\frac{3}{2}\frac{P}{2}\frac{\lambda_f}{L_s}\frac{v_d}{\omega_r}, \qquad (11)$$

where P is the number of the poles. Because the output torque is decided by i_q, torque control of the master motor is achieved by controlling i_{q1}. For the slave motor, however, current is indirectly controlled by output voltage of the master motor controller. The proposed indirect torque control method is described in Fig. 3. It shows the vector diagrams of the MIDP PMSM system where (a) and (b) are the master motor lagging cases and (c) is the master motor leading case. The positive and negative d axis currents move output voltage vectors upward and downward in (a) and (b), respectively. Because the d axis current of the master motor has no contribution to the torque generation, there is no change in v_{d1} while v_{d2} is altered by i_{d1}. For the master motor lagging cases, (a) and (b), positive i_{d1} decreases v_{d2} and negative i_{d1} increases v_{d2}. It means that the output torque

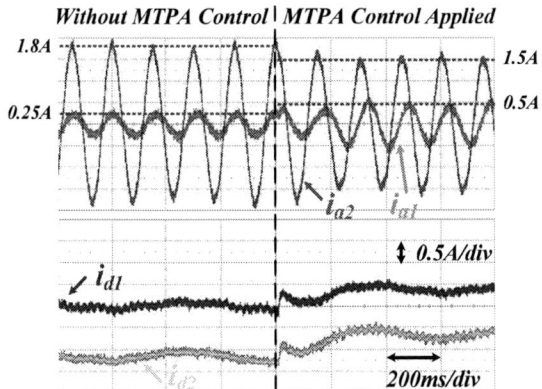

Fig. 5. Experimental results of proposed MTPA method for the MIDP PMSM drive system.

is also decreased or increased with v_{d2}. For the master motor leading case, (c), positive i_{d1} increases v_{d2} as opposed to the master motor lagging case. This phenomenon is contrary to the results of (a) and (b).

Fig. 4 shows the block diagram of proposed controller. Speed of the master motor is controlled by the conventional speed controller. To prevent the motors from diverging, slave motor is controlled to follow the speed of the master motor. Slave motor controller is composed of proportional controller to remove the steady state output. By removing the steady state output, d axis current can be controlled with MTPA current, $i_{d,MTPA}^*$ in the steady state. This proportional controller also can be thought as active damping. To prevent the transition shock caused by abrupt change of the d axis current reference at zero angle difference, ramp function to the

angle difference is applied where angle difference is low.

IV. EXPERIMENTAL RESULTS

To verify the proposed MTPA method and controller, experiments with 1kW motor of which the rated values are listed in Table 1. Fig. 5 shows the experimental results where master motor is under no load torque and slave motor is under 2Nm load torque condition. The motors are rotating with 400r/min speed. The result show that phase currents of the both motor is reduced from 1.8A, 0.25A to 1.5A and 0.5A by applying proposed MTPA current tracking method. The rms of the phase currents are reduced from 1.82A to 1.58A with the proposed method. Although the current of the master motor is increased, decreased slave motor current reduces more loss.

Fig. 6 and 7 show the dynamic response results of the proposed MIDP PMSM drive system. In this experiment the speeds of the motors are 1200r/min and the rated torques are loaded to each motors. In the steady state, 20% of step torque is unloaded and restored after the motors reach the steady state. For the Fig. 6, the currents of the two motors are presented and the speeds and d axis currents of the two motors are presented in the Fig. 7.

As shown in the Fig. 6, i_{q2} rapidly changes when load is changed thanks to the injected d axis current. As shown in the experimental results, the proposed system can stand the step load change and operate stably at different load condition. As shown in the Fig. 6 (a), the i_{d1} is slightly increased and i_{d2} is decreased when the load of the master motor is lower than that of the slave motor. In contrast, when the load of the master motor is higher than that of the slave motor, Fig. 6 (b), i_{d1} is slightly decreased

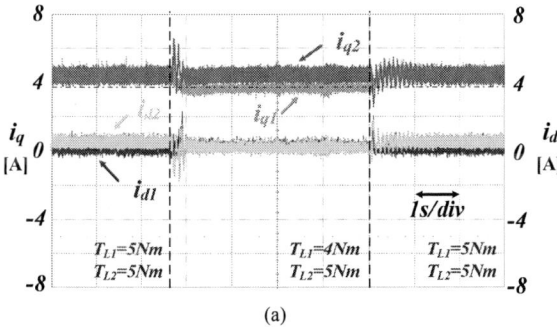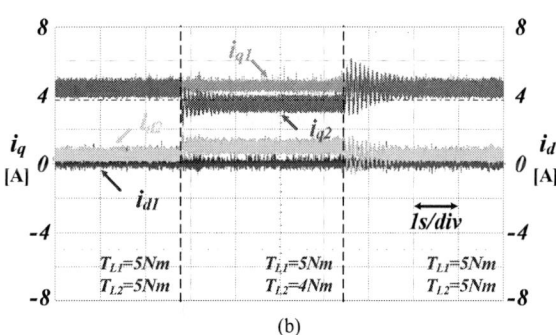

(a)　　　　　　　　　　　　　　　　　　(b)

Fig. 6. Experimental waveforms of d-q axis currents in the proposed control method for the MIDP PMSM drive system against the step load change. (a) Master motor load torque is varying. (b) Slave motor load torque is varying.

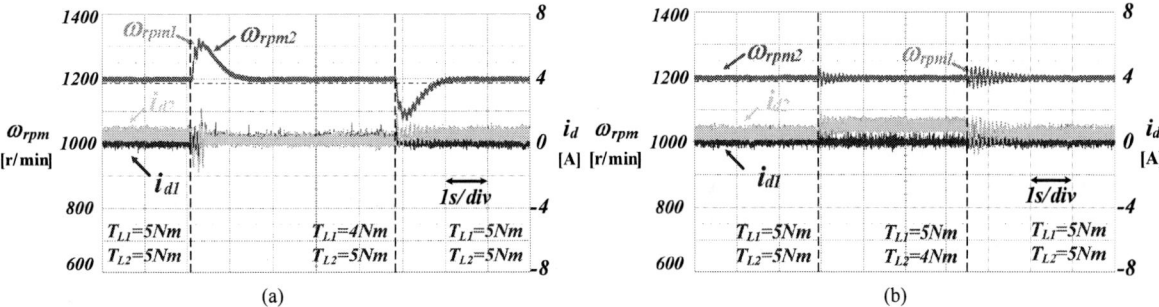

(a)　　　　　　　　　　　　　　　　　　(b)

Fig. 7. Experimental waveforms of d axis current and speed in the proposed control method for the MIDP PMSM drive system against the step load change. (a) Master motor load torque is varying. (b) Slave motor load torque is varying.

978-1-4799-2706-7/14 $31.00 © 2014 IEEE

and i_{d2} is increased for the MTPA operation. Although the optimal d axis currents at same load condition is zero, the d axis currents are not zero due to the different friction.

Fig. 7 shows the d axis currents and speeds of the two motors against step load change. The changes of the d axis currents of the Fig. 7 (a) and (b) are same with the results of the Fig. 6. The changes of the speeds in transient state shown in the Fig. 7 (a) and (b), however, are different according to the load in contrast with the currents. When the load of the master motor changed, Fig. 7 (a), the speeds of the two motors are increased or decreased according to the load change. This speed fluctuating mainly comes from the slow control dynamics of the speed controller. Due to the low bandwidth of the speed controller, torque of the master motor cannot quickly follow the step load change. On the other hand, when the load of the slave motor changed, Fig. 7 (b), the speeds does not changed significantly because the speed controller is not included to the slave motor speed control. Instead, the proposed controller for the slave motor quickly stabilizes the slave motor. The damping controller also synchronizes the speeds of the two motors for both cases as shown in the Fig. 7.

V. CONCLUSIONS

In this paper, MTPA and active damping control methods for MIDP PMSM drive system are proposed. MTPA currents for the steady state condition are calculated and analyzed for the SPM. Because the MTPA current is not based on the table but calculated in real time, it is more convenient than the conventional methods. Novel controller for the MIDP PMSM drive system which can control the motors with calculated MTPA currents is also proposed. The proposed method also includes active damping which uses the d axis current of the master motor to control the torque of the slave motor. Thanks to the proposed control method, not only the currents flowing in the machines are minimized but also the dynamic characteristics of the MIDP PMSM drive system is improved. Proposed methods are verified by the experiments with the 1kW SPMs.

ACKNOWLEDGMENTS

The authors want to acknowledge the assistance and financial support provided by Samsung Electronics DMC R&D Center.

REFERENCES

[1] V. Khanna, Insulated Gate Bipolar Transistor IGBT Theory and Design, Wiley-IEEE Press, 2003.

[2] J. –I. Ha, K. Ide, T. Sawa, and S. –K. Sul, "Sensorless rotor position estimation of an interior permanent-magnet motor from initial states," *IEEE Trans. Ind. Appl.*, vol. 39, no. 3, pp. 761–767, May/Jun. 2003.

[3] K. –W. Lee, and J. –I. Ha, "Evaluation of back-EMF estimators for sensorless control of permanent magnet synchronous motors", *Journal of Power Electronics*, vol. 12, no. 4, pp.604-614, 2012.

[4] J. –I. Ha, "Current prediction in vector-controlled PWM inverters using single DC-link current sensor," *IEEE Trans. Ind. Electron.*, vol. 57, no. 2, pp. 716–726, Feb. 2010.

[5] P. M. Kelecy and R. D. Lorenz, "Control methodology for single inverter, parallel connected dual induction motor drives for electric vehicles," *in Proc. IEEE PESC'94*, 1994, pp. 987–991.

[6] A. Bouscayrol, M. Pietrzak-David, P. Delarue, R. Pena-Eguiluz, P. E Vidal, and X. Kestelyn, "Weighted control of traction drives with parallel-connected AC machines," *IEEE Trans. Ind. Electron.*, vol. 53, no. 6, pp. 1799–1806, Dec. 2006.

[7] K. Matsuse, Y. Kouno, H. Kawai, and S. Yokomizo, "A speed-sensorless vector control method of parallel-connected dual induction motor fed by a single inverter," *IEEE Trans. Ind. Appl.*, vol. 38, no. 6, pp. 1566–1571, Nov./Dec. 2002.

[8] D. Bidart, M. Pietrzak-David, P. Maussion, and M. Fadel, "Mono inverter multi-parallel permanent magnet synchronous motor: structure and control strategy," *IET Journal on Elec. Power Appl.*, vol. 5, no. 3, pp. 288–294, Mar. 2011.

[9] N. L. Nguyen, M. Fadel, and A. Llor, "A new approach to predictive torque control with dual parallel PMSM system," *in Proc. IEEE ICIT*, Cape Town, South Africa, Feb. 25-27, 2013.

[10] J. M. Lazi, Z. Ibrahim, M. Sulaiman, I. W. Jamaludin, and M. Y. Lada, "Performance comparison of SVPWM and hysteresis current control for dual motor drives," *in IEEE International Conference on Power and Energy (PECon)*, Kuala Lumpur, Malaysia, 2011, pp. 75-80.

[11] A. Del Pizzo, D. Iannuzzi, and I. Spina, "High performance control technique for unbalanced operations of single-VSI dual-PM brushless motor drives," *2010 IEEE International Symposium on Industrial Electronics (ISIE)*, pp. 1302-1307, Jul 2010.

[12] P. Janssens, W. Vanloock, G. Pipeleers, and J. Swevers, "An efficient algorithm for solving time-optimal point-to-point motion control problems," *in Proc. IEEE Int. Conf. Mechatronics*, pp. 682–687, Feb. 2013.

Performance Comparison of Inverter and Drive Configurations with Open-End and Star-Connected Windings

Markus Neubert, Stefan Koschik, Rik W. De Doncker
Institute for Power Electronics and Electrical Drives (ISEA), RWTH Aachen University, Germany

Abstract—**Electrical drives with open-end winding configurations offer certain advantages over drives with star-connected windings such as higher dc-link voltage utilization and lower inverter losses. However, there are a couple of drawbacks like circulating currents, diminishing the performance of the drive. This paper compares the performance of inverter and drive configurations with star-connected windings and open-end windings regarding efficiency, THD and torque ripple over the entire operation range of the drive. The most widely used B6C topology is compared to the H-bridge topology for open-end winding machines using different modulation schemes. Inverter losses as well as losses due to switching harmonics are taken into account. The comparison is carried out for an interior permanent magnet synchronous machine (IPMSM) with a nominal power of 60 kW.**

I. INTRODUCTION

In the scope of the german national funded research project EMiLE, interior permanent magnet synchronous machines (IPMSM) with open-end windings have been considered a viable choice for highly integrated drives. They offer certain advantages over drives with star-connected windings such as easier inverter integration, higher dc-link voltage utilization, potentially higher switching frequencies and lower inverter losses. However, there are a couple of issues described in publications such as circulating currents or bearing currents in the machine caused by zero-sequence components of the voltage [1]–[3]. Various modulation schemes have been described to overcome these problems [4]–[6]. However, most of the publications focus either on the performance of the machine or on the performance of the inverter. Obviously, the modulation scheme and the control strategy of the machine and the inverter have a significant influence on the overall system performance. Especially in open-end winding configurations, a larger total harmonic distortion (THD) of the current can lead to additional inverter and machine losses. This paper presents simulation results for the whole drive system consisting of an electrical machine and an inverter. The simulation is built in MATLAB/Simulink using the PLECS toolbox [7] to allow a detailed loss comparison of the different inverter topologies and modulation schemes.

II. INVERTER MODELLING

A. B6C Inverter

The B6C inverter is the most widely used topology for electrical machines. It consists of three half bridges connected

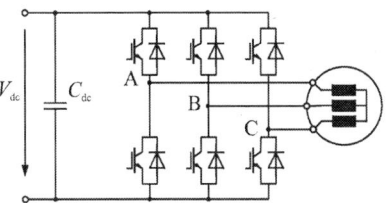

Fig. 1.   Schematic of the B6C inverter setup

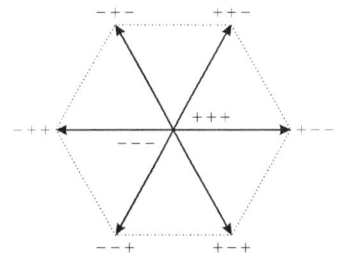

Fig. 2.   Space vector representation of the B6C inverter

to the machine (Fig. 1). The B6C inverter is controlled using space vector modulation (SVM). Figure 2 depicts the corresponding space vector representation. The switching states are given by the symbols next to the space vectors, where "+" denotes the on-state of the high-side IGBT and "-" the on-state of the low-side IGBT of the respective phase. The maximum peak value of the fundamental frequency of the phase voltage using space vector modulation without overmodulation equals $V_{dc}/\sqrt{3}$. At each instant in time only one semiconductor per phase is conducting. Due to the star connected windings of the machine, the sum of the phase currents is zero at every instant. Nevertheless, the semiconductors have to switch the whole dc-link voltage V_{dc}.

B. H-bridge Inverter

The H-bridge inverter is often used for drives with open-end windings. Figure 3 shows the schematic of the H-bridge inverter. Two half bridges are connected to each phase of the machine and thus twice the amount of switches is required compared to the B6C configuration. The maximum peak value of the phase voltage without overmodulation is equal to the dc-link voltage V_{dc}. This reduces the maximum phase current by a factor of $\sqrt{3}$ compared to the B6C inverter assuming equal electric input power. However, there are always two

978-1-4799-2706-7/14 $31.00 © 2014 IEEE

semiconductors conducting per phase. Hence, the switch utilization ratio (SUR) [8] of the H-bridge inverter is lower by a factor of $\sqrt{3}/2$ in comparison to the B6C inverter. The SUR is defined as the inverter's rms volt-ampere output at the fundamental frequency divided by the number of switches and their peak voltage and current ratings [8].

The H-bridge configuration offers the possibility to apply a zero voltage vector independently for each phase by turning on both high-side or low-side switches of one phase. The corresponding space vector representation is given in Fig. 4, where "+" denotes a positive and "-" a negative phase voltage. The zero voltage vectors are denoted by "0".

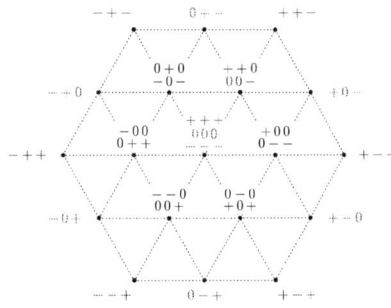

Fig. 3.   Schematic of the H-bridge inverter setup

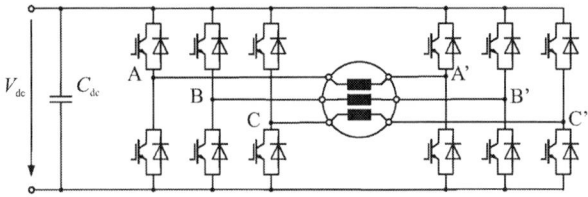

Fig. 4.   Space vector representation of the H-bridge inverter

To utilize the zero voltage vectors and to reduce the switching losses a 3-level sine-triangle modulation with flipped carrier can be used. Thereby, only one inverter leg of the H bridge is switched each period while the other is clamped dependent on the output voltage reference V_{ref}. Table I summarizes the switching states of this so-called clamped modulation, where T_{on} is the duty cycle of the high-side switch. The corresponding phase voltages of one exemplary switching period are shown in Fig. 5.

TABLE I
CLAMPED MODULATION FOR THE H-BRIDGE INVERTER

$V_{\text{ref}} \geq 0$	right inverter leg clamped to $0\,\text{V}$ left inverter leg switching, $T_{\text{on}} = V_{\text{ref}}/V_{\text{dc}}$
$V_{\text{ref}} < 0$	left inverter leg clamped to $0\,\text{V}$ right inverter leg switching, $T_{\text{on}} = V_{\text{ref}}/V_{\text{dc}}$

This simple modulation strategy reduces the switching losses of the H-bridge inverter. However, space vectors with zero-voltage sequences are used, and thus the condition for

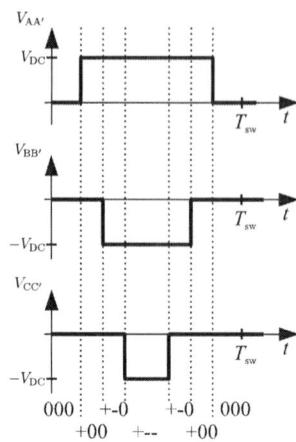

Fig. 5.   Phase voltages of H bridge using clamped modulation

the zero-voltage component V_0,

$$V_0 = \frac{V_{\text{AA}'} + V_{\text{BB}'} + V_{\text{CC}'}}{3} = 0, \tag{1}$$

is no longer applicable. Zero-voltage components cause circulating currents and an increased THD. To adress this issue, different modulation strategies have been proposed [4]–[6]. Theses modulation strategies avoid zero-voltage sequences by using space vectors without zero-voltage components only. The according space vectors are marked red in Fig. 4.

One of these modulation strategies is the zero common mode modulation (ZCMM). Exemplary voltage vectors as well as the according space vectors are given in Fig. 6(a). Using this modulation strategy the common mode voltages and thus the circulating currents can be cancelled, increasing the efficiency of the machine. However, one of the inverter phases is switched twice per switching period increasing the amount of switching instants by a factor of $1/3$.

To reduce zero-voltage sequences while maintaining the same amount of switching instants, a third modulation strategy, in the following called left aligned zero common mode modulation (LAZCMM), is implemented. Exemplary phase voltages and space vectors are given in Fig. 6(b).

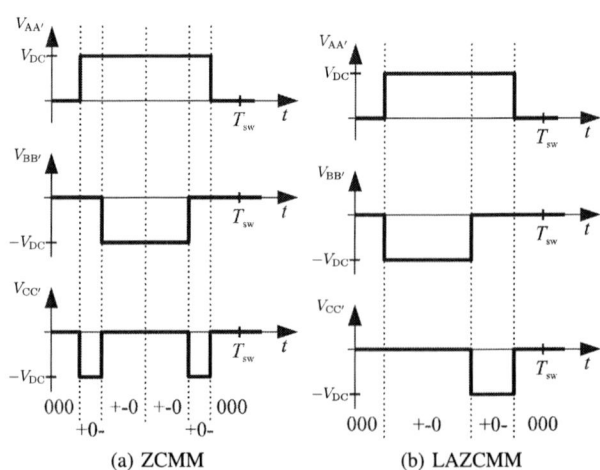

(a) ZCMM   (b) LAZCMM

Fig. 6.   Phase voltages of H bridge using zero common mode modulations

The 2014 International Power Electronics Conference

While the amount of switching instants is reduced, the phase voltages are no longer symmetric to $T_{sw}/2$. Assuming that the currents are measured at $T_{sw}/2$, the sampled values do not necessarily equal the average current of the respective switching period.

Beside the modulation strategy, the deadtime also contributes to common mode voltage [3]. Therefore, algorithms to compensate the deadtime are implemented separately for each modulation strategy.

The parameters of the different inverter topologies are given in Table II, where f_{sw} is the switching frequency of the inverter, $V_{s,rms,max}$ is the maximum rms phase voltage and $I_{s,rms,max}$ the maximum rms phase current.

TABLE II
INVERTER PARAMETERS

Topology	$V_{s,rms,max}$	$I_{s,rms,max}$	f_{sw}	Deadtime
B6C	230.9 V	360.8 A	10 kHz	1 µs
H bridge	400 V	208.4 A	10 kHz	1 µs

III. DRIVE MODELING

The drive is modeled as a three-phase machine with mutual coupling to allow independent control of each phase as described in [9]. The model is based on fundamental waveforms and does not include flux harmonics or cogging torque. The machine parameters for the reference design of the B6C inverter are given in Table III.

TABLE III
MACHINE PARAMETERS FOR REFERENCE DESIGN WITH B6C

Parameter	Value
Nominal voltage	400 V
Nominal power	60 kW
Peak power	80 kW
Peak torque	250 Nm
Number of pole pairs	4
$R_{s,B6C}$	0.015 Ω
$L_{d,B6C}$	0.384 mH
$L_{q,B6C}$	0.561 mH

For the H-bridge configuration, the characteristic values of the machine are scaled with a factor of $k = \sqrt{3}$ to accommodate the different maximum rated phase voltages. Parameters that have to be adapted are nominal phase current $I_{s,max}$, phase inductance L_{dq} in dq-axis, phase resistance R_s and flux linkage Ψ_f from permanent magnet:

$$I_{s,max,H} = \frac{I_{s,max,B6C}}{k} \qquad (2)$$

$$L_{dq,H} = L_{dq,B6C} \cdot k^2 \qquad (3)$$

$$R_{s,H} = R_{s,B6C} \cdot k^2 \qquad (4)$$

$$\Psi_{f,H} = \Psi_{f,B6C} \cdot k \qquad (5)$$

The control strategy of the machine is synthesized from the inductance values and the permanent magnet flux of the machine. Figure 7 shows the current locus of the IPMSM. In the base speed region the machine is operated at maximum torque per ampere (MTPA). Above base speed, constant torque operation is achieved by keeping the current on the maximum flux (MF) linkage ellipse [10]. For higher speeds the machine is operated along the maximum ampere (MA) circle and the maximum torque per flux (MTPF) line, respectively. This results in the operation area defined by the points P_0 to P_3 given in Fig. 7. The drive is operated in field weakening mode at speeds above 3000 rpm.

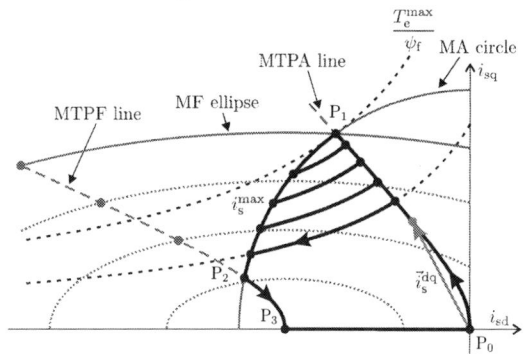

Fig. 7. Current locus of the IPMSM

IV. SIMULATION

When simulating electrical machines and power electronics as a whole the simulation time increases significantly due to their different time constants.

To address this issue, the simulation is carried out in two steps: at start up, the input signals of the model of the electrical machine, i.e. the voltage waveforms, are assumed to be ideal. Using ideal waveforms requires less computation time and allows the simulation to quickly reach the steady state. Once the steady state is reached, the inverter loss model built in PLECS [7] is enabled and substitutes the idealized waveforms at the input of the machine model. Thus, the controller only needs to adjust the differences between the idealized and the simulated waveforms. Consequently, the steady state is reached after a few periods which significantly reduces the simulation time. To further reduce the simulation time and to limit the amount of data only the last period of the fundamental waveform is evaluated.

Furthermore, a feed forward control was implemented to compensate dead time effects and zero current clamping [11]. The input values of the controller, i.e. the phase currents and the rotor position θ_m, are sampled at the center of each switching period. The calculated duty-cycles are forwarded to the inverter with a time delay of half the switching period.

A. Zero Current Ripple

Figure 8 shows the maximum peak to peak value of the zero current ripple for the different inverter setups and modulation strategies. The zero current ripple $I_{s,0}$ is defined as

$$I_{s,0} = I_{s,A} + I_{s,B} + I_{s,C}, \qquad (6)$$

where I_s is the phase current of the corresponding phase.

The 2014 International Power Electronics Conference

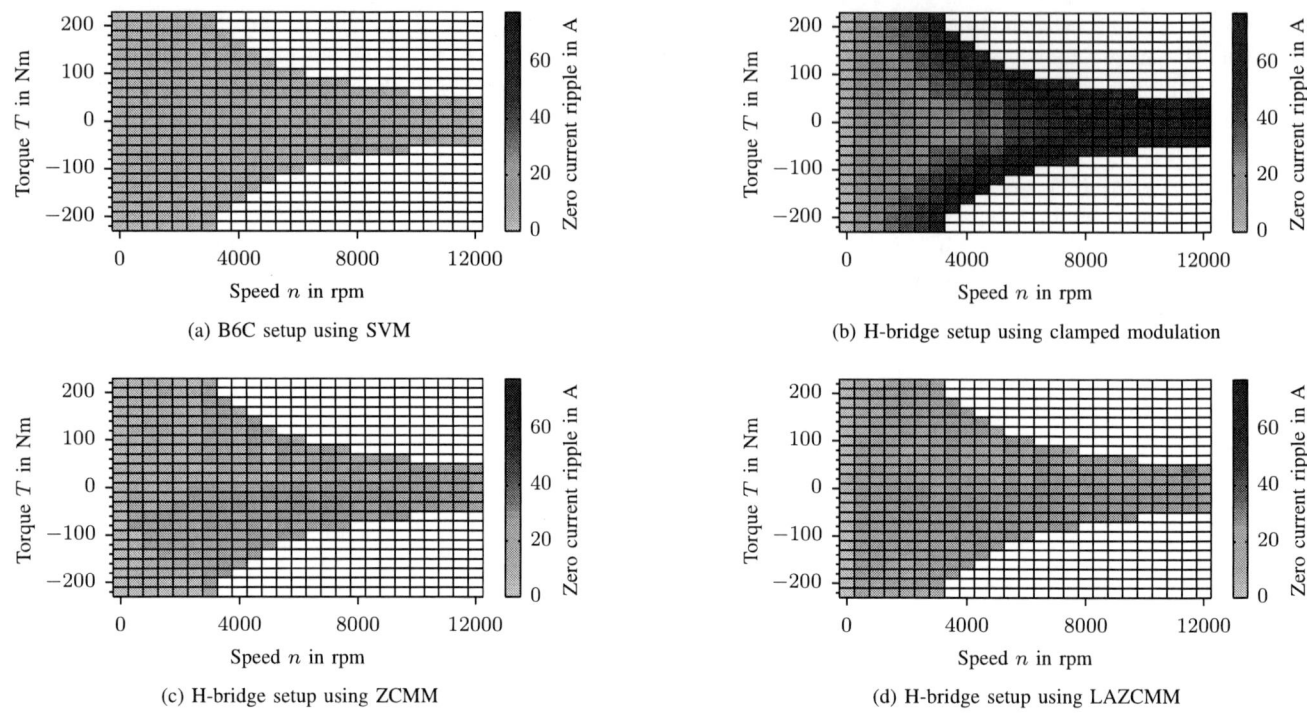

(a) B6C setup using SVM

(b) H-bridge setup using clamped modulation

(c) H-bridge setup using ZCMM

(d) H-bridge setup using LAZCMM

Fig. 8.   Maximum peak to peak value of the zero current ripple for the whole operating range at $V_{dc} = 400\,\text{V}$, $f_{sw} = 10\,\text{kHz}$

For the B6C inverter the zero current component is zero for the whole operating range, see Fig. 8(a). Due to the missing star-connection of the H-bridge inverter, the condition $I_{s,0} = 0$ cannot be guaranteed. A zero current regulator has to actively balance the phase currents in such a way that on average the zero current is zero. The ripple amplitude of the zero currents solely depends on the stray inductance of the machine. The stray inductance in PMSM machines is generally between 20% to 40% of the main inductance, leading to large current ripples.

As shown in Fig. 8(b) the zero current ripple of the H-bridge inverter using the clamped modulation is up to 75 A. Naturally, the maximum peak to peak value of the zero current increases with the duty-cycle and thus the output power. This results in an increased rms value of the phase currents, especially at high speed and low torque.

From Fig. 8(c) and 8(d) the influence of the zero common mode modulation becomes obvious: the zero current in the H-bridge inverter is reduced significantly. However, the zero current components cannot be eliminated completely. One of the main reasons are undesired deadtime effects, e.g. as a consequence of zero crossings of the current. These zero crossings lead to errors in the deadtime compensation and thus result in zero currents. Therefore, a zero current regulator is needed in addition to the zero common mode modulations. Due to the uniformly distributed phase voltages of the ZCMM (cf. Fig. 6), the ZCMM is less prone to deadtime effects than the LAZCMM resulting in slightly less zero currents.

B. THD

The THD of the phase currents in the electrical machine has a large effect on the conduction losses of the inverter. Higher

harmonics in the current waveform produce increased losses without contributing to the torque generation in the machine. The total harmonic distortion is defined as

$$\text{THD} = \frac{\sqrt{\sum_{m=2}^{\infty} I_{s,m}^{2}}}{I_{s,1}}, \qquad (7)$$

where $I_{s,m}$ is the amplitude of the harmonic m in the phase current [8]. The resulting THD for the entire operating range is shown in Fig. 9. At $n = 0$ the THD is not defined and thus is left out.

The THD of the H-bridge topology using the clamped modulation is significantly higher compared to the B6C configuration, especially for low torque. The reason for the higher THD can be found in the circulating currents due to zero-sequence components of the voltage (cf. Fig. 8).

Using the ZCMM to avoid the zero-sequence currents (cf. Fig. 8), the THD of the H-bridge inverter can be reduced significantly, see Fig. 9(c). However, the THD for the LAZCMM shown in Fig. 9(d) is higher than for the clamped modulation. The reason has been given earlier: the sampled values of the currents are undefined and lead to distortions in the phase currents. At high speeds the THD of the LAZM modulation is even higher than the THD of the clamped modulation, which compensates the advantage of the reduced zero currents.

C. Inverter Losses

The losses of the semiconductors are analyzed based on datasheet parameters [12]. To create a fair basis for the comparison, the same total silicon area is used for both topologies,

978-1-4799-2706-7/14 $31.00 © 2014 IEEE   3148

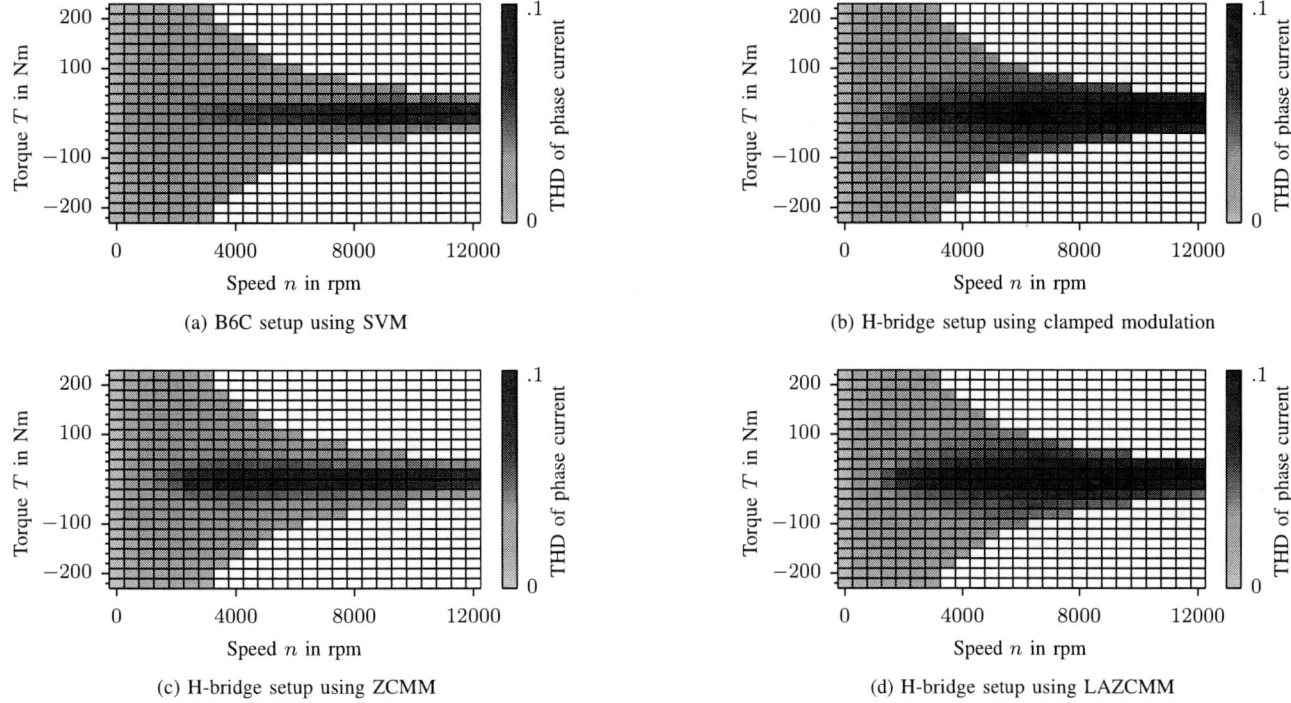

(a) B6C setup using SVM

(b) H-bridge setup using clamped modulation

(c) H-bridge setup using ZCMM

(d) H-bridge setup using LAZCMM

Fig. 9. THD of phase current for the entire operating range at $V_{dc} = 400\,V$, $f_{sw} = 10\,kHz$

the H bridge and the B6C. Thus, every switch of the B6C utilizes twice the silicon area compared to the switches of the H bridge. This apparently results in an oversizing of the B6C bridge regarding the currents and the SUR, respectively (cf. Table II). Nevertheless, the increased silicon area is needed due to loss dissipation as shown in the following. Figure 10 shows the efficiency for the entire range for both inverter setups and the different modulation strategies. For a better readability, the efficiency is restricted to a minimum of 90 %. In case of positive torque, the efficiency is referred to the electrical input power P_{el}, $\eta = {}^{P_{mech}}/P_{el}$. In contrast, the efficiency at negative torque, i.e. generator mode, is referred to the mechanical output power P_{mech}, $\eta = {}^{P_{el}}/P_{mech}$. The lower electric loading results in a decreased inverter efficiency of related operating points.

As expected, the efficiencies increase towards higher torque and higher speed. Comparing Fig. 10(a) and Fig. 10(b), an advantage of the H-bridge setup with open-end windings becomes obvious: the efficiency is higher in every operating point. This applies especially for light load condition ($n = 500\,rpm$, $T = 20\,Nm$), where the efficiency of the H-bridge setup is 7.4 % higher compared to the B6C setup. According to the datasheet, the conduction losses at high load increase faster than the switching losses, which diminishes the efficiency of the H-bridge inverter. Moreover, the H-bridge inverter suffers from a higher THD at high speeds (cf. Fig. 9). Consequently, the difference of efficiency of the B6C and the H-bridge inverter becomes smaller. At $n = 12000\,rpm$ the difference is negligible.

The inverter efficiencies of the H-bridge setup using zero common mode modulations are given in Fig. 10(c) and

Fig. 10(d) for the ZCMM and the LAZCMM, respectively. As a result of the increased amount of switching instants the efficiency of the ZCMM decreases significantly compared to the other H-bridge modulation schemes. Nevertheless, the efficiency at low load ($n = 500\,rpm$, $T = 20\,Nm$) is still 4.2 % higher than the efficiency of the B6C setup. According to the other H-bridge modulation schemes, the efficiency at higher loads converges towards the one of the B6C setup. At maximum load the efficiency is slightly lower compared to the B6C setup.

From Fig. 10(d) and Fig. 10(b) the efficiency of the H-bridge setup using LAZCMM is almost equal to the efficiency using the clamped modulation. For low loads, the zero currents within the inverter and thus the efficiency are roughly the same for both modulations (cf. Section IV-A). At high load, where the circulating currents become more predominant, the efficiency of the H-bridge inverter using the LAZCMM is expected to be higher. Nevertheless, the undefined sampling points of the phase currents of the LAZCMM increase the THD at high speeds. Both effects compensate each other. At high loads and low speeds, where the THD of both modulations is the same, the lower circulating currents result in an increased efficiency of 0.5 % when using the LAZCMM. However, the difference in the efficiency is relatively small. Thus, as a result of the simulation principle, where only the last fundamental period is evaluated (cf. Section IV), zero crossings of the currents may lead to small deviations. For a more detailed analysis, the loss distribution at the boundary to the field weakening operation at $n = 3000\,rpm$ is depicted in Fig. 11. For all setups the losses continuously rise with increasing torque and thus increasing current. The

The 2014 International Power Electronics Conference

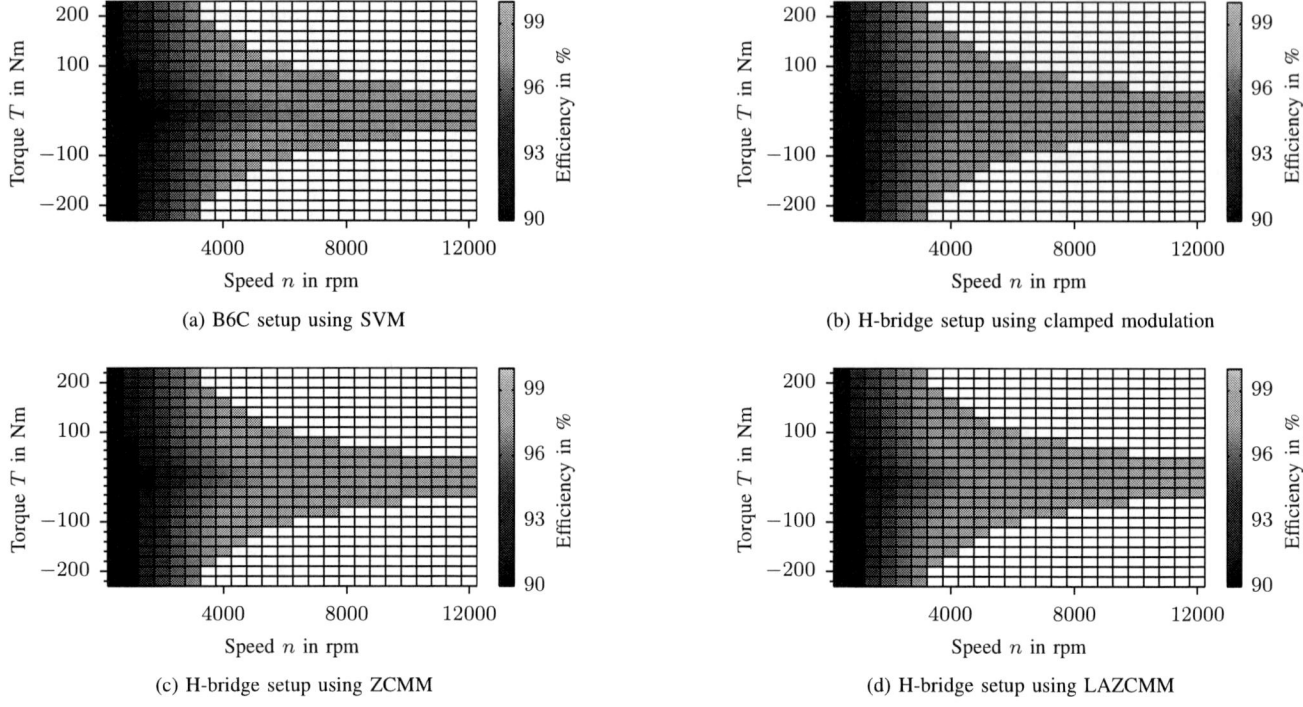

Fig. 10. Efficiency for the entire operating range at $V_{\mathrm{dc}} = 400\,\mathrm{V}$, $f_{\mathrm{sw}} = 10\,\mathrm{kHz}$

distribution of the conduction losses among the diodes and switches depends on the power factor of the electrical drive. Furthermore, the diodes conduct during the deadtime and thus the diode conduction losses depend on the deadtime. At negative torque, the conduction time of the diodes is higher, increasing their conduction losses.

With regard to the B6C inverter setup, the maximum losses of $728\,\mathrm{W}$ correspond to an efficiency of $96.7\,\%$. For both, diodes and switches, the turn-off losses significantly contribute to the overall losses. The loss distribution using an open winding configuration and an H-bridge inverter is given in Fig. 11(b). Obviously, the switching losses of the H bridge are significantly lower than those of the B6C (cf. Fig. 11(a)). Switching losses are proportional to voltage and current:

$$E_{\mathrm{sw}} \sim V_{\mathrm{sw}}, I_{\mathrm{sw}} \tag{8}$$

Thus, the switching losses of the H-bridge inverter are lower by a factor of approximately $1/\sqrt{3}$ (cf. Table II), making the H bridge an attractive choice for high speed drives. Nevertheless, the conduction losses are higher than the ones of the B6C as each semiconductor utilizes only half the silicon area but conducts $1/\sqrt{3}$ of the current. Hence, the conduction losses are increased by a factor of $2/\sqrt{3}$. Even though the B6C exhibits a higher SUR compared to the H bridge, the overall losses of the B6C setup are higher. With respect to an equal loss dissipation, this would require an even higher silicon area for the B6C setup. According to the loss distribution, a major difference in the design process of both setups becomes apparent: the H-bridge inverter should be optimized for conduction losses while the design process of the B6C inverter should focus on reducing the switching losses.

From Fig. 11(c), the maximum losses of the ZCMM are $748\,\mathrm{W}$, being higher than those of the B6C inverter. According to the modulation scheme, the amount of switching instants of the ZCMM and consequently the switching losses are increased. However, the switching losses do not increase by a factor of $4/3$ as one might expect. Considering the fact that the inverter phase with the smallest duty-cycle, i.e. the lowest voltage reference, is switched twice, the increase in switching losses strongly depend on the power factor of the machine. At $n = 3000\,\mathrm{rpm}$, $T = 220\,\mathrm{Nm}$ the switching losses are increased by a factor of 1.42.

Despite the reduction of the zero currents, the conduction losses of the ZCM and the clamped modulation are the same. On the one hand, the zero currents at $n = 3000\,\mathrm{rpm}$ only slightly contribute to the overall rms value of the phase currents. At $n = 3000\,\mathrm{rpm}$, $T = 220\,\mathrm{Nm}$ the rms value of the zero current is around $11\,\mathrm{A}$. The corresponding reference value of the phase current equals $200\,\mathrm{A}$. Assuming that both currents are orthogonal the total rms value of the phase current i_{rms} can be calculated by geometrically adding the rms values:

$$i_{\mathrm{rms}} = \sqrt{(200\,\mathrm{A})^2 + (11\,\mathrm{A})^2} \approx 200.3\,\mathrm{A} \tag{9}$$

Obviously, the influence of the zero current on the rms value of the phase current is negligible at $n = 3000\,\mathrm{rpm}$ and high torque. On the other hand, unpredicted zero crossings of the currents and the transient behavior of the regulator may slightly distort the waveforms and thus the rms currents. At higher speeds, the zero currents of the clamped modulation are up to one third of the rms value of the phase currents, and hence, significantly increase the conduction losses. Indeed, there are a couple of effects, like the transient switching

The 2014 International Power Electronics Conference

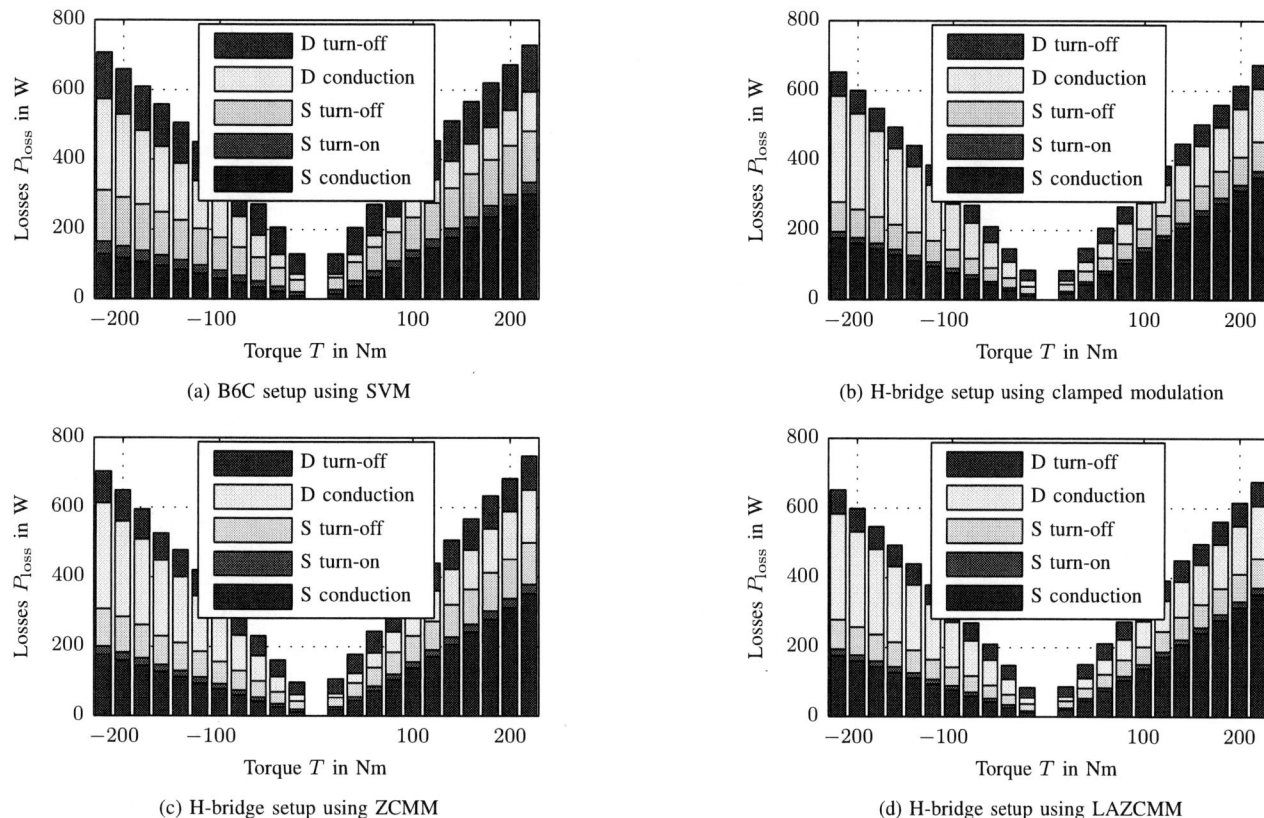

(a) B6C setup using SVM

(b) H-bridge setup using clamped modulation

(c) H-bridge setup using ZCMM

(d) H-bridge setup using LAZCMM

Fig. 11. Loss distribution per phase at $n = 2000\,\mathrm{rpm}$, $f_{\mathrm{sw}} = 10\,\mathrm{kHz}$, $V_{\mathrm{dc}} = 400\,\mathrm{V}$

behavior of semiconductors or the jitter of (gate-)signals, that contribute to the zero currents. Those effects cannot be sufficiently covered by the simulation. Therefore, the actual contribution of the zero currents to the rms value of the phase current is expected to be higher than given by the simulation. The LAZCMM offers lower switching losses than the ZCMM, see Fig. 11(d). Moreover, the above mentioned distortions in the phase currents due to the undefined sample point are not distinctive at $n = 3000$, cf. Fig. 9(d). Hence, the overall losses at $n = 3000$ are slightly lower than those of the clamped modulation. Nevertheless, the differences are neglectible.

D. Torque Ripple

Figure 12 shows the torque ripple for the entire operating range. The torque ripple is caused by harmonics in the flux linkage, i.e. in the phase currents. A fundamental waveform based model of the drive is used. Therefore, torque ripple components are solely produced by inverter switching actions and the current regulator.

Overall, switching induced air gap torque ripple is much lower in the B6C configuration. There, all three phases are coupled electrically and magnetically while the phases in the H-bridge inverter are coupled magnetically only. The electric coupling in the B6C inverter due to the star connection of the windings leads to five voltage levels $\pm^{1}/_{3}V_{\mathrm{dc}}$, $\pm^{2}/_{3}V_{\mathrm{dc}}$ and $0\,\mathrm{V}$ which are switched during one cycle. Using center aligned modulations, the voltage time areas are divided into

seven parts. This results in a low current ripple. For the H-bridge inverter, only full dc-link voltage and zero voltage can be applied to the phase. For each phase, only one voltage time area is applied per switching cycle, Fig. 5, leading to higher current ripple. With increasing load and therefore increasing duty cycles, also the current ripple increases. However, at higher machine speeds, the current waveforms around the zero crossing become smoother due to the increased back EMF of the machine. This in turn improves the current ripple at higher speeds for the same load level.

Using the H-bridge setup, the modulation strategy and the arrangement of the space vectors within one switching cycle have a significant influence on the torque ripple. The clamped modulation and the ZCMM perform equally well. For the ZCMM one of the phases is switched twice per cycle, decreasing the current ripple amplitude. However, the phase with the smallest duty cycle is switched twice. Therefore, the improvement in current ripple amplitude is negligible.

For the LAZCMM, torque ripple increases towards higher loads and higher speeds. While the currents of the LAZCMM are sampled center aligned, the left aligned voltage areas according to Fig. 6(b) result in an undefined sample point regarding the rms value of the according switching cycle. Therefore, the current regulator operates on current values which oscillate around the average. This infers a ripple in the equivalent dq currents, which results in a periodic torque ripple.

978-1-4799-2706-7/14 $31.00 © 2014 IEEE

The 2014 International Power Electronics Conference

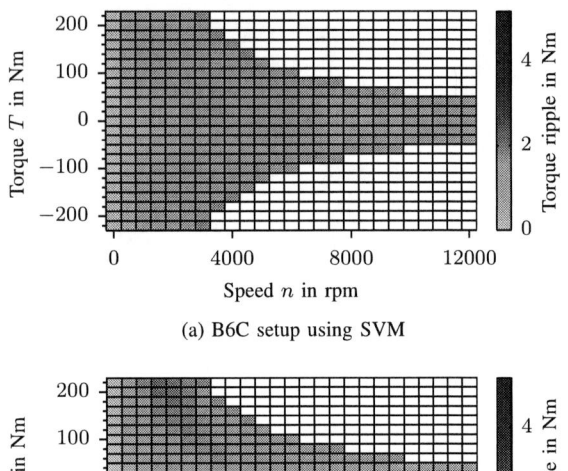

(a) B6C setup using SVM

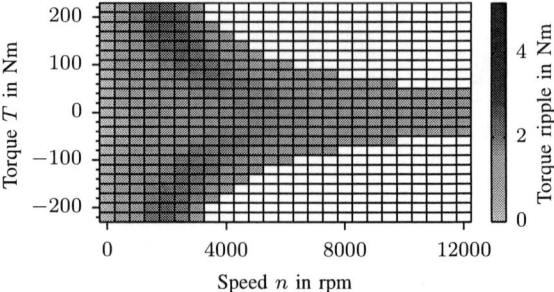

(b) H-bridge setup using clamped modulation

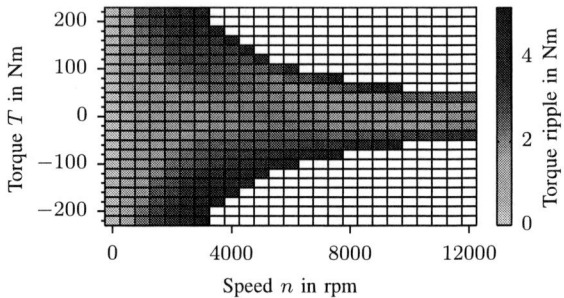

(c) H-bridge setup using ZCMM

(d) H-bridge setup using LAZCMM

Fig. 12. Torque ripple for the entire operating range at $V_{\mathrm{dc}} = 400\,\mathrm{V}$, $f_{\mathrm{sw}} = 10\,\mathrm{kHz}$

V. CONCLUSION

This paper gives a detailed analysis of the common B6C and H-bridge setup for electrical drives. For the H-bridge setup an electrical machine with open-end windings is required, resulting in different drive configurations. Moreover, several H-bridge modulation schemes have been evaluated to address the issue of zero-voltage sequences. To achieve a fair comparison, the same silicon area has been used for both inverter setups.

It has been found that the H bridge suffers from an increased zero current ripple and THD due to zero-voltage sequences. Despite using zero common mode modulations, the zero currents could not be avoided completely. Moreover, there are a couple of drawbacks using these zero common mode modulations: depending on the modulation strategy either the inverter losses or the THD and the torque ripple are increased significantly. However, it has been shown, that the open end winding configuration is superior in terms of inverter efficiency.

The project underlying this report was funded by the Federal Ministry of Education and Research, project number 16EMO0009. The responsibility for the content of this publications lies with the authors.

SPONSORED BY THE

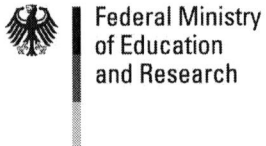

Federal Ministry
of Education
and Research

REFERENCES

[1] S. Chen, T. A. Lipo, and D. Fitzgerald, "Source of induction motor bearing currents caused by pwm inverters," *Energy Conversion, IEEE Transactions on*, vol. 11, no. 1, pp. 25–32, 1996.

[2] J. M. Erdman, R. J. Kerkman, D. W. Schlegel, and G. L. Skibinski, "Effect of pwm inverters on ac motor bearing currents and shaft voltages," *Industry Applications, IEEE Transactions on*, vol. 32, no. 2, pp. 250–259, 1996.

[3] A. Somani, R. K. Gupta, K. K. Mohapatra, and N. Mohan, "On the causes of circulating currents in pwm drives with open-end winding ac machines," *Industrial Electronics, IEEE Transactions on*, vol. 60, no. 9, pp. 3670–3678, 2013.

[4] Ratnayake, K. R. M. N. and Y. Murai, "A novel pwm scheme to eliminate common-mode voltage in three-level voltage source inverter," in *Power Electronics Specialists Conference, 1998. PESC 98 Record. 29th Annual IEEE*, vol. 1, 1998, pp. 269–274 vol. 1.

[5] H. Zhang, A. Von Jouanne, S. Dai, A. Wallace, and F. Wang, "Multilevel inverter modulation schemes to eliminate common-mode voltages," *Industry Applications, IEEE Transactions on*, vol. 36, no. 6, pp. 1645–1653, Nov 2000.

[6] M. R. Baiju, K. Mohapatra, R. S. Kanchan, and K. Gopakumar, "A dual two-level inverter scheme with common mode voltage elimination for an induction motor drive," *Power Electronics, IEEE Transactions on*, vol. 19, no. 3, pp. 794–805, May 2004.

[7] Plexim GmbH, "Plecs," http://www.plexim.com.

[8] Mohan, Undeland, Robbins, *Power Electronics Converters, Applications, and Design*. Wiley, 2003.

[9] Novotny D.W., Lipo T.A., *Vector Control and Dynamics of AC Drives*. Oxford University Press, 1996.

[10] R. U. Lenke, R. W. De Doncker, Mu-Shin Kwak, Tae-Suk Kwon, and Seung-Ki Sul, "Field weakening control of interior permanent magnet machine using improved current interpolation technique," in *Power Electronics Specialists Conference, 2006. PESC '06. 37th IEEE*, 2006, pp. 1–5.

[11] Jong-Woo Choi and Seung-Ki Sul, "New dead time compensation eliminating zero current clamping in voltage-fed pwm inverter," in *Industry Applications Society Annual Meeting, 1994. Conference Record of the 1994 IEEE*, 1994, pp. 977–984 vol. 2.

[12] Infineon Technologies AG, "Datasheet: F4-75R06W1E3."

978-1-4799-2706-7/14 $31.00 © 2014 IEEE   3152

Input Current Harmonics Reduction Control for Electrolytic Capacitor Less Inverter Based IPMSM Drive System

Kodai Abe
Nagaoka University of Technology
1603-1 Kamitomioka, Nagaoka
Niigata, JAPAN, 940-2188
abe_nagaoka@stn.nagaokaut.ac.jp

Kiyoshi Ohishi
Nagaoka University of Technology
1603-1 Kamitomioka, Nagaoka
Niigata, JAPAN, 940-2188
ohishi@vos.nagaokaut.ac.jp

Hitoshi Haga
Nagaoka University of Technology
1603-1 Kamitomioka, Nagaoka
Niigata, JAPAN, 940-2188
hagah@vos.nagaokaut.ac.jp

Abstract—**This paper proposes a current harmonics reduction method to improve the input current waveform of electrolytic capacitor less single-phase to three-phase power converters. Typically, the back back electromotive force (EMF) on interior permanent magnet (IPM) motor is not sinusoidal and contains harmonics caused by the rotor structure; therefore, it causes harmonic distortion in the motor control system and generates harmonics in the input current. This paper analyzes the cause of the input current harmonics aims to the back EMF and the d-q axis current controller. This paper proposes two control methods to reduce the current harmonics distortion. The first method filters out the harmonics of feedback d-q axis current by using harmonics filter. The second method compensates the d-q axis voltage references to reduce the input current harmonics. The d-q axis compensation voltages are obtained by feed forward controller. The maximum power factor of the proposed method obtains 98.47 %. The experimental results confirm that the proposed control method clears the guideline EN61000-3-2.**

Keywords—*electrolytic capacitor less inverter, interior permanent magnet synchronous motor, input current harmonic, compensation voltage*

I. INTRODUCTION

In a residential air-conditioner application, controlling the compressor with a variable-speed motor drive would allow overall system optimization that could significantly reduce the energy consumption [1]-[3]. However, the inverter driven variable-speed motor system requires improving the power quality of the AC sources. Recently, the home electrical appliances apply power factor correction (PFC) circuit in the rectifier. The purpose of PFC is to obtain high power factor, and to achieve the sinusoidal current control in the source side.

The conventional PFC technique requires the use of electrolytic capacitors, reactors and switching devices. The electrolytic capacitors in the converter circuit occupy a large volume; this prevents miniaturization of the system and limits its lifetime. Reactors and switching devices significantly contribute toward an increase in power loss, weight of the system, and its cost. In order to further conserve energy and resources of the compressor drive system, PFC circuits has demanded the reduction of size, weight, and high efficiency [4]-[9].

The authors have proposed electrolytic capacitor less single-phase to three-phase power converter for the compressor driven system. The proposed system consists of a single-phase diode rectifier, small film capacitor, three-phase voltage source inverter, IPM motor, and the converter does not have huge energy storage. The previous report in [10], the high power factor operation has obtained at any load condition. However, there are many current harmonics at the source side; the system could not clear the guideline EN61000-3-2. The previous report in [11], the harmonics current reduction method has proposed. However, the control method also could not clear the guideline EN61000-3-2.

This paper proposes two new control method to reduce the input current harmonics of the proposed power converter. Firstly, this paper analysis a cause of the increase of the input current harmonics in the proposed power converter. In order to overcome this problem, this paper proposes control methods to reduce the current harmonics distortion. The first method filters out the harmonics of feedback d-q axis current by using harmonics filter. The second method compensates the d-q axis voltage references to reduce the input current harmonics. The d-q axis compensation voltages are obtained by feed forward controller. The experimental results confirm that the proposed method improves the input current waveform and keeps the high power factor control.

II. ELECTROLYTIC CAPACITOR LESS INVERTER AND INVERTER OUTPUT POWER CONTROL

Fig. 1 shows a proposed power converter, which consists of a single-phase diode rectifier, small film capacitor, a three-phase voltage-source inverter, and IPM motor [10]. It consists of a few energy storage elements, and the source side power ripple is smoothed by the moment of inertia of the IPM motor. The motor speed is regulated averagely because of the moment of inertia in the motor. There are many torque ripples at the IPM motor. Hence, the proposed power converter applies to the compressor drive system. Small film-capacitor is used to absorb the DC-link current ripple due to the PWM of the three-phase inverter.

Fig. 2 shows the control block diagram of the inverter output power controller. As the high power factor

978-1-4799-2706-7/14 $31.00 © 2014 IEEE

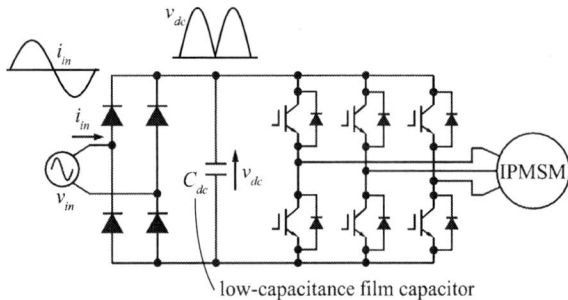

Fig. 1. Proposed control method

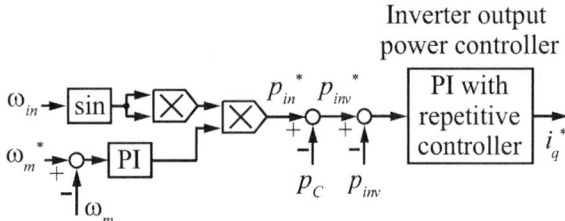

Fig. 2. Control block diagram of inverter output power control

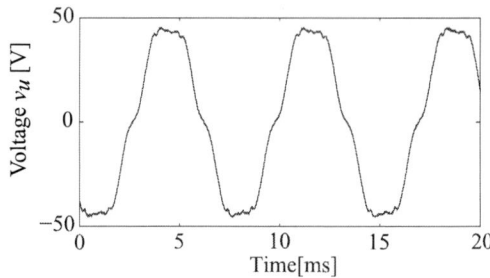

Fig. 3. Back EMF waveform of tested IPM motor

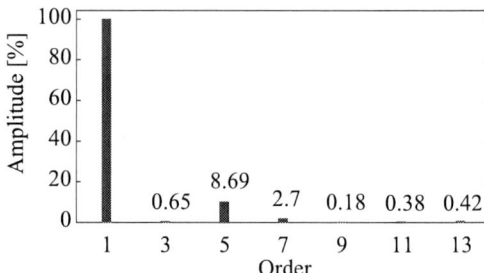

Fig. 4. Frequency analysis of back EMF

operation for the proposed power converter, the inverter output power control is available. In order to obtain high power factor, the inverter output power controls to ripple with twice the synchronized the source voltage frequency. The inverter output power reference p^*_{inv} is calculated by subtracting the compensation power of the DC-link capacitor from the input power reference p^*_{in}. The input power reference p^*_{in} is calculated by multiplying the output of the speed PI controller with $\sin 2\omega_{in}t$ generated from the input voltage. p_c is DC-link capacitor power calculated by v_{in}. The difference between p^*_{inv} and $Ppinv$ is the inverter output power error. This error is controlled by the inverter output-power controller, and its output is the q-axis current command i^*_q.

III. CURRENT HARMONICS OF PROPOSED POWER CONVERTER

Typically, a conventional inverter system having large electrolytic capacitor and a PFC circuit can control input current waveform. The PWM switching pattern in the inverter and the motor side harmonics does not influence to the source side current waveform because of the large electrolytic capacitor. However, the proposed electrolytic capacitor less power converter is affected by the motor side harmonics because of small DC-link capacitor. This chapter clarifies the influence of motor side harmonics on the input current harmonics.

A. Spatial Harmonics of IPM Motor

A back EMF of the IPM motor has spatial harmonics with the amplitude and frequency, proportional to motor speed, because the shape of the rotor has not a sine wave. Fig. 3 shows the back EMF waveform of the tested IPM motor. Fig. 4 shows FFT analysis result of the back EMF.

The largest harmonic content is the fifth followed by the seventh.

B. Harmonics current in Motor Control System

The fifth harmonics included in the three-phase current i_{u5}, i_{v5}, i_{w5} can be expressed as,

$$\begin{bmatrix} i_{u5} \\ i_{v5} \\ i_{w5} \end{bmatrix} = \omega_e \psi_c \begin{bmatrix} \sin 5\theta_e \\ \sin(5(\theta_e - \frac{2}{3}\pi)) \\ \sin(5(\theta_e + \frac{2}{3}\pi)) \end{bmatrix} \quad (1)$$

Where, ω_e is the rotor electrical angular speed [rad/s], θ_e is the motor position [rad], and ψ_c is back EMF coefficient. The current harmonics in the d-q axis frame can be expressed as

$$\begin{bmatrix} i_{dh} \\ i_{qh} \end{bmatrix} = \sqrt{\frac{2}{3}}[C] \begin{bmatrix} i_{u5} \\ i_{v5} \\ i_{w5} \end{bmatrix}$$

$$= \sqrt{\frac{3}{2}}\omega_e \psi_c \begin{bmatrix} \sin 6\theta_e \\ \cos 6\theta_e \end{bmatrix} \quad (2)$$

Where,

$$[C] = \begin{bmatrix} \cos \theta_e & \cos(\theta_e - \frac{2}{3}\pi) & \cos(\theta_e + \frac{2}{3}\pi) \\ -\sin \theta_e & -\sin(\theta_e - \frac{2}{3}\pi) & -\sin(\theta_e + \frac{2}{3}\pi) \end{bmatrix} \quad (3)$$

Hence, the harmonics current appears as the sixth harmonics in the d-q axis current controller.

IV. PROPOSED CONTROL METHOD

A. Harmonics Reduction Method using Harmonics Filter

Fig. 5 shows the control block diagram of the first method. The proposed method filters out the harmonics of feedback d-q axis current by notch filter. In order to apply the filter to the d-q axis frame, the filter is designed for the (n+1)th harmonics as follows.

$$f_{h(n+1)}[\text{Hz}] = \frac{(n+1)p}{60}\omega_m \tag{4}$$

Where, n is the targeted harmonic order number, ω_m is the motor speed in rpm, and p is pole number. In this paper, the largest harmonic content of tested motor is the fifth. Hence, the filter is designed to sixth order.

B. Harmonics Reduction Method using Compensation Voltages

Fig. 6 shows the control block diagram of the second method. The proposed method compensates the d-q axis voltage references to reduce the input current harmonics. The d-q axis compensation voltages v_{dcmp} and v_{qcmp} are obtained by feed forward controller. The compensation voltages v_{dcmp} and v_{qcmp} are calculated by using fol-

lowing equations.

$$v_{dcmp} = K_d \cdot \sin((n+1)\theta_e + \phi_d) \tag{5}$$
$$v_{qcmp} = K_q \cdot \cos((n+1)\theta_e + \phi_q) \tag{6}$$

Where, n is the targeted harmonic order number. In this paper, the compensation voltage is designed for the fifth harmonics shown in Fig. 4. Hence, n is set at 5. ϕ_d and ϕ_q are phase differences between electrical rotor position θ_e obtained from the encoder. K_d and K_q are the gain of the compensation voltages. In this paper, ϕ_d, ϕ_q, K_d and K_q are obtained by off-line experimental test.

C. Proposed Control Block Diagram and Design of d-q axis Current Controller

Fig. 7 shows the diagram of the entire circuit with the proposed control method using the harmonics filter and compensation voltages. The currents i_d and i_q are obtained by coordinate conversion from the detected currents i_u and i_w, respectively. The harmonics filter of the proposed method filters out harmonics current of i_d and i_q. These currents are subtracted from the reference currents i_d^* and i_q^*, and each current is regulated by the PI controller. The compensation voltages of the proposed method are added to the output of the PI current regulator.

In this paper, the d-q axis proportional gain K_{dp} and K_{qp} in the PI current regulator are designed as follows,

$$K_{dp} = L_d\omega_c \tag{7}$$
$$K_{qp} = L_q\omega_c \tag{8}$$

Where, ω_c is cut-off frequency of the current controller. K_{dp} is not same as K_{qp} because of L_d and L_q. The harmonic content of the motor current passes this current controller and becomes the inverter voltage references. The harmonics of d-q axis voltage references v_{dh}^*, v_{qh}^*

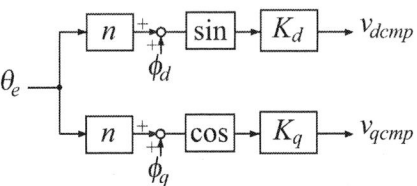

Fig. 5.   Control block diagram of harmonics filter

Fig. 6.   Control block diagram of compensation voltage

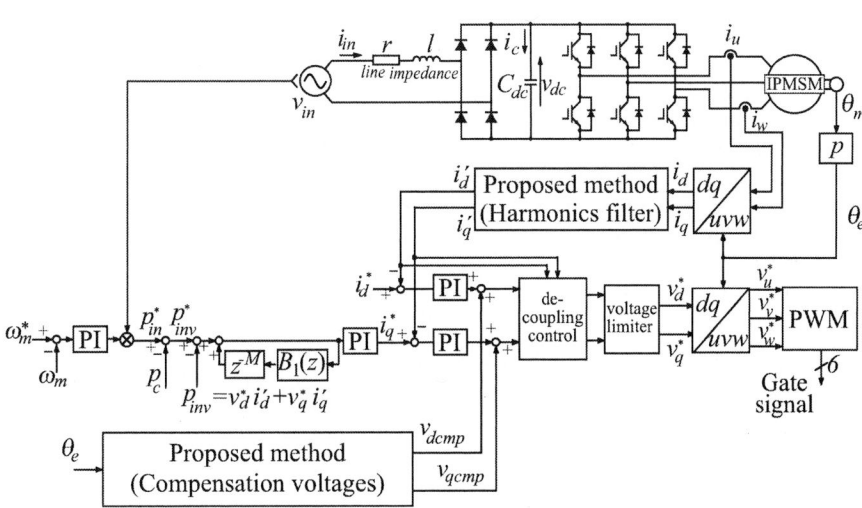

Fig. 7.   System configuration of proposed control method

are calculated by using Eq. (2), Eq. (7) and Eq. (8).

$$\begin{bmatrix} v_{dh}^* \\ v_{qh}^* \end{bmatrix} = \sqrt{\frac{3}{2}}\omega_e\psi_c \begin{bmatrix} K_{dp}\sin 6\theta_e \\ K_{qp}\cos 6\theta_e \end{bmatrix}$$

$$= \sqrt{\frac{3}{2}}\omega_e\omega_c\psi_c \begin{bmatrix} L_d\sin 6\theta_e \\ L_q\cos 6\theta_e \end{bmatrix} \quad (9)$$

Then, the harmonics in the u-v-w axis frame can be expressed as,

$$\begin{bmatrix} v_{uh}^* \\ v_{vh}^* \\ v_{wh}^* \end{bmatrix} = \sqrt{\frac{2}{3}}[D] \begin{bmatrix} v_{dh}^* \\ v_{qh}^* \end{bmatrix}$$

$$= \frac{1}{2}a \begin{bmatrix} (L_d+L_q)\sin 5\theta_e \\ (L_d+L_q)\sin(5\theta_e+\frac{2}{3}) \\ (L_d+L_q)\sin(5\theta_e-\frac{2}{3}) \end{bmatrix}$$

$$+ \frac{1}{2}a \begin{bmatrix} (L_d-L_q)\sin 7\theta_e \\ (L_d-L_q)\sin(7\theta_e-\frac{2}{3}) \\ (L_d-L_q)\sin(7\theta_e+\frac{2}{3}) \end{bmatrix} \quad (10)$$

Where,

$$[D] = \begin{bmatrix} \cos\theta_e & -\sin\theta_e \\ \cos(\theta_e-\frac{2}{3}\pi) & -\sin(\theta_e-\frac{2}{3}\pi) \\ \cos(\theta_e+\frac{2}{3}\pi) & -\sin(\theta_e+\frac{2}{3}\pi) \end{bmatrix} \quad (11)$$

$$a = \omega_e\omega_c\psi_c \quad (12)$$

Eq. (10) shows that there are harmonics of the fifth and the seventh. The seventh harmonics caused by the difference between K_{dp} and K_{qp}. The harmonic content of the inverter output voltage increases with increase of the cut-off frequency of d-q axis current controller ω_c. The decoupling controller also causes the harmonics except the fifth. Hence, the proposed electrolytic capacitor less power converter should design the cut-off frequency of d-q axis current controller at low frequency.

V. Experimental Results

Table I lists the tested motor parameter. The experimental results confirm the proposed control method. The input voltage and frequency are 200 Vrms and 50 Hz, and the inverter carrier frequency is 16 kHz. The bandwidth of the current regulator is set to 1500 rad/s. The DC-link capacitance and the line impedance are set at 14 μF, 0.2 mH and 0.5 Ω, respectively. The motor speed is set at 4500 rpm, and the load torques is set at 1.8 Nm. In this paper, the inverter output voltages are synchronized to the input voltage. The targeted harmonics order is the fifth harmonics content, and the targeted harmonics order in d-q axis frame is the sixth. Therefore, the harmonics filter and compensation voltages are designed to the sixth harmonics content; when the motor speed is set at 4500rpm, the sixth harmonics is 900 Hz.

Fig. 8-10 show the experimental results of the conventional control method and the proposed control method. Fig. 11 shows the FFT analysis result of the input current in the proposed power converter.

Fig. 8 shows experimental results of the conventional control method. The control method uses only the inverter output power controller in Fig. 2, and the d-axis current

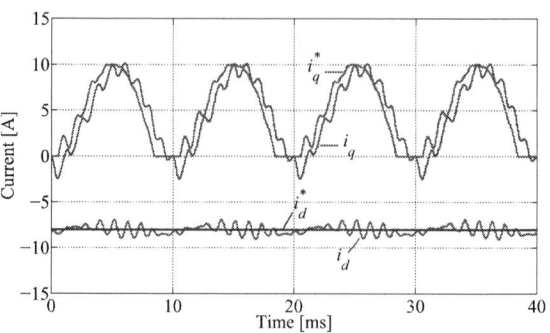

(a) d-q axis current

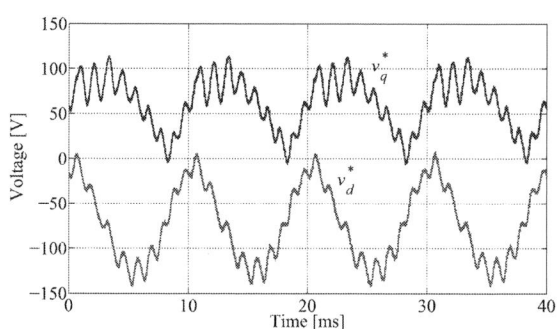

(b) d-q axis voltage references

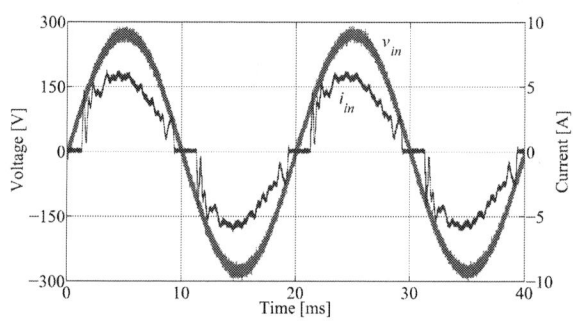

(c) Voltage and current waveform at source side

Fig. 8.   Conventional control method

reference is constant value. The conventional method causes the harmonics distortion on voltage references v_d^*, v_q^* and the input current.

TABLE I.   MOTOR PARAMETERS

Stator resistance R_a	0.615[Ω]
d-axis inductance L_d	7.1[mH]
q-axis inductance L_q	11.3[mH]
Linkage flux ϕ_a	0.124[Wb]
Pole number	4[pole]
Rated speed	4200[rpm]
Rated torque	1.8[Nm]

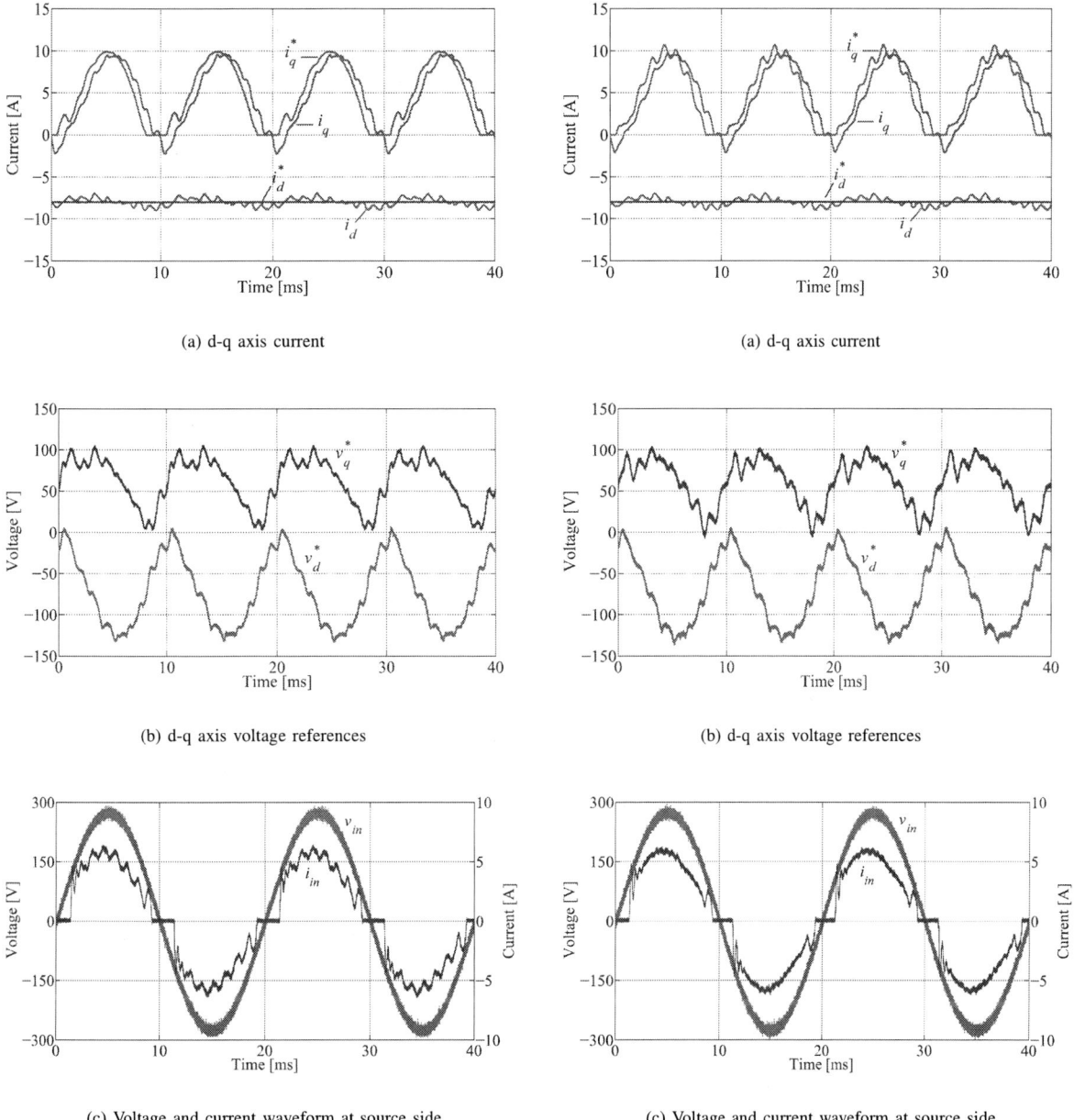

(a) d-q axis current

(b) d-q axis voltage references

(c) Voltage and current waveform at source side

Fig. 9. Proposed control method with harmonics filter

(a) d-q axis current

(b) d-q axis voltage references

(c) Voltage and current waveform at source side

Fig. 10. Proposed control method with harmonics filter and compensation voltages

Fig. 9 shows experimental results of the proposed control method with the harmonics filter. The harmonics filter is designed to reduce the target sixth order harmonic component in d-q axis frame. The first method reduces harmonic distortion of voltage references in Fig. 9. Therefore, the twenty-first harmonic of the input current due to voltage references distortion is reduced in Fig. 11. However, the influence of the harmonics in the motor side is clear, and the seventeenth and nineteenth harmonics of the input current are increased as shown Fig. 11.

Fig. 10 shows results of the proposed control method with the harmonics filter and the compensation voltages. The compensation voltages are designed for target sixth

order harmonic to reduce the input current harmonics due to the motor side harmonics. The second method reduces the input current harmonics in Fig. 10. These control methods clear the guideline EN61000-3-2 in Fig. 11.

Table II summarizes the input power factor and THD of the input current waveform. When the harmonics filter is applied, the twenty-first harmonic is reduced, but the seventeenth and nineteenth harmonics are increase as shown Fig. 11. Therefore, the power factor and THD with harmonics filter are worse than the conventional method. By combining the harmonics filter and compensation voltages, proposed methods improve the power factor and

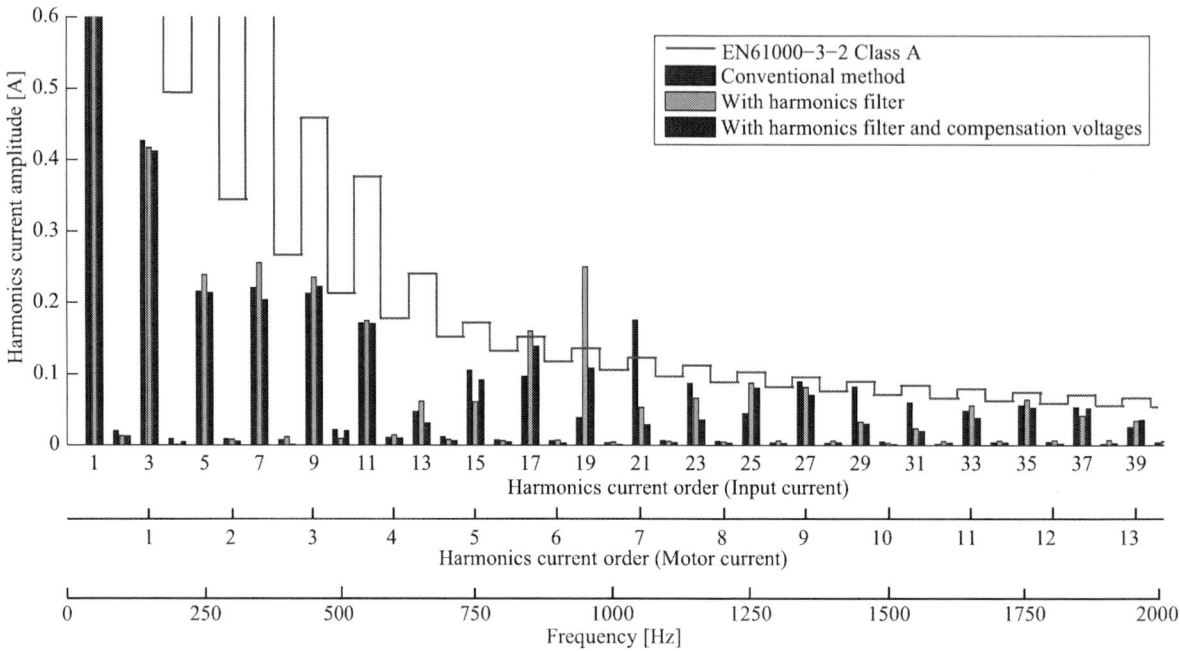

Fig. 11. FFT analysis result of the input current

TABLE II. INPUT POWER FACTOR AND THD OF THE INPUT CURRENT

	P.F. [%]	THD [%]
Conventional method	98.06	18.02
Proposed method with filter	98.01	18.79
Proposed method with filter, v_{dcmp} and v_{qcmp}	98.47	16.65

THD of the input current effectively.

VI. CONCLUSION

This paper proposes two new control methods to improve the input current waveforms of the electrolytic capacitor less power converter. Firstly, this paper analyzed a cause of the input current harmonics aims to the back EMF of the IPM motor and the d-q axis current controller. The proposed method reduces the harmonic content by the spatial harmonics of the motor. The proposed methods filter out the harmonics of feedback d-q axis current and compensate the d-q axis voltage references to reduce the input current harmonics. The d-q axis compensation voltages are obtained by feed forward controller. The effectiveness of the proposed method has been verified by experiments in comparison with the conventional meethod. These two control methods clear the guideline EN61000-3-2. The maximum power factor of the proposed method is 98.47 % and the THD of the input current is 16.65 %.

REFERENCES

[1] J. Gao and Y. Hu, "Direct Self-Control for BLDC Motor Drives Based on Three-Dimensional Coordinate System", *IEEE Trans. on Ind. Electron.*, vol. 57, no. 8, pp. 2836-2844, 2010.

[2] S. Shao, E. Abdi, and R. McMahon, "Low-Cost Variable Speed Drive Based on a Brushless Doubly-Fed Motor and a Fractional Unidirectional Converter", *IEEE Trans. Ind. Electron.*, vol. 59, no. 1, pp. 317-325 (2012)

[3] K. Kondo and H. Kubota, "Innovative Application Technologies of AC Motor Drive Systems", *IEEJ Journal of Industry Applications*, vol. 1, no. 3, pp. 132-140, 2012.

[4] M. K. H. Cheung, M. H. L. Chow, and C. K. Tse, "Design and Performance Considerations of PFC Switching Regulators Based on Noncascading Structures", *IEEE Trans. on Ind. Electron.*, vol. 57, no. 11, pp. 3730-3745, 2010.

[5] H. L. Cheng, Y. C. Hsieh, and C. S. Lin, "A Novel Single-Stage High-Power-Factor AC/DC Converter Featuring High Circuit Efficiency", *IEEE Trans. Ind. Electron.*, vol.58, no.2, pp.524-532 (2011)

[6] Z. Li, C. Y. Park, J. M. Kwon, and B. H. Kwon, "High-Power-Factor Single-Stage LCC Resonant Inverter for Liquid Crystal Display Backlight", *IEEE Trans. Ind. Electron.*, vol.58, no.3, pp.1008-1015 (2011)

[7] C. Larouci, T. Azib, A. Chaibet and M. Boukhnifer, "Control of a Flyback Converter in Mixed Conduction Mode: Influence on the Converter Design Using Optimization under Constraints", *IEEJ J. Ind. Appl.*, vol. 2, no. 3, pp. 132-140 (2013)

[8] M. Daniele, P. K. Jain, and G. Joos, "A single-stage power-factor corrected AC/DC converter", *IEEE Trans. Power Electron.*, vol. 14, no. 6, pp. 1046-1055, Nov. 1999.

[9] P. J. Grbovic, P. Delarue, P. Le Moigne, and P. Bartholomeus, "A Three-Terminal Ultracapacitor-Based Energy Storage and PFC Device for Regenerative Controlled Electric Drives", *IEEE Trans. on Ind. Electron.*, vol. 59, no. 1, pp. 301-316 (2012)

[10] K. Inazuma, H. Utsugi, K. Ohishi, and H. Haga, "High Power Factor Single-phase Diode Rectifier Driven by Repetitive Controlled IPM Motor", *IEEE Trans. on Industrial Electron.*, vol. 60, no. 10, pp. 4427-4437, 2013.

[11] H. Utsugi, K. Ohishi, and H. Haga., "Reduction in Current Harmonics of Electrolytic Capacitor-less Diode Rectifier using Inverter-Controlled IPM Motor", *The 38rd Annual Conference of the IEEE Industrial Electronics Society* (IECON), pp. 6210-6215, 2012.

Noncontact Guide System for Traveling Elastic Steel Plates
(Theoretical Study on the Shape of Traveling Steel Plate)

Kouichi Sakaba
Tokai University
Undergraduate, Dep. of Prime Mover Eng.
4-1-1, Kitakaname,Hiratsuka-shi,Kanagawa,Japan

Takayoshi Narita
Tokai University
Graduate Student, Course of Science & Tech.
4-1-1, Kitakaname,Hiratsuka-shi,Kanagawa,Japan

Shinya Hasegawa
Tokai University
Dep. of Prime Mover Eng.
4-1-1, Kitakaname,Hiratsuka-shi, Kanagawa,Japan

Yasuo Oshinoya
Tokai University
Dep. of Prime Mover Eng.
4-1-1, Kitakaname,Hiratsuka-shi,Kanagawa,Japan
ossy@keyaki.cc.u-tokai.ac.jp

Abstract- **In the factory, the continuous thin steel plates are subjected to iron and steel processes are supported by a series of rollers during processes such as plating and rolling. However, because the rollers come in contact with the steel plates, the problem of surface quality deterioration arises. To solve this problem, we developed a non-contact guide system for parts of the steel plate at which its traveling direction changes in high-speed traveling by applying an electromagnetic force from the direction of the edge of the thin steel plate, and experimentally examined the effectiveness of the system. However, the asymmetry traveling steel plate during uncontrolled is not calculated theoretically. In this study, the shape of traveling steel plate is calculated with numerical analysis. It is confirmed that, as we compared analysis result with experiment result, both have similar tendency.**

Keywords— Analysis, Electromagnet, Traveling steel plate

I. INTRODUCTION

These days, users demand high quality, high added value as well as high grade product manufactured by the continuous steel plates process. In the factory, the continuous thin steel plates are subjected to iron and steel processes are supported by a series of rollers during processes such as plating and rolling. However, because the rollers come in contact with the steel plates, the problem of surface quality deterioration arises. To find methods to solve this problem, research and development on the noncontact conveyance system of belt steel using a floater, whereby a belt steel is floated using a fluid, have been carried out. However, problems in the occurrence of wind ripples and irregularities on the surface of steel plates in processes such as plating are unavoidable. Therefore, there is a solving method for these problems. H. Matsumoto and T. Nakagawa performed a research on

vibration suppression using electromagnet of a stationary long steel plate [1,2]. It is an application of the magnetism technology where fluid is inessential. Researchers in our laboratory have examined noncontact edge control by applying an electromagnetic force near the edge of a continuous steel plate traveling along a straight line in order to suppress its vibration [3]. We also confirmed the effectiveness of the noncontact guide that is used to change the traveling direction of the traveling continuous steel plate and that has functions similar to those of sink rolls, on the vibration suppression of the steel plate. However, to improve the surface quality, a more stable vibration suppression performance of the actual equipment is required in processes such as drying and plating. In the previous studies, we have confirmed that the loop shape of the traveling steel plate, whose thickness is 0.2mm, under the condition without control changes at the section where the traveling direction of the steel plate changes. Then we constructed the guideway using electromagnets taking into consideration the loop shape, and examined the effect of the difference in the guideway shape on vibration suppression [4]. However, it is not yet possible to calculate the shape of the traveling steel plate at the time of uncontrolled theoretically. In this study, our aimed is to present the basic data in order to determine the placement position of the electromagnet which could anticipate control on vibration effect by obtaining numerical analysis of the shape of traveling steel plate.

II. NONCONTACT GUIDE SYSTEM

As shown in Fig. 1(a), in the plating process for continuous steel plates in a production line, a steel plate is contact-supported by a deflector or sink rolls. In this study, we aim to form a noncontact guide path using electromagnets at the roll sections as shown in Fig. 1(b). We prepared an experimental setup that simulates the plating bath through which the continuous steel plate passes, as shown in Fig. 2. Fig. 3 is the photograph of noncontact guide system. A continuous steel plate produced by welding a quenched steel plate (SUS632) with the dimensions of 6,894 mm (length) × 150 mm (width) × 0.2 mm (thickness) into a belt shape is suspended by a pulley with a diameter of 700 mm and a width of 154 mm. The width of the steel plate is set at 150 mm in this experiment to reduce deformation and elastic vibration in the direction perpendicular to the traveling direction on the plane of the steel plate and to clarify the effects of the edge control system on loop-shaped sections. The pulley is driven by a DC servomotor, enabling endless traveling of the continuous steel plate. The electromagnets were placed at the three positions (Nos. 1, 2, and 3) indicated in the figure.

x_1, x_2, and x_3 indicate the coordinates of the traveling direction of the steel plate at the three electromagnet positions. Similarly, y_1, y_2, and y_3 indicate the coordinates perpendicular to the traveling direction and on the same plane as the steel plate at the three electromagnet positions (hereafter, this direction is called the y direction). z_1, z_2, and z_3 indicate the coordinates perpendicular to the plane of the steel plate at the three electromagnet positions (hereafter, this direction is called the z direction). Electromagnets are pressed in place from inside the system, using two acrylic plates and anti-vibration pads against the two plates outside the continuous steel plate. The electromagnets can be placed arbitrarily along the surface of the acrylic plates, as shown in Fig. 4.

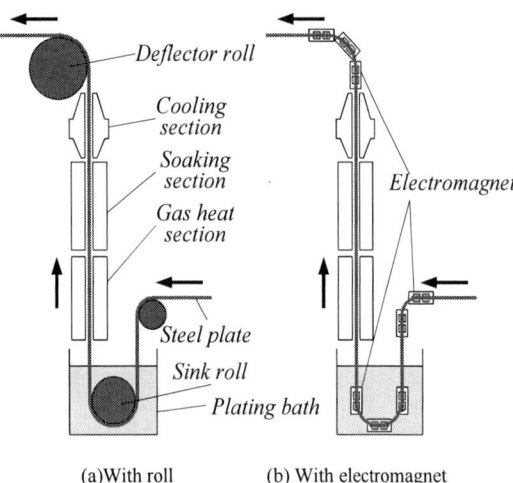

Fig. 3. Photograph of noncontact guide system.

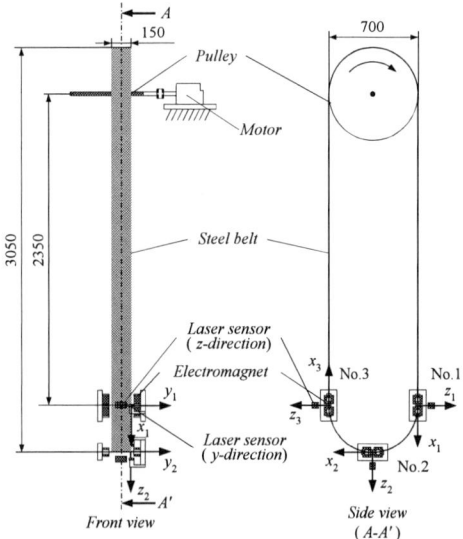

Fig. 1. Manufacturing process of continuous steel plates.

Fig. 4. Photograph of arrangement of electromagnets.

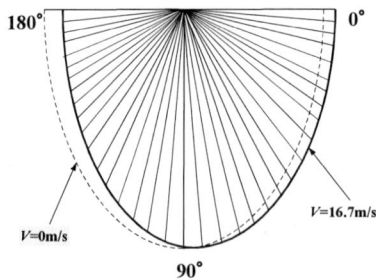

Fig. 2. Schematic illustration of noncontact guide system.

Fig. 5. Experimental result of steel plate shape.

III. THIN STEEL SHAPE ANALYSIS

. If a steel plate was traveling continuously, the form of the loop portion of a steel plate became asymmetrical by right, as shown in Fig. 5 [5]. If an electromagnet was arranged according to this traveling form, it was confirmed that the vibration suppression effect of a traveling steel plate improved [4]. Here, the form of a traveling steel plate is computed in finite-element-method analysis using Solidworks Simulation of Dassault System.

According to the friction theory of Euler [6], the formula for the effective tension T_e of a belt is as follows.

$$T_e = T_t - T_s \tag{1}$$

T_e : Effective tension[N] T_t : Strain side tention[N]
T_s : Slack side tention[N]

The formula which asks for the strain side tension T_t of a belt is as follows.

$$T_t = \frac{T_e e^{\mu\theta}}{e^{\mu\theta} - 1} + mV^2 \tag{2}$$

The formula which asks for the slack side tension T_s of a belt is as follows.

$$T_s = \frac{T_e}{e^{\mu\theta} - 1} + mV^2 \tag{3}$$

V : Steel plate traveling speed[m/s] μ : The coefficient of friction between a steel plate and a pulley θ : The angle to twine[rad] m : Mass per 1m of the steel plate[kg/m]

In this study, T_s and T_t are calculated using a formula (2) and (3). After inputting T_s and T_t into model of the stillness steel plate form acquired by experiment, steel plate form was computed in analysis. Analysis result is shown in Fig. 6. Graph in Fig. 7 shows the analysis result and experimental value of the loop form portion when the steel plate is traveling at a speed of 0m/s and 16.7m/s. The loop portion moved to the right and became asymmetrical form by right and left. In Fig. 7, the dotted line which represents the experiment result displaced around 10mm from the standstill shape. On the other hand, solid line which represent analysis result displaced around 40mm. There was an error between an experiment result and analysis result, but we were able to confirm that both tendency becoming similar.

IV. CONCLUSIONS

In this study, verification regarding changes of asymmetrical traveling form in the loop portion of a traveling continuation steel plate was performed from numerical analysis. As a result, we could assume that after changes, there is a possibility that the form becoming close to the form

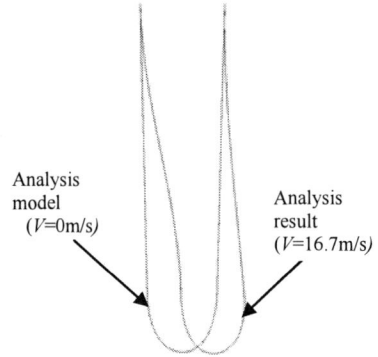

Fig. 6. Analysis result of steel plate shape.

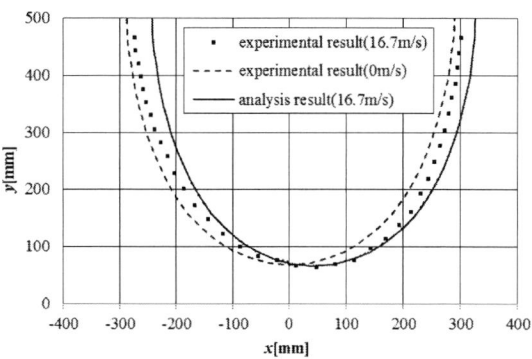

Fig. 7. Compare experimental result with analysis result.

acquired by experiment. From the provided study analysis, as we arranged electromagnet for the traveling shape of the traveling steel plate, improvement of vibration control can be expected. For our future task, we will perform a control experiment using traveling shape obtained by the analysis and check the effectiveness of this proposal.

REFERENCES

[1] H. Matsumoto, Y. Okada, et al., "Active Vibration Control of Thin Steel Sheet," *Transactions of the Japan Society of Mechanical Engineers*, Series C, vol.65, no. 630, pp. 96-102, 1999.

[2] T. Nakagawa, M. Hama and T. Furukawa, "Study of Magnetic Levitation Technique Applied to Steel Plate Production Line," *IEEE Transactions on Magnetics*, vol. 36, no. 5, pp. 3686-3689, 2000.

[3] K. Kashiwabara, et al., "Noncontact Guide System for Elastic Steel Plate Process (Fundamental Considerations on High-Speed Traveling)," *Journal of Magnetics Society of Japan*, 29-3, pp. 326-331, 2005.

[4] T. Narita, et al., "Noncontact Guide System for Traveling Elastic Steel Plates(Fundamental Considerations on Asymmetric Placement of Electromagnets)," *ICEE*, 2012.

[5] T. Hashimoto, et al., "Noncontact Guide System for Traveling Elastic Steel Plate (Fundamental Considerations on Shape of Traveling Steel Plate)," *MAGDA Conference in Sendai*, pp. 321-323, 2012.

[6] T. Koyama, Practical design of belt transmission, YOKENDO, 2006, pp. 21-25.

Active Seat Suspension for Ultra-Compact Vehicle (Fundamental Consideration on Electromyogram When Fall from the Bump)

Masahiro Mashino
Tokai University
Undergraduate, Dep. of Prime Mover Eng.
4-1-1 Kitakaname, Hiratsuka-shi, Kanagawa, Japan

Masaki Ishida
Tokai University
Graduate Student, Course of Mech. Eng.
4-1-1 Kitakaname, Hiratsuka-shi, Kanagawa, Japan

Keita Sunaga
Tokai University
Graduate Student, Course of Mech. Eng.
4-1-1 Kitakaname, Hiratsuka-shi, Kanagawa, Japan

Hideaki Kato
Tokai University
Graduate Student, Course of Science & Tech.
4-1-1 Kitakaname, Hiratsuka-shi, Kanagawa, Japan

Shinya Hasegawa
Tokai University
Dep. of Prime Mover Eng.
4-1-1 Kitakaname, Hiratsuka-shi, Kanagawa, Japan

Yasuo Oshinoya
Tokai University
Dep. of Prime Mover Eng.
4-1-1 Kitakaname, Hiratsuka-shi, Kanagawa, Japan
ossy@keyaki.cc.u-tokai.ac.jp

Abstract— In these days automobile industry, the request for ultra-compact electric vehicles has been increasing in Japan. Unlike conventional automobiles, such ultra-compact electric vehicles may frequently travel on roads of poor condition, such as narrow and unpaved roads. These roads have bumps and many small obstacles, and therefore, the ride comfort of the vehicles is expected to deteriorate. Therefore, it is concerned regarding lack of visibility and the head protection functions. We suggest active seat suspension to improve the ride comfort by the disorder road run as a high value-added product. The active seat suspension carried voice coil motor (VCM) which is a kind of a linear motor as actuator. The objective of this study is to inspect utility and the effectiveness of the active seat suspension. In this paper, we focused on the neck electricmyogram and seat acceleration when fall from the bump.

Keywords— Active Seat Suspension, Ride Comfort, Ultra Compact Electric Vehicle, Electromyogram

I. INTRODUCTION

In the current automobile industry, the demand for ultra-compact electric vehicles has been increasing in Japan. Unlike conventional automobiles, such ultra-compact electric vehicles may frequently travel on roads of poor condition, such as narrow and unpaved roads. These roads have bumps and many small obstacles, and therefore, the ride comfort of the vehicles is expected to deteriorate. Therefore, even in bumpy roads and gravels, system construction for safe traveling and comfortable drive is essential. We suggest active seat suspension to

improve the ride comfort by the bad road traveling as a high value-added product. The active seat suspension carried voice coil motor (VCM) which is a kind of a linear motor as actuator. We have previously examined the practicality of our active seat suspension, focusing on ride comfort [1], evaluated the ride comfort by sensor test, and examined the psychological factors that affect the ride comfort [2]. Moreover, to carry out traveling experiments outdoors, the control system and other units were installed into a vehicle, and the effect of the active seat suspension on the suppression of vibration of the traveling on stone paths was confirmed [3]. We examined the effect of the vibration restraint performance by the change in driver's weight with an ultra-compact electric car in a gravel road [4], [5]. We examined the effect of the vibration restraint performance by the change in driver's weight such as car sharing with an ultra-compact electric vehicle in a gravel road.

Furthermore, the elderly population has been increasing in Japan. Therefore, it has become increasingly important to ensure self-supported daily and social lives for the elderly. To this end, improving the availability and safety of transportation for the elderly has become increasingly important. Kamata *et al.* have reported a series of studies on transportation devices that can be easily used by the elderly, and they concluded that ultra-compact electric vehicles are more suitable transportation devices for the elderly [5]. However, Akamatsu also mentioned about ride contentment and its quality concerning some issues of new mobility technology such as ultra-compact electric vehicles as we

978-1-4799-2706-7/14 $31.00 © 2014 IEEE   3162

imply especially for the elderly regarding their body disability such as their lack in stability visual field and head support function. In the terms of cognitive function, the new mobility for the elderly ought to have vibration reduction preferably. Corresponding to this, the necessity for vehicle body construction and seat structure are mentioned [6]. Thus, if elderly people continue to use the ultra-compact electric vehicle, it is important to reduce driver's discomfort. Until now, Koizumi *et al.* have reported that the behavior of the head and upper body in the low frequency band is associated with feeling [7]. However, although subjective evaluation of the head behavior and ride comfort was made, the objective evaluation using biological signals is not yet performed. Therefore, we conduct the objective evaluation by using biological signals. Since we noticed that the stable condition of the head is supported by the neck muscles, we considered that reduction of burden in neck muscles helps reducing driver's discomfort. In addition, we confirmed that the bump of roadway boundary is one of the worst situations when traveling on road. In this paper, we focused on the electromyogram (EMG) of sternocleidomastoid (SCM) muscle and seat acceleration when fall from bump roadway boundary.

II. Experimental Setup

Figure 1 shows the ultra-compact electric vehicle used in our experiment. Figure 2 shows the active seat suspension installed into the seat of the automobile. An aluminum plate was used in the driver's seat, and the seat was supported by four coil springs and allowed to vibrate only in the vertical direction via a linear slider. A voice coil motor (VCM) that enables high-accuracy and high-speed control was adopted for the control actuator. This provides the merit of direct-drive maintenance-free control. Table I shows specifications for the vehicle and VCM.

III. System

Figure 3 shows the control system of the active seat suspension. Absolute displacements and velocities of the seat surface used for control are detected by the digital integration of the signals from the accelerometer, shown in the figure, using a computer. Furthermore, the current flowing through the VCM is detected, the control voltage is calculated from the measured values by a computer, and a control force is generated by driving the VCM. As explained above, the active seat is supported by coil springs, linear sliders, and a VCM installed in parallel to these components. No cushion is used, and it is assumed that the vibration of the seat surface is directly transmitted to the driver and that the driver and the seat move as one. The resonance frequencies of the chassis and seat suspensions in this setup are relatively close to each other. Therefore, the coupled vibration of these suspensions is considered to be negligible in this study. Thus, the active seat suspension is modeled as a one-degree-of-freedom system, in which the part above the floor is the target of control, as shown in Fig. 4, rather than a multiple-degree-of-freedom system that comprises the chassis suspension

Fig. 1. Photograph of ultra-compact electric vehicle.

Fig. 2. Photograph of active seat suspension.

and other components. The equation of motion is given as

$$m\ddot{y} + c\dot{y} + ky = Ki \tag{1}$$

Moreover, the VCM used in this experiment consists of two parts: a magnetic circuit including a magnet, and a coil, which generates a force. The VCM does not include a linear bearing to support the coil's motion. The circuit equation of the VCM is as follows.

$$L\dot{i} + Ri + K\dot{y} = v \tag{2}$$

Equations (1) and (2) are written as the following state equations;

$$\dot{\boldsymbol{y}}_s = \boldsymbol{A}_s \boldsymbol{y}_s + \boldsymbol{B}_s v \tag{3}$$

where

$$\boldsymbol{y}_s = \begin{bmatrix} y & \dot{y} & i \end{bmatrix}^{\mathrm{T}} \tag{4}$$

$$\boldsymbol{A}_s = \begin{bmatrix} 0 & 1 & 0 \\ -\dfrac{k}{m} & -\dfrac{c}{m} & \dfrac{K}{m} \\ 0 & -\dfrac{K}{L} & -\dfrac{R}{L} \end{bmatrix} \tag{5}$$

$$\boldsymbol{B}_s = \begin{bmatrix} 0 & 0 & \dfrac{1}{L} \end{bmatrix}^{\mathrm{T}} \tag{6}$$

Here, m is the sum of the masses of the driver (weight applied to the seat when the driver sits on the seat, excluding the weight of his/her legs or other parts) and the seat [kg], k is the spring constant (sum of four springs) [N/m], c is the apparent damping coefficient considering the friction of the linear slider and other factors [N/m], y is the absolute displacement of the seat[m], i is the control current [A], L is the coil inductance of the VCM [H], R is the coil resistance [Ω], K is the thrust constant of the actuator [N/A], and v is the input voltage [V].

IV. CONTROL METHOD AND CONTROL EXPERIMENT

A. Control method

In this study, a control system is constructed using a discrete time system; therefore, the evaluation function of a continuous system is digitized, and the optimal control law is obtained based on the optimal control theory of the discrete time system. Here, the following discrete time system is considered.

$$y_d(j+1) = \boldsymbol{\Phi}\, y_d(j) + \boldsymbol{\Gamma} v_d(j) \tag{7}$$

$$\boldsymbol{\Phi} = \exp(A_s T_s), \quad \boldsymbol{\Gamma} = \int_0^{T_s} \left[\exp(A_s \tau)\right] d\tau\, B_s$$

Here, the evaluation function of the discrete time system is expressed as follows.

$$J_d = \sum_{i=0}^{\infty} \left[y_d(j)^{\mathrm{T}} \boldsymbol{Q}_d y_d(j) + r_d v_d(j)^2 \right] \tag{8}$$

$$\boldsymbol{M} = \boldsymbol{\Phi}^{\mathrm{T}} \boldsymbol{M}\boldsymbol{\Phi} + \boldsymbol{Q}_d \\ - \boldsymbol{\Phi}^{\mathrm{T}} \boldsymbol{M}\boldsymbol{\Gamma}\left(r_d + \boldsymbol{\Gamma}^{\mathrm{T}} \boldsymbol{M}\boldsymbol{\Gamma}\right)^{-1} \boldsymbol{\Gamma}^{\mathrm{T}} \boldsymbol{M}\boldsymbol{\Phi} \tag{9}$$

$$v_d = -\boldsymbol{F}_d y_d \tag{10}$$

$$\boldsymbol{F}_d = \left(r_d + \boldsymbol{\Gamma}^{\mathrm{T}} \boldsymbol{M}\boldsymbol{\Gamma}\right)^{-1} \boldsymbol{\Gamma}^{\mathrm{T}} \boldsymbol{M}\boldsymbol{\Phi} \tag{11}$$

Where $\boldsymbol{Q}_d$ and r_d are weighting coefficients, $\boldsymbol{M}$ is the solution of the algebraic matrix, the Riccati equation, and T_s is a sampling interval. MATLAB command "lqrd" was used to solve eq. (9) and the digital controller was designed by using SIMULINK in the DSP.

In this study, the weight matrix of optimal control when determining feedback gain was formed as follows.

$$\boldsymbol{Q}_d = \begin{bmatrix} 10^{11} & 0 & 0 \\ 0 & 10^{9.2} & 0 \\ 0 & 0 & 10^5 \end{bmatrix} \tag{12}$$

$$r_d = 10^5 \tag{13}$$

TABLE I
SPECIFICATIONS FOR THE VEHICLE AND VOICE COIL MOTOR

	EVERYDAY COMS BASIC (Toyota Auto Body Co.,Ltd)	
Ultra-compact elctric vehicle	Total weight of vehicle	325 kg
	Total length	1935 mm
	Whole width	955 mm
	Total height	1600 mm
	Wheelbase	1280 mm
	Tread	840 mm (Front wheel)
		815 mm (Rear wheel)
Voice coil motor	Aoyama Special Steel Co., Ltd.	
	Effective stroke	20 mm
	Thrust constant	110 N/A
	Nominal thrust	160 N
	Maximum thrust	320 N or more
	Rated current	1.46 A

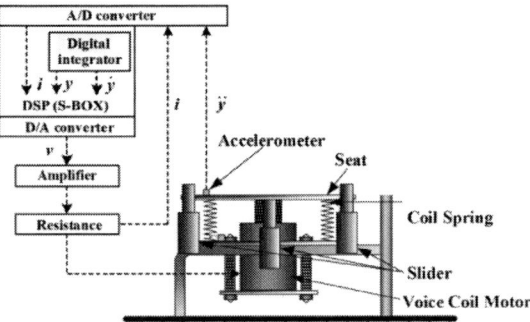

Fig. 3. Active seat suspension control system.

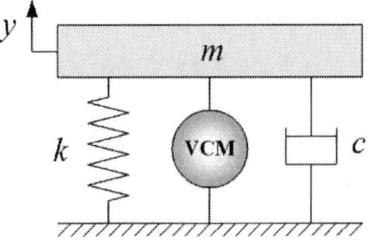

Fig. 4. Analytical model.

$$T_s = 1.0 \times 10^{-3} \tag{14}$$

B. Control experiment

We confirmed that the bump of roadway boundary is one of the worst situations when traveling on road. Figure 5 shows boundary roadway. In this study, during driving experiments, hard rubber plate 50mm height were used, assuming bump of the roadway boundary is as shown in Fig. 6. In addition, taking into account of the safe passage pedestrian and bicycle determined by the Ministry of Land, Infrastructure and Transport, the bump of bump roadway boundary should be less than 50mm [8]. Figure 7 shows the hard rubber. Since the position holding the steering wheel, the bending angle of the arm, the position to place the feet on the footrest to the vehicle ceiling of the driver's head and the depth of the sitting position of the seat distance was all match at each bump experiment, the driving position remain the same as show in Fig. 8. In the

Fig. 5. The bump of roadway boundary.

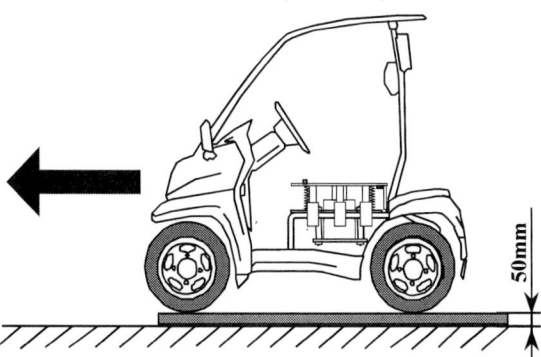

Fig. 6. Test course of bump road using hard rubber.

Fig. 7. Photograph of hard rubber.

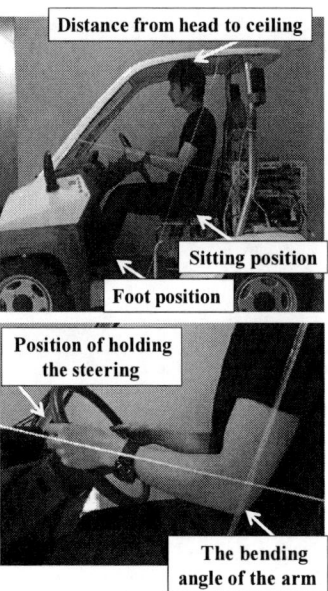

Fig. 8. Driving posture.

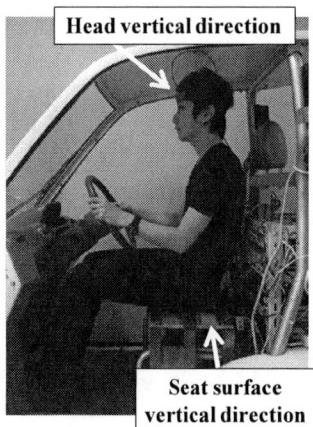

Fig. 9. Accelerometers of head vertical direction and seat surface vertical direction.

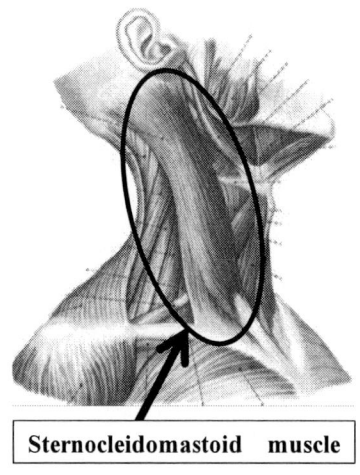

Fig. 10. Muscle of the neck [10].

experiments, the speed of the vehicle as it fell from the bump is about 2km / h. As the vehicle fell from the bump, the falling time of the driver was not taking into account. To evaluate the behavior of the head during falling from the bump, in addition to the measurement of the seat vertical direction absolute acceleration, we measured the vertical absolute acceleration of the head. As shown in Fig. 9, method of measuring the vertical. Experiments were performed on 3 healthy male subjects aged 21-23 years. As preparation experiment, the air pressure of the tires of the vehicle is set to the rated value.

We also measured surface EMG of SCM muscle which is a muscle that supports the head. The SCM muscles conduct muscle activity in order to restrain vibration of the head when maintenance of the head becomes unstable to maintain. Figure 10 shows a schematic diagram of the SCM muscle. Surface EMG measurement using is

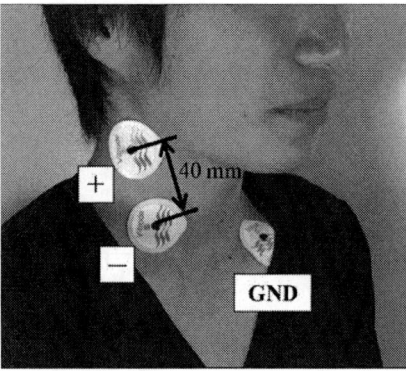

Fig. 11. Photograph of the position of the electrode for measuring EMG.

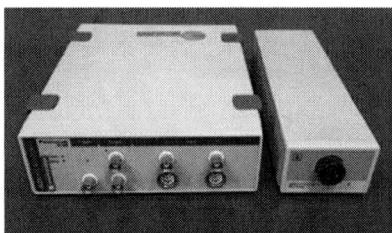

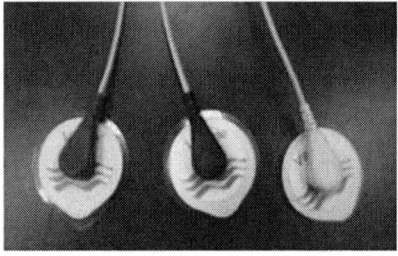

Fig. 12. Electromyogram measuring device.

disposable electrode J Bitorodo (Nihon Kohden Corporation) as a skin surface electrode was used. The sampling frequency of the EMG signal was 1000Hz. EMG which was amplified (high-pass filterat 500Hz, low-pass filter at 10Hz) by Bio Amp ML132 (AD Instruments company), was recorded through the A/D converter (Power Lab ML825 2125, AD Instruments company). Figure 11 shows the position of the electrode for measuring EMG. Figure 12 shows Electromyogram measuring device. Recorded EMG was calculated as the root mean square (RMS) of 0.1 second interval per 0.1 second [9].

We obtained approval from the Ethics Committee for Research on Human Subjects of Tokai University. We also explained to the subject about the contents of the research and obtained signed consent (using a form approved by the Committee) from the subject who participated in this research.

V. EXPERIMNTAL RESULT

Figures 13 and 14 show examples of time histories of seat surface accelerations and RMS of SCM muscle without control (a) [actuator installed control circuit

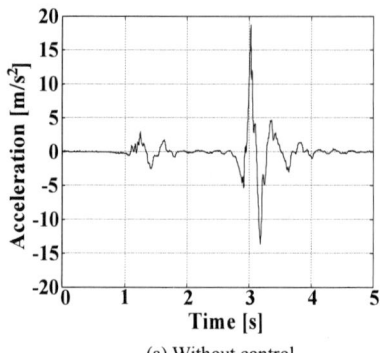

(a) Without control

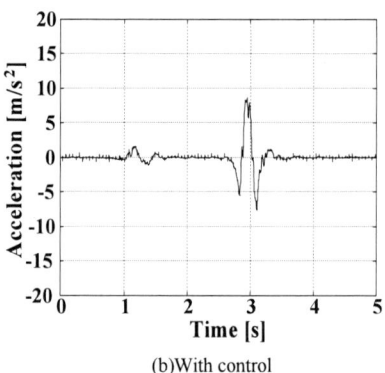

(b)With control

Fig. 13. Time histories of seat acceleration.

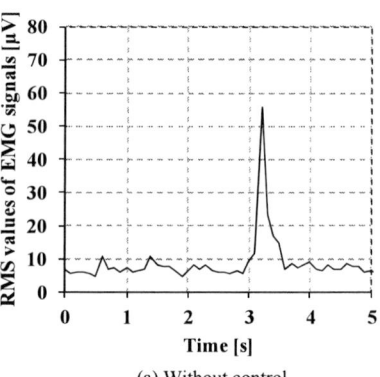

(a) Without control

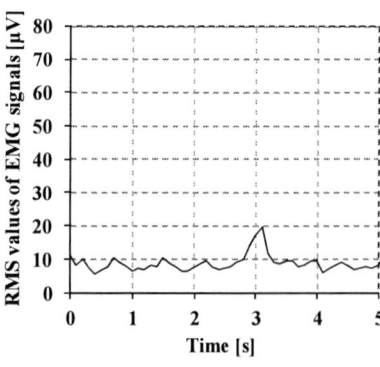

(b) With control

Fig. 14. Time histories of RMS values of EMG signals.

TABLE II
SUBJECTS DATA AND RESULT OF ACCELERATION AND RMS OF EMG SIGNAL (MEAN±SD)

Subject	Height [cm]	Weight [kg]	Age	Maximum amplitude value of seat acceleration [m/s^2]		Maximum RMS value of EMG of SCM muscle [μV]		Feeling of burden by RMS of EMG [μV]
				Without control	With control	Without control	With control	
A	166.2	56.3	21	28.56±0.89	15.87±0.40	61.84±9.56	31.88±4.78	48.38
B	171.0	60.4	21	28.02±0.53	13.95±0.22	93.61±42.43	34.18±11.7	57.41
C	178.5	57.1	23	30.01±2.37	14.31±0.59	63.82±19.93	19.32±2.43	45.91

release] and with control (b). As the rear wheels fell about 2 seconds after the front wheels, the seat acceleration of the rear wheels were larger than the front wheels. Rotation moment around center of gravity of the vehicle's longitudinal direction increased on the front wheels after they fell. Thenceforth, we can consider the falling acceleration of the rear wheels increased by balancing the rotation moment to its original position after the rear wheels fell. Figure 13 shows that maximum amplitude value of seat acceleration without control was assumed to be 100%, and as with control was reduced by approximately 47%. As in Fig. 14, it is shows that maximum RMS value of EMG of SCM muscle without control was assumed to be 100%, and as with control was reduced by approximately 65%. As a result, by lowering the seat acceleration, it was possible to reduce the EMG of SCM muscle which is a muscle that supports the head.

Table II shows result of maximum amplitude value of seat acceleration, maximum value of RMS of SCM muscle and feeling of burden by RMS of EMG from each 3 subjects in experiment data. These data represent the averages and standard deviation of 5 times traveling experiments. We assumed the 5 times results of driving experiment by conducting preliminary experiment in the same condition. Therefore, we confirmed that the 5 times driving experiment results show a proper value. Moreover, we carried out subjective evaluation for muscle burden for each 3 subjects. As for muscle burden assessment, we measured the EMG of SCM muscle for 5 seconds by the strain of subject's neck which subjects feel a burden and calculated as RMS of 0.1 second interval. We have confirmed that there was no discomfort or pain due to muscle fatigue in the neck since we provided time-out for the subjects before performing each experiment. From the table, as we compare the feeling of burden with the average of maximum RMS, we can assume that, as for vehicle with control, drivers did not feel burden on the neck. On the other hand, there was some muscle strains felt during falling time in vehicle without control. From all subjects' result, by lowering the seat acceleration using optimal control, it was possible to reduce the RMS value which is calculated by EMG. When using control, we confirmed that the RMS value decrease the degree that subjects are not feeling burden.

VI. CONCLUSION

In this paper, in order to verify the effectiveness of the active seat suspension in ultra-compact electric vehicles, traveling experiment using optimal control was carried

out. Observation on seat acceleration and EMG of SCM muscle when fall from bump road way boundary was performed. As a result, by lowering the acceleration of the seat, it was possible to reduce the head acceleration and the RMS of EMG. Therefore, we confirmed that it is possible to reduce impact on the neck. In addition, we confirmed that all subjects did not feel the line burden of the SCM muscle in a control vehicle.

Our future plan is to compare optimal control system with new control systems which have high utility, in order to verify the effectiveness of the head impact reduction by reducing seat acceleration. In addition, we should consider the %MVC which is possible to evaluate the feeling of burden of muscle to maximum muscular.

REFERENCES

[1] Y. Abe, H. Arai, Y. Oshinoya, and K. Ishibashi, "Improvement on Riding Comfort with an Active Seat Suspension (Basic Examination Using Small Electric Vehicle)", *JSME annual meeting*, vol.2002, no.7, pp.175-176, 2009, (in Japanese).

[2] H. Kato, Y. Oshinoya, S. Hasegawa, and H. Kasuya, "Psychological and Physiological Effects of Active Seat Suspension on Ride Comfort of Small Vehicles", *Proc. Sch. Eng. Tokai Univ., Ser. E*, vol.35, pp.47-51, 2010.

[3] H. Kato, Y. Oshinoya, S. Hasegawa, H. Kasuya and H. Kasuya, "Use of Active Seat Suspension to Improve Ride Comfort during Travel on Gravel Road", *Proc. Sch. Eng. Tokai Univ., Ser. E*, vol.35, pp.53-57, 2010.

[4] H. Kato, Y. Oshinoya, and S. Hasegawa, "Active Control of Ultra-Compact Vehicle Seat With Voice Coil Motor (Examination of the Control Performances during Driving on a Bad Road)," *J.Magn. Soc. Jpn.*, vol.37, no.3-1, pp.95-101 (2013), (in Japanese).

[5] H. Kato, K. Nakashima, Q. Lan, S Hasegawa, and Y. Oshinoya, "Active Seat Suspension for a Small Vehicle (Effect of Weight Change on Vibration Suppression Performance)", *ICEE*, 2012, (CD-ROM).

[6] M. Kamata, N. Fujii, and T. Akiyama, "Study on Human Friendly Vehicle Usable for Elderly People (Investigation on Physical Characteristics and Life Style of Elderly People, and Discussion of User Needs for Vehicle Design)", *Transactions of the JSME, Series C*, vol.68, no.665, pp.220-227, 2002, (in Japanese).

[7] M. Akamatsu, "Aging of Human Functions and Mobility Technologies for Elderly", JSAE, Vol. 67, No.3, pp. 49-54, 2013.

[8] T. Koizumi, N. Tsujiuchi, H. Abe, J. Ninomiya, K. Yamazaki: Establishment of Evaluation Method for Improving Ride Quality Based on Human Dynamics, *JSME Dynamics and Design Conference*, 2007, CD-ROM, (in Japanese).

[9] Ministry of Land, Infrastructure and Transport, "A Standard Regarding General Structure of the Sidewalk", 2005, (in Japanese).

[10] T. Asai, Y. Tagawa,"Alteration in Muscle Fatigue Cased by Load-Type Differences During Isometric Flexion of Human Fingers", *Transactions of the JSME, Series C*, vol.78, no.790, pp. 2152-2161, 2012, (in Japanese).

[11] T. Fujita, An Introduction to Human Anatomy, Nankodo, 2012, (in Japanese).

978-1-4799-2706-7/14 $31.00 © 2014 IEEE

Adaptive Current Tracking of Three-Phase Active Power Filter Using Backstepping Control

Yunmei Fang[1], Juntao Fei[2], Shixi Hou[2], Weili Dai[2]

[1] College of Mechanical and Electrical Engineering, Hohai University

[2] College of IOT Engineering, Hohai University
Changzhou, P.R.China
e-mail: jtfei@ieee.org

Abstract—**A backstepping controller (BC) is developed to eliminate harmonic contamination and improve the quality of power supply for three-phase active power filter (APF) in this paper. To deal with the nonlinearity of APF, the backstepping method is applied in the design of current tracking control system. The proposed backstepping controllers can ensure proper tracking of the reference current, and impose a desired dynamic behavior, giving robustness and insensitivity to parameter variations. Simulation studies in the MATLAB/ SimPowerSystems Toolbox demonstrate the high performance of the proposed backstepping control strategy.**

Keywords—*Backstepping control, harmonic compensation, total harmonic distrotion, active power fitler.*

I. INTRODUCTION

Power electronic technology brings great convenience to our daily life, however nonlinear loads often bring harmonic-related problems to the industrial power systems including low power factor, phase distortion, waveform surges and so on. Shunt active power filters are the most widely used solution because they can efficiently eliminate current distortion and the reactive power. The APF operates by injecting compensation current which is of the same magnitudes and opposite phases with the harmonic currents into the power system to eliminate harmonic contamination and improve the power factor. Compared with conventional current control methods including hysteresis control, single cycle control, and space vector control, many new control strategies have been designed to improve the dynamic response, such as sliding control, and adaptive control.

In recent years, scholars have in-depth studied the application of different methods in APF control system, including topology, harmonic detection, AC side current control and DC side voltage control. Braiek et al. [1] utilized feedback linearization technique to improve power balance in source side and APF sides. Komucugil et al. [2] presented a new control strategy for single-phase shunt active power filters (SAPF) based on Lyapunov stability theory. Rahmani et al. [3] proposed a nonlinear control technique for a three-phase SAPF and tested it on a laboratory prototype of an SAPF. Shyu et al. [4] introduced a model reference adaptive control for a single-phase SAPF. Matas et al. [5] developed the feedback linearization technique for a single-phase SAPF. Montero et al. [6] compared different methods for extracting the reference currents for SAPF in three-phase four-wire systems. Valdez et al. [7] showed an adaptive controller for a single-phase APF to compensate the current harmonic distortion. Marconi et al.[8] designed a robust nonlinear controller for SAPF to absorb harmonics. Hu et al.[9] introduced a multiresolution control method for an APF which is controlled by digital signal processor to meet the requirements for reducing the real-time computation. Advanced controllers have been investigated in APF to improve the power quality [10-13].

Backstepping [14-16] is similarly a recursive Lyapunov-based design procedure which breaks a design problem for the full system into a sequence of design problems for lower order systems. Nevertheless, Backstepping design has two advantages over sliding mode control: One is that the matching condition appeared in the design of sliding mode control can be relaxed for a class of systems satisfying the so called strict feedback form; The other is that backstepping designs can avoid cancellation of useful nonlinearities. The strategy of backstepping is to develop a controller recursively by considering some of the state variables as "virtual controls" and designing intermediate control laws. Adaptive neural network was also incorporated into backstepping design for nonlinear control system [17]. Backstepping technique was applied to the tracking control of industrial system in [18]. Backstepping can achieve the goals of stabilization and tracking. However, so far, adaptive backstepping method has not been employed to eliminate harmonics of active power filter. Our work will explore an adaptive backstepping control for active power filter.

There are no relevant research works combining adaptive backstepping control to APF before. Backstepping control is applied to a three-phase active power filter in this paper. For the APF, the design of backstepping controller contains two steps. First, a virtual control function is proposed by a Lyapunov function and then real controller is designed. This makes the control law design simpler and easier to be implemented. So the proposed control system has important theoretical and practical significance for promoting the application of APF, improving total harmonic distortion (THD) and strengthening the quality of power supply. This paper combines the adaptive control, and backstepping control to improve the robust design of control law. The adaptive backstepping control improves the power dynamic performance such as current tracking and THD performance.

This work is partially supported by National Science Foundation of China under Grant No. 61374100; Natural Science Foundation of Jiangsu Province under Grant No. BK20131136. The Fundamental Research Funds for the Central Universities under Grant No. 2013B19314.

Combination of these methods has a general sense and can be extended to other power electronic converter topology.

This paper is organized as follows. In section 2, dynamic model of APF is established. In section 3, backstepping controller is designed. Simulation results are shown in section 4 to demonstrate the performance of the proposed method. In the last section conclusions are given.

II. PRINCIPLE OF ACTIVE POWER FILTER

Shunt active power filters are usually applied to three-phase system where a large capacity is required. The most widely used parallel voltage type of APF is mainly discussed. In consideration of the practical application, the SAPF is most applied in the three-phase systems because of its excellent performance and easy implementation. A dynamic analytical model of the APF will be developed.

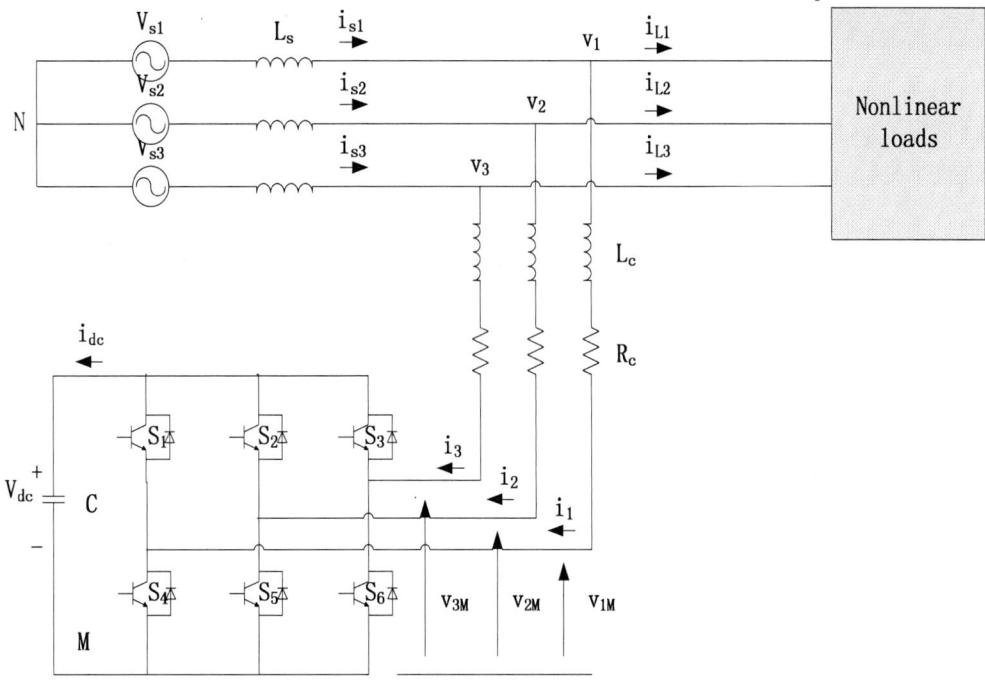

Fig. 1: Block diagram for APF

In practical operation, APF is equivalent to a flow control current source. The APF contains three sections, harmonic current detection module, control system and main circuit. The rapid detection of harmonic current based on instantaneous reactive power theory is most widely used in harmonic current detection module. The main circuit consisting of power switching devices produces compensation currents according to the control signal from the control system. For the sake of absorbing the harmonics created by the nonlinear loads, the compensation currents should be the same magnitudes and opposite phases with the harmonic currents. Fig. 1 shows the structure of three-phase three-wire APF.

The dynamic analytical model of APF is proposed in the following step. According to Kirchhoff's voltage and current laws we can get following circuit equations:

$$
\begin{cases}
v_1 = L_c \dfrac{di_1}{dt} + R_c i_1 + v_{1M} + v_{MN} \\[2mm]
v_2 = L_c \dfrac{di_2}{dt} + R_c i_2 + v_{2M} + v_{MN} \\[2mm]
v_3 = L_c \dfrac{di_3}{dt} + R_c i_3 + v_{3M} + v_{MN}
\end{cases}
\tag{1}
$$

The parameter of L_c and R_c are the inductance and resistance of the APF respectively, V_{MN} is the voltage between point M and N.

By summing the three equations in (1), taking into account the absence of the zero-sequence in the three wire system currents, and assuming that the AC supply voltages are balanced, we can obtain:

$$
v_{MN} = -\frac{1}{3} \sum_{m=1}^{3} v_{mM}
\tag{2}
$$

The switching function c_k denotes the ON/OFF status of the devices in the two legs of the IGBT bridge. We can define c_k as

$$
c_k =
\begin{cases}
1 & \text{if } S_k \text{ is on and } S_{k+3} \text{ is off} \\
0 & \text{if } S_k \text{ is off and } S_{k+3} \text{ is on}
\end{cases}
\tag{3}
$$

where $k=1,2,3$.

Hence, by writing $v_{kM} = c_k v_{dc}$, then (1) becomes

978-1-4799-2706-7/14 $31.00 © 2014 IEEE

$$\begin{cases} \dfrac{di_1}{dt} = -\dfrac{R_c}{L_c}i_1 + \dfrac{v_1}{L_c} - \dfrac{v_{dc}}{L_c}\left(c_1 - \dfrac{1}{3}\sum_{m=1}^{3}c_m\right) \\[2mm] \dfrac{di_2}{dt} = -\dfrac{R_c}{L_c}i_2 + \dfrac{v_2}{L_c} - \dfrac{v_{dc}}{L_c}\left(c_2 - \dfrac{1}{3}\sum_{m=1}^{3}c_m\right) \\[2mm] \dfrac{di_3}{dt} = -\dfrac{R_c}{L_c}i_3 + \dfrac{v_3}{L_c} - \dfrac{v_{dc}}{L_c}\left(c_3 - \dfrac{1}{3}\sum_{m=1}^{3}c_m\right) \end{cases} \quad (4)$$

Then we define the switching state function d_{nk} which is given as

$$d_{nk} = \left(c_k - \frac{1}{3}\sum_{m=1}^{3}c_m\right)_n \quad (5)$$

So d_{nk} depends on the switching function c_k, and based on the eight permissible switching states of the IGBT, we can obtain that

$$\begin{bmatrix} d_{n1} \\ d_{n2} \\ d_{n3} \end{bmatrix} = \frac{1}{3}\begin{bmatrix} 2 & -1 & -1 \\ -1 & 2 & -1 \\ -1 & -1 & 2 \end{bmatrix}\begin{bmatrix} c_1 \\ c_2 \\ c_3 \end{bmatrix} \quad (6)$$

Equation (4) becomes

$$\begin{cases} \dfrac{di_1}{dt} = -\dfrac{R_c}{L_c}i_1 + \dfrac{v_1}{L_c} - \dfrac{v_{dc}}{L_c}d_{n1} \\[2mm] \dfrac{di_2}{dt} = -\dfrac{R_c}{L_c}i_2 + \dfrac{v_2}{L_c} - \dfrac{v_{dc}}{L_c}d_{n2} \\[2mm] \dfrac{di_3}{dt} = -\dfrac{R_c}{L_c}i_3 + \dfrac{v_3}{L_c} - \dfrac{v_{dc}}{L_c}d_{n3} \end{cases} \quad (7)$$

Furthermore, we can define

$$\begin{cases} x_1 = i_k \\ x_2 = \dot{x}_1 = \dot{i}_k \end{cases} \quad (8)$$

The derivative of x_1 and x_2 with respect to time yields:

$$\dot{x}_1 = \dot{i}_k = -\frac{R_c}{L_c}i_k + \frac{v_k}{L_c} - \frac{v_{dc}}{L_c}d_k \quad (9)$$

$$\begin{aligned} \dot{x}_2 = \ddot{x}_1 = \ddot{i}_k &= \frac{d\left(-\dfrac{R_c}{L_c}i_k + \dfrac{v_k}{L_c} - \dfrac{v_{dc}}{L_c}d_k\right)}{dt} \\[2mm] &= -\frac{R_c}{L_c}\dot{i}_k + \frac{1}{L_c}\frac{dv_k}{dt} - \frac{1}{L_c}\frac{dv_{dc}}{dt}d_k \\[2mm] &= -\frac{R_c}{L_c}\left(-\frac{R_c}{L_c}i_k + \frac{v_k}{L_c} - \frac{v_{dc}}{L_c}d_k\right) + \frac{1}{L_c}\frac{dv_k}{dt} - \frac{1}{L_c}\frac{dv_{dc}}{dt}d_k \\[2mm] &= \frac{R_c^2}{L_c^2}i_k - \frac{R_c}{L_c^2}v_k + \frac{1}{L_c}\frac{dv_k}{dt} + \left(\frac{R_c}{L_c^2}v_{dc} - \frac{1}{L_c}\frac{dv_{dc}}{dt}\right)d_k \end{aligned} \quad (10)$$

Then the dynamic model of APF can be written as

$$\begin{cases} \dot{x}_1 = x_2 \\ \dot{x}_2 = f(x) + bu \end{cases} \quad (11)$$

where $f(x) = \dfrac{R_c^2}{L_c^2}i_k - \dfrac{R_c}{L_c^2}v_k + \dfrac{1}{L_c}\dfrac{dv_k}{dt}$, $b = \dfrac{R_c}{L_c^2}v_{dc} - \dfrac{1}{L_c}\dfrac{dv_{dc}}{dt}$, $u = d_k$.

III. DESIGN OF ADAPTIVE BACKSTEPPING CONTROLLER

The backstepping method is an efficient control approach for nonlinear systems, which uses the virtual control to simplify the design of control laws. In general, we transform the complex nonlinear systems into several subsystems through backstepping design, then design virtual control for each subsystem based on Lyapunov functions. So in each step, what we need is just to cope with an easier control law.

In this section, a backstepping controller is designed for APF to ensure the proper tracking of reference current and the stability of the closed-loop system is guaranteed based on Lyapunov analysis.

For the APF, the design of backstepping controller contains two steps. At first step, a virtual control function is proposed by a Lyapunov function V_1. The real control law is designed at the second step. In the following, the design procedure of the backstepping controller is given.

Step1: Define the reference current is y_d, and y_d has continuous second order derivatives. And the tracking error can be defined as

$$e_1 = x_1 - y_d \quad (12)$$

Then

$$\dot{e}_1 = \dot{x}_1 - \dot{y}_d = x_2 - \dot{y}_d \quad (13)$$

We select the virtual control as

$$\alpha_1 = -c_1e_1 + \dot{y}_d \quad (14)$$

where c_1 is a non-zero positive constant.

Define the error as

$$e_2 = x_2 - \alpha_1 \quad (15)$$

We consider the first Lyapunov function as

$$V_1 = \frac{1}{2}e_1^2 \quad (16)$$

and the derivative of V_1 becomes

$$\begin{aligned} \dot{V}_1 &= e_1\dot{e}_1 = e_1(x_2 - \dot{y}_d) \\ &= e_1(e_2 + \alpha_1 - \dot{y}_d) \\ &= e_1(e_2 - c_1e_1 + \dot{y}_d - \dot{y}_d) \\ &= -c_1e_1^2 + e_1e_2 \end{aligned} \quad (17)$$

If $e_2 = 0$, then $\dot{V}_1 = -c_1e_1^2 \le 0$. So we must construct the next step.

Step2: Based on (15), we obtain

978-1-4799-2706-7/14 $31.00 © 2014 IEEE

$$\dot{e}_2 = \dot{x}_2 - \dot{\alpha}_1$$
$$= f(x) + bu - \dot{\alpha}_1 \qquad (18)$$
$$= f(x) + bu - \ddot{y}_d + c_1\dot{e}_1$$

Define the second Lyapunov function as

$$V_2 = V_1 + \frac{1}{2}e_2^2 \qquad (19)$$

and the derivative of V_2 is

$$\dot{V}_2 = \dot{V}_1 + e_2\dot{e}_2$$
$$= -c_1 e_1^2 + e_1 e_2 + e_2[f(x) + bu - \ddot{y}_d + c_1\dot{e}_1] \qquad (20)$$

In order to obtain $\dot{V}_2 \leq 0$, the desired backstepping control law is designed as

$$u = \frac{1}{b}[-f(x) + \ddot{y}_d - c_1\dot{e}_1 - c_2 e_2 - e_1] \qquad (21)$$

where c_2 is a positive scalar.
Then

$$\dot{V}_2 = -c_1 e_1^2 - c_2 e_2^2 \leq 0 \qquad (22)$$

Define $c = \min\{c_1, c_2\}$, one obtains from (22)

$$\dot{V}_2 \leq -c e_1^2 - c e_2^2 = -2cV_2 \qquad (23)$$

Based on the same arguments in [15], the semi-globally uniformly ultimately bounded of the signals can be proved in the closed loop system via the proposed backstepping controller. Moreover, the tracking error will be as small as desired through the suitable parameters.

IV. SIMULATION STUDY

In order to validate the effectiveness and advantage of the proposed control strategies , the designed APF control system is implemented using Matlab/Simulink with SimPower Toolbox. In the simulation, the behavior of each method and its performances during steady and transient state are analyzed to verify the effectiveness of the proposed backstepping control. In the backstepping control, $c_1 = 120000$, $c_2 = 100000$. Other parameters are: $V_{s1} = V_{s2} = V_{s3} = 220V$, $f = 50Hz$ .The resistance in the nonlinear load is 10Ω and the inductance is $2mH$ .The inductance in the compensation circuit is $10mH$ and the capacitance is $100\mu F$. When $t = 0.04s$, the switch of compensation circuit is closed and APF begins to work. In practice, nonlinear loads are usually time varying, so it is necessary to study the dynamic performance of the APF when variations in the nonlinear loads are considered. When $t = 0.12s$, the same nonlinear load is added into the circuit.

Fig.2, Fig.4 and Fig.5 show the load current (only one phase current is represented for the clearness) and its harmonics spectrum. It is clearly shown that there is serious distortion of load current and the Total Harmonic Distortion (THD) is

relatively high(24.71% and 22.24%). Fig.3, Fig.6 and Fig.7 plot the source current and its harmonics spectrum using backstepping approach. It is observed that the source currents are close to sinusoidal wave and become balanced after compensation even with loads changing. The THD is reduced to 1.47% , 1.66% , 1.69% and 1.55% all within the limit of the harmonic standard of IEEE of 5%.The results confirm the capability of the control strategy to cancel the harmonics.

Fig. 8 shows the instruction current and compensation current, and compensation current tracking error is drawn in Fig. 9. It is shown that compensation current can properly track the instruction current using the proposed backstepping controllers. It indicates that the proposed backstepping control can ensure the proper tracking of the reference current. Fig.9 demonstrates that the DC capacitor voltage can track the reference voltage, and it tends to be steady state quickly when loads change.

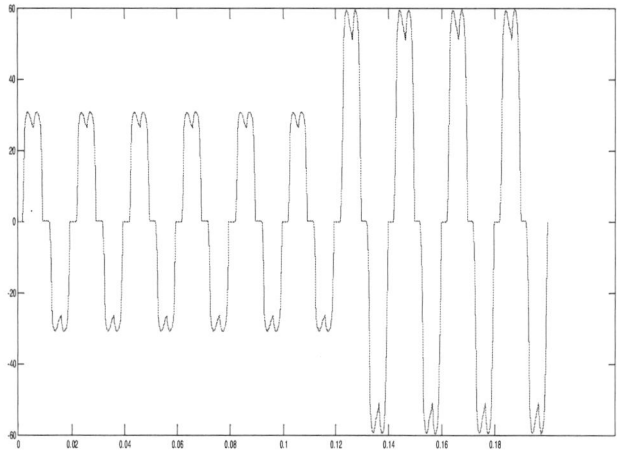

Fig. 2: A phase load current

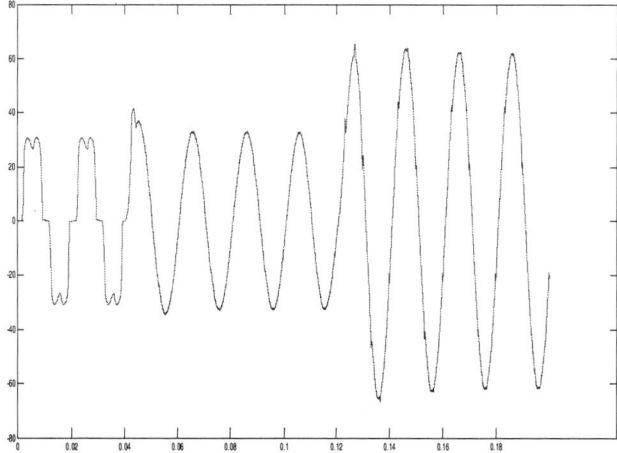

Fig. 3: A phase source current suing backstepping control

978-1-4799-2706-7/14 $31.00 © 2014 IEEE

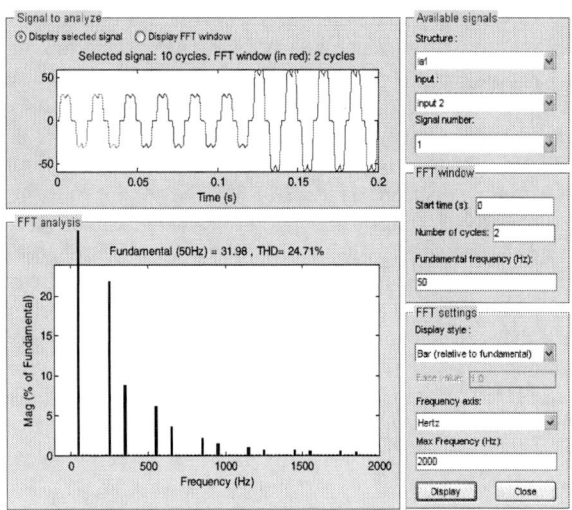

Fig. 4: Load current harmonic analysis when $t=0s$

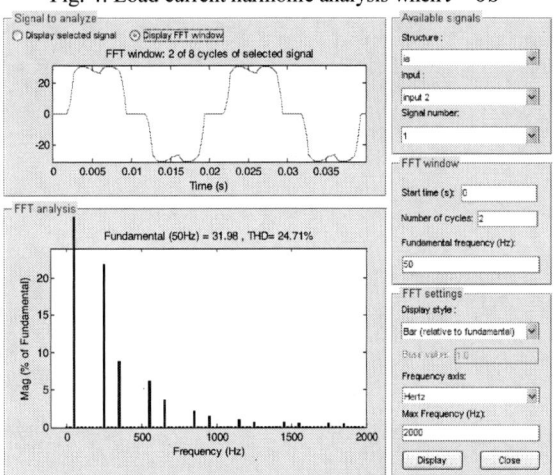

Fig. 5: Source current harmonic analysis when $t=0s$

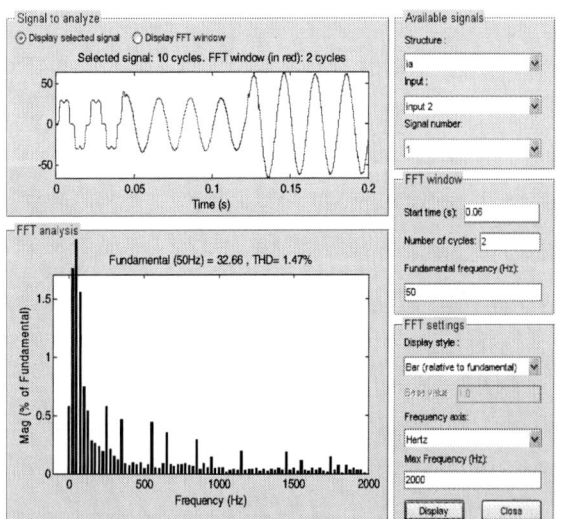

Fig. 6: Source current harmonic analysis when $t=0.06s$ using backstepping control

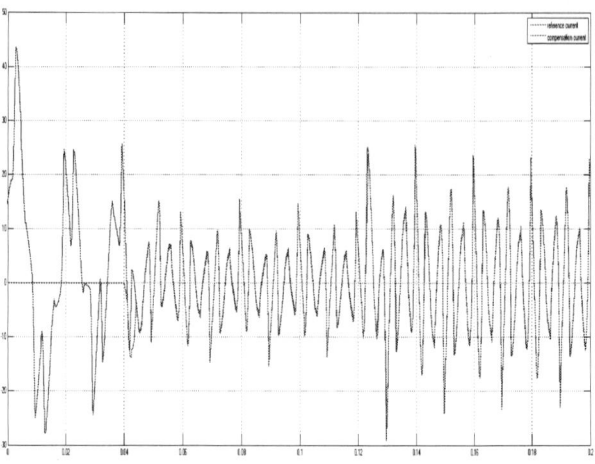

Fig. 7: Instruction current and compensation current using backstepping control

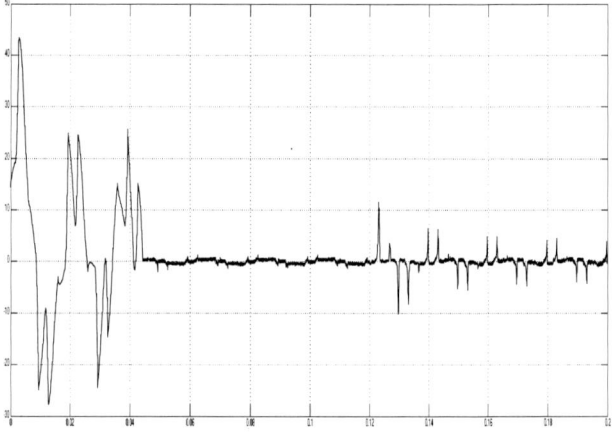

Fig. 8: Compensation current tracking error using backstepping control

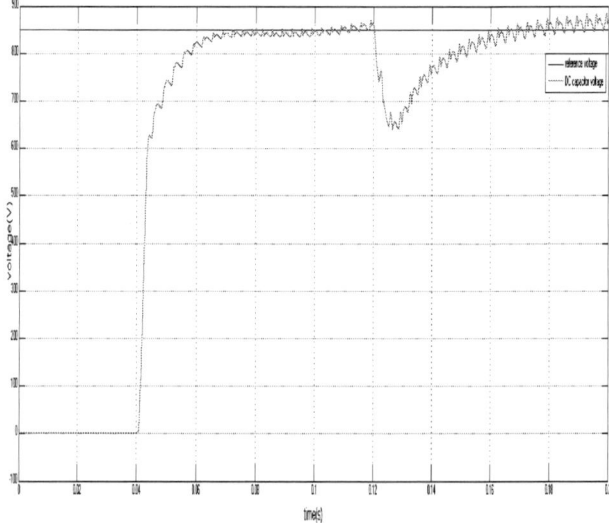

Fig. 9: DC capacitor voltage using backstepping control

TABLE 1. Performance for variation in filter inductance
and DC capacitor with BC when $t=0.16s$

L(mH)	C(uF)	THD(%)
		BC
10	100	1.69
10	200	1.88
10	500	2.37
8	100	1.19
12	100	1.11

In order to illustrate that the BC can achieve satisfactory performance, we testify the APF with the parameters variation. Table 1 shows the performance at $t=0.16s$ .. The results prove that the dynamic response for the APF can be improved using the proposed control strategies in the presence of variations in filter inductance and DC capacitor

V. CONCLUSIONS

In this paper, an adaptive backstepping control strategy for the APF is presented. The stability of the closed-loop system can be guaranteed with the proposed control strategy. The proposed controllers are able to keep the THD of the supply current below the limits specified by the IEEE-519 standard, and impose a desired dynamic behavior. The obtained results have demonstrated the high performance of the APF under both dynamic and steady state operations.

REFERENCES

[1] M. Braiek, F. Fnaiech, and K. Haddad, "Adaptive controller based on a feedback linearization technique applied to a three-phase shunt active power filter, " *Proceedings of IEEE Ind. Elec. Society Conf.*, pp. 975-980, 2005.

[2] H. Komucugil, O. Kukrer, "A new control strategy for single-phase shunt active power filters using a Lyapunov function, " *IEEE Trans. on Industrial Electronics*, vol.53, no. 1, pp.305-312, 2006.

[3] S. Rahmani, N. Mendalek, K. Haddad, "Experimental design of a nonlinear control technique for three-phase shunt active power filter, " *IEEE Trans. Industrial Electronics*, vol.57, no.10, pp. 3364 – 3375, 2010.

[4] K. Shyu, M. Yang, Y. Chen, Y. Lin, "Model reference adaptive control design for a shunt active-power-filter system , " *IEEE Trans. on Industrial Electronics*, vol.55, no. 1, pp. 97– 106, 2008.

[5] J. Matas, L. Vicuna, J. Miret, "Feedback linearization of a single-phase active power filter via sliding mode control, "*IEEE Trans. on Power Electronics,* vol.23, no. 1 , pp. 116 – 125, 2008.

[6] M. Montero, E. Cadaval, F. Gonzalez, "Comparison of control strategies for shunt active power filters in three-phase four-wire systems, " *IEEE Trans. on Power Electronics*, vol.22, no. 1, pp. 229 – 236, 2007.

[7] A. Valdez, G. Escobar, and R. Ortega, "An adaptive controller for shunt active filter considering a dynamic load and the line impedance, " *IEEE Trans. Control System Technolog*, vol.17, no. 2, pp. 458-464, 2009.

[8] L. Marconi, F. Ronchi, A. Tilli, "Robust nonlinear control of shunt active filters for harmonic current compensation," *Automatica*, vol. 43, no. 2, pp. 252-263, 2007.

[9] H. Hu, W. Shi, Y.Lu, Y.Xing, "Design considerations for DSP-Controlled 400 Hz shunt active power filter in an aircraft power system, " *IEEE Trans on Industrial Electronics*, vol.59, no. 9, pp.3624-3634, 2012.

[10] J. Fei, S. Hou, "Adaptive fuzzy control with fuzzy sliding switching for active power filter, " *Transactions of the Institute of Measurement and Control*, vol. 35, no. 8, pp. 1094-1103, 2013.

[11] M. Angulo, D. Caballero, J. Lago, M.Heldwein, S.Mussa, "Active power filter control strategy with implicit closed-loop current control and resonant controller, " *IEEE Trans. on Industrial Electronics*, vol. 6, no. 7, pp. 272–2730, 2013.

[12] Z. Shuai, A. Luo, J. Shen, X. Wang, "Double closed-loop control method for injection-type hybrid active power filter, "*IEEE Trans. on Power Electronics,* vol.26, no. 9, pp. 2393–2403, 2011.

[13] S. Litrán, P. Salmerón, "Analysis and design of different control strategies of hybrid active power filter based on the state model, *IET Power Electronics*, vol. 5, no.8, pp.1341-1350, 2012.

[14] M. Krstic, M. Kanellakopoulos, P. Kokotovic, "Nonlinear and Adaptive Control Design, John Willey & Sons, INC., 1995.

[15] C. Chen, Backstepping Control Design and Its Applications to Vehicle Lateral Control in Automated Highway Systems, Doctoral Dissertation, University of California, Berkeley, 1996.

[16] J. Zhou, C. Wen, Adaptive Backstepping Control of Uncertain Systems, Springer, Verlag Berlin Heidelberg, 2008.

[17] O. Kuljaca, N. Swamy, F. Lewis, "Design and implementation of industrial neural network controller using backstepping, " *IEEE Trans on Industrial Electronics*, vol. 50, no.1, pp.193-201, 2003.

[18] K. Kim, Y. Kim, "Robust backstepping control for slew maneuver using nonlinear tracking function, " *IEEE Trans on Control Systems Technology*, vol. 11, no.6, pp. 822-829, 2003.

Fast Identification of Resonance Characteristic for 2-Mass System with Elastic Load

Ming Yang, Liang Hao, Dianguo Xu
Harbin Institute of Technology
Electrical Engineering and Automation
Harbin, China
yangming@hit.edu.cn

Abstract—**Servo system is greatly influenced by mechanical resonance. The offline suppression of mechanical resonance is determined by the identification of mechanical resonance characteristic. The resonance characteristic can be obtained from the Bode diagram of current closed-loop. This paper discusses two methods of fast acquirement to get resonant characteristic based on pseudo random binary sequence (PRBS) and Chirp signals. Power spectrum method is used to process data. According to the identified frequency, the parameter of notch filter is determined. The results of simulation and experiment demonstrate the accuracy of this method and the suppression effect based on this method.**

Keywords— *Servo System; Identification of Resonance Characteristic; Pseudo Random Binary Sequence; Chirp Signals.*

I. INTRODUCTION

Servo drive system has some mechanical transmission devices to connect motor and load. Especially in the case of slender shaft, the factor of elasticity which cannot be ignored will engender mechanical resonance.

At present, various research plans are aimed at measuring the position information of the drive motor only, and there is no sensors increased at the transmission device and load side. Suppressing resonance research strategies can be roughly divided into two categories: active mode and passive mode.

Active mode is to take the initiative to change the controller parameters or controller structures to eliminate harmonic effects. Active mode can be divided into pure PI control (two degree of freedom PI control, RRC) [1-3], state feedback control based on PI control [4-6], and other advanced algorithms.

Passive mode is that notch filter will been inserted between the output of speed loop and current loop given, and other control system design unchanged. Notch filter can attenuate the amplitude of resonance frequency meanwhile has few effect on other frequency characteristics. Its parameter-design which is simple and practicable has explicit physical concepts. Many commercial servo systems generally adopt the precept that multiply notch filters are inserted to suppress multiply resonance frequencies [7-9]. But it requires an efficient resonant spectrum identification method to

identify the resonant frequency precisely. The offline suppression way firstly obtains system's Bode diagram by sweeping frequency or other methods, and then analyzes its characteristic of resonant frequencies to set filter parameters. This way takes much time and needs additional run at a time. Fixed-point arithmetic has some digital signal processing problems like delay, quantization error in [10]. Wider pass-band range of notch filter brings larger lag angle of phase. For increasing suppressed depth, narrow pass-band design will make resonant spectrum identification worse. Therefore, the design scheme adopting adaptive notch filter can identify resonance frequency and adjust filter parameter quickly [11]. Even through the equal numbers of notch filter's pole-zero will bring smaller lag angle of phase to pass-band range, the lag is still attached. In [12], using phase angle compensation to approach zero-lag angle achieves a certain effect. In [13], traditional notch filter is replaced by FIR filter which is achieved by changing just one parameter. But this method is limited by resonance identification and can't eliminate multiply resonance frequency. So the actual effect is modest.

Traditional method for acquiring frequency characteristic is sweeping frequency which is reliable and high-precision. But the operating time is long and it will bring frequency drift. In [14], PRBS and linear frequency modulation is considered. A power spectrum estimation method based on Welch is proposed in [15].

The offline suppression way needs the identification of characteristic. From the result of frequency characteristic, passive mode based on notch filter can be set intuitionistic and reasonable. And long operating time will increase the damage of system. So fast-identification of resonance characteristic is essential. This paper analyzes two methods based on PRBS signal and Chirp signal and uses the power spectrum method to process data. Accurate resonance characteristic can be acquired by offline resonance characteristic which is obtained by fast identification. It provides accurate resonant parameters for suppression of mechanical resonance in passive mode.

II. INFORMATION

Typical two-inertia mechanical drive system has the following differential equations, shown as (1). The shaft

978-1-4799-2706-7/14 $31.00 © 2014 IEEE

has certain torsion stiffness K and damping coefficient C_w. Motor has inertia J_1 and damping C_1, and Actuators have inertia J_2 and damping coefficient C_2. Electromagnetic torque T_e and load torque T_l are related by the shaft torque T_w. According to equation (1), we can obtain the mechanical transmission device model as shown in Fig. 1 (damping coefficient is neglected).

$$\begin{cases} J_1\ddot{\theta}_1 = T_e - C_1\dot{\theta}_1 - T_w \\ J_2\ddot{\theta}_2 = T_w - C_2\dot{\theta}_2 - T_l \\ T_w = C_w\left(\dot{\theta}_1 - \dot{\theta}_2\right) + K\left(\theta_1 - \theta_2\right) \\ \omega_1 = \dot{\theta}_1 \\ \omega_2 = \dot{\theta}_2 \end{cases} \quad (1)$$

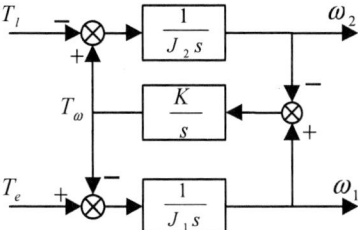

Fig. 1 Block diagram of mechanical transmission.

The transfer function between the motor speed and motor torque is shown as (2) when damping coefficient is neglected. The conjugate zero point refers to Anti-Resonance Frequency (ARF) and the conjugate pole point refers to Natural Torsional Frequency (NTF). The existence of these points makes system response more intense at specific frequency.

$$G_1(s) = \frac{\omega_1}{T_e} = \frac{J_2 s^2 + K}{J_1 J_2 s^3 + (J_1 + J_2) Ks} \quad (2)$$

High performance servo system often need high stiffness, the gain of speed controller K_p is generally large. The resonance frequency has two cases that ARF or NTF frequency and mainly NTF frequency in discrete system. Suppression of resonance in passive mode uses notch filter to filtering the resonance. If system sustained oscillates at steady state caused by lack of stability margin, filter frequency is NTF frequency; conversely, there is damped oscillation instead of sustained oscillation at steady state, filter frequency is ARF frequency.

This paper researches that obtaining Bode diagram of speed open-loop system to get resonance characteristic by offline mode. Fig. 2 is structure block diagram of resonance characteristic identification in discrete system. The reference of signal shouldn't too small; otherwise it will impact on identification accuracy. The signal amplitude of q-axis reference is one times of rated current value in this paper.

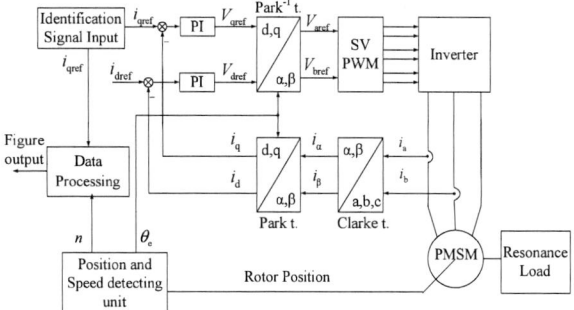

Fig. 2 Structure block diagram of resonance characteristic identification in discrete system.

A.   Identification signal

Basic sweeping frequency mainly provides continuous frequency signal with repetitive periodicity records the characteristics in different frequencies, and on this account, draws the curve of frequency characteristic. But this method is time-consuming, may loss the details of particular frequency characteristic and has DC offset when identifying open-loop system. A long time test on resonant system may even lead to mechanical damage, so it is necessary to study the other method that resonance characteristic can be obtained fast and accuracy.

The main feature of reference signal in fast identification is that the signal contains all the frequency components, has wide range of spectrum, and presents aperiodic change severely in time domain. White noise is the most typical signal according to these features, but white noise signal is not easy to implement in actual digital system and the noise variance is large. So it should be replaced by other signals conforming to these features.

a.   pseudo random sequence signal

Pseudo random sequence can fully embody the feature of fast identification signal. This paper uses M-sequence for pseudo random sequence. M-sequence is also called the maximum linear shift register sequence, which is generated by feedback shift register. There is a relation (3) between each element in sequence $x_1 x_2 \ldots x_p x_{p+1} \ldots$:

$$x_i = a_1 x_{i-1} \oplus a_2 x_{i-2} \oplus \cdots \oplus a_p x_{i-p}, i = p+1, p+2, \cdots \quad (3)$$

where a_i is 0 or 1, $\oplus$ express modulo 2 operation after summation.

Binary M-sequence has good pseudo random characteristic, which has the following features:

1. Balance: The difference of numbers of 0 and 1 is 1 in a cycle of binary M-sequence.

2. Binary autocorrelation function.

For a ±1 level M-sequence with a period of $T=2n-1$, n is the order of M-sequence, its autocorrelation function is (4):

$$R(\tau) = \sum_{k=0}^{T-1} x_k x_{k+\tau} = \begin{cases} T & \tau = 0 \\ -1 & \tau \neq 0 \end{cases} \quad (4)$$

where $u(k)$ is level value of sequence signal, τ is integer.

The autocorrelation characteristic is: $R(\tau)=T$ when τ is integral multiple of 0 or T, and weak correlation $R(\tau)= -1$

when τ is other value. When T is large, autocorrelation function of M-sequence is similar with white noise. This sharp autocorrelation characteristic is the reason that M-sequence can replace white noise as a current reference.

The number of 1 is one more than the one of -1 in a cycle of pseudo random binary sequence. The mean value of sequence is small, avoids the problem of too large accumulated speed in low-speed open-loop actual system. Fig. 3 is waveform graph of PRBS as current reference.

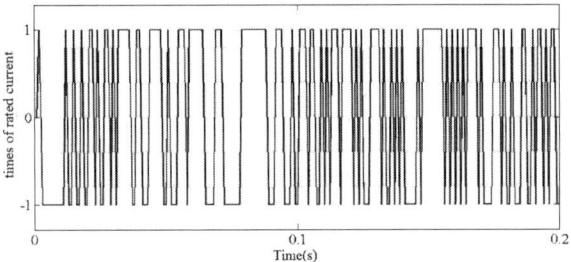

Fig. 3 Waveform graph of Pseudo random sequence signal.

b. Chirp signal

Except of PRBS, Chirp signal also has wide range of spectrum and constant value of frequency components. Chirp signal which is frequency modulated pulse sweep signal is a kind of continuous cosine sweep signal. The following (5) is used in obtaining the frequency characteristic.

$$u(t) = A\cos(2\pi(\beta t^2 + f_0 t)) \tag{5}$$

where A is amplitude of sweeping frequency, β is variation rate of frequency, f_0 is initial frequency. From (5), Chirp signal is the cosine function varies linearly with time. Fig. 4 is waveform graph of Chirp signal as current reference.

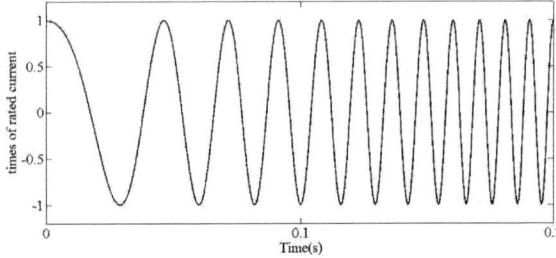

Fig. 4 Waveform graph of Chirp signal.

This signal has uniform frequency component and wide range of spectrum which is similar with white noise. The signal band of Chirp has cut-off range which depends on end-frequency of high frequency, higher end-frequency and larger band.

The above two signals have rapidity and accuracy, only need 1s complete the reference signal in discrete system that sampling frequency is 1 kHz.

B. Data process

The basic data processing method is FFT calculation. The excitation signal $y(t)$ caused by reference $x(t)$ has Fourier transform $y(j\omega)$ which equals to the product of system transfer function $G(j\omega)$ and $x(j\omega)$, shown as (6):

$$y(j\omega) = G(j\omega) \cdot x(j\omega) \tag{6}$$

From FFT results, in-phase quadrature component of the same frequency ω_r of the given and excitation signal is obtained, shown as (7). A is the amplitude of a certain phase. R and Q are the real part and the image part of the FFT result.

$$
\begin{aligned}
R_i(\omega_r) &= A_i \cos\phi_i \\
Q_i(\omega_r) &= A_i \sin\phi_i \\
R_o(\omega_r) &= A_o \cos\phi_o \\
Q_o(\omega_r) &= A_o \sin\phi_o
\end{aligned}
\tag{7}
$$

The frequency characteristic of system under test can be obtained by calculating the amplitude ratio and phase difference of every frequency. The calculation of amplitude-frequency and phase-frequency characteristic is shown in (8), and Bode figure can be draw in this way.

$$|G(j\omega_r)| = \frac{A_o}{A_i} = \sqrt{\frac{R_o^2 + Q_o^2}{R_i^2 + Q_i^2}} \tag{8}$$

$$\phi(\omega_r) = \phi_o - \phi_i = \tan^{-1}\frac{Q_o}{R_o} - \tan^{-1}\frac{Q_i}{R_i}$$

The method of FFT calculation is still need to calculate the characteristic of every frequency. It will cost much resource of system. In contrast, power spectrum method is fast and fit fast signals well.

The frequency characteristic $H(\omega)$ can be obtained by the cross-power spectrum from input and output $G_{xy}(\omega)$ and the auto-power spectrum from input $G_x(\omega)$, shown as (9).

$$H(\omega) = G_{xy}(\omega)/G_x(\omega) \tag{9}$$

If the result has much noise component, the improved method with spectral window, like Bartlett and Welch, will have desired result.

III. SIMULATION AND EXPERIMENT

A. Simulation results

The parameters of servo system used in simulation which is in MATLAB/Simulink are listed in Table I.

TABLE I
MAIN PARAMETER OF PMSM SERVO SYSTEM

Parameter	Value
Rated power	750 W
Rated torque	2.39 N·m
Rated speed	300 r/min
Rated current	4.4A
Number of pole-pairs	4
Inertia of motor	1.1×10^{-3} N·m²
Inertia of load	3×10^{-3} N·m²
Elasticity of shaft	626 N·m/rad

From this parameter, ARF=72 Hz and NTF=140 Hz. Fig. 5 is simulation contrast figure between M-sequence and Chirp signals. From the results, two reference signals

978-1-4799-2706-7/14 $31.00 © 2014 IEEE

can both acquire the characteristic of open-loop system (only amplitude-frequency in this paper), but the result of Chirp signal is better than the one of M-sequence, and noise variance is smaller.

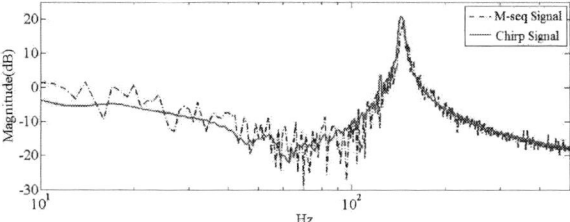

Fig. 5 Simulation results between two signals with power spectrum method.

B. Experiment results

Fig. 6 is real photo of 2-mass system. There is a transmission shaft between motor and load, which produces the mechanical resonance. The resonance frequency can be changed by attaching inertia load to any sides or by changing other shafts.

Fig. 6 Real plant of 2-mass system in laboratory. The left side is drive motor, and right side is load motor. The inertia of drive motor is 1.1×10^{-3} N·m². Rated parameters meet with the simulation.

Use the above methods to obtain mechanical resonance characteristics on physical platform. The speed sampling frequency of two reference signals is 1 kHz, the length of signals are1024, so it only need 1.07 s which is the total time with FFT operation to get the Bode figure of resonance characteristic. Fig. 7 is motor speed of two methods. Two methods both have the advantages of short operation time, no accumulated speed, avoiding the mechanical damage of long operation time and high speed on resonance system.

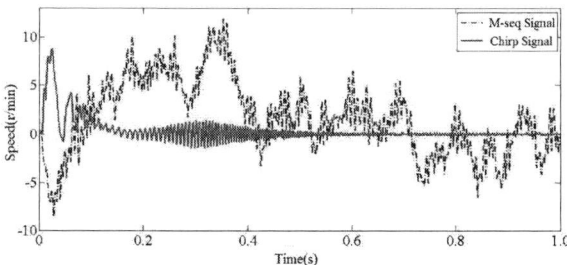

Fig. 7 Waveform of motor speed between two methods.

Fig. 8 is contrastive result of two signals with power spectrum method. Two methods can also obtain the resonance characteristic. The feasibility and practicability are proved in this experiment. And Chirp signal which has less noise than M-sequence is more practical.

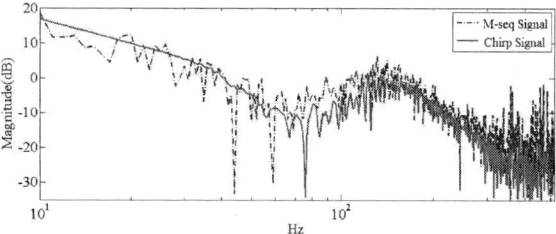

Fig. 8 Real results between two signals with power spectrum method.

Because of the sustained oscillation phenomenon in discrete system, proving deficiency of stability margin, the resonance frequency is NTF frequency when discrete system is oscillation at high stiffness state. Notch filter is set after the output of speed controller. The filter frequency is set by NTF frequency read from amplitude-frequency characteristic. The additional pole can be changed by adjusting the bandwidth of filter. The additional pole can be set to real axis when the parameter of bandwidth is 2 if the damping is ignored. Fig. 9 is motor speed step contrast before and after filtering. Experiment shows that the parameter of notch filter can be determined according to the identification characteristic, and suppression of mechanical resonance can be achieve by this method.

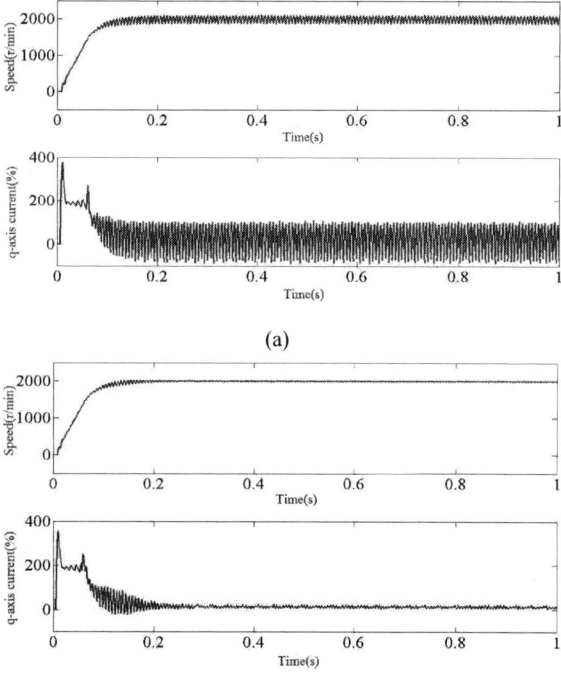

Fig. 9 Real comparison before and after setting notch filter according to identification of characteristics. (a) Notch filter is not set. (b) Notch filter is set according to identification result.

IV. CONCLUSIONS

The method of offline suppressing mechanical resonance is determined by the identification of mechanical resonance characteristic. This paper proposes two practical fast identification methods. Bode figure of resonance characteristic can be obtained by the reference signal of PRBS and Chirp signal. This identification method has a feature of rapidity and small rotation which can reduce the mechanical damage to resonant system and improves the safety of test.

ACKNOWLEDGMENT

The paper is supported by National Natural Science Foundation of China (Project 61273147).

This paper and its related research are supported by grants from the Power Electronics Science and Education Development Program of Delta Environmental & Educational Foundation.

REFERENCES

[1] G. Zhang, "Speed control of two-interia system by PI/PID control", *IEEE Trans. Ind. Electron.*,vol.47, no. 3, pp. 603-609, Jun. 2000.

[2] S. Katsura and K. Ohnishi, "Force servoing by flexible manipulator based on resonance ratio control," *IEEE Trans. Ind. Electron.*, vol. 54, no. 1,pp. 539–547, Feb. 2007.

[3] Y. Hori, H. Sawada, and Y. Chun, "Slow resonance ratio control for vibration suppression and disturbance rejection in torsional system," *IEEE Trans. Ind. Electron.*, vol. 46, no. 1, pp. 162–168, Feb. 1999

[4] K. Szabat and T. Orlowska-Kowalska, "Vibration suppression in a two-mass drive system using PI speed controller and additional feedbacks—Comparative study," *IEEE Trans. Ind. Electron.*, vol.54,no.2, pp.1193–1206, Apr. 2007.

[5] G. Shahgholian, P. Shadaghi, M. Zinali and S. Moalem, "State space analysis and control design of two-mass resonant system", *Computer and Electrical Engineering*, vol. 1,pp. 97-101,2007.

[6] T. Orlowska-Kowalska and M. Kaminski, "Effectiveness of Saliency-Based Methods in Optimization of Neural State Estimators of the Drive System With Elastic Couplings", *IEEE Trans. Ind. Electron.*, vol.56, no.10, pp.4043-4051, Oct.2009.

[7] G. Ellis and R. D. Lorenz, "Resonant Load Control Methods for Industrial Servo Drives", *2000 Industry Applications Conference*, vol. 3, pp.1438-1445, 2000.

[8] Wei-Chih Hsu, Chien-Liang Lai, and Pau-Lo Hsu, "A Novel Design for Vibration Suppression for Lightly-Damped Servo Control Systems", *Proceedings of 2011 8th Asian Control Conference (ASCC)*, pp.251-256, May 2011.

[9] P. Schmidt and T. Rehm, "Notch Filter Tuning for Resonant Frequency Reduction in Dual Inertia Systems", *IAS Annual Meeting (IEEE Industry Applications Society)*, vol. 3 pp.1730-1734, 1999.

[10] Lennart Harnefors, "Implementation of Resonant Controllers and Filters in Fixed-Point Arithmetic", *IEEE Trans. Ind. Electron.*, vol.56, no.4, pp.1273-1281, Apr.2009.

[11] C.-I. Kang and C.-H. Kim, "An Adaptive Notch Filter for Suppressing Mechanical Resonance in High Track Density Disk Drives," *Microsystem Technologies*, vol. 11, pp. 638-652, 2005.

[12] S. Kumagai K.Ohishi and T. Miyazaki, "High Performance Robot Motion Control Based on Zero Phase Error Notch Filter and D-PD Control", *Proceedings of the 2009 IEEE International Conference on Mechatronics*, pp. Apr. 2009.

[13] Vukosavic SN; Stojic MR, "Suppression of torsional oscillations in a high-performance speed servo drive", *IEEE Trans. Ind. Electron.*, Vol.45, no.1, pp.108-117, FEB 1998.

[14] Jennison B.K, "Performance of a linear frequency modulated signal detection algorithm". *The Record of the IEEE 2000 International*. 2000: 447-450.

[15] Villwock S,Pacas M. "Application of the Welch-Method for the Identification of Two- and Three-Mass-System". *IEEE Trans on Industrial Electronics*, 2008:457-466.

Autonomous Navigation System Based on Collision Danger-degree for Unmanned Ground Vehicle

Takashi Yasuno, Daiki Tanaka and Akinobu Kuwahara

Department of Electrical and Electronic Engineering, Faculty of Engineering

The University of Tokushima

Tokushima, Japan

yasuno@ee.tokushima-u.ac.jp

Abstract—This paper proposes an autonomous navigation system for an unmanned ground vehicle (UGV) using a collision danger-degree on outdoor environment. The proposed autonomous navigation system mainly consists of three kinds of control module, a self-tuning fuzzy (STF) path tracking control module, an obstacle avoidance control module, and a two degree-of-freedom speed control module. The STF path tracking control module estimates a steering angle for reference path tracking. On the other hand, the obstacle avoidance control module estimates a steering angle for realizing effective avoidance action by considering the reference path tracking. To solve a tradeoff problem between these control modules, estimated two steering angles are weighted by the collision danger-degree, and final steering angle command is determined. Some experimental results using the developed UGV demonstrate the usefulness of the proposed autonomous navigation system.

Keywords—Unmanned ground vehicle, Autonomous navigation, Collision danger-degree

I. INTRODUCTION

In recent years, autonomous mobile robots have been introduced into various scenes such as home and office cleaning, transportation, guarding, farm working, and so on. These scenes can be divided into two environments; indoor and outdoor environments. Indoor environment, we can easily set up landmarks or guide lines to recognize the self-position and direction of the mobile robot. On the other hand, in outdoor environment, we will take a lot of effort and cost to make a suitable workspace for the mobile robot in much the same way as indoor environment. In addition, there is a disadvantage that the outdoor environment is easily changed and complex. Therefore, if the autonomous control technique for the indoor environment is directly applied to the outdoor environment, various problems often cause.

Fundamental tasks of the autonomous mobile robot are a path planning for generating efficiently reference path from the current position to the goal, an obstacle avoidance control, a path tracking control for getting good tracking performances to the generated reference path, a moving speed planning for reaching the goal at scheduled time. However, these tasks has the tradeoff relationship each other. In order to solve these problems, it is well known that the subsumption architecture[1], [2] is one of the very useful approaches.

Until now, we have proposed the self-tuning fuzzy[3] path tracking control system and the feedforward type two d.o.f. speed control system[4], and independently confirmed the usefulness of these systems by several experiments using our developed unmanned ground vehicle (UGV)[5], [6].

In this paper, we propose a novel autonomous navigation system based on the subsumption architecture, which the priority of each task with tradeoff relationship is decided by using the collision danger-degree[7]. In following sections, we explain the proposed autonomous navigation system and show the usefulness of it from experimental results.

II. DEVELOPED AUTONOMOUS UGV

Figure 1 shows the appearance of our developed autonomous UGV (called here, RV-SCOT2) (overall length 1.35m, overall width 0.78m, overall height 1.3m) driven by a gasoline engine. Self-position and direction of the RV-SCOT2 can obtain by a GPS receiver (San Jose Navigation FV-M8) mounted on the roof of the RV-SCOT2 and a digital compass (HMC6352). In addition, a laser range finder (LRF, SICK LMS100) installing in front of the RV-SCOT2 can measure related positions between the RV-SCOT2 and obstacles.

The laptop PC (Panasonic CF-R8GC1AAS) mounted on the RV-SCOT2 obtains environmental information mentioned above and calculates control signals of the steering angle and the accelerator level according to the internal and external sensors data and the given reference path. In order to maintain safety during experiments, start and stop commands from the base station to the RV-SCOT2 are sent by the wireless communication unit (CIRCUIT DESIGN MU-1 RIK-429S). A sampling time of sensor data acquisition, steering control and moving speed control is set to 0.2 s.

III. PROBLEM STATEMENT

The mission of the RV-SCOT2 is to follow the given reference path and to continue traveling without collision

Fig. 1. Developed UGV (RV-SCOT2) with the proposed autonomous navigation system.

for static obstacles as shown in Fig. 2. Here, (x, y, t, v) indicates an absolute position (x, y), a traveling elapsed time t, and a traveling speed of the RV-SCOT2 v. a is a traveling direction angle of the RV-SCOT2 that is expressed in values relative to north. d_i is the minimum distance error between the current position of the RV-SCOT2 (x, y) and the reference straight path. ϕ_i is the direction error between the current speed vector and the current reference straight path. l_i is the distance error between the current position of the RV-SCOT2 (x, y) and the current subgoal position (X_i, Y_i). δ_i is the direction error between a vector from current position of the RV-SCOT2 to the subgoal and the vector of the reference straight path. Here it is assumed that the subgoal (X_i, Y_i) and the desired arrival time T_i are given in advance.

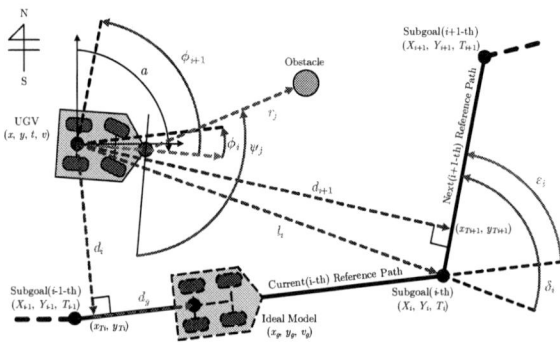

Fig. 2. Coordinates of the UGV, the obstacle, and the reference path with subgoals.

IV. AUTONOMOUS NAVIGATION SYSTEM UNDER COLLISION DANGER-DEGREE

Figure 3 shows the proposed autonomous navigation system which consists of three kinds of control modules

such as the STF path tracking control module, the obstacle avoidance control module, and the two d.o.f. speed control module. To achieve the given mission effectively, it is very important to maintain suitable balance of the tradeoff relationship between each control module in all cases.

First, an operator generates the reference straight path by setting subgoal positions (X_n, Y_n). In addition, target arrival time T_n for each subgoal are set.

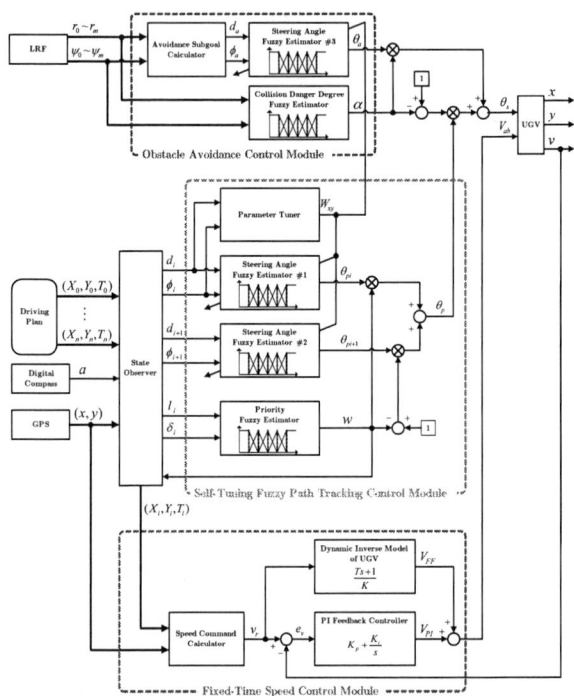

Fig. 3. Configuration of the proposed autonomous navigation system under collision danger-degree.

A. Self-tuning fuzzy path tracking control module

The STF path tracking control module which consists of three simplified fuzzy estimators calculates the suitable steering angle θ_p for the reference path on the basis of (d_i, ϕ_i), (d_{i+1}, ϕ_{i+1}), and (l_i, δ_i). The steering angle fuzzy estimator #1 and #2 estimate suitable steering angles θ_{pi} and θ_{pi+1} for current and next reference straight path, respectively.

The fuzzy rules (singleton parameter in consequent part W_{ab} as shown in Table I) of these fuzzy estimators are adjusted by the following tuning law.

$$\frac{dW_{ab}(t)}{dt} = \gamma e_f(t) \mu_{ab} \tag{1}$$

where $\gamma > 0$ is a tuning gain. $\mu_{ab} = \mu_a \cdot \mu_b$ (μ_a and μ_b correspond to the grade for input variables d_i and ϕ_i obtained from triangular-shaped membership functions) is a grade for the selected fuzzy rules. $e_f(t)$ is a tracking error as defined by

$$e_f(t) = \beta_1 d_i(t) + \beta_2 \phi_i(t) \tag{2}$$

where β_1 and β_2 are weighted parameters that have direct effects upon path tracking control performances. In Table I, negative big (NB), negative small (NS), zero (ZO), positive small (PS) and positive big (PB) denote five linguistic variables of triangular membership functions.

TABLE I.   RULES FOR STEERING ANGLE FUZZY ESTIMATER

		ϕ_i				
		NB	NS	ZR	PS	PB
d_i	NB	W_{00}	W_{01}	W_{02}	W_{03}	W_{04}
	ZR	W_{10}	W_{11}	W_{12}	W_{13}	W_{14}
	PB	W_{20}	W_{21}	W_{22}	W_{23}	W_{24}

To realize suitable tracking performances for changes of the reference straight path, the priority fuzzy estimator decides the value of the priority parameter w ($0 \leq w \leq 1$) for the current reference straight path by a simplified fuzzy reasoning based on the tuned fuzzy rules listed in Table II.

TABLE II.   RULES FOR PRIORITY FUZZY ESTIMATER

		δ_i				
		NB	NS	ZR	PS	PB
l_i	ZR	0.00	0.00	0.00	0.00	0.00
	PS	0.27	0.41	0.58	0.41	0.27
	PM	0.50	0.99	1.00	0.99	0.50
	PB	1.00	1.00	1.00	1.00	1.00

Finally, the steering angle command θ_p for path tracking action is calculated by

$$\theta_p = w \cdot \theta_{pi} + (1 - w) \cdot \theta_{pi+1}. \tag{3}$$

B. Obstacle avoidance control module

The obstacle avoidance control module consists of the avoidance subgoal calculator to set new subgoals for avoiding static obstacles based on Dijkstra's algorithm [8] as shown in Figs. 4(a) ~ 4(d), the steering angle fuzzy estimator, and the collision danger-degree fuzzy estimator. The steering angle θ_a to avoid the collision for obstacles based on the collision danger-degree α ($0 \leq \alpha \leq 1$) calculated by the collision danger degree fuzzy estimator. Estimated two steering angle commands θ_p and θ_a are weighted by the danger-degree α as follows and decide the final steering angle θ_s.

$$\theta_s = (1 - \alpha) \cdot \theta_p + \alpha \cdot \theta_a. \tag{4}$$

C. Traveling speed control module

The feedforward type two d.o.f. speed control module with a fixed gain PI feedback controller calculates the suitable acceleration level V_{ab} includes the brake command so that speed control performances are satisfy the given driving schedule (X_i, Y_i, T_i) sufficiently.

First, we consider the ideal RV-SCOT2 model that can reach in each subgoal (X_i, Y_i) at the scheduled arrival

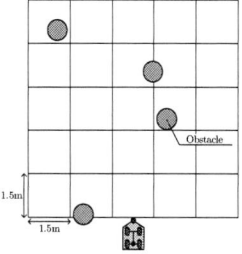 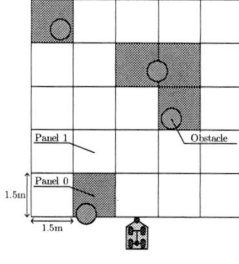

(a) setting virtual 5×5 panels (b) deviding into panels with or without static obstacles

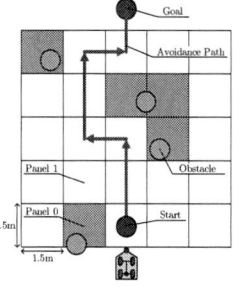 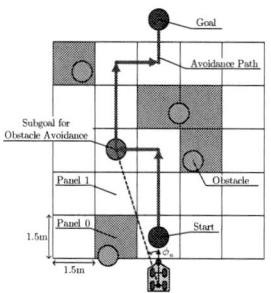

(c) searching the most shortest path from start to goal   (d) setting avoidance subgoals on the shortest path searched in (c)

Fig. 4.   Calculation procedure of distance error d_a and angle error ϕ_a based on Dijkstra's algorithm

time T_i. The traveling speed of the ideal RV-SCOT2 model v_g is calculated by

$$v_g = \frac{\sqrt{(X_i - X_{i-1})^2 + (Y_i - Y_{i-1})^2}}{T_i - t} K_k \tag{5}$$

where $K_k = 1.9438 \frac{\text{kt}}{\text{m/s}}$ is a conversion factor of unit. t is elapsed time.

Next, the traveling speed reference of the actual RV-SCOT2 v_r is decided by Eq.(6) so that the distance error d_g between the ideal RV-SCOT2 model and the actual RV-SCOT2 decreases.

$$v_r = v_g + K_g d_g \tag{6}$$

where K_g is a gain constant.

Irregular changes of traveling speed of the actual RV-SCOT2 are caused by load surface conditions and widely changes of the steering angle. Therefore, we assumed that these factors degenerating control performances are equivalent disturbances, and gains of the PI feedback controller K_p and K_i are designed to restrain these effects. On the other hand, the feedforward controller is set to be the dynamic inverse model of the RV-SCOT2 $G_{UGV}(s)^{-1}$ so that the tracking performance is improved as follow.

$$G_{UGV}(s)^{-1} = \frac{Ts + 1}{K} \tag{7}$$

where $K = 16.183 \frac{\text{kt}}{\text{m/s}}$ is a gain, $T = 0.9$s is a time constant.

V. COLLISION DANGER-DEGREE

As mentioned above, the collision danger-degree is the key parameter for resolving tradeoff relationship between each control module in this paper.

The LRF mounted on the RV-SCOT2 can measure the related distance r_j and direction ψ_j between the RV-SCOT2 and obstacles. However, it is notice that the measured data by the LRF includes uncertainty. Then, we consider how to understand the dangerous of collision for static obstacles quantitatively, and propose the collision danger-degree α that is calculated by using the simplified fuzzy reasoning.

In the antecedent part of the fuzzy rule, input data of distance r_j and direction ψ_j for the static obstacles are converted by the triangular-shaped membership functions crossing at the grade 0.5 as shown in Figs.5(a) and 5(b). When the input variables r_j and ψ_j are given to the fuzzy estimator, two fuzzy sets are selected for each input variable. Here, μ_{ij} is the grade for the selected fuzzy rules given by

$$\mu_{ij} = \mu_i \cdot \mu_j \qquad (8)$$

where μ_i and μ_j correspond to the grade for input variables r_j and ψ_j obtained from Figs.5(a) and 5(b). As the results, four fuzzy rules are selected.

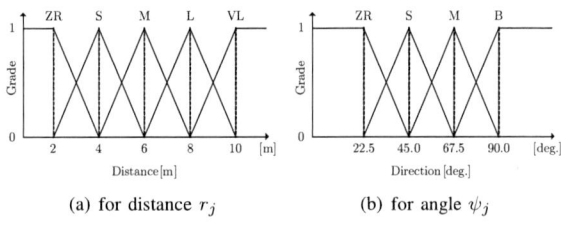

(a) for distance r_j (b) for angle ψ_j

Fig. 5. Membership functions of collision danger degree fuzzy estimator

In consequent part of the fuzzy rule, singleton variables W_{ij} of the selected fuzzy rules shown in Table III are decided by trial and error.

The collision danger-degree α_{Sn} for eight directions range every 22.5 deg. can be obtained by using the following equations:

$$\alpha_{Sn} = \frac{\sum \mu_{ij} \cdot W_{ij}}{\mu_{ij}} \qquad (9)$$

where the subscript $S(= R \text{ or } L)$ denotes the left or right side. $n(=1\sim4)$ is the direction range number. Finally, the collision danger-degree α is decided to the maximum value of the eight collision danger-degrees α_{Sn} as follow;

$$\alpha = \max_n \{\alpha_{Sn}\} \qquad (10)$$

TABLE III. DESIGNED FUZZY RULES OF COLLISION DANGER-DEGREE FUZZY ESTIMATOR

		ψ_j				
		ZR	S	M	L	VL
r_j	ZR	0.8	0.2	0.0	0.0	0.0
	S	0.9	0.4	0.1	0.0	0.0
	M	0.9	0.6	0.2	0.1	0.0
	B	1.0	0.8	0.5	0.2	0.0

VI. EXPERIMENTAL RESULTS

To ensure the validity of the proposed autonomous navigation system, several experiments using our developed RV-SCOT2 were conducted in intramural athletic ground. Here, we set $K_p = 0.2$, $K_i = 0.01$, $\gamma = 0.2$, and $K_g = 0.15$.

First, Figs. 6(a) and 6(b) show path tracking control results for the reference path linked between subgoals by a straight line without static obstacles. In this case, traveling speed command v_r is set to 3.8 kt constant. In early stage of tuning, tracking error is larger under tuning error of steering angle fuzzy rules. However, after tuning, the tracking performance is improved by the tuning law in Eq.(1). Moreover, the trajectory of the RV-SCOT2 is located in inner side or online of the reference path since the RV-SCOT2 shifts the reference straight path at the suitable timing.

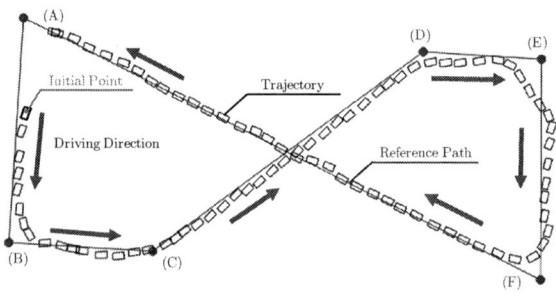

Fig. 6. Tracking control results for the reference path without static obstacles

Next, Fig.7 shows the path tracking control result for the reference path with some static obstacles. The RV-SCOT2 can achieve the path tracking control successfully without collision for static obstacles. In particular, we can confirm that the RV-SCOT2 selects suitable obstacle avoidance actions considering the reference path tracking from the subgoal (F) to (A).

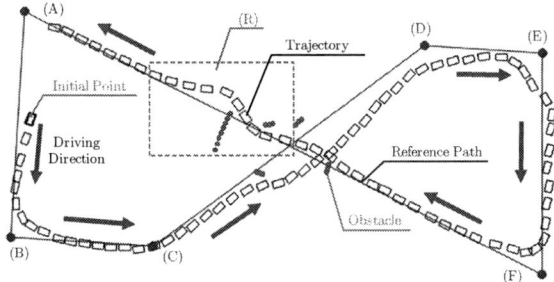

Fig. 7.   Tracking control result for reference path with static obstacles

Figs. 8(a) ∼ 8(f) show the obstacle avoidance performances in the inside of the area (R) shown in Fig.7. In Fig.8(a), the subgoal for the obstacle avoidance is set to the left front of the RV-SCOT2, and the RV-SCOT2 can avoid the static obstacles successfully by controlling the steering so that the RV-SCOT2 follows its subgoal. In Figs.8(b) and 8(c), although the static obstacles exists in front of the RV-SCOT2, the subgoal for the obstacle avoidance is set to the right front of the RV-SCOT2, because the free panel exists in the right front of the RV-SCOT2. On the other hand, in Figs.8(e) and 8(f), the RV-SCOT2 carried out the reference path tracking control with first priority since there are no static obstacles around the RV-SCOT2.

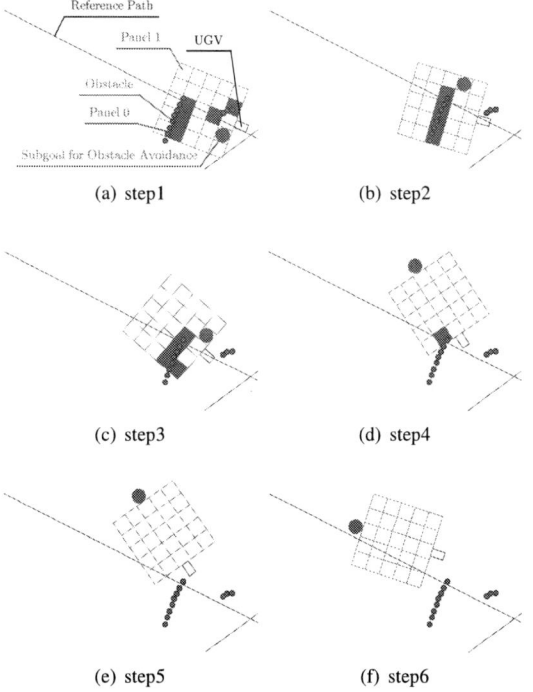

Fig. 8.   States of the obstacle avoidance action in the area (R).

Finally, Fig.9 shows the traveling trajectory for the traveling speed control. In this experiment, scheduled arrival times are set to as listed in Table IV. Although the error for the scheduled arrival time at first lap is

large because of a lack of enough fuzzy rule tuning, the RV-SCOT2 could pass each subgoal at scheduled arrival time certainly while maintaining the tracking control performance.

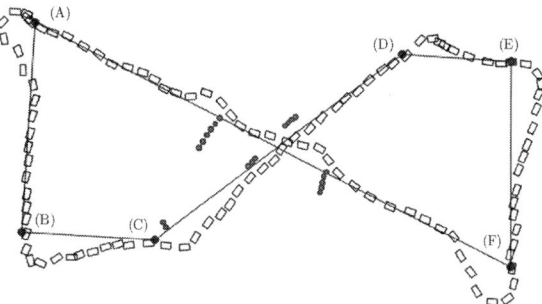

Fig. 9.   Traveling trajectory under traveling speed control (2nd laps)

TABLE IV.   SETTING SCHEDULED ARRIVAL TIME AND ERROR FOR EACH SUBGOAL

Subgoal	Scheduled arrival time [s]	Error [s]
(B)	16.0	+1.0
(C)	27.0	+1.4
(D)	48.0	±0.0
(E)	57.0	+0.8
(F)	73.0	±0.0
(A)	110.0	−0.2
(B)	126.0	+0.4
(C)	137.0	+0.8
(D)	158.0	+0.4
(E)	167.0	+0.4
(F)	183.0	+0.4
(A)	220.0	+0.4

VII.   CONCLUSIONS

In this paper, we proposed an autonomous navigation system for an unmanned ground vehicle (RV-SCOT2). From the experimental results using our developed RV-SCOT2, the effective path tracking and obstacle avoidance control performances can be obtained by solving tradeoff relationship between each control modules using the collision danger-degree.

Future works is to improve a path tracking accuracy and to apply the proposed autonomous navigation system to outdoor environment with static and dynamic obstacles.

REFERENCES

[1]   Brooks, R.A., "A robust layered control system for a mobile robot", *IEEE Journal of Robotis and Automation*, Vol.2, No.1, pp.14-23, 1986

[2]   Takashi Gomi, "Subsumption Architecture", *Journal of Japan Society for Fuzzy Theory and Systems*, Vol.7, No.5, pp.909-930, 1995

[3]   Takuya Kamano, Junji Fukumi, Takayuki Suzuki, Hironobu Harada, and Yu Kataoka "High Speed Synchronization of Two Axes System under Self-Tuning Fuzzy Control", *Journal of Japan Society for Fuzzy Theory and Systems*, Vol.8, No.1, pp.47-56, 1996

[4]   Takashi Yasuno, Takuya Kamano, Takayuki Suzuki, Hironobu Harada, and Yu Kataoka, "Design and responses of the two degree-of-freedom speed servo system under adaptive control", *IEEJ Journal of Industry Applications*, Vol.115-D, No.11, pp.1357-1364, 1995

[5] Shiro Ozaki, et al., "Self-Tuning Fuzzy Tracking Control for Autonomous Unmanned Ground Vehicle Using Preview of Reference Path", *2010 RISP Int. Workshop on NCSP*, pp.580-583, 2010

[6] Daiki Tanaka, et al., "Autonomous Navigation System for Unmanned Ground Vehicle Using Collision Danger-degree", *2011 RISP Int. Workshop on NCSP*, pp.40-43, 2011

[7] Maeda Y, Takegaki M. Collision avoidance control among moving obstacles for a mobile robot on the fuzzy reasoning. Journal of The Robotics Society of Japan, Vol.6, pp.518-522, 1988

[8] Dijkstra, E.W., "A note on two problems in connexion with graphs", *Numerische Mathematik*, 1, pp. 269 ∼ 271, 1959.

The 2014 International Power Electronics Conference

A High-Performance Bidirectional DC-DC Converter For DC Micro-Grid System Application

Shu-Wei Kuo, Yu-Kang Lo, and Huang-Jen Chiu
Department of Electronic Engineering
National Taiwan University of Science and Technology
Taipei City, Taiwan, ROC
Email: D9902210@mail.ntust.edu.tw

Shih-Jen Cheng, Chung-Yi Lin, and CS Yang
Flextronics Power
New Taipei City, Taiwan, ROC
Email: sj.cheng1@flextronics.com

Abstract-**This paper presents a high efficiency dual-half-bridge bidirectional DC-DC converter for DC micro-grid system applications. The operating principle and design consideration of the studied bidirectional converter are discussed. The zero-voltage-switching features and an adaptive phase-shift control method can be achieved under wide-range load variations. When power flow through from DC_BUS side to the battery side. A three-stage charging method is designed to meet the fast-charging demand and prevent battery overcharging to prolong the life-time of LiFePO$_4$ batteries. Otherwise, a digital-signal-processing (DSP) control IC is used to realize the power flow control, DC_BUS voltage regulation and battery charging/ discharging of the studied bidirectional DC-DC converter. A 10kW laboratory prototype converter with E-CAN communication function is built and tested for DC micro-grid system applications. The experimental results are shown to verify the feasibility of the proposed scheme. High efficiency performance can be achieved by the studied adaptive phase-shift control method and zero-voltage switching features. The measured Peak efficiency of the prototype converter expect to higher than 98% under discharging mode and charging mode.**
KEYWORDS: bidirectional converter, micro-grid, and zero-voltage-switching.

I. INTRODUCTION

Recent years, renewable energies such as fuel-cell, wind energy, and solar energy have more demand [1]-[4]. In a distributed electric power system, an independent power generation system or large-scale power system in parallel is a trend. Compared to the centralized power system that needs to fulfil different load requirements from different areas, a more reliable and economic DC micro-grid system is introduced in Fig. 1. DC-load power requirements can be supplied from renewable energy sources such as solar-cells or wind-energy, otherwise from parallel mains-grid for AC-load demands. The excessive DC power can also be fed back to the AC mains through an inverter, and re-sell the energy to power company. However, each renewable energy have different problems in energy capacity and response speed, and seldom features the energy storage function. Therefore, rechargeable batteries are usually needed for a DC micro-grid system. In a DC micro-grid system, the

excessive energy from renewable energy resources will transfer to the batteries as DC-load power demand by the bidirectional DC-DC converter. On the contrary, if the renewable energy resources cannot fully supply the load demand, then the bidirectional DC-DC converter will provide the energy from batteries to DC-load.

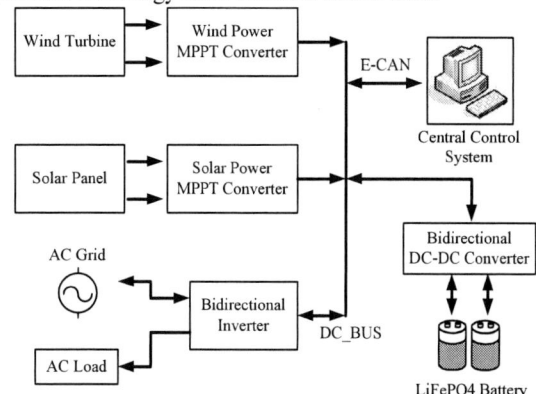

Fig. 1. Block diagram of a DC micro-grid system.

II. SYSTEM ARCHITECTURE ANALYSIS

Fig. 2 shows the studied bidirectional DC-DC converter that consists of an inductor, four switches, and two capacitors. There are two operation modes according to the power flow direction: discharging mode and charging mode for the studied bidirectional DC-DC converter [5]-[8]. During the discharging mode, renewable energy resource is not enough to provide output load, so that the battery is then discharged to meet load requirements. From the theoretical waveforms shown in Fig. 3(a), Q_A、 Q_B and Q_C、 Q_D are both complementary signals with phase-shift of ϕ. While the duty-cycle of Q_A is higher than 50% and the duty-cycle of Q_C and Q_D are fixed at 50%. During the charging mode, the power flow provided by the renewable energy is higher than the load requirement such that the abundant power is charged to battery for storage. From the theoretical waveforms shown in Fig. 3(b), the duty-cycle of Q_C is less than 50% while the duty-cycle of Q_A and Q_B is fixed at 50%.

978-1-4799-2706-7/14 $31.00 © 2014 IEEE

The 2014 International Power Electronics Conference

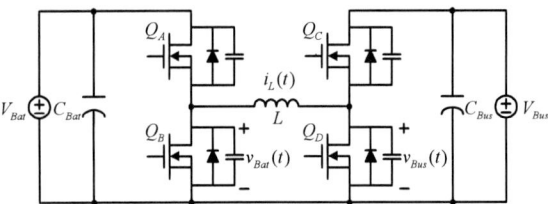

Fig. 2. Bidirectional DC-DC converter.

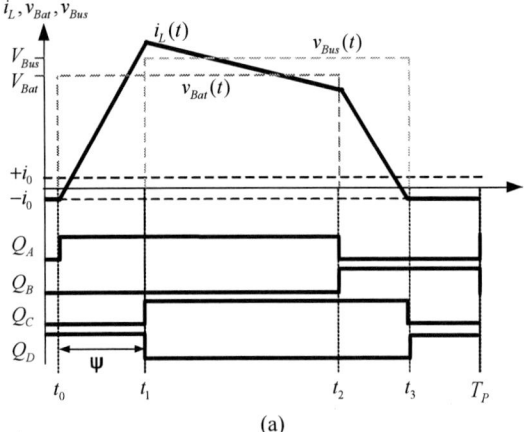

(a)

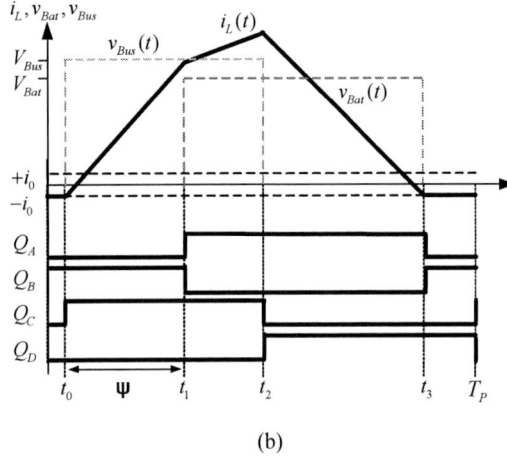

(b)

Fig. 3. Switching Waveforms during (a) Discharging, and (b) Charging Operation Modes .

III. DESIGN CONSIDERATION

A charging strategy is proposed in this paper as shown in Fig. 4.

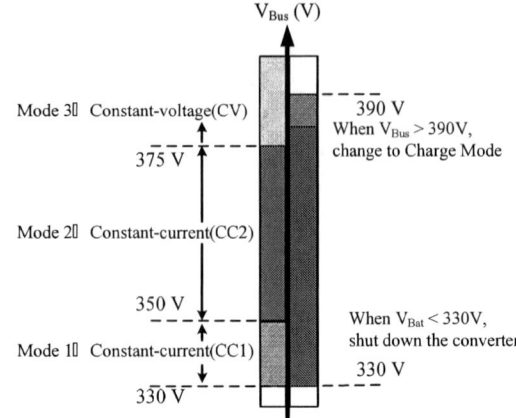

Fig. 4. The Control Strategy for the Bidirectional DC-DC Converter

The DC_BUS voltage (V_{Bus}) is regulated at 380V at normal operation. If the renewable energy is more than the load power, the DC_BUS voltage is increasing. The converter will transform to charge mode as DC_BUS voltage higher than 390V. In the Mode1, CC charging proceeds for the battery voltage between 330V and 350V. In the Mode2, CC charging is for the battery voltage between 350V and 375V. If the battery voltage grows up to 375V, the Mode3 will start to CV charging to avoid over-charging problem. Otherwise, when detected the DC_BUS voltage is over than 390V, it should be change to the charging mode. But if the battery voltage is less than 330V, the bidirectional DC-DC converter will shut down to prevent battery damage.

Fig. 5 shows the theoretical waveforms for average inductor current (i_L) under different phase-shift angles ($\phi_C > \phi_B > \phi_A$) at the same load condition. For a higher phase-shift angle, the peak-to-peak value of inductor current is higher ($i_{LC} > i_{LB} > i_{LA}$). It may cause different condition losses on power devices and reduce the conversion efficiency. Therefore, an adaptive phase-shift control is proposed and designed to reduce the reverse inductor current and then improve the efficiency performance of the studied converter.

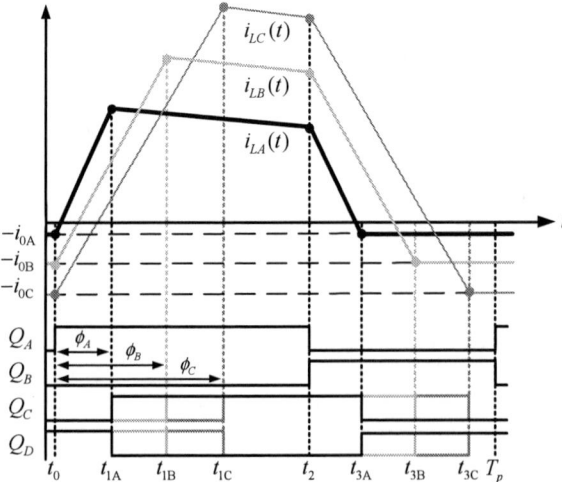

Fig. 5. The Average Inductor Current under Different Phase-shift Conditions ($\phi_C > \phi_B > \phi_A$)

Fig. 6 shows the control block diagram of the studied bidirectional DC-DC converter. This paper use the Texas Instruments DSP chip TMS320F28035 to realize the control scheme.

During the discharging mode operation, the duty cycle of power switches Q_A and Q_B is adjusted to regulate the DC_BUS voltage and the duty cycle of power switches Q_C and Q_D is fixed at 0.5. The central control system captured information of load power condition by sensing device form the DC_BUS voltage and current. One of leg that provides energy operates with 50% duty cycle while

the other leg operates with a duty cycle control to regulate the DC_BUS voltage. When the DC_BUS current reaches to the maximum value (26.5A), the converter is operated as a constant current source with a fixed phase-shift of 126°. Until the DC_BUS current is less than the preset maximum value, the phase-shift angle is adaptive in accordance with the load current variation. Through phase-shift angel control between Q_A and Q_C is varied to reduce the reverse inductor current and reach a high efficiency performance.

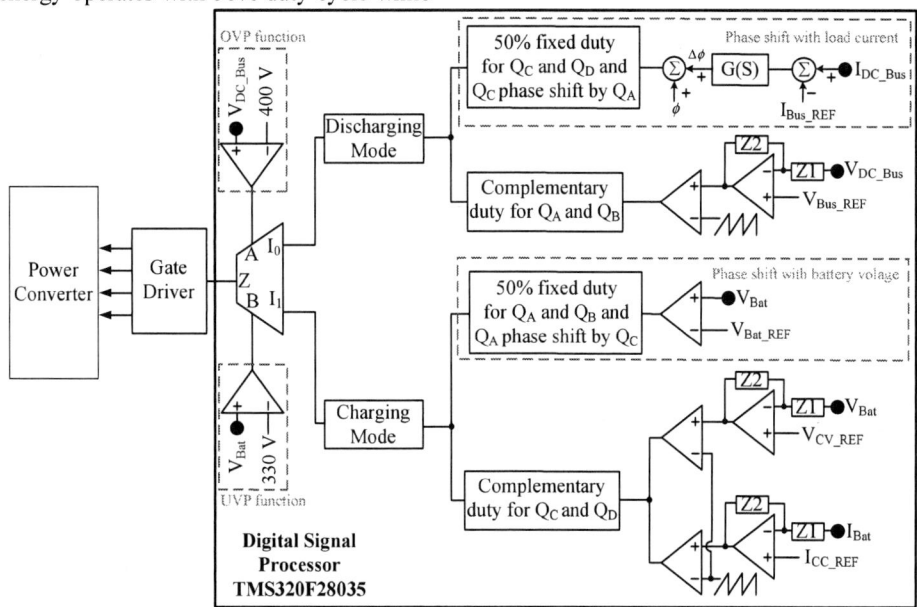

Fig. 6. Control Block Diagram for the Bidirectional DC-DC Converter

When during the charging mode operation, the duty cycle of power switches Q_C and Q_D is regulated by sensing the battery voltage and battery current and the duty cycle of power switches Q_A and Q_B is fixed at 0.5. The central control system captured information of load power condition by sensing device form the battery voltage and current. One leg that provides energy operates with 50% duty cycle while the other leg operates with a varying duty cycle to control the battery charging voltage and current. As mentioned above in Fig. 4, three-phase (CC-CC-CV) charging method is realized to satisfy the battery charging requirements. The adaptive phase-shift angle between Q_C and Q_A is 10° for Mode 1 CC charging (330V<V_{Bat}<350V), 85° for Mode 2 CC charging (350V<V_{Bat}<375V) and 10° for Mode 3 CV charging (375V<V_{Bat}) to achieve a high efficiency performance.

IV. EXPERIMENTAL VERIFICATIONS

In the final manuscripts, a 10-kW laboratory prototype was built and tested to verify the feasibility of the studied bidirectional DC-DC converter for DC micro-grid system applications. The system specifications are listed as in Table I.

TABLE I

PROTOTYPE SPECIFICATIONS

Description	Specification
Battery voltage (V_{Bat})	330 ~ 375 V
Battery capacity (C)	27 Ah
DC_BUS voltage (V_{DC_BUS})	380 V
Rated power (P_{rated})	10 kW
Maximum DC_BUS current ($I_{DC\_BUS, max}$)	26.5 A
Mode 1 charging Current (I_{CC1})	10 A (0.4C)
Mode 2 charging Current (I_{CC2})	27 A (1C)
Mode 3 charging voltage (CV)	375 V
Switching frequency (f_s)	20 kHz
Power inductor (L)	60 μH

Fig. 7 and 8 shows the measured circuit waveforms at 10 kW output load during discharging mode and charging mode. From the result, it can be observed that the experimental waveforms are agreed with the theoretical waveforms.

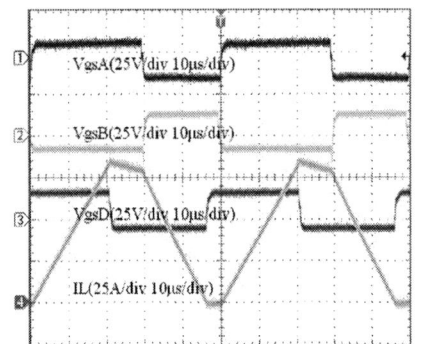

Fig. 7. Measured Circuit Waveforms under Discharging Operation Modes

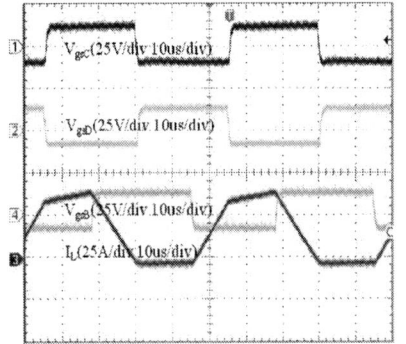

Fig. 8. Measured Circuit Waveforms under Charging Operation Modes

Fig. 9 and 10 shows the measured waveforms under different phase-shifts at 2.5 kW power output under 340V battery voltage and 380V DC-bus voltage conditions. It can be observed that the measured results are agreed with the theoretical waveforms shown in Fig. 5. For the higher phase-shift angle, the peak-to-peak value of inductor current is higher. It could cause higher conduction losses on power devices and reduce the conversion efficiency. Fig. 11 shows the measured efficiency comparison of the bidirectional DC-DC converter with and without adaptive phase-shift control. At the light-load efficiency can be effectively improved by varying the phase-shift angle in accordance with load current variations.

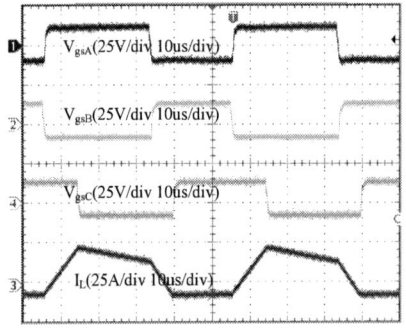

Fig. 9. Measured Waveforms without Phase-shift Control

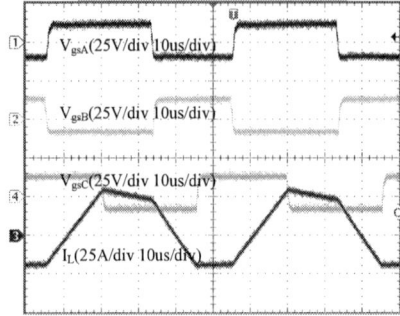

Fig. 10. Measured Waveforms with Phase-shift Control

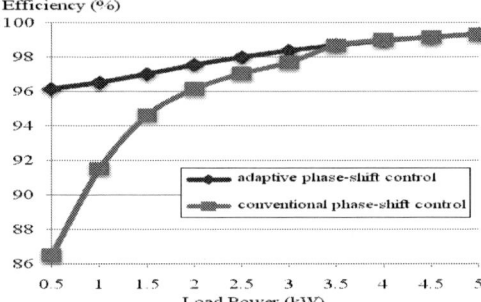

Fig. 11. Efficiency Comparison of the Converter with and without the Adaptive Phase-shift Control

Fig. 12 shows the measured waveforms for verifying the studied three-mode charging strategy. In the experiment, a 5.2kW DC electronic-load CHROMA 63204 is used as the simulated load at battery side for testing convenience. Considering the power rating of DC electronic load, the maximum load condition is set around 3kW. For Mode 1 constant current (CC1) and Mode 2 constant current (CC2) charging stages, the DC electronic-load is operated at CV mode. It can be observed that the output current is regulated at 3A for 330V (Mode 1) condition and 8.6A for 350V (Mode 2) condition. For Mode 3 constant voltage (CV) charging stage, the DC electronic load is operated at CC mode. At 0.5A constant-current load condition, the output voltage is regulated at 375V. In Fig. 8, a high current ripple presents on the inductor because the 10kW-rated prototype converter is operated at light-load condition for a 3kW testing.

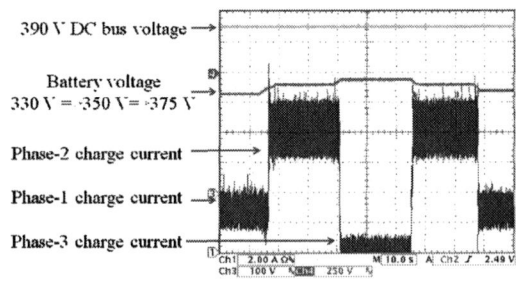

Fig. 12. Measured Waveforms for Three-phase Battery Charging

Fig. 13 shows the measured ZVS turned-on waveforms for power switches of the studied control method.

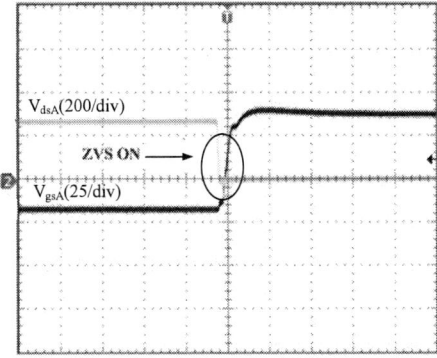

Fig. 13. Measured Waveforms for ZVS condition

The measured efficiencies under discharging and charging operation modes are listed in TABLE II and TABLE III. The peak efficiency can be up to 98% under discharging mode and 98.44% under charging mode.

TABLE II

MEASURED EFFICIENCY UNDER DISCHARGING OPERATION MODE AT V_{BAT}=340V

I_{Bat} (A)	V_{DC_BUS} (V)	P_O (W)	η(%)
3.17	387	1042	96.64
15.03	385	5029	98.42
30.12	384	10035	98

TABLE III

MEASURED EFFICIENCY AT CHARGING OPERATION MODE AT V_{DC_BUS}=390V

I_{DC_BUS} (A)	V_{Bat} (V)	P_O (W)	η(%)
2.75	355	1038	96.78
13.08	353	5026	98.53
26.11	352	10024	98.44

V. CONCLUSION

This paper presents a high performance bidirectional DC-DC converter for DC micro-grid system applications. Unlike the traditional control with hard switching, zero-voltage-switching (ZVS) is fulfilled with a novel control method to improve the conversion efficiency. The operating principles and design considerations of the studied bidirectional converter are analyzed and discussed. To satisfy the demands for fast charging and long lifetime of batteries, a three mode charging strategy is realized to prevent battery overcharging. Finally a digital signal processor (DSP) is used to implement a 10kW laboratory prototype is built and tested.Power flow control on a DC micro-grid system is also demonstrated for system safety considerations. The experimental results are shown to verify the feasibility of the proposed scheme. High efficiency can be achieved by the studied adaptive phase-shift control method and zero-voltage switching features. Peak efficiency of the prototype converter can be up to 98% under discharging mode and 98.44 % under charging mode.

REFERENCES

[1] S. Jiang, D. Cao, D. Y. Li, and F. Peng, "Grid-Connected Boost-Half-Bridge Photovoltaic Micro Inverter System Using Repetitive Current Control and Maximum Power Point Tracking," IEEE Transactions on Power Electronics, 2012. (In Press)

[2] W. S. Liu, J. F. Chen, T. J. Liang, and R. L. Lin , "Multicascoded Sources for a High-Efficiency Fuel-Cell Hybrid Power System in High-Voltage Application," IEEE Transactions on Power Electronics, vol.26, no.3, pp.931-942, March 2011.

[3] N. Lakshminarasamma, M. Masihuzzaman, and V. Ramanarayanan, "Steady-State Stability of Current-Mode Active-Clamp ZVS DC–DC Converters," IEEE Transactions on Power Electronics, vol.26, no.5,pp.1295-1304, May 2011

[4] H. Wu, J. Lu, W. Shi, and Y. Xing, "Nonisolated Bidirectional DC–DC Converters with Negative-Coupled Inductor," IEEE Transactions on Power Electronics , vol.27, no.5,pp.2231-2235, May 2012.

[5] S. Waffler and J. W. Kolar, "A Novel Low-loss Modulation Strategy for High-power Bidirectional Buck + Boost Converters," IEEE Trans. on Power Electronics, vol. 22, no. 6, pp. 1589– 1599, June 2009

[6] M. Pahlevaninezhad, P. Das, J. Drobnik, P. K. Jain, and A. Bakhshai, "A Novel ZVZCS Full-Bridge DC/DC Converter Used for Electric Vehicles," IEEE Transactions on Power Electronics , vol.27, no.6, pp.2752-2769, June 2012

[7] F. A. Himmelstoss and M. E. Ecker, "Analysis of a Bidirectional DC-DC Half-bridge Converter with Zero Voltage Switching," in Proc. International Symposium on Signals, Circuits and Systems, ISSCS' 05, July 2005, vol. 2, pp. 449-452.

[8] H. Qian, J. Zhang, J.-S. Lai, and W. Yu; , "A High-Efficiency Grid-Tied Battery Energy Storage System," IEEE Transactions on Power Electronics , vol.26, no. 3, pp.886-896, Mar. 2011

AUTHOR INDEX

Abe, Kodai 3153
Abe, Seiya 177, 1179, 2216, 2222, 3652
Abe, Shigeru 1109, 1115
Abe, T. 3007
Abe, Takashi...............................2183, 2189, 3024
Abe, Tomohiko 1575
Abiko, Hiroshi 634
Achara, Pichetjamroen....................................3687
Adachi, Mitsuo 92
Adhikari, Jeevan 1775
Agelidis, Vassilios G. 1458, 3758, 3764, 3933
Agelidis, Vassilios Georgios....................................640
Aguglia, Davide 3371
Ahmed, Furqan 480, 790
Ahssanuzzaman, S. M.....................................3582
Aiso, Kohei...1141
Ajima, Toshiyuki..................................383, 682
Akagi, Hirofumi 750, 1586, 1761, 2290, 2323, 3742
Akagi, Masataka 629
Akahane, Masashi...2302
Akatsu, Kan.................. 1128, 1141, 1234, 2673, 3828
Aketa, M. 2074
Akira 3784
Akiyama, Satoru 2285
Alemi, Payam 1201
Alipoor, Jaber...3298
Aljankawey, A. S. 2156
Allen, Scott...3447
Almer, Stefan..3563
Ama, Naji Rajai Nasri 2413, 2988, 3278
Amanci, Adrian Z. 1303
Amano, Yuki 1824
Amma, Ryosuke 2027
Anazawa, Yoshihisa 3801
Andersen, Michael A. E......... 78, 506, 2842, 3352, 3905
Ando, Itaru 1516
Ando, Masato 1317
Anthon, Alexander 78
An-Yeol Ko 796
Aoki, Mutsumi...2400
Aoyagi, Shigehisa 2451
Aoyama, Fumio 2644
Aoyama, Kohei 2266
Aoyama, Masahiro...1405
Aoyama, Tomohiro 3823
Ara, Takahiro..3044
Arai, Haruki 403
Arai, Manabu 3440
Araki, Jun 1728
Araki, Takahiro 1613
Arata, Masanori...1874
Arikawa, S. 3007
Arimatsu, Kenji 415
Arita, Hideaki...2673
Asai, Inami..123

Asakimori, Koki 1567
Asama, Junichi 988
Asano, Katsunori 3440
Asano, Yoshinari 1997
Asano, Yuji 3872
Ashikaga, Tadashi 1886
Athab, Hussain S. 3695
Atsushi, Manabe 2745
Awaji, Sosuke...3194
Ayano, Hideki 2385
Azuma, M. 1892
Baba, Jumpei 1849
Babasaki, Tadatoshi 1567
Bac Xuan Nguyen 2722
Bafleur, M. 707
Bahman, A. S. 2862
Bahrani, Behrooz 1386
Bak, Claus Leth 3320
Bakran, Mark-M. 2113, 3255
Bang, Deok-Je 2427
Bani Shamseh, Mohammad 2794
Baoquan Liu 1155, 3546
Barater, Davide 433
Barrade, Philippe 1081
Barth, Henry..2881
Basari, Amat A. 3194
Basu, Kaushik 3061
Bauer, Florian 3898
Bauer, Pavol 1193, 3200
Baumgartner, Thomas 1707
Beczckowski, Szymon 2547
Beczkowski, S. M. 2862
Beczkowski, Szymon 2850
Belanger, Jean 2644
Ben Guo 3129
Ben, Hongqi 2318
Beres, Remus 3320
Berhouet, S. 707
Berkouk, El Madjid..560
Bessegato, Luca 1087
Besselmann, Thomas.......................................3563
Bhat, Ashoka K. S. 1721
Bhattacharya, Subhashish 651, 656,
758, 1626, 2562, 3225, 3286, 3447, 3726
Bianda, Enea 3432
Biela, J. 868
Biela, Jurgen 1788
Bilal, Akin 230
Bin Wu 3482, 3695
Binbin Li 3680
Bizen, Yosio..2983
Blaabjerg, F. 548, 1912, 2862
Blaabjerg, Frede........................... 216, 857, 1529,
1634, 1801, 2610, 3320
Blank, Frederic...264

AUTHOR INDEX

Bo Wen...944
Bocker, J..2887
Bocker, Joachim.......................346, 1501, 1508
Boehm, Andreas.......................................283
Boillat, David O......................................1073
Boitier, V...707
Boroyevich, Dushan....................944, 2626, 3850
Bortis, D...1291
Bortis, Dominik........................1309, 2079, 3864
Bosshard, R...2167
Bosshard, Roman......................................1904
Boyu Wang..2893
Braz Cardoso, F.......................................3225
Burger, Niklaus.......................................1386
Burgos, Rolando.......................944, 2626, 3850
Burkart, Ralph M......................................891, 3460
Buticchi, Giampaolo...................................433
Byoungchang Jung.....................................1185
Byung Moon Han.......................................937
Byung-Geuk Cho.......................................2802
Byung-Gyu Yu...3784
Cai, Zheng-Xiu..429
Canales, Francisco....................1043, 3432
Cao, Guoen...2587
Cao, Wei...567
Cao, Yuan..647
Cardoso, Braz J.......................................3270
Caris, M. L. A..2954
Carvalho, Eden Luiz...................................1276
Casolari, Ronaldo Pedro...............................1276
Castellazzi, A..2503
Castellazzi, Alberto...........433, 2920, 3718
Ceballos, Salvador....................3758, 3764
Cha, Honnyong.......................110, 480, 790
Chai Feng..3129
Chang, C.-H..2050
Chang, Chien-Hsuan...................2523, 3333
Chang, Hsiu-Feng.....................................330
Chang, Kai-Chi.......................................105
Chang, L...2156
Chang, Yuan-Chih.....................330, 1832
Changsheng Hu..782
Changwoo Kim..1646
Chao Wang...2950
Chao-Fu Wang..2758
Chattopadhyay, Ritwik.................................3225
Chen, Ching-Guo......................................1734
Chen, H..1471
Chen, Hsin-Chih......................................1639
Chen, Hung-Chi.......................................2580
Chen, Jiann-Jong.....................................2910
Chen, Jung-Chieh.....................................677
Chen, Min..485
Chen, Qianhong.......................................1425
Chen, Shen-Li..236

Chen, Wei..66, 72
Chen, Wenjie...2950
Chen, Yaow-Ming......................................3592
Chen, Ying-Zuo.......................................351
Chen, Zhe..3538
Chen-Feng Chuang.....................................3379
Cheng Deng...782
Cheng, Chun-An.......................2523, 3333
Cheng, Hung-Liang....................2523, 3333
Cheng, Po-Tai........................1261, 1639
Cheng, Shih-Jen......................199, 2593
Cheng, Stone...3425
Cheng-Chieh Yu.......................................2910
Cheng-Wei Chen.......................................3592
Cheol-O Yeon...1738
Cheon, Jun P...3358
Chia-Chi Chu...3379
Chiang, Hsin-Wei.....................................2100
Chiba, Akira.........................982, 988, 3513
Chiba, Yoshinori.....................................634
Chien-Yu Lin...2758
Chih Wei Chen..3938
Ching-Hsiang Yang....................................1639
Ching-Tasi Pan.......................................3379
Ching-Wei Wang.......................................1639
Chiu, Chian-Song.....................................440
Chiu, Huang-Jen..........172, 199, 2593, 3328
Chiu, Tse-Wei..440
Cho, Bo-Hyung.......................2272, 2575
Choi, Bo H..........................2232, 3358
Choi, Byungcho.......................................3638
Choi, Hangseok.......................................2575
Choi, Seong-Chon.....................................409
Choi, Sewan.........................1394, 2247
Choi, Su Y...1103
Chokchai, Chuenwattanapraniti.........................3789
Chou, Tzu-Han.......................208, 421
Chow, T. Paul..2208
Chu, Xi..1322
Chun, Chang Yoon.....................................2272
Chung, Tsung-Yuan....................................2523
Chung-Chuan Hou......................................2821
Chung-Yi Lin...3185
Chung-Yuen Won......................796, 3532
Chunkag, Viboon......................................694
Chu-Shen Chang.......................................3928
Ciftci, Baris..3734
Coldevin, Grete H....................................1861
Colmenares, Juan.....................................3712
Concari, Carlo.......................................433
Cortes, Patricio.....................................3864
Cortizio, Porfirio C.................................3225
Cosovic, Mirsad......................................1148
Daesu Han...1185
Dahidah, Mohamed S. A...............................1283

AUTHOR INDEX

Dahono, Pekik Argo.................................3893
Dai, Wei-Fu 330, 1832
Daikoku, A...1892
Daikoku, Akihiro 2011, 2673
Dan Chen 3938
Darba, Araz 718
D'Arco, Salvatore..................................1544
Darus, Rosheila 3758, 3764
Daskalos, Mike 2330
Davoodnezhad, R....................................1482
Dawson, Francis P...................................1303
De Belie, Frederik 718
De Carvalho, Kelly Caroline
Mingorancia 2413, 2618, 2988, 3278
De Doncker, R. W....................................3898
De Doncker, Rik W.................736, 2729, 3145
De Haan, Sjoerd 2787
De Mallac, Louis 3371
De Miranda, Rubens Domingos 1276
De Paula, Helder 3225
De S. Brito, Jose A.................................3225
De Vega, Angel Ruiz 2547, 2850
De, Ankan 651, 2562, 3286, 3447
De, D..2503
De, Dipankar..433
Deguchi, Tadayoshi..................................3440
Dehong Xu 782
Dekka, Apparao 3468
Demetriades, Georgios D.............................1220
Deng, Lirong..465
Dianguo Xu 3174, 3680
Dianguo, Xu 341
Diduch, C. P..2156
Dilhac, J-M...707
Diniz, Rogerio Azevedo..............................3270
Doki, Shinji 907, 2445, 3079, 3823
Domoto, Kazuhide 3652
Dong Le 1837
Dong-Hee Lee 994, 2693
Dong-Jing Lee 1452
Dongkook Son 2914
Dongouk Kim 925
Dongwook Kim 2914
Dou, Qinyun 3604
Dowaki, Kiyoshi 1207
Do-Yun Kim 796
Drofenik, Uwe.......................................1043
Du Yan 2668
Du, Yimian 1721
Duarte, J. L..2954
Dujic, Drazen.......................................3476
Durand Estebe, P. 707
Dutta, R..2679
Endo, Takahisa...............................2541, 2977
Eni, E..1912

Enomoto, Toshio.....................................2421
Enomoto, Yuji.......................................1997
Erturk, Feyzullah 3734
Eui-Cheol Nho 2763
Fang Zheng 1342
Fang Zhuo 1155, 3546
Fang, Xiaocun 335
Fassler, Lukas 3864
Fei Lin 807
Fei Meng..2815
Fei Zhang 3857
Fernandes, B. G.....................................2433
Ferrari, Bruno Augusto 3278
Ferreau, Joachim 3563
Ferreira, J. A......................................1935
Ferreira, Jan A.....................................2787
Figueredo, Ricardo Souza......................2413, 2618
Fletcher, J. 2679
Fletcher, John 2926, 2932
Foo, Gilbert..2722
Fosso, Olav B. 1861
Foureaux, Nicole C. 3225
Fournier-Bidoz, Sebastien 3496
Franca, Gleisson J. 3270
Franceschini, Giovanni 433
Franke, Toke..78
Fritz, Dominik 3476
Frohleke, N. 2887
Fujii, Junji 1654
Fujii, Kansuke 1748
Fujii, Toshiyuki 2663
Fujimoto, Hiroshi 1671, 2421
Fujimoto, Takafumi 3857
Fujimoto, Yasutaka 1685, 1968
Fujisaki, Keisuke 289, 2856, 2874
Fujisawa, Hiroyuki 3440
Fujita, Hideaki...........1006, 1160, 1350, 2027, 2042
Fujitsuna, Masami...................................3079
Fuketa, Hiroshi 2228
Fukuda, Kenji 3440
Fukuhara, Shuhei 289
Fukumoto, Hisao 724, 730, 3067, 3249
Fukuoka, Hiroki 3341
Fukushima, Kentaro 2189
Fulin Zhou 1050
Funabiki, Shigeyuki 2470
Funaki, Tsuyoshi 1621
Funato, Hirohito 1728, 2517
Furukawa, Kimihisa..................................383
Furukawa, Tatsuya 724, 730, 3067, 3249
Furukawa, Yutaka 2252
Furuta, R...2120
Gaing, Zwe-Lee 278
Gao Qiang 3050
Gao, Qiang 614

AUTHOR INDEX

Geng, Hua 543
Gerling, Dieter 774
Ghimire, Pramod 2547, 2850
Giaretta, Antonio Ricardo..............1276
Goehler, Lutz...............................2554
Goh Teck Chiang 1028
Gohara, Hiromichi..........................671
Goto, Akira 130
Goto, Yasuyuki..............................1490
Goto, Yuichi 1671
Graus, Johannes.............................270
Grider, Dave 3726
Gruber, Wolfgang 1691, 1701
Gu, Beom W. 1103
Gueldner, Henry 2554
Guidi, Giuseppe 1544
Gunasekaran, Deepak........................1342
Guo, Wei 160, 475
Gurpinar, Emre 433, 3718
Ha, Jung-Ik 3140
Hafner, Jurgen 3667
Haga, Hitoshi.......................415, 3153
Haghbin, Saeid 1373
Hagiwara, Makoto 1586, 1761, 2323, 3742
Hahn, Ingo 270, 283
Haining Wang 3702
Haitao Yang 782
Hak-Soo Kim 2763
Hakutou, Takuma 2297
Hama, Ryota 2470
Hamasaki, Shin-Ichi 2775, 3674
Hamasaki, Sin-Ichi 3093
Hamazaki, Yasuhiro 2126
Hanada, T. 2074
Hanamoto, Tsuyoshi 538, 1811
Han-Shin Youn 1743
Hao Huang...................................2967
Hao Yi 1155, 2960, 3546
Hao, Xiang 2950
Hao-Chien Cheng 3379
Hara, Hidenori 1654, 1898
Harada, Shingo..............................1671
Harada, Shinsuke 3440
Harakawa, Masaya 2638
Hariya, Akinori 3630
Hasegawa, Isamu 1365
Hasegawa, Kohei............................3707
Hasegawa, Masaru...............183, 907, 2445
Hasegawa, Masataka 2212
Hasegawa, Shinya 294, 299, 2972, 3055, 3159, 3162
Hasegawa, Tomonori 2126
Hashimoto, Kento 1974
Hashimoto, S 1471
Hashimoto, Seiji...........................3194
Hashimoto, Shizuka 3018

Hassanpoor, Arman 3667
Hatanaka, Ayumu 2285
Hattori, Fumiya 811
Hatua, Kamalesh 758, 1626
Hau-Chen Yen 3928
Hava, Ahmet M. 498, 2034, 3734
Hayase, Masanori 1207
Hayashi, Makoto 1950
Hayashi, Toshihiko 3440
Hayashi, Yusuke 1560
Hayashiya, Hitoshi 1062
Hazeyama, M. 1892
Hazra, Samir...............758, 1626, 3447
He, Guofeng 485
Hee-Jun Lee 3532
Hei, Xinhong 647
Hella, Mona 2208
Hermansson, Willy 1220
Hernandez, Juan C. 3352
Heung-Geun Kim 790, 2763
Hibino, Shinya 2638
Hidaka, Akira 2216
Hidayat, Nabil M 573, 2529
Higuchi, Shinichi 1522
Higuchi, T. 3007
Higuchi, Tsuyoshi 3024
Hijikata, Hiroki 2673, 3828
Hikita, Masayuki 689
Hinata, Toshifumi 919
Hinkkanen, Marko 2489
Hino, Wataru 3525
Hintz, Andrew 2343
Hira, Yuki 730
Hirahara, Hideaki 3044
Hirai, Junji 1974
Hirakawa, Yuki 3024
Hiraki, Eiji 3292
Hirano, Yosei 1956
Hirao, Kuniaki 191, 1365
Hirase, Yuko 1552
Hirokado, K. 146
Hirose, Toshiro 2252
Hirota, Yukitsugu 1728
Hisada, Yoshihiro 3292
Hisato, Hosoyama 2745
Ho, Kung-Min 3942
Hoene, Eckart 2366
Hoffmann, Stefan 2366
Hofmann, Wilfried 2881
Hojo, Masahide.............................2152
Hojo, Toshiaki 1276
Hokazono, Hiroaki 2870
Holm, Toni 3432
Holmes, D. G 1482, 2019, 2372, 3306
Homma, Hiroshi 1880

AUTHOR INDEX

Hong Li 3314
Hong, Ki-Nam 2598
Hong-Hee Lee 1013
Hongqi Ben 3213
Hori, Yoichi 2421
Horiguchi, Takeshi 2290
Horita, Yasuhisa 1317
Hosaka, Tatsuya 1350
Hoshi, Nobukazu 1242
Hosoyamada, Yu 801
Hou, Chih-Hao 1796
Hou, Jiaxin 526
Hou, Lixiang 577
Hredzak, Branislav 1458
Hsieh, Guan-Cyun 526
Hsieh, Hung-I 526, 2380
Hsieh, Min-Fu 278
Hsieh, Yao-Ching 429, 1796
Hsin-Chih Chen 1261
Hsin-Ping Su 2821
Hu, Jia-Sheng 278
Hu, Shang-Hung 2606
Hu, Sheng 555
Hu, Taiyuan 335
Huang, Hsin-Wei 421
Huang, Jia-Wei 3233
Huang, Lang 2950
Huang, Min 2610
Huang, T. D. 2195
Huang, Wen-Nan 1734
Huang, Zhenhui 647
Huang-Jen Chiu 2758, 2810, 3185, 3913
Huber, Jonas E. 766
Huber, Tobias 1508
Hui Liu 1634
Hui Zhang 1365, 3455
Huisman, H. 2954
Hull, Brett 3447
Huu-Nhan Nguyen 1013
Hwang, Seon-Hwan 2427
Hwang, Yuh-Shyan 2910
Hwu, K. I. 204, 2754, 3190, 3392
Hyoyol Yoo 1646
Ichihara, Junichi 2189
Ichiya, Takahiro 370
Ichiyanagi, Katsuhiro 1490
Ide, Kozo 933
Ieda, Jun 2663
Iga, Yuichi 3341
Igarashi, Kazunori 2983
Igarashi, S. 3702
IIda, Mikiya 2977
IIjima, Ryuji 117
IIjima, Yukihia 2095
Ikawa, O. 2569, 3702

Ikeda, Hidehiro 2476
Ikeda, Masahiro 2183
Ikeda, Tomohiko 1575
Ikeda, Y. 2569
Ikeda, Yoshinari 2870
Il-Kuen Won 796
Ilves, Kalle 1087
Imai, Jun 2470
Imakiire, Akihiro 689
Imamura, Yasutaka 863
Imanishi, Takao 2663
Imaoka, Jun 811, 883, 2497
Inamori, Mamiko 3509
In-Dong Kim 2763
Inomata, Kentaro 1654
Inoue, Kaoru 3872
Inoue, Keita 130
Inoue, M. 1892
Inoue, Tatsuki 363
Inoue, Y. 246, 258, 312, 390
Inoue, Yukinori 324, 356, 363, 370, 2183, 3018, 3519
Irokawa, Shoichi 1357
Ise, Tomofumi 1430
Ise, Toshifumi 1536, 1560, 2632, 3298, 3687
Ishibashi, Makoto 724
Ishida, Koichi 2228
Ishida, M. 146
Ishida, Masaaki 3707
Ishida, Masaki 3162
Ishida, Takahito 634
Ishigami, Takashi 1880
Ishigma, Satoru 403
Ishihara, Chio 1984
Ishihara, Yuji 1135
Ishii, Hirotaka 294
Ishikawa, Hiroki 1135, 2183, 2189
Ishikawa, Katsumi 2140, 2285
Ishikawa, Takeo 252, 1697
Ishimaru, Yusuke 92
Ishimori, Hitoshi 3440
Ishitobi, Manabu 811
Ishizuka, Tomotsugu 2644
Ishizuka, Yoichi 2222, 2252, 2737, 3630, 3652
Isida, Takashi 1950
Itako, Kazutaka 3244
Ito, Yasuhide 3823
Ito, Yoichi 403
Itoh, Hideaki 724, 730, 3067, 3249
Itoh, Jum-Ichi 1943
Itoh, Jun-Chi 1253
Itoh, Junichi 130
Itoh, Jun-Ichi 84, 138, 152, 191, 682, 1021, 1028, 1095, 1613, 2277, 3659, 3815
Itoh, Tomomichi 850
Itoh, Youichi 415

AUTHOR INDEX

Itoh, Yuki 883, 2497
Iwaji, Yoshitaka 2451
Iwakami, Tetsuro 817
Iwasaki, Makoto 1665
Iwasaki, Shinya 2663
Iwata, Tetsuki 403
Iyer, Kartik V 3037, 3061
Iyer, Shivkumar 3482
Izumi, Toru 3440
Jacobson, Bjorn 3667
Jae-Bum Lee 1738
Jaeho Choi 2656
Jae-Hun Jung 2763
Jae-Hyun Kim 1738
Jaesig Kim 2656
Jang, Jinhaeng 3638
Jang, Young-Jin 664
Jang-Hwan Kim 925
Jardini, Jose Antonio 1276
Jauch, Felix 1788
Javed, Riffat 624
Jayoon Kang 1185
Jen-Hao Teng 1452
Jenn-Jong Shieh 3190
Jeon, Jin-Yong 166
Jeong, Seog Y 1103
Jeong, Seon-Yeong 2406
Jeongjoong Kim 1185
Jhen-Yu Jian 3928
Jia Liu 1536
Jia, Y. 1594
Jianfeng Li 3718
Jiang, Dawang 647
Jiang, Maoh-Chin 105
Jiang, W. 1471
Jiang, W. Z. 204, 3190
Jiang, Yongjie 458
Jianhui Su 2668
Jiann-Fuh Chen 2714
Jianwen Zhang 3124
Jih-Hua Yu 2910
Jin Miaoxin 3050
Jin, Miaoxin 614
Jin, Xu 341
Jin, Yasuhiro 3207
Jing Bian 3314
Jing-Hsiao Chen 3233
Jing-Yuan Lin 2758
Jinjun Liu 835
Jinno, Masahito 1781, 3333
Jin-Woo Ahn 994, 2693
Jinyong Zhang 3213
Ji-Shiang Lee 3346
Joebges, Philipp 2729
Jokipii, J. 514, 2240
Jokipii, Juha 1466
Jong Kyou Jeong 937
Jonghyung Park 1185
Jonishi, Akihiro 2302
Jou, Sung-Tak 224
Jung, Hochang 1990
Jung, Jae-Jung 1268
Jung, Sang-Yong 1990
Jung, Yong-Chae 166, 409
Junghum Lee 2656
Junjie Feng 835
Juntao Fei 3168
Juyoung Jang 2656
Kabasawa, Yuichiro 2175
Kabiri, R. 3306
Kadavelugu, Arun 758, 1626, 3726
Kai, Masahiko 1054
Kai-Hui Chen 2750, 3346
Kaipia, T. 587
Kajiwara, Kazuhiro 3950
Kakishima, Takeo 3513
Kalogera, Maria 1193, 3200
Kameshiro, Norifumi 2140
Kamikura, Mamoru 2064
Kamnarn, Uthen 694
Kanagawa, Kinji 2983
Kanai, Yasuyuki 1567
Kanamori, Masaki 2541, 2977
Kaneko, Junji 2745
Kaneko, Yasuyoshi 1109, 1115
Kanematsu, Masato 2421
Kanemoto, Daisuke 2737
Kang, Feel-Soon 2260
Kang, Yong 555
Kanno, Hiroshi 2302
Kano, Yoshiaki 2004, 2457
Kanoda, Akihiko 3920
Kanouda, Akihiko 2058
Kantar, Emre 2034
Kanthaphayao, Yutthana 694
Kari, Mat Nasir 573
Karki, Ujjwal 1342
Karvonen, Andreas 1373
Kasai, Makoto 3194, 3194
Kashihara, Yugo 1943
Kasper, Matthias 2079
Katade, Motohumi 130
Katakami, Shuji 3440
Kataoka, Yasuhiro 3801
Katayama, Noboru 1207, 1227
Kato, Hideaki 2972, 3162
Kato, Koji 403, 415
Kato, Shinji 2175
Kato, Takashi 3828
Kato, Taro 2972

AUTHOR INDEX

Kato, Tomohisa 3440
Kato, Toshiji 2183, 2189, 3872
Kato, Yutaka 2644
Katoh, Kaoru 2285
Katoh, Shuji 850
Katsuki, Akihiko 1575, 3624
Katsura, Seiichiro 1679
Kawachi, Konosuke 863
Kawaguchi, Shinichi 3959
Kawahara, Keiji 1062
Kawakami, Noriko 2095
Kawamura, Atsuo 801, 2266, 2794, 3403
Kawamura, Mitsuhiro 3012
Kawamura, Wataru 3742
Kawano, Daisuke 1671
Kawano, Kenji 883
Kawazoe, Yosuke 2011
Kazuya, Ogura 452
Kempen, S. 2887
Kenji, Matsumoto 3218
Kern, Ansgar 712
Khan, Ashraf Ali 110
Khan, Faisal H. 2161
Khant, Hlaing Kyi Pyar 183
Khomfoi, Surin 2392
Kicin, Slavo 3432
Kihyun Lee 1646
Kikuchi, Takuya 1328
Kim, Bong C. 3358
Kim, Chong-Eun 1738, 1743
Kim, Dong-Hun 790
Kim, Dong-Rak 409
Kim, Hee-Jun 2587
Kim, Heung-Geun 110, 480
Kim, Hyejin 2575
Kim, Jae-Hyun 1743
Kim, Jang-Mok 2406
Kim, Ji H. 2232
Kim, Ji-Won 2427
Kim, Jonghoon 619
Kim, Minjae 2247
Kim, Seonghye 2260
Kim, Su-Han 480
Kim, Sungmin 1268
Kim, Yong-Jae 1990
Kimoto, Tsunenobu 3440
Kimura, Hiroshi 1920
Kimura, Noriyuki...................1299, 1806, 2183, 3341
Kimura, Shota 883, 2497
Kinouchi, Shin-Ichi 750, 2290
Kish, Gregory J. 951
Kitabayashi, Tatsuaki 2517
Kitagawa, Wataru 2310, 3809
Kitajima, Jun 1247
Kitazawa, Satochi 1438

Kiyota, Kyohei 3513
Kleinecke, John 2330
Kluge, Andreas 2554
Knott, Arnold 506
Kobayashi, H. 2569
Kobayashi, Hiroya 2517
Kobayashi, Ryota 1115
Kobayashi, Takenori 1868
Kobayashi, Y. 2569
Kodama, Takashi 1365
Kogi, Ryosuke 2874
Kogoshi, Sumio 1207, 1227
Kohama, Teruhiko 522, 2781
Kohno, Yusuke 2183
Koiwa, Kazuhiro 84, 130, 1028
Kolar, J. W. 1291, 2167
Kolar, Johann W. 766, 821, 891, 899, 975,
    1073, 1309, 1707, 1904, 2079, 2834, 3365,
    3460, 3864
Komada, Satoshi 1974
Komatsu, Wilson 1276, 2413, 2988
Komeda, Shohei 1160
Komiya, Hiroshi 2421
Kon, Saytaro 3263
Kondo, Keiichiro 1438, 2126
Kondo, Seiji 415
Kondo, Takeshi 1365
Kondou, Masahiko 2421
Kono, Y. 2120
Kono, Yasuhiko 2140
Konoto, Masaaki 2189
Konstantinou, Georgios 1458, 3758, 3764
Korner, Olaf 2113
Kosaka, T. 2438
Kosaka, Takashi 1984, 1997
Koschik, Stefan 3145, 3898
Koseki, Takafumi 1334, 2126
Kotegawa, Ryo 1317
Kotera, Keito 3872
Kouki, Matsuse 3134
Kouno, Yusuke 2189
Kounoto, Masaaki 2175
Koyama, Masato 750
Krafft, Eberhard 2113
Krismer, Florian 2834
Kuan-Hsien Chou 3346
Kubo, H. 395, 1594
Kubo, Hajime 1601
Kubo, Yuji 3134
Kubota, Hisao 919, 1929, 3119, 3134
Kubota, Yutaka 2183
Kudo, Takahiro 1109
Kuga, Shotaro 3955
Kukita, Akio 1444, 2351
Kumagai, Shunji 3194

AUTHOR INDEX

Kumakura, Yoshito 1715
Kume, Tsuneo 1898
Kumsuwan, Yuttana 3417
Kun-Hung Chen 3592
Kunomura, Ken 1054
Kuo, Kuan-Yi 278
Kuperman, A. 2240
Kurabayashi, Toshiyuki 1962
Kuribayashi, H. 2569
Kurihara, Takeshi 299
Kurihara, Yoshihiro 1874
Kurita, Nobuyuki 252, 1697
Kuroda, Y. 1892
Kurokawa, Fujio 2108, 3611, 3950
Kusaka, Keisuke 191
Kusukawa, Jumpei 2904
Kusunoki, Hironobu 2330
Kutsuki, Tomohiro 2064
Kuwahara, Akinobu 3179
Kuzumaki, Atsuhiko 1929
Kwasinski, Alexis 2649
Kwasinski, Andres 2649
Kwon, Soon-Kurl 2359
Kyungbae Lim 2656
Kyungmin Sung 744
Kyungsub Jung 1646
Lai, Yen-Shin 3942
Lamantia, A. 2503
Lana, A. 587
Lang, Klaus-Dieter 2366
Larsson, Tomas 1220
Laska, Bernd 2113
Law, Kah Haw 1283
Le Hoai Nam 3659
Lee, Chia-Tse 1639
Lee, Dong-Choon 1201, 2406
Lee, Eun S. 1103, 2232, 3358
Lee, Hong-Hee 2826
Lee, Jae-Bum 1743
Lee, June-Hee 493, 532
Lee, June-Seok 493, 532
Lee, Kyo-Beum 224, 493, 532
Lee, Min-Hua 236
Lee, Seong Ryong 3292
Lee, Shiu-Hui 1734
Lee, Sung W. 1103
Lee, Taeck-Kie 595
Lee, Tzung-Lin 2606
Lee, Woo-Cheol 595
Lee, Ya-Ting 440
Lee, Yuang-Shung 208, 421
Lehmann, Oliver 3085
Lehn, Peter W. 951
Lei, Wanjun 160, 475
Leibl, Michael 899

Lelie, Markus 2729
Leslie, Scott 3726
Leuenberger, D. 868
Leuer, Michael 346
Li Yan 2899
Li, Ding 341
Li, Haiqing 2095
Li, Hong 2893
Li, Ning 160, 475
Li, Qian 2161
Li, Yanxiang 3002
Lian, K. L. 2195
Liang Hao 3174
Liangyi Tang 3695
Liao, Jhen-Yu 2580
Lie Guo 3489
Lin Cheng 3447
Lin, Chiao-Chien 3072
Lin, Chia-Yu 1832
Lin, Chien-Yu 172
Lin, Chung-Yi 199, 2593
Lin, Fei 335, 1322, 2133
Lin, Jing-Yuan 172
Lin, L.-C. 2050
Lin, Z. Y. 1471
Lindberg-Poulsen, Kristian 2842
Liping Zheng 1837
Liserre, Marco 857, 3320
Liu, Baoquan 577
Liu, Fang 567
Liu, Fangcheng 3604
Liu, Fuxin 458, 2768
Liu, Hanchao 967
Liu, Jianyu 614
Liu, Jilong 66, 72
Liu, Jinjun 624, 2815, 3604
Liu, Kangzhi 3568
Liu, Ning 2156
Liu, Rongqiang 3099
Liu, Tai-Chun 105
Liu, Xiankai 647
Liu, Xiaosheng 3002
Liu, Yi-Hua 3233
Liu, Yu-Chen 199, 2593
Liuchen Chang 1476, 2668, 3842
Lo, Yu-Kang 172, 199, 2593
Lobsiger, Yanick 1309
Loh, Poh Chiang 216, 1529, 1634, 1801, 2610
Longlong Zhang 782
Lopez-Arevalo, Saul 3718
Lovatt, Howard 2679
Low, K. S. 446
Lu, Dylan D. C. 3553
Lu, Kao-Yi 105
Luo, Guomin 2145

AUTHOR INDEX

Luthardt, Sven 3029
Ma, K. 548, 2862
Ma, Weigang 647
Madawala, Udaya K. 2722
Madhusoodhanan, Sachin 656, 1626
Maekawa, Sari 919, 1929
Maemura, Akihiko 1898
Maeyama, Shigetaka 1575, 3624
Maezono, Paulo Koiti 1276
Mahdavikhah, Behzad 3582
Mainali, Krishna 758, 1626
Makaino, Yuki 914
Makita, Shinji 3823
Mamun, Mostafa 97
Manias, Stefanos N 1606
Mannen, Tomoyuki 2042
Manolas, Iakovos 1606
Mao, Meiqin 2156
Maret, C. 3239
Marrero Sosa, Juan Alberto 3476
Martinz, Fernando Ortiz 2413, 2988, 3278
Maru, Naoki 2285
Marukawa, Yasuhiro 1984
Marumori, Hiroki 3055
Maruta, Hidenori 3611
Marz, Andreas 2113
Marzouk, Ahmad Diab 3496
Masaki, Kenji 2663
Mashino, Masahiro 3162
Masic, Semsudin 1148
Maskell, D. L. 3598
Masuda, Hiroyuki 92
Masui, Takeshi 1317
Masutomo, Kazufumi 3624
Masuzawa, Hiroshi 1054
Masuzawa, Takashi 2366
Matakas, Lourenco 2413, 2618, 2988, 3278
Matsubara, Masakatsu 1874
Matsuda, Katsuhiro 415
Matsuhashi, Daiki 1886
Matsuhashi, Masataka 1516
Matsui, Hitoshi 1586
Matsui, Keiju 183
Matsui, Mikihiko 3489
Matsui, N. 2438
Matsui, Ryota 1128
Matsui, Yoshihiro 2385
Matsui, Yoshinobu 2745
Matsumoto, Akira 1560
Matsumoto, Atsushi 2445
Matsumoto, Kazushi 3440
Matsumoto, Satoshi 2216
Matsumoto, Shuhei 1929
Matsumoto, Yasushi 1920
Matsuo, Hirofumi 1781

Matsuo, Keisuke 1886
Matsuo, Yusuke 1671
Matsuoka, Kazumasa 3207
Matsuoka, Yuji 744
Matsushima, Yoshitarou 3801
Matsushita, Makoto 3012
Matsuura, Kei 1516
Matsuura, Ken 3630
Matsuzaki, Ryohei 1978
Mattavelli, Paolo 3850
Mattsson, A. 587
Mauerer, M. 1291
McGrath, B. P. 1482, 2019, 2372, 3306
McLean, Kenneth 3496
Meiqin Mao 1476, 2668, 3842
Mekhilef, Saad 560, 3574
Melkebeek, Jan 718
Meng, Fei 624
Meng, Tao 2318
Merahi, Farid 560
Messo, T. 514, 2240
Messo, Tuomas 1466
Mihara, Teruyoshi 1728
Mii, Kenji 2737
Miiura, Yushi 1430
Mikihiko 3784
Mills, Liam 3718
Ming Yang 3174
Mingfei Wu 3553
Mingyan Wang 3129
Mino, Kazuaki 1920
Minoshima, N. 2438
Minowa, Masanao 3828
Minsoo Jang 3933
Mira, Maria C. 506
Mishima, Tomoakzu 2533
Mishra, Santanu 2707
Mishra, Santanu Kumar 3587
Misu, Daisuke 1874, 3012
Mitterhofer, Hubert 1701
Miura, Yushi 1536, 3298
Miyajima, Hiroki 1054
Miyajima, Takayuki 2421
Miyakawa, Takayuki 2421
Miyama, Yoshihiro 2673
Miyashita, S. 3702
Miyawaki, Satoshi 84
Miyazaki, Hideki 383
Miyazaki, Kensuke 601
Miyazaki, Toshimasa 1956
Miyazaki, Yuji 750
Mizoguchi, Takahiro 1660
Mizukami, Makoto 3440
Mizuki, Tatsuya 1575, 3624
Mizuma, Takeshi 2126

AUTHOR INDEX

Mizuno, Takayuki 1886
Mizusaki, Hiroshi 3093
Moballegh, Shiva 656
Mochikawa, Hiroshi 1929
Mochizuki, Eiji 671, 2870
Mochizuki, K. 2569
Mohamed, Essam Ebaid 3877
Mohamed, Tarek Hassan 3877
Mohan, Ned 1036, 1412, 3037, 3061, 3750
Mohd Arif, Mohd Johari 573
Molinas, Marta 1861
Momose, Fumihiko 671
Moo, Chin-Sien 1796, 3796, 3928
Moon, Dongok 1394
Moon, Gun-Woo 1738, 1743
Moon, Sang-Ho 224
Moorthy, Radha Sree Krishna 2087, 3616
Moraes, Lenin 3225
Mori, Tomohiro 2983
Morikawa, R. 258
Morimoto, Masayuki 3509
Morimoto, S. 246, 258, 312, 390
Morimoto, Shigeo 324, 356, 363, 370,
    1997, 3018, 3519
Morimoto, Shinya 1654
Morishita, Shin 130
Morita, Hiroshi 1490
Morita, Kazunori 191, 582
Morita, Kosuke 3624
Morita, M. 1892
Morizane, Toshimitsu 1299, 1806
Morizane, Tosimitsu 3341
Moroi, Takayuki 3134
Morozumi, Akira 671
Mory, David 2554
Motizuki, Shun 2745
Motoi, Naoki 801, 2266
Motomura, Masashi 3611
Mouri, Masayuki 1728
Mrak, Branimir 1701
Mukai, Ryosuke 2775
Mukunoki, Makoto 97, 1950
Munk-Nielsen, Stig 2547, 2850
Murai, Kensuke 1567
Murai, Toshiaki 1122
Murakami, Daichi 1728
Murakami, Kouhei 2385
Murakami, Toshiyuki 1962
Murata, Koji 2108
Murata, Munehiro 1173
Murata, Yuichiro 2064
Musing, Andreas 821
Mustapa, Rijalul Fahmi 2529
Muta, Shoichiro 3067
Nag, Soumya Shubhra 3587

Nagai, Shinichiroh 811
Nagano, Tetsuaki 2638
Nagano, Tsuyoshi 1253
Nagano, Y. 146
Nagashima, Tomohiro 2175
Nagata, Shun 2252
Nagatomo, Yoshinobu 2663
Nagel, Andreas 2113
Nagura, Hirokazu 2451
Nagy, Istvan 2700
Naitoh, Haruo 1135
Nakagawa, Hidehiko 1552
Nakagawa, Yuki 2533
Nakahara, Mizuki 744, 2511
Nakajima, Yoichiro 403
Nakamura, M. 376
Nakamura, Ritaka 92
Nakamura, Sota 2400
Nakamura, T. 2074
Nakamura, Tatsuya 1575
Nakamurame, Fuminori 2632
Nakanishi, Toshiki 1095
Nakano, Y. 2074
Nakao, Hiroshi 2745
Nakao, Noriya 1128, 1141
Nakaoka, Mutsuo 2359, 2533
Nakaoka, Mutuo 3341
Nakashima, Yoshiyasu 2745, 3386
Nakata, Yuki 138
Nakatsu, Kinya 2904
Nakatsugawa, Junnosuke 2451
Nakayama, Koji 3440
Nakayama, Naoyuki 3857
Nakayama, Yasushi 2290
Nakazawa, Yosuke 1357
Nam, Kwang-Hee 664
Narita, Takayoshi 294, 299, 3055, 3159
Nashida, N. 2569
Nayanasiri, D. R. 3598
Nee, Hans-Peter 3712
Neubert, Markus 3145
Nguyen, D. 2679
Nguyen, Quoc Khanh 318
Nguyen, Thanh Hai 2406
Nha, Quang Trong 3913
Nho Van Nguyen 2826
Nian Heng 843
Nicolae, Ileana-Diana 2996
Nicolae, Marian-Stefan 2996
Nicolae, Petre-Marian 2996
Niijima, Koji 1299, 1806
Nilssen, Robert 1412
Nimura, Tomohiro 3079
Ning Liu 2668

AUTHOR INDEX

Ninomiya, Tamotsu 177, 1179, 2216, 2222, 3630, 3652

Nishida, Katsumi 2359

Nishida, Yasuyuki 2189

Nishikata, Shoji 959

Nishimura, T. 3702

Nishimura, Tomohiro 2870

Nishimura, Yoshitaka 671

Nishio, Haruhiko 2302

Nishioka, Tomoya 2152

Nishisu, Koji 2285

Nishiyama, Noriyoshi 2011

Nishizawa, Shinichi 117, 744

Niu, Ruigen 160, 475

Noda, Koji 2541

Noda, Taku 2175

Noguchi, Kenji 2277

Noguchi, S. 2569

Noguchi, Toshihiko 1173, 1405

Noh, Yong-Su 166

Nomura, Naofumi 1522

Nomura, Shinichi 3218

Nonaka, Hirotaka 2737

Norigoe, Isami 117

Noro, Osamu 1552

Norrga, Staffan 1087

Noto, Yasuo 682

Nozaki, Takahiro 1660

Nozawa, Ryosuke 1115

Nussbaumer, Thomas 975, 3365

Nuutinen, P. 587

Oboe, Roberto 1679

O'Byrne, Sean 2926, 2932

Oda, Yoshinori 829

Odawara, Shunya 289, 2856, 2874

Ogasawara, Satoshi 1728, 2977, 3525

Ogashi, Yoshihiro 92

Ogawa, Kazutoshi 2140, 2285

Ogawa, Takashi 2285

Ogura, Kazuya 582, 3455

Ogura, Tsuneo 2068

Oh, Min-Seok 166

Ohara, Shinya 850

Ohashi, Hiromichi 117, 744

Ohashi, Shunsuke 3410

Ohchi, Masashi 724, 730, 3067, 3249

Ohishi, Kiyoshi 1247, 1516, 1956, 3153

Ohnishi, Kouhei 1660, 2483

Ohnuma, Takumi 914

Ohnuma, Yoshiya 84

Ohse, Naoyuki 3440

Ohtake, Asuka 3857

Oi, Kazunobu 452

Oi, Takeshi 2290

Oishi, K. 376

Oishi, Koji 3012

Oiwa, Takaaki 988

Ojika, Satoshi 1430

Oka, T. 376

Oka, Toshiaki 2330

Okamoto, Dai 3440

Okamoto, Masayuki 3292

Okamoto, Shoji 2290

Okamura, Kazuki 3674

Okazaki, Fumihiro 1728

Okazaki, Yuhei 1586

Okitsu, Takashi 1886

Okubo, Toshikazu 2058

Okuma, Jun 1978

Okuma, Yasuhiro 2834

Okumura, Hajime 3440

Okuyama, Yoshihiro 1811

Omata, Shinpei 2944

Omi, Masataro 1317

Omori, Hideki 1299, 1806, 3341

Omote, Kenichiro 1950

Omura, Mototsugu 1685

Ong, Andrew 2722

Onishi, Mitsuru 1054

Ooishi, Eiji 183

Ooshima, Masahide 1715

Orikawa, Koji 191, 1613, 2277, 3659

Ortiz, G. 1291

Ortiz, Gabriel 1309

Oshima, Ryo 1021

Oshinoya, Yasuo 294, 299, 2972, 3055, 3159, 3162

Oso, Hiroshi 629

Ota, Chiharu 3440

Ota, Satoru 2095

Otsuki, Midori 1054

Ouchi, Takayuki 3920

Ouyang, Shaodi 624

Ouyang, Ziwei 2842

Ozaki, Takayuki 2638

Ozkan, Ziya 498

Pala, Vipindas 2208

Palmour, John 3447

Pan, Miao 582

Panda, S K 1775

Panda, Sanjib Kumar 1580

Pansier, F. 1935

Papafotiou, Georgios 1606

Papastergiou, Konstantinos 1220

Park, Gyeong-Jae 1990

Park, Junsung 1394

Park, Yongsoon 2598

Parker, S. G. 2019, 2372

Partanen, J. 587

Patel, Dhaval 758, 1626

Pedersen, Kristian Bonderup 2547

AUTHOR INDEX

Peftitsis, Dimosthenis 3712
Peltoniemi, P. 587
Peng Gao 2926, 2932
Peng Wang 3124
Peng Wen 782
Peng, Han 2208
Peretti, L. 3111
Peters, A. 2887
Peters, Wilhelm 1508
Petersen, Lars P. 3352
Petersen, Lars Press 2842
Petrich, Matthias 318
Pettersson, Sami 3432
Pham Phu Hieu 3913
Pidaparthy, Syam Kumar 3638
Piepenbreier, Bernhard 1816, 3029
Ping-Heng Wu 1261
Pires, Igor A. 3225, 3270
Pittet, Serge 3371
Pittini, Riccardo 3905
Po-Chien Chou 3425
Poh Chiang Loh 857
Po-Jung Tseng 2810, 3328
Popa, Lucian-Dinut 2996
Popova, L. 548
Popovic, J. 1935
Poshtkouhi, Shahab 2336
Pou, Josep 3758, 3764
Prasanna, I. V. 1580
Prasanna, U. R. 230, 395, 1594
Prasanna, Udupi. R. 2343
Prodic, Aleksandar 3582
Pyrhonen, J. 548
Qi Zhang 3489
Qu, Lizhi 609
Qunzhan Li 1050
Rabkowski, Jacek 3712
Radic, Aleksandar 3582
Radman, Karlo 1691
Rae-Sung Yu 925
Rahman, F. 2679
Rahman, M. A. 982
Rahman, M. F. 2686
Rajashekara, K. 395, 1594
Rajashekara, Kaushik 230, 2343, 3134
Raju, Siddharth 1036
Ramadan, Husam A. 863
Rambetius, Alexander 270, 3029
Rannestad, Bjorn 2547
Rannested, Bjorn 2850
Rathore, Akshay K. 1775
Rathore, Akshay Kumar 2087, 3616
Ray, Olive 2707
Razik, H. 3239
Reiter, Tomas 774

Ren, Kangle 465
Riffat, Javid 2815
Rikitake, Jungo 2216
Rim, Chun T. 1103, 2232, 3358
Rivera, Marco 3574
Robbins, William P 3037, 3061
Rodriguez, Jose 3574
Rongfeng Yang 3680
Rosekeit, Martin 2729
Roth-Stielow, Jorg 264, 318, 3085
Roy, Sudhin 651, 2562, 3286
Ruan, Xinbo 458, 2768
Ruda, Harry E. 1303
Ruderman, Michael 1665
Rufer, Alfred 1081, 1386
Ryo, Mina 3440
Ryu, Sei-Hyung 3726
Saarakkala, Seppo E. 2489
Saga, Yasunao 1748
Sahoo, Ashish Kumar 3750
Saikusa, H. 3007
Saito, Eiichi 1679
Saito, Katsuhiko 2064
Saito, Ryo 3397
Saito, Ryoji 2189
Saito, Takashi 671
Saitoh, Ryoh 914
Sakaba, Kouichi 3159
Sakai, Kazuto 240
Sakai, Tomoyasu 1490
Sakai, Toshifumi 2451
Sakaino, Sho 1978
Sakimoto, Kenichi 1552
Sakurai, Naoki 2297
Sakurai, Takayasu 2228
Sampath, Prasad K. 2722
Sanada, M. 246, 258, 312, 390
Sanada, Masayuki 324, 356, 363, 370, 3018, 3519
Sand, Kjell 1861
Sariyildiz, Emre 2483
Sasaki, Tomotake 2745
Sasongko, Firman 1761
Sato, Daisuke 3815
Sato, Koji 1671
Satria, Andri 3893
Sauer, Dirk Uwe 2729
Sawada, Tadashi 1122
Sayed, Khairy 2359
Sayed, Mahmoud A. 3877
Schob, Reto. T. 1691
Schon, Andre 3255
Schrittwieser, L. 1291
Schupbach, Marcelo 3447
Schuster, Johannes 3085
Segaran, D. S. 2372

AUTHOR INDEX

Segsa, Karl-Heinz 2554
Seilmeier, Markus 1816
Sekiba, Yoichi 2175
Sekisue, Takayuki 2175, 2189
Seo, Gab-Su 2272
Seok-Jin Hong 3532
Seunghoo Song 1646
Seung-Ki Sul 925, 2802
Severson, Eric 1412
Shah, Shahil 843, 967
Shao Zhang 1342
Shaodi Ouyang 2815
Shaofeng Xie 1050
Shaohua Sun 3213
Shaohui Zhong 1155
Shen, Na 582
Shen, Zhiyu 3850
Shenghui Cui 1268
Sheng-Kai Kao 2714
Shi, Hongtao 577
Shi, Rongliang 567
Shibahara, Kohei 1575, 3624
Shibahara, Ryota 2222
Shibanuma, Kenichi 634
Shibata, Yuichiro 3950
Shieh, Hsin-Jang 351
Shieh, Jenn-Jong 204
Shigematsu, Koichi 2183, 2189
Shih, Bing-Jyun 105
Shih, Sheng-Fang 2380
Shih-Jen Cheng 3185
Shimada, Takae 2058
Shimamori, Hiroshi 2745
Shimao, Toshihiro 415
Shimatou, T. 2939
Shimizu, Kyohei 1968
Shimizu, Toshihisa ... 876, 1166, 2944, 2983, 3044, 3771
Shimizu, Toshimasa 1054
Shimode, Daisuke 1122
Shimomura, Junichi 2183
Shimono, Tomoyuki 1685
Shin Shiung Wang 3938
Shin, Hyunhak 110
Shin, Yesl 493
Shinagawa, Syuhei 252
Shinbo, Mitsuo 634
Shindo, Yuji 1552
Shinnaka, Shinji 1824
Shinohara, Atsushi 324
Shinozaki, Ikki 1728
Shinozuka, Yasuhiro 2228
Shioda, Masashi 130
Shirakawa, Kazuhiro 304, 1379
Shiraki, N. 2120
Shirasawa, Koki 3106

Shishida, Yasuhiro 2644
Shiting Weng 1476
Shixi Hou 3168
Shoeiby, B. 1482
Shoji, Hiroyuki 2058
Shoyama, Masahito 863, 3386
Shu-Hung Liao 1452
Shuitao Yang 1342
Shun-Chung Wang 3233, 3778
Shunke Sui 3680
Shuren Wang 3194
Shu-Wei Kuo 3185
Siemaszko, Daniel 3371
Silva, Marcelo 3864
Silva, Sidelmo M. 3225
Silventoinen, P. 587
Sin, Min-Ho 409
Singh, B. N. 3482
Sintamarean, C. 1912
Sitbon, M. 2240
Sivakumar K 1400
Siwakoti, Yam P. 1801
Skuriat, Robert 2920
Smadi, Issam 1968
Smaka, Senad 1148
Smiththisomboon, Somrat 3885
Sogawa, Yuki 522, 2781
Solomon, Adane Kassa 2920
Sone, Kodai 3525
Song Kejian 640
Song, Z. Q. 2686
Sonoda, Hideki 634
Soo-Cheol Shin 3532
Specht, Andreas 1501
Srirattanawichaikul, Watcharin 3417
Steinert, Daniel 975
Steinke, Gina 1081
Stumpf, Peter 2700
Su, Bonan 614
Su, Hong-Wei 1781
Su, Jianhui 2156
Suetsugu, Tadashi 3955
Sugao, Kazumi 403
Sugimoto, Hiroya 982
Sugiura, Makoto 1135
Suh, Yongsug 1185, 1646
Su-Han Kim 790
Sul, Seung-Ki 1268, 2598
Sumida, Hitoshi 2302
Sun, Jian 843, 967, 2202
Sun, Shaohua 2318
Sun, Wei 609
Sunaga, Keita 3162
Sung, Kyungmin 117, 829
Sunsoon Park 1646

AUTHOR INDEX

Suntio, T. 514, 2240
Suntio, Teuvo 1466
Suryadevara, Rohit 2433
Suto, Kenji 3194, 3194
Suul, Jon Are 1544
Suwankawin, Surapong 3885
Suzuki, Genri 1697
Suzuki, Hirokazu 3503
Suzuki, Katsumi 959
Suzuki, Michiaki 883
Suzuki, Nobuyuki 919, 2541
Suzuki, Ryosuke 2972
Suzuki, Shun 1166
Suzuki, Takashi 1062
Suzuki, Toshiki 907
Svensson, Jan R 1220
Tabira, K. 2939
Tadano, Yugo 1242
Tadokoro, D. 390
Taeck-Kie Lee 3532
Taekyun Kim 3933
Tae-Won Chun 2763
Taga, Hironori 1328
Tajima, G. 2438
Takada, Hiromu 1697
Takagi, Ryo 1328
Takahashi, Akiko 2470
Takahashi, Hiroki 152, 1021
Takahashi, Hirotaka 1068
Takahashi, Hisashi 3106
Takahashi, K. 2569
Takahashi, Naoya 3920
Takahashi, Nobuhiro 3207
Takahashi, Osamu 2285
Takahashi, Takehiro 3207
Takahashi, Yoshikazu 671, 2870
Takamiya, Makoto 2228
Takao, Kazuto 744, 3440, 3707
Takasaki, Mika 2252
Takashita, Haruomi 3386
Takasu, Shinji 3440
Takatsuka, Yushi 1898
Takayama, Masakazu 3801
Takayanagi, Atsushi 2794
Takeda, Kotaro 1654
Takeda, Masashi 801
Takeda, Takashi 1490
Takei, Manabu 3440
Takemoto, Masatsugu 1000, 3525
Takenaka, Kensuke 3440
Takenami, Fumiaki 2737
Takenoiri, S. 2939
Takeshita, Takaharu 123, 601, 2310, 3809, 3877
Takeuchi, Katsutoku 3012
Takeuchi, Shun 3646

Takezaki, Kenichi 3525
Taki, Hiroshi 876, 1379
Takino, Toshiaki 817
Tam Khanh Tu Nguyen 2826
Tamada, Shunsuke 1357
Tamura, Hiroshi 682
Tan, Nadia M. L. 750
Tanabe, Ryo 1234
Tanai, Masanobu 829
Tanaka, Daiki 3179
Tanaka, Junya 240
Tanaka, Kiminori 2222
Tanaka, Koutaro 1006
Tanaka, Seiyu 982
Tanaka, Takahide 2302
Tanaka, Toshihiko 3292
Tanaka, Yasunori 3440
Tanaka, Yuichiro 1880
Tanifuji, Hikaru 1115
Taniguchi, Shun 2465
Tao Meng 3213
Tatsuta, Fujio 959
Tauchi, Yuki 3119
Teng, Jen-Hao 677
Teodorescu, R. 1912
Tera, Takahiro 304, 876
Terabe, Ryosuke 2638
Terao, Yutaka 2644
Teshima, Masato 1068
Thiringer, Torbjorn 1373
Thogersen, Paul 2547
Thogersen, Paul Bach 2850
Tian, Yanjun 3538
Ting, Pangan 677
Ting, Yeh 2787
Tint Soe Win 3292
Toba, Akio 2011
Toda, Hiroaki 1984
Togashi, Ryo 356
Toi, Takahiro 1109
Tokiwa, Tsuyoshi 2977
Tokuda, Hirokazu 2175
Tokumasu, Akira 1379
Tokuyama, Takeshi 2904
Tominaga, Shinji 2290
Tomioka, Satoshi 3630
Tomita, Mutuwo 907
Tonogi, K. 2438
Toru 3784
Tosaka, Shuhei 1207, 1227
Town, Graham E. 1801
Toyoda, Hajime 1560
Tran, Q. V. 446
Trescases, Olivier 2336, 3496
Trillion Zheng 2893

AUTHOR INDEX

Trintis, Ionut 2547
Tripathi, Awneesh 758, 1626
Trompa, Thomas...2554
Tsai, Jiung-Lin 1781
Tsai, Ming-Hsiao 278
Tsan Chen 2810
Tse, Chi K. 1425
Tseng, K. J. 2145
Tsorng-Juu Liang 2750, 3346
Tsubakidani, Takashi 3674
Tsuboi, Yoshiki 1160
Tsuboi, Yuichi 92
Tsuchida, Kazuo...3397
Tsuda, Junichi 1929
Tsuji, Mineo.........................2775, 3093, 3674
Tsuji, Satoshi 522, 2781
Tsuji, Toshiaki 1978
Tsukakoshi, Masahiko 92, 2330
Tsuruma, Yoshinori 1054, 2644
Tsuruta, Hironori...629
Tsuruta, Ryoji 1350
Tsuruta, Yukinori 2266, 3403
Tsuyoshi, Hanamoto 2476
Tu, Yunwu...465
Tukiman, Rahayu 2529
Turpin, Santiago 1974
Tuysuz, Arda 1904
Tzou, Ying-Yu 3072
Uddin, Muslem 3574
Ueda, K. 312
Ueda, Tetsuya 403
Ueda, Tetsuzo 2075
Ueda, Yoshinobu 1855
Uemura, Hirofumi 891, 2834
Ukai, Hiroyuki 2400
Umeda, Nobuhiro 2183
Umeno, Masayoshi 183
Umesh B S 1400
Umetani, Kazuhiro 304
Undeland, Tore 1412
Uno, Masatoshi 1444, 2351
Urushibata, Hiroaki 2290
Urushibata, Shota 1365, 3455
Ushiro, Nobumasa..3106
Vaisanen, V. 587
Vajk, Istvan 2700
Van Brunt, Edward 3726
Van Wyk, J. D. 1935
Van-Long Tran 1214
Vasiladiotis, Michail 1386
Vasquez-Arnez, Ricardo Leon 1276
Veerasamy, Balaji 2310
Vieto, Ignacio..843
Viinamaki, J. 2240
Vilathgamuwa, D. M. 2722, 3598

Vogt, T. 2887
Wada, Keiji 744, 1379, 2511, 3646
Wahlstroem, Jonas 3476
Wajima, Kiyoshi 1984
Wallscheid, Oliver 1501
Wang Hui 640
Wang, Bin 2133
Wang, Chao-Fu 172
Wang, Fei 470
Wang, Fusheng 465
Wang, H. 1912
Wang, Hengli 72
Wang, Jun 944
Wang, Lingxiang 465
Wang, Lipeng 458
Wang, Xiaojian 2815
Wang, Xinyu 624, 2815
Wang, Xiongfei 216, 1529, 3320, 3538
Wang, Yanbo 3538
Wang, Yong 470
Wang, Yue 160, 475
Wang, Zhao'An 160, 475
Watanabe, Daisuke 988
Watanabe, Hiroki 84
Watanabe, S. 2939
Watanabe, Shoichiro 1334, 2126
Watashima, T. 2939
Wei Jiang 3194
Wei Liu 1050
Wei Wang...3680
Wei Yan 3455
Wei, Guo 2318
Wei, Sun 341
Weili Dai 3168
Weirong Chen 3695
Wei-Ting Hsu 2910
Wen, Bo 3850
Wen, Chao-Kai 677
Wen, Huiqing 702
Wen-Chien Hsu 2714
Wenjie Chen 2967
Wen-Tai Li 677
Wheeler, Pat...2920
Won, Chung-Yuen 166, 409
Wong, Siu-Chung 1425
Wonsuk Choi 2914
Woojin Choi 1214
Wu Mingli 640
Wu, Bin 3468
Wu, Chun-Wei..330, 1832
Wu, G. F. 1471
Wu, Gwo-Bin 3796
Wu, T.-F. 2050
Wu, Tsung-Hsi 1796
Wu, Weimin 2610

AUTHOR INDEX

Wu, Weiyang 582
Wu, Wenlong 470
Wu, Wen-Zhe 429
Wunsch, B. 2167
Xia, Huan 2133
Xiangdong Sun 3489
Xiang-Dong Sun 3784
Xiao, D. 2686
Xiao, Fei 66, 72
Xiao, Shuai 543
Xiaojie You 2893
Xiaojie Zhuang 2638
Xiaolong Ma 835
Xiaomei Song 2967
Xie, Ruiliang 2950
Xiong, Li 230
Xiuqin Wei 3955
Xu Cai 1842, 3124
Xu Dianguo 3050
Xu Yang 2967
Xu, David 3099
Xu, Dehong 485
Xu, Dianguo 609, 614, 3002
Xu, Haizhen 567
Xu, Rong 609
Xue, Danhong 3604
Xuling Chen 2768
Yablecki, Jessica 3496
Yachi, Toshiaki 3959
Yakabe, Seichiro 730
Yamada, Hiroaki 538, 1811
Yamada, Kenji 1898
Yamada, Ryuji 1920
Yamada, Takatoshi 2212
Yamada, Tatsuji 3263
Yamagata, Shinichi 829
Yamagishi, Tatsuya 750
Yamaguchi, Shota 3771
Yamaguchi, Takashi 1242
Yamaichi, Katsuya 2517
Yamaji, Masaharu 2302
Yamamoto, Eiji 1654
Yamamoto, Junichi 177, 1179
Yamamoto, Kenji 3106
Yamamoto, Kichiro 689
Yamamoto, Kohei 1438
Yamamoto, Masayoshi 811, 883, 2497
Yamamoto, Shu 3044
Yamamoto, Takashi 3707
Yamamoto, Yasuhiro 1601
Yamamura, N. 146
Yamanaka, Kenji 2152
Yamanaka, Tatsuya 1207, 1227
Yamanoi, Takashi 1062
Yamashita, Nobuyuki 2983

Yamashita, Shigeharu 2745
Yamazaki, Akira 933
Yan Li 3124
Yan Zhang 835
Yanagi, Hiroshige 3630
Yang Chuan 1962
Yang, Cs 199, 2593, 3185
Yang, Daeki 2247
Yang, Geng 543, 582
Yang, Guorun 66
Yang, Hong-Tzer 2100
Yang, Rongfeng 609
Yang, Shih-Sian 2606
Yang, Sihun 863
Yang, Xu 2950
Yang, Zhongping 335, 1322, 2133
Yanhong, Zhang 452
Yano, Yoshihiro 2775
Yanru Zhong 3489
Yaramasu, Venkata 3695
Yashiro, Daisuke 1974
Yashun Li 782
Yasubayashi, Mikio 183
Yasui, Kazuya 2465
Yasumura, Yuji 3568
Yasuno, Takashi 3179
Yau, Y. T. 2754, 3392
Yazdkhasti, Pegah 2156
Yi, Hao 577
Yi-Chun Lin 2714
Yi-Hsun Chiu 3778
Yi-Hua Liu 3778
Yixin Zhu 1155, 3546
Yizhanyi Tang 2977
Yoda, Kazuyuki 1748
Yokoi, Y. 3007
Yokoi, Yuichi 3024
Yokokura, Yuki 1956
Yokoyama, H. 2120
Yokoyama, Natsuki 2285
Yokoyama, Tomoki 3397, 3410
Yonemori, Ryo 689
Yonezawa, Hikaru 3055
Yonezawa, Yoshiyuki 3440
Yonezawa, Yu 2745, 3386
Yong Ding 1476, 3842
Yong, Yu 341
Yong-Cheol Kwon 925
Yongdong Tan 3538
Yongjae Lee 3140
Yoon, Sung Hyun 2272
Yoshida, Morito 3410
Yoshida, S. 2569
Yoshida, T. 2438
Yoshida, Yoshiaki 3503

AUTHOR INDEX

Yoshikawa, Yuichi 2011
Yoshimoto, Kantaro................................2421
Yoshimura, Eiji 1552
Yoshino, Teruo 2644, 3834
Yoshino, Yukio 2745
Yoshioka, S. 246
Yoshioka, Takashi....................................1956
Yoshizawa, Daisuke 97, 1950
Young-Do Kim 1738, 1743
Youngjoon Choi 1185
Young-Ryul Kim 796
Yu, Changzhou 567
Yu, F. Y. 1471
Yu, Ling-Chia 208
Yu, Shuai 2318
Yu, Weikai 647
Yu, Yifan 1458
Yu, Yong 567, 609
Yu-Chen Liu 2810, 3328
Yue Chen 2768
Yue, Xiaolong 2960
Yu-Jen Wang 1420
Yu-Kang Lo 2758, 2810, 3185, 3328, 3913
Yuki, Kazuaki 2465
Yukita, Kazuto 1490
Yukutake, Seigo 2140
Yunchang Kwak...2693
Yun-Chu Chiu 3328
Yung-Ching Huang 1452
Yunmei Fang 3168
Yunwei Li 3482
Yura, Masashi 2297
Yu-Shan Cheng 3233, 3778
Yuzurihara, Itsuo 2794
Zaitsu, Toshiyuki 177, 1179
Zanma, Tadanao 3568
Zargari, Navid R. 3468
Zehelein, Matthias.....................................3085
Zeliang Shu 1050
Zeljkovic, Sandra 774
Zhang Wei 3050
Zhang Yajing...2899
Zhang, Haodong 3604
Zhang, Huiguo...2202
Zhang, Tao..485
Zhang, Wei 614, 1425
Zhang, Xing 465, 567
Zhang, Xuning 2626
Zhang, Yuzhuo 647
Zhang, Zhe...78
Zhao, Wei..567
Zhao-Qin Guo 1580
Zhe Wang 3129
Zhe Zhang...3905
Zheng Dong 3842

Zheng Li 1842
Zheng, T. Q. 807
Zheng, Trillion Q. 2899, 3314
Zhengzhi Han 3124
Zhenyao Xu 994
Zhongping Yang 807
Zhu, B. 395, 1594
Zhu, Honglin 3002
Zhuo, Fang 577, 2960
Zian Qin 857
Zingerli, Claudius M. 3365
Zitouni, Y. 3239
Zong-Zhen Yang..3778
Zou, Xudong 555
Zwyssig, Christof.......................................1707